THE COGNITIVE NEUROSCIENCES

THE COGNITIVE NEUROSCIENCES

Sixth Edition

David Poeppel, George R. Mangun, and Michael S. Gazzaniga, *Editors-in-Chief*

Section Editors:

Danielle Bassett
Marina Bedny
Sarah-Jayne Blakemore
Alfonso Caramazza
Maria Chait
Anjan Chatterjee
Stanislas Dehaene
Mauricio Delgado
Karen Emmorey
Kalanit Grill-Spector
Richard B. Ivry
Sabine Kastner
John W. Krakauer
Nikolas Kriegeskorte
Steven J. Luck
Ulman Lindenberger
Josh McDermott
Elizabeth Phelps
Liina Pylkkänen
Charan Ranganath
Adina Roskies
Tomás J. Ryan
Wolfram Schultz
Daphna Shohamy

THE MIT PRESS
CAMBRIDGE, MASSACHUSETTS
LONDON, ENGLAND

This book was set in ITC New Baskerville Std by Westchester Publishing Services. Printed and bound in the United States of America.

Library of Congress Cataloging-in-Publication Data

Names: Poeppel, David, editor. | Mangun, G. R. (George Ronald), 1956- editor. | Gazzaniga, Michael S., editor.
Title: The cognitive neurosciences / edited by David Poeppel, George R. Mangun, and Michael S. Gazzaniga.
Description: Sixth edition. | Cambridge, MA : The MIT Press, [2020] | Includes bibliographical references and index.
Identifiers: LCCN 2019007816 | ISBN 9780262043250 (hardcover : alk. paper)
Subjects: LCSH: Cognitive neuroscience.
Classification: LCC QP360.5 .C63952 2020 | DDC 612.8/233--dc23 LC record available at https://lccn.loc.gov/2019007816

10 9 8 7 6 5 4 3 2 1

CONTENTS

II AUDITORY AND VISUAL PERCEPTION

III MEMORY

IX CONCEPTS AND CORE DOMAINS

X LANGUAGE

PREFACE

How compelling is cognitive neuroscience? So compelling that in the summer of 2018, 30 people were willing to spend three weeks in a windowless seminar room when they could have been enjoying one of the most beautiful places on Earth, Lake Tahoe. The seductive allure of the science held sway. A cohort of brilliant fellows showed up to every single session, listening to over 80 talks and peppering the speakers with candid, probing, and even, for the most part, polite questions. To cut to the chase: the state of the field is good.

The first edition of this book—both the inventory of our field and a critical perspective on it—was published in 1995, following the first meeting of the Cognitive Neuroscience Summer Institute, fondly known as "brain camp," in Tahoe. At that point, electroencephalography (EEG) was a well-established technique; magnetoencephalography (MEG) existed but was not widely appreciated beyond a group of aficionados; and positron emission tomography (PET) was around and available, though rapidly becoming supplanted by functional magnetic resonance imaging (fMRI), which was only a few years old at the time. Twenty-five years later, the technical advances in and the wide availability of fMRI constitute the most dramatic changes in the field. Cognitive neuroscience has also benefited from many additional experimental approaches and novel measurement methodologies. From single-unit, depth-electrode, and grid-based recordings in neurosurgical patients to near-infrared spectroscopy (NIRS) in newborns, no technique has been left unexploited and unexplored. This edition reflects the healthy and exciting methodological pluralism of our field. But mapping with MRI—both structural and functional—is the aspect of the human brain sciences that has most captured the public and professional imaginations.

Maps and Explanations

What is it about making maps that draws us in so powerfully? For starters, scientific success! Identifying a systematic spatial layout at any scale of neural organization suggests we are on the right track to understand function. Nobody questions the utility of knowing the retinotopic layout of visual areas, the sheer coolness of identifying 180 distinct cortical fields per hemisphere, or the payoff of being able to identify a brain region robustly implicated in speaking, or throwing, or remembering. Yet something is missing. To caricature the problem: *localization is not explanation*. Indeed, even spatial organization is not explanation.

The last 25 years have yielded incredible insights into how neural real estate is partitioned. Detailed descriptions have been generated across measurement methods and brain systems (perceptual, motor, cognitive, affective). Yet descriptions they remain, and while "descriptively adequate" is hardly an insult, our yearning should be higher. Our field should strive to achieve "explanatory adequacy."

David Marr famously argued that the study of a complex phenomenon or system can be profitably pursued by breaking it into three tightly linked levels of analysis: implementational, algorithmic, and computational. Whatever one thinks of Marr's framework, he identified a serious conceptual challenge. We are, unsurprisingly, enamored of our remarkable new measurement tools, and we use the tools for mapmaking (a) because we can and (b) because the findings can be so intuitively pleasing. But too often (either for practical reasons or due to our own epistemic insufficiency) we stay in the descriptive safety of the implementational level, without taking the critical additional step of linking to other levels of analysis. The current approaches can and do yield first-rate neuroscience, but do they yield satisfactory *cognitive* neuroscience?

One might point out that we are studying *networks*, not circumscribed areas. But the fundamental issue remains. It may be descriptively more adequate to suggest that a neurocognitive phenomenon is executed by network X-Y-Z rather than, say, area Y. However, invoking a "network" for explanation is no more satisfactory than invoking a single area. One might also argue that we are now using sophisticated computational models in the study of some neural and cognitive phenomena and can even make predictions that illustrate remarkable model-data fits. However, this descriptive use does not meet the standards of computational and algorithmic levels of *understanding*. The goal is to incorporate nuanced and theoretically motivated accounts of behavior that stand a chance of generating persuasive explanations. *Regression is not explanation.*

In short, we should celebrate our considerable successes while remaining fascinated by the even more substantial challenges that lie before us. The notion that we have achieved a deep understanding of any cognitive system "in neural terms" is misleading, counterproductive, and, frankly, no fun at all.

A Postscience Cognitive Neuroscience?

A recurring topic at the Summer Institute—both explicitly and implicitly present in many of the lectures and discussions—was the tension between big data versus deep data. The suite of techniques we now have at our disposal can generate enormous amounts of data. It is, however, reasonable to ask whether colossal amounts of data acquired from a large number of participants are likely to yield the kinds of *answers* we seek. Historically, the very intensive study of individual participants has yielded great success.

Obviously, the approach depends on the nature of the specific question. Some research questions may be best methodologically approached with big data; for others, deep data are likely to generate better insight. Nevertheless, it is appropriate to raise serious questions about our epistemological stance. One thing is certain: there is now an intense (though typically unspoken) compulsion to have truly enormous data sets. The attraction may be due to the computational tools we now have available and, in part, to very real issues with replication that different areas of the psychological sciences are grappling with.

Nobody, or at least nobody rational, is in principle against having a large number of observations for any kind of study. But the orgy of data has not always been accompanied by equally passionate theory building. As it stands, there is a clear and present mandate to obtain and work with data sets of unprecedented size. This unspoken code of conduct necessitates data-analytic approaches that capitalize on approaches such as machine-learning. However, while we as a field have embraced and enthusiastically pursued big and sometimes deep data, we have not pursued "big theory." Is this a problem? We are data-rich but theory-poor. Have we thereby maneuvered ourselves into an epistemological local minimum?

Perhaps the cognitive neurosciences have reached a stage at which engineering has supplanted science. In our modern scientific era, do scientists no longer matter? If research questions and approaches are driven not by understanding or mechanism or theory but by prediction or data-driven model fitting, our data *are* our theory. In this case, our science might uncharitably be characterized as the "mother of all regressions." When a field becomes increasingly theoretically sterile, we are in acute danger of prioritizing approaches that are commensurate with the kind of data and data analyses we currently prefer.

Certainly, science progresses by way of exciting new instruments. However, science also progresses by way of exciting new ideas. As we continue to drown in data sets, we should feel vindicated in emphasizing and prioritizing the value of theory, behavior, observation, and other such old-school scientific habits. A well-motivated hypothesis is a terrible thing to waste.

The Future of the Cognitive Neuroscientist

The Summer Institute was bookended by two terrific lectures on morality: the first on moral cognition in infants (see section I, Brain Circuits over a Lifetime), the last on morality in adults (see section XII, Neuroscience and Society). This was not a happy coincidence but a reflection on the cohort of lecturers. This year's fellows rightly insisted on discussing the importance of scientific and personal integrity, open science, and transparent and fair processes. Indeed, the fellows wrote a joint chapter for this book. It is noteworthy and heartening that scientific ethics is not an afterthought but a central issue for young researchers.

Science as a social activity and scientific achievement, the fruit of our collective labors, cannot help but reflect our social values. As such, the integrity of all participants is as important as technical competence. Too many examples of problems have existed on both the ethical and the technical sides. It is a welcome, sensible, necessary, and inspiring response of young researchers to refuse to choose one over the other. As a community, we can demand appropriate social interaction *and* appropriate technical validation. We can prioritize careful attention to replication, high ethical standards, and open access, all at once. It is not only scientific output that needs to be considered carefully. The scientific process should be subject to the same level of scrutiny.

Desiderata

In the first edition of this book in 1995, Michael Gazzaniga wrote: "The future of the field … is in working toward a science that truly relates brain and cognition in a mechanistic way." As outlined above, that desideratum is still number one on the list. Obtaining it requires the best ideas and theories to be paired with the best techniques and analyses, in large part through open-minded interaction across disciplines. Gazzaniga points out, in the second edition of 2000, that "interdisciplinary cross-pollination seems inevitable in a field whose subject is itself both a coherent whole and a motley conglomerate of components."

Two decades later, the motley conglomerate is perhaps even more motley. The dangers of interdisciplinary cross-sterilization over cross-fertilization remain. The flavors of the moment include the ubiquity of deep learning, the fascinating data from electrocorticography (ECoG), the utility of big data, the explanatory power of predictive coding, the insights into neurobiology and perception derived from oscillation-based frameworks, and other promising domains and techniques. Cognitive neuroscience memes such as these penetrate across areas. The spread is in itself exciting, but it is up to this new generation of cognitive neuroscientists—intellectually fearless, technically brilliant, and socially responsible—to take the best ideas and best techniques and address the deepest problems at the heart of the intersection of biology and the mind.

To succeed, the new leaders of the cognitive neurosciences will have to develop and examine hypotheses that plausibly *link* levels of description and yield an *understanding* of the system. For example, in the last few years, neural oscillations and their perceptual and cognitive correlates have been considered by some to be critical ingredients for certain linking hypotheses. However, building bridges and links is really, truly hard. Correlational relations will not suffice, and we, the editors, are relieved that this enormous burden is now also on the shoulders of an energetic new crop of researchers.

As pointed out in the preface to the first edition of this book by one of us (MSG), the field faces "the most fundamental problem of modern science—the problem of the explanatory gap. The gap here is the one between biologic process and the processes of mind." The gap remains and we look forward to seeing it closed.

Thanks

Huge thanks are due to Jayne Kelly and Marin Gazzaniga. Without Jayne, the Summer Institute would not have happened. Without Marin, this book would not exist. Their professionalism, competence, relentless flexibility, sense of humor, and deep tolerance of our idiosyncrasies allow this kind of scientific exchange to begin with. We are all extremely grateful. The section editors—all international scientific leaders—deserve our deep appreciation. They selected and curated excellent lectures and chapters, for which the entire field thanks them; their section overviews provide enlightenment on the topics in this volume. Enterprises of this scale require considerable support, and we thank the National Institute of Mental Health and the National Institute on Drug Abuse, the Kavli Foundation, and the University of California. As co–principal investigators, Michael Miller's and Barry Giesbrecht's contributions to the Summer Institute cannot be overstated, and we are also grateful for the support of staff members at the UC Davis Center for Mind and Brain, the UCSB Sage Center for the Study of the Mind, and the Max Planck Institute for Empirical Aesthetics. Finally, we are indebted to Philip Laughlin and his team and everyone at the MIT Press for the work they dedicated not only to this volume but to the field of cognitive neuroscience. We hope this book provides as much enjoyment and stimulation in the reading as it has to us in the writing. It is a pleasure and a privilege to be part of so heady and fun an intellectual and social enterprise.

David Poeppel
Max Planck Institute and New York University

George R. Mangun
University of California, Davis

Michael S. Gazzaniga
University of California, Santa Barbara

I

BRAIN CIRCUITS OVER A LIFETIME

Introduction

SARAH-JAYNE BLAKEMORE AND
ULMAN LINDENBERGER

The present section of *The Cognitive Neurosciences* spans human behavioral and neural development from infancy to old age. This change to the section theme is welcome, as it signals that human development does not end with adolescence but continues into advanced old age.

Individuals organize their exchange with the physical and social environment through behavior. On the one hand, the changing brain and the changing physical and cultural environment shape behavioral development. On the other hand, behavior alters both the brain and the environment. Hence, environments and brains act not only as antecedents but also as consequences of moment-to-moment variability and long-term changes in patterns of behavior. The dynamics of this system give rise to the diversity of individuals' trajectories through life (Molenaar, 2012; Nesselroade, 1991).

The general goal of developmental cognitive neuroscience is to identify neural mechanisms that generate invariance and variability in behavioral repertoires from infancy to old age. By identifying the commonalities, differences, and interrelations in the ontogeny of sensation, motor control, cognition, affect, social processing, and motivation, both within and across individuals, the field can move toward providing more comprehensive theories of behavioral development across different periods of the lifespan.

In attempts to explain the age-related evolution of this system, *maturation* and *senescence* (i.e., aging-related decline) denote the operation of developmental brain mechanisms and their effects on changes in behavior, which are especially pronounced during early childhood and late adulthood, respectively. In addition, *learning*, at any point during ontogeny, denotes changes in brain states induced by behavior-environment interactions. Note, however, that maturation cannot take place without learning and that some forms of learning cannot take place without maturation. Similarly, the ways in which senescence takes its toll on the aging brain depend on an individual's past and present learning and maturational histories. To complicate matters even more, processes commonly associated with maturation are not confined to early ontogeny, and processes related to senescence are not restricted to old and very old age. For instance, neurogenesis and synaptogenesis, which qualify as maturational mechanisms promoting plasticity, continue to exist in the adult brain; conversely, declines in dopaminergic neuromodulation, which indicate senescence-related changes in brain chemistry, commence in early adulthood. Thus, maturation, senescence, and learning mutually enrich and constrain each other throughout the entire lifespan and should preferably be understood and studied as interacting forces constituting and driving the brain-behavior-environment system (Benasich & Ribary, 2018; Lindenberger, Li, & Backman, 2006).

Thus, developmental cognitive neuroscientists are faced with three challenging tasks. First, there is the need to integrate theoretical and empirical research across functional domains to attain a comprehensive picture of individual development. For instance, sensorimotor and cognitive functioning are more interdependent in early childhood (e.g., Diamond, 2000) and old age (e.g., Lindenberger, Marsiske, & Baltes, 2000) than during middle portions of the lifespan, and developmental changes in either domain are better understood if studied in conjunction.

Second, there is a need to understand the mechanisms that link short-term variations to long-term change. Short-term variations are often reversible and transient, whereas long-term changes are often cumulative, progressive, and permanent. Establishing links between short-term variations and long-term changes is of eminent heuristic value, as it helps to identify mechanisms that drive development in different directions.

Third, to arrive at mechanistic explanations of behavioral change requires the integration of behavioral and neural levels of analysis. At any given point in the lifespan, one-to-one mappings between brain states and behavioral states are the exception rather than the rule, as the brain generally offers more than one implementation of an adaptive behavioral outcome. Therefore, ontogenetic changes in behavioral repertoires are accompanied by continuous changes in multiple brain-behavior mappings. Some of these remapping gradients may be relatively universal and age-graded, whereas others may be more variable, reflecting genetic differences, person-specific learning histories, the path-dependent nature of developmental dynamics, or a combination of all three. The resulting picture underscores the diversity and malleability of the organization of the brain and behavior as well as the constraints on diversity and malleability brought about by (1) universal age-related mechanisms associated with maturation and senescence, (2) general laws of neural and behavioral organization, and (3) cultural-social as well as physical regularities of the environment.

Research on brain development in the second half of the 20th century focused almost entirely on nonhuman animals and revealed a great deal about early neuronal and synaptic development (Wiesel & Hubel, 1965). These advances in animal research followed pioneering research in developmental psychology, particularly by Vygotsky and Piaget (Chapman, 1988). Their studies, which involved observing and analyzing children's behavior in meticulous detail, changed contemporary thinking about children's minds. Today, theory-guided series of behavioral experiments strongly support the claim that the foundational capacities of very young children are organized by guiding principles in physical, psychological, and sociomoral core domains. In this vein, Buyukozer Dawkins, Ting, Stavans, and Baillargeon propose in the opening chapter to this section that early sociomoral reasoning is guided by the principles of fairness, harm avoidance, in-group support, and authority.

While developmental psychology made great progress in the last century, it remained relatively removed from developmental neuroscience. Research on human neural development was heavily constrained by the technical challenges of studying the living human brain and, until fairly recently, was limited to the study of postmortem brains. In the past decades, however, the field of developmental cognitive neuroscience has undergone unprecedented expansion, at least in part due to technological advances. In particular, the increased and concerted use of various MRI techniques in children has created new opportunities to track structural and functional changes in the developing human brain. The use of these imaging methods has propelled our knowledge of how the human brain develops, and the data from developmental imaging studies have in turn spurred new interest in the changing structure and functions of the brain over the entire lifespan. Fifty years ago, who would have imagined that scientists would eventually be able to look inside the brains of living humans of all ages and track changes in brain structure and function from intrauterine development into old age?

Age-graded changes in the structure of the human brain from childhood to early adulthood are addressed in the chapter by Tamnes and Mills. They focus on measurements of brain morphometry and measurements derived from diffusion tensor imaging (DTI) while also discussing novel measures and approaches to examine structural brain development. Whereas structural MRI has enriched our knowledge of age-related changes in regional volume and structural connectivity, functional magnetic resonance imaging (fMRI), in concert with electroencephalography (EEG) and near-infrared spectroscopy (NIRS), has revealed developmental changes in regional brain activity and functional connectivity. Today many labs around the world use fMRI to investigate how neural systems associated with particular cognitive processes change with age. Crone and van Duijvenvoorde report the neural correlates of cognitive and affective decision-making in school-aged children, adolescents, and adults. They show that the development of basic to complex levels of cognitive control follows a pattern of specialization with age in the prefrontal cortex and the posterior parietal cortex, such that these areas are more strongly and more selectively recruited for specific tasks. Kilford and Blakemore trace the development of the social brain in adolescence and provide rich evidence for the substantial and protracted development of multiple aspects of social cognition, as well as the structural and functional development of the social brain network, during this period of life.

Inquiries into the plasticity of the brain and behavior are a rich source of developmental information; by assessing "changes in change," they offer the promise to observe the operation and proximal consequences of developmental mechanisms. Taking a lifespan and phylogenetic perspective, Walhovd and Lövdén review the evidence for age-graded differences in human plasticity from infancy to old age. This sets the stage for the chapter by Raz, who takes a systemic look at senescent changes in the brain and behavior, with particular emphasis on the role of vascular and metabolic factors. The aging brain is notorious for detrimental changes, but some older adults appear to display brain maintenance, defined as a widespread lack of senescent brain changes and age-related brain pathology. Nyberg and Lindenberger focus on the structural and functional maintenance of the hippocampus and argue that it is the primary determinant of preserved episodic-memory functioning in old age. Finally, Mather directs our attention to the role of the locus coeruleus norepinephrine system and provides evidence that the integrity of this system is crucial for maintaining cognitive functions in old age.

REFERENCES

Benasich, A. A., & Ribary, U. (Eds.). (2018). *Emergent brain dynamics: Prebirth to adolescence.* Strüngmann Forum Reports No. 25. Cambridge, MA: MIT Press.

Chapman, M. (1988). *Constructive evolution: Origins and development of Piaget's thought.* New York: Cambridge University Press.

Diamond, A. (2000). Close interrelation of motor development and cognitive development and of the cerebellum and prefrontal cortex. *Child Development, 71,* 44–56.

Lindenberger, U., Li, S.-C., & Bäckman, L. (2006). Delineating brain-behavior mappings across the lifespan: Substantive and methodological advances in developmental neuroscience. *Neuroscience & Biobehavioral Reviews, 30,* 713–717.

Lindenberger, U., Marsiske, M., & Baltes, P. B. (2000). Memorizing while walking: Increase in dual-task costs from young adulthood to old age. *Psychology and Aging, 15,* 417–436.

Molenaar, P. C. M. (2012). Stagewise development, behavior genetics, brain imaging, and a "Aha Erlebnis." *International Journal of Development Science, 6,* 45–49.

Nesselroade, J. R. (1991). The warp and the woof of the developmental fabric. In R. M. Downs, L. S. Liben, & D. S. Palermo (Eds.), *Visions of aesthetics, the environment and development: The legacy of Joachim F. Wohlwill* (pp. 213–240). Hillsdale, NJ: L. Erlbaum.

Wiesel, T. N., & Hubel, D. H. (1965). Extent of recovery from the effects of visual deprivation in kittens. *Journal of Neurophysiology, 28,* 1060–1072.

1 Early Moral Cognition: A Principle-Based Approach

MELODY BUYUKOZER DAWKINS, FRANSISCA TING, MAAYAN STAVANS, AND RENÉE BAILLARGEON

ABSTRACT There is considerable evidence that beginning early in life, abstract principles guide infants' reasoning about the displacements and interactions of objects (physical reasoning) and about the intentional actions of agents (psychological reasoning). Recently, developmental researchers have begun to explore whether early emerging principles also guide infants' reasoning about individuals' actions toward others (sociomoral reasoning). Investigations over the past few years suggest that at least four principles may guide early sociomoral reasoning: fairness, harm avoidance, in-group support, and authority. In this chapter, we review some of the evidence for these principles. In particular, we report findings that infants expect individuals to distribute windfall resources and rewards fairly; they expect individuals in a social group to help in-group members in need, to limit unprovoked and retaliatory harm toward in-group members, to prefer and align with in-group members, and to favor in-group members when distributing limited resources; and they expect an authority figure in a group to rectify transgressions among subordinate members of the group. Together, these findings support prior claims by a broad cross-section of social scientists that a small set of universal principles shapes the basic foundation of human moral cognition, a foundation that is then extensively revised by experience and culture.

Beginning in the first year of life, infants attempt to make sense of the world around them. How do they do so? A major hypothesis in developmental research has long been that in each core domain of causal reasoning, a skeletal framework of abstract principles and concepts guides how infants reason about events (Gelman, 1990; Leslie, 1995; Spelke, 1994). Initial investigations focused on infants' *physical reasoning* and found that principles of gravity, inertia, and persistence (with its corollaries of solidity, continuity, cohesion, boundedness, and unchangeableness) constrain early reasoning about objects' displacements and interactions (Baillargeon, 2008; Luo, Kaufman, & Baillargeon, 2009; Spelke, Phillips, & Woodward, 1995). Thus, even young infants realize that an inert object cannot remain suspended when released in midair (gravity); cannot spontaneously reverse course (inertia); cannot occupy the same space as another object (solidity); and cannot spontaneously disappear (continuity), break apart (cohesion), fuse with another object (boundedness), or change into a different object (unchangeableness).

Next, researchers turned to infants' *psychological reasoning* (also referred to as mental-state reasoning or theory of mind). Investigations revealed that when infants observe an agent act in a scene, they attempt to infer the agent's mental states; these can include motivational states (e.g., intentions), epistemic states (e.g., ignorance), and counterfactual states (e.g., false beliefs) (Gergely, Nádasdy, Csibra, & Bíró, 1995; Luo & Baillargeon, 2007; Onishi & Baillargeon, 2005). Infants then use these mental states, together with a principle of rationality (and its corollaries of consistency and efficiency), to predict and interpret the agent's subsequent actions (Baillargeon, Scott, & Bian, 2016; Gergely et al., 1995; Woodward, 1998). Thus, if an agent wants a toy and sees someone place it in one of two containers, infants expect the agent to reach for the correct container (consistency) and to retrieve the toy without expending unnecessary effort (efficiency).

More recently, researchers have begun to study infants' *sociomoral reasoning*. Initially, it appeared as though the skeletal framework in this domain, unlike those in the previous two domains, might involve no principles. In particular, infants seemed to hold no expectations about whether individuals would refrain from harming others or would help others in need of assistance. In a series of experiments, infants ages 3–19 months were presented with various scenarios depicting interactions among nonhuman individuals (e.g., different blocks with eyes; Hamlin, 2013, 2014; Hamlin & Wynn, 2011; Hamlin, Wynn, & Bloom, 2007, 2010; Hamlin, Wynn, Bloom, & Mahajan, 2011). Each scenario involved two events: a *positive* event, in which a nice character acted positively toward a protagonist (e.g., rolled a dropped ball back to the protagonist or helped the protagonist reach the top of a steep hill), and a *negative* event, in which a mean character acted negatively toward the same protagonist (e.g., stole the ball or knocked the protagonist down to the bottom of the hill). Across ages and scenarios, infants looked equally at the two events, suggesting that they detected no violations in the negative events and hence did not expect the mean character to either refrain from harming the protagonist or help it achieve its goal. These results did not stem from infants' inability to understand the scenarios presented: when encouraged to choose one of the two characters, infants 3–10 months old consistently preferred the nice one over the mean one (Hamlin et al., 2007, 2010; Hamlin & Wynn, 2011). Together, these results suggested that infants possess abstract concepts of welfare and harm, distinguish between positive and negative actions, and hold affiliative attitudes consistent with these valences. Nevertheless, infants seemed to lack principle-based expectations about individuals' actions toward others, suggesting that the skeletal framework for sociomoral reasoning included moral concepts but not moral principles (e.g., infants held no expectations as to whether the characters would harm or help the protagonist, but they did recognize harm or help when they saw it).

This characterization of early morality began to change, however, as researchers went on to explore other scenarios. It is now becoming clear that the skeletal framework that guides early sociomoral reasoning does include a small set of principles. However, because most of these principles apply only when specific preconditions are met, expectations related to the principles can be observed only with scenarios that satisfy these preconditions. For example, if infants view helping as expected only among in-group members, then they will expect an individual to aid another only when the two are clearly identified as members of the same social group.

Over the past few years, evidence has slowly been accumulating for at least four sociomoral principles (Baillargeon et al., 2015). The most general is *fairness*, which applies broadly to all individuals: all other things being equal, individuals are expected to treat others fairly, according to their just deserts. At the next level of generality is *harm avoidance*: when individuals belong to the same moral circle (e.g., humans), they are expected not to cause significant harm to each other. At the next level of generality is *in-group support*: when individuals in a moral circle belong to the same social group (e.g., teammates), additional expectations of in-group care and in-group loyalty are brought to bear. Finally, at the fourth and most specific level is *authority*: when individuals in a social group are identified as authority figures or subordinates, further expectations related to these group roles come into play (e.g., rectifying transgressions for the authority figures or obeying directives for the subordinates). Thus, each new structure in the social landscape—moral circle, social group, group roles—brings forth new expectations about how individuals will act toward others.

This emerging characterization of early morality supports long-standing claims, by a broad cross-section of social scientists, that the basic structure of human moral cognition includes a small set of universal foundations or principles (Baumard, André, & Sperber, 2013; Brewer, 1999; Cosmides & Tooby, 2013; Dawes et al., 2007; Dupoux & Jacob, 2007; Graham et al., 2013; Jackendoff, 2007; Pinker, 2002; Rai & Fiske, 2011; Shweder, Much, Mahapatra, & Park, 1997; Tyler & Lind, 1992; Van Vugt, 2006). Although details about the nature and contents of these principles vary across accounts, it is commonly assumed that the principles evolved during the millions of years our ancestors lived in small groups of hunter-gatherers, where survival depended on cooperation within groups and, to a lesser extent, between groups; that the principles interact in various ways and must be rank ordered when they suggest distinct courses of action; and that different cultures implement, stress, and rank order the principles differently, resulting in the diverse moral landscape that exists in the world today. Graham et al. (2013) aptly described this view as "a theory about the universal first draft of the moral mind and about how that draft gets revised in variable ways across cultures" (p. 65).

In the remainder of this chapter, we review some of the recent evidence that principles of fairness, harm avoidance, in-group support, and authority are included in the "first draft" of moral cognition.

Fairness

According to the principle of fairness, all other things being equal, individuals are expected to treat others fairly when allocating windfall resources, dispensing rewards, or meting out punishments (Baillargeon et al., 2015; Dawes et al., 2007; Graham et al., 2013; Rai & Fiske, 2011). Traditionally, investigations of fairness in preschoolers have used *first-party* tasks, in which the children tested are potential recipients, and *third-party* tasks, in which they are not. Perhaps not surprisingly, given young children's pervasive difficulty in curbing their self-interest, a concern for fairness has typically been observed only in third-party tasks (Baumard, Mascaro, & Chevallier, 2012; Olson & Spelke, 2008). Building on these results, investigations with infants have also used third-party tasks to examine early expectations about fairness.

Equality Do infants expect a distributor to divide windfall resources equally between similar recipients?

In a series of experiments (Buyukozer Dawkins, Sloane, & Baillargeon, 2019; Sloane, Baillargeon, & Premack, 2012), 4-, 9-, and 19-month-olds were tested using the violation-of-expectation method (this method takes advantage of infants' natural tendency to look longer at events that violate, as opposed to confirm, their expectations). Infants faced a puppet-stage apparatus and saw live events in which an experimenter brought in two identical items (e.g., two cookies) and divided them between two identical animated puppets (e.g., two penguins). In one event, the experimenter gave one item to each puppet (*equal* event); in the other, she gave both items to the same puppet (*unequal* event; figure 1.1*A*). At all ages, infants looked significantly longer if shown the unequal as opposed to the equal event, and this effect was eliminated if the puppets were inanimate (i.e., neither moved nor spoke).

Consistent with the claim that fairness applies broadly, positive results were also obtained when a monkey puppet divided items between two giraffe puppets (Bian, Sloane, & Baillargeon, 2018) and when an orange circle with eyes divided items between two yellow triangles with eyes (Meristo, Strid, & Surian, 2016). At the same time, however, other findings revealed that when the number of items allocated was increased to four, infants under 12 months of age failed to detect a violation when one recipient was given three items and the other recipient was given one item (Schmidt & Sommerville, 2011; Ziv & Sommerville, 2017). Thus, while a concern for fairness emerges early in life, there are initially sharp limits to the fairness violations young infants can detect, for reasons that are currently being explored.

Equity The preceding findings demonstrate that even young infants possess an expectation of fairness. But how should this expectation be construed? Do infants possess a simple concept of *equality* and expect all individuals to be treated similarly, or do they possess a richer notion of *equity* and expect individuals to receive their just deserts? One way to examine this issue is to present infants with scenarios in which treating individuals the same way would violate fairness. For example, would infants expect a worker, but not a slacker, to receive a reward? To find out, 21-month-olds were shown events in which an experimenter asked two assistants to put away a pile of toys and then left; next to each assistant was a clear lidded box (Sloane et al., 2012). In the *both-help* event, each assistant placed about

FIGURE 1.1 Infants detect a fairness violation when (*A*) an experimenter fails to divide windfall resources equally between two similar recipients or (*B*) fails to dispense rewards equitably between a worker, who put away toys as instructed, and a slacker, who did no work.

half of the toys in her box and then closed it. The experimenter then returned, inspected both boxes, and rewarded each assistant with a sticker. The *one-helps* event was similar except that one assistant put away all the toys in her box while the other assistant continued to play. Nevertheless, as before, the experimenter gave each assistant a reward (figure 1.1*B*). Infants looked significantly longer if shown the one-helps as opposed to the both-help event. This effect was eliminated if the boxes were opaque so the experimenter could no longer determine who had worked in her absence.

Additional experiments indicated that 10-month-olds detected a violation when an experimenter praised two assistants equally even though she could see that only one had performed the assigned task (Buyukozer Dawkins, Sloane, & Baillargeon, 2017); 21-month-olds detected a violation when an experimenter punished two assistants equally even though she could see that only one had not performed the assigned task (Buyukozer Dawkins et al., 2017); and 17-month-olds detected a violation when two workers shared a resource in a manner inconsistent with their respective efforts in obtaining this resource (Wang & Henderson, 2018).

Together, the preceding results suggest that infants' concern for fairness is equity-based: infants expect individuals to get their just deserts, be it an equal share of a windfall resource, a reward commensurate with their efforts, or a punishment that befits their actions.

In-Group Support

According to the principle of in-group support, members of a social group are expected to act in ways that sustain the group (Baillargeon et al., 2015; Brewer, 1999; Graham et al., 2013; Rai & Fiske, 2011; Shweder et al., 1997). The principle has two corollaries, *in-group care* and *in-group loyalty*, each of which carries a rich set of expectations. With respect to in-group care, for example, one is expected (a) to provide help and comfort to in-group members in need and (b) to limit harm to in-group members by refraining from unprovoked harm and by curbing retaliation. Similarly, with respect to in-group loyalty, one is expected (c) to prefer in-group members over out-group members and (d) to reserve limited resources for the in-group. Below, we report evidence that infants already hold these expectations.

Helping the in-group Do infants view helping as expected with an in-group individual but as optional otherwise? In one experiment, 17-month-olds watched events involving three female experimenters, E1–E3, who sat around three sides of an apparatus and announced their group memberships via novel labels (Jin & Baillargeon, 2017). In the *in-group* condition, E1 (on the right) and E2 (in back) belonged to the same group (e.g., "I'm a bem!"; "I'm a bem too!"), while E3 (on the left) belonged to a different group ("I'm a tig!"). In the *out-group* condition, E2 belonged to the same group as E3 instead of E1 (E1: "I'm a bem!"; E2: "I'm a tig!"; E3: "I'm a tig too!"). Finally, in the *no-group* condition, the Es used phrases that provided incidental information about objects they had seen, rather than inherent information about their social groups (E1: "I saw a bem!"; E2: "I saw a bem too!"; E3: "I saw a tig!"). In the test trial, E3 was absent (her main role was to help establish group affiliations), and while E2 watched, E1 selected discs of decreasing sizes from a clear box and stacked them on a base. The final, smallest disc rested across the apparatus from E1, out of her reach (but within E2's reach). E1 tried in vain to reach the disc until a bell rang; at that point, E1 said, "Oh, I have to go. I'll be back!" and left. E2 then picked up the smallest disc, inspected it, and either placed it in E1's box so that she could complete her stack when she returned (*help* event) or returned it to its same position on the apparatus floor, out of E1's reach (*ignore* event; figure 1.2*A*).

Infants in the in-group condition looked significantly longer if shown the ignore as opposed to the help event, whereas infants in the out-group and no-group conditions looked equally at the events. Thus, in accordance with the principle of in-group care, infants detected a violation when E2 chose not to help in-group E1. In additional experiments (Jin, Houston, Baillargeon, Groh, & Roisman, 2018), 4-, 8-, and 12-month-olds were shown videotaped events in which a woman was performing a household chore when a baby (who presumably belonged to the same group as the woman) began to cry. The woman either attempted to comfort the baby (*comfort* event) or ignored the baby and continued her work (*ignore* event). At all ages, infants detected a violation in the ignore event, and this effect was eliminated if the baby laughed instead.

Limiting harm toward the in-group If infants' sense of in-group care modulates their expectations about harm avoidance, they might expect individuals to direct *less* unprovoked and retaliatory harm at in-group members than at out-group members. To examine these predictions, 18-month-olds were first tested in a baseline *out-group* experiment (Ting, He, & Baillargeon, 2019a). Three female experimenters, E1–E3, sat around three sides of an apparatus, and their group memberships were marked by salient outfits: E1 (on the right) wore one outfit, while E2 (in back) and E3 (on the left) wore a different outfit. While E2 and E3 watched, E1 used small blocks to build two towers of four blocks. In the next trial, E3 was absent and E2 ate crackers from a

FIGURE 1.2 Infants detect an in-group–support violation when (*A*) an individual fails to help an in-group member in need of assistance, (*B*) fails to curb retaliation against an in-group member who stole and ate a cracker, (*C*) fails to prefer an in-group member over an out-group member, and (*D*) fails to favor the in-group when distributing limited resources.

small plate in front of her while watching E1 build a third tower. After completing this tower, E1 either simply left the scene (*no-provocation* condition) or first stole a cracker from E2 and then left the scene (*provocation* condition). In both conditions, E2 then knocked down one block from one tower (*one-block* event), one tower (*one-tower* event), or two towers (*two-tower* event). In the no-provocation condition, infants looked significantly longer if shown the one- or two-tower event as opposed to the one-block event; in the provocation event, in contrast, infants looked significantly longer if shown the one-block or one-tower event as opposed to the two-tower event. Thus, when no provocation had occurred, infants detected a violation in all but the one-block event: mild unprovoked harm to out-group E1 was acceptable, but more significant harm was not. Following provocation, however, infants detected a violation in all but the two-tower event, suggesting that they viewed knocking down at least two of out-group E1's towers as an appropriate retaliatory response for her theft of one cracker (perhaps in a sort of "two-for-one" accounting).

Would infants show similar expectations if E1 and E2 belonged to the same group, or would considerations of in-group care modulate these expectations, leading infants to expect both less unprovoked harm and less retaliatory harm? To find out, infants were tested in an *in-group* experiment identical to that above except that E2 wore the same outfit as E1 and hence belonged to the same group. Across conditions, infants now detected a violation in all but the one-block event of the provocation condition. Thus, when no provocation had occurred, infants expected E2 to refrain from knocking down *any* of in-group E1's blocks; following provocation, knocking down one block became permissible in retaliation for in-group E1's theft—but no more than one block and certainly not two towers, as in the out-group experiment (figure 1.2*B*).

Together, the preceding results make clear that from an early age, considerations of in-group care modulate expectations about harm avoidance: infants expect stricter limits on unprovoked and retaliatory harm when directed at in-group members. In line with these results, recent experiments have found that infants also expect individuals to *punish* harm to in-group members, at least indirectly, through the withholding of help (Ting, He, & Baillargeon, 2019b). When a bystander saw a wrongdoer harm a victim, and the wrongdoer subsequently needed help to complete a task, 13-month-olds expected the bystander to refrain from providing help if the victim was an in-group member, but not if she was an out-group member. Infants' concern for in-group care thus leads them to expect individuals both to limit harm to in-group members and to punish such harm, at least indirectly, when perpetrated by others.

Preferring the in-group Do infants expect individuals in a group to prefer in-group members over out-group members, in accordance with the principle of in-group loyalty? In one experiment (Bian & Baillargeon, 2016), 12-month-olds again saw events involving three female experimenters, E1–E3, whose group memberships were marked by salient outfits. In one familiarization trial, E2 (in back) sat alone; she picked up two-dimensional toys on the apparatus floor and placed them in a box near her, thus giving infants the opportunity to observe her outfit. In the next familiarization trial, E2 was absent and E1 (on the right) and E3 (on the left) read identical books; one E wore the same outfit as E2, and the other E wore a different outfit. In the test trial, E1 and E3 were joined by E2, who approached either the E from the same group (*approach-same* event) or the E from the other group (*approach-different* event; figure 1.2*C*) to read along. Infants looked significantly longer at the approach-different than at the approach-same event, suggesting that they expected E1 to approach her in-group member, in accordance with in-group loyalty, and detected a violation when she did not. This effect was eliminated when the first familiarization trial was modified to reveal that E2's outfit served an instrumental role: she now placed the toys she picked up in the large kangaroo pocket on her shirt, instead of in the box near her. Infants looked equally at the approach-different and approach-same events, suggesting that they no longer viewed the Es' outfits as providing information about their group memberships (in the same way, adults would not view pedestrians holding black umbrellas in the rain on a busy street, or travelers pulling black suitcases in a busy airport, as belonging to the same groups).

Similar results have been obtained in tasks using other cues to group memberships. After watching nonhuman adult characters soothe baby characters, 16-month-olds detected a violation if one baby preferred a baby who had been soothed by a different adult (and hence presumably belonged to a different group) over a baby who had been soothed by the same adult (and hence presumably belonged to the same group) (Spokes & Spelke, 2017). After watching two groups of nonhuman characters (identified by both physical and behavioral cues) perform distinct novel conventional actions, infants 7–12 months old detected a violation if a member of one group chose to imitate the other group's conventional action (Powell & Spelke, 2013). Finally, when faced with a native speaker of their language and a foreign speaker, infants 10–14 months old were more likely to prefer the native speaker (Kinzler, Dupoux, & Spelke, 2007), to select snacks endorsed by the native speaker (Shutts, Kinzler, McKee, & Spelke, 2009), and to imitate novel conventional actions modeled by the native speaker (Buttelmann, Zmyj, Daum, & Carpenter, 2013). One interpretation of these last results is that in this minimal setting contrasting two unfamiliar individuals, language served as a natural group marker, leading infants to prefer and align with the native speaker, in accordance with in-group loyalty.

Favoring the in-group when resources are limited If infants' sense of in-group loyalty modulates their expectations about fairness, they might expect a distributor to favor in-group over out-group recipients, particularly when resources are scarce or otherwise valuable. To examine this prediction, 19-month-olds saw resource-allocation events involving two groups of animated puppets, monkeys and giraffes (Bian et al., 2018). A puppet distributor (e.g., a monkey) brought in either three (*three-item* condition) or two (*two-item* condition) items and faced two potential recipients, an in-group puppet (another monkey) and an out-group puppet (a giraffe). In each condition, the distributor allocated two items: she gave one item each to the in-group and out-group puppets (*equal* event; figure 1.2*D*), she gave both items to the in-group puppet (*favors-in-group* event), or she gave both items to the out-group puppet (*favors-out-group* event). In the three-item condition, the third item was not distributed and was simply taken away by the distributor when she left. Infants in the three-item condition looked significantly longer if shown the favors-in-group or favors-out-group event than if shown the equal event, suggesting that when there were as many items as puppets, infants expected fairness to prevail: they detected a violation if the distributor chose to give two items to one recipient and none to the other, regardless of which recipient was advantaged. In contrast, infants in the two-item condition looked significantly longer if shown the equal or favors-out-group event rather than the favors-in-group event, suggesting that when the distributor had only enough items for the group to which she belonged (e.g., two items and two monkeys), infants expected in-group loyalty to prevail: they detected a violation if the distributor gave any of the items to the out-group puppet.

Together, these results suggest two conclusions. First, the "first draft" of moral cognition includes not only principles of fairness and in-group support but also a context-sensitive ordering of these principles that befits their contents: one is expected to adhere to fairness except in contexts where doing so would be detrimental to one's group. Second, a shortage of resources is one such context: when there is not enough to go around, the group must come first.

Authority

According to the principle of authority, when a social group accepts an individual in the group as a legitimate leader, rich expectations come into play that reflect this power asymmetry (Baillargeon et al., 2015; Graham et al., 2013; Rai & Fiske, 2011; Tyler & Lind, 1992; Van Vugt, 2006). On the one hand, the leader is expected to maintain order, provide protection, and facilitate cooperation toward group goals. On the other hand, the subordinates are expected to obey, respect, and defer to the leader. Do infants already possess authority-based expectations about the behaviors of leaders toward their subordinates or about the behaviors of subordinates toward their leaders?

Before addressing this question, developmental researchers first had to determine whether infants could represent power asymmetries. Over the past decade, evidence has steadily accumulated that by the second year of life, infants (a) can detect differences in social power (Pun, Birch, & Baron, 2016; Thomsen, Frankenhuis, Ingold-Smith, & Carey, 2011), (b) expect such differences to both endure over time and extend across situations (Enright, Gweon, & Sommerville, 2017; Mascaro & Csibra, 2012), and (c) distinguish between powerful individuals with respect-based as opposed to fear-based power (Margoni, Baillargeon, & Surian, 2018). Building on these results, recent experiments examined whether infants might also hold expectations about one specific type of respect-based power, the legitimate power of an authority figure (Stavans & Baillargeon, 2019). Specifically, these experiments asked whether infants would expect a powerful individual in a group to rectify a transgression perpetrated by one subordinate against another. The rationale was that positive results would suggest that infants cast the powerful individual in the role of legitimate leader and hence expected this leader to restore order in the group, in accordance with the principle of authority.

In these experiments, 17-month-olds watched live interactions among a group of three bear puppets (Stavans & Baillargeon, 2019). One puppet (at the back of the apparatus) served as the leader, and the other two puppets (on the left and right sides) served as the subordinates; in front of each subordinate was a place mat. In different scenarios, the leader was identified either by its larger size (physical cue) or by the subordinates' compliance with its instructions (behavioral cue); results

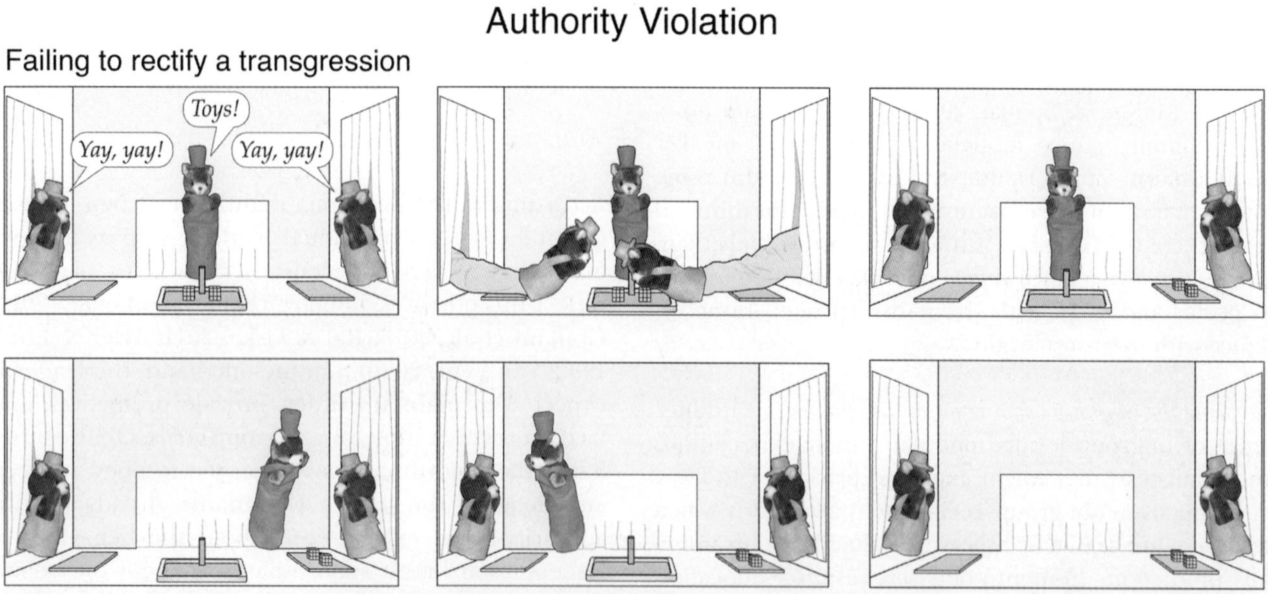

FIGURE 1.3 Infants detect an authority violation when a leader (*here marked by its larger size*) in a group fails to rectify a transgression between subordinate members of the group.

were identical across scenarios, so the size-based scenario is used here. To start, the leader brought in a tray with two identical toys for the subordinates to share. However, one subordinate (the perpetrator) quickly grabbed both toys and deposited them on its place mat so that the other subordinate (the victim) did not get a toy. In one event, the leader rectified this transgression by taking one of the toys away from the perpetrator and giving it to the victim (*rectify* event). In the other event, the leader again approached each subordinate in turn but did nothing to correct the transgression (*ignore* event; figure 1.3). Infants looked significantly longer if shown the ignore as opposed to the rectify event. This effect was eliminated if the leader was replaced by another member of the group who gave no evidence of being a leader (e.g., another bear of the same size as the two subordinates).

Together, these results suggest that when infants identify an individual as a legitimate leader in a group, they expect this leader to restore order if one subordinate transgresses against another, in accordance with the authority principle.

Conclusions

The evidence reviewed in this chapter suggests that from a very young age a skeletal framework of abstract principles guides infants' sociomoral reasoning. These principles include fairness, harm avoidance, in-group support (with its corollaries of in-group care and in-group loyalty), and authority. Although considerable research is needed to fully understand the "first draft" of human moral cognition and how experience and culture revises it (Graham et al., 2013), available findings indicate that this "first draft" makes possible surprisingly sophisticated moral expectations, evaluations, and attitudes.

Acknowledgments

This chapter was supported by a Graduate Fellowship from the National Science Foundation to Melody Buyukozer Dawkins, a Fulbright Postdoctoral Fellowship to Maayan Stavans, and a grant from the John Templeton Foundation to Renée Baillargeon.

REFERENCES

Baillargeon, R. (2008). Innate ideas revisited: For a principle of persistence in infants' physical reasoning. *Perspectives on Psychological Science, 3*, 2–13.

Baillargeon, R., Scott, R. M., & Bian, L. (2016). Psychological reasoning in infancy. *Annual Review of Psychology, 67*, 159–186.

Baillargeon, R., Scott, R. M., He, Z., Sloane, S., Setoh, P., Jin, K., & Bian, L. (2015). Psychological and sociomoral reasoning in infancy. In M. Mikulincer & P. R. Shaver (Eds.), E. Borgida & J. A. Bargh (Assoc. Eds.), *APA handbook of personality and social psychology: Vol. 1. Attitudes and social cognition* (pp. 79–150). Washington, DC: American Psychological Association.

Baumard, N., André, J. B., & Sperber, D. (2013). A mutualistic approach to morality: The evolution of fairness by partner choice. *Behavioral and Brain Sciences, 36*, 59–78.

Baumard, N., Mascaro, O., & Chevallier, C. (2012). Preschoolers are able to take merit into account when distributing goods. *Developmental Psychology, 48*, 492–498.

Bian, L., & Baillargeon, R. (2016, May). *Toddlers and infants expect individuals from novel social groups to prefer and align with ingroup members.* Poster presented at the International Conference on Infant Studies, New Orleans, LA.

Bian, L., Sloane, S., & Baillargeon, R. (2018). Infants expect ingroup support to override fairness when resources are limited. *Proceedings of the National Academy of Sciences, 115*(11), 2705–2710.

Brewer, M. B. (1999). The psychology of prejudice: Ingroup love or outgroup hate? *Journal of Social Issues, 55,* 429–444.

Buttelmann, D., Zmyj, N., Daum, M., & Carpenter, M. (2013). Selective imitation of in-group over out-group members in 14-month-old infants. *Child Development, 84,* 422–428.

Buyukozer Dawkins, M., Sloane, S., & Baillargeon, R. (2017, August). *Evidence for an equity-based sense of fairness in infancy.* Poster presented at the Dartmouth Workshop on Action Understanding, Hanover, NH.

Buyukozer Dawkins, M., Sloane, S., & Baillargeon, R. (2019). Do infants in the first year of life expect equal resource allocations? In J. Sommerville. K. Lucca, & J. K. Hamlin (Eds.), *Frontiers in Psychology, 10,* article 116 (special issue on "Early Moral Cognition and Behavior").

Cosmides, L., & Tooby, J. (2013). Evolutionary psychology: New perspectives on cognition and motivation. *Annual Review of Psychology, 64,* 201–229.

Dawes, C. T., Fowler, J. H., Johnson, T., McElreath, R., & Smirnov, O. (2007). Egalitarian motives in humans. *Nature, 466,* 794–796.

Dupoux, E., & Jacob, P. (2007). Universal moral grammar: A critical appraisal. *Trends in Cognitive Sciences, 9,* 373–378.

Enright, E. A., Gweon, H., & Sommerville, J. A. (2017). 'To the victor go the spoils': Infants expect resources to align with dominance structures. *Cognition, 164,* 8–21.

Gelman, R. (1990). First principles organize attention to and learning about relevant data: Number and the animate-inanimate distinction as examples. *Cognitive Science, 14,* 79–106.

Gergely, G., Nádasdy, Z., Csibra, G., & Bíró, S. (1995). Taking the intentional stance at 12 months of age. *Cognition, 56,* 165–193.

Graham, J., Haidt, J., Koleva, S., Motyl, M., Iyer, R., Wojcik, S. P., & Ditto, P. H. (2013). Moral foundations theory: The pragmatic validity of moral pluralism. *Advances in Experimental Social Psychology, 47,* 55–130.

Hamlin, J. K. (2013). Failed attempts to help and harm: Intention versus outcome in preverbal infants' social evaluations. *Cognition, 18,* 451–474.

Hamlin, J. K. (2014). Context-dependent social evaluation in 4.5-month-old human infants: The role of domain-general versus domain-specific processes in the development of social evaluation. *Frontiers in Psychology, 5,* 614.

Hamlin, J. K., & Wynn, K. (2011). Young infants prefer prosocial to antisocial others. *Cognitive Development, 26,* 30–39.

Hamlin, J. K., Wynn, K., & Bloom, P. (2007). Social evaluation by preverbal infants. *Nature, 450,* 557–559.

Hamlin, J. K., Wynn, K., & Bloom, P. (2010). Three-month-olds show a negativity bias in their social evaluations. *Developmental Science, 13,* 923–929.

Hamlin, J. K., Wynn, K., Bloom, P., & Mahajan, N. (2011). How infants and toddlers react to antisocial others. *Proceedings of the National Academy of Sciences, 108,* 19931–19936.

Jackendoff, R. (2007). *Language, consciousness, culture: Essays on mental structure.* Cambridge, MA: MIT Press.

Jin, K., & Baillargeon, R. (2017). Infants possess an abstract expectation of ingroup support. *Proceedings of the National Academy of Sciences, 114,* 8199–8204.

Jin, K., Houston, J. L., Baillargeon, R., Groh, A. M, & Roisman, G. I. (2018). Young infants expect an unfamiliar adult to comfort a crying baby: Evidence from a standard violation-of-expectation task and a novel infant-triggered-video task. *Cognitive Psychology, 102,* 1–20.

Kinzler, K. D., Dupoux, E., & Spelke, E. S. (2007). The native language of social cognition. *Proceedings of the National Academy of Sciences, 104,* 12577–12580.

Leslie, A. M. (1995). A theory of agency. In D. Sperber, D. Premack, & A. J. Premack (Eds.), *Causal cognition: A multidisciplinary debate* (pp. 121–149). Oxford: Clarendon Press.

Luo, Y., & Baillargeon, R. (2007). Do 12.5-month-old infants consider what objects others can see when interpreting their actions? *Cognition, 105,* 489–512.

Luo, Y., Kaufman, L., & Baillargeon, R. (2009). Young infants' reasoning about events involving inert and self-propelled objects. *Cognitive Psychology, 58,* 441–486.

Margoni, F., Baillargeon, R., & Surian, L. (2018). Infants distinguish between leaders and bullies. *Proceedings of the National Academy of Sciences, 115*(38), E8835–E8843.

Mascaro, O., & Csibra, G. (2012). Representation of stable social dominance relations by human infants. *Proceedings of the National Academy of Sciences, 109,* 6862–6867.

Meristo, M., Strid, K., & Surian, L. (2016). Preverbal infants' ability to encode the outcome of distributive actions. *Infancy, 21*(3), 353–372.

Olson, K. R., & Spelke, E. S. (2008). Foundations of cooperation in young children. *Cognition, 108,* 222–231.

Onishi, K. H., & Baillargeon, R. (2005). Do 15-month-old infants understand false beliefs? *Science, 308,* 255–258.

Pinker, S. (2002). *The blank slate: The modern denial of human nature.* New York: Viking.

Powell, L. J., & Spelke, E. S. (2013). Preverbal infants expect members of social groups to act alike. *Proceedings of the National Academy of Sciences, 110,* 3965–3972.

Pun, A., Birch, S. A., & Baron, A. S. (2016). Infants use relative numerical group size to infer social dominance. *Proceedings of the National Academy of Sciences, 113*(9), 2376–2381.

Rai, T. S., & Fiske, A. P. (2011). Moral psychology is relationship regulation: Moral motives for unity, hierarchy, equality, and proportionality. *Psychological Review, 118,* 57–75.

Schmidt, M. F. H., & Sommerville, J. A. (2011). Fairness expectations and altruistic sharing in 15-month-old human infants. *PLoS One, 6,* e23223.

Shutts, K., Kinzler, K. D., McKee, C. B., & Spelke, E. S. (2009). Social information guides infants' selection of foods. *Journal of Cognition and Development, 10,* 1–17.

Shweder, R. A., Much, N. C., Mahapatra, M., & Park, L. (1997). The "big three" of morality (autonomy, community and divinity) and the "big three" explanations of suffering. In A. M. Brandt & P. Rozin (Eds.), *Morality and health* (pp. 119–169). New York: Routledge.

Sloane, S., Baillargeon, R., & Premack, D. (2012). Do infants have a sense of fairness? *Psychological Science, 23,* 196–204.

Spelke, E. S. (1994). Initial knowledge: Six suggestions. *Cognition, 50,* 431–445.

Spelke, E. S., Phillips, A., & Woodward, A. L. (1995). Infants' knowledge of object motion and human action. In D. Sperber, D. Premack, & A. J. Premack (Eds.), *Causal cognition: A multidisciplinary debate* (pp. 44–78). Oxford: Clarendon Press.

Spokes, A. C., & Spelke, E. S. (2017). The cradle of social knowledge: Infants' reasoning about caregiving and affiliation. *Cognition, 159,* 102–116.

Stavans, M., & Baillargeon, R. (2019). Infants expect leaders to right wrongs. Manuscript under review.

Thomsen, L., Frankenhuis, W., Ingold-Smith, M., & Carey, S. (2011). Big and mighty: Preverbal infants mentally represent social dominance. *Science, 331,* 477–480.

Ting, F., He, Z., & Baillargeon, R. (2019a, March). *Group membership modulates early expectations about retaliatory harm.* Paper presented at the Society for Research in Child Development Biennial Meeting, Baltimore, MD.

Ting, F., He, Z., & Baillargeon, R. (2019b). Toddlers and infants expect individuals to refrain from helping an ingroup victim's aggressor. *Proceedings of the National Academy of Sciences, 116,* 6025–6034.

Tyler, T. R., & Lind, A. (1992). A relational model of authority in groups. *Advances in Experimental Social Psychology, 25,* 115–191.

Van Vugt, M. (2006). Evolutionary origins of leadership and followership. *Personality and Social Psychology Review, 10,* 354–371.

Wang, Y., & Henderson, A. M. (2018). Just rewards: 17-month-old infants expect agents to take resources according to the principles of distributive justice. *Journal of Experimental Child Psychology, 172,* 25–40.

Woodward, A. L. (1998). Infants selectively encode the goal object of an actor's reach. *Cognition, 69,* 1–34.

Ziv, T., & Sommerville, J. A. (2017). Developmental differences in infants' fairness expectations from 6 to 15 months of age. *Child Development, 88*(6), 1930–1951.

2 Imaging Structural Brain Development in Childhood and Adolescence

CHRISTIAN K. TAMNES AND KATHRYN L. MILLS

ABSTRACT The human brain undergoes a remarkably protracted development. Magnetic resonance imaging (MRI) has allowed us to capture these changes through longitudinal investigations. In this chapter we describe the typical developmental trajectories of the human brain structure between childhood and early adulthood. We focus on measurements of brain morphometry and measurements derived from diffusion tensor imaging (DTI). By integrating findings from multiple longitudinal investigations with seminal cellular studies, we describe the neurotypical patterns of structural brain development and the possible underlying biological mechanisms. Finally, we highlight several new measures and approaches to examine structural brain development.

Since the 1990s, several longitudinal investigations have examined brain development using MRI. Through these studies we have learned that the human brain undergoes a particularly protracted development, with some aspects of our brain maturing into the third decade of life. This chapter will review the current literature on the development of brain structure as measured through MRI. We will focus on aspects of brain morphometry, as well as tissue microstructure as measured through DTI. We will then discuss the biological mechanisms underlying the developmental changes in brain structure and highlight new imaging and analytic approaches to the study of structural brain development. While there has been a recent concerted effort to understand aspects of brain development during infancy and early childhood using MRI, we will focus primarily on brain development during later childhood and adolescence.

Brain Structural and Microstructural Development

MRI, based on the principles of nuclear magnetic resonance, detects proton signals from water molecules and allows us to produce high-quality images of the internal structure of living organs. MRI protocols designed to create anatomical images of the brain rely on signal intensities and contrasts to distinguish between gray matter, white matter, and cerebrospinal fluid (CSF), while other protocols can create, for example, images to probe tissue microstructural properties (figure 2.1).

Global volumes It is important to note that the cranial cavity itself continues to grow into the second decade of life. An investigation of four longitudinal developmental data sets presented evidence that intracranial volume increases around 1% annually between late childhood and midadolescence, when it begins to stabilize (Mills et al., 2016). In this regard, the growth of intracranial volume resembles the growth trajectories of other physical measures, such as height and bone density, although changes in body growth do not fully account for these changes in intracranial volume. In contrast to the steady increase in intracranial volume into midadolescence, whole-brain volume (the sum of the gray and white matter) reduces in size during adolescence before stabilizing in the early 20s (Mills et al., 2016). When these findings are considered alongside those from a large meta-analysis of longitudinal studies, it appears that whole-brain volume increases until around age 13 and then decreases until some point in the early 20s, after which it remains relatively stable until around age 40, when it begins to decrease again (Hedman, van Haren, Schnack, Kahn, & Hulshoff Pol, 2012). These findings go beyond early assertions that the brain is close to adult volume by childhood, as it is now clear that the overall size of the human brain continues to change across the first two decades of life.

Critically, volumetric growth of the two main subcomponents of the brain, gray matter and white matter, follows distinct developmental trajectories. Gray matter—that is, the cerebral and cerebellar cortex and distinct subcortical structures—is composed of neuronal bodies, glial cells, dendrites, blood vessels,

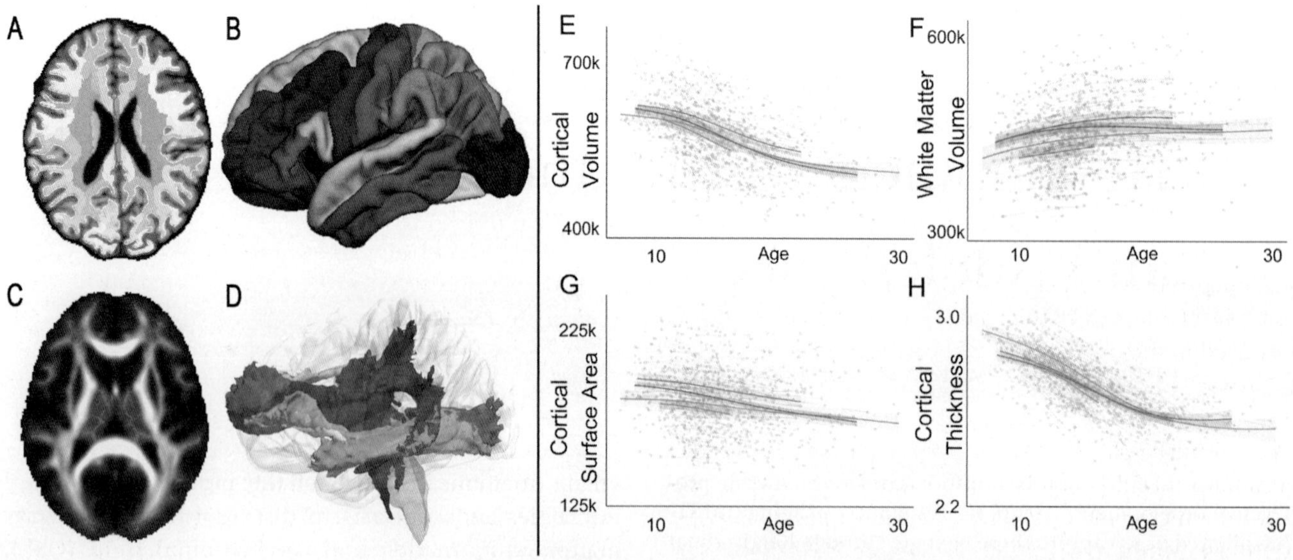

FIGURE 2.1 Illustration of key MRI methods and findings discussed in this review. *A*, Horizontal slice of T1 image showing a whole-brain segmentation used for volumetric analyses. *B*, A left-lateral view of an averaged parcellated cerebral cortex used for surface-based analyses, both from FreeSurfer. *C*, Horizontal slice of Tract-Based Spatial Statistics (TBSS) mean FA white matter skeleton overlaid on a mean FA map. *D*, a left-lateral view of a three-dimensional rendering of probabilistic fiber tracts from the Mori atlas. Developmental trajectories for global cortical measures from four independent samples for cortical volume (*E*), total white matter volume (*F*), cortical surface area (*G*), and mean cortical thickness (*H*). The colored lines represent the generalized additive mixed model (GAMM) fitting while the lighter-colored areas correspond to the 95% confidence intervals. *Note*: *Pink*, Child Psychiatry Branch (CPB); *purple*, Pittsburgh (PIT); *blue*, Neurocognitive Development (NCD); *green*, Braintime (BT). (See color plate 1.)

extracellular space, and both unmyelinated and myelinated axons. Cortical gray matter increases rapidly after birth, approximately doubling in volume in the first year of life (Gilmore et al., 2012). It then reaches its greatest volume in childhood and begins to decrease in late childhood and throughout adolescence before stabilizing in the third decade of life (Lebel & Beaulieu, 2011; Mills et al., 2016). In a study of four longitudinal data sets, we calculated that cortical volume decreases by (on average) 1.4% annually between late childhood and early adulthood, with the sharpest decline in volume occurring in early to midadolescence (Tamnes, Herting, et al., 2017). In contrast, cerebral white matter, which occupies almost half of the human brain and consists largely of organized myelinated axons, continues to increase in volume into at least the second decade of life but begins to decelerate at some point in midadolescence to late adolescence (Lebel & Beaulieu, 2011; Mills et al., 2016). In addition to these tissue-specific patterns, component-specific and regional differences in brain developmental timing and pace have been linked to adolescent-specific changes in behavior.

The cerebral cortex Because the cerebral cortex is a layer of tissue enveloping the cerebrum, it is often measured in terms of *thickness*, *surface area*, or their product, *volume*. Cortical thickness and surface area are influenced by various evolutionary, genetic, and cellular processes and show unique developmental changes (Mills, Lalonde, Clasen, Giedd, & Blakemore, 2014; Tamnes, Herting, et al., 2017; Vijayakumar et al., 2016; Wierenga, Langen, Oranje, & Durston, 2014). While brain size varies substantially in both mature and developing humans, there is less interindividual variability in cortical thickness than in surface area.

Average cortical thickness follows a similar nonlinear, decreasing trajectory as cortical volume (around 1% annually), although the decline in average cortical thickness is more pronounced across the second decade of life, before flattening out around age 20 (Tamnes, Herting, et al., 2017). Cortical thickness begins to decrease much earlier than gray matter volume or cortical surface area, with this process observed as early as 4 years of age (Walhovd, Fjell, Giedd, Dale, & Brown, 2017). In contrast, total cortical surface area increases in early development and begins to decrease in an almost linear fashion (around 0.5% annually) from late childhood to early adulthood (Tamnes, Herting, et al., 2017).

The cerebral cortex does not develop uniformly. Investigations of structural brain development starting in middle childhood have consistently found decelerating change in posterior cortical regions and accelerating change in anterior regions, in line with the posterior-anterior theory of cortical maturation (Yakovlev & Lecours, 1967). For example, the parietal lobes and lateral occipital cortices (involved in sensory processing) show larger volumetric reductions in late childhood and early adolescence, whereas the medial frontal cortex and the anterior temporal cortex pick up the pace in the teen years (Tamnes et al., 2013). The more pronounced changes in brain structure that occur during the second decade of life are likely related to cognitive processes involved in the developmental tasks of this period of life.

Notably, not all cortical regions undergo significant macrostructural changes between late childhood and early adulthood. Studies of several longitudinal data sets have found evidence for little to no change in the central sulcus, medial temporal, and medial occipital cortices (Mutlu et al., 2013; Tamnes et al., 2013). Given that the central sulcus and the medial occipital cortices are involved in primary sensory processes, they likely undergo more rapid change at earlier ages. Certain cortical regions also show a relatively greater surface area expansion between childhood and young adulthood (Hill et al., 2010). Between ages 4 to 20, this includes the lateral and medial temporal, cingulate, lateral orbitofrontal, superior and inferior frontal, insular, temporoparietal, cuneus, and lingual cortices (Fjell et al., 2015).

Cortical topography The human cortex is highly convoluted, with approximately one-third of the cortical surface exposed on gyri and two-thirds buried within sulci. The gyrification index of the whole brain is defined as the ratio of the total folded cortical surface over the total perimeter of the brain (Zilles, Armstrong, Schleicher, & Kretschmann, 1988), whereas the local gyrification index measures the degree of cortical folding at specific points of the cortical surface (Schaer et al., 2008). The gyrification index of the human brain decreases between childhood and young adulthood, whereas the amount of exposed cortical surface increases from childhood to late adolescence (Alemán-Gómez et al., 2013; Raznahan, Shaw, et al., 2011). One longitudinal study demonstrated that the cortex "flattens" during adolescence, mostly due to decreases in sulcal depth and increases in sulcal width (Alemán-Gómez et al., 2013). The developmental changes in local gyrification vary across the cortex, with regions in the medial prefrontal cortex, occipital cortex, and temporal cortex undergoing little to no change between ages 6 to 30 (Mutlu et al., 2013). However, similar to what has been found in whole-brain (Raznahan, Shaw, et al., 2011) and lobar-level (Alemán-Gómez et al., 2013) analyses, Mutlu et al. (2013) observed linear decreases in the local gyrification index across the majority of the cortex.

Subcortical structures Several subcortical structures and cortical infolds show substantial structural change between childhood and young adulthood, although generally at a lower rate than observed in the cortex (Tamnes et al., 2013). Longitudinal studies have found that the thalamus, pallidum, amygdala, caudate, putamen, and nucleus accumbens all show significant changes in volume across the second decade of life (Goddings et al., 2014; Herting et al., 2018; Wierenga, Langen, Ambrosino, et al., 2014). The caudate, putamen, and nucleus accumbens undergo linear reductions during this time, whereas the amygdala, thalamus, and pallidum follow nonlinear increases. These findings contradict with hypotheses and developmental models that assume that subcortical structures are mature by adolescence, as it is now clear that subcortical regions undergo structural development throughout the second decade of life.

White matter microstructure Diffusion MRI (dMRI) has over the last two decades grown in popularity as a method to study brain development, particularly that of white matter. dMRI uses the phenomenon of naturally moving water molecules in the brain to indirectly obtain information about the underlying tissue microstructure (Le Bihan & Johansen-Berg, 2012). This is possible since water diffusion in biological tissue is not free and uniform (isotropic) but reflects interactions with obstacles, such as cell membranes and myelin, and is therefore not necessarily the same in all directions (anisotropic). The diffusion patterns can reveal details about tissue architecture at a micrometer scale well beyond the usual millimetric resolution of morphometric MRI.

Typical quantification of dMRI is achieved in a tensor model, and this is referred to as DTI. Several indices can be derived; fractional anisotropy (FA) is used as a measure of the directionality of diffusion, mean diffusivity (MD) reflects the overall magnitude of diffusion, and axial diffusivity (AD) and radial diffusivity (RD) are diffusivity along and across the longest axis of the diffusion tensor, respectively. These indices can be analyzed on a voxelwise basis or in regions of interest. Tractography techniques can be used to reconstruct long-range connections, yielding possibilities for inferring patterns of structural connectivity (Jbabdi & Behrens, 2013). However, current techniques also have known limitations (see, e.g., Maier-Hein et al., 2017).

The major white matter fiber pathways in the brain are present and identifiable at birth, but very rapid changes in DTI indices are seen across infancy (Qiu, Mori, & Miller, 2015). For example, a large longitudinal study of young children indicated that during the first two years of life, FA in ten major tracts increases by 16%–55%, RD decreases by 24%–46%, and AD decreases by 13%–28%, with faster changes in the first year than in the second for all tracts investigated (Geng et al., 2012). Such massive changes are perhaps not surprising given the enormous behavioral and psychological developments in this period of life.

As for later childhood and adolescence, many cross-sectional, and an increasing number of longitudinal, DTI studies document consistent patterns of continued development in white matter microstructure. With increasing age, FA increases, while MD and RD decrease, in most white matter regions, but the results for AD are less consistent (Tamnes, Roalf, Goddings, & Lebel, 2018). For example, Krogsrud et al. (2016) focused on the preschool and early school years and found that for most white matter regions, FA showed a linear increase over time, while MD and RD showed a linear decrease. Lebel and Beaulieu (2011) studied a much broader age range, 5–32 years, and used tractography for ten major white matter tracts. Almost all showed nonlinear developmental trajectories, with decelerating increases for FA and decelerating decreases for MD, primarily due to decreasing RD (see also Simmonds, Hallquist, Asato, & Luna, 2014).

The timing and rates of the DTI developmental changes vary regionally in the brain. A pattern of maturation in which major tracts with frontotemporal connections develop more slowly than other tracts has emerged (Lebel, Walker, Leemans, Phillips, & Beaulieu, 2008). Lebel and Beaulieu (2011) also found a pattern in which changes in DTI parameters were mostly complete by late adolescence for projection and commissural tracts, while postadolescent development was indicated for both FA and MD in association tracts. Of the major fiber bundles, the cingulum, implicated in, for example, cognitive control, and the uncinate fasciculus, implicated in emotion and episodic memory, are among those shown to have particularly prolonged development (Lebel et al., 2012; Olson, Heide, Alm, & Vyas, 2015).

Relating Structural Brain Development to Biological Development

While most studies assess human brain development in relation to chronological age, other developmental processes, such as body growth and puberty, occur during the first two decades of life that likely have an impact on brain development. For several reasons, age might not be the most appropriate measure against which to judge brain development. For one, age only explains a certain proportion of the variance in modeled trajectories. Further, age provides little information about the possible cellular and molecular mechanisms underlying observed changes. During late childhood and adolescence, individuals undergo physical changes such as a growth spurt in height and puberty, which happen at different ages across individuals, and, on average, at different times for girls and boys.

Sex differences Although males, on average, show larger global and regional brain volumes than females and sex differences have been reported for many other imaging measures (Ruigrok et al., 2014), the findings are much less clear for sex differences in developmental changes and trajectories across childhood and adolescence (Herting et al., 2018; Mutlu et al., 2013; Vijayakumar et al., 2016). It has also proven difficult to clearly describe how puberty and related hormonal changes affect brain structural and microstructural development, and there are few longitudinal studies (Herting & Sowell, 2017). However, one large longitudinal study found that age and pubertal development had both independent and interactive influences on volume for the amygdala, hippocampus, and putamen in both sexes and the caudate in females (Goddings et al., 2014). The relatively subtle (or inconclusive) evidence for mean sex differences in brain development might suggest that we need to move beyond mean level differences. Robust sex differences in the variability in brain measures have recently been shown in both developmental (Wierenga, Sexton, Laake, Giedd, & Tamnes, 2018) and adult samples (Ritchie et al., 2018), with males showing greater variance at both upper and lower extremities of the distributions, which might have functional and clinical implications.

Cellular and molecular mechanisms underlying structural changes What do developmental changes, as assessed by, for example, T1-weighted MRI or DTI scans, reflect on a cellular and molecular level? To date, studies that have directly tested these relationships in humans are scarce. However, several hypotheses concern the mechanisms underlying these observed developmental changes (Paus, 2013). One hypothesis is that reductions in the gray matter volume during adolescence partly reflect synaptic pruning. However, synaptic boutons are very small and comprise only a fraction of gray matter volume. Even when synapses are particularly dense, they are estimated to represent only 2% of a cubic millimeter of neuropil, or less than 1.5% cortical volume (Bourgeois & Rakic, 1993). Given this small percentage, it is unlikely that the marked decreases in cortical volume observed across adolescence mainly reflect synaptic pruning. The reduction in the number of synapses might, however, in addition to a reduction in neuropil, also be accompanied by a reduction in the number of cortical glial cells or other processes. These events could together account for more of the cortical structural changes observed during development, although this remains a speculation. The encroachment of subcortical white matter, and/or continued intracortical myelination, likely affects the measurements of cortical gray matter by changing the signal intensity values and contrasts so that the boundary between white and gray matter moves outward with increasing age. Undoubtedly, there is a myriad of both parallel and interacting neurobiological processes underlying the macrostructural changes observed during childhood and adolescence in MRI studies.

Similarly, many factors, including axon caliber, myelin content, fiber density, water content, crossing or diverging fibers, and partial voluming, influence DTI indices (Beaulieu, 2009). Developmental changes in DTI indices are thought to mainly relate to increasing axon caliber and continued myelination, as well as changes in fiber-packing density (Paus, 2010). Animal studies indicate that axonal membranes are the primary determinants of FA, while myelin has a modulating role (Beaulieu, 2009; Concha, Livy, Beaulieu, Wheatley, & Gross, 2010). For example, rodent dysmyelination models show that FA values still indicate anisotropy and only reduce by approximately 15% in the complete absence of myelin (Beaulieu, 2009). Further, a rare study comparing human in vivo DTI with subsequent microscopy in patients with epilepsy found a robust positive correlation between FA and axonal membranes (Concha et al., 2010). Animal studies do, however, consistently indicate that RD is particularly sensitive to de- and dysmyelination (e.g., Song et al., 2005), and correlations between DTI and myelin content and, to a lesser degree, axon count have also been shown in the postmortem brains of human patients with multiple sclerosis (Schmierer et al., 2007). The myelin content interpretation has, because of these and other findings, often been stressed. Although myelination, a process that begins between weeks 20 and 28 of gestation, has been shown to continue throughout childhood and adolescence (Benes, 1989; Benes, Turtle, Khan, & Farol, 1994; Yakovlev & Lecours, 1967), it does not logically follow from these rodent and postmortem studies that healthy developmental changes in RD in humans reliably indicate myelination (Paus, 2010). DTI parameters are sensitive to the general diffusion properties of brain tissue and are not selective markers of specific biological properties.

The relative roles of specific cellular and molecular processes for developmental changes in brain structure and microstructure are likely also age-dependent, with different contributions, for instance, in infancy, during adolescence, and in old age. Precise interpretations of the underlying mechanisms of morphometric or DTI developmental changes are thus challenging and should be done with great caution. However, investigating multiple imaging indices concurrently might yield additional information to better characterize tissue properties, and new imaging techniques, as well as studies combining imaging and histology, can hopefully increase our understanding of the cellular and molecular developmental processes.

Future Directions

Beyond well-established morphometric approaches and DTI, imaging acquisition and analysis techniques are ever evolving, promising to provide more sensitive or specific measures. In this section, we briefly present a few selected emerging imaging and analytic approaches and discuss their application to structural brain development in childhood and adolescence.

A small but increasing number of studies use surface-based methods and examine age-related differences in specific signal intensity contrasts, such as cortical gray/white matter contrast (Lewis, Evans, & Tohka, 2018; Norbom et al., 2019) or the T1-weighted/T2-weighted ratio, also referred to as cortical *myelin mapping* (Glasser & Van Essen, 2011; Grydeland, Walhovd, Tamnes, Westlye, & Fjell, 2013; see also Geeraert et al., [2017] for a comparison of other neuroimaging markers of myelin content in children). In relation to the more widely used measures, these approaches appear to provide partly independent and possibly more specific biomarkers of brain structural alterations in development, but further studies are needed to test this.

More recent and advanced dMRI methods compared to DTI, including diffusion kurtosis imaging (DKI) and neurite orientation dispersion and density imaging (NODDI), also aim to provide biologically more specific measures. Developmental studies using these methods are becoming more common, yet only cross-sectional studies are so far available (for a review, see Tamnes, Roalf, Goddings, & Lebel, 2018). NODDI studies indicate that the FA increase during childhood and adolescence is dominated by an increasing neurite density index (NDI), which points to increasing myelin and/or axonal packing but negligible changes in axon coherence during development (Chang et al., 2015; Mah, Geeraert, & Lebel, 2017). Moreover, results indicate that NODDI metrics predict chronological age better than DTI metrics (Chang et al., 2015; Genc, Malpas, Holland, Beare, & Silk, 2017). These initial applications of these methods thus demonstrate utility in studying brain development. However, they currently require relatively long scan times, a hurdle for developmental studies.

An increasingly popular analytic approach is structural covariance, which refers to correlations across individuals in the properties of pairs of brain regions and aims to inform us about structural connectivity (Alexander-Bloch, Giedd, & Bullmore, 2013). A few longitudinal studies have used the approach of *maturational coupling*—that is, covariance in longitudinal changes across subjects. Frontotemporal association cortices show the strongest and most widespread maturational coupling with other cortical areas, while lower-order sensory cortices show the least (Raznahan, Lerch, et al., 2011). Another study looked at cortico-subcortical structural change relationships and found that these partly correspond to known functional networks; for example, a longitudinal change in hippocampal volume was found to be associated with longitudinal changes in the cortical areas involved in episodic memory (Walhovd et al., 2015). Maturational covariance, presumably reflecting coordinated development between brain regions, may also be responsible for cross-sectional structural covariance (Alexander-Bloch, Raznahan, Bullmore, & Giedd, 2013). Finally, a recent study indicates links between verbal intelligence and the strength of structural couplings of cortical regions in children and adolescents (Khundrakpam et al., 2017).

Beyond these measures, graph theoretical analyses are opening up new perspectives on the development of brain networks, potentially across imaging modalities and scales (Betzel & Bassett, 2017). Although many features of complex networks, such as small worldness, highly connected hubs (together forming a rich club), and modularity, are already established at birth, they are thought to mature across childhood and adolescence (Vértes & Bullmore, 2015; Wierenga et al., 2018). Few longitudinal studies have so far been performed, but one such study found that the efficiency of structural networks, as measured from DTI, changes in a nonlinear fashion from late childhood to early adulthood and that such development of network efficiency is related to intelligence (Koenis et al., 2017).

Conclusion

The human brain undergoes considerable changes in structure across the first two decades of life. Cortical gray matter increases into childhood and decreases steadily across adolescence before stabilizing in the early 20s, whereas white matter increases. The cortex thins around 1% annually throughout the second decade of life, and surface area decreases at approximately half this rate. Crucially, cortical and subcortical changes do not proceed uniformly. Rather, there are regional differences in timing and tempo, with a general trend for posterior regions to develop earlier than anterior regions of the brain. White matter microstructure also continues to change into the third decade of life, with frontotemporal connections developing more slowly than other tracts. Both morphometric properties and diffusion measures derived from MRI cannot currently be mapped to specific cellular processes. Our understanding of the underlying mechanisms driving structural changes in the brain will continue to improve as new measures and approaches become more widely applied to longitudinal investigations.

Acknowledgments

We thank Nandita Vijayakumar and Lara M. Wierenga for comments on earlier drafts of the manuscript. Christian K. Tamnes is funded by the Research Council of Norway, and Kathryn L. Mills is funded by the National Institutes of Health R01 MH107418.

REFERENCES

Alemán-Gómez, Y., Janssen, J., Schnack, H., Balaban, E., Pina-Camacho, L., Alfaro-Almagro, F., … Desco, M. (2013). The human cerebral cortex flattens during adolescence. *Journal of Neuroscience, 33*(38), 15004–15010. https://doi.org/10.1523/JNEUROSCI.1459-13.2013

Alexander-Bloch, A., Giedd, J. N., & Bullmore, E. (2013). Imaging structural co-variance between human brain regions. *Nature Reviews. Neuroscience, 14*(5), 322–336. https://doi.org/10.1038/nrn3465

Alexander-Bloch, A., Raznahan, A., Bullmore, E., & Giedd, J. N. (2013). The convergence of maturational change and structural covariance in human cortical networks. *Journal of Neuroscience, 33*(7), 2889–2899. https://doi.org/10.1523/JNEUROSCI.3554-12.2013

Beaulieu, C. (2009). The biological basis of diffusion anisotropy. In H. Johansen-Berg & T. E. J. Behrens (Eds.), *Diffusion MRI* (pp. 105–126). San Diego: Academic Press.

Benes, F. (1989). Myelination of cortical-hippocampal relays during late adolescence. *Schizophrenia Bulletin, 15*(4), 585–593.

Benes, F., Turtle, M., Khan, Y., & Farol, P. (1994). Myelination of a key relay zone in the hippocampal formation occurs in the human brain during childhood, adolescence, and adulthood. *Archives of General Psychiatry, 51*(6), 477–484.

Betzel, R. F., & Bassett, D. S. (2017). Generative models for network neuroscience: Prospects and promise. *Journal of the Royal Society Interface, 14*(136), 20170623. https://doi.org/10.1098/rsif.2017.0623

Bourgeois, Jean-P., & Rakic, P. (1993). Changes of synaptic density in the primary visual cortex of the macaque monkey from fetal to adult stage. *Journal of Neuroscience, 13*(7), 2801–2820.

Chang, Y. S., Owen, J. P., Pojman, N. J., Thieu, T., Bukshpun, P., Wakahiro, M. L. J., … Mukherjee, P. (2015). White matter changes of neurite density and fiber orientation dispersion during human brain maturation. *PLoS One, 10*(6), e0123656. https://doi.org/10.1371/journal.pone.0123656

Concha, L., Livy, D. J., Beaulieu, C., Wheatley, B. M., & Gross, D. W. (2010). In vivo diffusion tensor imaging and histopathology of the fimbria-fornix in temporal lobe epilepsy. *Journal of Neuroscience, 30*(3), 996–1002. https://doi.org/10.1523/JNEUROSCI.1619-09.2010

Fjell, A. M., Westlye, L. T., Amlien, I., Tamnes, C. K., Grydeland, H., Engvig, A., … Walhovd, K. B. (2015). High-expanding cortical regions in human development and evolution are related to higher intellectual abilities. *Cerebral Cortex, 25*(1), 26–34. https://doi.org/10.1093/cercor/bht201

Geeraert, B. L., Lebel, R. M., Mah, A. C., Deoni, S. C., Alsop, D. C., Varma, G., & Lebel, C. (2017). A comparison of inhomogeneous magnetization transfer, myelin volume fraction, and diffusion tensor imaging measures in healthy children. *NeuroImage, 182,* 343–350. https://doi.org/10.1016/j.neuroimage.2017.09.019

Genc, S., Malpas, C. B., Holland, S. K., Beare, R., & Silk, T. J. (2017). Neurite density index is sensitive to age related differences in the developing brain. *NeuroImage, 148*(Suppl. C), 373–380. https://doi.org/10.1016/j.neuroimage.2017.01.023

Geng, X., Gouttard, S., Sharma, A., Gu, H., Styner, M., Lin, W., … Gilmore, J. H. (2012). Quantitative tract-based white matter development from birth to age 2 years. *NeuroImage, 61*(3), 542–557. https://doi.org/10.1016/j.neuroimage.2012.03.057

Gilmore, J. H., Shi, F., Woolson, S. L., Knickmeyer, R. C., Short, S. J., Lin, W., … Shen, D. (2012). Longitudinal development of cortical and subcortical gray matter from birth to 2 years. *Cerebral Cortex, 22*(11), 2478–2485. https://doi.org/10.1093/cercor/bhr327

Glasser, M. F., & Van Essen, D. C. (2011). Mapping human cortical areas in vivo based on myelin content as revealed by T1- and T2-weighted MRI. *Journal of Neuroscience, 31*(32), 11597–11616. https://doi.org/10.1523/JNEUROSCI.2180-11.2011

Goddings, A.-L., Mills, K. L., Clasen, L. S., Giedd, J. N., Viner, R. M., & Blakemore, S.-J. (2014). The influence of puberty on subcortical brain development. *NeuroImage, 88,* 242–251. https://doi.org/10.1016/j.neuroimage.2013.09.073

Grydeland, H., Walhovd, K. B., Tamnes, C. K., Westlye, L. T., & Fjell, A. M. (2013). Intracortical myelin links with performance variability across the human lifespan: Results from T1- and T2-weighted MRI myelin mapping and diffusion tensor imaging. *Journal of Neuroscience, 33*(47), 18618–18630. https://doi.org/10.1523/JNEUROSCI.2811-13.2013

Hedman, A. M., van Haren, N. E. M., Schnack, H. G., Kahn, R. S., & Hulshoff Pol, H. E. (2012). Human brain changes across the life span: A review of 56 longitudinal magnetic resonance imaging studies. *Human Brain Mapping, 33*(8), 1987–2002. https://doi.org/10.1002/hbm.21334

Herting, M. M., Johnson, C., Mills, K. L., Vijayakumar, N., Dennison, M., Liu, C., … Tamnes, C. K. (2018). Development of subcortical volumes across adolescence in males and females: A multisample study of longitudinal changes. *NeuroImage, 172,* 194–205. https://doi.org/10.1016/j.neuroimage.2018.01.020

Herting, M. M., & Sowell, E. R. (2017). Puberty and structural brain development in humans. *Frontiers in Neuroendocrinology, 44,* 122–137. https://doi.org/10.1016/j.yfrne.2016.12.003

Hill, J., Inder, T., Neil, J., Dierker, D., Harwell, J., & Van Essen, D. (2010). Similar patterns of cortical expansion during human development and evolution. *Proceedings of the National Academy of Sciences of the United States of America, 107*(29), 13135–13140. https://doi.org/10.1073/pnas.1001229107

Jbabdi, S., & Behrens, T. E. (2013). Long-range connectomics. *Annals of the New York Academy of Sciences, 1305,* 83–93. https://doi.org/10.1111/nyas.12271

Khundrakpam, B. S., Lewis, J. D., Reid, A., Karama, S., Zhao, L., Chouinard-Decorte, F., … Brain Development Cooperative Group. (2017). Imaging structural covariance in the development of intelligence. *NeuroImage, 144* (Pt. A), 227–240. https://doi.org/10.1016/j.neuroimage.2016.08.041

Koenis, M. M. G., Brouwer, R. M., Swagerman, S. C., van Soelen, I. L. C., Boomsma, D. I., & Hulshoff Pol, H. E. (2017). Association between structural brain network efficiency and intelligence increases during adolescence. *Human Brain Mapping, 39*(2), 822–836. https://doi.org/10.1002/hbm.23885

Krogsrud, S. K., Fjell, A. M., Tamnes, C. K., Grydeland, H., Mork, L., Due-Tønnessen, P., … Walhovd, K. B. (2016). Changes in white matter microstructure in the developing brain—a longitudinal diffusion tensor imaging study of children from 4 to 11 years of age. *NeuroImage, 124*(Pt. A), 473–486. https://doi.org/10.1016/j.neuroimage.2015.09.017

Lebel, C., & Beaulieu, C. (2011). Longitudinal development of human brain wiring continues from childhood into adulthood. *Journal of Neuroscience, 31*(30), 10937–10947. https://doi.org/10.1523/JNEUROSCI.5302-10.2011

Lebel, C., Gee, M., Camicioli, R., Wieler, M., Martin, W., & Beaulieu, C. (2012). Diffusion tensor imaging of white matter tract evolution over the lifespan. *NeuroImage, 60*(1), 340–352. https://doi.org/10.1016/j.neuroimage.2011.11.094

Lebel, C., Walker, L., Leemans, A., Phillips, L., & Beaulieu, C. (2008). Microstructural maturation of the human brain from childhood to adulthood. *NeuroImage, 40*(3), 1044–1055. https://doi.org/10.1016/j.neuroimage.2007.12.053

Le Bihan, D., & Johansen-Berg, H. (2012). Diffusion MRI at 25: Exploring brain tissue structure and function. *NeuroImage, 61*(2), 324–341. https://doi.org/10.1016/j.neuroimage.2011.11.006

Lewis, J. D., Evans, A. C., & Tohka, J. (2018). T1 white/gray contrast as a predictor of chronological age, and an index of cognitive performance. *NeuroImage, 173,* 341–350. https://doi: 10.1016/j.neuroimage.2018.02.050

Mah, A., Geeraert, B., & Lebel, C. (2017). Detailing neuroanatomical development in late childhood and early adolescence using NODDI. *PLoS One, 12*(8), e0182340. https://doi.org/10.1371/journal.pone.0182340

Maier-Hein, K. H., Neher, P. F., Houde, J.-C., Côté, M.-A., Garyfallidis, E., Zhong, J., … Descoteaux, M. (2017). The challenge of mapping the human connectome based on diffusion tractography. *Nature Communications, 8*(1), 1349. https://doi.org/10.1038/s41467-017-01285-x

Mills, K. L., Goddings, A.-L., Herting, M. M., Meuwese, R., Blakemore, S.-J., Crone, E. A., … Tamnes, C. K. (2016). Structural brain development between childhood and adulthood: Convergence across four longitudinal samples. *NeuroImage, 141,* 273–281. https://doi.org/10.1016/j.neuroimage.2016.07.044

Mills, K. L., Lalonde, F., Clasen, L. S., Giedd, J. N., & Blakemore, S.-J. (2014). Developmental changes in the structure of the social brain in late childhood and adolescence. *Social Cognitive and Affective Neuroscience, 9(1),* 123–131. https://doi: 10.1093/scan/nss113

Mutlu, A. K., Schneider, M., Debbané, M., Badoud, D., Eliez, S., & Schaer, M. (2013). Sex differences in thickness, and folding developments throughout the cortex. *NeuroImage, 82,* 200–207. https://doi.org/10.1016/j.neuroimage.2013.05.076

Norbom, L. B., Doan, N. T., Alnæs, D., Kaufmann, T., Moberget, T., Rokocki, J., Andreassen, O. A., Westlye, L. T., & Tamnes, C. K. (2019). Probing brain development patterns of myelination and associations with psychopathology in youth using gray/white matter contrast. *Biological Psychiatry, 85*(5), 389–398. https://doi: 10.1016/j.biopsych.2018.09.027

Olson, I. R., Heide, R. J. V. D., Alm, K. H., & Vyas, G. (2015). Development of the uncinate fasciculus: Implications for theory and developmental disorders. *Developmental Cognitive Neuroscience, 14*(Suppl. C), 50–61. https://doi.org/10.1016/j.dcn.2015.06.003

Paus, T. (2010). Growth of white matter in the adolescent brain: Myelin or axon? *Brain and Cognition, 72*(1), 26–35. https://doi.org/10.1016/j.bandc.2009.06.002

Paus, T. (2013). How environment and genes shape the adolescent brain. *Hormones and Behavior, 64*(2), 195–202. https://doi.org/10.1016/j.yhbeh.2013.04.004

Qiu, A., Mori, S., & Miller, M. I. (2015). Diffusion tensor imaging for understanding brain development in early life. *Annual Review of Psychology, 66,* 853–876. https://doi.org/10.1146/annurev-psych-010814-015340

Raznahan, A., Lerch, J. P., Lee, N., Greenstein, D., Wallace, G. L., Stockman, M., … Giedd, J. N. (2011). Patterns of coordinated anatomical change in human cortical development: A longitudinal neuroimaging study of maturational coupling. *Neuron, 72*(5), 873–884. https://doi.org/10.1016/j.neuron.2011.09.028

Raznahan, A., Shaw, P., Lalonde, F., Stockman, M., Wallace, G. L., Greenstein, D., … Giedd, J. N. (2011). How does your cortex grow? *Journal of Neuroscience, 31*(19), 7174–7177. https://doi.org/10.1523/JNEUROSCI.0054-11.201.1

Ritchie, S. J., Cox, S. R., Shen, X., Lombardo, M. V., Reus, L. M., Alloza, C., … Deary, I. J. (2018). Sex differences in the adult human brain: Evidence from 5216 UK Biobank participants. *Cerebral Cortex, 28*(8), 2959–2975 . https://doi: 10.1093/cercor/bhy109

Ruigrok, A. N. V., Salimi-Khorshidi, G., Lai, M.-C., Baron-Cohen, S., Lombardo, M. V., Tait, R. J., & Suckling, J. (2014). A meta-analysis of sex differences in human brain structure. *Neuroscience and Biobehavioral Reviews, 39*(100), 34–50. https://doi.org/10.1016/j.neubiorev.2013.12.004

Schaer, M., Cuadra, M. B., Tamarit, L., Lazeyras, F., Eliez, S., & Thiran, J. (2008). A surface-based approach to quantify local cortical gyrification. *IEEE Transactions on Medical Imaging, 27*(2), 161–170. https://doi.org/10.1109/TMI.2007.903576

Schmierer, K., Wheeler-Kingshott, C. A. M., Boulby, P. A., Scaravilli, F., Altmann, D. R., Barker, G. J., … Miller, D. H. (2007). Diffusion tensor imaging of post mortem multiple sclerosis brain. *NeuroImage, 35*(2), 467–477. https://doi.org/10.1016/j.neuroimage.2006.12.010

Simmonds, D., Hallquist, M. N., Asato, M., & Luna, B. (2014). Developmental stages and sex differences of white matter and behavioral development through adolescence: A longitudinal diffusion tensor imaging (DTI) study. *NeuroImage, 92,* 356–368. https://doi.org/10.1016/j.neuroimage.2013.12.044

Song, S.-K., Yoshino, J., Le, T. Q., Lin, S.-J., Sun, S.-W., Cross, A. H., & Armstrong, R. C. (2005). Demyelination increases radial diffusivity in corpus callosum of mouse brain. *NeuroImage, 26*(1), 132–140. https://doi.org/10.1016/j.neuroimage.2005.01.028

Tamnes, C. K., Herting, M. M., Goddings, A.-L., Meuwese, R., Blakemore, S.-J., Dahl, R. E., … Mills, K. L. (2017). Development of the cerebral cortex across adolescence: A multisample study of inter-related longitudinal changes in cortical volume, surface area, and thickness. *Journal of Neuroscience, 37*(12), 3402–3412. https://doi.org/10.1523/JNEUROSCI.3302-16.2017

Tamnes, C. K., Roalf, D. R., Goddings, A.-L., & Lebel, C. (2018). Diffusion MRI of white matter microstructure development in childhood and adolescence: Methods, challenges and progress. *Developmental Cognitive Neuroscience, 33,* 161–175. https://doi: 10.1016/j.dcn.2017.12.002

Tamnes, C. K., Walhovd, K. B., Dale, A. M., Østby, Y., Grydeland, H., Richardson, G., … Fjell, A. M. (2013). Brain development and aging: Overlapping and unique patterns of change. *NeuroImage, 68C,* 63–74. https://doi.org/10.1016/j.neuroimage.2012.11.039

Vértes, P. E., & Bullmore, E. T. (2015). Annual research review: Growth connectomics—the organization and reorganization of brain networks during normal and abnormal development. *Journal of Child Psychology and Psychiatry, and Allied Disciplines, 56*(3), 299–320. https://doi.org/10.1111/jcpp.12365

Vijayakumar, N., Allen, N. B., Youssef, G., Dennison, M., Yücel, M., Simmons, J. G., & Whittle, S. (2016). Brain development during adolescence: A mixed-longitudinal investigation of cortical thickness, surface area, and volume. *Human Brain Mapping.* https://doi.org/10.1002/hbm.23154

Walhovd, K. B., Fjell, A. M., Giedd, J., Dale, A. M., & Brown, T. T. (2017). Through thick and thin: A need to reconcile contradictory results on trajectories in human cortical development. *Cerebral Cortex, 27,* 1472–1481. https://doi.org/10.1093/cercor/bhv301

Walhovd, K. B., Tamnes, C. K., Bjørnerud, A., Due-Tønnessen, P., Holland, D., Dale, A. M., & Fjell, A. M. (2015). Maturation of cortico-subcortical structural networks-segregation and overlap of medial temporal and fronto-striatal systems in development. *Cerebral Cortex, 25*(7), 1835–1841. https://doi.org/10.1093/cercor/bht424

Wierenga, L. M., Langen, M., Ambrosino, S., van Dijk, S., Oranje, B., & Durston, S. (2014). Typical development of basal ganglia, hippocampus, amygdala and cerebellum from age 7 to 24. *NeuroImage, 96,* 67–72. https://doi.org/10.1016/j.neuroimage.2014.03.072

Wierenga, L. M., Langen, M., Oranje, B., & Durston, S. (2014). Unique developmental trajectories of cortical thickness and surface area. *NeuroImage, 87,* 120–126. https://doi.org/10.1016/j.neuroimage.2013.11.010

Wierenga, L. M., Sexton, J. A., Laake, P., Giedd, J. N., & Tamnes, C. K. (2018). A key characteristic of sex differences in the developing brain: Greater variability in brain structure of boys than girls. *Cerebral Cortex, 28*(8), 2741–2751. https://doi.org/10.1093/cercor/bhx154

Wierenga, L. M., van den Heuvel, M. P., Oranje, B., Giedd, J. N., Durston, S., Peper, J. S., … Pediatric Longitudinal Imaging, Neurocognition, and Genetics Study. (2018). A multisample study of longitudinal changes in brain network architecture in 4–13-year-old children. *Human Brain Mapping, 39*(1), 157–170. https://doi.org/10.1002/hbm.23833

Yakovlev, P. A., & Lecours, I. R. (1967). The myelogenetic cycles of regional maturation of the brain. In A. Minkowski (Ed.), *Regional Development of the brain in early life* (pp. 3–70). Oxford: Blackwell.

Zilles, K., Armstrong, E., Schleicher, A., & Kretschmann, H.-J. (1988). The human pattern of gyrification in the cerebral cortex. *Anatomy and Embryology, 179*(2), 173–179. https://doi.org/10.1007/BF00304699

3 Cognitive Control and Affective Decision-Making in Childhood and Adolescence

EVELINE A. CRONE AND ANNA C. K. VAN DUIJVENVOORDE

ABSTRACT Childhood and adolescence are periods of pronounced cognitive and emotional advancement accompanied by significant changes in brain maturation. This chapter describes the development of cognitive control abilities, including working memory, response inhibition, feedback monitoring, and relational reasoning, vis-à-vis developmental changes in brain maturation. It also discusses the neurocognitive development of affective decision-making, highlighting the role of risk and reward in adolescents' decision-making. These findings are integrated and discussed in relation to neurodevelopmental models of brain development. These models highlight not only the potential vulnerabilities in adolescent development but also the opportunities for adolescents' exploration and learning.

Cognitive and Affective Decision-Making in Adolescence

Fast improvement in cognitive control One of the most consistent observations in the development of cognitive capacities across childhood and adolescence is the rapid increase in cognitive control functions, otherwise referred to as *executive functions*. Cognitive control functions are the capacities that enable us to keep relevant information in mind in order to obtain a future goal (Diamond, 2013; Miyake & Friedman, 2012). Cognitive control functions have been demarcated into correlated yet distinct factors. Such factors include the ability to store and manipulate information in one's mind to inhibit responses, to filter irrelevant information, and to switch between tasks (Friedman et al., 2016). In early development there is a marked improvement in these cognitive control functions that continues during school-aged development, with adult levels of functioning achieved around mid-adolescence (Davidson, Amso, Anderson, & Diamond, 2006). These cognitive control functions are crucial for all kinds of daily activities and central to academic attainment. For example, working memory—a key component of cognitive control—predicts future academic performance in areas such as reading and arithmetic (Peters, van der Meulen, Zanolie, & Crone, 2017; St. Clair-Thompson & Gathercole, 2006).

Social affective sensitivities At the onset of adolescence, the influence of social and affective context starts to have an impact on the decisions that adolescents make (see chapters 2 and 4). Adolescence begins with the onset of puberty (approximately 10–11 years of age), although there is individual variability (Goddings et al., 2014). During pubertal development there are substantial changes in terms of hormone release; these changes propagate alterations both in terms of bodily characteristics and social-affective sensitivities. These include an increased tendency toward risk-taking and a greater sensitivity to peer group influence (Crone & Dahl, 2012). Most changes in social sensitivity and increases in risk-taking behavior are adaptive and stimulate explorative learning, thus contributing to mature social functioning. However, in some cases such changes can have serious consequences, including accidents, drug abuse, and in extreme cases suicide attempts (Dahl & Gunnar, 2009).

These developmental patterns have inspired neuroscientists to investigate how different brain regions work together when children, adolescents, and adults make decisions.

The Neurocognitive Development of Cognitive Control

Basic cognitive control: working memory and response inhibition The basic components of cognitive control consist of several processes. Most developmental cognitive control research has focused on the processes of working memory and response inhibition.

Working memory Drawing from the adult literature (D'Esposito, 2007), several developmental neuroimaging studies have examined the role of the lateral prefrontal cortex (Brodmann [BA] 44 and BA 9/46) and posterior parietal cortex (BA 7) in working memory performance and development. These studies have reported that when 8- to 12-year-old children and adults are performing a visuospatial working memory task, adults' brains showed more activation in the lateral prefrontal cortex and posterior parietal cortex than children's (Klingberg, Forssberg, & Westerberg, 2002; Kwon, Reiss, & Menon, 2002; Scherf, Sweeney, & Luna, 2006; Thomason et al., 2009). Age-related increases in the recruitment of these areas were also found for other domains of working memory, such as verbal working memory (Thomason et al., 2009) and object working memory (Ciesielski, Lesnik, Savoy, Grant, & Ahlfors, 2006; Crone, Wendelken, Donohue, van Leijenhorst, & Bunge, 2006; Jolles, van Buchem, Rombouts, & Crone, 2011). This increase in activation in the lateral prefrontal and posterior parietal cortex correlates with performance in both adults and children (Crone et al., 2006; Finn, Sheridan, Kam, Hinshaw, & D'Esposito, 2010; Olesen, Nagy, Westerberg, & Klingberg, 2003), suggesting that both age and performance have partly independent contributions to activation levels in these areas.

More recently, researchers have highlighted the importance of having large samples, and thus variability in terms of age and task performance, in order to better understand the interplay between these factors and brain development. A large-scale neuroimaging study including 951 participants aged 8–22 years aimed to disentangle the relationship between age and working memory performance (Satterthwaite et al., 2013). Age was associated not only with greater activation in the lateral prefrontal and posterior parietal cortex but also with lower activation in regions of the default network, such as the medial prefrontal and temporal cortex. A similar, but stronger, pattern emerged for task performance and remained even when controlling for age. Finally, activation in the lateral prefrontal cortex mediated the relationship between age and performance improvement. Together, these results suggest that (1) frontal and parietal brain regions are important for successful working memory performance, and (2) greater independence of frontal parietal and default model networks contributes to performance improvements across development.

Demands in working memory tasks also influence age-related changes in prefrontal cortex recruitment. For instance, some studies have reported that adults are more responsive in terms of neural activity to specific task demands (e.g., load dependency or modality dependency) than 7- to 13-year-old children (Brahmbhatt, White, & Barch, 2010; Libertus, Brannon, & Pelphrey, 2009; O'Hare, Lu, Houston, Bookheimer, & Sowell, 2008), which may benefit performance.

Taken together, a comparison between children, adolescents, and adults shows that as age increases participants recruit the lateral prefrontal cortex and posterior parietal cortex and deactivate other regions of the cortex in a way that is helpful for the successful performance of a given task. In addition, children and adolescents seem less sensitive to different task demands.

Response inhibition A second basic control process contributing to cognitive control is the ability to inhibit inappropriate responses. In order to investigate this, most studies have made use of either stop-signal tasks, in which an already initiated response needs to be inhibited, or a go/no-go task, which involves the inhibition of a response to a rarely presented specific stimulus (e.g., letter *X*) that is presented in a series of other more prevalent stimuli requiring a response (e.g., other letters of the alphabet).

Neuroimaging studies in adults and patients have consistently reported that the right inferior frontal gyrus (BA 45/47) is important for successful inhibitory control (Aron & Poldrack, 2005) and for the greater attention demands associated with response inhibition (Hampshire, Chamberlain, Monti, Duncan, & Owen, 2010). Developmental neuroimaging studies have shown that the right inferior frontal gyrus is recruited more in adults than in children and adolescents (aged 8–17) and that adults perform better on inhibition tasks, suggesting that the right inferior frontal gyrus is important for successful inhibition (Durston et al., 2006; Rubia et al., 2006; Tamm, Menon, & Reiss, 2002). Indeed, performance on the stop-signal task correlates with activity in the right inferior frontal gyrus in children, adolescents, and adults (Cohen et al., 2010). Besides the age-related increase in activation in the right inferior frontal gyrus, some studies have reported more widespread activation in other parts of the lateral and medial prefrontal cortex in 8- to 12-year-old children than adults when inhibiting responses (Booth et al., 2003; Velanova, Wheeler, & Luna, 2008). Finally, a longitudinal functional Magnetic Resonance Imaging (fMRI) study used an antisaccade task to dissect components of inhibition. This study found that the lateral and medial prefrontal cortex, regions that are important for adjusting performance and signaling errors, showed a protracted developmental change into mid-adolescence and late adolescence (Ordaz, Foran, Velanova, & Luna, 2013).

This evidence suggests that better response inhibition is accompanied by greater activation of the right inferior frontal gyrus and that both inhibition performance and age drive activation differences in the lateral and medial prefrontal cortex. In addition, children rely on a wider network of areas for successful inhibition.

Adaptive control: feedback monitoring Whereas working memory and response inhibition require the implementation of specific task rules, most of our cognitive control requires us to respond to changing task demands. The ability to demonstrate adaptive behavior in response to changing task demands is referred to as *adaptive control*, a process required when, for example, feedback cues inform us that we need to change our behavior on a subsequent occasion.

Feedback monitoring has been widely studied in neuropsychological literature, using classic tasks such as the Wisconsin Card Sorting Task (WCST). Patient studies have found that several regions of the lateral and medial prefrontal cortex are important for monitoring negative feedback cues that inform participants that a previously applied rule (for example, sorting cards according to color) is no longer correct. This feedback cue (such as a minus sign, or the word *incorrect*) instructs the participant to switch to a new rule (for example, sorting cards according to shape) (Barcelo & Knight, 2002).

It was observed that while learning a certain sorting rule in a rule-learning and application task, both striatal and prefrontal regions were more active during the learning than applying phase. Longitudinal changes between ages 8 and 28 showed that these regions were more engaged when participants were older, with increases in neural activity until late adolescence (Peters, van Duijvenvoorde, Koolschijn, & Crone, 2016). Moreover, the dorsal striatum was most strongly engaged during late adolescence (16–18 years), and stronger activity predicted better learning performance at the testing session, as well as two years later (Peters & Crone, 2017). Together, these findings suggest that enhanced striatal activity is associated with an upregulation of cognitive control regions and, consequently, an increase in cognitive performance.

Other learning tasks focus on differences in valence, comparing positive or negative feedback ensuing from the response to certain task rules. Neuroimaging analyses reveal that in adults, receiving negative feedback results in activation in the same frontal parietal network and medial prefrontal cortex as is activated in working memory and inhibition studies (Zanolie, van Leijenhorst, Rombouts, & Crone, 2008). The negative feedback–related activity was greater for adults and 13- to 17-year-old adolescents than for 8- to 12-year-old children, specifically in the lateral prefrontal and posterior parietal cortices (Crone, Zanolie, van Leijenhorst, Westenberg, & Rombouts, 2008; van den Bos, Guroglu, van den Bulk, Rombouts, & Crone, 2009; van Duijvenvoorde, Zanolie, Rombouts, Raijmakers, & Crone, 2008). This activation increase correlated with successful performance independent of age, suggesting that these areas are important for updating behavior following negative feedback (Crone, Zanolie, et al., 2008).

In children aged 8–10 years however, the lateral prefrontal cortex and posterior parietal cortex are typically more active in the reverse contrast; that is to say, more activation is reported following positive compared to negative feedback, with a shift occurring in adolescence (Peters, Braams, Raijmakers, Koolschijn, & Crone, 2014; van den Bos et al., 2009; van Duijvenvoorde et al., 2008). This developmental difference is specific to situations in which participants learn new rules and not when applying rules that are already learned (van den Bos et al., 2009).

Together, this suggests that late adolescents may be particularly sensitive to feedback providing the potential for learning a new rule. In addition, there may be valence differences in feedback processing in which adults show more activation in the lateral prefrontal cortex and posterior parietal cortex when updating behavior following negative feedback, while children recruit these same areas more following positive feedback, with a transition occurring in adolescence. These valence differences are, however, specific to more complex rule-learning tasks that require a goal-directed choice.

Complex cognitive control: relational reasoning The ability to interpret problems from multiple perspectives, to integrate knowledge, or to infer new solutions from presently available information probably lies at the highest level of cognitive control. This type of complex reasoning often involves the combination of different control processes for successful performance.

Relational reasoning Previous research in adults and adolescents has demonstrated that this ability to integrate information relies on the most anterior part of the prefrontal cortex, the rostrolateral prefrontal cortex (Christoff et al., 2001; Dumontheil, 2014), which is modality independent (Magis-Weinberg, Blakemore, & Dumontheil, 2017). In a series of developmental neuroimaging studies, an adaptation of the Raven's Progressive Matrices task was used to study how neural activation differs when individuals need to integrate one dimension (e.g., follow a horizontal line of reasoning) or two dimensions (e.g., follow and integrate a horizontal and a vertical line of reasoning; see figure 3.1). The rostrolateral prefrontal cortex was more

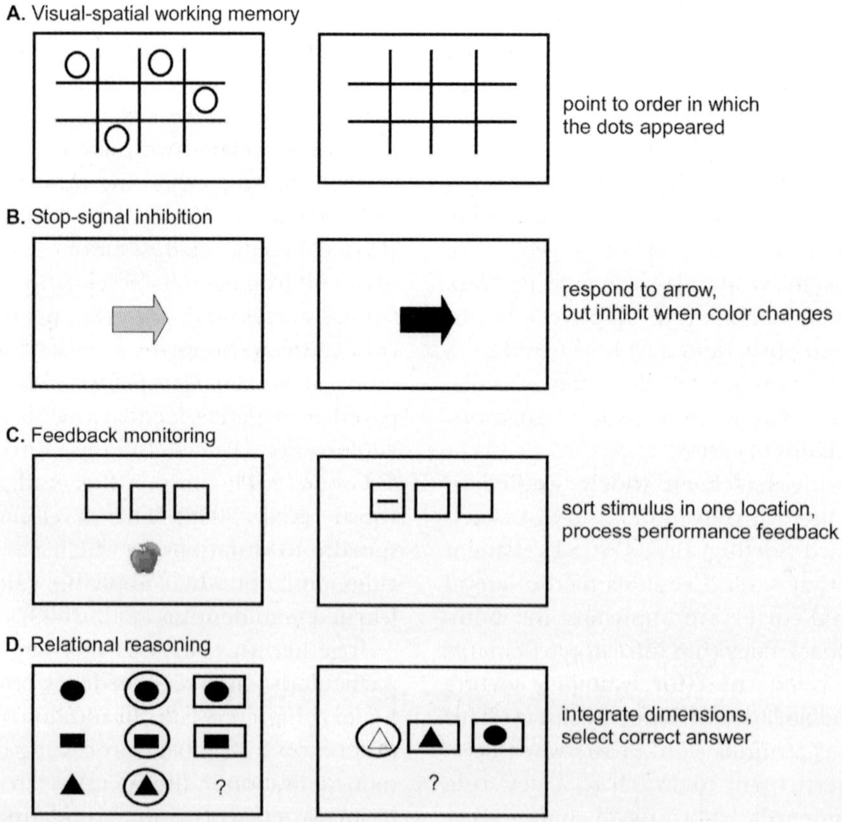

FIGURE 3.1 Examples of cognitive control paradigms. *A*, A visuospatial working memory typically involves the presentation of a grid in which dots are consecutively presented and need to be reproduced on the next trial. More dots will make the task more difficult. *B*, A stop-signal paradigm involves the presentation of a stimulus, which requires a left- or a right-hand response. In some trials, the arrow quickly changes color, which informs the participant that he/she should inhibit responding. *C*, A feedback-learning task typically involves a stimulus, which needs to be sorted in a specific location. The feedback screen informs the participant whether the response was correct or incorrect. *D*, Relational reasoning requires the participant to integrate dimensions of a presented stimulus. One-dimensional trials are those in which only one direction needs to be followed (e.g., a horizontal line), and two-dimensional trials are those in which more dimensions (e.g., a horizontal line and a vertical line) need to be integrated to come to the correct solution. (See color plate 2.)

active in 8- to 12-year-old children at the onset of stimulus presentation but failed to show sustained activation during problem solving. In contrast, in adults this region showed sustained activation throughout the problem-solving period (Crone et al., 2009). In addition, a study including children, adolescents, and adults showed that children aged 7–10 years recruited the rostrolateral prefrontal cortex for both one- and two-dimensional problems, whereas adolescents aged 11–14 years showed a small differentiation in activation patterns, and adolescents aged 15–18 years recruited the rostrolateral prefrontal cortex for two-dimensional, but not for one-dimensional, problems (Wendelken, O'Hare, Whitaker, Ferrer, & Bunge, 2011). Finally, a study including 95 children and adolescents ages 6–18 years including a pictorial propositional analogy task showed that the left anterior prefrontal cortex (BA 47/45), a brain region important for semantic retrieval, correlated positively with age and performance (Whitaker, Vendetti, Wendelken, & Bunge, 2018).

Together, these studies suggest that the specialization of the rostrolateral and anterior prefrontal cortex with regard to age is related to relational and semantic integration, respectively.

The Neurocognitive Development of Affective Decision-Making

Risks and rewards In order to understand how affective context influences the way we control our actions and make decisions, research has often focused on how children, adolescents, and adults process rewards. Reward processing has been examined in the context of risk-taking, based on the observation that adolescents are more prone than children and adults to take risks in daily life (Steinberg, 2011). Laboratory studies have demonstrated an age-related reduction in risk-taking (Crone, Bullens, van der Plas, Kijkuit, & Zelazo, 2008; van Duijvenvoorde, Jansen, Bredman, & Huizenga, 2012), but also nonlinear age effects, suggesting that adolescents take more risks than children and adults when there is a strong affective context (Burnett, Bault, Coricelli, & Blakemore, 2010; Figner, Mackinlay, Wilkening, & Weber, 2009). The specificity of these developmental differences has been studied in more detail using neuroimaging.

One line of research tested risk-taking using experimental laboratory tasks, in which adolescents needed to decide between a certain chance of getting a small reward (safe bet) and an uncertain chance of getting a high reward (risky bet). The value of the choice at hand is reflected in the activation of a number of key brain regions, such as the ventral medial prefrontal cortex, posterior cingulate cortex, and ventral striatum (Bartra, McGuire, & Kable, 2013; Clithero & Rangel, 2014). In addition, it has been suggested that separate brain regions, such as the insula and dorsal medial prefrontal cortex, assess the level of risk during choice (Mohr, Biele, & Heekeren, 2010). In one study, adolescents showed a higher sensitivity to value compared to adults, as reflected in greater ventral striatum activation (Barkley-Levenson & Galvan, 2014). In addition, 16- to 19-year-old adolescents, compared to 9- to 12-year-old children and 25- to 35-year-old adults, also show a greater neural sensitivity to risk, which was reflected in insula and dorsal medial prefrontal cortex activation (van Duijvenvoorde et al., 2015). Similarly, when taking risks a more ventral part of the anterior cingulate cortex (subgenual ACC) was more active in 12- to 17-year-olds compared to 8- to 10-year-olds and 18- to 25-year-olds (van Leijenhorst, Gunther Moor, et al., 2010). Together, these findings suggest that adolescents rely more on striatal and affective prefrontal cortex regions in assessing choices and risks than do adults.

In these laboratory decision-making tasks, each decision to take a risk (or not) will typically lead to a reward or a loss outcome. Several studies have reported that this reward response in the ventral striatum is higher in 13- to 17-year-old adolescents compared to children and adults (Ernst et al., 2005; Galvan et al., 2006; Padmanabhan, Geier, Ordaz, Teslovich, & Luna, 2011; van Leijenhorst, Zanolie, et al., 2010). Adolescents' higher ventral striatum activation response to rewarding outcomes has been confirmed by a formal meta-analysis of neuroimaging studies comparing adolescents and adults (Silverman, Jedd, & Luciana, 2015), as well as by a three-wave longitudinal study testing individuals between ages 8 and 29 (Schreuders et al., 2018). This heightened reward response indicates greater sensitivity to affective learning signals in adolescence.

An additional line of research used neuroimaging outcomes to predict self-reported risk-taking in daily life. Results showed positive associations between reward activation in the ventral striatum and the self-reported reward drive. That is, participants who indicated they were willing to exert more effort for a reward also showed greater striatum responses to rewards (Schreuders et al., 2018). Other studies showed that reward-related neural activation in the striatum and lateral prefrontal cortex was related to a variety of risky real-life behaviors, including risky sexual behavior, illicit drug use, and binge drinking (Blankenstein, Schreuders, Peper, Crone, & van Duijvenvoorde, 2018; Braams, Peper, van der Heide, Peters, & Crone, 2016; Galvan, Hare, Voss, Glover, & Casey, 2007). Finally, functional coupling between subcortical limbic structures and the orbitofrontal cortex (OFC) has been related to risky real-life behavior. That is, self-reported rule-breaking behavior has been related to a functional coupling of the striatum and OFC (Qu, Galvan, Fuligni, Lieberman, & Telzer, 2015).

Together, this suggests that functional activity and connectivity, within and between the striatal and prefrontal reward network, may be a marker for the propensity to display risk-taking behaviors.

Short-term and long-term consequences How individuals weigh up short- versus long-term consequences is often examined using delay-discounting tasks, where the option of obtaining an immediate (e.g., one dollar now) or a delayed reward (e.g., two dollars later) is presented with variable delays. Individuals tend to opt for the immediate reward more often when the delay for the larger reward is longer. In a delay-discounting task, preference for a delayed reward is associated with activation in the lateral prefrontal cortex, whereas preference for an immediate reward is associated with activation in the ventral striatum in both adolescents and adults (Christakou, Brammer, & Rubia, 2011). In adults, the ventromedial prefrontal cortex is more active than in 11- to 17-year-old adolescents when they choose immediate rewards, compared to 18- to 31-year-old adults. This region also shows age-related increases in functional connectivity within the ventral striatum, suggesting that the ventromedial prefrontal cortex works together with the striatum when selecting or inhibiting impulsive choices. Consistent with this line of reasoning, a diffusion tensor imaging (DTI) study reported that stronger structural connectivity between the (ventromedial) prefrontal cortex and the striatum relates to fewer short-term choices in adults (Peper et al., 2013). A longitudinal study showed that this brain connectivity increases with age and is predictive for the tendency to balance immediate and delayed rewards two year later (Achterberg, Peper, van Duijvenvoorde, Mandl, & Crone, 2016). These findings suggest that the prefrontal cortex plays a regulatory role when making impulsive choices. Moreover, the development of this connectivity seems to underlie the developmental improvements in the inhibition of impulsive choices between adolescence and adulthood.

Models of Neurocognitive Development

Several models have been introduced to explain how the differential development of various brain regions is important for control and thus influences decision-making in adolescence. These dual-processing models (Casey, 2015; Ernst, 2014; Strang, Chein, & Steinberg, 2013) suggest that affective limbic brain regions, such as the ventral striatum, develop at a faster pace than brain regions important to control and regulation, such as the prefrontal cortex, the dorsal ACC, and the posterior parietal cortex. This imbalance makes adolescence a sensitive time for risk-taking but also brings opportunities for exploration and adaptive learning.

Crone and Dahl (2012) suggest that puberty could be a driving force for heightened affective sensitivity in adolescence. Extensive animal research has pinpointed the timing of puberty in rodents and has reported specific effects of puberty on brain function and structure (Spear, 2011). Furthermore, pubertal hormones have been found to have a steering influence on the structural development of the human brain (Ladouceur, Peper, Crone, & Dahl, 2011). Finally, puberty influences affective responses to reward in the ventral striatum, independent of age (Forbes et al., 2010; Op de Macks et al., 2011). Puberty strongly influences the way we process affective and social information, preparing adolescents to obtain independence and adapt quickly to changing social contexts. Therefore, pubertal development may be an important contributor to the increased sensitivity to affective information, which, together with flexibility in recruitment of the prefrontal cortex, may facilitate explorative learning (figure 3.2).

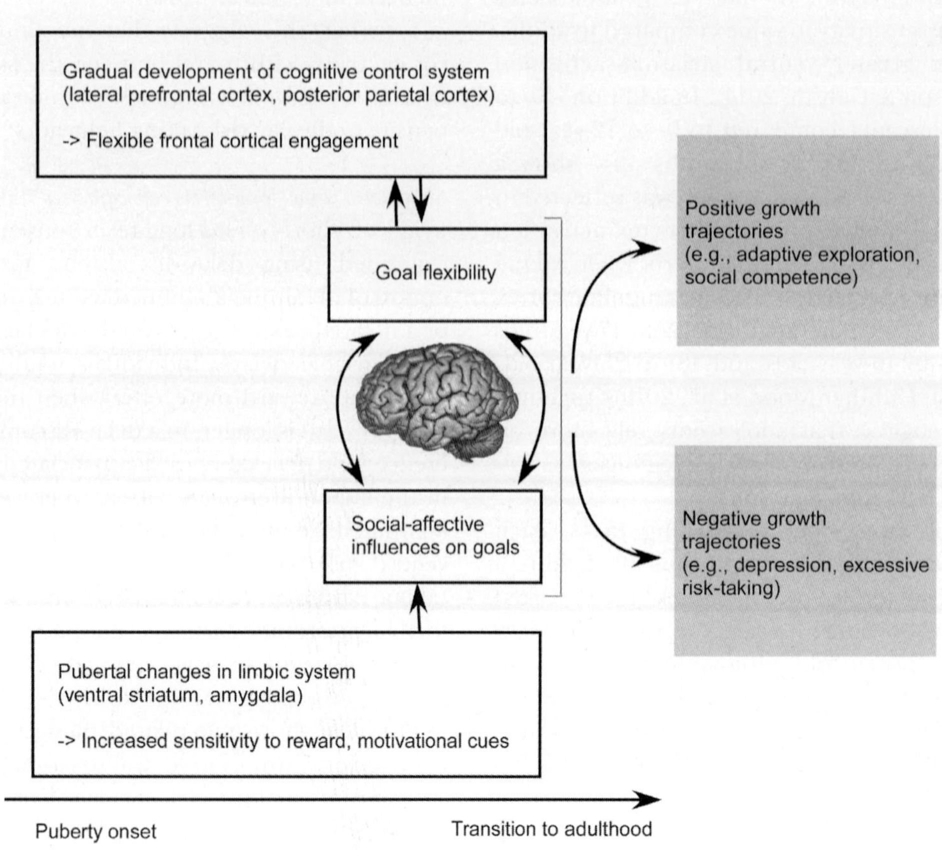

FIGURE 3.2 This model explains the slow developmental trajectory and flexible recruitment of the prefrontal and parietal cortex in adolescence, in combination with puberty-specific changes in the limbic system. This combination leads to positive growth trajectories in adolescence, as this is a natural time of exploration and social learning. However, in some cases the imbalance between these systems can cause negative growth trajectories, which can result in depression or excessive risk-taking. Adapted from Crone & Dahl (2012).

More formal models have been suggested to help disentangle the process of risk-taking and learning. These types of models rely heavily on a more computational framework. Reinforcement-learning (RL) models are one such example. RL models specify prediction errors as key learning signals. That is, prediction errors signal the difference between expected and observed outcomes and are used to update learning behavior. These prediction errors can be calculated from a behavioral-computational model and combined with neuroimaging methods to find brain regions that track variations in prediction error. The striatum and medial prefrontal cortex are found to signal prediction errors across development—that is, the difference between expected and observed outcomes (van den Bos, Cohen, Kahnt, & Crone, 2012). These models can advance the field by quantifying how learning changes across development. Moreover, they can be applied to learning across contexts, such as learning in social situations. Computational models of social learning may rely on a social Prediction Error (PE) signal that describes how we learn from and about our social world. This could occur through interaction with others or by observing others (vicarious rewards). A growing body of evidence suggests substantial overlap between nonsocial (individual) and social learning (e.g., Ruff & Fehr, 2014).

Conclusion and Future Directions

This chapter has described the neural correlates of cognitive and affective decision-making in school-aged children, in adolescents, and in adults. Literature relating to cognitive control shows that the development of basic to complex levels of control follows a pattern of specialization with age in the prefrontal cortex and the posterior parietal cortex, such that these areas are more strongly and more selectively recruited for specific tasks. The transition from widespread to focused networks takes place during adolescence, which is a period of explorative learning. This development coincides with increased affective sensitivity in mid-adolescence to affective cues, pinpointing nonlinear contributions of control and affective brain regions in development (van Duijvenvoorde & Crone, 2013). This integrative approach, in which the development of cognitive control and decision-making are studied in combination, is expected to allow for a richer description of adolescent brain development.

Acknowledgments

The authors of this chapter are supported by the European Research Council (ERC CoG PROSOCIAL 681632 to E.A.C.) and the Netherlands Organization for Scientific Research (NWO-VICI 453-14-001 E.A.C.) (NWO-ORA ASTA 464-15-176 to A.C.K.D.).

REFERENCES

Achterberg, M., Peper, J. S., van Duijvenvoorde, A. C., Mandl, R. C., & Crone, E. A. (2016). Fronto-striatal white matter integrity predicts development in delay of gratification: A longitudinal study. *Journal of Neuroscience, 36*(6), 1954–1961.

Aron, A. R., & Poldrack, R. A. (2005). The cognitive neuroscience of response inhibition: Relevance for genetic research in attention-deficit/hyperactivity disorder. *Biological Psychiatry, 57*(11), 1285–1292. doi:10.1016/j.biopsych.2004.10.026

Barcelo, F., & Knight, R. T. (2002). Both random and perseverative errors underlie WCST deficits in prefrontal patients. *Neuropsychologia, 40*(3), 349–356. Retrieved from http://www.ncbi.nlm.nih.gov/pubmed/11684168.

Barkley-Levenson, E., & Galvan, A. (2014). Neural representation of expected value in the adolescent brain. *Proceedings of the National Academy of Sciences of the United States of America, 111*(4), 1646–1651. doi:10.1073/pnas.1319762111

Bartra, O., McGuire, J. T., & Kable, J. W. (2013). The valuation system: A coordinate-based meta-analysis of BOLD fMRI experiments examining neural correlates of subjective value. *NeuroImage, 76,* 412–427. doi:10.1016/j.neuroimage.2013.02.063

Blankenstein, N. E., Schreuders, E., Peper, J. S., Crone, E. A., & van Duijvenvoorde, A. C. K. (2018). Individual differences in risk-taking tendencies modulate the neural processing of risky and ambiguous decision-making in adolescence. *NeuroImage, 172,* 663–673. doi:10.1016/j.neuroimage.2018.01.085

Booth, J. R., Burman, D. D., Meyer, J. R., Lei, Z., Trommer, B. L., Davenport, N. D., … Mesulam, M. M. (2003). Neural development of selective attention and response inhibition. *NeuroImage, 20*(2), 737–751. doi:10.1016/S1053-8119(03)00404-X

Braams, B. R., Peper, J. S., van der Heide, D., Peters, S., & Crone, E. A. (2016). Nucleus accumbens response to rewards and testosterone levels are related to alcohol use in adolescents and young adults. *Developmental Cognitive Neuroscience, 17,* 83–93. doi:10.1016/j.dcn.2015.12.014

Brahmbhatt, S. B., White, D. A., & Barch, D. M. (2010). Developmental differences in sustained and transient activity underlying working memory. *Brain Research, 1354,* 140–151. doi:10.1016/j.brainres.2010.07.055

Burnett, S., Bault, N., Coricelli, G., & Blakemore, S. J. (2010). Adolescents' heightened risk-seeking in a probabilistic gambling task. *Cognitive Development, 25*(2), 183–196. doi:10.1016/j.cogdev.2009.11.003

Casey, B. J. (2015). Beyond simple models of self-control to circuit-based accounts of adolescent behavior. *Annual Review of Psychology, 66,* 295–319. doi:10.1146/annurev-psych-010814-015156

Christakou, A., Brammer, M., & Rubia, K. (2011). Maturation of limbic corticostriatal activation and connectivity associated with developmental changes in temporal discounting. *NeuroImage, 54*(2), 1344–1354. doi:10.1016/j.neuroimage.2010.08.067

Christoff, K., Prabhakaran, V., Dorfman, J., Zhao, Z., Kroger, J. K., Holyoak, K. J., & Gabrieli, J. D. (2001). Rostrolateral prefrontal cortex involvement in relational integration during reasoning. *NeuroImage, 14*(5), 1136–1149. doi:10.1006/nimg.2001.0922

Ciesielski, K. T., Lesnik, P. G., Savoy, R. L., Grant, E. P., & Ahlfors, S. P. (2006). Developmental neural networks in children performing a categorical n-back task. *NeuroImage, 33*(3), 980–990. doi:10.1016/j.neuroimage.2006.07.028

Clithero, J. A., & Rangel, A. (2014). Informatic parcellation of the network involved in the computation of subjective value. *Social Cognitive and Affective Neuroscience, 9*(9), 1289–1302. doi:10.1093/scan/nst106

Cohen, J. R., Asarnow, R. F., Sabb, F. W., Bilder, R. M., Bookheimer, S. Y., Knowlton, B. J., & Poldrack, R. A. (2010). Decoding developmental differences and individual variability in response inhibition through predictive analyses across individuals. *Frontiers in Human Neuroscience, 4*, 47. doi:10.3389/fnhum.2010.00047

Crone, E. A., Bullens, L., van der Plas, E. A., Kijkuit, E. J., & Zelazo, P. D. (2008). Developmental changes and individual differences in risk and perspective taking in adolescence. *Development and Psychopathology, 20*(4), 1213–1229. doi:10.1017/S0954579408000588

Crone, E. A., & Dahl, R. E. (2012). Understanding adolescence as a period of social-affective engagement and goal flexibility. *Nature Reviews. Neuroscience, 13*(9), 636–650. doi:10.1038/nrn3313

Crone, E. A., Wendelken, C., Donohue, S., van Leijenhorst, L., & Bunge, S. A. (2006). Neurocognitive development of the ability to manipulate information in working memory. *Proceedings of the National Academy of Sciences of the United States of America, 103*(24), 9315–9320. doi:10.1073/pnas.0510088103

Crone, E. A., Wendelken, C., van Leijenhorst, L., Honomichl, R. D., Christoff, K., & Bunge, S. A. (2009). Neurocognitive development of relational reasoning. *Developmental Science, 12*(1), 55–66. doi:10.1111/j.1467-7687.2008.00743.x

Crone, E. A., Zanolie, K., van Leijenhorst, L., Westenberg, P. M., & Rombouts, S. A. (2008). Neural mechanisms supporting flexible performance adjustment during development. *Cognitive, Affective, & Behavioral Neuroscience, 8*(2), 165–177. Retrieved from http://www.ncbi.nlm.nih.gov/pubmed/18589507.

Dahl, R. E., & Gunnar, M. R. (2009). Heightened stress responsiveness and emotional reactivity during pubertal maturation: Implications for psychopathology. *Development and Psychopathology, 21*(1), 1–6. Retrieved from http://www.ncbi.nlm.nih.gov/entrez/query.fcgi?cmd=Retrieve&db=PubMed&dopt=Citation&list_uids=19144219.

Davidson, M. C., Amso, D., Anderson, L. C., & Diamond, A. (2006). Development of cognitive control and executive functions from 4 to 13 years: Evidence from manipulations of memory, inhibition, and task switching. *Neuropsychologia, 44*(11), 2037–2078. doi:10.1016/j.neuropsychologia.2006.02.006

D'Esposito, M. (2007). From cognitive to neural models of working memory. *Philosophical Transactions of the Royal Society of London. Series B, Biological Sciences, 362*(1481), 761–772. doi:10.1098/rstb.2007.2086

Diamond, A. (2013). Executive functions. *Annual Review of Psychology, 64*, 135–168. doi:10.1146/annurev-psych-113011-143750

Dumontheil, I. (2014). Development of abstract thinking during childhood and adolescence: The role of rostrolateral prefrontal cortex. *Developmental Cognitive Neuroscience, 10*, 57–76. doi:10.1016/j.dcn.2014.07.009

Durston, S., Davidson, M. C., Tottenham, N., Galvan, A., Spicer, J., Fossella, J. A., & Casey, B. J. (2006). A shift from diffuse to focal cortical activity with development. *Developmental Science, 9*(1), 1–8. doi:10.1111/j.1467-7687.2005.00454.x

Ernst, M. (2014). The triadic model perspective for the study of adolescent motivated behavior. *Brain and Cognition, 89*, 104–111. doi:10.1016/j.bandc.2014.01.006

Ernst, M., Nelson, E. E., Jazbec, S., McClure, E. B., Monk, C. S., Leibenluft, E., ... Pine, D. S. (2005). Amygdala and nucleus accumbens in responses to receipt and omission of gains in adults and adolescents. *NeuroImage, 25*(4), 1279–1291. doi:10.1016/j.neuroimage.2004.12.038

Figner, B., Mackinlay, R. J., Wilkening, F., & Weber, E. U. (2009). Affective and deliberative processes in risky choice: Age differences in risk taking in the Columbia Card Task. *Journal of Experimental Psychology: Learning, Memory, and Cognition, 35*(3), 709–730. doi:10.1037/a0014983

Finn, A. S., Sheridan, M. A., Kam, C. L., Hinshaw, S., & D'Esposito, M. (2010). Longitudinal evidence for functional specialization of the neural circuit supporting working memory in the human brain. *Journal of Neuroscience, 30*(33), 11062–11067. doi:10.1523/JNEUROSCI.6266-09.2010

Forbes, E. E., Ryan, N. D., Phillips, M. L., Manuck, S. B., Worthman, C. M., Moyles, D. L., ... Dahl, R. E. (2010). Healthy adolescents' neural response to reward: Associations with puberty, positive affect, and depressive symptoms. *Journal of the American Academy of Child and Adolescent Psychiatry, 49*(2), 162–172, e161–165. Retrieved from http://www.ncbi.nlm.nih.gov/pubmed/20215938.

Friedman, N. P., Miyake, A., Altamirano, L. J., Corley, R. P., Young, S. E., Rhea, S. A., & Hewitt, J. K. (2016). Stability and change in executive function abilities from late adolescence to early adulthood: A longitudinal twin study. *Developmental Psychology, 52*(2), 326–340. doi:10.1037/dev0000075

Galvan, A., Hare, T. A., Parra, C. E., Penn, J., Voss, H., Glover, G., & Casey, B. J. (2006). Earlier development of the accumbens relative to orbitofrontal cortex might underlie risk-taking behavior in adolescents. *Journal of Neuroscience, 26*(25), 6885–6892. doi:10.1523/JNEUROSCI.1062-06.2006

Galvan, A., Hare, T., Voss, H., Glover, G., & Casey, B. J. (2007). Risk-taking and the adolescent brain: Who is at risk? *Developmental Science, 10*(2), F8–F14. doi:10.1111/j.1467-7687.2006.00579.x

Goddings, A. L., Mills, K. L., Clasen, L. S., Giedd, J. N., Viner, R. M., & Blakemore, S. J. (2014). The influence of puberty on subcortical brain development. *NeuroImage, 88*, 242–251. doi:10.1016/j.neuroimage.2013.09.073

Hampshire, A., Chamberlain, S. R., Monti, M. M., Duncan, J., & Owen, A. M. (2010). The role of the right inferior frontal gyrus: Inhibition and attentional control. *NeuroImage, 50*(3), 1313–1319. doi:10.1016/j.neuroimage.2009.12.109

Jolles, D. D., van Buchem, M. A., Rombouts, S. A., & Crone, E. A. (2011). Developmental differences in prefrontal activation during working memory maintenance and manipulation for different memory loads. *Developmental Science, 14*(4), 713–724.

Klingberg, T., Forssberg, H., & Westerberg, H. (2002). Increased brain activity in frontal and parietal cortex underlies the development of visuospatial working memory capacity during childhood. *Journal of Cognitive Neuroscience, 14*(1), 1–10. doi:10.1162/089892902317205276

Kwon, H., Reiss, A. L., & Menon, V. (2002). Neural basis of protracted developmental changes in visuo-spatial working memory. *Proceedings of the National Academy of Sciences of the United States of America, 99*(20), 13336–13341. doi:10.1073/pnas.162486399

Ladouceur, C. D., Peper, J. S., Crone, E. A., & Dahl, R. E. (2011). White matter development in adolescence: The influence of puberty and implications for affective disorders. *Developmental Cognitive Neuroscience, 2*(1), 36–54.

Libertus, M. E., Brannon, E. M., & Pelphrey, K. A. (2009). Developmental changes in category-specific brain responses to numbers and letters in a working memory task. *NeuroImage, 44*(4), 1404–1414. doi:10.1016/j.neuroimage.2008.10.027

Magis-Weinberg, L., Blakemore, S. J., & Dumontheil, I. (2017). Social and nonsocial relational reasoning in adolescence and adulthood. *Journal of Cognitive Neuroscience, 29*(10), 1739–1754. doi:10.1162/jocn_a_01153

Miyake, A., & Friedman, N. P. (2012). The nature and organization of individual differences in executive functions: Four general conclusions. *Current Directions in Psychological Science, 21*, 8–14.

Mohr, P. N., Biele, G., & Heekeren, H. R. (2010). Neural processing of risk. *Journal of Neuroscience, 30*(19), 6613–6619. doi:10.1523/JNEUROSCI.0003-10.2010

O'Hare, E. D., Lu, L. H., Houston, S. M., Bookheimer, S. Y., & Sowell, E. R. (2008). Neurodevelopmental changes in verbal working memory load-dependency: An fMRI investigation. *NeuroImage, 42*(4), 1678–1685. doi:10.1016/j.neuroimage.2008.05.057

Olesen, P. J., Nagy, Z., Westerberg, H., & Klingberg, T. (2003). Combined analysis of DTI and fMRI data reveals a joint maturation of white and grey matter in a fronto-parietal network. *Brain Research. Cognitive Brain Research, 18*(1), 48–57. Retrieved from http://www.ncbi.nlm.nih.gov/pubmed/14659496.

Op de Macks, Z., Gunther Moor, B., Overgaauw, S., Guroglu, B., Dahl, R. E., & Crone, E. A. (2011). Testosterone levels correspond with increased ventral striatum activation in response to monetary rewards in adolescents. *Developmental Cognitive Neuroscience, 1*(4), 506–516.

Ordaz, S. J., Foran, W., Velanova, K., & Luna, B. (2013). Longitudinal growth curves of brain function underlying inhibitory control through adolescence. *Journal of Neuroscience, 33*(46), 18109–18124. doi:10.1523/JNEUROSCI.1741-13.2013

Padmanabhan, A., Geier, C. F., Ordaz, S. J., Teslovich, T., & Luna, B. (2011). Developmental changes in brain function underlying the influence of reward processing on inhibitory control. *Developmental Cognitive Neuroscience, 1*(4), 517–529. doi:10.1016/j.dcn.2011.06.004

Peper, J. S., Mandl, R. C., Braams, B. R., de Water, E., Heijboer, A. C., Koolschijn, P. C., & Crone, E. A. (2013). Delay discounting and frontostriatal fiber tracts: A combined DTI and MTR study on impulsive choices in healthy young adults. *Cerebral Cortex, 23*(7), 1695–1702. doi:10.1093/cercor/bhs163

Peters, S., Braams, B. R., Raijmakers, M. E., Koolschijn, P. C., & Crone, E. A. (2014). The neural coding of feedback learning across child and adolescent development. *Journal of Cognitive Neuroscience, 26*(8), 1705–1720. doi:10.1162/jocn_a_00594

Peters, S., & Crone, E. A. (2017). Increased striatal activity in adolescence benefits learning. *Nature Communications, 8*(1), 1983. doi:10.1038/s41467-017-02174-z

Peters, S., van der Meulen, M., Zanolie, K., & Crone, E. A. (2017). Predicting reading and mathematics from neural activity for feedback learning. *Developmental Psychology, 53*(1), 149–159. doi:10.1037/dev0000234

Peters, S., van Duijvenvoorde, A. C., Koolschijn, P. C., & Crone, E. A. (2016). Longitudinal development of frontoparietal activity during feedback learning: Contributions of age, performance, working memory and cortical thickness. *Developmental Cognitive Neuroscience, 19*, 211–222. doi:10.1016/j.dcn.2016.04.004

Qu, Y., Galvan, A., Fuligni, A. J., Lieberman, M. D., & Telzer, E. H. (2015). Longitudinal changes in prefrontal cortex activation underlie declines in adolescent risk taking. *Journal of Neuroscience, 35*(32), 11308–11314. doi:10.1523/JNEUROSCI.1553-15.2015

Rubia, K., Smith, A. B., Woolley, J., Nosarti, C., Heyman, I., Taylor, E., & Brammer, M. (2006). Progressive increase of frontostriatal brain activation from childhood to adulthood during event-related tasks of cognitive control. *Human Brain Mapping, 27*(12), 973–993. doi:10.1002/hbm.20237

Ruff, C. C., & Fehr, E. (2014). The neurobiology of rewards and values in social decision making. *Nature Reviews Neuroscience, 15*(8), 549–562. doi:10.1038/nrn3776

Satterthwaite, T. D., Wolf, D. H., Erus, G., Ruparel, K., Elliott, M. A., Gennatas, E. D., ... Gur, R. E. (2013). Functional maturation of the executive system during adolescence. *Journal of Neuroscience, 33*(41), 16249–16261. doi:10.1523/JNEUROSCI.2345-13.2013

Scherf, K. S., Sweeney, J. A., & Luna, B. (2006). Brain basis of developmental change in visuospatial working memory. *Journal of Cognitive Neuroscience, 18*(7), 1045–1058. doi:10.1162/jocn.2006.18.7.1045

Schreuders, E., Braams, B. R., Blankenstein, N. E., Peper, J. S., Guroglu, B., & Crone, E. A. (2018). Contributions of reward sensitivity to ventral striatum activity across adolescence and early adulthood. *Child Development*. doi:10.1111/cdev.13056

Silverman, M. H., Jedd, K., & Luciana, M. (2015). Neural networks involved in adolescent reward processing: An activation likelihood estimation meta-analysis of functional neuroimaging studies. *NeuroImage, 122*, 427–439. doi:10.1016/j.neuroimage.2015.07.083

Spear, L. P. (2011). Rewards, aversions and affect in adolescence: Emerging convergences across laboratory animal and human data. *Developmental Cognitive Neuroscience, 1*, 390–403.

St. Clair-Thompson, H. L., & Gathercole, S. E. (2006). Executive functions and achievements in school: Shifting, updating, inhibition, and working memory. *Quarterly Journal of Experimental Psychology, 59*(4), 745–759. doi:10.1080/17470210500162854

Steinberg, L. (2011). *The science of adolescent risk-taking*. Washington, DC: National Academies Press.

Strang, N. M., Chein, J. M., & Steinberg, L. (2013). The value of the dual systems model of adolescent risk-taking. *Frontiers in Human Neuroscience, 7*, 223. doi:10.3389/fnhum.2013.00223

Tamm, L., Menon, V., & Reiss, A. L. (2002). Maturation of brain function associated with response inhibition. *Journal of the American Academy of Child and Adolescent Psychiatry, 41*(10), 1231–1238. doi:10.1097/00004583-200210000-00013

Thomason, M. E., Race, E., Burrows, B., Whitfield-Gabrieli, S., Glover, G. H., & Gabrieli, J. D. (2009). Development of spatial and verbal working memory capacity in the human brain. *Journal of Cognitive Neuroscience, 21*(2), 316–332. doi:10.1162/jocn.2008.21028

van den Bos, W., Cohen, M. X., Kahnt, T., & Crone, E. A. (2012). Striatum-medial prefrontal cortex connectivity predicts developmental changes in reinforcement learning. *Cerebral Cortex, 22*(6), 1247–1255. doi:10.1093/cercor/bhr198

van den Bos, W., Guroglu, B., van den Bulk, B. G., Rombouts, S. A., & Crone, E. A. (2009). Better than expected or as bad as you thought? The neurocognitive development of probabilistic feedback processing. *Frontiers in Human Neuroscience, 3*, 52. doi:10.3389/neuro.09.052.2009

van Duijvenvoorde, A. C., & Crone, E. A. (2013). A neuroeconomic approach to adolescent decision making. *Current Directions in Psychological Science,22*(2), 108–113.

van Duijvenvoorde, A. C., Huizenga, H. M., Somerville, L. H., Delgado, M. R., Powers, A., Weeda, W. D., ... Figner, B. (2015). Neural correlates of expected risks and returns in risky choice across development. *Journal of Neuroscience, 35*(4), 1549–1560. doi:10.1523/JNEUROSCI.1924-14.2015

van Duijvenvoorde, A. C., Jansen, B. R., Bredman, J. C., & Huizenga, H. M. (2012). Age-related changes in decision making: Comparing informed and noninformed situations. *Developmental Psychology, 48*(1), 192–203. doi:10.1037/a0025601

van Duijvenvoorde, A. C., Zanolie, K., Rombouts, S. A., Raijmakers, M. E., & Crone, E. A. (2008). Evaluating the negative or valuing the positive? Neural mechanisms supporting feedback-based learning across development. *Journal of Neuroscience, 28*(38), 9495–9503. doi:10.1523/JNEUROSCI.1485-08.2008

van Leijenhorst, L., Gunther Moor, B., Op de Macks, Z. A., Rombouts, S. A., Westenberg, P. M., & Crone, E. A. (2010). Adolescent risky decision-making: Neurocognitive development of reward and control regions. *NeuroImage, 51*(1), 345–355. doi:10.1016/j.neuroimage.2010.02.038

van Leijenhorst, L., Zanolie, K., van Meel, C. S., Westenberg, P. M., Rombouts, S. A., & Crone, E. A. (2010). What motivates the adolescent? Brain regions mediating reward sensitivity across adolescence. *Cerebral Cortex, 20*(1), 61–69. doi:10.1093/cercor/bhp078

Velanova, K., Wheeler, M. E., & Luna, B. (2008). Maturational changes in anterior cingulate and frontoparietal recruitment support the development of error processing and inhibitory control. *Cerebral Cortex, 18*(11), 2505–2522. doi:10.1093/cercor/bhn012

Wendelken, C., O'Hare, E. D., Whitaker, K. J., Ferrer, E., & Bunge, S. A. (2011). Increased functional selectivity over development in rostrolateral prefrontal cortex. *Journal of Neuroscience, 31*(47), 17260–17268. doi:10.1523/JNEUROSCI.1193-10.2011

Whitaker, K. J., Vendetti, M. S., Wendelken, C., & Bunge, S. A. (2018). Neuroscientific insights into the development of analogical reasoning. *Developmental Science, 21*(2). doi:10.1111/desc.12531

Zanolie, K., van Leijenhorst, L., Rombouts, S. A., & Crone, E. A. (2008). Separable neural mechanisms contribute to feedback processing in a rule-learning task. *Neuropsychologia, 46*(1), 117–126. doi:10.1016/j.neuropsychologia.2007.08.009

4 Social Cognition and Social Brain Development in Adolescence

EMMA J. KILFORD AND SARAH-JAYNE BLAKEMORE

ABSTRACT Adolescence is a time of pronounced social, affective, and cognitive development, during which the social world becomes increasingly nuanced and dynamic. Social cognitive processes are critical in navigating complex social interactions and are associated with a network of brain areas termed the *social brain*. Neuroimaging and behavioral studies have demonstrated that the social brain undergoes significant and protracted structural and functional development during human adolescence, as do social cognitive abilities such as face-processing, mentalizing, perspective-taking, and social decision-making. The development of the social brain and social cognition does not occur in isolation but in the context of developments in other neurocognitive systems, such as those implicated in cognitive control and motivational-affective processes. Social contexts are a key source of motivational-affective responses, particularly during adolescence, when social factors increase in salience and value. The successful transition to adulthood requires the rapid refinement and integration of these processes, and many adolescent-typical behaviors, such as peer influence and sensitivity to social exclusion, involve dynamic interactions between these systems.

Adolescence can be defined as the period of life between the biological changes of puberty and the achievement of self-sufficiency and the individual attainment of a stable, independent role in society (Blakemore & Mills, 2014). While the concept of adolescence is recognized across cultures and throughout history, the nature of its biopsychosocial definition can make it challenging to define chronologically, as the timing of both pubertal onset and adult role transition varies both between and within cultures (Sawyer, Azzopardi, Wickremarathne, & Patton, 2018). This transitional period of development has long been associated with physical, social, behavioral, and cognitive changes. More recently, advances in brain imaging technology have enabled an increased understanding of structural and functional changes in the human brain during this developmental period and how they relate to social, affective, and cognitive development.

Many social changes occur during adolescence. These include the fact that, compared with children, adolescents form more complex and hierarchical peer relationships and are more sensitive to acceptance and rejection by their peers (Brown, 2004; Steinberg & Morris, 2001). Although the factors that underlie these social changes are likely to be multifaceted, one possible contributing factor is the development of the *social brain*, the network of brain areas involved in social perception and cognition (Frith & Frith, 2007). In this chapter we focus on the development of social cognition and the social brain in adolescence.

Social Cognition and the Social Brain

Social cognition refers to the ability to make sense of the world through processing signals generated by other members of the same species and encompasses a wide range of cognitive processes that enable individuals to understand and interact with one another (Frith & Frith, 2007). These include social perceptual processes, such as face processing, biological motion detection, and joint attention, in addition to more complex social cognitive processes involving inference and reasoning, such as mentalizing—the process of mental state attribution. Such social cognitive processes enable us to understand and predict the mental states, intentions, and actions of others and to modify our own accordingly (Frith & Frith, 2007). Social cognition thus plays a critical role in the successful negotiation of complex social interactions and decisions (Crone, 2013).

A wide network of brain areas, referred to as the *social brain network*, is involved in social perception and cognition. Regions within the social brain network include the posterior superior temporal sulcus (pSTS), the temporoparietal junction (TPJ), the dorsomedial prefrontal cortex (dmPFC; medial aspects of Brodmann area 10; mBA10), and the anterior temporal cortex (ATC) (Frith & Frith, 2007).

Structural Development of the Social Brain

Areas within the social brain network are among the regions that undergo the most protracted development

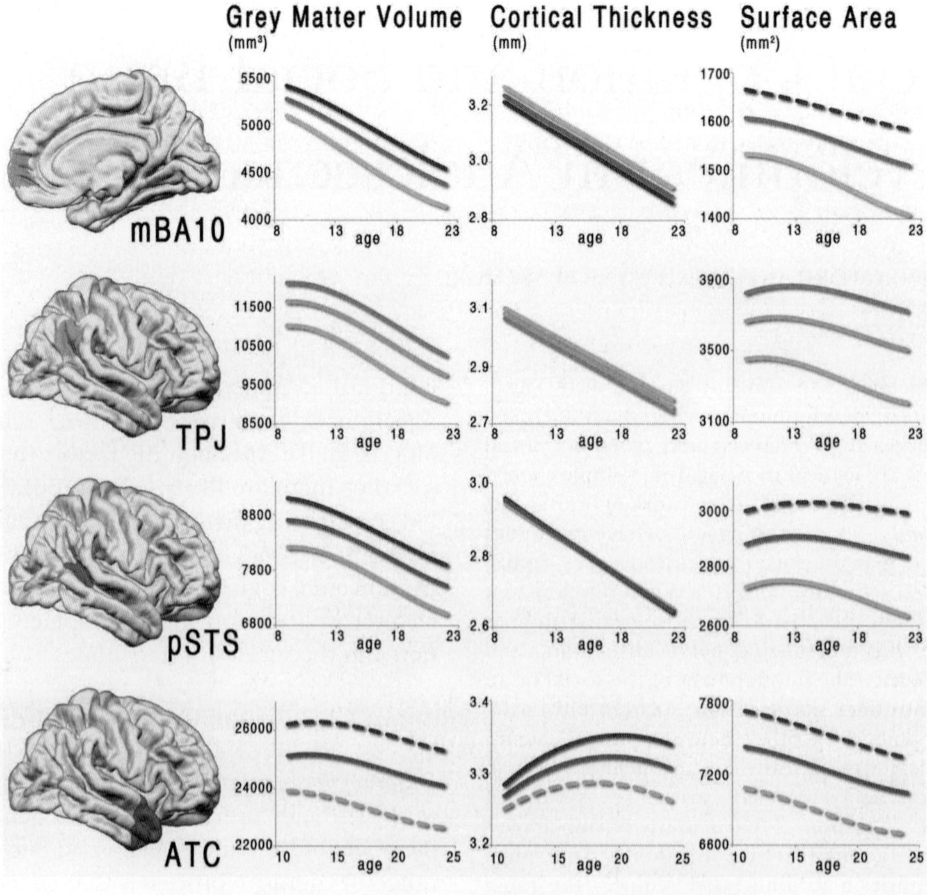

FIGURE 4.1 Structural development of the social brain. Structural developmental trajectories of brain areas associated with mentalizing across adolescence (gray matter volume, cortical thickness, surface area). The best-fitting models for all participants are shown for each region of interest (combined hemispheres). Models are fitted to the middle 80% of the sample (ages 9–22 years for mBA10, TPJ, and pSTS; ages 11–24 years for ATC). The lighter lines show the fitted models applied to females only, and the darker lines show the fitted models applied to males only. Solid lines indicate the fitted model was significant ($p < 0.05$), whereas dashed lines indicate the fitted model was not significant ($p \geq 0.05$). Reproduced from Mills et al. (2014). (See color plate 3.)

in humans, showing changes throughout adolescence before relatively stabilizing in the early to mid-20s. Mills, Lalonde, Clasen, Giedd, and Blakemore (2014) examined the structural developmental trajectories (gray matter volume, cortical thickness, and surface area) of brain areas associated with mentalizing. In a sample of 288 individuals with at least two brain scans between the ages of 7 and 30, they found that gray matter volume and cortical thickness decreased in the dmPFC, the TPJ, and the pSTS from childhood into the early 20s. In contrast, the ATC increased in gray matter volume until early adolescence (~12 years), decreasing thereafter, whereas cortical thickness increased until early adulthood (~19 years). Surface area in all four regions followed a cubic trajectory, reaching a peak in late childhood or early adolescence before decreasing in the early 20s (figure 4.1).

Development of Social Cognition and Social Brain Function

While certain social cognitive processes are present from an early age (see chapter 1), behavioral and neuroimaging studies have shown developments in a number of more complex social cognitive abilities and functional changes in associated brain networks across adolescence.

Face-processing A vast array of social information can be extracted from the faces of those around us, including identity, emotional expression, and direction of eye-gaze. In adolescence, new aspects of face-processing become increasingly important as more complex social cognitive abilities develop, such as judgments of attractiveness or social status, which rely on the successful detection and interpretation of communicative signals from face-processing. A small number of studies have suggested that face-processing abilities continue to improve during adolescence, with some studies observing a dip in face-processing proficiency in early adolescence. While the core face-processing network appears to be present from around 7 years of age, face-processing abilities and the brain systems that support them continue to show age-related changes between adolescence and adulthood (reviewed in Blakemore, 2008).

Evidence indicates that face-processing systems become increasingly specialized, particularly in relation to the more complex aspects of face-processing, such as detecting socially relevant information. Cohen Kadosh, Cohen Kadosh, Dick, and Johnson (2011) and Cohen Kadosh, Johnson, Dick, Cohen Kadosh, and Blakemore (2013) used a series of functional magnetic resonance imaging (fMRI) target detection tasks (detection of facial identity, expression, or eye-gaze direction) to investigate the developmental changes in the functional connectivity of the face-processing network when performing tasks involving face property-specific processes. Children, adolescents, and young adults activated the core face-processing network across all different tasks; however, the extent to which network connectivity was selectively modulated according to face-processing task demands increased between the ages of 7 and 11 (Cohen Kadosh et al., 2011). In adults, detecting a facial identity was associated with selective strengthening of the functional connectivity between the inferior occipital gyrus (IOG) and the fusiform gyrus, whereas detecting an emotional expression modulated connectivity between the IOG and the STS. No such modulation was found in children. A follow-up study indicated that with age there were also increases in task-specific activations that were additionally associated with changes in gray and white matter volume. These changes were distinct from more widespread activations that varied as a function of individual task performance (Cohen Kadosh et al., 2013). Thus, changes in face-processing during the first three decades of life appear to be related both to the continued structural development of underlying brain areas and to individual task performance, which might reflect changing cognitive strategies across development and experience (Cohen Kadosh et al., 2013).

Mentalizing Mentalizing describes the ability to make attributions about the mental states of others, including their beliefs, thoughts, desires, intentions, and feelings. There is a rich literature on the development of mentalizing in childhood, pointing to changes in the ability to understand others' mental states during the first five years of life (Frith & Frith, 2007). While certain aspects of mentalizing are present in infancy (Baillargeon, Scott, & He, 2010), it is not until around age 4 that children begin to explicitly understand that someone else can hold a belief that differs from their own and that this can be false. Until fairly recently, there was a shortage of studies looking into mentalizing after childhood, as it was generally assumed that these abilities were already mature by mid-childhood in typically developing children. However, adolescence is marked by substantial changes in social competence and social behavior, as well as structural development within the social brain (Mills et al., 2014). These changes may be paralleled by changes in the neural processing of mentalizing.

Neuroimaging studies of mentalizing have consistently found associations with activity within the dmPFC, TPJ, pSTS, and ATC, suggesting these regions are key to the process of mental state attribution. A number of developmental fMRI studies of mentalizing report decreases in dmPFC recruitment between adolescence and adulthood (reviewed in Blakemore, 2008). These studies have used a variety of tasks that require mental state attribution, such as thinking about social emotions such as guilt (Burnett, Bird, Moll, Frith, & Blakemore, 2009), understanding intentions (Blakemore, den Ouden, Choudhury, & Frith, 2007), thinking about the preferences and dispositions of oneself or a fictitious story character (Pfeifer et al., 2009), and making attributions about the emotional states of others (Gunther Moor et al., 2012). An example of such a task is the Reading the Mind in the Eyes Test (RMET; Baron-Cohen, Wheelwright, Hill, Raste, & Plumb, 2001), which assesses the ability to perceive, categorize, and make attributions about other people's mental and affective states based only on photographs of their eyes. Gunther Moor et al. (2012) used fMRI to compare brain activation while performing the RMET between early adolescents (10–12 years), mid-adolescents (14–16 years), and young adults (19–23 years) relative to a control condition (age and gender categorizations of the same stimuli). Whereas participants of all ages showed increased activation in the pSTS during the task, participants in the youngest group exhibited additional engagement of the dmPFC (Gunther Moor et al., 2012). A follow-up study that retested the same participants two years later indicated that these cross-sectional differences reflected longitudinal changes within individuals. Specifically, dmPFC activation during the RMET followed a quadratic developmental trajectory, being lowest during mid-adolescence (Overgaauw, van Duijvenvoorde, Moor, & Crone, 2015).

In some studies, higher activity in more posterior regions of the social brain, such as the pSTS/TPJ

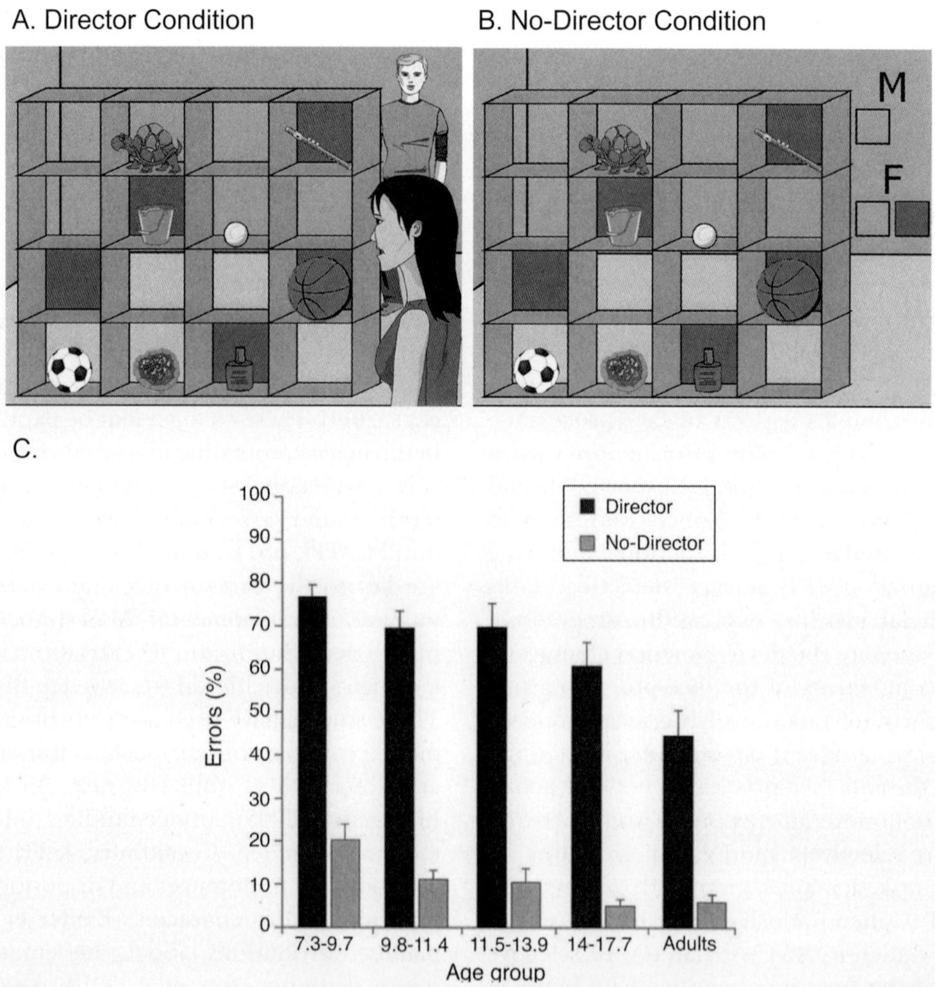

FIGURE 4.2 The Director task. Example stimuli from the Director task, including a social (A: Director) and nonsocial control condition (B: No-Director). In this example, in both conditions participants hear the instruction "Move the large ball up" in either a male or a female voice. In both examples, if the voice is female the correct object to move would be the basketball because in the Director condition (*A*) the female director is standing in front of the shelves and can see all the objects, and in the No-Director condition (*B*), the two boxes below the *F* (for "female") indicate that all objects can be moved by the participant. If the voice is male, the correct object to move would be the football (soccer ball) because in the Director condition (*A*) the male director is standing behind the shelves and therefore cannot see the larger basketball in the occluded slot, whereas in the No-Director condition (*B*) the single clear box below the *M* (for "male") indicates that only objects in open shelves can be moved/that no objects in front of a gray background can be moved. *C*, Adolescents (14–17 years) make more errors than adults in the Director condition, whereas in the No-Director condition no difference is found between these age groups. Adapted from Dumontheil, Apperly & Blakemore (2010); Dumontheil, Hillebrandt, Apperly & Blakemore (2012).

(Blakemore et al., 2007), and in the ATC (Burnett et al., 2009), was observed in adults as compared to adolescents.

Perspective-taking The ability to take another person's point of view into account, known as perspective-taking, is an important determinant of successful social functioning in everyday life. Fundamental aspects of perspective-taking develop during childhood. However, the ability to use these social competencies online continues to develop throughout adolescence.

The Director task can be used to investigate the ability to use perspective-taking to guide decisions in a referential communicative context. Participants are instructed to move objects around a set of shelves by a director, who cannot see some of the objects that the participant can see (figure 4.2; Apperly et al., 2010; Dumontheil, Apperly, & Blakemore, 2010). Correct interpretation of the instructions requires participants to use the director's perspective and only move objects that the director can see. Adult participants frequently make errors in this task, suggesting that despite possessing the ability to use mentalizing to understand that the director's perspective differs from their own, they often fail to use this information to guide decisions (Keysar, Lin, & Barr, 2003).

In a sample of participants aged 7–27 years, Dumontheil, Apperly, et al. (2010) used a computerized adaptation of the Director task. To differentiate between the general impact of cognitive control demands on task performance and effects that have a specific impact on the social components of the task, a control condition was included. In this condition the director was absent, and participants had to use a nonsocial rule to guide their decisions while following the (otherwise) identical instructions as in the Director condition. Although accuracy improved until mid-adolescence in both conditions, accuracy in the Director condition continued to improve after mid-adolescence. Similar findings were also observed in a more recent study (Symeonidou, Dumontheil, Chow, & Breheny, 2015). These findings suggest that the ability to use another's perspective to guide decisions continues to develop in late adolescence, over and above developmental improvements in more general cognitive control processes recruited by both conditions.

However, the improved integration of social cognition and cognitive control systems in adolescence may also contribute to developmental advances in perspective-taking (Dumontheil, Apperly, et al., 2010). Using a variant of the Director task adapted for fMRI (Apperly et al., 2010; Dumontheil, Küster, Apperly & Blakemore, 2010), in which the cognitive demands of the task were also manipulated, Dumontheil et al. (2012) demonstrated developmental differences in brain areas associated with both social cognitive and more domain-general cognitive control processes. In both social and nonsocial conditions, adults recruited frontoparietal regions associated with cognitive control more than adolescents. When social cues were needed to accurately perform the task, both adults and adolescents (11–16 years) recruited the dmPFC; however, adolescents also recruited the dmPFC when social cues were not needed. The authors suggest that this engagement of the dmPFC for irrelevant social stimuli may reflect the use of social brain regions even when they are not necessary, consistent with a pattern of increasing specialization within networks supporting social cognition.

Social decision-making Another line of research into the dynamic and interactive aspects of social cognition has employed paradigms from the field of behavioral economics to study the development of the cognitive and neural mechanisms that underlie more complex aspects of social exchanges, such as social preferences for fairness, trustworthiness, or cooperation (van Duijvenvoorde & Crone, 2013). Even though a basic sense of fairness in bargaining is observed in young children (Güroğlu, van den Bos, & Crone, 2009), the understanding of intentionality in social interactions develops gradually over the course of adolescence and early adulthood (Güroğlu, van den Bos, & Crone, 2009; van den Bos, Westenberg, van Dijk, & Crone, 2010). Age-related changes in social behavior beyond childhood, such as increases in trust and reciprocity during social interactions (Belli, Rogers, & Lau, 2012; van den Bos et al., 2010), may be associated with the increasing tendency to consider others' viewpoints and intentions.

In an fMRI study, young adolescents (12–14 years), older adolescents (15–17 years), and young adults (18–22 years) played the role of the second player in the Trust Game (van den Bos, van Dijk, Westenberg, Rombouts, & Crone, 2011). An anonymous first player would give an amount of money as an investment, which the participant could either divide equally between herself and the investor or keep the majority for herself. Participants' tendency to take the perspective of the first player into account was investigated by examining their sensitivity to the degree of risk (i.e., the amount of money that could be lost) taken by the investor. Whereas older adolescents and young adults were more likely to reciprocate when the investor trusted them and therefore risked losing larger amounts of money, the younger adolescents did not differentiate. These findings suggest that adolescence is not necessarily characterized by general increases in prosocial behavior but by an increase in sensitivity to the perspective of others.

These forms of perspective-taking behavior were associated with increased involvement of the left TPJ and the right dorsolateral PFC (dlPFC), which the authors suggest indicates a role for both social cognitive and cognitive control systems in the development of social behavior in adolescence. When participants observed that the first player trusted them, recruitment of the left TPJ increased with age, and this level of activation correlated with participants' sensitivity to the first player. Participants also showed increased engagement of the right dlPFC with age when receiving trust, which the authors speculate may indicate a regulatory role of the right dlPFC in social exchange—for example, in the inhibition of more egocentric behavior—although the precise psychological mechanisms cannot be ascertained from neural activity patterns (van den Bos et al., 2011).

Fett, Shergill, et al. (2014) investigated the relationship between perspective-taking and social processes—such as trust and reciprocity—in adolescence, using two variants of the Trust Game and the Director task.

Adolescents (13–18 years) with a higher perspective-taking tendency (measured as accuracy on the Director task) demonstrated greater trust toward others (initial investment in the Trust Game) and higher levels of trust during cooperative interactions (higher investments). While all adolescents modified their behavior in response to unfair interactions (decreased investments and more malevolent reciprocity) when they were treated unfairly, high perspective-takers did so more drastically, suggesting a greater decrease in trust. The authors propose that increases in perspective-taking tendencies in adolescence are therefore associated with specific developmental changes in trust and reciprocity, as opposed to simply generalized increases in prosocial behavior. Although this study did not include adult participants, the behavioral patterns of high perspective-takers were similar to those observed in adults in another study using this paradigm (Fett, Gromann, Giampietro, Shergill, & Krabbendam, 2014).

Behavioral findings add support to this conclusion. In a series of allocation games in which the identity of the interaction partner was manipulated (friends, antagonists, neutral classmates, anonymous peers), younger adolescents (9 and 12 years) showed similar levels of prosocial behavior to all interaction partners (Güroğlu, van den Bos, & Crone, 2014). However, older adolescents (15 and 18 years) showed increasing differentiation in prosocial behavior according to their relation with the partner, displaying the most prosocial behavior (both costly and noncostly) toward friends. This suggests that with age, *who* you are interacting with becomes more important. The age-related increase in noncostly prosocial behavior toward friends was mediated by self-reported perspective-taking skills.

Interactions between Social Cognition, Cognitive Control, and Motivational-Affective Processing in Adolescence

While there is a growing body of work investigating the interplay between cognitive control and motivational-affective processing, less is known about the way in which these processes interact with social cognitive processes, social contexts, and stimuli. Social interactions are a key source of elicited motivational-affective responses: social cues can elicit robust affective responses, and those around us can be a salient source of potential rewards and punishments. Socioaffective context, such as the heightened motivational salience of peers or the affective appraisal of the value of an outcome, appears to exert a great influence on the extent to which cognitive control systems are recruited in adolescence. This context can be external—for example, one's social context—or internal, such as one's affective state.

Social cognition and affective processing The perception, understanding, and interpretation of others' emotions is a fundamental aspect of social interaction and requires the integration of a range of perceptual, social cognition, and affective skills (Garcia & Scherf, 2015). These include basic aspects of affective processing, such as emotion perception and recognition, and more complex social cognitive processes, such as the ability to understand the affective states of others, sometimes referred to as *affective mentalizing*.

Affective mentalizing requires the integration of both social cognitive and affective processing networks. Sebastian et al. (2012) examined the development of affective (understanding emotions) and cognitive mentalizing (understanding thoughts, perspectives, and intentions) and their neural substrates during adolescence using cartoon vignettes. Both types of mentalizing were associated with activation in social brain network regions (including dmPFC, pSTS/TPJ, and ATC); however, affective mentalizing also elicited activation in the ventromedial PFC, which was greater in adolescents compared to adults. These findings extend the pattern of decreased dmPFC activation between childhood and adulthood to aspects of medial PFC more typically associated with affective processing and highlight the importance of considering the integration, overlap, and interplay of multiple developing brain regions and networks when investigating the development of complex social skills and behaviors during adolescence.

Affective consequences of social interactions *Social affect* refers to the interaction between our emotions and our behavior in the context of communication with others. The highly rewarding nature of peer interactions during adolescence is believed to increase the impact of both positive and negative aspects of such interactions. Social situations can evoke strong emotional responses, and there is evidence that this is particularly great in adolescence.

Studies of peer rejection in adolescence, using a range of experimental paradigms, repeatedly find that peer rejection is associated with worsened mood, increased distress, and increased anxiety in adolescents compared to child and adult groups, particularly in younger adolescents (reviewed in Platt, Cohen Kadosh, & Lau, 2013). Similarly, Silvers, McRae, Gabrieli, and Gross (2012) found that when viewing socioaffective stimuli, as opposed to nonsocial affective stimuli, younger adolescents had greater difficulty regulating their emotions than did older adolescents and adults.

Studies of the neural bases of emotional regulation in the context of social rejection have implicated prefrontal regions—notably, the right ventrolateral PFC (vlPFC). Compared to adults, adolescents show reduced activation of this region during experimental manipulations of social rejection, such as the Cyberball game (Bolling et al., 2011; Masten et al., 2009; Sebastian et al., 2011). It has been suggested that developmental increases in vlPFC activation may be associated with an increased regulation of social distress following exclusion. Consistent with this hypothesis, within adolescents, greater right vlPFC activation during exclusion has been associated with higher levels of parent-reported interpersonal competence, lower self-reported distress (Masten et al., 2009), and lower self-reported resistance to peer influence (Sebastian et al., 2011). Furthermore, Bolling et al. (2011) found age-related increases in functional connectivity between the right vlPFC and the ventral anterior cingulate cortex, an effect found only during social exclusion and not during a similar task in which social expectancies were violated but participants were not excluded.

Social context and peer influence Social context can have an impact on decision-making, such as the propensity to engage in prosocial or risky behaviors. Adolescents are often characterized as prone to engage in risky decision-making, which, although probably adaptive in many circumstances (Blakemore & Mills, 2014; Willoughby, Good, Adachi, Hamza, & Tavernier, 2013), can sometimes result in negative real-life outcomes, such as unsafe sexual behavior, substance abuse, and dangerous driving (Viner et al., 2011; Willoughby et al., 2013). As a result of these epidemiological findings, experimental studies of social context in adolescence have largely focused on the impact of peer influence on adolescent risk-taking behavior.

The presence of peers affects how likely adolescents are to take risks in a driving-simulation game. While adolescents (13–16 years), young adults (18–22 years), and adults (24 years and older) have been shown to take around the same number of driving risks when alone, in the presence of their friends adolescents took significantly more risks, whereas peer presence had no impact on risk-taking in adults and an intermediate effect in young adults (Gardner & Steinberg, 2005). Adolescents are also sensitive to the presence of peers when performing other experimental tasks involving risky and reward-related decisions (O'Brien, Albert, Chein, & Steinberg, 2011; Smith, Chein, & Steinberg, 2014).

The increased sensitivity to the presence of peers found in risky and reward-related decision-making appears to extend to other aspects of cognition. One study found that the presence of peers had a detrimental effect on female adolescents' performance on a high-level cognitive task (relational reasoning; Wolf, Bazargani, Kilford, Dumontheil, & Blakemore, 2015). Pairs of female friends were randomly assigned as either a participant or an observer. The participant then performed the task in three social contexts: alone, observed by a friend, or observed by an experimenter. Social context affected adolescent, but not adult, performance, an effect that was also influenced by the participants' age and the task difficulty. Older adolescents (14.9–17.8 years) exhibited poorer performance when being observed by their friend relative to the experimenter, independent of task difficulty, while younger adolescents (10.6–14.2 years) only showed this effect for easier reasoning trials (Wolf et al., 2015).

Social context can also modulate adolescents' attitudes toward risk (Engelmann, Moore, Capra, & Berns, 2012). A study investigated the development of social influence on risk perception from late childhood to adulthood by asking participants to rate the riskiness of everyday situations (Knoll, Magis-Weinberg, Speekenbrink, & Blakemore, 2015; figure 4.3). After recording their rating, individuals were informed about the ratings of a social-influence group (teenagers or adults) before rating each situation again. All age groups showed a significant social-influence effect, changing their risk ratings in the direction of the provided ratings, and this social-influence effect decreased with age. Most age groups adjusted their ratings more to conform to the ratings of the adult social-influence group than to the ratings of the teenage social-influence group. Only young adolescents (aged 12–14) were more strongly influenced by the teenage social-influence group than by the adult social-influence group, which suggests that to early adolescents, the opinions of other teenagers about risk matter more than the opinions of adults.

In contrast to the presence of peers, parents can have a protective effect on risk-taking in adolescence. Telzer, Ichien, and Qu (2015) showed that adolescents demonstrated reduced risk-taking behavior when their mothers were present compared with when alone. Safe decision-making was associated with greater recruitment of the vlPFC and greater functional coupling between the ventral striatum (VS) and the vlPFC, while risky decision-making was associated with decreased activation in the VS. The authors propose that heightened adolescent sensitivity in neural circuitry that is associated with greater risk-taking can also be redirected toward making thoughtful, more deliberative, and safer decisions.

Peer influence is largely associated with negative outcomes; however, peers can also have a positive influence

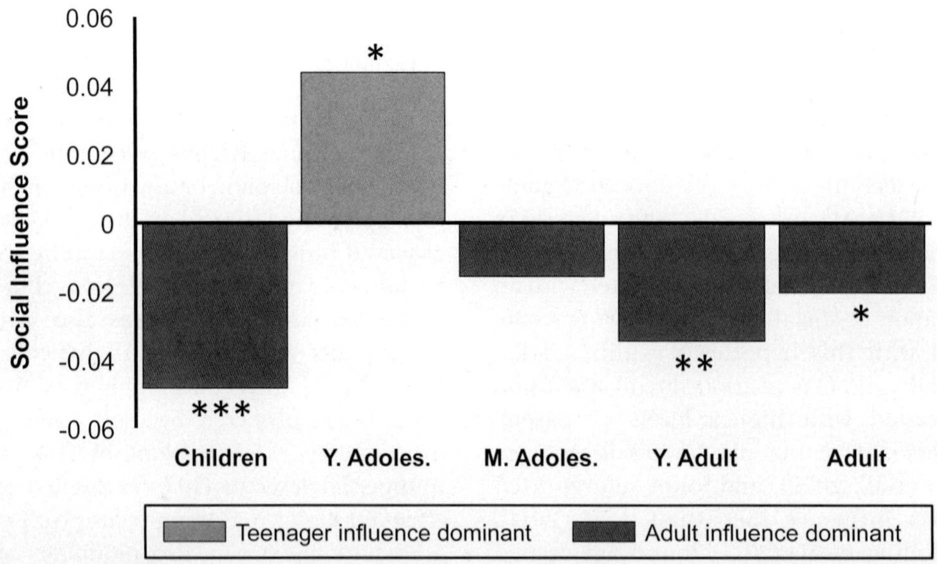

FIGURE 4.3 Effect of social influence on risk ratings. Participants (N=563) rated the riskiness of everyday situations before and after they were informed about the ratings of a social influence group (teenagers or adults). Social influence score, an index of conformity to other people's ratings, is shown relative to the source of the social influence for five age groups: children (aged 8–11 years), young adolescents (Y. Adoles., aged 12–14 years), mid-adolescents (M. Adoles., aged 15–18 years), young adults (Y. Adult, aged 19–25 years), and adults (aged 26–59 years). The difference in social influence effect between social influence groups (adults vs. teenagers) is shown for each age group (***$p<0.001$; **$p<0.01$; *$p<0.05$). Adapted from Fuhrmann, Knoll, & Blakemore, 2015; data published in Knoll et al. (2015).

on behavior. A study demonstrated that prosocial feedback from peers was associated with increased prosocial behavior compared to either no feedback or to antisocial feedback, which was associated with decreased prosocial behavior (van Hoorn, van Dijk, Meuwese, Rieffe, & Crone, 2014). The tendency to moderate behavior in line with the values of the people we are with likely involves both regulatory and social cognitive processes. Greater neural activity within cognitive control regions during a response inhibition task predicted safer decisions in the driving-simulation game, specifically when in the presence of a peer who expressed risk-averse attitudes, as compared to a risk-promoting peer (Cascio et al., 2015). The authors suggest that the ability to override risky tendencies in the presence of cautious peers may therefore be associated with individual differences in the systems involved in top-down cognitive control. In a follow-up fMRI study of the influence of peers on prosocial behavior, van Hoorn, van Dijk, Güroğlu, and Crone (2016) demonstrated that the presence of peers during decision-making resulted in increased activity in several social brain regions, including the dmPFC, TPJ, and pSTS, and that TPJ activity was positively associated with the degree to which peer feedback increased prosocial behavior. These findings may suggest that individual differences in social cognitive processes, such as perspective-taking, result in variation in the influence of peers on prosocial behavior.

In everyday life, the attitudes and values of those around us are not always explicitly expressed but must instead be inferred using social cognitive processes. Studies investigating the relationship between risk-taking and social exclusion in adolescence suggest that in addition to cognitive control processes, social cognition processes may also be uniquely implicated in adolescents' vulnerability to peer influence on risk-taking. Falk et al. (2014) used the Cyberball game to examine whether neural activation during simulated social exclusion predicted peer influence on risky decisions in the driving-simulation game one week later in 16- to 17-year-olds. Activity in the social brain network (dmPFC, right TPJ, posterior cingulate cortex) during social exclusion was positively associated with increased risk-taking when playing the game in the presence of a peer, relative to alone. In a further study, adolescent participants (14–17 years) completed the driving-simulation game—while in the implied presence of two online peers—before and after being socially excluded by these peers (Peake, Dishion, Stormshak, Moore, & Pfeifer, 2013). Exclusion was associated with greater behavioral risk-taking among adolescents with low self-reported resistance to peer influence (RPI). When making risky decisions after social exclusion, adolescents who had lower RPI exhibited higher levels of activity in the right TPJ, and this response was a significant mediator of the relationship between RPI and greater risk-taking after social exclusion. Lower RPI was also associated with lower levels of activity in the lateral PFC during risks following social exclusion, but this did not mediate the relationship between RPI and greater risk-taking after social exclusion.

Conclusion

Research over the past decade has demonstrated the substantial and protracted development of multiple aspects of social cognition, as well as the structural and functional development of the social brain network, across human adolescence.

Acknowledgments

Emma J. Kilford received support from a UK Medical Research Council (MRC) studentship. Sarah-Jayne Blakemore received a Royal Society University Research Fellowship, and her research is currently funded by the Wellcome Trust and the Jacobs Foundation.

REFERENCES

Apperly, I. A., Carroll, D. J., Samson, D., Humphreys, G. W., Qureshi, A., & Moffitt, G. (2010). Why are there limits on theory of mind use? Evidence from adults' ability to follow instructions from an ignorant speaker. *Quarterly Journal of Experimental Psychology, 63*(6), 1201–1217.

Baillargeon, R., Scott, R. M., & He, Z. (2010). False-belief understanding in infants. *Trends in Cognitive Sciences, 14*(3), 110–118.

Baron-Cohen, S., Wheelwright, S., Hill, J., Raste, Y., & Plumb, I. (2001). The "Reading the Mind in the Eyes" test revised version: A study with normal adults, and adults with Asperger syndrome or high-functioning autism. *Journal of Child Psychology and Psychiatry, 42*(2), 241–251.

Belli, S. R., Rogers, R. D., & Lau, J. Y. F. (2012). Adult and adolescent social reciprocity: Experimental data from the Trust Game. *Journal of Adolescence, 35*(5), 1341–1349.

Blakemore, S.-J. (2008). The social brain in adolescence. *Nature Reviews Neuroscience, 9*(4), 267–277.

Blakemore, S.-J., den Ouden, H., Choudhury, S., & Frith, C. (2007). Adolescent development of the neural circuitry for thinking about intentions. *Social Cognitive and Affective Neuroscience, 2*(2), 130–139.

Blakemore, S.-J., & Mills, K. L. (2014). Is adolescence a sensitive period for sociocultural processing? *Annual Review of Psychology, 65*(1), 187–207.

Bolling, D. Z., Pitskel, N. B., Deen, B., Crowley, M. J., Mayes, L. C., & Pelphrey, K. A. (2011). Development of neural systems for processing social exclusion from childhood to adolescence. *Developmental Science, 14*(6), 1431–1444.

Brown, B. B. (2004). Adolescents' relationships with peers. *Handbook of Adolescent Psychology, 2*, 363–394.

Burnett, S., Bird, G., Moll, J., Frith, C., & Blakemore, S.-J. (2009). Development during adolescence of the neural processing of social emotion. *Journal of Cognitive Neuroscience, 21*(9), 1736–1750.

Cascio, C. N., Carp, J., O'Donnell, M. B., Tinney, F. J., Bingham, C. R., Shope, J. T., ... Falk, E. B. (2015). Buffering social influence: Neural correlates of response inhibition predict driving safety in the presence of a peer. *Journal of Cognitive Neuroscience, 27*(1), 83–95.

Cohen Kadosh, K., Cohen Kadosh, R., Dick, F., & Johnson, M. H. (2011). Developmental changes in effective connectivity in the emerging core face network. *Cerebral Cortex, 21*(6), 1389–1394.

Cohen Kadosh, K., Johnson, M. H., Dick, F., Cohen Kadosh, R., & Blakemore, S.-J. (2013). Effects of age, task performance, and structural brain development on face processing. *Cerebral Cortex, 23*(7), 1630–1642.

Crone, E. A. (2013). Considerations of fairness in the adolescent brain. *Child Development Perspectives, 7*(2), 97–103.

Dumontheil, I., Apperly, I. A., & Blakemore, S.-J. (2010). Online usage of theory of mind continues to develop in late adolescence. *Developmental Science, 13*(2), 331–338.

Dumontheil, I., Hillebrandt, H., Apperly, I. A., & Blakemore, S.-J. (2012). Developmental differences in the control of action selection by social information. *Journal of Cognitive Neuroscience, 24*(10), 2080–2095.

Dumontheil, I., Küster, O., Apperly, I. A., & Blakemore, S.-J. (2010). Taking perspective into account in a communicative task. *NeuroImage, 52*(4), 1574–1583.

Engelmann, J. B., Moore, S., Capra, C. M., & Berns, G. S. (2012). Differential neurobiological effects of expert advice on risky choice in adolescents and adults. *Social Cognitive and Affective Neuroscience, 7*(5), 557–567.

Falk, E. B., Cascio, C. N., O'Donnell, M. B., Carp, J., Tinney, F. J., Bingham, C. R., ... Simons-Morton, B. G. (2014). Neural responses to exclusion predict susceptibility to social influence. *Journal of Adolescent Health, 54*(5 Suppl), S22–31.

Fett, A. K. J., Gromann, P. M., Giampietro, V., Shergill, S. S., & Krabbendam, L. (2014). Default distrust? An fMRI investigation of the neural development of trust and cooperation. *Social Cognitive and Affective Neuroscience, 9*(4), 395–402.

Fett, A. K. J., Shergill, S. S., Gromann, P. M., Dumontheil, I., Blakemore, S.-J., Yakub, F., & Krabbendam, L. (2014). Trust and social reciprocity in adolescence—a matter of perspective-taking. *Journal of Adolescence, 37*(2), 175–184.

Frith, C. D., & Frith, U. (2007). Social cognition in humans. *Current Biology, 17*(16), R724–732.

Fuhrmann, D., Knoll, L. J., & Blakemore, S.-J. (2015). Adolescence as a sensitive period of brain development. *Trends in Cognitive Sciences, 19*(10), 558–566.

Garcia, N. V., & Scherf, K. S. (2015). Emerging sensitivity to socially complex expressions: A unique role for adolescence? *Child Development Perspectives, 9*(2), 84–90.

Gardner, M., & Steinberg, L. (2005). Peer influence on risk taking, risk preference, and risky decision making in adolescence and adulthood: An experimental study. *Developmental Psychology, 41*(4), 625–635.

Gunther Moor, B., Op de Macks, Z. A., Güroğlu, B., Rombouts, S. A. R. B., van der Molen, M. W., & Crone, E. A. (2012). Neurodevelopmental changes of reading the mind in the eyes. *Social Cognitive and Affective Neuroscience, 7*(1), 44–52.

Güroğlu, B., van den Bos, W., & Crone, E. A. (2009). Fairness considerations: Increasing understanding of intentionality during adolescence. *Journal of Experimental Child Psychology, 104*(4), 398–409.

Güroğlu, B., van den Bos, W., & Crone, E. A. (2014). Sharing and giving across adolescence: An experimental study examining the development of prosocial behavior. *Frontiers in Psychology, 5*, 291.

Keysar, B., Lin, S., & Barr, D. J. (2003). Limits on theory of mind use in adults. *Cognition, 89*(1), 25–41.

Knoll, L. J., Magis-Weinberg, L., Speekenbrink, M., & Blakemore, S.-J. (2015). Social influence on risk perception during adolescence. *Psychological Science, 26*(5), 583–592.

Masten, C. L., Eisenberger, N. I., Borofsky, L. A., Pfeifer, J. H., McNealy, K., Mazziotta, J. C., & Dapretto, M. (2009). Neural correlates of social exclusion during adolescence: Understanding the distress of peer rejection. *Social Cognitive and Affective Neuroscience, 4*(2), 143–157.

Mills, K. L., Lalonde, F., Clasen, L. S., Giedd, J. N., & Blakemore, S.-J. (2014). Developmental changes in the structure of the social brain in late childhood and adolescence. *Social Cognitive and Affective Neuroscience, 9*(1), 123–131.

O'Brien, L., Albert, D., Chein, J., & Steinberg, L. (2011). Adolescents prefer more immediate rewards when in the presence of their peers. *Journal of Research on Adolescence, 21*(4), 747–753.

Overgaauw, S., van Duijvenvoorde, A. C. K., Moor, B. G., & Crone, E. A. (2015). A longitudinal analysis of neural regions involved in reading the mind in the eyes. *Social Cognitive and Affective Neuroscience, 10*(5), 619–627.

Peake, S. J., Dishion, T. J., Stormshak, E. A., Moore, W. E., & Pfeifer, J. H. (2013). Risk-taking and social exclusion in adolescence: Neural mechanisms underlying peer influences on decision-making. *NeuroImage, 82*, 23–34.

Pfeifer, J. H., Masten, C. L., Borofsky, L. A., Dapretto, M., Fuligni, A. J., & Lieberman, M. D. (2009). Neural correlates of direct and reflected self-appraisals in adolescents and adults: When social perspective-taking informs self-perception. *Child Development, 80*(4), 1016–1038.

Platt, B., Cohen Kadosh, K., & Lau, J. Y. F. (2013). The role of peer rejection in adolescent depression. *Depression and Anxiety, 30*(9), 809–821.

Sawyer, S. M., Azzopardi, P. S., Wickremarathne, D., & Patton, G. C. (2018). The age of adolescence. *Lancet Child & Adolescent Health, 2*(3), 223–228.

Sebastian, C. L., Fontaine, N. M. G., Bird, G., Blakemore, S.-J., De Brito, S. A., McCrory, E. J. P., & Viding, E. (2012). Neural processing associated with cognitive and affective theory of mind in adolescents and adults. *Social Cognitive and Affective Neuroscience, 7*(1), 53–63.

Sebastian, C. L., Tan, G. C. Y., Roiser, J. P., Viding, E., Dumontheil, I., & Blakemore, S.-J. (2011). Developmental influences on the neural bases of responses to social rejection: Implications of social neuroscience for education. *NeuroImage, 57*(3), 686–694.

Silvers, J., McRae, K., Gabrieli, J., & Gross, J. (2012). Age-related differences in emotional reactivity, regulation, and rejection sensitivity in adolescence. *Emotion, 12*(6), 1235–1247.

Smith, A. R., Chein, J., & Steinberg, L. (2014). Peers increase adolescent risk taking even when the probabilities of negative outcomes are known. *Developmental Psychology, 50*(5), 1564–1568.

Steinberg, L., & Morris, A. S. (2001). Adolescent development. *Annual Review of Psychology, 52*, 83–110.

Symeonidou, I., Dumontheil, I., Chow, W.-Y., & Breheny, R. (2015). Development of online use of theory of mind during adolescence: An eye-tracking study. *Journal of Experimental Child Psychology, 149*, 81–97.

Telzer, E. H., Ichien, N. T., & Qu, Y. (2015). Mothers know best: Redirecting adolescent reward sensitivity toward safe behavior during risk taking. *Social Cognitive and Affective Neuroscience, 10*(10), 1383–1391.

van den Bos, W., van Dijk, E., Westenberg, M., Rombouts, S. A. R. B., & Crone, E. A. (2011). Changing brains, changing perspectives: The neurocognitive development of reciprocity. *Psychological Science, 22*(1), 60–70.

van den Bos, W., Westenberg, M., van Dijk, E., & Crone, E. A. (2010). Development of trust and reciprocity in adolescence. *Cognitive Development, 25*(1), 90–102.

van Duijvenvoorde, A. C. K., & Crone, E. A. (2013). The teenage brain: A neuroeconomic approach to adolescent decision making. *Current Directions in Psychological Science, 22*(2), 108–113.

van Hoorn, J., van Dijk, E., Güroğlu, B., & Crone, E. A. (2016). Neural correlates of prosocial peer influence on public goods game donations during adolescence. *Social Cognitive and Affective Neuroscience, 11*(6), 923–933.

van Hoorn, J., van Dijk, E., Meuwese, R., Rieffe, C., & Crone, E. A. (2014). Peer influence on prosocial behavior in adolescence. *Journal of Research on Adolescence, 26*(1), 90–100.

Viner, R. M., Coffey, C., Mathers, C., Bloem, P., Costello, A., Santelli, J., & Patton, G. C. (2011). Fifty-year mortality trends in children and young people: A study of 50 low-income, middle-income, and high-income countries. *Lancet, 377*(9772), 1162–1174.

Willoughby, T., Good, M., Adachi, P. J. C. C., Hamza, C., & Tavernier, R. (2013). Examining the link between adolescent brain development and risk taking from a social-developmental perspective. *Brain and Cognition, 89*(3), 315–323.

Wolf, L. K., Bazargani, N., Kilford, E. J., Dumontheil, I., & Blakemore, S.-J. (2015). The audience effect in adolescence depends on who's looking over your shoulder. *Journal of Adolescence, 43*, 5–14.

5 A Lifespan Perspective on Human Neurocognitive Plasticity

KRISTINE BEATE WALHOVD AND MARTIN LÖVDÉN

ABSTRACT *Plasticity* can be defined as the ability of the brain to change its structure in response to a mismatch between its capacity and environmental demands. Taking a phylogenetic perspective, we review the evidence for qualitative and quantitative differences in human plasticity across the lifespan. In this perspective, humans, relative to other animals, possess low plasticity but develop greater flexibility. Current evidence emphasizes heightened vulnerability, rather than high experience-dependent plasticity, in early human life, with little evidence of qualitative age differences (e.g., critical periods) in plasticity other than those involved in the development of species-mandatory functions. Quantitative differences in plasticity, evident as decreases in higher ages, are likely to be more pronounced, although age-comparative studies of human structural brain plasticity are rare. Emerging evidence on the time course of brain changes in sensorimotor learning suggests that human data follow predictions from an exploration-selection-refinement model that have arisen in work on animal models. However, our phylogenetic considerations underscore that it cannot be automatically assumed that animal models directly inform the principles of human plasticity and that future work must address whether principles of plasticity for basic functions apply to other forms of human cognition.

A Phylogenetically Informed Perspective on Human Plasticity throughout the Lifespan

In this chapter we review the evidence on the lifespan development of human plasticity. This task requires precision in our definition of plasticity and, we argue, a phylogenetically informed perspective that can lend nuance to interpretations of extant findings. We define *plasticity* as the ability of the brain to undergo structural (i.e., anatomical) changes that are driven by a mismatch between environmental requirements (demand) and the organism's current capacity (supply). Altered environmental demands, but also changes in the organism's capacity (e.g., after injury), can induce such mismatches (Lövdén, Backman, Lindenberger, Schaefer, & Schmiedek, 2010). With this definition, human beings may not be particularly plastic compared to many other species (Paredes, Sorrells, Garcia-Verdugo, & Alvarez-Buylla, 2016; Rakic, 1985; Walhovd, Westerhausen, et al., 2016). The relative number of new neurons formed in life is much larger in fish and crustaceans than in mammals (Simoes & Rhiner, 2017; Zupanc, 2006, 2008). Compared to many "lower" vertebrates, the rate of neurogenesis in adult mammals also decreases more with age (Amrein, Isler, & Lipp, 2011). Many major structural changes in "lower" animals are arguably at least partly driven by mismatches between functional supply and environmental demands. The changes thus, by our definition, qualify as manifestations of plasticity. For instance, crustaceans, such as crabs, self-amputate legs and claws in response to being held in a fight and then grow new ones (Fleming, Muller, & Bateman, 2007), necessitating neural reorganization, too. Zebra fish regenerate functional nervous system tissue after injury (Mokalled et al., 2016). Among mammals there are similarities but also important differences. Neurogenesis might, for instance, be manifested differently in the brains of mice and men (Lazarov & Marr, 2013), and most Alzheimer's disease mouse models do not exhibit neuronal loss (Ameen-Ali et al., 2017), which is a characteristic of this disease that yields decreased plasticity in humans.

Humans appear to have evolved limitations of plasticity, along with the ability to rely heavily on accumulated experience (Paredes et al., 2016; Rakic, 1985, 2004). Though acquiring knowledge requires plasticity, maintaining it requires stability. Efficient synaptic plasticity and limits to larger-scale plasticity (e.g., neural turnover) may strike a balance that enables humans to expand their functional capacity during their protracted life-course (Hubener & Bonhoeffer, 2014; Meyer, Bonhoeffer, & Scheuss, 2014; Takesian & Hensch, 2013). Hence, humans may possess relatively constrained plasticity but have developed all the more *flexibility*. Flexibility-based reactions to environmental demands are based on existing functional supply (Lövdén et al., 2010). Flexibility is thus synonymous with functional capacity. Flexibility increases in childhood development, while a commonly held view is that plasticity decreases (Kühn & Lindenberger, 2016). It may be tempting to see increased flexibility as a result of the

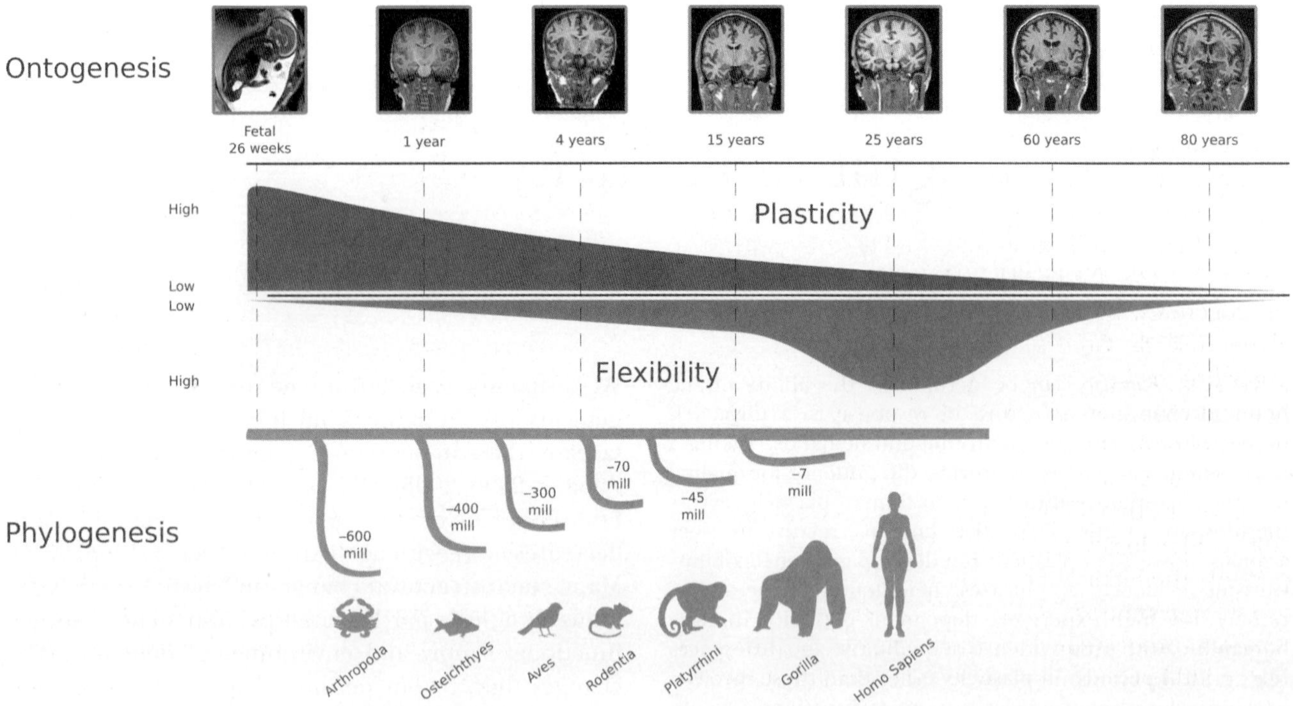

FIGURE 5.1 Schematic figure depicting plasticity (i.e., the capacity for anatomical brain changes induced by environmental demands) and flexibility (i.e., functional capacity) as a function of ontogenesis and phylogenesis. Note that early plasticity is linked to vulnerability, which for a number of functions and their neural substrates appears highest when flexibility is at its lowest in ontogenesis. (See color plate 4.)

brain being increasingly "completed" and less malleable to change. In this regard, plasticity and flexibility may be seen as opposing forces in phylogenesis and ontogenesis alike (figure 5.1). However, flexibility is not continuously increasing throughout the human lifespan (Baltes, Staudinger, & Lindenberger, 1999), and it cannot a priori be assumed that we have continuously decreasing plasticity. A number of mechanisms for altering functional supply remain also in adult human beings (Zatorre, Fields, & Johansen-Berg, 2012).

In the remainder of the chapter, we first briefly review the ontogenesis of the human brain. This is a necessary background to appreciate what anatomical changes in the brain may be influenced by experience at which times. Next, we turn to reviewing the evidence on lifespan differences in plasticity. The phylogenetic differences in plasticity, combined with differences in the timescales of ontogeny (Austad, 2010; Jones et al., 2014) and the uniquely human nature of some cognitive functions (MacLean, 2016), limit the knowledge to be directly inferred from animal models of human neurocognitive plasticity. Animal models yield invaluable insights that cannot be obtained otherwise (Hensch, 2005; Hubener & Bonhoeffer, 2014), and a better coordination of animal and human research on plasticity is needed (Lindenberger, 2018). However, in our view, especially given phylogenetic differences in plasticity, a challenge is that it cannot be assumed, without empirical evidence, that findings on lower animals can always be generalized to human beings. We therefore focus on and argue for the consideration of possible unique features of the *human* condition with regard to neurocognitive plasticity across the lifespan. Taking the same phylogenetic perspective, we will also be fully open to the possibility that mechanisms of plasticity may not generalize from sensorimotor functions to other cognitive functions (e.g., language, episodic memory, executive functions). In line with our definition of plasticity, our review is mainly restricted to brain plasticity, with a limited review of cognitive changes. We discuss whether there is evidence for variations in human plasticity across the lifespan, qualitatively (in terms of mechanisms) and quantitatively (in terms of magnitude), and before concluding, we review the emerging evidence on the maintenance and time course of experience-dependent structural brain changes.

Ontogenesis of the Human Brain

While the brain is made mainly in the womb, it does continue to change throughout life, and more so in early childhood than later. During the first year, the brain more than doubles its size, and it reaches 90% of its maximum before school age (Dekaban, 1977, 1978). The birth of new neurons, *neurogenesis*, takes place in fetal development (Rakic, 1988). While new neurons have been found in the dentate gyrus of adult humans (Eriksson et al., 1998), this has been debated (Boldrini et al., 2018; Sorrells et al., 2018). There is mixed evidence for the functional significance of postnatal neurogenesis, but it may contribute to new learning (Opendak & Gould, 2015). Uncertainties prevail regarding techniques for evaluating cellular age, but new neurons may also be added in the striatum and olfactory bulb, though not in the neocortex (Bhardwaj et al., 2006; Ernst et al., 2014; Lotsch et al., 2014). There is consensus that adult human neurogenesis is severely restricted in location and amount (Paredes et al., 2016).

While the birth of new neurons in fetal life is genetically programmed (Rakic, 1988), there is no evidence they are programmed to die in aging (Kirkwood & Melov, 2011). Contrary to previous estimates of 1% neocortical neuron loss per year (Anderson, Hubbard, Coghill, & Slidders, 1983; Brody, 1955), methodologically advanced studies indicate that humans do not lose a great proportion of neurons with age (Haug, Kuhl, Mecke, Sass, & Wasner, 1984; von Bartheld, 2017). However, there is neuronal shrinkage (Haug et al., 1984) and average neocortical neuron losses of 2%–4% from early to late life (von Bartheld, 2017). When neurons are born, they *migrate* to the places in the brain where they will reside, largely during fetal life. Neuronal *differentiation* is also defined early, by the place and time the neuron is born (Rakic, 1988). Many neurons undergo apoptosis, programmed cell death. This too is genetically determined, largely belongs to early development, and is not a candidate for later plastic changes. One can increase apoptosis, though—for instance, by exposing human fetal brain cells to drugs such as opioids (Hu, Sheng, Lokensgard, & Peterson, 2002).

While *synaptogenesis* starts at fetal stages, the peak amount of synapses in the human brain is reached postnatally (Huttenlocher, 1979), and synaptic changes are also seen in adulthood. Through development, including adolescence, there is considerable *pruning*, or elimination of synapses, as indicated by human cross-sectional data (Huttenlocher, 1984; Petanjek et al., 2011). While in part genetically programmed, synaptic remodeling can be affected by species-mandatory experience, such as visual input (Greenough, Black, & Wallace, 1987). In contrast, the evidence for variation in pruning related to normal variation in experience is weaker and in human beings, nonexistent. While animal models have shown variation in spine remodeling to covary with learning (Yang, Pan, & Gan, 2009), this cannot be generalized to indicate that normal variation in the experience of children determines whether a massive amount of their synapses survive.

Finally, *myelination*, the insulation of nerve connections by glial cells, while also beginning in fetal life, chiefly takes place after birth, in infancy, and continues well into adulthood (Yeung et al., 2014). While the number of glial cells does not differ much in young versus older adults (Pakkenberg et al., 2003), cross-sectional data indicate a considerable loss of myelinated fibers in normal aging (Marner, Nyengaard, Tang, & Pakkenberg, 2003). Characteristics of myelinated nerves can be influenced by experience (Walhovd, Johansen-Berg, & Karadottir, 2014) at different ages. There is considerable variation across cortical regions with regard to the ontogenetic timing, extent, and intraindividual variability of myelination, and these differences may relate to differences in plastic potential (Flechsig, 1901; Nieuwenhuys, 2013; Walhovd, Westerhausen, et al., 2016; Yakovlev & Lecours, 1967). In addition to the neural processes, changes in vasculature may take place at all ages.

In principle, the above structural changes may be detectable by modern in vivo imaging of human brains. In practice, the achievable resolution limits the possibility. The number of glial cells and axons, for instance, that can reside in one image voxel (e.g., in T1 or diffusion-weighted magnetic resonance imaging (MRI) scans) is enormous. By counting contents as visualized by electron micrograph images in a rodent 100 μm^3 white matter volume (1 mm^3 would be 1,000,000,000 μm^3), we found a median of 3,000 axons and on average 86 oligodendrocytes and 22 astrocytes. If axon density were equal in human beings, we estimate that a 2 mm^3 voxel could contain about 0.5 to 5 million axons (Walhovd et al., 2014). Thus, detecting true volume changes of the size of a few voxels requires growth or retractions of neural tissue of a magnitude not necessarily expected from short-term variations in motor learning or knowledge acquisition. However, alterations in the density and the composition of the tissue within a voxel may be detectable. For example, changes in dendritic spine density can account for experience-induced signal changes observed with MRI of the mouse brain (Keifer et al., 2015). Unfortunately, reverse inferences, from MRI changes to biological changes, are not readily possible. This is a weakness of our operational definition: human plasticity in terms of structural brain changes may not be readily detected with MRI, their interpretation will depend on sequence parameters, and we do not know exactly what the changes may entail. Work that includes the coordination of human and animal research data on plasticity may provide insight into the mechanisms involved (Sagi et al., 2012), and further such models are needed. However, we need to be aware that animal data do not necessarily provide accurate representations of the human condition. This makes human MRI studies with direct histological confirmation crucial for our understanding of tissue changes (Concha, 2014).

Lifespan Differences in Plasticity

The human brain is thus mostly made during fetal life, which is also when the environment can affect it the most. Toxins, drugs, and a lack of nutrients and oxygen at this stage can cause profound damage or restrictions (Barnes & Ozanne, 2011; Walhovd, Krogsrud, et al., 2016) that cannot be mended with postnatal intervention. The rapid brain growth after fetal life has been used to argue that this is when the brain responds best to training. This is flawed logic (Thompson & Nelson, 2001)—none would argue that we should impose the most weight training in infancy because that is when the muscles grow most rapidly. Early brain development is genetically programmed (Rakic, 1988), and though this program expects environmental stimulation, the maturational processes cannot be initiated or stopped, massively sped up, or slowed down by *normal* variation in environment and experience. That does not mean these processes cannot be influenced. For instance, there is evidence to suggest that relatively more deprived pre- or postnatal environments can act to lower cognitive scores as well as genetic variance (i.e., the heritability suppression hypothesis; see Scarr-Salapatek, 1971; Tucker-Drob & Bates, 2016; Turkheimer, Haley, Waldron, D'Onofrio, & Gottesman, 2003). However, we need sobriety in the conceptualization of plastic potential. By normal variation, we mean variations in the environment that the human species has been selected through evolution to operate in (for a broader account of behavioral genetics and sources of variation, see, e.g., Griffiths, Gelbart, Miller, & Lewontin, 1999). In other words, experience-expectant plasticity, the dependency of brain maturation on ubiquitous and species-mandatory environmental information (e.g., visual stimulation) must be separated from experience-dependent plasticity—that is, the incorporation of idiosyncratic environmental information (e.g., specific memories and skills) that is relatively more unique to the individual (Greenough, Black, & Wallace, 1987). Greenough and colleagues indicated different neural mechanisms for the two, with emphasis on the selection among numerous early synaptic connections for experience-expectant plasticity and the formation of new synaptic connections for experience-dependent plasticity (Greenough, Black, & Wallace, 1987). Early brain development is, in our opinion, a period of heightened plasticity mainly in the sense that it is primarily vulnerable to variations in the environment that are in conflict with that expected (such as sensory deprivation). In this sense, experience-expectant plasticity and vulnerability are invariably coupled. (For a perspective on individual differences in genetic vulnerability and plasticity, see, e.g., Belsky et al., 2009.) In contrast, variations within the normal environment have comparably less impact on brain development. Below, we review the evidence supporting our position on this issue, which is critically linked to the question of whether plasticity differs qualitatively with age.

Qualitative differences in plasticity with age The question of whether there are qualitative age differences in plasticity is intertwined with the concepts of critical or sensitive periods. A critical period denotes a specific time window within which the central nervous system is especially sensitive to environmental stimuli and must encounter that type of stimulus for an ability to develop or develop optimally. Hence, it is the functions that are species-mandatory and important to survival that likely have evolved to develop early and in critical periods through experience-expectant plasticity. Examples of such fundamental functions would be sight and hearing, while memorizing pi with a tremendous number of decimals would not usually be important to survival. It follows that the issue of whether we can elicit plastic changes with targeted training does not pertain to the brain's adaptation to species-mandatory and essential information, as in experience-expectant plasticity, but rather to the incorporation of unique information, as in experience-dependent plasticity. Yet critical periods and the importance of early experience are being mentioned in connection with anything from mathematics to music (Nelwan & Kroesbergen, 2016; Steele, Bailey, Zatorre, & Penhune, 2013). Do kids miss out forever on anything else other than fun, if they are not given huge amounts of homework?

There is unequivocal evidence for the existence of a critical period, and hence qualitative age differences, in the effect of sensory deprivation. Torsten Wiesel and David Hubel (1963) were among the first to show this, studying the functional and cortical consequences of different types of visual deprivation in newborn and older animals (Wiesel, 1982). They found that monocular deprivation (suturing one eye) during specific time spans early in life yielded permanent functional blindness in the closed eye, along with marked changes in the primary visual cortex (striate cortex). Here, cells are organized in ocular dominance columns, with alternating groups of cells responding primarily to stimulation of each eye. In animals monocularly deprived at a young age, connections with signals from the nondeprived eye will permanently occupy the space of connections from the deprived eye. This will not happen in mature animals deprived for long periods. We know that critical periods for perception also hold true for human beings, based on the phylogenetic similarity of these brain areas across mammals and the observed permanent consequences of untreated strabism or cataracts in early childhood. The fine-grained mechanisms of critical periods for sensory experience are still being researched. Importantly, studies of visual impairments in sight-recovery individuals with congenital, relative to late-onset, cataracts have yielded new knowledge suggesting that sight impairments may predominantly arise at higher cortical-processing stages (Sourav, Bottari, Kekunnaya, & Roder, 2018). Intriguingly, there is also evidence to suggest ways to alter or extend ocular dominance plasticity for longer time periods (Greifzu et al., 2014), which will further enhance the understanding of plasticity.

Upon receiving the Nobel Prize in 1981, Wiesel concluded his Nobel speech by saying that "it may well be that other aspects of brain function, such as language, complex perceptual tasks, learning, memory and personality, have different programs of development. Such sensitivity of the nervous system to the effects of experience may represent the fundamental mechanism by which the organism adapts to its environment during the period of growth and development."

For all his genius, we would say Wiesel has not yet been proven right about programs of development for various functions, other than sensory perceptual, with regard to such specific sensitivity of the nervous system. Nearly four decades later, the kind of exact and permanent neural mechanism for the effect of early experience so eloquently identified for sensory deprivation and cortical columns has yet to be shown for other types of environmental variation or experience.

There is no true experimental controlled study of human developmental deprivation, yet children are sometimes the victims of untreated conditions, cruelty, and neglect, the consequences of which have been studied. Based on this, it has, for example, been concluded that there is a critical period for language exposure, as select cases have not seemed able to acquire a first language fluently after entering the teenage years (Fromkin, Krashen, Curtiss, Rigler, & Rigler, 1974; Grimshaw, Adelstein, Bryden, & MacKinnon, 1998). A lack of language exposure rarely occurs amid otherwise optimized conditions, and there is uncertainty as to the brain mechanism, insofar as language is uniquely human, and no animal model can be created. Natural lesion studies do, however, support some early plasticity of the neural substrates of language in that left-hemisphere lesions in small children less often yield substantial aphasia or language disorders than equivalent lesions in adults (Chilosi et al., 2008). Note that this cannot be generalized to less serious effects of childhood brain damage in general. Lesions in the language and motor areas may, to some extent, yield lesser function loss at younger ages (Chilosi et al., 2008; Kennard & Fulton, 1942). However, this should not be taken to mean that early lesions cannot also yield serious language consequences (Chilosi et al., 2008), and one study found that for some language tests, stroke suffered prenatally or during the first nine months of life yielded poorer performance than stroke suffered between the ages of 1 and 14 years (Max, Bruce, Keatley, & Delis, 2010). Notably, for general cognitive ability and psychosocial functions, cerebral lesions at younger ages have been associated with worse outcomes than equivalent lesions at older ages (Anderson, Damasio, Tranel, & Damasio, 2000; Duval et al., 2008). So, while evidence suggests a sensitive period for first-language acquisition, an exact definition of a critical period for language exposure and its possible brain substrate remains unclear and cannot readily be likened to that for sight, where the exact qualitatively distinct cerebral mechanism has been established.

Beyond language, early cognitive and emotional deprivation effects in institutionalized children have been studied extensively, and a history of institutionalization has been associated with both cognitive, psychosocial, and cerebral differences relative to never institutionalized or foster-care children (Berens, Jensen, & Nelson, 2017; Rutter, 2012; Rutter, O'Connor, & English and Romanian Adoptees Study, 2004; Sheridan, Fox, Zeanah, McLaughlin, & Nelson, 2012). However, it is difficult to isolate physical and psychological environmental effects here. Furthermore, there is substantial heterogeneity in outcomes after such adversities, so the effects are not deterministic (Rutter, 2012; Rutter et al., 2004). Considerable variation is masked in the average group differences reported, where differences in brain characteristics are more of degree than kind. Again, an exact cerebral mechanism for permanent early change as an effect of experience of the kind reported for deviant early visual stimulation has not been identified for other functions, and this pertains also to cognitive and emotional neglect in childhood.

The concept of critical periods is also sometimes used as a way of giving preponderance to variations in *normal* cognitive experience in early childhood rather than later (Nelwan & Kroesbergen, 2016; Steele et al., 2013) and rather in childhood than in adulthood. The reasoning rests on the time course of synaptic pruning described for childhood and adolescence (Huttenlocher, 1984), with the idea that whether synapses persist and perish depends on how they are used. This is currently impossible to investigate in humans. Furthermore, experience-dependent pruning has also been shown to be a mechanism in adult mice (Chen et al., 2012; Xu et al., 2009; Yang, Pan, & Gan, 2009).

In summary, we have established that convincing evidence of qualitative age differences is currently restricted to sensory-perceptual functions and mainly to a dependency of brain maturation on normal, species-mandatory, environmental information. In contrast, there is no evidence for clearly defined age periods wherein the brain can and cannot change as a result of normal variations in experience. Thus, the massive campaigns for the importance of "early stimulation" (Thompson & Nelson, 2001) for children's development are misguided. There is little evidence suggesting that any additional early stimulation is necessary or should be undertaken. However, we need to consider whether there may be quantitative age differences in plasticity that are of degree more than kind.

Quantitative differences in plasticity with age Few studies have systematically compared brain plasticity in younger and older individuals. However, skill acquisition and the behavioral effects of cognitive training at different ages have been extensively studied. Training effects on performing trained tasks are seen throughout the lifespan. However, there is no indication that cognitive training serves to even out interindividual differences. Either proportional differences stay the same or those who have more to begin with gain more from intervention. This also goes for the typical adult age differences in cognition (Kliegl, Smith, & Baltes, 1990). Older adults, on average, do less well at baseline and do not "catch up" to young adults who also receive training. At best, they may improve proportionately as much as younger adults at some tasks (Brehmer, Shing, Heekeren, Lindenberger, & Backman, 2016; Carretti, Borella, & De Beni, 2007; Lövdén et al., 2012), but more often training appears to widen individual differences (Baltes & Kliegl, 1992; de Lange et al., 2017; Kliegl, Smith, & Baltes, 1990), with lesser gains in older age (Baltes & Kliegl, 1992; Burki, Ludwig, Chicherio, & de Ribaupierre, 2014; Dahlin, Nyberg, Backman, & Neely, 2008; de Lange et al., 2017; Nyberg et al., 2003). There is some evidence to suggest that children, although performing similar to older adults at baseline, tend to gain more from mnemonic training than older adults but not as much as higher-performing younger adults (Brehmer, Li, Muller, von Oertzen, & Lindenberger, 2007).

While training effects are found in most studies, the far transfer of effects to tasks that do not overlap with the trained tasks in terms of perceptual and motor expertise or strategic cognitive processing is relatively seldom seen (Melby-Lervag, Redick, & Hulme, 2016; Nelwan & Kroesbergen, 2016; Sala & Gobet, 2017). Does the potential for such transfer differ across age groups? Given the controversies surrounding the existence of transfer effects to begin with, this is especially hard to answer. For far transfer, meta-analytic studies suggest that any such effects are at best small (max ≈ 0.2 Standard Deviation) and may even be nonexistent when training groups are compared to active control groups. Comparing small effects across age groups requires massive sample sizes to achieve sufficient statistical power, and few such studies exist. However, in one study, roughly 100 younger and 100 older participants practiced working memory, episodic memory, and perceptual speed tasks for about 100 days. Compared to controls, younger adults showed the transfer of training effects on reasoning ability, but no transfer was observed in older adults (Schmiedek, Lövdén, & Lindenberger, 2010). Meta-analysis, however, shows no statistically significant differences between younger and older adults (Karbach & Verhaeghen, 2014). The evidence concerning the transfer of training effects, and potential age differences in such effects, thus remains inconclusive.

With respect to quantitative differences in brain plasticity, we know of no studies directly comparing children with adults, but some studies serve to illuminate adult age differences. Sensorimotor skill acquisition, such as juggling, and cognitive experience, such as spatial navigation training, have been shown to affect gray matter in both younger and older adults (Boyke, Driemeyer, Gaser, Buchel, & May, 2008; Draganski et al., 2004), either by cortical increases (Boyke et al., 2008) or relatively less hippocampal atrophy (Lövdén et al., 2012), compared to controls. With juggling training, the cortical changes were smaller in older adults, but they also showed increases in medial temporal lobe areas not evident in younger adults (Boyke et al., 2008; Draganski et al., 2004). However, for navigation training, regional cortical thickening was observed in younger, but not older, adults (Wenger et al., 2012). With strategy-based mnemonic training, though, both hippocampal and cortical gray matter increases have been observed in older adults (Engvig et al., 2010; Engvig et al., 2014), such as has been seen with juggling training (Boyke et al., 2008), but the extent of these relative to younger adults are unknown.

As for white matter, there is ample evidence for training effects at both younger and older ages. Increases in corpus callosum microstructural markers of integrity (in terms of decreased mean diffusivity [MD] and increased fractional anisotropy [FA]), as well as the anterior collosal area, have been observed in both younger and older adults training for a set of different cognitive tasks (Lövdén, Bodammer, et al., 2010). With mnemonic strategy training, regional changes in white matter microstructure, with decreased MD and increased FA, encompassing in part the anterior corpus callosum, have been observed in older and not younger adults (de Lange, Brathen, Rohani, Fjell, & Walhovd, 2018; de Lange et al., 2017). Of note, although the cognitive task-training study adjusted task difficulty to performance at baseline, it did not apply an adaptive training regimen with task difficulty adjusted to performance increases (Lövdén, Bodammer, et al., 2010). While the mnemonic study did (de Lange et al., 2017, 2018), there may be difficulties in making the training equally subjectively challenging to younger and older adults, and this may affect the neural changes. It is methodologically challenging to make sure that training programs are equally demanding to individuals of different ages. There are systematic differences in performance, brain structure, and function across the lifespan, making objectively identical interventions differentially demanding for different age groups. Because individual adjustments to ensure environmental demands exceed the current capacity are pivotal to maximize training gains (Brehmer, Westerberg, & Backman, 2012) and elicit plastic changes (Lövdén, Backman, et al., 2010), differential demands may thus bias the assessment of plasticity (i.e., capacity for changes).

The reviewed evidence does not imply qualitative differences in neural plastic potential in younger and older adults and remains inconclusive with respect to quantitative age differences in brain plasticity. However, the known age changes in brain and cognition imply that there may be quantitative differences. Age differences in gray and white matter characteristics (Sexton et al., 2014; Storsve et al., 2014) are indicated to affect plasticity: baseline hippocampal volumes are predictive of learning and retention across minutes and days in development (Ostby, Tamnes, Fjell, & Walhovd, 2012; Tamnes et al., 2014), recall over weeks and months in adulthood (Walhovd et al., 2004), and cognitive-training gain across weeks in aging (Engvig et al., 2012). Furthermore, white matter microstructure in areas of established age differences is predictive of cognitive training gains (de Lange et al., 2016). Hence, the conclusion at this time is that cognitive training gains and neural plasticity are seen across the lifespan, but the differences in neural integrity seen with normal aging are likely to, if anything, impact plasticity in a manner of degree but not kind. This is not proven but implied by what we know of how brains are built and maintained, together with the limited evidence from training studies so far.

The Time Course of Plastic Changes

What is the fate of training-induced changes of brain and cognition over time? Behaviorally, some extent of maintenance of training effects over prolonged periods is indicated (Schmiedek, Lövdén, & Lindenberger, 2014), but there is relatively little evidence at the neural level. On the one hand, cross-sectional studies reveal differences in gray and white matter characteristics between individuals with various degrees of expertise, suggesting that experience-dependent human brain changes may be relatively long lasting (de Manzano & Ullen, 2018; Maguire et al., 2000). These studies are, however, confounded by potential innate difference between groups and can rarely separate between past and present differences in behavior. There is mixed evidence on the maintenance of experience-dependent brain changes, with decreases, increases, or stability of training-related changes in gray matter structure observed after the termination of training (Draganski et al., 2004, 2006; Lövdén et al., 2012). A recent study investigated the temporal dynamics of brain plasticity across periods on and off memory training in older adults during a 40-week program, including two training periods separated by periods with no intervention (de Lange et al., 2018). A general decline in white matter microstructure was observed across the duration of the study but was moderated by the training periods, indicating that cognitive training may mitigate age-related brain deterioration. The training-related improvements were estimated to subside over time, indicating that continuous training may be a premise for the enduring attenuation of neural decline. Memory improvements were, however, largely maintained after the initial training period and thus may not rely on continuous training to the same degree as the neural changes. The results may also suggest that the structural changes observed with MRI are relatively transiently supporting learning or relate to use rather than learning, while representations of acquired skills (i.e., memories) are stored in a small subset of the structural changes.

Theoretical work and evidence from animal models suggest that learning sensorimotor skills involves several neural stages (Changeux & Dehaene, 1989; Edelman, 1987; Kilgard, 2012). Early in learning, the brain explores numerous candidate circuits that may execute the behavior. Probably based on reinforcement (Dhawale, Smith, & Olveczky, 2017; Makino, Hwang, Hedrick, & Komiyama, 2016), the best candidate circuit is then selected for further local refinement. This exploration-selection-refinement model is consistent with evidence indicating that rodents exhibit expanded cortical maps early in behavioral training but that these expansions later subside despite continued training and stable behavioral performance (Molina-Luna, Hertler, Buitrago, & Luft, 2008; Reed et al., 2011). The model is also consistent with findings that novel dendritic spines rapidly form in the deeper layers of the motor cortex of rodents during motor skill acquisition, but with continued training the phase of growth is followed by stabilization of the new spines and elimination of old spines, almost returning the overall number of spines to pre-training levels (Xu et al., 2009; Yang, Pan, & Gan, 2009). Thus, in the exploration-selection-refinement model, initial activity-related net structural growth may be helpful for learning, but not necessary for continued learning and maintenance of skill.

The dynamics of dendritic spines and their morphology and density are probably key to understanding the acquisition of sensorimotor skills (Hofer & Bonhoeffer, 2010; Kassem et al., 2013; Keifer et al., 2015). Experience-dependent alterations in the morphology of dendritic spines have been reported to account for a substantial amount of variance in changes of gray matter structure observed with MRI of the rodent brain, suggesting that these changes may also be possible to detect with imaging of the human brain (Kassem et al., 2013; Keifer et al., 2015). At the level of macroscale measurements of the brain structure with MRI, the microstructural processes related to the exploration-selection-refinement model may be reflected in expansion followed by partial renormalization of the gray matter during acquisition of a sensorimotor skill. A recent study tested this hypothesis by acquiring 18 structural (T1-weighted) MR images over a seven-week period during which young adults practiced left-hand writing and drawing every day. Results showed that the volume of the human primary motor cortices increased during the first weeks of learning and then partially renormalized during continued practice (Wenger et al., 2017). Thus, human structural gray matter changes during the acquisition of sensorimotor skills may be relatively transiently observable during an initial exploration phase. Adaptive skill acquisition, with repeated novel exploration phases for acquiring new subskills (e.g., learning complex skills such as playing musical instruments), may still result in larger and more permanent structural changes. However, in the study of left-hand writing and drawing, other brain regions (e.g., the cerebellum) than the primary motor cortices displayed more linear changes. In addition, it is difficult to interpret changes observed on T1-weighted images in terms of their microstructural nature. Notably, the T1-weighted image is also highly sensitive to myelin (Tardif et al., 2015), and the deeper layers of the cortex, in particular in the motor cortex, are densely myelinated (Whitaker et al., 2016). Experience-dependent myelination (Sampaio-Baptista & Johansen-Berg, 2017) of the deeper cortical layers in the later phases of skill acquisition could therefore result in the estimated decreases of gray matter thickness. Myelin content is influencing apparent cortical thickness, as derived from anatomical MRI scans (Glasser & Van Essen, 2011), and myelin proliferation is assumed to be one factor causing apparent cortical thinning during development (Walhovd, Fjell, Giedd, Dale, & Brown, 2017). For instance, this has been indicated by multimodal imaging of the observed "decreases" of the thickness of the visual cortex in human childhood development combined with adult postmortem histology (Natu et al., 2018). It is thus unknown whether the skill-related renormalization of the gray matter structure reflects the true retraction of tissue (e.g., of dendritic spine density) or a late phase of myelination. Much work remains to elucidate the temporal dynamics of human structural changes during skill acquisition.

Conclusions and Outlook

A major challenge for future studies is to identify which factors constrain and augment plasticity in human neurocognitive function and thus when and how they can be influenced. It should be clear from the above that one needs to employ a life-course perspective. Importantly, as the augmentation of human plasticity is often desired for complex neurocognitive functions, there is a need to acknowledge that principles derived from research on sensory and perceptual functions (the early critical periods included) may not be directly translatable to other aspects of human cognition and to the effects of allowable variation in human environment and experience across the life course. More basic and higher-order cognitive functions may be seen along a dimension, rather than as distinct categories, and functions usually thought of as forms of higher-order cognition are typically complex and draw on multiple capacities, including those that are more basic. However, we do not know if plasticity observed in the sensory and the motor cortices in the context of perceptual and motor skill acquisition offers a viable analogy of the role of the association cortices in the context of higher-order or complex cognitive abilities such as episodic memory, working memory, task set switching, and fluid intelligence (Lindenberger, 2018). We believe that in view of the protracted development of human higher-order functions and their reliance on the incorporation of unique information, as in experience-dependent plasticity (Greenough, Black, & Wallace, 1987), the critical-period plasticity framework and animal and human models of sensory and motor plasticity are likely to be insufficient to understand the breadth of human neurocognitive plasticity across the life span. This poses substantial challenges to experimental designs. While there is little to suggest major qualitative or quantitative differences in plasticity at different ages, there may still be critical age differences in when our capacity for change can be changed. To further test this, we suggest systematically comparing the effects of sensorimotor and higher-order cognitive training with repeated multimodal imaging and testing to delineate their temporal and neural dynamics. To partly mend the inherent confounds associated with age, such as differences in novelty, one may utilize new technologies such as apps and virtual reality paradigms, enabling rich enough settings for the systematic manipulation of sensory, motor, working, and episodic memory training. This may require the creation of alternative virtual worlds, where other rules apply, for the systematic training of different domains, which is an exciting future possibility.

Acknowledgments

We are grateful to a number of people for helping us create figure 5.1: Athanasia Monika Mowinckel for her invaluable help with setting up the graphics, Tomáš Paus for providing the fetal MRI image, Lilla Zöllei for providing the processed brain image for the 1-year-old, and Mikkel Mowinckel for assistance with the creation of the phylogenetic tree with breaks from the human evolutionary time line, which is based on data from the Tree of Life project (http://tolweb.org/tree/). The 12-month-old brain data set is one member of the data set described in Zöllei, Ou, Iglesias, Grant, and Fischl, "FreeSurfer image processing pipeline for infant clinical MRI images," OHBM2017. The volumetric segmentation was manually done and the surfaces were reconstructed using an automated infant-specific pipeline as described in de Macedo Rodrigues et al., "A FreeSurfer-compliant consistent manual segmentation of infant brains spanning the 0–2 year age range," *Frontiers in Human Neuroscience*, February 2015. https://www.frontiersin.org/articles/10.3389/fnhum.2015.00021/fu.

REFERENCES

Ameen-Ali, K. E., Wharton, S. B., Simpson, J. E., Heath, P. R., Sharp, P., & Berwick, J. (2017). Review: Neuropathology and behavioural features of transgenic murine models of Alzheimer's disease. *Neuropathology and Applied Neurobiology, 43*(7), 553–570. doi:10.1111/nan.12440

Amrein, I., Isler, K., & Lipp, H. P. (2011). Comparing adult hippocampal neurogenesis in mammalian species and orders: Influence of chronological age and life history stage. *European Journal of Neuroscience, 34*(6), 978–987. doi:10.1111/j.1460-9568.2011.07804.x

Anderson, J. M., Hubbard, B. M., Coghill, G. R., & Slidders, W. (1983). The effect of advanced old age on the neurone content of the cerebral cortex. Observations with an automatic image analyser point counting method. *Journal of the Neurological Sciences, 58*(2), 235–246.

Anderson, S. W., Damasio, H., Tranel, D., & Damasio, A. R. (2000). Long-term sequelae of prefrontal cortex damage acquired in early childhood. *Developmental Neuropsychology, 18*(3), 281–296. doi:10.1207/S15326942O2Anderson

Austad, S. N. (2010). Methusaleh's Zoo: How nature provides us with clues for extending human health span. *Journal of Comparative Pathology, 142*(Suppl 1), S10–21. doi:10.1016/j.jcpa.2009.10.024

Baltes, P. B., & Kliegl, R. (1992). Further testing of limits of cognitive plasticity: Negative age differences in a mnemonic skill are robust. *Developmental Psychology, 28*(1). doi:10.1037/0012-1649.28.1.121

Baltes, P. B., Staudinger, U. M., & Lindenberger, U. (1999). Lifespan psychology: Theory and application to intellectual functioning. *Annual Review of Psychology, 50*, 471–507. doi:10.1146/annurev.psych.50.1.471

Barnes, S. K., & Ozanne, S. E. (2011). Pathways linking the early environment to long-term health and lifespan. *Progress in Biophysics & Molecular Biology, 106*(1), 323–336. doi:10.1016/j.pbiomolbio.2010.12.005

Belsky, J., Jonassaint, C., Pluess, M., Stanton, M., Brummett, B., & Williams, R. (2009). Vulnerability genes or plasticity genes? *Molecular Psychiatry, 14*(8), 746–754. doi:10.1038/mp.2009.44

Berens, A. E., Jensen, S. K. G., & Nelson III, C. A. (2017). Biological embedding of childhood adversity: From physiological mechanisms to clinical implications. *BMC Medicine, 15*(1), 135. doi:10.1186/s12916-017-0895-4

Bhardwaj, R. D., Curtis, M. A., Spalding, K. L., Buchholz, B. A., Fink, D., Bjork-Eriksson, T., ... Frisen, J. (2006). Neocortical neurogenesis in humans is restricted to development. *Proceedings of the National Academy of Sciences of the United States of America, 103*(33), 12564–12568. doi:10.1073/pnas.0605177103

Boldrini, M., Fulmore, C. A., Tartt, A. N., Simeon, L. R., Pavlova, I., Poposka, V., ... Mann, J. J. (2018). Human hippocampal neurogenesis persists throughout aging. *Cell Stem Cell, 22*(4), 589–599, e585. doi:10.1016/j.stem.2018.03.015

Boyke, J., Driemeyer, J., Gaser, C., Buchel, C., & May, A. (2008). Training-induced brain structure changes in the elderly. *Journal of Neuroscience, 28*(28), 7031–7035. doi:10.1523/JNEUROSCI.0742-08.2008

Brehmer, Y., Li, S. C., Muller, V., von Oertzen, T., & Lindenberger, U. (2007). Memory plasticity across the life span: Uncovering children's latent potential. *Developmental Psychology, 43*(2), 465–478. doi:10.1037/0012-1649.43.2.465

Brehmer, Y., Shing, Y. L., Heekeren, H. R., Lindenberger, U., & Backman, L. (2016). Training-induced changes in subsequent-memory effects: No major differences among children, younger adults, and older adults. *NeuroImage, 131,* 214–225. doi:10.1016/j.neuroimage.2015.11.074

Brehmer, Y., Westerberg, H., & Backman, L. (2012). Working-memory training in younger and older adults: Training gains, transfer, and maintenance. *Frontiers in Human Neuroscience, 6,* 63. doi:10.3389/fnhum.2012.00063

Brody, H. (1955). Organization of the cerebral cortex. III. A study of aging in the human cerebral cortex. *Journal of Comparative Neurology, 102*(2), 511–516.

Burki, C. N., Ludwig, C., Chicherio, C., & de Ribaupierre, A. (2014). Individual differences in cognitive plasticity: An investigation of training curves in younger and older adults. *Psychological Research, 78*(6), 821–835. doi:10.1007/s00426-014-0559-3

Carretti, B., Borella, E., & De Beni, R. (2007). Does strategic memory training improve the working memory performance of younger and older adults? *Journal of Experimental Psychology, 54*(4), 311–320. doi:10.1027/1618-3169.54.4.311

Changeux, J. P., & Dehaene, S. (1989). Neuronal models of cognitive functions. *Cognition, 33*(1–2), 63–109.

Chen, J. L., Villa, K. L., Cha, J. W., So, P. T., Kubota, Y., & Nedivi, E. (2012). Clustered dynamics of inhibitory synapses and dendritic spines in the adult neocortex. *Neuron, 74*(2), 361–373. doi:10.1016/j.neuron.2012.02.030

Chilosi, A. M., Cipriani, P., Pecini, C., Brizzolara, D., Biagi, L., Montanaro, D., ... Cioni, G. (2008). Acquired focal brain lesions in childhood: Effects on development and reorganization of language. *Brain and Language, 106*(3), 211–225. doi:10.1016/j.bandl.2007.12.010

Concha, L. (2014). A macroscopic view of microstructure: Using diffusion-weighted images to infer damage, repair, and plasticity of white matter. *Neuroscience, 276,* 14–28. doi:10.1016/j.neuroscience.2013.09.004

Dahlin, E., Nyberg, L., Backman, L., & Neely, A. S. (2008). Plasticity of executive functioning in young and older adults: Immediate training gains, transfer, and long-term maintenance. *Psychology and Aging, 23*(4), 720–730. doi:10.1037/a0014296

Dekaban, A. S. (1977). Tables of cranial and orbital measurements, cranial volume, and derived indexes in males and females from 7 days to 20 years of age. *Annals of Neurology, 2*(6), 485–491. doi:10.1002/ana.410020607

Dekaban, A. S. (1978). Changes in brain weights during the span of human life: Relation of brain weights to body heights and body weights. *Annals of Neurology, 4*(4), 345–356. doi:10.1002/ana.410040410

de Lange, A. G., Brathen, A. C., Grydeland, H., Sexton, C., Johansen-Berg, H., Andersson, J. L., ... Walhovd, K. B. (2016). White matter integrity as a marker for cognitive plasticity in aging. *Neurobiology and Aging, 47,* 74–82. doi:10.1016/j.neurobiolaging.2016.07.007

de Lange, A. G., Brathen, A. C. S., Rohani, D. A., Fjell, A. M., & Walhovd, K. B. (2018). The temporal dynamics of brain plasticity in aging. *Cerebral Cortex, 28,* 1857–1865. doi:10.1093/cercor/bhy003

de Lange, A. G., Brathen, A. C. S., Rohani, D. A., Grydeland, H., Fjell, A. M., & Walhovd, K. B. (2017). The effects of memory training on behavioral and microstructural plasticity in young and older adults. *Human Brain Mapping, 38*(11), 5666–5680. doi:10.1002/hbm.23756

de Macedo Rodrigues, K., Ben-Avi, E., Sliva, D. D., Choe, M. S., Drottar, M., Wang, R., Fischl, B., Grant, P. E., & Zollei, L. (2015). A FreeSurfer-compliant consistent manual segmentation of infant brains spanning the 0-2 year age range. *Frontiers in Human Neuroscience, 9,* 21. doi:10.3389/fnhum.2015.00021

de Manzano, O., & Ullen, F. (2018). Same genes, different brains: Neuroanatomical differences between monozygotic twins discordant for musical training. *Cerebral Cortex, 28*(1), 387–394. doi:10.1093/cercor/bhx299

Dhawale, A. K., Smith, M. A., & Olveczky, B. P. (2017). The role of variability in motor learning. *Annual Review of Neuroscience, 40,* 479–498. doi:10.1146/annurev-neuro-072116-031548

Draganski, B., Gaser, C., Busch, V., Schuierer, G., Bogdahn, U., & May, A. (2004). Neuroplasticity: Changes in grey matter induced by training. *Nature, 427*(6972), 311–312. doi:10.1038/427311a

Draganski, B., Gaser, C., Kempermann, G., Kuhn, H. G., Winkler, J., Buchel, C., & May, A. (2006). Temporal and spatial dynamics of brain structure changes during extensive learning. *Journal of Neuroscience, 26*(23), 6314–6317. doi:10.1523/JNEUROSCI.4628-05.2006

Duval, J., Braun, C. M., Montour-Proulx, I., Daigneault, S., Rouleau, I., & Begin, J. (2008). Brain lesions and IQ: Recovery versus decline depends on age of onset. *Journal of Child Neurology, 23*(6), 663–668. doi:10.1177/0883073808314161

Edelman, G. M. (1987). *Neural Darwinism: The theory of neuronal group selection.* New York: Basic Books.

Engvig, A., Fjell, A. M., Westlye, L. T., Moberget, T., Sundseth, O., Larsen, V. A., & Walhovd, K. B. (2010). Effects of memory training on cortical thickness in the elderly. *NeuroImage, 52*(4), 1667–1676. doi:10.1016/j.neuroimage.2010.05.041

Engvig, A., Fjell, A. M., Westlye, L. T., Skaane, N. V., Dale, A. M., Holland, D., ... Walhovd, K. B. (2014). Effects of cognitive training on gray matter volumes in memory clinic patients with subjective memory impairment. *Journal of Alzheimer's Disease, 41*(3), 779–791. doi:10.3233/JAD-131889

Engvig, A., Fjell, A. M., Westlye, L. T., Skaane, N. V., Sundseth, O., & Walhovd, K. B. (2012). Hippocampal subfield volumes correlate with memory training benefit in subjective memory impairment. *NeuroImage, 61*(1), 188–194. doi:10.1016/j.neuroimage.2012.02.072

Eriksson, P. S., Perfilieva, E., Bjork-Eriksson, T., Alborn, A. M., Nordborg, C., Peterson, D. A., & Gage, F. H. (1998). Neurogenesis in the adult human hippocampus. *Nature Medicine, 4*(11), 1313–1317. doi:10.1038/3305

Ernst, A., Alkass, K., Bernard, S., Salehpour, M., Perl, S., Tisdale, J., ... Frisen, J. (2014). Neurogenesis in the striatum of the adult human brain. *Cell, 156*(5), 1072–1083. doi:10.1016/j.cell.2014.01.044

Flechsig, P. (1901). Developmental (myelogenetic) localisation of the cerebral cortex in the human subject. *Lancet, 158*(4977), 1027–1030.

Fleming, P. A., Muller, D., & Bateman, P. W. (2007). Leave it all behind: A taxonomic perspective of autotomy in invertebrates. *Biological Reviews of the Cambridge Philosophical Society, 82*(3), 481–510. doi:10.1111/j.1469-185X.2007.00020.x

Fromkin, V., Krashen, S., Curtiss, S., Rigler, D., & Rigler, M. (1974). The development of language in Genie: A case of language acquisition beyond the "critical period." *Brain and Language, 1,* 81–107.

Glasser, M. F., & Van Essen, D. C. (2011). Mapping human cortical areas in vivo based on myelin content as revealed by T1- and T2-weighted MRI. *Journal of Neuroscience, 31*(32), 11597–11616. doi:10.1523/JNEUROSCI.2180-11.2011

Greenough, W. T., Black, J. E., & Wallace, C. S. (1987). Experience and brain development. *Child Development, 58*(3), 539–559.

Greifzu, F., Pielecka-Fortuna, J., Kalogeraki, E., Krempler, K., Favaro, P. D., Schluter, O. M., & Lowel, S. (2014). Environmental enrichment extends ocular dominance plasticity into adulthood and protects from stroke-induced impairments of plasticity. *Proceedings of the National Academy of Sciences of the United States of America, 111*(3), 1150–1155. doi:10.1073/pnas.1313385111

Griffiths, A. F. J., Gelbart, W. M., Miller, J. H., & Lewontin, R. C. (1999). *Modern Genetic Analysis.* New York: W.H. Freeman.

Grimshaw, G. M., Adelstein, A., Bryden, M. P., & MacKinnon, G. E. (1998). First-language acquisition in adolescence: Evidence for a critical period for verbal language development. *Brain and Language, 63*(2), 237-255. doi:10.1006/brln.1997.1943

Haug, H., Kuhl, S., Mecke, E., Sass, N. L., & Wasner, K. (1984). The significance of morphometric procedures in the investigation of age changes in cytoarchitectonic structures of human brain. *Journal für Hirnforschung, 25*(4), 353–374.

Hensch, T. K. (2005). Critical period plasticity in local cortical circuits. *Nature Reviews Neuroscience, 6*(11), 877–888. doi:10.1038/nrn1787

Hofer, S. B., & Bonhoeffer, T. (2010). Dendritic spines: The stuff that memories are made of? *Current Biology, 20*(4), R157–159. doi:10.1016/j.cub.2009.12.040

Hu, S., Sheng, W. S., Lokensgard, J. R., & Peterson, P. K. (2002). Morphine induces apoptosis of human microglia and neurons. *Neuropharmacology, 42*(6), 829–836.

Hubener, M., & Bonhoeffer, T. (2014). Neuronal plasticity: Beyond the critical period. *Cell, 159*(4), 727–737. doi:10.1016/j.cell.2014.10.035

Huttenlocher, P. R. (1979). Synaptic density in human frontal cortex—developmental changes and effects of aging. *Brain Research, 163*(2), 195–205.

Huttenlocher, P. R. (1984). Synapse elimination and plasticity in developing human cerebral cortex. *American Journal of Mental Deficiency, 88*(5), 488–496.

Jones, O. R., Scheuerlein, A., Salguero-Gomez, R., Camarda, C. G., Schaible, R., Casper, B. B., ... Vaupel, J. W. (2014). Diversity of ageing across the tree of life. *Nature, 505*(7482), 169–173. doi:10.1038/nature12789

Karbach, J., & Verhaeghen, P. (2014). Making working memory work: A meta-analysis of executive-control and working memory training in older adults. *Psychological Science, 25*(11), 2027–2037. doi:10.1177/0956797614548725

Kassem, M. S., Lagopoulos, J., Stait-Gardner, T., Price, W. S., Chohan, T. W., Arnold, J. C., ... Bennett, M. R. (2013). Stress-induced grey matter loss determined by MRI is primarily due to loss of dendrites and their synapses. *Molecular Neurobiology, 47*(2), 645–661. doi:10.1007/s12035-012-8365-7

Keifer Jr., O. P., Hurt, R. C., Gutman, D. A., Keilholz, S. D., Gourley, S. L., & Ressler, K. J. (2015). Voxel-based morphometry predicts shifts in dendritic spine density and morphology with auditory fear conditioning. *Nature Communications, 6,* 7582. doi:10.1038/ncomms8582

Kennard, M., & Fulton, J. F. (1942). Age and reorganization of central nervous system. *Journal of Mount Sinai Hospital New York, 9,* 594–606.

Kilgard, M. P. (2012). Harnessing plasticity to understand learning and treat disease. *Trends in Neurosciences, 35*(12), 715–722. doi:10.1016/j.tins.2012.09.002

Kirkwood, T. B., & Melov, S. (2011). On the programmed/non-programmed nature of ageing within the life history. *Current Biology, 21*(18), R701–707. doi:10.1016/j.cub.2011.07.020

Kliegl, R., Smith, J., & Baltes, P. B. (1990). On the locus and process of magnification of age differences during mnemonic training. *Developmental Psychology, 26*(6). doi:10.1037/0012-1649.26.6.894

Kühn, S., & Lindenberger, U. (2016). Research on human plasticity in adulthood: A lifespan agenda. In K. W. Schaie & S. Willis (Eds.), *Handbook of the psychology of aging* (8th ed., pp. 105–123). San Diego: Academic Press.

Lazarov, O., & Marr, R. A. (2013). Of mice and men: Neurogenesis, cognition and Alzheimer's disease. *Frontiers in Aging Neuroscience, 5,* 43. doi:10.3389/fnagi.2013.00043

Lindenberger, U. (2018). Plasticity beyond early development: Hypotheses and questions. In A. A. Benasich & U. Ribary (Eds.), *Manifestations and mechanisms of dynamic brain coordination over development* (pp. 207–223). Cambridge, MA: MIT Press.

Lotsch, J., Schaeffeler, E., Mittelbronn, M., Winter, S., Gudziol, V., Schwarzacher, S. W., ... Ultsch, A. (2014). Functional genomics suggest neurogenesis in the adult human olfactory bulb. *Brain Structure and Function, 219*(6), 1991–2000. doi:10.1007/s00429-013-0618-3

Lövdén, M., Backman, L., Lindenberger, U., Schaefer, S., & Schmiedek, F. (2010). A theoretical framework for the study of adult cognitive plasticity. *Psychological Bulletin, 136*(4), 659–676. doi:10.1037/a0020080

Lövdén, M., Bodammer, N. C., Kuhn, S., Kaufmann, J., Schutze, H., Tempelmann, C., ... Lindenberger, U. (2010). Experience-dependent plasticity of white-matter microstructure extends into old age. *Neuropsychologia, 48*(13), 3878–3883. doi:10.1016/j.neuropsychologia.2010.08.026

Lövdén, M., Schaefer, S., Noack, H., Bodammer, N. C., Kuhn, S., Heinze, H. J., ... Lindenberger, U. (2012). Spatial navigation training protects the hippocampus against age-related changes during early and late adulthood. *Neurobiology of Aging, 33*(3), 620.e9-620.e22. doi:10.1016/j.neurobiolaging.2011.02.013

MacLean, E. L. (2016). Unraveling the evolution of uniquely human cognition. *Proceedings of the National Academy of Sciences of the United States of America, 113*(23), 6348–6354. doi:10.1073/pnas.1521270113

Maguire, E. A., Gadian, D. G., Johnsrude, I. S., Good, C. D., Ashburner, J., Frackowiak, R. S., & Frith, C. D. (2000). Navigation-related structural change in the hippocampi of taxi drivers. *Proceedings of the National Academy of Sciences of the United States of America, 97*(8), 4398–4403. doi:10.1073/pnas.070039597

Makino, H., Hwang, E. J., Hedrick, N. G., & Komiyama, T. (2016). Circuit mechanisms of sensorimotor learning. *Neuron, 92*(4), 705–721. doi:10.1016/j.neuron.2016.10.029

Marner, L., Nyengaard, J. R., Tang, Y., & Pakkenberg, B. (2003). Marked loss of myelinated nerve fibers in the human brain with age. *Journal of Comparative Neurology, 462*(2), 144–152. doi:10.1002/cne.10714

Max, J. E., Bruce, M., Keatley, E., & Delis, D. (2010). Pediatric stroke: Plasticity, vulnerability, and age of lesion onset. *Journal of Neuropsychiatry and Clinical Neurosciences, 22*(1), 30–39. doi:10.1176/appi.neuropsych.22.1.3010.1176/jnp.2010.22.1.30

Melby-Lervag, M., Redick, T. S., & Hulme, C. (2016). Working memory training does not improve performance on measures of intelligence or other measures of "far transfer": Evidence from a meta-analytic review. *Perspectives on Psychological Science, 11*(4), 512–534. doi:10.1177/1745691616635612

Meyer, D., Bonhoeffer, T., & Scheuss, V. (2014). Balance and stability of synaptic structures during synaptic plasticity. *Neuron, 82*(2), 430–443. doi:10.1016/j.neuron.2014.02.031

Mokalled, M. H., Patra, C., Dickson, A. L., Endo, T., Stainier, D. Y., & Poss, K. D. (2016). Injury-induced ctgfa directs glial bridging and spinal cord regeneration in zebrafish. *Science, 354*(6312), 630–634. doi:10.1126/science.aaf2679

Molina-Luna, K., Hertler, B., Buitrago, M. M., & Luft, A. R. (2008). Motor learning transiently changes cortical somatotopy. *NeuroImage, 40*(4), 1748–1754. doi:10.1016/j.neuroimage.2007.11.018

Natu, S. V., Gomez, J., Barentt, M., Jeska, B., Kirilina, E., Jaeger, C., … Grill-Spector, K. (2018). Apparent thinning of visual cortex during childhood is associated with myelination, not pruning. *bioRxiv.* Retrieved from https://www.biorxiv.org/content/early/2018/07/13/368274.doi:10.1101/368274

Nelwan, M., & Kroesbergen, E. H. (2016). Limited near and far transfer effects of jungle memory working memory training on learning mathematics in children with attentional and mathematical difficulties. *Frontiers in Psychology, 7*, 1384.

Nieuwenhuys, R. (2013). The myeloarchitectonic studies on the human cerebral cortex of the Vogt-Vogt school, and their significance for the interpretation of functional neuroimaging data. *Brain Structure and Function, 218*(2), 303–352. doi:10.1007/s00429-012-0460-z

Nyberg, L., Sandblom, J., Jones, S., Neely, A. S., Petersson, K. M., Ingvar, M., & Backman, L. (2003). Neural correlates of training-related memory improvement in adulthood and aging. *Proceedings of the National Academy of Sciences of the United States of America, 100*(23), 13728–13733. doi:10.1073/pnas.1735487100

Opendak, M., & Gould, E. (2015). Adult neurogenesis: A substrate for experience-dependent change. *Trends in Cognitive Sciences, 19*(3), 151–161. doi:10.1016/j.tics.2015.01.001

Ostby, Y., Tamnes, C. K., Fjell, A. M., & Walhovd, K. B. (2012). Dissociating memory processes in the developing brain: The role of hippocampal volume and cortical thickness in recall after minutes versus days. *Cerebral Cortex, 22*(2), 381–390. doi:10.1093/cercor/bhr116

Pakkenberg, B., Pelvig, D., Marner, L., Bundgaard, M. J., Gundersen, H. J., Nyengaard, J. R., & Regeur, L. (2003). Aging and the human neocortex. *Experimental Gerontology, 38*(1–2), 95–99.

Paredes, M. F., Sorrells, S. F., Garcia-Verdugo, J. M., & Alvarez-Buylla, A. (2016). Brain size and limits to adult neurogenesis. *Journal of Comparative Neurology, 524*(3), 646–664. doi:10.1002/cne.23896

Petanjek, Z., Judas, M., Simic, G., Rasin, M. R., Uylings, H. B., Rakic, P., & Kostovic, I. (2011). Extraordinary neoteny of synaptic spines in the human prefrontal cortex. *Proceedings of the National Academy of Sciences of the United States of America, 108*(32), 13281–13286. doi:10.1073/pnas.1105108108

Rakic, P. (1985). Limits of neurogenesis in primates. *Science, 227*(4690), 1054–1056.

Rakic, P. (1988). Specification of cerebral cortical areas. *Science, 241*(4862), 170–176.

Rakic, P. (2004). Neuroscience: Immigration denied. *Nature, 427*(6976), 685–686. doi:10.1038/427685a

Reed, A., Riley, J., Carraway, R., Carrasco, A., Perez, C., Jakkamsetti, V., & Kilgard, M. P. (2011). Cortical map plasticity improves learning but is not necessary for improved performance. *Neuron, 70*(1), 121–131. doi:10.1016/j.neuron.2011.02.038

Rutter, M. (2012). Achievements and challenges in the biology of environmental effects. *Proceedings of the National Academy of Sciences of the United States of America, 109*(Suppl 2), 17149–17153. doi:10.1073/pnas.1121258109

Rutter, M., O'Connor, T. G., & English and Romanian Adoptees Study Team (2004). Are there biological programming effects for psychological development? Findings from a study of Romanian adoptees. *Developmental Psychology, 40*(1), 81–94. doi:10.1037/0012-1649.40.1.81

Sagi, Y., Tavor, I., Hofstetter, S., Tzur-Moryosef, S., Blumenfeld-Katzir, T., & Assaf, Y. (2012). Learning in the fast lane: New insights into neuroplasticity. *Neuron, 73*(6), 1195–1203. doi:10.1016/j.neuron.2012.01.025

Sala, G., & Gobet, F. (2017). Does far transfer exist? Negative evidence from chess, music, and working memory training. *Current Directions in Psychological Science, 26*(6), 515–520. doi:10.1177/0963721417712760

Sampaio-Baptista, C., & Johansen-Berg, H. (2017). White matter plasticity in the adult brain. *Neuron, 96*(6), 1239–1251. doi:10.1016/j.neuron.2017.11.026

Scarr-Salapatek, S. (1971). Race, social class, and IQ. *Science, 174*(4016), 1285–1295.

Schmiedek, F., Lövdén, M., & Lindenberger, U. (2010). Hundred days of cognitive training enhance broad cognitive abilities in adulthood: Findings from the COGITO study. *Frontiers in Aging Neuroscience, 2.* doi:10.3389/fnagi.2010.00027

Schmiedek, F., Lövdén, M., & Lindenberger, U. (2014). Younger adults show long-term effects of cognitive training on broad cognitive abilities over 2 years. *Developmental Psychology, 50*(9). doi:10.1037/a0037388 25019946

Sexton, C. E., Walhovd, K. B., Storsve, A. B., Tamnes, C. K., Westlye, L. T., Johansen-Berg, H., & Fjell, A. M. (2014). Accelerated changes in white matter microstructure during aging: A longitudinal diffusion tensor imaging study. *Journal of Neuroscience, 34*(46), 15425–15436. doi:10.1523/JNEUROSCI.0203-14.2014

Sheridan, M. A., Fox, N. A., Zeanah, C. H., McLaughlin, K. A., & Nelson III, C. A. (2012). Variation in neural development as a result of exposure to institutionalization early in childhood. *Proceedings of the National Academy of Sciences of the United States of America, 109*(32), 12927–12932. doi:10.1073/pnas.1200041109

Simoes, A. R., & Rhiner, C. (2017). A cold-blooded view on adult neurogenesis. *Frontiers of Neuroscience, 11*, 327. doi:10.3389/fnins.2017.00327

Sorrells, S. F., Paredes, M. F., Cebrian-Silla, A., Sandoval, K., Qi, D., Kelley, K. W., … Alvarez-Buylla, A. (2018). Human hippocampal neurogenesis drops sharply in children to undetectable levels in adults. *Nature, 555*, 377–381. doi:10.1038/nature25975

Sourav, S., Bottari, D., Kekunnaya, R., & Roder, B. (2018). Evidence of a retinotopic organization of early visual cortex but impaired extrastriate processing in sight recovery individuals. *Journal of Visualized Experiments, 18*(3), 22. doi:10.1167/18.3.22

Steele, C. J., Bailey, J. A., Zatorre, R. J., & Penhune, V. B. (2013). Early musical training and white-matter plasticity in the corpus callosum: Evidence for a sensitive period. *Journal of Neuroscience, 33*(3), 1282–1290. doi:10.1523/JNEUROSCI.3578-12.2013

Storsve, A. B., Fjell, A. M., Tamnes, C. K., Westlye, L. T., Overbye, K., Aasland, H. W., & Walhovd, K. B. (2014). Differential longitudinal changes in cortical thickness, surface area and volume across the adult life span: Regions of accelerating and decelerating change. *Journal of Neuroscience, 34*(25), 8488–8498. doi:10.1523/JNEUROSCI.0391-14.2014

Takesian, A. E., & Hensch, T. K. (2013). Balancing plasticity/stability across brain development. *Progress in Brain Research, 207*, 3–34. doi:10.1016/B978-0-444-63327-9.00001-1

Tamnes, C. K., Walhovd, K. B., Engvig, A., Grydeland, H., Krogsrud, S. K., Ostby, Y., … Fjell, A. M. (2014). Regional hippocampal volumes and development predict learning and memory. *Developmental Neuroscience, 36*(34), 161–174. doi:10.1159/000362445

Tardif, C. L., Schafer, A., Waehnert, M., Dinse, J., Turner, R., & Bazin, P. L. (2015). Multi-contrast multi-scale surface registration for improved alignment of cortical areas. *NeuroImage, 111*, 107–122. doi:10.1016/j.neuroimage.2015.02.005

Thompson, R. A., & Nelson, C. A. (2001). Developmental science and the media: Early brain development. *American Psychologist, 56*(1), 5–15.

Tucker-Drob, E. M., & Bates, T. C. (2016). Large cross-national differences in gene x socioeconomic status interaction on intelligence. *Psychological Science, 27*(2), 138–149. doi:10.1177/0956797615612727

Turkheimer, E., Haley, A., Waldron, M., D'Onofrio, B., & Gottesman, I. (2003). Socioeconomic status modifies heritability of IQ in young children. *Psychological Science, 14*(6), 623–628. doi:10.1046/j.0956-7976.2003.psci_1475.x

von Bartheld, C. S. (2017). Myths and truths about the cellular composition of the human brain: A review of influential concepts. *Journal of Chemical Neuroanatomy, 93*, 2–15. doi:10.1016/j.jchemneu.2017.08.004

Walhovd, K. B., Fjell, A. M., Giedd, J., Dale, A. M., & Brown, T. T. (2017). Through thick and thin: A need to reconcile contradictory results on trajectories in human cortical development. *Cerebral Cortex, 27*(2), 1472–1481. doi:10.1093/cercor/bhv301

Walhovd, K. B., Fjell, A. M., Reinvang, I., Lundervold, A., Fischl, B., Quinn, B. T., & Dale, A. M. (2004). Size does matter in the long run: Hippocampal and cortical volume predict recall across weeks. *Neurology, 63*(7), 1193–1197.

Walhovd, K. B., Johansen-Berg, H., & Karadottir, R. T. (2014). Unraveling the secrets of white matter—bridging the gap between cellular, animal and human imaging studies. *Neuroscience, 276*, 2–13. doi:10.1016/j.neuroscience.2014.06.058

Walhovd, K. B., Krogsrud, S. K., Amlien, I. K., Bartsch, H., Bjornerud, A., Due-Tonnessen, P., … Fjell, A. M. (2016). Neurodevelopmental origins of lifespan changes in brain and cognition. *Proceedings of the National Academy of Sciences of the United States of America, 113*(33), 9357–9362. doi:10.1073/pnas.1524259113

Walhovd, K. B., Westerhausen, R., de Lange, A. M., Brathen, A. C., Grydeland, H., Engvig, A., & Fjell, A. M. (2016). Premises of plasticity—and the loneliness of the medial temporal lobe. *NeuroImage, 131*, 48–54. doi:10.1016/j.neuroimage.2015.10.060

Wenger, E., Kuhn, S., Verrel, J., Martensson, J., Bodammer, N. C., Lindenberger, U., & Lövdén, M. (2017). Repeated structural imaging reveals nonlinear progression of experience-dependent volume changes in human motor cortex. *Cerebral Cortex, 27*(5), 2911–2925. doi:10.1093/cercor/bhw141

Wenger, E., Schaefer, S., Noack, H., Kuhn, S., Martensson, J., Heinze, H. J., … Lövdén, M. (2012). Cortical thickness changes following spatial navigation training in adulthood and aging. *NeuroImage, 59*(4), 3389–3397. doi:10.1016/j.neuroimage.2011.11.015

Whitaker, K. J., Vertes, P. E., Romero-Garcia, R., Vasa, F., Moutoussis, M., Prabhu, G., … Bullmore, E. T. (2016). Adolescence is associated with genomically patterned consolidation of the hubs of the human brain connectome. *Proceedings of the National Academy of Sciences of the United States of America, 113*(32), 9105–9110. doi:10.1073/pnas.1601745113

Wiesel, T. N. (1982). Postnatal development of the visual cortex and the influence of environment. *Nature, 299*(5884), 583–591.

Wiesel, T. N., & Hubel, D. H. (1963). Single-cell responses in striate cortex of kittens deprived of vision in one eye. *Journal of Neurophysiology, 26*, 1003–1017. doi:10.1152/jn.1963.26.6.1003

Xu, T., Yu, X., Perlik, A. J., Tobin, W. F., Zweig, J. A., Tennant, K., … Zuo, Y. (2009). Rapid formation and selective stabilization of synapses for enduring motor memories. *Nature, 462*(7275), 915–919. doi:10.1038/nature08389

Yakovlev, P. I., & Lecours, A. (1967). The myelogenetic cycles of regional maturation of the brain. In Minkowski (Ed.), *Regional development of the brain in early life* (pp. 3–70). Oxford: Blackwell.

Yang, G., Pan, F., & Gan, W. B. (2009). Stably maintained dendritic spines are associated with lifelong memories. *Nature, 462*(7275), 920–924. doi:10.1038/nature08577

Yeung, M. S., Zdunek, S., Bergmann, O., Bernard, S., Salehpour, M., Alkass, K., … Frisen, J. (2014). Dynamics of oligodendrocyte generation and myelination in the human brain. *Cell, 159*(4), 766–774. doi:10.1016/j.cell.2014.10.011

Zatorre, R. J., Fields, R. D., & Johansen-Berg, H. (2012). Plasticity in gray and white: Neuroimaging changes in brain structure during learning. *Nature Neuroscience, 15*(4), 528–536. doi:10.1038/nn.3045

Zöllei, L., Ou, Y., Iglesias, J. E., Grant, P. E., & Fischl, B. (2017). FreeSurfer image processing pipeline for infant clinical MRI images. Organization for Human Brain Mapping (OHBM).

Zupanc, G. K. (2006). Neurogenesis and neuronal regeneration in the adult fish brain. *Journal of Comparative Physiology A: Neuroethology, Sensory, Neural, and Behavioral Physiology, 192*(6), 649–670. doi:10.1007/s00359-006-0104-y

Zupanc, G. K. (2008). Adult neurogenesis and neuronal regeneration in the brain of teleost fish. *Journal of Physiology-Paris, 102*(4–6), 357–373. doi:10.1016/j.jphysparis.2008.10.007

6 Brains, Hearts, and Minds: Trajectories of Neuroanatomical and Cognitive Change and Their Modification by Vascular and Metabolic Factors

NAFTALI RAZ

ABSTRACT Aging is accompanied by substantial reductions in brain volume, alterations in structural properties of the white matter, shifts in iron metabolism, and selective cognitive declines. Although it is assumed that brain changes precede and mediate cognitive aging, empirical support for this assumption is sparse. Cross-sectional studies cannot provide an adequate solution to the age-brain-cognition (ABC) triangle, and multiwave longitudinal investigations that can reveal mediation and moderation relationships among these variables are rare. Moreover, the trajectories of neural and cognitive aging and the distribution of variance in the ABC triad can be influenced by multiple factors reflecting vascular and metabolic risk and the propensity for neuroinflammatory response. To complicate matters further, the BC part of the trilateral relationship is not necessarily described by the brain's unidirectional effects on cognition. Rather, empirical evidence suggests that those are frequently connected by bidirectional paths. In addition, vascular, metabolic, and inflammatory risk factors modify and moderate the trajectories of age-related change. Although the extant literature is still scant, certain trends can be discerned, and the paths to elucidating the neural mechanisms of age-related changes through studying neuroanatomical changes in the context of the brain energy crisis can be outlined.

I list not prophecy; but let Time's news
Be known when 'tis brought forth.
—Shakespeare, *Winter's Tale*

Act IV, Scene 1

Whoever came up with the "time takes its toll" cliché knew what they were doing, for it is incontrovertible that in adult mammals, changes in virtually all systems and organs accompany the passage of time. Advanced age is associated with reduced lean mass and bone density, limited respiratory fitness, increased insulin resistance, low-grade chronic inflammation, lowered energy expenditure, and elevated vascular risk factors (Manini, 2010; Margolick & Ferrucci, 2015). Conceptualization of the multifarious aging process as a slow-rolling energy crisis (de la Torre, 2008) aligns with the dominant views of modern gerontology: the free radical accumulation model (Garaschuk, Semchyshyn, & Lushchak, 2018; Harman, 1956) and the disposable soma theory (Kirkwood, 1977). These theories converge on the conceptualization of postreproductive aging as an unprogrammed stochastic wear-and-tear process, driven by shifting of the dynamic equilibrium between the benefits of reactive oxygen (ROS), nitrogen (RNS), and carbonyl (RCS) species—RONCS (Garaschuk et al., 2018)—and the declines they cause through damaging mitochondria and inducing energetic deficit.

The inability to meet energy demands may underpin the differential age-related changes, and despite its universality, aging preferentially affects organs and systems with the highest energy requirements—the muscles, the liver, and the brain (Manini, 2010). The latter, while constituting only 2% of body weight, draws more than 20% of the organism's total energy (Attwell & Laughlin, 2001), most of which supports the brain's information-processing capability through maintaining neurotransmission (Howarth et al., 2012), while the remainder is allocated to building and maintaining structural components.

At any moment, the size and structural composition of the brain is determined by multiple forces and actors that make a veritable Robinson Crusoe list of things good and evil. Nature is a mighty seamstress toiling over alterations in the never-quite-perfect fabric of the brain, shortening and extending and widening and narrowing to meet the demands, within genetic constraints, of many biological processes and environmental experiences. On the one hand, neuropil volume grows through the expansion of dendritic arborization, synaptogenesis, myelin synthesis, angiogenesis, astrocyte proliferation, and neurogenesis. On the other hand, attrition and shrinkage of neurons, synapse stripping by microglia, dendritic spine retraction, demyelination, and loss of astrocytes offset the expansion. The complexity of forces at play makes inferring neurobiological mechanisms of cognitive aging from gross structural changes extremely challenging and calls for evaluating their molecular and metabolic underpinnings. Although animal models exploring the fundamental physiology of aging are plentiful, their translation to noninvasive studies of the human brain is lagging.

The Free-Radicals-Induced Energetic and Neural Decline in Senescence (FRIENDS; Raz & Daugherty, 2018; figure 6.1) model inspired by the free-radical (Harman, 1956) and the energy crisis (de la Torre, 2008) theories casts the gradual decline in the availability of energy resources as the core phenomenon of cognitive aging. The age-related waning of mitochondria is exacerbated by endothelial dysfunction that impairs the delivery of energy substrates and the removal of byproducts and debris of cellular decay (e.g., Singer et al., 2014). Whereas maintaining ion pumps is critical for neurotransmission and is unlikely to slow down and accommodate the dwindling energy supply, the upkeep of the metabolically expensive myelin sheath (Bartzokis, 2011), cellular membranes, and dendritic spines (Morrison & Baxter, 2012) is affected by energetic scarcity. Moreover, endothelial dysfunction may directly impair the maintenance of the myelin sheath (Rajani et al., 2018). Thus, brain connectivity is disrupted (Damoiseaux, 2017), and it becomes a progressively noisier system. Noise propagates in the information-processing networks (Myerson et al., 1990) and is likely to affect them in proportion to their complexity. Aging, therefore, is bound to induce a differential failure of those cognitive skills that depend on novel connections and the reconfiguration of existing networks rather than those that rely on existing "greased paths."

Despite significant advances in the understanding of the mechanisms that translate brain cellular activity into cognitive operations undergirding complex reasoning, the specifics of the associations between brain aging and age-related cognitive changes remain unclear, and elucidating the mechanisms of brain aging is a critical priority for understanding and mitigating senescent cognitive declines. Yet, because it is an integral part of the organism, clarifying the brain's role in age-related cognitive changes is impossible without describing the contributions of other organismic components that affect its properties and the trajectories of change (Fjell et al., 2014; Kennedy & Raz, 2015; Raz & Rodrigue, 2006). Below, I survey the current literature on age-related changes in brain structure and their relation to cognitive changes, as well as the role of various risk factors in shaping individual trajectories of neurocognitive aging. Functional brain changes are reviewed in chapters 5 and 7 of this volume.

Postmortem Studies

Neuroanatomical postmortem studies have demonstrated that the brains of older adults are smaller and lighter than those of their younger counterparts and exhibit significantly larger cerebrospinal fluid cavities (Kemper, 1994). Older brains evidence shrinkage, dysmorphology and the selective attrition of neurons (Pakkenberg et al., 2003), the impairment of neurogenesis and synaptogenesis, sparser dendritic spines (Morrison & Baxter, 2012), and lower myelin content in the cortex and subcortical white matter (Kaes, 1907). Iron content is elevated in older brains, especially in the basal ganglia, the substantia nigra, and the red nucleus (Hallgren & Sourander, 1958). With iron as a key player in mitochondrial energy production (Zecca et al., 2004), its excessive deposits may signal the disruption of energy metabolism with age as well as set off multiple paths to age-related neurodegeneration (Ashraf, Clark, & So, 2018). Postmortem studies, though indispensable in elucidating the natural history of mammalian aging, represent only a cross-sectional snapshot of the brain state. They reveal nothing about the temporal dynamics and individual differences in change, which became accessible only with noninvasive neuroimaging techniques.

Aging of the Brain Structure: In Vivo Evidence

For the past three decades, noninvasive neuroimaging has produced a substantial body of cross-sectional studies of age differences in the brain's gross structural properties. The findings from in vivo studies agree with the postmortem record. They document age-related differences in almost all brain regions, including smaller total and regional volumes, a thinner cortex, and larger cerebral ventricles in persons of advanced age compared to their younger counterparts (Raz, 2000). Meta-analyses of the extant studies and mega-analyses of aggregated samples show differential age-related alterations across brain regions: stronger

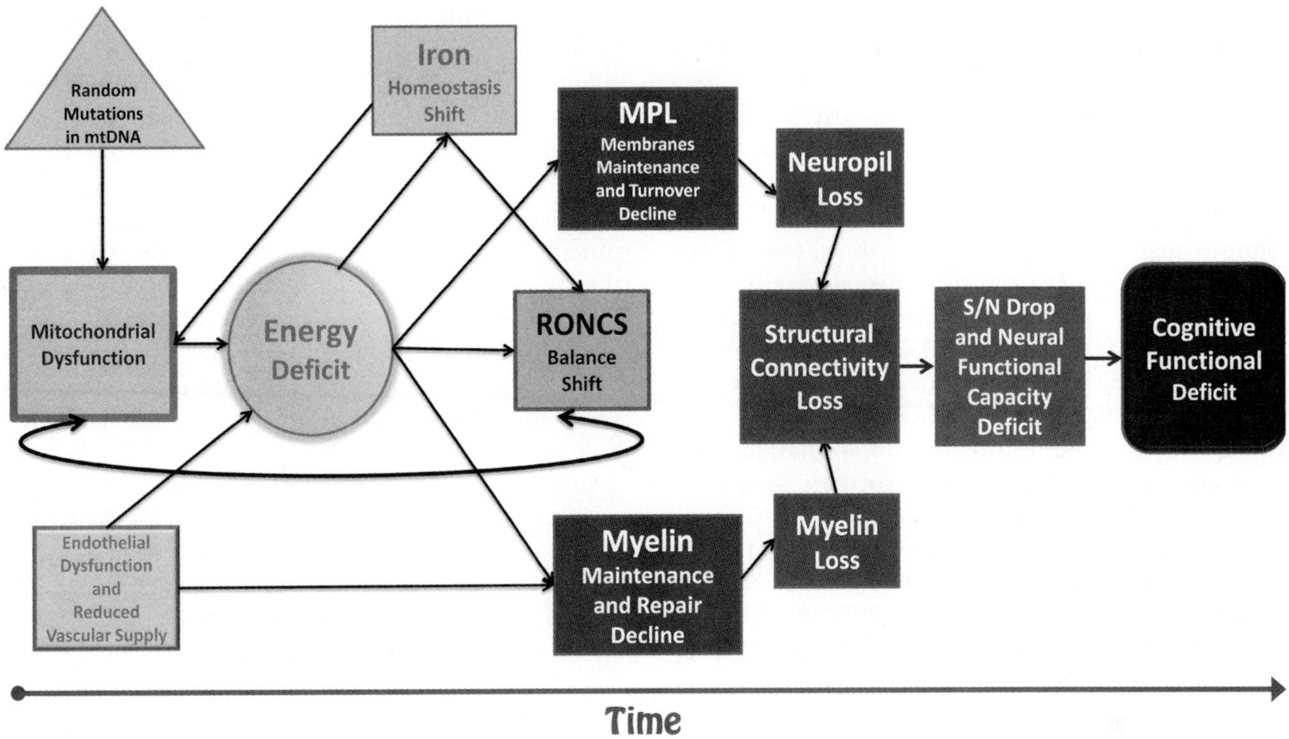

FIGURE 6.1 The FRIENDS model of cognitive aging. Random mutations accumulate in the mitochondrial DNA, and mitochondrial function declines with age. That decline results in the gradual reduction of energy substrates (ATP and NADH). The sequestration and transport of iron are affected, and the accumulation of iron, in turn, exacerbates mitochondrial declines. Endothelial dysfunction impairs the delivery of glucose and oxygen to the end organs in the brain and may directly stunt the regeneration of myelin. An energy crisis and further mitochondrial dysfunction ensue. Impaired mitochondrial respiration and the accumulation of iron lead to a shift in the homeostasis of RONCS and their excess further damages the mitochondria. This cascade of events further reduces the energy available for the maintenance of cell membranes and myelin and for supporting synaptic plasticity. Myelin and neuropil are gradually diminished, and connectivity among the brain regions is lost. That loss impairs the fidelity of neural communication, and the information-processing capabilities of the brain networks are degraded; the system becomes progressively noisier. Noise in the information- processing system affects all cognitive operations, with the more energy-dependent of them faring the worst of all. (See color plate 5.)

age associations in the prefrontal cortex, cerebellum, and neostriatum compared to the primary sensory cortices (e.g., Driscoll et al., 2009; Fjell et al., 2009; Raz et al., 1993, 1997, 2004; Walhovd et al., 2011; see Raz, 2000 for a review).

Despite their multiple advantages—for example, the feasibility of large-sample data collection and the potential for covering a lifespan age range—cross-sectional studies have an irreparable flaw: they yield no true estimates of change and individual differences in change trajectories (Hofer & Sliwinski, 2001; Lindenberger et al., 2011) due to the violation of ergodicity assumption (McArdle, 2009; Molenaar, 2004). These limitations are particularly severe in studies that evaluate the brain's role in mediating age influence on cognition (Raz & Lindenberger, 2011). Therefore, I limit the discussion to longitudinal studies and call upon cross-sectionally sourced findings only when longitudinal data are unavailable.

Longitudinal studies of the cerebral ventricles and the whole brain reveal expansion of the former (e.g., McArdle et al., 2004) and shrinkage of the latter time (Raz & Rodrigue, 2006). Age-related shrinkage happens even in selected samples of healthy adults (Anblagan et al., 2018; Callisaya et al., 2013; Crivello et al., 2014; Fjell et al., 2013; Hua et al., 2016; Leong et al., 2017; Opfer et al., 2018; Pfefferbaum et al., 2013; Persson et al., 2014; Rast et al., 2018; Raz et al., 2005, 2010, 2013; Squarzoni et al., 2018; Yuan, Voelkle, & Raz, 2018) within time windows as narrow as six months (Raz et al., 2013). Brain volume change over at least two occasions has been estimated at 0.15% per annum in middle-age adults (Opfer et al., 2018) and .21% (Takao et al., 2012) and .51% in a broad-age sample (Storvse et al., 2014), steadily increasing to .56% (Leong et al., 2017), .61% (Opfer et al., 2018), .55%–.77% (Sigurdsson et al., 2012), .80% (Aljondi et al., 2018), and up to .83% (Crivello et al., 2014) in samples of older participants. Notably, however, in some highly selected samples the total brain atrophy rates were modest even at the tenth decade of life (Mueller et al., 1998; Pressman et al., 2016).

Despite a general agreement on trends, specific findings vary across studies. The discrepancies may stem from differences in brain measurement methods, sample age ranges, and an admixture of individuals with vascular and metabolic risk factors, as well as frequency of assessments and the time elapsing between measurements. Most longitudinal studies of regional brain volumes reveal significant shrinkage of the neostriatum, cerebellum, posterior cingulate cortex, and tertiary association cortices. Lesser but still significant volume loss is noted in the fusiform, insular, anterior cingulate, and parahippocampal cortices, with minimal changes or no decline observed in the primary visual and somatosensory cortices (Crivello et al., 2014; Driscoll et al., 2009; Jiang et al., 2014; Kim et al., 2018; Leong et al., 2017; Persson et al., 2014; Pfefferbaum et al., 2013; Shaw et al., 2016; Storsve et al., 2014, see Kennedy & Raz for a review). Shrinkage of the hippocampus and the entorhinal cortex may follow a nonlinear course, with faster shrinkage in older participants (Crivello et al., 2014; Doré et al., 2013; Gorbatch et al., 2017; Jiang et al., 2014; Leong et al., 2017; Persson et al., 2014; Pfefferbaum et al., 2013; Raz et al., 2005, 2010; Storvse et al., 2014; see Kennedy & Raz, 2015, for a review). Other regions (e.g., the prefrontal white matter) exhibit age-accelerated changes in some samples (Raz et al., 2008), whereas in other studies a mixed pattern of acceleration and deceleration is noted, with age-accelerating changes in the temporal and the occipital and deceleration in the prefrontal and the anterior cingulate cortices (Storsve et al., 2014). Notably, cross-sectional designs underestimate age-related changes in cortical volume and thickness (Fjell et al., 2014; Kim et al., 2018; Pfefferbaum & Sullivan, 2015; Rast et al., 2017; Raz et al., 2005; Sigurdsson et al., 2012).

The magnitude of regional declines varies even across studies on the samples drawn from the same population, and some age trends are less replicable than others. Importantly, however, significant individual variability of change is observed in virtually all regions and in all studies that statistically modeled individual differences. For example, substantial shrinkage of the prefrontal cortex and the white matter (Raz et al., 2005) was not observed in other samples studied in the same lab with the same methods (Persson et al., 2014; Raz et al., 2010), whereas shrinkage of the cerebellum, the hippocampus, and the orbitofrontal cortex, accompanied by a relative stability of the primary visual cortex, was consistently replicated (Persson et al., 2014; Raz et al., 2005, 2010). Cortical thickness trajectories approximate the course of age-related reduction in regional volumes (Storsve et al., 2014; Thambisetti et al., 2010; Yuan et al., 2018; but see Pacheco et al. 2015 for prominent thinning of the primary motor cortex), with the measures of cortical surface yielding lower estimates of age-related change (Storsve et al., 2014). Cortical thinning and volume shrinkage are tightly correlated, but the cortical surface changes at a slower pace (Storvse et al., 2014). The shape of age-related change—linear, quadratic, or accelerated/decelerated exponential—is still unclear, as very few studies included more than one follow-up assessment, thus precluding the trajectory evaluation.

In summary, structural brain aging is heterochronic and individualized. The fact that some persons show minimal change, whereas others exhibit steep declines, spurs the search for influential factors mitigating age-related change and shaping individual trajectories of aging with minimal decline (Lindenberger, 2014). Determining the factors that induce individual variability in the brain aging of healthy adults is a challenge, as by definition, healthy samples are selected for the absence of detectable pathology and have restricted variance in negative modifiers of the brain structure. Nonetheless, several factors have been implicated in moderating the trajectories of brain aging. Physiological and genetic variations associated with vascular, metabolic, and neuroinflammatory risks and affecting endothelial function and energy expenditure while increasing the RONCS burden—the *vascular health triad* (Wadley, Veldhuijzen van Zanten, & Aldred, 2013)—appear particularly important. Arterial hypertension, even when medically controlled, is linked to the accelerated shrinkage of several brain regions, such as the hippocampus (Raz et al., 2005), raising a possibility that this detrimental effect stems not from elevated blood pressure per se but from the concomitant vascular changes. In many samples, gray matter shrinkage is exacerbated by an age-related elevation in vascular risk (Bobb et al., 2014; Kim et al., 2018; Raz et al., 2005, 2007, 2008),and in inflammation (Metti et al., 2015), as well as by genetic variants associated with an increased risk for Alzheimer's disease (AD; Donix et al., 2010; Harrison et al., 2016; Hua et al., 2016; Li et al., 2016; Moffat et al., 2000; but see Persson et al., 2014; Raz et al., 2010; Squarzoni et al., 2018 for negative findings), inflammation (Persson et al., 2014), and endothelial dysfunction (Zannas et al., 2014).

Carriers of the ApoE ε4 allele, a risk factor for AD, show faster shrinkage of the medial-temporal (Moffat et al., 2000), prefrontal, and cingulate (Rast et al., 2017) regions. Amyloid deposits are associated with faster rates of regional shrinkage, especially in the

7 Brain Maintenance and Cognition in Old Age

LARS NYBERG AND ULMAN LINDENBERGER

ABSTRACT The aging brain is notorious for detrimental changes. However, some older adults display *brain maintenance*, defined as a lack of senescent brain changes and age-related brain pathology. While general brain maintenance seems possible, the selective maintenance of some brain systems in the presence of age-related decline in others appears more likely. Here we focus on the structural and functional maintenance of the hippocampus. We propose that hippocampal maintenance is the primary determinant of preserved episodic-memory functioning in old age. We discuss several potential neural and nonneural mechanisms promoting hippocampal maintenance, including neuronal survival and generation, intact neuronal morphology, and vascular integrity. We provide evidence suggesting that correlated genetic and environmental factors cumulatively influence the operation of these mechanisms, in part through lifestyle choices. Future work should study the contributions of specific connections, transmitter systems, and subregions to hippocampal maintenance; further explore its etiology, prevalence, and behavioral relevance; and examine the generalizability of the maintenance concept for understanding individual differences in human cognitive aging to other regions and circuits.

More than 30 years ago, Rowe and Kahn (1987) introduced the notion of *successful aging* as a complement to the distinction between pathological and normal aging. While defining the criteria for success has stirred some debate, the preservation of broad cognitive abilities, such as episodic memory, is generally considered a hallmark of successful aging. Hence, the sizeable heterogeneity and malleability of human cognitive aging, including the potential for brain maintenance, have attracted increasing attention in recent years (Lindenberger, 2014; Nyberg, Lövdén, Riklund, Lindenberger, & Bäckman, 2012).

Cognitive abilities such as episodic memory, perceptual speed, reasoning, and semantic memory are characterized by large individual differences in age-related longitudinal change (e.g., Ghisletta, Rabbitt, Lunn, & Lindenberger, 2012; Lindenberger, 2014; Tucker-Drob, Brandmaier, & Lindenberger, 2019). Some individuals show a precipitous decline with advancing adult age, whereas others display high and stable levels of performance well into old age. In the following, we focus on the preservation of episodic memory as a particularly important aspect of successful cognitive aging (i.e., successful memory aging; see also Nyberg & Pudas, 2019.

Complex interactions among genetic and environmental factors underlie successful memory aging by influencing brain integrity. Major brain changes are linked to pathological memory aging—for example, in Alzheimer's dementia, as reviewed in detail elsewhere (e.g., Jack, Barrio, & Kepe, 2013). Also, while not of a pathological nature, brain changes at multiple levels (e.g., cellular, regional, and large-scale networks) are normative in aging (Fjell, McEvoy, Holland, Dale, & Walhovd, 2013) and underlie normal memory decline. Even in the presence of normative age-related brain changes, memory aging might be ameliorated through compensatory processes, as, for example, proposed by Reuter-Lorenz and Park (2014). Hence, successful memory aging is likely to proceed along multiple paths (Cabeza et al., 2018; Nyberg & Pudas, 2019; figure 7.1).

We have previously argued that the primary determinant of preserved cognition in older age is *brain maintenance*, or a relative lack of senescent and pathological brain changes in aging (Lindenberger, Burzynska, & Nagel, 2013; Nyberg et al., 2012). Despite its face validity and conceptual appeal, the relative maintenance of structural and functional brain integrity in aging remains an underexplored path toward successful memory aging. In this chapter we summarize relevant findings from studies of brain maintenance and cognition in old age and offer some thoughts on conceptual and theoretical refinements of the maintenance concept.

Brain Maintenance—General or Specific?

Will a group of older individuals who are characterized by, say, minimal cortical and subcortical atrophy, also display other kinds of brain maintenance (e.g., intact neurotransmitter systems, preserved structural connectivity, or low amyloid retention)? Alternatively, in the spirit of a multiple-factor framework (Buckner, 2004; Gabrieli, 1996), will different groups of individuals show

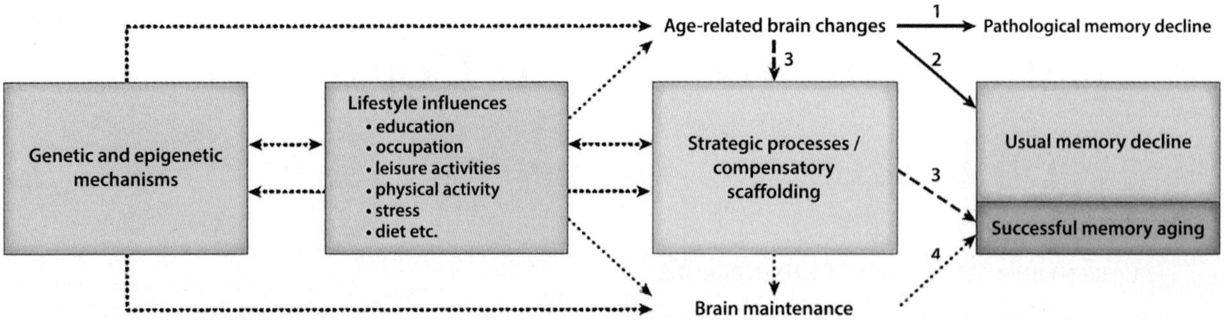

FIGURE 7.1 Model of life course paths to pathological, usual, and successful memory aging. Path 1 (not the focus of this chapter) represents trajectories characterized by pathological memory decline (e.g., dementia). Paths 2, 3, and 4 lead to normal memory aging (usual and successful). Path 2 represents usual memory decline in response to age-related brain changes (e.g., hippocampus atrophy). Path 3 (*dashed line*) is an indirect path to successful memory aging via compensation for age-related brain changes. Path 4 leads directly to successful memory aging via brain maintenance. The dotted lines reflect the influences of genetics, epigenetics, and lifestyle ("what" factors) as well as strategies ("how" factors) on brain integrity (maintenance versus decline) and memory outcomes. From Nyberg & Pudas (2019). (See color plate 6.)

distinct cascades of age-related degenerative changes that target different brain systems and vary in the rate of progression? In the latter view, it is conceivable that some individuals would display brain maintenance of certain major systems (such as the hippocampal/medial temporal lobe (MTL) system) while showing normal age-related changes in other systems (e.g., the fronto-striatal system). These competing accounts make distinct predictions for patterns of age-related cognitive changes (cf., Rabbitt, 1993).

Some findings from studies comparing the magnitude of age-related changes among different cognitive abilities, which should tax distinct brain circuits, at least in part, are in line with the generality of brain maintenance. For example, Ghisletta and colleagues (2012) found that about two-thirds of the age-based changes in memory are shared with those of other cognitive abilities, including reasoning, vocabulary, and perceptual speed (see also Dekhtyar et al., 2017; Tucker-Drob, Reynolds, Finkel, & Pedersen, 2014). These results were recently confirmed in a meta-analysis comprising 89 effect sizes representing shared variance in cognitive change from 22 unique data sets composed of over 30,000 individuals (Tucker-Drob, Brandmaier, & Lindenberger, 2019).

Broad cognitive abilities require multiple brain circuits and regions for their operation. Hence, individual differences in aging-induced cognitive change may covary across different abilities in the presence of distinct patterns of cerebral deterioration and maintenance. Longitudinal studies of brain aging do not portray a coherent picture as to the generality or specificity of brain maintenance in aging. For instance, analyses of individual five-year changes in the volumes of specific regions revealed correlations for some of the examined brain volumes, but not others (Raz et al., 2005). The highest change-change correlations were observed for the prefrontal white matter and the hippocampus ($r = .71$) and for the prefrontal white matter and the lateral prefrontal cortex ($r = .70$).

Some cross-sectional studies point to a certain degree of specificity of maintenance (Hedden et al., 2016; Lövdén et al., 2018; Salami, Pudas, & Nyberg, 2014). In a recent report from the COBRA Study, Salami and colleagues used a latent-profile analysis of data from 181 older adults aged 64 to 68 years who performed an n-back working-memory task during functional magnetic resonance imaging (fMRI) scanning. They identified one subgroup with higher performance ($n = 113$; 63%) and a second subgroup ($n = 55$; 31%) with lower performance. Analyses of data from an extensive off-line cognitive test battery confirmed that the lower-performing group had lower working-memory performance but nevertheless relatively intact episodic-memory performance. Moreover, analyses of different markers of brain integrity (cortical and subcortical volumes; functional connectivity; dopamine D2 system integrity) revealed specific alterations in the second group in frontoparietal regions, along with no group differences for other markers, such as hippocampal/default-mode regions.

Taken together, the degree of generality in brain maintenance remains a significant unresolved issue. Future large-scale longitudinal studies with multiple indicators of brain integrity, along with reliable measures of different cognitive functions, are needed to more conclusively address this topic. Awaiting such evidence, we focus on brain maintenance of the hippocampal system in relation to episodic memory, which is known to be hippocampus-dependent. We will draw on findings from longitudinal studies when available, as they provide more direct evidence on maintenance than cross-sectional comparisons. Select examples of nonhippocampal forms of brain maintenance will be provided in a separate section.

Hippocampus and Successful Episodic Memory

There is evidence that the amount of age-related atrophy, and by inference likely also maintenance, is not uniform across subregions of the hippocampal formation (Leal & Yassa, 2015; Pini et al., 2016; Shing et al., 2011). However, with only a few exceptions that will be noted, the available evidence relevant to brain maintenance refers to the whole hippocampus.

Mounting evidence indicates that hippocampal atrophy in aging underlies normal episodic-memory decline—notably, from change-change analyses (e.g., Gorbach et al., 2016). Hippocampal atrophy has been observed even in normal individuals who, according to genetic, biomarker, and cognitive criteria, are at minimal risk for showing early signs of pathological brain aging (Fjell et al., 2013; Raz et al., 2005). Importantly, there is evidence that some older individuals suffer minimal hippocampal atrophy with advancing adult age (Fjell et al., 2009; Raz et al., 2005), along with intact episodic memory (see Nyberg & Pudas, 2019). For example, Sun and colleagues (2016) reported that a group of "memory super-agers" had hippocampal volumes comparable to those of younger adults, in contrast to typically performing older adults, who had smaller volumes than younger individuals (see also Dekhtyar et al., 2017).

In addition to structure, functional integrity of the hippocampal formation matters as well. An fMRI study from the Betula project (Pudas et al., 2013) compared the brain activity of a group of individuals with maintained memory, as defined by baseline level and change over two to four test waves covering up to 15 years (Josefsson, De Luna, Pudas, Nilsson, & Nyberg, 2012), with individuals with normal memory development (i.e., moderate decline). The successful older adults were found to have a higher activity in the left hippocampus and prefrontal cortex during encoding of face-name pairs compared to the normal participants, and this activity was as high as the average activity observed in a young reference group (see also Düzel, Schütze, Yonelinas, & Heinze, 2011). Hippocampal activity correlated positively with memory performance, and this association continued to be present after statistically controlling for group differences in hippocampal volume. Further evidence for the importance of hippocampal functional integrity was obtained in a study showing that Betula participants who maintained their memory over up to 20 years had better-preserved hippocampal resting-state connectivity than older individuals with typical memory decline (Salami, Pudas, & Nyberg, 2014).

Additional aspects of hippocampal maintenance include dopaminergic neurotransmission (e.g., Nyberg et al., 2016) and neurovascular functions (Düzel, van Praag, & Sendtner, 2016), as well as minimal tau protein deposition (Schöll et al., 2016). Schöll and colleagues noted that tau is abundant in normal aging, in particular in the medial temporal lobes, and they observed that the amount of tau deposition across individuals was negatively related to episodic memory.

Not all of the above-mentioned studies used a longitudinal design. It has been acknowledged that cross-sectional data cannot serve as proxies for estimating associations between age changes in brain maintenance and age changes and cognitive performance (Lindenberger, von Oertzen, Ghisletta, & Hertzog, 2011; Nyberg et al., 2010; Raz & Lindenberger, 2011). Nevertheless, a cross-sectional data pattern in which older adults with more "youth-like" brain structures and functions show higher levels of cognitive performance is at least consistent with the brain maintenance notion. Hence, the longitudinal and cross-sectional findings taken together give some credence to the maintenance hypothesis as far as the hippocampal formation and episodic memory are concerned.

Brain Maintenance beyond the Hippocampus

Going beyond the focus on the hippocampal formation, other brain characteristics have also been related to well-preserved episodic memory in aging. One prominent example is greater thickness of the cingulate cortex, particularly its anterior segments (Gefen et al., 2015; Sun et al., 2016). Another example is stronger and more youth-like functional connectivity between the hippocampus and anterior brain regions, which have been found to correlate with better memory performance in old age (Fandakova, Lindenberger, & Shing, 2015; Lin, Ren, et al., 2017).

Minimal β-amyloid deposition might also be a salient, more general characteristic of brain maintenance, and it has been found to relate to high episodic-memory functioning in aging (Dekhtyar et al., 2017; Farrell et al., 2017). Also, recent findings suggest that integrity of the locus coeruleus and associated noradrenergic pathways may play an important role in maintaining memory performance in old age (Dahl et al., 2019; Hämmerer et al., 2018; Mather, this volume).

Possibly, minimal cortical pathology could also have contributed to observations of a youth-like pattern of functional brain activity in high-performing older adults while performing tasks assessing processing speed (Waiter et al., 2008), executive functioning (Burzynska et al., 2012), and working memory (Burzynska et al., 2013; Nagel et al., 2009, 2011). For instance, Burzynska et al. (2013) found that a more intact white matter microstructure was associated with less signal increase from rest to task in task-relevant brain regions and better performance on a working-memory task (n-back), regardless of adult age.

Moderators of Brain Maintenance

In view of global population aging, a pressing question for individuals as well as society is whether it is possible to influence successful aging in general and successful memory aging in particular (cf. Lindenberger, 2014; Nyberg & Pudas, 2019). In this chapter we have argued that the primary path to well-preserved episodic-memory functions in older age is via brain maintenance—notably, hippocampal integrity (Nyberg et al., 2012). Both genetic and environmental factors influence brain integrity (maintenance and change) in aging. The age-related increase in heritability for most cognitive abilities from childhood to adulthood suggests that genetic and environmental factors are correlated across ontogeny (Beam & Turkheimer, 2013). In particular, differences in lifestyle choices are likely to reflect, in part, genetic differences (Pedersen et al., 2013). Hence, genetic and environmental factors contribute synergistically and cumulatively to individual differences in brain maintenance.

Among the genetic factors, *APOE* stands out in relation to an accelerated cognitive decline in aging (e.g., Davies et al., 2014; Josefsson et al., 2012). APOE has also been identified as a characteristic of older adults displaying successful memory aging (Lin, Wang, et al., 2017). In addition, APOE has repeatedly been linked to hippocampal volume and atrophy (e.g., Davies et al., 2014; Li et al., 2016), though large-scale genome-wide association studies (GWAS) also point to other genetic loci in determining individual differences in hippocampal volume (Hibar et al., 2017) and atrophy (e.g., Mather et al., 2015).

Some lifestyle features may also promote brain maintenance, in part through epigenetic mechanisms (e.g., Mather, Kwok, Armstrong, & Sachdev, 2014; Spiegel, Sewal, & Rapp, 2014). In line with observational longitudinal data (Valenzuela, Sachdev, Wen, Chen, & Brodaty, 2008), intervention studies have linked cognitively stimulating activities, such as spatial navigation, to less hippocampal atrophy (e.g., Lövdén et al., 2012). Physical activity is also a crucial lifestyle factor, and there is evidence from epidemiological and intervention studies in older age for positive associations among physical exercise, hippocampal integrity, and memory (e.g., Düzel, van Praag, & Sendtner, 2016; Maass et al., 2015).

Specifically, physical activity may contribute to hippocampal maintenance by reducing the risk for hypertension. There is evidence that midlife hypertension increases the risk for hippocampal atrophy (e.g., Korf, White, Scheltens, & Launer, 2004), and it has been shown that accelerated hippocampal atrophy in older age is selective for individuals with hypertension (Raz et al., 2005). Also, hypertension has been linked to impaired hippocampus-related neurogenesis (Shih et al., 2016) and to smaller hippocampal subfield volumes and a greater propensity for false memories among older adults (Shing et al., 2011). Thus, the prevention and treatment of hypertension may be crucial for promoting hippocampal maintenance in aging, and physical exercise may be relevant in this regard.

In sum, genetic and environmental factors interactively and cumulatively contribute to brain maintenance, in part through individual differences in lifestyle. Next, we turn to the more mechanistic issue of what is maintained, and how.

Etiology and Expressions of Brain Maintenance

Evolutionary theories propose that human senescence reflects evolved limitations in somatic maintenance, resulting in a buildup of damage (Kirkwood, 2005). In light of the brain's disproportionately high energy needs, general theories of human brain aging stress the importance of bioenergetic mechanisms (Lindenberger, 2014; Raz, this volume; Raz & Daugherty, 2018). By this account, the "core cause of cognitive aging is a generalized and chronic decline in the availability of energy resources" (Raz & Daugherty, 2018, p. 55). As a corollary of this view, maintenance processes that are particularly expensive metabolically, such as myelin repair and regeneration (Hill, Li, & Grutzendler, 2018), are especially likely to suffer from normal aging. Brain maintenance, from this point of view, can be seen as the postponement and attenuation of a looming energy crisis.

In relation to the hippocampus, several possible etiologies of atrophy and functional impairment have been put forward (e.g., Dawe, Bennett, Schneider, & Arfanakis, 2011). These include (1) neuronal loss, (2) the loss of axons and dendrites, and (3) the loss of non-neural elements (e.g., glia, vessels). Each of these can be considered in relation to brain maintenance.

Starting with neuronal loss, postmortem MRI quantification of hippocampal volume was found to correlate with histology-based measurements in a sample of 11 patients with Alzheimer's disease and 4 healthy older controls (Bobinski et al., 2000). Specifically, in the study by Babinski and colleagues, a strong correlation between neuronal counts and MRI volumes was found (cf. Zarow et al., 2005). Hence, by inference, in vivo MRI observations of hippocampal atrophy in aging might reflect neuronal loss. Cell loss is a characteristic feature of pathological aging in Alzheimer's disease, but it has been argued that a marked loss of neurons in the hippocampus is not typical for normal aging (e.g., Dickstein, Weaver, Luebke, & Hof, 2013; Lister & Barnes, 2009; Wilson, Gallagher, Eichenbaum, & Tanila, 2006). Still, although the rate is higher in pathological aging, aging neurons seem to become more vulnerable to cell death (e.g., Mattson & Magnus, 2006), in particular in the CA1 region (Šimić, Kostović, Winblad, & Bogdanović, 1997). Moreover, it remains an open possibility that neuronal loss is more abundant in usual than in successful aging, resulting in a nonlinear loss gradient (successful < usual << pathological aging).

A related mechanism is age-related alteration in neurogenesis. New hippocampal neurons are generated throughout development, and perhaps throughout life (Spalding et al., 2013). Spalding and colleagues reported only a modest reduction in neurogenesis in aging, with marked interindividual differences from about age 20 to 80. Other studies have observed age-related reductions in neurogenesis and related such reductions to memory impairment in normal aging (see Seib & Martin-Villalba, 2015). Collectively, these past studies suggest that maintained neurogenesis could contribute to hippocampal integrity and preserved memory in certain older individuals, and some recent results further underscore this interpretation (Boldrini et al., 2018). It should be noted, though, that findings from another recent study suggested that neurogenesis in the dentate gyrus is rare or nonexistent in adult humans (Sorrells et al., 2018). Thus, at present, the evidence that effective neurogenesis in older age is a vital mechanism underlying brain maintenance is inconclusive (Snyder, 2019). Still, a role for neurogenesis in brain maintenance, even in view of the recent findings by Sorrells and colleagues, is bolstered by the suggestion that high degrees of neurogenesis at younger ages might protect against memory decline in old age (Seib & Martin-Villalba, 2015).

We turn now to the second class of possible etiologies of hippocampal atrophy, the loss of axons or dendrites (Dawe et al., 2011). Dendritic spine changes in aging

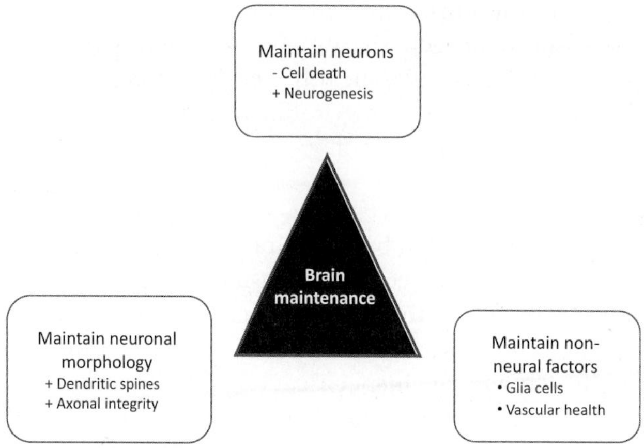

FIGURE 7.2 Putative underlying mechanisms of (hippocampal) brain maintenance.

have been suggested to potentially distinguish unsuccessful from successful cognitive aging (Dickstein et al., 2013). Alterations of myelin and nerve fibers in normal aging have been linked to cognitive decline (Peters, 2002). Morphological neuronal changes may also relate to decreased brain connectivity and disrupted neuronal synchrony in aging (Pannese, 2011). Relatedly, subtle synaptic alterations, including in the hippocampus, have been linked to cognitive impairment in normal aging (Morrison & Baxter, 2012).

In the third class, nonneural contributions to hippocampal atrophy, a recent study indicated that genes specific for astrocytes and oligodendrocytes, but not neurons, shift their expression patterns in aging—notably, in the hippocampus (Soreq et al., 2017)—thus pointing to a key role of glia cells in aging. Another example of a relevant nonneural contribution was revealed in a study of the neuropathological basis of age-associated brain atrophy (Erten-Lyons et al., 2013). Multiple antemortem MRI measurements of brain volumes were related to several neuropathological markers. Accelerated hippocampal volume decline trajectories were found to be significantly associated with the presence of amyloid angiopathy, and it was suggested that subtle chronic ischemia may contribute to elevated atrophy. More generally, there is much evidence that vascular pathology in normal and pathological aging negatively affects the brain and cognition in old age (see Farkas & Luiten, 2001).

In summary, we have reviewed some potential neural and nonneural mechanisms underlying brain maintenance (figure 7.2)—in particular, the maintenance of hippocampal integrity. Each of these mechanisms is likely influenced by the individual differences in lifestyles summarized above. Perhaps the most prominent example is physical exercise, which has been linked to each of the three domains in figure 7.1, such as increased neurogenesis (van Praag, Shubert, Zhao, & Gage, 2005), vascular health (see Düzel et al., 2016), and dendritic branching (Redila & Christie, 2006).

Future Directions

We have argued that brain maintenance is the primary path to preserved cognition in later adulthood, and more specifically that hippocampal maintenance is a key determinant of intact episodic-memory functioning in old age. Such specification of the general concept of brain maintenance has been called for (Cabeza et al., 2018; Nilsson & Lövdén, 2018). In future work, the underlying mechanisms of hippocampal maintenance should be explored in greater detail. This work will benefit from considering specific connections and subregions of the hippocampal formation, which in human MRI studies may require MR research at high field strengths (e.g., Maass et al., 2014).

At a more general level, the etiology, dimensionality, and promotion of brain maintenance needs to be addressed in large-scale studies that combine long-term longitudinal observations with interventions targeting behaviorally relevant brain regions and neural circuitry. At the same time, more detailed insights into the metabolic preconditions and molecular mechanisms of brain maintenance are needed to gauge their contributions to successful cognitive aging.

REFERENCES

Beam, C. R., & Turkheimer, E. (2013). Phenotype-environment correlations in longitudinal twin models. *Development and Psychopathology, 25*(1), 7–16. doi:10.1017/S0954579412000867

Bobinski, M., De Leon, M. J., Weigiel, J., Desanti, S., Convit, A., Saint Louis, L. A., … Wisniewski, H. M. (2000). The histological validation of post mortem magnetic resonance imaging-determined hippocampal volume in Alzheimer's disease. *Neuroscience, 95*(3), 721–725. doi:10.1016/s0306-4522(99)00476-5

Boldrini, M., Fulmore, C. A., Tartt, A. N., Simeon, L. R., Pavlova, I., Poposka, V., … Mann, J. J. (2018). Human hippocampal neurogenesis persists throughout aging. *Cell Stem Cell, 22*(4), 589–599.e585. doi:10.1016/j.stem.2018.03.015

Buckner, R. L. (2004). Memory and executive function in aging and AD: Multiple factors that cause decline and reserve factors that compensate. *Neuron, 44*(1), 195–208. doi:10.1016/j.neuron.2004.09.006

Burzynska, A., Garrett, D. D., Preuschhof, C., Nagel, I. E., Li, S.-C., Bäckman, L., … Lindenberger, U. (2013). A scaffold for efficiency in the human brain. *Journal of Neuroscience, 33*(43), 17150–17159. doi:10.1523/jneurosci.1426-13.2013

Burzynska, A. Z., Nagel, I. E., Preuschhof, C., Gluth, S., Bäckman, L., Li, S.-C., … Heekeren, H. R. (2012). Cortical thickness is linked to executive functioning in adulthood and aging. *Human Brain Mapping, 33*(7), 1607–1620. doi:10.1002/hbm.21311

Cabeza, R., Albert, M., Belleville, S., Craik, F. I. M., Duarte, A., Grady, C. L., … Rajah, M. N. (2018). Maintenance, reserve, and compensation: the cognitive neuroscience of healthy aging. *Nature Reviews Neuroscience, 19*(11), 701–710. doi:10.1038/s41583-018-0068-2

Dahl, M. J., Mather, M., Düzel, S., Bodammer, N. C., Lindenberger, U., Kühn, S., & Werkle-Bergner, M. (2019). Locus coeruleus integrity preserves memory performance across the adult life span.

Davies, G., Harris, S. E., Reynolds, C. A., Payton, A., Knight, H. M., Liewald, D. C., … Deary, I. J. (2014). A genome-wide association study implicates the APOE locus in nonpathological cognitive ageing. *Molecular Psychiatry, 19*(1), 76–87. doi:10.1038/mp.2012.159

Dawe, R. J., Bennett, D. A., Schneider, J. A., & Arfanakis, K. (2011). Neuropathologic correlates of hippocampal atrophy in the elderly: A clinical, pathologic, postmortem MRI Study. *PLoS One, 6*(10), e26286. doi:10.1371/journal.pone.0026286

Dekhtyar, M., Papp, K. V., Buckley, R., Jacobs, H. I. L., Schultz, A. P., Johnson, K. A., … Rentz, D. M. (2017). Neuroimaging markers associated with maintenance of optimal memory performance in late-life. *Neuropsychologia, 100,* 164–170. doi:10.1016/j.neuropsychologia.2017.04.037

Dickstein, D. L., Weaver, C. M., Luebke, J. I., & Hof, P. R. (2013). Dendritic spine changes associated with normal aging. *Neuroscience, 251,* 21–32. doi:10.1016/j.neuroscience.2012.09.077

Düzel, E., Schütze, H., Yonelinas, A. P., & Heinze, H.-J. (2011). Functional phenotyping of successful aging in long-term memory: Preserved performance in the absence of neural compensation. *Hippocampus, 21*(8), 803–814. doi:10.1002/hipo.20834

Düzel, E., van Praag, H., & Sendtner, M. (2016). Can physical exercise in old age improve memory and hippocampal function? *Brain, 139*(3), 662–673. doi:10.1093/brain/awv407

Erten-Lyons, D., Dodge, H. H., Woltjer, R., Silbert, L. C., Howieson, D. B., Kramer, P., & Kaye, J. A. (2013). Neuropathologic basis of age-associated brain atrophy. *JAMA Neurology, 70*(5), 616–622. doi:10.1001/jamaneurol.2013.1957

Fandakova, Y., Lindenberger, U., & Shing, Y. L. (2015). Episodic memory across the lifespan: General trajectories and modifiers. In A. Duarte, D. R. Addis, & M. D. Barense (Eds.), *The cognitive neuroscience of human memory* (pp. 309–325). Hoboken, NJ: Wiley-Blackwell.

Farkas, E., & Luiten, P. G. M. (2001). Cerebral microvascular pathology in aging and Alzheimer's disease. *Progress in Neurobiology, 64*(6), 575–611. doi:10.1016/S0301-0082(00)00068-X

Farrell, M. E., Kennedy, K. M., Rodrigue, K. M., Wig, G., Bischof, G. N., Rieck, J. R., … Park, D. C. (2017). Association of longitudinal cognitive decline with amyloid burden in middle-aged and older adults: Evidence for a dose-response relationship. *JAMA Neurology, 74*(7), 830–838. doi:10.1001/jamaneurol.2017.0892

Fjell, A. M., McEvoy, L., Holland, D., Dale, A. M., & Walhovd, K. B. (2013). Brain changes in older adults at very low risk for Alzheimer's disease. *Journal of Neuroscience, 33*(19), 8237–8242. doi:10.1523/JNEUROSCI.5506-12.2013

Fjell, A. M., Walhovd, K. B., Fenneman-Notestine, C., McEvoy, L. K., Hagler, D. J., Holland, D., … Dale, A. M. (2009). One-year brain atrophy evident in healthy aging. *Journal of*

Neuroscience, 29(48), 15223–15231. doi:10.1523/JNEURO SCI.3252-09.2009

Gabrieli, J. D. E. (1996). Memory systems analyses of mnemonic disorders in aging and age-related diseases. *Proceedings of the National Academy of Sciences of the United States of America, 93*(24), 13534–13540.

Gefen, T., Peterson, M., Papastefan, S. T., Martersteck, A., Whitney, K., Rademaker, A., … Geula, C. (2015). Morphometric and histologic substrates of cingulate integrity in elders with exceptional memory capacity. *Journal of Neuroscience, 35*(4), 1781–1791. doi:10.1523/JNEUROSCI.2998-14.2015

Ghisletta, P., Rabbitt, P., Lunn, M., & Lindenberger, U. (2012). Two thirds of the age-based changes in fluid and crystallized intelligence, perceptual speed, and memory in adulthood are shared. *Intelligence, 40*(3), 260–268. doi:10.1016/j.intell.2012.02.008

Gorbach, T., Pudas, S., Lundquist, A., Orädd, G., Josefsson, M., Salami, A., … Nyberg, L. (2016). Longitudinal association between hippocampus atrophy and episodic-memory decline. *Neurobiology of Aging, 51,* 167–176. doi:10.1016/j.neurobiolaging.2016.12.002

Hämmerer, D., Callaghan, M. F., Hopkins, A., Kosciessa, J., Betts, M., Cardenas-Blanco, A., … Düzel, E. (2018). Locus coeruleus integrity in old age is selectively related to memories linked with salient negative events. *Proceedings of the National Academy of Sciences of the United States of America, 115*(9), 2228–2233. doi:10.1073/pnas.1712268115

Hedden, T., Schultz, A. P., Rieckmann, A., Mormino, E. C., Johnson, K. A., Sperling, R. A., & Buckner, R. L. (2016). Multiple brain markers are linked to age-related variation in cognition. *Cerebral Cortex, 26*(4), 1388–1400. doi:10.1093/cercor/bhu238

Hibar, D. P., Adams, H. H. H., Jahanshad, N., Chauhan, G., Stein, J. L., Hofer, E., … Ikram, M. A. (2017). Novel genetic loci associated with hippocampal volume. *Nature Communications, 8,* 13624. doi:10.1038/ncomms13624

Hill, R. A., Li, A. M., & Grutzendler, J. (2018). Lifelong cortical myelin plasticity and age-related degeneration in the live mammalian brain. *Nature Neuroscience, 21,* 683–695. doi:10.1038/s41593-018-0120-6

Jack, C. R., Barrio, J. R., & Kepe, V. (2013). Cerebral amyloid PET imaging in Alzheimer's disease. *Acta Neuropathologica, 126*(5), 643–657. doi:10.1007/s00401-013-1185-7

Josefsson, M., De Luna, X., Pudas, S., Nilsson, L.-G., & Nyberg, L. (2012). Genetic and lifestyle predictors of 15-year longitudinal change in episodic memory. *Journal of the American Geriatrics Society, 60*(12), 2308–2312. doi:10.1111/jgs.12000

Kirkwood, T. B. L. (2005). Understanding the odd science of aging. *Cell, 120*(4), 437–447. doi:10.1016/j.cell.2005.01.027

Korf, E. S. C., White, L. R., Scheltens, P., & Launer, L. J. (2004). Midlife blood pressure and the risk of hippocampal atrophy. *Hypertension, 44*(1), 29–34. doi:10.1161/01.HYP.0000132475.32317.bb

Leal, S. L., & Yassa, M. A. (2015). Neurocognitive aging and the hippocampus across species. *Trends in Neurosciences, 38*(12), 800–812. doi:10.1016/j.tins.2015.10.003

Li, B., Shi, J., Gutman, B. A., Baxter, L. C., Thompson, P. M., Caselli, R. J., … Alzheimer's Disease Neuroimaging Initiative. (2016). Influence of APOE genotype on hippocampal atrophy over time: An N=1925 surface-based ADNI study. *PLoS One, 11*(4), e0152901. doi:10.1371/journal.pone.0152901

Lin, F. V., Ren, P., Mapstone, M., Meyers, S. P., Porsteinsson, A., & Baran, T. M. (2017). The cingulate cortex of older adults with excellent memory capacity. *Cortex, 86,* 83–92. doi:10.1016/j.cortex.2016.11.009

Lin, F. V., Wang, X., Wu, R., Rebok, G. W., Chapman, B. P., & Alzheimer's Disease Neuroimaging Initiative. (2017). Identification of successful cognitive aging in the Alzheimer's Disease Neuroimaging Initiative study. *Journal of Alzheimer's Disease, 59*(1), 101–111. doi:10.3233/JAD-161278

Lindenberger, U. (2014). Human cognitive aging: Corriger la fortune? *Science, 346*(6209), 572–578. doi:10.1126/science.1254403

Lindenberger, U., Burzynska, A. Z., & Nagel, I. E. (2013). Heterogeneity in frontal-lobe aging. In D. T. Stuss & R. T. Knight (Eds.), *Principles of frontal lobe functions* (2nd ed., pp. 609–627). New York: Oxford University Press.

Lindenberger, U., von Oertzen, T., Ghisletta, P., & Hertzog, C. (2011). Cross-sectional age variance extraction: What's change got to do with it? *Psychology and Aging, 26*(1), 34–47. doi:10.1037/a0020525

Lister, J. P., & Barnes, C. A. (2009). Neurobiological changes in the hippocampus during normative aging. *Archives of Neurology, 66*(7), 829–833. doi:10.1001/archneurol.2009.125

Lövdén, M., Karalija, N., Andersson, M., Wåhlin, A., Jan, A., Köhncke, Y., … Lindenberger, U. (2018). Latent-profile analysis reveals behavioral and brain correlates of dopamine-cognition associations. *Cerebral Cortex, 28*(11), 3894–3907. doi:10.1093/cercor/bhx253

Lövdén, M., Schaefer, S., Noack, H., Bodammer, N. C., Kühn, S., Heinze, H.-J., … Lindenberger, U. (2012). Spatial navigation training protects the hippocampus against age-related changes during early and late adulthood. *Neurobiology of Aging, 33*(3), 620.e629–620.e622. doi:10.1016/j.neurobiolaging.2011.02.013

Maass, A., Düzel, S., Goerke, M., Becke, A., Sobieray, U., Neumann, K., … Düzel, E. (2015). Vascular hippocampal plasticity after aerobic exercise in older adults. *Molecular Psychiatry, 20,* 585–593. doi:10.1038/mp.2014.114

Maass, A., Schütze, H., Speck, O., Yonelinas, A., Tempelmann, C., Heinze, H.-J., … Düzel, E. (2014). Laminar activity in the hippocampus and entorhinal cortex related to novelty and episodic encoding. *Nature Communications, 5,* 5547. doi:10.1038/ncomms6547

Mather, K. A., Armstrong, N. J., Wen, W., Kwok, J. B., Assareh, A. A., Thalamuthu, A., … Sachdev, P. S. (2015). Investigating the genetics of hippocampal volume in older adults without dementia. *PLoS One, 10*(1), e0116920. doi:10.1371/journal.pone.0116920

Mather, K. A., Kwok, J. B., Armstrong, N., & Sachdev, P. S. (2014). The role of epigenetics in cognitive ageing. *International Journal of Geriatric Psychiatry, 29*(11), 1162–1171. doi:10.1002/gps.4183

Mattson, M. P., & Magnus, T. (2006). Ageing and neuronal vulnerability. *Nature Reviews Neuroscience, 7*(4), 278–294. doi:10.1038/nrn1886

Morrison, J. H., & Baxter, M. G. (2012). The ageing cortical synapse: Hallmarks and implications for cognitive decline. *Nature Reviews Neuroscience, 13,* 240–250. doi:10.1038/nrn3200

Nagel, I. E., Preuschhof, C., Li, S.-C., Nyberg, L., Bäckman, L., Lindenberger, U., & Heekeren, H. R. (2009). Performance level modulates adult age differences in brain activation during spatial working memory. *Proceedings of the National Academy of Sciences of the United States of America, 106*(52), 22552–22557. doi:10.1073/pnas.0908238106

Nagel, I. E., Preuschhof, C., Li, S.-C., Nyberg, L., Bäckman, L., Lindenberger, U., & Heekeren, H. R. (2011). Load modulation of BOLD response and connectivity predicts working memory performance in younger and older adults. *Journal of Cognitive Neuroscience, 23*(8), 2030–2045. doi:10.1162/jocn.2010.21560

Nilsson, J., & Lövdén, M. (2018). Naming is not explaining: Future directions for the "cognitive reserve" and "brain maintenance" theories. *Alzheimer's Research & Therapy, 10*(1), 34. doi:10.1186/s13195-018-0365-z

Nyberg, L., Karalija, N., Salami, A., Andersson, M., Wahlin, A., Kaboovand, N., … Bäckman, L. (2016). Dopamine D2 receptor availability is linked to hippocampal-caudate functional connectivity and episodic memory. *Proceedings of the National Academy of Sciences of the United States of America, 113*(28), 7918–7923. doi:10.1073/pnas.1606309113

Nyberg, L., Lövdén, M., Riklund, K., Lindenberger, U., & Bäckman, L. (2012). Memory aging and brain maintenance. *Trends in Cognitive Sciences, 16*(5), 292–305. doi:10.1016/j.tics.2012.04.005

Nyberg, L., & Pudas, S. (2019). Successful memory aging. *Annual Review of Psychology, 70,* 2019–243. doi:10.1146/annurev-psych-010418-103052

Nyberg, L., Salami, A., Andersson, M., Eriksson, J., Kalpouzos, G., Kauppi, K., … Nilsson, L.-G. (2010). Longitudinal evidence for diminished frontal cortex function in aging. *Proceedings of the National Academy of Sciences of the United States of America, 107*(52), 22682–22686. doi:10.1073/pnas.1012651108

Pannese, E. (2011). Morphological changes in nerve cells during normal aging. *Brain Structure and Function, 216*(2), 85–89. doi:10.1007/s00429-011-0308-y

Pedersen, N. L., Christensen, K., Dahl, A. K., Finkel, D., Franz, C. E., Gatz, M., … Reynolds, C. A. (2013). IGEMS: The Consortium on Interplay of Genes and Environment across Multiple Studies. *Twin Research and Human Genetics, 16*(1), 481–489. doi:10.1017/thg.2012.110

Peters, A. (2002). The effects of normal aging on myelin and nerve fibers: A review. *Journal of Neurocytology, 31*(8–9), 581–593. doi:10.1023/A:1025731309829

Pini, L., Pievani, M., Bocchetta, M., Altomare, D., Bosco, P., Cavedo, E., … Frisoni, G. B. (2016). Brain atrophy in Alzheimer's disease and aging. *Ageing Research Reviews, 30,* 25–48. doi:10.1016/j.arr.2016.01.002

Pudas, S., Persson, J., Josefsson, M., de Luna, X., Nilsson, L.-G., & Nyberg, L. (2013). Brain characteristics of individuals resisting age-related cognitive decline over two decades. *Journal of Neuroscience, 33*(20), 8668–8677. doi:10.1523/JNEUROSCI.2900-12.2013

Rabbitt, P. (1993). Does it all go together when it goes? The Nineteenth Bartlett Memorial Lecture. *Quarterly Journal of Experimental Psychology A, 46*(3), 385–434. doi:10.1080/14640749308401055

Raz, N., & Daugherty, A. M. (2018). Pathways to brain aging and their modifiers: Free-radical-induced energetic and neural decline in senescence (FRIENDS) model—a mini-review. *Gerontology, 64*(1), 49–57. doi:10.1159/000479508

Raz, N., & Lindenberger, U. (2011). Only time will tell: Cross-sectional studies offer no solution to the age-brain-cognition triangle: Comment on Salthouse (2011). *Psychological Bulletin, 137*(5), 790–795. doi:10.1037/a0024503

Raz, N., Lindenberger, U., Rodrigue, K. M., Kennedy, K. M., Head, D., Williamson, A., … Acker, J. D. (2005). Regional brain changes in aging healthy adults: General trends, individual differences and modifiers. *Cerebral Cortex, 15*(11), 1676–1689. doi:10.1093/cercor/bhi044

Redila, V. A., & Christie, B. R. (2006). Exercise-induced changes in dendritic structure and complexity in the adult hippocampal dentate gyrus. *Neuroscience, 137*(4), 1299–1307. doi:10.1016/j.neuroscience.2005.10.050

Reuter-Lorenz, P. A., & Park, D. C. (2014). How does it STAC up? Revisiting the scaffolding theory of aging and cognition. *Neuropsychology Review, 24*(3), 355–370. doi:10.1007/s11065-014-9270-9

Rowe, J. W., & Kahn, R. L. (1987). Human aging: Usual and successful. *Science, 237*(4811), 143–149. doi:10.1126/science.3299702

Salami, A., Pudas, S., & Nyberg, L. (2014). Elevated hippocampal resting-state connectivity underlies deficient neurocognitive function in aging. *Proceedings of the National Academy of Sciences of the United States of America, 111*(49), 17654–17659. doi:10.1073/pnas.1410233111

Schöll, M., Lockhart, S. N., Schonhaut, D. R., O'Neil, J. P., Janabi, M., Ossenkoppele, R., … Jagust, W. J. (2016). PET imaging of tau deposition in the aging human brain. *Neuron, 89*(5), 971–982. doi:10.1016/j.neuron.2016.01.028

Seib, D. R. M., & Martin-Villalba, A. (2015). Neurogenesis in the normal ageing hippocampus: A mini-review. *Gerontology, 61*(4), 327–335. doi:10.1159/000368575

Shih, Y. H., Tsai, S. F., Huang, S. H., Chiang, Y. T., Hughes, M. W., Wu, S. Y., … Kuo, Y. M. (2016). Hypertension impairs hippocampus-related adult neurogenesis, CA1 neuron dendritic arborization and long-term memory. *Neuroscience, 322,* 346–357. doi:10.1016/j.neuroscience.2016.02.045

Shing, Y. L., Rodrigue, K. M., Kennedy, K. M., Fandakova, Y., Bodammer, N., Werkle-Bergner, M., … Raz, N. (2011). Hippocampal subfield volumes: Age, vascular risk, and correlation with associative memory. *Frontiers in Aging Neuroscience, 3,* 2. doi:10.3389/fnagi.2011.00002

Šimić, G., Kostović, I., Winblad, B., & Bogdanović, N. (1997). Volume and number of neurons of the human hippocampal formation in normal aging and Alzheimer's disease. *Journal of Comparative Neurology, 379*(4), 482–494.

Snyder, J. S. (2019). Recalibrating the relevance of adult neurogenesis. *Trends in Neurosciences, 42*(3), 164–178. doi:10.1016/j.tins.2018.12.001

Soreq, L., Rose, J., Soreq, E., Hardy, J., Trabzuni, D., Cookson, M. R., … Ule, J. (2017). Major shifts in glial regional identity are a transcriptional hallmark of human brain aging. *Cell Reports, 18*(2), 557–570. doi:10.1016/j.celrep.2016.12.011

Sorrells, S. F., Paredes, M. F., Cebrian-Silla, A., Sandoval, K., Qi, D., Kelley, K. W., … Alvarez-Buylla, A. (2018). Human hippocampal neurogenesis drops sharply in children to undetectable levels in adults. *Nature, 555,* 377–381. doi:10.1038/nature25975

Spalding, K. L., Bergmann, O., Alkass, K., Bernard, S., Salehpour, M., Huttner, Hagen B., … Frisén, J. (2013). Dynamics of hippocampal neurogenesis in adult humans. *Cell, 153*(6), 1219–1227. doi:10.1016/j.cell.2013.05.002

Spiegel, A. M., Sewal, A. S., & Rapp, P. R. (2014). Epigenetic contributions to cognitive aging: Disentangling mindspan and lifespan. *Learning & Memory, 21*(10), 569–574. doi:10.1101/lm.033506.113

Sun, F. W., Stepanovic, M. R., Andreano, J., Barrett, L. F., Touroutoglou, A., & Dickerson, B. C. (2016). Youthful brains in older adults: Preserved neuroanatomy in the default mode and salience networks contributes to youthful memory in superaging. *Journal of Neuroscience, 36*(37), 9659–9668. doi:10.1523/JNEUROSCI.1492-16.2016

Tucker-Drob, E. M., Brandmaier, A. M., & Lindenberger, U. (2019). Coupled cognitive change in adulthood: A meta-analysis. *Psychological Bulletin, 145*(3), 273–301. doi: 10.1037/bul0000179

Tucker-Drob, E. M., Reynolds, C. A., Finkel, D., & Pedersen, N. L. (2014). Shared and unique genetic and environmental influences on aging-related changes in multiple cognitive abilities. *Developmental Psychology, 50*(1), 152–166. doi:10.1037/a0032468

Valenzuela, M. J., Sachdev, P., Wen, W., Chen, X., & Brodaty, H. (2008). Lifespan mental activity predicts diminished rate of hippocampal atrophy. *PLoS One, 3*(7), e2598. doi:10.1371/journal.pone.0002598

van Praag, H., Shubert, T., Zhao, C., & Gage, F. H. (2005). Exercise enhances learning and hippocampal neurogenesis in aged mice. *Journal of Neuroscience, 25*(38), 8680–8685. doi:10.1523/JNEUROSCI.1731-05.2005

Waiter, G. D., Fox, H. C., Murray, A. D., Starr, J. M., Staff, R. T., Bourne, V. J., ... Deary, I. J. (2008). Is retaining the youthful functional anatomy underlying speed of information processing a signature of successful cognitive ageing? An event-related fMRI study of inspection time performance. *NeuroImage, 41*(2), 581–595. doi:10.1016/j.neuroimage.2008.02.045

Wilson, I. A., Gallagher, M., Eichenbaum, H., & Tanila, H. (2006). Neurocognitive aging: Prior memories hinder new hippocampal encoding. *Trends in Neurosciences, 29*(12), 662–670. doi:10.1016/j.tins.2006.10.002

Zarow, C., Vinters, H. V., Ellis, W. G., Weiner, M. W., Mungas, D., White, L., & Chui, H. C. (2005). Correlates of hippocampal neuron number in Alzheimer's disease and ischemic vascular dementia. *Annals of Neurology, 57*(6), 896–903. doi:10.1002/ana.20503

8 The Locus Coeruleus-Norepinephrine System's Role in Cognition and How It Changes with Aging

MARA MATHER

ABSTRACT With its extensive projections that release norepinephrine (NE), the locus coeruleus (LC) has a major influence on the rest of the brain and influences cognitive processes. It aggregates signals regarding arousal from various other brain systems and modulates neural processing to become more focused under arousal. Because of challenges in measuring its structure and function in living humans, there has historically been little focus on the LC's role in determining the fate of cognition in aging. However, in the past few years a growing number of animal and human findings have suggested that the integrity of the LC-NE system plays a key role in maintaining late life cognition. In particular, age declines in selective attention, episodic memory, working memory, and cognitive flexibility have been linked to the LC. Furthermore, LC degeneration can accelerate the progression of Alzheimer's disease, whereas an intact LC may contribute to cognitive reserve.

The locus coeruleus (LC) consists of two bilateral, small brain stem nuclei that are the source of most of the brain's norepinephrine (NE; Berridge & Waterhouse, 2003). Although small, LC neurons have axons that project through much of the rest of the brain (Schwarz et al., 2015), typically releasing NE into the extrasynaptic space rather than at specific synapses (Vizi, Fekete, Karoly, & Mike, 2010). Given this anatomy and its central role in arousal, the LC has been considered ideal for broadcasting the message to pay extra attention to goal-relevant or emotionally salient information under arousing circumstances, such as when under pressure or threat. However, findings that arousal affects representations differently depending on their salience or priority suggest that the LC does something more sophisticated than simply providing a one-size-fits-all broadcast message (Mather, Clewett, Sakaki, & Harley, 2016). As reviewed in this chapter, local cortical conditions help control how NE release modulates attention and cognition, allowing LC activity to increase cognitive selectivity and attentional focus. This chapter also covers emerging evidence that LC integrity plays a more important role in late life cognition than previously appreciated and that the LC is one of the earliest sites of Alzheimer's-related tau pathology in the brain. In particular, changes in the LC-NE system in aging are likely to contribute to declines in attention and memory.

The Locus Coeruleus Integrates Information about Brain Arousal State

Although the LC projects widely to much of the rest of the brain, in most cases this communication is not reciprocated (figure 8.1). Instead, inputs to the LC come from a select group of brain regions that signal different types of arousal, including stress from the amygdala and paraventricular nucleus of the hypothalamus; movement and motivation from dopaminergic nuclei; wakefulness and sleep from serotonergic, cholinergic, and dopaminergic nuclei, as well as the hypothalamus and medulla; and additional arousal signals from histamine and orexin neurons (for a review, see Samuels & Szabadi, 2008). These inputs allow the LC to be a hub region for integrating information about the current state of arousal. There is little cortical input to the LC, but a notable exception comes from the prefrontal cortex, which provides signals regarding effort-related arousal.

LC activity stimulates pupil dilation (Joshi, Li, Kalwani, & Gold, 2015), and indeed each of these various types of arousal are associated with pupil dilation. For example, experiencing threat of monetary loss (Clewett, Huang, Velasco, Lee, & Mather, 2018) and engaging in cognitive effort (figure 8.2) both briefly increase pupil size. Being drowsy has a more complex relationship with pupil size, as when people get sleepy they show slow waves of pupil constrictions (e.g., Lüdtke, Wilhelm, Adler, & Schaeffel, 1998). These oscillations in pupil size are associated with low frequency rhythms of

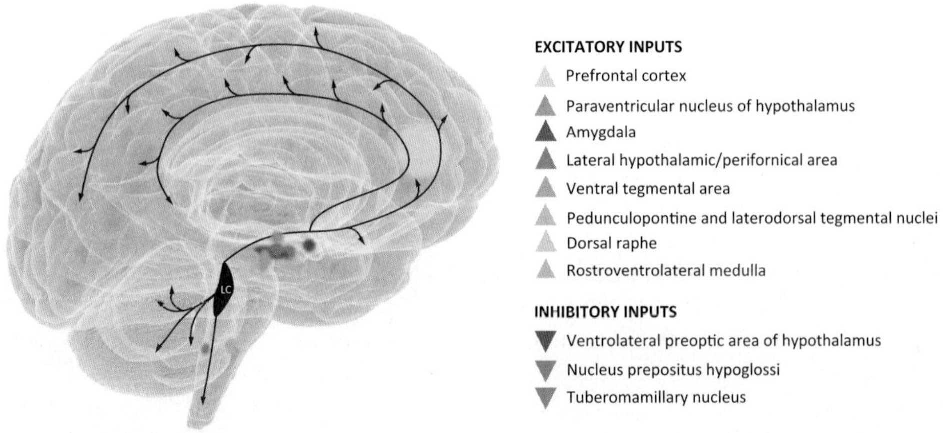

FIGURE 8.1 LC axons project to much of the rest of the brain, but the LC receives direct input from only a subset of these regions. These regions provide information about different aspects of arousal states and include excitatory input from the amygdala, paraventricular nucleus of the hypothalamus, lateral hypothalamus/perifornical area, ventral tegmental area, dorsal raphe, pedunculopontine and laterodorsal tegmental nuclei, and rostroventrolateral medulla and inhibitory input from the ventrolateral preoptic area of the hypothalamus, nucleus prepositus hypoglossi, and tuberomammillary nucleus (for a review, see Samuels & Szabadi, 2008). (See color plate 7.)

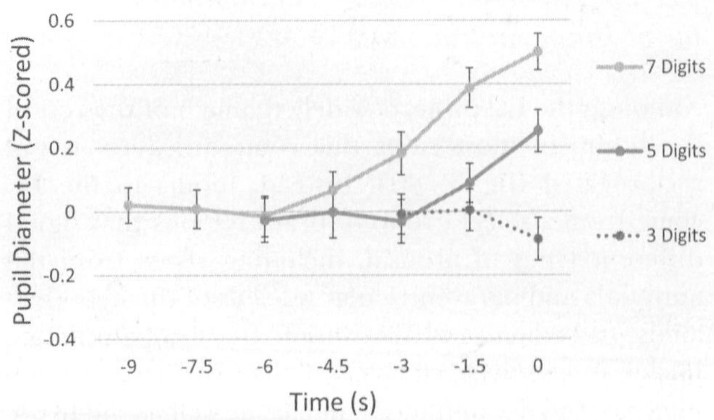

FIGURE 8.2 Participants heard one digit each 1.5 s and heard either three, five, or seven digits in a row that they retained for an immediate test. The *0* point on the right is the end of the auditory list presentation. After three digits were heard, pupil diameter increased for each additional digit added to their working-memory load.

cortical activity (McGinley, Vinck, Reimer, et al., 2015), which are suppressed by LC activity, as well as being modulated by other arousal-related nuclei (Neske, 2016; Steriade, Amzica, & Nunez, 1993).

Local Fluctuations in Glutamate Signal Priority to the Locus Coeruleus-Norepinephrine System

Having one brain region that integrates all these disparate signals about arousal provides the foundation to organize neural processing during many different types of arousal in a similar way. A common need during high arousal situations that involve high effort, threat, stress, or motivation is to focus resources on what really matters at that moment and avoid being distracted by low-priority information. But without direct input from most of the cortex, how can the LC "know" which neural representations are currently high priority?

Behaviorally, it is clear that arousal makes attention more selective. In particular, whatever would already stand out due to its bottom-up or top-down priority (i.e., perceptual salience or goal relevance) stands out even more, whereas lower-priority stimuli receive even less attention (Mather & Sutherland, 2011). For instance, when letters are flashed quickly on a white screen and people are asked to report which letters they saw, they are more likely to report letters that are more perceptually salient than others. However, if they hear an arousing sound seconds before seeing the letters, they are even more likely to report the more salient letters and less likely to report the less salient letters (Sutherland & Mather, 2012).

Given that NE is typically released into the extrasynaptic space, how can specific synaptic networks that are currently high priority be targeted for enhancement while currently low-priority networks are suppressed? One potential solution to this problem of the "priority-blind" LC is to have some local cortical signal of priority that modulates the effect of NE within the neural network generating the signal. In the brain, a simple signal of the current priority of a particular stimulus (for instance, the letter *P* flashing on a screen) is the level of excitatory activity at the neurons representing that stimulus. As outlined in the glutamate amplifies noradrenergic effects (GANE) model (Mather et al., 2016), glutamate (the brain's primary excitatory neurotransmitter) spills over from synapses, allowing it to stimulate receptors on nearby LC axons (figure 8.3*A*). These glutamatergic receptors stimulate more local release of NE (figure 8.3*B*), which in turn activates excitatory β-adrenergic receptors that lead to more glutamate release (figure 8.3*C*). This feedback mechanism stimulates hot spots of even higher activity among strongly activated neurons.

Having different types of noradrenergic receptors with different thresholds for activation plays a key role in allowing this increase in cortical gain. There are three classes of noradrenergic receptors: α2-, α1-, and β-adrenergic receptors, listed in order of their affinity for NE, as α2-adrenergic receptors require the lowest levels of NE to be activated, whereas β-adrenergic receptors require many times more NE to be activated (Ramos & Arnsten, 2007; Salgado, Kohr, & Trevino, 2012). The most common type of α2-adrenergic receptors (α2a) generally have inhibitory effects, serving as autoreceptors on LC neurons or on glutamatergic or other neurons that reduce neurotransmitter release (Starke, 2001). In contrast, β-adrenergic receptors typically have excitatory effects (Berridge, 2008; McCormick, Pape, & Williamson, 1991) and promote synaptic plasticity (Hagena, Hansen, Manahan-Vaughan, 2016; O'Dell, Connor, Gelinas, & Nguyen, 2010).

In rodents, stimulating LC and measuring NE levels via microdialysis indicate that stimulating LC activity can increase NE release in the cortex or hippocampus between one and five times the baseline NE levels (Berridge & Abercrombie, 1999; Fernández-Pastor & Meana, 2002; Florin-Lechner, Druhan, Aston-Jones, & Valentino, 1996; Pudovkina, Kawahara, de Vries, & Westerink, 2001). This is a much smaller range than the difference in NE levels needed to activate α2-adrenergic versus β-adrenergic receptors (see Mather et al., 2016), suggesting that differences in LC activity are not in themselves enough to account for the full range of NE levels that occur elsewhere in the brain. Thus, local cortical conditions, such as glutamate levels, likely play a key role in allowing for a greater range of NE levels, as well as levels that are high enough to activate β-adrenergic receptors that stimulate synaptic change.

In summary, the LC-NE system integrates many signals related to the brain's arousal state and has a wide network of afferent projections where it typically releases NE into extrasynaptic space. Yet despite relying on volume transmission, it can interact with local cortical excitation levels to modulate its impact on neural activity to promote even more selective processing during states of arousal.

The Nature of the Locus Coeruleus-Norepinephrine System Makes It Especially Vulnerable in Aging

The features that allow the LC to be a centralized aggregator of arousal signals that can broadcast messages throughout the brain yet have quite specific targeted effects make the LC more vulnerable in aging (Mather & Harley, 2016). In particular, the LC faces a uniquely high set of risk factors for tau pathology. One consists of the long, thin, and only sparsely myelinated nature of LC axons. Tau pathology tends to occur in neurons with long and thin-caliber axons with little or no myelin (Braak & Del Tredici, 2015). In contrast, short-axoned or highly myelinated projection neurons resist tau pathology. Highly myelinated neurons also transmit signals faster. So why would the LC-NE system, a key system involved in arousal and therefore involved in responding to threat, have slow axonal transmission instead of being optimized for speed?

In general, cells involved in sensing internal states and maintaining homeostasis are mostly unmyelinated, which in turn may enhance axons' ability to gather information about their milieu, via the exchange of ions or extracellular current flow (Damasio & Carvalho, 2013). This is consistent with the GANE model's postulate that the LC is sensitive to local cortical excitation levels, releasing more NE where there are already high levels of excitation and allowing highly active mental representations to get even more highly activated under arousal (Mather et al., 2016). To calibrate local NE release levels to local cortical conditions requires extensive two-way communication along the far-reaching LC axons. The cost of this may come in vulnerability to toxin exposure.

Furthermore, no other neuronal system has as extensive an exposure to circulating blood as the LC (Braak & Del Tredici, 2015; Pamphlett, 2014). LC neurons regulate the majority of the brain's microvasculature, with each LC neuron (with a tiny diameter of 45 μm) estimated to innervate nearly a tennis-court-length worth of capillaries (20 m; Pamphlett, 2014). Thus, a small amount of blood-borne toxic material could pose a relatively high burden for LC neurons.

In addition to blood, cranial nerves such as the trigeminal and vagal provide another route for toxic

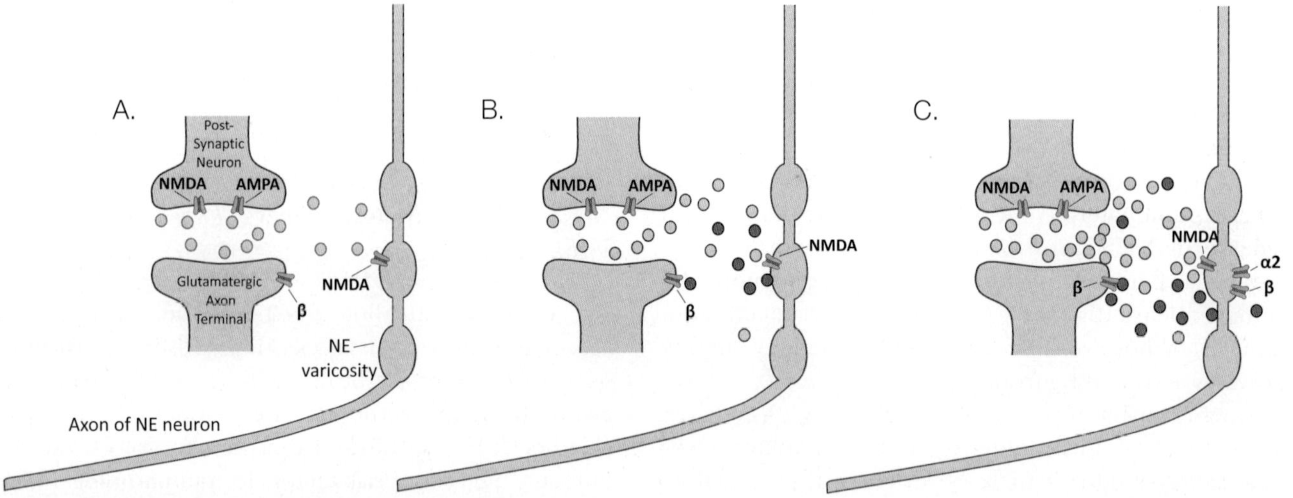

FIGURE 8.3 A graphic depiction of the proposed mechanisms of the GANE model. *A*, Glutamate (*gray circles*) spills over from glutamatergic synapses at the sites of excited representations. *B*, If the LC happens to be activated (that is, depolarized) at the same time as glutamate reaches the N-methyl D-aspartate (NMDA) receptors on the LC axons, this triggers more local release of NE (*black circles*) from those LC varicosities. *C*, Elevated local levels of NE activate β-adrenergic receptors that further stimulate glutamate release, leading to a local hot spot of high excitation. Autoreceptors at LC varicosities also contribute to increasing neural gain by inhibiting the release of noradrenaline when low levels of noradrenaline activate α-adrenergic receptors but increasing the release of noradrenaline when high levels of noradrenaline activate β-adrenergic receptors. Reprinted from Lee et al. (2018). (See color plate 8.)

substances to enter the brain. For instance, in mice, the avian flu virus travels from the gut and the peripheral nervous system to the brain stem solitary nucleus and from there to the LC, substantia nigra, and other brain stem nuclei, where the virus causes protein phosphorylation and aggregation that persists long after infection (Jang et al., 2009). Particulate matter in the air also leads to vagal nerve inflammation (Villarreal-Calderon et al., 2010), and postmortem examination of Mexico City residents with high exposures to air pollution show accelerated rates of Alzheimer's pathology in infants' and children's brain stems (Calderón-Garcidueñas et al., 2018).

Another risk factor for the LC is its high energy need. LC axons' sparse myelination increases the energy needed to transmit signals (Braak & Del Tredici, 2015), making the neurons vulnerable to activity-dependent oxidant stress (Matschke et al., 2015). In addition, LC neurons have intrinsic pacemaker mechanisms that maintain firing even when glutamate and gamma-aminobutyric acid (GABA) inputs are blocked and are modulated by calcium channels (Matschke et al., 2015). A comparison across neurogenerative disorders suggests that the types of neurons that are most vulnerable are those that exhibit highly regulated firing properties that rely heavily on calcium regulation, which in turns depends on proper cellular homeostasis (Roselli & Caroni, 2015). Thus, a number of features make the LC more vulnerable than most brain regions to degeneration over time.

Alzheimer's-Related Tau Pathology Is Found in the Locus Coeruleus Early in Adulthood

Recent findings suggest that Alzheimer's-related tau pathology first occurs in brain stem nuclei—in particular, in the LC and in most people by early adulthood. In a sample of 2,332 autopsied brains of people ranging from 1 to 100 years old at death (with a diverse set of precipitating causes of death), only 10 brains (all younger than 24 years old) had no indication of abnormal tau anywhere in the brain (Braak, Thal, Ghebremedhin, & Del Tredici, 2011). Another 58 brains had no abnormal cortical tau in the cortex but did have abnormal tau in the brain stem, predominantly in the LC. From the LC, the path of tau pathology appeared to spread to the transentorhinal cortex and then to the hippocampus and other cortical regions, with cortical amyloid β-peptide (Aβ) plaques emerging only after brain stem tau pathology. In samples with cognitive test data, neocortical tau pathology is more strongly associated with cognitive decline than amyloid pathology (Giannakopoulos, Gold, von Gunten, Hof, & Bouras, 2009; Nelson et al., 2012; Suemoto et al., 2017). In addition, in older adults, greater postmortem LC neuronal density and a higher LC neuron number are associated with better cognitive performance before death (Kelly et al., 2017; Wilson et al., 2013). Thus, understanding the LC's role in cognition in aging and how it relates to the spread of tau pathology throughout the brain is relevant even for "normal" aging.

The Locus Coeruleus-Norepinephrine System's Role in Various Cognitive Functions in Aging and Alzheimer's Disease

Selective attention. As already outlined, the LC integrates information about arousal states, and as arousal increases, the release of NE modulates cortical activity to increase the gain on processing and make it more selective (Mather et al., 2016). A recent study that manipulated arousal via fear-conditioned sounds and the salience of place versus object images revealed a number of interesting findings about how arousal and LC activity affect attention differently in older than in younger adults (Lee et al., 2018). For younger adults, playing a tone that predicted a shock increased activity in a scene-selective brain region (the parahippocampal place area or PPA) when the place image was more salient but decreased PPA activity when the place image was less salient than the object image (see also Lee, Sakaki, Cheng, Velasco, & Mather, 2014). In contrast, for older adults, arousal increased PPA activity in response to place images regardless of whether they were more or less salient than the competing object images. Thus, arousal increased the selectivity of attention for younger but not older adults.

The GANE model posits that greater excitation associated with processing salient stimuli under arousal is driven by local cortical control over how much NE is released when the LC is activated (Mather et al., 2016). Regions that are highly active and therefore have high levels of glutamate stimulate more local release of NE, which in turn activates β-adrenergic receptors that stimulate more glutamate release, increasing excitation in that region. This model predicts that the LC and any particular region of cortex should be most coordinated in their activation levels when arousal is high (activating the LC) and that cortical region is highly active (for instance, the PPA during viewing of a highly salient picture of a place). Consistent with this prediction, both younger and older adults showed the greatest functional connectivity between LC and PPA in trials that were arousing and included a salient place image (Lee et al., 2018). Thus, the role of the LC in increasing the excitation of highly active representations seems to be intact in older adults.

But what can explain the failure of arousal to suppress nonsalient stimuli in older adults? The coordination of stimuli salience or priority across different cortical regions relies on the frontoparietal network, which provides a priority map of the environment (Ptak, 2012). Lee et al. (2018) found that whereas for younger adults arousal increased activation levels within this frontoparietal network and increased its functional connectivity with the LC, arousal had a significantly diminished impact on older adults' frontoparietal activation levels and frontoparietal LC functional connectivity.

Perhaps the most obvious interpretation of these findings is that the LC is less effective at modulating frontoparietal activity in older adults. But findings that older adults show significantly *more* LC frontoparietal functional connectivity at rest than do younger adults (Jacobs, Müller-Ehrenberg, Priovoulos, & Roebroeck, 2018; Zhang, Hu, Chao, & Li, 2016) raise the possibility that LC-NE system activity may be upregulated to compensate for age-related decline within the system (Mather, forthcoming), leading to a higher tonic level of NE activity within the frontoparietal regions, making it more difficult to further upregulate activity in these regions. The issue of potential upregulated tonic NE activity is addressed more fully in a later section. In any case there are clear age-related changes in how the LC interacts with the frontoparietal brain regions involved in attentional selectivity.

Episodic memory NE plays an important role in episodic memory, especially for highly salient arousing events and stimuli (Hagena et al., 2016; Harley, 1987; McGaugh, 2004; Tully & Bolshakov, 2010). These beneficial effects of NE on episodic memory are typically driven by β-adrenergic receptors that can be activated by the glutamate-NE hot spot mechanism (Mather et al., 2016). Some have even argued that neuromodulators such as NE and dopamine are necessary for spike-timing-dependent long-term potentiation supporting memory consolidation (Huang et al., 2014; Johansen et al., 2014). Rodent research in which the infusion of NE into the hippocampus or hippocampal slices eliminates age-related memory deficits suggests that decreased basal extracellular NE levels in the hippocampus in older rodents account for a substantial component of their impaired spatial and emotional memory (Luo et al., 2015; Mei et al., 2015).

Because extracellular NE levels cannot be measured without invasive techniques, we have less information regarding the LC-NE system and its role in memory in humans. However, MRI sequences have been developed that indicate the location of the LC (Betts, Cardenas-Blanco, Kanowski, Jessen, & Düzel, 2017; Chen et al., 2014; Keren et al., 2015; Keren, Lozar, Harris, Morgan, & Eckert, 2009; Hämmerer et al., 2018; Langley, Huddleston, Liu, & Hu, 2017; Priovoulos et al., 2017; Sasaki et al., 2006; Watanabe et al., 2019). Postmortem examination in humans has confirmed that T1-weighted imaging can show hyperintensities in the locations corresponding with LC neuromelanin-containing neurons (Keren et al., 2015), although it is not yet clear if neuromelanin or some combination of properties of the LC cause its unique magnetic properties compared with the surrounding brainstem (Watanabe et al., 2019).

In any case, the measure does seem to reflect the integrity of the LC-NE system, as older adults with greater LC contrast ratios tend to have better episodic memory. Among 22 older adults, MRI LC contrast ratios were positively associated with memory for pictures appearing before losses (Hämmerer et al., 2018). Furthermore, among 233 older adults, LC contrast ratios were positively associated with a verbal learning task (Dahl et al., under review). These findings suggest that the structural integrity of the LC in late life helps predict episodic memory performance.

Working memory Animal research indicates that the activation of α2-adrenergic receptors in the dorsolateral prefrontal cortex enhances working memory processes (Arnsten & Goldman-Rakic, 1985; Franowicz et al., 2002). In humans, links between the LC and working memory function have been identified using pupil dilation as a window into LC-NE function (Granholm et al., 2017). Typically, on a digit-span task, pupil dilation increases as people hear and try to maintain each additional digit in working memory (Kahneman & Beatty, 1966; figure 8.2). Older adults with memory deficits show greater pupil dilation at low loads and reduced pupil dilation at high loads than nonimpaired older adults (Granholm et al., 2017), which may relate to impairments in LC function.

Cognitive flexibility Rodent studies indicate a key role for the LC in cognitive flexibility. For instance, reversibly silencing the LC using optogenetic techniques selectively impaired learning that was dependent on cognitive flexibility, such as reversal learning and extradimensional set shifting (Janitzky et al., 2015). Furthermore, a transgenic rat model of Alzheimer's disease (expressing disease-causing mutant amyloid precursor protein and presenilin-1) showed hyperphosphorylated tau in the LC before it spread to the entorhinal cortex, and these rats were worse at reversal learning compared to their wild-type littermates (Rorabaugh et al., 2017). This reversal learning impairment was eliminated by activating the LC via designer receptors.

Is the Locus Coeruleus-Norepinephrine System the Neural Mechanism of Cognitive Reserve?

Education and working in more cognitively challenging professions are associated with a reduced risk of being diagnosed with Alzheimer's disease but a faster decline once diagnosed (Stern, 2012). One possibility is that the LC-NE system is a key neural mechanism underlying this phenomenon, known as *cognitive reserve* (Mather & Harley, 2016; Robertson, 2013, 2014). Experiencing mental challenges and novel situations stimulate the LC and, in turn, should increase the protective effects of NE release outlined earlier. In particular, the anti-inflammatory effects of NE rely on β-adrenergic receptor activation. Based on the GANE model outlined earlier, the high levels of NE required to activate β-adrenergic receptors are most likely to occur at cortical sites processing salient or high-priority stimuli during phasic LC activation. This combination is especially likely to occur during the sort of mentally stimulating situations encountered while learning new things or navigating challenging social situations.

This fits with education delaying the onset of Alzheimer's disease, but how can it be reconciled with a decline, once diagnosed, that is faster among the higher educated? Stern (2012) argued that those with a higher cognitive reserve due to education or high professional attainment are at later stages of the disease when it is detected because they are better able to compensate for the pathology. In addition, by the time someone is in the symptomatic phases of Alzheimer's disease, the anti-inflammatory benefits of β-adrenergic receptor activity may be outweighed by the fact that β-adrenergic activity also stimulates Aβ production (Ni et al., 2006). Aβ is a by-product of mental activity that is typically flushed out during sleep (Shokri-Kojori et al., 2018), but it becomes toxic in Alzheimer's disease when it is no longer cleared out as effectively and aggregates into Aβ plaques.

Do Aging and Alzheimer's Disease Increase or Deplete Norepinephrine?

To assess NE levels in living Alzheimer's disease patients, researchers typically rely on cerebral spinal fluid and blood samples. These samples reveal that

Alzheimer's disease patients have normal or elevated levels of NE. The other standard approach to quantify NE is to examine homogenized tissue from postmortem brains. In these studies, brain samples from Alzheimer's disease patients have depleted NE levels (for a review, see Mather & Harley, 2016).

How can these opposite findings be reconciled? One possibility is that in Alzheimer's disease the overall NE levels may be lower due to fewer NE-containing LC neurons, but the circulating levels of NE in extracellular space remain constant or even increase (Gannon & Wang, 2018). Indeed, measurements of NE levels and dynamics in fresh brain tissue from patients who had diagnostic craniotomies indicated a maintained ability of neurons to release NE but a decreased varicosity uptake of NE and decreased tissue levels of NE in Alzheimer's patients compared with controls (Palmer et al., 1987).

As Alzheimer's disease slowly diminishes the number of functioning LC neurons, surviving neurons may upregulate their NE release (Gannon & Wang, 2018). Consistent with this possibility, Alzheimer's disease patients have a greater mRNA expression per remaining LC neuron of tyrosine hydroxylase, the rate-limiting enzyme in the synthesis of NE or NE transporter (NET) (Szot et al., 2000, 2006). As elevated levels of this enzyme can increase NE production, this suggests an upregulation in the capacity of each LC neuron to produce more NE. Alzheimer's patients also show increased sprouting of remaining LC neurons in the region right around the LC and in the hippocampus (Booze, Mactutus, Gutman, & Davis, 1993; Nelson, Kolasa, & McMahon, 2014; Szot et al., 2006), as well as in the prefrontal cortex (Szot et al., 2007). Also contributing to greater extracellular levels of NE, NET levels in the LC decrease in Alzheimer's disease (Gulyás et al., 2010; Szot et al., 2006). NET is a presynaptic pump that returns NE to the neuron to be metabolized, so decreases in NET leave more NE in the synapse.

Does the Locus Coeruleus-Norepinephrine System Protect against Alzheimer's Disease?

Animal research suggests that LC degeneration is not simply associated with Alzheimer's disease pathology elsewhere in the brain but in fact exacerbates progression of the disease. Using a neurotoxin to selectively lesion LC in rodent transgenic models of Alzheimer's disease accelerates Alzheimer's-like pathology and cognitive deficits (Chalermpalanupap et al., 2018; Heneka et al., 2006, 2010; Jardanhazi-Kurutz et al., 2010; Rey et al., 2012).

An intact LC may protect against Alzheimer's-related neuropathology and cognitive decline in a variety of ways. For instance, extracellular NE has protective effects due to its free radical scavenging, which reduces oxidative stress (Feinstein, Kalinin, & Braun, 2016). NE also reduces glial cell inflammatory responses via β-adrenergic receptor activation (Frohman, Vayuvegula, Gupta, & Van Den Noort, 1988; Galea, Heneka, Russo, & Feinstein, 2003; Hetier, Ayala, Bousseau, & Prochiantz, 1991; Minghetti et al., 1997). Also via β-adrenergic receptors and not α-adrenergic receptors, NE reduces the cytokine-dependent induction of type 2 nitric oxide synthase in astrocytes (Feinstein, Galea, & Reis, 1993). β-adrenergic receptors require high concentrations of NE (i.e., about 25 times higher than for α1-adrenergic receptors; Salgado, Kohr, & Trevino, 2012). LC activation does not lead to enough NE release at most downstream sites to activate these high-threshold receptors, but high levels of local activity—as might occur when a set of neurons representing a particular object are highly active because that object is the focus of attention—stimulate greater NE release at those sites (Mather et al., 2016). These specific mechanisms through which NE diminishes inflammation may contribute to the cognitive reserve that results from education and mental stimulation.

Conclusions

The LC is a small brain nucleus that has not been considered in most theories of cognitive aging. In part, the LC has been ignored because its small size, lack of clear borders, and pulsatile motion make its structure and activity hard to measure, especially in living humans. However, growing evidence suggests that it plays an important role in memory and cognition and that its large network of axons tuned to local cortical conditions are especially vulnerable to toxins and viruses, which leads to pathology. Initial LC tau pathology typically appears in early adulthood and slowly spreads to the entorhinal cortex and beyond. LC contributions to cognitive aging have likely been underappreciated due to a lack of ability to measure LC structure or activity in living humans. However, recent advances in neuroimaging have begun to make it possible to assess LC integrity and functioning and have confirmed the suggestions of postmortem findings, as well as yielded new insights.

Acknowledgments

I thank Christine Cho for assistance with figures 8.1 and 8.3 and Ringo Huang for assistance with figure 8.2. This work was supported by National Institute on Aging grant R01AG025340.

REFERENCES

Arnsten, A., & Goldman-Rakic, P. S. (1985). Alpha 2-adrenergic mechanisms in prefrontal cortex associated with cognitive decline in aged nonhuman primates. *Science, 230*(4731), 1273–1276.

Berridge, C. W. (2008). Noradrenergic modulation of arousal. *Brain Research Reviews, 58*(1), 1–17.

Berridge, C. W., & Abercrombie, E. D. (1999). Relationship between locus coeruleus discharge rates and rates of norepinephrine release within neocortex as assessed by in vivo microdialysis. *Neuroscience, 93*(4), 1263–1270.

Berridge, C. W., & Waterhouse, B. D. (2003). The locus coeruleus-noradrenergic system: Modulation of behavioral state and state-dependent cognitive processes. *Brain Research Reviews, 42*(1), 33–84.

Betts, M. J., Cardenas-Blanco, A., Kanowski, M., Jessen, F., & Düzel, E. (2017). In vivo MRI assessment of the human locus coeruleus along its rostrocaudal extent in young and older adults. *NeuroImage, 163*, 150–159.

Booze, R. M., Mactutus, C. F., Gutman, C. R., & Davis, J. N. (1993). Frequency analysis of catecholamine axonal morphology in human brain: II. Alzheimer's disease and hippocampal sympathetic ingrowth. *Journal of the Neurological Sciences, 119*(1), 110–118.

Braak, H., & Del Tredici, K. (2015). *Neuroanatomy and pathology of sporadic Alzheimer's disease.* Vol. 215. New York: Springer International.

Braak, H., Thal, D. R., Ghebremedhin, E., & Del Tredici, K. (2011). Stages of the pathologic process in Alzheimer disease: Age categories from 1 to 100 years. *Journal of Neuropathology and Experimental Neurology, 70*(11), 960–969.

Calderón-Garcidueñas, L., Gónzalez-Maciel, A., Reynoso-Robles, R., Delgado-Chávez, R., Mukherjee, P. S., Kulesza, R. J., … Villarreal-Ríos, R. (2018). Hallmarks of Alzheimer disease are evolving relentlessly in metropolitan Mexico City infants, children and young adults. APOE4 carriers have higher suicide risk and higher odds of reaching NFT stage V at ≤ 40 years of age. *Environmental Research, 164*, 475–487.

Chalermpalanupap, T., Schroeder, J. P., Rorabaugh, J. M., Liles, L. C., Lah, J. J., Levey, A. I., & Weinshenker, D. (2018). Locus coeruleus ablation exacerbates cognitive deficits, neuropathology, and lethality in P301S tau transgenic mice. *Journal of Neuroscience, 38*(1), 74–92.

Chen, X., Huddleston, D. E., Langley, J., Ahn, S., Barnum, C. J., Factor, S. A., … Hu, X. (2014). Simultaneous imaging of locus coeruleus and substantia nigra with a quantitative neuromelanin MRI approach. *Magnetic Resonance Imaging, 32*(10), 1301–1306.

Clewett, D. V., Huang, R., Velasco, R., Lee, T.-H., & Mather, M. (2018). Locus coeruleus activity strengthens prioritized memories under arousal. *Journal of Neuroscience, 38*(6), 1558–1574.

Dahl, M. J., Düzel, S., Mather, M., Lindenberger, U., Kühn, S., & Werkle-Bergner, M. (under review). Locus coeruleus integrity preserves memory performance across the adult lifespan.

Damasio, A., & Carvalho, G. B. (2013). The nature of feelings: Evolutionary and neurobiological origins. *Nature Reviews Neuroscience, 14*(2), 143.

Feinstein, D. L., Galea, E., & Reis, D. J. (1993). Norepinephrine suppresses inducible nitric oxide synthase activity in rat astroglial cultures. *Journal of Neurochemistry, 60*(5), 1945–1948.

Feinstein, D. L., Kalinin, S., & Braun, D. (2016). Causes, consequences, and cures for neuroinflammation mediated via the locus coeruleus: Noradrenergic signaling system. *Journal of Neurochemistry, 139*(S2), 154–178.

Fernández-Pastor, B., & Meana, J. J. (2002). In vivo tonic modulation of the noradrenaline release in the rat cortex by locus coeruleus somatodendritic α2-adrenoceptors. *European Journal of Pharmacology, 442*(3), 225–229.

Florin-Lechner, S. M., Druhan, J. P., Aston-Jones, G., & Valentino, R. J. (1996). Enhanced norepinephrine release in prefrontal cortex with burst stimulation of the locus coeruleus. *Brain Research, 742*(1–2), 89–97.

Franowicz, J. S., Kessler, L. E., Borja, C. M. D., Kobilka, B. K., Limbird, L. E., & Arnsten, A. F. (2002). Mutation of the α2A-adrenoceptor impairs working memory performance and annuls cognitive enhancement by guanfacine. *Journal of Neuroscience, 22*(19), 8771–8777.

Frohman, E. M., Vayuvegula, B., Gupta, S., & Van Den Noort, S. (1988). Norepinephrine inhibits gamma-interferon-induced major histocompatibility class II (Ia) antigen expression on cultured astrocytes via beta-2-adrenergic signal transduction mechanisms. *Proceedings of the National Academy of Sciences, 85*(4), 1292–1296.

Galea, E., Heneka, M. T., Russo, C. D., & Feinstein, D. L. (2003). Intrinsic regulation of brain inflammatory responses. *Cellular and Molecular Neurobiology, 23*(4–5), 625–635.

Gannon, M., & Wang, Q. (2018). Complex noradrenergic dysfunction in Alzheimer's disease: Low norepinephrine input is not always to blame. *Brain Research, 1702*, 12–16.

Giannakopoulos, P., Gold, G., von Gunten, A., Hof, P. R., & Bouras, C. (2009). Pathological substrates of cognitive decline in Alzheimer's disease. In *Dementia in clinical practice* (Vol. 24, pp. 20–29). Basel, Switzerland: Karger.

Granholm, E. L., Panizzon, M. S., Elman, J. A., Jak, A. J., Hauger, R. L., Bondi, M. W., … Kremen, W. S. (2017). Pupillary responses as a biomarker of early risk for Alzheimer's disease. *Journal of Alzheimer's Disease, 56*(4), 1419–1428.

Gulyás, B., Brockschnieder, D., Nag, S., Pavlova, E., Kása, P., Beliczai, Z., … Halldin, C. (2010). The norepinephrine transporter (NET) radioligand (S,S)-[18F]FMeNER-D2 shows significant decreases in NET density in the human brain in Alzheimer's disease: A post-mortem autoradiographic study. *Neurochemistry International, 56*(6), 789–798.

Hagena, H., Hansen, N., & Manahan-Vaughan, D. (2016). β-adrenergic control of hippocampal function: subserving the choreography of synaptic information storage and memory. *Cerebral Cortex, 26*(4), 1349–1364.

Hämmerer, D., Callaghan, M. F., Hopkins, A., Kosciessa, J., Betts, M., Cardenas-Blanco, A., … Dolan, R. J. (2018). Locus coeruleus integrity in old age is selectively related to memories linked with salient negative events. *Proceedings of the National Academy of Sciences, 115*(9), 2228–2233.

Harley, C. W. (1987). A role for norepinephrine in arousal, emotion and learning?: Limbic modulation by norepinephrine and the Kety hypothesis. *Progress in Neuro-Psychopharmacology and Biological Psychiatry, 11*, 419–458.

Heneka, M. T., Nadrigny, F., Regen, T., Martinez-Hernandez, A., Dumitrescu-Ozimek, L., Terwel, D., … Hanisch, U.-K.

(2010). Locus ceruleus controls Alzheimer's disease pathology by modulating microglial functions through norepinephrine. *Proceedings of the National Academy of Sciences, 107*(13), 6058–6063.

Heneka, M. T., Ramanathan, M., Jacobs, A. H., Dumitrescu-Ozimek, L., Bilkei-Gorzo, A., Debeir, T., ... Staufenbiel, M. (2006). Locus ceruleus degeneration promotes Alzheimer pathogenesis in amyloid precursor protein 23 transgenic mice. *Journal of Neuroscience, 26*(5), 1343–1354.

Hetier, E., Ayala, J., Bousseau, A., & Prochiantz, A. (1991). Modulation of interleukin-1 and tumor necrosis factor expression by β-adrenergic agonists in mouse ameboid microglial cells. *Experimental Brain Research, 86*(2), 407–413.

Huang, R., & Mather, M. (2018). Pupil dilation increases in both younger and older adults under increasing working memory load. Unpublished manuscript.

Huang, S., Rozas, C., Treviño, M., Contreras, J., Yang, S., Song, L., ... Kirkwood, A. (2014). Associative Hebbian synaptic plasticity in primate visual cortex. *Journal of Neuroscience, 34*(22), 7575–7579.

Jacobs, H. I., Müller-Ehrenberg, L., Priovoulos, N., & Roebroeck, A. (2018). Curvilinear locus coeruleus functional connectivity trajectories over the adult lifespan: A 7T MRI study. *Neurobiology of Aging, 69*, 167–176.

Jang, H., Boltz, D., Sturm-Ramirez, K., Shepherd, K. R., Jiao, Y., Webster, R., & Smeyne, R. J. (2009). Highly pathogenic H5N1 influenza virus can enter the central nervous system and induce neuroinflammation and neurodegeneration. *Proceedings of the National Academy of Sciences, 106*(33), 14063–14068.

Janitzky, K., Lippert, M. T., Engelhorn, A., Tegtmeier, J., Goldschmidt, J., Heinze, H.-J., & Ohl, F. W. (2015). Optogenetic silencing of locus coeruleus activity in mice impairs cognitive flexibility in an attentional set-shifting task. *Frontiers in Behavioral Neuroscience, 9*, 286.

Jardanhazi-Kurutz, D., Kummer, M. P., Terwel, D., Vogel, K., Dyrks, T., Thiele, A., & Heneka, M. T. (2010). Induced LC degeneration in APP/PS1 transgenic mice accelerates early cerebral amyloidosis and cognitive deficits. *Neurochemistry International, 57*(4), 375–382.

Johansen, J. P., Diaz-Mataix, L., Hamanaka, H., Ozawa, T., Ycu, E., Koivumaa, J., ... Boyden, E. S. (2014). Hebbian and neuromodulatory mechanisms interact to trigger associative memory formation. *Proceedings of the National Academy of Sciences, 111*(51), E5584–E5592.

Joshi, S., Li, Y., Kalwani, Rishi M., & Gold, Joshua I. (2015). Relationships between pupil diameter and neuronal activity in the locus coeruleus, colliculi, and cingulate cortex. *Neuron, 89*, 1–14.

Kahneman, D., & Beatty, J. (1966). Pupil diameter and load on memory. *Science, 154*(3756), 1583–1585.

Kelly, S. C., He, B., Perez, S. E., Ginsberg, S. D., Mufson, E. J., & Counts, S. E. (2017). Locus coeruleus cellular and molecular pathology during the progression of Alzheimer's disease. *Acta Neuropathologica Communications, 5*(1), 8.

Keren, N. I., Lozar, C. T., Harris, K. C., Morgan, P. S., & Eckert, M. A. (2009). In vivo mapping of the human locus coeruleus. *NeuroImage, 47*(4), 1261–1267.

Keren, N. I., Taheri, S., Vazey, E. M., Morgan, P. S., Granholm, A.-C. E., Aston-Jones, G. S., & Eckert, M. A. (2015). Histologic validation of locus coeruleus MRI contrast in post-mortem tissue. *NeuroImage, 113*, 235–245.

Langley, J., Huddleston, D. E., Liu, C. J., & Hu, X. (2017). Reproducibility of locus coeruleus and substantia nigra imaging with neuromelanin sensitive MRI. *Magnetic Resonance Materials in Physics, Biology and Medicine, 30*(2), 121–125.

Lee, T.-H., Greening, S. G., Ueno, T., Clewett, D., Ponzio, A., Sakaki, M., & Mather, M. (2018). Arousal increases neural gain via the locus coeruleus-noradrenaline system in younger adults but not in older adults. *Nature Human Behaviour, 2*(5), 356.

Lee, T. H., Sakaki, M., Cheng, R., Velasco, R., & Mather, M. (2014). Emotional arousal amplifies the effects of biased competition in the brain. *Social Cognitive and Affective Neuroscience, 9*(12), 2067–2077.

Luo, Y., Zhou, J., Li, M. X., Wu, P. F., Hu, Z. L., Ni, L., ... Wang, F. (2015). Reversal of aging-related emotional memory deficits by norepinephrine via regulating the stability of surface AMPA receptors. *Aging Cell, 14*(2), 170–179.

Mather, M. (forthcoming). Does the aging brain suffer from overarousal or underarousal? How arousal-related neurotransmitter systems compensate for age-related decline In A. Thomas & A. Gutchess (Eds.), *Handbook of cognitive aging: A life course perspective*. Cambridge: Cambridge University Press.

Mather, M., Clewett, D., Sakaki, M., & Harley, C. W. (2016). Norepinephrine ignites local hotspots of neuronal excitation: How arousal amplifies selectivity in perception and memory. *Behavioral and Brain Sciences, 39*, e200.

Mather, M., & Harley, C. W. (2016). The locus coeruleus: Essential for maintaining cognitive function and the aging brain. *Trends in Cognitive Sciences, 20*, 214–226.

Mather, M., & Sutherland, M. R. (2011). Arousal-biased competition in perception and memory. *Perspectives on Psychological Science, 6*, 114–133.

Matschke, L. A., Bertoune, M., Roeper, J., Snutch, T. P., Oertel, W. H., Rinné, S., & Decher, N. (2015). A concerted action of L- and T-type Ca2+ channels regulates locus coeruleus pacemaking. *Molecular and Cellular Neuroscience, 68*, 293–302.

McCormick, D., Pape, H.-C., & Williamson, A. (1991). Actions of norepinephrine in the cerebral cortex and thalamus: Implications for function of the central noradrenergic system. *Progress in Brain Research, 88*, 293–305.

McGaugh, J. L. (2004). The amygdala modulates the consolidation of memories of emotionally arousing experiences. *Annual Review of Neuroscience, 27*, 1–28.

McGinley, M. J., Vinck, M., Reimer, J., Batista-Brito, R., Zagha, E., Cadwell, C. R., ... McCormick, D. A. (2015). Waking state: rapid variations modulate neural and behavioral responses. *Neuron, 87*(6), 1143–1161.

Mei, Y., Jiang, C., Wan, Y., Lv, J., Jia, J., Wang, X., ... Tong, Z. (2015). Aging-associated formaldehyde-induced norepinephrine deficiency contributes to age-related memory decline. *Aging Cell, 14*(4) 659–668.

Minghetti, L., Nicolini, A., Polazzi, E., CRéminon, C., Maclouf, J., & Levi, G. (1997). Inducible nitric oxide synthase expression in activated rat microglial cultures is downregulated by exogenous prostaglandin E2 and by cyclooxygenase inhibitors. *Glia, 19*(2), 152–160.

Nelson, A. R., Kolasa, K., & McMahon, L. L. (2014). Noradrenergic sympathetic sprouting and cholinergic reinnervation maintains non-amyloidogenic processing of AβPP. *Journal of Alzheimer's Disease, 38*(4), 867–879.

Nelson, P. T., Alafuzoff, I., Bigio, E. H., Bouras, C., Braak, H., Cairns, N. J., ... Del Tredici, K. (2012). Correlation of Alzheimer disease neuropathologic changes with cognitive status: A review of the literature. *Journal of Neuropathology and Experimental Neurology, 71*(5), 362.

Neske, G. T. (2016). The slow oscillation in cortical and thalamic networks: mechanisms and functions. *Frontiers in Neural Circuits, 9*, 88.

Ni, Y., Zhao, X., Bao, G., Zou, L., Teng, L., Wang, Z., ... Pei, G. (2006). Activation of β 2-adrenergic receptor stimulates γ-secretase activity and accelerates amyloid plaque formation. *Nature Medicine, 12*(12), 1390.

O'Dell, T. J., Connor, S. A., Gelinas, J. N., & Nguyen, P. V. (2010). Viagra for your synapses: Enhancement of hippocampal long-term potentiation by activation of beta-adrenergic receptors. *Cellular Signalling, 22*(5), 728–736.

Palmer, A., Francis, P., Bowen, D., Benton, J., Neary, D., Mann, D., & Snowden, J. (1987). Catecholaminergic neurons assessed ante-mortem in Alzheimer's disease. *Brain Research, 414*(2), 365–375.

Pamphlett, R. (2014). Uptake of environmental toxicants by the locus ceruleus: A potential trigger for neurodegenerative, demyelinating and psychiatric disorders. *Medical Hypotheses, 82*(1), 97–104.

Priovoulos, N., Jacobs, H. I., Ivanov, D., Uludag, K., Verhey, F. R., & Poser, B. A. (2017). High-resolution in vivo imaging of human locus coeruleus by magnetization transfer MRI at 3T and 7T. *NeuroImage, 168*, 427–436.

Ptak, R. (2012). The frontoparietal attention network of the human brain action, saliency, and a priority map of the environment. *Neuroscientist, 18*(5), 502–515.

Pudovkina, O. L., Kawahara, Y., de Vries, J., & Westerink, B. H. C. (2001). The release of noradrenaline in the locus coeruleus and prefrontal cortex studied with dual-probe microdialysis. *Brain Research, 906*(1), 38–45.

Ramos, B. P., & Arnsten, A. F. T. (2007). Adrenergic pharmacology and cognition: Focus on the prefrontal cortex. *Pharmacology and Therapeutics, 113*(3), 523–536.

Rey, N. L., Jardanhazi-Kurutz, D., Terwel, D., Kummer, M. P., Jourdan, F., Didier, A., & Heneka, M. T. (2012). Locus coeruleus degeneration exacerbates olfactory deficits in APP/PS1 transgenic mice. *Neurobiology of Aging, 33*(2), 426.e421–426.e411.

Robertson, I. H. (2013). A noradrenergic theory of cognitive reserve: Implications for Alzheimer's disease. *Neurobiology of Aging, 34*(1), 298–308.

Robertson, I. H. (2014). A right hemisphere role in cognitive reserve. *Neurobiology of Aging, 35*(6), 1375–1385.

Rorabaugh, J. M., Chalermpalanupap, T., Botz-Zapp, C. A., Fu, V. M., Lembeck, N. A., Cohen, R. M., & Weinshenker, D. (2017). Chemogenetic locus coeruleus activation restores reversal learning in a rat model of Alzheimer's disease. *Brain, 140*(11), 3023–3038.

Roselli, F., & Caroni, P. (2015). From intrinsic firing properties to selective neuronal vulnerability in neurodegenerative diseases. *Neuron, 85*(5), 901–910.

Salgado, H., Kohr, G., & Trevino, M. (2012). Noradrenergic "tone" determines dichotomous control of cortical spike-timing-dependent plasticity. *Scientific Reports, 2*, 7.

Samuels, E. R., & Szabadi, E. (2008). Functional neuroanatomy of the noradrenergic locus coeruleus: Its roles in the regulation of arousal and autonomic function part I: Principles of functional organisation. *Current Neuropharmacology, 6*(3), 235–253.

Sasaki, M., Shibata, E., Tohyama, K., Takahashi, J., Otsuka, K., Tsuchiya, K., ... Sakai, A. (2006). Neuromelanin magnetic resonance imaging of locus ceruleus and substantia nigra in Parkinson's disease. *NeuroReport, 17*(11), 1215–1218.

Schwarz, L. A., Miyamichi, K., Gao, X. J., Beier, K. T., Weissbourd, B., DeLoach, K. E., ... Kremer, E. J. (2015). Viral-genetic tracing of the input-output organization of a central noradrenaline circuit. *Nature, 524*(7563), 88.

Shokri-Kojori, E., Wang, G.-J., Wiers, C. E., Demiral, S. B., Guo, M., Kim, S. W., ... Volkow, N. D. (2018). β-amyloid accumulation in the human brain after one night of sleep deprivation. *Proceedings of the National Academy of Sciences, 115*(17), 4483–4488.

Starke, K. (2001). Presynaptic autoreceptors in the third decade: Focus on α2-adrenoceptors. *Journal of Neurochemistry, 78*(4), 685–693.

Stern, Y. (2012). Cognitive reserve in ageing and Alzheimer's disease. *Lancet Neurology, 11*(11), 1006–1012.

Suemoto, C. K., Ferretti-Rebustini, R. E., Rodriguez, R. D., Leite, R. E., Soterio, L., Brucki, S. M., ... Chiavegatto Filho, A. (2017). Neuropathological diagnoses and clinical correlates in older adults in Brazil: A cross-sectional study. *PLoS Medicine, 14*(3), e1002267.

Sutherland, M. R., & Mather, M. (2012). Negative arousal amplifies the effects of saliency in short-term memory. *Emotion, 12*, 1367–1372.

Szot, P., Leverenz, J. B., Peskind, E. R., Kiyasu, E., Rohde, K., Miller, M. A., & Raskind, M. A. (2000). Tyrosine hydroxylase and norepinephrine transporter mRNA expression in the locus coeruleus in Alzheimer's disease. *Molecular Brain Research, 84*(1), 135–140.

Szot, P., White, S. S., Greenup, J. L., Leverenz, J. B., Peskind, E. R., & Raskind, M. A. (2006). Compensatory changes in the noradrenergic nervous system in the locus ceruleus and hippocampus of postmortem subjects with Alzheimer's disease and dementia with Lewy bodies. *Journal of Neuroscience, 26*(2), 467–478.

Szot, P., White, S. S., Greenup, J. L., Leverenz, J. B., Peskind, E. R., & Raskind, M. A. (2007). Changes in adrenoreceptors in the prefrontal cortex of subjects with dementia: Evidence of compensatory changes. *Neuroscience, 146*(1), 471–480.

Tully, K., & Bolshakov, V. Y. (2010). Emotional enhancement of memory: How norepinephrine enables synaptic plasticity. *Molecular Brain, 3*(1), 15.

Villarreal-Calderon, R., Torres-Jardon, R., Palacios-Moreno, J., Osnaya, N., Perez-Guille, B., Maronpot, R. R., ... Calderon-Garciduenas, L. (2010). Urban air pollution targets the dorsal vagal complex and dark chocolate offers neuroprotection. *International Journal of Toxicology, 29*(6), 604–615.

Vizi, E., Fekete, A., Karoly, R., & Mike, A. (2010). Non-synaptic receptors and transporters involved in brain functions and targets of drug treatment. *British Journal of Pharmacology, 160*(4), 785–809.

Watanabe, T., Tan, Z., Wang, X., Martinez-Hernandez, A., & Frahm, J. (2019). Magnetic resonance imaging of noradrenergic neurons. *Brain Structure and Function*, 1–17.

Wilson, R. S., Nag, S., Boyle, P. A., Hizel, L. P., Yu, L., Buchman, A. S., ... Bennett, D. A. (2013). Neural reserve, neuronal density in the locus ceruleus, and cognitive decline. *Neurology, 80*(13), 1202–1208.

Zhang, S., Hu, S., Chao, H. H., & Li, C.-S. R. (2016). Resting-state functional connectivity of the locus coeruleus in humans: In comparison with the ventral tegmental area/substantia nigra pars compacta and the effects of age. *Cerebral Cortex, 26*(8), 3413–3427.

II

AUDITORY AND VISUAL PERCEPTION

Introduction

KALANIT GRILL-SPECTOR AND MARIA CHAIT

This section about visual and auditory perception summarizes present knowledge about the neural mechanisms underlying vision and hearing. This is an exciting time for the cognitive neuroscience of vision and audition, as recent methodological and computational advancements have facilitated significant progress in understanding the cortical mechanisms that enable us to see, hear, and integrate visual and auditory information. While there have been huge leaps in both vision and audition, it will become apparent to the reader that the focus of these advancements varies across the two fields. In vision, there has been remarkable progress in understanding structural–functional relations in the brain, developing precise generative models of spatial and temporal computations, and leveraging these advancements to understand complex visual behavior, such as face recognition. In audition, the major breakthroughs have involved advances in signal processing and the development of decoding paradigms that now allow us to extract, and track with precise temporal resolution, brain activity specifically evoked by each sound source in a busy scene. This freed the field from the limitation inherent in the classic ERP (event-related potential) measurement and produced new insights into the nature of the computations that underpin everyday listening and their relationship to perception.

The perception section starts with four chapters on the visual system, followed by a chapter on audiovisual integration, and ends with three chapters on auditory perception. In line with the rich tradition in vision science, the chapters on vision cover the core components necessary for understanding brain function as proposed by David Marr: implementation, computations, and behavioral goals. Thus, these chapters discuss anatomical implementation (Weiner & Yeatman), computations (Winawer & Benson), face recognition behavior (Rossion & Retter), and scene/object categorization (Oliva). An emerging theme from these chapters is that individual subject analyses and generative models of brain function are critical for precisely understanding the functional neuroanatomy of the visual system. In other words, these chapters highlight that substantial information about brain function may be lost if neuroimaging data are only considered at the group level instead of the individual person level.

In chapter 9 Weiner and Yeatman highlight the utility of concurrent investigations of structure and function within individual brains. Individual subject analyses uncovered differences among people in both the patterns of their cortical folding and their functional representations. Critically, however, Weiner and Yeatman underscore that integrating brain function and structure within an individual brain leads to the remarkable discovery that across individuals there is a tight correspondence among cortical folding, functional representations, cytoarchitectonic areas, and white matter tracts. Two important take-home messages emerge from their chapter. First, understanding functional–structural relationships provides key insights about the computational strategies of the brain that enable the achievement of behavioral goals. Second, they discuss how their new approach—cognitive neuroanatomy—has the potential to revolutionize cognitive neuroscience broadly.

In chapter 10 Winawer and Benson detail how the population receptive field (pRF) model, a generative, stimulus-referred, computational model of neural populations, and its application to functional magnetic resonance imaging (fMRI) have advanced the understanding of the visual system. Specifically, the pRF approach has revealed key differences between visual areas in terms of the spatial and temporal scale over which brain areas analyze visual input. Two important insights emerge from the work detailed in the chapter. First, that areas higher in the visual-processing hierarchy pool visual information over larger extents of space and time and exhibit more pronounced nonlinearities than earlier visual areas. Second, the pRF approach presents an exceptional method to mechanistically unveil how sensory inputs are transformed into neuronal responses across stages of cortical-processing hierarchy within and beyond vision.

In chapter 11 Rossion and Retter suggest that recognizing people from their faces reflects domain-specific expertise characterized by accurate, rapid, and automatic perception as well as recognition of both the face and its identity. They propose that this expertise to faces is enabled and constrained by genetics and a domain-specific brain network. At the same time, however, experience during childhood development is vital for typical face recognition in adulthood. Like Weiner and Yeatman, they suggest that face-selective cortical regions should be considered anatomically on an individual subject basis. Like the chapters on audition, they underscore the importance of investigating the temporal characteristics of neural responses to understand the brain mechanisms of face recognition.

In chapter 13 Oliva provides an overview on deep neural networks and how the computations they enable support object recognition and scene perception. She proposes an interesting framework for using deep neural networks not only to predict responses in the human visual system but also to study representations at different levels of the visual-processing hierarchy.

The auditory chapters discuss the challenges of listening in crowded environments and the brain mechanisms that underpin these abilities. An emerging theme across all of the chapters is that our perceptual experiences in busy scenes are determined by an intricate interplay between bottom-up sensory processes and top-down priors: what we perceive at any given moment is determined not only by the input to our ears but, to a large extent, by our intention (i.e., what we are listening for), our implicit knowledge of the statistics of auditory objects, and our prior experiences in the environment.

In chapter 12 Noppeney discusses how the brain accumulates information from the visual and auditory senses to form perceptual decisions about the causal structure of the environment. She describes how human observers integrate sensory signals weighted by their a priori and task-induced reliabilities, though not always in the manner expected from a Bayesian observer. She goes on to argue, based on evidence from electroencephalography (EEG) and fMRI, that multisensory integration is refined along a hierarchy of cortical-processing stages, but surprisingly, Bayesian-like inference emerges only rather late in this process.

In chapter 14 Shinn-Cunningham reviews the building blocks of auditory perceptual organization, emphasizing that the process by which the brain constructs a representation of the acoustic scene is based on an interplay between goal-directed behavior and feedforward sensory processing. She discusses how attention shapes scene analysis not only by prioritizing the processing of the attended source but also by fundamentally affecting how other objects in the scene are represented. Interestingly, the brain networks that support auditory attention appear to overlap with the frontoparietal network associated with visual attention. This hints at the possibility that different sensory modalities may engage the same functional regions within the intraparietal sulcus (IPS) depending on the calculations performed—specifically, that timing-based calculations involve what are classically thought of as auditory regions, and spatial information–based calculations involve what are thought to be visual areas. Interestingly, some of the IPS regions highlighted in this chapter appear to overlap with those discussed in chapter 12, raising intriguing questions regarding the nature of the multisensory computations they perform.

In chapter 15 Obleser and Erb continue the theme of auditory attention but with an emphasis on how the interplay between target enhancement and distractor suppression supports listening in crowded environments. They highlight the different roles oscillatory brain activity plays in this process, with slow oscillations in the 0.5–8 Hz range supporting entrainment-based enhancement/suppression and 8–12 Hz oscillations supporting spatiotemporal inhibitory mechanisms of task-irrelevant information. They review how recent advances in precise, noninvasive measurements of attention-related activity can reveal, with great temporal resolution, a listener's perceptual state at any given moment. That is, from these measurements one can infer which auditory sources participants are tracking, which ones they are trying to ignore, and how much effort they are expending on the listening experience. Beyond basic research, these advancements are currently driving a paradigm shift in the field of brain–computer interface-based hearing aid design.

In chapter 16 Davis and Sohoglu propose that fast and accurate speech perception, as well as our ability to withstand the variety of interruptions and variability inherent to natural environments, are supported by predictive coding mechanisms. They suggest that listeners use their prior knowledge of the statistics of the speech signal, at the acoustic, phonetic, and semantic levels, to derive the most probable upcoming signal. They describe accumulating evidence consistent with the idea that predictions are generated in the inferior frontal gyrus and are fed to lower stages of the processing hierarchy, such that only prediction errors—the mismatch between the expected—are coded at lower levels.

While the chapters on visual and auditory perception highlight different aspects of brain computations enroute to perception, they also illustrate important commonalities. Of note, the IPS may be a key region that is involved in allocating attention as well as feature binding in both vision and audition and during multisensory integration. Nonetheless, it remains an open question as to whether the IPS is a general purpose region for attentional control across sensory domains or whether different functional regions within the IPS are involved in different aspects of top-down processing.

Despite the fact that our environments are fundamentally multisensory, research in hearing and vision has progressed largely in parallel. But is this parcellation of sensory perception useful? As discussed in several chapters here (and in other sections of this book), emerging work reveals fundamental similarities between the senses across all three of Marr's levels of analysis. Future research on computational principles that may be shared across sensory modalities, as well as aligning measurement and analysis approaches across systems (e.g., individual subject analyses are standard in vision but not yet in hearing), will pave the way toward a better understanding of the neural mechanisms of sensory perception.

9 The Cognitive Neuroanatomy of Human Ventral Occipitotemporal Cortex

KEVIN S. WEINER AND JASON D. YEATMAN

ABSTRACT Variability in human brain structure and function contributes to variability in perception and cognition. Nevertheless, in addition to the structural-functional variability of human ventral occipitotemporal cortex (VOTC), recent research reveals an unprecedented correspondence among cortical folding, functional representations, cytoarchitectonic areas, and white matter tracts in VOTC. This organization is remarkably stable across individuals and has been causally linked to domain-specific aspects of perception. In this chapter we provide a comprehensive review of this growing body of work. We use VOTC as a model system because it (1) plays an integral role in high-level visual perception and (2) displays an elegant correspondence between structure and function across spatial scales—from cells to networks. Specifically, we highlight that the correspondence between the anatomical and functional organization of VOTC can be used to (1) motivate and test computational models, (2) understand how neural responses in VOTC change during different cognitive tasks, and (3) relate to the perception of specific stimulus categories. Together, the findings reviewed in the present chapter lay the foundation for a new subdiscipline of cognitive neuroscience, which we refer to as *cognitive neuroanatomy* (CNA). We formally define CNA and end this chapter by proposing future applications of CNA.

A central question in cognitive neuroscience asks *How is the anatomy of the human brain linked to computations underlying neural responses and perception?* Let us consider the complexity of this question with one anatomical feature—cortical folding. Neuroanatomists write extensively about the "bewildering diversity" (Bailey & von Bonin, 1951) of cortical folding and how this variability or "diversity" corresponds to individual differences in cognition (Van Essen & Dierker, 2007). This variability extends to additional anatomical features across spatial scales—from microns to centimeters—which then add further complexity to understanding correspondences among brain structure, brain function, and perception. Contemporarily, Van Essen and Dierker (2007) state that "individual variability of the human cerebral cortex is a source of both fascination and frustration. The fascination arises because variability in cortical structure and function may account for many aspects of our unique personalities and cognitive capabilities. For neuroimagers, the frustration arises because variability presents serious obstacles when attempting to assign particular functional activation patterns to specific cortical areas" (p. 1050).

However, despite variability in brain structure, brain function, and perceptual abilities among individuals, recent evidence indicates that there are also stable features in brain structure across spatial scales and consistencies in the structural-functional coupling of cortical systems from one person to the next—especially within *human ventral occipitotemporal cortex* (VOTC)—that are perceptually meaningful. As such, the two main goals of this chapter are to (1) provide a comprehensive review of this stability and the orderliness it produces within human VOTC and (2) explain how this orderliness is not epiphenomenal. Indeed, as this section of the book is dedicated to perception and cognition, VOTC is an ideal model system to show how an understanding among cortical microstructure, cortical macrostructure, and white matter connectivity informs computational models that leverage this anatomical knowledge to provide insights into the underpinnings of high-level perception (e.g., word and face recognition).

In line with these goals, this chapter is divided into four sections. First, we review studies demonstrating a strong anatomical-functional coupling between cortical folding and functional representations spanning large-scale functional gradients and fine-scale functional regions within VOTC. In the second and third sections, we discuss two anatomical constraints that contribute to this structural-functional coupling: cytoarchitecture (i.e., the size, shape, and organization of cells across the layers of the cortical ribbon) and long-range white matter tracts connecting cortical regions. We then discuss how the consistent structural-functional organization of human VOTC motivates a new class of mechanistic models that leverages anatomical features to predict neural responses and perceptual functions. In the first three sections, we modify Van Essen and Dierker's (2007) forms of variability to consistency. We consider *functional representation-versus-folding consistency*, *cytoarchitectonic area-versus-folding consistency*, *cytoarchitectonic area-versus-functional region consistency*, and *connectional-versus-folding consistency*. In the fourth section, we propose that the ability to build mechanistic models that relate microanatomical features to macroanatomical (and functional) structures contributing to perception is the foundation of an emerging new subdiscipline of cognitive neuroscience: cognitive neuroanatomy (CNA). We formally define CNA and describe how it complements other subdisciplines in cognitive neuroscience. We end by providing examples of new questions and applications that emerge from CNA.

Consistency of Cortical Folding and Functional Representations

The folding of the cortex consists of both gyri (i.e., ridges) and sulci (i.e., furrows) in which the latter encompass more surface area than the former in the human brain (Destrieux et al., 2010; Van Essen, 2005; Zilles et al, 1997). Consequently, since more surface area of the human brain is sulcal as opposed to gyral in nature, accurate sulcal definitions are imperative for (1) quantifying the correspondence between cortical folding and functional representations and (2) motivating novel insights into the neurobiological foundations of perception. For example, related to the latter point, a recent study showed a surprising link between the fractionation of the occipitotemporal sulcus (OTS) and reading skills (Cachia et al., 2018). In addition to the OTS, modern methods accurately identify deep sulci within human VOTC such as the calcarine and collateral (CoS) sulci. However, four shallower VOTC sulci are also consistently identified in individual hemispheres (figure 9.1*A*): (1) the mid-fusiform sulcus (MFS), which bisects the fusiform gyrus (FG) into lateral and medial partitions, (2) the anterior (ALS) and posterior lingual sulci (PLS), which separate the lingual gyrus (LG) into anterior, mid, and posterior components, and (3) the posterior transverse collateral sulcus (ptCoS), which is a transverse component of the CoS. As discussed in subsequent sections, these sulci are landmarks identifying functional representations in VOTC.

Functional representation-versus-folding consistency Recent studies show that the relationship between cortical folding and functional representations in VOTC is much tighter than previously thought. We consider two scales of this relationship: large-scale functional gradients that span approximately 5,000 mm^3 (Weiner et al., 2014) and fine-scale functional regions, which are approximately an order of magnitude smaller (~200–500 mm^3; Weiner and Grill-Spector, 2010; Weiner et al., 2018).

Large scale As previously described (Grill-Spector & Weiner, 2014), the MFS identifies transitions in many large-scale functional maps, or gradients, that are composed of continuous or discrete stimulus dimensions, such as eccentricity (Malach, Levy, & Hasson, 2002), semantics (Huth, Nishimoto, Vu, & Gallant, 2012), animacy (Haxby et al., 2011), real-world object size (Konkle & Oliva, 2012), domain selectivity (Nasr et al., 2011), and conceptual knowledge (Martin, 2007). Despite differences in functional content, each map shares a common spatial layout in which representations positioned lateral to the MFS are functionally distinct from those positioned medial to the MFS. For example, lateral to the MFS, representations are biased toward foveal signals, while medial to the MFS, representations are biased toward peripheral signals (Malach, Levy, & Hasson, 2002); this same gradient predicts responsiveness to large versus small objects (Konkle & Oliva, 2012). We emphasize that some of these large-scale, lateral-medial gradients are not limited to visual perception, as they are present in blind individuals (van den Hurk, Van Baelen, & Op de Beeck, 2017) and can be produced in VOTC when sighted individuals simply imagine stimuli (Konkle & Oliva, 2012). These findings boast an impressive amount of predictive power: simply identifying the MFS in a person's brain predicts transition points in multiple functional gradients.

Fine scale While the MFS is a useful landmark to identify transitions in large-scale functional gradients, recent evidence indicates that particular sulcal features are useful for identifying the location of fine-scale clusters in VOTC in three particular ways. First, high-resolution functional magnetic resonance imaging (fMRI) shows that the traditional fusiform face area is composed of two spatially distinct clusters separated by 1–1.5 cm. The anterior component (mFus-faces) colocalizes with the anterior tip of the MFS. In fact, placing a 1 cm disk at the anterior tip of the MFS predicts 83% ± 7% of mFus-faces in the right hemisphere across individuals (Weiner et al., 2014). Second, the intersection between the ALS and the occipital component of the CoS predicts the location of the parahippocampal place area (CoS-places). This sulcal intersection is such a consistent landmark that a model built from one

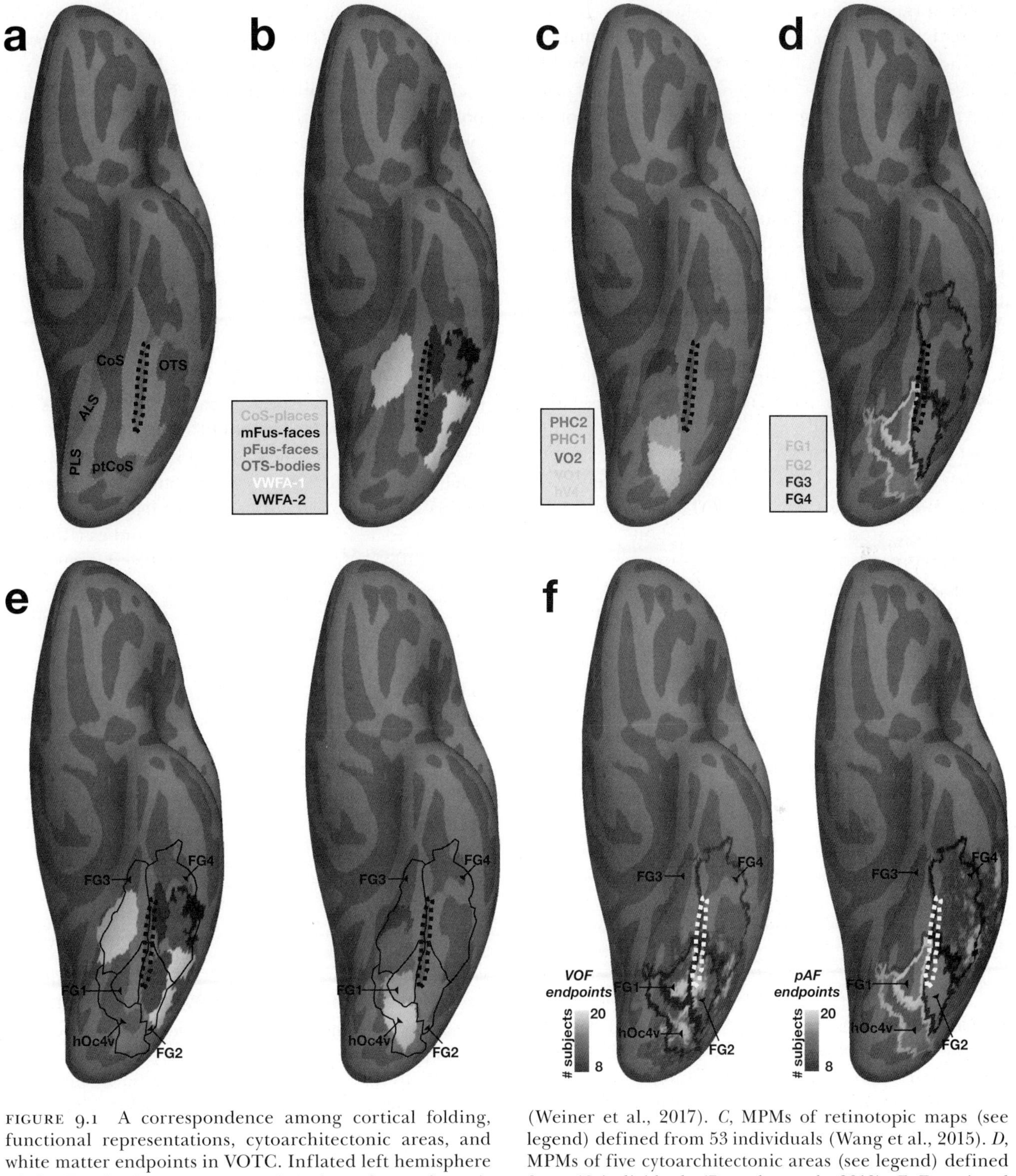

FIGURE 9.1 A correspondence among cortical folding, functional representations, cytoarchitectonic areas, and white matter endpoints in VOTC. Inflated left hemisphere of the FreeSurfer average template (www.freesurfer.net). *Dotted outline (black in A–E; white in F)*: Mid-fusiform sulcus. *A*, *Orange*: lingual gyrus; *blue*: fusiform gyrus; *ALS*: anterior lingual sulcus; *CoS*: collateral sulcus; *OTS*: occipitotemporal sulcus; *PLS*: posterior lingual sulcus; *ptCoS*: posterior transverse collateral sulcus. *B*, Maximum probability maps (MPMs) of domain-specific regions selective for faces, bodies, words, and places (see legend) from 12 individuals (Weiner et al., 2017). *C*, MPMs of retinotopic maps (see legend) defined from 53 individuals (Wang et al., 2015). *D*, MPMs of five cytoarchitectonic areas (see legend) defined from 10 individuals (Rosenke et al., 2018). *E*, Functional parcellations relative to cytoarchitectonic parcellations of VOTC. *Left*: domain-specific regions from (*B*). *Right*: retinotopic maps from (*C*). *F*, Probabilistic maps of cortical endpoints of the VOF (*left*) and pAF (*right*; *N* = 37; Weiner, Yeatman, & Wandell, 2017) relative to cytoarchitectonic areas from (*D*). (See color plate 9.)

group of participants can accurately predict the location of CoS-places in a separate group of participants (Weiner et al., 2018; figure 9.1*B*). Importantly, this structural-functional coupling between face- and place-selective regions and specific sulcal features is not epiphenomenal: electrical stimulation delivered to these locations produces domain-specific perceptual distortions (Megevand et al., 2014; Rangarajan et al., 2014). Third, the ptCoS identifies the boundary in the eccentricity maps distinguishing areas hV4 and VO-1 despite variability in cortical folding and map size (Witthoft et al., 2014; figure 9.1*C*). Of course, not all regions colocalize with sulcal features. In these cases, the topology of VOTC is useful. For example, body- and word-selective regions are consistently located lateral to face-selective regions within the OTS (Glezer & Riesenhuber, 2013; Weiner & Grill-Spector, 2010; Yeatman, Rauschecker, & Wandell, 2013).

The fact that there is functional representation-versus-folding consistency in VOTC is a major step forward for understanding the elegant correspondence between structure and function in high-level visual cortex. It also means that (1) anatomical axes reflect functional axes in VOTC and (2) fine-scale functional regions are embedded within large-scale functional gradients. Further, this nesting of functional representations is predicted by different scales of cortical folding: transitions in large-scale gradients are predicted by entire sulci, and local sulcal features predict the specific location of fine-scale functional regions.

Consistency of Cytoarchitectonics, Cortical Folding, and Functional Representations

Cytoarchitectonic area-versus-folding consistency Contrary to classic neuroanatomical studies, recent evidence supports a tight correspondence between sulci and cytoarchitectonic transitions in VOTC. Defined in the introduction, cytoarchitecture is considered the anatomical hardware that performs the computations underlying perception. Modern cytoarchitectonic approaches implement objective analyses in which algorithms traverse the cortical ribbon and statistically determine where adjacent pieces of tissue are cyotarchitectonically different (Amunts & Zilles, 2015). This algorithmic approach also allows the visualization of cytoarchitectonic areas using the same cortical surface-based methods that are used for visualizing in vivo MRI data. Employing these methods, eight areas have been identified in visual cortex, five of which are the most relevant for the present chapter: human occipital cytoarchitectonic area 4 ventral (hOc4v; Rottschy et al., 2007), as well as four cytoarchitectonic areas within the fusiform gyrus (FG1–FG4; Caspers et al., 2013; Lorenz et al., 2015; figures 9.1*D*, 9.2*A*). The MFS predicts cytoarchitectonic transitions between FG1 and FG2 in the posterior aspects of the FG, while the MFS also predicts cytoarchitectonic transitions between FG3 and FG4 in the middle portion of the FG. Additionally, the ptCoS predicts the transition between hOc4v and FG1 (Rosenke et al., 2018). Taken together, these findings also boast an impressive amount of predictive power: cellular-scale insight can be gleaned from centimeter-scale features of the cortical surface.

Cytoarchitectonic area-versus-functional region consistency In linking the previous two sections together, recent studies leveraged the fact that cytoarchitectonic areas and functional regions have a tight correspondence relative to cortical folding by developing tools to accurately project functional regions identified in living brains to cytoarchitectonic areas identified in postmortem tissue (and vice versa). These studies (Rosenke et al., 2018; Weiner et al., 2017) produced three main discoveries. First, each functional region is largely restricted to one cytoarchitectonic area. For instance, hV4 is largely restricted to hOc4v, while mFus-faces is located in a different cytoarchitectonic territory in FG4 (figure 9.1*E*). Second, regions selective for the same domain (e.g., faces) and located on the same macroanatomical structure (e.g., the FG), can be differentiated based on cytoarchitectonics. For example, mFus-faces is located within FG4 and displays different cytoarchitectonic features than pFus-faces, which is located in FG2. Third, and contrary to classic theories (Brodmann, 1909), there is not a one-to-one correspondence between a cytoarchitectonic area and a functional region. For example, face-, body-, and word-selective regions are all clustered within FG4. Understanding how computational and perceptual insight can be gleaned from the cytoarchitecture of a functional region (figure 9.2*A*) is discussed in the final section of this chapter.

Consistency of Connections in Ventral Occipitotemporal Cortex: From Models of Anatomy to Models of Neural Computations

Connectional-versus-folding consistency As the MFS is a functional and cytoarchitectonic landmark in VOTC, we briefly summarize recent findings showing that white matter tracts also have a consistent relationship relative to the MFS. Specifically, using a face-selective region and a place-selective region as separate seeds, Gomez et al. (2015) identified two longitudinal white

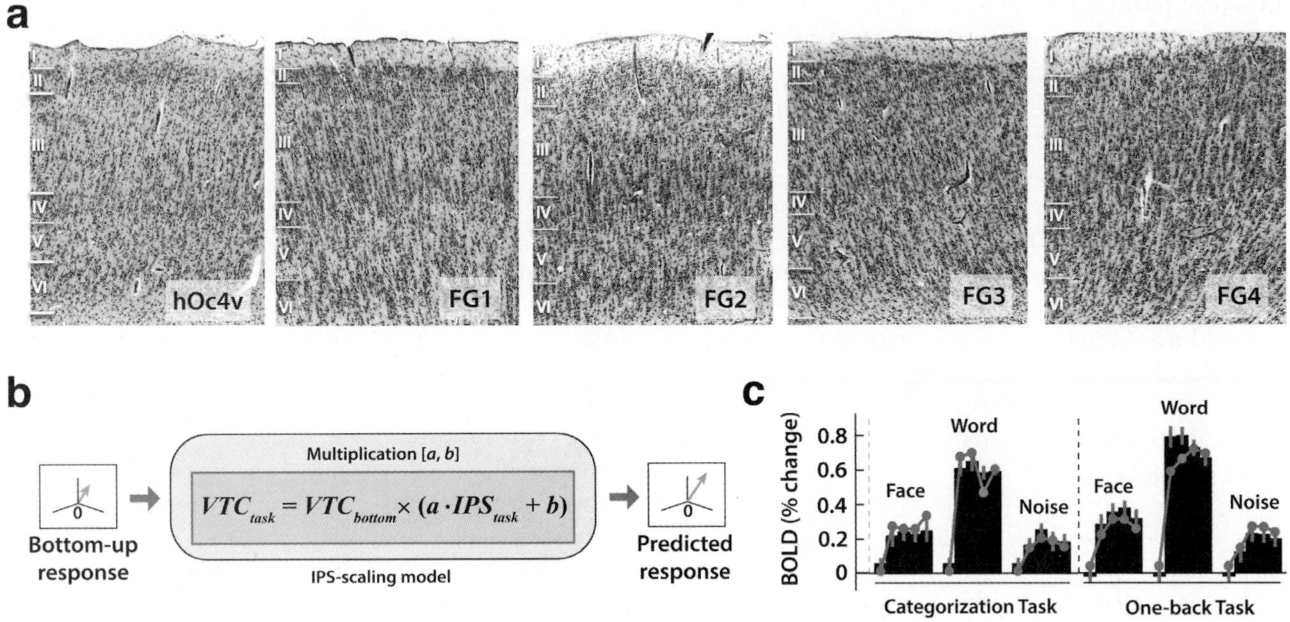

FIGURE 9.2 From cytoarchitecture to computational model architecture. *A*, Example histological slices from the five cytoarchitectonic areas in VOTC in figure 9.1 (modified from Rosenke et al., 2018). *B*, Architecture of the IPS-scaling model implemented by Kay and Yeatman (2017). The model begins with estimates of bottom-up neural signals in response to a stimulus (*left*). A top-down component is then added in which signals from the IPS modulate the neural responses within VWFA-1 using two free parameters (*middle*) to generate predicted responses, which are then compared to the actual measured responses (*right*). *C*, Blood oxygen level-dependent (BOLD) responses as a function of task (categorization; one-back) and stimulus (face, word, noise) during a contrast manipulation within VWFA-1. *Black bars*: measured BOLD responses; *gray lines*: model fits. There was a close correspondence between the model fits and the data (r=.96). As VWFA-1 is located within cytoarchitectonic area FG2, insights can now be made regarding how to incorporate cytoarchitectonic information into the model (see text for details).

matter tracts that were located (1) below the inferior longitudinal fasciculus and (2) lateral or medial to the MFS, respectively. Interestingly, the large-scale topology of these tracts was similar in typical adults and adults with prosopagnosia, or an inability to recognize faces. However, white matter properties within the tracts lateral, but not medial, to the MFS that intersected with face-selective regions were different between the patients compared to controls, indicating an important relationship between white matter properties and perception. Additionally, the endpoints of two vertical white matter tracts in VOTC, the vertical occipital fasciculus (VOF) and the posterior arcuate fasciculus (pAF), are also reliably positioned relative to the MFS: the anterior boundary of the VOF and the posterior boundary of the pAF are located near the midpoint of the MFS (figure 9.1*F*; Weiner et al., 2017; Yeatman et al., 2014). We emphasize that the cortical locations in which white matter tracts (1) start and end, as well as which (2) intersect with functional regions, guide both studies of perception and the construction of neurobiologically plausible computational models, which we expand on further in the next section.

A tight correspondence among cortical folding, functional representations, cytoarchitectonic areas, and white matter endpoints in VOTC Because there is (1) a consistent relationship between white matter endpoints and cortical folding and (2) a consistent relationship between functional regions and cortical folding, it is perhaps intuitive to the reader that white matter terminations in VOTC colocalize with functional regions. Indeed, the correspondence between functional responses and white matter anatomy in VOTC is so strong that a model of an individual's white matter connections can predict (1) the location of functional regions in adults (Saygin et al., 2012) and (2) where functional regions will emerge over the course of development (Saygin et al., 2016). A series of recent studies also show that category-selective regions—such as pFus-faces and the first visual word form area (VWFA-1)—are situated within the ventral terminations of the VOF (Kay & Yeatman, 2017; Yeatman, Rauschecker, & Wandell, 2013). As described in the previous section, both functional regions are also located within cytoarchitectonic area FG2. By visualizing probabilistic maps of white matter endpoints (Weiner, Yeatman, & Wandell, 2017) relative to probabilistic maps of cytoarchitectonic areas (Weiner et al., 2017), we can draw additional insights relating white matter transitions to cytoarchitectonic transitions (figure 9.1*F*). For example, VOF endpoints are largely restricted to FG1, FG2, and hOc4v. pAF endpoints are largely restricted to FG4. This suggests that there may be a relationship between large-scale white matter tracts and cytoarchitectonic boundaries, which can be examined further in future research. Taken together, there is a tight correspondence among cortical folding, functional representations, cytoarchitectonic transitions, and white matter endpoints in VOTC. We next illustrate how this precise anatomical knowledge can serve as the foundation for constraining models of cortical computation.

From models of anatomy to models of neural computations There are a wealth of studies in animals and humans in which knowledge of anatomical connections generates novel hypotheses about function—or lends support to a specific interpretation of information flow within a circuit (Noudoost & Moore, 2011; Saalmann, Pinsk, Wang, Li, & Kastner, 2012; and many others). Recently, anatomical connectivity was formally incorporated into a model of cortical computation within human VOTC (figure 9.2*B*), which linked task-driven modulations of word- and face-selective responses to the difficulty human observers experience when recognizing noisy images of words and faces (Kay & Yeatman, 2017). When attention is averted, the model showed that bottom-up response properties of pFus-faces and VWFA-1 could be computed by comparing a simple, V1-like representation of visual stimuli to a template of the preferred category. However, when viewers were prompted to make perceptual judgments of visual stimuli—such as categorizing an image as a word or a face—responses in category-selective regions were amplified by as much as 400%. This task-driven modulation of VOTC could not be explained by a bottom-up visual encoding model that simply predicted neural responses as a function of image features. However, when the authors considered that the VOF connected pFus-faces and VWFA-1 ventrally to regions of the intraparietal sulcus (IPS) that are involved in visual attention and perceptual decision-making, an intriguing question emerged: Could the observed, task-dependent modulation of VOTC responses reflect top-down signals that are related to the decision-making process within the IPS? Based on these anatomical connections, model modifications were made in which the IPS responses directly modulated the VOTC responses via signals conducted through the VOF. This top-down modulation introduced a gain to the bottom-up VOTC responses. This anatomically motivated model accurately predicted how responses in pFus-faces and VWFA-1 changed when the stimulus was held constant but the perceptual task changed (figure 9.2*B*). These results pave the way for new models that incorporate anatomical knowledge to predict how regions in human VOTC flexibly adapt to the demands of various perceptual tasks. For example, this model could serve as a general model explaining how top-down signals modulate visual encoding, which can be tested with additional perceptual tasks in future studies.

The Emergence of Cognitive Neuroanatomy

We have discussed how novel methodologies have generated insights into structural-functional correspondences in VOTC spanning spatial scales from cells to networks to computations underlying specific perceptual tasks. In this section we propose that these findings lay the groundwork for a new subdiscipline of cognitive neuroscience, cognitive neuroanatomy (CNA). In the sections below, we (1) provide a formal definition of CNA and (2) propose future applications of CNA based on the approaches reviewed in this chapter.

A formal definition of cognitive neuroanatomy CNA aims to explain how microanatomical structures generate macroanatomical functions contributing to measurable behaviors. At the heart of CNA is the goal to derive linking functions (i.e., models) that relate (1) behavioral output to (2) functional organization of the brain across spatial scales to (3) anatomical organization across spatial scales. The benefit of such functions is that once they are derived, specific aspects of human behavior can be explained as emergent properties of microarchitecture, macroarchitecture, and connectivity. As traditional anatomical measurements are made in static postmortem tissue, additional benefits are that once these linking functions are derived, variations in microanatomical features could predict variations in human behavioral performance. While this might seem far-fetched—for example, to predict behavioral performance in perceptual tasks from cellular or receptor measurements in postmortem tissue—in the next subsection, we show that this goal is not as far-fetched as it may seem.

Before doing so, we would like to clarify that *cognitive neuroanatomy* is a phrase that has appeared in the literature (Friston et al., 1996; Jäncke, 2003; Witelson, 1992) but without formal definition. The phrase was used interchangeably with *functional neuroanatomy*, *functional anatomy*, and *cognitive anatomy* (Friston & Price, 2003; Mechelli, Gorno-Tempini, & Price, 2003; Price &

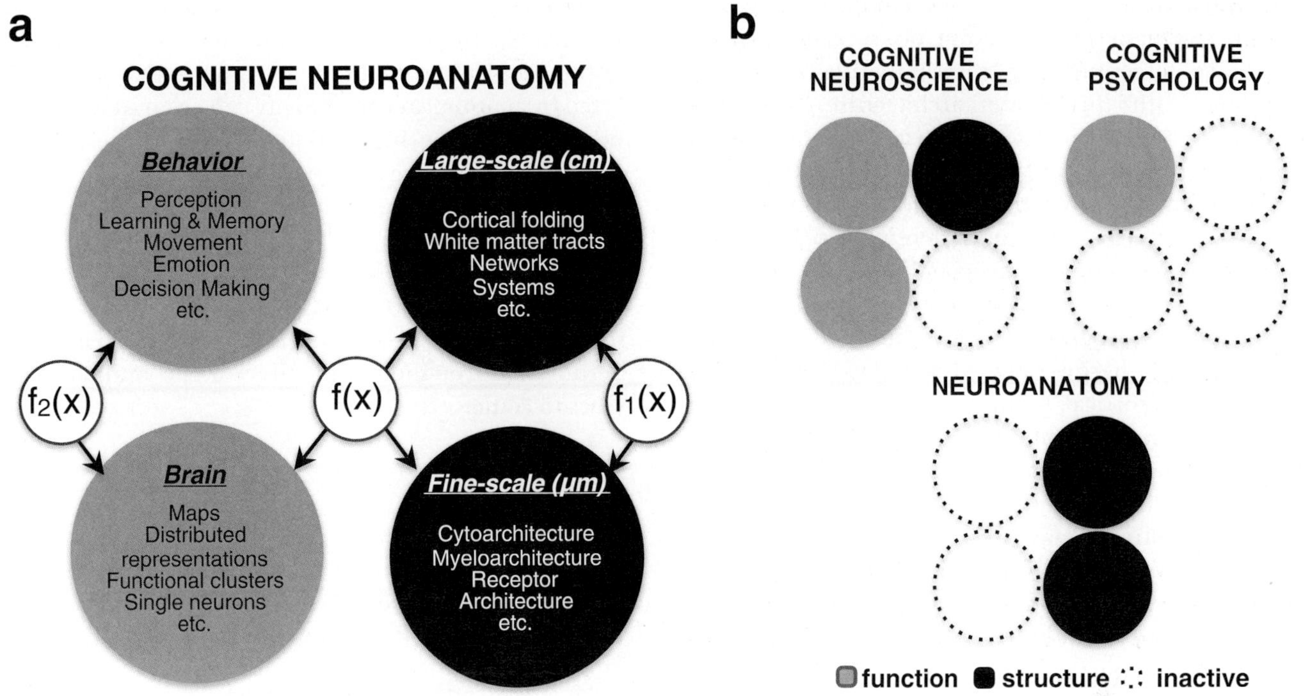

FIGURE 9.3 Cognitive neuroanatomy (CNA): a new subdiscipline of cognitive neuroscience. *A*, CNA works toward building mechanistic models explaining how microanatomical structures (*lower right*) generate macroanatomical functions (*lower left* and *upper right*) contributing to measurable behaviors (*upper left*) with an emphasis on deriving specific linking functions (depicted as $f_y(x)$). *B*, CNA integrates (1) cognitive neuroscience, which combines behavior with functional and structural features of the brain (three quadrants), (2) cognitive psychology, which incorporates function and behavior, as well as modeling (one quadrant), and (3) traditional neuroanatomy, which examines structural features of the brain across spatial scales (two quadrants). *Gray*: function; *black*: structure; *dotted line*: inactive.

Friston, 2002; Swanson et al., 1991). We do not see it as a phrase that is interchangeable with these other terms. Instead, CNA is a distinct subdiscipline of cognitive neuroscience emerging from a detailed understanding of anatomical principles that form the foundation of human brain computation and cognition. Subdisciplines become necessary as integration across fields increases (Naselaris et al., 2018). CNA can be considered the integration, or intersection, of cognitive psychology, cognitive neuroscience, and neuroanatomy (figure 9.3).

Future implementations of CNA based on the findings reviewed in the present chapter The model described in the previous section is an example of a linking function derived to explain how vertical white matter connections between VOTC and IPS carry signals that modulate processing in VOTC to accommodate the demands of face and word perception in the presence of noise. Progress can be gained by incorporating additional neuroanatomical and functional details to this (and future) model(s). Each successive step provides a new linking function—and with it, new insights into the architecture of human perception. For example, let us consider that VWFA-1 and pFus-faces are (1) already included in the model, (2) embedded within foveally biased cortex, (3) connected to the IPS through the VOF, and (4) located within FG2, which has a higher cell density than FG1 (Caspers et al., 2013). A logical hypothesis resulting from these structural-functional correspondences is that an increase in cell density is necessary to support the higher fidelity of foveal compared to peripheral processing. Previous neuroanatomical findings in nonhuman primates support this hypothesis by demonstrating that (1) cell density is highest in visual cortex compared to the rest of the brain, and (2) foveal representations in early visual cortex have higher neuronal densities than peripheral representations (Collins, Airey, Young, Leitch, & Kaas, 2010). Consequently, cell density across cortical layers serves a functional purpose and could be formalized into a model based on cytoarchitectonic data—for example, to predict differences in response properties between (1) posterior VWFA-1 and more anterior VWFA-2 (figure 9.1) and (2) mFus-faces and pFus-faces (figure 9.1). A perceptual component to the model could also be added by including psychophysical performance on different tasks related to word and face perception (e.g., discrimination versus recognition). With this model architecture, simulations titrating cell density could provide insight into how increased versus decreased cell density affects model performance in predicting neural responses, as well as behavioral performance on different perceptual tasks.

Additional components could also be integrated from models that already exist in the extended fields of neuroscience, cognitive science, and neuroanatomy. For example, there are powerful models that accurately predict (1) cortical folding across species (Tallinen, Chung, Biggins, & Mahadevan, 2014) and (2) the topological organization of maps in visual cortex (Kohonen, 1990). Incorporating mechanistic models of both folding and functional topography into the infrastructure described in figures 9.1 and 9.2 would provide a flexible model that could provide cellular insights into perception beyond the extensively studied regions of VOTC. Simulations could examine how the combination of cell density and area size influence perception. For instance, previous research indicates that (1) the surface area of functionally defined V1 predicts variability in conscious experience (Schwarzkopf, Song, & Rees, 2011), (2) a greater proportion of V1 is devoted to foveal compared to peripheral processing (Dougherty et al., 2003), and (3) there is an increased cell density devoted to foveal compared to peripheral processing. Thus, it is likely that there is a relationship among (1) cell density, (2) the surface area of eccentricity representations within V1, and (3) perception. Finally, cortical thickness and the surface area of functional regions would also be valuable anatomical factors to consider, as prior research shows that thin cortex with an enlarged surface area is linked to neural tuning and is perceptually advantageous (Song et al., 2015). Importantly, such models for visual perception may serve as the foundation for CNA models in other parts of the brain. As such, the details of brain anatomy, which are often considered tedious, will incrementally improve evolving CNA models that provide mathematical formulas defining the computations underlying the architecture of human perception and cognition.

Conclusion

Individual differences in the structure and function of VOTC have been described as "fascinating and frustrating" (Van Essen and Dierker, 2007) and linked to individual differences in perception and cognition. Nevertheless, in this chapter we have summarized a broad body of work that empirically supports the orderliness of VOTC to the point where (1) we know how cells are organized in multiple functionally defined regions and (2) computational models that are motivated by anatomical connectivity can accurately predict neural responses in relation to specific perceptual tasks. These findings are built on the systematic progress that has leveraged the centuries of work leading to the foundational knowledge of CNA, which is a new subdiscipline of cognitive neuroscience. While we have derived the definition of CNA from our field's understanding of VOTC, we look forward to applying this definition throughout the study of the brain in the years to come.

Acknowledgment

This work was supported by start-up funds from the University of California, Berkeley (KSW), and NSF SBE/BSF BCS-1551330 (JDY). We would like to thank Melina Uncapher for foundational ideas and discussions about cognitive neuroanatomy.

REFERENCES

Allison, T., Puce, A., Spencer, D. D., & McCarthy, G. (1999). Electrophysiological studies of human face perception. I: Potentials generated in occipitotemporal cortex by face and non-face stimuli. *Cerebral Cortex, 9*(5), 415–430.

Amunts, K., & Zilles, K. (2015). Architectonic mapping of the human brain beyond Brodmann. *Neuron, 88*(6), 1086–1107.

Bailey, P., & von Bonin, G. (1951). *The isocortex of man.* Urbana: University of Illinois Press.

Brodmann, K. (1909). *Vergleichende Lokalisationslehre der Großhirnrinde in ihren Prinzipien dargestellt auf Grund des Zellbaues.* Leipzig: Johann Ambrosius Barth Verlag.

Cachia, A., Roell, M., Mangin, J. F., Sun, Z. Y., Jobert, A., Braga, L., Houde, O., Dehaene, S., & Borst, G. (2018). How interindividual differences in brain anatomy shape reading accuracy. *Brain Structure and Function, 223*(2), 701–712.

Caspers, J., Zilles, K., Eickhoff, S. B., Schleicher, A., Mohlberg, H., & Amunts, K. (2013). Cytoarchitectonical analysis and probabilistic mapping of two extrastriate areas of the human posterior fusiform gyrus. *Brain Structure and Function, 218*(2), 511–526.

Collins, C. E., Airey, D. C., Young, N. A., Leitch, D. B., & Kaas, J. H. (2010). Neuron densities vary across and within cortical areas in primates. *Proceedings of the National Academy of Sciences of the United States of America, 107*(36), 15927–15932.

Destrieux, C., Fischl, B., Dale, A., & Halgren, E. (2010). Automatic parcellation of human cortical gyri and sulci using standard anatomical nomenclature. *Neuroimage, 53*(1), 1–15.

Dougherty, R. F., Koch, V. M., Brewer, A. A., Fischer, B., Modersitzki, J., & Wandell, B. A. (2003). Visual field representations and locations of visual areas V1/2/3 in human visual cortex. *Journal of Visualized Experiments, 10,* 586–598.

Friston, K. J., & Price, C. J. (2003). Degeneracy and redundancy in cognitive anatomy. *Trends in Cognitive Sciences, 7*(4), 151–152.

Friston, K. J., Price, C. J., Fletcher, P., Moore, C., Frackowiak, R. S., & Dolan, R. J. (1996). The trouble with cognitive subtraction. *NeuroImage, 4*(2), 97–104.

Glezer, L. S., & Riesenhuber, M. (2013). Individual variability in location impacts orthographic selectivity in the "visual word form area." *Journal of Neuroscience, 33*(27), 11221–11226.

Gomez, J., Pestilli, F., Witthoft, N., Golarai, G., Liberman, A., Poltoratski, S., ... Grill-Spector, K. (2015). Functionally defined white matter reveals segregated pathways in human ventral temporal cortex associated with category-specific processing. *Neuron, 85*(1), 216–227.

Grill-Spector, K., & Weiner, K. S. (2014). The functional architecture of the ventral temporal cortex and its role in categorization. *Nature Reviews Neuroscience, 15*(8), 536–548.

Haxby, J. V., Guntupalli, J. S., Connolly, A. C., Halchenko, Y. O., Conroy, B. R., Gobbini, M. I., ... Ramadge, P. J. (2011). A common, high-dimensional model of the representational space in human ventral temporal cortex. *Neuron, 72*(2), 404–416.

Huth, A. G., Nishimoto, S., Vu, A. T., & Gallant, J. L. (2012). A continuous semantic space describes the representation of thousands of object and action categories across the human brain. *Neuron, 76*(6), 1210–1224.

Jäncke, L. (2003). Current methods for cognitive neuroanatomy. In K. Hugdahl (Ed.), *Experimental Methods in Neuropsychology* (pp. 197–222). Dordrecht, the Netherlands: Kluwer Academic.

Kay, K. N., & Yeatman, J. D. (2017). Bottom-up and top-down computations in word- and face-selective cortex. *eLife, 6*, e22341.

Kohonen, T. (1990). Cortical maps. *Nature, 346*(6279), 24.

Konkle, T., & Oliva, A. (2012). A real-world size organization of object responses in occipitotemporal cortex. *Neuron, 74*(6), 1114–1124.

Lorenz, S., Weiner, K. S., Caspers, J., Mohlberg, H., Schleicher, A., Bludau, S., ... Amunts, K. (2015). Two new cytoarchitectonic areas on the human mid-fusiform gyrus. *Cerebral Cortex, 27*(1), 373–385.

Malach, R., Levy, I., & Hasson, U. (2002). The topography of high-order human object areas. *Trends in Cognitive Sciences, 6*(4), 176–184.

Mangin, J. F., Auzias, G., Coulon, O., Sun, Z. Y., Riviere, D., & Regis, J. (2015). Sulci as landmarks. In A. W. Toga (Ed.), *Brain Mapping: An Encyclopedic Reference* (pp. 45–52). New York: Academic Press.

Martin, A. (2007). The representation of object concepts in the brain. *Annual Review of Psychology, 58*, 25–45.

Mechelli, A., Gorno-Tempini, M. L., & Price, C. J. (2003). Neuroimaging studies of word and pseudoword reading: Consistencies, inconsistencies, and limitations. *Journal of Cognitive Neuroscience, 15*(2), 260–271.

Megevand, P., Groppe, D., Goldfinger, M. S., Hwang, S. T., Kingsley, P. B., Davidesko, I., & Mehta, A. D. (2014). Seeing scenes: Topographic visual hallucinations evoked by direct electrical stimulation of the parahippocampal place area. *Journal of Neuroscience, 34*(16), 5399–5405.

Naselaris, T., Bassett, D. S., Fletcher, A. K., Kording, K., Kriegeskorte, N., Nienborg, H., ... Kay, K. (2018). Cognitive computational neuroscience: A new conference for an emerging discipline. *Trends in Cognitive Sciences, 22*(5), 365–367.

Nasr, S., Liu, N., Devaney, K. J., Yue, X., Rajimehr, R., Ungerleider, L. G., & Tootell, R. B. (2011). Scene-selective cortical regions in human and nonhuman primates. *Journal of Neuroscience, 31*(39), 13771–13785.

Noudoost, B., & Moore, T. (2011). Control of visual cortical signals by prefrontal dopamine. *Nature, 474*(7351), 372–375.

Price, C. J., & Friston, K. J. (2002). Degeneracy and cognitive anatomy. *Trends in Cognitive Sciences, 6*(10), 416–421.

Rangarajan, V., Hermes, D., Foster, B. L., Weiner, K. S., Jacques, C., Grill-Spector, K., & Parvizi, J. (2014). Electrical stimulation of the left and right fusiform gyrus causes different effects of conscious face perception. *Journal of Neuroscience, 34*(38), 12828–12836.

Rosenke, M., Weiner, K. S., Barnett, M. A., Zilles, K., Amunts, K., Goebel, R., & Grill-Spector, K. (2018). A cross-validated cytoarchitectonic atlas of the human ventral visual stream. *NeuroImage, 170*, 257–270.

Rottschy, C., Eickhoff, S. B., Schleicher, A., Mohlberg, H., Kujovic, M., Zilles, K., & Amunts, K. (2007). Ventral visual cortex in humans: Cytoarchitectonic mapping of two extrastriate areas. *Human Brain Mapping, 28*(10), 1045–1059.

Saalmann, Y. B., Pinsk, M. A., Wang, L., Li, X., & Kastner, S. (2012). The pulvinar regulates information transmission between cortical areas based on attention demands. *Science, 337*(6095), 753–756.

Saygin, Z. M., Osher, D. E., Koldewyn, K., Reynolds, G., Gabrieli, J. D., & Saxe, R. R. (2012). Anatomical connectivity patterns predict face selectivity in the fusiform gyrus. *Nature Neuroscience, 15*(2), 321–327.

Saygin, Z. M., Osher, D. E., Norton, E. S., Youssoufian, D. A., Beach, S. D., Feather, J., ... Kanwisher, N. (2016). Connectivity precedes function in the development of the visual word form area. *Nature Neuroscience, 19*(9), 1250–1255.

Schwarzkopf, D. S., Song, C., & Rees, G. (2011). The surface area of human V1 predicts the subjective experience of object size. *Nature Neuroscience, 14*, 28–30.

Song, C., Schwarzkopf, D. S., Kanai, R., & Rees, G. (2015). Neural population tuning links visual cortical anatomy to human visual perception. *Neuron, 85*, 641–656.

Swanson, J. M., Posner, M., Potkin, S., Bonforte, S., Youpa, D., Fiore, C., ... Crinella, F. (1991). Activating tasks for the study of visual-spatial attention in ADHD children: A cognitive anatomic approach. *Journal of Child Neurology, 6*, S119–127.

Takemura, H., Rokem, A., Winawer, J., Yeatman, J. D., Wandell, B. A., & Pestilli, F. (2015). A major human white matter pathway between dorsal and ventral visual cortex. *Cerebral Cortex, 26*(5), 2205–2214.

Tallinen, T., Chung, J. Y., Biggins, J. S., & Mahadevan, L. (2014). Gyrification from constrained cortical expansion. *Proceedings of the National Academy of Sciences of the United States of America, 111*(35), 12667–12672.

Van Essen, D. C. (2005). A population-average, landmark- and surface-based (PALS) atlas of human cerebral cortex. *NeuroImage, 28*, 635–662.

Van Essen, D. C., & Dierker, D. (2007). On navigating the human cerebral cortex: Response to "in praise of tedious anatomy." *NeuroImage, 37*(4), 1050–1054; discussion 1066–1058.

van den Hurk, J., Van Baelen, M., & Op de Beeck, H. P. (2017). Development of visual category selectivity in ventral visual cortex does not require visual experience. *Proceedings of the National Academy of Sciences of the United States of America, 114*(22), E4501–E4510.

Wang, L., Mruczek, R. E .B., Arcaro, M. J., & Kastner, S. (2015). Probabilistic maps of visual topography in human cortex. *Cerebral Cortex, 25*(10), 3911–3931.

Weiner, K. S., Barnett, M. A., Lorenz, S., Caspers, J., Stigliani, A., Amunts, K., ... Grill-Spector, K. (2017). The cytoarchitecture of domain-specific regions in human high-level visual cortex. *Cerebral Cortex, 27*(1), 146–161.

Weiner, K. S., Barnett, M. A., Witthoft, N., Golarai, G., Stigliani, A., Kay, K. N., ... Grill-Spector, K. (2018). Defining the most probable location of the parahippocampal place area using cortex-based alignment and cross-validation. *NeuroImage, 170*, 373–384.

Weiner, K. S., Golarai, G., Caspers, J., Chuapoco, M. R., Mohlberg, H., Zilles, K., ... Grill-Spector, K. (2014). The mid-fusiform sulcus: A landmark identifying both cytoarchitectonic and functional divisions of human ventral temporal cortex. *NeuroImage, 84*, 453–465.

Weiner, K. S., & Grill-Spector, K. (2010). Sparsely-distributed organization of face and limb activations in human ventral temporal cortex. *Neuroimage, 52*(4), 1559–1573.

Weiner, K. S., Yeatman, J. D., & Wandell, B. A. (2017). The posterior arcuate fasciculus and the vertical occipital fasciculus. *Cortex, 97*, 274–276.

Weiner, K. S., & Zilles, K. (2016). The anatomical and functional specialization of the fusiform gyrus. *Neuropsychologia, 83*, 48–62.

Witelson, S. F. (1992). Cognitive neuroanatomy: A new era. *Neurology, 42*(4), 709–713.

Witthoft, N., Nguyen, M. L., Golarai, G., LaRocque, K. F., Liberman, A., Smith, M. E., & Grill-Spector, K. (2014). Where is human V4? Predicting the location of hV4 and VO1 from cortical folding. *Cerebral Cortex, 24*(9), 2401–2408.

Yeatman, J. D., Rauschecker, A. M., & Wandell, B. A. (2013). Anatomy of the visual word form area: Adjacent cortical circuits and long-range white matter connections. *Brain and Language, 125*(2), 146–155.

Yeatman, J. D., Weiner, K. S., Pestilli, F., Rokem, A., Mezer, A., & Wandell, B. A. (2014). The vertical occipital fasciculus: A century of controversy resolved by in vivo measurements. *Proceedings of the National Academy of Sciences of the United States of America, 111*(48), E5214–23.

Zilles, K., Schleicher, A., Langemann, C., Amunts, K., & Morosan, P. (1997). Quantitative analysis of sulci in the human cerebral cortex: Development, regional heterogeneity, gender difference, asymmetry, intersubject variability and cortical architecture. *Human Brain Mapping, 5*, 218–221.

10 Population Receptive Field Models in Human Visual Cortex

JONATHAN WINAWER AND NOAH C. BENSON

ABSTRACT An important goal in visual neuroscience is to understand how images are transformed into neuronal responses at different points in the visual pathway. Studies of activity in the living human brain, primarily with functional MRI, reflect neuronal responses pooled across large cell populations. The quantitative characterization of these population responses in terms of the visual stimulus is called *population receptive field modeling*. In recent years, the population receptive field model for fMRI has been used to study how the visual system combines inputs over space and time. This computational approach has revealed systematic differences in encoding properties across cortical location, task, development, and health. We review the basic principles behind the modeling, the connections to perception, and recent advances in extending population receptive field models to new domains.

Functional MRI: Localization and Computation

The ability to noninvasively measure the human brain increased dramatically following the advent of functional magnetic resonance imaging (fMRI) a quarter-century ago (Ogawa et al., 1992). The stimulus- or task-evoked Blood Oxygen Level Dependent (BOLD) fMRI response has a high signal-to-noise ratio in many parts of the brain and can be measured at the millimeter scale. For these reasons, fMRI was rapidly adopted by scientists and clinicians to study many topics spanning perception, cognition, emotion, health, development, and aging. Not everyone was impressed. In the *London Review of Books*, the philosopher Jerry Fodor asked, "Why, why, does everyone go on so about the brain?" with particular criticism aimed at questions of *where* in the brain responses correlate with a stimulus or behavior (Fodor, 1999). After all, Fodor complained, if we already know *that* something happens in the brain, who cares *where* it happens?

While localization continues to be important in human brain research and in medicine (a neurosurgeon might care where a language area is located), a computational approach serves a different and complementary function. Rather than asking where in the brain a stimulus (or task) is represented, one can instead ask *what* is computed or represented within a known brain area and *how* to explain patterns of brain responses as a function of stimulus and task. The purpose of this chapter is to explain methods for, and recent findings from, computational modeling of the human visual system, with a particular emphasis on the *population receptive field method*, a useful tool for quantifying how a neuronal population encodes visual information distributed over space and time.

A Brief History of the Receptive Field

The *receptive field* is a core principle in sensory neuroscience, unifying measurements across electrophysiology, fMRI, and behavior. Sherrington (1906) coined the phrase *receptive field* to describe a behavior, the scratch reflex in a dog. In his words, "The whole area of skin from whose points the scratch-reflex can be elicited may be conveniently termed the receptive field of that reflex." The key concept is that the behavior of interest is not described in units of the *output*, such as the speed or intensity of the muscle contraction, but in terms of *inputs*—the skin locations that elicit the behavior. This change of units is widely used in engineering and neuroscience and is called a *stimulus-referred* or *input-referred* measurement (Wandell, 1995). The change of units enabled Sherrington to ask questions such as, What is the extent of the skin patch that elicits the reflex; are all skin locations equally potent, or do some locations require more pressure than others to elicit the reflex; if two regions are lightly stimulated such that neither alone would elicit a reflex, does the simultaneous combined stimulation produce the reflex? Sherrington used the receptive field concept to reason about different types of reflexes, comparing, for example, the spatial extent of the flexion reflex versus the extensor thrust of the hind limb. (The flexion receptive field was much larger.) Using the receptive field concept, Sherrington explained behavior in terms of simple, generalizable units, such as the size and location of skin patches, rather than tracing the detailed neurophysiological pathways.

The receptive field concept was adapted from behavior to the electrophysiology of optic nerve fibers by Adrian and Matthews (1927) in the eel and Hartline (1938) in the bullfrog. Hartline, like Sherrington, emphasized the location and spatial extent of the inputs that lead to an output. Using the receptive field approach, Hartline asked quantitative questions, such as whether the size of the receptive field depends on the ambient illumination and how stimulation within the receptive field sums. Further, Hartline could compare results from diverse animals such as turtles, alligators, and frogs. Although the neurobiological pathways differ between these species, the input-referred measure showed that the general pattern of spatially extended, circular receptive fields on the retina held for optic nerve fibers across these species.

Even for single optic nerve fibers, receptive fields summarize network activity. The receptive field is a property of the entire sequence of events from light absorption to phototransduction, through all of the circuitry in the retina, and up to the nerve fiber that is actually measured. Thus, while the measurement may sample a single cell, the underlying computations are implemented by populations. A receptive field takes a village.

The Linear Population Receptive Field in Functional MRI

Just as Hartline extended the term *receptive field* from behavior to single-unit electrophysiology, the concept is naturally extended to neuronal population measures, including the local field potential (Victor, Purpura, Katz, & Mao, 1994) and the fMRI BOLD signal (Dumoulin & Wandell, 2008). The receptive field applied to these measures is sometimes called a *population receptive field* (pRF). Neuronal cell densities are estimated at approximately 20,000–40,000 per cubic millimeter in human gray matter and the glial-to-neuron ratio at about 1.5 (von Bartheld, Bahney, & Herculano-Houzel, 2016). Therefore, an fMRI voxel that is 1.6 mm on each side, as in the data used in figure 10.1*B*, likely contains about 100,000 to 200,000 neurons and comparable numbers of glial cells. The BOLD signal from this voxel results from the vascular response triggered by the activity of these large neuronal and glial populations. The voxel's response, like the action potentials of a single neuron, can be described in terms of locations in the visual field. This description is the voxel's pRF. The same logic holds for field potentials, which also pool signals over large populations of neurons. Despite the relatively coarse scale of neuronal pooling, these measures are nonetheless highly sensitive to stimulus properties such as position, contrast, and pattern.

The pRF has been most extensively developed for fMRI. Qualitatively, the pRF of a voxel measured with fMRI is the set of locations in the visual field that, when occupied by a contrast pattern, elicit a BOLD response. There are various ways to quantify the pRF. The simplest is the linear, circularly symmetric pRF (Dumoulin & Wandell, 2008). For this model the stimulus is converted to a binary contrast mask at each time point. The experiment is designed with the contrast mask in mind. In a typical experiment, a contrast pattern, such as a texture or checkerboard, is windowed within a slowly moving aperture, such as a bar, wedge, or ring. For each stimulus presentation, all the pixels inside the aperture are labeled *1* and all the pixels outside as *0* (figure 10.1*A*). The binary contrast mask is then multiplied point-wise by a circularly symmetric two-dimensional Gaussian and summed, yielding one number per time point. The two-dimensional Gaussian is the voxel's pRF. Finally, the time series is convolved with a hemodynamic response function, yielding the predicted fMRI time series for that voxel. The parameters of the Gaussian—its center (x, y) and size (σ)—are determined by a search algorithm to minimize differences between the predicted and observed time series. These parameters are expressed in degrees of visual angle, a stimulus-referred quantity.

The pRF model, even in its simplest form, goes beyond earlier retinotopy measurements, such as the traveling wave method (Engel et al., 1994). The advance is that the pRF specifies the spatial extent, not just the peak location, that elicits a BOLD response for a voxel. This additional dimension reveals interesting properties of the visual cortex. For example, pRF size increases systematically with eccentricity in each visual area and increases in later compared to earlier areas (figure 10.1*B*). These effects can be appreciated by visualizing schematic pRF solutions across the visual field (figure 10.1*C*). The increase in size with eccentricity approximately matches the granularity of visual analysis: larger pRFs in the periphery reflect the lower resolution of peripheral vision. Differences in pRF size between areas arise from other aspects of neuronal computations. For example, later visual areas are increasingly tolerant to changes in stimulus size and position, and this is partly reflected in the larger pRFs.

Just as the receptive field for a single neuron requires careful interpretation, so does the pRF of an fMRI voxel. Both the sizes and the scatter of neuronal receptive fields contribute to the voxel's pRF. A voxel might have a large pRF because it contains neurons with large receptive fields or because it contains neurons with

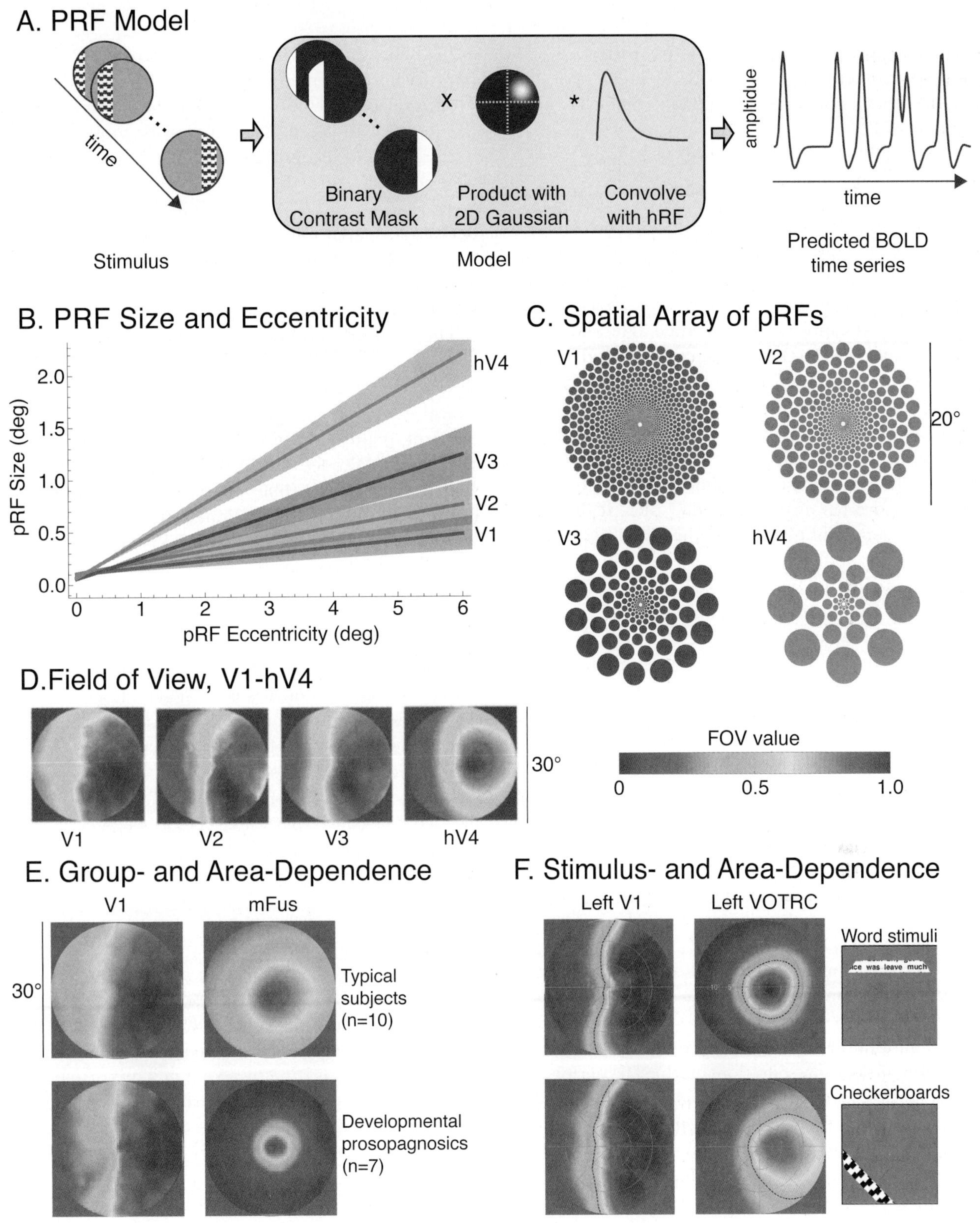

FIGURE 10.1 The fMRI population receptive field. *A*, The pRF model replaces the stimulus at each time point with a binary contrast mask, multiplied point-wise by the pRF, followed by convolution with a hemodynamic response function. *B*, pRF size increases with eccentricity and across visual areas. *C*, A cartoon showing the viewpoint of four visual areas based on pRF models. Circles represent one standard deviation of the pRF at each location (derived from the line plots in *B*). Data in *B* and *C* are from the Human Connectome Project (Benson et al., 2018; Ugurbil et al., 2013; Van Essen et al., 2013). *D*, The FOV for V1–V4, averaged across 10 subjects. pRFs from the right hemisphere are flipped across the y-axis. *E*, The FOV for bilateral V1 and a face-responsive region, mFus, for 10 typical subjects and 7 developmental prosopagnosics. *D* and *E* adapted from Witthoft et al., (2016) as per CC-BY-NC-ND 4.0 International license (https://creativecommons.org/licenses/by-nc-nd/4.0/). *F*, Group average (n=20) FOV for left V1 and VOTRC. The pRFs were derived from experiments with word stimuli or checkerboards within the bar apertures. Adapted from Le et al. (2017) as per CC-BY-NC-ND 4.0 International license (https://creativecommons.org/licenses/by-nc-nd/4.0/). (See color plate 10.)

small receptive fields scattered across different locations. Researchers have begun to model these two contributing factors to pRF size (Haak, Cornelissen, & Morland, 2012).

A Visual Area's Field of View

By combining pRFs from many locations, one can visualize the field of view of a visual area. The field of view is the portion of the visual field that elicits a response in any part of a region of interest. It is an aggregate measure of pRFs and can be computed as follows. First, exclude noisy voxels (low variance explained by the pRF model). Second, normalize the height of each pRF to 1. Third, at each point in the visual field take the maximum across the pRFs. Bootstrapping this procedure across voxels can reduce sensitivity to outliers. Because of the max operation, the field of view includes any region overlapping at least one pRF. It does not reflect the density of pRFs across the visual field. Alternatively, one could plot the *average* pRF value at each location in the visual field (Mackey, Winawer, & Curtis, 2017) or the fraction of voxels whose pRFs overlap each location (Sheremata & Silver, 2015), rather than the maximum value. These metrics reflect the density of pRFs across the visual field, emphasizing a different component of the representation than the field of view.

The field of view concisely summarizes the properties of a visual area. For V1–V3 the field of view is essentially the contralateral hemifield, with a slight bleeding into the ipsilateral field (figure 10.1*D*). (PRFs centered near the vertical midline extend into the ipsilateral field.) Some cortical areas with larger pRFs, like TO-2, show much more spillover into the ipsilateral field (Amano, Wandell, & Dumoulin, 2009; Huk, Dougherty, & Heeger, 2002).

It is not surprising that the field of view of V1 spans the visual field, since anywhere you can see requires a V1 representation. This is not true for all visual areas. Ventrotemporal areas tend to have a smaller field of view, with little representation of the far periphery. This is consistent with the proposal that much of the ventral visual pathway is specialized for recognition processes that depend on inputs from the central visual field (Levy, Hasson, Avidan, Hendler, & Malach, 2001). The earliest area to show this central bias is hV4 (figure 10.1*D*, right). Its field of view falls to its half-maximum value at about 10° to 15°. The smaller field of view of hV4 compared to V1–V3 arises not because the pRFs are smaller (in fact they are bigger); it is because the pRF centers are closer to the fovea.

A similar central bias is observed for face-responsive regions, such as mFus (figure 10.1*E*) (Witthoft et al., 2016). The limited field of view corresponds to the region where face recognition is good. The field of view in face-selective regions appears to be especially small in subjects with developmental prosopagnosia, a deficit in the ability to recognize faces. This decreased field of view occurs because the developmental prosopagnosics have smaller pRFs. An intriguing proposal is that decreased spatial integration in the ventral visual cortex, as reflected in smaller pRFs and a more central field of view, contributes to a reduced ability to recognize faces.

The ventral occipitotemporal reading circuitry (VOTRC) shows a similar foveal bias. In the VOTRC, the field of view extends only about 9° into the contralateral visual field, approximately the region used for efficient reading (figure 10.1*F*) (Le, Witthoft, Ben-Shachar, & Wandell, 2017). Interestingly, the field of view in the VOTRC (but not V1) shows some stimulus dependence: when the mapping stimuli are checkerboards rather than words, the VOTRC extends further into the periphery, suggesting that there are multiple routes by which visual information can reach these cortical regions. An important question yet to be resolved is whether individual differences in the VOTRC field of view predict individual differences in skilled reading.

Because the field of view is an aggregate measure, its properties can vary for multiple reasons. For example, hV4 has a smaller field of view but larger pRFs compared to V1. In contrast, face-selective areas of developmental prosopagnosics have a smaller field of view and smaller pRFs compared to typical subjects (on average).

Beyond the Linear Population Receptive Field

The pRF model developed by Dumoulin and Wandell pools contrast linearly over space. Linearity is implicit in the point-wise multiplication of the contrast mask and the pRF. The linear pRF model accurately fits the BOLD time series data for experiments with certain classes of stimuli. However, when spatial summation of contrast patterns is directly tested in a mixture experiment, the BOLD response is subadditive, meaning that linearity fails. For example, the response to a large contrast pattern tends to be less than the sum of the responses to parts of the pattern presented separately (figure 10.2*A*) (Kay, Winawer, Mezer, & Wandell, 2013). The linear prediction is somewhat too high in V1 and increasingly inaccurate in more anterior visual field maps. This subadditivity across space has been widely observed at the level of single neurons (e.g., Britten & Heuer, 1999) and fMRI (Kastner et al., 2001).

Subadditive spatial summation is an empirical observation not accounted for by the linear pRF model. The

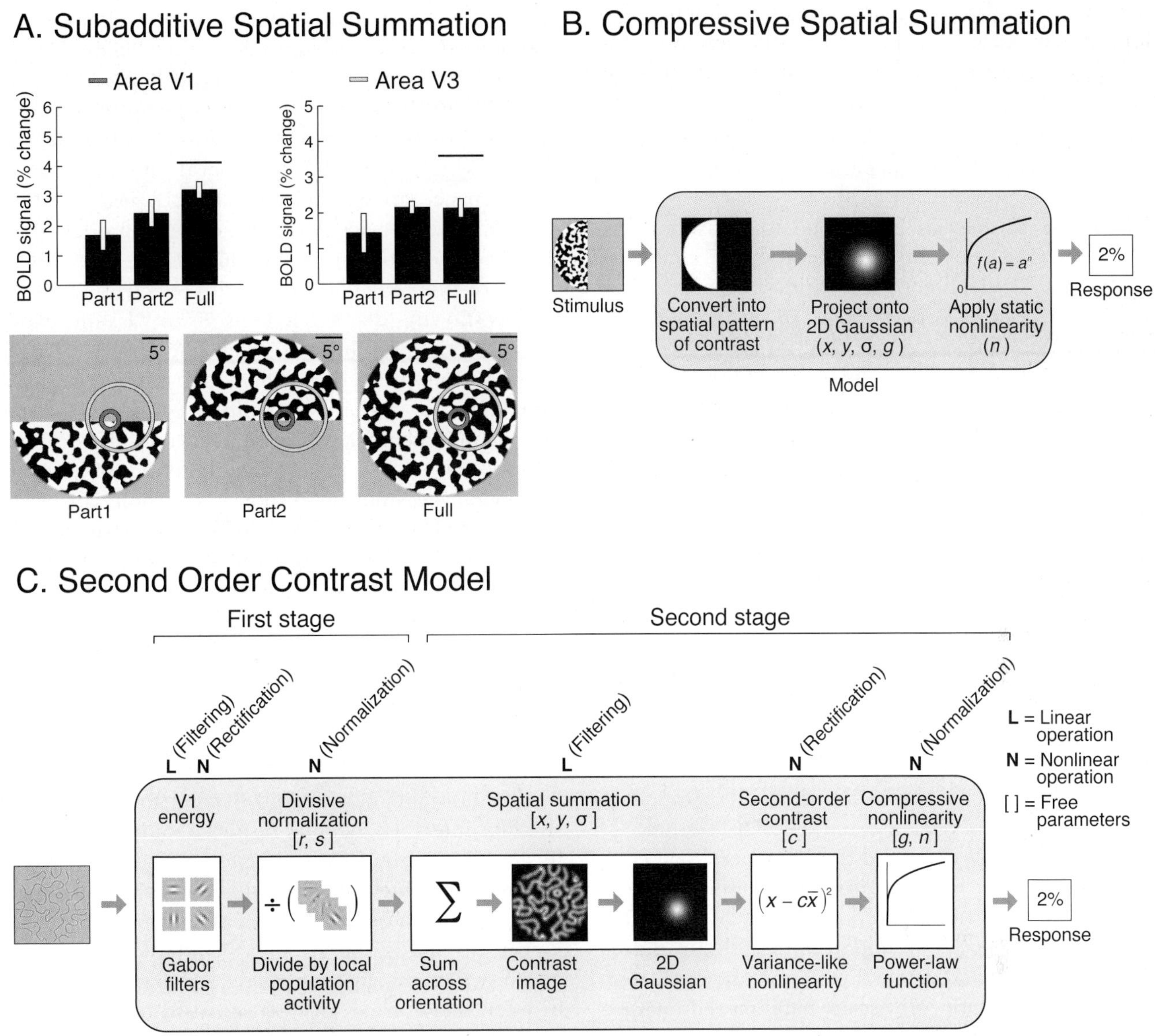

FIGURE 10.2 pRF model variants. *A*, The BOLD amplitude for a V1 and V3 voxel to stimulus parts 1 and 2, measured separately and together (*Full*). The pRFs are centered near the border between the two stimuli, indicated by the dark (*V1*) and light (*V3*) circles. The BOLD response for the full stimulus is less than the sum of the responses to the two parts, indicated by the horizontal line over the *full* bars in each plot. *B*, The compressive spatial summation pRF model is similar to the linear model in figure 10.1*A* but with an output static nonlinearity. Adapted from Kay, Winawer, Mezer, et al., (2013). *C*, The second-order contrast model operates directly on images rather than on binary contrast masks. The model includes a number of canonical neuronal calculations (filtering, divisive normalization, summation, second-order contrast, and a compressive nonlinearity) organized into two similar stages, each with three steps: linear (*L*), nonlinear (*N*), nonlinear (*N*). Adapted from Kay, Winawer, Rokem, et al. (2013).

limitation is effectively addressed by a compressive spatial summation (CSS) variant of the pRF model (figure 10.2*B*). In this model, after point-wise multiplication of the contrast mask and the pRF the output is passed through a power-law nonlinearity. The result of the nonlinearity is that the predicted response increases more gradually than the linear prediction. The form of the nonlinearity was shown to be compatible with divisive normalization, a calculation often used to model nonlinear responses in single neurons.

Kay and colleagues (2013) found that the CSS model was more accurate than the linear pRF model. The improvement was small for V1 and large for later areas. More generally, the authors observed that increasingly compressive spatial summation led to an increased tolerance for changes in stimulus size and position, an important property of visual responses and object recognition. A number of subsequent studies have fit the CSS model to fMRI data. It is especially useful for higher-level cortical areas, such as word- and face-responsive regions (Kay, Weiner, & Grill-Spector, 2015; Le et al., 2017) and for the parietal and frontal cortex (Mackey, Winawer, & Curtis, 2017).

The CSS model can be applied to measurements other than fMRI. Several groups have fit pRF models to intracranial field potentials in human subjects measured with electrocorticography, or ECoG (Harvey et al., 2013; Winawer et al., 2013; Yoshor, Bosking, Ghose, & Maunsell, 2007). The field potential includes components arising from multiple neural sources. Winawer and colleagues (2013) separated the ECoG time series into two sources, one broadband and one stimulus locked. They found that the two components of the signal both had time series well fit by pRF models. The broadband but not the stimulus-locked signal showed subadditive spatial summation properties matching fMRI. They concluded that the neuronal response giving rise to the ECoG broadband signal is largely responsible for the stimulus-evoked BOLD signal in the visual cortex. It also provided further support to the conclusion from the fMRI studies that the nonlinearity arose (at least in part) from nonlinear neural responses, rather than from the coupling between the neural response and the BOLD signal.

The pRF models discussed thus far operate on contrast masks rather than the original stimulus. Models that predict the BOLD response directly from the image pixel values are an important test of our understanding of the visual system. Even for a restricted set of images, such as static, gray-scale, band-passed images, relatively complex models are needed to make accurate predictions (figure 10.2*C*). One such model consists of a two-stage cascade (Kay, Winawer, Rokem, Mezer, & Wandell, 2013), including contrast energy, divisive normalization, spatial summation, second-order contrast, and a compressive nonlinearity. In V1–V4, the model accurately predicts responses to a wide range of images, varying in orientation, contrast, spatial location, and pattern. Importantly, the model solutions help to clarify and quantify differences between visual areas. For example, compared to V1, V2–V4 generally had larger receptive field sizes, stronger levels of normalization, and greater selectivity for second-order contrast.

From Space to Time

The visual system must pool signals over time as well as space. Recently, pRF models have been developed to quantify how visual areas encode information distributed over time. In one such effort, the CSS model was adapted from space to time (Zhou, Benson, Kay, & Winawer, 2018). Zhou and colleagues employed a compressive model because mixture tests revealed that the cortex exhibits subadditive temporal summation similar to space. For example, the BOLD response to a stimulus viewed twice in rapid succession was less than double the response to a stimulus viewed once. These fMRI results are consistent with electrophysiology studies showing that single neurons in V1 sum subadditively in time (Tolhurst, Walker, Thompson, & Dean, 1980). In adapting the CSS model to the temporal domain (CTS, compressive temporal summation), several changes were made: the input is a one-dimensional time course rather than a two-dimensional contrast envelope, the impulse response function is a one-dimensional gamma function rather than a two-dimensional Gaussian, and the static nonlinearity is implemented as divisive normalization rather than a power law (figure 10.3*A*). Similar to the spatial domain, the subadditive temporal model was more accurate than a linear model.

The CTS model captures some interesting temporal phenomena (figure 10.3*B*). First, responses to long or repeated stimuli were less than the linear prediction from brief stimuli (adaptation). Second, for longer interstimulus intervals, the total response to the two stimuli increased (recovery from adaptation) (figure 10.3*C*). The patterns of temporal summation in the visual hierarchy parallel those in space: temporal windows become longer in more anterior areas, and the summation becomes increasingly compressive. As a result, responses in later visual areas are less sensitive to the precise duration of a viewed stimulus. This might be thought of as tolerance for timing or duration, analogous to the position and size tolerance in visual responses.

Overall, the CTS model provides a simple and interpretable set of computations—temporal summation followed by a compressive nonlinearity—that operates on arbitrary temporal inputs and can be used to characterize neural response properties across the visual hierarchy. The similarity of the findings between spatial and temporal summation suggest that the visual cortex may use a similar processing strategy to achieve higher-level and increasingly invariant representations of the visual world.

The fMRI response to temporally modulated stimuli has also been modeled using a different formulation, in which neuronal responses are thought to reflect the summed output of two temporal channels, one more sustained and one more transient. Models like these have

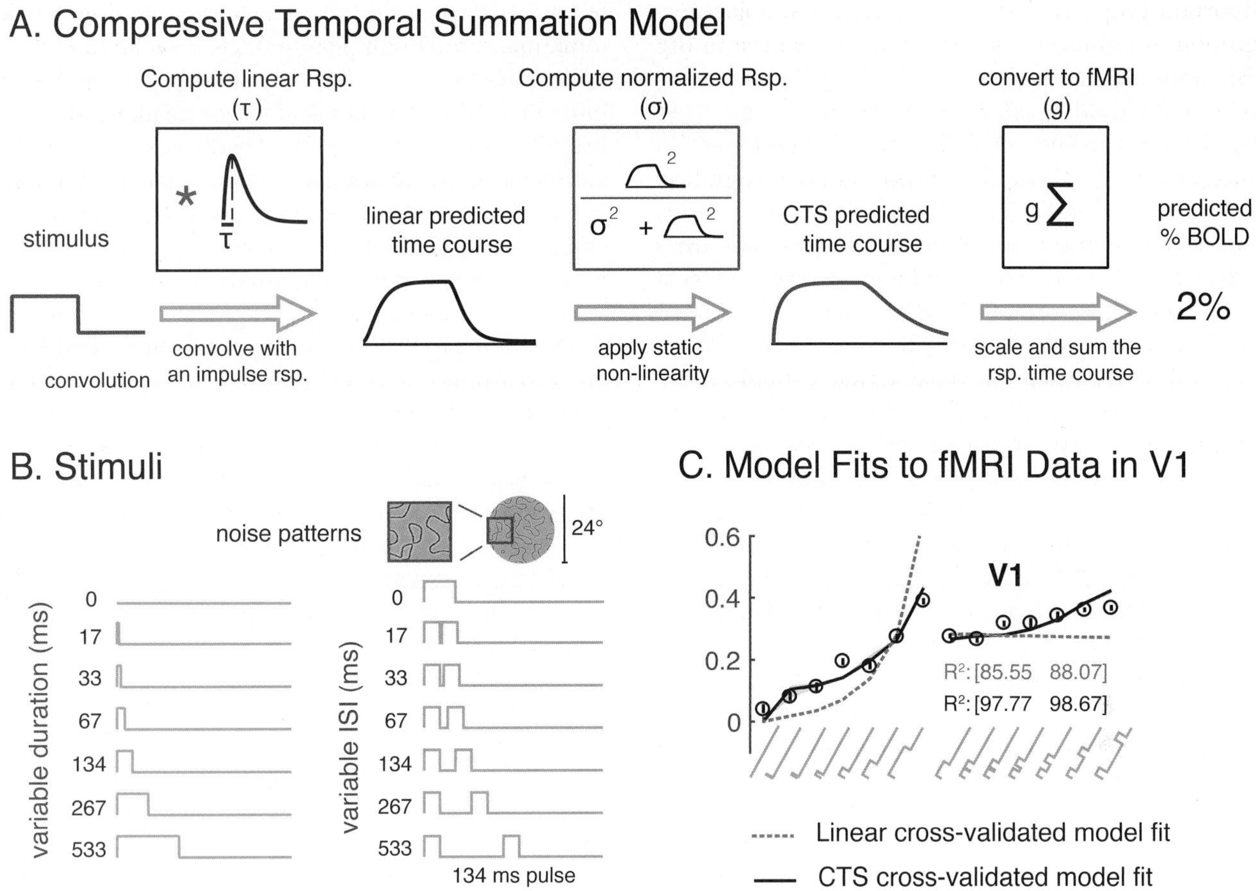

FIGURE 10.3 A temporal pRF model. *A*, The model takes the stimulus time course as input, which is convolved with an impulse response function. The impulse response function is parameterized by a time constant, τ, on the order of approximately 100 ms. The linear output is then passed through a static nonlinearity, implemented as divisive normalization, to yield the CTS output. This output is summed over the trial to predict the percent BOLD response. *B*, Stimuli for testing a temporal pRF model were large-field static texture patterns. The patterns were viewed either as one pulse (0 to 533 ms) or two pulses of 134 ms each, with a variable interstimulus interval. *C*, Cross-validated model predictions in V1. The linear model fit omits the divisive normalization step. The nonlinear (CTS) model predictions are more accurate. The R^2 values are the 50% confidence intervals from bootstrapped, cross-validated fits. Adapted from Zhou et al. (2018).

been used to explain psychophysical data (Watson & Robson, 1981), the V1 fMRI response to luminance modulations (Horiguchi, Nakadomari, Misaki, & Wandell, 2009) and the fMRI response in many cortical areas to contrast modulations (Stigliani, Jeska, & Grill-Spector, 2017). These models have been especially useful at capturing differences between foveal visual representations, which are most sensitive to sustained inputs, and peripheral representations, which are most sensitive to visual transients.

It is likely that a more complete model of temporal processing will involve elements of both models: a compressive summation over extended temporal windows and two or more channels to capture differential sensitivity to transient and sustained inputs.

Applications and Future Directions

The pRF model was introduced in fMRI to quantify the spatial extent to which cortical locations are sensitive. This model built on earlier fMRI findings demonstrating sensitivity to neuronal receptive field size (Smith, Singh, Williams, & Greenlee, 2001; Tootell et al., 1997). Early pRF studies largely measured human analogs to known results from animal studies, such as the increase in receptive field size with eccentricity. Having firmly established the methods, pRF modeling has now been substantially extended to make new discoveries and quantifications of the visual pathways, including applications to health, plasticity, development, and task. Many of these results were reviewed recently (Wandell &

Winawer, 2015). Here, we point to a few illustrative examples to demonstrate the type of new findings made with pRF models.

First, pRF models can characterize the field of view of a visual area. These results quantify the manner in which many ventral visual areas involved in recognition have a more centrally limited field of view. The results also show that the field of view of some visual areas depends on stimulus pattern (Le et al., 2017), subject population (Witthoft et al., 2016), and development (Gomez, Natu, Jeska, Barnett, & Grill-Spector, 2018). The method is sufficiently reliable that it should prove possible to test hypotheses linking field-of-view measures to visual performance in individual subjects.

Second, a number of studies have found differences in pRF properties in groups as a function of health. The interpretation of these differences varies across studies. For example, in early visual areas both pRF size and the depth of surround suppression were found to be reduced in people with schizophrenia (Anderson et al., 2017). The authors linked these findings to some of the visual deficits reported with schizophrenia and suggested they arise from imbalances of excitation and inhibition in the visual cortex. Another recent study showed that following cataract surgery, the average V1 pRF is larger in those patients with decreased contrast sensitivity, reflecting increased optical blur (Miranda et al., 2018). These findings highlight the fact that receptive field properties depend on the function of the entire visual pathway, from input (optics) to intracortical circuitry.

Third, a number of studies have quantified how attention affects pRF properties, including the size, position, and gain of pRFs (Kay, Weiner, & Grill-Spector, 2015; Sheremata & Silver, 2015; Sprague & Serences, 2013). These effects result in a change in the spatial uncertainty with which the pRFs encode the stimulus (Kay, Weiner, & Grill-Spector, 2015).

Fourth, pRF models are increasingly used to study particular stimulus properties, such as wavelength sensitivity (Welbourne, Morland, & Wade, 2018). Comparing luminance-modulated stimuli and chromatically modulated stimuli, these authors find that sensitivity to spatial frequency is lower for chromatic stimuli, but pRF size is unchanged. By showing a dissociation between receptive field size and sensitivity, the results highlight the importance of making other measurements (here, spatial frequency tuning) in tandem with pRF modeling, leading to new insights about how these two visual properties are represented in neuronal populations.

Fifth, anatomical templates of retinotopic organization bridge the gap between the pRF parameters of individual voxels and the organization of entire retinotopic maps. Such template models can predict retinotopic parameters in V1, V2, and V3 based on an individual brain's anatomical structure alone (Benson, Butt, Brainard, & Aguirre, 2014) or in combination with an individual's pRF measurements (Benson & Winawer, 2018). Moreover, by combining a template with a more complex pRF model, such as the two-stage cascade model described above, one can derive a standard cortical observer model—a model capable of predicting the distribution of cortical responses within several brain areas to arbitrary images (Benson, 2017).

Finally, there has been substantial progress in extending pRF measures to new domains. We reviewed studies in one new domain, temporal pRF models. In addition, pRF modeling has recently been used to study auditory (Thomas et al., 2015) and somatosensory cortex (Schellekens, Petridou, & Ramsey, 2017).

Together, the developments reviewed here demonstrate the power of the computational modeling approach. Even the relatively simple linear pRF model, when measured and interpreted carefully, can lead to new insights about visual function, health, development, and cognition. And new methods, first checked against known findings, can be further refined for appropriate uses, such as the nonlinear pRF models for space and time.

REFERENCES

Adrian, E. D., & Matthews, R. (1927). The action of light on the eye: Part I. The discharge of impulses in the optic nerve and its relation to the electric changes in the retina. *Journal of Physiology, 63*(4), 378–414.

Amano, K., Wandell, B. A., & Dumoulin, S. O. (2009). Visual field maps, population receptive field sizes, and visual field coverage in the human MT+ complex. *Journal of Neurophysiology, 102*(5), 2704–2718. doi:10.1152/jn.00102.2009

Anderson, E. J., Tibber, M. S., Schwarzkopf, D. S., Shergill, S. S., Fernandez-Egea, E., Rees, G., & Dakin, S. C. (2017). Visual population receptive fields in people with schizophrenia have reduced inhibitory surrounds. *Journal of Neuroscience, 37*(6), 1546–1556. doi:10.1523/JNEUROSCI.3620-15.2016

Benson, N. C. (2017). Toward a standard cortical observer. *Journal of Vision, 17*(15), 11–12. doi:10.1167/17.15.11a

Benson, N. C., Butt, O. H., Brainard, D. H., & Aguirre, G. K. (2014). Correction of distortion in flattened representations of the cortical surface allows prediction of V1–V3 functional organization from anatomy. *PLoS Computational Biology, 10*(3), e1003538. doi:10.1371/journal.pcbi.1003538

Benson, N. C., Jamison, K. W., Arcaro, M. J., Vu, A. T., Glasser, M. F., Van Essen, D. C., … Kay, K. N. (2018). The Human Connectome Project 7 Tesla retinotopy dataset: Description and population receptive field analysis. *Journal of Vision, 18*(13), 23. doi: 10.1167/18.13.23

Benson, N. C., & Winawer, J. (2018). Bayesian analysis of retinotopic maps. *Elife 7.* pii: e40224. doi:10.7554/eLife.40224

Britten, K. H., & Heuer, H. W. (1999). Spatial summation in the receptive fields of MT neurons. *Journal of Neuroscience, 19*(12), 5074–5084.

Dumoulin, S. O., & Wandell, B. A. (2008). Population receptive field estimates in human visual cortex. *NeuroImage, 39*(2), 647–660. doi:10.1016/j.neuroimage.2007.09.034

Engel, S. A., Rumelhart, D. E., Wandell, B. A., Lee, A. T., Glover, G. H., Chichilnisky, E. J., & Shadlen, M. N. (1994). fMRI of human visual cortex. *Nature, 369*(6481), 525. doi:10.1038/369525a0

Fodor, J. (1999). Let your brain alone. *London Review of Books, 21*(19), 68–69.

Gomez, J., Natu, V., Jeska, B., Barnett, M., & Grill-Spector, K. (2018). Development differentially sculpts receptive fields across early and high-level human visual cortex. *Nature Communications, 9*(1), 788. doi:10.1038/s41467-018-03166-3

Haak, K. V., Cornelissen, F. W., & Morland, A. B. (2012). Population receptive field dynamics in human visual cortex. *PLoS One, 7*(5), e37686. doi:10.1371/journal.pone.0037686

Hartline, H. K. (1938). The response of single optic nerve fibers of the vertebrate eye to illumination of the retina. *American Journal of Physiology—Legacy Content, 121*(2), 400–415.

Harvey, B. M., Vansteensel, M. J., Ferrier, C. H., Petridou, N., Zuiderbaan, W., Aarnoutse, E. J., ... Dumoulin, S. O. (2013). Frequency specific spatial interactions in human electrocorticography: V1 alpha oscillations reflect surround suppression. *NeuroImage, 65*, 424–432. doi:10.1016/j.neuroimage.2012.10.020

Horiguchi, H., Nakadomari, S., Misaki, M., & Wandell, B. A. (2009). Two temporal channels in human V1 identified using fMRI. *NeuroImage, 47*(1), 273–280. doi:10.1016/j.neuroimage.2009.03.078

Huk, A. C., Dougherty, R. F., & Heeger, D. J. (2002). Retinotopy and functional subdivision of human areas MT and MST. *Journal of Neuroscience, 22*(16), 7195–7205. doi:20026661

Kastner, S., De Weerd, P., Pinsk, M. A., Elizondo, M. I., Desimone, R., & Ungerleider, L. G. (2001). Modulation of sensory suppression: Implications for receptive field sizes in the human visual cortex. *Journal of Neurophysiology, 86*(3), 1398–1411.

Kay, K. N., Weiner, K. S., & Grill-Spector, K. (2015). Attention reduces spatial uncertainty in human ventral temporal cortex. *Current Biology, 25*(5), 595–600. doi:10.1016/j.cub.2014.12.050

Kay, K. N., Winawer, J., Mezer, A., & Wandell, B. A. (2013). Compressive spatial summation in human visual cortex. *Journal of Neurophysiology, 110*(2), 481–494. doi:10.1152/jn.00105.2013

Kay, K. N., Winawer, J., Rokem, A., Mezer, A., & Wandell, B. A. (2013). A two-stage cascade model of BOLD responses in human visual cortex. *PLoS Computational Biology, 9*(5), e1003079. doi:10.1371/journal.pcbi.1003079

Le, R., Witthoft, N., Ben-Shachar, M., & Wandell, B. (2017). The field of view available to the ventral occipito-temporal reading circuitry. *Journal of Visualized Experiments, 17*(4), 6. doi:10.1167/17.4.6

Levy, I., Hasson, U., Avidan, G., Hendler, T., & Malach, R. (2001). Center-periphery organization of human object areas. *Nature Neuroscience, 4*(5), 533–539. doi:10.1038/87490

Mackey, W. E., Winawer, J., & Curtis, C. E. (2017). Visual field map clusters in human frontoparietal cortex. *eLife, 6.* doi:10.7554/eLife.22974

Miranda, A. S. C., Martins Rosa, A. F., Patricio Dias, M. J., Harvey, B. M., Loureiro da Silva, M. F., de Sa, E. S. C.-B. M., & Murta, J. C. N. (2018). Optical properties influence visual cortical functional resolution after cataract surgery and both dissociate from subjectively perceived quality of vision. *Investigative Ophthalmology & Visual Science, 59*(2), 986–994. doi:10.1167/iovs.17-22321

Ogawa, S., Tank, D. W., Menon, R., Ellermann, J. M., Kim, S. G., Merkle, H., & Ugurbil, K. (1992). Intrinsic signal changes accompanying sensory stimulation: Functional brain mapping with magnetic resonance imaging. *Proceedings of the National Academy of Sciences, 89*(13), 5951–5955.

Schellekens, W., Petridou, N., & Ramsey, N. F. (2017). Mapping the human homunculus with receptive field analysis. Paper presented at the Annual Meeting of the Society for Neuroscience, Washington, DC.

Sheremata, S. L., & Silver, M. A. (2015). Hemisphere-dependent attentional modulation of human parietal visual field representations. *Journal of Neuroscience, 35*(2), 508–517. doi:10.1523/JNEUROSCI.2378-14.2015

Sherrington, C. S. (1906). Observations on the scratch-reflex in the spinal dog. *Journal of Physiology, 34*(1–2), 1–50.

Smith, A. T., Singh, K. D., Williams, A. L., & Greenlee, M. W. (2001). Estimating receptive field size from fMRI data in human striate and extrastriate visual cortex. *Cerebral Cortex, 11*(12), 1182–1190.

Sprague, T. C., & Serences, J. T. (2013). Attention modulates spatial priority maps in the human occipital, parietal and frontal cortices. *Nature Neuroscience, 16*(12), 1879–1887. doi:10.1038/nn.3574

Stigliani, A., Jeska, B., & Grill-Spector, K. (2017). Encoding model of temporal processing in human visual cortex. *Proceedings of the National Academy of Sciences.* doi:10.1073/pnas.1704877114

Thomas, J. M., Huber, E., Stecker, G. C., Boynton, G. M., Saenz, M., & Fine, I. (2015). Population receptive field estimates of human auditory cortex. *NeuroImage, 105*, 428–439. doi:10.1016/j.neuroimage.2014.10.060

Tolhurst, D. J., Walker, N. S., Thompson, I. D., & Dean, A. F. (1980). Non-linearities of temporal summation in neurons in area 17 of the cat. *Experimental Brain Research, 38*(4), 431–435.

Tootell, R. B., Mendola, J. D., Hadjikhani, N. K., Ledden, P. J., Liu, A. K., Reppas, J. B., ... Dale, A. M. (1997). Functional analysis of V3A and related areas in human visual cortex. *Journal of Neuroscience, 17*(18), 7060–7078.

Ugurbil, K., Xu, J., Auerbach, E. J., Moeller, S., Vu, A. T., Duarte-Carvajalino, J. M., ... Yacoub, E. (2013). Pushing spatial and temporal resolution for functional and diffusion MRI in the Human Connectome Project. *NeuroImage, 80*, 80–104. doi:10.1016/j.neuroimage.2013.05.012

Van Essen, D. C., Smith, S. M., Barch, D. M., Behrens, T. E., Yacoub, E., Ugurbil, K., & WU-Minn HCP Consortium (2013). The WU-Minn Human Connectome Project: An overview. *NeuroImage, 80*, 62–79. doi:10.1016/j.neuroimage.2013.05.041

Victor, J. D., Purpura, K., Katz, E., & Mao, B. (1994). Population encoding of spatial frequency, orientation, and color in macaque V1. *Journal of Neurophysiology, 72*(5), 2151–2166.

von Bartheld, C. S., Bahney, J., & Herculano-Houzel, S. (2016). The search for true numbers of neurons and glial cells in the human brain: A review of 150 years of cell counting. *Journal of Comparative Neurology, 524*(18), 3865–3895. doi:10.1002/cne.24040

Wandell, B. A. (1995). *Foundations of vision.* Sunderland, MA: Sinauer Associates.

Wandell, B. A., & Winawer, J. (2015). Computational neuroimaging and population receptive fields. *Trends in Cognitive Sciences, 19*(6), 349–357. doi:10.1016/j.tics.2015.03.009

Watson, A. B., & Robson, J. G. (1981). Discrimination at threshold: Labelled detectors in human vision. *Vision Research, 21*(7), 1115–1122.

Welbourne, L. E., Morland, A. B., & Wade, A. R. (2018). Population receptive field (pRF) measurements of chromatic responses in human visual cortex using fMRI. *NeuroImage, 167,* 84–94. doi:10.1016/j.neuroimage.2017.11.022

Winawer, J., Kay, K. N., Foster, B. L., Rauschecker, A. M., Parvizi, J., & Wandell, B. A. (2013). Asynchronous broadband signals are the principal source of the BOLD response in human visual cortex. *Current Biology, 23*(13), 1145–1153. doi:10.1016/j.cub.2013.05.001

Witthoft, N., Poltoratski, S., Nguyen, M., Golarai, G., Liberman, A., LaRocque, K. F., ... Grill-Spector, K. (2016). Reduced spatial integration in the ventral visual cortex underlies face recognition deficits in developmental prosopagnosia. *bioRxiv,* 051102.

Yoshor, D., Bosking, W. H., Ghose, G. M., & Maunsell, J. H. (2007). Receptive fields in human visual cortex mapped with surface electrodes. *Cerebral Cortex, 17*(10), 2293–2302. doi:10.1093/cercor/bhl138

Zhou, J., Benson, N. C., Kay, K. N., & Winawer, J. (2018). Compressive temporal summation in human visual cortex. *Journal of Neuroscience, 38*(3), 691–709. doi:10.1523/JNEUROSCI.1724-17.2017

11 Face Perception

BRUNO ROSSION AND TALIA L. RETTER

ABSTRACT Neurotypical human adults categorize faces—highly complex, socially informative visual stimuli—accurately, rapidly, and automatically, at both generic and individual levels. This defines *expertise* in individual face perception, which is domain-specific and may be restricted to neurotypical human adults. Individual face perception expertise is lost in prosopagnosia—that is, a rare category-specific deficit at face individuation following brain damage. Relying on holistic rather than analytic face perception, humans are experts for both familiar and unfamiliar faces, with repeated, meaningful, and variable exposure to an identity strengthening its unique, high-level percept. This human expertise is broadly enabled and constrained by genetics, but active experience is vital throughout development. The combination of these two factors leads to an extensive, right-lateralized ventral occipitotemporal cortical network subtending human-specific expertise at individual face perception. A full understanding of human face perception will require the inclusion of key aspects of expertise beyond accuracy—that is, processing time and automaticity—tested with a wide variety of discriminations and generalizations (e.g., fast periodic visual stimulation, a key paradigm to probe this function).

Faces emerged through evolution about 500 million years ago, as bridges between the central nervous system and the environment (McNeill, 2000). A face can be described as a collection of sensory receptors positioned together in proximity to the brain, enabling appropriate and rapid reactions to the external world. In humans, the face evolved to have importance for social functioning, conveying both quickly changeable information—for example, emotion from the expression, the direction of attention from the position of the head and gaze, and the state of health from the skin tone—and relatively invariant information, such as age, sex, race, attractiveness, and personality traits (Bruce & Young, 1998; Sheehan & Nachman, 2014; Todorov, 2017).

The face, undoubtedly, holds a special place in cognitive neuroscience research, having become probably the most widely studied stimulus (Calder, Rhodes, Johnson, & Haxby, 2011). Why? Besides being ecologically meaningful, the face is a highly frequent and familiar stimulus, especially in modern societies and the age of (social) media. Further, faces form a rich visual category, including a wide range of physically variable face exemplars (figure 11.1*A*). At the individual level, the face is also complex, containing multiple nameable features (eyes, mouth, nose …) and sources of variation in terms of shape, texture, and color. Hence, the face is an ideal stimulus to investigate the bases of perception in a modality that is dominant in the human species: vision.

At a neural level, human face perception has been associated with *specific* visual recognition processes since the proposal of a face-specific recognition deficit following brain damage, prosopagnosia (Bodamer, 1947). Lesion analysis of prosopagnosia (Meadows, 1974; Sergent & Signoret, 1992), as well as neuroimaging and intracranial recordings of human face perception (since Sergent, Ohta, & Macdonald, [1992] and Allison et al., [1994], respectively), have shown that a large, right-lateralized expanse of the cerebral cortex is involved in this function. Face perception research is therefore important for understanding the general organization of human brain function, including hemispheric lateralization.

Face Perception Emerges from Recognition

In (neuro)cognitive models of vision, perception is commonly considered to be the processing that leads to the formation of visual representations—that is, neural images of the external world. Perception is thought to precede *recognition,* when correctly formed perceptual representations are associated with representations stored in memory. Although this dissociation between perception and recognition is widely used in face perception research (e.g., the distinction between a *structural encoding stage* and *face recognition units* in the influential Bruce & Young [1986] model; see also Duchaine & Yovel, 2015; Haxby, Hoffman, & Gobbini, 2000; Jiang et al., 2006), it is not supported by empirical evidence (e.g., by a clear dissociation between perceptual and mnesic forms of prosopagnosia; Farah, 1990/2004; Gainotti, 2013) and causes a series of difficulties, such as understanding why *perceptual tasks* (e.g., the simultaneous matching of individual face images) are performed better with familiar than unfamiliar faces (Bruce et al., 2001).

Most importantly, it causes a fracture between *different levels* of face perception: discerning that a visual stimulus

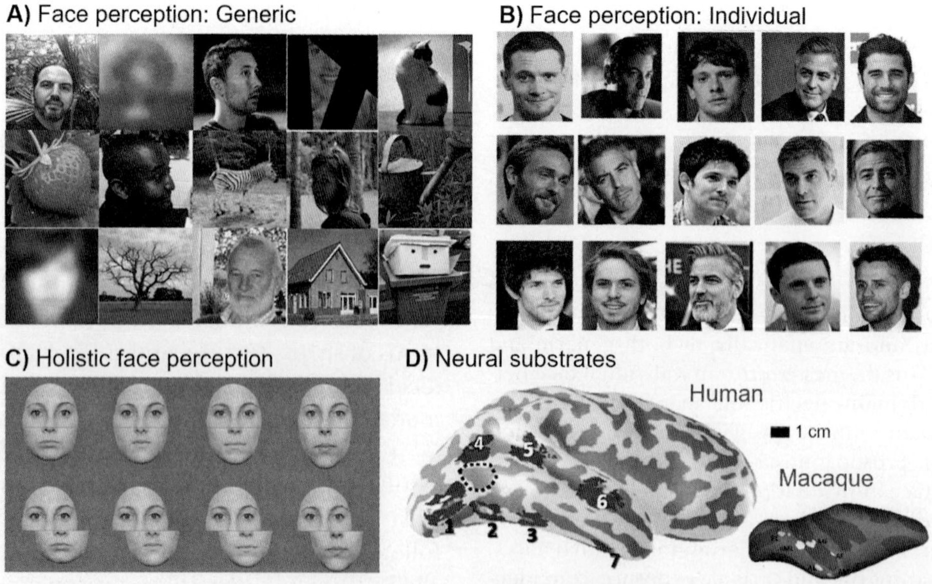

FIGURE 11.1 *A*, The human brain perceives stimuli as faces under a wide variety of viewing conditions, sometimes with little or no shared features. Understanding the nature of such generic face categorization (i.e., discrimination from nonface objects and generalization across variable face exemplars) requires going beyond sensory, quantifiable, low-level information from the visual stimulus and incorporating memory of the category in the process. *B*, Individual face perception can be understood as a more specific function than generic face perception but similarly requiring categorization: discrimination—here from various unfamiliar faces—and generalization across variable viewing conditions. *C*, The composite face illusion (adapted from Rossion, 2013) illustrates how face perception goes beyond the given sensory information, taking advantage of matches to holistic facial memories derived through experience. All four top halves (*above the narrow gap*) are physically identical. Yet aligned (*top row*; compare with misaligned below) with distinct bottom halves (all of different identities), they are perceived as being different. *D*, Inflated segmented brains showing typical locations of face-selective regions in humans (from Weiner & Grill-Spector, 2015) and macaques (from Tsao, Moeller, & Freiwald, 2008), plotted to a common scale. Up to six regions have been labeled in the macaque STS, whereas humans have regions in both the STS (labeled *4*, *5*, and *6*) and the VOTC (*1*, *2*, *3*, *7*). (See color plate 11.)

is a face is considered a *perceptual* task (face detection), while identifying somebody's face as that of a specific individual is considered a *recognition* task (figure 11.1*A* and 11.1*B*). These two levels are often taken to be neurofunctionally distinct and even to have opposing demands: that face detection requires extracting "what is common to faces," while individual recognition requires extracting "differences between faces" (Tsao & Livingstone, 2008). However, being able to tell that a face is a face (generic face perception) requires both *discrimination* (i.e., providing a selective response to faces) and *generalization* (i.e., providing the same response to different faces). In the same vein, recognizing someone's identity (individual face perception) requires both *discrimination* (from other faces) and *generalization* (across various views of the same face). Therefore, in a coherent theoretical framework, a facial identity should be considered as a category (Burton et al., 2016; figure 11.1*B*).

In this theoretical framework, perception arises through matching low-level (i.e., sensory, unconscious) information to high-level (categorical) representations in memory. This matching corresponds to *categorization*—that is, *recognition*—of the visual environment. In this view, *perception* does not precede recognition but is the *subjective experience that emerges from this process.* We consider this framework essential for understanding the nature of human face perception and its neural basis.

Human Expertise

Behavioral, eye movement, and electrophysiological studies in humans have shown that widely variable natural stimuli are perceived as faces extremely accurately, rapidly (shortly after 100 ms), and automatically (without intention to do so; Crouzet, Kirchner, & Thorpe; Hershler et al., 2010; Rossion, 2014a). These characteristics, applied across a wide range of discriminations and generalizations, define the *expertise* of neurotypical human adults at generic face perception. However, human's level of expertise is arguably best expressed when considering the challenge of *individual* face perception.

Every human face is unique in its morphological characteristics, and modern human societies are characterized by the presence of numerous familiar and unfamiliar individuals, augmented with (social) media. Since individual faces are similar in their overall visual configuration, the human visual system is formidably challenged to successfully individuate people by their faces. Nevertheless, a combination of this social (over) exposure and biological factors, as discussed below, allow people to recognize familiar individuals from their faces with astounding accuracy, even across changes of viewpoint and expression or under degraded conditions (occlusion, distortion, blurring: V. Bruce, 1982; Burton et al., 1999; Hole et al., 2002; Sinha, 2002). Familiar faces can also be discriminated from unfamiliar faces within a few hundred milliseconds (Caharel, Ramon, & Rossion, 2014; Visconti di Oleggio Catsello & Gobbini, 2015). Additionally, humans naturally (i.e., without formal training) recognize and encode a seemingly unlimited number of individual faces in memory, for short or long periods of time, throughout their lives (Bahrick, Bahrick, & Wittlinger, 1975).

In sum, although the capacity of the human brain to individuate faces must be better evaluated and documented by systematic quantitative research (see "Summary and Perspectives"), these observations suggest that neurotypical human adults can indeed be considered *experts* at individual face perception (Carey, 1992; McKone, Kanwisher, & Duchaine, 2007; Rossion, 2018b; Tanaka, 2001; Young & Burton, 2018). To illustrate this level of expertise, we compare neurotypical adults' performance to two kinds of nonexperts, other animal species and patients with prosopagnosia.

Nonhuman species Behavioral studies have shown that nonhuman species, such as sheep (Peirce, Leigh, & Kendrick, 2001), fish (Wang & Takeuchi, 2017), or even wasps (Sheehan & Tibbetts, 2011), can discriminate pictures of (conspecific) individual faces, drawing parallels to human performance. However, the face individuation performance of nonhuman species often follows long operant-conditioning procedures and, despite a small number of individual exemplars, remains relatively modest (Rossion & Taubert, 2019). Moreover, successful performance at such tasks appears to be based on low-level image characteristics (with limited across-image generalization) and/or external traits (e.g., antennae for wasps or ears for sheep; see respective references above).

Importantly, this limited ability also holds for the species often considered the best model for the human visual system: the macaque monkey. Macaques, as well as other monkey species (Pokorny & de Waal, 2009), are in fact quite poor at behavioral face individuation tasks (Parr, Heintz, & Pradhan, 2008; Rossion & Taubert, 2019). Moreover, monkeys do not rely on qualitatively similar processes as humans. That is, they fail to show a decrease of performance with stimulus (picture-plane) inversion (C. Bruce, 1982; Rossion & Taubert, 2019), one of the most compelling effects in human studies (Rossion, 2008), and rely primarily on external features (Parr, Heintz, & Pradhan, 2008). These observations raise questions as to the appropriateness of the macaque as a model for the neural basis of human (individual) face perception (Rossion & Taubert, 2019). More generally, humans may be the only species to naturally use the face as the primary source of information for individuation (as opposed to smell, vocalization, body appearance, and so on; Sheehan & Nachman, 2014), reaching an expertise at this function that is unparalleled in the animal world.

Prosopagnosia: a category-specific individuation impairment Another nonexpert system is demonstrated by prosopagnosia, a rare inability to individuate faces following brain damage in a neurotypically developed face perception system (not to be confused with *prosopdysgnosia* or *developmental/congenital prosopagnosia*, which is a difficulty individuating faces of developmental origin; Behrmann & Avidan, 2005; Duchaine & Nakayama, 2006; McConachie, 1976). Prosopagnosia is a massive impairment in recognizing both faces encountered before and after brain damage (i.e., it is both a retrograde and anterograde recognition deficit), which cannot be accounted for by low-level sensory deficits and/or intellectual deficiencies. Most importantly, it is a *category-specific* individual recognition disorder: cases of prosopagnosia are able to recognize nonface objects correctly and rapidly (Bodamer, 1947; Busigny et al., 2010; Rossion, 2018a).

Prosopagnosia appears to be caused by a failure to match sensory inputs to damaged, high-level neural regions responding selectively to faces (see "Neurofunctional Substrates"). Prosopagnosia is therefore a (visual) *recognition* disorder, which necessarily implies impairment at face *perception*. This is why, despite decades of research, there is no evidence for a clear-cut distinction between perceptual (apperceptive) and mnesic (associative) forms of prosopagnosia (Farah, 1990/2004; Gainotti, 2013).

Interestingly, despite their impaired individual face perception, patients with prosopagnosia are generally capable of generic face perception, which does not require detailed internal representations (Liu-Shuang, Torfs, & Rossion, 2016). They can also discriminate pictures of unfamiliar faces well above the chance level (e.g., Busigny et al., 2010; Sergent & Signoret, 1992) and even learn to individuate face pictures (Sergent & Signoret, 1992); in fact, they perform better than nonhuman species at such tasks. However, unlike neurotypical human adults, their performance is slow, effortful, and limited in generalization. This shows the importance of going beyond a "can/can't do" criterion for measuring observers' levels of expertise at (individual) face perception.

Holistic Face Perception

How are faces perceived? According to an *analytic view*, the face percept is built from the perception of independent face parts, such as the eyes, nose, mouth, and so on. The definition of parts is problematic, as it can be performed in a myriad of ways: feature-based (e.g., by cutting out rectangles around the eyes, the nose, and the mouth), halves (top/bottom or left/right), surface/shape, Gabor jets, or apertures or "bubbles" in response classification techniques and more. Although these parts are thought to be ultimately integrated into a global face percept, this face percept is thought to be equivalent to the sum of its parts (Gold, Mundy, & Tjan, 2012) so that the parts' respective contributions can be studied and defined independently.

However, visual illusions such as the composite face illusion, in which identical top face halves are aligned with differing bottom face halves, produce the impression of different top face halves (figure 11.1*C*).

Such illusions therefore provide compelling evidence that facial parts are mandatorily perceived interactively or interdependently of one another—that is, *holistically* (Rossion, 2013; Tanaka & Farah, 1993; Young, Hellawell, & Hay, 1987). The importance of holistic face perception is supported by its disruption following stimulus inversion (Tanaka & Farah, 1993; Young, Hellawell, & Hay, 1987) and in cases of prosopagnosia (Rossion, 2014b; Van Belle et al., 2010). Despite such observations, holistic perception is not fully embraced in scientific research, probably because it is very difficult to understand in neurocomputational terms (Spillman, 1999; but see Biederman & Kalocsai, 1997; Tan & Poggio, 2016).

Perception of Familiar and Unfamiliar Faces

Individual faces encoded in (long-term) memory—that is, familiar faces—are more easily individuated than unfamiliar faces, both in old/new recognition (V. Bruce, 1982) and simultaneous matching tasks (Bruce et al., 2001). Thus, although pictures of unfamiliar faces offer experimental advantages in terms of stimulus control, findings from these studies need to be carefully interpreted and may not fully extend to familiar face perception. Nevertheless, behavioral experiments show that without any task training, typical human adults remain very good at simultaneous or delayed unfamiliar face–matching tasks, even in difficult tasks requiring high levels of generalization across substantial changes in size, head orientation, and lighting conditions with many similar-looking distractors (Megreya & Burton, 2006; Rossion & Michel, 2018). Moreover, the perception of unfamiliar facial identities is significantly affected by picture-plane inversion (Rossion, 2008), against the claim of reliance on low-level information for unfamiliar face perception (Megreya & Burton, 2006). The composite face illusion is also best illustrated with pictures of unfamiliar faces (figure 11.1*C*). In addition, the "other-race" face effect, reflecting a level of expertise specific to experienced facial morphologies, is readily demonstrated with unfamiliar faces (Rossion & Michel, 2011). Electrophysiological studies also show that discrimination of individual unfamiliar faces is fast and automatic across changes—for example, in size and head orientation (Caharel, d'Arripe, & Ramon, 2009; Liu-Shuang, Norcia, & Rossion, 2014).

Thus, contrary to a recently expressed view limiting expertise to familiar faces (Young & Burton, 2018), we posit that neurotypical human adults are experts even with unfamiliar faces. However, an initial individual face percept is labile and will be registered as a distinct (dynamic) category in the brain only when it is seen repeatedly, variably, and meaningfully and is thus associated with contextual, affective, and semantic information in the brain of the observer. This exposure will also greatly expand the range of generalization ability for individuation of that face. Interestingly, the resulting advantage in individuation tasks for familiar over unfamiliar faces is found neither in patients with prosopagnosia (for obvious reasons) nor in macaques (Micheletta et al., 2015; Rossion & Taubert, 2019), so it may be considered an additional marker of human expertise at individual face perception.

Genetics and Experience throughout Development

To what extent is human face perception genetically predetermined, and to what extent does it arise due to visual experience? The complex interplay between these factors prevents them from being neatly disentangled, especially in the case of this naturally developed expertise in the human species. For example, developmental difficulties at individual face recognition may run in families (Susilo & Duchaine, 2013), and monozygotic twins' performance at face recognition tests are more strongly correlated than dizygotic twins' performance

(Wilmer et al., 2010), leading to claims that genetics largely determines face recognition. However, genetics could affect both specific and general functions essential for learning to individuate faces throughout development. Thus, while genetics certainly influences the neural development (from epigenetics through nervous system development to cytoarchitectonic and functional neural organization) required for face recognition, experience appears to be necessary or instrumental throughout the developmental time course.

The extent to which the visually unexperienced human newborn's brain exhibits selective responses to faces is debated (Johnson et al., 1991; Simion et al., 2003). Perhaps supported by an innate bias to look at the general structural characteristics present in faces (e.g., "top-heavy patterns": Simion et al., 2003), infants receive an enormous amount of exposure to faces (Sugden, Mohamed-Ali, & Moulson, 2014) and look preferentially at face stimuli per se after a few months of age (Chien, 2011). Four-to-six-month-old infants already show a specific neural response to variable natural images of faces compared to objects, even for single-glanced faces (de Heering & Rossion, 2015). Experience also leads to the *fine-tuning* of individual facial discrimination abilities; at nine months of age, infants show lower performance at discriminating nonexperienced than frequently experienced faces (e.g., other-species faces: Pascalis, de Haan, & Nelson, 2002; "other-race" human faces: Kelly et al., 2007). Other studies have illustrated that the type of experience that infants have with faces matters: active experience increases performance more than passive experience (Scott & Monesson, 2010). Nevertheless, even after nine years of age a predominant exposure to "other-race" faces can reverse such an early selection in discrimination ability (Sangrigoli & de Schonen, 2004). Thus, experience as well as genetics appears to be necessary and influential.

Overall, these studies show that the early encounter between a biological system and the environment *selects* a specific relevant type of facial morphology for perceptual *learning* (rather than *narrowing*). Indeed, individual face perception continues to develop until adulthood, but whether this is due to the progressive maturation of face perception specifically or is explained by more general cognitive development is debated (Crookes & McKone, 2009). Neuroimaging investigations show that the topology of face-selective regions in the posterior ventral occipitotemporal cortex (VOTC) is mature by age seven, but the spatial extent and degree of face selectivity continue to develop into adulthood, supporting a selective development of the function (Golarai, Liberman, & Grill-Spector, 2017).

To summarize, although the notion of expertise has been around for decades in the field of face perception (Carey, 1992; Diamond & Carey, 1986), it is generally conflated with the view that expertise arises from visual experience without genetic constraints and that it could be developed in a similar way for various kinds of nonface stimuli learned during development or even in adulthood (Diamond & Carey, 1986; Gauthier & Tarr, 1997). In this vein, "acquired" domain-general expertise has often been radically opposed to "innate" domain-specific expertise (McKone, Kanwisher, & Duchaine, 2007). Instead, current evidence suggests that neurotypical human adults possess a *domain-specific expertise* in (individual) face perception, emerging from both genetic factors and experience.

Neurofunctional Substrates

The neural bases of human individual face perception were first hinted at by patients presenting with prosopagnosia after lesions to the VOTC—in particular, the right fusiform gyrus (Meadows, 1974; Sergent & Signoret, 1992). Since the mid-1990s, functional magnetic resonance imaging (fMRI) studies have identified clusters of voxels (of a few cubic millimeters) that respond significantly more to pictures of faces than nonface objects, especially along the VOTC (in particular, the lateral fusiform gyrus) and the superior temporal sulcus, or STS (e.g., for reviews, see Duchaine & Yovel, 2015; Grill-Spector et al., 2017; Haxby, Hoffman, & Gobbini, 2000; figure 11.1*D*). More broadly distributed face-selective neural populations have been suggested from direct, invasive intracranial electroencephalographic recordings performed for clinical purposes (Allison et al., 1994). However, spatial clustering of face-selective responses, with a peak of activity in the right lateral fusiform gyrus, have also been reported in electrocorticography (Jacques et al., 2016) as well as with depth (i.e., intracerebral) electrodes with a frequency-tagging approach (for a review, see Jonas et al., 2016; Rossion et al., 2018; figure 11.2*D*).

While the human VOTC network has been associated with invariant aspects of faces—in particular, individual face perception—the more dorsal STS network is thought to be critical for the perception of dynamic aspects of faces, such as eye-gaze direction, head orientation, and facial expression (as in monkeys: Heywood & Cowey, 1992; humans: Allison, Puce, & McCarthy, 2000). In humans, more recent models have emphasized the dissociation between form and motion as the primary functional division between the ventral system along the VOTC and the dorsal system that goes through the STS, respectively (Bernstein & Yovel, 2015).

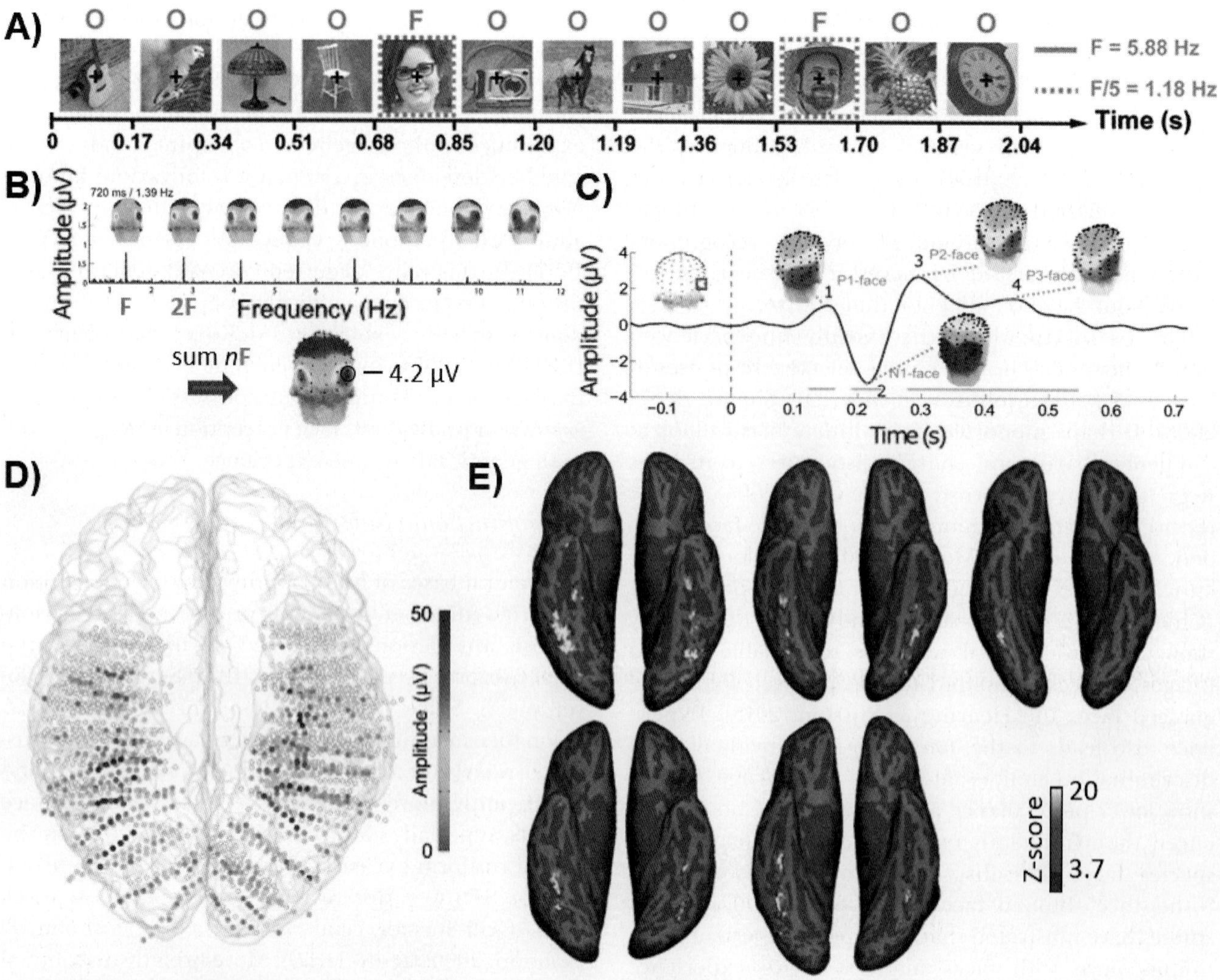

FIGURE 11.2 *A*, A fast periodic visual stimulation (FPVS) categorization paradigm (adapted from Rossion et al., 2015): stimuli are presented at a fast rate, 5.88 images per second (Hz), with faces appearing here as every fifth stimulus—that is, at 5.88 Hz / 5 = 1.18 Hz. Selective responses to faces are thus "tagged" at 1.18 Hz, under severe time constraints (one fixation per image), automatically (i.e., without an explicit perception task), and with variable stimuli ensuring ecological validity. *B*, Scalp electroencephalogram responses recorded in the amplitude spectrum at the presentation rate of faces (*F*) and its harmonics—that is, *2F*, *3F*, and so on, plotted from an occipitotemporal region indicated below. To quantify the response, these harmonics are baseline corrected and summed (Retter & Rossion, 2016). *C*, These selective responses to faces can also be represented in the time domain. This reveals at least four face-selective deflections, onsetting after about 100 ms and persisting for over 400 ms. *D*, Intracerebral depth electrode contacts' face-selective amplitudes are shown from a large sample of participants (*N* = 28; from Jonas et al., 2016). *E*, The ventral surfaces of individual human brains showing the wide variability, across five individuals, in their face-selective fMRI activation obtained with the same type of paradigm (data unsmoothed; Gao, Gentile, & Rossion, 2018). While these differences in terms of the spatial extent and number of clusters may be due to individual anatomy interacting with fMRI signal capture, they may reflect genuine (behaviorally relevant) interindividual differences.

In terms of time course, event-related potential (ERP) studies have reported several face-selective electrophysiological components, the most prominent and well-studied of which is undoubtedly the N170 (Bentin et al., 1996), a marker of conscious (individual) face perception (Rossion, 2014a). However, more recent studies isolating face-selective responses in a rapid train of nonface stimuli have shown that electrophysiological face-selective responses devoid of low-level visual confounds emerge about 100 ms after stimulus onset and persist for over 400 ms (Retter & Rossion, 2016; figure 11.2).

A hallmark of human face perception is that the right hemisphere is more functionally active than the left; again, this has been evidenced by both direct (Bentin et al., 1996; Jonas et al., 2016; Rossion, 2014a;) and indirect (neuroimaging) neural measurements (Kanwisher, McDermott, & Chun, 1997; Sergent, Ohta, & Macdonald, 1992), as well as by behavioral studies with divided visual field presentations in split-brain or normal observers (Parkin & Williamson, 1987) and with transcranial magnetic stimulation (Pitcher et al., 2007). Importantly, a right hemisphere lesion is both necessary and sufficient to cause prosopagnosia in right-handed individuals (Sergent & Signoret, 1992). The factors causing this right lateralization, which is specific to the human species (e.g., it is absent in macaques) remain largely unknown. While it has been proposed that this is due to competition from visual language acquisition (reading) taking place primarily in the left VOTC (Behrmann & Plaut, 2015), generic face perception is right lateralized as early as 4–6 months of age, well before reading acquisition (de Heering & Rossion, 2015), casting doubt on this view.

Neurofunctional models of face perception have been generated with modular roles attributed to clusters of voxels in the VOTC and STS, referred to as *face-selective areas* (Duchaine & Yovel, 2015; Haxby, Hoffman, & Gobbini, 2000). In the original models, such clusters were thought to be hierarchically spatially organized, with perceptual and mnesic information ordered along a posterior to anterior axis along VOTC areas (Haxby, Hoffman, & Gobbini, 2000). However, neuroimaging studies of brain-damaged patients with prosopagnosia (Rossion et al., 2003) and time-resolved fMRI (Jiang et al., 2011) have instead shown a nonhierarchical organization, forcing the revision of these models (Duchaine & Yovel, 2015). Studies using fMRI with a higher spatial resolution have dissociated the face-selective cluster in the fusiform gyrus ("Fusiform Face Area" [FFA]: Kanwisher, McDermott, & Chun, 1997) into several subclusters, such as the pFus-faces/FFA-1 and mFus-faces/FFA-2, and have linked them to cytoarchitectonically distinct regions (Grill-Spector et al., 2017; Weiner & Grill-Spector, 2012). However, in general there is great interindividual variability in the number of spatially dissociable face-selective regions and their extent throughout the cortex (Gao, Gentile, & Rossion, 2018; figure 11.2*E*). Whether this interindividual variability in the number and exact localization of face-selective regions is due to the measurement noise of functionally defined regions that are generalizable anatomically across people or is genuine (and potentially related to behavioral differences in face perception) remains unknown.

Recent years have seen a rise in fMRI studies attempting to "decode" the representations of (unfamiliar) individual faces in the cortical face network in order to localize and define the nature of individual face representations. However, the results of these studies are often inconsistent and describe relatively weak effects potentially driven by image-based cues (Guntupalli, Wheeler, & Gobbini, 2017; Kriegeskorte et al., 2007; Nestor, Plaut, & Behrmann, 2011). In contrast, fMR-adaptation studies (Grill-Spector & Malach, 2001), which are based on reduced fMRI signal to repeated facial identities as compared to different stimuli, have measured more consistent changes of activity during individual face discrimination in face-selective VOTC regions (Davies-Thompson, Gouws, & Andrews, 2009; Gauthier et al., 2000). Their findings are generally in agreement with studies of prosopagnosia and electrical brain stimulation in these regions (Jonas et al., 2012; Parvizi et al., 2012; Rossion, 2014b).

In line with the present theoretical framework, we suggest that a number of tightly interconnected face-selective clusters in the VOTC holds memories of (individual) faces, carved from the lifetime brain-environment interaction. Specific activity in these regions, triggered from low-level sensory inputs, leads to both generic and individual face recognition/perception, with the former temporally preceding the latter.

Neuroimaging studies show inconsistent differences in the amplitude (or pattern) of responses to familiar and unfamiliar faces across VOTC face-selective regions (Gobbini & Haxby, 2007; Natu & O'Toole, 2011; Visconti di Oleggio Castello et al., 2017), which is surprising given the role of these regions in causing prosopagnosia (Jonas et al., 2012; Rossion, 2014b; Sergent & Signoret, 1992). This could be due to contrasting effects in neuroimaging (e.g., larger fMR-adaptation effects for repeated familiar than unfamiliar faces, compensated by much richer associations to semantic information for familiar faces). It is also likely that familiarity effects will be revealed at smaller levels of organization (e.g., the strengthening of connections among groups of neurons, such as cortical columns or minicolumns) as well as shorter time scales in future studies, providing neural mechanisms for individual recognition of familiar faces.

Summary and Perspectives

In this chapter, we proposed a theoretical framework considering the perception of faces to *arise* from recognition, that is, from matching sensory inputs with high-level, face-selective, holistic representations, mainly in the VOTC. Moreover, we advocate reintegrating the notion of neurotypical human adult's *expertise* in the study of face perception, as a domain-specific expertise arising from both genetic and experiential factors.

This expertise in individual face perception—even for unfamiliar faces—is striking in comparison with non-expert populations, that is, other animal species and patients with prosopagnosia; additional non-expert populations could include infants, children and, arguably, artificial systems (if they are probed with valid measures of face recognition; see O'Toole et al., 2018; Philips & O'Toole, 2014; Scheirer et al., 2014). Importantly, such an emphasis on expertise highlights neurotypical human adults' extraordinary ability at (individual) face perception and encourages research into this complex function's origins and substrates.

We suggest operationalizing expertise not only in terms of accuracy, but also processing time and automaticity, across variable discriminations/generalizations of face stimuli. To this end, face perception could be measured with fast periodic visual stimulation, or FPVS (Retter & Rossion, 2016; Rossion, 2014a). In a recent FPVS categorization paradigm, a within-category stimulus (e.g., faces) appears as a "frequency-tagged" proportion of periodically-presented across-category stimuli (e.g., nonface objects; Rossion et al., 2015; figure 11.2*A*). Importantly for capturing *categorization* responses, that is, *recognition* responses in our theoretical framework, this approach can readily incorporate a large amount of variable discriminations and generalizations, with ecologically valid natural stimuli (see figure 11.2*A*). Automaticity is required when the process of interest is measured implicitly, e.g., while naïve participants perform an orthogonal task. Additionally, when stimuli are presented in succession at relatively fast rates (e.g., about 6 Hz in figure 11.2*A*), processing time of each stimulus is restricted. Moreover, FPVS can be advantageous applied (i.e., with high sensitivity and objectivity) to a variety of recording techniques, including human (intracerebral) electrophysiology, and fMRI (figure 11.2*B–D*). We thus propose that FPVS will allow the isolation of high-level processes from low-level sensory biases, while incorporating the key notions of accuracy, processing time, and automaticity, that define human face perception expertise.

REFERENCES

Allison, T., Ginter, H., McCarthy, G., Nobre, A. C., Puce, A., Luby, M., & Spencer, D. D. (1994). Face recognition in human extrastriate cortex. *Journal of Neurophysiology*, 71, 821–825.

Allison, T., Puce, A., & McCarthy, G. (2000). Social perception from visual cues: Role of the STS region. *Trends in Cognitive Sciences*, 4, 267–278.

Bahrick, H. P., Bahrick, P. O., & Wittlinger, R. P. (1975). Fifty years of memory for names and faces: A cross-sectional approach. *Journal of Experimental Psychology, General*, 104, 54–75.

Behrmann, M., & Avidan, G. (2005). Congenital prosopagnosia: Face-blind from birth. *Trends in Cognitive Sciences*, 9, 180–187.

Behrmann, M., & Plaut, D. C. (2015). A vision of graded hemispheric specialization. *Annals of the New York Academy of Sciences*, 1359, 30–46.

Bentin, S., Allison, T., Puce, A., Perez, E., & McCarthy, G. (1996). Electrophysiological studies of face perception in humans. *Journal of Cognitive Neuroscience*, 8, 551–565.

Bernstein, M., & Yovel, G. (2015). Two neural pathways of face processing: A critical evaluation of current models. *Neuroscience & Biobehavioral Reviews*, 55, 536–546.

Biederman, I., & Kalocsai, P. (1997). Neurocomputational bases of object and face recognition. *Philosophical Transactions of the Royal Society of London. Series B, Biological Sciences*, 352, 1203–1219.

Bodamer, J. (1947). Die Prosopagnosie. *Archiv fur Psychiatrie und Zeitschrift fur Neurologie*, 179, 6–54.

Bruce, C. (1982). Face recognition by monkeys: Absence of an inversion effect. *Neuropsychologia*, 20, 515–521.

Bruce, V. (1982). Changing faces: Visual and non-visual coding processes in face recognition. *British Journal of Psychology*, 73, 105–116.

Bruce, V., Henderson, Z., Newman, C., & Burton, A. M. (2001). Matching identities of familiar and unfamiliar faces caught on CCTV images. *Journal of Experimental Psychology: Applied*, 7, 207–218.

Bruce, V., & Young, A. (1986). Understanding face recognition. *British Journal of Psychology*, 77, 305–327.

Bruce, V., & Young, A. (1998). *In the eye of the beholder: The science of face perception*. New York: Oxford University Press.

Burton, A. M., Kramer, R. S., Ritchie, K. L., & Jenkins, R. (2016). Identity from variation: Representations of faces derived from multiple instances. *Cognitive Science*, 40, 202–223.

Burton, A. M., Wilson, S., Cowan, W., & Bruce, V. (1999). Face recognition in poor-quality video: Evidence from security surveillance. *Psychological Science*, 10, 243–248.

Busigny, T., Graf, M., Mayer, E., & Rossion, B. (2010). Acquired prosopagnosia as a face-specific disorder: Ruling out the general visual similarity account. *Neuropsychologia*, 48, 2051–2067.

Caharel, S., d'Arripe, O., Ramon, M., Jacques, C., & Rossion, B. (2009). Early adaptation to unfamiliar faces across viewpoint changes in the right hemisphere: Evidence from the N170 ERP component. *Neuropsychologia*, 47, 639–643.

Caharel, S., Ramon, M., & Rossion, B. (2014). Face familiarity decisions take 200 ms in the human brain: Electrophysiological evidence. *Journal of Cognitive Neuroscience*, 26, 81–95.

Calder, A. J., Rhodes, G., Johnson, M. H., & Haxby, J. V. (Eds.). (2011). *The Oxford handbook of face perception*. Oxford: Oxford University Press, 2011.

Carey, S. (1992). Becoming a face expert. *Philosophical Transactions of the Royal Society of London. Series B, Biological Sciences*, 335, 95–103.

Chien, S. H. (2011). No more top-heavy bias: Infants and adults prefer upright faces but not top-heavy geometric or face-like patterns. *Journal of Vision*, 11(6).

Crookes, K., & McKone, E. (2009). Early maturity of face recognition: No childhood development of holistic processing, novel face encoding, or face-space. *Cognition*, 111, 219–247.

Crouzet, S. M., Kirchner, H., & Thorpe, S. J. (2010). Fast saccades toward faces: Face detection in just 100 ms. *Journal of Vision*, 16, 1–17.

Davies-Thompson, J., Gouws, A., & Andrews, T. J. (2009). An image-dependent representation of familiar and unfamiliar faces in the human ventral stream. *Neuropsychologia*, 47, 1627–1635.

De Heering, A., & Rossion, B. (2015). Rapid categorization of faces in the infant right hemisphere. *eLife*, 4, e06564.

Diamond, R., & Carey, S. (1986). Why faces are and are not special: An effect of expertise. *Journal of Experimental Psychology, General*, 115, 107–117.

Duchaine, B. C., & Nakayama, K. (2006). Developmental prosopagnosia: A window to content-specific face processing. *Current Opinion in Neurobiology*, 16, 166–173.

Duchaine, B., & Yovel, G. (2015). A revised neural framework for face processing. *Annual Review of Vision Science*, 1, 293–416.

Farah, M. J. (2004). *Visual agnosia: Disorders of object recognition and what they tell us about normal vision*. Cambridge, MA: MIT Press. (Original work published in 1990).

Gainotti, G. (2013). Is the right anterior temporal variant of prosopagnosia a form of "associative prosopagnosia" or a form of "multimodal person recognition disorder"? *Neuropsychology Review*, 23, 99–110.

Gao, X., Gentile, F., & Rossion, B. (2018). Fast periodic stimulation (FPS): A highly effective approach in fMRI rain mapping. *Brain Structure and Function*, 223, 2433–2454.

Gauthier, I., & Tarr, M. J. (1997). Becoming a "Greeble" expert: Exploring mechanisms for face recognition. *Vision Research*, 37, 1673–1682.

Gauthier, I., Tarr, M. J., Moylan, J., Skudlarski, P., Gore, J. C., & Anderson, A. W. (2000). The FFA is part of a network that processes faces on an individual level. *Journal of Cognitive Neuroscience*, 12, 495–504.

Gobbini, M. I., & Haxby, J. V. (2007). Neural systems for recognition of familiar faces. *Neuropsychologia*, 45, 32–41.

Golarai, G., Liberman, A., & Grill-Spector, K. (2017). Experience shapes the development of neural substrates of face processing in human ventral temporal cortex. *Cerebral Cortex*, 27, 1229–1244.

Gold, J. M., Mundy, P. J., & Tjan, B. S. (2012) The perception of a face is no more than the sum of its parts. *Psychological Science*, 23, 427–434.

Grill-Spector, K., & Malach, R. (2001). FMR-adaptation: A tool for studying the functional properties of human cortical neurons. *Acta Psychologica*, 107, 293–321.

Grill-Spector, K., Weiner, K. S., Kay, K., & Gomez, J. (2017). The functional neuroanatomy of human face perception. *Annual Review of Vision Science*, 3, 167–196.

Guntupalli, J. S., Wheeler, K. G., & Gobbini, M. I. (2017). Disentangling the representation of identity from head view along the human face processing pathway. *Cerebral Cortex*, 27, 46–53.

Haxby, J. V., Hoffman, E. A., & Gobbini, M. I. (2000). The distributed human neural system for face perception. *Trends in Cognitive Sciences*, 4, 223–233.

Hershler, O., Golan, T., Bentin, S., & Hochstein, S. (2010). The wide window of face detection. *Journal of Vision*, 10, 21.

Heywood, C. A., & Cowey, A. (1992). The role of the "face-cell" area in the discrimination and recognition of faces by monkeys. *Philosophical Transactions of the Royal Society of London B*, 335, 31–37.

Hole, G. J., George, P. A., Eaves, K., & Rasek, A. (2002). Effects of geometric distortions on face-recognition performance. *Perception*, 31, 1221–1240.

Jacques, C., Witthoft, N., Weiner, K. S., Foster, B. L., Rangarajan, V., Hermes, D., Miller, K. J., Parvizi, J., & Grill-Spector, K. (2016). Corresponding ECoG and fMRI category-selective signals in human ventral temporal cortex. *Neuropsychologia*, 83, 14–28.

Jiang, F., Dricot, L., Weber, J., Righi, G., Tarr, M. J., Goebel, R., & Rossion, B. (2011). Face categorization in visual scenes may start in a higher order area of the right fusiform gyrus: Evidence from dynamic visual stimulation in neuroimaging. *Journal of Neurophysiology*, 106, 2720–2736.

Jiang, X., Rosen, E., Zeffiro, T., Vanmeter, J., Blanz, V., & Riesenhuber, M. (2006). Evaluation of a shape-based model of human face discrimination using FMRI and behavioral techniques. *Neuron*, 50, 159–172.

Johnson, M. H., Dziurawiec, S., Ellis, H., & Morton, J. (1991). Newborns' preferential tracking of face-like stimuli and its subsequent decline. *Cognition*, 40(1–2), 1–19.

Jonas, J., Descoins, M., Koessler, L., Colnat-Coulbois, S., Sauvee, M., Guye, M., Vignal, J-P., Vespignani, H., Rossion, B., & Maillard, L. (2012). Focal electrical intracerebral stimulation of a face-sensitive area causes transient prosopagnosia. *Neuroscience*, 222, 281–288.

Jonas J., C. Jacques, J. Liu-Shuang, Brissart, H., Colnat-Coulbois, S., Maillard, L., & Rossion, B. (2016). A face-selective ventral occipito-temporal map of the human brain with intracerebral potentials. *Proceedings of the National Academy of Sciences of the United States of America*, 113, E4088–E4097.

Kanwisher, N., McDermott, J. & Chun, M. (1997). The fusiform face area: A module in human extrastriate cortex specialized for face perception. *Journal of Neuroscience*, 17, 4302–4311.

Kelly, D. J., Liu, S., Ge, L., Quinn, P. C., Salater, A. M., Lee, K., Liu, Q., & Pascalis, O. (2007). Cross-race preferences for same-race faces extend beyond the African versus Caucasian contrast in 3-month-old infants. *Infancy*, 11(1), 87–95.

Kriegeskorte, N., Formisano, E., Sorger, B., & Goebel, R. (2007). Individual faces elicit distinct response patterns in human anterior temporal cortex. *Proceedings of the National Academy of Sciences of the United States of America*, 104(51), 20600–20605.

Liu-Shuang, J., Norcia, A. M., & Rossion, B. (2014). An objective index of individual face discrimination in the right occipito-temporal cortex by means of fast periodic oddball stimulation. *Neuropsychologia*, 52, 57–72.

Liu-Shuang, J., Torfs, K., & Rossion, B. (2016). An objective electrophysiological marker of face individualisation impairment in acquired prosopagnosia with fast periodic visual stimulation. *Neuropsychologia*, 83, 100–113.

McConachie, H. R. (1976). Developmental prosopagnosia. A single case report. *Cortex*, 12, 76–82.

McKone, E., Kanwisher, N., & Duchaine, B. C. (2007). Can generic expertise explain special processing for faces? *Trends in Cognitive Sciences*, 11, 8–15.

McNeill, D. (2000). *The Face: A Natural History*. Boston: Little, Brown and Company.

Meadows, J. C. (1974). The anatomical basis of prosopagnosia. *Journal of Neurology, Neurosurgery, and Psychiatry*, 37, 489–501.

Megreya, A. M., & Burton, A. M. (2006). Unfamiliar faces are not faces: Evidence from a matching task. *Memory & Cognition*, 34, 865–876.

Micheletta, J., Whitehouse, J., Parr, L. A., Marshman, P., Engelhardt, A., & Waller, B. M. (2015). Familiar and unfamiliar face recognition in crested macaques (Macaca nigra). *Royal Society Open Science*, 2(5), 150109.

Natu, V., & O'Toole, A. J. (2011). The neural processing of familiar and unfamiliar faces: A review and synopsis. *British Journal of Psychology*, 102, 726–747.

Nestor, A., Plaut, D. C., & Behrmann, M. (2011). Unraveling the distributed neural code of facial identity through spatiotemporal pattern analysis. *Proceedings of the National Academy of Sciences of the United States of America*, 108(24), 9998–10003.

O'Toole, A. J., Castillo, C. D., Parde, C. J., Hill, M. Q., & Chelappa, R. (2018). Face space representations in deep convolutional neural networks. *Trends in Cognitive Sciences*, 22, 794–809.

Parkin, A. J., & Williamson, P. (1987). Cerebral lateralisation at different stages of facial processing. *Cortex*, 23, 99–110.

Parr, L. A., Heintz, M., & Pradhan, G. (2008). Rhesus monkeys (Macaca mulatta) lack expertise in face processing. *Journal of Comparative Psychology*, 122, 390–402.

Parvizi, J., Jacques, C., Foster, B. L., Withoft, W., Rangarajan, V., Weiner, K. S., & Grill-Spector, K. (2012). Electrical stimulation of human fusiform face-selective regions distorts face perception. *Journal of Neuroscience*, 32, 14915–14920.

Pascalis, O., de Haan, M., & Nelson, C. A. (2002). Is face processing species-specific during the first year of life? *Science*, 296(5571), 1321–1323.

Peirce, J. W., Leigh, A. E., & Kendrick, K. M. (2001). Configurational coding, familiarity and the right hemisphere advantage for face recognition in sheep. *Neuropsychologia*, 38, 475–483.

Phillips, P. J., & O'Toole, A. J. (2014). Comparison of human and computer performance across face recognition experiments. *Image and Vision Computing*, 32, 74–85.

Pitcher, D., Walsh, V., Yovel, G., & Duchaine, B. (2007). TMS evidence for the involvement of the right occipital face area in early face processing. *Current Biology*, 17, 1568–1573.

Pokorny, J. J., & de Waal, F. B. (2009). Face recognition in capuchin monkeys (Cebus apella). *Journal of Comparative Psychology*, 123, 151–160.

Retter, T. L., & Rossion, B. (2016). Uncovering the neural magnitude and spatio-temporal dynamics of natural image categorization in a fast visual stream. *Neuropsychologia*, 91, 9–28.

Rilling, J. K., & Seligman, R. A. (2002): A quantitative morphometric comparative analysis of the primate temporal lobe. *Journal of Human Evolution*, 42, 505–533.

Rossion, B. (2008). Picture-plane inversion leads to qualitative changes of face perception. *Acta Psychologica*, 128, 274–289.

Rossion, B. (2013). The composite face illusion: A window to our understanding of holistic face perception. *Visual Cognition*, 121, 139–253.

Rossion, B. (2014a). Understanding face perception by means of human electrophysiology. *Trends in Cognitive Sciences*, 18, 310–318.

Rossion, B. (2014b). Understanding face perception by means of prosopagnosia and neuroimaging. *Frontiers in Bioscience, Elite*, 6, 258–307.

Rossion, B. (2018a). Damasio's error—prosopagnosia with intact within-category object recognition. *Journal of Neuropsychology*, 12(3), 357–388.

Rossion, B. (2018b). Humans are visual experts at unfamiliar face recognition. *Trends in Cognitive Sciences*, 22, 471–472.

Rossion, B., Caldara, R., Seghier, M., Schuller, A.-M., Lazeyras, F., & Mayer, E. (2003). A network of occipito-temporal face-sensitive areas besides the right middle fusiform gyrus is necessary for normal face processing. *Brain*, 126, 2381–2395.

Rossion, B., Jacques, C., & Jonas, J. (2018). Mapping face categorization in the human ventral occipito-temporal cortex with direct neural intracranial recordings. *Annals of the New York Academy of Science*, 1426(1), 5–24.

Rossion, B., & Michel, C. (2011). An experience-based holistic account of the other-race face effect. In A. Calder, G. Rhodes, M. Johnson, & J. Haxby (Eds.), *The Oxford Handbook of Face Perception*. Oxford: Oxford University Press.

Rossion, B., & Michel, C. (2018). Normative accuracy and response time data for the computerized Benton Facial Recognition Test (BFRT-c). *Behavioral Research Methods*, 50(6), 2442–2460.

Rossion, B., & Taubert, J. (2019). What can we learn about human individual face recognition from experimental studies in monkeys? *Vision Research*. doi:10.1016/j.visres.2018.03.012

Rossion, B., Torfs, K., Jacques, C., & Liu-Shuang, J. (2015). Fast periodic presentation of natural images reveals a robust face-selective electrophysiological response in the human brain. *Journal of Vision*, 15(1), 18: 1–18.

Sangrigoli, S., & de Schonen, S. (2004). Effect of visual experience on face processing: A developmental study of inversion and non-native effects. *Developmental Science*, 7(1), 74–87.

Scheirer, W. J. & Anthony, S., Nakayama, K., & Cox, D. (2014). Perceptual annotation: Measuring human vision to improve computer vision. *Pattern Analysis and Machine Intelligence, IEEE Transactions*, 36, 1679–1686.

Scott, L. S., & Monesson, A. (2010). Experience-dependent neural specialization during infancy. *Neuropsychologia*, 48(6), 1857–1861.

Sergent, J., Ohta, S., & Macdonald, B. (1992). Functional neuroanatomy of face and object processing—a positron emission tomography study. *Brain*, 115, 15–36.

Sergent, J., & Signoret, J.-L. (1992). Varieties of functional deficits in prosopagnosia. *Cerebral Cortex*, 2, 375–388.

Sheehan, M. J., & Nachman, M. W. (2014). Morphological and population genomic evidence that human faces have evolved to signal individual identity. *Nature Communications*, 5(4800), 1–10.

Sheehan, M. J., & Tibbetts, E. A. (2011). Specialized face learning is associated with individual recognition in paper wasps. *Science*, 334, 1272–1275.

Simion, F., Macchi Cassia, V., Turati, C., & Valenza, E. (2003). Non-specific perceptual biases at the origins of face processing. In A. Slater & O. Pascalis (Eds.), *The development of face processing in infancy and early childhood* (pp. 13–25). New York: Nova Science.

Sinha, P. (2002). Qualitative representations for recognition. In H. Bulthoff, C. Wallraven, S. W. Lee, & T. A. Poggio (Eds.), *Biologically motivated computer vision* (pp. 249–262). Berlin: Springer-Verlag.

Spillmann, L. (1999). From elements to perception: Local and global processing in visual neurons. *Perception*, 28, 1461–1492.

Sugden, N. A., Mohamed-Ali, M. I., & Moulson, M. C. (2014). I spy with my little eye: Typical, daily exposure to faces documented from a first-person infant perspective. *Developmental Psychobiology*, 56(2), 249–261.

Susilo T., & Duchaine, B. (2013). Advances in developmental prosopagnosia research. *Current Opinion in Neurobiology*, 23, 423–429.

Tan, C., & Poggio, T. (2016). Neural tuning size in a model of primate visual processing accounts for three key markers of holistic face processing. *PLoS One*, 11(3), e0150980.

Tanaka, J. W. (2001). The entry point of face recognition: Evidence for face expertise. *Journal of Experimental Psychology: General*, 130, 534–543.

Tanaka, J. W., & Farah, M. J. (1993). Parts and wholes in face recognition. *Quarterly Journal of Experimental Psychology*, 46A(2), 225–245.

Todorov, A. (2017). *Face value: The irresistible influence of first impressions*. Princeton, NJ: Princeton University Press.

Tsao, D. Y., & Livingstone, M. S. (2008). Mechanisms of face perception. *Annual Review of Neuroscience*, 31, 411–437.

Tsao, D. Y., Moeller, S., & Freiwald, W. A. (2008). Comparing face patch systems in macaques and humans. *Proceedings of the National Academy of Sciences of the United States of America*, 105, 19514–19519.

Van Belle, G., de Graef, P., Verfaillie, K., Busigny, T., & Rossion, B. (2010). Whole not hole: Expert face recognition requires holistic perception. *Neuropsychologia*, 48, 2609–2620.

Visconti di Oleggio Castello, M., & Gobbini, M. I. (2015). Familiar face detection in 180 ms. *PLoS One*, 10(8), e0136548.

Visconti di Oleggio Castello, M., Halchenko, Y. O., Gors, J. D., & Gobbini, M. I. (2017). The neural representation of personally familiar and unfamiliar faces in the distributed system for face perception. *Scientific Reports*, 7, 12237.

Wang, M.-Y., & Takeuchi, H. (2017). Individual recognition and the "face inversion effect" in medaka fish (Oryzias latipes). *eLife*, 6(e24728), 1–18.

Weiner, K. S., & Grill-Spector, K. (2012). The improbable simplicity of the fusiform face area. *Trends in Cognitive Sciences*, 16(5), 251–254.

Weiner, K. S., & Grill-Spector, K. (2015). The evolution of face processing networks. *Trends in Cognitive Sciences*, 19(5), 240–241.

Wilmer, J. B., Germine, L., Chabris, C. F., Chatterjee, G., Williams, M., Loken, E., Nakayama, K., & Duchaine, B. (2010). Human face recognition ability is specific and highly heritable. *Proceedings of the National Academy of Sciences of the United States of America*, 107, 5238–5241.

Wilmer, J. B., Germine, L., Chabris, C. F., Chatterjee, G., Gerbasi, M., & Nakayama, K. (2012). Capturing specific abilities as a window into human individuality: The example of face recognition. *Cognitive Neuropsychology*, 29, 360–392.

Young, A. W., & Burton, M. (2018). Are we face experts? *Trends in Cognitive Sciences*, 22, 100–110.

Young, A. W., Hellawell, D., & Hay, D. C. (1987). Configural information in face perception. *Perception*, 16, 747–759.

12 Multisensory Perception: Behavior, Computations, and Neural Mechanisms

UTA NOPPENEY

ABSTRACT Our senses are constantly bombarded with a myriad of diverse signals. Transforming this sensory cacophony into a coherent percept of our environment relies on solving two computational challenges: First, we need to solve the causal inference problem—deciding whether signals come from a common cause and thus should be integrated or come from different sources and should be treated independently. Second, when there is a common cause, we should integrate signals across the senses weighted in proportion to their sensory reliabilities. We discuss recent research at the behavioral, computational, and neural systems level that investigates how the brain addresses these two computational challenges in multisensory perception.

Computational Challenges in Multisensory Perception

In our natural environment, our senses are constantly bombarded with a myriad of diverse signals—for example, the sight of a rapidly approaching truck, the sound of its looming motor noise, and the smell of traffic fumes. The effortless ease with which we merge these signals into a seamless percept masks the complexity of the computations and neural mechanisms involved. While all of our senses continuously provide the brain with uncertain information about our environment, they are specialized for different sorts of information and environmental conditions. For instance, vision provides us with precise spatial information under optimal lighting conditions, but it is impaired in darkness and blind to sources behind us. Conversely, audition tends to be inferior to vision in its spatial precision during the daytime, yet it can exceed vision in darkness and provide spatial information about sources outside our current field of view. These considerations highlight the extraordinary benefits that we gain from collaborative interactions across all our senses, combining their complementary strengths and mitigating their individual limitations. Multisensory integration is one of the brain's key strategies to reduce its uncertainty and resolve ambiguities about the current state of the world. It is thus critical for effective interactions with our environment in everyday life.

More specifically, multisensory integration enhances our performance on two sorts of tasks. First, combining inputs from multiple senses increases the salience of an event, enabling faster and more accurate detection performance (Diederich & Colonius, 2004; Vroomen & de Gelder, 2000). We detect a potential threat, such as a wild boar in the forest, faster and more accurately when we combine the sight of its body obscured by foliage with the rustling noises it makes. Second, information integration increases the robustness and precision of a percept, leading to improved discrimination and estimation performance (Ernst & Bülthoff, 2004).

Within a normative Bayesian framework, the observer is thought to form a probabilistic generative model of the sensory inputs that is inverted during perceptual inference (Kersten, Mamassian, & Yuille, 2004). Bayesian probability theory offers a precise formulation of how an "ideal observer" should combine uncertain information as a benchmark of "optimal behavior" against which to compare human performance. It thus forms a starting point to study how the brain computes a perceptual interpretation from the noisy sensory inputs.

With respect to multisensory perception, the observer needs to solve two fundamental computational challenges. First, we need to solve the *binding* or *causal inference* problem—deciding whether signals originate from a common cause and should be integrated or come from different sources and should be treated independently. Formally, this goes beyond estimating an environmental property (i.e., a parameter such as the size of an object) and involves inferring a causal model of the world (i.e., causal structure inference; Kording et al., 2007). Second, when there is a common cause, we should integrate the multiple streams of sensory information along with prior knowledge, weighted by their uncertainties, giving more emphasis to more reliable information (Alais & Burr, 2004; Ernst & Banks, 2002; Fetsch, Pouget, DeAngelis, & Angelaki, 2012). The key to multisensory perceptual inference is thus the estimation and representation of different types of ignorance, including uncertainty about perceptual parameters and the world's causal structure.

In the following, we will review research at the behavioral, computational, and neural systems level that investigates how observers address these two computational challenges in multisensory perception. Our review will focus mainly on audiovisual perception in humans, though we also refer to the most relevant research involving other sensory modalities or species (e.g., nonhuman primates). First, we discuss the special *forced-fusion* case that ignores observers' uncertainty about the world's causal structure. In a second step, we discuss how the brain solves the causal inference problem and arbitrates between sensory integration and segregation. In each section we first describe the probabilistic generative model as a starting point and then review the extent to which behavioral and neural responses are in accordance with those computational principles.

Forced Fusion: Integrating Information from a Common Cause

Maximum likelihood estimation under forced-fusion assumptions Many events, objects, or properties can be perceived concurrently by multiple senses that provide complementary or redundant information about the external world. We can view an object from the front and concurrently explore it from the back via manipulative hand movements. Combining complementary information from vision and touch provides a more complete description of the object's three-dimensional shape. Alternatively, sensory channels may provide redundant information about an environmental property. For instance, at dawn we may spot a little robin sitting on a branch and hear its song. In this situation vision and audition provide redundant information about the bird's location. Importantly, all sensory inputs are corrupted by various sources of external environmental and internal neural noise. Sensory integration therefore provides a more precise (i.e., less noisy) estimate of an environmental property.

The forced-fusion model (e.g., for audiovisual spatial localization) assumes one single cause S_{AV} that generates the visual (x_V) and the auditory (x_A) signals, each corrupted by independent noise. We refer to this scenario as a *forced-fusion case* because the generative model does not include the option for two signals to come from independent sources. Hence, in the absence of any causal uncertainty an optimal Bayesian observer should integrate auditory and visual information mandatorily into one single unified percept (for further details, see Meijer & Noppeney, forthcoming).

During perceptual inference the observer is thought to compute the posterior probability (over spatial location) by combining the conditionally independent likelihoods of the auditory and visual signal with the prior (over space) according to Bayes' theorem.

$$1.\ P(S_{AV} \mid x_A, x_V) \propto P(x_A \mid S_{AV}) * P(x_V \mid S_{AV}) * P(S_{AV})$$

Assuming uniform priors (i.e., all spatial locations are a priori equally likely), Bayes' optimal integration reduces to maximum likelihood estimation (MLE):

$$2.\ P(S_{AV} \mid x_A, x_V) \propto P(x_A \mid S_{AV}) * P(x_V \mid S_{AV})$$

Under Gaussian assumptions, we can then derive the two key equations that relate estimates and variances of the audiovisual conditions to their unisensory counterparts.

First, the optimal (i.e., minimum variance) audiovisual estimate $\hat{S}_{AV}$ is a reliability-weighted (i.e., reliability as an inverse of variance $r = \frac{1}{\sigma^2}$) average of the auditory $\hat{S}_A$ and visual ($\hat{S}_V$) estimates:

$$3.\ \hat{S}_{AV} = w_A \hat{S}_A + w_V \hat{S}_V \text{ with } w_A = \frac{r_A}{r_A + r_V} \text{ and}$$

$$w_V = \frac{r_V}{r_A + r_V} = 1 - w_A.$$

An ideal observer should thus integrate auditory and visual information weighted by their relative reliabilities, giving more weight to the more reliable information.

Second, the reliability of the audiovisual estimate is equal to the sum of the unisensory reliabilities:

$$4.\ r_{AV} = r_A + r_V = \frac{\sigma_A^2 + \sigma_V^2}{\sigma_A^2 \sigma_V^2}.$$

As a consequence the uncertainty of the integrated percept should be reduced (or equal) relative to the uncertainty of the least variable unisensory percept. The greatest reduction in uncertainty (i.e., variance reduction by a factor of 2) occurs when the two signals are equally reliable (i.e., $\sigma_A^2 = \sigma_V^2$) (Alais & Burr, 2004; Ernst & Banks, 2002).

Multisensory perception—a prime example of human optimal behavior? Multisensory integration under forced-fusion conditions has been widely regarded as a prime example of human optimality—that is, with behavior consistent with normative Bayesian models (Fetsch, DeAngelis, & Angelaki, 2013; Hillis, Ernst, Banks, & Landy, 2002; Pouget, Beck, Ma, & Latham, 2013). Yet evidence has never been unequivocal. While statistically optimal integration has been shown for visuohaptic shape and audiovisual rate discrimination (Ernst & Banks, 2002; Helbig & Ernst, 2007; Raposo, Sheppard, Schrater, & Churchland, 2012), suboptimal multisensory variance reduction and/or sensory weights have been reported for audiovisual duration or visuovestibular self-motion discrimination (Burr, Banks, & Morrone, 2009; Butler, Smith, Campos, & Bülthoff, 2010; Fetsch et al., 2012). Further, audiovisual spatial discrimination was shown to be statistically optimal in one study (Alais & Burr, 2004) but associated with visual overweighting in another study (Battaglia, Jacobs, & Aslin, 2003). To resolve this controversy, we have recently reinvestigated whether human observers integrate audiovisual signals optimally for spatial discrimination in a large-scale preregistration study. Our results suggest that the majority of human observers integrate audiovisual signals into spatial estimates more precisely than any unisensory estimates, as predicted by the MLE model (Meijer & Noppeney, 2019). Yet, surprisingly, we replicated the visual overweighting previously reported by Battaglia, Jacobs, and Aslin (2003). How can we explain these moderate deviations from the predictions of the MLE model?

First, the structure of the MLE model ignores observers' uncertainty about whether or not sensory signals come from a common source. Observers may experience causal uncertainty and even misperceive signals arising from a common source as originating from different causes because of sensory noise, differences in perceptual biases between sensory modalities (Hillis et al., 2002), or intersensory conflicts introduced by experimental setups. For instance, Butler, Campos, Bülthoff, & Smith (2011) showed that optimal visuovestibular integration in self-motion perception depended critically on whether stereoscopic or simple binocular vision was provided. Moreover, the sensory weights (see equation 3) can only be estimated by introducing a small, unnoticeable intersensory conflict (e.g., audiovisual spatial disparity) in so-called perturbation experiments.

Second, the MLE model assumes that the perceptual estimates are corrupted by sensory noise prior to integration but do not accommodate additional noise, be it decisional or memory related, after integration (Burr, Banks, & Morrone, 2009; Hartcher-O'Brien, Di Luca, & Ernst, 2014). This "unmodeled" decisional noise can mimic suboptimal integration (see equation 4) even in cases in which observers integrate signals weighted by their reliabilities.

Third, the forced-fusion model focuses selectively on response choices and ignores response times. Yet in everyday life observers often need to make accurate responses under time pressure. If observers try to maximize response speed in addition to accuracy, this will have a significant impact on the definition of Bayesian optimality. Indeed, a recent study has demonstrated that putative suboptimal visuovestibular integration for heading discrimination can be considered optimal when the decisional dynamics and speed-accuracy trade-off is taken into account in diffusion models of multisensory decision-making (Drugowitsch, DeAngelis, Klier, Angelaki, & Pouget, 2014).

Fourth, the forced-fusion model implicitly assumes that the sensory uncertainty is instantaneously estimated or known by the observer. Yet sensory reliability may be learned over time. In line with this conjecture, we have recently shown that observers estimate sensory reliability by combining a reliability prior based on recent sensory signals with current sensory inputs (Beierholm, Rohe, Ferrari, Stegle, & Noppeney, 2019). As a result, observers' sensory weights lag behind the true sensory reliability. At a longer time scale, Battaglia, Jacobs, and Aslin (2003) suggested that observers may impose a prior on sensory reliabilities based on their lifelong experiences. For instance, because vision is usually more reliable than audition for spatial localization, observers may overweight vision for spatial localization tasks. In line with this conjecture, both human and nonhuman primates have been shown to overweight vestibular inputs in visuovestibular heading and auditory inputs in audiovisual duration discrimination tasks.

Fifth, the MLE model is based on the assumption that the sensory noise is Gaussian and independent across the sensory modalities. Violations of these assumptions may also account for deviations from the MLE predictions (Burr, Banks, & Morrone, 2009; Ernst, 2012).

In summary, human observers integrate sensory signals weighted by their reliabilities into estimates that tend to be more precise than their corresponding unisensory estimates. However, the sensory weights are often not exactly in proportion to the reliabilities of the sensory signals, and the variance reduction is smaller than predicted by the standard MLE model. These modest deviations may arise if various model assumptions do not hold in experimental settings. For instance, observers may experience causal uncertainty, impose priors on the sensory reliability, use cost functions that trade off speed versus accuracy, or deal with additional sources of noise. In more complex situations, observers may not even perform Bayesian inference but use approximate algorithms or simple heuristics (Ma, 2012; Rahnev & Denison, 2018; Shen & Ma, 2016).

Neural systems and mechanisms of reliability-weighted integration Traditionally, it was thought that the information gathered by distinct sensory organs and entering the brain via parallel pathways is integrated late in association cortices as well as in subcortical circuits. Accumulating evidence from neuroanatomy, neurophysiology, and neuroimaging has shown that multisensory influences are near ubiquitous in primate neocortex—occurring as early as primary cortices and increasing progressively across the cortical hierarchy (Ghazanfar & Schroeder, 2006; Werner & Noppeney, 2010). This raises the question of how brain areas integrate sensory signals across the cortical hierarchy. Where are signals integrated and weighted in line with their behavioral responses?

Recent elegant neurophysiological research has addressed this question for visuovestibular heading discrimination in nonhuman primates. Behaviorally, macaques integrated visual and vestibular signals weighted by their relative reliabilities on a trial to trial basis, though, as previously discussed, a modest vestibular overweighting was observed. At the neural level, population decoding of neural responses in the dorsal medial superior temporal area (MSTd) revealed sensory weights that closely correlated with monkeys' behavioral cue weighting and even showed the characteristic vestibular overweighting (Fetsch et al., 2012). Likewise, single MSTd neurons with congruent visual and tactile heading preferences integrated visual and vestibular signals weighted by their relative reliabilities (Fetsch et al., 2012), leading to increases in sensitivity paralleling behavior (Gu, Angelaki, & DeAngelis, 2008). Additional electrical microstimulation and chemical inactivation of MSTd further supported a causal link between the neural computations in MSTd and behavioral performance in a heading discrimination task (Gu, DeAngelis, & Angelaki, 2012). Collectively, this extensive body of elegant neurophysiological research in nonhuman primates suggests that neuronal populations and single neurons in MSTd integrate visual and vestibular signals weighted in proportion to their relative reliabilities into representations of heading direction approximately consistent with the MLE model. These computations have been suggested to be mediated by mechanisms of divisive normalization (Ohshiro, Angelaki, & DeAngelis, 2011), a canonical neural computation that has previously been proposed for visual processing and attentional modulation (Carandini & Heeger, 2012).

In humans, a series of neuroimaging studies have shown that both regional Blood Oxygen Level Dependent (BOLD) responses and connectivity between sensory areas are modulated by sensory reliability in multisensory integration paradigms (Boyle, Kayser, & Kayser, 2017; Nath & Beauchamp, 2011). Focusing on visuotactile integration in shape discrimination, an initial human functional magnetic resonance imaging (fMRI) study presented observers with ellipses in the visual and visuotactile modalities at multiple levels of visual reliability (Helbig et al., 2012). The study showed that the responses to tactile input increased in primary somatosensory cortices and the superior parietal lobe for unreliable visual shape information. Conversely, tactile inputs suppressed activations in the right posterior fusiform gyrus for unreliable blurred visual signals. Critically, the modulation of the BOLD response in the parietal and the occipitotemporal cortices was consistent with the weights that observers applied to the sensory signals at the behavioral level. These results suggest that higher-order association areas, such as the parietal cortices, integrate signals from vision and touch weighted by their reliabilities into representations of shape that are relevant for observers' behavioral responses.

In a more recent study, we investigated how auditory and visual signals are integrated into spatial representations across the cortical hierarchy by decoding multivariate fMRI patterns evoked by auditory, visual, and audiovisual stimuli with and without a small, unnoticeable spatial conflict (Rohe & Noppeney, 2018). On each trial, observers located either the auditory or the visual stimulus. We estimated observers' behavioral weights by fitting psychometric functions to participants' localization responses. Likewise, we estimated the neural weights by fitting neurometric functions to spatial locations decoded from regional fMRI activation patterns in auditory, visual, and parietal areas. Our results demonstrate that early auditory regions predominantly encoded the location of the auditory signal and visual areas the location of the visual signal. By contrast, a region in the intraparietal sulcus formed spatial representations that integrated auditory and visual signals weighted by their reliabilities approximately in line with observers' behavioral weights. Critically, the neural and behavioral weights and the variance of the spatial representations depended not only on the sensory reliabilities as predicted by the MLE model but also on observers' modality-specific attention (i.e., visual vs. auditory report). These results suggest that audiovisual integration in parietal areas is not exclusively determined by bottom-up sensory reliabilities but is also flexibly modulated by the task-relevance of the sensory modalities to guide spatial orienting and localization responses. Critically, the effects of modality-specific report cannot be accommodated by the MLE model because auditory and visual estimates should be identical under the forced-fusions assumption. They are better accounted for in Bayesian causal inference models, which will be the focus of the next section.

Multisensory Perception: Implicit and Explicit Bayesian Causal Inference

Bayesian causal inference model The second computational challenge the brain faces in our multisensory world is solving the causal inference, or binding, problem. To transform the myriad of sensory inputs into a coherent percept, we need to determine whether signals have a common cause and should be integrated or whether they originate from different causes and should be segregated. Because we do not have direct access to the causal structure of the world, we need to infer the world's causal structure from the noisy sensory signals themselves, such as the spatiotemporal and higher-order statistical correspondences.

The *Bayesian causal inference model* accounts for this binding problem and observers' causal uncertainty by explicitly modeling the potential causal structures that could have generated the sensory signals—that is, whether auditory and visual signals arise from one common or two independent sources (Acerbi, Dokka, Angelaki, & Ma, 2018; Kording et al., 2007; Rohe & Noppeney, 2015b; Wozny, Beierholm, & Shams, 2010). The generative model of Bayesian causal inference assumes that common ($C = 1$) or independent ($C = 2$) causes are sampled from a binomial distribution defined by the common cause prior P_{common}. For a common cause, one single-source location S_{AV} is drawn from the spatial prior distribution. For two independent causes, the auditory (S_A) and visual (S_V) locations are drawn independently. Finally, the spatial sources generate visual and auditory signals corrupted by independent noise (see figure 12.1*A*)

During inference, this generative process is thought to be inverted by the observer. The observer computes the posterior probability over causal structures (i.e., common or separate sources) by combining the common cause prior with the sensory evidence according to Bayes' rule

$$P(C = 1 \mid x_A, x_V) = \frac{p(x_A, x_V \mid C = 1)P_{common}}{p(x_A, x_V)}.$$

Under the assumption of a common signal source (i.e., the forced-fusion case), the two unisensory estimates of a physical property (e.g., spatial location) should be combined weighted according to their reliabilities. Under the hypothesis of two independent sources, separate spatial estimates are computed for auditory and visual signals (i.e., the segregation case). To account for the uncertainty about the world's causal structure, the observer obtains a final estimate of the environmental property (e.g., auditory location) by combining the forced-fusion and segregation estimates using various decisional strategies, such as model averaging, model selection, or probability matching (Wozny, Beierholm, & Shams, 2010). For instance, the *model-averaging* strategy combines the forced-fusion estimate with the segregated spatial estimate of the task-relevant sensory modality weighted by the respective posterior probabilities of the underlying causal structures:

$$\hat{S}_A = P(C = 1 \mid x_A, x_V) * \hat{S}_{AV, C=1} + (1 - P(C = 1 \mid x_A, x_V)) * \hat{S}_{A, C=2}.$$

$$\hat{S}_V = P(C = 1 \mid x_A, x_V) * \hat{S}_{AV, C=1} + (1 - P(C = 1 \mid x_A, x_V)) * \hat{S}_{V, C=2}.$$

Bayesian causal inference accommodates the fact that multisensory integration breaks down for large intersensory conflicts when it is unlikely that two signals originate from a common cause. It also explains that observers do not report identical perceptual estimates for auditory and visual signals as a result of their uncertainty about the world's causal structure. Forced fusion with a single unified audiovisual percept emerges only in the special case in which observers are absolutely certain (i.e., $P(C = 1|x_A, x_V) = 1$) that two signals arise from one common cause.

Previous modeling approaches have accounted for situations in which signals are not bound into a single unified percept by introducing a so-called coupling prior (Ernst, 2012) that specifies the joint probability over auditory and visual sources. When coupling priors are specified as mixture distributions, they can, in principle, model identical posterior distributions over environmental properties (e.g., spatial location) as Bayesian causal inference (Shams & Beierholm, 2010). Yet the Bayesian causal inference model goes beyond coupling priors by explicitly specifying the underlying generative process. It thereby enables predictions jointly for (1) observers' explicit causal inference—that is, deciding whether signals arise from common or independent causes—and (2) their implicit causal inference involved in forming a perceptual (e.g., spatial) estimate from multiple sensory signals (Acerbi et al., 2018; Rohe & Noppeney, 2015b). We will review the extent to which human observers' behavioral and neural responses are consistent with the predictions of Bayesian causal inference in explicit and implicit causal inference tasks.

Human behavior: multisensory perception and causal judgments A wealth of behavioral research has shown that

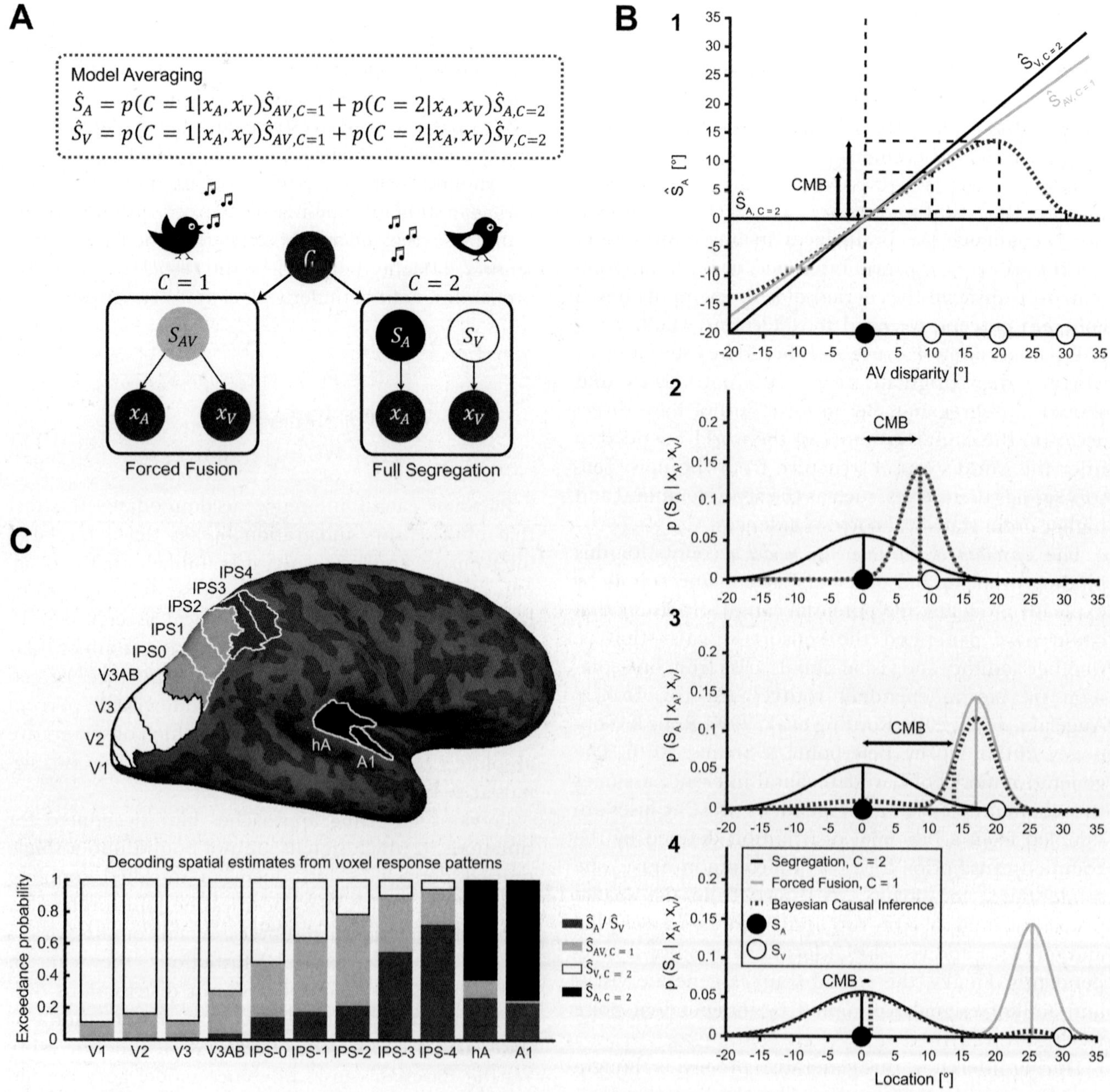

FIGURE 12.1 Bayesian causal inference.

A, Generative model. The generative model of Bayesian causal inference explicitly models the potential causal structures that could have generated the sensory signals. For the audiovisual spatial localization example, it models whether the "sight of the bird" and the bird's "singing" are generated by common (C=1) or independent (C=2) sources (Körding et al., 2007). In the case of a common source, the "true" audiovisual location (S_{AV}) is drawn from the prior spatial distribution. In the case of independent sources, the "true" auditory (S_A) and "true" visual (S_V) locations are drawn independently from this prior spatial distribution. Auditory (x_A) and visual (x_V) inputs are generated by adding independent sensory noise.

B, Posterior probability distribution and auditory spatial estimate. (1) The auditory spatial estimate ($\widehat{S_A}$) is shown as a function of audiovisual disparity, with auditory location set to 0° and visual location varying along the x-axis. According to the segregation model, $\widehat{S_A}$ is equal to the auditory location (black line, $\hat{S}_{A,C=2}$). In the visual capture scenario, $\widehat{S_A}$ is equal to the visual location (black line, $\hat{S}_{V,C=2}$). According to the forced-fusion model, auditory and visual inputs are integrated weighted by their relative reliabilities, thereby biasing $\widehat{S_A}$ toward the visual location (light gray line, $\hat{S}_{AV,C=1}$). According to Bayesian causal inference, the bias of $\widehat{S_A}$ depends nonlinearly on audiovisual disparity. For small disparities, the observer integrates auditory and visual inputs approximately weighted by their relative reliabilities. For large spatial disparities, audiovisual biases are attenuated (Rohe & Noppeney, 2015b). *Note:* CMB = crossmodal bias.

2–4. Posterior probability distributions for the segregation $P(S_A|x_A, x_V, C=2)$ (*black*), the forced fusion $P(S_A|x_A, x_V, C=1)$

multisensory integration is largely consistent with the principles of Bayesian causal inference. In explicit causal inference tasks, the size of the intersensory conflict determines whether observers perceive signals as coming from common or independent sources. As predicted by Bayesian causal inference, the percentage of common-cause responses or synchrony judgments declines with the size of the intersensory conflict with an inverted U-shaped function (Laurienti, Kraft, Maldjian, Burdette, & Wallace, 2004; Magnotti, Ma, & Beauchamp, 2013; Rohe & Noppeney, 2015b).

Likewise, perceptual illusions such as the McGurk (McGurk & Macdonald, 1976) or the ventriloquist effect as indices of audiovisual interactions in implicit causal inference tasks diminish with increasing intersensory conflict in time and/or space (Rohe & Noppeney, 2015b; van Wassenhove, Grant, & Poeppel, 2007; Wallace et al., 2004). In a spatial ventriloquist paradigm, observers were presented with auditory signals and visual signals in synchrony but with variable audiovisual spatial disparities. In each trial, observers reported the location of the auditory or visual stimulus. The results show that an observer's perceived sound location shifts toward a spatially displaced but synchronous visual stimulus, and vice versa, depending on the relative auditory and visual reliabilities (see figure 12.1*B*). In line with Bayesian causal inference, these audiovisual spatial biases—referred to as *spatial ventriloquism*—are attenuated for large audiovisual spatial disparities when it is unlikely that auditory and visual signals come from a common source (Kording et al., 2007; Wallace et al., 2004).

The results discussed so far have highlighted that audiovisual conflicts such as temporal asynchrony or spatial disparity are critical cues for observers to arbitrate between sensory integration and segregation both in explicit and implicit causal inference tasks. Critically, as a result of external sensory and internal neural noise, identical audiovisual signals can induce variable neural representations and behaviors across trials. This enables us to investigate the relationship between observers' explicit causal inference judgment and their spatial estimates and biases that rely on implicit causal inference. Models of Bayesian causal inference predict that the direction of the crossmodal bias in a particular trial depends on observers' explicit causal inference in a characteristic fashion (Kording et al., 2007): When audiovisual signals are perceived as coming from a common cause, the perceived sound location should be shifted toward the visual signal (i.e., *attractive* crossmodal bias). By contrast, when signals are perceived as arising from independent causes, the perceived sound location is shifted away from the visual signal, particularly for small spatial disparities (i.e., *repulsive* crossmodal bias). This repulsive crossmodal bias emerges because the sensory noise affects both observers' causal judgment (i.e., two causes) and their perceived sound location (i.e., sound is perceived as shifted away from the visual signal). Indeed, a previous study (Wallace et al., 2004) has shown attractive biases for common-source judgments and repulsive biases for independent-source judgments (although see Rohe & Noppeney, 2015b).

Collectively, this body of research demonstrates that human observers arbitrate between sensory integration and segregation consistent with the principles of Bayesian causal inference. However, future research is needed to better understand the qualitative and quantitative deviations of human behavior from the model's predictions.

Neural systems involved in Bayesian causal inference Human neuroimaging research has started to characterize how the human brain performs Bayesian causal inference for spatial localization (Rohe & Noppeney, 2015a, 2016). Inside the MRI scanner, observers were presented with audiovisual signals that varied in their spatial disparity and visual reliability. In each trial they either located the sound or the visual stimulus. Combining psychophysics, multivariate decoding, and Bayesian modeling, the study showed that the brain accomplishes Bayesian causal inference by encoding multiple spatial estimates across the auditory- and visual-processing hierarchy (see figure 12.1*C*). At the bottom of the hierarchy, early auditory areas predominantly encoded the location of the sound, and early visual areas predominantly encoded the location of the visual stimulus (= segregation). In the posterior intraparietal sulcus, location was estimated under the

(*light gray*), and the Bayesian causal inference model $P(S_A|x_A, x_V)$, with model averaging as a decision strategy (*gray dotted*) for (2) $S_A=0°$, $S_V=10°$, (3) $S_A=0°$, $S_V=20°$, and (4), $S_A=0°$, $S_V=30°$. The auditory location is indicated by the green dot and the visual location by the red dot. For the simulations we set $x_A=S_A$ and $x_V=S_V$ and the parameters for visual noise: $\sigma_V=3°$; auditory noise: $\sigma_A=7°$; spatial prior distribution: $\mu_{prior}=0°$; $\sigma_{prior}=1000°$; and common cause prior: $P_{common}=0.5$. *C*, Bayesian causal inference within the cortical hierarchy. Auditory areas represent predominantly the auditory location; early visual areas, the visual location. The posterior intraparietal cortex integrates sensory signals approximately according to forced-fusion principles. The anterior intraparietal sulcus computes the final Bayesian causal inference estimate that takes into account the observer's uncertainty about the world's causal structure (Rohe & Noppeney, 2015a). (See color plate 12.)

assumption that the two signals were from a common source (= forced fusion). Only at the top of the hierarchy, in the anterior intraparietal sulcus (IPS), was the uncertainty about the world's causal structure taken into account. As predicted by Bayesian causal inference, the anterior IPS estimated the location by combining the forced-fusion and the task-relevant full segregation estimates weighted by the posterior probabilities of common and independent sources (Rohe & Noppeney, 2015a, 2016). Likewise, a frontoparietal circuitry plays a key role in arbitrating between integration and segregation in explicit causal inference tasks. Thus, the IPS may form a spatial estimate that gracefully transitions from integration to segregation, informed by the causal inference in the dorsolateral prefrontal cortex (Mihalik & Noppeney, 2019).

This hierarchical organization of the neural processes underlying Bayesian causal inference raises the intriguing question of how these computations evolve dynamically over time in the human brain. How and when does the brain merge auditory and visual spatial information that is initially coded in different reference frames and representational formats? A recent electroencephalogram (EEG) study characterized the neural dynamics of Bayesian causal inference in spatial localization and showed that the brain initially encodes estimates based mainly on sensory segregation. Only relatively late, at about 350–450 ms, did it form a perceptual estimate that takes into account the uncertainty about the world's causal structure and combines signals as predicted by Bayesian causal inference (Aller & Noppeney, 2019). This late emergence of the Bayesian causal inference estimate was also replicated in a related EEG study focusing on the estimation of the number of audiovisual events (Rohe, Ehlis, & Noppeney, 2019). Collectively, this fMRI and EEG research suggests that Bayesian causal inference relies on the dynamic encoding of multiple perceptual estimates across the audiovisual cortical hierarchies.

Acknowledgment

This work was funded by the European Research Council, ERC multsens.

REFERENCES

Acerbi, L., Dokka, K., Angelaki, D. E., & Ma, W. J. (2018). Bayesian comparison of explicit and implicit causal inference strategies in multisensory heading perception. *PLOS Computational Biology, 14*(7), e1006110.

Alais, D., & Burr, D. (2004). Ventriloquist effect results from near-optimal bimodal integration. *Current Biology, 14*(3), 257–262.

Aller, M., & Noppeney, U. (2019). To integrate or not to integrate: Temporal dynamics of hierarchical Bayesian causal inference. *PLoS Biology, 17*(4), e3000210.

Battaglia, P. W., Jacobs, R. A., & Aslin, R. N. (2003). Bayesian integration of visual and auditory signals for spatial localization. *Journal of the Optical Society of America: A, Optics, Image Science, and Vision, 20*(7), 1391–1397.

Beierholm, U., Rohe, T., Ferrari, A., Stegle, O., & Noppeney, U. (2019). Using the past to predict sensory uncertainty. Manuscript submitted for publication. https://www.biorxiv.org/content/10.1101/589275v1.

Boyle, S. C., Kayser, S. J., & Kayser, C. (2017). Neural correlates of multisensory reliability and perceptual weights emerge at early latencies during audio-visual integration. *European Journal of Neuroscience, 46*(10), 2565–2577.

Burr, D., Banks, M. S., & Morrone, M. C. (2009). Auditory dominance over vision in the perception of interval duration. *Experimental Brain Research, 198*(1), 49–57.

Butler, J. S., Campos, J. L., Bülthoff, H. H., & Smith, S. T. (2011). The role of stereo vision in visual-vestibular integration. *Seeing and Perceiving, 24*(5), 453–470.

Butler, J. S., Smith, S. T., Campos, J. L., & Bülthoff, H. H. (2010). Bayesian integration of visual and vestibular signals for heading. *Journal of Vision, 10*(11), 23.

Carandini, M., & Heeger, D. J. (2012). Normalization as a canonical neural computation. *Nature Reviews Neuroscience, 13*(1), 51–62.

Diederich, A., & Colonius, H. (2004). Bimodal and trimodal multisensory enhancement: Effects of stimulus onset and intensity on reaction time. *Perception & Psychophysics, 66*(8), 1388–1404.

Drugowitsch, J., DeAngelis, G. C., Klier, E. M., Angelaki, D. E., & Pouget, A. (2014). Optimal multisensory decision-making in a reaction-time task. *ELife, 3.*

Ernst, M. O. (2012). Optimal multisensory integration: Assumptions and limits. In *The New Handbook of Multisensory Processes* (pp. 527–544).

Ernst, M. O., & Banks, M. (2002). Humans integrate visual and haptic information in a statistically optimal fashion. *Nature, 415*(6870), 429–433.

Ernst, M. O., & Bülthoff, H. H. (2004). Merging the senses into a robust percept. *Trends in Cognitive Sciences, 8*(4), 162–169.

Fetsch, C. R., DeAngelis, G. C., & Angelaki, D. E. (2013). Bridging the gap between theories of sensory cue integration and the physiology of multisensory neurons. *Nature Reviews Neurosciences, 14*(6), 429–442.

Fetsch, C. R., Pouget, A., DeAngelis, G. C., & Angelaki, D. E. (2012). Neural correlates of reliability-based cue weighting during multisensory integration. *Nature Neuroscience, 15*(1), 146–154.

Ghazanfar, A., & Schroeder, C. (2006). Is neocortex essentially multisensory? *Trends in Cognitive Sciences, 10*(6), 278–285.

Gu, Y., Angelaki, D. E., & DeAngelis, G. C. (2008). Neural correlates of multisensory cue integration in macaque MSTd. *Nature Neuroscience, 11*(10), 1201–1210.

Gu, Y., DeAngelis, G. C., & Angelaki, D. E. (2012). Causal links between dorsal medial superior temporal area neurons and multisensory heading perception. *Journal of Neuroscience, 32*(7), 2299–2313.

Hartcher-O'Brien, J., Di Luca, M., & Ernst, M. O. (2014). The duration of uncertain times: Audiovisual information

about intervals is integrated in a statistically optimal fashion. *PLoS One, 9*(3), e89339.
Helbig, H. B., & Ernst, M. O. (2007). Optimal integration of shape information from vision and touch. *Experimental Brain Research, 179*(4), 595–606.
Helbig, H. B., Ernst, M. O., Ricciardi, E., Pietrini, P., Thielscher, A., Mayer, K. M., & Noppeney, U. (2012). The neural mechanisms of reliability weighted integration of shape information from vision and touch. *NeuroImage, 60*(2), 1063–1072.
Hillis, J. M., Ernst, M. O., Banks, M. S., & Landy, M. S. (2002). Combining sensory information: Mandatory fusion within, but not between, senses. *Science, 298*(5598), 1627–1630.
Kersten, D., Mamassian, P., & Yuille, A. (2004). Object perception as Bayesian inference. *Annual Review of Psychology, 55*, 271–304.
Kording, K. P., Beierholm, U., Ma, W. J., Quartz, S., Tenenbaum, J. B., & Shams, L. (2007). Causal inference in multisensory perception. *PLoS One, 2*(9).
Laurienti, P. J., Kraft, R. A., Maldjian, J. A., Burdette, J. H., & Wallace, M. T. (2004). Semantic congruence is a critical factor in multisensory behavioral performance. *Experimental Brain Research, 158*(4), 405–414.
Ma, W. J. (2012). Organizing probabilistic models of perception. *Trends in Cognitive Sciences, 16*(10), 511–518.
Magnotti, J. F., Ma, W. J., & Beauchamp, M. S. (2013). Causal inference of asynchronous audiovisual speech. *Frontiers in Psychology, 4*, 798.
McGurk, H., & Macdonald, J. (1976). Hearing lips and seeing voices. *Nature, 264*, 691–811.
Meijer, D., Veselič, S., Calafiore, C., & Noppeney, U. 2019. Integration of audiovisual spatial signals is not consistent with maximum likelihood estimation. *Cortex, 119*, 74–88.
Meijer, D., & Noppeney, U. (forthcoming). Computational models of multisensory integration. In V. S. Sathian & K. Ramachandran (Eds.), *Multisensory perception: From Laboratory to Clinic.* Amsterdam: Academic Press, Elsevier.
Mihalik, A., & Noppeney, U. (2019). Causal inference in multisensory perception. Manuscript in preparation.
Nath, A. R., & Beauchamp, M. S. (2011). Dynamic changes in superior temporal sulcus connectivity during perception of noisy audiovisual speech. *Journal of Neuroscience, 31*(5), 1704–1714.
Ohshiro, T., Angelaki, D. E., & DeAngelis, G. C. (2011). A normalization model of multisensory integration. *Nature Neuroscience, 14*(6), 775–782.
Pouget, A., Beck, J. M., Ma, W. J., & Latham, P. E. (2013). Probabilistic brains: Knowns and unknowns, *16*(9).
Rahnev, D., & Denison, R. N. (2018). Suboptimality in perceptual decision making. *Behavioral and Brain Sciences*, 1–107.
Raposo, D., Sheppard, J. P., Schrater, P. R., & Churchland, A. K. (2012). Multisensory decision-making in rats and humans. *Journal of Neuroscience, 32*(11), 3726–3735.
Rohe T., Ehlis, A. C., & Noppeney, U. 2019. The neural dynamics of hierarchical Bayesian causal inference in multisensory perception. *Nature Communications, 10*(1), 1907.
Rohe, T., & Noppeney, U. (2015a). Cortical hierarchies perform Bayesian causal inference in multisensory perception. *PLoS Biology, 13*(2).
Rohe, T., & Noppeney, U. (2015b). Sensory reliability shapes perceptual inference via two mechanisms. *Journal of Vision, 15*(5), 22.
Rohe, T., & Noppeney, U. (2016). Distinct computational principles govern multisensory integration in primary sensory and association cortices. *Current Biology, 26*(4), 509–514.
Rohe, T., & Noppeney, U. (2018). Reliability-weighted integration of audiovisual signals can be modulated by top-down attention. *ENeuro, 5*(1), ENEURO.0315-17.2018.
Shams, L., & Beierholm, U. R. (2010). Causal inference in perception. *Trends in Cognitive Sciences.*
Shen, S., & Ma, W. J. (2016). A detailed comparison of optimality and simplicity in perceptual decision making. *Psychological Review, 123*(4), 452–480.
van Wassenhove, V., Grant, K. W., & Poeppel, D. (2007). Temporal window of integration in auditory-visual speech perception. *Neuropsychologia, 45*(3), 598–607.
Vroomen, J., & de Gelder, B. (2000). Sound enhances visual perception: Cross-modal effects of auditory organization on vision. *Journal of Experimental Psychology: Human Perception and Performance, 26*(5), 1583–1590.
Wallace, M. T. T., Roberson, G. E. E., Hairston, W. D. D., Stein, B. E. E., Vaughan, J. W. W., & Schirillo, J. A. A. (2004). Unifying multisensory signals across time and space. *Experimental Brain Research, 158*(2), 252–258.
Werner, S., & Noppeney, U. (2010). Distinct functional contributions of primary sensory and association areas to audiovisual integration in object categorization. *Journal of Neuroscience, 30*(7), 2662–2675.
Wozny, D. R., Beierholm, U. R., & Shams, L. (2010). Probability matching as a computational strategy used in perception. *PLoS Computational Biology, 6*(8).

13 Computational Models of Human Object and Scene Recognition

AUDE OLIVA

ABSTRACT To solve object and scene recognition tasks, the human brain has developed a particular cortical topology within its ventral and dorsal streams, recruiting regions in cascades to build a representation of what we see. A class of models, termed *artificial deep neural networks*, has been shown to learn a hierarchical representation of images, akin to the primate visual system, revealing internal representations that resemble the hierarchical topography in both the ventral and the dorsal visual streams of the human brain. Deep neural network models provide a hypothesis-testing framework to predict human brain responses as well as to give insights into how a network, natural or artificial, can learn and represent the visual world.

The High-Level Brain Regions Involved in Object and Scene Recognition

Brains are optimized to compute meaningful patterns from sensory inputs and to solve tasks fitted to their environment: while echolocation in dolphins is powerful under water, vision is the most dominant sense in primates. Yet visual object and scene recognition are difficult computational problems to solve given an almost infinite space of variation within our environment: Objects appear in different places, with different orientations, shapes, colors, and textures, and many can be made of different materials. In the real world, objects can be embedded in clutter, are often occluded, and can be observed from different distances and viewpoints. Despite these challenges, when we look at the world, a "feat of neural engineering" delivers a representation of what we see within only a few hundred milliseconds. How does the human brain instantiate visual recognition?

Entering the retina, visual information initially reaches the primary visual cortex (area V1, in the calcarine sulcus) before being transmitted to a series of retinotopically organized regions in visual occipital areas (areas V2, V3, human V4) and then higher-level regions of the cortex (for a review, see Grill-Spector and Weiner, 2014), which are responsible for visual perception and recognition. The types of computations and image features represented in different cortical sites have been extensively studied in primates in the past decades. For instance, low-level image properties, such as contrast, edges, and orientations, are processed within V1, V2, and V3; object form, shape, and physical size elicit stronger responses in the lateral occipital cortex, or LOC (Grill-Spector et al., 1999; Konkle & Oliva, 2012; Malach et al., 1995); images representing a layout, such as a scene or space, elicit stronger responses in the parahippocampal gyrus (Epstein & Kanwisher, 1998).

The temporal dynamics reflecting how these regions respond to an object or place suggest that the neural representation quickly passes through serial computations (Grill-Spector & Weiner, 2014). Neural responses first emerge in the occipital pole (V1, V2, V3) within 80–100 ms of viewing. Responses then spread rapidly and progressively in the anterior direction along the ventral stream (i.e., the LOC, ventral occipital cortex, temporal occipital cortex, and parahippocampal cortex) and the dorsal stream (intraparietal sulcus regions) within 110–170 ms after image onset (Cichy, Pantazis, & Oliva, 2016a).

These processing stages, particularly those in the ventral stream pathway, have inspired a class of computational models named deep (convolutional) neural networks (deep CNNs or DNNs). These models are quickly gaining popularity as a tool for the hypothesis testing of brain computations by providing simplified *artificial network streams.* What are DNNs and how can they be used to evaluate the computations performed within a biological brain?

Artificial Neural Networks

What is an artificial DNN? Artificial neural networks are a class of models that learn to recognize patterns from input data. One of their main properties is that they learn to progressively improve performance on a specific task without being explicitly programmed (for a review, see LeCun, Bengio, & Hinton, 2015). A given network trained in natural images is taught to achieve high performance on a specific task (e.g., a *detection* task, like finding the location of an object, or a *recognition* task, like identifying a place). Note that training these networks for a particular task, such as object classification (Krizhevsky, Sutskever, & Hinton, 2012) or scene classification (Zhou, Lapedriza, Xiao, Torralba, & Oliva, 2014), generally requires thousands of examples per class, due to a very large space of parameters that define the inner workings of the network.

While the basic operators of these networks are performed by interconnected units (i.e., artificial neurons), the entire network can be described as a high-dimensional mathematical function with many parameters (i.e., weights connecting units), which are tuned during a training (i.e., a learning) phase. In a supervised mode of training, an artificial network is taught associations between an input (i.e., an image) and an output (i.e., a label describing that image). It can adjust its parameters by iteratively reducing the errors between an output-input pair (i.e., the word *orange* and a specific picture of an orange fed, for instance, to the first layer as pixel values over a small zone of the image) across many presentations of the pair. The error is calculated based on the difference between the network's output value and the desired target value. It is then sent backward through the network to iteratively calculate error values for each layer, which are then used to update the network parameters.

This *back-propagation algorithm* (LeCun & Bengio, 1995) is able to find a solution to high-dimensional discrimination tasks—that is, learning the most useful features that recur across all the examples of pictures labeled *orange* in order to distinguish an orange from 1,000 other object categories. While it remains contentious whether a biological network like the brain uses a back-propagation function to learn feature matching between sensory inputs and classes of concepts, supervision signals can be triggered from different external and internal sources of reinforcement (e.g., context provided by different sensory modalities, direct verbal or tactile supervision, or comparison with mnemonic traces; for a review, see Kriegeskorte & Douglas, 2018).

How do artificial DNN models work? Let us look at a typical example of an artificial neural network, a CNN. A CNN is made of multiple layers, each implementing signal- and image-processing functions. Figure 13.1 illustrates the original architecture of a CNN for visual object recognition, nicknamed AlexNet and proposed by Krizhevsky, Sutskever, & Hinton (2012). AlexNet is composed of eight layers, some implementing one function only (e.g., layers three and four) and some implementing different functions in succession (e.g., layers one and two). The key function of a CNN is its convolution, which is implemented in most of the layers through the whole network (five convolutional layers in AlexNet): each layer is made by feature detection received through input from only a small part of the image. The convolution operation performs a summarization of each region of the image. For example, imagine a flashlight covering a region of five by five pixels directed over the top-left corner of an image. The flashlight can slide across the entire image, covering a new region with each movement. In this example, the flashlight is known as a filter (i.e., a neuron or kernel), which is an array of numbers (i.e., weights), and the region it is projecting light over is called a *receptive field*. The sliding motion of the flashlight filter over the whole image is called a *convolution*: as the filter is convolving, it is multiplying the values in the filter with the original pixel values of the image and summing these values. Therefore, for example, for each five-by-five-pixel image region, a single number is produced. With the flashlight filter moving through all regions in the image, a new feature map, of a smaller size than the original image, is created. Therefore, as more convolutional layers are added, the region of the input image analyzed by a given neuron (i.e, its receptive field) increases, giving to the CNN an important property for visual recognition: tolerance to the spatial position of a feature, or pattern, in the image (see the HMAX original model; Riesenhuber and Poggio, 1999; Serre, Wolf, Bileschi, Riesenhuber, & Poggio, 2007).

What are the advantages of DNNs? CNNs have been shown to yield several desirable outcomes observed in biological networks (for a review, see Bengio, 2009; Kriegeskorte, 2015). Akin to the observation that biological neurons exhibit receptive fields that increase in size between early visual areas and ventral stream regions, the "neurons" in later layers of an artificial neural network also have larger receptive fields. Larger receptive fields provide tolerance to the spatial translation of feature or shape in the image. This means that later layers will be able to discriminate a pattern (i.e., a circle or a corner shape) independently of the location of that shape in the original image input. As an added benefit, the shared parameters of the receptive fields that feed into each feature map allow for the same kernel function to be performed in different areas of the input to each layer. This results in increased robustness of a CNN to shifts and distortions in the data. This robustness is then compounded by the greater expressive power of deep architectures (Bengio & LeCun, 2007) where the features of each layer are combined to form a higher level of abstraction in the succeeding layer. This increasing abstraction from layer to layer allows deep CNNs to produce strong generalizations for highly variable functions. Additionally, this architecture supports the hypothesis that the network can disentangle at each step the factors of variation underlying the input–output data. As a result, the neurons in the network learn to

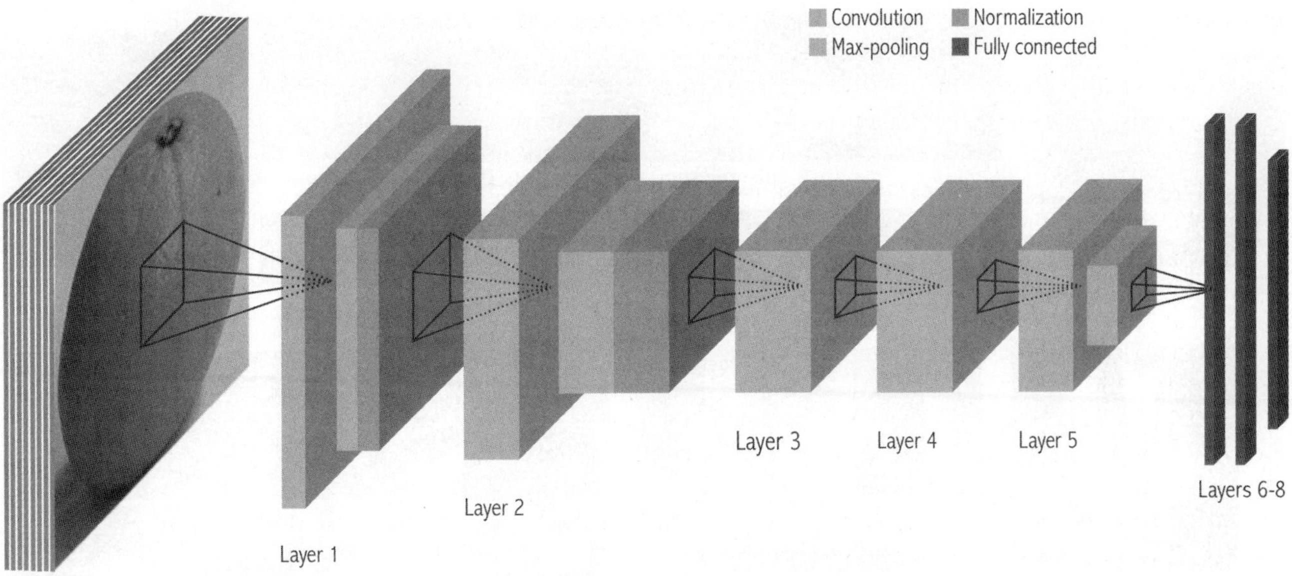

FIGURE 13.1 Deep CNN architecture AlexNet comprises eight layers. Each of layers one through five contains a convolution, some layers with max-pooling and normalization stages. The last three layers are fully connected, with the last layer acting as the output label. The network takes pixel values as inputs and propagates information in a feedforward manner through the layers, activating artificial units with particular weight values successively at each layer. (See color plate 13.)

represent increasingly complex visual patterns as layers deepen (see figure 13.2*B*). For example, in layer one of AlexNet, neurons are tuned to features such as contours and edges, while in layer two, neurons are tuned to curvature and repetitive patterns. In deeper layers, such as layers four and five, neurons become tuned to parts of objects and images, such as shape and form (Zeiler & Fergus, 2014; Zhou, Khosla, Oliva, & Torralba, 2015). This architecture of layers affords the artificial network the ability to learn many variations of a given object class, allowing the network to start resolving one of the most difficult challenges of object recognition: its diversity of views. In the real world, objects of the same category can be seen from different viewpoints, in different backgrounds, with different levels of clutter and occlusion, and at different retinal image sizes. While these models do not yet reach levels of human object recognition, CNNs have solved some of these issues better than models that are not hierarchically organized.

What do DNNs learn? Different methods for visualizing the artificial units' receptive fields and their selectivity to specific patterns have been proposed as a means toward understanding what a CNN has learned after training for a particular task (Bau, Zhou, Khosla, Oliva, & Torralba, 2017; Simonyan, Vedaldi, & Zisserman, 2014; Zeiler and Fergus, 2014).

In Zhou et al. (2015), the authors describe a method to estimate and quantify the selectivity of every single unit within the five convolutional layers of AlexNet as a function of the task (recognizing objects or scenes) the network was trained to perform. Specifically, a first AlexNet CNN (Object-Net) was trained to discriminate between classes of objects (is it a dog, a cat, or another animal, using the ImageNet data set; Deng et al., 2009). A second AlexNet (Scene-Net) was trained to discriminate between classes of scenes (is it a kitchen, a highway, or another object, using the Places data set; Zhou et al., 2014, 2017). To check what each artificial unit learned, Zhou et al. (2015) designed a pipeline inspired by neurophysiological experiments (illustrated in figure 13.2*A*): the two trained networks were shown many versions of the same set of testing images in which a small patch of random pixels occluded each image version. For each tested image, 5,000 stimulus versions were created by sliding the localized occluder over the image. Feeding all these occluded stimulus versions into a network and recording the change in activation compared to the original image (without any occluded region) allowed the calculation of a "discrepancy map" for each unit in the network. Examples of discrepancy maps for a particular unit for the ten images to which the unit was most responsive are shown in figure 13.2*A*. To consolidate the information, the discrepancy maps are centered around the spatial location of the unit that provided the maximum activation (see figure 13.2*A*, *bottom row*) and then averaged to build a final receptive field for that unit (figure 13.2*A*, *right*).

Figure 13.2*B* shows a visualization of the discrepancy map of a few such units: specifically, the visualization

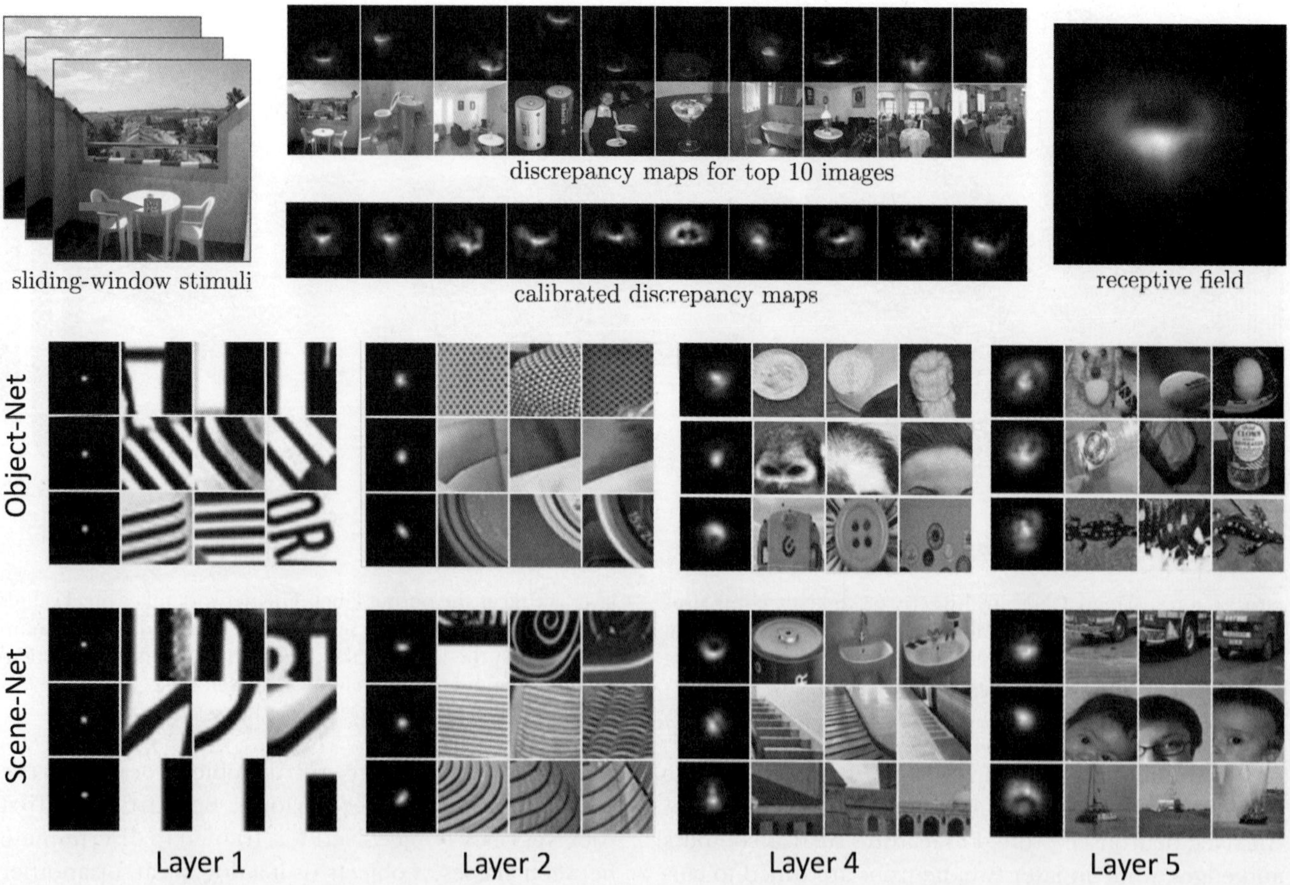

FIGURE 13.2 *A*, The pipeline for estimating the receptive field of each artificial unit of a DNN. Each sliding-window stimulus contains a small, randomized occluding patch (*red arrow*) at different spatial locations. By comparing the activation response of each sliding-window occluder with the activation response of the original image, a discrepancy map for each image can be built (*middle top*). By summing up the calibrated discrepancy maps (*middle bottom*) for the top-ranked images (i.e. 10), we can visualize the receptive field of that unit (*right*). *B*, The receptive fields of three units of pool1, pool2, conv4, and pool5 layers, respectively, for Object-Net (AlexNet trained with the ImageNet data set) and Scene-Net (AlexNet trained with Places data set), along with three images that correspond to particular images that activated these units the most. As the layers go deeper, the receptive field size gradually increases. (See color plate 14.)

shows the particular image patterns that drive the response of selected units in layers one, two, four, and five of AlexNet. For each unit's receptive field, three examples are shown that correspond to particular images that activated this neuron the most. Note that to robustly evaluate the true selectivity of a particular artificial unit, many images need to be passed through the network (Zhou et al., 2015).

As illustrated in figure 13.2*B*, as the layers deepen we can see that the units are tuned to increasingly complex features and sensitive to larger regions of the visual field—that is, they exhibit a larger receptive field, akin to the neurons found in the ventral visual-processing stream of primate brains.

Furthermore, as illustrated in figure 13.2*B*, the Object network and the Scene network learn highly similar patterns of oriented edges, curves, and textures in their two first layers. These results are most likely due to the similar distribution of image statistics across the two image data sets, as both are made of natural images (Torralba & Oliva, 2003). However, the properties of units in subsequent, higher layers of the network differ (see also Zeiler & Fergus, 2014): while a network trained to categorize objects (Object-Net in figure 13.2*B*) showed a predominance of units tuned to shapes and object parts, the network trained to classify scenes (Scene-Net in figure 13.2*B*) showed units selective to patterns that resemble whole objects, as well as more elaborate spatial layout patterns. This is expected given the hierarchical structure and the cost function the CNNs are programmed to solve: finding the discriminant or diagnostic information that maximizes performance on a specific task. As scene classes are mainly differentiated from the objects they are made of (i.e., a sofa makes a living room, a stove makes a kitchen), the network learned that patterns that we call "objects" are diagnostic features of scene classes. An emerging property is that both Object-Net and Scene-Net show units that seem specific to faces, people, and body parts, even as these networks did not explicitly learn to associate images to labels of faces, people, or body parts. Yet the emergence of such units suggests that the networks may have implicitly learned faces and body parts as diagnostic features that are contextually related to objects—for example, that baby faces are contextually related to *nursery*, or heads are contextually related to *hats*.

Human-interpretable concepts (i.e., a hat, a face, or an oval shape) emerge as *latent* variables in DNNs trained to solve detection or recognition tasks (Bau et al., 2017). These networks are not forced to decompose the classification problem in any interpretable way: the units do not have to specialize for particular patterns nor specialize for patterns that make sense to us (like object shapes, parts of objects, and more). However, these networks show interpretable units (figure 13.2*B*), which suggests they are able to spontaneously learn representations that are *disentangled* (Bau et al., 2017). A disentangled representation learns separate variables for separate meaningful features (Bengio, Courville, & Vincent, 2013)—for example, a concept or word that a person would use to describe a scene. While a network can learn an efficient encoding that makes economical use of hidden variables to distinguish between inputs, an internal structure with disentangled representation is an interesting property for comparing deep network representation to neural brain representations.

Applications of Deep Neural Networks to Investigate Cortical Regions of Object and Scene Processing

One can derive an algorithmically informed view on visual processing in the human brain by comparing, for the same set of images, the responses of a cortical region to the representation calculated from models. Because DNNs are made of a series of layers, a representation can be extracted from each layer for each image input. As shown in figure 13.2*B*, units in early layers of the DNN are more responsive to simple features like lines and corners, whereas units in later layers can be selective to complex patterns like shapes and object parts. This hierarchical representation of objects within the DNN resembles the hierarchical neural representation found in the primate visual brain (Grill-Spector & Weiner, 2014).

Using different neural-imaging techniques (i.e., electrophysiology, functional magnetic resonance imaging [fMRI], magneto/electroencephalography [M/EEG]), several recent studies have shown a systematic hierarchical relationship between convolutional neural network layers, such as in AlexNet (figure 13.1) and the processing cascade of information in both ventral and dorsal visual pathways in the human brain (Cadieu et al., 2014; Cichy, Khosla, Pantazis, Torralba, & Oliva, 2016b; Khaligh-Razavi, Cichy, Pantazis, & Oliva, 2018; Khaligh-Razavi & Kriegeskorte, 2014; Scholte, Losch, Ramakrishnan, de Haan, & Bohte, 2017; Seeliger et al., 2017; Yamins et al., 2014, among others). The relationship between model and brain data can be estimated using a technique termed *representational similarity analysis*, or RSA (Kriegeskorte, Mur, & Bandettini, 2008). Briefly, RSA builds dissimilarity distance matrices (referred to as a representational dissimilarity matrices, or RDMs) between every pair of images for which data are produced. For instance, for a data set of 100 images, an RDM can be built as a 100-by-100 matrix of distances between each pair of images. The RDMs can be built from the artificial units of a DNN layer, or from fMRI voxels activity in a brain region, or from the neuromagnetic signals captured from MEG sensors. The distributed nature of neural network data, where the response to an image input is best characterized as a pattern across artificial units within a layer, makes RSA an efficient analysis framework for relating the brain and models.

In the case of object processing, most studies have focused on comparing neural networks with early visual cortical (EVC) areas and inferior-temporal (IT) regions of the brain (Cadieu et al., 2014; Cichy et al. 2016b; Khaligh-Razavi & Kriegeskorte, 2014; Khaligh-Razavi et al., 2018; Yamins et al., 2014). Figure 13.3 illustrates results from such a study by Cichy et al. (2016b): the authors compared the RDMs from fMRI responses to object images to those from Object DNN (specifically, an AlexNet architecture trained on hundreds of object categories). This yielded eight brain maps (one for each of the eight layers of AlexNet; see figure 13.1) identifying the cortical regions where representations in the object DNN most correlated with cortical brain responses. As expected, for early model layers, similarities of visual representations were confined to the occipital lobe (i.e., the low-level and midlevel visual regions), and for late model layers, similarities of visual representations were found to correlate with more anterior regions in both the ventral and dorsal visual streams.

The comparison of DNNs for scene processing follows a similar pattern of results, with studies focusing on comparing scene models with brain regions previously identified to be more selective to images of places and scene layout than to other images (i.e., the parahippocampal place area, or PPA; Epstein & Kanwisher, 1998; the occipital place area, or OPA; Dilks, Julian, Paunov, & Kanwisher, 2013). For instance, Bonner and Epstein (2018)

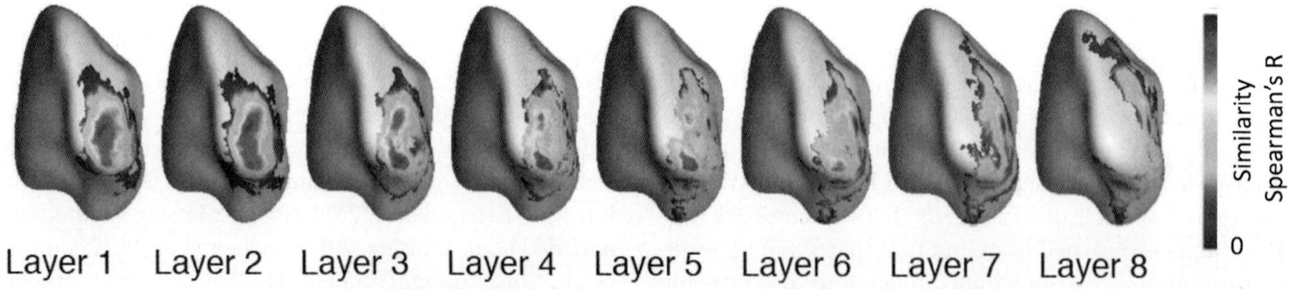

FIGURE 13.3 Spatial maps of visual representations common to human brain and object DNNs. The colors and the color bar show the strength of the brain-DNN similarity (similarity measured as Spearman's correlation, from 0 in *blue*, to a significant value in *red*, the maximum correlation varying between 0.07 and 0.14 depending on the layers). Low layers (one through two) have significant representational similarities with the occipital lobe of the brain—that is, low-level and midlevel visual regions. Higher layers have significant representational similarities with more anterior regions in the temporal and parietal lobe, with layers seven and eight reaching far into the inferior temporal cortex and inferior parietal cortex (from Cichy et al., 2016b). (See color plate 15.)

found that responses of a scene CNN to place images (Zhou et al., 2014, 2017) were highly predictive of human fMRI responses in the OPA, a scene-selective region of the dorsal occipitoparietal cortex. A visualization (as in figure 13.2*B*) of the feature selectivities in the scene network showed layout-type patterns, junctions, and large surfaces diagnostic of pictures of environments (see also Cichy, Khosla, Pantazis, & Oliva, 2017).

Conclusion

The isomorphism found between the human brain and DNNs suggests that hierarchical architectures are well suited for processing complex visual stimuli like real-world visual objects and scenes. This isomorphism also suggests that hierarchical systems of visual object and scene representations may emerge in both the human ventral and dorsal visual streams as the result of task constraints posed in everyday life (i.e., object categorization, place recognition for navigation).

While this chapter focused on the inner workings of only one class of DNN, neural network models simulating the neural dynamics of sensory information processing should have a better chance at representing the true spatiotemporal dynamics of vision. Decades of work in neuroscience suggest that vision starts as an initial feedforward phase, with signals going through the ventral stream (feedforward networks), followed by recurrent, long-range reverberating interactions between cortical regions (recurrent networks). Conceptually, feedforward and recurrent networks are sibling architectures, with the layered feedforward framework being a special case of a recurrent network in which some connections are missing (i.e., weights of connections between units are put at zero). Many neuroscience experiments suggest that performance at more attention-demanding tasks, such as object recognition in clutter, under occlusion, or with segmentation, are better approximated by a recurrent network. The basic assumption is that recurrent connections from higher to lower layers allow for looping back output to input, providing a temporal context to process the current input. Given that the state of the art in DNNs is evolving at an exponential pace, recurrent DNN models have a tremendous potential for explaining and operationalizing how biological brain networks learn diagnostic features, develop expertise, represent sensory information, solve recognition tasks, forecast the future, and never stop learning.

Acknowledgments

The author would like to thank Radoslaw Cichy, Mathew Lowe, Mathew Monfort, and Yalda Mohsenzadeh for thoughtful comments and feedback. This chapter was supported by the Vannevar Bush Faculty Fellowship program funded by the ONR grant No. N00014-16-1-3116 to A. Oliva.

REFERENCES

Bau, D., Zhou, B., Khosla, A., Oliva, A., & Torralba, A. (2017). Network dissection: Quantifying interpretability of deep visual representations. Proceedings of the 30th IEEE Conference on *Computer Vision and Pattern Recognition*, 3319–3327.

Bengio, Y. (2009). Learning deep architectures for AI. Hanover, MA: Now.

Bengio, Y., Courville, A., & Vincent, P. (2013). Representation learning: A review and new perspectives. *IEEE Transactions on Pattern Analysis and Machine Intelligence*, 35, 1798–1828.

Bengio, Y., & Lecun, Y. (2007). Scaling learning algorithms towards AI. In L. Bottou, O. Chapelle, D. DeCoste, &

J. Weston (Eds.), *Large-scale kernel machines.* Cambridge, MA: MIT Press.

Bonner, M. F., & Epstein, R. A. (2018). Computational mechanisms underlying cortical responses to the affordance properties of visual scenes. *Plos Computational Biology,* 14(4), e1006111.

Cadieu, C. F., Hong, H., Yamins, D. L. K., Pinto, N., Ardila, D., Solomon, E. A., Majaj, N. J., & DiCarlo, J. (2014). Deep neural networks rival the representation of primate IT cortex for core visual object recognition. *Plos Computational Biology,* 10(12), e1003963.

Cichy, R. M., Khosla, A., Pantazis, D., & Oliva, A. (2017). Dynamics of scene representations in the human brain revealed by magnetoencephalography and deep neural networks. *NeuroImage,* 153, 346–358.

Cichy, R. M., Khosla, A., Pantazis, D., Torralba, A., & Oliva, A. (2016b). Comparison of deep neural networks to spatio-temporal cortical dynamics of human visual object recognition reveals hierarchical correspondence. *Scientific Reports,* 6, 27755.

Cichy, R. M., Pantazis, D., & Oliva, A. (2016a). Similarity-based fusion of MEG and fMRI reveals spatio-temporal dynamics in human cortex during visual object recognition. *Cerebral Cortex,* 26(8), 3563–3579.

Deng, J., Dong, W., Socher, R., Li, L.-J., Lim, K., & Fei-Fei, L. (2009). ImageNet: A large-scale hierarchical image database. *IEEE Proceedings of the 22nd Conference on Computer Vision and Pattern Recognition,* 248–255.

Dilks, D. D., Julian, J. B., Paunov, A. M., & Kanwisher, N. (2013). The occipital place area is causally and selectively involved in scene perception. *Journal of Neuroscience,* 33(4), 1331–1336.

Epstein, R., & Kanwisher, N. (1998). A cortical representation of the local visual environment. *Nature,* 392(6676), 598.

Grill-Spector, K., Kushnir, T., Edelman, S., Avidan, G., Itzchak, Y., & Malach, R. (1999). Differential processing of objects under various viewing conditions in the human lateral occipital complex. *Neuron,* 24(1), 187–203.

Grill-Spector, K., & Weiner, K. S. (2014). The functional architecture of the ventral temporal cortex and its role in categorization. *Nature Neuroscience Review,* 15, 536–548.

Khaligh-Razavi, S. M., Cichy, R. M., Pantazis, D., & Oliva, A. (2018). Tracking the spatiotemporal neural dynamics of object properties in the human brain. *Journal of Cognitive Neuroscience,* 30(11), 1–18.

Khaligh-Razavi, S. M., & Kriegeskorte, N. (2014). Deep supervised, but not unsupervised, models may explain IT cortical representation. *Plos Computational Biology,* 10(11), e1003915.

Konkle, T., & Oliva, A. (2012). A real-world size organization of object responses in occipito-temporal cortex. *Neuron,* 74(6), 1114–1124.

Kriegeskorte, N. (2015). Deep neural networks: A new framework for modeling biological vision and brain information processing. *Annual Review of Vision Science,* 1, 417–446.

Kriegeskorte, N., & Douglas, P. K. (2018). Cognitive computational neuroscience. *Nature Neuroscience,* 21, 1148–1160.

Kriegeskorte, N., Mur, M., & Bandettini, P. (2008). Representational similarity analysis—connecting the branches of systems neuroscience. *Frontiers in Systems Neuroscience,* 24(2), 4.

Krizhevsky, A., Sutskever, I., & Hinton, G. E. (2012). ImageNet classification with deep convolutional neural networks. *Advances in Neural Information Processing Systems,* 25, 1097–1105.

LeCun, Y., & Bengio, Y. (1995). Convolutional networks for images, speech and time series. In M. A. Arbib (Ed.), *The Handbook of Brain Theory and Neural Networks* (pp. 255–258). Cambridge, MA: MIT Press.

LeCun, Y., Bengio, Y., & Hinton, G. E. (2015). Deep learning. *Nature,* 521, 436–444.

Malach, R., Reppas, J. B., Benson, R. R., Kwong, K. K., Jiang, H., Kennedy, W. A., Ledden, P. J., Brady, T. J., Rosen, B. R., & Tootell, R. B. (1995). Object-related activity revealed by functional magnetic resonance imaging in human occipital cortex. *Proceedings of the National Academy of Sciences,* 92(18), 8135–8139.

Riesenhuber, M., & Poggio, T. (1999). Hierarchical models of object recognition in cortex. *Nature Neuroscience,* 2(11), 1019–1025.

Scholte, S., Losch, M. M., Ramakrishnan, K., de Haan, E. H. F., & Bohte, S. M. (2017). Visual pathways from the perspective of cost functions and multi-task deep neural networks. *Cortex,* 98, 249–261.

Seeliger, K., Fritsche, M., Güçlü, U., Schoenmakers, S., Schoffelen, J. M., Bosch, S. E., & van Gerven, M. A. J. (2017). Convolutional neural network-based encoding and decoding of visual object recognition in space and time. *NeuroImage,* 180, 243–256.

Serre, T., Wolf, L., Bileschi, S., Riesenhuber, M., & Poggio, T. (2007). Robust object recognition with cortex-like mechanisms. *IEEE Transactions on Pattern Analysis and Machine Intelligence,* 29(3), 411–426.

Simonyan, K., Vedaldi, A., & Zisserman, A. (2014). Deep inside convolutional networks: Visualising image classification models and saliency maps. Paper presented at the International Conference on Learning Representation, April 14–16, 2014, Banff, Canada.

Torralba, A., & Oliva, A. (2003). Statistics of natural images categories. *Network: Computation in Neural Systems,* 14, 391–412.

Yamins, D. L. K., Hong, H., Cadieu, C. F., Solomon, E. A., Seibert, D., & Dicarlo, J. J. (2014). Performance-optimized hierarchical models predict neural responses in higher visual cortex. *Proceedings of the National Academy of Sciences,* 111, 8619–8624.

Zeiler, M. D., & Fergus, R. (2014). Visualizing and understanding convolutional networks. In *Proceedings of the 13th European Conference on Computer Vision* (pp. 818–833). New York: Springer.

Zhou, B., Khosla, A., Lapedriza, A., Oliva, A., & Torralba, A. (2015). Object detectors emerge in deep scene CNNs. Paper presented at the International Conference on Learning Representations, May 7–9, 2015, San Diego, USA.

Zhou, B., Lapedriza, A., Khosla, A., Oliva, A., & Torralba, A. (2017). Places: A 10 million image database for scene recognition. *IEEE Transactions on Pattern Analysis and Machine Intelligence,* 40(6), 1452–1464.

Zhou, B., Lapedriza, A., Xiao, J., Torralba, A., & Oliva, A. (2014). Learning deep features for scene recognition using places database. Advances in Neural Information Processing Systems (NIPS 2014), 27, Montreal, Canada.

14 Brain Mechanisms of Auditory Scene Analysis

BARBARA G. SHINN-CUNNINGHAM

ABSTRACT When listening to a mixture of sound sources, the brain relies on prior knowledge of statistical regularities in natural sound to estimate what sound energy comes from which source in the external world. This process of perceptual *object formation*, also known as auditory scene analysis, has been a focus of research for decades, most recently for rich, natural sounds like speech. Because the brain can only analyze one perceptual object at a time, object formation influences how we perceive and interpret the sounds in our environment. This chapter reviews the acoustic features that drive object formation and then considers neural correlates of auditory scene analysis and attention.

Our ears usually receive a mixture of sound sources overlapping in time and frequency content. Yet we perceive distinct perceptual "objects," or estimates of the acoustic energy emitted by physical sound sources in the external world (Darwin, 1997; Griffiths & Warren, 2004; Shinn-Cunningham, 2008). The process of separating sound mixtures into perceptual objects is known as *auditory scene analysis*, or ASA. This chapter provides an introduction to both behavioral and neurophysiological studies of ASA in natural settings (see also Darwin, 1997; Kondo, van Loon, Kawahara, & Moore, 2017).

Determining what sound energy comes from what physical sound source is a mathematically ill-posed problem: there are an infinite number of sound mixtures that could have produced the signals reaching your left and right ears. For instance, imagine the word *boot*. It could be from a single source or from one source, *boo*, temporally abutting another source *oot*, or the low frequencies of *boot* could be from one source and the high frequencies from a different source. How does your brain estimate the sources in the scene given that the problem itself is underconstrained?

The process of ASA matters practically because how you parse the acoustic scene has a direct impact on your ability to understand the meaning of the sources around you (Shinn-Cunningham, 2008). Cognitive processes are limited; you cannot process everything in the scene in detail. Instead, you use attention to focus on one object and suppress the neural responses from competing objects. Think about what this means: your ability to make sense of the world depends on attention, which suppresses the representation of sounds you ignore, but for attention to be effective, you must successfully segregate the sources in the scene. Thus, your ability to understand sound sources depends on your ability to parse the scene into auditory objects that represent the physical sound sources around you.

At first blush, there seems to be a hierarchy of processing: the scene is first analyzed into objects and then one object is selected and attended. In truth, these processes form a heterarchy: the processes occur in parallel, feeding back upon one another. Your decision about what to attend in turn influences what sound features are emphasized, which influences how objects are formed (akin to the reverse hierarchy theory; see Nahum, Nelken, & Ahissar, 2008). Conversely, sounds that are particularly salient are more likely to grab attention even in the absence of a conscious decision to focus on them (see Kaya & Elhilali, 2014a; Kayser, Petkov, Lippert, & Logothetis, 2005). Once a sound is the focus of attention, it influences object formation—whether attentional selection was the result of volitional effort or of stimulus-driven salience.

Although there is a large body of research exploring ASA and attention, the brain mechanisms realizing these processes are still poorly understood. In part, this is because object formation and attention involve a wide range of brain areas, working together. Moreover, how sound is organized into objects and how a particular object rises to be the focus of attention are both accomplished gradually across stages of the auditory pathway and as sound content evolves through time. Thus, one cannot pinpoint a specific neural place or a time at which an object is formed or becomes the focus of attention.

This chapter first discusses how the brain forms objects from sound mixtures and briefly reviews how object formation and attention allow us to interpret complex sound mixtures. It then considers different neural correlates of object formation and attention.

Statistical Regularities in Natural Sounds Guide Auditory Object Formation

The intellectual father of ASA, Al Bregman, wrote a tome that lays out most of its fundamental principles (Bregman, 1990). In general, auditory objects, the output of ASA, are the brain's "best guess" of what belongs together, based on statistical regularity in the spectrotemporal structure of natural sounds. Much of Bregman's seminal work focused on artificial sounds, such as mixtures of simple tones combined in different sequences and rates. Indeed, to most psychoacousticians, the shorthand *A-B-A paradigm* immediately brings to mind a sequence of two tones with different frequencies (A and B) that are each repeated, with the A tones isochronous at one rate, the B tones isochronous at half that rate, and each B tone falling midway between a pair of A tones (see figure 14.1). Listeners may perceive either one stream (of A-B-A tones, together, forming a triplet rhythm) or two streams (one of fast-repeating A tones and another of slow-repeating B tones, each with an isochronous rhythm). Listeners are more likely to hear a galloping stream when the presentation rate is faster or the two frequencies are closer together. The similarity of the A and B tones in other perceptual dimensions also can influence how the scene is parsed.

While simple stimuli allow one to isolate how specific features of sound influence ASA, they are quite impoverished compared to natural soundscapes, where most sources are broadband, time varying, and statistically unrelated to every other source. This makes it hard to extrapolate from experimental results of ABA studies to understand what occurs for richer acoustic stimuli like those encountered in everyday settings. Below, we focus on how the brain copes with more natural sound mixtures, such as independent streams of speech or temporally uncorrelated melodies.

Bregman broke down ASA mechanisms into *simultaneous* grouping (rules governing whether tones that are present at the same moment are heard in the same object) and *sequential* grouping (rules governing whether events that occur in a sequence are heard as part of the same, or a different, object). While these are useful concepts, the words can be misleading when considering natural sounds. To extend these concepts to common, everyday sounds, we consider statistical regularities that arise at a "local" level to form *syllables* (akin to Bregman's simultaneous grouping) and higher-order statistical regularities that can perceptually link together syllables separated by silent gaps (generalizing Bregman's concept of sequential grouping).

Syllables form from spectrotemporally structured, contiguous sound Many real physical objects generate energy in bursts, often broadly spread across the audible spectrum. Such bursts are not typically static; their content changes continuously through time. For instance, imagine a person saying the syllable *oy*, which has a spectrum that changes smoothly from one vowel to another (see figure 14.2). These changes are generally structured in

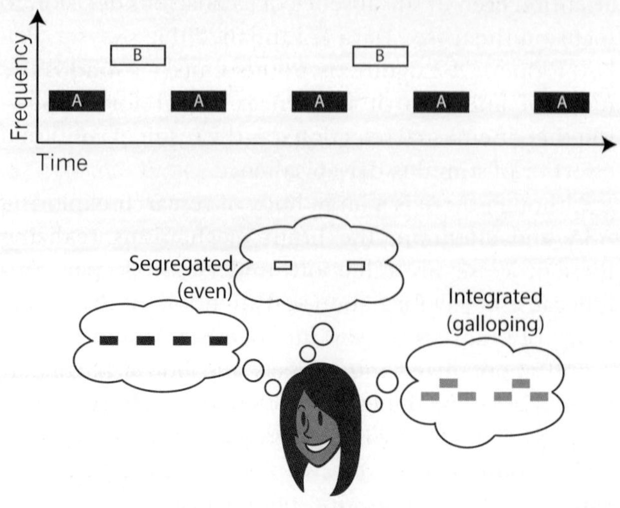

FIGURE 14.1 Schematic of the well tested A-B-A paradigm used to study auditory scene analysis. Two different frequency tones (denoted by *A* and *B*) are presented, each at a regular interval; however, the B tones repeat at half the rate of and in between the A tones. If listeners perceive the sound mixture as segregated, they report two separate streams, one of A tones and one of B tones, each of which has an "even" rhythm (*left side, bottom of figure*). If they perceive all of the tones as one single stream, they report a galloping rhythm.

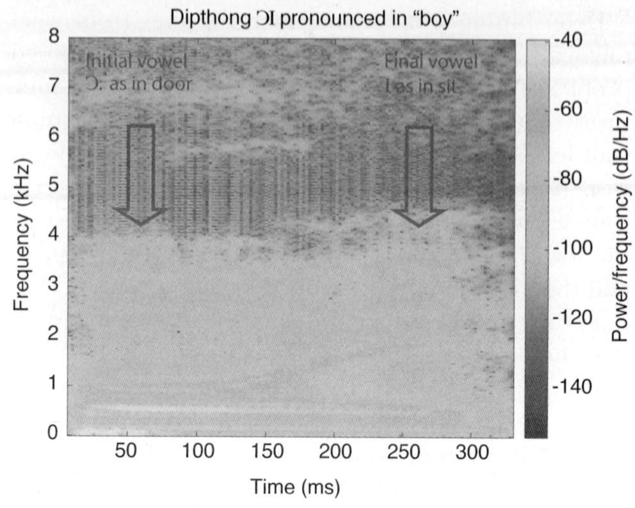

FIGURE 14.2 Spectrogram of the syllable *boy*, showing how dynamic changes in spectral content are nonetheless structured and continuous in time frequency. (See color plate 16.)

time frequency; as a result, the entire syllable is perceived automatically as a single unit, despite the dynamic changes in content over time.

In speech the unit of a syllable (like *oy*) has spectrotemporal continuity and a predictable structure that allows its widespread spectral content to nonetheless be grouped together unambiguously. In many nonspeech sounds, the same kind of spectrotemporal structure also arises. Throughout this chapter we use the term *syllable* to refer to any continuous burst of sound whose "local" spectrotemporal structure groups it together, perceptually, such as a note from an oboe, the crash of a glass smashing against a tile floor, or the rattle of a coin rolling to a stop on a counter.

In each example there are multiple (often temporally contiguous) spectral components making up individual syllables. Various spectral components, however, group together not because they are contiguous in frequency but because of other shared structure. For example, the music note from an oboe comprises harmonically related spectral elements with a common fundamental frequency. For the less structured smash of a breaking glass, spectral elements are not harmonically related but have similar, correlated temporal envelopes, with a sudden common onset that decays rapidly back to nothing. Although spatial cues are relatively weak in influencing the formation of spectrotemporally local objects, they still have some perceptual weight, especially when there is ambiguity in a sound mixture, and one spectral element could otherwise belong to two competing objects (e.g., see Darwin, 2006; Middlebrooks, 2017; Schwartz & Shinn-Cunningham, 2010). In general, the statistical structure of the spectrotemporal content, including harmonicity, common amplitude modulation, and spatial cues like interaural time or level differences, influence how spectral elements are grouped into syllables.

Sequences of syllables group together into streams based on feature similarity Just as typical sources produce syllables made up of statistically structured spectral components, they tend to produce sequences of syllables with common attributes. Each syllable has higher-level perceptual features that link together the spectral elements making up that syllable, including pitch, timbre, and location. For instance, a sequence of oboe notes is heard as a single melody. For syllables produced by the same source, these high-level syllabic features tend to change slowly over time, allowing the brain to connect the syllables, perceptually. When perceived as a single sequence, the syllables are said to form a *stream*.

Expectations and predictions guide object formation Both syllable and stream formation depend upon expectations: statistical regularities in low-level acoustic features that indicate spectral components belong to the same syllable and in higher-level perceptual features that indicate that distinct syllables are part of the same stream. Some of these expectations are learned, while others may be "hardwired" through evolution. For instance, common onsets strongly drive syllabic-level grouping; octopus cells in the brain stem are hardwired to detect and respond to common onsets across broad frequency ranges. On the other hand, if an acoustic signal in a mixture repeats, the brain rapidly learns (over the course of two to three repetitions) to expect this sequence, increasing the likelihood that it is heard as a coherent stream, even in the absence of attention (e.g., Masutomi, Barascud, Kashino, McDermott, & Chait, 2016).

Attention and Object Formation Are Intricately Intertwined

The brain processes one sound object in detail at any one time; it is not capable of attending to two sounds simultaneously (e.g., see Ihlefeld & Shinn-Cunningham, 2008). In those moments when you believe you are "sharing" attention among simultaneous objects, you are probably switching rapidly between them—and may be able to extract the gist of each if each is predictable, allowing you to fill in the unattended information. However, attention is being multiplexed, rather than truly divided.

What object we attend at a given moment depends partially on how we direct attention. This volitional focus is effected through feedback from control areas in the frontal cortex that modulate auditory sensory responses, suppressing inputs that do not match what we wish to attend. However, salient auditory objects may "get through" in spite of volitional attention. For instance, if you are trying to attend to a woman on your right, top-down attention will suppress a man on your left; however, if a gunshot goes off behind your head, it will interrupt attention to allow your brain to decide how to act (Fight? Flight? Or is it your mischievous cousin with his cap gun, which you can ignore?; see Dalton & Fraenkel, 2012). In vision, this battle between top-down, conscious attention and bottom-up, salience-driven attention is described as a *biased-competition*, in which volitional focus biases the incoming sensory representation but where this biasing may be overcome by salient inputs (Desimone & Duncan, 1995). This same model of biased competition accounts well for how auditory attention operates, consistent with the two sensory modalities sharing either common or similar attentional circuitry (Kaya & Elhilali, 2017; McDermott, 2009; Shinn-Cunningham, 2008).

Unambiguous objects form automatically, independent of attention In the natural world, auditory objects often are distinct, leading to strong interobject competition. The sounds of a phone ringing, music playing from your computer, the computer fan humming, and your colleague telling you about his weekend are mutually dissimilar; your brain has no trouble estimating the acoustic content of each source. However, because these objects are perceptually distinct, the automatic competition between them is strong, making it nearly impossible to analyze more than one (e.g., Best et al., 2006). Indeed, when distinct ongoing streams are present, an attended stream will tend to remain the focus of attention, biasing attention to remain on it as it continues (Best, Ozmeral, Kopco, & Shinn-Cunningham, 2008; Bressler, Masud, Bharadwaj, & Shinn-Cunningham, 2014; Woods & McDermott, 2015). This is not to say that top-down attentional focus (object selection) does not waver over time (e.g., oscillating between remaining focused on auditory target sounds versus *lapsing*, as shown recently in a cross-modal selective attention task in monkeys; Lakatos et al., 2016). Rather, while attentional selection may fail due to such top-down lapses of attentional control, top-down attention is not necessary to form streams from a sound mixture when the statistical structures of competing streams make each distinct. In these conditions, the objects are formed automatically, based on the statistical structure of the input sound mixture; top-down attentional failures may lead to selection of the wrong stream but will not lead to failures of object segregation (e.g., see Ruggles & Shinn-Cunningham, 2011).

Attention influences object formation when sound mixtures are ambiguous In some conditions, auditory object formation is ambiguous. Ambiguity may occur at the level of syllabic formation—such as when a flute and an oboe play a melody in unison—or at the level of stream formation—such as when two similar-sounding women speak at roughly the same intensity from roughly the same direction. In these cases, top-down focus affects what object will be attended and the formation of the objects themselves. If you focus on the sound of the flute within the flute-oboe mixture, your brain may, over time, isolate it and analyze it; however, if you focus on the flute and oboe as a single melody with an unusual timbre, you can perceive it as a single entity. As you focus on the woman who is slightly to the right and has a lower-pitched, raspier voice, you will be able to pull her speech out of the mixture; however, if you do not focus attention, you are likely to hear a gibberish of intermingled words spoken by both women (e.g., Best et al., 2006). Although syllables from each talker form automatically due to local structure, the stream corresponding to a speaker will not form without top-down, focused attention, and the specificity of attentional focus will tend to increase through time (e.g., see Best et al., 2008; Dai, Best, & Shinn-Cunningham, 2018). These examples illustrate two important principles of ambiguous mixtures. First, when objects are not particularly distinct, it takes time for objects to be segregated, perceptually. Second, in these cases attention can directly influence object formation, not just which object becomes the focus of detailed analysis.

In the above examples, object ambiguity arises because distinct physical sound sources produce sounds that are perceptually similar, either at the syllabic or the stream level. However, sometimes, ambiguity in a sound mixture arises because the scene is noisy. Imagine trying to listen to a talker in a cathedral. Echoes and reverberation smear out the sound, muddying both low-level attributes (temporal envelope, harmonic structure, location cues, and more) and higher-order perceptual features (pitch, location, and more), thereby interfering with object formation. In such situations, attention helps to isolate a desired sound, influencing both object selection and formation.

For listeners with peripheral hearing deficits, object formation may be ambiguous even in conditions for which a normal-hearing listener has no difficulty. Hearing impairment reduces spectral and temporal resolution, leading to less distinct acoustic features. The reduced spectral resolution of the hearing-impaired listener also increases the amount of acoustic overlap of competing, simultaneous sources. In a normal-hearing listener, peripheral neural channels are frequency-specific; therefore, each individual frequency channel is likely to be dominated by a single source at any one point in time. However, for a hearing-impaired listener with broader than normal peripheral filtering, each frequency channel is more likely to contain a mixture of sound sources, making it impossible to separate which energy is coming from which source (Best, Mason, & Kidd, 2011; Shinn-Cunningham & Best, 2008). These effects help explain the subjective reports of many hearing-impaired listeners (Dai, Best, & Shinn-Cunningham, 2018; Roverud, Best, Mason, Swaminathan, & Kidd, 2016). For instance, hearing aids can make perceptible sound that would otherwise be inaudible; however, many users report that in a noisy setting, improving audibility increases perceptual interference ("When I put in my hearing aid, it just makes the clattering dishes louder!"). These symptoms are consistent with failures of object formation: if the peripheral representation is degraded to the point that objects cannot form, then attentional selection will fail.

Neuroimaging Reveals Neural Correlates of Auditory Scene Analysis

The majority of neurophysiological studies of neural responses to sound mixtures have employed Bregman-like, A-B-A tone sequences (e.g., Itatani & Klump, 2018; Mehta, Jacoby, Yasin, Oxenham, & Shamma, 2017). More recent noninvasive studies in humans, however, have embraced the use of richer sound mixtures made up of speech (e.g., Ding & Simon, 2012), competing melodies (e.g., Xiang, Simon, & Elhilali, 2010), or complex "textures" of ongoing random events (e.g., Barascud, Pearce, Griffiths, Friston, & Chait, 2016). In exploring neural correlates of ASA, here we focus on neuroelectric (electro- and magnetoencephalography, or EEG and MEG) and functional magnetic resonance (fMRI) imaging studies in human listeners.

These two "views" of neural processing afford different strengths and weaknesses (Lee, Larson, Maddox, & Shinn-Cunningham, 2014). Whereas neuroelectric imaging provides exceptional temporal resolution, the neural locus of activity is difficult to determine. Conversely, fMRI provides good spatial precision in defining the brain regions involved in neural processing; however, its temporal resolution is too poor to track responses to individual sound events. While new approaches have been successfully used to wed the temporal resolution of neuroelectric imaging with the spatial resolution of fMRI in visual studies, these approaches have yet to be applied widely in studies of auditory processing (see Cichy & Teng, 2017). Still, the results of neuroelectric and fMRI studies converge to reveal that ASA and attention are controlled by interactions across disperse brain networks.

Neuroelectric imaging tracks neural responses to objects and streams EEG and MEG can both be used to measure the mismatch negativity (MMN), a larger than normal negative response to a syllable that is unexpected in a given stimulus context compared to when it is expected (e.g., Horvath, Czigler, Sussman, & Winkler, 2001; Näätänen, Paavilainen, Rinne, & Alho, 2007). For instance, in a sequence of repeated syllables ("dah, dah, dah"), a different syllable ("bah," known as a *deviant*) will elicit a stronger response than if it occurs in a sequence of "bahs." MMN responses occur relatively early in the neural cascade, typically within 200 ms of the deviant, consistent with a relatively early sensory response in the cortex. Importantly, MMN responses are preattentive (they can be elicited in a listener who is asleep) but tend to be enhanced when a listener is attending to the stream containing the deviant (Horvath et al., 2001; Näätänen et al., 2007).

The MMN provides evidence that syllables form coherent streams when they are expected and predictable and fit with the ongoing stream context. Deviants are inherently more salient than when the same syllable is expected. By deviating from built-up expectations about a stream's content, a deviant is more likely to be heard as a new object and to interrupt top-down processing in the biased competition governing attention. Thus, although rarely discussed in this way, the MMN is an early sensory marker of auditory stream formation, signaling an automatically detected deviation that may constitute a new object (see, for instance, Kaya & Elhilali, 2014b; Sohoglu & Chait, 2016; Southwell et al., 2017).

Early neuroelectric responses evoked by competing streams in a mixture are strongly modulated by attentional focus. In particular, when competing streams contain isolated syllables, the auditory N1 (a negativity in frontocentral electrodes, evoked by early auditory sensory responses), in response to a syllable of greater amplitude when the stream containing the syllable is attended compared to when it is ignored, provides a quantitative index of the strength of top-down attention (Choi, Rajaram, Varghese, & Shinn-Cunningham, 2013). Indeed, the strength of attentional modulation of the N1 response often correlates with how well listeners perform in challenging selective auditory attention tasks (Dai, Best, & Shinn-Cunningham, 2018). Attentional modulation also alters the strength of the correlation between the low-frequency portion of the measured neuroelectric waveform and the envelope of an attended speech stream (Ding, Melloni, Zhang, Tian, & Poeppel, 2016; Ding & Simon, 2013; O'Sullivan et al., 2014; Zion Golumbic et al., 2013). Consistent with behavioral studies demonstrating that it can take time to segregate an attended stream from competing sounds, these neural signatures of attentional focus show a buildup in the brain's ability to lock onto an attended sound stream amid competing sounds (Best, & Shinn-Cunningham, 2018; Riecke, Sack, & Schroeder, 2015; Dai, Best, & Shinn-Cunningham, 2018).

Neuroelectric studies comparing different forms of attentional control reveal that spatial attention tends to recruit activity in the superior prefrontal sulcus (sPCS; see Bharadwaj, Lee, & Shinn-Cunningham, 2014; Hill & Miller, 2010; Lee et al., 2013), while focusing attention on nonspatial features evokes activity in the temporal lobe (Hill & Miller, 2010; Lee et al., 2013). These results show that task demands change how control networks in the brain are activated. Neuroelectric studies of induced responses (looking at changes in the power of neural energy in different frequency bands that is not phase locked) also reveal differences in neural processing that depend on task demands, such as changes in energy in the alpha band (8–12 Hz) over the parietal cortex (see chapter 15). These studies suggest that IPS is engaged by auditory spatial selective attention, presumably by top-down signaling from the prefrontal control regions of the cortex (e.g., see Buschman & Miller, 2007).

Functional magnetic resonance imaging defines brain networks involved in auditory perception fMRI studies have been used to explore which brain regions are involved in different aspects of ASA and the control of auditory cognitive processes. These studies support the view that preattentive, primitive processes, largely in auditory sensory and multisensory regions of the brain, represent the fundamental features that drive object formation (see Nelken, 2004, for a review). Meanwhile, top-down cognitive control networks involving sensory, multisensory, and frontal regions modulate how auditory information is processed (Michalka, Rosen, Kong, Shinn-Cunningham, & Somers, 2016; Noyce, Cestero, Michalka, Shinn-Cunningham, & Somers, 2017; Osher, Tobyne, Congden, Michalka, & Somers, 2015; Shomstein & Yantis, 2004, 2006; Tobyne, Osher, Michalka, & Somers, 2017; Westerhausen et al., 2010). IPS is known to contain multisensory representations. In ASA experiments, IPS is active in the automatic, bottom-up processing that causes objects to form from complex acoustic scenes (Sohoglu & Chait, 2016; Teki et al., 2016).

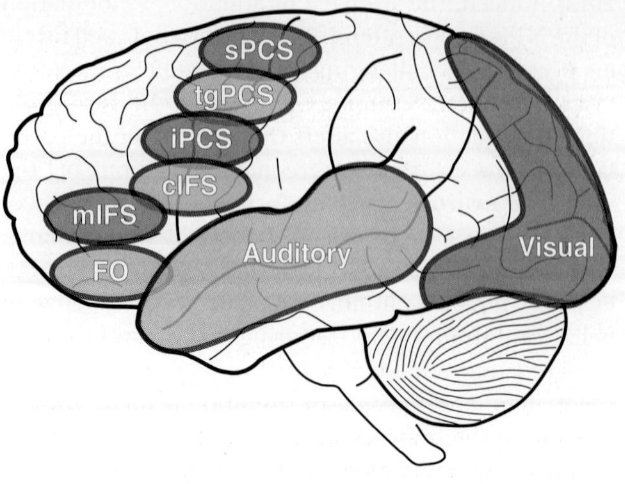

FIGURE 14.3 Schematic illustrating interleaved, discrete executive control regions that are preferentially recruited for visual (*dark gray*) or auditory (*light gray*) attention and working memory. Provided by Dr. Abigail Noyce, Boston University. The regions span from the superior precentral sulcus (*sPCS*, a visually biased region) through to the frontorbital cortex (*FO*, an auditory biased region). Other visually biased regions include the inferior precentral sulcus (*iPCS*) and the medial inferior frontal sulcus (*mIFS*), while auditory biased regions include the transverse gyrus of the precentral sulcus (*tgPCS*) and the caudo inferior frontal sulcus (*cIFS*).

fMRI studies show that auditory spatial selective attention engages areas in the frontoparietal network associated with visual spatial attention (Hill & Miller, 2010; Kong et al., 2014; Michalka et al., 2016). While these results hint that visual and auditory tasks use the same executive function regions, within-subject contrasts of activation in auditory versus in visual tasks reveal alternating sensory-biased regions (see schematic in figure 14.3; Michalka et al., 2016; Noyce et al., 2017). Importantly, if typical methods are used to coregister and average the activity of these caudolateral frontal cortex regions across subjects, the result is a large swath of multisensory control regions, smearing out the consistent alternating pattern (Noyce et al., 2017). Moreover, although these regions are biased toward either visual (figure 14.3, *blue regions*) or auditory (14.3, *orange regions*) inputs, the "more visual" areas are engaged when listeners process spatial, rather than temporal, aspects of an auditory input; conversely, the "more auditory" regions are more active when observers process temporal, rather than spatial, aspects of a visual input. Thus, while previous studies have shown that the caudolateral frontal cortex is part of a multiple-demand network (e.g., see Duncan, 2010; Fedorenko, Duncan, & Kanwisher, 2013), we see more structured, differentiated activity. Together, these studies suggest that multiple, broad brain networks with different processing specializations influence ASA and attention.

Summary

Behavioral and neuroimaging results show that ASA depends on both automatic and attentionally driven processes. Simple, unambiguous mixtures are parsed automatically into distinct perceptual objects, whereas noisier scenes may require both time and focused attention for auditory objects to emerge. In forming auditory objects, the brain relies not only on a prior knowledge of the statistical regularities of natural sounds but on an iterative estimation of ongoing sources, effected through multiple, distributed brain networks.

REFERENCES

Barascud, N., Pearce, M. T., Griffiths, T. D., Friston, K. J., & Chait, M. (2016). Brain responses in humans reveal ideal observer-like sensitivity to complex acoustic patterns. *Proceedings of the National Academy of Sciences, 113*(5), E616–E625.

Best, V., Gallun, F. J., Ihlefeld, A., & Shinn-Cunningham, B. G. (2006). The influence of spatial separation on divided listening. *Journal of the Acoustical Society of America, 120*(3), 1506–1516.

Best, V., Mason, C. R., & Kidd Jr., G. (2011). Spatial release from masking in normally hearing and hearing-impaired listeners as a function of the temporal overlap of competing talkers. *Journal of the Acoustical Society of America, 129*(3), 1616–1625. doi:10.1121/1.3533733

Best, V., Ozmeral, E. J., Kopco, N., & Shinn-Cunningham, B. G. (2008). Object continuity enhances selective auditory attention. *Proceedings of the National Academy of Science, 105*(35), 13174–13178. doi:10.1073/pnas.0803718105

Bharadwaj, H. M., Lee, A. K. C., & Shinn-Cunningham, B. G. (2014). Measuring auditory selective attention using frequency tagging. *Frontiers in Integrative Neuroscience, 8*. doi:10.3389/fnint.2014.00006

Bregman, A. S. (1990). *Auditory Scene Analysis: The Perceptual Organization of Sound.* Cambridge, MA: MIT Press.

Bressler, S., Masud, S., Bharadwaj, H., & Shinn-Cunningham, B. (2014). Bottom-up influences of voice continuity in focusing selective auditory attention. *Psychological Research, 78*(3), 349–360. doi:10.1007/s00426-014-0555-7

Buschman, T. J., & Miller, E. K. (2007). Top-down versus bottom-up control of attention in the prefrontal and posterior parietal cortices. *Science, 315*(5820), 1860–1862.

Choi, I., Rajaram, S., Varghese, L. A., & Shinn-Cunningham, B. G. (2013). Quantifying attentional modulation of auditory-evoked cortical responses from single-trial electroencephalography. *Frontiers in Human Neuroscience, 7*, 115. doi:10.3389/fnhum.2013.00115

Cichy, R. M., & Teng, S. (2017). Resolving the neural dynamics of visual and auditory scene processing in the human brain: A methodological approach. *Philosophical Transactions of the Royal Society B, 372*(1714), 20160108.

Dai, L., Best, V., & Shinn-Cunningham, B. G. (2018). Sensorineural hearing loss degrades behavioral and physiological measures of human spatial selective auditory attention. *Proceedings of the National Academy of Science, 115*(14), E3286–E3295. doi:10.1073/pnas.1721226115

Dalton, P., & Fraenkel, N. (2012). Gorillas we have missed: Sustained inattentional deafness for dynamic events. *Cognition, 124*(3), 367–372.

Darwin, C. J. (1997). Auditory grouping. *Trends in Cognitive Sciences, 1*(9), 327–333.

Darwin, C. J. (2006). Contributions of binaural information to the separation of different sound sources. *International Journal of Audiology, 45*(Suppl 1), S20–24.

Desimone, R., & Duncan, J. (1995). Neural mechanisms of selective visual attention. *Annual Review of Neuroscience, 18*, 193–222.

Ding, N., Melloni, L., Zhang, H., Tian, X., & Poeppel, D. (2016). Cortical tracking of hierarchical linguistic structures in connected speech. *Nature Neuroscience, 19*(1), 158–164. doi:10.1038/nn.4186

Ding, N., & Simon, J. Z. (2012). Emergence of neural encoding of auditory objects while listening to competing speakers. *Proceedings of the National Academy of Science, 109*(29), 11854–11859. doi:10.1073/pnas.1205381109

Ding, N., & Simon, J. Z. (2013). Adaptive temporal encoding leads to a background-insensitive cortical representation of speech. *Journal of Neuroscience, 33*(13), 5728–5735. doi:10.1523/JNEUROSCI.5297-12.2013

Duncan, J. (2010). The multiple-demand (MD) system of the primate brain: Mental programs for intelligent behaviour. *Trends in Cognitive Sciences, 14*(4), 172–179.

Fedorenko, E., Duncan, J., & Kanwisher, N. (2013). Broad domain generality in focal regions of frontal and parietal cortex. *Proceedings of the National Academy of Sciences, 110*(41), 16616–16621.

Griffiths, T. D., & Warren, J. D. (2004). What is an auditory object? *Nature Review Neuroscience, 5*(11), 887–892. doi:10.1038/nrn1538

Hill, K. T., & Miller, L. M. (2010). Auditory attentional control and selection during cocktail party listening. *Cerebral Cortex, 20*(3), 583–590. doi:10.1093/cercor/bhp124

Horvath, J., Czigler, I., Sussman, E., & Winkler, I. (2001). Simultaneously active pre-attentive representations of local and global rules for sound sequences in the human brain. *Brain Research. Cognitive Brain Research, 12*(1), 131–144. doi:S0926-6410(01)00038-6

Ihlefeld, A., & Shinn-Cunningham, B. (2008). Spatial release from energetic and informational masking in a divided speech identification task. *Journal of the Acoustical Society of America, 123*(6), 4380–4392. doi:10.1121/1.2904825

Itatani, N., & Klump, G. M. (2018). Interaction of spatial and non-spatial cues in auditory stream segregation in the European starling. *European Journal of Neuroscience.* doi:10.1111/ejn.13716

Kaya, E. M., & Elhilali, M. (2014a). Investigating bottom-up auditory attention. *Frontiers in Human Neuroscience, 8*, 327. doi:10.3389/fnhum.2014.00327

Kaya, E. M., & Elhilali, M. (2014b). Investigating bottom-up auditory attention. *Frontiers in Human Neuroscience, 8*, 327.

Kaya, E. M., & Elhilali, M. (2017). Modelling auditory attention. *Philosophical Transactions of the Royal Society B, 372*(1714), 20160101.

Kayser, C., Petkov, C. I., Lippert, M., & Logothetis, N. K. (2005). Mechanisms for allocating auditory attention: An auditory saliency map. *Current Biology, 15*(21), 1943–1947.

Kondo, H. M., van Loon, A. M., Kawahara, J.-I., & Moore, B. C. (2017). Auditory and visual scene analysis: An overview. *Philosophical Transactions of the Royal Society B, 372* (1714), 20160099.

Kong, L., Michalka, S. W., Rosen, M. L., Sheremata, S. L., Swisher, J. D., Shinn-Cunningham, B. G., & Somers, D. C. (2014). Auditory spatial attention representations in the human cerebral cortex. *Cerebral Cortex, 24*(3), 773–784. doi:10.1093/cercor/bhs359

Lakatos, P., Barczak, A., Neymotin, S. A., McGinnis, T., Ross, D., Javitt, D. C., & O'Connell, M. N. (2016). Global dynamics of selective attention and its lapses in primary auditory cortex. *Nature Neuroscience, 19*(12), 1707–1717. doi:10.1038/nn.4386

Lee, A. K. C., Larson, E., Maddox, R. K., & Shinn-Cunningham, B. G. (2014). Using neuroimaging to understand the cortical mechanisms of auditory selective attention. *Hearing Research, 307*, 111–120. doi:10.1016/J.Heares.2013.06.010

Lee, A. K. C., Rajaram, S., Xia, J., Bharadwaj, H., Larson, E., Hamalainen, M., & Shinn-Cunningham, B. G. (2013). Auditory selective attention reveals preparatory activity in different cortical regions for selection based on source location and source pitch. *Frontiers in Neuroscience, 6*. doi:10.3389/fnins.2012.00190

Masutomi, K., Barascud, N., Kashino, M., McDermott, J. H., & Chait, M. (2016). Sound segregation via embedded repetition is robust to inattention. *Journal of Experimental Psychology: Human Perception and Performance, 42*(3), 386.

McDermott, J. H. (2009). The cocktail party problem. *Current Biology, 19*(22), R1024–R1027.

Mehta, A. H., Jacoby, N., Yasin, I., Oxenham, A. J., & Shamma, S. A. (2017). An auditory illusion reveals the role of streaming in the temporal misallocation of perceptual objects. *Philosophical Transactions of the Royal Society B, 372*(1714), 20160114.

Michalka, S. W., Kong, L., Rosen, M. L., Shinn-Cunningham, B. G., & Somers, D. C. (2015). Short-term memory for space and time flexibly recruit complementary sensory-biased frontal lobe attention networks. *Neuron, 87*(4), 882–892. doi:10.1016/j.neuron.2015.07.028

Michalka, S. W., Rosen, M. L., Kong, L., Shinn-Cunningham, B. G., & Somers, D. C. (2016). Auditory spatial coding flexibly recruits anterior, but not posterior, visuotopic parietal cortex. *Cerebral Cortex, 26*(3), 1302–1308. doi:10.1093/cercor/bhv303

Middlebrooks, J. C. (2017). Spatial stream segregation. In J. C. Middlebrooks, J. Z. Simon, A. N. Popper, & R. R. Fay (Eds.), *The auditory system at the cocktail party* (pp. 137–168). New York: Springer.

Näätänen, R., Paavilainen, P., Rinne, T., & Alho, K. (2007). The mismatch negativity (MMN) in basic research of central auditory processing: A review. *Clinical Neurophysiology, 118*(12), 2544–2590.

Nahum, M., Nelken, I., & Ahissar, M. (2008). Low-level information and high-level perception: The case of speech in noise. *PLoS Biology, 6*(5), 978–991. doi:e12610.1371/journal.pbio.0060126

Nelken, I. (2004). Processing of complex stimuli and natural scenes in the auditory cortex. *Current Opinion in Neurobiology, 14*(4), 474–480.

Noyce, A. L., Cestero, N., Michalka, S. W., Shinn-Cunningham, B. G., & Somers, D. C. (2017). Sensory-biased and multiple-demand processing in human lateral frontal cortex. *Journal of Neuroscience, 37*(36), 8755–8766.

Osher, D., Tobyne, S., Congden, K., Michalka, S., & Somers, D. (2015). Structural and functional connectivity of visual and auditory attentional networks: Insights from the Human Connectome Project. *Journal of Vision, 15*(12), 223. doi:10.1167/15.12.223

O'Sullivan, J. A., Power, A. J., Mesgarani, N., Rajaram, S., Foxe, J. J., Shinn-Cunningham, B. G., Slaney, M., Shamma, S. A., & Lalor, E. C. (2014). Attentional selection in a cocktail party environment can be decoded from single-trial EEG. *Cerebral Cortex, 25*(7), 1697–1706. doi:10.1093/cercor/bht355

Riecke, L., Sack, A. T., & Schroeder, C. E. (2015). Endogenous delta/theta sound-brain phase entrainment accelerates the buildup of auditory streaming. *Current Biology, 25*(24), 3196–3201. doi:10.1016/j.cub.2015.10.045

Roverud, E., Best, V., Mason, C. R., Swaminathan, J., & Kidd Jr., G. (2016). Informational masking in normal-hearing and hearing-impaired listeners measured in a nonspeech pattern identification task. *Trends in Hearing, 20.* doi:10.1177/2331216516638516

Ruggles, D., & Shinn-Cunningham, B. (2011). Spatial selective auditory attention in the presence of reverberant energy: Individual differences in normal-hearing listeners. *Journal of the Association for Research in Otolaryngology, 12*(3), 395–405. doi:10.1007/s10162-010-0254-z

Schwartz, A. H., & Shinn-Cunningham, B. G. (2010). Dissociation of perceptual judgments of "what" and "where" in an ambiguous auditory scene. *Journal of the Acoustical Society of America, 128*(5), 3041–3051. doi:10.1121/1.3495942

Shinn-Cunningham, B. G. (2008). Object-based auditory and visual attention. *Trends in Cognitive Sciences, 12*(5), 182–186. doi:10.1016/j.tics.2008.02.003

Shinn-Cunningham, B. G., & Best, V. (2008). Selective attention in normal and impaired hearing. *Trends in Amplification, 12*(4), 283–299. doi:10.1177/1084713808325306

Shomstein, S., & Yantis, S. (2004). Control of attention shifts between vision and audition in human cortex. *Journal of Neuroscience, 24*(47), 10702–10706.

Shomstein, S., & Yantis, S. (2006). Parietal cortex mediates voluntary control of spatial and nonspatial auditory attention. *Journal of Neuroscience, 26*(2), 435–439.

Sohoglu, E., & Chait, M. (2016). Detecting and representing predictable structure during auditory scene analysis. *eLife, 5.*

Southwell, R., Baumann, A., Gal, C., Barascud, N., Friston, K., & Chait, M. (2017). Is predictability salient? A study of attentional capture by auditory patterns. *Philosophical Transactions of the Royal Society B, 372*(1714), 20160105.

Teki, S., Barascud, N., Picard, S., Payne, C., Griffiths, T. D., & Chait, M. (2016). Neural correlates of auditory figure-ground segregation based on temporal coherence. *Cerebral Cortex, 26*(9), 3669–3680.

Tobyne, S. M., Osher, D. E., Michalka, S. W., & Somers, D. C. (2017). Sensory-biased attention networks in human lateral frontal cortex revealed by intrinsic functional connectivity. *NeuroImage, 162*, 362–372.

Westerhausen, R., Moosmann, M., Alho, K., Belsby, S.-O., Hämäläinen, H., Medvedev, S., Specht, K., & Hugdahl, K. (2010). Identification of attention and cognitive control networks in a parametric auditory fMRI study. *Neuropsychologia, 48*(7), 2075–2081.

Woods, K. J., & McDermott, J. H. (2015). Attentive tracking of sound sources. *Current Biology, 25*(17), 2238–2246. doi:10.1016/j.cub.2015.07.043

Xiang, J., Simon, J., & Elhilali, M. (2010). Competing streams at the cocktail party: Exploring the mechanisms of attention and temporal integration. *Journal of Neuroscience, 30*(36), 12084–12093. doi:10.1523/JNEUROSCI.0827-10.2010

Zion Golumbic, E. M., Ding, N., Bickel, S., Lakatos, P., Schevon, C. A., McKhann, G. M., Goodman, R. R., Emerson, R., Mehta, A. D., Simon, J. Z., Poeppel, D., & Schroeder, C. E. (2013). Mechanisms underlying selective-neuronal tracking of attended speech at a "cocktail party." *Neuron, 77*(5), 980–991. doi:10.1016/j.neuron.2012.12.037

15 Neural Filters for Challenging Listening Situations

JONAS OBLESER AND JULIA ERB

ABSTRACT The listening brain cannot be studied without asking how attention in general operates and how attention manifests on a neural level. In this chapter we focus on the challenges that arise for human listeners as they choose what to attend and what to ignore in busy acoustic scenes. We introduce and discuss a set of nonexclusive neural filtering mechanisms in the temporal, spatial, and spectral domain. A special role here is suggested for such filter mechanisms that operate based on neural oscillations, with complementary roles for the phase alignment of slow neural oscillations in the approximate 1–8 Hz range versus the magnitude of alpha neural oscillations (approximately 8–12 Hz). We end on integrating these with knowledge on the functional neuroanatomy of auditory cortex as well as the key roles that domain-general networks play in auditory attention.

Introduction

To listen equals to attend, with our auditory senses, to an object in an auditory scene. Thus, *to listen* does of course not match *to see* but rather *to watch* in the much better understood visual domain. The listening brain cannot be studied without considering how attention operates in general; how attention might be implemented neurally; how attention in the auditory modality spells out; and how domain-specific, auditory processes interface with domain-general, attention-related neural processes.

In this chapter we will adopt a perspective that asks, in a Marr'ian sense (Marr, Ullman, & Poggio, 2010): If a listener's behavioral goal is to achieve successful listening in the face of acoustic and cognitive adversity, what are the necessary computations to achieve this goal? We will introduce a filter metaphor, which holds a special place in attention research and which has proven useful in the auditory, entirely time-dependent, domain, and then present potential neural implementations.

What is signal and what is noise is determined only in the instant by the listener and her goals. There is thus hardly a more "top-down" process fathomable in human cognitive neuroscience than listening in the presence of a concurrent sound. Hence, studying the brain in its ability to implement and maintain such neural filters in time, space, and frequency (spectrum) is fundamental to our understanding of the neurocognitive phenomenon of listening and communication as a whole.

What Characterizes a "Neural Filter" Suitable for Attentive Listening?

For successful listening under adverse conditions, we require (1) a set of mechanisms establishing and maintaining the behavioral goal and the listening target, (2) a mechanism that prioritizes the target signal (i.e., *sensory gain*), and (3) a gating, controlling, or inhibiting mechanism that supports the suppression of interfering sound sources or, more generally, distracting information.

Let us now look at suggested neural implementations for these key ingredients. Studying the listening brain calls for the neuroimaging methods best suited to match and capture acoustic communication signals in terms of temporal and spectral resolution. We will thus focus on neural oscillations, mainly through the lens of magneto- and electroencephalography (MEG, EEG). Where the functional neuroanatomical organization is key, we will emphasize results from nonhuman animal electrophysiology and from human functional magnetic resonance imaging (fMRI).

A Temporal Filter for Attentive Listening: Realignment of Neural Excitability to an Attended Sound Source (Neural Entrainment)

Neural oscillations are often quasisinusoidal (figure 15.1*A*) and are detectable at many frequencies and on many levels of granularity in the neural signal (Buzsaki & Draguhn, 2004). It is reasonable to assume that low-frequency neural oscillations in a range between approximately 0.5 to 1 up to 8 to 10 Hz (i.e., the classic delta and theta bands of the brain's frequency spectrum; Buzsaki & Draguhn, 2004) hold significance in selecting and prioritizing to-be-attended auditory input. Why is this? First, the long-term average modulation spectra of speech (Rosen, 1992) and music (Ding

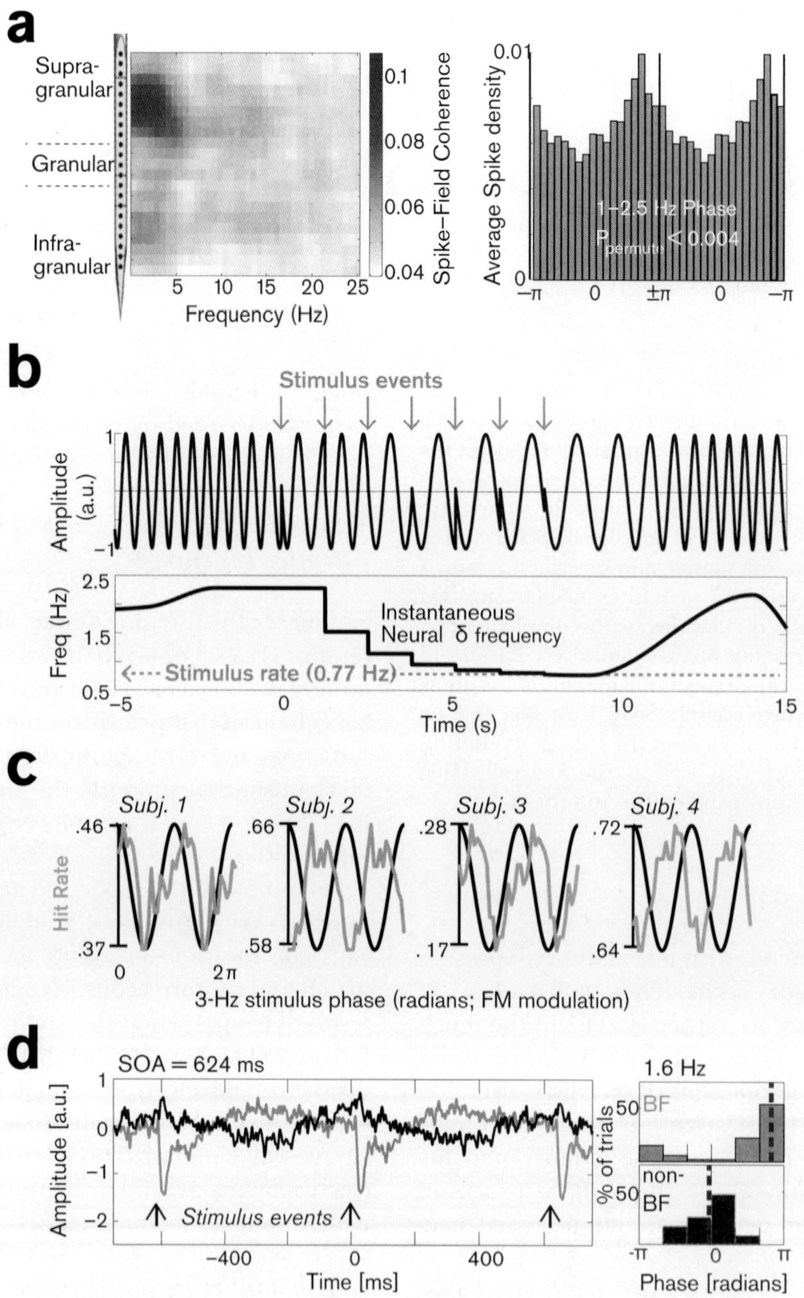

FIGURE 15.1 *A*, Slow neural oscillatory phase in supragranular layers of auditory cortex governs excitability (i.e., spike rate) in a granular layer. Spontaneous data recorded in A1 of an awake macaque. *Left panel*, spike-LFP (local field potential) coherence as a function of frequency (x-axis) and cortical depth (y-axis). *Right panel*, the modulation of granular-layer spike rate by supragranular 1–2.5 Hz (delta) LFP phase. Unpublished data by Jonas Obleser and Peter Lakatos. *B*, Neural oscillatory frequencies can become entrained to stimulus regularities. Here, an approximately 2 Hz neural oscillation adjusts its instantaneous phase (*top panel*; note the "phase slips" in the black waveform when stimulus events, *light gray on top*, occur) and frequency (*black line, bottom panel*) to the entraining 0.77 Hz stimulus rate (*light gray arrow and dashed line, bottom panel*). After stimulus-train offset, the endogenous frequency of approximately 2 Hz is reinstated. Adapted from Obleser, Henry, and Lakatos (2017). *C*, Listening success (target detection) becomes entrained by stimulus regularity. Four individuals from the Henry and Obleser (2012) data set are replotted. Irrespective of average gap-detection performance, detection ("Hit rate," *gray trace*) is a function of the phase of the 3 Hz stimulus modulation (*black trace*) that was employed. Note the individual differences in phase lag: some participants' performance is in phase with the stimulus modulation, while others' is out of phase. *D*, Slow-frequency neural entrainment in A1 as a spectrotemporal filter. Attended (BF, best-frequency, *gray*) and unattended (non-BF, *black*) sites in A1 entrain counterphasic at the stimulation rate (here, 1/624 ms or 1.6 Hz), resulting in amplified, sharpened responses at attended time points. Adapted from Lakatos et al. (2013). (See color plate 17.)

et al., 2017) hold most of their energy in this range. Second, there is now good evidence that inherent fluctuations of neural excitability in sensory cortices can reset or readjust (figure 15.1*B*) such that a stable alignment of the *duty cycle* (i.e., the excitable phase of a neural oscillation) with the dominant "rhythm" of the stimulus is achieved (Kayser, Ince, Gross, & Kayser, 2015; Obleser, Henry, & Lakatos, 2017)—a process conventionally labeled *neural entrainment.* Prima facie evidence for this phase alignment was obtained in the primary auditory cortex (A1) of the macaque (Lakatos, Karmos, Mehta, Ulbert, & Schroeder, 2008) while the general principle has been amply demonstrated at multiple timescales in multiple systems and has been elicited by auditory, visual, and somatosensory, as well as electrical, stimulus trains (for a review, see, e.g., Haegens & Zion Golumbic, 2018).

Neural entrainment and, more broadly defined, *speech-tracking* have proven to be game changers in studying auditory attention. Compared to the envelope of an ignored speech signal, the envelope of attended speech receives larger-amplitude tracking in auditory cortical areas (e.g., Ding & Simon, 2012; Fiedler et al., 2017). These attentional speech-tracking differences can be used to infer the attentional focus of listeners (e.g., O'Sullivan et al., 2015; see Forte, Etard, & Reichenbach [2017] for applying this rationale even to much earlier brain stem responses) or to establish an attentional hierarchy, with higher-order areas defined as almost exclusively representing the attended signal (e.g., Mesgarani & Chang, 2012; Zion Golumbic et al., 2013).

The power of the entrainment concept manifests in the flexibility of the listener to adjust such neural oscillations in the presence of multiple concurrent auditory streams. Accordingly, the notion of a "spectrotemporal filter" was introduced in a study by Lakatos et al. (2013; figure 15.1*D*). The authors recorded along the depth of macaque primary auditory cortex at two sites that, critically, differed in their best-frequency tuning by about two octaves. They then played two simultaneous streams with distinct time signatures consisting of pure tones that either matched or did not match the best-frequency tuning of the recording site. The authors did not only observe selective neural entrainment to the attended stream (i.e., to a particular frequency). Also, "optimal" alignment to the excitable ("up") neural phase only occurred in the *receptive field* (tonotopic area) selective for the attended frequency. Areas rather selective for other nonattended frequencies also became entrained to their "down" phase, effectively inhibiting—or so the interpretation goes—further cortical processing of nonattended stimulus information (figure 15.2*D*).

Although in human electrophysiology we are for most accounts lacking the concurrent temporal and spatial fidelity, there is compelling evidence that ongoing neural oscillatory activity becomes realigned to some degree by stimulus regularities: human behavior—namely, the detection and the discrimination of auditory sensory input near the perceptual threshold—can be biased depending on the phase of the neural excitability cycle it meets (Henry & Obleser, 2012; Rice & Hagstrom, 1989; Wilsch, Neuling, Obleser, & Herrmann, 2018).

In a set of studies (figure 15.1*C*), Molly Henry and colleagues showed that stimulation with frequency-modulated tones at a rate of 3 Hz first lead to increased phase coherence (i.e., increased trial-to-trial consistency in the neural phase), a process commensurate with neural entrainment. Second, the psychoacoustic detection of faint targets (gaps) close to the perceptual threshold was strongly modulated by phase: the hit rate appeared as a sinusoidal function of the phase of the acoustic 3 Hz modulation cycle at which a target occurred.

This leaves us with the appealing idea that regularities in our acoustic environment, especially those we attend to (Henry, Herrmann, Kunke, & Obleser, 2017), entrain endogenous neural activity and thus bias, shape, or even "optimize" our perception. However, note that any "good" or "optimal" neural phase corresponding to a particularly receptive state (Weisz et al., 2014) will be inevitably followed by a "poor" or "suboptimal" neural state (Ng, Schroeder, & Kayser, 2012). Thus, neural entrainment to an external rhythm is not always helpful. For example, if a listener's goal is to attend to an auditory signal or await an upcoming event as vigilantly as possible, a "rhythmic" neural filter might not be ideal. Instead, a suppression of slow neural oscillations altogether with their optimal but also suboptimal phases might be desirable. Such a mode of suppressed slow neural oscillations has been suggested, labeled a vigilant or continuous-processing mode (Herrmann, Henry, Haegens, & Obleser, 2016; Schroeder & Lakatos, 2009). Enter our second implementation of a neural filter for listening: a regime of high neural alpha/beta oscillatory power (8–12 and 15–25 Hz, respectively).

Alpha (8–12 Hz) Power: A Spatiotemporal Filter for Inhibition and Selection

In a 1944 paper with the disarmingly simple title "Brain Rhythms," Lord Adrian presented a lone EEG trace with one obvious, high-amplitude burst of an approximately 10 Hz waveform, which Adrian labeled as *hearing* (figure 15.2*A*). Relabeling this as *listening,* we can

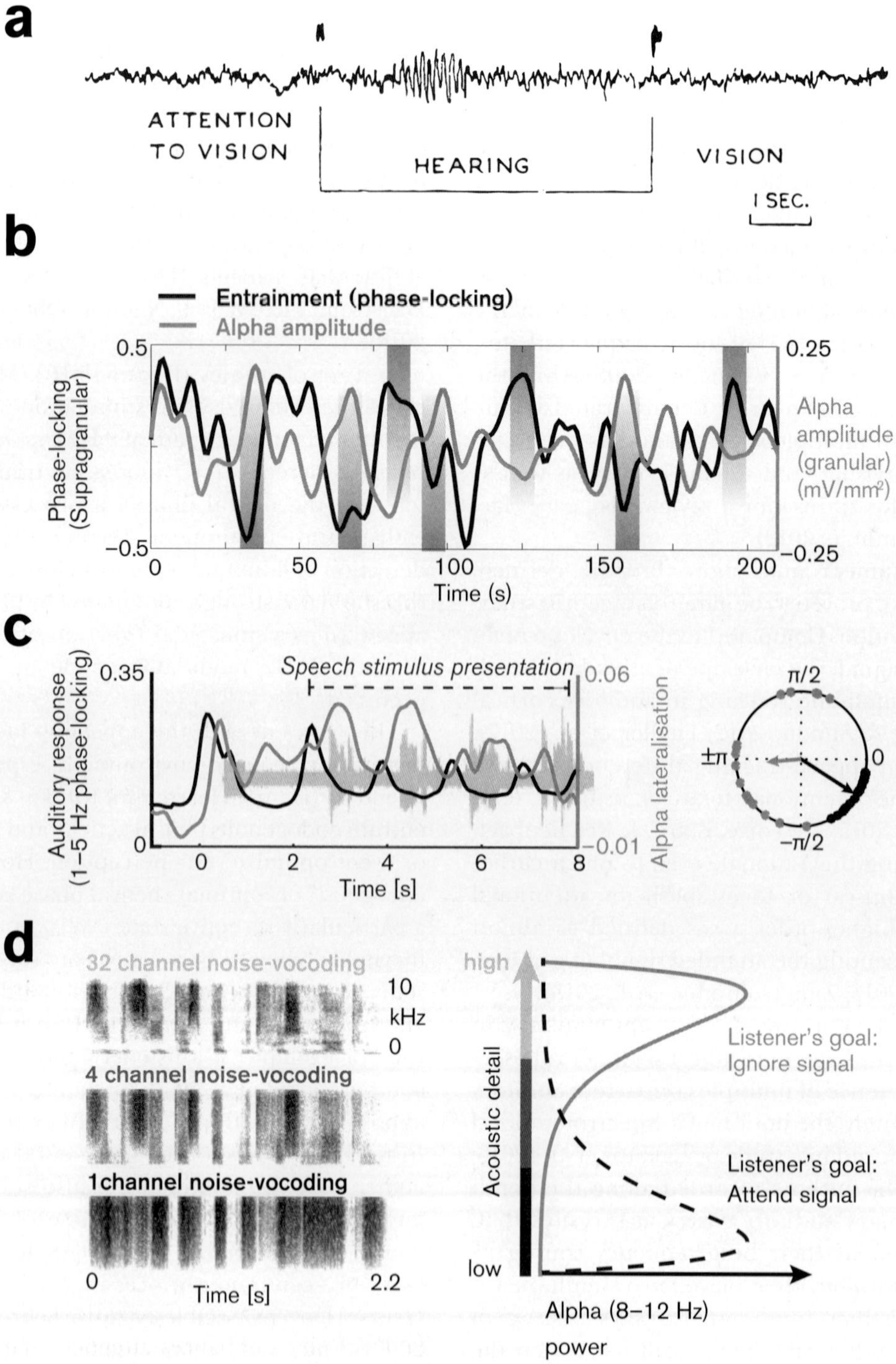

FIGURE 15.2 *A*, Listening (attention to sound) increases 8–12 Hz occipitoparietal alpha power as recorded at scalp. "During the middle section of the record the attention is concentrated on the tick of a watch" (Adrian, 1944, p. 361). Note that *listening* instead of *hearing* would have been more fitting. Reprinted with permission from Adrian (1944). *B*, Locally, in auditory cortex (macaque A1), states of high alpha power trade off with states of slow-oscillatory neural entrainment, taken to reflect neurally incommensurate brain states. Adapted from Lakatos et al. (2016). *C*, Globally, states of alpha-power lateralization trade off with states of slow-oscillatory neural entrainment during spatial selective attention to speech in lieu of a distracting speech stream. The neural entrainment response in auditory cortex (*black lines, dots, arrows*) is out of phase and temporally precedes the alpha lateralization response (*gray lines, dots, arrows*) from temporal as well as parietal cortex during a challenging dichotic listening task. Adapted from Wöstmann et al. (2016). *D*, Alpha power depends entirely on a listener's goals. While acoustic degradation (i.e., less acoustic detail) increases centroparietal alpha power in the EEG (illustrative data only) when the signal is attended, this effect can reverse completely when the signal is ignored. Adapted from Wöstmann, Lim, and Obleser (2017) and Obleser and Weisz (2012).

take Adrian's observation as the first of many to come that high-powered alpha oscillations, most likely emerging from occipitoparietal cortex, can indicate a listener's attention to sound (Foxe, Simpson, & Ahlfors, 1998; Strauss, Wöstmann, & Obleser, 2014; Weisz, Hartmann, Muller, Lorenz, & Obleser, 2011). A currently prevailing idea is that high-powered alpha oscillations are a neural means of *functional inhibition* (Jensen & Mazaheri, 2010), suppressing interfering or distracting neural processing in a target area (e.g., in visuospatial cortex in the case of auditory attention). In fact, the more intelligible a speech signal and thus the less effortful attentive listening becomes, the less prominent the occipitoparietal alpha power change will be (Obleser & Weisz, 2012).

More generally, we assume that alpha oscillations also pose a modulatory signal to auditory activity: first, alpha (and to a degree, beta, ~15–25 Hz) oscillations have been identified as top-down signals that enter cortex (e.g., primary visual cortex V1) predominantly in efferent layers (van Kerkoerle et al., 2014). Second, high-power alpha oscillations reduce the flow of "information," expressed as spike rate, in a neuronal population (e.g., Spaak, Bonnefond, Maier, Leopold, & Jensen, 2012) or in a Shannon/information-theoretic sense (Hanslmayr, Staudigl, & Fellner, 2012). Third, and most relevant to audition, states of high alpha power can occur independent from states of solid neural entrainment (figure 15.2; see also Zoefel & VanRullen, 2017). There might even be a trade-off or mutual exclusivity between states of high alpha power and states of auditory entrainment, as observed in nonhuman primate auditory cortex over the time course of minutes (Lakatos et al., 2016; figure 15.2*B*). Commensurate with the idea of functional inhibition, a state of high alpha power in auditory cortex might indicate the inhibition of neural communication or of the encoding of auditory input. For example, human listeners over the course of a stream of dichotically presented speech showed a stimulation-driven fluctuation of alpha-power lateralization in temporal (auditory) as well as in parietal cortex that peaked anticyclically to stimulus-induced auditory entrainment (Wöstmann et al., 2016; figure 15.2*C*). Not least, across human listeners of varying ages in a gap-detection task (as outlined previously), the individual degree of occipitoparietal alpha power suppressed or counteracted the effect that neural entrainment had on behavior (Henry et al., 2017).

Summing the available evidence, alpha power at the scalp level can be surprisingly informative about a listener's behavioral goals. Recently, Wöstmann, Lim, & Obleser (2017) instructed listeners to ignore some acoustically degraded speech in an auditory scene; the actual task was to retain in memory a sequence of digits. Strikingly, the above-mentioned pattern of *decreasing* occipitoparietal alpha power with increasing target-speech intelligibility (Obleser & Weisz, 2012) completely reversed. Thus, the alpha response is a surprisingly clear-cut proxy for a listener's goal (i.e., to attend to vs. to ignore speech; figure 15.2*D*; Wöstmann, Lim, & Obleser, 2017).

Nowadays, brain-computer-interface (BCI) researchers, hearing-aid engineers, and cognitive neuroscientists all study the challenges a listener is confronted with in real-world scenarios. There is hope that the temporally, spatially, and spectrally selective up- and down-regulations of alpha power as well as auditory neural entrainment (or *speech tracking*; Wöstmann, Fiedler, & Obleser, 2017) will ultimately serve as useful indicators of a listener's attentional control.

The Functional Neuroanatomy of Attentive Listening

Moving beyond primarily *temporal* filter mechanisms as physiological implementations of selective attention, let us now review evidence for the influence of attention on the *spatial* layout or topography of acoustic feature representations. Attention to auditory stimuli modulates sound representations along the entire auditory system up to the prefrontal and parietal cortex.

A well-established organizational principle of the auditory system is tonotopy (i.e., the spatially ordered representation of frequency; see figure 15.3*A*; e.g., Baumann et al., 2015; Formisano et al., 2003; Morel, Garraghty, & Kaas, 1993). More recently, an ordered spatial distribution of spectrotemporal modulations in auditory cortex has been additionally suggested (Baumann et al., 2015; Santoro et al., 2014). But how can attentive listening affect such topographical representations?

At the level of A1, the top-down demands of a behavioral task can rapidly change the gain or shape of receptive fields (see figure 15.3*C*). Frequency-selective attention acts as a spectrotemporal filtering mechanism that selectively enhances attended frequencies in tonotopically organized primary auditory cortex, and responses to least-preferred frequencies become suppressed (e.g., Fritz, Elhilali, & Shamma, 2007). This is akin to the spectrotemporal filter mechanism illustrated above (Lakatos et al., 2013). Long-term selective attention to stimulus features (here, extensive frequency discrimination training) further manifests in an increase of cortical surface dedicated to the processing of those features (Recanzone, Schreiner, & Merzenich, 1993).

In humans, few fMRI studies have investigated frequency-specific attentional effects on cortical sound representations. Here also, frequency-selective attention to tones in a multitone scene induced measurable changes in the tonotopic representation of the attended frequencies in primary auditory cortex (Da Costa, van der Zwaag, Miller, Clarke, & Saenz, 2013; Riecke et al., 2017).

Effects of attention are theorized to be specific to cortical layers (see above): bottom-up feedforward information enters at middle cortical layers, whereas top-down feedback information arrives at superficial and deep layers (figure 15.3*B*). Layer-specific sharpening of tuning width has been observed in macaque A1 (O'Connell, Barczak, Schroeder, & Lakatos, 2014). Recent advances in laminar fMRI at high field strengths (≥7 Tesla) provide insight into the selective activation of cortical layers noninvasively in humans (Lawrence, Formisano, Muckli, & de Lange, 2017). So far, one human laminar fMRI study has shown that an auditory pitch direction categorization task (compared to a visual change detection task) sharpened the frequency tuning in superficial (i.e., supragranular) more than in granular or infragranular layers (De Martino et al., 2015).

An unsolved question remains which, if any, brain area would show "first," "originating," or "causal" attention signals—that is, which brain area could be ascribed an attention-allocating role. From visual attention research, a particular role of prefrontal cortex and the frontal eye field is plausible, both anatomically and conceptually (Buschman & Kastner, 2015). The same holds for the intraparietal sulcus, but the relative hierarchy of the prefrontal and parietal cortex is a matter of ongoing debate. Closely related, it is all too unclear how far the recent progress on correlates of (i.e., not necessarily to be equated with causes of) attention in the visual domain (e.g., Fiebelkorn, Saalmann, & Kastner, 2013; Landau et al., 2015) will translate to auditory attention.

For visual attention, the frontal and parietal cortices are thought to constitute maps of prioritized space: retinotopic maps in higher-order areas of parietal cortex are primarily sensitive to attentional rather than sensory input (Jerde, Merriam, Riggall, Hedges, & Curtis, 2012). Are there attentional priority maps for auditory processing analogous to the visual system? If we consider frequency to be to audition what space is to vision, we would expect tonotopic organization to be retrievable from prefrontal cortex. However, the topographical organization of acoustic features appears to break down beyond auditory cortex. Instead, higher-order nonlinear transformations appear to take place in prefrontal cortex (Averbeck & Romanski, 2006; Plakke & Romanski, 2014). There is, however, preliminary evidence for attentional priority implemented in frontal cortex. For example, frontal cortex was found, using human fMRI, to flexibly represent any attended sound in a scene regardless of its category (and thus its acoustic features; Hausfeld, Riecke, & Formisano, 2018).

What about sensory degradation, which presents listeners with a "systems-level challenge that requires the allocation of executive cognitive resources" (Peelle, 2018)? The cingulo-opercular system has been consistently implicated in what has been termed *listening effort* (Alain, Du, Bernstein, Barten, & Banai, 2018). Most likely, it represents a domain-general system for executive function in difficult perceptual tasks (e.g., Vaden et al., 2013). For instance, both the recognition of spectrally degraded (vocoded) speech and increasingly difficult temporal modulation discrimination evoke enhanced fMRI responses in the insula and anterior cingulate cortex (Erb, Henry, Eisner, & Obleser, 2013; see figure 15.3*D*; vocoded speech is shown in figure 15.2*D*). Recruitment of this cingulo-opercular network is altered in older adults: it is active already under clear listening conditions, and hearing loss leads to higher dependence on the cingulo-opercular system (Erb & Obleser, 2013; see figure 15.3*E*). In the absence of any stimulation and task, the functional magnetic resonance imaging (fMRI) signal in the cingulo-opercular network is positively correlated with spontaneous alpha oscillations as measured in simultaneous EEG-fMRI recordings (Sadaghiani et al., 2010), where high alpha power has been taken as an index for "alertness" (see above). In a working-memory task under conditions of high cognitive load, increased alpha power was localized to the insular cortex using MEG (Wilsch, Henry, Herrmann, Maess, & Obleser, 2015). Together, these findings suggest a role of the cingulo-opercular system in the expression of alpha oscillations, likely serving the maintenance of auditory attention.

Conclusions

In this chapter we have discussed different neural filter mechanisms for auditory attention. We have emphasized that signal and noise, or target and distractor, can swap swiftly depending on a listener's behavioral goals and that listeners uphold this flexibility to switch under quite adverse acoustic conditions. We have also seen how the "central" stage of auditory attention, primary auditory cortex, cannot be viewed in isolation. Both electrophysiology and functional neuroimaging show clearly that interactions with the frontoparietal

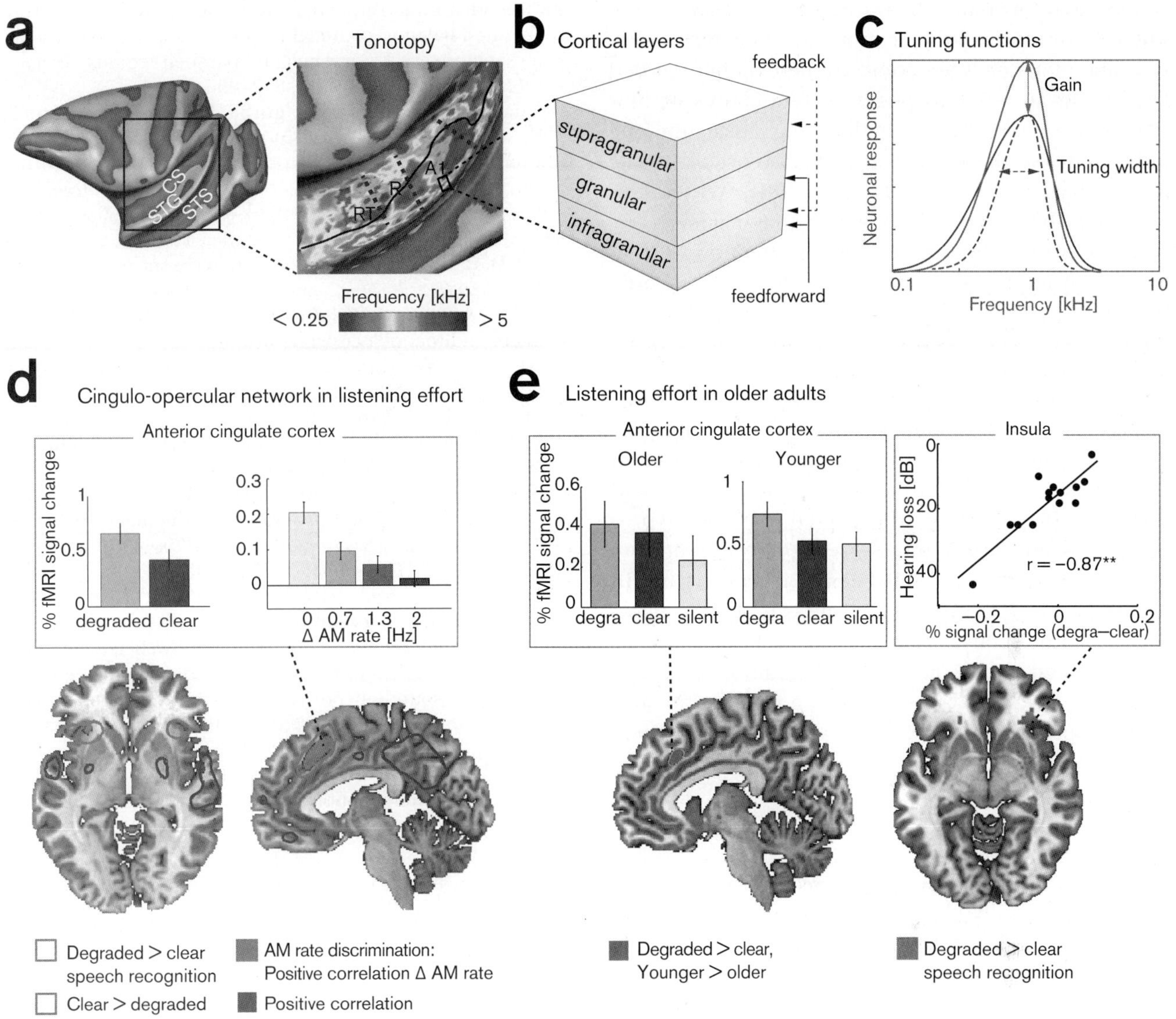

FIGURE 15.3 *A*, Tonotopy mapped with natural sounds. Tonotopic map is shown on the surface of the inflated left hemisphere of one macaque. Modified from Erb et al. (2018). *B*, Schematic of cortical layers in A1 and their inputs: bottom-up sensory feedforward information enters at deep and middle cortical layers; top-down feedback information arrives at superficial and deep layers (see also figure 15.1*A*). *C*, Task demands shape the gain or tuning width of neuronal (population) frequency response functions in a layer-dependent manner (De Martino et al., 2015; O'Connell et al., 2014). *D*, Attentive listening to spectrally degraded compared to clear speech evokes enhanced fMRI responses in insula and anterior cingulate cortex (*top panel, left; bottom panel: contours of the map of the speech degradation effects*). For amplitude modulation (AM) rate discrimination, activity levels parametrically increase in the same areas with decreasing AM rate difference between standard and deviant (Δ AM rate; note that this corresponds to an increasing difficulty level, *top panel, right*). Modified from Erb et al. (2013). *E*, An age-by-degradation interaction in the anterior cingulate cortex is driven by a decreased dynamic range in the older listeners who show an enhanced fMRI signal both in clear and degraded conditions (*left*). Hearing loss correlates with the fMRI signal difference between clear and degraded speech in the insula (*right*). Modified from Erb and Obleser (2013). *Note:* CS: circular sulcus; STG/STS: superior temporal gyrus/sulcus; AM: amplitude modulation; ** $p < 0.001$. (See color plate 18.)

and cingulo-opercular networks need consideration when we want to build a more mechanistic understanding of auditory attention. We here all but ignored fascinating attempts to push the boundaries of how attentional filters shape our listening down into the brain stem and the auditory periphery (Forte, Etard, & Reichenbach, 2017; Gruters et al., 2018; Wittekindt, Kaiser, & Abel, 2014). It is thus likely that in a few years' time the story to be told of auditory attention in challenging listening situations will have complicated further.

Acknowledgments

We are indebted to numerous colleagues who, over many years and after many guest lectures on this and closely related topics, have helped shape the ideas laid out here. In particular, the invited speakers at two workshops on Signal and Noise along the Auditory Pathway (SNAP) in 2013 in Leipzig, Germany, and in 2017 in Lübeck, Germany, have helped develop many of the leading thoughts of this chapter. Special thanks are extended to Maria Chait, Alexandra Bendixen, Molly Henry, Ingrid Johnsrude, and Peter Lakatos. This chapter is dedicated to the memory of Richard J. S. Wise.

REFERENCES

Adrian, E. D. (1944). Brain rhythms. *Nature, 153,* 360–362.

Alain, C., Du, Y., Bernstein, L. J., Barten, T., & Banai, K. (2018). Listening under difficult conditions: An activation likelihood estimation meta-analysis. *Human Brain Mapping, 39*(7), 2695–2709.

Arcaro, M. J., Pinsk, M. A., & Kastner, S. (2015). The anatomical and functional organization of the human visual pulvinar. *Journal of Neuroscience, 35*(27), 9848–9871.

Averbeck, B. B., & Romanski, L. M. (2006). Probabilistic encoding of vocalizations in macaque ventral lateral prefrontal cortex. *Journal of Neuroscience, 26*(43), 11023–11033.

Baumann, S., Joly, O., Rees, A., Petkov, C. I., Sun, L., Thiele, A., & Griffiths, T. D. (2015). The topography of frequency and time representation in primate auditory cortices. *eLife, 4.*

Bregman, A. S. (1994). *Auditory scene analysis: The perceptual organization of sound.* Cambridge, MA: MIT Press, Bradford Books.

Buschman, T. J., & Kastner, S. (2015). From behavior to neural dynamics: An integrated theory of attention. *Neuron, 88*(1), 127–144.

Buzsaki, G., & Draguhn, A. (2004). Neuronal oscillations in cortical networks. *Science, 304*(5679), 1926–1929.

Corbetta, M., & Shulman, G. L. (2002). Control of goal-directed and stimulus-driven attention in the brain. *Nature Reviews Neuroscience, 3*(3), 201–215.

Da Costa, S., van der Zwaag, W., Miller, L. M., Clarke, S., & Saenz, M. (2013). Tuning in to sound: Frequency-selective attentional filter in human primary auditory cortex. *Journal of Neuroscience, 33*(5), 1858–1863.

de la Mothe, L. A., Blumell, S., Kajikawa, Y., & Hackett, T. A. (2012). Thalamic connections of auditory cortex in marmoset monkeys: Lateral belt and parabelt regions. *Anatomical Record (Hoboken), 295*(5), 822–836.

De Martino, F., Moerel, M., Ugurbil, K., Goebel, R., Yacoub, E., & Formisano, E. (2015). Frequency preference and attention effects across cortical depths in the human primary auditory cortex. *Proceedings of the National Academy of Sciences of the United States of America, 112*(52), 16036–16041.

Ding, N., Patel, A. D., Chen, L., Butler, H., Luo, C., & Poeppel, D. (2017). Temporal modulations in speech and music. *Neuroscience & Biobehavioral Reviews, 81*(Part B), 181–187.

Ding, N., & Simon, J. Z. (2012). Emergence of neural encoding of auditory objects while listening to competing speakers. *Proceedings of the National Academy of Sciences of the United States of America, 109*(29), 11854–11859.

Dosenbach, N. U., Fair, D. A., Cohen, A. L., Schlaggar, B. L., & Petersen, S. E. (2008). A dual-networks architecture of top-down control. *Trends in Cognitive Sciences, 12*(3), 99–105.

Duncan, J. (2010). The multiple-demand (MD) system of the primate brain: Mental programs for intelligent behavior. *Trends in Cognitive Sciences, 14*(4), 172–179.

Elliott, T. M., & Theunissen, F. E. (2009). The modulation transfer function for speech intelligibility. *PLoS Computational Biology, 5*(3), e1000302.

Erb, J., Armendariz, M., De Martino, F., Goebel, R., Vanduffel, W., & Formisano, E. (2018). Homology and specificity of natural sound-encoding in human and monkey auditory cortex. *Cerebral Cortex.* doi:10.1093/cercor/bhy243. Advance online publication

Erb, J., Henry, M. J., Eisner, F., & Obleser, J. (2012). Auditory skills and brain morphology predict individual differences in adaptation to degraded speech. *Neuropsychologia, 50*(9), 2154–2164.

Erb, J., Henry, M. J., Eisner, F., & Obleser, J. (2013). The brain dynamics of rapid perceptual adaptation to adverse listening conditions. *Journal of Neuroscience, 33*(26), 10688–10697.

Erb, J., & Obleser, J. (2013). Upregulation of cognitive control networks in older adults' speech comprehension. *Frontiers in Systems Neuroscience, 7,* 116.

Fiebelkorn, I. C., Saalmann, Y. B., & Kastner, S. (2013). Rhythmic sampling within and between objects despite sustained attention at a cued location. *Current Biology, 23*(24), 2553–2558.

Fiedler, L., Wöstmann, M., Graversen, C., Brandmeyer, A., Lunner, T., & Obleser, J. (2017). Single-channel in-ear-EEG detects the focus of auditory attention to concurrent tone streams and mixed speech. *Journal of Neural Engineering, 14*(3), 036020.

Formisano, E., Kim, D. S., Di Salle, F., van de Moortele, P. F., Ugurbil, K., & Goebel, R. (2003). Mirror-symmetric tonotopic maps in human primary auditory cortex. *Neuron, 40*(4), 859–869.

Forte, A. E., Etard, O., & Reichenbach, T. (2017). The human auditory brainstem response to running speech reveals a subcortical mechanism for selective attention. *eLife, 6.*

Foxe, J. J., Simpson, G. V., & Ahlfors, S. P. (1998). Parieto-occipital approximately 10 Hz activity reflects anticipatory state of visual attention mechanisms. *Neuroreport, 9*(17), 3929–3933.

Fritz, J., Elhilali, M., & Shamma, S. A. (2007). Adaptive changes in cortical receptive fields induced by attention to complex sounds. *Journal of Neurophysiology, 98*(4), 2337–2346.

Garrod, S., & Pickering, M. (2016). *Language processing.* Abingdon, UK: Taylor & Francis.

Gruters, K. G., Murphy, D. L. K., Jenson, C. D., Smith, D. W., Shera, C. A., & Groh, J. M. (2018). The eardrums move when the eyes move: A multisensory effect on the mechanics of hearing. *Proceedings of the National Academy of Sciences of the United States of America, 115*(6), E1309–E1318.

Haegens, S., & Zion Golumbic, E. (2018). Rhythmic facilitation of sensory processing: A critical review. *Neuroscience & Biobehavioral Reviews, 86,* 150–165.

Hanslmayr, S., Staudigl, T., & Fellner, M. C. (2012). Oscillatory power decreases and long-term memory: The information via desynchronization hypothesis. *Frontiers in Human Neuroscience, 6,* 74.

Hausfeld, L., Riecke, L., & Formisano, E. (2018). Acoustic and higher-level representations of naturalistic auditory scenes in human auditory and frontal cortex. *NeuroImage, 173,* 472–483.

Henry, M. J., Herrmann, B., Kunke, D., & Obleser, J. (2017). Aging affects the balance of neural entrainment and top-down neural modulation in the listening brain. *Nature Communications, 8,* 15801.

Henry, M. J., & Obleser, J. (2012). Frequency modulation entrains slow neural oscillations and optimizes human listening behavior. *Proceedings of the National Academy of Sciences of the United States of America, 109*(49), 20095–20100.

Herrmann, B., Henry, M. J., Haegens, S., & Obleser, J. (2016). Temporal expectations and neural amplitude fluctuations in auditory cortex interactively influence perception. *NeuroImage, 124*(Pt A), 487–497.

Jensen, O., & Mazaheri, A. (2010). Shaping functional architecture by oscillatory alpha activity: Gating by inhibition. *Frontiers in Human Neuroscience, 4,* 186.

Jerde, T. A., Merriam, E. P., Riggall, A. C., Hedges, J. H., & Curtis, C. E. (2012). Prioritized maps of space in human frontoparietal cortex. *Journal of Neuroscience, 32*(48), 17382–17390.

Kayser, C., Ince, R. A., & Panzeri, S. (2012). Analysis of slow (theta) oscillations as a potential temporal reference frame for information coding in sensory cortices. *PLoS Computational Biology, 8*(10), e1002717.

Kayser, S. J., Ince, R. A., Gross, J., & Kayser, C. (2015). Irregular speech rate dissociates auditory cortical entrainment, evoked responses, and frontal alpha. *Journal of Neuroscience, 35*(44), 14691–14701.

Klimesch, W. (2012). Alpha-band oscillations, attention, and controlled access to stored information. *Trends in Cognitive Sciences, 16*(12), 606–617.

Lakatos, P., Barczak, A., Neymotin, S. A., McGinnis, T., Ross, D., Javitt, D. C., & O'Connell, M. N. (2016). Global dynamics of selective attention and its lapses in primary auditory cortex. *Nature Neuroscience, 19*(12), 1707–1717.

Lakatos, P., Karmos, G., Mehta, A. D., Ulbert, I., & Schroeder, C. E. (2008). Entrainment of neuronal oscillations as a mechanism of attentional selection. *Science, 320*(5872), 110–113.

Lakatos, P., Musacchia, G., O'Connel, M. N., Falchier, A. Y., Javitt, D. C., & Schroeder, C. E. (2013). The spectrotemporal filter mechanism of auditory selective attention. *Neuron, 77*(4), 750–761.

Lakatos, P., O'Connell, M. N., & Barczak, A. (2016). Pondering the pulvinar. *Neuron, 89*(1), 5–7.

Landau, A. N., Schreyer, H. M., van Pelt, S., & Fries, P. (2015). Distributed attention is implemented through theta-rhythmic gamma modulation. *Current Biology, 25*(17), 2332–2337.

Lawrence, S. J. D., Formisano, E., Muckli, L., & de Lange, F. P. (2017). Laminar fMRI: Applications for cognitive neuroscience. *NeuroImage.* http://dx.doi.org/10.1016/j.neuroimage.2017.07.004. Advance online publication.

Marr, D., Ullman, S., & Poggio, T. A. (2010). *Vision: A computational investigation into the human representation and processing of visual information.* Cambridge, MA: MIT Press.

Mesgarani, N., & Chang, E. F. (2012). Selective cortical representation of attended speaker in multi-talker speech perception. *Nature, 485*(7397), 233–236.

Miller, G. A. (2010). Mistreating psychology in the decades of the brain. *Perspectives on Psychological Science, 5*(6), 716–743.

Morel, A., Garraghty, P. E., & Kaas, J. H. (1993). Tonotopic organization, architectonic fields, and connections of auditory cortex in macaque monkeys. *Journal of Comparative Neurology, 335*(3), 437–459.

Ng, B. S., Schroeder, T., & Kayser, C. (2012). A precluding but not ensuring role of entrained low-frequency oscillations for auditory perception. *Journal of Neuroscience, 32*(35), 12268–12276.

O'Connell, M. N., Barczak, A., Schroeder, C. E., & Lakatos, P. (2014). Layer specific sharpening of frequency tuning by selective attention in primary auditory cortex. *Journal of Neuroscience, 34*(49), 16496–16508.

O'Sullivan, J. A., Power, A. J., Mesgarani, N., Rajaram, S., Foxe, J. J., Shinn-Cunningham, B. G., ... Lalor, E. C. (2015). Attentional selection in a cocktail party environment can be decoded from single-trial EEG. *Cerebral Cortex, 25*(7), 1697–1706.

Obleser, J., Henry, M. J., & Lakatos, P. (2017). What do we talk about when we talk about rhythm? *PLoS Biology, 15*(9), e2002794.

Obleser, J., & Weisz, N. (2012). Suppressed alpha oscillations predict intelligibility of speech and its acoustic details. *Cerebral Cortex, 22*(11), 2466–2477.

Peelle, J. E. (2018). Listening effort: How the cognitive consequences of acoustic challenge are reflected in brain and behavior. *Ear and Hearing, 39*(2), 204–214.

Pikovsky, A., Rosenblum, M., & Kurths, J. (2003). *Synchronization: A Universal Concept in Nonlinear Sciences.* Cambridge: Cambridge University Press.

Plakke, B., & Romanski, L. M. (2014). Auditory connections and functions of prefrontal cortex. *Frontiers in Neuroscience, 8,* 199.

Raz, A., & Buhle, J. (2006). Typologies of attentional networks. *Nature Reviews Neuroscience, 7*(5), 367–379.

Recanzone, G. H., Schreiner, C. E., & Merzenich, M. M. (1993). Plasticity in the frequency representation of primary auditory cortex following discrimination training in adult owl monkeys. *Journal of Neuroscience, 13*(1), 87–103.

Rice, D. M., & Hagstrom, E. C. (1989). Some evidence in support of a relationship between human auditory signal-detection performance and the phase of the alpha cycle. *Perceptual and Motor Skills, 69*(2), 451–457.

Riecke, L., Peters, J. C., Valente, G., Kemper, V. G., Formisano, E., & Sorger, B. (2017). Frequency-selective attention in auditory scenes recruits frequency representations throughout human superior temporal cortex. *Cerebral Cortex, 27*(5), 3002–3014.

Rosen, S. (1992). Temporal information in speech: Acoustic, auditory and linguistic aspects. *Philosophical Transactions of the Royal Society of London B: Biological Sciences, 336*(1278), 367–373.

Saalmann, Y. B., Pinsk, M. A., Wang, L., Li, X., & Kastner, S. (2012). The pulvinar regulates information transmission between cortical areas based on attention demands. *Science, 337*(6095), 753–756.

Sadaghiani, S., Scheeringa, R., Lehongre, K., Morillon, B., Giraud, A. L., & Kleinschmidt, A. (2010). Intrinsic connectivity networks, alpha oscillations, and tonic alertness: A simultaneous electroencephalography/functional magnetic resonance imaging study. *Journal of Neuroscience, 30*(30), 10243–10250.

Santoro, R., Moerel, M., De Martino, F., Goebel, R., Ugurbil, K., Yacoub, E., & Formisano, E. (2014). Encoding of natural sounds at multiple spectral and temporal resolutions in the human auditory cortex. *PLoS Computational Biology, 10*(1), e1003412.

Schroeder, C. E., & Lakatos, P. (2009). Low-frequency neuronal oscillations as instruments of sensory selection. *Trends in Neurosciences, 32*(1), 9–18.

Spaak, E., Bonnefond, M., Maier, A., Leopold, D. A., & Jensen, O. (2012). Layer-specific entrainment of gamma-band neural activity by the alpha rhythm in monkey visual cortex. *Current Biology, 22*(24), 2313–2318.

Strauss, A., Wöstmann, M., & Obleser, J. (2014). Cortical alpha oscillations as a tool for auditory selective inhibition. *Frontiers in Human Neuroscience, 8,* 350.

Szymanski, F. D., Rabinowitz, N. C., Magri, C., Panzeri, S., & Schnupp, J. W. (2011). The laminar and temporal structure of stimulus information in the phase of field potentials of auditory cortex. *Journal of Neuroscience, 31*(44), 15787–15801.

Vaden Jr., K. I., Kuchinsky, S. E., Cute, S. L., Ahlstrom, J. B., Dubno, J. R., & Eckert, M. A. (2013). The cingulo-opercular network provides word-recognition benefit. *Journal of Neuroscience, 33*(48), 18979–18986.

van Kerkoerle, T., Self, M. W., Dagnino, B., Gariel-Mathis, M. A., Poort, J., van der Togt, C., & Roelfsema, P. R. (2014). Alpha and gamma oscillations characterize feedback and feedforward processing in monkey visual cortex. *Proceedings of the National Academy of Sciences of the United States of America, 111*(40), 14332–14341.

Weisz, N., Hartmann, T., Muller, N., Lorenz, I., & Obleser, J. (2011). Alpha rhythms in audition: Cognitive and clinical perspectives. *Frontiers in Psychology, 2,* 73.

Weisz, N., & Obleser, J. (2014). Synchronisation signatures in the listening brain: A perspective from non-invasive neuroelectrophysiology. *Hearing Research, 307,* 16–28.

Weisz, N., Wuhle, A., Monittola, G., Demarchi, G., Frey, J., Popov, T., & Braun, C. (2014). Prestimulus oscillatory power and connectivity patterns predispose conscious somatosensory perception. *Proceedings of the National Academy of Sciences of the United States of America, 111*(4), E417–425.

Wilsch, A., Henry, M. J., Herrmann, B., Maess, B., & Obleser, J. (2015). Alpha oscillatory dynamics index temporal expectation benefits in working memory. *Cerebral Cortex, 25*(7), 1938–1946.

Wilsch, A., Neuling, T., Obleser, J., & Herrmann, C. S. (2018). Transcranial alternating current stimulation with speech envelopes modulates speech comprehension. *Neuroimage, 172,* 766–774.

Wittekindt, A., Kaiser, J., & Abel, C. (2014). Attentional modulation of the inner ear: A combined otoacoustic emission and EEG study. *Journal of Neuroscience, 34*(30), 9995–10002.

Wöstmann, M., Fiedler, L., & Obleser, J. (2017). Tracking the signal, cracking the code: Speech and speech comprehension in non-invasive human electrophysiology. *Language, Cognition and Neuroscience, 32*(7) 855–869.

Wöstmann, M., Herrmann, B., Maess, B., & Obleser, J. (2016). Spatiotemporal dynamics of auditory attention synchronize with speech. *Proceedings of the National Academy of Sciences of the United States of America 113* (14), 3873–3878.

Wöstmann, M., Lim, S. J., & Obleser, J. (2017). The human neural alpha response to speech is a proxy of attentional control. *Cerebral Cortex, 27*(6), 3307–3317.

Zhou, H., Schafer, R. J., & Desimone, R. (2016). Pulvinar-cortex interactions in vision and attention. *Neuron, 89*(1), 209–220.

Zion Golumbic, E. M., Ding, N., Bickel, S., Lakatos, P., Schevon, C. A., McKhann, G. M., … Schroeder, C. E. (2013). Mechanisms underlying selective neuronal tracking of attended speech at a "cocktail party." *Neuron, 77*(5), 980–991.

Zoefel, B., & VanRullen, R. (2017). Oscillatory mechanisms of stimulus processing and selection in the visual and auditory systems: State-of-the-art, speculations and suggestions. *Frontiers in Neuroscience, 11,* 296.

16 Three Functions of Prediction Error for Bayesian Inference in Speech Perception

MATTHEW H. DAVIS AND EDIZ SOHOGLU

ABSTRACT Spoken language is one of the most important sounds that humans hear yet also one of the most difficult sounds for nonhuman listeners or machines to identify. In this chapter we explore different neurocomputational implementations of Bayesian inference for speech perception. We propose, in line with predictive coding (PC) principles, that Bayesian inference is based on neural computations of the difference between heard and expected speech segments (prediction error). We will review three functions of these prediction error representations: (1) combining prior knowledge and degraded speech for optimal word identification, (2) supporting rapid learning processes so that perception remains optimal despite perceptual degradation or variation, and (3) ensuring that listeners detect instances of lexical novelty (previously unfamiliar words) so as to learn new words over the lifespan. Evidence from magneto-encephalography (MEG) and multivariate functional magnetic resonance imaging (fMRI) studies suggest computations of prediction error in the superior temporal gyrus (STG) during these three processes.

Speech is the most important sound we hear every day. Despite the many technologies that have changed our lives, speech remains our primary means of communication. Effective speech perception is critical for all human society, including educational, economic, and social functioning. Yet the processes by which we hear sounds, recognize words, and understand meanings remain largely invisible to listeners who are unaware of the neural processes that mediate between sound and meaning.

In this chapter we set out a computational and neural framework to explain the mechanisms by which listeners achieve certainty in what they are hearing despite the uncertainty of the perceptual signal. This framework is based around a predictive coding implementation of Bayesian perceptual inference for speech perception. We will provide a more precise motivation of this Bayesian approach based on the observed behavior of human listeners when the speed and accuracy of speech perception is measured. This experimental evidence outlines the functional goals of human listeners when understanding speech. We will then set out a predictive-processing framework that aims at achieving these goals through computations of prediction error—that is, the difference between heard and expected speech. We will focus on three critical perceptual functions supported by prediction error signals and review brain-imaging evidence (from fMRI and EEG/MEG) that supports this proposal.

Bayesian Inference in Speech Perception

In everyday conversation we hear speech at between 150 and 200 words per minute, though listeners can adapt to even faster speech (Dupoux and Green, 1997). Given the speed with which speech signals change over time and our fleeting sensory impressions (Crowder and Morton, 1969), listeners must be quick to comprehend speech correctly (Christiansen & Chater, 2016). For example, listeners can distinguish vocal and nonvocal sounds after just 4 ms of sound; less than a single pitch period (Suied et al., 2014). Listeners respond optimally to incoming speech; they can detect nonwords after hearing a single speech segment that mismatches the words they know (Marslen-Wilson, 1984). Listeners are similarly quick in making sense of connected speech, illustrated by the ability of *close shadowers* to repeat speech with a delay of less than 200 ms while also correcting for mispronunciations (Marslen-Wilson, 1973).

Listeners must also accommodate substantial variability, noise, and ambiguity in the sounds of speech. In addition to variation between speakers (Nusbaum & Magnuson, 1997), speech is often degraded or modified in many other ways yet remains comprehensible. As reviewed by Mattys et al. (2012), degradation can be due to the intrinsic properties of speech (e.g., accented, dysfluent, or apraxic speech) or due to the properties of the auditory environment, such as background noise or competing speakers. Coping with this variability and degradation is particularly challenging for listeners with sensory or neural impairments. Yet speech perception is largely successful for healthy listeners in a wide range of challenging situations; more so than even for state-of-the art machine speech recognition systems (Davis & Scharenborg, 2016).

We propose that listeners achieve this combination of speed, accuracy, and reliability by applying Bayesian inference to speech perception. That is, they compute the most probable interpretation of the current speech signal based on Bayes' theorem (see Kleinschmidt & Jaeger, 2015; Norris & McQueen, 2008). Bayesian perceptual inference is ideally suited to achieving fast and accurate speech identification for two reasons. First, it is probabilistic—that is, listeners compute the conditional probability of different meanings, words, or speech segments given the speech that has been heard, which allows for graded identification. Depending on the quality and uncertainty in the sensory signal, listeners generate a small set of high-probability interpretations (for clear, less ambiguous speech) or a larger set of less probable interpretations for speech that is noisier, degraded, or ambiguous. A second characteristic of Bayesian inference is that identification is guided by knowledge of the prior probability of specific interpretations. That is, an ambiguous sensory signal will be interpreted in line with the most likely speech sounds, words, or meanings. These two properties help ensure that the identification of speech is as fast and accurate as possible. Computational and cognitive models of speech perception have long sought to explain fast and accurate speech identification. However, we here suggest that this is best achieved by implementing Bayesian perceptual inference (see Norris & McQueen, 2008).

Implementations of Bayesian Perceptual Inference for Word Recognition

Despite agreement that Bayesian inference can account for the speed and accuracy of perceptual identification in a range of domains, it remains unclear how this process is neurally implemented (see Aitchison & Lengyel, 2017). We propose that predictive coding mechanisms (hereafter, PC) support perceptual inference for speech. PC describes a system for hierarchical perceptual processing in which successive levels of representation (for speech comprehension these might be acoustic, phonemic, lexical, and semantic levels) communicate by passing predictions from higher, more abstract levels down the hierarchy (e.g., lexical knowledge makes predictions for expected phonemes), and prediction errors (i.e., the difference between expected and heard speech) are passed back up the perceptual hierarchy. By this view, predictions act as hypotheses about the causes of sensory experience, and the perceptual hierarchy embodies a generative model of human speech and language. Prediction errors provide a neural signal to update higher-level perceptual and linguistic hypotheses, which thereby compute the posterior probability of different interpretations of current speech signals (see Rao & Ballard, 1999 for illustrative simulations in vision).

We depict this PC mechanism in figure 16.1*C* as a variant of two other hierarchical accounts of speech perception (MERGE, figure 16.1*A*: Norris, McQueen, & Cutler, 2000; TRACE, figure 16.1*B*: McClelland & Elman, 1986). In MERGE, information flow is purely bottom-up, and the system uses both excitatory and inhibitory connections so that phoneme representations can both activate matching words and suppress mismatching words. Word recognition is further achieved using a competitive selection mechanism, implemented by within-layer inhibitory connections. In contrast to this bottom-up theory, TRACE combines lexical inhibition with interactive activation—that is, top-down excitatory connections. This allows for knowledge of words to support identification of their constituent phonemes. This balance of interactive activation and mutual inhibition operates such that TRACE will settle into a stable state with appropriate intermediate representations (phonemes, or similar) being active. That is, those phonemes that receive bottom-up activation from speech and that are consistent with top-down activation from lexical expectations will be active. This allows TRACE to simulate the Ganong effect, whereby the recognition of phonemes is supported by the occurrence of familiar words (i.e., a single, ambiguous phoneme is heard as /d/ or /t/ in lexical contexts like *dark* and *task*; Ganong, 1980). These findings are simulated in MERGE using an additional decision mechanism (not shown in figure 16.1*A*).

In contrast, in a PC system, inhibitory connections express competing predictions for the expected form of sensory signals. These top-down predictions lead to the inhibition of expected sensory representations at lower levels of the system. Thus, in a PC system, activation at intermediate levels represents sensory prediction errors; these provide a means to update higher-level predictions. In a PC system, such as Rao and Ballard (1999), we can distinguish between positive and negative prediction errors that arise at the same time during the perception of a sensory signal and that have opposite effects on higher-level perceptual hypotheses. Positive prediction errors arise when sensory signals are stronger than predicted and act to strengthen relevant perceptual hypotheses. Negative prediction errors

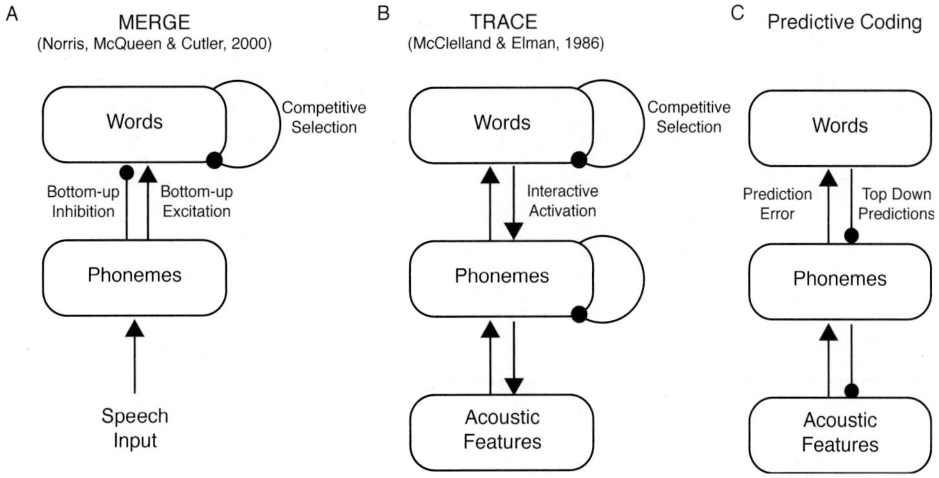

FIGURE 16.1 Three models of speech perception and spoken word recognition. *A*, MERGE (Norris, McQueen, & Cutler, 2000) includes only bottom-up information flow with phoneme representations supporting (excitation) or suppressing (inhibition) matching and mismatching words. Separate decision nodes (*not shown*) combine lexical and sublexical information for phoneme decisions. *B*, TRACE (McClelland & Elman, 1986) uses interactive activation mechanisms with a bidirectional, excitatory information flow between phoneme and word units. Within-level inhibition between words and phonemes support competitive selection processes (also in MERGE). *C*, Predictive coding (PC) uses higher-level representations to send inhibitory, top-down predictions for expected signals to lower levels. Prediction error is computed by subtracting a representation of expected signals from the current input. This generates positive prediction errors that support the identification of matching words and negative prediction errors that suppress activation of mismatching words. The PC account thereby implements excitatory and inhibitory lexical selection processes (like MERGE) through top-down predictions and bottom-up prediction error.

arise when sensory signals are weaker than predicted and act to suppress incorrect perceptual hypotheses. Thus, PC accounts provide an alternative mechanism for the bottom-up activation of matching perceptual interpretations and the bottom-up suppression of mismatching perceptual interpretations that achieves many of the same functional goals as MERGE or TRACE.

In this chapter, as in Gagnepain, Henson, and Davis (2012), we represent positive and negative prediction errors directly for clarity. However, we acknowledge that there are several ways in which prediction errors could be neurally implemented. For example, positive and negative prediction errors could be signaled by increases and decreases from baseline firing rates (Schultz, Dayan, & Montague, 1997) or by increased firing on separate positive and negative channels (Spratling, 2010). In a later section of the chapter and in figure 16.3, we will present a more detailed illustration of how these positive and negative prediction errors operate during the identification of specific words. However, we will first consider behavioral and neural evidence that favors a PC view of word identification over alternative accounts, such as MERGE and TRACE.

To assess evidence for these contrasting implementations of Bayesian perceptual inference, we will break the PC scheme down into two components. The first component is a reverse-hierarchical flow of information in which predictions are projected top-down (from higher to lower levels) to change prediction errors flowing bottom-up (from lower to higher levels) in the processing hierarchy. This contrasts with alternative accounts in which information flow is strictly bottom-up (e.g., Norris, McQueen, & Cutler, 2000; see figure 16.1*A*). The second component concerns the precise nature of the computations performed using top-down signals in PC. In PC, top-down predictions act to suppress (rather than support) bottom-up prediction error signals at lower hierarchical levels (contrast the PC model in figure 16.1*C* with TRACE, shown in figure 16.1*B*). Experimental evidence consistent with both these components would therefore provide unique support for PC over alternative computational accounts.

Our focus here is to review studies that have explored the influence of prior knowledge on speech perception. For example, fMRI studies have shown the effects of semantic context on auditory and higher-level responses to speech (Davis et al., 2011; Obleser & Kotz, 2011). While this is consistent with top-down information flow, to directly test for this requires imaging modalities with high temporal resolution. If prior knowledge or predictions are conveyed top-down as in TRACE

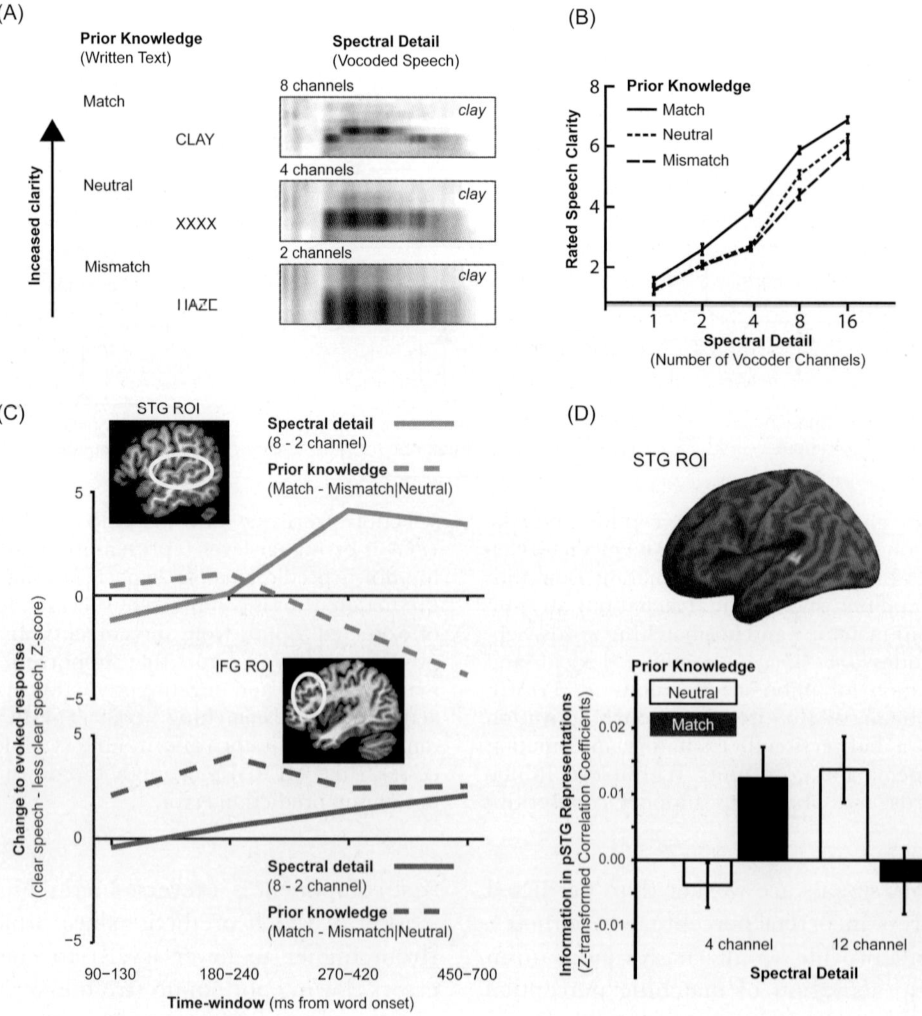

FIGURE 16.2 Experimental findings that illustrate the contrasting behavioral and neural effects of sensory and knowledge-driven manipulations on speech clarity. *A*, Experimental design from Sohoglu et al. (2012, 2014) in which speech clarity is influenced by prior knowledge (matching written text) for vocoded speech that varies in spectral detail (a varying number of channels). *B*, Clarity ratings are enhanced by prior knowledge (matching text) and increased spectral detail (vocoder channels; from Sohoglu et al., 2014, expt. 1). *C*, Differential neural effects of these two manipulations are seen in MEG responses in the superior temporal gyrus (STG, *blue*) and the inferior frontal gyrus (IFG, *red*), consistent with top-down predictions on match trials that reduce STG responses (redrawn from Sohoglu et al., 2012). *D*, Multivariate fMRI responses measured in the STG show interactions between prior knowledge and spectral detail that are in line with prediction error representations. Increased spectral detail enhances STG representations of speech following neutral text but reduces the information content of STG representations after matching text (redrawn from Blank & Davis, 2016). (See color plate 19.)

or PC, then manipulations of listeners' prior knowledge should change brain responses in higher-level regions (which represent higher-level context, or predictions) before those in lower-levels (representing current speech content or prediction errors). We therefore require careful experimental control of when and how prior knowledge of speech content is provided to listeners. To do this we have used experiments in which written text provides prior knowledge of the expected form of degraded (noise vocoded) speech (Shannon et al., 1995), which varies in physical clarity (Sohoglu et al., 2012, 2014; see figure 16.2*A*).

There are several important advantages of this method: first, the effect of prior knowledge on the subjective experience of speech is striking and very reliable. Degraded speech sounds dramatically clearer when it matches previously presented written text and not if text follows speech (Sohoglu et al., 2014). Effects of presentation order have also been observed for ratings of speech in noise (Goldinger, Kleider, & Shelley,

1999) and for objective measures of speech in noise or competing speech (e.g., same/different judgments; Freyman et al., 2017). These findings confirm that prior knowledge of speech content enhances perceptual and not decision processes (see Sohoglu et al., 2014 for discussion). The effect of prior knowledge on clarity ratings is equivalent to doubling the spectral detail of vocoded speech (see figure 16.2*B*) or a large change in the signal-to-noise ratio for speech in noise (>10 dB; Goldinger et al., 1999). This provides the second advantage of this method: we can contrast enhanced speech clarity due to prior knowledge with equivalent changes due to sensory manipulations. Despite equivalent perceptual outcomes, distinct neural consequences are observed consistent with the differences between knowledge-driven and sensory processes. This comparison helps rule out alternative explanations of behavioral and neural observations, such as listening effort or intelligibility, which would be similarly changed by both manipulations.

A third advantage of methods combining written text and degraded speech is that prior knowledge comes from a nonauditory source (written text). The neural consequences of prior knowledge in sensory cortex can only be due to top-down influences of higher-level knowledge on lower-level processes rather than local habituation, adaptation, or repetition suppression effects (Grill-Spector, Henson, & Martin, 2006; see Wild, Davis, & Johnsrude, 2012 for similar arguments).

One study to combine written text and degraded spoken words with fast brain imaging measures was reported by Sohoglu et al. (2012; figure 16.2*C*). In a combined MEG/EEG study, listeners were presented with distorted (noise-vocoded) words that varied in spectral detail after the presentation of written text that matched, mismatched, or was uninformative for spoken words. As expected, rated speech clarity was enhanced by matching prior text, and this was accompanied by increased brain responses to speech in a region associated with higher-level phonological processing, the inferior frontal gyrus (IFG). Importantly, frontal responses were modulated before responses in a lower-level acoustic-phonetic region, the superior temporal gyrus (STG). As outlined above, this temporal sequence is strongly suggestive of a top-down information flow from higher (IFG) to lower levels (STG) of perceptual processing.

Further evidence of top-down information flow underpinning the influence of prior knowledge on speech perception came from testing individuals with selective neurodegeneration in inferior frontal regions (progressive nonfluent aphasia, or PNFA; Cope et al., 2017). Aphasic listeners showed a delayed influence of prior knowledge on neural activity in the STG, despite normal gray matter volume in the temporal cortex and normal STG responses to manipulations of sensory detail in speech. Interestingly, patients' subjective reports showed that rather than a reduced influence of prior knowledge on perceptual outcomes, PNFA leads to an increased reliance on prior knowledge—patients underestimate the clarity of speech that mismatches or is not informed by prior written text. These findings provide causal evidence for top-down influences from higher-level (inferior frontal) to lower-level (superior temporal) regions during speech perception that plays a functional role in integrating prior knowledge and sensory signals. Evidence from causal connectivity analysis of MEG (Di Liberto, Lalor, & Millman, 2018; Gow et al., 2008; Park et al., 2015) and electrocorticography (Leonard et al., 2016) further suggests top-down mechanisms by which higher-level knowledge (e.g., phonological predictions in the IFG) modulate activity at lower processing levels (e.g., acoustic-phonetic processes in the STG).

Evidence that top-down signals influence lower-level neural responses during speech perception challenges purely bottom-up models. But, what evidence is there that these top-down signals contribute to computations of prediction error (i.e., the second component of the PC account)? Critical to distinguishing this possibility from other top-down accounts, such as TRACE, is comparing manipulations of top-down predictions and bottom-up speech content. In TRACE, both bottom-up and top-down influences on perceptual processing are excitatory—prior knowledge and sensory detail will facilitate perception in the same way and should have an equivalent influence on neural activity. In contrast, in PC accounts, neural responses do not represent sensory outcomes (as in TRACE) but represent the degree to which the sensory input diverges from expectations. The magnitude of prediction error signals will decrease if the divergence is small (as when matching prior knowledge is available) and increase if it is large (if clearer speech is heard without an accompanying improvement in prediction). Hence, the PC account proposes that when acoustic-phonetic clarity is increased, so too is the magnitude of acoustic-phonetic prediction error and hence neural responses (at least if listeners lack informative prior knowledge). The PC account further proposes that matching prior knowledge leads to top-down suppression of lower-level prediction errors and that this will be more pronounced for physically clearer speech. Both these effects have been demonstrated experimentally when measuring the magnitude of evoked MEG responses to spoken words in STG regions (see figure 16.2*C* from

Sohoglu et al., 2012; replications reported by Sohoglu & Davis, 2016; Cope et al., 2017). Computational simulations of a simple PC model with two levels of representation (acoustic-phonetic and phonological) provide a good fit to observed neural responses and perceptual outcomes for MEG studies with healthy listeners (Sohoglu & Davis, 2016) and PNFA patients (Cope et al., 2017).

While these neural findings are compatible with a PC model, this work does not entirely rule other explanations (such as interactive-activation models like TRACE). As noted by Aitchison and Lengyel (2017), reductions in the magnitude of neural responses for expected stimuli may be equally consistent with other neural implementations of Bayesian inference—including accounts in which prior expectations are added to or multiplied with perceptual representations. Whether neural differences between bottom-up and top-down manipulations of speech clarity (effects of prior knowledge and sensory detail) are challenging for models like TRACE is best assessed by directly comparing neural observations with explicit computational simulations.

In testing these accounts, we draw on the information contained in spatiotemporal patterns of neural activity and analyses of representational content (Kriegeskorte & Kievit, 2013). The key idea here is to distinguish between accounts—such as PC—in which neural activity represents the difference between heard and expected signals (prediction error) and accounts in which neural activity more directly represents the current perceptual experience (in Bayesian terms, the posterior). While perceptual experience (and hence the posterior) is enhanced similarly by prior knowledge or improvements in sensory clarity, prediction error representations will be less informative for speech that clearly matches prior expectations. Thus, analyses of representational content such as multivoxel pattern analysis (MVPA; Blank & Davis, 2016) or multivariate encoding/decoding (Holdgraf et al., 2016) can distinguish between PC and these alternative implementations of Bayesian inference (Aitchison & Lengyel, 2017; for similar arguments).

Two such studies were reported by Blank and colleagues (Blank & Davis, 2016; Blank, Spangenberg, & Davis, 2018), who combined informative/uninformative written text with degraded speech while measuring the representational content of STG responses with fMRI. In a first study, Blank and Davis (2016) showed that while word report was enhanced in an additive way by increased sensory detail and informative prior knowledge, fMRI data shows a striking interaction for neural representations in the posterior STG. Increased signal quality of speech presented after neutral text enhances neural representations, whereas the same change in signal quality after matching text reduces the information content of fMRI multivoxel patterns (see figure 16.2*D* from Blank & Davis, 2016). This interaction rules out bottom-up accounts (since in MERGE or Shortlist B, perceptual representations should not be modified by prior knowledge; see Norris, McQueen, & Cutler, 2000). They also rule out "sharpening" theories of speech perception (such as the interactive activation TRACE model; McClelland & Elman, 1986) since by these accounts, effects of prior knowledge and sensory quality should combine additively. Simulations of different computational mechanisms for degraded speech perception confirm that this interaction of sensory detail and prior knowledge for neural representations is uniquely consistent with STG representations of prediction error.

A follow-up study (Blank, Spangenberg, & Davis, 2018) further explored neural representations for degraded speech heard after written text that matches, partially matches, or fully mismatches. Reading and then hearing similar-sounding words (like *kick* followed by *pick*, or *kit* followed by *kitsch*) leads to frequent misperception, which is accompanied by reduced fMRI responses in the STG. These findings again suggest top-down influences on STG responses, and representational similarity analysis allows us to specify the underlying mechanism. Critically, this same STG region preferentially represented the sounds that differed between prior expectations and heard speech (like the initial /k../-/p../ in *kick-pick*), rather than the shared sounds (the rhyme segments /.Ik/ for *kick-pick*), and representations of these deviating sounds predicted perceptual outcomes—written-spoken pairs that evoked a clearer representation of the deviating sounds were more accurately perceived by listeners.

These two MVPA fMRI studies converge in showing representations of prediction error in the STG. These findings are incompatible with models like TRACE, in which top-down expectations enhance representations of heard segments, and in line with PC accounts in which the STG signals the discrepancy between heard and expected speech. Despite compelling evidence for a PC account, however, many questions concerning the functional and ecological significance of these mechanisms for everyday listening remain. We will explore these questions in the final section of the paper. Before that, however, we will propose two other functions of prediction error computations in adapting and learning from exposure to variable or novel speech.

Perceptual Learning by Minimizing Prediction Error

Prediction error computations in the STG provide an implementation of Bayesian perceptual inference that is compatible with neuroimaging observations when the perception of degraded speech is guided by prior knowledge. By this view, perceptual identification involves updating higher-level predictions when identifying speech sounds. However, this is not the only function of prediction error computations during perception. After identification, the system should also adapt its computations such that in the future, speech with similar acoustic, phonetic, or linguistic properties will also be optimally identified. This process, post-identification perceptual learning, helps ensure that human speech perception remains optimal despite longer-term changes in the linguistic environment.

Perceptual learning is apparent in a variety of experimental situations that are reviewed elsewhere (see Kleinschmidt & Jaeger, 2015; Samuel & Kraljic, 2009). However, we will argue that in a PC framework these processes can all be explained as resulting from long-term modifications to connection weights that convey top-down predictions from higher-level to lower-level representations. Any acoustic prediction error that remains once listeners have generated their best interpretation of the current speech signal should be used to modify longer-term predictions for how that word or segment should sound when it is heard next. In this way, today's posterior becomes tomorrow's prior.

In the literature on the categorical perception of speech, two different forms of perceptual learning have been described—selective adaptation and phonetic recalibration (Kleinschmidt & Jaeger, 2015). The distinction between these is that selective adaptation arises from repeated presentations of unambiguous tokens from a single category and leads to a shift in the category boundary away from the repeated item (i.e., a reduced likelihood of reporting the repeated category; see Samuel, 1986). Conversely, phonetic recalibration occurs when an ambiguous segment is presented in contexts that favor one interpretation (based on lexical information, visual speech, or other cues; Norris, McQueen, & Cutler, 2000; van Linden & Vroomen, 2007). Phonetic recalibration leads to an opposite shift to category boundaries such that ambiguous segments are reported as belonging to the repeated category. In line with the ideal adapter model of Kleinschmidt and Jaeger (2015), both these processes can arise from updating the distribution of acoustic features that signal specific categories based on recent experience. We will next review the behavioral and neural evidence consistent with the proposal that this form of perceptual learning (as well as perceptual inference, described previously) is implemented by a neural process that minimizes prediction error.

In PC, top-down predictions are iteratively updated by bottom-up prediction errors during perceptual inference. Perceptual identification occurs when prediction errors are minimized by activating appropriate higher-level representations. However, some residual prediction error may remain even after identification is complete (whenever the expected form of the most likely word does not perfectly match current sensory signals). Under the PC account, these residual prediction errors are used to modify the connection weights that link this higher-level interpretation with sensory predictions, resulting in perceptual learning. Prior knowledge that decreases perceptual uncertainty (such as the prior presentation of informative text) should lead to a reduction in prediction error (since the correct perceptual interpretation is more strongly predicted) but should also enhance learning—since residual prediction errors will arise from uncertainty concerning the acoustic realization of heard words and not from uncertainty in higher-level interpretations.

In line with this proposal, prior knowledge of perceptual content enhances the perceptual learning of speech. This is apparent for lexically guided phonetic recalibration of ambiguous speech sounds. Perceptual learning is shown for ambiguous sounds at word offset (when lexical predictions provide prior knowledge) but not for ambiguous speech sounds at word onset (when lexical knowledge is only available subsequently; Jesse & McQueen, 2011). Enhanced perceptual learning due to prior knowledge of speech content is also shown for noise vocoded speech. Listeners show more rapid improvements in word report for vocoded sentences (Davis et al., 2005) or words (Hervais-Adelman et al., 2008) if they have accurate prior knowledge of degraded speech content. These effects of prior knowledge on perceptual learning closely parallel the effects on speech clarity reported by Sohoglu et al. (2014) and reviewed in the previous section. These findings therefore suggest that perceptual outcomes due to short-term and long-term influences of prior knowledge depend on the same time-limited process that operates during the predictive processing of speech. One such mechanism that we have argued explains this time-limited behavior is auditory echoic memory. In the PC account, top-down and bottom-up signals must be compared to derive prediction errors. Hence, for learning to take place, top-down influences from higher-level representations (e.g., phonological representations that can be maintained in working memory for seconds or longer) must be available before the rapid decay of bottom-up auditory representations in echoic memory occurs (see Davis & Johnsrude, 2007; Sohoglu et al., 2014).

Further evidence to link perceptual learning to the behavioral and neural consequences of updating predictions during speech processing comes from an MEG/EEG study that combined the manipulations of prior knowledge and sensory detail described previously (see figure 16.2) with perceptual learning (Sohoglu & Davis, 2016). As before, listeners heard noise-vocoded spoken words preceded by either matching or mismatching text supplying informative or uninformative prior knowledge. Before and after this "training," phase listeners also performed a word report task on degraded speech during which their ability to report spoken words (without accompanying written text) was assessed. Word report accuracy significantly improved after training, and in line with enhanced predictions, this perceptual learning was associated with a reduction in the STG response that colocalizes with the immediate reduction that occurs with matching prior knowledge. Furthermore, the magnitude of both these neural reductions (due to prior knowledge and perceptual learning) were correlated across listeners with the behavioral manifestation of perceptual learning (i.e., improvements in word report accuracy).

These results therefore support the idea that the process by which prediction errors update predictions online for optimal Bayesian inference also supports longer-term perceptual learning. In Sohoglu and Davis (2016), we propose that both prior knowledge and perceptual learning act to change the distribution of expected acoustic cues represented in the STG, although in different ways. Matching written text (prior knowledge) increases the specificity of acoustic predictions by suppressing alternative perceptual hypotheses, which reduces prediction error. Perceptual learning also reduces prediction error, not due to the suppression of alternative perceptual hypotheses but rather because acoustic predictions for the realization of higher-level categories become more accurate (better matched to the acoustic feature distributions of the degraded speech signals).

Prediction Error and the Detection of Linguistic Novelty

We have thus far described the mechanisms by which higher-level knowledge (of words, meanings, and sentence structure) supports lower-level perceptual identification and guides longer-term perceptual learning of speech. However, the account presented so far has a significant and important flaw. Neither accurate identification nor perceptual learning will be possible if listeners hear unfamiliar words. We might naïvely suggest that the PC account be considered an account of adult identification but not of language development or childhood acquisition. However, lifespan analysis of vocabulary size shows that word learning is a near-daily experience, even for adults. For example, Brysbaert et al. (2016) compared the median vocabulary size of 20- and 60-year-old English speakers and inferred that, on average, adults learn a new lemma word every 2.4 days and a new base word every 6.3 days during the intervening 40 years. Looking at the new words that have entered all our vocabularies in the last decade (*emoji, selfie, vape*; see https://en.oxforddictionaries.com/word-of-the-year) makes clear that adult word learning is not confined to formal education.

Adult listeners therefore continue to detect new or previously unfamiliar spoken or written words. They must encode word form and possible meaning in order to add these new words to the lexicon. There is now considerable laboratory evidence exploring the cognitive and neural basis of the detection and encoding of newly heard unfamiliar words and their integration into the lexicon (see Davis & Gaskell, 2009; James et al., 2017 for reviews). Given the prevalence of word learning in adulthood, however, we argue that these processes must also be included in theories of speech perception. While space prohibits reviewing this work in detail, behavioral studies with adults and children document a dissociation between the rapid encoding of new word forms and meanings (which is apparent immediately after learning; see Gaskell & Dumay, 2003; Havas et al., 2018; Leach & Samuel, 2007) and slower consolidation, which appears necessary for new vocabulary items to function like other familiar words in the lexicon (competing for identification with existing words: Dumay & Gaskell, 2007; showing rapid generalization: Tamminen et al., 2012; or automatic semantic access: Tham, Lindsay, & Gaskell, 2015).

Yet it remains unclear how a Bayesian system for speech perception should operate when speech contains new or unfamiliar words. Using Bayes' theorem to determine which word or words are most probable becomes ill-defined when words with a prior probability of zero (i.e., unfamiliar words) are heard. Without some additional mechanism for detecting unfamiliar words—and implicitly assigning them a probability—machine speech recognition systems fail to correctly identify familiar words within such utterances (Hermansky, 2013). Additional mechanisms for processing unfamiliar words (*pseudowords*) have also been added to models of human spoken word recognition, such as the possible-word constraint (Norris et al., 1997) or adding a "dummy" pseudoword unit with a nonzero probability (Norris & McQueen, 2008). These ad hoc modifications permit the recognition of speech sequences that include pseudowords, at the cost of some parsimony. However, in the PC account, computation of lexical probabilities for familiar words involves generating a neural signal, a prediction error, that allows the detection of lexical novelty and supports the encoding of unfamiliar words. We will argue that this provides a unique advantage of the PC account in comparison with other implementations of Bayesian perceptual inference.

Central to the PC explanation is that the computation of prediction error contributes to the identification of familiar words (building on Gagnepain, Henson, & Davis, 2012). To illustrate this, we first describe the recognition of higher- and lower-frequency neighboring words like *captain* (/k{ptIn/; see figure 16.3*A*) and *captive* (/k{ptIv/; figure 16.3*B*) before considering the perception of a pseudoword neighbor of these words, *captick* (/k{ptIk/; figure 16.3*C*). The PC account can explain the behavioral observations reviewed earlier that listeners will use knowledge of prior probabilities to recognize familiar words quickly and detect pseudowords after hearing a single segment that mismatches with all familiar words (Marslen-Wilson, 1984). We will explain both these processes using PC mechanisms and computations of positive and negative prediction error.

The presentation of the speech sequence /k{ptI/ (i.e., the words *captain* or *captive* prior to the final segment) leads to the activation of these two matching words with contrasting predictions for upcoming speech. Given the greater frequency of occurrence of *captain*, segment /n/ is more strongly predicted than /v/ (figure 16.3, *white bars*). When the final segment is heard, these predictions are compared against the current speech input (*black bars*). The resulting prediction error distribution (*gray bars*) supports word identification by generating negative prediction errors for segments that are expected but absent from the input (such as /v/ in figure 16.3*A*, which is predicted for the word *captive*) and positive prediction errors for segments that are somewhat expected but clearly present in the input (such as the segment /n/ in figure 16.3*A* for *captain*). These two components of prediction error are of different magnitudes—due to differences in the prior probability of *captain* and *captive*—and support computations of lexical mismatch and lexical match.

Hearing a segment that mismatches with lexical expectations generates a negative predictive error signal that (when used to update lexical representations) will reduce the probability of previously active lexical candidates. The negative prediction error for /v/ when hearing *captain* (figure 16.3*A*) serves to suppress predictions from the lexical item *captive*. Conversely, the negative prediction error for /n/ when hearing *captive* (figure 16.3*B*) serves to suppress predictions from the lexical item *captain*. As is apparent by comparison with figure 16.3*A*, the presentation of /v/ elicits a larger prediction error since this segment was less expected. Recognition depends on generating a large, positive prediction error to signal a lexical match, which leads to additional difficulty for word identification (shown, for example, by cross-modal priming data in Gaskell and Marslen-Wilson [1998]).

Prediction error computations also contribute directly to pseudoword detection, as shown in figure 16.3*C*. Hearing the segment /k/ at the end of /k{ptIk/ generates a negative prediction error that suppresses word representations for both *captain* and *captive*. This effect of word-final mismatch has been shown in cross-modal priming studies reported by Marslen-Wilson (1993); hearing a nonword like *fleek* blocks semantic access for words that are related to a neighboring word (e.g., *ship*, related to *fleet*), just like the word *streak* blocks access to the meaning of *street*. Models like TRACE, which eschew bottom-up mismatch, may find this result challenging to explain, whereas this finding can be explained by negative prediction errors in the PC account.

The positive prediction error generated on hearing *captick* also serves an important function. Rather than increasing lexical activity for matching words (as in figure 16.3*A*, 16.3*B*), it contributes to the process of nonword detection and encoding. Since there is no familiar word that is compatible with the word-final /k/ in /k{ptIk/,[1] the absolute summed magnitude of prediction error will be maximal (only pseudowords can elicit a summed prediction error of 2). The magnitude of the prediction error response provides a signal of lexical novelty that can explain the speed of no responses in lexical decisions (see Marslen-Wilson, 1984) and triggers the encoding of the new word *captick*.

If, in line with the PC account discussed earlier, prediction error computations are performed by the STG, we would expect overlapping neural responses for word identification difficulty (since difficult to identify words elicit larger prediction error) and for nonwords

[1] One possible source of the final /k/ would be misidentification of the low-frequency word *haptic*. We speculate that detection and encoding of the nonword *captick* relies on hearing sufficiently clear speech that the lexical hypothesis *haptic* has a low probability. This proposal has—to our knowledge—not yet been tested.

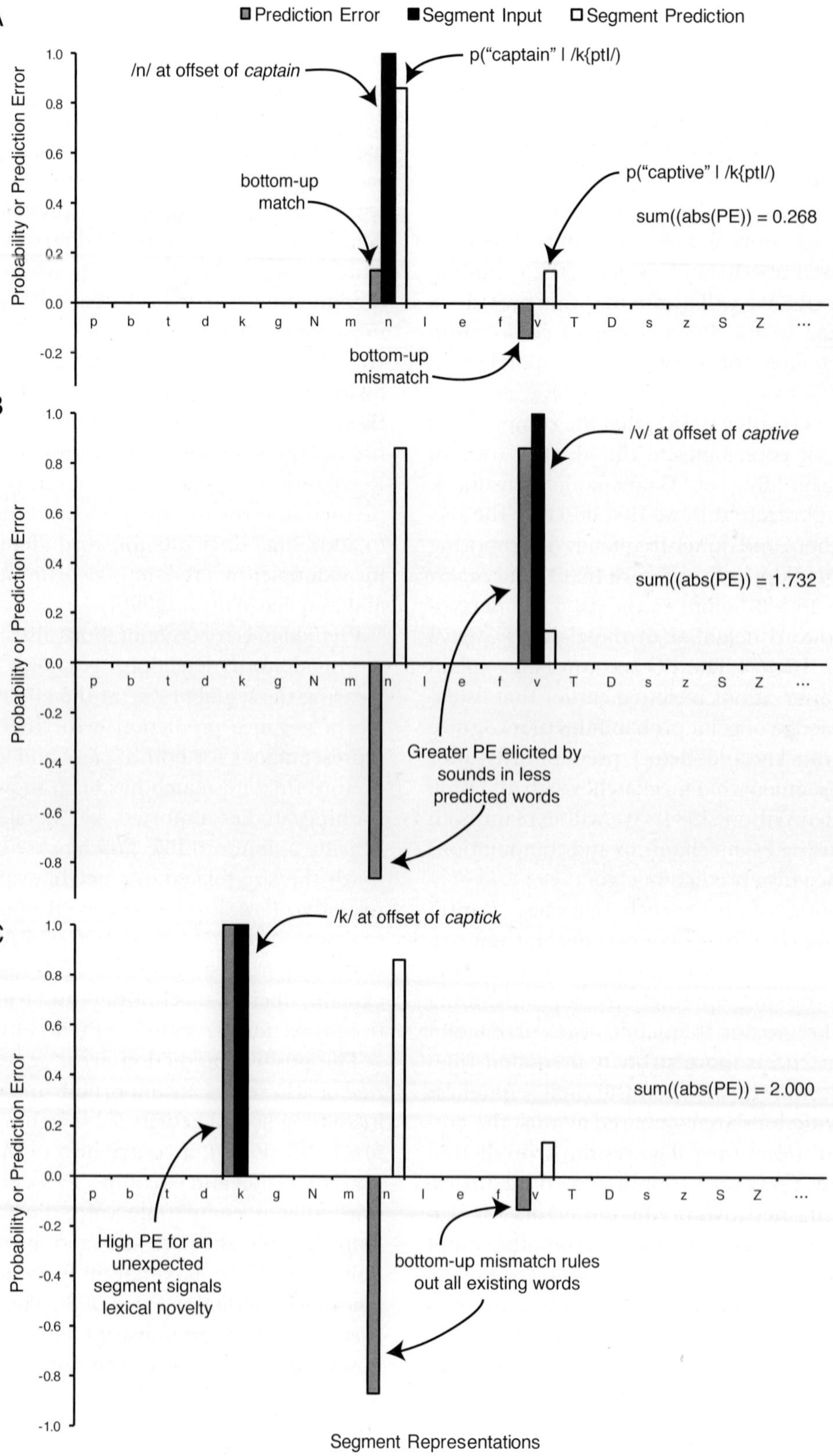

FIGURE 16.3 Representations of segment input (*black-filled bars*) and prediction probability (*white bars*) and prediction error (*gray bars*) on hearing: *A*, /n/ at the offset of *captain*, *B*, /v/ at the offset of *captive*, and *C*, /k/ at the offset of *captick*. As marked, segment prediction error provides a bottom-up signal of lexical match and mismatch (positive and negative prediction errors) that support word recognition in (*A*) and (*B*) and provide a signal of lexical novelty to drive new word learning, *C*.

compared to real words. This overlap is apparent in the functional imaging literature: a meta-analysis of PET and fMRI studies comparing word and nonword identification showed additional responses to spoken nonwords compared to real words in the STG (Davis & Gaskell, 2009), exactly the same region as shown to elicit additional activity for more difficult to identify words (see Davis, 2015 for a review of relevant fMRI data).

Further evidence to link both lexical competition and novelty detection processes with STG computations of prediction error comes from an MEG study reported by Gagnepain, Henson, and Davis (2012). This study explored the time course of STG responses during the identification of new words (e.g., *formuty*), how these responses differ from responses to familiar neighboring words (e.g., *formula*), and how both these responses change due to the learning and consolidation of neighboring novel words (e.g., *formubo*, for participants who were extensively trained on this word on the previous day). This study confirms the neural overlap shown by fMRI. Neural effects of recognition difficulty for real words with and without additional lexical competitors, additional neural responses to novel words compared to familiar words, and changes to both these effects for pseudowords that were learned prior to overnight consolidation all overlap in the STG. Furthermore, these neural responses are time-locked to the onset of the segments that deviate between familiar and unfamiliar words (i.e., /b5/, /l@/, or /t#/ in *formubo, formula, and formuty*). Computational simulations show that this pattern is exactly consistent with the PC account and inconsistent with other accounts in which these effects arise from lexical uncertainty (as in TRACE) or other mechanisms.

Of course, to adequately account for vocabulary acquisition we also need to explain the computations that are performed after detecting a novel word. Here we build on previous proposals in domain-general complementary learning systems theories (CLS; e.g., McClelland, McNaughton, & O'Reilly, 1995), extended to spoken word learning (Davis & Gaskell, 2009) and linked to predictive coding (Henson & Gagnepain, 2010). Specifically, prediction error responses on the detection of a nonword in the STG trigger neural activity and plasticity in medial temporal lobe regions, such as the hippocampus. These allow listeners to encode the form, meaning, and context of new words for later off-line consolidation. While the MEG study reported by Gagnepain, Henson, and Davis (2012) cannot speak to this proposal (since MEG has only limited sensitivity in medial temporal regions), similar fMRI studies have shown medial temporal lobe activity associated with the learning of new spoken nonwords (Breitenstein et al., 2005; Davis & Gaskell, 2009). This dissociation of cortical regions involved in recognizing familiar words and detecting new words and of medial temporal regions supporting the initial learning of new words is exactly in line with these CLS theories. Further behavioral and neural experiments to update this CLS account so that it operates in line with PC principles are ongoing.

Summary and Future Directions

In this chapter we offered evidence for a PC account in which computations of prediction error support three important aspects of Bayesian perceptual inference for speech. We first showed how prediction error signals permit accurate word identification by combining perceptual signals with prior knowledge. The magnitude, timing, and local representational content of neural responses measured in the STG are uniquely consistent with prediction error computations. However, the generality of this account remains to be established. During natural listening, multiple weak sources of prediction must be combined to provide optimal prior knowledge of upcoming speech. Even in combination these syntactic, semantic, and pragmatic cues provide weaker constraints than in experiments using matching written text. A second aspect of our experimental paradigm is the use of degraded speech. Behavioral and neural influences of prior knowledge on perception (in Bayesian accounts) should be most apparent when speech signals are degraded or ambiguous. In combination these might lead us to question whether similar mechanisms contribute to the identification of clear speech in more ecological listening situations.

The second section of this chapter proposed that weight changes after identification that minimize prediction error lead to longer-term improvements in speech perception (perceptual learning). This conclusion was supported by MEG evidence that showed how a simplified PC model could account for the neural effects of prior knowledge, spectral detail and perceptual learning. However, these simulations lack detail, and further evidence concerning how the neural representation of degraded and preserved sensory features—before and after perceptual learning—would increase our confidence in the validity of the PC account.

Finally, the third section proposed an account of novelty detection for spoken words that follows directly from prediction error representations. While existing MEG and fMRI evidence suggests a common neural correlate of lexical competition and novelty detection which is consistent with the PC account, a more detailed characterization of neural representations during word and pseudoword perception would lend substantial further support to this computational account. Furthermore, we lack a detailed account of how cortical and medial temporal/hippocampal learning mechanisms should combine. This is critical for understanding when listeners encode novel words (novelty detection) rather than modify existing higher-level representations (perceptual learning). Elevated prediction error signals (compared to clearly spoken familiar words) will be apparent for familiar speech that sounds unfamiliar (due to perceptual degradation) and for unfamiliar words that are clearly spoken. Further investigations are required if we are to understand the different forms of learning that are critical for effective speech perception in these circumstances.

Acknowledgments

The preparation of this chapter was supported by the UK Medical Research Council (RG91365/SUAG008). We are grateful to Helen Blank, Thomas Cope, Pierre Gagnepain, and Rik Henson for helping to advance the predictive coding account and to Maria Chait and Benjamin Gagl for comments and suggestions on a previous version of this chapter.

REFERENCES

Aitchison, L., & Lengyel, M. (2017). With or without you: Predictive coding and Bayesian inference in the brain. *Current Opinion of Neurobiology, 46,* 219–227.

Blank, H., & Davis, M. H. (2016). Prediction errors but not sharpened signals simulate multivoxel fMRI patterns during speech perception. *PLoS Biology, 14,* e1002577.

Blank, H., Spangenberg, M., & Davis, M. H. (2018). Neural prediction errors distinguish perception and misperception of speech. *Journal of Neuroscience, 38*(27), 6076–6089.

Breitenstein, C., Jansen, A., Deppe, M., Foerster, A. F., Sommer, J., Wolbers, T., & Knecht, S. (2005). Hippocampus activity differentiates good from poor learners of a novel lexicon. *NeuroImage, 25,* 958–968.

Brysbaert, M., Stevens, M., Mandera, P., & Keuleers, E. (2016). How many words do we know? Practical estimates of vocabulary size dependent on word definition, the degree of language input and the participant's age. *Frontiers in Psychology, 7,* 1116.

Christiansen, M. H., & Chater, N. (2016). The now-or-never bottleneck: A fundamental constraint on language. *Behavioral and Brain Sciences, 39,* e62.

Cope, T. E., Sohoglu E., Sedley, W., Patterson, K., Jones, P. S. S., Wiggins, J., Dawson, C., Grube, M., Carlyon, R. P. P., Griffiths, T. D. D., Davis, M. H., & Rowe, J. B. (2017). Evidence for causal top-down frontal contributions to predictive processes in speech perception. *Nature Communications, 8,* 2154.

Crowder, R. G., & Morton, J. (1969). Precategorical acoustic storage (PAS). *Perception & Psychophysics, 5,* 365–373.

Davis, M. H. (2015). The neurobiology of lexical access. In G. Hickok and S. Small (Eds.), *Neurobiology of language* (pp. 541–555). Amsterdam: Elsevier .

Davis, M. H., Ford, M. A., Kherif, F., & Johnsrude, I. S. (2011). Does semantic context benefit speech understanding through "top-down" processes? Evidence from time-resolved sparse fMRI. *Journal of Cognitive Neuroscience, 23,* 3914–3932.

Davis, M. H., & Gaskell, M. G. (2009). A complementary systems account of word learning: Neural and behavioural evidence. *Philosophical Transactions of the Royal Society of London B: Biological Sciences, 364,* 3773–3800.

Davis, M. H., & Johnsrude, I. S. (2007). Hearing speech sounds: Top-down influences on the interface between audition and speech perception. *Hearing Research, 229,* 132–147.

Davis, M. H., Johnsrude, I. S., Hervais-Adelman, A., Taylor, K., & McGettigan, C. (2005). Lexical information drives perceptual learning of distorted speech: Evidence from the comprehension of noise-vocoded sentences. *Journal of Experimental Psychology: General, 134,* 222–241.

Davis, M. H., & Scharenborg, O. (2016). Speech perception by humans and machines. In M. G. Gaskell & J. Mirković (Eds.), *Speech perception and spoken word recognition* (1st ed.) Abingdon, UK: Psychology Press.

Di Liberto, G. M., Lalor, E. C., & Millman, R. E. (2018). Causal cortical dynamics of a predictive enhancement of speech intelligibility. *NeuroImage, 166,* 247–258.

Dumay, N., & Gaskell, M. G. (2007). Sleep-associated changes in the mental representation of spoken words. *Psychological Science, 18,* 35–39.

Dupoux, E., & Green, K. (1997). Perceptual adjustment to highly compressed speech: Effects of talker and rate changes. *Journal of Experimental Psychology: Human Perception and Performance, 23,* 914–927.

Freyman, R. L., Terpening, J., Costanzi, A. C., & Helfer, K. S. (2017). The effect of aging and priming on same/different judgments between text and partially masked speech. *Ear and Hearing, 38,* 672–680.

Gagnepain, P., Henson, R. N., & Davis, M. H. (2012). Temporal predictive codes for spoken words in auditory cortex. *Current Biology, 22,* 1–7.

Ganong, W. F. (1980). Phonetic categorization in auditory word perception. *Journal of Experimental Psychology: Human Perception and Performance, 6,* 110–125.

Gaskell, M. G., & Dumay, N. (2003). Lexical competition and the acquisition of novel words. *Cognition, 89,* 105–132.

Gaskell, M. G., & Marslen-Wilson, W. D. (1998). Mechanisms of phonological inference in speech perception. *Journal of Experimental Psychology: Human Perception and Performance, 24,* 380–396.

Goldinger, S. D., Kleider, H. M., & Shelley, E. (1999). The marriage of perception and memory: Creating two-way illusions with words and voices. *Memory & Cognition, 27,* 328–338.

Gow, D. W., Segawa, J. A., Ahlfors, S. P., & Lin, F-H. H. (2008). Lexical influences on speech perception: A Granger causality analysis of MEG and EEG source estimates. *NeuroImage, 43,* 614–623.

Grill-Spector, K., Henson, R., & Martin, A. (2006). Repetition and the brain: Neural models of stimulus-specific effects. *Trends in Cognitive Sciences, 10,* 14–23.

Havas, V., Taylor, J., Vaquero, L., de Diego-Balaguer, R., Rodríguez-Fornells, A., & Davis, M. H. (2018). Semantic and phonological schema influence spoken word learning

and overnight consolidation. *Quarterly Journal of Experimental Psychology (Hove), 71,* 1469–1481.
Henson, R. N., & Gagnepain, P. (2010). Predictive, interactive multiple memory systems. *Hippocampus, 20,* 1315–1326.
Hermansky, H. (2013). Multistream recognition of speech: Dealing with unknown unknowns. *Proceedings of the IEEE, 101,* 1076–1088.
Hervais-Adelman, A., Davis, M. H., Johnsrude, I. S., & Carlyon, R. P. (2008). Perceptual learning of noise vocoded words: Effects of feedback and lexicality. *Journal of Experimental Psychology: Human Perception and Performance, 34,* 460–474.
Holdgraf, C. R., de Heer, W., Pasley, B., Rieger, J., Crone, N., Lin, J. J., Knight, R. T., & Theunissen, F. E. (2016). Rapid tuning shifts in human auditory cortex enhance speech intelligibility. *Nature Communications, 7,* 13654.
James, E., Gaskell, M. G., Weighall, A., & Henderson, L. (2017). Consolidation of vocabulary during sleep: The rich get richer? *Neuroscience & Biobehavioral Reviews, 77,* 1–13.
Jesse, A., & McQueen, J. M. (2011). Positional effects in the lexical retuning of speech perception. *Psychonomic Bulletin & Review, 18,* 943–950.
Kleinschmidt, D. F., & Jaeger, T. F. (2015). Robust speech perception: Recognize the familiar, generalize to the similar, and adapt to the novel. *Psychological Review, 122,* 148–203.
Kriegeskorte, N., & Kievit, R. A. (2013). Representational geometry: Integrating cognition, computation, and the brain. *Trends in Cognitive Sciences, 17,* 401–412.
Leach, L., & Samuel, A. G. (2007). Lexical configuration and lexical engagement: When adults learn new words. *Cognitive Psychology, 55,* 306–353.
Leonard, M. K., Baud, M. O., Sjerps, M. J., & Chang, E. F. (2016). Perceptual restoration of masked speech in human cortex. *Nature Communications, 7,* 13619.
Marslen-Wilson, W. (1973). Linguistic structure and speech shadowing at very short latencies. *Nature, 244,* 522–523.
Marslen-Wilson, W. (1984). Function and process in spoken word-recognition: A tutorial review. In H. Bouma & D. Bouwhuis (Eds.), *Attention and performance X: Control of language processes* (pp. 125–150). Hillsdale, NJ: Erlbaum.
Marslen-Wilson, W. (1993). Issues of process and representation in lexical access. In *Cognitive models of speech processing: The second Sperlonga meeting* (pp. 187–210). Mahwah, NJ: Lawrence Erlbaum.
Mattys, S. L., Davis, M. H., Bradlow, A. R., & Scott, S. K. (2012). Speech recognition in adverse conditions: A review. *Language and Cognitive Processes, 27*(7–8), 37–41.
McClelland, J. L., & Elman, J. L. (1986). The TRACE model of speech perception. *Cognitive Psychology, 18,* 1–86.
McClelland, J. L., McNaughton, B. L., & O'Reilly, R. C. (1995). Why there are complementary learning systems in the hippocampus and neocortex: Insights from the successes and failures of connectionist models of learning and memory. *Psychological Review, 102,* 419–457.
Norris, D., & McQueen, J. M. (2008). Shortlist B: A Bayesian model of continuous speech recognition. *Psychological Review, 115,* 357–395.
Norris, D., McQueen, J. M., & Cutler, A. (2000). Merging information in speech recognition: Feedback is never necessary. *Behavioral and Brain Sciences, 23,* 299–325, 325–370.
Norris, D., McQueen, J. M., & Cutler, A. (2003). Perceptual learning in speech. *Cognitive Psychology, 47,* 204–238.
Norris, D., McQueen, J. M., Cutler, A., & Butterfield, S. (1997). The possible-word constraint in the segmentation of continuous speech. *Cognitive Psychology, 34,* 191–243.
Nusbaum, H. C., & Magnuson, J. S. (1997). Talker normalization: Phonetic constancy as a cognitive process. In K. Johnson & J. W. Mullenix. (Eds.), *Talker variability in speech processing* (pp. 109–132). San Diego: Academic Press.
Obleser, J., & Kotz, S. A. (2011). Multiple brain signatures of integration in the comprehension of degraded speech. *NeuroImage, 55,* 713–723.
Park, H., Ince, R. A. A., Schyns, P. G., Thut, G., & Gross, J. (2015). Frontal top-down signals increase coupling of auditory low-frequency oscillations to continuous speech in human listeners. *Current Biology, 25,* 1649–1653.
Rao, R. P., & Ballard, D. H. (1999). Predictive coding in the visual cortex: A functional interpretation of some extra-classical receptive-field effects. *Nature Neuroscience, 2,* 79–87.
Samuel, A. G. (1986). Red herring detectors and speech perception: In defense of selective adaptation. *Cognitive Psychology, 18,* 452–499.
Samuel, A. G., & Kraljic, T. (2009). Perceptual learning for speech. *Attention, Perception, Psychophysics, 71,* 1207–1218.
Schultz, W., Dayan, P., & Montague, P. R. (1997). A neural substrate of prediction and reward. *Science, 275,* 1593–1599.
Shannon, R. V., Zeng, F-G., Kamath, V., Wygonski, J., & Ekelid, M. (1995). Speech recognition with primarily temporal cues. *Science, 270*(5234), 303–304.
Sohoglu, E., & Davis, M. H. (2016). Perceptual learning of degraded speech by minimizing prediction error. *Proceedings of the National Academy of Sciences of the United States of America, 113,* E1747–E1756.
Sohoglu, E., Peelle, J. E., Carlyon, R. P., & Davis, M. H. (2012). Predictive top-down integration of prior knowledge during speech perception. *Journal of Neuroscience, 32,* 8443–8453.
Sohoglu, E., Peelle, J. E., Carlyon, R. P., & Davis, M. H. (2014). Top-down influences of written text on perceived clarity of degraded speech. *Journal of Experimental Psychology: Human Perception and Performance, 40,* 186–199.
Spratling, M. W. (2010). Predictive coding as a model of response properties in cortical area V1. *Journal of Neuroscience, 30,* 3531–3543.
Suied, C., Agus, T. R., Thorpe, S. J., Mesgarani, N., & Pressnitzer, D. (2014). Auditory gist: Recognition of very short sounds from timbre cues. *Journal of the Acoustical Society of America, 135,* 1380–1391.
Tamminen, J., Davis, M. H., Merkx, M., & Rastle, K. (2012). The role of memory consolidation in generalisation of new linguistic information. *Cognition, 125,* 107–112.
Tham, E. K. H., Lindsay, S., & Gaskell, M. G. (2015). Markers of automaticity in sleep-associated consolidation of novel words. *Neuropsychologia, 71,* 146–157.
van Linden, S., & Vroomen, J. (2007). Recalibration of phonetic categories by lipread speech versus lexical information. *Journal of Experimental Psychology: Human Perception and Performance, 33,* 1483–1494.
Wild, C. J., Davis, M. H., & Johnsrude, I. S. (2012). Human auditory cortex is sensitive to the perceived clarity of speech. *NeuroImage, 60,* 1490–1502.

III
MEMORY

Introduction

TOMÁS J. RYAN AND CHARAN RANGANATH

Memory is the process by which the brain changes as a consequence of experience. Just as memory is defined by change, we have seen a massive change in our thinking about memory since the publication of the first *Cognitive Neurosciences* tome. In the beginning our understanding of the cognitive neuroscience of memory was like an archipelago of islands of knowledge—a collection of subfields in which inquiries were limited by theories, tools, and model systems. In the ensuing years, new generations of scientists sparked a rapid progression of technology and tools as well as paradigms and theories. As we assembled this section, we were struck by the confluence of new ideas, new technologies, and new approaches that span levels of analysis, from molecular changes that drive experience-dependent change, experimental strategies that explore the behavioral consequences of manipulating specific cell populations, and neuroimaging studies that reveal information about the representations that underlie complex cognitive processes. We have moved from seeing a collection of memory systems providing a static record of experience to a dynamic, adaptive process emerging from the complex interplay of molecular, cellular, and circuit-level interactions.

Lasting memory must involve a persistent change in the material structure of the brain. Whatever this change is, if it accounts for a specific memory, it can be referred to as a memory *engram*. Studying brain activity can greatly inform our understanding of learning and recall, but the engram itself necessarily exists in a nascent state. Genetic tools have given us new opportunities to narrow in on the physical basis of the engram in both mammalian and invertebrate organisms. Habituation, one of the most fundamental forms of memory, enables animals to modify their perception and behavior according to recent experience. By employing modern methodologies and informed by a long history of research from experimental psychology, Cooke and Ramaswami have developed a novel theory for sensory habituation that builds on their respective experimental research of habituation memory in *Drosophila* (fruit flies) and mice. They propose that the memory engrams for habituation experience may in fact be ensembles of inhibitory neurons forming to provide a "negative image" of the excitatory ensembles that mediate the perception of environmental stimuli. This theory potentially provides a novel and general way of understanding the organization of memory engrams in higher brain regions. Ryan's chapter describes the recent development and application of engram cell-labeling technology. The combination of immediate early gene (IEG) transgenics with optogenetics allows investigators to label and manipulate specific engram cells in awake, behaving rodents. Ryan discusses how this methodology, when used to test hypotheses directly informed by cognitive perspectives of memory, can lead to new insights into the plasticity mechanism(s) that underlie the long-term storage of specific memories. This line of recent research indicates that memory lies in stable changes in the fine-scale microanatomical structure of the brain. Ryan then proposes that the information storage mechanisms of memory and instinct may be essentially the same, and a novel proposition on the origin of innate information is offered.

How do the circuit-level mechanisms of memory relate to the kinds of events that we remember? This topic is taken up in the chapters by Julian and Doeller and Ranganath and Ekstrom. Both chapters draw on the fact that much of the work on memory in rodent models has focused on spatial learning, whereas work in humans has traditionally focused on memory for events, or episodic memory. Despite the substantial differences between these paradigms, both chapters highlight remarkable parallels. Tulving defined the idea of episodic memories as occurring in a particular spatiotemporal context. O'Keefe and Nadel, in turn, proposed that the hippocampus is the center of a memory system that organizes experiences according to their spatiotemporal context. Whereas Tulving drew from a combination of careful behavioral experiments and introspection, O'Keefe and Nadel drew from a large body of evidence on the functions of the rodent hippocampus in tests of spatial memory, as revealed by studies of place cells and lesion studies.

Central to both the Julian/Doeller and Ranganath/Ekstrom chapters is the idea that work in animal models provides key insights into our understanding of human episodic memory and, conversely, that our understanding of cognition provides key insights into spatial cognition across species.

Julian and Doeller focus on the concept of context as a means to think about how memories are formed and how hippocampal remapping in rodents indicates the critical role of context in episodic memory. They consider a range of work across species to provide an operational definition of how an organism constructs a representation of context. They then provide an overview of hippocampal anatomy and review converging evidence from human neuroimaging studies that lead them to conclude that the hippocampus is central to representing the contextual information that forms the basis for episodic memory.

Ranganath and Ekstrom tackle similar topics from the reverse direction. Starting with the evidence for the central importance of the hippocampus to episodic memory, they lay out several theories of hippocampal function and focus on the key issues of both consensus and disagreement. After considering a wide range of evidence from rodents, monkeys, and humans, they conclude that although each theory has some merit, the available evidence is best accounted for by a model that lays out key principles for understanding the central role of *time* and *space* in hippocampal function and the importance of its position as a central hub that can index sequences of states in multiple semimodular cortical networks.

Beyond understanding the circuits that support memory, it is essential to consider why some memories seem to be inaccessible, while others remain vivid long after the event has passed. The chapters by Meyer/Pattwell and Gruber/Ritchey take on this topic, focusing on the link between motivation, emotion, and memory. Although it is common for psychologists and neuroscientists to differentiate between emotion and cognition, it is highly likely that these processes are intimately linked with basic motivational processes critical to survival. In particular, for memory to be adaptive it is necessary to prioritize the retention of information that is of high significance for future behavior.

From a lifespan perspective, memory can be thought of as an extension of development. Genetically determined developmental processes construct our brains, and environmentally induced plasticity processes further develop this circuitry to form memories. Meyer and Pattwell take a developmental perspective to understand memory across the lifespan, considering how emotion, stress, and motivation affect the developing brain and, in turn, how the influence of these variables changes over the course of development. They provide a comprehensive summary and a unique synthesis of cognitive, behavioral, and molecular approaches to memory across development and shed much light on how memory must adapt to a developing body as well as a changing environment.

Ritchey and Gruber, in turn, focus on the role of motivation in prioritizing the events that will be remembered and how these events are remembered. Like Meyer and Pattwell, they consider both reward motivation and the effects of arousal elicited by aversive motivation. They consider the well-established evidence of the role of norepinephrine and dopamine in promoting the lasting retention of aversive and appetitive experiences, respectively. However, rather than seeing these neuromodulators as mechanisms of stabilizing learning through a relatively simple consolidation process, Ritchey and Gruber envision consolidation as a process that prioritizes the retention of the most important aspects of the most salient experiences. They also point out that many of the same neurobiological factors that influence consolidation also influence how a memory will be learned, such that positive and negative experiences shape both what is attended to during learning and how well those experiences will be retained.

Just as memory is an index of change, our memories themselves are dynamic, changing as they are activated during off-line and online states. Paller, Antony, Mayes, and Norman consider how memories are reactivated during both sleep and waking states and how reactivation can influence the fate of both reactivated memories and memories for competing events. Paller et al. review findings from detailed neurophysiological studies of single-unit and oscillatory activity in rats, suggesting that reactivation during slow-wave sleep depends on an orchestrated relationship between hippocampal firing sequences, hippocampal sharp wave ripples, thalamocortical sleep spindles, and cortical slow oscillations. To understand how reactivation could influence memory, Paller et al. focus on an innovative approach by which memories of specific experiences can be reactivated by providing cues during sleep. The reviewed research shows, remarkably, that targeted reactivation can significantly improve the retention of a wide variety of memories, including forms of learning previously thought of as independent of the hippocampus. The reviewed work demonstrates that rather than providing a crystallized record of past experiences, memories are dynamic, with many changes happening even during off-line states.

Continuing with this theme, Orederu and Schiller draw on both human and animal studies to provide a current perspective on the memory reconsolidation field. Reconsolidation, the general concept that consolidated long-term memories can be destabilized and modified or updated, has represented a paradigm shift in how memory is understood in behavioral neuroscience. It also provides a bridge for integrating seemingly disparate cognitive and behavioral perspectives of memory by understanding that memory engrams are never crystallized and can be altered with new experience. Orederu and Schiller detail the history of this topic and how it has led to a fundamental revision in how we understand the molecular and cellular basis of memory storage. The chapter details the nuances and criteria for investigating reconsolidation processes and brings us up to the frontier findings and questions of the field. It also describes current strategies to target individual memories for post-traumatic stress disorder (PTSD)treatment and drug addiction in humans.

In surveying these contributions, it is clear that the memory field itself is in a stage of change and transition. Molecular and systems neuroscience approaches are having a transformative effect on memory research, including the mechanistic neurobiology required for memory as well as our understanding of the cognitive processes that characterize memory function itself. The study of memory is developing into a mechanistic cognitive neuroscience that is ready for new concepts and investigative strategies. Younger researchers are approaching memory more expansively, combing perspectives and ideas from what is known about cognition, development, and evolution. We will not attempt to predict what the future of memory research will look like, but it certainly will not be boring.

17 Ignoring the Innocuous: Neural Mechanisms of Habituation

SAMUEL F. COOKE AND MANI RAMASWAMI

ABSTRACT Habituation is a form of learning that reduces behavioral responses to stimuli experienced repeatedly without reward or punishment. This fundamental form of learning is exhibited by a wide range of organisms. Habituation enables energy and attention to be devoted to stimuli that have already been established as meaningful, as well as to novel stimuli that may merit exploration or avoidance due to their potential to deliver reward or punishment. The detection of novelty requires memory for all things familiar, a lasting neural imprint revealed as behavioral habituation. Great difficulties arise for organisms that are unable to ignore familiar and innocuous elements of the environment due to the failure of habituation. Significantly, such difficulties are apparent across a range of psychiatric disorders. Early studies of habituation, which focused on accessible sensorimotor circuits, have recently been extended through several direct studies of how habituation processes are implemented via neural plasticity in the central nervous system. Together, these indicate that patterns of neural excitation triggered by novel stimuli can be attenuated with familiarity through the buildup of matching patterns of inhibition. Here we provide an integrated summary of the current understanding of habituation, familiarity, and novelty detection and discuss the questions that remain to be answered.

Consider a countryside denizen moving, for the first time, from a small, quiet rural village to a large, busy metropolis in pursuit of fame and fortune. That person's first visit to the city's downtown area would likely prove highly memorable but also discombobulating, as her sensory systems are bombarded with a wide array of intense, novel stimuli: gaudy streetlights, blaring traffic noise, and the malodor of exhaust fumes. Before she can efficiently engage in goal-directed behavior, such as crossing the street to find a place for lunch while avoiding oncoming cars, she must quickly habituate to the elements of her environment that are irrelevant to those goals. This process of *short-term habituation* filters out unnecessary cues, facilitating the attainment of immediate goals, which are to find the reward of food while avoiding the punishment of being run over by a car. When she next returns to that same setting, those features she previously habituated to may become relevant to new immediate goals. Therefore, habituation will occur to a separate set of stimuli that are now irrelevant to these new goals, which may include going to the theater or escaping from the rain. However, a second form of plasticity will occur as the person returns repeatedly to the same context, perhaps as she passes through it every day on her commute to work. This *long-term habituation*, in which generally innocuous stimuli that never predict impending reward or punishment become *familiar* over repeated experience, allows the person to disengage from sensory input and devote her brain to analysis or planning. Importantly, this habituated state does not prevent the person from responding swiftly to the emergence of a *novel* and potentially critical element of the world around her, such as a *Tyrannosaurus rex* walking down Main Street! You might also like to imagine what life might be like for this new urban inhabitant if she did not possess this amazing but apparently simple faculty of habituation. How might these remarkable abilities, which we often take for granted, be implemented in our central nervous system?

Habituation allows organisms to suppress *behavioral responses* to familiar stimuli that consistently fail to signal reward or punishment. This form of learning enables organisms to focus energy and attention on meaningful or novel elements of their environment that may predict reward or punishment. The fundamental importance of habituation is apparent in its conservation across a wide range of organisms, from those that do not possess nervous systems, such as paramecia (Jennings 1906), to simple nervous systems, such as those of nematodes (Rose and Rankin 2001) and sea slugs (Castellucci et al. 1970), to the progressively more complex nervous systems of fruit flies (Twick, Lee, and Ramaswami 2014), zebra fish (Marsden and Granato 2015), rabbits (Horn and Hill 1964), cats (Thompson and Spencer 1966), and humans (Barry and Sokolov 1993). This indicates that habituation can be implemented in various ways, supported by many different signaling systems and circuits, and suggests that multiple mechanisms operate in parallel in more evolved nervous systems. However, it is also possible that in complex systems such as the vertebrate brain, only a few mechanisms are commonly and efficiently implemented. The clear importance of habituation indicates that there is the potential to study a process that is as equally critical to humans as it is to the wide range of species that we use to model them.

Habituation is typically described as a nonassociative form of learning because in the experimental setting it occurs to stimuli that are explicitly not associated with reward or punishment (Pinsker et al. 1970). However, this simple form of learning serves as a gateway to higher-order cognition, which may involve reward or punishment or the formation of associations between neutral stimuli (Schmid, Wilson, and Rankin 2014). Deficits in habituation are apparent in a range of psychiatric conditions, including autism, schizophrenia, and intellectual disability, and likely contribute to characteristic higher cognitive and, perhaps, noncognitive aspects of these disorders (McDiarmid, Bernardos, and Rankin 2017; Ramaswami 2014). While a substantial body of investigative work has been conducted on habituation in a range of simple and sensorimotor preparations, the central mechanisms of behavioral habituation have historically been largely ignored. Focused attention on the mechanisms that underlie cognitive habituation is essential, not only for a deep understanding of this foundational process but also because such understanding may elucidate cellular mechanisms that generally operate for information storage and retrieval in higher-order forms of learning and memory.

Several factors make habituation an attractive form of learning to study. First, it occurs reliably in all possible animal models without the need for pretraining or shaping. Second, because it occurs to even the simplest of sensory stimuli, it can be studied with great experimental precision. Third, although it may be supported by plasticity occurring throughout the central nervous system, underlying neural events can be studied in regions of the brain proximal to sensory input where experimental access is relatively easy, where form and function are relatively well understood, and, critically, where information remains relatively unprocessed. In this chapter we will discuss what is known about habituation across different timescales. In so doing we will cover a range of experimental systems that have been used to gain insight, the knowledge of underlying circuitry and molecular mechanisms that has been acquired, and the major models that exist to explain these phenomena.

Fundamental and Defining Features of Habituation

In an influential article, Thompson and Spencer (1966) outlined what they regarded as nine fundamental features of behavioral habituation:

1. Given that a particular stimulus elicits a response, repeated applications of the stimulus result in decreased response (habituation). The decrease is usually a negative exponential function of the number of stimulus presentations.
2. If the stimulus is withheld, the response tends to recover over time (spontaneous recovery).
3. If repeated series of habituation training and spontaneous recovery are given, habituation becomes successively more rapid (this phenomenon might be called potentiation of habituation).
4. Other things being equal, the more rapid the frequency of stimulation, the more rapid and/or more pronounced is habituation.
5. The weaker the stimulus, the more rapid and/or more pronounced is habituation. Strong stimuli may yield no significant habituation.
6. The effects of habituation training may proceed beyond the zero or asymptotic response level.
7. Habituation of response to a given stimulus exhibits stimulus generalization to other stimuli.
8. Presentation of another (usually strong) stimulus results in recovery of the habituated response (dishabituation).
9. Upon repeated application of the dishabituatory stimulus, the amount of dishabituation produced habituates (this phenomenon might be called habituation of dishabituation).

All these criteria were proposed with short-term habituation in mind. An important tenth criterion was added in a recent revision of the defining features of habituation by Thompson and other influential colleagues (Rankin et al. 2009) to acknowledge the phenomenon of long-term habituation:

10. Some stimulus repetition protocols may result in properties of the response decrement (e.g. more rapid rehabituation than baseline, smaller initial responses than baseline, smaller mean responses than baseline, less frequent responses than baseline) that last hours, days or weeks. This persistence of aspects of habituation is termed long-term habituation. (p. 137)

Several of these features also apply to other forms of memory, such as associative memory—for instance, feature 2 (*spontaneous recovery*) and feature 7 (*generalization*). Some other features, while intellectually useful, are almost never established experimentally for most studied forms of habituation—for instance, feature 9 (*habituation of dishabituation*). While we refer keen students of this subject to the original deep discussions by Thompson and colleagues, we choose to focus here on three features, which we identify as the *defining properties* of habituation (figure 17.1), to evaluate how the brain is modified during habituation:

1. Habituation always manifests as a *reduced behavioral response* to a stimulus following repeated or sustained exposure. While this has been commented on in the literature for more than 3,000 years (e.g., Aesop's fable of the camel; [Townsend 1867]), it was experimentally perhaps most clearly documented by observations on spiders learning to ignore vibrations (Peckham and Peckham 1887) and by Ivan Pavlov, who described the "conditioning of the orienting reflex" in dogs, a phenomenon in which animals show reduced orientation toward familiar, repeated stimuli (Sechenov 1863).
2. Habituation is *gated*: it occurs less efficiently if reward, punishment, or strong emotional engagement occurs together with stimulus exposure. For example, Pavlov's dogs did not habituate to the sound of a bell that was a harbinger of food (instead, they developed a strong response to it). Similarly, Thompson and Spencer (1966) demonstrated that decerebrated cats habituate relatively easily to weak foot shocks when compared to shocks of higher intensity. This phenomenon is not restricted to reflexes. It is also seen for more complex exploratory behavior (Welker 1956).
3. Most interestingly, habituation is *subject to dishabituation or override*: for instance, the sudden loud noise of a car backfiring from the side of a street may cause the hypothetical country denizen, whom we met earlier, to abruptly attend to the surroundings she had previously habituated to. This ability to volitionally reengage with habituated elements is critically important, indicating that the process of habituation is mediated by a mechanism that allows it to be transiently suppressed when required. The phenomenon of dishabituation is particularly important because it distinguishes habituation from sensory adaptation, in which sensory epithelia are temporarily modified to optimize sensation, or motor fatigue, in which muscular output is temporarily reduced by a drain on metabolic resources. The instant reinstatement of the response that occurs in dishabituation could not occur if sensory adaptation or motor fatigue was a contributory factor.

Although in many cases, such as long-term olfactory habituation in *Drosophila* (Das et al. 2011), it may not be possible for the experimenter to easily identify a dishabituating novel stimulus, it is often still possible to show that behavioral habituation is rapidly reversible by coaxing animals into attending and responding to the familiar stimulus. For instance, mice habituated to a certain tonal frequency following days of passive exposure to the same tone quickly reengage with and respond to the familiar tone when it results in a food reward (Kato, Gillet, and Isaacson 2015). Thus, a key feature of habituation is that it is subject to override. As we will see below, some reported instances of dishabituation may in fact arise through a parallel but distinct process of behavioral sensitization (Groves et al. 1970; Pinsker et al. 1973), which increases general arousal and may nonspecifically reduce the sensory thresholds required for a broad range of sensory stimuli.

Before discussing the proposed models for habituation that account for these defining features with varying levels of success, it is also important to note that there are many time scales of habituation, which we here term as *fast*, with onset and recovery within seconds or less; *short term*, with onset and recovery with timescales of minutes; and *long term*, with onset and recovery with timescales of days and weeks. The latter form of habituation, like long-term forms of memory,

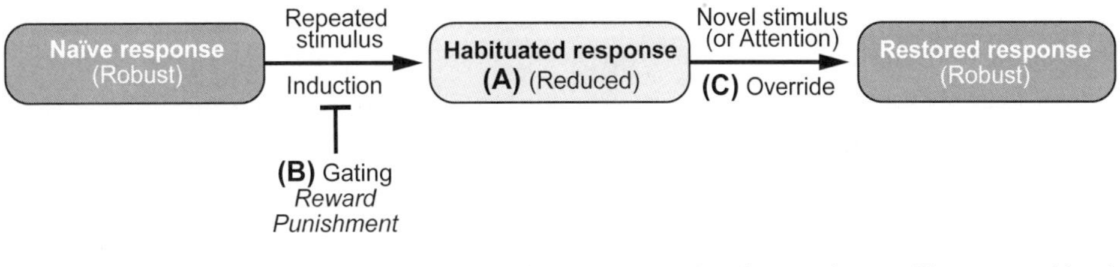

FIGURE 17.1 Defining features of habituation. Of the 10 defining criteria that have been proposed for habituation (Groves and Thompson 1970; Rankin et al. 2009; Thompson and Spencer 1966), we focus on those we consider the three most reliable and critical: *A*, Habituation always leads to a reduction in behavioral response. *B*, Habituation is gated by other factors. In the absence of reward, punishment, or intense arousal, habituation occurs, but in the presence of any of these factors, habituation will likely not occur. *C*, Habituation to one stimulus can be readily reversed by the presentation of an arousing stimulus through a process known as *dishabituation*. This, particularly for long-term forms of habituation, may not be easy to demonstrate experimentally due to the difficulty in determining an appropriate stimulus and intensity. However, attention can override habituation, showing that even after habituation animals retain the capacity to respond robustly to a familiar stimulus.

requires new gene expression and protein synthesis (Ezzeddine and Glanzman 2003). Most of our current understanding of habituation is based on the explicit study of short-term habituation in invertebrate species such as the sea slug *Aplysia californica* (Castellucci et al. 1970; Pinsker et al. 1970) and reflex pathways in vertebrates (Farel and Thompson 1972; Teyler, Roemer, and Thompson 1972). However, work undertaken on long-term habituation in these systems indicates that they are supported by different physiological mechanisms (Rankin et al. 2009; Sanderson and Bannerman 2011).

There has also been a great deal of work conducted on long-term habituation in the well-known rodent assay of familiar object recognition (FOR) or novel object detection (NOD), but this work has tended to focus on the process of retrieving an established memory (recognition and detection) rather than the process of learning itself (habituation) (Bevins and Besheer 2006). For the purposes of this chapter, we will not dwell on this work, although it is certainly important to acknowledge the relevance of familiarity memory in relation to understanding habituation.

Mechanisms of Habituation

Proposed mechanisms for habituation fall into two broad classes (figure 17.2). One class of mechanisms, which is also most frequently included in textbooks, posits that familiar inputs trigger a *weaker excitation* of the neurons that mediate behavioral outputs. A second class of mechanisms posits instead that familiar inputs trigger *stronger inhibition* onto downstream neurons that drive behavior. The two classes differ most crucially in the implied mechanism of dishabituation or habituation override: the first proposing a process of overlying sensitization; the second, an independent disinhibitory mechanism.

Excitatory Depression Models The pervading model of habituation remains one in which feedforward neuronal pathways connecting sensory neurons and response neurons are weakened by repeated stimulation (figure 17.2*A*). It is a common theme in excitatory depression models, such as self-generated depression (Horn 1967) or stimulus-response decrement (Groves and Thompson 1970), that the synaptic depression underlying habituation arises through reduced neurotransmitter release (Castellucci et al. 1970; Farel and Thompson 1976). However, alternative means of weakening excitatory drive, such as reducing the postsynaptic response (Wickelgren 1977) or a change in dendritic excitability (Marsden and Granato 2015), have also been implicated.

Homosynaptic depression A major motivating force behind the synaptic depression model is that it invokes Occam's razor, being the simplest of all possibilities (Horn 1967). A further benefit exists if we consider that depression could be synapse-specific, or homosynaptic, because neurons that integrate inputs from two modalities exhibit a response decrement during habituation to one modality without transfer to the other (Bell et al. 1964), indicating input specificity that could only be achieved by synaptic modification or some very localized change in excitation. However, the simplest version of the excitatory depression model cannot explain how to satisfy the key criterion of dishabituation, in which the behavioral response to the initial stimulus is immediately reinstated by the presentation of a second novel stimulus. If there is only a simple decrement in a purely feedforward system, then the presentation of a second novel stimulus, which must drive activity through a separate set of neurons and synapses to the original stimulus, would not be expected to immediately reinstate a response to the original stimulus unless some cross talk exists between the two pathways. Moreover, how could weakened synapses be instantaneously returned to their original state by the presentation of the second novel stimulus to mediate dishabituation to the original stimulus? Such immediate cued recovery of synaptic strength or whole-cell excitability is not a phenomenon that has ever been described under physiological conditions. Two key modifications to this model have been proposed to reconcile homosynaptic depression with dishabituation.

The dual-process model Thompson and Spencer (1966; Groves and Thompson 1970) put forward the dual-process model of habituation, which is somewhat similar to the proposed explanation of dishabituation by Gabriel Horn (1967). In this model, the overall behavioral output through reflex pathways is determined by a balance of two processes: first, habituation, which is a stimulus-selective phenomenon mediated by the feedforward depression of the stimulus-response pathway, and second, a counteracting form of nonassociative learning known as sensitization (Carew, Castellucci, and Kandel 1979). Sensitization, which has a generalized effect across stimuli and results from an enhancement of the neuronal "state" due to a sensitizing stimulus, could result from a variety of arousing stimuli, but one possible driver would be the sudden emergence of a novel feature in the environment, increasing the output of the entire nervous system and promoting the behavioral response through an unspecified positive neuromodulation of activity. Such an arousal system, possibly incorporating elements of the reticular

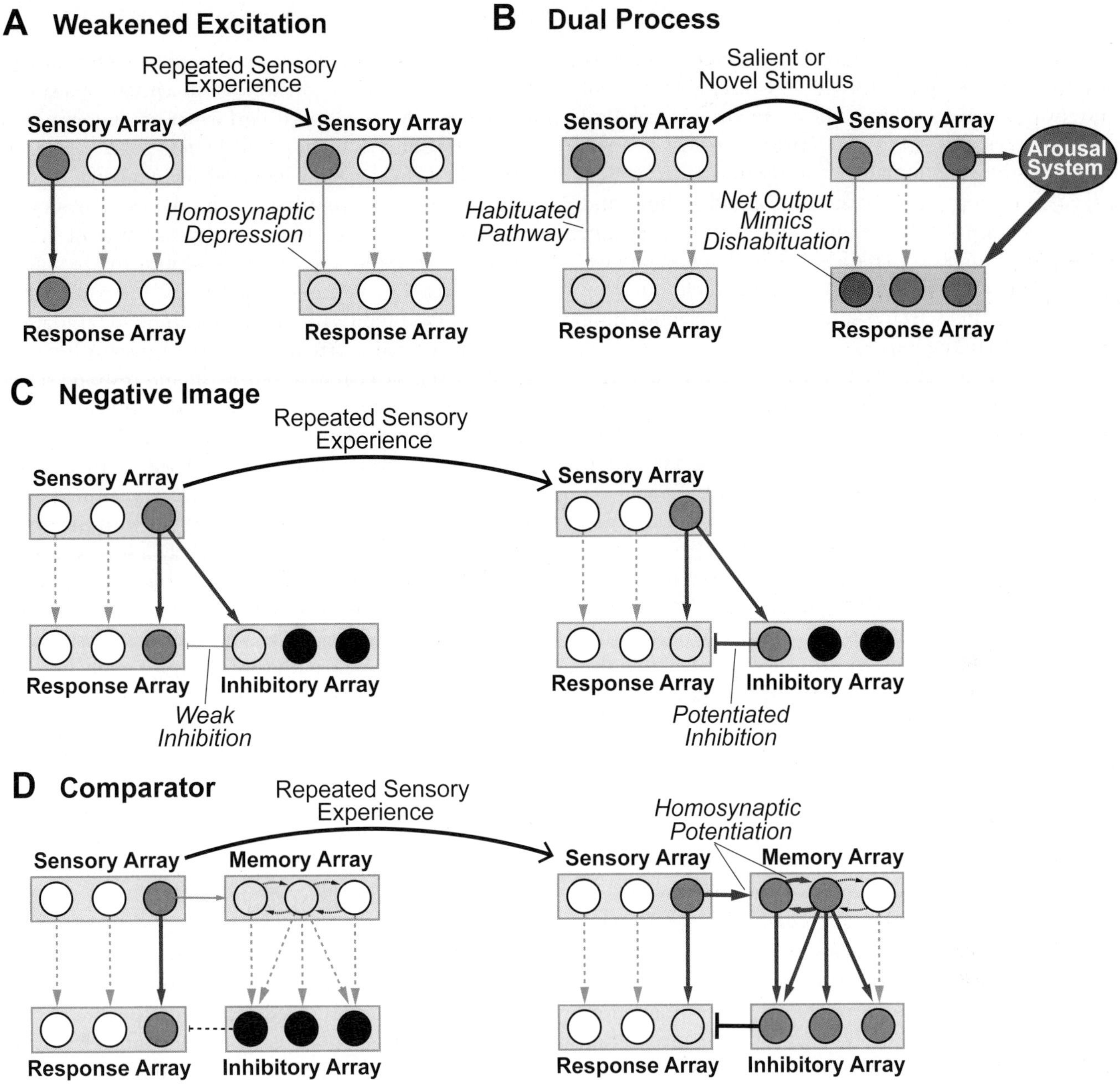

FIGURE 17.2 Models of habituation. Models are presented here with simplified arrays of neurons representing each conceptual stage. Thus, the stimulus-response pathway is modeled as a sensory array connected through one or more intermediates to a response array. *A*, The most parsimonious explanation of behavioral habituation is *weakened excitation* via feedforward synaptic connections between the sensory array, which first responds to sensory input, and an array of unidirectionally connected response neurons, which are responsible for driving behavioral response. Shown on the left of this panel are the arrays prior to habituation, when an innocuous stimulus (*dark blue*) drives a pronounced response through existing feedforward inputs. After repeated presentation of the same innocuous stimulation, the behavioral response is selectively weakened (*light blue*) by homosynaptic depression within these feedforward stimulus-response connections. *B*, An important add-on to this feedforward depression model is included in the *dual process model*, largely to explain the important phenomenon of dishabituation, in which response returns immediately to the habituated stimulus (*blue*) after the presentation of a novel or salient stimulus (*red*). This model regards behavioral output as a net effect (*purple*) of depressed response through habituation, which is highly stimulus-specific, and increased response through sensitization, which is not stimulus-specific and is mediated by a modulatory arousal system. Thus, while stimulus-response synapses remain weakened, a generalized increase in the output of the response array returns the response output to approximately prehabituation levels. *C*, An alternative model contains an added layer of complexity, which is an array of inhibitory interneurons. In this model the primary modification is a selective potentiation of inhibitory neurons that form a *negative image* to suppress the output of the response array. Although not depicted here, dishabituation is proposed to be mediated by disinhibiting the response array, meaning that the previously habituated response truly returns to basal levels after dishabituation. *D*, Finally, *comparator* models have been proposed in which a memory array is an additional intermediary between the stimulus and the response. This memory array, formed by initial experience, is an internal representation of the familiar stimulus. If sensory input subsequently matches this internal representation after comparison, the memory array acts on the response array through the inhibitory intermediaries to suppress its output. If a novel stimulus is presented for which no internal representation exists, then this "top-down" inhibition cannot be applied, leading to behavioral output. The advantages of memory arrays are explained in more detail in figure 17.3. (See color plate 20.)

activating system in vertebrate brain stems, does seem plausible, as it is a phylogenetically conserved region of the brain that contains nuclei mediating general arousal in response to threatening, rewarding, or novel stimuli through modulatory transmitters such as noradrenaline, dopamine, serotonin, acetylcholine, and histamine (Jones 2003). Thus, there is still an elegant simplicity to the idea that dishabituation could, in fact, reflect a generalized sensitization that ameliorates the still implemented selective stimulus-response weakening imposed by habituation.

However, this dual-process model makes clear predictions about the stimulus-specificity and generalization that should be observed during the dishabituation process. Although some observations are consistent with these predictions (Groves, Lee, and Thompson 1969; Thompson and Spencer 1966) many others are not: Notably, disinhibition does not always produce the generalized effect that one would expect if it were mediated by sensitization (Marcus et al. 1988), suggesting that dishabituation may be a unique process in its own right. Also, in the *Aplysia*, at least, the phenomenon emerges at a different time during development than sensitization (Rankin and Carew 1988), indicating that the two processes have separable underlying mechanisms. In the *Aplysia* there is also striking evidence that the decrement in synaptic release that demonstrably mediates stimulus-selective, short-term behavioral habituation (Castellucci et al. 1970) is directly reversed by the presentation of a dishabituating novel stimulus (Carew, Castellucci, and Kandel 1979). Though broadly in keeping with Groves and Thompson's dual-process theory because the dishabituation behavioral effect is mediated by an interaction between the stimulus-response pathway and an overall state effect, this observation is also at odds with that theory because the presentation of the different *apparently* sensitizing stimulus results in the *specific* reversal of plasticity invoked in the habituated stimulus-response pathway *without a general effect on other synapses.* Thus, the key difference between dishabituation and sensitization may be in whether they truly reverse the physiological consequences of habituation or instead compensate for the effect by potentiating neural output via a different target.

These observations highlight one limitation of excitatory depression models for habituation: if synaptic weakening through any mechanism were the sole basis of habituation, it would be difficult to envisage biophysical mechanisms at the synapse that would allow a dishabituating stimulus to immediately reverse such weakening. A second conceptual problem lies in an intrinsic limitation of the excitatory depression model. While it appears to be a reasonable solution for habituation to *hedonic stimuli,* in which the activity of a single neuron encodes meaningful information, it does not as effectively address habituation to *perceptual stimuli,* in which the information is encoded by neuronal assemblies. If objects are represented in the brain, as images are on a monitor, by the specific assembly of active neurons (or pixels), then a mechanism for habituation that invokes the dimming of each pixel that contributes to an object would result in the substantial degradation of all images that utilize the same pixel, greatly limiting the ability of the system to represent and respond selectively to different objects. Thus, at a theoretical level, it would be preferable to conceive of models in which habituation can act at the level of the entire object image, rather than at the level of each constituent neuron/pixel. Both of the above difficulties with the excitatory depression model are effectively addressed by a second class of habituation models that relies not on changes in feedforward excitatory synaptic strength but rather on changes in the strength of inhibition in the stimulus-response pathway as major factors in driving behavioral habituation.

Inhibitory potentiation models The potential role for increasing inhibition was appreciated in classical neuropsychological writings on habituation, perhaps most clearly by Clark Hull (1943), who referred to the buildup of "residual inhibition" as a potential underlying mechanism. However, this initial premise was not supported by studies of habituation in experimentally accessible sensorimotor reflex circuits. Here, electrophysiological recordings from neurons that mediate behavioral reflexes provided data that supported excitatory depression as the underlying mechanism, at least for rapid forms of habituation (Castellucci et al. 1970; Thompson and Spencer 1966). The wide acceptance of this model ignored the lack of evidence for excitatory depression in longer-lasting forms of habituation as well as the absence of information on central mechanisms supporting perceptual habituation (Ramaswami 2014; Rankin et al. 2009). However, with the emergence of experimentally accessible central circuits that encode percepts and behavior, central mechanisms of habituation have recently begun to be explored. These studies of brain systems now indicate key roles for inhibitory circuits in driving behavioral habituation (Kaplan et al. 2016; Kato, Gillet, and Isaacson 2015; Ramaswami 2014).

The negative image model Neural excitation is typically paired with inhibition in most organisms. Excitatory arrays transmit not only excitation to downstream neurons but also feedforward and feedback (or recurrent) inhibition. In addition, excitatory arrays often receive descending inhibition from downstream brain regions. Within this conserved architectural framework, the negative-image model proposes that habituation arises from the selective strengthening of inhibitory inputs onto excitatory arrays.

The simplest version of this model emerged from studies of olfactory habituation. This model was enabled by pioneering studies that detailed the conserved neurons and circuits involved in olfactory coding in insect and mammalian nervous systems (Joseph and Carlson 2015; Wilson 2013). In the insect olfactory system, an odor-activated sensory neuron array excites a corresponding array of projection neurons (PNs). Crucially, PNs also receive feedforward and feedback inhibition from local inhibitory interneurons. *Drosophila* olfactory habituation appears to occur through the specific and selective potentiation of inhibitory synapses made on the PN array (Das et al. 2011; Ramaswami 2014). This matching inhibitory pattern, termed a *negative image*, may be created through the implementation of a local synaptic learning rule: the strength of inhibitory synapses increases selectively on postsynaptic PNs that show sustained elevated levels of activity.

In addition to explaining how olfactory habituation occurs and is implemented through a simple underlying synaptic mechanism, this model proposes that gating and override of habituation occur, respectively, through the modulatory control of inhibitory synaptic plasticity and disinhibition (the inhibition of inhibitory neurons mediating habituation; Barron et al. 2017; Ramaswami 2014). In rodents, long-term auditory habituation to specific tonal frequencies also occurs through the potentiation of inhibition onto pyramidal cells tuned to respond to the familiar frequency. In this study the predicted role for disinhibition in the override of habituation has been directly observed (Kato, Gillet, and Isaacson 2015). Taken together, negative images formed through a homeostatic inhibitory potentiation mechanism, wherein inhibition is tuned to match the level of postsynaptic excitation within a specific time window, offer a potentially satisfying and empirically supported mechanism to explain forms of cognitive habituation in insect and mammalian brains.

However, the simplest version of the negative image model, wherein locally created, matched inhibitory patterns filter sensory input and selectively reduce the ability of familiar stimuli to excite downstream brain regions, does not explain a subjectively obvious feature of some forms of habituation. Quite simply, while familiar stimuli may be less salient, they are usually also accompanied by a memory, meaning that we actively recognize them as previously encountered stimuli. For instance, though thoroughly habituated to our office, we still explicitly recognize it as our office. Furthermore, a familiar stimulus in one context may appear novel or salient in a different one, as highlighted in O'Keefe and Nadel's (1978) observation that "the novelty of the spouse in the best friend's bed lies neither in the spouse, nor the friend, nor the bed, but in the unfamiliar conjunction of the three" (p. 241).

Thus, the details of many normally inconsequential but occasionally important stimuli are not only filtered but also stored as familiar memories in the brain that can be retrieved for a variety of purposes. Therefore, in addition to simple filtration, the brain must actively store and access the information of familiar people, places, objects, and events. This was recognized in the late 1950s by Yevgeny Sokolov, who proposed that inhibition of the response array came not via feedforward inhibition from sensory arrays but from a process that compared the current sensory input with the bank of stored memories so that matched stimuli would trigger inhibition from the memory center to the response array, while novel stimuli would provide a mismatch that would drive an uninhibited behavioral response. Others proposed similar models (Konorski 1967; Wagner 1979). Recent observations have resurrected interest in this previously influential but now less acknowledged set of models described collectively as comparator models.

Comparator models of habituation Comparator models propose the formation of an engram for familiar stimuli that can suppress the output of behavioral or arousal systems through feedforward inhibition (Konorski 1967; Sokolov 1960a; Wagner 1979). When a stimulus is familiar, the stored model is activated and the output suppressed, but when a stimulus is novel, no such model exists to be activated, and output is no longer suppressed. Thus, they interpose a memory system between the sensory array and inhibitory output onto the response array (figure 17.2*D*). In doing so, comparator models presciently articulate and invoke what is now considered to be the core feature of contemporary *predictive coding* theory (Rao and Ballard 1999). Of many advantages, one obvious desirable feature of comparator models is that they explain habituation while also providing a framework that supports the storage and volitional recall of familiar memories. Additionally, and most pertinently to this chapter, the comparator model is also necessary to explain recent experimental observations that cannot be rationalized based on either a purely feedforward depression model or a simple local inhibitory filtration model. Behavioral habituation to visual stimuli in the mouse is associated

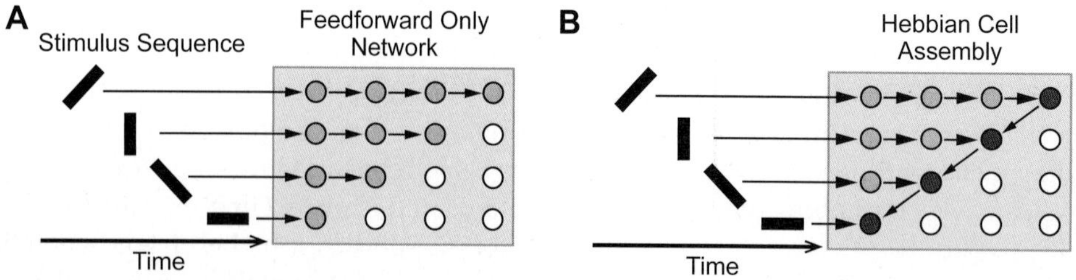

FIGURE 17.3 The advantages of memory arrays. What are the advantages of including the extra complication of a memory array in models of habituation? Memory arrays allow habituation to reflect all the complexities associated with memories, including context specificity, pattern completion for partial cues, and sensitivity to the spatiotemporal features of stimuli. Key to implementation of these features are not only feedforward excitation but also recurrent lateral and feedback circuitry. This figure, which is based on recent experiments (Gavornik and Bear 2014), shows how a Hebbian assembly (Hebb 1949) encoding spatiotemporal sequence information can form and be compared to corresponding features of incoming inputs. *A*, A sequence of oriented lines initially drives weak cortical responses that are not connected. Each orientation activates a distinct neuronal array. *B*, The potentiation of lateral connections between sensory elements that impact the array at different delays, however, "teaches" the network to encode both sequence, by providing strong preparatory excitatory inputs to the neurons (*arrows between dark gray elements*) that stimulate the next element of the sequence, and time, which could be encoded through synaptic delay lines defined by the number of synapses traversed by activity elicited from each stimulus. Habituation, depending on inhibitory inputs deriving from these memory arrays, would show all these distinctive features of memory.

with and requires pronounced increases in excitatory synaptic and neural activity in the superficial layers of the visual cortex (Cooke et al. 2015). Thus, an intermediate step in this habituation process involves the potentiation of excitatory transmission in the cortex.

The comparator-driven inhibitory mechanism for habituation can not only account for stored familiarity memory but can potentially also explain some other features of habituation that are difficult to justify by a local inhibitory filtering mechanism alone. For instance, habituation can be selective for the temporal frequency of stimulus presentations, even if all other features of the stimulus are maintained. Also, for familiar sensory sequences the omission of an element of a habituated sequence can trigger an active physiological and behavioral response (Bernstein 1969; Sokolov 1960b; Zimny and Schwabe 1965). Comparator models can explain the temporal specificity of habituation or the detection of novelty when a sequence element is omitted because they include the lateral and feedback connectivity within a memory array that can encode spatiotemporal sequences. One way in which this could be accomplished is depicted in figure 17.3, where a laterally connected memory array learns a sequence of visual inputs. Here, different oriented line stimuli produce activity in different polysynaptic feedforward pathways, forming synaptic delay lines. Because it takes some time for activity to pass from entry point to end point, neural activity is evoked at different points in each pathway for each element of the sequence. However, these pathways are also weakly connected to each other with both lateral and feedback inputs, providing an opportunity for Hebbian synaptic potentiation to strengthen the connections between coactive pathways. Selective strengthening between the relative delay point in each synaptic pathway forms a Hebbian cell assembly, which has the capacity to store not only spatiotemporal memories but also complete stored patterns by partially depolarizing a neuron preemptive to it being activated by sensory input. Just this kind of observation of sequence learning has been reported in V1 (Gavornik and Bear 2014), to the extent that phantom responses are produced by cortex to missing sequence elements. Without lateral/feedback connectivity, it is very hard to explain this phenomenon or related examples of pattern completion.

Hybrid models It is likely that elements of each model, with their respective advantages, operate for different forms of habituation in different nervous systems or that a hybrid model could be in operation for all forms of habituation. It is our contention that, as has been shown experimentally, short-term habituation probably relies upon excitatory depression in most cases, while longer-term forms of habituation likely rely upon the formation of engrams that incorporate some aspects of negative image and comparator models. Much work is required to understand these processes more deeply.

Conclusions

Here we have discussed the fundamentally important cognitive phenomenon of habituation. We have described its cardinal features and what is understood about its implementation in the central nervous system. We hope it is clear to the reader that much is yet to be understood about habituation and the related phenomena of familiarity and novelty detection. We have a wonderful opportunity to gain deep insight, given how reliable and pervasive these phenomena are compared to so many other higher-order forms of learning and memory. Many outstanding questions remain. Some of the more intriguing ones include: (1) Is habituation a default state for the nervous system that is bound to occur in the absence of reward and punishment? (2) Do similar neural processes support the formation of latent memory, which allows memories to be stored in silent, quiescent form; extinction, in which learned responses are diminished; and habituation, in which instinctual responses are diminished (Barron et al. 2017)? (3) Are higher-order cognitive deficits in psychiatric disorders a consequence of deficient habituation, or are they a reflection of shared underlying deficits?

Acknowledgments

The authors acknowledge collective insights from past and current colleagues and collaborators. Samuel F. Cooke acknowledges generous support from the Wellcome Trust and the Biotechnology and Biological Sciences Research Council (BBSRC). Mani Ramaswami acknowledges generous support from the Wellcome Trust, the Science Foundation Ireland, and the National Centre for Biological Sciences, Bangalore.

REFERENCES

Barron, H. C., T. P. Vogels, T. E. Behrens, and M. Ramaswami. (2017). Inhibitory engrams in perception and memory. *Proceedings of the National Academy of Sciences of the United States of America, 114*(26): 6666–6674.

Barry, R. J., and E. N. Sokolov. (1993). Habituation of phasic and tonic components of the orienting reflex. *International Journal of Psychophysiology, 15*(1): 39–42.

Bell, C., G. Sierra, N. Buendia, and J. P. Segundo. (1964). Sensory properties of neurons in the mesencephalic reticular formation. *Journal of Neurophysiology, 27*:961–987.

Bernstein, A. S. (1969). To what does the orienting response respond? *Psychophysiology, 6*(3): 338–350.

Bevins, R. A., and J. Besheer. (2006). Object recognition in rats and mice: A one-trial non-matching-to-sample learning task to study "recognition memory." *Nature Protocols, 1*(3): 1306–1311.

Carew, T., V. F. Castellucci, and E. R. Kandel. (1979). Sensitization in Aplysia: Restoration of transmission in synapses inactivated by long-term habituation. *Science, 205*(4404): 417–419.

Castellucci, V., H. Pinsker, I. Kupfermann, and E. R. Kandel. (1970). Neuronal mechanisms of habituation and dishabituation of the gill-withdrawal reflex in Aplysia. *Science, 167*(3926): 1745–1748.

Cooke, S. F., R. W. Komorowski, E. S. Kaplan, J. P. Gavornik, and M. F. Bear. (2015). Visual recognition memory, manifested as long-term habituation, requires synaptic plasticity in V1. *Nature Neuroscience, 18*(2): 262–271.

Das, S., M. K. Sadanandappa, A. Dervan, A. Larkin, J. A. Lee, I. P. Sudhakaran, R. Priya, R. Heidari, E. E. Holohan, A. Pimentel, A. Gandhi, K. Ito, S. Sanyal, J. W. Wang, V. Rodrigues, and M. Ramaswami. (2011). Plasticity of local GABAergic interneurons drives olfactory habituation. *Proceedings of the National Academy of Sciences of the United States of America, 108*(36): E646–654.

Ezzeddine, Y., and D. L. Glanzman. (2003). Prolonged habituation of the gill-withdrawal reflex in Aplysia depends on protein synthesis, protein phosphatase activity, and postsynaptic glutamate receptors. *Journal of Neuroscience, 23*(29): 9585–9594.

Farel, P. B., and R. F. Thompson. (1972). Habituation and dishabituation to dorsal root stimulation in the isolated frog spinal cord. *Behavioral Biology, 7*(7): 37–45.

Farel, P. B., and R. F. Thompson. (1976). Habituation of a monosynaptic response in frog spinal cord: Evidence for a presynaptic mechanism. *Journal of Neurophysiology, 39*(4): 661–666.

Gavornik, J. P., and M. F. Bear. (2014). Learned spatiotemporal sequence recognition and prediction in primary visual cortex. *Nature Neuroscience, 17*(5): 732–737.

Groves, P. M., D. L. Glanzman, M. M. Patterson, and R. F. Thompson. (1970). Excitability of cutaneous afferent terminals during habituation and sensitization in acute spinal cat. *Brain Research, 18*(2): 388–392.

Groves, P. M., D. Lee, and R. F. Thompson. (1969). Effects of stimulus frequency and intensity on habituation and sensitization in acute spinal cat. *Physiology & Behavior, 4*:383–388.

Groves, P. M., and R. F. Thompson. (1970). Habituation: A dual-process theory. *Psychological Review, 77*(5): 419–450.

Hebb, D. O. (1949). *The organization of behavior: A neuropsychological theory*. New York: Wiley.

Horn, G. (1967). Neuronal mechanisms of habituation. *Nature, 215*(5102): 707–711.

Horn, G., and R. M. Hill. (1964). Habituation of the response to sensory stimuli of neurones in the brain stem of rabbits. *Nature, 202*:296–298.

Hull, C. (1943). *Principles of behavior: An introduction to behavior theory*. Oxford: Appleton-Century.

Jennings, H. S. (1906). *Behavior of lower organisms*. New York: Columbia University Press.

Jones, B. E. (2003). Arousal systems. *Frontiers in Bioscience, 8*:s438–451.

Joseph, R. M., and J. R. Carlson (2015). Drosophila chemoreceptors: A molecular interface between the chemical world and the brain. *Trends in Genetics, 31*(12): 683–695.

Kaplan, E. S., S. F. Cooke, R. W. Komorowski, A. A. Chubykin, A. Thomazeau, L. A. Khibnik, J. P. Gavornik, and M. F. Bear. (2016). Contrasting roles for parvalbumin-expressing inhibitory neurons in two forms of adult visual cortical plasticity. *eLife, 5*.

Kato, H. K., S. N. Gillet, and J. S. Isaacson (2015). Flexible sensory representations in auditory cortex driven by behavioral relevance. *Neuron, 88*(5): 1027–1039.

Konorski, J. (1967). *Integrative activity of the brain*. Chicago: University of Chicago Press.

Marcus, E. A., T. G. Nolen, C. H. Rankin, and T. J. Carew. (1988). Behavioral dissociation of dishabituation, sensitization, and inhibition in Aplysia. *Science, 241*(4862): 210–213.

Marsden, K. C., and M. Granato (2015). In Vivo Ca(2+) imaging reveals that decreased dendritic excitability drives startle habituation. *Cell Reports, 13*(9): 1733–1740.

McDiarmid, T. A., A. C. Bernardos, and C. H. Rankin. (2017). Habituation is altered in neuropsychiatric disorders—a comprehensive review with recommendations for experimental design and analysis. *Neuroscience & Biobehavioral Reviews, 80*:286–305.

O'Keefe, J., and L. Nadel. (1978). *The hippocampus as a cognitive map.* Oxford: Oxford University Press.

Peckham, G. W., and E. G. Peckham. (1887). Some observations on the mental powers of spiders. *Journal of Morphology, 1*:383–419.

Pinsker, H. M., W. A. Hening, T. J. Carew, and E. R. Kandel. (1973). Long-term sensitization of a defensive withdrawal reflex in Aplysia. *Science, 182*(4116): 1039–1042.

Pinsker, H., I. Kupfermann, V. Castellucci, and E. Kandel. (1970). Habituation and dishabituation of the gill-withdrawal reflex in Aplysia. *Science, 167*(3926): 1740–1742.

Ramaswami, M. (2014). Network plasticity in adaptive filtering and behavioral habituation. *Neuron, 82*(6): 1216–1229.

Rankin, C. H., T. Abrams, R. J. Barry, S. Bhatnagar, D. F. Clayton, J. Colombo, G. Coppola, M. A. Geyer, D. L. Glanzman, S. Marsland, F. K. McSweeney, D. A. Wilson, C. F. Wu, and R. F. Thompson. (2009). Habituation revisited: An updated and revised description of the behavioral characteristics of habituation. *Neurobiology of Learning and Memory, 92*(2): 135–138.

Rankin, C. H., and T. J. Carew. (1988). Dishabituation and sensitization emerge as separate processes during development in Aplysia. *Journal of Neuroscience, 8*(1): 197–211.

Rao, R. P., and D. H. Ballard. (1999). Predictive coding in the visual cortex: A functional interpretation of some extra-classical receptive-field effects. *Nature Neuroscience, 2*(1): 79–87.

Rose, J. K., and C. H. Rankin. (2001). Analyses of habituation in Caenorhabditis elegans. *Learning & Memory, 8*(2): 63–69.

Sanderson, D. J., and D. M. Bannerman. (2011). Competitive short-term and long-term memory processes in spatial habituation. *Journal of Experimental Psychology. Animal Behavior Processes, 37*(2): 189–199.

Schmid, S., D. A. Wilson, and C. H. Rankin. (2014). Habituation mechanisms and their importance for cognitive function. *Frontiers in Integrative Neuroscience, 8*:97.

Sechenov, I. (1863/1965). *Reflexes of the brain.* Cambridge, MA: MIT Press. (Original work published in Russia in 1863.)

Sokolov, E. N. (1960a). *The central nervous system and behaviour.* Edited by M. A. Brazier. New York: Macy Foundation, 187.

Sokolov, E. N. (1960b). Neuronal models and the orienting influence. *The central nervous system and behavior III.* Edited by M. A. Brazier. New York: Macy Foundation.

Teyler, T. J., R. A. Roemer, and R. F. Thompson. (1972). Habituation of the pyramidal response in unanesthetized cat. *Physiology & Behavior, 8*(2): 201–205.

Thompson, R. F., and W. A. Spencer. (1966). Habituation: A model phenomenon for the study of neuronal substrates of behavior. *Psychological Review, 73*(1): 16–43.

Townsend, G. F. (1867). *Three hundred Aesop's fables.* Abingdon, UK: G. Routledge and Sons.

Twick, I., J. A. Lee, and M. Ramaswami. (2014). Olfactory habituation in Drosophila—odor encoding and its plasticity in the antennal lobe. *Progress in Brain Research, 208*:3–38.

Wagner, A. R. (1979). Habituation and memory. In A. Dickinson and R. A. Boakes (Eds.), *Mechanisms of learning and motivation: A memorial volume for Jerry Konorski* (pp. 53–82). Mahway, NJ: Lawrence Erlbaum.

Welker, W. I. (1956). Variability of play and exploratory behavior in chimpanzees. *Journal of Comparative and Physiological Psychology, 49*(2): 181–185.

Wickelgren, W. O. (1977). Post-tetanic potentiation, habituation and facilitation of synaptic potentials in reticulospinal neurones of lamprey. *Journal of Physiology, 270*(1): 115–131.

Wilson, R. I. (2013). Early olfactory processing in Drosophila: Mechanisms and principles. *Annual Review of Neuroscience, 36*:217–241.

Zimny, G. H., and L. W. Schwabe. (1965). Stimulus change and habituation of the orienting response. *Psychophysiology, 2*(2): 103–115.

18 Memory and Instinct as a Continuum of Information Storage

TOMÁS J. RYAN

ABSTRACT Memory engrams are the hypothetical storage sites of learned information. Learning induces material changes in specific groups of brain cells that retain information and are subsequently reactivated upon appropriate conditions, resulting in memory recall. Though the engram concept has intuitive appeal, experimental limitations have long prevented it from being directly tested. Over the past decade the ability to label, observe, and manipulate specific neuronal ensembles in an activity-dependent manner has allowed us to identify components of specific memory engrams in the rodent brain. This technology enables us to label sparse populations of brain cells that contribute to the storage of individual memories. Applying this methodology has resulted in novel insights into the kind(s) of plasticity that underlie various aspects of memory function. Though this line of research is in an early stage, a novel working theory of long-term memory has developed in which stable information storage is accomplished through the formation of distributed and hierarchical circuits composed of specific engram cell connectivity patterns. A hypothesis arising from this perspective is that memory engrams may influence the evolution of innate, instinctual information representations (ingrams). Understanding any equivalencies that exist between memories and instincts may aid in understanding the fundamental nature of information coding in the brain.

Memory is a fundamental cognitive property of all animals and is essential for adaptive behavior. It allows past experience to modulate present behavior in an uncertain environment, thereby increasing an organism's likelihood of survival and fitness. A behavioral definition of memory might describe a change in an animal's behavior due to a specific experience. A cognitive definition might focus on the formation and retention of an internal representation that affects how the animal perceives and interacts with the world. What is central to either perspective of memory is that it is *information* acquired by the organism through a process of learning. The ultimate goal of memory research in neuroscience is to understand how this information is encoded and decoded in the brain. What decides which features of an experience are encoded as memory? At what biological level(s) is the information represented? What are the plasticity mechanisms that enable information encoding? How is the information recalled?

But memory is not the only kind of information stored in our brains. The other form is instinct. Memories are not formed on a blank slate but on the preexisting information that all individuals of a given species possess. All animals have innate, genetically encoded instincts that are adaptive for the environments in which they evolved (Tinbergen, 1951). Mice know that cat urine signals a threat, even if they have never seen a cat. Sea turtles navigate without guidance back to the beach where they hatched. Orangutans can swing on trees without training. Primates recognize faces and humans smile at one another. Within an individual animal, the formation of an instinct is determined through the interaction of the genome with the developing brain. Through learning we form memories, and those memories build on instincts to enable beneficial behavior. Crucially, our instincts must interact with our memories in real psychological time in order to ensure appropriate behavioral action. Our brains must be able to draw on memory and whatever biological manifestation underlies instinct simultaneously. Therefore, if we are to understand memory function in an individual, we need to understand it within the context of the species' ancestral instincts. But what if, from the perspective of the brain, memories and instincts are essentially the same thing? What if instincts are actually descendant from memories? What if learned memory engrams give rise to genetically encoded information?

In this chapter I will examine the recent progress that has been made in understanding the biology of memory *engram* formation. I will draw a parallel of this and recent insights into the basis of instinct drawn from studies using the same, or similar, methodology. I will then propose a framework for considering memory and instinct as isomorphic forms of biological information that have distinct origins and mechanisms for formation but may share a common mode of storage and coding.

Memory Research

Empirical investigations into the nature of memory necessarily anchor themselves in phenomena, processes, and mechanisms that correlate with, or are necessary for, learning and recall as assayed at a behavioral level. The space in between—the stored information itself—is the essential property of memory that we aim to understand. Learning, the process of memory encoding, must involve a material change in the brain. Whatever material change is attributed to a specific memory can be referred to as a memory *engram*. The engram was originally defined and developed by the German zoologist Richard Semon (Schacter, 2001; Semon, 1904, 1909). Approaching memory from a biological perspective, Semon proposed that learning induces persistent biological changes (plasticity) in specific brain cells, allowing the brain to retain information and retrieve it through activation of these cells. He described the engram as "the enduring though primarily latent modification in the irritable substance produced by a stimulus (from an experience) upon appropriate retrieval conditions" (Semon, 1904). This abstract conception of a memory engram poses no problem for neuroscientists, or any scientific materialist, because it is a truism that memories must involve some kind of changes, or plasticity, in the brain. But problems arise when definitions of memory engrams become more concrete, which is a necessity of experimental research.

Following Karl Lashley's thorough but inconclusive searches for the location of a specific engram for a maze environment in the rat brain, experiments designed to localize engrams in the animal brain fell out of favor (Lashley, 1950). What emerged instead was a tradition of searching for the plasticity mechanism that enables the formation of engrams in general. Largely influenced by Donald Hebb's monograph and earlier hypotheses developed by others, investigators searched for plasticity of the synaptic connections in the animal brain and implicated it as a mechanism of memory. Hebb (1949) developed the theory that memory resides in specific *cell assemblies* formed through the strengthening of synaptic connections between cells. He further hypothesized that it was the coincident activation of connected cells that lead to this synaptic plasticity. While cell assemblies as a means of information coding are a feature of Hebb's theory reminiscent of Semon's engram cells, due to technical limitations most experimental investigations inspired by Hebb have focused exclusively on the mechanism of the synaptic plasticity that would effectively glue the assemblies together (Bliss, Collingridge, & Morris, 2003). Changes in synaptic weight (strength) due to suprathreshold synaptic activity have been discovered in both invertebrate and vertebrate organisms (Bliss & Lomo, 1973; Markram, Lubke, Frotscher, & Sakmann, 1997; Mulkey & Malenka, 1992; Tauc & Kandel, 1964). Synaptic weight can be strengthened by a process of long-term potentiation (LTP) or weakened through a process of long-term depression (LTD). Changes in synaptic weight can be induced by high- or low-frequency presynaptic stimulation or by paired pre- and postsynaptic stimulation (which can be noncontiguous in the cases of spike timing-dependent plasticity). Plasticity of synaptic weight has also been reported *in vivo* in brain regions receptive to particular kinds of behavioral experience (Clem & Huganir, 2010; McKernan & Shinnick-Gallagher, 1997; Rogan, Staubli, & LeDoux, 1997; Whitlock, Heynen, Shuler, & Bear, 2006). While this approach has been immensely productive in informing us about the biology of certain plasticity mechanisms and their importance for learned behavior, it has fallen short of providing us with insight into how information itself is stored as an engram.

Instinct Research

Relative to the literature on learning and memory, the neurobiological investigation of instinct has received much less attention. This emphasis on learned over innate behavior is a bias that is historically reflected in the experimental psychology literature, which assumed the primacy of learned behavior while instinct was traditionally the domain of ethology (Domjan, 2013; Lorenz, 1973; Mandler, 2007; Tinbergen, 1951). It is generally accepted that the neural circuits underlying instinctual behavior are programmed by the genome and sculpted through a process of brain development that may be modulated by general perceptual and motor activity during critical periods (Anderson, 2016). Brain development results in the construction of species-invariant *label lines* that connect specific perceptual stimuli with appropriate motor outputs, allowing an animal to innately respond to specific environmental features.[1] The circuit basis of many instinctual behaviors has been elegantly delineated using modern neuroscience techniques in *Drosophila* and rodent models. We now understand the circuitry behind a number of instinctual behaviors in invertebrate and vertebrate model organisms, including attraction or repulsion to specific odors and tastes of positive or negative valence,

[1] While perception, emotion, motivation, and central pattern generation can all be considered cognitive capacities that are necessarily innate, I do not categorize them as instincts here because they do not contain informational specificity about the animal's environment or how to react to it. The ability to sense an odor may be innate, but it's not an instinct. But the innate tendency to be afraid of a particular environmental odor is an instinct because it gives the animal specific information about its environment.

grooming, pheromone sensing in social contexts and courtship, and escape from predation (Choi et al., 2005; Evans et al., 2018; Han et al., 2017; Hong, Kim, & Anderson, 2014; Ishii et al., 2017; Kunwar et al., 2015; Manoli, Meissner, & Baker, 2006; Suh et al., 2007; Suh et al., 2004; Wang et al., 2018).

Plasticity

Based on the above summary, it would seem that while memories may be stored through the plasticity of synaptic strength, instinctual information is embedded in the hard-wired structure of the brain's neural circuit anatomy. However, recent research has demonstrated that the plasticity mechanisms underlying various aspects of mnemonic function, such as learning, consolidation, maintenance, retrievability, and recall, may be more diverse and nuanced than previously thought. Simply identifying and characterizing an enduring form of plasticity in the brain is not sufficient to establish it as a bona fide memory information substrate. This is because plasticity is ubiquitous and fundamental to biology. Every cell type, without exception, displays numerous forms of plasticity. Some of these are specific to the proper function of that cell type, such as the generation of antibody diversity by T lymphocytes in the immune system or the hypertrophy of muscle cells following repeated exercise. Other forms of plasticity are homeostatic in nature and serve to maintain the cells' metabolism and equilibrium. Brain cells, including neurons, astrocytes, oligodendrocytes, and microglia, all employ numerous forms of cellular plasticity. Moreover, plasticity across neuronal circuits can be observed at the molecular, synaptic, cellular, microcircuit, and brain systems levels. The empirical challenge therefore is to identify what kind(s) of plasticity are associated with learning and memory, which are attributable to different forms of memory (motor, perceptual, associative, habitual, episodic, semantic, and more), which underlie which timescales of memory (long term, short term, working memory, and more), and what plasticity mechanisms can underlie the formation of individual engrams.

In the case of long-term memory, the kind that can often last a lifetime, there has been good reason to attribute it to strengthened synaptic weight (enduringly induced by learning through some form of LTP or related physiological induction process). This is because interventions that disrupt the induction or maintenance of synaptic plasticity in physiological preparations also disrupt memory function in behavioral studies, resulting in experimental amnesia. For example, anterograde interventions that disrupt memory encoding, such as the antagonism of N-methyl-D-aspartate receptor (NMDA) receptors, also prevent the induction of synaptic plasticity (Morris, 2013; Park et al., 2014). Moreover, both memory formation and LTP have early and late phases that seem to require the same cell biological mechanisms (Frey, Huang, & Kandel, 1993). Short-term memory function (minutes to hours after training in rodents) does not require new gene expression or protein synthesis and neither does the early phase of LTP (E-LTP) (Poo et al., 2016). However, the administration of protein synthesis inhibitors after training results in retrograde amnesia for the trained behavior at long-term points (one day or more posttraining; McGaugh, 2000). Similarly, protein synthesis inhibition is known to prevent the maintenance of late-phase LTP (L-LTP) (Fonseca, Nagerl, & Bonhoeffer, 2006). These mechanistic parallels, first of learning and plasticity induction and second of memory consolidation and plasticity maintenance, have also been documented following a growing list of other more specific pharmacological and genetic manipulations (Kandel, 2001; Kandel, Dudai, & Mayford, 2014; McGaugh, 2000; Poo et al., 2016; Sweatt, 2016). Based on these studies, it seems almost self-evident that changes in synaptic strength is the plasticity mechanism that underlies long-term memory. Nevertheless, this standard model of memory storage has been challenged for numerous phenomenological and conceptual inconsistencies between associative learning and synaptic plasticity, but these are outside the scope of this chapter (Gallistel & Matzel, 2013; Miller & Matzel, 2000). What is directly relevant here is that behavioral and physiological studies of memory have been conducted almost entirely in distinct experimental paradigms—that is, behaving animals versus physiological slice preparations—and this has led to two limitations. First, these experiments have dealt with the behavior of the animal as a criterion for memory, the capacity for learned behavior or the capacity for recall, and not with the existence or persistence of the memory engram itself. Therefore, any behavioral case of apparent memory loss (amnesia) may in principle be due to a degraded or damaged memory engram or an inability to access a surviving engram. Second, these approaches do not allow for an investigation of which kind(s) of plasticity may be attributed to different latent or active properties of a specific memory engram. Recently, the standard model of memory storage has been challenged by experimental studies of memory engram cells (Poo et al., 2016; Queenan, Ryan, Gazzaniga, & Gallistel, 2017; Tonegawa, Liu, Ramirez, & Redondo, 2015; Tonegawa, Pignatelli, Roy, & Ryan, 2015).

Engram Technology

We know from many decades of ongoing research in the field of in vivo physiology that sparse populations of cells seem to represent certain features of an animal's perceived spatial environment and context (Hartley, Lever, Burgess, & O'Keefe, 2014). The fact that subsets of brain cells show specific patterns of activity that often correlate with specific experiences is a rubric of modern neuroscience and is consistent with both Semon's theory of engram cells and Hebbian cell assemblies. But crucial to an engram is "the enduring though primarily *latent* modification in the irritable substance produced by a stimulus" that Semon described. An engram is an engram even when it is silent. A memory can last a lifetime even when rarely recalled, so to empirically demonstrate engram cells we need to be able to observe and manipulate them even in a quiescent state. Moreover, if we are to demonstrate that specific subgroups of cells carry a particular memory, then it is necessary to move beyond correlative studies and show that these cells are sufficient and necessary for its recall. The fusion of optogenetics with transgenic immediate early gene (IEG) labeling has resulted in the development of an engram-labeling technology that allows for the specific tagging and in vivo reversible manipulation of putative engram cells with channelrhodopsin-2 (ChR2) and other opsins that permit the artificial light-induced activation of labeled neurons in awake, behaving rodents. IEGs are expressed in active cells, and if the promoter of an IEG, such as *c-fos* or *arc*, is used to express a temporally inducible transactivator, the expression of ChR2 can be controlled in a small population of experience-activated cells (figure 18.1) (Mayford, 2014).

The first demonstration of engram technology involved labeling active cells in the hippocampal dentate gyrus (DG) in a Pavlovian contextual fear-conditioning task (Liu et al., 2012; Ramirez, Tonegawa, & Liu, 2013). The direct stimulation of DG engram cells resulted in light-induced conditioned freezing behavior in a neutral context. Crucially, the optogenetic stimulation of engram cells for a different neutral context did not induce freezing behavior in animals that also possessed encoded unlabeled engrams of fear conditioning. Therefore, engram technology allows for the crucial criterion of *information specificity*—the ability to manipulate engrams of specific isolated experiences. The

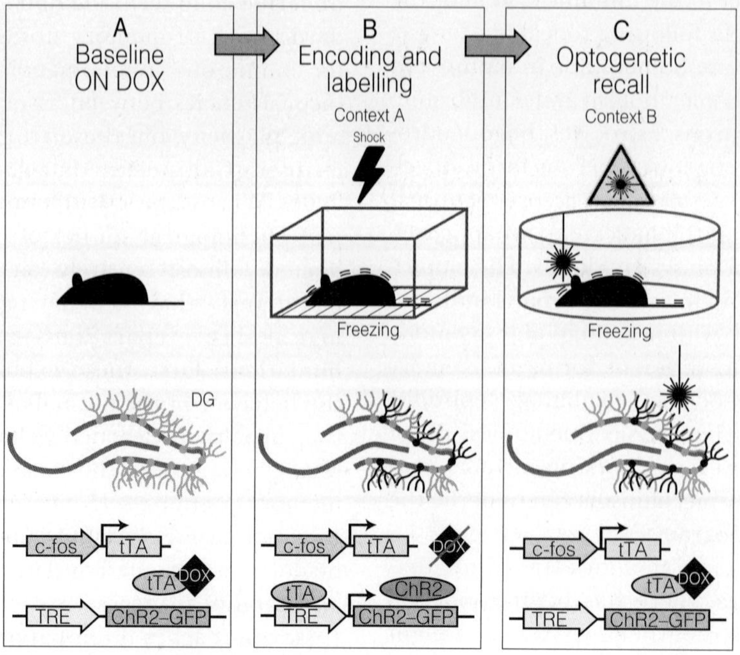

FIGURE 18.1 Engram cell-labeling technology. Schematic to describe engram-labeling technology, applied here to hippocampal dentate gyrus (DG) neurons. *A*, The promoter of the immediate early gene (IEG), in this case *c-fos*, drives the expression of tTA transactivator in an activity-dependent manner. Doxycycline (DOX), which is embedded in the mouse's food, prevents tTA from binding to the TRE element of the target transgene, in this case channelrhodopsin-2 (ChR2). *B*, Contextual fear conditioning causes fear to be associated with a new contextual memory (Context A), where animals would subsequently and specifically elicit a freezing response. In the absence of DOX, DG neurons that are active during the encoding of the Context A memory express ChR2. *C*, The optogenetic activation of engram neurons in novel, neutral Context B induces the recall of a distributed and context-specific fear response.

light-induced freezing behavior is not a general fear or anxiety effect because subsequent experiments showed that purely contextual engrams for neutral contexts could be labeled with ChR2 and artificially associated with shock information, resulting in permanent and natural conditioned freezing behavior in response to the context for which the engram was originally tagged but not other contextual engrams in the same brain (Ramirez et al., 2013). The optogenetic inhibition of engram cells in various hippocampal regions has shown that these cells are also necessary for the natural recall of specific engrams (Denny et al., 2014; Tanaka et al., 2014; Trouche et al., 2016; Yokose et al., 2017). The field of engram manipulation is rapidly growing, and the technology has been adapted to multiple brain regions and for diverse behavioral assays, such as place preference, object memory, social memory, stress assays, and operant conditioning (Nomoto et al., 2016; Ramirez et al., 2015; Redondo et al., 2014; Ryan, Roy, Pignatelli, Arons, & Tonegawa, 2015; Suto et al., 2016).

Engram Plasticity

One useful strategy for using engram technology to learn about memory storage mechanisms has been to investigate the nature of amnesia (Miller & Matzel, 2006; Ortega-de San Luis & Ryan, 2018). As discussed earlier, any case of amnesia (experimental or clinical) can be *a priori* due to a loss of the information (storage deficit) or a loss of access to the information (access deficit). Engram technology allows us to resolve this ambiguity in certain kinds of amnesia by the direct optogenetic stimulation of engram cells and to parallel the outcomes with studies of engram cell plasticity. Early investigation showed that the direct stimulation of engram cells in cases of retrograde amnesia induced by the pharmacological inhibition of protein synthesis resulted in normal memory recall in a range of experimental conditions (Ryan et al., 2015). These findings provided clear evidence that the apparently lost memories can be due to impaired access to the engram but that the learned information itself remains intact. These outcomes were subsequently corroborated and extended to various other forms of memory loss, including the transgenic induction of early Alzheimer's disease and infantile amnesia due to development (Abdou et al., 2018; Guskjolen et al., 2018; Roy et al., 2016). One of the main methodological strengths of engram labeling in vivo is that it allows for the study of both the behavioral functionality and the physiological properties of a particular engram in the same experimental preparation (Ryan et al., 2015; Tonegawa, Pignatelli, et al., 2015). Therefore, the plasticity of engram cell physiological measurements can be correlated with learning, memory, and recall. Through intracellular patch clamp recordings of engram cells, it was established that dentate gyrus engram cells show enhanced synaptic input strength, measured as an increased magnitude of excitatory postsynaptic currents (EPSCs) relative to nonengram cells (Ryan et al., 2015). This represents a learning-induced potentiation of engram cell synapses and was corroborated by the analysis of spontaneous excitatory postsynaptic currents (sEPSCs), intrinsic capacitance, as well as the dendritic spine density of engram cells relative to nonengram cells. All of these forms of engram cell–specific plasticity were abolished when consolidation of the target memory was disrupted through the administration of the protein synthesis inhibitor anisomycin immediately after learning.

Based on these findings, learning-dependent changes in synaptic strength may be crucial for normal memory retrieval (and also possibly for memory encoding) but are dispensable for the storage of memory information itself (Poo et al., 2016; Tonegawa, Pignatelli, et al., 2015). What else survived? Engram cells for a particular experience are distributed across brain regions, but engram cells tagged by the same experience in different regions are specifically connected to one another. This feature of engram circuit neurobiology survives amnesia and remains intact even when the memory seems inaccessible (Roy et al., 2016; Ryan et al., 2015). Directly stimulating these connections enables the retrieval of learned information in these circuits. Memory information can thus be viewed as *engravings* within the brain's microanatomy, initiated by salient events and resulting in newly formed synaptic connections between ensembles of brain cells. In this sense learned information would be stored not at a synaptic level per se but at a neuronal ensemble level, where basal synaptic connectivity naturally forms the connections necessary for an ensemble to exist. Memories, like instincts, might never really be forgotten (Kitamura et al., 2017; Tonegawa, Pignatelli, et al., 2015). A particular memory, like an instinct, might be represented as a new microanatomical pathway in a particular set of relevant brain areas.

Origin of Instinct

The conservative Darwinian perspective holds that instincts originate from random mutations that alter brain structure or function and result in useful behavioral phenotypes that are selected for in a population and thereby come to fixation in a species. On the other extreme, it has been speculated that instincts may be direct descendants of learned memories through epigenetic mechanisms that have yet to be identified but would require a Lamarckian mode of inheritance (Robinson & Barron, 2017). There is growing evidence that the experience of one generation can have an epigenetic effect on the homeostatic regulation of descendent generations—for example, stress response (Yeshurun & Hannan, 2018). But this kind of transgenerational epigenetic effect on behavior should not be confused with the transmission of learned information. A demonstration of the epigenetic transfer of memory would require evidence that a specific memory formed by individuals is transferred to their offspring.

The standard Darwinian paradigm can certainly account for the origin of innate behaviors, but it is very slow and requires that the mutant behavioral trait be of a fortuitous advantage to individuals in their population. If a mutant phenotype does not improve the fitness of an organism in its environment, then it is unlikely to come to fixation in the population. On the other hand, the epigenetic paradigm is attractive from a naïve perspective because it would mean that our own learning might directly influence the behavior of our offspring and help to direct the evolution of our species. Such speculations go back at least as far as Lamarck and tend to resurface whenever a new biological mechanism is characterized that can be imagined to somehow carry learned information from the brain of an individual animal to its germ cells, through the developing offspring, and into their brains, ultimately resulting in very specific effects on brain development.[2] As well as being biologically implausible, Lamarckian inheritance would be highly unstable because instincts would change with new incidental experience in each passing generation. This idea is at odds with the essential stability of innate behaviors across generations and the fact of conservation of similar instincts across related species.

Here I present a more parsimonious working hypothesis for how instincts may be evolutionarily descendant from memories, not through direct epigenetic transfer of a molecular substrate but by imitation of the informational content of an ancestral memory by an independently formed instinct. Over a century ago, the Baldwin effect described how learned behaviors may facilitate the evolution of similar innate behaviors by creating an environment or niche where hard-wired versions of those behaviors would have a selective advantage (Baldwin, 1896; Morgan, 1896; Osborn, 1896). Without such niche construction, a random mutation that leads to a new innate behavior is unlikely to have adaptive value in the population in which it emerges. But if that innate behavior can substitute for a valuable or necessary learned behavior that already exists in the population, then the new instinct will instill a competitive advantage on the mutant individuals relative to their wild-type peers in that ecosystem (figure 18.2). Learning is hard work—it is imperfect and acquiring the information by experience is fraught with risk. Instinct is free, consistent, and built in to the structure of the brain. Mutant organisms born with genetically encoded instincts will outcompete their less privileged peers who must learn the information for themselves.

While more biologically plausible than Lamarckian inheritance and supported by computational analysis (Hinton & Nowlan, 1987), the Baldwin effect has not been empirically demonstrated, and no concrete mechanism has been proposed. However, a conceptual synthesis of recent research in the neurobiology of memories and instincts may provide novel evidence for a continuity between memory and instinct. In neurobiology it is understood that instinct is information embedded in genetically determined brain structures formed by developmental processes that originate through biological evolution (Anderson, 2016). Instincts are a product of evolution, while memories are a product of learning. Clearly, instincts and memories are *encoded* through very different mechanisms (mutation and neural plasticity, respectively), and it has also been tacitly assumed that they are coded or represented as different neurobiological substrates (neuroanatomy and synaptic plasticity). But owing to studies of engram ensembles, it now seems likely that long-term memories, like instincts, are embedded in the brain as changes in the connectivity patterns between distributed engram cells (Poo et al., 2016; Tonegawa, Liu, et al., 2015; Tonegawa, Pignatelli, et al., 2015).

Instinctual behavior is dependent on the brain's hard-wired label lines, from perception to action (Anderson, 2016; Root, Denny, Hen, & Axel, 2014). When the same activity-dependent labeling techniques were used to tag subpopulations of olfactory sensory neurons responsive to specific odors innately perceived as attractive or aversive (e.g., that can be derived from rose oil or bobcat urine, respectively), optogenetic stimulation of these neurons elicited instinctive avoidance or attraction behavioral responses in untrained mice through the activation of specific regions of the cortical amygdala (Root et al., 2014). Equivalent findings have been reported for the brain's innate representations of bitter and sweet tastes in the gustatory system (Peng et al., 2015). Activity-dependent labeling has also

[2] While such proposals are radical, there have been striking reports of olfactory conditioning in mice promoting glomerular plasticity for similar odors in mice offspring (Dias & Ressler, 2014). While such cases may be valid, they seem to represent the exception rather than the rule.

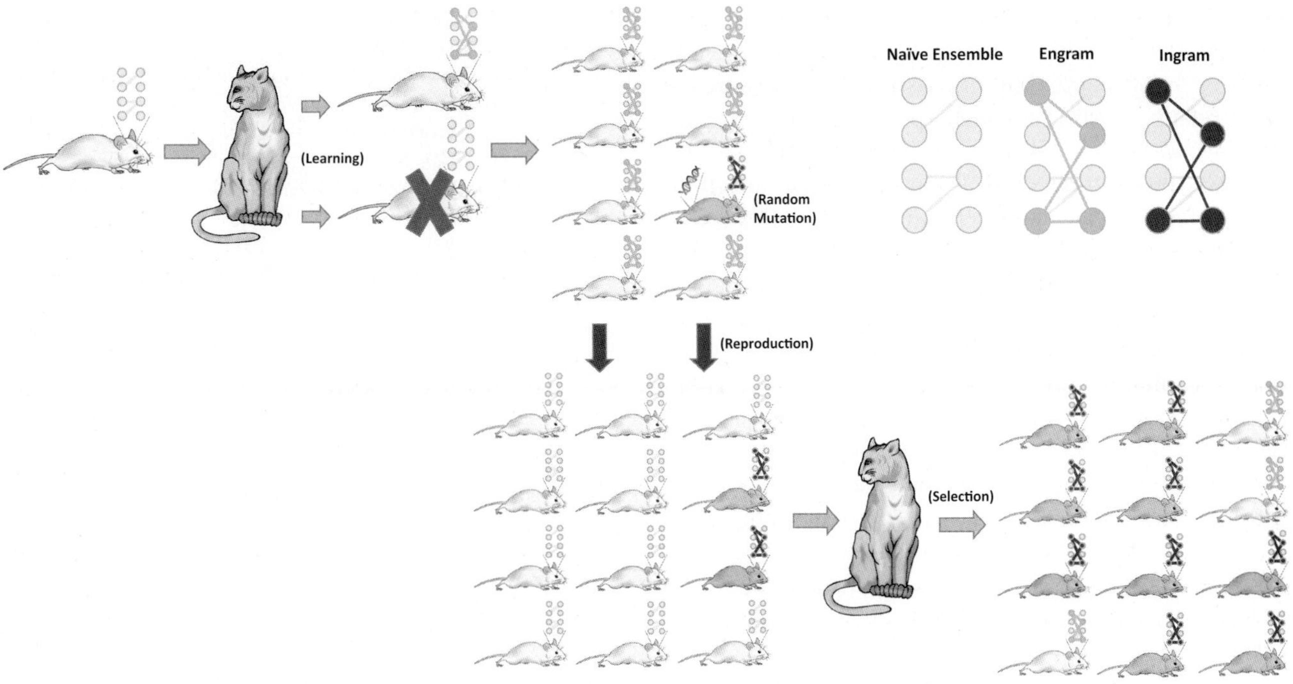

FIGURE 18.2 The innate fear response as an example of the Baldwin effect. *Top left,* Prey animals that encounter a predator must rapidly learn to regard it as a threat and avoid it for survival. All surviving individuals will possess an engram to help negotiate future encounters with the predator. *Top center,* As long as predators are a common part of the ecosystem, a niche will be constructed where information about negotiating the predator will be valuable to individuals. Relevant engrams will become essential for survival in the environment. Within the context of this niche, an individual who acquires a random genetic mutation (originating in the germ cells of one of its parents) experiences a developmental alteration of its brain structure that mimics the engrams of the population. *Bottom center,* The progeny of animals with engrams will begin life as naïve individuals who must learn the crucial memory by experience or social interaction. In contrast, the progeny of animals with ingrams will innately possess information pertaining to the predator. *Bottom right,* During future encounters with predators, animals with innate ingrams will outcompete the naïve individuals who must form engrams to adapt to the environment. Over many generations, animals with ingrams will be selected for, and the ingram will be driven to fixation in the population. *Top right,* Schematic of naïve neuronal ensemble (*gray*), a learning-informed engram ensemble (*blue*), and a genetically informed ingram ensemble (*red*). (See color plate 21.)

allowed the identification of ensembles within the amygdala that mediate the innate positive or negative valence of a given stimulus associated with both learned and innate perceptual stimuli (Gore et al., 2015; Kim, Pignatelli, Xu, Itohara, & Tonegawa, 2016; Redondo et al., 2014). Instincts and memories seem to both be coded as specific ensembles that have targeted and hard-wired connections to downstream brain regions that integrate the stimuli with other forms of innate information, such as an emotional response or motivated action.

If long-term memories are stored as permanent changes in the brain's hard-wired connectivity, then instinct and memory can plausibly interact with each using the same neurophysiological "language." Furthermore, memory engrams can provide an environment in which adaptive instincts can originate via a nonepigenetic mechanism of convergently evolving toward an equivalent ensemble structure of the relevant engram. Some instincts may originate simply by random mutation and the subsequent selection of the resulting phenotypic traits that happen to increase the fitness of the mutant organisms. But this process is slow, stochastic, and will only occur within a population and environment where individuals without the instinct can still survive, and those with the instinct will thrive. However, the ability to form engrams about how to navigate the world allows for populations to create such an environment by test-driving different engrams for their adaptive utility. Naïve individuals rapidly learn what is useful and what is dangerous by experience or social interaction and thereby can survive (figure 18.2). When a particular piece of information is valuable enough—say, for example, that a particular predator is bad and must be avoided—every individual in the population is forced to form that kind of engram for survival. Within such a population, an individual may arise with a random mutation that happens to alter the developmental system just enough for the brain to mimic or phenocopy the structure of the memory engram. This genetically encoded *ingram* then increases the fitness of its hosts relative to the rest of the population. Over generations, individuals with the innate ingram will outcompete those individuals who must form original engrams every time. Thus, the ingram becomes fixed in the population as an endowed instinct of the species.

Engrams and ingrams are clearly encoded by very different mechanisms, but the two resultant types of information content could be neurobiologically isomorphic once stored. Conceiving memory and instinct as a continuum of information allows us to consider the evolution of the information itself. Diversity of experience within a population results in innumerable engrams, the most useful of which become prevalent by a process of selection, and the selected ones may then be amplified across generations through descendant genetically encoded ingrams. A unified theory of memory and instinct may bring us closer to understanding the nature of the encoded information (Dennett, 2017).

Acknowledgments

My thanks go to Clara Ortega de San Luis for providing figure 18.1 and Lydia Marks for proofreading.

REFERENCES

Abdou, K., Shehata, M., Choko, K., Nishizono, H., Matsuo, M., Muramatsu, S. I., & Inokuchi, K. (2018). Synapse-specific representation of the identity of overlapping memory engrams. *Science, 360*(6394), 1227–1231. doi:10.1126/science.aat3810

Anderson, D. J. (2016). Circuit modules linking internal states and social behaviour in flies and mice. *Nature Reviews Neuroscience, 17*(11), 692–704. doi:10.1038/nrn.2016.125

Baldwin, J. M. (1896). Heredity and instinct. *Science, 3*(64), 438–441. doi:10.1126/science.3.64.438

Bliss, T. V., Collingridge, G. L., & Morris, R. G. (2003). Introduction. Long-term potentiation and structure of the issue. *Philosophical Transactions of the Royal Society of London B: Biological Sciences, 358*(1432), 607–611. doi:10.1098/rstb.2003.1282

Bliss, T. V., & Lomo, T. (1973). Long-lasting potentiation of synaptic transmission in the dentate area of the anaesthetized rabbit following stimulation of the perforant path. *Journal of Physiology, 232*(2), 331–356.

Choi, G. B., Dong, H. W., Murphy, A. J., Valenzuela, D. M., Yancopoulos, G. D., Swanson, L. W., & Anderson, D. J. (2005). Lhx6 delineates a pathway mediating innate reproductive behaviors from the amygdala to the hypothalamus. *Neuron, 46*(4), 647–660. doi:10.1016/j.neuron.2005.04.011

Clem, R. L., & Huganir, R. L. (2010). Calcium-permeable AMPA receptor dynamics mediate fear memory erasure. *Science, 330*(6007), 1108–1112.

Dennett, D. C. (2017). *Bacteria to bach and back: The evolution of minds.* New York: W. W. Norton.

Denny, C. A., Kheirbek, M. A., Alba, E. L., Tanaka, K. F., Brachman, R. A., Laughman, K. B., ... Hen, R. (2014). Hippocampal memory traces are differentially modulated by experience, time, and adult neurogenesis. *Neuron, 83*(1), 189–201. doi:S0896-6273(14)00404-8 [pii] 10.1016/j.neuron.2014.05.018

Dias, B. G., & Ressler, K. J. (2014). Parental olfactory experience influences behavior and neural structure in subsequent generations. *Nature Neuroscience, 17*(1), 89–96. doi:10.1038/nn.3594

Domjan, M. P. (2013). *The principles of learning and behavior* (7th ed.). Boston: Cengage.

Evans, D. A., Stempel, A. V., Vale, R., Ruehle, S., Lefler, Y., & Branco, T. (2018). A synaptic threshold mechanism for computing escape decisions. *Nature, 558*(7711), 590–594. doi:10.1038/s41586-018-0244-6

Fonseca, R., Nagerl, U. V., & Bonhoeffer, T. (2006). Neuronal activity determines the protein synthesis dependence of long-term potentiation. *Nature Neuroscience, 9*(4), 478–480. doi:nn1667 [pii] 10.1038/nn1667

Frey, U., Huang, Y. Y., & Kandel, E. R. (1993). Effects of cAMP simulate a late stage of LTP in hippocampal CA1 neurons. *Science, 260*(5114), 1661–1664.

Gallistel, C. R., & Matzel, L. D. (2013). The neuroscience of learning: Beyond the Hebbian synapse. *Annual Review of Psychology, 64*, 169–200. doi:10.1146/annurev-psych-113011-143807

Gore, F., Schwartz, E. C., Brangers, B. C., Aladi, S., Stujenske, J. M., Likhtik, E., ... Axel, R. (2015). Neural representations of unconditioned stimuli in basolateral amygdala mediate innate and learned responses. *Cell, 162*(1), 134–145. doi:10.1016/j.cell.2015.06.027

Guskjolen, A., Kenney, J. W., de la Parra, J., Yeung, B. A., Josselyn, S. A., & Frankland, P. W. (2018). Recovery of "lost" infant memories in mice. *Current Biology, 28*(14), 2283–2290, e2283. doi:10.1016/j.cub.2018.05.059

Han, W., Tellez, L. A., Rangel Jr., M. J., Motta, S. C., Zhang, X., Perez, I. O., ... de Araujo, I. E. (2017). Integrated control of predatory hunting by the central nucleus of the amygdala. *Cell, 168*(1–2), 311–324, e318. doi:10.1016/j.cell.2016.12.027

Hartley, T., Lever, C., Burgess, N., & O'Keefe, J. (2014). Space in the brain: How the hippocampal formation supports spatial cognition. *Philosophical Transactions of the Royal Society of London B: Biological Sciences, 369*(1635), 20120510. doi:10.1098/rstb.2012.0510

Hebb, D. O. (1949). *The organization of behavior: A neuropsychological theory.* New York: Wiley.

Hinton, G., & Nowlan, S. J. (1987). How learning can guide evolution. *Complex Systems, 1*, 495–502.

Hong, W., Kim, D. W., & Anderson, D. J. (2014). Antagonistic control of social versus repetitive self-grooming behaviors by separable amygdala neuronal subsets. *Cell, 158*(6), 1348–1361. doi:10.1016/j.cell.2014.07.049

Ishii, K. K., Osakada, T., Mori, H., Miyasaka, N., Yoshihara, Y., Miyamichi, K., & Touhara, K. (2017). A labeled-line neural circuit for pheromone-mediated sexual behaviors in mice. *Neuron, 95*(1), 123–137, e128. doi:10.1016/j.neuron.2017.05.038

Kandel, E. R. (2001). The molecular biology of memory storage: A dialogue between genes and synapses. *Science, 294*(5544), 1030–1038.

Kandel, E. R., Dudai, Y., & Mayford, M. R. (2014). The molecular and systems biology of memory. *Cell, 157*(1), 163–186. doi:S0092-8674(14)00290-6 [pii] 10.1016/j.cell.2014.03.001

Kim, J., Pignatelli, M., Xu, S., Itohara, S., & Tonegawa, S. (2016). Antagonistic negative and positive neurons of the basolateral amygdala. *Nature Neuroscience, 19*(12), 1636–1646. doi:10.1038/nn.4414

Kitamura, T., Ogawa, S. K., Roy, D. S., Okuyama, T., Morrissey, M. D., Smith, L. M., ... Tonegawa, S. (2017). Engrams and circuits crucial for systems consolidation of a memory. *Science, 356*(6333), 73–78. doi:10.1126/science.aam6808

Kunwar, P. S., Zelikowsky, M., Remedios, R., Cai, H., Yilmaz, M., Meister, M., & Anderson, D. J. (2015). Ventromedial hypothalamic neurons control a defensive emotion state. *eLife, 4.* doi:10.7554/eLife.06633

Lashley, K. (1950). In search of the engram. *Symposia of the Society for Experimental Biology, 4,* 454–482.

Liu, X., Ramirez, S., Pang, P. T., Puryear, C. B., Govindarajan, A., Deisseroth, K., & Tonegawa, S. (2012). Optogenetic stimulation of a hippocampal engram activates fear memory recall. *Nature, 484*(7394), 381–385. doi:nature11028 [pii] 10.1038/nature11028

Lorenz, K. (1973). *Die Rückseite des Spiegels* [Behind the mirror]. Munich: Piper Verlag.

Mandler, G. (2007). *A history of modern experimental psychology: From James and Wundt to cognitive science.* Cambridge, MA: MIT Press.

Manoli, D. S., Meissner, G. W., & Baker, B. S. (2006). Blueprints for behavior: Genetic specification of neural circuitry for innate behaviors. *Trends in Neurosciences, 29*(8), 444–451. doi:10.1016/j.tins.2006.06.006

Markram, H., Lubke, J., Frotscher, M., & Sakmann, B. (1997). Regulation of synaptic efficacy by coincidence of postsynaptic APs and EPSPs. *Science, 275*(5297), 213–215.

Mayford, M. (2014). The search for a hippocampal engram. *Philosophical Transactions of the Royal Society of London B: Biological Sciences, 369*(1633), 20130161. doi:rstb.2013.0161 [pii] 10.1098/rstb.2013.0161

McGaugh, J. L. (2000). Memory—a century of consolidation. *Science, 287*(5451), 248–251. doi:8182 [pii]

McKernan, M. G., & Shinnick-Gallagher, P. (1997). Fear conditioning induces a lasting potentiation of synaptic currents in vitro. *Nature, 390*(6660), 607–611. doi:10.1038/37605

Miller, R. R., & Matzel, L. D. (2000). Memory involves far more than "consolidation." *Nature Reviews Neuroscience, 1*(3), 214–216. doi:10.1038/35044578

Miller, R. R., & Matzel, L. D. (2006). Retrieval failure versus memory loss in experimental amnesia: Definitions and processes. *Learning & Memory, 13*(5), 491–497. doi:13/5/491 [pii] 10.1101/lm.241006

Morgan, C. L. (1896). On modification and variation. *Science, 4*(99), 733–740. doi:10.1126/science.4.99.733

Morris, R. G. (2013). NMDA receptors and memory encoding. *Neuropharmacology, 74,* 32–40. doi:10.1016/j.neuropharm.2013.04.014

Mulkey, R. M., & Malenka, R. C. (1992). Mechanisms underlying induction of homosynaptic long-term depression in area CA1 of the hippocampus. *Neuron, 9*(5), 967–975.

Nomoto, M., Ohkawa, N., Nishizono, H., Yokose, J., Suzuki, A., Matsuo, M., ... Inokuchi, K. (2016). Cellular tagging as a neural network mechanism for behavioural tagging. *Nature Communications, 7,* 12319. doi:10.1038/ncomms12319

Ortega-de San Luis, C., & Ryan, T. J. (2018). United states of amnesia: Rescuing memory loss from diverse conditions. *Disease Models & Mechanisms, 11*(5). doi:10.1242/dmm.035055

Osborn, H. F. (1896). Oytogenic and phylogenic variation. *Science, 4*(100), 786–789. doi:10.1126/science.4.100.786

Park, P., Volianskis, A., Sanderson, T. M., Bortolotto, Z. A., Jane, D. E., Zhuo, M., ... Collingridge, G. L. (2014). NMDA receptor-dependent long-term potentiation comprises a family of temporally overlapping forms of synaptic plasticity that are induced by different patterns of stimulation. *Philosophical Transactions of the Royal Society of London B: Biological Sciences, 369*(1633), 20130131. doi:rstb.2013.0131 [pii] 10.1098/rstb.2013.0131

Peng, Y., Gillis-Smith, S., Jin, H., Trankner, D., Ryba, N. J., & Zuker, C. S. (2015). Sweet and bitter taste in the brain of awake behaving animals. *Nature, 527*(7579), 512–515. doi:10.1038/nature15763

Poo, M. M., Pignatelli, M., Ryan, T. J., Tonegawa, S., Bonhoeffer, T., Martin, K. C., ... Stevens, C. (2016). What is memory? The present state of the engram. *BMC Biology, 14,* 40. doi:10.1186/s12915-016-0261-6

Queenan, B. N., Ryan, T. J., Gazzaniga, M. S., & Gallistel, C. R. (2017). On the research of time past: The hunt for the substrate of memory. *Annals of the New York Academy of Sciences, 1396*(1), 108–125. doi:10.1111/nyas.13348

Ramirez, S., Liu, X., Lin, P. A., Suh, J., Pignatelli, M., Redondo, R. L., ... Tonegawa, S. (2013). Creating a false memory in the hippocampus. *Science, 341*(6144), 387–391. doi:341/6144/387 [pii] 10.1126/science.1239073

Ramirez, S., Liu, X., MacDonald, C. J., Moffa, A., Zhou, J., Redondo, R. L., & Tonegawa, S. (2015). Activating positive memory engrams suppresses depression-like behaviour. *Nature, 522*(7556), 335–339. doi:10.1038/nature14514

Ramirez, S., Tonegawa, S., & Liu, X. (2013). Identification and optogenetic manipulation of memory engrams in the hippocampus. *Frontiers in Behavioral Neuroscience, 7,* 226. doi:10.3389/fnbeh.2013.00226

Redondo, R. L., Kim, J., Arons, A. L., Ramirez, S., Liu, X., & Tonegawa, S. (2014). Bidirectional switch of the valence associated with a hippocampal contextual memory engram. *Nature, 513*(7518), 426–430. doi:nature13725 [pii] 10.1038/nature13725

Robinson, G. E., & Barron, A. B. (2017). Epigenetics and the evolution of instincts. *Science, 356*(6333), 26–27. doi:10.1126/science.aam6142

Rogan, M. T., Staubli, U. V., & LeDoux, J. E. (1997). Fear conditioning induces associative long-term potentiation in the amygdala. *Nature, 390*(6660), 604–607. doi:10.1038/37601

Root, C. M., Denny, C. A., Hen, R., & Axel, R. (2014). The participation of cortical amygdala in innate, odour-driven behaviour. *Nature, 515*(7526), 269–273. doi:10.1038/nature13897

Roy, D. S., Arons, A., Mitchell, T. I., Pignatelli, M., Ryan, T. J., & Tonegawa, S. (2016). Memory retrieval by activating engram cells in mouse models of early Alzheimer's disease. *Nature, 531*(7595), 508–512. doi:10.1038/nature17172

Ryan, T. J., Roy, D. S., Pignatelli, M., Arons, A., & Tonegawa, S. (2015). Engram cells retain memory under retrograde amnesia. *Science, 348*(6238), 1007–1013. doi:10.1126/science.aaa5542

Schacter, D. L. (2001). *Forgotten ideas, neglected pioneers: Richard Semon and the story of memory.* New York: Routledge.

Semon, R. (1904). *Die mneme* [The mneme]. Leipzig: Wilhelm Engelmann.

Semon, R. (1909). *Die nmemischen empfindungen* [Mnemic psychology]. Leipzig: Wilhelm Engelmann.

Suh, G. S., Ben-Tabou de Leon, S., Tanimoto, H., Fiala, A., Benzer, S., & Anderson, D. J. (2007). Light activation of an innate olfactory avoidance response in Drosophila. *Current Biology, 17*(10), 905–908. doi:10.1016/j.cub.2007.04.046

Suh, G. S., Wong, A. M., Hergarden, A. C., Wang, J. W., Simon, A. F., Benzer, S., ... Anderson, D. J. (2004). A single population of olfactory sensory neurons mediates an innate avoidance behaviour in Drosophila. *Nature, 431*(7010), 854–859. doi:10.1038/nature02980

Suto, N., Laque, A., De Ness, G. L., Wagner, G. E., Watry, D., Kerr, T., ... Weiss, F. (2016). Distinct memory engrams in the infralimbic cortex of rats control opposing environmental actions on a learned behavior. *eLife, 5*. doi:10.7554/eLife.21920

Sweatt, J. D. (2016). Neural plasticity and behavior—sixty years of conceptual advances. *Journal of Neurochemistry, 139*(Suppl 2), 179–199. doi:10.1111/jnc.13580

Tanaka, K. Z., Pevzner, A., Hamidi, A. B., Nakazawa, Y., Graham, J., & Wiltgen, B. J. (2014). Cortical representations are reinstated by the hippocampus during memory retrieval. *Neuron, 84*(2), 347–354. doi:S0896-6273(14)00895-2 [pii] 10.1016/j.neuron.2014.09.037

Tauc, L., & Kandel, E. R. (1964). Heterosynaptic transfer of facilitation. [In French.] *Journal of Physiology (Paris), 56*, 446.

Tinbergen, N. Z. (1951). *The study of instinct.* Oxford: Clarendon Press.

Tonegawa, S., Liu, X., Ramirez, S., & Redondo, R. (2015). Memory engram cells have come of age. *Neuron, 87*(5), 918–931. doi:10.1016/j.neuron.2015.08.002

Tonegawa, S., Pignatelli, M., Roy, D. S., & Ryan, T. J. (2015). Memory engram storage and retrieval. *Current Opinion in Neurobiology, 35*, 101–109. doi:10.1016/j.conb.2015.07.009

Trouche, S., Perestenko, P. V., van de Ven, G. M., Bratley, C. T., McNamara, C. G., Campo-Urriza, N., ... Dupret, D. (2016). Recoding a cocaine-place memory engram to a neutral engram in the hippocampus. *Nature Neuroscience, 19*(4): 564–567.

Wang, L., Gillis-Smith, S., Peng, Y., Zhang, J., Chen, X., Salzman, C. D., ... Zuker, C. S. (2018). The coding of valence and identity in the mammalian taste system. *Nature, 558*(7708), 127–131. doi:10.1038/s41586-018-0165-4

Whitlock, J. R., Heynen, A. J., Shuler, M. G., & Bear, M. F. (2006). Learning induces long-term potentiation in the hippocampus. *Science, 313*(5790), 1093–1097. doi:10.1126/science.1128134

Yeshurun, S., & Hannan, A. J. (2018). Transgenerational epigenetic influences of paternal environmental exposures on brain function and predisposition to psychiatric disorders. *Molecular Psychiatry, 24*(4): 536–548.

Yokose, J., Okubo-Suzuki, R., Nomoto, M., Ohkawa, N., Nishizono, H., Suzuki, A., ... Inokuchi, K. (2017). Overlapping memory trace indispensable for linking, but not recalling, individual memories. *Science, 355*(6323), 398–403. doi:10.1126/science.aal2690

19 Context in Spatial and Episodic Memory

JOSHUA B. JULIAN AND CHRISTIAN F. DOELLER

ABSTRACT In this chapter we discuss the role of context in shaping spatial and episodic memories. We first survey the psychological literature on the types of cues that define context and offer an inclusive definition that focuses on the adaptive role of contextual representations for guiding behavioral and mnemonic outputs. Using observations from both humans and nonhuman animals, we then review the neural basis of contextual memory, focusing in particular on the hippocampus. We show that contextual representations in the hippocampus are organized by those same cues that define context cognitively. Finally, we characterize the inputs to the hippocampus mediating the recognition of context-defining cues. Together, our review supports the hypothesis that a function of the hippocampus and its primary inputs is to form the holistic context representations that shape memory.

Theories of memory suggest that encoding and retrieval are facilitated or hindered by *context* (Davies & Thomson, 1988; Smith & Vela, 2001). For example, it is easier to recognize someone when that person is in the same setting as when you initially encountered her. Context plays a particularly important role in shaping spatial and episodic memories. Spatial memory reflects memory for spatial information defined relative to a particular contextual frame of reference (e.g., my memory of the location of my seat in a movie theater). Episodic memories are detailed representations of the what, where, and when of past experiences (Tulving, 2002), and thus the ability to reinstate contextual information is one of the defining features of episodic memory (e.g., my memory of finding my seat in the movie theater). By contrast, other types of memory require no contextual information, such as knowledge of facts in the absence of memory for the context in which they were learned, or the recognition of stimuli based on a feeling of familiarity. A major scientific challenge has been to understand how the brain processes contextual information and how this information shapes spatial and episodic memories. In this chapter we review the cognitive role that context plays in memory and elucidate how contextual information is processed by the brain in service of such memories.

What Cues Define Contexts?

Despite the ubiquity of context in our lives and its clear importance for shaping memory, *context* has proven to be a surprisingly difficult concept to define (Nadel & Willner, 1980). Confusion around the definition of *context* is not new; Smith (1979) argued in the 1970s that context "is a kind of conceptual garbage ... that denotes a great variety of intrinsic and extrinsic characteristics of the presentation or test" of stimuli. Indeed, across studies purporting to interrogate contextual memory, *context* has been operationalized in terms as nearly anything associated with items or locations in an event, ranging from something as simple as the color of text in a word list to cues as complex as the physical environment. This ongoing lack of definitional clarity is due in part to the fact that general rules governing when cues do or do not define a context are unclear. Moreover, the type of context referred to in studies of memory is often underspecified, and it is not empirically clear that all types of cues used to operationalize context play identical mnemonic roles. To provide a handle for understanding the neural basis of context-dependent memory, it is thus critical to start by surveying the possible types of context-defining cues:

Spatial cues Everything we do occurs *somewhere*. The external sensory cues (visual, olfactory, auditory, and tactile) that denote this "somewhere" form the spatial context relative to which memories are encoded and retrieved. Early research using *interference reduction* paradigms demonstrated that confusion between two lists of items to remember is reduced if the lists are learned in different spatial environments rather than the same environment (Canas & Nelson, 1986; Emmerson, 1986; Godden & Baddeley, 1975; Smith & Vela, 2001). In other words, people exhibit better memory when tested in the presence of the same external sensory cues as those experienced during learning compared to people tested in new spatial contexts. Studies with both rodents and nonhuman primates have likewise found that changes to spatial cues strongly influence memory

(Bachevalier, Nemanic, & Alvarado, 2015; Bouton, 2002; Curzon, Rustay, & Browman, 2009; Dellu, Fauchey, Le Moal, & Simon, 1997; Pascalis, Hunkin, Bachevalier, & Mayes, 2009). For example, although animals are able to recognize objects after moving from one experimental chamber to another, memory is stronger when the familiar environment is used during both learning and retrieval (Dix & Aggleton, 1999). Any external sensory cue could theoretically constitute a spatial contextual cue, though for reasons that will become clear in the next section, *landmarks*—stable and salient environmental features—are particularly critical.

Situational cues Everything we do occurs in *some way*, and this state of affairs, or "situation," surrounding an event is often an important contextual cue. For instance, a wedding and a funeral are vastly different experiences even if they occur in the presence of the same spatial cues. Early reports noted that simple physical disruption between two lists of items to remember caused as much interference reduction as changes in spatial cues (Strand, 1970), and contextual interference is eliminated when participants tested in a new spatial context are instructed to recall the original learning environment just prior to recall (Smith, 1979). Such results show that situational cues, often operationalized in terms of task or motivational demands, influence memory independent of spatial cues. Moreover, memories are best retrieved if the brain state at encoding and retrieval are similar. *Brain state* refers to the internal state of the individual, which we include as a kind of situational cue, such as mood (Bower, 1981; Eich, 1995), hormonal state (McGaugh, 1989), or feelings associated with the administration of drugs (Overton, 1964). Whether external situational cues, such as the normative rules surrounding an event, and internal situational cues, such as the brain state, have qualitatively different influences on contextual representations remains an open question.

Temporal cues Everything we do occurs at *some time*, and it is possible to remember that different events that occurred in the presence of similar spatial or situational cues occurred at different times. Two kinds of temporal cues influence memory. First, an internal representation of the time of day at which learning occurs, tightly linked to an individual's circadian rhythm, has an influence on retrieval (Mulder, Gerkema, & Van Der Zee, 2013). Time of day can serve as an important mnemonic cue in spatial memory tasks (Boulos & Logothetis, 1990). Time-of-day effects are also observed in contextual fear-conditioning experiments that interrogate episodic memory, in which animals learn to fear a spatial context in which shock was previously experienced. Rodents display their strongest context-dependent fear response during their *inactive phase* (the light period; Chaudhury & Colwell, 2002). The second kind of temporal cue is the relative sequence in which learning takes place. Events experienced closer together in time are more similar than events experienced further apart. As a result, if a person experiences an event and her memory is later assessed, the ability to recall that event will decrease as the time between learning and retrieval increases (Rubin & Wenzel, 1996). Similarly, items encountered in close temporal proximity are more likely to be recalled sequentially than items encountered further apart (Howard & Kahana, 2002).

This brief taxonomy of context-defining cues suggests that context is characterized by factors external to the agent, including the set of environmental cues that define a place or the situation that characterizes an event, and the internal factors (e.g., temporal, cognitive, hormonal, affective) against which mnemonic processes operate. The cinema provides an apt metaphor for summarizing these context-defining cues: it contains multiple movie theaters (spatial cues) playing different movies (situational cues) at different times (temporal context) (figure 19.1*A*).

When Do Cues Not Define Contexts?

For context to be a useful scientific construct, there must be factors that differentiate contexts from other types of mnemonic cues. We suggest three important properties that limit the appropriate application of the term *context*.

First, for the brain to form contextual representations from statistical cue regularities, the cues that characterize context must be reliably present over time, or *stable* (Biegler & Morris, 1993; Robin, 2018; Stark, Reagh, Yassa, & Stark, 2017). For instance, the location of seats that define a movie theater context must not change often for the seat locations to form an integral part of that context. In contextual fear-conditioning experiments, if animals are briefly (e.g., less than 27 s) exposed to a context and shocked, they later show little fear of the context (Fanselow, 1990) (figure 19.1*B*). However, if they are first preexposed to the context, the shock elicits a fear response when the animal is subsequently returned to the conditioned context. Contextual conditioning thus only occurs if animals have an opportunity to learn the reliability of contextual cues through prolonged or repetitive exposure, indicating that the experience of cue stability is critical for the formation of contextual representations that organize memory.

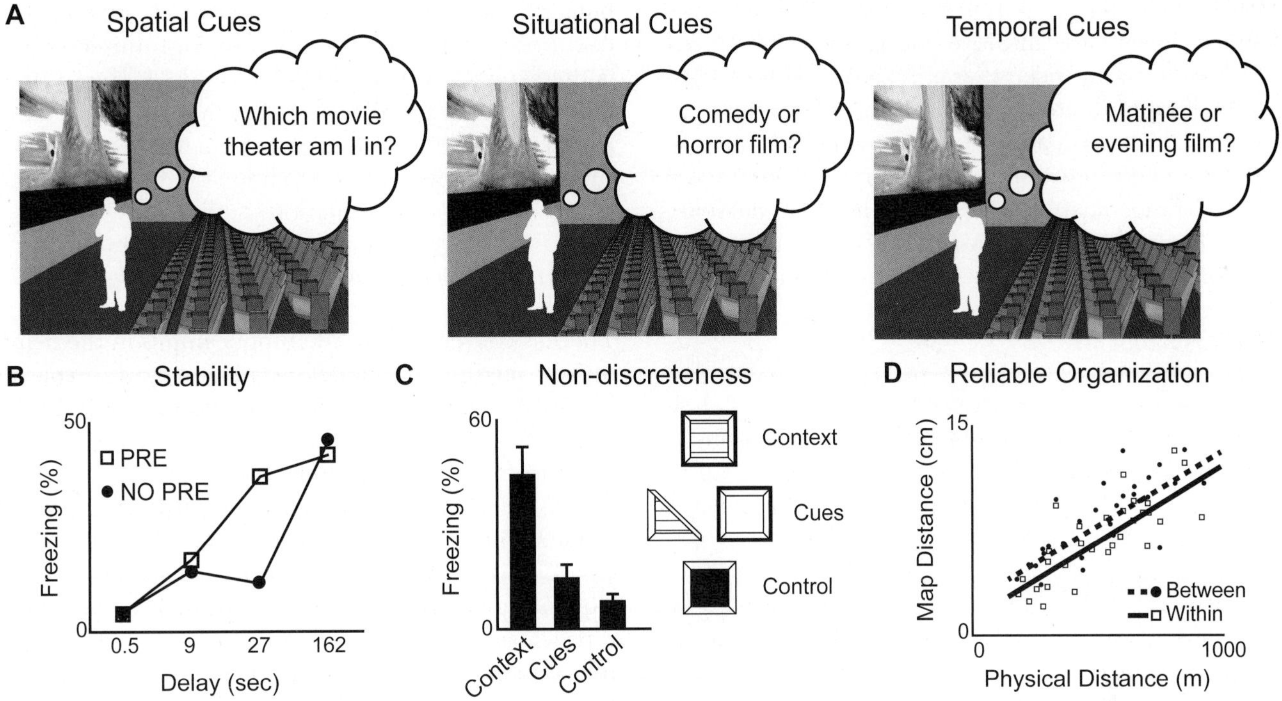

FIGURE 19.1 What is context? *A*, Contexts are defined by three cue types: spatial, situational, and temporal. *B*, Cues must be experienced as stable to form an integral part of context. The longer rodents experienced a context prior to fear conditioning, the more likely they were to show contextual conditioning (% freezing). Context preexposure (PRE) also resulted in stronger conditioning than no preexposure (Fanselow, 1990). *C*, Contexts are not defined by single discrete cues. When rodents were preexposed to either a spatial context (*Context*) separately from each of the cues that conjointly define that context (*Cues*) or a completely different context (*Control*), they subsequently displayed a fear response to the context only when initially exposed to the context itself (Rudy & O'Reilly, 1999). *D*, Contextual cues are represented as reliably organized. When participants recalled locations of landmarks in a city, their recall patterns showed evidence of hierarchical clustering into multiple smaller local contexts. Landmarks were drawn closer together on a map when recalled as being in similar local contexts (*Within*) than in different local contexts (*Between*) (Hirtle & Jonides, 1985).

Second, just as eating popcorn does not define being in a cinema (one can also eat popcorn at home), contexts are *not defined by any single discrete cue* (Robin, 2018). In other words, contexts are not the same as cues that serve as discrete signals for other events. Unlike contexts, increased time spent with a discrete cue does not alter conditioning to that cue (Fanselow, 1990). As a corollary, contexts are tolerant to changes in any one discrete cue. The context of your local movie theater could be recalled as such independent of whether you have popcorn, or are seeing a horror or a comedy film, or have consumed caffeine beforehand. This corollary suggests that context is not simply the set of cues associated with a particular event but rather a holistic representation of those cues. Consistent with this idea, rodents do not exhibit a typical contextual fear-conditioning response when exposed only to the cues individually that conjunctively form the conditioned context (Rudy & O'Reilly, 1999; figure 19.1*C*). Therefore, context is a neural construct, rather than something that exists in the world (Anderson, Hayman, Chakraborty, & Jeffery, 2003). As an illustration of this point, suppose the locations of the seats in your local movie theater are moved in your absence. When you later return to the theater, did you return to the same context or not? The answer to this question is not knowable a priori, but you could easily answer this question about your own memory.

Third, because contexts are not defined by any one discrete cue, different context-defining cues must have a *reliable organization* that allows them to be unified in a contextual representation. A common cue organization used by the brain to represent contexts is a hierarchy (Jeffery, Anderson, Hayman, & Chakraborty, 2004; Pearce & Bouton, 2001). There is an extensive literature demonstrating that the spatial environment is encoded as multiple hierarchically organized contexts, varying in spatial scale, instead of a single environmental context, and performance on memory tasks is influenced by this hierarchical structure (Han & Becker, 2014; Hirtle & Jonides, 1985; Holding, 1994; Marchette, Ryan, & Epstein, 2017; McNamara, 1986; McNamara, Hardy, & Hirtle, 1989; Montello & Pick, 1993; Wiener &

Mallot, 2003; figure 19.1*D*). Situational and temporal contexts also have intuitive hierarchical structures. Purchasing movie tickets or purchasing movie snacks are both subordinate to the larger class of transactional situational contexts, and the relative sequence of events can be organized over minutes or days. Beyond hierarchical arrangements, the set of possible relational structures between cues necessary for such cues to be associated in a contextual representation is unknown.

What Is Context?

Based on this survey of context-defining cues and their boundary conditions, we offer the following inclusive definition of context:

> *Context is a holistic representation of the internal and external (stable, nondiscrete, and reliably organized) cues that predict particular behavioral or mnemonic outputs.*

This definition unifies the contextual cues by placing emphasis on the adaptive function of contextual representations, rather than on any one specific cue type (Mizumori, 2013; Stachenfeld, Botvinick, & Gershman, 2017). Note that although this definition runs the risk of circularity, we have proposed three boundary conditions that limit the correct application of the context construct—stability, nondiscreteness, and reliable organization—and immunize against circularity. Insofar as the role of context is concerned, this definition is consistent with theories of memory that do not place particular importance on any one contextual cue type but rather focus on the function of contextual representations (Eichenbaum, 1993, 1996; Howard & Kahana, 2002; Mensink & Raaijmakers, 1988; Schacter, 2012; Schacter, Addis, & Buckner, 2007; Ranganath, 2010).

By contrast, others argue that spatial cues play a particularly special role in memory by serving as an ineluctable component of all memories (Burgess, Becker, King, & O'Keefe, 2001; Hassabis & Maguire, 2007; Maguire & Mullally, 2013; Nadel & Moscovitch, 1997; Robin, Buchsbaum, & Moscovitch, 2018). There is empirical evidence in favor of this position. For instance, when recalling previously read scenarios, participants spontaneously generate spatial contexts for the scenarios, even when the scenarios did not include any spatial cues (Robin, Buchsbaum, & Moscovitch, 2018; see also Hebscher, Levine, & Gilboa, 2017). However, as eluded to above, the situational and the temporal context can also strongly influence memory if they are behaviorally relevant. Our definition suggests that spatial cues may be strong determinants of contextual representations because they are often experienced as the most stable and thereby the most predictive of context-appropriate behaviors, even if they do not necessarily have unique cognitive status. An important area for future research is the extent to which different context-defining cues, matched in terms of their behavioral relevance—not just in an experimental situation but also over the lifetime of an individual or evolution—are incorporated into contextual representations.

The Hippocampal Basis of Contextual Memory

There is consensus that the hippocampus in the mammalian medial temporal lobe plays a crucial role in spatial and episodic memory, and neurobiological studies of contextual processing have focused on this brain area (for reviews, see Maren, Phan, & Liberzon, 2013; Myers & Gluck, 1994; Ranganath, 2010; Rudy, 2009; Rugg & Vilberg, 2013; Smith & Mizumori, 2006; Winocur & Olds, 1978). In the 1970s, Hirsch (1974) first explicitly proposed that the hippocampus mediates the retrieval of information in response to contextual cues that refer to the retrieved information. Since then, a wide variety of studies in both human and nonhuman animals have reinforced the importance of the hippocampus for context-dependent memory. Indeed, an automated meta-analysis (www.neurosynth.org) of functional magnetic resonance imaging (fMRI) studies of human context-dependent memory revealed common activation across these studies largely localized to the hippocampus (figure 19.2*A*).

Consistent with these neuroimaging findings, lesion studies have shown that the hippocampus is necessary for maintaining context-dependent memories (Anagnostaras, Gale, & Fanselow, 2001; Maren, 2001). When rodents are conditioned in one spatial context, for instance, they typically show a reduction of conditioned responses when tested in a new context, but animals with hippocampal damage continue to respond as if they failed to notice the spatial context change (Bachevalier, Nemanic, & Alvarado, 2015; Butterly, Petroccione, & Smith, 2012; Corcoran & Maren, 2001; Honey & Good, 1993; Penick & Solomom, 1991). Hippocampal damage also impairs memory for situational contexts (Ainge, van der Meer, Langston, & Wood, 2007); for example, hippocampal lesions disrupt the ability of rats to approach different goal objects depending on the rats' internal motivational state (hunger or thirst), even though object and motivational state discrimination are preserved (Kennedy & Shapiro, 2004). Finally, hippocampal lesions impair the ability to recall the biological time of day at which an event occurred (Cole et al., 2016) and remember the temporal sequence of events (i.e., the relative temporal context; Agster, Fortin, & Eichenbaum, 2002; Fortin, Agster, & Eichenbaum,

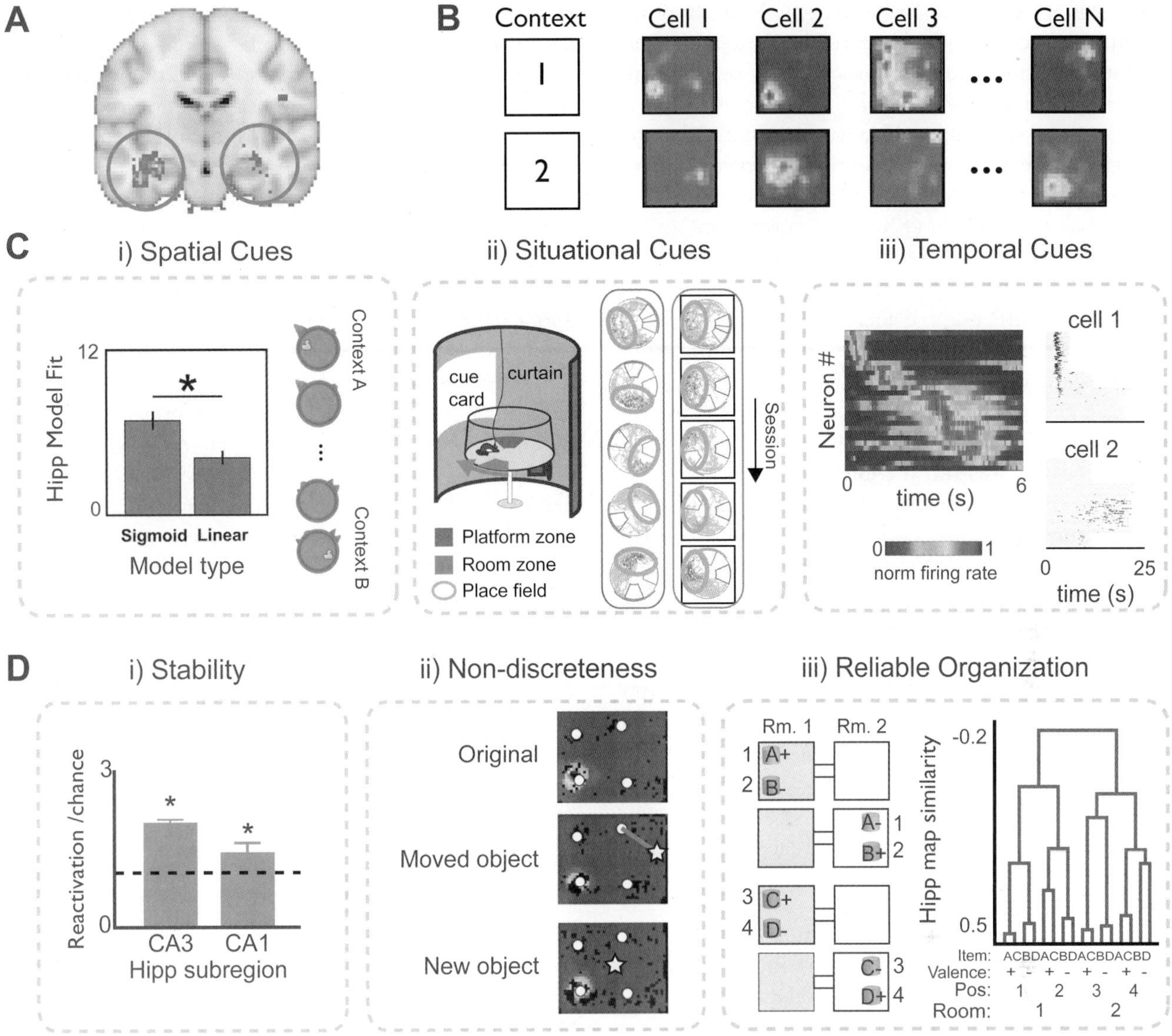

FIGURE 19.2 The hippocampal basis of contextual memory. *A*, Reverse inference meta-analysis (Yarkoni, Poldrack, Nichols, Van Essen, & Wager, 2011) of 36 context-dependent memory human fMRI studies. Overlapping activation across studies was largely localized to the hippocampus (threshold p < 0.01, false discovery rate (FDR)-corrected). *B*, Contextual memory is indexed by hippocampal remapping, in which all simultaneously recorded neurons alter their firing patterns across contexts (Alme et al., 2015). *C*, Remapping is induced by contextual cue changes: (1) Spatial cues. As visual cues (mountains) were gradually morphed from Context A to B during a spatial memory task, a rapid remapping of fMRI response patterns (*Sigmoidal*) better characterized hippocampal activity than a gradual change (*Linear*) (Steemers et al., 2016); (2) Situational cues. Hippocampal neurons represented locations in two different situational contexts, one relative to a moving platform (*left*) and another relative to the stable room (*right*; Keleman & Fenton, 2010); (3) Temporal cues. *Left*, Hippocampal neurons modulated by time. *Right*, Neurons changed firing patterns when the task's temporal parameters (*yellow bars*) were altered (MacDonald et al., 2011). *D*, Remapping reflects contextual boundary conditions: (1) Stability. The same hippocampal neurons (in subfields CA3 and CA1) reactivated two weeks later after mice were placed in the same context as initial exposure (Tayler et al., 2013); (2) Nondiscreteness. Example hippocampal neuron that did not remap when a discrete object (*white circles*) was moved (*magenta line to star*) or a novel object was added (*star*) (Deshmuch & Knierim, 2013); (3) Reliable organization. When rodents explored two chambers containing objects in different positions associated with different valences, hierarchical cue structure was reflected in hippocampal population activity patterns (McKenzie et al., 2014). (See color plate 22.)

2002; Kesner, Gilbert, & Barua, 2002). Thus, the hippocampus is necessary for the retrieval of memories associated with contexts characterized by the full range of context-defining cues.

At the cellular level, context is represented by the population activity of hippocampal neurons that fire whenever a navigator occupies particular environmental locations (*place fields*; O'Keefe & Dostrovsky, 1971). Within a context, different neurons have different place fields and thus, as a population, are thought to reflect a cognitive map of locations within the local context (O'Keefe & Nadel, 1978). Neuroimaging studies in humans likewise support the idea that the hippocampus represents a map of local context (Epstein, Patai, Julian, & Spiers, 2017). Beyond distinguishing between locations within a context, however, the hippocampus also stores multiple maps that allow it to represent multiple contexts (Bostock, Muller, & Kubie, 1991; Muller & Kubie, 1987). The hippocampus' ability to distinguish between contexts is indexed by a process known as *remapping* (figure 19.2*B*). During remapping, when an animal changes contexts, all simultaneously recorded neurons shift their relative place fields to new locations or stop firing altogether, quickly forming a new map-like representation (Bostock, Muller, & Kubie, 1991; Save, Nerad, & Poucet, 2000).[1] Current evidence suggests that a distinct ensemble of hippocampal neurons represents each different context (Alme et al., 2014; Anderson et al., 2003; Leutgeb et al., 2005; Leutgeb, Leutgeb, Treves, Moser, & Moser, 2004). If remapping mediates contextual memory, then remapping should occur between contexts defined by all contextual cue types and should be constrained by the same factors that limit when cues do not define contexts. As we will now review, this is indeed the case.

What Contextual Cues Induce Hippocampal Remapping?

Spatial cues Remapping is induced by spatial cue changes, such as when the walls of a familiar testing arena are replaced with walls of a different color (Bostock et al., 1991) or when the shape of the environment is altered (Lever, Wills, Cacucci, Burgess, & O'Keefe, 2002). For example, Wills et al. (2005) observed that incremental changes in the squareness or circularity of the walls of an experimental chamber produced no change in hippocampal activity until the cumulative changes became sufficiently great, at which point all neurons suddenly remapped to the other pattern. Human fMRI studies provide convergent evidence for the idea that the hippocampus represents spatial context as well (Alvarez, Biggs, Chen, Pine, & Grillon, 2008; Chadwick, Hassabis, & Maguire, 2011; Copara et al., 2014; Kyle, Stokes, Lieberman, Hassan, & Ekstrom, 2015; Steemers et al., 2016; Stokes, Kyle, & Ekstrom, 2015; figure 19.2*C*). Interestingly, rapid remapping following spatial cue changes is not always observed but rather depends on several factors, including prior learning experience (Bostock et al., 1991; Leutgeb et al., 2005) and the extent of differences between cues. Moreover, if there are sudden shifts from one spatial context to another, the hippocampus spontaneously "flickers" back to the original context representation (Jezek, Henriksen, Treves, Moser, & Moser, 2011). Remapping thus does not simply reflect changes to the perceived spatial cue constellation but rather reflects contextual memory.

Situational cues Task and motivational demands strongly influence the firing of hippocampal neurons (Frank, Brown, & Wilson, 2000; Gothard, Skaggs, & McNaughton, 1996; Hampson, Simeral, & Deadwyler, 1999; Kobayashi, Nishijo, Fukuda, Bures, & Ono, 1997; Lee, LeDuke, Chua, McDonald, & Sutherland, 2018; Markus et al., 1995; Redish, Rosenzweig, Bohanick, McNaughton, & Barnes, 2000; Smith & Mizumori, 2006; Wible et al., 1986). For instance, hippocampal neurons remap depending on the behavioral strategy used to solve a spatial memory task (Eschenko & Mizumori, 2007), or when navigators explore the same spatial context using different modes of transport (Song, Kim, Kim, & Jung, 2005), or when an animal's future goal changes (Skaggs & McNaughton, 1998; Wood, Dudchenko, Robitsek, & Eichenbaum, 2000). In an even more striking demonstration of the impact of situational context cues, Kelemen and Fenton (2010) trained rats to avoid two shock zones in a rotating disk-shaped arena; one zone was stationary relative to the larger room frame and the other rotated with the arena. Some neurons had place fields that were stationary relative to the broader room framework, while other fields rotated along with the local cues of the rotating arena (figure 19.2*C*). Thus, the hippocampus held distinct representations of two situational contexts in the same spatial context, one

[1] In contrast to remapping, in some cases the same neurons fire in the same locations across contexts but with reliably different firing rates, a process termed *rate remapping* (Leutgeb et al., 2005). The conditions under which remapping (sometimes called *global* or *complex* remapping) versus rate remapping are observed are not currently well understood, but whereas global remapping may relate more to contextual changes, rate remapping may reflect noncontextual, nonspatial influences on hippocampal representations (Leutgeb et al., 2005).

defined by the stable shock zone and the other defined by the rotating shock zone, and alternated between them when the situational contexts were placed in conflict. Human fMRI experiments provide convergent evidence for the hippocampal coding of situational contexts (Milivojevic, Varadinov, Grabovetsky, Collin, & Doeller, 2016). Changes in the affective brain state can induce remapping as well (Moita, Rosis, Zhou, LeDoux, & Blair, 2004; Wang, Yuan, Keinath, Álvarez, & Muzzio, 2015).

Temporal cues Circadian rhythms modulate the firing rates of hippocampal neurons (Munn & Bilkey, 2012), but whether changes in behaviorally relevant biological times of day induce remapping is less well studied. Greater evidence supports the idea that the hippocampus encodes the relative temporal context in which stimuli are learned and remaps between event sequences with different temporal structures. Temporal sequence information is represented by hippocampal cells that encode successive moments during a temporal gap between events (MacDonald, Lepage, Eden, & Eichenbaum, 2011; Sakon, Naya, Wirth, & Suzuki, 2014), even for sequences devoid of specific discrete cues (Farovik, Dupont, & Eichenbaum, 2010; Hales & Brewer, 2010; Meck, Church, & Olton, 1984; Moyer, Deyo, & Disterhoft, 1990; Staresina & Davachi, 2009). Critically, many hippocampal neurons sensitive to temporal information remap (or "retime") when the main temporal parameter of a task is altered (figure 19.2*C*), suggesting that such neural populations encode temporal context. Human fMRI studies have likewise found that temporal sequence-structure learning is associated with the hippocampus (Lehn et al., 2009; Schapiro, Turk-Browne, Norman, & Botvinick, 2016) and that the hippocampus generalizes across different sequences with similar temporal structures but not random sequences (Hsieh, Gruber, Jenkins, & Ranganath, 2014).

Effects of Contextual Boundary Conditions on Hippocampal Codes

Hippocampal context representations are stable Repeated visits to the same context reliably elicit activity in similar hippocampal populations (Cacucci, Wills, Lever, Giese, & O'Keefe, 2007; Guzowski, McNaughton, Barnes, & Worley, 1999; Kentros et al., 1998; Muller, Kubie, & Ranck, 1987; Thompson & Best, 1990). For example, Tayler and colleagues (2013) used genetically engineered mice that express a long-lasting marker of neural activity to compare the hippocampal population active at the time of initial exposure to a context with an active population in that same context two weeks later (figure 19.2*D*). Many neurons were active at both time points but not reactivated in a different context, indicating that hippocampal context representations remain stable over weeks. Inactivation of the hippocampus prior to context preexposure also eliminates the effect of preexposure in contextual fear-conditioning paradigms (Matus-Amat, Higgins, Barrientos, & Rudy, 2004), suggesting that preexposure allows the hippocampus to form a contextual representation reflecting stable cues. Likewise, spatial cues that are previously experienced as unstable have little control over place fields (Knierim, Kudrimoti, & McNaughton, 1995).

Despite the stability of hippocampal context representations, hippocampal population activity changes over time in the presence of the same spatial and situational cues (Mankin et al., 2012). Ziv and colleagues (2013) used calcium imaging to monitor the activity of hundreds of hippocampal neurons in mice over a 45-day period. Although many neurons had a place field on any given day, only 15%–25% were present on any other given day. Indeed, the overlap between hippocampal populations activated by two distinct spatial contexts acquired within a day is higher than when separated by a week (Cai et al., 2016). Therefore, in addition to forming stable contextual representations, hippocampal neurons change firing patterns over time in a manner that may reflect gradually shifting temporal context information, an idea also supported by human fMRI and intracranial recording studies (Copara et al., 2014; Deuker, Bellmund, Schröder, & Doeller, 2016; Manning, Polyn, Baltuch, Litt, & Kahana, 2011; Nielson, Smith, Sreekumar, Dennis, & Sederberg, 2015).

Hippocampal contextual representations do not reflect discrete cues Hippocampal lesions selectively impair context-dependent learning in rodents, but not conditioned responses to discrete cues such as a tone, during both episodic (Kim & Fanselow, 1992; Phillips & LeDoux, 1992; Selden, Everitt, Jarrard, & Robbins, 1991) and spatial (Pearce, Roberts, & Good, 1998) memory tasks. Human patients with hippocampal damage likewise have greater deficits in memory for contextual associations compared to recall or recognition of discrete cues and events (Giovanello, Verfaellie, & Keane, 2003; Holdstock, Mayes, Gong, Roberts, & Kapur, 2005; Mayes, Holdstock, Isaac, Hunkin, & Roberts, 2002; Turriziani, Fadda, Caltagirone, & Carlesimo, 2004). Human fMRI studies have also found that the hippocampus is more sensitive to contextual cues than information about the discrete cues learned within those contexts (Copara et al., 2014; Davachi, Mitchell, & Wagner, 2003; Hsieh et al., 2014; Ross & Slotnick, 2008).

Importantly, consistent with these lesion and neuroimaging results, changes to discrete spatial cues do not always elicit remapping (Cressant, Muller, & Poucet, 1997; Deshmukh & Knierim, 2013; figure 19.2*D*).

Hippocampal representations reflect reliable organization of contextual cues When spatial and episodic cues are hierarchically structured, hippocampal neurons differentiate between such cues using a hierarchical coding scheme (Takahashi, 2013). McKenzie and colleagues (2014) recorded hippocampal neurons while rats explored two rooms containing two objects (A and B) located in either of two positions (figure 19.2*D*). In one room, object A was rewarded and in the other, object B was rewarded. The rats subsequently learned new room-object-reward contingencies using a second object set (C and D) within the same rooms. At the most general level, hippocampal activity encoded room identity. At the next level, the population responded similarly to objects at similar positions independent of the valence, and so forth. Thus, the hippocampus can represent cues using a hierarchical coding scheme in which each kind of response represents a subset of the responses at the next highest level of coding. Broadly, this suggests that the hippocampus represents contextual cues in a manner that reflects the reliable organization of those cues. Interestingly, rather than a distinct hippocampal ensemble representing each different context, this would imply that hippocampal neurons do not remap randomly across contexts; rather, the similarity between different hippocampal context representations may reflect the similarity in across-context relational cue structure, thus enabling across-context behavioral predictions. Consistent with this idea, when only a subset of cues change across contexts, partial remapping can occur in which the place fields of only a proportion of neurons remap (Anderson & Jeffery, 2003).

Hippocampal Context Representations and Behavior

If the hippocampus mediates contextual memory, we would expect a link between hippocampal population activity and context-dependent behavior. Striking demonstrations of this link come from studies using optogenetics to stimulate hippocampal populations (Liu et al., 2012; Tanaka et al., 2014). In one recent example, mice were exposed to a spatial context, and the hippocampal neurons active in that context genetically labeled (Ramirez et al., 2013). The next day the mice were shocked in a different context while the labeled neurons from the original context were reactivated. When the mice were subsequently tested in the original context with no stimulation, they exhibited a fear response. Thus, the mice learned to fear an artificially reactivated representation of the original context even though they had never been shocked there. Since hippocampal activity elicited by stimulation acted as a serviceable substitute for contextual cues—akin to how recalling the original learning context at retrieval eliminates contextual interference effects—hippocampal context representations mediate context-dependent behavior.

Despite this growing evidence that hippocampal activity is sufficient to induce context-dependent behavior, there is conflicting evidence regarding whether remapping is necessary for contextual memory under more naturalistic conditions. On the one hand, Kennedy and Shapiro (2009) observed remapping due to changes in motivational state (hunger vs. thirst) only when such situational cues were required to select among goal-directed actions but not during random foraging when the situational cues were incidental to behavior. On the other hand, a consistent relationship between remapping and context-dependent behavior is not always found. Jeffery and colleagues (2003) trained rats to locate a reward in a chamber with black walls. When the wall color was changed to white, the rats still accurately chose the rewarded location despite the fact that the change in wall color induced remapping. This disconnect could have been due to the fact that behavior in this case was guided by discrete cues (i.e., behavior did not actually reflect contextual memory), even though the hippocampus remapped. Understanding the link between remapping and contextual memory is a critical area for future research.

Context Recognition Inputs to the Hippocampus

For context to influence memory, an agent must first recognize the cues that denote the current context. This context recognition process is cognitively dissociable from other aspects of spatial memory (Julian, Keinath, Muzzio, & Epstein, 2015). Since the hippocampus mediates both the contextual memory, as well as the recall, of locations, events, or items within a single context (Eichenbaum, Yonelinas, & Ranganath, 2007; Keinath, Julian, Epstein, & Muzzio, 2017; Redish & Touretzky, 1998; Ranganath, 2010), this raises the possibility that context recognition is performed upstream of the hippocampus itself.

The primary inputs to the hippocampus originate in entorhinal cortex (EC; Witter & Amaral, 2004), which has medial (MEC) and lateral (LEC) subdivisions. There is mixed evidence for the idea that EC supports context recognition. On the one hand, lesions of the entire entorhinal region produce contextual memory deficits that mirror those caused by hippocampal damage (Ji & Maren, 2008; Majchrzak et al., 2006). The perturbation of hippocampal inputs from MEC also induces spontaneous hippocampal remapping (Miao et al., 2015; figure 19.3*A*), suggesting that MEC in particular may be the source of hippocampal context representations. The MEC contains several types of place-modulated neurons (Hafting, Fyhn, Molden, Moser, & Moser, 2005; Sargolini et al., 2006), a subset of which are strongly contextually modulated (Kitamura et al., 2015). When contextually modulated MEC neurons change firing patterns across different spatial contexts (Barry, Hayman, Burgess, & Jeffery, 2007; Fyhn, Hafting, Treves, Moser, & Moser, 2007; Marozzi, Ginzberg, Alenda, & Jeffery, 2015), coincident remapping is found in the hippocampus (Fyhn et al., 2007). MEC sensitivity to behaviorally relevant situational cues has not been extensively explored, but some MEC neurons are modulated by temporal sequence information (Kraus et al., 2015). On the other hand, lesions specifically targeting MEC or LEC do not cause selective contextual memory deficits (Hales et al., 2014; Wilson et al., 2013), and lesions localized to MEC do not eliminate hippocampal remapping (Schlesiger, Boublil, Hales, Leutgeb, & Leutgeb, 2018). Thus, although EC is critical for transmitting contextual information to the hippocampus, it is unlikely to serve as a context recognition system itself.

In rodents, one of the primary MEC inputs is postrhinal cortex (POR; Ho & Burwell, 2014), which also projects directly to the hippocampus (Agster & Burwell, 2013). Cytoarchitectonic characteristics and anatomical connectivity suggest that POR is homologous to the primate posterior parahippocampal cortex (Burwell, 2001; Furtak, Wei, Agster, & Burwell, 2007), including a functionally defined region known as the *parahippocampal place area* (PPA) in humans (Aguirre, Zarahn, & D'Esposito, 1998; Epstein & Kanwisher, 1998; figure 19.3*B*). Growing evidence suggests that the POR/PPA plays an important role in context recognition (Julian, Keinath, Marchette, & Epstein, 2018). Damage to the human posterior parahippocampal cortex from stroke causes context recognition impairments (Aguirre & D'Esposito, 1999; Takahashi & Kawamura, 2002). Animal lesion studies have also confirmed the importance of the posterior parahippocampal/POR region for context-dependent memory (Bucci, Phillips, & Burwell, 2000; Bucci, Saddoris, & Burwell, 2002; Burwell, Bucci, Sanborn, & Jutras, 2004; Norman & Eacott, 2005; Peck & Taube, 2017; figure 19.3*C*). The magnitude of contextual memory deficits following POR lesions is not delay dependent, suggesting that the POR serves a context recognition function, rather than retrieving contextual memories per se (Liu & Bilkey, 2002). POR lesions have little effect on the stability of hippocampal representations in a single context (Nerad, Liu, & Bilkey, 2009), but whether POR damage disrupts hippocampal remapping is unknown.

Recent human fMRI studies provide convergent evidence for the role of the PPA in processing contextual information. The PPA response pattern is similar for visual scenes depicting different views of the same spatial context but only in participants who have learned that these views depict the same context (Marchette, Vass, Ryan, & Epstein, 2015; figure 19.3*C*), and posterior parahippocampal cortex is activated when participants process cues with strong contextual associations (Aminoff, Kveraga, & Bar, 2013; Bar & Aminoff, 2003; Bar, Aminoff, & Schacter, 2008; Davachi, Mitchell, & Wagner, 2003; Diana, 2017; Hayes, Nadel, & Ryan, 2007; Ross & Slotnick, 2008). The PPA is particularly sensitive to landmark cues that could serve as useful indicators of context (Epstein, 2014; Troiani, Stigliani, Smith, & Epstein, 2012), such as environmental boundaries (Epstein & Kanwisher, 1998; Kamps, Julian, Kubilius, Kanwisher, & Dilks, 2016; Kravitz, Peng, & Baker, 2011; Park, Brady, Greene, & Oliva, 2011) and large, stable objects (Julian, Ryan, & Epstein, 2016; Konkle & Oliva, 2012). The PPA is also modulated by the temporal sequence in which items are experienced (Turk-Browne, Simon, & Sederberg, 2012). However, one study found that the PPA is less strongly activated when participants identify scenes based on situational rather than spatial cues (Epstein & Higgins, 2006). Future studies are needed to resolve whether the POR/PPA is equally sensitive to all types of context-defining cues and to determine whether contextual representations in this region are constrained by all contextual cue boundary conditions.

Concluding Remarks

Based on a survey of the cues critical for shaping contextual representations and their boundary conditions, we propose that context is a holistic representation of the spatial, situational, and temporal cues that reliably predict particular behavioral and mnemonic outputs. Extensive research supports the idea that context-dependent memory is mediated by the hippocampus. At a mechanistic level, context is represented by the hippocampus through remapping, driven by parahippocampal context recognition inputs. Together, our chapter shows that the brain learns in a dynamic world by forming holistic representations of the stable and reliably structured cue constellations (i.e., contexts) that in turn make it possible to generate precise predictions about the future.

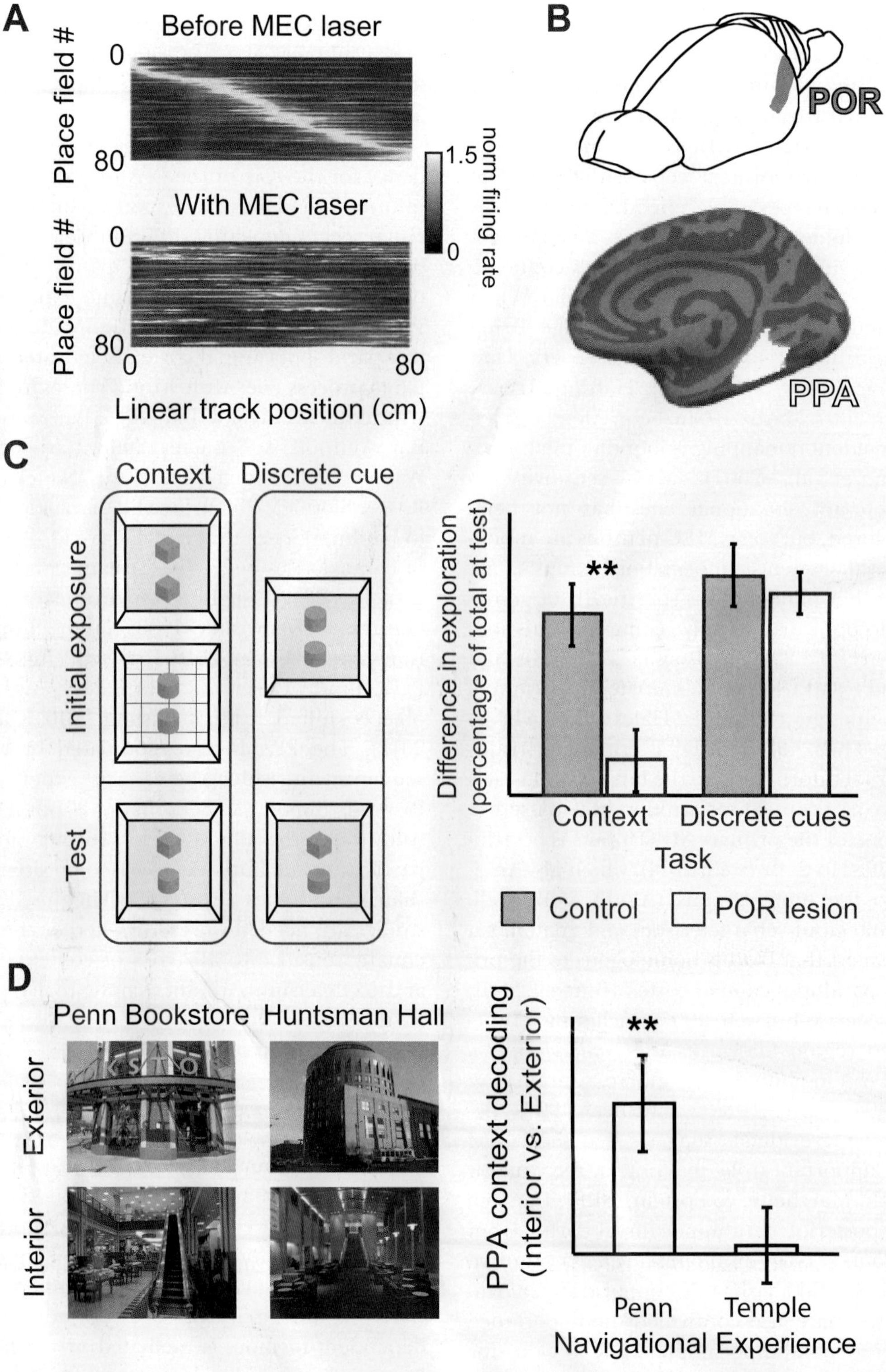

FIGURE 19.3 Parahippocampal context recognition inputs to the hippocampus. *A*, When rodents walked along a linear track, optogenetic (laser) inactivation of the MEC induced hippocampal remapping (Miao et al., 2015). *B*, A primary input to the rodent MEC is POR, which may be homologous to human PPA (shown on the inflated cortical surface; Julian, Fedorenko, Webster, & Kanwisher, 2012). *C*, POR lesions cause context recognition impairments. Control rats explore familiar discrete objects more when those objects appear in a different familiar context than when initially encountered, but POR lesions eliminate this object-context novelty preference. POR lesions had no effect in a comparable discrete cue object recognition task (Norman & Eacott, 2005). *D*, PPA mediates context recognition in humans. fMRI activity patterns in the PPA were similar for images of the interior and exterior of the same buildings, which share the same spatial context, but only in students who have experience with those buildings (*Penn*) and not in students who do not (*Temple*) (Marchette et al., 2015).

Acknowledgments

We acknowledge the support to Christian F. Doeller from the Max Planck Society; the European Research Council (ERC-CoG GEOCOG 724836); the Kavli Foundation, Centre of Excellence scheme of the Research Council of Norway–Centre for Neural Computation, Egil and Pauline Braathen and Fred Kavli Centre for Cortical Microcircuits, National Infrastructure scheme of the Research Council of Norway–NORBRAIN; and the Netherlands Organisation for Scientific Research (NWO-Vidi 452-12-009; NWO-Gravitation 024-001-006; NWO-MaGW 406-14-114; NWO-MaGW 406-15-291).

REFERENCES

Agster, K. L., & Burwell, R. D. (2013). Hippocampal and subicular efferents and afferents of the perirhinal, postrhinal, and entorhinal cortices of the rat. *Behavioural Brain Research, 254*, 50–64.

Agster, K. L., Fortin, N. J., & Eichenbaum, H. (2002). The hippocampus and disambiguation of overlapping sequences. *Journal of Neuroscience, 22*(13), 5760–5768.

Aguirre, G. K., & D'Esposito, M. (1999). Topographical disorientation: A synthesis and taxonomy. *Brain, 122*, 1613–1628.

Aguirre, G. K., Zarahn, E., & D'Esposito, M. (1998). An area within human ventral cortex sensitive to building stimuli: Evidence and implications. *Neuron, 21*, 373–383.

Ainge, J. A., van der Meer, M. A., Langston, R. F., & Wood, E. R. (2007). Exploring the role of context-dependent hippocampal activity in spatial alternation behavior. *Hippocampus, 17*(10), 988–1002.

Alme, C. B., Miao, C., Jezek, K., Treves, A., Moser, E. I., & Moser, M.-B. (2014). Place cells in the hippocampus: Eleven maps for eleven rooms. *Proceedings of the National Academy of Sciences, 111*(52), 18428–18435.

Alvarez, R. P., Biggs, A., Chen, G., Pine, D. S., & Grillon, C. (2008). Contextual fear conditioning in humans: Cortical-hippocampal and amygdala contributions. *Journal of Neuroscience, 28*(24), 6211–6219.

Aminoff, E. M., Kveraga, K., & Bar, M. (2013). The role of the parahippocampal cortex in cognition. *Trends in Cognitive Sciences, 17*, 379–390.

Anagnostaras, S. G., Gale, G. D., & Fanselow, M. S. (2001). Hippocampus and contextual fear conditioning: Recent controversies and advances. *Hippocampus, 11*(1), 8–17.

Anderson, M. I., Hayman, R., Chakraborty, S., & Jeffery, K. J. (2003). The representation of spatial context. In K. J. Jeffery (Ed.), *The neurobiology of spatial behaviour* (pp. 274–294). Oxford: Oxford University Press.

Anderson, M. I., & Jeffery, K. J. (2003). Heterogeneous modulation of place cell firing by changes in context. *Journal of Neuroscience, 23*, 8827–8835.

Bachevalier, J., Nemanic, S., & Alvarado, M. C. (2015). The influence of context on recognition memory in monkeys: Effects of hippocampal, parahippocampal and perirhinal lesions. *Behavioural Brain Research, 285*, 89–98.

Bar, M., & Aminoff, E. (2003). Cortical analysis of visual context. *Neuron, 38*, 347–358.

Bar, M., Aminoff, E., & Schacter, D. L. (2008). Scenes unseen: The parahippocampal cortex intrinsically subserves contextual associations, not scenes or places per se. *Journal of Neuroscience, 28*, 8539–8544.

Barry, C., Hayman, R., Burgess, N., & Jeffery, K. J. (2007). Experience-dependent rescaling of entorhinal grids. *Nature Neuroscience, 10*(6), 682.

Biegler, R., & Morris, R. G. (1993). Landmark stability is a prerequisite for spatial but not discrimination learning. *Nature, 361*, 631–633.

Bostock, E., Muller, R. U., & Kubie, J. L. (1991). Experience-dependent modifications of hippocampal place cell firing. *Hippocampus, 1*, 193–205.

Boulos, Z., & Logothetis, D. E. (1990). Rats anticipate and discriminate between two daily feeding times. *Physiology & Behavior, 48*(4), 523–529.

Bouton, M. E. (2002). Context, ambiguity, and unlearning: Sources of relapse after behavioral extinction. *Biological Psychiatry, 52*(10), 976–986.

Bower, G. H. (1981). Mood and memory. *American Psychologist, 36*(2), 129.

Bucci, D. J., Phillips, R. G., & Burwell, R. D. (2000). Contributions of postrhinal and perirhinal cortex to contextual information processing. *Behavioral Neuroscience, 114*(5), 882.

Bucci, D. J., Saddoris, M. P., & Burwell, R. D. (2002). Contextual fear discrimination is impaired by damage to the postrhinal or perirhinal cortex. *Behavioral Neuroscience, 116*(3), 479.

Burgess, N., Becker, S., King, J. A., & O'Keefe, J. (2001). Memory for events and their spatial context: Models and experiments. *Philosophical Transactions of the Royal Society B: Biological Sciences, 356*(1413), 1493–1503.

Burwell, R. D. (2001). Borders and cytoarchitecture of the perirhinal and postrhinal cortices in the rat. *Journal of Comparative Neurology, 437*, 17–41.

Burwell, R. D., Bucci, D. J., Sanborn, M. R., & Jutras, M. J. (2004). Perirhinal and postrhinal contributions to remote memory for context. *Journal of Neuroscience, 24*(49), 11023–11028.

Butterly, D. A., Petroccione, M. A., & Smith, D. M. (2012). Hippocampal context processing is critical for interference free recall of odor memories in rats. *Hippocampus, 22*(4), 906–913.

Cacucci, F., Wills, T. J., Lever, C., Giese, K. P., & O'Keefe, J. (2007). Experience-dependent increase in CA1 place cell spatial information, but not spatial reproducibility, is dependent on the autophosphorylation of the α-isoform of the calcium/calmodulin-dependent protein kinase II. *Journal of Neuroscience, 27*(29), 7854–7859.

Cai, D. J., Aharoni, D., Shuman, T., Shobe, J., Biane, J., Song, W., … Lou, J. (2016). A shared neural ensemble links distinct contextual memories encoded close in time. *Nature, 534*(7605), 115.

Canas, J. J., & Nelson, D. L. (1986). Recognition and environmental context: The effect of testing by phone. *Bulletin of the Psychonomic Society, 24*(6), 407–409.

Chadwick, M. J., Hassabis, D., & Maguire, E. A. (2011). Decoding overlapping memories in the medial temporal lobes using high-resolution fMRI. *Learning & Memory, 18*(12), 742–746.

Chaudhury, D., & Colwell, C. S. (2002). Circadian modulation of learning and memory in fear-conditioned mice. *Behavioural Brain Research, 133*(1), 95–108.

Cole, E., Mistlberger, R. E., Merza, D., Trigiani, L. J., Madularu, D., Simundic, A., & Mumby, D. G. (2016). Circadian time-place (or time-route) learning in rats with hippocampal lesions. *Neurobiology of Learning and Memory, 136*, 236–243.

Collin, S. H., Milivojevic, B., & Doeller, C. F. (2015). Memory hierarchies map onto the hippocampal long axis in humans. *Nature Neuroscience, 18*(11), 1562.

Copara, M. S., Hassan, A. S., Kyle, C. T., Libby, L. A., Ranganath, C., & Ekstrom, A. D. (2014). Complementary roles of human hippocampal subregions during retrieval of spatiotemporal context. *Journal of Neuroscience, 34*(20), 6834–6842.

Corcoran, K. A., & Maren, S. (2001). Hippocampal inactivation disrupts contextual retrieval of fear memory after extinction. *Journal of Neuroscience, 21*(5), 1720–1726.

Cressant, A., Muller, R. U., & Poucet, B. (1997). Failure of centrally placed objects to control the firing fields of hippocampal place cells. *Journal of Neuroscience, 17*, 2531–2542.

Curzon, P., Rustay, N. R., & Browman, K. E. (2009). Cued and contextual fear conditioning for rodents. In J. J. Buccafusco (Ed.), *Methods of behavior analysis in neuroscience* (pp. 1–12). Boca Raton: FL: CRC Press.

Davachi, L., Mitchell, J. P., & Wagner, A. D. (2003). Multiple routes to memory: Distinct medial temporal lobe processes build item and source memories. *Proceedings of the National Academy of Sciences, 100*(4), 2157–2162.

Davies, G. M., & Thomson, D. M. (1988). *Memory in context: Context in memory*. Hoboken, NJ: John Wiley & Sons.

Dellu, F., Fauchey, V., Le Moal, M., & Simon, H. (1997). Extension of a new two-trial memory task in the rat: Influence of environmental context on recognition processes. *Neurobiology of Learning and Memory, 67*(2), 112–120.

Deshmukh, S. S., & Knierim, J. J. (2013). Influence of local objects on hippocampal representations: Landmark vectors and memory. *Hippocampus, 23*(4), 253–267.

Deuker, L., Bellmund, J. L., Schröder, T. N., & Doeller, C. F. (2016). An event map of memory space in the hippocampus. *eLife, 5*, e16534.

Diana, R. A. (2017). Parahippocampal cortex processes the nonspatial context of an event. *Cerebral Cortex, 27*(3), 1808–1816.

Dix, S. L., & Aggleton, J. P. (1999). Extending the spontaneous preference test of recognition: Evidence of object-location and object-context recognition. *Behavioural Brain Research, 99*(2), 191–200.

Eich, E. (1995). Searching for mood dependent memory. *Psychological Science, 6*(2), 67–75.

Eichenbaum, H. (1993). *Memory, amnesia, and the hippocampal system*. Cambridge, MA: MIT Press.

Eichenbaum, H. (1996). Is the rodent hippocampus just for "place"? *Current Opinion in Neurobiology, 6*(2), 187–195.

Eichenbaum, H., Yonelinas, A. P., & Ranganath, C. (2007). The medial temporal lobe and recognition memory. *Annual Review of Neuroscience, 30*, 123–152.

Ekstrom, A. D., & Bookheimer, S. Y. (2007). Spatial and temporal episodic memory retrieval recruit dissociable functional networks in the human brain. *Learning & Memory, 14*(10), 645–654.

Emmerson, P. G. (1986). Effects of environmental context on recognition memory in an unusual environment. *Perceptual and Motor Skills, 63*(3), 1047–1050.

Epstein, R., & Kanwisher, N. (1998). A cortical representation of the local visual environment. *Nature, 392*, 598–601.

Epstein, R. A. (2014). Neural systems for visual scene recognition. In Kestutis Kveraga & Moshe Bar (Eds.), *Scene vision: Making sense of what we see* (pp. 105–134). Cambridge, MA: MIT Press.

Epstein, R. A., & Higgins, J. S. (2006). Differential parahippocampal and retrosplenial involvement in three types of visual scene recognition. *Cerebral Cortex, 17*(7), 1680–1693.

Epstein, R. A., Patai, E. Z., Julian, J. B., & Spiers, H. J. (2017). The cognitive map in humans: Spatial navigation and beyond. *Nature Neuroscience, 20*(11), 1504.

Eschenko, O., & Mizumori, S. J. (2007). Memory influences on hippocampal and striatal neural codes: Effects of a shift between task rules. *Neurobiology of Learning and Memory, 87*(4), 495–509.

Fanselow, M. S. (1990). Factors governing one-trial contextual conditioning. *Animal Learning & Behavior, 18*(3), 264–270.

Farovik, A., Dupont, L. M., & Eichenbaum, H. (2010). Distinct roles for dorsal CA3 and CA1 in memory for sequential nonspatial events. *Learning & Memory, 17*(1), 12–17.

Fortin, N. J., Agster, K. L., & Eichenbaum, H. B. (2002). Critical role of the hippocampus in memory for sequences of events. *Nature Neuroscience, 5*(5), 458.

Frank, L. M., Brown, E. N., & Wilson, M. (2000). Trajectory encoding in the hippocampus and entorhinal cortex. *Neuron, 27*(1), 169–178.

Furtak, S. C., Wei, S. M., Agster, K. L., & Burwell, R. D. (2007). Functional neuroanatomy of the parahippocampal region in the rat: The perirhinal and postrhinal cortices. *Hippocampus, 17*, 709–722.

Fyhn, M., Hafting, T., Treves, A., Moser, M.-B., & Moser, E. I. (2007). Hippocampal remapping and grid realignment in entorhinal cortex. *Nature, 446*, 190–194.

Giovanello, K. S., Verfaellie, M., & Keane, M. M. (2003). Disproportionate deficit in associative recognition relative to item recognition in global amnesia. *Cognitive, Affective, & Behavioral Neuroscience, 3*(3), 186–194.

Godden, D. R., & Baddeley, A. D. (1975). Context-dependent memory in two natural environments: On land and underwater. *British Journal of Psychology, 66*(3), 325–331.

Gothard, K. M., Skaggs, W. E., & McNaughton, B. L. (1996). Dynamics of mismatch correction in the hippocampal ensemble code for space: Interaction between path integration and environmental cues. *Journal of Neuroscience, 16*, 8027–8040.

Guzowski, J. F., McNaughton, B. L., Barnes, C. A., & Worley, P. F. (1999). Environment-specific expression of the immediate-early gene Arc in hippocampal neuronal ensembles. *Nature Neuroscience, 2*(12), 1120.

Hafting, T., Fyhn, M., Molden, S., Moser, M.-B., & Moser, E. I. (2005). Microstructure of a spatial map in the entorhinal cortex. *Nature, 436*, 801–806.

Hales, J. B., & Brewer, J. B. (2010). Activity in the hippocampus and neocortical working memory regions predicts successful associative memory for temporally discontiguous events. *Neuropsychologia, 48*(11), 3351–3359.

Hales, Jena B., Schlesiger, M. I., Leutgeb, J. K., Squire, L. R., Leutgeb, S., & Clark, R. E. (2014). Medial entorhinal cortex lesions only partially disrupt hippocampal place cells and hippocampus-dependent place memory. *Cell Reports, 9*(3), 893–901.

Hampson, R. E., Simeral, J. D., & Deadwyler, S. A. (1999). Distribution of spatial and nonspatial information in dorsal hippocampus. *Nature, 402*(6762), 610.

Han, X., & Becker, S. (2014). One spatial map or many? Spatial coding of connected environments. *Journal of Experimental Psychology: Learning, Memory, and Cognition, 40*(2), 511.

Hassabis, D., & Maguire, E. A. (2007). Deconstructing episodic memory with construction. *Trends in Cognitive Sciences, 11*(7), 299–306.

Hayes, S. M., Nadel, L., & Ryan, L. (2007). The effect of scene context on episodic object recognition: Parahippocampal cortex mediates memory encoding and retrieval success. *Hippocampus, 17*(9), 873–889.

Hebscher, M., Levine, B., & Gilboa, A. (2017). The precuneus and hippocampus contribute to individual differences in the unfolding of spatial representations during episodic autobiographical memory. *Neuropsychologia,* 110, 123–133.

Hirsh, R. (1974). The hippocampus and contextual retrieval of information from memory: A theory. *Behavioral Biology, 12*(4), 421–444.

Hirtle, S. C., & Jonides, J. (1985). Evidence of hierarchies in cognitive maps. *Memory & Cognition, 13*(3), 208–217.

Ho, J. W., & Burwell, R. D. (2014). Perirhinal and postrhinal functional inputs to the hippocampus. In D. Derdikman & J. J. Knierim (Eds.), *Space, time and memory in the hippocampal formation* (pp. 55–81). New York: Springer.

Holding, C. S. (1994). Further evidence for the hierarchical representation of spatial information. *Journal of Environmental Psychology, 14*(2), 137–147.

Holdstock, J. S., Mayes, A. R., Gong, Q. Y., Roberts, N., & Kapur, N. (2005). Item recognition is less impaired than recall and associative recognition in a patient with selective hippocampal damage. *Hippocampus, 15*(2), 203–215.

Honey, R. C., & Good, M. (1993). Selective hippocampal lesions abolish the contextual specificity of latent inhibition and conditioning. *Behavioral Neuroscience, 107*(1), 23.

Howard, M. W., & Kahana, M. J. (2002). A distributed representation of temporal context. *Journal of Mathematical Psychology, 46*(3), 269–299.

Hsieh, L.-T., Gruber, M. J., Jenkins, L. J., & Ranganath, C. (2014). Hippocampal activity patterns carry information about objects in temporal context. *Neuron, 81*(5), 1165–1178.

Hyman, J. M., Ma, L., Balaguer-Ballester, E., Durstewitz, D., & Seamans, J. K. (2012). Contextual encoding by ensembles of medial prefrontal cortex neurons. *Proceedings of the National Academy of Sciences, 109*(13), 5086–5091.

Jeffery, K. J., Anderson, M. I., Hayman, R., & Chakraborty, S. (2004). A proposed architecture for the neural representation of spatial context. *Neuroscience & Biobehavioral Reviews, 28,* 201–218.

Jeffery, K. J., Gilbert, A., Burton, S., & Strudwick, A. (2003). Preserved performance in a hippocampal-dependent spatial task despite complete place cell remapping. *Hippocampus, 13,* 175–189.

Jezek, K., Henriksen, E. J., Treves, A., Moser, E. I., & Moser, M.-B. (2011). Theta-paced flickering between place-cell maps in the hippocampus. *Nature, 478*(7368), 246.

Ji, J., & Maren, S. (2008). Lesions of the entorhinal cortex or fornix disrupt the context-dependence of fear extinction in rats. *Behavioural Brain Research, 194*(2), 201–206.

Julian, J. B., Fedorenko, E., Webster, J., & Kanwisher, N. (2012). An algorithmic method for functionally defining regions of interest in the ventral visual pathway. *NeuroImage, 60,* 2357–2364.

Julian, Joshua B., Keinath, A. T., Marchette, S., & Epstein, R. A. (2018). The neurocognitive basis of spatial reorientation. *Current Biology,* 28, R1059–R1073.

Julian, Joshua B., Keinath, A. T., Muzzio, I. A., & Epstein, R. A. (2015). Place recognition and heading retrieval are mediated by dissociable cognitive systems in mice. *Proceedings of the National Academy of Sciences, 112,* 6503–6508.

Julian, Joshua B., Ryan, J., & Epstein, R. A. (2016). Coding of object size and object category in human visual cortex. *Cerebral Cortex,* doi.org/10.1093/cercor/bhw150

Kamps, F. S., Julian, J. B., Kubilius, J., Kanwisher, N., & Dilks, D. D. (2016). The occipital place area represents the local elements of scenes. *NeuroImage, 132,* 417–424.

Keinath, A. T., Julian, J. B., Epstein, R. A., & Muzzio, I. A. (2017). Environmental geometry aligns the hippocampal map during spatial reorientation. *Current Biology, 27,* 309–317.

Kelemen, E., & Fenton, A. A. (2010). Dynamic grouping of hippocampal neural activity during cognitive control of two spatial frames. *PLoS Biology, 8*(6), e1000403.

Kennedy, P. J., & Shapiro, M. L. (2004). Retrieving memories via internal context requires the hippocampus. *Journal of Neuroscience, 24*(31), 6979–6985.

Kennedy, P. J., & Shapiro, M. L. (2009). Motivational states activate distinct hippocampal representations to guide goal-directed behaviors. *Proceedings of the National Academy of Sciences, 106*(26), 10805–10810.

Kentros, C., Hargreaves, E., Hawkins, R. D., Kandel, E. R., Shapiro, M., & Muller, R. V. (1998). Abolition of long-term stability of new hippocampal place cell maps by NMDA receptor blockade. *Science, 280,* 2121–2126.

Kesner, R. P., Gilbert, P. E., & Barua, L. A. (2002). The role of the hippocampus in memory for the temporal order of a sequence of odors. *Behavioral Neuroscience, 116*(2), 286.

Kim, J. J., & Fanselow, M. S. (1992). Modality-specific retrograde amnesia of fear. *Science, 256*(5057), 675–677.

Kitamura, T., Ogawa, S. K., Roy, D. S., Okuyama, T., Morrissey, M. D., Smith, L. M., … Tonegawa, S. (2017). Engrams and circuits crucial for systems consolidation of a memory. *Science, 356*(6333), 73–78.

Kitamura, T., Sun, C., Martin, J., Kitch, L. J., Schnitzer, M. J., & Tonegawa, S. (2015). Entorhinal cortical ocean cells encode specific contexts and drive context-specific fear memory. *Neuron, 87*(6), 1317–1331.

Kjelstrup, K. B., Solstad, T., Brun, V. H., Hafting, T., Leutgeb, S., Witter, M. P., … Moser, M.-B. (2008). Finite scale of spatial representation in the hippocampus. *Science, 321*(5885), 140–143.

Knierim, J. J., Kudrimoti, H. S., & McNaughton, B. L. (1995). Place cells, head direction cells, and the learning of landmark stability. *Journal of Neuroscience, 15,* 1648–1659.

Kobayashi, T., Nishijo, H., Fukuda, M., Bures, J., & Ono, T. (1997). Task-dependent representations in rat hippocampal place neurons. *Journal of Neurophysiology, 78*(2), 597–613.

Konkle, T., & Oliva, A. (2012). A real-world size organization of object responses in occipitotemporal cortex. *Neuron, 74,* 1114–1124.

Kornblith, S., Cheng, X., Ohayon, S., & Tsao, D. Y. (2013). A network for scene processing in the macaque temporal lobe. *Neuron, 79,* 766–781.

Kraus, B. J., Brandon, M. P., Robinson, R. J., Connerney, M. A., Hasselmo, M. E., & Eichenbaum, H. (2015). During running in place, grid cells integrate elapsed time and distance run. *Neuron, 88*(3), 578–589.

Kravitz, D. J., Peng, C. S., & Baker, C. I. (2011). Real-world scene representations in high-level visual cortex: It's the spaces more than the places. *Journal of Neuroscience, 31*, 7322–7333.

Kyle, C. T., Stokes, J. D., Lieberman, J. S., Hassan, A. S., & Ekstrom, A. D. (2015). Successful retrieval of competing spatial environments in humans involves hippocampal pattern separation mechanisms. *eLife, 4*, e10499.

Lee, J. Q., LeDuke, D. O., Chua, K., McDonald, R. J., & Sutherland, R. J. (2018). Relocating cued goals induces population remapping in CA1 related to memory performance in a two-platform water task in rats. *Hippocampus, 8*(6), 431–440.

Lehn, H., Steffenach, H.-A., van Strien, N. M., Veltman, D. J., Witter, M. P., & Håberg, A. K. (2009). A specific role of the human hippocampus in recall of temporal sequences. *Journal of Neuroscience, 29*(11), 3475–3484.

Leutgeb, J. K., Leutgeb, S., Treves, A., Meyer, R., Barnes, C. A., McNaughton, B. L., ... Moser, E. I. (2005). Progressive transformation of hippocampal neuronal representations in morphed environments. *Neuron, 48*, 345–358.

Leutgeb, S., Leutgeb, J. K., Barnes, C. A., Moser, E. I., McNaughton, B. L., & Moser, M.-B. (2005). Independent codes for spatial and episodic memory in hippocampal neuronal ensembles. *Science, 309*, 619–623.

Leutgeb, S., Leutgeb, J. K., Treves, A., Moser, M.-B., & Moser, E. I. (2004). Distinct ensemble codes in hippocampal areas CA3 and CA1. *Science, 305*(5688), 1295–1298.

Lever, C., Wills, T., Cacucci, F., Burgess, N., & O'Keefe, J. (2002). Long-term plasticity in hippocampal place-cell representation of environmental geometry. *Nature, 416*, 90–94.

Liu, P., & Bilkey, D. K. (2002). The effects of NMDA lesions centered on the postrhinal cortex on spatial memory tasks in the rat. *Behavioral Neuroscience, 116*, 860.

Liu, X., Ramirez, S., Pang, P. T., Puryear, C. B., Govindarajan, A., Deisseroth, K., & Tonegawa, S. (2012). Optogenetic stimulation of a hippocampal engram activates fear memory recall. *Nature, 484*(7394), 381.

MacDonald, C. J., Lepage, K. Q., Eden, U. T., & Eichenbaum, H. (2011). Hippocampal "time cells" bridge the gap in memory for discontiguous events. *Neuron, 71*(4), 737–749.

Maguire, E. A., & Mullally, S. L. (2013). The hippocampus: A manifesto for change. *Journal of Experimental Psychology: General, 142*(4), 1180.

Majchrzak, M., Ferry, B., Marchand, A. R., Herbeaux, K., Seillier, A., & Barbelivien, A. (2006). Entorhinal cortex lesions disrupt fear conditioning to background context but spare fear conditioning to a tone in the rat. *Hippocampus, 16*(2), 114–124.

Mankin, E. A., Sparks, F. T., Slayyeh, B., Sutherland, R. J., Leutgeb, S., & Leutgeb, J. K. (2012). Neuronal code for extended time in the hippocampus. *Proceedings of the National Academy of Sciences, 109*(47), 19462–19467.

Manning, J. R., Polyn, S. M., Baltuch, G. H., Litt, B., & Kahana, M. J. (2011). Oscillatory patterns in temporal lobe reveal context reinstatement during memory search. *Proceedings of the National Academy of Sciences, 108*(31), 12893–12897.

Marchette, S. A., Ryan, J., & Epstein, R. A. (2017). Schematic representations of local environmental space guide goal-directed navigation. *Cognition, 158*, 68–80.

Marchette, S. A., Vass, L. K., Ryan, J., & Epstein, R. A. (2015). Outside looking in: Landmark generalization in the human navigational system. *Journal of Neuroscience, 35*, 14896–14908.

Maren, S. (2001). Neurobiology of Pavlovian fear conditioning. *Annual Review of Neuroscience, 24*(1), 897–931.

Maren, S., Phan, K. L., & Liberzon, I. (2013). The contextual brain: Implications for fear conditioning, extinction and psychopathology. *Nature Reviews Neuroscience, 14*(6), 417.

Markus, E. J., Qin, Y.-L., Leonard, B., Skaggs, W. E., McNaughton, B. L., & Barnes, C. A. (1995). Interactions between location and task affect the spatial and directional firing of hippocampal neurons. *Journal of Neuroscience, 15*(11), 7079–7094.

Marozzi, E., Ginzberg, L. L., Alenda, A., & Jeffery, K. J. (2015). Purely translational realignment in grid cell firing patterns following nonmetric context change. *Cerebral Cortex, 25*(11), 4619–4627.

Matus-Amat, P., Higgins, E. A., Barrientos, R. M., & Rudy, J. W. (2004). The role of the dorsal hippocampus in the acquisition and retrieval of context memory representations. *Journal of Neuroscience, 24*(10), 2431–2439.

Mayes, A. R., Holdstock, J. S., Isaac, C. L., Hunkin, N. M., & Roberts, N. (2002). Relative sparing of item recognition memory in a patient with adult-onset damage limited to the hippocampus. *Hippocampus, 12*(3), 325–340.

McGaugh, J. L. (1989). Involvement of hormonal and neuromodulatory systems in the regulation of memory storage. *Annual Review of Neuroscience, 12*(1), 255–287.

McKenzie, S., Frank, A. J., Kinsky, N. R., Porter, B., Rivière, P. D., & Eichenbaum, H. (2014). Hippocampal representation of related and opposing memories develop within distinct, hierarchically organized neural schemas. *Neuron, 83*(1), 202–215.

McNamara, T. P. (1986). Mental representations of spatial relations. *Cognitive Psychology, 18*(1), 87–121.

McNamara, T. P., Hardy, J. K., & Hirtle, S. C. (1989). Subjective hierarchies in spatial memory. *Journal of Experimental Psychology: Learning, Memory, and Cognition, 15*(2), 211.

Meck, W. H., Church, R. M., & Olton, D. S. (1984). Hippocampus, time, and memory. *Behavioral Neuroscience, 98*(1), 3.

Mensink, G.-J., & Raaijmakers, J. G. (1988). A model for interference and forgetting. *Psychological Review, 95*(4), 434.

Miao, C., Cao, Q., Ito, H. T., Yamahachi, H., Witter, M. P., Moser, M.-B., & Moser, E. I. (2015). Hippocampal remapping after partial inactivation of the medial entorhinal cortex. *Neuron, 88*(3), 590–603.

Milivojevic, B., Varadinov, M., Grabovetsky, A. V., Collin, S. H., & Doeller, C. F. (2016). Coding of event nodes and narrative context in the hippocampus. *Journal of Neuroscience, 36*(49), 12412–12424.

Mizumori, S. J. (2013). Context prediction analysis and episodic memory. *Frontiers in Behavioral Neuroscience, 7*, 132.

Moita, M. A., Rosis, S., Zhou, Y., LeDoux, J. E., & Blair, H. T. (2004). Putting fear in its place: Remapping of hippocampal place cells during fear conditioning. *Journal of Neuroscience, 24*(31), 7015–7023.

Montello, D. R., & Pick Jr., H. L. (1993). Integrating knowledge of vertically aligned large-scale spaces. *Environment and Behavior, 25*(3), 457–484.

Moyer, J. R., Deyo, R. A., & Disterhoft, J. F. (1990). Hippocampectomy disrupts trace eye-blink conditioning in rabbits. *Behavioral Neuroscience, 104*(2), 243.

Mulder, C. K., Gerkema, M. P., & Van Der Zee, E. A. (2013). Circadian clocks and memory: Time-place learning. *Frontiers in Molecular Neuroscience, 6*, 8.

Muller, R. U., & Kubie, J. L. (1987). The effects of changes in the environment on the spatial firing of hippocampal complex-spike cells. *Journal of Neuroscience, 7,* 1951–1968.

Muller, R. U., Kubie, J. L., & Ranck, J. B. (1987). Spatial firing patterns of hippocampal complex-spike cells in a fixed environment. *Journal of Neuroscience, 7*(7), 1935–1950.

Munn, R. G., & Bilkey, D. K. (2012). The firing rate of hippocampal CA1 place cells is modulated with a circadian period. *Hippocampus, 22*(6), 1325–1337.

Myers, C. E., & Gluck, M. A. (1994). Context, conditioning, and hippocampal rerepresentation in animal learning. *Behavioral Neuroscience, 108*(5), 835.

Nadel, L., & Moscovitch, M. (1997). Memory consolidation, retrograde amnesia and the hippocampal complex. *Current Opinion in Neurobiology, 7*(2), 217–227.

Nadel, L., & Willner, J. (1980). Context and conditioning: A place for space. *Physiological Psychology, 8,* 218–228.

Nasr, S., Liu, N., Devaney, K. J., Yue, X., Rajimehr, R., Ungerleider, L. G., & Tootell, R. B. (2011). Scene-selective cortical regions in human and nonhuman primates. *Journal of Neuroscience, 31,* 13771–13785.

Nerad, L., Liu, P., & Bilkey, D. K. (2009). Bilateral NMDA lesions centered on the postrhinal cortex have minimal effects on hippocampal place cell firing. *Hippocampus, 19,* 221–227.

Nielson, D. M., Smith, T. A., Sreekumar, V., Dennis, S., & Sederberg, P. B. (2015). Human hippocampus represents space and time during retrieval of real-world memories. *Proceedings of the National Academy of Sciences, 112,* 11078–11083.

Norman, G., & Eacott, M. J. (2005). Dissociable effects of lesions to the perirhinal cortex and the postrhinal cortex on memory for context and objects in rats. *Behavioral Neuroscience, 119,* 557.

O'Keefe, J., & Dostrovsky, J. (1971). The hippocampus as a spatial map: Preliminary evidence from unit activity in the freely-moving rat. *Brain Research, 34,* 171–175.

O'Keefe, J., & Nadel, L. (1978). *The hippocampus as a cognitive map* (Vol. 3). Oxford: Clarendon Press.

Overton, D. A. (1964). State dependent or "dissociated" learning produced with pentobarbital. *Journal of Comparative and Physiological Psychology, 57*(1), 3.

Park, S., Brady, T. F., Greene, M. R., & Oliva, A. (2011). Disentangling scene content from spatial boundary: Complementary roles for the parahippocampal place area and lateral occipital complex in representing real-world scenes. *Journal of Neuroscience, 31,* 1333–1340.

Pascalis, O., Hunkin, N. M., Bachevalier, J., & Mayes, A. R. (2009). Change in background context disrupts performance on visual paired comparison following hippocampal damage. *Neuropsychologia, 47*(10), 2107–2113.

Pearce, J. M., & Bouton, M. E. (2001). Theories of associative learning in animals. *Annual Review of Psychology, 52*(1), 111–139.

Pearce, J. M., Roberts, A. D., & Good, M. (1998). Hippocampal lesions disrupt navigation based on cognitive maps but not heading vectors. *Nature, 396*(6706), 75.

Peck, J. R., & Taube, J. S. (2017). The postrhinal cortex is not necessary for landmark control in rat head direction cells. *Hippocampus, 27,* 156–168.

Penick, S., & Solomom, P. R. (1991). Hippocampus, context, and conditioning. *Behavioral Neuroscience, 105*(5), 611.

Phillips, R. G., & LeDoux, J. E. (1992). Differential contribution of amygdala and hippocampus to cued and contextual fear conditioning. *Behavioral Neuroscience, 106*(2), 274.

Ramirez, S., Liu, X., Lin, P.-A., Suh, J., Pignatelli, M., Redondo, R. L., ... Tonegawa, S. (2013). Creating a false memory in the hippocampus. *Science, 341*(6144), 387–391.

Ranganath, C. (2010). Binding items and contexts: The cognitive neuroscience of episodic memory. *Current Directions in Psychological Science, 19*(3), 131–137.

Ranganath, C., & Ritchey, M. (2012). Two cortical systems for memory-guided behaviour. *Nature Reviews Neuroscience, 13*(10), 713.

Redish, A. D., Rosenzweig, E. S., Bohanick, J. D., McNaughton, B. L., & Barnes, C. A. (2000). Dynamics of hippocampal ensemble activity realignment: Time versus space. *Journal of Neuroscience, 20*(24), 9298–9309.

Redish, A. D., & Touretzky, D. S. (1998). The role of the hippocampus in solving the Morris water maze. *Neural Computation, 10,* 73–111.

Robin, J. (2018). Spatial scaffold effects in event memory and imagination. *Wiley Interdisciplinary Reviews: Cognitive Science.* doi: 10.1002/wcs.1462

Robin, J., Buchsbaum, B. R., & Moscovitch, M. (2018). The primacy of spatial context in the neural representation of events. *Journal of Neuroscience, 38*(11), 2755–2765.

Ross, R. S., & Slotnick, S. D. (2008). The hippocampus is preferentially associated with memory for spatial context. *Journal of Cognitive Neuroscience, 20*(3), 432–446.

Rubin, D. C., & Wenzel, A. E. (1996). One hundred years of forgetting: A quantitative description of retention. *Psychological Review, 103*(4), 734.

Rudy, J. W. (2009). Context representations, context functions, and the parahippocampal–hippocampal system. *Learning & Memory, 16*(10), 573–585.

Rudy, J. W., & O'Reilly, R. C. (1999). Contextual fear conditioning, conjunctive representations, pattern completion, and the hippocampus. *Behavioral Neuroscience, 113*(5), 867.

Rugg, M. D., & Vilberg, K. L. (2013). Brain networks underlying episodic memory retrieval. *Current Opinion in Neurobiology, 23*(2), 255–260.

Sakon, J. J., Naya, Y., Wirth, S., & Suzuki, W. A. (2014). Context-dependent incremental timing cells in the primate hippocampus. *Proceedings of the National Academy of Sciences, 111*(51), 18351–18356.

Sargolini, F., Fyhn, M., Hafting, T., McNaughton, B. L., Witter, M. P., Moser, M.-B., & Moser, E. I. (2006). Conjunctive representation of position, direction, and velocity in entorhinal cortex. *Science, 312,* 758–762.

Save, E., Nerad, L., & Poucet, B. (2000). Contribution of multiple sensory information to place field stability in hippocampal place cells. *Hippocampus, 10,* 64–76.

Schacter, D. L. (2012). Adaptive constructive processes and the future of memory. *American Psychologist, 67*(8), 603.

Schacter, D. L., Addis, D. R., & Buckner, R. L. (2007). Remembering the past to imagine the future: The prospective brain. *Nature Reviews Neuroscience, 8*(9), 657.

Schapiro, A. C., Turk-Browne, N. B., Norman, K. A., & Botvinick, M. M. (2016). Statistical learning of temporal community structure in the hippocampus. *Hippocampus, 26*(1), 3–8.

Schlesiger, M. I., Boublil, B. L., Hales, J. B., Leutgeb, J. K., & Leutgeb, S. (2018). Hippocampal global remapping can occur without input from the medial entorhinal cortex. *Cell Reports, 22*(12), 3152–3159.

Selden, N. R. W., Everitt, B. J., Jarrard, L. E., & Robbins, T. W. (1991). Complementary roles for the amygdala and hippocampus in aversive conditioning to explicit and contextual cues. *Neuroscience, 42*(2), 335–350.

Skaggs, W. E., & McNaughton, B. L. (1998). Spatial firing properties of hippocampal CA1 populations in an environment containing two visually identical regions. *Journal of Neuroscience, 18*, 8455–8466.

Smith, S. M. (1979). Remembering in and out of context. *Journal of Experimental Psychology: Human Learning and Memory, 5*(5), 460–471.

Smith, D. M., & Mizumori, S. J. (2006). Hippocampal place cells, context, and episodic memory. *Hippocampus, 16*(9), 716–729.

Smith, S. M., & Vela, E. (2001). Environmental context-dependent memory: A review and meta-analysis. *Psychonomic Bulletin & Review, 8*(2), 203–220.

Song, E. Y., Kim, Y. B., Kim, Y. H., & Jung, M. W. (2005). Role of active movement in place-specific firing of hippocampal neurons. *Hippocampus, 15*(1), 8–17.

Stachenfeld, K. L., Botvinick, M. M., & Gershman, S. J. (2017). The hippocampus as a predictive map. *Nature Neuroscience, 20*(11), 1643.

Staresina, B. P., & Davachi, L. (2009). Mind the gap: Binding experiences across space and time in the human hippocampus. *Neuron, 63*(2), 267–276.

Steemers, B., Vicente-Grabovetsky, A., Barry, C., Smulders, P., Schröder, T. N., Burgess, N., & Doeller, C. F. (2016). Hippocampal attractor dynamics predict memory-based decision making. *Current Biology, 26*, 1750–1757.

Stokes, J., Kyle, C., & Ekstrom, A. D. (2015). Complementary roles of human hippocampal subfields in differentiation and integration of spatial context. *Journal of Cognitive Neuroscience, 27*(3), 546–559.

Strand, B. Z. (1970). Change of context and retroactive inhibition. *Journal of Verbal Learning and Verbal Behavior, 9*(2), 202–206.

Takahashi, N., & Kawamura, M. (2002). Pure topographical disorientation: The anatomical basis of landmark agnosia. *Cortex, 38*, 717–725.

Takahashi, S. (2013). Hierarchical organization of context in the hippocampal episodic code. *eLife, 2*. e00321.

Tanaka, K. Z., Pevzner, A., Hamidi, A. B., Nakazawa, Y., Graham, J., & Wiltgen, B. J. (2014). Cortical representations are reinstated by the hippocampus during memory retrieval. *Neuron, 84*(2), 347–354.

Tayler, K. K., Tanaka, K. Z., Reijmers, L. G., & Wiltgen, B. J. (2013). Reactivation of neural ensembles during the retrieval of recent and remote memory. *Current Biology, 23*(2), 99–106.

Thompson, L. T., & Best, P. J. (1990). Long-term stability of the place-field activity of single units recorded from the dorsal hippocampus of freely behaving rats. *Brain Research, 509*(2), 299–308.

Troiani, V., Stigliani, A., Smith, M. E., & Epstein, R. A. (2012). Multiple object properties drive scene-selective regions. *Cerebral Cortex*, doi: 10.1093/cercor/bhs364

Tulving, E. (2002). Episodic memory: From mind to brain. *Annual Review of Psychology, 53*(1), 1–25.

Turk-Browne, N. B., Simon, M. G., & Sederberg, P. B. (2012). Scene representations in parahippocampal cortex depend on temporal context. *Journal of Neuroscience, 32*(21), 7202–7207.

Turriziani, P., Fadda, L., Caltagirone, C., & Carlesimo, G. A. (2004). Recognition memory for single items and for associations in amnesic patients. *Neuropsychologia, 42*(4), 426–433.

Wang, M. E., Yuan, R. K., Keinath, A. T., Álvarez, M. M. R., & Muzzio, I. A. (2015). Extinction of learned fear induces hippocampal place cell remapping. *Journal of Neuroscience, 35*(24), 9122–9136.

Wible, C. G., Findling, R. L., Shapiro, M., Lang, E. J., Crane, S., & Olton, D. S. (1986). Mnemonic correlates of unit activity in the hippocampus. *Brain Research, 399*(1), 97–110.

Wiener, J. M., & Mallot, H. A. (2003). "Fine-to-coarse" route planning and navigation in regionalized environments. *Spatial Cognition and Computation, 3*(4), 331–358.

Wills, T. J., Lever, C., Cacucci, F., Burgess, N., & O'Keefe, J. (2005). Attractor dynamics in the hippocampal representation of the local environment. *Science, 308*, 873–876.

Wilson, D. I., Langston, R. F., Schlesiger, M. I., Wagner, M., Watanabe, S., & Ainge, J. A. (2013). Lateral entorhinal cortex is critical for novel object-context recognition. *Hippocampus, 23*(5), 352–366.

Winocur, G., & Olds, J. (1978). Effects of context manipulation on memory and reversal learning in rats with hippocampal lesions. *Journal of Comparative and Physiological Psychology, 92*(2), 312.

Witter, M. P., & Amaral, D. G. (2004). The hippocampal region. In G. Paxinos (Ed.), *The rat nervous system*, (pp. 637–703). San Diego: Elsevier/Academic Press.

Wood, E. R., Dudchenko, P. A., Robitsek, R. J., & Eichenbaum, H. (2000). Hippocampal neurons encode information about different types of memory episodes occurring in the same location. *Neuron, 27*(3), 623–633.

Yarkoni, T., Poldrack, R. A., Nichols, T. E., Van Essen, D. C., & Wager, T. D. (2011). Large-scale automated synthesis of human functional neuroimaging data. *Nature Methods, 8*, 665–670.

Ziv, Y., Burns, L. D., Cocker, E. D., Hamel, E. O., Ghosh, K. K., Kitch, L. J., … Schnitzer, M. J. (2013). Long-term dynamics of CA1 hippocampal place codes. *Nature Neuroscience, 16*(3), 264.

20 Maps, Memories, and the Hippocampus

CHARAN RANGANATH AND ARNE D. EKSTROM

ABSTRACT Converging evidence in rats, monkeys, and humans has shown that the hippocampus plays a critical role in forming and accessing memories of past events, or episodic memory, and in spatial learning and memory-guided navigation. Several different theories have been proposed to explain how the hippocampus contributes to spatial and episodic memory. Whereas some theories suggest that the hippocampus simply stores associations between all kinds of stimuli, others emphasize the idea that spatial, temporal, and/or situational context plays a privileged role in hippocampal representations. In this chapter, we review the relevant evidence and conclude that the hippocampus is indeed disproportionately necessary for expressions of memory that require contextual information or for associations between people and things and the context in which they are encountered. Although the hippocampus does not appear to represent object information per se, in specific task contexts, it does appear to map dimensional information about task-relevant stimuli that are not obviously driven by spatial or temporal cues. To explain these findings, we propose that discrete neocortical networks compete for access to hippocampal processing and that the hippocampus maps sequences of activity states in the currently prioritized network. In novel environments, sequences of activity patterns in networks that signal spatial information will generally be prioritized, and as a result, spatiotemporal information will be prominent in hippocampal representations. However, during the processing of complex events or goal-directed cognitive tasks, the hippocampus will index sequential states across multiple networks, thereby representing multiple dimensions of experience.

Since the first studies of patient H. M. in the 1950s (Scoville & Milner 1957), there is now almost universal consensus that the human hippocampus is necessary for episodic memory and, in particular, its contextual components (i.e., "When and where did I eat dinner last night?") (Eacott & Gaffan 2005; Eichenbaum, Yonelinas, & Ranganath 2007; Ranganath 2010). Starting with the discovery of *place cells* by O'Keefe and Dostrovsky (1971), research with nonhuman animals has also highlighted the idea that the hippocampus plays a central role in the representation of, and memory for, spatial relationships. Beyond these findings, recent studies have suggested that the hippocampus might have an even broader reach than previously imagined. In this chapter we consider this evidence and propose a set of core functions that might explain hippocampal function across a wide range of domains.

Theories of Hippocampal Function and Memory: Areas of Convergence and Debate

Cognitive map theory (CMT), proposed by O'Keefe and Nadel (1978), was one of the first comprehensive theories of hippocampal function along these lines. Inspired by the work of Tolman (1948) and the first findings of place cells in rodents (O'Keefe & Dostrovsky 1971), they proposed that "the hippocampus is the core of a neural memory system providing an objective spatial framework within which the items and events of an organism's experience are located and interrelated." Although this sentence is sometimes interpreted to suggest that the hippocampus literally represents distances and angles between points in space (e.g., McNaughton, Battaglia, Jensen, Moser, & Moser 2006), CMT was actually much broader in scope. Central to the theory is that the hippocampus supports "memory for items or events within a spatio-temporal context" (p. 381). This idea was heavily influenced by Tulving's (1972) definition of episodic memory, which proposed that events occur "at a particular spatial location and in a particular temporal relation to other events that already have occurred" and that "these temporal relations amongst experienced events are also somehow represented as properties of items in the episodic memory system."

Whereas CMT (O'Keefe & Nadel 1978) and related theories (e.g., McNaughton et al. 2006) relied heavily on research on spatial processing in rodents, Neal Cohen and Howard Eichenbaum's (1993) relational memory theory (RMT) placed greater emphasis on the idea that the hippocampus plays a primary role in memory. Cohen and Eichenbaum reframed findings on the hippocampal processing of space in rodents as a subset of its more general role in supporting memory for the "capacity for relational representation, supporting both memory for relationships among perceptually distinct items and flexible expression of memories in novel contexts."

The binding of items and contexts (BIC) theory (Eacott & Gaffan 2005; Eichenbaum, Yonelinas, & Ranganath 2007) emphasized that neocortical areas are sufficient for the representation of certain kinds of relationships (i.e., between pairs of items [Haskins et al. 2008] or between features of an integrated scene context [Epstein 2008a]), whereas the hippocampus is disproportionately critical for associating information about specific items relative to a contextual framework that is specified by spatial, temporal, and situational features (Ranganath 2010). The distinction between item cues and contextual cues can be operationalized in terms of temporal stationarity (i.e., contexts are stable in time, items change more rapidly), spatial scale (i.e., contexts are large and can contain items that are small), and attentional focus (i.e., because of their temporal stability, contexts tend to be backgrounded, whereas items tend to capture attentional focus). A different, but complementary idea is the temporal context model (TCM) which proposes that the hippocampus associates incoming information about items and relatively stationary contextual elements with a neural context representation that gradually changes over time (Howard & Eichenbaum 2015). A key element of TCM is that, even when the environment and the situation are held constant, memories for items are differentiated from one another based on their relative proximity in time.

Whereas the CMT, RMT, and TCM focus on explaining what is represented by the hippocampus (i.e., space, relations, or time), models by David Marr (1971) and others (O'Reilly & Norman 2002; Rolls & Kesner 2006) propose that the hippocampus is uniquely specialized for certain computations. These models propose that sparse coding in the dentate gyrus differentiates overlapping inputs from the entorhinal cortex (*pattern separation*) and that CA3 neurons enable the network to reconstruct a previously learned pattern (*pattern completion*) from noisy or partial input. Although pattern separation and pattern completion are sometimes portrayed as opposing processes (Yassa & Stark 2011), the computational models emphasize the idea that the two processes actually work hand in hand (Norman 2010). For instance, if you need to recall where you parked your car, pattern separation enables the most recent parking event to be represented separately from previous parking events. As a result, a context cue can trigger pattern completion such that you recollect the location of your parked car. Without pattern separation, competition between different parking events would make it difficult to recover the current parking place.

Almost every model of hippocampal function described above proposes that the hippocampus is needed to support episodic memory and aspects of spatial memory, but they differ in terms of the kinds of information represented by the hippocampus. The CMT, BIC theory, and TCM propose a privileged role for the hippocampus in the representation of information about spatiotemporal context (When and where?), and the BIC theory extends this concept to include the situational context (How?). In contrast, the RMT and variants of the Marr model generally propose that the hippocampus represents information about specific items, contexts, and relationships with equal importance. Below, we consider how well these theories stack up against the extant data from humans and nonhuman animals.

TABLE 20.1 Effects of hippocampal lesions on different kinds of tasks, based on studies of humans or nonhuman animals.

The hippocampus is critical for	The hippocampus is not essential for
Context fear conditioning	Cued fear conditioning
Conditioned place preference	Pavlovian conditioning or reinforcement learning
Recollection-based recognition of words, objects, or scenes	Familiarity-based word, object, or scene recognition
Temporal order memory	Conceptual or perceptual priming
Source memory	Coarse spatial memory in humans
Trace conditioning	
Context-specific extinction of cued fear	
Water maze retention (in rodents)	
Free recall	
Place recognition and object-location associative learning in rodents	
High-precision odor and object recognition	

Space, Time, and Context Representation in the Hippocampus

There is a vast literature on the effects of hippocampal lesions of the hippocampus on memory in rodents, monkeys, and humans. Although there are some conflicting findings in the literature, it is clear that certain tasks tend to be relatively impaired and others tend to be relatively spared following hippocampal lesions (Eichenbaum, Yonelinas, & Ranganath 2007; Kesner 2018; Yonelinas et al. 2010). As summarized in table 20.1, hippocampal lesions affect memory tasks that require the representation of spatial, temporal, or situational context, as well as the association of items with contextual information and other tasks that require precise memory judgments.

Given these generalities in the literature, we can turn to the question of *how* the spatial, temporal, and situational context might be represented by the hippocampus. In rodents, the most relevant findings come from studies of place cells that fire at specific spatial locations (O'Keefe & Dostrovsky 1971) and *time cells* that fire at specific time points in a predictable sequence of events, even when the animal remains in the same location (Eichenbaum 2014). The two populations appear to overlap, in that many time cells also show spatial selectivity, and many place cells show temporal selectivity (Eichenbaum 2014). Notably, the spatial and temporal selectivity of individual place cells and/or populations can dramatically change (or *remap*) if spatial context (Muller & Kubie 1987), temporal structure of the task (Kraus et al. 2013; MacDonald et al. 2011, 2013), or currently relevant behavioral goals (Ferbinteanu & Shapiro 2003; Wood et al. 2000) are changed. Consistent with the physiology data, evidence from human fMRI studies has also shown that spatial contexts can be decoded during virtual navigation (Kyle et al. 2015), and that the position of an object in a temporal sequence (Hsieh et al. 2014) can be decoded from hippocampal activity patterns. Although time cells appear to represent relatively short timescales relative to predictable events, considerable evidence suggests that the same spatial context may be represented by different cell populations (Mankin et al. 2012; Mau et al. 2018) over the course of several days. Moreover, hippocampal ensembles can form associations between different spatial contexts that were explored in close temporal succession (Cai et al. 2016). These findings are consistent with the idea that hippocampal ensembles carry a multiplexed representation of time and space, across short and long timescales (Eichenbaum 2017).

Studies of human episodic memory support a similar conclusion. During item recognition, hippocampal activity is enhanced during the successful recollection of the spatial, temporal, or situational context (e.g., memory for the encoding task) associated with the test item (Diana, Yonelinas, & Ranganath 2007). Moreover, hippocampal activity patterns during item recognition carry information about the spatial location, task context, and temporal context in which an item was previously encountered (Bellmund et al. 2016; Jonker et al. 2018; Ritchey et al. 2015; Stokes, Kyle, and Ekstrom 2015). Hippocampal representations can carry spatial and temporal information either independently (Copara et al. 2014; Nielson et al. 2015) or in an integrated fashion (Dimsdale-Zucker et al. 2018), and the retrieval of an item can elicit the reinstatement of a temporally and contextually linked event representation in the hippocampus (Jonker et al. 2018).

In contrast to the robust coding of time, place, or situational context, object coding is relatively weak. Hippocampal activity patterns in rodents (McKenzie et al. 2014) and humans (Dimsdale-Zucker et al. 2018; Hsieh et al. 2014; Libby, Hannula, & Ranganath 2014; Libby et al. 2018; Ritchey et al. 2015) carry little information about objects in and of themselves, but they do carry information about the context associated with a particular item. One exception to this rule is that the hippocampus can encode information about how specific items vary along dimensions that are relevant to a particular task. For example, in rats trained to perform complex mappings between sounds and manual responses, hippocampal neurons formed discrete firing fields at particular sound frequencies (Aronov, Nevers, & Tank 2017). Functional imaging studies in humans have likewise indicated that the hippocampus can encode one's relative position in a social hierarchy (Tavares et al. 2015) or features that differentiate items in a category learning task (Davis, Love, & Preston 2012). These findings demonstrate that space and time are not the only dimensional variables encoded by the hippocampus (Ekstrom & Ranganath 2017; Mack, Love, & Preston 2018; Schiller et al. 2015).

Our review shows that the hippocampus represents information about spatial context (even when it is not task-relevant), sequences of experiences that form the basis for episodic memories, and information about nonspatial stimulus dimensions that are relevant in a particular task context. None of the theories proposed so far can explain all of this evidence. Perhaps a more significant shortcoming of the models described so far is that they either do not say much about the neocortical or subcortical connections of the hippocampus or they focus on a few medial temporal lobe regions known to contribute to specific tasks. In actuality, the hippocampus interacts with specific neocortical networks beyond the medial temporal lobes (Aggleton

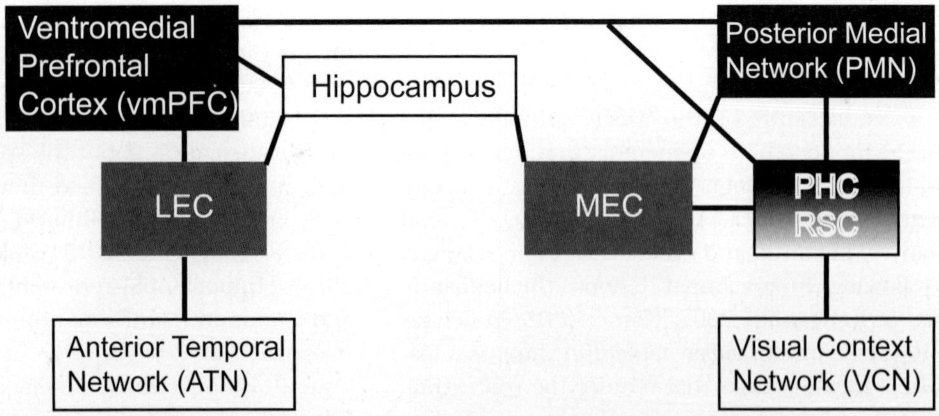

FIGURE 20.1 Schematic depiction of network-level connectivity of the hippocampus.

2011). Accordingly, we will digress a bit in the next section and briefly cover what is known about network-level connectivity in the hippocampus.

Networks That Interact with the Hippocampus

The network-level connectivity of the hippocampus has been described in many previous studies (Aggleton 2011; Kravitz et al. 2011; Nadel & Peterson 2013; Ranganath & Ritchey 2012), and we summarize this evidence, as well as the functions of different corticohippocampal networks. As shown in figure 20.1, the medial entorhinal cortex (MEC) is positioned as a hub for networks that are classically considered to provide "spatial information." Movement-based information (e.g., information about velocity or head and body position) thought to be critical for path integration is conveyed from a subcortical pathway that includes the anterior thalamus and mammillary bodies to the hippocampus, the MEC, the parahippocampal cortex (PHC), and the retrosplenial cortex (RSC; Aggleton 2011; Kahn et al. 2008; Libby et al. 2012; Maass et al. 2015; Witter et al. 2000). Sections of PHC and RSC appear to be at the apex of a hierarchical pathway along the ventral visual stream that includes medial occipital and posterior medial temporal areas. Because activity patterns in these areas are highly sensitive to characteristics of visual scenes, landmarks, and objects that have strong associations with particular spatial contexts (Epstein 2008b), we collectively refer to these areas as a *visual context network* (VCN). Other sections of PHC and RSC are more closely affiliated with a *posterior medial network* (PMN) consisting of the precuneus (i.e., medial parietal cortex), posterior cingulate, ventrolateral parietal cortex, and lateral temporal cortex. Whereas visual information appears to predominate in the VCN, processing in the PMN is not restricted to any particular modality (Baldassano et al. 2017; Bird et al. 2015; Chen et al. 2017). The PMN, like the hippocampus, is extensively engaged in spatial navigation and episodic memory. Ranganath and colleagues proposed that the PMN encodes structured knowledge (*schemas*) that specify the spatial, temporal, and causal relationships that generally apply within a particular event context (Cohn-Sheehy & Ranganath 2017; Inhoff & Ranganath 2017; Ranganath & Ritchey 2012). Similar proposals regarding a network basis for spatial navigation have also been put forth (Ekstrom & Ranganath 2017; Watrous & Ekstrom 2014). Zooming out, the MEC is at a critical juncture, positioned to provide a compressed representation of concurrent activity within the VCN and PMN (Behrens et al. 2018; Mok & Love 2018).

The lateral entorhinal cortex (LEC) is extensively interconnected with an *anterior temporal network* (ATN) that includes the perirhinal cortex, amygdala, anterior-lateral inferior temporal cortex, and ventral temporopolar cortex. As reviewed elsewhere, available evidence indicates that the ATN represents organized knowledge about the people and things that largely remain constant across events. Additionally, the LEC is directly interconnected with the ventromedial prefrontal cortex (vmPFC). Notably, vmPFC also receives a large direct projection from the hippocampus and a reciprocal projection via the nucleus reuniens of the thalamus. There is considerable evidence (Gruber et al. 2018; Navawongse & Eichenbaum 2013; Place et al. 2016; Young & Shapiro 2009) suggesting that the vmPFC, possibly via the nucleus reuniens (Ito et al. 2015), conveys input to the hippocampus and the LEC that relates particular contexts and event types (e.g., a dinner, a wedding, or a birthday party) to abstract rules that are relevant to particular goals (e.g., getting food, avoiding embarrassment, or others).

How Does the Hippocampus Map Experiences?

Based on concepts proposed in previous theories, we assert that the functions of the hippocampus emerge from a set of core principles:

1. *Intrinsic sequence generation* (Buzsáki & Tingley 2018; Levy 1996; Wallenstein, Eichenbaum, & Hasselmo 1998): Our review suggests that core functions of the hippocampus may emerge from the fact that the hippocampal cell assemblies tend to fire in sequential order over short intervals (Eichenbaum 2014) and that single-neuron coding of contextual information gradually drifts across days or longer (Mau et al. 2018). One explanation of this phenomenon is that randomly connected neural ensembles in the hippocampus may be excitable at different moments in time (Cai et al. 2016), and at the network level, this manifests as a drifting change in the neural population vector (i.e., the relative pattern of firing rates across different cells in the population at any given time). Due to Hebbian plasticity (i.e., the ability to link together inputs that come in at the same time), inputs from MEC and LEC can be rapidly associated with the currently active subset of neurons. Because overlapping sets of neurons are active across contiguous time points, cell assemblies associated with inputs at any given moment are linked synaptically with the overlapping neuronal populations associated with previous and future inputs (Levy 1996; Wallenstein, Eichenbaum, & Hasselmo 1998).
2. *Dynamic connectivity*: As reviewed above, the hippocampus, MEC, and LEC interface with multiple semimodular neocortical networks (see figure 20.1). Interactions with neocortical network "hubs" might be a mechanism for the hippocampus to preferentially emphasize any dimension of information coding (Schedlbauer et al. 2014; Zhang & Ekstrom 2013). Under some circumstances, these networks can actively interact with one another in a coordinated fashion, but under other circumstances, inputs from some networks may be prioritized at the expense of others (Inhoff & Ranganath 2017; Ranganath 2018).
3. *Prediction and error-driven learning*: MEC and LEC can be described as encoding a compressed representation of the current pattern of activity across currently prioritized networks (e.g., the conjunction of active ensembles of neurons in the VCN and PMN). Due to intrinsic sequence generation, hippocampal firing sequences enable the association of past inputs—that is, representations of activity states in currently prioritized networks—and future inputs. In other words, the hippocampus is optimized to link sequences of activity patterns in the neocortical networks that emerge as one experiences an event (Levy 1996; Wallenstein, Eichenbaum, & Hasselmo 1998). Later, if a hippocampal input—that is, an activity state in the VCN, PMN, ATN, and/or vmPFC—activates a significant subset of the cell assemblies that were part of a previously mapped experience, hippocampal pattern completion will reinstate the past activity sequence, thereby reinstating the sequence of activity states in the cortical networks activated during learning (Ranganath 2018). In other words, hippocampal ensembles take in a current pattern of activity in the prioritized network (e.g., the currently active ensemble of neurons in the PMN) and generate predictions of future states of the network (Gershman 2017; Lisman & Otmakhova 2001). When hippocampal predictions do not match up with future inputs, new information is linked synaptically to the existing ensemble via error-driven learning (Ketz, Morkonda, & O'Reilly 2013; Lisman & Otmakhova 2001). In this way, specific temporal trajectories can be updated to reflect the environment in a probabilistic manner (Gershman 2017; Stachenfeld, Botvinick, & Gershman 2017).
4. *Spatiotemporal scaffold*: We assume that, in a novel environment, the hippocampus preferentially prioritizes subcortical path integration-based information and information about environmental borders, landmarks from the VCN. These inputs, when associated with hippocampal cell sequences, enable a continuous representation of spatial and temporal context—a spatiotemporal scaffold (Ekstrom & Ranganath 2017). The spatiotemporal scaffold is a context representation or a segment of experience that can be associated with other salient inputs via Hebbian learning (in a novel environment) or error-driven learning in a familiar environment (O'Keefe & Nadel 1978). In the latter case, hippocampal context representations will be modified over time to reflect the statistical properties of a particular environment (Gershman, Blei, & Niv 2010).
5. *Context-specific coding*: As noted above, we argue that hippocampal cell assemblies generate predictions about future states in the prioritized neocortical network(s) (Ranganath 2018). If the hippocampus generates a predicted state that deviates substantially from the actual state of the

neocortex (i.e., a large prediction error), the new input should trigger a significant change in the currently active neural ensemble (*remapping*). This could occur if the input triggers the activation of a cell assembly sequence previously associated with a different context. If there is no strong match, the input is associated with an entirely new cell assembly sequence. Remapping need not be limited to changes in spatial context, however. For instance, a change in the goal state can trigger the activation of a different corticohippocampal sequence (MacDonald et al. 2013; Wood et al. 2000).

These principles can explain many aspects of spatial coding in the hippocampus. During exploration of a novel environment, the MEC integrates sequential input from the VCN—reflecting a continuous stream of incoming sensory information about environmental borders, landmarks, and visual context (O'Keefe & Nadel 1978). The sequence of MEC inputs, in turn, is associated with hippocampal cell assembly sequences—overlapping populations of neurons that are active across successive time points (Kraus et al. 2015). In other words, hippocampal cell assemblies encode a visuospatial sequence that can relate past self-motion and visual context inputs to future input conjunctions. Over the course of exploration, as one takes overlapping paths to different points in the same context (e.g., traveling to the same door from two different corners of the room), hippocampal predictions of future activity states will be violated, and error-driven learning will rapidly transform representations of specific movement sequences to representations of one's current position and upcoming positions relative to contextual boundaries.

The hippocampal representation of a broader range of experiences can be explained by accounting for neocortical networks that are likely to be prioritized in certain behavioral contexts. During a relatively novel event, the hippocampus builds an episodic memory representation by mapping the sequence of activity states across the PMN, the ATN, and the VCN. Later, the activation of a subset of the cell assemblies that mapped the past event leads to reinstatement of the previously encoded activity sequence, thereby resulting in recollection of the event as it unfolded over time (Ranganath 2018).

In addition to explaining spatial learning and episodic memory, our account is compatible with findings showing that the hippocampus (Aronov, Nevers, & Tank 2017; Davis, Love, & Preston 2012; Tavares et al. 2015) encodes dimensions of nonspatial stimuli that are task-relevant. For example, in one experiment, animals were placed in an operant-conditioning chamber and exposed to a pure auditory tone (Aronov, Nevers, & Tank 2017). The animals gradually learned to push a joystick in order to gradually raise the frequency of the tone, and they were rewarded when the tone matched a target frequency. As the animals used the joystick to approach the target sound, hippocampal cells exhibited sequential firing patterns so that different cells appeared to encode different auditory frequencies.

These findings can be explained in terms of learning-related changes in corticohippocampal indexing (Ranganath 2018; Teyler & DiScenna 1986). In a completely naïve animal, inputs from the ATN, PMN, and PFC might be mapped by hippocampal cell assemblies, but subcortical and cortical inputs about the spatial context will be relatively prioritized. Through exploration and error-driven learning, the animal acquires a corticohippocampal representation of the spatial context. Next, during training, the animal learns that it can receive rewards by manipulating the joystick, and the relative order of sound frequencies is always constant. As such, the sequence of neocortical activity states elicited as the animal manipulates the joystick will be mapped to a hippocampal cell assembly sequence. In other words, the initial spatiotemporal scaffold enables other relevant variables to be mapped to the environment. A critical prediction of our framework, however, is that a large prediction error, such as a sudden change of context or task rules, should trigger remapping so that a different or new context representation is activated and associated with the current inputs (Axmacher et al. 2010).

The critical point in these examples is that the hippocampus initially associates intrinsic cell assembly sequences with sequences of inputs from the currently prioritized cortical network. In a largely novel situation that involves exploration with body, head, or eye movements, subcortical path integration inputs and inputs from the VCN will be prioritized. As a result, the sequence of states in the VCN will be mapped to hippocampal cell assembly sequences, and over the course of learning, hippocampal neurons will signal current, past, and likely future locations—a predictive map (Stachenfeld, Botvinick, & Gershman 2017) of the environment. If salient inputs from other networks are prioritized, then state sequences from these networks would either be incorporated into the current predictive map or associated with a different map.

Concluding Remarks

Our review highlights the challenges inherent in accounting for the vast literature on hippocampal function. Most theories to date do a good job of explaining at least some of the evidence, but none appear to be completely sufficient. We suggest that a possible solution to the problem is to consider the hippocampus as a flexible hub, tracking changes in states of activity over time, both within and across neocortical networks. Although this theory is undoubtedly incomplete, we hope that it will initiate a fruitful sequence of new research studies for the field to encode in the years to come.

Acknowledgments

Charan Ranganath acknowledges funding from a Vannevar Bush Fellowship (Office of Naval Research Grant N00014-15-1-0033) and a Multi-University Research Initiative Grant (Office of Naval Research Grant N00014-17-1-2961). Arne D. Ekstrom acknowledges funding from National Institutes of Health/National Institute of Neurological Disorders and Stroke grants NS076856, NS093052 (ADE), and NSF BCS-1630296. Any opinions, findings, and conclusions or recommendations expressed in this material are those of the author(s) and do not necessarily reflect the views of the Office of Naval Research or the US Department of Defense.

REFERENCES

Aggleton, J. P. 2011. Multiple anatomical systems embedded within the primate medial temporal lobe: Implications for hippocampal function. *Neuroscience & Biobehavioral Reviews.* 36(7): 1579–1596. doi: 10.1016/j.neubiorev.2011.09.005

Aronov, D., Nevers, R., & Tank, D. W. 2017. Mapping of a nonspatial dimension by the hippocampal-entorhinal circuit. *Nature* 543(7647): 719–722.

Axmacher, N., Cohen, M. X., Fell, J., Haupt, S., Dümpelmann, M., et al. 2010. Intracranial EEG correlates of expectancy and memory formation in the human hippocampus and nucleus accumbens. *Neuron* 65(4): 541–549.

Baldassano, C., Chen, J., Zadbood, A., Pillow, J. W., Hasson, U., & Norman, K. A. 2017. Discovering event structure in continuous narrative perception and memory. *Neuron* 95(3): 709–721.e5. doi: 10.1016/j.neuron.2017.06.041

Behrens, T. E. J., Muller, T. H., Whittington, J. C. R., Mark, S., Baram, A., et al. 2018. What is a cognitive map? Organising knowledge for flexible behaviour. *Neuron* 100(2): 490–509. doi: 10.1016/j.neuron.2018.10.002

Bellmund, J. L. S., Deuker, L., Schröder, T. N., & Doeller, C. F. 2016. Grid-cell representations in mental simulation. *eLife* 5:e17089. doi: 10.7554/eLife.17089

Bird, C. M., Keidel, J. L., Ing, L. P., Horner, A. J., & Burgess, N. 2015. Consolidation of complex events via reinstatement in posterior cingulate cortex. *Journal of Neuroscience* 35(43): 14426–14434.

Buzsáki, G., & Tingley, D. 2018. Space and time: The hippocampus as a sequence generator. *Trends in Cognitive Sciences* 22(10): 853–869.

Cai, D. J., Aharoni, D., Shuman, T., Shobe, J., Biane, J., et al. 2016. A shared neural ensemble links distinct contextual memories encoded close in time. *Nature* 534(7605): 115–118.

Chen, J., Leong, Y. C., Honey, C. J., Yong, C. H., Norman, K. A., & Hasson, U. 2017. Shared memories reveal shared structure in neural activity across individuals. *Nature Neuroscience* 20(1): 115–125. doi: 10.1038/nn.4450

Cohen, N. J., & Eichenbaum, H. 1993. *Memory, amnesia, and the hippocampal system.* Cambridge, MA: MIT Press.

Cohn-Sheehy, B. I., & Ranganath, C. 2017. Time regained: How the human brain constructs memory for time. *Current Opinion in Behavioral Sciences* 17, 169–177. doi: 10.1016/j.cobeha.2017.08.005

Copara, M. S., Hassan, A. S., Kyle, C. T., Libby, L. A., Ranganath, C., & Ekstrom, A. D. 2014. Complementary roles of human hippocampal subregions during retrieval of spatiotemporal context. *Journal of Neuroscience* 34(20): 6834–6842.

Davis, T., Love, B. C., & Preston, A. R. 2012. Learning the exception to the rule: Model-based fMRI reveals specialized representations for surprising category members. *Cerebral Cortex* 22(2): 260–273.

Diana, R. A., Yonelinas, A. P., & Ranganath, C. 2007. Imaging recollection and familiarity in the medial temporal lobe: A three-component model. *Trends in Cognitive Sciences* 11(9): 379–386.

Dimsdale-Zucker, H. R., Ritchey, M., Ekstrom, A. D., Yonelinas, A. P., & Ranganath, C. 2018. CA1 and CA3 differentially support spontaneous retrieval of episodic contexts within human hippocampal subfields. *Nature Communications*, 9(1): 294.

Eacott, M. J., & Gaffan, E. A. 2005. The roles of perirhinal cortex, postrhinal cortex, and the fornix in memory for objects, contexts, and events in the rat. *Quarterly Journal of Experimental Psychology Section B* 58(3–4): 202–217.

Eichenbaum, H. 2014. Time cells in the hippocampus: A new dimension for mapping memories. *Nature Reviews Neuroscience* 15(11): 732–744.

Eichenbaum, H. 2017. On the integration of space, time, and memory. *Neuron* 95(5): 1007–1018. doi: 10.1016/j.neuron.2017.06.036

Eichenbaum, H., Yonelinas, A. P., & Ranganath, C. 2007. The medial temporal lobe and recognition memory. *Annual Review of Neuroscience* 30:123–152.

Ekstrom, A. D., & Ranganath, C. 2017. Space, time, and episodic memory: The hippocampus is all over the cognitive map. *Hippocampus* 28(9): 680–687. doi: 10.1002/hipo.22750

Epstein, R. A. 2008a. Parahippocampal and retrosplenial contributions to human spatial navigation. *Trends in Cognitive Sciences* 12(10): 388–396. doi: 10.1016/j.tics.2008.07.004

Epstein, R. A. 2008b. Parahippocampal and retrosplenial contributions to human spatial navigation. *Trends in Cognitive Sciences* 12(10): 388–396.

Ferbinteanu, J., & Shapiro, M. L. 2003. Prospective and retrospective memory coding in the hippocampus. *Neuron* 40(6): 1227–1239.

Gershman, S. J. 2017. Predicting the past, remembering the future. *Current Opinion in Behavioral Sciences* 17, 7–13.

Gershman, S. J, Blei, D. M., & Niv, Y. 2010. Context, learning, and extinction. *Psychological Review* 117(1): 197–209. doi: 10.1037/a0017808

Gruber, M. J., Hsieh, L. T., Staresina, B. P., Elger, C. E., Fell, J., et al. 2018. Theta phase synchronization between the human hippocampus and prefrontal cortex increases during encoding of unexpected information: A case study. *Journal of Cognitive Neuroscience* 30(11): 1646–1656.

Haskins, A. L., Yonelinas, A. P., Quamme, J. R., & Ranganath, C. 2008. Perirhinal cortex supports encoding and familiarity-based recognition of novel associations. *Neuron* 59(4): 554–560. doi: 10.1016/j.neuron.2008.07.035

Howard, M. W., & Eichenbaum, H. 2015. Time and space in the hippocampus. *Brain Research* 1621: 345–354.

Hsieh, L. T., Gruber, M. J., Jenkins, L. J., & Ranganath, C. 2014. Hippocampal activity patterns carry information about objects in temporal context. *Neuron* 81(5): 1165–1178.

Inhoff, M. C., & Ranganath, C. 2017. Dynamic cortico-hippocampal networks underlying memory and cognition: The PMAT framework. In D. E. Hannula & M. C. Duff (Eds.), *The hippocampus from cells to systems: Structure, connectivity, and functional contributions to memory and flexible cognition* (pp. 559–589). Cham, Switzerland: Springer International.

Ito, H. T., Zhang, S. J., Witter, M. P., Moser, E. I., & Moser, M. B. 2015. A prefrontal-thalamo-hippocampal circuit for goal-directed spatial navigation. *Nature* 522(7554): 50–55. doi: 10.1038/nature14396

Jonker, T. R., Dimsdale-Zucker, H., Ritchey, M., Clarke, A., & Ranganath, C. 2018. Neural reactivation in parietal cortex enhances memory for episodically linked information. *Proceedings of the National Academy of Sciences* 115(43): 11084–11089. doi: 10.1073/pnas

Kahn, I., Andrews-Hanna, J. R., Vincent, J. L., Snyder, A. Z., & Buckner, R. L. 2008. Distinct cortical anatomy linked to subregions of the medial temporal lobe revealed by intrinsic functional connectivity. *Journal of Neurophysiology* 100(1): 129–139.

Kesner, R. P. 2018. Exploration of the neurobiological basis for a three-system, multiattribute model of memory. *Current Topics in Behavioral Neurosciences* 37, 325–359.

Ketz, N., Morkonda, S. G., & O'Reilly, R. C. 2013. Theta coordinated error-driven learning in the hippocampus. *PLoS Computational Biology* 9(6): e1003067.

Kraus, B. J., Brandon, M. P., Robinson, R. J., Connerney, M. A., Hasselmo, M. E., & Eichenbaum, H. 2015. During running in place, grid cells integrate elapsed time and distance run. *Neuron* 88(3): 578–589. doi: 10.1016/j.neuron.2015.09.031

Kraus, B. J., Robinson, R. J., White, J. A., Eichenbaum, H., & Hasselmo, M. E. 2013. Hippocampal "time cells": Time versus path integration. *Neuron* 78(6): 1090–1101. doi: 10.1016/j.neuron.2013.04.015

Kravitz, D. J., Saleem, K. S., Baker, C. I., & Mishkin, M. 2011. A new neural framework for visuospatial processing. *Nature Reviews Neuroscience* 12(4): 217–230.

Kyle, C. T., Stokes, J. D., Lieberman, J. S., Hassan, A. S., & Ekstrom, A. D. 2015. Successful retrieval of competing spatial environments in humans involves hippocampal pattern separation mechanisms. *eLife* 4(November).

Levy, W. B. 1996. A sequence predicting CA3 is a flexible associator that learns and uses context to solve hippocampal-like tasks. *Hippocampus* 6(6): 579–590.

Libby, L. A., Ekstrom, A. D., Ragland, J. D., & Ranganath, C. 2012. Differential connectivity of perirhinal and parahippocampal cortices within human hippocampal subregions revealed by high-resolution functional imaging. *Journal of Neuroscience* 32(19): 6550–6560.

Libby, L. A., Hannula, D. E., & Ranganath, C. 2014. Medial temporal lobe coding of item and spatial information during relational binding in working memory. *Journal of Neuroscience* 34(43): 14233–14242.

Libby, L. A., Reagh, Z. M., Bouffard, N. R., Ragland, J. D., & Ranganath, C. 2018. The hippocampus generalizes across memories that share item and context information. *Journal of Cognitive Neuroscience*, 21, 1–12. doi: 10.1162/jocn_a_01345

Lisman, J. E., & Otmakhova, N. A. 2001. Storage, recall, and novelty detection of sequences by the hippocampus: Elaborating on the SOCRATIC model to account for normal and aberrant effects of dopamine. *Hippocampus* 11(5): 551–568.

Maass, A., Berron, D., Libby, L., Ranganath, C., & Düzel, E. 2015. Functional subregions of the human entorhinal cortex. *eLife* 4:1–20.

MacDonald, C. J., Carrow, S., Place, R., & Eichenbaum, H. 2013. Distinct hippocampal time cell sequences represent odor memories in immobilized rats. *Journal of Neuroscience* 33(36): 14607–14616.

MacDonald, C. J., Lepage, K. Q., Eden, U. T., & Eichenbaum, H. 2011. Hippocampal "time cells" bridge the gap in memory for discontiguous events. *Neuron* 71(4): 737–749.

Mack, M. L., Love, B. C., & Preston, A. R. 2018. Building concepts one episode at a time: The hippocampus and concept formation. *Neuroscience Letters* 680: 31–38.

Mankin, E. A., Sparks, F. T., Slayyeh, B., Sutherland, R. J., Leutgeb, S., & Leutgeb, J. K. 2012. Neuronal code for extended time in the hippocampus. *Proceedings of the National Academy of Sciences of the United States of America* 109(47): 19462–19467.

Marr, D. 1971. Simple memory: A theory for archicortex. *Philosophical Transactions of the Royal Society of London* 262:23–81.

Mau, W., Sullivan, D. W., Kinsky, N. R., Hasselmo, M. E., Howard, M. W., & Eichenbaum, H. 2018. The same hippocampal CA1 population simultaneously codes temporal information over multiple timescales. *Current Biology* 28(10): 1499–1508.e4.

McKenzie, S., Frank, A. J., Kinsky, N. R., Porter, B., Riviere, P. D., & Eichenbaum, H. 2014. Hippocampal representation of related and opposing memories develop within distinct, hierarchically organized neural schemas. *Neuron* 83(1): 202–215.

McNaughton, B. L., Battaglia, F. P., Jensen, O., Moser, E. I., & Moser, M. B. 2006. Path integration and the neural basis of the "cognitive map." *Nature Reviews Neuroscience* 7(8): 663–678.

Mok, R. M., & Love, B. C. 2018. A non-spatial account of place and grid cells based on clustering models of concept learning. *bioRxiv*. doi: 10.1101/421842

Muller, R. U., & Kubie, J. L. 1987. The effects of changes in the environment on the spatial firing of hippocampal complex-spike cells. *Journal of Neuroscience* 7(7): 1951–1968.

Nadel, L., & Peterson, M. A. 2013. The hippocampus: Part of an interactive posterior representational system spanning perceptual and memorial systems. *Journal of Experimental Psychology: General* 142(4): 1242–1254.

Navawongse, R., & Eichenbaum, H. 2013. Distinct pathways for rule-based retrieval and spatial mapping of memory representations in hippocampal neurons. *Journal of Neuroscience* 33(3): 1002–1013.

Nielson, D. M., Smith, T. A., Sreekumar, V., Dennis, S., & Sederberg, P. B. 2015. Human hippocampus represents space and time during retrieval of real-world memories. *Proceedings of the National Academy of Sciences of the United States of America* 112(35): 11078–11083. doi: 10.1073/pnas.1507104112

Norman, K. A. 2010. How hippocampus and cortex contribute to recognition memory: Revisiting the complementary learning systems model. *Hippocampus* 20(11): 1217–1227.

O'Keefe, J., & Dostrovsky, J. 1971. The hippocampus as a spatial map: Preliminary evidence from unit activity in the freely-moving rat. *Brain Research* 34(1): 171–175.

O'Keefe, J., & Nadel, L. 1978. *The hippocampus as a cognitive map*. Oxford: Oxford University Press.

O'Reilly, R. C., & Norman, K. A. 2002. Hippocampal and neocortical contributions to memory: Advances in the complementary learning systems framework. *Trends in Cognitive Sciences* 6(12): 505–510.

Place, R., Farovik, A., Brockmann, M., & Eichenbaum, H. 2016. Bidirectional prefrontal-hippocampal interactions support context-guided memory. *Nature Neuroscience* 19(8): 992–994.

Ranganath, C. 2010. A unified framework for the functional organization of the medial temporal lobes and the phenomenology of episodic memory. *Hippocampus* 20(11): 1263–1290.

Ranganath, C. 2018. Time, memory, and the legacy of Howard Eichenbaum. *Hippocampus* 29(3): 146–161.

Ranganath, C, & Ritchey, M. 2012. Two cortical systems for memory-guided behavior. *Nature Reviews Neuroscience* 13(10): 713–726.

Ritchey, M., Montchal, M. E., Yonelinas, A. P., & Ranganath, C. 2015. Delay-dependent contributions of medial temporal lobe regions to episodic memory retrieval. *eLife* 13;4. doi: 10.7554/eLife.05025

Rolls, E. T., & Kesner, R. P. 2006. A computational theory of hippocampal function, and empirical tests of the theory. *Progress in Neurobiology* 79(1): 1–48.

Schedlbauer, A. M., Copara, M. S., Watrous, A. J., & Ekstrom, A. D. 2014. Multiple interacting brain areas underlie successful spatiotemporal memory retrieval in humans. *Scientific Reports* 4: 6431.

Schiller, D., Eichenbaum, H., Buffalo, E. A., Davachi, L., Foster, D. J., et al. 2015. Memory and space: Towards an understanding of the cognitive map. *Journal of Neuroscience* 35(41): 13904–13911.

Scoville, W. B., & Milner, B. 1957. Loss of recent memory after bilateral hippocampal lesions. *Journal of Neurology, Neurosurgery, and Psychiatry* 20:11–21.

Stachenfeld, K. L., Botvinick, M. M., & Gershman S. J. 2017. The hippocampus as a predictive map. *Nature Neuroscience* 20, 1643–1653.

Stokes, J., Kyle, C., & Ekstrom, A. D. 2015. Complementary roles of human hippocampal subfields in differentiation and integration of spatial context. *Journal of Cognitive Neuroscience* 27(3): 546–559.

Tavares, R. M., Mendelsohn, A., Grossman, Y., Williams, C. H., Shapiro M., et al. 2015. A map for social navigation in the human brain. *Neuron* 87(1): 231–243.

Teyler, T. J., & DiScenna, P. 1986. The hippocampal memory indexing theory. *Behavioral Neuroscience* 100(2): 147–154.

Tolman, E. C. 1948. Cognitive maps in rats and men. *Psychological Review* 55, 189–208.

Tulving, E. 1972. Episodic and semantic memory. In E. Tulving & W. Donaldson (Eds.), *Organization of memory* (pp. 382–402). New York: Academic Press.

Wallenstein, G. V., Eichenbaum, H., & Hasselmo, M. E. 1998. The hippocampus as an associator of discontiguous events. *Trends in Neurosciences* 21(8): 317–323.

Watrous, A. J., & Ekstrom, A. D. 2014. The spectro-contextual encoding and retrieval theory of episodic memory. *Frontiers in Human Neuroscience* 8: 75. doi: 10.3389/fnhum.2014.00075

Witter, M. P., Wouterlood, F. G., Naber, P. A., & Van Haeften, T. 2000. Anatomical organization of the parahippocampal-hippocampal network. *Annals of the New York Academy of Sciences* 911: 1–24.

Wood, E. R., Dudchenko, P. A., Robitsek, R. J., & Eichenbaum, H. 2000. Hippocampal neurons encode information about different types of memory episodes occurring in the same location. *Neuron* 27(3): 623–633.

Yassa, M. A., & Stark, C. E. 2011. Pattern separation in the hippocampus. *Trends in Neurosciences* 34(10): 515–525.

Yonelinas, A. P., Aly, M., Wang, W. C., & Koen, J. D. 2010. Recollection and familiarity: Examining controversial assumptions and new directions. *Hippocampus* 20(11): 1178–1194.

Young, J. J., & Shapiro, M. L. 2009. Double dissociation and hierarchical organization of strategy switches and reversals in the rat PFC. *Behavioral Neuroscience* 123(5): 1028–1035.

Zhang, H., & Ekstrom, A. 2013. Human neural systems underlying rigid and flexible forms of allocentric spatial representation. *Human Brain Mapping* 34(5): 1070–1087.

21 Memory across Development, with Insights from Emotional Learning: A Nonlinear Process

HEIDI C. MEYER AND SIOBHAN S. PATTWELL

ABSTRACT While many traits associated with normative development traverse via a linear trajectory, properties of learning and memory, particularly emotional learning, exhibit dynamic changes across the lifespan. Nonlinear changes associated with the capacity for both aversive and appetitive learning are associated with underlying changes in the neural circuitry regulating these unique types of memories. By studying the behavioral and neural changes across development as they relate to both fear learning and memory, as well as appetitive learning and memory, insight can be gained into typical neurodevelopment, as well as atypical changes associated with psychiatric disorders and psychopathology unique to particular age groups, such as children or adolescents. Here, we review the neural circuits and behavioral manifestations associated with emotionally salient learning and memory tasks across development to provide a context for better understanding the brain under both normative and atypical trajectories.

The understanding of learning and memory remains one of the central goals of modern neuroscience. The study of emotional memory, in particular, has garnered significant interest in recent years for its inherent role in various psychiatric disorders. The dysregulation of emotional memory systems is a principle component in many affective disorders, including depression, specific phobias, generalized anxiety disorder, agoraphobia, and post-traumatic stress disorder (PTSD). Specifically, alterations in memory processing for aversive or traumatic experiences lie at the heart of many clinical psychiatric disorders, which often trace their roots to the early childhood and adolescent years. Reinforcement-processing abnormalities have also been implicated in a variety of psychiatric disorders and are linked to drastic and long-term effects on behavior. Indeed, blunted signaling in reinforcement-related brain regions is apparent in major depression (e.g., anhedonia) and the negative symptoms of schizophrenia, while elevated signaling manifests during manic episodes in bipolar disorder. By studying the neural circuitry of emotional memory, insight can be gained into not only how these systems function normally but also how they may go awry in the case of psychiatric disorders.

The focus of this chapter is to provide an overview of the neurobiological substrates underlying emotional learning and memory across development throughout early life. By exploring the behavioral, neural, and molecular properties of both aversive and appetitive learning as a function of age, this chapter highlights the recent developmental advances of memory systems implicated in emotional psychopathology.

Aversive Learning and Memory

Under normal circumstances, fear learning is an adaptive, evolutionarily conserved process that allows one to respond appropriately to cues predictive of danger. In the case of psychiatric disorders, however, fear may persist long after a threat has passed. This unremitting fear is a core component of many anxiety disorders, including PTSD, and often involves exaggerated or inappropriate fear responses, as well as a lack of reappraisal once a stimulus switches from a cue of threat to a cue of safety. It is estimated that 18.1% of Americans, or 40 million people, are living with a diagnosable anxiety disorder, accounting for nearly US$58 billion in health-care costs (AHRQ/NIMH, 2006; Merikangas et al., 2011). Globally, the World Health Organization (WHO) estimates that more than 260 million people suffer from anxiety disorders, and along with depression, anxiety disorders are estimated to cost the global economy over US$1 trillion per year in productivity loss.

Experimental methods for studying aversive memory Behavioral paradigms relying on Pavlovian principles have become standard for studying fear in both humans and nonhuman animals (Pavlov, 1927). Through associative learning techniques based on these classical-conditioning principles, long-lasting, aversive memories can be formed in the rodent (Maren, 2001), and animal models of fear learning are frequently relied on and held in high regard due to their ease and experimental control. Adult studies have exploited the finely tuned adult brain to identify key regions in fear memory acquisition, retrieval, expression, extinction, and erasure, including the medial prefrontal cortex (PFC), amygdala, and hippocampus (Krabbe, Grundemann, & Luthi, 2018; Maren & Quirk, 2004; Sotres-Bayon & Quirk, 2010).

Developmental influences on fear learning and memory While existing therapies and medications offer significant benefit to adult patients, a comparative knowledge gap surrounding the dynamic fear neural circuitry across early development may prohibit similarly successful treatment outcomes in children and adolescents (Liberman, Lipp, Spence, & March, 2006). It is estimated that 10–20% of the world's 2.2 billion children and adolescents suffer from neuropsychiatric disorders, highlighting the need to further tease apart how the neurobiological substrates of emotional memory change across development (Kieling et al., 2011).

Infant and juvenile fear memories Studies investigating aversive learning in infants and juveniles have uncovered key developmental windows involving both critical and sensitive periods (Marin, 2016). For clarity, a critical period is associated with molecular or genetic brakes/accelerators and is defined as a time of extreme interdependence between experience and development, after which there is a *decrease* in neural plasticity. These resultant behavioral changes are typically irreversible, as is seen with amblyopia of the visual system (Nabel & Morishita, 2013). Conversely, a sensitive period is a window during which a functional process and its underlying brain circuit temporarily experience *heightened* plasticity. Neural development is especially receptive to particular types of experience during this time (Nabel & Morishita, 2013).

Fear learning in rodents emerges very early in postnatal development and coincides with amygdala maturation. During this early developmental window (within 10 postnatal days [P10]), rodents develop a seemingly paradoxical Pavlovian fear response to odor/tone shock pairings (Camp & Rudy, 1988; Sullivan, Landers, Yeaman, & Wilson, 2000) during a sensitive period for attachment learning, in which maternal presence serves to block the acquisition of fear (Landers & Sullivan, 2012). Coinciding with the onset of learning-induced synaptic plasticity in the amygdala after P10, rodents begin to exhibit more traditional cued fear learning to odor-shock pairs (Thompson, Sullivan, & Wilson, 2008), yet this can be modified by maternal presence up until about P15 (Moriceau & Sullivan, 2006). Fear memories acquired prior to P10 are not as robust or as persistent as those acquired later in life and remain susceptible to forgetting through a process known as infantile amnesia, which is highly influenced by exposure to early life stress, such as maternal deprivation (Alberini & Travaglia, 2017; Callaghan & Richardson, 2012; Campbell & Spear, 1972; Kim & Richardson, 2007; Pattwell & Bath, 2017). Contextual fear conditioning in rodents emerges later (P23 in rats) than cued fear learning (P18 in rats) (Akers, Arruda-Carvalho, Josselyn, & Frankland, 2012; Rudy, 1993), which may reflect the maturation of hippocampal-amygdala connectivity or hippocampal activity (Raineki et al., 2010) and can be dissociated from cued fear learning in infant-juvenile rodents.

In addition to new learning associated with fear conditioning, the capacity for fear extinction learning also changes across early juvenile periods. Prior to P24 (circa weaning age), rodent pups display a normal decrease in fear expression when undergoing classical extinction paradigms, yet this learning differs from that of the adult, as the fear neither reemerges with reinstatement or renewal nor exhibits a spontaneous recovery, which is potentially indicative of infantile amnesia (Gogolla, Caroni, Luthi, & Herry, 2009; Kim, Hamlin, & Richardson, 2009; Yap & Richardson, 2007), although notable differences in female rats have been observed (Park, Ganella, & Kim, 2017). The closing of this window for memory erasure coincides with changes in extracellular matrix chondroitin sulfate proteoglycans within the juvenile amygdala, after which fear memories are protected from erasure by perineuronal nets (PNNs), representing a critical period in fear memory across development associated with structural changes and altered gamma-aminobutyric acid (GABAergic) signaling.

Adolescent fear memories Adolescence, in particular, is a period of increased prevalence of emotional psychopathology (Monk et al., 2003), and it is estimated that over 75% of adults with fear-related disorders met diagnostic criteria as children and adolescents (Kim-Cohen et al., 2003; Pollack et al., 1996), yet fewer than one in five children or adolescents are estimated to receive treatment for their anxiety disorders (Merikangas et al., 2010). Adolescence also coincides with a period of significant cortical rearrangement that is normatively accompanied by drastic cognitive and behavioral changes (Spear, 2000). Longitudinal studies of brain maturation illustrate a nonlinear process that is not complete until early adulthood (Giedd et al., 1999; Gogtay et al., 2004), with regionally specific, age-dependent, linear increases in white matter and nonlinear increases in gray matter indicative of increased axonal myelination and synaptic pruning. Prefrontal cortical regions, such as those implicated in top-down control, response inhibition, executive function, and fear extinction learning, undergo protracted development relative to subcortical structures, including the amygdala (Casey, Jones, & Somerville, 2011; Casey, Glatt, & Lee, 2015; Casey, Tottenham, Liston, & Durston, 2005). During tasks involving self-regulation and reappraisal, children show a greater and more diffuse activation of prefrontal loci compared to adults, suggestive of regional immaturity (Galvan et al., 2006; Levesque et al., 2004). It is of clinical interest to examine whether diffuse patterns of PFC activity, observed in adolescents during tasks requiring the control of subcortical structures, will also influence the precise interactions between inhibitory and excitatory hippocampal-prefrontal-amygdala circuits during fear regulation. Regardless of the type of task being performed, healthy adolescent humans display increased activity in frontal amygdala circuits, which may alter the balance in the excitation and inhibition of finely tuned glutamatergic/GABAergic bidirectional projections to the amygdala (Monk et al., 2003). Converging evidence from human and rodent studies suggests that insufficient top-down regulation of subcortical structures, such as the amygdala, may coincide with impairments in prototypical extinction learning. In addition, recent work highlights distinct patterns of amygdaloid and medial temporal lobe activation between children and adolescents when learning about neutral versus fearful faces (Pinabiaux et al., 2013).

Sensitive periods and critical periods have been the focus of infant and juvenile models for some time, yet throughout the past decade, rodent models have started incorporating the older, more intermediate adolescent ages between P23 and P42 (Hefner & Holmes, 2007; J. H. Kim, Li, & Richardson, 2011; McCallum, Kim, & Richardson, 2010; Pattwell, Bath, Casey, Ninan, & Lee, 2011; Pattwell et al., 2012; Pattwell et al., 2016; Shen et al., 2010). By examining fear conditioning as mice transitioned through adolescence, recent research has uncovered an aspect of fear learning in which contextual fear expression is suppressed during adolescence (Pattwell et al., 2011). This lack of contextual fear expression did not result from global impairments in fear memory acquisition or consolidation, as amygdala-dependent cued fear remained intact at all developmental ages examined and correlated with electrophysiological recordings in their amygdalae. Interestingly, despite a suppression of contextual fear expression and corresponding blunted synaptic activity in the basal amygdala and hippocampus during adolescence, mice were able to retrieve and express the contextual fear memory as they transitioned out of adolescence and into adulthood. This transition occurred in concordance with a delayed increase in basal amygdala synaptic potentiation as measured by field excitatory postsynaptic potentials (fEPSPs), highlighting the importance of this developmental transition on behavioral, neural, and molecular outcomes. Despite a lack of contextual fear expression, mice given contextual extinction during this adolescent window did not exhibit the fear later as adults, suggesting prophylactic extinction—when behavior was otherwise absent—may prevent fear memory expression in adulthood (Pattwell et al., 2011). Despite the suppression of contextual fear expression in adolescent mice, cued fear expression appeared to be not only enhanced but also highly resistant to extinction in both adolescent rodents and humans (Drysdale et al., 2014; Johnson & Casey, 2015; McCallum, Kim, & Richardson, 2010; Pattwell et al., 2012).

The period of diminished capacity for cue-specific extinction learning coincides with a time when the PFC is undergoing maturational changes in the dynamic interaction between the ventromedial PFC and the amygdala (Gee et al., 2013) and correlates with blunted infralimbic (IL) activity in rodents on fear extinction tasks (Cruz et al., 2015; Pattwell et al., 2012). Converging evidence from human and rodent studies suggests that insufficient top-down regulation of subcortical structures (Casey et al., 2010), such as the amygdala, may coincide with impairments in prototypical extinction learning. Studies utilizing retrograde tracers revealed enhanced structural connectivity between the ventral hippocampus and the prelimbic (PL) cortex during adolescence, compared to juvenile and adult mice, and this surge is to the PL (Pattwell et al., 2016), while two-photon imaging of medial PFC shows a surge in the formation of excitatory postsynaptic dendritic spines in the medial PFC occurring during adolescence. Dense populations of PL-projecting cell bodies within the basolateral amygdala also significantly increased from the juvenile period to adolescence and subsequently decreased by adulthood, which may maintain a positive feedback loop during enhanced extinction-resistant cued fear expression. The optogenetic examination of PFC-amygdala circuitry across development also revealed an adolescent surge in feedforward inhibition with increased spontaneous inhibitory currents in excitatory neurons (Arruda-Carvalho, Wu, Cummings, & Clem, 2017). Given the well-established role of hippocampal-PL inputs for suppressing fear expression and a surge in vCA1-PL connectivity, studies designed for maximally targeting the contextual component of a prior conditioned fear showed that combinatorial context-cue extinction sessions offered significant benefits over cued extinction alone during this adolescent sensitive period (Pattwell et al., 2016). See figure 21.1*A* and 21.1*B*

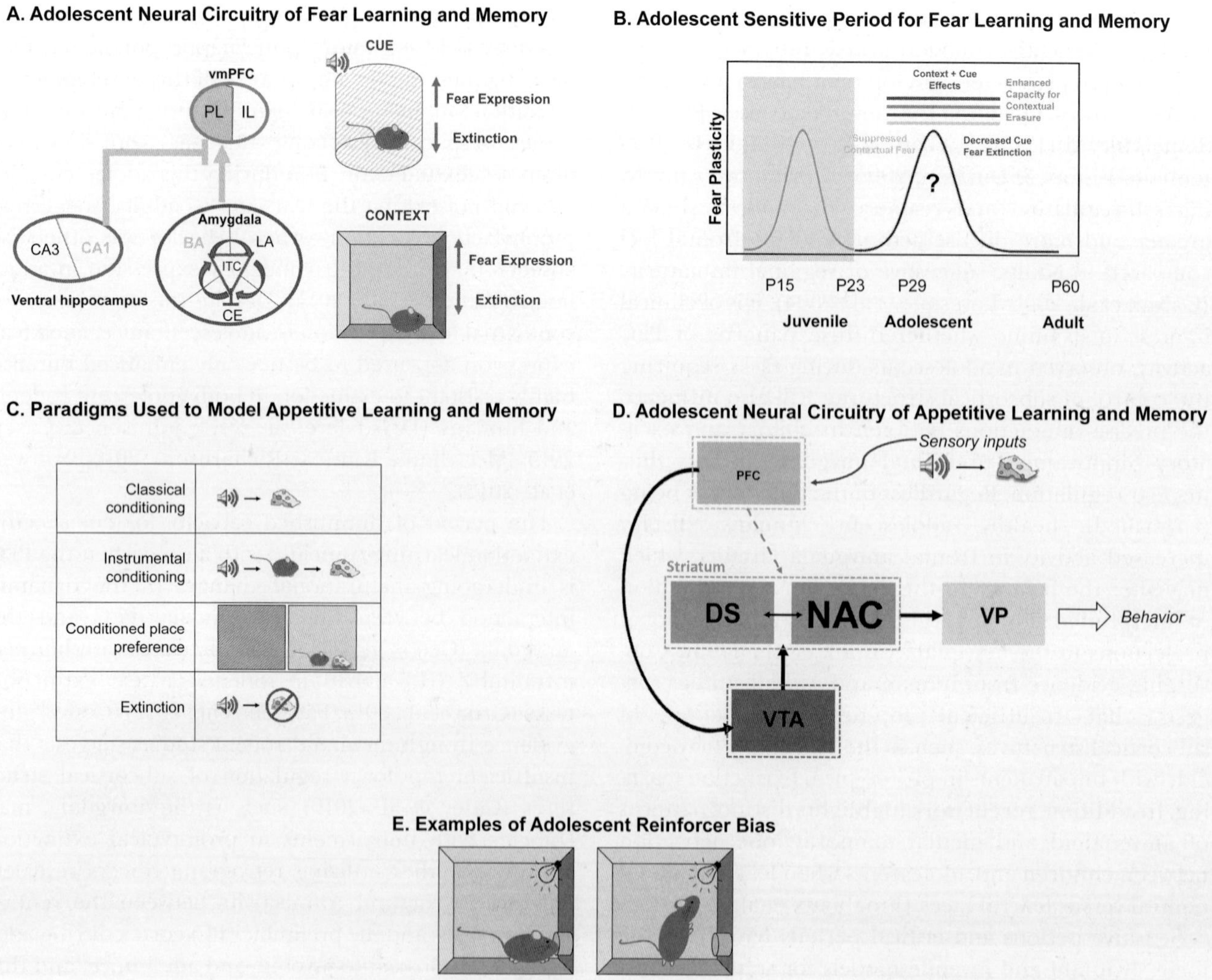

FIGURE 21.1 Emotional memory-formation processes during adolescence. *A*, A schematic of the neural circuitry of adolescent cued fear as simplified from retrograde tracer studies (Pattwell et al., 2016) shows an adolescent surge in connectivity between vCA1 and PL, as well as BA and PL. *Abbreviations*: basal amygdala, BA; central amygdala, CE; infralimbic, IL; intercalated cells, ITC; lateral amygdala, LA; prelimbic, PL. *B*, Developmental sensitive periods for fear learning and memory—insights into adolescent fear memory and behavior. *C*, Paradigms used to model appetitive learning and memory. *D*, A schematic of the neural circuitry of appetitive memory during adolescence. Line weight and font size indicate relative contributions to appetitive memory. Dashed lines indicate notable differences from adult circuitry. *Abbreviations*: dorsal striatum, DS; nucleus accumbens, NAC; prefrontal cortex, PFC; ventral pallidum, VP; ventral tegmental area, VTA. *E*, Appetitive memory strength during adolescence may be influenced by a higher salience of reinforcers, a higher salience of reinforcer-associated cues, or a combination of both. *Left*, An adolescent rodent perseverates on the delivery of a reinforcer by spending more time in the reinforcer receptacle. *Right*, An adolescent rodent perseverates on a visual stimulus associated with reinforcer delivery. (See color plate 23.)

for a summary of adolescent retrograde tracer findings and corresponding sensitive periods of fear behavior.

Of particular importance for the vulnerable adolescent age group are the deleterious effects that psychiatric disorders can have on social and academic contexts (Ginsburg, La Greca, & Silverman, 1998), when peer relationships are paramount, as well as the enhanced potential for persisting disorders in adulthood (Foulkes & Blakemore, 2018). As adolescence is also a time associated with prototypical increases in risky behavior, stress, thrill seeking, impulsivity, and heightened reward sensitivity, seeking more effective treatments for anxiety and affective disorders in this population may also indirectly lead to reductions in substance abuse and the other maladaptive behaviors often employed as forms of anxiolytic self-medication.

Appetitive Learning and Memory

The core purpose of fear learning and memory is to facilitate the avoidance of aversive outcomes. In contrast, appetitive learning and memory provide information about the reinforcement-predictive properties of a cue, as well as the circumstances that modulate these properties. In turn, this facilitates the fine-tuning of behavioral patterns that will maximize the opportunity for an appetitive outcome (i.e., a reinforcer).

Early in development, appetitive memory is critical for the ability to establish beneficial social networks, initially with caregivers and later with peers. Subsequently, an elevated focus on appetitive stimuli and outcomes can contribute to enhanced learning and flexibility (McCormick & Telzer, 2017). This is particularly important in late childhood (i.e., the juvenile stage in a rodent) and throughout adolescence, as an individual encounters novel settings and situations during the transition to independence from the caregiver and home environment.

Unfortunately, the pursuit of appetitive outcomes can in some cases lead to risky and impulsive behaviors that increase the possibility of harm or even premature death. Moreover, the altered processing of reinforcement has been implicated in a variety of clinical psychiatric disorders, many of which emerge during development, and has been associated with an increased vulnerability to substance use and abuse (Cardinal & Everitt, 2004; Chambers, Taylor, & Potenza, 2003). Thus, an understanding of how appetitive memories are encoded will inform the underpinnings of goal-directed behavior, reveal how a disruption of this process can manifest in psychiatric disorders, and further advise psychiatric treatments as well as interventions for pathological reinforcer-seeking behaviors.

Experimental methods for studying appetitive memory In the laboratory, appetitive conditioning, not unlike fear conditioning, trained through repeated pairings of an initially neutral stimulus with an appetitive outcome will provide value to an initially neutral cue in the environment, thus increasing the salience of the cue (figure 21.1*C*). In turn, the salience of a cue is included in the information encoded about the cue and upon subsequent recall can be used to guide behavior. Quantifiable measures of the strength of the reinforcing properties include the number of head entries *during the cue* (i.e., preceding the actual reinforcer delivery, a Pavlovian measure) into a port where the reinforcer is delivered or the increased performance of behavioral response over time as it is learned that this will, in many cases, increase the total amount of reinforcer (an instrumental measure). The strength of the appetitive memory can also be measured by how long it takes to update the memory once the cue is no longer paired with reinforcement (i.e., extinction). Appetitive-conditioning processes can also be applied to diffuse contexts, rather than discrete cues, when the presence of a reinforcer in a given context results in a preference for that context relative to a similar context in which no reinforcer has been presented.

Developmental influences on appetitive memory in infancy One of the earliest examples of appetitive memory in development is the attachment to a caregiver. This attachment promotes the survival of an infant by facilitating access to resources and protection (Bowlby, 1969). Neonatal mice as young as P3 can form an appetitive memory for an odor predicting access to the mother (Armstrong, DeVito, & Cleland, 2006). Similarly, rat pups exhibit learned preferences for odors paired with tactile stimulation comparable to that received from the dam (Sullivan & Leon, 1987). To date, no cortical regions for attachment have been found in the neonatal mammalian brain.

In humans, infants are capable of encoding appetitive memories that underlie the subsequent expectation of reinforcement. Indeed, in the *mobile conjugate reinforcement paradigm* (Rovee & Rovee, 1969), infants learn the contingency between the instrumental response of kicking their legs and the movement of a mobile hanging above their crib. A high specificity of a cue necessary for the associative recall of the appetitive memory is apparent until three months (Rovee-Collier & Hayne, 1987), diminishing thereafter alongside increases in the ability to generalize across stimuli and experiences. The retention for the appetitive association also shows a gradual increase across infancy. Notably, the ability to learn that a cue itself is representative of an appetitive outcome is limited in early infancy, the first year of life, despite intact cue recognition memory (Diamond, Churchland, Cruess, & Kirkham, 1999).

Childhood and adolescence Reinforcement learning and appetitive memory formation during childhood in humans occur similarly to that observed in adulthood (Galvan et al., 2006; Somerville, Hare, & Casey, 2011), although children have been shown to differ in their capacity to differentiate behavioral responses between cues predictive of differing reinforcer magnitude (Galvan et al., 2006). Strikingly, subsequent changes in components of the appetitive memory circuitry during adolescence in both humans and animals have been shown to greatly influence the utilization of appetitive memory in service of guiding behavior (figure 21.1*D*).

Because the appetitive properties of environmental cues directly influence how they are encoded, sensitivity to reinforcement may account for a great deal of observable differences in adolescent behavior (figure 21.1*E*). In humans, adolescents have been shown to exhibit hypersensitivity to primary reinforcers (Fareri, Martin, & Delgado, 2008; Steinberg, 2008), with similar patterns observed in mice (Adriani, Chiarotti, & Laviola, 1998). Research in rats has also indicated that the appetitive qualities of drugs and alcohol are elevated during adolescence (Pautassi, Myers, Spear, Molina, & Spear, 2008; Vastola, Douglas, Varlinskaya, & Spear, 2002). Notably, both human and rodent adolescents display increased responsiveness to environmental cues *signaling* a potential reinforcer (Hare et al., 2008; Meyer & Bucci, 2016), and evidence from rats has highlighted a greater effort to obtain a reinforcer than adults (Friemel, Spanagel, & Schneider, 2010; Stolyarova & Izquierdo, 2015).

In line with this, the presence of an appetitive stimulus produces a drastically different pattern of performance in inhibitory control tasks relative to tests with neutral stimuli, with both humans and rodents specifically exhibiting difficulty suppressing responses to appetitive cues during adolescence compared to younger or older ages (Galván, 2013; Hare et al., 2008; Meyer & Bucci, 2017; Somerville, Hare, & Casey, 2011).

Furthermore, across species, appetitive memories appear to be more resistant to updating with new information during adolescence (Levin et al., 1991; Newman & McGaughy, 2011). Particularly notable examples of this effect have been shown in studies considering the extinction of either a Pavlovian appetitive cue or an instrumental reinforcer-eliciting response in rats (Andrzejewski et al., 2011; Meyer & Bucci, 2016; Sturman, Mandell, & Moghaddam, 2010). Perseveration on the reinforcing properties of appetitive cues, even in the absence of the expected reinforcer, has been taken to indicate increased strength of the appetitive cue memory specifically during adolescence.

The neural circuitry of adolescent appetitive memory During the initial encoding of an appetitive memory, dopaminergic projections from the ventral tegmental area (VTA) communicate information about the predictive value of an appetitive outcome to the nucleus accumbens (NAC) via the mesolimbic pathway. In turn, the NAC promotes reinforcer-seeking behaviors through connectivity with the ventral pallidum (VP; Leung & Balleine, 2013; Smith, Tindell, Aldridge, & Berridge, 2008). Notably, although robust differences in dopaminergic neurotransmission are apparent during adolescence in rats (Matthews, Bondi, Torres, & Moghaddam, 2013; Robinson, Zitzman, Smith, & Spear, 2011) sensitivity to reinforcement during adolescence does not appear to be driven by hyperactivity of VTA dopaminergic neurons. Indeed, while adolescents and adults show similar increases in appetitive cue-evoked VTA activity as learning progresses, adolescent dopamine neurons exhibit an attenuated response preceding the delivery of a reinforcer (i.e., reinforcer anticipation) relative to adults, along with a reduced response to reinforcer delivery (Kim, Simon, Wood, & Moghaddam, 2015). Conversely, during extinction, while VTA activity associated with reinforcer-predictive cues decreases over time in adults, persistent VTA responding is observed in adolescents (Kim et al., 2015). Furthermore, activity remained higher in adolescence even when behavioral measures of extinction learning (i.e., reduced reinforcer-seeking behavior) matched that seen in adults. Thus, persistent appetitive cue-related activity may contribute to an increased susceptibility to both generalization and spontaneous recovery of the original appetitive memory, despite subsequent learning about the decreased likelihood of reinforcement.

In humans, substantial evidence has shown that subcortical limbic regions (including the NAC) mature earlier than cortical control areas, indicating a potential explanation for the differing incentive salience attribution processes apparent during the adolescent period. As a result, activity in subcortical regions is disproportionately higher than in PFC during adolescence (Casey, Jones, & Hare, 2008). Moreover, evidence in humans has shown stronger signaling to reinforcement in NAC during adolescence relative to adulthood (Ernst et al., 2005; Galvan et al., 2006). However, separate studies have shown the opposite, with adolescents mounting a weaker NAC response to reinforcement (Bjork et al., 2004; Bjork, Smith, Chen, & Hommer, 2010) or, alternatively, more complex context-dependent patterns (Geier, Terwilliger, Teslovich, Velanova, & Luna, 2010). Thus, differences in NAC signaling may be sensitive to nuances of reinforcement contingencies and vary with the component of behavior such as anticipation versus the receipt of a reinforcer.

Interestingly, age differences in reinforcement processing may also be attributable to altered signaling in dorsal striatum. Dorsal striatal circuitry is recruited both earlier and to a greater degree in adolescents relative to adults during the retrieval of a reinforcer (Sturman & Moghaddam, 2012). Interactions between the mesolimbic system and the nigrostriatal system, extending between substantia nigra and dorsal striatum, are of great importance for mediating the interface between motivation and action (Mogenson, Jones, & Yim, 1980; Nauta, Smith, Faull, & Domesick, 1978), indicating a possible mechanism underlying the heightened approach of appetitive cues observed during adolescence.

Finally, within the PFC, apparent immaturities in the orbitofrontal cortex (OFC) likely influence the ability of an adolescent to appropriately reconcile appetitive information in the context of long-term goals (Ernst et al., 2005; Galvan et al., 2006). Similar age differences in OFC activity specific to reinforcement processing have also been observed in rats (Sturman & Moghaddam, 2011).

Discussion

Sources of information relevant to an individual can differ greatly depending on one's developmental stage. Here, we have outlined examples of how the individual stimulus representations composing the memory of an environment can have great impact on subsequently manifesting behaviors. We have discussed a range of dynamic neurobiological changes in circuitries for both aversive and appetitive learning and memory that offers context for understanding how individuals at varying developmental stages utilize alternative processes in the generation of behavioral goals and the influence of memories on overt behavior.

Importantly, many of the age-specific features of emotional memory we have discussed promote behavioral patterns that are adaptive for the developmental period during which they manifest, highlighting evolutionary biases in the context of brain development that allow one to meet the environmental demands of each stage of life and acquire the skills necessary to progress through subsequent stages. Moreover, striking parallels in the developmental features of both aversive and appetitive memory systems indicate that despite differences in underlying circuitry these memory systems are coordinated in their ability to recognize the most salient features of an environment and subsequently use this information in the service of goal-directed behavior targeted to discrete developmental stages.

For example, both fear and appetitive memory during infancy are biased toward forming an attachment to a caregiver, which maximizes the chances of survival (Brown, 1986). Interactions between the oxytocin and dopamine systems allow an infant to distinguish social from nonsocial cues and promote reinforcement learning specifically for the caregiver (Nelson & Panksepp, 1996). Moreover, a maternal presence serves as a buffer, modifying cued fear learning in rodents (Moriceau & Sullivan, 2006).

In addition, emotional memory processes apparent during adolescence can facilitate the acquisition of the skills and experiences necessary for the maturation to adulthood (Spear, 2010). Adolescence (especially in rodents) is a time when heightened exploratory behaviors facilitate the transition away from parental dependence to relative independence. This is reflected in fear response patterns that promote not only the exploration of new environments but the generalization of fear toward cues that predict a threat (Fanselow, 1994). Decreased exploration as a result of contextual fear could result in the depletion of food in the home environment and a failure to mate. Similarly, heightened sensitivity to cues of threat in novel environments contributes to vigilance to threats and is similarly adaptive as an evolutionary measure. Thus, heightened cued fear expression combined with attenuated contextual fear expression during adolescence (McCallum, Kim, & Richardson, 2010; Pattwell et al., 2011; Pattwell et al., 2012) allows the adolescent to remain both exploratory and cautious. Likewise, characteristics of appetitive memory during adolescence are well suited for forms of learning that occur in uncertain or changing environments (Johnson & Wilbrecht, 2011; Qin et al., 2004). Indeed, the contingencies defining when and how much of an appetitive outcome will be available can be highly variable in different environments. Thus, during the transition to independence, as an adolescent is likely to experience increased exposure to new environments, hypersensitivity to reinforcers and the perseveration on reinforcer-associated behaviors may actually increase the likelihood of attaining reinforcement, until such a time when sufficient information about contingencies in discrete environments can be established.

Despite these evolutionarily advantageous developmental changes in emotional memory, a multitude of psychiatric conditions emerge during development as the brain is undergoing complex and dynamic changes. Unfortunately, the earlier emergence of emotional disorders has in some cases been associated with an increased severity of symptoms as well as comorbidities (Andersen & Teicher, 2008; Gutman & Nemeroff, 2003). Thus, there is significant interest in understanding the interplay between the specific neurobiological and behavioral factors that characterize developmental stages and in identifying why particular individuals are susceptible to negative outcomes. While this chapter provides an overview of the behavioral, neural, and molecular properties of both aversive and appetitive learning as a function of age, various factors, including but not limited to gender, early life stress, the environment, and genetic differences, may also influence the properties outlined here and should be considered in the developmental landscape of learning and memory (Pattwell & Bath, 2017).

As more is uncovered about the brain through the modern technologies associated with basic neuroscience research, the field of developmental memory is on the verge of great advances. Still to uncover are many answers surrounding not just how memories are acquired or expressed but how they change across the lifespan in both declarative form and content and also in the emotional and age-specific salience unique to one's developmental state at any given time. A body of literature has begun to probe these questions for various types of emotional memory, investigating whether memories depend on the age at which they are encoded or the age at which they are retrieved (Barnet & Hunt, 2006; Richardson & Fan, 2002; Simcock & Hayne, 2002). How retrieval processes, such as those outlined in chapter 23 on reconsolidation, may strengthen or weaken memories across development is also of great interest for understanding just how the brain forms, maintains, and alters aversive and appetitive memories across the formative years of childhood and adolescence and how this sets the stage for the adult memory processing of similar or related experiences.

REFERENCES

Adriani, W., Chiarotti, F., & Laviola, G. (1998). Elevated novelty seeking and peculiar d-amphetamine sensitization in periadolescent mice compared with adult mice. *Behavioral Neuroscience, 112*(5), 1152–1166.

AHRQ/NIMH [Agency for Healthcare Research and Quality]. (2006). Total expenses and percent distribution for selected conditions by type of service. Medical expenditure panel survey household component data. United States. Retrieved from http://www.meps.ahrq.gov/mepsweb/data_stats/tables_compendia_hh_interactive.jsp?_SERVICE=MEPSSocket0&_PROGRAM=MEPSPGM.TC.SAS&File=HCFY2006&Table=HCFY2006%5FCNDXP%5FC&_Debug=.

Akers, K. G., Arruda-Carvalho, M., Josselyn, S. A., & Frankland, P. W. (2012). Ontogeny of contextual fear memory formation, specificity, and persistence in mice. *Learning & Memory, 19*(12), 598–604. doi:10.1101/lm.027581.112

Alberini, C. M., & Travaglia, A. (2017). Infantile amnesia: A critical period of learning to learn and remember. *Journal of Neuroscience, 37*(24), 5783–5795. doi:10.1523/JNEUROSCI.0324-17.2017

Andersen, S. L., & Teicher, M. H. (2008). Stress, sensitive periods and maturational events in adolescent depression. *Trends in Neurosciences, 31*(4), 183–191. doi:10.1016/j.tins.2008.01.004 10.1016/j.tins.2008.01.004

Andrzejewski, M. E., Schochet, T. L., Feit, E. C., Harris, R., McKee, B. L., & Kelley, A. E. (2011). A comparison of adult and adolescent rat behavior in operant learning, extinction, and behavioral inhibition paradigms. *Behavioral Neuroscience, 125*(1), 93–105. doi:10.1037/a0022038 10.1037/a0022038

Armstrong, C. M., DeVito, L. M., & Cleland, T. A. (2006). One-trial associative odor learning in neonatal mice. *Chemical Senses, 31*(4), 343–349. doi:10.1093/chemse/bjj038

Arruda-Carvalho, M., Wu, W. C., Cummings, K. A., & Clem, R. L. (2017). Optogenetic examination of prefrontal-amygdala synaptic development. *Journal of Neuroscience, 37*(11), 2976–2985. doi:10.1523/JNEUROSCI.3097-16.2017

Barnet, R. C., & Hunt, P. S. (2006). The expression of fear-potentiated startle during development: Integration of learning and response systems. *Behavioral Neuroscience, 120*(4), 861–872. doi:10.1037/0735-7044.120.4.861

Bjork, J. M., Knutson, B., Fong, G. W., Caggiano, D. M., Bennett, S. M., & Hommer, D. W. (2004). Incentive-elicited brain activation in adolescents: Similarities and differences from young adults. *Journal of Neuroscience, 24*(8), 1793–1802. doi:10.1523/jneurosci.4862-03.2004 10.1523/JNEUROSCI.4862-03.2004

Bjork, J. M., Smith, A. R., Chen, G., & Hommer, D. W. (2010). Adolescents, adults and rewards: Comparing motivational neurocircuitry recruitment using fMRI. *PLoS One, 5*(7), e11440. doi:10.1371/journal.pone.0011440 10.1371/journal.pone.0011440

Bowlby, J. (1969). *Attachment and loss* (Vol. 1). New York: Basic Books.

Brown, R. E. (1986). Paternal behavior in the male Long-Evans rat (rattus norvegicus). *Journal of Comparative Psychology, 100*(2), 162.

Callaghan, B. L., & Richardson, R. (2012). Early-life stress affects extinction during critical periods of development: An analysis of the effects of maternal separation on extinction in adolescent rats. *Stress, 15*(6), 671–679. doi:10.3109/10253890.2012.667463

Camp, L. L., & Rudy, J. W. (1988). Changes in the categorization of appetitive and aversive events during postnatal development of the rat. *Developmental Psychobiology, 21*(1), 25–42. doi:10.1002/dev.420210103

Campbell, B. A., & Spear, N. E. (1972). Ontogeny of memory. *Psychological Review, 79*(3), 215–236.

Cardinal, R. N., & Everitt, B. J. (2004). Neural and psychological mechanisms underlying appetitive learning: Links to drug addiction. *Current Opinion in Neurobiology, 14*(2), 156–162. doi:10.1016/j.conb.2004.03.004 10.1016/j.conb.2004.03.004

Casey, B. J., Glatt, C. E., & Lee, F. S. (2015). Treating the developing versus developed brain: Translating preclinical mouse and human studies. *Neuron, 86*(6), 1358–1368. doi:10.1016/j.neuron.2015.05.020

Casey, B. J., Jones, R. M., & Hare, T. A. (2008). The adolescent brain. *Annals of the New York Academy of Sciences, 1124*, 111–126. doi:10.1196/annals.1440.01010.1196/annals.1440.010

Casey, B. J., Jones, R. M., Levita, L., Libby, V., Pattwell, S. S., Ruberry, E. J., ... Somerville, L. H. (2010). The storm and stress of adolescence: Insights from human imaging and mouse genetics. *Developmental Psychobiology, 52*(3), 225–235.

Casey, B. J., Jones, R. M., & Somerville, L. H. (2011). Braking and accelerating of the adolescent brain. *Journal of Research on Adolescence, 21*(1), 21–33.

Casey, B. J., Tottenham, N., Liston, C., & Durston, S. (2005). Imaging the developing brain: What have we learned about cognitive development? *Trends in Cognitive Sciences, 9*(3), 104–110.

Chambers, R. A., Taylor, J. R., & Potenza, M. N. (2003). Developmental neurocircuitry of motivation in adolescence: A critical period of addiction vulnerability. *American Journal of Psychiatry, 160*(6), 1041–1052. doi:10.1176/appi.ajp.160.6.1041

Cruz, E., Soler-Cedeno, O., Negron, G., Criado-Marrero, M., Chompre, G., & Porter, J. T. (2015). Infralimbic EphB2 modulates fear extinction in adolescent rats. *Journal of Neuroscience, 35*(36), 12394–12403. doi:10.1523/JNEUROSCI.4254-14.2015

Diamond, A., Churchland, A., Cruess, L., & Kirkham, N. Z. (1999). Early developments in the ability to understand the relation between stimulus and reward. *Developmental Psychobiology, 35*(6), 1507–1517.

Drysdale, A. T., Hartley, C. A., Pattwell, S. S., Ruberry, E. J., Somerville, L. H., Compton, S. N., … Walkup, J. T. (2014). Fear and anxiety from principle to practice: Implications for when to treat youth with anxiety disorders. *Biological Psychiatry, 75*(11), e19–20. doi:10.1016/j.biopsych.2013.08.015

Ernst, M., Nelson, E. E., Jazbec, S., McClure, E. B., Monk, C. S., Leibenluft, E., … Pine, D. S. (2005). Amygdala and nucleus accumbens in responses to receipt and omission of gains in adults and adolescents. *Neuroimage, 25*(4), 1279–1291. doi:10.1016/j.neuroimage.2004.12.038 10.1016/j.neuroimage.2004.12.038

Fanselow, M. S. (1994). Neural organization of the defensive behavior system responsible for fear. *Psychonomic Bulletin & Review, 1*(4), 429–438.

Fareri, D. S., Martin, L. N., & Delgado, M. R. (2008). Reward-related processing in the human brain: Developmental considerations. *Development and Psychopathology, 20*(4), 1191–1211. doi:10.1017/s0954579408000576 10.1017/S0954579408000576

Foulkes, L., & Blakemore, S. J. (2018). Studying individual differences in human adolescent brain development. *Nature Neuroscience, 21*(3), 315–323. doi:10.1038/s41593-018-0078-4

Friemel, C. M., Spanagel, R., & Schneider, M. (2010). Reward sensitivity for a palatable food reward peaks during pubertal developmental in rats. *Frontiers in Behavioral Neuroscience, 4*. doi:10.3389/fnbeh.2010.00039 10.3389/fnbeh.2010.00039

Galván, A. (2013). The teenage brain: Sensitivity to rewards. *Current Directions in Psychological Science, 22*(2), 88–93.

Galvan, A., Hare, T. A., Parra, C. E., Penn, J., Voss, H., Glover, G., & Casey, B. J. (2006). Earlier development of the accumbens relative to orbitofrontal cortex might underlie risk-taking behavior in adolescents. *Journal of Neuroscience, 26*(25), 6885–6892.

Gee, D. G., Gabard-Durnam, L. J., Flannery, J., Goff, B., Humphreys, K. L., Telzer, E. H., … Tottenham, N. (2013). Early developmental emergence of human amygdala-prefrontal connectivity after maternal deprivation. *Proceedings of the National Academy of Sciences of the United States of America, 110*(39), 15638–15643. doi:10.1073/pnas.1307893110

Geier, C. F., Terwilliger, R., Teslovich, T., Velanova, K., & Luna, B. (2010). Immaturities in reward processing and its influence on inhibitory control in adolescence. *Cerebral Cortex, 20*(7), 1613–1629. doi:10.1093/cercor/bhp225 10.1093/cercor/bhp225

Giedd, J. N., Blumenthal, J., Jeffries, N. O., Castellanos, F. X., Liu, H., Zijdenbos, A., … Rapoport, J. L. (1999). Brain development during childhood and adolescence: A longitudinal MRI study. *Nature Neuroscience, 2*(10), 861–863.

Ginsburg, G. S., La Greca, A. M., & Silverman, W. K. (1998). Social anxiety in children with anxiety disorders: Relation with social and emotional functioning. *Journal of Abnormal Child Psychology, 26*(3), 175–185.

Gogolla, N., Caroni, P., Luthi, A., & Herry, C. (2009). Perineuronal nets protect fear memories from erasure. *Science, 325*(5945), 1258–1261. doi:10.1126/science.1174146 325/5945/1258 [pii]

Gogtay, N., Giedd, J. N., Lusk, L., Hayashi, K. M., Greenstein, D., Vaituzis, A. C., … Thompson, P. M. (2004). Dynamic mapping of human cortical development during childhood through early adulthood. *Proceedings of the National Academy of Sciences of the United States of America, 101*(21), 8174–8179. doi:10.1073/pnas.0402680101 0402680101 [pii]

Gutman, D. A., & Nemeroff, C. B. (2003). Persistent central nervous system effects of an adverse early environment: Clinical and preclinical studies. *Physiology & Behavior, 79*(3), 471–478.

Hare, T. A., Tottenham, N., Galvan, A., Voss, H. U., Glover, G. H., & Casey, B. J. (2008). Biological substrates of emotional reactivity and regulation in adolescence during an emotional go-nogo task. *Biological Psychiatry, 63*(10), 927–934. doi:10.1016/j.biopsych.2008.03.015 S0006-3223(08)00359-4 [pii]

Hefner, K., & Holmes, A. (2007). Ontogeny of fear-, anxiety- and depression-related behavior across adolescence in C57BL/6J mice. *Behavioural Brain Research, 176*(2), 210–215.

Johnson, C., & Wilbrecht, L. (2011). Juvenile mice show greater flexibility in multiple choice reversal learning than adults. *Developmental Cognitive Neuroscience, 1*(4), 540–551. doi:10.1016/j.dcn.2011.05.008 10.1016/j.dcn.2011.05.008

Johnson, D. C., & Casey, B. J. (2015). Extinction during memory reconsolidation blocks recovery of fear in adolescents. *Scientific Reports, 5*, 8863. doi:10.1038/srep08863

Kieling, C., Baker-Henningham, H., Belfer, M., Conti, G., Ertem, I., Omigbodun, O., … Rahman, A. (2011). Child and adolescent mental health worldwide: Evidence for action. *Lancet, 378*(9801), 1515–1525. doi:10.1016/S0140-6736(11)60827-1

Kim, J. H., Hamlin, A. S., & Richardson, R. (2009). Fear extinction across development: The involvement of the medial prefrontal cortex as assessed by temporary inactivation and immunohistochemistry. *Journal of Neuroscience, 29*(35), 10802–10808. doi:10.1523/JNEUROSCI.0596-09.2009 29/35/10802 [pii]

Kim, J. H., Li, S., & Richardson, R. (2011). Immunohistochemical analyses of long-term extinction of conditioned fear in adolescent rats. *Cerebral Cortex, 21*(3), 530–538.

Kim, J. H., & Richardson, R. (2007). A developmental dissociation of context and GABA effects on extinguished fear in rats. *Behavioral Neuroscience, 121*(1), 131–139. doi:10.1037/0735-7044.121.1.131

Kim, Y., Simon, N. W., Wood, J., & Moghaddam, B. (2015). Reward anticipation is encoded differently by adolescent ventral tegmental area neurons. *Biological Psychiatry, 79*(11), 878–886. doi:10.1016/j.biopsych.2015.04.026 10.1016/j.biopsych.2015.04.026

Kim-Cohen, J., Caspi, A., Moffitt, T. E., Harrington, H., Milne, B. J., & Poulton, R. (2003). Prior juvenile diagnoses in adults with mental disorder: Developmental follow-back of a prospective-longitudinal cohort. *Archives of General Psychiatry, 60*(7), 709–717.

Krabbe, S., Grundemann, J., & Luthi, A. (2018). Amygdala inhibitory circuits regulate associative fear conditioning. *Biological Psychiatry, 83*(10), 800–809. doi:10.1016/j.biopsych.2017.10.006

Landers, M. S., & Sullivan, R. M. (2012). The development and neurobiology of infant attachment and fear. *Developmental Neuroscience, 34*(2–3), 101–114. doi:000336732

Leung, B. K., & Balleine, B. W. (2013). The ventral striato-pallidal pathway mediates the effect of predictive learning on choice between goal-directed actions. *Journal of Neuroscience, 33*(34), 13848–13860. doi:10.1523/jneurosci.1697-13.2013 10.1523/JNEUROSCI.1697-13.2013

Levesque, J., Joanette, Y., Mensour, B., Beaudoin, G., Leroux, J. M., Bourgouin, P., & Beauregard, M. (2004). Neural basis of emotional self-regulation in childhood. *Neuroscience, 129*(2), 361–369.

Levin, H. S., Culhane, K. A., Hartmann, J., Evankovich, K., Mattson, A. J., Harward, H., ... Fletcher, J. M. (1991). Developmental changes in performance on tests of purported frontal lobe functioning. *Developmental Neuropsychology, 7*(3), 377–395.

Liberman, L. C., Lipp, O. V., Spence, S. H., & March, S. (2006). Evidence for retarded extinction of aversive learning in anxious children. *Behaviour Research and Therapy, 44*(10), 1491–1502.

Maren, S. (2001). Neurobiology of Pavlovian fear conditioning. *Annual Review of Neuroscience, 24*, 897–931. doi:10.1146/annurev.neuro.24.1.897 24/1/897 [pii]

Maren, S., & Quirk, G. J. (2004). Neuronal signalling of fear memory. *Nature Reviews Neuroscience, 5*(11), 844–852.

Marin, O. (2016). Developmental timing and critical windows for the treatment of psychiatric disorders. *Nature Medicine, 22*(11), 1229–1238. doi:10.1038/nm.4225

Matthews, M., Bondi, C., Torres, G., & Moghaddam, B. (2013). Reduced presynaptic dopamine activity in adolescent dorsal striatum. *Neuropsychopharmacology, 38*(7), 1344–1351. doi:10.1038/npp.2013.32 10.1038/npp.2013.32

McCallum, J., Kim, J. H., & Richardson, R. (2010). Impaired extinction retention in adolescent rats: Effects of D-cycloserine. *Neuropsychopharmacology, 35*(10), 2134–2142. doi:10.1038/npp.2010.92 npp201092 [pii]

McCormick, E. M., & Telzer, E. H. (2017). Adaptive adolescent flexibility: Neurodevelopment of decision-making and learning in a risky context. *Journal of Cognitive Neuroscience, 29*(3), 413–423. doi:10.1162/jocn_a_01061

Merikangas, K. R., He, J. P., Burstein, M., Swanson, S. A., Avenevoli, S., Cui, L., ... Swendsen, J. (2010). Lifetime prevalence of mental disorders in U.S. adolescents: Results from the National Comorbidity Survey Replication—Adolescent Supplement (NCS-A). *Journal of the American Academy of Child and Adolescent Psychiatry, 49*(10), 980–989. doi:10.1016/j.jaac.2010.05.017 S0890-8567(10)00476-4 [pii]

Merikangas, K. R., He, J. P., Burstein, M., Swendsen, J., Avenevoli, S., Case, B., ... Olfson, M. (2011). Service utilization for lifetime mental disorders in U.S. adolescents: Results of the National Comorbidity Survey-Adolescent Supplement (NCS-A). *Journal of the American Academy of Child and Adolescent Psychiatry, 50*(1), 32–45. doi:10.1016/j.jaac.2010.10.006 S0890-8567(10)00783-5 [pii]

Meyer, H. C., & Bucci, D. J. (2014). The ontogeny of learned inhibition. *Learning & Memory, 21*(3), 143–152. doi:10.1101/lm.033787.113 10.1101/lm.033787.113

Meyer, H. C., & Bucci, D. J. (2016). Age differences in appetitive Pavlovian conditioning and extinction in rats. *Physiology & Behavior, 167*, 354–362.

Meyer, H. C., & Bucci, D. J. (2017). Negative occasion setting in juvenile rats. *Behavioural Processes, 137*, 33–39. doi:10.1016/j.beproc.2016.05.003 10.1016/j.beproc.2016.05.003

Mogenson, G. J., Jones, D. L., & Yim, C. Y. (1980). From motivation to action: Functional interface between the limbic system and the motor system. *Progress in Neurobiology, 14*(2–3), 69–97.

Monk, C. S., McClure, E. B., Nelson, E. E., Zarahn, E., Bilder, R. M., Leibenluft, E., ... Pine, D. S. (2003). Adolescent immaturity in attention-related brain engagement to emotional facial expressions. *Neuroimage, 20*(1), 420–428. doi:S1053811903003550 [pii]

Moriceau, S., & Sullivan, R. M. (2006). Maternal presence serves as a switch between learning fear and attraction in infancy. *Nature Neuroscience, 9*(8), 1004–1006.

Nabel, E. M., & Morishita, H. (2013). Regulating critical period plasticity: Insight from the visual system to fear circuitry for therapeutic interventions. *Frontiers in Psychiatry, 4*, 146. doi:10.3389/fpsyt.2013.00146

Nauta, W. J., Smith, G. P., Faull, R. L., & Domesick, V. B. (1978). Efferent connections and nigral afferents of the nucleus accumbens septi in the rat. *Neuroscience, 3*(4–5), 385–401.

Nelson, E., & Panksepp, J. (1996). Oxytocin mediates acquisition of maternally associated odor preferences in preweanling rat pups. *Behavioral Neuroscience, 110*(3), 583–592.

Newman, L. A., & McGaughy, J. (2011). Adolescent rats show cognitive rigidity in a test of attentional set shifting. *Developmental Psychobiology, 53*(4), 391–401. doi:10.1002/dev.20537 10.1002/dev.20537

Park, C. H. J., Ganella, D. E., & Kim, J. H. (2017). Juvenile female rats, but not male rats, show renewal, reinstatement, and spontaneous recovery following extinction of conditioned fear. *Learning & Memory, 24*(12), 630–636. doi:10.1101/lm.045831.117

Pattwell, S. S., & Bath, K. G. (2017). Emotional learning, stress, and development: An ever-changing landscape shaped by early-life experience. *Neurobiology of Learning and Memory, 143*, 36–48. doi:10.1016/j.nlm.2017.04.014

Pattwell, S. S., Bath, K. G., Casey, B. J., Ninan, I., & Lee, F. S. (2011). Selective early-acquired fear memories undergo temporary suppression during adolescence. *Proceedings of the National Academy of Sciences of the United States of America, 108*(3), 1182–1187. doi:10.1073/pnas.1012975108 1012975108 [pii]

Pattwell, S. S., Duhoux, S., Hartley, C. A., Johnson, D. C., Jing, D., Elliott, M. D., ... Lee, F. S. (2012). Altered fear learning across development in both mouse and human. *Proceedings of the National Academy of Sciences of the United States of America, 109*(40), 16318–16323. doi:10.1073/pnas.1206834109 1206834109 [pii]

Pattwell, S. S., Liston, C., Jing, D., Ninan, I., Yang, R. R., Witztum, J., ... Lee, F. S. (2016). Dynamic changes in neural circuitry during adolescence are associated with persistent attenuation of fear memories. *Nature Communications, 7,* 11475. doi:10.1038/ncomms11475

Pautassi, R. M., Myers, M., Spear, L. P., Molina, J. C., & Spear, N. E. (2008). Adolescent but not adult rats exhibit ethanol-mediated appetitive second-order conditioning. *Alcoholism: Clinical and Experimental Research, 32*(11), 2016–2027. doi:10.1111/j.1530-0277.2008.00789.x 10.1111/j.1530-0277.2008.00789.x

Pavlov, I. P. (1927). *Conditioned reflexes: An investigation of the physiological activity of the cerebral cortex* (G. Anrep, Trans.). London: Oxford University Press.

Pinabiaux, C., Hertz-Pannier, L., Chiron, C., Rodrigo, S., Jambaque, I., & Noulhiane, M. (2013). Memory for fearful faces across development: Specialization of amygdala nuclei and medial temporal lobe structures. *Frontiers in Human Neuroscience, 7,* 901. doi:10.3389/fnhum.2013.00901

Pollack, M. H., Otto, M. W., Sabatino, S., Majcher, D., Worthington, J. J., McArdle, E. T., & Rosenbaum, J. F. (1996). Relationship of childhood anxiety to adult panic disorder: Correlates and influence on course. *American Journal of Psychiatry, 153*(3), 376–381.

Qin, Y., Carter, C. S., Silk, E. M., Stenger, V. A., Fissell, K., Goode, A., & Anderson, J. R. (2004). The change of the brain activation patterns as children learn algebra equation solving. *Proceedings of the National Academy of Sciences of the United States of America, 101*(15), 5686–5691. doi:10.1073/pnas.0401227101 10.1073/pnas.0401227101

Raineki, C., Holman, P. J., Debiec, J., Bugg, M., Beasley, A., & Sullivan, R. M. (2010). Functional emergence of the hippocampus in context fear learning in infant rats. *Hippocampus, 20*(9), 1037–1046.

Richardson, R., & Fan, M. (2002). Behavioral expression of learned fear in rats is appropriate to their age at training, not their age at testing. *Animal Learning & Behavior, 30*(4), 394–404.

Robinson, D. L., Zitzman, D. L., Smith, K. J., & Spear, L. P. (2011). Fast dopamine release events in the nucleus accumbens of early adolescent rats. *Neuroscience, 176,* 296–307. doi:10.1016/j.neuroscience.2010.12.016 10.1016/j.neuroscience.2010.12.016

Rovee, C. K., & Rovee, D. T. (1969). Conjugate reinforcement of infant exploratory behavior. *Journal of Experimental Child Psychology, 8*(1), 33–39.

Rovee-Collier, C., & Hayne, H. (1987). Reactivation of infant memory: Implications for cognitive development. *Advances in Child Development and Behavior, 20,* 185–238.

Rudy, J. W. (1993). Contextual conditioning and auditory cue conditioning dissociate during development. *Behavioral Neuroscience, 107*(5), 887–891.

Shen, H., Sabaliauskas, N., Sherpa, A., Fenton, A. A., Stelzer, A., Aoki, C., & Smith, S. S. (2010). A critical role for alpha-4betadelta GABAA receptors in shaping learning deficits at puberty in mice. *Science, 327*(5972), 1515–1518.

Simcock, G., & Hayne, H. (2002). Breaking the barrier? Children fail to translate their preverbal memories into language. *Psychological Science, 13*(3), 225–231. doi:10.1111/1467-9280.00442

Simon, N. W., & Moghaddam, B. (2014). Neural processing of reward in adolescent rodents. *Developmental Cognitive Neuroscience, 11,* 145–154. doi:10.1016/j.dcn.2014.11.001

Smith, K. S., Tindell, A. J., Aldridge, J. W., & Berridge, K. C. (2008). Ventral pallidum roles in reward and motivation. *Behavioural Brain Research, 196*(2), 155–167. doi:10.1016/j.bbr.2008.09.038 10.1016/j.bbr.2008.09.038

Somerville, L. H., Hare, T., & Casey, B. J. (2011). Frontostriatal maturation predicts cognitive control failure to appetitive cues in adolescents. *Journal of Cognitive Neuroscience, 23*(9), 2123–2134. doi:10.1162/jocn.2010.21572 10.1162/jocn.2010.21572

Sotres-Bayon, F., & Quirk, G. J. (2010). Prefrontal control of fear: More than just extinction. *Current Opinion in Neurobiology, 20*(2), 231–235. doi:10.1016/j.conb.2010.02.005 S0959-4388(10)00027-9 [pii]

Spear, L. P. (2000). The adolescent brain and age-related behavioral manifestations. *Neuroscience & Biobehavioral Reviews, 24*(4), 417–463.

Spear, L. P. (2010). *The behavioral neuroscience of adolescence.* New York: W. W. Norton.

Steinberg, L. (2008). A social neuroscience perspective on adolescent risk-taking. *Developmental Review, 28*(1), 78–106. doi:10.1016/j.dr.2007.08.002 10.1016/j.dr.2007.08.002

Stolyarova, A., & Izquierdo, A. (2015). Distinct patterns of outcome valuation and amygdala-prefrontal cortex synaptic remodeling in adolescence and adulthood. *Frontiers in Behavioral Neuroscience, 9,* 115. doi:10.3389/fnbeh.2015.00115 10.3389/fnbeh.2015.00115

Sturman, D. A., Mandell, D. R., & Moghaddam, B. (2010). Adolescents exhibit behavioral differences from adults during instrumental learning and extinction. *Behavioral Neuroscience, 124*(1), 16–25. doi:10.1037/a0018463 10.1037/a0018463

Sturman, D. A., & Moghaddam, B. (2011). Reduced neuronal inhibition and coordination of adolescent prefrontal cortex during motivated behavior. *Journal of Neuroscience, 31*(4), 1471–1478. doi:10.1523/jneurosci.4210-10.2011 10.1523/JNEUROSCI.4210-10.2011

Sturman, D. A., & Moghaddam, B. (2012). Striatum processes reward differently in adolescents versus adults. *Proceedings of the National Academy of Sciences of the United States of America, 109*(5), 1719–1724. doi:10.1073/pnas.1114137109 10.1073/pnas.1114137109

Sullivan, R. M., Landers, M., Yeaman, B., & Wilson, D. A. (2000). Good memories of bad events in infancy. *Nature, 407*(6800), 38–39. doi:10.1038/35024156

Sullivan, R. M., & Leon, M. (1987). One-trial olfactory learning enhances olfactory bulb responses to an appetitive conditioned odor in 7-day-old rats. *Brain Research, 432*(2), 307–311.

Thompson, J. V., Sullivan, R. M., & Wilson, D. A. (2008). Developmental emergence of fear learning corresponds with changes in amygdala synaptic plasticity. *Brain Research, 1200,* 58–65. doi:10.1016/j.brainres.2008.01.057

Vastola, B. J., Douglas, L. A., Varlinskaya, E. I., & Spear, L. P. (2002). Nicotine-induced conditioned place preference in adolescent and adult rats. *Physiology & Behavior, 77*(1), 107–114.

Yap, C. S., & Richardson, R. (2007). Extinction in the developing rat: An examination of renewal effects. *Developmental Psychobiology, 49*(6), 565–575.

22 Episodic Memory Modulation: How Emotion and Motivation Shape the Encoding and Storage of Salient Memories

MATTHIAS J. GRUBER AND MAUREEN RITCHEY

ABSTRACT Emotion and reward motivation are key factors in shaping the contents of memory. In this chapter we review evidence from two parallel literatures revealing the influence of emotion and reward motivation on episodic memory processes, mediated by the amygdala and the dopaminergic system, respectively. Taking an adaptive-memory perspective, we argue that emotion- and reward-related information is prioritized in memory from the earliest stages of encoding, leading to targeted effects on memory for salient information as well as spillover effects that affect memory for other information encoded around the same time. We distinguish these effects at encoding from the modulation of consolidation processes, which may serve to further prioritize memory for emotion- and reward-related information. Importantly, across the different stages of memory formation, emotion- and reward-related memories appear to share several key principles. These parallels shed light on the similar adaptive impact of two distinct neuromodulatory systems on memory.

Throughout our lives we forget more than we remember. The selectivity of memory has been an enduring puzzle: Why do we easily remember some information for years but quickly forget most information that we encounter? In this chapter we review evidence that memory systems are adaptive, protecting memories for information that could be useful in the future, such as events that signal potential threats or rewards, while discarding the rest. We focus on the effects of emotionally negative and rewarding events on encoding and consolidation processes that shape episodic memory.

Negative emotions and rewards are thought to influence episodic memory through separable neural circuits. Enhancements in memory for emotional experiences have been linked to noradrenergic activity in the amygdala (reviewed by LaBar & Cabeza, 2006; McGaugh, 2004). The amygdala is strongly interconnected with other structures in the medial temporal lobes (MTL), including the hippocampus, entorhinal cortex, and perirhinal cortex (Stefanacci, Suzuki, & Amaral, 1996), which are necessary for encoding new experiences into long-term memory. Amygdala activity and concomitant changes in stress hormone levels are thought to modulate the consolidation of new memories, thereby protecting memories for arousing experiences. The amygdala is also positioned to influence the quality of memory encoding through its connections with the multiple brain systems involved in attention and perception (Price, 2006).

In contrast, reward-based memories are thought to depend on the mesolimbic dopaminergic circuit (for current reviews, see Miendlarzewska, Bavelier, & Schwartz, 2016; Murty & Dickerson, 2017). Theories and recent findings have suggested that the hippocampus, together with two critical regions within the dopaminergic circuit, the nucleus accumbens and the substantia nigra/ventral tegmental area (SN/VTA) complex, are highly interconnected (Düzel, Bunzeck, Guitart-Masip, & Düzel, 2010; Shohamy & Adcock, 2010). The models suggest that the three regions are thought to be key to forming a functional loop that prioritizes learning and memory for rewarded information by enhancing plasticity (Lisman & Grace, 2005; Lisman, Grace, & Düzel, 2011).

Targeted Effects of Emotion and Reward on Encoding

Encoding processes supporting memory for emotional content From the earliest stages of neural processing, emotionally evocative stimuli compete for prioritized neural representation (Dolan & Vuilleumier, 2006; Mather & Sutherland, 2011). Emotional content influences perceptual processes as early as 100 to 200 ms after stimulus onset (Pizzagalli et al., 2002), resulting

The authors contributed equally to this work.

in enhanced activity in perceptual-processing areas (Lane, Chua, & Dolan, 1999; Vuilleumier, Armony, Driver, & Dolan, 2001). Emotional information is also more likely to reach conscious awareness under conditions of reduced attentional resources (Anderson & Phelps, 2001). Such effects have been shown to depend on the integrity of the amygdala (Anderson & Phelps, 2001; Vuilleumier, Richardson, Armony, Driver, & Dolan, 2004), which has direct projections back to primary sensory cortex (Amaral, Behniea, & Kelly, 2003).

The early biasing of perception and attention has direct implications for the quality of memory encoding. For instance, divided attention has a smaller effect on emotional memory encoding compared to neutral (Kensinger & Corkin, 2004; Talmi, Schimmack, Paterson, & Moscovitch, 2007), suggesting that reflexive orienting toward arousing information facilitates memory encoding. Although arousal appears to drive these effects (Kensinger & Corkin, 2004), emotional valence may influence which features are attended and encoded. It has been suggested that negative memories include more perceptual details, whereas positive memories include more semantic details. Negative objects are remembered with greater visual detail, and negative memory encoding elicits greater activity in visual cortex than neutral encoding (Kensinger, Garoff-Eaton, & Schacter, 2007). Positive memory encoding, on the other hand, elicits greater activity in lateral prefrontal areas (Mickley & Kensinger, 2008) and stronger prefrontal-hippocampal interactions supporting encoding (Ritchey, LaBar, & Cabeza, 2011).

Emotion-related changes in encoding processes have also been observed within MTL subregions, including the perirhinal cortex, which is seated at the apex of the ventral visual stream (Murray & Bussey, 1999), and the hippocampus, which is thought to bind item and context information in memory (Davachi, 2006; Eichenbaum, Yonelinas, & Ranganath, 2007). Compared to neutral item encoding, negative item encoding is associated with greater activity in the amygdala and perirhinal cortex (e.g., (Ritchey, Wang, Yonelinas, & Ranganath, 2018). Enhancements in emotional item recollection have not necessarily been tied to improvements in memory for source context (Yonelinas & Ritchey, 2015). Some studies have suggested that emotional arousal might actually interfere with associative memory encoding, leading to diminished hippocampal activity and worse memory for associations including emotional items (Bisby, Horner, Hørlyck, & Burgess, 2016; Madan, Fujiwara, Caplan, & Sommer, 2017).

Encoding processes supporting memory for rewarding information Similar to emotional material, cues that signal a future reward have been shown to enhance early perceptual and attentional processes (Bunzeck, Guitart-Masip, Dolan, & Düzel, 2011; Gruber & Otten, 2010; Yeung & Sanfey, 2004). In the last decade, evidence has accumulated of how reward anticipation facilitates encoding via the mesolimbic dopaminergic circuit. In one study, activity elicited by high-reward cues—but not low-reward cues—was predictive of whether the upcoming image would be remembered later (Adcock, Thangavel, Whitfield-Gabrieli, Knutson, & Gabrieli, 2006). This anticipatory effect of reward on memory was evident in the hippocampus, along with the nucleus accumbens and the SN/VTA (i.e., the critical areas that have previously been shown to code reward anticipation (Knutson, Adams, Fong, & Hommer, 2001). Furthermore, functional connectivity between the SN/VTA and the hippocampus during high-reward cues was also predictive of reward-related memory enhancements, illustrating that activity and communication within the mesolimbic dopaminergic circuit *prior* to the encoding of upcoming reward-related information benefits later memory. In addition, evidence from electroencephalography (EEG) studies that take advantage of the higher temporal resolution compared to functional Magnetic Resonance Imaging (fMRI) confirms that reward-related memory enhancements are driven by anticipatory processes (Gruber & Otten, 2010; Gruber, Watrous, Ekstrom, Ranganath, & Otten, 2013).

More specifically, hippocampal activity predicting memory enhancements could be pinpointed to the hippocampal subfields DG/CA2,3 (but not CA1 and the subiculum) and functional connectivity between the DG/CA2,3 subfields and the SN/VTA (Wolosin, Zeithamova, & Preston, 2012). Furthermore, findings on multivoxel activity patterns suggest that the hippocampus codes the value of information, thereby leading to enhanced memory for high-value information (Gruber, Ritchey, Wang, Doss, & Ranganath, 2016; Wolosin, Zeithamova, & Preston, 2013). In another seminal study, participants *incidentally* encoded scene images that served as reward cues (Wittmann et al., 2005). Consistent with prominent theories on dopamine and hippocampus-dependent consolidation (Lisman & Grace, 2005), a reward effect on memory emerged for high-reward compared to low-reward scene cues in a three-week delayed memory test. In line with these behavioral findings, brain activity in the SN/VTA and hippocampus during the encoding of high-reward scene cues predicted the three-week delayed memory enhancement.

In summary, although there is increasing evidence of how reward enhances incidental and intentional hippocampus-dependent learning via the mesolimbic dopaminergic circuit, more research is needed to better understand how reward affects memory. For example, future research would need to delineate the neural effects of reward-related anticipation compared to the effects of reward feedback and outcome (Mather & Schoeke, 2011). In addition to the dopaminergic modulation on memory, reward/value motivation can also lead to the strategic engagement of semantic processes supported by a frontotemporal network (Cohen, Rissman, Suthana, Castel, & Knowlton, 2014). Future research would need to address how interactions between reward- and semantic-related processes (e.g., via prefrontal cortex functions; Ballard et al., 2011) affect later memory.

Spillover Effects of Emotion and Reward during Encoding

Spillover effects of emotion during encoding Studies of emotional memory have primarily focused on enhancements for the emotional information itself. However, a growing literature has documented the existence of emotional spillover effects: changes in memory for intrinsically neutral information that is encoded around the same time as an emotional stimulus or while in a state of arousal. For instance, enhancements in memory for emotional items tend to be accompanied by impairments in memory for their neutral background scenes (Waring & Kensinger, 2009, 2011). This effect has been associated with enhanced activity in temporoparietal regions associated with attention (Waring & Kensinger, 2011). Interestingly, both emotion-related trade-offs (Hurlemann et al., 2005) and memory enhancements (Anderson, Wais, & Gabrieli, 2006) have been observed for neutral information encoded shortly before or after an emotional stimulus. It has been argued that this apparent discrepancy can be explained by differences in prioritization during encoding—that is, emotional arousal gives way to memory enhancements for prioritized information and memory impairments for everything else, due to arousal-biased competition for encoding resources (Mather & Sutherland, 2011).

Sustained states of arousal can also influence the efficacy of encoding. For instance, one study has shown that prolonged periods of emotional encoding "carried over" into a neutral encoding block so that neutral items encoded in an experimental block after a block of emotional items were remembered better compared to those studied after a block of neutral items (Tambini, Rimmele, Phelps, & Davachi, 2016). Under these circumstances, neutral encoding elicited patterns of neural activity similar to those observed during emotional encoding. Other studies have shown memory enhancements for information interleaved with emotionally arousing videos (Henckens, Hermans, Pu, Joëls, & Fernández, 2009), an effect that is counterintuitively associated with reductions in hippocampal activity.

Finally, memories are enhanced for items that are intrinsically neutral yet signal threat. Recent investigations of the mnemonic consequences of fear conditioning have shown that conditioned stimuli (CS+) are remembered better than their safe (CS–) counterparts (Dunsmoor, Murty, Davachi, & Phelps, 2015). Threatening outcomes need not be experienced during encoding to secure this benefit: the mere threat of an aversive outcome has been shown to enhance memory for those items tied to the outcome (Clewett, Huang, Velasco, Lee, & Mather, 2018; Murty, LaBar, & Adcock, 2012). These memory benefits have been linked to activations in the amygdala (Murty, LaBar, & Adcock, 2012) and locus coeruleus (Clewett et al., 2018), the latter of which has been specifically tied to changes in pupil diameter, a putative marker of noradrenergic tone.

Spillover effects of reward during encoding In contrast to the emotion literature, studies that investigate the spillover effects of reward have typically revealed enhancing rather than impairing effects. For example, reward-related memory enhancements spread from rewarded to neighboring nonrewarded information (Mather & Schoeke, 2011). Furthermore, neutral images showed memory enhancements when preceded by an unrelated rewarded reaction-time task (Murayama & Kitagami, 2014). In addition to these temporal proximity effects on memory, two recent studies have shown that memory enhancements for rewarded information can also "spill over" to semantically related, nonrewarded information that is not part of the same study phase as the reward information (Oyarzún, Packard, de Diego-Balaguer, & Fuentemilla, 2016; Patil, Murty, Dunsmoor, Phelps, & Davachi, 2017).

Importantly, one study that showed that curiosity states depend on the mesolimbic dopaminergic circuit in a way similar to reward anticipation investigated the neural mechanisms underlying salient spillover effects on neutral information (Gruber, Gelman, & Ranganath, 2014). In this study, participants encoded a series of high- and low-curiosity trivia questions and anticipated their associated answers. Critically, during the anticipation period participants also incidentally encoded neutral faces. In line with the above findings on reward-related spillover effects, faces presented during high- compared to low-curiosity states were better remembered in immediate and 24-hour-delayed memory tests. Importantly, individual variations in SN/VTA and hippocampal activity and functional connectivity between the two regions predicted the subsequent spillover effect on incidental face images, providing evidence for mesolimbic dopaminergic involvement with regard to salient spillover effects.

Recent findings have suggested that reward-related spillover effects depend on the exact presentation time of an incidental image during reward anticipation and on the reward probability (Stanek, Dickerson, Chiew, Clement, & Adcock, 2019). The findings suggest that phasic dopamine responses (elicited by a reward cue) and the sustained levels of or ramping up of dopamine (during reward anticipation) might be two separate mechanisms that enhance reward-related spillover effects.

The Influence of Emotion and Reward on Memory Consolidation Processes

Emotion effects on consolidation The standard account of enhanced emotional memory holds that arousal-mediated mechanisms that promote consolidation into long-term memory protect emotional memories (McGaugh, 2004; Roozendaal & McGaugh, 2011). Emotional memory enhancements have been shown to depend on the integrity of the amygdala and noradrenergic transmission (see Roozendaal & Hermans, 2017) for a comparison of rodent and human findings). These findings provide indirect support for the modulatory consolidation account of emotional memory (LaBar & Cabeza, 2006). Although it is challenging to directly study consolidation processes in humans, certain lines of evidence have been used to infer emotion effects on human memory consolidation.

First, emotion effects on memory are time-dependent. Emotional memories tend to be forgotten more slowly than neutral memories (Kleinsmith & Kaplan, 1963; LaBar & Phelps, 1998; Sharot & Yonelinas, 2008), leading to emotional memory enhancements that emerge after a delay, compared to immediately. This effect has been linked to amygdala engagement during encoding (Mackiewicz, Sarinopoulos, Cleven, & Nitschke, 2006; Ritchey, Dolcos, & Cabeza, 2008), and patients with amygdala damage do not show a time-dependent enhancement in emotional memory (LaBar & Phelps, 1998).

Second, postencoding arousal influences episodic memory. Several researchers have examined the effects of stress manipulations (e.g., the cold-pressor task) on memory for recently learned information. Across studies, postencoding stress appears to have a protective effect (Shields, Sazma, McCullough, & Yonelinas, 2017), suggesting the enhancement of postencoding memory consolidation processes. The effects of stress are not uniform, however, and seem to vary in a dose-dependent way (Andreano & Cahill, 2006; McCullough, Ritchey, Ranganath, & Yonelinas, 2015). Factors at encoding, including the emotional valence of the memoranda (Cahill, Gorski, & Le, 2003; Smeets, Otgaar, Candel, & Wolf, 2008; but see McCullough et al., 2015; Preuss & Wolf, 2009) and the amount of MTL engagement observed during encoding (Ritchey, McCullough, Ranganath, & Yonelinas, 2017), have been shown to moderate the effects of postencoding stress. Together, these results suggest that the arousal modulation of memory consolidation, like encoding, is not a simple on-off switch—it interacts with priorities and representations laid down at encoding and reorganizes them in light of new information.

Finally, researchers have recently begun to examine how emotion shapes neural activity during the postencoding consolidation period. Emotional arousal has been associated with functional connectivity changes that persist into rest periods following an arousal induction (van Marle, Hermans, Qin, & Fernández, 2010). Individual differences in functional connectivity changes between the amygdala and the hippocampus predicted arousal-related enhancements in memory for recently learned information (de Voogd, Klumpers, Fernández, & Hermans, 2017). Related results were obtained for rest periods following fear learning (Hermans, Kanen, & Tambini, 2016). In this study, multivoxel patterns corresponding to fear learning were also shown to be reinstated during postlearning rest.

Reward effects on consolidation Theoretical models have highlighted how dopamine affects cellular consolidation processes in the hippocampus (Düzel et al., 2010; Lisman & Grace, 2005; Shohamy & Adcock, 2010). Central to these models, VTA dopaminergic neurons are thought to enhance hippocampal late long-term potentiation, thereby prioritizing memory consolidation for dopamine-related memories. Consistent with these ideas, the reviewed studies have shown that SN/VTA and hippocampal activity predicted reward-related memory enhancements in memory tests delayed by at least 24 hours (Adcock et al., 2006; Wittmann et al., 2005).

Similar to emotion-related memories, studies have also suggested a time dependency of the effects of reward on memory. For example, some studies showed reward-related memory enhancements in 24-hour-delayed—but not immediate—memory tests (Murayama & Kitagami, 2014; Murayama & Kuhbandner, 2011). This time dependency of reward-related effects is also evident for reward-related spillover effects on semantically related information, suggesting a consolidation-dependent memory enhancement (Oyarzún et al., 2016; Patil et al., 2017). These recent studies suggest that reward-related spillover effects might be consolidation-dependent.

Nevertheless, several neuroimaging studies have also shown reward-related effects in immediate memory tests (Cohen et al., 2014; Gruber et al., 2013; Murty & Adcock, 2014). Consistent with this evidence, it has been suggested that dopamine could potentially affect dopamine-related *encoding* mechanisms via different dopaminergic properties (e.g., extracellular dopamine release; Floresco, West, Ash, Moore, & Grace, 2003; Shohamy & Adcock, 2010). A recent study might reconcile the ideas of distinct encoding- and consolidation-dependent dopamine mechanisms (Stanek et al., 2019), suggesting that different physiological dopaminergic properties enhance memory on different timescales.

Postencoding manipulations have also been used to infer reward effects on consolidation, particularly how reward interacts with postencoding sleep or interference. For example, administering a dopamine agonist during postencoding sleep boosted later memory for low-reward information up to the level of high-reward information, suggesting dopamine-dependent consolidation mechanisms (Feld, Besedovsky, Kaida, Münte, & Born, 2014). Furthermore, in line with the evidence that postencoding wakeful rest enhances consolidation (Dewar, Alber, Butler, Cowan, & Della Sala, 2012), a recent unpublished study from our laboratory demonstrated that wakeful rest during a postencoding period was necessary to show the effects of reward on memory in an immediate memory test (Gruber & Ranganath, in preparation). Conforming with the idea that different dopaminergic properties can enhance memory on different timescales, these latter findings suggest that wakeful rest might facilitate early consolidation effects on salient memories.

In order to directly address the neural mechanisms during reward-related memory consolidation, two recent fMRI studies targeted the neural dynamics during postencoding rest periods. During postencoding rest periods, individual variation in resting-state functional connectivity between the hippocampus and the representational cortical areas of the encoded material correlated with the magnitude of reward-related memory enhancements for such material (Murty, Tompary, Adcock, & Davachi, 2017). The findings suggest a potential mechanism of prioritized systems consolidation for rewarded material (Murty et al., 2017). Furthermore, consistent with the results that dopamine affects cellular hippocampal consolidation processes in rodents (McNamara, Tejero-Cantero, Trouche, Campo-Urriza, & Dupret, 2014; Singer & Frank, 2009), individual variability in postencoding increases in resting-state functional connectivity between the SN/VTA and the hippocampus predicted later reward-related memory enhancements (Gruber et al., 2016). In addition, using multivoxel pattern analyses, postencoding increases in the spontaneous reactivation of high-reward hippocampal representations correlated with the magnitude of later reward-related memory enhancements (Gruber et al., 2016). The findings are in line with prioritized hippocampal consolidation mechanisms for high-reward information.

Concluding Remarks

We reviewed the current evidence on how emotion- and reward-related information is prioritized during different stages of memory formation. Several models have been proposed to explain how neuromodulators, such as norepinephrine and dopamine, contribute to the prioritization of salient information in memory. One dominant model—synaptic tag-and-capture—proposes that new memory tags capture plasticity-related products that are available during or shortly after encoding (Redondo & Morris, 2011; Viola, Ballarini, Martinez, & Moncada, 2014), resulting in memory benefits for salient information and other information encoded around the same time. This model can explain recent behavioral evidence in humans documenting spillover effects that enhance memory in the context of rewarding events (Gruber, Gelman, & Ranganath, 2014; Loh, Deacon, de Boer, Dolan, & Düzel, 2015; Mather & Schoeke, 2011; Murayama & Kitagami, 2014; Stanek et al., 2019) and threatening experiences (Dunsmoor et al., 2015). It remains to be seen whether synaptic tag-and-capture models can also explain some of the memory-impairing (i.e., competitive) effects of emotional arousal. Another model has recently been developed to explain such competitive effects. The Glutamate Amplifies Noradrenergic Effects, or GANE model (Mather, Clewett, Sakaki, & Harley, 2015), argues that norepinephrine influences neural activity as a function of local glutamatergic activity, leading to enhanced plasticity for prioritized representations and reduced plasticity for other representations. This can lead to changes in the efficacy of encoding or consolidation for prioritized over nonprioritized information. Another model that bridges the findings on reward and negative emotion (Murty & Adcock, 2017) explains how dopamine and norepinephrine modulate different aspects of MTL function, resulting in distinct profiles of memory expression. In line with the reviewed evidence, the model suggests that reward enhances associative memory via SN/VTA-hippocampal mechanisms, whereas emotionally negative events enhance item memory via mechanisms in the amygdala and cortical MTL (Murty & Adcock, 2017).

Affect and motivation are intertwined in their effects on cognition (c.f., Chiew & Braver, 2011). Both affective and motivational states involve changes in arousal that could engage both the noradrenergic and dopaminergic pathways. Disentangling these contributions to episodic memory modulation is a key challenge for future research. Another open question is how the effects of neuromodulators on encoding processes interact with neuromodulatory effects on consolidation processes. Most studies suggest that the observed consolidation effects are independent of encoding-related processes (Gruber et al., 2016; Murty et al., 2017; Tambini et al., 2016). However, other evidence indicates that the effects of postencoding arousal depend on processes engaged during encoding (Bennion, Mickley Steinmetz, Kensinger, & Payne, 2013; Dunsmoor et al., 2015; Ritchey et al., 2017), consistent with the idea that encoding "tags" lead to enhanced consolidation. It remains to be seen whether similar interactions support the memory prioritization of rewarding events. Finally, although we focused only on the effects of negative emotion and reward on encoding and consolidation, these factors may have an additional impact on memory retrieval (Bowen, Kark, & Kensinger, 2017; Wolosin, Zeithamova, & Preston, 2013). Future research must consider the cumulative and interacting effects of neuromodulators on multiple memory processes.

Acknowledgments

Matthias J. Gruber was supported by a COFUND Fellowship from the European Commission and the Welsh government. Maureen Ritchey was supported by National Institutes of Health grant R00MH103401.

REFERENCES

Adcock, R. A., Thangavel, A., Whitfield-Gabrieli, S., Knutson, B., & Gabrieli, J. D. E. (2006). Reward-motivated learning: Mesolimbic activation precedes memory formation. *Neuron, 50*(3), 507–517.

Amaral, D. G., Behniea, H., & Kelly, J. L. (2003). Topographic organization of projections from the amygdala to the visual cortex in the macaque monkey. *Neuroscience, 118*(4), 1099–1120.

Anderson, A. K., & Phelps, E. A. (2001). Lesions of the human amygdala impair enhanced perception of emotionally salient events. *Nature, 411*(6835), 305–309.

Anderson, A. K., Wais, P. E., & Gabrieli, J. D. E. (2006). Emotion enhances remembrance of neutral events past. *Proceedings of the National Academy of Sciences of the United States of America, 103*(5), 1599–1604.

Andreano, J. M., & Cahill, L. (2006). Glucocorticoid release and memory consolidation in men and women. *Psychological Science, 17*(6), 466–470.

Ballard, I. C., Murty, V. P., Carter, R. M., MacInnes, J. J., Huettel, S. A., & Adcock, R. A. (2011). Dorsolateral prefrontal cortex drives mesolimbic dopaminergic regions to initiate motivated behavior. *Journal of Neuroscience, 31*(28), 10340–10346.

Bennion, K. A., Mickley Steinmetz, K. R., Kensinger, E. A., & Payne, J. D. (2013). Sleep and cortisol interact to support memory consolidation. *Cerebral Cortex, 25*(3), 646–657.

Bisby, J. A., Horner, A. J., Hørlyck, L. D., & Burgess, N. (2016). Opposing effects of negative emotion on amygdalar and hippocampal memory for items and associations. *Social Cognitive and Affective Neuroscience, 11*(6), 981–990.

Bowen, H. J., Kark, S. M., & Kensinger, E. A. (2017). NEVER forget: Negative emotional valence enhances recapitulation. *Psychonomic Bulletin & Review, 25*(3), 870–891.

Bunzeck, N., Guitart-Masip, M., Dolan, R. J., & Düzel, E. (2011). Contextual novelty modulates the neural dynamics of reward anticipation. *Journal of Neuroscience, 31*(36), 12816–12822.

Cahill, L., Gorski, L., & Le, K. (2003). Enhanced human memory consolidation with post-learning stress: Interaction with the degree of arousal at encoding. *Learning & Memory, 10*(4), 270–274.

Chiew, K. S., & Braver, T. S. (2011). Positive affect versus reward: Emotional and motivational influences on cognitive control. *Frontiers in Psychology, 2,* 279.

Clewett, D., Huang, R., Velasco, R., Lee, T.-H., & Mather, M. (2018). Locus coeruleus activity strengthens prioritized memories under arousal. *Journal of Neuroscience, 38*(6), 1558–1574.

Cohen, M. S., Rissman, J., Suthana, N. A., Castel, A. D., & Knowlton, B. J. (2014). Value-based modulation of memory encoding involves strategic engagement of fronto-temporal semantic processing regions. *Cognitive, Affective & Behavioral Neuroscience, 14*(2), 578–592.

Davachi, L. (2006). Item, context and relational episodic encoding in humans. *Current Opinion in Neurobiology, 16*(6), 693–700.

de Voogd, L. D., Klumpers, F., Fernández, G., & Hermans, E. J. (2017). Intrinsic functional connectivity between amygdala and hippocampus during rest predicts enhanced memory under stress. *Psychoneuroendocrinology, 75,* 192–202.

Dewar, M., Alber, J., Butler, C., Cowan, N., & Della Sala, S. (2012). Brief wakeful resting boosts new memories over the long term. *Psychological Science, 23*(9), 955–960.

Dolan, R. J., & Vuilleumier, P. (2006). Amygdala automaticity in emotional processing. *Annals of the New York Academy of Sciences, 985*(1), 348–355.

Dunsmoor, J. E., Murty, V. P., Davachi, L., & Phelps, E. A. (2015). Emotional learning selectively and retroactively strengthens memories for related events. *Nature, 520*(7547), 345–348.

Düzel, E., Bunzeck, N., Guitart-Masip, M., & Düzel, S. (2010). NOvelty-related motivation of anticipation and exploration by dopamine (NOMAD): Implications for healthy aging. *Neuroscience and Biobehavioral Reviews, 34*(5), 660–669.

Eichenbaum, H., Yonelinas, A. P., & Ranganath, C. (2007). The medial temporal lobe and recognition memory. *Annual Review of Neuroscience, 30,* 123–152.

Feld, G. B., Besedovsky, L., Kaida, K., Münte, T. F., & Born, J. (2014). Dopamine D2-like receptor activation wipes out preferential consolidation of high over low reward memories during human sleep. *Journal of Cognitive Neuroscience, 26*(10), 2310–2320.

Floresco, S. B., West, A. R., Ash, B., Moore, H., & Grace, A. A. (2003). Afferent modulation of dopamine neuron firing differentially regulates tonic and phasic dopamine transmission. *Nature Neuroscience, 6*(9), 968–973.

Gruber, M. J., Gelman, B. D., & Ranganath, C. (2014). States of curiosity modulate hippocampus-dependent learning via the dopaminergic circuit. *Neuron, 84*(2), 486–496.

Gruber, M. J., & Otten, L. J. (2010). Voluntary control over prestimulus activity related to encoding. *Journal of Neuroscience, 30*(29), 9793–9800.

Gruber, M. J., & Ranganath, C. Wakeful rest prioritizes associative memory for high-reward information. Manuscript in preparation.

Gruber, M. J., Ritchey, M., Wang, S.-F., Doss, M. K., & Ranganath, C. (2016). Post-learning hippocampal dynamics promote preferential retention of rewarding events. *Neuron, 89*(5), 1110–1120.

Gruber, M. J., Watrous, A. J., Ekstrom, A. D., Ranganath, C., & Otten, L. J. (2013). Expected reward modulates encoding-related theta activity before an event. *NeuroImage, 64*, 68–74.

Henckens, M. J. A. G., Hermans, E. J., Pu, Z., Joëls, M., & Fernández, G. (2009). Stressed memories: How acute stress affects memory formation in humans. *Journal of Neuroscience, 29*(32), 10111–10119.

Hermans, E. J., Kanen, J. W., & Tambini, A. (2016). Persistence of amygdala-hippocampal connectivity and multivoxel correlation structures during awake rest after fear learning predicts long-term expression of fear. *Cerebral Cortex, 27*(5), 3028–3041.

Hurlemann, R., Hawellek, B., Matusch, A., Kolsch, H., Wollersen, H., Madea, B., ... Dolan, R. J. (2005). Noradrenergic modulation of emotion-induced forgetting and remembering. *Journal of Neuroscience, 25*(27), 6343–6349.

Kensinger, E. A., & Corkin, S. (2004). Two routes to emotional memory: Distinct neural processes for valence and arousal. *Proceedings of the National Academy of Sciences of the United States of America, 101*(9), 3310–3315.

Kensinger, E. A., Garoff-Eaton, R. J., & Schacter, D. L. (2007). How negative emotion enhances the visual specificity of a memory. *Journal of Cognitive Neuroscience, 19*(11), 1872–1887.

Kleinsmith, L. J., & Kaplan, S. (1963). Paired-associate learning as a function of arousal and interpolated interval. *Journal of Experimental Psychology, 65*, 190–193.

Knutson, B., Adams, C. M., Fong, G. W., & Hommer, D. (2001). Anticipation of increasing monetary reward selectively recruits nucleus accumbens. *Journal of Neuroscience, 21*(16), RC159.

LaBar, K. S., & Cabeza, R. (2006). Cognitive neuroscience of emotional memory. *Nature Reviews. Neuroscience, 7*(1), 54–64.

LaBar, K. S., & Phelps, E. A. (1998). Arousal-mediated memory consolidation: Role of the medial temporal lobe in humans. *Psychological Science, 9*(6), 490–493.

Lane, R. D., Chua, P. M.-L., & Dolan, R. J. (1999). Common effects of emotional valence, arousal and attention on neural activation during visual processing of pictures. *Neuropsychologia, 37*, 989–997.

Lisman, J. E., & Grace, A. A. (2005). The hippocampal-VTA loop: Controlling the entry of information into long-term memory. *Neuron, 46*(5), 703–713.

Lisman, J., Grace, A. A., & Düzel, E. (2011). A neoHebbian framework for episodic memory; role of dopamine-dependent late LTP. *Trends in Neurosciences, 34*(10), 536–547.

Loh, E., Deacon, M., de Boer, L., Dolan, R. J., & Düzel, E. (2015). Sharing a context with other rewarding events increases the probability that neutral events will be recollected. *Frontiers in Human Neuroscience, 9*, 683.

Mackiewicz, K. L., Sarinopoulos, I., Cleven, K. L., & Nitschke, J. B. (2006). The effect of anticipation and the specificity of sex differences for amygdala and hippocampus function in emotional memory. *Proceedings of the National Academy of Sciences of the United States of America, 103*(38), 14200–14205.

Madan, C. R., Fujiwara, E., Caplan, J. B., & Sommer, T. (2017). Emotional arousal impairs association-memory: Roles of amygdala and hippocampus. *NeuroImage, 156*, 14–28.

Mather, M., Clewett, D., Sakaki, M., & Harley, C. W. (2015). Norepinephrine ignites local hot spots of neuronal excitation: How arousal amplifies selectivity in perception and memory. *Behavioral and Brain Sciences, 39*, e200.

Mather, M., & Schoeke, A. (2011). Positive outcomes enhance incidental learning for both younger and older adults. *Frontiers in Neuroscience, 5*, 129.

Mather, M., & Sutherland, M. R. (2011). Arousal-biased competition in perception and memory. *Perspectives on Psychological Science: A Journal of the Association for Psychological Science, 6*(2), 114–133.

McCullough, A. M., Ritchey, M., Ranganath, C., & Yonelinas, A. (2015). Differential effects of stress-induced cortisol responses on recollection and familiarity-based recognition memory. *Neurobiology of Learning and Memory, 123*, 1–10.

McGaugh, J. L. (2004). The amygdala modulates the consolidation of memories of emotionally arousing experiences. *Annual Review of Neuroscience, 27*(1), 1–28.

McNamara, C. G., Tejero-Cantero, Á., Trouche, S., Campo-Urriza, N., & Dupret, D. (2014). Dopaminergic neurons promote hippocampal reactivation and spatial memory persistence. *Nature Neuroscience, 17*(12), 1658–1660.

Mickley, K. R., & Kensinger, E. A. (2008). Emotional valence influences the neural correlates associated with remembering and knowing. *Cognitive, Affective & Behavioral Neuroscience, 8*(2), 143–152.

Miendlarzewska, E. A., Bavelier, D., & Schwartz, S. (2016). Influence of reward motivation on human declarative memory. *Neuroscience and Biobehavioral Reviews, 61*, 156–176.

Murayama, K., & Kitagami, S. (2014). Consolidation power of extrinsic rewards: Reward cues enhance long-term memory for irrelevant past events. *Journal of Experimental Psychology. General, 143*(1), 15–20.

Murayama, K., & Kuhbandner, C. (2011). Money enhances memory consolidation—but only for boring material. *Cognition, 119*(1), 120–124.

Murray, E. A., & Bussey, T. J. (1999). Perceptual-mnemonic functions of the perirhinal cortex. *Trends in Cognitive Sciences, 3*(4), 142–151.

Murty, V. P., & Adcock, R. A. (2014). Enriched encoding: Reward motivation organizes cortical networks for hippocampal detection of unexpected events. *Cerebral Cortex, 24*(8), 2160–2168.

Murty, V. P., & Adcock, R. A. (2017). Distinct medial temporal lobe network states as neural contexts for motivated memory formation. In D. E. Hannula & M. C. Duff (Eds.), *The hippocampus from cells to systems* (pp. 467–501). New York: Springer.

Murty, V. P., & Dickerson, K. C. (2017). Motivational influences on memory. In *Recent developments in neuroscience research on human motivation* (pp. 203–227). Bingley, UK: Emerald Group Publishing.

Murty, V. P., LaBar, K. S., & Adcock, R. A. (2012). Threat of punishment motivates memory encoding via amygdala, not midbrain, interactions with the medial temporal lobe. *Journal of Neuroscience, 32*(26), 8969–8976.

Murty, V. P., Tompary, A., Adcock, R. A., & Davachi, L. (2017). Selectivity in postencoding connectivity with high-level visual cortex is associated with reward-motivated memory. *Journal of Neuroscience, 37*(3), 537–545.

Oyarzún, J. P., Packard, P. A., de Diego-Balaguer, R., & Fuentemilla, L. (2016). Motivated encoding selectively promotes memory for future inconsequential semantically-related events. *Neurobiology of Learning and Memory, 133,* 1–6.

Patil, A., Murty, V. P., Dunsmoor, J. E., Phelps, E. A., & Davachi, L. (2017). Reward retroactively enhances memory consolidation for related items. *Learning & Memory, 24*(1), 65–69.

Pizzagalli, D. A., Lehmann, D., Hendrick, A. M., Regard, M., Pascual-Marqui, R. D., & Davidson, R. J. (2002). Affective judgments of faces modulate early activity (approximately 160 ms) within the fusiform gyri. *NeuroImage, 16*(3), 663–677.

Preuss, D., & Wolf, O. T. (2009). Post-learning psychosocial stress enhances consolidation of neutral stimuli. *Neurobiology of Learning and Memory, 92*(3), 318–326.

Price, J. L. (2006). Comparative aspects of amygdala connectivity. *Annals of the New York Academy of Sciences, 985*(1), 50–58.

Redondo, R. L., & Morris, R. G. M. (2011). Making memories last: The synaptic tagging and capture hypothesis. *Nature Reviews. Neuroscience, 12*(1), 17–30.

Ritchey, M., Dolcos, F., & Cabeza, R. (2008). Role of amygdala connectivity in the persistence of emotional memories over time: An event-related FMRI investigation. *Cerebral Cortex, 18*(11), 2494–2504.

Ritchey, M., LaBar, K. S., & Cabeza, R. (2011). Level of processing modulates the neural correlates of emotional memory formation. *Journal of Cognitive Neuroscience, 23*(4), 757–771.

Ritchey, M., McCullough, A. M., Ranganath, C., & Yonelinas, A. P. (2017). Stress as a mnemonic filter: Interactions between medial temporal lobe encoding processes and post-encoding stress. *Hippocampus, 27*(1), 77–88.

Ritchey, M., Wang, S.-F., Yonelinas, A. P., & Ranganath, C. (2018). Dissociable medial temporal pathways for encoding emotional item and context information. *Neuropsychologia, 124,* 66–78.

Roozendaal, B., & Hermans, E. J. (2017). Norepinephrine effects on the encoding and consolidation of emotional memory: Improving synergy between animal and human studies. *Current Opinion in Behavioral Sciences, 14,* 115–122.

Roozendaal, B., & McGaugh, J. L. (2011). Memory modulation. *Behavioral Neuroscience, 125*(6), 797–824.

Sharot, T., & Yonelinas, A. P. (2008). Differential time-dependent effects of emotion on recollective experience and memory for contextual information. *Cognition, 106*(1), 538–547.

Shields, G. S., Sazma, M. A., McCullough, A. M., & Yonelinas, A. P. (2017). The effects of acute stress on episodic memory: A meta-analysis and integrative review. *Psychological Bulletin, 143*(6), 636–675.

Shohamy, D., & Adcock, R. A. (2010). Dopamine and adaptive memory. *Trends in Cognitive Sciences, 14*(10), 464–472.

Singer, A. C., & Frank, L. M. (2009). Rewarded outcomes enhance reactivation of experience in the hippocampus. *Neuron, 64*(6), 910–921.

Smeets, T., Otgaar, H., Candel, I., & Wolf, O. T. (2008). True or false? Memory is differentially affected by stress-induced cortisol elevations and sympathetic activity at consolidation and retrieval. *Psychoneuroendocrinology, 33*(10), 1378–1386.

Stanek, J. K., Dickerson, K. C., Chiew, K. S., Clement, N. J., & Adcock, R. A. (2019). Expected reward value and reward uncertainty have temporally dissociable effects on memory formation. *Journal of Cognitive Neuroscience.* doi: 10.1162/jocn_a_01411

Stefanacci, L., Suzuki, W. A., & Amaral, D. G. (1996). Organization of connections between the amygdaloid complex and the perirhinal and parahippocampal cortices in macaque monkeys. *Journal of Comparative Neurology, 375*(4), 552–582.

Talmi, D., Schimmack, U., Paterson, T., & Moscovitch, M. (2007). The role of attention and relatedness in emotionally enhanced memory. *Emotion, 7*(1), 89–102.

Tambini, A., Rimmele, U., Phelps, E. A., & Davachi, L. (2016). Emotional brain states carry over and enhance future memory formation. *Nature Neuroscience, 20*(2), 271–278.

van Marle, H. J. F., Hermans, E. J., Qin, S., & Fernández, G. (2010). Enhanced resting-state connectivity of amygdala in the immediate aftermath of acute psychological stress. *NeuroImage, 53*(1), 348–354.

Viola, H., Ballarini, F., Martinez, M. C., & Moncada, D. (2014). The tagging and capture hypothesis from synapse to memory. *Progress in Molecular Biology and Translational Science, 122,* 391–423.

Vuilleumier, P., Armony, J. L., Driver, J., & Dolan, R. J. (2001). Effects of attention and emotion on face processing in the human brain: An event-related fMRI study. *Neuron, 30*(3), 829–841.

Vuilleumier, P., Richardson, M. P., Armony, J. L., Driver, J., & Dolan, R. J. (2004). Distant influences of amygdala lesion on visual cortical activation during emotional face processing. *Nature Neuroscience, 7*(11), 1271–1278.

Waring, J. D., & Kensinger, E. A. (2009). Effects of emotional valence and arousal upon memory trade-offs with aging. *Psychology and Aging, 24*(2), 412–422.

Waring, J. D., & Kensinger, E. A. (2011). How emotion leads to selective memory: Neuroimaging evidence. *Neuropsychologia, 49*(7), 1831–1842.

Wittmann, B. C., Schott, B. H., Guderian, S., Frey, J. U., Heinze, H.-J., & Düzel, E. (2005). Reward-related FMRI activation of dopaminergic midbrain is associated with enhanced hippocampus-dependent long-term memory formation. *Neuron, 45*(3), 459–467.

Wolosin, S. M., Zeithamova, D., & Preston, A. R. (2012). Reward modulation of hippocampal subfield activation during successful associative encoding and retrieval. *Journal of Cognitive Neuroscience, 24*(7), 1532–1547.

Wolosin, S. M., Zeithamova, D., & Preston, A. R. (2013). Distributed hippocampal patterns that discriminate reward context are associated with enhanced associative binding. *Journal of Experimental Psychology. General, 142*(4), 1264–1276.

Yeung, N., & Sanfey, A. G. (2004). Independent coding of reward magnitude and valence in the human brain. *Journal of Neuroscience, 24*(28), 6258–6264.

Yonelinas, A. P., & Ritchey, M. (2015). The slow forgetting of emotional episodic memories: An emotional binding account. *Trends in Cognitive Sciences, 19*(5), 259–267.

23 Replay-Based Consolidation Governs Enduring Memory Storage

KEN A. PALLER, JAMES W. ANTONY, ANDREW R. MAYES, AND KENNETH A. NORMAN

ABSTRACT The human ability to remember unique experiences from many years ago comes so naturally that we often take it for granted. It depends on three stages: (1) encoding, when new information is initially registered, (2) storage, when encoded information is held in the brain, and (3) retrieval, when stored information is used. Historically, cognitive neuroscience studies of memory have emphasized encoding and retrieval. Yet the intervening stage may hold the most intrigue and has become a major research focus in the years since the last edition of this book. Here we describe recent investigations of postacquisition memory processing in relation to enduring storage. This evidence of memory processing belies the notion that memories stored in the brain are held in stasis, without changing. Various methods for influencing and monitoring brain activity have been applied to study off-line memory processing. In particular, memories can be reactivated during sleep and during resting periods, with distinctive physiological correlates. These neural signals shed light on the contribution of hippocampal-neocortical interactions to memory consolidation. Overall, results converge on a framework whereby memory reactivation is a critical determinant of systems-level consolidation, and thus of future remembering, which in turn facilitates future planning and problem solving.

How do we acquire new knowledge? Not easily! We often fail to retain important information, even when we try to forestall forgetting by rehearsing what we wish to keep. Indeed, repeated retrieval may be the key to enduring memory storage. Yet a deep conundrum remains in that intentional retrieval alone cannot explain the seemingly unpredictable way that some memories drift away while others are retained.

This chapter explores the idea that memory storage also depends on rehearsal that occurs unintentionally and implicitly, including while we sleep. A key driving force behind consolidation, according to our view, is the regular reactivation of memories without our awareness. This view goes beyond the first-person sense of rehearse-to-remember. When rehearsal is hidden, the consequences may go unnoticed. Whereas speculations about consolidation have largely been derived from behavioral and neural studies of memory change over time, particularly in retrograde amnesia, the incremental improvements in storage due to consolidation have been difficult to observe. The additional consideration that we emphasize here, with implications for making such observations, is that memories change in fundamental ways in conjunction with unconscious rehearsal.

The journey of a memory, such as the memory of a unique life event like reading this sentence, begins with encoding and concurrent neural plasticity. The journey may be a long one; a single event may be remembered many years later. If so, one might say that such a memory existed for the duration of that multiyear period, like a file secured away in a file drawer. This commonplace notion—that "the memory" per se lasts from encoding until retrieval—reifies it as existing in a static manner, independently, set apart from other memories. This view is misleading.

Somehow, neural substrates of memory storage must traverse the entire storage interval for a memory to ultimately be retrieved. However, if memories are not static entities, how should we characterize memory storage during this interval? Changes in storage are not a simple matter of the memory transitioning from a labile state to a stable one, such as when a newly created ceramic object is heated. A progression of neural restructuring seems more likely, particularly for an episode from long ago.

Such progressive changes are widely acknowledged as fundamental to the neurobiology of consolidation, now being intensively investigated on many fronts. Through neural restructuring, the informational content of memories can also change. Memories are subject to gradual integration with other stored knowledge; emergence of a theme or interpretation; stabilization of certain features; stripping away of details; gist formation; generalization; forming novel associations among features; producing creative new ideas; and, ultimately, the crystallization of a set of memories that form the fabric of one's life story.

Whereas our thesis is that memory reactivation is a critical determinant of memory storage, one classic memory phenomenon—the *flashbulb memory*—seems in direct opposition. A classical flashbulb memory is found when a person can recount, in detail, learning of some momentous public event, such as an assassination. The metaphorical flashbulb would illuminate everything in view at that instant; that singular moment would be frozen in time, preserved in a permastore to remain forever available. Livingston (1967) proposed that the emotional impact engaged a "now-print" mechanism that permanently preserved the event and all concurrent details. However, flashbulb memories become distorted just like ordinary episodic memories (Schmolck, Buffalo, & Squire, 2000). Repeatedly retelling a story is a common way to introduce distortions. So our view is that these momentous events are not immediately etched into memory. In place of the classic view of flashbulb memories, we attribute their dramatic persistence to repeated memory reactivation.

Likewise, we may carry some memories with us throughout our lives, thanks to consolidation rather than to superior encoding. The most decisive memory process could be repeated reactivation, some of which occurs implicitly. Off-line reactivation and concomitant plasticity may even be a necessity for enduring memory storage, ultimately determining which memories we keep. In this account of memory preservation, how should we now conceptualize the "replay" of a memory?

Defining "Replay" in the Context of Memory Categories

The prime directive of a *Star Trek* expedition to an alien planet is to avoid undue interference with another culture. The prime directive of an expedition in memory research is to acknowledge that different types of memory depend on distinct mechanisms.

What type of memory are we talking about? William James' (1890) classic distinction between *primary memory* and *secondary memory* is an appropriate starting point. The former comprises the content of our moment-to-moment train of thought, whereas the latter concerns information brought back to mind after departing from awareness. James' terms were supplanted by the contrast between *short-* and *long-term memory* (STM and LTM), but this distinction is problematic because it emphasizes time span. As long as active rehearsal continues, information can be kept alive. In place of STM, with time span as the defining feature, *immediate memory* and *working memory* adequately designate information kept in mind.

Time span is nevertheless essential to consider. Memory research typically emphasizes acquisition-to-retrieval delays not longer than a few minutes. In contrast, here we strive to explain *enduring memory storage*—memories that somehow last days, weeks, even years in the face of the daily trudge of new learning, wherein forgetting seems to be the rule.

Declarative memory is defined as the type of memory used in recalling and recognizing episodes and facts. Patients with circumscribed amnesia have difficulty with recent episodic and factual knowledge. Their capabilities on tests designed to assess other types of memory—such as skills, procedures, priming, conditioning, and habits—can be entirely preserved. These other types of memory have been categorized collectively as *nondeclarative memory*. Although replay is certainly relevant for nondeclarative memory, here we focus on declarative memory.

The fundamental distinctiveness of declarative memory likely arises in relation to (1) storage across multiple neocortical regions and (2) the potential for conscious recollection. For example, the components of a specific event, including relevant causes and repercussions, are represented in multiple neocortical regions specialized for processing different informational features. Recollecting an enduring declarative memory relies on combining such assorted elements. Because the cortical fragments are spatially separated in the brain, they must be linked to form a cohesive unit, requiring what at a neural level can be called cross-cortical storage (Paller, 1997, 2002) or, at a cognitive level, relational representations (Eichenbaum & Cohen, 2001; Shimamura, 2002).

Another fundamental characteristic of enduring declarative memories is that storage is altered gradually via consolidation (Squire, Cohen, & Nadel, 1984). Which pathway will a newly formed memory take—stabilization, integration, corruption, forgetting? Optimally, an initial stage of rapid plasticity involving the formation of new hippocampal connections with various cortical representations is followed by a gradual process involving further hippocampal-neocortical interaction (McClelland, McNaughton, & O'Reilly, 1995). Postacquisition processing may promote cross-cortical storage by gradually and thoroughly binding together a memory's distinct representational components. Synaptic consolidation involves molecular changes at individual synapses shortly after learning; systems consolidation concerns changes in storage that take place over a prolonged period of time and that involve multiple brain regions. Systems consolidation can include restructuring, and this restructuring may continue indefinitely (Dudai, 2012).

A pivotal physiological bond between consolidation and the hippocampus comes from reports of *hippocampal replay* in rodent place cells (reviewed by Foster, 2017). Firing patterns during sleep mirrored those previously exhibited during exploratory behavior in a new environment (Pavlides & Winson, 1989; Wilson & McNaughton, 1994). Replay is also found during wake, in cortical regions, in the striatum, and in various forms in multiple species. Although the term *replay* is sometimes restricted to repeated firing sequences in hippocampal place cells, here we use the term *replay* to encompass the notion of any neural recapitulation of stored information and *hippocampal replay* to denote this specific example.

If replay is at the heart of declarative memory consolidation, the opportunity may arise each and every time a memory is reactivated, online or off-line. Online reactivation would be when one knowingly recalls a memory, intentionally or otherwise. The canonical example of an off-line period is when we sleep.

Memory Processing during Sleep

The notion that memories change during sleep has not always been on the radar of memory researchers. Our view is that declarative memories change both during waking and during sleep and that such changes contribute to the gradual process of consolidation (Paller, 1997; Paller & Voss, 2004). Substantial empirical support has accrued for sleep-based memory processing (Rasch & Born, 2013). According to this view, memories do not just lie dormant during sleep but instead receive regular exercise that changes what is stored.

Sleep has a complex physiological architecture. The classic staging of sleep into just four stages is deceptive in its apparent simplicity. Electroencephalographic (EEG) signals differ markedly between *slow-wave sleep* (SWS, also known as N3) and *rapid eye movement sleep* (REM). Non-REM sleep includes three stages—N1, N2, and N3—going from light sleep to deep sleep. Current thinking is that SWS and REM have complementary memory functions.

In prior decades before the recent waves of empirical support, many theories on memory and sleep were entertained (e.g., Cartwright, 1977; Marr, 1971; Winson, 1985). An intuitively reasonable idea was that sleep supports adaptive mechanisms for evaluating recent experiences and relating them to current goals. Hippocampal replay connects with these ideas, although early studies of hippocampal replay lacked suitable behavioral measures that might show improved spatial memory following sleep, so hippocampal replay could not be directly linked with consolidation.

A good case can now be made to link consolidation with both hippocampal replay and hippocampal *sharp-wave ripples* (SWRs; ripples are high-frequency bursts in field-potential recordings, 100–250 Hz, lasting approximately 50 ms). For example, hippocampal replay can occur during SWRs, which increase as a function of learning (Dupret, O'Neill, Pleydell-Bouverie, & Csicsvari, 2010; O'Neill, Senior, Allen, Huxter, & Csicsvari, 2008; Peyrache, Khamassi, Benchenane, Wiener, & Battaglia, 2009). More telling, hippocampal replay is specific to learning-related ensembles and correlates with retention (Dupret et al., 2010). Furthermore, manipulating SWRs alters memory (Barnes & Wilson, 2014; Ego-Stengel & Wilson, 2009; Girardeau, Benchenane, Wiener, Buzsáki, & Zugaro, 2009).

Additional evidence brings in cortical activity, as neocortical SWRs and hippocampal SWRs can be observed together with thalamocortical *sleep spindles* (Khodagholy, Gelinas, & Buzsáki, 2017; Siapas & Wilson, 1998). Spindles are brief (0.5–3 s) oscillations at approximately 11–16 Hz. Spindles may both be temporally guided by cortical slow waves and help to synchronize hippocampal SWRs with cortical activity.

In humans, ample results demonstrate superior memory after a period of sleep compared to a period of wake (Rasch & Born, 2013). In an extreme way, sleep deprivation can produce such a result, but this can be problematic because of memory difficulties arising from excessive sleepiness or nonspecific effects of deprivation, such as stress. In any such sleep/wake comparison, wakefulness can entail more memory interference than sleep, calling into question whether sleep necessarily made a specific contribution. Thus, this sort of evidence provides only tentative support for the notion that sleep after learning improves memory.

To get a better handle on how the physiology of sleep might map onto processing pertaining to consolidation, we will need to better specify connections between specific signals in sleep EEG and specific aspects of memory processing. One way to reach for this goal, while also avoiding the problem of differential memory interference that plagues sleep/wake comparisons, is to use subtle but systematic sensory stimulation during sleep.

Manipulating memory during sleep The literature on presenting a sleeper with cues to information recently learned while awake has grown considerably in the last few years (Cellini & Capuozzo, 2018; Oudiette & Paller, 2013; Schouten, Pereira, Tops, & Louzada, 2017). Note that gaining new knowledge presented only during sleep was ostensibly ruled out by Emmons and Simon (1956), who investigated presenting spoken facts during sleep. Their subjects showed no evidence of learning as long as no signs of arousal were present in EEG recordings. Many studies on this topic up to that point did not include physiological verification of sleep state, which came to be deemed essential. The work of Emmons and Simon led to widespread skepticism in the scientific community about the validity of so-called sleep learning, impeding workers from pursuing many adjacent research directions (Paller & Oudiette, 2018). However, recent findings show that some implicit learning during sleep may indeed be possible (Arzi et al., 2012; Andrillon et al., 2017).

Here we focus instead on the use of sensory stimulation to study brain mechanisms, whereby memories formed while awake can be consolidated during sleep. Among the early studies on this topic were classical-conditioning studies in rats trained to fear a tone repeatedly paired with a shock during wakefulness; conditioning was enhanced by a mild shock during sleep (Hars, Hennevin, & Pasques, 1985; Hennevin, Hars, Maho, & Bloch, 1995). Smith and Weeden (1990) trained people in a complex finger-tapping task while listening to a ticking sound, and performance was improved by playing the sound during sleep. In the landmark study of Rasch and colleagues (2007), a rose odor was presented while subjects learned spatial locations of objects. Presenting the rose odor again during SWS improved cued recall of all the learned locations (relative to several control conditions in other subjects) and functional magnetic resonance imaging (fMRI) showed hippocampal activation, a putative correlate of the memory reactivation.

In 2009 we took the further step of showing that specific memories could be strengthened using sounds during sleep (Rudoy, Voss, Westerberg, & Paller, 2009; figure 23.1). *Targeted memory reactivation* (TMR) refers to this method for selectively manipulating memory during sleep. Whereas memory comparisons following a period of sleep versus wake can be confounded by indirect effects of alertness or interference, TMR studies are immune from this problem. TMR studies generally rely on within-subject contrasts of postsleep performance for cued versus uncued material. Selectively improved recall performance after TMR during sleep thus demonstrated that specific memories were changed, an effect replicated in subsequent studies (e.g., Creery, Oudiette, Antony, & Paller, 2014; Vargas, Schechtman, & Paller, 2019).

Auditory processing may be reduced during sleep, but it is not eliminated. Van Dongen and colleagues (2012) examined TMR while subjects slept during fMRI scanning. Subjects were motivated to suppress auditory processing, given the exceedingly loud scanning noise. Supporting the idea of sensory gating operative at the level of the thalamus, the degree of memory benefit, which was not reliable overall, was correlated with brain activation in the thalamus across subjects. The degree of memory benefit was also correlated with activity in the medial temporal lobe and the cerebellum, as well as with parahippocampal-precuneus connectivity, thus identifying several measures of brain activity associated with sound-cued memory reactivation (see also Berkers, Ekman, van Dongen, Takashima, Paller, & Fernandez, 2018; Shanahan, Gjorgieva, Paller, Kahnt, & Gottfried, 2018). In another study with the same spatial recall task, we showed that sleep without sounds favored high-value information (Oudiette, Antony, Creery, & Paller, 2013) recall for low-value items was brought up to the level of high-value items when low-value sound cues were presented during SWS. In a variation on these procedures with rodents, Bendor and Wilson (2012) used TMR to link reactivation with hippocampal replay. Tones previously associated with spatial learning were played during sleep, and a systematic bias in hippocampal place cell firing was found as a function of which tone was presented.

With TMR during sleep, memory can be manipulated by surreptitiously presenting part of what has been learned prior to sleep. In addition to influencing learning of spatial locations, TMR can influence a variety of other types of learning, including learning complex skills (Antony, Gobel, O'Hare, Reber, & Paller, 2012), foreign vocabulary (Schreiner & Rasch, 2014), conditioning (Hauner, Howard, Zelano, & Gottfried, 2013), body-ownership changes (Honma et al., 2016), and words in locations (Fuentemilla et al., 2013). In this last study, the degree of word recall benefit after TMR was inversely correlated with the degree of medial temporal damage in epileptic patients.

Another way to manipulate sleep that can provide clues about the relevant physiology is to entrain brain oscillations. Slow waves and sleep spindles have been linked with memory consolidation on the basis of correlative findings, along with direct manipulations, that strongly suggest a causal link. Disrupting SWS can produce memory difficulties (e.g., Landsness et al., 2009), but the disruption could affect memory either directly or indirectly. Therefore, sleep-memory connections can more convincingly be derived by facilitating SWS. Marshall and colleagues (2006) were the first to show that transcranial stimulation with slow oscillatory electrical currents can enhance slow waves and thereby benefit word-pair learning. Precisely timed auditory stimulation can have similar effects (e.g., Ngo, Martinetz, Born, & Mölle, 2013). Thus, there is convincing evidence that slow waves play a causal role in sleep-based memory consolidation.

Slow-wave entrainment often produces a concomitant increase in spindles as well. Spindles can also be entrained electrically (Lustenberger et al., 2016) or with auditory stimulation (Antony & Paller, 2017). A pharmacological approach, using Ambien, produced both an increase in spindles and an improvement in memory (Mednick et al., 2013). Spindle timing relative to slow-wave phase may be critical (Helfrich, Mander, Jagust, Knight, & Walker, 2018; Niknazar, Krishnan, Bazhenov, & Mednick, 2015). Although the precise role of sleep spindles in memory consolidation remains to be elucidated, recent studies have made significant headway (Antony et al., 2018; Cairney, Guttesen, El Marj, & Staresina, 2018; Schreiner, Lehmann, & Rasch, 2015; figure 23.2).

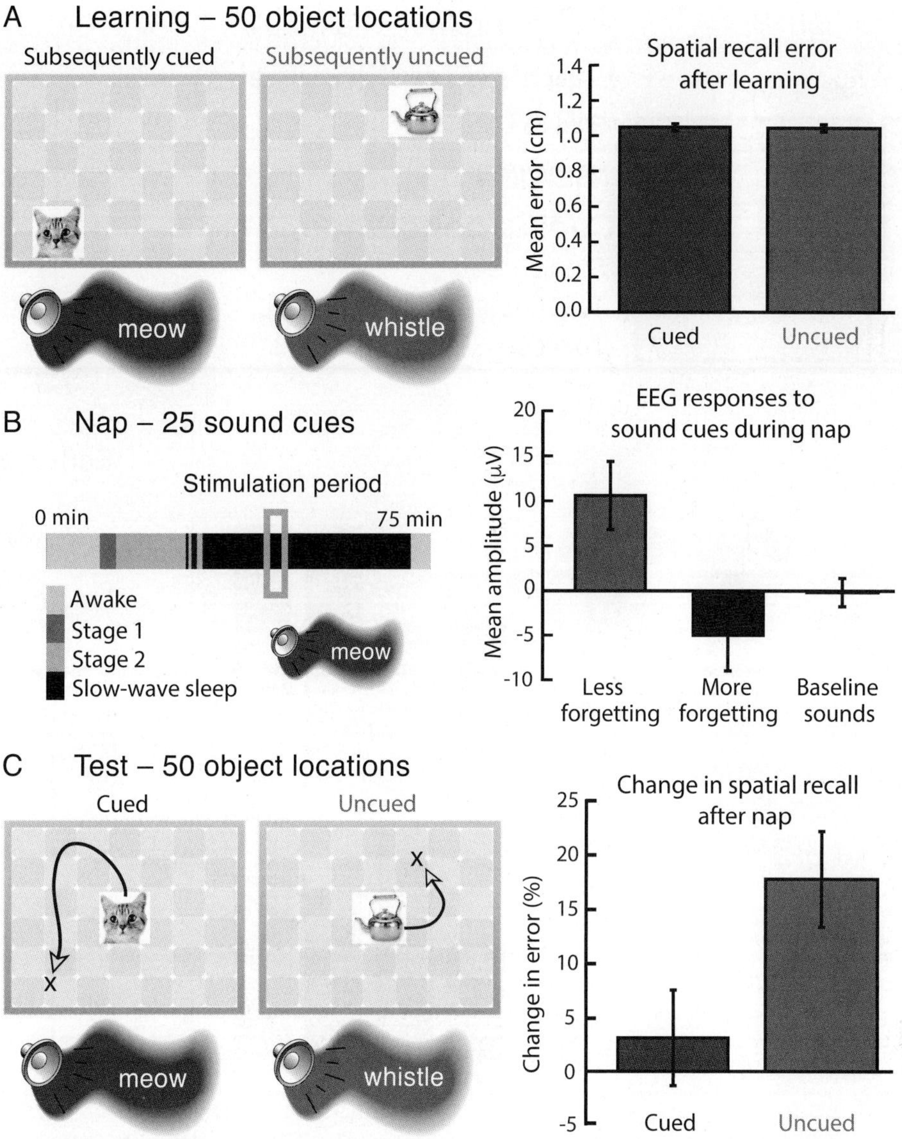

FIGURE 23.1 Targeted memory reactivation (TMR). *A*, Subjects in the study by Rudoy and colleagues (2009) first learned 50 object-location associations. Each object was presented with its characteristic sound. Following an interactive learning procedure, location recall was tested. Half of the objects were assigned to be cued during sleep such that recall accuracy was matched for cued and uncued objects. *B*, Next, subjects slept with EEG monitoring. When signs of SWS were evident, 25 of the sounds were presented at a low intensity. These sounds influenced memory storage without waking people up. *C*, Recall of locations was tested again after the nap. Subjects moved each object from the center to where they thought it belonged (*arrows*). Recall was more accurate for cued versus uncued objects. Mean EEG responses from 400–800 ms following the onset of each sound presented during sleep were found to be more positive for those objects with less decline in recall (*Less forgetting* in *B*) compared to the remaining objects or to baseline sounds. These responses resembled typical event-related potentials predictive of later memory (Dm effects; Paller et al., 1987), suggesting that spatial memory reactivation occurred as a consequence of cue presentation, leading to improved spatial recall after awakening. Reprinted from Rudoy et al. (2009). (See color plate 24.)

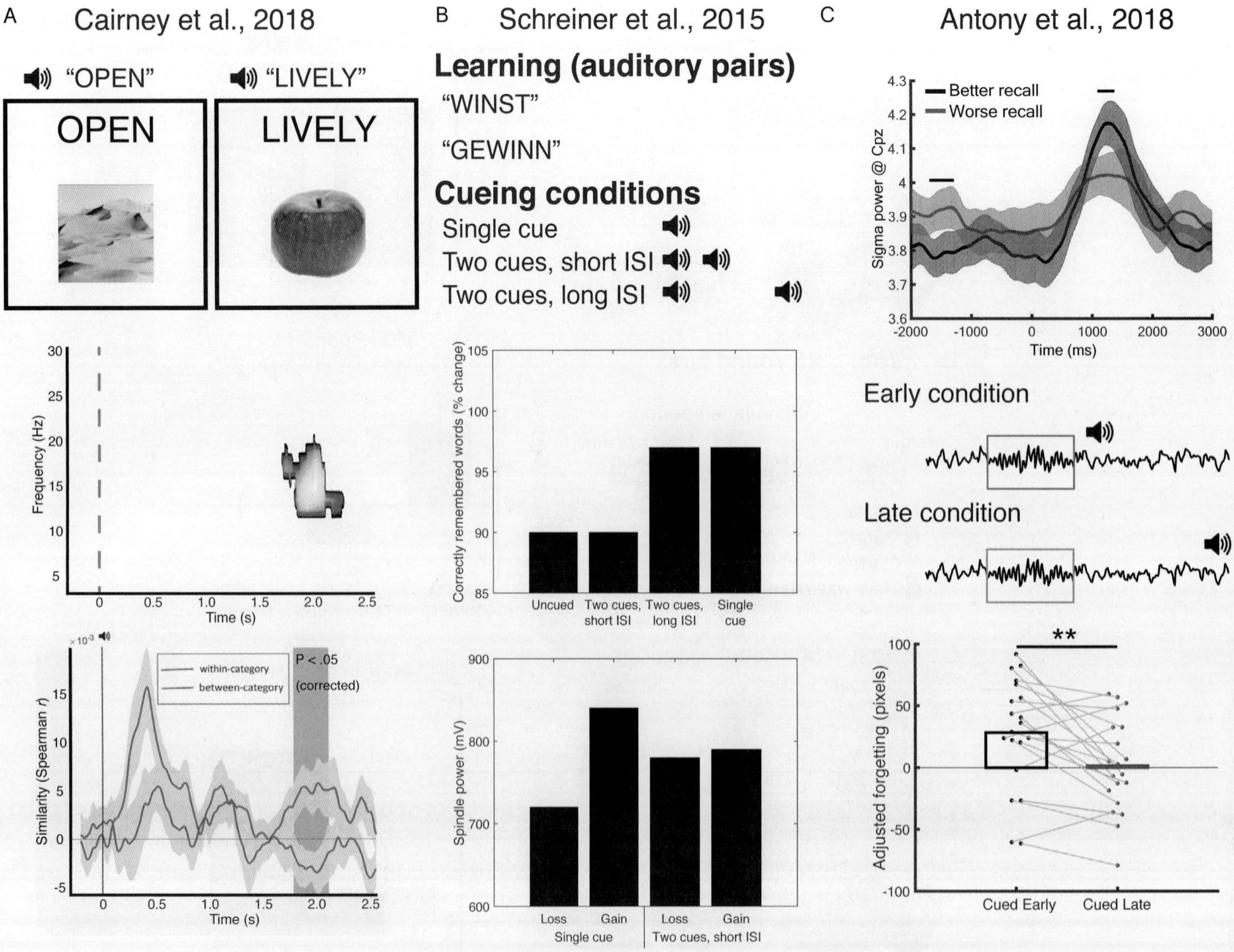

FIGURE 23.2 Sleep spindles and memory as studied in three experiments. *A*, Subjects in Cairney et al. (2018) learned adjective-scene and adjective-object associations. A subset of spoken adjectives were then presented during postlearning sleep. These cues elicited higher EEG power in the spindle band (sigma, ~15 Hz) for learned than for nonlearned words (1.7–2.3 s after cue onset). Additionally, within-category neural similarity (object vs. scene) exceeded between-category similarity at roughly the same time, suggesting that spindles mediate relevant memory reactivation. *B*, Subjects in Schreiner, Lehmann, & Rasch (2015) learned auditory word pairs. Cues presented during sleep included single words, two words separated by 200 ms, or two words separated by 1,500 ms (i.e., a long or short interstimulus interval [ISI]). Subsequent recall was best with single cues or two cues (long ISI), and spindle power within the immediate postcue period predicted memory change with single cues only. *C*, Antony et al. (2018) similarly found that postcue sigma power predicted memory improvement for spatial recall. Additionally, precue sigma power negatively predicted memory, suggesting that precue spindles impede reactivation in that a well-timed postcue spindle is unlikely in these cases. Spindles were found to be most likely to reoccur after about 4–6 s. Using software to track spindles in real time, TMR benefits were better for sounds presented late (long ISI after prior spindle) versus early (short ISI after prior spindle). These results suggest that memory reactivation is linked with spindles, which also means that there may be pauses in reactivation corresponding with the normal pauses between spindles. (See color plate 25.)

In sum, evidence from TMR and from direct manipulation of neural oscillations strongly favors the view that memory storage can be enhanced during sleep. Slow waves may set the stage for the drama of intricate interactions manifested by neural oscillations and their cross-frequency coupling. Furthermore, spindles can be taken as a prime example of neural sleep signals that have a causal impact in enhancing specific memories due to replay-based consolidation. A neuropsychological perspective may have intriguing relevance, given the literature on diencephalic amnesia (e.g., Aggleton & Saunders, 1997). That is, we speculate that the central role of the thalamus in generating spindles and corresponding replay events may be at the heart of *both* sleep-based consolidation and the classic symptoms of amnesia after diencephalic damage.

Memory Processing during Wake

Many electrophysiological and behavioral findings implicate memory reactivation during wake. Rodent hippocampal replay can be observed during or just after learning (Diba & Buzsáki, 2007), as well as more remotely during both wake and sleep (Karlsson & Frank, 2009). Likewise, SWRs occur during waking immobility (Buzsáki, Lai-Wo, & Vanderwolf, 1983) and contain replay content (Davidson, Kloosterman, & Wilson, 2009; Karlsson & Frank, 2009). These wake SWRs correlate with retention (Dupret et al., 2010), and their disruption impairs performance on a working memory task (Jadhav, Kemere, German, & Frank, 2012).

In human studies, fMRI data acquired shortly after learning have shown increases in connectivity between the hippocampus and cortical regions (e.g., Schlichting & Preston, 2014). In addition, specific patterns of hippocampal activity associated with what was just learned can appear spontaneously shortly after learning and can correlate with retention (Gruber, Ritchey, Wang, Doss, & Ranganath, 2016; Schapiro, McDevitt, Rogers, Mednick, & Norman, 2018; Tambini & Davachi, 2013). Moreover, a brief rest after encoding can apparently aid retention (e.g., Craig & Dewar, 2018).

Memory reactivation engaged when relevant information is encountered commonly leads to improved subsequent memory. This observation borders on the territory of standard methods to improve learning. Restudying material strengthens memories, but recall provides a superior benefit (Roediger & Karpicke, 2006). Likewise, cued recall in a spatial task one day after initial learning improves recall accuracy the following day (Bridge & Paller, 2012). Additionally, TMR during wake can improve memory when delivered with subliminal cues (Tambini, Berners-Lee, & Davachi, 2017) or during an engaging task that likely limited rehearsal (Oudiette et al., 2013). Furthermore, reactivation of learning-related neural patterns occurs during restudy (Xue et al., 2010), during successful retrieval (Karlsson Wirebring et al., 2015; Ritchey, Wing, LaBar, & Cabeza, 2013), and even during subliminal wake reactivation (Henke et al., 2003).

Finally, both retrieval (relative to restudy) and sleep (relative to wake) were found to improve consolidation (Antony & Paller, 2018; Bäuml, Holterman, & Abel, 2014). These similar effects of retrieval during wake and sleep support a recent idea that retrieval may naturally engender online consolidation (Antony, Ferreira, Norman, & Wimber, 2017). In sum, consolidation may proceed during sleep and during wake, in conjunction with reactivation that can be intentional, unintentional, with awareness of retrieval, or without awareness of retrieval.

Consolidation and Interference

Whereas research on sleep and memory has largely focused on memory strengthening via replay, a limitation of this approach is that it typically neglects interactions between memories. These interactions may be crucial for shaping retention. Decades of memory research have established that interference from other similar memories can cause forgetting (Underwood, 1957). To predict whether memories will be retained in the long term, we need to understand both how reactivation can cause interference and how it might mitigate interference.

Numerous studies have found, during wake, that retrieving a memory can lead to forgetting competing memories (e.g., Anderson, Bjork, & Bjork, 2000; Lewis-Peacock & Norman, 2014; Norman, Newman, & Detre, 2007). Recent studies using TMR have found that these forgetting effects can also occur when memories are reactivated during sleep (Antony, Cheng, Brooks, Paller, & Norman, 2018; Oyarzún, Moris, Luque, Diego-Balaguer, & Fuentemilla, 2017). In addition to causing interference, reactivation-related learning might restructure memories in a way that mitigates interference. Generally speaking, there are two ways to reduce interference between two memories while still preserving the retrievability of both memories: integrating them into a single, cohesive memory or differentiating them so one memory does not trigger retrieval of the other. Intuitively, this corresponds to the two main ways to prevent enemies from fighting—you can make them friends (integration) or you can separate them (differentiation). Drawing on prior studies showing that strong activation leads to strengthening of memory associations but moderate activation leads to weakening of these associations (e.g., Detre, Natarajan, Gershman, & Norman, 2013), Antony and colleagues (2017) describe how retrieval-driven learning could lead to integration and differentiation. If two memories strongly coactivate during retrieval, this will lead to strengthened connections between the memories, integrating them. Conversely, if two memories show a moderate level of coactivation during retrieval (such that one tends to moderately activate when the other is retrieved and vice versa), this will lead to weakened connections between the memories, differentiating them.

Further progress will require studies that link three measures: neural measures of reactivation during sleep (or wake/rest), neural measures of memory restructuring (e.g., from fMRI pattern analysis; Kim, Norman, & Turk-Browne, 2017), and behavioral measures of memory interference. At present, some data speak to pieces of this puzzle, but no extant studies connect all three. For example, a reduction in memory interference has been observed after sleep (Baran, Wilson, & Spencer, 2010; McDevitt, Duggan, & Mednick, 2015), but these studies did not include neural measures of memory restructuring. Other studies have shown memory integration or differentiation effects with fMRI pattern analysis after a delay that includes sleep, but they did not relate this restructuring to neural activity during the intervening sleep period (Favila, Chanales, & Kuhl, 2016; Kim, Norman, & Turk-Browne, 2017; Tompary & Davachi, 2017).

A related challenge is understanding the role of specific sleep stages in restructuring memories. Prior neural network modeling has found that interleaved learning—repeatedly looping through a playlist of memories marked as important, doing incremental learning each time—is the most effective way to force the brain to reconcile competing representations (McClelland, McNaughton, & O'Reilly, 1995). One intriguing hypothesis is that REM sleep provides a focused period of interleaved learning of competing memories, thereby driving representational change that helps the memories coexist, either through integration or differentiation (Norman, Newman, & Perotte, 2005). The idea that REM is especially important for restructuring representations has the potential to explain results from a wide range of studies, including studies showing that REM leads to improved performance when multi-item integration is required (Cai, Mednick, Harrison, Kanady, & Mednick, 2009; Schapiro et al., 2017; Stickgold & Walker, 2013); studies showing that REM helps to reduce interference between similar memories, potentially through differentiation of representations (Baran, Wilson, & Spencer, 2010; McDevitt, Duggan, & Mednick, 2015); and studies showing that REM plays a role in gaining new insights (Cai et al., 2009; Fosse, Stickgold, & Hobson, 2001; Nishida, Pearsall, Buckner, & Walker, 2009; Payne, Stickgold, Swanberg, & Kensinger, 2008; Wagner, Gais, Haider, Verleger, & Born, 2004).

Future Directions

The results surveyed here convincingly document sleep's relevance for memory storage. Still, many outstanding questions remain about the neurocognitive mechanisms that support sleep-based memory consolidation and offline consolidation generally (figure 23.3).

Whereas memories may be reactivated throughout the sleep-wake cycle, the divergent physiological signals apparent during sleep versus wake suggest different mechanisms of memory change. Future research should seek to elucidate these mechanisms. In particular, deciphering the significance of signals such as slow waves and spindles for memory reactivation could be a big step in advancing our understanding of consolidation.

Various neuroscience techniques will likely provide future insights into these mechanisms. Recent optogenetic work provides a glimpse into how systems-level interactions can be revealed; for example, plasticity in cortical neurons may begin early and then change gradually (e.g., Kitamura et al., 2017; Lesburguères et al., 2011). The hypothetical progression of neural restructuring thought to underlie consolidation may entail a complex set of neural interactions across regions. Prolonged hippocampal-neocortical interactions (e.g., Goshen et al., 2011; Rothschild, 2019) could mediate consolidation in conjunction with memory reactivation. Although few experimental studies have examined long retention delays, there is evidence supporting the importance of repeatedly revising memories (e.g., Cepeda, Vul, Rohrer, Wixted, & Pashler, 2008).

The notion that repeated reactivation is at the core of declarative memory consolidation is consonant with various theories of consolidation. For example, Squire, Cohen, and Nadel (1984) pointed out that "the neural elements participating in memory storage can undergo reorganization with the passage of time after learning" (p. 201). More ideas about the complexities of reorganization were added in subsequent theoretical conceptions (e.g., Moscovitch et al., 2005). Competition has also long been recognized as relevant—"loss of connectivity among elements due to forgetting is accompanied by, causes, or results from a process of reorganization of that which remains" (Squire, Cohen, & Nadel, 1984, p. 201). Whereas concepts of reorganization and

FIGURE 23.3
Outstanding questions for future research.

- **What is the physiology of memory reactivation, and how does reactivation lead to changes in memory storage?**
- **In what ways does consolidation progress differently during wake reactivation and sleep reactivation?**
- **In what ways does consolidation progress differently during reactivation with awareness of retrieval versus reactivation without awareness of retrieval?**
- **How can studies of human memory consolidation best connect with fine-grained neurobiological analyses (e.g., two-photon microscopy and optogenetics)?**
- **Does the principle of expanding retrieval practice hold for sleep reactivation, such that consolidation is best with repeated reactivation after progressively longer delays?**

competition have been acknowledged within theoretical frameworks for consolidation, what happens to engender progressive memory changes over the course of consolidation has usually not been fleshed out. Going back even to Burnham's (1903) early view citing both "a physical process of [re]organization and a psychological process of repetition and association," consolidation theories usually allow for neural changes to progress without necessarily being tied to replay. The current view proposes a shift in emphasis from prior views: repeated memory reactivation is here explicitly conceived as the motive force behind progressive changes in memory storage, which, along with intermemory competition, ultimately determines what information is available for retrieval.

Memory—what is it good for? This question has become a focal point of the overarching orientation to contemporary memory research and has alerted us to the importance of memory for future planning and problem-solving in particular. In this chapter we have zeroed in on enduring memories of episodes and facts. These long-enduring memories have the greatest potential for influencing our future actions. We have a lot to learn about how all types of memories persevere in the brain and manage to remain operative months and years after they are initially acquired. What we eventually can retrieve after long delays is not a pure record of the initial experience but rather a function of a progression of changes in memory storage resulting from intervening retrieval, an idea that has been evident in memory research since Bartlett (1932). Understanding the progressive changes that underlie consolidation will help us gain a fuller conception of learning, and may also provide insights into the fundamental forces that determine the biographical story line and identity that we each carry with us.

Acknowledgements

We gratefully acknowledge research support from the National Science Foundation (BCS-1461088, BCS-1533511, BCS-1829414), the National Institutes of Health (F31-MH100958, T32-AG020506), and the Mind Science Foundation. We also thank Monika Schönauer, Elizabeth McDevitt, and Charan Ranganath for helpful input.

REFERENCES

Aggleton, J. P., & Saunders, R. C. (1997). Relationships between temporal lobe and diencephalic structures implicated in anterograde amnesia. *Memory, 5*(1/2), 49–71.

Anderson, M. C., Bjork, E. L., & Bjork, R. A. (2000). Retrieval-induced forgetting: Evidence for a recall-specific mechanism. *Psychonomic Bulletin & Review, 7*(3), 522–530.

Andrillon, T., Pressnitzer, D., Léger, D., & Kouider, S. (2017). Formation and suppression of acoustic memories during human sleep. *Nature Communications, 8*, 179.

Antony, J. W., Cheng, L. Y., Brooks, P. P., Paller, K. A., & Norman, K. A. (2018). Competitive learning modulates memory consolidation during sleep. *Neurobiology of Learning and Memory, 155*, 216–230.

Antony, J. W., Ferreira, C. S., Norman, K. A., & Wimber, M. (2017). Retrieval as a fast route to memory consolidation. *Trends in Cognitive Sciences, 21*(8), 573–576.

Antony, J. W., Gobel, E. W., O'Hare, J. K., Reber, P. J., & Paller, K. A. (2012). Cued memory reactivation during sleep influences skill learning. *Nature Neuroscience, 15*(8), 1114–1116.

Antony, J. W., & Paller, K. A. (2017). Using oscillating sounds to manipulate sleep spindles. *Sleep, 40*(3), 1–8.

Antony, J. W., & Paller, K. A. (2018). Retrieval and sleep both counteract the forgetting of spatial information. *Learning & Memory, 25*(6), 258–263.

Antony, J. W., Piloto, L., Wang, M., Brooks, P. P., Norman, K. A., & Paller, K. A. (2018). Sleep spindle refractoriness segregates periods of memory reactivation. *Current Biology, 28*(11), 1736–1743.e4.

Arzi, A., Shedlesky, L., Ben-Shaul, M., Nasser, K., Oksenberg, A., Hairston, I. S., & Sobel, N. (2012). Humans can learn new information during sleep. *Nature Neuroscience, 15*(10), 1460–1465.

Baran, B., Wilson, J., & Spencer, R. M. C. (2010). REM-dependent repair of competitive memory suppression. *Experimental Brain Research, 203*(2), 471–477.

Barnes, D. C., & Wilson, D. A. (2014). Slow-wave sleep-imposed replay modulates both strength and precision of memory. *Journal of Neuroscience, 34*(15), 5134–5142.

Bartlett, F. C. (1932). *Remembering.* Cambridge: Cambridge University Press.

Bäuml, K.-H. T., Holterman, C., & Abel, M. (2014). Sleep can reduce the testing effect: It enhances recall of restudied items but can leave recall of retrieved items unaffected. *Journal of Experimental Psychology: Learning, Memory, and Cognition, 40*(6), 1568–1581.

Bendor, D., & Wilson, M. A. (2012). Biasing the content of hippocampal replay during sleep. *Nature Neuroscience, 15*(10), 1439–1444.

Berkers, R. M. W. J., Ekman, M., van Dongen, E. V., Takashima, A., Paller, K. A., & Fernandez, G. (2018). Cued reactivation during slow-wave sleep induces connectivity changes related to memory stabilization. *Scientific Reports, 8,* 16958.

Bridge, D. J., & Paller, K. A. (2012). Neural correlates of reactivation and retrieval-induced distortion. *Journal of Neuroscience, 32*(35), 12144–12151.

Burnham, W. H. (1903). Retroactive amnesia: Illustrative cases and a tentative explanation. *American Journal of Psychology, 14,* 382–396.

Buzsáki, G., Lai-Wo, S., & Vanderwolf, C. H. (1983). Cellular bases of hippocampal EEG in the behaving rat. *Brain Research Reviews, 6*(2), 139–171.

Cai, D. J., Mednick, S. A., Harrison, E. M., Kanady, J. C., & Mednick, S. C. (2009). REM, not incubation, improves creativity by priming associative networks. *Proceedings of the National Academy of Sciences of the United States of America, 106*(25), 10130–10134.

Cairney, S. A., Guttesen, A. á. V., El Marj, N., & Staresina, B. P. (2018). Memory consolidation is linked to spindle-mediated information processing during sleep. *Current Biology, 28*(6), 948–954.e4.

Cartwright, R. (1977). *Night life: Explorations in dreaming.* Englewood Cliffs, NJ: Prentice Hall.

Cellini, N., & Capuozzo, A. (2018). Shaping memory consolidation via targeted memory reactivation during sleep. *Annals of the New York Academy of Sciences, 1426*(1), 52–71.

Cepeda, N. J., Vul, E., Rohrer, D., Wixted, J. T., & Pashler, H. (2008). Spacing effects in learning. *Psychological Science, 19*(11), 1095–1102.

Craig, M., & Dewar, M. (2018). Rest-related consolidation protects the fine detail of new memories. *Scientific Reports, 8*(1), 1–9.

Creery, J. D., Oudiette, D., Antony, J. W., & Paller, K. A. (2014). Targeted memory reactivation during sleep depends on prior learning. *Sleep, 38*(5), 755–763.

Davidson, T. J., Kloosterman, F., & Wilson, M. A. (2009). Hippocampal replay of extended experience. *Neuron, 63*(4), 497–507.

Detre, G. J., Natarajan, A., Gershman, S. J., & Norman, K. A. (2013). Moderate levels of activation lead to forgetting in the think/no-think paradigm. *Neuropsychologia, 51*(12), 2371–2388.

Diba, K., & Buzsáki, G. (2007). Forward and reverse hippocampal place-cell sequences during ripples. *Nature Neuroscience, 10*(10), 1241–1242.

Dudai, Y. (2012). The restless engram: Consolidations never end. *Annual Review of Neuroscience, 35,* 227–247.

Dupret, D., O'Neill, J., Pleydell-Bouverie, B., & Csicsvari, J. (2010). The reorganization and reactivation of hippocampal maps predict spatial memory performance. *Nature Neuroscience, 13*(8), 995–1002.

Ego-Stengel, V., & Wilson, M. A. (2009). Disruption of ripple-associated hippocampal activity during rest impairs spatial learning in the rat. *Hippocampus, 20*(1), 1–10.

Eichenbaum, H. B., & Cohen, N. J. (2001). *From conditioning to conscious recollection: Memory systems of the brain.* New York: Oxford University Press.

Emmons, W., & Simon, C. (1956). The non-recall of material presented during sleep. *American Journal of Psychology, 69*(1), 76–81.

Favila, S. E., Chanales, A. J. H., & Kuhl, B. A. (2016). Experience-dependent hippocampal pattern differentiation prevents interference during subsequent learning. *Nature Communications, 7,* 11066.

Fosse, R., Stickgold, R., & Hobson, J. A. (2001). The mind in REM sleep: Reports of emotional experience. *Sleep, 24*(8), 947–955.

Foster, D. J. (2017). Replay comes of age. *Annual Review of Neuroscience, 40*(1), 581–602.

Fuentemilla, L., Miró, J., Ripollés, P., Vilà-Balló, A., Juncadella, M., Castañer, S., Salord, N., Monasterio, C., Falip, M., & Rodríguez-Fornells, A. (2013). Hippocampus-dependent strengthening of targeted memories via reactivation during sleep in humans. *Current Biology, 23*(18), 1769–1775.

Girardeau, G., Benchenane, K., Wiener, S. I., Buzsáki, G., & Zugaro, M. B. (2009). Selective suppression of hippocampal ripples impairs spatial memory. *Nature Neuroscience, 12*(10), 1222–1223.

Goshen, I., Brodsky, M., Prakash, R., Wallace, J., Gradinaru, V., Ramakrishnan, C., & Deisseroth, K. (2011). Dynamics of retrieval strategies for remote memories. *Cell, 147*(3), 678–689.

Gruber, M. J., Ritchey, M., Wang, S. F., Doss, M. K., & Ranganath, C. (2016). Post-learning hippocampal dynamics promote preferential retention of rewarding events. *Neuron, 89*(5), 1110–1120.

Hars, B., Hennevin, E., & Pasques, P. (1985). Improvement of learning by cueing during postlearning paradoxical sleep. *Behavioural Brain Research, 18,* 241–250.

Hauner, K. K., Howard, J. D., Zelano, C., & Gottfried, J. A. (2013). Stimulus-specific enhancement of fear extinction during slow-wave sleep. *Nature Neuroscience, 16*(11), 1553–1555.

Helfrich, R. F., Mander, B. A., Jagust, W. J., Knight, R. T., & Walker, M. P. (2018). Old brains come uncoupled in sleep: Slow wave-spindle synchrony, brain atrophy, and forgetting. *Neuron, 97*(1), 221–230.

Henke, K., Mondadori, C. R., Treyer, V., Nitsch, R. M., Buck, A., & Hock, C. (2003). Nonconscious formation and reactivation of semantic associations by way of the medial temporal lobe. *Neuropsychologia, 41*(8), 863–876.

Hennevin, E., Hars, B., Maho, C., & Bloch, V. (1995). Processing of learned information in paradoxical sleep: Relevance for memory. *Behavioural Brain Research, 69*(1–2), 125–135.

Honma, M., Plass, J., Brang, D., Florczak, S. M., Grabowecky, M., & Paller, K. A. (2016). Sleeping on the rubber-hand illusion: Memory reactivation during sleep facilitates multisensory recalibration. *Neuroscience of Consciousness, 2016*(1), niw020.

Jadhav, S. P., Kemere, C., German, P. W., & Frank, L. M. (2012). Awake hippocampal sharp-wave ripples support spatial memory. *Science, 336*(6087), 1454–1458.

James, W. (1890). *The Principles of psychology.* New York: Henry Holt.

Karlsson, M. P., & Frank, L. M. (2009). Awake replay of remote experiences in the hippocampus. *Nature Neuroscience, 12*(7), 913–918.

Karlsson Wirebring, L., Wiklund-Hornqvist, C., Eriksson, J., Andersson, M., Jonsson, B., & Nyberg, L. (2015). Lesser neural pattern similarity across repeated tests is associated with better long-term memory retention. *Journal of Neuroscience, 35*(26), 9595–9602.

Khodagholy, D., Gelinas, J. N., & Buzsáki, G. (2017). Learning-enhanced coupling between ripple oscillations in association cortices and hippocampus. *Science, 358*(6361), 369–372.

Kim, G., Lewis-Peacock, J. A., Norman, K. A., & Turk-Browne, N. B. (2014). Pruning of memories by context-based prediction error. *Proceedings of the National Academy of Sciences of the United States of America, 111*(24), 8997–9002.

Kim, G., Norman, K. A., & Turk-Browne, N. B. (2017). Neural differentiation of incorrectly predicted memories. *Journal of Neuroscience, 37*(8), 2022–2031.

Kitamura, T., Ogawa, S. K., Roy, D. S., Okuyama, T., Morrissey, M. D., Smith, L. M., … Tonegawa, S. (2017). Engrams and circuits crucial for systems consolidation of a memory. *Science, 356*(6333), 73–78.

Landsness, E. C., Crupi, D., Hulse, B. K., Peterson, M. J., Huber, R., Ansari, H., … Tononi, G. (2009). Sleep-dependent improvement in visuomotor learning: A causal role for slow waves. *Sleep, 32*(10), 1273–1284.

Lesburguères, E., Gobbo, O. L., Alaux-Cantin, S., Hambucken, A., Trifilieff, P., & Bontempi, B. (2011). Early tagging of cortical networks is required for the formation of enduring associative memory. *Science, 331*(6019), 924–928.

Lewis-Peacock, J. A., & Norman, K. A. (2014). Competition between items in working memory leads to forgetting. *Nature Communications, 5*, 5768.

Livingston, R. (1967). Reinforcement. In G. C. Quarton, T. Melnechuk, & F. O. Schmitt (Eds.), *The neurosciences: A study program* (pp. 568–576). New York: Rockefeller Press.

Lustenberger, C., Boyle, M. R., Alagapan, S., Mellin, J. M., Vaughn, B. V., & Fröhlich, F. (2016). Feedback-controlled transcranial alternating current stimulation reveals a functional role of sleep spindles in motor memory consolidation. *Current Biology, 26*(16), 2127–2136.

Marr, D. (1971). Simple memory: A theory for archicortex. *Philosophical Transactions of the Royal Society of London. Series B, Biological Sciences, 262*, 23–81.

Marshall, L., Helgadóttir, H., Mölle, M., & Born, J. (2006). Boosting slow oscillations during sleep potentiates memory. *Nature, 444*(7119), 610–613.

McClelland, J. L., McNaughton, B. L., & O'Reilly, R. C. (1995). Why there are complementary learning systems in the hippocampus and neocortex: Insights from the successes and failures of connectionist models of learning and memory. *Psychological Review, 102*(3), 419–457.

McDevitt, E. A., Duggan, K. A., & Mednick, S. C. (2015). REM sleep rescues learning from interference. *Neurobiology of Learning and Memory, 122*, 51–62.

Mednick, S., McDevitt, E., Walsh, J., Wamsley, E., Paulus, M., Kanady, J., & Drummond, S. (2013). The critical role of sleep spindles in hippocampal-dependent memory: A pharmacology study. *Journal of Neuroscience, 33*(10), 4494–4504.

Moscovitch, M., Rosenbaum, R. S., Gilboa, A., Addis, D. R., Westmacott, R., Grady, C., … Nadel, L. (2005). Functional neuroanatomy of remote episodic, semantic and spatial memory: A unified account based on multiple trace theory. *Journal of Anatomy, 207*(1), 35–66.

Nadel, L., & Moscovitch, M. (1997). Memory consolidation, retrograde amnesia and the hippocampal complex. *Current Opinion in Neurobiology, 7*(2), 217–227.

Newman, E. L., & Norman, K. A. (2010). Moderate excitation leads to weakening of perceptual representations. *Cerebral Cortex, 20*(11), 2760–2770.

Ngo, H. V, Martinetz, T., Born, J., & Mölle, M. (2013). Auditory closed-loop stimulation of the sleep slow oscillation enhances memory. *Neuron, 78*(3), 545–553.

Niknazar, M., Krishnan, G. P., Bazhenov, M., & Mednick, S. C. (2015). Coupling of thalamocortical sleep oscillations are important for memory consolidation in humans. *PloS One, 10*(12), 1–14.

Nishida, M., Pearsall, J., Buckner, R. L., & Walker, M. P. (2009). REM sleep, prefrontal theta, and the consolidation of human emotional memory. *Cerebral Cortex, 19*, 1158–1166.

Norman, K. A., Newman, E. L., & Detre, G. (2007). A neural network model of retrieval-induced forgetting. *Psychological Review, 114*(4), 887–953.

Norman, K. A., Newman, E. L., & Perotte, A. J. (2005). Methods for reducing interference in the Complementary Learning Systems model: Oscillating inhibition and autonomous memory rehearsal. *Neural Networks, 18*(9), 1212–1228.

O'Neill, J., Senior, T. J., Allen, K., Huxter, J. R., & Csicsvari, J. (2008). Reactivation of experience-dependent cell assembly patterns in the hippocampus. *Nature Neuroscience, 11*(2), 209–215.

Oudiette, D., Antony, J. W., Creery, J. D., & Paller, K. A. (2013). The role of memory reactivation during wakefulness and sleep in determining which memories endure. *Journal of Neuroscience, 33*(15), 6672–6678.

Oudiette, D., & Paller, K. A. (2013). Upgrading the sleeping brain with targeted memory reactivation. *Trends in Cognitive Sciences, 17*(3), 142–149.

Oyarzún, J., Moris, J., Luque, D., de Diego-Balaguer, R., & Fuentemilla, L. (2017). Targeted memory reactivation during sleep adaptively promotes the strengthening or weakening of overlapping memories. *Journal of Neuroscience, 37*(32), 7748–7758.

Paller, K. A. (1997). Consolidating dispersed neocortical memories: The missing link in amnesia. *Memory, 5*(1/2), 73–88.

Paller, K. A. (2002). Cross-cortical consolidation as the core defect in amnesia. In L. R. Squire & D. L. Schacter (Eds.), *Neuropsychology of memory* (3rd ed., pp. 73–87). New York: Guilford Press.

Paller, K. A., Kutas, M., & Mayes, A. R. (1987). Neural correlates of encoding in an incidental learning paradigm. *Electroencephalography and Clinical Neurophysiology, 67*(4), 360–371.

Paller, K. A., & Oudiette, D. (2018). Sleep learning gets real: Experimental techniques demonstrate how to strengthen memories when our brains are off-line. *Scientific American, 319*, 26–31.

Paller, K. A., & Voss, J. L. (2004). Reactivation and consolidation of memory during sleep. *Learning & Memory, 11*(6), 664–670.

Pavlides, C., & Winson, J. (1989). Influences of hippocampal place cell firing in the awake state on the activity of these cells during subsequent sleep episodes. *Journal of Neuroscience, 9*(8), 2907–2918.

Payne, J. D., Stickgold, R., Swanberg, K., & Kensinger, E. A. (2008). Sleep preferentially enhances memory for emotional components of scenes. *Psychological Science, 19*(8), 781–788.

Peyrache, A., Khamassi, M., Benchenane, K., Wiener, S. I., & Battaglia, F. P. (2009). Replay of rule-learning related neural patterns in the prefrontal cortex during sleep. *Nature Neuroscience, 12*(7), 919–926.

Rasch, B., & Born, J. (2013). About sleep's role in memory. *Physiological Reviews, 93*(2), 681–766.

Rasch, B., Büchel, C., Gais, S., & Born, J. (2007). Odor cues during slow-wave sleep prompt declarative memory consolidation. *Science, 315*(5817), 1426–1429.

Ritchey, M., Wing, E. A., LaBar, K. S., & Cabeza, R. (2013). Neural similarity between encoding and retrieval is related to memory via hippocampal interactions. *Cerebral Cortex, 23*(12), 2818–2828.

Roediger, H. L., & Karpicke, J. D. (2006). The power of testing memory: Basic research and implications for educational practice. *Perspectives on Psychological Science, 1*(3), 181–210.

Rothschild, G. (2019). The transformation of multi-sensory experiences into memories during sleep. *Neurobiology of Learning and Memory, 160*, 58–66.

Rudoy, J. D., Voss, J. L., Westerberg, C. E., & Paller, K. A. (2009). Strengthening individual memories by reactivating them during sleep. *Science, 326*(5956), 1079.

Schapiro, A. C., McDevitt, E. A., Chen, L., Norman, K. A., Mednick, S. C., & Rogers, T. T. (2017). Sleep benefits memory for semantic category structure while preserving exemplar-specific information. *Scientific Reports, 7*(1), 1–13.

Schapiro, A. C., McDevitt, E. A., Rogers, T. T., Mednick, S. C., & Norman, K. A. (2018). Human hippocampal replay during rest prioritizes weakly-learned information and predicts memory performance. *Nature Communications, 9*, 3920.

Schlichting, M. L., & Preston, A. R. (2014). Memory reactivation during rest supports upcoming learning of related content. *Proceedings of the National Academy of Sciences, 111*(44), 15845–15850.

Schmolck, H., Buffalo, E. A., & Squire, L. R. (2000). Memory distortions develop over time: Recollections of the O. J. Simpson trial verdict after 15 and 32 months. *Psychological Science, 11*(1), 39–45.

Schouten, D. I., Pereira, S. I. R., Tops, M., & Louzada, F. M. (2017). State of the art on targeted memory reactivation: Sleep your way to enhanced cognition. *Sleep Medicine Reviews, 32*, 123–131.

Schreiner, T., Lehmann, M., & Rasch, B. (2015). Auditory feedback blocks memory benefits of cueing during sleep. *Nature Communications, 6*, 8729.

Schreiner, T., & Rasch, B. (2014). Boosting vocabulary learning by verbal cueing during sleep. *Cerebral Cortex, 25*(11), 4169–4179.

Shanahan, L. K., Gjorgieva, E., Paller, K. A., Kahnt, T., & Gottfried, J. A. (2018). Odor-evoked category reactivation in human ventromedial prefrontal cortex during sleep promotes memory consolidation. *eLife, 7, e39681.*

Shimamura, A. (2002). Relational binding theory and the role of consolidation in memory retrieval. In L. R. Squire & D. L. Schacter (Eds.), *Neuropsychology of memory* (3rd ed., pp. 61–72). New York: Guilford Press.

Siapas, A. G., & Wilson, M. A. (1998). Coordinated interactions between hippocampal ripples and cortical spindles during slow-wave sleep. *Neuron, 21*, 1123–1128.

Smith, C., & Weeden, K. (1990). Post training REMs coincident auditory stimulation enhances memory in humans. *Psychiatric Journal of the University of Ottawa, 15*(2), 85–90.

Squire, L., Cohen, N., & Nadel, L. (1984). The medial temporal region and memory consolidation: A new hypothesis. In H. Weingartner & E. Parder (Eds.), *Memory consolidation* (pp. 185–210). Hillsdale, NJ: Erlbaum.

Stickgold, R., & Walker, M. P. (2013). Sleep-dependent memory triage: Evolving generalization through selective processing. *Nature Neuroscience, 16*(2), 139–145.

Tambini, A., Berners-Lee, A., & Davachi, L. (2017). Brief targeted memory reactivation during the awake state enhances memory stability and benefits the weakest memories. *Scientific Reports, 7*(1), 1–17.

Tambini, A., & Davachi, L. (2013). Persistence of hippocampal multivoxel patterns into postencoding rest is related to memory. *Proceedings of the National Academy of Sciences of the United States of America, 110*(48), 19591–19596.

Tompary, A., & Davachi, L. (2017). Consolidation promotes the emergence of representational overlap in the hippocampus and medial prefrontal cortex. *Neuron, 96*(1), 228–241.e5.

Underwood, B. J. (1957). Interference and forgetting. *Psychological Review, 64*(1), 49–60.

van Dongen, E. V, Takashima, A., Barth, M., Zapp, J., Schad, L. R., Paller, K. A., & Fernández, G. (2012). Memory stabilization with targeted reactivation during human slow-wave sleep. *Proceedings of the National Academy of Sciences of the United States of America, 109*(26), 10575–10580.

Vargas, I. M., Schechtman, E., & Paller, K. A. (2019). Targeted memory reactivation during sleep to strengthen memory for arbitrary pairings. *Neuropsychologia, 124*, 144–150.

Wagner, U., Gais, S., Haider, H., Verleger, R., & Born, J. (2004). Sleep inspires insight. *Nature, 427*(22), 352–355.

Wilson, M., & McNaughton, B. (1994). Reactivation of hippocampal ensemble memories during sleep. *Science, 265*(5172), 676–679.

Winson, J. (1985). *Brain and psyche: The biology of the unconscious.* Garden City, NY: Anchor Press/Doubleday.

Xue, G., Dong, Q., Chen, C., Lu, Z., Mumford, J. A, & Poldrack, R. A. (2010). Greater neural pattern similarity across repetitions is associated with better memory. *Science, 330*(6000), 97–101.

24 The Dynamic Memory Engram Life Cycle: Reactivation, Destabilization, and Reconsolidation

TEMIDAYO OREDERU AND DANIELA SCHILLER

ABSTRACT Recent discoveries on memory demand a reconsideration of core beliefs in favor of a new view. For most of history, neuroscientists believed that memories are initially unstable but stabilize into permanent fixtures through a process called consolidation. New evidence shows that consolidated memories can return to their unstable states and, once destabilized, can be diminished, enhanced, or modified. This chapter examines the factors facilitating shifts between stable and unstable memory states, the paths available to memories occupying each state, and the therapeutic promise of the continuing research investigating memory modification.

When visualizing a metaphor for memory, you might picture a box you can open at any time to reveal its original contents. This model of memory is not only overly simplistic but also factually incorrect. A memory may differ at each retrieval. Why? While a memory is protected when in its "box," once removed it becomes susceptible to change.

An engram (the memory "box") is a hypothesized aggregate of synaptic changes thought to encode a memory. New information enters short-term memory (STM) but degrades if not converted to long-term memory (LTM) through a process termed *synaptic*, or cellular, consolidation. Memories were previously thought to be stable and protected from alteration post consolidation (McGaugh, 1966). Neuroscientists now believe memories shift between unstable and stable states throughout their lifetimes (Nader & Hardt, 2009). This chapter explores recent discoveries on memory and the implications of a dynamic engram.

Dynamic Perspectives on Memory Dynamics

The concept of varying memory states is only a few decades old among neuroscientists, but psychologists have acknowledged memory dynamics since 1932. Frederic Bartlett described memories as reconstructions integrated with present-day knowledge (Bartlett, 1932). In 1968 a landmark study prompted neuroscientists to adopt perspectives long held by psychologists (Misanin, Miller, & Lewis, 1968). It was already known that manipulations such as electroconvulsive shock (ECT) or amnestic pharmacological agents could induce amnesia by disrupting the consolidation of an unstable STM. Identical manipulations did not similarly affect older memories, revealing a temporal gradient of retrograde amnesia (memory for recent events disproportionately impaired compared to memory for remote events). Misanin, Miller, & Lewis (1968) tested whether temporally graded amnesia would emerge for older memories if a reminder was administered before ECT. They thought a reminder might "reactivate" the memory and trigger conversion from its stable to unstable state. The authors trained rats to associate an acoustic tone (conditioned stimulus [CS]) with an electric shock (unconditioned stimulus [US]). After 24 hours, the tone was presented as a reminder of the tone-shock association, followed by ECT. Associative memory was tested the next day by measuring rates of lick suppression (interrupting water licking is akin to freezing) during tone presentation. Rats that received ECT following a reminder tone demonstrated amnesia. The control group, however, received ECT without the reminder and retained the associative memory. The authors used the term *cue-induced amnesia* to describe their discovery that amnesia for a consolidated memory could be induced if a reminder trial is presented prior to the amnestic manipulation.

Today, the phenomenon is most commonly referred to as *reconsolidation*, due to the supposition that a consolidated memory returns to its unstable state and must be consolidated again to persist in LTM (Przybyslawski & Sara, 1997; Spear, 1973). This discovery caused excitement among scientists and motivated subsequent studies that adopted the three-day framework (memory acquisition, memory reminder + interference, and memory test, respectively; figure 24.1) to replicate and expound on cue-induced amnesia. Within a few years, however, enthusiasm for cue-induced amnesia tapered, and the phenomenon fell out of the spotlight.

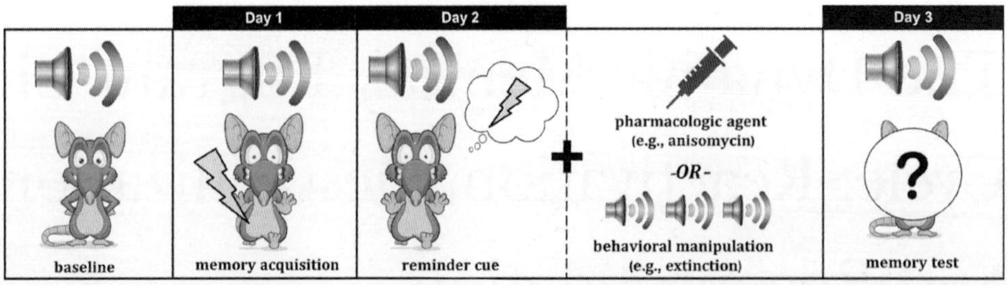

FIGURE 24.1 A standard reconsolidation protocol. A conditioned stimulus (CS; tone) does not originally elicit noteworthy behavioral responding. During conditioning, the animal learns to associate the CS with an unconditioned stimulus (US; shock), and the CS alone then elicits a response. After 24 hours, the CS is presented to reactivate the associative memory. This is followed by a pharmacologic or behavioral manipulation thought to disrupt reconsolidation. Lastly, a memory test is conducted by measuring the behavior indicative of memory retention. (See color plate 26.)

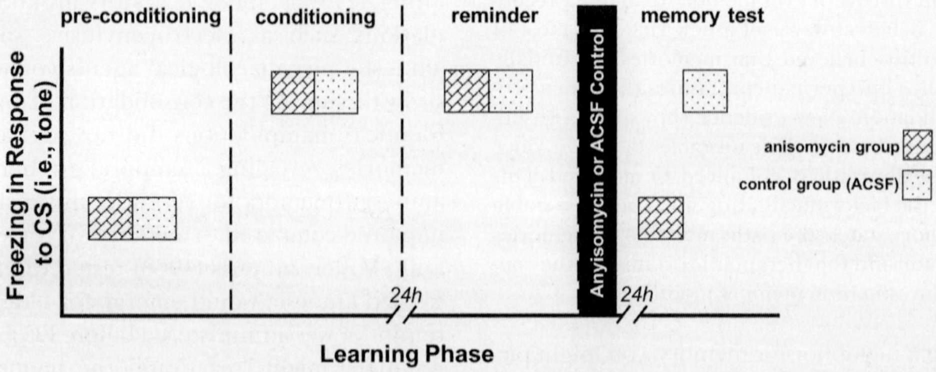

FIGURE 24.2 Threat memories undergo protein-synthesis-dependent reconsolidation. Nader et al. (2000) conditioned rats to associate a tone with shock. Freezing in response to the tone was the index of memory strength. Freezing was initially low but rose during conditioning. One day later, rodents exhibited high freezing during a reminder trial, indicating that the associative memory persisted overnight. Following the reminder, the authors injected anisomycin into the amygdala. Twenty-four hours postinjection, freezing rates among rats that received anisomycin returned to their preconditioning levels, indicating an attenuated or erased memory. This can be compared to the control rats that received a reminder cue plus artificial cerebrospinal fluid (ACSF) and maintained high rates of freezing during the memory test.

Interest in cue-induced amnesia resurfaced decades later, when Nader, Schafe, & LeDoux (2000) tested whether reactivation could destabilize consolidated memories. They used a threat-conditioning paradigm in rats and paired an acoustic tone (CS) with an electric shock (US). They then utilized a protein synthesis inhibitor (PSI; anisomycin) to induce amnesia, following previous findings that PSIs, which block new gene expression, disrupt memory consolidation. They infused anisomycin into the amygdala (a core storage site for emotional memory), which elicited amnesia for the CS-US association (figure 24.2). This finding revived interest in cue-induced amnesia or, by its better-known name, *reconsolidation.*

What Is Reconsolidation Theory?

The reconsolidation effect can be subdivided into three distinct stages of reactivation, destabilization, and reconsolidation (figure 24.3). During reactivation, an inactive memory becomes active via firing of the ensemble of neurons that initially encoded the memory (Tayler, Tanaka, Reijmers, & Wiltgen, 2013). Reactivation can lead to any combination of (1) retrieval, or the conscious process of "remembering" in humans, (2) expression, or a behavioral response to the memory, and/or (3) destabilization, or the return of the memory to its unstable state. In this chapter we focus on destabilization, which is the only fate that necessarily precedes reconsolidation. Reconsolidation theory is named after the final stage, when the engram is restabilized to LTM. This is why many refer to the entire sequence (reactivation, destabilization, and restabilization/reconsolidation) collectively as reconsolidation. Going forward, we use the terms *reactivation, destabilization,* and *restabilization* to differentiate between processes often clustered under *memory reconsolidation.*

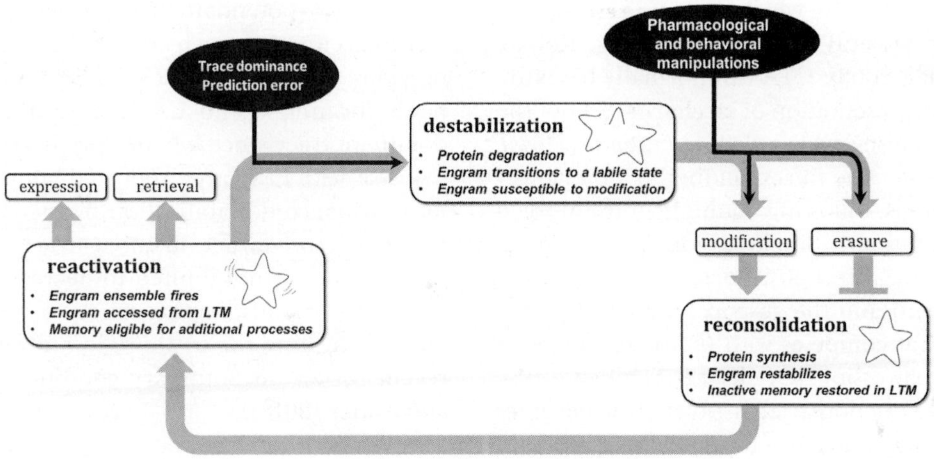

FIGURE 24.3 The life cycle of memory dynamics. Memories are usually inactive but can reactivate upon engram firing. Reactivation makes the memory eligible for retrieval, behavioral expression, and/or destabilization. Trace dominance and prediction error are two of several boundary conditions that increase the likelihood of destabilization. If destabilization occurs, a cascade of neural processes (including protein degradation) initiates a transition from the engram's stable to unstable state, where it is susceptible to modification. If protein synthesis and other reconsolidation processes follow, the memory is restabilized and re-stored in LTM. However, manipulations that interfere with reconsolidation (amnestic agents, memory enhancers, or behavioral interference) can reroute the memory toward erasure, augmentation, or updating. If the memory is not erased, it may cycle into another sequence of reactivation.

Reactivation: Awakening the Engram

Countless memories are encoded in our synapses, yet only a subset is "active" at any given moment. New memories are active by default and trigger a unique pattern of neural ensemble firing before consolidation into an inactive LTM. The dormant engram is awakened when its unique ensemble refires. Such "reactivation" qualifies the memory for additional neural processes (see Gisquet-Verrier & Riccio, 2012 for review). For example, researchers have harnessed reactivation to inject emotional content into a neutral contextual memory within mice by stimulating the hippocampal engram encoding the context while administering a foot shock. The mice later exhibited a freezing response to the context, even though no aversive events occurred there (Ramirez et al., 2013). A more recent study conducted in mice (Khalaf et al., 2018) demonstrated that a 28-day-old memory could be dampened with training but only when the hippocampal engram specifically reactivated during training. The memory was unresponsive to training when engram cells were chemically silenced, exemplifying the prerequisite role of reactivation in various neural processes, such as memory destabilization and subsequent modification, as in the above examples.

Destabilization: Unraveling the Engram

Destabilization converts a consolidated memory back to its initially unstable state, where it is labile and susceptible to influence. Destabilization is only possible if a memory is first reactivated but is not a necessary destination for all reactivated memories (Lee & Flavell, 2014). Research in animal models has elucidated a host of molecular, cellular, and genetic events that must occur in engram cells for a reactivated memory to destabilize; these include protein degradation, glutamatergic and dopaminergic signaling, microRNA expression, and chromatin modifications (for a review, see Flavell, Lambert, Winters, & Bredy, 2013).

Neural destabilization markers have largely been identified using procedures in animal models that are unsafe for human use. The consequent lack of destabilization markers in humans impedes the interpretation of many reconsolidation studies. Until researchers can confirm that an experimental protocol successfully elicited destabilization, they cannot definitively ascertain that observed effects are related to reconsolidation. In light of this, some scientists have taken to examining the behavioral parameters that influence memory destabilization (also called *boundary conditions*) to better understand what determines if a reactivated memory will destabilize and what accounts for divergent findings in experiments (Haubrich & Nader, 2018). Since boundary conditions are often manipulated behaviorally, they can be studied in humans. Here, we will discuss trace dominance and prediction error (PE), which are two of several boundary conditions often cited in the literature.

Trace dominance was first described following a study in which rats underwent conditioned taste aversion training (CTA), in which the taste of saccharin was paired with visceral malaise (Eisenberg, Kobilo, Berman, & Dudai, 2003). CTA can normally be extinguished with one presentation of saccharin without the induction of malaise, also called an *extinction trial.* Extinction is the decline in responding to a previously reinforced stimulus following multiple unreinforced stimulus presentations. Single-session extinction is thought to produce a second association between the CS (e.g., saccharin) and the absence of a US (e.g., visceral malaise) that competes with the original CS-US association for expression (Orederu & Schiller, 2018). When a single extinction trial is isolated, it becomes operationally identical to a reactivation trial. Eisenberg et al. (2003) administered such an extinction trial to rats possessing a CTA memory followed by an injection of anisomycin into the insular (taste) cortex. Our knowledge of anisomycin as a PSI might lead us toward two opposing predictions: (1) anisomycin will prevent the new CS-no US association from consolidating into a stable, long-term memory, leading to *sustained* CTA, or 2) the reactivation trial will elicit memory destabilization, but restabilization will be prevented by anisomycin and cause *diminished* CTA. Indeed, both instances occurred. CTA was sustained when the aversive memory was trained using one CS-US pairing and diminished when training was intensified with a second CS-US pairing. The authors attribute the divergent results to trace dominance, or the idea that a retrieved memory will only destabilize and become eligible for modification if it is the dominant memory trace in the brain at that moment. Standard training results in a weak CTA memory that cannot dominate the newly forming extinction memory, while intensified training creates a CTA memory that can dominate, destabilize, and be disrupted by anisomycin. By manipulating a target memory's control over behavior prior to reactivation, the authors uncovered the importance of trace dominance in influencing a memory's eligibility for destabilization.

The presence and magnitude of PEs also influence whether a memory will destabilize (Sevenster, Beckers, & Kindt, 2013). PE is the discrepancy between an expected outcome and what actually occurs. While large PEs initiate new memory formation, small PEs indicate that an existing memory should be slightly updated and destabilizes the memory to allow for modification (Gershman, Monfils, Norman, & Niv, 2017). PEs are often regarded as a distinct boundary condition for memory destabilization, but they may indirectly affect destabilization by influencing trace dominance. In studies examining reminder duration (e.g., Hu et al., 2018), short reminders destabilize the CS-US memory, while long reminders create a separate CS-no US association—potentially because long reminders allow PEs to accumulate such that current observations and previous memories are excessively incongruent. The newly forming CS-no US memory then becomes the dominant trace since it better predicts observations.

Just as with PEs, other reported boundary conditions may influence destabilization by increasing the dominance of the target memory trace. Such methods include but are not limited to decreasing the number of reminder trials, increasing the age of the memory, and reactivating the memory in a novel context (for a recent review on boundary conditions, see Haubrich and Nader [2018]).

Restabilization: Reassembling the Engram

In order for a destabilized memory to remain an entity in the brain, it must be restabilized through the process of reconsolidation. The term *reconsolidation* suggests that restabilization is a recapitulation of the consolidating processes seen when a new memory is converted to LTM. Indeed, both consolidation and restabilization depend on RNA synthesis and de novo protein synthesis in the brain regions implicated in memory, such as the amygdala, hippocampus, and nonspecific motor areas. Additionally, hippocampal mitogen-activated protein kinase (MAPK), amygdala protein kinase A, and cAMP response element-binding protein (CREB) are required for both consolidation and restabilization in recognition memory, CTA, and contextual threat conditioning, respectively (for an extensive review on the neurobiological mechanisms of restabilization, see Besnard, Caboche, and Laroche [2012]).

Despite these similarities and others, some researchers have raised issues with the verbiage that suggests consolidation and restabilization are one and the same. Alberini (2005) summarizes the metabolic, epigenetic, and proteomic differences between reconsolidation and consolidation, concluding that the two processes are mechanistically distinct. Nader, Hardt, and Wang (2005) argues against this, noting that reported differences most likely result from experimental variations (e.g., a CS-US during consolidation vs. a CS only during restabilization; novelty during consolidation vs. expectation during restabilization; Nader, Hardt, & Wang, 2005). Such a disagreement between scientific groups does not detract from the powerful implications of reconsolidation theory but does highlight the need for continued efforts to understand it.

The "Why?" of reconsolidation As we move through our environment and identify relevant observations, we encounter information that may later be useful and store it within an engram. However, when we encounter new information with some degree of familiarity, do we always store it in a brand-new engram? By now you might imagine a more efficient approach: identify an existing memory containing similar information, reactivate and destabilize its contents, and update it with the new observations. Reconsolidation may be the process that endows animals with this precise ability to selectively update individual memories (for review, see Lee, Nader, & Schiller, 2017), maximizing efficiency by integrating new information into existing memory traces. Not all new observations, though, should trigger memory updating. The requirement for a small PE serves as a gatekeeper, ensuring that memories are only modified when observations are similar to a previously encoded memory. Likewise, the requirement for trace dominance limits updating to only the most relevant memories. First observed in animal models, this concept of memory updating uprooted core neuroscience beliefs, prompting researchers to examine the extent to which the phenomenon occurs in varying memory subtypes and in humans. The following subsections will discuss reconsolidation as an update mechanism of motor, declarative, and emotional memories, with an emphasis on human work.

Motor memory The first experiment to translate rodent reconsolidation findings to humans did so in motor memory, or the memory for a procedural skill (Walker, Brakefield, Hobson, & Stickgold, 2003). Scientists trained participants on a target sequence of finger taps and shortly after trained the same participants on a second sequence, expecting the second motor memory to interfere with consolidation of the first. Indeed, learning the second interference sequence impaired performance of the target sequence one day later. By contrast, when participants learned the interference sequence 24 hours after the target sequence, performance of the target sequence did not suffer, suggesting that consolidation stabilized the memory. Given the literature on reconsolidation, the researchers next investigated whether learning the interference sequence would disrupt performance of the target sequence if the memory of the target sequence was reactivated prior to interference training. In support of reconsolidation theory, the consolidated target memory was destabilized and disrupted by a reactivation plus interference procedure. Memory impairment was not immediate but emerged one day later, suggesting the effect was not a result of immediate memory reversal but was instead due to restabilization impairments.

Declarative memory Declarative memory, or the memory for facts (semantic memory) and autobiographical events (episodic memory), has been observed to undergo reconsolidation in hippocampal-dependent object recognition memory in rats (Rossato et al., 2007) and memories of words or narratives in humans (Chan & LaPaglia, 2013; Hupbach, Gomez, Hardt, & Nadel, 2007). In one human study (Hupbach et al., 2007), participants were prompted to memorize the contents of a basket filled with objects on day 1. On day 2 an experimenter presented the basket without the objects to trigger memory reactivation without directly prompting memory recall. Control participants did not encounter this reminder cue. Subsequently, both groups of participants memorized a second batch of objects. During a day 3 recall test, participants who encountered the reminder cue incorrectly incorporated a higher number of items from list 2 into list 1. The intrusion of list 2 items into list 1 suggests that the memory for list 1 destabilized following reactivation, was updated with list 2 information, and reconsolidated after memory modification. An alternative explanation for this finding is that both memories remained intact but participants had trouble deciphering whether objects belonged to list 1 or list 2. This would be a possibility if list 1 items had also intruded into list 2, but this was not the case, supporting the notion that the destabilized memory was selectively susceptible to memory updating via reconsolidation.

The above study assumed that the reminder cue successfully elicited memory reactivation but did not explicitly test for it. In a more recent study (Chan & LaPaglia, 2013), researchers verified memory reactivation by eliciting recall. The study employed two experiments whereby participants viewed a movie about a fictional terrorist attack followed by memory reactivation via a recall test either 20 minutes or 48 hours later. Control participants performed a distractor task (a computer game) in lieu of memory reactivation. Postreactivation, or control, participants listened to an audio recount of the terrorist attack, but the recording misrepresented several details. During a memory test either 20 minutes or 24 hours later, participants showed impaired memory for details that were misrepresented, but only if reactivation of the movie preceded the audio recording. This it yet another example of memory updating that demonstrates the malleability of a reactivated memory in the face of new information.

Within their second experiment, the authors assessed the degree of specificity needed for new information to update a reactivated memory. To this end, they presented the postreactivation misinformation as part of a story line unrelated to the original movie. This manipulation did not affect the memory of the initial account, suggesting that declarative memory may only become eligible for update if the new information is highly specific to the original memory. We would expect such selectivity from our knowledge of trace dominance. The authors further explain that this requirement for specificity is the reason our declarative memories are not constantly modified by new pieces of information encountered during daily life.

Emotional memory While many manipulations that target emotional memory reconsolidation in animal models are not suitable for use in humans, there is a pharmacological agent that modifies memories in animals and is safe for human use: propranolol. Propranolol acts through beta-receptor antagonism to regulate the noradrenergic system, which is involved in the consolidation and reconsolidation of emotional memories. In rodents, propranolol has varying influence on memory modification, depending on the memory subtype. Propranolol with reactivation reduced the response to a CS in cued threat-conditioning studies, but the effect was only modest within contextual threat conditioning. In appetitive-conditioning tasks, propranolol with reactivation decreased the self-administration of cocaine and sucrose, with modest effects on reducing alcohol administration. In humans, propranolol with memory reactivation decreased emotional responses to threat memories in healthy controls as well as anxiety patient populations. Similarly, in tasks of appetitive drug-cue associations, recall for emotional memory components was impaired in participants who received propranolol with reactivation, indicating that beta-receptor antagonism may specifically reduce the emotional affect associated with a memory. These results and others illustrate the therapeutic promise for using propranolol to modify maladaptive memories, although the specific clinical applications might be more complex. Some studies have found no effect of propranolol in patient populations, while others have demonstrated efficacy with multiple doses and prereactivation administration. Aside from propranolol, several other agents have been implicated in memory modification, including methylenedioxymethamphetamine (MDMA), ketamine, cortisol, glucose, and cannabinoids (for reviews, see Agren, 2014; Elsey, Van Ast, & Kindt, 2018; Fattore, Piva, Zanda, Fumagalli, & Chiamulera, 2018).

Alongside discoveries using pharmacological agents, scientists have also found noninvasive means to update emotional memories. Conditioned threat memory can be diminished with a behavioral extinction paradigm applied during the reconsolidation window in both rats (Monfils, Cowansage, Klann, & LeDoux, 2009) and humans, with humans showing attenuated threat responding even one year later (Schiller et al., 2010). Extinction during reconsolidation may be regarded as a form of updating the initial memory with the "safe" association conveyed during extinction. Similar threat response attenuation was demonstrated using counterconditioning (replacing a negative cue association with a positive one) during reconsolidation and when participants played a computer game following the reminder, which is thought to funnel cognitive resources away from restabilization—thereby disrupting it. These findings support a model of therapeutic reconsolidation with the potential to offer lasting treatment options to patients with anxiety-based psychiatric conditions rooted in maladaptive emotional memories. As appetitive associations are also susceptible to noninvasive interventions during reconsolidation, psychiatric disorders rooted in dysfunctional reward circuitry, such as addiction, are also likely to benefit from reconsolidation-based therapeutics (for a review, see Lee, Nader, & Schiller, 2017).

Potentiating reconsolidation Future therapies that target reconsolidation must be careful to modulate memories in the appropriate direction, as experimental manipulations to impair reconsolidation coexist with manipulations that can enhance it. Memory enhancement, though, has therapeutic potential in its own right, as it would be desirable to enhance adaptive memories (e.g., memory for nondrug cues or safe contexts) that are difficult to acquire or maintain. Stress has repeatedly been found to enhance hippocampus-dependent memory in animal models (Maroun & Akirav, 2008), as well as in humans. Coccoz, Maldonado, and Delorenzi (2011), for example, utilized cold pressor stress (CPS; stress induced by a protocol in which participants submerge their arms in an ice-cold water bath) to demonstrate that a mild acute stressor during reconsolidation improved memory for cue-syllable associations. Another study (Coccoz, Sandoval, Stehberg, & Delorenzi, 2013) tested whether declarative memory could be enhanced during the reconsolidation of a forgotten memory. Coccoz et al. (2013) again utilized CPS to enhance destabilized memories but also administered oral glucose, which had previously been shown to enhance memory in healthy adults, adults with Down syndrome, and adults with Alzheimer's disease. Six days after training in a cue-syllable associative task, a control group of participants showed poor recall for the memory. Participants who did not receive the memory test on day 6 and instead underwent either (1) *reactivation plus glucose* or (2) *reactivation plus CPS* showed enhanced memory for the cue-associations the following day. When declarative memory was tested 20 days after learning, *reactivation plus glucose* was still able to enhance declarative memory, but *reactivation plus CPS* was not. The authors note that their employed stressor was milder than that of other studies and that a more intense stressor may have enhanced memory even at day 20. The specific type of stress may also have an impact on the direction of reconsolidation effects, as both the elevated platform task (the rat is placed on an elevated platform in a brightly lit room) and context unfamiliarity (the rat is not exposed to the training context prior to training) induce increased glucocorticoid secretion, but the two tasks enhance and impair the reconsolidation of object recognition memory, respectively (Maroun & Akirav, 2008). Stress, though, is not the only mechanism that can enhance the reconsolidation of declarative memory. For example, low, but not high, doses of nicotine administered during the reconsolidation of object recognition enhanced memory in rats (Tian, Pan, & You, 2015), covert variations in sensorimotor demands enhanced motor memory in humans (Wymbs, Bastian, & Celnik, 2016), and transcranial direct current stimulation (a noninvasive method of electrically stimulating the brain using electrodes placed on the scalp) enhanced declarative memory when applied during consolidation and reconsolidation in humans (Javadi & Cheng, 2013).

Life of the Engram Postreconsolidation

In a typical reconsolidation study, a memory is acquired on day 1 then reactivated and manipulated on day 2. Between 6 and 24 hours should pass to allow the memory to reconsolidate. What happens next?

To assess whether day 2 had a lasting effect on the target memory, researchers determine the strength and accessibility of the memory trace by presenting a reminder cue. Probing for recall is a logical method for memory testing, but it is important to keep in mind that the seemingly simple act of stimulating memory retrieval requires reactivation, which makes the memory again susceptible to a number of fates, including destabilization. The life cycle of the dynamic engram is exactly that—a cycle. Each reactivation, even those that occur during memory testing, can initiate a cascade of events.

In the days, weeks, and months following memory acquisition, consolidation, reactivation, destabilization, and restabilization, even more still happens to the engram. Thus far we have discussed *synaptic*, or cellular, consolidation and reconsolidation, which refer to changes at the level of the synapse occurring minutes to hours after learning. *Systems* consolidation is a process driven by synaptic consolidation but specifically refers to circuit-level changes that convert a memory from an initial hippocampus-dependent state to a hippocampus-independent state. Systems consolidation was discovered when researchers found that lesioning the hippocampus 24 hours postlearning disrupted a contextual threat memory, showing that intact hippocampus function is necessary for memory retrieval. Lesioning the hippocampus 28 days after memory acquisition, however, did not affect memory recall (Kim & Fanselow, 1992). Thus, the hippocampus was determined to be involved in initial synaptic consolidation but with time, the memory is distributed to a range of cortical memory storage sites. In short, the hippocampus is necessary for the acquisition of STM and its consolidation to LTM. However, a distinction can be drawn between recent LTM that still depends on the hippocampus and remote LTM that has to be redistributed throughout the cortex (Kim & Fanselow, 1992).

In the tradition of reconsolidation mechanisms mirroring those of consolidation, scientists have additionally uncovered evidence for systems reconsolidation. In 2000, Land et al. challenged the idea that hippocampal dependence is contingent on the age of a memory. Though many studies report the hippocampal independence of older memories, the authors noted, those studies conflate the memory state and age and fail to account for the fact that older memories are more likely to be in an inactive state. The researchers dissociated hippocampal involvement in active memories from incidental associations with memory age by reactivating remote memories in rats prior to lesioning their hippocampi. They found that hippocampal lesions caused amnesia only if the memory was reactivated prior to the lesion, indicating that reactivation caused the memory to become hippocampus-dependent (Land, Bunsey, & Riccio, 2000). Debiec and colleagues (2002) later used a contextual threat-conditioning paradigm to directly probe systems reconsolidation using a task known to rely on the hippocampus for initial memory encoding and consolidation. Their results again revealed that a hippocampus-independent consolidated contextual threat memory could be made hippocampus-dependent by reactivating the memory, supporting the notion that hippocampal dependence is a function of memory state (i.e., active vs. inactive) rather than memory age.

Therapeutic Reconsolidation: Fact or Science Fiction?

Memory researchers have uncovered several pharmacological and behavioral manipulations that relieve the symptoms of psychopathologies rooted in maladaptive memory processing. Patient studies in reconsolidation aim to repurpose these manipulations to go deeper than symptom relief and modify the maladaptive memory itself. A handful of studies have directly assessed the ability to harness reconsolidation to modify pathological memory associations (for reviews, see Exton-McGuinness & Milton, 2018; Kroes, Schiller, LeDoux, & Phelps, 2016; Lee, Nader, & Schiller, 2017). Alcohol craving, for example, was diminished in a study in which researchers triggered PE in patients with alcohol use disorder by instructing them to consume an alcoholic beverage but interrupting before each participant could take a first sip. After this reactivating and destabilizing procedure, participants viewed alcohol cues paired with disgusting images in a counterconditioning protocol that lead to a later reduction in cue-induced craving. Cravings also diminished in two other studies examining participants with heroin use disorder and participants who smoke cigarettes. A retrieval-extinction procedure led to reduced craving 24 hours later and at a six-month follow-up among patients with heroin use disorder and a one-month follow-up among patients with tobacco use disorder. Participants with a spider phobia also experienced lasting clinical improvements in response to a retrieval-extinction protocol and a reactivation-propranolol protocol, as evidenced by increased approach behavior toward spiders 24 hours after the extinction session as well as six months and one year later, respectively. Two other studies using retrieval-extinction protocols to modify behavioral expression in spider phobics did not show conclusive evidence of memory modification resulting from reconsolidation manipulation. Together, these experiments demonstrate the potential for therapeutic reconsolidation but also indicate the necessity for clarification of the parameters that reliably correspond to significant clinical improvements.

Challenges to Reconsolidation Theory

The validity of any scientific theory must be challenged by considering alternative explanations for experimental observations. Accordingly, some scientists argue that the changes in behavioral expression thought to reflect memory modification during reconsolidation could be attributed to other processes that do not modify the memory. The question of whether retrograde amnesia constitutes a storage failure or retrieval failure is at the heart of this reconsolidation debate. If perceived memory modification results from a storage failure, amnesia occurs because a destabilized memory cannot be successfully restabilized. However, if the memory engram remains intact and does not undergo modification, amnesia must occur because the participant no longer has access to the engram, constituting a retrieval failure. In the case of a retrieval failure, the manipulation does not modify the memory itself but modifies the ability for a retrieval cue to successfully access the memory. The support for retrieval failure stems largely from studies that have reversed retrograde amnesia. One such study found that after anisomycin-induced amnesia for a consolidated memory, the direct stimulation of hippocampal engram cells reactivated a contextual threat memory, as evidenced by an increase in threat responding to a CS. This memory restoration occurred despite the reversal of synaptic plasticity in engram cells (increased potentiation and dendritic spine density) following blocked reconsolidation (Ryan, Roy, Pignatelli, Arons, & Tonegawa, 2015).

Retrograde amnesia generated by a PSI can also be reversed by readministering the drug prior to memory testing. This was also observed in a CTA memory acquired in the presence of lithium chloride, which induces gastric malaise and enhances CTA without affecting protein synthesis. This suggests that retrograde amnesia may be the result of *state-dependent learning* rather than a failure of memory restorage. In state-dependent learning, an animal's internal state during memory reactivation (e.g., drug state) becomes linked with the memory, and subsequently, the memory can only be retrieved when the animal enters that state again (Gisquet-Verrier et al., 2015). State-dependent learning and reconsolidation theory, however, are not necessarily mutually exclusive. A destabilized memory could conceivably be updated with the neural representation of the animal's physiological state and subsequently reconsolidated so that future recall of the memory is most effectively triggered by reactivating the drug state. Additionally, retrograde amnesia may be a shared end point for several neural processes, including disrupted reconsolidation and state-dependent learning.

Summary

Though memory was once thought to be immutable following consolidation, neuroscientists have found that memory fluctuates between active and inactive states that differentially permit modification. Only an active memory, whether newly acquired or reactivated, can undergo memory destabilization, a neural process that returns the memory to its unstable state through a cascade of molecular, cellular, and genetic events. Once destabilized, the memory can be diminished if restabilization is interrupted, enhanced by potentiating manipulations, or updated with new information. Though a reactivated memory is subject to multiple fates, the last few decades have been marked by increased interest in the specific sequence of memory reactivation, destabilization, and restabilization—namely, due to the immense potential it poses for the treatment of psychopathologies marked by maladaptive memory processing. However, the invasive nature of many experimental manipulations used in studying memory modification prohibits their use in humans, complicating the translation of animal findings to humans. Researchers are currently developing strategies to circumvent this obstacle and have already made strides in uncovering knowledge on memory modifications in humans.

Acknowledgments

Funding was provided by NIMH grant R01MH105535 and a Klingenstein-Simons Fellowship Award in the Neurosciences to Daniela Schiller; and NIMH grant R01MH05535-04S1 to Daniela Schiller and Temidayo Orederu.

REFERENCES

Agren, T. (2014). Human reconsolidation: A reactivation and update. *Brain Research Bulletin, 105,* 70–82. https://doi.org/10.1016/j.brainresbull.2013.12.010

Alberini, C. M. (2005). Mechanisms of memory stabilization: Are consolidation and reconsolidation similar or distinct processes? *Trends in Neurosciences, 28*(1), 51–56. https://doi.org/10.1016/j.tins.2004.11.001

Bartlett, F. (1932). *Remembering: An experimental and social study.* Cambridge: Cambridge University.

Besnard, A., Caboche, J., & Laroche, S. (2012). Reconsolidation of memory: A decade of debate. *Progress in Neurobiology, 99*(1), 61–80. https://doi.org/10.1016/j.pneurobio.2012.07.002

Chan, J. C. K., & LaPaglia, J. A. (2013). Impairing existing declarative memory in humans by disrupting reconsolidation. *Proceedings of the National Academy of Sciences of the United States of America, 110*(23), 9309–9313. https://doi.org/10.1073/pnas.1218472110

Coccoz, V., Maldonado, H., & Delorenzi, A. (2011). The enhancement of reconsolidation with a naturalistic mild stressor improves the expression of a declarative memory in humans. *Neuroscience, 185,* 61–72. https://doi.org/10.1016/j.neuroscience.2011.04.023

Coccoz, V., Sandoval, A. V., Stehberg, J., & Delorenzi, A. (2013). The temporal dynamics of enhancing a human declarative memory during reconsolidation. *Neuroscience, 246,* 397–408. https://doi.org/10.1016/j.neuroscience.2013.04.033

Debiec, J., LeDoux, J. E., & Nader, K. (2002). Cellular and systems reconsolidation in the hippocampus. *Nature, 36*(3), 527–538. https://doi.org/10.1016/S0896-6273(02)01001-2

Eisenberg, M., Kobilo, T., Berman, D. E., & Dudai, Y. (2003). Stability of retrieved memory: Inverse correlation with trace dominance. *Science, 301*(5636), 1102–1104. https://doi.org/10.1126/science.1086881

Elsey, J. W. B., Van Ast, V. A., & Kindt, M. (2018). Human memory reconsolidation: A guiding framework and critical review of the evidence. *Psychological Bulletin, 144*(8), 797–848. https://doi.org/10.1037/bul0000152

Exton-McGuinness, M. T. J., & Milton, A. L. (2018). Reconsolidation blockade for the treatment of addiction: Challenges, new targets, and opportunities. *Learning & Memory, 25*(9), 492–500. https://doi.org/10.1101/lm.046771.117

Fattore, L., Piva, A., Zanda, M. T., Fumagalli, G., & Chiamulera, C. (2018). Psychedelics and reconsolidation of traumatic and appetitive maladaptive memories: Focus on cannabinoids and ketamine. *Psychopharmacology, 235*(2), 433–445. https://doi.org/10.1007/s00213-017-4793-4

Flavell, C. R., Lambert, E. A., Winters, B. D., & Bredy, T. W. (2013). Mechanisms governing the reactivation-dependent destabilization of memories and their role in extinction. *Frontiers in Behavioral Neuroscience, 7.* https://doi.org/10.3389/fnbeh.2013.00214

Gershman, S. J., Monfils, M.-H., Norman, K. A., & Niv, Y. (2017). The computational nature of memory modification. *eLife, 6.* https://doi.org/10.7554/eLife.23763

Gisquet-Verrier, P., Lynch, J. F., Cutolo, P., Toledano, D., Ulmen, A., Jasnow, A. M., & Riccio, D. C. (2015). Integration of new information with active memory accounts for retrograde amnesia: A challenge to the consolidation/reconsolidation hypothesis? *Journal of Neuroscience, 35*(33), 11623–11633. https://doi.org/10.1523/JNEUROSCI.1386-15.2015

Gisquet-Verrier, P., & Riccio, D. C. (2012). Memory reactivation effects independent of reconsolidation. *Learning & Memory, 19*(9), 401–409. https://doi.org/10.1101/lm.026054.112

Haubrich, J., & Nader, K. (2018). Memory reconsolidation. *Current Topics in Behavioral Neurosciences, 37,* 151–176. https://doi.org/10.1007/7854_2016_463

Hu, J., Wang, W., Homan, P., Wang, P., Zheng, X., & Schiller, D. (2018). Reminder duration determines threat memory modification in humans. *Scientific Reports, 8*(1), 8848. https://doi.org/10.1038/s41598-018-27252-0

Hupbach, A., Gomez, R., Hardt, O., & Nadel, L. (2007). Reconsolidation of episodic memories: A subtle reminder triggers integration of new information. *Learning & Memory, 14*(1–2), 47–53. https://doi.org/10.1101/lm.365707

Javadi, A. H., & Cheng, P. (2013). Transcranial direct current stimulation (tDCS) enhances reconsolidation of long-term memory. *Brain Stimulation, 6*(4), 668–674. https://doi.org/10.1016/j.brs.2012.10.007

Khalaf, O., Resch, S., Dixsaut, L., Gorden, V., Glauser, L., & Gräff, J. (2018). Reactivation of recall-induced neurons contributes to remote fear memory attenuation. *Science, 360*(6394), 1239–1242. https://doi.org/10.1126/science.aas9875

Kim, J. J., & Fanselow, M. S. (1992). Modality-specific retrograde amnesia of fear. *Science, 256*(5057), 675–677.

Kroes, M. C. W., Schiller, D., LeDoux, J. E., & Phelps, E. A. (2016). Translational approaches targeting reconsolidation. *Current Topics in Behavioral Neurosciences, 28,* 197–230. https://doi.org/10.1007/7854_2015_5008

Land, C., Bunsey, M., & Riccio, D. C. (2000). Anomalous properties of hippocampal lesion-induced retrograde amnesia. *Psychobiology, 28*(4), 476–485. https://doi.org/10.3758/BF03332005

Lee, J. L. C., & Flavell, C. R. (2014). Inhibition and enhancement of contextual fear memory destabilization. *Frontiers in Behavioral Neuroscience, 8,* 144. https://doi.org/10.3389/fnbeh.2014.00144

Lee, J. L. C., Nader, K., & Schiller, D. (2017). An update on memory reconsolidation updating. *Trends in Cognitive Sciences, 21*(7), 531–545. https://doi.org/10.1016/j.tics.2017.04.006

Maroun, M., & Akirav, I. (2008). Arousal and stress effects on consolidation and reconsolidation of recognition memory.

Neuropsychopharmacology: Official Publication of the American College of Neuropsychopharmacology, 33(2), 394–405. https://doi.org/10.1038/sj.npp.1301401

McGaugh, J. L. (1966). Time-dependent processes in memory storage. *Science, 153*(3742), 1351–1358.

Misanin, J. R., Miller, R. R., & Lewis, D. J. (1968). Retrograde amnesia produced by electroconvulsive shock after reactivation of a consolidated memory trace. *Science, 160*(3827), 554–555.

Monfils, M.-H., Cowansage, K. K., Klann, E., & LeDoux, J. E. (2009). Extinction-reconsolidation boundaries: Key to persistent attenuation of fear memories. *Science, 324*(5929), 951–955. https://doi.org/10.1126/science.1167975

Nader, K., & Hardt, O. (2009). A single standard for memory: The case for reconsolidation. *Nature Reviews Neuroscience, 10*(3), 224–234. https://doi.org/10.1038/nrn2590

Nader, K., Hardt, O., & Wang, S.-H. (2005). Response to Alberini: Right answer, wrong question. *Trends in Neurosciences, 28*(7), 346–347. https://doi.org/10.1016/j.tins.2005.04.011

Nader, K., Schafe, G. E., & LeDoux, J. E. (2000). Fear memories require protein synthesis in the amygdala for reconsolidation after retrieval. *Nature, 406*(6797), 722–726. https://doi.org/10.1038/35021052

Orederu, T., & Schiller, D. (2018). Fast and slow extinction pathways in defensive survival circuits. *Current Opinion in Behavioral Sciences, 24*, 96–103. https://doi.org/10.1016/j.cobeha.2018.06.004

Przybyslawski, J., & Sara, S. J. (1997). Reconsolidation of memory after its reactivation. *Behavioural Brain Research, 84*(1–2), 241–246.

Ramirez, S., Liu, X., Lin, P.-A., Suh, J., Pignatelli, M., Redondo, R. L., … Tonegawa, S. (2013). Creating a false memory in the hippocampus. *Science, 341*(6144), 387–391. https://doi.org/10.1126/science.1239073

Rossato, J. I., Bevilaqua, L. R. M., Myskiw, J. C., Medina, J. H., Izquierdo, I., & Cammarota, M. (2007). On the role of hippocampal protein synthesis in the consolidation and reconsolidation of object recognition memory. *Learning & Memory, 14*(1–2), 36–46. https://doi.org/10.1101/lm.422607

Ryan, T. J., Roy, D. S., Pignatelli, M., Arons, A., & Tonegawa, S. (2015). Engram cells retain memory under retrograde amnesia. *Science, 348*(6238), 1007–1013. https://doi.org/10.1126/science.aaa5542

Schiller, D., Monfils, M.-H., Raio, C. M., Johnson, D. C., LeDoux, J. E., & Phelps, E. A. (2010). Preventing the return of fear in humans using reconsolidation update mechanisms. *Nature, 463*(7277), 49–53. https://doi.org/10.1038/nature08637

Sevenster, D., Beckers, T., & Kindt, M. (2013). Prediction error governs pharmacologically induced amnesia for learned fear. *Science, 339*(6121), 830–833. https://doi.org/10.1126/science.1231357

Spear, N. E. (1973). Retrieval of memory in animals. *Psychological Review, 80*(3), 163.

Tayler, K. K., Tanaka, K. Z., Reijmers, L. G., & Wiltgen, B. J. (2013). Reactivation of neural ensembles during the retrieval of recent and remote memory. *Current Biology, 23*(2), 99–106. https://doi.org/10.1016/j.cub.2012.11.019

Tian, S., Pan, S., & You, Y. (2015). Nicotine enhances the reconsolidation of novel object recognition memory in rats. *Pharmacology, Biochemistry, and Behavior, 129*, 14–18. https://doi.org/10.1016/j.pbb.2014.11.019

Walker, M. P., Brakefield, T., Hobson, J. A., & Stickgold, R. (2003). Dissociable stages of human memory consolidation and reconsolidation. *Nature, 425*(6958), 616–620. https://doi.org/10.1038/nature01930

Wymbs, N. F., Bastian, A. J., & Celnik, P. A. (2016). Motor skills are strengthened through reconsolidation. *Current Biology, 26*(3), 338–343. https://doi.org/10.1016/j.cub.2015.11.066

IV

ATTENTION AND WORKING MEMORY

Introduction

SABINE KASTNER AND STEVEN J. LUCK

Our section focuses on attention, working memory, and their interactions—and this is an exciting new development for the sixth edition of this book. Previous editions focused on attention in isolation, but the focus of research has shifted over recent years. The cognitive neuroscience of working memory has become a large and relatively mature field, and working memory is strongly intertwined with attention, so it made sense to combine attention and working memory in the same section. Interestingly, although we invited the chapter authors to contribute a chapter on attention *or* working memory, most of the authors wrote chapters on attention *and* working memory.

A second exciting innovation for our section is that we include, for the first time, a chapter on the development of attention and working-memory functions (by Scerif). The field of development was grounded in behavioral psychology and has now become an integral part of the field of cognitive neuroscience.

A third and final innovation is that for the first time our section includes a chapter on the role of the thalamus in selective attention (by Usrey and Kastner). Whereas most neural accounts of cognitive processing have focused on cortical systems, the involvement of the thalamus and its significance for the healthy and pathologic brain have become increasingly apparent. Particularly, the study of thalamocortical interactions holds great promise in leading to a more complete understanding of cognition.

We will start our section overview with a brief account of terminology to clarify the terms *attention* and *working memory*, which are broad and have multiple definitions that can lead to substantial confusion.

In cognitive neuroscience the term *attention* most commonly refers to *selective attention*, the set of mechanisms by which we select a subset of the available sensory inputs or tasks for enhanced processing. Selective attention is important for avoiding information overload and for dealing with competition between stimuli or tasks. The chapter by Rosenberg and Chun describes three additional types of attention: alerting (the general state of arousal), executive attention (engaging in controlled processing and overriding automatic responses), and sustained attention (maintaining a goal over time and avoiding mind wandering). Although the term *attention* is used to refer to all of these processes, they are very different in terms of both cognitive mechanisms and neural substrates.

The term *working memory* does not have such distinctly different meanings, yet there is still quite a bit of variation in how the term is used. Virtually all definitions refer to some kind of relatively brief *memory* (on the scale of seconds for some researchers and minutes for others) with a limited storage capacity and some kind of *work* (a cognitive process that makes use of this memory). However, some researchers stress the *memory* part, whereas others stress the *work* part. That is, for some researchers, working memory is mainly a temporary storage buffer, whereas for other researchers, working memory is mainly a system that protects and manipulates the information in this buffer. Cognitive neuroscientists have focused mainly (although not exclusively) on the storage aspect rather than the manipulation aspect, and this can be seen in the present volume in the chapters by Awh and Vogel, by Jensen and Hanslmayr, by Nobre and Stokes, and by Scerif.

Cognitive neuroscience research on attention and working memory has progressed rapidly since the last edition of this volume. We now highlight some important emerging trends, which the chapters in this section cover in detail.

Interactions between attention and working memory Much recent research has focused on the bidirectional interactions between working memory and attention. Indeed, these cognitive processes are so densely interactive, and overlap so much neuroanatomically, that some researchers have proposed them to be a single system (see, for example, the idea that working memory can be considered *internally focused attention* in the chapter by Rosenberg and Chun). However, it is probably more accurate to think of attention and working memory as analogous to the heart and the lungs, which work toward a set of common goals but are nonetheless distinct organs. The chapter by Nobre and Stokes does an excellent job of summarizing the interactions between attention and working memory (and long-term memory, as well). Because working memory capacity is highly limited, attention plays an essential gatekeeper role, ensuring that only the most relevant information is stored in working memory (and ultimately in long-term memory). Attention can also be used to strengthen and protect information that has already been stored in working memory. Working memory, in turn, plays a key role in controlling attention: by storing a goal in working memory, attention will be directed to items that match that goal. As described in the chapter by Scerif, these bidirectional interactions between attention and working memory develop from infancy through adolescence and into adulthood. The chapter by Moore, Jonikaitis, and Pettine discusses the neural mechanisms of these interactions, describing how working-memory representations of locations can be maintained by means of sustained neural activity in the frontal eye fields, which produces feedback signals in the visual cortex that boost the neural coding of objects presented at the corresponding locations.

Nature of working-memory representations A great deal of empirical and theoretical work in cognitive neuroscience currently focuses on the mechanisms underlying working-memory storage. The kind of sustained neural activity discussed by Moore, Jonikaitis, and Pettine has been studied for several decades, but two new trends are worth noting. First, as described by Nobre and Stokes and by Buschman and Miller, working memory representations may also be stored by means of short-term changes in synaptic plasticity, without sustained firing (*activity-silent* representations). Second, as described by Awh and Vogel, working memory can be described in terms of both the number of representations that can be maintained (*capacity*) and the precision of the representations (*resolution*).

Individual differences Most research in cognitive neuroscience seeks to explain how the "average" brain carries out cognitive functions, ignoring the obvious fact that people vary enormously in their experiences, their abilities, their motivations, and other factors. Cognitive psychologists started taking these individual differences seriously many years ago, and the study of individual differences is now common in cognitive neuroscience as well. This is beautifully exemplified in Rosenberg and Chun's chapter, which focuses on individual differences in patterns of functional connectivity as revealed by functional MRI (fMRI). These individual differences in functional network properties predicted individual differences in the ability of people to sustain their attention, to suppress salient-but-irrelevant distractors, and to maintain precise representations in working memory.

Large-scale networks and graph theory Rosenberg and Chun also highlight another important trend, the use of graph theory and related methods to characterize large-scale patterns of information flow in the brain. Whereas Rosenberg and Chun focus on applying these methods to fMRI data, the chapter by Jensen and Hanslmayr discusses the application of network-level methods to electroencephalography (EEG) and magnetoencephalography (MEG).

Oscillations in attention and working memory The study of the neural mechanisms of attention and working memory has shifted during the last years from characterizing the correlations of local neural activity and behavioral outcome to the relations of large-scale network activity and behavior. Electrophysiologists have recently turned to the important question of how these large-scale networks are organized to allow their participating hubs to contribute to the network function and output. One important mechanism that has been identified is the task-dependent synchronization of neural activity in different frequency bands. Jensen and Hanslmayr illustrate this effort by summarizing the relevant MEG/EEG literature on alpha-band (8–13 Hz) oscillations. Alpha oscillations provide region-specific functional inhibition to suppress hubs that are not engaged in the task at hand and thereby indirectly maximize the allocation of computational resources to the hubs that are the most task-relevant (see also the chapter by Buschman and Miller). Usrey and Kastner show, in their chapter, how the cortical attention network is temporally organized through thalamocortical interactions that modulate neuronal synchronization across interconnected cortical hubs. These chapters provide examples of emerging work from the growing field of cognitive network science.

Subcortical contributions The thalamus has been traditionally viewed as a slave system to the cortex. For example, the lateral geniculate nucleus (LGN) is best known for its function as a relay station between the retina and the cortex. In contrast, neural mechanisms of cognitive processing—such as those related to attention and working memory—have traditionally been associated with the cortex. This corticocentric view of cognition was largely based on early negative findings when exploring the thalamus in attention tasks in nonhuman primates and later in difficulties obtaining high-resolution functional images from the human thalamus. This view has begun to change, and an increasing amount of research is being directed at the role of the primate (and rodent) thalamus in attention. Usrey and Kastner summarize the findings for both first-order (e.g., LGN) and higher-order thalamic nuclei (e.g., pulvinar). Examining the role of the thalamus in attention and other processes will lead to a more complete understanding of the fundamental mechanistic operations underlying cognition.

25 Memory and Attention: The Back and Forth

A. C. (KIA) NOBRE AND M. S. STOKES

ABSTRACT Memory and attention are two core domains of our psychological functions. Accordingly, they anchor two major fields of inquiry within cognitive neuroscience. These have developed relatively independently, with each field focusing on the attributes that distinguish the two functions. However, as this chapter highlights, memory and attention have much in common and often work together in a mutually supportive way toward a common purpose: to guide flexible and adaptive behavior.

Memory Back and Attention Forth

Our folk psychological intuitions tell us that memory is about what has passed and that attention is about what is to come. Memory retrieves and attention anticipates. Perhaps unsurprisingly, research has largely followed these intuitions in separating memory and attention into the "back" and the "forth." However, these arrows of time are misleading. When we take an ecological, functional view and ask what purpose memory and attention serve, the arrows of time break down, and the two cognitive domains come much closer together. The core purpose of both memory and attention is to guide adaptive behavior in a flexible way that takes into account what is relevant within a given context. As elaborated in the rest of the chapter, the brain draws on experience from multiple timescales to anticipate and prepare for incoming stimulation and guide adaptive action. Within this framework it becomes more difficult to separate memory from attention. Memory ceases to be just about the past, and its prospective nature comes to light. In turn, attention is recognized to rely heavily on previous experience. Thus, a better way to define each of these interrelated functions is to consider the role each plays in this process of linking the past to the future. In guiding flexible and adaptive behavior, memory provides the informational content, and attention prioritizes and selects what is likely to be important.

Memory Forth

The traces left behind through experience are the essence of memory. Some types of traces support conscious recollection, while others do not; however, all types of traces can interact with incoming stimulation to change behavior. That is the fundamental purpose of memory—collecting relevant past experience to anticipate future demands and guide behavior.

These prospective properties of memory are increasingly recognized. The field of attention has given particular importance to working-memory traces. These are thought to maintain a template of stimulus attributes that are relevant for current goals and thus to constitute an important source of top-down, attention-related signals that bias the analysis of incoming sensory stimulation (Desimone & Duncan, 1995). Accordingly, the current chapter will focus on the relation between working memory and attention; however, it is important to appreciate that more remote traces from long-term memory also influence the processing of incoming stimulation (see Aly & Turk-Browne, 2017; Awh, Belopolsky, & Theeuwes, 2012; Nobre & Mesulam, 2014; see figure 25.1).

Working memory: from retrospective representational states to prospective functional states Working memory (WM) refers to the ability to store and manipulate recently acquired information independently of continuous sensory stimulation. A stable internal cognitive state is needed for integrating information over sensory discontinuities (e.g., eye movements), performing cognitive operations such as mental arithmetic or object rotations, and, more generally, guiding behavior over the short term (Baddeley, 2003). As such, WM provides the functional backbone to high-level cognition, allowing us to perform complex actions based on time-extended goals and contextual contingencies (Fuster, 2001). We argue that WM is not simply a *representational state* of *past experience* but is better conceived as a *functional state* for guiding *future behavior* (Myers, Stokes, & Nobre, 2017). WM traces are adaptive, dynamic, and proactive, bridging previous contexts and sensations to anticipated actions and outcomes.

Tonic delay activity Single-unit neurophysiology in the awake, behaving monkey provided influential

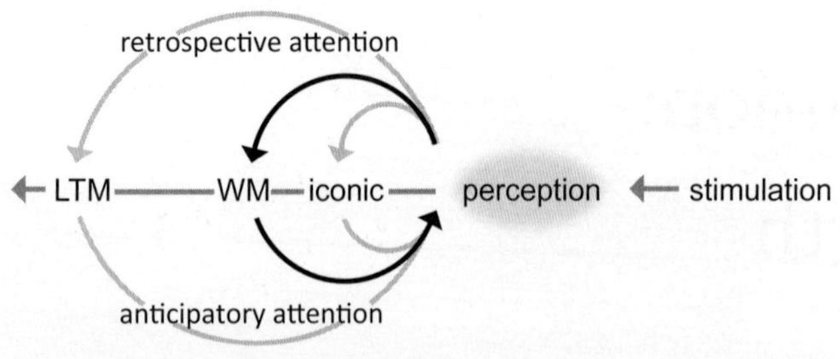

FIGURE 25.1 Mutual interactions between memory and attention. Attention draws on past experience from multiple timescales to anticipate and prepare for incoming stimulation and guide adaptive action. Conversely, attention is not only forward looking but can select and bias information in memory. These mutual interactions feed a virtuous cycle that tunes our minds to the most relevant features of the environment. Although multiple mnemonic timescales are important for attention, we focus on the interactions with working memory in this chapter.

breakthroughs in WM research. Recordings from the prefrontal cortex (PFC) discovered so-called memory cells that are persistently active during the delay period in delayed-response tasks (Fuster & Alexander, 1971; Kubota & Niki, 1971). WM delay activity was subsequently replicated in PFC (e.g., Funahashi, Bruce, & Goldman-Rakic, 1989; Miller, Erickson, & Desimone, 1996) and later also identified in the parietal cortex (Chaffee & Goldman-Rakic, 1998) and in visual areas such as the inferior temporal cortex (IT; Fuster & Jervey, 1981). Importantly, delay activity is selective for the content of WM—cells are more active when their preferred stimulus is held in mind (e.g., Miller, Erickson, & Desimone, 1996), which, at the population level, gives rise to a decodable signal for downstream systems. Brain-imaging studies in humans provided converging evidence (see Christophel, Klink, Spitzer, Roelfsema, & Haynes, 2017) by revealing WM-related activity in the PFC (Courtney, Petit, Maisog, Ungerleider, & Haxby, 1998), the parietal cortex (Todd & Marois, 2004), and the visual cortex (Awh et al., 1999).

Together, single-unit and imaging studies have contributed to the prevailing view that tonically maintained neuronal activity is the representational state supporting WM (Goldman-Rakic, 1987; Zylberberg & Strowbridge, 2017). This view emphasizes the retrospective aspect of WM function—in preserving a record of previous stimulation—and paints it as a rather static and inert record. However, it is important to recognize the importance of the prospective nature of WM—to guide future behavior.

Selective and adaptive traces When viewing WM from its prospective perspective, a much more adaptive, dynamic, and proactive set of mechanisms emerges. Findings from the classic single-unit *delay-activity* studies become more nuanced. For example, activity tends to increase during the delay in expectation of the probe (Watanabe & Funahashi, 2007) and can disappear altogether to reemerge at the anticipated time of the probe stimulus without compromising performance (Watanabe & Funahashi, 2014). Importantly, it has been noted that the PFC and parietal cortex do not equally represent all aspects of ongoing stimulation but instead pick up the dimensions of stimulation that are specifically relevant, given the current task goals (see Duncan, 2001). For example, by using stimuli morphed along multiple dimensions, Freedman, Riesenhuber, Poggio, and Miller (2001) showed that neurons were selectively sensitive to the dimensions that monkeys were required to discriminate in the task. Similar effects were found in the parietal cortex when monkeys were required to discriminate between arbitrary categorical boundaries along continuous feature dimensions (Freedman & Assad, 2006). Even when monkeys view the same memory stimuli but are trained to expect different kinds of memory probes, the activity in PFC adapts prospectively to the expected task demands (Rainer, Rao, & Miller, 1999; Warden & Miller, 2010).

In humans, similar prospective signals have been observed in the visual cortex. For example, using multivariate methods to derive population response properties from functional magnetic resonance imaging (fMRI) data, Serences and colleagues (2009) found that decoding during a WM delay depended as much on the memory stimulus as the expected demands during recall (see figure 25.2). Specifically, they told participants that either the color or the orientation of a visual stimulus would be probed at the end of a memory delay. Patterns of activity in the visual cortex selectively maintained the task-relevant feature, consistent with a prospective memory code for guiding future behavior.

The prospective use of WM traces has also been highlighted in neural studies of attentional top-down biasing signals. When monkeys guide visual search on the basis of an object or location in WM, the delay activity in visual areas representing the relevant object (Chelazzi, Duncan, Miller, & Desimone, 1998) or location (Luck, Chelazzi, Hillyard, & Desimone, 1997) remains elevated in anticipation of the search array. Human fMRI studies have also reported elevated levels of activity for the spatial location (Kastner, Pinsk, De Weer, Desimone, & Ungerleider, 1999) or identity (Stokes, Thompson, Nobre, & Duncan, 2009) of relevant, anticipated objects based on WM templates. Such findings have supported the influential idea that persistent activity associated with maintenance in WM

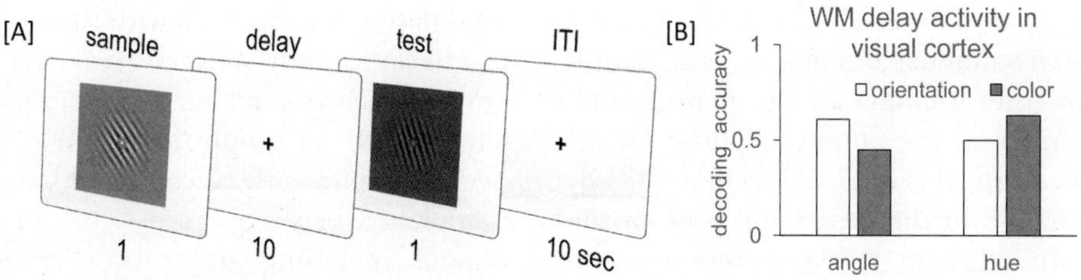

FIGURE 25.2 Working memory is prospective, representing the information most likely to be relevant for behavior. *A*, In this example, Serences and colleagues (2009) used fMRI to show that working memory maintains sensory information that is most relevant to behavior. *B*, Decoding patterns of activity in early visual cortex, they found that activity in the memory delay carried orientation-angle information when orientation was relevant for future decision-making or the color-hue information when color was relevant.

provides the major neurophysiological mechanism for top-down attentional modulation by effectively biasing the subsequent activation of matching sensory input (Desimone & Duncan, 1995).

Dynamic traces Most studies to date have highlighted the persistence of item-specific information that can be decoded during WM delays. However, finer-grained analysis of the qualitative patterns of brain activity coding for specific items reveals a much more dynamic picture (Stokes, 2015). The basic logic of machine-learning approaches to neural decoding can be extended to track qualitative changes in coding format. Rather than comparing the accuracy of decoding between two independent but equivalent sets of data, decoding can be compared among data drawn from different contexts. For example, to test how neural coding evolves over time, decoding can be performed in a way that tests the generalizability (or specificity) of discriminative patterns at different time points by training a classifier at one time point and testing performance at different time points (e.g., cross-temporal generalization; see King & Dehaene, 2014; Stokes, 2015).

This general approach suggests that neural activity patterns are constantly changing (Sreenivasan, Curtis, & D'Esposito, 2014). Analyses exploiting the high-temporal resolution of neurophysiological recordings from nonhuman primates reveal dynamic patterns of neuronal activity in the PFC (Meyers, Freedman, Kreiman, Miller, & Poggio, 2008; Stokes et al., 2013) and parietal cortex (Crowe, Averbach, & Chafee, 2010), even when the cognitive state remains stable (Murray et al., 2013; Spaak, Watanabe, Funahashi, & Stokes, 2017). Similar dynamics are also seen with noninvasive electrophysiological methods in humans (Myers et al., 2015; Wolff, Jochim, Akyurek, & Stokes, 2017).

In addition to their intrinsically dynamic neural nature, WM can also be utilized flexibly and proactively over time, according to the temporal regularities within a given context (see Nobre & van Ede, 2018). We showed this using multivariate analysis methods in a magnetoencephalogram (MEG) task in which participants matched visual orientation stimuli against a memorized template to detect infrequent matches (Myers et al., 2015). Information related to the template was associated with a dynamically evolving pattern of neural activity. Rather than being tonically elevated, the pattern became manifest around the predicted time of stimulus appearance. These results highlight how WM information can be used in a temporally structured proactive fashion to guide behavioral performance.

Functional states In light of the evidence that WM traces are adaptive and prospective, we can reframe WM as a flexible shift in how the brain processes new information (Stokes, 2015). Rather than acting as a representational state that preserves the past as persistent activity, it makes more sense to consider it as a functional neural state that shifts the coding properties of the system to anticipate future task demands. It is the functionality of the neural state that is most important, not merely its decodability. From a mechanistic perspective, decodability is only a minimal requirement. To understand how memories are used for recall, attention, or anything else, it is necessary to understand how the mnemonic states interact with subsequent input to produce context-dependent output.

Recent developments provide an expanding tool box for exploring the functional properties of mnemonic states. For example, we developed an impulse-response approach to probe how WM states change the input-output behavior of the neural system (Wolff, Ding, Myers, & Stokes, 2015). The logic borrows from *active sonar*, in which a well-characterized impulse (ping) is emitted toward a hidden landscape, and the contours are inferred from distortions in the reflected signal. In the case of *neural sonar*, we present a sensory impulse (i.e., a neutral visual stimulus) and measure the neural response. We can infer changes in the neural landscape from distortions in the output response (Wolff et al., 2017). Importantly, this approach is theoretically sensitive to any change in the functional state of the targeted system. In addition to the manifest delay-activity states that have been the focus of most studies of WM, it can also reveal latent, activity-silent neural states (see also Rose et al., 2016).

At the neural level, it is possible to maintain a functional state in persistent activity patterns (Machens, Romo, & Brody, 2005; Mante, Sussillo, Senoy, & Newsome, 2013). However, this is not the only way to maintain a functional state. A great diversity of alternative neurophysiological mechanisms could also play important roles (Barak & Tsodyks, 2014; Buonomano & Maass, 2009). For example, numerous computational models propose that short-term synaptic plasticity (STSP; see Zucker & Regehr, 2002) plays an important role in maintaining functional states in WM networks (e.g., Mongillo, Barak, & Tsodyks, 2008). Activity-dependent STSP has been observed in the frontal cortex of rodents (Hempel et al., 2000) and has been correlated to performance in memory-guided tasks (Fujisawa, Amarasingham, Harrison, & Buzsaki, 2008).

Another useful approach to study the functional state of WM is to explore the context-dependent response to the stimulus used to probe the memory. Previous studies have found evidence for a match-filter response, which signals the degree of match between the memory probe and the previous memory item (e.g., Hayden & Gallant, 2013). Such a signal could be used to guide performance in a delayed-match-to-sample task (Miller & Desimone, 1993) and could be implemented by delay activity (Machens, Romo, & Brody, 2005) or activity-silent mechanisms (Sugase-Miyamoto, Wiener, Optican, & Richmond, 2008). In a recent MEG study requiring orientation judgments against a memorized template, we showed that a synaptic model of a match filter storing parametrically varying stimulus orientation could be used to infer the direction as well as the magnitude of orientation change (Myers et al., 2015). Such flexibility suggests the same coding scheme could be used for guiding different types of WM-dependent behaviors.

Attention Back

Attention is clearly future serving, prioritizing and selecting useful information to guide adaptive behavior. This is often taken to mean that attention biases necessarily act upon the incoming sensory stream. Indeed, the vast majority of research on attention-related modulation is concerned with the presence and nature of effects in early sensory areas. Yet attention has its effects much beyond early sensory processing. Having left behind the simplistic notion of a unitary bottleneck, we currently recognize that attention-related modulatory biases operate across a multitude of brain regions, including early subcortical nuclei, numerous sensory cortices, sensorimotor and association areas, and regions involved in motor control (Nobre & Kastner, 2014). However, that is only one side of the story. In addition to biasing "forth" along the sensorimotor axis associated with incoming information, attention also acts upon mnemonic content to prioritize and select relevant information from WM (and long-term memory). By biasing mnemonic information, the brain can use memories more flexibly and effectively in the service of adaptive future behavior.

The ability of attention to point "back" to influence internal representations is recognized in the most classic definition of attention, by William James (1890), who stated that attention "is the taking possession by the mind, in clear and vivid form, of one out of what seem several simultaneously possible objects or *trains of thought*" (pp. 403–404). Yet, surprisingly, the ability of attention to modulate representations in WM was not demonstrated until relatively recently.

Initial studies in the 1960s had indicated that attention-directing cues were ineffective at improving the reporting of items in visual short-term memory (Sperling, 1960). When participants viewed a large array of items, the proportion of those they could report improved significantly if an immediate postcue (i.e., within a few hundred milliseconds) prompted them to report items from one row only. If the postcue was delayed by more than 1 s after the memory array, however, it conferred no benefit. These findings were interpreted to suggest that although visual memories over very brief periods (iconic memory; Neisser, 1967) had greater capacity than suspected, rapid forgetting ensued, leaving only a limited number of items in a more robust form of short-term memory. Similar findings were obtained with visual material that could not easily be transferred into verbal codes (Averbach & Coriell, 1961). For about 40 years thereafter, visual WM was studied as an inflexible, short-term store of limited capacity in which items were resistant to interference and accessible through serial search.

Cueing attention in working memory In the early 2000s, two independent research groups, including ours, upgraded this view of WM (Griffin & Nobre, 2003; Landman, Spekreijse, & Lamme, 2003; figure 25.3). Both groups showed the significant benefits of cues presented

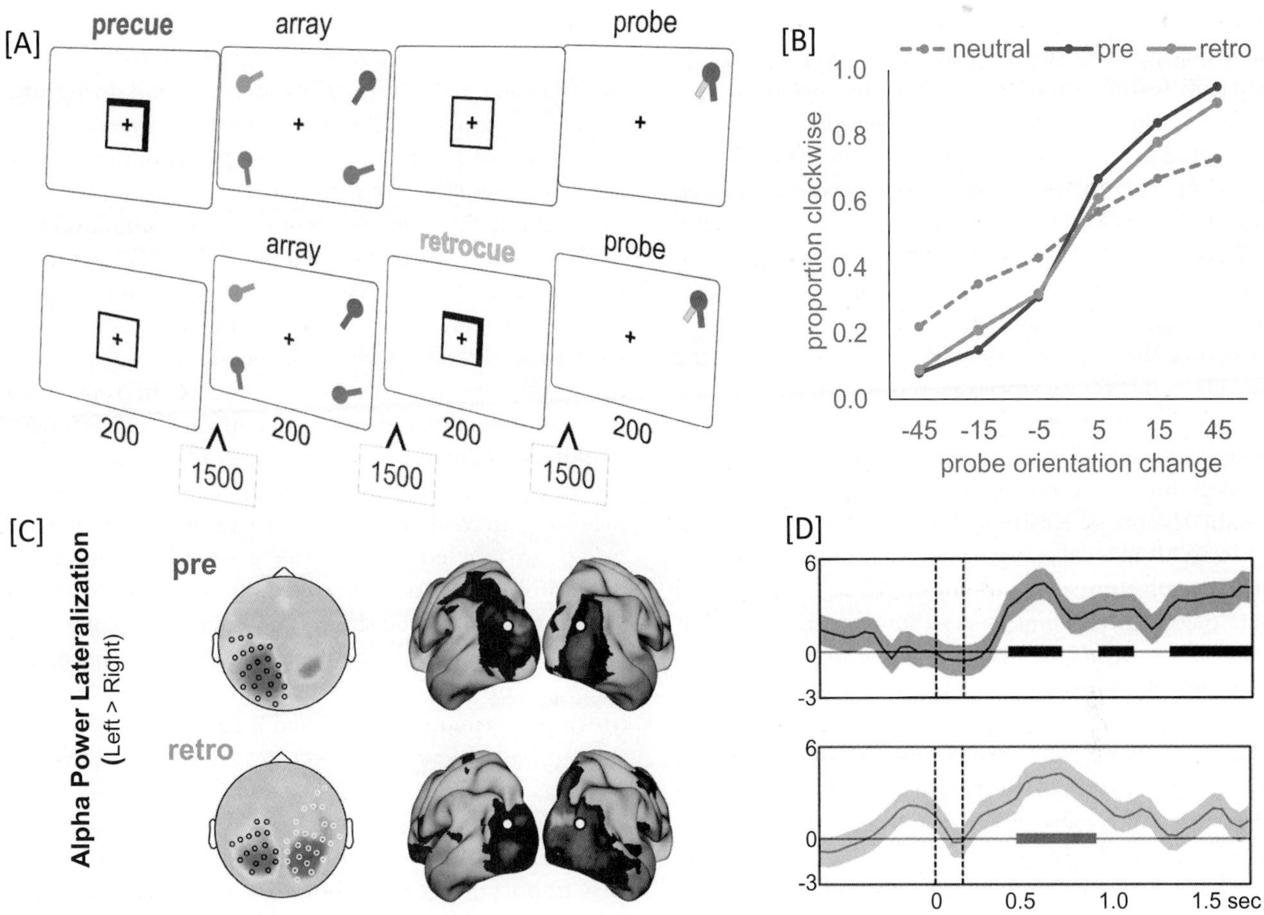

FIGURE 25.3 Attention is also retrospective, operating on the content of working memory. *A*, In this example, Wallis and colleagues (2015) used MEG to compare the neural dynamics underlying prospective and retrospective attention. *B*, Spatial cues presented either before encoding (precue) or during the memory delay (retro-cue) were both effective for optimizing memory performance. *C*, Both cue types also elicited a classic signature of spatial attention (contralateral desynchronization of posterior alpha oscillations; see topological plots). *D*, Time course analysis further showed that preparatory attention involves sustained alpha lateralization, but contralateral desynchronization was relatively transient following the retro-cue. (See color plate 27.)

during WM retention that indicated which memorized item would be relevant for subsequent task performance. These cues provided retroactively predictive information (*retro-cues*) about the relevance of items encoded into WM. The initial reports were met with some degree of skepticism, given the long-standing dogma about the inflexible nature of WM. However, since these original studies, the benefits of retro-cues have been replicated numerous times by laboratories around the world (for a review see Souza & Oberauer, 2016).

The immediate question that comes to mind is why retro-cues succeeded when the original postcues failed. Some technical reasons and task-specific parameters may contribute, but one fundamental difference is "time." Time is required for the information carried by retro-cues to influence neural activity associated with the memoranda according to their predicted relevance. The postcues in early studies prompted immediate recall. They left no time for attention-related modulation to operate, and they may have even interfered with the storage of and/or the access to the relevant memoranda. Retro-cues are followed by an interval before the final imperative memory prompt. Take away that interval and effective retro-cues become ineffective postcues. Studies directly comparing the consequences of retro-cues and postcues illustrate this difference well (Makovski, Sussman, & Jiang, 2008; Murray, Nobre, Clark, Cravo, & Stokes, 2013; Sligte, Scholte, & Lamme, 2008).

Early retro-cue studies manipulated spatial attention in visual WM, but subsequent research has shown that retro-cueing is also effective in different WM modalities and when using different types of attentional cues (Stokes & Nobre, 2012; Souza & Oberauer, 2016). For example, retro-cueing has been reported for spatial information in audition (Backer & Alain, 2012), for visual object categories (Lepsien & Nobre, 2006), and for visual feature dimensions (Niklaus, Nobre, & van Ede, 2017). Similar facilitation of WM performance has been noted in tasks using *refresh* cues, which prompt participants simply to "think back" to a previously viewed item (Johnson, Mitchell, Raye, D'Esposito, & Johnson, 2007), or by *incidental* cues, which prompt participants to perform an unrelated task on one of the memoranda (Zokaei, Manohar, Husain, & Feredoes, 2014).

Much of the current research concerns pinpointing how retro-cues act on stored memories. However, looking for one general mechanism may be naïve. Analogously to the plurality of sites and modes of modulation revealed for attention operating in the perceptual domain (Nobre & Kastner, 2014), orienting attention within WM may also involve multiple mechanisms. Some putative effects of attending to an item in WM include activating latent traces (Sprague, Ester, & Serences, 2016; Wolff et al., 2017), highlighting active traces of constitutive features (Griffin & Nobre, 2003), protecting from decay or interference (Matsukara, Luck, & Vecera, 2007), reducing interference from competing items (Kuo, Stokes, & Nobre, 2012), prioritizing retrieval (Nobre, Griffin, & Rao, 2008), and activating associated response codes (Chatham, Frank, & Badre, 2014). The exact type of modulation will inevitably depend on stimulus parameters and task goals.

We have proposed that retro-cues do more than create a sustained focus of internal attention (Wallis, Stokes, Cousijn, Woolrich, & Nobre, 2015; Myers et al., 2017). The main reason is that the WM versus perceptual domains have different affordances. In perception, focusing neural receptors and processing on a subset of locations or features necessarily compromises how other competing items are processed and encoded (Carrasco, 2014). However, this need not be the case in WM. In principle, at least, prioritizing and selecting a given memorandum does not have to compromise other traces that have been encoded. Furthermore, compared to orienting attention in perception, prioritization and selection within WM can benefit more readily from task and action goals and thus directly support output gating (see Chatham, Frank, & Badre, 2014).

Flexible updating of attention in working memory Behavioral studies illustrate the flexible nature of retro-cueing. Evidence that retro-cues do more than just foster the continued maintenance of a cued item comes from studies showing superior performance to a retro-cued item than to an uncued item retrieved much earlier, at the time of retro-cue presentations (Makovski, Sussman, & Jiang, 2008; Murray et al., 2013; Sligte, Scholte, & Lamme, 2008). Accounts based on retro-cues only acting in a way to protect items from decay or interference therefore fall short of explaining the results. Retro-cues confer active performance benefits. The ability of retro-cues to confer performance advantages without compromising other competing traces is highlighted by a set of experiments using multiple probes after a retro-cue (Myers et al., 2017). In these experiments, spatial retro-cues indicate one of four orientation stimuli that will be probed at the end of the trial. In addition to probing the cued location, a second probe assesses performance for one of the remaining uncued items. Spatial retro-cues in these experiments conferred reliable performance benefits compared to uninformative neutral retro-cues. However, critically, orienting in WM did not significantly impair performance to uncued items. After a retro-cued item has been probed, performance to spatially uncued items at the subsequent probe showed no decrement compared to performance to neutrally cued items. Furthermore, there was no indication of any trade-off between benefits in performance at the cued location versus costs at the probed uncued location across trials. Thus, the findings challenge accounts of WM as a unitary finite resource, which propose that gains conferred to a given item should come with correlated costs to other items. It is important to note, however, that invalidity costs have occasionally been reported in a number of retro-cueing studies comparing performance on uncued versus neutral items (see Myers et al., 2017; Rerko, Souza, & Oberauer, 2014). Whether invalidity costs arise is likely to depend on specific task factors. Completely dropping uncued items can be advantageous in some task contexts—for example, with fully or highly predictive cues and only one probe—and is therefore more likely to occur (e.g., Berryhill, Richmond, Shay, & Olson, 2012; Gunseli, van Moorselaar, Meeter, & Olivers, 2015).

A recent WM study of ours, in which attention-orienting cues were internalized (van Ede, Niklaus, & Nobre, 2017), demonstrated the flexible and temporally dynamic updating of item prioritization in WM. Participants viewed two peripheral colored, oriented bars and, at the end of the trial, were prompted to reproduce the orientation of one. No cues were presented, but participants learned that one of the colored items was more likely to be probed at an early interval, while the other was more likely to be probed later (i.e., after the early interval lapsed). These purely endogenous, internalized "retro-cues" were highly effective at modulating behavioral performance. During the early interval, participants were more accurate and faster to report the orientation of the predicted item. Critically, similar performance and benefit levels occurred for stimuli probed at the late interval. Thus, items that had been relatively deprioritized and had yielded poorer performance when probed earlier during the WM delay became reprioritized over the passage of time to yield optimal performance. EEG recordings during task performance showed that neural markers of attention-related selection in WM covaried with the flexible orienting and reorienting of spatial attention in the task.

Output gating Evidence that retro-cues lead to output gating is mounting. Brain-imaging studies show that retro-cues engage the dorsal frontoparietal network involved in orienting attention in the perceptual domain as well as a cingulo-opercular network additionally implicated in top-down, action-related control (e.g., Nobre et al., 2004; Nee & Jonides, 2009; Nelissen, Stokes, Nobre, & Rushworth, 2013). We replicated this pattern of findings in a MEG study comparing spatial retro-cues and precues (Wallis et al., 2015). Additionally, the temporal resolution of MEG showed earlier engagement of the frontoparietal network followed by subsequent engagement of the cingulo-opercular network. Our MEG study also showed that, contrary to spatial precues, spatial retro-cues modulate visual excitability in a dynamic and short-lived way (see figure 25.1*C*). Replicating numerous findings in visual spatial attention (e.g., Worden, Foxe, Wang, & Simpson, 2000; Rihs, Michel, & Thut, 2007; Foster, Sutterer, Serences, Vogel, & Awh, 2017), spatial precues in the MEG study led to sustained changes in the level of alpha-band lateralization in anticipation of the item array. However, when spatial retro-cues were presented during the WM delay, alpha lateralization was brief and followed a temporally dynamic pattern (Wallis et al., 2015; see also Poch, Carretie, & Campo, 2017). We speculated that, rather than eliciting a state of sustained spatial focus, retro-cues operate by reactivating relevant sensory information, as evidenced by the transient pattern of alpha lateralization, therefore placing it in a prioritized state to guide action (Olivers, Peters, Houtkamp, & Roelfsema, 2011), akin to the process of output gating (Chatham, Frank, & Badre, 2014). The frontoparietal and cingulo-opercular networks may mediate these different stages of input and output gating, though more research will be needed to verify the relative contribution of these control processes. A follow-up MEG study measuring neural modulation by spatial retro-cues in older participants replicated the transient modulation of alpha-band lateralization and further showed that greater benefits conferred by spatial retro-cueing were correlated with more transient modulations of alpha lateralization (Mok, Myers, Wallis, & Nobre, 2016).

Closing the Loop

This chapter has departed from the traditional treatment of working memories concerning the past and attention concerning the future to highlight how working memories also concern the future and how attention can operate on traces from the past. Closing the loop, we can see how the past is constantly informing our interface with the incoming future and how the selective products of perception come to occupy our memory banks. Memories from multiple timescales, shaped by attention, carry the most important information into the future to guide adaptive behavior. The results of these biases then continue to shape the mnemonic landscape, which in turn influences attention, which again biases memories, and so on. This positive-feedback loop between attention and memory feeds a virtuous cycle that tunes our minds to the most relevant features of the environment.

Acknowledgments

This work was funded by a Wellcome Trust Senior Investigator Award (104571/Z/14/Z) and a James S. McDonnell Foundation Understanding Human Cognition Collaborative Award (#220020448) to A. C. Nobre and by a James S. McDonnell Foundation Scholar Award (#220020405) to M. S. Stokes and was supported by the National Institute for Health Research Oxford Health Biomedical Research Centre. The Wellcome Centre for Integrative Neuroimaging is supported by core funding from the Wellcome Trust (203139/Z/16/Z).

REFERENCES

Aly, M., & Turk-Browne, N. B. (2017). How hippocampal memory shapes, and is shaped by, attention. In *The hippocampus from cells to systems* (pp. 369–403). New York: Springer.

Averbach, E., & Coriell, A. S. (1961). Short-term memory in vision. *Bell Labs Technical Journal, 40*(1), 309–328.

Awh, E., Belopolsky, A. V., & Theeuwes, J. (2012). Top-down versus bottom-up attentional control: A failed theoretical dichotomy. *Trends in Cognitive Sciences, 16*(8), 437–443.

Awh, E., Jonides, J., Smith, E. E., Buxton, R. B., Frank, L. R., Love, T., ... & Gmeindl, L. (1999). Rehearsal in spatial working memory: Evidence from neuroimaging. *Psychological Science, 10*(5), 433–437.

Backer, K. C., & Alain, C. (2012). Orienting attention to sound object representations attenuates change deafness. *Journal of Experimental Psychology: Human Perception and Performance, 38*(6), 1554.

Baddeley, A. (2003). Working memory: Looking back and looking forward. *Nature Reviews Neuroscience, 4*(10), 829.

Barak, O., & Tsodyks, M. (2014). Working models of working memory. *Current Opinion Neurobiology, 25*, 20–24.

Berryhill, M. E., Richmond, L. L., Shay, C. S., & Olson, I. R. (2012). Shifting attention among working memory representations: Testing cue type, awareness, and strategic control. *Quarterly Journal of Experimental Psychology, 65*(3), 426–438.

Buonomano, D. V., & Maass, W. (2009). State-dependent computations: Spatiotemporal processing in cortical networks. *Nature Reviews Neuroscience, 10*(2), 113–125.

Carrasco, M. (2014). Spatial covert attention: Perceptual modulation. In *The Oxford handbook of attention* (pp. 183–230). New York: Oxford University Press.

Chafee, M. V., & Goldman-Rakic, P. S. (1998). Matching patterns of activity in primate prefrontal area 8a and parietal area 7ip neurons during a spatial working memory task. *Journal of Neurophysiology, 79*(6), 2919–2940.

Chatham, C. H., Frank, M. J., & Badre, D. (2014). Corticostriatal output gating during selection from working memory. *Neuron, 81*(4), 930–942.

Chelazzi, L., Duncan, J., Miller, E. K., & Desimone, R. (1998). Responses of neurons in inferior temporal cortex during memory-guided visual search. *Journal of Neurophysiology, 80*(6), 2918–2940.

Christophel, T. B., Klink, P. C., Spitzer, B., Roelfsema, P . R., & Haynes, J . D. (2017). The distributed nature of working memory. *Trends in Cognitive Sciences, 21*(2), 111–124.

Courtney,S .M., Petit, L., Maisog, J. M., Ungerleider, L. G., & Haxby, J . V. (1998). An area specialized for spatial working memory in human frontal cortex. *Science, 279*(5355), 1347–1351.

Crowe, D. A., Averbeck, B. B., & Chafee, M. V. (2010). Rapid sequences of population activity patterns dynamically encode task-critical spatial information in parietal cortex. *Journal of Neuroscience, 30*(35), 11640–11653.

Curtis, C. E., & D'Esposito, M. (2003). Persistent activity in the prefrontal cortex during working memory. *Trends in Cognitive Sciences, 7*(9), 415–423.

Desimone, R., & Duncan, J. (1995). Neural mechanisms of selective visual attention. *Annual Review of Neuroscience, 18*(1), 193–222.

Duncan, J. (2001). An adaptive coding model of neural function in prefrontal cortex. *Nature Reviews Neuroscience, 2*(11), 820–829.

Foster, J. J., Sutterer, D. W., Serences, J. T., Vogel, E. K., & Awh, E. (2017). Alpha-band oscillations enable spatially and temporally resolved tracking of covert spatial attention. *Psychological Science, 28*(7), 929–941.

Freedman, D. J., & Assad, J. A. (2006). Experience-dependent representation of visual categories in parietal cortex. *Nature, 443*(7107), 85.

Freedman, D. J., Riesenhuber, M., Poggio, T., & Miller, E. K. (2001). Categorical representation of visual stimuli in the primate prefrontal cortex. *Science, 291*(5502), 312–316.

Fujisawa, S., Amarasingham, A., Harrison, M. T., & Buzsaki, G. (2008). Behavior-dependent short-term assembly dynamics in the medial prefrontal cortex. *Nature Neuroscience, 11*(7), 823–833.

Funahashi, S., Bruce, C. J., & Goldman-Rakic, P. S. (1989). Mnemonic coding of visual space in the monkey's dorsolateral prefrontal cortex. *Journal of Neurophysiology, 61*(2), 331–349.

Fuster, J. M. (2001). The prefrontal cortex—an update: Time is of the essence. *Neuron, 30*(2), 319–333.

Fuster, J. M., & Alexander, G. E. (1971). Neuron activity related to short-term memory. *Science, 173*(3997), 652–654.

Fuster, J. M., & Jervey, J. P. (1981). Inferotemporal neurons distinguish and retain behaviorally relevant features of visual stimuli. *Science, 212*(4497), 952–955.

Goldman-Rakic, P. S. (1987). Circuitry of primate prefrontal cortex and regulation of behavior by representational memory. In *Handbook of Physiology, The Nervous System, Higher Functions of the Brain* (pp. 373–417). Bethesda: American Physiological Society.

Griffin, I. C., & Nobre, A. C. (2003). Orienting attention to locations in internal representations. *Journal of Cognitive Neuroscience, 15*(8), 1176–1194.

Gunseli, E., van Moorselaar, D., Meeter, M., & Olivers, C. N. (2015). The reliability of retro-cues determines the fate of noncued visual working memory representations. *Psychonomic Bulletin & Review, 22*(5), 1334–1341.

Hayden, B. Y., & Gallant, J. L. (2013). Working memory and decision processes in visual area V4. *Frontiers in Neuroscience, 7*, 18.

Hempel, C. M., Hartman, K, H., Wang, X. J., Turrigiano, G, G., & Nelson, S. B. (2000). Multiple forms of short-term plasticity at excitatory synapses in rat medial prefrontal cortex. *Journal of Neurophysiology, 83*, 3031–3041.

James, W. (1890). *The principles of psychology.* New York: Henry Holt.

Johnson, M. R., Mitchell, K. J., Raye, C. L., D'Esposito, M., & Johnson, M. K. (2007). A brief thought can modulate activity in extrastriate visual areas: Top-down effects of refreshing just-seen visual stimuli. *NeuroImage, 37*(1), 290–299.

Kastner, S., Pinsk, M. A., De Weerd, P., Desimone, R., & Ungerleider, L. G. (1999). Increased activity in human visual cortex during directed attention in the absence of visual stimulation. *Neuron, 22*(4), 751–761.

King, J. R., & Dehaene, S. (2014). Characterizing the dynamics of mental representations: The temporal generalization method. *Trends in Cognitive Sciences, 18*(4), 203–210.

Kubota, K., & Niki, H. (1971). Prefrontal cortical unit activity and delayed alternation performance in monkeys. *Journal of Neurophysiology, 34*(3), 337–347.

Kuo, B. C., Stokes, M. G., & Nobre, A. C. (2012). Attention modulates maintenance of representations in visual short-term memory. *Journal of Cognitive Neuroscience, 24*(1), 51–60.

Landman, R., Spekreijse, H., & Lamme, V. A. (2003). Large capacity storage of integrated objects before change blindness. *Vision Research, 43*(2), 149–164.

Lepsien, J., & Nobre, A. C. (2006). Attentional modulation of object representations in working memory. *Cerebral Cortex, 17*(9), 2072–2083.

Luck, S. J., Chelazzi, L., Hillyard, S. A., & Desimone, R. (1997). Neural mechanisms of spatial selective attention in areas V1, V2, and V4 of macaque visual cortex. *Journal of Neurophysiology, 77*(1), 24–42.

Machens,C. K., Romo, R., & Brody, C. D. (2005). Flexible control of mutual inhibition: A neural model of two-interval discrimination. *Science,* 307(5712), 1121–1124.

Makovski, T., Sussman, R., & Jiang, Y. V. (2008). Orienting attention in visual working memory reduces interference from memory probes. *Journal of Experimental Psychology: Learning, Memory, and Cognition, 34*(2), 369.

Mante, V., Sussillo, D., Shenoy, K. V., & Newsome, W. T. (2013). Context-dependent computation by recurrent dynamics in prefrontal cortex. *Nature, 503*(7474), 78–84.

Matsukura, M., Luck, S. J., & Vecera, S. P. (2007). Attention effects during visual short-term memory maintenance: Protection or prioritization? *Perception & Psychophysics, 69*(8), 1422–1434.

Meyers, E. M., Freedman, D. J., Kreiman, G., Miller, E. K., & Poggio, T. (2008). Dynamic population coding of category information in inferior temporal and prefrontal cortex. *Journal of Neurophysiology, 100*(3), 1407–1419.

Miller, E. K., Li, L., & Desimone, R. (1993). Activity of neurons in anterior inferior temporal cortex during a short-term memory task. *Journal of Neuroscience, 13*(4), 1460–1478.

Miller, E. K., Erickson, C. A., & Desimone, R. (1996). Neural mechanisms of visual working memory in prefrontal cortex of the macaque. *Journal of Neuroscience, 16*(16), 5154–5167.

Mok, R. M., Myers, N. E., Wallis, G., & Nobre, A. C. (2016). Behavioral and neural markers of flexible attention over working memory in aging. *Cerebral Cortex, 26*(4), 1831–1842.

Mongillo G., Barak, O., & Tsodyks, M. (2008). Synaptic theory of working memory. *Science, 319*, 1543–1546.

Murray, A. M., Nobre, A. C., Clark, I. A., Cravo, A. M., & Stokes, M. G. (2013). Attention restores discrete items to visual short-term memory. *Psychological Science, 24*(4), 550–556.

Myers, N. E., Chekroud, S. R., Stokes, M. G., & Nobre, A. C. (2017). Benefits of flexible prioritization in working memory can arise without costs. *Journal of Experimental Psychology: Human Perception and Performance, 44*(3), 398–411.

Myers, N. E., Rohenkohl, G., Wyart, V., Woolrich, M. W., Nobre, A. C., & Stokes, M. G. (2015). Testing sensory evidence against mnemonic templates. *eLife, 4*.

Myers, N. E., Stokes, M. G., & Nobre, A. C. (2017). Prioritizing information during working memory: Beyond sustained internal attention. *Trends in Cognitive Sciences, 21*(6), 449–461.

Nee, D. E., & Jonides, J. (2009). Common and distinct neural correlates of perceptual and memorial selection. *NeuroImage, 45*(3), 963–975.

Neisser, U. (1967). *Cognitive psychology*. New York: Appleton-Century-Crofts.

Nelissen, N., Stokes, M., Nobre, A. C., & Rushworth, M. F. (2013). Frontal and parietal cortical interactions with distributed visual representations during selective attention and action selection. *Journal of Neuroscience, 33*(42), 16443–16458.

Nelson, K. (2003) Self and social functions: Individual autobiographical memory and collective narrative. *Memory, 11*(2), 12536.

Niklaus, M., Nobre, A. C., & Van Ede, F. (2017). Feature-based attentional weighting and spreading in visual working memory. *Scientific Reports, 7*, 42384.

Nobre, A. C., Coull, J. T., Maquet, P., Frith, C. D., Vandenberghe, R., & Mesulam, M. M. (2004). Orienting attention to locations in perceptual versus mental representations. *Journal of Cognitive Neuroscience, 16*(3), 363–373.

Nobre, A. C., Griffin, I. C., & Rao, A. (2008). Spatial attention can bias search in visual short-term memory. *Frontiers in Human Neuroscience, 2*, 4.

Nobre, A. C., & Kastner, S. (2014). Attention: Time capsule 2013. In *Oxford handbook of attention* (pp. 1201–1222). New York: Oxford University Press.

Nobre, A. C., & Mesulam, M. M. (2014). Large-scale networks for attentional biases. In *Oxford handbook of attention* (pp. 105–151). New York: Oxford University Press.

Nobre, A. C., & van Ede, F. (2018). Anticipated moments: Temporal structure in attention. *Nature Reviews Neuroscience, 19*(1), 34.

Olivers, C. N., Peters, J., Houtkamp, R., & Roelfsema, P. R. (2011). Different states in visual working memory: When it guides attention and when it does not. *Trends in Cognitive Sciences, 15*(7), 327–334.

Poch, C., Carretie, L., & Campo, P. (2017). A dual mechanism underlying alpha lateralization in attentional orienting to mental representation. *Biology Psychology*, 128, 63–70.

Rainer, G., Rao, S. C., & Miller, E. K. (1999). Prospective coding for objects in primate prefrontal cortex. *Journal of Neuroscience, 19*(13), 5493–5505.

Rerko, L., Souza, A. S., & Oberauer, K. (2014). Retro-cue benefits in working memory without sustained focal attention. *Memory & Cognition, 42*(5), 712–728.

Rihs, T. A., Michel, C. M., & Thut, G. (2007). Mechanisms of selective inhibition in visual spatial attention are indexed by α-band EEG synchronization. *European Journal of Neuroscience, 25*(2), 603–610.

Rose, N. S., LaRocque, J. J., Riggall, A. C., Gosseries, O., Starrett, M. J., Meyering, E. E., & Postle, B. R. (2016). Reactivation of latent working memories with transcranial magnetic stimulation. *Science, 354*, 1136–1139.

Serences, J. T., Ester, E. F., Vogel, E. K., & Awh, E. (2009). Stimulus-specific delay activity in human primary visual cortex. *Psychological Science, 20*(2), 207–214.

Serences, J. T., Saproo, S., Scolari, M., Ho, T., & Muftuler, L. T. (2009). Estimating the influence of attention on population codes in human visual cortex using voxel-based tuning functions. *NeuroImage, 44*(1), 223–231.

Shimi, A., Nobre, A. C., Astle, D., & Scerif, G. (2014). Orienting attention within visual short-term memory: Development and mechanisms. *Child Development, 85*(2), 578–592.

Sligte, I. G., Scholte, H. S., & Lamme, V. A. (2008). Are there multiple visual short-term memory stores? *PLoS One, 3*(2), e1699.

Souza, A. S., & Oberauer, K. (2016). In search of the focus of attention in working memory: 13 years of the retro-cue effect. *Attention, Perception, & Psychophysics, 78*(7), 1839–1860.

Spaak, E., Watanabe, K., Funahashi, S., & Stokes, M. G. (2017). Stable and dynamic coding for working memory in primate prefrontal cortex. *Journal of Neuroscience, 37*(27), 6503–6516.

Sperling, G. (1960). The information available in brief visual presentations. *Psychological Monographs: General and Applied, 74*(11), 1.

Sprague, T. C., Ester, E. F., & Serences, J. T. (2016). Restoring latent visual working memory representations in human cortex. *Neuron, 91*(3), 694–707.

Sreenivasan, K. K., Curtis, C. E., & D'Esposito, M. (2014). Revisiting the role of persistent neural activity during working memory. *Trends in Cognitive Sciences, 18*(2), 82–89.

Stokes, M. G. (2015). "Activity-silent" working memory in prefrontal cortex: A dynamic coding framework. *Trends in Cognitive Sciences, 19*(7), 394–405.

Stokes, M. G., Kusunoki, M., Sigala, N., Nili, H., Gaffan, D., & Duncan, J. (2013). Dynamic coding for cognitive control in prefrontal cortex. *Neuron, 78*(2), 364–375.

Stokes, M. G., & Nobre, A. C. (2012). Top-down biases in visual short-term memory. In G. R. Mangun (Ed.), *The neuroscience of attention: Attentional control and selection* (pp. 209–228). Oxford: Oxford University Press.

Stokes, M., Thompson, R., Nobre, A. C., & Duncan, J. (2009). Shape-specific preparatory activity mediates attention to targets in human visual cortex. *Proceedings of the National Academy of Sciences, 106*(46), 19569–19574.

Sugase-Miyamoto, Y., Liu, Z., Wiener, M . C., Optican, L. M., & Richmond, B. J. (2008). Short-term memory trace in rapidly adapting synapses of inferior temporal cortex. *PLoS Computation Biology, 4*(5).

Todd, J. J., & Marois, R. (2004). Capacity limit of visual short-term memory in human posterior parietal cortex. *Nature, 428*(6984), 751.

van Ede, F., Niklaus, M., & Nobre, A. C. (2017). Temporal expectations guide dynamic prioritization in visual working memory through attenuated α oscillations. *Journal of Neuroscience, 37*(2), 437–445.

van Moorselaar, D., Olivers, C. N., Theeuwes, J., Lamme, V. A., & Sligte, I. G. (2015). Forgotten but not gone: Retro-cue costs and benefits in a double-cueing paradigm suggest multiple states in visual short-term memory. *Journal of Experimental Psychology: Learning, Memory, and Cognition, 41*(6), 1755.

Wallis, G., Stokes, M., Cousijn, H., Woolrich, M., & Nobre, A. C. (2015). Frontoparietal and cingulo-opercular networks play dissociable roles in control of working memory. *Journal of Cognitive Neuroscience, 27*(10), 2019–2034.

Warden, M. R., & Miller, E. K. (2010). Task-dependent changes in short-term memory in the prefrontal cortex. *Journal of Neuroscience, 30*(47), 15801–15810.

Watanabe, K., & Funahashi, S. (2007). Prefrontal delay-period activity reflects the decision process of a saccade direction during a free-choice ODR task. *Cerebral Cortex, 17*(Suppl 1), i88–i100.

Watanabe, K., & Funahashi, S. (2014). Neural mechanisms of dual-task interference and cognitive capacity limitation in the prefrontal cortex. *Nature Neuroscience, 17*(4), 601.

Wolff, M. J., Ding, J., Myers, N. E., & Stokes, M. G. (2015). Revealing hidden states in visual working memory using electroencephalography. *Frontiers in Systems Neuroscience, 9*, 123.

Wolff, M. J., Jochim, J., Akyürek, E. G., & Stokes, M. G. (2017). Dynamic hidden states underlying working-memory-guided behavior. *Nature Neuroscience, 20*(6), 864.

Worden, M. S., Foxe, J. J., Wang, N., & Simpson, G. V. (2000). Anticipatory biasing of visuospatial attention indexed by retinotopically specific-band electroencephalography increases over occipital cortex. *Journal of Neuroscience, 20*(RC63), 1–6.

Zokaei, N., Manohar, S., Husain, M., & Feredoes, E. (2014). Causal evidence for a privileged working memory state in early visual cortex. *Journal of Neuroscience, 34*(1), 158–162.

Zucker, R. S., & Regehr, W. G. (2002). Short-term synaptic plasticity. *Annual Review of Physiology,* 64, 355–405.

Zylberberg, J., & Strowbridge, B. W. (2017). Mechanisms of persistent activity in cortical circuits: Possible neural substrates for working memory. *Annual Review of Neuroscience, 40*, 603–627.

26 The Developmental Dynamics of Attention and Memory

GAIA SCERIF

ABSTRACT Attentional control plays a crucial role in biasing incoming information in favor of what is relevant to further information processing, action selection, and long-term goals. A cognitive neuroscience approach illustrates how attentional processes are best understood not simply as a control homunculus; rather, they are bidirectionally influencing and influenced by prior experience. It therefore becomes very useful to place attention and memory dynamics into a developmental context. From very early in infancy, we are equipped with exquisite attentional skills whose improvement is coupled with the increased effectiveness of control networks. Later in childhood, both behavioral and neural indices suggest similarities and differences in how children and young adults deploy attentional control to optimize maintenance in short-term memory. Influences of attention on encoding into memory are also apparent through the effects that highly salient social, attentional biases have on learning and later recall from longer-term memory. At the same time, attentional effects on memory are not unidirectional: previously learned information and resistance to distraction during learning guide later attentional deployment, both in adulthood and in childhood. In conclusion, assessing attentional development and its dynamics points to the bidirectional influences between attention and memory.

Placing Interactions between Attention and Memory into a Developmental Time Frame

Multiple attentional control mechanisms influence processing by the adult attentive brain, within the remit of perception and short-term memory all the way to encoding into and recall from long-term memory. Starting from influences on perception, classic neurocognitive models of adult attention detail the mechanisms by which top-down biases from ongoing task goals play a key role in resolving the competition arising in complex visual input (Desimone & Duncan, 1995; Kastner & Ungerleider, 2000). Other classic neurocognitive models also emphasize both interactions and distinctions between goal-driven and input-driven influences on attentional selection in the adult brain (Corbetta & Shulman, 2002), as well as how overlapping but separable attention mechanisms govern behavior in space through spatial orienting, in time through alerting processes and over goals through executive attention (Petersen & Posner, 2012; Posner & Petersen, 1990). Despite differences in the level at which each of these proposals operate and their many exciting new mechanistic foci (Buschman & Kastner, 2015; Halassa & Kastner, 2017), core to these neurocognitive models is the concept of attention as a set of biases resolving competition in a complex visual environment and therefore constraining further processing into memory. Increasingly, views of how the adult attentive brain operates have been modified to incorporate influences on attention by the contents of working goals or long-term memories (Chun, Golomb, & Turk-Browne, 2011; Gazzaley & Nobre, 2012). It is, in particular, the interface between attention and these internally held representations that will be the focus of the current chapter. In the first section, I detail the role of attention in shaping short- and long-term memory from infancy into childhood, with a focus on both changing and stable mechanisms, whereas the second section highlights growing evidence of how the contents of short-term and longer-term representations influence attention deployment across development.

Attentional Influences on Short-Term and Long-Term Memory over Development

Before delving into attentional influences on memory, it is worth describing, briefly, the amazing changes that characterize attention mechanisms from infancy into adulthood. From the first months of life, changes in attention are indexed by the way in which infants increasingly control their eye movements. While referring the interested reader to fuller reviews on the neural basis of attention development in infancy (e.g., Richards, Reynolds, & Courage, 2010) or early childhood (Amso & Scerif, 2015), it is key to note here, first, that eye movements are very powerful mechanisms through which all observers, from infancy, select relevant information in their environment. Second, even though attention orienting can dissociate from eye movements (*covert attention*), even in adults there is a high degree of overlap in neural correlates supporting overt and covert orienting (e.g., Nobre, Gitelman, Dias, & Mesulam, 2000). However, and finally, it is very difficult to study covert attention in infants, as this normally requires observers to follow explicit instructions (e.g., "orient your attention to the periphery while fixating in the center"; see Johnson, Posner, and Rothbart [1994] for an infant covert-orienting paradigm), and therefore most infant studies focus on rapid changes in eye-movement control over the first year of life.

Indeed, many aspects of oculomotor control show dramatic improvements between birth and 4 months (Johnson, 1994). The engagement and efficiency of these circuits improves staggeringly and steadily from infancy into adulthood. For example, the ability to inhibit overt orienting toward salient peripheral stimuli emerges from 3 or 4 months of age (Johnson, 1995), but it continues to develop over early childhood and well into adulthood, as indexed by the increasing accuracy in producing antisaccades (Luna, Velanova, & Geier, 2008). Alongside the control of overt eye movements, infants between 4 and 6 months of age become increasingly able to orient covert attention to stimuli in the environment, as indexed by the benefits that peripheral visual cues accrue to their orienting (Hood, 1993; Johnson, Posner, & Rothbart, 1994).

In neural terms these gradual changes in the control of the overt and covert orienting of attention have long been accounted for by increasing frontoparietal control on subcortical mechanisms (e.g., Atkinson et al., 1992; Johnson, 1990), a suggestion bolstered by more recent infant work (Richards, 2010). Early electrophysiological evidence pertaining to eye movements indicated that the infant brain before 1 year of age deploys frontoparietal mechanisms when preparing eye movements (e.g., Csibra, Tucker, & Johnson, 1998). Developments in methods such as near infrared spectroscopy have more recently also pinpointed a role for classic control nodes in frontal and parietal cortex from early during the first year of life, when young infants direct attention to higher-level representations that might guide their actions (Werchan, Collins, Frank, & Amso, 2016). Later in childhood and into adolescence, attentional mechanisms continue to develop, with increasing control over the orienting of attention in space, over the temporal alerting of attention, and over competing responses (Amso & Scerif, 2015; Rueda et al., 2004; Rueda, Posner, & Rothbart, 2005). These changes are supported by the maturation of the cognitive control regions and, most importantly, by strengthened effective connectivity across the frontoparietal areas and their partners across the brain (Fair et al., 2007, 2009).

Of note, initial neurocognitive models of infant and childhood attention development treated attentional processes as relatively independent from other developing processes, as they were keenly focused on tracing the onset and maturation of attention in and of itself. In contrast, recent work has investigated how attention influences short-term and long-term memory in differentiable ways that distinguish infants, children, and adults, to which we now turn.

Influences of attention development on short-term memory Given the protracted changes in attentional circuitry described above, it is not surprising that the effects of attentional cues on memory also show protracted change over infancy and into childhood. Although traditions differ in whether they use the term *working memory* interchangeably with *short-term memory* or distinguish between the two (see Cowan, 2017 for a recent review), perhaps one of the most robust findings in developmental science is the fact that in both infants (Ross-Sheehy, Oakes, & Luck, 2003) and young children (visual but also auditory), short-term memory spans index lower capacity than those of older children and adults (Cowan et al., 2005; Gathercole, Pickering, Ambridge, & Wearing, 2004). For example, Ross-Sheehy, Oakes, and Luck (2003) used a simple change-detection paradigm to show that visual short-term memory (VSTM) capacity increases significantly from 4 to 13 months of age. Adapting this change-detection paradigm, Ross-Sheehy, Oakes, and Luck (2011) investigated the role of attentional cues on memory for 5- and 10-month-old infants, who experienced changes in arrays composed of three differently colored squares. In each trial one square changed color, and one square was cued. Sometimes the cued item was the changing item and sometimes it was not. Older infants detected changes for the cued item when the cue was spatial (a peripheral flash preceding the onset of the item at its location), but even younger infants could exhibit this enhanced memory, although the necessary cue here was motion. These data showed that, although limited, young infants' encoding into VSTM can benefit from attention cues.

Although primarily cognitive in nature, this literature inspired developmental cognitive neuroscientists to ask questions about the neural mechanisms by which attention may bolster children's ability to maintain information in short-term memory. Indeed, attention may influence how well children and adults remember in different ways: by dynamically preparing to encode information better or by refreshing it while it is held in memory. As the attentional networks that support adaptive cognitive control are slow to develop, their maturation may also constrain the efficiency with which memories are encoded and maintained. Let us take, for example, a very simple memory task, such as being presented with four items that then disappear and then asked if a memory probe item was part of the initial array. Using a version of this task with both 9- to 12-year-olds and adults, Astle et al. (2015) found that children in particular are highly variable in how they manage to recruit cognitive control in service of memory (see figure 26.1).

The authors recorded oscillatory brain activity using magnetoencephalography (MEG) while children and adults performed the VSTM task. By combining temporal independent component analysis (ICA) with general linear modeling, they tested whether frontoparietal activity correlated with VSTM performance on a trial-by-trial basis. In children, but not in adults, slow frequency theta (4–7 Hz) activity within a right lateralized frontoparietal network, specifically in anticipation of the memoranda, predicted the accuracy with which those memory items were subsequently retrieved, suggesting that the inconsistent use of anticipatory control mechanisms during encoding contributes to trial-to-trial variability in children's VSTM maintenance.

In addition to the general involvement of attentional control networks at encoding, spatially selective attention mechanisms seem to play an even more specific role in the maintenance of visual information. Cognitive neuroscientists have long demonstrated that spatially directed cues presented during the maintenance period facilitate adults' accurate recall from memory (Griffin & Nobre, 2003). As discussed extensively in other chapters for this section, benefits accrued from cues presented in anticipation of encoding information into memory (*precues*) and those presented in the maintenance period (*retro-cues*) have very interesting behavioral similarities in adults, although they are also characterized by a growing set of neural differences

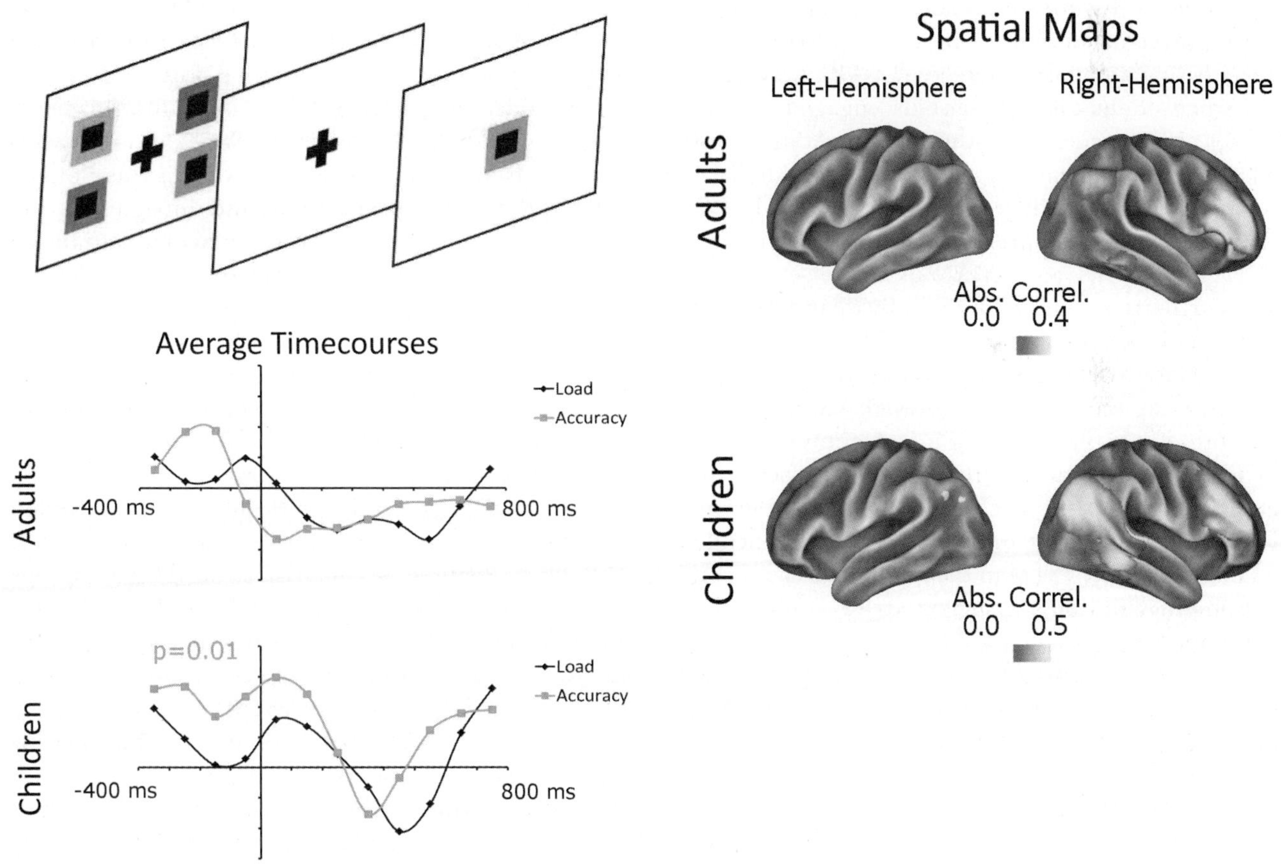

FIGURE 26.1 *A*, Graphical representation of the memory task employed here. *B*, Activity in frontoparietal network (slow frequency theta 4–7 Hz) oscillations predicted accuracy of memory at the end of the trial in children and similarly, but not significantly so, in adults. The map shows the spatial extent of the component networks (in terms of the absolute Pearson correlation values between each brain location and this component). *C*, The time course of the regressor (*black line*) shows that accuracy is predicted by oscillations for this network at the time of encoding of the memoranda. The time course also presents another regressor as a comparison (load—2 vs. 4 items, *cyan line*) to show that this network was not differentially recruited by just any demand, like task difficulty. Adapted with permission from Astle et al. (2014). (See color plate 28.)

(Myers, Walther, Wallis, Stokes, & Nobre, 2015). Exploiting the retro-cueing paradigm, Shimi, Nobre, Astle, and Scerif (2014) asked whether the interactions between spatial attentional cues and memory show age-related dissociations. They found that although children as young as 7 years of age are as capable as adults at drawing benefits from spatial attentional precues to better remember information encoded into short-term memory, their ability to use retro-cues is less well developed. Extending this work to younger children, Guillory, Gliga, and Kaldy (2018) found an increasing refinement in short-term memory capacity in 4- to 7-year-olds such that precues were more effective than retro-cues in benefiting their short-term memory capacity. Furthermore, electroencephalographic (EEG) data have already provided further insights into the mechanisms of potential differences in attentional recruitment by children and adults when they use retro-cues (Shimi, Kuo, Astle, Nobre, & Scerif, 2014). Known neural markers of spatial orienting—that is, early-directing attention negativity (EDAN), anterior-directing attention negativity (ADAN), and late-directing attention positivity (LDAP), were examined when adults and 10-year-olds engaged in using precues or retro-cues to aid their VSTM. Adults exhibited a set of neural markers that were broadly similar in preparation for encoding and maintenance. In contrast, in children these processes dissociated, with little evidence of EDAN and ADAN in response to retro-cues. Furthermore, in children, individual differences in the amplitude of neural markers of prospective orienting related to individual differences in VSTM capacity, suggesting that children with high VSTM capacity are more efficient at selecting information for encoding into VSTM. Drawing from these behavioral and neural findings, it seems clear that spatial attentional processes control what information will be encoded and maintained in VSTM in the face of increased competition. In children, as suggested for adults, these attentional refreshment mechanisms may operate by reactivating and strengthening the signal of visual representations associated with memoranda (Astle et al., 2015). As a whole, the emerging developmental literature on attentional cues and their benefits to VSTM suggests that developing spatial attentional control skills contribute to young children's ability to maintain items in VSTM. This is not to say that spatial attentional biases are the sole, or even independent, contributor to the development of VSTM capacity. Other key contributing factors (such as memory load itself, decay of information over time, and the nature of the memoranda) also deserve further investigation by developmental cognitive neuroscientists, as they have, in the main, been studied only through behavioral indices by developmental psychologists (see Shimi and Scerif [2017] for a review and integrative proposal).

Evidence that not all attentional mechanisms play equivalent roles in the interaction between attention and memory over development comes from other recent electroencephalographic evidence. A candidate mechanism contributing to individual differences in VSTM capacity in adults has been the ability to filter out distracting information while maintaining potential target items (Fukuda & Vogel, 2009). Is this a factor underpinning developmental capacity differences? Astle, Harvey, et al. (2014) presented participants with arrays of to-be-remembered items containing two targets, four targets, or two targets and two distracters. Participants consisted of high-VSTM capacity adults, low-VSTM capacity adults, and typically developing children. Children's performance on the VSTM task was poor and equivalent to that of the low-capacity adults. Using electroencephalography, as expected, a relative negativity in the maintenance delay (called *contralateral delay activity*, or *CDA*) was measured over the scalp contralateral to the original locations of the memoranda, and in the low-capacity adults, this negativity was modulated similarly by target and distracter items, indicative of poor selectivity. This was not the case for the high-capacity adults and, intriguingly, the children: the response to memory arrays containing two target items and two distracters was equivalent to the response elicited by arrays containing only two target items. Importantly, despite their obvious differences in capacity, children were not specifically impaired at filtering out distracters, a characteristic of low-capacity adults. Indeed, these findings are consistent with cognitive work by Cowan and colleagues, especially when the number of items to be encoded into memory is small (e.g., two items; Cowan, Morey, AuBuchon, Zwilling, & Gilchrist, 2010). These findings suggest that while the activity of attentional control networks may contribute to efficient recall, not all attentional mechanisms seem to contribute equally to developmental differences in VSTM.

Of note, the development of the mechanisms by which distracters are suppressed deserves further investigation with multiple imaging modalities in addition to EEG: using functional imaging, resistance to distraction during maintenance had previously differentiated adults and young adolescents' VSTM (Olesen, Macoveanu, Tegner, & Klingberg, 2007). This study measured brain activity with functional magnetic resonance imaging in adults and 13-year-olds using a paradigm in which participants were provided information to maintain in memory. During the delay period, they were also presented with irrelevant distracter stimuli. Adults were more accurate and less distractible than children. Distraction during the delay evoked activation in the parietal and occipital cortices in both adults and children, whereas it activated frontal cortex only in children, suggesting overlapping and yet distinct cortical recruitment while suppressing competing distracter information.

In summary, developing attentional mechanisms result in differential attention benefits at distinguishable points over the timeline, leading to successful recall from VSTM, and they involve the recruitment of frontoparietal networks whose coordination is critical to selective encoding and maintenance in VSTM. Resistance to distracters competing for attentional resources seems to recruit overlapping but also differing networks over development, with neural signatures that deserve further investigation, as they have been studied in the context of attentional influences on longer-term memory, to which we now turn.

Attention development and its influence on long-term memory A parallel body of work suggests that basic attentional mechanisms influence long-term memory from infancy onward. For example, Markant and Amso (2013) found that visual selection mechanisms limit distracter interference during item encoding for infants, a process they found to be key to successfully retaining information in long-term memory. In a modified spatial cueing task, 9-month-old infants encoded multiple objects following orienting cues that required them to inhibit distracter information, as opposed to a condition that did not. When their memory was tested, infants in the distracter-suppression condition retrieved item-specific information from memory (by discriminating items that were old from new). These data suggested that developing selective attention (and, more precisely, the suppression of distracting information) enhances the efficacy of memory encoding for subsequent retrieval. The effects of these attentional biases on the encoding of information in long-term memory span beyond infancy and into childhood and adolescence. Markant and Amso (2014) used a similar spatial-cueing paradigm geared to engage distracter suppression, while also incidentally presenting participants with unique line drawings of objects, across a large sample spanning 6 to 16 years of age. Across the full sample, distracter suppression resulted in long-term benefits for a surprise memory recognition test that followed the cueing phase of the study. Functional-imaging evidence in adults indeed also suggests that engaging distracter-suppression mechanisms may result in better long-term memory encoding. fMRI analyses revealed that this memory benefit was driven by the attention modulation of visual cortex activity, as increased suppression of the previously attended location in visual cortex during target object encoding predicted better subsequent recognition memory performance (Markant, Worden, & Amso, 2015).

The mechanisms underpinning the role of attentional cueing and distracter-processing effects on long-term memory relate to the growing literature on memory-guided attention (Stokes, Atherton, Patai, & Nobre, 2012; Summerfield, Lepsien, Gitelman, Mesulam, & Nobre, 2006). As reviewed in depth in this section (see chapter 25), memory-guided attention paradigms ask participants to search repeatedly for unique targets in scenes. Repeated searching engenders learning, after which long-term memory for target locations is assessed. In a final memory-guided attention-orienting phase, the speed of target detection is assessed for targets that are presented at locations consistent with their locations in memory, as opposed to locations inconsistent with memory. Attention allocation is faster at locations consistent with memory and recruits both frontoparietal and hippocampal circuits (Summerfield et al., 2006). Like the cueing paradigms by Amso and colleagues above, memory-guided attention paradigms therefore offer the opportunity to test both the effects of attentional allocation during learning and the role of distracters competing for attention while encoding information in long-term memory, in both adults and children. First, in adults, Doherty, Patai, Duta, Nobre, and Scerif (2017) asked participants to search for targets in scenes containing social or nonsocial distracters. The subsequent memory precision for target locations was tested. Eye tracking revealed significantly more attentional capture to social compared to nonsocial distracters matched for low-level visual salience. Critically, memory precision for target locations was poorer for social scenes, suggesting a role for differential attentional allocation to competing distracters on long-term memory. In a recent extension to younger children, Doherty, Fraser, et al. (under review) found that children directed first looks to the social distracter even more than adults and that memory precision was lower, for both children and adults, when a social distracter was present (see figure 26.2). The powerful effects of social distracters alert us to the fact that attentional biases influencing later memory do not operate equivalently across stimuli of all types but that preexisting preferences for certain stimuli also guide attention.

Attentional influences on long-term memory are robust from infancy and into childhood. Distracter effects, albeit far from fully understood, also suggest that the nature of the items to which attention is directed (e.g., preexisting strong social biases) have a

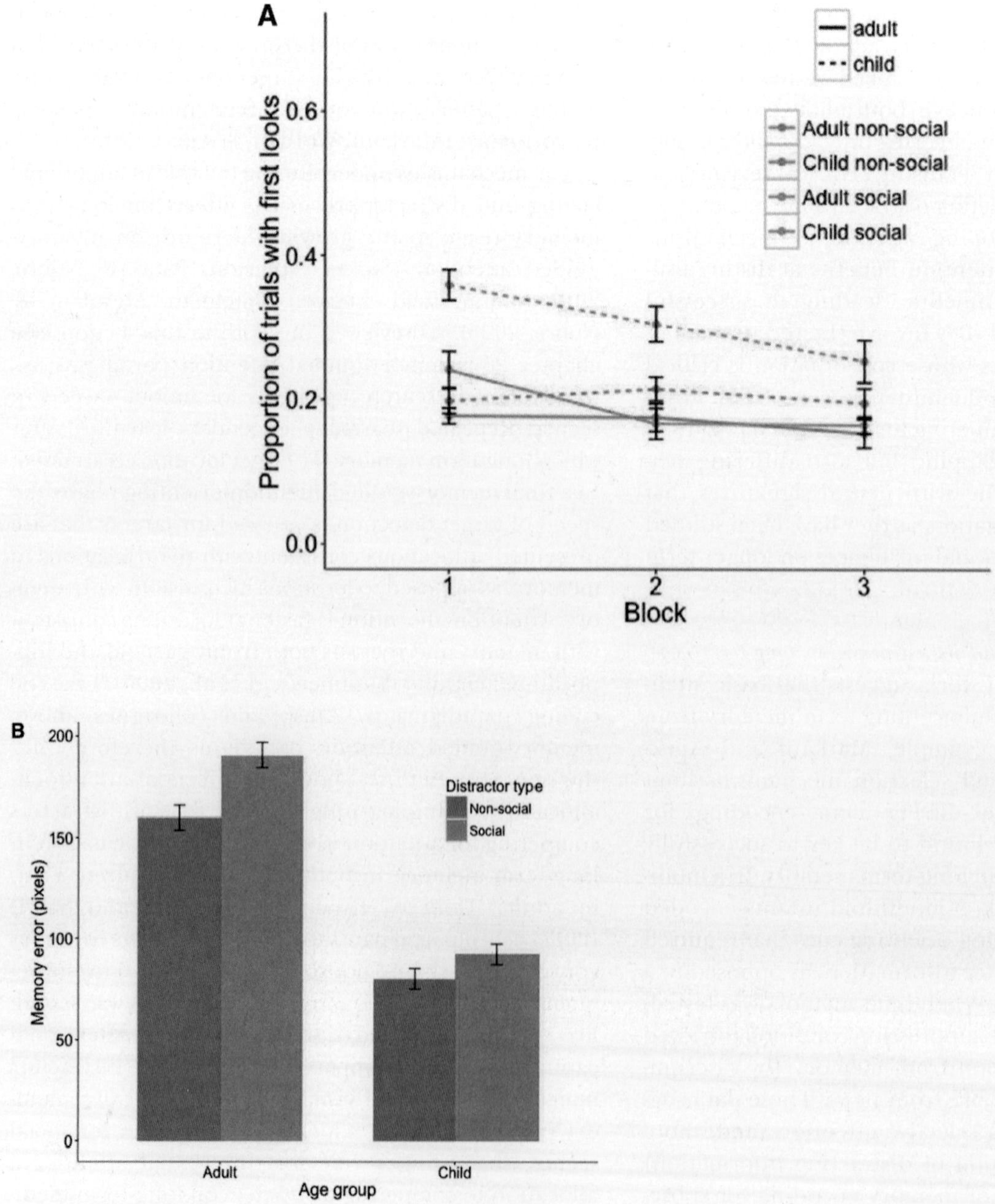

FIGURE 26.2 *A*, First looks were more likely to be directed to social compared to nonsocial distracters by both adults and children, and differentially more so for children. *B*, Subsequent memory precision was lower for social compared to nonsocial distracters for both children and adults. Note that, intriguingly, children's memory precision was higher than that of adults. A possible interpretation is that slower and less efficient attentional orienting may paradoxically result in a longer or qualitatively different exploration of complex natural scenes in children compared to adults and therefore, in the longer run, better encoding of the context and location at which targets were places. Error bars indicate standard errors. Adapted with permission from Doherty, Fraser, et al. (under review).

strong influence on attention. We therefore now turn to how developmental studies can begin to investigate the mechanisms by which these preexisting representations influence attention.

Influences of Short-Term and Long-Term Memory Representations on Attention Deployment

In this section I overview developmental data suggesting that the contents of memory have a powerful influence on attention. Starting from the realm of short-term memory representations, an open question is how attentional biases interact with the nature of the internal memory codes on which they operate. In infancy, recent work has shown the influences of VSTM on attention (Mitsven, Cantrell, Luck, & Oakes, 2018). Later in childhood, the influence of short-term memory representations on attentional deployment has also been studied. Shimi and Scerif (2015) asked 7-year-olds, 11-year-olds, and adults to complete the retro-cueing paradigm described above: spatial cues guided participants' attention to the likely location of a to-be-probed item during maintenance. The memoranda contained either highly familiar items or unfamiliar abstract shapes. Replicating earlier findings, all participants benefited from cues during maintenance, although benefits were smaller for 7-year-olds than for older participants. Critically, attentional benefits interacted with the nature of the memoranda: better VSTM maintenance was obtained for cued familiar items—and differentially more so for children compared to adults. These data suggest that attentional biases during maintenance operate more efficiently on memory representations that are more familiar and can therefore be retrieved more easily, pointing to the need to consider the influence of memory representations themselves on attention orienting.

Work investigating memory-guided attention orienting most directly tackles the influence of memory traces onto attention. These paradigms were developed for use with adults (Stokes et al., 2012; Summerfield et al., 2006), but they have been recently adapted for use in children. Nussenbaum, Scerif, and Nobre (forthcoming) pitted against each other the effects of salient visual cues and of memory-guided cues on attention orienting in children and in adults. Over three complementary experiments, children demonstrated faster reaction times to targets both when they were cued by sudden visual events and by memories (see figure 26.3). These findings suggest that memories may be a particularly robust source of influence on attention in children.

Returning to the critical role of the nature of memory traces themselves, Doherty, van Ede et al. (under review) asked whether the differential effects of social scenes on memory alter subsequent memory-guided attention orienting and the corresponding anticipatory dynamics of 8–12 Hz alpha-band oscillations as measured with EEG. After searching for targets in scenes that contained either social or nonsocial distracters, young adults' reaction time was measured as participants oriented to targets appearing in those scenes at either valid (previously learned) locations or invalid (different) locations. Poorer memory performance for scenes with social distracters was marked by reduced anticipatory dynamics of spatially lateralized 8–12 Hz alpha-band oscillations during the orienting phase. But do the effects of distracters influence memory-guided attention differently in children compared to adults? After the learning and memory phases, Doherty, Fraser et al. (under review) asked participants to perform a speeded target-detection task. Intriguingly, although both children and adults were less precise in remembering targets that had appeared in social versus nonsocial scenes, children demonstrated overall better memory precision than adults. Furthermore, when participants detected previously learned targets within visual scenes, adults were slower for targets appearing at unexpected (invalid) locations within social scenes compared to nonsocial scenes, whereas children did not show this cost, suggesting that social memory traces may play a different role for them than for adults.

In summary, therefore, the contents of short- and long-term memory guide attention across development. The differential effects of memoranda and distracters point to the possibility that one's prior learning history or strong attentional bias for certain stimuli could influence memory-guided attention orienting, a bidirectional chain that may further reinforce attentional biases.

Conclusion and Future Directions—Attention and Memory Interactions over Development

A growing body of evidence suggests that developmental changes in attentional control constrain co-occurring changes in short-term memory and long-term memory skills from infancy and into childhood. The efficiency of a frontoparietal network engaged in attentional control seems critical to these increasingly adult-like interactions. I have also described how early goal- and memory-related activity bias attention from very early on in infancy and therefore how the interactions between attention, memory, and learning are the target of much recent work in the developmental cognitive neuroscience of this area. As a whole, these findings suggest that the interplay between attentional

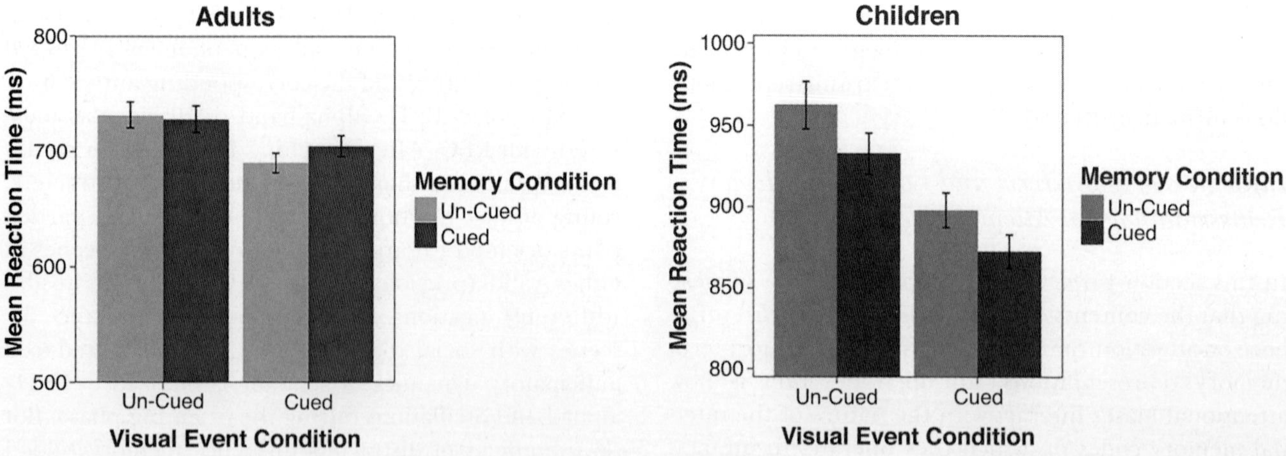

FIGURE 26.3. After learning about the specific locations of objects within scenes over repeated learning blocks, participants were presented with an orienting task in which they had to respond as quickly as possible to targets that appeared either at the location cued by their memory, at a location that was inconsistent with that memory, at a location cued by the sudden presentation of a visual event (a flash), or at a location inconsistent with the visual event. (*A*) Adults and (*B*) children both demonstrated faster reaction times when the visual event cued the target location. However, only children benefited significantly in response to memories, demonstrating faster reaction times when the memory cued the target location. Error bars indicate standard errors. Adapted with permission from Nussenbaum, Scerif, and Nobre (forthcoming).

biases, differential memory traces, and memory-guided attention is complex and modulated by age-related differences. Of note, interactions between attention and short-term and longer-term memory over developmental time have only recently been tackled with methods that are complementary to behavioral data: eye tracking and electro- and magnetoencephalography, as well as functional neuroimaging methods, are increasingly being used in this field and will yield many needed insights. Complementary methodologies in developmental cognitive neuroscience will be needed to shed further light on the mechanisms through which attention and memory interact over development.

Acknowledgments

I am very grateful to too many colleagues and students to acknowledge all in full as I should, but I dedicate this chapter to Annette Karmiloff-Smith and Jon Driver, two scientists and mentors who influenced me a great deal and who are sorely missed.

REFERENCES

Amso, D., & Scerif, G. (2015). The attentive brain: Insights from developmental cognitive neuroscience. *Nature Reviews Neuroscience, 16*(10), 606–619. doi:10.1038/nrn4025

Astle, D. E., Harvey, H., Stokes, M., Mohseni, H., Nobre, A. C., & Scerif, G. (2014). Distinct neural mechanisms of individual and developmental differences in VSTM capacity. *Developmental Psychobiology, 56*(4), 601–610. doi:10.1002/dev.21126

Astle, D., Luckhoo, H., Woolrich, M., Kuo, B.-C., Nobre, A. C., & Scerif, G. (2015). Electrophysiological measures of fronto-parietal networks in typically developing children using magnetoencephalography. *Cerebral Cortex, 25*(10), 3868–3876. doi:10.1093/cercor/bhu271

Atkinson, J., Hood, B., Wattambell, J., & Braddick, O. (1992). Changes in infants' ability to switch visual-attention in the 1st 3 months of life. *Perception, 21*(5), 643–653.

Buschman, T. J., & Kastner, S. (2015). From behavior to neural dynamics: An integrated theory of attention. *Neuron, 88*(1), 127–144. doi:10.1016/j.neuron.2015.09.017

Chun, M. M., Golomb, J. D., & Turk-Browne, N. B. (2011). A taxonomy of external and internal attention. *Annual Review of Psychology, 62*, 73–101. doi:10.1146/annurev.psych.093008.100427

Corbetta, M., & Shulman, G. L. (2002). Control of goal-directed and stimulus-driven attention in the brain. *Nature Reviews Neuroscience, 3*(3), 201–215. doi:10.1038/nrn755

Cowan, N. (2017). The many faces of working memory and short-term storage. *Psychonomic Bulletin & Review, 24*(4), 1158–1170. doi:10.3758/s13423-016-1191-6

Cowan, N., Elliott, E. M., Saults, J. S., Morey, C. C., Mattox, S., Hismjatullina, A., & Conway, A. R. A. (2005). On the capacity of attention: Its estimation and its role in working memory and cognitive aptitudes. *Cognitive Psychology, 51*(1), 42–100. doi:10.1016/j.cogpsych.2004.12.001

Cowan, N., Morey, C. C., AuBuchon, A. M., Zwilling, C. E., & Gilchrist, A. L. (2010). Seven-year-olds allocate attention like adults unless working memory is overloaded. *Developmental Science, 13*(1), 120–133. doi:10.1111/j.1467-7687.2009.00864.x

Crone, E. A. (2009). Executive functions in adolescence: Inferences from brain and behavior. *Developmental Science, 12*(6), 825–830. doi:10.1111/j.1467-7687.2009.00918.x

Csibra, G., Tucker, L. A., & Johnson, M. H. (1998). Neural correlates of saccade planning in infants: A high-density ERP study. *International Journal of Psychophysiology, 29*(2), 201–215. doi:10.1016/s0167-8760(98)00016-6

Desimone, R., & Duncan, J. (1995). Neural mechanisms of selective visual-attention. *Annual Review of Neuroscience, 18,* 193–222.

Doherty, B. R., Fraser, A., Nobre, A. C. N., & Scerif, G. (under review). The functional consequences of social attention on memory precision and on memory-guided orienting in development.

Doherty, B. R., Patai, E. Z., Duta, M., Nobre, A. C., & Scerif, G. (2017). The functional consequences of social distraction: Attention and memory for complex scenes. *Cognition, 158,* 215–223. doi:10.1016/j.cognition.2016.10.015

Doherty, B. R., van Ede, F., Fraser, A., Patai, E. Z., Nobre, A. C. N., & Scerif, G. (under review). The functional consequences of social attention for memory-guided attention orienting and anticipatory neural dynamics.

Fair, D. A., Cohen, A. L., Power, J. D., Dosenbach, N. U. F., Church, J. A., Miezin, F. M., ... Petersen, S. E. (2009). Functional brain networks develop from a "local to distributed" organization. *PloS Computational Biology, 5*(5). doi:10.1371/journal.pcbi.1000381

Fair, D. A., Dosenbach, N. U. F., Church, J. A., Cohen, A. L., Brahmbhatt, S., Miezin, F. M., ... Schlaggar, B. L. (2007). Development of distinct control networks through segregation and integration. *Proceedings of the National Academy of Sciences of the United States of America, 104*(33), 13507–13512.

Fan, J., McCandliss, B. D., Sommer, T., Raz, A., & Posner, M. I. (2002). Testing the efficiency and independence of attentional networks. *Journal of Cognitive Neuroscience, 14*(3), 340–347. doi:10.1162/089892902317361886

Fukuda, K., & Vogel, E. K. (2009). Human variation in overriding attentional capture. *Journal of Neuroscience, 29*(27), 8726–8733. doi:10.1523/jneurosci.2145-09.2009

Gathercole, S. E., Pickering, S. J., Ambridge, B., & Wearing, H. (2004). The structure of working memory from 4 to 15 years of age. *Developmental Psychology, 40*(2), 177–190. doi:10.1037/0012-1649.40.2.177

Gazzaley, A., & Nobre, A. C. (2012). Top-down modulation: Bridging selective attention and working memory. *Trends in Cognitive Sciences, 16*(2), 129–135. doi:10.1016/j.tics.2011.11.014

Griffin, I. C., & Nobre, A. C. (2003). Orienting attention to locations in internal representations. *Journal of Cognitive Neuroscience, 15*(8), 1176–1194. doi:10.1162/089892903322598139

Guillory, S. B., Gliga, T., & Kaldy, Z. (2018). Quantifying attentional effects on the fidelity and biases of visual working memory in young children. *Journal of Experimental Child Psychology, 167,* 146–161. doi:10.1016/j.jecp.2017.10.005

Halassa, M. M., & Kastner, S. (2017). Thalamic functions in distributed cognitive control. *Nature Neuroscience, 20*(12), 1669–1679. doi:10.1038/s41593-017-0020-1

Hood, B. M. (1993). Inhibition of return produced by covert shifts of visual-attention in 6-month-old infants. *Infant Behavior & Development, 16*(2), 245–254. doi:10.1016/0163-6383(93)80020-9

Johnson, M. H. (1990). Cortical maturation and the development of visual attention in early infancy. *Journal of Cognitive Neuroscience, 2*(2), 81–95. doi:10.1162/jocn.1990.2.2.81

Johnson, M. H. (1994). Visual-attention and the control of eye-movements in early infancy. In *Attention and performance XV: Conscious and nonconscious information processing* (Vol. 15, pp. 291–310). Cambridge, MA: MIT Press.

Johnson, M. H. (1995). The inhibition of automatic saccades in early infancy. *Developmental Psychobiology, 28*(5), 281–291. doi:10.1002/dev.420280504

Johnson, M. H., Posner, M. I., & Rothbart, M. K. (1994). Facilitation of saccades toward a covertly attended location in early infancy. *Psychological Science, 5*(2), 90–93. doi:10.1111/j.1467-9280.1994.tb00636.x

Kannass, K. N., Oakes, L. M., & Shaddy, D. J. (2006). A longitudinal investigation of the development of attention and distractibility. *Journal of Cognition and Development, 7*(3), 381–409. doi:10.1207/s15327647jcd0703_8

Kastner, S., & Ungerleider, L. G. (2000). Mechanisms of visual attention in the human cortex. *Annual Review of Neuroscience, 23,* 315–341. doi:10.1146/annurev.neuro.23.1.315

Luna, B., Velanova, K., & Geier, C. F. (2008). Development of eye-movement control. *Brain and Cognition, 68*(3), 293–308. doi:10.1016/j.bandc.2008.08.019

Markant, J., & Amso, D. (2013). Selective memories: Infants' encoding is enhanced in selection via suppression. *Developmental Science, 16*(6), 926–940. doi:10.1111/desc.12084

Markant, J., & Amso, D. (2014). Leveling the playing field: Attention mitigates the effects of intelligence on memory. *Cognition, 131*(2), 195–204. doi:10.1016/j.cognition.2014.01.006

Markant, J., Worden, M. S., & Amso, D. (2015). Not all attention orienting is created equal: Recognition memory is enhanced when attention orienting involves distractor suppression. *Neurobiology of Learning and Memory, 120,* 28–40. doi:10.1016/j.nlm.2015.02.006

Mitsven, S. G., Cantrell, L. M., Luck, S. J., & Oakes, L. M. (2018). Visual short-term memory guides infants' visual attention. *Cognition, 177,* 189–197. doi:10.1016/j.cognition.2018.04.016

Myers, N. E., Walther, L., Wallis, G., Stokes, M. G., & Nobre, A. C. (2015). Temporal dynamics of attention during encoding versus maintenance of working memory: Complementary views from event-related potentials and alpha-band oscillations. *Journal of Cognitive Neuroscience, 27*(3), 492–508. doi:10.1162/jocn_a_00727

Nobre, A. C., Gitelman, D. R., Dias, E. C., & Mesulam, M. M. (2000). Covert visual spatial orienting and saccades: Overlapping neural systems. *NeuroImage,* 11(3), 210–216.

Nussenbaum, K., Scerif, G., & Nobre, A. C. N. (forthcoming). Differential effects of salient visual events on memory-guided attention in adults and children. *Child Development.* doi: 10.1111/cdev.13149. [Epub ahead of print]

Oakes, L. M., Kannass, K. N., & Shaddy, D. J. (2002). Developmental changes in endogenous control of attention: The role of target familiarity on infants' distraction latency. *Child Development, 73*(6), 1644–1655. doi:10.1111/1467-8624.00496

Olesen, P. J., Macoveanu, J., Tegner, J., & Klingberg, T. (2007). Brain activity related to working memory and distraction in children and adults. *Cerebral Cortex, 17*(5), 1047–1054. doi:10.1093/cercor/bhl014

Petersen, S. E., & Posner, M. I. (2012). The attention system of the human brain: 20 years after. *Annual Review of Neuroscience, 35,* 73–89. doi:10.1146/annurev-neuro-062111-150525

Posner, M. I., & Petersen, S. E. (1990). The attention system of the human brain. *Annual Review of Neuroscience, 13,* 25–42.

Richards, J. E. (2010). The development of attention to simple and complex visual stimuli in infants: Behavioral and psychophysiological measures. *Developmental Review, 30*(2), 203–219. doi:10.1016/j.dr.2010.03.005

Richards, J. E., Reynolds, G. D., & Courage, M. L. (2010). The neural bases of infant attention. *Current Directions in Psychological Science, 19*(1), 41–46. doi:10.1177/0963721409360003

Ross-Sheehy, S., Oakes, L. M., & Luck, S. J. (2003). The development of visual short-term memory capacity in infants. *Child Development, 74*(6), 1807–1822. doi:10.1046/j.1467-8624.2003.00639.x

Ross-Sheehy, S., Oakes, L. M., & Luck, S. J. (2011). Exogenous attention influences visual short-term memory in infants. *Developmental Science, 14*(3), 490–501. doi:10.1111/j.1467-7687.2010.00992.x

Rueda, M. R., Fan, J., McCandliss, B. D., Halparin, J. D., Gruber, D. B., Lercari, L. P., & Posner, M. I. (2004). Development of attentional networks in childhood. *Neuropsychologia, 42*(8), 1029–1040. doi:10.1016/j.neuropsychologia.2003.12.012

Rueda, M. R., Posner, M. I., & Rothbart, M. K. (2005). The development of executive attention: Contributions to the emergence of self-regulation. *Developmental Neuropsychology, 28*(2), 573–594. doi:10.1207/s15326942dn2802_2

Shimi, A., Kuo, B.-C., Astle, D. E., Nobre, A. C., & Scerif, G. (2014). Age group and individual differences in attentional orienting dissociate neural mechanisms of encoding and maintenance in visual STM. *Journal of Cognitive Neuroscience, 26*(4), 864–877. doi:10.1162/jocn_a_00526

Shimi, A., Nobre, A. C., Astle, D., & Scerif, G. (2014). Orienting attention within visual short-term memory: Development and mechanisms. *Child Development, 85*(2), 578–592. doi:10.1111/cdev.12150

Shimi, A., & Scerif, G. (2015). The interplay of spatial attentional biases and mental codes in VSTM: Developmentally informed hypotheses. *Developmental Psychology, 51*(6), 731–743. doi:10.1037/a0039057

Shimi, A., & Scerif, G. (2017). Towards an integrative model of visual short-term memory maintenance: Evidence from the effects of attentional control, load, decay, and their interactions in childhood. *Cognition, 169*, 61–83. doi:10.1016/j.cognition.2017.08.005

Stokes, M. G., Atherton, K., Patai, E. Z., & Nobre, A. C. (2012). Long-term memory prepares neural activity for perception. *Proceedings of the National Academy of Sciences of the United States of America, 109*(6), E360–E367. doi:10.1073/pnas.1108555108

Summerfield, J. J., Lepsien, J., Gitelman, D. R., Mesulam, M. M., & Nobre, A. C. (2006). Orienting attention based on long-term memory experience. *Neuron, 49*(6), 905–916. doi:10.1016/j.neuron.2006.01.021

Werchan, D. M., Collins, A. G. E., Frank, M. J., & Amso, D. (2016). Role of prefrontal cortex in learning and generalizing hierarchical rules in 8-month-old infants. *Journal of Neuroscience, 36*(40), 10314–10322. doi:10.1523/jneurosci.1351-16.2016

27 Network Models of Attention and Working Memory

MONICA D. ROSENBERG AND MARVIN M. CHUN

ABSTRACT Attention and working memory, critical for navigating everyday life, are dominant topics of study in cognitive psychology and neuroscience. Despite major theoretical advances, there is not yet a comprehensive ontology that describes their component processes and the interactions between them. Here we suggest that new techniques in network neuroscience, which conceptualizes the brain as a system of interacting units, can inform taxonomies of attention and working memory. In particular, these approaches can reveal common and unique brain systems that underlie attention- and memory-related processes. We begin with a bird's-eye view of network neuroscience before focusing on network models of attention and working memory measured with functional magnetic resonance imaging (fMRI), distinguishing descriptive models that characterize cognitive processes at the group level from predictive models that forecast behavior in single individuals. We highlight the theoretical and practical benefits of predictive network models, which have so far provided evidence for interactions between sustained attention, other attentional components, and memory.

Network Neuroscience

At every spatial scale, the brain is a network of interacting components (Bassett & Sporns, 2017). At the molecular level, genes and proteins interact to regulate gene expression; at the cellular level, neurons and glia form circuits to process and transmit information; and at the systems level, brain regions interact via structural and functional connections to guide behavior.

Network neuroscience, an emerging field at the intersection of graph theory, cognitive neuroscience, and neurobiology, offers a new conceptual framework for understanding the principles of brain function at multiple levels of organization (Bassett & Sporns, 2017). From a network neuroscientific perspective, parts of the brain represent nodes, and interactions between them form connections, or edges. Edges can be either structural connections (e.g., white matter tracts) or functional connections (e.g., correlations between neuroimaging signals in spatially distinct regions). Brain networks can be directed, comprising edges that begin at one node and end at another, or undirected, comprising bidirectional edges. Networks also vary according to the information they carry about individual edges: whereas edges in unweighted (binary) networks are either present or absent, edges in weighted networks are associated with a value that indicates their strength (Boccaletti, Latora, Moreno, Chavez, & Hwang, 2006). Once brain data are represented as networks, graph theoretical tools can be applied to reveal previously unappreciated organizational features of the brain.

In cognitive neuroscience, network analyses are applied primarily to structural connectivity networks measured with diffusion tensor imaging or functional connectivity networks measured with techniques such as fMRI to describe features of the human connectome common to healthy individuals or different in individuals with a disease or disorder. Characterizing the "typical" structural and functional human brain connectomes has uncovered principles of large-scale brain organization. Early functional connectivity analyses identified a set of networks whose nodes coactivate during task engagement and remain functionally connected (i.e., show correlated activity over time) in the absence of an explicit task (Fox et al., 2005; Smith et al., 2009). These canonical networks include subcortical (e.g., cerebellar, basal gangliar), sensorimotor (e.g., visual, motor, and auditory), and association (e.g., default mode, dorsal attention, ventral attention, frontoparietal, cingulo-opercular, salience) networks and are thought to comprise the gross functional architecture of the brain (Bressler & Menon, 2010; Power et al., 2011; Yeo et al., 2011).

In parallel, graph theoretical approaches have demonstrated that human brain networks show features common to other complex systems. For example, brains exhibit the property of small-worldness—that is, like individuals in social networks, most structural and functional brain nodes are not directly connected to each other but are indirectly connected by only a small number of steps, and paths from one node to another often traverse highly connected *hub* regions (Bassett & Bullmore, 2016; van den Heuvel, Stam, Boersma, & Hulshoff Pol, 2008). These hubs form a neural *rich club*, meaning they tend to connect to other hub regions rather than to more sparsely connected nodes (Grayson et al., 2014; van den Heuvel & Sporns, 2011, 2013). Damage to hub regions disproportionally disrupts network structure and cognitive function (Buckner et al., 2009; Crossley et al., 2014; Fornito, Zalesky, & Breakspear, 2015). Interestingly, functional hubs overlap with regions of the default mode network, a system implicated in neurological and psychiatric disorders (Buckner, Andrews-Hanna, & Schacter, 2008; Whitfield-Gabrieli & Ford, 2012).

Recent work in cognitive network neuroscience has focused not just on describing large-scale brain systems at the group level, but also on characterizing how brain connectivity differs across individuals and how these differences predict interindividual variability in behavior (Finn et al., 2015; Medaglia, Lynall, & Bassett, 2015). Such individual differences approaches offer scientific and practical benefits (Dubois & Adolphs, 2016). From a basic science perspective, linking individual differences in brain features and behavior offers a new way to identify neural mechanisms of cognition. Characterizing connectivity-behavior relationships can also shed light on the functional organization of the mind—for example, by identifying common and specific brain networks that support processes such as attention and working memory (Rosenberg, Finn, Scheinost, Constable, & Chun, 2017). Practically, predicting traits, behavior, and clinical symptoms at the individual level can improve health and education outcomes by providing early, objective diagnoses and assessments and identifying those who may benefit from a particular treatment, training, or intervention (Gabrieli, Ghosh, & Whitfield-Gabrieli, 2015; Rosenberg et al., 2018; Woo et al., 2017).

Network Models of Attention

Attention, while a useful catchall concept, is not a single process. Rather, *attention* is an umbrella term that encompasses the selection and enhancement of relevant information, inhibition of distraction, and maintenance of vigilance over time (Chun, Golomb, & Turk-Browne, 2011). Further complicating the definition, attentional processes operate along a number of different dimensions. Attention can be directed to one's outside surroundings or inner thoughts (Chun, Golomb, & Turk-Browne, 2011), to features or objects (Maunsell & Treue, 2006), and to space or time (Nobre & van Ede, 2017). Attention can be guided by current goals, selection history, or stimulus salience (Awh, Belopolsky, & Theeuwes, 2012); focused on a single target or divided between multiple (Treisman, 1969); and deployed briefly or maintained over time (Egeth & Yantis, 1997).

In distinguishing different types of attention—external versus internal, object-based versus feature-based, top-down versus bottom-up, spatial versus temporal, selective versus divided, transient versus sustained—researchers are attempting to "carve it at its joints" by uncovering its underlying architecture. A subsequent issue of clear importance is whether joints in the mind are reflected in the brain. Here we review current neuroanatomical models of attention components, emphasizing distinctions between *descriptive models* that characterize large-scale brain systems at the group level and *predictive models* that characterize attentional abilities at the level of the individual. In doing so, we suggest that predictive network models, in addition to their practical benefits, can inform an ontology of attention.

Descriptive Models of Attention

Alerting, orienting, and executive control In contrast to current methods that define functional networks using correlation-based or signal decomposition approaches (e.g., principal or independent component analysis), initial network models of attention were based on univariate fMRI contrasts that identified regions coactivated during specific attention challenges. Using this approach, Posner and Petersen argued that three independent processes comprise attention (Fan, McCandliss, Fossella, Flombaum, & Posner, 2005; Petersen & Posner, 2012; Posner & Petersen, 1990). In this model, a largely right-lateralized alerting network that includes regions of the norepinephrine system in thalamic, frontal, and parietal areas supports our ability to respond to cues and maintain vigilance. A distinct orienting network, responsible for directing attention to internal or external stimuli, includes the posterior parietal lobe, lateral pulvinar nucleus of the thalamus, superior colliculus, frontal eye fields, and temporoparietal junction (Petersen & Posner, 2012). Two executive control networks support our ability to detect and resolve conflicting information. The frontoparietal control network, which spans lateral frontal and parietal regions distinct from those of the orienting network, is related to task initiation and switching, while the cingulo-opercular network, which includes midline and anterior insular regions, is related to the maintenance of task performance (Dosenbach, Fair, Cohen, Schlaggar, & Petersen, 2008; Petersen & Posner, 2012).

Top-down versus bottom-up attention A dual-network model subdivides attentional orienting into an endogenous top-down system that "pushes" attention toward goal-relevant stimuli and an exogenous bottom-up system that "pulls" attention to stimuli with low-level salience (Corbetta & Shulman, 2002; Desimone & Duncan, 1995). In this model, a bilateral dorsal frontoparietal system supports top-down control. This dorsal attention network includes the intraparietal sulci and frontal eye fields and activates in response to cues about the features or location of upcoming stimuli. Regions of the dorsal attention network contain topographic maps relevant for covert and overt spatial attention and are presumably responsible for selecting goal-relevant stimuli and linking them to appropriate behavioral responses (Corbetta & Shulman, 2011; Vossel, Geng, & Fink, 2014). A right-lateralized ventral attention system is involved in bottom-up processing. The ventral attention network includes temporoparietal and ventral frontal cortices and activates in response to behaviorally relevant but unexpected stimuli—essentially acting as a "circuit breaker" for the dorsal system (Corbetta, Patel, & Shulman, 2008; Corbetta & Shulman, 2002). Functional connectivity studies show that even during rest the dorsal and ventral attention systems are reflected in the brain's functional organization (Fox, Corbetta, Snyder, Vincent, & Raichle, 2006).

Internal versus external attention Activation and functional connectivity analyses of task-based and resting-state (task-free) fMRI data have revealed distinct networks associated with internal and external attention. The default mode network, which includes ventral and dorsal medial prefrontal cortex, medial and lateral parietal and temporal cortex, and posterior cingulate cortex, is more active during rest than task performance (Buckner, Andrews-Hanna, & Schacter, 2008; Raichle, 2015). Although the default network has been related primarily to internally directed attention, such as that observed during mind wandering and task-irrelevant or self-referential thought (Buckner et al., 2008; Christoff, Gordon, Smallwood, Smith, & Schooler, 2009; Mason et al., 2007), it may also support environment monitoring (Hahn, Ross, & Stein, 2007) and "in-the-zone" task performance (Esterman et al., 2013). In contrast, a "task-positive" network includes the intraparietal sulci and frontal eye fields (the dorsal attention system) as well as dorsolateral and ventral prefrontal cortex, the insula, and the supplementary motor area (Fox et al., 2005). Activity in this network increases during task engagement and is anticorrelated with that of the default network during task performance and rest (Fox et al., 2005; Kelly et al., 2008).

PREDICTIVE MODELS OF ATTENTION Although canonical network models characterize the neural correlates of attention at the group level, they do not capture individual differences in the ability to focus. Recent work, however, has emphasized the importance of models that account for individual variability in attention function (Rosenberg et al., 2017). Models that describe (1) how brain regions coordinate to support attentional processes on average and (2) how differences in the integrity of these systems relate to differences in attention function go a step further in characterizing neural mechanisms of attention than models that do not account for individual differences. A hypothetical set of models, or neuromarkers, that predicts each individual's unique pattern of attentional abilities from that person's brain data can help refine proposed taxonomies of attention by identifying specific and general attention factors, and may benefit personalized medicine and education (Rosenberg et al., 2017). Here we review recent advances in the predictive modeling of attention, highlighting functional network models of sustained attention, distractor suppression, and alerting.

Sustained attention In contrast to descriptive models that summarize a set of observations, predictive models forecast outcomes from previously unseen data (Shmueli, 2010). Connectome-based predictive modeling, or CPM, is a recently developed technique for building predictive models from brain features (Finn et al., 2015; Shen et al., 2017; Yoo et al., 2018; figure 27.1). The CPM method identifies functional connections that are related to behavior in a group of individuals (the training set) and examines the strength of these connections in novel individuals (the test set) to predict their behavior. Of note, CPM and other regression modeling approaches generate continuous predictions, offering greater precision than classification models, which categorize individuals into discrete groups.

Given that maintaining focus over time is a central feature of attention, CPM was applied to predict individual differences in sustained attention. During fMRI, 25 healthy adult participants performed the gradual-onset continuous performance task (gradCPT; Esterman et al., 2013; Rosenberg et al., 2013), which engaged attention circuitry and presumably magnified associated individual differences in functional connectivity (Rosenberg, Finn, et al., 2016). Models were defined to relate connectivity patterns to task performance (sensitivity, or d') using data from n–1 participants and then applied to data from the left-out individual to generate a predicted d' score. Demonstrating that functional connectivity observed during task engagement can provide an objective index of sustained attention, predicted and observed d' scores were significantly correlated across individuals. Furthermore, models generalized to predict performance from resting-state functional connectivity alone, demonstrating for the first time that we can measure attention without a task challenge (Rosenberg, Finn, et al., 2016).

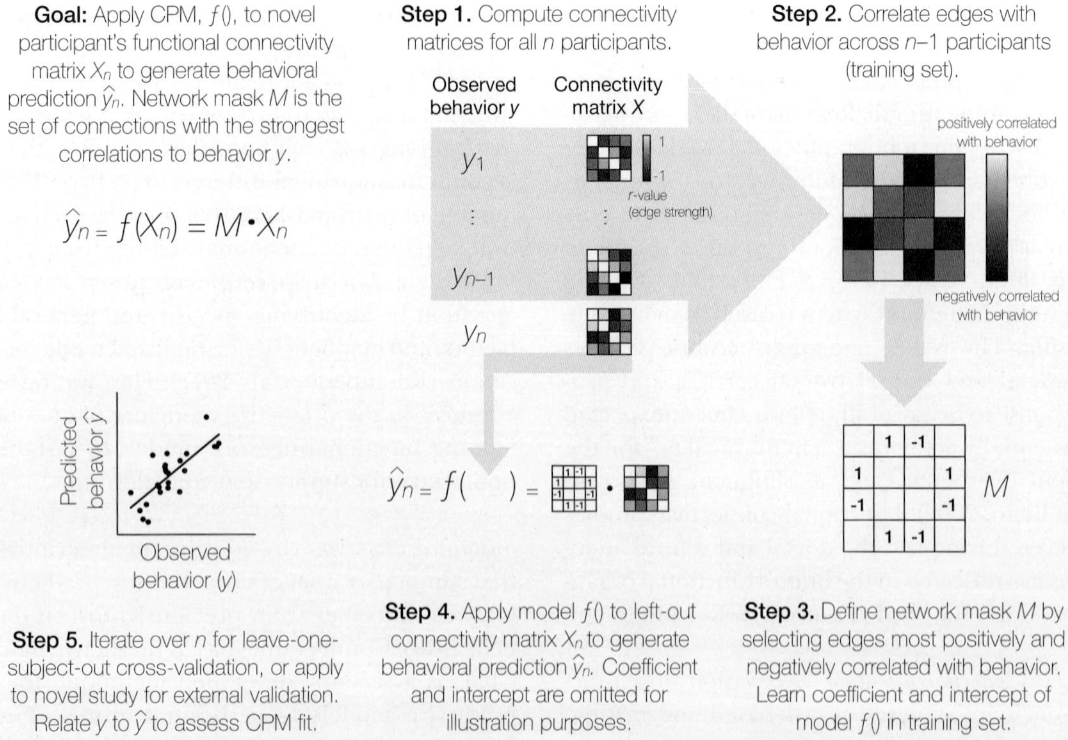

FIGURE 27.1 Schematic of the connectome-based predictive modeling (CPM) pipeline (Finn et al., 2015; Shen et al., 2017). The CPM approach identifies behaviorally relevant functional connections in a training set of individuals and measures their strength in a novel test set to predict behavior.

To test whether the sustained attention CPM predicts gradCPT performance in particular or sustained attentional abilities in general, the model was applied to an external validation sample. This independent data set included resting-state fMRI data and clinician-rated attention deficit hyperactivity disorder (ADHD) symptom scores from individuals aged 8–16. Even controlling for IQ, predictions of the sustained attention CPM were inversely correlated with symptom scores, meaning that the model predicted that children with fewer ADHD symptoms would have higher *d'* scores if they were to perform the gradCPT (Rosenberg, Finn, et al., 2016). Furthermore, this same network model generalized to predict stop-signal task performance in a third independent group of individuals and was sensitive to attention changes resulting from pharmacological intervention (Rosenberg, Zhang, et al., 2016). These results suggest that a common functional network underlies variation in sustained attention in adulthood and attention dysfunction in development.

The sustained attention CPM generates predictions from the strength of a high-attention network of edges positively correlated with sustained attention and a low-attention network of edges inversely correlated with attention (figure 27.2). These networks include prefrontal, parietal, and cerebellar nodes implicated in attention (Castellanos & Proal, 2012), but do not rely on these regions to make predictions. Instead, variance in behavior is captured by connections that span the cortex, subcortex, and cerebellum, and models are not reducible to a single structure, lobe, or canonical network (Rosenberg, Finn, et al., 2016).

Complementary functional connectivity models support the finding that distributed systems underlie interindividual differences in sustained attention. Using resting-state connectivity data from 519 individuals, Kessler, Angstadt, and Sripada (2016) developed a maturational growth chart to predict children's ADHD diagnoses and success on a continuous performance task. They found that complex interactions within and between nodes of the default mode, frontoparietal, and dorsal and ventral attention networks predicted attention. O'Halloran et al. (2018) used task-based functional connectivity in a sample of 758 adolescents to predict response time variability on a stop-signal task, and found that lower cerebellar-motor, cerebellar- prefrontal, and occipitomotor connectivity predicted better sustained attention, whereas greater intramotor, motor-parietal, motor-prefrontal, and motor-limbic connectivity predicted worse attention. Although the sustained attention CPM, connectivity growth chart, and response time variability model have not been

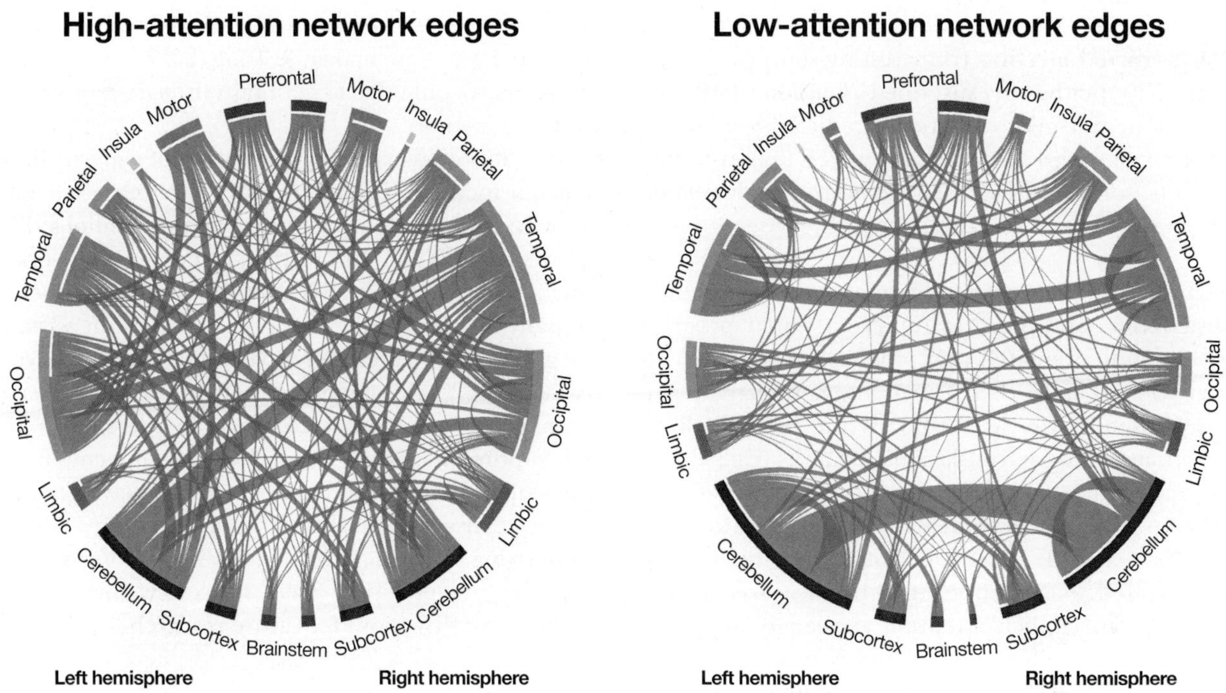

FIGURE 27.2 Functional connections (edges) in the high-attention and low-attention networks (Rosenberg, Finn, et al., 2016). Network nodes are grouped into macroscale brain regions; lines between them represent edges. Line width corresponds to the number of edges between region pairs. (See color plate 29.)

compared directly, integrating their predictive features or identifying their overlap could help refine a maximally generalizable model of sustained attention.

Distractor suppression Closely related to sustained attention is the ability to resist internal distraction (mind wandering) and external distraction (attention capture by task-irrelevant stimuli). To characterize individual differences in reactive control, or the ability to disengage from a stimulus after it has captured attention, Poole and colleagues (2016) analyzed resting-sate functional connectivity patterns from 32 adults who later performed a singleton task. In this task, participants were instructed to identify a unique shape in an eight-item array. On half of the trials, a unique distractor color was also present. Attention capture was measured as the difference in correct-trial response time between trials with and without irrelevant color distractors. Using leave-one-subject-out cross-validation, Poole et al. (2016) trained models to predict attention capture scores from functional connectivity within and between the default mode and the dorsal and ventral attention networks. Models successfully predicted left-out participants' attentional capture scores, revealing that participants with stronger within-default connectivity but weaker default mode to dorsal and ventral attention network connectivity were less disrupted by task-irrelevant distractors. Thus, in addition to sustained attention, the brain's intrinsic functional architecture contains a signature of the ability to disengage from a visual distractor.

Alerting, orienting, and executive control In the three-component model of attention, sustained attention falls under the umbrella of *alerting*, a subsystem encompassing both phasic alerting (changing attention in response to a signal or cue) and tonic alerting (maintaining alertness or vigilance; Posner & Petersen, 1990). To test the relationship between sustained attention and phasic alerting, the sustained attention CPM was applied to functional connectivity data measured as novel participants performed the Attention Network Task (ANT), which uses the difference in response time to trials with and without warning cues to measure a person's ability to prepare to respond to upcoming stimuli (Fan et al., 2005). Evidencing a perhaps unappreciated distinction between sustained attention and alerting, the sustained attention CPM predicted *overall* ANT performance (accuracy and response time variability), but not alerting scores (Rosenberg et al., 2018). Instead, model predictions were more closely related to individuals' executive control abilities, measured in the ANT as the difference in response time on trials with target-congruent and target-incongruent distractors. Furthermore, whereas a new data-driven CPM predicted alerting from resting-state functional connectivity, neither the sustained attention CPM nor a new data-driven network model predicted spatial orienting. Intriguingly, these results suggest that sustained attention (tonic alerting) may be more closely related to executive control than phasic alerting.

Looking ahead Network models of sustained attention, distractibility, and alerting represent initial progress toward a suite of models that predicts a person's attentional abilities from their functional connectivity patterns (Rosenberg et al., 2017). While individualized predictions can have translational benefits (e.g., identifying individuals at risk for future attention deficits), they can also inform what we know about attention itself. For example, predictive network models have demonstrated that attention can be measured in the absence of an explicit attention challenge, and have provided evidence for relationships between sustained attention and executive control but not phasic alerting. In the future, predictive modeling approaches may be applied to other attention factors and cognitive processes to elucidate relationships between them and, together with behavioral individual differences studies (Huang, Mo, & Li, 2012), contribute to a data-driven taxonomy of attention.

Network Models of Working Memory

Working memory is a capacity-limited system that enables the storage and manipulation of information (Baddeley, 1992). Like attention, working memory is not a single process, but rather is best characterized as a collection of mechanisms related to information maintenance and modulation. Cognitive psychological theories posit that capacity, approximately three to four items on average, arises from a fixed number of memory slots (Luck & Vogel, 2013) or a fixed amount of attentional resources (Ma, Husain, & Bays, 2014). Examining working memory precision (the quality of a memory representation) has provided evidence for both views. As predicted by the slots model, increasing the number of to-be-remembered items from three to six decreases the probability that any one item will be held in memory but does not affect the precision of the information that is maintained (Zhang & Luck, 2008). As predicted by the resource view, a model allowing memory precision to vary across items and trials better fits behavioral data than a slot-based model (van den Berg, Shin, Chou, George, & Ma, 2012). However, more recent findings suggest that these results are partly explained by guessing (Adam, Vogel, & Awh, 2017) and, for some stimuli, a reliance on categorical representations (Pratte, Park, Rademaker, & Tong, 2017), and that participants are only able to maintain three to four items in working memory.

In addition to exploring the nature of capacity limits, a major focus of working memory research has been to explain how and why working memory abilities differ across individuals. Individual differences in working memory capacity are stable over time and consequential in daily life, explaining more than 40% of the variance in global fluid intelligence (Fukuda, Vogel, Mayr, & Awh, 2010). Working memory deficits are also observed in a range of neuropsychiatric disorders, including schizophrenia (Luck & Vogel, 2013). Approaches in cognitive neuroscience, and, more recently, network neuroscience, have revealed large-scale brain systems underlying individual differences in working memory capacity and precision. Here we review these models and suggest directions for future research.

Descriptive Models of Working Memory

Capacity Working memory representations are maintained with sustained activity and *activity-silent* mechanisms—functional connectivity patterns or dynamic population codes (Stokes, 2015)—spanning prefrontal, parietal, and sensory cortices (for a recent review, see D'Esposito & Postle, 2015). Whereas the prefrontal cortex is thought to support top-down control by representing current goals (D'Esposito & Postle, 2015), converging evidence suggests that capacity limits are related to activity in the inferior parietal sulcus (IPS). For example, the fMRI signal in IPS scales with working memory load until working memory capacity is reached (McNab & Klingberg, 2007; Todd & Marois, 2004; Xu & Chun, 2006), and this change point varies with capacity across individuals (Todd & Marois, 2005). Resting-state functional connectivity analyses suggest that IPS centrality, a measure of the most important nodes in a network, is also related to individual differences in capacity: in individuals with higher capacity limits, the IPS is less influential in the whole-brain network (Markett et al., 2018). Furthermore, changes in parietal activity and frontoparietal functional connectivity have been observed following working memory training (Constantinidis & Klingberg, 2016), and these connectivity increases appear to track post-training behavioral improvements (Thompson, Waskom, & Gabrieli, 2016). Corroborating findings from fMRI, a magnetoencephalography (MEG) study found that synchrony in a large-scale brain network was related to individual differences in working memory capacity and that the central hub of this network was the intraparietal sulcus (Palva, Monto, Kulashekhar, & Palva, 2010).

fMRI functional connectivity studies also point to relationships between the function of distributed brain networks and working memory capacity. One study found relationships between better working memory performance, decreased connectivity in the task-positive network, and decreased anticorrelation between the task-positive and default mode networks (Magnuson et al., 2015). Another observed relationships between working memory capacity and whole-brain network small-worldness and modularity (a measure of a network's community structure) during rest (Stevens, Tappon, Garg, & Fair, 2012). Recent work, however, found an inverse relationship between working memory function, modularity, and local efficiency (measures of network segregation) during task performance (Cohen & D'Esposito, 2016). Cohen and D'Esposito (2016) also reported positive relationships between working memory, global efficiency, and the number of connector hubs (measures of network integration), suggesting that communication between large-scale networks during task engagement underlies successful working memory performance.

Findings in electroencephalography (EEG) suggest that, in parallel, sustained voltage changes during working memory retention, known as contralateral delay activity, track working memory load and capacity differences across individuals. In the hemisphere contralateral to the visual field location of items maintained in working memory, EEG signal amplitude during retention tracks memory load until set size exceeds capacity. This asymptote is related to capacity differences across individuals, such that the contralateral delay activity scales with higher set sizes in people with higher capacity limits (Luck & Vogel, 2013).

Evidence from fMRI, MEG, and EEG suggests that brain networks involving parietal cortex in particular are related to an individual's working memory capacity. Although these individual differences approaches provide valuable insight into the neural mechanisms of working memory, models have not yet been applied to predict behavior in novel individuals. In the future, validating models on unseen data can help identify the most reliable predictors of working memory at the level of single individuals.

Precision Although the majority of individual differences studies of working memory have focused on capacity, people also differ in their working memory precision. Curtis, Rao, and D'Esposito (2004) first investigated the neural mechanisms of working memory precision by looking at differences in representational fidelity over time rather than across individuals. They found that, within subjects, fMRI activity in the frontal eye fields reflected the accuracy of memory-guided saccades in an oculomotor delayed-response task. Using a similar approach, Emrich, Riggall, LaRocque, and Postle (2013) measured patterns of fMRI activity in sensory cortex as participants performed a working memory task. Increases in set size were accompanied by performance decrements and lower pattern classification accuracy for the remembered stimuli, a measure of representational precision. In one individual differences design, Ester, Anderson, Serences, and Awh (2013) applied forward-encoding models to fMRI data collected during a task requiring participants to remember the orientation of line gratings. Estimates of orientation selectivity in visual cortex were correlated with differences in representational acuity across participants, also suggesting links between working memory precision and sustained neural activity in sensory cortex. Finally, Galeano Weber, Peters, Hahn, Bledowski, and Fiebach (2016) reported that participants with more stable working memory performance (i.e., less variable representational precision) in conditions of high memory load showed greater load-dependent increases in IPS activity. Based on these findings, they argue that the IPS supports working memory by decreasing the variability of memory precision under conditions of high load.

Predictive Models of Working Memory To date, predictive network models have characterized individual differences in the precision, but not capacity, of working memory. Asking whether interactions between perceptual and attentional systems affect working memory precision, Galeano Weber, Hahn, Hilger, and Fiebach (2017) scanned participants while they performed a visual working memory and a visual attention task. They fit participants' behavioral data with a model that assumed fixed working memory capacity but variable memory precision over time and across items, providing an estimate of each individual's working memory capacity and precision. For each participant, they also calculated functional connectivity between the occipital and parietal regions activated during both tasks. Using leave-one-subject-out cross-validation, Galeano Weber et al. (2017) found that functional connectivity observed during the working memory task, but not the visual attention task, predicted memory precision but not capacity. Participants with better working memory precision showed higher connectivity between occipital and parietal regions during encoding. Mirroring findings with attention, these results suggest that engaging memory-related circuits magnifies individual differences in memory-related functional connections.

However, unlike aspects of attention, working memory precision may not be reflected in the brain's intrinsic functional architecture (that is, because predictions were not significant when working memory was not engaged during the visual attention task). Nonetheless, these results leave open the possibility that models based on whole-brain functional connectivity, rather than a circumscribed set of regions of interest, could predict individual differences in working memory capacity.

Attention-memory interactions Although attention and working memory are often studied in isolation, they are intimately intertwined (Engle, 2002). For example, attentional mechanisms can gate entry into our capacity-limited working memory (Awh, Vogel, & Oh, 2006) and manipulate stored information (Myers, Stokes, & Nobre, 2017), the contents of working memory can influence how we focus our attention and resist distraction (de Fockert, Rees, Frith, & Lavie, 2001; Downing, 2000), and working memory itself can be considered a form of internally directed attention (Chun, Golomb, & Turk-Browne, 2011).

Interactions between attention and memory are also evident at the level of large-scale brain networks. As one example, the sustained attention CPM was applied to functional connectivity data collected while participants read a Greek history lecture transcript during fMRI. The model significantly predicted memory-test performance, such that individuals with stronger high-attention networks and weaker low-attention networks during reading better comprehended and remembered what they had read (Jangraw et al., 2018). These results demonstrate links between sustained attention and short-term memory and suggest that cross-task prediction approaches can elucidate relationships between the constituent processes of attention and working memory.

Current work explores relationships between aspects of attentional control and memory. In particular, Avery et al. (2018) used task-based and resting-state functional connectivity data from 502 adults in the Human Connectome Project sample to build predictive models of 2-back task performance, a measure reflecting working memory capacity, memory-based discrimination abilities, attentional control, and executive function (Jaeggi, Buschkuehl, Perrig, & Meier, 2010). These models generalized to predict visual and verbal memory in 157 older adults from a Samsung Medical Center data set, highlighting relationships between processes underlying attention, working memory, and short-term memory across the lifespan (Avery et al., 2018).

Limitations of Predictive Network Models

Although this chapter has focused on the benefits of predictive network models, there are several limitations associated with the approach. First, individual differences studies provide correlational (rather than causal) evidence of brain-behavior relationships and are limited by sample size and composition, the reliability of single-subject data, and the degree to which data reflect state-like versus trait-like influences (Braver, Cole, & Yarkoni, 2010). Confounds such as head motion can also induce spurious relationships between functional connectivity and behavior, undermining model validity if not appropriately controlled. Finally, translating brain-based predictive models to clinical settings requires the careful consideration of issues related to implementation and patient privacy (Rosenberg, Casey, & Holmes, 2018).

Conclusions

A driving question in psychology is how the mind is organized into distinct processes. Proposed taxonomies of attention and working memory have suggested that attention comprises three independent systems (alerting, orienting, and executive control), that these components vary along a number of dimensions (e.g., top-down vs. bottom-up orienting, tonic vs. phasic alerting, internal vs. external focus), and that attention and working memory rely on common processes (Chun, Golomb, & Turk-Browne, 2011). Predictive network models, which forecast an individual's abilities and behavior from their unique pattern of brain connectivity (Finn et al., 2015), can help advance proposed ontologies by identifying general and specific models of cognitive performance (Rosenberg et al., 2017). Thus, moving forward, cognitive network neuroscientific approaches may not only shed light on the functional organization of the brain, but may also inform the organization of the mind.

Acknowledgment

This work was supported by National Institutes of Health grant MH108591 and National Science Foundation grant BCS1558497 to Marvin M. Chun.

REFERENCES

Adam, K. C. S., Vogel, E. K., & Awh, E. (2017). Clear evidence for item limits in visual working memory. *Cognitive Psychology, 97*, 79–97.

Avery, E. W., Yoo, K., Rosenberg, M. D., Na, D. L., Greene, A. S., Gao, S., Scheinost, D., Constable, R. T., & Chun,

M. M. (2018). Whole-brain functional connectivity predicts working memory performance in novel healthy and memory-impaired individuals. Program No. 426.16. 2018 Neuroscience Meeting Planner. San Diego, CA: Society for Neuroscience, 2018. Online.

Awh, E., Belopolsky, A. V., & Theeuwes, J. (2012). Top-down versus bottom-up attentional control: A failed theoretical dichotomy. *Trends in Cognitive Sciences, 16*(8), 437–443.

Awh, E., Vogel, E. K., & Oh, S.-H. (2006). Interactions between attention and working memory. *Neuroscience, 139*(1), 201–208.

Baddeley, A. (1992). Working memory. *Science, 255*, 556–559.

Bassett, D. S., & Bullmore, E. T. (2016). Small-world brain networks revisited. *Neuroscientist, 23*(5), 499–516.

Bassett, D. S., & Sporns, O. (2017). Network neuroscience. *Nature Neuroscience, 20*(3), 353–364.

Boccaletti, S., Latora, V., Moreno, Y., Chavez, M., & Hwang, D.-U. (2006). Complex networks: Structure and dynamics. *Physics Reports, 424*(4), 175–308.

Braver, T. S., Cole, M. W., & Yarkoni, T. (2010). Vive les differences! Individual variation in neural mechanisms of executive control. *Current Opinion in Neurobiology, 20*(2), 242–250.

Bressler, S. L., & Menon, V. (2010). Large-scale brain networks in cognition: Emerging methods and principles. *Trends in Cognitive Sciences, 14*(6), 277–290.

Buckner, R. L., Andrews-Hanna, J. R., & Schacter, D. L. (2008). The brain's default network: Anatomy, function, and relevance to disease. *Annals of the New York Academy of Sciences*, 1124, 1–38.

Buckner, R. L., Sepulcre, J., Talukdar, T., Krienen, F. M., Liu, H., Hedden, T., ... Johnson, K. A. (2009). Cortical hubs revealed by intrinsic functional connectivity: Mapping, assessment of stability, and relation to Alzheimer's disease. *Journal of Neuroscience, 29*(6), 1860–1873.

Castellanos, F. X., & Proal, E. (2012). Large-scale brain systems in ADHD: Beyond the prefrontal-striatal model. *Trends in Cognitive Sciences, 16*(1), 17–26.

Christoff, K., Gordon, A. M., Smallwood, J., Smith, R., & Schooler, J. W. (2009). Experience sampling during fMRI reveals default network and executive system contributions to mind wandering. *Proceedings of the National Academy of Sciences of the United States of America, 106*(21), 8719–8724.

Chun, M. M., Golomb, J. D., & Turk-Browne, N. B. (2011). A taxonomy of external and internal attention. *Annual Review of Psychology, 62*(1), 73–101.

Cohen, J. R., & D'Esposito, M. (2016). The segregation and integration of distinct brain networks and their relationship to cognition. *Journal of Neuroscience, 36*(48), 12083–12094.

Constantinidis, C., & Klingberg, T. (2016). The neuroscience of working memory capacity and training. *Nature Reviews Neuroscience, 17*(7), 438–449.

Corbetta, M., Patel, G., & Shulman, G. L. (2008). The reorienting system of the human brain: From environment to theory of mind. *Neuron, 58*(3), 306–324.

Corbetta, M., & Shulman, G. L. (2002). Control of goal-directed and stimulus-driven attention in the brain. *Nature Reviews Neuroscience, 3*, 201–215.

Corbetta, M., & Shulman, G. L. (2011). Spatial neglect and attention networks. *Annual Review of Neuroscience, 34*, 569–599.

Crossley, N. A., Mechelli, A., Scott, J., Carletti, F., Fox, P. T., McGuire, P., & Bullmore, E. T. (2014). The hubs of the human connectome are generally implicated in the anatomy of brain disorders. *Brain, 137*(8), 2382–2395.

Curtis, C. E., Rao, V. Y., & D'Esposito, M. (2004). Maintenance of spatial and motor codes during oculomotor delayed response tasks. *Journal of Neuroscience, 24*(16), 3944–3952.

D'Esposito, M., & Postle, B. R. (2015). The cognitive neuroscience of working memory. *Annual Review of Psychology, 66*(1), 115–142.

de Fockert, J. W., Rees, G., Frith, C. D., & Lavie, N. (2001). The role of working memory in visual selective attention. *Science, 291*(5509), 1803–1806.

Desimone, R., & Duncan, J. (1995). Neural mechanisms of selective visual attention. *Annual Review of Neuroscience, 18*, 193–222.

Dosenbach, N. U. F., Fair, D. A., Cohen, A. L., Schlaggar, B. L., & Petersen, S. E. (2008). A dual-networks architecture of top-down control. *Trends in Cognitive Sciences, 12*(3), 99–105.

Downing, P. E. (2000). Interactions between visual working memory and selective attention. *Psychological Science, 11*(6), 467–473.

Dubois, J., & Adolphs, R. (2016). Building a science of individual differences from fMRI. *Trends in Cognitive Sciences, 20*(6), 1–19.

Egeth, H. E., & Yantis, S. (1997). Visual attention: Control, representation, and time course. *Annual Review of Psychology, 48*, 269–297.

Emrich, S. M., Riggall, A. C., LaRocque, J. J., & Postle, B. R. (2013). Distributed patterns of activity in sensory cortex reflect the precision of multiple items maintained in visual short-term memory. *Journal of Neuroscience, 33*(15), 6516–6523.

Engle, R. W. (2002). Working memory capacity as executive attention. *Current Directions in Psychological Science, 11*(1), 19–23.

Ester, E. F., Anderson, D. E., Serences, J. T., & Awh, E. (2013). A neural measure of precision in visual working memory. *Journal of Cognitive Neuroscience, 25*(5), 754–761.

Esterman, M., Noonan, S. K., Rosenberg, M., & Degutis, J. (2013). In the zone or zoning out? Tracking behavioral and neural fluctuations during sustained attention. *Cerebral Cortex, 23*(11), 2712–2723.

Fan, J., McCandliss, B. D., Fossella, J., Flombaum, J. I., & Posner, M. I. (2005). The activation of attentional networks. *NeuroImage, 26*, 471–479.

Finn, E. S., Shen, X., Scheinost, D., Rosenberg, M. D., Huang, J., Chun, M. M., Papademetris, X., & Constable, R. T. (2015). Functional connectome fingerprinting: Identifying individuals using patterns of brain connectivity. *Nature Neuroscience, 18*(11), 1664–1671.

Fornito, A., Zalesky, A., & Breakspear, M. (2015). The connectomics of brain disorders. *Nature Reviews Neuroscience, 16*, 159.

Fox, M. D., Corbetta, M., Snyder, A. Z., Vincent, J. L., & Raichle, M. E. (2006). Spontaneous neuronal activity distinguishes human dorsal and ventral attention systems. *Proceedings of the National Academy of Sciences of the United States of America, 103*(26), 10046–10051.

Fox, M. D., Snyder, A. Z., Vincent, J. L., Corbetta, M., Van Essen, D. C., & Raichle, M. E. (2005). The human brain is intrinsically organized into dynamic, anticorrelated functional networks. *Proceedings of the National Academy of Sciences of the United States of America, 102*(27), 9673–9678.

Fukuda, K., Vogel, E., Mayr, U., & Awh, E. (2010). Quantity not quality: The relationship between fluid intelligence and working memory capacity. *Psychonomic Bulletin & Review, 17*(5), 673–679.

Gabrieli, J. D. E., Ghosh, S. S., & Whitfield-Gabrieli, S. (2015). Prediction as a humanitarian and pragmatic contribution from human cognitive neuroscience. *Neuron, 85*(1), 11–26.

Galeano Weber, E. M., Hahn, T., Hilger, K., & Fiebach, C. J. (2017). Distributed patterns of occipito-parietal functional connectivity predict the precision of visual working memory. *NeuroImage, 146*, 404–418.

Galeano Weber, E. M., Peters, B., Hahn, T., Bledowski, C., & Fiebach, C. J. (2016). Superior intraparietal sulcus controls the variability of visual working memory precision. *Journal of Neuroscience, 36*(20), 5623–5635.

Grayson, D. S., Ray, S., Carpenter, S., Iyer, S., Dias, T. G. C., Stevens, C., Nigg, J. T., & Fair, D. A. (2014). Structural and functional rich club organization of the brain in children and adults. *PLOS One, 9*(2), e88297.

Hahn, B., Ross, T. J., & Stein, E. A. (2007). Cingulate activation increases dynamically with response speed under stimulus unpredictability. *Cerebral Cortex, 17*(7), 1664–1671.

Huang, L., Mo, L., & Li, Y. (2012). Measuring the interrelations among multiple paradigms of visual attention: An individual differences approach. *Journal of Experimental Psychology: Human Perception and Performance, 38*(2), 414–28.

Jaeggi, S. M., Buschkuehl, M., Perrig, W. J., & Meier, B. (2010). The concurrent validity of the N-back task as a working memory measure. *Memory, 18*(4), 394–412.

Jangraw, D. C., Gonzalez-Castillo, J., Handwerker, D. A., Ghane, M., Rosenberg, M. D., Panwar, P., & Bandettini, P. A. (2018). A functional connectivity-based neuromarker of sustained attention generalizes to predict recall in a reading task. *NeuroImage, 166*, 99–109.

Kelly, C. A. M., Uddin, L. Q., Biswal, B. B., Castellanos, F. X., & Milham, M. P. (2008). Competition between functional brain networks mediates behavioral variability. *NeuroImage, 39*(1), 527–537.

Kessler, D., Angstadt, M., & Sripada, C. (2016). Growth charting of brain connectivity networks and the identification of attention impairment in youth. *JAMA Psychiatry, 73*(5), 481–489.

Luck, S. J., & Vogel, E. K. (2013). Visual working memory capacity: From psychophysics and neurobiology to individual differences. *Trends in Cognitive Sciences, 17*(8), 391–400.

Ma, W. J., Husain, M., & Bays, P. M. (2014). Changing concepts of working memory. *Nature Neuroscience, 17*(3), 347–356.

Magnuson, M. E., Thompson, G. J., Schwarb, H., Pan, W.-J., McKinley, A., Schumacher, E. H., & Keilholz, S. D. (2015). Errors on interrupter tasks presented during spatial and verbal working memory performance are linearly linked to large-scale functional network connectivity in high temporal resolution resting state fMRI. *Brain Imaging and Behavior, 9*(4), 854–867.

Markett, S., Reuter, M., Heeren, B., Lachmann, B., Weber, B., & Montag, C. (2018). Working memory capacity and the functional connectome—insights from resting-state fMRI and voxelwise centrality mapping. *Brain Imaging and Behavior, 12*(1), 238–246.

Mason, M. F., Norton, M. I., Van Horn, J. D., Wegner, D. M., Grafton, S. T., & Macrae, C. N. (2007). Wandering minds: The default network and stimulus-independent thought. *Science, 315*(5810), 393–395.

Maunsell, J. H. R., & Treue, S. (2006). Feature-based attention in visual cortex. *Trends in Neurosciences, 29*(6), 317–322.

McNab, F., & Klingberg, T. (2007). Prefrontal cortex and basal ganglia control access to working memory. *Nature Neuroscience, 11*, 103–107.

Medaglia, J. D., Lynall, M.-E., & Bassett, D. S. (2015). Cognitive network neuroscience. *Journal of Cognitive Neuroscience, 27*(8), 1471–1491.

Myers, N. E., Stokes, M. G., & Nobre, A. C. (2017). Prioritizing information during working memory: Beyond sustained internal attention. *Trends in Cognitive Sciences, 21*(6), 449–461.

Nobre, A. C., & van Ede, F. (2017). Anticipated moments: Temporal structure in attention. *Nature Reviews Neuroscience, 19*, 34–48.

O'Halloran, L., Cao, Z., Ruddy, K., Jollans, L., Albaugh, M. D., Aleni, A., ... Whelan, R. (2018). Neural circuitry underlying sustained attention in healthy adolescents and in ADHD symptomatology. *NeuroImage, 169*, 395–406.

Palva, J. M., Monto, S., Kulashekhar, S., & Palva, S. (2010). Neuronal synchrony reveals working memory networks and predicts individual memory capacity. *Proceedings of the National Academy of Sciences of the United States of America, 107*(16), 7580–7585.

Petersen, S. E., & Posner, M. I. (2012). The attention system of the human brain: 20 years after. *Annual Review of Neuroscience, 35*(1), 73–89.

Poole, V. N., Robinson, M. E., Singleton, O., DeGutis, J., Milberg, W. P., McGlinchey, R. E., Salat, D. H., & Esterman, M. (2016). Intrinsic functional connectivity predicts individual differences in distractibility. *Neuropsychologia, 86*, 176–182.

Posner, M. I., & Petersen, S. E. (1990). The attention system of the human brain. *Annual Review of Neuroscience, 13*, 25–42.

Power, J. D., Cohen, A. L., Nelson, S. M., Wig, G. S., Barnes, K. A., Church, J. A., ... Petersen, S. E. (2011). Functional network organization of the human brain. *Neuron, 72*(4), 665–678.

Pratte, M. S., Park, Y. E., Rademaker, R. L., & Tong, F. (2017). Accounting for stimulus-specific variation in precision reveals a discrete capacity limit in visual working memory. *Journal of Experimental Psychology: Human Perception and Performance, 43*(1), 6–17.

Raichle, M. E. (2015). The brain's default mode network. *Annual Review of Neuroscience, 38*, 433–447.

Rosenberg, M., Noonan, S., DeGutis, J., & Esterman, M. (2013). Sustaining visual attention in the face of distraction: A novel gradual-onset continuous performance task. *Attention Perception Psychophysics, 75*(3), 426–439.

Rosenberg, M. D., Casey, B. J., & Holmes, A. J. (2018). Prediction complements explanation in understanding the developing brain. *Nature Communications, 9*(1), 589.

Rosenberg, M. D., Finn, E. S., Scheinost, D., Constable, R. T., & Chun, M. M. (2017). Characterizing attention with predictive network models. *Trends in Cognitive Sciences, 21*(4), 290–302.

Rosenberg, M. D., Finn, E. S., Scheinost, D., Papademetris, X., Shen, X., Constable, R. T., & Chun, M. M. (2016). A neuromarker of sustained attention from whole-brain functional connectivity. *Nature Neuroscience, 19*(1), 165–171.

Rosenberg, M. D., Hsu, W.-T., Scheinost, D., Constable, R. T., & Chun, M. M. (2018). Connectome-based models predict separable components of attention in novel individuals. *Journal of Cognitive Neuroscience, 30*(2), 160–173.

Rosenberg, M. D., Zhang, S., Hsu, W.-T., Scheinost, D., Finn, E. S., Shen, X., Constable, R. T., Li, C.-S. R., & Chun, M. M. (2016). Methylphenidate modulates functional network connectivity to enhance attention. *Journal of Neuroscience, 36*(37), 9547–9557.

Shen, X., Finn, E. S., Scheinost, D., Rosenberg, M. D., Chun, M. M., Papademetris, X., & Constable, R. T. (2017). Using connectome-based predictive modeling to predict individual behavior from brain connectivity. *Nature Protocols, 12*(3), 506–518.

Shmueli, G. (2010). To explain or to predict? *Statistical Science, 25*(3), 289–310.

Smith, S. M., Fox, P. T., Miller, K. L., Glahn, D. C., Fox, P. M., Mackay, C. E., ... Beckmann, C. F. (2009). Correspondence of the brain's functional architecture during activation and rest. *Proceedings of the National Academy of Sciences of the United States of America, 106*(31), 13040–13045.

Stevens, A. A., Tappon, S. C., Garg, A., & Fair, D. A. (2012). Functional brain network modularity captures inter- and intra-individual variation in working memory capacity. *PLOS One, 7*(1), e30468.

Stokes, M. G. (2015). "Activity-silent" working memory in prefrontal cortex: A dynamic coding framework. *Trends in Cognitive Sciences, 19*(7), 394–405.

Thompson, T. W., Waskom, M. L., & Gabrieli, J. D. E. (2016). Intensive working memory training produces functional changes in large-scale frontoparietal networks. *Journal of Cognitive Neuroscience, 28*(4), 575–588.

Todd, J. J., & Marois, R. (2004). Capacity limit of visual short-term memory in human posterior parietal cortex. *Nature, 428*(6984), 751–754.

Todd, J. J., & Marois, R. (2005). Posterior parietal cortex activity predicts individual differences in visual short-term memory capacity. *Cognitive, Affective, & Behavioral Neuroscience, 5*(2), 144–155.

Treisman, A. M. (1969). Strategies and models of selective attention. *Psychological Review, 76*(3), 282–299.

van den Berg, R., Shin, H., Chou, W.-C., George, R., & Ma, W. J. (2012). Variability in encoding precision accounts for visual short-term memory limitations. *Proceedings of the National Academy of Sciences of the United States of America, 109*(22), 8780–8785.

van den Heuvel, M. P., & Sporns, O. (2011). Rich-club organization of the human connectome. *Journal of Neuroscience, 31*(44), 15775–15786.

van den Heuvel, M. P., & Sporns, O. (2013). Network hubs in the human brain. *Trends in Cognitive Sciences, 17*(12), 683–696.

van den Heuvel, M. P., Stam, C. J., Boersma, M., & Hulshoff Pol, H. E. (2008). Small-world and scale-free organization of voxel-based resting-state functional connectivity in the human brain. *NeuroImage, 43*(3), 528–539.

Vossel, S., Geng, J. J., & Fink, G. R. (2014). Dorsal and ventral attention systems distinct neural circuits but collaborative roles. *Neuroscientist, 20*, 150–159.

Whitfield-Gabrieli, S., & Ford, J. M. (2012). Default mode network activity and connectivity in psychopathology. *Annual Review of Clinical Psychology, 8*, 49–76.

Woo, C.-W., Chang, L. J., Lindquist, M. A., & Wager, T. D. (2017). Building better biomarkers: Brain models in translational neuroimaging. *Nature Neuroscience, 20*(3), 365–377.

Xu, Y., & Chun, M. M. (2006). Dissociable neural mechanisms supporting visual short-term memory for objects. *Nature, 440*(7080), 91–95.

Yeo, B. T. T., Krienen, F. M., Sepulcre, J., Sabuncu, M. R., Lashkari, D., Hollinshead, M., ... Buckner, R. L. (2011). The organization of the human cerebral cortex estimated by intrinsic functional connectivity. *Journal of Neurophysiology, 106*(3), 1125–1165.

Yoo, K., Rosenberg, M. D., Hsu, W.-T., Zhang, S., Li, C.-S. R., Scheinost, D., Constable, R. T., & Chun, M. M. (2018). Connectome-based predictive modeling of attention: Comparing different functional connectivity features and prediction methods across datasets. *NeuroImage, 167*, 11–22.

Zhang, W., & Luck, S. J. (2008). Discrete fixed-resolution representations in visual working memory. *Nature, 453*, 233–235.

28 The Role of Alpha Oscillations for Attention and Working Memory

OLE JENSEN AND SIMON HANSLMAYR

ABSTRACT Selective attention and working memory are key functions supporting human cognition—namely, the allocation of neurocomputational resources and the active retention of newly arrived information. Research using electroencephalography (EEG) and magnetoencephalography (MEG) has demonstrated that the human alpha rhythm (8–13 Hz) is strongly modulated in such tasks. The modulation is regional-specific and serves to dynamically allocate resources in the network constituting the working brain. We will explain how the functional inhibition by the alpha rhythm serves to support attention and working-memory operations. In particular, functional inhibition serves to suppress the regions not required for the task at hand, thus allocating neurocomputational resources to regions supporting the required computations. While an increase in alpha power reflects functional inhibition, a decrease allows for the representation of information and working-memory maintenance. The modulation of alpha oscillations is under top-down control. We are now beginning to get a good handle on the frontostriatal network involved in this control, as well as the possible pathways by which the control is exercised. In sum, it is now clear that alpha oscillations play a crucial role supporting the network dynamics required for attention and working memory. Future research endeavors would further serve to uncover the neurocomputational role contributed by the phasic modulation of the alpha oscillations.

Alpha Oscillations and the Allocation of Computational Resources: A Physiological Perspective

Cognitive neuroscientists have investigated selective attention and working memory for decades. This is mainly because these functions rely on key mechanisms supporting human cognition—namely, prioritization and the maintenance of recent information. How does the brain network implement the mechanisms supporting such functions? When performing attention and working-memory tasks, some regions are task-relevant, whereas other regions are task-irrelevant. Therefore, mechanisms are required that support the engagement and communication between task-relevant regions. Such mechanisms can be implemented by simply shutting down the task-irrelevant regions (figure 28.1*A*), which then leaves the task-relevant regions to communicate and process.

Furthermore, this shutting down—the functional inhibition—is achieved by brain oscillations in the alpha band (8–13 Hz; Jensen & Mazaheri, 2010; Klimesch, Sauseng, & Hanslmayr, 2007). Brain oscillations, such as the alpha rhythm, are generated by large ensembles of neurons activating in synchrony. This results in a population signal that can be detected at the scalp level using EEG and MEG in individuals performing attention- and working-memory tasks. We will explain how synchronization in the alpha band serves to allocate resources in the working brain by inhibiting specific regions.

There is converging experimental support that alpha oscillations reflect regional-specific functional inhibition. Direct evidence comes from intracranial recordings in monkeys in which single-unit firing is related to alpha oscillations observed in local field potentials. It was demonstrated that an increase in the magnitude of the alpha oscillations is associated with a decrease in firing rate in sensorimotor regions (Haegens, Nacher, Luna, Romo, & Jensen, 2011). Furthermore, neuronal firing is strongly modulated in a phasic manner by alpha oscillations; that is, firing is blocked in every cycle, resulting in neuronal pulsing approximately every 100 ms. A similar relationship has been demonstrated in early visual regions (Buffalo, Fries, Landman, Buschman, & Desimone, 2011; Spaak, Bonnefond, Maier, Leopold, & Jensen, 2012; van Kerkoerle et al., 2014). Also, gamma band activity (40–100 Hz) is modulated by the phase of ongoing alpha oscillations (Khan et al., 2013; Osipova, Hermes, & Jensen, 2008; Park et al., 2011; Spaak, Bonnefond, Maier, Leopold, & Jensen, 2012). The general finding is that as alpha power goes up, both neuronal firing and gamma activity are diminished. Other studies combining functional magnetic resonance imaging (fMRI) and EEG have demonstrated that an increase in alpha power is associated with a decrease in the blood oxygen level-dependent (BOLD) signal, which is thought to index neuronal activity (Goldman, Stern, Engel, & Cohen, 2002; Laufs et al., 2003; Scheeringa et al., 2009). Combined EEG

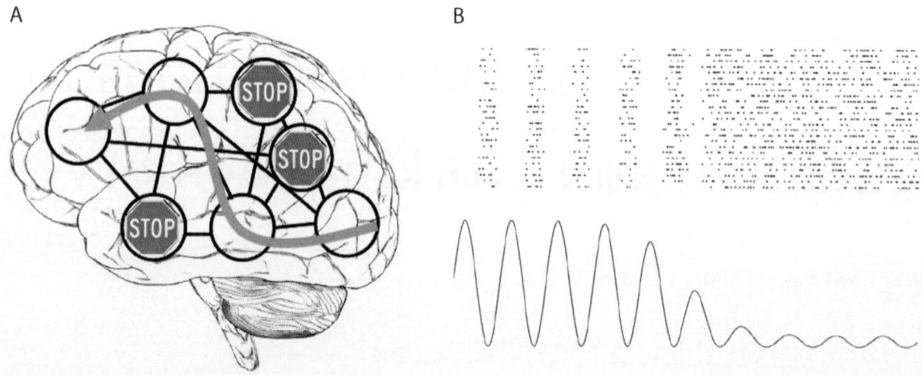

FIGURE 28.1 *A*, Routing by alpha inhibition. It has been proposed that task-relevant regions are left to communicate by selectively inhibiting task-irrelevant regions. This inhibition is reflected by an increase in alpha oscillations in the task-irrelevant regions. This mechanism supports the routing of information at the network level in attention and working-memory tasks. *B*, A schematic illustration of the firing of 25 example neurons explaining how the measured alpha power increases as neuronal firing decreases. *Top*, neurons are firing continuously, resulting in a direct current (DC) signal in the field potential conceptualized as the summed activity. *Bottom*, the firing of the neurons are repeatedly inhibited every 100 ms. This produces a rhythmic signal in the group activity at ~10 Hz while the firing rate decreases. This mechanism is at play when engaging and disengaging regions in attention and working-memory tasks.

and transcranial magnetic stimulation (TMS) demonstrate the inhibition by alpha oscillations: the perception of phosphenes evoked by TMS pulses over visual cortex is reduced during periods of increased alpha power (Romei et al., 2008) but also phasically modulated by the alpha oscillations (Dugue, Marque, & VanRullen, 2011).

In sum, converging evidence demonstrates that the magnitude of alpha oscillations are inversely related to neuronal firing and that neuronal firing is modulated phasically by the alpha oscillations. This does, however, pose an apparent paradox. Why is the strongest signal measured from the brain—the alpha rhythm—associated with reduced neuronal processing? Figure 28.1*B* provides a compelling explanation. It posits that alpha band oscillations emerge from the rhythmic inhibition of ongoing neuronal firing. Without this rhythmic inhibition, no oscillatory signal can be measured from the brain at the scalp level (figure 28.1*B*, *bottom*). Rather, the rhythmic inhibition serves to break the firing of a large cell assembly, thus producing a highly robust oscillatory signal that can be readily detected (figure 28.1*B*, *top*). This simple scheme explains why functional inhibition is associated with an increase in the magnitude of alpha oscillations. As we will outline below, the regional specific inhibition by alpha band oscillations plays a crucial role for the allocation of neurocomputational resources in attention and working-memory tasks.

It deserves mentioning that until the early 2000s the dominant view was that alpha oscillations reflected a state of "idling" or rest rather than regional-specific functional inhibition (Pfurtscheller, Stancak, & Neuper, 1996). The idling notion was based on the observation that alpha oscillations become strong when subjects are at rest but still vigilant (Berger, 1929). The revised view on the inhibitory role of alpha oscillations has resulted in a revived appreciation for the role of alpha oscillations, particularly for attention and working-memory operations.

Selective Attention

Cross-modal allocation of attention One of the first reports on alpha oscillations in relation to attention comes from (Adrian, 1944). He asked participants to attend to either visual or auditory streams of stimuli presented simultaneously while recording the ongoing EEG. When attention was allocated to the auditory modality, he observed a relative increase in posterior alpha power (figure 28.2*A*). These findings have later been replicated in more comprehensive studies using both EEG and MEG (Fu et al., 2001; Mazaheri et al., 2014). The findings can be explained by the functional inhibition of visual regions by alpha oscillations; this inhibition serves to reduce interference from visual stimuli when attending to auditory input. This interpretation is confirmed by results from a combined TMS/EEG study showing that attention to auditory stimuli leads to increased TMS-induced alpha responses in the visual system (Herring, Thut, Jensen, & Bergmann, 2015). Intriguingly, these early findings of Adrian are inconsistent with the idling notion of alpha oscillations, as the allocation of auditory attention requires considerable effort.

Selective spatial attention A large number of studies have investigated brain oscillations in relation to visual

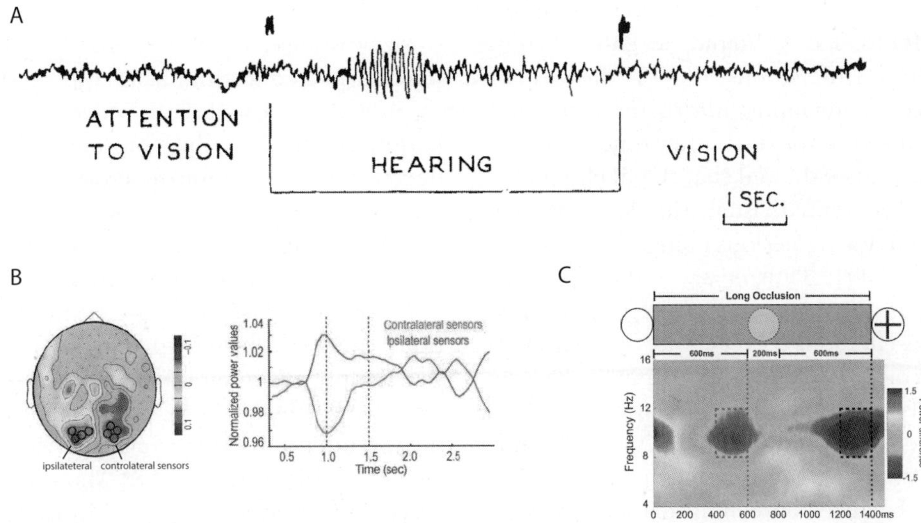

FIGURE 28.2 Alpha and the allocation of selective attention. *A*, In an EEG study, a subject was asked to attend to either visual or auditory input. As attention was allocated to the auditory input stream, the alpha power increased. This observation is consistent with the functional inhibition of visual areas when auditory information is attended to. Reproduced from Adrian (1944). *B*, In a spatial attention task, subjects were asked to attend to items continuously presented in the left or the right visual field. This resulted in an alpha power decrease contralateral to the attended direction, which reflects the engagement of this hemisphere. Importantly, the ipsilateral alpha power prevents the processing of unattended stimuli. Reproduced from Händel, Haarmeier, and Jensen (2011). *C*, In a temporal attention task, an occluded visual item could reappear either at time 800 ms or 1,400 ms. Just prior to these anticipated time points, the alpha power decreased. Reproduced from Rohenkohl and Nobre (2011). (See color plate 30.)

attention. Many of these studies have focused on the allocation of attention to stimuli anticipated to appear in either the left or the right visual hemifield. A common finding is that when spatial attention is allocated to, for example, the left, alpha oscillations in the contralateral right hemisphere decrease (and vice versa). Importantly, the alpha oscillations in the hemisphere ipsilateral to the attended location—for example, the left hemisphere (and vice versa)—remain relatively strong (Worden, Foxe, Wang, & Simpson, 2000). As such, posterior alpha oscillations are hemispherically lateralized with respect to the allocation of attention (figure 28.2*B*). These findings are consistent with the notion that a decrease in alpha power reflects the engagement of the visual hemisphere processing the attended incoming information. The stronger ipsilateral alpha power reflects a relative disengagement of the visual areas processing unattended—that is, irrelevant—information. The hemispheric lateralization of alpha band activity correlates with behavior both in terms of reaction times and accuracy (Noonan et al., 2016; Okazaki, De Weerd, Haegens, & Jensen, 2014; Popov, Kastner, & Jensen, 2017; Thut, Nietzel, Brandt, & Pascual-Leone, 2006). Importantly, it has been demonstrated that the ipsilateral alpha band power reflects the inhibition of unattended items (Handel, Haarmeier, & Jensen, 2011), although the generality of this finding has been questioned (Noonan, Crittenden, Jensen, & Stokes, 2017). More recently, the role of alpha oscillations have been investigated with a better spatial resolution, using intracranial recordings in monkeys to replicate the human findings. The allocation of spatial attention was associated with a decrease in the alpha power recorded directly in early visual cortex. Removing attention, on the other hand, resulted in an increase of alpha oscillations and a decrease in neuronal firing (Buffalo et al., 2011). As such, the modulation of alpha activity with attention is not specific to humans, and it is a local phenomenon that can be observed using intracranial recordings.

The modulation of oscillatory brain activity with respect to spatial visual attention generalizes to the somatosensory system. The same hemispheric lateralization of alpha band oscillations is observed when attention is allocated to either the left- or right-hand receiving somatosensory input (Haegens, Osipova, Oostenveld, & Jensen, 2010; van Ede, Szebenyi, & Maris, 2014). Importantly, also in the somatosensory system, the ipsilateral alpha oscillations are associated with the inhibition of distracting sensory input (Haegens, Luther, & Jensen, 2012).

Temporal fluctuations of attention Alpha oscillations are all but stationary. Instead, they fluctuate considerably over time and fluctuations can be observed in attention tasks as well (Monto, Palva, Voipio, & Palva, 2008). These spontaneous fluctuations render the system highly susceptible to incoming information at some points in time and less sensitive at other times. A series of EEG and MEG studies showed that the likelihood of detecting a briefly presented visual stimulus decreases when it is presented during periods of high alpha power (Ergenoglu et al., 2004; Thut et al., 2006; van Dijk, Schoffelen, Oostenveld, & Jensen, 2008). Importantly, decreased alpha oscillations correlate not only with better hit rates (i.e., correctly perceiving a stimulus that was presented) but also with higher false alarm rates (i.e., erroneously perceiving a stimulus that actually was not presented). This finding is consistent with alpha oscillations, reflecting a balance between inhibitory and excitable neuronal activity in visual cortex (Iemi, Chaumon, Crouzet, & Busch, 2017). Thus, spontaneous fluctuations in alpha power that have an impact on neuronal excitation can produce false percepts.

Consistent with the pulsed notion of brain oscillations, it has been demonstrated that an instantaneous oscillatory phase predicts visual perception. Specifically, the phase of 5–12 Hz oscillations at which a stimulus arrives is predictive of perception (Busch & VanRullen, 2010; Hanslmayr, Volberg, Wimber, Dalal, & Greenlee, 2013; Mathewson, Gratton, Fabiani, Beck, & Ro, 2009). Alpha oscillations also fluctuate over time in terms of interregional phase consistency (Hanslmayr et al., 2007). Interregional phase consistency is a measure of oscillatory synchronization thought to reflect interregional communication (Varela, Lachaux, Rodriguez, & Martinerie, 2001). Similarly to the findings obtained on alpha power, increased alpha synchrony between regions is negatively related to the likelihood of correctly identifying a briefly presented visual stimulus (Hanslmayr et al., 2007). A combined EEG-fMRI study demonstrated that the alpha phase as measured in the EEG modulates visual perception by supporting the communication flow between lower and higher visual areas, as measured by connectivity measures of the BOLD signal (Hanslmayr et al., 2013). In conclusion, visual perception is modulated by both the alpha phase and power. The influence of the alpha phase has led to the hypothesis of *perceptual snapshots*. In an analogy to the shutter mechanism of a video camera, visual perception is not continuous but is formed by snapshots at a rate of approximately 10 Hz (VanRullen, 2016).

While there is a random element to the temporal fluctuations of alpha oscillations, they are also under top-down control. As such, if alpha oscillations are a mechanism for routing information, they should be modulated when visual attention is allocated at a particular moment. This prediction was confirmed in a study by Rohenkohl and Nobre (2011), in which participants expected a stimulus to appear at a certain point in time. Alpha oscillations indeed decreased in anticipation of the stimulus. Moreover, this alpha power decrease was rhythmically modulated by the slow frequency (~1 Hz) in which the stimuli were presented. Therefore, alpha power decreases closely followed the time course of the stimulus stream presentation. It has also been demonstrated that the phase of alpha oscillations can be adjusted in anticipation of a predicted visual input (Bonnefond & Jensen, 2012; Samaha, Gosseries, & Postle, 2017). While the generality of this finding was not replicated in an audiovisual EEG study (van Diepen, Cohen, Denys, & Mazaheri, 2015), it was recently reproduced (Solis-Vivanco, Jensen, & Bonnefond, 2018). In sum, these findings demonstrate that alpha oscillations can be top-down controlled with respect to temporal attention both in terms of power and phase. This adds to the computational versatility of the alpha rhythms in terms of resource allocation.

Alpha oscillations have also been linked to another hallmark of temporal attention: the *attentional blink* (Raymond, Shapiro, & Arnell, 1992). The attentional blink is typically observed when target stimuli need to be detected within a stream of distracter stimuli, which are presented sequentially at a rate of 7–13 Hz. Subjects typically have no problem identifying a target; however, the likelihood of correctly identifying a second target drops dramatically when it is presented approximately 300 ms after the first target. This attentional blink, elicited by the processing of the first target, lasts for about 500 ms. Interestingly, the frequency at which visual stimuli are presented in order to create the strongest attentional blink effect matches the frequency of human alpha oscillations (~10 Hz). Accordingly, a neurophysiological explanation based on alpha oscillations was proposed (Mazaheri et al., 2014; Shapiro & Hanslmayr, 2014) in which the combination of externally driving the visual system at alpha frequency and processing a target in working memory leads to high alpha power and high corticocortical alpha connectivity (Hanslmayr, Gross, Klimesch, & Shapiro, 2011). This increase in alpha activity protects the system from interference, and while it promotes the processing of the first target, it prevents perception of the second target. Thus, alpha oscillations might be intimately connected to the mechanism generating the attentional blink (Kranczioch, Debener, Maye, & Engel, 2007; Zauner et al., 2012). If true, then the attentional blink should only be observed in the frequency range of human alpha oscillations. This critical prediction was confirmed in a behavioral study (Shapiro, Hanslmayr, Enns, & Lleras, 2017).

To summarize, alpha oscillations serve to route information processing not only in space but also in time. Alpha oscillations are subject to top-down modulation in order to facilitate information processing at particular time points (Rohenkohl & Nobre, 2011) or to promote the internal processing of information (Shapiro & Hanslmayr, 2014).

The Network Control of Alpha Oscillations in Attention Tasks

When a cue directs attention to upcoming targets in the left or right visual hemifield, posterior alpha oscillations are strongly modulated even when the screen is blank. These posterior oscillations are therefore under top-down control. Multiple studies have made progress on identifying the network involved in this top-down control. Not surprisingly, several studies suggest a role of the dorsal attention network. In particular, converging work points to the involvement of the frontal eye field (FEF). Temporally lesioning the FEF with repetitive transcranial magnetic stimulation (rTMS) results in a reduced ability to modulate posterior alpha oscillations in spatial attention tasks (Marshall, O'Shea, Jensen, & Bergmann, 2015). Using TMS, the intraparietal sulcus (IPS) has been implicated in the control of alpha oscillations (Capotosto, Corbetta, Romani, & Babiloni, 2012). Combining EEG with fMRI has demonstrated that the magnitude of visual alpha oscillations are negatively correlated with the BOLD signal in the dorsal attention network, including the intraparietal sulcus, and the right FEF (Zumer, Scheeringa, Schoffelen, Norris, & Jensen, 2014). These findings are consistent with the notion that the dorsal attention network suppresses alpha band activity in visual cortex in a regional-specific manner. A recent MEG study aimed to identify the dynamics associated with top-down control using Granger causality (Popov, Kastner, & Jensen, 2017). By asking which regions were driving the visual oscillations, it was found that, in particular, the FEF exercised top-down control. This suggests that the FEF is controlling the alpha oscillations in terms of phase and magnitude. This then begs the question: How is this control mediated? The superior longitudinal fascicular (SLF) are white matter fibers connecting the frontal and posterior brain regions. The so-called SLF-I denotes the dorsal fibers connecting regions overlapping with the dorsal attention network. A recent study combined MEG and MR diffusion tensor imaging to identify the white matter fibers of the SLF-I (Marshall, Bergmann, & Jensen, 2015). MEG was used to quantify the ability of individuals to modulate alpha oscillations in the right versus the left hemisphere in a spatial attention task. The key finding was that individuals with larger right compared to left hemisphere fiber bundles in the SLF-I were better at modulating their right hemisphere, compared to left hemisphere, alpha oscillations (and vice versa). This suggests that the top-down control of the alpha activity is—at least partly—mediated by the white matter tracts in SLF-I. The detailed neuronal mechanism implementing the top-down control remains largely unknown. Undoubtedly, it involves a complex interplay between neuronal dynamics and neurotransmitters as well as neuromodulators. A recent study found that the cholinergic agonist physostigmine enhanced posterior alpha and beta oscillations in a spatial attention task (Bauer et al., 2012). More work is required to identify the neuromodulators involved in top-down control.

While neocortical regions are clearly involved in the top-down control supporting the allocation of attention, subcortical regions are likely to play a role as well. For instance, an fMRI study suggests that the striatum and associated regions are involved in cognitive control, modulating the engagement of extrastriate visual areas (van Schouwenburg, O'Shea, Mars, Rushworth, & Cools, 2012). Direct recordings from the nucleus accumbens (NAc) combined with scalp EEG recordings have demonstrated an oscillatory coupling in both the theta and alpha band between the NAc and prefrontal cortical areas (Horschig et al., 2015). A recent study combined structural MRI measures and MEG from subjects performing spatial attention tasks. It was found that subjects with a larger right hemisphere globus pallidus, compared to the left, were better at modulating their right hemisphere alpha oscillation, compared to the left (and vice versa). This was particularly the case in tasks in which the visual input was associated with rewards or losses (Mazzetti et al., submitted). The extended striatal regions modulate the posterior sensory regions via the anterior thalamus, which connects not only to neocortical areas but also to posterior thalamic regions like the pulvinar. Indeed, intracranial recordings in the thalami of monkeys and dogs have demonstrated the role of the pulvinar (Lopes da Silva, Vos, Mooibroek, & Van Rotterdam, 1980). In particular, the pulvinar exercises a phasic drive that synchronizes regions in the ventral visual stream in a spatial attention task (Saalmann, Pinsk, Wang, Li, & Kastner, 2012).

In sum, the dorsal attention network seems to be causally involved in the top-down control of visual alpha oscillations in spatial attention tasks. This control is likely to be mediated via neocortical pathways, implicating the SLF as well as subcortical regions, including the thalamus and the extended striatal network. In future work it would be of great interest to further uncover the mechanisms implementing the top-down control of posterior alpha oscillations in attention tasks.

Brain Oscillations during Working-Memory Maintenance

Beyond attention, neuronal oscillations have been implicated in working-memory maintenance. These oscillations reflect both resource allocation implemented by local inhibition and the dynamics serving to sustain the memory trace. In terms of resource allocation, several EEG studies on working memory have contributed to refuting the idling or resting state notion of alpha oscillations in favor of a much more active role. In particular, it was found that alpha oscillations increase in power when several items are held in memory, compared to zero items (Klimesch, Doppelmayr, Schwaiger, Auinger, & Winkler, 1999). This finding was complemented by a study applying the Sternberg task (Sternberg, 1966). In this task, up to six letters are presented sequentially. After a retention interval of a few seconds, a letter is presented probing the content of the working memory. Several EEG and MEG studies have quantified the alpha power during the retention interval. The key finding is that alpha power parametrically increases with the number of items to be held in working memory (Jensen, Gelfand, Kounios, & Lisman, 2002). In a replication using MEG and the maintenance of faces, the increase in alpha power was identified to early visual areas (Tuladhar et al., 2007). A study combining EEG and fMRI demonstrated that the alpha power increase with memory load was produced in early visual regions (Scheeringa et al., 2009). The increase in alpha power with memory load suggests that alpha oscillations reflect the functional inhibition of early visual regions not required for the task (Klimesch, Sauseng, & Hanslmayr, 2007; Mazaheri & Jensen, 2010). The alpha power increase was hypothesized to reflect the suppression of potentially distracting visual information. An MEG study directly tested this hypothesis by presenting distracting items during the maintenance interval. Distracter items were presented at a fixed time so that participants could anticipate their appearance. The key finding was that the alpha power increased just prior to distracter onset. Intriguingly, this increase in alpha power was predictive of performance (Bonnefond & Jensen, 2012) such that a strong increase in alpha power resulted in faster reaction times when identifying the probe (see Payne, Guillory, & Sekuler, 2013 for related findings).

The allocation of resources by alpha oscillations has been investigated in working-memory studies in which items were presented in the left or the right hemifield. In these studies visuospatial configurations were maintained, relying on the engagement of visual areas. Importantly, alpha power in the ipsilateral hemisphere remained strong, whereas alpha power decreased in the hemisphere contralateral to the to-be-remembered items (Leenders, Lozano-Soldevilla, Roberts, Jensen, & De Weerd, 2018; Sauseng et al., 2009). Note that this is in contrast to the results for the above-mentioned Sternberg task in which items such as letters and faces were maintained without necessarily relying on visuospatial representations. Taken together, the above reviewed findings suggest that alpha oscillations ensure that resources are allocated to task-relevant regions via inhibiting regions that are either irrelevant or even interfering. Therefore, alpha oscillations ensure that initially fragile working-memory traces can be maintained and processed by reducing the "noise" from other systems; alpha oscillations thus would achieve a function similar to tuning out the sound of a radio when reading a complex book chapter on oscillations and cognitive neuroscience.

However, another function of alpha oscillations is associated with actively maintaining the memory trace. As mentioned above, *decreases* in alpha power contralateral to the to-be-maintained information indicate the active engagement of areas that internally maintain the representations. This begs the question: Which mechanism do alpha power decreases use to allow for information representation? We discussed earlier in this chapter that alpha power decreases are associated with increased firing rates (Haegens et al., 2011). Therefore, a sustained decrease of alpha power, arguably, allows individual neurons to fire in a sustained manner, which is a classic mechanism supporting the online maintenance of information in working memory (Funahashi, Bruce, & Goldman-Rakic, 1989). One simple perspective is that alpha power decreases during working-memory maintenance reflect the increased firing rates of neurons that hold on to internally represented information. However, there might be more to it. Specifically, there might be a computational utility in the desynchronization itself, other than just allowing for more spiking to occur. For instance, a decrease in alpha power lets neurons spike less regularly—that is, in a less stereotypic manner. From a purely mathematical point of view, less regularity means less predictability, which means less redundancy and hence more information. In other words, the less predictably spikes occur, the more information is carried in these events (Shannon & Weaver, 1949). Desynchronized firing on a population level is therefore necessary to allow neurons to code

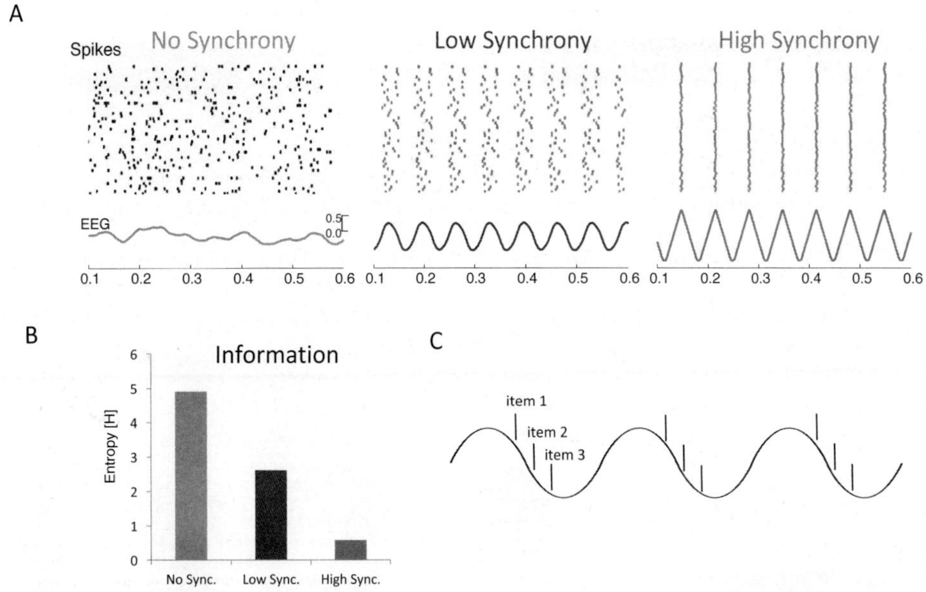

FIGURE 28.3 The computational utility of alpha power decreases. *A*, Raster plots for a simulated population of neurons (*rows*) are shown in different synchronization regimes, ranging from no synchrony (*left*) to very high synchrony (*right*). *B*, Information, as measured with entropy decreases with increasing synchronization. *C*, *Phase coding* refers to the notion that different representations (items) activate at different phases of the oscillatory cycle. The scheme is consistent with the low-synchrony scenario in panel *A*.

complex messages, via a synergistic code (Schneidman et al., 2011), as described in the information via desynchronization hypothesis (figure 28.3*A*; Hanslmayr, Staudigl, & Fellner, 2012). Notably, this information via desynchronization hypothesis is compatible with the notion of alpha phase coding (Jensen et al., 2014), which suggests that the computational utility of decreases in alpha power is that they prolong the *duty cycle*—that is, the opportunity for neurons to fire (figure 28.3*B*). Therefore, alpha power decreases lead to (1) increased firing and (2) more flexible firing. If alpha power decreases allow for the coding of information, we should be able to decode information from desynchronized EEG/MEG traces. Indeed, recent MEG and EEG studies confirmed this assumption by showing that the identity of maintained stimuli can be decoded from desynchronized alpha oscillations, as shown in figure 28.4 (Michelmann, Bowman, & Hanslmayr, 2016, 2018).

These findings are consistent with the notion that the hemisphere contralateral to the items to be remembered carries the memory trace while the ipsilateral hemisphere is being disengaged. This principle also generalizes to the dorsal and ventral stream. It is well established that the ventral stream is dedicated to object-specific processing, such as faces, whereas the dorsal stream is involved in spatial operations. An MEG study revealed that when face identity, as compared to face orientation, was maintained in working memory, alpha power increased in the dorsal stream (Jokisch & Jensen, 2007). When the face orientation was maintained, alpha power increased in the ventral stream, as revealed by an electrocorticography (ECoG) study (Leszczynski, Fell, Jensen, & Axmacher, 2017). More recently, working memory maintenance in relation to brain oscillations has been investigated using retro-cuing paradigms. In these paradigms, items are presented simultaneously in the left and right visual field. The items then have to be maintained until a probe appears. The probe directs participants to focus on items previously presented in either the left or the right hemisphere. The probe presentation resulted in a robust hemispheric lateralization of alpha power that was predictive of performance in terms of precision (Myers, Walther, Wallis, Stokes, & Nobre, 2015). These findings are consistent with an updating of working memory in response to the cue in which the alpha increase serves to suppress working memory representations not required for the task.

Phase Coding

While we have so far focused on the magnitude of brain oscillations, some theories elaborate on the role of phase (see figure 28.3*C*). Recordings from the hippocampus of behaving rats have demonstrated that different spatial information is represented at different

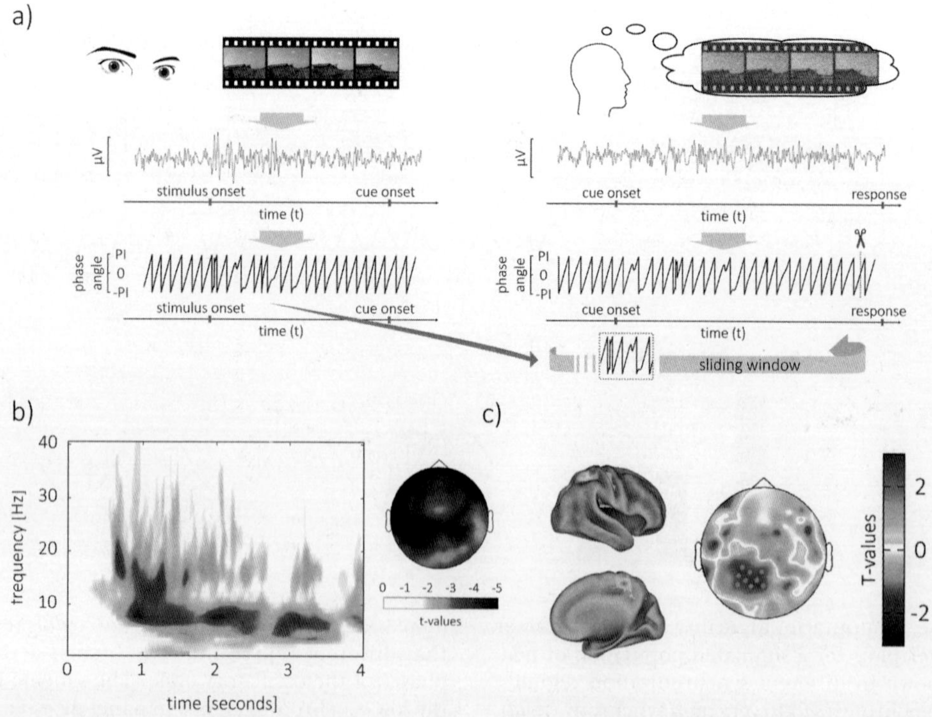

FIGURE 28.4 Alpha power decreases during memory maintenance code stimulus-specific information. *A*, Subjects first encoded a video (*left*) and then maintained a vivid imagination of that video at a later time point (*right*). The phase time course was extracted from the EEG during encoding and retrieval in order to calculate a similarity measure between encoding and retrieval. *B*, During maintenance, strong and sustained alpha power decreases were observed. *C*, Reactivation of stimulus-specific information, as measured with phase similarity, could be detected in the alpha frequency band with a maximum in parietal regions. Reproduced from Michelmann, Bowman, and Hanslmayr (2018). (See color plate 31.)

phases of the theta cycle (O'Keefe & Recce, 1993). This finding inspired a proposal put forward by Lisman and Idiart (1995). They suggested a computational model based on coupled theta and gamma oscillations (Lisman & Idiart, 1995). The basic idea is that a set of working memory items is sequentially activated—one item per gamma cycle—within a theta cycle. Depending on the frequency of the gamma activity, about five to seven items can be activated within one theta cycle (for an elaborate review, see Lisman & Jensen, 2013). Theories have also been put forward regarding the role of alpha phase. It has been suggested that competing visual items are activated sequentially along an alpha cycle as a pulse of inhibition ramps down (Jensen, Bonnefond, Marshall, & Tiesinga, 2015). Recent empirical evidence was established for the Lisman and Idiart model of working memory maintenance (Bahramisharif, Jensen, Jacobs, & Lisman, 2018). This study using ECoG recordings showed that different memory items (*consonants*) were associated with high-frequency gamma power at different electrodes. This demonstrated that different memory items activated sequentially within the oscillations in the alpha band. In sum, the notion of phase coding is gaining ground, but more empirical work is required to uncover the generality of the principle.

Conclusion

Numerous human studies using EEG, MEG, and intracranial recordings have demonstrated that brain oscillations are strongly modulated in attention and working-memory tasks. This modulation is particularly strong in the alpha band. The findings suggest that the oscillations are involved in the temporal coordination of the neuronal activity supporting core functions such as routing and the temporary maintenance of information. A good understanding has emerged for what these oscillations are doing in terms of power. In particular, it is clear that alpha oscillations serve to allocate neurocomputational resources by inhibiting regions not required for a given task. The field is now headed toward understanding the phasic role of these ongoing oscillations.

Acknowledgments

This work was supported by the James S. McDonnell Foundation Understanding Human Cognition Collaborative

Award (grant number 220020448) to O.J.; Wellcome Trust Investigator Award in Science (grant number 207550) to O.J.; Royal Society Wolfson Research Merit Award to O.J. and S.H.; European Research Council (ERC) Code-4Memory Consolidator Grant (grant number 647954) to S.H.; Economic and Social Sciences Research Council (ESRC) TIME (grant number ES/R010072/1) to S.H.

REFERENCES

Adrian, E. D. (1944). Brain rhythms. *Nature, 153,* 360–362.

Bahramisharif, A., Jensen, O., Jacobs, J., & Lisman, J. (2018). Serial representation of items during working memory maintenance at letter-selective cortical sites. *PLoS Biology, 16*(8), e2003805. doi:10.1371/journal.pbio.2003805

Bauer, M., Kluge, C., Bach, D., Bradbury, D., Heinze, H. J., Dolan, R. J., & Driver, J. (2012). Cholinergic enhancement of visual attention and neural oscillations in the human brain. *Current Biology, 22*(5), 397–402. doi:10.1016/j.cub.2012.01.022

Berger, H. (1929) Über das Elektrenkephalogramm des Menschen. *Archiv für Psychiatrie und Nervenkrankheiten,* 87, 527–570.

Bonnefond, M., & Jensen, O. (2012). Alpha oscillations serve to protect working memory maintenance against anticipated distracters. *Current Biology, 22*(20), 1969–1974. doi:10.1016/j.cub.2012.08.029

Buffalo, E. A., Fries, P., Landman, R., Buschman, T. J., & Desimone, R. (2011). Laminar differences in gamma and alpha coherence in the ventral stream. *Proceedings of the National Academy of Sciences of the United States of America, 108*(27), 11262–11267. doi:10.1073/pnas.1011284108

Busch, N. A., & VanRullen, R. (2010). Spontaneous EEG oscillations reveal periodic sampling of visual attention. *Proceedings of the National Academy of Sciences of the United States of America, 107*(37), 16048–16053. doi:10.1073/pnas.1004801107

Capotosto, P., Corbetta, M., Romani, G. L., & Babiloni, C. (2012). Electrophysiological correlates of stimulus-driven reorienting deficits after interference with right parietal cortex during a spatial attention task: A TMS-EEG study. *Journal of Cognitive Neuroscience, 24*(12), 2363–2371. doi:10.1162/jocn_a_00287

Dugue, L., Marque, P., & VanRullen, R. (2011). The phase of ongoing oscillations mediates the causal relation between brain excitation and visual perception. *Journal of Neuroscience, 31*(33), 11889–11893. doi:10.1523/JNEUROSCI.1161-11.2011

Ergenoglu, T., Demiralp, T., Bayraktaroglu, Z., Ergen, M., Beydagi, H., & Uresin, Y. (2004). Alpha rhythm of the EEG modulates visual detection performance in humans. *Brain Research. Cognitive Brain Research, 20*(3), 376–383. doi:10.1016/j.cogbrainres.2004.03.009

Fu, K. M., Foxe, J. J., Murray, M. M., Higgins, B. A., Javitt, D. C., & Schroeder, C. E. (2001). Attention-dependent suppression of distracter visual input can be cross-modally cued as indexed by anticipatory parieto-occipital alpha-band oscillations. *Brain Research. Cognitive Brain Research, 12*(1), 145–152.

Funahashi, S., Bruce, C. J., & Goldman-Rakic, P. S. (1989). Mnemonic coding of visual space in the monkey's dorsolateral prefrontal cortex. *Journal of Neurophysiology, 61*(2), 331–349. doi:10.1152/jn.1989.61.2.331

Goldman, R. I., Stern, J. M., Engel Jr., J., & Cohen, M. S. (2002). Simultaneous EEG and fMRI of the alpha rhythm. *Neuroreport, 13*(18), 2487–2492. doi:10.1097/01.wnr.0000047685.08940.d0

Haegens, S., Luther, L., & Jensen, O. (2012). Somatosensory anticipatory alpha activity increases to suppress distracting input. *Journal of Cognitive Neuroscience, 24*(3), 677–685. doi:10.1162/jocn_a_00164

Haegens, S., Nacher, V., Luna, R., Romo, R., & Jensen, O. (2011). Alpha-oscillations in the monkey sensorimotor network influence discrimination performance by rhythmical inhibition of neuronal spiking. *Proceedings of the National Academy of Sciences of the United States of America, 108*(48), 19377–19382. doi:10.1073/pnas.1117190108

Haegens, S., Osipova, D., Oostenveld, R., & Jensen, O. (2010). Somatosensory working memory performance in humans depends on both engagement and disengagement of regions in a distributed network. *Human Brain Mapping, 31*(1), 26–35. doi:10.1002/hbm.20842

Händel, B. F., Haarmeier, T., & Jensen, O. (2011). Alpha oscillations correlate with the successful inhibition of unattended stimuli. *Journal of Cognitive Neuroscience, 23*(9), 2494–2502. doi:10.1162/jocn.2010.21557

Hanslmayr, S., Aslan, A., Staudigl, T., Klimesch, W., Herrmann, C. S., & Bauml, K. H. (2007). Prestimulus oscillations predict visual perception performance between and within subjects. *NeuroImage, 37*(4), 1465–1473. doi:10.1016/j.neuroimage.2007.07.011

Hanslmayr, S., Gross, J., Klimesch, W., & Shapiro, K. L. (2011). The role of alpha oscillations in temporal attention. *Brain Research Reviews, 67*(1–2), 331–343. doi:10.1016/j.brainresrev.2011.04.002

Hanslmayr, S., Staudigl, T., & Fellner, M. C. (2012). Oscillatory power decreases and long-term memory: The information via desynchronization hypothesis. *Frontiers in Human Neuroscience, 6,* 74. doi:10.3389/fnhum.2012.00074

Hanslmayr, S., Volberg, G., Wimber, M., Dalal, S. S., & Greenlee, M. W. (2013). Prestimulus oscillatory phase at 7 Hz gates cortical information flow and visual perception. *Current Biology, 23*(22), 2273–2278. doi:10.1016/j.cub.2013.09.020

Herring, J. D., Thut, G., Jensen, O., & Bergmann, T. O. (2015). Attention modulates TMS-locked alpha oscillations in the visual cortex. *Journal of Neuroscience, 35*(43), 14435–14447. doi:10.1523/JNEUROSCI.1833-15.2015

Horschig, J. M., Smolders, R., Bonnefond, M., Schoffelen, J. M., van den Munckhof, P., Schuurman, P. R., ... Jensen, O. (2015). Directed communication between nucleus accumbens and neocortex in humans is differentially supported by synchronization in the theta and alpha band. *PLoS One, 10*(9), e0138685. doi:10.1371/journal.pone.0138685

Iemi, L., Chaumon, M., Crouzet, S. M., & Busch, N. A. (2017). Spontaneous neural oscillations bias perception by modulating baseline excitability. *Journal of Neuroscience, 37*(4), 807–819. doi:10.1523/JNEUROSCI.1432-16.2016

Jensen, O., Bonnefond, M., Marshall, T. R., & Tiesinga, P. (2015). Oscillatory mechanisms of feedforward and feedback visual processing. *Trends in Neurosciences, 38*(4), 192–194. doi:10.1016/j.tins.2015.02.006

Jensen, O., Gips, B., Bergmann, T. O., & Bonnefond, M. (2014). Temporal coding organized by coupled alpha and gamma oscillations prioritize visual processing. *Trends in Neurosciences, 37*(7), 357–369. doi:10.1016/j.tins.2014.04.001

Jensen, O., Gelfand, J., Kounios, J., & Lisman, J. E. (2002). Oscillations in the alpha band (9–12 Hz) increase with memory load during retention in a short-term memory task. *Cerebral Cortex, 12*(8), 877–882.

Jensen, O., & Mazaheri, A. (2010). Shaping functional architecture by oscillatory alpha activity: Gating by inhibition. *Frontiers in Human Neuroscience, 4,* 186. doi:10.3389/fnhum.2010.00186

Jokisch, D., & Jensen, O. (2007). Modulation of gamma and alpha activity during a working memory task engaging the dorsal or ventral stream. *Journal of Neuroscience, 27*(12), 3244–3251. doi:10.1523/JNEUROSCI.5399-06.2007

Khan, S., Gramfort, A., Shetty, N. R., Kitzbichler, M. G., Ganesan, S., Moran, J. M., ... Kenet, T. (2013). Local and long-range functional connectivity is reduced in concert in autism spectrum disorders. *Proceedings of the National Academy of Sciences of the United States of America, 110*(8), 3107–3112. doi:10.1073/pnas.1214533110

Klimesch, W., Doppelmayr, M., Schwaiger, J., Auinger, P., & Winkler, T. (1999). "Paradoxical" alpha synchronization in a memory task. *Brain Research. Cognitive Brain Research, 7*(4), 493–501.

Klimesch, W., Sauseng, P., & Hanslmayr, S. (2007). EEG alpha oscillations: The inhibition-timing hypothesis. *Brain Research Reviews, 53*(1), 63–88. doi:10.1016/j.brainresrev.2006.06.003

Kranczioch, C., Debener, S., Maye, A., & Engel, A. K. (2007). Temporal dynamics of access to consciousness in the attentional blink. *NeuroImage, 37*(3), 947–955. doi:10.1016/j.neuroimage.2007.05.044

Laufs, H., Kleinschmidt, A., Beyerle, A., Eger, E., Salek-Haddadi, A., Preibisch, C., & Krakow, K. (2003). EEG-correlated fMRI of human alpha activity. *NeuroImage, 19*(4), 1463–1476.

Leenders, M. P., Lozano-Soldevilla, D., Roberts, M. J., Jensen, O., & De Weerd, P. (2018). Diminished alpha lateralization during working memory but not during attentional cueing in older adults. *Cerebral Cortex, 28*(1), 21–32. doi:10.1093/cercor/bhw345

Leszczynski, M., Fell, J., Jensen, O., & Axmacher, N. (2017). Alpha activity in the ventral and dorsal visual stream controls information flow during working memory. *BioRxiv,* doi:10.1101/180166

Lisman, J. E., Idiart, M. A. (1995). Storage of 7 +/− 2 short-term memories in oscillatory subcycles. *Science,* 267(5203), 1512–1515.

Lisman, J. E., & Jensen, O. (2013). The theta-gamma neural code. *Neuron, 77*(6), 1002–1016. doi:10.1016/j.neuron.2013.03.007

Lopes da Silva, F. H., Vos, J. E., Mooibroek, J., & Van Rotterdam, A. (1980). Relative contributions of intracortical and thalamo-cortical processes in the generation of alpha rhythms, revealed by partial coherence analysis. *Electroencephalography and Clinical Neurophysiology, 50*(5–6), 449–456.

Marshall, T. R., Bergmann, T. O., & Jensen, O. (2015). Frontoparietal structural connectivity mediates the top-down control of neuronal synchronization associated with selective attention. *PLoS Biology, 13*(10), e1002272. doi:10.1371/journal.pbio.1002272

Marshall, T. R., O'Shea, J., Jensen, O., & Bergmann, T. O. (2015). Frontal eye fields control attentional modulation of alpha and gamma oscillations in contralateral occipitoparietal cortex. *Journal of Neuroscience, 35*(4), 1638–1647. doi:10.1523/JNEUROSCI.3116-14.2015

Mathewson, K. E., Gratton, G., Fabiani, M., Beck, D. M., & Ro, T. (2009). To see or not to see: Prestimulus alpha phase predicts visual awareness. *Journal of Neuroscience, 29*(9), 2725–2732. doi:10.1523/JNEUROSCI.3963-08.2009

Mazaheri, A., & Jensen, O. (2010). Rhythmic pulsing: Linking ongoing brain activity with evoked responses. *Frontiers in Human Neuroscience, 4,* 177. doi:10.3389/fnhum.2010.00177

Mazaheri, A., van Schouwenburg, M. R., Dimitrijevic, A., Denys, D., Cools, R., & Jensen, O. (2014). Region-specific modulations in oscillatory alpha activity serve to facilitate processing in the visual and auditory modalities. *NeuroImage, 87,* 356–362. doi:10.1016/j.neuroimage.2013.10.052

Mazzetti, C., Staudigl, T., Marshall, T. R., Zumer, J. M., Fallon, S. J., & Jensen, O. (submitted). Hemispheric asymmetry of globus pallidus predicts reward-related posterior alpha modulation.

Michelmann, S., Bowman, H., & Hanslmayr, S. (2016). The temporal signature of memories: Identification of a general mechanism for dynamic memory replay in humans. *PLoS Biology, 14*(8), e1002528. doi:10.1371/journal.pbio.1002528

Michelmann, S., Bowman, H., & Hanslmayr, S. (2018). Replay of stimulus-specific temporal patterns during associative memory formation. *Journal of Cognitive Neuroscience, 30*(11), 1577–1589. doi:10.1162/jocn_a_01304

Monto, S., Palva, S., Voipio, J., & Palva, J. M. (2008). Very slow EEG fluctuations predict the dynamics of stimulus detection and oscillation amplitudes in humans. *Journal of Neuroscience, 28*(33), 8268–8272. doi:10.1523/JNEUROSCI.1910-08.2008

Myers, N. E., Walther, L., Wallis, G., Stokes, M. G., & Nobre, A. C. (2015). Temporal dynamics of attention during encoding versus maintenance of working memory: Complementary views from event-related potentials and alpha-band oscillations. *Journal of Cognitive Neuroscience, 27*(3), 492–508. doi:10.1162/jocn_a_00727

Noonan, M. P., Adamian, N., Pike, A., Printzlau, F., Crittenden, B. M., & Stokes, M. G. (2016). Distinct mechanisms for distractor suppression and target facilitation. *Journal of Neuroscience, 36*(6), 1797–1807. doi:10.1523/JNEUROSCI.2133-15.2016

Noonan, M. P., Crittenden, B. M., Jensen, O., & Stokes, M. G. (2017). Selective inhibition of distracting input. *Behavioural Brain Research.* doi:10.1016/j.bbr.2017.10.010

Okazaki, Y. O., De Weerd, P., Haegens, S., & Jensen, O. (2014). Hemispheric lateralization of posterior alpha reduces distracter interference during face matching. *Brain Research, 1590,* 56–64. doi:10.1016/j.brainres.2014.09.058

O'Keefe, J., & Recce, M. L. (1993). Phase relationship between hippocampal place units and the EEG theta rhythm. *Hippocampus, 3*(3), 317–330. doi:10.1002/hipo.450030307

Osipova, D., Hermes, D., & Jensen, O. (2008). Gamma power is phase-locked to posterior alpha activity. *PLoS One, 3*(12), e3990. doi:10.1371/journal.pone.0003990

Park, H., Kang, E., Kang, H., Kim, J. S., Jensen, O., Chung, C. K., & Lee, D. S. (2011). Cross-frequency power correlations reveal the right superior temporal gyrus as a hub region during working memory maintenance. *Brain Connect, 1*(6), 460–472. doi:10.1089/brain.2011.0046

Payne, L., Guillory, S., & Sekuler, R. (2013). Attention-modulated alpha-band oscillations protect against intrusion of irrelevant information. *Journal of Cognitive Neuroscience, 25*(9), 1463–1476. doi:10.1162/jocn_a_00395

Pfurtscheller, G., Stancak Jr., A., & Neuper, C. (1996). Event-related synchronization (ERS) in the alpha band—an electrophysiological correlate of cortical idling: A review. *International Journal of Psychophysiology, 24*(1–2), 39–46.

Popov, T., Kastner, S., & Jensen, O. (2017). FEF-controlled alpha delay activity precedes stimulus-induced gamma-band activity in visual cortex. *Journal of Neuroscience, 37*(15), 4117–4127. doi:10.1523/JNEUROSCI.3015-16.2017

Raymond, J. E., Shapiro, K. L., & Arnell, K. M. (1992). Temporary suppression of visual processing in an RSVP task: An attentional blink? *Journal of Experimental Psychology: Human Perception and Performance, 18*(3), 849–860.

Rohenkohl, G., & Nobre, A. C. (2011). Alpha oscillations related to anticipatory attention follow temporal expectations. *Journal of Neuroscience, 31*(40), 14076–14084. doi:10.1523/JNEUROSCI.3387-11.2011

Romei, V., Brodbeck, V., Michel, C., Amedi, A., Pascual-Leone, A., & Thut, G. (2008). Spontaneous fluctuations in posterior alpha-band EEG activity reflect variability in excitability of human visual areas. *Cerebral Cortex, 18*(9), 2010–2018. doi:10.1093/cercor/bhm229

Saalmann, Y. B., Pinsk, M. A., Wang, L., Li, X., & Kastner, S. (2012). The pulvinar regulates information transmission between cortical areas based on attention demands. *Science, 337*(6095), 753–756. doi:10.1126/science.1223082

Samaha, J., Gosseries, O., & Postle, B. R. (2017). Distinct oscillatory frequencies underlie excitability of human occipital and parietal cortex. *Journal of Neuroscience, 37*(11), 2824–2833. doi:10.1523/JNEUROSCI.3413-16.2017

Sauseng, P., Klimesch, W., Heise, K. F., Gruber, W. R., Holz, E., Karim, A. A., … Hummel, F. C. (2009). Brain oscillatory substrates of visual short-term memory capacity. *Current Biology, 19*(21), 1846–1852. doi:10.1016/j.cub.2009.08.062

Scheeringa, R., Petersson, K. M., Oostenveld, R., Norris, D. G., Hagoort, P., & Bastiaansen, M. C. (2009). Trial-by-trial coupling between EEG and BOLD identifies networks related to alpha and theta EEG power increases during working memory maintenance. *NeuroImage, 44*(3), 1224–1238. doi:10.1016/j.neuroimage.2008.08.041

Schneidman, E., Puchalla, J. L., Segev, R., Harris, R. A., Bialek, W., & Berry, M. J. (2011). Synergy from silence in a combinatorial neural code. *Journal of Neuroscience, 31*(44), 15732–15741. doi:10.1523/JNEUROSCI.0301-09.2011

Shannon, C. E., & Weaver, W. (1949). *The mathematical theory of communication.* Urbana: University of Illinois Press.

Shapiro, K. L., & Hanslmayr, S. (2014). The role of brain oscillations in the temporal limits of attention. In A. C. Nobre & S. Kastner (Eds.), *The Oxford handbook of attention* (pp. 620–650). Oxford: Oxford University Press.

Shapiro, K. L., Hanslmayr, S., Enns, J. T., & Lleras, A. (2017). Alpha, beta: The rhythm of the attentional blink. *Psychonomic Bulletin & Review, 24*(6), 1862–1869. doi:10.3758/s13423-017-1257-0

Solis-Vivanco, R., Jensen, O., & Bonnefond, M. (2018). Top-down control of alpha phase adjustment in anticipation of temporally predictable visual stimuli. *Journal of Cognitive Neuroscience, 30*(8), 1157–1169. doi:10.1162/jocn_a_01280

Spaak, E., Bonnefond, M., Maier, A., Leopold, D. A., & Jensen, O. (2012). Layer-specific entrainment of gamma-band neural activity by the alpha rhythm in monkey visual cortex. *Current Biology, 22*(24), 2313–2318. doi:10.1016/j.cub.2012.10.020

Sternberg, S. (1966). High-speed scanning in human memory. *Science, 153*(3736), 652–654.

Thut, G., Nietzel, A., Brandt, S. A., & Pascual-Leone, A. (2006). Alpha-band electroencephalographic activity over occipital cortex indexes visuospatial attention bias and predicts visual target detection. *Journal of Neuroscience, 26*(37), 9494–9502. doi:10.1523/JNEUROSCI.0875-06.2006

Tuladhar, A. M., ter Huurne, N., Schoffelen, J. M., Maris, E., Oostenveld, R., & Jensen, O. (2007). Parieto-occipital sources account for the increase in alpha activity with working memory load. *Human Brain Mapping, 28*(8), 785–792. doi:10.1002/hbm.20306

van Diepen, R. M., Cohen, M. X., Denys, D., & Mazaheri, A. (2015). Attention and temporal expectations modulate power, not phase, of ongoing alpha oscillations. *Journal of Cognitive Neuroscience, 27*(8), 1573–1586. doi:10.1162/jocn_a_00803

van Dijk, H., Schoffelen, J. M., Oostenveld, R., & Jensen, O. (2008). Prestimulus oscillatory activity in the alpha band predicts visual discrimination ability. *Journal of Neuroscience, 28*(8), 1816–1823. doi:10.1523/JNEUROSCI.1853-07.2008

van Ede, F., Szebenyi, S., & Maris, E. (2014). Attentional modulations of somatosensory alpha, beta and gamma oscillations dissociate between anticipation and stimulus processing. *NeuroImage, 97*, 134–141. doi:10.1016/j.neuroimage.2014.04.047

van Kerkoerle, T., Self, M. W., Dagnino, B., Gariel-Mathis, M. A., Poort, J., van der Togt, C., & Roelfsema, P. R. (2014). Alpha and gamma oscillations characterize feedback and feedforward processing in monkey visual cortex. *Proceedings of the National Academy of Sciences of the United States of America, 111*(40), 14332–14341. doi:10.1073/pnas.1402773111

VanRullen, R. (2016). Perceptual cycles. *Trends in Cognitive Sciences, 20*(10), 723–735. doi:10.1016/j.tics.2016.07.006

van Schouwenburg, M. R., O'Shea, J., Mars, R. B., Rushworth, M. F., & Cools, R. (2012). Controlling human striatal cognitive function via the frontal cortex. *Journal of Neuroscience, 32*(16), 5631–5637. doi:10.1523/JNEUROSCI.6428-11.2012

Varela, F., Lachaux, J. P., Rodriguez, E., & Martinerie, J. (2001). The brainweb: Phase synchronization and large-scale integration. *Nature Reviews Neuroscience, 2*(4), 229–239. doi:10.1038/35067550

Worden, M. S., Foxe, J. J., Wang, N., & Simpson, G. V. (2000). Anticipatory biasing of visuospatial attention indexed by retinotopically specific alpha-band electroencephalography increases over occipital cortex. *Journal of Neuroscience, 20*(6), RC63.

Zauner, A., Fellinger, R., Gross, J., Hanslmayr, S., Shapiro, K., Gruber, W., … Klimesch, W. (2012). Alpha entrainment is responsible for the attentional blink phenomenon. *NeuroImage, 63*(2), 674–686. doi:10.1016/j.neuroimage.2012.06.075

Zumer, J. M., Scheeringa, R., Schoffelen, J. M., Norris, D. G., & Jensen, O. (2014). Occipital alpha activity during stimulus processing gates the information flow to object-selective cortex. *PLoS Biology, 12*(10), e1001965. doi:10.1371/journal.pbio.1001965

29 A Role for Gaze Control Circuitry in the Selection and Maintenance of Visual Spatial Information

TIRIN MOORE, DONATAS JONIKAITIS, AND WARREN PETTINE

ABSTRACT Human behavioral studies indicate that spatial attention and spatial working memory may be interdependent in complex ways. Within the visual domain, past neurophysiological studies in animal models and neuroimaging studies in humans have revealed neural correlates of both cognitive functions within similar structures within the visual and prefrontal cortex. However, only recently has evidence emerged of how a common neural circuitry may give rise both to the spatial selection of visual information and the persistence of that information during working memory. Here, we summarize this evidence and describe how identifying a role of the gaze-control circuits in spatial attention seems to have revealed an accompanying role in spatial working memory.

The selection and maintenance of sensory information is essential to goal-directed behavior. The information available from the sensory stimuli most relevant to behavior must be adequately extracted from the environment and retained sufficiently long to guide decisions and actions. Evidence both from neurophysiological studies in animal models and from neuroimaging studies in humans has revealed that selective attention heightens the sensory processing of relevant stimuli by neurons throughout the brain (Kastner & Ungerleider, 2000; Noudoost et al., 2010). Similarly, other studies have demonstrated that working memory (WM) involves the persistent signaling of relevant information by neurons in a distributed set of brain areas (Ester, Sprague, & Serences, 2015; Fuster, 1973; Goldman-Rakic, 1995; Srimal & Curtis, 2008). In vision, our dominant sense, the value of selecting and maintaining sensory information is perhaps best exemplified by visual exploration via scanning eye movements. The restriction of high-acuity vision to the fovea necessitates the use of saccadic eye movements (*saccades*), which are executed roughly every few hundred milliseconds. Through these gaze shifts, information from the visual environment is accumulated across multiple fixations in order to achieve a complete perception of objects or scenes. This process necessarily requires that the preparation of each gaze shift selects enough information about the target to accurately fixate it. The process also requires target information to be preserved at least long enough to integrate pre- and post-movement stimuli, and thus a relationship between the mechanisms controlling this basic sensorimotor function to attention and WM bears consideration. In fact, human psychophysical studies have long noted an influence of eye movement planning and/or execution both on visual spatial attention (Deubel & Schneider, 1996; Hoffman & Subramanian,1995) and on visual spatial WM (Baddeley, 1986; Bays & Husain, 2008; Lawrence et al., 2001; Postle et al., 2006). To date, the neural circuit bases of these influences have primarily focused on the role of gaze-control mechanisms in visual spatial attention (Moore, Armstrong, & Fallah, 2003), but more recently, evidence of a similar basis for visual spatial WM has been emerging. Below, we describe both sets of evidence.

Control of Visual Spatial Attention by Gaze-Control Networks

The role of gaze-control mechanisms in visual spatial attention has been appreciated for more than a century (Moore & Zirnsak, 2017). In particular, gaze-control neurons within parietal and prefrontal cortex, as well as within the midbrain of both birds and mammals (Knudsen, 2007), have been implicated as playing a causal role in directing attention within visual space, even when attention is directed covertly. The evidence appears to be particularly strong for the prefrontal cortex. During the 20th century, lesion studies identified the specific involvement of a small cortical area within the prefrontal cortex—namely, the frontal eye field (FEF; Latto & Cowey, 1971; Welch & Stuteville, 1958). The FEF is appropriately situated for a role in visually guided saccades. FEF neurons receive projections from most of the functionally defined areas within visual cortex (Schall et al., 1995) and also send feedback projections to much of the visual cortex (Schall et al., 1995; Stanton, Bruce, & Goldberg, 1995). In addition, FEF neurons project both to the brain stem saccade generator and to the superior colliculus (SC; Stanton, Goldberg, & Bruce, 1988), a midbrain structure with a known involvement in saccade production (Wurtz & Goldberg, 1971). The visually driven responses of some classes of FEF neurons (visual and visuomovement) are enhanced when the stimulus inside the neuron's receptive field (RF) is used as a saccade target compared to when no saccade is made to the stimulus (Bruce & Goldberg, 1985; Goldberg & Bushnell, 1981; Wurtz, Goldberg, & Robinson, 1982). Initial studies suggested that activity within the FEF, as well as the SC, was only enhanced prior to the execution of saccades (overt attention; Goldberg & Bushnell, 1981; Wurtz, Goldberg, & Robinson, 1982) and thus that perhaps these areas are not involved in covert attention. However, a wealth of more recent evidence has overturned this view, confirming Ferrier's (1890) 19th-century hypothesis that this area directly contributes to the "faculty of attention." Examples of this evidence are summarized in figure 29.1.

Motivated by the early lesion evidence and by human psychophysics (e.g., Deubel & Schneider, 1996) and neuroimaging studies (e.g., Kastner & Ungerleider, 2000), Moore and Fallah (2001, 2004) demonstrated that the electrical microstimulation of sites within the FEF could augment monkeys' performance on a covert attention task. They found that when sites within the FEF were stimulated using currents too low to evoke saccades (subthreshold), they could nonetheless enhance covert attentional deployment in a spatially specific manner (figure 29.1*A*). Subsequent studies revisiting the attentional modulation of FEF activity found that it is robustly enhanced during covert attention (Armstrong, Chang, & Moore, 2009; Thompson, Biscoe, & Sato, 2005). In addition, other studies reported similar spatially specific enhancements in covert spatial attention following subthreshold microstimulation of the SC (Cavanaugh & Wurtz, 2004; Müller, Philiastides, & Newsome, 2005), consistent with newly emerging evidence of SC modulation during covert attention (e.g., Ignashchenkova et al., 2004). A later study examined the effect of subthreshold FEF microstimulation on the metrics of voluntarily evoked saccades made to visual stimuli (Schafer & Moore, 2007; figure 29.1*B*). In control trials, the end points of saccades made to drifting gratings are biased in the direction of grating drift in spite of the fact that the grating aperture is stationary. Subthreshold FEF microstimulation augments this motion-induced saccadic bias for gratings positioned at locations represented by neurons at the stimulated FEF site. This result provides evidence of how sensory and motor (and covert and overt) processes are integrated within gaze-control circuits. Specifically, it shows that the activation of FEF neurons drives the selection of retinotopically corresponding visual stimuli and the integration of visual stimulus properties into an appropriately guided movement.

For a particular set of neurons to have a role in the top-down control of attention, as opposed to bottom-up attention, it should follow that their activity is under some degree of voluntary, or operant, control not solely determined by external (e.g., sensory) input. To test this, Schafer and Moore (2011) employed an operant-training paradigm to examine the extent to which FEF neurons could be voluntarily controlled (figure 29.1*C*). Monkeys were provided with real-time auditory feedback based on the firing rate of FEF neurons and rewarded for either increasing or decreasing that activity to some threshold (in alternating Up and Down blocks of trials) while remaining fixated. Overall, monkeys were able to alter the average firing rate of FEF neurons in Up versus Down operant trials and maintained that firing rate for several seconds. Schafer and Moore also probed the consequences of the voluntary control of FEF activity on behavior. They introduced probe trials during the voluntary control paradigm to assess the monkeys' performance on a visual search task. They observed that when the target appeared within the neuronal RF, failures to detect the target (*misses*) were more frequent on the Down trials than on the Up trials. In contrast, the frequency of such errors for targets appearing outside the RF was unaffected by voluntary control. Furthermore, the selectivity of FEF neurons to the target stimulus, versus the distracter, was significantly increased during Up trials, compared to Down trials. These results indicate that the portion of FEF response variability subject to operant control is correlated both with attentional performance and FEF target selectivity.

In addition to producing perceptual benefits, the voluntary deployment of covert attention is known to modulate the visual responses of neurons in visual cortex (Noudoost et al., 2010). The observation that FEF microstimulation produced benefits in attentional performance in monkeys suggested that perhaps such stimulation would also modulate the activity of neurons within visual cortex. To test this, Moore and colleagues measured the effects of subthreshold FEF microstimulation on the visually driven responses of extrastriate area V4 neurons with RFs that corresponded retinotopically to the stimulated FEF site (Moore & Armstrong, 2003; Armstrong, Fitzgerald, &

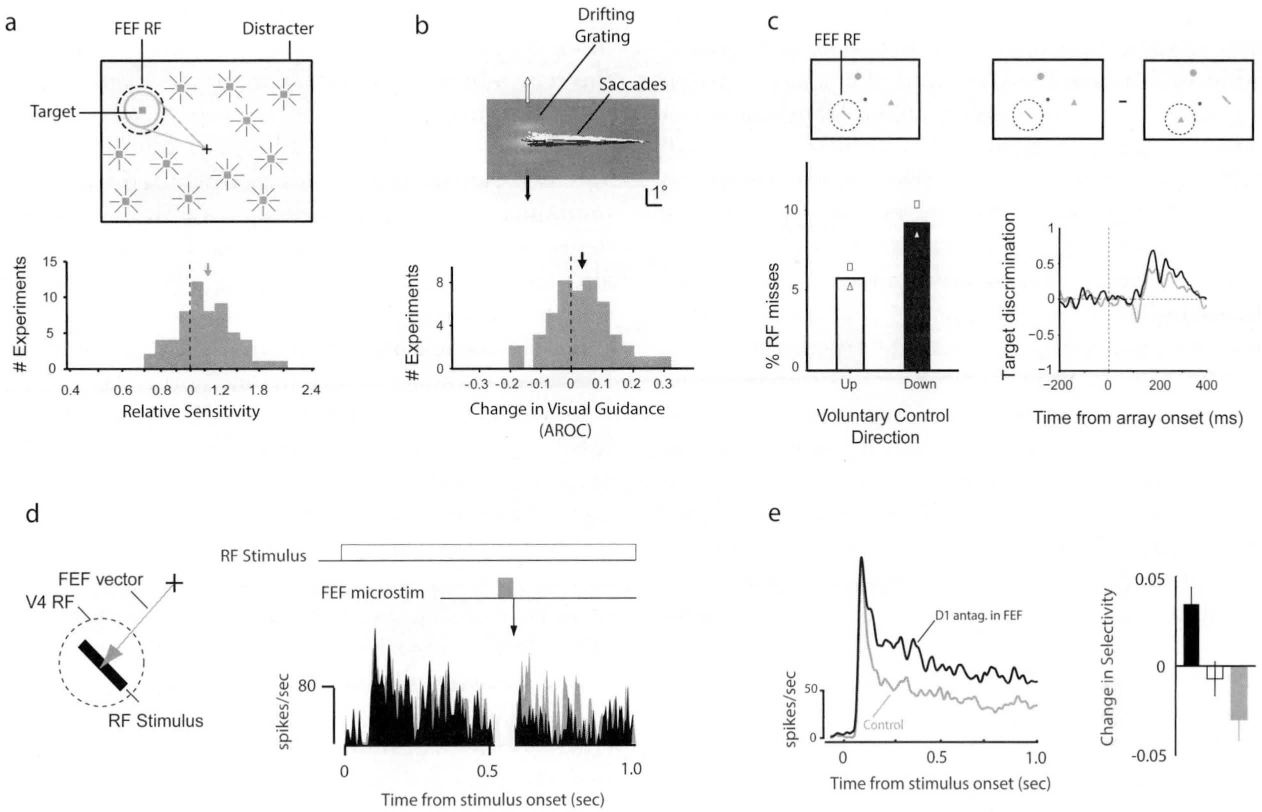

FIGURE 29.1 Perceptual and neurophysiological benefits elicited by perturbations of neural activity in the FEF. *A*, Electrical microstimulation of the FEF improves spatial attention performance. *Top*, Monkeys covertly attended (spotlight icon) a peripheral target stimulus and detected luminance changes in the target while ignoring flashing distracter stimuli. *Bottom*, The microstimulation of sites within the FEF improved the detection of luminance changes compared to control (nonstimulation) trials ($\text{sensitivity}_{\text{microstim}}/\text{sensitivity}_{\text{control}}$). *B*, FEF microstimulation increases the visual guidance of saccades made to visual stimuli. *Top*, Saccades made to drifting gratings are biased in the direction of grating drift (*white traces*, upward motion; *black traces*, downward motion). FEF microstimulation increased the influence of motion on saccadic end points. *C*, Voluntary control of FEF neuronal activity affects visual search errors and FEF target selectivity. Monkeys were operantly conditioned to increase or decrease neuronal activity at a site within the FEF. Upward changes in FEF activity led to fewer visual search errors in the FEF receptive field (RF; % RF misses), compared to downward changes in activity. During upward voluntary changes in FEF activity, the selectivity of FEF neurons for the searched target (*diagonal bar*) was increased compared to downward voluntary changes. All of the behavioral effects above (*A–C*) are spatially specific; effects are only observed at the part of visual space corresponding to the FEF recording/stimulation sites. *D*, Brief microstimulation of the FEF enhances visually driven responses of neurons in visual cortex (area V4). *Shown in black*, The average spike density histogram of a single V4 neuron following the onset of a bar stimulus in the RF. *Gray histogram*, The same response but on trials in which a 50 ms train of microstimulation was delivered to the FEF. *E*, Perturbation of D1-mediated activity within the FEF increases the visual responses and stimulus selectivity of V4 neurons. *Right bar plot*, The change in selectivity following an infusion of a d1 antagonist into an FEF site (black), compared to infusions of a d2 agonist (*white*), and inactivation of FEF activity with a GABAa agonist (*gray*). Both of the above V4 effects (*D* and *E*) are spatially specific and observed only when the FEF and V4 cortical sites correspond retinotopically.

Moore, 2006; Armstrong & Moore, 2007; figure 29.1*D*). Indeed, they found that microstimulation of the FEF enhanced the responses of V4 neurons to visual stimuli and that this enhancement depended critically on the overlap of the V4 RF and the end point of saccades evoked from the FEF stimulation site. Later studies demonstrated that microstimulation of the FEF evoked widespread modulation of sensory responses in the visual cortex in monkeys (Ekstrom et al., 2008) as well as humans (Ruff et al., 2006). Furthermore, whereas inactivation of the SC in monkeys fails to alter attentional modulation within the visual cortex (Zenon & Krauzlis, 2012), damage to prefrontal cortex appears to result in reductions in that modulation (Gregoriou et al., 2014). Thus, inputs from the FEF may be necessary and sufficient both for attentional deployment and for driving selective modulation in visual cortex.

Although it appears that attention-related modulation of visual cortex is in part driven by the FEF, this influence seems to be under neuromodulatory control.

Noudoost and Moore (2011) demonstrated that the manipulation of dopamine (DA)-mediated activity within FEF sites was sufficient to alter visually driven responses in area V4 (figure 29.1*E*). Manipulation of D1R-mediated FEF activity was achieved via small (≤microliter) infusions of a selective D1 antagonist (SCH23390) into sites within the FEF. Behaviorally, the drug manipulation increased the tendency of monkeys to make saccades to visual targets appearing in the part of visual space affected by the drug infusion. In addition, the manipulation also enhanced the visual responses of area V4 neurons with RFs within the drug-affected part of visual space. The enhanced visual responses also became more selective to stimulus orientation, as well as less variable across trials, compared to controls. Similar infusions of a D2R agonist, which produced nearly identical behavioral effects on saccadic choice, failed to alter V4 visual responses. Infusions of the gamma-aminobutyric acid subtype A (GABAa) agonist muscimol reduced the visual selectivity of V4 neurons. Notably, the observed changes in V4 visual activity with the D1R manipulation are also known effects of visual spatial attention (Noudoost et al., 2010). Thus, dopamine D1Rs appear to mediate the FEF's influence on sensory responses in the visual cortex.

Coincident Representations of Attended and Remembered Stimuli

At a coarse level, evidence implicating the prefrontal cortex in the control of spatial attention seems consistent with the notion of common mechanisms for spatial WM and spatial attention, if only because of the strong evidence that prefrontal areas also contribute to spatial WM. For example, neurons in area 46 are classically known to exhibit persistent activity during the delay period of spatial delayed-response tasks (Fuster, 1973; Goldman-Rakic, 1995), and activity in this area appears to be necessary for the performance of spatial WM tasks (Sawaguchi & Iba, 2001). But more recent evidence demonstrates that neurons in this area also robustly signal the direction of top-down spatial attention (Buschman & Miller, 2007). Furthermore, area 46 neurons also appear able to simultaneously signal both the direction of attention and the location of remembered stimuli. Lebedev and colleagues (2004) examined the activity of area 46 neurons during the performance of a task that engaged both spatial attention and WM simultaneously, at separate locations. They trained monkeys to remember one location while attending to a second location and found that during the execution of this task, a majority of neurons signaled the remembered location, the attended location, or both (Lebedev et al., 2004). Although significantly more neurons represented the attended location than the remembered location, approximately one-third of those showing any modulation were affected by both WM and attention. Thus, sources of robust attention and WM signals appear to be colocalized within the nonhuman primate brain. Importantly, similar evidence of that colocalization in the human brain has also emerged from neuroimaging studies (Srimal & Curtis, 2008).

Similar to neurons within area 46, neurons in the FEF also exhibit persistent memory-delay activity (Clark, Noudoost, & Moore, 2012; Hasegawa, Pterson, & Goldberg, 2004). Indeed, it remains unclear whether persistent activity in area 46 and the FEF differs in any significant way, either in terms of its origin or its function in spatial WM. As mentioned above, in spite of earlier reports to the contrary, FEF neurons are clearly modulated during covert spatial attention (Bushmann & Miller, 2007; Gregoriou et al., 2009; Thompson, Biscoe, & Sato, 2005) and directly contribute to attentional deployment and its modulation of activity within the visual cortex (Moore, Armstrong, & Fallah, 2003). Yet how the attention and memory-related functions of neurons in this area (or within other prefrontal areas) relate to one another remains unclear. To investigate the relationship between attentional modulation and sustained memory activity within the FEF, Moore and colleagues recorded FEF activity during a *change-blindness task*. In change-blindness tasks, observers have difficulty detecting localized changes between two visual scenes when they are flashed in quick succession (Cavanaugh & Wurtz, 2004; Rensink, 2002). Directing spatial attention to a particular location can greatly increase the ability of observers to correctly detect changes (Rensink, 2002). Armstrong, Chang, and Moore (2009) recorded the activity of FEF neurons in monkeys performing a change-blindness task. Monkeys indicated a change in one of six stimuli by releasing a lever while maintaining fixation. The activity of FEF neurons with RFs at the cued location was elevated during the delay immediately following the cue, during the presentation of the visual stimuli themselves, and in the interval between the flashed stimulus array (IFI; figure 29.2*A*). FEF neurons thus signaled the remembered cue location and distinguished the target from distracters during visual stimulation (array flashes). Most interestingly, neurons with persistent delay-period activity were considerably better at signaling the target stimulus during the array flash and the IFI than neurons without delay-period activity (figure 29.2*B*). Classifiers trained from populations of delay-period neurons grossly outperformed nondelay neurons at

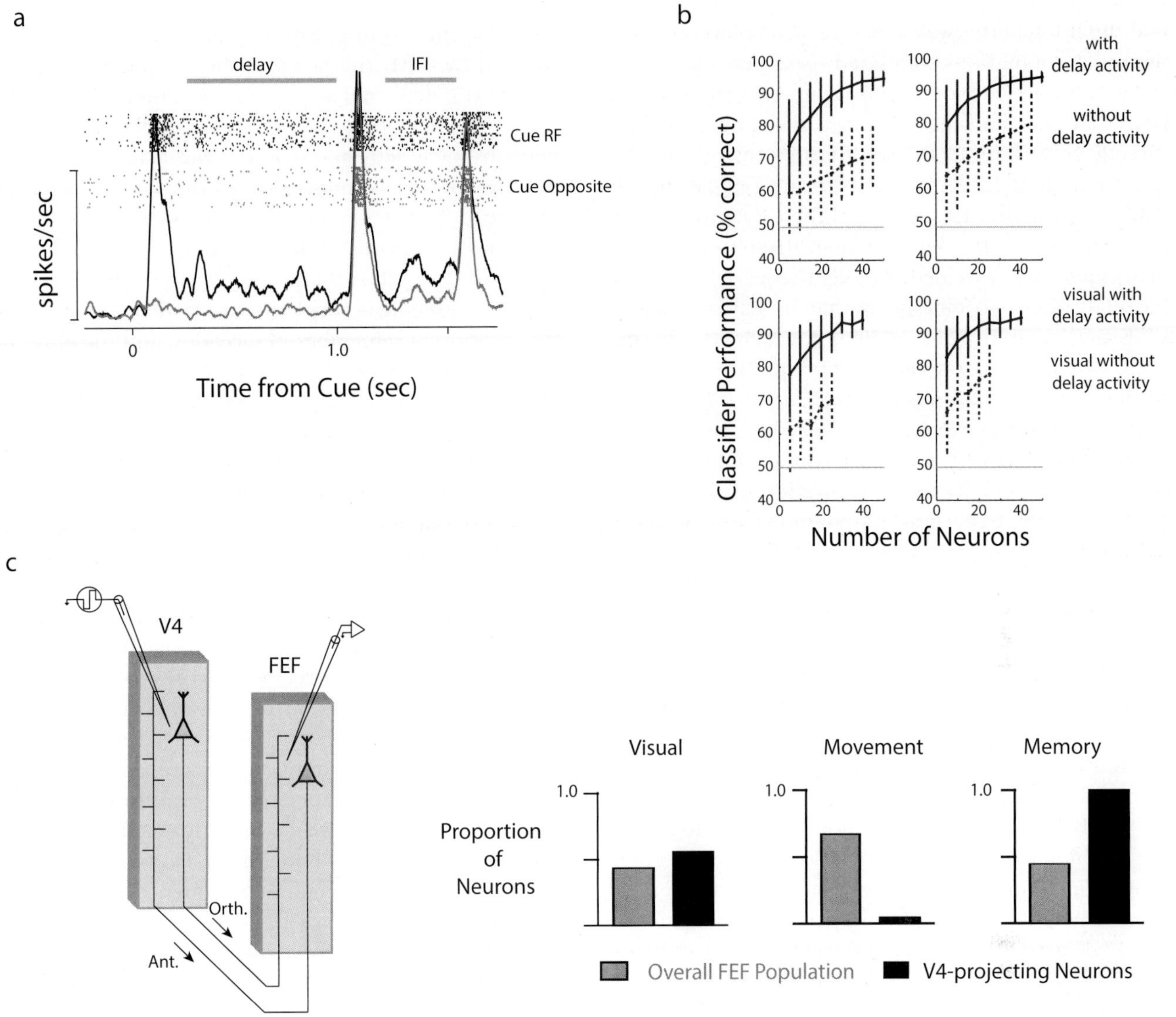

FIGURE 29.2 Evidence of a direct role of persistent neuronal activity in attentional selection. *A*, Persistent activity of an FEF neuron during sustained attention; spike density functions and spike rasters during trials in which a monkey was cued to attend to the RF location (*black*) versus a non-RF location (*gray*). The response to the brief (120–270 ms) cue is transient, but activity remains elevated during the delay period, relative to the Cue Opposite condition, as the monkey awaits a flashing six-item array. During the flash, the item at the cued location either does or does not change orientation, and the monkey must detect the change for a reward. Activity after 1 s reflects the neuron's response to two array flashes and the interflash interval (IFI). Visual stimulation is identical across cue conditions up until the second flash. Note the larger response of the neuron during the IFI in the Cue RF condition. *B*, Population decoding of the cued/attended location from FEF spiking activity during the response to flash 1 and the IFI of the task shown in *A* reveals greater performance of neurons with activity during the delay period, whether those neurons are visually responsive or not. Classifier performance was determined from a support vector machine trained to distinguish between Cue RF and Cue-Opposite locations in the presence (flash 1) or absence of (IFI) visual stimulation, and is plotted as a function of population size. *C*, FEF neurons preferentially transmit persistent, memory-related signals to visual cortex. The functional properties of FEF neurons were determined using a standard delayed-saccade task, and antidromic stimulation of area V4 was used to identify which types of signals are projected from the FEF to V4. Neurons with visual activity were equally likely to project to V4, while neurons with movement activity were significantly less likely to project to V4. In contrast, all identified neurons exhibited persistent memory-delay activity.

localizing the target stimulus during the delay flash and the IFI, and this was true regardless of whether the neuronal populations contained visual activity.

The above evidence is consistent with the speculation that mechanisms holding information in WM directly contribute to the selection of current sensory representations (Desimone & Duncan, 1995; Knudsen, 2007). Yet whether this is indeed the case and how it is implemented at the level of neural circuitry is only beginning to be revealed. Among the set of significant recent findings to emerge is that in humans spatial WM, as measured by an oculomotor delayed-response task, depends not on dorsolateral prefrontal cortex (dlPFC) but instead on precentral cortex (PC; Mackey et al., 2016; Mackey & Curtis, 2017). Motivated by the observation that imaging studies have often failed to demonstrate clear persistent activation of dlPFC during WM in spite of clear activation in PC (e.g., Srimal & Curtis, 2008), Mackey and Curtis (2017) tested neurological patients with damage to either the dlPFC or PC on an oculomotor delayed-saccade task. They found that although PC patients had clear deficits on this task, dlPFC patients were largely normal (Mackey & Curtis, 2017). Moreover, similar effects were observed in subjects receiving transcranial magnetic stimulation of the PC or dlPFC. These results call into question a common dogma of the dominant role of the dlPFC in WM and also raise questions about the homology with the nonhuman primate brain. As described above, neurons in monkey dlPFC, most notably area 46, exhibit robust, persistent delay-period activity during oculomotor delayed-saccade tasks (Fuster, 1973; Goldman-Rakic, 1995) that appears to be necessary for the performance of this task, as demonstrated by the effects of reversible inactivation (Sawaguchi & Iba, 2001). As noted above, neurons in the monkey FEF also exhibit robust delay-period activity (Armstrong, Chang, & Moore, 2009; Clark, Noudoost, & Moore, 2012; Hasegawa, Peterson, & Goldberg, 2004), and the reversible inactivation of FEF activity dramatically disrupts memory-guided saccades (Clark, Noudoost, & Moore, 2012; Dias & Segraves, 1999). Thus, the relative roles of the dlPFC and the FEF in spatial WM may differ to some significant degree between human and nonhuman primates (Mackey & Curtis, 2017).

Modulation of Sensory Signals by Persistently Active Neurons

The discovery of a causal role of gaze-control structures in visual spatial attention raised a number of important questions, most crucially how this role is achieved by neurons within these structures given the heterogeneity of neuronal properties there. Within all three of the oft-implicated structures in attentional control, the FEF, the SC, and the lateral intraparietal area (LIP), neuronal activity is associated with a broad range of behaviorally relevant factors. All three contain neurons solely activated by either the visual stimulation of their RFs (visual neurons) or prior to the execution of saccades of a particular direction and amplitude (movement neurons), but often both (visuomovement neurons; Wurtz, Goldberg, & Robinson, 1982). In addition, all three structures contain neurons that signal the location of remembered saccades, and the elimination of this activity in either structure is sufficient to impair the performance of monkeys on a memory-guided saccade task (Dias & Segraves, 1999; Hikosaka & Wurtz, 1985; Li, Mazzoni, & Andersen, 1999). Yet until recently it was unclear which of these signals is used to control spatial attention. Based on evidence of separable contributions of the FEF to saccadic programming and attention deployment (e.g., Juan, Shorter-Jacobi, & Schall, 2004), it seemed plausible that perhaps attentional control was achieved largely via the outputs of visual neurons. Moreover, it was observed that the increased synchronization of FEF and visual cortical activity within the gamma frequency band that occurs during covert spatial attention (Gregoriou et al., 2009) is most robust when specifically examining the synchronization of FEF visual neurons with local field activity in visual cortex (Gregoriou, Gotts, & Desimone, 2012). While not a direct line of evidence, this observation suggests that, indeed, FEF visual neurons are uniquely responsible for driving attentional modulation within the visual cortex. However, more recent work appears to indicate otherwise.

As described above, dopamine neuromodulation through D1 receptors appears to play a key role in the influence that FEF neurons exert on visual cortical activity (Noudoost & Moore, 2011a; figure 29.1*E*). On the face of it, this observation may seem to have little to do with the question of which class of FEF neurons contributes to spatial attention. However, it is important to note that dopamine D1 receptors are well known as a key mechanism in the maintenance of persistent, delay-period activity within the prefrontal cortex. The iontophoretic application of D1 agonists and antagonists within the dlPFC can selectively enhance or reduce delay-period activity (Williams & Goldman-Rakic, 1995), and local infusions of similar drugs impair performance on delayed saccade tasks (Sawaguchi & Goldman-Rakic, 1991). This evidence, in addition to the results described in figure 29.2*B*, prompted the speculation that perhaps FEF delay neurons uniquely contribute to the modulation of visual cortical signals and that this control is modulated by dopaminergic inputs (Noudoost & Moore, 2011b). Yet direct evidence of this was missing until recently.

Using antidromic stimulation, Merrikkhi et al. (2017) directly addressed the contribution of different functional classes of FEF neurons to the top-down modulation of the visual cortex (figure 29.2*C*). FEF neurons were classified into standard functional groups using a delayed-saccade task and identified as area V4-projecting if they could be activated antidromically by the electrical stimulation of retinotopically corresponding sites within V4. Three key observations were made. First, V4-projecting FEF neurons were equally likely to be visually responsive compared to the overall population of FEF neurons. Second, a significantly lower proportion of V4-projecting neurons exhibited movement-related activity, indicating a relative absence of perisaccadic movement signals projecting to V4 from the FEF. This result appears to be consistent with the observation that inactivation of the FEF fails to reduce the presaccadic enhancement of visual responses in V4, in spite of the reduction stimulus selectivity it produces there (Noudoost, Clark, & Moore, 2014). Third, and most importantly, it was observed that all of the identified V4-projecting FEF neurons exhibited persistent delay-period activity. Thus, the FEF appears to project disproportionately strong memory-related delay signals to visual cortex, and therefore the modulation of sensory activity in visual cortex by the FEF, modulation associated with visual spatial attention, appears to derive predominantly from FEF memory delay neurons. Note that this finding is consistent with the report that the magnitude of the impairments of visual attention resulting from inactivation of the FEF are correlated with the

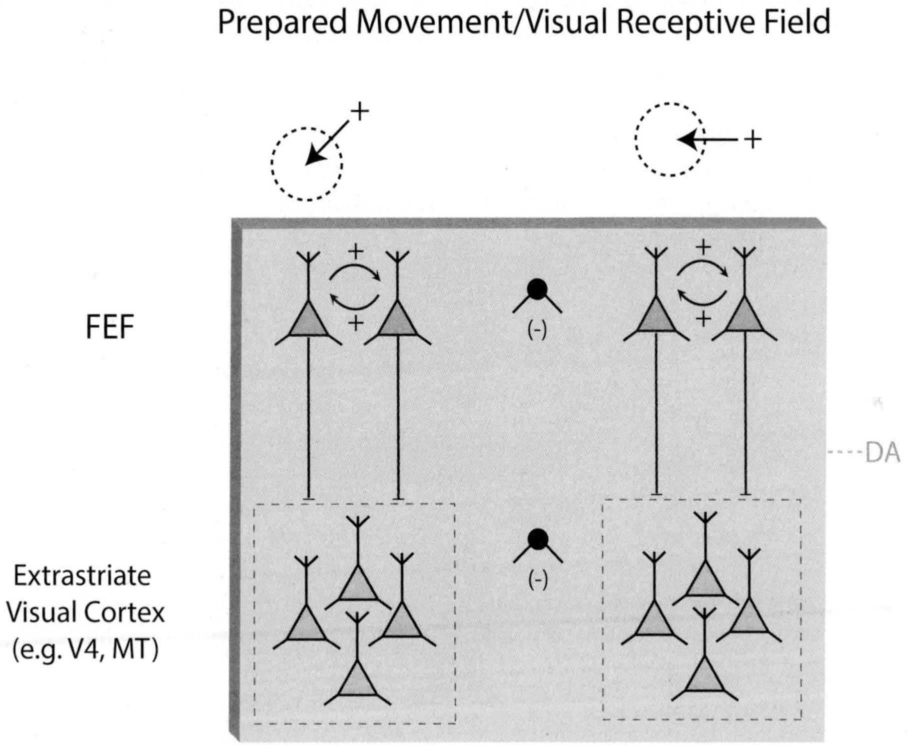

FIGURE 29.3 A simplified circuit model of the FEF's influence on visual cortex and its modulation by dopamine innervation. The diagram depicts the top-down projection of layer II–III pyramidal neurons in the FEF to neurons within extrastriate cortex—for example, area V4 or MT. Evidence shows that most of the FEF inputs to V4 synapse onto the spines of pyramidal neurons across layers II–VI. Two adjacent columns are shown to illustrate the projection of a retinotopically organized FEF, where neurons have visual RFs and coordinate saccades of particular direction and amplitudes (*top cartoon*) to corresponding columns in retinotopically organized visual areas. The adjacent columns of both areas are shown to competitively interact via mutual inhibition (middle inhibitory neuron), consistent with evidence. In addition, the dominance of persistent delay-period activity in the signals transmitted to visual cortex is shown in the FEF as a recurrent excitatory circuit. In this model, delay-period activity, which is dependent on the level of dopamine (DA) release throughout cortex, maintains saccadic plans to particular locations in space and effectively amplifies feedforward visual inputs to extrastriate cortical neurons. In the absence of visual input, activity in this circuit reflects remembered locations and planned movements to those locations; in the presence of visual input, activity in this circuit reflects the attentional priority of visual stimuli.

magnitude of the impairments in memory-guided saccades (Monosov & Thompson, 2009).

The evidence that delay signals dominate FEF inputs to the visual cortex raises a key question about its generalizability to other instances of projections from premotor to sensory areas of cortex. A number of recent studies in rodents reveal potent influences of motor cortical feedback on feedforward sensory responses of sensory cortical neurons. For example, neurons in mouse vibrissal cortex receive somatotopically specific excitatory inputs from the vibrissal motor cortex, inputs that alter sensory processing and increase the reliability of responses to complex whisker stimulation (Lee, Carvell, & Simons, 2008; Zagha et al., 2013). Similar to primates, neurons in mouse visual cortex are modulated by inputs from frontal cortex that can increase the selectivity of visual cortical neurons (Zhang et al., 2014). In both these examples, improvements in sensory processing are affected by spatially specific inputs from motor and premotor networks of neurons in spite of differences in modality and apparent differences in precise circuitry. However, it will be important to know if there are similarities in the functional properties of the sensory cortex projecting motor and premotor neurons. For example, do these neurons themselves tend to exhibit sensory responses or premovement bursts? Or, perhaps more enticingly, do they exhibit persistent delay-period activity as observed among FEF neurons projecting to the visual cortex in primates?

Models of Working Memory and Their Relation to Attentional Selection

Attentional selection and WM have been a major focus of theoretical models. Although there are notable exceptions, neuronal attractor states have been the primary framework studied. Attractor states are stable patterns of neural activity that can represent a memory maintained during a delay or the choice of saccade direction. These stable patterns of activity can also represent a source of the attentional selection signal transmitted from the prefrontal cortex to a sensory map—for example, within the visual cortex.

In one of the earliest examples of these models, Amit and Brunel (1997) used populations of excitatory and inhibitory integrate-and-fire neurons with a biophysically realistic learning rule to study delay activity. They showed how stable regimes of neuron firing during delay periods differ between familiar and novel stimuli. The physiology of the circuit was then further characterized by adding realistic glutamate receptor channel dynamics and showing how the proportion of those receptor types strongly shape sustained delay activity in prefrontal areas (Wang, 1999). The activity of units in these models reproduces the behavior of single neurons recorded from the prefrontal cortex of nonhuman primates (Compte et al., 2000). Furthermore, dopaminergic modulation of units in these sustained activity networks produces effects similar to those observed experimentally (Brunel & Wang, 2001; Durstewitz, Kelc, & Güntürkün, 1999; Durstewitz, Seamans, & Sejnowski, 2000). Interestingly, these same kinds of WM models reproduce the nonhuman primate saccade statistics in perceptual decision-making tasks (Amit et al., 2003; Wang, 2002), suggesting an important role of these mechanisms in the allocation of attention.

Since this early work, single-area models of attractor states have been used to study specific problems such as multi-item storage (Dempere-Marco, Melcher, & Deco, 2012; Edin et al., 2009; Rolls, Dempere-Marco, & Deco, 2013; Wei, Wang, & Wang, 2012) or the advantages of random versus structured unit connectivity (Maass, Natschläger, & Markram, 2002; Rigotti et al., 2010). Alternatives to persistent activity have also been proposed, such as synaptic facilitation (Mongillo, Barak, & Tsodyks, 2008) or feedforward chains (Goldman, 2009; Murphy & Miller, 2009). More recently, models have been developed that unify the attentional effects of normalization and surround suppression, with the maintenance of stimulus representation through a delay (Ahmadian et al., 2013; Kraynyukova & Tchumatchenko, 2018; Persi et al., 2011; Rubin, Van Hooser, & Miller, 2015). These *stabilized supralinear networks* rely on strong feedback inhibition, whose source may be local or long range.

Multiarea models of WM and attention, however, are still in their early stages. One approach has been to use connectivity data produced by tract tracing in nonhuman primates (Markov et al., 2014) to construct a whole-cortex simulation (Chaudhuri et al., 2015; Joglekar et al., 2018; Mejias et al., 2016). The structure of each area within these models is constrained by biophysical gradients across cortex, such as the strength of recurrent excitation, or the relative balance of AMPA and NMDA receptors. Feedforward and feedback connections are explicitly implemented using laminar projection profiles or bidirectional weights. This means that activity in a visual area like V4 provides feedforward inputs to a prefrontal area like the FEF, which are then transformed and sent as feedback to V4. Importantly, these models reproduce gross effects sensory-modality attention—for example, stimulation of auditory cortex versus visual (Mejias et al., 2016). Furthermore, the same framework can maintain stimulus representation throughout a delay (Chaudhuri et al., 2015). Thus, large-scale theoretical modeling also demonstrates the shared mechanisms of attention and WM.

Conclusions

Regardless of the precise circuitry employed to persistently encode information over brief periods, the persistently encoded information likely interacts with the processing of incoming sensory information. As we have described, recent studies indicating the role of gaze-control structures in the control of visual spatial attention in nonhuman (and human) primates have also brought forth evidence of the direct role of neurons that maintain visual spatial signals during WM in the selection of visual stimuli during spatial attention. This evidence appears consistent with observations from a broad range of human psychophysical studies, demonstrating an influence of remembered information on the perception of visual stimuli (Awh & Jonides, 2001). Many other studies show that the content and precision of visual WM is heavily dependent upon the preparation and/or execution of eye movements (Bays & Husain, 2008; Hanning et al., 2016; Lawrence et al., 2004; Tas, Luck, & Hollingworth, 2016). Thus, the prevalence of persistent activity within motor-related structures, such as those involved in gaze control, might suggest that spatial attention and spatial WM emerge from the preparation of sensory-guided movements and that the persistence of premovement network states carries with it both the maintenance of recently associated sensory stimuli and the gating of subsequent sensory events. Nonetheless, evidence from across a range of experimental approaches suggests a fundamental relationship between visual spatial attention, visual spatial WM, and gaze control and that much of the neural circuitry underlying this relationship awaits discovery.

REFERENCES

Ahmadian, Y., Rubin, D. B., and Miller, K. D. (2013). Analysis of the stabilized supralinear network. *Neural Computation, 25*, 1994–2037.

Amit, D. J., Bernacchia, A., & Yakovlev, V. (2003). Multiple-object working memory—A model for behavioral performance. *Cerebral Cortex 13*, 435–443.

Amit, D. J., & Brunel, N. (1997). Model of global spontaneous activity and local structured activity during delay periods in the cerebral cortex. *Cerebral Cortex*, 7(3), 237–252.

Armstrong, K. M., Chang, M. H., & Moore, T. (2009). Selection and maintenance of spatial information by frontal eye field neurons. *Journal of Neuroscience, 29*, 15621–15629.

Armstrong, K. M., Fitzgerald, J. F., & Moore, T. (2006). Changes in visual receptive fields with microstimulation of frontal cortex. *Neuron*, 50, 791–798.

Armstrong, K. M., & Moore, T. (2007). Rapid enhancement of visual cortical response discriminability by microstimulation of the frontal eye field. *Proceedings of the National Academy of Sciences, 104*(22), 9499–9504.

Awh, E., & Jonides, J. (2001). Overlapping mechanisms of attention and spatial working memory. *Trends in Cognitive Sciences, 5*, 119–126.

Baddeley, A. D. (1986). *Working memory*. London: Oxford University Press.

Bays, P. M., & Husain, M. (2008). Dynamic shifts of limited working memory resources in human vision. *Science, 321*, 851–854.

Bruce, C. J., & Goldberg M. E. (1985). Primate frontal eye fields. I. Single neurons discharging before saccades. *Journal of Neurophysiology, 53*, 603–635.

Brunel, N., & Wang, X. J. (2001). Effects of neuromodulation in a cortical network model of object working memory dominated by recurrent inhibition. *Journal of Computational Neuroscience, 11*(1), 63–85.

Burrows, B. E., Zirnsak, M., Akhlaghpour, H., & Moore, T. (2014). Global selection of saccadic target features by neurons in area V4. *Journal of Neuroscience, 34*, 6700–6706.

Buschman, T. J., & Miller, E. K. (2007). Top-down versus bottom-up control of attention in the prefrontal and posterior parietal cortices. *Science, 315*, 1860–1862.

Cavanaugh, J., & Wurtz, R. H. (2004). Subcortical modulation of attention counters change blindness. *Journal of Neuroscience, 24*, 11236–11243.

Chaudhuri, R., Knoblauch, K., Gariel, M.-A., Kennedy, H., & Wang, X.-J. (2015). A large-scale circuit mechanism for hierarchical dynamical processing in the primate cortex. *Neuron, 88*, 419–431.

Clark, K. L., Noudoost, B., & Moore, T. (2012). Persistent spatial information in the frontal eye field during object-based short-term memory. *Journal of Neuroscience, 32*, 10907–10914.

Compte, A., Brunel, N., Goldman-Rakic, P. S., & Wang, X.-J. (2000). Synaptic mechanisms and network dynamics underlying spatial working memory in a cortical network model. *Cerebral Cortex, 10*, 910–923.

Dempere-Marco, L., Melcher, D. P., & Deco, G. (2012). Effective visual working memory capacity: An emergent effect from the neural dynamics in an attractor network. *PLoS One, 7*, e42719.

Desimone, R., & Duncan, J. (1995). Neural mechanisms of selective visual attention. *Annual Review of Neuroscience, 18*, 193–222.

Deubel, H., & Schneider, W. X. (1996). Saccade target selection and object recognition: Evidence for a common attentional mechanism. *Vision Research, 36*, 1827–1837.

Dias, E. C., & Segraves, M. A. (1999). Muscimol-induced inactivation of monkey frontal eye field: Effects on visually and memory-guided saccades. *Journal of Neurophysiology, 81*(5), 2191–2214.

Druckmann, S., & Chklovskii, D. B. (2012). Neuronal circuits underlying persistent representations despite time varying activity. *Current Biology, 22*, 2095–2103.

Durstewitz, D., Kelc, M., & Güntürkün, O. (1999). A neurocomputational theory of the dopaminergic modulation of working memory functions. *Journal of Neuroscience, 19*, 2807–2822.

Durstewitz, D., Seamans, J. K., & Sejnowski, T. J. (2000). Dopamine-mediated stabilization of delay-period activity in a network model of prefrontal cortex. *Journal of Neurophysiology, 83*, 1733–1750.

Edin, F., Klingberg, T., Johansson, P., McNab, F., Tegnér, J., & Compte, A. (2009). Mechanism for top-down control of working memory capacity. *Proceedings of the National Academy of Sciences, 106*, 6802–6807.

Ekstrom, L. B., Roelfsema, P. R., Arsenault, J. T., Bonmassar, G., & Vanduffel, W. (2008). Bottom-up dependent gating of frontal signals in early visual cortex. *Science, 321*, 414–417.

Ester, E. F., Sprague, T. C., & Serences, J. T. (2015). Parietal and frontal cortex encode stimulus-specific mnemonic representations during visual working memory. *Neuron, 87*, 893–905.

Ferrier, D. 1890. *Cerebral localisation*. London: Smith, Elder.

Fuster, J. M. (1973). Unit activity in prefrontal cortex during delayed-response performance: Neuronal correlates of transient memory. *Journal of Neurophysiology, 36*, 61–78.

Goldberg, M. E., & Bushnell, M. C. (1981). Behavioral enhancement of visual responses in monkey cerebral cortex. II. Modulation in frontal eye fields specifically related to saccades. *Journal of Neurophysiology, 46*, 773–787.

Goldman, M. S. (2009). Memory without feedback in a neural network. *Neuron, 61*, 621–634.

Goldman-Rakic, P. S. (1995). Cellular basis of working memory. *Neuron, 14*, 477–485.

Gregoriou, G. G., Gotts, S. J., and Desimone, R. (2012). Cell-type-specific synchronization of neural activity in FEF with V4 during attention. *Neuron* 73, 581–594.

Gregoriou, G. G., Gotts, S. J., Zhou, H., & Desimone, R. (2009). High-frequency, long-range coupling between prefrontal and visual cortex during attention. *Science, 29*, 1207–1210.

Gregoriou, G. G., Rossi, A. F., Ungerleider, L. G., & Desimone, R. (2014). Lesions of prefrontal cortex reduce attentional modulation of neuronal responses and synchrony in V4. *Nature Neuroscience, 17*, 1003–1011.

Hanning, N. M., Jonikaitis, D., Deubel, H., & Szinte, M. (2016). Oculomotor selection underlies feature retention in visual working memory. *Journal of Neurophysiology, 115*, 1071–1076.

Hasegawa, R. P., Peterson, B. W., & Goldberg, M. E. (2004). Prefrontal neurons coding suppression of specific saccades. *Neuron, 43*, 415–425.

Hikosaka, O., & Wurtz, R. H. (1985). Modification of saccadic eye movements by GABA-related substances. I. Effect of muscimol and bicuculline in monkey superior colliculus. *Journal of Neurophysiology, 53*, 266–291.

Hoffman, J. E., & Subramaniam, B. (1995). The role of visual attention in saccadic eye movements. *Perception & Psychophysics, 57*, 787–795.

Hopfield, J. J. (1982) Neural networks and physical systems with emergent collective computational abilities. *Proceedings of the National Academy of Sciences, 79*, 2554–2558.

Ignashchenkova, A., Dicke, P. W., Haarmeier, T., & Thier, P. (2004). Neuron-specific contribution of the superior colliculus to overt and covert shifts of attention. *Nature Neuroscience, 7*, 56–64.

Joglekar, M. R., Mejias, J. F., Yang, G. R., & Wang, X.-J. (2018). Inter-areal balanced amplification enhances signal propagation in a large-scale circuit model of the primate cortex. *Neuron, 98*, 222–234.e8.

Juan, C. H., Shorter-Jacobi, S. M., & Schall, J. D. (2004). Dissociation of spatial attention and saccade preparation. *Proceedings of the National Academy of Sciences of the United States of America, 101*, 15541–15544.

Jüttner, M., & Röhler, R. (1993). Lateral information transfer across saccadic eye movements. *Perception & Psychophysics, 53*, 210–220.

Kastner, S., & Ungerleider, L. G. (2000). Mechanisms of visual attention in the human cortex. *Annual Review of Neuroscience, 23*, 315–341.

Knudsen, E. I. (2007). Fundamental components of attention. *Annual Review of Neuroscience, 30*, 57–78.

Kraynyukova, N., & Tchumatchenko, T. (2018). Stabilized supralinear network can give rise to bistable, oscillatory, and persistent activity. *Proceedings of the National Academy of Sciences, 115*, 3464–3469.

Latto, R., & Cowey, A. (1971).Visual field defects after frontal eye-field lesions in monkeys. *Brain Research, 30*, 1–24.

Lawrence, B. M., Myerson, J., Oonk, H. M., & Abrams, R. A. (2001). The effects of eye and limb movements on working memory. *Memory, 9*, 433–444.

Lebedev, M. A., Messinger, A., Kralik, J. D., & Wise, S. P. (2004). Representation of attended versus remembered locations in prefrontal cortex. *PLoS Biology, 2*, e365.

Lee, S., Carvell, G. E., & Simons, D. J. (2008). Motor modulation of afferent somatosensory circuits. *Nature Neuroscience, 11*, 1430–1438.

Li, C. S., Mazzoni, P., & Andersen, R. A. (1999). Effect of reversible inactivation of macaque lateral intraparietal area on visual and memory saccades. *Journal of Neurophysiology, 81*, 1827–1838.

Maass, W., Natschläger, T., & Markram, H. (2002). Real-time computing without stable states: A new framework for neural computation based on perturbations. *Neural Computation, 14*, 2531–2560.

Mackey, W. E., & Curtis, C. E. (2017). Distinct contributions by frontal and parietal cortices support working memory. *Scientific Reports, 7*, 6188.

Mackey, W. E., Devinsky, O., Doyle, W. K., Meager, M. R., & Curtis, C. E. (2016). Human dorsolateral prefrontal cortex is not necessary for spatial working memory. *Journal of Neuroscience, 36*, 2847–2856.

Markov, N. T., Vezoli, J., Chameau, P., Falchier, A., Quilodran, R., Huissoud, C., Lamy, C., Misery, P., Giroud, P., Ullman, S., et al. (2014). Anatomy of hierarchy: Feedforward and feedback pathways in macaque visual cortex. *Journal of Comparative Neurology, 522*, 225–259.

Mejias, J. F., Murray, J. D., Kennedy, H., & Wang, X. J. (2016). Feedforward and feedback frequency-dependent interactions in a large-scale laminar network of the primate cortex. *Science Advances, 2*(11), e1601335.

Merrikhi, Y., Clark, K. L., Albarran, E., Mohammadbagher, P., Zirnsak, M., Moore, T., & Noudoost, B. (2017). Spatial working memory alters the efficacy of input to visual and prefrontal cortex. *Nature Communications, 8*, 15041.

Mongillo, G., Barak, O., & Tsodyks, M. (2008). Synaptic theory of working memory. *Science, 319*, 1543–1546.

Monosov, I. E., & Thompson, K. G. (2009). Frontal eye field activity enhances object identification during covert visual search. *Journal of Neurophysiology, 102*, 3656–3672.

Moore, T., & Armstrong, K. M. (2003). Selective gating of visual signals by microstimulation of frontal cortex. *Nature, 421*(6921), 370–373.

Moore, T., Armstrong, K. M., & Fallah, M. (2003). Visuomotor origins of covert spatial attention. *Neuron, 40*(4), 671–683.

Moore, T., & Fallah, M. (2001). Control of eye movements and spatial attention. *Proceedings of the National Academy of Sciences of the United States of America, 98*, 1273–1276.

Moore, T., & Fallah, M. (2004). Microstimulation of frontal eye fields and its effects on covert spatial attention. *Journal of Neurophysiology, 91,* 152–162.

Moore, T., & Zirnsak, M. (2017). Neural mechanisms of selective visual attention. *Annual Review of Psychology, 68*(1), 47–72.

Müller, J. R., Philiastides, M. G., & Newsome, W. T. (2005). Microstimulation of the superior colliculus focuses attention without moving the eyes. *Proceedings of the National Academy of Sciences, 102*(3), 524–529.

Murphy, B. K., & Miller, K. D. (2009). Balanced amplification: A new mechanism of selective amplification of neural activity patterns. *Neuron, 61,* 635–648.

Noudoust, B., Chang, M. H., Steinmetz, N. A., & Moore, T. (2010). Top-down control of visual attention. *Current Opinion in Neurobiology, 20,* 183–190.

Noudoost, B., Clark, K. L., & Moore, T. (2014). Distinct contribution of the frontal eye field to the representation of saccadic targets. *Journal of Neuroscience, 34,* 3687–3698.

Noudoost, B., & Moore, T. (2011a). The control of visual cortical signals by prefrontal dopamine. *Nature,* 474, 372–375.

Noudoost, B., & Moore, T. (2011b). The role of neuromodulators in selective attention. *Trends in Cognitive Sciences, 15*(12), 585–591.

Persi, E., Hansel, D., Nowak, L., Barone, P., & van Vreeswijk, C. (2011). Power-law input-output transfer functions explain the contrast-response and tuning properties of neurons in visual cortex. *PLOS Computational Biology, 7,* e1001078.

Postle, B. R., Idzikowski, C., Sala, S. D., Logie, R. H., & Baddeley, A. D. (2006). The selective disruption of spatial working memory by eye movements. *Quarterly Journal of Experimental Psychology (Hove), 59,* 100–120.

Rensink, R. A. 2002. Change detection. *Annual Review of Psychology, 53,* 245–277.

Rigotti, M., Rubin, D. B. D., Wang, X.-J. & Fusi, S. (2010). Internal representation of task rules by recurrent dynamics: The importance of the diversity of neural responses. *Frontiers in Computational Neuroscience, 4* eCollection. https://www.ncbi.nlm.nih.gov/pubmed/?term=Internal+representation+of+task+rules+by+recurrent+dynamics%3A+The+importance+of+the+diversity+of+neural+responses

Rolls, E. T., Dempere-Marco, L., & Deco, G. (2013). Holding multiple items in short term memory: A neural mechanism. *PloS One, 8,* e61078.

Rubin, D. B., Van Hooser, S. D., & Miller, K. D. (2015). The stabilized supralinear network: A unifying circuit motif underlying multi-input integration in sensory cortex. *Neuron, 85,* 402–417.

Ruff, C. C., Blankenburg, F., Bjoertomt, O., Bestmann, S., Freeman, E., Haynes, J. D., Rees, G., Josephs, O., Deichmann, R., & Driver, J. (2006). Concurrent TMS-fMRI and psychophysics reveal frontal influences on human retinotopic visual cortex. *Current Biology, 16,* 1479–1488.

Sawaguchi, T., & Goldman-Rakic, P. S. (1991). D1 dopamine receptors in prefrontal cortex: Involvement in working memory. *Science, 251,* 947–950.

Sawaguchi, T., & Iba, M. (2001). Prefrontal cortical representation of visuospatial working memory in monkeys examined by local inactivation with muscimol. *Journal of Neurophysiology, 86,* 2041–2053.

Schafer, R. J., & Moore, T. (2007). Attention governs action in the primate frontal eye field. *Neuron, 56,* 541–551.

Schafer, R. J., & Moore, T. (2011). Selective attention from voluntary control of prefrontal neurons. *Science, 332,* 1568–1571.

Schall, J. D., Morel, A., King, D. J., & Bullier, J. (1995). Topography of visual cortex connections with frontal eye field in macaque: Convergence and segregation of processing streams. *Journal of Neuroscience, 15,* 4464–4487.

Squire, R. F., Noudoost, B., Schafer, R. J., & Moore, T. (2013). Prefrontal contributions to visual selective attention. *Annual Review of Neuroscience, 36,* 451–466.

Srimal, R., & Curtis, C. E. (2008). Persistent neural activity during the maintenance of spatial position in working memory. *NeuroImage, 39,* 455–468.

Stanton, G. B., Bruce, C. J., & Goldberg, M. E. (1995). Topography of projections to posterior cortical areas from the macaque frontal eye fields. *Journal of Comparative Neurology, 353,* 291–305.

Stanton, G. B., Goldberg, M. E., & Bruce, C. J. (1988). Frontal eye field efferents in the macaque monkey: II. Topography of terminal fields in midbrain and pons. *Journal of Comparative Neurology, 271,* 493–506.

Steinmetz, N. A., & Moore, T. (2014). Eye movement preparation modulates neuronal responses in area V4 when dissociated from attentional demands. *Neuron, 83,* 496–506.

Tas, A. C., Luck, S. J., & Hollingworth, A. (2016). The relationship between visual attention and visual working memory encoding: A dissociation between covert and overt orienting. *Journal of Experimental Psychology-Human Perception and Performance, 42*(8), 1121–1138.

Thompson, K. G., Biscoe, K. L., & Sato, T. R. (2005). Neuronal basis of covert spatial attention in the frontal eye field. *Journal of Neuroscience, 25,* 9479–9487.

Wang, X.-J. (1999). Synaptic basis of cortical persistent activity: The importance of NMDA receptors to working memory. *Journal of Neuroscience, 19,* 9587–9603.

Wang, X.-J. (2002). Probabilistic decision making by slow reverberation in cortical circuits. *Neuron, 36,* 955–968.

Wei, Z., Wang, X.-J., & Wang, D.-H. (2012). From distributed resources to limited slots in multiple-item working memory: A spiking network model with normalization. *Journal of Neuroscience, 32,* 11228–11240.

Welch, K., & Stuteville, P. (1958). Experimental production of unilateral neglect in monkeys. *Brain, 81,* 341–347.

Williams, G. V., & Goldman-Rakic, P. S. (1995). Modulation of memory fields by dopamine D1 receptors in prefrontal cortex. *Nature, 376,* 572–575.

Wurtz, R. H., & Goldberg, M. E. (1971). Superior colliculus cell responses related to eye movements in awake monkeys. *Science, 171,* 82–84.

Wurtz, R. H., Goldberg, M. E., & Robinson, D. L. (1982). Brain mechanisms of visual attention. *Scientific American, 246,* 124–135.

Zagha, E., Casale, A. E., Sachdev, R. N., McGinley, M. J., & McCormick, D. A. (2013). Motor cortex feedback influences sensory processing by modulating network state. *Neuron, 79,* 567–578.

Zénon, A., & Krauzlis, R. J. (2012). Attention deficits without cortical neuronal deficits. *Nature, 489,* 434–437.

Zhang, S., Xu, M., Kamigaki, T., Do, J. P. H., Chang, W. C., Jenvay, S., Miyamichi K., Luo, L., & Dan, Y. (2014). Long-range and local circuits for top-down modulation of visual cortex processing. *Science, 345,* 660–665.

30 Online and Off-Line Memory States in the Human Brain

EDWARD AWH AND EDWARD K. VOGEL

ABSTRACT Working memory (WM) allows us to hold information "in mind" to support virtually all forms of complex cognition. Embedded process models of WM refer to a highly restricted set of representations that can be held in the *focus of attention* and distinguished from the passively stored representations in long-term memory, or *activated* long-term memory. Here, we review recent work that has identified neural signals that track the online components of memory, including the number of items stored and the content of those representations, as well as individual differences in WM capacity. These studies suggest that the focus of attention is not a monolithic process but depends on a collaboration between at least two distinct processes that support item-based memory and the spatial indexing of the prioritized items. Because of their tight link with behavioral indices of the focus of attention, we suggest that these components of WM delay activity may provide a powerful tool for characterizing the complex interplay between the online and off-line components of memory, both of which are critical for intelligent behavior.

Working memory (WM) is an "online" memory system where information can be readily accessed in the service of ongoing cognitive tasks. While the centrality of WM for intelligent behaviors is well accepted, modern conceptions of WM acknowledge that a satisfying model of this process requires an explicit characterization of how it interacts with other forms of memory (e.g., Cowan, 1999). For example, Cowan proposed an embedded process perspective in which the online contents of WM are restricted to three or four items—referred to as the focus of attention—that comprise a subset of the activated portion of long-term memory (LTM). Thus, his model asserts three distinct states of memory: first, all of the representations stored in LTM; second, the "activated" portion of LTM, where representations are latent but more readily accessible because of recency or contextual priming; and finally, a handful of representations that can be maintained online in the focus of attention. Critically, the performance of virtually any complex task engages all three aspects of memory. Other variations of this embedded process perspective (e.g., Ericsson & Delaney, 1999; Jonides et al., 2008; Oberauer, 2002) differ in terms of the number of "layers" of memory that are distinguished and the capacity limits implied for each, but the broader perspective has stood the test of time.

Although embedded process models have provided a productive theoretical platform, they also highlight an important challenge for the interpretation of both behavioral and neural signatures of memory function. Given that representations can move fluidly between activated LTM and the focus of attention, the mere fact that a subject can recall or use a piece of information does not diagnose which aspect of memory was guiding behavior. Adding to this challenge, verbal definitions of WM highlight how WM representations are readily accessible and important for guiding ongoing cognition, but a growing body of work makes it clear that representations in activated LTM have all the same properties. For example, Ericsson and Kintsch (1995) introduced the concept of long-term WM, in which information stored in LTM is made readily available by the maintenance of efficient retrieval cues. They showed that these long-term working memories can be rapidly accessed and demonstrated how they could support complex cognitive activities, such as reading comprehension and chess. Thus, long-term WM is essentially the same thing as activated LTM, which in turn shares many properties with the focus of attention. These similarities pose a challenge for distinguishing between the systems on the basis of behavioral data. Indeed, the observation of similar empirical patterns for short-term and long-term memory tasks has been used to challenge whether it is productive to maintain the distinction between WM and LTM (Crowder, 1982; Öztekin, Davachi, & McElree, 2010).

We believe there are strong reasons to maintain this theoretical distinction. First, we'll review compelling evidence that the focus of attention is subject to a relatively strict capacity limit. While controversy has arisen over the nature of these limits, most models agree that WM is much more restricted to small amounts of information compared to the vast capacity for storage in LTM. Decades of work have left little doubt that individual differences in WM capacity are strong predictors of broad cognitive ability (Fukuda et al., 2010; Unsworth et al., 2014). Critically, studies of individual differences have also revealed that WM and LTM ability are best modeled with separate latent variables that explain distinct variances in fluid intelligence (e.g., Unsworth & Engle, 2007). Thus, lumping together WM and LTM constructs undermines the goal of characterizing the unique components of intellectual function. Finally, representations in WM are associated with sustained patterns of neural activity that track both the number (Todd & Marois, 2004; Vogel & Machizawa, 2004) and content (Harrison & Tong, 2009; Serences et al., 2009) of stored items. By contrast, storage in LTM is mediated by changes in synaptic connectivity that enable the reinstantiation of latent memories into an online state. Thus, even though the experimental paradigms focused on LTM and WM elicit activity in similar cortical and subcortical regions (Jonides et al., 2008), there is still a clear neural distinction between active and passive representations of past experience. Indeed, our view is that focusing on the neural substrates of these processes may provide better traction for determining when and how each memory system is contributing to ongoing cognition.

Focus of Attention: Capacity Limits

Within these embedded process models of memory, the focus of attention construct is thought to determine one of WM's most notable features: its sharply limited capacity. It has long been known that only a small amount of information can be accurately held in WM at a given moment (reviewed in Cowan, 2001). For example, Luck and Vogel (1997) found that observers were nearly perfect at remembering the color of arrays of up to three items, but that performance systematically declined for larger arrays. This result is consistent with a capacity limit of three items, but the same pattern is also consistent with the storage of all items with reduced fidelity as the number of items stored increases. Thus, while the Luck and Vogel findings pointed to a sharp capacity limit, they did not establish whether a limit exists on the number of items that can be stored. Zhang and Luck (2008) helped advance this debate by developing an analytical approach to separately measure the probability that an item is stored, as well as the precision of the stored representations. This work provided some of the first clear evidence that subjects failed to store more than about three items and were reduced to random guesses when the number of items exceeded this relatively low item limit. However, Zhang and Luck's findings could also be well fit by a model that proposed that all items were stored but with wide variations in the precision of the memories (van den Berg et al., 2012). From this view, some items from an array are precisely stored, and others are imprecisely stored in memory; critically, however, all items are stored regardless of their number. In a meta-analysis, van den Berg et al. (2014) found that while models asserting an item limit had a numerical advantage over models that denied storage failures, the difference was not large enough to provide clear evidence for one over the other.

Recently, Adam, Vogel, and Awh (2017) attempted to break this theoretical stalemate using a whole report procedure that tested memory for all items on each trial. This whole-report procedure provides a richer picture of performance across all items in a trial than the typical procedures that randomly probe a single item. They found that for arrays of six items, a strong majority of subjects exhibited random guessing distributions for three of the six items (indicating that these three items were completely absent from WM). Moreover, this empirical pattern was clear enough to break the deadlock between models, providing compelling evidence against models that deny item limits in visual WM tasks. Interestingly, the leading model that denies item limits still provided a tight fit to the aggregate data in this experiment, but a closer inspection revealed that this model posits a high prevalence of "memories" that are literally indistinguishable from random guesses. In other words, if subjects actually have more than three to four items in memory, the representations of these items are so imprecise that they cannot be distinguished from completely random guesses. Finally, subjects' self-reports of whether they felt they were guessing closely tracked the guessing rates estimated by Zhang and Luck's analytical procedure. Thus, both quantitative modeling of subjects' responses, as well as their own reports of whether or not they had information, suggest that a strictly limited number of representations, rather than low-precision representations, best explain limits in WM performance. However, because these studies relied exclusively on behavioral responses, a critical ambiguity still persists: At what stage are these item capacity limits imposed? While many models propose a limit to the number of items that can be stored, a prominent class of models suggest that these limits arise only when the information in memory is being accessed at test (Oberauer & Lin, 2017). With behavior alone it is difficult to discern what stage of processing yielded these capacity limits, which is one reason why there has been strong motivation to develop neural measures that can track the online representations in WM throughout the processing stages that lead up to a behavioral response.

Neural Evidence for Sustained Activity during Working Memory

Characterizing the mechanics of WM in the brain has been a challenging exercise over the past 45 years. We have long known that various measures of neural activity show what appears to be sustained activity during the retention interval of WM tasks. For example, many cells in parietal and prefrontal cortical areas show what is often referred to as *delay activity*, in which cells show above-baseline firing rates during the maintenance phase of delayed match to sample tasks (Fuster & Alexander, 1971). Often this delay activity is observed only for memoranda that match the selectivity of the recorded cell, such as its position (Chaffee & Goldman-Rakic, 1998) or visual identity (Miller, Li, & Desimone, 1993). In other words, neurons that produce a sensory response to a stimulus also show sustained activity when the item is being maintained in WM.

Recent theoretical and empirical work, however, has questioned whether this activity is truly persistent and sustained. In particular, many neurons that contribute to WM performance are heterogeneous with regard to both their stimulus selectivity and time course. While some show clear patterns of sustained firing, many others show sporadic bursts of activity throughout the retention period. These results have been argued to support the notion that WM activity may not actually be sustained and persistent but instead supported by brief "ripples" of neural activity. This view is generally consistent with models that argue for activity-silent changes in synaptic connectivity that mediate WM storage (Lundqvist, Herman, & Miller, 2018; Stokes, 2015).

However, WM activity encompasses much of the cortex (e.g., Ester, Sprague, & Serences, 2015), and individual neuron activity may provide a too-limited view to characterize whether item-specific delay activity is sustained in this large-scale system. Much recent progress has been made when examining activity pooled across many heterogeneous individual cells, which gives the opportunity to characterize *population-level* responses. For example, Murray et al. (2017) used a dynamical *bump-attractor* model of WM that produced sustained and highly stable population responses despite being based on data from a large number of highly heterogeneous individual cells, many of which did not exhibit sustained activity. This work suggests that stable and persistent WM representations may be an emergent property of a large-scale population response with heterogeneous neural inputs. Thus, sporadic or dynamic representations that are observed within a small subset of neurons may not provide compelling evidence against the hypothesis that online representations in WM are supported by sustained delay period activity.

Neural Evidence for the Focus of Attention Construct

Most of human neuroscience relies on population level signals such as blood oxygen level dependent (BOLD) observed in functional magnetic resonance imaging (fMRI) and electroencephalogram (EEG) activity measured at the human scalp. These methods have provided many demonstrations of sustained neural responses during WM tasks. Numerous areas in inferior temporal, parietal, and prefrontal cortex show increased BOLD activation during WM retention periods. This set of cortical areas expands to include many more regions, such as V1, when sensitive multivariate analyses are used to decode the actual feature value of the memoranda rather than the mean amplitude of BOLD signals within each region (Ester et al., 2013; Harrison & Tong, 2009; Serences et al., 2009). These new analyses provide content-specific evidence for maintained representations held in WM. However, because fMRI has a poor temporal resolution, it is difficult to discern whether activity at a given moment reflects actively represented information or the lingering trace of information recently in the focus of attention. Considering this limitation, EEG recordings that reveal storage-related neural activity offer important advantages for characterizing the nature of WM delay activity because the excellent temporal resolution of the method is better able to reveal the time course of an ephemeral memory trace.

Initial EEG work by Ruchkin and colleagues (1990) reported a sustained negative-voltage slow wave during the retention period of WM tasks. While the activity showed distinct scalp topographies from visual and verbal memoranda, the nonspecific nature of the activity made it difficult to distinguish from other nonmnemonic activity general to most tasks, such as perceptual responses, arousal, and response anticipation. Vogel and Machizawa (2004) developed a lateralized version of a change detection WM paradigm that allowed them to better isolate the neural activity generated by WM-related processes (see figure 30.1*A*). Stimuli are presented bilaterally while subjects hold central fixation and are instructed to remember only the objects in a single visual hemifield. Shortly following the onset of the memory items, a sustained negative-going voltage is observed at posterior electrode sites over the hemisphere contralateral to the to-be-remembered items. A difference wave subtracting the ipsilateral activity from the contralateral activity can be used to observe the

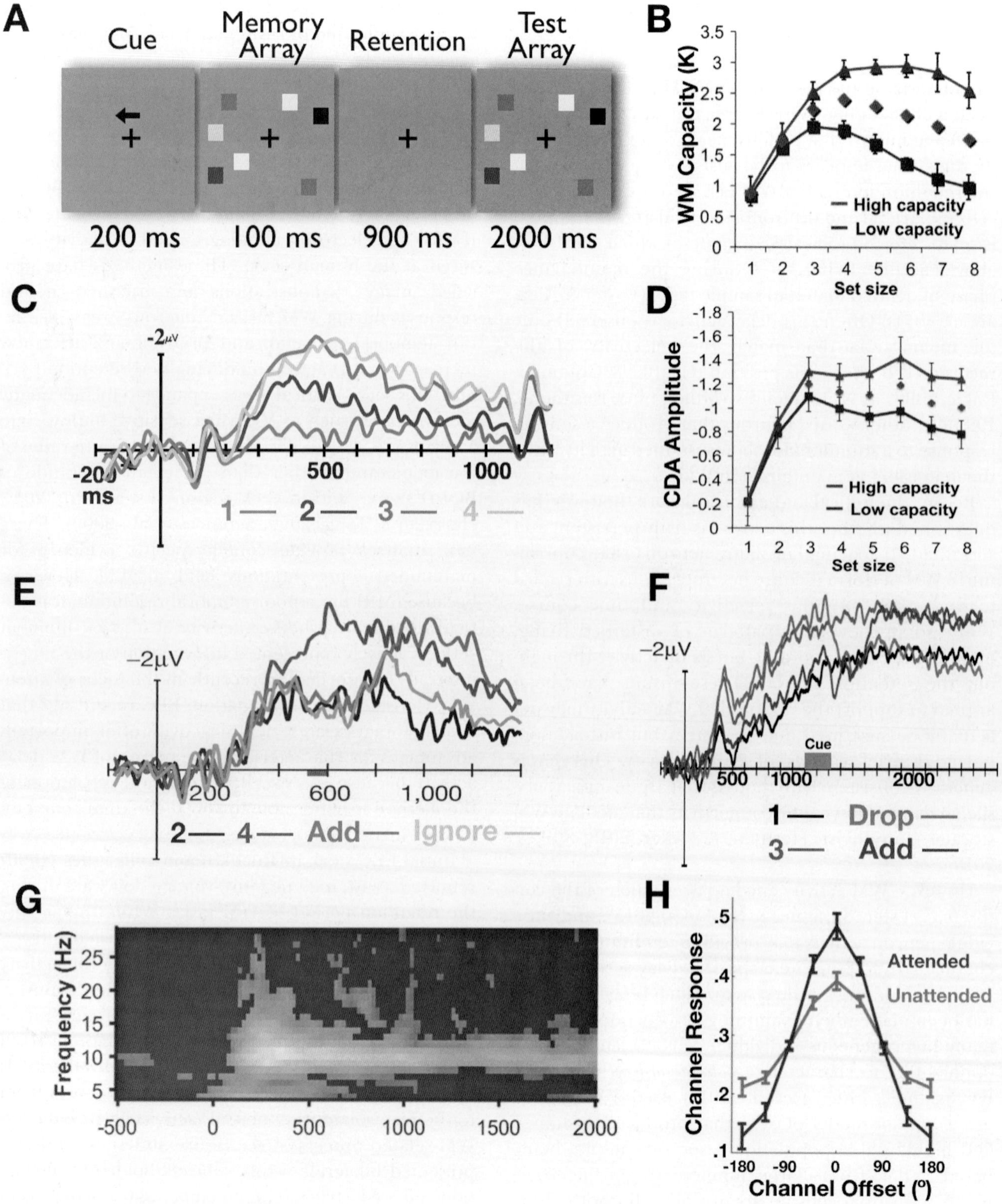

FIGURE 30.1 *A*, Stimuli and procedure for a typical CDA WM paradigm. *B*, Behavioral capacity estimates (K) across set sizes for high- and low-WM-capacity individuals. *C*, CDA amplitudes as a function of the number of memory items. *D*, CDA amplitude across set size for high- and low-WM-capacity individuals. *E*, CDA for sequential displays. In the Add condition, a two-item array is followed by another two-item array that must be stored (i.e., 2+2). In the Ignore condition, a two-item array is followed by another two-item array that must be ignored. *F*, CDA for dynamic load changes. In the Add condition, subjects initially tracked one item and then, following a cue, began tracking two additional items (i.e., 1+3). In the Drop condition, subjects tracked three items but were instructed to drop two of those items (i.e., 3−2). *G*, Sustained location selectivity about remembered position is concentrated in the alpha band (8–12 Hz). *H*, Alpha channel-tuning functions (CTF) show a graded profile of channel activity-tracking locations in WM for attended and unattended items. (See color plate 32.)

properties of this component, often referred to as the contralateral delay activity (CDA). This procedure isolates the activity specific to the selection and storage of the memoranda while controlling for the general arousal and sensory stimulation equated between the two hemispheres.

Contralateral Delay Activity as an Index of the Number of Objects in the Focus of Attention

The CDA has proven to be a useful tool for studying WM and attention-related phenomena across a wide range of task contexts, such as perceptual monitoring (Tsubomi et al., 2013), mental rotation (Prime & Jolicoeur, 2010), filtering and attentional capture (Vogel, McCollough, & Machizawa, 2005), visual search (Emrich et al., 2009), and multiple object tracking (Drew & Vogel, 2008). Across these contexts, the CDA has shown itself to be a robust signal that tracks the currently relevant items that the subject must represent to perform the task. This is primarily because the CDA shares several characteristic properties with those attributed to the focus of attention construct within WM models. Notably, the CDA is primarily sensitive to how many items are currently being attended. CDA amplitude increases as a function of the number of items currently held in memory (see figure 30.1*C*, *D*). Critically, the activity reaches a limit at three items, which is comparable to the typically assumed capacity limit. Furthermore, the activity is highly sensitive to individual differences in behaviorally measured capacity: high-capacity individuals show stable amplitudes at large array sizes, while low-capacity individuals show a decrease in CDA amplitude (Fukuda, Mance, & Vogel, 2015; Vogel & Machizawa, 2004). In addition to its sensitivity to between-subject variability in WM capacity, its amplitude also tracks trial-to-trial fluctuations in the number of successfully maintained items (Adam, Robison, & Vogel, 2018).

The CDA has also been shown to respect the presence of grouping cues, which effectively reduce the set of memoranda by allowing them to be "chunked" into fewer total items (Luria & Vogel, 2011; Peterson & Berryhill, 2013). When gestalt factors such as similarity, connectedness, and common fate can be utilized to decrease the effective number of items that must be maintained, the CDA shows predictable reductions in amplitude. Importantly, CDA amplitude is generally insensitive to manipulations of the visual arrays that are not expected to change the current memory load but will affect other task processes, such as the effort and difficulty of discrimination. For example, reducing the visual contrast of the memoranda does not affect CDA amplitude, even though it makes the task more difficult and reduces accuracy (Ikkai, McCollough, & Vogel, 2010; Luria et al., 2010). Furthermore, manipulations of the spatial extent of the attended memory items (i.e., near or far spacing) have a negligible impact on the CDA, further suggesting that it is primarily modulated by the number of items per se rather than the size of the attended region (Drew & Vogel, 2008; McCollough, Machizawa, & Vogel, 2007). Taken together, these findings point to an item-based interpretation of CDA activity, such that CDA amplitude reflects the number of individuated representations in memory, rather than details about the number of feature values or the number of elements within a visual chunk.

Contralateral Delay Activity Quickly Responds to Dynamic Changes in Current Focus

In many task contexts, the current contents of the focus are presumed to rapidly change as the trial progresses over time. Likewise, CDA amplitude rapidly responds to changes in what is currently being held in the focus rather than statically representing what was initially encoded. This property was initially observed in WM tasks in which the items are presented sequentially across two separate arrays (i.e., two items + two items), compared to simultaneously presented displays of equivalent total set size. As shown in figure 30.1*E*, following the first array, the amplitude reaches a two-item level but then quickly rises to a four-item level following presentation of the second array. Importantly, this rise does not occur obligatorily for all object onsets but is primarily for items that the subject is attempting to encode (Vogel, McCollough, & Machizawa, 2005). This property can also be observed in task contexts in which subjects are cued to update the contents of the focus by switching which items must be attended in the middle of the trial. For example, Drew et al. (2012) found that CDA amplitude rapidly changed to reflect the new current number of attended items when cues instructed subjects to either add new items or drop existing items from being attended (figure 30.1*F*). Recent work from Luria and colleagues (Balaban & Luria, 2017; Balaban, Drew, & Luria 2018) has extended this demonstration to contexts in which the set of attended items must be reinterpreted because of dynamic changes to the objects themselves. When a single attended object moves about the screen and then splits into two independently moving objects, the CDA quickly "resets" to the new set size because the prior interpretation of the item in the focus is no longer valid (Balaban & Luria, 2017). Together, these results highlight the orderly changes in CDA amplitude to dynamic changes in the number of attended items.

Contralateral Delay Activity, Alpha, and Spatial Attention

Another EEG-based candidate for the focus of attention construct is oscillatory activity in the alpha band (8–12 Hz), which similarly shows sustained modulations during the retention period and sensitivity to ongoing task demands (see chapter 28 for a detailed discussion of alpha-band EEG oscillations). Indeed, the similarity between the CDA and alpha has led to proposals that they may be related or even the same activity. Specifically, van Dijk et al. (2010) proposed that the CDA is an averaging artifact of trial-level modulations of alpha activity, which they showed in simulations could produce a sustained slow wave similar to the CDA. Because alpha has primarily been viewed as a spatial signal, they argued that the CDA reflects attention to the positions of the items rather than the item representations themselves. Fukuda et al. (2015a) tested this possibility by measuring the alpha power response across manipulations of set size, which are known to produce characteristic responses in the CDA. Consistent with the initial proposal, alpha power was reduced as the number of items increased, reaching an asymptote around three to four items. Further, the difference in alpha power between the low and high set sizes predicted the subject's WM capacity. Although this empirical pattern is nearly identical to that typically observed for the CDA, the two measures of sustained activity were clearly dissociated. First, the two components were uncorrelated with each other, and while both predicted WM capacity, they explained distinct variance in WM scores. Second, in an experiment manipulating retention interval, CDA and alpha indices of set size persisted for different durations. These two results support the provocative suggestion that the focus of attention may not simply be a monolithic process applied to attended items. It may instead comprise at least two complementary but distinct facets of neural activity.

Hakim et al. (2018) bolstered this hypothesis by testing whether sustained spatial attention alone is sufficient to drive the CDA response without item-based storage. They compared neural activity in WM and attention tasks that employed the same displays but in which only the WM task encouraged storage of the items in the sample display. In the WM task, subjects stored two or four items from one side of space. In the attention task, subjects instead attended to the positions of the colors in anticipation of an occasional brief target whose orientation had to be discriminated. The parameters of the target discrimination task were such that it was necessary to sustain attention to the precise positions of the cue items, thereby matching the requirement to maintain location information across the WM and attention tasks. In line with the expectation that both tasks would recruit spatial attention to the relevant side, both tasks produced highly reliable modulations of sustained contralateral alpha power. In the WM task, a large set-size-dependent CDA was produced as expected. However, in the attention task, virtually no CDA was observed. This indicates that despite evidence for sustained spatial attention and the maintenance of precise location information, the CDA was not engaged when attention was directed for the purpose of apprehending new items instead of storing the objects in the sample display. These results provide initial evidence that these two neural measures of the focus of attention may play distinct roles: one that represents objects in active memory and another that provides a map of currently prioritized space (see also Bae & Luck, 2018).

Alpha and Prioritized Space

The modulations of contralateral alpha power in the Hakim et al. (2018) study fall in line with a longstanding body of work showing that the scalp topography of alpha oscillations tracks currently attended positions (Fox & Snyder, 2011). Moreover, recent work has demonstrated that alpha topography precisely tracks the relevant position in a hemifield, not just the attended side of space (e.g., Foster, Sutterer, et al., 2017; Rihs et al., 2007). In line with past work that has shown strong links between spatial attention and WM (Awh & Jonides, 2001), Foster et al. (2016) used a multivariate encoding model to show that alpha topography precisely tracked locations stored in spatial WM (see figure 30.1*G*). A highlight of this analytic approach is that the encoding model provided a visualization of the full distribution of activity associated with all possible positions in the task, yielding a *channel-tuning function* (CTF) that peaked at the stored position and declined at positions farther away. Thus, the spatial information encoded in alpha activity has the graded character that is a hallmark of sensory representations of space. Moreover, it is straightforward to quantify the basic tuning properties of these CTFs, providing new insights about how the precision of spatially selective neural activity is affected by various experimental factors. For instance, as shown in figure 30.1*H*, Foster, Bsales, et al. (2017) showed that the amplitude of CTFs was substantially higher for voluntarily stored items compared to distracters (see also Ester et al., 2018). Likewise, Foster,

Sutterer, et al. (2017) used CTFs to demonstrate that the timing of spatially selective responses in the alpha band predicts visual search latencies, an analysis that required moment-by-moment quantification of the spatial selectivity of alpha activity. Finally, this alpha index of spatial position is also robustly observed during nonspatial WM tasks in which position is irrelevant to the behavioral task (Foster, Bsales, et al., 2017), suggesting that alpha activity may be integral to storage in visual WM even when position is not behaviorally relevant.

These findings suggest that at least two distinct neural signals track items within the focus of attention. The CDA indexes the number of items in WM, even when the number of relevant items in displays shifts from one moment to the next. By contrast, alpha activity—while sensitive to the number of items stored—explains distinct variance in WM capacity and often follows a time course that is distinct from that of the CDA. Our working hypothesis is that alpha activity reflects a spatial-indexing signal that tracks the position of prioritized items in WM and may facilitate the rehearsal and access of information from visual WM. Thus, the neural activity supporting the focus of attention reflects a collaboration between multiple processes that play distinct roles in online memory.

Characterizing the Collaboration between Online and Off-Line Memory Processes

While we have emphasized the active neural signals that track information in the focus of attention, recent *activity-silent* conceptions of WM storage have challenged whether persistent delay activity is integral to storage in WM (e.g., Lewis-Peacock et al., 2012; Rose et al. 2016; Stokes, 2015). A central motivation for activity-silent models of WM storage is the finding that neurally active delay signals are not always sustained throughout the time between encoding and testing the memory. For instance, Lewis-Peacock et al. (2012) showed that when subjects were cued to pay attention to a subset of the items they had encoded into WM, neural activity tracking the unattended memory item dropped to baseline. When attention returned to that item, the neural activity tracking that item returned. Thus, the authors argued that active neural signals are not integral to storage in WM because behavioral performance remains intact despite the waxing and waning of those signals. In line with this interpretation, Stokes (2015) has proposed that WM storage is accomplished primarily via rapid changes in the pattern of synaptic weights that maintain information in a manner similar to that posited for LTM. Here, information is stored in a passive manner that enables the rapid reactivation of recently attended information. This mode of storage is less metabolically demanding and may be particularly well suited for guiding comparisons between new inputs and recently attended ones. Indeed, more recent studies have shown that transcranial magnetic stimulation (Rose et al., 2016) or irrelevant visual stimulation (Wolff et al., 2017) can elicit a reactivation of neural signals that track information recently encoded into WM, supporting the hypothesis that latent representations can be brought back into mind by nonspecific input signals that reactivate potentiated neural connections.

On the one hand, the recent work on activity-silent memory has provided an exciting new window into the neural mechanisms that can support the retention of information over brief delays (e.g, Rose et al., 2016; Wolff et al., 2017). This perspective underscores the importance of passive memory processes in the brain, much like the activated LTM component of embedded process models. On the other hand, there is room for debate regarding the most productive way to position these activity-silent phenomena within a taxonomy of memory. Is a rapid shift of synaptic weights—in the absence of active neural signals—best understood as *working memory*? One might presume so, given that behavioral tests show that subjects can still access the target information following the short delay. But this interpretation presumes that WM is the only memory system that maintains information across short delays when it has been understood for decades (e.g., Atkinson & Shiffrin, 1968) that multiple memory systems, including activated LTM and LTM, can guide behavior in short-delay tasks. If both WM and LTM contribute to performance in such tasks, activity-silent periods may simply reflect a temporary off-loading of information from WM so that limited resources can be directed elsewhere (Rhodes & Cowan, 2018).

In our view the distinction between WM and LTM is well motivated, and the presence of active neural representations of the memoranda may be a productive way to draw a line between the two. There are multiple arguments for this taxonomy. First, this scheme has high face validity because it dovetails with the common conception of WM as an "online" memory system in which information is held "in mind." Indeed, a common thread in recent demonstrations of activity-silent memory is that subjects—either by instruction or because of the demands of an intervening task—are forced to direct attention away from initially encoded information. Thus, activity-silent memories have typically been referred to as *unattended memory items* (e.g., Lewis-Peacock et al., 2012; Rose et al., 2016; Wolff et al., 2017). Second, individual difference studies using very similar tasks show that a person's ability to retrieve those unattended memory items is well predicted by standard measures of LTM retrieval, such as the free recall of word lists (Unsworth et al., 2014). Thus, associating WM with active neural signals captures the common conception of WM as the subset of information currently in mind, as well as the structure of memory abilities as revealed by individual differences. Of course, this argument does not minimize the importance of activated LTM for ongoing cognition. A central virtue of the embedded process models is their acknowledgment that both active and passive aspects of memory are critical for virtually any complex cognitive task.

That said, this discussion underscores an important area for future research. What are the key functional differences between representations stored online in WM and the contents of activated LTM? For instance, subjects have voluntary control over encoding into WM, and when information is no longer needed, it can be dropped (Williams & Woodman, 2012). While it has been postulated that the sustained maintenance of activity-silent representations may be contingent on current behavioral relevance (e.g., Rose et al., 2016), other work has shown that this may occur even for recently attended but currently irrelevant representations (e.g., Bae and Luck, 2019). Thus, more work is needed to determine the boundary conditions for reactivation. Many studies have shown that recently attended or rewarded events can elicit subsequent attentional capture even when it is contrary to the subject's current goals. Likewise, the contents of past trials shape the responses to items in the present, even though past trials are completely irrelevant. Thus, given that recently attended items often exert influence when they are behaviorally irrelevant (Awh, Belopolsky, & Theeuwes, 2012), more work is needed to determine the relationship between activity-silent representations and voluntary control.

Our intent is not to promote endless debate over how to label various memory phenomena. Our goal is to consider how different conceptions of WM and LTM may provide the most productive platform for understanding how these memory systems interact to guide intelligent behaviors. Even amid any ongoing controversy regarding the best way to categorize different memory phenomena, there is nevertheless a consensus that we should push forward with the effort to link robust behavioral indices of memory function with clear models of the underlying neural processes. Thus, no matter where one might choose to draw the line between WM and LTM, the effort to connect brain and behavior will be critical for understanding this core cognitive process.

Acknowledgment

This research was supported by National Institute of Mental Health Grant 5R01 MH087214-08 and Office of Naval Research Grant N00014-12-1-0972.

REFERENCES

Adam, K. C., Robison, M. K., & Vogel, E. K. (2018). Contralateral delay activity tracks fluctuations in working memory performance. *Journal of Cognitive Neuroscience, 30*(9), 1229–1240.

Adam, K. C., Vogel, E. K., & Awh, E. (2017). Clear evidence for item limits in visual working memory. *Cognitive Psychology, 97*, 79–97.

Atkinson, R. C., & Shiffrin, R. M. (1968). Human memory: A proposed system and its control processes. In *Psychology of learning and motivation* (Vol. 2, pp. 89–195). Academic Press.

Awh, E., Belopolsky, A. V., & Theeuwes, J. (2012). Top-down versus bottom-up attentional control: A failed theoretical dichotomy. *Trends in Cognitive Sciences, 16*(8), 437–443.

Awh, E., & Jonides, J. (2001). Overlapping mechanisms of attention and spatial working memory. *Trends in Cognitive Sciences, 5*(3), 119–126.

Bae, G. Y., & Luck, S. J. (2018). Dissociable decoding of spatial attention and working memory from EEG oscillations and sustained potentials. *Journal of Neuroscience*, 38, 409–422.

Bae, G. Y., & Luck, S. J. (2019). Reactivation of Previous Experiences in a Working Memory Task. *Psychological Science*, 0956797619830398.

Balaban, H., Drew, T., & Luria, R. (2018). Delineating resetting and updating in visual working memory based on the object-to-representation correspondence. *Neuropsychologia, 113*, 85–94.

Balaban, H., & Luria, R. (2017). Neural and behavioral evidence for an online resetting process in visual working memory. *Journal of Neuroscience, 37*(5), 1225–1239.

Chafee, M. V., & Goldman-Rakic, P. S. (1998). Matching patterns of activity in primate prefrontal area 8a and parietal area 7ip neurons during a spatial working memory task. *Journal of Neurophysiology, 79*(6), 2919–2940.

Cowan, N. (1999). An embedded-processes model of working memory. *Models of Working Memory: Mechanisms of Active Maintenance and Executive Control, 20*, 506.

Crowder, R. G. (1982). The demise of short-term memory. *Acta Psychologica, 50*(3), 291–323.

Drew, T., Horowitz, T. S., Wolfe, J. M., & Vogel, E. K. (2012). Neural measures of dynamic changes in attentive tracking load. *Journal of Cognitive Neuroscience, 24*(2), 440–450.

Drew, T., & Vogel, E. K. (2008). Neural measures of individual differences in selecting and tracking multiple moving objects. *Journal of Neuroscience, 28*(16), 4183–4191.

Emrich, S. M., Al-Aidroos, N., Pratt, J., & Ferber, S. (2009). Visual search elicits the electrophysiological marker of visual working memory. *PloS One, 4*(11), e8042.

Ericsson, K. A., & Delaney, P. F. (1999). Long-term working memory as an alternative to capacity models of working memory in everyday skilled performance. In A. Miyake & P. Shah (Eds.), *Models of working memory: Mechanisms of active maintenance and executive control* (pp. 257–297). New York: Cambridge University Press.

Ericsson, K. A., & Kintsch, W. (1995). Long-term working memory. *Psychological Review, 102*(2), 211.

Ester, E. F., Anderson, D. E., Serences, J. T., & Awh, E. (2013). A neural measure of precision in visual working memory. *Journal of Cognitive Neuroscience, 25*(5), 754–761.

Ester, E. F., Nouri, A., & Rodriguez, L. (2018). Retrospective cues mitigate information loss in human cortex during working memory storage. *Journal of Neuroscience, 38*(40), 8538–8548.

Ester, E. F., Sprague, T. C., & Serences, J. T. (2015). Parietal and frontal cortex encode stimulus-specific mnemonic representations during visual working memory. *Neuron, 87*(4), 893–905.

Foster, J. J., Bsales, E. M., Jaffe, R. J., & Awh, E. (2017). Alpha-band activity reveals spontaneous representations of spatial position in visual working memory. *Current Biology, 27*(20), 3216–3223.

Foster, J. J., Sutterer, D. W., Serences, J. T., Vogel, E. K., & Awh, E. (2016). The topography of alpha-band activity tracks the content of spatial working memory. *Journal of Neurophysiology, 115*(1), 168–177.

Foster, J. J., Sutterer, D. W., Serences, J. T., Vogel, E. K., & Awh, E. (2017). Alpha-band oscillations enable spatially and temporally resolved tracking of covert spatial attention. *Psychological Science, 28*(7), 929–941.

Foxe, J. J., & Snyder, A. C. (2011). The role of alpha-band brain oscillations as a sensory suppression mechanism during selective attention. *Frontiers in Psychology, 2,* 154.

Fukuda, K., Mance, I., & Vogel, E. K. (2015a). α power modulation and event-related slow wave provide dissociable correlates of visual working memory. *Journal of Neuroscience, 35*(41), 14009–14016.

Fukuda, K., Vogel, E., Mayr, U., & Awh, E. (2010). Quantity, not quality: The relationship between fluid intelligence and working memory capacity. *Psychonomic Bulletin & Review, 17*(5), 673–679.

Fukuda, K., Woodman, G. F., & Vogel, E. K. (2015b). Individual differences in visual working memory capacity: Contributions of attentional control to storage. *Mechanisms of Sensory Working Memory: Attention and Performance, 25,* 105.

Fuster, J. M., & Alexander, G. E. (1971). Neuron activity related to short-term memory. *Science, 173*(3997), 652–654.

Gao, Z., Ding, X., Yang, T., Liang, J., & Shui, R. (2013). Coarse-to-fine construction for high-resolution representation in visual working memory. *PloS One, 8*(2), e57913.

Hakim, N., Adam, K. C., Gunseli, E., Awh, E., & Vogel, E. K. (2018). Dissecting the neural focus of attention reveals distinct processes for spatial attention and object-based storage in visual working memory. *Psychological Science,* 0956797619830384.

Harrison, S. A., & Tong, F. (2009). Decoding reveals the contents of visual working memory in early visual areas. *Nature, 458*(7238), 632.

Ikkai, A., McCollough, A. W., & Vogel, E. K. (2010). Contralateral delay activity provides a neural measure of the number of representations in visual working memory. *Journal of Neurophysiology, 103*(4), 1963–1968.

Jensen, O., & Mazaheri, A. (2010). Shaping functional architecture by oscillatory alpha activity: Gating by inhibition. *Frontiers in Human Neuroscience,* 4, 186.

Jonides, J., Lewis, R. L., Nee, D. E., Lustig, C. A., Berman, M. G., & Moore, K. S. (2008). The mind and brain of short-term memory. *Annual Review of Psychology, 59,* 193–224.

Lewis-Peacock, J. A., Drysdale, A. T., Oberauer, K., & Postle, B. R. (2012). Neural evidence for a distinction between short-term memory and the focus of attention. *Journal of Cognitive Neuroscience, 24*(1), 61–79.

Luck, S. J., & Vogel, E. K. (1997). The capacity of visual working memory for features and conjunctions. *Nature, 390*(6657), 279.

Lundqvist, M., Herman, P., & Miller, E. K. (2018). Working memory: Delay activity, yes! Persistent activity? Maybe not. *Journal of Neuroscience, 38*(32), 7013–7019.

Luria, R., Sessa, P., Gotler, A., Jolicœur, P., & Dell'Acqua, R. (2010). Visual short-term memory capacity for simple and complex objects. *Journal of Cognitive Neuroscience, 22*(3), 496–512.

Luria, R., & Vogel, E. K. (2011). Shape and color conjunction stimuli are represented as bound objects in visual working memory. *Neuropsychologia, 49*(6), 1632–1639.

McCollough, A. W., Machizawa, M. G., & Vogel, E. K. (2007). Electrophysiological measures of maintaining representations in visual working memory. *Cortex, 43*(1), 77–94.

Miller, E. K., Li, L., & Desimone, R. (1993). Activity of neurons in anterior inferior temporal cortex during a short-term memory task. *Journal of Neuroscience, 13*(4), 1460–1478.

Murray, J. D., Bernacchia, A., Roy, N. A., Constantinidis, C., Romo, R., & Wang, X. J. (2017). Stable population coding for working memory coexists with heterogeneous neural dynamics in prefrontal cortex. *Proceedings of the National Academy of Sciences, 114*(2), 394–399.

Oberauer, K. (2002). Access to information in working memory: exploring the focus of attention. *Journal of Experimental Psychology: Learning, Memory, and Cognition, 28*(3), 411.

Oberauer, K., & Lin, H. Y. (2017). An interference model of visual working memory. *Psychological Review, 124*(1), 21.

Öztekin, I., Davachi, L., & McElree, B. (2010). Are representations in working memory distinct from representations in long-term memory? Neural evidence in support of a single store. *Psychological Science, 21*(8), 1123–1133.

Peterson, D. J., & Berryhill, M. E. (2013). The gestalt principle of similarity benefits visual working memory. *Psychonomic Bulletin & Review,* 20(6), 1282–1289.

Prime, D. J., & Jolicoeur, P. (2010). Mental rotation requires visual short-term memory: Evidence from human electric cortical activity. *Journal of Cognitive Neuroscience, 22*(11), 2437–2446.

Rhodes, S., & Cowan, N. (2018). Attention in working memory: Attention is needed but it yearns to be free. *Annals of the New York Academy of Sciences, 1424*(1), 52–63.

Rihs, T. A., Michel, C. M., & Thut, G. (2007). Mechanisms of selective inhibition in visual spatial attention are indexed by α-band EEG synchronization. *European Journal of Neuroscience, 25*(2), 603–610.

Rose, N. S., LaRocque, J. J., Riggall, A. C., Gosseries, O., Starrett, M. J., Meyering, E. E., & Postle, B. R. (2016). Reactivation of latent working memories with transcranial magnetic stimulation. *Science, 354*(6316), 1136–1139.

Ruchkin, D. S., Johnson Jr., R., Canoune, H., & Ritter, W. (1990). Short-term memory storage and retention: An event-related brain potential study. *Electroencephalography and Clinical Neurophysiology, 76*(5), 419–439.

Serences, J. T., Ester, E. F., Vogel, E. K., & Awh, E. (2009). Stimulus-specific delay activity in human primary visual cortex. *Psychological Science, 20*(2), 207–214.

Stokes, M. G. (2015). 'Activity-silent' working memory in prefrontal cortex: a dynamic coding framework. *Trends in Cognitive Sciences, 19*(7), 394–405.

Todd, J. J., & Marois, R. (2004). Capacity limit of visual short-term memory in human posterior parietal cortex. *Nature, 428*(6984), 751.

Tsubomi, H., Fukuda, K., Watanabe, K., & Vogel, E. K. (2013). Neural limits to representing objects still within view. *Journal of Neuroscience, 33*(19), 8257–8263.

Unsworth, N., & Engle, R. W. (2007). The nature of individual differences in working memory capacity: Active maintenance in primary memory and controlled search from secondary memory. *Psychological Review, 114*(1), 104.

Unsworth, N., Fukuda, K., Awh, E., & Vogel, E. K. (2014). Working memory and fluid intelligence: Capacity, attention control, and secondary memory retrieval. *Cognitive Psychology, 71,* 1–26.

Van den Berg, R., Awh, E., & Ma, W. J. (2014). Factorial comparison of working memory models. *Psychological Review, 121*(1), 124.

Van den Berg, R., Shin, H., Chou, W. C., George, R., & Ma, W. J. (2012). Variability in encoding precision accounts for visual short-term memory limitations. *Proceedings of the National Academy of Sciences, 109*(22), 8780–8785.

van Dijk, H., van der Werf, J., Mazaheri, A., Medendorp, W. P., & Jensen, O. (2010). Modulations in oscillatory activity with amplitude asymmetry can produce cognitively relevant event-related responses. *Proceedings of the National Academy of Sciences, 107*(2), 900–905.

Vogel, E. K., & Machizawa, M. G. (2004). Neural activity predicts individual differences in visual working memory capacity. *Nature, 428*(6984), 748.

Vogel, E. K., McCollough, A. W., & Machizawa, M. G. (2005). Neural measures reveal individual differences in controlling access to working memory. *Nature, 438*(7067), 500.

Williams, M., Pouget, P., Boucher, L., & Woodman, G. F. (2013). Visual-spatial attention aids the maintenance of object representations in visual working memory. *Memory & Cognition, 41*(5), 698–715.

Williams, M., & Woodman, G. F. (2012). Directed forgetting and directed remembering in visual working memory. *Journal of Experimental Psychology: Learning, Memory, and Cognition, 38*(5), 1206.

Wolff, M. J., Jochim, J., Akyürek, E. G., & Stokes, M. G. (2017). Dynamic hidden states underlying working-memory-guided behavior. *Nature Neuroscience, 20*(6), 864.

Zhang, W., & Luck, S. J. (2008). Discrete fixed-resolution representations in visual working memory. *Nature, 453*(7192), 233.

31 How Working Memory Works

TIMOTHY J. BUSCHMAN AND EARL K. MILLER

ABSTRACT Working memory (WM) is the ability to hold things "in mind." It lies at the core of cognition, a mental sketch pad on which thoughts are held, transformed, and then used to guide actions. WM has a severely limited capacity—we can only hold a few items in mind at once. To compensate, WM is tightly controlled. Here, we review the neural mechanisms of WM. First, we review how information is maintained in WM. Next, we discuss why WM has a limited capacity. Finally, we discuss how the contents of WM are controlled.

Working memory (WM) is the contents of our thoughts, a mental sketch pad where we can hold information "in mind." We "think" by manipulating this information. For example, WM allows us to remember our coffee order or do mental arithmetic. However, despite its importance, WM has a severely limited capacity—we can only hold a few thoughts simultaneously. To compensate, the contents of WM are tightly controlled—access is regulated and unnecessary items are discarded. Here we review the neural mechanisms of WM, discuss why it may have a limited capacity, and examine how it is controlled.

Representation of Working Memory

Working memory is distributed In the past, WM was predominantly associated with the prefrontal cortex (PFC). The first discoveries of the neural correlates of WM were neurons in lateral prefrontal cortex showing elevated spiking that maintained task-relevant information over brief (1 s or more) memory delays (Fuster, 2015; Goldman-Rakic, 1995).

More recent work has shown that WM is distributed across the cortex (figure 31.1*A*; Christophel, Klink, Spitzer, Roelfsema, & Haynes, 2017). In addition to PFC, neurons in parietal and sensory cortex carry WM information, as do several subcortical regions (particularly regions connected with the PFC, such as basal ganglia and the thalamus; Passingham, 1993).

Given that WM is distributed across many cortical and subcortical areas, a key question is how all those distributed representations are organized into a seamless unitary experience. Synchronization of the brain's rhythms may play a role.

Rhythms organize working memory What mechanism could organize such scattered WM representations? It would need to be flexible and able to quickly form (and disperse) neural ensembles as items move in and out of WM. Rhythmic synchrony could serve this purpose (Fries, 2015).

The brain oscillates at different frequencies from below 1 Hz to over 100 Hz. These oscillations are synchronized across thousands to millions of neurons, allowing them to be easily detected in local field potentials (LFPs; the summed electrical activity of neurons within a few millimeters of cortex). Synchronizing the activity of neural populations can facilitate communication within the ensemble. When synchronized in phase with one another, neurons are excitable (or not) at the same time. When they are both in an excitable state, spikes from one neuron will have a greater impact on the other, facilitating communication. On the other hand, if neurons are out of sync or anticorrelated, one set of neurons may be spiking when another set is in a low state of excitement, hindering the impact of spikes and thus limiting communication between them.

Several lines of evidence suggest synchrony is involved in WM. First, areas involved in WM—frontal, parietal, sensory, and temporal cortex—become synchronized during WM tasks (figure 31.1*A*; Palva, Monto, Kulashekhar, & Palva, 2010). Second, synchrony forms memory-specific ensembles, linking together a group of neurons representing an item in WM. Evidence for this comes from observations of different patterns of synchrony between LFPs at different recording sites depending on the information being held in WM (Antzoulatos & Miller, 2014; Buschman, Denovellis, Diogo, Bullock, & Miller, 2012; Salazar, Dotson, Bressler, & Gray, 2012). The advantage of forming ensembles via rhythmic synchrony is that they are flexible. Ensembles can be formed, discarded, and then reformed, all by changing the pattern of synchrony without needing to change the physical structure. Such cognitive flexibility is a hallmark of higher cognitive functioning and of WM.

Sustained versus dynamic representations The first neurophysiological observations of WM suggested that memories were maintained by the persistent activity of neurons in response to a stimulus (figure 31.1*B*; Funahashi, Bruce, & Goldman-Rakic, 1989). The idea was

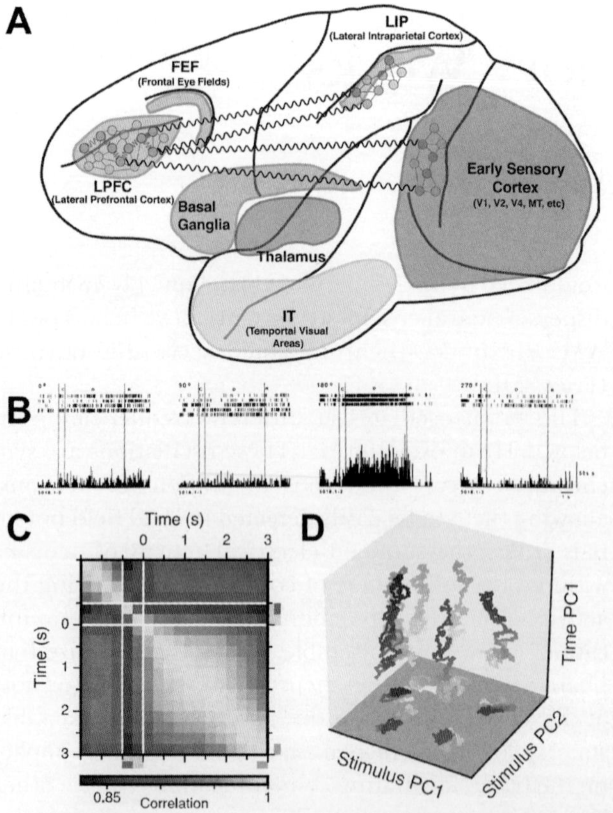

FIGURE 31.1 The neurophysiological basis of working memory. *A*, Working memory representations are distributed across the brain, including in sensory regions, parietal regions, and prefrontal regions, as well as subcortical regions, such as the basal ganglia and the thalamus. Synchrony within and between different brain regions is thought to help organize the distributed representation into a cohesive representation. *B*, Working memory is represented in the sustained neural activity of prefrontal cortex neurons. For example, a prefrontal cortex neuron persistently responds when a monkey remembers a stimulus presented to the left of fixation (*third column*) compared to when the same stimulus was presented to the right, up, or down (*other columns, from left to right*). Adapted from Funahashi, Bruce, and Goldman-Rakic (1989). *C*, Working memory representations are dynamic. Cross-temporal correlation shows that, across a population of prefrontal cortex neurons, neural activity at one time point (*x-axis*) is not well correlated with activity at other time points (*y-axis*). In particular, correlation is low between the response to the stimulus presentation (*shaded gray on x-axis*) and memory delay. Adapted from Murray et al. (2017). *D*, Dynamics are orthogonal to mnemonic subspace. Despite the dynamics seen in *C*, a mnemonic subspace exists in which different memories (*indicated by different colors*) can be stably decoded. Instead, dynamics appear to track time (*z-axis*). Adapted from Murray et al. (2017).

that an ensemble representing a stimulus is activated when that stimulus is seen. That ensemble is then held in WM by keeping it "online" in an active state. This is thought to be due to recurrent connections between the neurons that belong to the same ensemble. The idea is that once activity passes a threshold, there is enough recurrence to sustain its activity.

A common version of such a model is the *bump attractor*. In this model, neurons are topographically arranged around a ring according to their selectivity—nearby neurons share similar selectivity. Local recurrent connections then sustain initial inputs into the ring, leading to a "bump" of activity, while more distal inhibitory connections stabilize the memory in place. This leads to a persistent *attractor state*, corresponding to a specific pattern of activity. This type of model has been the dominant view of the neurobiology of WM. However, recent work is beginning to challenge that (reviewed in Lundqvist, Herman, & Miller, 2018; Stokes, 2015).

First, WM spiking is not as persistent as once thought. Much of the prior evidence for persistent spiking comes from studies that averaged spiking across time and trials. While this shows that the average spike rate of neurons increases over the delay, it masks the details of the spiking itself. When examined in "real time" (i.e., in single trials), spiking is typically sparse, with gaps of time hundreds of milliseconds between bursts of spikes.

Second, persistent activity may not be necessary for WM. Watanabe and Funahashi (2014) trained monkeys on an oculomotor delayed-saccade task that required memory for the location of a saccade target. During part of the memory delay, animals had to attend to a different location. During this time, there was little or no delay activity in the PFC even though the monkeys could later still demonstrate memory for the saccade location. This suggests that persistent spiking per se may not be necessary for WM maintenance.

Third, WM activity does not seem to be simple maintenance of a previously activated ensemble. Instead, it changes over time. When the delay duration is fixed, robust spiking may only emerge late in the delay. The resulting "ramp" in neural activity may reflect preparation for the upcoming memory probe, suggesting the activity spiking is a readout, not memory, mechanism.

The pattern of activity across neurons (the population code) also changes. This can be evaluated by testing if a decoder trained on activity at one time in the trial can decode memories at other times. If not, there has been a change in code. Cross-temporal decoding fails soon into the memory delay (figure 31.1*C*; Stokes, 2015). It is possible, however, to find a linear combination of neurons that will maintain a stable code, "a stable subspace" (figure 31.1*D*; Murray et al., 2017). However, it is important to note that this has been demonstrated with "empty" delays without additional inputs or distractions. In contrast, WM in the real world is rarely held over empty delays. Decoders trained before additional inputs do not perform well following them. This change in code is consistent with mixed selectivity—individual neurons sensitive to the combination of multiple behavioral conditions and items (Rigotti et al., 2013).

These results argue against the notion that persistent activity is the only neural representation of WM. Instead, it suggests that WM is complex and dynamic. Some investigators have taken note of this and have proposed new models of WM.

Activity-silent models of working memory An alternative type of model proposes that WM is *activity silent* (see reviews by Miller, Lundqvist, & Bastos, 2018; Stokes, 2015). Rather than persistent activity, spiking is sparse. The spiking temporarily changes synaptic weights through short-term synaptic plasticity (STSP), leaving behind a stimulus trace that preserves its memory between spiking. In other words, the spikes leave an "impression" in networks that preserve the memory of the activity. Indeed, spiking activity can produce fast synaptic enhancement that last hundreds of milliseconds (Wang et al., 2006). Memories can be maintained over a longer time scale by "refreshing" the synaptic weight changes with occasional spiking.

Such activity-silent representations have functional advantages over persistent spiking. Memories held by persistent spiking alone can be labile because they are lost when activity is disrupted. Models of persistent spiking have trouble holding more than one memory at a time. If there is any overlap in the ensembles/attractor states, they tend to meld into one. Plus, neurons optimize information when they spike sparsely and in bursts, not persistently.

Activity-silent models predict content-dependent changes in network connections. It is difficult to directly test this prediction, as it is difficult to record from a pair of monosynpatically connected neurons. However, Fujisawa et al. (2008) used multicontact silicon probes to record from a handful of theoretically connected neurons in rat PFC during WM (~1%–2% of all pairs). They found spiking-related changes in effective synaptic connectivity.

A further prediction is that neural responses to a new input should depend on the information already encoded in synaptic weights. To test this, Stokes et al. (2013) trained monkeys to perform a delayed-association WM task. During the memory delay, there was presentation of a *null* (irrelevant) stimulus. The neural response to the null stimulus depended on the current contents of WM. Rose et al. (2016) found that the decoding of the contents of WM using functional magnetic resonance imaging (fMRI) decreased to chance early in the memory delay, but after a pulse of transcranial magnetic stimulation (TMS), the memory could once again be decoded. This was no longer possible after the item had been "cleared" from WM. These results are consistent with the idea that WM can be stored in a latent form (e.g., via synaptic weights).

The need to refresh weight changes may explain the limited capacity of WM. If too many items are simultaneously held, the requirement to refresh the synapses causes a buildup of interference due to competition for the limited time available for the refresh. This limited capacity is a hallmark of WM storage. Unlike long-term memory, which has enough capacity to hold a lifetime of experiences and knowledge, we can only hold very few items "in mind" simultaneously. This is discussed next.

The Limited Capacity of Working Memory

WM has a severely limited capacity. The average adult human can only hold about four items in memory at a time. This is obvious in our lives (e.g., restaurant servers write down orders). Individual capacity varies from one to seven and is highly correlated with fluid intelligence, reflecting that capacity limits are a fundamental restriction in cognition (Fukuda, Vogel, Mayr, & Awh, 2010). This makes sense: the more thoughts that can be simultaneously held and manipulated, the more associations, connections, and relationships can be made and therefore the more sophisticated a thought can be. We begin our discussion of limited capacity on a general level.

Slots versus pools: behavioral evidence for limited working-memory capacity What accounts for the limited capacity of WM? Do we simply miss new items once we have filled our thoughts? Or do we try to take in as much information as possible, eventually spreading ourselves too thin? Both may be true.

Some models posit that WM has a limited number of discrete "slots" (figure 31.2*A*, *top row*), and therefore, you stop storing items once you've filled all of the slots. Alternative models predict that WM is a flexible resource that can be subdivided among objects and that the limited capacity of WM is due to spreading it too thin to support behavior (figure 31.2*A*, *bottom row*). Buschman et al. (2011) found an intriguing possibility: both the slot and flexible-resource models are correct, albeit for different reasons.

Visual WM capacity is typically studied using change-detection tasks. In these tasks, subjects must remember a screen with a variable number of objects (such as colored squares). Then, after a delay of a few seconds, the subjects see a second "test" screen of objects. Subjects must detect

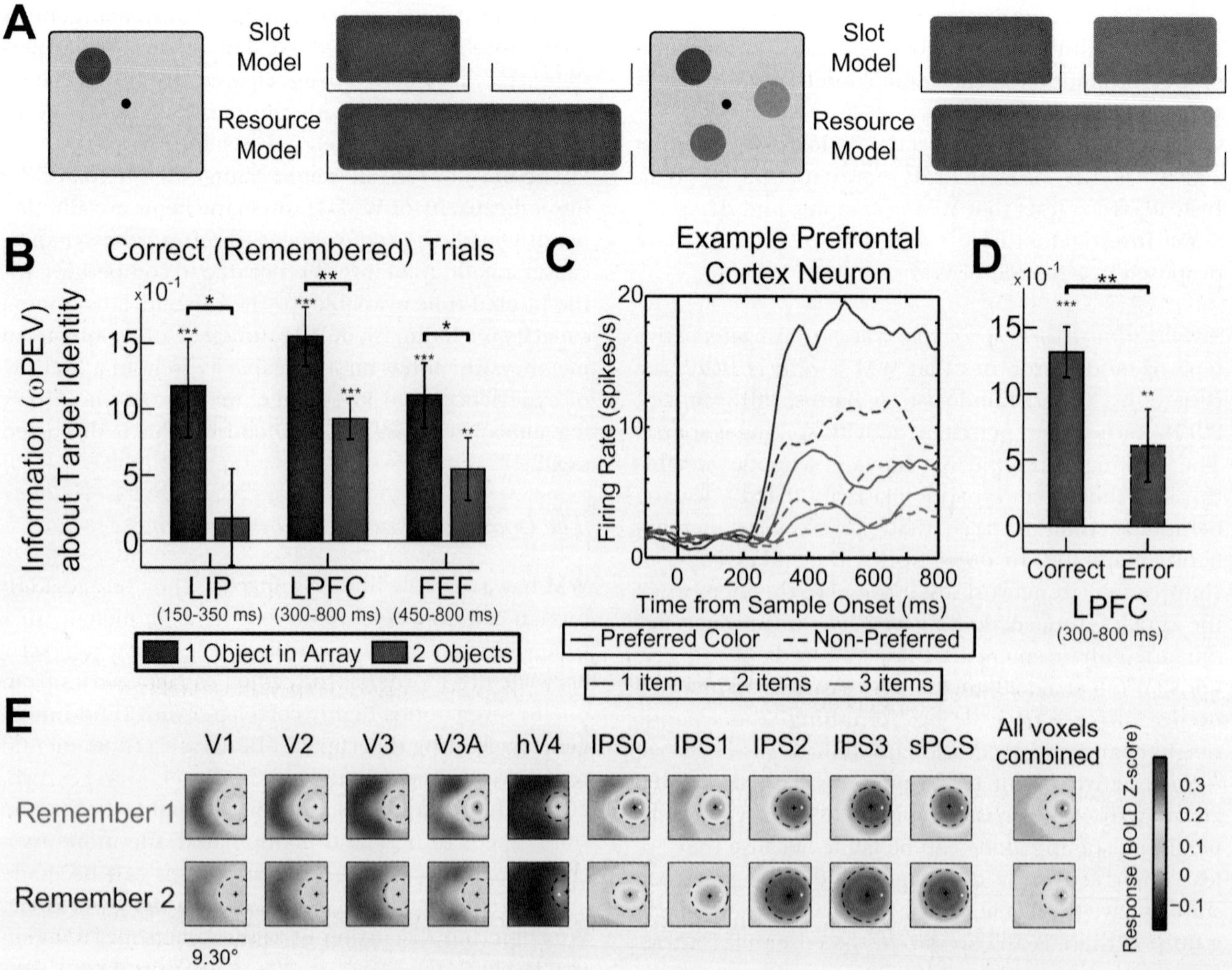

FIGURE 31.2 The slot versus pools models of working-memory capacity limits. *A*, Capacity limits in working memory have been modeled either as the result of a limited number of slots (*top row*) or limited resources (*bottom row*). The slot model predicts that increasing the memory load (*right*) leads to failure to maintain certain memories (e.g., light gray is not stored). In contrast, the resource model predicts that increasing memory load should reduce the information about any single item. *B–C*, Neurophysiological evidence for the resource model. Information about a memory item is reduced in prefrontal and parietal cortex when working-memory load is increased (*B, dark vs. light gray bars*). In addition, prefrontal cortex neurons carry information about a to-be-remembered stimulus even when the animal is unable to report it (i.e., it is forgotten). Adapted from Buschman et al. (2012). *D–E*, Reduced information about a stimulus is thought to be due to the divisive normalization of responses. This is seen at the level of single neurons (*D*) and in the neural population (using blood-oxygen-level dependent activity, BOLD, *E*). Firing rate and decodability are reduced when the number of items to be remembered is increased. Adapted from Buschman et al. (2012) and Sprague, Ester, and Serences (2014). (See color plate 33.)

the object that changed from the previous screen (if any did). When the subjects' WM capacity is exceeded, they make errors (by missing changes). Monkeys, like humans, showed a decline in performance above four items. However, closer investigation revealed that the monkeys' overall capacity was actually two independent capacities: two objects in the right visual hemifield and two in the left visual hemifield (to the right and left of the vision). WM in the right hemifield was unaffected by the objects in the left hemifield (and vice versa), but adding even one object on the same side of the gaze as another decreased performance. Hemifield independence has also been seen during attention tasks (Alvarez & Cavanagh, 2005; Umemoto, Drew, Ester, & Awh, 2010) and so may influence the encoding of items into WM (Delvenne & Holt, 2012). Hemifield independence has not always been observed in studies of human WM. However, much of the human work did not monitor the subjects' eye position to ensure they maintained central fixation, as we did in our studies in animals. Any spurious eye movements could mask or attenuate hemifield independence by bringing stimuli into the other hemifield.

The independence between visual hemifields is consistent with the slot model (i.e., a right slot and a left slot). However, within each hemifield's "slot," we found that WM was a flexible resource. In other words, within each hemifield, information was shared and spread among objects. This was revealed by a closer look at how neurons encoded the contents of WM.

The slot model predicts that encoding is all or none; an object is encoded or not. However, we found that even when an object was successfully encoded, neural information about that specific object was reduced when another object was added to the same visual hemifield (figure 31.2*B*), as if a limited amount of neural information was spread between the objects on one side of vision.

The slot model also predicts that if a subject misses an object, no information about it should be encoded. The flexible-resource model suggests that some information about the object could be encoded, just not enough to support behavior. We found the latter within each visual hemifield. Even when the change was unnoticed, there was still significant, albeit reduced, information (figure 31.2*D*). In sum, the two cerebral hemispheres (visual hemifields) act like discrete resource slots, but within them, information is divided among objects in a graded fashion (like a flexible resource).

The division of information between objects within a hemifield appeared to be due to the normalization of neural activity. Neurons that were selective for a stimulus at one location were inhibited when a second or third item was added to the display (figure 31.2*C*). Similar effects have been seen in humans (figure 31.2*E*; Sprague, Ester, & Serences, 2014). This reduction in response is similar to the divisive normalization seen with the crowding of receptive fields during perception (Buschman & Kastner, 2015), suggesting that WM capacity limitations reflect a fundamental limit for all cognition.

But why is there a limitation? It seems unlikely to be the number of neurons—the average adult human has around 100 billion neurons. Energy constraints also seem unlikely—the brain already consumes more energy than any other part of the body, so a few more kilocalories seem a small burden. One explanation may be a limitation in the coding scheme.

One possibility was discussed above. Activity-silent models explain it by a build-up of interference between different items in WM, which is consistent with the flexible-resource model. Another, not incompatible, explanation involves the role of synchronized oscillations. Different items could be separated in WM by multiplexing them at different phases of an oscillation (Lisman & Idiart, 1995). In this model each item is represented in a single cycle of a high-frequency gamma oscillation (~50 Hz). To maintain item order, gamma oscillations are nestled within *theta* oscillations (~4–8 Hz). The capacity limit is because only four to seven gamma oscillations (each ~20 ms long) can fit in the well of a theta oscillation (~100 ms long).

In partial support, there is evidence that information is multiplexed across oscillatory phases. Siegel, Warden, and Miller (2009) found that PFC neurons encode objects at different phases of an approximately 32 Hz oscillation. Phase-based coding has an inherent capacity limitation because WM contents have to fit within an oscillatory cycle.

This sounds like a slot model, with each phase representing a different slot. However, if information about each object is maximal in, but not limited to, each phase, it can also be compatible with the flexible-resource model or a hybrid of the two.

Given both the limited capacity of WM and its importance for cognition, the brain should have mechanisms to optimize its use. Next, we highlight two potential mechanisms: compression of items (chunking) and judicious control over access to WM (executive control).

Optimizing Working Memory

Compressing items in working memory *Chunking* is the combination of multiple items into a single "chunk" that requires less space in WM than the sum of the constituent parts. This is an approach we often use—we remember phone numbers as two groups of numbers (three plus four) rather than a string of seven individual digits.

Psychophysics suggests that chunked items are formed based on statistical regularities in the world. Brady, Konkle, and Alvarez (2009) had subjects perform a classic change-detection task (as described above). The stimuli had two parts, an inner and an outer ring, each of a different color that was independent and random for most stimuli. However, for a subset of stimuli, there were statistical regularities between the inner and outer colors. This allowed these stimuli to be chunked—the inner and outer color could be combined into a single object (e.g., labeling a common red outer/green inner stimulus as X), reducing the amount of information needed to specify the stimulus.

This is what was seen in humans. Although subjects were not aware of the color combinations, they were better able to remember these stimuli compared to others. Furthermore, the effect extended beyond the common stimuli—the existence of a chunked stimulus in the display improved memory performance for *other* stimuli because the chunked stimulus required fewer

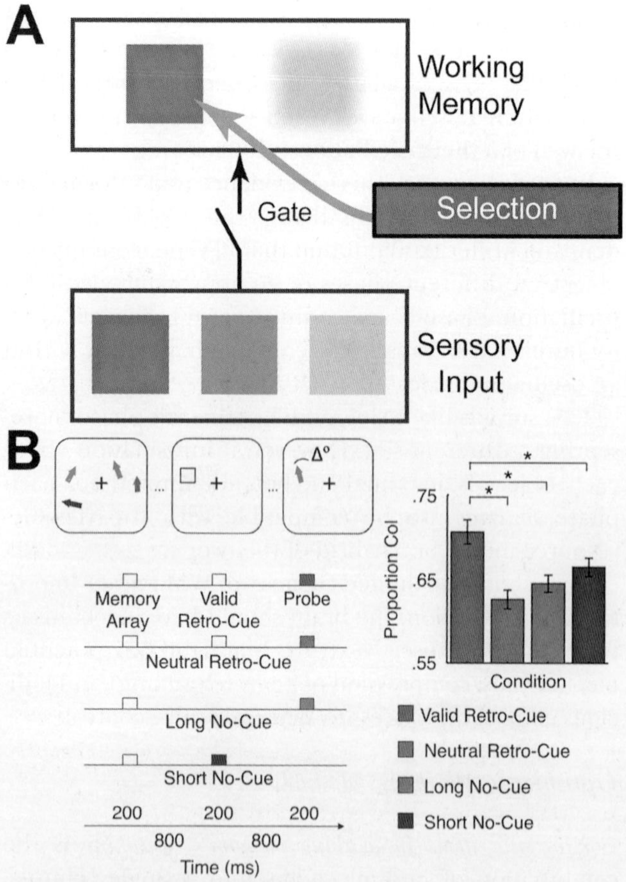

FIGURE 31.3 Working memory is tightly controlled. *A*, Given the importance of working memory and its limited capacity, the contents of working memory must be tightly controlled. Information must be "gated" into working memory (*arrow*) and, once in working memory, a memory can be "selected" and used to guide behavior. *B*, Subjects were asked to remember the direction of a set of arrows. After a memory delay, they reported the direction of a cued stimulus. Subjects were cued as to *which* stimulus they would report either halfway through the long memory delay (valid) or at the end of a short or long delay (no cue). Receiving the cue earlier in the delay improved the accuracy of memory recalls even more than testing at the shorter delay. Adapted from Murray et al. (2013).

resources, allowing resources to be allocated to other stimuli.

Controlling the contents of working memory To compensate for its limited capacity, WM is a highly dynamic resource. This requires a 'central executive' to control and manipulate the contents of WM "sketch pads" (figure 31.3A).

WM limits vary between individuals and are highly correlated with measures of fluid intelligence. However, experimental evidence suggests that the true variability across individuals is their ability to *control* the contents of WM. Fukuda and Vogel (2011) had subjects perform a change-detection task. However, instead of memorizing the entire visual display, a cue indicated whether subjects should remember stimuli on the left or the right. They found a strong correlation between an individual's ability to filter out distracters and their overall "capacity" (measured in an independent test). Thus, everyone may have a similar WM capacity; what differs is how well they control access to it. Indeed, disrupting one's ability to control WM can be pathological. Such disruptions may partly underlie intrusive thoughts in anxiety (Brewin & Beaton, 2002) and may contribute to schizophrenia (Braver, Barch, & Cohen, 1999), although other evidence suggests there is an actual reduction in WM capacity associated with schizophrenia (Erickson et al., 2015). WM can be controlled in two primary ways (figure 31.3A). First, one must control access. Then, once items are in memory, one must select them for use in behavior.

A *gating* signal is thought to control access to WM. Without it, WM is susceptible to noise and cannot be flexibly updated (Hochreiter & Schmidhuber, 1997). Braver, Barch, and Cohen (1999) propose that gating occurs when neurons in PFC are transiently activated by dopaminergic innervations. Dopamine modulates active afferent synapses, changing the dynamics such that a stimulus input is maintained in memory. Alternatively, Frank, Loughry, and O'Reilly (2001) proposed that the basal ganglia gate memories into PFC. The activation of striatal neurons in the basal ganglia disinhibit neurons in the thalamus, engaging recurrent prefrontal-thalamic loops and sustaining memories.

Selection is the process by which one memory, from a set of remembered items, can be activated and used to guide behavior. It is like attention, except attention selects one stimulus from a field of stimuli, improving its perception (Buschman & Kastner, 2015). Instead, selection retrieves one item from a set of items held in WM.

We previously showed that WM capacity limitations are due to interference among the neural representations of remembered stimuli, at least within a visual hemifield (Buschman et al., 2011). This is similar to the competitive interference among visible stimuli that is thought to underlie limitations in perception. In perception, attention compensates for these limitations by selecting a specific stimulus for greater neural representation. This biases the competition between stimuli, resolving interference and improving perceptual accuracy for the attended stimulus (at the cost of losing accuracy for unattended stimuli).

Selection plays a similar role for WM (figure 31.3B, see chapter 25). Selecting a stimulus leads to improvements in memory accuracy for the selected stimulus (e.g., Sprague, Ester, & Serences, 2014). In these studies, subjects are asked to hold two items in WM. After a short delay, a *retro-cue* indicates which of the two items the subjects should report. These studies add a second memory delay after the retro-cue and before the final report. This allows the stimulus to be "selected" and then maintained in WM alone. They have found that if a retro-cue occurs earlier in the trial, performance improves. This makes sense—if interference between memories causes memory representations to decay over time, then selection acts to reduce this interference. Indeed, valid retro-cues improve the accuracy of human WM.

A neural infrastructure for working-memory control We noted that WM is associated with LFPs rhythms in the alpha/beta (10–30 Hz) and gamma (>30 Hz) bands. The interplay between these rhythms in different cortical layers may be an infrastructure for WM gating and selection (see review by Miller, Lundqvist, & Bastos, 2018).

Bottom-up (sensory) information held in WM has been associated with brief bursts of spiking linked to bursts of gamma in LFPs (Lundqvist et al., 2016). The gamma bursts are interleaved with bursts of alpha/beta in a push-pull fashion. If gamma is up, beta is down and vice versa. Alpha/beta has been associated with top-down functions such as volitional shifts of attention (Buschman & Miller, 2007) and top-down information such as task rules (Buschman et al., 2012). Importantly, alpha/beta has also been associated with inhibition. Alpha/beta increases, for example, when a motor response must be inhibited. Thus, top-down information associated with alpha/beta can inhibit the bottom-up gamma/spiking that holds stimuli in WM.

Support for this came from observations that the gamma (30–100 Hz) bursts and spikes carrying WM contents are stronger in the superficial feedforward cortical layers that carry bottom-up sensory information (layers 2 and 3; Bastos, Loonis, Kornblith, Lundqvist, & Miller, 2018). By contrast, alpha/beta (10–30 Hz) is stronger in the deep feedback cortical layers associated with top-down information (layers 5 and 6). The deep-layer alpha/beta is coupled to superficial-layer gamma, and their power is anticorrelated. This all suggests that top-down deep-layer alpha/beta can regulate the expression of superficial-layer gamma and thus gate bottom-up information into WM. It may also clear out the contents of WM when it is no longer needed. When memories become irrelevant, increases in PFC beta power can result in a corresponding decrease in gamma and in spiking, discarding the contents of WM (Lundqvist, Herman, Warden, Brincat, & Miller, 2018). Thus, this interplay between different rhythms in distinct cortical layers may underlie the executive, volitional control over WM.

Conclusions

WM is the fundamental function by which we break free from reflexive input-output reactions and gain control over our own thoughts. Early models of its neurobiology focused on how it maintains information over short delays. This was thought to depend on persistent spiking. Recent studies have examined this on a more granular level. They indicate that there is more going on than a simple persistence of spiking. Instead, brief bursts of spiking and associated gamma bursting reflect activation and reactivation of the neural ensembles for the WM memoranda. The spiking could cause temporary changes in synaptic weights—impressions—that carry the memories between spiking. This solves many of the problems with persistent spiking. It makes the memories more robust to interference. It allows multiple items to be held in WM by "juggling" their activations in time. This new perspective is part of mounting evidence that the neural basis of cognition is not continuous but discrete and periodic (reviewed in Buschman & Miller, 2010). Sparse spiking also leaves room for rhythmic interplay between oscillations of different bands, gamma and alpha/beta, which are observed during WM tasks. Beta is associated with top-down information and seems to have an inhibitory role. It has a push-pull relationship with gamma (when beta is up, gamma is down and vice versa), suggesting that beta could be a gating signal for WM. In other words, this may be the infrastructure for controlling WM storage, with beta turning on and off the "faucet" of gamma/spike-based WM storage (Miller, Lundqvist, & Bastos, 2018).

Acknowledgment

This work was supported by National Institute of Mental Health grant R56MH115042 and Office of Naval Research grant N00014-14-1-0681 to Timothy J. Buschman and National Institute of Mental Health grant R37MH087027, Office of Naval Research Multidisciplinary University Research Initiative grant N00014-16-1-2832, and the MIT Picower Innovation Fund to Earl K. Miller.

REFERENCES

Alvarez, G. A., & Cavanagh, P. (2005). Independent resources for attentional tracking in the left and right visual hemifields. *Psychological Science, 16*(8), 637–643. https://doi.org/10.1111/j.1467-9280.2005.01587.x

Antzoulatos, E. G., & Miller, E. K. (2014). Increases in functional connectivity between prefrontal cortex and striatum during category learning. *Neuron, 83*(1), 216–225. https://doi.org/10.1016/j.neuron.2014.05.005

Bastos, A. M., Loonis, R., Kornblith, S., Lundqvist, M., & Miller, E. K. (2018). Laminar recordings in frontal cortex suggest distinct layers for maintenance and control of working memory. *Proceedings of the National Academy of Sciences, 115*(5), 1117–1122. https://doi.org/10.1073/pnas.1710323115

Brady, T. F., Konkle, T., & Alvarez, G. A. (2009). Compression in visual working memory: Using statistical regularities to form more efficient memory representations. *Journal of Experimental Psychology: General, 138*(4), 487–502. https://doi.org/10.1037/a0016797

Braver, T. S., Barch, D. M., & Cohen, J. D. (1999). Cognition and control in schizophrenia: A computational model of dopamine and prefrontal function. *Biological Psychiatry, 46*(3), 312–328. https://doi.org/10.1016/S0006-3223(99)00116-X

Brewin, C. R., & Beaton, A. (2002). Thought suppression, intelligence, and working memory capacity. *Behaviour Research and Therapy, 40*(8), 923–930. https://doi.org/10.1016/S0005-7967(01)00127-9

Buschman, T. J., Denovellis, E. L., Diogo, C., Bullock, D., & Miller, E. K. (2012). Synchronous oscillatory neural ensembles for rules in the prefrontal cortex. *Neuron, 76*(4), 838–846. https://doi.org/10.1016/j.neuron.2012.09.029

Buschman, T. J., & Kastner, S. (2015). From behavior to neural dynamics: An integrated theory of attention. *Neuron, 88*(1), 127–144. https://doi.org/10.1016/j.neuron.2015.09.017

Buschman, T. J., & Miller, E. K. (2007). Top-down versus bottom-up control of attention in the prefrontal and posterior parietal cortices. *Science, 315*, 1860–1862. https://doi.org/10.1126/science.1138071

Buschman, T. J., & Miller, E. K. (2010). Shifting the spotlight of attention: Evidence for discrete computations in cognition. *Frontiers in Human Neuroscience, 4*. https://doi.org/10.3389/fnhum.2010.00194

Buschman, T. J., Siegel, M., Roy, J. E., & Miller, E. K. (2011). Neural substrates of cognitive capacity limitations. *Proceedings of the National Academy of Sciences, 108*(27), 11252–11255. https://doi.org/10.1073/pnas.1104666108

Christophel, T. B., Klink, P. C., Spitzer, B., Roelfsema, P. R., & Haynes, J.-D. (2017). The distributed nature of working memory. *Trends in Cognitive Sciences, 21*(2), 111–124. https://doi.org/10.1016/j.tics.2016.12.007

Delvenne, J.-F., & Holt, J. L. (2012). Splitting attention across the two visual fields in visual short-term memory. *Cognition, 122*(2), 258–263. https://doi.org/10.1016/j.cognition.2011.10.015

Erickson, M. A., Hahn, B., Leonard, C. J., Robinson, B., Gray, B., Luck, S. J., & Gold, J. (2015). Impaired working memory capacity is not caused by failures of selective attention in schizophrenia. *Schizophrenia Bulletin, 41*(2), 366–373. https://doi.org/10.1093/schbul/sbu101

Erickson, M., Hahn, B., Leonard, C., Robinson, B., Luck, S., & Gold, J. (2014). Enhanced vulnerability to distraction does not account for working memory capacity reduction in people with schizophrenia. *Schizophrenia Research: Cognition, 1*(3), 149–154. https://doi.org/10.1016/j.scog.2014.09.001

Frank, M. J., Loughry, B., & O'Reilly, R. C. (2001). Interactions between frontal cortex and basal ganglia in working memory: A computational model. *Cognitive, Affective, & Behavioral Neuroscience, 1*(2), 137–160. https://doi.org/10.3758/CABN.1.2.137

Fries, P. (2015). Rhythms for cognition: Communication through coherence. *Neuron, 88*(1), 220–235. https://doi.org/10.1016/j.neuron.2015.09.034

Fujisawa, S., Amarasingham, A., Harrison, M. T., & Buzsáki, G. (2008). Behavior-dependent short-term assembly dynamics in the medial prefrontal cortex. *Nature Neuroscience, 11*(7), 823–833. https://doi.org/10.1038/nn.2134

Fukuda, K., & Vogel, E. K. (2011). Individual differences in recovery time from attentional capture. *Psychological Science, 22*(3), 361–368. https://doi.org/10.1177/0956797611398493

Fukuda, K., Vogel, E., Mayr, U., & Awh, E. (2010). Quantity, not quality: The relationship between fluid intelligence and working memory capacity. *Psychonomic Bulletin & Review, 17*(5), 673–679.

Funahashi, S., Bruce, C. J., & Goldman-Rakic, P. S. (1989). Mnemonic coding of visual space in the monkey's dorsolateral prefrontal cortex. *Journal of Neurophysiology, 61*(2), 331–349. https://doi.org/10.1152/jn.1989.61.2.331

Fuster, J. (2015). *The prefrontal cortex*. Cambridge, MA: Academic Press.

Goldman-Rakic, P. (1995). Cellular basis of working memory. *Neuron, 14*(3), 477–485. https://doi.org/10.1016/0896-6273(95)90304-6

Hochreiter, S., & Schmidhuber, J. (1997). Long short-term memory. *Neural Computation, 9*(8), 1735–1780. https://doi.org/10.1162/neco.1997.9.8.1735

Lisman, J. E., & Idiart, M. A. (1995). Storage of 7 +/− 2 short-term memories in oscillatory subcycles. *Science, 267*(5203), 1512–1515.

Lundqvist, M., Herman, P., & Miller, E. K. (2018). Working memory: Delay activity, yes! Persistent activity? Maybe not. *Journal of Neuroscience, 38*(32), 7013–7019. https://doi.org/10.1523/JNEUROSCI.2485-17.2018

Lundqvist, M., Herman, P., Warden, M. R., Brincat, S. L., & Miller, E. K. (2018). Gamma and beta bursts during working memory readout suggest roles in its volitional control. *Nature Communications, 9*(1), 394. https://doi.org/10.1038/s41467-017-02791-8

Lundqvist, M., Rose, J., Herman, P., Brincat, S. L., Buschman, T. J., & Miller, E. K. (2016). Gamma and beta bursts underlie working memory. *Neuron, 90*(1), 152–164. https://doi.org/10.1016/j.neuron.2016.02.028

Miller, E. K., Lundqvist, M., & Bastos, A. M. (2018). Working memory 2.0. *Neuron, 100*(2), 463–475. https://doi.org/10.1016/j.neuron.2018.09.023

Murray, A. M., Nobre, A. C., Clark, I. A., Cravo A. M., & Stokes, M. G. (2013) Attention restores discrete items to visual short-term memory. *Psychological Science, 24*(4), 550–556. https://doi.org/10.1177/0956797612457782

Murray, J. D., Bernacchia, A., Roy, N. A., Constantinidis, C., Romo, R., & Wang, X.-J. (2017). Stable population coding for working memory coexists with heterogeneous neural dynamics in prefrontal cortex. *Proceedings of the National Academy of Sciences, 114*(2), 394–399. https://doi.org/10.1073/pnas.1619449114

Palva, J. M., Monto, S., Kulashekhar, S., & Palva, S. (2010). Neuronal synchrony reveals working memory networks and predicts individual memory capacity. *Proceedings of the National Academy of Sciences, 107*(16), 7580–7585. https://doi.org/10.1073/pnas.0913113107

Passingham, R. E. (1993). *The frontal lobes and voluntary action*. New York: Oxford University Press.

Rigotti, M., Barak, O., Warden, M. R., Wang, X.-J., Daw, N. D., Miller, E. K., & Fusi, S. (2013). The importance of mixed selectivity in complex cognitive tasks. *Nature, 497*(7451), 585–590. https://doi.org/10.1038/nature12160

Rose, N. S., LaRocque, J. J., Riggall, A. C., Gosseries, O., Starrett, M. J., Meyering, E. E., & Postle, B. R. (2016). Reactivation of latent working memories with transcranial magnetic stimulation. *Science, 354*(6316), 1136–1139. https://doi.org/10.1126/science.aah7011

Salazar, R. F., Dotson, N. M., Bressler, S. L., & Gray, C. M. (2012). Content-specific fronto-parietal synchronization during visual working memory. *Science, 338*(6110), 1097–1100. https://doi.org/10.1126/science.1224000

Siegel, M., Warden, M. R., & Miller, E. K. (2009). Phase-dependent neuronal coding of objects in short-term memory. *Proceedings of the National Academy of Sciences, 106*(50), 21341–21346. https://doi.org/10.1073/pnas.0908193106

Sprague, T. C., Ester, E. F., & Serences, J. T. (2014). Reconstructions of information in visual spatial working memory degrade with memory load. *Current Biology, 24*(18), 2174–2180. https://doi.org/10.1016/j.cub.2014.07.066

Stokes, M. G. (2015). "Activity-silent" working memory in prefrontal cortex: A dynamic coding framework. *Trends in Cognitive Sciences, 19*(7), 394–405. https://doi.org/10.1016/j.tics.2015.05.004

Stokes, M. G., Kusunoki, M., Sigala, N., Nili, H., Gaffan, D., & Duncan, J. (2013). Dynamic coding for cognitive control in prefrontal cortex. *Neuron, 78*(2), 364–375. https://doi.org/10.1016/j.neuron.2013.01.039

Umemoto, A., Drew, T., Ester, E. F., & Awh, E. (2010). A bilateral advantage for storage in visual working memory. *Cognition, 117*(1), 69–79. https://doi.org/10.1016/j.cognition.2010.07.001

Wang, Y., Markram, H., Goodman, P. H., Berger, T. K., Ma, J., & Goldman-Rakic, P. S. (2006). Heterogeneity in the pyramidal network of the medial prefrontal cortex. *Nature Neuroscience, 9*(4), 534–542. https://doi.org/10.1038/nn1670

Watanabe, K., & Funahashi, S. (2014). Neural mechanisms of dual-task interference and cognitive capacity limitation in the prefrontal cortex. *Nature Neuroscience, 17*(4), 601–611. https://doi.org/10.1038/nn.3667

32 Functions of the Visual Thalamus in Selective Attention

W. MARTIN USREY AND SABINE KASTNER

ABSTRACT Selective attention is a cognitive process that allows an organism to direct processing resources preferentially to behaviorally relevant stimuli. This is important since attention is a limited resource, and stimulus detection and discrimination are improved with selective attention. Although the neural mechanisms for selective attention have traditionally been thought to reside solely within the cortex, emerging evidence indicates that this view should be reassessed, as subcortical structures, including the thalamus, also play a significant role. This chapter focuses on thalamocortical network interactions and how they contribute to selective attention.

The thalamus and cerebral cortex are inseparable and essential partners for vision. In primates, the cerebral cortex contains more than 20 visual cortical areas, and each area receives input from and projects to the thalamus (Jones, 2007). This close association allows the thalamus and cortex to work together dynamically to process visual signals that are necessary for behavior and cognition. Selective attention, the ability to direct visual attention to specific stimulus features (e.g., red versus green), objects, or specific spatial locations without moving the eyes, is a cognitive activity known to improve both the detection and discrimination of visual stimuli (Nobre & Kastner, 2014). Although most studies of selective attention have focused on effects in the cortex, results from an increasing number of experiments indicate that attention also enhances subcortical activity and thalamocortical network interactions. This chapter examines the role of the primate thalamus in selective visual attention.

The two major thalamic nuclei that process visual signals and communicate with the visual cortex are the dorsal lateral geniculate nucleus (LGN) and the pulvinar nucleus. Although both nuclei have important roles in vision, they have distinct circuitry and serve different functions. As shown in figure 32.1*A*, the LGN receives visual signals directly from the retina and relays these signals to primary visual cortex (V1). In contrast, as illustrated in figure 32.1*B*, the many divisions of the pulvinar nucleus collectively receive feedforward input from every visual cortical area and project, in turn, back to the cortex, perhaps to facilitate corticocortical communication (Sherman & Guillery, 2013). Based on the source of their feedforward input, retina versus cortex, the LGN and the pulvinar are referred to as first-order and higher-order thalamic nuclei, respectively. In the sections below, we compare and contrast the cells and circuits that comprise the primate LGN and pulvinar and examine their contributions to selective attention. Specifically, we will focus on the role of the visual thalamus in spatial attention, given that most studies thus far have explored this particular selection mechanism, and little is known about other selection mechanisms studied at the cortical level, such as feature- and object-based attention.

The Lateral Geniculate Nucleus: More than a Relay Station between the Retina and Cortex

Anatomical and functional organization Anatomically and functionally distinct parallel-processing streams are particularly prominent in the retinogeniculocortical pathway of primates (see Casagrande & Xu, 2004; Jones, 2007; Usrey & Alitto, 2015). In Old World monkeys and humans, the LGN contains four parvocellular layers, two magnocellular layers, and six koniocellular layers (figure 32.1*A*). Relay neurons in the parvocellular layers receive input from midget retinal ganglion cells and send axons to V1 neurons in layer 4Cβ, whereas neurons in the magnocellular layers receive input from parasol retinal ganglion cells and send axons to neurons in layer 4Cα. Neurons in the koniocellular layers receive input from a variety of additional retinal ganglion cell types, including the small and large bistratified cells, and send axons that pass through layer 4C to terminate in the more superficial layers of V1. While neural computations can occur more rapidly when conducted in parallel, parallel-processing streams also provide a substrate for selectively processing specific aspects of the visual scene (e.g., color, form, motion, and texture).

The response properties of neurons in the magnocellular, parvocellular, and koniocellular layers typically match those of their retinal input (reviewed in

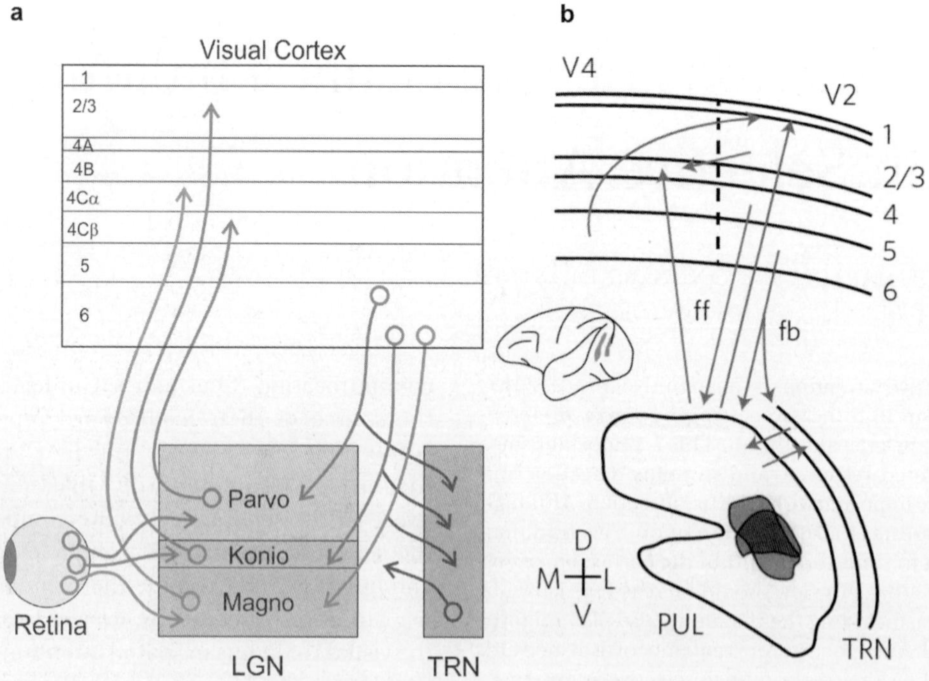

FIGURE 32.1 Thalamocortical connectivity. *A*, The retinogeniculocortical pathway is composed of three distinct streams—the parvocellular, magnocellular, and koniocellular streams—that arise from distinct cell classes in the retina, remain segregated in the LGN, and terminate in different layers of V1. The parallel feedforward streams are matched with similarly specific streams of corticogeniculate feedback. Feedback axons provide monosynaptic excitation to LGN neurons as well as disynaptic inhibition via local interneurons and neurons in the thalamic reticular nucleus. *B*, Pulvinar; direct corticocortical connections (*top*) and indirect corticopulvinocortical loops exemplified by V2-pulvino-V4 circuitry. Tracer injections into V2 (*blue*) and V4 (*pink; inset*) showed overlapping (*purple*) projection zones in the pulvinar (*bottom*). Adapted with permission from Adams et al. (2000). (See color plate 34.)

Usrey & Alitto, 2015). Thus, parvocellular LGN neurons have small receptive fields, produce sustained responses to stationary visual stimuli, and often display chromatic selectivity. In contrast, magnocellular neurons have larger receptive fields, produce transient responses, and have little selectivity for the chromatic properties of a stimulus. Magnocellular neurons also have greater response gain to low-contrast stimuli and greater extraclassical surround suppression than parvocellular neurons. Less is known about the response properties of koniocellular neurons; however, unlike magnocellular and parvocellular neurons, which respond exclusively to one eye, some koniocellular neurons have binocular responses (Cheong et al., 2013).

Given the similarity in receptive field properties between the retina and the LGN, the question arises as to what purpose the LGN serves. One answer to this question involves the diversity of extraretinal inputs to LGN neurons that serve to modulate the gain of LGN responses to incoming retinal signals (reviewed in Jones, 2007; Sherman & Guillery, 2013; Usrey & Alitto, 2015). Nonvisual, extraretinal sources of input to LGN neurons include noradrenergic input from the reticular formation, cholinergic input from the parabrachial nucleus, and serotonergic input from the dorsal raphe nucleus. Although these extraretinal inputs do not directly evoke LGN responses, they play an important role in adjusting LGN activity levels as a function of the sleep-wake cycle and alertness (Bereshpolova et al., 2011; Livingstone & Hubel, 1981; McCormick, McGinley, & Salkoff, 2015; Steriade, 2004). In addition to these nonvisual inputs, LGN neurons also receive visually evoked, extraretinal glutamatergic feedback input from the visual cortex and gamma-aminobutyric acid (GABA)ergic input from the thalamic reticular nucleus (TRN), a neighboring nucleus with neurons that integrate feedback input from cortex and feedforward input from the LGN (figure 32.1*A*; reviewed in Guillery, Feig, & Lozsádi, 1998). If, as discussed below, LGN activity is modulated by attention, then it seems likely that the effects of attention include the involvement of the corticogeniculate feedback pathway and/or the TRN.

Attentional response modulation Covert spatial attention, the ability to direct visual attention to specified retinotopic locations, has been shown to improve the

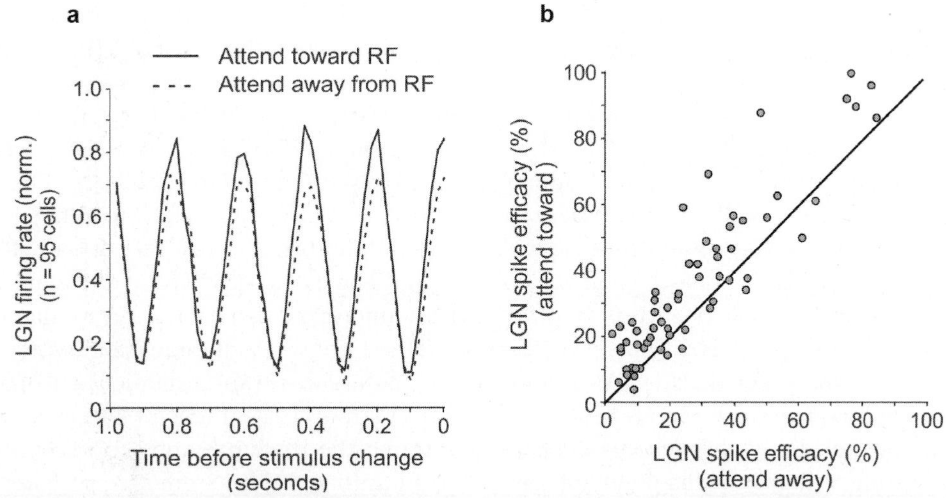

FIGURE 32.2 Influence of attention on LGN activity and geniculocortical communication. *A*, The firing rate of LGN neurons is greater when attention is directed toward their receptive fields (RFs) than when attention is directed away. The plot shows the average firing rate of 95 LGN neurons in the macaque monkey performing a contrast-change-detection task. In this task, the animal maintains fixation on a central point while two drifting grating stimuli (5 Hz) are presented on a computer screen; one stimulus is located over the recorded cell's RF and the other at a different location. Based on the color of the fixation point, the animal attends to one or the other grating in preparation for a change in the stimulus contrast (time = 0). Animals are rewarded for reporting the contrast change. Adapted with permission from Alitto and Usrey (2015). *B*, Synaptic communication between LGN neurons and target neurons in layer 4C of V1 is enhanced with spatial attention. Here, animals perform a similar attention task to that described for (*A*); however, a stimulating electrode placed into the LGN evokes spikes at specific times while animals attend toward or away from the RF of a synaptically connected cortical layer 4C neuron. The efficacy of shock-evoked geniculate spikes to evoke a cortical response (i.e., percentage of successful shocks) is shown when animals attend toward and away from the RF of the recorded cortical neuron. Adapted with permission from Briggs, Mangun, and Usrey (2013).

detection and discrimination of visual stimuli at attended locations, compared with unattended locations (Carrasco, 2011). Within the cortex, spatial attention has been shown to increase neuronal responses to visual stimuli at attended locations (reviewed in Maunsell, 2015; Reynolds & Chelazzi, 2004) and increase the coherence between single-unit activity and the local field potential (LFP) in specific frequency bands (reviewed in Buschman & Kastner, 2015; Fries, 2015).

Although the effects of spatial attention are typically strongest in extrastriate cortical areas (e.g., V4, MT, VIP), attention has been found to influence neuronal activity in subcortical areas, including the LGN. For instance, spatial attention increases the single-unit activity of LGN neurons in macaque monkeys (figure 32.2*A*) and the blood oxygenation level-dependent (BOLD) response in the LGN in humans (McAlonan, Cavanaugh, & Wurtz, 2008; O'Connor et al., 2002; Schneider & Kastner, 2009). Moreover, the mechanisms contributing to the effects of attention on LGN neurons appear to include the release of inhibition from the TRN, as spatial attention decreases the activity levels of TRN neurons (McAlonan, Brown, & Bowman, 2000; McAlonan, Cavanaugh, & Wurtz, 2006; see also Wimmer et al., 2015). Because the influence of attention on TRN neurons is more transient than that for LGN neurons, it seems likely that additional pathways and mechanisms contribute to attentional effects in the LGN. Although untested, feedback from the cortex is a likely candidate for the extended effects of attention on LGN neurons. Along these lines, it is important to note that the corticogeniculate feedback pathway comprises stream-specific projections that selectively innervate the magnocellular, parvocellular, and koniocellular layers of the LGN (Briggs et al., 2016; Briggs & Usrey, 2009; Fitzpatrick et al., 1994; Ichida, Mavity-Hudson, & Casagrande, 2014). Thus, it is possible that cortical feedback may be able to exert stream-specific attentional effects on visual signals traveling from the LGN to cortex.

Functional interactions between the lateral geniculate nucleus and V1 Spatial attention also modulates the strength of geniculocortical communication. By pairing the electrical stimulation of LGN neurons with recordings from synaptically coupled target neurons in macaque V1, researchers have shown that spatial attention increases the percentage of electrically evoked spikes that successfully drive postsynaptic responses in V1 (figure 32.2*B*; Briggs, Mangun, & Usrey, 2013). Thus, attention not only increases the firing rate of LGN neurons but also increases the efficacy, or likelihood, that LGN spikes will be successful in evoking postsynaptic cortical responses.

Rhythmic (also called *oscillatory*) activity patterns are common in the brain and have been proposed to play a role in facilitating the communication of signals between brain regions that are oscillating in phase with each other (Fries, 2005). With respect to this idea, it is interesting to note that oscillatory phase synchronization between the LGN and V1 has been reported for neural activity in the alpha (8–14 Hz) and beta (15–30 Hz) frequency bands (Bastos et al., 2014). Moreover, an analysis of directed connectivity reveals that beta-band interactions are mediated by geniculocortical feedforward processing, whereas alpha-band interactions are mediated by corticogeniculate feedback processing. Given the presence of oscillatory activity in the LGN and V1, and the phase synchronization seen between the two structures, an open and important question to answer is whether or not attention serves to modulate the strength of oscillatory interactions between the two structures, as has been shown to occur with the pulvinar and cortex (see below).

The Pulvinar: Attention Control from the Center of the Brain

Anatomical and functional organization The pulvinar is the largest nucleus in the primate thalamus and is considered a higher-order thalamic nucleus because it forms input-output loops almost exclusively with the cortex. The pulvinar has undergone a significant expansion during evolution, which is on the order of that observed in prefrontal cortex (Jones, 2007). This in itself suggests that corticopulvinar interactions may play an important role in the increasingly flexible mechanisms underlying perception, action, and cognition that parallel this evolutionary expansion of brain structures.

Several different schemes may be used to subdivide the pulvinar based on connectivity, neurochemistry, or electrophysiological properties (Adams et al., 2000; Gutierrez, Yaun, & Cusick, 1995; Stepniewska & Kaas, 1997). For reasons of simplicity, we will not adopt any specific scheme but will broadly refer to the medial (PM), lateral (PL), and inferior (PI) pulvinar. PI and PM are located ventrally and dorsally, respectively, whereas PL has both a ventral and a dorsal part. Each part can be further subdivided into regions that receive distinct sets of inputs and project differentially to a distinct set of cortical regions. Briefly, the PI and PL divisions contain the highest number of visually responsive neurons, and each contains one or more retinotopic maps (Arcaro, Pinsk, & Kastner, 2015; Kaas & Lyon, 2007). The PI map is based on inputs from early visual cortex (V1–V3), whereas the PL map, located in its ventral subdivision, receives dense projections from extrastriate areas V2–V4. The dorsal subdivision of PL (sometimes referred to as *Pdm*) is preferentially targeted by parietal and frontal inputs. Finally, PM receives a diverse set of inputs that include temporal, frontal, parietal, limbic, and insular cortices (Romanski et al., 1997). The dorsal subdivision of PL and PM contain the least amount of visually responsive neurons.

There are two well-established types of corticopulvinar pathways: a transthalamic corticopulvinar feedforward pathway that connects two cortical areas indirectly through the thalamus and a corticopulvinar feedback pathway that projects from a cortical area to its thalamic projection zone (Sherman & Guillery, 2013; Shipp, 2003). As for the indirect transthalamic pathway, a general anatomical principle appears to apply such that directly connected cortical areas form indirect loops through the pulvinar (figure 32.1*B*). Specifically, the direct corticocortical feedforward connections originating in layer 3 of cortical area A and terminating in layer 4 of cortical area B (Felleman & Van Essen, 1991) are paralleled by a putative indirect feedforward pathway through the pulvinar that originates in cortical layer 5 of cortical area A and terminates in layers 3 and 4 of cortical area B (see figure 32.1*B* for an example circuit linking areas V2 and V4). In contrast, the feedback pathway to the pulvinar originates in cortical layer 6 of a given area and projects to an area-specific zone, which itself projects to layer 1 of the same cortical area (e.g., Shipp, 2003). Interestingly, the direct corticocortical feedback connections commonly project from layer 6 to layer 1 of the lower cortical area. Thus, direct and indirect pathways terminate in similar cortical layers, thereby providing an opportunity for the two pathways to interact. Due to the overall connectivity pattern, the pulvinar is positioned to play multiple functional roles, such as routing information from one cortical area to the next (Theyel, Llano, & Sherman, 2010) or regulating corticocortical information transmission according to behavioral context (Saalmann, Pinsk, Wang, Li, & Kastner, 2012; Zhou, Schafer, & Desimone, 2016).

Pulvinar neurons in the ventral parts of PL and in PI reflect the response properties of early visual cortex, such as orientation tuning, directional preference, or color selectivity, including color-opponent responses; however, their tuning properties are generally much broader than those observed in the cortical areas providing input to these pulvinar regions (e.g., Petersen, Robinson, & Keys, 1985; reviewed in Saalmann & Kastner, 2011). Intriguingly, the ventral pulvinar also responds to high-level visual information. For example, in human functional magnetic resonance imaging (fMRI) studies, a posterior medial region of the ventral pulvinar responded preferentially to face stimuli (versus scenes) and was functionally coupled with the fusiform face area at rest (Arcaro, Pinsk, & Kastner, 2018). These results are consistent with anatomical connectivity studies in nonhuman primates demonstrating projections from the medial ventral pulvinar to the cortical face patch network (Grimaldi, Saleem, & Tsao, 2016). Despite this broad reflection of neural response properties of the ventral pathway, it is not clear to which extent pulvinar neurons encode visual information that is essential for computation. For example, lesions of ventral PL and PI do not lead to deficits in the visual discrimination of patterns or color.

Similarly, pulvinar neurons in the dorsal parts of PL reflect response properties of the dorsal visual pathway, such as eye movements (e.g., Vargas et al., 2017). In the human, the dorsal pulvinar also reflects human-specific adaptations, such as tool responses, and is functionally interconnected with the parietal tool network (Arcaro, Pinsk, & Kastner, 2018). Generally, dorsal pulvinar responses depend more strongly on behavior rather than physical properties of the external environment.

Effects of pulvinar lesions The most compelling evidence for the pulvinar playing an important role in visual attention comes from lesion studies in humans and monkeys that can lead to deficits in the orienting of attention, or the filtering of distracter information, among others.

Cortical lesions involving the posterior parietal cortex (PPC) may lead to profound attentional deficits, such as visuospatial hemineglect, a syndrome associated with a failure to direct attention to contralesional space. Neglect is not only associated with cortical lesions but can also occur after thalamic lesions that include the pulvinar. More specifically, the PPC is interconnected with the dorsal pulvinar, and accordingly, inactivation of the dorsal pulvinar in monkeys leads to deficits in directing attention to contralateral space (Wilke et al., 2010). Even though thalamic neglect in humans is rare and severe attentional deficits that occur as a consequence of pulvinar lesions typically do not persist, a milder deficit that may be a residual form of thalamic neglect has been observed as a slowing of orienting responses to contralesional space. This deficit has been specifically related to an impairment in engaging attention at a cued location (Rafal & Posner, 1987).

Patients with pulvinar lesions also show deficits in filtering distracter information. While these patients have no difficulty discriminating target stimuli when shown alone, discrimination performance is impaired when salient distracters that compete with the target for attentional resources are present, which is consistent with a difficulty in filtering out the unwanted information present in the visual display (e.g., Snow et al., 2009). Similar filtering deficits have been observed after PPC lesions in humans (Friedman-Hill et al., 2003) and after extrastriate cortex lesions that include area V4 in humans (Gallant, Shoup, & Mazer, 2000) and monkeys (De Weerd et al., 1999), suggesting that the pulvinar is part of a distributed network of brain areas that subserve visuospatial attention.

Attentional response modulation The findings from lesion studies are corroborated by electrophysiology and neuroimaging studies showing that neural responses in the pulvinar reflect the behavioral relevance of visual input. In human neuroimaging studies, the modulation of responses has been shown in several different parts of the pulvinar, including dorsomedial, lateral, and inferior parts, using selective attention tasks that emphasized directing attention to a spatial location (e.g., Arcaro, Pinsk, & Kastner, 2018), filtering distracter information (e.g., Fischer & Whitney, 2012), and shifting attention across the visual field (e.g., Yantis et al., 2002). Interestingly, some of these functions, such as distracter filtering, may also extend to working memory (Rotshtein et al., 2011). In monkey physiology studies, it has been demonstrated that spatial attention modulates the response magnitude of neurons in dorsal, lateral, and inferior parts of the pulvinar (Petersen, Robinson, & Keys, 1985; Saalmann et al., 2012; Zhou, Schafer, & Desimone, 2016). In a typical attention study, a location in the visual field at which a target stimulus will occur after a variable delay period is cued, and neural responses are compared when attention is directed to a neuron's receptive field or when attention is directed away from it. It has been shown in visual cortex that neural responses to attended visual stimuli typically increase by up to 25% or more, compared to when the same stimuli are ignored. Pulvinar neurons show similar attentional response enhancement (figure 32.3*A*). Remarkably, and again similar to cortical neurons, baseline activity also increased during delay periods after an animal was cued to deploy and sustain attention at a spatial location (figure 32.3*A*; Saalmann et al., 2012; Zhou, Schafer, & Desimone, 2016). Such elevated delay activity is obtained during a pure cognitive state and is not contaminated by sensory input from the environment. In addition to response magnitude, the timing and variability of pulvinar responses are likely to influence information transmission to the cortex. Accordingly, pulvinar neurons show reduced response variability during peripheral attention and

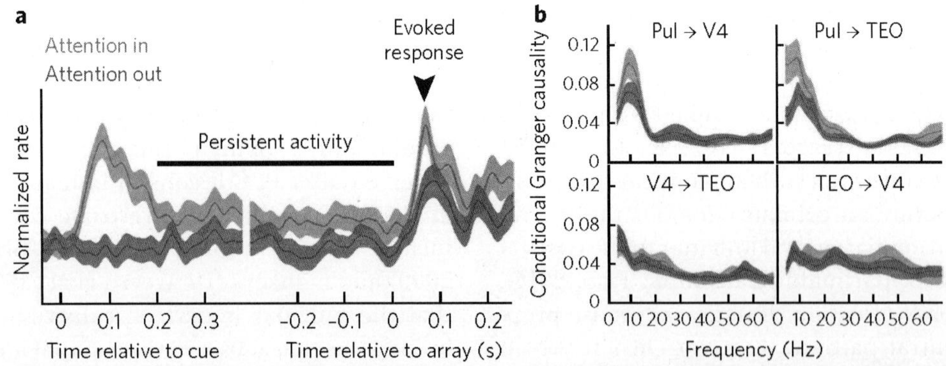

FIGURE 32.3 Attentional modulation in the pulvinar. *A*, When attention is directed to a neuron's RF by a visual cue (attention in) as compared to when attention is directed away from it (attention out), there is elevated persistent activity during a delay and moderate attentional enhancement in response to an array. *B*, Conditional Granger causality analysis suggests a role for the pulvinar in increasing coherence in an alpha frequency band between V4 and TEO during the delay period. The two cortical areas do not appear to interact using the direct corticocortical pathways during that period, and interareal interactions go mainly through the indirect thalamocortical pathways. Adapted with permission from Saalmann et al. (2012).

saccade tasks (Petersen, Robinson, & Keys, 1985; Saalmann et al., 2012).

Functional interactions with cortex The direct corticocortical pathways are commonly thought to be the major routes for the transmission of visual information between cortical areas (but see, e.g., Sherman & Guillery, 2013), whereas the functional roles of the indirect pathways through the pulvinar have been less clear. In vitro studies demonstrated that microstimulation of the indirect pathway between the primary and secondary sensory cortical areas strongly activated the interconnected cortical areas (Theyel, Llano, & Sherman, 2010). Moreover, inactivation of the thalamic projection zone that these cortical areas share led to a failure of corticocortical communication, raising the possibility that all corticocortical information transmission may depend strongly on thalamic loops (Theyel, Llano, & Sherman, 2010). These results were corroborated by in vivo studies, on anesthetized prosimian primates, exploring the thalamocortical interactions between V1 and PI, including pharmacological interventions. Muscimol inactivation diminished visually evoked responses of V1 neurons to their preferred orientation but enhanced their relative responses to other orientations (Purushothaman et al., 2012). Thus, it is possible that pulvinar inputs are required for augmenting synaptic connections among similarly tuned V1 neurons, and in the absence of thalamocortical signals, weak inputs (such as from neurons with opposite orientation preferences) would be abnormally strengthened. Both studies suggest that cortical computation in early sensory cortex strongly depends on normally functioning pulvinocortical interactions. It is not clear whether such dependence of cortical function on an intact thalamus also extends to higher-order cortex (but see Zhou, Schafer, & Desimone, 2016).

Studies on corticocortical functional interactions suggest that the selective routing of behaviorally relevant information across the attention network depends on the degree of synchrony between cortical areas (reviewed in Buschman & Kastner, 2015; Fries, 2015). Researchers tested whether the pulvinar synchronized oscillations between interconnected cortical areas according to attentional demands, thereby modulating the efficacy of corticocortical information transfer. To do this, simultaneous recordings were obtained from two interconnected cortical areas along the ventral visual pathway, V4 and TEO, as well as from the corresponding projection zone in the pulvinar of macaques performing a spatial attention task (Saalmann et al., 2012). While monkeys maintained spatial attention, cortical areas V4 and TEO synchronized in the alpha frequency range and to a smaller extent in the gamma frequency range. At the same time, the pulvinar causally influenced oscillatory activity in both V4 and TEO predominantly in the alpha frequency range, suggesting that the pulvinar controlled the alpha frequency synchrony between cortical areas (figure 32.3*B*). Pulvinar influence on the cortex may also extend to gamma frequencies through a cross-frequency coupling mechanism. Pulvinar-controlled alpha oscillations in the cortex modulated gamma frequency activity in both V4 and TEO, likely contributing to the synchrony observed between these cortical areas in the gamma frequency range. Thus, the pulvinar may be able to regulate information transfer between cortical areas based on attentional demands.

These results were corroborated and extended in a study by Zhou, Schafer, and Desimone (2016) performing simultaneous recordings from areas V4 and IT and the ventral part of PL. Critical support for a causal role of the pulvinar having an impact on cortex was obtained through pharmacological inactivation. Muscimol infusion into the pulvinar resulted in local effects on V4 neurons, including diminished visually evoked responses and increased baseline firing rates and, presumably, decreased synchrony between V4 and IT as a consequence. These effects were associated with impaired behavioral performance, such as a significant spatial bias away from the site of inactivation consistent with a neglect syndrome. The elevated baseline responses may indicate that the pulvinar regulates synaptic gain within and possibly across visual cortical regions and as a consequence functional connectivity between interconnected areas.

The pulvinar control of cortical processing challenges the common conceptualizing of cognitive functions as restricted to cortex. During maintained spatial attention in the delay period between a cue and a subsequent target, pulvinocortical influences were strong, whereas direct corticocortical influences were weak (Saalmann et al., 2012). This suggests that internal processes such as the maintenance of attention in expectation of visual stimuli and short-term memory rely heavily on pulvinocortical interactions. Because of common cellular mechanisms and thalamocortical connectivity principles across sensorimotor domains, a general function of higher-order thalamic nuclei may be the regulation of cortical synchrony to selectively route information across cortex. Thus, one of the functional roles of pulvinar may be to organize cognitive cortical networks in time. Such a timekeeper function is essential to the control of attentional selection. In this view, attentional control emerges in a distributed fashion with specific roles for the cortex and thalamus (see Halassa & Kastner, 2017).

Conclusions

Selective attention is one of the best-understood cognitive operations and serves as a role model to gain a deeper understanding of cognition in the primate brain. Traditional views have emphasized a top-down model, in which a distributed frontoparietal network of brain regions generates attention signals that are then fed back to visual cortex to modulate ongoing processing. In this corticocentric view, the thalamus mainly serves to relay visual signals to cortex. More recent evidence, reviewed in this chapter, has begun to change this view quite substantially. First, it has become clear that selective attention modulates neural gain at the level of the LGN through corticogeniculate feedback and interactions with the TRN. Such modulation could even bypass most of the cortex via direct interactions of the frontal cortex and the TRN converging onto the LGN, as shown in rodents. However, such a direct influence remains to be demonstrated in the primate brain. Also, the exact mechanisms of gain control achieved at the LGN level will need thorough characterization through further empirical study and computational models. It is possible that the LGN-TRN system serves as a thalamic gatekeeper of sensory input to the cortex, as originally proposed by Crick (1984). Second, even though it has long been known that pulvinar lesions impair attention function and that pulvinar neurons are modulated during spatial attention, functions of the vast interconnectivity between the pulvinar and cortex remained elusive until recently. The emerging evidence suggests that pulvinocortical interactions serve to temporally coordinate interconnected cortical areas in order to optimize signal transfer between them. Such a timekeeper function contributes to the control of the attentional selection process, thereby undermining the corticocentric top-down model and suggesting a distributed attentional control function. It is unclear whether such function is unique to spatial attention or will also apply to other aspects of selection, such as feature- or object-based attention. Further, it remains to be shown what kind of functions (if any) pulvinocortical interactions play in other cognitive domains.

Acknowledgments

We thank the National Eye Institute, the National Institute of Mental Health, and the James S. McDonnell Foundation for the support of our studies.

REFERENCES

Adams, M. M., Hof, P. R., Gattass, R., Webster, M. J., & Ungerleider, L. G. (2000). Visual cortical projections and chemoarchitecture of macaque monkey pulvinar. *Journal of Comparative Neurology, 419,* 377–393.

Alitto, H. J., & Usrey, W. M. (2015). Behavioral modulation of visual responses and network dynamics in the lateral geniculate nucleus. *Society for Neuroscience Abstracts, 148,* 24.

Arcaro, M. J., Pinsk, M. A., & Kastner, S. (2015). The anatomical and functional organization of the human visual pulvinar. *Journal of Neuroscience, 35,* 9848–9871.

Arcaro, M. J., Pinsk, M. A., & Kastner, S. (2018). Organizing principles of pulvino-cortical connectivity in humans. *Nature Communications* 9, 5382.

Bastos, A. M., Briggs, F., Alitto, H. J., Mangun, G. R., & Usrey, W. M. (2014). Simultaneous recordings from the primary visual cortex and lateral geniculate nucleus reveal rhythmic interactions and a cortical source for gamma-band oscillations. *Journal of Neuroscience, 34*(22), 7639–7644.

Bereshpolova, Y., Stoelzel, C. R., Zhuang, J., Amitai, Y., Alonso, J. M., & Swadlow, H. A. (2011). Getting drowsy? Alert/nonalert transitions and visual thalamocortical network dynamics. *Journal of Neuroscience, 31*(48), 17480–17487.

Briggs, F., Kiley, C. W., Callaway, E. M., & Usrey, W. M. (2016). Morphological substrates for parallel streams of corticogeniculate feedback originating in both V1 and V2 of the macaque monkey. *Neuron, 90*(2), 388–399.

Briggs, F., Mangun, G. R., & Usrey, W. M. (2013). Attention enhances synaptic efficacy and the signal-to-noise ratio in neural circuits. *Nature, 499*(7459), 476–480.

Briggs, F., & Usrey, W. M. (2009). Parallel processing in the corticogeniculate pathway of the macaque monkey. *Neuron, 62*(1), 135–146.

Buschman, T. J., & Kastner, S. (2015). From behavior to neural dynamics: An integrated theory of attention. *Neuron, 88*, 127–144.

Carrasco, M. (2011). Visual attention: The past 25 years. *Vision Research, 51*, 1484–1525.

Casagrande, V. A., & Xu, X. (2004). Parallel visual pathways: A comparative perspective. In L. M. Chalupa, & J. S. Werner (Eds.), *The visual neurosciences* (pp. 494–506). Cambridge, MA: MIT Press.

Cheong, S. K., Tailby, C., Solomon, S. G., & Martin, P. R. (2013). Cortical-like receptive fields in the lateral geniculate nucleus of marmoset monkeys. *Journal of Neuroscience, 33*, 6864–6876.

Crick, F. (1984). Function of the thalamic reticular complex: The searchlight hypothesis. *Proceedings of the National Academy of Sciences of the United States of America, 81*, 4586–4590.

De Weerd, P., Peralta III, M. R., Desimone, R., & Ungerleider, L. G. (1999). Loss of attentional stimulus selection after extrastriate cortical lesions in macaques. *Nature Neuroscience, 2*, 753–758.

Felleman, D. J., & Van Essen, D. C. (1991). Distributed hierarchical processing in the primate cerebral cortex. *Cerebral Cortex, 1*, 1–47.

Fischer, J., & Whitney, D. (2012). Attention gates visual coding in the human pulvinar. *Nature Communications, 3*, 1051.

Fitzpatrick, D., Usrey, W. M., Schofield, B. R., & Einstein, G. (1994). The sublaminar organization of corticogeniculate neurons in layer 6 of macaque striate cortex. *Visual Neuroscience, 11*, 307–315.

Friedman-Hill, S. R., Robertson, L. C., Desimone, R., & Ungerleider, L. G. (2003). Posterior parietal cortex and the filtering of distractors. *Proceedings of the National Academy of Sciences of the United States of America, 100*, 4263–4268.

Fries, P. (2005). A mechanism for cognitive dynamics: Neuronal communication through neuronal coherence. *Trends in Cognitive Science, 9*, 474–480.

Fries, P. (2015). Rhythms for cognition: Communication through coherence. *Neuron, 88*, 220–235.

Gallant, J. L., Shoup, R. E., & Mazer, J. A. (2000). A human extrastriate area functionally homologous to macaque V4. *Neuron, 27*, 227–235.

Grimaldi, P., Saleem, K. S., & Tsao, D. (2016). Anatomical connections of the functionally defined "face patches" in the macaque monkey. *Neuron, 90*, 1325–1342.

Guillery, R. W., Feig, S. L., & Lozsádi, D. A. (1998). Paying attention to the thalamic reticular nucleus. *Trends in Neuroscience, 21*, 28–32.

Gutierrez, C., Yaun, A., & Cusick, C. G. (1995). Neurochemical subdivisions of the inferior pulvinar in macaque monkeys. *Journal of Comparative Neurology, 363*, 545–562.

Halassa, M. M., & Kastner, S. (2017). Thalamic functions in distributed cognitive control. *Nature Neuroscience, 20*, 1669–1679.

Ichida, J. M., Mavity-Hudson, J. A., & Casagrande, V. A. (2014). Distinct patterns of corticogeniculate feedback to different layers of the lateral geniculate nucleus. *Eye and Brain, 6*(Suppl. 1), 57.

Jones, E. G. (2007). *The thalamus*. 2nd ed. Cambridge: Cambridge University Press.

Kaas, J. H., & Lyon, D. C. (2007). Pulvinar contributions to the dorsal and ventral streams of visual processing in primates. *Brain Research Reviews, 55*, 285–296.

Livingstone, M. S., & Hubel, D. H. (1981). Effects of sleep and arousal on the processing of visual information in the cat. *Nature, 291*, 554–561.

Maunsell, J. H. R. (2015). Neuronal mechanisms of visual attention. *Annual Review of Vision Science, 1*, 373–391.

McAlonan, K., Brown, V. J., & Bowman, E. M. (2000). Thalamic reticular nucleus activation reflects attentional gating during classical conditioning. *Journal of Neuroscience, 20*(23), 8897–8901.

McAlonan, K., Cavanaugh, J., & Wurtz, R. H. (2006). Attentional modulation of thalamic reticular neurons. *Journal of Neuroscience, 26*(16), 4444–4450.

McAlonan, K., Cavanaugh, J., & Wurtz, R. H. (2008). Guarding the gateway to cortex with attention in visual thalamus. *Nature, 456*, 391–394.

McCormick, D. A., McGinley, M. J., & Salkoff, D. B. (2015). Brain state dependent activity in the cortex and thalamus. *Current Opinion in Neurobiology, 31*, 133–140.

Nobre, A. C., & Kastner, S. (2014). *The Oxford handbook of attention*. Oxford: Oxford University Press.

O'Connor, D. H., Fukui, M. M., Pinsk, M. A., & Kastner, S. (2002). Attention modulates responses in the human lateral geniculate nucleus. *Nature Neuroscience, 5*, 1203–1209.

Petersen, S. E., Robinson, D. L., & Keys, W. (1985). Pulvinar nuclei of the behaving rhesus monkey: Visual responses and their modulation. *Journal of Neurophysiology, 54*, 867–886.

Purushothaman, G., Marion, R., Li, K., & Casagrande, V. A. (2012). Gating and control of primary visual cortex by pulvinar. *Nature Neuroscience, 15*, 905–912.

Rafal, R. D., & Posner, M. I. (1987). Deficits in human visual spatial attention following thalamic lesions. *Proceedings of the National Academy of Sciences of the United States of America, 84*, 7349–7353.

Reynolds, J. H., & Chelazzi, L. (2004). Attentional modulation of visual processing. *Annual Review of Neuroscience, 27*, 611–647.

Romanski, L. M., Giguere, M., Bates, J. F., & Goldman-Rakic, P. S. (1997). Topographic organization of medial pulvinar connections with the prefrontal cortex in the rhesus monkey. *Journal of Comparative Neurology, 379*, 313–332.

Rotshtein, P., Soto, D., Gregucci, A., Geng, J. J., & Humphreys, G. W. (2011). The role of the pulvinar in resolving competition between memory and visual selection: A functional connectivity study. *Neuropsychologia, 49*, 1544–1552.

Saalmann, Y. B., & Kastner, S. (2011). Cognitive and perceptual functions of the visual thalamus. *Neuron, 71*, 209–223.

Saalmann, Y. B., Pinsk, M. A., Wang, L., Li, X., & Kastner, S. (2012). The pulvinar regulates information transmission between cortical areas based on attention demands. *Science, 337*, 753–756.

Schneider, K. A., & Kastner, S. (2009). Effects of sustained spatial attention in the human lateral geniculate nucleus and superior colliculus. *Journal of Neuroscience, 29*, 1784–1795.

Sherman, S. M., & Guillery, R. W. (2013). *Thalamocortical processing: Understanding the messages that link the cortex to the world.* Cambridge, MA: MIT Press.

Shipp, S. (2003). The functional logic of cortico-pulvinar connections. *Philosophical Transactions of the Royal Society of London B: Biological Sciences, 358*, 1605–1624.

Snow, J. C., Allen, H. A., Rafal, R. D., & Humphreys, G. W. (2009). Impaired attentional selection following lesions to human pulvinar: Evidence for homology between human and monkey. *Proceedings of the National Academy of Sciences of the United States of America, 106*, 4054–4059.

Stepniewska, I., & Kaas, J. H. (1997). Architectonic subdivisions of the inferior pulvinar in New World and Old World monkeys. *Visual Neuroscience, 14*, 1043–1060.

Steriade, M. (2004). Acetylcholine systems and rhythmic activities during the waking-sleep cycle. *Progress in Brain Research, 145*, 179–196.

Theyel, B. B., Llano, D. A., & Sherman, S. M. (2010). The corticothalamocortical circuit drives higher-order cortex in the mouse. *Nature Neuroscience, 13*, 84–88.

Usrey, W. M., & Alitto, H. J. (2015) Visual functions of the thalamus. *Annual Review of Vision Science, 1*, 351–371.

Vargas, A. U., Schneider, L., Wilke, M., & Kagan, I. (2017). Electrical microstimulation of the pulvinar biases saccade choices and reaction times in a time-dependent manner. *Journal of Neuroscience, 37*, 2234–2257.

Wilke, M., Turchi, J., Smith, K., Mishkin, M., & Leopold, D. A. (2010). Pulvinar inactivation disrupts selection of movement plans. *Journal of Neuroscience, 30*, 8650–8659.

Wimmer, R. D., Schmitt, L. I., Davidson, T. J., Nakajima, M., Deisseroth, K., & Halassa, M. M. (2015). Thalamic control of sensory selection in divided attention. *Nature, 526*(7575), 705–709.

Yantis, S., Schwarzbach, J., Serences, J. T., Carlson, R. L., Steinmetz, M. A., Pekar, J. J., & Courtney, S. M. (2002). Transient neural activity in human parietal cortex during spatial attention shifts. *Nature Neuroscience, 5*, 995–1002.

Zhou, H., Schafer, R. J., & Desimone, R. (2016). Pulvinar-cortex interactions in vision and attention. *Neuron, 89*, 209–220.

V

NEUROSCIENCE, COGNITION, AND COMPUTATION: LINKING HYPOTHESES

Introduction

STANISLAS DEHAENE AND JOSH MCDERMOTT

This is the sixth edition of *The Cognitive Neurosciences*. With it we bring to you a new section of the book, in which we aim to survey work that links cognitive science and neuroscience via computation. Cognitive neuroscience has, of course, always aimed to create bridges between fundamental neuroscience and cognition. However, the field is increasingly shaped by the power of computational models to instantiate theories and generate predictions for both behavior and brain responses. Models continue to expand in scope due to advances in theory, engineering, and computing resources, as does the ability to use them to make and evaluate predictions. And classic ideas from computational neuroscience are being extended to new problems. Each of the seven chapters in this section highlights examples of the ways in which computation can help to bridge neuroscience with perception and cognition.

The section begins with two chapters that describe different approaches to harnessing the recent advances in artificial intelligence research in order to build explicit models of challenging computational problems in perception. Yamins describes the use of artificial neural networks to develop new models of the ventral visual stream. The key theoretical claim is that tasks place significant constraints on neural systems, such that optimizing for them in a distributed multistage model might generate representations like those in the brain. Current methods for training deep neural networks enable human-level recognition performance on some real-world recognition tasks, and the resulting models produce quantitatively accurate predictions of responses deep in the visual system and approximate the hierarchical structure of the ventral stream.

Yildirim, Siegel, and Tenenbaum take a complementary approach, proposing that humans use physically realistic internal models of objects for perceiving and thinking about the world. They adopt the Helmholtzian notion that perception consists of inverting the process by which sensory signals are generated from causes in the world, leveraging recent advances in machine learning to make this inference process tractable for some classes of realistic three-dimensional objects. Their framework incorporates feedforward neural networks but uses them to initialize inference in generative models, yielding both psychophysical and neural predictions that the authors have begun to confirm.

From perception we turn to decision-making and two chapters that apply some classic tools to new domains. Rossi-Pool, Vergara, and Romo exploit the somatosensory system as a model for decision-making, presenting a detailed comparison of psychophysical behavior, concurrent neural recordings, and the effects of microstimulation in awake, behaving monkeys. They find that primary somatosensory cortex faithfully represents sensory information but that most other stages of the presumptive behavior-generating pathway reflect aspects of the animal's decision.

Summerfield and Tsetsos consider decision-making in both perceptual and economic contexts. They ask the normative question of whether human decisions can be understood as the solution to a constrained optimization problem. Extending the framework of efficient coding (commonly applied to explain sensory representations), they argue that human decisions that might not be traditionally defined as rational can nonetheless be interpreted as normative given a constraint on processing costs.

The two chapters on decision-making are followed by an examination of abnormal decision-making in mental illness. Bennett and Niv survey the growing field of computational psychiatry, which attempts to understand different forms of mental illness as abnormalities in specific components of the decision-making process in healthy individuals. Computational models help to specify processes or variables that might be altered in mental illness, thus leading to precise predictions that can be tested. The best explored example thus far lies in reinforcement learning, models of which have been adapted to explain depression and bipolar disorder.

Reinforcement learning is also discussed by Koechlin, but as a starting point for a broad theory of prefrontal cortex evolution and function. Koechlin argues that the limitations of reinforcement learning in simple organisms necessitated an expansion of cognitive resources, both to recognize new situations that require new patterns of action and to store multiple task sets. He proposes that these demands drove the expansion of the frontal cortex in rodents, monkeys, and humans, and he speculates on the novel prefrontal architectures that might underlie human-specific abilities for language and other recursive structures.

The final chapter confronts the representation of meaning in the brain. Semantic representations in the brain have traditionally been studied using focal contrasts between small numbers of categories of stimuli. Gallant and Popham discuss a new approach in which brain responses are measured to natural stimuli such as movies and stories. Semantic descriptors of the stimulus content are then regressed against the responses of voxels measured with functional magnetic resonance imaging (fMRI). Here the role of computation is to define a semantic feature space and to relate it to brain responses. Gallant and Popham describe evidence for a vast array of semantic representations distributed throughout most of the human cortex.

These contributions provide a glimpse of the ways in which computation is bridging the gap between the brain and cognition. The chapters are diverse but exhibit some common themes. New computational methods derived from the latest innovations in engineering are being used alongside decades-old methods and ideas that continue to stand the test of time. And bridges are being built at all scales of measurement, from single neurons to whole-brain maps, and at all levels of computational analysis—from top-level descriptions of the problem being solved to specific neural circuit components. Perhaps the most exciting development documented here is the increasing ability to characterize and study realistic modes of behavior and cognition using new developments in artificial intelligence, engineering, and computing combined with real-world tasks and stimuli. This trend seems likely to continue in the coming years and to make important contributions to the next generation of brain-behavior models.

33 An Optimization-Based Approach to Understanding Sensory Systems

DANIEL YAMINS

ABSTRACT Recent results have shown that deep neural networks (DNN) may have significant potential to serve as quantitatively precise models of sensory cortex neural populations. However, the implications these results have for our conceptual understanding of neural mechanisms are subtle. This is because many modern DNN brain models are best understood as the products of task-constrained optimization processes, unlike the intuitively simpler handcrafted models from earlier approaches. In this chapter we illustrate these issues by first discussing the nature of information processing in the primate ventral visual pathway and review results comparing the response properties of units in goal-optimized DNN models to neural responses found throughout the ventral pathway. We then show how DNN visual system models are just one instance of a more general optimization framework whose logic may be applicable to understanding the underlying constraints that shape neural mechanisms throughout the brain.

Nothing in biology makes sense except in light of evolution.
—Theodosius Dobzhansky

Nothing in neurobiology makes sense except in light of behavior.
—Gordon Shepherd

An important part of a scientist's job is to answer "why" questions. For cognitive neuroscientists, a core objective is to uncover the underlying reasons why the structures of the human brain are as they are. Since brains are biological systems, answering such questions is ultimately a matter of identifying the evolutionary and developmental constraints that shape brain structure and function. Such constraints are in part architectural: What large-scale brain structures are put in place genetically to enable a brain to help its host organism better meet evolutionary challenges? In light of the centrality of behavior in understanding the brain, an ethological investigation is also indicated: What behavioral goals most strongly constrain a given neural system? And since many complex behaviors in higher organisms are not entirely genetically determined and must instead be partly derived through experience of the world, a core question of learning is also involved: How do learning rules that absorb experiential data constrain what brains look like?

The interactions between architectural structure, behavioral goals, and learning rules suggest a *quantitative optimization framework* as one route toward answering these "why" questions. Put simply, this means postulating one or several goal behavior(s) as driving the evolution and/or development of a neural system of interest, finding architecturally plausible computational models that (attempt to) optimize for the behavior, and then quantitatively comparing the internal structures arrived at in the optimized models to measurements from large-scale neuroscience experiments. To the extent that there is a match between the optimized models and the real data that is very substantially better than that found for various controls (e.g., models designed by hand or optimized for other tasks), this is evidence that something important has been understood about the underlying constraints that shape the brain system under investigation. Though it might sound challenging to put this approach into practice, recent successes suggest we might add to our list of maxims the observation that *nothing in computational cognitive neuroscience makes sense except in light of optimization.*

Case Study: The Primate Ventral Visual Stream

The most thoroughly developed example of these optimization-based ideas is the visual system—in particular, the ventral visual stream in humans and nonhuman primates. While a complete review of the work that lead to the present understanding of the primate ventral stream is beyond the scope of this chapter (see DiCarlo, Zoccolan, and Rust [2012] for a summary), discussing key computational aspects of the ventral stream in some detail will lay the groundwork for the optimization approach more generally.

The computational crux of the vision problem The human brain effortlessly reformats the "blooming, buzzing confusion" of unstructured visual data streams into powerful abstractions that serve high-level behavioral

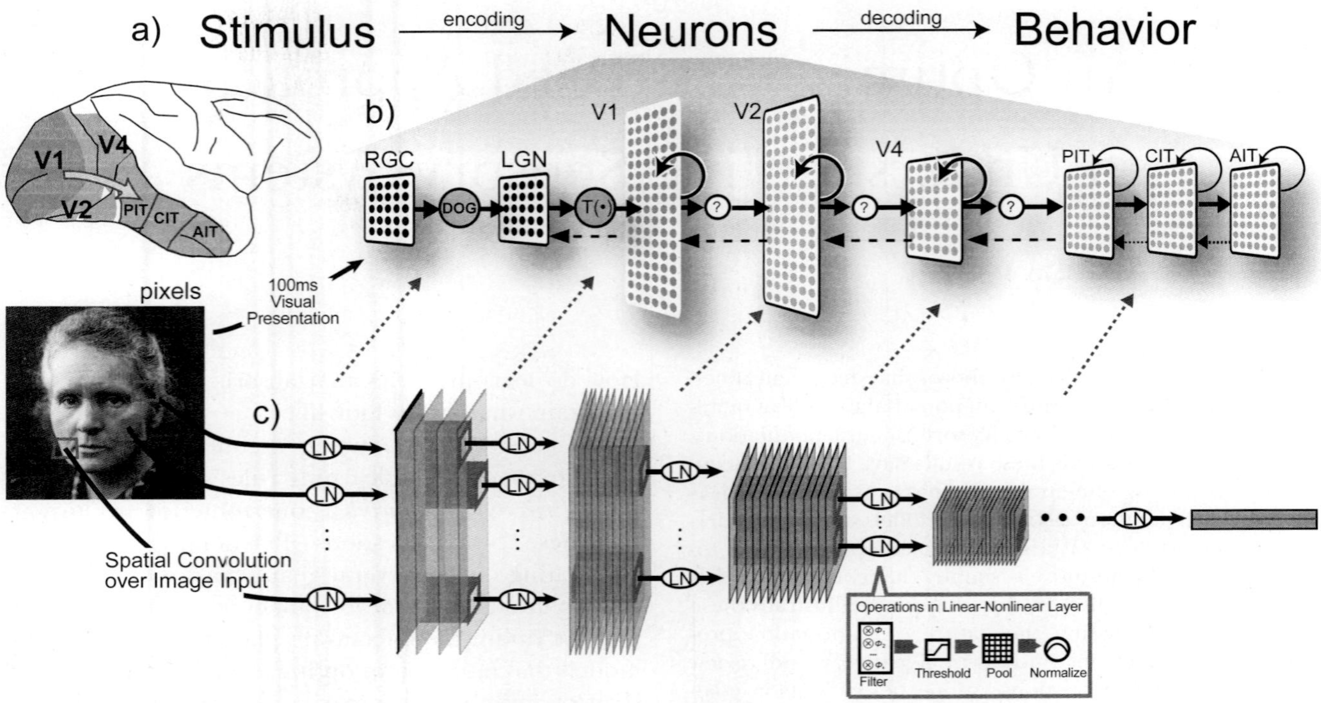

FIGURE 33.1 Hierarchical convolutional neural networks as models of sensory cortex. *A*, The basic framework in which sensory cortex is studied is one of encoding, the process by which stimuli are transformed into patterns of neural activity, and decoding, the process by which neural activity generates behavior. *B*, The ventral visual pathway of humans and nonhuman primates is one of the most comprehensively studied sensory systems in neuroscience. It consists of a series of connected cortical brain areas that are thought to operate in a sensory cascade, from early visual areas such as V1 to later visual areas such as inferior temporal (IT) cortex. Neural responses in the ventral pathway are believed to encode an abstract representation of objects in visual images. *C*, Hierarchical convolutional neural networks (HCNNs) are multilayer neural networks that have been proposed as models of the ventral pathway. Each layer of an HCNN is made up of a linear-nonlinear (LN) combination of simple operations such as filtering, thresholding, pooling, and normalization. The filter bank in each layer consists of a set of weights analogous to synaptic strengths. Each filter in the filter bank corresponds to a distinct template, analogous to Gabor wavelets with different frequencies and orientations (the image shows a model with four filters in layer 1, eight in layer 2, and so on). The operations within a layer are applied locally to spatial patches within the input, corresponding to simple limited-size receptive fields (*red boxes*). The composition of multiple layers leads to a complex nonlinear transform of the original input stimulus. At each layer, retinotopy decreases and effective receptive field size increases. (See color plate 35.)

goals, such as scene understanding, navigation, and action planning (James 1890). But parsing retinal input into rich object-centric scene descriptions is a major computational challenge. The crux of the problem is that the axes of the low-level input space (i.e., light intensities at each retinal "pixel") don't correspond to the natural axes along which high-level constructs vary. For example, translation, rotation in depth, deformation, or relighting of a single object (e.g., one person's face) can lead to large and complex nonlinear transformations of the original image. Conversely, images of two ecologically quite distinct objects—for example, different individuals' faces—may be very close in pixel space. Behaviorally relevant dimensions are thus highly "tangled" in the original input space (DiCarlo and Cox 2007), and to recognize objects and understand scenes, the brain must rapidly and accurately accomplish the complex and often ill-posed nonlinear untangling process (DiCarlo, Zoccolan, and Rust 2012).

Hierarchy and retinotopy in the ventral pathway Sparked by the seminal ideas of Hubel and Wiesel, six decades of work in visual systems neuroscience have shown that the homologous visual system in humans and nonhuman primates generates robust object recognition behavior via a series of anatomically distinguishable cortical areas known as the *ventral visual stream* (figure 33.1*A–B*; Connor, Brincat, and Pasupathy 2007; DiCarlo, Zoccolan, and Rust 2012; Felleman and Van Essen 1991; Malach, Levy, and Hasson 2002; Rust and DiCarlo 2010). Two basic principles of architectural organization emerging from this work are that the ventral stream is

1. *hierarchical*, with visual information passing along a cascade of processing stages embodied by distinct cortical areas, and
2. *retinotopic*, composed of structurally similar operations with spatially local receptive fields tiling the overall visual field, with decreasing spatial resolution in each subsequent stage of the hierarchy.

Visual areas early in the hierarchy, such as V1 cortex, capture low-level features, including edges and center-surround patterns (Carandini et al. 2005; Movshon, Thompson, and Tolhurst 1978). Neural population responses in the highest ventral visual area, the anterior inferior temporal (AIT) cortex, can be used to decode object category, robust to significant variations present in natural images (Hung et al. 2005; Majaj et al. 2015; Yamane et al. 2008). Midlevel visual areas such as V2, V3, V4, and posterior IT (PIT) are less well characterized by such "word models" than higher or lower visual areas closer to the sensorimotor periphery. Nonetheless, these intermediate areas appear to contain computations at an intermediate level of complexity between simple edges and complex objects, along a pipeline of increasing receptive field size (Brincat and Connor 2004; DiCarlo and Cox 2007; DiCarlo, Zoccolan, and Rust 2012; Freeman and Simoncelli 2011; Gallant et al. 1996; Lennie and Movshon 2005; Schiller 1995; Schmolesky et al. 1998; Yau et al. 2012).

Linear-nonlinear cascades A core hypothesis is that the ventral stream employs sensory cascades because (1) the overall stimulus-to-neuron transforms required to support complex behaviors are extremely complicated—after all, since the original input tangling is highly nonlinear, the inverse untangling process is also highly nonlinear; but (2) the capacities of any single stage of neural processing are limited to comparatively simple operations, such as weighted sums of inputs, thresholding nonlinearities, and local normalization (Carandini et al. 2005). To build up a sufficiently complex end-to-end transform with a reasonable number of neurons, a cascade of stages is needed. Complex nonlinear transformations arise from multiple such stages applied in series (Sharpee, Kouh, and Reyholds 2012). Such cascades are not only present in the visual system but are common in a wide variety of sensory areas (Hegner, Lindner, and Braun 2017; Petersen 2007; Pickles 2008; Romanski and LeDoux 1993).

A very simplified version of the feedforward component of the multistage sensory cascade may thus be represented symbolically by:

$$\text{stimulus} \overset{T_1}{\mapsto} \mathbf{n}_1 \overset{T_2}{\mapsto} \mathbf{n}_2 \ldots \overset{T_{\text{top}}}{\mapsto} \mathbf{n}_{\text{top}} \qquad (33.1)$$

where the $\mathbf{n}_i$ represent neural responses in brain area i, and T_i is the transform computed by the neurons in area i based on input from area $i-1$. In the macaque ventral stream, this will (at least) include several subcortical stages prior to the ventral stream (e.g., the retinal ganglion and lateral geniculate nucleus (LGN)), followed by cortical areas V1, V2, V4, PIT, and AIT. The homologous structure in humans is similar but likely to be substantially more complex (Wang et al. 2014).

Robust empirical observations (Carandini et al. 2005) suggest that the transforms T_i can be reasonably well modeled as *linear-nonlinear* (LN) blocks of the form:

$$T_i = N_i \circ L_i.$$

Biologically, the linear transforms L_i are inspired by the observation that neurons are admirably suited for taking dot products—that is, summing up their inputs on each incoming dendrite, weighted by synaptic strengths. The transforms L_i formalize the synaptic strengths as numerical matrices. Mathematically, the L_i map the input feature space output by one area to an intermediate feature space in the next. In the case of L_1 (the transform between the input image and the first visual area, taken to be either subcortical or in V1), the input space is the three-channel RGB-like representation of pixels, while the output space is substantially higher dimensional, corresponding to the number of different neural projections computed at each retinotopic location. An extensive line of research characterizing V1 responses (Carandini et al. 2005; Hubel and Wiesel 1959; Ringach, Shapley, and Hawken 2002) yielded the realization that the linear transforms early on in the cascade can be reasonably well characterized as spatial convolution with a filter bank of Gabor wavelets in a range of frequencies and orientations (Willmore et al. 2008).

The nonlinear component N_i has been shown to involve combinations of very basic transforms, including rectification, pooling, and normalization operations (Brincat and Connor 2004; Carandini et al. 2005). While the T_i s are simple, it is critical that they are at least somewhat nonlinear: the composition of linear operations is linear, so additional complexity can't be built up by a sequence of linear operations, and there would be no evolutionary point to allocating multiple brain areas for them in the first place.

It is tempting to ascribe specific functional roles for each of the constituent operations within an LN block, described in terms of features of the original input stimulus. While this may be possible early in the sensory cascade, the compounding of multiple nonlinearities makes it unlikely that this type of description is adequate for intermediate or higher-sensory areas. Instead, it is probably more effective to think of the LN block as combining a dimension-expanding component (the linear-filtering step), a dimension-reducing aggregation component (the pooling operation), and a range-centering component to ensure the cascade can be effectively extended hierarchically (the normalization operations). These features allow LN cascades to cover a wide range of complex nonlinear functions in an efficient manner (Bengio 2012; Poole et al. 2016), consistent with the idea that good LN cascade architectures can be discovered by evolutionary and developmental processes.

A common visual feature basis The features computed by the sensory cascade are often thought of as constituting a *visual representation.* One way to interpret this idea is that the output from area n_{top}—which is considerably *upstream* of highly task-modulated decision-making or motor areas—is able to support observed organism output behaviors via simple decoders. Symbolically, the pipeline in diagram (1) can be extended to this observation:

$$\text{stimulus} \ldots \mapsto \mathbf{n}_{top} \overset{D}{\mapsto} \text{behavior} \tag{33.2}$$

where D is a population decoder. The requirement that D be "simple" just means that it can also be cast in the form of a single LN block rather than requiring many stages of nonlinearity. In the case of the macaque visual system, the role of n_{top} seems to be played by anterior IT cortex, where it has been robustly shown that simple decoders, such as linear classifiers or linear regressors, operating on neural responses in IT cortex can support patterns of visual behavior at a high degree of behavioral resolution (DiCarlo and Cox 2007; Hung et al. 2005; Majaj et al. 2015; Rajalingham, Schmidt, and DiCarlo 2015; Rust and DiCarlo 2010). The linear classifiers embody a computational description of the stimulus-driven component of hypothetical decoding circuits downstream of the ventral visual representation (Freedman et al. 2001; Pagan et al. 2013).

The representation concept is enhanced by the observation that IT cortex can provide useful support for *many different* visual behaviors. In addition to object category, attributes such as fine-grained within-category identification, object position, size, pose, and complex lighting and material properties can be decoded from IT neural activity (Hong et al. 2016; Nishio et al. 2014). Symbolically, this might be represented by the diagram

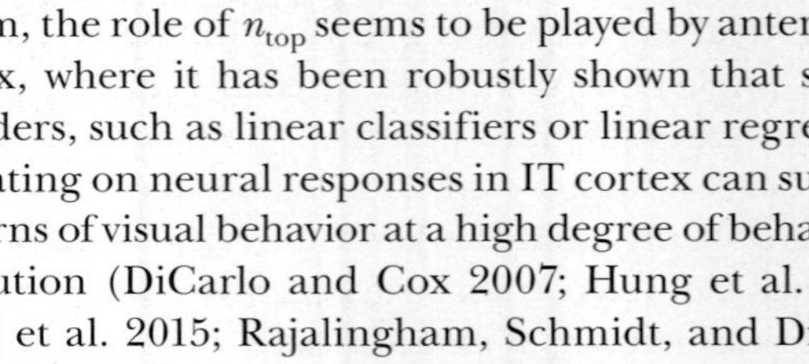

in which D_1, D_2,... are different readout decoders for the various possible visually driven behaviors.

A key observation is that for naturalistic scenes with realistically high levels of image variability, these same visual properties cannot be robustly read out from the visually evoked neural responses in earlier areas such as the retina, V1, or V2 using simple decoders and can be read out only partially in intermediate areas such as V4 (Hong et al. 2016; Majaj et al. 2015). Of course, the information *must* in some way be present in these areas since the properties can be determined by looking at the image. However, as alluded to earlier, these properties are "tangled up" in the representations in early areas and so cannot be easily decoded. The nonlinear operations of the ventral stream cascade culminating in the IT representation have reformatted the information in the input image stimuli into a *common basis,* from which it is possible to generate many different behaviorally relevant readouts.

Not just an information channel These considerations suggest that the ventral stream is *not* best thought of as a "channel" in the sense of Shannon information theory. As a result of (converse of) Shannon's famous channel-coding theorem, with every step of the cascade, the system can only lose information in an information-theoretic sense (Cover and Thomas 2012). The more stages in the case, the less good it will be as a pure information channel. The existence of a many-stage LN cascade in the ventral pathway suggests that the evolutionary constraint on the system is *not* the veridical preservation of information about the stimulus. Rather, the constraining evolutionary goal of the sensory cascade is more likely to be making behaviorally relevant information—such as the identity of a face present in the image—much more *explicitly* available for easy access by downstream brain areas while discarding other information about the stimuli—such as pixel-level details—that is less behaviorally relevant.

Neural Network Models of the Ventral Stream

In this section we will discuss how the neurophysiological observations described above can be formalized mathematically. But before diving into models of the ventral stream, it is worth briefly considering why we might want to make quantitative neural network models of the ventral stream in the first place. After all, neuroscientists did not need such models to discover the important insights described in the previous section.

Two convergent problems, however, strongly motivate the building of large-scale formal models. First, the simpler word-model approach useful for characterizing the shape of visual feature tuning curves in earlier cortical areas, such as the retina or V1, was found to be difficult to generalize to intermediate- and higher-level visual areas (Pinto, Cox, and DiCarlo 2008). Though some progress has been made using intuition to find visual features to which intermediate- and higher-area neurons would respond (Connor, Brincat, and Pasupathy 2007; Tanaka 2003; Yau et al. 2012), a more systematic approach is needed to organize and generalize these disparate observations. Second, the most naïve implementations of multilayer hierarchical retinotopic models performed very poorly on tests of performance generalization in real-world settings (Pinto, DiCarlo, Doukhan, and Cox 2009). Although hierarchy and retinotopy appeared to be important high-level principles, they were insufficiently detailed to actually produce operational algorithms with anything like the visual abilities of a macaque or a human. Echoing Feynman's famous dictum that "what I cannot create, I do not understand," the inability to create from scratch a truly working visual recognition system meant that some key feature of understanding was missing.

Hierarchical convolutional neural networks Hierarchical convolutional neural networks (HCNNs) are a broad generalization of Hubel and Wiesel's ideas that has been developed over the past 40 years by researchers in biologically inspired computer vision (Fukushima 1980; LeCun and Bengio 1995; Yamins and DiCarlo 2016). HCNNs consist of cascades of layers containing simple neural circuit motifs repeated retinotopically across the sensory input (figure 33.1*C*). Each layer is simple, but a deep network composed of such layers computes a complex transformation of the input data roughly analogous to the organization of the ventral stream. The specific operations comprising a single HCNN layer were inspired directly by the LN neural motif (Carandini et al. 2005), including convolutional filtering, a linear operation that takes the dot product of local patches in the input stimulus with a set of templates, typically followed by rectified activation, mean or maximum pooling (Serre, Oliva, and Poggio 2007), and some form of normalization (Carandini and Heeger 2012). All the basic operations exist within a single HCNN layer, which is designed to be analogous to a single cortical area within the visual pathway.

A key feature of HCNNs is that all operations are applied locally, over a fixed-size input zone that is smaller than the full spatial extent of the input. HCNNs employ *convolutional weight sharing*, meaning that the same filter templates are applied at all spatial locations. Since identical operations are applied everywhere, spatial variation in the output arises entirely from spatial variation in the input stimulus. The brain is unlikely to literally implement weight sharing, since the physiology of the ventral stream appears to rule out the existence of a single "master" location in which shared templates could be stored. However, the natural visual statistics of the world are themselves largely shift invariant in space (or time), so experience-based learning processes in the brain should tend to cause weights at different spatial locations to converge. Shared weights are therefore likely to be a reasonable approximation, at least within the central visual field.

Although the local fields seen by units in a single HCNN layer have a fixed small size, the effective receptive field size relative to the original input increases with succeeding layers in the hierarchy. Like the brain's ventral pathway, multilayer HCNNs typically become less retinotopic with each succeeding layer, consistent with empirical observations (Malach, Levy, and Hasson 2002). However, the number of filter templates used in each layer typically increases. Thus, the dimensionality changes through the layers from being dominated by spatial extent to being dominated by more abstract feature dimensions. After many layers the spatial component of the output may be so reduced that convolution is no longer meaningful, whereupon networks may be extended using one or more fully connected layers that further process information without explicit retinotopic structure. The last layer is usually used for *readout*—for example, for each of several visual categories, the likelihood of the input image containing an object of the given category might be represented by one output unit.

Learning modern deep HCNNs The earliest HCNNs were not particularly effective at either solving vision tasks or quantitatively describing neurons. Arbitrary hierarchical retinotopic nonlinear functions do not appear to compute useful representations (Pinto et al. 2009), and hand-designed filter banks in multilayer networks were also not performant (Pinto, Cox, and DiCarlo 2008; Pinto et al. 2009). It was realized early on, however, that the parameters of the HCNNs could be *learned*—that is, optimized so that the network output maximized performance. Parameters subject to optimization include discrete choices about the particular architecture to be used (How many layers? How many features per layer? What local receptive field should be used at a given layer?), as well as the continuous parameters of the linear transforms L_i at each layer.

Initial attempts to learn HCNNs led to intriguing and suggestive results (LeCun and Bengio 1995) but were not entirely satisfactory either in terms of neural similarity or task performance. However, recent work in computer vision and artificial intelligence has sought to use advances in hardware-accelerated computing to optimize the parameters of DNNs to maximize their performance on more challenging large-scale visual tasks (Deng, Li, et al. 2009). Leveraging computer vision and machine-learning techniques, together with large amounts of real-world labeled images used as supervised training data (Bergstra, Yamins, and Cox 2013; Krizhevsky, Sutskever, and Hinton 2012), HCNNs have, arguably, achieved human-level performance on several challenging object categorization tasks (He et al. 2016; Zoph et al. 2018).

In fact, the power of HCNNs trained on large data sets goes beyond merely doing well on training sets. Unlike small data sets that are prone to severe overfitting, large highly variable data sets, such as ImageNet, have yielded networks that can serve as useful bases for solving a variety of other visual tasks (Girshick 2015; Simonyan and Zisserman 2014). State-of-the-art solutions to ImageNet categorization often exhibit especially good transfer capabilities (Zoph et al. 2018). In other words, training HCNNs in a supervised manner has at least some power to produce robust visual representations.

Quantitative matches between HCNNs and ventral pathway areas A core result linking the deep HCNNs used in modern computer vision to ideas from visual systems neuroscience is that an HCNN's ability to predict neural responses in visual cortex is strongly correlated with its performance on challenging object categorization tasks (Yamins et al. 2013, 2014). Such correlations have been investigated by high-throughput studies comparing tens of thousands of distinct HCNN model instantiations to neural data from large-scale array electrophysiology experiments in macaques (Yamins et al. 2014), as well as human fMRI (Khaligh-Razavi and Kriegeskorte 2014). While the correlation is present for HCNNs with randomly chosen architectures, it is especially high when architectures are optimized for task performance (figure 33.2*A*).

Inferior temporal cortex Tighter relationships between HCNNs and neural data are observed on a per-area basis. Model responses from hidden layers near the top of HCNNs optimized for ImageNet categorization performance are highly predictive of neural responses in IT cortex, both in electrophysiological (figure 33.3*A*; Cadieu et al. 2014; Yamins et al. 2014) and fMRI data (Güçlü and Gerven 2015; Khaligh-Razavi and Kriegeskorte 2014). These deep, goal-optimized neural networks (*red squares*, figure 33.3*A*) have thus yielded the first quantitatively accurate, predictive model of

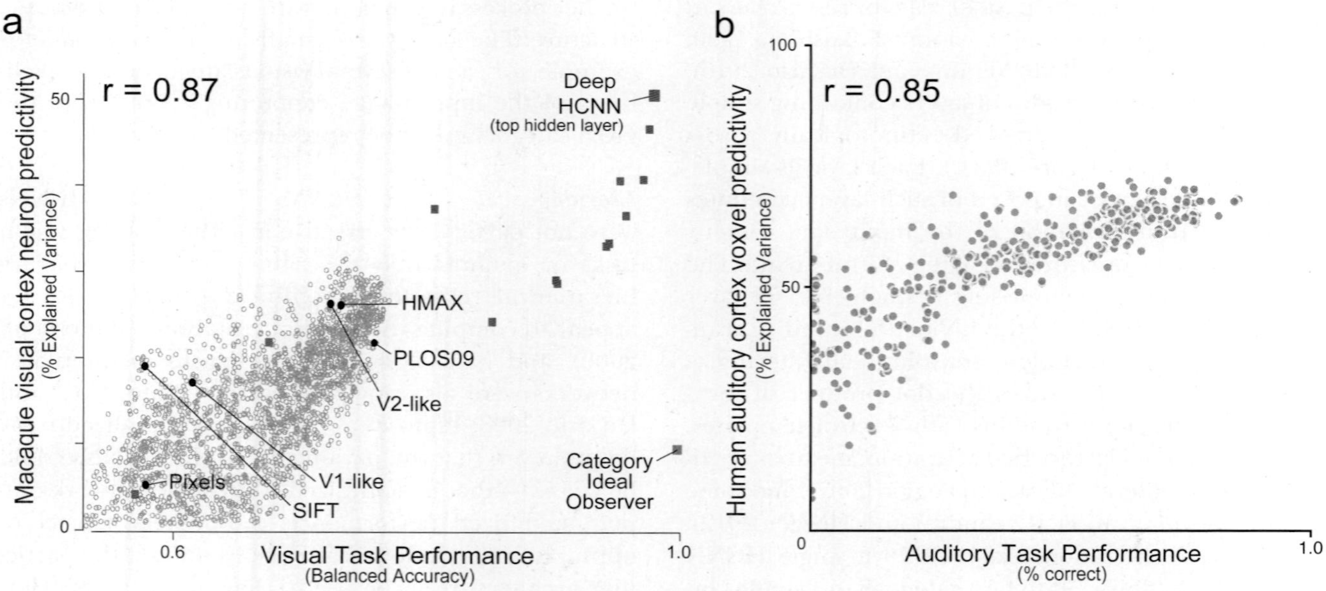

FIGURE 33.2 *A*, Visual object categorization task performance (*x*-axis) is highly correlated with the ability to predict IT cortex neural responses (*y*-axis). Adapted from Yamins et al. (2014). *Blue dots*, Various three-layer (shallow) HCNN models, either with random weights or optimized for categorization performance or to predict IT responses. *Black squares*, A variety of previous models. *Red dots*, Increasing performance and predictivity over time as a deep HCNN is trained. ***r*** value is for red and black points. *Green square*, A category ideal observer with perfect semantic category knowledge, to control for how much neural variance is explained just by categorical features alone. *B*, Analogous result for neural networks optimized for auditory tasks. Adapted from Kell et al. (2018). (See color plate 36.)

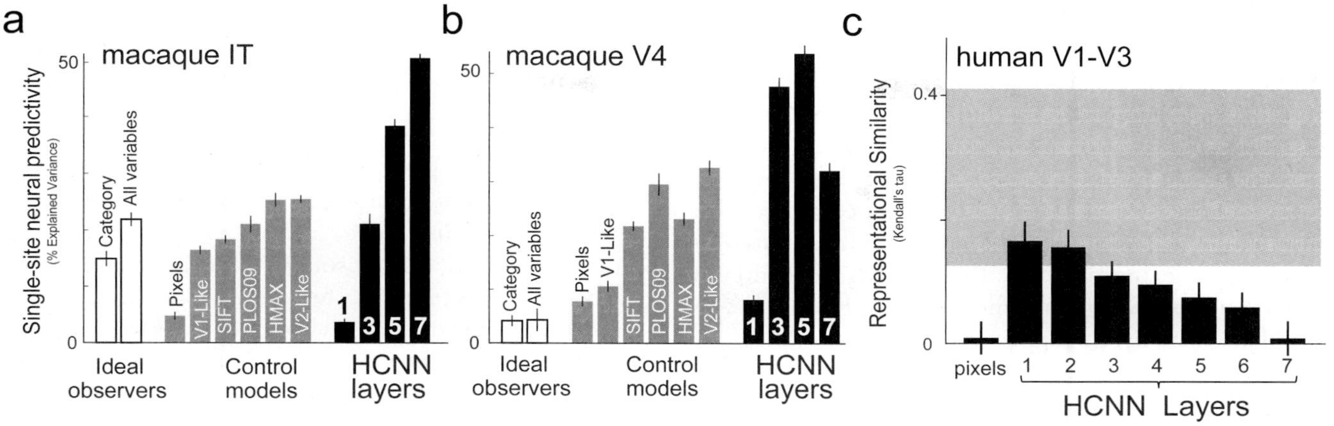

FIGURE 33.3 *A*, Based on Yamins et al. (2014), a comparison of the ability of various computational models to predict neural responses of populations of macaque IT neurons (*right*). The HCNN model (*black bars*) is a significant improvement in neural response prediction compared to previous models (*gray bars*) and task ideal observer controls (*open bars*). The top HCNN layer 7 best predicts IT responses. *B*, Similar to *A*, but for macaque V4 neurons. Note that intermediate layer 5 best predicts V4 responses. *C*, Representational similarity between visual representations in HCNN model layers and human V1–V3, based on fMRI data. Adapted with permission from Khaligh-Razavi and Kriegeskorte (2014). *Horizontal gray bar*, The inherent noise ceiling of the data. Note that earlier HCNN model layers most resemble early visual areas.

population responses in a higher cortical brain area. These quantitative models are also substantially better at predicting neural response variance in IT than semantic models based on word-level descriptions of object category or other attributes (*green square*, figure 33.3*A*; Yamins et al. 2014). Recent high-performing ImageNet-trained architectures also appear to provide the best matches to the visual behavioral patterns of primates (Rajalingham et al. 2018).

Intermediate visual areas Intermediate layers of the same HCNNs whose higher layers match IT neurons also turn out to yield state-of-the-art predictions of neural responses in V4 cortex (figure 33.3*B*; Güçlü and Gerven 2015; Yamins et al. 2014), the dominant cortical input to IT. Similarly, recent models with especially good performance have distinct layers clearly segregating late-intermediate visual area PIT neurons from downstream central IT (CIT) and AIT neurons (Nayebi et al. 2018). These results are important because they show that high-level, ecologically relevant constraints on network function—that is, the categorization task imposed at the network's output layer—are strong enough to inform upstream visual features in a nontrivial way. In other words, HCNN models suggest that the computations performed by the circuits in V4 and PIT are structured so that downstream computations in AIT can support high-variation robust categorization tasks. Thus, even though there may be no simple word model describing *what* the features in an intermediate cortical area such as V4 are, HCNNs can provide a principled description of *why* the area's neural responses might be as they are.

Early visual cortex Results in early visual cortex are equally striking. The filters emergent in HCNNs' early layers from the learning process naturally resemble Gabor wavelets without having to build this structure (Krizhevsky, Sutskever, and Hinton 2012). Extending the correspondence between HCNN layers and ventral stream layers down further, it has been shown that lower HCNN layers match neural responses in early visual cortex areas, such as V1 (figure 33.3*C*; Güçlü and Gerven 2015; Khaligh-Razavi and Kriegeskorte 2014; Seibert et al. 2016). In fact, recent high-resolution results show that early-intermediate layers of performance-optimized HCNNs are substantially better models of macaque V1 neural responses to natural images than previous state-of-the-art models hand-designed to replicate qualitative neuroscience observations (Cadena et al. 2019).

Taken together, these results indicate that combining two general biological constraints—the behavioral constraint of object recognition performance and the architectural constraint imposed by the HCNN model class—leads to improved models of multiple areas through the visual pathway hierarchy.

A contrast to curve fitting A key feature of these results is that the parameters of the HCNN models are optimized to solve a visual performance goal that is ethologically plausible for the organism, rather than being directly fit to neural data. Yet the resulting neural network effectively models the biology as well or better than direct curve fits (Cadena et al. 2019; Yamins et al. 2014). This is the idea of *goal-driven* modeling (Yamins and DiCarlo 2016). Goal-driven modeling is attractive as a method for building quantitative cortical models for several reasons. Practically speaking, it does not require the collection of the unrealistically massive amounts of neurophysiological data that would be needed to fit deep networks to such data. Second, because model validity is assessed on a completely different metric (and different data set) than that used to choose model parameters, the results are comparatively free from overfitting and/or multiple-comparison problems. Finally, the approach posits an evolutionally plausible functional *reason* for choices of model parameters throughout the hierarchy.

A Tripartite Optimization Framework

While the results described in the previous section are in some ways specific to the primate ventral pathway, they are based on a more general underlying logic that can apply to neural network-modeling problems throughout computational neuroscience. Specifically, three fundamental components underlie all functionally optimized neural network models:

- An *architecture class* $\mathcal{A}$ containing potential neural network structures from which the real system is drawn. $\mathcal{A}$ captures the structural constraints on the network drawn from knowledge about a brain system's anatomical and functional connectivity.
- A *computational goal* that the system seeks to accomplish, mathematically expressed as a loss target function

$$\mathcal{L}:\mathcal{A}\rightarrow\mathbb{R}$$

to be minimized by parameter choices within the set $\mathcal{A}$. For any potential network $\mathbf{a}\in\mathcal{A}$, the value $\mathcal{L}(\mathbf{a})$ represents the error that network incurs in attempting to solve the computational goal. $\mathcal{L}$ captures the functional constraints on the network drawn from hypotheses about the organism's behavioral repertoire.

- A *learning rule* by which optimization for $\mathcal{L}$ occurs within the architecture class $\mathcal{A}$. This is a function such that, at least statistically, for any nonoptimal network $\mathbf{A}\in\mathcal{A}$,

$$\mathcal{R}_{\mathcal{L}}:\mathcal{A}\rightarrow\mathcal{A}$$

$$\mathcal{L}(\mathcal{R}_{\mathcal{L}}(\mathbf{A}))<\mathcal{L}(\mathbf{A}). \tag{33.3}$$

Biologically, the learning rule captures the way that the error signal from mismatches between the system's current output and the correct outputs (as defined by the computational goal) is used to identify better parameter choices, over evolutionary and developmental timeframes.

This framework predicts that, statistically, the actual biological system is approximated by the optimal solution within $\mathcal{A}$ to the goal posed by $\mathcal{L}$—that is,

$$\mathbf{A}^*=argmin_{\mathbf{A}\in\mathcal{A}}\,\mathcal{L}(\mathbf{A}). \tag{33.4}$$

Of course, biological systems produced by evolution and development are not guaranteed to be optimal for their evolutionary niche, so this prediction is really more an informed heuristic for hypothesis generation rather than a candidate for natural law. In fact, any practically implementable learning rule will not perfectly meet the criterion in inequality (3), being subject to the same problem that evolution/development faces: failures to achieve the optimum due to incomplete optimization or capture by local minima. Insofar as the model of the learning rule and initial condition distribution is itself biologically accurate, the same patterns of performance failures should be observed in both the model and the real behavioral data (Rajalingham et al. 2018).

Returning to the example of the primate ventral stream, the model architecture class $\mathcal{A}$ has been taken to include feedforward HCNNs, broadly capturing aspects of the known neuroanatomical structure of the ventral visual pathway. The parameters describing this class of models include (1) discrete choices about (e.g.) the number of layers in the cascade, the specific nonlinear operations to employ at each layer, and the sizes of local receptive fields (see Yamins and DiCarlo [2016] for more details on these parameters) and (2) the continuous-valued filter templates embodied by the linear transforms L_i at each layer. The loss target $\mathcal{L}$ has typically been chosen as a categorization error on the 1,000-way object recognition task in the ImageNet data set (Deng, Li, et al. 2009), capturing the fact that primates have especially strong invariant object recognition capacities.

The learning rule used for optimizing HCNNs to solve categorization problems is composed of two pieces, corresponding to the two types of model parameters: (1) an "outer loop" of metaparameter optimization used for selecting the discrete parameters, typically either just random choice (Pinto et al. 2009) or a simple evolutionary algorithm (Yamins et al. 2014) and (2) an "inner loop" of smooth optimization of the synaptic strength parameters L_i, typically involving *gradient descent*:

$$\frac{dL_i}{dt}=-\lambda(t)\cdot\nabla_{L_i}[\mathcal{L}].$$

This expression formalizes the idea that learning modifies the synaptic strengths L_i of the visual system over time—the derivative $dL_i\,/\,dt$—by greedily following the local gradient of the loss target, scaled in magnitude by the *learning rate* $\lambda(t)$.

Many variants of gradient descent have been explored in the machine-learning literature, some of which scale better or achieve faster or better optimization (Bottou 2010; Kingma and Ba 2014; Zeiler 2012). Though Hebbian learning rules have been proposed many times in neuroscience (Montague, Dayan, and Sejnowski 1996; Song, Miller, and Abbott 2000) and have attractive theoretical properties (Gerstner and Kistler 2002), explicit error-based rules such as gradient descent have proven substantially more computationally effective. There is much debate about the biological realism of gradient descent (Stork 1989), and an ongoing area of research seeks to discover more biologically plausible versions of explicit error-driven learning rules (Bengio et al. 2015; Lillicrap et al. 2014).

While a vast oversimplification, the relationship between optimizing discrete architecture parameters and synaptic strength parameters is somewhat analogous to the relationship between evolutionary and developmental learning. Changes to synaptic strengths are continuous and can occur without modifying the overall system architecture, and thus could support experience-driven optimization during the lifetime of the organism. Changes in the discrete parameters, in contrast, restructure the computational primitives, the number of sensory areas (model layers) and the number of neurons in each area, and thus are more likely to be selected over evolutionary time.

Mapping models to data A goal-optimized model generates computationally precise hypotheses for how data collected from the real system will look. Testing these hypotheses involves assessing metrics of similarity between the model and the brain system, both for the output behaviors of the system and for internal responses of the system's neural components. Several commonly used metrics for assessing the mapping of models to empirical data include (from coarsest to finest resolution):

- *Behavioral consistency* Even before any neural data is collected, high-throughput systematic measurements of psychophysical data can be used to obtain a "fingerprint" of human behavioral responses across a wide variety of task conditions (Rajalingham et al. 2018). This fingerprint can then be compared to output behavior on these tasks as generated by neural network models. For example, Rajalingham et al. (2018) show that achieving consistency with high-resolution human error patterns in visual categorization tasks is a very strong test of correctness for models of the primate visual system.
- *Population-level neural comparison* The representation dissimilarity matrix (RDM) is a convenient tool for comparing two neural representations at a population level (Kriegeskorte et al. 2008). Each entry in the RDM corresponds to one stimulus pair, with high/low values indicating that the population as a whole treats the pair stimuli as very different/similar. Taken over the whole stimulus set, the RDM characterizes the layout of the images in the high-dimensional neural population space. A measure of how similar the representations are between real neural populations and those produced by a neural network can be obtained by assessing the correlations between the RDMs from each layer of a neural network model and the RDMs from real neural populations. This technique, which is called representational similarity analysis (RSA), has been effectively used for comparing visual representations in human fMRI data to HCNN models (Khaligh-Razavi and Kriegeskorte 2014).
- *Single-neuron regression* Linear regression is a convenient method for mapping units from neural network models to individual neural-recording sites (Yamins et al. 2014). For each neural site, this technique seeks to identify the linear weighting of neural network model output units (typically from one network layer) that is most predictive of that neural site's actual output on a fixed set of sample images. The "synthetic neuron" then produces response predictions on novel stimuli not used in the regression training, which are then compared to the actual neural site's output. Accuracy in regression prediction has shown to be a useful tool for achieving finer-grained model-brain mappings when higher resolution (e.g., electrophysiological) data are available (Nayebi et al. 2018; Yamins et al. 2014).

See Yamins and DiCarlo (2016) for a more detailed description and evaluation of these and other mapping procedures.

Properly assessing model complexity When comparing any two models of data, it is important to ensure that model complexity is taken into account: a complex model with many parameters may not be an improvement over a simple model with fewer parameters, even if the former fits the data somewhat better. However, even though goal-optimized DNNs have many parameters *before* task optimization, those parameters are *determined* by the optimization process in attempting to solve the computational goal itself. Thus, when the optimized networks are subsequently mapped to brain data, these parameters are no longer available for free modification to fit the neurons. Hence, although it may at first be somewhat counterintuitive, these predetermined parameters cannot be counted when assessing model complexity—for example, when computing scores such as the Akaike or Bayesian information criteria (Schwarz et al. 1978). Instead, once the optimized network has been produced, the only free parameters used when comparing to neural data are just those required by the mapping procedure itself. For example, when using RSA, *no* free parameters are needed at all since building the RDM matrix is a parameter-free procedure. Thus, if a larger goal-optimized neural network achieves a match between its RDMs and those in neural populations, it has done so fairly—that is, *not* by using those parameters to better (over)fit the neural data but because the bigger network has (presumably) achieved better performance on the computational goal, and the computational goal is itself highly relevant to the real biological constraints on the neural mechanism. Similarly, when performing single-neuron regression, the number of free parameters is equal to the number of model neurons used as linear regressor dimensions. In this case it is necessary (but easy) to ensure fair comparisons between models with different numbers of features by simply subsampling a fixed number of model units as regressors (as done in, e.g., Yamins et al. 2014) or using some unsupervised dimension-reduction procedure (such as principal components analysis) prior to regression.

Relationship to previous work in visual modeling Other approaches to modeling the visual system can be placed in the context of the optimization framework. Efficient coding hypotheses seek to generate efficient, low-dimensional representations of natural input statistics. This corresponds to a choice of architecture class $\mathcal{A}$ -containing "hourglass-shaped" networks (Hinton and Salakhutdinov 2006) composed of a compressive intermediate encoding followed by a decoding that produces an image-like output. The loss target is then (roughly) of the form

$$\mathcal{L}(x) = \|x - \mathbf{D}(\mathbf{E}(x))\| + \mathbf{Regularization}(\mathbf{E}(x))$$

where $\mathbf{E}(x)$ is the network encoding of image x, and $\mathbf{D}$ is the corresponding decoding. The first term of $\mathcal{L}$ is the reconstruction error, measuring the ability of the decoded representation to reproduce the original input, while the second term prevents overfitting by imposing a "simpleness prior" on the encoder. Efficient coding is an attractive idea because it combines functional requirements and biophysical constrains (e.g., metabolic efficiency). Early versions of this idea, such as sparse autoencoders (Olshausen and Field 1996), have shown promise in training shallow (one-layer) convolutional networks that naturally discover the Gabor-like filter patterns seen in V1 cortex. More recent methods such as variational autoencoders, generative adversarial networks (GANs), and BiGANs (Donahue, Krähenbühl, and Darrell 2016; Goodfellow et al. 2014; Kingma and Welling 2013) essentially correspond to improvements in the choice of regularization functions and have shown promise in training deeper networks. While such ideas have been effective in limited visual domains, improving their applicability to unrestricted visual image space is an open question and an important area for innovation (Karras et al. 2017).

Another line of work has attempted to fit neural networks directly to data from V1 (Klindt et al. 2017), V2 (Vintch et al. 2012), and V4 (Cadieu et al. 2007) cortex. These results are consistent with the optimization framework insofar as they involve finding parameters that optimize a loss function—in this case, the mismatch between network output and the measured neural data. Such investigations can be very informative, as they contribute to the discovery of which classes of neural architectures best capture the data. However, unlike the goal-driven-modeling approach or the efficient coding ideas, these direct curve fits do not generate a normative explanation underlying *why* the neural responses are as they are.

An interesting approach combines neural fits and normative explanations. In McIntosh et al. (2016), comparatively shallow HCNNs were fit to responses in retinal ganglion cells (RGCs). A key finding in this work was that characteristic properties of bipolar cells, which are upstream of the RGCs, naturally emerge in the networks' first layers just by forcing the network's last layer to correctly emulate RGC response patterns. While this work does not explain why the RGCs are as they are, it does suggest a kind of conditional normative explanation for why the bipolar cell patterns are as they are, given the RGCs as output. Understanding whether this holds for other parts of the retinal circuit (e.g, the intermediate cells in the amacrine layer) and whether the RGC patterns themselves arise from a higher-level downstream computational goal are exciting open questions.

Beyond the visual system The goal-driven optimization approach has also had success building quantitatively accurate models of the human auditory system (Güçlü et al. 2016; Kell et al. 2018). Using HCNNs as the architecture class but substituting a computational goal defined by speech and music genre recognition, this work finds a strong correlation between auditory task performance and auditory cortex neural response predictivity (figure 33.2*B*). A representational hierarchy is also found in auditory cortex, suggesting interesting similarities to the visual system, in that the robustness to variability (e.g., position, size, and pose tolerance) that makes convolutional networks useful for visual object recognition may have rough equivalents in the auditory domain that make convolution useful for parsing auditory "objects." However, the work of Kell et al. (2018) goes beyond models of a single processing stream, exhibiting multistream networks that solve several auditory tasks simultaneously with an initial common architecture that subsequently splits into multiple task-specific pathways. The different pathways of the network differentially explain neural variance in different parts of the auditory cortex, illustrating how task-optimized neural networks can help further our understanding of large-scale functional organization in the brain. Recent work along similar lines has begun to tackle somatosensory systems (Zhuang et al. 2017).

A functionally driven optimization approach has also been effective at driving progress in modeling the motor system (Lillicrap and Scott 2013; Sussillo et al. 2015). This work shows how imposing the computational goal of creating behaviorally useful motor output constrains internal neural network components to match otherwise nonobvious features of neurons in motor cortex, and provides a modern computational basis for earlier work on movement efficiency (Flash and Hogan 1985). Unlike work on sensory systems, the goals in motor networks are not representational but instead focus on the generation of dynamic patterns of motor preparation and movement (Churchland et al. 2012). For this reason, the models involved in these efforts are typically recurrent neural networks (RNNs) rather than feedforward HCNNs. These results show that the goal-driven optimization idea has power across a wide range of network architectures and behavioral goal types.

Analyzing constraints rather than optima A classic approach to analyzing a population of (in most cases, sensory) neurons is to classify the shape of their tuning curves in response to systematically changing input stimuli along certain characteristic axes that are key drivers of the populations' variability. This approach has been successful in a variety of brain areas—most notably, in early visual cortex (Hubel and Wiesel 1959), where tuning curves illustrating the orientation and frequency selectivity of V1 neurons laid the groundwork for Gabor wavelet–based models.

Relative to the optimization framework described above, the analysis of tuning curves is essentially an attempt to characterize optimal networks $\mathbf{A}^*$ in non-optimization-based terms. When a small number of mathematically simple stimulus-domain axes can be found in which the tuning curves of $\mathbf{A}^*$ have a mathematically simple shape, $\mathbf{A}^*$ can largely be constructed by a simple closed-form procedure without any reference to learning through iterative optimization. This is to some extent feasible for V1 neurons and perhaps in early cortical areas in other domains, such as primary auditory cortex (Chi, Ru, and Shamma 2005). It is possible that this type of simplification is most helpful for understanding neural responses that arise largely from highly constrained stereotyped genetic developmental programs, rather than those that depend heavily on experience-driven learning (Espinosa and Stryker 2012), or where biophysical constraints—such as metabolic cost or noise reduction—might also impose "simplicity priors" on the neural architecture (Olshausen and Field 1996; Sussillo et al. 2015).

In general, however, it is not guaranteed that closed-form expressions describing the response properties of task-optimized models can be found. Evolution and development are under no general constraint to make their products conform to simple mathematical shapes, especially for intermediate and higher cortical areas removed from the sensory or motor periphery. However, even if such analytical simplifications do not exist, the optimization framework nonetheless, provides a method for generating meta-understanding via characterizing the constraints on the system, rather than analyzing the specific outcome network itself. By varying the architectural class, the computational goal, or the learning rule, and identifying which choices lead to networks that best match the observed neural data, it is possible to learn much about the brain system of interest even if its tuning curves are inscrutable.

Understanding multiple optima What happens when multiple optimal network solutions exist? For many architecture classes, there may be infinitely many qualitatively very similar networks with the same or substantially similar outputs—for example, those created by applying orthonormal rotations to linear transforms present in the network. Sometimes, however, qualitatively very distinct networks might achieve similar performance levels on a task. For example, very deep residual network architectures (He et al. 2016) and comparatively shallower (but much more locally complex) architectures arising from a neural architecture search (Zoph et al. 2018) achieve roughly similar performance on ImageNet categorization despite key structural differences.

The optimization framework does not require a unique best solution to the computational goal to make useful predictions. If several subclasses of high-performing solutions to a given task are identified, this is equivalent to formulating multiple very qualitatively distinct hypotheses for the neural circuits underlying function in a given brain area. Recent work in modeling rodent whisker trigeminal cortex, in which similar task performance on whisker-driven shape recognition can be achieved by several distinct neural architecture classes, illustrates this idea (Zhuang et al. 2017). Comparison of the distinct model types to experimental results, either from detailed behavioral or neural experiments, is then likely to point toward one of these hypotheses as explaining the data better than others. Techniques similar to those used to create the models in the first place can be deployed to generate *optimal stimuli* for separating the predictions of the multiple models as widely as possible, which would in turn directly inform experimental design. In these cases, the optimization framework serves as an efficient generator of strong hypotheses.

In contrast, if most high-performing solutions to a computational goal fall into a comparatively narrower band of variability, the set of model solutions may correspond to actual variability in the real subject population. For some brain regions, especially those in intermediate or higher cortical areas, the particular collection of neural circuits present in any one subject's brain may vary considerably between conspecifics (Baldassarre et al. 2012). The optimization framework naturally supports at least two potential sources of such variation, including the following:

- Variation of *initial conditions*, described as a probability distribution over the starting point models $\mathbf{A}_0$ to which the learning rule is applied. For example, different random draws of initial values for linear filters L_i will lead to distinct final optimized HCNNs. While many high-level representational properties are shared between these networks, meaningful differences can exist (Li et al. 2015) and may explain aspects of the variation between real visual systems.
- Variation of *computational goal*, described as a distribution over stimuli in the data set defining the goal task. This idea captures the concept that different individuals will experience somewhat different stimulus diets during development and learning.

Understanding the computational sources of intraspecific variation is itself an important modeling question for future work (Van Horn, Grafton, and Miller 2008).

A contravariance principle Though it may at first seem counterintuitive, the *harder* the computational goal, the *easier* the model-to-brain matching problem is likely to be. This is because the set of architectural solutions to an easy goal is large, while the set of solutions to a challenging goal is comparatively smaller. In mathematical terms, the size of the set of optima is *contravariant* in the difficulty of the optimization problem.

A simple thought experiment makes this clear: Imagine if, instead of trying to solve 1,000-way object classification in the real-world ImageNet data set, one simply asked a network to solve the binary discrimination between two simple geometric shapes shown on uniform gray backgrounds. The set of networks that can solve the latter task is much less narrowly constrained than that which solves the former. And given that primates actually do exhibit robust object classification, the more strongly constrained networks that pass the same hard performance tests are more likely to be homologous to the real primate visual system. A detailed example of how optimizing a network to achieve high performance on a low-variation training set can lead to poor performance generalization and neurally inconsistent features is illustrated in Hong et al. (2016).

The contravariance principle makes a strong prescription for using the optimization framework to design effective computationally driven experiments. Unlike the typical practice in experimental neuroscience but echoing recent theoretical discussions of task dimensionality (Gao et al. 2017), it does *not* make sense from the optimization perspective to choose the most reduced version of a given task domain and then seek to thoroughly understand the mechanisms that solve the reduced task before attempting to address more realistic versions of the task. In fact, this sort of highly reductive approach is likely to lead to confusing results precisely because the reduced task may admit many spurious solutions. It is more effective to impose the challenging real-world task from the beginning, both in designing training sets for optimizing the neural network models and in designing experimental stimulus sets for making model-data comparisons. Even if the absolute performance numbers of networks on the harder computational goal are lower, the resulting networks are likely to be better models of the real neural system.

There is a natural balance between network size and capacity. In general, the optimization-based approach is likely to be most efficient when the network sizes are just large enough to solve the computational task. Thus, another way to constrain networks while still using a comparatively simple computational goal is to reduce the network size. This idea is consistent with results from experiments measuring neural dynamics in the fruit fly, where a small but apparently near-optimal circuit has been shown to be responsible for the fly's simple but robust navigational control behaviors (Turner-Evans et al. 2017). It remains unknown whether the specific architectural principles discovered in such simplified settings will prove useful for understanding the larger networks needed for achieving more sophisticated computational goals in higher organisms.

Major Future Directions

The optimization framework suggests a wide variety of important future directions to be explored.

Better sensory models Within the domain of the visual system, many substantial differences remain between state-of-the-art models and the real neural system. For neurons throughout the macaque ventral visual stream, the best neural network models are able to explain only approximately 65% of the reliable time-averaged neural responses to static natural stimuli. This neural result is echoed by the fact that while the models are behaviorally consistent with primate and human visual error patterns at the category or object level (Rajalingham, Schmidt, and DiCarlo 2015), they fail to entirely account for error patterns at the finest image-by-image grain (Rajalingham et al. 2018), especially in the context of adversarially created stimuli (Kurakin, Goodfellow, and Bengio 2016). Closing the explanatory gap will require a next generation of improved models.

Another major open direction involves understanding recurrence and feedback in visual (and other sensory) processing and the corresponding modeling of neurons' temporal dynamics. While some recent progress has been made on functionally driven neural models of temporal dynamics that integrate RNN motifs into HCNNs (Nayebi et al. 2018; Spoerer, McClure, and Kriegeskorte 2017), it is unlikely that a full understanding of the functional role of feedback has been achieved. While most modeling efforts have so far focused on the ventral visual pathway, understanding the functional demands that lead to the emergence of multiple visual pathways, or combining constraints at multiple levels (e.g., behavioral and biophysical), is another key direction for future work. Likewise, little attention has been paid to understanding the physical layout of brain areas. While some of the most robust results in human cognitive neuroscience involve identifying the subregions of visual cortex that selectively respond to certain classes of stimuli—for example, the well-known face, body, and place areas (Downing et al. 2001; Epstein and Kanwisher 1998; Kanwisher, McDermott, and Chun 1997)—the computational-level constraints leading to these topographical features are poorly understood.

Learning Though the optimization framework has shown exciting progress at the intersection of machine learning and computational neuroscience, there is a fundamental problem confronting the approach. Typical neural network training uses heavily supervised methods involving huge numbers of high-level semantic labels—for example, category labels for thousands of examples in each of thousands of categories (Deng, Dong, et al. 2009; Mahajan et al. 2018). Viewed as technical tools for tuning algorithm parameters, such procedures can be acceptable, although they limit the purview of the method to situations with large existing labeled data sets. As real models of learning in the brain, they are highly unrealistic because, among other reasons, human infants and nonhuman primates simply do not receive millions of category labels during development. There has been a substantial amount of research on unsupervised, semisupervised, and self-supervised visual-learning methods (Goodfellow et al. 2014; Kingma and Welling 2013; Olshausen and Field 1996; Sener and Savarese 2017; Settles 2011; Tarvainen and Valpola 2017). Despite these advances, the gap between supervised and unsupervised approaches still remains significant. The discovery of procedures that are computationally powerful but use substantially less labeled data is a key challenge for understanding real biological learning.

Modeling integrated agents rather than isolated systems Cognition is not just about the passive parsing of sensory streams or the disembodied generation of motor commands. Humans are agents, interacting with and modifying their environment via a tight visuomotor loop. Effective courses of action based both on sensory input and the agent's goals afford the agent the opportunity to restructure its surroundings to better pursue those goals. By the same token, however, constructing and evaluating a complex action policy imposes a substantial additional computational challenge for the agent that goes considerably beyond "mere" sensory processing. Applying the optimization framework to modeling full agents is an exciting possibility, and some recent speculative work in deep reinforcement learning has made progress in this direction (Wayne et al. 2018; Yang et al. 2018). However, fully fleshing out neural network models of memory, decision-making, and higher cognition that have the resolution and completeness to be quantitatively compared to experimental data will require substantial improvements at the algorithmic level.

The problem of learning becomes especially acute in the context of interactive systems. Human infants employ an active learning process that builds representations underlying sensory judgments and motor planning (Begus et al. 2014; Goupil, Romand-Monnier, and Kouider 2016; Kidd et al.2012). Children exhibit a wide range of interesting, apparently spontaneous, visuomotor behaviors—including navigating their environment, seeking out and attending to novel objects, and engaging physically with these objects in novel and surprising ways (Begus 2014; Fantz 1964; Goupil, Romand-Monnier, and Kouider 2016; Hurley, Kovack-Lesh, and Oakes 2010; Hurley and Oakes 2015; Gopnik, Meltzoff, and Kuhl 2009; Twomey and Westermann 2017). Modeling these key behaviors, and the brain systems underlying them, is a formidable challenge for computational cognitive neuroscience (Haber et al. 2018).

REFERENCES

Baldassarre, Antonello, Christopher M. Lewis, Giorgia Committeri, Abraham Z. Snyder, Gian Luca Romani, and Maurizio Corbetta. 2012. Individual variability in functional connectivity predicts performance of a perceptual task. *Proceedings of the National Academy of Sciences, 109*(9), 3516–3521.

Begus, Katarina, Teodora Gliga, and Victoria Southgate. 2014. Infants learn what they want to learn: Responding to infant pointing leads to superior learning. *PLoS One, 9*(10), 1–4. https://doi.org/10.1371/journal.pone.0108817

Bengio, Y. 2012. Deep learning of representations for unsupervised and transfer learning. In *Proceedings of ICML Workshop on Unsupervised and Transfer Learning,* 17–36.

Bengio, Y., Dong-Hyun Lee, Jorg Bornschein, Thomas Mesnard, and Zhouhan Lin. 2015. Towards biologically plausible deep learning. *arXiv.* Retrieved from 1502.04156.

Bergstra, James, Daniel Yamins, and David Cox. 2013. Making a science of model search: Hyperparameter optimization in hundreds of dimensions for vision architectures. *Proceedings of the 30th International Conference on Machine Learning,* 115–123.

Bottou, Léon. 2010. Large-scale machine learning with stochastic gradient descent. *Proceedings of COMPSTAT'2010,* 177–186.

Brincat, S. L., and C. E. Connor. 2004. Underlying principles of visual shape selectivity in posterior inferotemporal cortex. *Nature Neuroscience, 7*(8), 880–886.

Cadena, Santiago A., George H. Denfield, Edgar Y. Walker, Leon A. Gatys, Andreas S. Tolias, Matthias Bethge, and Alexander S. Ecker. 2019. Deep convolutional models improve predictions of macaque V1 responses to natural images. *PLoS Computational Biology 15*(4), e1006897.

Cadieu, Charles F., Ha Hong, Daniel L. K. Yamins, Nicolas Pinto, Diego Ardila, Ethan A. Solomon, Najib J. Majaj, and James J. DiCarlo. 2014. Deep neural networks rival the representation of primate IT cortex for core visual object recognition. *PLoS Computational Biology, 10*(12), e1003963.

Cadieu, C., M. Kouh, A. Pasupathy, C. E. Connor, M. Riesenhuber, and T. Poggio. 2007. A model of V4 shape selectivity and invariance. *Journal of Neurophysiology, 98*(3), 1733–1750.

Carandini, M., J. B. Demb, V. Mante, D. J. Tolhurst, Y. Dan, B. A. Olshausen, J. L. Gallant, and N. C. Rust. 2005. Do we know what the early visual system does? *Journal of Neuroscience, 25*(46), 10577–10597.

Carandini, M., and David J. Heeger. 2012. Normalization as a canonical neural computation. *Nature Reviews Neuroscience, 13*(1), 51–62.

Chi, Taishih, Powen Ru, and Shihab A. Shamma. 2005. Multiresolution spectrotemporal analysis of complex sounds. *Journal of the Acoustical Society of America, 118*(2), 887–906.

Churchland, Mark M., John P. Cunningham, Matthew T. Kaufman, Justin D. Foster, Paul Nuyujukian, Stephen I. Ryu, and Krishna V. Shenoy. 2012. Neural population dynamics during reaching. *Nature, 487*(7405), 51.

Connor, C. E., S. L. Brincat, and A. Pasupathy. 2007. Transformation of shape information in the ventral pathway. *Current Opinion in Neurobiology, 17*(2), 140–147.

Cover, Thomas M., and Joy A. Thomas. 2012. *Elements of information theory.* Hoboken, NJ: John Wiley & Sons.

Deng, J., W. Dong, R. Socher, L.-J. Li, K. Li, and L. Fei-Fei. 2009. ImageNet: A large-scale hierarchical image database. In *IEEE Computer Society Conference on Computer Vision and Pattern Recognition CVPR09,* 248–255.

DiCarlo, J. J., and D. D. Cox. 2007. Untangling invariant object recognition. *Trends in Cognitive Sciences, 11*(8), 333–341.

DiCarlo, J. J., D. Zoccolan, and N. C. Rust. 2012. How does the brain solve visual object recognition? *Neuron, 73*(3), 415–434.

Donahue, Jeff, Philipp Krähenbühl, and Trevor Darrell. 2016. Adversarial feature learning. *arXiv.* Retrieved from 1605.09782.

Downing, P. E., Y. Jiang, M. Shuman, and N. Kanwisher. 2001. A cortical area selective for visual processing of the human body. *Science, 293,* 2470–2473.

Epstein, R., and N. Kanwisher. 1998. A cortical representation of the local visual environment. *Nature, 392(*6676), 598–601.

Espinosa, J. Sebastian, and Michael P. Stryker. 2012. Development and plasticity of the primary visual cortex. *Neuron, 75*(2), 230–249.

Fantz, R. L. 1964. Visual experience in infants: Decreased attention to familiar patterns relative to novel ones. *Science, 146*(3644), 668–670.

Felleman, D. J., and D. C. Van Essen. 1991. Distributed hierarchical processing in the primate cerebral cortex. *Cerebral Cortex, 1*(1), 1–47.

Flash, Tamar, and Neville Hogan. 1985. The coordination of arm movements: An experimentally confirmed mathematical model. *Journal of Neuroscience, 5*(7), 1688–1703.

Freedman, David J., Maximilian Riesenhuber, Tomaso Poggio, and Earl K. Miller. 2001. Categorical representation of visual stimuli in the primate prefrontal cortex. *Science, 291*(5502), 312–316.

Freeman, J., and E. Simoncelli. 2011. Metamers of the ventral stream. *Nature Neuroscience, 14*(9), 1195–1201.

Fukushima, K. 1980. Neocognitron: A self-organizing neural network model for a mechanism of pattern recognition unaffected by shift in position. *Biological Cybernetics, 36*(4), 193–202.

Gallant, J. L., C. E. Connor, S. Rakshit, J. W. Lewis, and D. C. Van Essen. 1996. Neural responses to polar, hyperbolic, and Cartesian gratings in area V4 of the macaque monkey. *Journal of Neurophysiology, 76*(4), 2718–2739.

Gao, Peiran, Eric Trautmann, M. Yu Byron, Gopal Santhanam, Stephen Ryu, Krishna Shenoy, and Surya Ganguli. 2017. A theory of multineuronal dimensionality, dynamics and measurement. *bioRxiv*, 214262.

Gerstner, Wulfram, and Werner M. Kistler. 2002. Mathematical formulations of Hebbian learning. *Biological Cybernetics, 87*(5–6), 404–415.

Girshick, Ross. 2015. Fast R-Cnn. In *Proceedings of the IEEE International Conference on Computer Vision*, 1440–1448. Vancouver, BC.

Goodfellow, Ian, Jean Pouget-Abadie, Mehdi Mirza, Bing Xu, David Warde-Farley, Sherjil Ozair, Aaron Courville, and Yoshua Bengio. 2014. Generative adversarial nets. *Advances in Neural Information Processing Systems*, 2672–2680.

Gopnik, A., A. N. Meltzoff, and P. K. Kuhl. 2009. *The scientist in the crib: Minds, brains, and how children learn.* New York: HarperCollins. https://books.google.com/books?id=ui6KAniUJfsC.

Goupil, Louise, Margaux Romand-Monnier, and Sid Kouider. 2016. Infants ask for help when they know they don't know. *Proceedings of the National Academy of Sciences, 113*(13), 3492–3496. https://doi.org/10.1073/pnas.1515129113

Güçlü, Umut, Jordy Thielen, Michael Hanke, and Marcel Van Gerven. 2016. Brains on beats. *Advances in Neural Information Processing Systems*, 2101–2109.

Güçlü, Umut, and Marcel A. J. van Gerven. 2015. Deep neural networks reveal a gradient in the complexity of neural representations across the ventral stream. *Journal of Neuroscience, 35*(27), 10005–10014.

Haber, Nick, Damian Mrowca, Li Fei-Fei, and Daniel L. K. Yamins. 2018. Learning to play with intrinsically-motivated self-aware agents. *Advances in Neural Information Processing Systems.*

He, Kaiming, Xiangyu Zhang, Shaoqing Ren, and Jian Sun. 2016. Deep residual learning for image recognition. *Proceedings of the IEEE Conference on Computer Vision and Pattern Recognition*, 770–778.

Hegner, Yiwen Li, Axel Lindner, and Christoph Braun. 2017. A somatosensory-to-motor cascade of cortical areas engaged in perceptual decision making during tactile pattern discrimination. *Human Brain Mapping, 38*(3), 1172–1181.

Hinton, Geoffrey E., and Ruslan R. Salakhutdinov. 2006. Reducing the dimensionality of data with neural networks. *Science, 313*(5786), 504–507.

Hong, Ha, Daniel L. K. Yamins, Najib J. Majaj, and James J. DiCarlo. 2016. Explicit information for category-orthogonal object properties increases along the ventral stream. *Nature Neuroscience, 19*(4), 613–622.

Hubel, David H., and Torsten N. Wiesel. 1959. Receptive fields of single neurons in the cat's striate cortex. *Journal of Physiology, 148*(3), 574–591.

Hung, C. P., G. Kreiman, T. Poggio, and J. J. DiCarlo. 2005. Fast readout of object identity from macaque inferior temporal cortex. *Science, 310*(5749), 863–866.

Hurley, K. B., K. A. Kovack-Lesh, and L. M. Oakes. 2010. The influence of pets on infants' processing of cat and dog images. *Infant Behavior and Development, 33*(4), 619–628.

Hurley, K. B., and L. M. Oakes. 2015. Experience and distribution of attention: Pet exposure and infants' scanning of animal images. *Journal of Cognition and Development, 16*(1), 11–30.

James, William. 1890. *The principles of psychology* (Vol. 1). New York: Henry Holt, 474.

Kanwisher, N., J. McDermott, and M. M. Chun. 1997. The fusiform face area: A module in human extrastriate cortex specialized for face perception. *Journal of Neuroscience, 17*(11), 4302–4311.

Karras, Tero, Timo Aila, Samuli Laine, and Jaakko Lehtinen. 2017. Progressive growing of gans for improved quality, stability, and variation. *arXiv*. Retrieved from 1710.10196.

Kell, Alexander J. E., Daniel L. K. Yamins, Erica N. Shook, Sam V. Norman-Haignere, and Josh H. McDermott. 2018. A task-optimized neural network replicates human auditory behavior, predicts brain responses, and reveals a cortical processing hierarchy. *Neuron, 98*(3), 630–644.

Khaligh-Razavi, S. M., and N. Kriegeskorte. 2014. Deep supervised, but not unsupervised, models may explain it cortical representation. *PLoS Computational Biology, 10*(11).

Kidd, Celeste, Steven T. Piantadosi, and Richard N. Aslin. 2012. The Goldilocks effect: Human infants allocate attention to visual sequences that are neither too simple nor too complex. *PLoS One, 7*(5), 1–8. https://doi.org/10.1371/journal.pone.0036399

Kingma, Diederik P., and Jimmy Ba. 2014. Adam: A method for stochastic optimization. *arXiv,* 1412.6980.

Kingma, Diederik P., and Max Welling. 2013. Auto-encoding variational bayes. *arXiv,* 1312.6114.

Klindt, David, Alexander S. Ecker, Thomas Euler, and Matthias Bethge. 2017. Neural system identification for large populations separating "what" and "where." *Advances in Neural Information Processing Systems*, 3506–3516.

Kriegeskorte, N., M. Mur, D. A. Ruff, R. Kiani, J. Bodurka, H. Esteky, K. Tanaka, and P. A. Bandettini. 2008. Matching categorical object representations in inferior temporal cortex of man and monkey. *Neuron, 60*(6), 1126–1141.

Krizhevsky, Alex, Ilya Sutskever, and Geoffrey E. Hinton. 2012. Imagenet classification with deep convolutional neural networks. In *Advances in Neural Information Processing Systems,* 1097–1105.

Kurakin, Alexey, Ian Goodfellow, and Samy Bengio. 2016. Adversarial examples in the physical world. *arXiv*. Retrieved from 1607.02533.

LeCun, Y., and Y. Bengio. 1995. Convolutional networks for images, speech, and time series. *The handbook of brain theory and neural networks.* Cambridge, MA: MIT Press, 255–258.

Lennie, P., and J. A. Movshon. 2005. Coding of color and form in the geniculostriate visual pathway. *Journal of the Optical Society of America. A, Optics, Image Science, and Vision, 22*(10), 2013–2033.

Li, Yixuan, Jason Yosinski, Jeff Clune, Hod Lipson, and John E. Hopcroft. 2015. Convergent learning: Do different neural networks learn the same representations? In *Feature Extraction: Modern Questions and Challenges,* 196–212.

Lillicrap, Timothy P., Daniel Cownden, Douglas B. Tweed, and Colin J. Akerman. 2014. Random feedback weights support learning in deep neural networks. *arXiv*. Retrieved from 1411.0247.

Lillicrap, Timothy P., and Stephen H. Scott. 2013. Preference distributions of primary motor cortex neurons reflect control solutions optimized for limb biomechanics. *Neuron, 77*(1), 168–179.

Mahajan, Dhruv, Ross Girshick, Vignesh Ramanathan, Kaiming He, Manohar Paluri, Yixuan Li, Ashwin Bharambe, and Laurens van der Maaten. 2018. Exploring the limits of weakly supervised pretraining. In *Proceedings of the European Conference on Computer Vision (ECCV),* 181–196.

Majaj, Najib J., Ha Hong, Ethan A. Solomon, and James J. DiCarlo. 2015. Simple learned weighted sums of inferior temporal neuronal firing rates accurately predict human core object recognition performance. *Journal of Neuroscience, 35*(39), 13402–13418.

Malach, R., I. Levy, and U. Hasson. 2002. The topography of high-order human object areas. *Trends in Cognitive Sciences, 6*(4), 176–184.

McIntosh, Lane, Niru Maheswaranathan, Aran Nayebi, Surya Ganguli, and Stephen Baccus. 2016. Deep learning models of the retinal response to natural scenes. *Advances in Neural Information Processing Systems*, 1369–1377.

Montague, P. Read, Peter Dayan, and Terrence J. Sejnowski. 1996. A framework for mesencephalic dopamine systems based on predictive Hebbian learning. *Journal of Neuroscience, 16*(5), 1936–1947.

Movshon, J. Anthony, Ian D. Thompson, and David J. Tolhurst. 1978. Spatial summation in the receptive fields of simple cells in the cat's striate cortex. *Journal of Physiology, 283*(1), 53–77.

Nayebi, Aran, Daniel Bear, Jonas Kubilius, Kohitij Kar, Surya Ganguli, David Sussillo, James J. DiCarlo, and Daniel L. K. Yamins. 2018. Task-driven convolutional recurrent models of the visual system. *Advances in Neural Information Processing Systems.*

Nishio, Akiko, Takeaki Shimokawa, Naokazu Goda, and Hidehiko Komatsu. 2014. Perceptual gloss parameters are encoded by population responses in the monkey inferior temporal cortex. *Journal of Neuroscience, 34*(33), 11143–11151.

Olshausen, Bruno A., and David J. Field. 1996. Emergence of simple-cell receptive field properties by learning a sparse code for natural images. *Nature, 381*(6583), 607–609.

Pagan, Marino, Luke S. Urban, Margot P. Wohl, and Nicole C. Rust. 2013. Signals in inferotemporal and perirhinal cortex suggest an untangling of visual target information. *Nature Neuroscience, 16*(8), 1132–1139.

Petersen, Carl C. H. 2007. The functional organization of the barrel cortex. *Neuron, 56*(2), 339–355. https://doi.org/10.1016/j.neuron.2007.09.017

Pickles, James O. 2008. *An introduction to the physiology of hearing.* Bingley, UK: Emerald Insight.

Pinto, Nicolas, David Doukhan, James J. DiCarlo, and David D. Cox. 2009. A high-throughput screening approach to discovering good forms of biologically inspired visual representation. *PLoS Computational Biology, 5*(11), e1000579.

Pinto, Nicolas, David D. Cox, and James J. DiCarlo. 2008. Why is real-world visual object recognition hard? *PLoS Computational Biology, 4*(1), e27.

Poole, Ben, Subhaneil Lahiri, Maithra Raghu, Jascha Sohl-Dickstein, and Surya Ganguli. 2016. Exponential expressivity in deep neural networks through transient chaos. *Advances in Neural Information Processing Systems*, 3360–3368.

Rajalingham, Rishi, Elias B. Issa, Pouya Bashivan, Kohitij Kar, Kailyn Schmidt, and James J. DiCarlo. 2018. Large-scale, high-resolution comparison of the core visual object recognition behavior of humans, monkeys, and state-of-the-art deep artificial neural networks. *Journal of Neuroscience, 38*(33), 7255–7269.

Rajalingham, R., K. Schmidt, and J. J. DiCarlo. 2015. Comparison of object recognition behavior in human and monkey. *Journal of Neuroscience, 35*(35), 12127–12136.

Ringach, Dario L., Robert M. Shapley, and Michael J. Hawken. 2002. Orientation selectivity in macaque V1: Diversity and laminar dependence. *Journal of Neuroscience, 22*(13), 5639–5651.

Romanski, Lizabeth M., and Joseph E. LeDoux. 1993. Information cascade from primary auditory cortex to the amygdala: Corticocortical and corticoamygdaloid projections of temporal cortex in the rat. *Cerebral Cortex, 3*(6), 515–532.

Rust, N. C., and J. J. DiCarlo. 2010. Selectivity and tolerance ("invariance") both increase as visual information propagates from cortical area V4 to it. *Journal of Neuroscience, 30*(39), 12978–12995.

Schiller, P. H. 1995. Effect of lesion in visual cortical area V4 on the recognition of transformed objects. *Nature, 376* (6538), 342–344.

Schmolesky, M. T., Y. Wang, D. P. Hanes, K. G. Thompson, S. Leutgeb, J. D. Schall, and A. G. Leventhal. 1998. Signal timing across the macaque visual system. *Journal of Neurophysiology, 79*(6), 3272–3278.

Schwarz, Gideon. 1978. Estimating the dimension of a model. *Annals of Statistics, 6*(2), 461–464.

Seibert, Darren, Daniel L. Yamins, Diego Ardila, Ha Hong, James J. DiCarlo, and Justin L. Gardner. 2016. A performance-optimized model of neural responses across the ventral visual stream. *bioRxiv.* https://doi.org/10.1101/036475

Sener, Ozan, and Silvio Savarese. 2017. A geometric approach to active learning for convolutional neural networks. *Computing Research Repository.* Retrieved from abs/1708.00489. http://dblp.uni-trier.de/db/journals/corr/corr1708.html#abs-1708-00489.

Serre, T., A. Oliva, and T. Poggio. 2007. A feedforward architecture accounts for rapid categorization. *Proceedings of the National Academy of Sciences of the United States of America, 104*(15), 6424–6429.

Settles, Burr. 2011. *Active learning* (Vol. 18). Williston, VT: Morgan & Claypool.

Sharpee, T. O., M. Kouh, and J. H. Reyholds. 2012. Trade-off between curvature tuning and position invariance in visual area V4. *Proceedings of the National Academy of Sciences of the United States of America, 110*(28), 11618–11623.

Simonyan, Karen, and Andrew Zisserman. 2014. Very deep convolutional networks for large-scale image recognition. *arXiv.* Retrieved from 1409.1556.

Song, Sen, Kenneth D. Miller, and Larry F. Abbott. 2000. Competitive Hebbian learning through spike-timing-dependent synaptic plasticity. *Nature Neuroscience, 3*(9), 919.

Spoerer, Courtney J., Patrick McClure, and Nikolaus Kriegeskorte. 2017. Recurrent convolutional neural networks: A better model of biological object recognition. *Frontiers in Psychology, 8*, 1551.

Stork, David G. 1989. Is backpropagation biologically plausible. *International Joint Conference on Neural Networks, 2*, 241–246.

Sussillo, David, Mark M. Churchland, Matthew T. Kaufman, and Krishna V. Shenoy. 2015. A neural network that finds a naturalistic solution for the production of muscle activity. *Nature Neuroscience, 18*(7), 1025–1033.

Tanaka, Keiji. 2003. Columns for complex visual object features in the inferotemporal cortex: Clustering of cells with similar but slightly different stimulus selectivities. *Cerebral Cortex, 13*(1), 90–99.

Tarvainen, Antti, and Harri Valpola. 2017. Mean teachers are better role models: Weight-averaged consistency targets improve semi-supervised deep learning results. *Advances in Neural Information Processing Systems*, 1195–1204.

Turner-Evans, Daniel, Stephanie Wegener, Herve Rouault, Romain Franconville, Tanya Wolff, Johannes D. Seelig, Shaul Druckmann, and Vivek Jayaraman. 2017. Angular velocity integration in a fly heading circuit. *eLife 6*, e23496.

Twomey, K. E., and G. Westermann. 2018. Curiosity-based learning in infants: A neurocomputational approach. *Developmental Science, 21*(4), e12629.

Van Horn, John Darrell, Scott T. Grafton, and Michael B. Miller. 2008. Individual variability in brain activity: A nuisance or an opportunity? *Brain Imaging and Behavior, 2*(4), 327–334.

Vintch, Brett, Andrew Zaharia, J. Movshon, and Eero P. Simoncelli. 2012. Efficient and direct estimation of a neural subunit model for sensory coding. *Advances in Neural Information Processing Systems*, 3104–3112.

Wang, Liang, Ryan E. B. Mruczek, Michael J. Arcaro, and Sabine Kastner. 2014. Probabilistic maps of visual topography in human cortex. *Cerebral Cortex, 25*(10), 3911–3931.

Wayne, Greg, Chia-Chun Hung, David Amos, Mehdi Mirza, Arun Ahuja, Agnieszka Grabska-Barwinska, Jack Rae, et al. 2018. Unsupervised predictive memory in a goal-directed agent. *arXiv*. Retrieved from 1803.10760.

Willmore, Ben, Ryan J. Prenger, Michael C-K. Wu, and Jack L. Gallant. 2008. The Berkeley wavelet transform: A biologically inspired orthogonal wavelet transform. *Neural Computation, 20*(6), 1537–1564.

Yamane, Y., E. T. Carlson, K. C. Bowman, Z. Wang, and C. E. Connor. 2008. A neural code for three-dimensional object shape in macaque inferotemporal cortex. *Nature Neuroscience, 11*, 1352–1360.

Yamins, Daniel L. K., and James J. DiCarlo. 2016. Using goal-driven deep learning models to understand sensory cortex. *Nature Neuroscience, 19*(3), 356–365.

Yamins, Daniel L., Ha Hong, Charles Cadieu, and James J. DiCarlo. 2013. Hierarchical modular optimization of convolutional networks achieves representations similar to macaque IT and human ventral stream. In *Advances in Neural Information Processing Systems*, 3093–3101.

Yamins, Daniel L. K., Ha Hong, Charles F. Cadieu, Ethan A. Solomon, Darren Seibert, and James J. DiCarlo. 2014. Performance-optimized hierarchical models predict neural responses in higher visual cortex. Proceedings of the *National Academy of Sciences, 111*(23), 8619–8624.

Yang, Guangyu Robert, Igor Ganichev, Xiao-Jing Wang, Jonathon Shlens, and David Sussillo. 2018. A dataset and architecture for visual reasoning with a working memory. *arXiv*. Retrieved from 1803.06092.

Yau, Jeffrey M., Anitha Pasupathy, Scott L. Brincat, and Charles E. Connor. 2012. Curvature processing dynamics in macaque area V4. *Cerebral Cortex, 23*(1), 198–209.

Zeiler, Matthew D. 2012. ADADELTA: An adaptive learning rate method. *arXiv*. Retrieved from 1212.5701.

Zhuang, Chengxu, Jonas Kubilius, Mitra J. Z. Hartmann, and Daniel L. Yamins. 2017. Toward goal-driven neural network models for the rodent whisker-trigeminal system. *Advances in Neural Information Processing Systems*, 2555–2565.

Zoph, Barret, Vijay Vasudevan, Jonathon Shlens, and Quoc V. Le. 2018. Learning transferable architectures for scalable image recognition. In *Proceedings of the IEEE Conference on Computer Vision and Pattern Recognition*, 8697–8710.

34 Physical Object Representations for Perception and Cognition

ILKER YILDIRIM, MAX SIEGEL, AND JOSHUA TENENBAUM

ABSTRACT Theories of perception typically assume that the goal of sensory processing is to output simple categorical labels or low-dimensional quantities, such as the identities and locations of objects in a scene. But humans perceive much more in a scene: we perceive rich and detailed three-dimensional shapes and surfaces, substance properties of objects (such as whether they are light or heavy, rigid or soft, solid or liquid), and relations between objects (such as which objects support, contain, or are attached to other objects). These physical targets of perception support flexible and complex action as the substrate of planning, reasoning, and problem-solving. In this chapter we introduce and argue for a theory of how people perceive, learn, and reason about objects in our sensory environment in terms of what we call *physical object representations* (PORs). We review recent work showing how this explains many human judgments in intuitive physics, provides a basis for object shape perception when traditional visual cues are not available, and, in one domain of high-level vision, suggests a new way to interpret multiple stages of hierarchical processing in the primate brain.

Consider the scenes in figure 34.1*A* and *B*. In each case we see a set of apples in a certain geometric arrangement (figure 34.1*C, D*). But we also see so much more: We see fine-grained details of their three-dimensional (3-D) shapes. We infer their physical properties and relationships: which objects are supporting which others and how heavy or light or hard or soft they would feel if we picked them up. We can predict whether the stack would topple if the middle apple on the bottom row were removed, and we can plan how to pick the designated apple without making the rest unstable. We can also "see" that picking the apple in figure 34.1*B* is much easier and can be achieved with just one action using just one hand (as opposed to the two hands or a more complex sequence of actions needed for the stack in figure 34.1*A*). These abilities are present even early in childhood (figure 34.1*E*) and are likely shared with other species, particularly nonhuman primates (figure 34.1*F*). They are general purpose and can be used to think about many different kinds of physical scenarios and judgments: For instance, can you arrange a set of objects into a stable tower using wooden blocks or Lego bricks (as in figure 34.1*E*)? What about using stones or bricks or cups or even apples?

How might we explain these flexible, seemingly effortless judgments? This chapter presents an answer centered at the notion of physical object representations (PORs), a basic system of knowledge that supports perceiving, learning, and reasoning about all the objects in our environment—their shapes, appearances, affordances, substances, and the way they react to forces applied to them. Our goal here is to outline a computational framework for studying the form and content of PORs in the mind and brain. PORs can be considered an interface between perception and cognition, linking what we perceive to how we plan our actions and talk about the world. Despite their fundamental role in perception, many important questions about object representations remain open. What kind of information formats or data structures underlie PORs so as to support the many ways in which humans flexibly and creatively interact with the world? How can properties of objects be inferred from sensory inputs, and how are they represented in neural circuits? How can these representations integrate sense data across vision, touch, and audition?

After introducing the computational ingredients of POR theory from a reverse-engineering perspective, we review recent work that is beginning to answer some of these questions. We focus on three case studies: (1) how PORs can explain human judgments in intuitive physics, across a broad range of physical outcome prediction scenarios; (2) how PORs provide a substrate for physically mediated object shape perception in scenarios where traditional visual cues fail and a natural substrate for multimodal (visual-haptic) perception and crossmodal transfer; and (3) how in one domain of high-level vision—face perception—PORs might be computed by neural circuits, and how thinking in terms of PORs suggests a new way to interpret multiple stages of processing in the primate brain.

Physical Object Representations

How, in engineering terms, can we formalize PORs? There are two main aspects to our proposal. The first is

FIGURE 34.1 *A* and *B*, How would you pick up the apples indicated while maintaining a stable arrangement of the other objects? It is easy to see that you will likely need to touch more objects (and probably use two hands) in panel (*A*), while the apple in panel (*B*) can be removed on its own with just one hand. *C* and *D*, What is where? Semantic segmentation maps showing class labels and locations of objects from panels (*A* and *B*). *E*, A child playing with stacking cups. Screenshot from https://www.youtube.com/watch?v=dEnDjyWHN4A. *F*, An orangutan building a tower with large Lego-like blocks. Screenshot from https://www.youtube.com/watch?v=MxRJjzSY_JE&t=21s. (See color plate 37.)

a working hypothesis about the contents of PORs. We draw on tools developed for video game engines (Gregory, 2014), including graphics (Blender Online Community, 2015) and physics engines (Coumans, 2010; Macklin, Müller, Chentanez, & Kim, 2014) and planning engines from robotics for grasping and other humanoid motions (Miller & Allen, 2004; Todorov, Erez, & Tassa, 2012; Toussaint, 2015). These tools instantiate simplified but algorithmically tractable models of reality that capture our basic knowledge of how objects work and how our bodies interact with them. In these systems, objects are described by just those attributes needed to simulate natural-looking scenes and motion over short timescales (~2 seconds): 3-D geometry, substance or mechanical material properties (e.g., rigidity), optical material properties (e.g., texture), and dynamical properties (e.g., mass). Video game engines provide causal models in the sense that the process by which the data (i.e., natural-looking scenes) are generated has some abstract level of resemblance to its corresponding real-world process in a form efficient enough to support real-time interactive simulation.

Second, we embed these simulation engines within probabilistic generative models. Physical properties of an object are not directly observable in the raw signals arriving at our sensory organs. These properties, including 3-D shape, mass, or support relations, are latent variables that need to be inferred given sense inputs; they are products of perception. Probabilistic modeling provides the mathematical language to rigorously and unambiguously specify the domain and task being studied, and to explain how, given sensory inputs, latent properties and relations in the underlying physical scene can be reliably inferred through some form of approximate Bayesian inference (see Kersten and Schrater [2002] for an in-depth treatment of this perspective). The probabilistic models we build to capture PORs can be seen as a special case of *probabilistic programs*, or generalizations of directed graphical models (Bayesian networks) that define random variables and conditional probability distributions relating variables using more general data structures and algorithms than simply graphs and matrix algebra (see Ghahramani [2015] and Goodman and Tenenbaum [2016] for an introduction).

The POR framework is closely related to *analysis-by-synthesis* (A×S) accounts of perception: the notion that perception is fundamentally about inverting the causal process of image formation (Helmholtz & Southall, 1924; Rock, 1983). In this view, perceptual systems model the causal processes by which natural scenes are constructed, as well as the process by which images are formed from scenes; this is a mechanism for the hypothetical "synthesis" of natural images, in the style of computer graphics, by using a graphics engine. Perception (or "analysis") is then the search for or inference to the best explanation (or plausible explanations) of an observed image in terms of this synthesis, which in the POR framework can be implemented using Bayesian inference.

Most mechanisms for approximating Bayesian inference that have traditionally been proposed in analysis by synthesis (e.g., Markov chain Monte Carlo, or MCMC) seem implausible when considered as an algorithmic account of perception: they are inherently iterative and almost always far too slow relative to the dynamics of perception in the mind or brain. We draw on recent advances in machine learning and probabilistic programming (including deep neural networks, particle filters or sequential importance samplers, data-driven MCMC, approximate Bayesian computation, and hybrids of these methods) to construct efficient and neurally plausible approximate algorithms for the physical inference tasks specified with our probabilistic models.

While our focus in this chapter is perception, the domain of the POR framework is more general. With a causal model of the world (including its state-space structure—i.e., object dynamics and interactions in a physics engine) and a planner based on a body model, the POR framework transforms the physical environment around us into something computable, naturally supporting many aspects of cognition, including reasoning, imagery, and planning for locomotion and object manipulation via simulation-based inference and control algorithms. In this sense, PORs express functionality somewhat analogous to the "emulators" of emulation theory (Grush, 2004), an earlier proposal for an integrated account of perception, imagery, and motor planning that also fits broadly within a Bayesian approach to inference and control. A key difference is the language of representation for state, dynamics, and observation. Emulation theory was formulated using classical ideas from estimation and control, such as the Kalman filter: body and environment state are represented as vectors, dynamics are linear, and observations are linear functions of the state with Gaussian added noise. The computations supported are simpler but much less expressive than in the POR framework, where state is represented with structured object and scene descriptions, dynamics using physics engines, and observation models using graphics engines. PORs can thus explain how cognitive and perceptual processes operate over a much wider range of physical scenarios, varying greatly in complexity and content, although they require more algorithmic machinery to do so.

Intuitive Physical Reasoning

Having overviewed the basic components of PORs, we now turn to recent computational and behavioral work exploring their application in several domains. We begin with intuitive physics, in the context of scene understanding. Recall the introductory example displayed in figure 34.1. The POR framework was first introduced to answer these kinds of questions, in a form similar to how we characterize it here, by Battaglia, Hamrick, and Tenenbaum (2013). They showed that approximate probabilistic inferences over simulations in a game-style physics engine could be used to perform many different tasks in blocks-world type scenes. While physics engines are designed to be deterministic, Battaglia, Hamrick, and Tenenbaum (2013) found that human judgments were best captured using a probabilistic model that combined the deterministic dynamics of the physics engine with probability distributions over the uncertain geometry of objects' initial configurations and/or shapes, their physical attributes (e.g., their masses), and perhaps the nature of the forces at work (e.g., friction or perturbations of the supporting surface).

In one version of this model (figure 34.2), input images comprised one or more static 2-D views of a tower of blocks in 3-D that might fall over under gravity, and the task was to make various judgments about what would or could happen in the near future. Object shapes and physical properties were assumed to be known, but the model had to estimate the 3-D scene configuration for the blocks. This inference step used A×S with a top-down stochastic search-based (MCMC) procedure: Block positions in 3-D are iteratively and randomly adjusted until the rendered (synthesized) 2-D images approximately match the input images; multiple runs of this procedure yield slightly different outputs, representing samples from an approximate Bayesian posterior distribution on scenes given images. Once these physical object representations are established, they support a wide range of dynamical inferences that go well beyond the purely static content in the perceptual input. How likely is the tower to fall? If it falls, how much of the tower will fall? In which direction will the blocks fall? How far will they fall? If the table supporting the tower were bumped, how many or which of the blocks would fall off the table? If the tower is unstable, what kind of applied force or other action could hold it stable?

To see how these judgments are computed, consider answering the questions: How likely is the tower to fall?

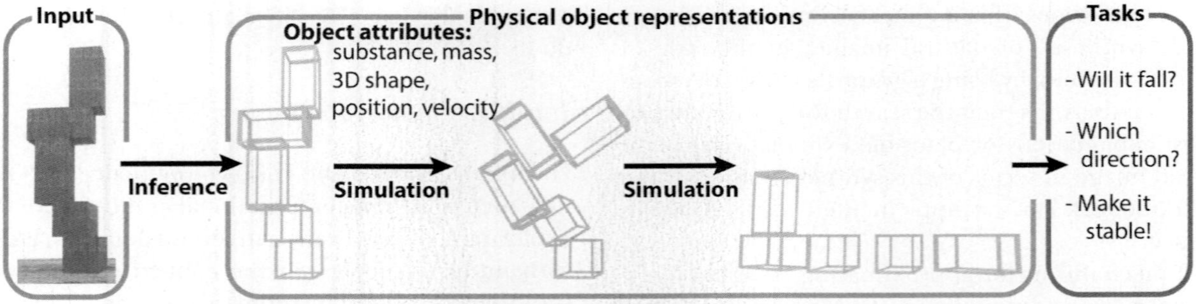

FIGURE 34.2 A schematic of the POR framework applied to intuitive physical reasoning with a tower of wooden blocks. *Left to right,* The input image; inference to recover the 3-D scene and physical properties of objects; physics engine simulation to predict near-future states given the inferred initial configuration; and questions that can be answered and tasks that can be performed based on such simulations.

How much of this tower is likely to fall? One way to make these judgments is to run a small number of forward simulations using a physics engine (implemented, e.g., using Bullet & Coumans, 2010), starting from the sample of configurations returned by the probabilistic 3-D scene inference procedure. These simulations run until all objects stop moving, or some short time limit has elapsed. The distribution of their outcomes represents a sample of the Bayesian posterior predictive distribution on future states, conditioned on the input image and the model's representation of physics. Predictive judgments such as those above can then be calculated by simply querying each sample and aggregating: for example, the model's judgment of "How likely is the tower to fall?" is calculated as the average number of simulations in which the tower fell (relative to the total number of simulations ran); "How much of the tower is likely to fall?" is calculated by averaging the proportion of blocks that fell in each simulation.

Strikingly, Battaglia, Hamrick, and Tenenbaum (2013) found that only a few such posterior samples (they estimated typically three to seven samples per participant, per trial), generated from the highly approximate simulations of video game physics engines under perceptual uncertainty, were sufficient to account for human judgments across a wide range of tasks with high quantitative accuracy. In the last several years, a growing number of behavioral and computational studies have developed approximate probabilistic simulation models of the PORs underlying our everyday physical reasoning abilities. Studies have examined intuitive judgments of mass from how towers do or don't fall (Hamrick, Battaglia, Griffiths, & Tenenbaum, 2016); predictions about future motions (Smith, Battaglia, & Vul, 2013b; Smith, Dechter, Tenenbaum, & Vul, 2013a); judgments of multiple physical properties (e.g., friction as well as mass) and latent forces such as magnetism from examining how objects move and collide in planar motion (Ullman, Stuhlmuller, Goodman, & Tenenbaum, 2018; see also the seminal earlier work on probabilistic inference in collisions by Sanborn, Mansinghka, and Griffiths [2013]); and predictions about the behavior of liquids such as water and honey (Bates, Yildirim, Battaglia, & Tenenbaum, 2015; Kubricht et al., 2016), and granular materials such as sand (Kubricht et al., 2017), falling under gravity. Taken together, these studies show how the POR framework provides a broadly applicable, quantitatively testable, and functionally powerful computational substrate for everyday intuitive physical scene understanding.

How might PORs and their associated computations be implemented in neural hardware? As a first step toward addressing this question, a recent functional magnetic resonance imaging (fMRI) study in humans aimed to localize cortical regions involved in many of the intuitive physics judgments discussed above (Fischer, Mikhael, Tenenbaum, & Kanwisher, 2016). Fischer et al. (2016) found a network of parietal and premotor regions that was differentially activated for physical reasoning tasks in contrast to difficulty-matched nonphysical tasks (such as color judgments, or social predictions) with the same or highly similar stimuli. These regions were consistent across multiple experiments controlling for different task demands and across different visual scenarios. A recent fMRI study in macaques found a similar brain network differentially recruited for analogous physical versus nonphysical stimulus contrasts, in a passive-viewing paradigm (Sliwa & Freiwald, 2017). These networks closely overlap with networks for action planning and tool use in humans (see Gallivan and Culham [2015] for a review) and the mirror neuron system in monkeys that is thought to be involved in action understanding (Rizzolatti & Craighero, 2004), consistent with the proposal that PORs provide a bridge between perception and cognitive functions of action

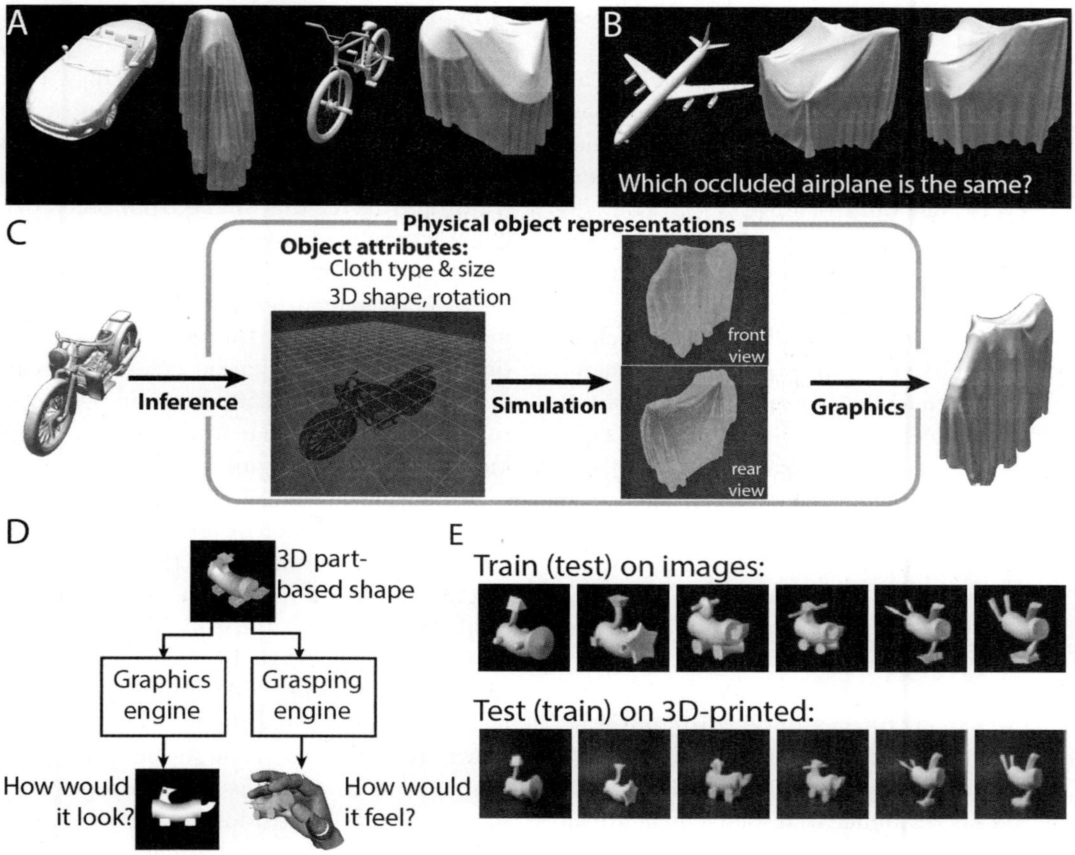

FIGURE 34.3 *A*, Example pairs of unoccluded objects and cloth-occluded matches in different poses. *B*, An example trial from Yildirim, Siegel, and Tenenbaum (2016), where the task is to match the unoccluded object to one of the two occluded objects. *C*, A schematic of the POR framework applied to the object-under-cloth task. *Left to right*, The input image; inference to recover the 3-D shape of the unoccluded object and imagining a cloth positioned above it; physics engine simulation to the predict dropping of the cloth on the object shown at two different angles; and graphics to predict what the resulting scene would look like. *D*, A multisensory causal model combining a graphics engine with a grasp-planning engine. *E*, Example novel objects from Yildirim and Jacobs (2013), rendered visually and photographed after 3-D printing using plastic.

planning, reasoning, and problem solving. Future experimental work using physiological recordings, informed by some of the more neurally grounded models discussed later in this chapter, can now target neural populations in these brain networks in order to elucidate the neural circuits underlying intuitive physics.

Physics-Mediated Object Shape Perception

We now turn to the role of PORs in a more purely perceptual task: perceiving object shape. Vision scientists traditionally study many cues as routes to 3-D shape, such as contours, shading, stereo disparity, or motion. But physics can also be an essential route to shape, especially when these traditional cues are unavailable or insufficient; such cues may be necessary for the correct recovery of a target shape but fail to capture all of the causal processes underlying the appearance of an image. Consider seeing an object that is heavily or even entirely occluded, as when draped by a cloth (figures 34.2*B* and 34.3*A*). It is likely you haven't seen airplanes or bicycles occluded under a cloth before, but it is still relatively easy to pair an unoccluded object with its randomly rotated and occluded counterpart. Of course, shading cues allow you to see the contours of the cloth as an occluding surface. Yet these cues alone do not explain how you perceive the shape of the underlying occluded object, which together with the physical properties of the cloth is the real cause of the shading patterns observed.

Most contemporary approaches to visual object perception emphasize learning to "untangle" or become invariant to sources of variation in the image (DiCarlo & Cox, 2007; Serre, Oliva, & Poggio, 2007). On this account, a processing hierarchy (such as a deep neural network) progressively transforms sensory inputs until reaching an encoding that is diagnostic for a particular object shape or identity and invariant to other factors (Riesenhuber & Poggio, 1999). These approaches can perform very well when trained to ignore a given class of variations, but to achieve optimal performance, they must be trained anew (or at least "fine-tuned") independently for every new kind of invariance. They do not show instantaneous (zero-shot) invariance for new ways an object might appear, such as those arising from an occluding cloth.

The POR framework provides a different approach in which the goal is not learning invariances but explaining variation in the image with respect to the causal process generating images from 3-D physical scenes (e.g., Mumford, 1997; Yuille & Kersten, 2006). For the object-under-cloth task, this process can be captured by composing (1) a physics engine simulating how cloth drapes over 3-D rigid shapes, (2) a graphics engine simulating how images look from the resulting scenes (occluded or unoccluded), and (3) a probabilistic inference engine. The inference engine inverts the graphics process to recover 3-D shapes from unoccluded images and then imagines likely images under different ways these shapes could be rotated and draped under cloth (figure 34.3*C*). Yildirim, Siegel, and Tenenbaum (2016) presented preliminary evidence that such a mechanism fits human judgments in a match-to-sample task, akin to figure 34.3*B*, across four difficulty levels. In contrast, a deep neural network trained for invariant object recognition, but not specifically for scenes involving cloth-based occlusion, could fit the easiest human judgments but failed to generalize above chance for the harder judgments. These results illustrate a key advantage of the POR framework: the ability to generalize to novel settings not by requiring further training but by combining or composing existing causal models.

The POR framework supports combining causal models not only across multiple visual cues but also across sensory modalities. This is because the contents of PORs are not specific to vision or any single modality but instead capture the physical properties of objects that are the root causes of sense data in every modality, via appropriate modality-specific "rendering" engines (such as a graphics engine in vision). Embedded in a framework for probabilistic inference to invert these renderers, PORs provide a basis for perceiving shape from any form of sense data, as well as for multisensory integration and cross-modal perception. Consider the POR-based model shown in figure 34.3*D*: Starting from a probabilistic generative model over part-based body shapes in 3-D, the multisensory causal model combines a visual graphics engine that generates the 2-D appearance of each shape viewed in a given pose with a touch or haptic rendering engine, based on a kinematic grasp planner, that generates the way a shape feels in the hand given a certain grasp trajectory. Bayesian inference then allows the model to estimate a 3-D shape that explains inputs from either visual or haptic channels, or both, as well as to automatically and without further training transfer that shape from objects first encountered in one modality (e.g., visually) to recognize how they would be perceived in another modality (e.g., haptically). Yildirim and Jacobs (2013) found that this model accounted for the performance of human participants in a visual-haptic crossmodal categorization task (example stimuli are shown in figure 34.3*E*). These results were extended to a visual-haptic shape similarity judgment task (Erdogan, Yildirim, & Jacobs, 2015).

The idea that shared neural representations support object perception across multiple sensory modalities is consistent with a number of fMRI studies (e.g., Amedi, Jacobson, Hendler, Malach, & Zohary, 2002; James et al., 2002; Lacey, Tal, Amedi, & Sathian, 2009; Lee Masson, Bulthé, Op de Beeck, & Wallraven, 2016; Tal & Amedi, 2009). The POR framework provides explicit hypotheses as to what the format of such multisensory neural representations might be. Erdogan, Chen, Garcea, Mahon, and Jacobs (2016) used fMRI to test one such hypothesis introduced in their earlier computational work (Erdogan, Yildirim, & Jacobs, 2015). In addition to finding that visual and haptic exploration of novel objects gave rise to similar patterns of neural activity in the lateral occipital cortex (LOC), they also found that this activity could be crossmodally decoded to the part-based 3-D object structure mentioned above (Erdogan, Yildirim, & Jacobs, 2015). This activity may be a result of visual imagery as opposed to haptic processing; however, other work suggests that imagery only minimally activates LOC (Amedi, Malach, Hendler, Peled, & Zohary, 2001; James et al., 2002). Further experimental work along these lines, aiming to quantitatively test specific POR models and ideally extending into physiological recordings from neural populations, could lead to a more precise understanding of the neurocomputational basis of multisensory perception and crossmodal transfer.

Reverse-Engineering Ventral Visual Stream Computations Using Physical Object Representations

We now turn to discussing how the POR framework can illuminate aspects of the neural circuits underlying perception. Even though traditional A×S methods can recover PORs from sense inputs, these algorithms (based on top-down, iterated stochastic search) do not

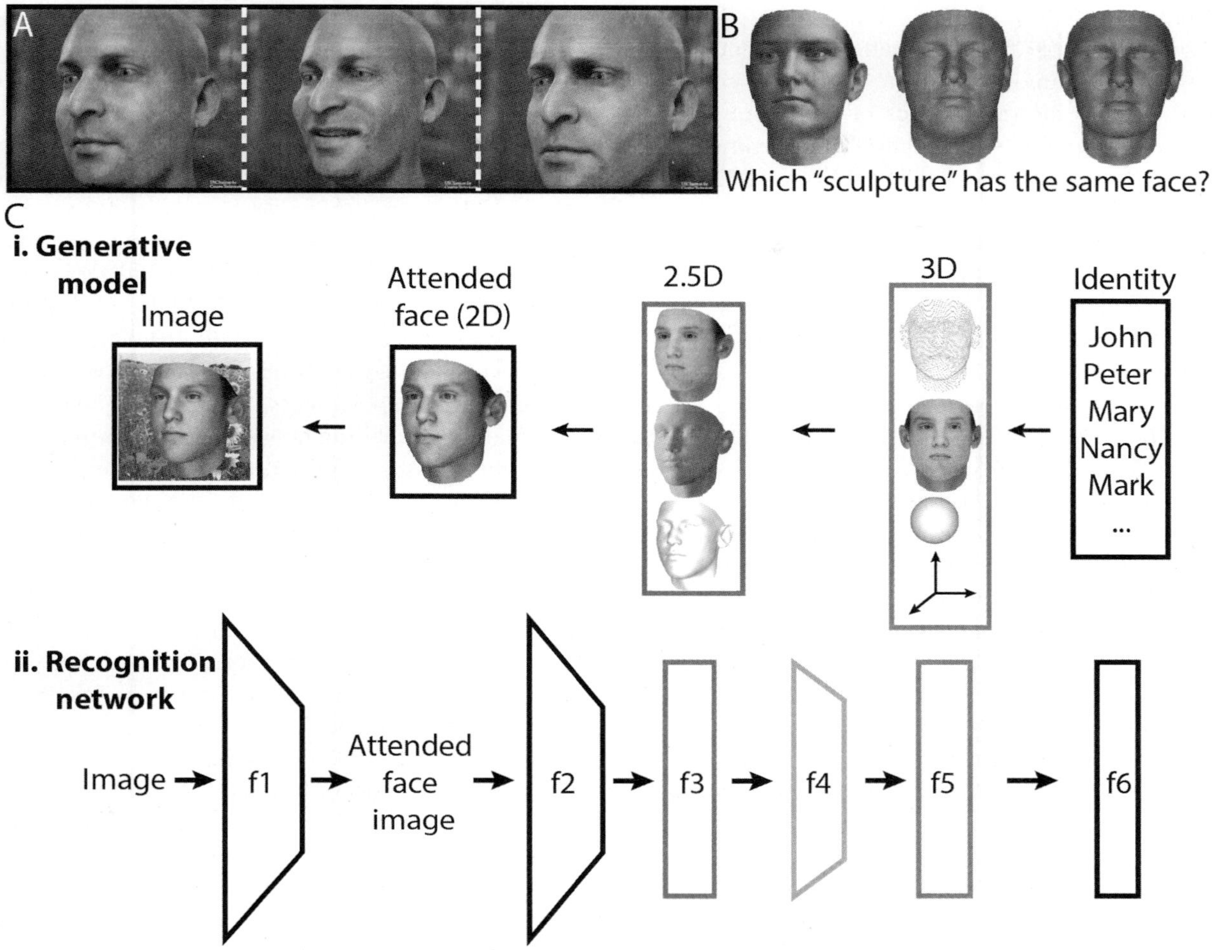

FIGURE 34.4 *A*, Samples from a modern 3-D graphics model of a human face, yielding near photorealistic images (Credit: NVIDIA and University of Southern California Institute for Creative Technologies). Across the three images of this face, in addition to knowing that identity is preserved, we can also appreciate the details of the face's 3-D shape and texture, the subtleties of expression, that vary or remain constant across images. *B*, Despite their unfamiliarity, most observers can match the identity of the naturalistic face on the left to one of the textureless faces ("sculptures"), which must rely on a sense of 3-D shape. *C*, Schematic of the efficient A×S approach, including a probabilistic generative model of face image formation (*panel i*) and the recognition network (*panel ii*). Layers f1 through f6 indicate the different components of the recognition network. Trapezoids show single or multiple layers of transformations where a layer can consist of convolution, normalization, and a nonlinear activation function. Yildirim et al. (2019) found that transformations across the model layers f3, f4, and f5 closely captured the transformations observed in the neural data from ML/MF (middle lateral and middle fundus areas) to AL (anterior lateral area) to AM (anterior medial area; Freiwald & Tsao, 2010). (See color plate 38.)

readily map onto neural computation. Many authors have thus preferred feedforward network models, most recently deep convolutional neural networks (CNNs), which are both more directly relatable to neural circuit-level mechanisms and more consistent with the fast bottom-up processing observed in perception. However, CNNs, typically trained for invariant object recognition or "untangling," do not explicitly address the question of how vision recovers the causal structure of scene and image formation. Therefore, neither traditional approaches to A×S nor modern CNNs really answer the challenge: How do our brains compute rich descriptions of scenes, with detailed 3-D shapes and surface appearances, in much less than a second?

A new class of computational models aim to combine the best aspects of these two approaches by using CNNs or recurrent networks to map images to their underlying scene descriptions, thereby accomplishing otherwise computationally costly inference in one or a few bottom-up passes on the image (Eslami et al., 2018; George et al., 2017; Kulkarni, Kohli, Tenenbaum, & Mansinghka, 2015; Yildirim, Kulkarni, Freiwald, &

Tenenbaum, 2015). Yildirim, Belledonne, Freiwald, and Tenenbaum (2019) developed one such approach using the POR framework and tested it as a computational theory of multiple stages of processing in the ventral visual stream, a hierarchy of processing stages in the visual brain (Conway, 2018). This model consists of two parts: a generative model based on a multistage 3-D graphics program for image synthesis (figure 34.4*C*) and a recognition model based on a CNN that approximately inverts the generative model, stage by stage (figure 34.4*C*). The recognition network is different from conventional CNNs for vision in two ways. First, it is trained to produce the inputs to a graphics engine, the latent or unobservable variables of the probabilistic model, instead of predicting class labels such as face identities. And second, it is trained in a self-supervised fashion, with inputs and targets internally synthesized by the probabilistic graphics component; no externally generated labels are needed. This approach differs from other recent efficient A×S approaches (Eslami et al., 2018; Kulkarni et al., 2015) and their earlier counterparts (Dayan, Hinton, Neal, & Zemel, 1995) in that it is based on a probabilistic graphics engine (instead of learning an unstructured generative model via a generic function approximator) and therefore more closely captures the causal structure of how 3-D scenes give rise to images.

Yildirim, Belledonne, Freiwald, and Tenenbaum (2019) tested their approach in one domain of high-level perception, the perception of faces. Faces give rise to a rich sense of 3-D shape in addition to percepts of a discrete individual's identity (see figure 34.4*A, B*), and face perception has been extensively studied in both psychology and neurophysiology, thus providing a rich source of data and constraints for modeling. The sense of a face's 3-D shape also crosses between visual and haptic modes of perception (Dopjans, Wallraven, & Bulthoff, 2009), as in the examples discussed above.

Yildirim, Belledonne, Freiwald, and Tenenbaum (2019) compared two broad classes of hypotheses for how we perceive the 3-D shape of a face and how these computations are implemented in the primate ventral stream: (1) the efficient A×S hypothesis implemented in their recognition network, which posits that the targets of ventral stream processing are latent variables in a probabilistic causal model of image formation, and (2) the untangling hypothesis implemented in standard deep CNNs for face recognition, which posits that the target of ventral stream processing is an embedding space optimized for discriminating among facial identities. Their recognition network implementing the A×S hypothesis recapitulated transformations across multiple stages of processing in inferio temporal (IT) cortex from middle lateral and middle fundus areas (ML/MF) to anterior lateral area (AL) to anterior medial area (AM)—the three sites in the monkey face patch system—with respect to the similarity structure of the population-level activity in each stage (Freiwald & Tsao, 2010). Both in the neural data and in the model, these similarity structures progressed from view-based to mirror-symmetric to view-invariant representations. Alternative models, including a number implementing the untangling hypothesis, did not capture these transformations. The efficient A×S model also accurately matched human error patterns in psychophysical experiments, including experiments designed to determine how flexibly humans can attend to either the shape or texture components of a face stimulus (figure 34.4*B*). Finally, the recognition model suggested an interpretable account of some intermediate representations in this hierarchy: in particular, population-level similarity structure of middle face patches (ML/MF) can be well accounted for by the similarity structure arising from intermediate surface representations, such as intrinsic images (normal maps or depth maps for surface geometry and albedos for surface color) or a 2.5-D sketch.

The efficient A×S approach thus offers a potential resolution to the issue of interpretability in systems neuroscience (Yamins & DiCarlo, 2016). In addition to assessing accounts of the brain in terms of how much variance in neural firing rates they explain, the efficient A×S approach suggests that computational neuroscientists could aim for "semi-interpretable" models of perception where the recognition network as a whole can be understood as inverting a causal generative model, and subpopulations of neurons in particular stages of the recognition network (such as ML/MF and AM) can be understood as inverting distinct, identifiable stages in the generative model, explicitly representing hypotheses about the corresponding aspects of scene structure encoded in those generative model stages. Other populations of neurons (such as AL) might be better explained as implementing valuable hidden-layer nonlinear transforms between more interpretable parts of the system.

Conclusion and Future Directions

We believe that there is promising, if preliminary, evidence for the centrality of PORs in the mind and brain. The strongest aspect of this proposal so far is theoretical: PORs offer a solution to problems both old (e.g., multimodal perception) and new (e.g., the cloth-draping task presented above), perceptual phenomena that are difficult to explain with alternative accounts in either cognitive neuroscience or artificial intelligence. There remain, however, significant challenges. Empirical work has only begun to test strong predictions of the POR framework; far more behavioral and physiological data are needed. As we have noted, PORs provide a rich foundation for structuring perception and behavior, but this comes with a heavy computational burden. The efficient A×S approach is one possible way the brain might handle this complexity, but again more study is needed, especially relating the dynamics of processing in these models to the dynamics of neural computation. Further theoretical work is also required to explore the origins of PORs: how an organism comes to possess an object-based causal model of the world around it.

The POR framework also offers new research directions for studying aspects of complex behavior production and object manipulation. An important advantage of the POR framework is that causal models of the world allow for flexible action planning, reasoning, and intelligent object manipulation. To illustrate, we revisit the grasping engine shown in figure 34.3*D* in its broader context. This grasping engine implements a planner based on a simulatable body model (similar to forward models typically invoked in models of motor control; Jordan & Rumelhart, 1992; Wolpert & Flanagan, 2009; Wolpert & Kawato, 1998). Such a model allows embodied agents to evaluate the consequences of their actions by simulating them internally before (or without ever) actually performing them. Many organisms likely use this approach—for example, performing simulations for making a judgment about the action "Can I jump?" Brecht (2017) suggested that the microcircuits in the mammal somatosensory cortex implement a simulatable body model that can be used for action planning and decision-making. The POR framework provides a toolkit to capture these computations in engineering terms using existing simulation engines (e.g., see Yildirim, Gerstenberg, Saeed, Toussaint, and Tenenbaum [2017] for a proof-of-concept implementation in the context of complex object manipulation).

Perhaps the most important open question is also the most challenging: How could simulations with richly structured generative models, such as graphics engines, physics engines, and body models, be implemented in neural mechanisms? Recent developments in machine learning and perception suggest intriguing possibilities based on deep learning systems that are trained to emulate a structured generative model in an artificial neural network architecture. Deep networks that emulate graphics engines were mentioned above; while they do not yet come close to the full functionality of traditional graphics engines, their performance in narrow domains can be surprisingly impressive and continues to improve. In intuitive physics, hybrids of discrete symbolic and distributed representations, such as neural physics engines (Chang, Ullman, Torralba, & Tenenbaum, 2016), interaction networks (Battaglia, Pascanu, Lai, & Rezende, 2016) and other graph networks (Battaglia et al., 2018), and hierarchical relation networks (Mrowca et al., 2018), have received much attention lately. These systems assume discrete symbolic representations for each object and its relation to other objects and vector representations for the rules of physical interactions between objects; this allows the dynamics of object motion and interaction (e.g., collisions) to be learned efficiently end-to-end from simulated data. Artificial neural networks such as these can be considered partial hypotheses for how graphics and physics might be implemented in biological neural circuits; they are almost surely wrong or at best incomplete, but they suggest a way forward. Further work is needed to test these models empirically and to develop their capacities; currently, they are very limited in the scope of physics they can learn (e.g., a limited class of rigid body interactions, such as billiard balls colliding on a table). Nevertheless, with these advances and building on the example of the efficient A×S approach and other research linking artificial neural networks to neural representations in the brain, we see promise in linking the POR framework to neural computation in perception and well beyond.

Acknowledgments

We thank Amir A. Soltani and Mario Belledonne for their help with the figures. We thank James Traer, Max Kleiman-Weiner, and our section editor Josh McDermott for their feedback on earlier versions of this chapter. This work was supported by the Center for Brains, Minds and Machines (CBMM) and funded by National Science Foundation STC award CCF-1231216; the Office of Naval Research Multidisciplinary University Research Initiatives grant N00014-13-1-0333; a grant from the Toyota Research Institute; and a grant from the Mitsubishi Electric Corporation.

REFERENCES

Amedi, A., Jacobson, G., Hendler, T., Malach, R., & Zohary, E. (2002). Convergence of visual and tactile shape processing in the human lateral occipital complex. *Cerebral Cortex, 12*(11), 1202–1212.

Amedi, A., Malach, R., Hendler, T., Peled, S., & Zohary, E. (2001). Visuo-haptic object-related activation in the ventral visual pathway. *Nature Neuroscience, 4*(3), 324.

Bates, C., Battaglia, P., Yildirim, I., & Tenenbaum, J. B. (2015). Humans predict liquid dynamics using probabilistic simulation. In *Proceedings of the 37th Annual Conference of the Cognitive Science Society*, 172–177.

Battaglia, P. W., Hamrick, J. B., Bapst, V., Sanchez-Gonzalez, A., Zambaldi, V., Malinowski, M., … Gulcehre, C. (2018). Relational inductive biases, deep learning, and graph networks. *arXiv*. Retrieved from 1806.01261.

Battaglia, P. W., Hamrick, J. B., & Tenenbaum, J. B. (2013). Simulation as an engine of physical scene understanding. *Proceedings of the National Academy of Sciences, 110* (45), 18327–18332.

Battaglia, P., Pascanu, R., Lai, M., & Rezende, D. J. (2016). Interaction networks for learning about objects, relations and physics. In *Advances in Neural Information Processing systems*, 4502–4510. Curran Associates, Inc.

Blender Online Community. (2015). Blender—a 3D modelling and rendering package [Computer software manual]. Amsterdam: Blender Institute. http://www.blender.org.

Brecht, M. (2017). The body model theory of somatosensory cortex. *Neuron, 94*(5), 985–992.

Chang, M. B., Ullman, T., Torralba, A., & Tenenbaum, J. B. (2016). A compositional object-based approach to learning physical dynamics. *arXiv*. Retrieved from 1612.00341.

Conway, B. R. (2018). The organization and operation of inferior temporal cortex. *Annual Review of Vision Science, 4, 381–402.*

Coumans, E. (2010). Bullet physics engine. [Open-source software]. http://bulletphysics.org.

Dayan, P., Hinton, G. E., Neal, R. M., & Zemel, R. S. (1995). The Helmholtz machine. *Neural Computation, 7*(5), 889–904.

DiCarlo, J. J., & Cox, D. D. (2007). Untangling invariant object recognition. *Trends in Cognitive Sciences, 11*(8), 333–341.

Dopjans, L., Wallraven, C., & Bulthoff, H. H. (2009). Crossmodal transfer in visual and haptic face recognition. *IEEE Transactions on Haptics, 2*(4), 236–240.

Erdogan, G., Chen, Q., Garcea, F. E., Mahon, B. Z., & Jacobs, R. A. (2016). Multisensory part-based representations of objects in human lateral occipital cortex. *Journal of Cognitive Neuroscience, 28*(6), 869–881.

Erdogan, G., Yildirim, I., & Jacobs, R. A. (2015). From sensory signals to modality-independent conceptual representations: A probabilistic language of thought approach. *PLoS Computational Biology, 11*(11), e1004610.

Eslami, S. A., Rezende, D. J., Besse, F., Viola, F., Morcos, A. S., Garnelo, M., … Reichert, D. P. (2018). Neural scene representation and rendering. *Science, 360*(6394), 1204–1210.

Fischer, J., Mikhael, J. G., Tenenbaum, J. B., & Kanwisher, N. (2016). Functional neuroanatomy of intuitive physical inference. *Proceedings of the National Academy of Sciences, 113*(34), E5072–E5081.

Freiwald, W. A., & Tsao, D. Y. (2010). Functional compartmentalization and viewpoint generalization within the macaque face-processing system. *Science, 330*(6005), 845–851.

Gallivan, J. P., & Culham, J. C. (2015). Neural coding within human brain areas involved in actions. *Current Opinion in Neurobiology, 33*, 141–149.

George, D., Lehrach, W., Kansky, K., Lázaro-Gredilla, M., Laan, C., Marthi, B., … Lavin, A. (2017). A generative vision model that trains with high data efficiency and breaks text-based CAPTCHAs. *Science, 358*(6368), eaag2612.

Ghahramani, Z. (2015). Probabilistic machine learning and artificial intelligence. *Nature, 521*(7553), 452.

Goodman, N. D., Tenenbaum, J. B., & The ProbMods Contributors. (2016). *Probabilistic models of cognition* (2nd ed.). Retrieved September 1, 2018, from https://probmods.org.

Gregory, J. (2014). *Game engine architecture*. Boca Raton, FL: CRC Press.

Grush, R. (2004). The emulation theory of representation: Motor control, imagery, and perception. *Behavioral and Brain Sciences, 27*(3), 377–396.

Hamrick, J. B., Battaglia, P. W., Griffiths, T. L., & Tenenbaum, J. B. (2016). Inferring mass in complex scenes by mental simulation. *Cognition, 157*, 61–76.

Helmholtz, H. V., & Southall, J. P. C. (1924). *Helmholtz's treatise on physiological optics*. Rochester, NY: Optical Society of America.

James, T. W., Humphrey, G. K., Gati, J. S., Servos, P., Menon, R. S., & Goodale, M. A. (2002). Haptic study of three-dimensional objects activates extrastriate visual areas. *Neuropsychologia, 40*(10), 1706–1714.

Jordan, M. I., & Rumelhart, D. E. (1992). Forward models: Supervised learning with a distal teacher. *Cognitive Science, 16*(3), 307–354.

Kersten, D., & Schrater, P. R. (2002). Pattern inference theory: A probabilistic approach to vision. In R. Mausfeld & D. Heyer (Eds.), *Perception and the physical world*, 191–228. Chichester, UK: John Wiley & Sons.

Kubricht, J. R., Holyoak, K. J., & Lu, H. (2017). Intuitive physics: Current research and controversies. *Trends in Cognitive Sciences, 21*(10), 749–759.

Kubricht, J., Jiang, C., Zhu, Y., Zhu, S. C., Terzopoulos, D., & Lu, H. (2016). Probabilistic simulation predicts human performance on viscous fluid-pouring problem. In *Proceedings of the 38th Annual Conference of the Cognitive Science Society*, 1805–1810.

Kubricht, J., Zhu, Y., Jiang, C., Terzopoulos, D., Zhu, S. C., & Lu, H. (2017). Consistent probabilistic simulation underlying human judgment in substance dynamics. In *Proceedings of the 39th Annual Meeting of the Cognitive Science Society*, 700–705.

Kulkarni, T. D., Kohli, P., Tenenbaum, J. B., & Mansinghka, V. (2015). Picture: A probabilistic programming language for scene perception. *Proceedings of the IEEE Conference on Computer Vision and Pattern Recognition*, 4390–4399.

Lacey, S., Tal, N., Amedi, A., & Sathian, K. (2009). A putative model of multisensory object representation. *Brain Topography, 21*(3–4), 269–274.

Le, T. A., Baydin, A. G., & Wood, F. (2016). Inference compilation and universal probabilistic programming. *arXiv*. Retrieved from 1610.09900.

Lee Masson, H., Bulthé, J., Op de Beeck, H. P., & Wallraven, C. (2016). Visual and haptic shape processing in the human brain: Unisensory processing, multisensory convergence, and top-down influences. *Cerebral Cortex, 26*(8), 3402–3412.

Macklin, M., Müller, M., Chentanez, N., & Kim, T. Y. (2014). Unified particle physics for real-time applications. *ACM Transactions on Graphics, 33*(4), 153.

Marr, D. (1982). *Vision: A computational investigation into the human representation and processing of visual information*. Cambridge, MA: MIT Press.

Miller, A. T., & Allen, P. K. (2004). Graspit! A versatile simulator for robotic grasping. *IEEE Robotics & Automation Magazine, 11*(4), 110–122.

Mrowca, D., Zhuang, C., Wang, E., Haber, N., Fei-Fei, L., Tenenbaum, J. B., & Yamins, D. L. (2018). Flexible neural representation for physics prediction. *arXiv.* Retrieved from 1806.08047.

Mumford, D. (1996). Pattern theory: A unifying perspective. In D. C. Knill & W. Richards (Eds.), *Perception as Bayesian inference,* 25–62. Cambridge: Cambridge University Press.

Pascual-Leone, A., & Hamilton, R. (2001). The metamodal organization of the brain. In C. Casanova & M. Ptito (Eds.), *Progress in brain research* (Vol. 134, pp. 427–445). New York: Elsevier.

Riesenhuber, M., & Poggio, T. (1999). Hierarchical models of object recognition in cortex. *Nature Neuroscience, 2*(11), 1019.

Rizzolatti, G., & Craighero, L. (2004). The mirror-neuron system. *Annual Review of Neuroscience, 27,* 169–192.

Rock, I. (1983). *The logic of perception.* Cambridge, MA: MIT Press.

Sanborn, A. N., Mansinghka, V. K., & Griffiths, T. L. (2013). Reconciling intuitive physics and Newtonian mechanics for colliding objects. *Psychological Review, 120*(2), 411.

Serre, T., Oliva, A., & Poggio, T. (2007). A feedforward architecture accounts for rapid categorization. *Proceedings of the National Academy of Sciences, 104*(15), 6424–6429.

Sliwa, J., & Freiwald, W. A. (2017). A dedicated network for social interaction processing in the primate brain. *Science, 356*(6339), 745–749.

Smith, K. A., Battaglia, P., & Vul, E. (2013b). Consistent physics underlying ballistic motion prediction. In *Proceedings of the 35th Annual Meeting of the Cognitive Science Society,* 426–3431.

Smith, K. A., Dechter, E., Tenenbaum, J. B., & Vul, E. (2013a). Physical predictions over time. In *Proceedings of the 35th Annual Meeting of the Cognitive Science Society,* 1342–1347.

Tal, N., & Amedi, A. (2009). Multisensory visual-tactile object related network in humans: insights gained using a novel crossmodal adaptation approach. *Experimental Brain Research, 198*(2), 165–182.

Todorov, E., Erez, T., & Tassa, Y. (2012). MuJoCo: A physics engine for model-based control. In *2012 IEEE/RSJ International Conference on Intelligent Robots and Systems,* 5026–5033. Vilamoura.

Toussaint, M. (2015). Logic-geometric programming: An optimization-based approach to combined task and motion planning. *International Joint Conferences on Artificial Intelligence,* 1930–1936.

Ullman, T. D., Spelke, E., Battaglia, P., & Tenenbaum, J. B. (2017). Mind games: Game engines as an architecture for intuitive physics. *Trends in Cognitive Sciences, 21*(9), 649–665.

Ullman, T. D., Stuhlmüller, A., Goodman, N. D., & Tenenbaum, J. B. (2018). Learning physical parameters from dynamic scenes. *Cognitive Psychology, 104,* 57–82.

Wolpert, D. M., & Flanagan, J. R. (2009). Forward models. In T. Bayne, A. Cleeremans, & P. Wilken (Eds.), *The Oxford Companion to Consciousness,* 294–296. New York: Oxford University Press.

Wolpert, D. M., & Kawato, M. (1998). Multiple paired forward and inverse models for motor control. *Neural Networks, 11*(7–8), 1317–1329.

Wu, J., Yildirim, I., Lim, J. J., Freeman, B., & Tenenbaum, J. (2015). Galileo: Perceiving physical object properties by integrating a physics engine with deep learning. *Advances in neural information processing systems,* 127–135. Curran Associates, Inc.

Yamins, D. L., & DiCarlo, J. J. (2016). Using goal-driven deep learning models to understand sensory cortex. *Nature Neuroscience, 19*(3), 356.

Yildirim, I., Belledonne, M., Freiwald, W., & Tenenbaum, J. (2019). Efficient inverse graphics in biological face processing. *bioRxiv,* 282798v2.

Yildirim, I., Gerstenberg, T., Saeed, B., Toussaint, M., & Tenenbaum, J. (2017). Physical problem solving: Joint planning with symbolic, geometric, and dynamic constraints. *arXiv.* Retrieved from 1707.08212.

Yildirim, I., & Jacobs, R. A. (2013). Transfer of object category knowledge across visual and haptic modalities: Experimental and computational studies. *Cognition, 126*(2), 135–148.

Yildirim, I., Kulkarni, T. D., Freiwald, W. A., & Tenenbaum, J. B. (2015). Efficient and robust analysis-by-synthesis in vision: A computational framework, behavioral tests, and modeling neuronal representations. In *Proceedings of the 35th Annual Conference of the Cognitive Science Society,* 2751–2756.

Yildirim, I., Siegel, M., & Tenenbaum, J. (2016). Perceiving fully occluded objects via physical simulation. In *Proceedings of the 36th Annual Conference of the Cognitive Science Society,* 1265–1270.

Yuille, A., & Kersten, D. (2006). Vision as Bayesian inference: Analysis by synthesis? *Trends in Cognitive Sciences, 10*(7), 301–308.

35 Constructing Perceptual Decision-Making across Cortex

ROMÁN ROSSI-POOL, JOSÉ VERGARA, AND RANULFO ROMO

ABSTRACT Here we review the neural computations involved in vibrotactile detection and discrimination tasks across cortex. A common feature for vibrotactile detection and discrimination is that primary somatosensory cortex (S1) is essential for feeding information to a large cortical network involved in perceptual decision-making. S1 generates a neural copy of the sensory input in these tasks. The S1 representation is then transformed across cortex, beginning in the secondary somatosensory cortex, and transformed again in the frontal lobe circuits into a neural signal consistent with the subject's decision report. Importantly, we discuss evidence that frontal lobe circuits represent current and remembered sensory inputs, their comparison, and the motor commands expressing the result—that is, the entire cascade linking the evaluation of sensory stimuli with a motor decision report. These findings provide a fairly complete panorama of the neural dynamics across cortex that underlies perceptual decision-making.

A fundamental issue in neurobiology is understanding precisely which component of the neuronal activity evoked by a sensory stimulus is meaningful for perception. Indeed, pioneering investigations in several sensory systems have shown how neural activity represents the physical parameters both in the periphery and central nervous system (Hubel and Wiesel, 1962; Mountcastle et al., 1967; Talbot et al., 1968). These investigations have paved the way for new questions more directly related to cognitive processing. For example, where and how in the brain do the neuronal responses that encode sensory stimuli translate into responses that encode a decision (Romo and de Lafuente, 2013; Romo and Salinas, 2003)? What components of the neuronal activity evoked by a sensory stimulus are directly related to perception (Romo et al., 1998; Salzman, Britten, and Newsome, 1990)? These questions have been investigated in behavioral tasks in which the sensory stimuli are under precise quantitative control, and the subjects' psychophysical performances are quantitatively measured (Hernández et al., 1997; Newsome, Britten, and Movshon, 1989). One of the main challenges of this approach is that even the simplest cognitive tasks engage a large number of cortical areas, and each one might encode the sensory information in a different way (Romo and de Lafuente, 2013; Romo and Salinas, 2003). Also, the sensory information might be combined in these cortical areas with other types of stored signals representing, for example, past experiences and future actions. Thus, an important issue is to decode from the neuronal activity all these processes that might be related to perceptual decision-making. Indeed, recent studies have provided new insights into this problem using highly simplified psychophysical tasks (de Lafuente and Romo, 2005; Hernández et al., 1997). In particular, these studies have shown the neural codes related to sensation, working memory, and decision reports in these tasks (Romo and de Lafuente, 2013; Romo and Salinas, 2003).

In this chapter we discuss the cortical representation of tactile stimuli, its relation to behavior and perception, its dependence on behavioral context, and its persistence in working memory, all crucial ingredients in decision-making. Notoriously, we describe neural responses found in cortical areas traditionally involved in motor behavior that, in our tasks, seem to reflect much more complex responses involved in the decision-making process. The results also illustrate population neural signals that condense the heterogeneity among the individual neuron response coding associated with the major components of the behavioral tasks. An important finding—using the somatosensory system as a model to investigate these processes—is that the primary somatosensory cortex (S1) drives higher cortical areas from the parietal and frontal lobes, which combine past and current sensory information, such that a comparison of the two evolves into a decision report. Another important finding is that quantifiable percepts can be triggered by directly activating the S1 circuit that drives cortical areas associated with perceptual decision-making (Romo et al., 1998, 2000). Finally, the direct activation of frontal lobe circuits can also produce quantifiable percepts (de Lafuente and Romo, 2005), suggesting the existence of facilitated circuits beyond S1 engaged in perceptual decision-making. This evidence favors the existence of distributed brain circuits engaged in perceptual decision-making.

Constructing Decision-Making during Sensory Detection

One of the simplest perceptual experiences that can be studied is the detection of sensory stimuli. Further, it is a requirement for more complex sensory processing. A singular feature of sensory detection is that near-threshold stimuli may or may not generate a percept. Consequently, a sensory-detection task represents a simple and appropriate design to study the neuronal processes by which the sensory information is analyzed and gives rise to perception. The intention in this task is to determine correlations between neuronal activity and the subject's perceptual report. In other words, which areas in the brain exhibit neuronal activity that correlates with the subject's perceptual decision reports?

In the last years, the detection of sensory stimuli has been studied using the somatosensory system as a model (de Lafuente and Romo, 2005, 2006). In these studies, monkeys were trained to perform a vibrotactile detection task. In each trial, the animal reported whether the tip of a mechanical stimulator vibrated or not (figure 35.1*A*). Stimuli were sinusoidal, of varied amplitude across trials, had a fixed frequency of 20 Hz, and were delivered to the glabrous skin of one fingertip of the restrained hand. Trials with stimulus-presence (stimulus amplitude higher than 0 μm) were combined randomly with an equal number of trials in which no mechanical vibration was delivered (stimulus amplitude equal to 0 μm). Stimulus detection thresholds were calculated from the animal's behavioral responses (*left panel*, figure 35.1*B*). In addition, monkeys' responses can be classified into four types: hits and misses (stimulus-present trials) and correct rejections and false alarms (stimulus-absent trials; *right panel*, figure 35.1*B*). The main goal of this experiment was to record simultaneously the behavioral responses together with the neuronal activity across cortex (*top panel*, figure 35.1*C*), in an attempt to explain the neuronal mechanisms involved in sensory detection.

Notably, the activity patterns of neurons recorded in S1 (areas 3b and 1) exquisitely encoded the physical properties of the vibratory stimuli but gave no information as to how the monkeys perceived the stimuli (de Lafuente and Romo, 2005). Remarkably, the psychophysical threshold for stimulus detection matches quite closely the sensitivity of single S1 neurons. Additionally, there is a high correspondence between the mean neurometric curve resulting from the activity of S1 neurons and the monkey's psychometric curve. Further, de Lafuente and Romo (2005) found no significant differences between the activity of S1 neurons either between hits and misses or between correct rejections and false alarms. They simply identified a gradual relationship between the stimulus amplitude and the evoked neuronal responses (*black line, lower panel*, figure 35.1*C*). Thus, the responses of S1 neurons did not predict the monkey's behavior; they only coded the stimulus intensity. These results fit well with the idea that central areas should be reading out the homogenous responses of S1 neurons to infer if the stimulus was present or not. Thus, S1 generates a neural representation of the sensory input for further processing in downstream areas in this task.

Conversely, activity recorded from neurons in the frontal lobe correlates closely with the animal's perception in the detection task (*top panel*, figure 35.1*C*). Specifically, neuron responses from the ventral (VPC), medial (MPC), and dorsal (DPC) premotor cortices closely covaried with the monkeys' behavioral reports. Premotor neurons responded in an all-or-none mode that was only weakly modulated by the stimulus amplitude (*light gray lines, lower panel*, figure 35.1*C*). Remarkably, this feature was observed even when those reports did not correctly reflect the stimulus characteristics (false alarms and misses). Consequently, the neuronal responses were clearly different between hits and misses and between false alarms and correct rejections. These results showed a close association between premotor neuronal activity and behavior, supporting the idea that frontal lobe neurons do not code the stimulus parameter but rather convey information about perceptual judgments (stimulus-presence or stimulus-absence).

The results described above raise the question of whether the neural correlate of perceptual judgments emerges abruptly in a particular cortical area or gradually builds up as information is transmitted and transformed across areas between S1 and the premotor cortex. To quantify the role of each area, the relationship between stimulus amplitude and firing rate was calculated (figure 35.1*C*; de Lafuente and Romo, 2006). The authors performed a linear regression on the normalized firing rate as a function of the logarithm of the stimulus amplitude. The semilogarithm slopes approximate increasingly to zero in neurons downstream to S1 (areas 3b and 1), areas 2 and 5, and second somatosensory cortex (S2). As a consequence, responses from downstream areas to somatosensory areas do not modulate their activity as functions of the stimulus amplitude, as early somatosensory areas do. Therefore, the stimulus encoding was transformed from a stimulus parametric code to an abstract representation. Thus, frontal lobe circuits that employ this abstract coding do not modulate their activity as a function of stimulus amplitude. This means that frontal neurons exhibit all-or-none responses, depending on whether the subject

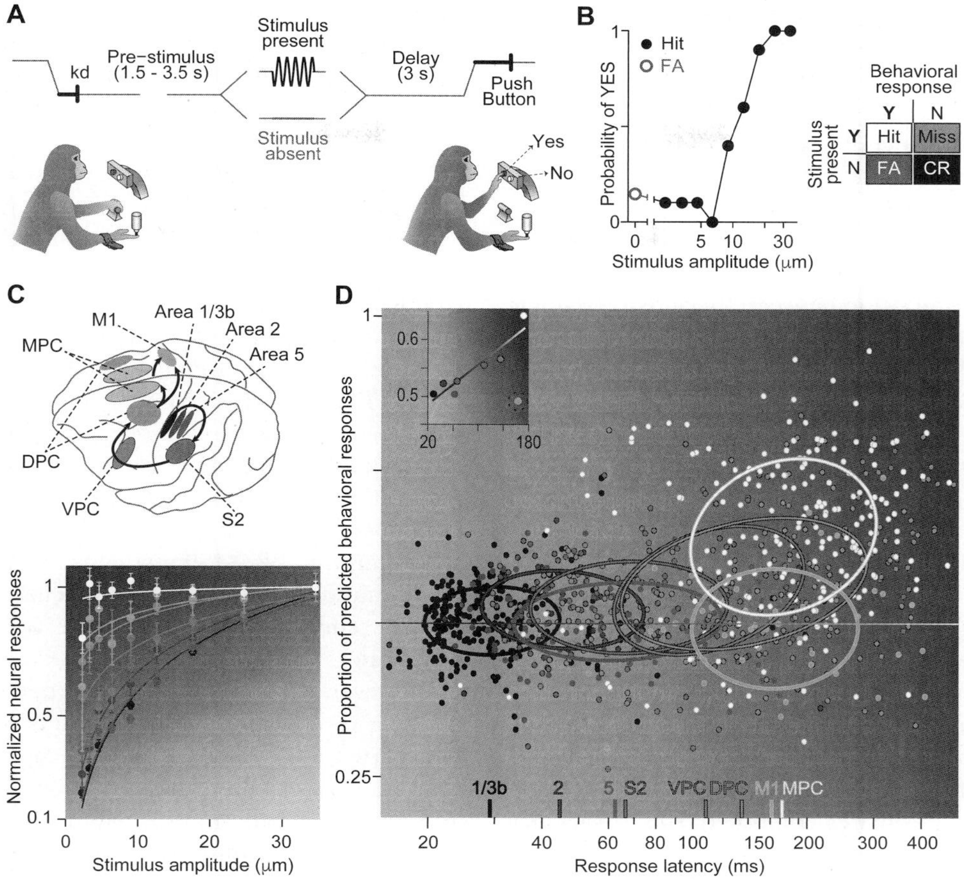

FIGURE 35.1 Vibrotactile detection task. *A*, Sequence of behavioral events during detection task. A trial began when the mechanical probe indented the glabrous skin of one fingertip of the right restrained hand, and the monkey reacted by placing its left free hand on an immovable key (key down [kd]). After a variable delay (prestimulus period, 1.5–3.5 s), a vibratory stimulus of variable amplitude (equal frequency and duration; 20 Hz, 0.5 s) was presented on one-half of the trials (stimulus-present); no stimulus was presented on the other half of the trials (stimulus-absent). Then the stimulator moved up after a fixed delay period (3 s), cueing the monkey to communicate its decision about stimulus-presence or stimulus-absence by pressing one of two push-buttons (yes-button; no-button). *B, Left panel,* The psychometric detection curve resulting from plotting the proportion of yes-button responses as a function of stimulus amplitude. *Right panel,* The four possible types of trials that can be obtained based on whether the stimulus was present or absent and the subject's behavioral reports: Hit (stimulus-present and yes-button), Miss (stimulus-present and no-button), False Alarm (FA; stimulus-absent and yes-button), and Correct Rejection (CR; stimulus-absent and no-button). *C, Upper panel,* the recorded areas. *Lower panel,* Mean normalized firing rate in stimulus-present trials across all the recorded cortical areas. Lines correspond to linear fitting of the firing rate as a function of the stimulus-amplitude logarithm. *D*, Timing and the ability to predict the behavioral response across cortical areas. Dots correspond to the choice probability indices (mean value: Hits vs. Misses and CR vs. FA; ROC [receiving operating characteristic] analysis) from each individual neuron as a function of their stimulus-response latencies. Ellipses are the 1σ contour for a two-dimensional Gaussian fit to the neurons from each recorded area. Grayscale vertical markers above the abscissa-axis indicate the mean response latency for each cortical region. The top left inset plot illustrates the increase of the mean choice probability as a function of the mean response latency ($r^2 = 0.87$; linear fit excluding M1 neurons [*lower dot surrounded by dotted black circle*]). Recorded areas include areas 1/3b, 2, 5, second somatosensory cortex (S2), and ventral premotor cortex (VPC) on the left hemisphere; dorsal and medial premotor cortices (DPC and MPC, respectively) recorded bilaterally; primary motor cortex (M1) recorded on the right hemisphere. The stimulus's response latencies and the ability to predict the subject's behavior show that vibrotactile information flows from sensory areas in the parietal cortex to premotor and motor areas in the frontal lobe (*black arrows*). Adapted from de Lafuente and Romo (2006).

felt or missed the stimulus. This evidence suggests that this task involves the conjoined activity of many brain areas. Hence, the vibrotactile stimulus evoked a distributed activity from S1 to premotor and motor areas.

Although neurons could respond during the detection task, they may or may not be part of the perceptual construction. To understand how the sensory percept emerges, it is necessary to define proper measures to quantify how neural responses covary with the perceptual behavior. Covariation between the activity of single neurons and the subject's choice is often quantified by the choice probability (CP) index (de Lafuente and Romo, 2006; Green and Swets, 1966). This quantity measures the average probability by which an external observer could predict the monkey's decision from the activity of a single neuron. On one side, S1 (areas 3b and 1) and area 2 neurons exhibited little predictive capacity regarding the animal's response to a near-threshold stimulus (CP ≈ 0.5 in figure 35.1*D*). As explained above, a near-threshold stimulus may (hit) or may not be detected (miss). However, S1 neurons were not associated with this perceptual behavior. In contrast, neurons from premotor cortices showed high values of choice probability (CP ≈ 0.75 in figure 35.1*D*). Interestingly, when the neural populations from these premotor areas optimally combined the covariance with behavior, they saturated the maximum predictive capacity (CP ≈ 1; Carnevale et al., 2013). Additionally, the activity of S2 neurons displayed intermediate CP values: correlation with behavioral outcomes was significantly above chance (CP > 0.5 in figure 35.1*D*). Notably, primary motor (M1) cortex neurons did not predict the animal's decision report. This evidence suggests that premotor areas seem more involved in perceptual judgments than in the motor responses during the detection task.

A notable feature is the response latency to the stimulus for each cortical area during the detection task. Indeed, de Lafuente and Romo (2005) sought to relate the response latency with the hierarchy of each area in sensory processing. To quantify this relationship between the predictive capacity of these neurons and the processing hierarchy, CP indexes were plotted as a function of the response latency (figure 35.1*D*; de Lafuente and Romo, 2006). Remarkably, neurons located in areas with longer mean latencies (higher-order areas downstream to S1) exhibited a large covariance with the monkeys' perceptual reports. To further show this phenomenon, these authors plotted the mean CP index as a function of the mean response latency for each cortical area (*left top inset,* figure 35.1*D*). Plainly, there is a linear dependence between these two quantities. As mentioned above, the activity from M1 neurons is excluded because their response is essentially involved in movement and displayed low CP values.

Additionally, CP indexes for neurons within each premotor area (VPC, DPC, and MPC) covaried with their response latency. This analysis further showed that even neurons within the same processing state (hierarchy) tended to correlate more with the monkey's perceptual outcome (figure 35.1*D*; de Lafuente and Romo, 2006). Recently, timescales of intrinsic fluctuations in spiking activity across areas were related to an analogous hierarchical ordering (Murray et al., 2014). These intrinsic timescales, measured with the autocorrelation function, revealed areal specialization for task-relevant computations. In particular, frontal areas exhibit much longer timescales (~200 ms) than somatosensory areas (~65 ms). Intermediate values were found in S2 (~150 ms). Future studies could help understand what underlying mechanisms contribute to the cortical areal hierarchy of these intrinsic timescales.

However, the construction of a perceptual decision-making process may involve circuits outside the cerebral cortex. Obvious candidates are the sensory thalamus and, particularly for the detection task, the ventral posterior nucleus (VPL). Neurons from the VPL behaved almost similarly to S1 during animals' task performance (Vázquez et al., 2012; Vázquez, Salinas, and Romo, 2013; Tauste et al., 2019). Interestingly, de Lafuente and Romo (2011) sought to determine other types of neurons not directly related to somatosensory processing during the detection task. Midbrain dopamine (DA) neurons were recorded to explore more about reward prediction, given the rich behavior during stimulus detection (hits and misses vs. correct rejections and false alarms). Some interesting features about reward were observed. Unexpectedly, these authors observed that DA neurons increased their firing rates as a function of the stimulus amplitude during the detection task when monkeys correctly detected its presence. Notoriously, when the subjects were instructed to communicate their decision (go cue signal), DA neurons modulated their firing activity according to the uncertainty associated with the perceptual judgment. In other words, the same go cue produced different DA responses according to the uncertainty level of a judgment made during stimulus-presence. This means that suprathreshold stimuli that are easy to detect elicit small DA responses. In contrast, stimulus-absent trials evoke large DA responses associated with the uncertainty after the go cue. For the subject in this task, it is impossible to differentiate between subthreshold stimulus-present trials and stimulus-absent trials. These results suggest that DA responses are not modulated by the sensory intensity but rather to the perceived intensity of the stimulus. This is in concordance with the fact that DA latency responses are much longer than those from the somatosensory areas, but they closely match the onset of MPC neurons (de Lafuente and Romo, 2012). Hence, DA neurons code not only the reward prediction but also the subjective sensory experience and uncertainty emerging internally from perceptual decisions in the detection task (de Lafuente and Romo, 2011; Sarno et al., 2017). These results show that cortical and subcortical structures encode several components of the detection task and should urge the neuroscience community to investigate the role of DA neurons beyond reward prediction (Romo and Schultz, 1990; Schultz, 1998).

Constructing Decision-Making across Cortex during Sensory Discrimination

Two important perceptual processes are impossible to study in the sensory-detection task. The first is the mechanism to store in working memory a previously transformed and encoded sensory input. This mnemonic process (Rossi-Pool, Vergara, and Romo, 2018), associated with an internal representation of the stimulus, cannot be addressed with the detection task. Another important missing step is the comparison of the current sensory input to a sensory referent, which could have been stored in working memory or in long-term memory. To understand the value of sensory transformation, working memory, and comparison in the generation of perceptual decision-making, Romo and colleagues (Hernández et al., 1997) designed a behavioral task in which monkeys were trained to discriminate (compare) the frequency of two vibratory stimuli applied sequentially to one fingertip. Monkeys had to indicate whether the frequency of the comparison stimulus (f2) was lower or higher (f2 < f1 or f2 > f1) than the frequency of a base stimulus (f1) that was stored in working memory during a fixed delay period (figure 35.2*A*). Furthermore, the key condition for a real discrimination is to vary the first stimulus frequency (f1) in each trial, such that each f1 value is followed by a higher or a lower comparison frequency (f2). Notice that these are scalar analog quantities on which the discrimination performance must be based.

As described previously for the detection task (figure 35.1*C*), neurons from several cortical areas were recorded during the discrimination task (Hernández et al., 2000, 2010; Hernández, Zainos, and Romo, 2002; Romo et al., 1999, 2002; Romo, Hernández, and Zainos, 2004; Salinas et al., 2000). Neurons in S1 respond with a fine temporal structure of spike trains, representing f1 and f2. In general, mean firing rate responses increase monotonically as a function of the increasing stimulus frequency. Thus, the S1 responses could be described reasonably well as a linear function of the stimulus frequency. In this model, coefficient a1 is the slope of the activity frequency function and is a measure of how strongly a neuron is driven by changes of f1 frequency (*top*, formula, figure 35.2*B*). Notably, S1 neurons exhibit only positive slope values (a1 > 0; *green dots*, figure 35.2*B*). The higher the stimulation frequency, the higher the firing rate of the response. Analogously, during f2, S1 neurons are also modulated as a function of f2, with positive linear functions (a2 > 0; *red dots*, figure 35.2*B*). Additionally, after the end of f1, S1 neurons almost immediately cease coding f1. This means that during the delay period between f1 and f2, no stimulus-modulation responses are found (figure 35.2*B*). Hence, S1 neurons code the stimulus quantities, f1 and f2, only during the stimulus periods in this task. Using this simple decoding method, in figure 35.2*B* we show the slope distributions derived from neural responses recorded in several cortical areas, from three different time intervals: f1, the period between f1 and f2, f2, and f2 > f1 or f2 < f1.

Notably, areas downstream from S1 show more complex and heterogenous responses. In addition to heterogeneity among responses, single neurons by themselves display complex dynamics often attributable to more than one task component (mixed selectivity; Rigotti et al., 2013). In particular, neurons from S2 exhibit in their activity a disappearance of the phase lock responses observed in S1 neurons. Even if there is strong evidence that S2 is directly driven by S1, a clear sensory transformation was recognized between these two areas (Salinas et al., 2000). Another important difference is that neurons in S2 have both positive and negative slopes during f1 (*green dots*, figure 35.2*B*). In negative neurons, the firing rate decreases as a function of the increasing stimulus frequency in an approximately linear manner. Except for M1, all areas downstream from S1 (S2, VPC, prefrontal cortex (PFC), DPC and MPC; *green dots*, figure 35.2*B*) show comparable monotonic responses and similar proportions of positive and negative slopes during f1 presentation (figure 35.2*B*). Importantly, analogous complementary populations of positive and negative tuning have also been observed in another task (Rossi-Pool et al., 2016). Such equalization may be useful for subtracting common noise and increasing coding efficiency (Carnevale et al., 2013; Romo et al., 2003).

Importantly, several S2 neurons display f1-dependent responses that continue for some hundreds of

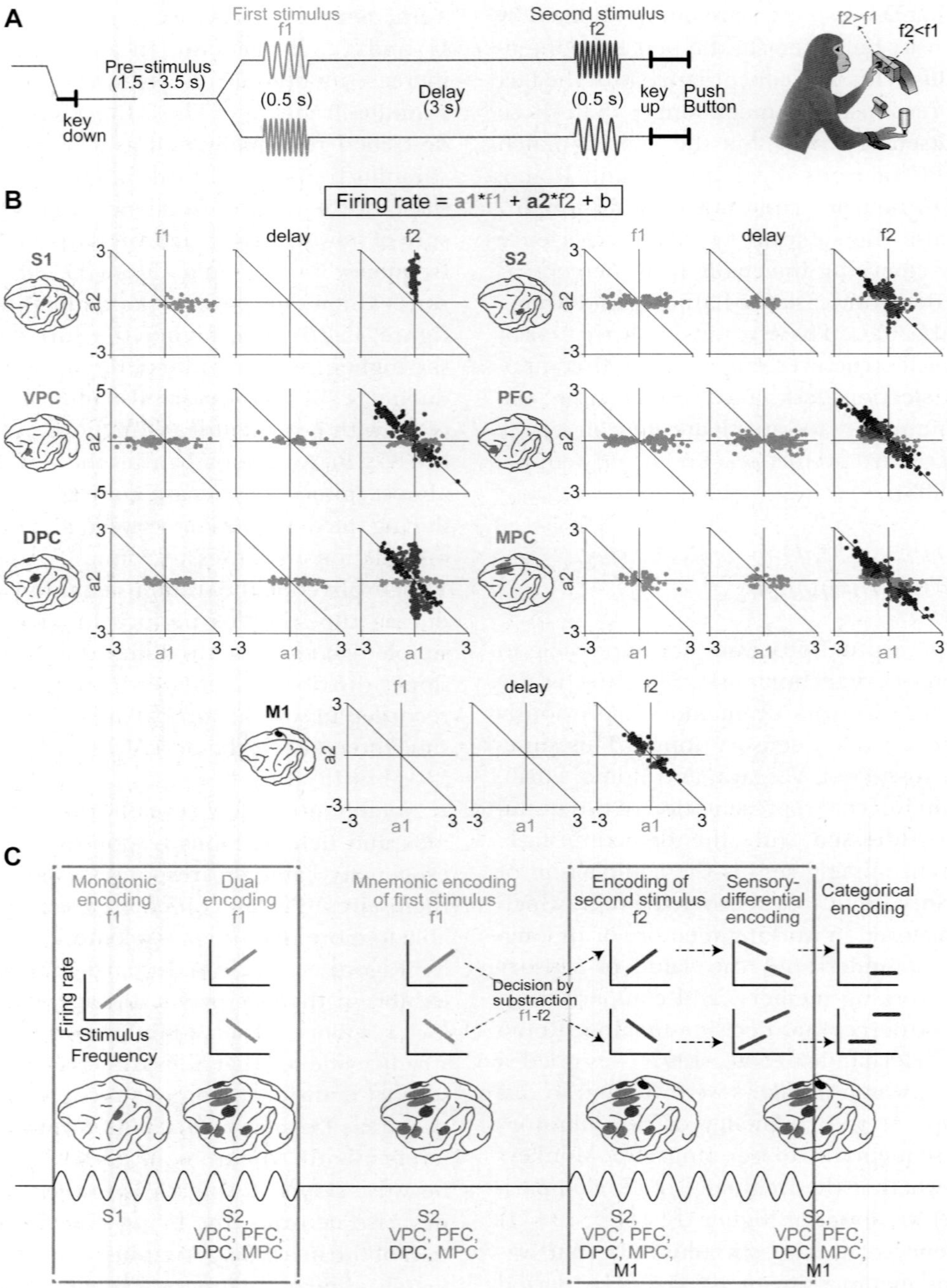

FIGURE 35.2 Vibrotactile discrimination task. *A*, Sequence of events during a frequency discrimination task. As for the detection task (figure 35.1*A*), a trial began when the mechanical probe indented the fingertip of the left restrained hand, triggering the monkey to place its right free hand on an immovable key (key down [kd]). After a variable prestimulus period (1.5–3.5), the probe oscillates vertically at the base stimulus frequency (f1, 10–34 Hz, 0.5 s), and following a delay of 3 s, a second mechanical vibration is delivered at the comparison frequency (f2, 10–34 Hz, 0.5 s). At the end of the second stimulus, the monkey releases the key (key up) and presses either the medial or lateral push-buttons to indicate whether the comparison frequency was lower (f2<f1) or higher (f2>f1) than the base frequency. *B*, Single-neuron dynamics across cortical areas during a discrimination task. For each neuron, responses were fitted to the equation: firing rate = a1 × f1 + a2 × f2 + b, where f1 is the base stimulus frequency, f2 is the comparison stimulus frequency, and a1, a2, and b are coefficients. Each data point corresponds to one neuron with at least one significant coefficient (a1 ≠ 0, a2 ≠ 0, or both are different from zero, p < 0.05) evaluated by 200 ms bins. Each panel shows the highest coefficients from each significant neuron coding during three different epochs: the first stimulus period (f1, 0.5 s), the delay between f1 and f2 (delay, 3 s), and the second stimulus period (f2, 0.5 s). Green and red circles correspond to those neurons

milliseconds immediately after the end of f1, into the working memory delay between f1 and f2 (*green dots*, figure 35.2*B*). Remarkably, no persistent f1 coding is observed in S2 neurons. In contrast, neurons in the frontal lobe (VPC, PFC, DPC, and MPC; *green dots*, figure 35.2*B*) carry information about f1 into the whole delay period between f1 and f2. Some neurons convey information during the early part, others only during the late part, and still others persistently throughout the entire delay period. This means that the mnemonic representation of f1 is not static, in the sense that the intensity of the coding activity varies across the delay. A comparison across areas shows a considerable overlap between the working memory coding, possibly reflecting interconnectivity between them.

Upon the presentation of f2, neuronal responses in areas downstream from S1 are no longer defined by one variable (f1) but by two (both f1 and f2). Therefore, the potential repertoire of responses increases greatly, and analysis of the neural data should take this into account. To quantify the simultaneous dependence of the firing rate on f1 and f2, a first-order approximation to a bilinear function of f1 and f2 was used (Romo et al., 2002). That is, neuronal firing rates were modeled as linear functions of both f1 and f2: firing rate = a1.f1 + a2.f2 + b, where b is a constant, and a1 and a2 are the coefficients that measure how strongly f1 and f2 modulate the neuron response. Over the course of the comparison period, a1 and a2 might change, indicating mixed selectivity. The right panels of each cortical area in figure 35.2*B* summarize the population coding during the comparison period. Except for S1, all the other cortical areas contain neurons with four different types of coding. Green dots correspond to neurons that had only significant f1 dependence, and red points correspond to neurons that have a significant f2 coding. Additionally, blue dots correspond to point cluster along the diagonal a2 = −a1, meaning that during that period the neurons respond as functions of the difference between f2 and f1. The neurons that encode this difference indicate the discrimination result, and they are interpreted as categorical decision coding. Additionally, gray dots indicate sensory differential encoding (intermediate decision coding), with significant but not equal values for a1 and a2. Notably, during the first 100 ms of f2, the activity of several neurons across cortical areas (except S1) was mainly a function of f1 frequency (*green dots*). This finding is consistent with a memory recall of the base stimulus frequency (f1). Further, some neurons initially code f1 or f2 frequencies and later code whether f2 is greater than f1 or f2 is less than f1 (*blue and gray dots*, figure 35.2*B*). Notoriously, during f2 presentation, the coding dynamics of S2, VPC, PFC, DPC, MPC, and M1 neurons are undistinguishable between them. Actually, just as in the neural representation of the sensory stimuli, decision-coding neurons were represented by two complementary (positive and negative) populations.

In brief, the decision of which of two stimuli has the higher vibration frequency engages multiple cortical areas on the parietal and frontal lobes (figure 35.2*B*). In figure 35.2*C*, representations of the monkey brain illustrate cortical areas with discrimination task activity. The vibrotactile information arrives to S1, assuming in this model that this is the initial representation of

whose responses depend on f1 only (a1 ≠ 0, a2 = 0; *dots on the abscissa axis*) or on f2 only (a1 = 0, a2 ≠ 0; *red dots on the ordinate axis*), respectively. Gray circles correspond to neurons with both significant coefficients of opposite signs (a1 > 0 and a2 < 0; a1 < 0 and a2 > 0) but significantly different magnitudes (a1 ≠ −a2); these responses are classified as partially differential neurons (*dots between the diagonal and the ordinate or abscissa axis*). Blue circles correspond to neurons with both significant coefficients (a1 ≠ 0 and a2 ≠ 0) but opposite signs and statistically equal magnitudes (a1 = −a2); these responses encode f2–f1 in a categorical or fully differential manner (*dots on the diagonal*). *C*, Schematic representation of the coding engagement across cortex during a flutter discrimination task. The macaque brain diagrams depict the activated cortical areas during the three main periods of the task: f1, working memory delay, and f2 (*bottom traces, not drawn to scale*). The primary somatosensory cortex (S1) encodes f1 only through positive monotonic responses (see also panel *B*). This sensory information is distributed to other cortical areas and transformed in such a way that opposite representations (dual encoding) arise in secondary somatosensory cortex (S2), ventral, dorsal, and medial premotor cortices (VPC, DPC, and MPC, respectively) and prefrontal cortex (PFC). Dual representations of the first stimulus (f1) are encoded during the delay period (mnemonic representation) by the sustained activity on the parietal and frontal lobe cortical areas. After the working memory period, the frequency of the second stimulus (f2) is encoded in all recorded areas of the frontal and parietal lobes, including the primary motor cortex (M1). Importantly, at the end of the second stimulus, neurons with decision-related activity in frontal and parietal cortical areas reflected the difference in frequency between the two stimuli frequencies (f2–f1). This decision-related activity could arise by subtracting the firing rates of neurons encoding f1 and f2 with opposite tuning. Note that with the exception of S1, neurons with opposite tuning are observed across all recorded areas. Adapted from Romo and de Lafuente (2013); Romo and Salinas (2003). (See color plate 39.)

the sensory input. Hence, S1 neurons display a faithful response to f1 and f2. Note that somatosensory neurons encode the stimulus frequencies (f1 and f2) only through positive monotonic responses (*red spot*, figure 35.2*C*). This sensory information is distributed to downstream cortical areas and transformed in such a way that opposite representation (positive and negative encoding) arises in S2 (pink spot), VPC (orange spot), PFC (green spot), DPC (violet spot), and MPC (*cyan spot*, figure 35.2*C*). Notably, this dual encoding emerges in S2 neurons earlier (~45ms) than in other frontal areas (~100 ms, VPC, PFC, DPC, and MPC; Hernández et al., 2010). These findings suggest that S2 may be involved in this sensory transformation to further distribute this processed information to downstream frontal areas. Importantly, during the delay between the two stimuli (f1 and f2), information of f1 is retained (mnemonic coding) by the sustained activity of frontal lobe areas. During this period the dual representation is preserved. Even if S2 neurons encode during the delay between f1 and f2, they are only engaged for some hundreds of milliseconds immediately after the end of f1. Notably, after the delay period the information of f2 is encoded in all recorded areas of the frontal and parietal lobes, including M1. Again, S1 neurons are the only ones to exhibit solely positive encoding. Remarkably, some M1 neurons encoded sensory information on which the decision is based (f1 and f2). Importantly, neurons with decision-related activity, reflecting the comparison of f1 against f2, emerge in all cortical areas except S1. This decision activity is coded as the difference in frequency between the two stimuli (sensory differential encoding), and they are different from the categorical, all-or-none motor signal. This categorical encoding appeared at the end of the f2 period in all the cortical areas (S2, VPC, PFC, DPC, MPC, and M1) except S1.

A general picture from figure 35.2 exhibits that the comparison process underlying the decision-making in this task is widely distributed across cortex. Excluding S1, decision-related activity could be recorded from S2, MPC, VPC, PFC, DPC, and M1 neurons. Notoriously, decision-encoding latency appeared almost simultaneously in all these cortical areas (Hernández et al., 2010). These results are evidence that decision-making is a distributed process that arises not from serial processing but from the cooperative activity of many cortical areas. In brief, in the frequency discrimination task S1 neurons represent the stimulus frequency both in the phase lock and mean firing-rate responses. Downstream from S1, neurons encode the stimulus frequencies in their firing rates, employing a dual manner, with positive and negative monotonic responses. This coding scheme is used as a subtraction mechanism to preserve information about f1 not only during the stimulus period, but also during the working memory and the comparison period to generate a neural computation consistent with the animal report (Hernández et al., 2010; Romo et al., 2003).

Do Frontal Lobe Circuits Code Motor Actions, Decision Processes, or Both?

Historically, premotor cortices have been associated with motor commands (Caminiti et al., 1991; Crammond and Kalaska, 2000; Wise and Mauritz, 1985). However, several studies have shown that these areas could be involved in more complex cognitive processes, such as sensorimotor associations, the conversion of working memory to motor sequences, the timing of sequential motor actions, and decision-making (Hernández et al., 2010; Ohbayashi, Ohki, and Miyashita, 2003; Ohbayashi, Picard, and Strick, 2016; Tanji, 1994; Thura and Cisek, 2014). The results presented above strongly question the limited premotor status assigned earlier to these cortical areas and rather support a role in more complex processes, such as perception, working memory, and decision-making. Considering the evidence presented here, the whole premotor network is actually associative and therefore not necessarily limited to generating motor actions.

Additionally, many frontal area neurons were tested in a task variant in which the same tactile stimuli were presented (with the same event sequence and identical frequency pairs), but an additional visual cue indicated the correct push-button to press to the monkeys. Under this control task, the animals should ignore (we assume) the tactile stimuli and easily answer according to the visual instruction. Note that the flutter stimuli and arm movements are the same as during the regular tasks, but the cognitive events—stimulus transformation, working memory, and stimulus comparison—were not. Remarkably, under this condition the coding and decision activity elicited during the active task disappeared (Hernández, Zainos, and Romo, 2002; Lemus et al., 2007; Ponce-Alvarez et al., 2012; Romo, Hernández, and Zainos, 2004; Rossi-Pool et al., 2016, 2017). Hence, the decision coding observed during the discrimination task is not simply related to a motor action but represents an associative process fully gated by the behavioral context of the task.

Another notable feature of premotor cortices is that they represent sensory inputs with an abstract code format. Recently, DPC responses were studied in a vibrotactile temporal pattern discrimination task, where monkeys were trained to discriminate whether the temporal structure of two stimuli of equal mean frequency was the same or different (Rossi-Pool et al., 2016). Subjects performed the task in blocks of trials in which the two patterns had a fixed mean frequency. Surprisingly, DPC neurons were able to respond independently of the mean stimulus frequency in such a way that they coded the identity of the stimulus pattern in a frequency-invariant and thus somewhat abstract format. Additionally, in a bimodal discrimination task, monkeys were trained to discriminate the frequency of either two vibrotactile flutter stimuli, two acoustic flutter stimuli, or two crossmodal stimuli. Notably, monkeys were able to store both tactile and acoustic information in working memory using the same code (Vergara et al., 2016). Remarkably, MPC neurons coded parametrically both acoustic and tactile information during working memory. This means that MPC neurons use the same abstract and amodal (bimodal) code for maintaining information during working memory.

Surprisingly, under the control task condition described above, M1 decision neurons either stopped responding altogether or fired at higher rates compared with the baseline but with the same intensity for both arm movements (Romo et al., 1997; Salinas and Romo, 1998). Overall, about three-quarters of the M1 decision neurons tested during visually guided movements for push-button presses stopped responding differentially. However, a motor plan is needed in the context of the vibrotactile tasks. One possibility is that during the control task, motor planning is maintained in other circuits—for example, in the spinal cord (Prut and Fetz, 1999). In this case weak signals sent from the cortical lobe circuits could activate the execution of the motor plan during the control tasks. This conjecture is supported by the fact that frontal lobe circuits are known to send their projections to the motor apparatus of the spinal cord (Dum and Strick, 1991; He, Dum, and Strick, 1993). However, few studies have explored the functional role of frontal lobe neurons that project to the spinal cord during cognitive tasks (Kraskov et al., 2009). Thus, whether spinal motor circuits receive an instruction signal to execute the motor plan in the vibrotactile and control tasks is an open question.

Provoking Percepts by Inserting Artificial Information in Circuits of the Cerebral Cortex

Even if the results presented above reveal close associations among the neural activity of a particular cortical area and sensory inputs and decision-making, they are not definitive proof that this activity provides sufficient basis for a specific cognitive function. It is essential to verify the impact of such activity on perception and subsequent behavior. Intracortical microstimulation has provided the most compelling evidence to establish a causal link between the responses of localized neurons and specific cognitive functions. Several studies in monkeys have revealed that it is possible to evoke behavioral responses activating sensory neurons (Britten and van Wezel, 1998; Graziano, Taylor, and Moore, 2002; de Lafuente and Romo, 2005; Murphey and Maunsell, 2007; Romo et al., 1998, 2000; Salzman, Britten, and Newsome, 1990). Recently, these results were extended to humans: tetraplegic patients were able to restore lost tactile and proprioceptive feeling by artificially activating the appropriate cortical neurons (Salas et al., 2018). In this chapter we discuss the results obtained with microstimulation for each of the two tasks discussed above.

To further study the association between MPC activity and the behavioral report during the stimulus-detection task, de Lafuente and Romo (2005) artificially microstimulated the MPC circuit of the frontal lobe. The authors injected a weak electric current into the cortex in both stimulus-present and stimulus-absent trials (*left panel*, figure 35.3*A*). In figure 35.3*A*, the resulting psychometric curve is shown (*gray line*), including the mechanical-only curve (*dark line*) for comparison. Even if the monkeys increase the probability of detecting the vibratory stimuli with microstimulation (>0 μm), the number of false alarms increases, too (0 μm). These results are consistent with the idea that MPC activity is involved in perceptual judgments. Additionally, if the mechanical stimuli are substituted by electric currents of varying strengths, the artificial activation of MPC gives rise to a detection curve that resembles that obtained during the detection task (*right panel*, figure 35.3*A*). Hence, detection behavior could be triggered with purely electrical stimuli (*gray line*) resembling that obtained with mechanical stimulation to the skin (*dark line*, figure 35.3*A*). These results give further evidence that psychometric performance based on the microstimulation of MPC neurons mimics that based on the vibrotactile stimuli delivered to the skin during the detection task.

Despite the fact that artificial injected currents elicit psychometric curves analogous to those obtained with mechanical stimulation, it is uncertain whether they evoke the same somatosensory sensation. Another feasible hypothesis is that injected current activates neurons associated with a task rule, such as "stimulus present." Under this hypothesis, increasing the microstimulation current could increase the number of neurons that code the stimulus-present detection. Note that MPC responses during a detection task are much more homogenous than during a discrimination task. Hence, in a more complex task the microstimulation approach appears unlikely in frontal areas because they show high heterogeneity in their neuronal responses.

Based on the hypothesis that S1 neurons are necessary to represent and transmit sensory information to downstream areas, Romo and colleagues microstimulated S1 neurons with receptive fields during the discrimination task (Romo et al., 1998, 2000). In the first step, the authors substituted the comparison stimulus with microstimulation in half of the trials (f1 mechanical pulse and f2 pulses substituting f2, *left top panel,* figure 35.3*B*). Artificial stimuli consisted of periodic current bursts injected at the same comparison frequencies as the mechanical stimuli (mechanical pulses during f1 and f2, *left top panel,* figure 35.3*B*). Notably, the subjects were able to discriminate the mechanical (f1) and electrical (f2) stimulus with performance profiles that resembled those obtained with only tactile stimuli (*right top panel,* figure 35.3*B*). Therefore, the artificially induced psychophysical performance could produce sensations in S1 that closely mimic the natural vibrotactile stimuli.

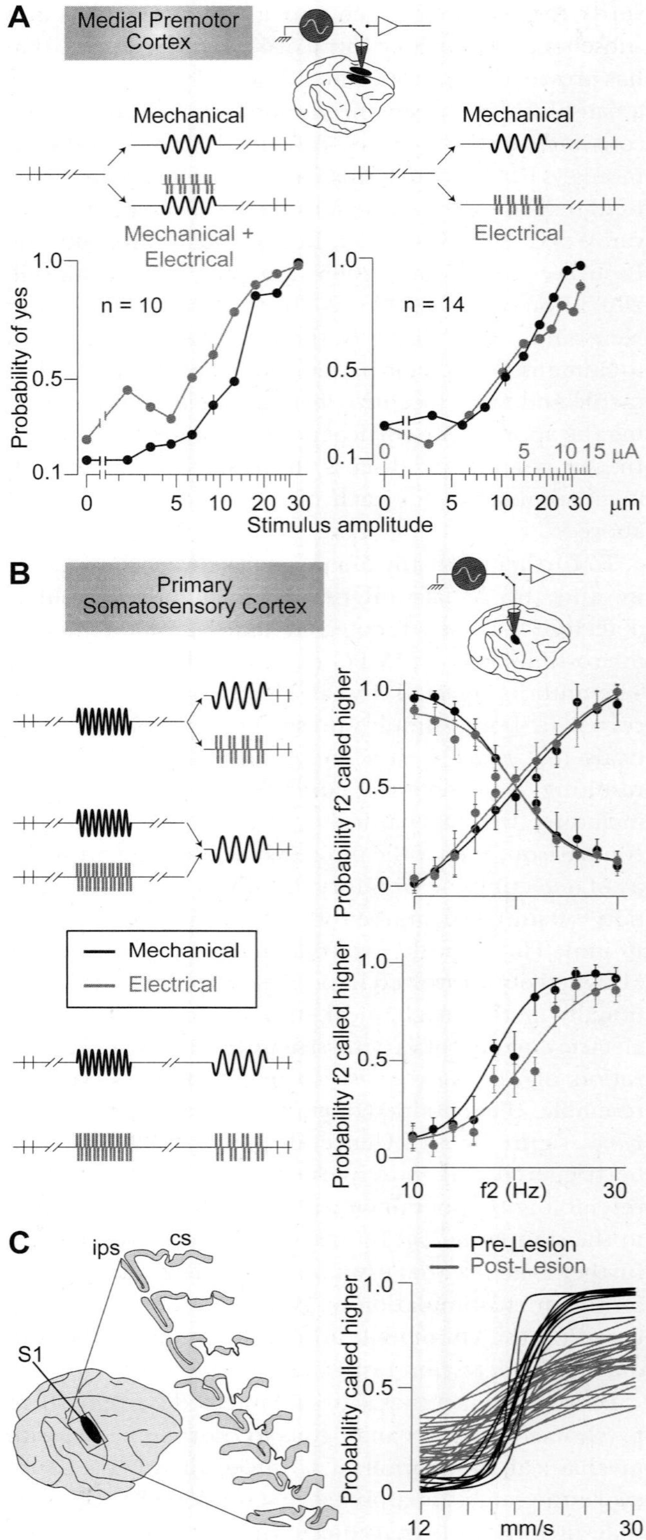

FIGURE 35.3 Psychophysical performance based on cortical microstimulation and after a cortical lesion. *A,* Detection curves during electrical microstimulation of MPC neurons. *Left panel,* Mean detection curves for mechanical stimuli (*black traces*) and for mechanical-plus-electrical stimuli (*gray traces*). Trials were randomly interleaved. *Right panel,* Mean detection curves for purely mechanical (*black traces*) and purely electrical stimuli (*gray traces*). Trials were randomly interleaved. Small vertical lines indicate SEM. n = number of sessions (each session consists of 10 repetitions of each kind of stimuli). *B,* Frequency discrimination task performed by mechanical stimulation of the skin or by direct electrical microstimulation of S1 neurons. In half of the trials, the monkeys compared two mechanical vibrations; in the other half, one or both stimuli pulses were replaced by two biphasic current pulses microinjected into clusters of quickly adapting neurons in area 3b. The mechanical and electrical trials were interleaved, and frequencies always change from trial to trial. *Right panels,* Show the psychophysical performances using the four protocols illustrated in the left panels. Dark and gray circles indicate mechanical and electrical performance, respectively; continuous lines are fits to the data points. The monkey's performance was practically the same with natural and electrical stimuli. *C,* Psychophysical performance after a lesion in S1 (*left panel;* IPS, intraparietal sulcus; CS, central sulcus). In this task the animal categorized a tactile moving stimulus across the skin of one fingertip as lower (12 mm/s) or higher (30 mm/s) by pressing with the free hand one of two push-buttons, as in the detection and discrimination task. *Left panel,* The top view of the brain with a black spot marking the lesion area, together with histological serial sections. *Right panel,* After the S1 lesion, categorization decreased at chance levels (*gray lines*), compared to the prelesion performance (*black traces*). Panel (*A*) was adapted from de Lafuente and Romo (2005); panel (*B*) was adapted from Romo et al. (1998, 2000); panel (*C*) was adapted from Zainos et al. (1997).

Moreover, in experiments in which f1 was substituted with electric injected current (*left top panel,* figure 35.3*B*), the monkey's psychometric curve was indistinguishable from that observed with only tactile stimuli (*right top panel,* figure 35.3*B*). This means that an artificial stimulus (f1) injected in S1 could be stored and recalled in working memory for use during the comparison period (f2) with roughly the same fidelity. Further, monkeys were able to execute the whole task (*lower left panel,* figure 35.3*B*), with little degradation in performance, using purely artificial (f1 and f2) stimuli (*right lower panel,* figure 35.3*B*). These results suggest that the S1 circuit distributes the representation of the flutter stimuli to more central structures to solve the discrimination task. In other words, neurons in S1 are sufficient to trigger all the cognitive processes of the discrimination task.

The results obtained in another tactile task support this interpretation (Zainos et al., 1997). In this task, the animal categorized the stimulus speed across the skin of one fingertip as low or high. However, after a lesion of S1 (*black spot, left,* on the brain figurine and serial sections in panel, figure 35.3*C*), the animal psychophysical performance decreased to chance level (*gray traces, right,* figure 35.3*C*). The categorization performance after the S1 lesion was followed for 60 daily sessions, but animals were unable to recuperate this capacity. Importantly, the reaction and movement times were not affected by the S1 lesion. This would indicate that animals detected the moving stimuli but were unable to extract sensory information for categorization. The authors concluded that S1 is essential for tactile perception. In other words, downstream areas require the S1 circuit for constructing perceptual decision-making.

Population Coding Approach during Perceptual Detection and Discrimination

Frontal neurons exhibit a baffling heterogeneity among their neuronal responses during the vibrotactile flutter tasks (de Lafuente and Romo, 2006; Romo et al., 1999, 2003). Historically, this heterogeneity has often been neglected, preselecting cells based on particular criteria. Actually, as we discussed above, most neurons in higher cortical areas typically encode several task parameters and therefore exhibit what has been denominated mixed selectivity (Rigotti et al., 2013). A reasonable approach to handle this heterogeneity and mixed selectivity is to use dimensionality reduction methods; the resulting responses describe population activity in a compact format and could convey clearer, hidden signals. The relevance of this approach is well supported by recent works showing the potentiality of these methods to decode population responses that cannot be inferred from single units (Chaisangmongkon et al., 2017; Mante et al., 2013; Rossi-Pool et al., 2017). In this section we focus on a couple of recent studies that apply this approach to population responses recorded in the frontal lobe during detection (Carnevale et al., 2015) and discrimination tasks (Barak, Tsodyks, and Romo, 2010; Kobak et al., 2016; Murray et al., 2017).

During the detection task (figure 35.1*A*), monkeys are able to predict neither the timing nor the presence of the stimulus. Carnevale et al. (2015) showed how monkeys exploit previous knowledge to cope with the uncertainty of stimulus arrival over time. Using a template-matching algorithm, the neural correlates of false-alarm events could be identified. Notably, neural correlates of false-alarm events occurred during the possible stimulation window. Hence, there is a neural mechanism by which previous information is intrinsically coded in the dynamics of the premotor neural population (figure 35.4*A*). This means that the optimal response criterion employed by the network is modulated according to the learned temporal structure of the task. In other words, the strength of the sensory evidence required to produce a stimulus-present response is modulated throughout the detection task. The authors proposed that this mechanism could be dynamically implemented by a separatrix, in the population neural space, dividing the two possible responses (yes and no attractor), stimulus-present and stimulus-absent (figure 35.4*A*).

Focusing on the discrimination task, single-neuron activity in frontal areas during working memory is heterogenous and strongly dynamic (Brody et al., 2003; Romo et al., 1999), raising questions about the stability and purpose of this representation. Despite this temporal dynamics, there is a population-level representation of the first (f1) stimulus frequency that is maintained stably during the delay between f1 and f2 (Barak et al., 2010; Murray et al., 2017). The high-dimensional state space of PFC population activity contains a low-dimensional subspace in which the stimulus representation is stable during working memory. Notably, this population coding is modulated in an approximately linear manner (figure 35.4*B*, frequency component; Kobak et al., 2016; Murray et al., 2017). These results fit well with the idea that parametric monotonic coding is used by the PFC population to maintain information during working memory. Additionally, a population decision component appeared during the comparison period (figure 35.4*B*). The population decision signals that correspond to the same answer (f1 > f2, *dashed*

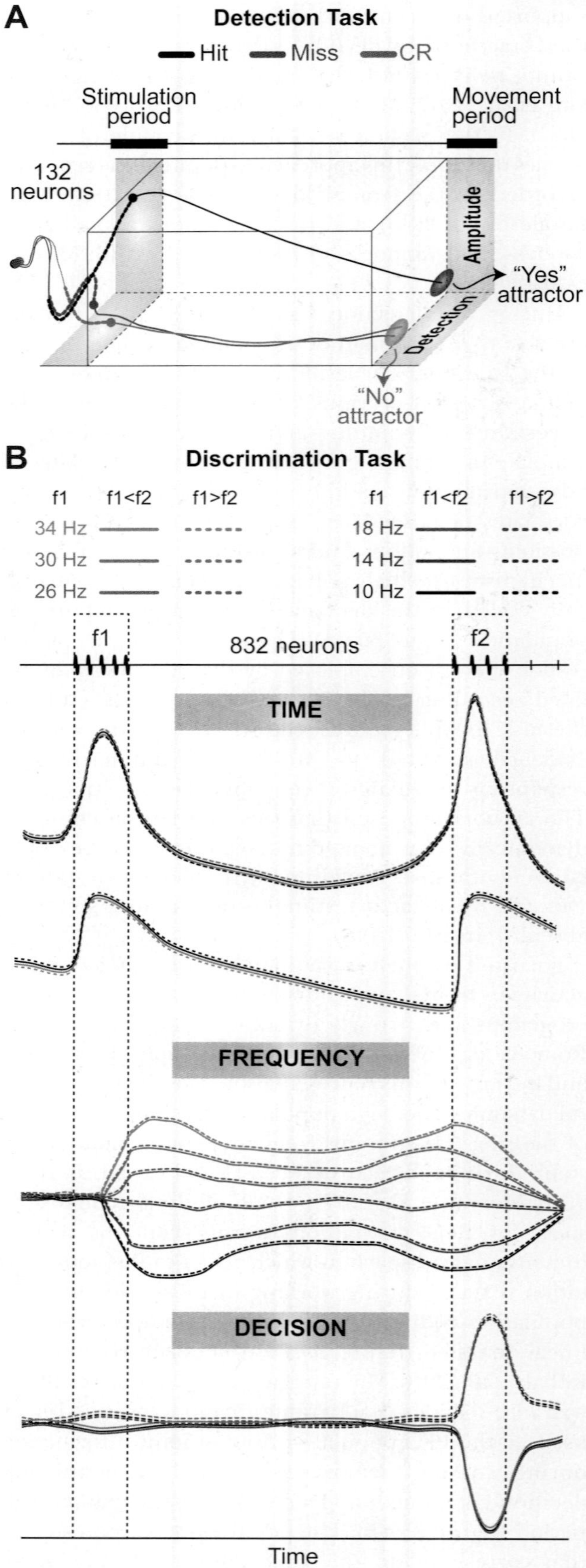

lines; f2 > f1, *solid lines*) closely overlapped. Notably, the population decision component emerged with a latency analogous to the single-neurons choice probability.

In Kobak et al. (2017), the authors propose a new methodological approach to split the contribution of different task parameters and time during the discrimination task. In particular, they found that purely temporal signals explained a high percentage of the total response variance (~70%, figure 35.4*B*, first two temporal components), suggesting that they are heavily involved in task execution. The first two components, shown in figure 35.4*B*, are involved in different aspects of task execution: sensory inputs and ramping activity during the delay. In addition, there is a large disparity between the whole-task variance explained by pure temporal signals versus first stimulus (f1) and decision components (f2 > f1 or f2 < f1). Notably, similar differences have been found in at least four other tasks (Kobak et al., 2016; Rossi-Pool et al., 2017, 2019), suggesting that this is a general feature. We propose that these temporal signals could be understood as a substrate necessary to provide an infrastructure on which the coding responses can develop, combine, and reach a decision during these tasks (Rossi-Pool et al., 2019).

Concluding Remarks

The somatosensory system is a model suitable for studying the neural mechanism involved in perceptual decision-making. This model has produced meaningful experimental and theoretical results for understanding processes ranging from detection to decision-making. In this chapter we showed evidence of how a sensory stimulus is coded across cortex and how

FIGURE 35.4 Population coding during vibrotactile detection and discrimination task. *A*, Three-dimensional population dynamics during the detection task. Traces correspond to the average neural trajectories from neurons recorded in the premotor cortex (VPC, DPC, and MPC) during hits, misses, and correct rejection trials (*upper legend*). All the trajectories were obtained by projecting the population activity onto two task-related axes or subspaces as a function of time (*x-axis*); one corresponds to the stimulus amplitude (*z-axis*) and the second to the decision report (stimulus detection, *y-axis*). *B*, Population dynamics of PFC neurons during a flutter discrimination task. The traces correspond to the projection of all the neural activity onto those subspaces that capture the highest variance related to the task parameters of time, frequency, and decision. The population activity was sorted by f1-decision identity (12 conditions, *upper legend*), and the respective neural trajectories onto each subspace were defined via demixed principal component analysis. Panel (*A*) was adapted from Carnevale et al. (2015); panel (*B*) was adapted from Kobak et al. (2016).

such representation relates to sensation, memory, and decision-making. This reveals that neuronal computations across cortex have provided an extended panorama of the neural activity engaged in both detection and discrimination tasks. Remarkably, a large number of cortical areas of the parietal and frontal lobe are engaged during both tasks. Specifically, S1 is essentially sensory, faithfully representing the information arriving from tactile receptive fields. The phase-lock stimulus representation is transformed by areas downstream from S1 into a simple firing-rate code, with a dual representation (positive and negative encoding) resulting in a subtraction operation consistent with the animal decision report. Thus, the sensory information is progressively converted into a subject's perceptual decision report. In addition to the contribution of several cortical areas, subcortical structures are also needed to generate a decision report. This could be a general processing principle not only for the tactile tasks discussed here but also for the other sensory modalities requiring the comparison between past and current sensory inputs (Lemus, Hernández, and Romo, 2009a, 2009b, 2010; Vergara et al., 2016).

The decision processes discussed here seem to evolve as if they were part of a network dynamic plan. Notably, this plan could be dynamically changed or reconfigured according to experience. In fact, the stronger decision coding signals are found in frontal lobe areas: PFC, VPC, MPC, and DPC. These results fit well with the interpretation that these circuits encode not only the planning of motor actions but also the information on which the motor action is based (Carpenter, Georgopoulos, and Pellizzer, 1999; Hoshi and Tanji, 2004; Ohbayashi, Ohki, and Miyashita, 2003; Shima et al., 2007).

To conclude, this chapter shows how distinct cortical circuits contribute to perceptual detection and discrimination. However, future experiments are needed to reveal how neuronal populations of distinct brain areas join the efforts, in real time, to solve perceptual decision-making in the tasks discussed here, as well as other modality tasks.

Acknowledgments

We thank H. Diaz, M. Alvarez, and A. Zainos for technical assistance. The research of Ranulfo Romo was partially supported by the Dirección General de Asuntos del Personal Académico de la Universidad Nacional Autónoma de México (UNAM; PAPIIT-IN202716 and PAPIIT-IN210819) and Consejo Nacional de Ciencia y Tecnología (CONACYT-240892).

REFERENCES

Barak, O., Tsodyks, M., & Romo, R. (2010). Neuronal population coding of parametric working memory. *Journal of Neuroscience, 30*(28), 9424–9430. doi:10.1523/JNEUROSCI.1875-10.2010

Britten, K. H., & van Wezel, R. J. (1998). Electrical microstimulation of cortical area MST biases heading perception in monkeys. *Nature Neuroscience, 1*(1), 59–63. doi:10.1038/259

Brody, C. D., Hernández, A., Zainos, A., & Romo, R. (2003). Timing and neural encoding of somatosensory parametric working memory in macaque prefrontal cortex. *Cerebral Cortex, 13*(11), 1196–1207. doi: 10.1093/cercor/bhg100

Caminiti, R., Johnson, P. B., Galli, C., Ferraina, S., & Burnod, Y. (1991). Making arm movements within different parts of space: The premotor and motor cortical representation of a coordinate system for reaching to visual targets. *Journal of Neuroscience, 11*(5), 1182–1197. doi:10.1523/JNEUROSCI.11-05-01182.1991

Carnevale, F., de Lafuente, V., Romo, R., Barak, O., & Parga, N. (2015). Dynamic control of response criterion in premotor cortex during perceptual detection under temporal uncertainty. *Neuron, 86*(4), 1067–1077. doi:10.1016/j.neuron.2015.04.014

Carnevale, F., de Lafuente, V., Romo, R., & Parga, N. (2013). An optimal decision population code that accounts for correlated variability unambiguously predicts a subject's choice. *Neuron, 80*(6), 1532–1543. doi:10.1016/j.neuron.2013.09.023

Carpenter, A. F., Georgopoulos, A. P., & Pellizzer, G. (1999). Motor cortical encoding of serial order in a context-recall task. *Science, 283*(5408), 1752–1757. doi:10.1126/science.283.5408.1752

Chaisangmongkon, W., Swaminathan, S. K., Freedman, D. J., & Wang, X. J. (2017). Computing by robust transience: How the fronto-parietal network performs sequential, category-based decisions. *Neuron, 93*(6), 1504–1517. e1504. doi:10.1016/j.neuron.2017.03.002

Crammond, D. J., & Kalaska, J. F. (2000). Prior information in motor and premotor cortex: Activity during the delay period and effect on pre-movement activity. *Journal of Neurophysiology, 84*(2), 986–1005. doi:10.1152/jn.2000.84.2.986

de Lafuente, V., & Romo, R. (2005). Neuronal correlates of subjective sensory experience. *Nature Neuroscience, 8*(12), 1698–1703. doi:10.1038/nn1587

de Lafuente, V., & Romo, R. (2006). Neural correlate of subjective sensory experience gradually builds up across cortical areas. *Proceedings of the National Academy of Sciences of the United States of America, 103*(39), 14266–14271. doi:10.1073/pnas.0605826103

de Lafuente, V., & Romo, R. (2011). Dopamine neurons code subjective sensory experience and uncertainty of perceptual decisions. *Proceedings of the National Academy of Sciences of the United States of America, 108*(49), 19767–19771. doi:10.1073/pnas.1117636108

de Lafuente, V., & Romo, R. (2012). Dopaminergic activity coincides with stimulus detection by the frontal lobe. *Neuroscience, 218*, 181–184. doi:10.1016/j.neuroscience.2012.05.026

Dum, R. P., & Strick, P. L. (1991). The origin of corticospinal projections from the premotor areas in the frontal lobe. *Journal of Neuroscience, 11*(3), 667–689. doi:10.1523/JNEUROSCI.11-03-00667.1991

Graziano, M. S., Taylor, C. S., & Moore, T. (2002). Complex movements evoked by microstimulation of precentral cortex. *Neuron, 34*(5), 841–851. doi:10.1016/S0896-6273(02)00698-0

Green, D. M., & Swets, J. A. (1966). *Signal detection theory and psychophysics.* New York: Wiley.

He, S. Q., Dum, R. P., & Strick, P. L. (1993). Topographic organization of corticospinal projections from the frontal lobe: Motor areas on the lateral surface of the hemisphere. *Journal of Neuroscience, 13*(3), 952–980. doi:10.1523/JNEUROSCI.13-03-00952.1993

Hernández, A., Nacher, V., Luna, R., Zainos, A., Lemus, L., Alvarez, M., Vázquez, Y., Camarillo, L., & Romo, R. (2010). Decoding a perceptual decision process across cortex. *Neuron, 66*(2), 300–314. doi:10.1016/j.neuron.2010.03.031

Hernández, A., Salinas, E., Garcia, R., & Romo, R. (1997). Discrimination in the sense of flutter: New psychophysical measurements in monkeys. *Journal of Neuroscience, 17*(16), 6391–6400. doi:10.1523/JNEUROSCI.17-16-06391.1997

Hernández, A., Zainos, A., & Romo, R. (2000). Neuronal correlates of sensory discrimination in the somatosensory cortex. *Proceedings of the National Academy of Sciences of the United States of America, 97*(11), 6191–6196. doi:10.1073/pnas.120018597

Hernández, A., Zainos, A., & Romo, R. (2002). Temporal evolution of a decision-making process in medial premotor cortex. *Neuron, 33*(6), 959–972. doi:10.1016/S0896-6273(02)00613-X

Hoshi, E., & Tanji, J. (2004). Differential roles of neuronal activity in the supplementary and presupplementary motor areas: From information retrieval to motor planning and execution. *Journal of Neurophysiology, 92*(6), 3482–3499. doi:10.1152/jn.00547.2004

Hubel, D. H., & Wiesel, T. N. (1962). Receptive fields, binocular interaction and functional architecture in the cat's visual cortex. *Journal of Physiology, 160,* 106–154. doi:10.1113/jphysiol.1962.sp006837

Kobak, D., Brendel, W., Constantinidis, C., Feierstein, C. E., Kepecs, A., Mainen, Z. F., Qi, X. L., Romo, R., Uchida, N., & Machens, C. K. (2016). Demixed principal component analysis of neural population data. *eLife, 5.* doi:10.7554/eLife.10989

Kraskov, A., Dancause, N., Quallo, M. M., Shepherd, S., & Lemon, R. N. (2009). Corticospinal neurons in macaque ventral premotor cortex with mirror properties: A potential mechanism for action suppression? *Neuron, 64*(6), 922–930. doi:10.1016/j.neuron.2009.12.010

Lemus, L., Hernández, A., Luna, R., Zainos, A., Nacher, V., & Romo, R. (2007). Neural correlates of a postponed decision report. *Proceedings of the National Academy of Sciences of the United States of America, 104*(43), 17174–17179. doi:10.1073/pnas.0707961104

Lemus, L., Hernández, A., Luna, R., Zainos, A., & Romo, R. (2010). Do sensory cortices process more than one sensory modality during perceptual judgments? *Neuron 67*(2), 335–348. doi:10.1016/j.neuron.2010.06.015

Lemus, L., Hernández, A., & Romo, R. (2009a). Neural codes for perceptual discrimination of acoustic flutter in the primate auditory cortex. *Proceedings of the National Academy of Sciences of the United States of America, 106*(23), 9471–9476. doi:10.1073/pnas.0904066106

Lemus, L., Hernández, A., & Romo, R. (2009b). Neural encoding of auditory discrimination in ventral premotor cortex. *Proceedings of the National Academy of Sciences of the United States of America, 106*(34), 14640–14645. doi:10.1073/pnas.0907505106

Mante, V., Sussillo, D., Shenoy, K. V., & Newsome, W. T. (2013). Context-dependent computation by recurrent dynamics in prefrontal cortex. *Nature, 503*(7474), 78–84. doi:10.1038/nature12742

Mountcastle, V. B., Talbot, W. H., Darian-Smith, I., & Kornhuber, H. H. (1967). Neural basis of the sense of flutter-vibration. *Science, 155*(3762), 597–600. doi:10.1126/science.155.3762.597

Murphey, D. K., & Maunsell, J. H. (2007). Behavioral detection of electrical microstimulation in different cortical visual areas. *Current Biology, 17*(10), 862–867. doi:10.1016/j.cub.2007.03.066

Murray, J. D., Bernacchia, A., Freedman, D. J., Romo, R., Wallis, J. D., Cai, X., Padoa-Schioppa, C., Pasternak, T., Seo, H., Lee, D., Wang, X. J. (2014). A hierarchy of intrinsic timescales across primate cortex. *Nature Neuroscience, 17*(12), 1661–1663. doi:10.1038/nn.3862

Murray, J. D., Bernacchia, A., Roy, N. A., Constantinidis, C., Romo, R., & Wang, X. J. (2017). Stable population coding for working memory coexists with heterogeneous neural dynamics in prefrontal cortex. *Proceedings of the National Academy of Sciences of the United States of America, 114*(2), 394–399. doi:10.1073/pnas.1619449114

Newsome, W. T., Britten, K. H., & Movshon, J. A. (1989). Neuronal correlates of a perceptual decision. *Nature, 341*(6237), 52–54. doi:10.1038/341052a0

Ohbayashi, M., Ohki, K., & Miyashita, Y. (2003). Conversion of working memory to motor sequence in the monkey premotor cortex. *Science, 301*(5630), 233–236. doi:10.1126/science.1084884

Ohbayashi, M., Picard, N., & Strick, P. L. (2016). Inactivation of the dorsal premotor area disrupts internally generated, but not visually guided, sequential movements. *Journal of Neuroscience, 36*(6), 1971–1976. doi:10.1523/JNEUROSCI.2356-15.2016

Ponce-Alvarez, A., Nacher, V., Luna, R., Riehle, A., & Romo, R. (2012). Dynamics of cortical neuronal ensembles transit from decision making to storage for later report. *Journal of Neuroscience, 32*(35), 11956–11969. doi:10.1523/JNEUROSCI.6176-11.2012

Prut, Y., & Fetz, E. E. (1999). Primate spinal interneurons show pre-movement instructed delay activity. *Nature, 401*(6753), 590–594. doi:10.1038/44145

Rigotti, M., Barak, O., Warden, M. R., Wang, X. J., Daw, N. D., Miller, E. K., & Fusi, S. (2013). The importance of mixed selectivity in complex cognitive tasks. *Nature, 497*(7451), 585–590. doi:10.1038/nature12160

Romo, R., Brody, C. D., Hernández, A., & Lemus, L. (1999). Neuronal correlates of parametric working memory in the prefrontal cortex. *Nature, 399*(6735), 470–473. doi:10.1038/20939

Romo, R., & de Lafuente, V. (2013). Conversion of sensory signals into perceptual decisions. *Progress in Neurobiology, 103,* 41–75. doi:10.1016/j.pneurobio.2012.03.007

Romo, R., Hernández, A., & Zainos, A. (2004). Neuronal correlates of a perceptual decision in ventral premotor cortex. *Neuron, 41*(1), 165–173. doi:10.1016/S0896-6273(03)00817-1

Romo, R., Hernández, A., Zainos, A., Brody, C. D., & Lemus, L. (2000). Sensing without touching: Psychophysical performance based on cortical microstimulation. *Neuron, 26*(1), 273–278. doi:10.1016/S0896-6273(00)81156-3

Romo, R., Hernández, A., Zainos, A., Lemus, L., & Brody, C. D. (2002). Neuronal correlates of decision-making in secondary somatosensory cortex. *Nature Neuroscience, 5*(11), 1217–1225. doi:10.1038/nn950

Romo, R., Hernández, A., Zainos, A., & Salinas, E. (1998). Somatosensory discrimination based on cortical microstimulation. *Nature, 392*(6674), 387–390. doi:10.1038/32891

Romo, R., Hernández, A., Zainos, A., & Salinas, E. (2003). Correlated neuronal discharges that increase coding efficiency during perceptual discrimination. *Neuron, 38*(4), 649–657. doi:10.1016/S0896-6273(03)00287-3

Romo, R., Merchant, H., Zainos, A., & Hernández, A. (1997). Categorical perception of somesthetic stimuli: Psychophysical measurements correlated with neuronal events in primate medial premotor cortex. *Cerebral Cortex, 7*(4), 317–326. doi:10.1093/cercor/7.4.317

Romo, R., & Salinas, E. (1999). Sensing and deciding in the somatosensory system. *Current Opinion in Neurobiology, 9*(4), 487–493. doi:10.1016/S0959-4388(99)80073-7

Romo, R., & Salinas, E. (2003). Flutter discrimination: Neural codes, perception, memory and decision making. *Nature Reviews Neuroscience, 4*(3), 203–218. doi:10.1038/nrn1058

Romo, R., & Schultz, W. (1990). Dopamine neurons of the monkey midbrain: Contingencies of responses to active touch during self-initiated arm movements. *Journal of Neurophysiology, 63*(3), 592–606. doi:10.1152/jn.1990.63.3.592

Rossi-Pool, R., Salinas, E., Zainos, A., Alvarez, M., Vergara, J., Parga, N., & Romo, R. (2016). Emergence of an abstract categorical code enabling the discrimination of temporally structured tactile stimuli. *Proceedings of the National Academy of Sciences of the United States of America, 113*(49), E7966–E7975. doi:10.1073/pnas.1618196113

Rossi-Pool, R., Vergara, J., & Romo, R. (2018). The Memory Map of Visual Space. *Trends in Neuroscience, 41*(3), 117–120. doi:10.1016/j.tins.2017.12.005

Rossi-Pool, R., Zainos, A., Alvarez, M., Zizumbo, J., Vergara, J., & Romo, R. (2017). Decoding a decision process in the neuronal population of dorsal premotor cortex. *Neuron, 96*(6), 1432–1446, e1437. doi:10.1016/j.neuron.2017.11.023

Rossi-Pool, R., Zizumbo, J., Alvarez, M., Vergara, J., Zainos, A., & Romo, R. (2019). Temporal signals underlying a cognitive process in the dorsal premotor cortex. *Proceedings of the National Academy of Sciences of the United States of America, 116*(15), 7523–7532. doi:10.1073/pnas.1820474116

Salas, M. A., Bashford, L., Kellis, S., Jafari, M., Jo, H., Kramer, D., Shanfield, K., Pejsa, K., Lee, B., Liu, C., & Andersen, R. A. (2018). Proprioceptive and cutaneous sensations in humans elicited by intracortical microstimulation. *eLife, 7*. doi:10.7554/eLife.32904

Salinas, E., Hernández, A., Zainos, A., & Romo, R. (2000). Periodicity and firing rate as candidate neural codes for the frequency of vibrotactile stimuli. *Journal of Neuroscience, 20*(14), 5503–5515. doi:10.1523/JNEUROSCI.20-14-05503.2000

Salinas, E., & Romo, R. (1998). Conversion of sensory signals into motor commands in primary motor cortex. *Journal of Neuroscience, 18*(1), 499–511. doi:10.1523/JNEUROSCI.18-01-00499.1998

Salzman, C. D., Britten, K. H., & Newsome, W. T. (1990). Cortical microstimulation influences perceptual judgements of motion direction. *Nature, 346*(6280), 174–177. doi:10.1038/346174a0

Sarno, S., de Lafuente, V., Romo, R., & Parga, N. (2017). Dopamine reward prediction error signal codes the temporal evaluation of a perceptual decision report. *Proceedings of the National Academy of Sciences of the United States of America, 114*(48), E10494–E10503. doi:10.1073/pnas.1712479114

Schultz, W. (1998). Predictive reward signal of dopamine neurons. *Journal of Neurophysiology, 80*(1), 1–27. doi:10.1152/jn.1998.80.1.1

Shima, K., Isoda, M., Mushiake, H., & Tanji, J. (2007). Categorization of behavioural sequences in the prefrontal cortex. *Nature, 445*(7125), 315–318. doi:10.1038/nature05470

Talbot, W. H., Darian-Smith, I., Kornhuber, H. H., & Mountcastle, V. B. (1968). The sense of flutter-vibration: Comparison of the human capacity with response patterns of mechanoreceptive afferents from the monkey hand. *Journal of Neurophysiology, 31*(2), 301–334. doi:10.1152/jn.1968.31.2.301

Tanji, J. (1994). The supplementary motor area in the cerebral cortex. *Neuroscience Research, 19*(3), 251–268. doi:10.1016/0168-0102(94)90038-8

Tauste Campo, A., Vázquez, Y., Álvarez, M., Zainos, A., Rossi-Pool, R., Deco, G., Romo, R. (2019). Feed-forward information and zero-lag synchronization in the sensory thalamocortical circuit are modulated during stimulus perception. *Proceedings of the National Academy of Sciences of the United States of America, 116*(15), 7513-7522. doi:10.1073/pnas.1819095116

Thura, D., & Cisek, P. (2014). Deliberation and commitment in the premotor and primary motor cortex during dynamic decision making. *Neuron, 81*(6), 1401–1416. doi:10.1016/j.neuron.2014.01.031

Vázquez, Y., Salinas, E., & Romo, R. (2013). Transformation of the neural code for tactile detection from thalamus to cortex. *Proceedings of the National Academy of Sciences of the United States of America, 110*(28), E2635–E2644. doi:10.1073/pnas.1309728110

Vázquez, Y., Zainos, A., Alvarez, M., Salinas, E., & Romo, R. (2012). Neural coding and perceptual detection in the primate somatosensory thalamus. *Proceedings of the National Academy of Sciences of the United States of America, 109*(37), 15006–15011. doi:10.1073/pnas.1212535109

Vergara, J., Rivera, N., Rossi-Pool, R., & Romo, R. (2016). A neural parametric code for storing information of more than one sensory modality in working memory. *Neuron, 89*(1), 54–62. doi:10.1016/j.neuron.2015.11.026

Wise, S. P., & Mauritz, K. H. (1985). Set-related neuronal activity in the premotor cortex of rhesus monkeys: Effects of changes in motor set. *Proceeding of the Royal Society B: Biological Sciences, 223*(1232), 331–354. doi:10.1098/rspb.1985.0005

Zainos, A., Merchant, H., Hernández, A., Salinas, E., & Romo, R. (1997). Role of primary somatic sensory cortex in the categorization of tactile stimuli: Effects of lesions. *Experimental Brain Research, 115*(2), 357–360. doi: 10.1152/jn.1995.73.2.525

36 Rationality and Efficiency in Human Decision-Making

CHRISTOPHER SUMMERFIELD AND KONSTANTINOS TSETSOS

ABSTRACT How should humans think and act? This question is relevant to a multitude of academic disciplines, from statistics to philosophy. In this chapter we consider this question from the standpoint of computation, cognition, and neurobiology. Our focus is the study of human decision-making. The chapter summarizes theoretical work that has sought to define normative (i.e., optimal or rational) principles for making decisions and empirical work that has asked whether humans make optimal choices about sensory signals (perceptual decision-making) and rational choices about economic prospects (value-based decision-making). We argue that human decisions are not always optimal or rational as traditionally defined. For example, humans exhibit biases that lead to inaccurate judgments about the sensory world or follow courses of action that fail to maximize potential reward. However, we argue that humans have evolved to make *efficient* decisions—those that mitigate processing costs by capitalizing on knowledge of the structure of the world. We support this argument with recent evidence from behavioral testing, computational modeling, and neural recordings in humans and other animals.

The goal of psychologists, cognitive neuroscientists, and other researchers in the behavioral sciences is to understand the determinants of human behavior. Decisions are the precursors to behavior, so understanding human decision-making is a prerequisite for this endeavor. Decisions occur whenever multiple potential courses of action are available, but only one can be followed at a time. This is the default case in natural environments. When arriving at a fork in the road, you can only take one of the two available routes. At a restaurant, the menu might contain many tasty dishes, but you usually only have the appetite to eat one. At the polling booth, your ballot will be spoiled if you vote for more than one candidate. This constraint ensures that noisy, continuous, high-dimensional signals from perception and memory have to be mapped onto a single, discrete course of action. When studying decision-making, this is the neurocognitive process that we are seeking to understand.

In this chapter our focus is on a question that lies at the heart of the decision sciences: Do humans make decisions as they should? We start with the traditional definitions of "rational" (or "optimal") decisions that focus on the maximization of accuracy or reward or the exhibition of consistent preferences. We go on to chart well-described decision biases that suggest departures from rationality in human decision processes, particularly where perceptual and economic choices are swayed by irrelevant contextual information. We then ask how these can be understood by considering the pressures that may have shaped the evolution of neural information-processing systems in natural environments. Our conclusion is that although human decisions deviate from traditional normative benchmarks, they are *efficient*—that is, they capitalize on the structure of natural environments in order to minimize the computational cost of information processing.

Optimality in Perceptual Decision-Making

To tackle the problem of how decisions are made, researchers tend to study very simple choice scenarios. One successful domain, known as perceptual decision-making, examines how humans and other animals discriminate or categorize sensory stimuli. For example, participants might be presented with a cloud of dots and asked whether they are moving to the left or right (Britten, Shadlen, Newsome, & Movshon, 1992). Perceptual decision experiments are usually crafted so that there is a clear correct or incorrect answer. For example, if you respond "left" when the dots are actually moving to the right, then you are wrong. It might be tempting to think that rational decisions are those that are correct, and irrational decisions are those that are erroneous. However, a foundational principle in the decision sciences is that choices are made under uncertainty (Glimcher, 2004). Errors can occur because of the intrinsic variability in sensory signals, noise arising during neural encoding, or limitations in subsequent computation. Deciding whether an individual has made a good decision or not depends on how these various sources of variability are characterized.

Normative theories of choice begin with the premise that neurons encode and represent stimuli in the local environment. For any stimulus in the external world x, we can posit a neural state $\hat{x}$ that is computed internally. The observer does not have access to x, so decisions are based on a learned mapping function (or policy) linking $\hat{x}$ to a motor output. When x is corrupted by noise, internal estimates $\hat{x}$ may favor an incorrect choice so that even an observer who uses the correct policy will misidentify the stimulus. For example, a stimulus might be erroneously categorized due to random signal fluctuations in the grating itself or during transduction—that is, directly at the level of $\hat{x}$. A canonical approach, known as signal-detection theory, assumes that decisions are corrupted by a single source of Gaussian noise. Thus, $\hat{x} = x + \mathcal{N}(0,\sigma)$, where σ is an estimate of stimulus variability. Signal-detection theory provides statistical tools for measuring the sensitivity of human judgment under this simple assumption (Green & Swets, 1966).

Often, multiple independent sources of information concerning the identity of x are available to an observer. Decisions that consider all of the relevant evidence are more likely to be correct. However, optimal choices require an observer to account for the relative reliability of different sources of information. For example, when deciding whether a defendant is guilty in a court of law, more credence should be given to the testimony of a reliable than an unreliable witness. When deciding if an individual is male or female, we might use both information about facial features (vision) and voice (audition), but to make the best decisions, we should rely more on audition when the room is darkened, and the face is hard to see. Our internal estimate $\hat{x}$ should thus be formed by combining the available noisy signals, each weighted by their reliability, to form a maximum likelihood estimate. If the noise is Gaussian distributed, then reliability is simply the reciprocal of σ for each sensory estimate (e.g., visual, auditory). An extensive literature has asked whether humans behave in this way, and a view has emerged that on average, they do. For example, in one study, participants given both haptic information and visual signals of variable quality were asked to judge the height of a bar. After independently measuring the sensory noise in each modality, the researchers were able to predict human psychophysical performance in the multimodal case, using a model that combined cues weighted by their reliability (Ernst & Banks, 2002). Similar results have been seen when observers integrate information from vision and audition (Kanitscheider, Brown, Pouget, & Churchland, 2015), or from the density and orientation of a texture (Blake, Bulthoff, & Sheinberg, 1993). On this basis, a canonical view has emerged that humans optimally weight sensory information by its reliability (Ernst & Bulthoff, 2004).

Decisions are also made in the context of information that occurred previously. The optimality of judgment will also depend on whether past information is appropriately factored into a decision. Bayesian decision theory begins with the assertion that optimal decisions are made by combining current evidence (concerning the likelihood of $\hat{x}\,|\,x$) with prior beliefs about the base rate probability of x. For example, imagine you are trying to decide whether your opponent in a tennis match will hit the ball long or short, given uncertain sensory information about her racquet stroke. If you have previously observed that she frequently plays drop shots, then optimal inference will be biased toward short. In psychophysical experiments, humans can learn the distributions of likely stimuli and use these to bias their sensorimotor behavior in an approximately optimal fashion (Kording, 2007)—for example, when reporting the location (Kording & Wolpert, 2004), duration (Jazayeri & Shadlen, 2010), or motion direction (Hanks, Mazurek, Kiani, Hopp, & Shadlen, 2011) of a sensory stimulus.

Optimal decisions depend on an observer's sensitivity to the sources of noise that corrupt information processing. Studies have demonstrated that when making perceptual decisions, participants show a striking sensitivity to the reliability of sensory information and that human decisions follow lawful statistical principles, as prescribed by Bayes' rule (Ma & Jazayeri, 2014). However, as we shall see below, human perceptual judgments can also show striking deviations from veridicality. These can be explained in part by accounting for learning about the structure of the world.

Natural Priors and Local Expectations

Our starting point is that decisions are shaped by learning from past experiences. As we encounter natural environments, we learn about the relative frequencies of different states of the world and their patterns of mutual covariation. Learning leads to the formation of stable representations that in turn specify the prior distribution over possible states of the world that guides decisions in the laboratory. Where the input states are highly structured, as in natural environments, the priors that guide decisions are informative. For example, real-world inputs are temporally autocorrelated so that an object present at time t will often be present at time $t+1$ and spatially autocorrelated so that if a point on the retina is stimulated by green light, it is more likely that adjacent regions will also be green. Observers should thus expect sensory signals to be relatively stable over time and to obey gestalt principles, such as proximity, similarity, and good continuation.

Deviations from veridical perception observed in the lab can be explained by considering the natural priors that humans may have formed in the real world (Geisler, 2008; Knill & Pouget, 2004). In natural scenes, objects that are farther away tend to have both lower contrast and to move more slowly due to parallax error, such as when a distant mountain is viewed from a moving train. Thus, when viewing two gratings moving with equal speed, humans will tend to report the lower-contrast grating as slower, as if otherwise optimal inference occurs under this prior (Weiss, Simoncelli, & Adelson, 2002). Another well-described bias is the tendency for judgments about sensory stimuli to be biased toward exemplars that are more familiar. For example, when reproducing a color that is a mixture of green and blue, participants will often judge it to be closer to green or blue than it really is. This ubiquitously observed phenomenon, known as categorical perception, can be understood if humans have learned a real-world prior that most textures are blue (such as the sky) or green (such as the grass), rather than a mixture of these two colors, and inference is biased by this knowledge (Tenenbaum & Griffiths, 2001). The same argument can be used to understand a range of canonical visual illusions as optimal inference, such as when we extract shape from shading under the long-term assumption that light comes from above (Ramachandran, 1988).

Natural priors may explain how decision biases are shaped by representation learning. However, sensory representations are acquired gradually during development and modified only after extensive new experience, whereas human decision biases can vary rapidly with the local stimulation context. One salient class of bias, known as sequential effects, occurs when a decision made about one event carries over to the next (Fischer & Whitney, 2014). This is exemplified by the popular misconception that good luck comes in streaks when playing sports or games of chance, implying an illusory benefit of repeated action, a phenomenon known as the *hot hand* fallacy (Gilovich, Tversky, & Vallone, 1985). When judging sensory stimuli, such as tilted gratings, numbers, or faces, humans are often biased to make consistent judgments on successive trials, and this effect is heightened if stimuli are perceptually ambiguous (Akaishi, Umeda, Nagase, & Sakai, 2014). A related bias occurs when humans make two judgments about the same noisy stimulus. When asked to first categorize a dot motion stimulus and then estimate its orientation, the estimation judgment is repulsed away from the category boundary in the direction of the reported category (Jazayeri & Movshon, 2007). These biases lead to reductions in accuracy in the lab, where conditions are deliberately randomized. However, they may be normative in the real world, where sensory stimulation is temporally autocorrelated, and so recent stimuli and responses carry predictive information that is relevant for current choices.

How is past information incorporated rapidly and flexibly into the neural variables that determine decisions? One possibility is that decision variables are simultaneously integrated over multiple timescales in higher association cortex (Bernacchia, Seo, Lee, & Wang, 2011), and sequential effects occur when neural signals relating to a past event are inappropriately factored into the decision variable for a current event (Mattar, Kahn, Thompson-Schill, & Aguirre, 2016). Natural environments have an intrinsically hierarchical temporal structure. For example, when visiting a restaurant, some visual signals remain constant (e.g., the décor), some change slowly (e.g., the food on your plate), and others change fast (e.g., the waitstaff rushing around). It is likely that biological systems have evolved mechanisms that integrate information over different windows of time, allowing real-world decisions to be modulated by both currently and recently available signals.

Single-cell neurophysiology and human brain imaging have been used to ask how prior information modulates current decisions over multiple timescales. One possible locus for this integration is the parietal cortex, which is known to be a key site for the short-term storage and accumulation of decision information. For example, when the prior probability of the occurrence of a given stimulus is experimentally manipulated, this is reflected in the responding of parietal neurons both at stimulus onset and during integration (Hanks et al., 2011). When comparing two successive stimuli, such as two auditory tones, humans and other animals display a *contraction* bias whereby estimates of the first stimulus drift toward the mean of recent stimulation, leading to lower discrimination performance (Ashourian & Loewenstein, 2011). In rodents, this bias can be removed after the optogenetic inactivation of posterior parietal neurons, increasing the accuracy of discrimination judgments (Akrami, Kopec, Diamond, & Brody, 2018). Higher regions, such as the parietal cortex, may also incorporate prior information into decisions by modulating activity in sensory regions via top-down connections. For example, when faces are conditionally probable given the recent stimulation sequence, both single-cell activity (Bell, Summerfield, Morin, Malecek, & Ungerleider, 2016) and blood-oxygen-level-dependent (BOLD) responses (Egner, Monti, & Summerfield, 2010) are modulated in the fusiform gyrus, a key extrastriate region for face perception. One popular model, known as predictive coding, has suggested that perceptual inference over multiple timescales is shaped by the dynamic interplay between higher and lower brain regions, with higher regions encoding long-term predictions modulating the response to punctate stimulation in lower regions, which in turn compute error signals that allow future predictions to be updated (Friston, 2005).

Irrationality in Economic Decision-Making

The phenomena described above pertain to decisions about the perceptual world. A different subfield, developed within psychology and economics, has investigated whether humans make rational choices about economic prospects. The normative principles on which this research is founded are rather different. This is because unlike the sensory properties of a stimulus (e.g., dots moving left or right), which are known to the experimenter, value is an inherently subjective quality. If offered the choice between red wine or white wine, I might prefer white wine, whereas you prefer red. But that does not mean that one of us is wrong, just that we have different preferences. One could argue that some stimuli, such as financial rewards, provide an objective standard for valuation that is not subject to the vagaries of preference. However, a difference in outcome of five dollars might be inconsequential to a millionaire but could mean the difference between life and death for an individual on the brink of starvation. Over and above any idiosyncratic risk attitudes, decisions about whether to forego a sure five dollars in favor of a risky but higher-valued sum might thus depend on the status quo wealth of the agent. In other words, values, unlike sensory signals, are inherently subjective, and this complicates the specification of normative principles for economic decision-making.

One assumption that allows normative economic principles to be defined is that human decisions follow a fixed value function. That is, I have learned a function that maps the value of external stimuli, such as red or white wine, onto an internal representation $u(\hat{x})$ that encodes its utility as a fixed quantity. Preferences may vary idiosyncratically between individuals, but rational decisions should be consistent with the dictates of this utility function—in my case that $u(\hat{x}_{white}) > u(\hat{x}_{red})$. This assumption allows the specification of a set of axioms that should be obeyed by a rational observer (Von Neumann & Morgenstern, 1944), such that preferences are internally consistent (or *menu invariant*). To illustrate, if I prefer white wine to red wine when only these two options are available, I should also prefer white to red when rosé appears on the menu (axiom of independence). A straightforward implication of menu invariance is that preferences should be well-ordered: if I choose white wine over red wine and red wine over rosé, then I should choose white wine over rosé (axiom of transitivity). Where choices are made among gambles—that is, sums of money that can be gained or lost with a given probability—it is possible to construct a choice set (known as a *Dutch book*) for which an agent that fails to respect these axiomatic principles is guaranteed to lose money, on average. This is one principle by which bookmakers seek to turn a profit—for example, when offering odds on a horse race.

A long tradition in psychology and behavioral economics has suggested that humans can be observed to systematically violate these rational principles (Kahneman, Slovic, & Tversky, 1982). The inconsistency of human preferences has been most vividly shown in experiments in which the exact same choice set is presented under different frames—for example, as a gain or a loss. For example, when offered a choice between (1) saving one-third of a population from a fictitious pandemic for sure or (2) a one-third chance of saving everyone, participants tend to prefer the first option. However, they prefer (2) if the gamble is framed as a choice between a sure loss of two-thirds of the population or a two-thirds chance of saving nobody, even though this choice set is identical (Tversky & Kahneman, 1981). In general, when presented with descriptive scenarios such as these, human preferences tend to reverse systematically, such that they are risk averse in the frame of gains and risk seeking in the frame of losses. Descriptive economic models can capture this finding by assuming that the function $u(\cdot)$, which maps objective values onto their subjective counterparts, can vary with contextual factors, such as status quo wealth or satisfaction. For example, in prospect theory, if the utility function has a steeper slope for that portion of the value space that is lower than the current status quo, then losses will "loom larger" than equivalent gains, leading to effects of the sort described above (Kahneman & Tversky, 1979).

Among the most ubiquitous violations of rationality are contrastive effects, which occur when a prospect occurs in the context of another item, even if that item is unavailable or unwanted. According to the axiom of independence, when deciding between a preferred item A and a dispreferred item B, the choice should not depend on whether a less preferred item C is available. For example, when choosing between a magazine subscription that is available in print ($50) or a print plus online form ($60), a consumer's decision should be unaffected by an additional, less-preferred offer of online only ($60). However, a large literature suggests that human preferences reverse in stereotypical ways in the presence of such "decoy" stimuli. For example, imagine you are buying a house and the relevant factors are price and size. Consider a choice between two equally valued properties: house A, which is large and expensive, and house B, which is smaller but more modestly priced. House A will be chosen more often in the presence of (1) house C_{sim}, which is equally valued but similar to B; (2) house C_{att}, which is overall equally inferior to both A and B but more similar to A (both more costly and smaller than A but larger than B); and (3) house $C_{extreme}$, which is overall equivalently valued but even larger and more expensive than A. Accounting for this complex pattern of irrational behaviors, known respectively as the *similarity*, *attraction*, and *compromise* effects, within a single model remains a major endeavor within psychology and behavioral economics (Tsetsos, Usher, & Chater, 2010).

While far from exhaustive, these examples are intended to reveal the consensus view that, unlike perceptual decisions, economic choices are biased and irrational and fail to maximize reward. This conclusion about the quality of human decisions is rather different from that typically made by researchers in psychophysics and sensory neuroscience, who tend to emphasize the optimality of human performance. Various explanations have been proposed to explain this discrepancy. For example, psychophysical studies typically use simple sensory stimuli and employ prolonged training accompanied by feedback. By contrast, many experiments in behavioral economics simply ask participants to imagine a single hypothetical scenario (e.g., via a written vignette) and respond as if it were real. It is possible, thus, that the differences between the domains arise from the nature or format of the experimental materials or the level of training and feedback provided (Jarvstad, Hahn, Rushton, & Warren, 2013; Wu, Delgado, & Maloney, 2009). However, this possibility is less viable in light of recent empirical reports showing psychophysical analogs of irrational behaviors—for example, that "decoy" effects occur when judging the perceptual properties of a stimulus, such as its height or width (Trueblood & Pettibone, 2017; Tsetsos, Chater, & Usher, 2012; Tsetsos et al., 2016).

Another possibility concerns the differing sources of uncertainty that corrupt perceptual and economic decisions (Juslin & Olsson, 1997). In psychophysical experiments, participants have to classify the stimulus on the basis of the sensory evidence (e.g., the relative masculinity or femininity of a face) but are not obliged to retrieve stored estimates of the value of the stimulus (e.g., whether they think the face is attractive or not). In economic decisions, such as choosing between two pieces of fruit, participants can easily recognize whether the stimulus is an apple or an orange but may be unclear about which they prefer. The locus of uncertainty thus lies not in sensory signals but in the value function itself. One possibility is that humans are more sensitive to the sources of uncertainty that corrupt perceptual judgments and can adjust their decision policy accordingly. However, in this article we emphasize a different perspective, and one that appeals to the commonalities, rather than the differences, between perceptual and economic decisions.

Efficient Coding in a Structured World

The quality of a decision varies with the level of expertise of the decision-maker. Humans make more sensitive judgments about stimuli with which they are familiar—for example, when discriminating or remembering faces from their own race compared to a different ethnicity (Meissner & Brigham, 2001). In a well-described visual phenomenon, known as the oblique effect, discrimination thresholds for cardinally oriented stimuli are lower than those for diagonally oriented stimuli (Appelle, 1972). This is consistent with the greater prevalence of horizontal and vertical lines in the natural world to which we are exposed (Girshick, Landy, & Simoncelli, 2011).

A principled understanding of these phenomena is provided by the theory that biological brains have evolved learning rules that allow the formation of efficient codes for sensory stimuli (Barlow, 1961; Simoncelli, 2003). Efficient coding systems can capitalize on the structure of the world to represent data in a compressed format, a fact exploited by the algorithms that produce zipped file formats on a modern computer. Efficient representations will emerge naturally from various biologically plausible classes of learning rules, such as Hebbian learning, which ensure that neural systems reduce the dimensionality of input data in a way similar to principal components analysis (Oja, 1982). The efficiency principle ensures that internal representations are distributed in a way that matches the statistics of the external environment. Thus, if stimulus x is drawn from a distribution with statistics ϕ, then the distribution of neural states $\hat{x}$ should also have statistics ϕ. This will ensure that those features or objects most commonly encountered are relatively overrepresented and can thus be discriminated and recognized with the highest accuracy, at the expense of sensitivity for less commonly occurring stimuli. Neuroscience has also provided evidence that representations are distributed to match the statistics of the external world. For example, cardinal orientations are overrepresented in early visual cortex, as indexed with both single-cell recordings (Li, Peterson, & Freeman, 2003) and functional neuroimaging (Furmanski & Engel, 2000), consistent with an efficient coding explanation of the oblique effect (Girshick, Landy, & Simoncelli, 2011).

In the lab, accurate decisions will be made when objective stimulus features or values are linearly transduced to objective decision values. For example, consider an observer who is attempting to reproduce the orientation of a grating by turning a wheel. If the function that maps external (true) orientation onto internal (subjective) orientation is nonlinear or otherwise distorted, then the observer will make less accurate estimation judgments. The same principle applies for value. Since Bernoulli, it has been known that some economically irrational behaviors can be described by assuming that the value function $u(\hat{x})$ exhibits a compressive nonlinearity. For example, most humans will care more about the difference between \$1 and \$11 than they do about the difference between \$101 and \$111, even though in both cases the difference is exactly \$10. This follows naturally from the assumption that the value function is steeper for low values than for high values—that is, we are more sensitive to values at the lower than upper end of the scale. Although this may appear suboptimal, it may be normative in the natural world, in which outcomes are approximately encountered according to a power-law distribution, such that prospects of low value (e.g., a coffee for \$2) are more commonly encountered or evaluated than prospects of high value (e.g., a car for \$20,000; Stewart, Chater, & Brown, 2006). In fact, this view can account for a range of scalar variability effects, by which stimulus sensitivity varies logarithmically with sensory magnitude across the human behavioral repertoire (Mackay, 1963). For example, it has been known since the 19th century that noticeable differences in lighter objects (say those of approximately 50 g) are smaller than for heavier objects (those of ~5 kg).

One way of understanding the idiosyncrasy of human decisions is that we have evolved efficient coding schemes for representation learning. The idea that sensory stimuli are encoded efficiently but decoded optimally predicts a lawful relationship between discriminability and bias, which states that bias should always be proportional to the slope of the square of the discrimination threshold (Wei & Stocker, 2012). Remarkably, this law has been found to hold over a variety of different directional estimates, including motion (Gros, Blake, & Hiris, 1998), heading (Crane, 2012), and pursuit (Krukowski & Stone, 2005), as well as orientation discrimination (Wei & Stocker, 2015), suggesting a general role for efficient coding in human decision-making.

Efficient Computation and Relative Coding

An efficient system will allocate neuronal resources in proportion to the prevalence of stimuli in the external world. However, when the world changes rapidly, this resource allocation needs to occur flexibly and dynamically and faster than is permitted by the gradual mechanisms that underlie representation learning. In other words, brains may have evolved mechanisms that economize on both neural resources (e.g., a fixed budget of cells for neural coding) and processing resources (e.g., a fixed number of spikes for neural signaling). This is consistent with the idea that capacity limitations in neural systems arise both through limits on cortical availability (Franconeri, Alvarez, & Cavanagh, 2013), which require *efficient coding*, and a need to keep metabolic expenditure low (Lennie, 2003), which requires *efficient computation*.

One likely substrate for efficient computation is *divisive normalization*, a ubiquitous feature of cortical circuits (Carandini & Heeger, 2012). The assumption that inputs are divisively normalized over time can explain the adaptive effects that occur when neuronal responsivity declines after prolonged exposure to a given context. For example, dark adaptation allows the retina to transduce effectively despite ambient light varying by some 14 orders of magnitude over the diurnal cycle (Bartlett, 1965). Other classic examples of normalization over space include the local inhibitory interactions that give rise to center-surround opponency in V1 cells, or the form of the contrast saturation function following exposure to an adapting stimulus or mask (Carandini & Heeger, 1994). However, more complex adaptive effects may occur during the computation of higher-order decision variables, explaining a number of key phenomena that characterize perceptual and economic choice behavior in humans and animals.

To illustrate, consider cells in the mammalian orbitofrontal cortex, which have been found to signal stimulus value with a rate code—that is, higher values elicit faster spiking (Padoa-Schioppa & Assad, 2006). For one such neuron, consider the challenge of simultaneously coding items with low value (e.g., two brands of pasta) and high value (e.g., two brands of laptop computer). A neuron's dynamic range is limited by biophysical constraints that set an upper bound on its firing rate (say, 100 Hz). If the gain function that maps values onto spikes is fixed, then the two brands of pasta will be coded with similar firing rates near the bottom of the 0–100 Hz range. However, due to stochasticity in neural firing, the spike rates generated by the two similarly valued stimuli will frequently overlap, and the agent will sometimes pick the dispreferred option when shopping at the supermarket. Now let us assume instead that the gain function can adapt, permitting the neuron to use its full dynamic range to encode the options in the choice set. For example, the upper reaches of the range can be used to represent one brand of pasta and the lower portion the other, minimizing confusion about the value of the two products. Alternatively, neurons would need a reduced dynamic range (say, 0–10 Hz) to represent the same variety of stimuli, thereby increasing neural efficiency (Rangel & Clithero, 2012).

The precise form of the normalization that might occur in cortical circuits remains a matter of debate (Louie, Glimcher, & Webb, 2015). In one form of normalization, known as range adaptation, firing rates evoked by a stimulus $r(A)$ are related to its value $v(A)$ scaled by the range of possible values across an experiment or block.

$$r(A) = \frac{v(A)}{v_{max} - v_{min}}. \tag{36.1}$$

Range adaptation effects have been observed in the lateral orbitofrontal neurons as a macaque chooses among rewarding stimuli, such as drops of a sweet fruit drink. When an offer occurs in the context of a block of low-valued offers, the gain function is steeper than when it occurs in a block of both high and low offers (Padoa-Schioppa, 2009; Tremblay & Schultz, 1999). Interestingly, when options A and B themselves vary systematically over different ranges, this range adaptation is corrected to avoid arbitrary choice biases (Rustichini, Conen, Cai, & Padoa-Schioppa, 2017). Behavioral data suggests that human value judgments are modulated by context in a similar fashion. For example, when making monetary payments to avoid painful shocks, humans will pay more to avoid medium-intensity shocks that occur in a block of mostly low-strength than mostly high-strength stimulation (Vlaev, Seymour, Dolan, & Chater, 2009). Accordingly, when performing intertemporal choice tasks for monetary value, human BOLD signals in the ventromedial cortex are scaled according to the range of values in the local context (Cox & Kable, 2014).

Range normalization divides all items by a common scalar term $v_{max} - v_{min}$, and so the resulting functions that map sensory signals onto decision values, although rescaled in slope, remain linear in the input space. Another possibility is that normalization varies with the intensity of recent items or the value of locally available alternatives.

$$r(A) = \frac{v(A)}{v(A) + v(B)} \tag{36.2}$$

One illustrative example of normalization by context comes from the measurement of neural responses in the auditory cortex of the ferret (Rabinowitz, Willmore, Schnupp, & King, 2011). During high-variance auditory stimulation, the gain function that maps stimulus contrast onto firing rates is attenuated compared to low-variance stimulation, meaning that sensitivity for low-contrast auditory stimuli is greater when they occur in the context of low-variance stimulation. These data fit extremely well with a divisive normalization model, and unlike the orbitofrontal data described above, the range of observed spike rates remained greater for the high-variance stimulation.

This form of divisive normalization also provides an explanation for some violations of menu invariance. For example, in a behavioral phenomenon dubbed the *distracter effect*, a dispreferred item B is more often chosen over a preferred item A in the presence of a decoy C that approaches A and B in value. Imagine that stimulus A is coded by a neuron with rates $r(A)$ that on average scales with $v(A)$ but is normalized in proportion to the sum of available values $v(A) + v(B) + v(C)$. The strength of this normalization term grows with $v(C)$, leading to greater compression of overall signals for higher average values of A, B, and C. This means that noisy signals for A and B are harder to distinguish when C is increased in value, providing a unidimensional violation of menu invariance (Louie, Khaw, & Glimcher, 2013). This effect is supported by evidence from neurophysiological recordings in the parietal cortex. When monkeys were rewarded for making a saccade to an instructed target within the response field of the neuron, firing rates were modulated not only by the value of the instructed target but also by the value of an irrelevant stimulus in the opposite hemifield. The form of the modulation was well captured by a divisive normalization model with the form described in equation 36.2 above (Louie, Grattan, & Glimcher, 2011). This model has also been found to account for the modulation exhibited by neurons in the medial orbitofrontal cortex (OFC) when monkeys choose between a safe and a risky option in blocks where the safe option has a different value (Yamada, Louie, Tymula, & Glimcher, 2018).

In a further form of normalization, the efficiency of computation is increased by explicitly calculating decision variables relative to a variable reference point, given by the average of a local context. In this case neuronal responses are modulated by prediction errors—that is, the difference between current and recent stimulation. For example, the response to a stimulus A might be computed as

$$r(A) = \frac{v(A)}{\left| v(A) - \mathbb{E}[v(A)] \right|} \tag{36.3}$$

where $E[v(A)]$ is the expectation of $v(A)$—for example, given the average of recent stimulation. There is emerging evidence from categorization studies that gain is allocated adaptively across features in a way that increases the efficiency of computation during perceptual decisions (Summerfield & Tsetsos, 2015). In one paradigm, participants are asked to average sequential information—for example, pertaining to the tilt of a grating, relative to a category boundary. Human participants display a bias to overweight information that is consistent with recent stimulation even within a single trial, leading to a suboptimal bias. However, this behavior can be explained by a model in which each item is evaluated by a gain function that is constantly updated according to recent stimulation, ensuring that the highest gain is allocated to expected information, as in equation 36.3. Indeed, a heightened gain of encoding for consistent samples is observed in electroencephalography (EEG) signals that peak over the parietal cortex (Cheadle et al., 2014). This form of adaptive gain control can unfold over very fast timescales, even within a single trial.

In a different variant of the task that involves spatial averaging—for example, when observers are asked to categorize the mean tilt in a ring of gratings as clockwise or counterclockwise with respect to a reference orientation, human observers give more weight to gratings that fall closer to the global mean feature, which by design lies near the reference (*robust averaging*; de Gardelle & Summerfield, 2011). In principle, robust averaging is suboptimal because from the experimenter's perspective, there is no reason why differential weight should be given to the available information when all gratings are equally reliable. However, robust averaging can be explained if observers allocate neural resources in proportion to the distribution of features in the experiment. Indeed, computational modeling shows that under explicit assumptions about the limited capacity of integration, a model that engages in robust averaging will outperform one that does not (Li, Herce Castanon, Solomon, Vandormael, & Summerfield, 2017). Furthermore, the gain-control model successfully predicts how changing the distribution of stimuli from trial to trial will affect performance. For example, when participants view an irrelevant "prime" array with high variance, the model suggests that gain should be allocated more broadly, facilitating performance when a subsequent "target" array also has high variance. This model-predicted "variance priming" phenomenon is observed in human observers (Michael, de Gardelle, & Summerfield, 2014).

Critically, the same principle provides a normative motivation for econometric models, such as prospect theory, which assume that stimuli are evaluated relative to a status quo reference point. Consider a canonical economic choice between a sure bet of $10 and a 50/50 chance of receiving $20. Faced with this choice, most humans prefer to take the safe option. However, if participants are first endowed with $20 and offered a sure bet of losing $10 or a 50/50 chance of keeping everything, they tend to prefer the risky option (De Martino, Kumaran, Holt, & Dolan, 2009). This preference reversal occurs despite the fact that the two choice sets are formally identical. The key innovation provided by prospect theory is that value functions adapt according to a reference point defined by the status quo wealth of the agent. In other words, the $20 endowment shifts the reference point (and so the value function) in a way that all new prospects are compared relative to the new status quo (Kahneman & Tversky, 1979). The nonlinear form of the prospect theory value predicts that near to the reference, subjective utility is inflated away from objective value most sharply, so this process acts very similarly to the gain-control mechanism described above. Indeed, it has been noted that prospect theory can be considered an efficient form of sensory distortion, akin to robust averaging and other phenomena from the perceptual decision-making literature (Woodford, 2012).

Here, we have summarized three candidate normalization schemes and discussed the empirical evidence that may support them. Each of these schemes potentially has a normative justification, depending on the putative cost function that organisms strive to minimize in ongoing behavior. A plausible starting assumption is that biological cost functions entail the joint minimization of metabolic expenses and of decision errors. The three schemes all reduce metabolic costs to different extents, at the expense of decision accuracy, and thus can all be normatively justified under different assumptions about an agent's willingness to sacrifice accuracy for computational efficiency.

In summary, neural mechanisms that promote efficient computation inflate the effective dynamic range of neurons or neuronal populations and thus facilitate downstream stimulus decoding. However, when sensory signals are computed on a relative (rather than an absolute) scale, stimuli will be evaluated differently according to the context in which they occur. This can give rise to the contrastive effects or other contextual biases typically observed in human economic decision-making. These decisions may appear suboptimal in the lab but can be understood as respecting the efficiency principles that have evolved to deal with the highly structured natural environment in which animals evolve.

Framing Effects and Selective Integration

Another classic violation of axiomatic rationality is observed when human choices are susceptible to the framing of a decision. Consider the choice between two holiday destinations, Bali and Baltimore. For a traveler from the United States, Bali might be more exotic, but Baltimore has the merit of being less expensive. Paradoxically, participants who are broadly indifferent to either of these options will be biased to choose Bali when asked which of these options they prefer but to reject Bali when asked which of the options they disprefer. According to one theory (Shafir, 1993), the framing of the question changes the relative salience of the positive and negative attributes of a multidimensional stimulus so that Bali is accepted in the positive frame because it is exotic and rejected in the negative frame because it is expensive. Psychophysical analogs of this task produce the same phenomenon. For example, when asked to choose between two simultaneously occurring streams of numbers with equivalent means but differing variance, participants are biased to choose the more variable stream—that is, that with the more salient or outlying values. This occurs both when asked which is higher and which is lower, equivalent to the "accept" and "reject" frames for the holiday destinations described above (Tsetsos, Chater, & Usher, 2012).

One model, known as selective integration, explains these findings by proposing that during evidence evaluation, humans give more weight to evidence that is frame-consistent. The model states that when (for example) averaging streams of numbers, participants neglect lower-valued samples when asked to report which stream has the larger average and neglect higher-valued samples when asked which stream has the smaller average. In other words, observers selectively discard some information (promoting efficiency), by allocating reduced gain to "locally losing" samples of information. Selective integration can explain the framing effects reported above, as well as other violations of axiomatic rationality. For example, it predicts that during multialternative choice, the probability of choosing the most valuable option will depend on the rank ordering of all the options, including those irrelevant to the choice (Tsetsos, Chater, & Usher, 2012; c.f., the attraction effect above). The "salience-driven" bias proposed by this model is similar to that proposed by models in which attention modulates the process of evidence accumulation during perceptual and economic choice (Busemeyer & Townsend, 1993; Krajbich, Armel, & Rangel, 2010).

A further study demonstrated that selective integration can explain the systematic intransitivity of human decisions, a canonical violation of rational choice theory (Tsetsos et al., 2016). The task involved a choice set A, B, and C (streams of bars of varying height) and observers were asked to make binary choices about the stream with the highest average height (e.g., A vs. B). The choice set was constructed so that A, B, and C had equivalent mean height, but A had more local winners when paired against B, and B had more winners when paired with C, while C had more winners when paired with A. Selective integration explains the pattern of intransitivity observed in human decisions because it predicts that choices depend on the local rank of the evidence between alternatives.

Interestingly, and related to the examples of "efficient" computation described above, it can be shown that (under simple and plausible assumptions) selective integration can paradoxically increase decision accuracy despite discarding part of the choice-relevant information. The authors modeled the data to include a biologically plausible "late" noise term that occurred during information integration (but beyond the sensory stage). This late noise term might be thought of as an explicit limit on the fidelity of information integration, akin to a bound on higher-processing capacity. In simulation, the selective integration model reaped more reward than the traditionally normative perfect averaging model. This occurred because selective integration exaggerates the differences among winners and losers, conferring robustness on decisions that are corrupted by late noise. Similar to the result from the robust averaging studies above, this shows that when psychologically and neurally plausible constraints are incorporated into decision models—such as the notion that processing capacity is not limitless—the reward-maximizing policy may differ from that proposed by the traditional model conceived under the Bayesian framework (Wald & Wolfowitz, 1949).

Theoretically, to reap the maximum levels of reward, selective integration needs to be employed in direct proportion to the levels of late noise that corrupt the decision. Further analyses of the bar height integration task suggested that indeed, humans demonstrated this proportional relationship between late noise and selective integration. In other words, as in the examples above, a policy that explicitly discards information can maximize reward when the imperfections of neural computation are realistically taken into account. The assumption that human performance is mainly limited by noise downstream from the sensory representation is plausible given the hierarchical and distributed nature of information processing in the brain. This opens up the possibility for a broader definition of optimality beyond the conventional decision theoretic framework.

Conclusions

This chapter began by asking whether human decisions should be described as "optimal" or "rational." However, whether a decision is optimal or not depends on what is being optimized. In machine learning, optimization begins with a cost function, a theoretical construct that specifies whether a given outcome is desired or not (Marblestone, Wayne, & Kording, 2016). When modeling optimal or rational behavior, psychologists, neuroscientists, and economists have traditionally considered only the behavioral cost (e.g., the need to maximize accuracy or reward), often without giving due consideration to the foundational principle from cognitive science that information-processing systems are limited by capacity and by hierarchically distributed processing noise. Here, we argue that normative models should also consider the neural cost—that is, the need for computation to be efficient (Gershman, Horvitz, & Tenenbaum, 2015), as well as the nature of neural noise. We have summarized a breadth of work that suggests that decision policies have evolved to place a strong premium on computational efficiency, both by learning representations that match the statistics of the external world (efficient coding) and by engaging in context-dependent normalization mechanisms that accentuate local differences among stimuli in space and time. These hallmarks of neural information-processing systems entail that the policies exhibited by biological agents may deviate from those that would be optimal if agents had limitless capacity, yielding what may appear—at first glance—to be irrational perceptual and economic choices. However, the theoretical arguments and computational simulations described above imply that these mechanisms can be adaptive and even reward maximizing for limited-capacity agents negotiating a world that is highly structured in space and time. Our article thus summarizes the neural coding schemes, and mechanisms, that promote efficient and reward-maximizing decisions in humans and other animals.

REFERENCES

Akaishi, R., Umeda, K., Nagase, A., & Sakai, K. (2014). Autonomous mechanism of internal choice estimate underlies decision inertia. *Neuron, 81*(1), 195–206. doi:10.1016/j.neuron.2013.10.018

Akrami, A., Kopec, C. D., Diamond, M. E., & Brody, C. D. (2018). Posterior parietal cortex represents sensory history and mediates its effects on behaviour. *Nature, 554*(7692), 368–372. doi:10.1038/nature25510

Appelle, S. (1972). Perception and discrimination as function of stimulus orientation. *Psychological Bulletin, 78*(4), 266–278.

Ashourian, P., & Loewenstein, Y. (2011). Bayesian inference underlies the contraction bias in delayed comparison tasks. *PLoS One, 6*(5), e19551. doi:10.1371/journal.pone.0019551

Barlow, H. (1961). Possible principles underlying the transformation of sensory messages. *Sensory communication.* Cambridge, MA: MIT Press.

Bartlett, N. R. (1965). Dark adaptation and light adaptation. In C. H. Graham (Ed.), *Vision and visual perception* (pp. 185–207). New York: John Wiley and Sons.

Bell, A. H., Summerfield, C., Morin, E. L., Malecek, N. J., & Ungerleider, L. G. (2016). Encoding of stimulus probability in macaque inferior temporal cortex. *Current Biology, 26*(17), 2280–2290. doi:10.1016/j.cub.2016.07.007

Bernacchia, A., Seo, H., Lee, D., & Wang, X. J. (2011). A reservoir of time constants for memory traces in cortical neurons. *Nature Neuroscience, 14*(3), 366–372. doi:10.1038/nn.2752

Blake, A., Bulthoff, H. H., & Sheinberg, D. (1993). Shape from texture: Ideal observers and human psychophysics. *Vision Research, 33*(12), 1723–1737.

Britten, K. H., Shadlen, M. N., Newsome, W. T., & Movshon, J. A. (1992). The analysis of visual motion: A comparison of neuronal and psychophysical performance. *Journal of Neuroscience, 12*(12), 4745–4765.

Busemeyer, J. R., & Townsend, J. T. (1993). Decision field theory: A dynamic-cognitive approach to decision making in an uncertain environment. *Psychological Review, 100*(3), 432–459.

Carandini, M., & Heeger, D. J. (1994). Summation and division by neurons in primate visual cortex. *Science, 264*(5163), 1333–1336.

Carandini, M., & Heeger, D. J. (2012). Normalization as a canonical neural computation. *Nature Reviews Neuroscience, 13*(1), 51–62. doi:nrn3136 [pii] 10.1038/nrn3136

Cheadle, S., Wyart, V., Tsetsos, K., Myers, N., de Gardelle, V., Herce Castanon, S., & Summerfield, C. (2014). Adaptive gain control during human perceptual choice. *Neuron, 81*(6), 1429–1441. doi:10.1016/j.neuron.2014.01.020

Cox, K. M., & Kable, J. W. (2014). BOLD subjective value signals exhibit robust range adaptation. *Journal of Neuroscience, 34*(49), 16533–16543. doi:10.1523/JNEUROSCI.3927-14.2014

Crane, B. T. (2012). Direction specific biases in human visual and vestibular heading perception. *PLoS One, 7*(12), e51383. doi:10.1371/journal.pone.0051383

de Gardelle, V., & Summerfield, C. (2011). Robust averaging during perceptual judgment. *Proceedings of the National Academy of Sciences of the United States of America, 108*(32), 13341–13346. doi:1104517108 [pii] 10.1073/pnas.1104517108

De Martino, B., Kumaran, D., Holt, B., & Dolan, R. J. (2009). The neurobiology of reference-dependent value computation. *Journal of Neuroscience, 29*(12), 3833–3842. doi:10.1523/JNEUROSCI.4832-08.2009

Egner, T., Monti, J. M., & Summerfield, C. (2010). Expectation and surprise determine neural population responses in the ventral visual stream. *Journal of Neuroscience, 30*(49), 16601–16608. doi:30/49/16601 [pii] 10.1523/JNEUROSCI.2770-10.2010

Ernst, M. O., & Banks, M. S. (2002). Humans integrate visual and haptic information in a statistically optimal fashion. *Nature, 415*(6870), 429–433. doi:10.1038/415429a 415429a [pii]

Ernst, M. O., & Bulthoff, H. H. (2004). Merging the senses into a robust percept. *Trends in Cognitive Sciences, 8*(4), 162–169. doi:10.1016/j.tics.2004.02.002

Fischer, J., & Whitney, D. (2014). Serial dependence in visual perception. *Nature Neuroscience, 17*(5), 738–743. doi:10.1038/nn.3689

Franconeri, S. L., Alvarez, G. A., & Cavanagh, P. (2013). Flexible cognitive resources: Competitive content maps for attention and memory. *Trends in Cognitive Sciences, 17*(3), 134–141. doi:10.1016/j.tics.2013.01.010

Friston, K. (2005). A theory of cortical responses. *Philosophical Transactions of the Royal Society of London B: Biological Sciences, 360*(1456), 815–836. doi:W5T4QMCP8T4K0UP8 [pii] 10.1098/rstb.2005.1622

Furmanski, C. S., & Engel, S. A. (2000). An oblique effect in human primary visual cortex. *Nature Neuroscience, 3*(6), 535–536. doi:10.1038/75702

Geisler, W. S. (2008). Visual perception and the statistical properties of natural scenes. *Annual Review of Psychology, 59*, 167–192. doi:10.1146/annurev.psych.58.110405.085632

Gershman, S. J., Horvitz, E. J., & Tenenbaum, J. B. (2015). Computational rationality: A converging paradigm for intelligence in brains, minds, and machines. *Science, 349*(6245), 273–278. doi:10.1126/science.aac6076

Gilovich, T., Tversky, A., & Vallone, R. (1985). The hot hand in basketball: On the misperception of random sequences. *Cognitive Psychology, 17*(3), 295–314.

Girshick, A. R., Landy, M. S., & Simoncelli, E. P. (2011). Cardinal rules: Visual orientation perception reflects knowledge of environmental statistics. *Nature Neuroscience, 14*(7), 926–932. doi:10.1038/nn.2831

Glimcher, P. W. (2004). *Decision, uncertainty and the brain: The science of neuroeconomics.* Cambridge, MA: MIT Press.

Green, D. M., & Swets, J. A. (1966). *Signal detection theory and psychophysics.* New York: Wiley & Sons.

Gros, B. L., Blake, R., & Hiris, E. (1998). Anisotropies in visual motion perception: A fresh look. *Journal of the Optical Society of America A, 15*(8), 2003–2011.

Hanks, T. D., Mazurek, M. E., Kiani, R., Hopp, E., & Shadlen, M. N. (2011). Elapsed decision time affects the weighting of prior probability in a perceptual decision task. *Journal of Neuroscience, 31*(17), 6339–6352. doi:31/17/6339 [pii] 10.1523/JNEUROSCI.5613-10.2011

Jarvstad, A., Hahn, U., Rushton, S. K., & Warren, P. A. (2013). Perceptuo-motor, cognitive, and description-based decision-making seem equally good. *Proceedings of the National Academy of Sciences of the United States of America, 110*(40), 16271–16276. doi:10.1073/pnas.1300239110

Jazayeri, M., & Movshon, J. A. (2007). A new perceptual illusion reveals mechanisms of sensory decoding. *Nature, 446*(7138), 912–915. doi:10.1038/nature05739

Jazayeri, M., & Shadlen, M. N. (2010). Temporal context calibrates interval timing. *Nature Neuroscience, 13*(8), 1020–1026. doi:10.1038/nn.2590

Juslin, P., & Olsson, H. (1997). Thurstonian and Brunswikian origins of uncertainty in judgment: A sampling model of confidence in sensory discrimination. *Psychological Review, 104*(2), 344–366.

Kahneman, D., Slovic, P., & Tversky, A. (1982). *Judgment under uncertainty: Heuristics and biases.* New York: Cambridge University Press.

Kahneman, D., & Tversky, A. (1979). Prospect theory: An analysis of decision under risk. *Econometrica, 47*(2), 263–291.

Kanitscheider, I., Brown, A., Pouget, A., & Churchland, A. K. (2015). Multisensory decisions provide support for probabilistic number representations. *Journal of Neurophysiology, 113*(10), 3490–3498. doi:10.1152/jn.00787.2014

Knill, D. C., & Pouget, A. (2004). The Bayesian brain: The role of uncertainty in neural coding and computation. *Trends in Neurosciences, 27*(12), 712–719. doi:10.1016/j.tins.2004.10.007

Kording, K. P. (2007). Decision theory: What "should" the nervous system do? *Science, 318*(5850), 606–610. doi:10.1126/science.1142998

Kording, K. P., & Wolpert, D. M. (2004). Bayesian integration in sensorimotor learning. *Nature, 427*(6971), 244–247. doi:10.1038/nature02169 nature02169 [pii]

Krajbich, I., Armel, C., & Rangel, A. (2010). Visual fixations and the computation and comparison of value in simple choice. *Nature Neuroscience, 13*(10), 1292–1298. doi:nn.2635 [pii] 10.1038/nn.2635

Krukowski, A. E., & Stone, L. S. (2005). Expansion of direction space around the cardinal axes revealed by smooth pursuit eye movements. *Neuron, 45*(2), 315–323. doi:10.1016/j.neuron.2005.01.005

Lennie, P. (2003). The cost of cortical computation. *Current Biology, 13*(6), 493–497.

Li, B., Peterson, M. R., & Freeman, R. D. (2003). Oblique effect: A neural basis in the visual cortex. *Journal of Neurophysiology, 90*(1), 204–217. doi:10.1152/jn.00954.2002

Li, V., Herce Castanon, S., Solomon, J. A., Vandormael, H., & Summerfield, C. (2017). Robust averaging protects decisions from noise in neural computations. *PLoS Computational Biology, 13*(8), e1005723. doi:10.1371/journal.pcbi.1005723

Louie, K., Glimcher, P. W., & Webb, R. (2015). Adaptive neural coding: From biological to behavioral decision-making. *Current Opinion in Behavioral Sciences, 5*, 91–99. doi:10.1016/j.cobeha.2015.08.008

Louie, K., Grattan, L. E., & Glimcher, P. W. (2011). Reward value-based gain control: Divisive normalization in parietal cortex. *Journal of Neuroscience, 31*(29), 10627–10639. doi:31/29/10627 [pii] 10.1523/JNEUROSCI.1237-11.2011

Louie, K., Khaw, M. W., & Glimcher, P. W. (2013). Normalization is a general neural mechanism for context-dependent decision making. *Proceedings of the National Academy of Sciences of the United States of America, 110*(15), 6139–6144. doi:1217854110 [pii] 10.1073/pnas.1217854110

Ma, W. J., & Jazayeri, M. (2014). Neural coding of uncertainty and probability. *Annual Review of Neuroscience, 37*, 205–220. doi:10.1146/annurev-neuro-071013-014017

Mackay, D. M. (1963). Psychophysics of perceived intensity: A theoretical basis for Fechner's and Stevens' laws. *Science, 139*(3560), 1213–1216.

Marblestone, A. H., Wayne, G., & Kording, K. P. (2016). Toward an integration of deep learning and neuroscience. *Frontiers in Computational Neuroscience, 10*, 94. doi:10.3389/fncom.2016.00094

Mattar, M. G., Kahn, D. A., Thompson-Schill, S. L., & Aguirre, G. K. (2016). Varying timescales of stimulus integration unite neural adaptation and prototype formation. *Current Biology, 26*(13), 1669–1676. doi:10.1016/j.cub.2016.04.065

Meissner, C. A., & Brigham, J. C. (2001). Thirty years of investigating the own-race bias in memory for faces: A meta-analytic review. *Psychology, Public Policy and Law, 7*(1), 3–35.

Michael, E., de Gardelle, V., & Summerfield, C. (2014). Priming by the variability of visual information. *Proceedings of the National Academy of Sciences of the United States of America, 111*(21), 7873–7878. doi:10.1073/pnas.1308674111

Oja, E. (1982). Simplified neuron model as a principal component analyzer. *Journal of Mathematical Biology, 15*(3), 267–273.

Padoa-Schioppa, C. (2009). Range-adapting representation of economic value in the orbitofrontal cortex. *Journal of Neuroscience, 29*(44), 14004–14014. doi:29/44/14004 [pii] 10.1523/JNEUROSCI.3751-09.2009

Padoa-Schioppa, C., & Assad, J. A. (2006). Neurons in the orbitofrontal cortex encode economic value. *Nature, 441*(7090), 223–226. doi:nature04676 [pii] 10.1038/nature 04676

Rabinowitz, N. C., Willmore, B. D., Schnupp, J. W., & King, A. J. (2011). Contrast gain control in auditory cortex. *Neuron, 70*(6), 1178–1191. doi:10.1016/j.neuron.2011.04.030

Ramachandran, V. S. (1988). Perception of shape from shading. *Nature, 331*(6152), 163–166. doi:10.1038/331163a0

Rangel, A., & Clithero, J. A. (2012). Value normalization in decision making: Theory and evidence. *Current Opinion in Neurobiology, 22*(6), 970–981. doi:10.1016/j.conb.2012.07.011

Rustichini, A., Conen, K. E., Cai, X., & Padoa-Schioppa, C. (2017). Optimal coding and neuronal adaptation in economic decisions. *Nature Communications, 8*(1), 1208. doi:10.1038/s41467-017-01373-y

Shafir, E. (1993). Choosing versus rejecting: Why some options are both better and worse than others. *Memory & Cognition, 21*(4), 546–556.

Simoncelli, E. P. (2003). Vision and the statistics of the visual environment. *Current Opinion in Neurobiology, 13*(2), 144–149.

Stewart, N., Chater, N., & Brown, G. D. (2006). Decision by sampling. *Cognitive Psychology, 53*(1), 1–26. doi:10.1016/j.cogpsych.2005.10.003

Summerfield, C., & Tsetsos, K. (2015). Do humans make good decisions? *Trends in Cognitive Sciences, 19*(1), 27–34. doi:10.1016/j.tics.2014.11.005

Tenenbaum, J. B., & Griffiths, T. L. (2001). Generalization, similarity, and Bayesian inference. *Behavioral and Brain Sciences, 24*(4), 629–640, discussion 652–791.

Tremblay, L., & Schultz, W. (1999). Relative reward preference in primate orbitofrontal cortex. *Nature, 398*(6729), 704–708.

Trueblood, J. S., & Pettibone, J. C. (2017). The phantom decoy effect in perceptual decision making. *Journal of Behavioral Decision Making, 30*(2), 157–167.

Tsetsos, K., Chater, N., & Usher, M. (2012). Salience driven value integration explains decision biases and preference reversal. *Proceedings of the National Academy of Sciences of the United States of America, 109*(24), 9659–9664. doi:10.1073/pnas.1119569109

Tsetsos, K., Moran, R., Moreland, J., Chater, N., Usher, M., & Summerfield, C. (2016). Economic irrationality is optimal during noisy decision making. *Proceedings of the National Academy of Sciences of the United States of America, 113*(11), 3102–3107. doi:10.1073/pnas.1519157113

Tsetsos, K., Usher, M., & Chater, N. (2010). Preference reversal in multiattribute choice. *Psychological Review, 117*(4), 1275–1293. doi:10.1037/a0020580

Tversky, A., & Kahneman, D. (1981). The framing of decisions and the psychology of choice. *Science, 211*(4481), 453–458.

Vlaev, I., Seymour, B., Dolan, R. J., & Chater, N. (2009). The price of pain and the value of suffering. *Psychological Science, 20*(3), 309–317. doi:10.1111/j.1467-9280.2009.02304.x

Von Neumann, J., & Morgenstern, O. (1944). *Theory of games and economic behavior.* Princeton, NJ: Princeton University Press.

Wald, A., & Wolfowitz, J. (1949). Bayes solutions of sequential decision problems. *Proceedings of the National Academy of Sciences of the United States of America, 35*(2), 99–02.

Wei, X., & Stocker, A. A. (2012). Efficient coding provides a direct link between prior and likelihood in perceptual Bayesian inference. *Advances in Neural Information Processing Systems, 25*(1), 1304–1312.

Wei, X. X., & Stocker, A. A. (2015). A Bayesian observer model constrained by efficient coding can explain "anti-Bayesian" percepts. *Nature Neuroscience, 18*(10), 1509–1517. doi:10.1038/nn.4105

Weiss, Y., Simoncelli, E. P., & Adelson, E. H. (2002). Motion illusions as optimal percepts. *Nature Neuroscience, 5*(6), 598–604. doi:10.1038/nn858

Woodford, M. (2012). Prospect theory as efficient perceptual distortion. *American Economic Review, 102,* 41–46.

Wu, S. W., Delgado, M. R., & Maloney, L. T. (2009). Economic decision-making compared with an equivalent motor task. *Proceedings of the National Academy of Sciences of the United States of America, 106*(15), 6088–6093. doi:10.1073/pnas.0900102106

Yamada, H., Louie, K., Tymula, A., & Glimcher, P. W. (2018). Free choice shapes normalized value signals in medial orbitofrontal cortex. *Nature Communications, 9*(1), 162. doi:10.1038/s41467-017-02614-w

37 Opening Burton's Clock: Psychiatric Insights from Computational Cognitive Models

DANIEL BENNETT AND YAEL NIV

ABSTRACT Computational psychiatry is a nascent field that seeks to use computational tools from neuroscience and cognitive science to understand psychiatric illness. In this chapter we make the case for computational cognitive models as a bridge between the cognitive and affective deficits experienced by those with a psychiatric illness and the neurocomputational dysfunctions that underlie these deficits. We first review the history of computational modeling in psychiatry and conclude that a key moment of maturation in this field occurred with the transition from qualitative comparison between computational models and human behavior to formal quantitative model fitting and model comparison. We then summarize current research at one of the most exciting frontiers of computational psychiatry: reinforcement-learning models of mood disorders. We review state-of-the-art applications of such models to major depression and bipolar disorder and outline important open questions to be addressed by the coming wave of research in computational psychiatry.

The brain must needs primarily be misaffected, as the seat of reason ... for our body is like a clock, if one wheel be amiss, all the rest are disordered; the whole fabric suffers.

—Robert Burton, *The Anatomy of Melancholy*

For a watch repairer, the first task in fixing a faulty watch is diagnosis: What is the dysfunctional mechanism that is responsible for the fault? If the watch is losing time, is it because the mainspring is insufficiently wound, or could dirt be causing the gears to stick? If the watch has stopped, could this be the result of a loose balance wheel, or does the battery simply need changing?

In his analogy between human mental illness and the faulty mechanics of a clock, Robert Burton captured the essence of one of the most durable problems of contemporary biological psychiatry. In a clock a given functional disturbance, such as running fast or running slow, may be the result of any number of mechanical faults, and it is typically impossible to determine which mechanism is primarily amiss by observing the timekeeping dysfunction alone. Moreover, this inverse problem grows in difficulty with the complexity of the mechanism inside the watch: a fault is easier to diagnose when the underlying mechanism is simpler (e.g., a vibrating quartz crystal in a modern analog watch) than when it is complex (e.g., the many gears and springs of a 17th-century watch). Analogously, it has long been understood that psychiatric symptoms such as thought disorder and mania are aberrant behaviors produced by dysfunctions within an exceedingly complex dynamical system, the human brain (Hoffman, 1987; Joseph, Frith, & Waddington, 1979). It is no surprise, then, that identifying the specific neural-processing deficits that cause a given psychiatric symptom is difficult.

In this chapter we argue that computational psychiatry should approach this problem using computational cognitive models, with a focus on testing specific behavioral predictions made by different candidate neurocomputational dysfunctions. Just as the ticking sounds of a clock can be decomposed with spectral analyses to diagnose a mechanical fault (He, Su, & Du, 2008), computational cognitive models can be used to infer the latent neurocomputational deficits that underlie psychiatric conditions as diverse as depression and psychosis. However, just as in the clock analogy, the utility of these inferences critically depends upon two factors: first, an accurate mechanistic model of how the system operates and second, a sensitive behavioral assay of its operations. To this end, computational psychiatry should seek to integrate normative and process models from computational neuroscience and biological psychiatry with behavioral tests from cognitive psychology, computer science, and economics. By applying computational cognitive models to sensitive measures of human behavior, we may make substantial progress in identifying the dysfunctions of neural computation that give rise to psychiatric illness.

This chapter first reviews the history of the computational-modeling paradigm in psychiatry through the cognitive revolution of the 1960s and 1970s and the rise of parallel distributed processing and reinforcement-learning models in the 1980s and 1990s. We then summarize the current state of the art of computational psychiatry in the study of mood disorders such as major depression and bipolar disorder using reinforcement-learning models.

The History of Computational Psychiatry

> Psychopathology has been rather a disappointment to the instinctive materialism of the doctors, who have taken the view that every disorder must be accompanied by actual lesions of some specific tissue involved... . This distinction between functional and organic disorders is illuminated by the consideration of the computing machine.
>
> —Norbert Wiener, *Cybernetics*

The idea that psychiatric illness might result from dysfunctions of neural or mental computation was proposed within 10 years of the invention of the modern digital computer. Writing in 1948 as part of a broader argument that the central nervous system ought to be treated as a self-regulating circuit, Norbert Wiener suggested a novel perspective on the 19th-century psychiatric distinction between organic and functional disorders (Fürstner, 1881, as cited by Beer, 1996). This dichotomy contrasts organic disorders caused by a purely biological pathology (such as a brain tumor or neurodegeneration) with functional disorders that cannot be diagnosed solely by the inspection of brain tissue. Wiener proposed that functional disorders—among which he included schizophrenia and bipolar disorder—could be best understood by analogy with the operations of a computer. This was, he proposed, because deficits in these disorders arose not from aberrations in the physical structure of the brain but from dysfunctions in the way the physical structure processed information (Wiener, 1948).

This information-processing paradigm was immensely influential in early cognitive psychology but gained traction much more slowly in psychiatry. Early research using computational models in psychiatry was rudimentary and consisted of little more than qualitative comparisons between simple computational models and aspects of contemporary psychiatric theory. For instance, Callaway (1970) pursued the analogy of a malfunctioning computer in an attempt to understand conceptual disorganization and the loosening of associations in schizophrenia. Drawing upon contemporary advances in cognitive science, Callaway posited that cognitive structures in schizophrenia could be represented as simple computational architectures called TOTE (test-operate-test-exit) units (Miller, Galanter, & Pribram, 1960). Deficits in schizophrenia were posited to result from interference in the test operations of these units by excessive neural noise. While the TOTE architecture has not proved durable, Callaway's notion that deficits in schizophrenia result from excessive levels of noise in neural computation has remained influential to the present day (e.g., Silverstein, Wibral, & Phillips, 2017; Winterer & Weinberger, 2004).

Separately, Colby (1964) used a computational dictionary seeded with quotations from human psychiatric patients to generate synthetic dialogues resembling those of a therapist with a psychiatric patient (e.g., "Father preferred sister. I avoid father." Colby, 1964, p. 221). Colby proposed that distorted beliefs in psychosis arose as a result of conflict between mutually exclusive impulses. Colby, Hilf, Weber, and Kraemer (1972) presented practicing psychotherapists with teletype printouts of a number of putative therapist/patient dialogues—half real and half generated by algorithm—and assessed the therapists' ability to distinguish real patients from simulated ones. It was found that therapists could not identify the real patients at an above-chance level and in some cases offered detailed psychoanalytic interpretations of the unconscious processes underlying algorithmically generated dialogues. The algorithm that generated the text engaged in dialogue by performing a rudimentary form of natural language processing with the intention of classifying its interlocutor's statements as either malevolent, benevolent, or neither. Depending on the values of the variables used to perform this classification, the algorithm then selected an internal response (e.g., anger or fear) and a corresponding utterance (e.g., verbal hostility in the case of high levels of anger). This algorithm can therefore be thought of as an early cognitive model of psychosis (albeit one that does not invoke unconscious processing, contrary to then-dominant theoretical ideas).

Other early work applying computational and mathematical methods to psychiatric illness did not adapt the computer metaphor directly. For instance, Rashevsky (1964) posited a rudimentary biophysical neural-processing system to explain the positive symptoms of schizophrenia in terms of the excessive reinforcement of endogenously generated responses. Houghton (1969) sought to specify a formal mathematical framework for understanding psychoanalysis by positing a negative feedback relationship between an "id module" and an "ego module," resulting in distortions of a topological space. Such theories have little empirical relevance for contemporary research; instead, they primarily reinforce the importance of grounding models of psychiatric illness in biologically principled models of neural computation.

The first computational models that are of more than historical interest to current research in computational psychiatry were made possible by advances in computational models of neural information processing. For instance, a computational theory of the distribution of attention among stimuli based on recurrent lateral inhibition between noisy processing channels (Walley & Weiden, 1973) gave rise directly to a computational model of attentional deficits in schizophrenia (Joseph, Frith, & Waddington, 1979). This model proposed that an excess of dopaminergic activity led to increased overall levels of mutual inhibition between sensory inputs in schizophrenia and thereby to a dysfunction in the system's ability to produce winner-take-all network dynamics.

The advent of more advanced neural network architectures in the 1980s stimulated the development of more sophisticated computational psychiatric models. For instance, Hopfield (1982) described a fully interconnected neural network that produced emergent properties resembling human recognition memory, categorization, and generalization. In turn, Ralph Hoffman showed how dysfunctions of computation within Hopfield nets led to aberrant dynamics resembling schizophrenia and mania (Hoffman, 1987) and linked the putative computational deficit in schizophrenia to aberrant patterns of cortical pruning in frontal cortex (Hoffman & Dobscha, 1989). At the same time, the immense influence of parallel distributed-processing connectionist architectures in cognitive science (Rumelhart & McClelland, 1987) led naturally to the adaptation of multilayer neural networks for psychiatric research (e.g., Ruppin, 1995; Spitzer, 1995; Stein & Ludik, 1998).

Of particular note, Cohen and Servan-Schreiber (1992) used a multilayer neural network to model a failure to maintain mental context in schizophrenia. This work demonstrated a *quantitative* correspondence between the behavior of trained neural network models and the behavior of patients with schizophrenia on three tasks: a Stroop task, a continuous performance task, and a lexical disambiguation task. The computational mechanism by which these deficits were produced in the model was a reduction of the gain of units in the network representing task context, and this computational dysfunction was linked by the authors to decreased dopaminergic activity in the prefrontal cortex in schizophrenia. This work marks a point of transition between qualitative and quantitative comparisons of models and behavior in computational psychiatry. As such, it stands in contrast to prior research that had proceeded after the fashion of Callaway (1970) by suggesting *qualitative* parallels between patterns of information processing in psychiatric illness and patterns of information processing in real or hypothetical computational architectures.

Arguably, this development—the quantitative fitting of computational models to behavior produced by individuals with a psychiatric illness—is responsible for much of the subsequent achievement, and much of the future promise, of computational methods in psychiatric research. The ability of computational models to make quantitative predictions about human behavior means that different psychiatric theories can be compared by instantiating each as a different model and determining which model provides the most accurate and parsimonious account of behavior. Once identified, a model serves at least two purposes: First, it provides a quantitative device for the measurement of cognitive-psychiatric symptoms that may aid in diagnosis and treatment selection in psychiatry in much the same way that a blood glucose test aids in diagnosing and treating diabetes. Second, a good correspondence between the predictions of a model and observed behaviors may offer a window into the functional causes of aberrant experiences in psychiatric illness, since it suggests mechanisms by which these symptoms may be produced.

As computational approaches to psychiatry have expanded in recent years, the behavioral model-fitting and model-comparison paradigm has grown to encompass computational models from disciplines including economic game theory (King-Casas et al., 2008), hierarchical probabilistic inference (Friston, Stephan, Montague, & Dolan, 2014), and Bayesian decision theory (Huys, Daw, & Dayan, 2015). In the remainder of this chapter, we review these developments with a specific focus on the state-of-the-art computational modeling of two mood disorders: major depression and bipolar disorder. In particular, we explore the extent to which dysfunctions in these conditions can be understood through the lens of reinforcement learning (see, e.g., Maia & Frank, 2011).

Reinforcement Learning Models of Mood Disorders

Below, we summarize the insights that reinforcement-learning models provide into the neurocomputational substrates of depression and bipolar disorder. Our intention is not to claim that mood disorders are disorders of learning narrowly defined. Instead, we argue that the mathematical formalisms of reinforcement learning provide a language that can describe how representations of the reinforcement value of the environment go astray in mood disorders.

Briefly, reinforcement learning describes a set of computational principles by which an agent in an uncertain or complex environment can act to maximize future expected reward (Dayan & Niv, 2008; Sutton & Barto, 1998). The framework relies on several relatively simple psychological primitives: Representations of different states of the environment, of the actions that can be taken by the agent in each state, and of the rewards that are received following each action. Reinforcement-learning algorithms then describe operations by which an agent can update its representations of the values of different actions as it interacts with the environment.

The foundational computational variable in reinforcement learning is the prediction error δ, calculated as the difference between the actual reward received after taking some action and the amount of reward an agent had expected to result from that action:

$$\delta = R_t - Q_t\,(s_t,\, a_t) \qquad (37.1)$$

Here, R_t denotes the reward (or, if negative, punishment) received on trial t, and $Q_t\,(s_t,\, a_t)$ denotes the expected value on trial t of taking action a_t in state s_t. δ takes a positive value when the received reward exceeds the expected reward amount (a positive reward prediction error) and a negative value when the reward received is less than expected. Given this prediction error, one can then update expectations for trial $t+1$ according to a simple Rescorla-Wagner learning rule (Rescorla & Wagner, 1972):

$$Q_{t+1}\,(s_t,\, a_t) = Q_t(s_t,\, a_t) + \eta \cdot \delta \qquad (37.2)$$

where η is a learning rate parameter controlling the speed with which action values are acquired. Equation 37.2 ensures that the expected value of actions will be incremented following positive reward prediction errors and decremented following negative reward prediction errors. Neurally, the prediction error signal δ (and, more precisely, its temporal difference cousin that accounts for the timing of prediction error signals within a trial; Schultz, Dayan, & Montague, 1997) is thought to be instantiated in the brain by the phasic release of dopamine in the basal ganglia.

From this foundation we can derive increasingly complex and sophisticated reinforcement-learning algorithms. For instance, the simple update rule described above is typically referred to as model-free reinforcement learning since it learns solely about the value of taking particular actions in particular states and not about the structure of the environment itself. This contrasts with model-based reinforcement learning, in which agents learn an internal model of the environment (possibly using prediction error signals) and use this model to plan actions through mental simulations of alternative options and their predicted outcomes (see, e.g., Doll, Simon, & Daw, 2012).

The domain of reinforcement learning is an agent's cognitive and behavioral responses to the affective feedback (i.e., rewards and punishments) that it receives from the environment. This domain is also a primary area of cognitive dysfunction in mood disorders, including major depression and bipolar disorder (Admon & Pizzagalli, 2015; Eshel & Roiser, 2010; Whitton, Treadway, & Pizzagalli, 2015). As such, reinforcement-learning models are well suited to the study of neurocomputational dysfunction in mood disorders. For instance, individuals with depression show a number of cognitive biases consistent with a reduced learned value of the environment and the preferential processing of negative information, such as pessimistic expectations regarding the value of future events (Showers & Ruben, 1990), an increased tendency to retrieve negatively valenced items from memory (Blaney, 1986), and decreased sensitivity to rewarding feedback (Henriques & Davidson, 2000). Similarly, a recent theory has suggested that oscillatory mood dynamics characteristic of bipolar disorder might be produced by an interaction between mood and the valuation of outcomes (Eldar & Niv, 2015; Eldar, Rutledge, Dolan, & Niv, 2016). As we will show, each of these phenomena can be described well in terms of dysfunctions of computation within a reinforcement-learning model.

Depression

Phenomenology and theories of depression The two most common diagnostic taxonomies of psychiatric illness, the Diagnostic and Statistical Manual of Mental Disorders (*DSM-5*) and the International Statistical Classification of Diseases (*ICD-10*), concur on two primary symptoms of major depression: persistent low mood or sadness and an inability to take pleasure in everyday events (anhedonia). The two taxonomies also concur on other secondary symptoms of depression, including fatigue or lack of energy (anergia), poor concentration, disturbances of sleep and appetite, thoughts of suicide or self-harm, feelings of guilt or worthlessness, and psychomotor disturbances (either agitation or motor slowing).

Cognitive theories of depression have posited a number of distinct information- processing biases that might underlie these symptoms (Gotlib & Joormann, 2010; Ingram, 1984). For instance, Beck (1967) proposed that preexisting representations (*schemas*) of oneself, other people, and the external world bias the processing of emotional information in a schema-congruent way. One example of a depressive schema, for instance, is a core belief that one is unlovable; this belief would lead to the interpretation of neutral or ambiguous social cues as consistent with the fact that one is unlovable, thereby reinforcing the schema. Other cognitive theories have emphasized the operation of different cognitive processes, but most agree that the biased processing of emotional information plays a crucial role in the onset and maintenance of depression. For instance, Bower (1981) and Ingram (1984) emphasized the role of disturbed semantic networks in depression, leading to the increased activation of negatively valenced nodes in an associative network. By contrast, Lewinsohn (1974) adopted a behaviorist perspective and emphasized the role of a lack of response-contingent reinforcement in depression, whereas Rehm (1977) emphasized the role of self-control in the selective processing of negative outcomes, and Seligman (1975) highlighted the role of learned helplessness (that is, the distorted belief that one's experiences of positive and negative events are not under one's own control).

Cognitive theories of depression have been highly influential, both in empirical research on cognition in depression and in the development of applied cognitive therapies for depression. However, these theories are persistently criticized because they merely redescribe known phenomena and do not offer any novel insights (Blaney, 1977; Ingram, 1984). The computational approach to psychiatry that we argue for in this chapter provides a tool to address this shortcoming. This is because the requirement that theories of psychiatric illness be embedded in a computational model means that quantitative behavioral predictions of different theories can be generated directly via model simulation. Empirical work can then test the extent to which these predictions are borne out by human behavior. Additionally, by mapping information-processing biases in depression onto putative neural computations—especially within the framework of reinforcement learning—computational models can flesh out cognitive theories of depression with reference to our understanding of how these computations are implemented in the human brain.

Computational modeling of depression The basic reinforcement-learning framework detailed in equations 37.1 and 37.2 can be extended to capture the cognitive phenomena of depression in a number of ways. One possibility proposed by Huys, Pizzagalli, Bogdan, and Dayan (2013) is that anhedonia represents a diminished hedonic response to rewarding outcomes in depression, which affects prediction errors as below:

$$\delta = \rho \cdot R_t - Q_t(s_t, a_t) \qquad (37.3)$$

where $0 \leq \rho \leq 1$ is a reward sensitivity parameter that describes the degree to which primary hedonic responses to rewarding outcomes are diminished in individuals with depression. The pattern of behavior produced by this model matches the phenomenological experience of anhedonia in the sense that since the effective reward value of outcomes is diminished, individuals with lower values of ρ will experience outcomes as subjectively less rewarding. Because reinforcement learning from prediction errors means they will also learn that the reward value of actions and options in the environment is lower, such individuals will form pessimistic expectations about future outcomes.

To provide evidence for this model, Huys et al. (2013) fit a version of the computational model described by equation 37.3 to the behavior of individuals with varying levels of anhedonia as they performed a simple learning task designed to measure reward sensitivity (Pizzagalli, Jahn, & O'Shea, 2005). Huys et al. (2013) found that across both healthy individuals and those with major depression, self-reported anhedonia was positively correlated with participants' estimated reward sensitivity ρ but not their estimated learning-rate parameter η.

However, further evidence complicates this view and suggests that anhedonia should not be simply viewed as a deficiency in hedonic responses to rewarding outcomes (Huys et al., 2015). If it were true that primary hedonic responses to rewards were diminished in depression, it would be expected that individuals with depression would report less enjoyment of pleasant primary rewards, such as sweet liquids. However, this is not the case: those with depression do not differ from healthy controls in the self-reported pleasantness of sucrose solutions (Amsterdam, Settle, Doty, Abelman, & Winokur, 1987). In addition, a recent study found no differences between those with depression and healthy controls in the strength of the relationship between reward prediction error magnitude and self-reported mood during a gambling task (Rutledge et al., 2017). This leads to the question: What computational mechanisms other than reduced hedonic response to rewards might explain an apparent reduction in reward sensitivity in depression?

A re-examination of cognitive theories of depression suggests asymmetric responses to positive and negative outcomes as one candidate. For instance, the self-control theory of Rehm (1977) proposes that depression is associated with selective attention to negative outcomes, as well as a tendency to make stronger inferences about the self from negative feedback than positive feedback. Similarly, the reinforcement theory of Lewinsohn (1974) posits that a reduction in the degree to which actions are reinforced by positive feedback is central to depression. From the perspective of reinforcement learning, one way of capturing this proposed information-processing bias is as an asymmetry in learning rates for positive versus negative reward prediction errors (Gershman, 2015; Mihatsch & Neuneier, 2002; Niv, Edlund, Dayan, & O'Doherty, 2012):

$$Q_{t+1}(s_t, a_t) = \begin{cases} Q_t(s_t, a_t) + \eta^+ \cdot \delta, & \delta > 0 \\ Q_t(s_t, a_t) + \eta^- \cdot \delta, & \delta < 0 \end{cases} \quad (37.4)$$

In equation 37.4, η^+ is the learning rate for positive reward prediction errors, and η^- is the learning rate for negative reward prediction errors. When $\eta^- > \eta^+$, value updates are affected more strongly by negative reward prediction errors, consistent with the proposed negative information-processing bias in major depression. This bias produces an underestimation of the value of uncertain rewards that is qualitatively similar to that produced by a reduction of the reward sensitivity parameter ρ in equation 37.3. However, deterministic rewards are learned correctly by this model (Niv et al., 2012).

Importantly, underestimations of reward value could be produced in equation 37.4 by hypersensitivity to negative reward prediction errors (increased η^-), by hyposensitivity to positive reward prediction errors (decreased η_+), or both. Empirical evidence from behavioral studies of depression is divided on this question. While there is consistent evidence that individuals with depression display diminished learning from positive feedback (Henriques & Davidson, 2000; Henriques, Glowacki, & Davidson, 1994; Korn, Sharot, Walter, Heekeren, & Dolan, 2014; Robinson, Cools, Carlisi, Sahakian, & Drevets, 2012; Vrieze et al., 2013), evidence for increased sensitivity to negative feedback is more equivocal. Some studies have shown that those with depression respond more to worse than expected outcomes than healthy controls, (Garrett et al., 2014; Nelson & Craighead, 1977) but others have found no difference (Henriques & Davidson, 2000; Henriques, Glowacki, & Davidson, 1994; Robinson et al., 2012; Santesso et al., 2008). This suggests, on balance, that aberrant reward processing in depression is more likely to result from hyposensitivity to positive reward prediction errors than from hypersensitivity to negative reward prediction errors. Further study of this question is required, however, and an important open question is whether different symptom profiles of depression are associated with different patterns of learning from positive and negative reward prediction errors. For instance, it is known that anxiety, a disorder highly comorbid with major depression (Sartorius, Üstün, Lecrubier, & Wittchen, 1996), is associated with hypersensitivity to punishment and increased attention to potentially threatening events (Bishop, 2007). This suggests the interesting possibility that low-level computational mechanisms of depression might differ between major depression with and without comorbid anxiety.

As a further prediction, asymmetric learning rates as per equation 37.4, but not changes in reward sensitivity as per equation 37.3, induce preferences with respect to the *risk* of outcomes (in the economic sense of risk, referring to outcome variance; Mihatsch & Neuneier, 2002). Learning rate asymmetry in depression would therefore also predict that individuals with depression should display increased risk aversion. This is because high-risk choice options are those associated with larger deviations, on average, between individual instances of reward and long-term reward averages, meaning larger absolute reward prediction errors. As a result, high-risk choice options will be more devalued when $\eta^- > \eta^+$ than low-risk choice options, resulting in risk aversion. This prediction is consistent with behavioral data showing increased risk aversion in individuals with depression performing the Iowa Gambling Task (Smoski et al., 2008), as well as greater self-reported risk aversion (Leahy, Tirch, & Melwani, 2012; Wiersma et al., 2011).

Separately, recent theories in computational psychiatry have also proposed a role for the dysfunction of model-based reinforcement learning in depression. As introduced above, model-based reinforcement learning applies to scenarios in which an agent's decisions are dependent upon a learned internal model of the environment (a *model* of the environment, hence *model-based* reinforcement learning). This is distinguished from model-free reinforcement learning, in which agents learn solely about the values of individual actions (Daw, Gershman, Seymour, Dayan, & Dolan, 2011). Two candidate model-based mechanisms for depression proposed by Huys et al. (2015) are biased attention toward negative possibilities in internal estimates of a current state and a failure to "prune" negative states from contemplation in planning future sequences of action.

The first of these, a bias in the internal representation of a state, reflects the fact that states of the world (s in the equations above) are not necessarily observable features; instead, a "state" represents an agent's inferences about the structure of rewards in the world at a given point in time and about the way that structure may change if different actions are taken (Schuck, Cai, Wilson, & Niv, 2016). For instance, while waiting at a bus stop, one can only estimate whether the state of the world is "the bus is shortly arriving" or "the bus already passed and I missed it." If the inferences used to construct this state are biased in a pessimistic way—such as because negative potential outcomes are weighted more strongly than positive outcomes—then an agent may believe itself to be in a worse state than is truly the case. Such a process might underlie the pessimistic representations of future outcomes in depression and might also provide an explanation for experiences of anergia, since low response vigor and reduced energy expenditure are rational strategies for an agent to adopt in states where few rewarding outcomes can result from action.

The second model-based mechanism is a failure to "prune" negative states from future planned actions in depression. In planned decision-making, nondepressed individuals typically avoid excessive focus upon the future possible states associated with large negative outcomes (Huys et al., 2012). This is an adaptive strategy since it means that cognitive resources can be directed instead toward plans that have a high a priori chance of reaching future states associated with a high reward value. Less pruning of negative states would be associated with a relatively greater focus on negative-valued paths in future planning, potentially leading to the patterns of ruminative thought characteristic of depression (Whitmer & Gotlib, 2013).

Open questions for the computational modeling of depression The literature reviewed above suggests several important open questions to be addressed via the computational modeling of behavior in depression.

First, to what extent can anhedonia in depression be characterized by asymmetric learning from positive and negative reward prediction errors, rather than reduced consummatory pleasure in reward receipt? Second, what combination of model-based and model-free reinforcement learning best describes the cognitive deficits observed in depression? On the one hand, depression may be associated with a low-level asymmetry in (model-free) learning. On the other hand, depression may be better characterized by model-based deficits in the construction of the present state and planning for future states. Or, depression may involve both deficits. Importantly, these questions can be answered using computational models and tasks specifically tailored to measure the parameters of these models in each individual.

Finally, how might the computational deficits underlying depression be expressed in different contexts? As Beck (1967) observed, inferences in depression are far more likely to be negatively biased when their object is one's own worth than when their object is an abstract statistical quantity. In the language of reinforcement learning, it is almost certainly not the case that learning rates for positive and negative prediction errors will be expressed equivalently in all domains. Instead, one possibility is that individual differences in the allocation of attention to positive and negative outcomes in different settings might provide a principled explanation for apparent differences in reinforcement sensitivity in depression. For instance, it is possible that attention to outcomes—and therefore learning rates—may fluctuate commensurate with the outcomes' congruency with prior beliefs regarding oneself. Designing sensitive measures of the context-dependence of reinforcement learning dysfunction in depression is therefore a crucial task for future research.

Bipolar Disorder

Phenomenology and subtypes of bipolar disorder In contrast to major depression, which is characterized solely by episodes of depression, bipolar disorder is characterized by episodes of both depression and mania. Under common definitions in the *DSM-5* and *ICD-10*, *mania* refers to a state in which mood is elevated (euphoria), and there is increased energy and goal-directed activity. Mania, and its less severe counterpart hypomania, are also typically characterized by increased risk-taking behavior, a decreased subjective need for sleep, and increased self-esteem, potentially leading to delusions of grandiosity (Goodwin & Jamison, 2007).

Typologies of bipolar disorder distinguish between two subtypes, bipolar I and bipolar II, which differ in the relative frequency and intensity of manic and depressed episodes. Bipolar I disorder is characterized by at least one episode of mania and often (but not necessarily) by other episodes of depression. By contrast, bipolar II disorder is typified by episodes of both major depression and hypomania (not meeting the full criteria for mania). Both forms of bipolar disorder are typified by a functional recovery between episodes of mania or depression to a mood in the normal range.

Whereas cognitive theories of depression have abounded since the 1960s, until recent years bipolar disorder was largely viewed through a psychopharmacological lens (Goodwin & Jamison, 2007), with a relative paucity of cognitive theorizing (but see, e.g., Alloy et al., 2008). One finding in this literature, however, is of mood-congruent information-processing biases in bipolar disorder. That is, individuals with bipolar disorder may display negative information-processing biases when in a low mood, as in depression, but positive information-processing biases when in a good mood (for reviews, see Alloy, Reilly-Harrington, Fresco, & Flannery-Schroeder, 2005; Whitton, Treadway, & Pizzagalli, 2015). This mood congruence is a critical feature of bipolar disorder that computational models must seek to account for; it also represents a significant point of contrast with cognitive theories of depression, which rather emphasize trait-level information-processing biases as a cognitive mechanism for the disorder.

Computational modeling of bipolar disorder A recent model has posited a set of computational mechanisms that may partly explain mood-congruent information-processing biases in bipolar disorder. Using a reinforcement-learning framework, Eldar and Niv (2015) proposed that mood oscillations and information-processing biases may be governed by a dynamic interaction between mood and outcome valuation. Specifically, their model proposed that the reward value of outcomes R_t is biased by a mood-dependent factor f^{m_t} in the calculation of prediction errors:

$$\delta = f^{m_t} \cdot R_t - Q_t(s_t, a_t) \tag{37.5}$$

Here, $-1 \leq m_t \leq 1$ represents mood at trial t, with negative values of m_t denoting negatively valenced moods and positive values of m_t denoting positively valenced moods. f is a parameter governing the strength of the interaction between mood and outcome valuation such that values of f greater than 1 indicate mood-congruent changes in outcome valuation (i.e., the overestimation of outcome value in good moods and the underestimation of outcome value in bad moods).

The model also proposes that mood changes over time according to a weighted average of recent reward prediction errors that is transformed to lie between −1 and 1 by a sigmoidal function:

$$h_{t+1} = h_{t+1} + \eta_\eta \cdot (\delta - h_t) \tag{37.6}$$

$$m_t = \tanh(\delta - h_t) \tag{37.7}$$

where η_h is a learning-rate parameter for this reward prediction error history. Together, equations 37.5–37.7 specify a dynamic system in which reward prediction errors trigger the mood-congruent processing of subsequent rewards. This, in turn, leads to escalatory mood dynamics that may explain the emergence of mania and depression in bipolar disorder.

There is an important parallel between this model of bipolar disorder and the models of depression reviewed above. Specifically, the form of equation 37.5 closely resembles that of the reward-sensitivity model of depression in equation 37.3, as posited by Huys et al. (2013). The difference between the two models is that Huys et al. (2013) posit a trait-level parameter ρ to govern blunted reward sensitivity in depression, whereas Eldar and Niv (2015) propose a mood-dependent term f^{m_t}.

This comparison may be instructive. In reviewing the models of depression above, we observed that the reward-sensitivity model of depression posited by Huys et al. (2013) made predictions similar to a model in which depression affected not the hedonic value of rewards (through ρ) but rather the asymmetry between the effects of positive and negative reward prediction errors (through η^+ and η^-). A similar principle applies to models of bipolar disorder. This means that an alternative model to that of Eldar and Niv (2015) is one in which mood affects not the hedonic value of rewards but the relative strength of learning from positive versus negative reward prediction errors:

$$Q_{t+1}(s_t, a_t) = \begin{cases} Q_t(s_t, a_t) + f^{m_t} \cdot \eta^+ \cdot \delta, & \delta > 0 \\ Q_t(s_t, a_t) + f^{-m_t} \cdot \eta^- \cdot \delta, & \delta < 0 \end{cases} \tag{37.8}$$

where δ is defined according to equation 37.3, not equation 37.5. The cognitive interpretation of equation 37.8 is that positive moods lead to increases in learning rate from positive reward prediction errors and decreases in learning from negative reward prediction errors and vice versa for negative moods.

Here, too, the reward-sensitivity model of Eldar and Niv (2015) and the model specified by equation 37.8 make different predictions concerning attitudes toward risk in bipolar disorder. This is because equation 37.8, but not the model of Eldar and Niv (2015), predicts that positive moods should be associated with decreased risk aversion (increased risk seeking). This is consistent with a large body of evidence suggesting that mania and hypomania are associated with increased risk-taking behavior (e.g., Mason, O'Sullivan, Montaldi, Bentall, & El-Deredy, 2014; Thomas, Knowles, Tai, & Bentall, 2007), as well as with diagnostic guidelines specifying risk-taking as a symptom of bipolar disorder in the *DSM-5*. Testing this prediction via behavioral model fitting in bipolar disorder is therefore a key task for future research.

Conclusion

In the 17th century, Robert Burton compared psychiatric illness to a clock in which one faulty gear interfered with the operation of the whole machine. In adapting this metaphor, we realize that in every age the brain has been likened to the most sophisticated contemporary machine—including clocks, steam locomotives, and now digital computers—none of which the brain is likely all that similar to. Nevertheless, the present chapter has considered how, given such a clock, we might apply computational methods to determine which gear is at fault. We have reviewed the history of a computational approach to psychiatric illness, with a focus on the current state of the art for reinforcement-learning models of major depression and bipolar disorder. Cutting-edge future research in this field will involve two lines of work: research to identify the algorithmic principles that govern human mood and affect and research to characterize how these algorithms go awry in psychiatric illness. Our contention is that these questions are best addressed by adapting computational cognitive models to human behavioral data.

A strong version of our behavioral argument holds that it is only by making distinct predictions about human behavior that psychiatric theories can meaningfully differ from one another. After all, if two different psychiatric theories made entirely equivalent predictions about behavior (and therefore about all phenomenological aspects of a patient's experience that are accessible to empirical inquiry), it would be reasonable to conclude that these two theories were functionally isomorphic, even if they proposed seemingly dissimilar theoretical constructs to explain psychiatric dysfunction (Putnam, 1975). A less strong, more pragmatic version of this same argument is that by adopting the quantitative prediction of behavior as the ground truth of psychiatric theory, it is relatively straightforward to reject theories that may seem conceptually sound while making no sensible predictions regarding behavior (e.g., Houghton, 1969). A focus on the prediction of behavior evaluates theories according to their empirical content and not the sophistication of their mathematical superstructures.

If it is true that scientific revolutions occur not necessarily because of serendipitous discovery but because certain scientists come to ask better questions, then the promise of computational psychiatry lies in the nature of the questions that it can ask about psychiatric illness. We propose that as a source for such questions, computational cognitive models are a critically important tool. Such models can be used to identify the nature of the computations employed by the brain, the role of aberrant computations in the production of psychiatric illness, and the potential biological and cognitive remedies for computational dysfunction.

Acknowledgment

This work was supported by a CJ Martin Early Career Fellowship (#1165010) to DB from the NHMRC.

REFERENCES

Admon, R., & Pizzagalli, D. A. (2015). Dysfunctional reward processing in depression. *Current Opinion in Psychology, 4,* 114–118.

Alloy, L. B., Abramson, L. Y., Walshaw, P. D., Cogswell, A., Grandin, L. D., Hughes, M. E., et al. (2008). Behavioral approach system and behavioral inhibition system sensitivities and bipolar spectrum disorders: Prospective prediction of bipolar mood episodes. *Bipolar Disorders, 10*(2), 310–322.

Alloy, L. B., Reilly-Harrington, N. A., Fresco, D. M., & Flannery-Schroeder, E. (2005). Cognitive vulnerability to bipolar spectrum disorders. In Lauren B. Alloy & John H. Riskind (Eds.), *Cognitive vulnerability to emotional disorders* (pp. 93–124). Hillsdale, NJ: Erlbaum.

Amsterdam, J. D., Settle, R. G., Doty, R. L., Abelman, E., & Winokur, A. (1987). Taste and smell perception in depression. *Biological Psychiatry, 22*(12), 1481–1485.

Beck, A. T. (1967). *Depression: Clinical, experimental, and theoretical aspects.* Philadelphia: University of Pennsylvania Press.

Beer, M. D. (1996). The dichotomies: Psychosis/neurosis and functional/organic: A historical perspective. *History of Psychiatry, 7*(26), 231–255.

Bishop, S. J. (2007). Neurocognitive mechanisms of anxiety: An integrative account. *Trends in Cognitive Sciences, 11*(7), 307–316.

Blaney, P. H. (1977). Contemporary theories of depression: Critique and comparison. *Journal of Abnormal Psychology, 86*(3), 203.

Blaney, P. H. (1986). Affect and memory: A review. *Psychological Bulletin, 99*(2), 229.

Bower, G. H. (1981). Mood and memory. *American Psychologist, 36*(2), 129.

Burton, R. (1847). *The anatomy of melancholy.* New York: Wiley and Putnam. (Original work published 1621.)

Callaway, E. (1970). Schizophrenia and interference: An analogy with a malfunctioning computer. *Archives of General Psychiatry, 22*(3), 193–208.

Cohen, J. D., & Servan-Schreiber, D. (1992). Context, cortex, and dopamine: A connectionist approach to behavior and biology in schizophrenia. *Psychological Review, 99*(1), 45–77.

Colby, K. M. (1964). Experimental treatment of neurotic computer programs. *Archives of General Psychiatry, 10*(3), 220–227.

Colby, K. M., Hilf, F. D., Weber, S., & Kraemer, H. C. (1972). Turing-like indistinguishability tests for the validation of a computer simulation of paranoid processes. *Artificial Intelligence, 3,* 199–221.

Daw, N. D., Gershman, S. J., Seymour, B., Dayan, P., & Dolan, R. J. (2011). Model-based influences on humans' choices and striatal prediction errors. *Neuron, 69*(6), 1204–1215.

Dayan, P., & Niv, Y. (2008). Reinforcement learning: The good, the bad and the ugly. *Current Opinion in Neurobiology, 18*(2), 185–196.

Doll, B. B., Simon, D. A., & Daw, N. D. (2012). The ubiquity of model-based reinforcement learning. *Current Opinion in Neurobiology, 22*(6), 1075–1081.

Eldar, E., & Niv, Y. (2015). Interaction between emotional state and learning underlies mood instability. *Nature Communications, 6,* 6149.

Eldar, E., Rutledge, R. B., Dolan, R. J., & Niv, Y. (2016). Mood as representation of momentum. *Trends in Cognitive Sciences, 20*(1), 15–24.

Eshel, N., & Roiser, J. P. (2010). Reward and punishment processing in depression. *Biological Psychiatry, 68*(2), 118–124.

Friston, K. J., Stephan, K. E., Montague, R., & Dolan, R. J. (2014). Computational psychiatry: The brain as a phantastic organ. *Lancet Psychiatry, 1*(2), 148–158.

Fürstner, C. (1881). Über delirium acutum. *Archiv für Psychiatrie und Nervenkrankheiten, 11,* 517–531.

Garrett, N., Sharot, T., Faulkner, P., Korn, C. W., Roiser, J. P., & Dolan, R. J. (2014). Losing the rose tinted glasses: Neural substrates of unbiased belief updating in depression. *Frontiers in Human Neuroscience, 8,* 639.

Gershman, S. J. (2015). Do learning rates adapt to the distribution of rewards? *Psychonomic Bulletin & Review, 22*(5), 1320–1327.

Goodwin, F. K., & Jamison, K. R. (2007). *Manic-depressive illness: Bipolar disorders and recurrent depression.* Oxford: Oxford University Press.

Gotlib, I. H., & Joormann, J. (2010). Cognition and depression: Current status and future directions. *Annual Review of Clinical Psychology, 6*, 285–312.

He, Q., Su, S., & Du, R. (2008). Separating mixed multi-component signal with an application in mechanical watch movement. *Digital Signal Processing, 18*(6), 1013–1028.

Henriques, J. B., & Davidson, R. J. (2000). Decreased responsiveness to reward in depression. *Cognition & Emotion, 14*(5), 711–724.

Henriques, J. B., Glowacki, J. M., & Davidson, R. J. (1994). Reward fails to alter response bias in depression. *Journal of Abnormal Psychology, 103*(3), 460.

Hoffman, R. E. (1987). Computer simulations of neural information processing and the schizophrenia-mania dichotomy. *Archives of General Psychiatry, 44*(2), 178–188.

Hoffman, R. E., & Dobscha, S. K. (1989). Cortical pruning and the development of schizophrenia: A computer model. *Schizophrenia Bulletin, 15*(3), 477–490.

Hopfield, J. J. (1982). Neural networks and physical systems with emergent collective computational abilities. *Proceedings of the National Academy of Sciences, 79*(8), 2554–2558.

Houghton, G. (1969). A lie group topology for normal and abnormal human behavior. *Bulletin of Mathematical Biophysics, 31*(2), 275–293.

Huys, Q. J., Daw, N. D., & Dayan, P. (2015). Depression: A decision-theoretic analysis. *Annual Review of Neuroscience, 38*, 1–23.

Huys, Q. J., Eshel, N., O'Nions, E., Sheridan, L., Dayan, P., & Roiser, J. P. (2012). Bonsai trees in your head: How the Pavlovian system sculpts goal-directed choices by pruning decision trees. *PLoS Computational Biology, 8*(3), e1002410.

Huys, Q. J., Pizzagalli, D. A., Bogdan, R., & Dayan, P. (2013). Mapping anhedonia onto reinforcement learning: A behavioral meta-analysis. *Biology of Mood & Anxiety Disorders, 3*(1), 12.

Ingram, R. E. (1984). Toward an information-processing analysis of depression. *Cognitive Therapy and Research, 8*(5), 443–477.

Joseph, M. H., Frith, C. D., & Waddington, J. L. (1979). Dopaminergic mechanisms and cognitive deficit in schizophrenia. *Psychopharmacology, 63*(3), 273–280.

King-Casas, B., Sharp, C., Lomax-Bream, L., Lohrenz, T., Fonagy, P., & Montague, P. R. (2008). The rupture and repair of cooperation in borderline personality disorder. *Science, 321*(5890), 806–810.

Korn, C., Sharot, T., Walter, H., Heekeren, H., & Dolan, R. (2014). Depression is related to an absence of optimistically biased belief updating about future life events. *Psychological Medicine, 44*(3), 579–592.

Leahy, R. L., Tirch, D. D., & Melwani, P. S. (2012). Processes underlying depression: Risk aversion, emotional schemas, and psychological flexibility. *International Journal of Cognitive Therapy, 5*(4), 362–379.

Lewinsohn, P. M. A. (1974). A behavioral approach to depression. In R. J. Friedman & M. M. Katz (Eds.), *The psychology of depression: Contemporary theory and research* (pp. 157–184). Washington, DC: V. H. Winston.

Maia, T. V., & Frank, M. J. (2011). From reinforcement learning models to psychiatric and neurological disorders. *Nature Neuroscience, 14*(2), 154.

Mason, L., O'Sullivan, N., Montaldi, D., Bentall, R. P., & El-Deredy, W. (2014). Decision- making and trait impulsivity in bipolar disorder are associated with reduced prefrontal regulation of striatal reward valuation. *Brain, 137*(8), 2346–2355.

Mihatsch, O., & Neuneier, R. (2002). Risk-sensitive reinforcement learning. *Machine Learning, 49*(2–3), 267–290.

Miller, G. A., Galanter, E., & Pribram, K. H. (1960). *Plans and the structure of behavior.* New York: Henry Holt.

Nelson, R. E., & Craighead, W. E. (1977). Selective recall of positive and negative feedback, self-control behaviors, and depression. *Journal of Abnormal Psychology, 86*(4), 379.

Niv, Y., Edlund, J. A., Dayan, P., & O'Doherty, J. P. (2012). Neural prediction errors reveal a risk-sensitive reinforcement-learning process in the human brain. *Journal of Neuroscience, 32*(2), 551–562.

Pizzagalli, D. A., Jahn, A. L., & O'Shea, J. P. (2005). Toward an objective characterization of an anhedonic phenotype: A signal-detection approach. *Biological Psychiatry, 57*(4), 319–327.

Putnam, H. (1975). Philosophy and our mental life. In *Philosophical papers vol. 2: Mind, language, and reality* (pp. 291–303). Cambridge: Cambridge University Press.

Rashevsky, N. (1964). A neurobiophysical model of schizophrenias and of their possible treatment. *Bulletin of Mathematical Biophysics, 26*(2), 167–185.

Rehm, L. P. (1977). A self-control model of depression. *Behavior Therapy, 8*(5), 787–804.

Rescorla, R. A., & Wagner, A. R. (1972). A theory of Pavlovian conditioning: Variations in the effectiveness of reinforcement and nonreinforcement. In A. H. Black & W. F. Prokasy (Eds.), *Classical conditioning II: Current research and theory* (Vol. 2, pp. 64–99). New York: Appleton-Century-Crofts.

Robinson, O. J., Cools, R., Carlisi, C. O., Sahakian, B. J., & Drevets, W. C. (2012). Ventral striatum response during reward and punishment reversal learning in unmedicated major depressive disorder. *American Journal of Psychiatry, 169*(2), 152–159.

Rumelhart, D. E., & McClelland, J. L. (1987). *Parallel distributed processing* (Vol. 1). Cambridge, MA: MIT Press.

Ruppin, E. (1995). Neural modelling of psychiatric disorders. *Network: Computation in Neural Systems, 6*(4), 635–656.

Rutledge, R. B., Moutoussis, M., Smittenaar, P., Zeidman, P., Taylor, T., Hrynkiewicz, L., … Dolan, R. J. (2017). Association of neural and emotional impacts of reward prediction errors with major depression. *JAMA Psychiatry, 74*(8), 790–797.

Santesso, D. L., Steele, K. T., Bogdan, R., Holmes, A. J., Deveney, C. M., Meites, T. M., & Pizzagalli, D. A. (2008). Enhanced negative feedback responses in remitted depression. *Neuroreport, 19*(10), 1045.

Sartorius, N., Üstün, T. B., Lecrubier, Y., & Wittchen, H.-U. (1996). Depression comorbid with anxiety: Results from the WHO study on "psychological disorders in primary health care." *British Journal of Psychiatry, 168*(S30), 38–43.

Schuck, N. W., Cai, M. B., Wilson, R. C., & Niv, Y. (2016). Human orbitofrontal cortex represents a cognitive map of state space. *Neuron, 91*(6), 1402–1412.

Schultz, W., Dayan, P., & Montague, P. R. (1997). A neural substrate of prediction and reward. *Science, 275*(5306), 1593–1599.

Seligman, M. E. P. (1975). *Helplessness: On depression, development, and death.* New York: W. H. Freeman.

Showers, C., & Ruben, C. (1990). Distinguishing defensive pessimism from depression: Negative expectations and positive coping mechanisms. *Cognitive Therapy and Research, 14*(4), 385–399.

Silverstein, S. M., Wibral, M., & Phillips, W. A. (2017). Implications of information theory for computational modeling of schizophrenia. *Computational Psychiatry, 1*, 82–101.

Smoski, M. J., Lynch, T. R., Rosenthal, M. Z., Cheavens, J. S., Chapman, A. L., & Krishnan, R. R. (2008). Decision-making and risk aversion among depressive adults. *Journal of Behavior Therapy and Experimental Psychiatry, 39*(4), 567–576.

Spitzer, M. (1995). A neurocomputational approach to delusions. *Comprehensive Psychiatry, 36*(2), 83–105.

Stein, D. J., & Ludik, J. (1998). *Neural networks and psychopathology: Connectionist models in practice and research.* Cambridge: Cambridge University Press.

Sutton, R. S., & Barto, A. G. (1998). *Reinforcement learning: An introduction.* Cambridge, MA: MIT Press.

Thomas, J., Knowles, R., Tai, S., & Bentall, R. P. (2007). Response styles to depressed mood in bipolar affective disorder. *Journal of Affective Disorders, 100*(1), 249–252.

Vrieze, E., Pizzagalli, D. A., Demyttenaere, K., Hompes, T., Sienaert, P., de Boer, P., ... Claes, S. (2013). Reduced reward learning predicts outcome in major depressive disorder. *Biological Psychiatry, 73*(7), 639–645.

Walley, R. E., & Weiden, T. D. (1973). Lateral inhibition and cognitive masking: A neuropsychological theory of attention. *Psychological Review, 80*(4), 284–302.

Whitmer, A. J., & Gotlib, I. H. (2013). An attentional scope model of rumination. *Psychological Bulletin, 139*(5), 1036.

Whitton, A. E., Treadway, M. T., & Pizzagalli, D. A. (2015). Reward processing dysfunction in major depression, bipolar disorder and schizophrenia. *Current Opinion in Psychiatry, 28*(1), 7.

Wiener, N. (1948). Cybernetics. *Scientific American, 179*(5), 14–19.

Wiersma, J. E., van Oppen, P., Van Schaik, D., Van der Does, A., Beekman, A., & Penninx, B. (2011). Psychological characteristics of chronic depression: A longitudinal cohort study. *Journal of Clinical Psychiatry, 72*(3), 288–294.

Winterer, G., & Weinberger, D. R. (2004). Genes, dopamine and cortical signal-to-noise ratio in schizophrenia. *Trends in Neurosciences, 27*(11), 683–690.

38 Executive Control and Decision-Making: A Neural Theory of Prefrontal Function

ETIENNE KOECHLIN

ABSTRACT In mammals, the prefrontal cortex is one of the brain regions that has evolved the most. The prefrontal cortex primarily subserves executive control and decision-making. In this chapter we describe how prefrontal function may have evolved from rodents to monkeys and humans by progressively implementing increasingly sophisticated inferential, selective, and creative processes that gradually optimize adaptive behavior in uncertain, changing, and open-ended environments. We outline how this evolution may have contributed to endowing humans with unique, high-level cognitive faculties like language and reasoning.

The prefrontal cortex (PFC) is one of the brain regions that has evolved the most in humans compared to nonhuman animals, and it prominently contributes to uniquely human cognitive abilities such as judgment, reasoning, and language. The PFC has appeared in mammalian brains in front of the (pre)motor cortex (Uylings, Groenewegen, & Kolb, 2003). In rodents, the PFC comprises the orbitofrontal cortex (OFC) and the anterior cingulate cortex (ACC; Uylings, Groenewegen, & Kolb, 2003). The PFC has further evolved in monkeys with the apparition of the lateral PFC (laPFC; Fuster, 1989). In humans, further evolutions are observed: (1) the development of a lateral region in the frontal pole, usually referred to as the frontopolar cortex (poPFC) and mainly connected with neighboring PFC regions and with no homologs in nonhuman (possibly nonhominoid) monkeys (Koechlin, 2011; Mansouri, Koechlin, Rosa, & Buckley, 2017; Neubert, Mars, Thomas, Sallet, & Rushworth, 2014; Semendeferi, Armstrong, Schleicher, Zilles, & Van Hoesen, 2001; Teffer & Semendeferi, 2012); (2) the emergence of left-right asymmetry, yielding to the notion of Broca's area in the left caudal laPFC (Schenker et al., 2010; Uylings, Jacobsen, Zilles, & Amunts, 2006), which plays a prominent role in language (Broca, 1861); (3) a decreased connectivity between the temporal cortex and the ACC accompanied by an increased connectivity between the caudal laPFC (including Broca's area in the left hemisphere) and superior temporal cortex (Neubert et al., 2014). A key feature is that all these PFC regions form parallel loop circuits with basal ganglia (Alexander, DeLong, & Strick, 1986). Basal ganglia are subcortical brain nuclei common to vertebrates that especially comprise the striatum subserving reinforcement learning (RL; Doya, 2007; Samejima, Ueda, Doya, & Kimura, 2005; Schultz, 1997; Stephenson-Jones, Samuelsson, Ericsson, Robertson, & Grillner, 2011). RL, and more specifically its temporal-difference algorithmic implementation (Sutton & Barto, 1998), is a basic adaptive behavior process that adjusts online stimulus-action associations according to the discrepancy between actual and expected rewards. RL is a very simple, robust, and efficient adaptive process that can learn complex tasks even in uncertain environments. In particular, when rewards only depend upon current external states and actions, RL potentially converges toward the behavioral strategy maximizing rewards (Sutton & Barto, 1998). Reinforcing signals in the ventral striatum, like reward prediction errors, serve to adjust stimulus-action associations, while the dorsal striatum along with the premotor cortex guides action selection based on learned stimulus-action associations (Atallah, Lopez-Paniagua, Rudy, & O'Reilly, 2007; Kahnt et al., 2009; O'Doherty et al., 2004; Samejima et al., 2005). However, RL has severe adaptive limitations, suggesting that the PFC has primarily evolved to overcome these limitations. Here, we propose a comprehensive theory of PFC function based on this premise. We first identity two major limitations in RL adaptive capabilities in view of the adaptive behavior problem the individual faces. We then describe the PFC evolution from rodents to humans as the gradual addition of new control and inferential capabilities that progressively tackle the adaptive behavior problem more efficiently. We next show how the resulting human PFC executive system guiding adaptive behavior may have contributed to the emergence of reasoning and language abilities. We conclude by notably discussing how the present theoretical framework potentially dismisses two central premises commonly used to conceptualize the PFC function—namely, the notion of goal-directed behavior and utility maximization.

Beyond Reinforcement Learning: The Adaptive-Planning Problem

RL has a first major limitation when an animal's internal state (e.g., needs) changes and alters rewards' subjective value. In RL, problematically, the strength of learned stimulus-action associations scales with rewards' subjective values when learning occurs and may subsequently become highly maladaptive when these values change (Balleine & Dickinson, 1998; Dickinson, 1985). For instance, consider two actions A and B, which in a given situation lead to water and food, respectively. If the animal is thirsty but replete, RL will reinforce action A relative to B. When the situation reoccurs, the animal will then select action A rather than B. However, this behavior is certainly maladaptive when the animal becomes hungry rather than thirsty. The problem arises because in basic RL (also referred to as *model-free* RL), only stimulus-action associations are learned. These associations form an internal model, referred to as a *selective model,* that guides behavior without learning and using action-outcome associations per se (e.g., A, water vs. B, food). Overcoming this limitation thus requires learning an internal model, referred to as a *predictive* model, that encodes action-outcome associations in response to stimuli. This model simply learns the statistical occurrences of actual outcomes given actions and current states.

Learning selective and predictive models in parallel allows for selecting actions based on stimuli and action outcomes, respectively. Moreover, predictive models enable the internal emulation of RL without physical action (Sutton & Barto, 1998): as predictive models predict outcomes from actions derived from selective models, the rewarding values of action outcomes may be internally experienced according to the agent's current motivational state (e.g., thirst or hunger), yielding stimulus-action associations in selective models to be adjusted accordingly through standard RL algorithms. This emulation that reflects covert planning is commonly referred to as *model-based* RL. Selective models thus adjust to the agent's motivational state before guiding overt behavior. Behavioral studies confirm that animal behavior comprises both a model-free and a model-based RL component (Gershman, Markman, & Otto, 2014; Otto, Gershman, Markman, & Daw, 2013; Simon & Daw, 2011). Some authors have proposed that model-free and model-based RL actually form two competitive instrumental systems guiding behavior. Arbitrating between the two systems would rely on the relative uncertainty/reliability about reward and outcome expectations drawn from selective and predictive models, respectively (Daw, Niv, & Dayan, 2005; Lee, Shimojo, & O'Doherty, 2014). However, recent behavioral results support the idea that model-free and model-based RL instead form two cooperative systems, with model-free RL guiding overt behavior while model-based RL covertly runs off-line to continuously adjust model-free RL (Gershman et al., 2014; Pezzulo, Rigoli, & Chersi, 2013; Sutton & Barto, 1998). This cooperative combination of model-free and model-based RL enables faster learning but still leaves open the problem of their relative contribution to behavior. As the OFC appears to encode predictive models (Jones et al., 2012; Wilson, Takahashi, Schoenbaum, & Niv, 2014), we argue here that the PFC has evolved to enable model-based RL—namely, to regulate when and how much covert model-based RL needs to invest before acting under the guidance of model-free RL.

Beyond Reinforcement Learning: The Adaptive Inference Problem

A second major limitation of model-free/model-based RL is that learning new contingencies occurs by gradually erasing previously learned contingencies. This limitation has little impact when previously learned situations never reoccur or even when the environment comprises only a constant number of recurrent situations, as RL processes easily generalize to such closed environments (see Doya, Samejima, Katagiri, & Kawato, 2002). The limitation becomes problematic when, in addition to presenting recurrent situations, the environment is *open-ended* by constantly featuring new situations that were never experienced in the past and may even become recurrent in the future. With no additional mechanisms identifying recurrent and new situations, learning new contingencies erases what was previously learned and consequently prevents the exploitation of the partially recurrent nature of the environment. Open-ended environments thus feature an infinite number of dimensions. As no physical systems can plausibly represent such an infinite-dimensional space in a parametric fashion, overcoming this RL limitation requires an animal to create new dimensions whenever it infers a new situation occurs. The animal will then gradually build an extended repertoire of discrete dimensions, or mental sets, that will ideally correspond to the various situations the animal has encountered. The problem then becomes how the animal infers that the current situation is new, in which case a new mental set should be created and learned (through RL), versus a recurrent situation, in which case previously created and learned mental sets should be retrieved to guide behavior.

An optimal solution to this adaptive probabilistic inference problem exists, which is usually referred as to *mixtures of Dirichlet processes* (MDP; Collins & Koechlin, 2012; Doshi-Velez, 2009; Gershman, Blei, & Niv, 2010; Teh, Jordan, Beal, & Blei, 2006). However, this mathematical solution is computationally intractable (Collins & Koechlin, 2012) because (1) probabilistic inferences bear upon a number of mental sets that indefinitely grow with time; (2) as creating mental sets is a nonparametric, discrete (all-or-none) event, optimality requires the flexibility to constantly revise, in a backward fashion, the history of set creation whenever new observations are made (in other words, reparameterize a nonparametric event). As a result, computational costs grow exponentially with time, which makes the MDP a biologically implausible solution to overcome the RL limitation. Here, we argue that the PFC has evolved from rodents to monkeys and humans by gradually adding new inferential capabilities approximating a better and better MDP solution to this adaptive inferential problem (Koechlin, 2014).

Task Sets as Basic Executive Blocks Driving Behavior

As noted above, overcoming major RL limitations requires considering selective and predictive models along with the creation of discrete mental sets for guiding adaptive behavior in open-ended environments. We therefore consider a mental set to primarily encompass the selective and predictive model that has learned the contingencies of the situation associated with the creation of this mental set. Such mental sets are thus fully equipped to drive adaptive behavior in a given situation and correspond to the psychological notion of *task sets* (Rogers & Monsell, 1995). We thus view the core PFC function as developing inferential processes to manage task sets guiding behavior (Sakai, 2008). Accordingly, task sets are abstract, discrete entities linking the selective and predictive models over which inferential processes in the PFC operate. Task sets instantiate situations deemed as *distinct latent states* through PFC inferential processes.

As the optimal solution to this management problem is computationally intractable (see above), the PFC function has presumably evolved to optimize task set management under some computational constraints. A first constraint is certainly the inability to monitor the whole repertoire of task sets created so far along the animal's life history. As behavioral results have confirmed (Collins & Koechlin, 2012; Donoso, Collins, & Koechlin, 2014), the PFC function is able to monitor and make inferences about only a limited number of task sets. This inferential buffer corresponds to the psychological notion of capacity-limited *working memory* (Cowan, 2005; Risse & Oberauer, 2010). The PFC function consequently has no access to the whole repertoire of previously created task sets to infer whether one task set or none fits the current situation or, equivalently, whether it faces a recurrent or new situation. This implies that the repertoire of task sets outside the inferential buffer is no longer within the scope of PFC function, and none of these task sets can be *directly* retrieved in a top-down fashion to guide behavior. A second constraint is likely the inability to make computationally costly MDP-like backward inferences (see above). The PFC function is likely based only on forward inference processes over task sets—that is, only inferring from past information the likely futures. Forward inference models indeed account better for subjects' adaptive performance than MDP models, which, when computable, largely outperform subjects' performances (Collins & Koechlin, 2012).

Accordingly, we assume that the PFC function has evolved to optimize task set management under these computational constraints. In rodents the emergence of the OFC and ACC is assumed to implement the *minimal* inferential capabilities required to overcome the two RL limitations outline above. In monkeys and humans, the inferential capabilities associated with the OFC and ACC are preserved while the development of the laPFC and poPFC provides additional inferential capabilities that further optimize task set management.

The Rodent Prefrontal Cortex: Executive Control as Factual Reactive Inference

The minimal inferential capability corresponds to an inferential buffer monitoring only one task set—that is, the one guiding ongoing behavior and learning current behavioral contingencies and referred to as the *actor*. And the minimal requirement to overcome RL limitations is to infer when the current situation changes to form a new actor. This inference relies on evaluating the actor ability to predict actual action outcomes. Our theory assumes that the development of paralimbic prefrontal regions (ACC and OFC) in lower mammals (rodents) implements these minimal inferential and executive processes guiding behavior (figure 38.1). Such inferences are *factual* as bearing only upon the actor and *reactive* as operating only after observing

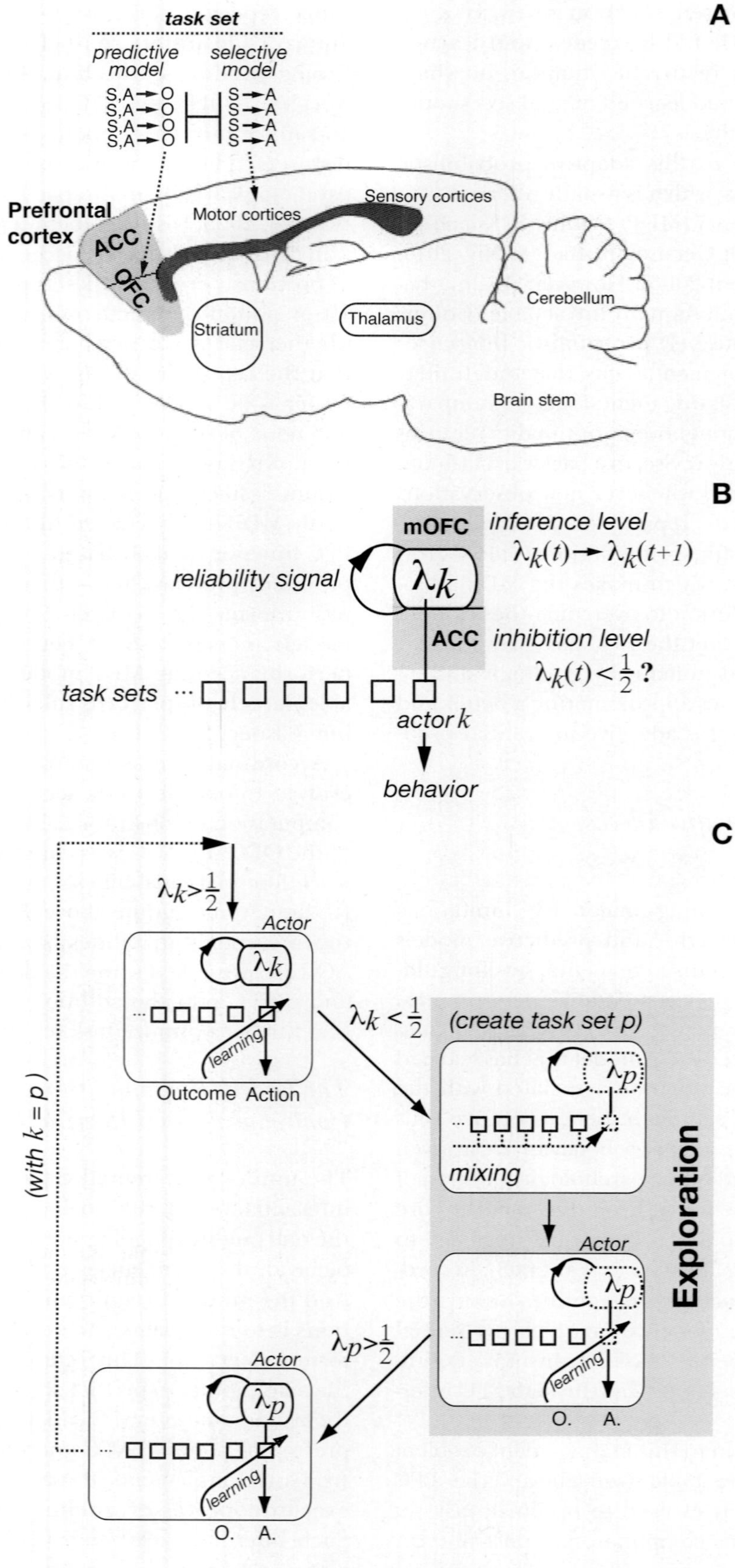
A
task set
predictive model
selective model
S,A→O
S,A→O
S,A→O
S,A→O
S→A
S→A
S→A
S→A
Prefrontal cortex
ACC
OFC
Motor cortices
Sensory cortices
Striatum
Thalamus
Cerebellum
Brain stem
B
mOFC
inference level
$\lambda_k(t) \rightarrow \lambda_k(t+1)$
reliability signal
λ_k
ACC
inhibition level
$\lambda_k(t) < \frac{1}{2}$?
task sets
actor k
behavior
C
$\lambda_k > \frac{1}{2}$
Actor
λ_k
learning
Outcome
Action
$\lambda_k < \frac{1}{2}$
(create task set p)
λ_p
mixing
Actor
λ_p
learning
O.
A.
Exploration
$\lambda_p > \frac{1}{2}$
Actor
λ_p
learning
O.
A.
(with k = p)

action outcomes. Adaptive behavior thus derives from either adjusting selective and predictive models while perseverating with the same actor or switching to a new actor for guiding subsequent behavior.

Arbitrating between these two alternatives is based on inferring actor *reliability*—that is, the posterior probability that the current situation remains the same or, equivalently, that the current external contingencies match those the actor has learned (Koechlin, 2014). Updating online actor reliability according to actual action outcomes involves forward Bayesian inferences comparing the likelihood of actual action outcomes according to the actor predictive model to their likelihood according to *any* potential predictive models (Koechlin, 2014). The latter cannot be exactly computed, but following the maximal entropy principle (Jaynes, 1957), this likelihood is estimated as the equiprobability of action outcomes *produced by* the actor (Koechlin, 2014). Actor reliability λ_t in every trial t serves to arbitrate between staying versus switching away from the current actor. While the actor remains more likely reliable than unreliable ($\lambda_t > 1 - \lambda_t$), the current situation is likely to remain unchanged. The same actor is then kept and continues to adjust to current external contingencies (notably, through RL). The system thus operates in an *exploitation* mode. When, conversely, the actor becomes unreliable ($\lambda_t < 1 - \lambda_t$) following the last action outcome, the situation has likely changed. A new actor is then optimally built from the whole repertoire of previously learned task sets stored in long-term memory (corresponding to all previously formed actors). The new actor selective (resp. predictive) model is thus formed as the mixture of selective (resp. predictive) models stored in the repertoire possibly weighted according to the last action outcome given task sets' predictive models (figure 38.1*B*).

As the new actor is created from the task sets' repertoire, its initial reliability corresponds to the repertoire reliability—that is, the posterior probability that the animal faces a previously encountered situation rather than a never-encountered situation. This reliability corresponds to the likelihood of the current action outcome in all previously encountered situations (deriving from the new actor predictive model mixing all previously learned predictive models) compared to that in a presumably new situation. The latter is actually equal to 1 because in this situation only one outcome has been observed so far. As a result, the new actor guiding behavior is initially inferred as being unreliable ($\lambda_0 < 1 - \lambda_0$). The system thus starts operating in an *exploration* mode with an unreliable actor. This promotes actor learning by preventing a switch away from the new actor *while* it remains unreliable. When the new actor becomes reliable, thanks to learning, the system then returns to the exploitation mode described above. Switching away from the actor and creating a new one may occur again when this actor becomes unreliable. Consequently, the lower the initial actor reliability, the more likely the situation was never encountered and the longer the exploration will last: initially, the new actor less likely matches the new situation and, consistently, more trials are required for the new actor to learn the new external contingencies.

Given its intrinsic computational constraints (forward, factual, and reactive inferences only), this executive system implements an optimal adaptive process in environments featuring both new and recurrent situations. The system especially exhibits two key functional properties consistent with empirical data. First, the model shows abrupt rather than gradual behavioral changes. When following variations in external contingencies, the actor becomes unreliable, and a new one is created. Such abrupt behavioral changes are routinely observed in rodent, monkey, and human experiments (Asaad, Rainer, & Miller, 1998; Karlsson, Tervo, & Karpova, 2012; Nassar, Wilson, Heasly, & Gold, 2010;

FIGURE 38.1 The proposed rodent prefrontal function. *A*, Schematic representation of the rodent brain. The prefrontal function involves the anterior cingulate cortex (ACC) and the orbitofrontal cortex (OFC) and manages the creation of actor task sets comprising selective and predictive models (i.e., stimulus-action associations and stimulus-action-outcome associations, respectively). Selective models are encoded in motor/premotor cortices, while the OFC encodes predictive models. *B*, Diagram showing inferential and inhibition processes composing the rodent prefrontal function (*square*: task sets stored in long-term memory). The OFC infers actor reliability λ_κ—that is, its ability to predict action outcomes to monitor when the situation changes. While the actor remains reliable ($\lambda_\kappa > 1 - \lambda_\kappa$), the actor drives behavior and adjusts its internal models (model-free RL; exploitation periods). The ACC detects when the actor becomes unreliable ($\lambda_\kappa < 1 - \lambda_\kappa$)—that is, the situation has presumably changed. The ACC inhibits the unreliable actor and triggers the creation of a new actor (initiating exploration periods). *C*, Diagram showing the transitions between exploitation and exploration periods corresponding to creating a new actor task set *p*. Actor creation results from mixing stored task sets yielding to forming an unreliable actor ($\lambda_p < 1 - \lambda_p$) and implementing model-based RL. While this newly created actor remains unreliable, it drives behavior and learns external contingencies (model-free RL; exploration period). When it subsequently becomes reliable, it is consolidated in long-term memory as an additional task set. A new exploitation period starts with again the possibility to subsequently create new actors. See text for explanation.

Pasupathy & Miller, 2005). Second, the long-term repertoire of task sets expands whenever new actors are created so that the *reoccurrence* frequency of external situations has a major influence on shaping new actors. Whenever external situations reoccur, indeed, new actors are created with selective and predictive models again learning the associated external contingencies, thereby replicating in the repertoire the selective and predictive models previously learned from previous occurrences: the more external situations reoccur, the more these models are then replicated in long-term memory and, consequently, the more they contribute to the formation of new actors. This executive system thus exhibits a basic feature of Dirichlet processes (Teh et al., 2006), whereby more recurrent situations are more likely to repeat in the future. Collins and Koechlin (2012) showed that this executive system accounts for increased human performance in recurrent situations.

Additionally, our theory logically assumes that actor formation triggers model-based RL before the new actor starts driving behavior and learning through model-free RL. Because new actors are initially built from previously learned task sets in long-term memory, this model-based RL directly adjusts off-line the new actor's selective model to animals' current needs before the latter drive behavior. Accordingly, when the executive system infers the situation has changed, animal adaptive behavior following this inference predominantly derives from model-based RL through actor creation, rather than model-free RL. By contrast, when the situation is subsequently inferred as perpetuating over time, animal adaptive behavior increasingly reflects model-free RL relative to model-based RL. Consistently, empirical studies provide evidence that behavioral performances predominantly reflect model-based RL after changes in external contingencies and subsequently reflect model-free RL when external contingencies remain stable (Lee, Shimojo, & O'Doherty, 2014). Finally, switching away from the current actor and creating a new actor that triggers model-based RL may also occur when the animal's motivational state (i.e,. its current needs) changes. This happens provided that in paralimbic regions, task sets' predictive models are assumed to encode action outcomes along the *ordinal* dimensions characterizing an animal's motivational states: the actor predictive model learns a collection of motivational values that are minimally expected from action outcomes experienced in the current situation. Actor reliability thus derives from the likelihood of actual action outcomes projected onto these ordinal dimensions (see above). Accordingly, when the animal's motivational state changes, actual action outcomes may be associated with motivational values below what is expected from the actor predictive model. The actor may then become unreliable—that is, the executive system infers that the situation has changed, thereby triggering the formation of a new actor that yields to switching from model-free to model-based RL. The present theory thus accounts for behavioral changes observed in rodents (e.g., extinction effects) reflecting transitions from model-free to model-based RL following experimental alterations of animals' motivational states (see, e.g., Daw, Niv, & Dayan, 2005).

As generally agreed, the premotor cortex encodes and stores stimulus-action associations learned through model-free RL (i.e., the selective models; Badre, Hoffman, Cooney, & D'Esposito, 2009; Hadj-Bouziane, Meunier, & Boussaoud, 2003; Histed, Pasupathy, & Miller, 2009; Koechlin, Ody, & Kouneiher, 2003; Packard & Knowlton, 2002), while the dorsal and ventral striatum implement model-free RL, adjusting the actor selective model (Doya, 2007; Samejima et al., 2005; Schultz, 1997; Stephenson-Jones et al., 2011). Empirical evidence further suggests that the OFC encodes task sets' predictive models and updates actor reliability according to action outcomes. Neural data in rodents, monkeys, and humans show that the lateral OFC is involved in learning motivational values associated with perceptual attributes of action outcomes, while the medial OFC in humans predicts and responds to action outcomes based on action outcome associations (Burke, Franz, Miller, & Schoenbaum, 2008; Hampton, Bossaerts, & O'Doherty, 2006; Jones et al., 2012; McDannald, Lucantonio, Burke, Niv, & Schoenbaum, 2011; Rouault, Drugowitsch, & Koechlin, 2019 Rudebeck & Murray, 2014; Walton, Behrens, Buckley, Rudebeck, & Rushworth, 2010). Accordingly, task sets appear as spatially long-range neural frames linking selective and predictive models encoded in premotor cortex and OFC, respectively. In humans, neuroimaging studies further show that medial OFC activations encode current latent states determining action outcome contingencies (Schuck, Cai, Wilson, & Niv, 2016; Stalnaker, Cooch, & Schoenbaum, 2015) and correlate with actor reliability (Donoso, Collins, & Koechlin, 2014). Additionally, actor reliability reflects the confidence in perseverating with the same task set—that is, repeating the same response to the same situation/stimuli. Human neuroimaging studies consistently show that subjects' confidence judgments are associated with medial OFC activations (De Martino, Fleming, Garrett, & Dolan, 2013; Lebreton, Abitbol, Daunizeau, & Pessiglione, 2015).

By contrast, empirical evidence suggests that the ACC is involved in detecting when the actor becomes unreliable and in triggering the creation of new actors.

Rodent studies show that in the ACC, neuronal activity exhibits abrupt phase transitions in neural ensembles in relation to behavioral switches (Durstewitz, Vittoz, Floresco, & Seamans, 2010; Hayden, Pearson, & Platt, 2011; Karlsson, Tervo, & Karpova, 2012; Tervo et al., 2014). Moreover, monkey electrophysiological and human neuroimaging studies indicate that the ACC is involved in monitoring when to switch from exploitation to exploration behaviors (Kolling, Behrens, Mars, & Rushworth, 2012; Quilodran, Rothe, & Procyk, 2008) and exhibits phasic activations, especially when the actor task set becomes unreliable (Donoso, Collins, & Koechlin, 2014). More generally, the ACC is involved in responding to surprising outcomes triggering behavioral switches (Alexander & Brown, 2011) and in starting the execution of new tasks (Dosenbach et al., 2006).

Anatomical and functional data (Alexander, DeLong, & Strick, 1986; Badre, Kayser, & D'Esposito, 2010; Donoso, Collins, & Koechlin, 2014; Liljeholm & O'Doherty, 2012) further suggest that frontal-striatal loop circuits involving the premotor cortex, ACC, and OFC may subserve actor creation. The ACC may prevent the striatum from filtering out nonactor task sets and allow stored task sets to mix for forming new actors: the mixture of selective and predictive models, given current action outcomes, may then occur in the premotor cortex and OFC, respectively. Actor creation is further assumed to rely on model-based RL that critically implicates predictive models. Consistent with the hypothesis, functional magnetic resonance imaging (fMRI) studies have shown the OFC and striatum to be implicated in model-based RL (McDannald et al., 2011; Wunderlich, Dayan, & Dolan, 2012).

The Monkey Prefrontal Cortex: Executive Control as Factual Proactive Inference

The rodent PFC system described above has two critical limitations. First, new actor task sets are created only in reaction to experiencing action outcomes, which may be detrimental with adverse outcomes. Second, and perhaps more problematically, actor creation ignores the context in which stored task sets were learned and mainly relies on the reoccurrence frequency of encountered situations. Accordingly, our theory assumes that the development of the laPFC in monkeys and humans implements an additional inferential capability overcoming these limitations—namely, using external cues (in addition to action outcomes) for inferring when creating a new actor (figure 38.2). In contrast to action outcomes, external cues occur regardless of subjects' behavior so that their occurrences may inform about changes in external contingencies before acting. The resulting executive system thus exhibits *proactive* inferences in arbitrating between adjusting versus creating actors.

For making such proactive inferences, the actor comprises and learns an additional internal model, which we refer to as the *contextual model*. The actor contextual model simply learns the occurrence and frequency of external cues and is stored in long-term memory along with selective and predictive models. Task sets thus comprise selective, predictive, and contextual models. When external cues occur, the actor reliability is then updated through forward Bayesian inferences comparing the likelihood of these external cues derived from the actor contextual model with their likelihood according to *any* potential contextual models (Koechlin, 2014). The latter is again not exactly computable. External cues are, however, presumed independent of subjects' behavior. This likelihood can therefore be estimated as the frequency of current external cues observed in the past, which simply derives from the mean of the contextual models stored in long-term memory (Koechlin, 2014). Consequently, external cues that are less likely to occur in the current than past situations degrade the actor reliability and may yield to proactively switching away from the current actor to create a new one, as described above.

When a new actor is created following reactive and/or proactive inferences, its selective, predictive, and contextual models are again computed as the respective mixtures of these models across the repertoire of task sets stored in long-term memory. These mixtures, however, are now additionally weighted according to current external cues given contextual models. Accordingly, actor creation now depends upon current external cues in addition to the reoccurrence frequency of previously encountered situations: task sets learned within more distant contexts to the current one contribute less to actor creation, while task sets learned within closer contexts contribute more to actor creation.

The initial reliability of new actors again corresponds to the repertoire reliability—that is, the posterior probability that the animal faces a previously encountered rather than a never-encountered situation. This probability is now evaluated according to current external cues (along with action outcomes when actor creation occurs in response to action outcomes). The probability thus corresponds to the likelihood of current external cues in all previously encountered situations (deriving from the new actor contextual model—that is, the weighted mixture of stored contextual models) compared to that in a presumably new situation. As external cues are presumably independent of subjects' behavior, the latter likelihood is simply estimated as the frequency

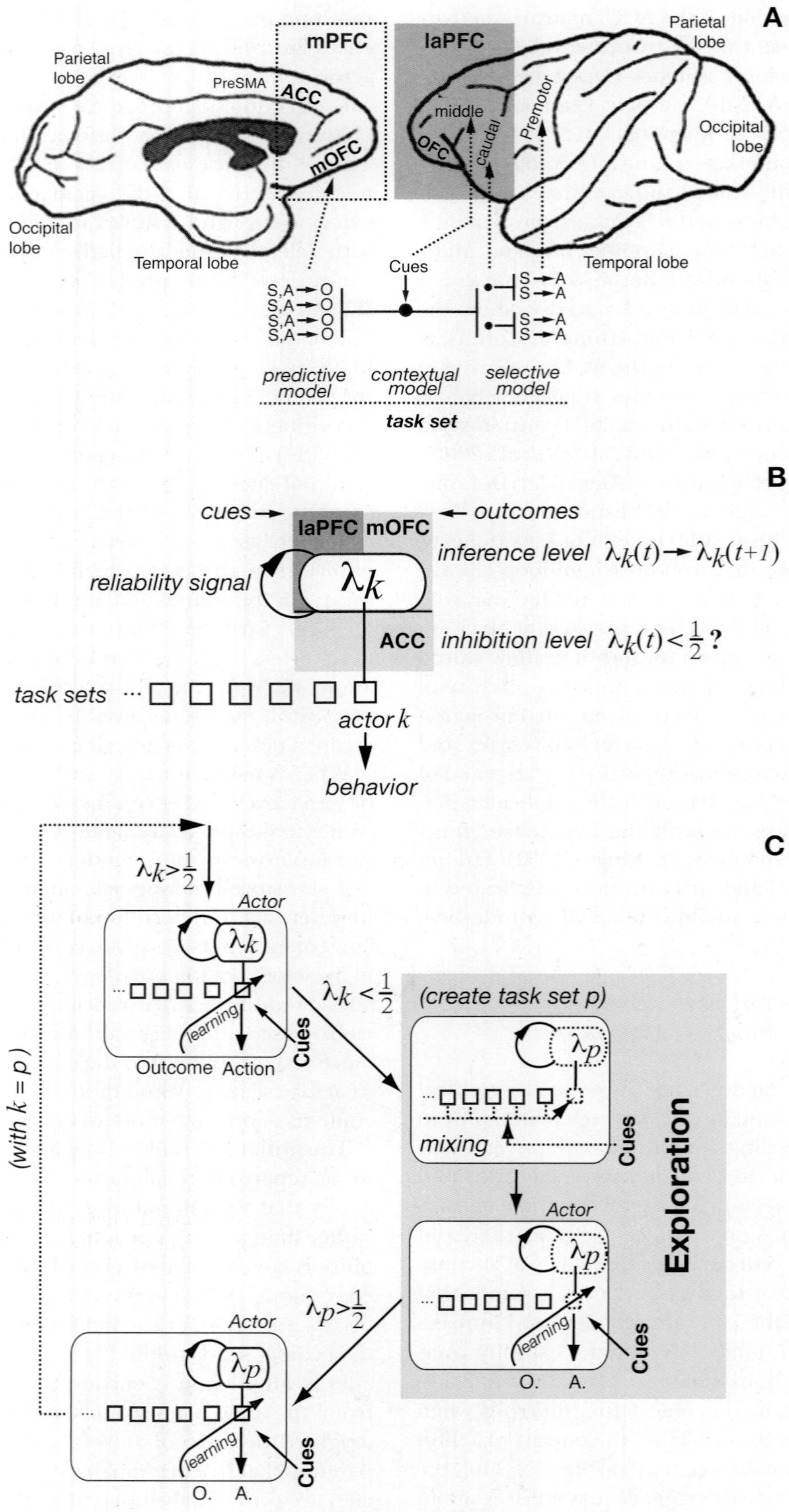
A
mPFC
laPFC
Parietal lobe
PreSMA
ACC
mOFC
middle
OFC
caudal
Premotor
Parietal lobe
Occipital lobe
Occipital lobe
Temporal lobe
Cues
Temporal lobe
S,A→O
S,A→O
S,A→O
S,A→O
S→A
S→A
S→A
S→A
predictive model
contextual model
selective model
task set
B
cues
outcomes
laPFC
mOFC
inference level $\lambda_k(t) \rightarrow \lambda_k(t+1)$
reliability signal
λ_k
ACC
inhibition level $\lambda_k(t) < \frac{1}{2}$?
task sets
actor k
behavior
C
$\lambda_k > \frac{1}{2}$
Actor
λ_k
learning
Cues
Outcome Action
$\lambda_k < \frac{1}{2}$
(create task set p)
λ_p
mixing
Cues
Exploration
Actor
λ_p
learning
Cues
O. A.
(with k = p)
$\lambda_p > \frac{1}{2}$
Actor
λ_p
learning
Cues
O. A.

of these external cues observed in the past, which corresponds with the mean rather than the weighted mixture of stored contextual models. Consequently, new actors may now be initially inferred as being immediately reliable ($\lambda_0 > 1 - \lambda_0$), thereby limiting exploration periods (figure 38.2*B*). This may happen when current external cues are highly specific to a given situation, and the repertoire contains task sets learned in the presence of such cues. In that event, actor creation resembles retrieving these task sets directly according to current external cues and may lead new actors to be rejected as soon as they guide behavior when becoming unreliable. This proactive executive system thus provides the ability to flexibly control behavior by rapidly recreating new actors yielding to switch across learned task sets according to external cues. This form of executive control has also been termed *episodic control*, in the sense that it enables the learning and maintenance of task sets guiding ongoing behavior over time, along with their retrieval (*through actor creation*) with respect to episodic occurrences of external cues (Koechlin, Ody, & Kouneiher, 2003; Koechlin & Summerfield, 2007).

Under its intrinsic computational constraints (forward and factual inferences only), this computational model optimally uses external cues in addition to action outcomes for adapting to environments featuring both new and recurrent situations. The present theory assumes that in monkeys, the laPFC and, more specifically, its middle sector (typically Brodmann's areas 46/9) learns and encodes task sets' contextual models for updating actor reliability in connection with the OFC and for triggering the creation of new actors through the ACC with respect to external cues. Accordingly, the middle laPFC represents task sets as abstract discrete nodes linked to external cues. Connected with both the premotor cortex and the OFC (Ongur & Price, 2000; Pandya & Yeterian, 1996; Tomassini et al., 2007), the middle laPFC is thus assumed to form a central hub of task set representations associated with external cues and linking selective models in premotor regions and predictive models in OFC. There is ample empirical evidence from both monkey electrophysiological recordings and human neuroimaging and lesion studies supporting the idea that the middle laPFC constitutes this central node subserving episodic control (Azuar et al., 2014; Badre, 2008; Badre & D'Esposito, 2007; Bahlmann, Aarts, & D'Esposito, 2015; Koechlin, Ody, & Kouneiher, 2003; Koechlin & Summerfield, 2007; Kouneiher, Charron, & Koechlin, 2009; Nee & D'Esposito, 2016, 2017; Passingham & Wise, 2012; Sakai & Passingham, 2003). In human neuroimaging experiments, furthermore, effective connectivity analyses measuring information flows across frontal regions provide evidence that episodic control in middle laPFC operates in a top-down fashion onto premotor regions for retrieving selective models guiding ongoing behavior (Koechlin, Ody, & Kouneiher, 2003; Nee & D'Esposito, 2016, 2017). The middle laPFC also has major reciprocal connections with the ACC (Beckmann, Johansen-Berg, & Rushworth, 2009; Medalla & Barbas, 2009, 2010). Consistently, the notion of episodic control described here still assumes that the ACC detects when the actor becomes unreliable, inhibits it, and triggers the *creation* of new task sets through the middle laPFC to serve as the actor. There is no actor *selection*, per se. In agreement with this view, dorsal ACC activations were observed (at least in humans) to influence middle laPFC activations irrespective of task set selection processes (Kouneiher, Charron, & Koechlin, 2009).

Contextual models associate task sets to external cues. As described above, contextual models are learned so that these cues reflect any stimuli acting as predictors of task set reliability. By contrast, selective models associate actions to stimuli so that through RL, these stimuli act as predictors of action values when the corresponding task set is the actor. This scheme leaves open the possibility that the same stimulus is involved in both contextual and selective models. Additionally, selective

FIGURE 38.2 The proposed monkey prefrontal function. *A*, Schematic medial and lateral representations of the monkey cerebral cortex. Compared to rodents, the monkey cortex additionally has a lateral prefrontal cortex (laPFC) comprising a middle and caudal sector. Task sets are thus assumed to further comprise contextual models (associating task set to external cues) encoded in the middle laPFC. Contextual models indexing task sets allow chunking processes in caudal laPFC to operate within task sets (see text). *B*, Diagram showing inferential and inhibition processes composing the monkey prefrontal function (*square*: task sets stored in long-term memory). Inferential and inhibition processes are similar to those in rodents (see figure 38.1) except that contextual models enable an update to actor reliability according to the occurrences of external cues (in addition to action outcomes). *C*, Diagram showing the transitions between exploitation and exploration periods corresponding to creating a new actor task set *p*. These transitions are similar to those in rodent prefrontal function (figure 38.1) except that contextual models allow actor creation to also occur proactively in response to external cues before acting. Contextual models also have a major role in shaping actor creation: the mixture of task sets in long-term memory is now weighted by current external cues according to contextual models. As a result, new actors may be created as immediately reliable ($\lambda_p(0) > 1 - \lambda_p(0)$; see text). In that event the exploration period is skipped, yielding to the ability to recreate new actors much more rapidly.

models are able to learn through RL combinations of stimuli predicting action values. However, empirical evidence is that in the presence of such predictive combinations (e.g., colors and shapes), subjects spontaneously form hierarchical rather than flat stimulus-action mappings (Badre, Kayser, & D'Esposito, 2010; Collins, Cavanagh, & Frank, 2014; Collins & Frank, 2013): one stimulus (e.g., shapes) is mapped onto responses while another (e.g., colors) is preferentially mapped onto this set of stimulus-action associations, which we refer to as *action chunks*. Such hierarchical structures in selective models are built even when there are no immediate behavioral advantages in forming these representations (Collins, Cavanagh, & Frank, 2014; Collins & Frank, 2013). Yet these hierarchical structures favor the generalization of subordinate stimulus-action mappings to new combinations (Collins & Frank, 2013). These structures are conditionally formed through hierarchical inferential processes upon the assumption that external contingencies remain stable over time (Collins & Frank, 2013)—that is, upon the inference that the situation remains unchanged and, consequently, that the same task set is maintained as the actor driving ongoing behavior. Such hierarchical selective models are thus learned and embedded within task sets and allow switching across action chunks according to immediate cues within the same actor task set. This hierarchical form of executive control thus forms an intermediate control level operating across *hierarchical levels* and embedded in the episodic control of task sets operating along the *temporal dimension* (Koechlin, 2007; Koechlin, Ody, & Kouneiher, 2003; Koechlin & Summerfield, 2007).

Human neuroimaging studies provide evidence that the caudal laPFC in the front of the premotor cortex forms hierarchical selective models within task sets. While the premotor cortex encodes stimulus-action associations (see above), the caudal laPFC is engaged in forming action chunks (corresponding to stimulus-action mappings or action sequences) associated with concomitant cues (Badre, Kayser, & D'Esposito, 2010; Koechlin, Danek, Burnod, & Grafman, 2002). Moreover, the caudal laPFC is engaged when subjects select responses to stimuli based on such hierarchical selective models or, equivalently, when subjects' responses to stimuli are contingent upon concomitant cues (Alamia et al., 2016; Azuar et al., 2014; Badre & D'Esposito, 2007; Badre et al., 2009; Balaguer, Spiers, Hassabis, & Summerfield, 2016; Dippel & Beste, 2015; Duverne & Koechlin, 2017; Koechlin, Ody, & Kouneiher, 2003). In neuroimaging studies, furthermore, effective connectivity analyses measuring information flows from middle laPFC to premotor cortex provide evidence that the middle laPFC representing the actor task set controls the selection of action chunks in caudal laPFC, which in turn control the selection of stimulus-action associations in the premotor cortex (Koechlin, Ody, & Kouneiher, 2003; Kouneiher, Charron, & Koechlin, 2009), thereby reflecting a top-down hierarchy of selection processes from the middle to caudal laPFC and premotor cortex.

In the same way the ACC is involved in inhibiting the actor task set represented in the middle laPFC when it is deemed unreliable, empirical evidence is that the presupplementary motor area (pre-SMA) posterior to the ACC in the dorsomedial PFC is engaged in inhibiting subordinate components within the actor's hierarchical selective model. The pre-SMA is activated at the onset of external cues, inducing switches across action chunks (Hikosaka & Isoda, 2010; Nachev, Kennard, & Husain, 2008) followed by caudal LaPFC activations (Jha et al., 2015; Neubert, Mars, Buch, Olivier, & Rushworth, 2010; Rae, Hughes, Anderson, & Rowe, 2015; Swann et al., 2012). Consistently, the pre-SMA and caudal laPFC are involved in inhibiting irrelevant responses to stimuli (Aron, Behrens, Smith, Frank, & Poldrack, 2007; Aron, Robbins, & Poldrack, 2014; Hikosaka & Isoda, 2010; Isoda & Hikosaka, 2007; Nachev et al., 2008; Nachev, Wydell, O'Neill, Husain, & Kennard, 2007).

The Human Prefrontal Cortex: Executive Control as Counterfactual Inferences

The monkey executive system described above has one major limitation. Inferences about the perpetuation versus termination of the current situation yielding to maintain the same actor or to create a new one are only factual: such inferences bear only upon the actor reliability, based on its predictive and contextual model. Accordingly, our theory assumes that the development of the frontopolar cortex (poPFC) in humans endows them with an additional inferential capability overcoming this limitation—namely, inferring *when* the current situation changes, as well as *which* alternative, previously encountered situations might reoccur instead. The human executive system is thus assumed to develop *counterfactual* inferences about the reliability of alternative task sets, which are *not* guiding ongoing behavior (figure 38.3). These counterfactual inferences are able to infer online concomitantly when to change the actor and which previously learned task sets might be selected as the new actor. Optimally, counterfactual inferences should bear upon the whole repertoire of stored task sets. This seems, however, computationally costly and biologically implausible. Accordingly, counterfactual inferences are assumed to develop only over a limited number of task sets, forming the *inferential buffer*.

One might consider the inferential buffer as forming a global actor guiding behavior by mixing online monitored task sets over the buffer with respect to their relative reliability (Doya et al., 2002). Collins and Koechlin (2012) showed that this hypothesis is inconsistent with human behavioral performance in sequential decision tasks. This is also theoretically suboptimal because the global actor may be inferred as reliable with only unreliable task sets, while another task set stored in long-term memory but outside the inferential buffer would be reliable. More optimally, the human executive system is assumed to concurrently infer the reliability of every monitored task set i, and when *none* are inferred as being reliable (more likely not applicable than applicable to the current situation, i.e., $\lambda_t^i < 1-\lambda_t^i$), a new task set is created from long-term memory to serve as actor and added to the inferential buffer (Collins & Koechlin, 2012; Koechlin, 2014; figure 38.3*B*). When conversely one (i_0) is inferred as being reliable ($\lambda_t^{i_0} > 1-\lambda_t^{i_0}$ or, equivalently, $\lambda_t^{i_0} > 1/2$), the others are necessary unreliable, even when considered collectively: by construction, indeed, inferred reliabilities sum up to 1 or less, as the current situation may match no monitored task sets (Collins & Koechlin, 2012; Koechlin, 2014). Accordingly, the reliable task set becomes *the* actor guiding behavior and learning external contingencies by adjusting its selective, predictive, and contextual model The inferential buffer is thus assumed to comprise the actor plus a number of alternative task sets, which we refer to as the *counterfactual* task sets.

The actor may thus be replaced rather than adjusted either by *retrieving* and *switching to* a reliable counterfactual task set or by *creating* a new task set from long-term memory, as described above (figure 38.3*B*). In the former case, the executive system continues to operate in the exploitation mode because the new actor is initially deemed reliable. In the latter case, the new actor may be created as unreliable, in which event the inferential system switches into the *exploration* mode. The executive system may then return to the exploitation mode in two ways (Collins & Koechlin, 2012; Koechlin, 2014): the newly created actor becomes reliable, thanks to learning, while the counterfactual task sets remain unreliable and the former is *confirmed* and stored in long-term memory with others, or a counterfactual task set becomes reliable, while the newly created actor remains unreliable. The former then becomes the actor, and the latter is *rejected* from the buffer and *disbanded*. Exploration periods thus correspond to probing newly created actors before storing them in long-term memory when they are deemed reliable (figure 38.3*B*). Accordingly, counterfactual task sets are the former actors that have been reliably assigned to an external situation that previously occurred. When newly created actors are confirmed, however, the number of task sets in the inferential buffer increases and possibly reaches the buffer capacity limit. In that event, the task set used the least recently as actor is simply assumed to leave the inferential buffer. The rationale is that older situations are potentially less frequent and, consequently, less likely to reoccur in the short run. The inferential buffer thus keeps monitoring counterfactual task sets, which are more likely to match the *next* external situation.

The computations implementing this counterfactual executive system are essentially the same as those described above. Reactive and proactive inferences are simply extended to counterfactual task sets. The differences are as follows (Collins & Koechlin, 2012): First, the reliability of every monitored task set is now inferred by comparing the likelihood of action outcomes/external cues derived from the task set predictive/contextual model with that derived from the predictive/contextual model of other monitored task sets, in addition to that derived from any alternative predictive/contextual models. Second, the action outcome likelihood derived from any predictive models is now better estimated as the equiprobability of outcomes registered by both the actor *and* counterfactual task sets. Collins and Koechlin (2012) show that the full executive system comprising factual, counterfactual, reactive, and proactive inferences reproduced human adaptive performances in environments featuring both recurrent and new situations associated with uncertain and variable contingencies along with possible occurrences of external cues. Moreover, they show that all the system components were necessary for accounting for human performances. Furthermore, the best account was found when the buffer capacity corresponds to two/three counterfactual task sets. This size matches the capacity previously proposed for human (declarative) working memory (Cowan, 2005).

There is converging evidence from human neuroimaging studies that the poPFC is involved in monitoring counterfactual task sets. The poPFC is engaged in *cognitive branching*, when subjects temporarily hold off executing one task to perform another task in response to unpredictable events (Charron & Koechlin, 2010; Koechlin, Basso, Pietrini, Panzer, & Grafman, 1999; Koechlin, Corrado, Pietrini, & Grafman, 2000; Koechlin & Hyafil, 2007). Furthermore, the poPFC is involved in monitoring the opportunity to switch back and forth between two alternative courses of action (Boorman, Behrens, & Rushworth, 2011; Boorman, Behrens,

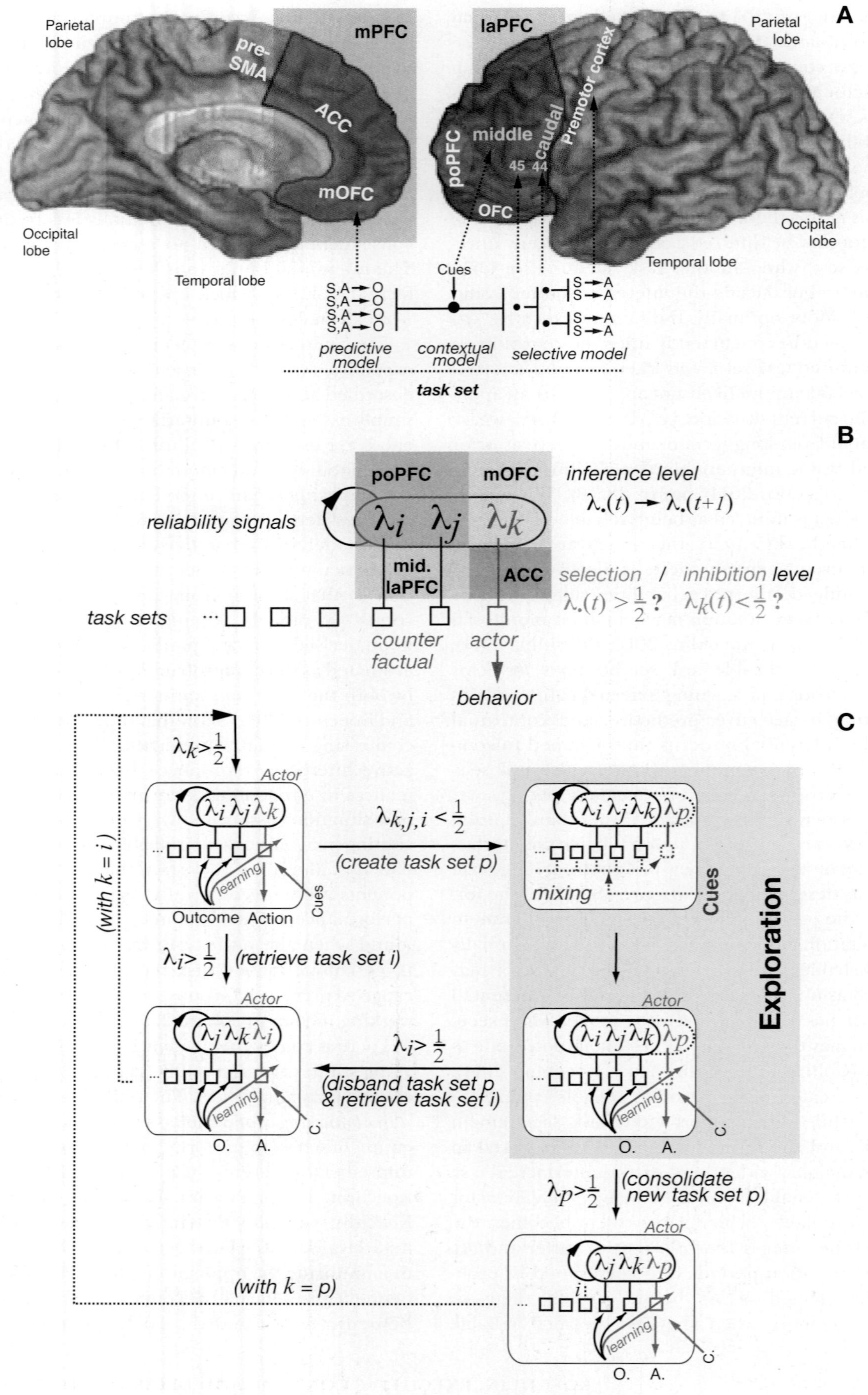

A
Parietal lobe
pre-SMA
mPFC
ACC
mOFC
Occipital lobe
Temporal lobe
laPFC
Premotor cortex
middle
caudal
poPFC
45
44
OFC
Parietal lobe
Occipital lobe
Temporal lobe
Cues
S,A→O
S→A
predictive model
contextual model
selective model
task set
B
poPFC
mOFC
inference level
$\lambda_{\cdot}(t) \rightarrow \lambda_{\cdot}(t+1)$
reliability signals
λ_i λ_j λ_k
mid. laPFC
ACC
selection / inhibition level
$\lambda_{\cdot}(t) > \frac{1}{2}$? $\lambda_k(t) < \frac{1}{2}$?
task sets
counter factual
actor
behavior
C
$\lambda_k > \frac{1}{2}$
(with $k = i$)
Actor
learning
Cues
Outcome Action
$\lambda_{k,j,i} < \frac{1}{2}$
(create task set p)
mixing
Cues
Exploration
$\lambda_i > \frac{1}{2}$ (retrieve task set i)
$\lambda_i > \frac{1}{2}$
(disband task set p & retrieve task set i)
O. A.
C.
$\lambda_p > \frac{1}{2}$ (consolidate new task set p)
(with $k = p$)

Woolrich, & Rushworth, 2009). More recent neuroimaging results even provide direct evidence that the poPFC monitors the reliability of two concurrent counterfactual task sets (Donoso, Collins, & Koechlin, 2014). By contrast, the middle laPFC is engaged when one counterfactual task set becomes reliable and is retrieved as the actor for guiding behavior (Donoso, Collins, & Koechlin, 2014). Consistent with its role as a central hub representing task sets as abstract entities linking together selective, predictive, and contextual models (see the section on the monkey PFC), the middle laPFC thus appears to detect when one counterfactual task set monitored in the poPFC becomes reliable for selecting it as the actor—that is, retrieving its embedded selective, predictive, and contextual models through laPFC projections to caudal laPFC and OFC to guide behavior. Finally, highly specific activations in the ventral striatum have been observed when newly created actors become reliable (Donoso, Collins, & Koechlin, 2014). As the ventral striatum is involved in reinforcement learning, this finding supports the idea that when newly created actors become reliable, they are consolidated in long-term memory as regular task sets.

It is worth noting that the human PFC executive system outlined here endows humans with the ability to switch between two learned stimulus-response mappings (i.e., action chunks), according to external cues, in three different ways. First, the two action chunks may belong to two distinct task sets monitored in the inferential buffer. Task switching then results from one task set (the actor) becoming unreliable while the other (the counterfactual one) becomes reliable. In that case, task switching presumably engages the poPFC and, in a top-down fashion, the middle and caudal laPFC. Second, the two action chunks may still belong to distinct task sets, but only one among these task sets is monitored in the inferential buffer and serving as actor. Task switching then results from the actor becoming unreliable, yielding to create a new actor from long-term memory that resembles the task set retrieval, as mentioned above (see the section on the monkey PFC). In that case, task switching presumably engages the middle and caudal laPFC. Third, the two action chunks belong to the same task set comprising a hierarchical selective model associating these chunks to external cues (see the monkey PFC section). Task switching then results from action chunk selection within the hierarchical selective model of this unique task set. In that case, task switching presumably engages the caudal laPFC only. Although the three cases result in apparently the same simple cognitive operation—namely, task switching (Rogers & Monsell, 1995)—they actually correspond to radically distinct inferential/control processes based on different subjective representations and constructs of the environment contingencies. While all cases involve the caudal laPFC, only the first and second case involve the middle laPFC, and only the first one involves the poPFC. These nesting activation effects have been reported in the same neuroimaging study (Koechlin et al., 1999). They may also explain discrepancies sometimes observed across studies investigating task-switching operations in differently administered and framed behavioral paradigms (Badre, 2008). Through extensive additional training, case 1 is likely to reduce to case 2 and then 3 distinct task sets may gradually merge into a unique task set driving behavior. Consistent with this prediction, prefrontal regions have been reported to gradually disengage from poPFC to caudal laPFC during learning sequences of action chunks (Koechlin et al., 2002; Sakai et al., 1998).

Finally, action chunk sequences are an example of *superordinate chunks*—that is, chunks of chunks. Neuroimaging studies provide evidence that in humans, hierarchical selective models driving action selection within task sets indeed comprise two hierarchical levels associated with distinct regions in caudal laPFC: (1) a lower level associated with the posterior sector of caudal laPFC (typically, BA 44) involved in selecting action

FIGURE 38.3 The proposed human prefrontal function. *A*, Schematic representations of the human cerebral cortex showing the prefrontal cortex (PFC). Compared to monkeys, the human cortex comprises a frontopolar region (poPFC) in the lateral forefront of the PFC with no known homolog regions in monkeys. Additionally, the caudal sector in the lateral PFC witnesses the development of Brodman's area (BA) 44 and 45, yielding to the notion of Broca's area in the left hemisphere. In humans compared to monkeys, task sets thus comprise two nested, abstract levels of chunking involving BA 44 and 45 and playing a major role in language. *B*, Inferential, inhibition, and selection processes forming the human PFC function. Compared to monkeys (see figure 38.2), the human poPFC allows for inferring and monitoring the reliability of a few task sets comprising the actor and three/four counterfactual (i.e., unreliable and not driving current behavior) task sets. This inferential buffer enables the middle lateral PFC to directly select/retrieve a counterfactual task set as an actor when it becomes reliable. *C*, Diagram showing the transitions between exploitation and exploration periods corresponding to creating a new actor task set *p* when no monitored task sets are reliable, to reject and disband newly created actor *p* during exploration periods when one counterfactual task set again becomes reliable to serve as actor, or to confirm and consolidate newly created actor *p* in long-term memory. These transitions realize hypothesis testing has a bearing upon task set creation. See the text for an explanation. (See color plate 40.)

chunks according to external cues or as elements of superordinate chunks; (2) a higher level associated with the anterior sector of caudal laPFC (typically BA 45) involved in selecting superordinate chunks according to cues (Badre, 2008; Koechlin & Jubault, 2006; Koechlin & Summerfield, 2007; figure 38.3*A*). Accordingly, task sets represented in middle laPFC comprise hierarchical selective models implementing a top-down hierarchy of selection processes operating from middle to anterior and posterior caudal laPFC and up to the premotor cortex that encode simple stimulus-action associations.

From Executive Control to Language and Reasoning

The human executive system outlined above may help to understand how the evolution of the PFC may have contributed to the emergence of human language. Language production is certainly the most advanced example of hierarchically organized behavior. As any behavior, first, speech primarily unfolds over time as a sequence of sentences, each forming a consistent temporal episode of words hierarchically organized according to syntactic rules. In that sense, sentences may be viewed as task sets comprising hierarchically organized selective models. We thus conceptualize speech as producing a series of task sets. This sequential production is based on inferential processes involving, as mentioned above, medial OFC and dorsal ACC, along with middle laPFC and poPFC, monitoring their successive reliability—that is, to which extent each task set/sentence is applicable to the ongoing discourse situation. Neuroimaging studies confirm that these PFC regions are involved in discourse generation (review in Bourguignon, 2014).

Second, sentence generation may be viewed as actor creation, which as indicated above primarily involves the caudal PFC (i.e., Broca's area and its right homolog) and premotor cortex bilaterally (Donoso, Collins, & Koechlin, 2014). Consistently, neuroimaging studies confirm the central role of Broca's area in sentence generation (see the review in Bourguignon, 2014). We have proposed above that actor creation and, more specifically, the creation of new selective models, consists in mixing previously stored selective models weighted by contextual cues according to associated contextual models. Mathematically, this operation is able to generate any selective models within the high-dimensional space comprising all combinations of previously learned selective models and, consequently, might account for sentence generation. By contrast, for poorly learned nonnative languages, processing complex multiutterance sentences was found to involve the middle laPFC and poPFC (Jeon & Friederici, 2015). In this case, sentence processing might simply require generating successive utterances as independent task sets and consequently engaging these anterior PFC regions.

Third, Broca's area and its right homolog implement selective models controlling action selection through two *nested, abstract* levels of chunking that we have referred to as action chunks and superordinate chunks (Koechlin & Jubault, 2005). Mathematically, such a two-level abstract chunking capability is sufficient to generate nested tree structures of unlimited depth, providing that, through a loop circuit, low-level chunks may instantiate high-level chunks in a recursive manner. Such nested tree structures are considered to be fundamental characteristics of the human faculty of language (Dehaene, Meyniel, Wacongne, Wang, & Pallier, 2015). The increased connectivity between posterior language areas (superior temporal cortex) and Broca's area in humans compared to monkeys (Neubert et al., 2014) might constitute this loop circuit (figure 38.3*B*) and, consequently, serve to generate such nested tree structures accounting for the evolution of language (Rouault & Koechlin, 2018). Recent studies support this view, as Broca's area is causally engaged in processing nested-tree structures (Udden, Ingvar, Hagoort, & Petersson, 2017). Beside production, language comprehension requires decoding the syntactic structure of sentences. This is a highly automatized process, at least for the native language, which also engages Broca's area (Jeon & Friederici, 2015; Udden et al., 2017). The same two-nested levels of abstract chunking that operate in Broca's area in connection with the superior temporal cortex may also be used to decode syntactic structures and, as a task-set creation process, to map complex sentences onto their semantic representation. In this view, the same neural circuit corresponding to the execution of a single task set is engaged in sentence production and comprehension with the activation and inactivation of motor outputs, respectively.

The proposed human executive system may also provide insights about the emergence of human reasoning (Donoso, Collins, & Koechlin, 2014; Oaksford & Chater, 2009). Inferring task-set reliability in medial OFC and poPFC is based on forward Bayesian inference processes regarding the possible latent causes, or *causal hypotheses*, determining the observed contingencies and instantiated through task sets. Reliability assessments in dorsal ACC and middle laPFC yield to binary judgments (reliable/unreliable) about the applicability of causal hypotheses to the current situation, which may be viewed as true/false judgments. The inferential buffer monitoring a few counterfactual task sets in the poPFC further endows humans with the ability to jointly consider several causal hypotheses simultaneously and consequently to realize *hypothesis testing* through actor creation: actor creation is equivalent to the formation of a new causal hypothesis from long-term memory when no monitored hypotheses are deemed reliable or true. This new hypothesis serving as actor is then tested as accounting for observed contingencies. The hypothesis is subsequently *confirmed* and consolidated in long-term memory as a subsequently recoverable hypothesis through activations in the ventral striatum when it is deemed reliable. Conversely, the hypothesis is *rejected* and disbanded when it remains unreliable while through middle laPFC activations one counterfactual hypothesis is finally deemed reliable. Hypothesis testing is the most basic form of backward inference, as the decision to form a new actor/hypothesis is subsequently revised according to the acquisition of subsequent information. Backward inferences are indeed critical in optimal inferential processes operating in open-ended environments for dealing with the intrinsic nonparametric nature of creating new latent causes (Teh et al., 2006). Through reliability judgments, accordingly, this PFC executive system crucially combines Bayesian inference and hypothesis-testing capabilities, which may constitute the foundations of human reasoning and creative abilities.

Concluding Remarks

The present theoretical framework provides a principled account of the PFC function as primarily optimizing adaptive behavior in uncertain, changing, and open-ended environments featuring both recurrent and new situations. Accordingly, the PFC function is described as implementing inferential and selection processes involved in: (1) creating task sets as instantiating inferred latent causes determining environment contingencies, (2) selecting and adjusting task sets as actors guiding behavior, and (3) storing task sets in long-term memory for subsequently contributing to the creation of new task sets. It is worth noting that this principled account dismisses two notions often viewed as central premises characterizing the PFC function: the notion of goal-directed behavior and utility maximization. The PFC function is indeed often conceptualized as guiding behavior according to internal goals, which problematically raises the issue of how goals are selected. The PFC function is also often conceptualized as maximizing action utility, which is computationally an intractable problem in uncertain environments featuring both new and recurrent situations. In the present theory, instead, actor task sets are selected or created based on reliability judgments assessing to which extent task sets are applicable to the current situation or, equivalently, to which extent task sets have learned enough of the current situation. Notably, actor task sets guide behavior through RL mechanisms that gradually converge to action selection processes maximizing action utility. As result, task selection may look like maximizing action utility, although the selection is actually based on task set reliability. Finally, the theoretical construct of goals as guiding behavior—and possibly as phenomenological experiences—might simply reflect that reliability judgments are primarily based, from rodents to humans, on the ability of medial OFC representations to predict action outcomes.

REFERENCES

Alamia, A., Solopchuk, O., D'Ausilio, A., Van Bever, V., Fadiga, L., Olivier, E., & Zenon, A. (2016). Disruption of Broca's area alters higher-order chunking processing during perceptual sequence learning. *Journal of Cognitive Neuroscience, 28*(3), 402–417.

Alexander, G. E., DeLong, M. R., & Strick, P. L. (1986). Parallel organization of functionally segregated circuits linking basal ganglia and cortex. *Annual Review of Neuroscience, 9*, 357–381.

Alexander, W. H., & Brown, J. W. (2011). Medial prefrontal cortex as an action-outcome predictor. *Nature Neuroscience, 14*(10), 1338–1344.

Aron, A. R., Behrens, T. E., Smith, S., Frank, M. J., & Poldrack, R. A. (2007). Triangulating a cognitive control network using diffusion-weighted magnetic resonance imaging (MRI) and functional MRI. *Journal of Neuroscience, 27*(14), 3743–3752.

Aron, A. R., Robbins, T. W., & Poldrack, R. A. (2014). Inhibition and the right inferior frontal cortex: One decade on. *Trends in Cognitive Sciences, 18*(4), 177–185.

Asaad, W. F., Rainer, G., Miller, E. K. (1998). Neural activity in the primate prefrontal cortex during associative learning. *Neuron, 21*(6), 1399–1407.

Atallah, H. E., Lopez-Paniagua, D., Rudy, J. W., & O'Reilly, R. C. (2007). Separate neural substrates for skill learning and performance in the ventral and dorsal striatum. *Nature Neuroscience, 10*(1), 126–131.

Azuar, C., Reyes, P., Slachevsky, A., Volle, E., Kinkingnehun, S., Kouneiher, F., … Levy, R. (2014). Testing the model of caudo-rostral organization of cognitive control in the human with frontal lesions. *NeuroImage, 84*(1), 1053–1060.

Badre, D. (2008). Cognitive control, hierarchy and the rostrocaudal organization of the frontal lobes. *Trends in Cognitive Sciences, 12*(5), 193–200.

Badre, D., & D'Esposito, M. (2007). Functional magnetic resonance imaging evidence for a hierarchical organization of the prefrontal cortex. *Journal of Cognitive Neuroscience, 19*(12), 2082–2099.

Badre, D., Hoffman, J., Cooney, J. W., & D'Esposito, M. (2009). Hierarchical cognitive control deficits following damage to the human frontal lobe. *Nature Neuroscience, 12*(4), 515–522.

Badre, D., Kayser, A. S., & D'Esposito, M. (2010). Frontal cortex and the discovery of abstract action rules. *Neuron, 66*(2), 315–326.

Bahlmann, J., Aarts, E., & D'Esposito, M. (2015). Influence of motivation on control hierarchy in the human frontal cortex. *Journal of Neuroscience, 35*(7), 3207–3217.

Balaguer, J., Spiers, H., Hassabis, D., & Summerfield, C. (2016). Neural mechanisms of hierarchical planning in a virtual subway network. *Neuron, 90*(4), 893–903.

Balleine, B. W., & Dickinson, A. (1998). Goal-directed instrumental action: Contingency and incentive learning and their cortical substrates. *Neuropharmacology, 37*(4–5), 407–419.

Beckmann, M., Johansen-Berg, H., & Rushworth, M. F. (2009). Connectivity-based parcellation of human cingulate cortex and its relation to functional specialization. *Journal of Neuroscience, 29*(4), 1175–1190.

Boorman, E. D., Behrens, T. E., & Rushworth, M. F. (2011). Counterfactual choice and learning in a neural network centered on human lateral frontopolar cortex. *PLoS Biology, 9*(6), e1001093.

Boorman, E. D., Behrens, T. E., Woolrich, M. W., & Rushworth, M. F. (2009). How green is the grass on the other side? Frontopolar cortex and the evidence in favor of alternative courses of action. *Neuron, 62*(5), 733–743.

Bourguignon, N. J. (2014). A rostro-caudal axis for language in the frontal lobe: The role of executive control in speech production. *Neuroscience & Biobehavioral Reviews, 47*, 431–444.

Broca, P. (1861). Remarques sur le siège de la faculté du langage articulé suivie d'une observation d'aphémie. *Bulletin de la Société d'Anatomie (Paris), 6*, 330.

Burke, K. A., Franz, T. M., Miller, D. N., & Schoenbaum, G. (2008). The role of the orbitofrontal cortex in the pursuit of happiness and more specific rewards. *Nature, 454*(7202), 340–344.

Charron, S., & Koechlin, E. (2010). Divided representation of concurrent goals in the human frontal lobes. *Science, 328*(5976), 360–363.

Collins, A. G., Cavanagh, J. F., & Frank, M. J. (2014). Human EEG uncovers latent generalizable rule structure during learning. *Journal of Neuroscience, 34*(13), 4677–4685.

Collins, A. G., & Frank, M. J. (2013). Cognitive control over learning: Creating, clustering, and generalizing task-set structure. *Psychological Review, 120*(1), 190–229.

Collins, A. G., & Koechlin, E. (2012). Reasoning, learning, and creativity: Frontal lobe function and human decision-making. *PLoS Biology, 10*(3), e1001293.

Cowan, N. (2005). Working-memory capacity limits in a theoretical context. In C. Izawa & N. Ohta (Eds.), *Human learning and memory: Advances in theory and applications* (pp. 155–175): Mahwah, NJ: Erlbaum.

Daw, N. D., Niv, Y., & Dayan, P. (2005). Uncertainty-based competition between prefrontal and dorsolateral striatal systems for behavioral control. *Nature Neuroscience, 8*(12), 1704–1711.

Dehaene, S., Meyniel, F., Wacongne, C., Wang, L., & Pallier, C. (2015). The neural representation of sequences: From transition probabilities to algebraic patterns and linguistic trees. *Neuron, 88*(1), 2–19.

De Martino, B., Fleming, S. M., Garrett, N., & Dolan, R. J. (2013). Confidence in value-based choice. *Nature Neuroscience, 16*(1), 105–110.

Dickinson, A. (1985). Actions and habits: The development of a behavioural autonomy. *Philosophical Transactions of the Royal Society of London B: Biological Sciences, 308*, 67–78.

Dippel, G., & Beste, C. (2015). A causal role of the right inferior frontal cortex in implementing strategies for multi-component behaviour. *Nature Communications, 6*, 6587.

Donoso, M., Collins, A. G., & Koechlin, E. (2014). Human cognition: Foundations of human reasoning in the prefrontal cortex. *Science, 344*(6191), 1481–1486.

Dosenbach, N. U., Visscher, K. M., Palmer, E. D., Miezin, F. M., Wenger, K. K., Kang, H. C., ... Petersen, S. E. (2006). A core system for the implementation of task sets. *Neuron, 50*(5), 799–812.

Doshi-Velez, F. (2009). The infinite partially observable Markov decision process. *Advances in Neural Information Processing Systems, 21*, 477–485.

Doya, K. (2007). Reinforcement learning: Computational theory and biological mechanisms. *Human Frontier Science Program Journal, 1*(1), 30–40.

Doya, K., Samejima, K., Katagiri, K., & Kawato, M. (2002). Multiple model-based reinforcement learning. *Neural Computation, 14*(6), 1347–1369.

Durstewitz, D., Vittoz, N. M., Floresco, S. B., & Seamans, J. K. (2010). Abrupt transitions between prefrontal neural ensemble states accompany behavioral transitions during rule learning. *Neuron, 66*(3), 438–448.

Duverne, S., & Koechlin, E. (2017). Rewards and cognitive control in the human prefrontal cortex. *Cerebral Cortex, 27*(10), 5024–5039.

Fuster, J. (1989). *The prefrontal cortex: Anatomy, physiology, and neuropsychology of the frontal lobes.* New York: Raven Press.

Gershman, S. J., Blei, D. M., & Niv, Y. (2010). Context learning, and extinction. *Psychological Review, 117*(1), 1997–1209.

Gershman, S. J., Markman, A. B., & Otto, A. R. (2014). Retrospective revaluation in sequential decision making: A tale of two systems. *Journal of Experimental Psychology: General, 143*(1), 182–194.

Hadj-Bouziane, F., Meunier, M., & Boussaoud, D. (2003). Conditional visuo-motor learning in primates: A key role for the basal ganglia. *Journal of Physiology, Paris, 97*(4–6), 567–579.

Hampton, A. N., Bossaerts, P., & O'Doherty, J. P. (2006). The role of the ventromedial prefrontal cortex in abstract state-based inference during decision making in humans. *Journal of Neuroscience, 26*(32), 8360–8367.

Hayden, B. Y., Pearson, J. M., & Platt, M. L. (2011). Neuronal basis of sequential foraging decisions in a patchy environment. *Nature Neuroscience, 14*(7), 933–939.

Hikosaka, O., & Isoda, M. (2010). Switching from automatic to controlled behavior: Cortico-basal ganglia mechanisms. *Trends in Cognitive Sciences, 14*(4), 154–161.

Histed, M. H., Pasupathy, A., & Miller, E. K. (2009). Learning substrates in the primate prefrontal cortex and striatum: Sustained activity related to successful actions. *Neuron, 63*(2), 244–253.

Isoda, M., & Hikosaka, O. (2007). Switching from automatic to controlled action by monkey medial frontal cortex. *Nature Neuroscience, 10*(2), 240–248.

Jaynes, E. T. (1957). Information theory and statistical mechanics. *Physical Review Serie II, 106*(4), 620–630.

Jeon, H. A., & Friederici, A. D. (2015). Degree of automaticity and the prefrontal cortex. *Trends in Cognitive Sciences, 19*(5), 244–250.

Jha, A., Nachev, P., Barnes, G., Husain, M., Brown, P., & Litvak, V. (2015). The frontal control of stopping. *Cerebral Cortex, 25*(11), 4392–4406.

Jones, J. L., Esber, G. R., McDannald, M. A., Gruber, A. J., Hernandez, A., Mirenzi, A., & Schoenbaum, G. (2012). Orbitofrontal cortex supports behavior and learning using inferred but not cached values. *Science, 338*(6109), 953–956.

Kahnt, T., Park, S. Q., Cohen, M. X., Beck, A., Heinz, A., & Wrase, J. (2009). Dorsal striatal-midbrain connectivity in humans predicts how reinforcements are used to guide decisions. *Journal of Cognitive Neuroscience, 21*(7), 1332–1345.

Karlsson, M. P., Tervo, D. G., & Karpova, A. Y. (2012). Network resets in medial prefrontal cortex mark the onset of behavioral uncertainty. *Science, 338*(6103), 135–139.

Koechlin, E. (2007). The cognitive architecture of human lateral prefrontal cortex. In P. Haggard, Y. Rossetti, & M. Kawato (Eds.), *Sensorimotor foundations of higher cognition: Attention & performance* (Vol. 22). Oxford: Oxford University Press.

Koechlin, E. (2011). Frontal pole function: What is specifically human? *Trends in Cognitive Sciences, 15*(6), 241.

Koechlin, E. (2014). An evolutionary computational theory of prefrontal executive function in decision-making. *Philosophical Transactions of the Royal Society of London B: Biological Sciences, 369*. doi:10.1098/rstb.2013.0474

Koechlin, E., Basso, G., Pietrini, P., Panzer, S., & Grafman, J. (1999). The role of the anterior prefrontal cortex in human cognition. *Nature, 399*(6732), 148–151.

Koechlin, E., Corrado, G., Pietrini, P., & Grafman, J. (2000). Dissociating the role of the medial and lateral anterior prefrontal cortex in human planning. *Proceedings of the National Academy of Sciences of the United States of America, 97*(13), 7651–7656.

Koechlin, E., Danek, A., Burnod, Y., & Grafman, J. (2002). Medial prefrontal and subcortical mechanisms underlying the acquisition of motor and cognitive action sequences in humans. *Neuron, 35*(2), 371–381.

Koechlin, E., & Hyafil, A. (2007). Anterior prefrontal function and the limits of human decision-making. *Science, 318*(5850), 594–598.

Koechlin, E., & Jubault, T. (2005). Broca's area and the hierarchical organization of human behavior. *Neuron, 15*(6), 963–974.

Koechlin, E., & Jubault, T. (2006). Broca's area and the hierarchical organization of human behavior. *Neuron, 50*(6), 963–974.

Koechlin, E., Ody, C., & Kouneiher, F. (2003). The architecture of cognitive control in the human prefrontal cortex. *Science, 302*(5648), 1181–1185.

Koechlin, E., & Summerfield, C. (2007). An information theoretical approach to prefrontal executive function. *Trends in Cognitive Sciences, 11*(6), 229–235.

Kolling, N., Behrens, T. E., Mars, R. B., & Rushworth, M. F. (2012). Neural mechanisms of foraging. *Science, 336*(6077), 95–98.

Kouneiher, F., Charron, S., & Koechlin, E. (2009). Motivation and cognitive control in the human prefrontal cortex. *Nature Neuroscience, 12*(7), 939–945.

Lebreton, M., Abitbol, R., Daunizeau, J., & Pessiglione, M. (2015). Automatic integration of confidence in the brain valuation signal. *Nature Neuroscience, 18*(8), 1159–1167.

Lee, S. W., Shimojo, S., & O'Doherty, J. P. (2014). Neural computations underlying arbitration between model-based and model-free learning. *Neuron, 81*(3), 687–699.

Liljeholm, M., & O'Doherty, J. P. (2012). Contributions of the striatum to learning, motivation, and performance: An associative account. *Trends in Cognitive Sciences, 16*(9), 467–475.

Mansouri, F. A., Koechlin, E., Rosa, M. G. P., & Buckley, M. J. (2017). Managing competing goals—a key role for the frontopolar cortex. *Nature Reviews Neuroscience, 18*(11), 645–657.

McDannald, M. A., Lucantonio, F., Burke, K. A., Niv, Y., & Schoenbaum, G. (2011). Ventral striatum and orbitofrontal cortex are both required for model-based, but not model-free, reinforcement learning. *Journal of Neuroscience, 31*(7), 2700–2705.

Medalla, M., & Barbas, H. (2009). Synapses with inhibitory neurons differentiate anterior cingulate from dorsolateral prefrontal pathways associated with cognitive control. *Neuron, 61*(4), 609–620.

Medalla, M., & Barbas, H. (2010). Anterior cingulate synapses in prefrontal areas 10 and 46 suggest differential influence in cognitive control. *Journal of Neuroscience, 30*(48), 16068–16081.

Nachev, P., Kennard, C., & Husain, M. (2008). Functional role of the supplementary and pre-supplementary motor areas. *Nature Reviews Neuroscience, 9*(11), 856–869.

Nachev, P., Wydell, H., O'Neill, K., Husain, M., & Kennard, C. (2007). The role of the pre-supplementary motor area in the control of action. *NeuroImage, 36*(Suppl. 2), T155–163.

Nassar, M. R., Wilson, R. C., Heasly, B., & Gold, J. I. (2010). An approximately Bayesian delta-rule model explains the dynamics of belief updating in a changing environment. *Journal of Neuroscience, 30*(37), 12366–12378.

Nee, D. E., & D'Esposito, M. (2016). The hierarchical organization of the lateral prefrontal cortex. *eLife, 5*. doi:10.7554/eLife.12112

Nee, D. E., & D'Esposito, M. (2017). Causal evidence for lateral prefrontal cortex dynamics supporting cognitive control. *eLife, 6*. doi:10.7554/eLife.28040

Neubert, F. X., Mars, R. B., Buch, E. R., Olivier, E., & Rushworth, M. F. (2010). Cortical and subcortical interactions during action reprogramming and their related white matter pathways. *Proceedings of the National Academy of Sciences of the United States of America, 107*(30), 13240–13245.

Neubert, F. X., Mars, R. B., Thomas, A. G., Sallet, J., & Rushworth, M. (2014). Comparison of human ventral cortex areas for cognitive control and language with areas in monkey frontal cortex. *Neuron, 81*(3), 700–713.

Oaksford, M., & Chater, N. (2009). Precis of Bayesian rationality: The probabilistic approach to human reasoning. *Behavioral and Brain Sciences, 32*(1), 69–84.

O'Doherty, J., Dayan, P., Schultz, J., Deichmann, R., Friston, K., & Dolan, R. J. (2004). Dissociable roles of ventral and dorsal striatum in instrumental conditioning. *Science, 304*(5669), 452–454.

Ongur, D., & Price, J. L. (2000). The organization of networks within the orbital and medial prefrontal cortex of rats, monkeys and humans. *Cerebral Cortex, 10*(3), 206–219.

Otto, A. R., Gershman, S. J., Markman, A. B., & Daw, N. D. (2013). The curse of planning: Dissecting multiple reinforcement-learning systems by taxing the central executive. *Psychological Science, 24*(5), 751–761.

Packard, M. G., & Knowlton, B. J. (2002). Learning and memory functions of the basal ganglia. *Annual Review of Neuroscience, 25*, 563–593.

Pandya, D. N., & Yeterian, E. H. (1996). Morphological correlations of the human and monkey frontal lobe. In A. R. Damasio, H. Damasio, & Y. Christen (Eds.), *Neurobiology of decision-making* (pp. 13–46). Berlin: Springer-Verlag.

Passingham, R. E., & Wise, S. P. (2012). *The neurobiology of the prefrontal cortex*. Oxford: Oxford University Press.

Pasupathy, A., & Miller, E. K. (2005). Different time courses of learning-related activity in the prefrontal cortex and striatum. *Nature, 433*(7028), 873–876.

Pezzulo, G., Rigoli, F., & Chersi, F. (2013). The mixed instrumental controller: Using value of information to combine habitual choice and mental simulation. *Frontiers in Psychology, 4*, 92.

Quilodran, R., Rothe, M., & Procyk, E. (2008). Behavioral shifts and action valuation in the anterior cingulate cortex. *Neuron, 57*(2), 314–325.

Rae, C. L., Hughes, L. E., Anderson, M. C., & Rowe, J. B. (2015). The prefrontal cortex achieves inhibitory control by facilitating subcortical motor pathway connectivity. *Journal of Neuroscience, 35*(2), 786–794.

Risse, S., & Oberauer, K. (2010). Selection of objects and tasks in working memory. *Quarterly Journal of Experimental Psychology, 63*(4), 784–804.

Rogers, R. D., & Monsell, S. (1995). Costs of predictable switch between simple cognitive tasks. *Journal of Experimental Psychology: General, 124*(2), 207–231.

Rouault, M., Drugowitsch, J., & Koechlin, E. (2019). Prefrontal mechanisms integrating rewards and beliefs in human decision-making. *Nature Communications, 10*, 301.

Rouault, M., & Koechlin, E. (2018). Prefrontal function and cognitive control: From action to language. *Current Opinion in Behavioral Sciences, 21*, 106–111.

Rudebeck, P. H., & Murray, E. A. (2014). The orbitofrontal oracle: Cortical mechanisms for the prediction and evaluation of specific behavioral outcomes. *Neuron, 84*(6), 1143–1156.

Sakai, K. (2008). Task set and prefrontal cortex. *Annual Review of Neuroscience, 31*, 219–245.

Sakai, K., Hikosaka, O., Miyauchi, S., Takino, R., Sasaki, Y., & Putz, B. (1998). Transition of brain activation from frontal to parietal areas in visuomotor sequence learning. *Journal of Neuroscience, 18*(5), 1827–1840.

Sakai, K., & Passingham, R. E. (2003). Prefrontal interactions reflect future task operations. *Nature Neuroscience, 6*(1), 75–81.

Samejima, K., Ueda, Y., Doya, K., & Kimura, M. (2005). Representation of action-specific reward values in the striatum. *Science, 310*(5752), 1337–1340.

Schenker, N. M., Hopkins, W. D., Spocter, M. A., Garrison, A. R., Stimpson, C. D., Erwin, J. M., … Sherwood, C. C. (2010). Broca's area homologue in chimpanzees (pan troglodytes): Probabilistic mapping, asymmetry, and comparison to humans. *Cerebral Cortex, 20*(3), 730–742.

Schuck, N. W., Cai, M. B., Wilson, R. C., & Niv, Y. (2016). Human orbitofrontal cortex represents a cognitive map of state space. *Neuron, 91*(6), 1402–1412.

Schultz, W. (1997). A neural substrate of prediction and reward. *Science, 275*(5306), 1593–1599.

Semendeferi, K., Armstrong, E., Schleicher, A., Zilles, K., & Van Hoesen, G. W. (2001). Prefrontal cortex in humans and apes: A comparative study of area 10. *American Journal of Physical Anthropology, 114*(3), 224–241.

Simon, D. A., & Daw, N. D. (2011). Neural correlates of forward planning in a spatial decision task in humans. *Journal of Neuroscience, 31*(14), 5526–5539.

Stalnaker, T. A., Cooch, N. K., & Schoenbaum, G. (2015). What the orbitofrontal cortex does not do. *Nature Neuroscience, 18*(5), 620–627.

Stephenson-Jones, M., Samuelsson, E., Ericsson, J., Robertson, B., & Grillner, S. (2011). Evolutionary conservation of the basal ganglia as a common vertebrate mechanism for action selection. *Current Biology, 21*(13), 1081–1091.

Sutton, R. S., & Barto, A. G. (1998). *Reinforcement learning*. Cambridge, MA: MIT Press.

Swann, N. C., Cai, W., Conner, C. R., Pieters, T. A., Claffey, M. P., George, J. S., … Tandon, N. (2012). Roles for the pre-supplementary motor area and the right inferior frontal gyrus in stopping action: Electrophysiological responses and functional and structural connectivity. *NeuroImage, 59*(3), 2860–2870.

Teffer, K., & Semendeferi, K. (2012). Human prefrontal cortex: Evolution, development, and pathology. *Progress in Brain Research, 195*, 191–218.

Teh, Y. W., Jordan, M. I., Beal, M. J., & Blei, D. M. (2006). Hierarchical Dirichlet processes. *Journal of the American Statistical Association, 101*(476), 1566–1581.

Tervo, D. G., Proskurin, M., Manakov, M., Kabra, M., Vollmer, A., Branson, K., & Karpova, A. Y. (2014). Behavioral variability through stochastic choice and its gating by anterior cingulate cortex. *Cell, 159*(1), 21–32.

Tomassini, V., Jbabdi, S., Klein, J. C., Behrens, T. E. J., Pozzilli, C., Matthews, P. M., … Johansen-Berg, H. (2007). Diffusion-weighted imaging tractography-based parcellation of the human lateral premotor cortex identifies dorsal and ventral subregions with anatomical and functional specializations. *Journal of Neuroscience, 27*(38), 10259–10269.

Udden, J., Ingvar, M., Hagoort, P., & Petersson, K. M. (2017). Broca's region: A causal role in implicit processing of grammars with crossed non-adjacent dependencies. *Cognition, 164*, 188–198.

Uylings, H. B., Groenewegen, H. J., & Kolb, B. (2003). Do rats have a prefrontal cortex? *Behavioural Brain Research, 146*(1–2), 3–17.

Uylings, H. B., Jacobsen, A. M., Zilles, K., & Amunts, K. (2006). Left-right asymmetry in volume and number of neurons in adult Broca's area. *Cortex, 42*(4), 652–658.

Walton, M. E., Behrens, T. E., Buckley, M. J., Rudebeck, P. H., & Rushworth, M. F. (2010). Separable learning systems in the macaque brain and the role of orbitofrontal cortex in contingent learning. *Neuron, 65*(6), 927–939.

Wilson, R. C., Takahashi, Y. K., Schoenbaum, G., & Niv, Y. (2014). Orbitofrontal cortex as a cognitive map of task space. *Neuron, 81*(2), 267–279.

Wunderlich, K., Dayan, P., & Dolan, R. J. (2012). Mapping value based planning and extensively trained choice in the human brain. *Nature Neuroscience, 15*(5), 786–791.

39 Semantic Representation in the Human Brain under Rich, Naturalistic Conditions

JACK L. GALLANT AND SARA F. POPHAM

ABSTRACT Conceptual understanding of the world is mediated by a broadly distributed network of brain areas that represent semantic information about our current experience and prior knowledge. Several decades of cognitive neuroscience research suggest that semantic processing in the natural world is supported by three distinct subsystems: modality-specific semantic representations are located in sensory and motor areas; amodal semantic representations are located in association areas; and the prefrontal cortex exercises the cognitive control required to understand rich semantic content in context. In this chapter we briefly review the large body of work on semantic representation. We then examine current views of semantic representation in light of a recent series of studies in which brain activity was recorded while individuals performed naturalistic tasks, such as listening to stories or watching movies. These studies revealed that semantic information is represented in an intricate mosaic of semantically selective regions that are mapped continuously across much of the human cerebral cortex and are highly consistent across individuals. These data have two profound implications for current views of semantic representation. First, they indicate that modal sensory information likely enters the amodal semantic system through multiple routes. Second, they suggest that current views that the prefrontal cortex does not directly represent semantic information need to be revised. These data suggest that the semantic system is a hybrid network in which connections between modal sensory areas and amodal semantic representations bind information about current experience, in parallel with a separate system for semantic memory access mediated by the anterior temporal lobes.

Natural human behavior is based on a complex interaction between immediate sensory experience, stored knowledge about the natural world, and continuous evaluation of the world relative to our own plans and goals. Even seemingly simple tasks, such as watching a movie or listening to a story, likely involve a range of different perceptual and cognitive processes whose underlying circuitry is broadly distributed across the brain. When watching a movie, we integrate visual and auditory information into a perceptual whole; we recognize the objects and actions in the movie and the intentions of the actors; and we understand the narrative arc of the story as it develops over time. When reading a book, we can still comprehend the story and its narrative arc even though the perceptual information available to us is greatly reduced compared with a film of the same story. A large body of research indicates that these remarkable capacities are underpinned by a broadly distributed network of brain areas that represents and processes information relevant to different parts of these tasks (Binder, Desai, Graves, & Conant, 2009; Huth, de Heer, Griffiths, Theunissen, & Gallant, 2016; Huth, Nishimoto, Vu, & Gallant, 2012). In this review we focus on one specific aspect of this system, the representation of conceptual information about the world: semantics (Binder et al., 2009; Martin & Chao, 2001; Patterson, Nestor, & Rogers, 2007; Ralph, Jefferies, Patterson, & Rogers, 2017).

The question of how the brain represents semantic information has been an intense topic of research in cognitive neuroscience for the past 40 years. Much of the early work on this topic involved neurological patients with temporal lobe degeneration, which causes a syndrome called semantic dementia (Hodges, Patterson, Oxbury, & Funnell, 1992; Snowden, 2015; Warrington, 1975; Wilkins & Moscovitch, 1978). About 25 years ago, researchers began to use neuroimaging to investigate this issue, first with positron emission tomography (PET; Damasio, Grabowski, Tranel, Hichwa, & Damasio, 1996; Diehl et al., 2004) and later with functional magnetic resonance imaging (fMRI; Mummery et al., 2000; Visser, Jefferies, & Lambon Ralph, 2010). These studies, and the subsequent research reviewed below, support the idea that semantic processing in the natural world is supported by three distinct subsystems. First, modality-specific semantic representations are located in sensory and motor areas. Second, amodal semantic representations are located in association areas, though the precise location and nature of these representations are more controversial. Third, prefrontal cortex appears to be involved in the cognitive control required to understand rich semantic content in context.

In this chapter we will first review the existing literature on each of these three aspects of semantic representation. Then we will summarize findings on semantic representation that have grown out of recent naturalistic experiments and evaluate how these data fit into existing theories.

Modality-Specific Semantic Representations

Both lesion studies and neuroimaging experiments support the view that modality-specific semantic representations are distributed in a network of distinct sensory and motor areas. Lesion studies have shown that individuals who have suffered stroke often exhibit modality-specific comprehension deficits, such as pure word deafness (Auerbach, Allard, Naeser, Alexander, & Albert, 1982; Kussmaul, 1877) or visual agnosia (Farah, 2004; Riddoch & Humphreys, 1987). Neuroimaging studies using positron emission tomography (PET; Damasio, Grabowski, Tranel, Hichwa, & Damasio, 1996) and functional magnetic resonance imaging (fMRI; Chao, Haxby, & Martin, 1999; Goldberg, Perfetti, & Schneider, 2006; Hauk, Johnsrude, & Pulvermüller, 2004) both indicate that modality-specific semantic information is represented in a network of brain areas broadly distributed across sensory and motor cortex. For example, watching a close-up of a Western gunfighter pulling his weapon out of its holster would produce activity in visual areas that represent body parts (Nishimoto et al., 2011) and in premotor areas that represent the hand (Hasson, Nir, Levy, Fuhrmann, & Malach, 2004). Modality-specific representations have been identified in the visual and auditory systems, around the precentral and postcentral gyri, and across much of the ventral temporal cortex.

These data have been used to support the view that semantic information is represented in a distributed form in the network of sensory and motor areas that serve as the source and sink for all human interactions with the world (Barsalou, 1999; Martin, 2007; Pulvermüller, 2013). According to this view, semantic concepts arise from connections between these distributed modality-specific representations (Meteyard, Cuadrado, Bahrami, & Vigliocco, 2012). This family of theories is usually called embodied or grounded cognition. While the theory of embodied cognition is broadly consistent with a large body of data, one area of contention concerns how such a system can represent abstract semantic concepts that have no direct sensory or motor correlates, such as truth, justice, and love (Meteyard et al., 2012; Vigliocco, Meteyard, Andrews, & Kousta, 2009).

Amodal Semantic Representations

Other lesion and imaging data suggest that semantic information is also represented in an amodal form that is not closely tied to sensory or motor representations. Most importantly, some neurodegenerative diseases or brain lesions appear to affect semantic judgment regardless of modality. The most profound of these disorders is semantic dementia, which causes a progressive bilateral atrophy of the anterior temporal lobes (ATL; Desgranges et al., 2007; Diehl et al., 2004; Galton et al., 2001; Hodges et al., 1992; Mummery et al., 2000; Nestor, Fryer, & Hodges, 2006; Snowden, 2015; Snowden et al., 2018; Snowden, Goulding, & Neary, 1989; Rosen et al., 2002; Warrington, 1975). ATL degeneration results in deficits in the amodal conceptual representations of words, pictures, sounds, smells, and actions (Bozeat, Lambon Ralph, Patterson, Garrard, & Hodges, 2000; Bozeat, Ralph, Patterson, & Hodges, 2002; Garrard & Carroll, 2006; Jefferies, Patterson, Jones, & Lambon Ralph, 2009; Luzzi et al., 2007; Schwartz, Marin, & Saffran, 1979; Wilkins & Moscovitch, 1978). Individuals with ATL degeneration also suffer from anomia and cannot name concepts based on the sensory evidence provided. For example, a patient with anomia might identify a zebra as a horse and express confusion about the presence of stripes (Patterson, Nestor, & Rogers, 2007). However, other aspects of cognition (syntax, numerical abilities, executive function) appear to be relatively spared (Jefferies, Patterson, Jones, Bateman, & Lambon Ralph, 2004; Hodges et al., 1992, 1999; Kramer et al., 2003). These profound semantic deficits are not observed in other neurodegenerative diseases that affect the hippocampus, parahippocampal cortex, and limbic structures, areas more closely involved with autobiographical memory than with semantic memory (Chan et al., 2001). In sum, degeneration of the temporal lobe is a key cause of semantic dementia. However, several aspects of this disorder are still in dispute.

First, there is some controversy about the organization of semantic representations along the temporal lobe. Some studies argue that degeneration of the most anterior regions of the temporal lobe produce the most profound deficits of semantic comprehension and that the degeneration of more posterior regions does not affect semantic judgment (Nestor, Fryer, & Hodges, 2006). Others have argued that the degradation of posterior regions is involved in semantic dementia (Galton et al., 2001) or rather that connections between posterior and anterior temporal lobe regions are in fact more critical for semantic judgment than the anterior regions (Martin & Chao, 2001; Mummery et al., 1999).

Another point of contention in studies of semantic dementia concerns whether this disease affects semantic comprehension in general (Lambon Ralph, Graham, Patterson, & Hodges, 1999) or whether it is mainly a deficit of lexical semantics (Lauro-Grotto, Piccini, & Shallice, 1997). The answer to this question has profound implications for any theory of semantic representation. The first case would indicate that the ATL is a critical hub for semantic comprehension, while the second would imply that the ATL is a critical interface mediating between perceptual and language systems. However, the evidence bearing on this issue is still mixed. Some studies have argued that this disorder impairs representations of categories of concrete objects but that verbs and abstract concepts are relatively spared (Breedin, Saffran, & Branch Coslett, 1994; Silveri, Brita, Liperoti, Piludu, & Colosimo, 2018). Others argue that representations of concrete categories, verbs, and abstract concepts are all degraded equally in semantic dementia if the base-rate frequencies for the exemplars used in testing are all equated (Bird, Lambon Ralph, Patterson, & Hodges, 2000; Ralph, Graham, Ellis, & Hodges, 1998). However, whether this impairment occurs at the level of concepts or the linguistic representations of those concepts is still unclear (Caramazza & Mahon, 2003; Kiefer & Pulvermüller, 2012).

Additionally, individuals with semantic dementia appear to lose finer categorical distinctions first and then coarser categorical distinctions at later stages of the disease (Ralph, Sage, Jones, & Mayberry, 2010; Lambon Ralph & Patterson, 2008). For example, someone with mild semantic dementia might be able to identify a picture of a robin as a bird but could be confused when presented with an ostrich (see Patterson, Nestor, & Rogers, 2007). Then, with further progression of the disease, the person would become unable to identify any bird. This pattern of deficits has been used to support the idea that semantic dementia impairs access to information about the hierarchical categorical structure of the world (Garrard, Ralph, Hodges, & Patterson, 2001; Laisney et al., 2011).

Much of the recent work on semantic dementia has proposed that the modality-specific semantic representations in sensory and motor areas serve as spokes that feed into a single semantic hub located in the ATL (Ralph et al., 2017). However, an older, alternative view suggests that multiple semantic convergence zones outside of the ATL serve as interfaces between different areas of unimodal semantic representations (A. R. Damasio, 1989; Damasio et al., 1996; Damasio, Tranel, Grabowski, Adolphs, & Damasio, 2004; Devereux, Clarke, Marouchos, & Tyler, 2013; Fairhall & Caramazza, 2013). This earlier idea proposes that different convergence zones mediate the interaction of different kinds of information, based on anatomical constraints and individual life experiences. A meta-analysis of over 120 studies of semantic representation in the brain identified a set of putative high-level convergence zones, including the angular gyrus; middle temporal gyrus; precuneus, fusiform, and parahippocampal gyri; and some portions of frontal cortex (Binder et al., 2009). When tested directly, the posterior middle temporal gyrus, angular gyrus, and precuneus were found to be responsive to both visual and linguistic stimuli of the same categories, lending support to the argument that they may function as high-level convergence zones (Fairhall & Caramazza, 2013). At this time it remains unclear whether these convergence zones support sensory integration or memory access and precisely how their functional properties differ from the ATL.

Control Processes for Semantic Comprehension

Substantial evidence suggests that regions of prefrontal cortex, particularly the inferior frontal gyrus (IFG), play a role in controlling the processes that mediate semantic judgments. Early PET and fMRI studies of semantic processing suggested that some prefrontal cortex areas are specifically involved in semantic retrieval, rather than serving as general-purpose cognitive-control regions (Demb et al., 1995; Martin, Haxby, Lalonde, Wiggs, & Ungerleider, 1995). This theory was further supported by reports that neurodegenerative diseases and lesions that affect the prefrontal cortex but leave the temporal cortex intact sometimes cause semantic deficits (Jefferies & Lambon Ralph, 2006). A more recent study argued that the IFG mediates decision-making only in semantic contexts but is not involved in other difficult decision-making processes (Whitney, Kirk, O'Sullivan, Lambon Ralph, & Jefferies, 2011). Finally, it has been argued that the prefrontal cortex contains specific regions that mediate semantic judgments but remain completely separate from the regions involved in cognitive control (Fedorenko, Behr, & Kanwisher, 2011).

In contrast, other studies of patients with lesions to the prefrontal cortex have reported that semantic deficits tend to be expressed only in tasks with relatively greater executive demands, such as comprehension of a complex narrative (Jefferies & Lambon Ralph, 2006). This suggests that prefrontal lesions do not affect semantic representations directly. Instead, they affect control processes that govern how semantic information is accessed, sequenced, and integrated (Jefferies & Lambon Ralph, 2006; Thompson-Schill, D'Esposito, Aguirre, & Farah, 1997). Consistent with this, in cognitively normal subjects the IFG is engaged during the comprehension of sentences that are semantically ambiguous

(Bedny & Thompson-Schill, 2006; Rodd, Davis, & Johnsrude, 2005), and its activity is modulated by the difficulty of a semantic decision-making task (Roskies, Fiez, Balota, Raichle, & Petersen, 2001). A meta-analysis also revealed that the IFG is recruited in language tasks that require nonsemantic judgments (Bookheimer, 2002). Finally, there is evidence that the left inferior prefrontal cortex (LIPC) is important for the retrieval of task-relevant information, regardless of whether the task requires semantic information (Wagner, Paré-Blagoev, Clark, & Poldrack, 2001). This is supported by the finding that the LIPC is more engaged when subjects are presented with semantic violations and violations of factual knowledge (Hagoort, Hald, Bastiaansen, & Petersson, 2004).

In sum, a wide variety of lesion and neuroimaging studies suggest that prefrontal cortex is involved in cognitive-control and selection processes rather than semantic representation per se (Badre, Poldrack, Paré-Blagoev, Insler, & Wagner, 2005; Gold et al., 2006). However, this interpretation has not received unanimous support (Nozari & Thompson-Schill, 2016).

Recent Studies of Semantic Representation

Until recently, much of the debate regarding semantic representation has focused on where semantic information is represented (Humphries, Binder, Medler, & Liebenthal, 2007; Patterson, Nestor, & Rogers, 2007; Visser, Jefferies, & Lambon Ralph, 2010), rather than precisely how semantic information is mapped across the cerebral cortex. Furthermore, the studies that have attempted to understand where some specific type of semantic information is represented have used classical experimental paradigms that manipulate a few semantic parameters under highly controlled and simplistic conditions (Binder, Westbury, McKiernan, Possing, & Medler, 2005; Epstein & Kanwisher, 1998; Kanwisher, McDermott, & Chun, 1997). While simple controlled studies have ample statistical power to identify specific semantic representations, they lack the power to support broad mapping of the semantic space. Our lab has taken a different approach to understanding semantic representations by using brain activity evoked by complex, naturalistic stimuli to create quantitative, high-dimensional models of semantic selectivity (Naselaris, Kay, Nishimoto, & Gallant, 2011; Wu, David, & Gallant, 2006). This approach allows us to create rich, high-dimensional maps of semantic selectivity across the entire cerebral cortex (Çukur, Nishimoto, Huth, & Gallant, 2013; Huth et al., 2012, 2016; Imamoglu, Huth, & Gallant, 2016; Naselaris et al., 2011; Popham, Huth, Bilenko, & Gallant, 2018).

Our experiments are based on a naturalistic, data-driven approach designed to reveal how semantic information is represented in individuals watching movies or listening to stories. Thus, our experiments are quite different from those usually used to study semantic representation, which often involve very reduced tasks such as naming pictures or defining words (Patterson, Nestor, & Rogers, 2007). We analyze these rich data by means of a powerful statistical approach called voxelwise modeling (Naselaris et al., 2011). The procedure proceeds in several steps (see figure 39.1). First, semantic features—objects and actions in movies and stories—are extracted from the stimuli and encoded in an appropriate semantic feature space. Each of the semantic features is used as a regressor in a regularized (ridge) regression procedure run separately for each of the approximately 50,000–100,000 voxels in each individual's brain. Our methods allow us to model thousands of semantic features simultaneously, providing a means to answer many questions about semantic representations in parallel. Second, the output of this procedure produces a separate weight vector for every voxel that describes how each semantic feature contributes to measured brain activity within that voxel. Features present in a movie or story that tend to elicit activity from a voxel will be given positive weights; features whose presence or absence has no effect on a voxel's response will be given zero weights; and features that tend to suppress a voxel's response when present will be given negative weights. Third, the semantic model of each voxel is tested using a separate data set reserved for this purpose. The model predicts how the voxel will respond to the new stimulus, and this prediction is compared to the voxel's actual response to the stimulus as measured by fMRI. Prediction accuracy is quantified by the correlation between the prediction and the observed response, and statistical significance is assessed by permutation testing. The end result is a list of semantic features that significantly modulate activity in each cortical voxel, ordered by the influence of each feature on voxel responses. This entire procedure is performed separately for each voxel in each subject. Finally, the fit voxelwise models are examined to understand how semantic features are represented across the cerebral cortex. The simplest method for this is to use principal component analysis to find a low-dimensional semantic space that best accounts for the data. An inspection of these principal components reveals the relative importance of each semantic feature within the semantic space. The principal components can also be visualized on the cortical surface to reveal how the dimensions of the semantic space are mapped across the surface of the cerebral cortex. Comparing these maps across

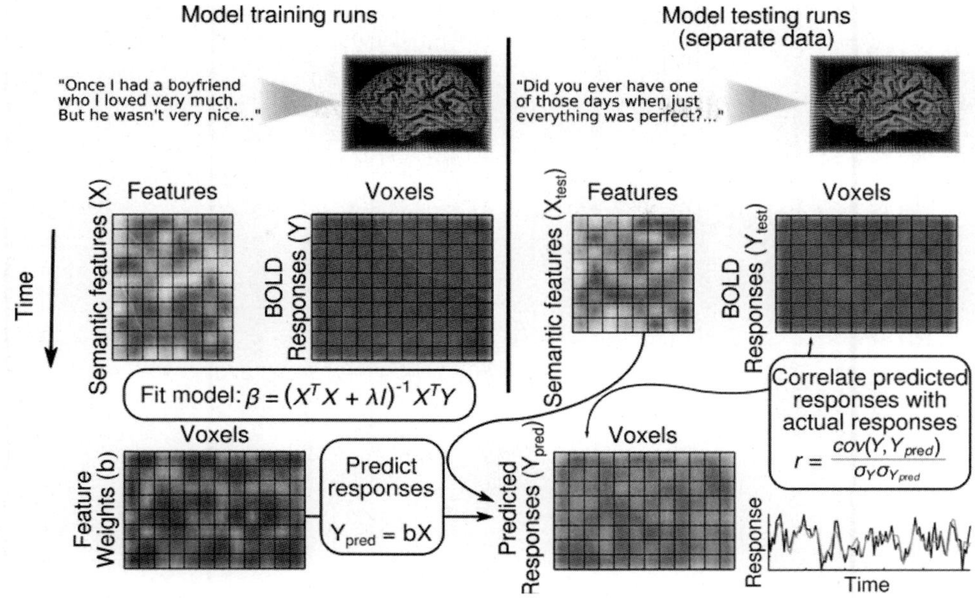

FIGURE 39.1 Voxelwise modeling procedure. Functional MRI data are recorded while subjects listen to natural stories or watch natural movies. These data are separated into two sets: a training set used to fit voxelwise models and a separate test set used to validate the fit models. Semantic features are extracted from the stimuli in each data set. *Left*, For each separate voxel, ridge regression is used to find a model that explains recorded brain activity as a weighted sum of the semantic features in the stories. *Right*, Prediction accuracy of the fit voxelwise models is assessed by using the model weights obtained in the previous step to predict voxel responses to the testing data and then comparing the predictions of the fit models to the obtained brain activity. Statistical significance of predictions and of specific model coefficients is assessed through permutation testing. (See color plate 41.)

subjects shows which aspects of semantic representation are common at the group level and which reflect individual differences.

We have used voxelwise modeling to recover semantic representations from brain activity recorded during several different naturalistic paradigms: while subjects were presented with a series of natural photographs (Naselaris et al., 2011); while they watched a series of very short (~20 seconds each) natural movie clips (Huth et al., 2012); while they listened to natural narrative short stories (Huth et al., 2016); while they read a text version of these same narrative stories (Imamoglu, Huth, & Gallant, 2016); while they watched natural short films with sound (Nunez-Elizalde, Deniz, Gao, & Gallant, 2018); and while they watched short films while attending to the presence of vehicles or humans (Çukur et al., 2013). All these studies show that semantic information is represented in an intricate mosaic of semantically selective regions that are mapped continuously across much of the human cerebral cortex and which are highly consistent across individuals. (For the purposes of this chapter, a semantic region is a patch of cortex with fairly uniform semantic tuning, whether unimodal or amodal.) For example, numbers appear to be represented in a collection of semantic regions distributed broadly across the cerebral cortex (*dark green patches*, figure 39.2). Social concepts appear to be represented in a different collection of semantic regions distributed broadly across the cerebral cortex (*bright red patches*, figure 39.2). However, there is no obvious systematic relationship between the distribution of the semantic regions pertaining to one domain versus another.

Furthermore, the semantic maps produced in these studies appear to be largely consistent regardless of whether they were acquired during listening to stories or during reading (Imamoglu, Huth, & Gallant, 2016). This consistency is found across a broadly distributed set of regions, including posterior cingulate cortex, parahippocampal cortex, the temporal lobes, posterior parietal cortex, the temporal-parietal junction, dorsolateral prefrontal cortex, ventromedial prefrontal cortex, and orbitofrontal cortex. The only regions that produce inconsistent maps across reading and listening are primary sensory and motor regions, an unsurprising result.

Finally, we find evidence for both modal and amodal semantic regions (Huth et al., 2012, 2016; Imamoglu, Huth, & Gallant, 2016). Modal regions appear to be located in higher-order sensory areas in the occipital and temporal lobes and in motor areas between the motor strip and prefrontal cortex (see figure 39.2). Amodal regions are located predominantly in the posterior parietal cortex, temporoparietal junction, dorsolateral

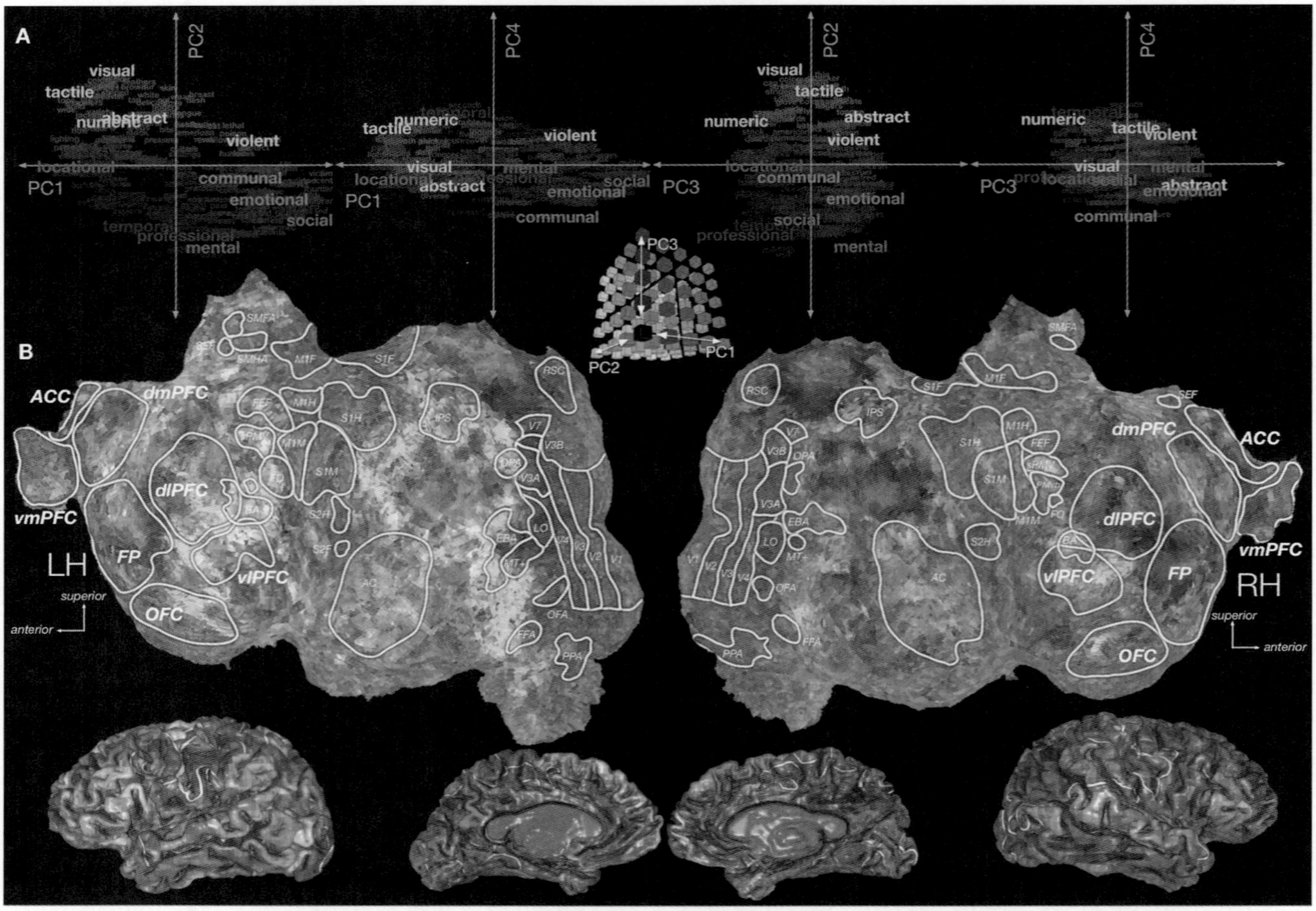

FIGURE 39.2 Semantic maps obtained from subjects who listened to narrative stories. Principal components analysis of voxelwise model weights reveals four important semantic dimensions in the brain. *A*, A Red, Green, Blue (RGB) color map was used to color both words and voxels based on the first three dimensions of the semantic space. Words that best matched the four semantic dimensions were found and then collapsed into 12 categories using k-means clustering. Each category was manually assigned a label. The 12 category labels (large words) and a selection of the 458 best words (small words) are plotted here along four pairs of semantic dimensions. The largest axis of variation lies roughly along the first dimension and separates perceptual and physical categories (tactile, locational) from human-related categories (social, emotional, violent). *B*, Voxelwise model weights were projected onto the semantic dimensions and then colored using the same RGB color map. Projections for one subject (S2) are shown on that subject's cortical surface. Semantic information seems to be represented in intricate patterns across much of the semantic system. White lines show conventional anatomical and/or functional ROIs. Labeled ROIs in prefrontal cortex reflect the typical anatomical parcellation into seven broad regions: dorsolateral prefrontal cortex (dlPFC), ventrolateral prefrontal cortex (vlPFC), dorsomedial prefrontal cortex (dmPFC), ventromedial prefrontal cortex (vmPFC), orbitofrontal cortex (OFC), anterior cingulate cortex (ACC), and the frontal pole (FP). Each of these conventional prefrontal ROIs contains multiple semantic domains, suggesting that the role of prefrontal cortex in semantic comprehension is more complicated than the current cognitive-control view would suggest. Reproduced and modified from Huth et al. (2016). (See color plate 42.)

prefrontal cortex, ventromedial prefrontal cortex, and orbitofrontal cortex.

As discussed earlier, many previous studies have identified semantically selective regions of interest (ROIs) in many different locations across the cerebral cortex, such as the fusiform face area (FFA; Kanwisher, McDermott, & Chun, 1997), the parahippocampal place area (PPA; Epstein & Kanwisher, 1998), and so on. These regions identified previously also appear in our functional maps. However, our studies also reveal a rich, continuous pattern of semantically selective regions that have not been identified previously. Furthermore, we find that many of the classical functional ROIs located within visual cortex are actually composed of several subdivisions. For example, the FFA contains three spatially segregated functional subregions that differ primarily in their responses for nonface categories, such as animals, vehicles, and communication verbs (Çukur et al., 2013). Three place-selective ROIs—the PPA, the retrosplenial cortex (RSC), and the occipital place area (OPA, also called the transverse occipital sulcus)—each contain two functional subregions, one selectively biased toward static stimuli and one biased toward dynamic stimuli (Çukur, Huth, Nishimoto, & Gallant, 2016). The temporoparietal junction (TPJ) is a broad region usually thought to represent information related to theory of mind and social meaning (Saxe & Kanwisher, 2003), but our data suggest that the TPJ encompasses many separate semantic regions that represent different aspects of social information (Huth et al., 2016). Cognitive-control regions within the prefrontal cortex, such as the dorsolateral prefrontal cortex (DLPFC), are quite large, but our data show that each of these ROIs may contain several distinct semantic regions (see figure 39.2*B*).

Implications of Recent Studies for Current Theories of Semantic Representation

Taken together, the results from our studies have important implications for two key aspects of current theories regarding semantic representation: the role of the ATL as a semantic hub and the role of prefrontal areas in semantic processing.

The anterior temporal lobe as a semantic hub As explained earlier, the current hub-and-spoke theory of semantic representation holds that the ATL serves as a hub that integrates distributed semantic representations. This view proposes that all information flowing between the unimodal and amodal semantic systems passes through the ATL. The studies from our laboratory do not offer much new information about semantic representation in the ATL itself. The ATL is difficult to image using fMRI (Binder et al., 2011; Visser, Jefferies, & Lambon Ralph, 2010), and correlations between ATL lesions and semantic deficits seen with PET are not readily apparent with fMRI (Devlin et al., 2000). Functional imaging of the ATL requires specialized protocols that can reveal ATL function but substantially lower image quality in the rest of the brain. Our laboratory chooses imaging protocols designed to optimize image quality across the entire cortex and thus the image quality in the ATL in our previous studies has been poor. For this reason, our data are agnostic about semantic representation within the ATL.

However, our data suggest that the ATL may not be the sole route for information flow through the semantic system and between modal and amodal representations. Instead, we suspect that there are multiple routes for modal semantic information to enter the amodal semantic system. In recent work we compared maps obtained when individual subjects watched brief movie clips versus when they listened to stories (Huth et al., 2012, 2016; Popham et al., 2018). We found that the representations of semantic information received through the visual modality and information received through the linguistic modality abut one another just anterior to occipital cortex (see figure 39.3). Furthermore, the arrangement of semantically selective regions along this border corresponds between vision and language. That is, for each patch of semantically selective visual cortex lying posterior to this border, there is another patch of semantically selective cortex immediately anterior to the border that responds to the same semantic content when it occurs in stories. It seems unlikely that this very specific arrangement would arise by chance; it seems more likely that some relationship exists between semantically selective regions on each side of this border. A well-known principle of cortical anatomy holds that nearby structures are relatively more likely to be anatomically connected than more distant structures. Therefore, we suspect that this arrangement is evidence of a direct parallel pathway that connects visual to lexical representations in the same semantic regions. This conflicts with a basic assumption of the hub-and-spoke model of the ATL, which holds that all modal semantic information must pass through the ATL in order to enter the amodal system (Ralph et al., 2017). Our result is more in line with the theory of multiple high-level convergence zones (Damasio & Damasio, 1994; Devereux et al., 2013; Fairhall & Caramazza, 2013).

Cognitive control of semantic access and use in prefrontal cortex As summarized earlier, it is well known that regions of prefrontal cortex become activated under conditions requiring the integration or use of complex semantic information but that prefrontal activation is much reduced under conditions requiring only simple semantic judgments. In contrast, lesions or degeneration of the ATL interferes with all semantic judgments, regardless of task complexity (Hodges et al., 1999). For these reasons, prefrontal cortex is not usually thought to be a primary site of semantic representation. Instead, it is thought to control the sequencing, ordering, access, and use of semantic information (Jefferies & Lambon Ralph, 2006). This idea is consistent with the common view of prefrontal cortex as a major site of cognitive control (Badre et al., 2005; Gold et al., 2006).

Several different lines of evidence from our studies suggest that this conventional view of the role of prefrontal cortex in semantic tasks may be oversimplified. The current view holds that the regions of prefrontal cortex responsible for cognitive control do not

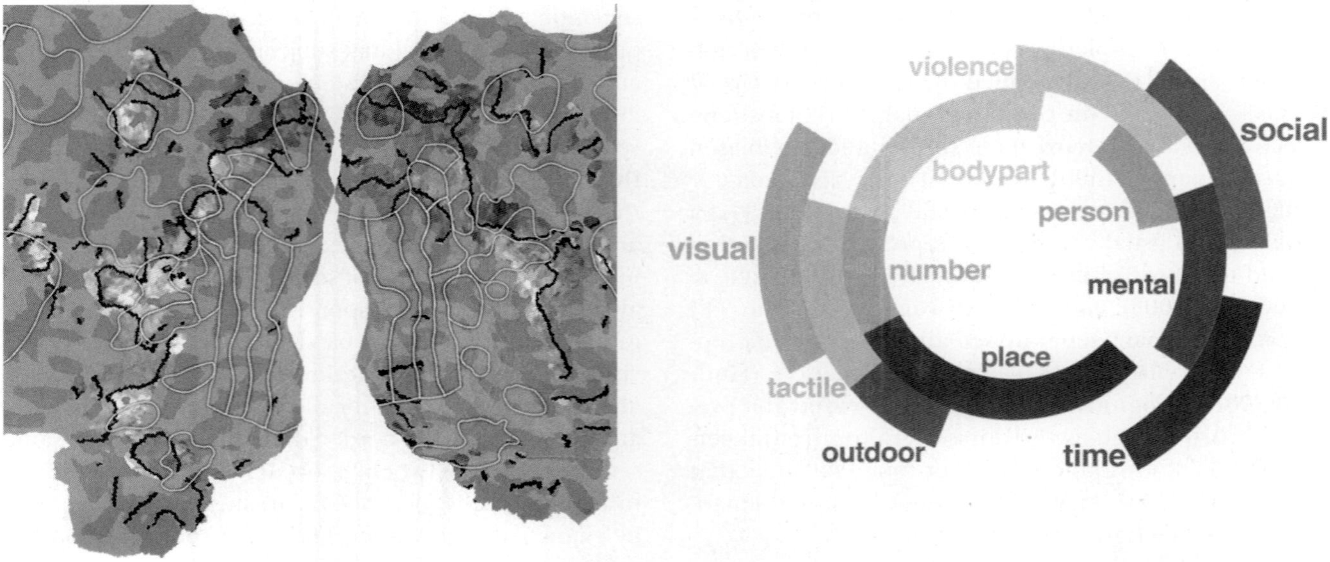

FIGURE 39.3 Relationship between visual and linguistic semantic representations along the boundary of visual cortex. The black boundary indicates the border between cortical regions activated by brief movie clips versus stories. Voxels posterior to the boundary (i.e., nearer the center of the figure) are activated by movie clips but not stories. Voxels anterior to the border are activated by stories but not movie clips. Each of the voxels activated by only one modality is colored based on fit model weights that indicate the semantic category for which it is selective (legend at right; data from Huth et al. [2012] and Huth et al. [2016]). For almost all semantic concepts, the semantic selectivity of voxels posterior to the boundary is similar to the semantic selectivity of voxels anterior to the boundary. The only exception seems to be "mental" concepts (purple voxels located in the dorsal region of the boundary in the right hemisphere), which appear to be represented only in the stories. However, these concepts were not labeled explicitly in the movies and therefore cannot be found in the visual semantic map. (See color plate 43.)

represent specific semantic information. However, our data show that prefrontal cortex is highly semantically selective during naturalistic semantic tasks (Huth et al., 2012, 2016; Imamoglu, Huth, & Gallant, 2016). The intricate pattern of semantic selectivity found in prefrontal cortex varies on a scale much finer than would be predicted based on the conventional parcellations of prefrontal cortex (see figure 39.2*B*). The current view predicts that activity in prefrontal cortex should depend only on the task requirements and not semantic content. In contrast, the semantic maps that we have obtained during reading and listening appear to be very similar (Imamoglu, Huth, & Gallant, 2016). Furthermore, unpublished preliminary data from our lab suggest that answering questions about specific semantic categories produces patterns of prefrontal activity that can be predicted by semantic selectivity during narrative comprehension. Finally, attention alters semantic selectivity in prefrontal cortex even under constant task conditions (Çukur et al., 2013). If prefrontal areas were involved in cognitive control exclusive of semantic content, then these results should not occur.

Taken together, our data suggest three different possibilities regarding the nature of semantic selectivity in prefrontal cortex. First, cognitive-control areas might be organized at a scale finer than currently believed so that each semantically selective region in prefrontal cortex has its own associated cognitive-control network. Second, cognitive-control areas might be interdigitated with semantically selective regions. Third, cognitive-control areas might be functionally distinct from, but overlap, semantically selective regions. Further studies will be required to determine which of these hypotheses is correct. One way to address this issue would be to obtain semantic maps simultaneously with cognitive-control localizers within the same set of subjects.

Summary and Conclusion

Data from naturalistic fMRI experiments in which subjects watch movies or listen to stories largely support a distributed view of semantic knowledge. Semantic comprehension appears to involve a large network of surprisingly specific semantic regions that are distributed broadly across most of the cerebral cortex (Huth et al., 2012, 2016). Areas located nearer primary sensory areas appear to represent semantic information within a specific sensory modality, while those located farther from primary sensory areas and in prefrontal cortex appear to represent amodal semantic information. However, our experiments reveal that the structure of these semantic maps is far richer and more detailed than previously suspected. This detail is most prominent in areas that represent amodal semantic information outside the ATL, such as the temporoparietal junction, parietal cortex, and prefrontal cortex.

Parietal areas are thought to be a key part of the network for directed attention (Farah, Wong, Monheit, & Morrow, 1989; Lynch, Mountcastle, Talbot, & Yin, 1977; Posner, Walker, Friedrich, & Rafal, 1987), and we speculate that perhaps semantic selectivity in parietal regions reflects semantically selective attentional demands of perception under natural conditions (Çukur et al., 2013). Semantic selectivity in prefrontal cortex is thought to reflect the operation of cognitive-control processes required for sequencing and organizing semantic information under natural conditions (Badre et al., 2005; Gold et al., 2006; Jefferies & Lambon Ralph, 2006). However, this explanation cannot account for the rich organization of semantic domains within prefrontal cortex.

Our data also show a close correspondence between semantic maps along the anterior border of the visual system and along the posterior border of the semantic system that is activated during naturalistic comprehension (Popham et al., 2018). This correspondence suggests that these areas may communicate directly along pathways that are independent of the ATL. Given the strong evidence that the ATL serves as a semantic hub, it seems unlikely that these direct connections are sufficient to provide semantic assignment to sensory experience. We propose that these connections provide the pathways necessary to bind information from different sensory modalities to each other, in parallel to the memory access processes mediated by the ATL. This explanation would reconcile the results found in support of both the ATL as a semantic hub and the existence of multiple high-level convergence zones. In other words, the hub-and-spoke model and the convergence zone model of semantic representation may merely describe different phases of semantic comprehension.

REFERENCES

Auerbach, S. H., Allard, T., Naeser, M., Alexander, M. P., & Albert, M. L. (1982). Pure deafness. *Brain, 105*(2), 271–300.

Badre, D., Poldrack, R. A., Paré-Blagoev, E. J., Insler, R. Z., & Wagner, A. D. (2005). Dissociable controlled retrieval and generalized selection mechanisms in ventrolateral prefrontal cortex. *Neuron, 47*(6), 907–918.

Barsalou, L. W. (1999). Perceptions of perceptual symbols. *Behavioral and Brain Sciences, 22*(04), 637–660.

Bedny, M., & Thompson-Schill, S. L. (2006). Neuroanatomically separable effects of imageability and grammatical class during single-word comprehension. *Brain and Language, 98*(2), 127–139.

Binder, J. R., Desai, R. H., Graves, W. W., & Conant, L. L. (2009). Where is the semantic system? A critical review and meta-analysis of 120 functional neuroimaging studies. *Cerebral Cortex, 19*(12), 2767–2796.

Binder, J. R., Gross, W. L., Allendorfer, J. B., Bonilha, L., Chapin, J., Edwards, J. C., ... Weaver, K. E. (2011). Mapping anterior temporal lobe language areas with fMRI: A multicenter normative study. *NeuroImage, 54*(2), 1465–1475.

Binder, J. R., Westbury, C. F., McKiernan, K. A., Possing, E. T., & Medler, D. A. (2005). Distinct brain systems for processing concrete and abstract concepts. *Journal of Cognitive Neuroscience, 17*(6), 905–917.

Bird, H., Lambon Ralph, M. A., Patterson, K., & Hodges, J. R. (2000). The rise and fall of frequency and imageability: Noun and verb production in semantic dementia. *Brain and Language, 73*(1), 17–49.

Bookheimer, S. (2002). Functional MRI of language: New approaches to understanding the cortical organization of semantic processing. *Annual Review of Neuroscience, 25*, 151–188.

Bozeat, S., Lambon Ralph, M. A., Patterson, K., Garrard, P., & Hodges, J. R. (2000). Non-verbal semantic impairment in semantic dementia. *Neuropsychologia, 38*(9), 1207–1215.

Bozeat, S., Ralph, M. A. L., Patterson, K., & Hodges, J. R. (2002). The influence of personal familiarity and context on object use in semantic dementia. *Neurocase, 8*(1–2), 127–134.

Breedin, S. D., Saffran, E. M., & Branch Coslett, H. (1994). Reversal of the concreteness effect in a patient with semantic dementia. *Cognitive Neuropsychology, 11*(6), 617–660.

Caramazza, A., & Mahon, B. Z. (2003). The organization of conceptual knowledge: The evidence from category-specific semantic deficits. *Trends in Cognitive Sciences, 7*(8), 354–361.

Chan, D., Fox, N. C., Scahill, R. I., Crum, W. R., Whitwell, J. L., Leschziner, G., ... Rossor, M. N. (2001). Patterns of temporal lobe atrophy in semantic dementia and Alzheimer's disease. *Annals of Neurology, 49*(4), 433–442.

Chao, L. L., Haxby, J. V., & Martin, A. (1999). Attribute-based neural substrates in temporal cortex for perceiving and knowing about objects. *Nature Neuroscience, 2*(10), 913–919.

Çukur, T., Huth, A. G., Nishimoto, S., & Gallant, J. L. (2016). Functional subdomains within scene-selective cortex: Parahippocampal place area, retrosplenial complex, and occipital place area. *Journal of Neuroscience, 36*(40), 10257–10273.

Çukur, T., Nishimoto, S., Huth, A. G., & Gallant, J. L. (2013). Attention during natural vision warps semantic representation across the human brain. *Nature Neuroscience, 16*(6), 763–770.

Damasio, A. R. (1989). The brain binds entities and events by multiregional activation from convergence zones. *Neural Computation, 1*(1), 123–132.

Damasio, A. R., & Damasio, H. (1994). Cortical systems for retrieval of concrete knowledge: The convergence zone framework. *Large-scale neuronal theories of the brain, 6174.*

Damasio, H., Grabowski, T. J., Tranel, D., Hichwa, R. D., & Damasio, A. R. (1996). A neural basis for lexical retrieval. *Nature, 380*(6574), 499–505.

Damasio, H., Tranel, D., Grabowski, T., Adolphs, R., & Damasio, A. (2004). Neural systems behind word and concept retrieval. *Cognition, 92*(1–2), 179–229.

Demb, J. B., Desmond, J. E., Wagner, A. D., Vaidya, C. J., Glover, G. H., & Gabrieli, J. D. (1995). Semantic encoding and retrieval in the left inferior prefrontal cortex: A functional MRI study of task difficulty and process specificity. *Journal of Neuroscience, 15*(9), 5870–5878.

Desgranges, B., Matuszewski, V., Piolino, P., Chételat, G., Mézenge, F., Landeau, B., ... Eustache, F. (2007). Anatomical and functional alterations in semantic dementia: A voxel-based MRI and PET study. *Neurobiology of Aging, 28*(12), 1904–1913.

Devereux, B. J., Clarke, A., Marouchos, A., & Tyler, L. K. (2013). Representational similarity analysis reveals commonalities and differences in the semantic processing of words and objects. *Journal of Neuroscience, 33*(48), 18906–18916.

Devlin, J. T., Russell, R. P., Davis, M. H., Price, C. J., Wilson, J., Moss, H. E., ... Tyler, L. K. (2000). Susceptibility-induced loss of signal: Comparing PET and fMRI on a semantic task. *NeuroImage, 11*(6 Pt. 1), 589–600.

Diehl, J., Grimmer, T., Drzezga, A., Riemenschneider, M., Förstl, H., & Kurz, A. (2004). Cerebral metabolic patterns at early stages of frontotemporal dementia and semantic dementia. A PET study. *Neurobiology of Aging, 25*(8), 1051–1056.

Epstein, R., & Kanwisher, N. (1998). A cortical representation of the local visual environment. *Nature, 392*(6676), 598–601.

Fairhall, S. L., & Caramazza, A. (2013). Brain regions that represent amodal conceptual knowledge. *Journal of Neuroscience, 33*(25), 10552–10558.

Farah, M. J. (2004). *Visual agnosia.* Cambridge, MA: MIT Press.

Farah, M. J., Wong, A. B., Monheit, M. A., & Morrow, L. A. (1989). Parietal lobe mechanisms of spatial attention: Modality-specific or supramodal? *Neuropsychologia, 27*(4), 461–470.

Fedorenko, E., Behr, M. K., & Kanwisher, N. (2011). Functional specificity for high-level linguistic processing in the human brain. *Proceedings of the National Academy of Sciences of the United States of America, 108*(39), 16428–16433.

Galton, C. J., Patterson, K., Graham, K., Lambon-Ralph, M. A., Williams, G., Antoun, N., ... Hodges, J. R. (2001). Differing patterns of temporal atrophy in Alzheimer's disease and semantic dementia. *Neurology, 57*(2), 216–225.

Garrard, P., & Carroll, E. (2006). Lost in semantic space: A multi-modal, non-verbal assessment of feature knowledge in semantic dementia. *Brain, 129*(Pt. 5), 1152–1163.

Garrard, P., Ralph, M. A., Hodges, J. R., & Patterson, K. (2001). Prototypicality, distinctiveness, and intercorrelation: Analyses of the semantic attributes of living and nonliving concepts. *Cognitive Neuropsychology, 18*(2), 125–174.

Gold, B. T., Balota, D. A., Jones, S. J., Powell, D. K., Smith, C. D., & Andersen, A. H. (2006). Dissociation of automatic and strategic lexical-semantics: Functional magnetic resonance imaging evidence for differing roles of multiple frontotemporal regions. *Journal of Neuroscience, 26*(24), 6523–6532.

Goldberg, R. F., Perfetti, C. A., & Schneider, W. (2006). Perceptual knowledge retrieval activates sensory brain regions. *Journal of Neuroscience, 26*(18), 4917–4921.

Hagoort, P., Hald, L., Bastiaansen, M., & Petersson, K. M. (2004). Integration of word meaning and world knowledge in language comprehension. *Science, 304*(5669), 438–441.

Hasson, U., Nir, Y., Levy, I., Fuhrmann, G., & Malach, R. (2004). Intersubject synchronization of cortical activity during natural vision. *Science, 303*(5664), 1634–1640.

Hauk, O., Johnsrude, I., & Pulvermüller, F. (2004). Somatotopic representation of action words in human motor and premotor cortex. *Neuron, 41*(2), 301–307.

Hodges, J. R., Patterson, K., Oxbury, S., & Funnell, E. (1992). Semantic dementia: Progressive fluent aphasia with temporal lobe atrophy. *Brain, 115*(Pt. 6), 1783–1806.

Hodges, J. R., Patterson, K., Ward, R., Garrard, P., Bak, T., Perry, R., & Gregory, C. (1999). The differentiation of semantic dementia and frontal lobe dementia (temporal and frontal variants of frontotemporal dementia) from early Alzheimer's disease: A comparative neuropsychological study. *Neuropsychology, 13*(1), 31–40.

Humphries, C., Binder, J. R., Medler, D. A., & Liebenthal, E. (2007). Time course of semantic processes during sentence comprehension: An fMRI study. *NeuroImage, 36*(3), 924–932.

Huth, A. G., de Heer, W. A., Griffiths, T. L., Theunissen, F. E., & Gallant, J. L. (2016). Natural speech reveals the semantic maps that tile human cerebral cortex. *Nature, 532*(7600), 453–458.

Huth, A. G., Nishimoto, S., Vu, A. T., & Gallant, J. L. (2012). A continuous semantic space describes the representation of thousands of object and action categories across the human brain. *Neuron, 76*(6), 1210–1224.

Imamoglu, F., Huth, A. G., & Gallant, J. L. (2016). The representation of semantic information in the human brain during listening and reading. Paper presented at the Society for Neuroscience, San Diego, CA.

Jefferies, E., & Lambon Ralph, M. A. (2006). Semantic impairment in stroke aphasia versus semantic dementia: A case-series comparison. *Brain, 129*(Pt. 8), 2132–2147.

Jefferies, E., Patterson, K., Jones, R. W., Bateman, D., & Lambon Ralph, M. A. (2004). A category-specific advantage for numbers in verbal short-term memory: Evidence from semantic dementia. *Neuropsychologia, 42*(5), 639–660.

Jefferies, E., Patterson, K., Jones, R. W., & Lambon Ralph, M. A. (2009). Comprehension of concrete and abstract words in semantic dementia. *Neuropsychology, 23*(4), 492–499.

Kanwisher, N., McDermott, J., & Chun, M. M. (1997). The fusiform face area: A module in human extrastriate cortex specialized for face perception. *Journal of Neuroscience, 17*(11), 4302–4311.

Kiefer, M., & Pulvermüller, F. (2012). Conceptual representations in mind and brain: Theoretical developments, current evidence and future directions. *Cortex, 48*(7), 805–825.

Kramer, J. H., Jurik, J., Sha, S. J., Rankin, K. P., Rosen, H. J., Johnson, J. K., & Miller, B. L. (2003). Distinctive neuropsychological patterns in frontotemporal dementia, semantic dementia, and Alzheimer disease. *Cognitive and Behavioral Neurology, 16*(4), 211–218.

Kussmaul, A. (1877). Word deafness and word blindness. In *Cyclopaedia of the practice of medicine.* New York: William Wood, 770–778.

Laisney, M., Giffard, B., Belliard, S., de la Sayette, V., Desgranges, B., & Eustache, F. (2011). When the zebra loses its stripes: Semantic priming in early Alzheimer's disease and semantic dementia. *Cortex, 47*(1), 35–46.

Lambon Ralph, M. A., Graham, K. S., Patterson, K., & Hodges, J. R. (1999). Is a picture worth a thousand words?

Evidence from concept definitions by patients with semantic dementia. *Brain and Language, 70*(3), 309–335.

Lambon Ralph, M. A., & Patterson, K. (2008). Generalization and differentiation in semantic memory: Insights from semantic dementia. *Annals of the New York Academy of Sciences, 1124*, 61–76.

Lauro-Grotto, R., Piccini, C., & Shallice, T. (1997). Modality-specific operations in semantic dementia. *Cortex, 33*(4), 593–622.

Luzzi, S., Snowden, J. S., Neary, D., Coccia, M., Provinciali, L., & Lambon Ralph, M. A. (2007). Distinct patterns of olfactory impairment in Alzheimer's disease, semantic dementia, frontotemporal dementia, and corticobasal degeneration. *Neuropsychologia, 45*(8), 1823–1831.

Lynch, J. C., Mountcastle, V. B., Talbot, W. H., & Yin, T. C. (1977). Parietal lobe mechanisms for directed visual attention. *Journal of Neurophysiology, 40*(2), 362–389.

Martin, A. (2007). The representation of object concepts in the brain. *Annual Review of Psychology, 58*, 25–45.

Martin, A., & Chao, L. L. (2001). Semantic memory and the brain: Structure and processes. *Current Opinion in Neurobiology, 11*(2), 194–201.

Martin, A., Haxby, J. V., Lalonde, F. M., Wiggs, C. L., & Ungerleider, L. G. (1995). Discrete cortical regions associated with knowledge of color and knowledge of action. *Science, 270*(5233), 102–105.

Meteyard, L., Cuadrado, S. R., Bahrami, B., & Vigliocco, G. (2012). Coming of age: A review of embodiment and the neuroscience of semantics. *Cortex, 48*(7), 788–804.

Mummery, C. J., Patterson, K., Price, C. J., Ashburner, J., Frackowiak, R. S., & Hodges, J. R. (2000). A voxel-based morphometry study of semantic dementia: Relationship between temporal lobe atrophy and semantic memory. *Annals of Neurology, 47*(1), 36–45.

Mummery, C. J., Patterson, K., Wise, R. J., Vandenberghe, R., Price, C. J., & Hodges, J. R. (1999). Disrupted temporal lobe connections in semantic dementia. *Brain, 122* (Pt. 1), 61–73.

Naselaris, T., Kay, K. N., Nishimoto, S., & Gallant, J. L. (2011). Encoding and decoding in fMRI. *NeuroImage, 56*(2), 400–410.

Nestor, P. J., Fryer, T. D., & Hodges, J. R. (2006). Declarative memory impairments in Alzheimer's disease and semantic dementia. *NeuroImage, 30*(3), 1010–1020.

Nishimoto, S., Vu, A. T., Naselaris, T., Benjamini, Y., Yu, B., & Gallant, J. L. (2011). Reconstructing visual experiences from brain activity evoked by natural movies. *Current Biology, 21*(19), 1641–1646.

Nozari, N., & Thompson-Schill, S. L. (2016). Left ventrolateral prefrontal cortex in processing of words and sentences. In G. Hickok & S. L. Small (Eds.), *Neurobiology of language* (pp. 569–584). San Diego: Academic Press.

Nunez-Elizalde, A. O., Deniz, F., Gao, J. S., & Gallant, J. L. (2018). Discovering brain representations across multiple feature spaces using brain activity recorded during naturalistic viewing of short films. Paper presented at the Society for Neuroscience, San Diego, CA.

Patterson, K., Nestor, P. J., & Rogers, T. T. (2007). Where do you know what you know? The representation of semantic knowledge in the human brain. *Nature Reviews Neuroscience, 8*(12), 976–987.

Popham, S. F., Huth, A. G., Bilenko, N. Y., & Gallant, J. L. (2018). Visual and linguistic semantic representations are aligned at the boundary of human visual cortex. Paper presented at the Computational and Systems Neuroscience Meeting, Denver, CO.

Posner, M. I., Walker, J. A., Friedrich, F. A., & Rafal, R. D. (1987). How do the parietal lobes direct covert attention? *Neuropsychologia, 25*(1A), 135–145.

Pulvermüller, F. (2013). How neurons make meaning: Brain mechanisms for embodied and abstract-symbolic semantics. *Trends in Cognitive Sciences, 17*(9), 458–470.

Ralph, M. A. L., Graham, K. S., Ellis, A. W., & Hodges, J. R. (1998). Naming in semantic dementia—what matters? *Neuropsychologia, 36*(8), 775–784.

Ralph, M. A. L., Jefferies, E., Patterson, K., & Rogers, T. T. (2017). The neural and computational bases of semantic cognition. *Nature Reviews Neuroscience, 18*(1), 42–55.

Ralph, M. A. L., Sage, K., Jones, R. W., & Mayberry, E. J. (2010). Coherent concepts are computed in the anterior temporal lobes. *Proceedings of the National Academy of Sciences, 107*(6), 2717–2722.

Riddoch, M. J., & Humphreys, G. W. (1987). A case of integrative visual agnosia. *Brain, 110*(Pt. 6), 1431–1462.

Rodd, J. M., Davis, M. H., & Johnsrude, I. S. (2005). The neural mechanisms of speech comprehension: fMRI studies of semantic ambiguity. *Cerebral Cortex, 15*(8), 1261–1269.

Rosen, H. J., Gorno-Tempini, M. L., Goldman, W. P., Perry, R. J., Schuff, N., Weiner, M., … Miller, B. L. (2002). Patterns of brain atrophy in frontotemporal dementia and semantic dementia. *Neurology, 58*(2), 198–208.

Roskies, A. L., Fiez, J. A., Balota, D. A., Raichle, M. E., & Petersen, S. E. (2001). Task-dependent modulation of regions in the left inferior frontal cortex during semantic processing. *Journal of Cognitive Neuroscience, 13*(6), 829–843.

Saxe, R., & Kanwisher, N. (2003). People thinking about thinking people: The role of the temporo-parietal junction in "theory of mind." *NeuroImage, 19*(4), 1835–1842.

Schwartz, M. F., Marin, O. S. M., & Saffran, E. M. (1979). Dissociations of language function in dementia: A case study. *Brain and Language, 7*(3), 277–306.

Silveri, M. C., Brita, A. C., Liperoti, R., Piludu, F., & Colosimo, C. (2018). What is semantic in semantic dementia? The decay of knowledge of physical entities but not of verbs, numbers and body parts. *Aphasiology, 32*(9), 989–1009.

Snowden, J. S. (2015). Semantic memory. In Wright, James D., *International encyclopedia of the social & behavioral sciences* (pp. 572–578). Elsevier.

Snowden, J. S., Goulding, P. J., & Neary, D. (1989). Semantic dementia: A form of circumscribed cerebral atrophy. *Behavioural Neurology, 2*(3), 167–182.

Snowden, J. S., Harris, J. M., Thompson, J. C., Kobylecki, C., Jones, M., Richardson, A. M., & Neary, D. (2018). Semantic dementia and the left and right temporal lobes. *Cortex, 107*, 188–203.

Thompson-Schill, S. L., D'Esposito, M., Aguirre, G. K., & Farah, M. J. (1997). Role of left inferior prefrontal cortex in retrieval of semantic knowledge: a reevaluation. *Proceedings of the National Academy of Sciences, 94*(26), 14792–14797.

Vigliocco, G., Meteyard, L., Andrews, M., & Kousta, S. (2009). Toward a theory of semantic representation. *Language and Cognition, 1*(2), 219–247.

Visser, M., Jefferies, E., & Lambon Ralph, M. A. (2010). Semantic processing in the anterior temporal lobes: A meta-analysis of the functional neuroimaging literature. *Journal of Cognitive Neuroscience, 22*(6), 1083–1094.

Wagner, A. D., Paré-Blagoev, E. J., Clark, J., & Poldrack, R. A. (2001). Recovering meaning: Left prefrontal cortex guides controlled semantic retrieval. *Neuron, 31*(2), 329–338.

Warrington, E. K. (1975). The selective impairment of semantic memory. *Quarterly Journal of Experimental Psychology, 27*(4), 635–657.

Whitney, C., Kirk, M., O'Sullivan, J., Lambon Ralph, M. A., & Jefferies, E. (2011). The neural organization of semantic control: TMS evidence for a distributed network in left inferior frontal and posterior middle temporal gyrus. *Cerebral Cortex, 21*(5), 1066–1075.

Wilkins, A., & Moscovitch, M. (1978). Selective impairment of semantic memory after temporal lobectomy. *Neuropsychologia, 16*(1), 73–79.

Wu, M. C.-K., David, S. V., & Gallant, J. L. (2006). Complete functional characterization of sensory neurons by system identification. *Annual Review of Neuroscience, 29*, 477–505.

VI
INTENTION, ACTION, CONTROL

Introduction

RICHARD B. IVRY AND JOHN W. KRAKAUER

The study of the motor system has a distinguished pedigree, evident in the writings of the ancient Greeks, prominent in the work of the first scientists to probe the brain, and a central concern in present-day research. From a cultural standpoint, motor skills are a source of massive fascination for the general public; over a billion people watched the 2018 World Cup final. Aristotle in his book *On the Motion of Animals* and in other works considers movement to be the "actualization" of being. At the turn of the 20th century, the Nobel laureate Charles Sherrington stated that "to move things is all that mankind can do whether it be the whisper of a syllable or the felling of a forest." Thus, from its very beginnings it is apparent that the study of action and of the motor system can make a dual contribution: it is of interest both in its particularity as a component of the functioning nervous system and in its generality as a model system for the study of cognition. The chapters in this section exemplify both of these themes. The experimental and theoretical tractability of motor behavior makes it particularly suitable for the generation of new principles that can then spread to other areas of neuroscience. Moreover, the authors bring a fresh perspective to areas of motor neuroscience that have been colonized of late with half-truths and somewhat stale ideas.

In direct lineage with Sherrington and his work on the spinal cord reflex, Monica A. Perez discusses the reorganization of the corticospinal tract (CST) after incomplete spinal cord injury. In contrast to the work of Sherrington, Perez is not interested in the behavior of the isolated spinal cord but in the more complex interaction between residual descending pathways and altered segmental circuitry. Her work reflects the recent shift in emphasis away from thinking that autonomous central pattern generators below the level of the lesion should be the main rehabilitative target. Rather, she emphasizes the importance of considering the influence of residual descending pathways through the lesioned territory. Perez goes on to discuss how noninvasive brain stimulation methods can provide insight into how the CST can reorganize, both spontaneously and in response to rehabilitation, presumably in a way causally related to recovery. Most excitingly, Perez discusses a new protocol inspired by classic cellular physiology work on spike-timing-dependent plasticity. Here repeated transcranial magnetic stimulation (TMS)-elicited corticospinal activity over a region of the primary motor cortex is timed to arrive a few milliseconds before antidromic activation of the motoneurons by supramaximal electrical stimulation of the peripheral nerve. This procedure induces an increase in the amplitude of motor-evoked potentials following cervicomedullary stimulation, highlighting a promising intervention to strengthen connections weakened by incomplete spinal cord injury.

There has been much excitement over the last 15 years or so with regard to the clinical implications of brain-machine interfaces (BMI) in the treatment of spinal cord injury, stroke, and a range of neurological disorders. With BMI, neural-recording technology, often in the form of implantable electrodes, serves as a conduit between the brain, or intentions of the person, and the external world. Two currents of research are discernible in the BMI world, one focusing on developing ever better biomimetic-decoding algorithms for more effective control of prosthetics and the other asking basic questions about how neural populations learn motor skills. Andrew Jackson shows how these two directions have converged of late. He focuses both on what is being discovered about the hierarchical structure present in high-dimensional neural state spaces and how it changes over the course of learning, and on how it can be exploited to improve biomimetic BMI decoding. Jackson posits that complex movements are constructed from muscle/joint synergies and submovement segments in the same way that complex sentences are built from phonemes and words. From this analogy, he segues to the interesting idea that the current state of BMI efforts can be compared to early speech recognition software in the 1970s before the advent of machine-learning approaches. Indeed, recent work suggests that machine learning by neural networks can yield decoders capable of considerable generalization to untrained behaviors.

Although this section of the book is focused on action, it has long been appreciated that accurate and purposeful movement cannot be achieved without ongoing sensory feedback, a fact apparent even in the monosynaptic reflex arc. Jeffrey Weiler and Andrew Pruszynski press this point by noting that "approximately 90% of the axons in the peripheral nerves of the upper limb transmit sensory information from the periphery into the central nervous system, while the remaining 10% of axons carry the motor commands from the central nervous system to muscles." After reviewing the devastating consequences of proprioceptive loss for motor control, Weiler and Pruszynski turn their attention to long-latency stretch reflexes. Even though these responses occur with latencies substantially shorter than voluntary reaction times, their expression is modulated by the subject's intent, sensitivity to task and limb structure, and they are engaged during decision-making and learning tasks. Work of this kind reveals that the spectrum from simple reflexes to voluntary movements can be seen as a hierarchy of feedback control loops of ever-increasing "intelligence." The chapter also reviews how sensory feedback from multiple modalities is integrated in real time and the relationship between somatosensory feedback for perception versus motor control. With regard to the latter, fascinating new data are described showing that people have far greater tactile acuity during motor control compared to when they are asked to make a perceptual report, underscoring the importance of studying sensory systems when embedded in motor behavior rather than in isolation.

A long-standing and cherished principle of organization in the sensory and motor cortices is the somatotopic map. Changes in cortical maps, either in response to use and learning or as a consequence of central and peripheral injury, have been thought to have significant behavioral implications. Tamar R. Makin, Jörn Diedrichsen, and John W. Krakauer take a critical look at sensorimotor cortical maps and in particular question whether reorganization, generally understood as a *qualitative* change in the input-output characteristics of a cortical area, ever happens. That is, does one representation invade or "take over" another? They examine this question by considering three putative triggers for reorganization: learning, loss of cortical inputs from amputation, and loss of cortical substrate following stroke. They conclude that changes in cortical maps from experience or injury are likely not due to reorganization but result from the unmasking of preexisting cortical connections or *subcortical* reorganization. They also argue that map changes, regardless of their causes, are not the causal factors in behavioral change.

The basal ganglia are a set of subcortical nuclei long implicated in motor control and motor learning in health and disease. There is, however, increasing awareness that these nuclei contribute to perception and cognition. Similar to current work on that other prominent subcortical structure, the cerebellum, the holy grail in basal ganglia research seems to be finding a universal computation, with regional differences attributable to this computation being performed on different variables—an idea that seems to be implied by the multiple parallel cortical-basal ganglionic loops. An important challenge for this endeavor is to reconcile what seem to be distinct learning versus performance functions of the basal ganglia. David Robbe and Joshua Tate Dudman review human and nonhuman animal data on the role of the striatum and its dopaminergic inputs with regard to action selection, motor control, decision-making, and learning. They favor an emphasis on the role of the basal ganglia in the selection of overlearned actions and their associated degree of vigor. It is less clear, in their view, whether the basal ganglia are needed for either learning or executing a skilled movement.

The idea that an action must be planned seems so obvious as to need no re-examination. Adrian Haith and Sven Bestmann show that this is clearly not the case. Indeed, they put forward a new view. They argue that movement preparation is a process of setting the state of the motor system once an action goal is identified, priming it to generate a single, task-appropriate movement. Contrary to traditional views, this preparatory process occurs very rapidly and is perhaps completed within approximately 50 ms. However, completing preparation does not directly trigger initiation of the movement; initiation is conceptualized as a separate, independent process. In addition, Haith and Bestmann provide alternative explanations for two prominent ideas in the literature: first, that several movements can be prepared in parallel and second, that the circuitry and mechanisms for decision-making and those for movement representation overlap. The authors argue instead that only one movement-control policy is present at any point in time and that this policy reflects the instantaneous state of decision uncertainty across goals. That is to say, there can be multiple goals but only one plan. They review recent physiology data from nonhuman primates that support this view.

There is a prevailing assumption, both in the cognitive neuroscience community and in the world at large, that there is something a bit undemanding, intellectually, about having a motor skill—the notion of the "dumb jock." Although claims to a distinction between "knowing what" and "knowing how" go back to the Greeks, it was given seeming intellectual respectability by the seminal findings in the patient H. M. using a mirror-drawing task. In their chapter, Jordan A. Taylor and Samuel D. McDougle question this simple dichotomous framework. They summarize a series of studies using visuomotor adaptation tasks to show that even simple motor-learning paradigms, like mirror drawing, do in fact comprise implicit learning mechanisms and explicit strategies that combine to accomplish the task. They conclude that, like all other cognitive tasks, motor learning recruits a full taxonomy of memory systems. Their position can be summarized as saying that skilled motor behaviors are far too important to leave to just one part of the brain.

Two abilities that lie right at the interface of cognition and movement are imitation and tool use. Humans, even compared to chimpanzees, our closest primate relative, are markedly superior at both. Fascinatingly, in humans both of these abilities are often lost when a left hemispheric lesion causes apraxia. It has been surprisingly difficult, however, to bring apraxia into some kind of conceptual and taxonomic order. For the most part, what we have been given instead are increasingly elaborate descriptions of the apraxic phenomena and a proliferation of terms for them. Laurel J. Buxbaum and Solène Kalénine have sought to rectify this situation by mapping behaviors onto putative computations and their associated left hemispheric anatomy. In particular, they delineate three major clusters of behaviors that reflect damage to conceptual, spatiotemporal, and selection-based components of tool use and imitation, which in turn are associated with posterior temporal, inferior parietal, and frontal network nodes, respectively.

It is to be hoped that the ambitious, interesting, and original chapters in this section demonstrate that the study of action can provide a fruitful terrain for deriving principles applicable to all of cognitive neuroscience.

40 The Physiology of the Healthy and Damaged Corticospinal Tract

MONICA A. PEREZ

ABSTRACT The corticospinal tract (CST) is a major descending motor pathway contributing to the control of voluntary movement in mammals. Anatomical and electrophysiological studies have shown significant reorganization in the CST following spinal cord injury (SCI) in humans. Noninvasive strategies that have targeted the CST have proven to be efficient to potentiate, at least to some extent, voluntary motor output after chronic, incomplete SCI. These approaches have used transcranial magnetic stimulation over the primary motor cortex and electrical stimulation over peripheral nerves as tools to induce plasticity in residual corticospinal synaptic connections, following the principles of spike-timing-dependent plasticity. The results of this work, together with information about the extent of the injury, provide a new framework for exploring the contribution of the CST to the recovery of function following SCI.

There are over 400,000 persons with spinal cord injury (SCI) in the United States and several million worldwide. These individuals have limited motor function, resulting in serious disability. Although experimental strategies ranging from neuroprotection to cell transplantation are designed to restore sensorimotor function following SCI, the efficacy of these treatments has been limited. At present, rehabilitation-based approaches are more common and are widely used to promote recovery after injury. These interventions likely depend on the recruitment of descending motor pathways, including the corticospinal tract (CST). The CST contributes significantly to the control of skilled movements in mammals (Lemon, 2008) and is a prominent target for investigating injury-induced plasticity and motor recovery after SCI (Oudega & Perez, 2012).

The first aim of this chapter is to review anatomical evidence of corticospinal reorganization after human SCI. Postmortem examination of spinal cord tissue has revealed anatomical changes in the CST and the presence of continuity of CNS parenchyma several segments below the injury. The second aim of this chapter is to highlight the main physiological features of cortical and corticospinal reorganization that can be observed at rest and during movement in people with SCI. Electrophysiological studies employing transcranial magnetic stimulation (TMS) have been used extensively to study the corticospinal system in humans since the output of the primary motor cortex can be easily assessed from the motor-evoked potentials (MEP) observed in electromyographic (EMG) recordings. TMS probes reveal reorganization in different aspects of corticospinal function after injury, including the threshold, amplitude, and latencies of MEPs. The utility of TMS as a tool for clinical diagnosis and clinical research studies will also be discussed.

Corticospinal Reorganization after Spinal Cord Injury: Anatomical Evidence

Early after an SCI, necrosis and apoptosis are responsible for the death of neurons and glia both near to and distant from the lesion. At later stages, the lesion commonly consists of a multilocular cavity traversed by vascular-glial bundles, accompanied by regenerated nerve roots (Kakulas, 2004). Postmortem examination of human spinal cord tissue and in vivo magnetic resonance imaging (MRI) analysis reveal Wallerian degeneration in the CST as early as a few days (Becerra et al., 1995; Buss et al., 2004) to a few weeks (Becerra et al., 1995; Quencer & Bunge, 1996) postlesion. The areas of Wallerian degeneration exhibit progressive astrogliosis (Bunge et al., 1993; Puckett et al., 1997). In the chronically injured human spinal cord, the number of reactive astrocytes around the lesion cavities is small (Bunge et al., 1993; Puckett et al., 1997) in comparison to that found in rodent models of SCI (Murray et al., 1990). This finding may have implications for the regenerative ability of axons in the injured human spinal cord, as they may not be exposed to the growth-inhibitory molecules expressed by reactive astrocytes to the same degree as in rodents.

Histological (Buss et al., 2004) and neuroimaging (Wrigley et al., 2009) data show that the loss of CST axons and/or myelin in humans with an SCI is gradual. Water diffusion changes are observed in tracts not damaged by the spinal injury, suggesting that in humans, as in animal models of SCI, uninjured tracts undergo reorganization after the lesion. Despite ample evidence for the presence of CST sprouting in animal models of SCI, comparable evidence in humans is sparse and indirect. A few studies have shown a reduced number of myelinated corticospinal axons and retrograde degeneration in postmortem material after chronic SCI (Bronson et al., 1978; Fishman, 1987; Hunt, 1904). A marked depletion of CST axons is observed at the injury site, whereas close to normal numbers of CST axons are seen at a distance from the injury, regardless of the injury duration. This suggests that degenerated axons are replaced by collateral sprouts of surviving axons (Fishman, 1987).

Based on postmortem analyses, approximately 75% of individuals with a diagnosis of clinically complete SCI exhibit evidence of some continuity of central nervous system (CNS) tissue across the injured segments (Kakulas, 1988). Histological analysis at the epicenter of the lesion revealed continuity of CNS parenchyma in approximately 62% of the tested spinal cord specimens (Bunge et al., 1993). These observations are in agreement with earlier neurophysiological studies that showed individuals with clinically complete SCI could present a tonic vibratory response (Dimitrijevic et al., 1977, 1984), voluntarily suppress responses to stimulation (Cioni et al., 1986), and respond to reinforcement maneuvers (Dimitrijevic et al., 1977, 1984). This indicates that some supraspinal control of muscles below the level of the injury was preserved, leading to the categorization of these individuals as discomplete (Dimitrijevic, 1988).

Contemporary evidence continues to support the view that a large number of individuals with clinically complete SCI are discomplete. For example, approximately 66% of individuals with a clinical diagnosis of no preserved motor function below the injury level were able to produce volitional EMG signals in muscles with motoneurons located below their injury level (Heald et al., 2017). Responses evoked by TMS over the primary motor cortex and/or voluntary muscle activity in muscles innervated below the lesion are also observed in most individuals with clinically complete SCI (Edwards et al., 2013; Squair et al., 2016). Behavioral evidence of the discomplete condition comes from studies using epidural or transcutaneous spinal cord stimulation, combined with motor training. This intervention can produce the recovery of some voluntary function in individuals with clinically complete SCI (Angeli et al., 2014; Donati et al., 2016; Harkema et al., 2011). Altogether, these studies suggest the presence of some residual CST connectivity both right after the injury and after an extended period of recovery.

Corticospinal Reorganization after Spinal Cord Injury: Physiological Evidence

TMS has emerged as an important noninvasive tool to investigate the contribution of the CST to human motor control. TMS has been used extensively for studying the corticospinal system since the output of the primary motor cortex can be easily assessed by measuring MEPs from EMG recordings. This is achieved using a short-lasting magnetic field that peaks after 0.2 ms and readily penetrates to the cortex due to the low impedance of the scalp. In contrast to electrical stimulation over the scalp, there is minimal discomfort with TMS since the magnetic field does not activate nociceptors. The short-lasting field of most available stimulators favors the excitation of axons over cell bodies, and the rapid decline in intensity with distance enables the excitation of superficial cortical layers. Corticospinal neurons are most likely activated where the axon bends away from the direction of the magnetic field (Amassian et al., 1993; Maccabee et al., 1993). Note that TMS can directly activate corticomotoneuronal cells as well as disynaptic pathways, both of which contribute to the size of MEPs (Petersen et al., 2010).

The first studies using TMS in humans with an SCI were published in the early 1990s (Brouwer, Bugaresti, & Ashby, 1992; Levy et al., 1990; Topka et al., 1991), offering the promise of this method for exploring the mechanisms involved in cortical and corticospinal reorganization after injury. Levy and collaborators (1990) used TMS with two quadriplegic individuals who had regained some voluntary control in proximal arm muscles, while the distal muscles remained paretic. They were able to elicit MEPs in proximal muscles from a much wider area of the scalp than in control subjects. Similarly, Topka and colleagues (1991) elicited MEPs from muscles in the abdominal wall rostral to the injury site from a larger number of scalp positions in individuals with SCI compared to control subjects. Brouwer, Bugaresti, and Ashby (1992) demonstrated that the short latency facilitation of MEPs in lower-limb muscles, reflecting activation of the fast corticospinal pathway, was present in individuals with acute and chronic SCI, although the latencies were delayed. Since these early publications, a large number of studies have provided evidence that TMS can be used to assess transmission along the corticospinal pathways, providing insights about reorganization and the presence of residual connectivity, as well as a tool to investigate clinical rehabilitation plasticity.

One of the important pathological processes affecting white matter tracts after an SCI is the chronic and progressive demyelination of long motor axons (Bunge et al., 1993; Griffiths & McCulloch, 1983; Totoiu & Keirstead, 2005). Histological examination in animal and human tissue has shown that after an SCI myelin loss is most pronounced in large-diameter fibers (Blight & Young, 1989; Quencer et al., 1992). In humans, transmission in large-diameter, fast-conducting fibers can be, to some extent, assessed by testing the effect of TMS on single motor unit recordings (Brouwer, Bugaresti, & Ashby, 1992). The majority of studies using TMS in humans with an incomplete SCI have reported delayed MEP latencies in partially paralyzed muscles. MEP latencies are delayed by approximately 2–10 ms in patients with cervical and thoracic SCI. These delays can be observed from the initial assessment on the day of injury to months and years after the injury (Alexeeva, Broton, & Calancie, 1998; Bunday & Perez, 2012a, 2012b; Curt, Keck, & Dietz, 1998). Resting and active motor thresholds also tend to increase in individuals with incomplete SCI. For example, a longitudinal study in individuals with incomplete SCI demonstrated that the motor thresholds tested at rest or during a small voluntary contraction significantly increased over the first year of injury (Smith et al., 2000). Similarly, in individuals with cervical SCI, resting and active motor thresholds were increased several years postinjury (Barry et al., 2013). The motor threshold may also be related to the degree of impairment; thus, individuals with a small amount of motor impairment can show thresholds similar to controls (Bunday & Perez, 2012a, 2012b).

A single TMS pulse over the primary motor cortex evokes temporally synchronized descending waves in the CST that can be recorded from the epidural space (Di Lazzaro et al., 2012). The shortest wave is likely due to direct stimulation of the corticospinal neuron (D wave) at some distance from the cell body, while the later indirect (I) waves (termed I1, I2, and I3) possibly arise from the transsynaptic activation of corticospinal neurons by intracortical circuits (Di Lazzaro et al., 2012). Notably, the duration and intensity of the field as well as the direction of the induced current in the brain affect the characteristics of MEPs elicited by TMS in healthy individuals (D'Ostilio et al., 2016; Hanna & Rothwell, 2017) and in people with SCI (Jo et al., 2018). TMS-induced electrical currents flowing from posterior to anterior (PA) across the central sulcus preferentially evoke highly synchronized corticospinal activity, while currents flowing from anterior to posterior (AP) preferentially evoke less synchronized activity, with their peaks partially matching the timing of the PA-evoked activity (Day et al., 1989; Sakai et al., 1997). The characteristics of PA and AP activity resemble the I waves recorded in animal studies (Kernell & Chien-Ping, 1967; Patton & Amassian, 1954), and the interval between I waves in primates (Maier et al., 1997) and humans (Di Lazzaro et al., 1998) is similar.

In controls, MEPs elicited with the coil in the AP orientation have longer latency, larger latency dispersion, and a higher threshold than MEPs elicited in the PA orientation (Di Lazzaro et al., 2012; Di Lazzaro, Rothwell, & Capogna, 2017). Orienting the coil to induce currents flowing from lateral to medial (LM) favors the direct activation of the corticospinal neurons. This results in MEPs with shorter latencies compared with PA and AP stimulation (Werhahn et al., 1994). MEP latencies in all coil orientations are prolonged in humans with SCI compared with control subjects (Jo et al., 2018; figure 40.1*A*). In addition, latencies of MEPs elicited by PA and AP stimulation, relative to those elicited by LM stimulation, are shorter in SCI compared with control subjects and larger for MEPs elicited by AP stimulation, suggesting that neural structures activated by AP-induced currents are more affected after SCI (Jo et al., 2018; figure 40.1*B*).

Another way of making inferences about descending corticospinal volleys in humans is by using paired-TMS paradigms. Paired-TMS pulses can be precisely timed to increase the amplitude of MEPs at interstimulus intervals of approximately 1.5 ms, compatible with the I waves recorded from the epidural space in control subjects (Tokimura et al., 1996; Ziemann et al., 1998) and in individuals with SCI (Cirillo et al., 2016). MEP peaks mimicking early and late I waves have decreased amplitude in SCI subjects compared with controls (figure 40.2*A*). The second and third peaks were delayed, with the third peak also showing an increased duration (figure 40.2*B*). A relationship was observed between the temporal and spatial aspects of the late peaks and MEP amplitude and hand voluntary motor output, suggesting that late corticospinal inputs on the spinal cord might be crucial for the recruitment of motoneurons after SCI.

A few studies have examined cortical and corticospinal reorganization in individuals with SCI during motor performance. Corticospinal reorganization associated with the recovery of motor function may be reflected by changes in the recruitment order of motoneurons. Davey and collaborators (1999) tested individuals with SCI rostral to C8–T1 segments and examined the effect of increasing levels of isometric voluntary contraction on the size of MEPs elicited in thenar muscles. The individuals showed a less pronounced increase in MEP size with increasing TMS stimulus intensity compared with control subjects. A similar decrease in corticospinal recruitment has also been reported in humans with SCI during functionally relevant motor tasks, such as a

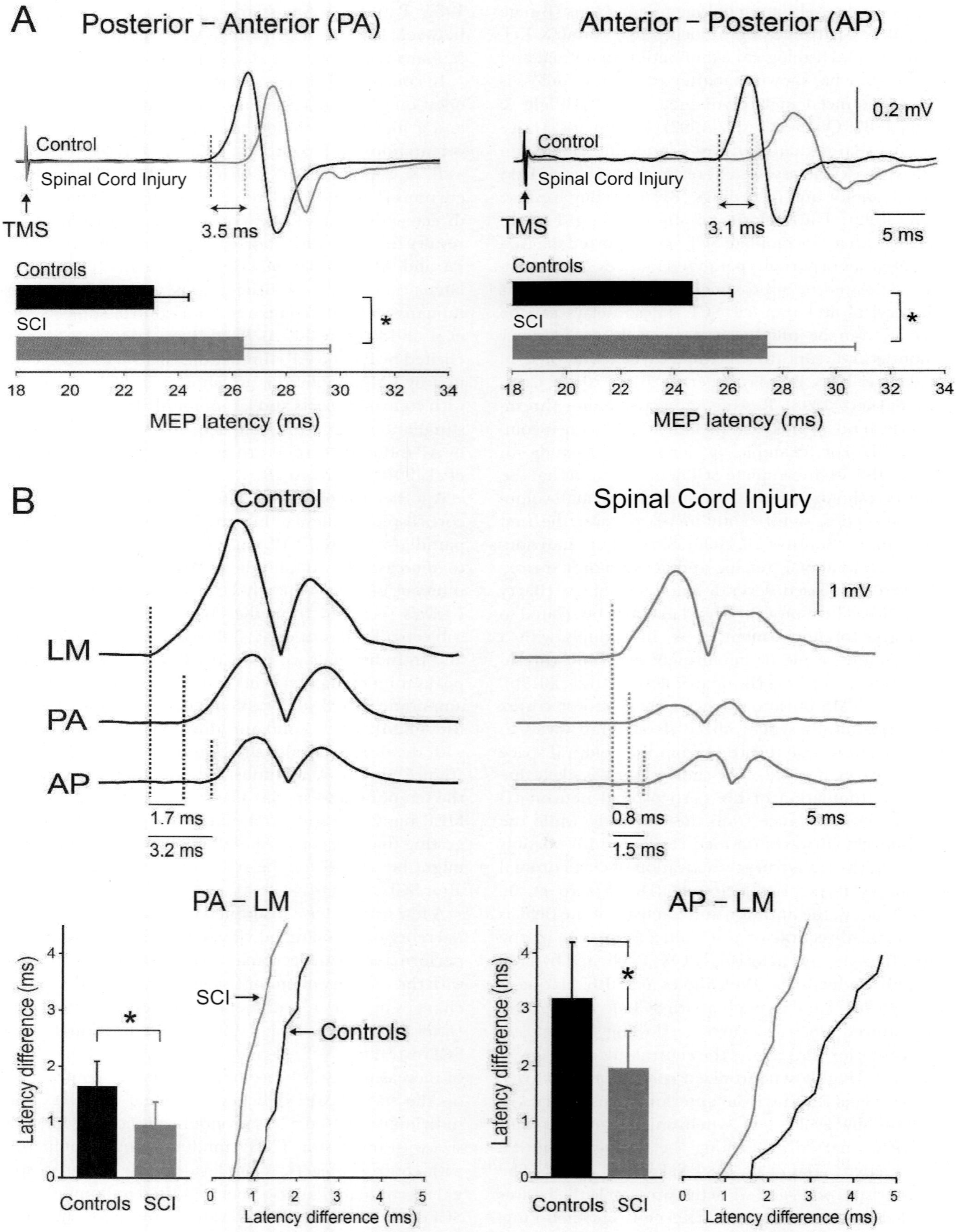
A
Posterior – Anterior (PA)
Anterior – Posterior (AP)
Control
Spinal Cord Injury
TMS
3.5 ms
3.1 ms
0.2 mV
5 ms
Controls
SCI
*
MEP latency (ms)
B
Control
Spinal Cord Injury
LM
PA
AP
1 mV
1.7 ms
3.2 ms
0.8 ms
1.5 ms
5 ms
PA – LM
AP – LM
Latency difference (ms)
Controls
SCI
Latency difference (ms)

precision grip (Bunday et al., 2014). It is possible that, after injury, changes in the reorganization of connections within the corticospinal system are needed for a muscle to function over its entire effective range. This might be accomplished by inputs from other descending or segmental inputs that contribute to increase the drive to spinal motoneurons, with the remaining corticospinal output helping modulate the voluntary contraction.

Another study used TMS during locomotion in individuals with chronic incomplete SCI. Parameters such as MEP amplitude at rest and MEP latency during a voluntary contraction correlated with the degree of foot drop (Barthélemy et al., 2010). This suggests that transmission in the corticospinal drive to lower-limb spinal motoneurons is of functional importance for lifting the foot during the early swing phase of the gait cycle. Importantly, these results demonstrate a linkage between electrophysiological measurements of corticospinal function and a behavioral deficit observed during locomotion after SCI. This is also in agreement with evidence showing that several months of locomotor training can enhance corticospinal excitability, measured by changes in the size of the maximal MEP and the slope of input-output excitability recruitment curves for lower-limb muscles (Thomas & Gorassini, 2005). The percentage change in MEP size in lower-limb muscles was correlated with the improvements in locomotor ability. This suggests that the recovery of locomotion may be mediated, in part, by changes in corticospinal function.

In another study, MEPs were measured in a resting hand muscle during increasing levels of isometric voluntary contraction by a contralateral finger muscle and a more proximal arm muscle (Bunday & Perez, 2012a). The size of the MEPs in the resting hand remained unchanged during increasing levels of voluntary contraction with a contralateral distal or proximal arm muscle in SCI participants. In contrast, MEP amplitude in a resting hand muscle increased during the same motor tasks in controls. To examine the mechanisms contributing to increases in MEP size, the authors examined short-interval intracortical inhibition (SICI), F waves, and cervicomedullary MEPs (CMEPs). SICI, F-wave amplitude and persistence, and MEP amplitude during contraction of the contralateral arm remained unchanged after cervical SCI, whereas in controls, SICI decreased, and the other measures increased (figure 40.3). The SCI effects may result from a lack of changes in the excitability of index finger motoneurons after chronic cervical SCI.

Overall, the results from these studies have increased our understanding of how the reorganized corticospinal pathway responds during voluntary movement. The work also makes clear that a better understanding of the involvement of the reorganized corticospinal pathways in functionally relevant tasks is important for elucidating the mechanisms underlying recovery after human SCI. Although insights have been gained about how to stimulate residual corticospinal tract connections following SCI, effective protocols that engage these connections to facilitate motor recovery remain limited (Tazoe & Perez, 2015).

Noninvasive Stimulation of the Corticospinal Tract

In the past 15 years, several noninvasive stimulation protocols have been used in an effort to restore function in humans with SCI (Tazoe & Perez, 2015). Here we focus on a protocol that targets the CST using the principles of spike-timing-dependent plasticity (STDP). *STDP* refers to a process by which synaptic strength changes as a function of the spiking activity in pre- and postsynaptic neurons (Bi & Poo, 1998). When presynaptic activation precedes postsynaptic antidromic somatic activation, synaptic connectivity increases, whereas the reverse order results in a decline in synaptic strength. As suggested by Hebb (1949), the temporal characteristics indicate a mechanism that depends on causal inference rather than mere coincidence, a process thought to be mediated by the N-methyl-D-aspartate (NMDA) glutamate receptor (Kampa et al., 2004). Both in vitro STDP (Carporale & Dan, 2008) and STDP-like changes in humans (Donges et al., 2017)

FIGURE 40.1 Motor-evoked potentials (MEPs). *A*, MEPs elicited in the first dorsal interosseous (FDI) muscle during index finger abduction when the current in the transcranial magnetic stimulator (TMS) coil was flowing in the posterior-anterior (PA) and anterior-posterior (AP) direction in a control subject (*black traces*) and a participant with SCI (*red traces*). Waveforms represent the average of 20 MEPs. Bar graphs show group data (control, $n=17$; SCI, $n=17$). MEP latency is plotted on the abscissa (control = *black bar*, SCI = *red bar*). *B*, Comparison of MEP latencies elicited with the coil in the lateromedial (LM), PA, and AP orientation during index finger abduction in the FDI muscle in a control and SCI subject. Waveforms represent the average of 20 trials. Group data (control, $n=17$; SCI, $n=17$) showing PA–LM and AP–LM MEP latency differences during index finger abduction in controls (*black bars*) and SCI (*red bars*). Error bars indicate SD. $^*p<0.05$; comparison between groups and lines represents MEP latency differences in individual subjects (control = *black lines*; SCI = *red lines*). Modified from Jo et al. (2018).

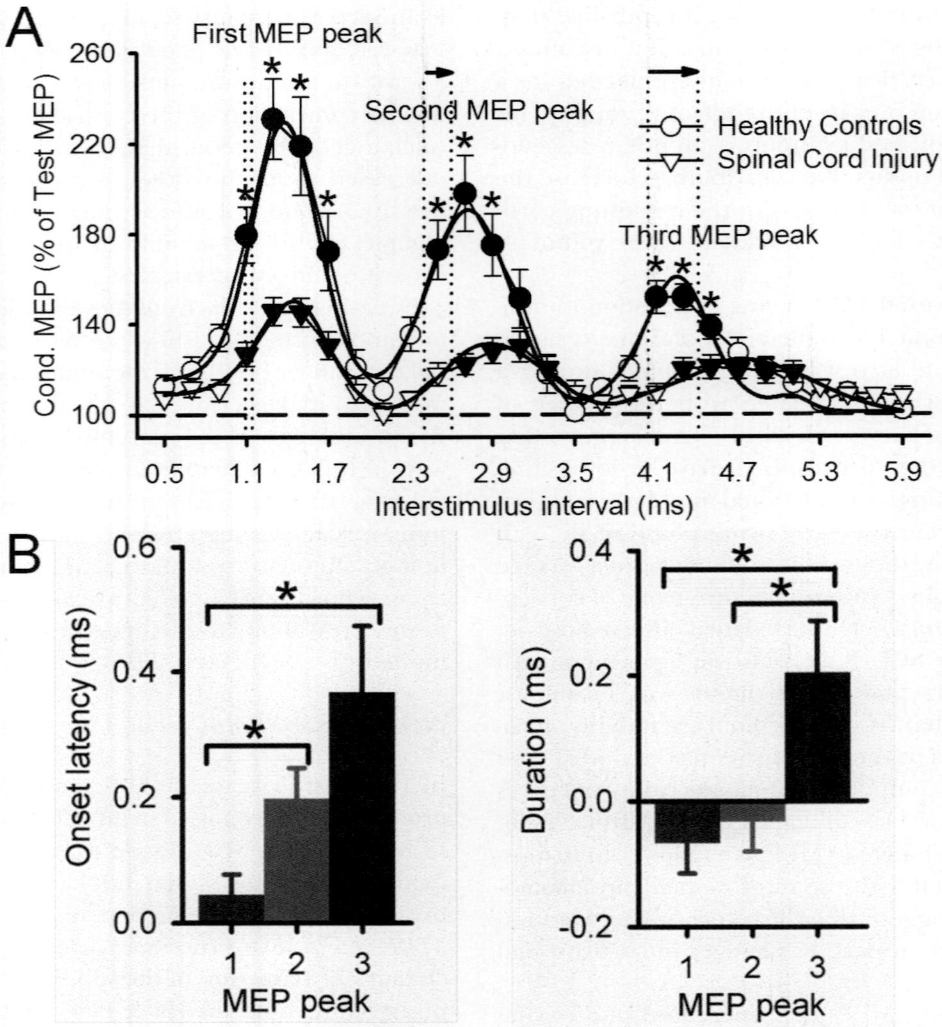

FIGURE 40.2 Latency of peaks in the MEP, estimated by a Gaussian model. *A*, Group data estimated by a Gaussian model for each MEP peak in controls (*circles*) and subjects with SCI (*triangles*). The interstimulus interval (ISI) between TMS pulses is plotted on the abscissa, and the ordinate shows the size of the conditioned MEPs (expressed as a percentage of the test MEP), with the solid line indicating the curve fit. Note that each MEP peak is constructed from the average of changes in conditioned MEPs expressed as a percentage of the test MEP and do not represent individual MEPs. The vertical dotted lines indicate the beginning of each MEP peak as determined by a model fit. Note the delay, as shown by the arrows, in SCI subjects compared with controls for the second and third MEP peak. The horizontal dashed line represents the size of the test MEP (control, $n=14$; SCI, $n=12$). *B*, Graphs show the group difference in each MEP peak in onset and duration for the first (*red*), second (*green*), and third (*blue*) MEP peaks between SCI subjects and controls. Each MEP peak is plotted on the abscissa (1, 2, and 3), with the ordinate depicting the onset (*left*) and duration (*right*) of each peak. Overall, subjects with SCI showed more pronounced changes in duration and onset latency in the third compared with the second and first MEP peaks. Error bars indicate SEs. $^*P<0.05$. Modified from Cirillo et al. (2016).

are thought to engage long-term potentiation (LTP) and long-term depression (LTD) mechanisms that depend on NMDA receptor activity. The timely arrival of the antidromic postsynaptic activation removes the Mg^{2+} receptor blocker, which enables Ca^{2+} to diffuse through the NMDA receptor (Kampa et al., 2004). The NMDA receptor thus detects the postsynaptic depolarization and the presynaptic release of glutamate. The resultant change in the Ca^{2+} influx leads to LTD or LTP induction, depending on the intracellular calcium concentration.

Noninvasive paired associative stimulation protocols using principles of STDP were first used in humans to target the primary motor cortex (Stefan et al., 2000). Here, STDP was evoked by timing the arrival of afferent volleys at the somatosensory cortex and the TMS pulse (Stefan et al., 2002; Wolters et al., 2003). Accordingly, the underlying mechanisms were thought to be

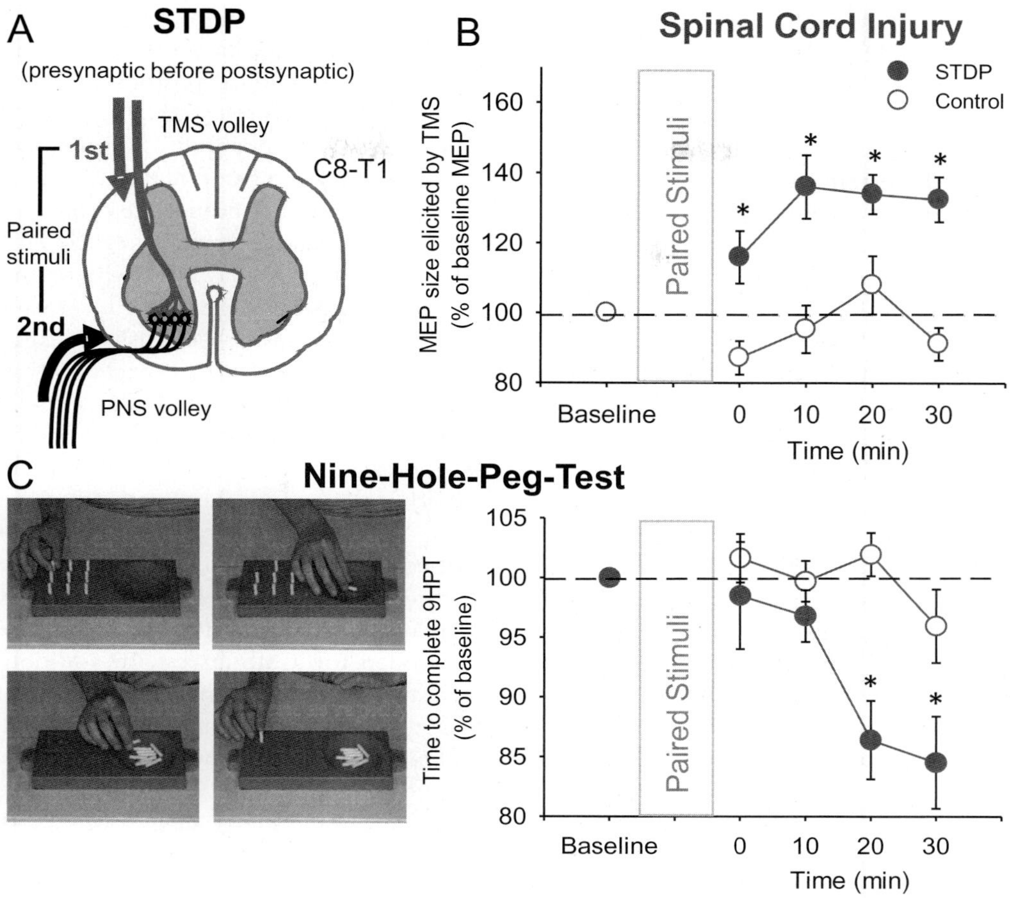

FIGURE 40.3 Paired corticomotoneuronal stimulation (PCMS). *A*, Illustration of the PCMS protocol used to enhance corticospinal function after SCI. Here, corticospinal neurons were activated at a cortical level by using TMS (TMS volley, first) delivered over the hand motor cortex, and spinal motoneurons were activated antidromically by peripheral nerve stimulation (PNS volley, second) delivered to the ulnar nerve. *B*, MEP amplitude increased after PCMS in which postsynaptic pulses were timed to arrive at the synapses 5 ms before presynaptic activation. *C*, Improvement in hand function as assessed with a nine-hold-peg test in 18 participants with chronic cervical SCI. Error bars, SEs, $^{*}p < 0.05$. Modified from Bunday and Perez (2012b). (See color plate 44.)

changes in synaptic efficacy due to LTP- and LTD-like processes. Since the observed effects arise from changes in populations of neurons on a much larger scale than cellular STDP, this noninvasive form of plasticity is often referred to as STDP-like changes.

In humans with SCI, protocols resulting in STDP-like changes have targeted the sensorimotor cortex (Roy, Yang, & Gorassini, 2010) and spinal cord (Bunday & Perez, 2012b; Bunday, Urbin, & Perez, 2018; Urbin et al., 2017). STDP-like changes in the spinal cord were first demonstrated by Taylor and Martin (2009) in control subjects. The authors showed that when repeated TMS-elicited corticospinal activity over a region of the primary motor cortex was timed to arrive a few milliseconds before antidromic activation of the motoneurons by supramaximal peripheral nerve electrical stimulation, the size of MEPs evoked by cervicomedullary stimulation (CMEPs) increased. In contrast, when postsynaptic depolarization preceded the presynaptic activation, the size of the CMEPs decreased. CMEPs measure corticospinal transmission and/or the excitability of spinal motoneurons (Taylor & Gandevia, 2004), suggesting the presence of plasticity in the spinal cord. Because this form of plasticity targeted synapses between corticospinal neurons and spinal motoneurons using pairs of stimuli, the protocol has been referred to as paired corticomotoneuronal stimulation (PCMS).

Note that the precise timing of the two stimuli is based on measurements of central and peripheral motor conduction times, which can be estimated from the latency of an MEP, the latency of a response elicited by cervical root stimulation, and the latency of the M wave and F wave. Bunday and Perez (2012b) reproduced those results in control subjects and, for the first time, used PCMS in humans with incomplete SCI. In their study the size of MEPs elicited by electrical and magnetic stimulation of the primary motor cortex, as well as the size of CMEPs, increased and decreased according to the timing of the arrival of volleys in the pre- and postsynaptic terminals. These changes were observed without corresponding changes in F waves (Bunday & Perez, 2012b). Although this pattern of results is consistent with the hypothesis that the arrival of presynaptic volleys prior to the depolarization of spinal motoneurons will strengthen corticospinal-motoneuronal synaptic transmission, it should be noted that there are limitations in the extent to which F-wave measurements can assess motoneuron excitability (Espiritu et al., 2003; Hultborn & Nielsen, 1995).

PCMS has also been used to target lower-limb muscles following SCI (Urbin et al., 2017). In agreement with the results found in arm (Fitzpatrick et al., 2016) and hand (Bunday & Perez, 2012b; Bunday, Urbin, & Perez, 2018) muscles, this work supports the view that PCMS effects are dose-dependent. Interestingly, in the lower limbs the excitatory effects, but not the inhibitory effects, of PCMS were similar to those found in upper-limb muscles. The inhibitory effect of PCMS in the tibialis anterior muscle was contingent on a suppressive effect of peripheral nerve stimulation on the MEP during the conditioning protocol (Urbin et al., 2017). The neurophysiological mechanisms underlying this inter-individual variability are yet to be determined, but the supramaximal nature of the peripheral stimulation suggests that both homonymous and heteronomous mixed nerve activation contribute, leading to complex integration at a spinal level.

PCMS can increase the recruitment of low-threshold motor units during voluntary movement, as evidenced by increases in low levels of electromyography (EMG) and force generation (Bunday & Perez, 2012b; Taylor & Martin, 2009) and, to a lesser extent, increases in the level of maximal voluntary contraction (D'Amico, Donges, & Taylor, 2017). In humans with SCI, changes in corticospinal transmission positively correlate with enhancements in voluntary motor output, suggesting an association between motor output and the strength of the induced plasticity (Bunday & Perez, 2012b). In addition, the clinical potential of this approach is highlighted by decreases in the time required to complete a nine-hole pegboard task.

The effect of PCMS on corticospinal excitability can be augmented when the pairs of stimuli are delivered during short-lasting, low-intensity isometric voluntary contractions (Bunday, Urbin, & Perez, 2018). Increasing the excitability of the motoneurons through voluntary activation augmented the increase in MEP amplitude; indeed, this augmentation procedure resulted in positive effects observed in some individuals who had not responded to standard protocols. Since spinal lesions are associated with reduced corticospinal inputs to the motoneurons, the possibility for potentiating the PCMS with voluntary activation makes the clinical applicability of this procedure even more promising.

The proposed mechanisms of the enhanced effects are unexplored but may involve increasing the size and number of descending volleys. Increasing motoneuronal excitability by a low-intensity background contraction also increases cortical excitability in control (Di Lazzaro et al., 1998) and SCI (Barry et al., 2013) participants. Voluntary contraction could constitute a method for increasing the size and number of descending volleys, thus decreasing the threshold of spinal motoneurons. The expected outcome would be increased corticospinal facilitation, which is consistent with the finding of an increased number of responders and increased facilitation for individuals with SCI who responded to PCMS at rest. In addition, similar augmentation in corticospinal excitability is observed when PCMS is administered in combination with acute intermittent hypoxia, a noninvasive strategy known to increase corticospinal plasticity (Christiansen et al., 2018). Taken together, these findings suggest that PCMS alone or in combination with other noninvasive therapeutic protocols can enhance CST function and the beneficial effects of sensorimotor rehabilitation after SCI.

Conclusions and Future Directions

To date, most of the information about CST reorganization after SCI in humans has come from electrophysiological studies. Although there is a consensus that MEP latencies and motor thresholds are increased after SCI and remain elevated for extended periods of time, emerging studies are focusing on better understanding the involvement of the reorganized corticospinal pathway during functionally relevant motor tasks. This may be especially important for elucidating the mechanisms of recovery after SCI.

Since spinal cord injuries are rarely complete, it is possible that the spared CST projections may prove useful for functional recovery. PCMS protocols have been employed to target physiological connections that have been weakened by SCI. Noninvasive stimulation protocols aim to strengthen the connections between corticospinal neurons and motoneurons to enhance motor output, thus supporting spinal plasticity. Note that the design here is based on central and peripheral conduction times, suggesting that electrophysiology can be successfully used to guide interventions designed to enhance motor output in humans with SCI. Several issues need to be considered in the translation of this protocol to a clinical environment. First, it will be important to assess the effectiveness of PCMS using lower intensities, given that current PCMS protocols in humans with SCI have entailed maximal stimulator output. In control subjects, submaximal stimulation intensities have been successful in inducing both physiological and behavioral plasticity. However, the higher motor thresholds in humans with SCI might require the use of higher TMS intensities to elicit MEPs in targeted muscles.

Second, for PCMS to be clinically useful, a clear understanding is required of the individual variation in the interstimulus intervals needed between TMS and peripheral nerve electrical stimulation, as well as between pairs of TMS pulses. If there is considerable variability between individuals, then the clinical utility is compromised to some degree. On the other hand, the electrophysiological times needed for PCMS can be incorporated into clinical diagnostic procedures. In addition, latencies of EMG responses, which depend on the generation of action potentials to the muscle, can be used for PCMS estimations. Thus, proper training and careful methodological considerations represent important steps for bringing these strategies to the clinic.

Overall, PCMS represents a new and promising tool. Further investigations are needed to identify the underlying mechanisms and determine the optimal dose, size, and duration of stimulation for employing the procedure outside laboratory settings. The effects also need to be examined in multicenter clinical trials, where the potential for cumulative effects of several sessions, along with interactions with motor practice, can be explored.

REFERENCES

Alexeeva, N., Broton, J. G., & Calancie, B. (1998). Latency of changes in spinal motoneuron excitability evoked by transcranial magnetic brain stimulation in spinal cord injured individuals. *Electroencephalography and Clinical Neurophysiology, 109,* 297–303.

Amassian V. E., Cracco R. Q., Maccabee P. J., Cracco J. B., Rudell A. P., & Eberle, L. (1993). Unmasking human visual perception with the magnetic coil and its relationship to hemispheric asymmetry. *Brain Research, 605*(2), 312–316.

Amassian, V. E., Eberle, L., Maccabee, P. J., & Cracco, R. Q. (1992). Modelling magnetic coil excitation of human cerebral cortex with a peripheral nerve immersed in a brain-shaped volume conductor: The significance of fiber bending in excitation. *Electroencephalography and Clinical Neurophysiology, 85*(5), 291–301.

*Angeli, C. A., Edgerton, V. R., Gerasimenko, Y. P., & Harkema, S. J. (2014). Altering spinal cord excitability enables voluntary movements after chronic complete paralysis in humans. *Brain, 137,* 1394–1409.

Barry, M. D., Bunday, K. L., Chen, R., & Perez, M. A. (2013). Selective effects of baclofen on use-dependent modulation of $GABA_B$ inhibition after tetraplegia. *Journal of Neuroscience, 33*(31), 12898–12907.

*Barthélemy, D., Willerslev-Olsen, M., Lundell, H., Conway, B. A., Knudsen, H., Biering-Sørensen, F., & Nielsen, J. B. (2010). Impaired transmission in the corticospinal tract and gait disability in spinal cord injured persons. *Journal of Neurophysiology, 104,* 1167–1176.

*Becerra, J. L., Puckett, W. R., Hiester, E. D., Quencer, R. M., Marcillo, A. E., Post, M. J., & Bunge, R. P. (1995). MR-pathologic comparisons of Wallerian degeneration in spinal cord injury. *American Journal of Neuroradiology, 16,* 125–133.

*Blight, A. R., & Young, W. (1989). Central axons in injured cat spinal cord recover electrophysiological function following remyelination by Schwann cells. *Journal of the Neurological Sciences, 91,* 15–34.

*Bronson, R., Gilles, F. H., Hall, J., & Hedley-Whyte, E. T. (1978). Long term post-traumatic retrograde corticospinal degeneration in man. *Human Pathology, 9,* 602–607.

Brouwer, B., Bugaresti, J., & Ashby, P. (1992). Changes in corticospinal facilitation of lower limb spinal motor neurons after spinal cord lesions. *Journal of Neurology, Neurosurgery, and Psychiatry, 55,* 20–24.

Bunday, K. L., & Perez, M. A. (2012a). Impaired crossed facilitation of the corticospinal pathway after cervical spinal cord injury. *Journal of Neurophysiology, 107*(10), 2901–2911.

Bunday, K. L., & Perez, M. A. (2012b). Motor recovery after spinal cord injury enhanced by strengthening corticospinal synaptic transmission. *Current Biology, 22*(24), 2355–2361.

*Bunday, K. L., Tazoe, T., Rothwell, J. C., & Perez, M.A. (2014). Subcortical control of precision grip after human spinal cord injury. *Journal of Neuroscience, 34,* 7341–7350.

Bunday, K. L., Urbin, M. A., & Perez, M. A. (2018). Potentiating paired corticospinal-motoneuronal plasticity after spinal cord injury. *Brain Stimulation, 11*(5), 1083–1092.

*Bunge, R. P., Puckett, W., Becerra, J., Marcillo, A., & Quencer, R. (1993). Observations on the pathology of human spinal cord injury: A review and classification of 22 new cases with details from a case of chronic cord compression with extensive focal demyelination. *Advances in Neurology, 59,* 75–89.

Buss, A., Brook, G. A., Kakulas, B., Martin, D., Franzen, R., Schoenen, J., Noth, J., & Schmitt, A. B. (2004). Gradual loss of myelin and formation of an astrocytic scar during Wallerian degeneration in the human spinal cord. *Brain, 127*(1), 34–44.

Caporale, N., & Dan, Y. (2008). Spike timing-dependent plasticity: A Hebbian learning rule. *Annual Review of Neuroscience, 31,* 25–46.

Christiansen, L., Urbin, M. A., Mitchell, G. S., & Perez, M. A. (2018). Acute intermittent hypoxia enhances corticospinal synaptic plasticity in humans. *eLife, 7,* e34304.

*Cioni, B., Dimitrijevic, M. R., McKay, W. B., & Sherwood, A. M. (1986). Voluntary supraspinal suppression of spinal reflex activity in paralyzed muscles of spinal cord injury patients. *Experimental Neurology, 93,* 574–583.

Cirillo, J., Calabro, F., & Perez, M. A. (2016). Impaired organization of TMS-induced I-waves after human spinal cord injury. *Cerebral Cortex, 26,* 2167–2177.

Curt, A., Keck, M. E., & Dietz, V. (1998). Functional outcome following spinal cord injury: Significance of motor-evoked potentials and ASIA scores. *Archives of Physical Medicine and Rehabilitation, 79*(1), 81–86.

D'Amico, J. M., Donges, S. C., & Taylor, J. L. (2017). Paired corticospinal-motoneuronal stimulation increases maximal voluntary activation of human adductor pollicis. *Journal of Neurophysiology,119*(1), 369–376.

Davey, N. J., Smith, H. C., Savic, G., Maskill, D. W., Ellaway, P. H., & Frankel, H. L. (1999). Comparison of input-output patterns in the corticospinal system of normal subjects and incomplete spinal cord injured patients. *Experimental Brain Research, 127*(4), 382–390.

Day, B. L., Dressler, D., Maertens de Noordhout, A., Marsden, C. D., Nakashima, K., Rothwell, J. C., & Thompson, P. D. (1989). Electric and magnetic stimulation of human motor cortex: Surface EMG and single motor unit responses. *Journal of Physiology, 412,* 449–473.

Debanne, D., & Poo, M.-M. (2010). Spike-timing dependent plasticity beyond synapse—pre- and post-synaptic plasticity of intrinsic neuronal excitability. *Frontiers in Synaptic Neuroscience, 2,* 21.

Di Lazzaro, V., Oliviero, A., Profice, P., Saturno, E., Pilato, F., Insola, A., Mazzone, P., Tonali, P., & Rothwell, J. C. (1998). Comparison of descending volleys evoked by transcranial magnetic and electric stimulation in conscious humans. *Electroencephalography and Clinical Neurophysiology, 109*(5), 397–401.

Di Lazzaro, V., Profice, P., Ranieri, F., Capone, F., Dileone, M., Oliviero, A., & Pilato, F. (2012). I-wave origin and modulation. *Brain Stimulation, 5*(4), 512–525.

Di Lazzaro, V., Rothwell, J. C., & Capogna, M. (2017). Noninvasive stimulation of the human brain: Activation of multiple cortical circuits. *Neuroscientist, 24*(3), 246–260.

Dimitrijević, M. R. (1988). Residual motor functions in spinal cord injury. *Advances in Neurology, 47,* 138–155

Dimitrijevic, M. R., Dimitrijevic, M., Faganel, M. J., & Sherwood, A. M. (1984). Suprasegmentally induced motor unit activity in paralyzed muscles of patients with established spinal cord injury. *Annals of Neurology, 16*(2), 216–221.

Dimitrijevic, M. R, Spencer, W. A., Trontelj, J. V., & Dimitrijevic, M. (1977). Reflex effects of vibration in patients with spinal cord lesions. *Neurology, 27*(11), 1078.

Donati, A. R. C., Shokur, S., Morya, E., Campos, D. S. F., Moioli, R. C., Gitti, C. M., ... Nicolelis, M. A. L. (2016). Long-term training with a brain-machine interface-based gait protocol induces partial neurological recovery in paraplegic patients. *Scientific Reports, 6,* 30383.

Donges, S. C., D'Amico, J. M., Butler, J. E., & Taylor, J. L. (2017). The involvement of N-methyl-D-aspartate receptors in plasticity induced by paired corticospinal-motoneuronal stimulation in humans. *Journal of Neurophysiology, 119*(2), 652–661.

D'Ostilio, K., Goetz, S. M., Hannah, R., Ciocca, M., Chieffo, R., Chen, J.-C. A., ... Rothwell, J. C. (2016). Effect of coil orientation on strength-duration time constant and I-wave activation with controllable pulse parameter transcranial magnetic stimulation. *Clinical Neurophysiology, 127*(1), 675–683.

Edwards, D., Cortes, M., Thickbroom, G., Rykman, A., Pascual-Leone, A., & Volpe, B. (2013). Preserved corticospinal conduction without voluntary movement after spinal cord injury. *Spinal Cord, 51*(10), 765–767.

Espiritu, M. G., Lin, C. S. Y., & Burke, D. (2003). Motoneuron excitability and the F wave. *Muscle & Nerve, 27,* 720–727.

Fishman, P. S. (1987). Retrograde changes in the corticospinal tract of posttraumatic paraplegics. *Archives of Neurology, 44*(10), 1082–1084.

Fitzpatrick, S. C., Luu, B. L., Butler, J. E., & Taylor, J. L. (2016). More conditioning stimuli enhance synaptic plasticity in the human spinal cord. *Clinical Neurophysiology, 127*(1), 724–731.

Griffiths, I. R., & McCulloch, M. C. (1983). Nerve fibers in spinal cord impact injuries. Part 1. Changes in the myelin sheath during the initial 5 weeks. *Journal of Neurological Sciences, 58*(3), 335–349.

Hannah, R., & Rothwell, J. C. (2017). Pulse duration as well as current direction determines the specificity of transcranial magnetic stimulation of motor cortex during contraction. *Brain Stimulation, 10*(1), 106–115.

Harkema, S., Gerasimenko, Y., Hodes, J., Burdick, J., Angeli, C., Chen, Y., ... Edgerton, V. R. (2011). Effect of epidural stimulation of the lumbosacral spinal cord on voluntary movement, standing, and assisted stepping after motor complete paraplegia: A case study. *Lancet, 377*(9781), 1938–1947.

Heald, E., Hart, R., Kilgore, K., & Peckham, P. H. (2017). Characterization of volitional electromyographic signals in the lower extremity after motor complete SCI. *Neurorehabilitation and Neural Repair, 31*(6), 583–591.

Hebb, D. (1949). *The organization of behavior.* New York: Wiley & Sons.

Hultborn, H., & Nielsen, J. B. (1995). H-reflexes and F-responses are not equally sensitive to changes in motoneuronal excitability. *Muscle & Nerve, 18*(12), 1471–1474.

Hunt, R. H. (1904). The retrograde atrophy of the pyramidal tracts. *Journal of Nervous and Mental Disease, 31,* 504–512.

Jo, H. J., & Perez, M. A. (2018). Effect of coil orientation on motor evoked potentials in humans with tetraplegia. *Journal of Physiology.* Advance online publication.

Kakulas, A. (1988). The applied neurobiology of human spinal cord injury: A review. *Spinal Cord, 26*(6), 371.

Kakulas, B. (2004). Neuropathology: The foundation for new treatments in spinal cord injury. *Spinal Cord, 42,* 549–563.

Kampa, B. M., Clements, J., Jonas, P., & Stuart, G. J. (2004). Kinetics of Mg2+ unblock of NMDA receptors: Implications for spike-timing dependent synaptic plasticity. *Journal of Physiology, 556,* 337–345.

Kernell, D., & Chien-Ping, W. (1967). Responses of the pyramidal tract to stimulation of the baboon's motor cortex. *Journal of Physiology, 191,* 653–672.

Lemon, R. N. (2008). Descending pathways in motor control. *Annual Review of Neuroscience, 31,* 195–218.

Levy, W. J., Amassian, V. E., Traad, M., & Cadwell, J. (1990). Focal magnetic coil stimulation reveals motor cortical system reorganized in humans after traumatic quadriplegia. *Brain Research, 510*(1), 130–134.

Maccabee, P. J., Amassian, V. E., Eberle, L. P., & Cracco, R. Q. (1993). Magnetic coil stimulation of straight and bent amphibian and mammalian peripheral nerve in vitro: Locus of excitation. *Journal of Physiology, 460,* 201–219.

Maier, M. A., Olivier, E., Baker, S. N., Kirkwood, P. A, Morris, T., & Lemon, R. N. (1997). Direct and indirect corticospinal control of arm and hand motoneurons in the squirrel monkey (*Saimiri sciureus*). *Journal of Neurophysiology, 78*, 721–733.

Murray, M., Wang, S. D., Goldberger, M. E., & Levitt, P. (1990). Modification of astrocytes in the spinal cord following dorsal root or peripheral nerve lesions. *Experimental Neurology, 110*(3), 248–257.

Oudega, M. & Perez, M. A. (2012). Corticospinal reorganization after spinal cord injury. *Journal of Physiology, 590*, 3647–3663.

Patton, H. D., & Amassian, V. E. (1954). Single and multiple-unit analysis of cortical stage of pyramidal tract activation. *Journal of Neurophysiology, 17*, 345–363.

Petersen, N. C., Butler, J. E., Taylor, J. L., & Gandevia, S. C. (2010). Probing the corticospinal link between the motor cortex and motoneurones: Some neglected aspects of human motor cortical function. *Acta Physiologica, 198*(4), 403–416.

Puckett, W. R, Hiester, E. D., Norenberg, M. D., Marcillo, A. E., & Bunge, R. P. (1997). The astroglial response to Wallerian degeneration after spinal cord injury in humans. *Experiments in Neurology, 148*, 424–432.

Quencer, R. M., & Bunge, R. P. (1996). The injured spinal cord: Imaging, histopathologic clinical correlates, and basic science approaches to enhancing neural function after spinal cord injury. *Spine, 21*(18), 2064–2066.

Quencer, R. M., Bunge, R., Egnor, M., Puckett, W., Naidich, T. P., & Green, M. J. D. (1992). Post MR/pathological correlations in the acute post-traumatic central cervical spinal cord syndrome.

Roy, F. D., Yang, J. F., & Gorassini, M. A. (2010). Afferent regulation of leg motor cortex excitability after incomplete spinal cord injury. *Journal of Neurophysiology, 103*(4), 2222–2233.

Sakai, K., Ugawa, Y., Terao, Y., Hanajima, R., Furubayashi, T., & Kanazawa I. (1997). Preferential activation of different I waves by transcranial magnetic stimulation with a figure-of-eight-shaped coil. *Experimental Brain Research, 113*(1), 24–32.

Squair, J. W., Bjerkefors, A., Inglis, J. T., Lam, T., & Carpenter, M. G. (2016). Cortical and vestibular stimulation reveal preserved descending motor pathways in individuals with motor-complete spinal cord injury. *Journal of Rehabilitation Medicine, 48*(7), 589–596.

Smith, H. C., Savic, G., Frankel, H. L., Ellaway, P. H., Maskill, D. W., Jamous, M. A., & Davey, N. J. (2000). Corticospinal function studied over time following incomplete spinal cord injury. *Spinal Cord, 38*, 292–300.

Stefan, K., Kunesch, E., Benecke, R., Cohen, L. G., & Classen, J. (2002). Mechanisms of enhancement of human motor cortex excitability induced by interventional paired associative stimulation. *Journal of Physiology, 543*(Pt. 2), 699–708.

Stefan, K., Kunesch, E., Cohen, L. G., Benecke, R., & Classen, J. (2000). Induction of plasticity in the human motor cortex by paired associative stimulation. *Brain, 123*(3), 572–584.

Taylor, J. L., & Gandevia, S. C. (2004). Noninvasive stimulation of the human corticospinal tract. *Journal of Applied Physiology, 96*(4), 1496–1503.

Taylor, J. L., & Martin, P. G. (2009). Voluntary motor output is altered by spike-timing-dependent changes in the human corticospinal pathway. *Journal of Neuroscience, 29*(37), 11708–11716.

Tazoe, T., & Perez, M. A. (2015). Effects of repetitive transcranial magnetic stimulation on recovery of function after spinal cord injury. *Archives of Physical Medicine and Rehabilitation, 96*(Suppl. 4), S145–S155.

Terae, S., Taneichi, H., & Abumi, K. (1993). MRI of Wallerian degeneration of the injured spinal cord. *Journal of Computer Assisted Tomography, 17*(5), 700–703.

Thomas, S. L., & Gorassini, M. A. (2005). Increases in corticospinal tract function by treadmill training after incomplete spinal cord injury. *Journal of Neurophysiology, 94*(4), 2844–2855.

Tokimura, H., Ridding, M. C., Tokimura, Y., Amassian, V. E., & Rothwell, J. C. (1996). Short latency facilitation between pairs of threshold magnetic stimuli applied to human motor cortex. *Electroencephalography and Clinical Neurophysiology, 101*, 263–272.

Topka, H., Cohen, L. G., Cole, R. A., & Hallett, M. (1991). Reorganization of corticospinal pathways following spinal cord injury. *Neurology, 41*(8), 1276–1283.

Totoiu, M. O., & Keirstead, H. S. (2005). Spinal cord injury is accompanied by chronic progressive demyelination. *Journal of Comparative Neurology, 486*(4), 373–383.

Urbin, M. A., Ozdemir, R. A., Tazoe, T., & Perez, M. A. (2017). Spike-timing-dependent plasticity in lower-limb motoneurons after human spinal cord injury. *Journal of Neurophysiology, 118*(4), 2171–2180.

Werhahn, K. J., Fong, J. K., Meyer, B. U., Priori, A., Rothwell, J. C., Day, B. L., & Thompson, P. D. (1994). The effect of magnetic coil orientation on the latency of surface EMG and single motor unit responses in the first dorsal interosseous muscle. *Electroencephalography and Clinical Neurophysiology, 93*(2), 138–146.

Wolters, A., F., Sandbrink, A., Schlottmann, E., Kunesch, K., Stefan, L. G., Cohen, R., & Benecke, J. (2003). Classen: A temporally asymmetric Hebbian rule governing plasticity in the human motor cortex. *Journal of Neurophysiology, 89*(5), 2339–2345.

Wrigley, P. J., Gustin, S. M., Macey, P. M., Nash, P. G., Gandevia, S. C., Macefield, V. G., Siddall, P. J., & Henderson, L. A. (2009). Anatomical changes in human motor cortex and motor pathways following complete thoracic spinal cord injury. *Cerebral Cortex, 19*(1), 224–232.

Ziemann, U., Corwell, B., & Cohen, L. G. (1998). Modulation of plasticity in human motor cortex after forearm ischemic nerve block. *Journal of Neuroscience, 18*, 1115–1123.

41 The Neuroscience of Brain-Machine Interfaces

ANDREW JACKSON

ABSTRACT Brain-machine interfaces (BMIs) act as direct communication channels between the central nervous system and external effectors to restore motor function after paralysis. Proof-of-principle preclinical and clinical BMI demonstrations have been made possible by technological developments in implantable multielectrode arrays and real-time signal-processing techniques, but performance improvements are also being driven by new scientific insights into how information is represented and processed by neural circuits. BMI research also offers a unique experimental window onto the acquisition of novel motor skills at a neural level. This chapter will explore the main neuroscientific themes of recent BMI research, with a focus on hierarchical levels of population structure in high-dimensional neural recordings. I will discuss how this structure can be exploited to improve biomimetic BMI decoding and consider how it may also constrain the new neural strategies that emerge during BMI learning.

The central nervous system represents and processes information using electrical signals. Thus, it is no surprise that the idea of communication between the brain and external electronics has recurred many times in science fiction. However, this century has witnessed considerable progress toward realizing BMI systems, with demonstrations in nonhuman primates of closed-loop neural control of two- and three-dimensional computer cursors (Carmena et al., 2003; Serruya et al., 2002; Taylor, Tillery, and Schwartz, 2002), followed swiftly by brain-controlled robotic arms (Velliste et al., 2008), muscle stimulation (Ethier et al., 2012; Moritz, Perlmutter, and Fetz, 2008), and spinal cord stimulation (Capogrosso et al., 2016; Zimmermann and Jackson, 2014). Clinical trials with paralyzed users have proceeded at a similarly impressive pace (Ajiboye et al., 2017; Bouton et al., 2016; Collinger et al., 2013; Hochberg et al., 2006, 2012), and in recent years, Silicon Valley entrepreneurs have announced ambitious visions of brain-controlled commercial and consumer technology. For some, the development of BMIs promises the next phase of our evolution, freeing our cognitive capabilities from neuroanatomical constraints much as the evolution of direct connections from the cortex to motoneurons of the hand afforded our primate ancestors the dexterity to first develop tool use. More pragmatically, the advent of brain-controlled devices could be of great practical use to individuals with paralysis resulting from neurological injury and disease.

While there remain considerable technical challenges to achieving reliable, robust, and clinically useful BMIs, high-density implantable electrode arrays and low-power electronics can now relay rich streams of neural spiking activity continuously from the nervous system. At the same time, advances in neuroimaging technologies promise to improve the performance capabilities of wearable, noninvasive interfaces exploiting field potentials or hemodynamic signals to infer brain activity. Importantly, however, the barriers to interfacing electronics with the brain do not lie solely within the domain of engineering. Decoding neural signals requires an understanding of how information is represented and processed by cortical circuits and how these networks change under altered sensorimotor contingences. At the same time, BMI experiments offer new paradigms to study scientific questions about neural coding, learning, and plasticity. In discussing these issues, this chapter will focus on invasive BMIs for restoring upper-limb movement, as this has been a major driver of research and remains a priority for those living with high spinal cord injuries.

Biomimetic Decoding Strategies

To operate an external device, a BMI must interpret (or "decode") high-dimensional neural features (e.g., the firing rates of hundreds of neurons) in real time to generate low-dimensional control signals (e.g., the velocity of a computer cursor or the end effector of a robotic arm). Usually, BMIs are initially trained using a labeled data set of neural patterns recorded during actual or imagined movements, such as *center-out* reaches with the arm to different locations in the workspace. During subsequent online operation, the volitional generation of the same neural patterns is decoded into appropriate movements of the BMI. Of course, this biomimetic scheme assumes the neural correlates of online control will match the corresponding off-line movements in the training data, which is unlikely to be the case. Arm movements rely heavily on continuous somatosensory and proprioceptive information, and until further progress is made with the artificial restoration of this feedback (Flesher et al., 2016; Tabot et al., 2013; Tomlinson and Miller, 2016), online BMI control will never fully mimic these natural sensorimotor loops. Therefore, researchers have attempted to train and modify decoders online, adapting to patterns of brain activity observed during BMI movements. Nevertheless, these approaches still rely on supervised learning based on data labeled with targets that are instructed (Gilja et al., 2012) or inferred, retrospectively, from user selections (Jarosiewicz et al., 2015).

A principle limitation of this biomimetic approach is the absence of causal principles that relate neuronal features to their associated movements. Since biomimetic decoding is based only on statistical correlations within the training data, there is no guarantee that the decoder will generalize (Jackson and Fetz, 2011). For example, while a neuron may exhibit cosine tuning around a preferred direction of movement from a central position, this tuning often varies when movements begin at different locations (Aflalo and Graziano, 2006); as such, a decoder trained only on center-out movements may not generalize across the workspace. Moreover, the tuning of neurons is also influenced by many other factors beyond the start/end location, such as limb posture and loading (Scott and Kalaska, 1997) or behavioral contexts—for example, goal-directed versus corrective movements (Archambault, Caminiti, and Battaglia-Mayer, 2009).

One might suppose that inferring the intended influence of neurons on muscle activity would provide a principled, generalizable approach to decoding movement. However, even for corticomotoneuronal neurons, output connectivity does not predict firing rate modulations during movement (Bennett and Lemon, 1996). Indeed, the 20th century saw extensive, unresolved debates over whether motor cortex represented "muscles or movements," with the diversity of neuronal responses ultimately leading many to question whether single neurons encoded meaningful (i.e., intuitive) parameters (Fetz, 1992; Scott, 2008; Shenoy, Sahani, and Churchland, 2013). It seems that, unlike early sensory processing, the output of motor cortex does not operate as a *labeled line*. Without a simplifying representational scheme that generalizes across behaviors, fitting decoders to the full repertoire of natural movements using a purely biomimetic approach will require extremely large, individualized training data sets covering a multiplicity of behavioral contexts.

Generously, we might compare the current state-of-the-art BMIs with early speech recognition software in the 1970s, systems that could recognize only a handful of words spoken slowly by a single user. Today we have commercial systems that interpret natural speech without individualized training, with error rates approaching human performance. Is it possible to conceive a similar trajectory of improvement for biomimetic BMI decoding? Since each implanted cortical array samples different elements of intermingled neural circuitry, it is impossible to predict in advance how any particular neuron will covary with behavior. However, natural speech is similarly confounded by variation between individual voices, so the mapping from low-level spectral features to phonemes is not straightforward. In addition, the meaning of these phonemes is context-dependent (e.g., *meat* vs. *meet*). Importantly, speech comprises multiple levels of organization (lexical, grammatical, semantic, and more), and progress in artificial recognition initially relied upon constructing hierarchical models of these structures to disambiguate variable acoustic data. At the risk of overextending the analogy, we can speculate whether comparable principles could apply to neural decoding for BMIs (figure 41.1).

Population Structure in the Motor Cortex

Recent neuroscientific findings are revealing structure within the activity of neuronal populations that can be exploited for BMI applications (Gallego et al., 2017). For example, activity is often constrained to low-dimensional neural manifolds within the *state space* (wherein the instantaneous activity of each recorded neuron defines a high-dimensional neural state). This population structure may reflect constraints afforded by the geometry and biomechanics of the limbs. For example, the distribution of neuronal preferred directions is skewed toward and away from the body, which is explained by mechanical anisotropies of the limb (Lillicrap and Scott, 2013). Similarly, biomechanics constrains the cocontraction of specific joint and muscle synergies, and this may be represented at the level of the cortex (Overduin et al., 2015). Identifying such global structure using unsupervised techniques confers advantages for BMI decoding in the presence of local variation (e.g., the loss of neurons during recordings). Kao et al. (2013) showed that reducing the dimensionality of the input neural state (via principal components analysis [PCA]) prior to decoding conferred robustness in the face of simulated neural instabilities. Effectively, decoders could reject the components of neural variability that fell outside the identified low-dimensional input manifold. Using a related approach, Wong et al. (2013) applied dimensionality reduction in the output (joint angle) space to

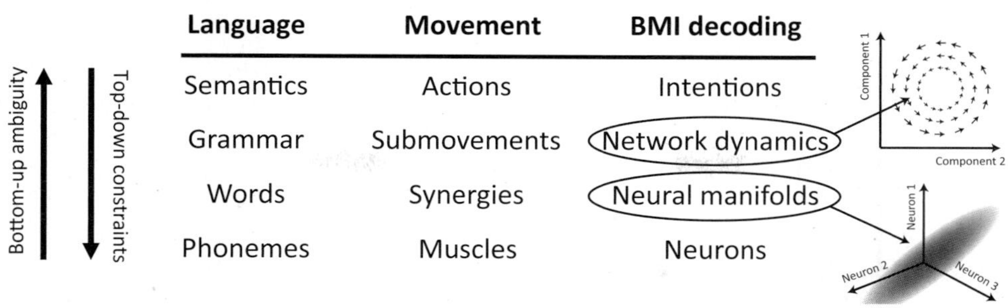

FIGURE 41.1 Hierarchical organization of behavior and neural correlates. Both language and movement are characterized by a hierarchical structure, whereby high-level organizational principles constrain low-level features. For example, grammatical rules constrain the ordering of words (*left*), while movements are composed from submovements and muscle synergies (*center*). In speech recognition this organization can be exploited to resolve ambiguities inherent in low-level features. Organizational constraints can similarly assist BMI decoding (*right*). For example, neural manifolds constrain activity to defined regions of the high-dimensional neural state space while consistent dynamic properties govern their evolution through time.

improve decoder performance. This can be understood as having the effect of rejecting the input noise decoded into movements outside of the identified effector manifold.

An additional level of organization is revealed when the evolution of the neural population is observed across time. State-space trajectories exhibit a conserved rotational structure at low frequencies during movement (Churchland et al., 2012). Hall, de Carvalho, and Jackson (2014) related state-space rotation to the well-known 2–3 Hz intermittency in continuous tracking movements, with each cycle of the rotation corresponding to a single submovement. Movement speed could readily be decoded from the areal velocity swept out by that cycle (figure 41.2*A*). Moreover, rotational structure was conserved across a wide range of behaviors and even evident during sleep, suggesting it arises from intrinsic patterns of connectivity that may be specific to particular cortical areas (Lara, Cunningham, and Churchland, 2018). As with the static neural manifolds described above, incorporating models of dynamic constraints on population activity (inferred using unsupervised techniques) can further improve decoder robustness (Pandarinath et al., 2018; Aghagolzadeh and Truccolo, 2016; Kao et al., 2015; figure 41.2*B*).

In summary, complex movements seem to be constructed from the simple building blocks of muscle/joint synergies and submovement segments in much the same way that complex sentences are built from phonemes and words. We are beginning to understand how this shapes the neural population structure and reflects computational principles under constraints imposed by the biomechanical properties of the limbs. Since these biomechanics are broadly conserved across different individuals, it is not surprising that structure is present in neural populations even without a one-to-one correspondence with low-level features. Increasingly, unsupervised statistical algorithms (e.g., Pandarinath et al., 2018; Aghagolzadeh and Truccolo, 2016; Vargas-Irwin et al., 2015) are providing methods to transform these populations to extract invariant high-level features for subsequent decoding.

Outlook for Biomimetic Decoding

Ultimately, the major fillip for advances in speech recognition was the availability of rich sets of training data that were amenable to powerful machine learning approaches. Datasets of this scale are not yet available for BMI research, but a recent study attempted to construct a larger-than-usual training set by combining data across multiple sessions (Sussillo et al., 2016). Machine learning by neural networks yielded decoders that could cope better with new variants encountered in future, without a significant decline in performance on trained data sets. In this case, the different datasets shared low-level commonalities (e.g., many of the same neurons were present in different recordings, and even when new cells appeared, they were presumably sampled from the same local region). Nevertheless, this study raises the intriguing prospect that conserved population structure in neural representations could be similarly exploited to build decoders that, like modern speech recognition systems, are robust in the face of day-to-day and perhaps even user-to-user variation.

Brain-Machine Interfaces as Abstract Tools

Despite advances in decoding strategies, seamlessly replacing a missing limb with a BMI remains a formidable challenge. It may be more productive to view BMIs instead as tools that can be used to restore function, comparable to a wheelchair that can be used for

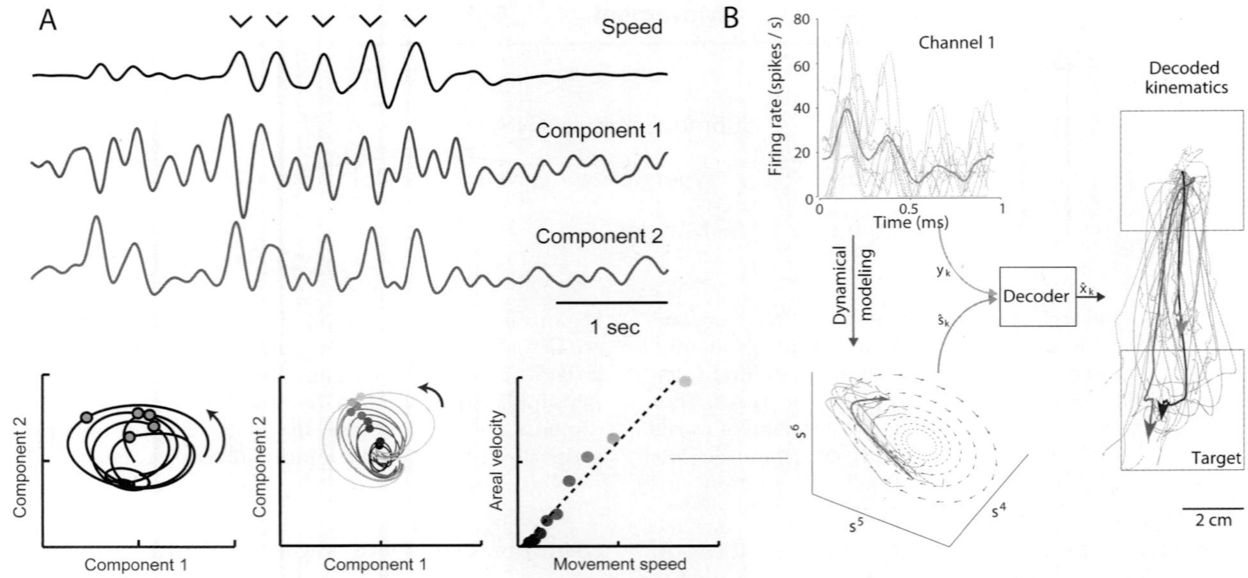

FIGURE 41.2 Neural decoding exploiting dynamic constraints. *A, Top panel*, Movement speed and principal components of associated field potentials recorded from primary motor cortex during an isometric tracking task. Note the regular submovements (indicated by ticks). *Bottom left*, When plotted as a trajectory in principal component space, each submovement is associated with a single cycle with a consistent direction and angular frequency. *Bottom right*, Faster submovements (*yellow traces*) are associated with larger cycles, so cursor speed can be decoded from the areal velocity of trajectories. Adapted from Hall, de Carvalho, and Jackson (2014). *B*, Decoding movement kinematics is improved by incorporating a model of consistent network dynamics. Neural signals (*light blue*) are considered to be noisy observations of a low-dimensional dynamical neural state (*orange*). Movements decoded from this inferred state provide a better match to the movements produced by the subject (*black lines*). Reproduced from Kao et al. (2015). (See color plate 45.)

transport even though it does not physically resemble a leg. In principle, future BMIs may even enhance normal capabilities, in the same way that motorized vehicles enable us to travel faster than walking. From this perspective, the key requirement is not that a BMI decoder can learn to interpret naturally-occurring neural patterns, but that the operator can learn to generate new patterns to effectively operate the BMI. Given we are adept tool-users, it is no surprise that BMI performance improves with practice, and this is associated with altered activity of neurons directly controlling output (Taylor, Tillery, and Schwartz, 2002) and within the wider network (Ganguly et al., 2011). Exploiting the brain's innate ability to learn will be key to improving BMI performance. However, the extent of, and constraints upon, neural reorganization during BMI learning remain important open questions.

Neurofeedback and Brain-Machine Interface Learning

Whether and which brain features are under volitional control has been the subject of *neurofeedback* experiments since the last century. In a pioneering study, Eberhard Fetz (1969) showed that monkeys could modulate individual motor cortex neurons when provided with visual feedback of their firing rate. Neurofeedback researchers have subsequently demonstrated volitional control of brain activity using a variety of different recording modalities and brain areas, and the field is enjoying a resurgence in recent years as a paradigm for studying learning processes at a neural level (Sitaram et al., 2017).

From a scientific perspective, several questions are uniquely amenable to neurofeedback experiments. Most simply, we may ask: What basic neural features are under volitional control? The current evidence suggests that signals from a wide range of cortical regions can be modulated, given appropriate real-time feedback (Cerf et al., 2010; Kobayashi, Schultz, and Sakagami, 2010; Moritz and Fetz, 2011; Neely et al., 2018). Within the motor cortex, both individual neurons and small ensembles can be brought under volitional control (Law, Rivlis, and Schieber, 2014). However, despite the apparent flexibility of single neurons, BMI operation requires activity patterns that are distributed over many neurons simultaneously.

A more subtle experimental approach within the neurofeedback paradigm is to examine how easily arbitrary combinations of different neurons can be controlled. Some insights into this question can be gleaned from comparable interfaces that use electromyogram signals from the muscles for real-time control. When learning an abstract myoelectric interface, the distal hand muscles can readily be combined into new patterns (Radhakrishnan, Baker, and Jackson, 2008), consistent with our capacity to produce a diverse repertoire of fractionated finger movements. Indeed, the synergist structure between arbitrary pairs of hand muscles can be changed on a trial-by-trial basis through differential modulation of divergent corticospinal pathways (Nazarpour, Barnard, and Jackson, 2012).

In contrast, arm muscles appear more constrained to habitual patterns (de Rugy, Loeb, and Carroll, 2012), and adaptation is faster to perturbations that preserve the low-dimensional structure of naturalistic synergies (Berger et al., 2013). It seems that BMI control by monkeys, at least using neurons from the arm area of motor cortex, may be similarly constrained to a low-dimensional neural space. After initiating a BMI task using a conventional biomimetic decoder, Sadtler et al. (2014) compared different perturbations applied during online control. Perturbations that stayed within a naturalistic 10-dimensional *intrinsic manifold* were learned quickly, while perturbations requiring neural strategies that lay outside of this manifold were not learned within a single session.

A further experimental approach within the neurofeedback paradigm exploits the redundancy inherent in the brain control (by many neurons) of (low-dimensional) effectors. Since multiple neural strategies could, in principle, be used to produce a desired movement, the particular choice that emerges through training can be informative of learning mechanisms. An early example of this type of study was performed by Jarosiewicz et al. (2008), who perturbed a previously learned BMI decoder by rotating the effect of half of the neurons on cursor movement. They examined possible adaptation strategies, considering solutions that ranged from the global adjustment of all neurons (reaiming) to local changes reflecting the neuron-specific nature of the perturbation (remapping). Notably, the global reaiming strategy contributed most to the behavioral improvements observed within each session, in effect reassigning the existing repertoire of neural patterns to a new set of rotated targets. A recent reanalysis (Golub et al., 2018) of the *within manifold* perturbations from Sadtler et al. (2014) similarly found that the learned strategies preserved the same overall neural repertoire. They proposed that learning, at least in the early stage, is driven predominantly by a reassociation of preexisting neural patterns with new behavioral outcomes (figure 41.3*A*, *B*).

These findings are also consistent with an earlier report that operant conditioning of individual parietal cortex neurons engaged ensemble patterns that were within the repertoire of natural behavior (Hwang, Bailey, and Andersen, 2013). Indeed, given that the monkeys in these experiments have no a priori way to assign credit to the particular neurons acting on the BMI cursor (or even know they are participating in a BMI experiment!), it seems inevitable that the initial exploration of a new problem will be informed by a set of naturalistic strategies that have been successful in the past. Nevertheless, the apparent flexibility with which the preexisting neuronal repertoire can be reassociated with new behavioral outcomes offers a powerful alternative to the biomimetic BMI approach.

In the field of human-computer interaction, *affordances* refer to action possibilities that are intuitively suggested

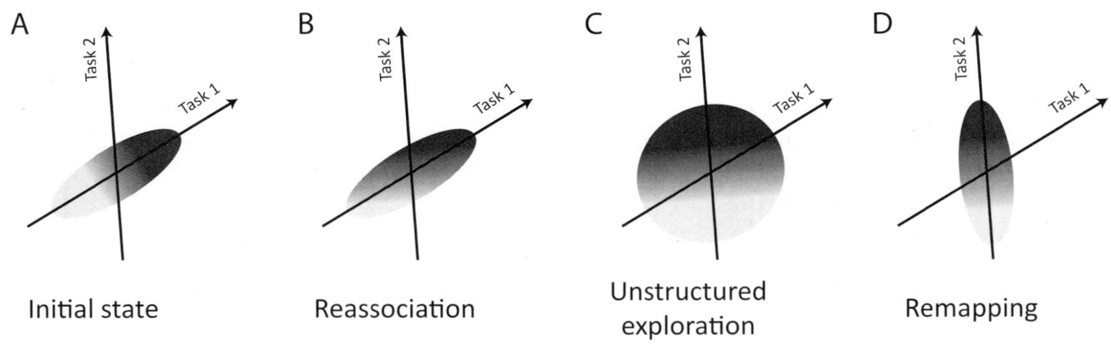

FIGURE 41.3 Possible mechanisms of BMI learning. *A*, Initially, the neural manifold is aligned to the output-potent axis of Task 1 (e.g., a naturalistic movement or a biomimetic BMI). Shading indicates movements associated with different regions of the manifold (e.g., left/right). Task 2 is an abstract BMI and thus introduces a new output-potent axis within the neural state space. *B*, Neural strategies within the original manifold may be reassociated with their new actions, indicated by the different pattern of shading. This solution is suboptimal since the dominant neural components do not efficiently drive movement. *C*, Unstructured exploration of the full neural space may eventually find new strategies that optimally drive movement. *D*, Over time, this may lead to a new manifold that is appropriately mapped onto the new output-potent axis.

by the design of an interface (e.g., a knob is designed to suggest turning). Similarly, we might envisage an intuitive BMI that, even if it does not mimic natural control, affords the exploration of effective control strategies. Unsupervised approaches that extract neural manifolds from large, unlabeled data sets could be used to predict how subjects will explore potential BMI control strategies and thus design decoders that can be learned through rapid *reassociation* processes. A nice example of this concept is provided by research into the body-machine interfaces operated by partially paralyzed individuals using residual motion measured by wearable sensors. Casadio et al. (2009) applied PCA to sensor data collected during free "dancing" movements to music to define the low-dimensional space that could readily be explored by users. Simply by projecting the first two dimensions of this space onto a computer screen, the users rapidly learned to control cursors within a single session, even though the mapping from sensor space to cursor movements was abstract.

Long-Term Reorganization during Brain-Machine Interface Learning

An important question is whether the constraints on early stages of BMI learning reflect hard-wired limitations of the motor cortex that cannot be overcome or simply a tendency for subjects to explore strategies that have previously been associated with rewarding outcomes. Due to the local nature of the BMI perturbations, global reaiming/reassociation solutions are generally suboptimal since individual neurons become maximally active for directions misaligned with their effect on cursor movements (figure 41.3*B*). However, finding optimal strategies requires a time-consuming search of the high-dimensional neural space. Therefore, it is interesting to ask whether, with sufficient training, local remapping could be achieved when solutions lie outside of the manifold associated with natural movements.

Experiments of this sort are difficult to perform, but Ganguly et al. (2011) reported that BMI learning over several days with the same set of neurons did indeed lead to the formation of new, stable ensemble patterns. However, it is not clear whether these reflect an entirely new manifold or simply different regions within a larger repertoire of naturalistic strategies. A related question is: What mechanisms guide this exploration for new neural strategies? A recent reanalysis of the same data set (Athalye et al., 2017) found that variability during the first few days of learning was apparently unstructured (figure 41.3*C*) before being shaped into a new manifold that reflected the current BMI task (figure 41.3*D*). One possibility is that high variability during the exploratory phase of BMI learning (Zacksenhouse et al., 2007) is driven by local sources of noise in individual neurons (Legenstein et al., 2010), with successful strategies reinforced by reward circuits involving the basal ganglia (Koralek et al., 2012).

There remains the possibility of intermediate levels of reorganization between a single constrained manifold and entirely unstructured exploration. Fast reaiming/reassociation effectively finds a new mapping at the highest level in figure 41.1, between preexisting global patterns of neural activity and new behavioral goals (e.g., rotated targets). Local remapping reflects reorganization at the lowest level in figure 41.1—individual neurons—over many days. Exploration at intermediate levels—for example, preserving the dynamic structure of neuronal populations or the synchronous ensembles of neurons projecting to the same muscles (Jackson et al., 2003)—may be faster than finding an entirely new neural representation. The hierarchical remapping hypothesis (Jackson and Fetz, 2011) suggests that reorganization should proceed through these levels over time, as behavior is optimized in progressively higher-dimensional spaces. Examining the evolution of neural strategies over time promises to be a fruitful line of future BMI research and should shed light on more general questions of how we learn new motor skills.

Summary

Over the last few decades, proof-of-principle BMIs have been demonstrated in both preclinical and clinical studies. As BMI technologies improve, we can expect to learn more about population structure in cortical activity and how this shapes the acquisition of new motor skills. This understanding should lead to more intuitive BMIs that require less training of both decoders and users. In combination with improved techniques to relay information back into the brain and spinal cord, BMIs have the potential to become powerful therapeutic tools for restoring motor function after neurological injury, as well as unique opportunities for investigating mechanisms of learning and plasticity in neural circuits.

REFERENCES

Aflalo, T. N., and Graziano, M. S. (2006). Partial tuning of motor cortex neurons to final posture in a free-moving paradigm. *Proceedings of the National Academy of Sciences of the United States of America, 103*(8), 2909–2914.

Aghagolzadeh, M., and Truccolo, W. (2016). Inference and decoding of motor cortex low-dimensional dynamics via latent state-space models. *IEEE Transactions on Neural Systems and Rehabilitation Engineering, 24*(2), 272–282.

Ajiboye, A. B., Willett, F. R., Young, D. R., Memberg, W. D., Murphy, B. A., Miller, J. P., Walter, B. L., Sweet, J. A., Hoyen, H. A., Keith, M. W., Peckham, P. H., Simeral, J. D., Donoghue, J. P., Hochberg, L. R., and Kirsch, R. F. (2017). Restoration of reaching and grasping movements through brain-controlled muscle stimulation in a person with tetraplegia: A proof-of-concept demonstration. *Lancet, 389*(10081), 1821–1830.

Archambault, P. S., Caminiti, R., and Battaglia-Mayer, A. (2009). Cortical mechanisms for online control of hand movement trajectory: The role of the posterior parietal cortex. *Cerebral Cortex, 19*(12), 2848–2864.

Athalye, V. R., Ganguly, K., Costa, R. M., and Carmena, J. M. (2017). Emergence of coordinated neural dynamics underlies neuroprosthetic learning and skillful control. *Neuron, 93*(4), 955–970.e5.

Bennett, K. M., and Lemon, R. N. (1996). Corticomotoneuronal contribution to the fractionation of muscle activity during precision grip in the monkey. *Journal of Neurophysiology, 75*(5), 1826–1842.

Berger, D. J., Gentner, R., Edmunds, T., Pai, D. K., and d'Avella, A. (2013). Differences in adaptation rates after virtual surgeries provide direct evidence for modularity. *Journal of Neuroscience, 33*(30), 12384–12394.

Bouton, C. E., Shaikhouni, A., Annetta, N. V., Bockbrader, M. A., Friedenberg, D. A., Nielson, D. M., Sharma, G., Sederberg, P. B., Glenn, B. C., Mysiw, W. J., Morgan, A. G., Deogaonkar, M., and Rezai, A. R. (2016). Restoring cortical control of functional movement in a human with quadriplegia. *Nature, 533*(7602), 247–250.

Capogrosso, M., Milekovic, T., Borton, D., Wagner, F., Moraud, E. M., Mignardot, J. B., Buse, N., Gandar, J., Barraud, Q., Xing, D., Rey, E., Duis, S., Jianzhong, Y., Ko, W. K., Li, Q., Detemple, P., Denison, T., Micera, S., Bezard, E., Bloch, J., and Courtine, G. (2016). A brain-spine interface alleviating gait deficits after spinal cord injury in primates. *Nature, 539*(7628), 284–288.

Carmena, J. M., Lebedev, M. A., Crist, R. E., O'Doherty, J. E., Santucci, D. M., Dimitrov, D. F., Patil, P. G., Henriquez, C. S., and Nicolelis, M. A. (2003). Learning to control a brain-machine interface for reaching and grasping by primates. *PLoS Biology, 1*(2), E42.

Casadio, M., Pressman, A., Danziger, Z., Tseng, H. Y., Fishbach, A., and Mussa-Ivaldi, F. A. (2009). Functional reorganization of upper-body movements for wheelchair control. *Conference Proceedings IEEE Engineering in Medicine and Biology Society, 10*(10), 4607–4610.

Cerf, M., Thiruvengadam, N., Mormann, F., Kraskov, A., Quiroga, R. Q., Koch, C., and Fried, I. (2010). On-line, voluntary control of human temporal lobe neurons. *Nature, 467*(7319), 1104–1108.

Churchland, M. M., Cunningham, J. P., Kaufman, M. T., Foster, J. D., Nuyujukian, P., Ryu, S. I., and Shenoy, K. V. (2012). Neural population dynamics during reaching. *Nature, 487*(7405), 51–56.

Collinger, J. L., Wodlinger, B., Downey, J. E., Wang, W., Tyler-Kabara, E. C., Weber, D. J., McMorland, A. J., Velliste, M., Boninger, M. L., and Schwartz, A. B. (2013). High-performance neuroprosthetic control by an individual with tetraplegia. *Lancet, 381*(9866), 557–564.

de Rugy, A., Loeb, G. E., and Carroll, T. J. (2012). Muscle coordination is habitual rather than optimal. *Journal of Neuroscience, 32*(21), 7384–7391.

Ethier, C., Oby, E. R., Bauman, M. J., and Miller, L. E. (2012). Restoration of grasp following paralysis through brain-controlled stimulation of muscles. *Nature, 485*(7398), 368–371.

Fetz, E. E. (1969). Operant conditioning of cortical unit activity. *Science, 163*(3870), 955–958.

Fetz, E. E. (1992). Are movement parameters recognizably coded in the activity of single neurons? *Behavioral and Brain Sciences, 15*(4), 154.

Flesher, S. N., Collinger, J. L., Foldes, S. T., Weiss, J. M., Downey, J. E., Tyler-Kabara, E. C., Bensmaia, S. J., Schwartz, A. B., Boninger, M. L., and Gaunt, R. A. (2016). Intracortical microstimulation of human somatosensory cortex. *Science Translational Medicine, 8*(361), 361ra141.

Gallego, J. A., Perich, M. G., Miller, L. E., and Solla, S. A. (2017). Neural manifolds for the control of movement. *Neuron, 94*(5), 978–984.

Ganguly, K., Dimitrov, D. F., Wallis, J. D., and Carmena, J. M. (2011). Reversible large-scale modification of cortical networks during neuroprosthetic control. *Nature Neuroscience, 14*(5), 662–667.

Gilja, V., Nuyujukian, P., Chestek, C.A., Cunningham, J. P., Yu, B. M., Fan, J. M., Churchland, M. M., Kaufman, M. T., Kao, J. C., Ryu, S. I., and Shenoy, K. V. (2012). A high-performance neural prosthesis enabled by control algorithm design. *Nature Neuroscience, 15*(12), 1752–1757.

Golub, M. D., Sadtler, P. T., Oby, E. R., Quick, K. M., Ryu, S. I., Tyler-Kabara, E. C., Batista, A. P., Chase, S. M., and Yu, B. M. (2018). Learning by neural reassociation. *Nature Neuroscience, 21*(4), 607–616.

Hall, T. M., de Carvalho, F., and Jackson, A. (2014). A common structure underlies low-frequency cortical dynamics in movement, sleep, and sedation. *Neuron, 83*(5), 1185–1199.

Hochberg, L. R., Bacher, D., Jarosiewicz, B., Masse, N. Y., Simeral, J. D., Vogel, J., Haddadin, S., Liu, J., Cash, S. S., van der Smagt, P., and Donoghue, J. P. (2012). Reach and grasp by people with tetraplegia using a neurally controlled robotic arm. *Nature, 485*(7398), 372–375.

Hochberg, L. R., Serruya, M. D., Friehs, G. M., Mukand, J. A., Saleh, M., Caplan, A. H., Branner, A., Chen, D., Penn, R. D., and Donoghue, J. P. (2006). Neuronal ensemble control of prosthetic devices by a human with tetraplegia. *Nature, 442*(7099), 164–171.

Hwang, E. J., Bailey, P. M., and Andersen, R. A. (2013). Volitional control of neural activity relies on the natural motor repertoire. *Current Biology, 23*(5), 353–361.

Jackson, A., and Fetz, E. E. (2011). Interfacing with the computational brain. *IEEE Transactions on Neural Systems and Rehabilitation Engineering, 19*(5), 534–541.

Jackson, A., Gee, V. J., Baker, S. N., and Lemon, R. N. (2003). Synchrony between neurons with similar muscle fields in monkey motor cortex. *Neuron, 38*(1), 115–125.

Jarosiewicz, B., Chase, S. M., Fraser, G. W., Velliste, M., Kass, R. E., and Schwartz, A. B. (2008). Functional network reorganization during learning in a brain-computer interface paradigm. *Proceedings of the National Academy of Sciences of the United States of America, 105*(49), 19486–19491.

Jarosiewicz, B., Sarma, A. A., Bacher, D., Masse, N. Y., Simeral, J. D., Sorice, B., Oakley, E. M., Blabe, C., Pandarinath, C., Gilja, V., Cash, S. S., Eskandar, E. N., Friehs, G., Henderson, J. M., Shenoy, K. V., Donoghue, J. P., and Hochberg,

L. R. (2015). Virtual typing by people with tetraplegia using a self-calibrating intracortical brain-computer interface. *Science Translational Medicine, 7*(313), 313ra179.

Kao, J. C., Nuyujukian, P., Ryu, S. I., Churchland, M. M., Cunningham, J. P., and Shenoy, K. V. (2015). Single-trial dynamics of motor cortex and their applications to brain-machine interfaces. *Nature Communications, 6*(7759), 7759.

Kao, J. C., Nuyujukian, P., Stavisky, S., Ryu, S. I., Ganguli, S., and Shenoy, K. V. (2013). Investigating the role of firing-rate normalization and dimensionality reduction in brain-machine interface robustness. *Conference Proceedings 35th Annual IEEE Engineering in Medicine and Biology Society (EMBC)*, Osaka, 293–298.

Kobayashi, S., Schultz, W., and Sakagami, M. (2010). Operant conditioning of primate prefrontal neurons. *Journal of Neurophysiology, 103*(4), 1843–1855.

Koralek, A. C., Jin, X., Long II, J. D., Costa, R. M., and Carmena, J. M. (2012). Corticostriatal plasticity is necessary for learning intentional neuroprosthetic skills. *Nature, 483*(7389), 331–335.

Lara, A. H., Cunningham, J. P., and Churchland, M. M. (2018). Different population dynamics in the supplementary motor area and motor cortex during reaching. *Nature Communications, 9*(1), 2754.

Law, A. J., Rivlis, G., and Schieber, M. H. (2014). Rapid acquisition of novel interface control by small ensembles of arbitrarily selected primary motor cortex neurons. *Journal of Neurophysiology, 112*(6), 1528–1548.

Legenstein, R., Chase, S. M., Schwartz, A. B., and Maass, W. (2010). A reward-modulated Hebbian learning rule can explain experimentally observed network reorganization in a brain control task. *Journal of Neuroscience, 30*(25), 8400–8410.

Lillicrap, T. P., and Scott, S. H. (2013). Preference distributions of primary motor cortex neurons reflect control solutions optimized for limb biomechanics. *Neuron, 77*(1), 168–179.

Moritz, C. T., and Fetz, E. E. (2011). Volitional control of single cortical neurons in a brain-machine interface. *Journal of Neural Engineering, 8*(2), 025017.

Moritz, C. T., Perlmutter, S. I., and Fetz, E. E. (2008). Direct control of paralysed muscles by cortical neurons. *Nature, 456*(7222), 639–642.

Nazarpour, K., Barnard, A., and Jackson, A. (2012). Flexible cortical control of task-specific muscle synergies. *Journal of Neuroscience, 32*(36), 12349–12360.

Neely, R. M., Koralek, A. C., Athalye, V. R., Costa, R. M., and Carmena, J. M. (2018). Volitional modulation of primary visual cortex activity requires the basal ganglia. *Neuron, 97*(6), 1356–1368.e4.

Overduin, S. A., d'Avella, A., Roh, J., Carmena, J. M., and Bizzi, E. (2015). Representation of muscle synergies in the primate brain. *Journal of Neuroscience, 35*(37), 12615–12624.

Pandarinath, C., O'Shea, D. J., Collins, J., Jozefowicz, R., Stavisky, S. D., Kao, J. C., Trautmann, E. M., Kaufman, M. T., Ryu, S. I., Hochberg, L. R., Henderson, J. M., Shenoy, K. V., Abbott, L. F., and Sussillo, D. (2018). Inferring single-trial neural population dynamics using sequential auto-encoders. *Nature Methods, 15*, 805–815.

Radhakrishnan, S. M., Baker, S. N., and Jackson, A. (2008). Learning a novel myoelectric-controlled interface task. *Journal of Neurophysiology, 100*(4), 2397–2408.

Sadtler, P. T., Quick, K. M., Golub, M. D., Chase, S. M., Ryu, S. I., Tyler-Kabara, E. C., Yu, B. M., and Batista, A. P. (2014). Neural constraints on learning. *Nature, 512*(7515), 423–426.

Scott, S. H. (2008). Inconvenient truths about neural processing in primary motor cortex. *Journal of Physiology, 586*(5), 1217–1224.

Scott, S. H., and Kalaska, J. F. (1997). Reaching movements with similar hand paths but different arm orientations. I. Activity of individual cells in motor cortex. *Journal of Neurophysiology, 77*(2), 826–852.

Serruya, M. D., Hatsopoulos, N. G., Paninski, L., Fellows, M. R., and Donoghue, J. P. (2002). Instant neural control of a movement signal. *Nature, 416*(6877), 141–142.

Shenoy, K. V., Sahani, M., and Churchland, M. M. (2013). Cortical control of arm movements: A dynamical systems perspective. *Annual Review of Neuroscience, 36*, 337–359.

Sitaram, R., Ros, T., Stoeckel, L., Haller, S., Scharnowski, F., Lewis-Peacock, J., Weiskopf, N., Blefari, M. L., Rana, M., Oblak, E., Birbaumer, N., and Sulzer, J. (2017). Closed-loop brain training: The science of neurofeedback. *Nature Reviews Neuroscience, 18*(2), 86–100.

Sussillo, D., Stavisky, S. D., Kao, J. C., Ryu, S. I., and Shenoy, K. V. (2016). Making brain-machine interfaces robust to future neural variability. *Nature Communications, 7*(13749), 13749.

Tabot, G. A., Dammann, J. F., Berg, J. A., Tenore, F. V., Boback, J. L., Vogelstein, R. J., and Bensmaia, S. J. (2013). Restoring the sense of touch with a prosthetic hand through a brain interface. *Proceedings of the National Academy of Sciences of the United States of America, 110*(45), 18279–18284.

Taylor, D. M., Tillery, S. I., and Schwartz, A. B. (2002). Direct cortical control of 3D neuroprosthetic devices. *Science, 296*(5574), 1829–1832.

Tomlinson, T., and Miller, L. E. (2016). Toward a proprioceptive neural interface that mimics natural cortical activity. *Advances in Experimental Medicine and Biology, 957*, 367–388.

Vargas-Irwin, C. E., Brandman, D. M., Zimmermann, J. B., Donoghue, J. P., and Black, M. J. (2015). Spike Train SIMilarity Space (SSIMS): A framework for single neuron and ensemble data analysis. *Neural Computation, 27*(1), 1–31.

Velliste, M., Perel, S., Spalding, M. C., Whitford, A. S., and Schwartz, A. B. (2008). Cortical control of a prosthetic arm for self-feeding. *Nature, 453*(7198), 1098–1101.

Wong, Y. T., Putrino, D., Weiss, A., and Pesaran, B. (2013). Utilizing movement synergies to improve decoding performance for a brain machine interface. *Conference Proceedings 35th Annual IEEE Engineering in Medicine and Biology Society (EMBC)*, Osaka, 289–292.

Zacksenhouse, M., Lebedev, M. A., Carmena, J. M., O'Doherty, J. E., Henriquez, C., and Nicolelis, M. A. (2007). Cortical modulations increase in early sessions with brain-machine interface. *PLoS One, 2*(7), e619.

Zimmermann, J. B., and Jackson, A. (2014). Closed-loop control of spinal cord stimulation to restore hand function after paralysis. *Frontiers in Neuroscience, 8*(87), 87.

42 Somatosensory Input for Real-World Hand and Arm Control

JEFFREY WEILER AND J. ANDREW PRUSZYNSKI

ABSTRACT In this chapter we consider how sensory feedback from the somatosensory periphery is integrated for the ongoing control of reaching, grasping, and object manipulation. We first review the peripheral neural basis of hand and arm control, specifically the signals arising from receptors embedded in muscles, tendons, and skin. Then, in the context of recent behavioral and neurophysiological studies, we describe several prominent factors that make dealing with feedback signals in the context of real-world hand control particularly interesting: the hierarchical organization of feedback control loops, the need to integrate sensory inputs from multiple modalities in real time, and the potential disconnect between perception and control.

At the age of 19, Ian Waterman developed an autoimmune disorder that selectively destroyed the large-diameter afferent neurons that convey sensory information from the muscles, tendons, and skin. This rare condition effectively rendered him unable to move, not because he couldn't plan his actions or produce muscle force but because his nervous system could not receive somatosensory feedback about the state of his body or about mechanical interactions with objects in the external world. Ian's story highlights the crucial role that somatosensory feedback plays in motor control, a role that is accentuated for dexterous actions like reaching, grasping, and object manipulation. Peripheral feedback is so essential for controlling the hand and arm that approximately 90% of the axons in the peripheral nerves of the upper limb transmit sensory information from the periphery into the central nervous system, while only 10% carry the motor commands from the central nervous system to the muscles (Gesslbauer et al., 2017).

Although sensory feedback was at the heart of Sherrington's classic work at the turn of the 20th century, the topic was brought back into the spotlight of modern motor control research by Emo Todorov and Michael Jordan (2002), who suggested that voluntary movement itself arises via the sophisticated manipulation of sensory feedback. The theoretical framework they introduced (but see Loeb et al., 1990), based on optimal feedback control, aims to mathematically describe a motor action—for example, a reach—via a formal cost function that incorporates the potential benefits and penalties associated with the execution of that action. The optimal reach, then, is the one that minimizes this cost function, which, in turn, is implemented by adjusting the gains of feedback control loops over time. With certain cost functions, the optimal feedback-control framework can account for many of the core features of reaching, as well as many other tasks. It also provides insight into how the system should best integrate sensory feedback in real time to achieve the task being performed. Because optimal feedback controllers only correct those errors that adversely affect task performance, observing how the nervous system responds to sensory inputs can provide clues to the cost function that governs a particular action. Thus, in this framework, understanding how the nervous system processes sensory feedback is key to understanding motor control.

Optimal feedback control offers a useful normative framework to understand motor behavior. Unfortunately, as it turns out, the mathematics of optimal feedback control are difficult even in simple scenarios, and establishing firm links between optimal feedback control and behavior is hard because there exist many cost functions that can reproduce the same behavioral predictions. Moreover, a general scheme to make predictions about underlying neural circuits, say, at the level of single neurons, remains elusive. Luckily, there are heuristics, as others have laid out in detail (Scott, 2004, 2016; Shadmehr and Krakauer, 2008). In this chapter we stick to the level of heuristics and examine how sensory feedback rapidly contributes to reaching, grasping, and object manipulation. Although all sensory modalities can help support these actions, we primarily detail how the nervous system uses somatosensory feedback provided by receptors embedded in muscles, tendons, and skin. We focus on real-world control in the sense that relatively naturalistic paradigms raise some specific issues, including (1) the hierarchical organization of feedback control loops and how this organization trades computational flexibility for temporal delays, (2) how sensory feedback from multiple modalities is integrated in real time, and (3) the relationship between somatosensory feedback for perception versus motor control.

The Peripheral Neural Basis of Somatosensory Inputs

Understanding anatomy is always a good place to start. We first describe the main somatosensory receptors for hand and arm control. These receptors signal information to the processing centers of the spine, brain stem, and cortex, which all generate feedback responses. As we next describe, the signals arising from these receptors are more complex than typically appreciated, and understanding how these signals contribute to hand and arm function remains a substantial challenge.

Proprioceptive Inputs from the Hand and Arm

Muscle spindles are receptors embedded within, and running parallel to, the large extrafusal muscle fibers that generate muscle force. They are comprised of two types of intrafusal fibers: nuclear bag fibers and nuclear chain fibers. Large-diameter afferent nerve fibers (type Ia) wrap around the noncontractile central portions of both types of intrafusal fibers, forming the primary sensory ending traditionally thought to signal information about dynamic and static muscle length. Smaller-diameter afferent nerve fibers (type II) selectively wrap around the nuclear chain fibers, forming the secondary sensory ending traditionally thought to encode static length information. The distal portions of the intrafusal fibers are able to contract and are innervated by gamma motor neurons. Gamma motor neuron activity controls the contractile state of the intrafusal fibers, which modulates the spindles' sensitivity to muscle stretch but does not appreciably contribute to muscle force. Understanding the precise role of gamma motor neurons remains elusive (Windhorst, 2007); the simplest idea is that their activation is yoked to the activation of alpha motor neurons (which innervate extrafusal muscle fibers) so that the muscle spindle remains in its working range when its parent muscle changes length.

Golgi tendon organs (GTOs) are sensory receptors that encapsulate 3–25 extrafusal muscles fibers and are arranged in series within the musculotendinous junction to signal the tension generated by a muscle. Each GTO is innervated by a large-diameter myelinated afferent fiber (type Ib) that branches through the tendon's loosely packed collagen fibers. When a muscle contracts, these collagen fibers straighten and compress the afferent axon, generating action potentials in numbers roughly proportional to muscle tension.

Understanding the static and dynamic configuration of the body in space—the sense of proprioception—is obviously important for planning and executing actions and is often linked to muscle spindle activity. The direct connection between muscle spindles and proprioception stems from studies in animals and ex vivo preparations, which showed that spindle activity is correlated with muscle length and velocity, features that are also correlated with the position and velocity of the joint(s) that the muscle spans. Indeed, muscle vibration, which stimulates the muscle spindles, can cause illusory joint rotations (Eklund, 1972; Goodwin, McCloskey, and Matthews, 1972).

Recent work casts some doubt on a simple link between muscle spindle activity and proprioception because these receptors appear to encode kinematic and kinetic variables besides length information. For example, in the cat, muscle spindle activity can encode the muscle forces that arise because of the mechanical properties of muscle cross bridges (Blum et al., 2017). Human experiments have long described a more complex mapping between muscle length/velocity and spindle activity in the context of active movement. For example, Dimitriou and Edin (2010) recorded from muscle spindle afferents while people grasped objects and pressed keys, two relatively natural hand actions, and found that spindle activity best predicts the future state of the parent muscle and not its instantaneous joint position or muscle length. They suggested that muscle spindles operate as a forward sensory model that predicts the consequences of intended movement, acting as a peripheral mechanism to minimize the negative impact of sensory delays on controlling actions.

Cutaneous Signals from the Glabrous Skin of the Hand

Four distinct tactile afferent neurons innervate the glabrous (i.e., hairless) skin of the human hand and fingers. These afferents collectively encode most of the critical information about the mechanical events that occur when interacting with and manipulating objects (Johansson and Flanagan, 2009). Fast-adapting-type-1 (FA-1) and slow-adapting-type-1 (SA-1) neurons innervate Meissner corpuscles and Merkel disks, respectively, both of which are located close to the skin's surface and are densely packed in the fingertips. Although both receptors respond to light touch, FA-1 neurons preferentially respond to high-frequency skin deformations, like the moment an edge is contacted. SA-1 neurons, in contrast, can respond to steady-state indentations.

Fast-adapting-type-2 (FA-2) neurons innervate Pacinian corpuscles, which are located in the deep dermal regions of the fingers and hands. FA-2 neurons are extremely sensitive and respond most vigorously to high-frequency transient events that occur when making or breaking mechanical contact with an object or surface. Slow-adapting-type-2 (SA-2) innervate Ruffini endings embedded in the collagenous tissues of the dermis and respond to specific directions of skin stretch.

Most work investigating the function of tactile afferent neurons has focused on perceptions that arise when tactile stimuli are passively applied to the skin (Goodwin and Wheat, 2004; Johnson, 2001). These studies typically associate the four classes of tactile afferents, and their distinct anatomy and physiology, with non-overlapping functions based on correlations with perceptual phenomena. However, there is substantial evidence that almost all real-world functions attributed to specific tactile afferents actually involve integration across multiple types of tactile afferent neurons (Saal and Bensmaia, 2014). For example, some textures, often those used experimentally, preferentially excite a particular class of tactile afferent neurons. However, the textures encountered during real-world actions typically recruit all types of afferents and may even engage unique coding schemes—from spatial-coding schemes that operate statically at low spatial frequencies to temporal-coding schemes the require movement at high spatial frequencies. Determining how the nervous system integrates inputs across these afferents at various levels of the neuraxis remains an underserviced area of study.

Studies investigating shape perception typically assume that tactile afferent neurons are distributed in a pixel-like array across the skin and that the spatial aspects of a stimulus, like edge orientation or curvature, are represented in a veridical way—that is, the spatial layout of activated neurons provides a "neural image" of the stimulus being indented into the skin. However, this conceptualization ignores a basic organizational principle of the somatosensory system—namely, that tactile afferent neurons branch and innervate many mechanosensitive end organs. For example, each FA-1 neuron branches 4–6 times and innervates approximately 30 Meissner corpuscles (Nolano et al., 2003). As a result, tactile afferent neurons critical for resolving the spatial details of touched objects, FA-1s and SA-1s, have spatially complex receptive fields that do not provide a veridical representation of a touched object or surface but rather emphasize some canonical features, such as edge orientation (Pruszynski and Johansson, 2014). Important questions in the field include determining the evolutionary pressures that cause this peripheral complexity to arise, the limits of peripheral feature extraction, and how upstream neurons decode this information (Zhao et al., 2018).

Muscle Feedback for Posture and Reaching: The Hierarchical Nature of Feedback Control

Real-world hand actions can be thrown off course because of noise internal to the motor system or by external forces (e.g., an unexpected encounter with someone or something). One way the nervous system overcomes these challenges is by processing sensory information arising from muscle receptors and generating a series of rapid feedback responses (i.e., stretch reflexes) that engage multiple levels of the neural hierarchy.

The hierarchical organization of stretch reflexes and their underlying neural circuits have been extensively studied, with much of the effort focusing on documenting functional differences between evoked responses at different latencies (for reviews, see Pruszynski, 2014; Scott, 2016; Shemmell, Krutky, and Perreault, 2010). A common way to experimentally probe this neural hierarchy is to mechanically stretch a muscle and record a rapid set of phasic responses in an electromyogram (EMG) aligned on perturbation onset. Although relatively large mechanical perturbations are used in these studies to evoke large reflex responses, the same set of responses are also evoked by perturbations on the much smaller scale of natural motor variability, suggesting that muscle feedback is continually processed in real time to control our actions (Crevecoeur, Kurtzer, and Scott, 2012).

The standard view is that the hierarchical organization of fast feedback responses trades temporal delays for computational sophistication: spinal circuits are quick but "dumb," whereas cortical circuits are slow but "smart" (figure 42.1). The earliest feedback response—called the short-latency stretch reflex—manifests as a phasic wave of EMG activity 20–50 ms after perturbation onset. It is this reflex that a doctor is testing with a tendon hammer. Because of its latency, the short-latency stretch reflex almost certainly engages purely spinal circuits, most prominently a monosynaptic pathway linking primary muscle spindle afferents to motor neurons that project back to their parent muscle but also oligosynaptic pathways that can target other functionally related muscles (Pierrot-Deseilligny and Burke, 2005). In terms of function, the short-latency stretch reflex is typically thought to simply resist changes in muscle length. Consistent with this role, it is influenced by relatively low-level features of the stimulus or sensorimotor apparatus, such as the amount and speed of muscle stretch or the state of the motor neuron pool.

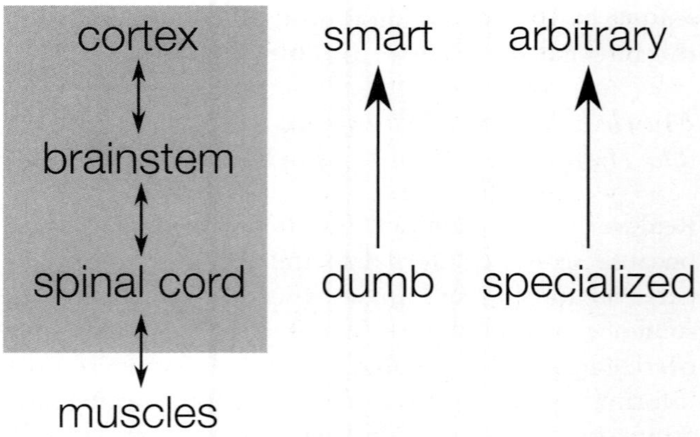

FIGURE 42.1 *Left,* Diagram of the nested-feedback pathways that contribute to hand and arm control. *Middle,* The traditional view is that the distinction between spinal and cortical feedback circuits relates to their functional capacity. *Right,* An alternative view is that the distinction relates to the flexibility of these circuits, the spinal cord providing complex but relatively fixed functions driven by long-term evolutionary pressures and the cerebral cortex allowing for relatively arbitrary input-output relationships.

The next feedback response—the long-latency stretch reflex—manifests as a phasic wave of EMG activity 50–100 ms after perturbation onset. Given its latency, the neural basis of the long-latency stretch reflex has been a point of substantial debate, with some evidence favoring a transcortical pathway mediated by fast (type Ia) muscle afferents and other evidence indicating a spinal pathway mediated by slower (type II) muscle afferents (Matthews, 1991). We now know that the long-latency stretch reflex reflects processing in spinal and cortical circuits and that there are likely additional contributions via the brain stem (Shemmell, Krutky, and Perreault, 2010). The key piece of evidence for a spinal circuit is that spinalized cats and monkeys still exhibit a clear long-latency stretch reflex (Ghez and Shinoda, 1978; Tracey, Walmsley, and Brinkman, 1980). There are two key pieces of evidence for a transcortical pathway. First, corticomotorneuronal cells in monkey primary motor cortex influence distal muscles of the limb and respond to mechanical perturbations at latencies early enough to contribute to the long-latency stretch reflex (Cheney and Fetz, 1984). In fact, cortical processing is not restricted to primary motor cortex, as mechanical perturbations quickly evoke vigorous, yet selective, responses in parietal area 5, dorsal premotor cortex, and the supplementary motor area (Hummelsheim et al., 1988; Omrani et al., 2016). Second, patients with undesired bilateral movements because of bifurcating corticospinal projections from primary motor cortex show long-latency stretch reflexes in both limbs when the muscles of one limb are stretched (Capaday et al., 1991; Matthews, Farmer, and Ingram, 1990).

A remarkable aspect of the long-latency stretch reflex is that, even though it occurs with latencies substantially shorter than typical measures of voluntary reaction time, it exhibits many sophisticated features typically attributed to voluntary control processes. These include modulation by subject intent, sensitivity to task demands, engagement during decision-making and motor learning, the routing of sensory information across muscles and effectors, and knowledge of the limb's dynamics and environment (for reviews, see Pruszynski and Scott, 2012; Scott, 2016). Perhaps the best real-world example of such functionality was demonstrated by Cole, Gracco, and Abbs (1984), who mechanically extended the thumb while participants were performing a pinching action, as if to grasp an object. Unlike the short-latency stretch reflex, which was evoked only in the stretched thumb muscle, the long-latency stretch reflex was coordinated across thumb and finger muscles in a way that supported accurate grasping. Although the presence of such sophisticated functional attributes is often attributed to the cortical circuits that underlie the long-latency stretch reflex, definitive links between particular attributes and specific neural circuits are rare, and more work is definitely warranted.

The functional dichotomy emphasized in the literature between short-latency and long-latency stretch reflexes makes it tempting to draw conclusions about how spinal and cortical circuits process somatosensory input for real-world control. But spinal circuits can drive many complex behaviors. A spinalized cat can be made to walk on a treadmill and can even change gait based on treadmill speed. A spinalized frog can swat noxious stimuli placed on its back and even account for different initial limb configurations. Indeed, the short-latency stretch reflex is not immutable. It is profoundly modulated at the transition from posture and movement, as well as over the course of cyclical actions like gait and hand cycling. It can even change sensitivity with long-term learning when its magnitude is directly reinforced, a promising intervention for improving gait after spinal cord injury (Thompson, Pomerantz, and Wolpaw, 2013). Thus, rather than contrasting spinal and cortical contributions according to their relative sophistication, a more accurate description of the hierarchy may relate to relative flexibility (figure 42.1). Whereas cortical circuits are able to transform sensory inputs onto motor responses in a seemingly arbitrary way, the spinal circuits support specific core functions of particular ethological relevance. This may be particularly true for hand control, as recent work has shown that the short-latency stretch reflex can help control the hand's position in external space by flexibly coordinating rotations of the elbow and wrist (Weiler, Gribble, and Pruszynski, 2019).

Taken together, fast feedback responses to mechanical disturbances are organized in a hierarchical fashion spanning the entire neuraxis. There are clear functional differences across this hierarchy. The long-latency stretch reflex, perhaps due to its transcortical contribution, is modulated by many, if not all, of the factors that influence voluntary control processes. The purely spinal short-latency stretch reflex is less flexible but can support some core functions or behaviors. One interesting aspect of this hierarchical organization that has received little attention is determining whether and how different levels of the hierarchy interact. Indeed, it is unknown whether and when longer feedback-control loops predict how the shorter control loops are going to act or whether each level of the hierarchy acts in isolation.

Reaching Is Automatically Guided by Visual and Tactile Feedback: Integrating Sensory Inputs in Real Time

People usually move their hands to touch and manipulate objects. In the real world, these objects can move, either because environmental forces are acting on them (e.g., someone bumps the table and a cup begins to fall) or because the object itself is an active agent (e.g., swatting a moving fly). Responding quickly to object motion is a critical component of real-world reaching. Because information about object movement is often visual in nature, a great deal of work has investigated how unexpected changes in the visual location of a target elicits rapid feedback corrections of the hand's trajectory (Day, 2014; Gaveau et al., 2014). Such corrections occur without conscious awareness, leading to the idea that there exists an "automatic pilot" of the hand with specialized neural resources—separate from those that initially generate the reaching action—acting to keep the hand moving toward its ultimate goal.

Vision is not the only modality that contributes information about the location of a reach target. In object manipulation and tool use, tactile inputs can be the main source of such information. Consider the many tasks in which one hand is reaching toward part of an object held by the other hand: the cap of a bottle, a berry on a branch. Consider also tasks in which an object held in one hand is obscured by the digits of that hand: the clasp of a piece of jewelry, the hole in a small screw nut. In these cases, changes in the position of the target can only be detected through touch.

Recent work has examined whether tactile information about a handheld object also elicits relatively automatic corrections to keep the reaching hand on target (Pruszynski, Johansson, and Flanagan, 2016). Using a task akin to berry picking, participants put the thumb of their left hand on a stick and made simple reaching movements with their right hand toward a ball mounted on the end of the stick. On a small proportion of trials, the stick unexpectedly rotated after the reach was initiated, and the participants had to alter their reach trajectory to succeed in touching the ball. Critically, on some trials vision was occluded. Here, participants had to rely on mechanical signals from the left hand as well as their memory of the length of the stick. In other trials, participants did not touch the stick with their left thumb but were given full vision, permitting a direct contrast of visually guided and touch-guided reach corrections. Participants were readily able to perform both the tactile and visual tasks immediately, meaning the process did not require substantial learning or training. Both sensory modalities evoked similar rapid corrective responses appropriate for bringing the hand to the target. Changes in muscle activation were evident 75 ms after the ball moved, and substantial kinematic deviations toward the new ball location started about 110 ms after the target jump.

What is the neural basis of such fast corrective responses, and do rapid tactile and visual corrections share a common neural substrate? The similarity of these responses, particularly in terms of timing, suggest that this might be the case. Though little is known about the rapid tactile responses, the neural machinery of rapid visual corrections has been investigated in various animal models as well as in studies involving neurological patients. In cats, midflight adjustments to a sudden target jump are likely mediated by subcortical structures (Alstermark and Isa, 2012). In humans, rapid corrections are often attributed to the posterior parietal cortex, based largely on the observation that patients with posterior parietal lesions are relatively normal at initiating actions but impaired at making fast corrective responses (Pisella et al., 2000). But work with a patient who has complete agenesis of the corpus callosum—a condition that eliminates direct communication between cerebral cortices—points to a brain stem pathway (Day and Brown, 2001). In this person, initiating a reach toward a target had greater latencies when the target was shown in the visual hemifield contralateral to the arm being used, a signature of a purely cortical process. However, the corrective action was the same regardless of how the target jumped or which hand was used, behavior suggestive of a subcortical contribution. Although the ultimate subcortical pathway is unknown, work in monkeys has shown that the superior colliculus receives direct visual inputs and has a substantial

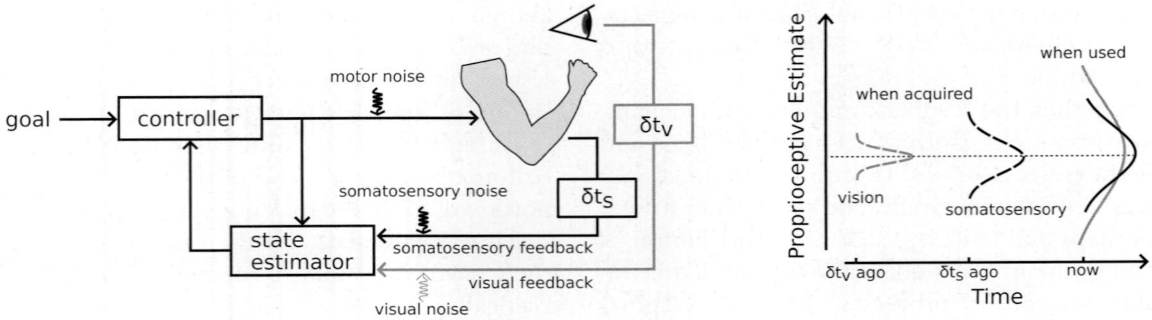

FIGURE 42.2 *Left,* Diagram of a feedback pathway integrating visual and somatosensory inputs at different delays. *Right,* Diagram showing the influence of sensory delays on the veracity of visual and somatosensory inputs. Note that, although visual inputs are more precise at the moment of their transduction, the increased latencies associated with this pathway mean they provide a less precise estimate of hand position by the time the transduced information is processed for the purposes of hand control. Reproduced with permission from Crevecoeur, Munoz, and Scott (2016).

population of neurons activated during reaching (Werner, Dannenberg, and Hoffmann, 1997). Indeed, when carefully examining muscle activity relative to target jumps, the fastest response, starting approximately 75 ms after the target jump, appears to be locked to the presentation of the target jump (Pruszynski et al., 2010), much like the neural activity in the superior colliculus that causes express saccades (Dorris, Paré, and Munoz, 1997).

An important consideration when trying to integrate multiple sources of feedback is that the different modalities provide feedback information with different acuity and at different latencies. The classical solution to this problem in the perceptual and decision-making literature is Bayesian estimation, whereby the input signals are combined in a manner inversely proportional to their variance (Fetsch, DeAngelis, and Angelaki, 2013). Since visual information is almost certainly more accurate than tactile information in providing an estimate of the tip of a handheld object, the simplest formulation of Bayesian integration would suggest that visual information should dominate the overall response when both vision and touch are available.

What does this extremely informative perceptual work tell us about real-world hand control? Perhaps not very much because the static Bayesian framework does not consider the fact that tactile information is available with substantially shorter latencies. Although these temporal differences can likely be ignored for perceptual processes that develop over seconds, in the context of real-world hand control tens of milliseconds matter. As such, there may be substantial benefit to getting crude but fast sensory inputs compared to precise sensory inputs with additional delay (figure 42.2). Indeed, recent work suggests that, in the context of visually guided reaching, the nervous system does not simply consider the relative acuity of the available sensory inputs but rather formulates a dynamic model that substantially penalizes delays (Crevecoeur, Munoz, and Scott, 2016). An important outstanding question relates to the underlying neural implementation of such a process: Can it be implemented via subcortical structures, or does it require the specialization of cortical machinery? And how does it relate to the neural mechanisms that underlie static Bayesian estimation?

Taken together, there exists a specialized system—perhaps subcortical in terms of neural implementation—that can update the details of an ongoing reach trajectory when the goal target position changes. Although this system is typically considered a specialization of vision and how visual inputs guide reaching, this seems unlikely to be the case, as tactile inputs can drive similar responses at similar latencies. On the other hand, whether rapid visual and tactile responses are generated by the same neural circuit, and thus reflect part of the same control process, remains unknown. Clues could come from behavioral and neurophysiological experiments looking for the multisensory integration of visual and tactile inputs, but such comparisons need to be cognizant of the inherent limitations associated with porting ideas about static Bayesian integration from perceptual studies into the domain of motor control.

Tactile Feature Extraction during Hand Control: Generalizing from Perception to Control

Putting on a necklace often requires pinching a small clasp and aligning it with an even smaller ring. In this and many other real-world manipulation tasks, executing fine control of the hands and digits requires that the nervous system has fast and accurate information about the orientation of each object. Such tasks are often done in the absence of vision, either because the action is performed out of sight (e.g., behind the head) or because the objects themselves are so small that direct vision is obscured by the digits. In these situations, information about an object's orientation is largely based on how its edges activate mechanoreceptors in the glabrous skin of the fingertips. Events that degrade the tactile signals generated via the mechanoreceptors, such as prolonged cold exposure or peripheral nerve injury, impair or even preclude fine manual dexterity.

Given its importance in the real world, it is not surprising that a wealth of studies have investigated edge-orientation processing. In most of these studies, an edge or a series of edges is applied to the participant's fingertip, and the participant is asked to judge the orientation of the edge or the relative orientation of multiple edges. Edge-orientation acuity, as measured in these perceptual identification and discrimination tasks, develops on the timescale of seconds and ranges between approximately 20° for long edges that span the whole contact surface of the finger (Bensmaia et al., 2008). For short edges on the order of a few millimeters (akin to the necklace clasp and ring) edge-orientation acuity is very poor, approximately 90° (Peters, Staibano, and Goldreich, 2015).

The speed and acuity of tactile edge-orientation processing reported in these perceptual studies is far too slow and crude to support many aspects of dexterous manipulation. Recent work has investigated edge-orientation acuity directly in the context of a simple object manipulation task (Pruszynski, Flanagan, and Flanagan, 2018). Rather than ask participants to report edge orientation directly, the experimenters asked participants to touch and rotate a randomly oriented dial toward a prespecified target position. Critically, information about the dial's initial orientation could only be extracted by touching a raised edge on the dial's surface. Thus, extracting edge-orientation information was not an explicit feature of the task, as in perceptual assays, but rather an implicit component of the object manipulation task being performed. Participants performed remarkably well in this task—nearly an order of magnitude better than in perceptual studies. On average, they were able to orient the dial within 3° of the target orientation for long edges spanning their entire fingertip, and this task remained impressive even for very short edges (e.g., 11° for a 2 mm long edge). Participants were not only exceptionally accurate but able to gather tactile edge-orientation information very quickly, within approximately 130 ms of initially contacting the edge and well before they reached a steady level of contact force. Using tactile information so quickly is consistent with the automaticity by which tactile signals are used in other aspects of manipulation to maintain stability when grasping an object. In fact, participants never used all of the available time nor did the fidelity of the sensory information (in this case, the edge length) influence the latency or timing of dial rotation. This points to a relatively low-level feedback-control scheme, contrasting with many perceptual or decision-making studies that have shown increases in reaction time as the veracity of the sensory input decreases.

Several factors likely account for the speed and acuity differences in tactile processing for perception versus motor control. One explanation is that there may exist more information when the participant is actively engaged with an object, perhaps due to subtle differences in contact dynamics. There are also differences in experimental timing that may put more stress on memory systems in the perceptual studies. For example, forced-choice paradigms present stimuli sequentially, requiring the participant to hold information about the first presented stimulus in memory in order to make a judgment about it relative to the second presented stimulus. Memory decay in many real-world manipulation tasks is minimized because the action is initiated quickly after the object is contacted and because contact is maintained during execution.

It is also possible that the difference between tactile processing for perception and control reflects substantive differences in their underlying neural mechanisms. Indeed, a large number of studies have shown that the processing of sensory information is sensitive to task and context. Perhaps the best example comes from the visual system, where it is well established that visual inputs are differentially processed for motor and perceptual tasks and that such differences reflect the engagement of specialized visual-processing pathways in the cerebral cortex (Milner and Goodale, 2006). The ventral visual pathway extracts detailed semantic information about objects, supporting perception, cognition, and memory. The dorsal visual pathway builds a pragmatic representation of objects in the environment, supporting real-time motor actions and guiding behavior. An equivalent segregation for the sense of touch has not been studied with equivalent protocols. For example, no one has directly tested whether the same tactile stimuli recruit different neural pathways depending on whether the acquired sensory information is used for perception or control.

Perception and control may also be served by different neural codes at the level of the peripheral apparatus. The currently accepted model supporting tactile acuity is built on a long tradition, starting with Mountcastle, of comparing neural responses to some set of stimuli in monkeys with perceptual reports from monkeys or, more often, from humans using similar stimuli (Johnson, 2001). This formulation starts with the notion that

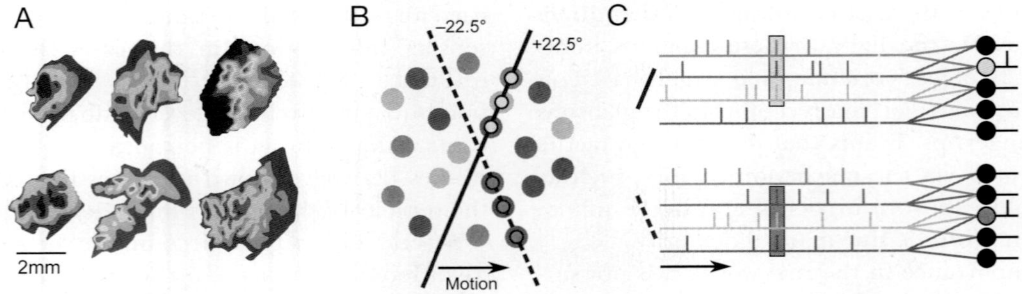

FIGURE 42.3 *A*, Exemplar receptive fields for first-order tactile neurons. The colors reflect sensitivity to a small indentation at that location on the skin. Note that receptive fields are complex, showing many zones of high sensitivity. *B*, Diagram showing the receptive field of four first-order tactile neurons (*color-coded ellipses*), each with multiple highly sensitive zones (*color-coded dots*). Solid and dashed lines illustrate differently oriented edges. *C*, Schematic of how the neurons in (*B*) will respond as the oriented edges move across the skin. Note that edge orientation influences how each neuron responds because of the distribution of its highly sensitive zones. At a given instance, one edge simultaneously stimulates one set of neurons, and the other edge stimulates another set of neurons (*yellow and orange rectangles*). These different sets of coincident action potentials cause responses in different higher-order neurons (*yellow and orange*), propagating edge-orientation information upstream. Reproduced with permission from Pruszynski and Johansson (2014). (See color plate 46.)

tactile afferent neurons have simple Gaussian sensitivity profiles. That is, they respond most robustly at the center of their receptive field and progressively less toward the margin. At the population level, the receptive fields of many tactile afferent neurons overlap, and spatial details are resolved based on the relative firing rates of neurons with nearby receptive fields via some neural interpolation scheme (Saal et al., 2017; Wheat, Goodwin, and Browning, 1995). Although this approach does well at explaining perceptual phenomena, a key weakness of an intensity-coding scheme is that estimating the firing rates of individual neurons takes a substantial amount of time and thus appears too slow to underlie hand control.

An alternative model that could support tactile acuity for hand control stems from the fact that human and monkey tactile afferent neurons branch in the skin and innervate many spatially distinct mechanoreceptors, yielding spatially complex receptive fields with multiple zones of high sensitivity (Johansson, 1978; Pruszynski and Johansson, 2014; figure 42.3*A*). These complex receptive fields are highly overlapping such that, at a population level, an edge at a given orientation will excite one subset of first-order tactile neurons, whereas an edge at a slightly different orientation will excite some other subset of first-order tactile neurons (figure 42.3*B*). The resulting coincidence code can yield high acuity and potentially operates at the timescale of the first incoming action potentials. Moreover, there exist established neural mechanisms for processing such a code based on the massive divergence and convergence of first-order tactile neurons onto second-order neurons in the cuneate nucleus (figure 42.3*C*).

Taken together, there is a functional and perhaps neurophysiological distinction between how tactile information is processed for perception and control. Determining the degree of these functional differences and the precise neural mechanisms that underlie them is an important avenue for future research. For example, appreciating these differences could improve brain-machine interfaces trying to reanimate hand function by "writing-in" somatosensory information into the brain (Bensmaia and Miller, 2014). The biomimetic stimulation protocols currently used are largely based on neural responses inspired by perceptual studies that, in the worst-case scenario, are orthogonal to the neural code that typically subserves motor control.

Conclusion

The goal of this chapter was to describe how somatosensory information is critical to real-world hand and arm control. We focused on three specific issues: the hierarchical organization of feedback-control loops, the ways in which sensory feedback from multiple modalities is integrated in real time, and the relationship between somatosensory feedback for perception versus motor control. These three topics were not meant as an exhaustive list but rather templates for thinking about how more naturalistic tasks reveal behavioral and neural insights unobtainable from more artificial laboratory paradigms.

REFERENCES

Alstermark, B., and Isa, T. (2012). Circuits for skilled reaching and grasping. *Annual Review of Neuroscience, 35*, 559–578.

Bensmaia, S. J., Hsiao, S. S., Denchev, P. V., Killebrew, J. H., and Craig, J. C. (2008). The tactile perception of stimulus orientation. *Somatosensory & Motor Research, 25*, 49–59.

Bensmaia, S. J., and Miller, L. E. (2014). Restoring sensorimotor function through intracortical interfaces: Progress and looming challenges. *Nature Reviews Neuroscience, 15*, 313–325.

Blum, K. P., Lamotte D'Incamps, B., Zytnicki, D., and Ting, L. H. (2017). Force encoding in muscle spindles during stretch of passive muscle. *PLoS Computational Biology, 13*, e1005767.

Capaday, C., Forget, R., Fraser, R., and Lamarre, Y. (1991). Evidence for a contribution of the motor cortex to the long-latency stretch reflex of the human thumb. *Journal of Physiology, 440*, 243–255.

Cheney, P. D., and Fetz, E. E. (1984). Corticomotoneuronal cells contribute to long-latency stretch reflexes in the rhesus-monkey. *Journal of Physiology-London, 349*, 249–272.

Cole, K. J., Gracco, V. L., and Abbs, J. H. (1984). Autogenic and nonautogenic sensorimotor actions in the control of multiarticulate hand movements. *Experimental Brain Research, 56*, 582–585.

Crevecoeur, F., Kurtzer, I., and Scott, S. H. (2012). Fast corrective responses are evoked by perturbations approaching the natural variability of posture and movement tasks. *Journal of Neurophysiology, 107*, 2821–2832.

Crevecoeur, F., Munoz, D. P., and Scott, S. H. (2016). Dynamic multisensory integration: Somatosensory speed trumps visual accuracy during feedback control. *Journal of Neuroscience, 36*, 8598–8611.

Day, B. L. (2014). Subcortical visuomotor control of human limb movement. *Advances in Experimental Medicine and Biology, 826*, 55–68.

Day, B. L., and Brown, P. (2001). Evidence for subcortical involvement in the visual control of human reaching. *Brain, 124*, 1832–1840.

Dimitriou, M., and Edin, B. B. (2010). Human muscle spindles act as forward sensory models. *Current Biology, 20*, 1763–1767.

Dorris, M. C., Paré, M., and Munoz, D. P. (1997). Neuronal activity in monkey superior colliculus related to the initiation of saccadic eye movements. *Journal of Neuroscience, 17*, 8566–8579.

Eklund, G. (1972). General features of vibration-induced effects on balance. *Upsala Journal of Medical Sciences, 77*, 112–124.

Fetsch, C. R., DeAngelis, G. C., and Angelaki, D. E. (2013). Bridging the gap between theories of sensory cue integration and the physiology of multisensory neurons. *Nature Reviews Neuroscience, 14*, 429–442.

Gaveau, V., Pisella, L., Priot, A.-E., Fukui, T., Rossetti, Y., Pélisson, D., and Prablanc, C. (2014). Automatic online control of motor adjustments in reaching and grasping. *Neuropsychologia, 55*, 25–40.

Gesslbauer, B., Hruby, L. A., Roche, A. D., Farina, D., Blumer, R., and Aszmann, O. C. (2017). Axonal components of nerves innervating the human arm. *Annals of Neurology, 82*, 396–408.

Ghez, C., and Shinoda, Y. (1978). Spinal mechanisms of functional stretch reflex. *Experimental Brain Research, 32*, 55–68.

Goodwin, A. W., and Wheat, H. E. (2004). Sensory signals in neural populations underlying tactile perception and manipulation. *Annual Review of Neuroscience, 27*, 53–77.

Goodwin, G. M., McCloskey, D. I., and Matthews, P. B. (1972). Proprioceptive illusions induced by muscle vibration: Contribution by muscle spindles to perception? *Science, 175*, 1382–1384.

Hummelsheim, H., Bianchetti, M., Wiesendanger, M., and Wiesendanger, R. (1988). Sensory inputs to the agranular motor fields: A comparison between precentral, supplementary-motor and premotor areas in the monkey. *Experimental Brain Research, 69*, 289–298.

Johansson, R. S. (1978). Tactile sensibility in the human hand: Receptive field characteristics of mechanoreceptive units in the glabrous skin area. *Journal of Physiology, 281*, 101–125.

Johansson, R. S., and Flanagan, J. R. (2009). Coding and use of tactile signals from the fingertips in object manipulation tasks. *Nature Reviews Neuroscience, 10*, 345–359.

Johnson, K. O. (2001). The roles and functions of cutaneous mechanoreceptors. *Current Opinion in Neurobiology, 11*, 455–461.

Loeb, G. E., Levine, W. S., and He, J. (1990). Understanding sensorimotor feedback through optimal control. *Cold Spring Harbor Symposia on Quantitative Biology, 55*, 791–803.

Matthews, P. B. C. (1991). The human stretch reflex and the motor cortex. *Trends in Neurosciences, 14*, 87–91.

Matthews, P. B. C., Farmer, S. F., and Ingram, D. A. (1990). On the localization of the stretch reflex of intrinsic hand muscles in a patient with mirror movements. *Journal of Physiology-London, 428*, 561–577.

Milner, D., and Goodale, M. (2006). *The visual brain in action*. Oxford: Oxford University Press.

Nolano, M., Provitera, V., Crisci, C., Stancanelli, A., Wendelschafer-Crabb, G., Kennedy, W. R., and Santoro, L. (2003). Quantification of myelinated endings and mechanoreceptors in human digital skin. *Annals of Neurology, 54*, 197–205.

Omrani, M., Murnaghan, C. D., Pruszynski, J. A., and Scott, S. H. (2016). Distributed task-specific processing of somatosensory feedback for voluntary motor control. *eLife, 5*.

Peters, R. M., Staibano, P., and Goldreich, D. (2015). Tactile orientation perception: An ideal observer analysis of human psychophysical performance in relation to macaque area 3b receptive fields. *Journal of Neurophysiology, 114*, 3076–3096.

Pierrot-Deseilligny, E., and Burke, D. (2005). The circuitry of the spinal cord: Its role in motor control and movement disorders. Cambridge: Cambridge University Press.

Pisella, L., Grea, H., Tilikete, C., Vighetto, A., Desmurget, M., Rode, G., Boisson, D., and Rossetti, Y. (2000). An "automatic pilot" for the hand in human posterior parietal cortex: Toward reinterpreting optic ataxia. *Nature Neuroscience, 3*, 729–736.

Pruszynski, J. A. (2014). Primary motor cortex and fast feedback responses to mechanical perturbations: A primer on what we know now and some suggestions on what we should find out next. *Frontiers in Integrative Neuroscience, 8*, 72.

Pruszynski, J. A., Flanagan, J. R., and Johansson, R. S. (2018). Fast and accurate edge orientation processing during object manipulation. *eLife, 7*.

Pruszynski, J. A., and Johansson, R. S. (2014). Edge-orientation processing in first-order tactile neurons. *Nature Neuroscience, 17*, 1404–1409.

Pruszynski, J. A., Johansson, R. S., and Flanagan, J. R. (2016). A rapid tactile-motor reflex automatically guides reaching toward handheld objects. *Current Biology, 26*, 788–792.

Pruszynski, J. A., King, G. L., Boisse, L., Scott, S. H., Flanagan, J. R., and Munoz, D. P. (2010). Stimulus-locked responses on human arm muscles reveal a rapid neural pathway linking visual input to arm motor output. *European Journal of Neuroscience, 32,* 1049–1057.

Pruszynski, J. A., and Scott, S. H. (2012). Optimal feedback control and the long-latency stretch response. *Experimental Brain Research, 218,* 341–359.

Saal, H. P., and Bensmaia, S. J. (2014). Touch is a team effort: Interplay of submodalities in cutaneous sensibility. *Trends in Neurosciences, 37,* 689–697.

Saal, H. P., Delhaye, B. P., Rayhaun, B. C., and Bensmaia, S. J. (2017). Simulating tactile signals from the whole hand with millisecond precision. *Proceedings of the National Academy of Sciences of the United States of America, 114,* E5693–E5702.

Scott, S. H. (2004). Optimal feedback control and the neural basis of volitional motor control. *Nature Reviews Neuroscience, 5,* 534–546.

Scott, S. H. (2016). A functional taxonomy of bottom-up sensory feedback processing for motor actions. *Trends in Neurosciences, 39,* 512–526.

Shadmehr, R., and Krakauer, J. W. (2008). A computational neuroanatomy for motor control. *Experimental Brain Research, 185,* 359–381.

Shemmell, J., Krutky, M. A., and Perreault, E. J. (2010). Stretch sensitive reflexes as an adaptive mechanism for maintaining limb stability. *Clinical Neurophysiology, 121,* 1680–1689.

Thompson, A. K., Pomerantz, F. R., and Wolpaw, J. R. (2013). Operant conditioning of a spinal reflex can improve locomotion after spinal cord injury in humans. *Journal of Neuroscience, 33,* 2365–2375.

Todorov, E., and Jordan, M. I. (2002). Optimal feedback control as a theory of motor coordination. *Nature Neuroscience, 5,* 1226–1235.

Tracey, D. J., Walmsley, B., and Brinkman, J. (1980). Long-loop reflexes can be obtained in spinal monkeys. *Neuroscience Letters, 18,* 59–65.

Weiler, J., Gribble, P. L., and Pruszynski, J. A. (2019). Spinal stretch reflexes support efficient hand control. *Nature Neuroscience, 22,* 529–533.

Werner, W., Dannenberg, S., and Hoffmann, K. P. (1997). Arm-movement-related neurons in the primate superior colliculus and underlying reticular formation: Comparison of neuronal activity with EMGs of muscles of the shoulder, arm and trunk during reaching. *Experimental Brain Research, 115,* 191–205.

Wheat, H. E., Goodwin, A. W., and Browning, A. S. (1995). Tactile resolution: Peripheral neural mechanisms underlying the human capacity to determine positions of objects contacting the fingerpad. *Journal of Neuroscience, 15,* 5582–5595.

Windhorst, U. (2007). Muscle proprioceptive feedback and spinal networks. *Brain Research Bulletin, 73,* 155–202.

Zhao, C. W., Daley, M. J., and Pruszynski, J. A. (2018). Neural network models of the tactile system develop first-order units with spatially complex receptive fields. *PloS One, 13,* e0199196.

43 Reorganization in Adult Primate Sensorimotor Cortex: Does It Really Happen?

TAMAR R. MAKIN, JÖRN DIEDRICHSEN, AND JOHN W. KRAKAUER

ABSTRACT The primary somatosensory (S1) and motor (M1) cortices are somatotopically organized, with distinct brain areas receiving information from—and controlling the movements of—specific body parts. But once this specialized organization has formed, can it be altered to support the changing needs of the organism? In this chapter, we reexamine the extent to which reorganization of the cortical hand representation occurs in the adult primate brain, with respect to both physiology and behavior. We review seminal findings reporting changes in the cortical maps of humans and nonhuman primates, which have been interpreted as evidence of brain reorganization: a qualitative change in the response of neurons. For example, a hand sensory area now responds to tactile stimulation of the face. We focus on three potential triggers for brain reorganization: changed input (due to training), input loss (due to amputation), and substrate loss (due to stroke). We review recent research demonstrating that the canonical functional organization in S1 and M1 is more resilient than originally assumed. Instead, experience or injury-dependent cortical map changes more likely result either from the unmasking of preexisting cortical connections or from subcortical reorganization, rather than true cortical reorganization. Finally, we examine whether cortical map changes, whatever their physiological origin, have any causal adaptive or maladaptive consequence on perception and action and conclude that they do not. Overall, our review suggests that although the adult brain can change based on experience or injury, cortical reorganization does not need to be invoked as the driving mechanism.

Representation is a much-debated concept in systems and cognitive neuroscience. The sensorimotor cortex has provided a compelling empirical model for studying how the brain represents the body and the world because, in addition to its well-characterized input-output structure, it contains a topographic map. Traditionally, topographic representation means a point-to-point correspondence between activated areas on the cortical surface and the object or space they represent (e.g., in primary visual and auditory cortices). In the sensorimotor system, neighboring parts of the body (or points on the skin) are represented in proximity on the cortical surface, a principle exemplified by the iconic sensorimotor homunculus (figure 43.1*A*). This relatively straightforward organizational principle, coupled with an accessible cortical location for various recording techniques, lends itself to studying how the brain maintains and updates sensorimotor representations. Over the years, changes in S1/M1 somatotopic maps have provided a key model system for studying cortical plasticity and a particular form of it termed *reorganization*.

Plasticity is a general term often used to describe how a neuron, circuit, or brain region quantitatively changes so that a novel output occurs in response to the same input. Crucially, this circuit change is driven by changes in behavior and, indeed, supports the development of these changes (e.g., long-term potentiation as a mechanism for learning and memory). Here we focus on an extreme case of plasticity termed *reorganization*, in which a specific neuron (and in fact an entire brain area) is assumed to change its input-output dynamics so that the reorganized neuron/area becomes responsive to a new input (retuning), or the circuit produces an entirely new motor output. These novel computations have frequently been inferred from qualitative changes in the spatial layout of cortical maps. Here, we critically evaluate whether such map changes in the adult primate brain really represent reorganization and whether they are behaviorally relevant. Before we can discuss evidence relating to putative reorganization, we first discuss native organization in the sensorimotor cortex.

While S1 is typically considered an input and M1 an output structure, this functional assignment is not clear-cut. For example, M1 is known to respond vigorously to somatosensory inputs (Maier, Bennett, Hepp-Reymond, & Lemon, 1993), whereas S1 has been shown to send disynaptic connections to the spinal cord (Rathelot, Dum, & Strick, 2017) and could potentially evoke hand movements. Even though the direct connections from M1 to S1 are not well described, it is clear that both structures

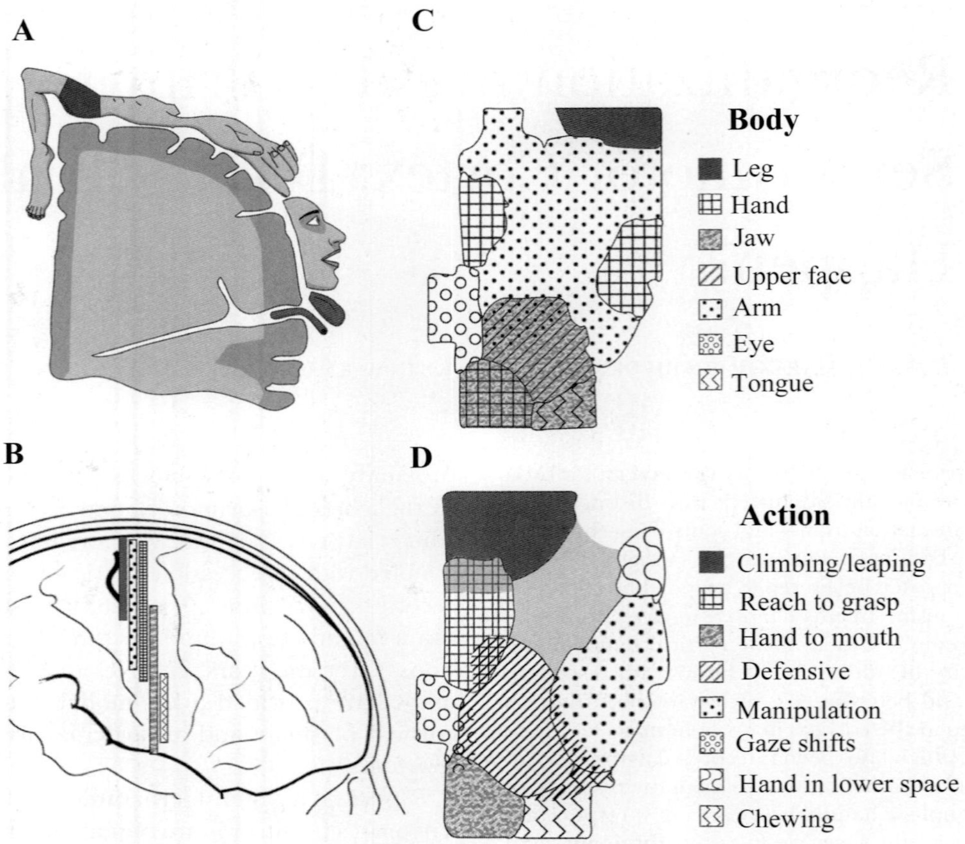

FIGURE 43.1 The sensorimotor homunculus based on a "winner takes all" caricature (*A*), microstimulation recordings in humans showing extensive overlap across body-part representations (*B*), or a computational model that optimizes local continuities based on body part and action representation (*C–D*). Note that the optimal arrangement based on body parts and actions results in a similar spatial layout. *A*, Image adapted from *The Homunculus Mapper* by Max Planck Florida Institute of Neuroscience, Creative Commons License (CC-BY-NC-SA). *B*, Modified from Catani (2017). *C–D*, Modified from Graziano and Aflalo (2007).

interact bidirectionally to contribute to sensorimotor control (Jones, Coulter, & Hendry, 1978). We will therefore discuss these two regions together throughout the chapter, emphasizing both similarities and differences.

Somatotopic Representation

The S1/M1 body representation is considered a form of map-like topographic organization, or specifically, a somatotopy. The relationship between the location of cortical activity and the represented body part, however, is not always as straightforward as the term *map* implies. For example, the spatial resolution of somatotopic maps differs across brain regions, with S1 showing highly detailed representations of individual hand digits (figure 43.2), whereas M1 only represents body parts crudely (figure 43.1*B*), with no clear organization of single digits (Schieber & Hibbard, 1993). It is also important to consider that the features of a cortical map are highly dependent on the experimental method and analysis approach used: a measurement of the most responsive aspect of the receptive field (RF; e.g., winner takes all) will reveal a crisp map, whereas a measurement of RF boundaries will show a blurrier map (figure 43.1). Finally, somatotopic maps contain discontinuities. For example, in the S1 and M1 maps the face area neighbors the hand area (figure 43.1). It is therefore important to note that topographic maps are likely only a superficial description of the system.

Further insight into how the body is represented in sensorimotor cortex can be gained from examining the relative similarities and dissimilarities across S1 and M1 activity patterns (see chapter 56). The similarity of two activity patterns may be determined not only by the physical proximity of two body parts but by the statistical relationship between two movements, suggesting that bodily information is represented in a more complex way related to behavior. For example, the relative overlap of activity patterns for individual digit movements is tightly related to the statistics of ordinary hand use (Ejaz, Hamada, & Diedrichsen, 2015). Two digits that move in a correlated fashion in everyday life have greater cortical overlap than two digits that move independently (figure 56.2 in chapter 56).

Thus, cortical maps do not just represent a single variable (body part) but rather contain multiple representational dimensions juxtaposed on top of each other (Graziano & Aflalo, 2007). Thought about this way, the full motor repertoire may be fit onto M1 to optimize local continuity across the cortical sheet to support specific behaviors (figure 43.1*D*). This alternative account for the organizing principle underlying S1/M1 can explain a somatotopic-like organization (neighboring digits are represented adjacently as they move in a more correlated fashion), as well as more complex features of the map (hand and face proximity, e.g., due to feeding behavior). As such, topographic maps may simply be an emergent property of a more fundamental underlying organizational principle based on a neural representation of behaviorally relevant movements. Here, we will nonetheless utilize the more classical view of a cortical map simply because it has been the main approach in analyzing S1/M1 cortical representations. We use the term *topographic maps* to imply an orderly relationship of activity patterns on the cortical sheet that reflects the relationships of stimuli/movements along a specific physical dimension (e.g., space and frequency). Moreover, for a representation to be considered topographic, the activity pattern relationship needs to be, to some degree, spatially invariant across different individuals.

If somatotopic maps are shaped by the statistical regularities or behavioral relevance of movements, then experience must play an important role in shaping these maps (please note this does not imply the converse). It has been long realized that body parts are represented disproportionately to their physical size. For example, the cortical hand area of the human homunculus is much larger than that of the foot. While cortical magnification may relate to the number of mechanoreceptors relaying information from the body part, it may also reflect increased usage in everyday life (termed *afferent magnification*). For example, spider monkeys who use their tails for manipulating objects show increased representation of their glabrous tail pads (Fulton & Dousser de Barenne, 1933), whereas rats that use their whiskers for palpating their near environment show increased representation of their whiskers. In summary, somatotopic organization may result from a dynamic learning process that attempts to spatially group neurons that fire in a coordinated fashion together on the cortical sheet. This process depends on a balance between (1) plasticity mechanisms that lead neurons with similar representational selectivity to reinforce each other and (2) local lateral inhibition, which introduces competitive interactions between neighboring neurons. This excitation/inhibition balance will ultimately shape the homunculus.

Modulations of the Cortical Hand Map

Altered inputs following training As stated above, cortical reorganization implies a qualitatively altered functional affiliation of the underlying tissue, as opposed to just quantitative modulation in response intensity. For example, brain area *A*, which previously responded to stimulus *a* and was nonresponsive to stimulus *b*, now becomes responsive to stimulus *b*; or a motor area that natively controls the arm now controls the digits. To conclude that such true reorganization occurred, however, it is necessary to rule out *gain modulation*—the process by which brain area *A*, which previously responded weakly to stimulus *b* or received sparse input from brain area *B*, becomes more responsive/connected to *b*/*B*. The critical point is that gain modulation can lead to changes in the extent of cortical maps, but such changes in somatotopy do *not* provide de facto evidence for functional reorganization. We will first describe what has been considered classical evidence for reorganization from studies in nonhuman primates and then discuss more recent studies from human research. We focus on two key questions: (1) Do the reported changes to native organization result from neuronal retuning to new inputs? (2) Are these neuronal changes, whether they represent true reorganization or not, necessary to support new behavior?

To the degree that there are givens in neuroscience, one of them is that somatotopic maps can be changed by experience; an idea that gained prominence in classic studies of the monkey somatosensory hand representation in the monkey (figure 43.2*A*). It has been shown that the boundaries of the hand map can be blurred when inputs are synchronized across the digits, such as by stitching two of them together, stimulating them synchronously, or performing repetitive and highly stereotypic hand movements (Wang, Merzenich, Sameshima, & Jenkins, 1995). Following prolonged periods of synchronized inputs, neurons previously showing greater selectivity to one digit expand their tuning to include the co-stimulated digit/s. Conversely, tactile training restricted to a single fingertip results in increased cortical representation of the stimulated digit. Originally, these studies were taken as evidence for behavior driving actual RF retuning. However, more recent research has emphasized the preexisting overlap in RF tuning across digits (Tamè et al., 2012; Tommerdahl, Favorov, & Whitsel, 2010). If the neurons along the digit boundaries of the hand map were *already* natively tuned to receive inputs from the neighboring digit, then these findings better fit our above definition of gain modulation rather than strict reorganization. Therefore, while the examples discussed here involve brain plasticity mechanisms, we believe that the modulation of RF tuning properties based on existing architecture does not meet the strong criteria of reorganization (see the introduction for our distinction between plasticity and reorganization).

In summary, patterns of synchronized sensory input, due to daily hand usage and learning, dictate the tuning properties of neurons comprising the hand map, likely through well-established processes of Hebbian plasticity and gain modulation. It is worth noting that these map changes can also be induced by passively generated inputs, which suggests that it is not a unique consequence of voluntary movement. Indeed, there is no clear evidence that these fine-scale brain organization changes have any implications on sensorimotor processing, in terms of perceptual or motor skill. Critically, changes in a map's boundaries could be induced by gain modulation without the need to invoke true reorganization. To address this potentially contentious issue more definitively, we next consider what happens when inputs are altered more dramatically—specifically, when the hand map is partially or completely deprived of synchronized inputs.

Reorganization following amputation What happens to the highly organized hand area in S1 (figure 43.2) if input to the map is lost? Seminal studies in monkeys have shown that the input-deprived brain territory becomes responsive to inputs from another body part. If, for example, digit 3 of the hand is amputated, over the course of weeks and months, the digit 3 part of the hand map becomes responsive to inputs from digits 2 and 4 (Merzenich et al., 1984; figure 43.2*A*). If input from the median nerve (innervating the glabrous skin of digits 1-3) is abolished due to nerve transection, then representation of the ulnar and radial nerves (innervating the dorsal skin of the hand) will seemingly expand into the deprived (median nerve) cortex (Merzenich et al., 1983). Since ulnar and radial responses were apparent in the deprived cortex immediately following deprivation, it is likely that these inputs were already native to this area but masked. While the newly observed representation in the deprived cortex of the dorsal skin is not initially topographically organized, over time it becomes so. This indicates that the synchronization of input, over the course of months, can promote topographic structure. Most strikingly, if input from the entire hand and arm is lost, due to deafferentation, neurons in the hand area become activated by inputs from the chin (Pons et al., 1991). Since the deprived hand area normally does not receive facial inputs, this is precisely the kind of result that had been interpreted as strong evidence for true reorganization. In all the cases of deafferentation discussed here, a new body-part representation, originally represented adjacently, is thought to "take over" the freed-up territory. We shall revisit these interpretations below.

Are there phenomena in M1 analogous to those in S1? Changes in M1 maps are most commonly studied using microstimulation to evoke muscle responses. Therefore, to study changes in M1 following input loss it is necessary to leave the efferent motor pathways preserved. When the dorsal column of the spinal cord is selectively injured (leaving the ventral column intact), S1 undergoes the above-mentioned facial remapping, yet movement representation in M1 is relatively unchanged (Kambi, Tandon, Mohammed, Lazar, & Jain, 2011). The preservation of M1 organization is probably due to the preservation of some motor function with the affected hand (e.g., climbing, locomotion, gross grasping), sufficient to maintain motor organization despite the loss of fine-scale motor control (due to input loss). Therefore, it seems that face-to-hand S1 reorganization does not necessitate a topographic change in M1. This unique dissociation between the S1 and M1 maps is intriguing, considering the strong bidirectional coupling between these two regions: If the former hand territory in S1 now supports facial processing, whereas its M1 counterpart still controls hand function, how would the functionality of the hand sensorimotor loop be maintained if true reorganization had occurred?

Persistent functional organization following amputation When aiming to characterize putative cortical reorganization in animals, scientists investigate the representation in the deprived sensory area of intact body parts—for example, facial responses following arm amputation. While this approach is suitable for documenting shifts in representational maps, it leaves unexplored the possibility that the original function of the deprived region may be preserved, although latent. Amputees experiencing phantom sensations provide a unique model to study what happens to the deprived hand area itself after sensory input loss. The phantom hand can be perceived as vividly as an intact hand. Many amputees report that they can elicit voluntary movements in their phantom hands. By utilizing reported phantom sensations, multiple studies have found that representations relating to the phantom hand persist in the peripheral and central nervous systems (Raffin, Mattout, Reilly, & Giraux, 2012). For

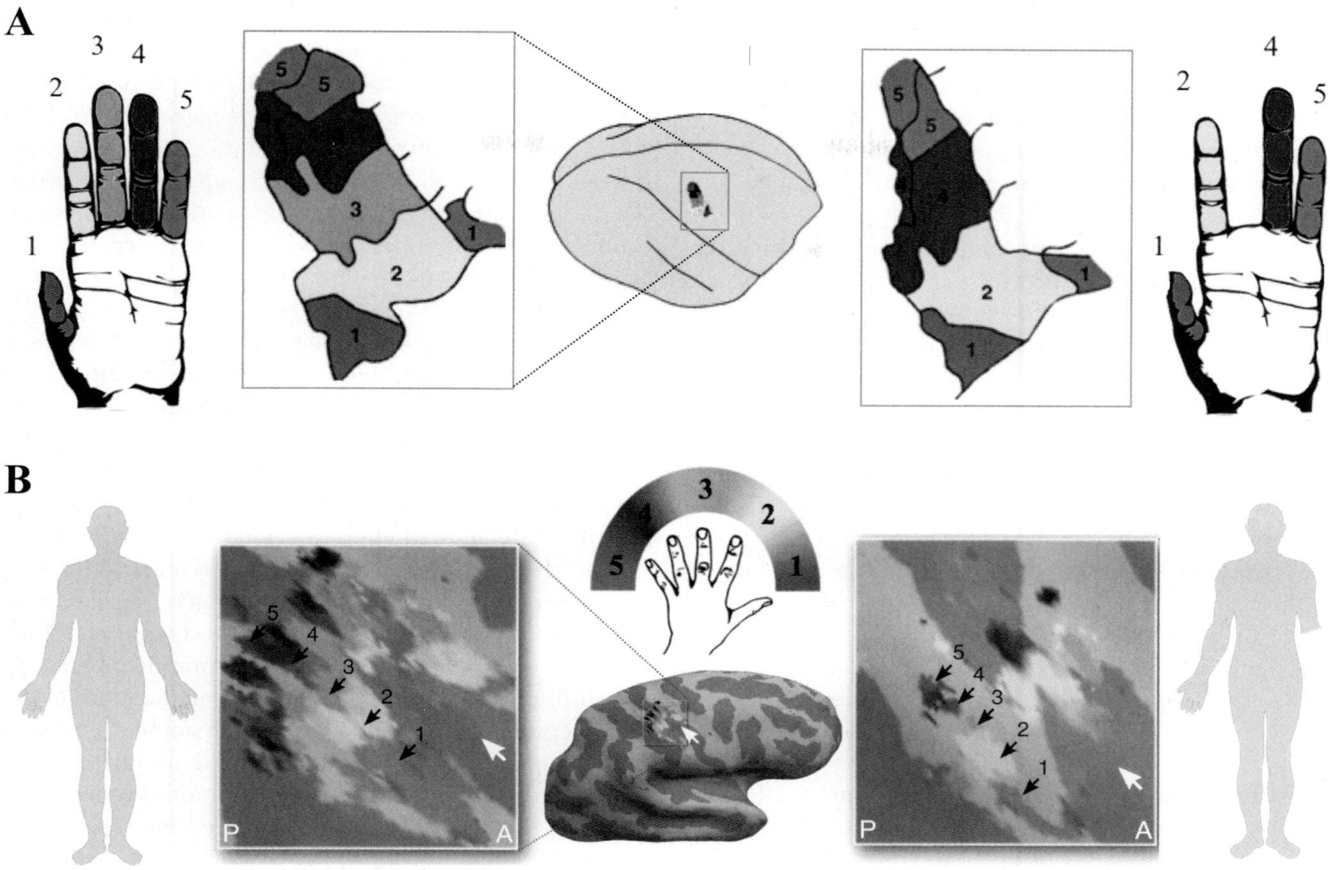

FIGURE 43.2 Digit maps in the monkey (*A*) and human (*B*). The left panels show characteristic maps of intact hands. The right panels show digit maps following the amputation of one digit (*A*) or the entire arm (*B*). *A*, Modified from Merzenich et al. (1984). *B*, Modified from Kikkert et al. (2016). (See color plate 47.)

example, functional magnetic resonance imaging (fMRI) data show that when amputees move their phantom fingers, they trigger a preserved and complete S1 topographic mapping of the missing hand, despite the physical absence of their hand (or organized sensory input) for several decades (Kikkert et al., 2016; figure 43.2*B*). Most strikingly, intracranial microstimulation in S1 of a tetraplegic patient elicited sensations of touch in specific hand locations (Flesher et al., 2016). Other studies show persistent communication between sensorimotor and higher-order motor areas relating to hand motor control. Collectively, these human studies clearly demonstrate that (1) the original functional layout of the sensory cortex is stable and does not disappear with afferent input loss; (2) representation in the deprived hand area is not altered—that is, the hand remains the target of S1/M1 processing; and (3) both upstream and downstream areas interpret input to or readout from the deprived hand area as relating to the missing hand.

How can we resolve the evidence from monkeys, showing extensive face remapping in the hand area of S1, with the human evidence suggesting persistent representation of the missing hand? The remapping observed in monkeys was originally thought to result from the widespread sprouting of intracortical connections, where new inputs from the face area travel to the hand area via lateral connections, formed following amputation (Pons et al., 1991). Alternatively, sparse widespread cortical connections that normally exist in the region may become unmasked following input loss, although anatomical evidence speaks against this possibility. Recent evidence suggests a third possibility: what appears to be cortical reorganization actually reflects *subcortical* reorganization, with subsequent changes in subcortical inputs to the cortex. For example, S1 facial remapping is abolished when the cuneate nucleus (normally receiving input from the hand and body) is inactivated (Kambi et al., 2014). This indicates that at the cortical level, the responses to subcortical inputs remain stable after amputation. The unchanged cortical hand area continues to receive input from the cuneate nucleus (via the thalamus) through the original pathways, but these are now being driven by sprouting from the trigeminal nucleus to the cuneate. These findings can resolve the apparent conflict between face-to-hand remapping in monkeys and the preservation of cortical hand organization in S1 of humans after amputation: while facial inputs may activate the hand area, due to changes subcortically, cortical hand architecture remains invariant.

In summary, classic studies in monkeys in which the main sensory input was removed from S1 were interpreted to mean that "freed-up" territory was invaded by new representations. Such findings were considered seminal and striking proof of the adult primate brain's capacity for cortical reorganization and formed the conceptual basis of many rehabilitative interventions. However, more recent studies challenge this reorganizational hypothesis. Instead, they demonstrate the immutability of native functional organization in the deprived cortex, with map changes being attributable either to the facilitation of *preexisting* cortical architecture or *subcortical* reorganization. Therefore, maps changes are either due to gain modulation or altered subcortical inputs, rather than a categorical change in the identity of a cortical neuron or network.

Behavioral Consequences of Map Changes

Regardless of what leads to cortical map changes, it still needs to be asked whether they are causally relevant to perception and action. That is to say, even if the cortical architecture has not qualitatively changed, the changes in cortical maps could still be necessary for *functional* reorganization, which we define as a neural change of any kind that is causally related to a change in behavior. We next ask if changes in sensorimotor maps are necessary for functional reorganization.

Map changes following training Motor training of a specific new movement—for example, learning how to rapidly extend the thumb—is associated with an apparent expansion of the motor map. This can manifest as an increase in activity (Shmuelof, Yang, Caffo, Mazzoni, & Krakauer, 2014), a larger cortical area devoted to that digit, or a lower threshold with which movement can be elicited by cortical stimulation (Classen, Liepert, Wise, Hallett, & Cohen, 1998). These map changes can be driven by gain changes, as defined above, but likely do not reflect true reorganization. This is because the map changes are only transient and disappear within hours or days after training stops, whereas the behavioral performance improvements persist (Molina-Luna, Hertler, Buitrago, & Luft, 2008). This dissociation clearly shows that increased map sizes are not a neural correlate of the new motor representation. Rather, it seems that it is the transiently shifted excitation/inhibition balance itself, rather than the map expansion it causes, that is crucial for learning, perhaps being required for the formation of new synapses. After these changes have occurred, the excitation/inhibition balance can return to normal and the map to its original shape, as demonstrated for the relationship between auditory perceptual learning and auditory cortex map changes (Reed et al., 2011).

Input loss following amputation It has long been suggested that the expanded representation of the spared input into deprived cortex following amputation results in perceptual gains. For example, Merzenich and colleagues (1984) proposed that reorganization following digit amputation (figure 43.2*A*) should lead to increased tactile acuity of the neighboring digits. Underlying this idea is the assumption that the brain is able to correctly interpret signals arising from the deprived area (missing digit territory) as relating to its newly assigned function (neighboring digit representation), thereby providing greater (or better) information about the new representation. However, tactile acuity in human digit amputees is not affected by the amputation. Similarly, the popular notion that cross-modal reorganization in the visual cortex of people with congenital blindness contributes to heightened tactile abilities has also been questioned. For example, activity in visual cortex in response to touch does not contain information that pertains to somatotopic digit representation (Wesselink et al., 2018). Instead, the improved perceptual abilities in populations suffering from sensory deprivation could be attributed to altered experience and training, which does not necessitate reorganization (Grant, Thiagarajah, & Sathian, 2000; Polley, Chen-Bee, & Frostig, 1999).

Referred sensations in amputees have famously also been regarded as the behavioral correlates of brain reorganization. Interestingly, some amputees report experiencing touch on their phantom hand when the skin of a different body part is touched. This "referred" sensation is most commonly experienced by touching the residual arm and thus it could be explained by aberrant reinnervation of the peripheral sensory nerves. In a few amputees, however, referred sensations have been reported to occur on the face, which have been suggested as even stronger evidence for reorganization—if touch on the face activates the hand area, the brain might interpret this touch as arising from the missing hand (Ramachandran, Stewart, & Rogers-Ramachandran, 1992). However, facial referred sensations are not common, and referred sensations can be triggered by touch applied to multiple body parts (e.g., feet, chest, and neck), as well as to both sides of the body. Given that referred sensations to the phantom hand are triggered by body parts that do not neighbor the missing hand territory, these findings are inconsistent with the idea of cortical reorganization.

Another highly influential model for the functional consequences of brain reorganization following amputation asserts that reorganization is harmful. While a phantom sensation per se is not necessarily bothersome, it can be highly debilitating when associated with phantom limb pain. Flor and colleagues (1995) explored the behavioral correlates of facial remapping in arm amputees and identified shifted cortical representation of the lower face. These shifts were greater in those amputees who reported experiencing the worst phantom limb pain. Based on these and similar findings, it was proposed that the mismatch between the invading facial inputs and the underlying infrastructure for the missing hand results in an "error" signal, consequentially interpreted by the brain as pain arising from the missing hand. This theory of maladaptive brain plasticity has been extremely influential, not only in the neuroscience community but also in the clinical literature: if pain is caused by maladaptive reorganization, then alleviating phantom pain could be achieved by reversing the reorganization. One common approach has been to "reinstate" the representation of the missing hand to its original territory, using illusory visual feedback of the missing hand (mirror box therapy). However, the evidence is, at best, mixed in terms of the success of these treatments (Richardson & Kulkarni, 2017). Furthermore, recent research has challenged the notion that the lower-face representation invades the missing hand cortex following amputation (Makin, Scholz, Henderson Slater, Johansen-Berg, & Tracey, 2015). Together with accumulating evidence emphasizing the role of persistent peripheral inputs after amputation in generating phantom pain, it appears that evidence for a relationship between brain reorganization and phantom limb pain is weak.

In summary, dramatic changes in activity patterns following amputation, interpreted as reorganization, have long been considered to support behavioral change. Most famously, facial activity in the cortical hand areas was considered to be the mechanism underlying referred sensations and phantom limb pain. Yet, despite the appeal of this view, especially in the clinical community, the current evidence does not support the hypothesis that reorganization has a causal role, either adaptive or maladaptive, on either perception or action.

Brain reorganization after stroke An applied area for which the idea of brain reorganization has had a lasting and problematic influence is the presumptive mechanism of motor recovery after stroke and other forms of focal brain injury. As in the case of S1, many studies of M1 have been interpreted as showing a takeover of motor function by undamaged areas of the motor cortex. To reiterate, reorganization in this context implies that following a lesion in region *A* of the motor cortex, region *B* (that moved effector *b* before the injury) can now move effector *a*. It is important to note that reorganization can be considered in both the negative and positive sense. In the positive, recovery-related sense, takeover pertains to when region *A* loses representation for effector *a* due to stroke, but region *B* for effector *b* reorganizes to move effector *a*. In the negative sense, to support effector a, region *B* loses some of its representation for effector *b*. We say this is negative because effector *b* loses representation due to the reorganization. This latter argument is directly analogous to the original claims made for S1 maladaptive reorganization after deafferentation that we reviewed above. It should be stated that sometimes these two forms of "reorganization" get conflated in the literature, confusing things even further.

To make this more concrete, we briefly describe a series of studies that are considered the most compelling for motor cortical reorganization after a focal lesion, although careful consideration leads to more nuanced conclusions. The studies were conducted by Randy Nudo and colleagues beginning in the 1990s. In the original studies, squirrel monkeys were pretrained on a hand dexterity task and then small subtotal infarcts were made in the M1 hand representation. In the first study, which sought to investigate spontaneous recovery (Nudo & Milliken, 1996), the monkeys took about two months to return to preoperative levels of hand dexterity. Notably, however, cortical mapping with intracortical microstimulation revealed that the small infarcts resulted in a widespread reduction in the areal extent of digit representations adjacent to the stroke (figure 43.3*A*, *B*). These results are not consistent with the idea that reorganization is behaviorally relevant: although there was a loss of digit representation from a map perspective, digit behavior nevertheless returned to normal levels. Interestingly, the authors concluded that "substantial functional reorganization occurs in primary motor cortex of adult primates following a focal ischemic infarct, but at least in the absence of post-infarct training, the movements formerly represented in the infarcted zone do not reappear in adjacent cortical regions." Thus, the idea of reorganization was inferred from the changes in map topography (i.e., remapping), even though they were not considered responsible for the recovery.

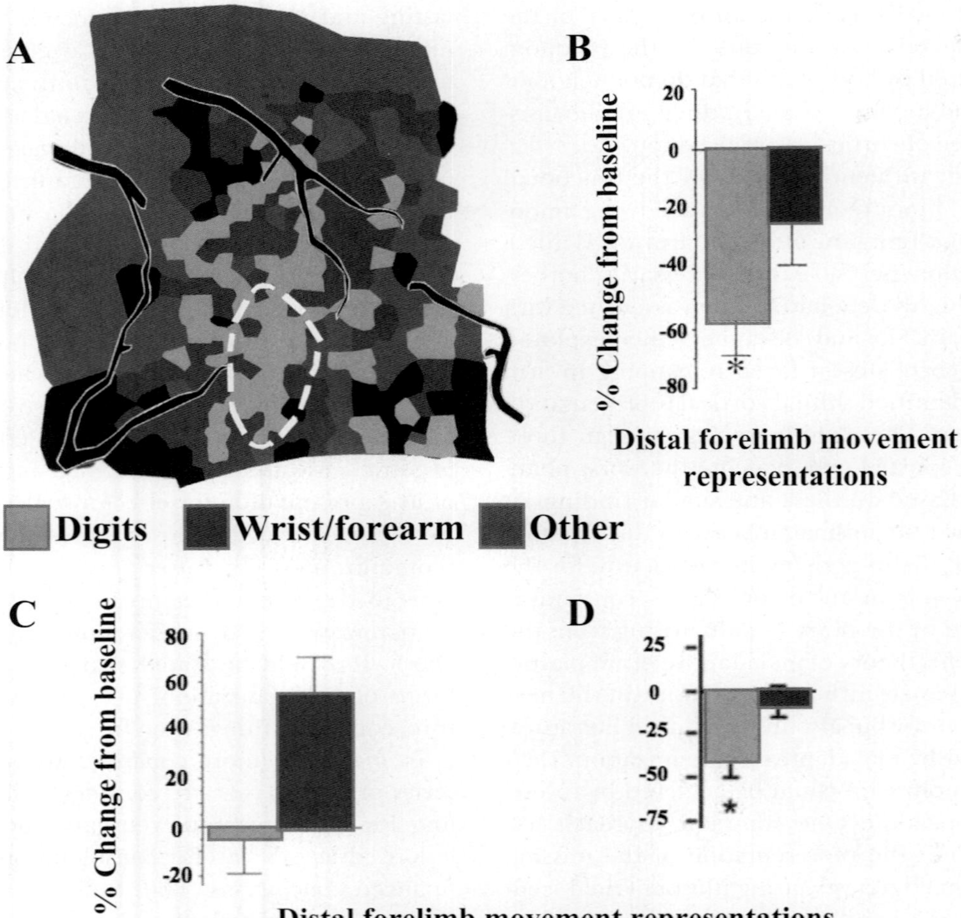

FIGURE 43.3 Remapping of the monkey hand and arm representation following focal M1 infarcts. *A*, Hand representation prior to M1 lesion (*shown by white line*). Remapping of hand representation after lesion to the digit area following (*B*) spontaneous recovery, (*C*) training-based recovery, and (*D*) delayed training. Despite the different remapping profiles, the monkeys regained similar digit function. The y-axis shows the change in motor map area devoted to the distal forelimb following the infarct and training. *A–B*, Modified from Nudo and Milliken (1996). *C*, Modified from Nudo et al. (1996). *D*, Modified from Barbay et al. (2006). (See color plate 48.)

In a classic follow-up study, intense daily training identical to the pretraining was resumed at poststroke day 5 and continued until the monkeys regained their former dexterity around 1 month poststroke. In this case, the hand and arm map expanded its representation (Nudo, Wise, Sifuentes, Milliken, & Millikent, 1996; figure 43.3*C*). Thus, the *opposite* mapping effect was observed compared to the first study, but with the *same* behavioral outcome. Finally, a decade later the authors investigated the effects of delayed retraining (Barbay et al., 2006), beginning at 1 month poststroke. Under this condition, the monkeys regained preoperative levels of behavior but without the maintenance of the spared peri-infarct hand representation seen with early-onset training (figure 43.3*D*).

Thus, when considered as a whole, these studies suggest that the notion of cortical reorganization as causally supporting recovery is highly problematic, even though these studies are frequently cited as strong evidence for functional reorganization. Instead, the most parsimonious conclusion is that map expansions and contractions are epiphenomena, perhaps use- and learning-related, but not the causal factors for behavioral recovery. In other words, processes relating to learning and adaptive behavior, likely necessary following injuries such as amputation and stroke, may be driving the map changes, but these changes are not driving behavioral changes. This conclusion is consistent with the above-mentioned studies of use-dependent plasticity in humans, showing that the relevance of cortical changes to voluntary motor control are minimal.

Concluding Remarks

A reconsideration of various studies of S1/M1 map changes in the adult primate brain leads to two primary conclusions: (1) The effects represent either the facilitation of preexisting connections or subcortical reorganization rather than categorical changes in cortical representation, and (2) there is currently no compelling evidence that they are the causal factor for behavioral change. Instead, the observed changes in the boundaries of cortical maps are likely to be markers for use and learning effects, with causal effects instead likely arising subcortically. Even in the extreme circumstance of amputation, representation of the hand is retained in the sensorimotor cortex for years and even decades following amputation. The effects of sensory deprivation and brain injury likely facilitate the appearance of training-induced cortical changes in two ways: First, they induce shifts in excitatory/inhibitory balance, potentially improving the neurophysiological profile for inducing long-term potentiation and depression (Polley, Chen-Bee, & Frostig, 1999). Second, the injury causing the amputation/stroke will also profoundly impair the motor abilities of the individual, promoting the learning of adaptive strategies and therefore changing input/output synchronization patterns. We therefore suggest that map changes should neither be attributed to categorical changes in cortical reorganization nor given causal behavioral relevance.

Acknowledgments

We thank Lisa Quarrell for artwork and Victoria Root and Andrew Pruszynski for helpful comments.

REFERENCES

Barbay, S., Plautz, E. J., Friel, K. M., Frost, S. B., Dancause, N., Stowe, A. M., & Nudo, R. J. (2006). Behavioral and neurophysiological effects of delayed training following a small ischemic infarct in primary motor cortex of squirrel monkeys. *Experimental Brain Research, 169*(1), 106–116. https://doi.org/10.1007/s00221-005-0129-4

Catani, M. (2017). A little man of some importance. *Brain, 140*(11), 3055–3061.

Classen, J., Liepert, J., Wise, S. P., Hallett, M., & Cohen, L. G. (1998). Rapid plasticity of human cortical movement representation induced by practice. *Journal of Neurophysiology, 79*(2), 1117–1123. https://doi.org/10.1152/jn.1998.79.2.1117

Ejaz, N., Hamada, M., & Diedrichsen, J. (2015). Hand use predicts the structure of representations in sensorimotor cortex. *Nature Neuroscience, 18*(7), 1034–1040.

Flesher, S. N., Collinger, J. L., Foldes, S. T., Weiss, J. M., Downey, J. E., Tyler-Kabara, E. C., … Gaunt, R. A. (2016). Intracortical microstimulation of human somatosensory cortex. *Science Translational Medicine, 8*(361), 1–11. https://doi.org/10.1126/scitranslmed.aaf8083

Flor, H., Elbert, T., Knecht, S., Wienbruch, C., Pantev, C., Birbaumer, N., … Taub, E. (1995). Phantom-limb pain as a perceptual correlate of cortical reorganization following arm amputation. *Nature, 375*(6531), 482–484.

Fulton, J. F., & Dousser de Barenne, J. G. (1933). The representation of the tail in the motor cortex of primates, with special reference to spider monkeys. *Journal of Cellular and Comparative Physiology, 2*(4), 399–426.

Grant, A. C., Thiagarajah, M. C., & Sathian, K. (2000). Tactile perception in blind braille readers: A psychophysical study of acuity and hyperacuity using gratings and dot patterns. *Perception and Psychophysics, 62*(2), 301–312. https://doi.org/10.3758/BF03205550

Graziano, M. S. A., & Aflalo, T. N. (2007). Mapping behavioral repertoire onto the cortex. *Neuron, 56*(2), 239–251. https://doi.org/10.1016/j.neuron.2007.09.013

Jones, E. G., Coulter, J. D., & Hendry, S. H. C. (1978). Intracortical connectivity of archtectonic fields in the somatic sensory, motor and parietal cortex of monkeys. *Journal of Comparative Neurology, 181*(2), 291–348.

Kambi, N., Halder, P., Rajan, R., Arora, V., Chand, P., Arora, M., & Jain, N. (2014). Large-scale reorganization of the somatosensory cortex following spinal cord injuries is due to brainstem plasticity. *Nature Communications, 5*, 3602. https://doi.org/10.1038/ncomms4602

Kambi, N., Tandon, S., Mohammed, H., Lazar, L., & Jain, N. (2011). Reorganization of the primary motor cortex of adult macaque monkeys after sensory loss resulting from partial spinal cord injuries. *Journal of Neuroscience, 31*(10), 3696–3707. https://doi.org/10.1523/JNEUROSCI.5187-10.2011

Kikkert, S., Kolasinski, J., Jbabdi, S., Tracey, I., Beckmann, C. F., Johansen-Berg, H., & Makin, T. R. (2016). Revealing the neural fingerprints of a missing hand. *eLife, 5*, e15292. https://doi.org/10.7554/eLife.15292

Maier, M. A., Bennett, K. M., Hepp-Reymond, M.-C., & Lemon, R. N. (1993). Contribution of the monkey corticomotoneuronal system to the control of force in precision grip. *Journal of Neurophysiology, 69*(3), 772–785. https://doi.org/8463818

Makin, T. R., Scholz, J., Henderson Slater, D., Johansen-Berg, H., & Tracey, I. (2015). Reassessing cortical reorganization in the primary sensorimotor cortex following arm amputation. *Brain, 138*(8), 2140–2146. https://doi.org/10.1093/brain/awv161

Merzenich, M. M., Kaas, J. H., Wall, J., Nelson, R. J., Sur, M., & Felleman, D. (1983). Topographic reorganization of somatosensory cortical areas 3b and 1 in adult monkeys following restricted deafferentation. *Neuroscience, 8*(1), 33–55. https://doi.org/10.1016/0306-4522(83)90024-6

Merzenich, M. M., Nelson, R. J., Stryker, M. P., Cynader, M. S., Schoppmann, A., & Zook, J. M. (1984). Somatosensory cortical map changes following digit amputation in adult monkeys. *Journal of Comparative Neurology, 224*(4), 591–605.

Molina-Luna, K., Hertler, B., Buitrago, M. M., & Luft, A. R. (2008). Motor learning transiently changes cortical somatotopy. *NeuroImage, 40*(4), 1748–1754. https://doi.org/10.1016/j.neuroimage.2007.11.018

Nudo, R. J., & Milliken, G. W. (1996). Reorganization of movement representations in primary motor cortex following focal ischemic infarcts in adult squirrel monkeys.

Journal of Neurophysiology, 75(5), 2144–2149. https://doi.org/10.1152/jn.1996.75.5.2144

Nudo, R. J., Wise, B. M., Sifuentes, F., Milliken, G. W., & Millikent, G. W. (1996). Neural substrates for the effects of rehabilitative training on motor recovery after ischemic infarct. *Science, 272*(June), 1791–1794.

Polley, D. B., Chen-Bee, C. H., & Frostig, R. D. (1999). Two directions of plasticity in the sensory-deprived adult cortex. *Neuron, 24*(3), 623–637. https://doi.org/10.1016/S0896-6273(00)81117-4

Pons, T. P., Garraghty, P. E., Ommaya, A. K., Kaas, J. H., Taub, E., & Mishkin, M. (1991). Massive cortical reorganization after sensory deafferentation in adult macaques. *Science, 252*(5014), 1857–1860.

Raffin, E., Mattout, J., Reilly, K. T., & Giraux, P. (2012). Disentangling motor execution from motor imagery with the phantom limb. *Brain, 135*(2), 582–595. https://doi.org/10.1093/brain/awr337

Ramachandran, V. S., Stewart, M., & Rogers-Ramachandran, D. C. (1992). Perceptual correlates of massive cortical reorganization. *Neuroreport, 3*(7), 583–586. https://doi.org/10.1126/science.1439826

Rathelot, J.-A., Dum, R. P., & Strick, P. L. (2017). Posterior parietal cortex contains a command apparatus for hand movements. *Proceedings of the National Academy of Sciences, 114*(16), 4255–4260. https://doi.org/10.1073/pnas.1608132114

Reed, A., Riley, J., Carraway, R., Carrasco, A., Perez, C., Jakkamsetti, V., & Kilgard, M. P. (2011). Cortical map plasticity improves learning but is not necessary for improved performance. *Neuron, 70*(1), 121–131. https://doi.org/10.1016/j.neuron.2011.02.038

Richardson, C., & Kulkarni, J. (2017). A review of the management of phantom limb pain: Challenges and solutions. *Journal of Pain Research, 10,* 1861–1870. https://doi.org/10.2147/JPR.S124664

Schieber, M. H., & Hibbard, L. S. (1993). How somatotopic is the motor cortex hand area? *Science, 261*(5120), 489–492. http://www.jstor.org/stable/2881942 JSTOR.

Shmuelof, L., Yang, J., Caffo, B., Mazzoni, P., & Krakauer, J. W. (2014). The neural correlates of learned motor acuity. *Journal of Neurophysiology, 112*(4), 971–980. https://doi.org/10.1152/jn.00897.2013

Tamè, L., Braun, C., Lingnau, A., Schwarzbach, J., Demarchi, G., Li Hegner, Y., … Pavani, F. (2012). The contribution of primary and secondary somatosensory cortices to the representation of body parts and body sides: An fMRI adaptation study. *Journal of Cognitive Neuroscience, 24*(12), 2306–2320. https://doi.org/10.1162/jocn_a_00272

Tommerdahl, M., Favorov, O. V., & Whitsel, B. L. (2010). Dynamic representations of the somatosensory cortex. *Neuroscience and Biobehavioral Reviews, 34*(2), 160–170. https://doi.org/10.1016/j.neubiorev.2009.08.009

Wang, X., Merzenich, M. M., Sameshima, K., & Jenkins, W. M. (1995). Remodelling of hand representation in adult cortex determined by timing of tactile stimulation. *Nature, 378,* 71–75.

Wesselink, D. B., Kikkert S., Bridge H., & Makin, T. R. (2018). Sensorimotor finger-specific information in the cortex of the congenitally blind. *Journal of Vision, 18*(10), 431–431.

44 The Basal Ganglia Invigorate Actions and Decisions

DAVID ROBBE AND JOSHUA TATE DUDMAN

ABSTRACT The basal ganglia are a group of interconnected subcortical nuclei well conserved across vertebrate species. They integrate multimodal information from the entire neocortex, thalamus, limbic regions, and dopaminergic midbrain nuclei. The basal ganglia output nuclei can powerfully regulate behavior either by modulating neuronal activity in frontal cortical regions indirectly through a thalamocortical pathway or directly by projections to midbrain/brain stem premotor regions. Dysfunction of the basal ganglia, occurring in diseases in humans or induced experimentally in animal models, leads to profound behavioral impairments, the most consistent being a reduction in speed and extent of movement. However, the specific computational aspect(s) of the basal ganglia that relate to the control of behavior remains the subject of considerable debate. Here, informed by features of the evolutionary conserved anatomy of the basal ganglia and the analysis of the behavioral deficits in human disease, we propose that a key computational function of the basal ganglia is to control low-level parameters of movements. As a result, the basal ganglia provide a critical circuit in translating both implicit and explicit aspects of voluntary decisions into action. We also review a series of neurophysiological studies across multiple vertebrate species that begin to elucidate the circuit-level implementation mechanisms of basal ganglia function.

Functional Anatomy of Basal Ganglia

The basal ganglia are a collection of interconnected subcortical nuclei that are found essentially unchanged in their pattern of connectivity, identified cell types, and neurochemical markers in all vertebrate species, ranging from the lamprey to the primate (Grillner, Robertson, & Stephenson-Jones, 2013). This conservation across species covers approximately 500 million years of vertebrate evolution. Over the last several years, the intersection of cell type–specific labeling, whole-brain anatomical-tracing techniques, and the development of refined molecular tools for anatomical tracing have led to substantial clarification of the cellular details of the inputs to the basal ganglia (Hintiryan et al., 2016; Hunnicutt et al., 2016). There are summaries of anatomical data in several mammalian species (e.g., Dudman & Gerfen, 2015); however, detailed quantitative comparisons across species are still lacking (but see Heilbronner, Rodriguez-Romaguera, Quirk, Groenewegen, & Haber, 2016). Our goal in this section is to provide a brief overview of the conserved functional architecture of the basal ganglia.

The basal ganglia transform descending inputs from the telencephalon and then project them via a reentrant pathway through the thalamus back to the cortex and through convergent, feedforward pathways to subcortical premotor structures, such as the reticular areas of the midbrain, the deep layers of the superior colliculus, and the mesencephalic locomotor region (figure 44.1). There is a dramatic reduction in cell number as one progresses through the internal circuitry of the basal ganglia (figure 44.1; Hardman et al., 2002; Oorschot, 1996). Although the reentrant pathway through the thalamus is often highlighted in descriptions that focus on the primate brain, the descending projection of the basal ganglia output is at least as prominent in all vertebrates so far studied (Cebrián, Parent, & Prensa, 2005; Deniau, Mailly, Maurice, & Charpier, 2007; Hikosaka, 2007) and appears to have arisen earlier in evolution (Robertson et al., 2014).

In mammals, afferent inputs to the basal ganglia arise from excitatory projection neurons of the neocortex, thalamus, amygdala, and hippocampus. Three primary classes of excitatory projection neurons in cortex provide input to basal ganglia: superficial intratelencephalic neurons of layer 2/3, deep intratelencephalic neurons of layer 5a, and collaterals from corticofugal projection neurons of layer 5b (Dudman & Gerfen, 2015). In mammals, the caudate and putamen (collectively referred to as the striatum in rodents due to its perforation by massive tracts of the internal capsule) receives all three types of cortical input. Deeper structures in the basal ganglia, such as the subthalamic nucleus and substantia nigra, receive corticofugal (layer 5b) input via collaterals but do not receive direct input from either intratelencephalic population. Compared to the subthalamic nucleus and substantia nigra, the striatum contains the greatest number of neurons by a few orders of magnitude (Hardman et al., 2002; Oorschot, 1996) As such, the striatum is often considered the primary input

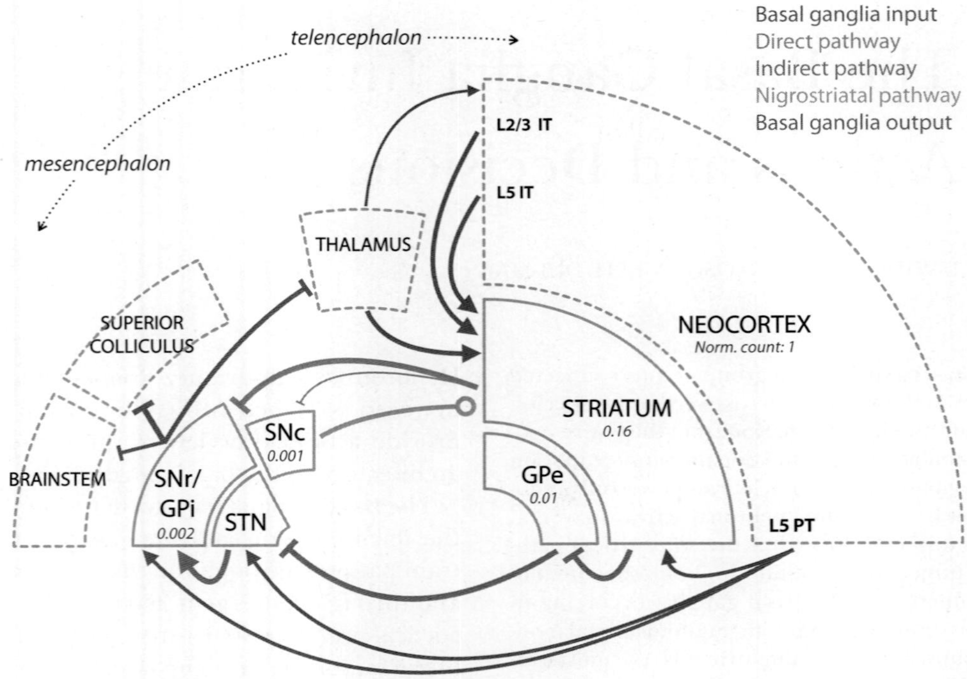

FIGURE 44.1 Functional anatomy of mammalian basal ganglia circuits. The organization of the major pathways of the dorsal basal ganglia is indicated by arrows for predominantly excitatory projections, lines for inhibitory projections, and an open circle for dopaminergic, modulatory projections. Detailed reviews of internal circuitry and topographical organization can be found in Dudman and Gerfen (2015). The weights of arrows reflect the rough density of projection. Numbers in italics reflect approximate normalized cell counts for several key nuclei normalized to the number of cortical inputs and derived from the best stereological counts taken in rats (Oorschot, 1996). *Abbreviations*: globus pallidus external, GPe; globus pallidus internal, GPi; intratelencephalic, IT; layer 5, L5; layer 2/3, L2/3; pyramidal tract, PT; substantia nigra pars compacta, SNc; substantia nigra pars reticulata, SNr; subthalamic nucleus, STN. (See color plate 49.)

nucleus of the basal ganglia and will be a focus of discussion in this chapter.

The striatum is composed of approximately 95% inhibitory projection neurons that fall into two distinct cell types distinguished by their neurochemical markers, dopamine receptor expression, neuropeptide expression, and efferent projections. In the dorsal striatum, these cell types and the functional pathways they constitute are generally referred to as the direct and indirect projection pathways to primary basal ganglia output nuclei (figure 44.1). Direct-projecting principal neurons of striatum, or spiny projection neurons (dSPNs), express D1-type dopamine receptors. Indirect projecting SPNs (iSPNs) express D2-type dopamine receptors. The output nuclei of the basal ganglia, the substantia nigra pars reticulata and the globus pallidus interna, are composed exclusively of GABAergic projection neurons with high baseline firing rates. dSPNs and iSPNs have opposite effects on activity in the output nuclei of the basal ganglia, with dSPNs reducing activity. Thus, dSPNs act to disinhibit downstream targets of basal ganglia output, whereas iSPNs act to enhance the inhibition of downstream targets.

Given this difference in the sign of the effects of dSPN and iSPN activity on basal ganglia output, it has long been suggested that the two pathways are antagonistic with respect to the control of movements (Albin, Young, & Penney, 1989; Mink, 1996). Consistent with this idea, the direct stimulation of dSPNs enhances locomotor behavior, whereas the stimulation of iSPNs suppresses it (Kravitz et al., 2010). However, the advent of methods that combine cell type identification with the recording of activity in behaving rodents (Cui et al., 2013; Markowitz et al., 2018; Parker et al., 2018) has revealed that dSPNs and iSPNs display largely concomitant activity. The apparent contradiction between these results might result from the inherent limitations of perturbation and correlative experiments. First, the behavioral effects triggered by the prolonged open-loop optogenetic stimulation of SPNs are likely caused by the supraphysiological modulation of basal ganglia output signals. Second, the algorithmic meaning of patterns of activity that correlate with certain aspects of behavior can remain obscure, even when complemented with perturbation experiments (Krakauer, Ghazanfar, Gomez-Marin, MacIver, & Poeppel, 2017). Insight into the function of the striatal circuits (and, more generally, of the basal ganglia) may be gained through a critical examination and comparison of the behavioral alterations associated with various forms of its dysfunction, in daily life activities or experimental setups.

The Role of the Basal Ganglia in Motor Control

In this section we are interested in the contribution of the basal ganglia in the performance of movements or actions that are already part of the subject's or animal's repertoire, because those movements have been learned much earlier during development or through instrumental conditioning as adults. The association of the basal ganglia with motor control is historically linked to the description of the behavioral deficits seen in Parkinson's disease (PD); deficits that were first reported by James Parkinson (1817) in his famous *Essay on the Shaking Palsy*. Three main deficits are considered typical of PD: (1) resting tremor, which is maybe the most apparent symptom and typically affects one hand or the head at disease onset; 2) rigidity—stiff muscles; and 3) bradykinesia, or slowness of movement. In addition, people with PD often display secondary symptoms such as a shuffling gait (also called *festination*), micrographia, and speech difficulties. In 1919, Tetriakoff performed postmortem examination on the brains of patients with PD and was the first to notice a systematic marked loss of pigmented nigral neurons. This discovery, largely overlooked at the time (Lees, Selikhova, Andrade, & Duyckaerts, 2008), was later explained by the fact that the aforementioned motor deficits are caused by dopamine depletion in the striatum following the progressive degeneration of dopaminergic nigrostriatal neurons (Hornykiewicz, 2006). The presence of resting tremor and rigidity in PD, in association with the erroneous assumption that this disease can be understood as a loss of function of the basal ganglia, led to the prevalent idea that the basal ganglia contribute to motor coordination and the smooth execution of movements. If this was the case, further perturbation or lesioning of the basal ganglia should worsen movement deficits. On the contrary, however, surgical interventions in PD that reduce the activity of the basal ganglia output nuclei, such as deep brain stimulation of the subthalamic nucleus or the surgical removal of the globus pallidus, dramatically suppress resting tremor and rigidity (Munhoz, Cerasa, & Okun, 2014; Obeso et al., 2009). This suggests that aberrant gain of function may account for some symptoms of PD. Additional evidence will be presented below showing that motor execution is indeed preserved in PD.

Vigor as a Common Motor Deficit in Patients Suffering from Basal Ganglia–Related Disorders

One fruitful approach to understanding the contribution of the basal ganglia to the performance of overlearned actions is to search for a common deficit across various pathologies of the basal ganglia such as PD, Huntington's disease (HD), and stroke-induced lesions. Individuals with bilateral lesions of the striatopallidal complex are characterized by an extreme lack of spontaneous mental activity (*psychic akinesia*) that can be reversed upon external stimulation (Laplane & Dubois, 2001). In addition, when tested with an incentivized handgrip squeezing task, these individuals display a complete incapacity to scale the vigor of their squeeze with the magnitude of a monetary reward (figure 44.2*A*; Schmidt et al., 2008). Importantly, their capacity to squeeze with increasing strength was identical to control patients in an instructed version of the task. These findings, along with the fact that patients do not display tremor or rigidity, suggest that the basal ganglia are not required for the execution of movements per se but for their invigoration. Interestingly, PD patients showed similar deficits (figure 44.2*A–C*). Thus, dopamine likely provides the motivational signal that is translated into invigorated output from the basal ganglia. In support of this notion, micrographia, which can be considered as a deficit in movement extent, is present in both PD patients (Oliveira, Gurd, Nixon, Marshall, & Passingham, 1997) and patients with focal striatal lesions (Barbarulo, Grossi, Merola, Conson, & Trojano, 2007). In addition, bradykinesia (slower and smaller movements) in PD patients is reversed by dopaminergic treatment. Why do individuals with PD or striatal lesions tend to make slower, shorter, and less vigorous movements? Mazzoni, Hristova, and Krakauer (2007) compared the performance of PD patients and healthy subjects in a self-paced speed-accuracy tradeoff (SAT) task comprising conditions of increasing difficulty (figure 44.2*D*). The results indicated that although PD patients were reluctant to perform fast movements, their accuracy was intact when they were forced to make them. Additional analysis suggested that bradykinesia in PD may arise from an implicit reluctance to perform fast movements due to an overestimation of their energetic cost. This hypothesis may also account for the often reported, but understudied, paradoxical kinesia, in which normal locomotion or movements are observed in PD patients in response to life-threatening events or in contexts in which additional sensory cues are provided. Improvement in movement speed/extent by external cues or by urgency occurs in

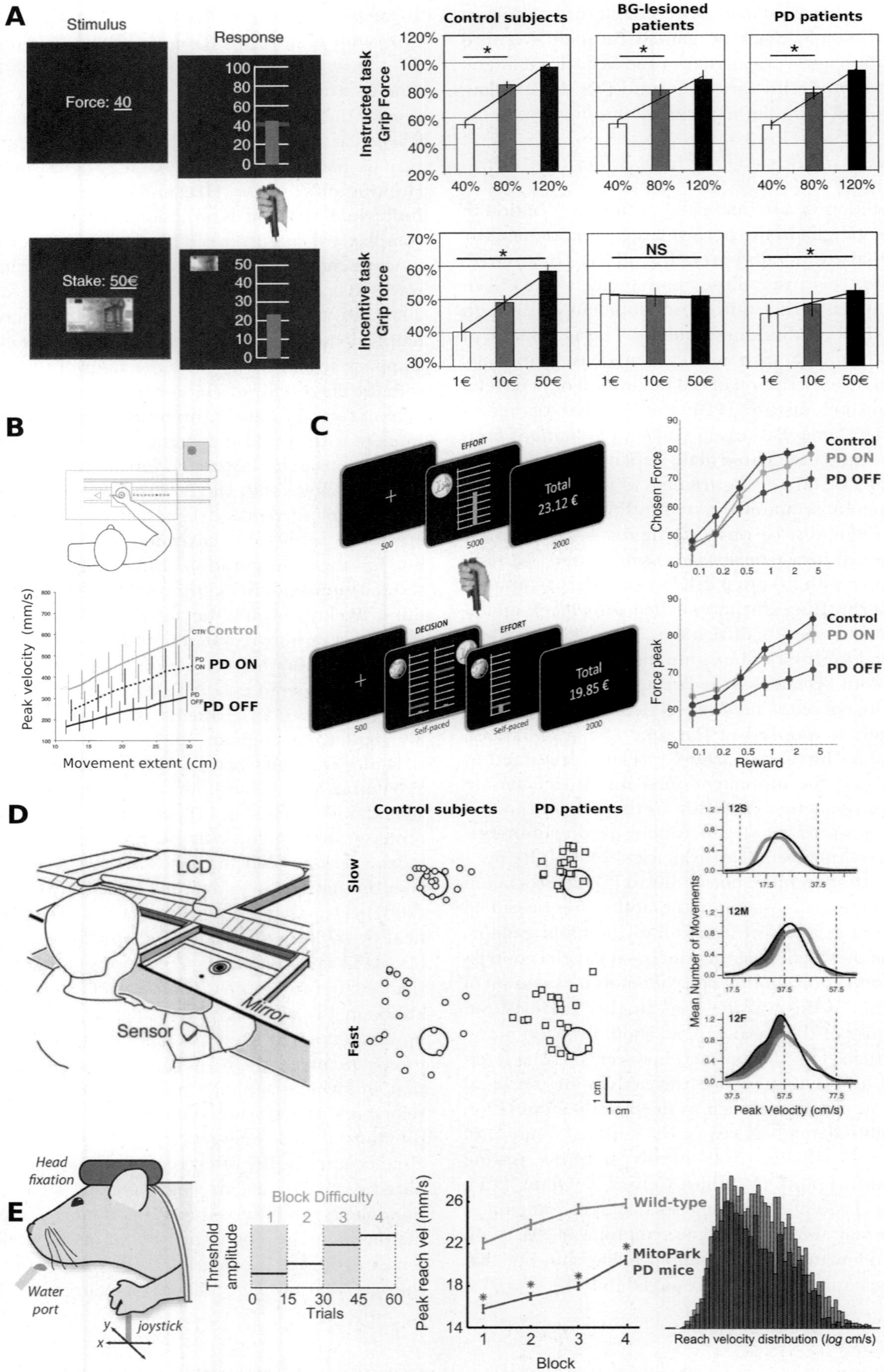

A
Stimulus
Response
Force: 40
Stake: 50€
Control subjects
BG-lesioned patients
PD patients
Instructed task Grip Force
Incentive task Grip force
NS
40% 80% 120%
1€ 10€ 50€
B
Peak velocity (mm/s)
Movement extent (cm)
Control
PD ON
PD OFF
C
EFFORT
DECISION
Total 23.12 €
Total 19.85 €
Self-paced
Chosen Force
Force peak
Reward
Control
PD ON
PD OFF
D
LCD
Mirror
Sensor
Control subjects
PD patients
Slow
Fast
1 cm
Mean Number of Movements
Peak Velocity (cm/s)
E
Head fixation
Water port
joystick
Block Difficulty
Threshold amplitude
Trials
Peak reach vel (mm/s)
Block
Wild-type
MitoPark PD mice
Reach velocity distribution (log cm/s)

PD patients, age-matched control subjects (Ballanger et al., 2006), and patients with striatopallidal lesions (see above). This supports the view that damaged basal ganglia do not lead to a fundamental deficit in movement control but a modulatory deficit due to blunted motivation or perhaps an implicit overestimation of the energetical cost of moving (Baraduc, Thobois, Gan, Broussolle, & Desmurget, 2013)

Reduced Vigor Is Also Present in Experimentally Induced Disorders of the Basal Ganglia

The idea that an important function of the dopaminergic projection to the striatum is to scale movement extent and speed according to motivational factors is supported by pharmacological manipulations of the dopaminergic system in the ventral striatum that alter the amount of work animals are willing to exert to obtain rewards (Salamone, Correa, Farrar, & Mingote, 2007). Panigrahi et al. (2015) reported that MitoPark mice, a model of PD, produce less vigorous movements (see also figure 44.2*E*). Recordings of spiking activity in the dorsal striatum of control mice showed that such activity in a majority of neurons was modulated at an early movement phase and correlated with movement speed. Such modulations, compatible with a representation of movement vigor, were markedly altered in MitoPark mice. Importantly, both the behavioral and neuronal abnormalities were reversed by dopamine replacement therapy. A set of studies investigating reaching movements in nonhuman primates showed that inactivation/lesioning of the globus pallidus consistently slowed down movements and induced hypometria but preserved reaction time and movement accuracy (Desmurget & Turner, 2008; Horak & Anderson, 1984), consistent with a selective contribution of the basal ganglia to movement vigor. The vigor viewpoint is also consistent with results obtained in rats performing a completely different form of motor sequence using a motorized treadmill that required the animals to learn to modulate their running speed (Rueda-Orozco & Robbe, 2015). Inactivation of the sensorimotor striatum did not affect the overall structure of the motor sequence but increased variability in the speed of execution. The spiking activity of striatal neurons correlated on a moment-to-moment basis with running speed and the animal's position on the treadmill, suggesting a role in representing and modulating the kinematic parameters of the motor sequence.

Altogether, the studies reviewed in this section suggest that the basal ganglia do not generate motor commands to muscles but instead modulate the vigor of ongoing movements (figure 44.2), consistent with a role of dopamine modulation on striatal-spiking activity . An important question is raised by the aforementioned studies: What are the respective contributions of dopamine acting on the dorsal and ventral striatum? One possibility is that the modulation of neuronal activity in these regions contributes to two parameters that usually covary: response vigor (How many times do I knock on the door to be heard?) and movement vigor (How hard is each knock?). Due to space constraints, this chapter is largely focused on the dorsal striatum, but we refer interested readers to a recent review on the possible functional specificity of neuronal processes occurring in the dorsal and ventral striatum (Hart, Leung, & Balleine, 2014). Finally, while the focus of this section is on movement, the fact that dopamine

FIGURE 44.2 Impaired movement vigor in basal ganglia–related disorders. *A*, Individuals with bilateral striatopallidal lesions lack the capacity to scale movement vigor with motivation (Schmidt et al., 2008). *Upper left*, The instructed task. Subjects were explicitly instructed what force to exert. *Lower left*, The incentive task. Subjects were shown a monetary incentive as an image. This was the trigger to exert a force on a hand grip. *Upper right*, Lesioned individuals and PD patients displayed moderate alteration of instructed grip force. *Lower right*, Lesioned individuals lacked any modulation of grip force by incentive cues, while PD patients showed reduced grip force. *B*, PD slows motor performance (Baraduc et al., 2013). *Top*, Experimental setup. Subjects held the handle and made linear movements toward the targets. *Bottom*, The relationship between movement amplitude and speed is altered in PD patients. Reduced vigor is less in PD patients with subthalamic nucleus deep brain stimulation (PD ON) than without (PD OFF). *C*, PD impairs the capacity to scale movement vigor and motivation (Le Bouc et al., 2016). *Upper left*, The incentive force task. After a fixation cross, subjects were shown the monetary incentive as a coin image (€0.1, €0.2, €0.5, €1, €2, €5). This was the trigger to exert a force on a hand grip. *Lower left*, The binary choice task. After a fixation cross, subjects were shown two options side by side, each corresponding to a potential monetary reward (*coin image*) associated with a required force level (*orange bar*). *Right*, Forces exerted by control and PD patients with (PD ON) or without (PD OFF) dopaminergic medication in response to cues in the incentive (*top*) and choice (*bottom*) tasks. *D*, PD slows reach movement velocity without altering the speed-accuracy trade-off (Mazzoni, Hristova, & Krakauer, 2007). *Left*, Experimental setup. *Middle*, Movement end points (*small circle*) relative to target center (*big circle*) for valid trials with slow (*up*) or fast (*down*) speed instruction. *Right*, Distributions of movement speeds for slow-, medium-, and fast-speed instructions revealed a preference for slow movements in PD patients. *E*, Mice models of PD display bradykinesia in an effort-based joystick task. *Left*, Experimental setup and task design. *Right*, Speed of reaches versus difficulty level and speed distributions in wild-type and PD mice.

depletion in mice is associated with the altered processing of sensory information in the striatum (Ketzef et al., 2017) suggests that the role of dopamine may extend beyond vigor control or that vigor is not only driven by motivation. In a constantly changing world and body, the role of the dorsal striatum in representing movements and their associated sensory consequences, coupled with dopaminergic teaching signals, could be important to maintain and update previously learned actions or to develop an adaptive repertoire of actions (Dudman & Krakauer, 2016; Robbe, 2018).

Basal Ganglia Contribution to Motor Learning

Neurons in the substantia nigra compacta and ventral tegmental area (VTA) are well known to generate transient signals thought to be critical for reinforcement learning (Schultz, Dayan, & Montague, 1997). The processing of such signals by the striatum, which also integrates multimodal information, mainly from the cortex and the thalamus, is believed to be critical to motor learning. We will start this section by defining motor learning as the execution of movements with increased precision and accuracy (motor acuity) and at shorter latency (Krakauer, Hadjiosif, Xu, Wong, & Haith, 2018). It is important to distinguish motor learning from the process by which humans and animals learn to select the "right" actions in a given context. Motor sequence tasks are based on the execution of one or several ordered movements according to specific rules. The elementary movements are usually easy (i.e., already part of the subject's motor repertoire), and the motor system is simply used to read out higher-order cognitive decisions that constitute the action selection process. In the field of animal operant conditioning, learning can take several weeks but, typically, this duration reflects the time it takes an animal to discover rules the experimenter could not verbally explain. At the end of this process, animals have learned to select movements (and sometimes their speed), but motor learning per se may not have occurred. Before conditioning, a rodent will not know how to press a lever or run in a particular manner in a maze. But it will often succeed in this type of task by assembling previously mastered elementary movements into an adaptive motor sequence. Importantly, a reduction in reaction and movement time in motor sequences following extensive practice has often been assumed to reflect implicit motor learning. However, it was recently shown in humans that such performance improvements can be driven by a gain in explicit knowledge acting on vigor (Wong, Lindquist, Haith, & Krakauer, 2015). A more suitable strategy is to use tasks that require unnatural movements and/or focus on motor accuracy (such as reaching a small target, performing dexterous movements to grab an object, or running on a narrow track) and examine the animal's (or subject's) capacity to execute these movements faster and faster while maintaining or improving accuracy. We will see in this section that, surprisingly, most studies investigating the role of the basal ganglia in motor learning have not done this.

Motor Learning and the Basal Ganglia in Humans

Despite the aforementioned limitations of motor sequence tasks, they have been widely used to examine the learning capacity of PD patients, with contradictory results (for a review, see Ruitenberg, Duthoo, Santens, Notebaert, & Abrahamse, 2015). In one study (Wilkinson, Khan, & Jahanshahi, 2009), subjects had to respond with four fingers of the right hand. Responding with independent movements of all four fingers (especially the ring and little finger) is atypical and may require some motor dexterity when trying to respond as quickly as possible. PD patients performed above chance level but failed to improve their performance across training blocks, compared to control subjects. Unfortunately, the quality of the finger movements was not quantified, leaving open the possibility that the longer reaction times were caused by vigor deficits (performance confound). Very few studies with human participants have directly examined how basal ganglia dysfunctions affect learning in tasks requiring motor acuity. One study compared the performance of HD patients and aged-match healthy subjects in a pursuit rotary task using increasing rotation speeds of the target. To account for group differences in initial performance, the rotation speed of the target was adjusted. HD patients exhibited a marked deficit compared to control participants, as well as patients with Alzheimer's disease (Gabrieli, 1997; Heindel, Butters, & Salmon, 1988). PD patients also displayed slower improvements in this task (Harrington, Haaland, Yeo, & Marder, 1990). The problem here is that normalization procedures do not solve the issue of how to properly compare motor learning across groups with different baselines. Finally, a more recent study using a reaching to target task under increasing speed constraints showed that HD patients failed to increase both the accuracy and speed of their reaching movements (i.e., shift their SAT function), which is the hallmark of motor skill learning in control subjects (Shabbott et al., 2013). Neurophysiological data that could provide a clue for how the basal ganglia might contribute to true motor learning are scarce. Activation of the caudate and putamen was reported during the dexterous four-fingers version of the serial reaction time (Jenkins, Brooks, Nixon, Frackowiak, & Passingham, 1994; Jueptner, Frith, Brooks, Frackowiak, & Passingham, 1997; Lehéricy et al., 2005). However, here, too, movement quality was not directly measured, thus additional studies will be required to delineate the contribution of the basal ganglia to motor skill learning.

Motor Learning and the Basal Ganglia: Insights from Studies in Animals

Similar to human studies, most work in other mammals has examined the contribution of the basal ganglia in the operant-conditioning framework, leaving improvements in motor acuity largely unexplored. In monkeys performing a sequential push-button task, inactivation of the anterior caudate and putamen slowed down the acquisition of a new sequence (Miyachi, Hikosaka, Miyashita, Kárádi, & Rand, 1997). The performance of previously learned sequences was not impaired, but because the movements themselves were simple, it is likely that the shorter response times in control animals did not result from better movements but from better knowledge acting through vigor (Wong et al., 2015). Mice with altered corticostriatal synaptic transmission displayed limited learning in a lever-press motor sequence task (Jin & Costa, 2010), and rats with a lesion of the striatum were impaired in learning navigational tasks (DeCoteau & Kesner, 2000; Moussa, Poucet, Amalric, & Sargolini, 2011). In addition, mice with lesions of the sensorimotor or associative striatum failed to properly learn to increase their running speed in a rotarod task (Yin et al., 2009). Again, in all these studies success did not necessarily require motor skill learning per se, and movement quality was not quantified. Whether the learning deficits were confounded by vigor deficits (performance confound) was not directly investigated. An interesting dissociation between neural bases of action learning and performance has been revealed by the study of songbirds that learn highly stereotyped motifs of sounds at specific frequencies interleaved with brief periods of silence. A lesion of Area X (a brain region that has homology with the mammalian striatum), or altering its function using RNA interference, prevented the development of songs (Haesler et al., 2007; Scharff & Nottebohm, 1991). Remarkably, this brain region appears to be dispensable for song production after learning (Bottjer, Miesner, & Arnold, 1984). Moreover, continuously blocking the anterior forebrain pathway (AFP), a series of nuclei equivalent to the basal ganglia-thalamocortical loop) during learning prevented the expression but not the learning of the song (Charlesworth, Warren, & Brainard, 2012). While these studies highlight a role of the avian basal ganglia in learning how to sing (using variation in intensity, speed, pitch ...), it is not clear whether such capacity can be related to the concept of speed/accuracy trade-off pertaining to motor skill learning. But they provide additional evidence for a role of the striatum in the acquisition of adaptive actions following operant conditioning (Packard & Knowlton, 2002). In addition to learning what action to execute in a given context, the striatum's role in selecting the kinematic parameters of actions is supported by a study in which brief closed-loop stimulation of dSPNs and iSPNs was delivered when mice pressed a bar at specified speeds (fast or slow presses; Yttri & Dudman, 2016). The stimulation of dSPNs induced a progressive sustained preference for the speeds that triggered stimulation, independently of whether that speed was fast or slow. In contrast, mice tended to avoid the range of speeds that triggered the stimulation of iSPNs. These effects persisted after cessation of the stimulation, suggesting that the modulation of both dSPNs and iSPNs contributes to the trial-and-error learning process that shapes the kinematic content of actions. Further studies will be required to investigate if such a process is at play during motor skill learning and/or can be generalized to learn to select adaptive actions (Kravitz, Tye, & Kreitzer, 2012)

In conclusion, running quickly versus slowly, or reaching left versus right, may not require any actual motor skill learning but rather learning to select such actions in the context provided by the experimenter. Very few studies have directly investigated the basal ganglia's contribution to true motor skill learning. In animals, there are as of yet no convincing experiments supporting a role of the basal ganglia in the simultaneous improvement in movement speed and accuracy. In rats, it has been reported that the dopaminergic projection from the VTA to the primary motor cortex is necessary to learn a precision forelimb-reaching task but is dispensable for skilled performance after learning (Hosp, Pekanovic, Rioult-Pedotti, & Luft, 2011). Thus, the striatum may not be the primary site where dopaminergic teaching signals act to produce lasting changes in neuronal activity that contribute to motor skill acquisition.

Basal Ganglia Function in Cognition

The striatum integrates diverse inputs from associative areas of the neocortex, from the intralaminar nuclei of the thalamus, and from the limbic areas. As a result, there has long been interest in the role of basal ganglia in more abstract, or cognitive, aspects of the control of purposive actions. For example, while the sensorimotor striatum appears to be involved in the learning and control of complex movements (see the section on reduced vigor), it has also been proposed to contribute to the selection among competing actions (Alexander & Crutcher, 1990; Gurney, Humphries, Wood, Prescott, & Redgrave, 2004; Lo & Wang, 2006; Mink, 1996) according to expected outcomes (Bogacz & Gurney, 2007; Dayan & Balleine, 2002; Niv, Joel, & Dayan, 2006). In addition, dorsal basal ganglia circuits have been implicated in aspects of decision-making that may reflect more abstract aspects of vigor—motivation (Averbeck & Costa, 2017), selective attention (Krauzlis, Bollimunta, Arcizet, & Wang, 2014), the urgency to act (Thura & Cisek, 2017), the representation of elapsed time (Paton & Buonomano, 2018), and the amount of integrated evidence required to render a decision (Bogacz, Wagenmakers, Forstmann, & Nieuwenhuis, 2010). Here we consider these functions from the perspective of the role of the basal ganglia in the control of movement and their anatomical organization.

Action Selection and Habits

The dorsolateral striatum, owing to its extensive inputs from primary motor and somatosensory cortical regions, is thought to be particularly important for movements and actions acquired through extensive training (Barnes, Kubota, Hu, Jin, & Graybiel, 2005; Jin, Tecuapetla, & Costa, 2014; Rueda-Orozco & Robbe, 2015; Thorn, Atallah, Howe, & Graybiel, 2010; Tricomi, Balleine, & O'Doherty, 2009; Yin et al., 2009). Such actions are relatively insensitive to the devaluation of the consequences of their execution (e.g. Rueda-Orozco & Robbe, 2015) and thus have been described as "habits" (Balleine & Dickinson, 1998). A large body of work, primarily in rodents, has established that extensively trained actions are both relatively slower to be extinguished following experimental changes in the contingency of outcomes and are preferentially sensitive to lesions of the dorsolateral striatum (Yin & Knowlton, 2006).

These behavioral observations are complemented by influential studies mapping the responsiveness of individual, putative SPNs in the dorsolateral striatum and/or putamen to movements of or the physical manipulation of specific body parts. In rats, mice, and primates, many neurons respond preferentially or even exclusively to passive and/or active movements of individual body parts (reviewed in Robbe, 2018). Moreover, response properties tend to cluster in small groups of a couple hundred microns in a "salt-and-pepper" arrangement. At the level of basal ganglia output, there also appears to be differential activity associated with distinct actions (Anderson & Turner, 1991; Barter et al., 2015; Gulley, Kosobud, & Rebec, 2002; Hikosaka & Wurtz, 1983). The salt-and-pepper somatotopic representation in the dorsolateral striatum and putamen has been a critical element for many models of basal ganglia. Alexander, DeLong, and Strick (1986) proposed that the basal ganglia were "composed of multiple, parallel subcircuits or channels concerned with movement of individual body parts." In the context of computational models, this has been interpreted as supporting a parallel architecture in which each action an animal performs is represented by a distinct neural population (e.g., Bogacz & Gurney, 2007). In such models, the basal ganglia (the striatum, in particular) play a critical role in the selection of actions by learning a relative "preference" for each action, expressed as a weight within each discrete neural population. While the concept of the parallel flow of information has been influential to support the action selection model, the extent to which corticostriatal channels representing distinct sensory or motor information are strictly parallel (independent) remains unclear (Reig & Silberberg, 2014). In addition, a distinct representation of distinct movements is not unique to action selection models and is a feature of other models, such as the regulation of vigor (Yttri & Dudman, 2018). It is also important to recall that the basal ganglia have properties that both serve to make representations of specific movements distinct (e.g., topography of afferent projections) and mix representations (e.g., intrinsic recurrent connectivity, somatotopic input to dorsomedial striatum). Outstanding questions turn on elaborating models that encompass these diverse aspects of functional organization in the basal ganglia.

Decision-Making

Numerous studies have implicated basal ganglia function (and pathological dysfunction) in decision-making processes that may be divorced from action selection. Animals are often confronted with the need to render a decision in the face of variable and uncertain information. In the context of the laboratory, this has often been studied in paradigms in which a subject needs to use ambiguous sensory information to make a decision given a well-known sensorimotor mapping. For example, if visual stimulus motion is to the right, the subject must reach over and touch a button to the right. Since the sensory evidence of motion can be manipulated by the experimenter, the difficulty of the task can be parametrically varied (Gold & Shadlen, 2007). Another important modification of such tasks is the ability to manipulate the demands for how quickly a decision must be rendered (Kiani, Hanks, & Shadlen, 2008). This leads to a well-known and fundamental trade-off between the speed of making a decision and the time required to integrate sensory evidence in favor of a decision, a form of an SAT (Heitz, 2014). SATs in the context of decision-making have often been argued to be computationally equivalent to the speed/accuracy trade-off characteristic of targeted movements (see the section on motor learning). However, PD patients are capable of producing movements as fast and accurate as those of control subjects, under the right conditions (Mazzoni, Hristova, & Krakauer, 2007). This has led to the proposal that the basal ganglia might play a role in determining the slope of the SAT (Herz et al., 2017; Manohar et al., 2015; Perugini, Ditterich, & Basso, 2016). Seen in this way, in the context of decision-making, the SAT could have a similar dependency on basal ganglia function as does the control of vigor. The altered performance in perceptual decision-making tasks seen in PD patients would thus result from a basal ganglia-dependent deficit in the integration of sensory evidence coupled with additional changes in the offset or the bound that do not correlate with PD progression and possibly may occur outside the basal ganglia (Perugini, Ditterich, Shaikh, Knowlton, & Basso, 2018).

The SAT, especially in the context of decision-making, is thought to be determined by dissociable computations (Bogacz et al., 2010; Ding & Gold, 2013; Heitz, 2014). A remarkably diverse set of behavioral observations can be explained by variants of models in which the integration of sensory evidence proceeds like a diffusion-to-bound process in which the strength of the sensory evidence biases the diffusion process, and boundary crossing determines the timing and identity of the decision (Gold & Shadlen, 2001). In these models, free parameters describe a prior bias in the integration, the height of the bound, the drift rate (a function of sensory evidence), and a "leak" in the integration process. Combinations of these parameters can be seen as spanning a space of models from those that yield accurate, deliberative decisions involving perfect integration (zero leaks) to those that yield more impulsive decisions resulting from a local running average of sensory evidence and a collapsing bound (Thura & Cisek, 2016). Distinguishing the regime in which a subject is operating in any given task/data set has been a subject of considerable debate.

While converging lines of evidence point to the basal ganglia's important role in modulating the SAT, there is relatively little consensus about which processes relevant to the SAT are most critically influenced by basal ganglia activity or disrupted when this structure is dysfunctional (Bogacz et al., 2010). In general, model-based analyses in human subjects point to a role for the basal ganglia in determining the integration starting point (offset; Perugini, Ditterich, & Basso, 2016); however, electrophysiological recordings and theoretical considerations have suggested a role for basal ganglia in multiple aspects, including evidence accumulation, setting prior biases in integration, and setting the boundary/threshold (Ding & Gold, 2010; Dunovan & Verstynen, 2016; Thura & Cisek, 2017; Yartsev, Hanks, Yoon, & Brody, 2018). Activity in cortical regions has generally been more closely associated with the integration process, although there is also evidence favoring a role in a cortical determinant of the bound (Hanks, Kiani, & Shadlen, 2014; Kiani, Hanks, & Shadlen, 2008). Finally, there is substantial evidence that the basal ganglia are critically involved in the representation of elapsed time (Paton & Buonomano, 2018), a feature that could suggest a more central role in time-dependent aspects (i.e., a collapsing integration bound).

Conclusion and Outlook

For an animal, survival is determined not only by its capacity to select the right action at the right time from an innate or acquired behavioral repertoire but also by how it executes voluntary actions. If a prey animal decides to run away from a predator but this decision is not executed correctly (e.g., through a vigorous escape), the benefit of the good decision (run versus fight versus freeze) can be lost. Natural selection presumably imposes a strong pressure to develop algorithm(s) endowing animals with the capacity to fine-tune the processes required for decision-making and the intensity of the response according to internal states (e.g., hunger) and external cues (e.g., genetically programmed or acquired knowledge about potential predators). In this chapter we have reviewed evidence in human and several other vertebrate species that support the view that the basal ganglia implement a low-level computation essential to scaling movement extent and speed according to extrinsic or intrinsic motivational factors (figure 44.2). In the formalism of David Marr, such a function must be supported by a biological algorithm (Krakauer et al., 2017) whose contours are only beginning to take shape (Yttri & Dudman, 2018). These computational ideas are consistent with basic features of the basal ganglia (figure 44.1) as a system that (1) integrates input carrying sensory, motor, emotional, and cognitive information; (2) has output that can modulate action execution, both via ascending and descending pathways; and (3) is composed of two distinct pathways whose activation has opposite effects on basal ganglia output activity and that are differentially modulated by dopamine.

The basal ganglia have also been proposed to contribute to higher-level cognitive functions such as decision-making. Here, we have (provocatively) argued that such contributions can often be understood as analogous to the role of the basal ganglia in controlling low-level kinematic parameters of learned actions. This claim is compatible with models that invoke functions such as urgency, effort, blocking initiated actions, monitoring, and vigor. It is also in agreement with anatomical considerations—for example, a substantial reduction in neuron number throughout the cortico-basal ganglia circuit and relatively weak recurrent and local connectivity (at least compared to neocortex). In contrast, models proposing that the basal ganglia play a critical role in action selection through the suppression of competing actions, selection of high-value actions, representation of the space of action values, or attentional selection would seem to call for higher dimensional representations difficult to reconcile with anatomical constraints and the frequent dissociation of effects on performance and selection in perturbation experiments (Turner & Desmurget, 2010). Nonetheless, recent work has highlighted the presence of recurrent connectivity within rodent basal ganglia (Brown, Pan, & Dudman, 2014). This raises the possibility that future work can identify how basal ganglia circuits reflect an evolutionarily early solution to harnessing the power of recurrent connectivity and dynamics for the control of behavior.

Acknowledgments

David Robbe's work is supported by a European Research Council grant (ERC-2013-CoG—615699_NeuroKinematics, D.R.). Joshua Dudman's work is funded through the Howard Hughes Medical Institute.

REFERENCES

Albin, R. L., Young, A. B., & Penney, J. B. (1989). The functional anatomy of basal ganglia disorders. *Trends in Neurosciences, 12*(10), 366–375.

Alexander, G. E., & Crutcher, M. D. (1990). Functional architecture of basal ganglia circuits: Neural substrates of parallel processing. *Trends in Neurosciences, 13*(7), 266–271.

Alexander, G. E., DeLong, M. R., & Strick, P. L. (1986). Parallel organization of functionally segregated circuits linking basal ganglia and cortex. *Annual Review of Neuroscience, 9*(1), 357–381. https://doi.org/10.1146/annurev.ne.09.030186.002041

Anderson, M. E., & Turner, R. S. (1991). A quantitative analysis of pallidal discharge during targeted reaching movement in the monkey. *Experimental Brain Research, 86*(3), 623–632.

Averbeck, B. B., & Costa, V. D. (2017). Motivational neural circuits underlying reinforcement learning. *Nature Neuroscience, 20*(4), 505–512. https://doi.org/10.1038/nn.4506

Ballanger, B., Thobois, S., Baraduc, P., Turner, R. S., Broussolle, E., & Desmurget, M. (2006). "Paradoxial kinesis" is not a hallmark of Parkinson's disease but a general property of the motor system. *Movement Disorders, 21*(9), 1490–1495. https://doi.org/10.1002/mds.20987

Balleine, B. W., & Dickinson, A. (1998). Goal-directed instrumental action: Contingency and incentive learning and their cortical substrates. *Neuropharmacology, 37*(4–5), 407–419.

Baraduc, P., Thobois, S., Gan, J., Broussolle, E., & Desmurget, M. (2013). A common optimization principle for motor execution in healthy subjects and parkinsonian patients. *Journal of Neuroscience, 33*(2), 665–677. https://doi.org/10.1523/JNEUROSCI.1482-12.2013

Barbarulo, A. M., Grossi, D., Merola, S., Conson, M., & Trojano, L. (2007). On the genesis of unilateral micrographia of the progressive type. *Neuropsychologia, 45*(8), 1685–1696. https://doi.org/10.1016/j.neuropsychologia.2007.01.002

Barnes, T. D., Kubota, Y., Hu, D., Jin, D. Z., & Graybiel, A. M. (2005). Activity of striatal neurons reflects dynamic encoding and recoding of procedural memories. *Nature, 437*(7062), 1158–1161. https://doi.org/10.1038/nature04053

Barter, J. W., Li, S., Sukharnikova, T., Rossi, M. A., Bartholomew, R. A., & Yin, H. H. (2015). Basal ganglia outputs map instantaneous position coordinates during behavior. *Journal of Neuroscience, 35*(6), 2703–2716. https://doi.org/10.1523/JNEUROSCI.3245-14.2015

Bogacz, R., & Gurney, K. (2007). The basal ganglia and cortex implement optimal decision making between alternative actions. *Neural Computation, 19*(2), 442–477. https://doi.org/10.1162/neco.2007.19.2.442

Bogacz, R., Wagenmakers, E. J., Forstmann, B. U., & Nieuwenhuis, S. (2010). The neural basis of the speed-accuracy tradeoff. *Trends in Neurosciences, 33*(1), 10–16. https://doi.org/10.1016/j.tins.2009.09.002

Bottjer, S. W., Miesner, E. A., & Arnold, A. P. (1984). Forebrain lesions disrupt development but not maintenance of song in passerine birds. *Science, 224*(4651), 901–903.

Brown, J., Pan, W. X., & Dudman, J. T. (2014). The inhibitory microcircuit of the substantia nigra provides feedback gain control of the basal ganglia output. *eLife, 3*, e02397. https://doi.org/10.7554/eLife.02397

Cebrián, C., Parent, A., & Prensa, L. (2005). Patterns of axonal branching of neurons of the substantia nigra pars reticulata and pars lateralis in the rat. *Journal of Comparative Neurology, 492*(3), 349–369.

Charlesworth, J. D., Warren, T. L., & Brainard, M. S. (2012). Covert skill learning in a cortical-basal ganglia circuit. *Nature, 486*(7402), 251–255. https://doi.org/10.1038/nature11078

Cui, G., Jun, S. B., Jin, X., Pham, M. D., Vogel, S. S., Lovinger, D. M., & Costa, R. M. (2013). Concurrent activation of striatal direct and indirect pathways during action initiation. *Nature, 494*(7436), 238–242. https://doi.org/10.1038/nature11846

Dayan, P., & Balleine, B. W. (2002). Reward, motivation, and reinforcement learning. *Neuron, 36*(2), 285–298.

DeCoteau, W. E., & Kesner, R. P. (2000). A double dissociation between the rat hippocampus and medial caudoputamen in processing two forms of knowledge. *Behavioral Neuroscience, 114*(6), 1096–1108. https://doi.org/10.1037/0735-7044.114.6.1096

Deniau, J.-M., Mailly, P., Maurice, N., & Charpier, S. (2007). The pars reticulata of the substantia nigra: A window to basal ganglia output. *Progress in Brain Research, 160*, 151–172. https://doi.org/10.1016/S0079-6123(06)60009-5

Desmurget, M., & Turner, R. S. (2008). Testing basal ganglia motor functions through reversible inactivations in the posterior internal globus pallidus. *Journal of Neurophysiology, 99*(3), 1057–1076. https://doi.org/10.1152/jn.01010.2007

Ding, L., & Gold, J. I. (2010). Caudate encodes multiple computations for perceptual decisions. *Journal of Neuroscience, 30*(47), 15747–15759. https://doi.org/10.1523/JNEUROSCI.2894-10.2010

Ding, L., & Gold, J. I. (2013). The basal ganglia's contributions to perceptual decision making. *Neuron, 79*(4), 640–649. https://doi.org/10.1016/j.neuron.2013.07.042

Dudman, J. T., & Gerfen, C. R. (2015). The basal ganglia. In G. Paxinos (Ed.), *The rat nervous system* (4th ed., pp. 391–440). New York: Elsevier. https://doi.org/10.1016/B978-0-12-374245-2.00017-6

Dudman, J. T., & Krakauer, J. W. W. (2016). The basal ganglia: From motor commands to the control of vigor. *Current Opinion in Neurobiology, 37*, 158–166. https://doi.org/10.1016/j.conb.2016.02.005

Dunovan, K., & Verstynen, T. (2016). Believer-skeptic meets actor-critic: Rethinking the role of basal ganglia pathways during decision-making and reinforcement learning. *Frontiers in Neuroscience, 10*, 106. https://doi.org/10.3389/fnins.2016.00106

Gabrieli, J. D. E. (1997). Intact mirror-tracing and impaired rotary-pursuit skill learning in patients with Huntington's disease: Evidence for dissociable memory systems in skill learning. *Neuropsychology, 11*(2), 272–281. https://doi.org/10.1037/0894-4105.11.2.272

Gold, J. I., & Shadlen, M. N. (2001). Neural computations that underlie decisions about sensory stimuli. *Trends in Cognitive Sciences, 5*(1), 10–16.

Gold, J. I., & Shadlen, M. N. (2007). The neural basis of decision making. *Annual Review of Neuroscience, 30*, 535–574. https://doi.org/10.1146/annurev.neuro.29.051605.113038

Grillner, S., Robertson, B., & Stephenson-Jones, M. (2013). The evolutionary origin of the vertebrate basal ganglia and its role in action selection. *Journal of Physiology, 591*(22), 5425–5431. https://doi.org/10.1113/jphysiol.2012.246660

Gulley, J. M., Kosobud, A. E. K., & Rebec, G. V. (2002). Behavior-related modulation of substantia nigra pars reticulata neurons in rats performing a conditioned reinforcement task. *Neuroscience, 111*(2), 337–349.

Gurney, K. N., Humphries, M., Wood, R., Prescott, T. J., & Redgrave, P. (2004). Testing computational hypotheses of brain systems function: A case study with the basal ganglia. *Network, 15*(4), 263–290.

Haesler, S., Rochefort, C., Georgi, B., Licznerski, P., Osten, P., & Scharff, C. (2007). Incomplete and inaccurate vocal imitation after knockdown of FOXP2 in songbird basal ganglia nucleus area X. *PLoS Biology, 5*(12), 2885–2897. https://doi.org/10.1371/journal.pbio.0050321

Hanks, T., Kiani, R., & Shadlen, M. N. (2014). A neural mechanism of speed-accuracy tradeoff in macaque area LIP. *eLife, 3*. https://doi.org/10.7554/eLife.02260

Hardman, C. D., Henderson, J. M., Finkelstein, D. I., Horne, M. K., Paxinos, G., & Halliday, G. M. (2002). Comparison of the basal ganglia in rats, marmosets, macaques, baboons, and humans: Volume and neuronal number for the output, internal relay, and striatal modulating nuclei. *Journal of Comparative Neurology, 445*(3), 238–255.

Harrington, D. L., Haaland, K. Y., Yeo, R. A., & Marder, E. (1990). Procedural memory in Parkinson's disease: Impaired motor but not visuoperceptual learning. *Journal of Clinical and Experimental Neuropsychology, 12*(2), 323–339. https://doi.org/10.1080/01688639008400978

Hart, G., Leung, B. K., & Balleine, B. W. (2014). Dorsal and ventral streams: The distinct role of striatal subregions in the acquisition and performance of goal-directed actions. *Neurobiology of Learning and Memory, 108*, 104–118. https://doi.org/10.1016/j.nlm.2013.11.003

Heilbronner, S. R., Rodriguez-Romaguera, J., Quirk, G. J., Groenewegen, H. J., & Haber, S. N. (2016). Circuit-based corticostriatal homologies between rat and primate. *Biological Psychiatry, 80*(7), 509–521. https://doi.org/10.1016/j.biopsych.2016.05.012.

Heindel, W. C., Butters, N., & Salmon, D. P. (1988). Impaired learning of a motor skill in patients with Huntington's disease. *Behavioral Neuroscience, 102*(1), 141–147. https://doi.org/10.1037/0735-7044.102.1.141

Heitz, R. P. (2014). The speed-accuracy tradeoff: History, physiology, methodology, and behavior. *Frontiers in Neuroscience, 8*, 150. https://doi.org/10.3389/fnins.2014.00150

Herz, D. M., Tan, H., Brittain, J. S., Fischer, P., Cheeran, B., Green, A. L., … Brown, P. (2017). Distinct mechanisms mediate speed-accuracy adjustments in cortico-subthalamic networks. *eLife, 6*, 1–25. https://doi.org/10.7554/eLife.21481

Hikosaka, O. (2007). GABAergic output of the basal ganglia. *Progress in Brain Research, 160*, 209–226. https://doi.org/10.1016/S0079-6123(06)60012-5

Hikosaka, O., & Wurtz, R. H. (1983). Visual and oculomotor functions of monkey substantia nigra pars reticulata. II. Visual responses related to fixation of gaze. *Journal of Neurophysiology, 49*(5), 1254–1267.

Hintiryan, H., Foster, N. N., Bowman, I., Bay, M., Song, M. Y., Gou, L., … Dong, H.-W. W. (2016). The mouse cortico-striatal projectome. *Nature Neuroscience, 19*(8), 1100–1114. https://doi.org/10.1038/nn.4332

Horak, F. B., & Anderson, M. E. (1984). Influence of globus pallidus on arm movements in monkeys. I. Effects of kainic acid-induced lesions. *Journal of Neurophysiology, 52*(2), 290–304. https://doi.org/10.1152/jn.1984.52.2.290

Hornykiewicz, O. (2006). *The discovery of dopamine deficiency in the parkinsonian brain.* Originally printed in 1957. *Journal of Neural Transmission. Supplementum*, (70), *9–15.*

Hosp, J. A., Pekanovic, A., Rioult-Pedotti, M. S., & Luft, A. R. (2011). Dopaminergic projections from midbrain to primary motor cortex mediate motor skill learning. *Journal of Neuroscience, 31*(7), 2481–2487. https://doi.org/10.1523/JNEUROSCI.5411-10.2011

Hunnicutt, B. J., Jongbloets, B. C., Birdsong, W. T., Gertz, K. J., Zhong, H., & Mao, T. (2016). A comprehensive excitatory input map of the striatum reveals novel functional organization. *eLife, 5*, e19103. https://doi.org/10.7554/eLife.19103

Jenkins, I. H., Brooks, D. J., Nixon, P. D., Frackowiak, R. S., & Passingham, R. E. (1994). Motor sequence learning: A study with positron emission tomography. *Journal of Neuroscience, 14*(June), 3775–3790.

Jin, X., & Costa, R. M. (2010). Start/stop signals emerge in nigrostriatal circuits during sequence learning. *Nature, 466*(7305), 457–462. https://doi.org/10.1038/nature09263

Jin, X., Tecuapetla, F., & Costa, R. M. (2014). Basal ganglia subcircuits distinctively encode the parsing and concatenation of action sequences. *Nature Neuroscience, 17*(3), 423–430. https://doi.org/10.1038/nn.3632

Jueptner, M., Frith, C. D., Brooks, D. J., Frackowiak, R. S., & Passingham, R. E. (1997). Anatomy of motor learning. II. Subcortical structures and learning by trial and error. *Journal of Neurophysiology, 77*(3), 1325–1337. https://doi.org/10.1152/jn.1997.77.3.1325

Ketzef, M., Spigolon, G., Johansson, Y., Bonito-Oliva, A., Fisone, G., & Silberberg, G. (2017). Dopamine depletion impairs bilateral sensory processing in the striatum in a pathway-dependent manner. *Neuron, 94*(4), 855–865.e5. https://doi.org/10.1016/j.neuron.2017.05.004

Kiani, R., Hanks, T. D., & Shadlen, M. N. (2008). Bounded integration in parietal cortex underlies decisions even when viewing duration is dictated by the environment. *Journal of Neuroscience, 28*(12), 3017–3029. https://doi.org/10.1523/JNEUROSCI.4761-07.2008

Krakauer, J. W., Ghazanfar, A. A., Gomez-Marin, A., MacIver, M. A., & Poeppel, D. (2017). Neuroscience needs behavior: Correcting a reductionist bias. *Neuron, 93*(3), 1–23. https://doi.org/10.1016/j.neuron.2016.12.041

Krakauer, J. W., Hadjiosif, A., Xu, J., Wong, A. L., & Haith, A. M. (2018). Motor learning. *Comprehensive Physiology, 9*(2), 613–663.

Krauzlis, R. J., Bollimunta, A., Arcizet, F., & Wang, L. (2014). Attention as an effect not a cause. *Trends in Cognitive Sciences, 18*(9), 457–464. https://doi.org/10.1016/j.tics.2014.05.008

Kravitz, A. V., Freeze, B. S., Parker, P. R. L., Kay, K., Thwin, M. T., Deisseroth, K., & Kreitzer, A. C. (2010). Regulation of parkinsonian motor behaviours by optogenetic control of basal ganglia circuitry. *Nature, 466*(7306), 622–626. https://doi.org/10.1038/nature09159

Kravitz, A. V., Tye, L. D., & Kreitzer, A. C. (2012). Distinct roles for direct and indirect pathway striatal neurons in reinforcement. *Nature Neuroscience, 15*(6), 816–818. https://doi.org/10.1038/nn.3100

Laplane, D., & Dubois, B. (2001). Auto-activation deficit: A basal ganglia related syndrome. *Movement Disorders, 16*(5), 810–814. https://doi.org/10.1002/mds.1185

Le Bouc, R., Rigoux, L., Schmidt, L., Degos, B., Welter, M.-L., Vidailhet, M., … Pessiglione, M. (2016). Computational dissection of dopamine motor and motivational functions in humans. *Journal of Neuroscience, 36*(25), 6623–6633. https://doi.org/10.1523/JNEUROSCI.3078-15.2016

Lees, A. J., Selikhova, M., Andrade, L. A., & Duyckaerts, C. (2008). The black stuff and Konstantin Nikolaevich Tretiakoff. *Movement Disorders, 23*(6), 777–783. https://doi.org/10.1002/mds.21855

Lehéricy, S., Benali, H., Van de Moortele, P.-F., Pélégrini-Issac, M., Waechter, T., Ugurbil, K., & Doyon, J. (2005). Distinct basal ganglia territories are engaged in early and advanced motor sequence learning. *Proceedings of the National Academy of Sciences of the United States of America, 102*(35), 12566–12571. https://doi.org/10.1073/pnas.0502762102

Lo, C.-C., & Wang, X.-J. (2006). Cortico-basal ganglia circuit mechanism for a decision threshold in reaction time tasks. *Nature Neuroscience, 9*(7), 956–963. https://doi.org/10.1038/nn1722

Manohar, S. G., Chong, T. T. J., Apps, M. A. J., Batla, A., Stamelou, M., Jarman, P. R., … Husain, M. (2015). Reward pays the cost of noise reduction in motor and cognitive control. *Current Biology, 25*(13), 1707–1716. https://doi.org/10.1016/j.cub.2015.05.038

Markowitz, J. E., Gillis, W. F., Beron, C. C., Neufeld, S. Q., Robertson, K., Bhagat, N. D., … Datta, S. R. (2018). The striatum organizes 3D behavior via moment-to-moment action selection. *Cell, 174*(1), 44–58 e17. https://doi.org/10.1016/j.cell.2018.04.019

Mazzoni, P., Hristova, A., & Krakauer, J. W. (2007). Why don't we move faster? Parkinson's disease, movement vigor, and implicit motivation. *Journal of Neuroscience, 27*(27), 7105–7116. https://doi.org/10.1523/JNEUROSCI.0264-07.2007

Mink, J. W. (1996). The basal ganglia: Focused selection and inhibition of competing motor programs. *Progress in Neurobiology, 50*(4), 381–425. https://doi.org/10.1016/S0301-0082(96)00042-1

Miyachi, S., Hikosaka, O., Miyashita, K., Kárádi, Z., & Rand, M. K. (1997). Differential roles of monkey striatum in learning of sequential hand movement. *Experimental Brain Research, 115*(1), 1–5.

Moussa, R., Poucet, B., Amalric, M., & Sargolini, F. (2011). Contributions of dorsal striatal subregions to spatial alternation behavior. *Learning & Memory, 18*(7), 444–451. https://doi.org/10.1101/lm.2123811

Munhoz, R. P., Cerasa, A., & Okun, M. S. (2014). Surgical treatment of dyskinesia in Parkinson's disease. *Frontiers in Neurology, 5*(April), 1–9. https://doi.org/10.3389/fneur.2014.00065

Niv, Y., Joel, D., & Dayan, P. (2006). A normative perspective on motivation. *Trends in Cognitive Sciences, 10*(8), 375–381. https://doi.org/10.1016/j.tics.2006.06.010

Obeso, J. A., Jahanshahi, M., Alvarez, L., Macias, R., Pedroso, I., Wilkinson, L., … Rothwell, J. C. (2009). What can man do without basal ganglia motor output? The effect of combined unilateral subthalamotomy and pallidotomy in a patient with Parkinson's disease. *Experimental Neurology, 220*(2), 283–292. https://doi.org/10.1016/j.expneurol.2009.08.030

Oliveira, R. M., Gurd, J. M., Nixon, P., Marshall, J. C., & Passingham, R. E. (1997). Micrographia in Parkinson's disease: The effect of providing external cues. *Journal of Neurology, Neurosurgery & Psychiatry, 63*(4), 429–433. https://doi.org/10.1136/jnnp.63.4.429

Oorschot, D. E. (1996). Total number of neurons in the neostriatal, pallidal, subthalamic, and substantia nigral nuclei of the rat basal ganglia: A stereological study using the cavalieri and optical disector methods. *Journal of Comparative Neurology, 366*(4), 580–599. https://doi.org/10.1002/(SICI)1096-9861(19960318)366:4<580::AID-CNE3>3.0.CO;2-0

Packard, M. G., & Knowlton, B. J. (2002). Learning and memory functions of the basal ganglia. *Annual Review of Neuroscience, 25*(1), 563–593. https://doi.org/10.1146/annurev.neuro.25.112701.142937

Panigrahi, B., Martin, K. A., Li, Y., Graves, A. R., Vollmer, A., Olson, L., … Dudman, J. T. (2015). Dopamine is required for the neural representation and control of movement vigor. *Cell, 162*(6), 1418–1430. https://doi.org/10.1016/j.cell.2015.08.014

Parker, J. G., Marshall, J. D., Ahanonu, B., Wu, Y. W., Kim, T. H., Grewe, B. F., … Schnitzer, M. J. (2018). Diametric neural ensemble dynamics in parkinsonian and dyskinetic states. *Nature, 557*(7704), 177–182. https://doi.org/10.1038/s41586-018-0090-6

Parkinson, J. (1817). *An essay on the shaking palsy* (Vol. 124). London. https://doi.org/10.1007/s00702-017-1684-0

Paton, J. J., & Buonomano, D. V. (2018). The neural basis of timing: Distributed mechanisms for diverse functions. *Neuron, 98*(4), 687–705. https://doi.org/10.1016/j.neuron.2018.03.045

Perugini, A., Ditterich, J., & Basso, M. A. (2016). Patients with Parkinson's disease show impaired use of priors in conditions of sensory uncertainty. *Current Biology, 26*(14), 1902–1910. https://doi.org/10.1016/j.cub.2016.05.039

Perugini, A., Ditterich, J., Shaikh, A. G., Knowlton, B. J., & Basso, M. A. (2018). Paradoxical decision-making: A framework for understanding cognition in Parkinson's disease. *Trends in Neurosciences, 41*(8), 512–525. https://doi.org/10.1016/j.tins.2018.04.006

Reig, R., & Silberberg, G. (2014). Multisensory integration in the mouse striatum. *Neuron, 83*(5), 1200–1212. https://doi.org/10.1016/j.neuron.2014.07.033

Robbe, D. (2018). To move or to sense? Incorporating somatosensory representation into striatal functions. *Current Opinion in Neurobiology, 52,* 123–130. https://doi.org/10.1016/j.conb.2018.04.009

Robertson, B., Kardamakis, A., Capantini, L., Pérez-Fernández, J., Suryanarayana, S. M., Wallén, P., … Grillner, S. (2014). The lamprey blueprint of the mammalian nervous system. *Progress in Brain Research, 212*(C), 337–349. https://doi.org/10.1016/B978-0-444-63488-7.00016-1

Rueda-Orozco, P. E., & Robbe, D. (2015). The striatum multiplexes contextual and kinematic information to constrain motor habits execution. *Nature Neuroscience, 18*(3), 453–460. https://doi.org/10.1038/nn.3924

Ruitenberg, M. F. L., Duthoo, W., Santens, P., Notebaert, W., & Abrahamse, E. L. (2015). Sequential movement skill in Parkinson's disease: A state-of-the-art. *Cortex, 65,* 102–112. https://doi.org/10.1016/j.cortex.2015.01.005

Salamone, J. D., Correa, M., Farrar, A., & Mingote, S. M. (2007). Effort-related functions of nucleus accumbens dopamine and associated forebrain circuits. *Psychopharmacology, 191*(3), 461–482. https://doi.org/10.1007/s00213-006-0668-9

Scharff, C., & Nottebohm, F. (1991). A comparative study of the behavioral deficits following lesions of various parts of the zebra finch song system: Implications for vocal learning. *Journal of Neuroscience, 11*(9), 2896–2913.

Schmidt, L., D'Arc, B. F., Lafargue, G., Galanaud, D., Czernecki, V., Grabli, D., … Pessiglione, M. (2008). Disconnecting force from money: Effects of basal ganglia damage on incentive motivation. *Brain, 131*(Pt. 5), 1303–1310. https://doi.org/10.1093/brain/awn045

Schultz, W., Dayan, P., & Montague, P. R. (1997). A neural substrate of prediction and reward. *Science, 275*(5306), 1593–1599.

Shabbott, B., Ravindran, R., Schumacher, J. W., Wasserman, P. B., Marder, K. S., & Mazzoni, P. (2013). Learning fast accurate movements requires intact frontostriatal circuits. *Frontiers in Human Neuroscience, 7*(November), 752. https://doi.org/10.3389/fnhum.2013.00752

Thorn, C. A., Atallah, H., Howe, M., & Graybiel, A. M. (2010). Differential dynamics of activity changes in dorsolateral and dorsomedial striatal loops during learning. *Neuron, 66*(5), 781–795. https://doi.org/10.1016/j.neuron.2010.04.036

Thura, D., & Cisek, P. (2016). On the difference between evidence accumulator models and the urgency gating model. *Journal of Neurophysiology, 115*(1), 622–623. https://doi.org/10.1152/jn.01050.2015

Thura, D., & Cisek, P. (2017). The basal ganglia do not select reach targets but control the urgency of commitment. *Neuron, 95*(5), 1160–1170.e5. https://doi.org/10.1016/j.neuron.2017.07.039

Tricomi, E., Balleine, B. W., & O'Doherty, J. P. (2009). A specific role for posterior dorsolateral striatum in human habit learning. *European Journal of Neuroscience, 29*(11), 2225–2232. https://doi.org/10.1111/j.1460-9568.2009.06796.x

Turner, R. S., & Desmurget, M. (2010). Basal ganglia contributions to motor control: A vigorous tutor. *Current Opinion in Neurobiology, 20*(6), 704–716. https://doi.org/10.1016/j.conb.2010.08.022

Wilkinson, L., Khan, Z., & Jahanshahi, M. (2009). The role of the basal ganglia and its cortical connections in sequence learning: Evidence from implicit and explicit sequence learning in Parkinson's disease. *Neuropsychologia, 47*(12), 2564–2573. https://doi.org/10.1016/j.neuropsychologia.2009.05.003

Wong, A. L., Lindquist, M. A., Haith, A. M., & Krakauer, J. W. (2015). Explicit knowledge enhances motor vigor and performance: Motivation versus practice in sequence tasks. *Journal of Neurophysiology, 114*(1), 219–232. https://doi.org/10.1152/jn.00218.2015

Yartsev, A. M. M., Hanks, T. D., Yoon, A. M., & Brody, C. D. (2018). Causal contribution and dynamical encoding in the striatum during evidence accumulation. *elife, 7.* https://doi.org/10.7554/eLife.34929

Yin, H. H., & Knowlton, B. J. (2006). The role of the basal ganglia in habit formation. *Nature Reviews. Neuroscience, 7*(6), 464–476. https://doi.org/10.1038/nrn1919

Yin, H. H., Mulcare, S. P., Hilário, M. R. F., Clouse, E., Holloway, T., Davis, M. I., … Costa, R. M. (2009). Dynamic reorganization of striatal circuits during the acquisition and consolidation of a skill. *Nature Neuroscience, 12*(3), 333–341. https://doi.org/10.1038/nn.2261

Yttri, E. A., & Dudman, J. T. (2016). Opponent and bidirectional control of movement velocity in the basal ganglia. *Nature, 533*(7603), 402–406. https://doi.org/10.1038/nature17639

Yttri, E. A., & Dudman, J. T. (2018). A proposed circuit computation in basal ganglia: History-dependent gain. *Movement Disorders, Official Journal of the Movement Disorder Society, 33*(5), 704–716. https://doi.org/10.1002/mds.27321

45 Preparation of Movement

ADRIAN M. HAITH AND SVEN BESTMANN

ABSTRACT It is widely believed that voluntary movements must be prepared before they are executed, but there is so far little consensus on exactly what mechanisms and computations must occur prior to a movement. Here, we argue that movement preparation is most usefully thought of as a process of setting an appropriate state to poise the motor system to generate a particular desired movement. This process, which can occur in as little as 50 ms, neither directly triggers movement initiation nor is required to initiate a movement. Instead, the initiation of movement is determined by a distinct process independent of the state of movement preparation. We suggest that during the course of determining a task goal, or making a decision, the prepared state is continually updated as the desired movement changes in light of new evidence and beliefs. This ensures that the motor system is able to generate an appropriate movement as rapidly as possible when necessary.

How does the brain generate movement? A longstanding idea in neuroscience is that movements must be "prepared" before they can be executed. This notion is inspired by the fact that the patterns of muscle activation needed to achieve even simple goals can be extremely complex and thus require equally complex, time-consuming computations. Numerous behavioral and physiological observations also support it. Over recent years, however, as our understanding of how movements are generated by the brain has advanced, the extent to which movement preparation is truly necessary has become less and less clear.

There is, in fact, surprisingly little consensus on what the term *preparation* should even mean. The term often collapses together multiple distinct processes that occur prior to movement (e.g., selecting what action to take versus specifying the details of how to execute that action). At other times, the term *preparation* is used without any explanation of what exactly it may or may not entail, or whether it even necessarily occurs. A dictionary definition of preparation is "the act of being made ready for use or consideration." We suggest that this definition aptly applies to movement preparation. It emphasizes several important features of movement preparation: that the act of preparation is distinct from the act of generating a movement, that it happens before we ever need to move, and that it is time-consuming. Furthermore, this definition suggests that preparation pertains to readiness to perform a *specific* action, and we will focus here on preparation in this sense, rather than in the more generic sense of readiness to transition from being at rest to moving.

In this chapter we will review conventional assumptions about movement preparation and outline a new emerging perspective. We start by reviewing behavioral studies examining the time course of movement preparation and its relationship to movement initiation. We then consider convergent evidence gained from neurophysiological investigations. Collectively, these lines of evidence point to a reappraisal of movement preparation as a relatively rapid process whereby specified high-level goals are translated into a policy suitable for movement and ready to be invoked when movement initiation is triggered.

Is Movement Preparation Time-Consuming?

It has long been argued that preparation is a time-consuming process, based on the general observation that when participants know beforehand which movement will be required, their reaction times are considerably faster (and movement accuracy is improved) than when the required movement is cued at the same time as a GO cue.

An example of this effect comes from the work of David Rosenbaum (1980), in which human participants were instructed, via a series of letters, which among eight possible targets they might have to move to (e.g., F = forward; B = backward). In some instances, certain letters were precued, narrowing down the range of possible targets ahead of the GO cue. The more information about the forthcoming movement was disclosed beforehand, the faster the reaction times. The interpretation of these findings, which has dominated much of the thinking about movement preparation and initiation, is that specifying each parameter of the movement is time-consuming. Precuing some aspect of the required movement enables the corresponding parameter to be specified in advance, allowing the movement to then be generated with a shorter reaction time.

As pointed out by Goodman and Kelso (1980), however, a major frailty in this logic is that specifying the movement may not have been the most challenging aspect of this task. Parsing the presented letters into a required movement goal is not straightforward and could well have been sufficiently time-consuming to account for much of the reaction time. Indeed, the reaction times exhibited by participants in tasks such as Rosenbaum's are relatively long, often around 500–700 ms. This is far longer than the typical reaction times of 200–250 ms for reaching movements toward a visually presented target location (Haith, Pakpoor, & Krakauer, 2016). It is unlikely, therefore, that the typical reductions in reaction time afforded by precues are attributable to movement preparation per se.

This raises an important point about movement preparation: that it should only encompass processes specific to generating the movement itself, not to deciding which movement to make. Although parsing the stimulus is clearly a prerequisite for making the right movement, it seems unreasonable to class this as *movement* preparation—just as planning your next move in a chess match should not qualify as such.

A more focused definition of movement preparation is a process that bridges the high-level goals of a movement (*what* you want to do, out of the many alternatives available) and the control policy for executing the movement (*how* you will achieve the goal). Only the specification of the policy prescribing the actual movement to be generated, contingent upon a selected goal, should be considered movement preparation. Processes required for specifying goals should not be.

With this in mind, movement preparation is perhaps best studied through tasks in which the required movement is cued by a highly salient spatial cue so that deciding what action to take should be trivial (though we discuss preparation in the context of decision-making further below). In this case, precuing the target location before a GO cue still leads to a reduction in reaction time, consistent with the theory that it enables some time-consuming preparation phase to be completed in advance (Goodman & Kelso, 1980). However, the reduction in reaction time enabled by direct spatial precues is only of the order of 50–100 ms and can be as little as 30 ms in monkeys (Churchland, Yu, Ryu, Santhanam, & Shenoy, 2006). The time consumed by preparation, therefore, seems to be quite modest and certainly much shorter than most classical theories postulate.

Movement Preparation versus Movement Initiation

Approaches to understanding movement preparation based on examining reaction times are inevitably indirect and rely on strong assumptions about what determines reaction times. In particular, it is commonly assumed that a movement is generated as soon as preparation is complete. As we shall see, this assumption does not withstand closer scrutiny of how we prepare and execute movements.

A more direct approach to characterizing movement preparation was devised by Claude Ghez and colleagues. Rather than treating the reaction time as a dependent variable, they instead employed a method to control the reaction time and measured the quality of movement as a dependent variable (Ghez et al., 1997). In their *timed-response* paradigm (figure 45.1*A*), participants were required to initiate a movement synchronously with a predictable auditory cue—the fourth tone in a sequence of four equally spaced tones. The allowed reaction time (and, consequently, the time available for movement preparation) could be varied by varying the moment at which the target was presented relative to this fixed time of movement initiation.

This approach reveals how movements transition from a default movement generated at very low allowed reaction times to a movement precisely tuned to the presented target at longer times—with appropriate movements able to be executed within as little as 150 ms following presentation of the target (Haith, Pakpoor, & Krakauer, 2016; figure 45.1*B*). This duration is scarcely longer than the signaling delays that would be necessary for information to travel from the eye, through the cortex, and to muscles. It can be even shorter for frequently repeated movements (Mawase, Lopez, Celnik, & Haith, 2018). Therefore, to the extent that movement preparation does occur, it does not seem to occupy the bulk of the reaction time.

Contrasting the time course of preparation from this timed-response approach with behavior in a more conventional "free" reaction time paradigm (in which participants react as quickly as possible to move toward a presented target) reveals that typical reaction times are much longer than apparently necessary (figure 45.1*B*). Whereas accurate movements could be reliably generated whenever at least 150 ms preparation time was allowed, typical "free" reaction times were delayed relative to this time by around 80 ms—around a third of the reaction time (Haith, Pakpoor, & Krakauer, 2016).

The substantial delay between the inferred timing of movement preparation and the timing of movement initiation suggests distinct mechanisms for these processes. This is perhaps unsurprising—after all, delay paradigms depend on the ability to prepare a movement without initiating it. However, it has also been noted that participants still exhibited a speed-accuracy trade-off even when acting under "free" reaction-time conditions (Haith, Pakpoor, & Krakauer, 2016). That is, they committed errors in which they apparently initiated movement before preparation had become complete. Critically, this suggests that movement initiation

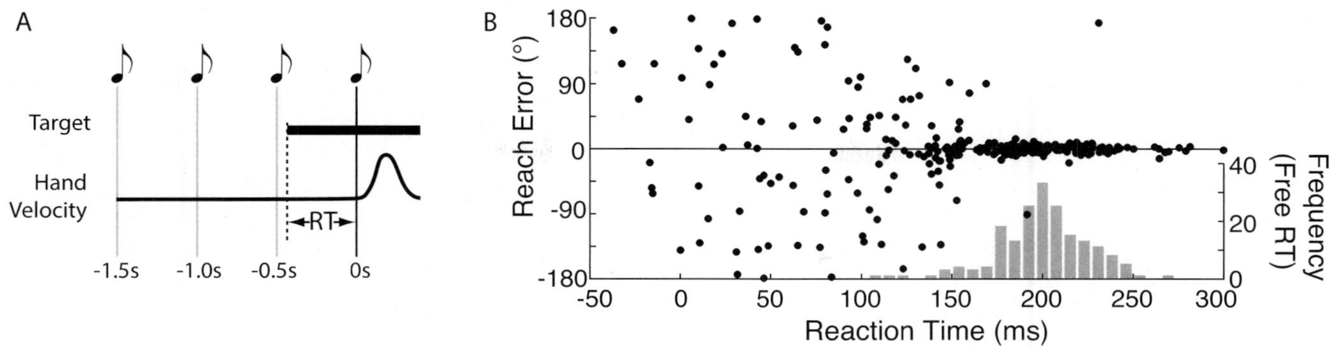

FIGURE 45.1 Relative timing of movement preparation versus movement initiation. Adapted from (Haith, Pakpoor, & Krakauer, 2016). *A*, Timed-response paradigm. In each trial, participants must initiate a response (a reach to one of eight possible targets) coincident with the fourth of four predictable tones. The time at which the true target is presented can be varied relative to the fourth tone to impose a particular reaction time. *B*, Behavioral results from a representative participant. The black dots show imposed reaction time and initial directional error for individual trials. This participant could reliably generate an accurate response whenever the imposed preparation time was greater than 150 ms, indicating that movement preparation was complete by that time. The gray histogram shows the distribution of reaction times for the same participant in a "free" reaction-time version of the task. Note that, despite a perfect performance above the imposed reaction times of 150 ms, typical free reaction times were around 200 ms and often even longer—illustrating the delay between the time required for preparation and the time at which movement is initiated.

was independent of, and not contingent on, movement preparation having become complete. This also offers a potential explanation as to why there should be a delay between movement preparation and initiation: it serves to minimize the risk of *fast errors*, in which a movement is initiated before the correct movement can be selected and prepared.

Movement Preparation and Freedom from Immediacy

An important implication of the independence of movement preparation and initiation is that, regardless of the time a movement is ultimately initiated, there is a transition from being unable to make the correct movement to being able to do so. This illustrates an important sense in which movements are indeed prepared: the movement is made ready for later execution, and this readied movement can be held in a prepared state until it is executed. Preparation in this sense differs from classical characterizations as a necessary and time-consuming computation, which triggers movement initiation as soon as it is complete (Allen, Kornblum, & Meyer, 1986).

Many of our movements are generated without being held in a readied state prior to their initiation. One trivial example is the stretch reflex, such as the patellar reflex, which might be induced by a doctor tapping on the tendon below your kneecap. This reflex is mediated by a simple monosynaptic circuit within the spinal cord that translates information about muscle length into motor neuron activity that causes the muscle to contract. In this case, movement happens as a direct consequence of the stimulus in a direct *stimulus-response association*, but this movement is not held in any "prepared" state prior to being initiated.

Such immediacy of movement also occurs when an ongoing movement is perturbed. For example, displacing the target of a movement elicits a rapid response that corrects for the effects of the perturbation (Day & Lyon, 2000; Pruszynski, Kurtzer, & Scott, 2008), often with considerable sophistication (Nashed, Crevecoeur, & Scott, 2014; see also chapter 42). These corrective movements are involuntary in that they occur even when they conflict with task goals—for example, an instruction to move in the opposite direction of a target jump.

The ability to prepare a movement without immediately initiating it thus seems critical to the "freedom from immediacy" that is a defining aspect of volitional control. It has long been appreciated, when it comes to internally generated, self-initiated movements, that decisions about *what to do* occur independently of decisions about *when to do it*. It should therefore perhaps not be surprising that voluntary movements cued by an external stimulus follow a similar organization.

Neural Mechanisms of Movement Preparation: Activity during Delay Periods

The behavioral evidence reviewed above suggests that voluntary movements are "prepared" before they are executed in the sense that a movement is made ready for execution. Neurophysiological investigations have yielded more or less the same conclusion.

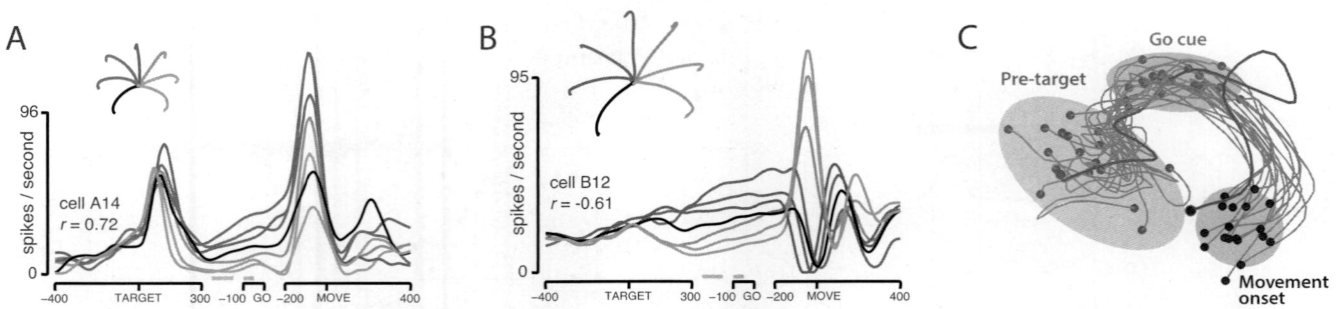

FIGURE 45.2 Neural mechanisms of movement preparation. *A, B,* Preparatory and movement-related activity in two example neurons from motor cortex. Reproduced from Churchland et al. (2010). Different colors indicate activity for different movements (*shown in inset*). *A,* A neuron whose preparatory activity (between TARGET and GO) is positively correlated (r = 0.72) with its activity during movement (MOVE). *B,* A neuron whose analogous correlation is negative (r = −0.61). *C,* At the population level (here, a two-dimensional view of overall population activity), neural activity is initially variable across trials before the target appears (*blue dots*) but converges to a specific state by the time of the GO cue (*green dots*). After the GO cue, neural activity follows a stereotyped activity profile. The thicker red curve indicates data from one outlying trial. Reproduced from Shenoy, Sahani, and Churchland (2013). (See color plate 50.)

Experiments seeking to characterize neural mechanisms of movement preparation have also typically employed a precuing approach, enabling preparatory activity to be distinguished from activity related to actually generating the movement. Many neurons active during movement exhibit similar tuning during the delay between a sensory cue and the ultimate initiation of the movement (Crammond & Kalaska, 2000; Kurata, 1989; Riehle & Requin, 1989; Tanji & Evarts, 1976; Weinrich & Wise, 1982; figure 45.2*A*). This premovement activity can also be influenced by the number of possible movement options, information about the timing of the movement, and various types of precues (e.g., Bastian, Schöner, & Riehle, 2003; Dekleva, Ramkumar, Wanda, Kording, & Miller, 2016). It has thus been suggested that movement preparation amounts to "priming" movement by bringing activity in these regions close to, but just below, some threshold required for generating a movement.

Implicit in such accounts is that individual neurons represent specific aspects of movement, such as the kinematics of the hands or joints, applied forces or torques, or the activity of specific muscles. Perhaps consistent with this view, microstimulation in either premotor cortex or M1 (henceforth PM/M1) can elicit movement of specific body parts (Dum & Strick, 2002; Leyton & Sherrington, 1917). It therefore seems plausible that premovement activity might serve to give the specification of these parameters a head start so that when the movement needs to ultimately be made, it is generated faster and more accurately (Hanes & Schall, 1996).

Recent work, however, challenges the view that movement preparation serves to simply hold movement-related activity in a sub–threshold state. The systematic examination of delay-period activity in hundreds of neurons in PM/M1 reveals that delay-period activity just as often shows an opposite pattern to activity during movement (see the example in figure 45.2*B*) or is entirely unrelated to it (Churchland, Cunningham, Kaufman, Ryu, & Shenoy, 2010).

Movement Preparation as State Setting for Subsequent Movement

Although there seems to be no straightforward relationship between preparatory activity and movement-related activity, what is clear is that premovement population activity in PM/M1 reliably converges to a specific state (figure 45.2*C*). For example, the variability in neural firing rates in premotor cortex decreases, while mean firing rates remain largely unaffected (Churchland et al., 2006). Similar decreases in variability occur in corticospinal excitability in humans before movement (Klein-Flügge, Nobbs, Pitcher, & Bestmann, 2013). This convergence appears to be functionally relevant; disrupting PM/M1 activity prior to movement erases the reaction-time advantage gained by precuing the target location (Churchland & Shenoy, 2007; Schluter, Rushworth, Passingham, & Mills, 1998). Thus, it appears that the overall population of neurons in PM/M1 must be active before movement with a very specific pattern.

A closer examination of neural activity during movement provides clues as to why this may be necessary. The joint population activity of neurons can be visualized and analyzed using dimensionality-reduction techniques, which have helped to reveal that premovement activity occurs in distinct patterns (or, to be more precise, different subspaces; Elsayed, Lara, Kaufman, Churchland, & Cunningham, 2016; Kaufman, Churchland,

Ryu, & Shenoy, 2014). Furthermore, during movement, the population-level activity across PM/M1 exhibits characteristic dynamics in which the future state of the neural population depends largely on its current state (Churchland et al., 2010, 2012). While different movements exhibit different patterns of activity, all movements exhibit stereotypical transitions from one state to another; such structure is not present in downstream muscle activity.

Movement preparation might therefore be conceived as attaining a specific state within PM/M1 that seeds the initial state of the dynamical system to generate a specific time-varying pattern of activity that will then drive muscle activity in the required way (Churchland et al., 2010, 2012). In this view the state of activity of neurons in PM/M1 prior to movement encodes the policy for how the subsequent movement will be generated. Indeed, trial-by-trial comparisons show an initially high variability of firing rates for individual neurons at the start of each trial, which rapidly diminishes after the target is presented (Churchland et al., 2006). Coupled with distinct changes in mean firing rate for different movements, this suggests a convergence of the population-level activity onto the specific state that is required to generate the desired movement (Churchland et al., 2006).

This perspective accords well with our understanding of movement preparation gained from behavioral investigations. Behaviorally, we determined that movement preparation must be rapid, requiring no more than 50–100 ms. Delay-period population activity typically converges to a stable state within around 50–100 ms. Behavioral evidence has also suggested that movement preparation is more than just a time-consuming process that precedes movement, instead reflecting an underlying transition from being unable to make the required movement to being able to do so. This is entirely consistent with the convergence of neural population activity onto a specific state.

One concern with the above proposal is that it is primarily founded on neural activity occurring during an artificially imposed delay period. It remains unclear to what extent patterns of neural activity in such paradigms can be generalized to more naturalistic scenarios in which action may be required immediately. However, the basic structure of preparatory events seems to be preserved even when no instructed delay is present (Lara, Elsayed, Zimnik, Cunningham, & Churchland, 2018).

Finally, neural data also suggest that movement initiation is distinct from and does not appear to be directly triggered by movement preparation. The state of PM/M1 at the time of the GO cue is predictive of reaction times, although it only accounts for around 20% of their variance (Afshar et al., 2011). Motor cortical population activity does, however, exhibit a rapid, stepwise change that is strongly predictive of reaction time. Unlike most other aspects of activity in PM/M1, this stepwise change occurs regardless of the specific movement being made (Kaufman et al., 2016). This is compatible with the conclusion, from behavioral investigations, that movement initiation is governed by a process distinct from that which is responsible for movement preparation.

Single versus Parallel Plans

One unresolved question that has occupied the field is whether several movements can be prepared in parallel, or whether only a single movement can be prepared. A popular view holds that the brain prepares multiple potential movements at the same time. For example, Cisek and Kalaska (2005) reported that if there is more than one potential target during the delay period, neurons tuned to each of these *specific options* appear to be active during this time. One interpretation of this observation is that multiple movements are prepared in parallel, presumably to allow the reaction-time savings secured through advance preparation to be gained for either potential target.

This interpretation, however, is somewhat at odds with the dynamical-systems account of movement generation. According to the latter, the premovement state of PM/M1 seeds the initial state of a dynamic system, leading to movement. Each possible state is therefore associated with generating a single movement, thus framing PM/M1 as a tuned population of neurons whose collective state encodes a single movement. This is in stark contrast to the idea underlying parallel preparation, in which different movements are represented by different subpopulations of neurons, instead framing PM/M1 as a population of tuned neurons.

How can these conflicting neural accounts of movement preparation be reconciled? Observed preparatory activity is generally presumed to reflect specification of a movement or control policy, especially when it occurs in an ostensibly "motor" region of the brain. However, this activity might alternatively reflect processes related to the specification of a more abstract movement goal, which would need to be elaborated into a policy elsewhere. It seems very plausible that we can entertain multiple potential *goals* in parallel, particularly if they are not mutually exclusive. But, given that we can only ever generate one movement at a time, it is not much of a restriction to only be able to prepare a single movement.

A number of lines of research have sought to establish behavioral evidence for the existence of parallel movement plans, based on the following logic: If two similar movement plans are indeed simultaneously activated, they may interfere with one another, leading to a movement that resembles a mixture of the features of the individual movements. For instance, two reaching movements planned to adjacent targets might ultimately give rise to a movement that reflects the average of the two trajectories, falling in an intermediate direction between the potential targets. This phenomenon of intermediate movements is indeed well established. Whenever there is ambiguity surrounding the required target location, participants are prone to generate movements between potential targets. This effect is most starkly revealed in "go-before-you-know" experiments that deliberately create ambiguity by presenting multiple potential targets and forcing participants to initiate movement before the true target location is disclosed (Chapman et al., 2010).

An alternative explanation for these intermediate movements, however, is that they are quite deliberate, goal-directed movements not generated through the inadvertent mixing of parallel plans. If the motor system is only able to prepare a single movement given an ambiguous target array, it should prepare the best possible movement for achieving all of the potential task goals. An intermediate movement between two potential targets brings the hand closer toward both of them—a sensible policy, since it makes it then easier to move toward either target if one emerges as more appealing later in the movement. A number of findings favor this optimization-based view: intermediate movements become less likely as the spatial separation between targets increases (Ghez et al., 1997); participants can avoid intermediate movements if these are penalized (Haith, Huberdeau, & Krakauer, 2015); and fast movements prohibit midmovement corrections, tipping the balance in favor of moving directly toward one target or the other (Wong & Haith, 2017). Collectively, these results suggest that parallel preparation of multiple movements is, at the very least, not an essential requirement for action, and they lend support to the idea that only a single movement is prepared by the motor system. This conclusion is supported by a recent study using large-scale neural recording in premotor cortex, which found that only single, and not multiple, plans are represented even when multiple potential targets are present (Dekleva, Kording, & Miller, 2018).

Relationship between Movement Preparation and Decision-Making

In many cases our movements are not unambiguously cued by a salient stimulus, and instead we must decide for ourselves what movement to make. The process of making such decisions is well studied. However, what remains less well understood is exactly how this process interacts with the generation of movement required to indicate one's decision.

A common paradigm for studying decision-making is to ask participants to make judgments about a stimulus. For instance, a series of dots may appear and disappear on a screen, resembling "snow" on an untuned television, except that some proportion of the dots move coherently in a particular direction. Participants must judge which direction the coherent dots are moving, and the difficulty of this judgment can be altered by varying the proportion of dots that move coherently. When, say, only 5% of the dots are moving coherently, it becomes necessary to observe the stimulus for some time, accumulating evidence from momentary observations, before deciding which way the dots are moving.

A simplistic serial viewpoint might suggest that such a decision is first made on the basis of visual evidence, and after commitment to a particular decision, the required action is prepared. It seems, however, that movements can be continuously modulated by the state of an inchoate decision, speaking against such an interpretation. For instance, in an experiment by Gold and Shadlen (2000), monkeys had to make a saccade in the appropriate direction of a random-dot-motion stimulus once they had judged the direction the dots were moving. At times, prior to monkeys having made up their minds about the direction of the dots, an involuntary eye movement would be elicited by stimulating the frontal eye field. Although the monkey had not yet completed its decision, the elicited eye movements deviated in the direction favored by the current evidence that had accumulated so far.

This phenomenon is not restricted to eye movements. In a study by Selen, Shadlen, and Wolpert (2012), human participants performed a similar task but had to generate an arm movement indicating the perceived motion direction. After various durations viewing the stimulus, the participant's arm was perturbed, and the strength of the low-latency feedback responses was closely tuned to the strength and direction of the evidence provided up to that point.

These findings suggest an overlap between the mechanisms by which a decision is formed and the mechanisms for representing a movement. At a neural level, the state of a decision (which can typically be inferred from the stimulus, often with the aid of a model) can be clearly decoded from neurons in motor areas (Thura & Cisek, 2014). In fact, the ubiquity with which the decision state has been found to be apparently represented in the motor system has even led to the suggestion that movement preparation may play a causal role in influencing the outcome of a decision (Cisek & Kalaska, 2005).

We propose an alternative interpretation. As we have just discussed, current evidence suggests that it is only possible to prepare a single movement at any point in time. During a developing decision, it would make sense for the motor system to specify the movement that corresponds to the best possible course of action, given the current available evidence. If the decision, and, consequently, the goal of movement, is ambiguous, prematurely initiated movements will reflect this ambiguity and be biased toward the movement corresponding to the current most likely goal. Indeed, this account can explain intermediate movements amid uncertainty; in these cases the optimal course of action among two neighboring targets is to move along an intermediate movement trajectory, anticipating that more evidence will be available later in the movement (Haith, Huberdeau, & Krakauer, 2015).

In support of this general view that the motor system may not encode a decision per se but rather a policy associated with the current evidence for one goal or another, a recent paper by Hanks and colleagues found differential tuning for decision-related variables in different brain regions in rats (Hanks et al., 2015). In posterior parietal cortex, neural activity was closely tuned to the current evidence available in the task, varying continuously with the available evidence. In prefrontal cortex, however, neural activity followed a more categorical tuning; population activity fell into one of two specific states depending on which response option was, on current evidence, more likely. As the authors stated, activity in these areas effectively answered the question: "If the go signal came now, which action should I take?" The response required by the animal in this case was rather more abstract than in the case of a saccade. However, we suggest that the distinction apparent in these data reflect a general principle of the gradient from evidence to action in decision-making. Within this gradient the bulk of the required computations specifies the goals of our actions, whereas the preparation and initiation of action is, comparatively, not especially time-consuming.

Summary

Classical views of movement preparation suggest that it is a time-consuming process entailing complex computations necessary for movement. However, we have argued that these ideas are difficult to uphold. Instead, a new view of movement preparation has emerged: movement preparation is a process of setting the state of the motor system once an action goal is identified, priming it to generate a single, specific required movement. Contrary to traditional views, this process occurs very rapidly, within around 50 ms, and does not directly trigger initiation of the movement upon completion.

REFERENCES

Afshar, A., Santhanam, G., Yu, B. M., Ryu, S. I., Sahani, M., & Shenoy, K. V. (2011). Single-trial neural correlates of arm movement preparation. *Neuron, 71*(3), 555–564. https://doi.org/10.1016/j.neuron.2011.05.047

Allen, O., Kornblum, S., & Meyer, D. E. (1986). The point of no return in choice reaction time: Controlled and ballistic stages of response preparation. *Journal of Experimental Psychology, 12*(3), 243–258.

Bastian, A., Schöner, G., & Riehle, A. (2003). Preshaping and continuous evolution of motor cortical representations during movement preparation. *European Journal of Neuroscience, 18*(7), 2047–2058. https://doi.org/10.1046/j.1460-9568.2003.02906.x

Chapman, C. S., Gallivan, J. P., Wood, D. K., Milne, J. L., Culham, J. C., & Goodale, M. A. (2010). Reaching for the unknown: Multiple target encoding and real-time decision-making in a rapid reach task. *Cognition, 116*(2), 168–176. https://doi.org/10.1016/j.cognition.2010.04.008

Churchland, M. M., Cunningham, J. P., Kaufman, M. T., Foster, J. D., Nuyujukian, P., Ryu, S. I., & Shenoy, K. V. (2012). Neural population dynamics during reaching. *Nature, 487*(7405), 51–56. https://doi.org/10.1038/nature11129

Churchland, M. M., Cunningham, J. P., Kaufman, M. T., Ryu, S. I., & Shenoy, K. V. (2010). Cortical preparatory activity: Representation of movement or first cog in a dynamical machine? *Neuron, 68*(3), 387–400. https://doi.org/10.1016/j.neuron.2010.09.015

Churchland, M. M., & Shenoy, K. V. (2007). Delay of movement caused by disruption of cortical preparatory activity. *Journal of Neurophysiology, 97*(1), 348–359. https://doi.org/10.1152/jn.00808.2006

Churchland, M. M., Yu, B. M., Ryu, S. I., Santhanam, G., & Shenoy, K. V. (2006). Neural variability in premotor cortex provides a signature of motor preparation. *Journal of Neuroscience, 26*(14), 3697–3712. https://doi.org/10.1523/JNEUROSCI.3762-05.2006

Cisek, P., & Kalaska, J. F. (2005). Neural correlates of reaching decisions in dorsal premotor cortex: Specification of multiple direction choices and final selection of action. *Neuron, 45*(5), 801–814. https://doi.org/10.1016/j.neuron.2005.01.027

Crammond, D. J., & Kalaska, J. F. (2000). Prior information in motor and premotor cortex: Activity during the delay period and effect on pre-movement activity. *Journal of Neurophysiology, 84*(2), 986–1005. https://doi.org/10.1152/jn.2000.84.2.986

Day, B. L., & Lyon, I. N. (2000). Voluntary modification of automatic arm movements evoked by motion of a visual target. *Experimental Brain Research, 130*(2), 159–168. https://doi.org/10.1007/s002219900218

Dekleva, B. M., Kording, K. P., & Miller, L. E. (2018). Single reach plans in dorsal premotor cortex during a two-target task. *Nature Communications, 9*(1), 3556. https://doi.org/10.1038/s41467-018-05959-y

Dekleva, B. M., Ramkumar, P., Wanda, P. A., Kording, K. P., & Miller, L. E. (2016). Uncertainty leads to persistent effects on reach representations in dorsal premotor cortex. *eLife*. https://doi.org/10.7554/eLife.14316

Dum, R. P., & Strick, P. L. (2002). Motor areas in the frontal lobe of the primate. *Physiology & Behavior, 77*(4), 677–682. https://doi.org/10.1016/S0031-9384(02)00929-0

Elsayed, G. F., Lara, A. H., Kaufman, M. T., Churchland, M. M., & Cunningham, J. P. (2016). Reorganization between preparatory and movement population responses in motor cortex. *Nature Communications, 7*, 13239. https://doi.org/10.1038/ncomms13239

Ghez, C., Favilla, M., Ghilardi, M. F., Gordon, J., Bermejo, R., & Pullman, S. (1997). Discrete and continuous planning of hand movements and isometric force trajectories. *Experimental Brain Research, 115*(2), 217–233. https://doi.org/10.1007/PL00005692

Gold, J. I., & Shadlen, M. N. (2000). Representation of a perceptual decision in developing oculomotor commands. *Nature, 404*(6776), 390–394. https://doi.org/10.1038/35006062

Goodman, D., & Kelso, J. A. (1980). Are movements prepared in parts? Not under compatible (naturalized) conditions. *Journal of Experimental Psychology. General, 109*(4), 475–495. https://doi.org/10.1037//0096-3445.109.4.475

Haith, A. M., Huberdeau, D. M., & Krakauer, J. W. (2015). Hedging your bets: Intermediate movements as optimal behavior in the context of an incomplete decision. *PLoS Computational Biology, 11*(3), e1004171. https://doi.org/10.1371/journal.pcbi.1004171

Haith, A. M., Pakpoor, J., & Krakauer, J. W. (2016). Independence of movement preparation and movement initiation. *Journal of Neuroscience, 36*(10), 3007–3015. https://doi.org/10.1523/JNEUROSCI.3245-15.2016

Hanes, D. P., & Schall, J. D. (1996). Neural control of voluntary movement initiation. *Science, 274*(5286), 427–430. https://doi.org/10.1126/science.274.5286.427

Hanks, T. D., Kopec, C. D., Brunton, B. W., Duan, C. A., Erlich, J. C., & Brody, C. D. (2015). Distinct relationships of parietal and prefrontal cortices to evidence accumulation. *Nature, 520*(7546), 220–223. https://doi.org/10.1038/nature14066

Kaufman, M. T., Churchland, M. M., Ryu, S. I., & Shenoy, K. V. (2014). Cortical activity in the null space: Permitting preparation without movement. *Nature Neuroscience, 17*(3), 440–448. https://doi.org/10.1038/nn.3643

Kaufman, M. T., Seely, J. S., Sussillo, D., Ryu, S. I., Shenoy, K. V., & Churchland, M. M. (2016). The largest response component in the motor cortex reflects movement timing but not movement type. *eNeuro, 3*(4), ENEURO.0085-16.2016. https://doi.org/10.1523/ENEURO.0085-16.2016

Klein-Flügge, M. C., Nobbs, D., Pitcher, J. B., & Bestmann, S. (2013). Variability of human corticospinal excitability tracks the state of action preparation. *Journal of Neuroscience, 33*(13), 5564–5572. https://doi.org/10.1523/JNEUROSCI.2448-12.2013

Kurata, K. (1989). Distribution of neurons with set- and movement-related activity before hand and foot movements in the premotor cortex of rhesus monkeys. *Experimental Brain Research, 77*(2), 245–256. https://doi.org/10.1007/BF00274982

Lara, A. H., Elsayed, G. F., Zimnik, A. J., Cunningham, J. P., & Churchland, M. M. (2018). Conservation of preparatory neural events in monkey motor cortex regardless of how movement is initiated. *eLife*, https://doi.org/10.7554/eLife.31826

Leyton, A. S. F., & Sherrington, C. S. (1917). Observations on the excitable cortex of the chimpanzee, orang-utan, and gorilla. *Quarterly Journal of Experimental Physiology, 11*(2), 135–222. https://doi.org/10.1113/expphysiol.1917.sp000240

Mawase, F., Lopez, D., Celnik, P. A., & Haith, A. M. (2018). Movement preparation facilitates response preparation. *Cell Reports, 24*(4), 801–808. https://www.cell.com/cell-reports/fulltext/S2211-1247(18)31039-8.

Nashed, J. Y., Crevecoeur, F., & Scott, S. H. (2014). Rapid online selection between multiple motor plans. *Journal of Neuroscience, 34*(5), 1769–1780. https://doi.org/10.1523/JNEUROSCI.3063-13.2014

Pruszynski, J. A., Kurtzer, I., & Scott, S. H. (2008). Rapid motor responses are appropriately tuned to the metrics of a visuospatial task. *Journal of Neurophysiology, 100*(1), 224–238. https://doi.org/10.1152/jn.90262.2008

Riehle, A., & Requin, J. (1989). Monkey primary motor and premotor cortex: Single-cell activity related to prior information about direction and extent of an intended movement. *Journal of Neurophysiology, 61*(3), 534–549.

Rosenbaum, D. A. (1980). Human movement initiation: Specification of arm, direction, and extent. *Journal of Experimental Psychology: General, 109*(4), 444–474. https://doi.org/10.1037/0096-3445.109.4.444

Schluter, N. D., Rushworth, M. F., Passingham, R. E., & Mills, K. R. (1998). Temporary interference in human lateral premotor cortex suggests dominance for the selection of movements. A study using transcranial magnetic stimulation. *Brain, 121*(5), 785–799. https://doi.org/10.1093/brain/121.5.785

Selen, L. P. J., Shadlen, M. N., & Wolpert, D. M. (2012). Deliberation in the motor system: Reflex gains track evolving evidence leading to a decision. *Journal of Neuroscience, 32*(7), 2276–2286. https://doi.org/10.1523/JNEUROSCI.5273-11.2012

Shenoy, K. V., Sahani, M., & Churchland, M. M. (2013). Cortical control of arm movements: A dynamical systems perspective. *Annual Review of Neuroscience, 36*(1), 337–359. https://doi.org/10.1146/annurev-neuro-062111-150509

Tanji, J., & Evarts, E. V. (1976). Anticipatory activity of motor cortex neurons in relation to direction of an intended movement. *Journal of Neurophysiology, 39*(5), 1062–1068. https://doi.org/10.1152/jn.1976.39.5.1062

Thura, D., & Cisek, P. (2014). Deliberation and commitment in the premotor and primary motor cortex during dynamic decision making. *Neuron, 81*(6), 1401–1416. https://doi.org/10.1016/j.neuron.2014.01.031

Weinrich, M., & Wise, S. P. (1982). The premotor cortex of the monkey. *Journal of Neuroscience, 2*(9), 1329–1345. https://doi.org/10.1523/JNEUROSCI.02-09-01329.1982

Wong, A. L., & Haith, A. M. (2017). Motor planning flexibly optimizes performance under uncertainty about task goals. *Nature Communications, 8*, 14624. https://doi.org/10.1038/ncomms14624

46 Visuomotor Adaptation Tasks as a Window into the Interplay between Explicit and Implicit Cognitive Processes

JORDAN A. TAYLOR AND SAMUEL D. McDOUGLE

ABSTRACT Motor learning is traditionally characterized as a function of *procedural memory*, distinct from episodic, semantic, and other forms of memory. Motor adaptation, in particular, describes the process of maintaining the calibration between motor commands and their desired sensory outcomes and has been characterized as a purely implicit, error-based learning process. Increasingly, however, evidence supports the idea that this common conception of motor adaptation is just one piece of the puzzle and should be expanded. Here we review recent evidence showing that when humans adapt visuomotor behaviors, they rely on multiple learning processes, with deliberate action selection helping to bootstrap the learning curve and flexibly maintain performance. We discuss the implications of these discoveries in short- and long-term motor learning.

Motor learning is often discussed in terms of the optimization of motor execution—that is, the specification of movement parameters that minimize noise and maximize accuracy, or the "how to do it" aspects of movement. This conception is the foundation of sensorimotor adaptation tasks, which are considered models for studying how the execution of actions is calibrated to align with desired goal states. But adaptive motor behavior needs to contend with another optimization problem: the intelligent selection of rewarding, attainable goals in the first place, or the "what to do" aspect of movement. This selection process is fundamental to the flexibility that is characteristic of a learned motor skill, though it is not known if or how it interacts with the calibration process. Recent research suggests that even in cases of learning very simple motor tasks, humans apply controlled, often explicit, reasoning processes to rapidly and flexibly select and adjust actions to improve motor performance. In this light, behavior in sensorimotor adaptation tasks involves improvements in action selection processes in addition to improvements in calibration processes.

Here we review recent psychophysical, neuropsychological, and neurophysiological evidence pertaining to how explicit selection processes are recruited throughout motor learning and how these processes may or may not interact with traditional implicit learning systems. Much of this evidence has come from sensorimotor adaptation tasks, which involve a perturbed relationship between motor commands and sensory feedback.

Although cognitive strategies were once central to theories of sensorimotor skill acquisition (Fitts & Posner, 1967), their role in motor learning has largely been overlooked in recent years. As Stanley and Krakauer (2013) have pointed out, one likely reason may be the seminal findings from patient H. M., which led to the distinction between explicit (declarative) and implicit (procedural) memory systems (Milner, Corkin, & Teuber, 1968). H. M. lacked the neural structures that support explicit memory formation, although he could apparently learn a new sensorimotor skill (e.g., mirror drawing). Thus, it would appear that explicit processes are unnecessary to sensorimotor learning. In the following sections, we hope to modify this interpretation.

The Sensorimotor Learning Curve

A number of behavioral tasks have been developed to study motor adaptation, including prism adaptation (Martin et al., 1996b), force field learning (Shadmehr & Mussa-Ivaldi, 1994), and visuomotor rotation (Krakauer et al., 2000; figure 46.1*A*). Regardless of the particular type of task, they all show similar learning curves (figure 46.1*B*) and have primarily been used as model tasks to study supervised, implicit error-based learning.

The canonical learning curve proceeds as follows (figure 46.1*A, B*): After a baseline phase, participants

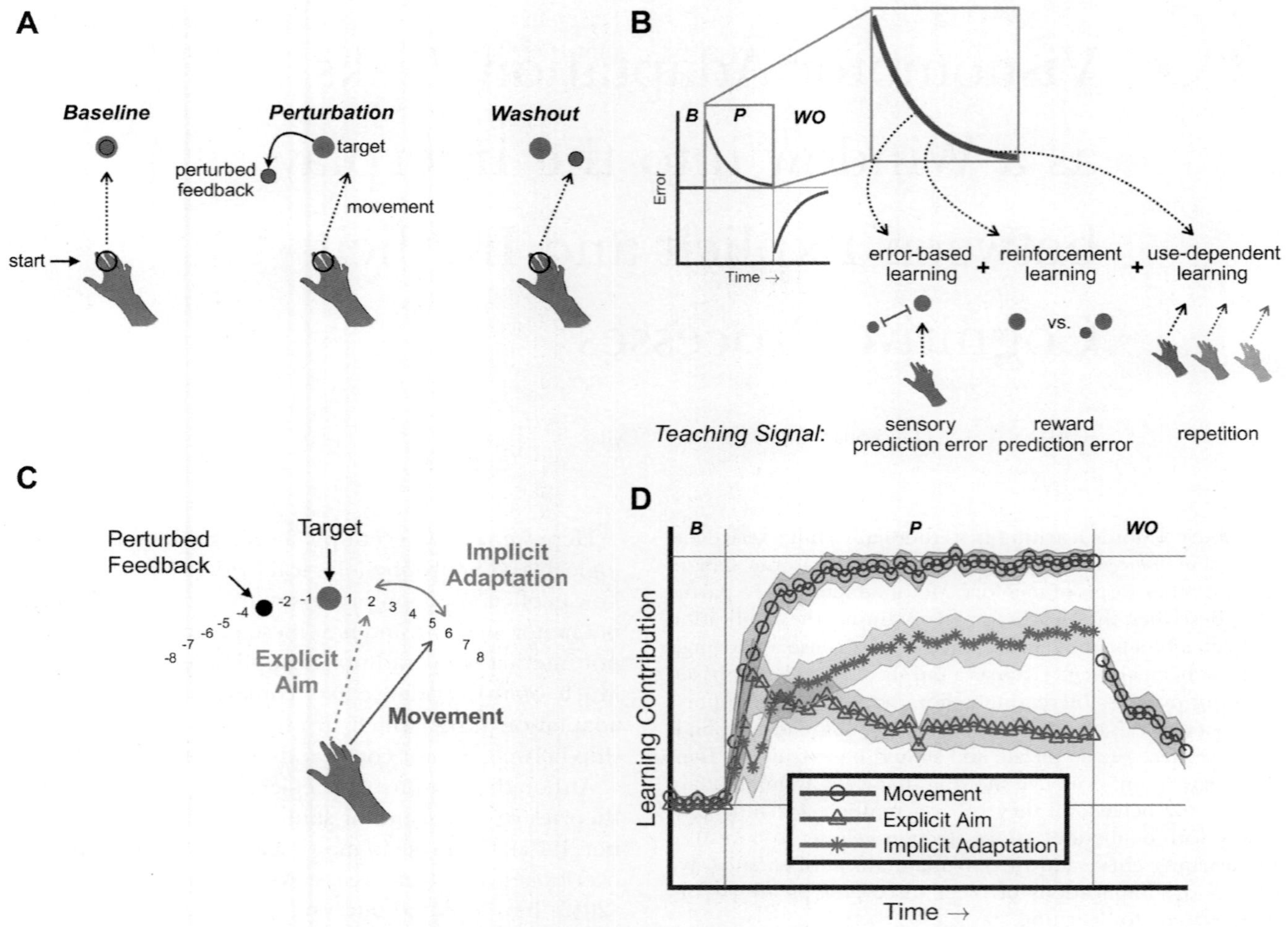

FIGURE 46.1 *A*, Standard adaptation task procedure (visuomotor rotation task is shown). Participants attempt to land a cursor (*red circle*) on a target (*green circle*), with vision of their limb occluded. *B*, Canonical adaptation learning curve over the baseline (B), perturbation (P), and washout (WO) phases depicted in panel (*A*), with a schematic of multiple learning processes making up the learning curve. Inspired by Huang et al. (2011) and Izawa and Shadmehr (2011). Learning processes rely on distinct teaching signals. *C*, Schematic of explicit report method used by Taylor, Krakauer, and Ivry (2014) and others, which separates explicit and implicit learning. *D*, Explicit and implicit learning contributions revealed by the reporting method depicted in (*C*). Adapted with permission from Taylor, Krakauer, and Ivry (2014).

experience large errors upon the introduction of a perturbation. Over time they adapt to these errors in a manner resembling a power function, eventually approaching a learning asymptote. Finally, in a washout phase, even though the perturbation is removed, adapted movements do not rapidly disappear—participants often show pronounced *aftereffects*, which are unlearned at a comparable rate to initial learning. Aftereffects are the gold standard of implicit adaptation, reflecting a new association between a goal (e.g., hitting a target at 90°) and a movement command (e.g., reaching at an angle of 120° instead of 90°). The dynamics of this learning function can be approximated by a Markovian *state-space* model of adaptation, where an internal state is updated trial by trial (Thoroughman & Shadmehr, 2000). Identifying the underlying processes and neural systems that give rise to this learning function has been a primary focus of research in the motor control field.

Error-Based Learning, Reinforcement Learning, and Use-Dependent Plasticity in Adaptation Tasks

One demonstration of the various processes underlying motor adaptation came from Huang et al. (2011), who dissociated multiple factors that appeared to drive different aspects of learning (figure 46.1*B*): sensory prediction error (e.g., a visual cursor's deviation from a movement direction), task success (e.g., a cursor hitting a target or not), and the direction of movement (e.g., the path of the hand). The authors were able to dissociate three distinct forms of learning that appeared to respond, respectively, to these different signals: (1) error-based adaptation, which responds to discrepancies between movement and feedback, (2) reinforcement learning, which reinforces movements that result in task success (i.e., reward), and (3) use-dependent learning, which merely reinforces repeated movement directions (figure 46.1*B*).

In another study, Izawa and Shadmehr (2011) showed how sensory prediction errors (the difference between expected and observed sensory feedback) and reward prediction errors (the difference between expected and observed reward feedback) drive different learning processes: the former was linked to a proprioceptive illusion, where at the end of learning participants misjudge the position of their hand in space. Moreover, while error-based learning led to broad generalization of the learned mapping to novel target locations, learning with reward prediction errors generalized more narrowly (and did not cause sensory recalibration). Taken together, these two studies showed how learning in a sensorimotor adaptation task is clearly multifaceted (figure 46.1*B*).

The cerebellum has been singled out as the primary locus of error-based recalibration via sensory prediction error (Tseng et al., 2007). The cerebellum is known to be vital for the learning and execution of coordinated movements (Thach, Goodkin, & Keating, 1992), and this learning is thought to rely on plasticity at the parallel fiber-Purkinje cell synapse (Albus, 1971; Marr, 1969): Parallel fibers carry a state representation that includes sensory context and current outgoing motor commands (e.g., efference copy) to the cerebellar cortex, while the second primary input to the Purkinje cells, the climbing fibers, provide a teaching signal when an unexpected sensory event occurs (i.e., a sensory prediction error). Long-term depression at the parallel fiber-Purkinje cell synapse induced by these teaching signals is thought to eventually lead to adapted motor behaviors that reduce error over time (Ito, 2006).

Consistent with a role for the cerebellum in error-based reach adaptation, cerebellar damage in humans leads to attenuated aftereffects in prism adaptation (Martin et al., 1996a; Weiner, Hallett, & Funkenstein, 1983). Similarly, patients with cerebellar damage show significant deficits in force field learning (Izawa, Criscimagna-Hemminger, & Shadmehr, 2012) and visuomotor rotation adaptation (Morehead et al., 2017; Tseng et al., 2007).

Positron emission topography (PET) and functional magnetic resonance imaging (fMRI) studies have revealed that a wide network of areas, including the cerebellum, prefrontal cortex, premotor and primary motor cortices, parietal cortex, and basal ganglia, are active during sensorimotor adaptation tasks (Krakauer et al., 2004; Seidler, Noll, & Chintalapati, 2006). As mentioned above, another process involved in sensorimotor learning is reward-based reinforcement: This process is thought to be homologous to standard model-free reinforcement learning (Daw et al., 2011), which involves reinforcing actions that directly lead to reward (e.g., hitting a target). This learning process is dubbed *model-free* because it does not require planning to future reward states but merely associates a running average of rewards with certain actions (Sutton & Barto, 1998).

This type of learning is likely reliant on dopamine-dependent reward prediction errors, which are computed in the midbrain and broadcast to a wide network that includes the basal ganglia and frontal cortex (Schultz, 1998). Patients with damage to the basal ganglia generally show a normal ability to adapt to sensory prediction errors (Smith & Shadmehr, 2005; Weiner, Hallett, & Funkenstein, 1983) but show diminished abilities, relative to controls, to rapidly relearn a perturbation days after initial training (Marinelli et al., 2009), suggesting a role for reinforcement learning in savings and relearning (Huang et al., 2011).

A third procedural process, use-dependent learning, is characterized as an "attractor" toward movements that are repeated (figure 46.1*B*; Huang et al., 2011): As a particular movement is repeated, similar movement commands are "pulled" toward the repeated direction (Verstynen & Sabes, 2011). This bias is primarily thought to result from plastic changes in primary motor cortex (Mawase et al., 2017), and the stimulation of motor cortex with transcranial magnetic stimulation (TMS) has been shown to enhance the effects of use-dependent learning (Bütefisch et al., 2004). This form of learning is essentially an acquired bias, described as a form of unsupervised learning (Doya, 2000).

We have just outlined three learning processes that are generally put under the "procedural" umbrella. Given the classic results of patient H.M. and the concomitant declarative/procedural divide, it has often been assumed that procedural processes fully account for motor learning. Do any nonprocedural processes subserve motor learning?

Deliberate Action Selection in Sensorimotor Adaptation

Various lines of evidence from prism adaptation studies suggest that *strategic control* plays a role in the rapid reduction of performance errors (Redding & Wallace, 1996). For instance, in a prism adaptation study by

Martin et al. (1996b), one individual appeared to immediately counteract the prism-induced shift after a single trial. Indeed, this individual reported using a deliberate "aiming" strategy. Upon explicit instruction to inhibit that strategy, the participant returned to an incremental learning function. How common are such deliberate strategies, and do they represent a fundamental aspect of motor learning?

Several studies in visuomotor rotation learning have provided indirect evidence for the concurrent operation of deliberate selection strategies and procedural learning. Benson et al. (2011) showed that explicitly describing a rotational perturbation to participants speeds their learning. Heuer and Hegele (2008) used verbal posttests to show that participants could provide a relatively accurate explicit spatial description of the perturbation at the end of learning, suggesting that they could have leveraged this knowledge during learning.

To directly probe explicit learning, Taylor, Krakauer, and Ivry (2014) developed a task in which participants explicitly reported their deliberate aiming strategy before every trial (figure 46.1*C*). This was accomplished by tiling the learning environment with numbered landmarks that corresponded to possible aiming directions. Participants' verbal reports could be converted to angular coordinates and subtracted from their actual movement directions on each trial, revealing the time course of implicit adaptation (figure 46.1*D*). Indeed, both explicit and implicit processes seemed to be active during learning, with the explicit selection of movements rapidly bootstrapping the learning curve and implicit learning slowly calibrating movements over time. Critically, the implicit learning curve was contiguous with observed aftereffects.

A follow-up study (McDougle, Bond, & Taylor, 2015) linked these two processes with a popular model of motor adaptation, the *two-state* model (Smith, Ghazizadeh, & Shadmehr, 2006). In this model, motor adaptation is shown to be parsimoniously explained by the concurrent operation of two subcomponents with different time constants: a "fast" process, which is fast and flexible, and a "slow" process, which is rigid and robust. Importantly, the assumption was that these two processes are qualitatively the same but quantitatively different. Results from McDougle, Bond, and Taylor (2015), however, suggest that the two proposed subcomponents of adaptation represent two fundamentally different learning systems: error-based implicit learning, which appears to map on to the slow process, and explicit action selection, which appears to map on to the fast process.

Explicit learning likely functions in force field adaptation as well: adding a cognitive load through a dual task causes decreased trial-by-trial adaptation to force perturbations (Taylor & Thoroughman, 2007), and this decrement is mediated by the difficulty of the secondary task (Taylor & Thoroughman, 2008). Awareness and instruction about force perturbations correlates with improved force adaptation (Hwang, Smith, & Shadmehr, 2006), and the fast process of force field learning appears to require access to declarative memory resources (Keisler & Shadmehr, 2010). Lastly, participants can verbally report using explicit strategies during force field adaptation, and the time course of these reports closely matches that seen in rotation tasks (McDougle, Bond, & Taylor, 2015).

The specific mechanisms driving explicit motor-learning processes are still being uncovered. Studies have shown that working-memory capacity correlates with faster rotation learning (Anguera et al., 2010), and this effect has been linked specifically to the explicit learning process (Christou et al., 2016). Eye gaze has been linked to explicit, but not implicit, learning (de Brouwer et al., 2018; Rand & Rentsh, 2014). Strategic processes are linked to high movement preparation times (Haith, Huberdeau, & Krakauer, 2015), implying the recruitment of executive control. Consistent with these behavioral results, regions in prefrontal cortex and premotor cortex are most active in the early phases of sensorimotor learning (Krakauer et al., 2004), when explicit strategies are especially important. As learning progresses, activity follows a rostrocaudal shift from frontal to parietal and subcortical areas (Krakauer et al., 2004), supporting a model where expensive computations may be slowly replaced by cached, overlearned responses (Haith & Krakauer, 2018).

The Interaction of Controlled and Automatic Motor Learning Processes

Given this multitude of learning processes, a fundamental open question concerns if and how they interact. In one clever study, Mazzoni and Krakauer (2006) found that explicit and implicit processes appear to be relatively independent: In a rotation task, after participants experienced two standard rotation trials, the experimenters intervened and instructed participants to aim at a landmark distal to the target that fully counteracted the rotation. Participants easily adopted this strategy and performed perfectly on the subsequent trial. However, after several trials participants' movements began to overcompensate and drift in the direction opposite the rotation, paradoxically making them less accurate.

This decidedly suboptimal behavior suggested that the error-based system was not responding to performance

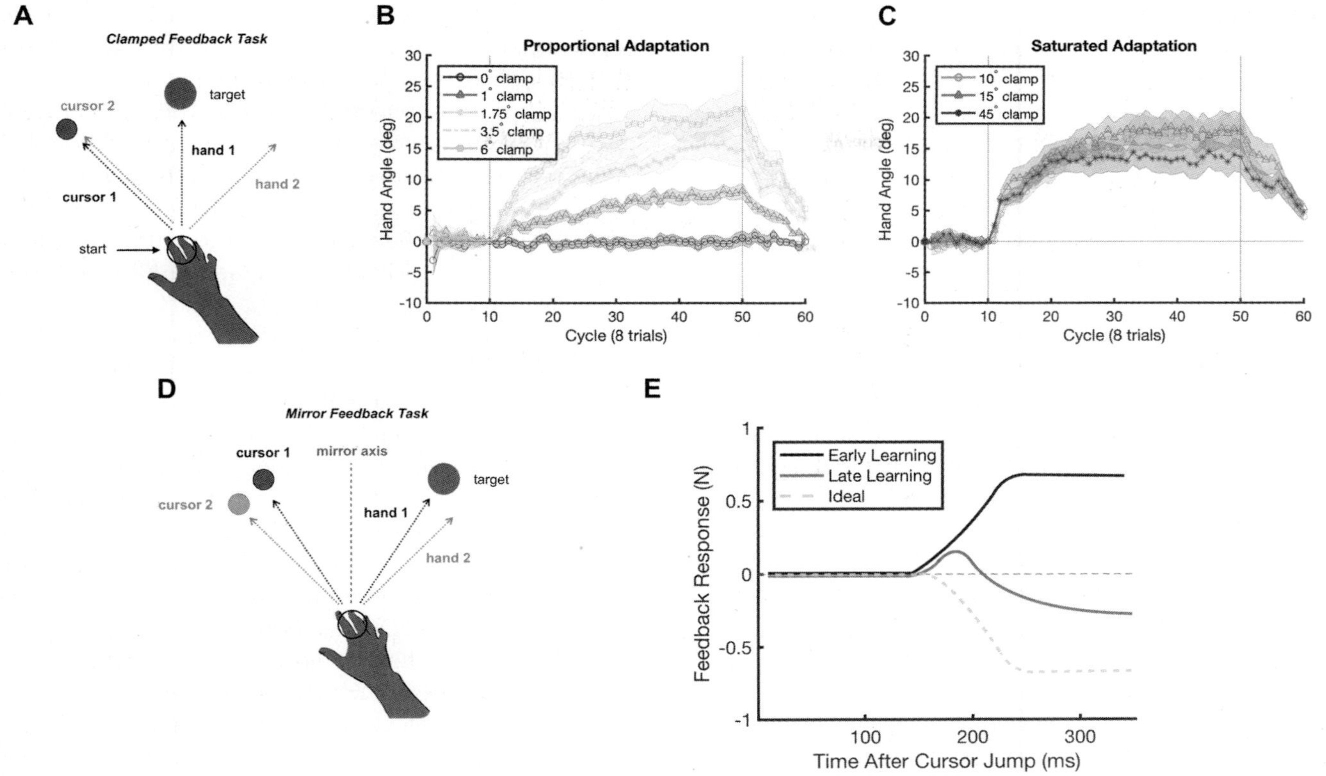

FIGURE 46.2 *A*, Clamped visual feedback used to isolate implicit learning processes. Adapted from Morehead et al. (2017). The hand will adapt in response to a sensory prediction, even though performance error is clamped. *B*, A small range of sensory prediction errors show scaling effects on the rate of learning, but (*C*) large sensory prediction errors do not scale learning rates. Adapted from Kim et al. (2018). *D*, Schematic of mirror task used in Telgen, Parvin, and Diedrichsen (2014). Standard adaptation causes an increase in error, suggesting that the task must be learned in other ways. *E*, Short-latency feedback responses to a displaced cursor do not show ideal adaptation even after prolonged exposure to a mirror perturbation. Adapted from Telgen, Parvin, and Diedrichsen (2014). (See color plate 51.)

errors in the task (i.e., the discrepancy between the observed feedback and the target) but rather to a task-irrelevant sensory prediction error (i.e., the discrepancy between the observed feedback and the predicted feedback, given the particular motor command). The rigidity of error-based learning was further highlighted by Morehead and colleagues (2017). They developed a task in which participants are instructed to aim directly at visual targets and ignore a consistent rotational error on every trial (figure 46.2*A*). Critically, the imposed error is of the same magnitude for every trial regardless of the participant's direction of movement. This *error-clamp* paradigm produced a surprising result: adaptation proceeded even though the error had no relevance to the participant's movement direction, and movements were adapted well past where they would need to go to fully compensate for the error (e.g., 25° of adaptation for a 7.5° rotational error). These results suggest that implicit adaptation occurs at a relatively fixed rate without regard to performance error, supporting the idea that the processes may not interact directly (Mazzoni & Krakauer, 2006). Furthermore, error-based learning is generally insensitive to the magnitude of sensory prediction errors (Bond & Taylor, 2015; Morehead et al., 2017), though it shows limited sensitivity when errors are small (figure 46.2*B*, *C*; Kim et al., 2018).

Given the above limitations, it is suggested that error-based cerebellar adaptation alone cannot fully solve the motor adaptation problem; rather, this process is best suited for retooling a well-trod internal model, or learned *control policy*. Consistent with this interpretation, Telgen, Parvin, and Diedrichsen (2014) further exposed the constraints on error-based learning using a mirror-reversed visuomotor transformation instead of a rotation. In mirror reversal, the solution to the task is not to adjust movements in the direction opposite the error direction but rather toward and then across the mirror axis (figure 46.2*D*). Here, a simple directionally sensitive algorithm would fail to counter the perturbation and would actually make performance worse by adapting movements away from the mirror axis. Indeed, after extensive exposure to a mirror reversal, appropriate motor commands for countering the mirror were rarely elicited—rather, they appear to be blunted—suggesting that participants had not formed an automatic, mirror-reversed internal model (figure 46.2*E*; Telgen, Parvin, & Diedrichsen, 2014). After rotation learning, however, adapted motor commands were highly reliable and were easily expressed at short reaction times, consistent with the updating of an existing control policy. These results suggest that when humans are required to learn a novel, complex visuomotor mapping, error-based adaptation alone cannot forge a new automatic control policy—that process likely requires weeks of training and perhaps other neural systems, like the basal ganglia. In this sense, mirror-reversal learning is similar to learning a novel motor skill from scratch.

Critically, participants in the mirror-reversal study (Telgen, Parvin, & Diedrichsen, 2014) were given limited reaction time to block time-consuming explicit planning processes. It is possible that performance under mirror reversal would be much improved given ample time to deliberate on each trial; indeed, explicit knowledge of the environment does affect how simple motor skills like the serial reaction time task (SRT) are learned and which brain networks are recruited during learning (Grafton, Hazeltine, & Ivry, 1995).

Recent results have put forward a potential interaction between deliberate action selection and implicit adaptation. Evidence has been provided that adaptation is maximal at the most visited aiming direction (Day et al., 2016)—that is, the actual parameter that undergoes implicit adaption may be tied to the volitional motor plan itself, rather than the downstream muscle commands (but see Gonzalez Castro, Monsen, & Smith, 2011). This can be tested using a generalization task, in which aftereffects are probed at various locations. The largest aftereffects are seen when, in the washout phase, participants reach in the direction they most commonly reported aiming toward during learning, not the direction they actually moved to (McDougle, Bond, & Taylor, 2017). Thus, the explicit planning stage may determine which internal representation is updated by subsequent sensory prediction errors.

This interpretation is echoed in interference studies, in which two opposing force field perturbations are induced over the same movement direction (on different trials). Sheahan, Franklin, and Wolpert (2016) showed that interfering perturbations can be learned if each is associated with a different movement plan, operationalized by different "follow-through" targets that participants would occasionally have to move toward at the end of the reach (i.e., after pushing through the force field). Without cueing distinct plans, massive interference is observed in such tasks. These results are consistent with previous work (Hirashima & Nozaki, 2012; Krakauer et al., 2006) showing that the context of learning a specific perturbation (e.g., the effector used during learning) may act as a cue. This earlier work led to a top-down cognitive interpretation for interference effects in visuomotor adaptation experiments (Krakauer, Ghez, & Ghilardi, 2005; Krakauer et al., 2006), an interpretation that fits with the Sheahan, Franklin, and Wolpert (2016) results.

The neural interpretation is that the plan sets the initial conditions of a neural trajectory in the downstream dynamic system that directly controls motor output: recent results using a brain-machine interface (BMI) in nonhuman primates revealed that training on a visuomotor rotation "covertly" (i.e., using the BMI but not eliciting actual arm movements) transfers to overt motor behavior on that same rotation and that this transfer is linked to corresponding motor cortical activity in the planning stages of both contexts (Vyas et al., 2018). It is likely that in many settings the planning stage encompasses explicit selection of a desired action.

Explicit action selection can rapidly increase the likelihood of arriving at a rewarding state, allowing implicit, model-free reinforcement processes to then reinforce movements that maintain that state. But what if the environment changes? Learning environments are often not static, undergoing changes over time and contexts. One phenomenon in motor learning is *savings*, whereby an association that appears to have been extinguished is learned more rapidly upon a second exposure.

Huang et al. (2011) suggested that savings is attributable to reinforcement learning, though whether this effect was driven by explicit strategizing or not was unclear. In a recent study, Morehead et al. (2015) showed that savings is attributable to explicit selection processes but not implicit processes. Moreover, participants can leverage contextual cues to "re-aim" and rapidly improve their performance when the environment changes (Morehead et al., 2015), highlighting the contextual sensitivity of selection strategies. In another study Huberdeau, Haith, and Krakauer (2015) showed that even very brief exposure to a rotation (e.g., as few as five trials) leads to faster relearning of that rotation the next day. This result is not parsimoniously explained by error-based adaptation, which acts on much slower timescales (McDougle, Bond, & Taylor, 2015).

Haith, Huberdeau, and Krakauer (2015) used a forced-response time paradigm to dissociate explicit planning processes (which require ample preparation time) and implicit adaptation (which requires very little preparation time; Telgen, Parvin, & Diedrichsen, 2014).

Savings was attributable to the process that required high preparation time, consistent with the role of deliberative planning in savings. Surprisingly, the importance of explicit processes for savings has been corroborated in gait adaptation (Roemmich & Bastian, 2015), which is generally considered an implicit learning process.

Explicit Learning and a Role for Factual Knowledge in Motor Learning

Why have so many learning processes? Intuitively, for an optimal learner, it seems useful to have multiple processes that act on different timescales and respond to different error signals (Huberdeau, Krakauer, & Haith, 2015); different systems could be suited to different points on the learning curve or to different tasks altogether. One relevant analogy is the dissociation of model-based and model-free reinforcement learning (Daw et al., 2011; Haith & Krakauer, 2013). Model-based learning requires an explicit model of transition probabilities between responses and sensory inputs, and model-free learning merely reinforces directly rewarded responses. One speculation is that responses reflecting model-based computations (which could initially rely on working memory) are themselves made automatic (Economides et al., 2015; Huberdeau, Krakauer, & Haith, 2018). Perhaps during learning there is a transition from explicit model-based planning to an intermediate form of working memory-based stimulus-response association (S-R) and finally to habitual, model-free S-R. In this model, cognitive strategies and implicit learning work cooperatively (Collins & Frank, 2018).

While direct interactions between explicit and implicit learning processes are poorly understood, explicit learning can counter the sometimes negative effects of implicit learning (Taylor & Ivry, 2011). Moreover, as mentioned above, results from generalization (McDougle, Bond, & Taylor, 2017) and interference studies (Hirashima & Nozaki, 2012; Sheahan, Franklin, & Wolpert, 2016) suggest that volitional planning carves out the neural representations that undergo implicit adaptation. In this view, explicit selection is the fundamental first step of learning, providing the foundation on which subtle, incremental performance improvements are built. This logic could be extended to the role of reinforcement as well, where an explicit strategy guides the learner to the subset of movements that produce rewarding states, and these movements are then incrementally reinforced (Holland et al., 2018).

The sequestering of *procedural* memory in cognitive neuroscience can be attributed in part to the seminal studies on hippocampal patient H. M. (Milner, 1962). In a classic experiment, H. M. appeared to show improvements in mirror tracing, a difficult task in which a complex shape must be traced under mirror-reversed vision of the hand. H. M. appeared to improve at this, and similar motor tasks, without explicit awareness that he had ever performed the task before (Milner, Corkin, & Teuber, 1968; Shadmehr, Brandt, & Corkin, 1998).

Critically, two facts about the studies on H. M. are relevant: First, while H. M. did show some improvements, his performance was often well below that of controls, suggesting that some aspect of performance was impeded by medial temporal lobe damage. If learning such tasks was purely procedural, why should he experience any deficits?

Second, and more importantly, for H. M. to properly perform the motor tasks each day he had to receive explicit instructions on both what the task required and how to do it (Stanley & Krakauer, 2013). Although H. M. could follow these instructions to set about doing the task, he would forget the instructions between sessions. Similar results have been found in studies of amnesiac patients learning to use novel tools: without daily explicit instruction, amnesiac patients cannot begin to solve such tasks (Roy & Park, 2010).

Access to episodic memory may play a key role in motor learning by setting the "initial state" of the learning process—that is, to rapidly reduce the dimensionality of the learning problem by selecting a manifold of correct actions to be performed. One possibility is that the medial temporal lobe helps the learner recognize the learning context and recall the appropriate subset of actions that have been useful in similar situations in the past. Indeed, *episodic control* has recently become an important concept in reinforcement learning (Gershman & Daw, 2017).

Moving forward, motor learning should be understood as recruiting a full taxonomy of memory systems, even in the context of simple adaptation tasks. That a wide network of neural systems serves motor adaptation should not be surprising; indeed, adaptive, precise movement is perhaps the most fundamental function of the central nervous system.

REFERENCES

Albus, J. S. (1971). A theory of cerebellar function. *Mathematical Biosciences, 10*, 25–61.

Anguera, J. A., Reuter-Lorenz, P. A., Willingham, D. T., & Seidler, R. D. (2010). Contributions of spatial working memory to visuomotor learning. *Journal of Cognitive Neuroscience, 22*, 1917–1930.

Benson, B. L., Anguera, J. A., & Seidler, R. D. (2011). A spatial explicit strategy reduces error but interferes with sensorimotor adaptation. *Journal of Neurophysiology, 105*, 2843–2851.

Bond, K. M., & Taylor, J. A. (2015). Flexible explicit but rigid implicit learning in a visuomotor adaptation task. *Journal of Neurophysiology,* 113, 3836–3849.

Bütefisch, C. M., Khurana, V., Kopylev, L., & Cohen, L. G. (2004). Enhancing encoding of motor memory in the primary motor cortex by cortical stimulation. *Journal of Neurophysiology, 91,* 2110–2116.

Christou, A. I., Miall, R. C., McNab, F., & Galea, J. M. (2016). Individual differences in explicit and implicit visuomotor learning and working memory capacity. *Scientific Reports, 6,* 36633.

Collins, A. G. E., & Frank, M. J. (2018). Within- and across-trial dynamics of human EEG reveal cooperative interplay between reinforcement learning and working memory. *Proceedings of the National Academy of Sciences of the United States of America, 115,* 2502–2507.

Daw, N. D., Gershman, S. J., Ben Seymour, Dayan, P., & Dolan, R. J. (2011). Model-based influences on humans' choices and striatal prediction errors. *Neuron, 69,* 1204–1215.

Day, K. A., Roemmich, R. T., Taylor, J. A., & Bastian, A. J. (2016). Visuomotor learning generalizes around the intended movement. *eNeuro,* 3(2), e0005–0016.

de Brouwer, A. J., Albaghdadi, M., Flanagan, R., & Gallivan, J. P. (2018). Using gaze behavior to parcellate the explicit and implicit contributions to visuomotor learning. *Journal of Neurophysiology, 120,* 1602–1615. doi:10.1152/jn.00113.2018

Doya, K. (2000). Complementary roles of basal ganglia and cerebellum in learning and motor control. *Current Opinion in Neurobiology, 10,* 732–739.

Economides, M., Kurth-Nelson, Z., Lubbert, A., Guitart-Masip, M., & Dolan, R. J. (2015). Model-based reasoning in humans becomes automatic with training. *PLoS Computational Biology, 11,* e1004463.

Fitts, P. M., & Posner, M. I. (1967). *Human performance: Basic concepts in psychology series.* Belmont, CA: Brooks/Cole.

Gershman, S. J., & Daw, N. D. (2017). Reinforcement learning and episodic memory in humans and animals: An integrative framework. *Annual Review of Psychology, 68,* 101–128.

Gonzalez Castro, L. N., Monsen, C. B., & Smith, M. A. (2011). The binding of learning to action in motor adaptation. *PLoS Computational Biology, 7,* e1002052.

Grafton, S. T., Hazeltine, E., & Ivry, R. B. (1995). Functional mapping of sequence learning in normal humans. *Journal of Cognitive Neuroscience, 7,* 497–510.

Haith, A. M., Huberdeau, D. M., & Krakauer, J. W. (2015). The influence of movement preparation time on the expression of visuomotor learning and savings. *Journal of Neuroscience, 35,* 5109–5117.

Haith, A. M., & Krakauer, J. W. (2013). Model-based and model-free mechanisms of human motor learning. In M. Richardson, M. Riley, & K. Shockley (Eds.), *Progress in motor control* (pp. 1–21). New York: Springer.

Haith, A. M., & Krakauer, J. W. (2018). The multiple effects of practice: Skill, habit and reduced cognitive load. *Current Opinion in Behavioral Sciences, 20,* 196–201.

Heuer, H., & Hegele, M. (2008). Adaptation to visuomotor rotations in younger and older adults. *Psychology and Aging, 23,* 190–202.

Hirashima, M., & Nozaki, D. (2012). Distinct motor plans form and retrieve distinct motor memories for physically identical movements. *Current Biology, 22,* 432–436.

Holland, P., Codol, O., & Galea, J. M. (2018). Contribution of explicit processes to reinforcement-based motor learning. *Journal of Neurophysiology, 119,* 2241–2255.

Huang, V. S., Haith, A., Mazzoni, P., & Krakauer, J. W. (2011). Rethinking motor learning and savings in adaptation paradigms: Model-free memory for successful actions combines with internal models. *Neuron, 70,* 787–801.

Huberdeau, D. M., Haith, A. M., & Krakauer, J. W. (2015). Formation of a long-term memory for visuomotor adaptation following only a few trials of practice. *Journal of Neurophysiology, 114,* 969–977.

Huberdeau, D. M., Krakauer, J. W., & Haith, A. M. (2015). Dual-process decomposition in human sensorimotor adaptation. *Current Opinion in Neurobiology, 33,* 71–77.

Huberdeau, D. M., Krakauer, J. W., & Haith, A. M. (2018). Practice induces a qualitative change in the memory representation for visuomotor learning. *bioRxiv.* doi:10.1101/226415

Hwang, E. J, Smith, M. A., & Shadmehr, R. (2006). Dissociable effects of the implicit and explicit memory systems on learning control of reaching. *Experimental Brain Research, 173,* 425–437.

Ito, M. (2006). Cerebellar circuitry as a neuronal machine. *Progress in Neurobiology, 78,* 272–303.

Izawa, J., Criscimagna-Hemminger, S. E., & Shadmehr, R. (2012). Cerebellar contributions to reach adaptation and learning sensory consequences of action. *Journal of Neuroscience, 32,* 4230–4239.

Izawa, J., & Shadmehr, R. (2011). Learning from sensory and reward prediction errors during motor adaptation. *PLoS Computational Biology, 7,* e1002012.

Keisler, A., & Shadmehr, R. (2010). A shared resource between declarative memory and motor memory. *Journal of Neuroscience, 30,* 14817–14823.

Kim, H. E., Morehead, J. R., Parvin, D. E., & Ivry, R. B. (2018). Invariant errors reveal limitations in motor correction rather than constraints on error sensitivity. *Communications Biology, 1,* 19.

Krakauer, J. W., Ghez, C., & Ghilardi, M. F. (2005). Adaptation to visuomotor transformations: Consolidation, interference, and forgetting. *Journal of Neuroscience, 25,* 473–478.

Krakauer, J. W., Ghilardi, M.-F., Mentis, M., Barnes, A., Veytsman, M., Eidelberg, D., & Ghez, C. (2004). Differential cortical and subcortical activations in learning rotations and gains for reaching: A PET study. *Journal of Neurophysiology, 91,* 924–933.

Krakauer, J. W., Mazzoni, P., Ghazizadeh, A., Ravindran, R., & Shadmehr, R. (2006). Generalization of motor learning depends on the history of prior action. *PLoS Biology, 4,* e316.

Krakauer, J. W., Pine, Z. M., Ghilardi, M. F., & Ghez, C. (2000). Learning of visuomotor transformations for vectorial planning of reaching trajectories. *Journal of Neuroscience, 20,* 8916–8924.

Marinelli, L., Crupi, D., Di Rocco, A., Bove, M., Eidelberg, D., Abbruzzese, G., & Ghilardi, M. F. (2009). Learning and consolidation of visuo-motor adaptation in Parkinson's disease. *Parkinsonism & Related Disorders, 15,* 6–11.

Marko, M. K., Haith, A. M., Harran, M. D., & Shadmehr, R. (2012). Sensitivity to prediction error in reach adaptation. *Journal of Neurophysiology, 108,* 1752–1763.

Marr, D. (1969). A theory of cerebellar cortex. *Journal of Physiology, 202,* 437–470.

Martin, T. A., Keating, J. G., Goodkin, H. P., Bastian, A. J., & Thach, W. T. (1996a). Throwing while looking through prisms: I. Focal olivocerebellar lesions impair adaptation. *Brain, 119*, 1183–1198.

Martin, T. A., Keating, J. G., Goodkin, H. P., Bastian, A. J., & Thach, W. T. (1996b). Throwing while looking through prisms: II. Specificity and storage of multiple gaze-throw calibrations. *Brain, 119*, 1199–1211.

Mawase, F., Uehara, S., Bastian, A., & Celnik, P. (2017). Motor learning enhances use-dependent plasticity. *Journal of Neuroscience, 37*, 2673–2685.

Mazzoni, P., & Krakauer, J. W. (2006). An implicit plan overrides an explicit strategy during visuomotor adaptation. *Journal of Neuroscience, 26*, 3642–3645.

McDougle, S. D., Bond, K. M., & Taylor, J. A. (2015). Explicit and implicit processes constitute the fast and slow processes of sensorimotor learning. *Journal of Neuroscience, 35*, 9568–9579.

McDougle, S. D., Bond, K. M., & Taylor, J. A. (2017). Implications of plan-based generalization in sensorimotor adaptation. *Journal of Neurophysiology, 118*, 383–393.

Milner, B. (1962). Les troubles de la memoire accompagnant des lesions hippocampiques bilaterales. In *Physiologie de l'hippocampe*, 257–272. Paris: Centre National de la Recherche Scientifique.

Milner, B., Corkin, S., & Teuber, H. L. (1968). Further analysis of the hippocampal amnesic syndrome: 14-year follow-up study of HM. *Neuropsychologia, 6*, 215–234.

Morehead, J. R., Qasim, S. E., Crossley, M. J., & Ivry, R. (2015). Savings upon re-aiming in visuomotor adaptation. *Journal of Neuroscience, 35*, 14386–14396.

Morehead, J. R., Taylor, J. A., Parvin, D. E., & Ivry, R. B. (2017). Characteristics of implicit sensorimotor adaptation revealed by task-irrelevant clamped feedback. *Journal of Cognitive Neuroscience, 29*, 1061–1074.

Rand, M. K., & Rentsch, S. (2014). Gaze locations affect explicit process but not implicit process during visuomotor adaptation. *Journal of Neurophysiology, 113*, 88–99.

Redding, G. M., & Wallace, B. (1996). Adaptive spatial alignment and strategic perceptual-motor control. *Journal of Experimental Psychology, 22*, 379.

Roemmich, R. T., & Bastian, A. J. (2015). Two ways to save a newly learned motor pattern. *Journal of Neurophysiology, 113*, 3519–3530.

Roy, S., & Park, N. W. (2010). Dissociating the memory systems mediating complex tool knowledge and skills. *Neuropsychologia, 48*, 3026–3036.

Schultz, W. (1998). Predictive reward signal of dopamine neurons. *Journal of Neurophysiology, 80*, 1–27.

Seidler, R. D., Noll, D. C., & Chintalapati, P. (2006). Bilateral basal ganglia activation associated with sensorimotor adaptation. *Experimental Brain Research, 175*, 544–555.

Shadmehr, R., Brandt, J., & Corkin, S. (1998). Time-dependent motor memory processes in amnesic subjects. *Journal of Neurophysiology, 80*, 1590–1597.

Shadmehr, R., & Mussa-Ivaldi, F. A. (1994). Adaptive representation of dynamics during learning of a motor task. *Journal of Neuroscience, 14*, 3208–3224.

Sheahan, H. R., Franklin, D. W., & Wolpert, D. M. (2016). Motor planning, not execution, separates motor memories. *Neuron, 92*, 773–779.

Smith, M. A., Ghazizadeh, A., & Shadmehr, R. (2006). Interacting adaptive processes with different timescales underlie short-term motor learning. *PLoS Biology, 4*, e179.

Smith, M. A., & Shadmehr, R. (2005). Intact ability to learn internal models of arm dynamics in Huntington's disease but not cerebellar degeneration. *Journal of Neurophysiology, 93*, 2809–2821.

Stanley, J., & Krakauer, J. W. (2013). Motor skill depends on knowledge of facts. *Frontiers in Human Neuroscience, 7*, 503.

Sutton, R. S., & Barto, A. G. (1998). *Reinforcement learning: An introduction.* Cambridge, MA: MIT Press.

Taylor, J. A., & Ivry, R. B. (2011). Flexible cognitive strategies during motor learning. *PLoS Computational Biology, 7*, e1001096.

Taylor, J. A., Krakauer, K. W., & Ivry, R. B. (2014). Explicit and implicit contributions to learning in a sensorimotor adaptation task. *Journal of Neuroscience, 34*, 3023–3032.

Taylor, J. A., & Thoroughman, K. A. (2007). Divided attention impairs human motor adaptation but not feedback control. *Journal of Neurophysiology, 98*, 317–326.

Taylor, J. A., & Thoroughman, K. A. (2008). Motor adaptation scaled by the difficulty of a secondary cognitive task. *PLoS One, 3*, e2485.

Telgen, S., Parvin, D., & Diedrichsen, J. (2014). Mirror reversal and visual rotation are learned and consolidated via separate mechanisms: Recalibrating or learning de novo? *Journal of Neuroscience, 34*, 13768–13779.

Thach, W. T., Goodkin, H. P., & Keating, J. G. (1992). The cerebellum and the adaptive coordination of movement. *Annual Review of Neuroscience, 15*, 403–442.

Thoroughman, K. A., & Shadmehr, R. (2000). Learning of action through adaptive combination of motor primitives. *Nature, 407*, 742–747.

Tseng, Y. W., Diedrichsen, J., Krakauer, J. W., Shadmehr, R., & Bastian, A. J. (2007). Sensory prediction errors drive cerebellum-dependent adaptation of reaching. *Journal of Neurophysiology, 98*, 54–62.

Verstynen, T., & Sabes, P. N. (2011). How each movement changes the next: An experimental and theoretical study of fast adaptive priors in reaching. *Journal of Neuroscience, 31*, 10050–10059.

Vyas, S., Even-Chen, N., Stavisky, S. D., Ryu, S. I., Nuyujukian, P., & Shenoy, K. V. (2018). Neural population dynamics underlying motor learning transfer. *Neuron, 97*, 1177–1186.

Weiner, M. J., Hallett, M., & Funkenstein, H. H. (1983). Adaptation to lateral displacement of vision in patients with lesions of the central nervous system. *Neurology, 33*, 766–766.

47 Apraxia: A Disorder at the Cognitive-Motor Interface

LAUREL J. BUXBAUM AND SOLÈNE KALÉNINE

ABSTRACT Limb apraxia is a complex disorder of skilled movement characterized by spatiotemporal errors in movement imitation and pantomime, difficulty selecting object-appropriate actions, and deficits in tool action knowledge. Despite its demonstrated relevance for our understanding of the cognitive neuroscience of skilled action, it has received relatively little study. Classical accounts of the disorder focused on a classification approach to apraxia's numerous patterns of symptom dissociation and association, usually without accompanying neuroanatomic data, hindering scientific progress. Recent advances in approaches to lesion analysis and corroborating data from functional neuroimaging studies of neurotypical subjects have enabled a new understanding of a specialized praxis network in the left hemisphere. After a brief discussion of traditional accounts, this chapter provides a review of our current neurocognitive understanding of this praxis system. In particular, three major clusters of behaviors reflecting damage to conceptual, spatiotemporal, and selection-based components of skilled action planning are discussed, attributable to posterior temporal, inferior parietal, and frontal network nodes and their interconnections. We conclude by offering suggestions for several future research directions that will benefit the study of apraxia, its rehabilitation, and our broad understanding of action imitation and tool use.

Limb apraxia is a poorly understood disorder of skilled action, imitation, action recognition, and (to a lesser degree) tool use, occurring in nearly 60% of individuals who have suffered a left-hemisphere stroke. It is also prominent in corticobasal degeneration, Alzheimer's disease, and behavioral variant frontotemporal dementia. Because much of our understanding of the disorder derives from studies of individuals with focal lesions, this chapter is concerned primarily with apraxia in left-hemisphere stroke and how apraxia—a disorder at the cognitive-motor interface—has elucidated the representation and planning of object-related actions in the neurotypical mind and brain.

Limb apraxia is a disorder of *action*, rather than motor planning or control. That is, apraxia occurs despite the relatively unimpaired planning and programming of specific muscle movements. We (and others) use the term *action* rather broadly to refer to *visual* memories (hereafter, *representations*) of what stored actions *look like*, *kinesthetic/proprioceptive* representations of how familiar actions *feel*, *abstract* representations of kinematic shape and motion trajectories, and, importantly, the processes that *transform* information from one of these formats to another and from these formats to motor plans.

The term *limb apraxia* refers to a cluster of related symptoms, some or all of which may be present in any given individual. Nearly every possible dissociation of clinical symptoms has been reported. Nonetheless, apraxia's primary clinical hallmarks are the presence of errors in the shape, amplitude, and/or timing of movements (hereafter, *spatiotemporal errors*), in the pantomime of manipulable object-related (hereafter, *tool)* actions to the sight of tools or upon command, and in imitation of either or both meaningful and novel movements. Clinically evident spatiotemporal errors may also be seen in actual tool use, though these tend to be more subtle. Critically, errors occur despite adequate strength and motor control and are prominent even in the ipsilesional, nonparetic left hand. Deficits in action recognition are also prominent in some patients, indicating that the disorder may affect representations of the appearance of skilled actions. Other apraxia patients may have no difficulty recognizing actions yet produce spatiotemporal errors when required to imitate meaningless actions. Some patients present a combination of recognition and production deficits, while still others are particularly prone to errors pantomiming the use of tools associated with competing action alternatives. The relationship of deficits in the prominent hallmarks of limb apraxia—imitation of meaningful and/or meaningless actions, pantomime to the sight of tools, "settling on" a single appropriate action, and gesture recognition—remains a perplexing challenge for many clinicians and researchers.

Apraxia significantly affects stroke survivors' disability and caregiver burden and is the single greatest predictor of disability in activities of daily living in left-hemisphere stroke. Despite this fact, it is the subject of surprisingly little research. Prior studies have often been conducted with small samples of fewer than 20 patients, and there is continued disagreement over apraxia's underlying mechanisms and whether there are distinguishable subtypes. This chapter will focus upon recent advances in the characterization of apraxic performance that have guided us to a new neurocognitive understanding of the praxis system.

Classic Apraxia Subtypes and Neuroanatomy

In the early 1900s, Hugo Liepmann advanced the hypothesis that conceptualizing goal-directed movements and converting concepts into motor commands could be disrupted at various processing stages. Two of the several apraxia subtypes he distinguished remain influential in modern taxonomies (Liepmann, 1988). The first, ideomotor apraxia, was posited to reflect difficulty retrieving kinematic patterns or integrating these patterns into a movement plan. Spatial errors, substitutions, and omissions were denoted as typical. The second type, ideational apraxia, was described as a loss of the concept underlying complex multistep actions, including the sequential and spatial "procedure." Deficiencies in this concept were proposed to result in errors of omission or the selection of incorrect tools. In contrast to ideomotor apraxia, the imitation of simple movements was proposed to be preserved. Another prominent account (DeRenzi & Lucchelli, 1988) advanced a similar view but noted that sequence errors did not occur; instead, errors of misuse, omission, and mislocation were said to be characteristic.

Note the similarities between DeRenzi's description of ideational apraxia and Liepmann's definition of ideomotor apraxia. Moreover, diagnostic clarity is further complicated by an additional apraxia designation—so-called frontal apraxia or action disorganization syndrome—focused on errors of action that occur primarily during complex, temporally extended, naturalistic tasks but that overlap with many descriptions of ideational apraxia (Luria, 1966). Due to inconsistencies of this sort in the proposed hallmarks of apraxia subtypes and the fact that errors of many types usually co-occur in the same patients, the classic syndromes have hindered theoretical and clinical understanding of the disorder.

The Cognitive Praxis Model

In the 1970s researchers began to focus on explaining the multiple patterns of association and dissociation that may be observed in apraxia syndromes (e.g., Geschwind, 1975). The most influential of the information-processing models of that time was the cognitive praxis model of Gonzalez Rothi et al. (1991), which contains numerous separate components that may be selectively impaired. On this model, memory traces of skilled movements are stored in an action lexicon located in the left inferior parietal lobe (IPL), which is translated into an *innervatory* pattern for actual gesture production. Further, gesture production may be achieved both directly and indirectly. Performing gestures via the indirect route involves accessing representations in the *action lexicon*—the stored repository of *space-time representations* of familiar actions. Alternatively, the imitation of gestures (whether meaningful or meaningless) can occur via a route that directly translates the visual stimulus into a motor plan, as evidenced by preserved gesture imitation with impaired gesture to the sight of tools. Deficits in the action lexicon were said to account for patterns of impairment that encompass both perception and production deficits. In contrast, impairments restricted to gesture production were posited to be due to deficits in the implementation of the motor plan. Finally, lesions of the praxis conceptual system subserving knowledge of the functions of tools and the relationships between tools and recipient objects were proposed to be the source of *ideational* apraxia.

Several refinements of the cognitive praxis model have been proposed in recent years (e.g., Randerath et al., 2011). Regardless of their details, the essential insight that there are both conceptual (or indirect) and nonconceptual (or direct) routes to praxis has persisted in some form in all of these and other models and, with some modifications, remains viable today.

Informative Dissociations and the Development of Neurocognitive Models of Praxis

The cognitive models previously described are not neuroanatomically grounded, and it has been difficult to extract the meaningful functional neuroanatomic organizational principles of the praxis system. In this context, however, there are several behavioral dissociations between preserved and deficient skills on ostensibly similar tasks that have fueled additional theoretical development.

Perhaps the most striking of these dissociations is apraxics' relatively preserved ability to produce reaching and grasping actions based on the currently visualized structure of objects despite flagrant errors in action tasks not entailing actual object interactions, including pantomime and the imitation of both meaningful and meaningless movements. This disparity may be related to the observation that the left IPL, frequently damaged in apraxia, is an important locus of processes involved in calculating predicted actions (perhaps through simulation) and in translating stored *visual* representations of action appearance into kinesthetic/proprioceptive and abstract (amodal) formats that can be used in motor planning.

A second dissociation of interest is that between the kinematic and conceptual aspects of action processing. A relatively "pure" test of kinematic processing is the imitation of meaningless movements. Meaningless movements cannot be retrieved from memory but must be computed in real time based on the target posture or trajectory to be imitated. On the other hand, a relatively "pure" test of conceptual action processing is gesture recognition, which requires no overt motor output but taxes the ability to link incoming visual information to aspects of cognition and language enabling access to the gestures' identities. Indeed, performance on these two tasks critically depends on the IPL or posterior temporal cortex (PTC), respectively, either (or both) of which may be lesioned in patients with apraxia.

These pairs of dissociations lay the groundwork for our understanding of two important functional gradients in the praxis system.

Dorsal-Ventral and Anterior-Posterior Functional Gradients in the Praxis System

Milner and Goodale (1993) described a modification of the classic two-visual-stream model (Ungerleider & Mishkin, 1982) that includes a ventral *What* stream projecting from the occipital to inferior temporal cortex, which constructs representations slowly via repeated perceptual exposure. In contrast, the dorsal *How* stream, which projects from the occipital cortex to the posterior parietal cortex, processes current perceptual information rapidly and online for the control of visually guided action.

Neuroanatomical interstream connectivity, as well as patterns of performance in patients, indicate that the two streams are highly interactive. Additionally, based both on patient data and monkey neuroanatomical studies, a third stream has been proposed, branching off ventrally from the dorsal stream (Rizzolatti & Matelli, 2003). In the monkey, this ventrodorsal stream starts in the posterior temporal lobule and projects to the IPL and has been characterized as important for understanding and organizing actions. In humans, lesions to the ventrodorsal stream with sparing of the dorsodorsal stream give rise to limb apraxia without deficits in visually guided reaching and grasping, thereby explaining the first of the common dissociations mentioned earlier.

Recently, a neurocognitive framework specifying two dorsal action systems was proposed (Binkofski & Buxbaum, 2013; Buxbaum & Kalenine, 2010). In this model the bilateral dorsodorsal stream represents the *grasp* system, responsible for goal-directed actions based on the processing of currently visualized structural (shape, size, and orientation) stimulus information. The left ventrodorsal stream is characterized as the *use* system, which is specialized for skilled, functional tool-related actions. These streams are both distinguished from the ventral stream, which subserves object recognition. Thus, when considered together, the three streams reflect a dorsal-ventral gradient from current visually guided actions (dorsodorsal) to nonaction aspects of object semantic processing (ventral), with information requiring the integration of stored and online information occupying the middle of the gradient (ventrodorsal).

A second organizational feature of the praxis system reflects regional specialization along the posterior-anterior axis. While the parietal lobe is critical for the kinesthetic/proprioceptive and abstract specification of potential actions, prefrontal regions are particularly critical in temporally extended, hierarchical action tasks (such as complex naturalistic actions or tasks requiring difficult selection among competing action alternatives; Grafton & Hamilton, 2007). Inferior parietal and prefrontal regions work together to enable the timely selection of task-relevant actions.

Set against the backdrop of these gradients, we focus next on three major nodes that play a prominent role in praxis.

Three Nodes of the Praxis Network

MIDDLE POSTERIOR TEMPORAL NODE

Tool and hand representations for the recognition and learning of tool use actions Neuroimaging studies indicate that execution of tool use actions recruits the posterior part of the middle and/or superior temporal cortices (hereafter posterior temporal cortex, or PTC). Performing tool-related gestures with the tool in view but without holding it tends to show even greater reliance on the PTC, suggesting a role in integrating tool representations with tool-specific actions. Consistent with this hypothesis, learning studies reveal activity in the PTC when participants discriminate novel tools they have recently learned to use. Lesion analyses in left-hemisphere stroke patients using voxel-based lesion-symptom mapping (VLSM) techniques indicate that both the production and recognition of tool use pantomimes depend on the integrity of the PTC.

The PTC is composed of several associative regions coding for shape and/or manipulation and/or motion tool properties (Fernandino et al., 2016). Area hMT+ in the PTC is selectively activated when participants observe actions and may encode specific rigid aspects of tool visual motion. A nearby region in the lateral occipitotemporal cortex has been identified as the tool/hand selective area. Moreover, these regions demonstrate a strong structural and functional connectivity with other nodes of the praxis system, in particular the IPL. Following from this, apraxia is associated with diminished connectivity between the PTC and these other nodes (Watson et al., 2019). These and other data suggest that the PTC provides a representation of the visual appearance of target actions that may inform action generation. In particular, the PTC is specialized for representations of the *hand shape* appropriate for skilled actions with tools (see Buxbaum, 2017).

The left posterior middle temporal gyrus, in particular, has been characterized as the brain's major hub for tool semantics (see van Elk, van Schie, & Bekkering, 2014). On several accounts, hand shape and the visual appearance of tool motions are key aspects of action semantics. That is, our explicit knowledge of how to position the hand for tool use and how tools move entails the reactivation (implicit simulation) of stored visual appearance information.

THEMATIC SEMANTIC REPRESENTATIONS The performance of pantomimed tool use in left-hemisphere-lesioned patients is correlated with the ability to identify tool-recipient relationships, such as hammer-nail. Tool-recipient relationships are a prominent example of thematic relationships, thought to conceptually bind objects that typically occur in the same spatiotemporal context or event. Thematic relationship coding is likely to rely on the action semantic system, which may encode tools and recipients *seen in use together*, and thus differs mechanistically from the anterior temporal region typically viewed as an abstract, amodal semantic hub (see Mirman, Landrigan, & Britt, 2017). Patients with left temporoparietal lesions and/or deficits in action recognition selectively fail to encode objects in terms of their action relationships to one another, and lesions to the PTC are associated with both poorer identification of thematic relations for tools and impaired recognition of functional tool use gestures, an aspect of apraxia.

The PTC cannot readily be characterized as a ventral stream or ventrodorsal stream structure. In action recognition tasks, the PTC plays a role akin to that ascribed to the ventral stream in the case of object recognition. On the other hand, the left PTC is densely interconnected with the IPL and frontal lobe via the superior longitudinal fasciculus, and its hypothesized role in informing tool use actions is more consistent with functions ascribed to the ventrodorsal stream. It is perhaps most accurate to consider the PTC a hub lying at the intersection of these two streams. A related role for PTC has been proposed in the language domain (Hickok & Poeppel, 2007).

INFERIOR PARIETAL LOBE KINESTHETIC

Proprioceptive and abstract components of use representations The IPL is greatly enlarged in humans compared to nonhuman primates and performs several functions critical to action imitation and to the production of actions in response to the sight of tools or upon command. One of these critical functions is the transformation of information: Imitation requires the transformation of visual input of complex movement into a format that the motor system can use. The production of actions in response to the sight of tools or upon command requires a similar transformation of stored visual representations of tool and body motion and posture shape, likely provided by the PTC. Our current understanding of the nature of the transformational process critical to praxis is limited. There are at least two possibilities that are not mutually exclusive. The first is that praxis requires the transformation of visual information into body-centered reference frames in which kinesthetic/proprioceptive information is critical. The second nonexclusive possibility is that praxis depends on the transformation of visual information into a more abstract kinematic trajectory shape representation (see Wong, Haith, & Krakauer, 2015).

More clear is the fact that the IPL is critical to both imitation and pantomime. When lesioned by stroke or disrupted by continuous theta burst stimulation, the IPL is associated with errors in imitating unfamiliar as well as familiar gestures. In addition, the particular error types associated with IPL damage in apraxia tend to be spatiotemporal and sequence-related (errors in amplitude and timing), in contrast to the postural (in particular, hand shape) aspects of tool action information associated with lesions to the left PTC. Moreover, patients with IPL damage also make errors in recognizing the appropriate spatiotemporal parameters for tool use and have trouble matching kinematic action trajectories to static shapes. Multivoxel pattern analyses have shown that the IPL and PTC are both sensitive to similarity in the kinematics or appearance of performed pantomimes (e.g., corkscrew and key).

Much as the PTC-computed visual appearance of hand-tool actions may provide a template against which produced gestures may be matched for the purpose of error correction, the IPL may contribute to a target or goal representation in the kinesthetic/proprioceptive domain—that is, a memory of how skilled postures and gestures are supposed to *feel*. In addition, the IPL participates in providing a dynamic estimate of the *current* state of the body by integrating tactile and proprioceptive feedback with the tactile and proprioceptive outcomes predicted from motor commands (Buneo & Andersen, 2006).

In tool-related pantomime and imitation tasks, connectivity between the PTC and the IPL mediated by arcuate fasciculus/superior longitudinal fasciculus white matter pathways may be critical in tuning abstracted representations of stereotypical action appearance stored in the PTC so that produced gestures are appropriate to current constraints. In other words, the PTC and IPL may work together to generate the visual appearances of canonical tool actions, "customized" online based on the current state of the body, and, perhaps, via interconnections between the IPL and superior parietal and intraparietal sulcus, based on current visual information.

SPATIOTEMPORAL ACTION PREDICTION Action prediction computed in part by the IPL allows for the anticipation of future states of the body, as well as upcoming states, in an action sequence or trajectory. Apraxics with IPL lesions are unable to prospectively judge the hand posture and wrist rotation they would adopt were they permitted to reach out and grasp different three-dimensional shapes presented at different orientations. Nevertheless, they perform normally when they actually reach out and grasp the shapes under visual guidance, suggesting that they may make use of visual information to perform adjustments of hand aperture and orientation based on target attributes (Buxbaum, Johnson-Frey, & Bartlett-Williams, 2005). The possibility of abnormal reliance on visual information is supported by evidence that apraxics' ability to pantomime and imitate tool-related movements is substantially disrupted when they are blindfolded. Moreover, in the face of IPL damage, apraxics produce *conduite d'approche*—that is, successive approximations to the correct action—and rely abnormally upon visual feedback to correct these errors. Interestingly, these correction attempts are only successful if action recognition is intact. It is thus likely that current visual information is used in the detection of mismatches between produced actions and stored visual representations of how actions should look, similar to an "external route" by which conduction aphasics use auditory feedback to repair errors.

Apraxia and IPL damage are also associated with the deficient predictive processing of movement trajectories, whether human or nonbiological (e.g., detecting whether a human's or a rolling ball's trajectory, when briefly occluded, is discontinuous in time with respect to its expected trajectory). These observations are consistent with numerous studies showing involvement of the IPL in the prediction of movement trajectories in neurotypical subjects. The involvement of the parietal lobe during prediction of others' movements may reflect the adaptive use of the same predictive processes used to plan one's own skilled movements.

On some accounts, anticipation of future body states depends upon kinesthetic/proprioceptive processing and serves as a signal to the motor system to correct errors early in production or even before they are produced (e.g., Kilner, Friston, & Frith, 2007). Deficits in the translation of visual frames of reference into kinesthetic/proprioceptive goals may underlie the inability of apraxics to imitate viewed actions or to predict future states of their bodies (which in turn may rely upon simulation—see Krakauer & Shadmehr, 2007). Another interpretation focuses on a deficit in transforming information from extrinsic (e.g., eye, head, hand-centered) coordinates into the intrinsic coordinates required for programming muscle forces or joint torques (Mutha, Sainburg, & Haaland, 2010; and see Jax, Rosa-Leyra, & Buxbaum, 2014). An alternative but not mutually exclusive framework suggests that at least some apraxic deficits may result from the inability to specify an abstract kinematic trajectory that can be used to inform motor planning (e.g., Wong, Haith, & Krakauer, 2015). Careful experimental studies are needed to adjudicate between these accounts.

REPRESENTATIONS OF BODY PART RELATIONSHIPS The IPL node of the praxis network is also important for processing the configuration and relative movements of the body—a dynamic representations often termed the *body schema*. Biomechanically possible (but not impossible) movements activate the IPL, and this region is also activated when instructions specify that movements should be imitated with the same limb as that used by a model. The IPL is activated when subjects judge the laterality of viewed hands, a task subject to biomechanical constraints. Critically, the same hand laterality task is uniquely associated with the imitation of meaningless movements in apraxia. An interpretation alternative to that of a body schema proposes that the IPL's role in imitation concerns the processing of categorical spatial relationships between the hand and body parts, the hand and tool, or tools and recipient objects (Goldenberg, 2009).

MECHANICAL PROBLEM-SOLVING The IPL node has also been proposed to support mechanical knowledge (Osiurak & Badets, 2016). Mechanical knowledge abilities are typically evaluated in situations involving the use of novel tools to achieve a certain action goal or the use of familiar tools in an unconventional way (e.g., screwing a screw with a knife). Left IPL lesions have an impact on the demonstration of both familiar and novel tool use actions on recipient objects. One possibility is that "mechanical knowledge" rests upon the ability to predict how a tool and/or the body should be moved to achieve a target visual or kinesthetic/proprioceptive state (see Buxbaum, 2017).

WORKING MEMORY AND BUFFERING OF MULTIPLE POTENTIAL ACTIONS The left IPL, along with the IFG, plays a critical role in the selection of actions appropriate to task goals (e.g., the instruction or self-generated goal to use a tool or to hand it to someone) and context (e.g., whether a mechanical kitchen timer is in a drawer, requiring transport, or on a kitchen counter, ready for turning of the rotary dial). Apraxics with IPL damage have long been noted to produce multiple responses and to have difficulty selecting appropriate actions. More recently, Cisek (2007) posited that visual information is transformed via the parietal lobe into parallel representations of potential actions, which compete with one another in the frontoparietal cortex. This competition is biased on the basis of task demands and goals by prefrontal regions, the basal ganglia, and the ventral stream, and a single response is selected when neural activity reaches a threshold emerging from the dynamics between competing neuronal populations (Cisek & Kalaska, 2010). While there is evidence that multiple *motor* plans are not concurrently activated (e.g., Wong & Haith, 2017), extensive evidence suggests that multiple visual, kinesthetic/proprioceptive, and/or abstract actions are indeed coactivated in parallel prior to the specification of a single motor plan.

Studies with healthy participants suggest that action plans for functional tool use may interfere with plans appropriate to picking up tools to move them. Consider, for example, a television remote control, which is picked up and held with a "clench" but used with a "poke" (we term such tools *conflict objects*). Long-lasting interference is observed when a task requiring the demonstration of how to use a conflict object precedes a task requiring the demonstration of how to move it. The long-lasting interference indicates that use representations may exert competition even when task-irrelevant. Additional data from eye-tracking studies indicate that when instructed to click on a picture of a target tool (e.g., a key) presented in an array with distracters, similar-use actions associated with a distractor object (e.g., screwdriver) cause competition (as measured by looks at the screwdriver) in healthy subjects. Peak latencies in looks toward action-similar distracters are abnormally late in apraxia, and the magnitude of this delay is significantly correlated with the degree of clinical apraxia and with IPL damage (Lee, Mirman, & Buxbaum, 2014; Lee et al., 2018). These and other data suggest that the implicit activation of (and competition between) tool-related actions is abnormal in parietal apraxia and that these implicit abnormalities inform overt action errors.

Additional data support the claim that abnormal response competition related to IPL damage is a substantial source of overt action errors in apraxia. A VLSM analysis showed that left supramarginal gyrus (SMG, a portion of the IPL) and IFG damage was associated with disproportionately poor performance in pantomiming the use of conflict objects, and a subsequent functional connectivity study showed that reduced connectivity of these same regions with other brain regions was related with an increased number of production errors (Watson et al., 2019; see figure 47.1). The SMG, in particular, was proposed to buffer potentially relevant actions in short-term memory until a single action is selected (see also Randerath, Martin, & Frey, 2013).

PREFRONTAL/PREMOTOR NODE: CONTEXT-, GOAL-, AND TASK-RELATED BIASING AND ACTION SELECTION The classical apraxia literature has long noted a subtype of apraxia characterized by difficulty selecting and sequencing actions, particularly as task complexity increases. The IFG may bias the activation of these action representations buffered primarily in the SMG in accordance with the context, goals, and task demands, making certain actions more likely to be selected at the appropriate time than others. In general, the frontal node of the praxis network appears critical for representing high-level action goals (e.g., turn out a light) independent of action means (e.g., whether the light is turned out with a clockwise or counterclockwise motion).

The frontal node plays a critical role in action selection during action observation as well. A recent neuroimaging study revealed increased activity in the bilateral ventrolateral premotor cortex during the observation of actions involving objects with numerous action representations compared to objects with few action representations (Schubotz et al., 2014). Additional behavioral data indicate that manipulable objects flexibly evoke different action representations. Contextual factors such as the visual scene surrounding the object, object position in space, and task constraints largely determine whether grasp-to-use or grasp-to-move gestures may be more rapidly accessed.

Recent data indicate that the left IFG and neighboring middle frontal gyrus (MFG) are also likely to be critical for detecting and correcting action errors when they occur. Patients with lesions to these regions make fewer attempts to correct errors during tool pantomime tasks, and their existing correction attempts are less

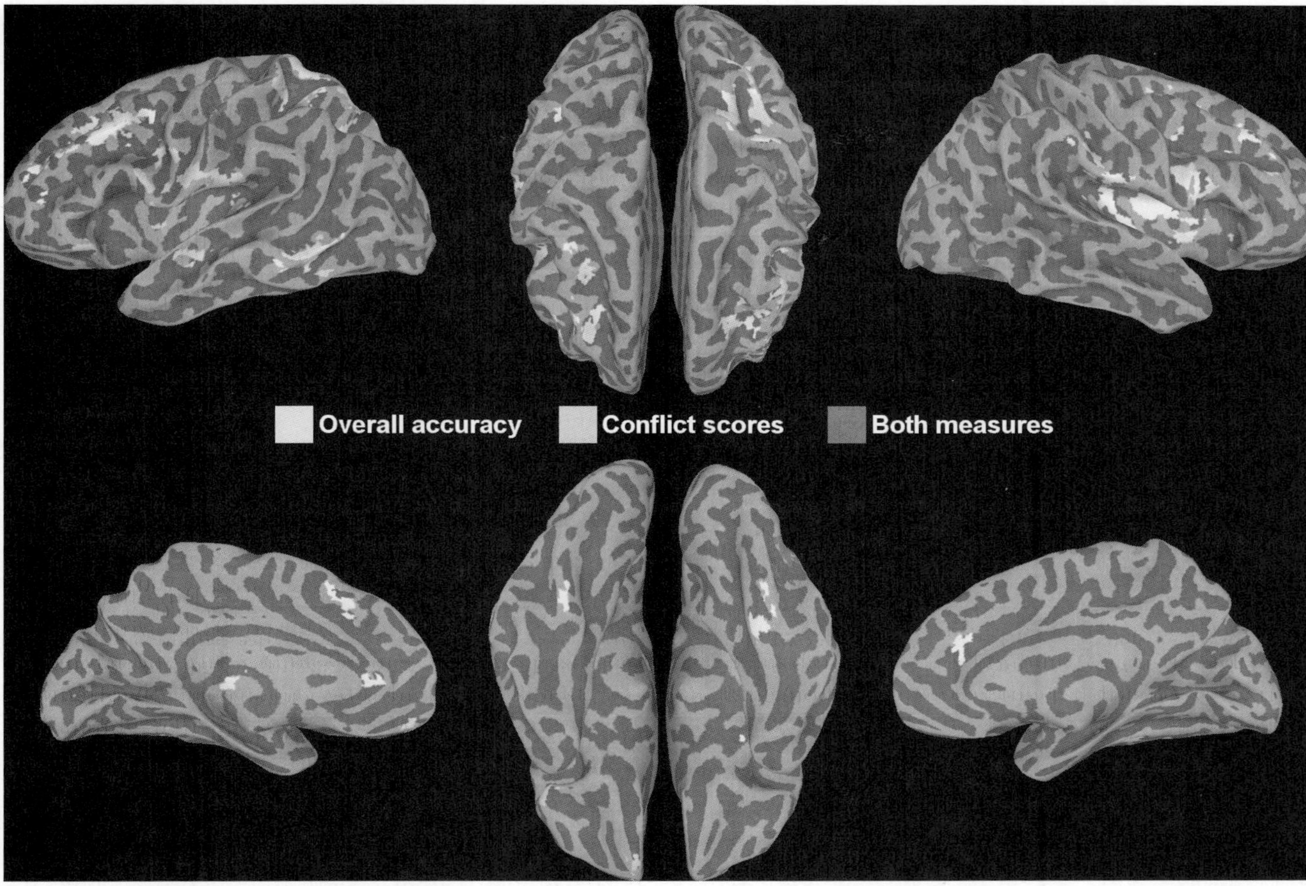

FIGURE 47.1 Brain regions (standardized Talaraich space) whose resting state functional connectivity in left-hemisphere stroke patients is (1) significantly reduced as compared with age-matched control subjects and (2) correlated with performance on tool gesture pantomime (*light gray*), pantomime of "conflict" tools associated with more than one action (*medium gray*), or both measures (*dark gray*). Reproduced from Watson et al. (2019). (See color plate 52.)

likely to be successful. Consistent with a much broader literature on cognitive control, the frontal node of the praxis network appears critical for monitoring action performance and detecting mismatches between high-level goals and produced outcomes.

Learning and Relearning of Actions in Apraxia

Despite the fact that apraxia has substantial consequences for caregiver burden and quality of life, there are surprisingly few studies of learning or relearning in patients with the disorder. One experimental study is relevant to the claim that apraxia is a disorder of the ventrodorsal stream. Left-hemisphere stroke patients (some with apraxia) were trained to learn novel gestures for using novel tools. Apraxics performed more poorly overall than nonapraxics, but apraxics whose lesions were more ventral, sparing the dorsodorsal system, performed better on both the production and matching tasks with tools for which the gesture was highly consistent with object structure (*afforded*). Patients whose lesions impinged on the dorsodorsal stream did not benefit from affordances. This supports our earlier suggestion that the integrity of representations of tool use gestures can be distinguished from the use of affordances derived from object structure (Barde, Buxbaum, & Moll, 2007).

In a more clinical vein, evidence-based rehabilitation approaches for patients with limb apraxia are rare (see Buxbaum et al., 2008). Gesture training and guided training of the activities of daily living are among the few most promising approaches. For example, apraxics can be trained to produce object-related gestures to tools, videos, and pictures over many repeated trials, and there is some evidence of transfer effects to untrained activities of daily living.

Neurostimulation may be another promising approach to the rehabilitation of apraxia. Transcranial direct current stimulation (tDCS) delivered to the IPL of left-hemisphere-lesioned patients may reduce errors in gesture imitation in some patients. Additional demonstrations of the effectiveness of neurostimulation in apraxia rehabilitation will be of considerable interest.

Future Directions and Conclusions

Traditional accounts of apraxia focused largely on the characterization of subtypes based on error types and patterns of performance with imitation, pantomime, single tools, and multiple objects. Increasing statistical sophistication in lesion analysis approaches as well as evidence from functional neuroimaging and neurostimulation studies has enabled a more nuanced appreciation of the distributed left-hemisphere network critical for skilled actions. Among the insights from this more recent work is a greater understanding of the dorsal-ventral and posterior-anterior functional gradients in the praxis system, the relationship of the left ventrodorsal stream to the classically defined (*dorso-*) dorsal and ventral visual pathways, and the integrated roles of the three major nodes of the praxis network.

With these points in mind, we will close with three considerations for future research. The first is that, rather than attempting to label subtypes of apraxia based purely on error types, the study of apraxic symptoms should be more closely aligned with increasingly detailed neurocognitive models of the roles of the temporal, parietal, and frontal regions in action representation and selection. The classical disorder associated with the loss of a movement "idea" is ideational apraxia. Confusingly, however, the diagnosis of ideational apraxia is often based on sequencing, omissions, and substitution errors on multistep, multiple objects tasks. Such errors are also very commonly observed in right-hemisphere stroke and bilateral pathologies and may result from executive deficits. We now understand that the left PTC plays an important role in representing visual tool motion and appropriate hand shape for tool use—specific aspects of distributed, multi-modal "ideas" of appropriate tool-related movements. We suggest that a combination of action recognition impairments, along with specific hand posture deficits (i.e., in the shape or aperture of the hand), in pantomiming to the sight of tools are more targeted potential indicators of a loss of tool action knowledge related to PTC damage.

Classical ideomotor apraxia is associated with a failure of intact "ideas" to inform motor planning. We have learned that the IPL is associated with the prediction of future actions, body-related spatial coding, buffering actions to enable selection in accordance with current task or environmental constraints, and the transformation from visual to kinesthetic/proprioceptive and abstract formats. Some of these functions, too, are associated with specific aspects of a distributed "idea" of movement. We suggest that the hallmarks of "pure" IPL apraxia are amplitude and timing errors in imitation of meaningless actions, with better performance in the imitation and production of meaningful actions. Finally, we understand that lesions to the IFG result in specific difficulties in selecting actions appropriate to task demands and in recognizing and correcting errors when they occur. Although broadly consistent with descriptions of impaired multistep action performance in classical frontal apraxia/action disorganization syndrome, recent evidence suggests that such failures occur even on tool pantomime tasks.

Critically, the regions of the praxis network described here are densely interconnected, such that even relatively discrete lesions to one region may be expected to have upstream and downstream consequences. Moreover, larger middle cerebral artery strokes rarely respect discrete architectonic boundaries. As a result, a mixture of deficits is the expected pattern in most patients. Additionally, functional connectivity analyses are beginning to reveal details of the white matter pathways connecting frontal, temporal, and parietal regions, along with a greater role for the right hemisphere than is usually assumed. Future studies of the neuroanatomic substrates of various components of the apraxia syndrome and changes in these substrates with learning and experience will increasingly benefit from network connectivity approaches.

The second consideration for future research is that characterizations of the computations of the ventrodorsal stream such as those outlined here parallel recent computational and theoretical models of the dorsal language pathway in the left hemisphere (e.g., Hickok & Poeppel, 2007). Approximately 70% of aphasics are also apraxic while approximately 90% of apraxics exhibit aphasia (Weiss et al., 2016). Like the ventrodorsal action stream, the dorsal language pathway projects from posterior temporal regions through the IPL and into the IFG. And like the ventrodorsal action stream, the dorsal language stream is specialized for the prediction and selection of the spatiotemporal aspects of action (in this case, speech-related actions), perhaps in a relatively abstract format. Moreover, lesions to this stream may give rise to conduction aphasia, characterized in part by deficits in phonological selection and phonological short-term memory, paralleling the selection and memory-buffering functions observed in the limb action domain. Consideration of the mechanisms that are shared and distinct to each syndrome continues to be an important direction for future research.

Finally, a third consideration for future research is that studies of the rehabilitation of apraxia lag far behind in number, sophistication, and rigor as compared to rehabilitation research in other domains. Given the overlap in the neural substrates of apraxia and aphasia, consideration of the "active ingredients" in successful treatments for aphasia may be useful. For example, a successful approach to the treatment of naming deficits attempts to strengthen the associative links between objects and actions as signified by nouns and verbs. A related approach in the apraxia domain might seek to strengthen the associative links between objects, people, locations, and actions. This could be contrasted with approaches emphasizing practice with the procedural, sensorimotor aspects of skilled action, which may be particularly relevant in patients who are relatively deficient in spatiotemporal processing (as evidenced, for example, by specific impairments in imitating meaningless actions). Compensatory strategies aiming at diminishing simultaneous action possibilities for both simple and complex actions might be best suited for apraxic patients who exhibit action selection deficits. Whether these approaches may be differentially beneficial for patients whose presentation is most consistent with one of the three major apraxia subtypes will be of great interest for the future.

Acknowledgments

Preparation of this chapter was supported by National Institutes of Health (NIH) grant R01 NS099061 to Laurel J. Buxbaum. We are grateful to our many current and former collaborators, postdoctoral fellows, and research assistants whose work is cited here.

REFERENCES

Barde, L., Buxbaum, L., & Moll, A. (2007). Abnormal reliance on object structure in apraxics' learning of novel actions. *Journal of the International Neuropsychological Society, 13*(6), 997–1008.

Binkofski, F., & Buxbaum, L. J. (2013). Two action systems in the human brain. *Brain and Language, 127*(2), 222–229. doi:10.1016/j.bandl.2012.07.007

Buneo, C. A., & Andersen, R. A. (2006). The posterior parietal cortex: sensorimotor interface for the planning and online control of visually guided movements. *Neuropsychologia, 44*(13), 2594–2606.

Buxbaum, L. J. (2017). Learning, remembering, and predicting how to use tools: Distributed neurocognitive mechanisms: Comment on Osiurak and Badets (2016). *Psychological Review, 124*(3), 346–360. doi:10.1037/rev0000051

Buxbaum, L. J., Haaland, K., Hallett, M., Wheaton, L., Heilman, K., Rodriguez, A., & Gonzalez-Rothi, L. (2008). Treatment of limb apraxia: Moving forward to improved action. *American Journal of Physical Medicine and Rehabilitation, 87* 149–161.

Buxbaum, L. J., Johnson-Frey, S. H., & Bartlett-Williams, M. (2005). Deficient internal models for planning hand-object interactions in apraxia. *Neuropsychologia, 43*(6), 917–929.

Buxbaum, L. J., & Kalenine, S. (2010). Action knowledge, visuomotor activation, and embodiment in the two action systems. *Annals of the New York Academy of Sciences, 1191*(1), 201–218.

Buxbaum, L. J., Shapiro, A. D., & Coslett, H. B. (2014). Critical brain regions for tool-related and imitative actions: A componential analysis. *Brain, 137*(7), 1971–1985.

Cisek, P. (2007). Cortical mechanisms of action selection: The affordance competition hypothesis. *Philosophical Transactions of the Royal Society of London B: Biological Sciences, 362*(1485), 1585–1599. doi:10.1098/rstb.2007.2054

Cisek, P., & Kalaska, J. F. (2010). Neural mechanisms for interacting with a world full of action choices. *Annual Review of Neuroscience, 33,* 269–298. doi:10.1146/annurev.neuro.051508.135409

Cubelli, R., Marchetti, C., Boscolo, G., & Della Sala, S. (2000). Cognition in action: Testing a model of limb apraxia. *Brain and Cognition, 44*(2), 144–165.

DeRenzi, E., & Lucchelli, F. (1988). Ideational apraxia. *Brain, 111*(5), 1173–1185.

Fernandino, L., Binder, J. R., Desai, R. H., Pendl, S. L., Humphries, C. J., Gross, W. L., Conant, L. L., Seidenberg, M. S. (2016). Concept representation reflects multimodal abstraction: a framework for embodied semantics. *Cerebral Cortex, 26*(5), 2018–2034.

Geschwind, N. (1975). The apraxias: Neural mechanisms of disorders of learned movement. *American Scientist, 63*(2), 188–195.

Goldenberg, G. (2009). Apraxia and the parietal lobes. *Neuropsychologia, 47*(6), 1449–1459. doi:S0028-3932(08)00298-4 [pii] 10.1016/j.neuropsychologia.2008.07.014

Gonzalez Rothi, L. J., Ochipa, C., & Heilman, K. M. (1991). A cognitive neuropsychological model of limb praxis. *Cognitive Neuropsychology, 8*(6), 443–458.

Grafton, S., & Hamilton, A. (2007). Evidence for a distributed hierarchy of action representation in the brain. *Human Movement Science, 26*(4), 590–616.

Hamilton, A., & Grafton, S. (2007). The motor hierarchy: From kinematics to goals and intentions. In P. Haggard, Y. Rossetti, & M. Kawato (Eds.), *Sensorimotor foundations of higher cognition* (pp. 381–408). Oxford: Oxford University Press.

Hickok, G., & Poeppel, D. (2007). The cortical organization of speech processing. *Nature Reviews Neuroscience, 8*(5), 393–402.

Jax, S. A., Rosa-Leyra, D. L., & Buxbaum, L. J. (2014). Conceptual- and production-related predictors of pantomimed tool use deficits in apraxia. *Neuropsychologia, 62,* 194–201. doi:10.1016/j.neuropsychologia.2014.07.014

Kilner, J. M., Friston, K. J., & Frith, C. D. (2007). Predictive coding: An account of the mirror neuron system. *Cognitive Processing, 8*(3), 159–166. doi:10.1007/s10339-007-0170-2

Krakauer, J. W., & Shadmehr, R. (2007). Towards a computational neuropsychology of action. *Progress in Brain Research, 165,* 383–394.

Lee, C. L., Huang, H. W., Federmeier, K. D., & Buxbaum, L. J. (2018). Sensory and semantic activations evoked by action attributes of manipulable objects: Evidence from ERPs. *NeuroImage, 167,* 331–341. doi:10.1016/j.neuroimage.2017.11.045

Lee, C. L, Mirman, D., Buxbaum, L. J. (2014). Abnormal dynamics of activation of object use information in apraxia: Evidence from eyetracking. *Neuropsychologia, 59,* 13–26.

Liepmann, H. (1988). The clinical picture of apraxia. In J. W. Brown (Ed.), *Agnosia and apraxia: Selected papers of Leipmann, Lange, and Potzl.* Hillsdale, NJ: Lawrence Erlbaum. Originally published in 1900.

Luria, A. R. (1966). *Higher cortical functions in man.* New York: Basic Books.

Milner, A. D., & Goodale, M. A. (1993). Visual pathways to perception and action. *Progress in Brain Research, 95,* 317–337.

Mirman, D., Landrigan, J. F., & Britt, A. E. (2017). Taxonomic and thematic semantic systems. *Psychological Bulletin, 143*(5), 499–520. doi:10.1037/bul0000092

Mutha, P. K., Sainburg, R. L., & Haaland, K. Y. (2010). Coordination deficits in ideomotor apraxia during visually targeted reaching reflect impaired visuomotor transformations. *Neuropsychologia, 48*(13), 3855–3867. doi:10.1016/j.neuropsychologia.2010.09.018

Osiurak, F., & Badets, A. (2016). Tool use and affordance: Manipulation-based versus reasoning-based approaches. *Psychological Review, 123*(5), 534–568.

Randerath, J., Goldenberg, G., Spijkers, W., Li, Y., & Hermsdorfer, J. (2011). From pantomime to actual use: How affordances can facilitate actual tool-use. *Neuropsychologia, 49*(9), 2410–2416. doi:10.1016/j.neuropsychologia.2011.04.017

Randerath, J., Martin, K. R., & Frey, S. H. (2013). Are tool properties always processed automatically? The role of tool use context and task complexity. *Cortex, 49*(6), 1679–1693. doi:10.1016/j.cortex.2012.08.016

Rizzolatti, G., & Matelli, M. (2003). Two different streams form the dorsal visual system: Anatomy and functions. *Experimental Brain Research, 153*(2), 146–157.

Schubotz, R. I., Wurm, M. F., Wittmann, M. K., & von Cramon, D. Y. (2014). Objects tell us what action we can expect: Dissociating brain areas for retrieval and exploitation of action knowledge during action observation in fMRI. *Frontiers in Psychology, 5,* 636. doi:10.3389/fpsyg.2014.00636

Stark, A., & Zohary, E. (2008). Parietal mapping of visuomotor transformations during human tool grasping. *Cerebral Cortex, 18*(10), 2358–2368.

Tarhan, L. Y., Watson, C. E., & Buxbaum, L. J. (2015). Shared and distinct neuroanatomic regions critical for tool-related action production and recognition: Evidence from 131 left-hemisphere stroke patients. *Journal of Cognitive Neuroscience, 27*(12), 2491–2511. doi:10.1162/jocn_a_00876

Ungerleider, L. G., & Mishkin, M. (1982). Two cortical visual systems. In D. J. Ingle, M. A. Goodale, & R. J. W. Manfield (Eds.), *Analysis of visual behavior* (pp. 549–586). Cambridge, MA: MIT Press.

van Elk, M., van Schie, H., & Bekkering, H. (2014). Action semantics: A unifying conceptual framework for the selective use of multimodal and modality-specific object knowledge. *Physics of Life Reviews, 11*(2), 220–250. doi:10.1016/j.plrev.2013.11.005

Watson, C. E., Gotts, S. J., Martin, A., & Buxbaum, L. J. (2019). Bilateral functional connectivity at rest predicts apraxic symptoms after left hemisphere stroke. *NeuroImage: Clinical, 21,* 101526.

Weiss, P. H., Ubben, S. D., Kaesberg, S., Kalbe, E., Kessler, J., Liebig, T., Fink, G. R. (2016). Where language meets meaningful action: A combined behavioral and lesion analysis of apraxia and aphasia. *Brain Structure and Function, 221*(1), 563–576.

Wolpert, D. M. (1997). Computational approaches to motor control. *Trends in Cognitive Sciences, 1*(6), 209–216.

Wong, A. L., & Haith, A. M. (2017). Motor planning flexibly optimizes performance under uncertainty about task goals. *Nature Communications, 8,* 14624.

Wong, A. L., Haith, A. M., Krakauer, J.W. (2015). Motor planning. *Neuroscientist, 21*(4), 385–398.

VII

REWARD AND DECISION-MAKING

Introduction

DAPHNA SHOHAMY AND WOLFRAM SCHULTZ

The cognitive neuroscience of decision-making and reward addresses the fundamental issue of how the brain supports motivated behaviors that allow an organism to pursue rewards, many of which (such as food) are necessary for survival. We need to acquire a number of substances contained in drinks and foods, and we need to reproduce in order to have our genes stay in the pool. Evolution has led to the emergence of brain systems that optimize this process of acquisition of primary rewards. Closely related brain processes have emerged that control the perception, selection, and acquisition of other stimuli, objects, and events that support the primary function of reward maximization, such as clothing, exchangeable goods, and social encounters, to name just a few.

By its very definition, reward-based decision-making involves a subjective evaluation of an individual's own preferences and needs. As such, the cognitive neuroscience of reward-based decision-making must address internal representations and goals, rather than external objective measures that can be derived simply from perceiving the physical parameters of rewards, such as milligrams of juice, numbers of chocolate bars, or amounts of dollars. A chocolate bar is physically the same even if I have eaten two chocolate bars already and no longer desire more chocolate at the moment; thus, the subjective value deviates from the physical parameter, and it is the subjective value that is critical in guiding motivated, reward-based behavior. This hallmark feature of subjectivity requires the development of cognitive and neural theories that conceptualize reward function from behavioral rather than from physical aspects, which dovetail with other aspects of internal representation and which can be linked to neural mechanisms.

Animal-learning theory, starting with Pavlov 100 years ago, provided one fundamental insight that has had a lasting influence on current studies of reward-based decision-making: that our subjective appetitive response to neutral cues can be heavily influenced by experience and learned associations. Reinforcement-learning theory, starting with Harry Klopf and Sutton and Barto more than 50 years ago, uses key concepts from animal-learning theory not only to understand reward learning but also to formalize these principles and to build artificial agents that learn to perform tasks faster and easier than humans using principles derived from natural biological behavior. Economic decision theory, starting with Bernoulli almost 300 years ago, makes the step from reward acquisition to the beneficial exchange of goods and formalizes the way decision-makers choose between such goods, and Kahnmenan and Tversky's prospect theory formalizes this subjectivity as following principled differentiation between losses and gains. All of these theories provide excellent and well-worked-out concepts that address distinct aspects of the functions of rewards and their influence on decisions.

The current collection of chapters tries to include representative examples of neurobiological studies based on these theories and address the possible ways our brains process rewards and losses and decide between them to obtain the best possible outcome. The selected chapters focus on a few key topics. Much attention across chapters is devoted to dopamine, given its well-established role in reward-based decision-making. Interestingly, though, much earlier investigations of the dopamine system started with the neurological deficits of parkinsonism, the psychiatric disorder of psychosis, and the observation that dopamine-enhancing drugs can be addictive. We now have a much more cohesive understanding of the role of dopamine across these highly complex dysfunctions and recognize that each of these pathologies may reveal only some aspects of dopamine's role in behavior. Indeed, subsequent investigations have confirmed, across a wide variety of functions, the important realization that dopamine plays a varied role and that it functions on different time courses, in different target regions, under different behavioral demands. These functions are revealed both by deficits arising from dopamine lesions or receptor alterations and by measurable dopamine changes in relation to specific forms of behavior, as discussed in the chapter by Westbrook, Cools, and Frank. The reward function concerns the most rapid changes of dopamine activity. Following the initial discovery of dopamine reward prediction error signaling, modeling studies revealed relationships to efficient reinforcement learning, such as temporal difference (TD) learning. Recent work has shown the suitability of a dopamine-like reinforcement signal for training computer models of the Asian game of *GO* that are capable of beating world champions. An update on the current state of TD modeling is provided in the chapter by Starkweather and Uchida. The field of reward neuroscience begins to incorporate basic concepts of economic decision-making in a relatively new field called *neuroeconomics*. Utility is one of the most basic and well-formulated economic decision variables; it is a mathematical representation of subjective reward value and, in being a mathematical function, predicts behavioral choices beyond those used to estimate the function. Thus, utility goes well beyond the notion of ad hoc determined subjective value that is unable to predict choices of options with other value. Utility is an internal decision variable that cannot be measured directly but needs to be inferred from choices. The underlying assumption is that in every choice the decision-maker considers internally the utility of each option and then selects that with the highest utility. Thus, as some economists note, utility is only inferred and might not really exist. The chapter by Stauffer and Schultz suggests, however, that the dopamine signal seems to reflect utility and thus constitutes a physical implementation that helps to interpret economic decision processes.

The dopamine reward signal does not exist alone in the brain but acts on other reward centers. The chapter by Conen and Padoa-Schioppa describes well-characterized types of reward signals that serve for economic decisions in the orbitofrontal cortex. Lesion studies in humans have long suggested this brain structure as a candidate for decision processes. The chapter is based on the initial discovery of the coding of specific decision variables. *Object value* refers to the value of a reward object irrespective of its being chosen. In a choice between two options, each object would have its specific value, and the decision process compares the value between the different objects, selects the object with the highest value, and feeds that decision to the motor apparatus. Thus, object value is supposedly an input variable to the decision process. By contrast, *chosen value* refers to the value of the object that has been attended to or has been selected by the decision mechanism. In the latter sense, chosen value reflects the result of the decision process. These notions reflect intuitive and straightforward notions of decision-making, and their implementation in a frontal brain structure involved in decisions attests to the power of these concepts.

A decision is a commitment to a proposition, among alternatives, based on evidence. Stine, Zylberberg, Ditterich, and Shadlen adopt this perspective in their chapter to address the core challenge of identification of the basic computations by which evidence is considered and how it guides decision-making—whether about internal subjective value or about external perceptual features. Their chapter views this process through the lens of simple perceptual decision-making in nonhuman primates, which provides a window into the basic process of operations such as inference, the integration of evidence, attention, and the selection of actions. This overview of neural and cognitive computations of perceptual decision-making offers an approach to understanding the basic mechanisms that underlie many different kinds of decisions, whether based on external or internal evidence.

Nonetheless, reward-guided decisions tend to be based on internal representations, raising questions about where the evidence for such decisions comes from. How is internal evidence generated and deliberated? At the most fundamental level, this process must depend on memory. Memory enables rapid associations between neutral cues and their associated value—how else do we know to associate a positive value with the yellow wrapper of peanut M&M's? The role of memory in reward-based decisions is discussed in the chapter by Duncan and Shohamy, which also looks at how memory allows the sort of flexible inferences essential for reward-based decision-making in a world full of uncertainty, sparse information, and novel situations. This chapter focuses on the balance between different forms of memory, on their complementary roles in different kinds of reward-based decisions, and on the essential and broad role that the hippocampus plays in the deliberation and consideration of future rewarding outcomes.

The amygdala has long been known to play a key role in emotional processing and has increasingly become considered to be a component of the brain's reward and decision system. Classically a prime structure for fear and aversive processing, a reevaluation of its neuronal responses has demonstrated a major role in reward processing. The chapter by Grabenhorst, Salzman, and Shultz describes this fast and interesting transition. The work began with Pavlovian conditioning as a benchmark test for controlled reward processing in C. Daniel Salzman's laboratory more than 10 years ago and was followed by a number of refinements, such as contingency, temporal processing, and the value of motivational states. Subsequent work confirmed the amygdala's major role in reward by showing an involvement in economic decision processes, including neuronal signals for input and output variables, the identification of distant reward goals, and the progress toward attaining them. Thus, amygdala neurons not only process reward events in a direct manner but also represent internal cognitive states. It still remains to be shown how amygdala neurons process social information that relates to the emotional function of this ancient brain structure.

The link between affective processing and reward and decision-making is also a central consideration in understanding how circuits for reward-based decisions develop over the life span. This topic is particularly relevant during adolescence, a phase of development that involves rapid growth toward independence, as well as extreme vulnerability to emotions, stress, and depression. In their chapter Galván, Delevich, and Wilbrecht review important recent discoveries regarding changes in corticostriatal circuitry and dopaminergic plasticity during adolescence and their implications for impulsive behavior, reward seeking, and learning. Studies across animals and humans have shown that the striatum undergoes changes during adolescence, and these influence the neural computations that arise from inputs to the striatum and are reflected in behavior.

Alterations in dopamine function and striatal circuitry are also related to the changes in affect and behavioral control typically associated with many psychiatric and neurological disorders. Westbrook, Cools, and Frank devote their chapter to discussing the complex relationship between dopamine imbalances, changes in the subcircuits of the basal ganglia, and behavior. As they emphasize, a full account of how reward-based decisions are supported by these circuits in health and disease cannot be explained by any one simple model. But substantial progress has been made in understanding the complex interactions between the role of dopamine in processing reward and loss, its effect on balances within corticostriatal circuits, and different aspects of behavior.

Altogether, the chapters span a wide range of topics and levels of analysis, from neurons to circuits, to human imaging and behavior. Across these levels of analysis, several of the crosscutting themes include learning, subjective internal representations, asymmetries in rewards and losses, and the importance of different timescales for reward. The selection here was intended to reflect current advances and areas of growth; this is a vibrant and large field, and the selection is necessarily limited by space and dependent on the editors. We apologize to all the authors whom we could not include in this short collection. We thank all the contributors to this volume for their effort and patience and wish the reader an enjoyable experience.

48 Dopamine Reward Prediction Errors: The Interplay between Experiments and Theory

CLARA K. STARKWEATHER AND NAOSHIGE UCHIDA

ABSTRACT Reinforcement-learning theories provide a normative perspective on learning and decision-making. In the 1990s, neurophysiology experiments revealed an exceptional correspondence between the activity of midbrain dopamine neurons and the reward prediction error (RPE) signal used to train computers in a reinforcement-learning algorithm called *temporal difference* (TD) learning. Studies of midbrain dopamine neurons play a pivotal role at the interface of empirical and theoretical studies. A theoretical framework for reinforcement learning has facilitated the interpretation of neurophysiology data and has guided the design of future studies. Here we discuss recent developments in the interplay between experimental findings and theories of dopamine signaling. In particular, recent studies emphasize the importance of state uncertainty in the neurobiological implementation of reinforcement learning.

An inexperienced shopper brings home a variegated bag of plums. After finding that the green and red plums are sour and the darker plums are sweet, she develops an eye for deep purple plums. She selects only ripe plums during future grocery trips. Animals learn to predict the value of outcomes in order to adapt successfully to their environments. The process of trial and error—and, crucially, learning from errors—reinforces the pairing of value with otherwise neutral sensory stimuli.

This chapter covers models and neural mechanisms of reinforcement learning. We will focus on the midbrain dopaminergic system, thought to be at the center of reinforcement learning in the mammalian brain. Research on the dopamine system occupies a unique space at the interface of empirical and theoretical studies. A theoretical framework for reinforcement learning provides twofold utility: it has aided in the interpretation of neurophysiology data and guides the design of future studies. In this spirit, we will frame this chapter using these normative models.

Pioneering Models and Neural Correlates of Reinforcement Learning

Certain outcomes (*unconditioned stimuli*, or US) trigger a biological response in the absence of learning, such as salivation accompanying a bite of tasty food. During Pavlovian conditioning, animals learn to predict a US, following a neutral sensory stimulus that does not evoke an automatic biological response (*conditioned stimulus*, or CS). Evidence of this prediction is seen in the animal's level of conditioned responding. Classical-conditioning tasks elicit conditioned responding that matches the automatic reaction to the anticipated US. The level of conditioned responding is measured during the time between the CS and US (the *interstimulus interval*, or ISI) and gradually increases across trials of learning.

An early theory describing the formation of a CS-US association (Rescorla & Wagner, 1972) set the change in associative strength between the CS and the US ($\Delta\hat{V}$) directly proportional to the discrepancy between the actual US level (V) and the predicted US level ($\hat{V}$):

$$\Delta\hat{V} = \alpha(V - \hat{V}) \tag{48.1}$$

In equation 48.1, α is a constant with a value between 0 and 1 that reflects the animal's learning rate. Prior to learning, $\hat{V}$ is 0. Thus, $(V - \hat{V})$ is large once an unexpected reward is received, driving large changes in $\hat{V}$ during initial CS-US pairings. As $\hat{V}$ increases over trials, smaller incremental changes in $\hat{V}$ occur on each subsequent trial. No additional associative strength is gained unless an existing prediction is violated. The Rescorla-Wagner model formalized the idea that an error between the actual and predicted outcome $(V - \hat{V})$ is needed in order to drive learning.

Subsequently, a learning theory was born in computer science (Sutton, 1988; Sutton & Barto, 1990). This theory, called temporal difference (TD) learning, also used the discrepancy between actual and expected outcome to drive learning, similar to the Rescorla-Wagner model. The model learns values, formally defined as the discounted sum of future reward (Sutton & Barto, 1990) associated with each "state" (figure 48.1*A*):

$$V(s_t) = \sum_{\tau=t}^{\infty} \gamma^{\tau-t}\, r(\tau) \tag{48.2}$$

$$= r(t) + \gamma r(t+1) + \gamma^2 r(t+2) + \gamma^3 r(t+3) + \cdots$$

where s_t is the state at time t, t is the current time, $r(\tau)$ is the reward at time τ, and γ is a discount factor ($0 \leq \gamma \leq 1$) that down weights future rewards. States represent the environment, or *task space,* in a way that is useful to performing the current task and here represent the world at different moments in time following some observable stimuli (although these may be used in other ways, as seen later in this chapter). States s, and transitions between these states, form a Markov process, allowing equation 48.2 to be written recursively using the Bellman equation:

$$V(s_t) = r(t) + \gamma V(s_{t+1}). \tag{48.3}$$

After transitioning from the state t to the state $t+1$, the agent obtains new feedback, $r(t)$. Both sides of equation 48.3 are the predicted value for s_t—that is, $\hat{V}(s_t)$. Due to the new information ($r(t)$) obtained by taking the state transition, the right-hand side can be seen as a more accurate prediction. The discrepancy between the left-hand side and the right-hand side is called the TD error because it corresponds to the discrepancy between the predictions at consecutive time points:

$$\delta(t) = r(t) + \gamma \hat{V}(s_{t+1}) - \hat{V}(s_t), \tag{48.4}$$

where $\delta(t)$ is the TD error at time t. The model then updates the predicted value for state s_t according to the following update rule:

$$\hat{V}(s_t) \leftarrow \hat{V}(s_t) + \alpha\delta(t), \tag{48.5}$$

where α is the learning rate. We will illustrate how TD learning would operate for a CS-US pairing, with a constant time delay. Before learning, the TD model has not learned any values. When the model experiences US (reward) delivery at time t following CS onset, this produces a positive prediction error $\delta(t)$. This positive prediction error increases the value associated with the corresponding state s_t, resulting in a positive value estimate at time t. During the subsequent CS-US pairing, there is a positive TD error at time $t-1$ because $\gamma\hat{V}(s_{t+1})$ is now positive, producing a positive prediction error $\delta(t-1)$ and increasing the value associated with the previous state s_{t-1}. In this way, the positive prediction errors propagate back in time over trials, increasing the values of states that fully tile the ISI, until the only positive prediction error occurs at the time of the CS (if CS onset is unpredictable). The TD algorithm eventually learns a sustained value estimate that precisely spans the time between the CS and US (figure 48.1*B*).

Although TD learning was originally developed for computer science applications, neurophysiology experiments in the 1990s by Schultz and colleagues revealed an exceptional correspondence between the error signals predicted by TD learning and those signaled by midbrain dopamine neurons (Bayer & Glimcher, 2005; Hollerman & Schultz, 1998; Mirenowicz & Schultz, 1994; Schultz, Dayan, & Montague, 1997; figure 48.1*B, C*). The authors trained monkeys on a classical-conditioning task in which a CS was followed by a US, roughly 1 s later. Before learning, dopamine neurons showed phasic activation only at the time of the US. After learning, this phasic dopamine response instead occurred at the time of the CS, and the signal at the time of the US was much smaller than before learning. Strikingly, if the US was unexpectedly omitted, dopamine neurons briefly paused their tonic firing exactly at the time of the usual US delivery. These signals, collectively referred to as reward prediction error (RPE), match the error signals proposed by the TD-learning algorithm (figure 48.1*B, C*):

1. Before learning, the only positive TD error $\delta(t)$ is at the time of the US due to positive excitation from reward $r(t)$.
2. After learning, the sustained value signal commences following the CS, resulting in a positive TD error and therefore a positive TD error at CS onset.
3. After learning, because the sustained value prediction drops to zero at the time of the predicted US, the TD error is negative at the time of the US and cancels out positive excitation from reward $r(t)$, resulting in a smaller response at the time of a predicted reward.
4. After learning, if the US is not delivered at the predicted time, the negative TD error is not canceled out by the arrival of a reward, resulting in a negative signal.

In these ways, dopamine signals recapitulate TD error signals. While the Rescorla-Wagner model predicts a smaller error signal at the time of a predicted US, it lacks a timing mechanism within a trial and therefore would not necessarily capture the CS response or the time-locked negative signal observed on reward omission with a nonzero CS-US delay.

The idea that dopamine neurons convey TD error signals has been substantiated by further experiments in a variety of species, including rats (Flagel et al., 2011; Hart, Rutledge, Glimcher, & Phillips, 2014; Pan, Schmidt, Wickens, & Hyland, 2005; Roesch, Calu, & Schoenbaum, 2007; Stuber et al., 2008), mice (Cohen, Haesler, Vong, Lowell, & Uchida, 2012; Eshel et al., 2015; Eshel, Tian,

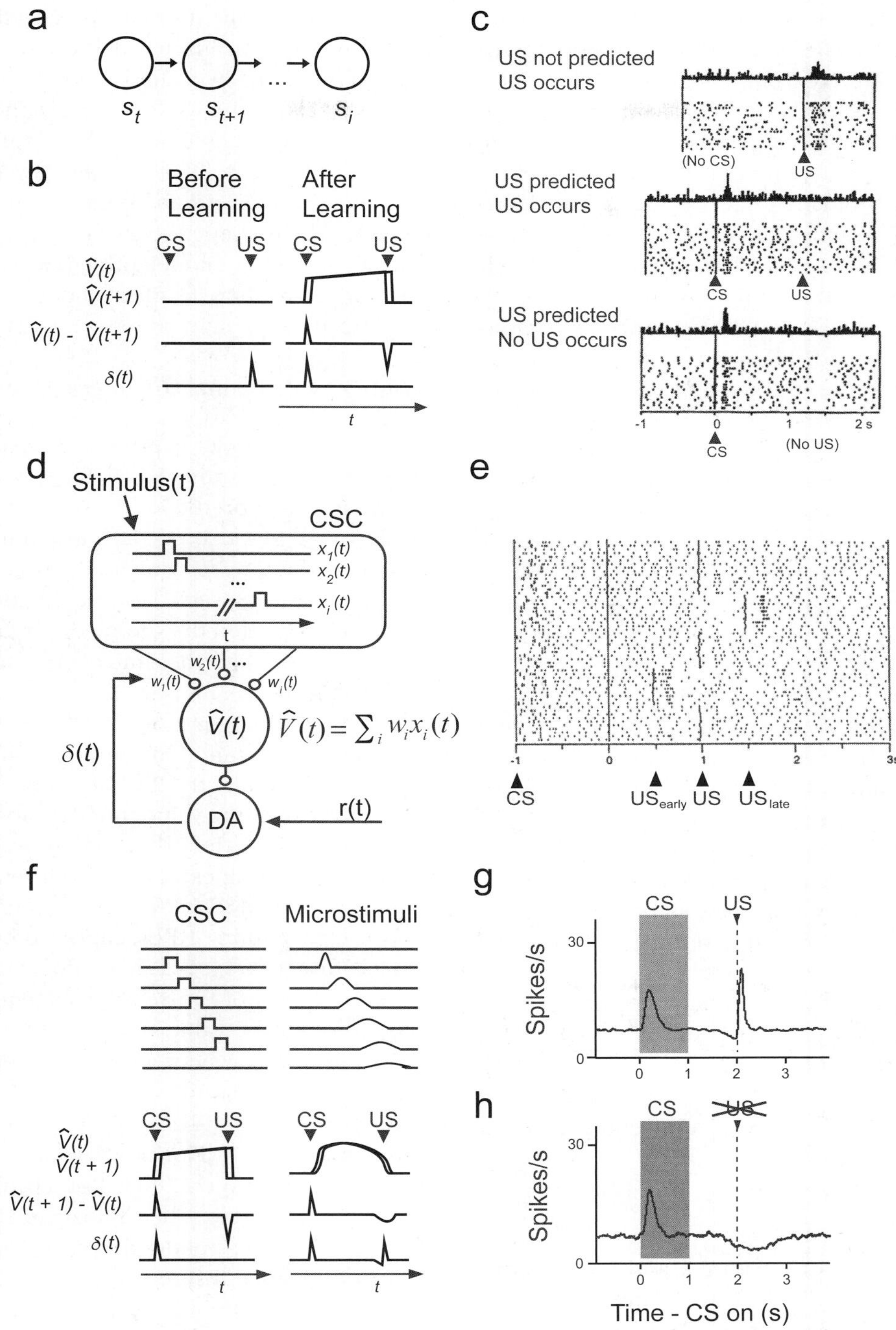

FIGURE 48.1 Temporal difference (TD) learning and dopamine signaling. *A*, States can be modeled as each corresponding to an arbitrary unit of time. *B*, Value signal, TD of value signal, and error signal produced by a simple TD model. *C*, Firing pattern of a putative dopaminergic neuron during a classical-conditioning task. From Schultz, Dayan, and Montague (1997). *D*, Neural circuit hypothesis for implementing TD learning in the brain, using the complete serial compound (CSC) feature representation. Value is computed as the linear sum of features (each "CSC" component, indexed from 1–i) multiplied by their weights. *E*, The timing of US delivery following CS onset was jittered (either earlier or later) by 500 ms in probe trials. Note there was no reward omission response if US was given early. From Hollerman and Schultz (1998). *F*, CSC versus microstimulus features. *G*, Averaged data from dopamine neurons showed a small dip in the tonic rate of dopamine firing prior to reward delivery. From Tian and Uchida (2015). *H*, Averaged data from dopamine neurons showed a temporally spread dip upon reward omission that lasted roughly 1 s. From Tian and Uchida (2015).

Bukwich, & Uchida, 2016; Menegas, Babayan, Uchida, & Watabe-Uchida, 2017; Parker et al., 2016), and humans (D'Ardenne, McClure, Nystrom, & Cohen, 2008). Importantly, recent studies applied optogenetic methods to unambiguously identify dopamine neurons while recording both in mice (Cohen et al., 2012; Eshel et al., 2015, 2016) and in monkeys (Stauffer et al., 2016). Eshel et al. (2015, 2016) demonstrated that reward expectation reduces dopamine reward responses in a purely subtractive fashion, and dopamine neurons in the lateral ventral tegmental area (VTA) exhibit very similar response properties, indicating that these neurons signal TD errors homogeneously. The correspondence between TD errors and dopamine signals supports the hypothesis that TD learning approximates how reinforcement learning is actually implemented in the brain. This hypothesis is further supported by optogenetic stimulation experiments, which show that the temporally precise manipulation of dopamine signaling at the time of expected outcomes leads to increased behavioral responding to preceding CSs, just as TD learning would predict (Steinberg et al., 2013).

A classic adaptation of TD learning to a biological circuit model utilizes a *complete serial compound* (CSC) feature representation that tracks elapsed time relative to observable stimuli (Schultz, Dayan, & Montague, 1997). Each CSC feature $x_i(t)$ is a vector of zeros, except for a value of 1 at timepoint i relative to cue onset (figure 48.1*D*). Each CSC feature can be conceived as representing the occupancy of a single state s described in the example above. $x_1(t)$ would have a value of 1 at time point 1, corresponding to occupancy of state s_1. A neuron corresponding to $x_1(t)$ would only fire at time point 1. A population of analogous neurons could show sequential activations and represent the entire interval between the CS and US. The value function estimate is modeled as a linear combination of these features (figure 48.1*D*):

$$\hat{V}(t) = \sum_i w_i x_i(t), \tag{48.6}$$

where $x(t)$ represents CSC features, and w_i is a predictive weight associated with feature i. The weights are updated according to the following learning rule:

$$\Delta w_i = \alpha x_i(t)\, \delta(t), \tag{48.7}$$

where α is a learning rate $(0 \le \alpha \le 1)$, and $\delta(t)$ is the prediction error in the value signal. The discrepancy between actual and predicted value is computed similar to that above (48.4), according to

$$\delta(t) = r(t) + \gamma \hat{V}(t+1) - \hat{V}(t), \tag{48.8}$$

where $r(t)$ represents reward at time t. The TD error increases the weights of sequential neural activations (modeled as CSC features) such that a downstream area can compute an accurate value prediction. In line with this algorithmic hypothesis, other theoretical works speculated that the cortex conveys the CSC temporal features into the striatum, where value is computed (Houk, Adams, & Barto, 1995). By modulating the weights of corticostriatal synapses, dopamine signals could plausibly shape the striatal value representation. The TD-learning model provided a cornerstone for understanding how reinforcement learning could be biologically implemented.

Timing in Temporal Difference Models

The CSC TD model captures the exquisite temporal specificity of dopamine RPEs. First, dopamine neurons show a negative prediction error exactly at the time of an expected reward. Second, dopamine responses to reward are suppressed only at the time of the expected reward. In a 1998 study, Hollerman and Schultz (1998) occasionally shifted the timing of reward by just 500 ms earlier or later than the time of the usual reward delivery (figure 48.1*E*). This small temporal jitter evoked a larger dopamine response than if the reward were delivered at its usual time. This finding is consistent with the errors produced by the TD model because the temporal difference of the value signal is most negative at the cue-reward delay time that the animal was trained on (thereby canceling excitation from reward only at that time; see figure 48.1*B*). Another timing-related characteristic of dopamine RPEs, captured by the TD model, is delay discounting. Because value corresponds to the discounted sum of future rewards (equation 48.2), the magnitude of the value signal at cue onset is inversely related to the delay between cue and reward. Consistent with this, several studies in both rodents and primates have observed delay discounting in the cue response of dopamine neurons (Fiorillo, Newsome, & Schultz, 2008; Kobayashi & Schultz, 2008; Roesch, Calu, & Schoenbaum, 2007; Starkweather, Babayan, Uchida, & Gershman, 2017). Cues followed by late rewards result in smaller dopamine responses than cues followed by early rewards. Thus, the TD framework matches several findings relating to timing and the dopamine system.

However, additional experimental observations suggest that the TD model—particularly, the CSC implementation—does not provide a complete account of timing in the dopamine system. The first observation is that dopamine responses are not thoroughly suppressed at the time of expected rewards, particularly for CS-US pairings involving long ISIs (Fiorillo et al., 2008; Kobayashi & Schultz, 2008). A CSC TD model would suppress the US responses equally irrespective of the delay time. The second observation is that the "dip" observed upon reward omission is temporally extended for up to 1 s (Cohen et al., 2012; Matsumoto & Hikosaka, 2007; Schultz, Dayan, & Montague, 1997; Tian & Uchida, 2015). In contrast, a CSC TD model produces a sharp dip only at the time of the expected reward (figure 48.1*B*).

Based on these experimental observations, Ludvig, Sutton, and Kehoe (2008) proposed a TD model that swaps the CSC representation for Gaussian distributions, whose widths increase over elapsed time, called *microstimuli* (figure 48.1*F*). The increasing widths of the microstimuli respect Weber's law, which stipulates that the variance of timing estimation increases linearly with elapsed time (Balsam & Gallistel, 2009). If a reward occurs at a long delay from the CS onset, the reward increases the weights of more than one microstimulus feature (contrasting with the CSC TD model) because the Gaussian distribution of multiple features will have nonzero values at the moment a reward is received. Accordingly, the TD model increases the weights of multiple microstimulus features that are active at that time. Over many trials this produces a value signal that is less sharply resolved in time than the CSC TD model (figure 48.1*F*). Specifically, the value function will be positive at more time points than just the exact moment that reward is received, and it will have a smaller amplitude (later microstimuli are modeled as having smaller heights). Therefore, the TD error, $\gamma\hat{V}(t+1)-\hat{V}(t)$, will be shallower and more spread out over time. As a consequence, when a reward is given, the response cannot be completely canceled out. The reward response is greater for longer delays because the model's ability to cancel out excitation from the reward becomes less precise for longer intervals. Furthermore, if a predicted reward is omitted, the "dip" is more spread out across time than predicted by the TD model with the CSC (figure 48.1*H*). Finally, a common observation from our lab is that the baseline tonic firing rate of dopamine neurons decreases slightly before a reward is received (figure 48.1*G*; Starkweather et al., 2017; Tian et al., 2016; Tian & Uchida, 2015). This prereward decrease in firing is reproduced by the microstimulus model because the temporal imprecision results in a negative prediction error commencing prior to the exact time of the reward. Therefore, the microstimulus TD model explains common observations of dopamine recordings, which were inconsistent with the CSC.

Other updates to the TD model have been proposed, which we will discuss in the next section. However, it is important to keep in mind that all these models possess a timekeeping mechanism. Dopamine RPEs are likely computed on an imperfect and noisy timing mechanism. Therefore, any modeling refinements proposed in future work should also contain a timing representation constrained by scalar timing uncertainty.

State Inference in the Temporal Difference Model

In 1998 Hollerman and Schultz made another important experimental observation in monkeys trained on a CS-US pairing separated by a constant delay time. In a small proportion of probe trials, rewards were given earlier than predicted. As predicted by the TD model, this early reward produced a large dopamine response because the timing of the reward was unexpected. However, there was no omission response (i.e., negative RPE) at the time of the usual reward (figure 48.1*E*). In contrast, the CSC TD model produced an omission response at the time of the usual reward (figure 48.2*A*). Based on this experimental observation, two main categories of modifications were proposed to the TD model.

The first proposal was that the TD model simply "resets" after a reward is received. Future errors could be fixed at zero after a reward is received (Suri & Schultz, 1998, 1999). Alternatively, a reward receipt could terminate the set of CSC features activated during the ISI and trigger a new set of CSC features activated during the intertrial interval, or ITI (Brown, Bullock, & Grossberg, 1999). Implicit in this reset model is that the animal knows (upon reward) that a reward is no longer expected. This means that the animal infers the "state" of the task based on observable stimuli (the reward). If one imagines the task is divided into two states—the ISI state during which a reward is expected and the ITI state during which a reward is not expected—the receipt of a reward initiates CSC features that correspond to the ITI state and terminates those that correspond to the ISI state. However, this is an ad hoc modification to the TD model with the CSC. It adds a simple form of state inference to the TD model to match the data but does not acknowledge other instances in which state inference comes into play. For example, what if reward were omitted? Would the ISI features simply continue on indefinitely because no reward is received? Or would the animal infer over time that a reward is not coming, thereby terminating the ISI features? These issues remain unresolved with the reset model.

The second proposed modification is that the TD model's features themselves represent the inferred state of the environment (Daw, Courville, & Touretzky, 2006; Rao, 2010). We will refer to this as a *belief-state* TD model. The temporal CSC or microstimulus representation is swapped with a belief state, which is an inferred probability distribution over possible states. Value is no longer proportional to the weight assigned to a particular active state. Rather, value is equal to the weight assigned

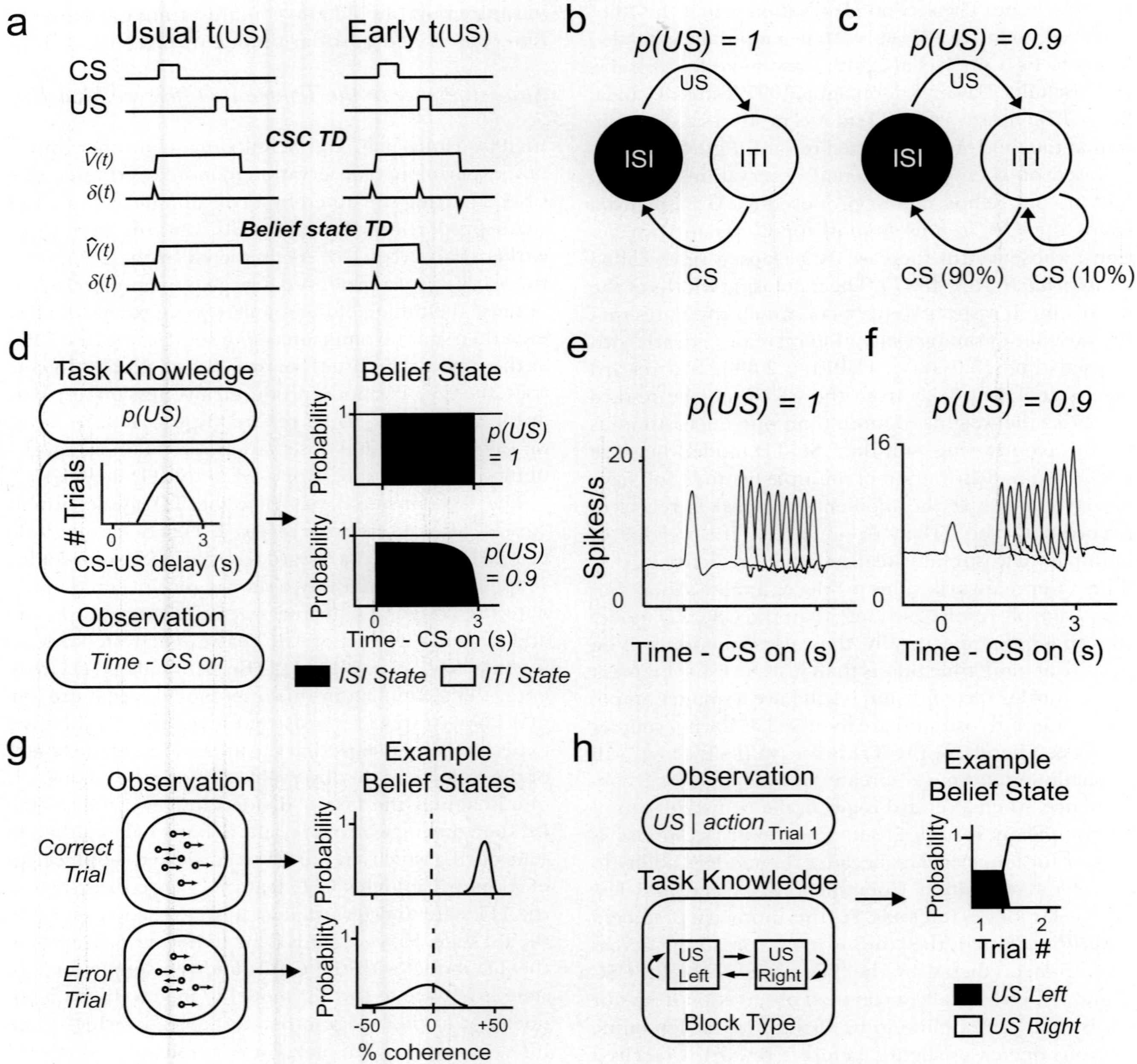

FIGURE 48.2 Belief-state features for reinforcement learning. *A*, Schematic for values and error signals produced by early reward delivery, in the belief-state TD model versus CSC TD model. CSC TD model produces a spurious reward omission "dip" after an early reward delivery, which is not observed in the data shown in figure 48.1*E*. *B*, Semi-Markov schematic for computing the belief state under fully observable task conditions. Adapted from Daw et al. (2006). *C*, Semi-Markov schematic for computing the belief state under partially observable task conditions. Adapted from Daw et al. (2006). *D*, In a task invoking variable delay times and particular reward contingencies (illustrated for 100%—and 90%–rewarded contingencies), the agent uses its observation of time passed since cue onset in addition to its knowledge of reward timing/contingency to compute a probability distribution over possible states from figure 48.2*B*, *C*. This probability distribution is called a *belief state*. *E*, Dopamine neurons produced smaller error signals as a function of time when reward is delivered if the task illustrated in figure 48.2*D* is 100% rewarded. *F*, Dopamine neurons produced larger error signals as a function of time when reward is delivered if the task illustrated in figure 48.2*D* is 90% rewarded. This can be explained by taking the belief state into account—as belief favors the possibility of a reward omission trial at later time points, reward evokes a larger error signal if it is actually delivered. *G*, Based on ambiguous sensory stimuli (random dot motion), the agent may compute a belief state over possible coherences and direction of movement. Adapted from Lak et al. (2017). *H* Based on observation of the first trial type within a block, an agent may use its knowledge of the task structure to compute a belief state. Schematized based on Bromberg-Martin et al. (2010).

to a state, scaled by the probability (from the belief state) allotted to that state, summed over all states:

$$\hat{V}(t) = \sum_i w_i b_i(t), \tag{48.9}$$

where $b(t)$ represents the belief state (i.e., the probability of being in each state i) at time t, with i indexing individual states, and w_i is a predictive weight associated with state i. Furthermore, weights are updated proportionally to the probability with which the agent "believes" it occupies a particular state:

$$\Delta w_i = \alpha b_i(t)\ \delta(t) \tag{48.10}$$

The belief-state TD model was originally conceived as a semi-Markov process involving just two states: the ISI and the ITI (figure 48.2*B*). Semi-Markov dynamics imply that the time spent in a state is probabilistic and is defined by a probability distribution called a *dwell time distribution* (contrasting with a Markov process, where each "state" represents one arbitrary unit of time). For that reason, rewards are discounted by however much time elapses within a state once received because this time interval varies from task to task: prediction errors should be larger, therefore driving up the value of a particular state more if the reward arrives early, whereas the opposite should be true if the reward arrives late.

The belief-state TD model accounts for the Hollerman and Schultz result and captures other experimental findings (Daw, Courville, & Touretzky, 2006). The Hollerman and Schultz experiment was rewarded in 100% of trials, meaning that the task is fully observable because the state of the task (ISI or ITI) can be deciphered based on sensory cues alone. Upon observing the CS, the belief state allots 100% probability into the ISI state and 0% probability into the ITI state. Upon observing the US, the belief state allots 100% probability into the ITI state and 0% probability into the ISI state (figure 48.2*D*). If the ISI accrues a larger weight (as it should, because rewards are only received when the belief state favors the ISI), a transition into the ITI upon receiving the US would eliminate future reward expectation. This would abolish the spurious reward omission response reported by the CSC TD model upon receiving an early reward (figure 48.2*A*). A second set of experimental results, compatible with the belief-state TD model, consists of dopamine responses recorded in 100%-rewarded task contingencies with variable delay times. In these experiments, on any given trial, the delay between the CS and US (or a reward-predicting CS) was drawn from a uniform distribution (Fiorillo et al., 2008; Nomoto, Schultz, Watanabe, & Sakagami, 2010; Pasquereau & Turner, 2015). Rewards delivered earliest in the variable delay interval produced the largest dopamine responses, and rewards delivered later in the interval produced smaller dopamine responses. While the authors argued that these results are explained by expectancy over time resembling a hazard function (the momentary likelihood that an event will occur, given that it hasn't occurred yet), this result is also compatible with the belief-state TD model (Starkweather et al., 2017). Because late rewards *within a particular state* are discounted more heavily in a semi-Markov framework, the belief-state TD model also captures this pattern of prediction errors across time. In contrast, a CSC TD model, which lacks an explicit representation of state space (ISI vs. ITI), would produce the same prediction error at each possible time of reward delivery because the uniform distribution of timings teaches the model to apply the same value prediction at each time point.

The belief-state TD model captures experimental findings from tasks that jitter the timing of reward relative to cue. However, these tasks do not implicate state uncertainty, which is a core tenant of the belief-state TD model. A simple modification that renders the task states partially observable is reducing the probability of reward from 100% to 90% (Starkweather et al., 2017; Starkweather, Gershman, & Uchida, 2018). Once the cue comes on, the animal no longer knows for certain whether it is in the ISI or the ITI. This is because in 10% of trials (the unrewarded trials), a cue onset leads to a hidden state transition from the ITI to the ITI (figure 48.2*C*). Whereas the belief state was uniform, following cue onset, under deterministic task conditions, the belief state evolves over time as the animal gathers evidence in favor of one hidden state over the other. After cue onset, the belief state is 90%–10%, with 90% allotted to the ISI. As time elapses and the animal does not receive a reward, the belief state shifts more probability into the ITI state, yielding to the possibility of a reward omission trial (figure 48.2*D*). Because the belief state shifts toward the ITI state, the value prediction is low, and thus, the prediction error is high if a reward is actually received at a later time point. This has been demonstrated experimentally and is a key prediction of the belief-state TD model: state uncertainty should dramatically affect how reward expectation evolves over time. In a 100%-rewarded deterministic scenario (figure 48.2*E*), later rewards evoke smaller prediction errors due to discounting over a lengthy dwell time, whereas in a 90%-rewarded nondeterministic scenario, later rewards evoke larger prediction errors due to the belief favoring the unrewarded state (figure 48.2*F*).

Experimental results in other tasks implicating state uncertainty are also explained by the belief-state TD model. One study used a belief-state TD model to capture dopamine signals observed in primates during a perceptual decision-making task (Lak, Nomoto, Keramati, Sakagami, & Kepecs, 2017; figure 48.2*G*). In this task a random-dot motion stimulus with variable coherence was presented following a fixation cue. Based on the perceived direction of the random dot motion, the animal made an action that if correct resulted in reward. The belief state was modeled as a probability distribution over a range of motion directions and coherences. On average the belief state should be centered at the true motion direction and coherence. However, the belief state would be different from trial to trial, as it was assumed there was sampling noise in the perception of the stimulus. Action value was computed based on the belief state, with the model choosing the action on the higher-valued side of the left-versus-right decision boundary. The value prediction was proportional to the probability of receiving a reward on a particular trial, given the belief state and corresponding choice—equivalent to a confidence signal—and would produce a difference in dopamine signals between correct (higher confidence) and error (lower confidence) trials. In contrast, the CSC TD model does not have access to a distribution representing the uncertainty in perceptual stimuli and would therefore predict dopamine signals of similar magnitude for "correct" and "error" trials. The authors found that dopamine signals were well matched by a belief-state TD model.

While both of the studies discussed above invoke a belief state that is computed on every trial, a belief state could pertain to more global aspects of task structure involving a block of trials. In one study (figure 48.2*H*), the cued side (left vs. right) on which a reward was presented switched between blocks (Bromberg-Martin, Matsumoto, Hong, & Hikosaka, 2010). So, if a formerly rewarded side went unrewarded, this signaled a block change. On a subsequent trial, animals should infer that the formerly unrewarded side was now rewarded. This inferred switch was reflected in dopamine signals. Dopamine responses to the inferred rewarded cue were larger than in the previous block, indicating a higher value prediction even if the animal had not yet experienced a reward on the newly rewarded side. This finding cannot be explained by a simple cached-value system (*cached*, meaning that the value must be assigned to a state based on direct experience) because the dopamine response reflected the new inferred value without the animal directly experiencing the new cue-outcome pairing. However, it may be explained by a belief state that shifts to reflect the rewarded side of the new block (Costa, Tran, Turchi, & Averbeck, 2015; Fuhs & Touretzky, 2007; Hampton, Bossaerts, & O'Doherty, 2006). Finally, a recent study similarly showed that dopamine responses reflect the inferred value of cues (Babayan, Uchida, & Gershman, 2018). In this study, blocks consisted of identical trials of a cue followed either by "big" or "small" rewards. On rare blocks, rewards were of intermediate size. After experiencing one of these intermediate rewards, the dopamine signal at the time of the reward on the subsequent trial appeared to subtract either the value of the "big" or "small" reward from the value of the received intermediate reward, as if the value prediction conveyed to dopamine neurons reflected whether the current block was inferred to be either "big" or "small," the two states in this task.

The belief-state TD model is needed to account for a growing number of empirical observations. In addition, belief-state TD models solve other existing controversies. For instance, in the original CSC model the value signal and the TD error signal propagate backward from the time of reward to the time of cue only gradually during learning. Neurophysiology data have not supported this gradual shift (Menegas et al., 2017; Pan et al., 2005). These results have been taken as evidence against TD models. However, belief-state models do not require this gradual shift because they must learn whether a cue is associated with a reward to begin with (in theoretical terms, learning about the state space), separately from when exactly the reward will occur.

Model-Based Feature Representation for Reinforcement Learning

Model-free systems predict discounted future reward without an explicit model of the environment. For instance, the CSC TD model lacks knowledge that a reward signifies a state transition into the ITI, hence the spurious "reward omission" dip after an early reward was delivered. Knowledge of the transition structure between states, and of the corresponding observations and probabilities of triggering these state occupancies, constitute *model-based* information. Therefore, the feature representation for the belief-state TD model is model-based. However, the TD-learning rules, including the computation of the TD error and the update of "states" weights, remain model-free. This model-free component means that the belief-state TD model stores, or "caches," weights for particular states, to be deployed every time a particular state is occupied (meaning that the belief state allots a nonzero probability to that state). These weights must be learned through the direct experience of occupying certain states.

The belief-state TD model differs from fully model-based reinforcement learning, which uses a forward model of the environment to compute values for various states. Consider a maze in which an animal receives different rewards at each possible exit (Niv, Joel, & Dayan, 2006). The animal may be hungry and therefore place greater value on food versus liquid rewards at the end of the maze. The animal simulates mental paths through the maze and computes values for states (each corresponding to various locations in the maze). States forming trajectories leading to food rewards would have higher values than other states perhaps leading to liquid rewards. If the animal was thirsty, a different set of state values would be simulated based on the higher utility of paths leading to liquid rewards. Importantly, the animal would not have to be trained while it was thirsty in order to use its model of the environment to reassign values for various states. This contrasts with the belief-state TD model, or any system that uses a model-free-learning rule, in which a state must be directly experienced in order to update its value (*cached value*). Work utilizing a two-step decision-making task has shown that humans, to varying degrees, combine model-based and model-free value computation (Daw, Gershman, Seymour, Dayan, & Dolan, 2011). Blood oxygen level-dependent signals in the ventral striatum, thought to correlate with dopamine activity, signal prediction errors in value predictions that reflect both model-based and model-free computation. Furthermore, while the computations that feed into the value prediction may involve a complex cognitive model of the environment, fully model-based accounts still endorse the idea that dopamine signals errors in value prediction.

Some recent studies argue that dopamine may not exclusively signal discrepancies in value prediction. In one experiment rats were trained to associate a cue with a reward. Characteristics of the reward (e.g., flavor of milk) were varied across blocks (Takahashi et al., 2017). Changes in milk flavor—even though rats preferred chocolate and vanilla equally—evoked a positive dopamine signal at the beginning of a block switch. In another experiment (Sharpe et al., 2017) an otherwise blocked sensory association between two cues could be "unblocked" by optogenetically activating dopamine neurons. These studies lead to the proposal that dopamine signals carry prediction errors along axes other than value (Langdon, Sharpe, Schoenbaum, & Niv, 2018). This differs from the belief-state TD model, which maintains the model-free computation of the TD error and update rules and places model-based, belief-state computations upstream of the value function. We chose to focus on the belief-state TD model, which contains model-free prediction errors, because it is widely accepted that dopamine signals are sensitive to errors in value prediction and drive learning about value. Furthermore, dopaminergic axons in the ventral striatum convey canonical model-free TD errors (Menegas et al., 2017; Parker et al., 2016). The unpredicted presentation of neutral stimuli caused no response from these axons (Menegas et al., 2017), incongruent with a pure "sensory" surprise signal. Furthermore, the activation or inactivation of dopamine signals exerts effects consistent with positive or negative RPEs (Chang et al., 2016; Steinberg et al., 2013). In these ways the model-free "value" axis of dopamine's role in learning is well established. By contrast, while it is possible that errors in sensory prediction (such as chocolate vs. vanilla milk) drive learning about state space, it is unclear how these errors could be separated along multiple axes by downstream circuitry. While recent work has shown that dopamine signals broadcast different types of information to subregions of the striatum (Howe & Dombeck, 2016; Menegas et al., 2017; Parker et al., 2016), it remains unknown whether dopamine signals containing state prediction errors can be separated downstream to drive learning about a world model. This is an important experimental hurdle that must be addressed before theories further integrate model-based prediction errors.

Implementing a Belief-State Temporal Difference Model

Neural implementations of a belief state may be modeled as partially observable Markov decision processes (POMDP; Daw et al., 2006; Lak et al., 2017; Rao, 2010; Starkweather et al., 2017). While the resulting belief state derived from a POMDP is equivalent to a belief state in the semi-Markov model, a benefit of a POMDP is that the belief state will propagate between nodes of the model as time elapses, potentially providing a closer analogy to neural activity. Rao (2010) harnessed this property to propose a simple implementation of a belief state using a recurrent neural network. The belief state is computed as follows:

$$b_i(t) \propto p(o(t)|i) \sum_j p(i|j)\, b_j(t-1), \qquad (48.11)$$

where $b_i(t)$ is the posterior probability that the animal is in substate i at time t, $p(o(t)|i)$ is the likelihood of the observation $o(t)$ under hypothetical substate i, and $p(i|j)$ is the probability of transitioning from substate j to substate i. It is possible to compute the belief state recursively from the belief state computed at $(t-1)$ because of the Markov property. In this way, a simple recurrent circuit that maintains feedback from the previous time point (in addition to incorporating feedforward information for new observations) could implement a belief state. We illustrated how the Starkweather et al. (2017) task could map onto this proposed implementation (figure 48.3*A*). Individual units of the feedforward layer would convey the likelihood of each observation,

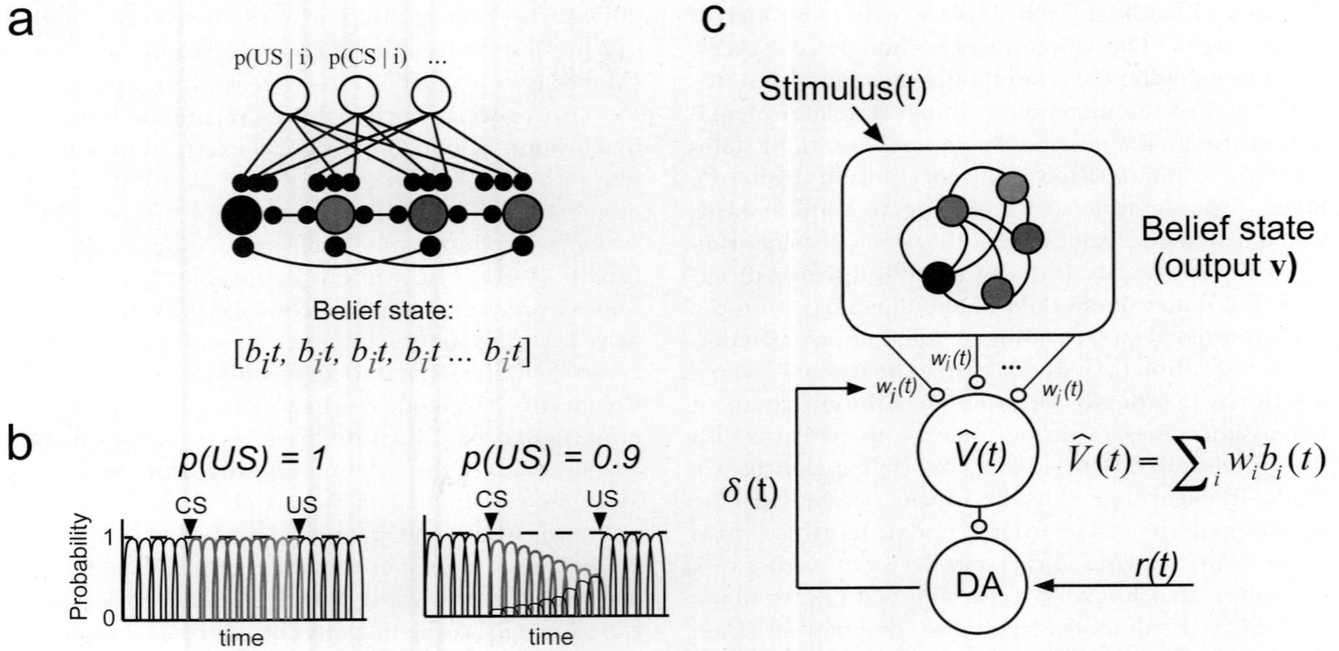

FIGURE 48.3 Recurrent network implementation of a belief state. *A*, A simple network with feedforward connections and a recurrently connected layer, adapted to computing a belief state for the task described in figure 48.2*D*. *B*, Predicted firing rates for output layer neurons in the task described in figure 48.2*D* if these neurons each signal one value within the belief-state vector. *C*, Revised neural circuit hypothesis of TD-learning implementation. Belief-state features could be represented by a recurrent network. The outputs from this recurrent network, multiplied by their weights, would be linearly summed to produce the value estimate. (See color plate 53.)

given a particular state occupancy; individuals units of the output recurrent layer would each correspond to a particular state and would each fire in proportion to the probability allotted to that state, collectively reading out the belief state (figure 48.3*B*). These output units would provide inputs to the striatum, where they could shape value predictions.

One important question in reinforcement learning is how an agent knows in the first place which features, or "states," to learn. Even in a simple TD model such as the CSC TD model, time could be tracked from the onset of any observable stimuli, meaning that the number of temporal "states" the TD model could erroneously assign weight is enormous. The same sort of problem exists when considering the belief-state TD model: there are an unconstrained number of possible states to compute inferences over in any given environment. How does the brain select the right belief state appropriate to maximizing reward in the current task? One possibility is that units projecting from cortex to striatum are maximally active when they represent higher belief in a particular state—and, critically, only in states that are relevant to the task. One way the cortex may be able to accomplish this is by using spike-timing-dependent synaptic mechanisms to hone a state representation based on temporal coincidences during a task. This was postulated as a mechanism for learning about state space during vocal learning in the songbird (Mackevicius & Fee, 2018). If the brain actively anticipates state transitions (e.g., tries to predict reward timing), temporal coincidences that commonly occur during the task may automatically hone the belief state representation into only that relevant for the current task. A second possibility is that the cortex uses the dopamine signal to compute a lower-dimensional belief state that helps the animal maximize discounted future reward. Rao proposed feeding the belief state outputs into a hidden layer containing fewer units and computing value using the outputs of this smaller hidden layer (Rao, 2010). The weights for inputs into this "hidden" layer would be tuned to the relevant belief-state representation by being trained on dopamine-like TD error signals. A third possibility is that the cortex simultaneously computes many belief states (some of these irrelevant for the current task) and feeds all of these into the striatum. Then, only those cortical inputs carrying belief states relevant to value prediction on the current task achieve synaptic potentiation with their striatal targets. One mechanism by which this could occur is dopamine-dependent modulation of the spike-timing-dependent plasticity (STDP) rule (Brzosko, Zannone, Schultz, Clopath, & Paulsen, 2017). The timing of dopamine release itself also modulates the strength of STDP-evoked dendritic spine enlargement in the striatum (Yagishita et al., 2014). In these ways, dopamine itself may play a role in strengthening only the belief-state inputs into the striatum that are temporally coincident with rewards, leading to greater weights for only the subset of belief-state inputs that allow the animal to maximize future rewards.

The belief-state TD model extends the theoretical framework for reinforcement learning in the brain. It connects the cortex's ability to represent probabilistic models of the environment with the goal of computing accurate future values. Further experiments should probe the neural implementation of a belief state and identify ways in which the brain efficiently knows which belief state to use in order to maximize expected future reward.

Acknowledgments

The authors thank Dr. Samuel Gershman for the development of the authors' studies. The authors thank all the members of the Uchida lab for discussions and for their contributions to various aspects of the studies cited in this article.

REFERENCES

Babayan, B. M., Uchida, N., & Gershman, S. J. (2018). Belief state representation in the dopamine system. *Nature Communications, 9*(1), 1891. https://doi.org/10.1038/s41467-018-04397-0

Balsam, P. D., & Gallistel, C. R. (2009). Temporal maps and informativeness in associative learning. *Trends in Neurosciences, 32*(2), 73–78. https://doi.org/10.1016/j.tins.2008.10.004

Bayer, H. M., & Glimcher, P. W. (2005). Midbrain dopamine neurons encode a quantitative reward prediction error signal. *Neuron, 47*(1), 129–141. https://doi.org/10.1016/j.neuron.2005.05.020

Bromberg-Martin, E. S., Matsumoto, M., Hong, S., & Hikosaka, O. (2010). A pallidus-habenula-dopamine pathway signals inferred stimulus values. *Journal of Neurophysiology, 104*(2), 1068–1076. https://doi.org/10.1152/jn.00158.2010

Brown, J., Bullock, D., & Grossberg, S. (1999). How the basal ganglia use parallel excitatory and inhibitory learning pathways to selectively respond to unexpected rewarding cues. *Journal of Neuroscience, 19*(23), 10502–10511.

Brzosko, Z., Zannone, S., Schultz, W., Clopath, C., & Paulsen, O. (2017). Sequential neuromodulation of Hebbian plasticity offers mechanism for effective reward-based navigation. *eLife, 6*. https://doi.org/10.7554/eLife.27756

Chang, C. Y., Esber, G. R., Marrero-Garcia, Y., Yau, H.-J., Bonci, A., & Schoenbaum, G. (2016). Brief optogenetic inhibition of dopamine neurons mimics endogenous negative reward prediction errors. *Nature Neuroscience, 19*(1), 111–116. https://doi.org/10.1038/nn.4191

Cohen, J. Y., Haesler, S., Vong, L., Lowell, B. B., & Uchida, N. (2012). Neuron-type-specific signals for reward and punishment in the ventral tegmental area. *Nature, 482*(7383), 85–88. https://doi.org/10.1038/nature10754

Costa, V. D., Tran, V. L., Turchi, J., & Averbeck, B. B. (2015). Reversal learning and dopamine: A Bayesian perspective. *Journal of Neuroscience, 35*(6), 2407–2416. https://doi.org/10.1523/JNEUROSCI.1989-14.2015

D'Ardenne, K., McClure, S. M., Nystrom, L. E., & Cohen, J. D. (2008). BOLD responses reflecting dopaminergic signals in the human ventral tegmental area. *Science, 319*(5867), 1264–1267. https://doi.org/10.1126/science.1150605

Daw, N. D., Courville, A. C., & Touretzky, D. S. (2006). Representation and timing in theories of the dopamine system. *Neural Computation, 18*(7), 1637–1677. https://doi.org/10.1162/neco.2006.18.7.1637

Daw, N. D., Gershman, S. J., Seymour, B., Dayan, P., & Dolan, R. J. (2011). Model-based influences on humans' choices and striatal prediction errors. *Neuron, 69*(6), 1204–1215. https://doi.org/10.1016/j.neuron.2011.02.027

Eshel, N., Bukwich, M., Rao, V., Hemmelder, V., Tian, J., & Uchida, N. (2015). Arithmetic and local circuitry underlying dopamine prediction errors. *Nature, 525*(7568), 243–246. https://doi.org/10.1038/nature14855

Eshel, N., Tian, J., Bukwich, M., & Uchida, N. (2016). Dopamine neurons share common response function for reward prediction error. *Nature Neuroscience, 19*(3), 479–486. https://doi.org/10.1038/nn.4239

Fiorillo, C. D., Newsome, W. T., & Schultz, W. (2008). The temporal precision of reward prediction in dopamine neurons. *Nature Neuroscience, 11*(8), 966–973. https://doi.org/10.1038/nn.2159

Flagel, S. B., Clark, J. J., Robinson, T. E., Mayo, L., Czuj, A., Willuhn, I., … Akil, H. (2011). A selective role for dopamine in stimulus-reward learning. *Nature, 469*(7328), 53–57. https://doi.org/10.1038/nature09588

Fuhs, M. C., & Touretzky, D. S. (2007). Context learning in the rodent hippocampus. *Neural Computation, 19*(12), 3173–3215. https://doi.org/10.1162/neco.2007.19.12.3173

Hampton, A. N., Bossaerts, P., & O'Doherty, J. P. (2006). The role of the ventromedial prefrontal cortex in abstract state-based inference during decision making in humans. *Journal of Neuroscience, 26*(32), 8360–8367. https://doi.org/10.1523/JNEUROSCI.1010-06.2006

Hart, A. S., Rutledge, R. B., Glimcher, P. W., & Phillips, P. E. M. (2014). Phasic dopamine release in the rat nucleus accumbens symmetrically encodes a reward prediction error term. *Journal of Neuroscience, 34*(3), 698–704. https://doi.org/10.1523/JNEUROSCI.2489-13.2014

Hollerman, J. R., & Schultz, W. (1998). Dopamine neurons report an error in the temporal prediction of reward during learning. *Nature Neuroscience, 1*(4), 304–309. https://doi.org/10.1038/1124

Houk, J. C., Adams, J. L., & Barto, A. G. (1995). A model of how the basal ganglia generate and use neural signals that predict reinforcement. In J. C. Houk, J. L. Davis, & D. G. Beiser (Eds.), *Computational neuroscience: Models of information processing in the basal ganglia* (pp. 249–270). Cambridge, MA: MIT Press.

Howe, M. W., & Dombeck, D. A. (2016). Rapid signalling in distinct dopaminergic axons during locomotion and reward. *Nature, 535*(7613), 505–510. https://doi.org/10.1038/nature18942

Kobayashi, S., & Schultz, W. (2008). Influence of reward delays on responses of dopamine neurons. *Journal of Neuroscience,*

28(31), 7837–7846. https://doi.org/10.1523/JNEUROSCI.1600-08.2008
Lak, A., Nomoto, K., Keramati, M., Sakagami, M., & Kepecs, A. (2017). Midbrain dopamine neurons signal belief in choice accuracy during a perceptual decision. *Current Biology, 27*(6), 821–832. https://doi.org/10.1016/j.cub.2017.02.026
Langdon, A. J., Sharpe, M. J., Schoenbaum, G., & Niv, Y. (2018). Model-based predictions for dopamine. *Current Opinion in Neurobiology, 49,* 1–7. https://doi.org/10.1016/j.conb.2017.10.006
Ludvig, E. A., Sutton, R. S., & Kehoe, E. J. (2008). Stimulus representation and the timing of reward-prediction errors in models of the dopamine system. *Neural Computation, 20*(12), 3034–3054. https://doi.org/10.1162/neco.2008.11-07-654
Mackevicius, E. L., & Fee, M. S. (2018). Building a state space for song learning. *Current Opinion in Neurobiology, 49,* 59–68. https://doi.org/10.1016/j.conb.2017.12.001
Matsumoto, M., & Hikosaka, O. (2007). Lateral habenula as a source of negative reward signals in dopamine neurons. *Nature, 447*(7148), 1111–1115. https://doi.org/10.1038/nature05860
Menegas, W., Babayan, B. M., Uchida, N., & Watabe-Uchida, M. (2017). Opposite initialization to novel cues in dopamine signaling in ventral and posterior striatum in mice. *eLife, 6.* https://doi.org/10.7554/eLife.21886
Mirenowicz, J., & Schultz, W. (1994). Importance of unpredictability for reward responses in primate dopamine neurons. *Journal of Neurophysiology, 72*(2), 1024–1027.
Niv, Y., Joel, D., & Dayan, P. (2006). A normative perspective on motivation. *Trends in Cognitive Sciences, 10*(8), 375–381. https://doi.org/10.1016/j.tics.2006.06.010
Nomoto, K., Schultz, W., Watanabe, T., & Sakagami, M. (2010). Temporally extended dopamine responses to perceptually demanding reward-predictive stimuli. *Journal of Neuroscience, 30*(32), 10692–10702. https://doi.org/10.1523/JNEUROSCI.4828-09.2010
Pan, W.-X., Schmidt, R., Wickens, J. R., & Hyland, B. I. (2005). Dopamine cells respond to predicted events during classical conditioning: Evidence for eligibility traces in the reward-learning network. *Journal of Neuroscience, 25*(26), 6235–6242. https://doi.org/10.1523/JNEUROSCI.1478-05.2005
Parker, N. F., Cameron, C. M., Taliaferro, J. P., Lee, J., Choi, J. Y., Davidson, T. J., … Witten, I. B. (2016). Reward and choice encoding in terminals of midbrain dopamine neurons depends on striatal target. *Nature Neuroscience, 19*(6), 845–854. https://doi.org/10.1038/nn.4287
Pasquereau, B., & Turner, R. S. (2015). Dopamine neurons encode errors in predicting movement trigger occurrence. *Journal of Neurophysiology, 113*(4), 1110–1123. https://doi.org/10.1152/jn.00401.2014
Rao, R. P. N. (2010). Decision making under uncertainty: A neural model based on partially observable Markov decision processes. *Frontiers in Computational Neuroscience, 4,* Article 146, 1–18. https://doi.org/10.3389/fncom.2010.00146
Rescorla, R. A., & Wagner, A. R. (1972). A theory of Pavlovian conditioning: Variations in the effectiveness of reinforcement and nonreinforcement. In A. Black & W. Prokasy (Eds.), *Classical conditioning II: Current research and theory* (pp. 64–99). New York: Appleton-Century-Crofts.
Roesch, M. R., Calu, D. J., & Schoenbaum, G. (2007). Dopamine neurons encode the better option in rats deciding between differently delayed or sized rewards. *Nature Neuroscience, 10*(12), 1615–1624. https://doi.org/10.1038/nn2013
Schultz, W., Dayan, P., & Montague, P. R. (1997). A neural substrate of prediction and reward. *Science, 275*(5306), 1593–1599.
Sharpe, M. J., Chang, C. Y., Liu, M. A., Batchelor, H. M., Mueller, L. E., Jones, J. L., … Schoenbaum, G. (2017). Dopamine transients are sufficient and necessary for acquisition of model-based associations. *Nature Neuroscience, 20*(5), 735–742. https://doi.org/10.1038/nn.4538
Starkweather, C. K., Babayan, B. M., Uchida, N., & Gershman, S. J. (2017). Dopamine reward prediction errors reflect hidden-state inference across time. *Nature Neuroscience, 20*(4), 581–589. https://doi.org/10.1038/nn.4520
Starkweather, C. K., Gershman, S. J., & Uchida, N. (2018). The medial prefrontal cortex shapes dopamine reward prediction errors under state uncertainty. *Neuron, 98*(3), 616–629.e6. https://doi.org/10.1016/j.neuron.2018.03.036
Stauffer, W. R., Lak, A., Yang, A., Borel, M., Paulsen, O., Boyden, E. S., & Schultz, W. (2016). Dopamine neuron-specific optogenetic stimulation in rhesus macaques. *Cell, 166*(6), 1564–1571.e6. https://doi.org/10.1016/j.cell.2016.08.024
Steinberg, E. E., Keiflin, R., Boivin, J. R., Witten, I. B., Deisseroth, K., & Janak, P. H. (2013). A causal link between prediction errors, dopamine neurons and learning. *Nature Neuroscience, 16*(7), 966–973. https://doi.org/10.1038/nn.3413
Stuber, G. D., Klanker, M., de Ridder, B., Bowers, M. S., Joosten, R. N., Feenstra, M. G., & Bonci, A. (2008). Reward-predictive cues enhance excitatory synaptic strength onto midbrain dopamine neurons. *Science, 321*(5896), 1690–1692. https://doi.org/10.1126/science.1160873
Suri, R. E., & Schultz, W. (1998). Learning of sequential movements by neural network model with dopamine-like reinforcement signal. *Experimental Brain Research, 121*(3), 350–354.
Suri, R. E., & Schultz, W. (1999). A neural network model with dopamine-like reinforcement signal that learns a spatial delayed response task. *Neuroscience, 91*(3), 871–890.
Sutton, R. S. (1988). Learning to predict by the methods of temporal differences. *Machine Learning, 3*(1), 9–44. https://doi.org/10.1007/BF00115009
Sutton, R. S., & Barto, A. G. (1990). Time-derivative models of Pavlovian reinforcement. In M. Gabriel & J. Moore (Eds.), *Learning and computational neuroscience: Foundations of adaptive networks* (pp. 497–537). Cambridge, MA: MIT Press.
Takahashi, Y. K., Batchelor, H. M., Liu, B., Khanna, A., Morales, M., & Schoenbaum, G. (2017). Dopamine neurons respond to errors in the prediction of sensory features of expected rewards. *Neuron, 95*(6), 1395–1405.e3. https://doi.org/10.1016/j.neuron.2017.08.025
Tian, J., Huang, R., Cohen, J. Y., Osakada, F., Kobak, D., Machens, C. K., … Watabe-Uchida, M. (2016). Distributed and mixed information in monosynaptic inputs to dopamine neurons. *Neuron, 91*(6), 1374–1389. https://doi.org/10.1016/j.neuron.2016.08.018
Tian, J., & Uchida, N. (2015). Habenula lesions reveal that multiple mechanisms underlie dopamine prediction errors. *Neuron, 87*(6), 1304–1316. https://doi.org/10.1016/j.neuron.2015.08.028
Yagishita, S., Hayashi-Takagi, A., Ellis-Davies, G. C. R., Urakubo, H., Ishii, S., & Kasai, H. (2014). A critical time window for dopamine actions on the structural plasticity of dendritic spines. *Science, 345*(6204), 1616–1620. https://doi.org/10.1126/science.1255514

49 Dopamine Prediction Error Responses Reflect Economic Utility

WILLIAM R. STAUFFER AND WOLFRAM SCHULTZ

ABSTRACT Rewards are objects, substances, or events that reinforce approach behavior, confer hedonic pleasure, and motivate action. Likewise, rewards guide decision-making. The principal goal of economic choices is to obtain the most valuable rewards. Dopamine neurons are critical for learning which rewards are the most valuable. Dopamine neurons code for reward prediction error, the difference between received and predicted rewards. This signal can be directly read out by downstream neurons to update behavior. The magnitude of dopamine prediction error responses scales with reward size, probability, and delay. Ultimately, it appears that dopamine neurons integrate these variables into a biological teaching signal for utility. *Utility* is defined in economics as a mathematical function for subjective value that predicts behavioral choices. Accordingly, dopamine reward prediction error responses reside at the interface between learning and decision-making. In this chapter we will focus on the coding of economic variables in the neural reward system and the role of dopamine neurons in updating the values used for choices.

Rewards include items such as energetically dense foods that provide fuel for cellular processes and financial benefits such as year-end bonuses that provide an incentive for hard work. Yet considered more broadly, many other objects, events, thoughts, and actions are rewarding as well. Classifying objects and events as rewards according to this broad perspective of behavioral effect—as exogenous and endogenous outcomes that are rewarding—serves to highlight the ubiquity of rewards and their fundamental role in behavioral function. Rewards are critically important for enjoyment and pleasure, but they are also critically important for adaptive behavioral functions, including learning and decision-making. In this regard, rewards drive learning as positive reinforcement, and they guide decision-making as outcomes endowed with value.

The ubiquity of rewards and the fundamental behaviors they support underscore the importance of neuronal reward processing. The proper appreciation and use of rewards requires significant neural resources because reward processing poses a difficult neurocomputational problem. For example, common, everyday rewards include delicious wine, good company, and excellent music. These rewards have very little in common. There is no common biophysical receptor that can detect all of these rewards, and there is no physical ruler that can compare them. Nevertheless, individuals instantaneously recognize rewards, rank them according to preference, and even exchange one for another. How does the brain manage these polysensory, high-dimensional problems? In fact, particular brain areas appear specialized for processing rewards and reward-related behaviors. These highly interconnected brain regions include the dopaminergic midbrain, the ventral striatum, the orbitofrontal cortex (OFC), and the amygdala. These (and possibly other) regions define a loosely organized brain network that we describe as the reward system.

Dopamine neurons are a critical part of the reward system. These neurons reside in the midbrain and send axons to the basal ganglia and frontal cortex (figure 49.1*A*). Dopamine neurons emit action potentials at low background frequencies (0–8 imp/s) and respond to behaviorally relevant events with phasic responses—short duration (150–300 ms) bursts or pauses of action potentials. These phasic responses code for reward prediction errors, the difference between received and predicted rewards. Prediction errors are a necessary condition for associative learning, and therefore dopamine reward prediction error responses are natural teaching signals. These signals can be read out to directly update behavioral responses. Increases in activity indicate the outcome was better than predicted, and the preceding behavior should be repeated or invigorated, whereas reductions in activity indicate that the outcome was worse than predicted and the preceding behavior should be lessened or avoided. Crucially, dopamine activations and suppressions indicate not just what direction to update, positive or negative. The magnitude of the response also indicates *by what degree* behaviors should be updated.

The magnitudes of dopamine responses correlate positively with reward amount and expected value (reward amount multiplied by the probability of delivery). These factors are important determinants of what decision-makers are most likely to choose. To properly

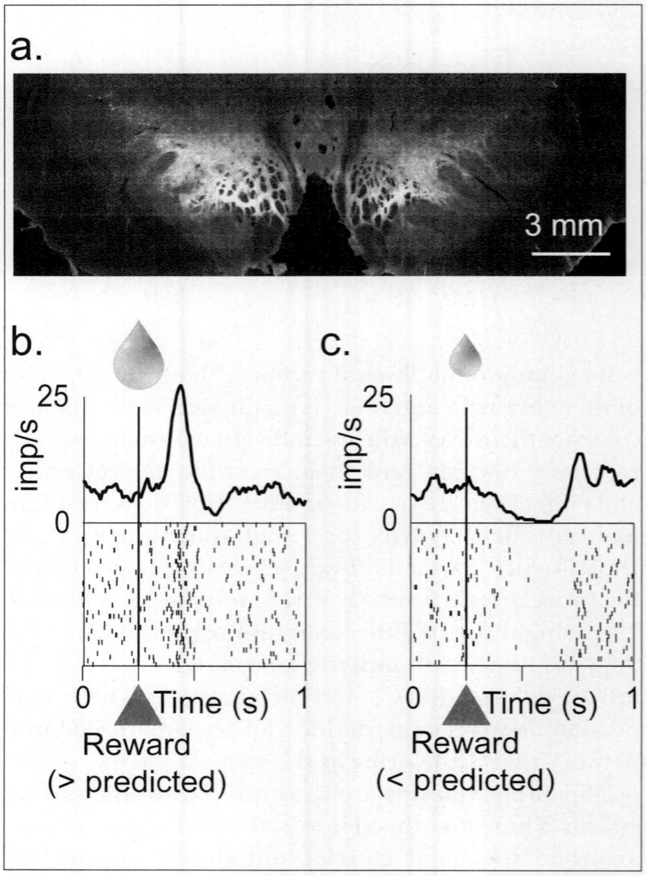

FIGURE 49.1 Dopamine neurons code for positive and negative prediction errors. *A*, Dopamine cell bodies are located in the midbrain. Bright regions are stained with an antibody for tyrosine hydroxylase, an enzyme specific to dopamine neurons in the midbrain. *B*, Raster plot (*bottom*) and peristimulus time histogram (PSTH) (*top*) of a single dopamine neuron shows the positive response (*left*) to rewards greater than predicted and negative responses to rewards smaller than predicted (*right*). Gray triangles indicate the time of the rewards.

quantify choice behavior, cognitive neuroscientists have recently been looking toward economic theory. Economic theory and experiments provide formal structures and behavioral tools to quantify the variables associated with individuals' choices. In the context of this chapter, we are particularly interested in economic theory's description of value: utility. *Utility* is defined in economics as a mathematical description of subjective value inferred from observed choice behavior. As we will see, psychophysical measurements of utility combined with neurophysiology have shown that dopamine teaching signals reflect the same values used for economic decisions. Thus, dopamine neurons constitute a critical neural interface between learning and decision-making. They teach downstream neurons what objects to choose. The purpose of this chapter is to examine reward coding in the brain and the neural interface between learning and decision-making.

Reward Prediction Error Responses

Some of the earliest indications that value is a quantity coded for by single neurons in the brain arose from dopamine neuron recordings. Unpredicted rewards briefly activated dopamine neurons, whereas unpredicted reward omissions caused dopamine activity to pause (Schultz, Apicella, & Ljungberg, 1993). This pattern of dopamine activity resembles the prediction error in the Rescorla-Wagner (R-W) reinforcement-learning rule, defined as *reward received—reward predicted.* Moreover, dopamine neurons are more readily activated by rewards than by motivationally equivalent aversive outcomes (Mirenowicz & Schultz, 1996). Accordingly, phasic dopamine responses are known as reward prediction error responses.

The explicit relationship between reward prediction error responses and value was brought to light by the formal application of temporal difference (TD) reinforcement learning to phasic dopamine signals (Montague, Dayan, & Sejnowski, 1996). Whereas the R-W model refers to associability between the CS and the US, the TD model consists of an explicit value function that reflects reward expectation through time (Sutton & Barto, 1998). TD model prediction errors constitute deviations from this value function, and it is customary to consider TD prediction errors as signaling rewards as *better or worse than predicted.* Thus, the TD model represents prediction errors as value differences—specifically, as $Value_{received} - Value_{predicted}$ ($Vr - Vp$). Like the TD model prediction error, dopamine neurons are activated by rewards that are better than predicted and silenced by rewards that are worse than predicted (figure 49.1*B, C*). Similarly, dopamine responses code for the prediction errors associated with higher-order conditional stimuli (Schultz, Apicella, & Ljungberg, 1993) and are sensitive to small temporal deviations in learned intervals (Hollerman & Schultz, 1998). These features and several others demonstrate the basic compatibility of dopamine responses with TD reinforcement learning.

Amount, Probability, and Expected Value

Biological evidence for value coding is found in the magnitude of dopamine responses to different amounts of unpredicted rewards. Reward prediction error responses scale positively with the amount of juice delivered. Dopamine neurons respond with the largest activation to the largest reward and the smallest activation to the smallest reward (Tobler, Fiorillo, & Schultz, 2005). Thus, dopamine responses show a positive monotonic relationship to reward amount, a primary factor that determines reward value.

Similar to reward amount, reward probability positively influences value. The predicted probability of reward delivery affects value in a manner proportional to the chance of getting the reward. The more frequently reward is delivered, the higher the predicted value. This concept is formalized by the expected value (EV) of the lottery. The EV is defined as the probability multiplied by amount, or $\sum_{i=1}^{n} p_i x_i$ for lotteries of size *n*. It is worth highlighting that EV is not a tangible object that has a physical manifestation such as a visual stimulus, a juice reward, or a movement. Platonic realism aside, EV is a concept that only exists in the space of thought. Accordingly, the demonstration of EV coding in the brain represents a neuroscientific milestone: the coding of a purely conceptual variable by single neurons (Platt & Glimcher, 1999). Subsequently, brain signals reflecting EV have been located in many brain regions, including dopamine neurons. The dopamine response to a particular reward is larger when the predicted probability of its delivery is smaller, and the reward responses are diminished as the predicted probability gets larger. The logic is consistent with the notion of prediction error responses. As the probability gets higher, *Vp* increases and *Vr*− *Vp* decreases. Likewise, the responses to cues predicting the reward increase with probability and EV (Fiorillo, Tobler, & Schultz, 2003). Thus, dopamine neurons reflect the integration of amount and probability into EV.

EV is a potential decision variable. Indeed, Pascal famously used EV as the decision variable to justify a belief in God (Pascal's wager). However, more modern thinkers have shown that EV is a poor normative guide (cf., the Saint Petersburg paradox) and fails to predict what decision-makers actually choose. These facts reveal that objective factors including amount and probability and their mathematical expectation are only partially responsible for decisions. As we will see, individuals transform objective value into subject-specific value (utility), and the probability multiplied by utility, or expected utility (EU), is far better than EV at predicting dopamine activity and observable choice behavior.

Subjective Value

Reward amount is a physical concept similar to contrast, duration, orientation, and other dimensions. These dimensions exist independently of individuals or behavior and thus are objective. Value, on the other hand, is subjective. Value is dependent on individuals and cannot be directly observed. For example, individuals will often consume too much of a highly valued reward, and the resulting state of satiety temporarily devalues that same reward. Thus, the same physical reward takes on different values based on the individual decision-maker. This demonstrates that value must be inferred from subjects' behavior, rather than the object itself, and therefore this quantity is commonly referred to as *subjective value.* The neural coding of subjective value can reveal brain regions involved in making decisions and has been observed in OFC, amygdala, dopamine neurons, and many other regions of the frontal cortex and limbic system.

Two reward parameters in particular, timing and risk, are critical for separating objective factors, such as amount, from subjective values. With regard to time, individuals often prefer to receive rewards sooner rather than later. A $10 reward is more valuable when it is delivered immediately, compared to when it is delivered in one week, despite the physical amount of reward being identical. This phenomenon is called *temporal discounting* and can be used to separate the coding of subjective value from the coding of the physical amount.

Temporal discounting has been used to demonstrate that dopamine neurons, basal ganglia, and frontal cortex regions code for the subjective value of rewards. The rate at which value is discounted decreases as a function of amount and time. Decreasing discount rate is often modeled with hyperbolic functions. Hyperbolic-discounting models—rather than exponential-discounting models that use a constant discount rate—predict well human and monkey preferences. Dopamine responses are diminished by the predicted reward delay, according to a hyperbolic-discounting model (Kobayashi & Schultz, 2008). This result provides strong evidence that dopamine neurons code for subjective value.

Risk

Like time, risk can be used to separate subjective value from objective properties. Risk has several definitions, and the one in everyday usage is distinctly different from the concept of risk in economics. In everyday language *risk* is defined as the probability of a bad outcome. For example, BASE jumping is riskier than golf or chess because strapping on a parachute and jumping off a building has a higher probability of mortal injury than swinging a golf club or moving a pawn. This definition is based primarily on the first moment of the statistical distribution, EV, as well as on psychological factors. In

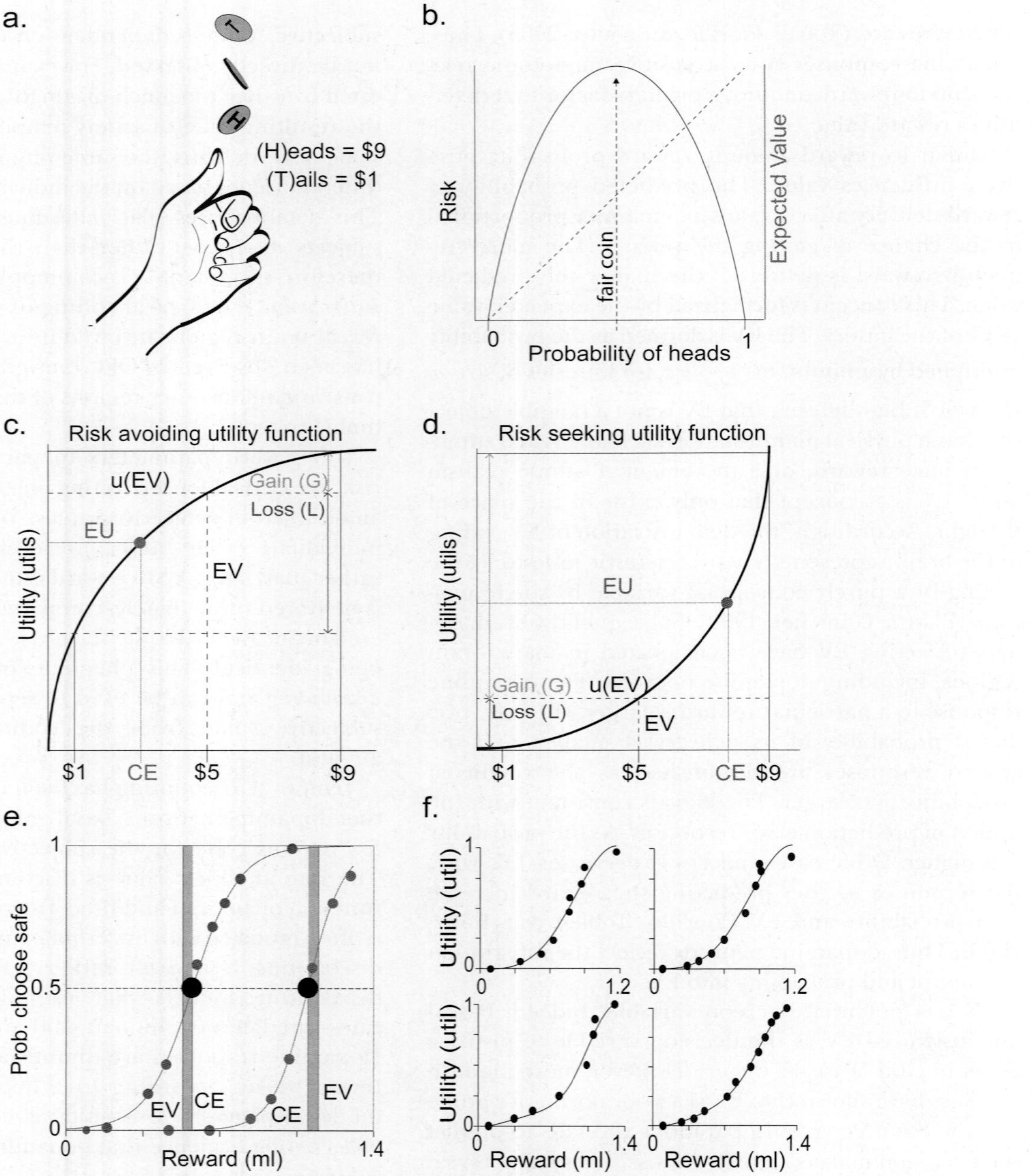

FIGURE 49.2 Risk, value, and utility. *A*, Flipping a coin, where heads pays a large reward ($9) and tails returns a small reward ($1), constitutes an ideal model of the binary gambles we describe. *B*, For a Bernoulli process such as a weighted coin flip, this graph illustrates the relationship between risk, probability, and value. The black curve plateaus at $p = 0.5$ (fair coin) and demonstrates the relationship between risk (formally: entropy $\sum_{i=1}^{n} -P(x_i)log_2P(x_i)$) and the probability that a coin comes up heads. A fair coin has the most unpredictable outcome and therefore has the highest risk. The red dotted line demonstrates that expected value (EV) is a linear function of probability. *C*, A risk-avoiding (concave) utility function describes the behavior of an individual who would prefer a sure $5 to the fair coin flip illustrated in (*A*). The potential utility loss (L) is greater than the potential utility gain (G). Accordingly, the certainty equivalent (CE) is smaller than the gamble EV, and therefore the expected utility (EU) is smaller than the utility of the EV u(EV). *D*, A risk seeking (convex) utility function describes the behavior of an individual who would prefer the gamble described in (*A*) to a sure $5 payout. The potential utility gain (G) outweighs the potential loss (L). The CE is larger than the gamble EV, and therefore EU > u(EV). *E*, Psychometric measurement of monkey risk seeking toward a small EV gamble (*blue*) and risk avoiding for a large EV gamble (*red*). The large black dots represent the certainty equivalents (CE)—the points of choice indifference between the gambles and safe options. Note that the CE is larger than the EV for the small gamble and less than the EV for the large gamble. The red and blue dots represent the probability of choosing a safe reward over the gamble. The red and blue curves represent the fitted psychometric functions. The shaded regions show the risk premiums—the differences between the CE and the gamble EV—for the two gambles. Data adapted with permission from Genest, Stauffer, and Schultz (2016). *F*, Utility functions for four different monkeys reflect risk seeking for small rewards and risk aversion for larger rewards. The black dots represent CE from iterative gambles within the reward range (as in [*E*]). Data adapted with permission from Genest, Stauffer, and Schultz (2016) and Stauffer, Lak, and Schultz (2014). (See color plate 54.)

contrast, economic theory defines risk as statistical variance in outcome distributions. Thus, risk is defined in economics with the second moment of the distribution. For example, all other factors being equal, a gamble with an equiprobable chance of gain or loss is riskier than a gamble with 9 in 10 chances of loss. Whether you consider this risky depends on your definition, yet economic logic dictates that there is little risk involved when the outcome is known with high certainty. The formal relationship between risk, probability, and EV is illustrated in figure 49.2*A*. For our purposes and in the context of this chapter, we will follow standard economics practice and define risk as the variance in the outcome distribution.

Perhaps surprisingly, given the narrow definition of risk employed in economic theory, neuronal signals predicted by this concept of risk can be observed in the several brain regions implicated in value and decision-making. Recordings from the OFC were made while monkeys observed cues that predicted a gamble between two outcomes. Nearly one in four neurons was sensitive to economic risk independently of value (O'Neill & Schultz, 2010). A similar risk signal was later found in the lateral septum. Neurons in the lateral septum responded most strongly to probabilities close to 0.5, and their responses were diminished as reward probability approached certainty (Monosov & Hikosaka, 2013). As we will see, risk attitudes are a fundamental aspect of economic value, and therefore the coding of such pure risk might constitute a neural antecedent of value computations.

Risk attitudes, alternatively called risk preferences, account for the subjective value added or subtracted from a gamble's EV. Risk attitudes are a fundamental aspect of economic decision-making (Bernoulli, 1954). Risk attitudes are derived from the preferences of individuals during choices under risk. Risk avoiders prefer a safe outcome to a gamble with the same mean value (same EV), whereas risk seekers prefer the gamble. Like temporal discounting, risk attitudes provide a means to separate objective from subjective value coding. Risk attitude–dependent neural signals have been observed in several reward-related brain areas, including the OFC. When a risk-seeking monkey chose gambles over safe options with the same EV, OFC neurons responded more strongly to the gamble than the safe option (Raghuraman & Padoa-Schioppa, 2014). Similar signals have been observed in the posterior cingulate cortex (area 23; McCoy & Platt, 2005), supplementary eye fields (SEF; So & Stuphorn, 2010), and dopamine neurons (Fiorillo, 2011; Lak, Stauffer, & Schultz, 2014).

Utility

Utility is defined in economics as subjective value derived from choice behavior. It is our "internal currency" for gains and losses (Luce, 1959). The fact that utility can only be measured from choices distinguishes the concept of utility from other measures of subjective value, including pleasantness ratings, reaction times, response rate, relative response rate, and force (Hursh & Silberberg, 2008). Within economics, utility is a purely theoretical construct used to describe the behavior of individuals. From the perspective of cognitive neuroscience, however, it is of great interest to understand whether utility is a variable coded for by single neurons and whether neuronal utility coding determines choice behavior. In this regard, economic theory provides a formal framework to measure utility functions.

All choices reveal utility rankings, but not all choices provide as much information about the underlying utility function. Choices between different types or amounts of reward reveal that the chosen option has a higher utility than the unchosen options. However, the only constraint this puts on the underlying utility function is one of rank. Any mathematical function with any shape that preserves the utility rankings—that is, that designates the chosen option as having a utility at least as high as the unchosen option—is equally as valid. In fact, an infinite number of utility functions satisfy this requirement. Thus, choices between reward amounts or types reveal utility functions uniquely defined up to a positive monotonic (rank-preserving) transformation (Mas-Colell, Whinston, & Green, 1995). As we will see, it is also true that an infinite number of utility functions fit risky choices, but crucially, every member of that infinite set has the same shape.

Risk attitudes are mathematically linked to the shape of the underlying utility functions. The concept that links risk attitude to the shape of the utility function is the *certainty equivalent* (CE)—the singular reward amount that has the same utility as a gamble. An individual is risk averse if CE < gamble EV, whereas an individual is risk seeking when CE > gamble EV. When the CE for a gamble is less than the gamble EV (risk aversion), only a concave utility function will connect the data (figure 49.2*C*). When the CE is greater than the gamble EV (risk seeking), only a convex utility function will connect the data (figure 49.2*D*). Thus, during risky choices, the shape of the utility function is nonarbitrary. Convex utility functions cannot explain risk aversion, and concave utility cannot account for risk-taking. These mathematical relationships signify that the shape of the utility function is meaningful. In particular, the

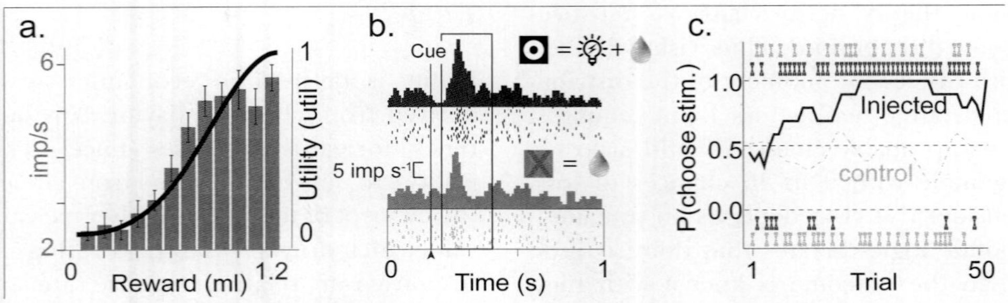

FIGURE 49.3 Dopamine responses code for utility. *A*, Dopamine responses were recorded to a range of rewards between 0.1 and 1.2 ml in 0.1 ml steps. The gray bars indicate the average responses of the population (n=16 dopamine neurons) as a function of reward size. The overlaid black line is the utility function measured from the choices of the same individual. Error bars are ±SEM (Standard Error of the Mean) across neurons. Data adapted with permission from Stauffer, Lak, and Schultz (2014). *B*, Dopamine neurons responded more strongly to a cue that predicts a juice reward and dopamine stimulation than to a cue that predicts juice alone. Black and gray PSTH (*top*) and raster plots (*bottom*) show responses from a single neuron to cues that predict light stimulation and reward (*top inset*) and reward alone (*bottom inset*). *C*, Choices between stimulation and no stimulation indicate that the monkeys learn to prefer the cue that predicts optical stimulation. The black and gray *X*s on the top and bottom show actual choices for the stimulated option (*top*) and the unstimulated option (*bottom*) when the stimulating probe was in the infected and control hemispheres, respectively. The black and gray lines indicate a running average of the 1–0 choices for the session with the stimulating probe in the infected hemisphere (*black*) and in the noninfected control hemisphere (*gray*). Data from (*B*) and (*C*) adapted with permission from Stauffer et al. (2016).

function is unique up to a positive affine (shape-preserving) transformation (Mas-Colell, Whinston, & Green, 1995).

Why are such intricate details from economic theory important for cognitive neuroscience? The reason is simple: To gain mechanistic insights into utility coding and economic decision-making, it is imperative that the behavioral and neuronal measures can be meaningfully correlated. If the shape of the utility function is arbitrary—if it is uniquely defined only up to a positive monotonic transformation—then the degree of correlation between a neuronal function and the utility function is meaningless. By contrast, choices under risk reveal a utility function with a nonarbitrary shape. Therefore, these choices can be used to define a utility function relatable to neural data. Using a chained psychophysical procedure, a utility function that is unique up to a positive affine transformation can be estimated from choices between gambles and certain values (Caraco, Martindale, & Whittam, 1980; Stauffer, Lak, & Schultz, 2014). The details of this psychometric chaining procedure can be found in Stauffer et al. (2014). A utility function measured in this way expresses utility as a function of reward size.

Standard utility functions are concave—as in figure 49.2*C*. Concave utility functions reflect the fact that human decision-makers are mostly risk averse (Holt & Laury, 2002). In contrast to the risk aversion associated with standard concave utility functions, numerous studies have reported that monkeys are risk seeking. What can account for this discrepancy? It is likely that the sizes of rewards in most studies of monkey behavior are too small to fully characterize behavior. Choices over a broader reward range reveal behavior more consistent with general economic principles. In fact, monkeys are risk seeking (CE > EV) for gambles with a low expected value (figure 49.2*E*, *blue*). However, as the range of rewards gets larger, monkeys make fewer risky choices and eventually demonstrate frank risk aversion (CE < EV; figure 49.2*E*, *red*). Thus, measured utility functions are convex at small reward magnitudes and become more linear before finally concaving as the reward magnitude increases (figure 49.2*F*). This shape reflects the reward-magnitude-dependent transition from risk-seeking choices to risk-avoiding ones.

The convex and then concave shape of the utility function is nonarbitrary, and thus it can be used as a psychometric function for meaningful comparison with neuronal data. To gather neuronal data for a neurometric function, dopamine responses were recorded while different-sized rewards spanning the range of the measured utility function were delivered at unpredictable intervals. The recorded dopamine responses reflected the shape of the utility functions measured under choices. Figure 49.3*A* shows an overlay of the psychometric and neurometric functions. Despite the fact that the functions were measured in entirely distinct behavioral contexts—the psychometric function was measured during choice behavior, whereas the neurometric function was collecting during passive reward-delivery trials—the correlation between these functions was greater than 0.9. Thus, dopamine prediction error responses code a neural signal suitable for teaching downstream neurons the utility of rewards.

Dopamine-dependent- (i.e., reward) learning studies provide even more evidence that dopamine learning acts as the interface between learning and decision-making. Strikingly, the amount that individuals learn from rewards can be predicted based on their wealth status, according to basic economic principles. Higher-wealth-status individuals learn less from a reward than lower-status individuals do from the same reward (Tobler, Fletcher, Bullmore, & Schultz, 2007). This learning behavior is consistent with a learning signal that is shaped by decreasing marginal utility, a concept at the heart of many economic theories. Decreasing marginal utility states that the same unit of reward will be worth less and less as an individual becomes wealthier. Thus, this behavioral result provides indirect evidence that dopamine signals act as a bridge between learning theory and economics.

Single-unit dopamine studies provide direct evidence that utility is a quantity coded by dopamine neurons. This finding is consistent with decades of observations from many laboratories showing that phasic dopamine responses scale with reward magnitude, expected value, and other factors that determine decisions. Learning studies in humans provide further indirect evidence for the role of dopamine as a bridge between learning and decision-making. Yet these studies are based on correlations between behavior and other variables. Strong evidence for the relationship between value and dopamine can be found in studies that directly stimulate dopamine neurons. Indeed, one of the earliest indications that dopamine was involved in value came from observations that electrical stimulation near dopamine neurons is rewarding (Corbett & Wise, 1980). New techniques based on the genetic regulation of protein expression permit the stimulation of dopamine neurons with light, rather than electric current, and this breakthrough enables the high-resolution interrogation of the function of specific cell types—notably, including dopamine neurons.

Optogenetic Stimulation of Dopamine Neurons

Correlations can reveal the underlying relationships between neuronal activity and behavior, but newer techniques such as optogenetics allow investigators to directly manipulate neuronal activity and observe changes in behavior. Optogenetics uses genetically coded optical actuators, opsins, to enable neurons to transduce light stimulation into action potentials. Using two viral vectors, one to define cell type specificity and the other to confer optical sensitivity, monkey dopamine neurons were selectively infected with channelrhodopsin (ChR2). Light flashes directed to ChR2-infected dopamine neurons caused the neurons to emit action potentials. To test for the relationship between dopamine activations and behavior, monkeys were trained that one cue predicted optical dopamine stimulation and a juice reward, whereas a different cue predicted a juice reward alone. Remarkably, after training, dopamine neurons responded more strongly to the cue that predicted juice and stimulation, compared to the cue predicting juice alone (figure 49.3*B*). This result provides direct evidence that dopamine action potentials are used to train the predictive neural responses to the cue. Furthermore, since dopamine response magnitudes scale with the utility of the cues, this result suggests that the monkey will prefer the cues predicting juice and stimulation. Indeed, given the choice between two cues they had never seen before, monkeys quickly learned to choose the option predicting juice and stimulation over the option promising juice alone (figure 49.3*C*). The behavioral and neuronal acquisition of conditioned responses (US-CS transfer) is the hallmark of all basic associative learning, and making decisions is the ultimate expression of value. Thus, the optogenetic stimulation of dopamine neurons demonstrates that dopamine activations teach animals what to choose (Stauffer et al., 2016).

Dopamine Contributions to Choices

Dopamine neurons send the majority of their axons to brain regions implicated in value learning and decision-making, including the striatum and the frontal cortex. Phasic dopamine responses and the associated dopamine release at the projection targets likely play several roles in decision-making that occur on multiple timescales. On the slowest timescale, dopamine release affects decision-making via learning. This has been one of the central points of this chapter. We have emphasized that dopamine neurons code for teaching signals (reward prediction errors) and scale with utility and that the artificial activation of dopamine neurons induces behaviors associated with value learning. At the cellular level, dopamine release plays a critical role in long-term potentiation (LTP)—the putative cellular mechanism for learning. Three-factor Hebbian learning involves presynaptic action potentials, postsynaptic action potentials, and dopamine release. It is thought that dopamine acts on preexisting synaptic traces to bridge the time interval between rewards and predictors.

Can dopamine release influence decisions on a faster timescale? For instance, can phasic dopamine responses that occur before a decision influence that decision?

The jury is still out. Dopamine responses reflect the chosen value even before the choice is indicated (Lak, Stauffer, & Schultz, 2016; Morris, Nevet, Arkadir, Vaadia, & Bergman, 2006). Chosen value is still considered a postdecision variable, and thus the decision appears to have been made internally before dopamine neurons respond. Optogenetic stimulation before choices might tell us more. In rodents, decisions are influenced by early optogenetic dopamine activation. However, these results seem to be explained by altered motivation, attention, and learning rather than as direct alterations of value-based calculations. Accordingly, more work is needed to fully understand the role of phasic dopamine responses in moment by moment behavioral control.

Conclusions

Dopamine neurons are critical for numerous behaviors. This chapter focused on the role of phasic dopamine signals as a neural interface between learning and decision-making. The key points to take away from this chapter are as follows: (1) The phasic activity of dopamine neurons constitutes a neuronal teaching signal. The reward prediction error nature of dopamine responding was the first firm evidence that phasic dopamine activity plays a role in learning. The evidence for a learning role has been repeatedly confirmed by neurophysiological experiments and, more recently, by optogenetic experiments. (2) Economic theory provides a critical framework for measuring behavior. Value cannot be directly observed; it must be inferred from well-controlled behavior. Economic theory has the tools and the techniques to assess the quality of the behavior and to estimate underlying functions. (3) Dopamine neuron reward prediction error responses code for utility as defined by economic theory. Thus, they code for utility prediction error. The responses of dopamine neurons to unpredicted rewards reflect the shape of the utility functions measured during choices. These are distinctly different behavioral contexts, and the neural evidence for utility coding in the former provides strong evidence that this is a key function for dopamine neurons. Furthermore, optogenetically stimulating dopamine neurons produces behavioral effects consistent with the notion that they drive utility learning. Together, these lines of evidence point to the role of dopamine prediction error responses as neural signals to teach the brain what to choose.

Acknowledgments

Our work has been supported by the Wellcome Trust and the European Research Council (ERC; Wolfram Schultz) and the National Institutes of Health 1DP2MH113095 (William R. Stauffer).

REFERENCES

Bernoulli, D. (1954). Exposition of a new theory on the measurement of risk. *Econometrica, 22*(1), 23–36. doi:10.2307/1909829

Caraco, T., Martindale, S., & Whittam, T. S. (1980). An empirical demonstration of risk-sensitive foraging preferences. *Animal Behaviour, 28*(3), 820–830.

Corbett, D., & Wise, R. A. (1980). Intracranial self-stimulation in relation to the ascending dopaminergic systems of the midbrain: A moveable electrode mapping study. *Brain Research, 185*(1), 1–15.

Fiorillo, C. D. (2011). Transient activation of midbrain dopamine neurons by reward risk. *Neuroscience, 197,* 162–171. doi:10.1016/j.neuroscience.2011.09.037

Fiorillo, C. D., Tobler, P. N., & Schultz, W. (2003). Discrete coding of reward probability and uncertainty by dopamine neurons. *Science, 299*(5614), 1898–1902.

Genest, W., Stauffer, W. R., & Schultz, W. (2016). Utility functions predict variance and skewness risk preferences in monkeys. *Proceedings of the National Academy of Sciences of the United States of America, 113*(30), 8402–8407. doi:10.1073/pnas.1602217113

Hollerman, J. R., & Schultz, W. (1998). Dopamine neurons report an error in the temporal prediction of reward during learning. *Nature Neuroscience, 1*(4), 304–309. doi:10.1038/1124

Holt, C. A., & Laury, S. K. (2002). Risk aversion and incentive effects. *American Economic Review, 92*(5), 1644–1655.

Hursh, S. R., & Silberberg, A. (2008). Economic demand and essential value. *Psychological Review, 115*(1), 186–198. doi:10.1037/0033-295X.115.1.186

Kobayashi, S., & Schultz, W. (2008). Influence of reward delays on responses of dopamine neurons. *Journal of Neuroscience, 28*(31), 7837–7846. doi:10.1523/JNEUROSCI.1600-08.2008

Lak, A., Stauffer, W. R., & Schultz, W. (2014). Dopamine prediction error responses integrate subjective value from different reward dimensions. *Proceedings of the National Academy of Sciences of the United States of America, 111*(6), 2343–2348. doi:10.1073/pnas.1321596111

Lak, A., Stauffer, W. R., & Schultz, W. (2016). Dopamine neurons learn relative chosen value from probabilistic rewards. *eLife, 5.* doi:10.7554/eLife.18044

Luce, D. (1959). *Individual choice behavior: A theoretical analysis.* Hoboken, NJ: Wiley.

Mas-Colell, A., Whinston, M. D., & Green, J. R. (1995). *Microeconomic theory.* Oxford: Oxford University Press.

McCoy, A. N., & Platt, M. L. (2005). Risk-sensitive neurons in macaque posterior cingulate cortex. *Nature Neuroscience, 8*(9), 1220–1227. doi:10.1038/nn1523

Mirenowicz, J., & Schultz, W. (1996). Preferential activation of midbrain dopamine neurons by appetitive rather than aversive stimuli. *Nature, 379*(6564), 449–451. doi:10.1038/379449a0

Monosov, I. E., & Hikosaka, O. (2013). Selective and graded coding of reward uncertainty by neurons in the primate anterodorsal septal region. *Nature Neuroscience, 16*(6), 756–762. doi:10.1038/nn.3398

Montague, P. R., Dayan, P., & Sejnowski, T. J. (1996). A framework for mesencephalic dopamine systems based on predictive Hebbian learning. *Journal of Neuroscience, 16*(5), 1936–1947.

Morris, G., Nevet, A., Arkadir, D., Vaadia, E., & Bergman, H. (2006). Midbrain dopamine neurons encode decisions for future action. *Nature Neuroscience, 9*(8), 1057–1063. doi:10.1038/nn1743

O'Neill, M., & Schultz, W. (2010). Coding of reward risk by orbitofrontal neurons is mostly distinct from coding of reward value. *Neuron, 68*(4), 789–800. doi:10.1016/j.neuron.2010.09.031

Platt, M. L., & Glimcher, P. W. (1999). Neural correlates of decision variables in parietal cortex. *Nature, 400*(6741), 233–238. doi:10.1038/22268

Raghuraman, A. P., & Padoa-Schioppa, C. (2014). Integration of multiple determinants in the neuronal computation of economic values. *Journal of Neuroscience, 34*(35), 11583–11603. doi:10.1523/JNEUROSCI.1235-14.2014

Schultz, W., Apicella, P., & Ljungberg, T. (1993). Responses of monkey dopamine neurons to reward and conditioned stimuli during successive steps of learning a delayed response task. *Journal of Neuroscience, 13*(3), 900–913.

So, N. Y., & Stuphorn, V. (2010). Supplementary eye field encodes option and action value for saccades with variable reward. *Journal of Neurophysiology, 104*(5), 2634–2653. doi:10.1152/jn.00430.2010

Stauffer, W. R., Lak, A., & Schultz, W. (2014). Dopamine reward prediction error responses reflect marginal utility. *Current Biology, 24*(21), 2491–2500. doi:10.1016/j.cub.2014.08.064

Stauffer, W. R., Lak, A., Yang, A., Borel, M., Paulsen, O., Boyden, E. S., & Schultz, W. (2016). Dopamine neuron-specific optogenetic stimulation in rhesus macaques. *Cell, 166*(6), 1564–1571 e1566. doi:10.1016/j.cell.2016.08.024

Sutton, R., & Barto, A. (1998). *Reinforcement learning: An introduction*. Cambridge, MA: MIT Press.

Tobler, P. N., Fiorillo, C. D., & Schultz, W. (2005). Adaptive coding of reward value by dopamine neurons. *Science, 307*(5715), 1642–1645. doi:10.1126/science.1105370

Tobler, P. N., Fletcher, P. C., Bullmore, E. T., & Schultz, W. (2007). Learning-related human brain activations reflecting individual finances. *Neuron, 54*(1), 167–175. doi:10.1016/j.neuron.2007.03.004

50 The Role of the Orbitofrontal Cortex in Economic Decisions

KATHERINE E. CONEN AND CAMILLO PADOA-SCHIOPPA

ABSTRACT Economic choice between goods entails the computation and comparison of subjective values. Neuroeconomics aspires to understand the neural mechanisms underlying these choices. The first generation of studies showed that subjective values are computed and explicitly represented at the neuronal level. More recently, the field has focused on the question of where and how values are compared to make a decision. In this chapter we review several lines of evidence suggesting that economic decisions are formed within the orbitofrontal cortex (OFC). We review results from lesion studies, neurophysiology experiments, and computational work, highlighting the following key findings: (1) OFC lesions impair value-guided behavior; (2) during economic choices, neurons in OFC encode both the input and the output of the decision process; (3) the neuronal representation of goods and values in OFC is contextually flexible but functionally stable; (4) activity fluctuations in OFC neurons are correlated with choice variability; and (5) computational models built on different principles recover the three groups of neurons identified in OFC. While several questions remain open, these results support the hypothesis that economic decisions may be generated in a neural circuit within the OFC.

Humans and animals often make decisions based on subjective preferences, without an intrinsically correct answer. This behavior, termed *economic choice*, takes place in a variety of contexts, ranging from trivial (*What socks should I wear today?*) to life changing (*What career should I pursue?*). Choice behavior has been a central interest for economists and experimental psychologists since the 18th century. More recently, economic choice has become a lively area of research in neuroscience. Neuroeconomics aims to understand the cognitive and neural mechanisms that underlie choice behavior. Research in this field has promoted an intense dialogue between economists, psychologists, and neuroscientists. Neurophysiology research often builds on constructs defined in economic theory and experimental psychology. Conversely, experimental observations on brain activity inform new economic models. Importantly, one long-term goal of the field is to use information about neural circuits to understand the choice deficits associated with mental and neurological disorders such as frontotemporal dementia, obsessive-compulsive disorder, and drug addiction.

Behavioral theories of choice have a cornerstone in the concept of value (Kreps 1990). The first generation of neuroeconomics studies focused on whether the value construct is valid at the neural level. Perhaps the most enduring result of the field has been the identification of explicit value signals in the brain during choice behavior. Subjective value signals reflect different dimensions along which goods may vary, including risk, delay, effort, ambiguity, and more (reviewed in Bartra, McGuire, and Kable 2013; Clithero and Rangel 2013; O'Doherty 2014; Padoa-Schioppa and Cai 2011; Wallis 2012). Building on these foundational results, research in the past few years has increasingly focused on the difficult question of where in the brain and how exactly subjective values are compared. Candidate regions have included ventromedial prefrontal cortex (vmPFC), posterior parietal cortex, and premotor regions (Hare et al. 2011; Hunt et al. 2012; Kable and Glimcher 2009; Strait, Blanchard, and Hayden 2014). Other groups advanced the hypothesis that economic decisions take place across multiple brain regions (Cisek 2012). While this remains an area of active research, converging lines of evidence suggest that economic decisions might be generated in a neural circuit within the orbitofrontal cortex (OFC).

This chapter presents our current understanding of the neural mechanisms underlying economic choices, focusing on the OFC. We begin by discussing the notion of value in neuroeconomics. Next, we describe the anatomical connectivity of OFC and review lesion studies linking OFC to economic choice behavior. In the following three sections, we describe distinctive features of the neuronal activity in OFC, focusing primarily on studies from nonhuman primates. First, we review early work on OFC and describe three classes of neurons identified in this area. Second, we describe properties of value encoding in OFC that provide a balance between stability and flexibility across contexts. Third, we describe how trial-by-trial variability in neuronal firing rates correlates with variability in choices. In the penultimate section of the chapter, we describe current neurocomputational models of choice and their relationship to the groups of neurons found in OFC. In the final section, we summarize the main points and indicate open questions for future research.

Subjective Value in Neuroeconomics

Economic choice may be conceptualized as a two-stage mental process: subjective values are assigned to the available options, and a decision is made by comparing values. The notion of value defined by this framework is closely tied to concepts defined in economics and learning theory.

The understanding of value in economics evolved from the early writings of Adam Smith and Jeremy Bentham (Niehans 1990). In standard or neoclassical economics, value is a weak concept (Kreps 1990). The theory asserts that choices are made *as if* based on subjective values. The standard theory is built only on revealed preferences and is completely agnostic about the mental processes underlying preference formation and choice (Ross 2005). This agnosticism might surprise the neophyte. To make sense of it, note that, at the behavioral level, the concept of value is somewhat circular. Choices allegedly maximize subjective values, but values cannot be measured independently of choices. Having no direct access to values, the standard theory builds on the only observables—namely, choices. A deliberate goal of neuroeconomics is to surpass this explanatory level. Thus, for the first generation of studies, the challenge was to assess whether choice behavior indeed entails an explicit neural representation of subjective values (it does). More recently, a major goal has been to dissociate the neural mechanisms of value assignment and value comparison (i.e., the decision).

The notion of value in neuroeconomics is also closely related to *goal-directed behavior*, discussed in learning theory (O'Doherty 2014; Padoa-Schioppa and Schoenbaum 2015). As the term suggests, goal-directed behavior refers to actions driven by the subject's motivation to achieve a specific outcome. It describes situations where subjects act with intent, informed by some understanding of the relationship between behavior and outcome. Goal-directed behavior is generally revealed through a reinforcer devaluation paradigm (Balleine and Dickinson 1998). In one version of this paradigm, subjects learn to perform a task to obtain a particular food. After training, subjects are divided into two groups: the experimental group can consume that food to satiation; the control group can consume some other food. When subjects are tested on the task, the experimental group performs at a lower level than the control group, reflecting the decreased value subjects place on the food after selective satiation. This decrease in performance indicates that the value of the food depends on the subject's motivational state. In other words, the value is subjective and computed "on the fly" (McDannald et al. 2014). Neuroeconomics embraces this concept: a neural signal may be said to encode subjective values only if it covaries with behavioral measures of value and is affected by the environmental and motivational factors that affect choices.

Anatomy and Lesion Studies of Orbitofrontal Cortex

The OFC is situated in the central part of a densely interconnected network of brain areas on the orbital surface of the frontal lobe, collectively referred to as the *orbital network* (Ongur and Price 2000). In this chapter the term *OFC* specifically refers to areas 13m/l and 11l in that network. Anatomical studies found that inputs from visual, somatosensory, olfactory, and gustatory regions converge in the OFC, along with connections from limbic regions and the dorsal raphe. This pattern of connectivity allows OFC to integrate information about sensory signals and the internal state to compute subjective values. Outputs from the OFC extend to the lateral prefrontal cortex, which in turn projects to motor and premotor regions. Through this pathway, the OFC can influence action planning and execution (figure 50.1; discussed in Padoa-Schioppa and Conen 2017).

Starting from the 19th-century case of Phineas Gage, numerous studies found that OFC lesions affect choice behavior (Fellows 2011; Rudebeck and Murray 2014). More recently, it has been observed that human patients with OFC damage make more inconsistent choices compared to patients with other prefrontal lesions and controls (Fellows and Farah 2007). Given three goods A, B, and C, OFC patients are more likely to choose A over B, B over C, and C over A—a pattern that violates preference transitivity. Experimental studies in rodents and primates have also shown that OFC lesions impair goal-directed behavior by reducing the effects of reinforcer devaluation (Gremel and Costa 2013; Rudebeck and Murray 2011). The loss of devaluation effects suggests that OFC lesions disrupt the ability to compute and/or use subjective values to guide behavior.

The link between OFC lesions and deficits in value-guided behaviors appears quite specific. Early studies suggested that OFC lesions also impaired reversal learning, but recent work showed that this deficit was caused by damage to white matter fibers passing immediately above the OFC rather than the OFC itself. When OFC lesions were procured using excitotoxic agents (which preserve white matter), reversal learning remained intact (Rudebeck et al. 2013). Furthermore, individuals

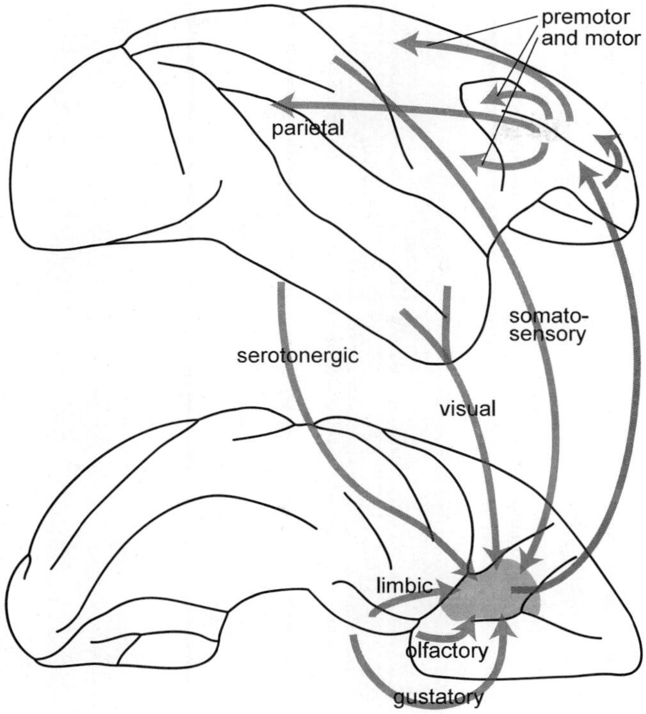

FIGURE 50.1 Anatomical connectivity of the orbitofrontal cortex. Lateral and ventral view of a monkey brain, with the front of the brain on the right. The figure shows the inputs and outputs of OFC considered most relevant to economic choice behavior. The OFC receives input across multiple sensory and limbic regions and sends outputs to lateral prefrontal cortex (LPFC), which in turn projects to several motor and premotor regions. Adapted with permission from Padoa-Schioppa and Conen (2017).

with OFC lesions can still perform accurate perceptual judgments and strategic, rule-based decisions (Baxter et al. 2009; Fellows and Farah 2007). Other studies showed that deficits in goal-directed behavior occur only after lesions of the OFC or the amygdala (e.g., Rhodes and Murray 2013). In contrast, lesions to vmPFC, lateral prefrontal cortex, prelimbic cortex, or the hippocampus do not affect these behaviors (discussed in Padoa-Schioppa and Conen 2017). Interestingly, OFC and amygdala lesions affect performance in reinforcer devaluation tasks in similar but not identical ways. Specifically, Machado and Bachevalier (2007) found that monkeys with amygdala lesions continued to select an object associated with a particular food even after the monkey had consumed that food to satiation. However, these monkeys did not take or eat the actual food placed underneath the object. In contrast, monkeys with OFC lesions took both the reward-associated object and the food itself. This distinction suggests that subjects with amygdala lesions may lose the ability to predict the reward associated with given cues or actions. In contrast, subjects with OFC lesions may have a general deficit in value-guided behavior, leading them to fall back on habitual behavior.

Neuronal Responses in Orbitofrontal Cortex

Starting in the 1980s, neurophysiology experiments showed that neurons in OFC encode information related to reward and punishment. Thorpe, Rolls, and Maddison (1983) recorded from the OFC of awake nonhuman primates while presenting animals with a wide array of foods and objects. They found that neurons responded specifically to rewarding or aversive stimuli in a way that could not be explained by the sensory properties of the stimuli. For example, many cells responded to both the sight and taste of a food, indicating that their activity was not specific to one sensory modality. Moreover, responses depended on the pleasantness or unpleasantness of a visual stimulus, not its physical appearance alone. One example neuron responded to the sight of a syringe, but only when the syringe was associated with saltwater (an aversive stimulus). When the researchers replaced the salt with sugar, the response was eliminated, despite the fact that the visual appearance of the syringe was identical. After switching back to salt, the response reappeared. Subsequent experiments found that reward-related responses depended on the behavioral context and the motivational state of the monkey. One study observed that the response to a particular juice was enhanced or reduced depending on the other juice being delivered in the current block of trials (Tremblay and Schultz 1999). Other experiments demonstrated that OFC responses depended on the subject's level of satiety and that a neuron's response to a given reward would be selectively reduced if the monkey had a chance to consume that reward to satiation (Critchley and Rolls 1996; Rolls, Sienkiewicz, and Yaxley 1989).

These studies established two key features of OFC: (1) neuronal responses in this area were not specific to any sensory modality; and (2) neuronal responses combined information about multiple internal and external factors such as motivation, stimulus magnitude, pleasantness, and delay. However, these experiments did not provide a quantitative measure of the monkeys' *subjective* preferences. To establish that a response encodes subjective value, one must record the signal during a choice task and analyze the neural activity in relation to the behaviorally measured value. Taking this approach, Padoa-Schioppa and Assad (2006) recorded neural responses while monkeys made a series of choices between two juices, A and B. Juice A was defined as the option the monkey preferred when the juices

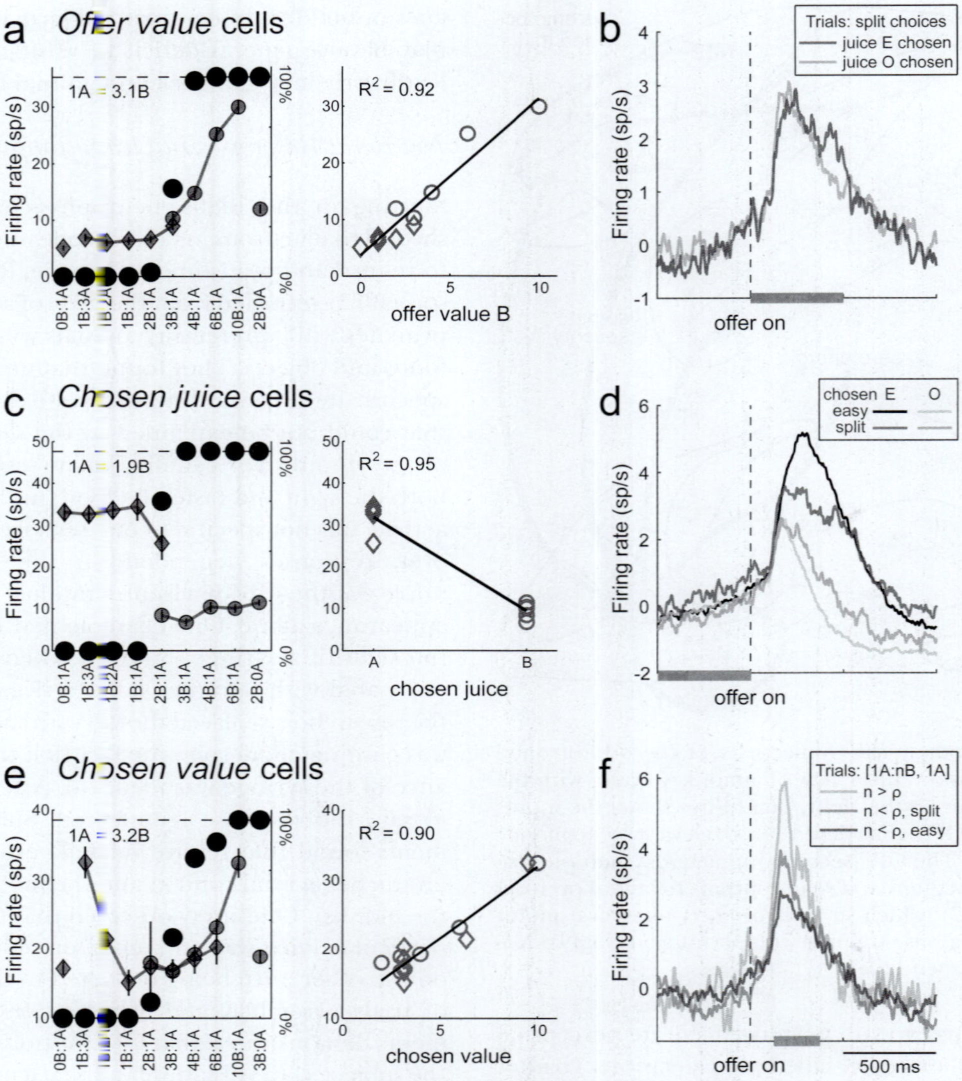

FIGURE 50.2 Three groups of neurons in OFC. *A, C, E,* Example neurons recorded from OFC during a juice choice task. *Left,* Neuronal responses and choice behavior. The x-axis shows the offer types available during the recording session, ranked by the increasing ratio of #B/#A. The black dots represent the proportion of trials for each offer type in which the monkey chose juice B (choice behavior). A sigmoid fit of this data was used to determine the relative value of the two juices. Gray symbols show neuronal activity, with diamonds and circles indicating trials in which the animal chose juice A and juice B, respectively. *Right,* Neuronal response as a function of the encoded variable. *Offer value* and *chosen value* neurons respond to value in a linear way. Neurons shown encode (*A*) offer value A, (*C*) chosen juice A, and (*E*) chosen value. *B, D, F,* The time course of neuronal activity for different choice types. *B,* Activity fluctuations in offer value neurons. Traces show the average baseline-subtracted activity of offer value neurons for offer types in which a monkey's choices were split between juice A and juice B. Traces are separated based on whether the monkey chose the juice encoded by the neuron (juice E) or the other juice (juice O). The juice E is slightly elevated compared to the juice O trace in the time window following the offer presentation. *D,* Predictive activity of *chosen juice* cells. Traces show the average baseline-subtracted activity of chosen juice neurons. Activity was divided into four groups depending on whether the animal chose the encoded juice (juice E) or the other juice (juice O) and whether the decisions were easy (all choices for one of the two juices) or hard (decisions split between the two juices). For offers with split decisions, neuronal activity was slightly elevated before offer onset in trials in which the monkey chose the encoded juice. Separation may reflect residual activity from the previous trial as well as random fluctuations in neuronal activity. *F,* Activity overshooting in chosen value neurons. Traces show the average baseline-subtracted activity of a large number of chosen value cells, including only trials in which the monkey chose 1A. Activity is divided into three groups depending on whether the quantity of the nonchosen juice (n) was greater or less than the relative value of the two juices (ρ). Cases with $n < \rho$ were also separated based on whether the decision was easy or split. During the decision window (~200–450 ms after the offer), chosen value neurons show the greatest peak activity when n is higher, which corresponds to more difficult decisions. Adapted with permission from Padoa-Schioppa (2013). (See color plate 55.)

were offered in equal quantities (offer 1B:1A). Offered quantities varied from trial to trial, inducing a quality/quantity trade-off in the animal's choices. For example, in the session shown in figure 50.2*A* (*black circles*), the monkey consistently chose juice A when the quantity ratio #B:#A was ≤2:1; it chose the two juices in roughly equal proportions when offered exactly 3B:1A; and it chose juice B consistently when the quantity ratio #B:#A was ≥4:1. In each session a sigmoid fit provided a measure for the relative value of the two juices. For the session in figure 50.2*A*, 1A = 3.1B (relative value = 3.1). Based on this measure, the authors defined a gamut of value-related variables and examined neuronal firing rates in relation to these variables. They found that neuronal responses in OFC encoded one of three variables: the value of one of the two juices (*offer value A* or *B*), the identity of the chosen option (*chosen juice*), and the value of the chosen option (*chosen value*) (figure 50.2*A, C, E*).

Subsequent analyses showed that these three variables were encoded by three distinct groups of neurons (Padoa-Schioppa 2013). Furthermore, chosen value cells truly encoded subjective value, as opposed to some objective property of the juices, such as sugar concentration or juice volume. These neurons integrated information about both the juice type and the quantity, and an analysis of their firing rates provided a neural measure for the relative value of the two juices that was statistically indistinguishable from behavioral measures (Padoa-Schioppa and Assad 2006). Another experiment showed that offer value responses also reflected the subjective nature of value, integrating information about probability and quantity in a way that reflected the animal's risk attitude. Thus, in sessions in which monkeys were more sensitive to risk, neural responses were also more strongly modulated by it (Raghuraman and Padoa-Schioppa 2014). These findings provided evidence of neural signals encoding subjective value. Moreover, they showed that different groups of neurons in OFC encoded the decision input (*offer values*) and the decision outcome (*chosen juice* and *chosen value*), suggesting that the decision might be formed within this area.

Numerous studies in human and nonhuman primates extended these results, finding value signals in OFC and other brain regions (for a review, see O'Doherty 2014; Padoa-Schioppa 2011; Schultz 2015). In OFC, a few features consistently stand out. First, the representation of value is generally independent of the spatial or sensorimotor features of the task. Second, the activity of neurons in OFC reflects a wide range of variables affecting choices, including reward quantity, probability, action cost, and even social information. Most studies found that these variables are integrated into a unified value signal. Importantly, single neurons in OFC respond to rewarding and aversive stimuli in opposite ways, consistent with a general representation of value (Morrison and Salzman 2011).

Stability and Versatility in the Decision Circuit

Any circuit responsible for economic decisions faces a challenge: it must be stable enough to compute and compare values in a reliable way, but it must be flexible enough to support choices in a variety of behavioral contexts. Three features of the neuronal representation in OFC reflect a balance between stability and versatility: menu invariance, range adaptation, and neuronal remapping.

Menu invariance refers to the fact that the activity of a neuron encoding the value of a given option does not depend on the value or identity of alternative options. Menu invariance was observed in a task in which monkeys chose between three types of juice offered pairwise: A:B, B:C, or A:C. Choices between the three juice pairs were interleaved, so the monkey might choose between A and B on one trial and then between A and C on the next trial. Offer value cells were consistently associated with one juice (A, B, or C). Furthermore, their activity was only affected by the value of the encoded juice, not the identity or value of the alternative option. For example, the tuning of offer value B cells was the same regardless of whether the alternative option was A or C. Importantly, menu invariance is closely related to preference transitivity. By definition, preferences are transitive if A > B and B > C imply A > C, where > means "is preferred to." From an ecological perspective, transitivity is vital. Intransitive preferences could lead a person who owns A to pay $1 to trade A for B, pay $1 to trade B for C, and then pay $1 again to trade C for A. At the end of this loop, that person would be in the same initial position (owning A), only $3 poorer. In most circumstances, human and animal preferences are indeed transitive (but see Tversky 1969). Notably, transitivity may be violated only if the value assigned to a particular option varies depending on the alternative (Tversky and Simonson 1993). In other words, if decisions are based on a menu-invariant representation, choices are necessarily transitive.

Where menu invariance reflects a certain stability, range adaptation illustrates the flexibility of the decision circuit. *Range adaptation* refers to the fact that value-encoding neurons change the gain of their response depending on the range of values available in a given context. Specifically, the gain of the encoding is lower when the range of available options is wider. In the juice choice experiments described above, range adaptation was observed in both offer value and chosen value cells (Padoa-Schioppa 2009). Within a session, the quantity of each juice varied from trial to trial within a fixed range. Across sessions, however, the value range varied. The activity of offer value cells and chosen value cells varied with the encoded value in a roughly linear way (linear tuning). However, across sessions the slope of encoding was inversely related to the range of values available in any given session. Subsequent studies confirmed this finding in individual cells (Kobayashi, Pinto de Carvalho, and Schultz 2010) and in the fMRI blood oxygen level dependent (BOLD) signal (Cox and Kable 2014). Theoretical and experimental work shows that range adaptation in offer value cells reduces choice variability, increasing expected payoff across trials (Rustichini et al. 2017). Interestingly, a recent study found that neurons only adapt partially to changes in value range, despite the fact that partial adaptation theoretically reduces the expected payoff (Conen and Padoa-Schioppa, 2019). Partial adaptation may reflect a tradeoff between stability and flexibility in the circuit.

Finally, neuronal remapping is a qualitative form of context adaptation, by which neurons in OFC become associated with different goods in different behavioral contexts. This property was observed in a study in which monkeys chose between different juice pairs in two blocks of trials (Xie and Padoa-Schioppa 2016). First, monkeys chose between juices A and B for approximately 200 trials. Then the juices were changed and the monkeys chose between two new juices, C and D, for approximately 200 trials. Strikingly, neurons maintained the same identity across blocks—for example, offer value cells remained offer value cells, chosen value cells remained chosen value cells, and the sign of the encoding was maintained. At the same time, when the context changed, each neuron remapped and became associated with one of the new juices available in the second trial block. Interestingly, two neurons associated with the same juice in the first block remapped together and became associated with the same juice again in the second block. In other words, the overall organization of the decision circuit remained stable across contexts. In that study, remapping appeared dictated by the preference ranking (i.e., neurons associated with juice A became associated with juice C). However, more work is needed to ascertain the rules governing neuronal remapping in general. In any case, the orderly remapping observed by Xie and Padoa-Schioppa (2016) shows how the neural circuit in OFC maintains a stable structure over time while also adapting to the current choice context.

Variability in Neurons and Behavior

When two similarly valued options are offered against each other multiple times, subjects typically split their choices. For example, in figure 50.2*E*, consider trials in which the monkey chose between 4B and 1A (4B:1A). In approximately 80% of trials, the animal chose juice B; in the remaining 20% of trials, it chose juice A. Presumably, this behavioral variability reflects some variability in the neural circuit. If decisions are indeed generated within the OFC, neuronal activity in that region should also explain choice variability across trials. Several studies examined this issue.

Offer value cells in OFC are thought to represent the input layer of the decision circuit. Thus, it is natural to examine whether fluctuations in their activity predict choice variability. However, measuring the behavioral effects of activity fluctuations in single neurons presents a challenge. There are approximately 50,000 neurons/mm^3 in the macaque orbitofrontal cortex (Dombrowski, Hilgetag, and Barbas 2001). Considering approximately 10 mm^3 of OFC in each hemisphere with roughly 20% offer value cells, about 100,000 neurons encode the offer value of each option during a juice choice task. For simplicity, we can assume that decisions emerge from the combined activity of this population and that every offer value cell contributes to the decision with equal weight. If the activity fluctuations of different neurons are independent of one another, the variability in any single neuron has a vanishingly small effect on the choice. However, trial-by-trial fluctuations in the activity of different cells present some degree of correlation. This correlation, termed *noise correlation*, is rather small—typically 0.1–0.2 in sensory regions (for a review, see Cohen and Kohn 2011) and even smaller (~0.01) in OFC (Conen and Padoa-Schioppa 2015). Nevertheless, the presence of noise correlation induces some relationship between activity fluctuations in individual offer value cells and the response of the overall neuronal population (i.e., choice behavior). The precise nature of this relationship depends on the pattern of noise correlation and on the way offer value signals are processed in the decision circuit (Kohn et al. 2016). However, under reasonable assumptions one can predict how the variability in individual neurons across trials relates to the animal's choices (Haefner et al. 2013). Specifically, if decisions are primarily based on the activity of offer value cells, given the noise correlation measured in OFC, there should be a weak but positive relation between activity fluctuations and the monkey's choices. In other words, when the same two options are offered repeatedly, the monkey will be slightly more likely to choose juice A when the typical offer value A cell has higher activity and slightly more likely to choose juice B when the typical offer value A cell has lower activity (for a more detailed explanation of this effect, see Britten et al. [1996]). Neuronal measures confirmed these predictions (figure 50.2*B*; Conen and Padoa-Schioppa, 2015; Padoa-Schioppa, 2013).

Another possible source of choice variability was found at a later stage in the decision circuit. In juice choice experiments, monkeys generally showed a slight bias toward the option they had received in the previous trial—a phenomenon termed *choice hysteresis* (Padoa-Schioppa, 2013). Choice hysteresis did not correspond to any variability in offer value cells, but it did correlate with trial-by-trial fluctuations in chosen juice cell activity. These neurons were frequently active at the end of the trial, upon juice delivery. Their average activity dropped in the intertrial interval, but it did not reach baseline levels before the beginning of the next trial. This tail activity appeared to induce a choice bias in the next trial. This effect can be observed in the activity of chosen juice cells in the 0.5 s preceding the offer (figure 50.2*D*). When the activity of chosen juice cells associated with a particular juice was slightly elevated, the monkey was more likely to choose that juice—a phenomenon termed *predictive activity* (Padoa-Schioppa, 2013). The presence of predictive activity suggests that choice variability arises not only from fluctuations in the decision input but also from within the decision circuit itself.

The precise relation between neuronal variability and choice can provide some insight into the organization of the decision circuit. As with offer value and chosen juice cells, the activity of chosen value cells varies systematically across different types of decisions. In particular, in trials in which the monkey chooses a particular option (e.g., 1A), the activity of chosen value neurons varies as a function of the decision difficulty (Padoa-Schioppa, 2013). Figure 50.2*F* illustrates this effect. When a monkey chooses 1A over a high quantity of juice B, chosen value neurons show transient *activity overshooting* shortly after the offer. When the quantity of B is lower but the decision is still difficult (split decisions), the activity overshoots to an intermediate level. Finally, when the quantity of B is so low that the monkey never chooses that option (easy decisions), the activity of chosen value cells is lowest. This effect may in part reflect variation in the subjective value across trials—when the quantity of B is particularly high, juice A can only beat the competing offer in trials when the monkey happens to assign an unusually high value to juice A. However, if this is the case, it is worth asking why the effect appears to be stronger in chosen value neurons than in offer value cells. Alternatively, the overshooting in chosen value activity may reflect the structure of the decision circuit. Notably, one neurocomputational model of choice naturally reproduces this effect (Rustichini and Padoa-Schioppa 2015). In this model, based on an attractor network originally developed by Wang (2002), chosen value signals arise from a pool of inhibitory interneurons that mediate competition between the two choice options. When a decision is more difficult, activity in these cells increases transiently, leading to overshooting that closely resembles the empirical data.

Neurocomputational Models of Economic Decisions

The attractor of Rustichini and Padoa-Schioppa (2015) is only one of several computational models recently proposed to account for economic decisions. In the past few years, research groups have shown that binary economic decisions can be implemented through a probabilistic generative model (Solway and Botvinick 2012), a spiking network that learns to optimize potential states (Friedrich and Lengyel 2016), and two recurrent neural networks trained on a wide variety of cognitive tasks (Song, Yang, and Wang 2017; Zhang et al. 2018). These models differ in fundamental ways, but remarkably, they all reproduce the three groups of neurons identified in the OFC. Furthermore, while differing in internal connectivity, these models have a common overall structure, whereby offer value cells provide inputs to a circuit of chosen value and chosen juice neurons, which process these inputs and generate a binary decision (figure 50.3).

These models establish offer value, chosen value, and chosen juice neurons as a common signature across many decision mechanisms. At the same time, the findings raise a question. If different networks make similar predictions, how can we adjudicate between competing hypotheses and eventually develop new and more accurate models? Looking forward, several strategies seem worth pursuing. First, it will be important to test whether computational models reproduce more detailed features of neuronal activity, such as correlation with choice variability, within-trial dynamics in neural populations, and context adaptation. Second, an accurate neurocomputational model of economic decisions should also reproduce choice anomalies observed in human and animal behavior. Among other experimental approaches, genetic tools available in rodents can help dissect the decision circuit and test predictions of biologically inspired models.

Summary and Future Directions

Neuroeconomics aims to uncover the neural and cognitive mechanisms underlying economic choice behavior.

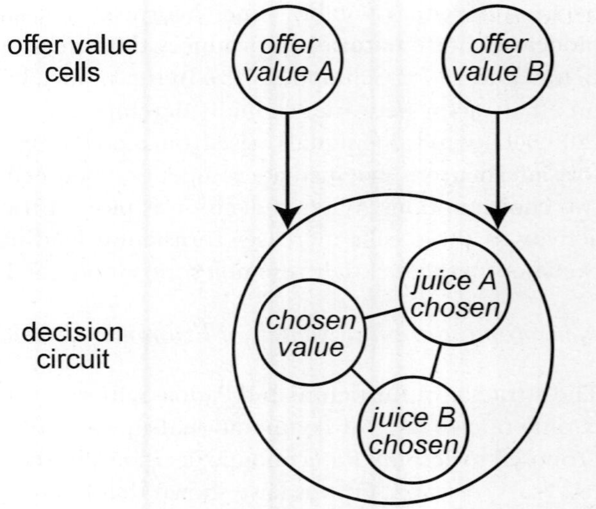

FIGURE 50.3 General schematic of an economic choice circuit. Offer value cells provide inputs to a neural circuit that includes chosen value cells and chosen juice cells and produces a binary choice output. Reprinted with permission from Padoa-Schioppa (2013).

The most important result so far has been to show that subjective values are explicitly represented by neurons during choice. Building on this foundational result, research in the field has increasingly focused on the difficult question of where in the brain and how exactly subjective values are compared. Several lines of evidence suggest that economic decisions between goods might be generated within the OFC. Experimental observations supporting this view can be summarized as follows: (1) OFC lesions specifically impair goal-directed behavior and economic choice; (2) during choice tasks, different groups of neurons in OFC encode the value of individual offers, the binary choice outcome and the chosen value. These groups of neurons capture both the input and the output of the decision process, suggesting that they are the building blocks of a decision circuit; (3) the neuronal representation in OFC presents a combination of stability and flexibility that is vital to make effective decisions in different behavioral contexts; (4) trial-by-trial fluctuations in the activity of OFC neurons correlate with variability in choice behavior; and (5) complementing these experimental results, computational models suggest that the groups of neurons identified in OFC are sufficient—and maybe necessary—to generate economic decisions.

Despite this evidence, the proposal that a neural circuit within the OFC underlies economic decisions remains a working hypothesis. Future research should shed light on several open questions. First, value-encoding neurons have been recorded in many brain regions, including the amygdala, vmPFC, parietal cortex, lateral prefrontal cortex, and premotor areas. While value signals may inform multiple cognitive functions—associative learning, perceptual attention, action planning, emotion, and more—some of these brain areas might also play a role in economic decisions. Second, fundamental features of the neural circuit within OFC remain poorly understood. For example, it is not clear whether the groups of cells described here correspond to different morphological cell types, whether they are preferentially excitatory or inhibitory, or whether they reside preferentially in different cortical layers. The connectivity between these cell groups and between them and other cortical regions is also unclear. Third, experiments to date have not established direct causal links between neuronal activity in OFC and decisions. Demonstrating such a link would provide unequivocal evidence for the working hypothesis put forth in this chapter. Fourth, in most studies to date, subjects made decisions between offers presented simultaneously. Yet offers in real-life decisions often appear sequentially. Thus, it is critical to examine whether and how current notions on the neural mechanisms underlying economic decisions generalize to choices under sequential offers. Research on many of these issues in ongoing, and the coming years are likely to witness new and exciting developments.

Acknowledgment

Our research is supported by the National Institutes of Health (grant numbers R01-MH104494 and R21-DA0 42882 to Camillo Padoa-Schioppa).

REFERENCES

Balleine, B. W., & Dickinson, A. (1998). Goal-directed instrumental action: Contingency and incentive learning and their cortical substrates. *Neuropharmacology, 37*, 407–419.

Bartra, O., McGuire, J. T., & Kable, J. W. (2013). The valuation system: A coordinate-based meta-analysis of BOLD fMRI experiments examining neural correlates of subjective value. *NeuroImage, 76*, 412–427.

Baxter, M. G., Gaffan, D., Kyriazis, D. A., & Mitchell, A. S. (2009). Ventrolateral prefrontal cortex is required for performance of a strategy implementation task but not reinforcer devaluation effects in rhesus monkeys. *European Journal of Neuroscience, 29*, 2049–2059.

Britten, K. H., Newsome, W. T., Shadlen, M. N., Celebrini, S., & Movshon, J. A. (1996). A relationship between behavioral choice and the visual responses of neurons in macaque MT. *Visual Neuroscience, 13*, 87–100.

Cisek, P. (2012). Making decisions through a distributed consensus. *Current Opinion in Neurobiology, 22*, 927–936.

Clithero, J. A., & Rangel, A. (2013). Informatic parcellation of the network involved in the computation of subjective value. *Social Cognitive and Affective Neuroscience, 9*, 1289–1302.

Cohen, M. R., & Kohn, A. (2011). Measuring and interpreting neuronal correlations. *Nature Neuroscience, 14*, 811–819.

Conen, K. E., & Padoa-Schioppa, C. (2015). Neuronal variability in orbitofrontal cortex during economic decisions. *Journal of Neurophysiology, 114*, 1367–1381.

Conen, K. E., & Padoa-Schioppa, C. (2019). Partial adaptation to the value range in the macaque orbitofrontal cortex. *Journal of Neuroscience, 39*, 3498–3513.

Cox, K. M., & Kable, J. W. (2014). BOLD subjective value signals exhibit robust range adaptation. *Journal of Neuroscience, 34*, 16533–16543.

Critchley, H. D., & Rolls, E. T. (1996). Hunger and satiety modify the responses of olfactory and visual neurons in the primate orbitofrontal cortex. *Journal of Neurophysiology, 75*, 1673–1686.

Dombrowski, S. M., Hilgetag, C. C., & Barbas, H. (2001). Quantitative architecture distinguishes prefrontal cortical systems in the rhesus monkey. *Cerebral Cortex, 11*, 975–988.

Fellows, L. K. (2011). Orbitofrontal contributions to value-based decision making: Evidence from humans with frontal lobe damage. *Annals of the New York Academy of Sciences, 1239*, 51–58.

Fellows, L. K., & Farah, M. J. (2007). The role of ventromedial prefrontal cortex in decision making: Judgment under uncertainty or judgment per se? *Cerebral Cortex, 17*, 2669–2674.

Friedrich, J., & Lengyel, M. (2016). Goal-directed decision making with spiking neurons. *Journal of Neuroscience, 36*, 1529–1546.

Gremel, C. M., & Costa, R. M. (2013). Orbitofrontal and striatal circuits dynamically encode the shift between goal-directed and habitual actions. *Nature Communications, 4*, 1–12.

Haefner, R. M., Gerwinn, S., Macke, J. H., & Bethge, M. (2013). Inferring decoding strategies from choice probabilities in the presence of correlated variability. *Nature Neuroscience, 16*, 235–242.

Hare, T. A., Schultz, W., Camerer, C. F., O'Doherty, J. P., & Rangel, A. (2011). Transformation of stimulus value signals into motor commands during simple choice. *Proceedings of the National Academy of Sciences of the United States of America, 108*, 18120–18125.

Hunt, L. T., Kolling, N., Soltani, A., Woolrich, M. W., Rushworth, M. F. S., & Behrens, T. E. J. (2012). Mechanisms underlying cortical activity during value-guided choice. *Nature Neuroscience, 15*, 470–476.

Kable, J. W., & Glimcher, P. W. (2009). The neurobiology of decision: Consensus and controversy. *Neuron, 63*, 733–745.

Kobayashi, S., Pinto de Carvalho, O., & Schultz, W. (2010). Adaptation of reward sensitivity in orbitofrontal neurons. *Journal of Neuroscience, 30*, 534–544.

Kohn, A., Coen-Cagli, R., Kanitscheider, I., & Pouget, A. (2016). Correlations and neuronal population information. *Annual Review of Neuroscience, 39*, 237–256.

Kreps, D. M. (1990). *A course in microeconomic theory*. Princeton, NJ: Princeton University Press.

Machado, C. J., & Bachevalier, J. (2007). The effects of selective amygdala, orbital frontal cortex or hippocampal formation lesions on reward assessment in nonhuman primates. *European Journal of Neuroscience, 25*, 2885–2904.

McDannald, M. A., Jones, J. L., Takahashi, Y. K., & Schoenbaum, G. (2014). Learning theory: A driving force in understanding orbitofrontal function. *Neurobiology of Learning and Memory, 108*, 22–27.

Morrison, S. E., & Salzman, C. D. (2011). Representations of appetitive and aversive information in the primate orbitofrontal cortex. *Annals of the New York Academy of Sciences, 1239*, 59–70.

Niehans, J. (1990). *A history of economic theory: Classic contributions, 1720–1980*. Baltimore: Johns Hopkins University Press.

O'Doherty, J. P. (2014). The problem with value. *Neuroscience & Biobehavioral Reviews, 43*, 259–268.

Ongur, D., & Price, J. (2000). The organization of networks within the orbital and medial prefrontal cortex of rats, monkeys and humans. *Cerebral Cortex, 10*, 206–219.

Padoa-Schioppa, C. (2009). Range-adapting representation of economic value in the orbitofrontal cortex. *Journal of Neuroscience, 29*, 14004–14014.

Padoa-Schioppa, C. (2011). Neurobiology of economic choice: A good-based model. *Annual Review of Neuroscience, 34*, 333–359.

Padoa-Schioppa, C. (2013). Neuronal origins of choice variability in economic decisions. *Neuron, 80*, 1322–1336.

Padoa-Schioppa, C., & Assad, J. A. (2006). Neurons in the orbitofrontal cortex encode economic value. *Nature, 441*, 223–226.

Padoa-Schioppa, C., & Cai, X. (2011). The orbitofrontal cortex and the computation of subjective value: Consolidated concepts and new perspectives. *Annals of the New York Academy of Sciences, 1239*, 130–137.

Padoa-Schioppa, C., & Conen, K. E. (2017). Orbitofrontal cortex: A neural circuit for economic decisions. *Neuron, 96*, 736–754.

Padoa-Schioppa, C., & Schoenbaum, G. (2015). Dialogue on economic choice, learning theory, and neuronal representations. *Current Opinion in Behavioral Sciences, 5*, 16–23.

Raghuraman, A. P., & Padoa-Schioppa, C. (2014). Integration of multiple determinants in the neuronal computation of economic values. *Journal of Neuroscience, 34*, 11583–11603.

Rescorla, R. A., & Wagner, A. R. (1972). A theory of Pavlovian conditioning: Variations in the effectiveness of reinforcement and nonreinforcement. *Classical Conditioning II: Current Research and Theory, 21*, 64–99.

Rhodes, S. E. V., & Murray, E. A. (2013). Differential effects of amygdala, orbital prefrontal cortex, and prelimbic cortex lesions on goal-directed behavior in rhesus macaques. *Journal of Neuroscience, 33*, 3380–3389.

Rolls, E. T., Sienkiewicz, Z. J., & Yaxley, S. (1989). Hunger modulates the responses to gustatory stimuli of single neurons in the caudolateral orbitofrontal cortex of the macaque monkey. *European Journal of Neuroscience, 1*, 53–60.

Ross, D. (2005). *Economic theory and cognitive science: Microexplanation*. Cambridge, MA: MIT Press.

Rudebeck, P. H., & Murray, E. A. (2011). Dissociable effects of subtotal lesions within the macaque orbital prefrontal cortex on reward-guided behavior. *Journal of Neuroscience, 31*, 10569–10578.

Rudebeck, P. H., & Murray, E. A. (2014). The orbitofrontal oracle: Cortical mechanisms for the prediction and evaluation of specific behavioral outcomes. *Neuron, 84*, 1143–1156.

Rudebeck, P. H., Saunders, R. C., Prescott, A. T., Chau, L. S., & Murray, E. A. (2013). Prefrontal mechanisms of behavioral flexibility, emotion regulation and value updating. *Nature Neuroscience, 16,* 1140–1145.

Rustichini, A., Conen, K. E., Cai, X., & Padoa-Schioppa, C. (2017). Optimal coding and neuronal adaptation in economic decisions. *Nature Communications, 8,* 1–14.

Rustichini, A., & Padoa-Schioppa, C. (2015). A neurocomputational model of economic decisions. *Journal of Neurophysiology, 114,* 1382–1398.

Schultz, W. (2015). Neuronal reward and decision signals: From theories to data. *Physiological Reviews, 95,* 853–951.

Solway, A., & Botvinick, M. M. (2012). Goal-directed decision making as probabilistic inference: A computational framework and potential neural correlates. *Psychological Review, 119,* 120–154.

Song, H. F., Yang, G. R., & Wang, X. J. (2017). Reward-based training of recurrent neural networks for cognitive and value-based tasks. *eLife, 6.*

Strait, C. E., Blanchard, T. C., & Hayden, B. Y. (2014). Reward value comparison via mutual inhibition in ventromedial prefrontal cortex. *Neuron, 82,* 1357–1366.

Thorpe, S. J., Rolls, E. T., & Maddison, S. (1983). The orbitofrontal cortex: Neuronal activity in the behaving monkey. *Experimental Brain Research, 49,* 93–115.

Tremblay, L., & Schultz, W. (1999). Relative reward preference in primate orbitofrontal cortex. *Nature, 398,* 704–708.

Tversky, A. (1969). Intransitivity of preferences. *Psychological Review, 76,* 31–48.

Tversky, A., & Simonson, I. (1993). Context-dependent preferences. *Management Science, 39,* 1179–1189.

Wallis, J. D. (2012). Cross-species studies of orbitofrontal cortex and value-based decision-making. *Nature Neuroscience, 15,* 13–19.

Wang, X. J. (2002). Probabilistic decision making by slow reverberation in cortical circuits. *Neuron, 36,* 955–968.

Xie, J., & Padoa-Schioppa, C. (2016). Neuronal remapping and circuit persistence in economic decisions. *Nature Neuroscience, 19,* 855–861.

Zhang, Z., Cheng, Z., Lin, Z., Nie, C., & Yang, T. (2018). A neural network model for the orbitofrontal cortex and task space acquisition during reinforcement learning. *PLoS Computational Biology, 14,* 1–24.

51 Neural Mechanisms of Perceptual Decision-Making

GABRIEL M. STINE, ARIEL ZYLBERBERG, JOCHEN DITTERICH, AND MICHAEL N. SHADLEN

ABSTRACT As we interact with the world, we must decide what to do next based on previously acquired and incoming information. The study of perceptual decision-making uses highly controlled sensory stimuli and exploits known properties of sensory and motor systems to understand the processes that occur between sensation and action. Even these relatively simple decisions often invoke operations like inference, integration of evidence, attention, appropriate action selection, and the assignment of levels of belief or confidence. Thus, the neurobiology of perceptual decision-making offers a tractable way to study mechanisms that play a role in higher cognitive function and reward-motivated behavior. This chapter provides a brief overview of the neural mechanisms that underlie decisions based on visual information, focusing on experiments in nonhuman primates and the principles they reveal. We then highlight some challenges the field faces—in particular, the identification of a subject's decision-making strategy from behavioral observations alone.

A decision is a commitment to a proposition, among alternatives, based on evidence. Many operations that would be characterized as cognitive—inference, integration of information, attention, appropriate action selection, the assignment of levels of belief to our inferences (i.e., confidence)—also play a central role in the decision process. Often, as in decisions based on subjective value, the evidence relevant to a decision is poorly understood. However, in perceptual decision-making, the experimenter has precise control over the source, reliability, and timing of the evidence that bears on a choice. This allows the experimenter to gain insight from quantitative relationships between the presented sensory evidence and different behavioral measures (e.g., accuracy, response time, confidence ratings) and to associate these behavioral measures with neural responses and perturbations of neural activity. In this way, perceptual decision-making offers a "sweet spot" in basic research on cognition: extracting an understanding of its neural mechanisms will contribute to our knowledge of higher brain function, yet it is simple enough to remain tractable.

The goal of this chapter is to provide a brief overview of the mechanisms that underlie perceptual decision-making and to highlight some of the gaps and limitations. Most of the work summarized in the first part of the chapter is from highly trained rhesus monkeys on tasks with reasonably well-established neural mechanisms from sensation to action. In the second part of the chapter, we will highlight some challenges, especially the importance and difficulty of determining an animal's decision-making strategy.

A Useful Task

The ability to control the reliability, or signal-to-noise ratio (SNR), of sensory evidence is critical to the study of perceptual decision-making. SNR is formally defined as the average magnitude of the signal divided by its standard deviation, and it is what determines how easily a signal can be discriminated from noise. Thus, the SNR is the ultimate arbiter of performance and response times, especially when signals are weak. With control of the SNR, the experimenter can quantify the relationship between the SNR and these behavioral measures. A low SNR regime is the main target of perceptual decision-making because if decisions are too easy, they are, effectively, instructed responses that do not require deliberation.

The stimulus used in most of the work we will discuss is a dynamic random-dot motion (RDM) movie, which is composed of flickering, moving dots. The subject's task is to judge in which of two possible directions these dots are moving and report their decision with an eye movement to a corresponding choice target (figure 51.1*A*). Each trial's difficulty (i.e., the SNR) is controlled by adjusting the *coherence*, or the expectation of the proportion of dots that are displaced coherently, as opposed to reappearing at a random location, in the viewing aperture. Critically, the specific dots that move coherently change within the trial, which imbues each dot with a "limited lifetime." This design discourages solving the task by waiting to observe a streak of motion. By forcing the animal to deliberate about the direction of motion in a low SNR regime, the RDM stimulus encourages the integration of motion information across time.

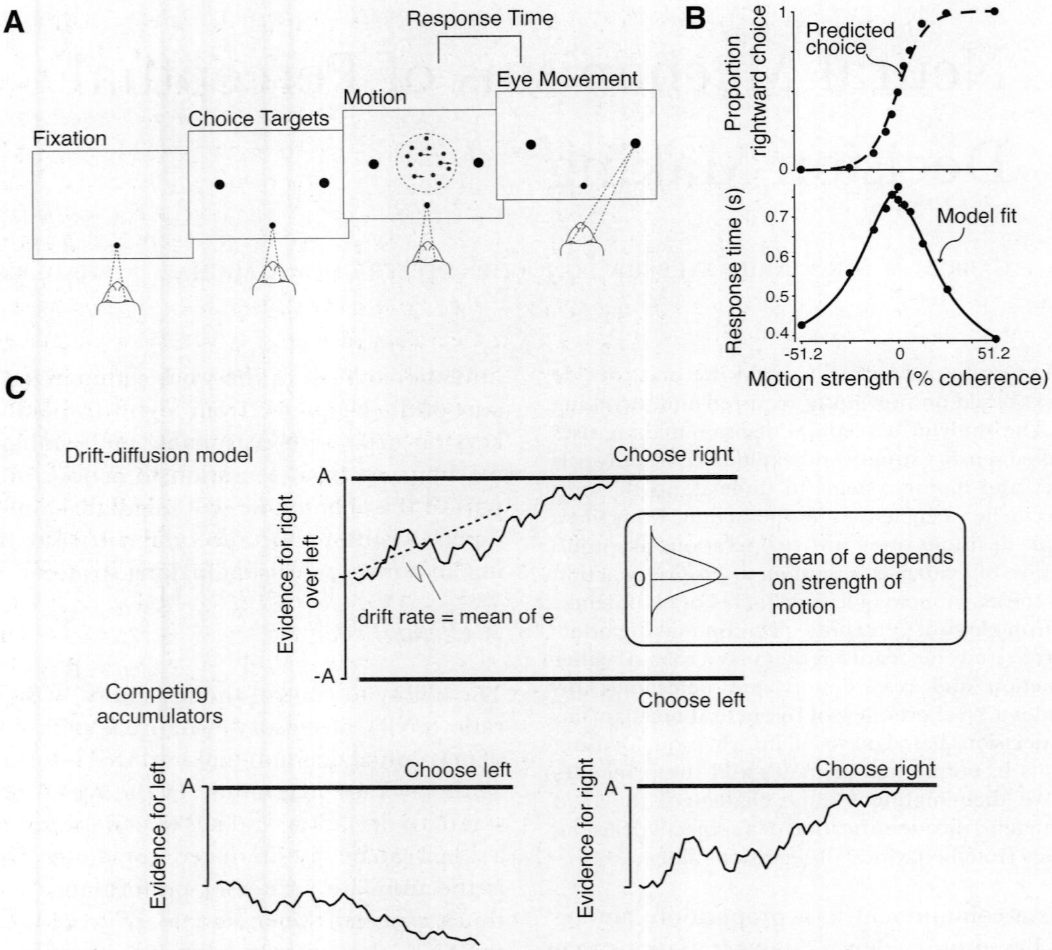

FIGURE 51.1 Bounded accumulation of noisy evidence (BANE) as a framework for understanding perceptual decisions. *A*, Choice-response time (RT) version of the RDM discrimination task. The subject judges in which of two possible directions the dynamic random dots are moving (in this case right or left). When ready, the subject indicates a decision with a saccade to the corresponding choice target. Difficulty is controlled by the motion strength—the fraction of coherently moving dots. *B*, The effect of motion strength on choice and RT. Positive and negative coherences correspond to rightward and leftward motion, respectively. Choices are faster and more accurate at higher motion strengths. The solid curve is the fit of a drift-diffusion model to the RT data. The dashed curve is the predicted choice data based on the RT fit. *C*, Models of BANE.. *Top*, Schematic of the drift-diffusion model. Evidence is accumulated until either the upper or lower bound is crossed. The drift rate, which is determined by the motion strength and direction, is the expectation of the slope of the random walk. The terminating thresholds at ±A control the trade-off between speed and accuracy. *Bottom*, Schematic of competing accumulators. The accumulation process can also be implemented as a race between competing accumulation processes. The first to reach its positive threshold at +A terminates the decision. If the accumulators are perfectly negatively correlated, this implementation is equivalent to the drift-diffusion model. Behavioral data is from Roitman and Shadlen (2002). *C*, Adapted with permission from Gold and Shadlen (2007).

Relationship between the Speed and Accuracy of a Decision

A fruitful version of the RDM task allows subjects to respond as soon as they are ready, which gives rise to two behavioral measures on each trial: the choice (e.g., left or right) and the response time (RT), defined as the elapsed time between stimulus onset and the indication of the choice. Models of the decision process provide a powerful tool for exploring quantitative relationships between choice, RT, and the SNR of the stimulus (determined by coherence). The most widely used of these models is bounded accumulation of noisy evidence (BANE), which posits that noisy sensory evidence is sampled sequentially and accumulated to a threshold level, at which point a commitment to a choice is made. The well-known drift-diffusion model in psychological literature formalizes such an accumulation mechanism

(Link, 1992; Laming, 1968; Ratcliff, 1978; figure 51.1*C*, *upper graph*). In the drift-diffusion model, evidence for each of the two choices is accumulated symmetrically until it exceeds either an upper or lower threshold. In other words, evidence for the two options is perfectly anticorrelated.

BANE models explain several common observations—for example, why harder decisions take longer to make (weaker evidence takes longer to be integrated to a threshold) and why speed and accuracy trade off with one another (decisions are faster when less evidence has to be accumulated). Impressively, BANE can also predict a subject's accuracy using only the model fits to RT measures (figure 51.1*B*). Finally, for many tasks used in perceptual decision-making—including the RDM task—BANE is a sensible and, in many cases, an optimal strategy. For these and other reasons, models of BANE have dominated the study of perceptual decision-making. Indeed, it is generally assumed that this is how animals under study (e.g., monkeys, rats, mice) and humans solve perceptual decision-making tasks. In the second part of the chapter, we will explore why this assumption is not automatically justified and might even lead to results being misinterpreted. In the meantime, however, we will exploit BANE as a general framework for how decisions are made and how different components of the decision process are represented and implemented in the brain.

Before discussing the mechanisms that underlie perceptual decisions, however, some additional qualifications are in order. Many, if not most, perceptual decisions are completed in less than 250 ms, roughly the amount of time that the gaze remains fixed on one location in the visual field before sampling elsewhere. By contrast, decisions in the RDM task often require many hundreds of milliseconds to several seconds—hence, the RDM task is more representative of cognitive deliberation than it is of perception. In fact, the direction judgment does not involve frank motion perception, which concerns the gradual displacement of an object or feature over time. At low SNR, the stimulus looks more like random snowflakes in a wind storm. The decision is an inference about the wind direction, not the movement of the snowflakes. Thus, we can use this highly controlled task to study how bits of noisy information are accumulated over time to construct beliefs about the world and to guide corresponding actions.

Neural Representation of Momentary Evidence

Decades of work using the RDM task have established that direction-selective (DS) neurons in visual cortex provide the sensory momentary evidence for the decision. DS neurons in the middle temporal area (MT; figure 51.2*A*, *upper plot*) respond with a fidelity that closely matches that of the animal's choices (Britten, Shadlen, Newsome, & Movshon, 1992). Specifically, the SNR associated with the response of pools of MT neurons with similar direction preferences is predictive of the monkey's error rates. MT's role in the RDM task has been supported by microstimulation (μStim) experiments in which electrical current is used to stimulate an approximately 100 μm radius sphere of neurons that share similar direction tuning (Ditterich, Mazurek, & Shadlen, 2003; Salzman, Britten, & Newsome, 1990). Stimulating rightward-preferring neurons causes monkeys to choose rightward more often, make rightward decisions in less time, and make leftward decisions more slowly. This last observation shows that rightward-preferring neurons contribute negatively to leftward choices; these neurons are not just ignored. Importantly, the specific effect on choice and RTs is consistent with a shift in the momentary evidence in a BANE framework (figure 51.2*C*)—as if μStim causes the monkeys to perceive stronger rightward motion in the stimulus.

From these studies and others, we deduce that DS neurons in areas like MT supply the evidence that a monkey uses to make decisions during the RDM task. The momentary evidence for one direction and against the other is the difference in firing rates between pools of DS neurons that prefer each of the opposing directions—for example, rightward- and leftward-preferring DS neurons with receptive fields that overlap the RDM stimulus. Notice that this difference signal is expressed as a spike rate (e.g., spikes per second); the time integral of this difference is therefore expressed in units of excess spikes favoring right over left.

The time integral of momentary evidence from visual cortex determines the choice, response time, and even confidence. Because the momentary evidence is noisy, its accumulation in any single trial is approximated by a drift-diffusion process—the accumulation of a deterministic, motion-strength-dependent response and unbiased noise. The deterministic drift component is an approximately linear function of the motion coherence and direction, as transformed by the DS neurons in area MT. The unbiased noise derives in part from the RDM stimulus and in part from the inherently variable discharge of DS neurons in visual cortex.

Neural Representation of Accumulated Evidence

The lateral intraparietal area (LIP) was the first area hypothesized to represent this time integral in the RDM task. LIP receives inputs from a variety of visual areas, including MT, and LIP's main projections are to the frontal eye field and the superior colliculus, which play a role in directing the gaze. This pattern of projections was one of two features of LIP that established it as a candidate for evidence accumulation in the RDM task. The second feature was that LIP neurons were shown to respond to visual stimuli when the stimuli are targets of the next eye movement; LIP neurons respond for up to seconds before that eye movement is made, even if the object has vanished during the delay, and the eye movement is guided entirely by memory (Gnadt & Andersen, 1988). Intriguingly, one way to get this step-like, persistent activity during the delay is to compute the integral of the pulse-like activity induced by the visual stimulus. So, it was hypothesized that if LIP represented the integral of a pulse, then, as long as decisions were about where to direct the gaze, maybe LIP would also reflect the integral of noisy sensory evidence (Shadlen & Newsome, 2001).

This hypothesis was not guaranteed to be correct, as LIP neurons might have responded only after the decision was made. But, LIP recordings during the RDM task suggest that they reflect the formation of the decision. Figure 51.2*A* (*lower plot*) shows the averaged firing rates from 54 neurons recorded during the RDM choice-RT task. For each recording session, one of the choice targets was placed in the neuron's response field (RF; for simplicity, we will refer to this choice target as the rightward target, even though the actual location of the choice targets varied). Beginning at about 180 ms after motion onset, the responses begin to increase or decrease gradually, depending not only on the motion's direction but also on its strength. The black curves, corresponding to easy decisions, ramp up or down quickly, whereas the lightest curve, corresponding to the 0% coherence condition, meanders. Intermediate motion strengths give rise to intermediate buildup rates. However, these curves do not exactly depict the accumulated noisy evidence that the monkey uses to make its choice on each trial—what we often call the *decision variable.* Because we have to average over many trials to estimate the firing rates of the neurons, we cannot directly observe the diffusion component of the decision variable—the accumulation of unbiased (i.e., zero mean) noise. The ramping signals we observe correspond to the drift, which can be thought of as the accumulation of the mean of the momentary evidence (or the signal component).

While we cannot directly observe the decision variable for each trial, several pieces of evidence tell us that the ramps are indeed the average of many drift-diffusion paths. First, we can look for signatures of a drift-diffusion process in the variance of the responses across trials, specifically in the way this variance evolves as a function of time. The sum of two independent random numbers, $x+y$, has a variance equal to the sum of the variances of x and y. Thus, a signal that reflects the accumulation of evidence should have a variance that increases linearly throughout the decision formation epoch. This prediction was confirmed in neural recordings from LIP and so was a related prediction about the evolution of covariance between firing rates sampled in two epochs from the same trial (Churchland et al., 2011). Second, a brief background pulse of motion during decision formation has a persistent effect on LIP activity (Huk & Shadlen, 2005), which is consistent with LIP representing the temporal integration of the sensory evidence. Finally, like μStim of MT, μStim of LIP affects choices and RTs but does so in a way consistent with a shift in the decision variable (figure 51.2*D*; Hanks, Ditterich, & Shadlen, 2006). Interestingly, however, the opposite effect is not seen when LIP is chemically inactivated—monkeys are somehow able to compensate so that the inactivation has no effect on choices (Katz, Yates, Pillow, & Huk, 2016), perhaps because LIP is not the only area that represents a decision variable.

LIP also reflects the decision threshold. The firing rate curves in figure 51.2*B* are aligned in time to the beginning of the rightward eye movements and separated by RT. Notice that the responses reach a common level of activity about 80 ms before the rightward eye movement begins. This suggests that the decision terminates when downstream areas detect that the firing rate has reached a threshold level, in this case about 60 spikes per second. It is the neural correlate of the bound in the model. The leftward choice trials are not shown, but the responses on those trials do not merge. That is because those decisions were terminated by neurons concerned with making leftward eye movements.

The emerging architecture is a representation of accumulating evidence for a proposition and against its alternative. Thus, it can be conceived of as a competition between evolving action plans, which is sometimes referred to as a race (figure 51.1*C*, *bottom*). The race architecture replaces the lower stopping threshold with the competing mechanism's upper stopping threshold. There are several virtues to this architecture: It naturally extends to more than two options. All that is required is expanding the number of accumulators participating in the race; and it simplifies termination, making it a threshold operation on a high firing rate. For example, rather than adjusting the threshold level of activity needed to terminate the decision, changes in speed accuracy setting are achieved by adding or subtracting an evidence-independent signal to the decision variable (Hanks, Kiani, & Shadlen, 2014). A similar mechanism is used to achieve a time-dependent collapse in

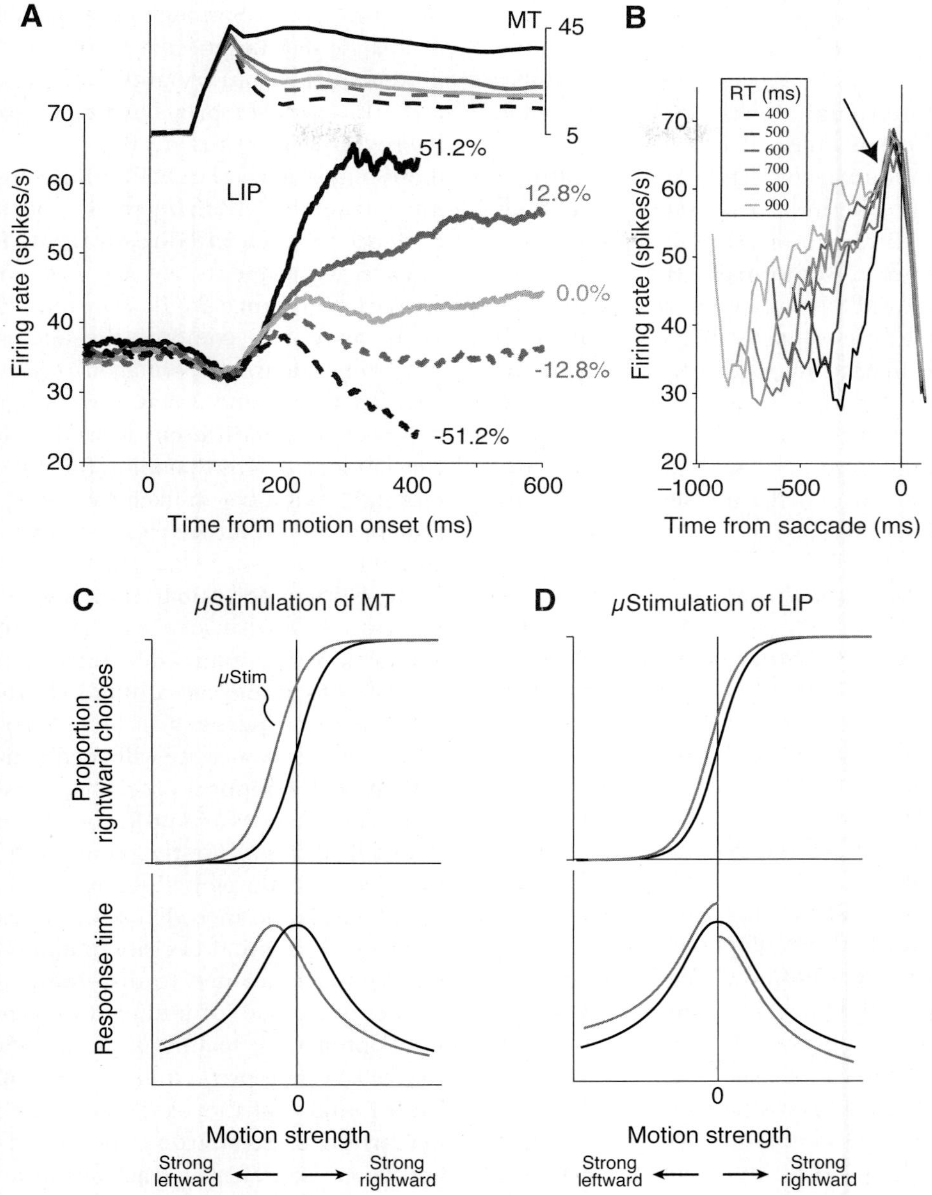

FIGURE 51.2 Neural correlates of the decision process. *A*, Averaged neural responses in MT and LIP during the RDM task. Responses are grouped by motion strength (*shading*) and direction (*solid/dashed*, toward/away from the RF; also indicated by the sign of coherence). LIP curves are truncated at the median RT. *B*, LIP responses grouped by RT and aligned to time of the saccade. Only choices toward the RF are shown. The arrow marks the coalesced firing rates approximately 80 ms preceding the saccade. This a neural correlate of the upper terminating threshold level in the competing accumulators (see figure 51.1). *C*, Theoretical predictions for the effect of MT μStim on choice and RT. The prediction is equivalent to a rightward shift in the momentary evidence distribution and is consistent with experimental results (Ditterich, Mazurek, & Shadlen, 2003). *D*, Same as (*C*) but for LIP. LIP μStim predicts an additive shift of the decision variable. This prediction is consistent with experimental results (Hanks, Ditterich, & Shadlen, 2006). LIP data is from Roitman and Shadlen (2002). *B*, Adapted with permission from Roitman and Shadlen (2002). Upper plot in (*A*) and (*C* and *D*) are adapted with permission from Gold and Shadlen (2007).

the decision bound—the mathematically optimal solution (Drugowitsch et al., 2012). An evidence-independent, time-dependent signal, termed an *urgency signal* (Churchland, Kiani, & Shadlen, 2008; Ditterich, 2006; Thura & Cisek, 2014), is added to all the racing accumulators, which leads to the acceptance of less accumulated evidence for terminating the decision as time passes.

It is important to stress that LIP is not the only area that represents a decision variable nor do these signals necessarily arise in LIP de novo. Similar signals have been observed in other areas concerned with directing gaze, like the frontal eye field (Kim & Shadlen, 1999), the dorsolateral prefrontal cortex (Kim & Shadlen, 1999), the superior colliculus (Horwitz & Newsome, 1999), and the caudate nucleus (Ding & Gold, 2013). A critical question is whether these signals are truly redundant or play different, but complementary, roles in the decision process. Additionally, these areas only reflect a decision variable because the decision is contrived to be about where to make an eye movement. If the decision were instead to require reaching to the choice targets, then we might expect a different group of brain areas to be involved, like the medial intraparietal area (de Lafuente, Jazayeri, & Shadlen, 2015) and the dorsal premotor cortex (Chandrasekaran, Peixoto, Newsome, & Shenoy, 2017).

Beyond Random Dots and Primates

The mechanisms we have discussed thus far are by no means specific to motion discrimination. Similar decision-making studies have used other tasks that involve the discrimination of stochastic stimuli. Some examples include discriminations of depth, color, orientation, and objects. Even value-based decisions are well predicted by models of bounded evidence accumulation (Krajbich, Lu, Camerer, & Rangel, 2012; Krajbich & Rangel, 2011), and their neural correlates have been shown to overlap with those of perceptual decisions (Polanía, Krajbich, Grueschow, & Ruff, 2014). Interestingly, while the general principles of computing a decision variable may be the same in these tasks, the source of evidence is different.

The generality of these principles is illustrated by a study that used a sequence of highly discriminable shapes (Kira, Yang, & Shadlen, 2015). The monkeys learned to associate each shape with a degree of reliability about which choice target would be rewarded and were able to sensibly combine information from each of the shapes to make the best decision. As in the RDM experiments, the monkeys indicated their decisions with an eye movement, and it was again possible to observe the formation of the decision in the neural responses of LIP neurons. What neurons supply the evidence in this task? Since the shapes themselves have no inherent meaning to the monkey, the evidence that they supply must come from memory. In other words, each shape is associated with a remembered action value, which gets integrated with those of previously seen shapes. This would seem to implicate hippocampal and striatal circuits. An intriguing idea is that similar circuits are involved even in simpler tasks such as RDM discrimination (Shadlen & Shohamy, 2016). Similar to the shapes, the association of a particular direction of motion with a particular eye movement is arbitrary and thus must be learned. In fact, although MT projects directly to LIP, there is an approximately 80 ms delay between the presence of motion information in MT and the impact of that same information on LIP firing rates—much too long for the evidence to reach LIP directly from MT.

These experiments beg the question of how different sources of evidence are flexibly routed to and operated on by areas that compute the decision variable. The source of evidence can switch on a millisecond-timescale, too fast to depend solely on changing the synaptic weights between different areas. Indeed, exploring the circuit mechanisms that allow for this immense computational flexibility will likely be one of the more important problems that neuroscientists will tackle in the coming decade.

More tractable animal models will be critical to this endeavor. Indeed, evidence accumulation has been studied not only in humans and monkeys but also in rodents, flies, worms, and other animals. To dissect the circuits and circuit properties that allow for the flexible accumulation of evidence, we will need to manipulate neural activity with temporal precision, record from neurons with known inputs and projections, and investigate specific cell classes within the circuit. While these tools are being actively developed for use in monkeys, they are readily available in rodents, which can be trained to perform perceptual decision-making tasks. For example, rats were trained to discriminate the overall frequency of a "cloud" of auditory tones and the researchers used optogenetic techniques to specifically explore the role of striatum-projecting neurons in auditory cortex. They found that they could bias the rat's decisions by perturbing these neurons (Znamenskiy & Zador, 2013) and were even able to track synaptic plasticity in the striatum while the rats learned to perform the task (Xiong, Znamenskiy, & Zador, 2015).

Excluding Alternatives to Evidence Accumulation

While the future is promising, the study of perceptual decision-making is not without its unique challenges. The decision process itself cannot be directly observed—its properties and timing must be inferred from behavioral measures like choice and RT. This is in stark contrast to work on sensory and motor systems, in which the variables of interest can be precisely controlled and/or measured. As cognitive neuroscientists, we are most interested in the kinds of decisions that involve deliberation over time—that is, when decision-making acts as a window on cognition (Shadlen &

Kiani, 2013). But there is no guarantee that subjects will accumulate evidence over time when making perceptual decisions, even if evidence accumulation is the optimal strategy. Nevertheless, it is often assumed that experimental subjects use a strategy that involves evidence accumulation. It could be highly problematic if this assumption were incorrect. For example, if a decision were actually made from the detection of short bursts of salient information, a neuroscientist might mistake step-like activity for a neural mechanism of integration. For computational psychiatrists, decision components would be incorrectly implicated in different disease phenotypes. Thus, characterizing each subject's decision process and verifying that evidence is accumulated over time is essential.

What, then, are the observations in behavioral data that confer such verification? One way to answer this question is to look for conditions in which BANE makes substantially different predictions compared to those of alternative strategies that do not involve evidence accumulation. An example strategy involves a subject waiting for the occurrence of a high SNR signal—an extremum—to instruct an action (Ditterich, 2006; Watson, 1979). Earlier, we discussed that this type of strategy might be encouraged in the RDM task if the dots didn't have a limited lifetime. This strategy is equivalent to extremely leaky evidence accumulation. Another strategy involves subjects picking an arbitrary, random time in the trial to pay attention to the stimulus—a snapshot—and basing their decisions solely on that snapshot of information. We call these two strategies *extrema detection* and *snapshot*, respectively. They replace deliberation with a momentary transition, based on one sample of evidence. Their dynamics are therefore step-like; more consistent with an instructed action than a process of deliberation. We will consider the predictions that these strategies make in three popular behavioral paradigms: fixed stimulus duration (FSD), variable stimulus duration (VSD), and response time (RT). We show that these alternative strategies are surprisingly difficult to rule out.

In a FSD paradigm, the stimulus is presented for a fixed amount of time in every trial, and the subject must wait for the stimulus to turn off before responding. Although this paradigm is widely used, it has several disadvantages. There is no way to estimate the decision time on each trial, and it is known that subjects often commit to a decision before the stimulus turns off (Kiani, Hanks, & Shadlen, 2008). At the level of single trials, it is not possible to know when this commitment occurs. Thus, any observed neural activity might have occurred after the decision was already made. Nevertheless, experimenters often use two observations to conclude that a subject is accumulating evidence: (1) the subject performs the task well, and (2) the choices appear to be informed by evidence obtained across most or all of the stimulus presentation epoch (i.e., flat positive psychophysical kernels; see figure 51.3*B*, *legend*). Both of these observations are predicted by BANE and, indeed, are commonly observed in subjects performing a FSD task. But simulations indicate that these two observations are not uniquely predicted by evidence accumulation. Extrema detection and snapshot strategies can produce psychometric functions that match those produced by BANE and also predict quantitatively similar psychophysical kernels (figure 51.3).

The models are also difficult to disentangle in a VSD paradigm, in which the stimulus duration in each trial is determined by a random variable and is unknown to the subject. This paradigm offers a number of benefits over a FSD paradigm, including the option of a flat hazard rate for stimulus duration, a richer set of behavioral measurements (e.g., accuracy as a function of stimulus duration), and the ability to constrain decision-making models that involve time, such as BANE, extrema detection, and snapshot. BANE predicts that accuracy should increase with increasing stimulus duration. This makes intuitive sense, as the longer a subject is able to accumulate evidence, the more accurate the subject will be. But this observation alone would not be enough to conclude that a subject is integrating—extrema detection and snapshot also predict this observation. The models make a guess on trials in which the stimulus turns off before a decision is reached; therefore, the proportion of guess trials decreases as stimulus duration increases, which leads to increasing accuracy.

Thankfully, a RT paradigm offers us some hope in disentangling these models. Because subjects can respond as soon as they are ready, a RT paradigm offers critical benefits: it illuminates the trade-off between speed and accuracy, and it delineates the epoch in which the decision is being formed but has not yet completed. This is not only extremely useful for studying the decision process at the neural level; it also offers much stronger constraints on behavioral models. A RT paradigm lets us easily rule out a snapshot strategy. Snapshot predicts no systematic relationship between stimulus strength and RT, a prediction we know is incorrect. Surprisingly, however, it is still not trivial to disentangle BANE from extrema detection. The key difference between the two models is their prediction for the nondecision time (i.e., the time needed for any decision-independent processes like sensory and motor delays). Because this difference manifests at the strongest stimulus strengths, a sufficiently wide range of stimulus strengths would constrain the nondecision

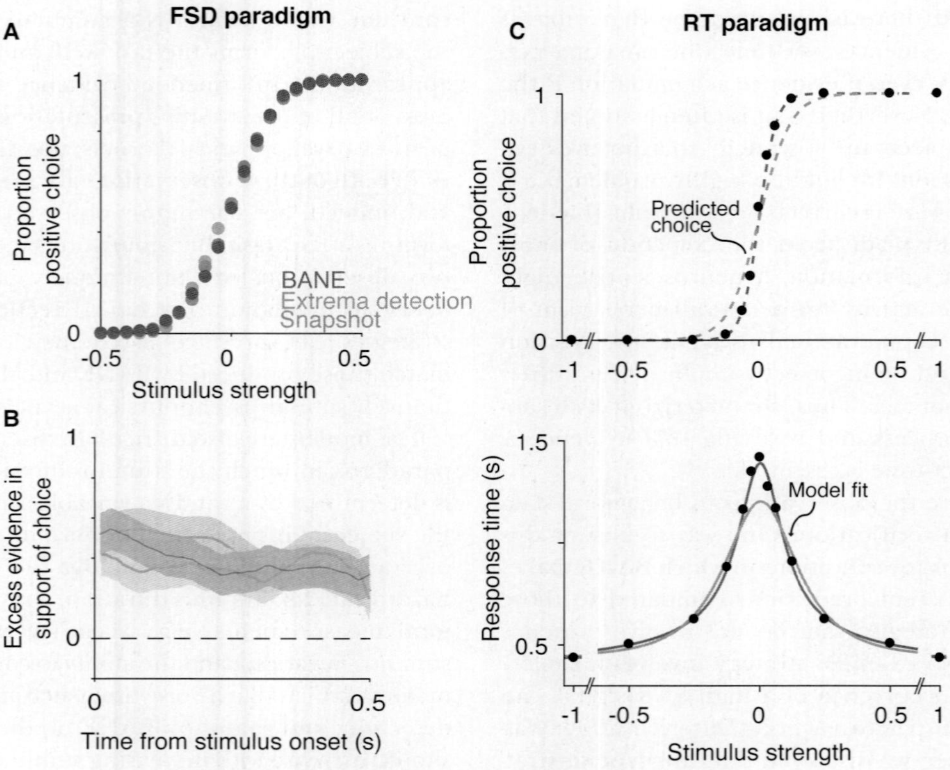

FIGURE 51.3 Alternative strategies to evidence accumulation. *A*, Simulated choice data from three decision-making strategies in a fixed stimulus duration (FSD) paradigm. Colors correspond to different strategies. *B*, Simulated psychophysical kernels using the same parameters as in (*A*). Kernels are calculated by computing the choice-conditioned average of stimulus fluctuations at several time points during stimulus presentation. Note that all three strategies make similar predictions in both (*A*) and (*B*). Stimulus fluctuations affect choice at different times in different trials in FSD and extrema detection. In BANE, stimulus fluctuations affect choices across time in any one trial. The kernel analysis fails to reflect this important distinction. *C*, The choice-response time paradigm can disentangle the models. Choice-RT data (*symbols*) are simulated from a BANE model. RT means are fit with both a BANE model and an extrema detection model (*solid curves*). Snapshot is not shown because it furnishes no explanation of stimulus strength-dependent RTs. The dashed lines show the predicted choice data using the parameters obtained from the RT fits. Note that the predictions and RT fits of the extrema detection strategy are substantially worse than those of BANE. (See color plate 56.)

time. This constraint, coupled with RTs in difficult trials that are substantially longer than the nondecision time, forces the two models to make different predictions. This is shown in figure 51.3*C*; BANE can predict choice accuracy using only the mean RTs, whereas extrema detection cannot.

Finally, it is critical to stress that different subjects might use different strategies, and even the same subject might vary strategies under different conditions or in different stages of training. This fact underscores the importance of assessing the strategy of each subject separately and designing tasks in a way that discourages unwanted strategies, especially when studying animals that cannot be given explicit task instructions. The important point is that inferring a subject's decision process from behavioral measurements alone is difficult, but nevertheless necessary, if we want to understand the neural underpinnings of perceptual decisions.

Conclusion

By studying the neurobiology of perceptual decision-making, we can begin to understand fundamental cognitive processes in a highly controlled and tractable manner. When experimental subjects deliberate about a stimulus in a low SNR regime, they invoke a variety of cognitive operations: allocating attention to relevant information, integrating evidence, weighing sources of evidence in accordance with reliability, pitting speed against accuracy, strategizing, prioritizing, choosing appropriate actions, and assigning levels of belief to their inferences. We have only touched on a few of these in this chapter, but the neuroscientific study of decision-making promises to elucidate many of these cognitive essentials. It seems likely that many of these operations fail in mental disorders, and successful treatments will have to somehow restore the disrupted brain functions. Through the use of contrived tasks and the careful quantification of behavioral measures, we can find correlates of these operations in the brain and begin to understand how they are implemented by neural circuits and networks. With the development of new tools and experimental paradigms, together with the careful identification and characterization of the decision process, the field is progressing toward a multilevel understanding of the neural mechanisms of deliberation and, thus, of higher brain function.

REFERENCES

Blatt, G. J., Andersen, R. A., & Stoner, G. R. (1990). Visual receptive field organization and cortico-cortical connections of the lateral intraparietal area (area LIP) in the macaque. *Journal of Comparative Neurology, 299*(4), 421–445. https://doi.org/10.1002/cne.902990404

Britten, K., Shadlen, M., Newsome, W., & Movshon, J. (1992). The analysis of visual motion: A comparison of neuronal and psychophysical performance. *Journal of Neuroscience, 12*(12), 4745–4765. https://doi.org/10.1523/JNEUROSCI.12-12-04745.1992

Chandrasekaran, C., Peixoto, D., Newsome, W. T., & Shenoy, K. V. (2017). Laminar differences in decision-related neural activity in dorsal premotor cortex. *Nature Communications, 8*(1), 614. https://doi.org/10.1038/s41467-017-00715-0

Churchland, A. K., Kiani, R., Chaudhuri, R., Wang, X.-J., Pouget, A., & Shadlen, M. N. (2011). Variance as a signature of neural computations during decision making. *Neuron, 69*(4), 818–831. https://doi.org/10.1016/j.neuron.2010.12.037

Churchland, A. K., Kiani, R., & Shadlen, M. N. (2008). Decision-making with multiple alternatives. *Nature Neuroscience, 11*(6), 693–702. https://doi.org/10.1038/nn.2123

de Lafuente, V., Jazayeri, M., & Shadlen, M. N. (2015). Representation of accumulating evidence for a decision in two parietal areas. *Journal of Neuroscience, 35*(10), 4306–4318. https://doi.org/10.1523/JNEUROSCI.2451-14.2015

Ding, L., & Gold, J. I. (2013). The basal ganglia's contributions to perceptual decision making. *Neuron, 79*(4), 640–649. https://doi.org/10.1016/j.neuron.2013.07.042

Ditterich, J. (2006). Stochastic models of decisions about motion direction: Behavior and physiology. *Neural Networks, 19*(8), 981–1012. https://doi.org/10.1016/j.neunet.2006.05.042

Ditterich, J., Mazurek, M. E., & Shadlen, M. N. (2003). Microstimulation of visual cortex affects the speed of perceptual decisions. *Nature Neuroscience, 6*(8), 891–898. https://doi.org/10.1038/nn1094

Drugowitsch, J., Moreno-Bote, R., Churchland, A. K., Shadlen, M. N., & Pouget, A. (2012). The cost of accumulating evidence in perceptual decision making. *Journal of Neuroscience, 32*(11), 3612–3628. https://doi.org/10.1523/JNEUROSCI.4010-11.2012

Fetsch, C. R., Kiani, R., Newsome, W. T., & Shadlen, M. N. (2014). Effects of cortical microstimulation on confidence in a perceptual decision. *Neuron, 83*(4), 797–804. https://doi.org/10.1016/j.neuron.2014.07.011

Gnadt, J. W., & Andersen, R. A. (1988). Memory related motor planning activity in posterior parietal cortex of macaque. *Experimental Brain Research, 70*(1), 216–220. https://doi.org/10.1007/BF00271862

Gold, J. I., & Shadlen, M. N. (2007). The neural basis of decision making. *Annual Review of Neuroscience, 30*(1), 535–574. https://doi.org/10.1146/annurev.neuro.29.051605.113038

Hanks, T. D., Ditterich, J., & Shadlen, M. N. (2006). Microstimulation of macaque area LIP affects decision-making in a motion discrimination task. *Nature Neuroscience, 9*(5), 682–689. https://doi.org/10.1038/nn1683

Hanks, T., Kiani, R., & Shadlen, M. N. (2014). A neural mechanism of speed-accuracy tradeoff in macaque area LIP. *eLife, 3.* https://doi.org/10.7554/eLife.02260

Horwitz, G. D., & Newsome, W. T. (1999). Separate signals for target selection and movement specification in the superior colliculus. *Science, 284*(5417), 1158–1161. https://doi.org/10.1126/science.284.5417.1158

Huk, A. C., & Shadlen, M. N. (2005). Neural activity in macaque parietal cortex reflects temporal integration of visual motion signals during perceptual decision making. *Journal of Neuroscience, 25*(45), 10420–10436. https://doi.org/10.1523/JNEUROSCI.4684-04.2005

Katz, L. N., Yates, J. L., Pillow, J. W., & Huk, A. C. (2016). Dissociated functional significance of decision-related activity in the primate dorsal stream. *Nature, 535*(7611), 285–288. https://doi.org/10.1038/nature18617

Kiani, R., Hanks, T. D., & Shadlen, M. N. (2008). Bounded integration in parietal cortex underlies decisions even when viewing duration is dictated by the environment. *Journal of Neuroscience, 28*(12), 3017–3029. https://doi.org/10.1523/JNEUROSCI.4761-07.2008

Kim, J.-N., & Shadlen, M. N. (1999). Neural correlates of a decision in the dorsolateral prefrontal cortex of the macaque. *Nature Neuroscience, 2*(2), 176–185. https://doi.org/10.1038/5739

Kira, S., Yang, T., & Shadlen, M. N. (2015). A neural implementation of Wald's sequential probability ratio test. *Neuron, 85*(4), 861–873. https://doi.org/10.1016/j.neuron.2015.01.007

Krajbich, I., Lu, D., Camerer, C., & Rangel, A. (2012). The attentional drift-diffusion model extends to simple purchasing decisions. *Frontiers in Psychology, 3,* 193. https://doi.org/10.3389/fpsyg.2012.00193

Krajbich, I., & Rangel, A. (2011). Multialternative drift-diffusion model predicts the relationship between visual fixations and choice in value-based decisions. *Proceedings of the National Academy of Sciences, 108*(33), 13852–13857. https://doi.org/10.1073/pnas.1101328108

Laming, D. R. J. (1968). *Information theory of choice-reaction times.* Oxford: Academic Press.

Link, S. W. (1992). *The wave theory of difference and similarity.* London: Psychology Press.

Polanía, R., Krajbich, I., Grueschow, M., & Ruff, C. C. (2014). Neural oscillations and synchronization differentially support evidence accumulation in perceptual and value-based decision making. *Neuron, 82*(3), 709–720. https://doi.org/10.1016/j.neuron.2014.03.014

Ratcliff, R. (1978). A theory of memory retrieval. *Psychological Review, 85*(2), 59–108. https://doi.org/10.1037/0033-295X.85.2.59

Roitman, J. D., & Shadlen, M. N. (2002). Response of neurons in the lateral intraparietal area during a combined cisual discrimination reaction time task. *Journal of Neuroscience, 22*(21), 9475–9489. https://doi.org/10.1523/JNEUROSCI.22-21-09475.2002

Salzman, C. D., Britten, K. H., & Newsome, W. T. (1990). Cortical microstimulation influences perceptual judgements of motion direction. *Nature, 346*(6280), 174–177. https://doi.org/10.1038/346174a0

Shadlen, M. N., & Kiani, R. (2013). Decision making as a window on cognition. *Neuron, 80*(3), 791–806. https://doi.org/10.1016/j.neuron.2013.10.047

Shadlen, M. N., & Newsome, W. T. (2001). Neural basis of a perceptual decision in the parietal cortex (area LIP) of the rhesus monkey. *Journal of Neurophysiology, 86*(4), 1916–1936. https://doi.org/10.1152/jn.2001.86.4.1916

Shadlen, M. N., & Shohamy, D. (2016). Decision making and sequential sampling from memory. *Neuron, 90*(5), 927–939. https://doi.org/10.1016/j.neuron.2016.04.036

Thura, D., & Cisek, P. (2014). Deliberation and commitment in the premotor and primary motor cortex during dynamic decision making. *Neuron, 81*(6), 1401–1416. https://doi.org/10.1016/j.neuron.2014.01.031

Watson, A. B. (1979). Probability summation over time. *Vision Research, 19*(5), 515–522. https://doi.org/10.1016/0042-6989(79)90136-6

Xiong, Q., Znamenskiy, P., & Zador, A. M. (2015). Selective corticostriatal plasticity during acquisition of an auditory discrimination task. *Nature, 521*(7552), 348–351. https://doi.org/10.1038/nature14225

Znamenskiy, P., & Zador, A. M. (2013). Corticostriatal neurons in auditory cortex drive decisions during auditory discrimination. *Nature, 497*(7450), 482–485. https://doi.org/10.1038/nature12077

52 Memory, Reward, and Decision-Making

KATHERINE DUNCAN AND DAPHNA SHOHAMY

ABSTRACT Decisions about preferences often use past experience to predict the likelihood of rewards in the future. Much work has focused on the role of the striatum and the habitual learning of stimulus-reward associations in decision-making. However, many facets of reward-based decisions do not depend on habits but on other forms of memory. A central challenge has been to understand the cognitive and neural mechanisms by which other forms of memory guide reward-based decisions and the circumstances under which different forms of memory contribute to reward-guided behaviors. Here we review recent advances in understanding the role of the hippocampus in episodic and relational memory, highlighting the different ways in which hippocampal-dependent memories support value-based decisions. Converging evidence suggests a role for the hippocampus in a broad range of memory-guided decisions, including sampling of one-shot episodes, integration across related events and their values, and imagining possible rewards in the future. We consider how these forms of memory complement existing theoretical, physiological, and cognitive accounts to provide a more complete understanding of how multiple forms of memory work together to support value-based decisions.

A fundamental challenge in cognitive neuroscience is to understand how the brain learns from experience to make adaptive decisions. Major progress has been made in understanding the neural and cognitive mechanisms by which the brain learns from repeated choices and their outcomes to guide decisions. The brief summary is that reward-based decisions often involve learning about cues or actions that repeatedly led to reward in the past. Extensive converging evidence suggests that this learning depends on dopaminergic inputs to the striatum, an idea supported by data from single-cell recordings, computational models of reinforcement learning, human functional magnetic resonance imaging (fMRI), and studies of patients with dopaminergic cell loss due to Parkinson's disease. Together, these studies suggest that the striatum and its dopaminergic inputs support decisions by learning the average reward value of candidate cues or actions (Barto, Mirolli, & Baldassarre, 2013; Daw & O'Doherty, 2013; Frank, 2005; Frank, Seeberger, & O'Reilly, 2004; Hare, Camerer, & Rangel, 2009; Houk & Adams, 1995; O'Doherty et al., 2004; O'Doherty, Dayan, Friston, Critchley, & Dolan, 2003; Pessiglione et al., 2008; Schonberg et al., 2010; Schultz, Dayan, & Montague, 1997; Shohamy et al., 2004). This learning is thought to take place incrementally, over many experiences, and is thought to underlie the formation of learned habits to automatically guide reward-seeking behavior.

This habit-learning system, by its very definition, supports only decisions that have an extensive reinforcement history. Yet in many situations, decisions must be made based on relatively sparse information, such as a single past event, or under novel circumstances in which past experience is not directly replicated but instead must be flexibly used to guide inferences, generalization, or deliberation about possible outcomes. Such decisions are not well served by a habit-learning system and likely depend on other forms of memory.

The notion that there are multiple complementary forms of memory that serve distinct functional roles has received extensive attention in cognitive and systems neuroscience (e.g., Eichenbaum & Cohen, 2001; Gabrieli, 1998; Knowlton, Mangels, & Squire, 1996; Squire & Zola, 1996). Yet the neural mechanisms by which rapidly acquired, flexible memories guide decisions have not been extensively studied. This is largely because the role of the hippocampus in memory has been examined almost exclusively in the context of memory itself, rather than how memory is used to guide behavior. But it is precisely the sorts of mnemonic functions ascribed to the hippocampus that have been missing from a more complete account of reward-based decision-making. Thus, the convergence between memory and decision-making serves to fill gaps in both fields.

In this chapter we survey these developments and discuss how understanding the convergence between these areas provides a framework for understanding the mechanisms by which memory guides decisions. After reviewing the evidence for classic reward-learning theories, we turn our focus to how the hippocampus' well-established role in *episodic memory* (rich memory for single events) could shape decision-making. Then, we discuss *relational models* of hippocampal memory representations and how they support the integration of features within an experience, as well as the integration across interrelated experiences. Finally, we discuss how the hippocampus supports the imagining of events in the future—referred to as *prospection*—and how prospection can guide decisions.

Dopamine, Reinforcement Learning, and Habits

Extensive research implicates the striatum and its dopaminergic inputs in reward learning and habit formation. Dopaminergic inputs to the striatum arise from neurons located in two midbrain nuclei, the substantia nigra pars compacta and the ventral tegmental area. Recordings from these dopaminergic neurons in behaving monkeys have revealed an important reward-related signal. These neurons have a low background-firing rate punctuated by brief, phasic excitations and inhibitions. Following a series of reports describing various circumstances under which these phasic firing modulations occur (Schultz, Dayan, & Montague, 1997; for an early review, see Schultz, 1992), seminal computational work pointed out that many of these responses could collectively be understood as signaling a *reward prediction error* (Houk & Adams, 1998; Montague, Dayan, & Sejnowski, 1996; Schultz, Dayan, & Montague, 1997). A reward prediction error is the difference between the reward received and the reward that was expected—in other words, a form of feedback that indicates how errant a choice was given its outcome. Indeed, dopamine neurons show a phasic excitation to reward that is proportional to the prediction error: largest for a completely unpredicted reward, virtually nonexistent for a fully predicted reward, and suppressed for rewards that fall short of expectations (Fiorillo, Tobler, & Schultz, 2003). In computer science and engineering, reward prediction errors are commonly used to implement reinforcement learning (Sutton & Barto, 1998). In particular, this signal underpins a class of reinforcement-learning algorithms—*model-free reinforcement learning*—that use unexpected rewards to "stamp in" preceding choices or actions. It is this form of learning that is thought to underlie automatic responses or habits.

The relationship between dopamine neurons and reward prediction error signaling has been replicated and extended in monkey and rodent studies. Convergent findings have also been revealed in humans with fMRI studies: the blood oxygen level-dependent (BOLD) signal at dopaminergic targets (principally ventral striatum) is correlated with reward prediction errors (McClure, Berns, & Montague, 2003; O'Doherty et al., 2003; Pessiglione, Seymour, Flandin, Dolan, & Frith, 2006). Of course, the metabolic activity detected by fMRI does not specifically reveal the activity of a particular neuromodulator such as dopamine. However, pharmacological studies in both healthy individuals and those with dopamine abnormalities (such as patients with Parkinson's disease) indicate that dopamine affects the prediction error-related BOLD signal (Pessiglione et al., 2006; Schmidt, Braun, Wager, & Shohamy, 2014; Schonberg et al., 2010).

Converging evidence supports the idea that the reward prediction error signal is not only found across species but is in fact important for learning. In humans, studies in patients with Parkinson's disease have revealed that the loss of dopaminergic transmission that characterizes the disease has a detrimental effect on reward-based learning mechanisms (Frank, Seeberger, & O'Reilly, 2004; Maia & Frank, 2011; Schonberg et al., 2010; Shohamy et al., 2004). Such studies typically use tasks involving a series of decisions for possible reward, with participants choosing between two options and the likelihood of reward given each option varying across trials. These tasks are often referred to as *probabilistic-learning tasks* (because the likelihood of reward given a choice is probabilistically determined) or as *two-armed-bandit tasks*, in reference to the gamble that the participant makes on each trial. Studies have found that learning to perform such tasks involves reward prediction error-related activity in the striatum in healthy participants and that patients with Parkinson's disease show both weaker striatal BOLD responses and less adaptive choices (Frank, Seeberger, & O'Reilly, 2004; Maia & Frank, 2011; Schonberg et al., 2010; Shohamy et al., 2004; for a review, see Foerde & Shohamy, 2011).

Habitual versus Goal-Directed Behavioral Control

Dopamine's involvement in model-free reinforcement learning is thus supported by both correlational and causal findings. But there are many other kinds of reward-based decisions not accounted for by this framework. If—as the model-free theories suggest—decisions result from the strengthened tendency to repeat previously rewarded actions, then the resulting behaviors are expected to have a hallmark inflexibility. This system for learning is not well suited for guiding behavior based on sparse experience, or for guiding flexible behaviors in abruptly changing environments.

For instance, habitual learning can take you back to a restaurant you've repeatedly enjoyed in the past, but it can't take you to a new restaurant you've just heard about, even if it is in a familiar neighborhood. Similarly, if you enjoy cake every afternoon but suddenly develop diabetes, a model-free reinforcement-learning mechanism would rigidly guide you to have the same habitual sweet cake each afternoon (it having always been rewarded in the past), rather than choosing a different snack appropriate to your new circumstances. This is because these model-free mechanisms learn only how well actions have turned out previously; because they do not explicitly encode the specific experienced outcomes, they cannot support prospective reasoning about the consequences of specific actions.

Indeed, a long tradition in psychology has aimed to distinguish between behaviors that are habit-like and others that are more informed or deliberative (Dickinson, 1985; Dickinson & Balleine, 2002; Tolman, 1948). The latter are called *goal-directed* actions because they are based on knowledge of a particular desirable goal (such as avoiding sugary foods) and knowledge of the action that will produce it. In contrast to the model-free-learning algorithms associated with the dopaminergic reward prediction error signal, deliberative behaviors are known as *model-based* decisions, after a family of reinforcement-learning algorithms that learn such knowledge (an *internal model* of the task or environment) and use it to evaluate options and guide decisions (Daw, Niv, & Dayan, 2005). A key insight of this research is that many behaviors can be ambiguous. A person ordering from a menu and receiving food might in principle be doing so because that action has been reinforced in the past or, alternatively, because she has knowledge about the predicted outcome and can flexibly choose it.

Multiple Memory and Control Systems

The notion of an internal model that guides decisions raises interesting questions about where this model comes from. For an internal model to adaptively guide flexible behaviors, it, too, must be learned from past experiences. But how? Like model-free learning, the internal model supporting model-based decisions is extracted from many experiences, averaging across them to represent probable outcomes and the steps required to achieve them (Daw, Niv, & Dayan, 2005). These forms of reinforcement learning are thus considered *parametric*, in that they estimate parameters that capture regularities across experience while discarding their idiosyncratic details (Gershman & Daw, 2017). But many real-world decisions are made despite a paucity of directly relevant experiences. From our most consequential choices, like voting for a politician or choosing a school to attend, to the more quotidian, like ordering a new dish at a favorite restaurant, we are often forced to choose between options with which we have had minimal experience, or none at all. A second challenge for parametric forms of reinforcement learning involves the difficulty in knowing how to attribute outcomes to cues and actions (Gershman, Blei, & Niv, 2010; Niv et al., 2015). In rich environments containing many elements, how does the brain know which one to associate with a desired outcome? Acing a test or having a baby sleep through the night, for example, are unambiguously positive outcomes, but deciphering which specific factors will solicit those same outcomes in the future may require a rich and multidimensional memory representation that encompasses many features of the context.

Addressing these challenges can be informed by incorporating theories from a separate literature investigating the cognitive and systems neuroscience mechanisms underlying learning and memory. The key insight is simple—the human brain can learn and remember the same experience in multiple ways (Eichenbaum & Cohen, 2004; Gabrieli, 1998; Poldrack & Packard, 2003; Squire & Dede, 2015; Squire & Zola, 1996). The different types of knowledge acquired by each memory system could, in turn, offer solutions to the challenges raised in the domain of reward-based decisions (Doll, Shohamy, & Daw, 2015; Foerde & Shohamy, 2011).

Bridging between these two literatures is facilitated by the remarkable parallels between the proposed organization of memory systems and value-based decision-making systems, which have mostly been studied independently. According to traditional memory systems theories, at the highest level memory is divided into declarative and procedural systems, distinguished by their accessibility to conscious awareness (Squire & Dede, 2015). Implicit procedural systems, dedicated to learning "how" to act, most closely parallel the type of learning described by model-free reinforcement learning. Despite being most often characterized in terms of skill-based habits, such as riding a bike, procedural memory shares central characteristics and mechanisms with model-free reinforcement learning, such as a reliance on striatal dopamine inputs (Knowlton, Mangels, & Squire, 1996), the extraction of statistical regularities (Knowlton, Squire, & Gluck, 1994), and the enabling of rapid, automatic actions (Cohen & Bacdayan, 1994).

By contrast, consciously experienced declarative memory, comprising hippocampus-dependent episodic (event) memory and cortical semantic (world knowledge) memory (Tulving, 1972), more closely parallels model-based reinforcement learning; this sort of memory is thought to represent outcomes and support flexible goal-directed behavior. As we review below, however, the behavioral control afforded by episodic memory, as well as other forms of hippocampal memory that don't fit as tidily into the traditional multiple memory system framework (Shohamy & Turk-Browne, 2013), extend beyond model-based control, potentially resolving some of its greatest challenges.

Episodic Sampling and Decisions

Episodic memory refers to rich, detailed memories of events or specific moments in time (Tulving, 1972). On one hand, the rich structure of an episodic memory could link states, actions, and outcomes—an ideal representation for model-based and flexible decision-making (Doll, Shohamy, & Daw, 2015; Palombo, Keane, & Verfaellie, 2015). Supporting this conjecture, rodent hippocampal neurons have been shown to track value expectations and outcomes, suggesting that value may be an integral aspect of episodic memories (Lee, Ghim, Kim, Lee, & Jung, 2012). Moreover, by supporting the rapid learning of an event—even something that happened only once—episodic memory is well positioned to guide decisions about options with which we have minimal experience (Gershman & Daw, 2017; Lengyel & Dayan, 2008; Santoro, Frankland, & Richards, 2016). On the other hand, these isolated snapshots are presumably less useful for making decisions that depend on knowledge of statistical regularities observed across many experiences, particularly in stable or slowly changing contexts (Gershman & Daw, 2017; Lengyel & Dayan, 2008; Santoro, Frankland, & Richards, 2016).

There is now extensive empirical data supporting the prevalent use of episodic memories across a variety of decision tasks in humans. In such experiments, each of a rich set of distinctive cues is typically associated with reward values, each exposed only in a single trial. Researchers can then assess what participants do when faced with a decision and the mechanisms underlying their use of "one-shot" memories to guide choices. This procedure is quite different from the "bandit" tasks used in standard reinforcement-learning studies, which associate single images or cues with reward outcomes across hundreds of trials. Studies of episodic sampling have found that participants prefer images associated with higher outcomes versus lower outcomes, a preference consistently observed across a variety of value-learning contexts, ranging from direct instructions to learn the concealed value of individual images (Duncan & Shohamy, 2016) to incidental pairings in which image identity was ostensibly unpredictive of outcomes (Bornstein, Khaw, Shohamy, & Daw, 2017; Bornstein & Norman, 2017; Wimmer & Buechel, 2016). Moreover, episodic memory has been shown to influence decisions in both social and nonsocial domains (Murty, FeldmanHall, Hunter, Phelps, & Davachi, 2016). Successful use of one-shot learning was also found to depend on having an accurate associative memory linking the image to its outcome, suggesting that one-shot value learning is, indeed, mediated by consciously available episodic memories (Murty et al., 2016; figure 52.1*A*). Thus, it appears that mnemonic records of events that only happened once encode value information that is retrieved and used to guide reward-based decisions.

Prioritized encoding of value-relevant memories The flexibility conferred by episodic sampling raises questions about *which* episodes from memory to sample when faced with a decision. Of course, the likelihood of any one memory being used to guide a decision is strongly influenced by the strength with which that memory was encoded to begin with. This encoding strength, in turn, is modulated by motivational relevance. For example, long-term potentiation in the hippocampus depends on neuromodulators, including dopamine (Lemon & Manahan-Vaughan, 2006; Li, Cullen, Anwyl, & Rowan, 2003), norepinephrine (Izumi & Zorumski, 1999; Stanton & Sarvey, 1985), and acetylcholine (Blitzer, Gil, & Landau, 1990; Huerta & Lisman, 1995). These neuromodulators are released during salient events, including reward, punishment, and expectancy violation (Lisman & Grace, 2005; Mather, Clewett, Sakaki, & Harley, 2016; Ruivo et al., 2017). By affecting the strength of memory encoding, neuromodulatory signals could adaptively favor the later retrieval of biologically important events, prioritizing their influence on behavior. This work highlights that common neuromodulatory mechanisms could underlie the prioritization of episodic encoding in the hippocampus while at the same time driving more habitual learning of repeated associations in the striatum. Memory encoding is also enhanced when people are in control of their environment (Murty et al., 2016; Voss, Gonsalves et al., 2011)—free to actively choose what happens next—and when they are motivated by potential rewards or punishments (Adcock, Thangavel, Whitfield-Gabrieli, Knutson, & Gabrieli, 2006; Murty, LaBar, & Adcock, 2012). Collectively, this work has recast episodic memory in an adaptive, potentially strategic light, as a memory system that stores the contents of those events best positioned to guide future actions (Duncan & Schlichting, 2018; Shohamy & Adcock, 2010).

The context of decisions influences the retrieval and use of episodic memories The context in which a decision is made also influences which memories are used, above and beyond the strength of any given memory. Episodic memories are inherently contextualized, containing information about where, when, and how an event unfolded. Thus, the specific episodic memories cued by the decision context will be more likely to influence

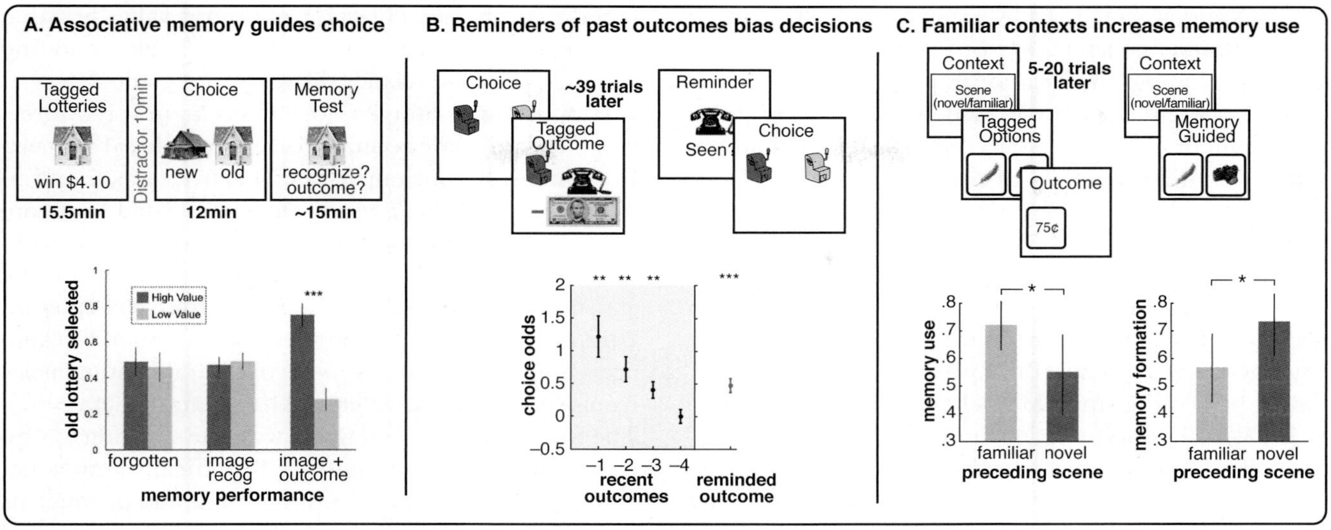

FIGURE 52.1 Evidence that episodic memories guide decisions. *A*, Participants first play many lotteries, each tagged with a distinctive house (*Tagged Lottery Phase*). Participants were then more likely to reengage with lotteries that resulted in higher outcomes (*Choice Phase*). This adaptive use of single experiences, however, was only seen for lotteries that were recognized and whose outcomes were remembered, as determined in the final *Memory Test Phase*. Adapted from Murty et al. (2016). *B*, Participants chose between two slot machines with outcomes that slowly varied across the experiment. The monetary outcome of each choice was tagged with a unique object "ticket." Tickets could then reappear as a reminder many trials later. Postticket choices were influenced by the choices made and outcomes experienced on the reminded trial. Adapted from Bornstein et al. (2017). *C*, Participants chose between pairs of cards tagged with unique objects. Two new cards were dealt on roughly half of the trials, but a previously selected card was dealt alongside a new card on the remaining trials. Participants were more likely to select the familiar card if it had resulted in a high outcome. This adaptive use of single experiences was heightened when participants made choices following the presentation of a familiar but unrelated contextual image, as compared to a novel image. By contrast, novel contextual images heightened the encoding and later use of memories. Adapted from Duncan and Shohamy (2016).

decisions, even if such memories store events from the distant past. For example, a friend's dessert recommendation given long ago at a trendy restaurant would be more likely to influence your order if her name comes up while you peruse the menu. Experimentally, this has been shown by tagging specific reward outcomes with a specific image. When later re-presented with an image, participants' choices were influenced by the outcome from the particular cued trial (Bornstein et al., 2017; figure 52.1*B*) or the context containing the cued trial (Bornstein & Norman, 2017). Moreover, the degree to which these prior outcomes influence choice was related to evidence for contextual neural reactivation (Bornstein & Norman, 2017). This work suggests that when cued at the time of choice, specific experiences from the distant past can carry as much weight as more recently experienced outcomes.

Context has also been shown to have an impact on the use of episodic memories by creating a *state of mind* that is more conducive to episodic memory retrieval (Duncan & Shohamy, 2016). This research was inspired by theoretical and empirical work proposing that context can adaptively bias the hippocampus toward either memory formation or retrieval; novel contexts facilitate memory formation, whereas familiar contexts facilitate memory retrieval (Duncan, Sadanand, & Davachi, 2012; Easton, Douchamps, Eacott, & Lever, 2012; Hasselmo, Wyble, & Wallenstein, 1996; Meeter, Murre, & Talamini, 2004; Patil & Duncan, 2018). This memory state hypothesis thus predicts that episodic memories would be most influential when choices are made in familiar contexts. Supporting this prediction, values learned in a single trial were found to carry more influence on decisions made after viewing an unrelated familiar, as compared to novel, image (Duncan & Shohamy, 2016). Thus, in addition to cuing particular memories, familiar contexts enhance the use of episodic memories by biasing people toward the process of memory retrieval. Conversely, memories formed after viewing a novel as compared to familiar image were more likely to be encoded well and later influence choices, underscoring the specificity of this contextual bias.

The Hippocampus and Relational Encoding

Think back on an episode from your life: perhaps this morning's breakfast or last year's birthday. What distinguishes that event from other similar experiences?

Events are rarely set apart by a single feature but rather by the unique constellation of features that comprise them, including the relationships between features. Indeed, there is extensive work demonstrating that episodic memory is, in its essence, *relational*: it depends on rapidly binding together pieces of an experience as it unfolds so that the pieces (and just the pertinent pieces) can be put back together again when the memory is retrieved.

Accordingly, many models of hippocampal function focus on its capacity to bind the pieces of experience together in relational (Eichenbaum, Otto, & Cohen, 1994) or configural (McClelland, McNaughton, & O'Reilly, 1995; Sutherland & Rudy, 1989) memory representations. The hippocampus has ideal anatomical connections for this binding. It sits atop the visual-processing hierarchy, receiving converging input about the identity and location of complex objects (Davachi, 2006; Lavenex & Amaral, 2000; Van Essen, Anderson, & Felleman, 1992). The hippocampus also receives, directly or indirectly, information from other modalities, such as audition and olfaction (Insausti & Amaral, 2012). On top of this sensory input, the hippocampus also receives modulatory input directly from the amygdala and indirectly from prefrontal regions (Insausti & Amaral, 2012; Vertes, Hoover, Do Valle, Sherman, & Rodriguez, 2006), which may reflect emotions and goals, respectively. These multimodal inputs are thoroughly intermixed both within the hippocampus proper and to some degree within medial temporal lobe (MTL) cortical regions, connecting disparate inputs (Insausti & Amaral, 2012). Hippocampal binding is also thought to be organized across time and space by neurons that reliably fire in particular locations (dubbed *place cells*; O'Keefe & Nadel, 1978) and at particular times (dubbed *time cells*; Eichenbaum, 2014). These neurons could bridge from one location and moment to another within an event.

These relational memories offer important insights into how past experience can be used to guide decisions. First, they could help to resolve the ambiguity in attributing outcomes to actions in complex environments. Just like life events, options under consideration are rarely distinguishable in terms of a single feature (at least outside the lab). These complex choice options could be evaluated in a piecemeal fashion by combining the learned values of each feature to derive the value of the whole. Conversely, relational memory prebinds features into configurations so that values can be directly associated with the complex option (Melchers, Shanks, & Lachnit, 2008). Critically, the configural approach allows the learned value of a complex option (e.g., sauerkraut-flavored ice cream) to be independent from the value of the parts that comprise it (e.g., the separate values of sauerkraut and ice cream). Configural value learning thus could increase decision flexibility by incorporating contingencies and relationships into preferences.

Hippocampal contributions to configural reinforcement learning have recently received empirical support. In humans, hippocampal BOLD activity was found to increase when people used values associated with configurations, as opposed to values associated with constituent features, to guide choice in a probabilistic classification task (Duncan, Daw, Doll, & Shohamy, 2018; figure 52.2*A*). Moreover, patients with MTL damage were impaired at learning configural contingencies from feedback in a related task (Kumaran et al., 2007). The hippocampus likely works with the striatum to support behavior in these contexts. Specifically, functional connectivity between the hippocampus and the nucleus accumbens has been related to learning the values of combinations of stimuli in both humans (Duncan, Daw, Doll, & Shohamy, 2018) and rats (Ito, Robbins, Pennartz, & Everitt, 2008). Together, this work suggests that hippocampal relational representations help us learn the values of previously experienced choice options when those options are made up of multiple pieces.

Relational memory representations could also guide choices that have not been directly reinforced in the past. Many decisions require incorporating information gained across multiple experiences, such as navigating a new route by piecing together familiar ones. Relational models propose that common elements of experiences are encoded by *node* neurons shared across hippocampal representations of related events (Eichenbaum et al., 1994). In this way, the intersections between different familiar routes are physically coded within the memory representation, fostering their integration. This integration is also thought to extend beyond spatial navigation, linking the contents of experiences that share people, places, or objects (Zeithamova, Schlichting, & Preston, 2012).

An intriguing consequence of this relational coding scheme is that some novel inferences could be precomputed during value learning, in anticipation of future choices. For a concrete example, consider a task dubbed *sensory preconditioning*. In the first phase of this task, two otherwise unrelated stimuli (*S1* and *S2*) are associated by repeatedly presenting them in close succession. Then, one stimulus is reinforced—for example, by pairing *S2* with a reward. In the critical test phase, subjects choose between *S1* and another equally familiar stimulus to determine whether the learned value of *S2* transferred to *S1*. Humans and other animals tend to prefer *S1* despite it never being directly rewarded.

A neuroimaging study showed that value transfer in the sensory-preconditioning task is related to

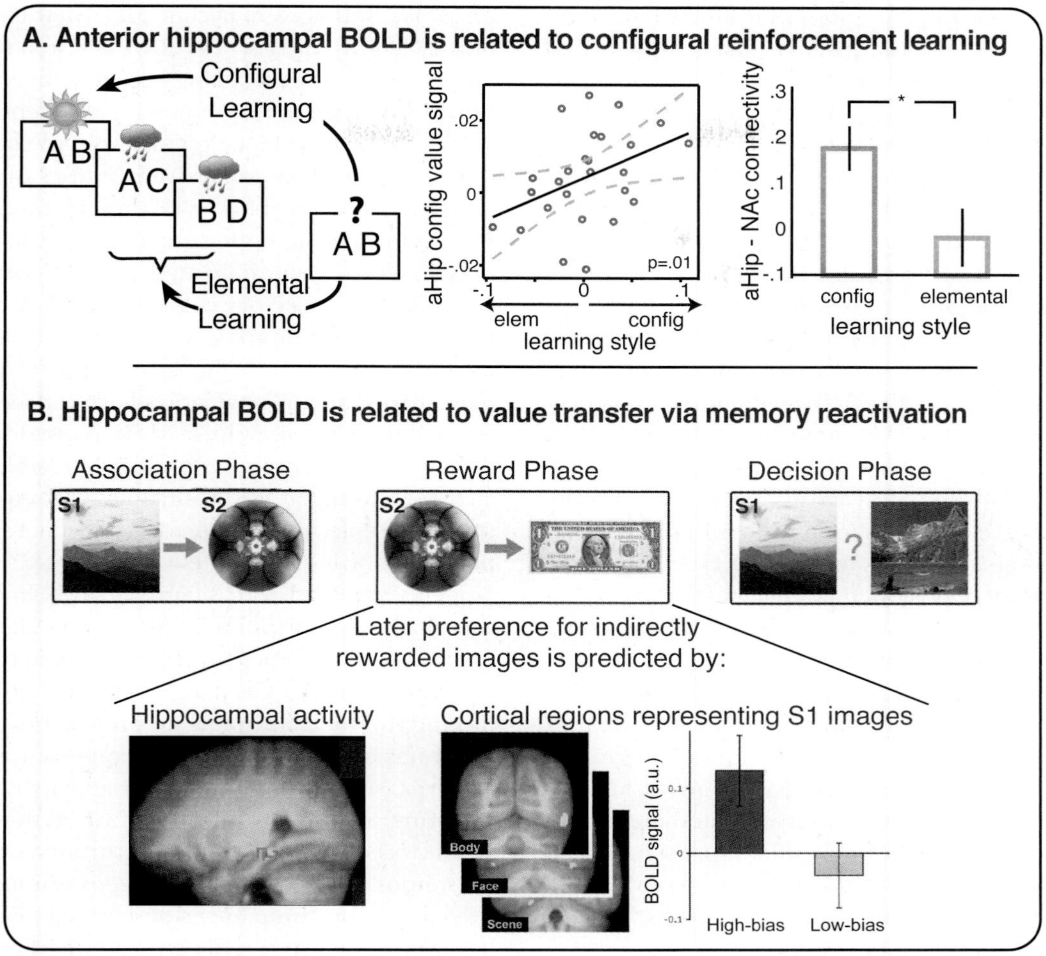

FIGURE 52.2 fMRI evidence for hippocampal involvement in decisions that depend on relational processing. *A*, Participants made weather predictions using pairs of abstract cues in a probabilistic classification task. Reinforcement-learning models quantified the likelihood that choices were made using experience with configurations (e.g., AB) versus individual elements (e.g., A). BOLD responses in the anterior hippocampus (aHip) and functional connectivity between the aHip and the nucleus accumbens tracked the degree to which participants used configural learning. Adapted from Duncan et al. (2018). *B*, Participants performed a sensory-preconditioning task in which multiple S1-S2 pairs were first associated with each other. S2 stimuli were then either rewarded or not rewarded, and preferences for indirectly rewarded S1 stimuli were measured in the final *Decision Phase*. The transfer of value to S1 stimuli was related to BOLD activity during the reward phase in both the hippocampus and the category-specific visual area corresponding to the S1 stimulus' class (scene, face, or body part). Adapted from Wimmer and Shohamy (2012). (See color plate 57.)

hippocampal BOLD activity during initial value learning (Wimmer & Shohamy, 2012; figure 52.2*B*). A clever feature of this task enabled additional insight into the mechanisms by showing that value transfer was related to reactivation of the specific categories of *S1* stimuli associated with the rewarded *S2* stimuli. This was accomplished by using *S1* stimuli from visual categories (face, place, and body part images) known to elicit activity in specific visual cortical areas (Reddy & Kanwisher, 2006). During the *S2* reward-pairing phase, participants who showed greater evidence of *S1* reactivation also showed greater value transfer during the later test. The link between neural reactivation and later transfer was also observed in an magnetoencephalography's (MEG) study using a similar paradigm, in which MEG greater temporal resolution isolated transfer-related reactivation to a few hundred milliseconds following reward (Kurth-Nelson, Barnes, Sejdinovic, Dolan, & Dayan, 2015). Conceptually related paradigms have also demonstrated a relationship between hippocampal activity during learning and later flexible decisions, such as making associative inference judgments (Schlichting, Zeithamova, & Preston, 2014; Zeithamova, Dominick, & Preston, 2012). Of note, these tasks involve *precomputing* the relationships between the stimuli themselves in the service of future decisions. Thus, integrated hippocampal representations might form the building blocks for the schemas, likely represented in the ventral medial prefrontal cortex (PFC) (Gilboa & Marlatte, 2017; Preston & Eichenbaum, 2013), that are ultimately used by model-based decisions.

Novel choices can also be made by integrating distinct memories at the time of decision. Returning to the example of sauerkraut ice cream, you are unlikely to have precomputed or stored its value in memory, having never tasted it before. Yet you can use separate past experiences with sauerkraut and ice cream to evaluate the dish. Indeed, fMRI adaptation shows that people access representations of each ingredient when evaluating novel (but somewhat more appealing) dishes, like "tea jelly" (Barron, Dolan, & Behrens, 2013). Neural processing at the time of decision also shapes inferential decisions, like those described above in the sensory-preconditioning task. For example, BOLD activity in the hippocampus increases when people make new choices that require inference, as compared to choices that were directly reinforced in the past (Heckers, Zalesak, Weiss, Ditman, & Titone, 2004; Preston, Shrager, Dudukovic, & Gabrieli, 2004). It is unclear, however, whether this activity reflects the retrieval and online integration of multiple distinct memories (Kumaran & McClelland, 2012) or the retrieval of preintegrated memories. Conversely, the rodent orbitofrontal cortex has been specifically linked to the online integration of values to support inferential choices (Jones et al., 2012).

In summary, hippocampal relational binding mechanisms, well studied for their memory contributions, may confer the flexibility needed to make decisions in everyday environments. First, binding within an experience allows configurations to take on values that are independent from their comprising features. Second, binding across related experiences could support novel inferential decisions. These "inferences" could be precomputed by either transferring values across associated options or directly encoding experienced relationships between options. Alternatively, these same inferential decisions could be supported by retrieving distinct memories and integrating their content at the time of decision.

Prospecting on Future States for Future Selves

Episodic memory and hippocampal processes can also shape decisions by supporting prospection—the representation of possible futures. Specifically, the hippocampus is thought to support the simulation of future scenarios, which can be used to make predictions, plan for the future, and set adaptive intentions or goals (Szpunar, Spreng, & Schacter, 2014). In this way, prospection is heavily intertwined with decision-making, as it represents the consequences of actions as well as our future selves, who receive those consequences.

Compelling evidence for the role of the hippocampus in prospection comes from studies of spatial navigation in rodents. This work capitalizes on the strong spatial tuning of hippocampal place cells, which fire maximally when animals are in particular locations, regardless of trajectory or orientation, and are argued to collectively form a *cognitive map* of the environment (O'Keefe & Nadel, 1978). Recordings from hippocampal neurons during navigation have revealed suggestive signals at decision points: while paused at junctures, sequences of place cells "preplay" possible spatial trajectories (Johnson & Redish, 2007). Moreover, the content of preplayed sequences has been related to the path that will be selected (Pfeiffer & Foster, 2013; Singer, Carr, Karlsson, & Frank, 2013), and disrupting the sharp wave ripples in which preplay events are embedded has been shown to impair spatial decisions (Jadhav, Kemere, German, & Frank, 2012), causally linking this mechanism to action selection. In humans, fMRI has also been used to decode future paths from patterns of hippocampal BOLD activity during spatial planning (Brown et al., 2016). Together, this work points to a concrete mechanism through which the hippocampus could support prospective simulation in the service of multistep decision-making. Further work, however, is required to determine whether this type of preplay extends beyond spatial planning in a manner that could support the more general-purpose cognitive simulations that have been described in humans. Notably, the short timescale of spatial navigation differs substantially from the long timescales involved in deciding about, for instance, which vacation to take, or which college to attend.

Nonetheless, there is some evidence that the hippocampus contributes to such longer-term prospection in humans. Amnesiac patients suffering from damage to the hippocampal region (Andelman, Hoofien, Goldberg, Aizenstein, & Neufeld, 2010; Hassabis, Kumaran, Vann, & Maguire, 2007; Race, Keane, & Verfaellie, 2011) show impaired prospection about future events, reflected in impoverished details of imagined personal experiences, such as sitting on a beach in the future. Converging neuroimaging evidence also demonstrates a striking overlap between the networks engaged during the successful recollection of memories for past events and the simulation of future events that never occurred (see Benoit & Schacter, 2015 for a recent meta-analysis). This literature suggests that decisions that depend on

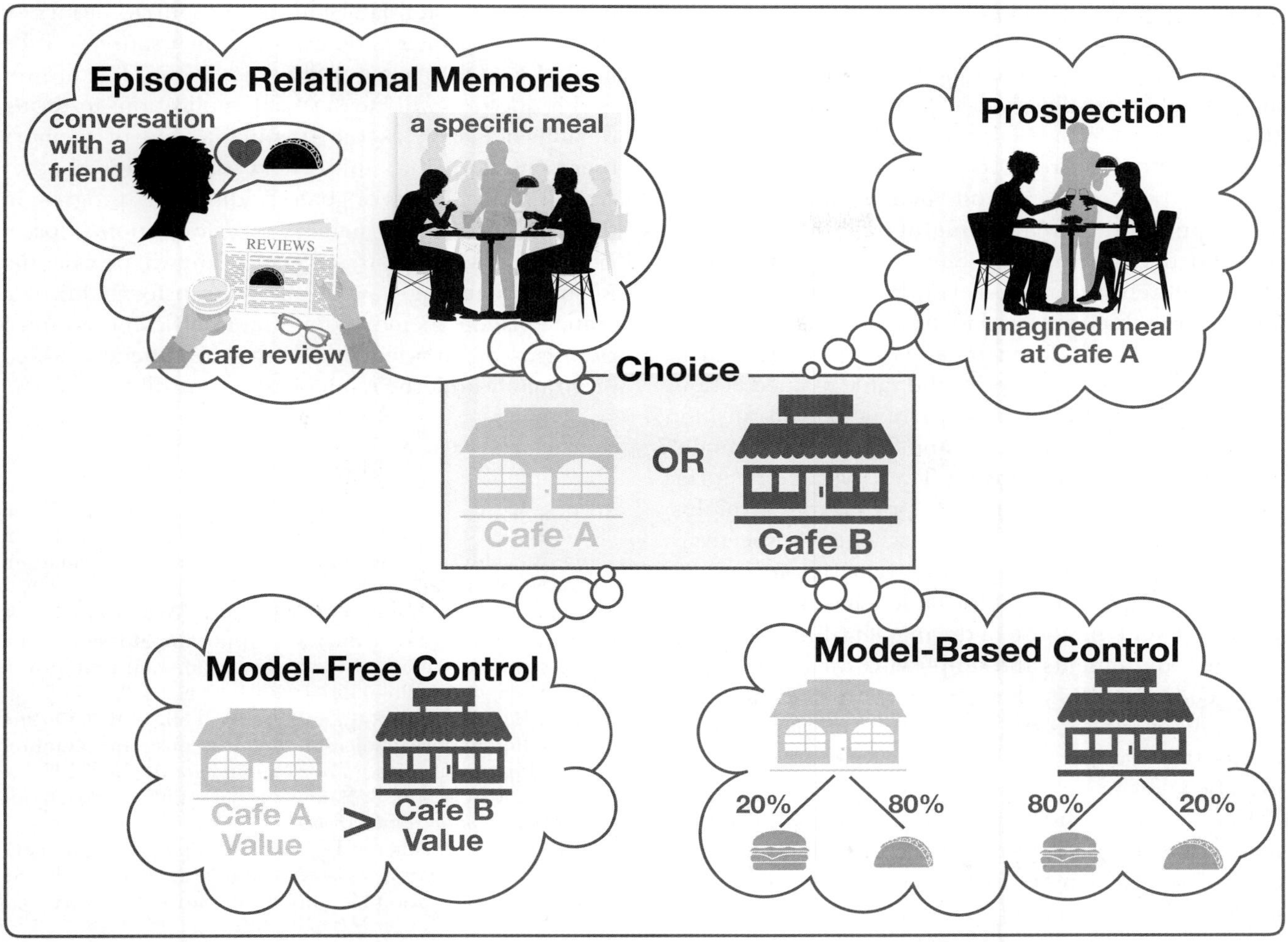

FIGURE 52.3 Multiple forms of memory can guide choices. With multiple control and memory systems, the same decision could be arrived at through different cognitive and neural processes. Take, for example, choosing between two cafés to meet a visiting friend. Much work has focused on *model-free* control, according to which an organism will habitually pick the option that resulted in greater reward across repeated past experiences. Conversely, *model-based* control would allow one to consider the plausible outcomes of each choice, permitting more flexible goal-directed decisions. While both model-free and model-based control depend on many experiences with each café, *episodic memories* could support decisions about less familiar options. You could recall your friend saying that she loves tacos, as well as the details of a recent review you read about a great Mexican café in your neighborhood. The *relational structure* of these memories allows you to recall contextual details that become important later on and to integrate across experiences to draw new inferences. Lastly, these memories can be used in complex planning via hippocampal-mediated *prospection*, supporting the ability to deliberate and imagine the potential future outcomes of each choice.

successfully simulating oneself in the future would also depend on hippocampal function.

These findings highlight the constructive nature of episodic memory, along with the flexibility it provides, broadening the scope of decisions on which episodic memory might bear. Accordingly, episodic memory has been found to influence counterfactual reasoning (Schacter, Benoit, De Brigard, & Szpunar, 2015), divergent thinking (Madore, Addis, & Schacter, 2015), open-ended problem-solving (Sheldon, McAndrews, & Moscovitch, 2011), and emotional reappraisal (Jing, Madore, & Schacter, 2016). Hippocampal-mediated prospection can also bias how we value immediate versus delayed rewards. When choosing between an immediate and a delayed but larger reward, people often take the immediate option, discounting the delayed option according to the wait time. But this tendency is reduced when people are encouraged to imagine particular future events, an effect that has been linked to both hippocampal BOLD activity in healthy individuals (Peters & Büchel, 2010) and MTL damage in amnesiac patients (Palombo, Keane, & Verfaellie, 2015).

Conclusions and Summary

The human brain has multiple ways to make decisions and multiple parallel ways to learn from experience (figure 52.3). Given their conservation in the face of evolutionary pressures, it is reasonable to assume that each of these learning and memory systems guides actions in distinct and meaningful ways, bridging memory and decision-making (Sherry & Schacter, 1987). Dominant reinforcement-learning and behavioral control theories, however, have focused on memory systems that incrementally learn by averaging across experiences, whether it be to derive the value of cued actions in model-free reinforcement learning or the transition probabilities between states and outcomes in model-based reinforcement learning. These parametric forms of learning closely parallel procedural stimulus-response learning and semantic schemas, respectively. Parametric memory has clear benefits for action control—the pertinent information has already been extracted and integrated during learning, reducing storage requirements and simplifying the decision process. There is a cost, though. They are only acquired across many experiences, and in new, complex environments the relevant features to average over may not even be known.

Here, we highlight nonparametric hippocampal memory and describe several ways in which episodic and relational memory might resolve important challenges in decision-making research. Our review focused on key features of hippocampal memory pertinent to decision-making. First, hippocampal memories capture single experiences, rather than averaging across experiences. This property could enable choices in relatively new contexts while learning the rules governing them—one can simply recall the most similar experience and repeat the action if the resulting outcome is desirable. Second, we provide evidence that the relational nature of hippocampal memories further resolves ambiguity in complex environments by associating configurations of features with outcomes, negating the need to identify and select specific relevant features in advance. Further, the relational structure could bridge interrelated events, supporting novel decisions via inference and value transfer. Last, we discussed the emerging role of the hippocampus in prospection and creative actions, extending beyond simple preferences.

Studying how hippocampal memory guides decisions is a young pursuit, in need of many avenues of empirical support. In addition to rigorous tests of the ideas put forth here, important extensions involve the interaction between memory systems in the service of decision-making. How does the brain arbitrate between multiple and potentially conflicting sources of memories, and what factors determine which source will be used (Lee, O'Doherty, & Shimojo, 2015)? The answer to this question will have crucial implications fostering flexible behavior, as the dominant type of memory representation may ultimately determine whether actions reflect habits or goals. Additionally, progress in the study of episodic memory transformation (Moscovitch, Cabeza, Winocur, & Nadel, 2016) will be essential for understanding how the brain transforms idiosyncratic episodic memories into more efficient parametric forms of knowledge, such as the schemas, which presumably underlie model-based control.

REFERENCES

Adcock, R. A., Thangavel, A., Whitfield-Gabrieli, S., Knutson, B., & Gabrieli, J. D. (2006). Reward-motivated learning: Mesolimbic activation precedes memory formation. *Neuron, 50*(3), 507–517.

Andelman, F., Hoofien, D., Goldberg, I., Aizenstein, O., & Neufeld, M. Y. (2010). Bilateral hippocampal lesion and a selective impairment of the ability for mental time travel. *Neurocase, 16*(5), 426–435.

Barron, H. C., Dolan, R. J., & Behrens, T. E. (2013). Online evaluation of novel choices by simultaneous representation of multiple memories. *Nature Neuroscience, 16*(10), 1492.

Barto, A., Mirolli, M., & Baldassarre, G. (2013). Novelty or surprise? *Frontiers in Psychology, 4*, 907.

Benoit, R. G., & Schacter, D. L. (2015). Specifying the core network supporting episodic simulation and episodic memory by activation likelihood estimation. *Neuropsychologia, 75*, 450–457.

Blitzer, R. D., Gil, O., & Landau, E. M. (1990). Cholinergic stimulation enhances long-term potentiation in the CA1 region of rat hippocampus. *Neuroscience Letters, 119*(2), 207–210.

Bornstein, A. M., Khaw, M. W., Shohamy, D., & Daw, N. D. (2017). Reminders of past choices bias decisions for reward in humans. *Nature Communications, 8*, 15958.

Bornstein, A. M., & Norman, K. A. (2017). Reinstated episodic context guides sampling-based decisions for reward. *Nature Neuroscience, 20*(7), 997.

Brown, T. I., Carr, V. A., LaRocque, K. F., Favila, S. E., Gordon, A. M., Bowles, B., … Wagner, A. D. (2016). Prospective representation of navigational goals in the human hippocampus. *Science, 352*(6291), 1323–1326.

Buckner, R. L., & Carroll, D. C. (2007). Self-projection and the brain. *Trends in Cognitive Sciences, 11*(2), 49–57.

Cohen, M. D., & Bacdayan, P. (1994). Organizational routines are stored as procedural memory: Evidence from a laboratory study. *Organization Science, 5*(4), 554–568.

Davachi, L. (2006). Item, context and relational episodic encoding in humans. *Current Opinion in Neurobiology, 16*(6), 693–700.

Daw, N. D., Niv, Y., & Dayan, P. (2005). Uncertainty-based competition between prefrontal and dorsolateral striatal systems for behavioral control. *Nature Neuroscience, 8*(12), 1704–1711.

Daw, N. D., & O'Doherty, J. P. (2013). Multiple systems for value learning. In P. W. Glimcher & Ernst Fehr (Eds.), *Neuroeconomics: Decision making, and the brain* (2nd ed., pp. 393–410). New York: Elsevier.

Dickinson, A. (1985). Actions and habits: The development of behavioural autonomy. *Philosophical Transactions of the Royal Society of London B: Biological Sciences, 308*(1135), 67–78.

Dickinson, A., & Balleine, B. (2002). The role of learning in the operation of motivational systems. In C. R. Gallistel (Ed.), *Stevens' handbook of experimental psychology: Learning, motivation and emotion* (3rd ed., Vol. 3, pp. 497–534). New York: John Wiley & Sons.

Doll, B. B., Shohamy, D., & Daw, N. D. (2015). Multiple memory systems as substrates for multiple decision systems. *Neurobiology of Learning and Memory, 117*, 4–13.

Duncan, K., Doll, B. B., Daw, N. D., & Shohamy, D. (2018). More than the Sum of its parts: A role for the hippocampus in configural reinforcement learning. *Neuron, 98*(3), 645–657.

Duncan, K. D., Sadanand, A., & Davachi, L. (2012). Memory's penumbra: Episodic memory decisions induce lingering mnemonic biases. *Science, 337*(6093), 485–487.

Duncan, K. D., & Schlichting, M. L. (2018). Hippocampal representations as a function of time, subregion, and brain state. *Neurobiology of Learning and Memory, 153*(Pt A), 40–56.

Duncan, K. D., & Shohamy, D. (2016). Memory states influence value-based decisions. *Journal of Experimental Psychology: General, 145*(11), 1420.

Easton, A., Douchamps, V., Eacott, M., & Lever, C. (2012). A specific role for septohippocampal acetylcholine in memory? *Neuropsychologia, 50*(13), 3156–3168.

Eichenbaum, H. (2014). Time cells in the hippocampus: A new dimension for mapping memories. *Nature Reviews Neuroscience, 15*(11), 732.

Eichenbaum, H., & Cohen, N. J. (2001). *From conditioning to conscious recollection: Memory systems of the brain.* Oxford Psychology Series no. 35. New York: Oxford University Press.

Eichenbaum, H., & Cohen, N. J. (2004). *From conditioning to conscious recollection: Memory systems of the brain.* Oxford: Oxford University Press.

Eichenbaum, H., Otto, T., & Cohen, N. J. (1994). Two functional components of the hippocampal memory system. *Behavioral and Brain Sciences, 17*(3), 449–472.

Fiorillo, C. D., Tobler, P. N., & Schultz, W. (2003). Discrete coding of reward probability and uncertainty by dopamine neurons. *Science, 299*(5614), 1898–1902.

Foerde, K., & Shohamy, D. (2011). Feedback timing modulates brain systems for learning in humans. *Journal of Neuroscience, 31*(37), 13157–13167. doi:10.1523/JNEUROSCI.2701-11.2011

Frank, M. J. (2005). Dynamic dopamine modulation in the basal ganglia: A neurocomputational account of cognitive deficits in medicated and nonmedicated parkinsonism. *Journal of Cognitive Neuroscience, 17*(1), 51–72.

Frank, M. J., Seeberger, L. C., & O'Reilly, R. C. (2004). By carrot or by stick: Cognitive reinforcement learning in parkinsonism. *Science, 306*(5703), 1940–1943. doi:10.1126/science.1102941

Gabrieli, J. D. (1998). Cognitive neuroscience of human memory. *Annual Review of Psychology, 49*(1), 87–115.

Gershman, S. J., Blei, D. M., & Niv, Y. (2010). Context, learning, and extinction. *Psychological Review, 117*(1), 197.

Gershman, S. J., & Daw, N. D. (2017). Reinforcement learning and episodic memory in humans and animals: An integrative framework. *Annual Review of Psychology, 68*, 101–128.

Gilboa, A., & Marlatte, H. (2017). Neurobiology of schemas and schema-mediated memory. *Trends in Cognitive Sciences, 21*(8), 618–631.

Hare, T. A., Camerer, C. F., & Rangel, A. (2009). Self-control in decision-making involves modulation of the vmPFC valuation system. *Science, 324*(5927), 646–648.

Hassabis, D., Kumaran, D., & Maguire, E. A. (2007). Using imagination to understand the neural basis of episodic memory. *Journal of Neuroscience, 27*(52), 14365–14374.

Hassabis, D., Kumaran, D., Vann, S. D., & Maguire, E. A. (2007). Patients with hippocampal amnesia cannot imagine new experiences. *Proceedings of the National Academy of Sciences, 104*, 1726–1731.

Hasselmo, M. E., Wyble, B. P., & Wallenstein, G. V. (1996). Encoding and retrieval of episodic memories: Role of cholinergic and GABAergic modulation in the hippocampus. *Hippocampus, 6*(6), 693–708.

Heckers, S., Zalesak, M., Weiss, A. P., Ditman, T., & Titone, D. (2004). Hippocampal activation during transitive inference in humans. *Hippocampus, 14*(2), 153–162.

Houk, J. C., Adams, J. L., & Barto A. G. (1995). A model of how the basal ganglia generate and use neural signals that predict reinforcement. In J. C. Houk, J. L. Davis, & D. G. Beiser (Eds.), *Models of information processing in the basal ganglia* (1st ed., pp. 249–270). Cambridge, MA: MIT Press.

Huerta, P. T., & Lisman, J. E. (1995). Bidirectional synaptic plasticity induced by a single burst during cholinergic theta oscillation in CA1 in vitro. *Neuron, 15*(5), 1053–1063.

Insausti, R., & Amaral, D. G. (2012). Hippocampal formation. In J. K. Mai & G. Paxinos (Eds.), *The human nervous system* (3rd ed., pp. 896–942). New York: Elsevier.

Ito, R., Robbins, T. W., Pennartz, C. M., & Everitt, B. J. (2008). Functional interaction between the hippocampus and nucleus accumbens shell is necessary for the acquisition of appetitive spatial context conditioning. *Journal of Neuroscience, 28*(27), 6950–6959.

Izumi, Y., & Zorumski, C. F. (1999). Norepinephrine promotes long-term potentiation in the adult rat hippocampus in vitro. *Synapse, 31*(3), 196–202.

Jadhav, S. P., Kemere, C., German, P. W., & Frank, L. M. (2012). Awake hippocampal sharp-wave ripples support spatial memory. *Science, 336*(6087), 1454–1458.

Jing, H. G., Madore, K. P., & Schacter, D. L. (2016). Worrying about the future: An episodic specificity induction impacts problem solving, reappraisal, and well-being. *Journal of Experimental Psychology: General, 145*(4), 402.

Jones, J. L., et al. (2012). Orbitofrontal cortex supports behavior and learning using inferred but not cached values. *Science, 80*(338), 953–956.

Johnson, A., & Redish, A. D. (2007). Neural ensembles in CA3 transiently encode paths forward of the animal at a decision point. *Journal of Neuroscience, 27*(45), 12176–12189.

Knowlton, B. J., Mangels, J. A., & Squire, L. R. (1996). A neostriatal habit learning system in humans. *Science, 273*(5280), 1399–1402.

Knowlton, B. J., Squire, L. R., & Gluck, M. A. (1994). Probabilistic classification learning in amnesia. *Learning & Memory, 1*(2), 106–120.

Kumaran, D., Hassabis, D., Spiers, H. J., Vann, S. D., Vargha-Khadem, F., & Maguire, E. A. (2007). Impaired spatial and non-spatial configural learning in patients with hippocampal pathology. *Neuropsychologia, 45*(12), 2699–2711.

Kumaran, D., & McClelland, J. L. (2012). Generalization through the recurrent interaction of episodic memories: A model of the hippocampal system. *Psychological Review, 119*(3), 573.

Kurth-Nelson, Z., Barnes, G., Sejdinovic, D., Dolan, R., & Dayan, P. (2015). Temporal structure in associative retrieval. *eLife, 4*, e04919.

Lavenex, P., & Amaral, D. G. (2000). Hippocampal-neocortical interaction: A hierarchy of associativity. *Hippocampus, 10*(4), 420–430.

Lee, H., Ghim, J.-W., Kim, H., Lee, D., & Jung, M. (2012). Hippocampal neural correlates for values of experienced events. *Journal of Neuroscience, 32*(43), 15053–15065.

Lee, S. W., O'Doherty, J. P., & Shimojo, S. (2015). Neural computations mediating one-shot learning in the human brain. *PLoS Biology, 13*(4), e1002137.

Lemon, N., & Manahan-Vaughan, D. (2006). Dopamine D1/D5 receptors gate the acquisition of novel information through hippocampal long-term potentiation and long-term depression. *Journal of Neuroscience, 26*(29), 7723–7729.

Lengyel, M., & Dayan, P. (2008). *Hippocampal contributions to control: The third way.* Paper presented at the Advances in Neural Information Processing Systems conference, Vancouver, BC.

Li, S., Cullen, W. K., Anwyl, R., & Rowan, M. J. (2003). Dopamine-dependent facilitation of LTP induction in hippocampal CA1 by exposure to spatial novelty. *Nature Neuroscience, 6*(5), 526.

Lisman, J. E., & Grace, A. A. (2005). The hippocampal-VTA loop: Controlling the entry of information into long-term memory. *Neuron, 46*(5), 703–713.

Madore, K. P., Addis, D. R., & Schacter, D. L. (2015). Creativity and memory: Effects of an episodic-specificity induction on divergent thinking. *Psychological Science, 26*(9), 1461–1468.

Maia, T. V., & Frank, M. J. (2011). From reinforcement learning models to psychiatric and neurological disorders. *Nature Neuroscience, 14*(2), 154.

Mather, M., Clewett, D., Sakaki, M., & Harley, C. W. (2016). Norepinephrine ignites local hotspots of neuronal excitation: How arousal amplifies selectivity in perception and memory. *Behavioral and Brain Sciences, 39*(200), 1–75.

McClelland, J. L., McNaughton, B. L., & O'Reilly, R. C. (1995). Why there are complementary learning systems in the hippocampus and neocortex: Insights from the successes and failures of connectionist models of learning and memory. *Psychological Review, 102*(3), 419.

McClure, S. M., Berns, G. S., & Montague, P. R. (2003). Temporal prediction errors in a passive learning task activate human striatum. *Neuron, 38*(2), 339–346.

Meeter, M., Murre, J., & Talamini, L. (2004). Mode shifting between storage and recall based on novelty detection in oscillating hippocampal circuits. *Hippocampus, 14*(6), 722–741.

Melchers, K. G., Shanks, D. R., & Lachnit, H. (2008). Stimulus coding in human associative learning: Flexible representations of parts and wholes. *Behavioural Processes, 77*(3), 413–427.

Montague, P. R., Dayan, P., & Sejnowski, T. J. (1996). A framework for mesencephalic dopamine systems based on predictive Hebbian learning. *Journal of Neuroscience, 16*(5), 1936–1947.

Moscovitch, M., Cabeza, R., Winocur, G., & Nadel, L. (2016). Episodic memory and beyond: The hippocampus and neocortex in transformation. *Annual Review of Psychology, 67*, 105–134.

Murty, V. P., FeldmanHall, O., Hunter, L. E., Phelps, E. A., & Davachi, L. (2016). Episodic memories predict adaptive value-based decision-making. *Journal of Experimental Psychology: General, 145*(5), 548.

Murty, V. P., LaBar, K. S., & Adcock, R. A. (2012). Threat of punishment motivates memory encoding via amygdala, not midbrain, interactions with the medial temporal lobe. *Journal of Neuroscience, 32*(26), 8969–8976.

Niv, Y., Daniel, R., Geana, A., Gershman, S. J., Leong, Y. C., Radulescu, A., & Wilson, R. C. (2015). Reinforcement learning in multidimensional environments relies on attention mechanisms. *Journal of Neuroscience, 35*(21), 8145–8157.

O'Doherty, J. P., Dayan, P., Friston, K., Critchley, H., & Dolan, R. J. (2003). Temporal difference models and reward-related learning in the human brain. *Neuron, 38*(2), 329–337.

O'Doherty, J. P., Dayan, P., Schultz, J., Deichmann, R., Friston, K., & Dolan, R. J. (2004). Dissociable roles of ventral and dorsal striatum in instrumental conditioning. *Science, 304*(5669), 452–454. doi:10.1126/science.1094285

O'Keefe, J., & Nadel, L. (1978). *The hippocampus as a cognitive map.* Oxford: Clarendon Press.

Palombo, D. J., Keane, M. M., & Verfaellie, M. (2015). The medial temporal lobes are critical for reward-based decision making under conditions that promote episodic future thinking. *Hippocampus, 25*(3), 345–353.

Patil, A., & Duncan, K. (2018). Lingering cognitive states shape fundamental mnemonic abilities. *Psychological Science, 29*(1), 45–55.

Pessiglione, M., Petrovic, P., Daunizeau, J., Palminteri, S., Dolan, R. J., & Frith, C. D. (2008). Subliminal instrumental conditioning demonstrated in the human brain. *Neuron, 59*(4), 561–567.

Pessiglione, M., Seymour, B., Flandin, G., Dolan, R. J., & Frith, C. D. (2006). Dopamine-dependent prediction errors underpin reward-seeking behaviour in humans. *Nature, 442*(7106), 1042–1045.

Peters, J., & Büchel, C. (2010). Episodic future thinking reduces reward delay discounting through an enhancement of prefrontal-mediotemporal interactions. *Neuron, 66*(1), 138–148.

Pfeiffer, B. E., & Foster, D. J. (2013). Hippocampal place-cell sequences depict future paths to remembered goals. *Nature, 497*(7447), 74.

Poldrack, R. A., & Packard, M. G. (2003). Competition among multiple memory systems: Converging evidence from animal and human brain studies. *Neuropsychologia, 41*(3), 245–251.

Preston, A. R., & Eichenbaum, H. (2013). Interplay of hippocampus and prefrontal cortex in memory. *Current Biology, 23*(17), R764–R773.

Preston, A. R., Shrager, Y., Dudukovic, N. M., & Gabrieli, J. D. (2004). Hippocampal contribution to the novel use of relational information in declarative memory. *Hippocampus, 14*(2), 148–152.

Race, E., Keane, M. M., & Verfaellie, M. (2011). Medial temporal lobe damage causes deficits in episodic memory and episodic future thinking not attributable to deficits in narrative construction. *Journal of Neuroscience, 31*(28), 10262–10269.

Reddy, L., & Kanwisher, N. (2006). Coding of visual objects in the ventral stream. *Current Opinion in Neurobiology, 16*(4), 408–414.

Ruivo, L. M. T.-G., Baker, K. L., Conway, M. W., Kinsley, P. J., Gilmour, G., Phillips, K. G., … Mellor, J. R. (2017). Coordinated acetylcholine release in prefrontal cortex and hippocampus is associated with arousal and reward on distinct timescales. *Cell Reports, 18*(4), 905–917.

Santoro, A., Frankland, P. W., & Richards, B. A. (2016). Memory transformation enhances reinforcement learning in dynamic environments. *Journal of Neuroscience, 36*(48), 12228–12242.

Schacter, D. L., Addis, D. R., & Buckner, R. L. (2007). Remembering the past to imagine the future: The prospective brain. *Nature Reviews Neuroscience, 8*(9), 657.

Schacter, D. L., Benoit, R. G., De Brigard, F., & Szpunar, K. K. (2015). Episodic future thinking and episodic counterfactual thinking: Intersections between memory and decisions. *Neurobiology of Learning and Memory, 117*, 14–21.

Schlichting, M. L., Zeithamova, D., & Preston, A. R. (2014). CA1 subfield contributions to memory integration and inference. *Hippocampus, 24*(10), 1248–1260.

Schmidt, L., Braun, E. K., Wager, T. D., & Shohamy, D. (2014). Mind matters: Placebo enhances reward learning in Parkinson's disease. *Nature Neuroscience, 17*(12), 1793–1797.

Schonberg, T., O'Doherty, J., Joel, D., Inzelberg, R., Segev, Y., & Daw, N. (2010). Selective impairment of prediction error signaling in human dorsolateral but not ventral striatum in Parkinson's disease patients: Evidence from a model-based fMRI study. *NeuroImage, 49*(1), 772–781. doi:S1053-8119(09)00873-8 [pii] 10.1016/j.neuroimage.2009.08.011

Schultz, W. (1992). Activity of dopamine neurons in the behaving primate. *Seminars in Neuroscience, 4*, 129–138.

Schultz, W., Dayan, P., & Montague, P. R. (1997). A neural substrate of prediction and reward. *Science, 275*(5306), 1593–1599.

Sheldon, S., McAndrews, M. P., & Moscovitch, M. (2011). Episodic memory processes mediated by the medial temporal lobes contribute to open-ended problem solving. *Neuropsychologia, 49*(9), 2439–2447.

Sherry, D. F., & Schacter, D. L. (1987). The evolution of multiple memory systems. *Psychological Review, 94*(4), 439.

Shohamy, D., & Adcock, R. A. (2010). Dopamine and adaptive memory. *Trends in Cognitive Sciences, 14*(10), 464–472.

Shohamy, D., Myers, C. E., Grossman, S., Sage, J., Gluck, M. A., & Poldrack, R. A. (2004). Cortico-striatal contributions to feedback-based learning: Converging data from neuroimaging and neuropsychology. *Brain, 127*(Pt. 4), 851–859. doi:10.1093/brain/awh100

Shohamy, D., & Turk-Browne, N. B. (2013). Mechanisms for widespread hippocampal involvement in cognition. *Journal of Experimental Psychology: General, 142*(4), 1159.

Singer, A. C., Carr, M. F., Karlsson, M. P., & Frank, L. M. (2013). Hippocampal SWR activity predicts correct decisions during the initial learning of an alternation task. *Neuron, 77*(6), 1163–1173.

Squire, L. R., & Dede, A. J. (2015). Conscious and unconscious memory systems. *Cold Spring Harbor Perspectives in Biology, 7*(3), a021667.

Squire, L. R., & Zola, S. M. (1996). Structure and function of declarative and nondeclarative memory systems. *Proceedings of the National Academy of Sciences, 93*(24), 13515–13522.

Stanton, P. K., & Sarvey, J. M. (1985). Depletion of norepinephrine, but not serotonin, reduces long-term potentiation in the dentate gyrus of rat hippocampal slices. *Journal of Neuroscience, 5*(8), 2169–2176.

Sutherland, R. J., & Rudy, J. W. (1989). Configural association theory: The role of the hippocampal formation in learning, memory, and amnesia. *Psychobiology, 17*(2), 129–144.

Sutton, R. S., & Barto, A. G. (1998). *Reinforcement learning: An introduction* (Vol. 1). Cambridge, MA: MIT Press.

Szpunar, K. K., Spreng, R. N., & Schacter, D. L. (2014). A taxonomy of prospection: Introducing an organizational framework for future-oriented cognition. *Proceedings of the National Academy of Sciences, 111*(52), 18414–18421.

Tolman, E. C. (1948). Cognitive maps in rats and men. *Psychological Review, 55*(4), 189–208. doi:10.1037/h0061626

Tulving, E. (1972). Episodic and semantic memory. *Organization of Memory, 1*, 381–403.

Van Essen, D. C., Anderson, C. H., & Felleman, D. J. (1992). Information processing in the primate visual system: An integrated systems perspective. *Science, 255*(5043), 419–423.

Vertes, R. P., Hoover, W. B., Do Valle, A. C., Sherman, A., & Rodriguez, J. (2006). Efferent projections of reuniens and rhomboid nuclei of the thalamus in the rat. *Journal of Comparative Neurology, 499*(5), 768–796.

Voss, J. L., Gonsalves, B. D., Federmeier, K. D., Tranel, D., & Cohen, N. J. (2011). Hippocampal brain-network coordination during volitional exploratory behavior enhances learning. *Nature Neuroscience, 14*(1), 115.

Wimmer, G. E., & Buechel, C. (2016). Reactivation of reward-related patterns from single past episodes supports memory-based decision making. *Journal of Neuroscience, 36*(10), 2868–2880.

Wimmer, G. E., & Shohamy, D. (2012). Preference by association: How memory mechanisms in the hippocampus bias decisions. *Science, 338*(6104), 270–273.

Zeithamova, D., Dominick, A. L., & Preston, A. R. (2012). Hippocampal and ventral medial prefrontal activation during retrieval-mediated learning supports novel inference. *Neuron, 75*(1), 168–179.

Zeithamova, D., Schlichting, M. L., & Preston, A. R. (2012). The hippocampus and inferential reasoning: Building memories to navigate future decisions. *Frontiers in Human Neuroscience, 6*, 70.

53 The Role of the Primate Amygdala in Reward and Decision-Making

FABIAN GRABENHORST, C. DANIEL SALZMAN, AND WOLFRAM SCHULTZ

ABSTRACT Rewards influence learning, attention, decision-making, emotion, and behavior. Long implicated in aversive processing, the amygdala is now recognized as a key component of the neural systems that process rewards. During reinforcement learning, distinct amygdala neurons encode positive and negative stimulus values in close correspondence with behavior. Amygdala neurons signal value across sequential presentations of different stimuli, representing global *state value*, a key concept in reinforcement-learning theory. Value representations in the amygdala are sensitive to parameters critical for learning, including reward contingency, relative reward quantity, and temporal reward structure. Amygdala reward signals are well suited to support economic decision-making. Recent data show that during reward-based decisions, amygdala neurons encode both the value inputs and the corresponding choice outputs of economic decision processes. Over sequential choices, amygdala "planning activities" signal internally set reward goals and progress toward obtaining these goals, thus reflecting the internal cognitive state. Consistent with this, amygdala neurons can encode the abstract conceptual information (*task sets*) needed to assess the value of upcoming stimuli and the spatial information in the service of allocating attention toward rewarding stimuli. Collectively, the amygdala's elaborate cognitive, reward, and decision signals provide a neuronal foundation for guiding primates' sophisticated behavioral repertoire toward the acquisition of the best rewards.

The amygdala, a nuclear complex in the anterior-medial temporal lobe, participates in a diversity of functions, including emotion, learning, memory, and reward-guided behavior. The amygdala receives inputs from all sensory systems, the prefrontal cortex, the hippocampus, and the rhinal cortices and typically returns these projections; additional outputs target the striatum, hypothalamus, midbrain, and brain stem (Amaral & Price, 1984; McDonald, 1998). These connections predispose the amygdala to link information about sensory stimuli with emotional and behavioral responses. Early lesion studies showed that amygdala damage in primates alters reinforcement-guided behaviors (Weiskrantz, 1956). Subsequent classical work in rodents established the amygdala as a critical structure for fear conditioning and revealed the underlying cellular and molecular mechanisms (LeDoux, 2000; Maren & Quirk, 2004). Human studies confirmed and elaborated the amygdala's role in emotion (Adolphs, 2013; Seymour & Dolan, 2008; Phelps & LeDoux, 2005). Recent reviews provide perspectives on amygdala functions in rodents (Janak & Tye, 2015; Krabbe, Grundemann, & Luthi, 2017) and humans (Adolphs, 2013; Rutishauser, Mamelak, & Adolphs, 2015; Seymour & Dolan, 2008). This chapter reviews the neuronal processes that mediate the more recently acknowledged functions of the primate amygdala in reward processing, reinforcement learning, and decision-making, focusing on the nature of the neural representations that mediate these functions.

Lesion studies in monkeys, as well as neuroimaging studies in humans, have provided evidence that the amygdala is involved not only in fear but also in reward processing (Amaral, 2016; Gottfried, O'Doherty, & Dolan, 2003; Grabenhorst et al., 2010; Murray & Rudebeck, 2013). Early neurophysiological investigations of the primate amygdala described neuronal responses to visual and other sensory stimuli, some of which were related to reinforcement (Nishijo, Ono, & Nishino, 1988; Rolls, 2000). However, it largely remained unclear if amygdala neural response properties were specifically related to either rewarding or aversive events. More recent studies reviewed here demonstrated that primate amygdala neurons preferentially represent either the positive or negative value of visual stimuli during learning (Belova et al., 2007, Belova, Paton, & Salzman, 2008; Paton et al., 2006). Studies in rodents have since established that distinct neural ensembles in the amygdala process appetitive and aversive information and that activity in these ensembles is causally related to valence-specific innate and learned emotional behavior (Gore et al., 2015; Redondo et al., 2014). Building on these studies, we will discuss recent advances in understanding the nature of neural representations of reward-related variables in the primate amygdala. Reward-related variables contribute to many different types of functions, including learning, attention, decision-making, and social behavior, all of which involve the amygdala.

Reinforcement Learning

In any given moment, a subject's current situation is defined by a set of internal and external variables. These variables include internal cognitive variables (e.g., conceptual knowledge or beliefs, plans, memories of recent events, and more) and internal physiological variables (e.g., thirst, hunger, physical pain), as well as external variables (e.g., the stimuli present and the stimuli recently experienced). Together these sets of variables define a subject's situation and are referred to in theories of reinforcement learning as a subject's *state* (Salzman & Fusi, 2010; Sutton & Barto, 1998). In any given state, a subject has a predisposition to act, where actions can be internal (e.g., cognitive or psychophysiological) or external (e.g., a physical action reflecting a decision; Salzman & Fusi, 2010). A central tenet of theories of reinforcement learning is that during learning, subjects assign values to states (Sutton & Barto, 1998). This process of updating the assignment of values to states is integral to decision-making since optimal decisions serve to maximize the value of a subject's state. Historically, the amygdala has been conceptualized as providing a neural substrate for linking neural representations of previously neutral conditioned stimuli (CSs) with motivationally significant unconditioned stimuli (USs). However, recent experiments show that amygdala neurons—across the population—do not merely represent values of CSs or USs but instead appear to modulate their activity in relation to manipulations of state value, which can be induced by a variety of experimental manipulations. In experimental settings, investigators have most commonly designed experiments inspired by animal-learning theory. Here the values assigned to states (*state values*) are manipulated by presenting to subjects conditioned (predictors) and unconditioned stimuli that have rewarding or aversive qualities. For example, the value of a state can be manipulated during experiments utilizing Pavlovian conditioning. This provides a means for previously neutral visual stimuli to induce a positive or negative state value through their association with rewarding and aversive unconditioned stimuli. The notion that the amygdala represents positive and negative state value was first suggested by experiments in which single neurons in primate amygdalae were recorded during a reversal learning task (Belova et al., 2007, 2008; Paton et al., 2006). In this task, monkeys learned that novel abstract images (conditioned stimuli, CSs) were linked to either positive or negative value through associations with rewarding or aversive unconditioned stimuli (USs). After learning, CS presentations instantiated a positive or negative state. Contingencies between CSs and USs were then reversed in order to determine if neural activity was related to the sensory properties of the CSs or to the value of states instantiated by CS presentation. Neural responses to CSs changed upon reversals to reflect the change in state value, and these changes in activity occurred fast enough to account for changing approach and defensive behaviors that reflected learning.

The reversal-learning task contained more states than those instantiated by CS presentations, as two other types of stimuli were presented during the experiment: a fixation point (FP) that appeared at the beginning of each trial and US presentations that appeared at the end of each trial. The FP induced a mildly positive state to monkeys because monkeys chose to foveate it to initiate trials. Regardless of whether presentations of FP, CSs, or USs caused state transitions, different populations of amygdala neurons tracked the positive or negative value of the current state (figure 53.1*A–D*; Belova et al., 2008). Positive value-coding neurons increase firing rates for positive states and negative value-coding neurons do the opposite. A recent study (Munuera, Rigotti, & Salzman, 2018) showed that amygdala value-coding neurons can also respond to social information (figure 53.1*E*), as described in more detail below. Subsequent studies have further demonstrated how amygdala neurons are sensitive to changes in state value by manipulating the rewards associated with other interleaved CSs during a contrast revaluation procedure (Saez et al., 2017). The role of the amygdala in representing the value of states helps explain how the amygdala can coordinate a range of physiological and behavioral responses constitutive of emotional behavior.

State Variables

Theories of reinforcement learning provide elegant algorithms that explain how values may be assigned to states and updated, but these theories do not provide an account for how representations of the states themselves are represented. Two recent studies demonstrate how the amygdala participates in the representation of state-related variables and how these representations may then be linked to reward-related variables to guide cognitive behaviors. In one study, amygdala neurons encoded information about the spatial location and reward associations of visual cues (Peck, Lau, & Salzman, 2013). Furthermore, fluctuating amygdala neural responses to these cues were correlated with trial-to-trial variability in behavioral measures of spatial attention in monkeys performing demanding visual tasks. Thus, the amygdala integrates spatial and motivational information, two different types of state variables, and this representation helps account for the allocation of a

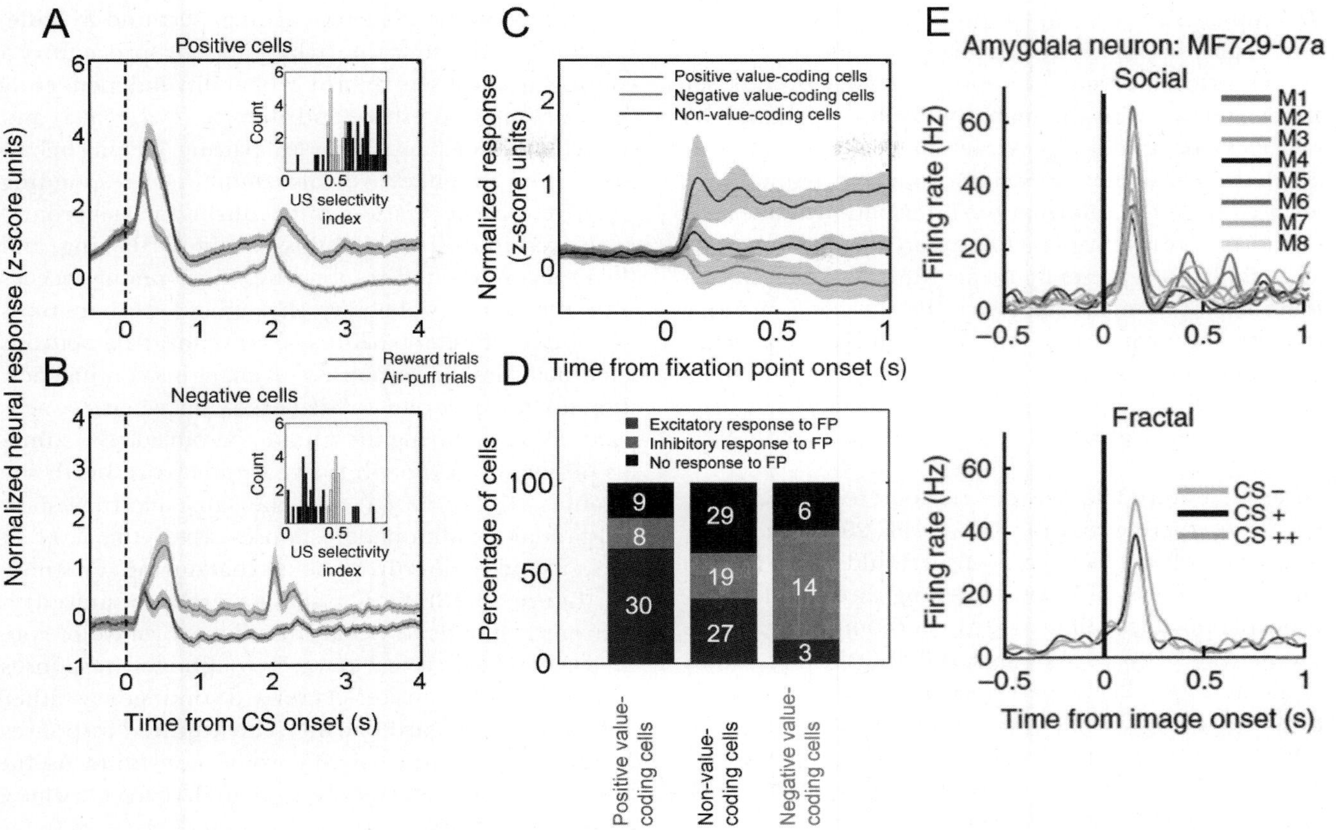

FIGURE 53.1 The amygdala represents the positive and negative value of stimuli. *A*, *B*, Normalized and averaged neural responses plotted as a function of time relative to CS onset for neurons in the amygdala that respond more strongly to CSs associated with rewards (*A*) or air puffs (*B*). *Blue traces*, Responses in rewarded trials; *red traces*, responses in trials in which subjects received an aversive air puff. *Inset histograms*, Selectivity index characterizing the preference for expected reward or air puff, where values > 0.5 indicate preference for reward and < 0.5 preference for the expected air puff. *C*, Normalized and averaged neural responses to the fixation point presentation for neurons classified according to their selectivity for the valence of CSs (*blue*, cells responding strongest to CSs predicting reward; *red*, cells that responded strongest to CSs predicting air puff; *black*, cells without valence selectivity). *D*, The percentage of cells that responded with increased (*blue*), decreased (*red*), or no change (*black*) activity upon fixation presentation classified according to the preference of the cell for CSs of different valence. *E*, Average activity of a single amygdala neuron recorded during passive viewing of either social images (images of faces of a monkey's social cohort, *top*) or nonsocial images that predicted different reward amounts (*bottom*). Monkeys within the social cohort formed a stable social hierarchy. Neural responses to the face images encoded the hierarchical rank of viewed monkey faces, with the strongest response observed when viewing the most dominant monkey (M1) and weaker responses the more submissive the viewed monkey (e.g., M8). The same neuron responded most strongly to a fractal image that predicted a large reward compared to smaller rewards. Since social rank is a motivationally significant parameter, these data indicate that this amygdala neuron signaled information about the value of both social and nonsocial stimuli. (See color plate 58.)

subsequent cognitive action—spatial attention toward emotionally relevant stimuli.

A second recent study (Saez et al., 2015) was built on the observation that the same sensory stimulus (or CS) can produce different emotional and behavioral responses depending upon the situation (or state). Traditionally, changing the motivational significance of a CS has often been accomplished by using different contextual cues to dictate how to interpret the reinforcement predicted by a CS. However, contextual information is often defined abstractly (e.g., consider different types of social situations or games with different rules for how to interpret playing cards). Thus, conceptual information alone is sufficient to define an "abstract" context that guides how to predict reward upon a CS presentation. By using a serial-reversal learning task, it was demonstrated that monkeys develop conceptual knowledge of task sets, the sets of CS-US mappings that occur in a block of trials. These task sets defined abstract contexts, and the same CS instantiated different state values in each context. Neural responses in the amygdala not only reflected state values that were updated rapidly and appropriately but also represented abstract conceptual information about task sets. A failure to maintain the representation of a task set was correlated with the inability to adjust assignments of state value upon CS presentation. This study emphasized, therefore, the critical importance of maintaining abstract cognitive information, information about state variables, in the amygdala to support flexible behavior. This ability to represent abstract information is consistent with the demonstration that the amygdala provides cognitive information related to planning sequential decisions (see below).

Reward Contingency

Given that amygdala neurons respond to conditioned, reward-predicting stimuli (Paton et al., 2006), the question arises whether such responses would follow fundamental principles of animal learning theory. Standard concepts postulate that reward conditioning relies on the pairing between a reward and an arbitrary stimulus. This is what Pavlov describes. However, later assessments of conditioning reveal that pairing is not the crucial aspect; even perfect, repeated, well-experienced pairing will not lead to behavioral conditioning or to conditioned neuronal responses. What is required is specific information about the reward, which can be acquired when the reward depends, or is contingent, on the stimulus. This occurs when more, or less, reward occurs during the stimulus compared to its absence. Only in this situation does the stimulus carry specific reward information. Of course, when the reward occurs, it becomes inadvertently paired with the stimulus—hence, the notion of stimulus-reward pairing. By contrast, when a reward occurs both without and during a stimulus, the stimulus is still well paired with the reward, but importantly, the stimulus carries no specific information about the reward; whether the stimulus occurs or not, the reward is the same. In other words, the value of the state is the same irrespective of the stimulus occurring. Carefully controlled experimental tests confirm this concept: the "truly random" procedure shows that noncontingent reinforcer occurrence does not lead to conditioning despite pairing (Rescorla, 1967). A stimulus that is paired with a noncontingent reward does not become conditioned (Delamater, 1995). Thus, the crucial variable explaining learning is contingency, not the inadvertent pairing.

These theoretical notions demand to ask whether conditioned reward processing in amygdala neurons follows reward contingency rather than reward pairing. Lesions in the amygdala make rats insensitive to background reward changes, thus suggesting a general role in contingency-dependent learning (Ostlund & Balleine, 2008). To follow the "truly random" procedure, a contingency test sets reward probability independently during stimulus presence and absence (*background*) and thus controls for stimulus-reward pairing. Licking behavior in rhesus monkeys reveals conditioned responding when reward occurs less often during a background compared to a visual stimulus but similar licking, and thus no conditioned responding, when reward occurs with the same probability throughout both periods (Bermudez & Schultz, 2010a). Many amygdala neurons alter their stimulus response with changing contingency; they do not respond to visual stimuli when rewards occur as often during the stimulus as outside the stimulus period, even though the reward is paired with the stimulus (figure 53.2*A*), but the same neurons show substantial conditioned responses when the level of background reward drops below that during the stimulus (figure 53.2*B*). Only a smaller group of amygdala neurons responds to visual stimuli irrespective of contingency. Thus, typical reward-conditioned responses in amygdala neurons reflect reward contingency rather than stimulus-reward pairing. The acquired responses to reward-predicting stimuli would constitute useful inputs to the neuronal mechanisms underlying informed economic choices.

Reward Amount

Given that amygdala neurons learn to respond to rewarding events (Paton et al., 2006) and that these responses reflect contingency (Bermudez & Schultz, 2010a), a participation in economic decision processes would require the graded coding of reward value. Reward amount is probably the most fundamental variable that determines reward value. Measurable choices reveal reward value usually on a subjective scale, often as a positive monotonic function of physical amount (as long as satiety is avoided); thus, subjective value reflects the reward amount in a ranked, ordinal fashion: more reward is better, and larger reward amounts are more frequently chosen than smaller amounts. This scaling is reflected in the responses of amygdala neurons to reward reception (Bermudez & Schultz, 2010b): they increase or, in separate neurons, decrease monotonically with more reward (figure 53.2*C*). Importantly, choices are made relative to what is on offer, and decision-makers tend to select the best reward within a given distribution of rewards, irrespective of the real, physical amount. A reward that is the larger of two rewards and, accordingly, is chosen more frequently than its alternative, may be the smallest when the originally smaller reward is replaced by a much larger one;

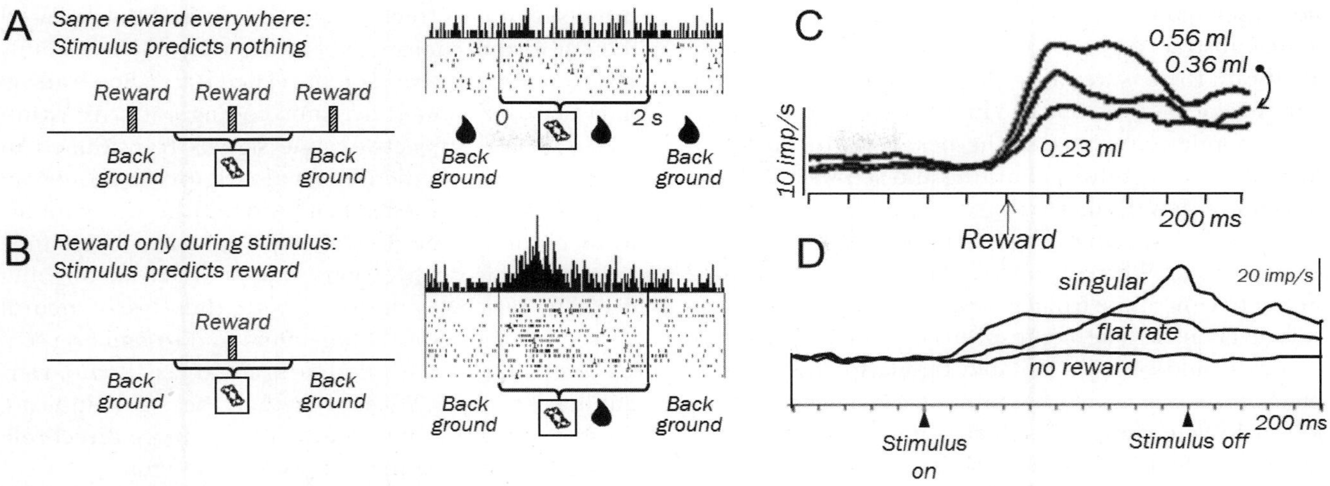

FIGURE 53.2 Signaling of fundamental reward parameters in the amygdala: contingency, amount, and timing. *A*, *B*, Contingency: animal-learning theory states that the important variable for acquiring reward prediction is contingency, not stimulus-reward pairing. *A*, No response in a single amygdala neuron to a fractal stimulus when the same reward also occurs with the same frequency in the absence of the stimulus (*background*). Despite being paired with the reward, this stimulus is not differentially informative about reward. *B*, Response in the same neuron to the same stimulus with less background reward. Only in this situation is the reward contingent on the stimulus; the stimulus predicts the reward. *C*, Reward value coding in a single amygdala neuron. Reward response increases monotonically with the liquid amount. *D*, Reward timing in a population of 86 amygdala neurons. Different temporal profiles of reward expectation-related activity reflect different instantaneous reward probabilities (indicated by three different stimuli: *top*, singular reward at stimulus end; *middle*, flat reward rate during stimulus, rewarded trials excluded from analysis; *bottom*, no reward).

that same first reward will now be chosen less frequently than the large reward, even though it has not changed physically. To be involved in economic choices, reward amount responses in at least some amygdala neurons should vary systematically depending on the other reward on offer. This is indeed the case in a good fraction of reward-sensitive amygdala neurons, both for responses to reward itself and for responses to reward-predicting stimuli. Responses to the smallest reward remain smallest and to the largest reward remain largest, irrespective of the physical reward amount (Bermudez & Schultz, 2010b). Responses to reward-predicting stimuli show the same phenomenon, being smaller when the reward is the smaller of the two but larger when the same reward is the largest (Saez et al., 2017). All these responses reflect the rank order of the rewards rather than their physical amounts. Such neurons faithfully identify by their response which reward is the best in any changing distribution of rewards. This quantitative reward code would allow amygdala neurons to participate in economic decisions.

Reward Timing

The predicted time of future reward has fundamental importance in behavior, as evidenced in the temporal difference model of reinforcement learning (Sutton & Barto, 1981) and the ramping activity in drift-diffusion and race models in decision-making (Stone, 1960; Vickers, 1970). Thus, for amygdala neurons to participate in these processes, they should have access to information about reward timing. When distinct visual stimuli predict particular instantaneous reward rates, the preparatory licking of monkeys suggests that animals have specific temporal reward expectations (Bermudez, Göbel, & Schultz, 2012). The activity in a group of amygdala neurons reflects these expectations; when a reward is delivered at a fixed time point at the end of the stimulus, the activity of these neurons ramps up gradually to stimulus end; however, when the same reward amount is spread out equally over the whole stimulus, their activity shows a smaller, tonic increase (figure 53.2*D*). Amygdala neurons also show time-sensitive responses to the reward itself. In one case, responses increase with increasing instantaneous reward rate, suggesting a positive, confirmatory relationship to expected reward reception. In the other, opposite, case, responses increase with decreasing instantaneous reward probability, which may reflect a surprise in reward occurrence or even positive reward prediction error and confirms responses seen earlier (Belova et al., 2007). These different forms of sensitivity to temporal reward structure would allow amygdala neurons to play an active role in the timing processes underlying learning and decision-making.

Economic Decisions

To obtain the best rewards, animals must not only evaluate sensory stimuli but also make choices among competing options. Moreover, the best rewards are often distant, which requires planning and stepwise, sequential choices toward internally set goals. These internally set goals are variables that in part define a subject's current state. Primate amygdala neurons seem well suited to contribute to such reward-based decision processes. Their flexible, context-sensitive value signals (see above) would provide suitable inputs to the decision mechanisms operating by winner-take-all competition. Reward-related responses during multistep behavioral schedules (Sugase-Miyamoto & Richmond, 2005) also suggest amygdala contributions to sequential reward pursuit. Traditionally, the amygdala's role in decision-making was seen as restricted to the evaluation of choice options and outcomes. However, recent data suggest that primate amygdala neurons contribute more directly to decision-making by encoding not only the value inputs but also the choice outputs of economic decision processes.

A series of studies examined primate amygdala neurons in a sequential economic decision-making task (Grabenhorst, Hernadi, & Schultz, 2012, 2016; Hernadi, Grabenhorst, & Schultz, 2015). In this "save-spend task," monkeys make consecutive choices to "save" (i.e., accumulate) liquid rewards for future trials until deciding to "spend" (i.e., consume) the saved amount. Reward amounts increased over consecutive save choices according to a variable, cued "interest rate." Higher interest rates resulted in exponential reward growths; accordingly, monkeys produced longer saving sequences when interest rates were high. A critical task feature was that by knowing the current interest rate, the animals could plan to obtain specific reward amounts through saving sequences of given lengths. This design allowed the animals to plan their choices over multiple trials and anticipate final rewards over 100 s in advance. Behavioral tests confirmed that the animals successfully tracked saved reward amounts and anticipated final rewards, consistent with internally planned behavior.

A significant number of amygdala neurons in this task predicted the monkeys' forthcoming trial-by-trial save-spend choices (Grabenhorst, Hernadi, & Schultz, 2012). These choice-predictive activities (figure 53.3*A*) neither explained left-right actions nor reflected the subjective values of save-spend choices or chosen value. Rather, amygdala neurons signaled the animal's abstract (action-independent) economic choice to spend a reward for immediate consumption or to save it for future trials. Choice-predictive activities were often specific to freely made choices and disappeared during forced choices, suggesting they did not simply reflect reward expectation or related states. Some amygdala neurons showed dynamic coding patterns within trials, in that subjective value signals transitioned to explicit choice-predictive signals (figure 53.3*A, dashed line*); these neurons explicitly linked the value inputs for decision-making to a choice output. Such dynamic value-to-choice conversions could reflect an ongoing decision process, consistent with theories of neural choice computation (Wang, 2002) and resembling sensory decision signals in other brain areas (Romo, Hernandez, & Zainos, 2004). Together, these data indicate that primate amygdala neurons play a more direct role in decision-making than previously thought.

Planning and Progress Tracking for Distant Rewards

Amygdala neurons exhibited additional, sophisticated decision activities that were critical for optimal performance in the save-spend task (Grabenhorst, Hernadi, & Schultz, 2016; Hernadi, Grabenhorst, & Schultz, 2015). One set of neurons showed prospective planning activities that signaled features of the animal's internally planned choice sequence (Hernadi, Grabenhorst, & Schultz, 2015). Specifically, neurons encoding *sequence value* signaled the subjective value of the current saving sequence (figure 53.3*B, solid line*), whereas neurons encoding *sequence length* directly signaled the number of forthcoming save choices (figure 53.3*B, dashed curve*). Sequence values were related to the animals' preferences for different choice sequences, reflecting both the benefits of specific rewards and the costs due to delay and effort. Importantly, sequence value was a nonmonotonic function of sequence length: depending on the interest rate, sequence value was highest for intermediate sequence lengths (figure 53.3*B, black bars*) that offered a compromise between large rewards (figure 53.3*B, thick curve*) and moderate delays. Notably, planning activities often disappeared during instructed behavior, despite comparable reward timing and anticipation, and in control analyses were unrelated to reward proximity and expectation. These results and the distinct activities related to sequence value or sequence length suggest that amygdala planning activities did not simply reflect motivation or state valuation. A recent imaging study translated the save-spend task to human economic saving behavior (Zangemeister, Grabenhorst, & Schultz, 2016) and identified corresponding activities in human amygdalae. Such reward-based planning activities in amygdala neurons could serve in the guidance of behavior toward internal goals.

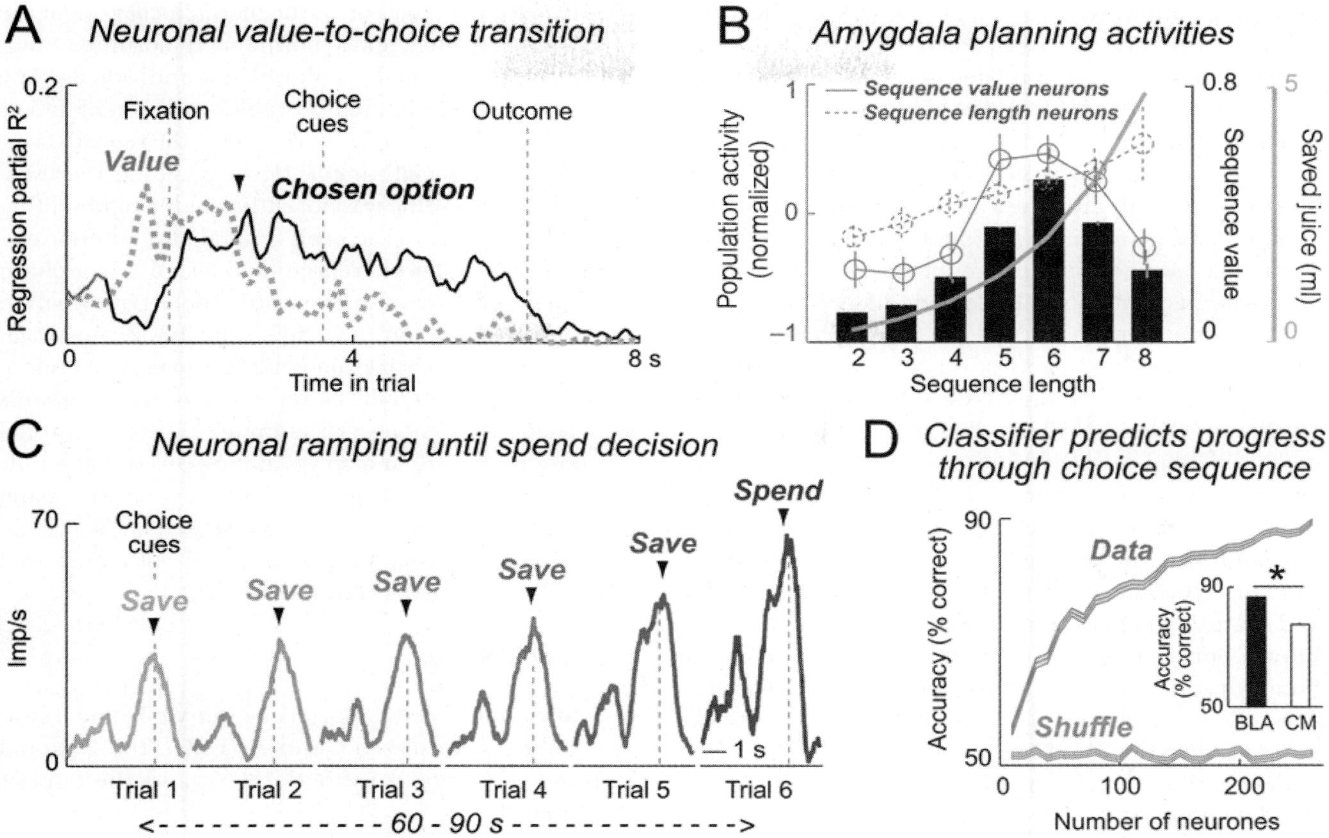

FIGURE 53.3 Signaling of economic decision processes in amygdala neurons. *A*, A single amygdala neuron dynamically coding value input (*dashed line*) and choice output (*solid*) of an economic decision process. Monkeys chose to save the liquid reward for later or spend (consume) in the current trial; they could plan save-spend choices (but not left-right actions) before choice cues. The neuron's activity transitioned from signaling value to predicting the forthcoming save-spend choice (partial R^2, sliding-window regression with value and choice regressors; *arrowhead*, onset of significant choice signal). *B*, Amygdala neurons with planning activity for distant rewards. Neurons signaled the length of the planned choice sequence (*dashed line*, population activity) or its subjective value (*solid line*). Value activity was highest during sequences lasting six trials, which had the highest subjective value (*black bars*)—that is, they were preferred by the animals because they offered a large reward (*thick curve*) for moderate delay. *C*, Single amygdala neuron tracking progress during sequential choices. Activity increased with consecutive save choices over 60–90 s until the monkey decided to spend the reward. *D*, Decoding progress from amygdala neurons. Cross-validated progress-decoding accuracy of nearest-neighbor classifier. *Inset*, Better decoding from basolateral than centromedial amygdala neurons.

A third type of neuronal activity observed in the save-spend task explicitly signaled the monkeys' progress through a saving sequence (Grabenhorst, Hernadi, & Schultz, 2016). Progress-tracking amygdala neurons showed gradually increasing "ramping" activity over consecutive save choices until the animal decided to spend (figure 53.3*C*). These responses occurred in the absence of external progress cues and often specifically during internally guided choices. The slope of these ramping activities depended on the forthcoming sequence length, with steeper neuronal ramping during shorter sequences, which suggested adaptation to an internal plan. Population decoding revealed a highly accurate neuronal progress code (figure 53.3*D*). Notably, neurons signaling sequence progress or the animal's internal plan (sequence length or value) were more prevalent in the basolateral amygdala compared to the centromedial amygdala. Basolateral neurons also encoded progress more accurately in predecision task periods (figure 53.3*D, inset*), suggesting particular importance for the basolateral amygdala in economic decisions. Progress evaluations in amygdala neurons (figure 53.3*C*) seem critical for linking amygdala planning activities (figure 53.3*B*) with choice signals in individual trials (figure 53.3*A*).

Given the amygdala's diverse projection targets (Amaral & Price, 1984), the neuronal representations of the decision variables described above could serve various functions. First, via amygdala outputs to physiological and attentional effectors, decision signals could serve to regulate attention, arousal, and affective state (*state variables*). Second, neurons projecting to frontal areas could influence decision-related activities in prefrontal cortex. A third, tantalizing possibility is the existence of a local decision mechanism within amygdala circuits (Grabenhorst, Hernadi, & Schultz, 2012). This suggestion is consistent with the observed explicit value-to-choice conversions in individual amygdala neurons (figure 53.3*A*) and with features of amygdala inhibitory microcircuits that could implement decision computations by mutual-inhibitory winner-take-all competition. Although amygdala inactivation does not necessarily cause choice deficits when reward values are stable (Wellman, Gale, & Malkova, 2005), the amygdala does make essential contributions in situations requiring dynamic valuations (Murray & Rudebeck, 2013). Consistently, amygdala lesions impair prefrontal value coding and behavioral choices during reinforcement learning (Rudebeck et al., 2017). The decision signals and planning activities in amygdala neurons may also inform our understanding of amygdala dysfunction in human psychiatric conditions, including mood disorders. These conditions have an impact on the motivation to plan for and pursue distant rewards. To better understand amygdala functions in health and disease, it will be important to establish whether amygdala neuronal circuits directly implement a local decision-making mechanism.

Reward and Social Information

Investigation of the role of the amygdala in decision-making is in its infancy, as is the study of how the amygdala participates in interactions between the emotional and cognitive variables that drive many forms of behavior. However, the amygdala does not only contribute to decision-making by virtue of its role in cognitive and emotional processing. Increasingly, the amygdala is recognized as playing a role in processing social stimuli, especially those of faces (Gothard et al., 2007; Rolls, 2000; Rutishauser, Mamelak, & Adolphs, 2015). The amygdala receives prominent input from the parts of the inferotemporal cortex that represent faces, and a long history of work has shown that amygdala neurons respond to images of faces (Gothard et al., 2007; Rolls, 2000; Rutishauser, Mamelak, & Adolphs, 2015). A key question is whether the amygdala processes social information in distinct neural circuits as compared to nonsocial stimuli that are related to rewards. A recent study showed that the majority of amygdala neurons that respond to face images of a monkey's social cohort also respond to abstract CSs that predict rewards (Munuera, Rigotti, & Salzman, 2018). Strikingly, when the responses to faces were examined in relation to the hierarchical rank of the individuals depicted, it was shown that the primate amygdala encodes the social rank of these individuals. The representation of hierarchical rank was observed in the same neuronal ensembles that encoded the rewards associated with nonsocial stimuli (figure 53.1*B*). A representation of reward value was not sufficient for representing hierarchical rank, as the orbitofrontal and anterior cingulate cortices lacked representations of hierarchical rank despite representing reward values. Information about hierarchical rank, which is intimately related to an assessment of the social value of individuals, is therefore linked in the amygdala to representations of rewards associated with nonsocial stimuli. These neuronal response properties and amygdala responses that jointly signal our own reward and a conspecific reward (Chang et al., 2015) position the amygdala to play a role in decision-making that integrates social with nonsocial information to guide behavior.

Acknowledgments

Our work has been supported by the Wellcome Trust, the European Research Council (ERC), the National Institute of Mental Health (NIMH), and the Simons Foundation.

REFERENCES

Adolphs, R. (2013). The biology of fear. *Current Biology, 23*, R79–93.

Amaral, D. G. (2016). A short history of the lesion technique for probing amygdala function. In D. G. Amaral & R. Adolphs (Eds.), *Living without an amygdala* (pp. 72–100). New York: Guilford Press.

Amaral, D. G., & Price, J. L. (1984). Amygdalo-cortical projections in the monkey (macaca fascicularis). *Journal of Comparative Neurology, 230*, 465–496.

Belova, M. A., Paton, J. J., Morrison, S. E., & Salzman, C. D. (2007). Expectation modulates neural responses to pleasant and aversive stimuli in primate amygdala. *Neuron, 55*, 970–984.

Belova, M. A., Paton, J. J., & Salzman, C. D. (2008). Moment-to-moment tracking of state value in the amygdala. *Journal of Neuroscience, 28*, 10023–10030.

Bermudez, M. A., Göbel, C., & Schultz, W. (2012). Sensitivity to temporal reward structure in amygdala neurons. *Current Biology, 22*, 1839–1844.

Bermudez, M. A., & Schultz, W. (2010a). Responses of amygdala neurons to positive reward predicting stimuli depend on background reward (contingency) rather than stimulus-reward pairing (contiguity). *Journal of Neurophysiology, 103*, 1158–1170.

Bermudez, M. A., & Schultz, W. (2010b). Reward magnitude coding in primate amygdala neurons. *Journal of Neurophysiology, 104*, 3424–3432.

Chang, S. W., Fagan, N. A., Toda, K., Utevsky, A. V., Pearson, J. M., & Platt, M. L. (2015). Neural mechanisms of social decision-making in the primate amygdala. *Proceedings of the National Academy of Sciences of the United States of America, 112,* 16012–16017.

Delamater, A. R. (1995). Outcome selective effects of intertrial reinforcement in a Pavlovian appetitive conditioning paradigm with rats. *Animal Learning and Behavior, 23,* 31–39.

Gore, F., Schwartz., E. C., Brangers, B. C., Aladi, S., Stujenske, J. M., Likhtik, E., Russo, M. J., Gordon, J. A., Salzman, C. D., & Axel., R. (2015). Neural representations of unconditioned stimuli in basolateral amygdala mediate innate and learned responses. *Cell, 162,* 134–145.

Gothard, K. M., Battaglia, F. P., Erickson, C. A., Spitler, K. M., & Amaral, D. G. (2007). Neural responses to facial expression and face identity in the monkey amygdala. *Journal of Neurophysiology, 97,* 1671–1683.

Gottfried, J. A., O'Doherty, J., & Dolan, R. J. (2003). Encoding predictive reward value in human amygdala and orbitofrontal cortex. *Science, 301,* 1104–1107.

Grabenhorst, F., Hernadi, I., & Schultz, W. (2012). Prediction of economic choice by primate amygdala neurons. *Proceedings of the National Academy of Sciences of the United States of America, 109,* 18950–18955.

Grabenhorst, F., Hernadi, I., & Schultz, W. (2016). Primate amygdala neurons evaluate the progress of self-defined economic choice sequences. *eLife, 5.*

Grabenhorst, F., Rolls, E. T., Parris, B. A., & D'Souza, A. (2010). How the brain represents the reward value of fat in the mouth. *Cerebral Cortex, 20,* 1082–1091.

Hernadi, I., Grabenhorst, F., & Schultz, W. (2015). Planning activity for internally generated reward goals in monkey amygdala neurons. *Nature Neuroscience, 18,* 461–469.

Janak, P. H., & Tye, K. M. (2015). From circuits to behaviour in the amygdala. *Nature, 517,* 284–292.

Krabbe, S., Grundemann, J., & Luthi, A. (2017). Amygdala inhibitory circuits regulate associative fear conditioning. *Biological Psychiatry, 83,* 800–809.

LeDoux, J. E. (2000). Emotion circuits in the brain. *Annual Review of Neuroscience, 23,* 155–184.

Maren, S., & Quirk, G. J. (2004). Neuronal signalling of fear memory. *Nature Reviews Neuroscience, 5,* 844–852.

McDonald, A. J. (1998). Cortical pathways to the mammalian amygdala. *Progress in Neurobiology, 55,* 257–332.

Munuera, J., Rigotti, M., & Salzman, C. D. (2018). Shared neural coding for social hierarchy and reward value in primate amygdala. *Nature Neuroscience, 21,* 415–423.

Murray, E. A., & Rudebeck, P. H. (2013). The drive to strive: Goal generation based on current needs. *Frontiers in Neuroscience, 7,* 112.

Nishijo, H., Ono, T., & Nishino, H. (1988). Topographic distribution of modality-specific amygdalar neurons in alert monkey. *Journal of Neuroscience, 8,* 3556–3569.

Ostlund, S. B., & Balleine, B. W. (2008). Differential involvement of the basolateral amygdala and mediodorsal thalamus in instrumental action selection. *Journal of Neuroscience, 28,* 4398–4405.

Paton, J. J., Belova, M. A., Morrison, S. E., & Salzman, C. D. (2006). The primate amygdala represents the positive and negative value of visual stimuli during learning. *Nature, 439,* 865–870.

Peck C. J., Lau, B., & Salzman C. D. (2013). The primate amygdala combines information about space and value. *Nature Neuroscience, 16,* 340–348.

Phelps, E. A., & LeDoux, J. E. (2005). Contributions of the amygdala to emotion processing: From animal models to human behavior. *Neuron, 48,* 175–187.

Redondo, R. L., Kim, J., Arons, A. L., Liu, X., & Tonegawa, S. (2014). Bidirectional switch of the valence associated with a hippocampal contextual memory engram. *Nature, 513,* 426–430.

Rescorla, R. A. (1967). Pavlovian conditioning and its proper control procedures. *Psychological Review, 74,* 71–80.

Rolls, E. T. (2000). Neurophysiology and functions of the primate amygdala, and the neural basis of emotion. In J. P. Aggleton (Ed.), *The amygdala: A functional analysis* (pp. 447–478). Oxford: Oxford University Press.

Romo, R., Hernandez, A., & Zainos, A. (2004). Neuronal correlates of a perceptual decision in ventral premotor cortex. *Neuron, 41,* 165–173.

Rudebeck, P. H., Ripple, J. A., Mitz, A. R., Averbeck, B. B., & Murray, E. A. (2017). Amygdala contributions to stimulus-reward encoding in the macaque medial and orbital frontal cortex during learning. *Journal of Neuroscience, 37,* 2186–2202.

Rutishauser, U., Mamelak, A. N., & Adolphs, R. (2015). The primate amygdala in social perception—insights from electrophysiological recordings and stimulation. *Trends in Neurosciences, 38,* 295–306.

Saez, A., Rigotti, M., Ostojic, S., Fusi, S., & Salzman, C. D. (2015). Abstract context representation in primate amygdala and prefrontal cortex. *Neuron, 87,* 869–881.

Saez, R. A., Saez, A., Paton, J. J., Lau, B., & Salzman, C. D. (2017). Distinct roles for the amygdala and orbitofrontal cortex in representing the relative amount of expected reward. *Neuron, 95,* 70–77.

Salzman, C. D., & Fusi, S. (2010). Emotion, cognition, and mental state representation in amygdala and prefrontal cortex. *Annual Review of Neuroscience, 33,* 173–202.

Seymour, B., & Dolan, R. (2008). Emotion, decision making, and the amygdala. *Neuron, 58,* 662–671.

Stone, M. (1960). Models for choice reaction time. *Psychometrika, 25,* 251–260.

Sugase-Miyamoto, Y., & Richmond, B. J. (2005). Neuronal signals in the monkey basolateral amygdala during reward schedules. *Journal of Neuroscience, 25,* 11071–11083.

Sutton, R. S., & Barto, A. G. (1981). Toward a modern theory of adaptive networks: Expectation and prediction. *Psychological Review, 88,* 135–170.

Sutton, R. S., & Barto, A. G. (1998). *Reinforcement learning.* Cambridge, MA: MIT Press.

Vickers, D. (1970). Evidence for an accumulator model of psychophysical discrimination. *Ergonomics, 13,* 37–58.

Wang, X. J. (2002). Probabilistic decision making by slow reverberation in cortical circuits. *Neuron, 36,* 955–968.

Weiskrantz, L. (1956). Behavioral changes associated with ablation of the amygdaloid complex in monkeys. *Journal of Comparative and Physiological Psychology, 49,* 381–391.

Wellman, L. L., Gale, K., & Malkova, L. (2005). GABAA-mediated inhibition of basolateral amygdala blocks reward devaluation in macaques. *Journal of Neuroscience, 25,* 4577–4586.

Zangemeister, L., Grabenhorst, F., & Schultz, W. (2016). Neural basis for economic saving strategies in human amygdala-prefrontal reward circuits. *Current Biology, 26,* 3004–3013.

54 Cortico-Striatal Circuits and Changes in Reward, Learning, and Decision-Making in Adolescence

ADRIANA GALVÁN, KRISTEN DELEVICH, AND LINDA WILBRECHT

ABSTRACT Reward processing, feedback learning, and decision-making evince significant changes during adolescence. This topic is of interest to scientists and stakeholders keen on identifying the mechanisms that underlie adolescent behavior in order to best support young people as they transition from childhood into adulthood. The study of cognitive development during this dynamic period also provides a unique opportunity to elucidate the relationship between neurobiology and behavior using a comparative approach. To these ends, our chapter reviews existing literature on the adolescent development of inputs to the dorsal and ventral striatum and develops three central ideas. First, the striatum is a structure that plays a central role in generating behavior by integrating glutamatergic and neuromodulatory inputs from other structures in a manner similar to a ballot box. Second, because cortical and limbic structures send convergent and coactive inputs to the striatum, the developmental changes in the relative ratio of these inputs will influence the neural computations that arise from the striatum. Third, the distinct developmental trajectories of inputs to the dorsal and ventral striatum likely contribute to a complex series of changes in behavioral patterns in learning and decision-making that are unique to adolescents.

Adolescents are characterized by their vast development across physical, neurobiological, psychological, and cognitive domains. Although normative and adaptive, these changes have significant implications for long-term outcomes, particularly as adolescents begin to explore their changing social and psychological landscape and make independent decisions. Over the past two decades, longitudinal imaging and computational neuroscience methods have provided increasingly sophisticated insights into the structure and function of the developing brain. Here we focus on how developmental changes in the convergence and relative influence of inputs to the basal ganglia can influence adolescent learning and decision-making.

This review is organized around three central tenets. The first is that cortical-striatal and limbic-striatal circuits play a central role in reward and decision-making (Averbeck & Costa, 2017; Friedman et al., 2015; Haber & Behrens, 2014; Lee et al., 2015; Yartsev et al., 2018). Second, inputs to the striatum undergo changes during adolescence that influence the neural computations that arise from this structure. Third, changes in striatal computations may orchestrate developmental changes in reward processing, feedback-based learning, and decision-making, forming part of a normative developmental process.

To date, much has been written about the potentially detrimental aspects of reward-based decision-making in adolescents. Here, we take the opportunity to examine positive aspects of this function at this time and introduce a framework by which interventions may be informed by developmental science.

Development of Inputs to the Dorsal Striatum and the Ballot Box Metaphor

The striatal regions of the basal ganglia (also referred to as the caudate and the putamen in primates) are critical hubs for learning, reward processing, and decision-making. The medial portions of the dorsal striatum (caudate) are implicated in flexible decision-making and action selection (Smith, Surmeier, Redgrave, & Kimura, 2011; Yartsev et al., 2018; Yin & Knowlton, 2006), while the lateral portions (putamen) are implicated in highly stereotyped learned motor sequences and inflexible stimulus-response habits (Yin, Knowlton, & Balleine, 2004; Yin et al., 2009; Yin & Knowlton, 2006). The ventral portions of the striatum are implicated in motivation, reward, and reinforcement (Averbeck & Costa, 2017; Haber, 2011; Roitman, Wheeler, & Carelli, 2005; Yin & Knowlton, 2006).

The majority of neurons in the striatum are inhibitory GABAergic medium spiny neurons (MSNs). MSNs, also referred to as spiny projection neurons (SPNs), can directly influence motor outputs through descending projections to the substantia nigra reticulata, superior colliculus, and downstream motor centers, as well as influence cortical motor and associative regions via basal ganglia projections to the thalamus (Redgrave, Prescott, & Gurney, 1999). Striatal MSNs sit at hyperpolarized resting potentials and require substantial depolarization (~20–30 mV) in order to fire an action potential, requiring hundreds to thousands of convergent glutamatergic inputs (Stern, Jaeger, & Wilson, 1998; Stern, Kincaid, & Wilson, 1997; Wilson & Kawaguchi, 1996). The striatal MSNs receive and integrate massively convergent input from across the overlying associative, limbic, and sensory neocortices; the thalamus; the amygdala; and the hippocampal systems (figure 54.1). It is important to note that the striatum is more than a mere relay station for these inputs. It is thought that the striatal neurons, MSNs in particular, play a key role in integrating these diverse sources of information to perform computations essential to decision-making, "on the fly" action selection, and the evaluation and reinforcement of actions for future decisions (Redgrave, Prescott, & Gurney, 1999; Smith, Surmeier, Redgrave, & Kimura, 2011; Tai et al., 2012; Yartsev et al., 2018). Some have likened the striatum and its neural computations to a "ballot box" in which different inputs to the striatum have a chance to "vote" for or against an action via inputs to different cell types within the striatum (Krauzlis, Bollimunta, Arcizet, & Wang, 2014; McHaffie, Stanford, Stein, Coizet, & Redgrave, 2005; Redgrave, Prescott, & Gurney, 1999). The ballot box model highlights the importance of (1) identifying the distinct afferent sources (voters) that drive activity in the striatum and (2) understanding how global activity in the striatum (e.g., as measured by functional imaging) relates to the neural activity of different striatal output populations (for example, the direct and indirect pathway MSNs).

Here we seek to apply this ballot box model to the developmental context. A prominent framework for understanding the maturation of cognitive control during adolescence is the dual-systems model, which suggests that the delayed maturation of prefrontal cortex relative to subcortical regions (including the striatum) produces an imbalance that promotes sensation seeking and risk-taking in adolescents (Shulman et al., 2016; Somerville, Jones, & Casey, 2010; Steinberg et al., 2008). While the dual-systems model separates the prefrontal cortex and subcortical striatal activity, a fundamental feature of the striatum is that it requires the convergence of many glutamatergic inputs (largely cortical) to drive activity. Therefore, rather than framing adolescent decision-making in terms of striatal "gas" versus prefrontal cortex "brakes," it may be more informative to consider "who" is driving striatal activation at distinct points in development (figure 54.1). Indeed, there are clear developmental shifts (with both gains and losses) in the strength of different inputs to striatal subregions: generally, more dorsal cortical regions wax in strength while more limbic and ventral connections wane and yet others exhibit U-shaped trajectories (outlined below). We can interpret these shifts in the framework of the ballot box metaphor: strengthening connections may be "gaining votes" while weakening connections are potentially "losing votes." In this way, shifts in the balance of the votes arising from different inputs adjust how different sources of information are weighted to guide active decision-making and action selection.

Recent developmental human brain-imaging work is beginning to make it possible to compare the relative strength of striatal inputs to determine which may play a dominant role in the selection of behavior at different ages. In one recent report using diffusion magnetic resonance-imaging (MRI) methodology, a striking shift in the relative strength of striatal inputs was observed by examining the cortical networks that converge on a dorsal striatal seed region. Here striatal inputs from a cortical "limbic network" associated with affective function were compared to a "frontoparietal network" associated with cognitive control (Larsen et al., 2017). Note that cortex to striatum directionality can be assumed in this case because the striatum does not project directly back to the neocortex. Over adolescence, the quantitative anisotropy (QA) of the two networks' projections to the striatum changed such that the convergence ratio of the limbic network versus the frontoparietal network significantly decreased from age 10 to 25 (Larsen et al., 2017). Moreover, the ratio of these connections in the striatum significantly mediated the relationship between age and incentive-based improvements in accuracy in an antisaccade task (Larsen et al., 2017). Similar general conclusions were drawn from a complementary study using resting-state connectivity analyses (Van Duijvenvoorde et al., 2015), and yet another study found that the increasing strength of frontostriatal white matter tracts correlated with reductions in temporal discounting rates across adolescence, consistent with decreased impatience (Van den Bos et al., 2015). The interpretation of these data in light of the ballot box metaphor suggests that changes in adolescent behavior likely emerge from the relative maturation of the spectrum of striatal inputs (with some strengthening and some weakening), rather than the independent maturation of the prefrontal cortex in isolation.

Future analyses of the maturation of the dorsal striatum should evaluate the relative influence of other glutamatergic inputs to this structure. Changes in the amygdala, cortical primary sensory inputs, and

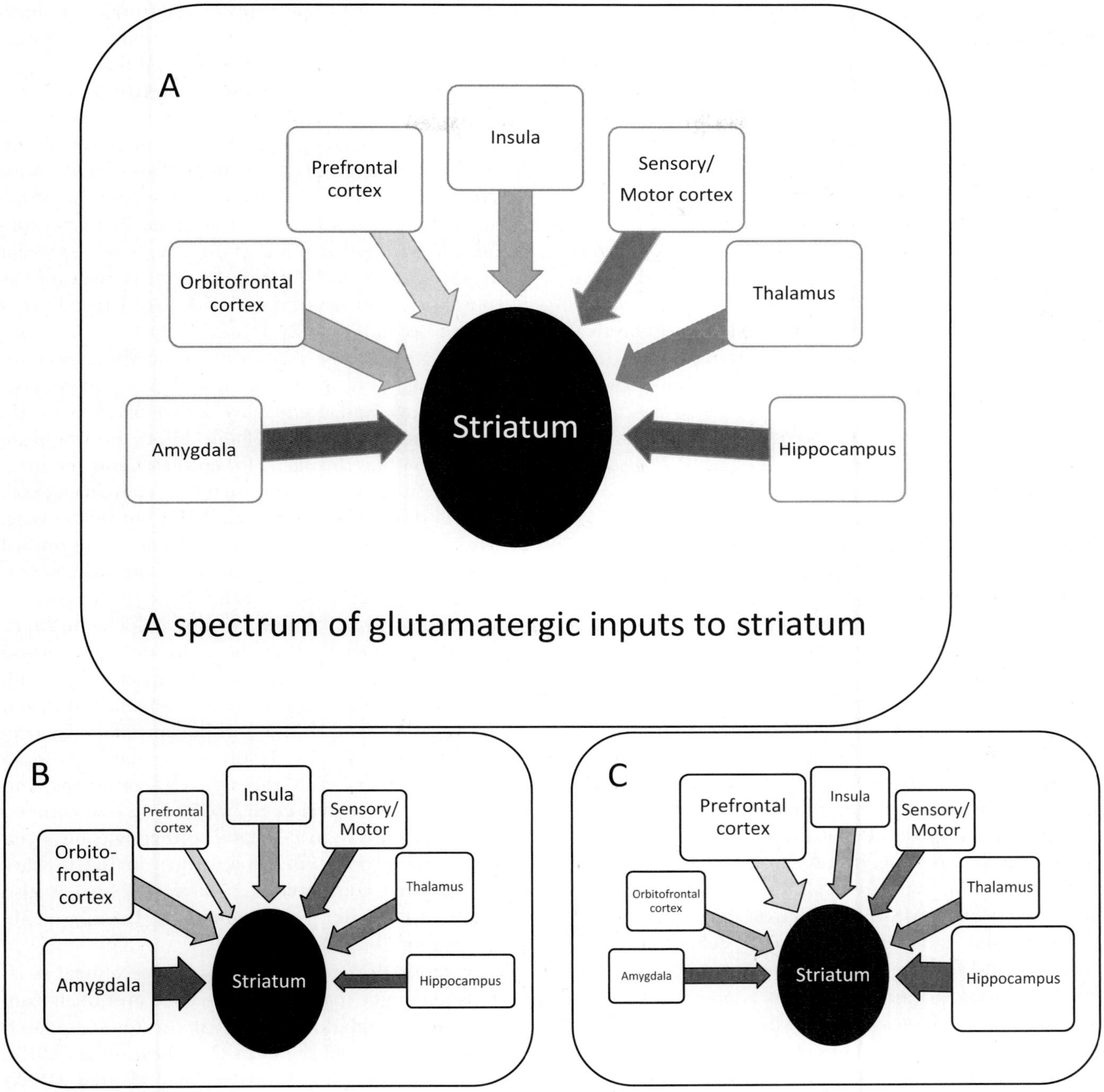

FIGURE 54.1 The striatum as a ballot box that undergoes change in the relative strength of "votes" coming from different regions through development. *A*, A schematic of the spectrum of glutamatergic inputs to the striatum. In theory, glutamatergic inputs all have the capacity to drive excitatory activity in the striatal medium spiny neurons and therefore have a "vote" in the context of the striatal ballot box metaphor (hence in [*A*] all inputs are shown as equal). Note, neuromodulatory inputs from dopamine or other neurotransmitters are not represented. *B*, In a hypothetical model, in early adolescence, inputs from the orbitofrontal cortex, amygdala, and insula to the striatum may dominate relative to other inputs. *C*, With adolescent development, prefrontal cortex inputs to the striatum, particularly from cortical cognitive control networks, may increase their relative strength compared to the amygdala, orbitofrontal cortex, and insula. Hippocampal inputs, although considered "limbic," may also increasingly contribute with age. These changes in relative influence over striatal activity may explain changes in striatal computation and behavior with age. (See color plate 59.)

glutamatergic thalamic projections to the striatum may be just as important as prefrontal cortex glutamatergic inputs in informing decision-making, action selection, and learning (Averbeck & Costa, 2017; Lee et al., 2015; Parker, Lalive, & Kreitzer, 2016), as well as play a role in alerting behavior (Smith, Surmeier, Redgrave, & Kimura, 2011). Changes in MSN excitability and neuromodulatory function also likely contribute to changing computation in the striatum through development (Lieberman et al., 2018). Dopamine innervation in particular may alter how information is processed within the striatum (Matthews, Bondi, Torres, & Moghaddam, 2013). The ballot box metaphor provides a theoretical framework in which to integrate findings about individual regions, pathways, and cell types in future work.

Adolescence and the Ventral Striatum: What Is the Significance and Source of an Adolescent Peak in Ventral Striatal Activation?

While the dorsal striatum is thought to play a causal role in current decision evaluation and action selection (Lee et al., 2015), the ventral striatum, which contains the nucleus accumbens (NAcc) region core and shell, is thought to be critical for motivation, Pavlovian conditioning, learning from reinforcement to anticipate reward, incentive salience, and informing future choice behavior. Crucially, lesions of the ventral striatum in nonhuman primates produce deficits in the ability to learn the value of stimuli but spare action-based reinforcement learning (Rothenhoefer et al., 2017), highlighting the importance of the ventral striatum in stimulus value learning. In informing future choice, the ventral striatum can interact with the dorsal striatum via ventral striatal MSN influence on the dopaminergic system (Haber, 2011; Haber, Fudge, & McFarland, 2000; Xia et al., 2011).

The ventral striatum has been a popular topic of investigation for adolescence research, but the data show a complex pattern of changes that do not easily align. A number of functional-imaging studies report that activation of the ventral striatum in response to rewards is highest in midadolescence (such that the developmental pattern forms an inverted U shape; Braams et al., 2015; Galván et al., 2006; Galván & McGlennen, 2013; Van Leijenhorst et al., 2010). These blood oxygen level dependent (BOLD) activity data have been tied to enhanced reward salience in adolescence and are often cited to explain risky and/or exploratory behavior in this midadolescent age group. However, it is not clear which changes in glutamatergic inputs or neuromodulatory inputs are responsible for this midadolescent peak in reward-related ventral striatal activation. Functional connectivity data of the ventral striatum of subjects measured in the "resting state" and in subjects performing tasks with feedback often show opposing changes across development. For example, prefrontal cortical connectivity to the ventral striatum shows decreases in connectivity over adolescence when analyzed using resting-state functional connectivity (Fareri et al., 2015; Van Duijvenvoorde et al., 2015). When examining functional connectivity after a correct versus incorrect outcome, functional coupling between the medial prefrontal cortex (mPFC) and the ventral striatum has been reported to strengthen from age 8 to 25 years (Van den Bos et al., 2012).

Insular cortex projections to the ventral striatum are of increasing interest due to their potential role in appetitive learning and compulsive behavior (Seif et al., 2013; Smith, Steinberg, & Chein, 2014). Resting-state analyses show that the insula to ventral striatum connection weakens by midadolescence and remains weaker than in childhood (Fareri et al., 2015; but see Stevens, Pearlson, & Calhoun, 2009). In a task context of reward anticipation, the connection between the insular cortex and the ventral striatum is adult-like in midadolescence (Cho et al., 2013; Smith, Steinberg, & Chein, 2014).

Notably, hippocampal connectivity with the ventral striatum using resting-state scans increases through adolescence into adulthood (Fareri et al., 2015; Van Duijvenvoorde et al., 2015), while amygdala ventral-striatal connectivity in resting-state data shows a U-shaped pattern, with the weakest connectivity in midadolescence (Fareri et al., 2015). In a task context, hippocampal and ventral-striatal connectivity has been shown to be stronger in adolescence during positive feedback when compared to adults, and this is also associated with differences in memory (Davidow et al., 2016; see more on this below).

Presynaptic dopamine function in the ventral striatum also changes through adolescence and likely contributes to a midadolescent peak in reward-related activity in the ventral striatum (Matthews et al., 2013). Ventral striatum and ventral tegmental area (VTA) functional coupling is highest in midadolescence when measured in the context of a motivated task (Murty et al., 2018). Also, the dorsal and ventral striatum show divergent developmental trajectories of T2*-weighted imaging (Larsen & Luna 2015), an indicator of the tissue-iron concentration, which may indirectly reflect differences in the dopaminergic system with age (Beard, Erikson, & Jones, 2003; Jellen et al., 2013).

In rodents under anesthesia, VTA dopamine neurons show faster firing in midadolescence compared to adults (McCutcheon et al., 2012). Dopamine D1 and D2 type receptor levels in the dorsal but not ventral striatum peak in adolescence and decline in adults (Teicher et al., 1995). Also, D1R expression in cortical neurons that project to the ventral striatum also peaks in adolescence and declines in adults (Brenhouse, Sonntag, & Andersen, 2008). These data suggest that adolescent rodents, too, may have enhanced dopamine signaling (Walker et al., 2017). In contrast, another study showed that when dopamine neurons are recorded in awake, behaving adolescent and adult rats performing a task, the dopamine neurons in adolescent and adult rats fire at similar rates during reward receipt, but adolescents actually show less dopamine neuron activity than adults during the epoch of reward anticipation (Kim, Simon, Wood, & Moghaddam, 2016). Recording data from dopamine neuron soma should be interpreted cautiously, as recent data have highlighted the potential disconnect between dopamine neuron firing and dopamine release, which can be regulated at the level of the axon terminals (Cachope & Cheer, 2014; Liu & Kaeser, 2019). Work in rodents also can remind the human subject research community that activity in the striatum measured using functional magnetic resonance imaging (fMRI) may arise from different MSN cell types (D1- and D2-expressing neurons), which have different projections and functions (Lee et al., 2015; Xia et al., 2011).

In light of this complexity, it is currently unclear how an inverted U pattern of activation emerges in the ventral striatum across adolescence during rewarding feedback in human fMRI studies. It is likely a result of the convergence of multiple factors in midadolescence that will require new research designs to disentangle.

How Do Changes in the Adolescent Brain Affect Learning, Reward Processing, and Decision-Making?

Learning is a major developmental task during adolescence, when youth must develop the knowledge and abilities they need to succeed in increasingly challenging academic and social domains and to transition into adult roles. Historical neurodevelopmental models highlighted the fact that the prefrontal cortex is relatively immature and that its development accompanies a more adult-like performance in learning and decision-making. This is often depicted as a linear trajectory in which prefrontal cortex development supports behavioral development until adulthood is reached.

In dual-systems models of development (mentioned above), the relative immaturity of the prefrontal cortex is contrasted with enhancement in motivation and sensation seeking, attributed to striatal and limbic structures (Shulman et al., 2016; Steinberg et al., 2008). A putative imbalance in these systems is thought to contribute to more risky behavior and potentially negative outcomes for youth. Alternate neurodevelopmental models also build on the evidence that adolescents have heightened sensation seeking and affect and/or exhibit heightened activity in striatal and limbic structures to emphasize that these changes could support the transition to independence and new skill learning (Crone & Dahl, 2012; Davidow et al., 2016; Larsen & Luna, 2018; Pfeifer & Allen, 2012).

These variant models of adolescent brain development can be reconciled within the ballot box metaphor. In midadolescence, limbic areas may have a stronger capacity to drive striatal activity compared to the prefrontal cortex, but in later adolescence the relative influence of prefrontal structures on the striatum grows relative to limbic structures (figure 54.1). Despite the potential for consilience in this framework, it is still currently difficult to summarize how learning and decision-making change through adolescence. The complexity of the findings may in part be due to differences in the neurobiological structures engaged by each task, their relative ability to engage neuromodulation, and last but not least, their relative strength of connectivity with the striatum at different stages of adolescent development.

The maturation of executive function during the teenage years is of great importance to pragmatic legal applications that establish when someone is able to become legally independent, can vote, or can be held culpable of a crime. A long list of behavioral studies have shown that behavioral inhibition, working memory, and cognitive flexibility improve during adolescence (Huizinga et al., 2006; Marek et al., 2015; Shulman et al., 2016; Steinberg et al., 2018). Reaction times also speed up with age, while variability in motor responses and brain activity decreases with age (Montez, Calabro, & Luna, 2017). Also, learning and cognition may become more complex. For example, one recent study found that adults use more complex "model-based" strategies when learning compared to children and that these complex strategies begin to be used in adolescence (Decker et al., 2016). A second recent study found that adults may incorporate more counterfactual information and absorb more from complex feedback information than adolescents (Palmintieri et al., 2016). A third recent study found that adults are better able to modulate their cognitive effort when stakes are high versus low when compared to adolescents (Insel et al., 2017).

Many of these studies attribute these gains in function to maturation of the prefrontal cortex; however, the functional explanation behind these changes may be even broader. Recent studies have argued that the gain of function as we reach adulthood may be due to greater engagement and integration of multiple learning systems, particularly the hippocampus and the striatum (Davidow et al., 2016; Jocham, Klein, & Ullsperger, 2011; Peters & Crone, 2017). For practical and scientific reasons, it is important and exciting to continue to clarify how children and adolescents learn and make decisions differently than adults and to pinpoint when and how learning and decision-making change.

To disentangle the complexities of how adolescents make decisions and learn from different outcomes, recent studies in adolescents have often incorporated neuroeconomic approaches and reinforcement-learning models. These approaches have often been applied with probabilistic-learning tasks to determine whether adolescents integrate feedback information differently in time and/or learn differentially from positive versus negative feedback. A crucial aspect of reinforcement learning is using feedback to estimate the expected value of the available options. First, the computation of a prediction error, calculating the difference between expected and experienced outcomes, is necessary. A positive prediction error indicates that outcomes are better than expected; a negative prediction error indicates outcomes are worse than expected (Sutton & Barto, 1998). These prediction errors are used to update the expected value associated with the chosen option over time. This updating of value, parameterized by the learning rate, can be done rapidly (in a single trial, for example) or can be done gradually, allowing greater integration of experience over time. In some reinforcement-learning models, learning from negative and positive outcomes can be assigned unique learning rates, reflecting different sensitivity to gains versus losses.

Fitting a canonical reinforcement-learning model to a simple probabilistic-learning task, Davidow and colleagues (2016) found that adolescents (aged 13–17 years) had a lower learning rate than adults (aged 24–30 years), which reflects the extent to which feedback in each trial is used to update future choices. It is important to consider that a lower learning rate is not necessarily a deficit. Davidow et al. (2016) point out that in a probabilistic task, a lower learning rate may be more optimal because it ensures that behavior is updated by evidence accumulated over trials rather than the outcome of any single trial (Davidow et al., 2016). These data also run counter to common sense expectation because they show that adolescents integrate their learning and decision-making more broadly in time than adults, rather than impulsively making a decision based on any one event. The different learning rate in Davidow et al. (2016) was also accompanied by greater episodic-memory performance and greater coactivation between the striatum and hippocampus in the adolescent group when compared with adults. These data are consistent with a separate longitudinal study in participants ages 8–29 that showed that enhanced activation in the striatum and NAcc during feedback-based learning was accompanied by enhanced concurrent and future learning performance (Peters & Crone, 2017).

Notably, there is still conflict in the literature in terms of the adolescent response to negative versus positive feedback. Using a probabilistic task and a reinforcement-learning model, Van den Bos and colleagues (2012) found that negative feedback had a stronger impact on how children learned than negative feedback given to adolescents and adults, suggesting that with increasing age, the impact of negative prediction errors on the future expected value decreases (Van den Bos et al., 2012). However, in a probabilistic task with a reversal, Hauser et al. (2015) reported greater learning from negative feedback in adolescents compared to adults (Hauser et al., 2015). Similarly, Van der Schaaf et al. (2011) also reported superior performance in adolescents (ages 13–17), when compared to adults and children, in a reversal task that used negative frowning faces as feedback. In this same study, when feedback signals were positive smiling faces, adults and adolescents showed similar reversal performance (Van der Schaaf et al., 2011). In contrast, studies using different tasks have found learning advantages when using positive valence feedback in adolescents. For example, Davidow et al. (2016) found that in adolescents, episodic memory was better for items presented with positive versus negative outcomes, but this valence-dependent difference in recall was not observed in adults. These contrasting results are difficult to summarize and also to build upon. Currently, if an educator is, for example, looking to the literature to design new educational strategies for adolescents, it is not simple to predict if positive or negative feedback is more effective. Progress in this area will likely come from dissecting the neurobiological basis of the tasks that produce different results and examining probabilistic and deterministic learning, updating, and recall performance separately.

Future Directions

Two directions we would like to highlight as interesting for future research into adolescent development are the effects of puberty and pubertal timing on neural circuit development and studies of the role of effort in decision-making.

Recent work has suggested a potentially causal role for gonadal hormones at puberty in scaling the adolescent peak in ventral striatum activation and individual differences in risk-taking behavior (Braams et al., 2015). Also, several studies on brain structure have found changes in cortical gray matter thickness, area, and white matter that scale with gonadal and adrenal hormones and/or self-report pubertal development scales (Chahal et al., 2018; Herting & Sowell, 2017; Piekarski et al., 2017). Recent work on the development of the anterior cingulate cortex in mice has revealed that gonadal hormones play an organizational role in the maturation of inhibitory neurotransmission onto layer 2/3 pyramidal neurons, a potentially important sensitive period mechanism (Piekarski et al., 2017; Piekarski, Boivin, & Wilbrecht, 2017). Also, hormonal injection to jump-start pubertal onset can speed the maturation of reversal learning performance in mice (Piekarski, Boivin, & Wilbrecht, 2017). These cellular and behavioral effects in mice could be mediated by the expression of gonadal hormone receptors in interneurons in the neocortex (Piekarski et al., 2017) and are consistent with the relationships reported between pubertal timing and the development of several white matter tracts, including the cingulum bundle in humans (Chahal et al., 2018). Future studies of the role of pubertal timing are ongoing and are likely to reveal further circuit and behavioral specificity, where puberty-dependent and independent effects need to be detangled using cell type-specific tools and/or clever study designs (Boivin et al., 2018; Paul et al., 2018; Walker et al., 2017).

A second horizon of interest is a deeper dive into motivation and effort in adolescence. Decision-making requires the consideration of both the potential benefits of a given choice and the potential costs, which can include risk, delay, and effort. Whereas the former has received considerable research attention, the effects of effort on adolescent decision-making are a growing focus. Using a task validated in adults (Sullivan-Toole, Richey, & Tricomi, 2017), DePasque and Galván (2019) identified developmental differences in the willingness to expend effort during goal pursuit in adolescents versus young adults. In this task, adults (18–23) preferred lower-effort options when outcome values were held constant (Sullivan-Toole, Richey, & Tricomi, 2017). Adolescents (aged 13–16 years) made choices that did not discriminate between physical effort costs. These effects did not appear to be driven by differences in subjective task motivation or awareness of the effort requirements (DePasque & Galván, 2019). These findings echo those of Insel et al. (2017), who found that adolescents did not modulate effort when the value of expected outcomes was varied, and the work of Castel et al. (2011) on the role of value in memory bias. These findings provide some evidence that adolescence may be a time of increased willingness to expend effort during goal pursuit and are consistent with a recent framework put forth to describe how value can facilitate decision-making and cognitive performance (Davidow, Insel, & Somerville, 2018).

Conclusion

Although adolescence has long been recognized as an important transitional period in our society, it has recently drawn more attention from policy-makers, public health officials, and cognitive neuroscientists (Dahl et al., 2018). Adolescence is now a forefront topic of health reports (United Nations, 2012), advocacy documents, and publications in academic journals. There is also worldwide attention paid to adolescents and their health, illustrated by several recent United Nations initiatives (Ki-moon, 2013). Insights from cognitive neuroscience have helped refine the traditional models of adolescent development by exemplifying the complexity of the neural circuits (and their continued maturation throughout adolescence and early adulthood) that underlie the developmental milestones that typify adolescents, including reward processing, learning, and decision-making.

Many of the studies reviewed here have identified developmental differences in the circuits that underlie these computations in adolescents as compared to children and adults. We propose that the ballot box metaphor and analyses of the relative strength and activation of different inputs to the basal ganglia have the potential to help detangle the source of these differences and inspire new research designs. The heightened reactivity of the adolescent striatum and the behavioral observations of more experientially driven learning in youth suggest that learning may be more closely tied to limbic-striatal mechanisms in adolescents than in adults, for whom later-maturing cortical and perhaps hippocampal inputs facilitate the use of additional learning strategies and enhanced executive functioning (Davidow et al., 2016; Peters et al., 2014; Wimmer & Shohamy, 2012).

Characterizing changes in the circuits that underlie reward, learning, and decision-making is important for identifying ways to support healthy adolescent development. This research has already helped trigger challenging but important conversations among scientists and policy-makers to advance policies and sanctions consonant with adolescent development. In contrast to early models of adolescent neurobiology that emphasized the protracted development of the prefrontal cortex in adolescent behavior, more contemporary research has placed emphasis on the striatum and changing inputs to the striatum from diverse circuits. The studies on learning, in particular, demonstrate that the brain is highly plastic well beyond the early postnatal years and that the adolescent brain may make unique computations potentially adapted to the series of challenges we face within this unique life stage.

Acknowledgments

We thank Dr. Anne G. E. Collins and Dr. Daphna Shohamy for discussion and comments.

REFERENCES

Averbeck, B. B., & Costa, V. D. (2017). Motivational neural circuits underlying reinforcement learning. *Nature Neuroscience, 20,* 505–512.

Beard, J., Erikson, K. M., & Jones, B. C. (2003). Neonatal iron deficiency results in irreversible changes in dopamine function in rats. *Journal of Nutrition, 133,* 1174–1179.

Boivin, J. R., Piekarski, D. J., Thomas, A. W., & Wilbrecht, L. (2018). Adolescent pruning and stabilization of dendritic spines on cortical layer 5 pyramidal neurons do not depend on gonadal hormones. *Developmental Cognitive Neuroscience, 30,* 100–107.

Braams, B., Van Duijvenvoorde, A. C., Peper, J. S., & Crone, E. A. (2015). Longitudinal changes in adolescent risk-taking: A comprehensive study of neural responses to rewards, pubertal development, and risk-taking behavior. *Journal of Neuroscience, 35,* 7226–7238.

Brenhouse, H. C., Sonntag, K. C., & Andersen, S. L. (2008). Transient D1 dopamine receptor expression on prefrontal cortex projection neurons: Relationship to enhanced motivational salience of drug cues in adolescence. *Journal of Neuroscience, 28,* 2375–2382.

Castel, A. D., Humphreys, K. L., Lee, S. S., Galván, A., Balota, D. A., & McCabe, D. P. (2011). The development of memory efficiency and value-directed remembering across the life span: A cross-sectional study of memory and selectivity. *Developmental Psychology, 47,* 1553–1564.

Cachope, R., & Cheer, J. F. (2014). Local control of striatal dopamine release. *Frontiers in Behavioral Neuroscience, 8,* 188.

Chahal, R., Vilgis, V., Grimm, K. J., Hipwell, A. E., Forbes, E. E., Keenan, K., & Guyer, A. E. (2018). Girls' pubertal development is associated with white matter microstructure in late adolescence. *NeuroImage, 181,* 659–669.

Cho, Y. T., Fromm, S., Guyer, A. E., Detloff, A., Pine, D. S., Fudge, J. L., et al. (2013). Nucleus accumbens, thalamus and insula connectivity during incentive anticipation in typical adults and adolescents. *NeuroImage, 66,* 508–521.

Crone, E. A., & Dahl, R. E. (2012). Understanding adolescence as a period of social-affective engagement and goal flexibility. *Nature Reviews Neuroscience, 13,* 636–650.

Dahl, R. E., Allen, N. B., Wilbrecht, L., & Suleiman, A. B. (2018). Importance of investing in adolescence from a developmental science perspective. *Nature, 554,* 441–450.

Davidow, J. Y., Foerde, K., Galván, A., & Shohamy, D. (2016). An upside to reward sensitivity. The hippocampus supports enhanced reinforcement learning in adolescence. *Neuron, 92,* 93–99.

Davidow, J. Y., Insel, C., & Somerville, L. H. (2018). Adolescent development of value-guided goal pursuit. *Trends in Cognitive Science, 18,* 30116.

Decker, J. H., Otto, A. R., Daw, N. D., & Hartley, C. A. (2016). From creatures of habit to goal-directed learners: Tracking the developmental emergence of model-based reinforcement learning. *Psychological Science, 27,* 848–858.

DePasque, S., & Galván, A. (2019). Neurobiological responses in the adolescent striatum to being "tested." *Social Cognitive, Affective Neuroscience, 14,* 3–12.

Fareri, D. S., Gabard-Durnam, L., Goff, B., Flannery, J., Gee, D. G., Lumian, D. S., Caldera, C., & Tottenham, N. (2015). Normative development of ventral striatal resting state connectivity in humans. *NeuroImage, 118,* 422–437.

Friedman, A., Homma, D. Gibb, L. G., Amemori, K., Rubin, S. J., Hood, A. S., Riad, M. H., & Graybiel, A. M. (2015). A corticostriatal path targeting striosomes controls decision-making under conflict. *Cell, 161,* 1320–1333.

Galván, A., Hare, T., Davidson, M. C., Spicer, J., Glover, G., & Casey, B. J. (2005). The role of ventral frontostriatal circuitry in reward-based learning in humans. *Journal of Neuroscience, 25,* 8650–8656.

Galván, A., Hare, T. A., Parra, C. E., Penn, J., Voss, H., Glover, G., & Casey, B. J. (2006). Earlier development of the accumbens relative to orbitofrontal cortex might underlie risk-taking behavior in adolescents. *Journal of Neuroscience, 26,* 6885–6892.

Galván, A., & McGlennen, K. M. (2013). Enhanced striatal sensitivity to aversive reinforcement in adolescents versus adults. *Journal of Cognitive Neuroscience, 25,* 284–296.

Haber, S. N. (2011). Neuroanatomy of reward: A view from the ventral striatum. In J. A. Gottfried (Ed.), *Neurobiology of sensation and reward* (chapter 11).Boca Raton, FL: CRC Press/Taylor & Francis.

Haber, S. N., & Behrens, T. E. (2014). The neural network underlying incentive-based learning: Implications for interpreting circuit disruptions in psychiatric disorders. *Neuron, 83,* 1019–1039.

Haber, S. N., Fudge, J. L., & McFarland, N. R. (2000). Striatonigrostriatal pathways in primates form an ascending spiral from the shell to the dorsolateral striatum. *Journal of Neuroscience, 20,* 2369–2382.

Hauser, T. U., Iannaccone, R., Walitza, S., Brandeis, D., & Brem, S. (2015). Cognitive flexibility in adolescence: Neural and behavioral mechanisms of reward prediction error processing in adaptive decision making during development. *NeuroImage, 104,* 347–354.

Herting, M. M., & Sowell, E. R. (2017). Puberty and structural brain development in humans. *Frontiers in Neuroendocrinology, 44,* 122–137.

Huizinga, M., Dolan, C. V., & Van der Molen, M. W. (2006). Age-related change in executive function: developmental trends and a latent variable analysis. *Neuropsychologia, 44,* 2017–2036.

Insel, C., Kastman, E. K., Glenn, C. R., & Somerville, L. H. (2017). Development of corticostriatal connectivity constrains goal-directed behavior during adolescence. *Nature Communications, 8,* 1605–1610.

Jellen, L. C., Lu, L., Wang, X., Unger, E. L., Earley, C. J., Allen, R. P., Williams, R. W., & Jones, B. C. (2013). Iron deficiency alters expression of dopamine-related genes in the ventral midbrain in mice. *Neuroscience, 252,* 13–23.

Jocham, G., Klein, T. A., & Ullsperger, M. (2011). Dopamine-mediated reinforcement learning signals in the striatum and ventromedial prefrontal cortex underlie value-based choices. *Journal of Neuroscience, 31*, 1606–1613.

Kim, Y., Simon, N. W., Wood, J., & Moghaddam, B. (2016). Reward anticipation is encoded differently by adolescent ventral tegmental area neurons. *Biological Psychiatry, 79*, 878–886.

Ki-moon, B. (2013). *Five-year action agenda: Office of the Secretary General's Envoy on Youth*. New York: United Nations.

Krauzlis, R. J., Bollimunta, A., Arcizet, F., & Wang, L. (2014). Attention as an effect not a cause. *Trends in Cognitive Science, 18*, 457–464.

Larsen, B., & Luna, B. (2015). In vivo evidence of neurophysiological maturation of the human adolescent striatum. *Developmental Cognitive Neuroscience, 12*, 74–85.

Larsen, B., & Luna, B. (2018). Adolescence as a neurobiological critical period for the development of higher-order cognition. *Neuroscience and Biobehavioral Reviews, 94*, 179–195.

Larsen, B., Verstynen, T. D., Yeh, F. C., & Luna, B. (2017). Developmental changes in the integration of affective and cognitive corticostriatal pathways are associated with reward-driven behavior. *Cerebral Cortex, 5*, 1–12.

Lee, A. M., Tai, L. H., Zador, A., & Wilbrecht, L. (2015). Between the primate and "reptilian" brain: Rodent models demonstrate the role of corticostriatal circuits in decision making. *Neuroscience, 296*, 66–74.

Lieberman, O. J., McGuirt, A. F., Mosharov, E. V., Pigulevskiy, I., Hobson, B. D., Choi, S., Frier, M. D., Santini, E., Borgkvist, A., & Sulzer, D. (2018). Dopamine triggers the maturation of striatal spiny projection neuron excitability during a critical period. *Neuron, 99*, 540–554.

Liu, C., & Kaeser, P. S. (2019). Mechanisms and regulation of dopamine release. *Current Opinion in Neurobiology, 57*, 46–53.

Marek, S., Hwang, K., Foran, W., Hallquist, M. N., & Luna, B. (2015). The contribution of network organization and integration to the development of cognitive control. *PLoS Biology, 13*, e1002328.

Matthews, M., Bondi, C., Torres, G., & Moghaddam, B. (2013). Reduced presynaptic dopamine activity in adolescent dorsal striatum. *Neuropsychopharmacology, 38*, 1344–1351.

McCutcheon, J. E., Conrad, K. L., Carr, S. B., Ford, K. A., McGehee, D. S., & Marinelli, M. (2012). Dopamine neurons in the ventral tegmental area fire faster in adolescent rats than in adults. *Journal of Neurophysiology, 108*, 1620–1630.

McHaffie, J. G., Stanford, T. R., Stein, B. E., Coizet, V., & Redgrave, P. (2005). Subcortical loops through the basal ganglia. *Trends in Neuroscience, 28*, 401–407.

Montez, D. F., Calabro, F. J., & Luna, B. (2017). The expression of established cognitive brain states stabilizes with working memory development. *elife, 6*, e25606.

Murty, V. P., Shah, H., Montez, D., Foran, W., Calabro, F., & Luna, B. (2018). Age-related trajectories of functional coupling between the VTA and nucleus accumbens depend on motivational state. *Journal of Neuroscience, 38*, 7420–7427.

Palminteri, S., Kilford, E. J., Coricelli, G., & Blakemore, S. J. (2016). The computational development of reinforcement learning during adolescence. *PLoS Computational Biology, 12*, 1–25.

Parker, P. R., Lalive, A. L., & Kreitzer, A. C. (2016). Pathway-specific remodeling of thalamostriatal synapses in parkinsonian mice. *Neuron, 89*, 734–740.

Paul, M. J., Probst, C. K., Brown, L. M., & de Vries, G. J. (2018). Dissociation of puberty and adolescent social development in a seasonally breeding species. *Current Biology, 28*, 1116–1123.e2.

Peters, S., & Crone, E. A. (2017). Increased striatal activity in adolescence benefits learning. *Nature Communications, 8*, 1983–1989.

Peters, S., Koolschijn, P. C., Crone, E. A., Van Duijvenvoorde, A. C., & Raijmakers, M. E. (2014). Strategies influence neural activity for feedback learning across child and adolescent development. *Neuropsychologia, 62*, 365–374.

Pfeifer, J. H., & Allen, N. B. (2012). Arrested development? Reconsidering dual-systems models of brain function in adolescence and disorders. *Trends in Cognitive Sciences, 16*, 322–329.

Piekarski, D. J., Boivin, J. R., & Wilbrecht, L. (2017). Ovarian hormones organize the maturation of inhibitory neurotransmission in the frontal cortex at puberty onset in female mice. *Current Biology, 27*, 1735–1745.e3.

Piekarski, D. J., Johnson, C. M., Boivin, J. R., Thomas, A. W., Lin, W. C., Delevich, K., Galarce, E. M., & Wilbrecht, L. (2017). Does puberty mark a transition in sensitive periods for plasticity in the associative neocortex? *Brain Research, 1654*(Pt. B), 123–144.

Redgrave, P., Prescott, T. J., & Gurney, K. (1999). The basal ganglia: A vertebrate solution to the selection problem? *Neuroscience, 89*, 1009–1023.

Roitman, M. F., Wheeler, R. A., & Carelli, R. M. (2005). Nucleus accumbens neurons are innately tuned for rewarding and aversive taste stimuli, encode their predictors, and are linked to motor output. *Neuron, 45*, 587–597.

Rothenhoefer, K. M., Costa, V. D., Bartolo, R., Vicario-Feliciano, R., Murray, E. A., & Averbeck, B. B. (2017). Effects of ventral striatum lesions on stimulus-based versus action-based reinforcement learning. *Journal of Neuroscience, 37*, 6902–6914.

Seif, T., Chang, S. J., Simms, J. A., Gibb, S. L., Dadgar, J., Chen, B. T., Harvey, B. K., Ron, D., Messing, R. O., Bonci, A., & Hopf, F. W. (2013). Cortical activation of accumbens hyperpolarization-active NMDARs mediates aversion-resistant alcohol intake. *Nature Neuroscience, 16*, 1094–1100.

Shulman, E. P., Smith, A. R., Silva, K., Icenogle, G., Duell, N., Chein, J., & Steinberg, L. (2016). The dual systems model: Review, reappraisal, and reaffirmation. *Developmental Cognitive Neuroscience, 17*, 103–117.

Smith, A. R., Steinberg, L., & Chein, J. (2014). The role of the anterior insula in adolescent decision making. *Developmental Neuroscience, 36*, 196–209.

Smith, Y., Surmeier, D. J., Redgrave, P., & Kimura, M. (2011). Thalamic contributions to basal ganglia-related behavioral switching and reinforcement. *Journal of Neuroscience, 31*, 16102–16106.

Somerville, L. H., Jones, R. M., & Casey, B. J. (2010). A time of change: Behavioral and neural correlates of adolescent sensitivity to appetitive and aversive environmental cues. *Brain and Cognition, 72*, 124–133.

Steinberg, L., Albert, D., Cauffman, E., Banich, M., Graham, S., & Woolard, J. (2008). Age differences in sensation seeking and impulsivity as indexed by behavior and self-report: Evidence for a dual systems model. *Developmental Psychology, 44*, 1764–1778.

Steinberg, L., Icenogle, G., Shulman, E. P., Breiner, K., Chein, J., Bacchini, D., Chang, L., Chaudhary, N., Giunta, L. D., Dodge, K. A., Fanti, K. A., Lansford, J. E., Malone, P. S., Oburu, P., Pastorelli, C., Skinner, A. T., Sorbring, E., Tapanya, S., Tirado, L. M. U., Alampay, L. P., Al-Hassan, S. M., & Takash, H. M. S. (2018). Around the world, adolescence is a time of heightened sensation seeking and immature self-regulation. *Developmental Science, 21,* 1–13.

Stern, E. A., Jaeger, D., & Wilson, C. J. (1998). Membrane potential synchrony of simultaneously recorded striatal spiny neurons in vivo. *Nature, 394,* 475–478.

Stern, E. A., Kincaid, A. E., & Wilson, C. J. (1997). Spontaneous subthreshold membrane potential fluctuations and action potential variability of rat corticostriatal and striatal neurons in vivo. *Journal of Neurophysiology, 77,* 1697–1715.

Stevens, M. C., Pearlson, G. D., & Calhoun, V. D. (2009). Changes in the interaction of resting-state neural networks from adolescence to adulthood. *Human Brain Mapping, 30,* 2356–2366.

Sullivan-Toole, H., Richey, J. A., & Tricomi, E. (2017). Control and effort costs influence the motivational consequences of choice. *Frontiers in Psychology, 8,* 675.

Sutton, R. S., & Barto, A. G. (1998). *Reinforcement learning: An introduction.* Cambridge, MA: MIT Press.

Tai, L. H., Lee, A. M., Benavidez, N., Bonci, A., & Wilbrecht, L. (2012). Transient stimulation of distinct subpopulations of striatal neurons mimics changes in action value. *Nature Neuroscience, 15,* 1281–1289.

Teicher, M. H., Andersen, S. L., & Hostetter, J. C., Jr. (1995). Evidence for dopamine receptor pruning between adolescence and adulthood in striatum but not nucleus accumbens. *Brain Research Developmental Brain Research, 89,* 167–172.

United Nations. (2012). World population monitoring: Adolescents and youth—A concise report. https://www.un.org/en/development/desa/population/publications/pdf/fertility/12_66976_adolescents_and_youth.pdf

Van den Bos, W., Cohen, M. X., Kahnt, T., & Crone, E. A. (2012). Striatum-medial prefrontal cortex connectivity predicts developmental changes in reinforcement learning. *Cerebral Cortex, 22,* 1247–1255.

Van den Bos, W., Rodriguez, C. A., Schweitzer, J. B., & McClure, S. M. (2015). Adolescent impatience decreases with increased frontostriatal connectivity. *Proceedings of the National Academy of Sciences of the United States of America, 112,* E3765–3774.

Van der Schaaf, M. E., Warmerdam, E., Crone, E. A., & Cools, R. (2011). Distinct linear and non-linear trajectories of reward and punishment reversal learning during development: Relevance for dopamine's role in adolescent decision making. *Developmental Cognitive Neuroscience, 1,* 578–590.

Van Duijvenvoorde, A. C., Huizenga, H. M., Somerville, L. H., Delgado, M. R., Powers, A., Weeda, W. D., Casey, B. J., Weber, E. U., & Figner, B. (2015). Neural correlates of expected risks and returns in risky choice across development. *Journal of Neuroscience, 35,* 1549–1560.

Van Leijenhorst, L., Gunther Moor, B., Op de Macks, Z. A., Rombouts, S. A., Westenberg, P. M., & Crone, E. A. (2010). Adolescent risky decision-making: Neurocognitive development of reward and control regions. *NeuroImage, 51,* 345–355.

Walker, D. M., Bell, M. R., Flores, C., Gulley, J. M., Willing, J., & Paul, M. J. (2017). Adolescence and reward: Making sense of neural and behavioral changes amid the chaos. *Journal of Neuroscience, 37,* 10855–10866.

Wilson, C. J., & Kawaguchi, Y. (1996). The origins of two-state spontaneous membrane potential fluctuations of neostriatal spiny neurons. *Journal of Neuroscience, 16,* 2397–2410.

Wimmer, G. E., & Shohamy, D. (2012). Preference by association: How memory mechanisms in the hippocampus bias decisions. *Science, 338,* 270–273.

Xia, Y., Driscoll, J. R., Wilbrecht, L., Margolis, E. B., Fields, H. L., & Hjelmstad, G. O. (2011). Nucleus accumbens medium spiny neurons target non-dopaminergic neurons in the ventral tegmental area. *Journal of Neuroscience, 31,* 7811–7816.

Yartsev, M. M., Hanks, T. D., Yoon, A. M., & Brody, C. D. (2018). Causal contribution and dynamical encoding in the striatum during evidence accumulation. *eLife, 7,* pii:e34929. doi:10.7554/eLife.34929

Yin, H. H., & Knowlton, B. J. (2006). The role of the basal ganglia in habit formation. *Nature Reviews Neuroscience, 7,* 464–476.

Yin, H. H., Knowlton, B. J. & Balleine, B. W. (2004). Lesions of dorsolateral striatum preserve outcome expectancy, but disrupt habit formation in instrumental learning. *European Journal of Neuroscience, 19,* 181–189.

Yin, H. H., Mulcare, S. P., Hilário, M. R., Clouse, E., Holloway, T., Davis, M. I., Hansson, A. C., Lovinger, D. M., & Costa, R. M. (2009). Dynamic reorganization of striatal circuits during the acquisition and consolidation of a skill. *Nature Neuroscience, 12,* 333–341.

55 Dopamine and Reward: Implications for Neurological and Psychiatric Disorders

ANDREW WESTBROOK, ROSHAN COOLS, AND MICHAEL J. FRANK

ABSTRACT Dopamine is implicated in both learning and performing actions to pursue rewards. While disorders have been variously characterized as either having too high or too low dopamine function, this simplistic narrative ignores much richer computational and functional roles. Namely, dopamine appears to mediate multiple opponent trade-offs for adapting behavior to our highly dynamic environment. Here, we articulate how dopamine mediates key trade-offs and describe how dopamine dysregulation and maladaptive trade-offs underlie numerous neurological and psychiatric disorders.

Originally studied for its role in motor facilitation, dopamine has long been implicated in mediating both the *learning* and *performance* of behaviors to maximize reward (Graybiel, Aosaki, Flaherty, & Kimura, 1994). Meanwhile, dopamine dysregulation has been linked with diverse pathologies ranging from Parkinson's disease (PD) to attention deficit hyperactivity disorder (ADHD), addiction, schizophrenia, Tourette's syndrome, problem gambling, and depression (figure 55.1; Cools, 2011; Hauser, Fiore, Moutoussis, & Dolan, 2016; Maia & Frank, 2011). In drug addiction, for example, amplified dopamine release during drug intake drives powerful habit learning and compulsive drug pursuit (Dalley, Everitt, & Robbins, 2011). On the other hand, a profound loss of dopamine signaling, as in PD, yields both gross motor disturbances and a diminished motivation to work for reward (Chong et al., 2015).

Too much or too little dopamine can be clearly detrimental, but from a theoretical perspective, the story about dopamine's function is much more interesting and can in turn enhance our understanding of disorders. Namely, dopamine appears to mediate essential trade-offs in adaptive behavior, including relative sensitivity to rewards versus punishments, the weighting of economic costs versus benefits, reliance on hardwired behaviors versus online action selection, and stability versus flexibility. Thus, for example, while the dopamine precursor levodopa remediates motivational deficits in PD (alongside other core cognitive and motor symptoms), it also promotes problem gambling and compulsive sexual behaviors (Moore, Glenmullen, & Mattison, 2014). In the cognitive domain, while levodopa enhances PD patients' ability to switch between alternating task rules (increased cognitive flexibility), it also makes them worse at ignoring irrelevant stimulus-response pairings (reduced cognitive stability; Cools, Miyakawa, Sheridan, & D'Esposito, 2010). But how can such a complex set of functional outcomes be linked with a single neurotransmitter? What determines whether dopamine medications help or hurt? What mechanisms could unify these diverse phenotypes?

In this chapter we review key evidence for dopamine's functional roles in learning and performing behaviors in pursuit of reward. We articulate central mechanisms of action selection, with a focus on corticostriatal synaptic effects. Along the way we describe how dopamine mediates essential trade-offs in adaptive behavior and how dysregulation gives rise to numerous neurological and psychiatric disorders. Note that by focusing on dopamine we neglect other equally critical neurotransmitter systems like norepinephrine and serotonin. Our intent is to highlight the numerous functions in which dopamine has been implicated, not to imply that dopamine is either uniquely or wholly responsible for such functions. Although dopamine has been central to the reward-processing literature, a complete picture involves rich and complex interactions among multiple neurotransmitter systems.

Dopamine and Reinforcement Learning

There is broad agreement that dopamine is central to reinforcement learning—adapting behavior to maximize rewards and minimize losses (Sutton & Barto, 1998). A revolution in theorizing about the computational role of dopamine signaling, and theories about adaptive learning, was spawned with the observation that midbrain dopamine firing reports the degree to

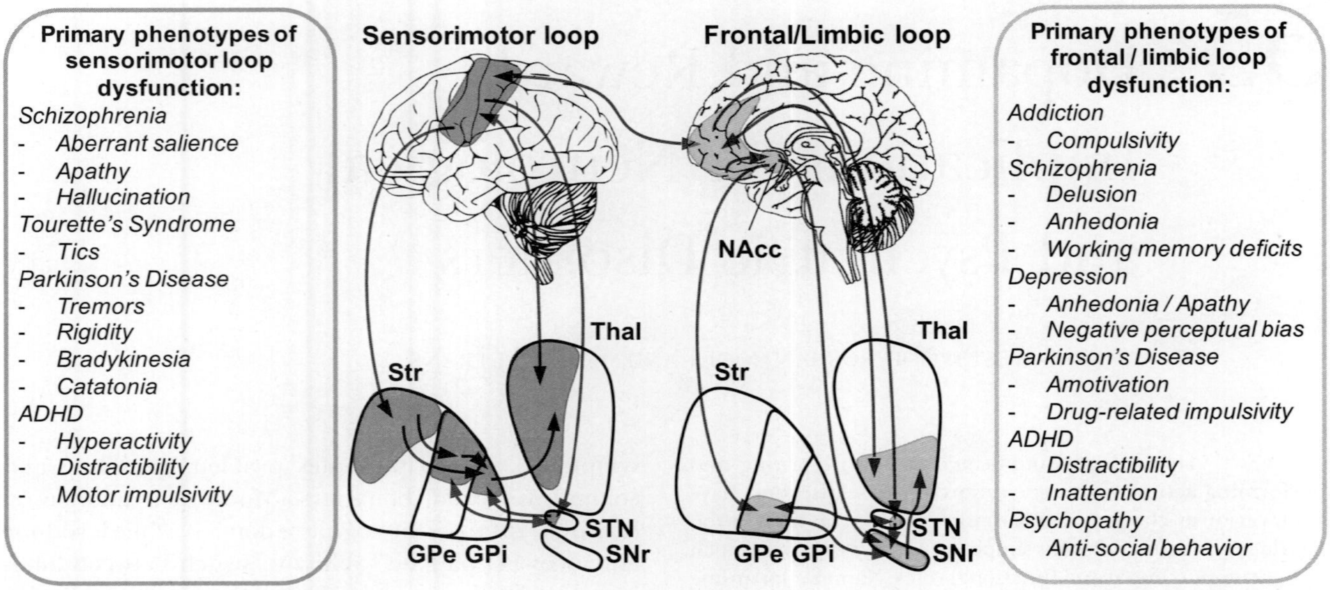

FIGURE 55.1 Diverse disorders can be linked, in part, to dysfunction in parallel corticostriatal loops where DA signaling shapes synaptic weights during learning and modulates circuit functionality during task performance. Phenotype lists are illustrative but not exhaustive. ADHD: attention deficit hyperactivity disorder; GPe: external globus pallidus; GPi: internal globus pallidus; NAcc: nucleus accumbens; STN: subthalamic nucleus; SNr: substantia nigra pars reticulata; Str: striatum; Thal: thalamus. Adapted from Hauser et al. (2016).

which reward outcomes are better or worse than expected, known as *reward prediction errors* (Montague, Dayan, & Sejnowski, 1996; Schultz, Dayan, & Montague, 1997). Dopamine dysregulation thus critically affects how quickly we adapt to changes in the outcomes of our actions. Moreover, it influences what we perceive as rewarding or punishing and, relatedly, how costly and beneficial we perceive our actions to be. In this section we discuss the implications of dopamine dysregulation on learning and action selection for multiple disorders.

Reward Versus Punishment Learning

Direct and indirect pathway synapses encode reward and punishment statistics In reinforcement learning, errors in predicted reward are used to update subsequent predictions about the value of states and actions. In terms of dopamine signaling, unexpected reward, or positive prediction errors, drive burst firing in dopamine neurons, while unexpected reward omissions, or negative prediction errors, transiently suppress dopamine neurons, yielding phasic dips in dopamine tone (Montague, Dayan, & Sejnowski, 1996; Schultz, Dayan, & Montague, 1997). As a neuromodulator, dopamine tunes excitatory or inhibitory drive and thus can, through modified Hebbian mechanisms, promote long-term potentiation (LTP) or long-term depression (LTD) at projection targets. Thus, dopamine has the capacity to "train" synapses to reflect reward and punishment statistics to make predictions about rewarding states and actions.

The impact of such signals is most established in the striatum, where two populations of *medium spiny neurons* (MSNs) facilitate action selection of the most rewarding choices. One population, which primarily expresses D1 receptors, sends inhibitory projections along the *direct pathway* (in corticostriatalthalamic loops; figure 55.2*A*) such that activity in this population increases the probability that rewarding actions are selected (by dis-inhibiting the thalamus). The other population primarily expresses D2 receptors and sends inhibitory projections along the *indirect pathway* such that their activity reduces the probability that less rewarding actions will be selected (by increasing inhibition of the thalamus). Phasic dopamine bursts and dips incrementally potentiate and depress these D1/D2 corticostriatal synapses in an opponent manner, making actions more or less likely in the future (Frank, 2005). Indeed, dopamine transients shape corticostriatal synapses with surgical precision—producing, for example, spine enlargement on D1-expressing MSNs specific to spines receiving afferent input from the cortex (i.e., those reflecting the state and action) and not those of other spines, even on the same neuron (Yagishita et al., 2014).

The net effect is that dopamine release reinforces specific actions by increasing the sensitivity of direct

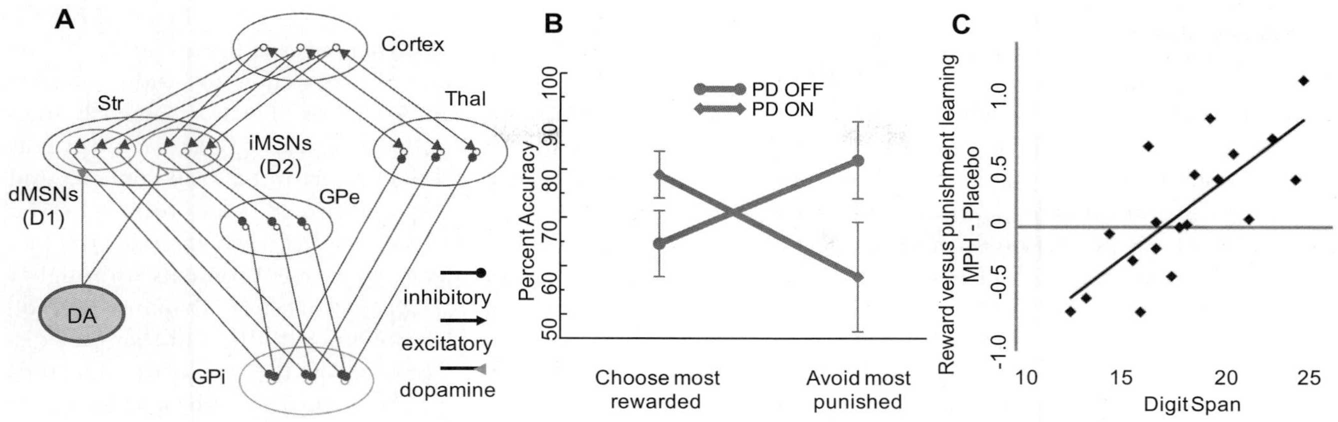

FIGURE 55.2 *A*, Phasic dopamine (DA) signals train corticostriatal synapses to reflect reward and punishment statistics for cortically represented actions. If these synapses facilitate enough activity in the D1-expressing direct versus the D2-expressing indirect pathway, actions are selected via disinhibition of the thalamus (Thal), which recurrently excites the evoking cortical representation. DA: dopamine; GPe: external globus pallidus; GPi: internal globus pallidus; MSN: medium spiny neuron; Str: Striatum. From Collins and Frank (2014). *B*, Medication effects on reinforcement learning in PD. Unmedicated PD patients are relatively impaired at learning to choose the most rewarded stimuli but are relatively better at avoiding the most punished stimuli. The pattern flips when patients are medicated with dopaminergic agents. From Frank, Seeberger, & O'Reilly (2004). Reprinted with permission from the AAAS. *C*, Effects of the dopamine-elevating drug methylphenidate (MPH) on learning vary by digit span, a putative index of individual differences in baseline striatal dopamine function (Cools & D'Esposito, 2011). MPH increases reward versus punishment learning for those with high span while decreasing it for those with low span. From Van der Schaaf et al. (2013). (See color plate 60.)

pathway MSNs (dMSNs) to cortical inputs while decreasing the sensitivity of indirect pathway MSNs (iMSNs). Conversely, dopamine dips punish actions by potentiating iMSN synapses and depressing dMSN synapses (Collins & Frank, 2014; Frank, 2005). Consistent with this account, potentiated activity in the direct and indirect pathways is both necessary and sufficient to reinforce and punish actions (Hikida, Kimura, Wada, Funabiki, & Nakanishi, 2010; Kravitz, Tye, & Kreitzer, 2012). Moreover, recent studies provide compelling evidence that the dopamine-mediated potentiation of dMSN/iMSN activity is specifically related to reinforcing or punishing the type of action selected and not simply modulating the overall likelihood of movement (Yttri & Dudman, 2016). Indeed, in this study, while dMSN stimulation following fast action promoted faster movements, stimulation following a slow action promoted slower movements. The reverse was true for iMSN stimulation, which promoted faster actions if stimulated after slow movements (i.e., slow movements were punished). These findings thus support the opponent reinforcement model of striatal dopamine and go beyond the classic model by which dopamine and D1 versus D2 activity are simply related to more or less movement.

A central prediction of the hypothesis that dopamine trains corticostriatal synapses, with major psychiatric implications, is that lesions to the dopamine system, as in PD, or dopamine medications, like levodopa, will modulate the propensity to learn from rewards and punishments (figure 55.2*B*). Since this prediction was first confirmed in PD patients (Cools, 2006; Frank, Seeberger, & O'Reilly, 2004), it has garnered robust support from more than 15 pharmacology and imaging studies with PD patients, recently including Piray et al. (2014) and Weismüller et al. (2018). Consistently replicated results include (1) that PD patients off dopamine medications (low dopamine) learn better from punishments but more poorly from rewards, and (2) patients on medications (high dopamine) learn better from rewards and more poorly from punishments. Impulse control disorders arising from dopamine therapies can be understood through this framework where medications make patients hypersensitive to rewards and hyposensitive to punishments by biasing striatal responses to prediction errors (Maia & Frank, 2011; Piray et al., 2014; Voon et al., 2010). Moreover, a model of amplified dopamine signaling biasing reward over punishment sensitivity may explain problem gambling in general, beyond PD (Rigoli et al., 2016; Van Holst et al., 2018). Conversely, hyposensitivity to rewards versus punishments, as resulting from an acute inflammation challenge, has been speculated to explain the link between chronic inflammation and depression (Harrison et al., 2016). Thus, different kinds of disorders may lie at the extremes of a dopamine-mediated trade-off between punishment and reward sensitivity.

The same patterns can also be seen in healthy subjects given dopamine drugs. A critical insight from this literature is that dopamine drugs do not have simple, unidirectional effects on reward-versus-punishment learning. Instead, drug effects appear to depend on individual differences in baseline dopamine function (Cools et al., 2009; Frank & O'Reilly, 2006). In one study, the stimulant methylphenidate (MPH) increased reward-versus-punishment learning, presumably by amplifying phasic prediction error signals, but it did so only for participants with a high working memory capacity, a putative marker of high baseline striatal dopamine function (figure 55.2*C*; Cools & D'Esposito, 2011; Van der Schaaf, Fallon, Huurne, Buitelaar, & Cools, 2013). For putatively low baseline dopamine function participants, the effect was reversed, consistent with an effect on presynaptic, autoregulatory D2 receptors that would paradoxically reduce phasic dopamine signals (Frank & O'Reilly, 2006). Though the exact mechanisms require further investigation, there is considerable evidence that dopamine drugs interact with baseline dopamine function in the trade-off between reward and punishment learning (Frank & Fossella, 2010). Indeed, many of the trade-offs described in this chapter are baseline dopamine-dependent—an insight that should give pause to casual "smart drug" users attempting to enhance cognitive function by pharmacological means.

COST VERSUS BENEFIT LEARNING While D2-receptor binding and iMSN activity are causally involved in the avoidance of risky gambles (Zalocusky et al., 2016), numerous pharmacology studies have also repeatedly implicated dopamine in effort learning and decision-making (Winstanley & Floresco, 2016). Thus, the notion that iMSNs and dMSNs encode punishment and reward statistics may be extended more broadly to encoding economic costs and benefits.

Animal and human studies have demonstrated that dopamine midbrain activity and dopamine release in the striatum mediate physical effort cost-benefit learning and predict effort expenditure. Optogenetic work has causally implicated striatal dopamine release while rodents learn about and decide whether to press levers for reward (Schelp et al., 2017), while functional magnetic resonance imaging (fMRI) has shown that the dopaminergic midbrain encodes prediction errors, tracking costs, and benefits while humans decide whether to squeeze handgrips for reward (Hauser, Eldar, & Dolan, 2017). Conversely, aberrant dopamine function has been hypothesized to explain the deficient effort expenditure symptomatic of multiple disorders, including depression, schizophrenia, and ADHD (Chong & Husain, 2016; Gold et al., 2013; Salamone et al., 2016).

Cognitive control is effortful, and, like physical effort, cognitive effort cost learning may also be dopamine-mediated (Cools, 2015; Westbrook & Braver, 2016). Cognitive control concerns the ability to pursue goals flexibly, using abstract rules to guide behavior rather than inflexible habits. Thus, subjectively high costs and a reduced willingness to exert control may partly explain deficient planning, self-control, and impulsivity in multiple disorders (Westbrook & Braver, 2015). Evidence that control is costly includes the fact that healthy adults learn to avoid contexts with higher cognitive control (task-switching) demands (Kool, McGuire, Rosen, & Botvinick, 2010) and that they discount monetary reward by the cognitive demands required to obtain the reward (Westbrook, Kester, & Braver, 2013). Moreover, discounting increases with negative symptom severity, suggesting that the lack of organized reward pursuit in schizophrenia may partly reflect subjectively high effort costs (Culbreth, Westbrook, & Barch, 2016).

There is evidence that cognitive effort learning is dopamine-dependent, just like physical effort. First, the anterior cingulate cortex (ACC), which is hypothesized to regulate cognitive control according to costs and benefits (Shenhav, Botvinick, & Cohen, 2013), tracks reinforcement learning prediction errors about cognitive effort demands (Nagase et al., 2018). In return, the ACC appears to reciprocally coordinate dopamine signaling. In one study, participants performed a cognitive conflict task while receiving probabilistic reward or punishment feedback (Cavanagh, Masters, Bath, & Frank, 2014). A subsequent transfer phase revealed that participants treated control demands as costly: rewarded stimuli were perceived as less rewarding and punishments more punishing when they had been paired with cognitive conflict. Critically, the degree to which conflict was punishing was related to an electroencephalography signal likely arising from the ACC and dopaminergic genes related to striatal plasticity, and was amplified by cabergoline, a D2 agonist. These results support the hypothesis that the ACC coordinates a punishment signal when faced with demands for cognitive control, conveyed by midbrain dopamine signals that, in turn, train striatal synapses to reflect cognitive effort costs.

LEARNING WHAT TO STORE IN WORKING MEMORY AND WHEN TO USE IT Corticostriatal loops govern cognitive action selection to maximize reward, just like motor actions (Frank, 2005; Frank & O'Reilly, 2006; Frank, Loughry, & O'Reilly, 2001). In the cognitive control domain, this means learning what rules to represent in working memory to maximize future reward in the same way the system might learn to press levers that predict reward. Thus, dopamine-trained corticostriatal synaptic weights determine the likelihood that task rules are *gated* into frontal working memory circuits to guide task execution. Critically, working memory gating can be hierarchical such that higher-level, more abstract rules (e.g., "I want to lose weight.") bias the gating of lower-level, more concrete rules (e.g., "When I see the doughnut, I will not eat it."), affording the powerful capacity to guide goal pursuit across complex, changing contexts (Collins & Frank, 2013).

If precise gating of the right information at the right time is essential to adaptive behavior, dysfunctional gating can be profoundly disruptive. For example, inappropriate rule representations may underlie psychosis and delusions in schizophrenia (Cannon, 2015). By one proposal, aberrant working memory gating in schizophrenia results from a combination of abnormally high striatal dopamine synthesis capacity and dysregulated striatal responses to rewards and punishments, where the wrong representations are imbued with high relative cached value (in corticostriatal synaptic weights), in the wrong contexts (Boehme et al., 2015). At the other end of the spectrum lies perseveration, where new rules are slow to be adopted when contingencies change. PD patients are particularly slow at adapting to changing contexts, as revealed by *reversal learning* tasks (Cools, Barker, Sahakian, & Robbins, 2001), described in the next section.

HABITS, REVERSAL LEARNING, AND WORKING MEMORY: SLOW VERSUS FAST BEHAVIORAL ADAPTATION During learning, dopamine subserves another essential trade-off in the speed of adaptation, with overly slow, perseverative behavior at one end of the spectrum and instability at the other. A central question concerns the development of *habits*, which are slow to acquire but robust to change and can be optimal in stable environments. However, habits also underlie compulsivity and addiction. Furthermore, when the environment changes, as in reversal-learning tasks, rapid adaptation is better.

Habit formation Reinforcement learners are fundamentally good at integrating reward statistics over many encounters yet fundamentally bad at adapting to rapid changes. Thus, strongly learned behaviors, or *habits*, can be difficult to override, even when they are maladaptive (Dalley, Everitt, & Robbins, 2011; Volkow, Wise, & Baler, 2017). Addiction-related habits may be more likely to form when stronger phasic dopamine signals more robust learning. In one study, for example, PET measures of dopamine release were larger to alcohol cues versus Gatorade, particularly for participants with alcoholic relatives, implying a genetic predisposition to alcoholism that is related to larger dopaminergic responses to alcohol cues (Oberlin et al., 2013).

Fast learning rates are advantageous when the context changes. Adaptation to changing contexts has been studied extensively in humans using reversal-learning paradigms (Clatworthy et al., 2009; Cools et al., 2009; Dodds et al., 2008). In a typical task, while participants learn to select between rewarding and punishing stimuli, reward contingencies switch unannounced, and participants must learn to switch choices. Reinforcement learners converge to the best answer over many trials following a change of context; however, their immediate behavior is dominated, in the short run, by past experience. The degree to which experience dominates appears to be determined by multiple facets of dopamine signaling. In a large human genetics study, variants of the dopamine transporter (DAT) gene associated with more dynamic dopamine responses to rewards predicted greater difficulty for participants to reverse their choice behavior when response contingencies changed (den Ouden et al., 2013). This result was interpreted as consistent with the account that more robust reward-related dopamine transients predicted more robust "stamping in" of prior statistics, which were subsequently more difficult to override. Thus, individual differences in the propensity for habit formation and compulsivity may partly reflect underlying DAT gene variants.

Working memory versus reinforcement learning Working memory gating is fast relative to reinforcement learning—indeed, it's immediate! On the other hand, unlike reinforcement learning, working memory is capacity-limited and subject to interference and decay. While participants, therefore, can remember what they did in their last trial, they cannot reliably track what they did and the outcomes they received over the last ten trials. Thus, the optimal approach involves balancing the relative merits of fast and flexible working memory with steady and reliable reinforcement learning. Indeed, participants use both systems when solving stimulus-response learning tasks in a dynamic manner, depending on task demands (Collins & Frank, 2012, 2018). An interesting consequence of coordination between systems is that better working memory–based predictions mean smaller prediction errors and thus slower reinforcement learning.

Imbalances in the degree to which working memory and reinforcement learning contribute to task performance have been linked with psychopathology: people with schizophrenia show reduced working memory contributions with relatively intact reinforcement learning (Collins, Albrecht, Waltz, Gold, & Frank, 2017).

The balance between working memory versus reinforcement learning also mirrors a classic distinction between flexible goal-directed action selection and incremental habit formation (Dolan & Dayan, 2013). Here, the constructs are connected in that goal states are represented (hierarchically) in working memory to bias behavior, often in opposition to habits acquired via reinforcement learning. Interestingly, an increased reliance on habitual versus goal-directed action appears to predict the formation of both eating disorders and drug addiction (Voon et al., 2014), highlighting that the balance between the two systems is critical for adaptive behavior. The distinction between goals and habits has also been mapped onto the more recent distinction between model-based versus model-free decision-making, which, as will be discussed in the second half of this chapter, also varies as a function of dopamine function.

Performance Effects of Dopamine

While dopamine signals are central to learning and adaptation, dopamine also modulates task performance, independent of its learning effects (Beeler, Daw, Frazier, & Zhuang, 2010; Cagniard et al., 2006). These performance effects also mediate key trade-offs in adaptive behavior, including instantaneous sensitivity to costs versus benefits, stability versus flexibility, and reliance on deeply ingrained behaviors versus online action selection. Thus, as with learning, dopamine's effects on performance are widespread, with implications for multiple disorders.

Action Selection

Instantaneous cost-benefit sensitivity During action selection, dopamine instantaneously modulates the expression of learned costs and benefits (Collins & Frank, 2014). As dopamine tone rises, action benefits are weighed more heavily than costs—and vice versa when dopamine falls (cf., *incentive salience*; Zhang, Berridge, Tindell, Smith, & Aldridge, 2009). In the effort domain, a robust rodent literature supports the hypothesis that striatal dopamine increases, while a striatal dopamine blockade decreases, high-effort, high-reward selection versus low-effort, low-reward alternatives (Hamid et al., 2016; Salamone et al., 2016). Importantly, a dopamine blockade not only reduces vigor and suppresses action but can specifically bias the avoidance of high costs.

If dopamine conveys information about economic context, dopamine depletion should reduce effort selection in humans as well. In fact, PD patients show diminished high-effort (handgrip squeezing), high-benefit selection versus low-effort, low-benefit options, and this effect is meliorated by dopamine medications (Chong et al., 2015). Clinical apathy, more broadly, also predicts reduced effort selection (Chong & Husain, 2016), and apathy has been specifically related to the degree of reward hyposensitivity in PD (Muhammed et al., 2016). Importantly, just like in the motor domain, higher dopamine tone may also promote high-cost, high-benefit cognitive control over low-cost, low-benefit habits (Manohar et al., 2015; Westbrook & Braver, 2016). Thus, reduced dopamine function should undermine control. Indeed, the lack of top-down control in ADHD may partly reflect the reduced motivational signaling associated with diminished striatal dopamine function (Volkow et al., 2010).

Vigor and economy of action More effort is not always better, particularly if it is metabolically costly. An influential, normative account proposes that dopamine mediates a strategic trade-off between faster movement with higher energetic costs and slower action with higher opportunity costs (Niv, Daw, Joel, & Dayan, 2007). Since opportunity costs increase with more reward in the environment, action should be more vigorous with increasing reward density. Striatal dopamine tone is proposed to play roles here, both in signaling the average rate of reward obtained—a proxy for reward density—and in invigorating behavior by lowering the threshold for action. Mechanistically, dopamine can lower the threshold for action by instantaneously making dMSN cells more excitable and iMSN cells less excitable so that all candidate actions more easily reach threshold dMSN-versus-iMSN activity (Collins & Frank, 2014).

The normative account implicating dopamine in action invigoration is well supported in humans. For example, action is systematically invigorated by average reward in humans (Cools, Nakamura, & Daw, 2010; Guitart-Masip, Beierholm, Dolan, Duzel, & Dayan, 2011), in a dopamine drug-dependent manner (Beierholm et al., 2013). A major clinical implication of deficient dopamine signaling, therefore, is diminished behavioral vigor. Indeed, *bradykinesia* (slowed and halting movements) is a central symptom of PD (Mazzoni, Hristova, & Krakauer, 2007) and may thus reflect altered cost-based decision-making. PD patients are also systematically worse at speeding up to obtain reward and better at slowing down to avoid punishment (Moustafa, Cohen, Sherman, & Frank, 2008).

Stability versus flexibility We have seen how dopamine likely mediates a dynamic balance between the opponent requirements of slow and fast learning in stable and changing environments, respectively. An analogous proposal has been put forward in the domain of working memory, where dopamine tone might mediate the trade-off between cognitive stability and flexibility (Cools & D'Esposito, 2011; Frank & O'Reilly, 2006). Here, rather than lowering the threshold to invigorate motoric action, striatal dopamine is thought to lower the threshold for working memory gating, increasing the chance that new information becomes represented in working memory circuits. On one end of the trade-off, insufficient gating may underlie overly stable, perseverative behaviors, as observed in PD patients off medication, who are significantly slower at gating-dependent working memory updating (Cools et al., 2001). On the other end, hyperflexibility and distractibility result from excessive gating, undermining the goal maintenance needed for protracted task engagement.

In contrast with the striatum, where dopamine tone is regulated by the fast reuptake by dopamine transporters, dopamine tone in the cortex is chiefly regulated by dopamine degradation by catechol-O-methyltransferase, with a much slower time constant (Cools & D'Esposito, 2011). This distinction implies that cortical dopamine governs relatively protracted, stable processes. Indeed, a core function of cortical dopamine is promoting working memory maintenance (Arnsten, 2011; Sawaguchi & Goldman-Rakic, 1991). Under typical conditions, increasing cortical dopamine tone promotes behavioral stability, via the robust maintenance of existing representations, at the cost of reduced behavioral flexibility, via insensitivity to new input (Cools & D'Esposito, 2011). This hypothesis is supported by evidence that MPH, which increases dopamine tone globally, has been shown to promote cognitive stability, via reliably increased cortical activity, and numerically larger striatal response during flexible working memory updating (Fallon, Van der Schaaf, Huurne, & Cools, 2016).

HARDWIRED VERSUS CONTEXT-SENSITIVE ACTION Although many behaviors are acquired via reinforcement learning, some, thankfully, are innate—like withdrawing a hand from a hot stove. These tendencies are evolutionarily endowed and are deeply ingrained. However, we might want to override hardwired behavior—for example, when we want to reach toward the hot stove to retrieve a pan—and thus we may prefer to select our behaviors online. The degree to which our behaviors are governed by ingrained tendencies versus online action selection is yet another key trade-off mediated by dopamine.

Pavlovian action biases An approach toward appetitive stimuli, or the avoidance of aversive stimuli, and acting under the prospect of appetitive outcomes while inhibiting behavior under the threat of aversive outcomes are two types of *Pavlovian biases*. Dopamine appears to underpin these biases, perhaps in coordination with serotonin (Boureau & Dayan, 2011; Cools, Nakamura, & Daw, 2010). For example, it is easier to learn to actively press a button to obtain a reward (a GO response) than it is to withhold a response to obtain a reward (a NoGO response), and the coupling of action and reward is modulated by dopamine signaling (Crockett, Clark, & Robbins, 2009; Guitart-Masip et al., 2012; Swart et al., 2017).

Interestingly, both the absence and exaggeration of Pavlovian biases have been linked with various forms of psychopathology. While healthy controls couple active avoidance reactions to angry face stimuli, violent offenders do not, suggesting that the lack of a key Pavlovian social bias contributes to violent psychopathy (Ly et al., 2016). Similarly, the absence of coupling between an approach to appetitive stimuli and a withdrawal from aversive stimuli both distinguishes depressed patients from healthy controls and predicts a lesser recovery from depression 4–6 months after a laboratory visit (Huys et al., 2016). Overactive Pavlovian biases can also be profoundly detrimental. For example, alcohol dependency and relapse can both be predicted by an abnormally strong bias of appetitive Pavlovian stimuli on instrumental button pressing and its associated signal in the ventral striatum (Garbusow et al., 2015). Aberrant Pavlovian control of behavior, stemming from dopamine dysregulation, may thus contribute to diverse disorders.

Model-based versus model-free action selection At the other end of the spectrum from hardwired Pavlovian control is online goal selection, or *model-based* control, with habitual or *model-free* control lying somewhere in between (Daw, Niv, & Dayan, 2005; Dolan & Dayan, 2013). *Model-based* refers to the fact that arbitrarily selected goals can be maintained and guide action selection based on an internal model of the environment. Note that such an internal model is not required in the case of habitual "model-free" actions in which one simply repeats actions that were rewarded in the past, without regard for their distal consequences. Deficient model-based decision-making is profoundly problematic, leading to an overreliance on habits or hardwired Pavlovian biases and a diminished sensitivity to future consequences. Thus, as noted above, reduced model-based decision-making has been linked with compulsivity disorders, including binge eating and methamphetamine addiction (Voon et al., 2014).

Model-based decision-making (Daw, Niv, & Dayan, 2005) involves the computationally intensive simulation of models of actions and their consequences. Thus,

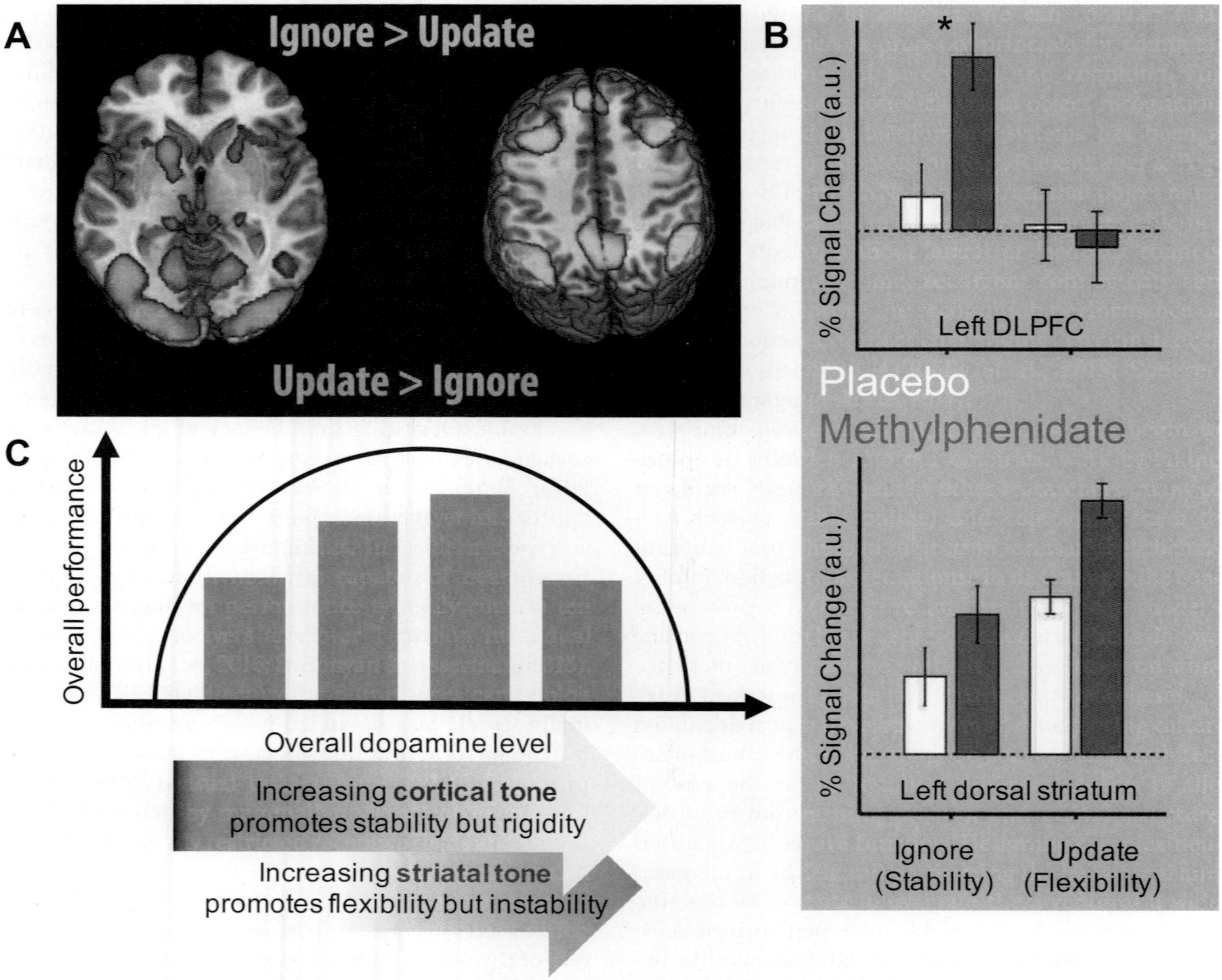

FIGURE 55.3 *A*, Brain regions recruited differentially as a function of demands for cognitive stability versus flexibility in a working memory task. *A*, The frontoparietal cortical network is more active for stable ignore/maintenance versus updating trials, while striatal regions are more active for flexible updating. *B*, Dopamine upregulation with MPH reliably increases the frontal cortical response when demands for stability are higher and numerically increases the striatal response when demands for flexibility are higher. *C*, Globally increased dopamine release has opposing effects in increasing cognitive stability via the cortex and increasing flexibility via the striatum, with optimal overall performance at the midrange of global dopamine function. *A* and *B* adapted from Fallon et al. (2016). (See color plate 61.)

model-based control appears to be subjectively costly in a way that can be offset by greater incentives (Kool, Gershman, & Cushman, 2017). The capacity of striatal dopamine to bias high-cost, high-benefit actions may partly explain why a higher striatal dopamine synthesis capacity (Deserno et al., 2015) and the administration of levodopa (Wunderlich, Smittenaar, & Dolan, 2012) both promote model-based control. Alternatively, a global dopamine agonist like levodopa may act to promote the stability of cortical working memory circuits, thus promoting model-based control by increasing the capacity for model-based computation. Future research is required to disentangle cortical and subcortical effects, but converging lines of evidence increasingly implicate dopamine function in model-based versus model-free control of behavior.

Conclusion

Dopamine signaling has profound consequences for organizing behavior to pursue reward and avoid punishment. It optimizes action selection in cases when the environment is stable and, alternately, when it is changing, facilitating the learning and expression of reward statistics that are relevant and overriding them when they are not. It promotes adaptation to changing contexts and mediates between working memory and reinforcement-learning systems to optimize learning as a function of experience. It couples action and valence in ways that are more often beneficial than not. It conveys instantaneous economic context and determines the weighting of benefits versus costs during action selection and economic choice. It mediates between behavioral stability and flexibility, via action in the cortex and striatum, to orient behavior toward immediate or delayed outcomes. Consequently, the disruption of dopamine signaling anticipates numerous and diverse disorders. Improving psychiatric treatment with respect to dopamine disruption will require grappling with complex mechanisms of action with multifaceted functional implications.

REFERENCES

Arnsten, A. F. T. (2011). Catecholamine influences on dorsolateral prefrontal cortical networks. *Biological Psychiatry, 69*(12), e89–e99. http://doi.org/10.1016/j.biopsych.2011.01.027

Beeler, J. A., Daw, N. D., Frazier, C. R. M., & Zhuang, X. (2010). Tonic dopamine modulates exploitation of reward learning. *Frontiers in Behavioral Neuroscience, 4.* http://doi.org/10.3389/fnbeh.2010.00170

Beierholm, U., Guitart-Masip, M., Economides, M., Chowdhury, R., Duzel, E., Dolan, R. J., & Dayan, P. (2013). Dopamine modulates reward-related vigor. *Neuropsychopharmacology, 38*(8), 1495–1503. http://doi.org/10.1038/npp.2013.48

Boehme, R., Deserno, L., Gleich, T., Katthagen, T., Pankow, A., Behr, J., et al. (2015). Aberrant salience is related to reduced reinforcement learning signals and elevated dopamine synthesis capacity in healthy adults. *Journal of Neuroscience, 35*(28), 10103–10111. http://doi.org/10.1523/JNEUROSCI.0805-15.2015

Boureau, Y.-L., & Dayan, P. (2011). Opponency revisited: Competition and cooperation between dopamine and serotonin. *Neuropsychopharmacology, 36*(1), 74–97. http://doi.org/10.1038/npp.2010.151

Cagniard, B., Beeler, J. A., Britt, J. P., McGehee, D. S., Marinelli, M., & Zhuang, X. (2006). Dopamine scales performance in the absence of new learning. *Neuron, 51*(5), 541–547. http://doi.org/10.1016/j.neuron.2006.07.026

Cannon, T. D. (2015). How schizophrenia develops: Cognitive and brain mechanisms underlying onset of psychosis. *Trends in Cognitive Sciences, 19*(12), 744–756. http://doi.org/10.1016/j.tics.2015.09.009

Cavanagh, J. F., Masters, S. E., Bath, K., & Frank, M. J. (2014). Conflict acts as an implicit cost in reinforcement learning. *Nature Communications, 5*(5394). http://doi.org/10.1038/ncomms6394

Chong, T. T. J., Bonnelle, V., Manohar, S., Veromann, K.-R., Muhammed, K., Tofaris, G. K., et al. (2015). Dopamine enhances willingness to exert effort for reward in Parkinson's disease. *Cortex, 69,* 40–46. http://doi.org/10.1016/j.cortex.2015.04.003

Chong, T. T. J., & Husain, M. (2016). The role of dopamine in the pathophysiology and treatment of apathy. In Bettina Studer & Stefan Knecht (Eds.), *Progress in brain research* (Vol. 229, pp. 389–426). New York: Elsevier. http://doi.org/10.1016/bs.pbr.2016.05.007

Clatworthy, P. L., Lewis, S. J. G., Brichard, L., Hong, Y. T., Izquierdo, D., Clark, L., et al. (2009). Dopamine release in dissociable striatal subregions predicts the different effects of oral methylphenidate on reversal learning and spatial working memory. *Journal of Neuroscience, 29*(15), 4690–4696. http://doi.org/10.1523/JNEUROSCI.3266-08.2009

Collins, A. G. E., Albrecht, M. A., Waltz, J. A., Gold, J. M., & Frank, M. J. (2017). Interactions between working memory, reinforcement learning and effort in value-based choice: A new paradigm and selective deficits in schizophrenia. *Biological Psychiatry, 82*(6) 431—439. http://doi.org/10.1016/j.biopsych.2017.05.017

Collins, A. G. E., & Frank, M. J. (2012). How much of reinforcement learning is working memory, not reinforcement learning? A behavioral, computational, and neurogenetic analysis. *European Journal of Neuroscience, 35*(7), 1024–1035. http://doi.org/10.1111/j.1460-9568.2011.07980.x

Collins, A. G. E., & Frank, M. J. (2013). Cognitive control over learning: Creating, clustering, and generalizing task-set structure. *Psychological Review, 120*(1), 190–229. http://doi.org/10.1037/a0030852

Collins, A. G. E., & Frank, M. J. (2014). Opponent actor learning (OpAL): Modeling interactive effects of striatal dopamine on reinforcement learning and choice incentive. *Psychological Review, 121*(3), 337–366. http://doi.org/10.1037/a0037015

Collins, A. G. E., & Frank, M. J. (2018). Within- and across-trial dynamics of human EEG reveal cooperative interplay between reinforcement learning and working memory. *Proceedings of the National Academy of Sciences, 115*(10), 2502–2507. http://doi.org/10.1214/aos/1176344136

Cools, R. (2006). Dopaminergic modulation of cognitive function-implications for l-DOPA treatment in Parkinson's disease. *Neuroscience and Biobehavioral Reviews, 30*(1), 1–23. http://doi.org/10.1016/j.neubiorev.2005.03.024

Cools, R. (2011). Dopaminergic control of the striatum for high-level cognition. *Current Opinion in Neurobiology, 21*(3), 402–407. http://doi.org/10.1016/j.conb.2011.04.002

Cools, R. (2015). The cost of dopamine for dynamic cognitive control. *Current Opinion in Behavioral Sciences, 4,* 152–159. http://doi.org/10.1016/j.cobeha.2015.05.007

Cools, R., Barker, R. A., Sahakian, B. J., & Robbins, T. W. (2001). Enhanced or impaired cognitive function in Parkinson's disease as a function of dopaminergic medication and task demands. *Cerebral Cortex, 11*(12), 1136–1143.

Cools, R., & D'Esposito, M. (2011). Inverted-U-shaped dopamine actions on human working memory and cognitive control. *Biological Psychiatry, 69*(12), e113–e125. http://doi.org/10.1016/j.biopsych.2011.03.028

Cools, R., Frank, M. J., Gibbs, S. E., Miyakawa, A., Jagust, W., & D'Esposito, M. (2009). Striatal dopamine predicts outcome-specific reversal learning and its sensitivity to dopaminergic drug administration. *Journal of Neuroscience, 29*(5), 1538–1543. http://doi.org/10.1523/JNEUROSCI.4467-08.2009

Cools, R., Miyakawa, A., Sheridan, M., & D'Esposito, M. (2010). Enhanced frontal function in Parkinson's disease. *Brain, 133*(1), 225–233. http://doi.org/10.1093/brain/awp301

Cools, R., Nakamura, K., & Daw, N. D. (2010). Serotonin and dopamine: Unifying affective, activational, and decision functions. *Neuropsychopharmacology, 36*(1), 98–113. http://doi.org/10.1038/npp.2010.121

Crockett, M. J., Clark, L., & Robbins, T. W. (2009). Reconciling the role of serotonin in behavioral inhibition and aversion: Acute tryptophan depletion abolishes punishment-induced inhibition in humans. *Journal of Neuroscience, 29*(38), 11993–11999. http://doi.org/10.1523/JNEUROSCI.2513-09.2009

Culbreth, A., Westbrook, A., & Barch, D. (2016). Negative symptoms are associated with an increased subjective cost of cognitive effort. *Journal of Abnormal Psychology, 125*(4), 528–536. http://doi.org/10.1037/abn0000153

Dalley, J. W., Everitt, B. J., & Robbins, T. W. (2011). Impulsivity, compulsivity, and top-down cognitive control. *Neuron, 69*(4), 680–694. http://doi.org/10.1016/j.neuron.2011.01.020

Daw, N. D., Niv, Y., & Dayan, P. (2005). Uncertainty-based competition between prefrontal and dorsolateral striatal systems for behavioral control. *Nature Neuroscience, 8*(12), 1704–1711. http://doi.org/10.1038/nn1560

den Ouden, H. E. M., Daw, N. D., Fernandez, G., Elshout, J. A., Rijpkema, M., Hoogman, M., et al. (2013). Dissociable effects of dopamine and serotonin on reversal learning. *Neuron, 80*(4), 1090–1100. http://doi.org/10.1016/j.neuron.2013.08.030

Deserno, L., Huys, Q. J. M., Boehme, R., Buchert, R., Heinze, H.-J., Grace, A. A., et al. (2015). Ventral striatal dopamine reflects behavioral and neural signatures of model-based control during sequential decision making. *Proceedings of the National Academy of Sciences, 112*(5), 1595–1600. http://doi.org/10.1073/pnas.1417219112

Dodds, C. M., Muller, U., Clark, L., Van Loon, A., Cools, R., & Robbins, T. W. (2008). Methylphenidate has differential effects on blood oxygenation level-dependent signal related to cognitive subprocesses of reversal learning. *Journal of Neuroscience, 28*(23), 5976–5982. http://doi.org/10.1523/JNEUROSCI.1153-08.2008

Dolan, R. J., & Dayan, P. (2013). Goals and habits in the brain. *Neuron, 80*(2), 312–325. http://doi.org/10.1016/j.neuron.2013.09.007

Fallon, S. J., Van der Schaaf, M. E., Ter Huurne, N., & Cools, R. (2016). The neurocognitive cost of enhancing cognition with methylphenidate: Improved distractor resistance but impaired updating. *Journal of Cognitive Neuroscience, 29*(4), 652–663. http://doi.org/10.1162/jocn_a_01065

Frank, M. J. (2005). Dynamic dopamine modulation in the basal ganglia: A neurocomputational account of cognitive deficits in medicated and nonmedicated parkinsonism. *Journal of Cognitive Neuroscience, 17*(1), 51–72.

Frank, M. J., & Fossella, J. A. (2010). Neurogenetics and pharmacology of learning, motivation, and cognition. *Neuropsychopharmacology, 36*(1), 133–152. http://doi.org/10.1038/npp.2010.96

Frank, M. J., Loughry, B., & O'Reilly, R. C. (2001). Interactions between frontal cortex and basal ganglia in working memory: A computational model. *Cognitive, Affective, & Behavioral Neuroscience, 1*(2), 137–160.

Frank, M. J., & O'Reilly, R. C. (2006). A mechanistic account of striatal dopamine function in human cognition: Psychopharmacological studies with cabergoline and haloperidol. *Behavioral Neuroscience, 120*(3), 497–517. http://doi.org/10.1037/0735-7044.120.3.497

Frank, M. J., Seeberger, L. C., & O'Reilly, R. C. (2004). By carrot or by stick: Cognitive reinforcement learning in parkinsonism. *Science, 306*(5703), 1940–1943. http://doi.org/10.1146/annurev.ento.50.071803.130456

Garbusow, M., Schad, D. J., Sebold, M., Friedel, E., Bernhardt, N., Koch, S. P., et al. (2015). Pavlovian-to-instrumental transfer effects in the nucleus accumbens relate to relapse in alcohol dependence. *Addiction Biology, 21*(3), 719–731. http://doi.org/10.1038/npp.2013.252

Gold, J. M., Strauss, G. P., Waltz, J. A., Robinson, B. M., Brown, J. K., & Frank, M. J. (2013). Negative symptoms of schizophrenia are associated with abnormal effort-cost computations. *Biological Psychiatry, 74*(2), 130–136. http://doi.org/10.1016/j.biopsych.2012.12.022

Graybiel, A. M., Aosaki, T., Flaherty, A. W., & Kimura, M. (1994). The basal ganglia and adaptive motor control. *Science, 265*(5180), 1826–1831.

Guitart-Masip, M., Beierholm, U. R., Dolan, R. J., Duzel, E., & Dayan, P. (2011). Vigor in the face of fluctuating rates of reward: An experimental examination. *Journal of Cognitive Neuroscience, 23*(12), 3933–3938.

Guitart-Masip, M., Chowdhury, R., Sharot, T., Dayan, P., Duzel, E., & Dolan, R. J. (2012). Action controls dopaminergic enhancement of reward representations. *Proceedings of the National Academy of Sciences of the United States of America, 109*(19), 7511–7516. http://doi.org/10.1073/pnas.1202229109

Hamid, A. A., Pettibone, J. R., Mabrouk, O. S., Hetrick, V. L., Schmidt, R., Vander Weele, C. M., et al. (2016). Mesolimbic dopamine signals the value of work. *Nature, 19*. http://doi.org/10.1038/nn.4173

Harrison, N. A., Voon, V., Cercignani, M., Cooper, E. A., Pessiglione, M., & Critchley, H. D. (2016). Archival report. *Biological Psychiatry, 80*(1), 73–81. http://doi.org/10.1016/j.biopsych.2015.07.018

Hauser, T. U., Eldar, E., & Dolan, R. J. (2017). Separate mesocortical and mesolimbic pathways encode effort and reward learning signals. *Proceedings of the National Academy of Sciences, 114*(35), E7395–E7404. http://doi.org/10.1155/2000/421719

Hauser, T. U., Fiore, V. G., Moutoussis, M., & Dolan, R. J. (2016). Computational psychiatry of ADHD: Neural gain impairments across Marrian levels of analysis. *Trends in Neurosciences, 39*(2), 63–73. http://doi.org/10.1016/j.tins.2015.12.009

Hikida, T., Kimura, K., Wada, N., Funabiki, K., & Nakanishi, S. (2010). Distinct roles of synaptic transmission in direct and indirect striatal pathways to reward and aversive behavior. *Neuron, 66*(6), 896–907. http://doi.org/10.1016/j.neuron.2010.05.011

Huys, Q. J. M., Gölzer, M., Friedel, E., Heinz, A., Cools, R., Dayan, P., & Dolan, R. J. (2016). The specificity of Pavlovian regulation is associated with recovery from depression. *Psychological Medicine, 46*(05), 1027–1035. http://doi.org/10.1017/S0033291715002597

Kool, W., Gershman, S. J., & Cushman, F. A. (2017). Cost-benefit arbitration between multiple reinforcement-learning systems. *Psychological Science, 28*(9), 1321–1333. http://doi.org/10.1177/0956797617708288

Kool, W., McGuire, J. T., Rosen, Z. B., & Botvinick, M. M. (2010). Decision making and the avoidance of cognitive demand. *Journal of Experimental Psychology-General, 139*(4), 665–682. http://doi.org/10.1037/a0020198

Kravitz, A. V., Tye, L. D., & Kreitzer, A. C. (2012). Distinct roles for direct and indirect pathway striatal neurons in reinforcement. *Nature Neuroscience, 15*(6), 816–818. http://doi.org/10.1038/nn.3100

Ly, V., Von Borries, A. K. L., Brazil, I. A., Bulten, B. H., Cools, R., & Roelofs, K. (2016). Reduced transfer of affective value to instrumental behaviors in violent offenders. *Journal of Abnormal Psychology, 125*(5), 657–663. http://doi.org/10.1037/abn0000166

Maia, T. V., & Frank, M. J. (2011). From reinforcement learning models to psychiatric and neurological disorders. *Nature Neuroscience, 14*(2), 154–162. http://doi.org/10.1038/nn.2723

Manohar, S. G., Chong, T. T. J., Apps, M. A. J., Batla, A., Stamelou, M., Jarman, P. R., et al. (2015). Reward pays the cost of noise reduction in motor and cognitive control. *Current Biology, 25*(13), 1707–1716. http://doi.org/10.1016/j.cub.2015.05.038

Mazzoni, P., Hristova, A., & Krakauer, J. W. (2007). Why don't we move faster? Parkinson's disease, movement vigor, and implicit motivation. *Journal of Neuroscience, 27*(27), 7105–7116. http://doi.org/10.1523/JNEUROSCI.0264-07.2007

Montague, P., Dayan, P., & Sejnowski, T. (1996). A framework for mesencephalic dopamine systems based on predictive Hebbian learning. *Journal of Neuroscience, 16*(5), 1936.

Moore, T. J., Glenmullen, J., & Mattison, D. R. (2014). Reports of pathological gambling, hypersexuality, and compulsive shopping associated with dopamine receptor agonist drugs. *Journal of the American Medical Association Internal Medicine, 174*(12), 1930. http://doi.org/10.1001/jamainternmed.2014.5262

Moustafa, A. A., Cohen, M. X., Sherman, S. J., & Frank, M. J. (2008). A role for dopamine in temporal decision making and reward maximization in parkinsonism. *Journal of Neuroscience, 28*(47), 12294–12304. http://doi.org/10.1523/JNEUROSCI.3116-08.2008

Muhammed, K., Manohar, S., Ben Yehuda, M., Chong, T. T. J., Tofaris, G., Lennox, G., et al. (2016). Reward sensitivity deficits modulated by dopamine are associated with apathy in Parkinson's disease. *Brain, 139*(10), 2706–2721. http://doi.org/10.1016/j.neuropsychologia.2015.11.019

Nagase, A. M., Onoda, K., Clifford Foo, J., Haji, T., Akaishi, R., Yamaguchi, S., et al. (2018). Neural mechanisms for adaptive learned avoidance of mental effort. *Journal of Neuroscience, 38*(10), 2631–2651. http://doi.org/10.1523/JNEUROSCI.1995-17.2018

Niv, Y., Daw, N. D., Joel, D., & Dayan, P. (2007). Tonic dopamine: Opportunity costs and the control of response vigor. *Psychopharmacology, 191*(3), 507–520.

Oberlin, B. G., Dzemidzic, M., Tran, S. M., Soeurt, C. M., Albrecht, D. S., Yoder, K. K., & Kareken, D. A. (2013). Beer flavor provokes striatal dopamine release in male drinkers: Mediation by family history of alcoholism. *Neuropsychopharmacology, 38*(9), 1617–1624. http://doi.org/10.1038/npp.2013.91

Piray, P., Zeighami, Y., Bahrami, F., Eissa, A. M., Hewedi, D. H., & Moustafa, A. A. (2014). Impulse control disorders in Parkinson's disease are associated with dysfunction in stimulus valuation but not action valuation. *Journal of Neuroscience, 34*(23), 7814–7824. http://doi.org/10.1523/JNEUROSCI.4063-13.2014

Rigoli, F., Rutledge, R. B., Chew, B., Ousdal, O. T., Dayan, P., & Dolan, R. J. (2016). Dopamine increases a value-independent gambling propensity. *Neuropsychopharmacology, 41*(11), 2658–2667. http://doi.org/10.1038/npp.2016.68

Salamone, J. D., Correa, M., Yohn, S., Cruz, L. L., Miguel, N. S., & Alatorre, L. (2016). The pharmacology of effort-related choice behavior: Dopamine, depression, and individual differences. *Behavioural Processes, 127*, 3–17. http://doi.org/10.1016/j.beproc.2016.02.008

Sawaguchi, T., & Goldman-Rakic, P. S. (1991). D1 dopamine receptors in prefrontal cortex: Involvement in working memory. *Science, 251*(4996), 947–950.

Schelp, S. A., Pultorak, K. J., Rakowski, D. R., Gomez, D. M., Krzystyniak, G., Das, R., & Oleson, E. B. (2017). A transient dopamine signal encodes subjective value and causally influences demand in an economic context. *Proceedings of the National Academy of Sciences, 367*(52), E11303–E11312. http://doi.org/10.1021/ac402884n

Schultz, W., Dayan, P., & Montague, P. R. (1997). A neural substrate of prediction and reward. *Science, 275*(5306), 1593–1599.

Shenhav, A., Botvinick, M. M., & Cohen, J. D. (2013). The expected value of control: An integrative theory of anterior cingulate cortex function. *Neuron, 79*(2), 217–240. http://doi.org/10.1016/j.neuron.2013.07.007

Sutton, R. S., & Barto, A. G. (1998). *Reinforcement learning: An introduction* (Vol. 1, pp. 1–398). Cambridge, MA: MIT Press.

Swart, J. C., Froböse, M. I., Cook, J. L., Guerts, D. E., Frank, M. J., Cools, R., & den Ouden, H. E. M. (2017). Catecholaminergic challenge uncovers distinct Pavlovian and instrumental mechanisms of motivated (in)action. *eLife.* http://doi.org/10.7554/eLife.22169.001

Van der Schaaf, M. E., Fallon, S. J., Ter Huurne, N., Buitelaar, J., & Cools, R. (2013). Working memory capacity predicts effects of methylphenidate on reversal learning. *Neuropsychopharmacology, 38*(10), 2011–2018. http://doi.org/10.1038/npp.2013.100

Van Holst, R. J., Sescousse, G., Janssen, L. K., Janssen, M., Berry, A. S., Jagust, W. J., & Cools, R. (2018). Increased striatal dopamine synthesis capacity in gambling addiction. *Biological Psychiatry, 83*(12), 1036–1043. http://doi.org/10.1016/j.biopsych.2017.06.010

Volkow, N. D., Wang, G.-J., Newcorn, J. H., Kollins, S. H., Wigal, T. L., Telang, F., et al. (2010). Motivation deficit in ADHD is associated with dysfunction of the dopamine reward pathway. *Molecular Psychiatry, 16*(11), 1147–1154. http://doi.org/10.1038/mp.2010.97

Volkow, N. D., Wise, R. A., & Baler, R. (2017). The dopamine motive system: Implications for drug and food addiction. *Nature Reviews Neuroscience, 18*(12), 741–752. http://doi.org/10.1080/08039488.2016.1176252

Voon, V., Derbyshire, K., Rück, C., Irvine, M. A., Worbe, Y., Enander, J., et al. (2014). Disorders of compulsivity: A common bias towards learning habits. *Molecular Psychiatry, 20*(3), 345–352. http://doi.org/10.1038/mp.2014.44

Voon, V., Pessiglione, M., Brezing, C., Gallea, C., Fernandez, H. H., Dolan, R. J., & Hallett, M. (2010). Mechanisms underlying dopamine-mediated reward bias in compulsive behaviors. *Neuron, 65*(1), 135–142. http://doi.org/10.1016/j.neuron.2009.12.027

Weismüller, B., Ghio, M., Logmin, K., Hartmann, C., Schnitzler, A., Pollok, B., et al. (2018). Effects of feedback delay on learning from positive and negative feedback in patients with Parkinson's disease off medication. *Neuropsychologia,*

117, 46–54. http://doi.org/10.1016/j.neuropsychologia.2018.05.010

Westbrook, A., & Braver, T. S. (2015). Cognitive effort: A neuroeconomic approach. *Cognitive, Affective, & Behavioral Neuroscience, 15*(2), 395–415. http://doi.org/10.3758/s13415-015-0334-y

Westbrook, A., & Braver, T. S. (2016). Dopamine does double duty in motivating cognitive effort. *Neuron, 89*(4), 695–710. http://doi.org/10.1016/j.neuron.2015.12.029

Westbrook, A., Kester, D., & Braver, T. S. (2013). What is the subjective cost of cognitive effort? Load, trait, and aging effects revealed by economic preference. *PloS One, 8*(7), e68210. http://doi.org/10.1371/journal.pone.0068210.t004

Winstanley, C. A., & Floresco, S. B. (2016). Deciphering decision making: Variation in animal models of effort- and uncertainty-based choice reveals distinct neural circuitries underlying core cognitive processes. *Journal of Neuroscience, 36*(48), 12069–12079. http://doi.org/10.1523/JNEUROSCI.1713-16.2016

Wunderlich, K., Smittenaar, P., & Dolan, R. J. (2012). Dopamine enhances model-based over model-free choice behavior. *Neuron, 75*(3), 418–424. http://doi.org/10.1016/j.neuron.2012.03.042

Yagishita, S., Hayashi-Takagi, A., Ellis-Davies, G. C. R., Urakubo, H., Ishii, S., & Kasai, H. (2014). A critical time window for dopamine actions on the structural plasticity of dendritic spines. *Science, 345*(6204), 1616–1620. http://doi.org/10.1126/science.1255514

Yttri, E. A., & Dudman, J. T. (2016). Opponent and bidirectional control of movement velocity in the basal ganglia. *Nature, 533*(7603), 402–406. http://doi.org/10.1038/nature17639

Zalocusky, K. A., Ramakrishnan, C., Lerner, T. N., Davidson, T. J., Knutson, B., & Deisseroth, K. (2016). Nucleus accumbens D2R cells signal prior outcomes and control risky decision-making. *Nature, 531*(7596), 642–646. http://doi.org/10.1038/nature17400

Zhang, J., Berridge, K. C., Tindell, A. J., Smith, K. S., & Aldridge, J. W. (2009). A neural computational model of incentive salience. *PLoS Computational Biology, 5*(7), e1000437. http://doi.org/10.1371/journal.pcbi.1000437.g004

VIII
METHODS ADVANCES

Introduction

DANIELLE BASSETT AND NIKOLAUS KRIEGESKORTE

Methodological advances are necessary to capitalize on the increasingly rich and diverse data afforded by modern imaging technologies and to address novel hypotheses regarding brain structure and function and their relation to behavior. While methodological advances are often technical in nature, they enable theoretical progress, allowing us to unwrap the intricacies of the brain with a swiftness and elegance that would otherwise be impossible. Without purposefully hiding the technicalities, we do take special care to ensure that the chapters in this section are accessible to a broad audience invested in the fundamental open questions of cognitive neuroscience. Citations to prior literature are liberal in an effort to ensure that interested students and practitioners alike can deepen their understanding by further reading.

The two focus areas of this section are (1) the modeling of representations, with the aim to reveal the information encoded in cortical areas; and (2) the modeling of networks, spanning cortical and subcortical structures. Conceptually, these two areas share a focus on information but operationalize that focus in different ways. Methods for the modeling of representations seek to offer descriptions, explanations, and mechanisms for the encoding of information in the brain. In contrast, methods for the modeling of networks seek to offer descriptions, explanations, and mechanisms for the transmission and processing of information in the brain. Together, the two types of modeling provide a richer context in which to pursue experiments and theory to reveal how the brain represents, transforms, and transmits information.

Information Encoding: Representational Models

At the outset of cognitive neuroscience, neuroimaging served primarily to measure the overall activation of entire functional regions in the human brain. The goal was to find the regions involved in a given task component. Since the turn of the century, the field has progressed toward a focus on the information represented within functional regions. Multivariate pattern analysis methods, including decoding, were used to detect the information encoded in fine-grained patterns of activity. This approach was complemented by advances in functional imaging, including the higher spatial resolution afforded by functional magnetic resonance imaging at higher magnetic field strengths.

Four chapters in this section present several exciting current directions of methodological development that build on multivariate pattern analyses. Diedrichsen describes the field of representational models. Whereas decoding analyses merely test whether particular information is present in a region's activity patterns, a representational model makes comprehensive predictions about the content and format of a region's representation. Encoding models, representational similarity analysis, and pattern component models are three variants of representational modeling. Carlson, Grootswagers, and Robinson describe how time-resolved decoding and representational similarity analysis can be used to study the dynamics of brain representations. They explain how to apply these pattern analyses to magneto- and electroencephalographic (M/EEG) time series using a sliding temporal window. King and colleagues build on these advances and combine encoding, decoding, and temporal cross-generalization analyses to get closer to revealing the algorithms of information processing from high-temporal-resolution electrophysiological data. Finally, Storrs and Kriegeskorte introduce deep neural network models, which fully define hypothetical algorithms of brain computation. Deep neural network models can serve as bases for representational models, which can be evaluated as described in the chapter by Diedrichsen. The chapter describes the various ways in which cognitive neuroscientists can engage deep neural network models—whether or not they build such models in their own labs.

Information Transmission: Network Models

Conceptualizing the brain as a network of units whose pattern of interunit connections allows for information transmission, manipulation, and computation is embedded in historic work in neuroanatomy and physiology. Current efforts stand apart from past work in the space due to the formalization of that conception in the mathematical terms of networks and graphs. The emerging field of *network science* comprises the study of complex systems in a manner predicated on two assumptions: (1) that such systems can be modeled as graphs or graph-based objects and (2) that the topology of these graphs has an impact on the system's dynamics, development, function, and evolution. Subdomains of the field include the development of statistics and other formal methods to characterize network topology, the investigation of network reconfiguration or the modeling of dynamics at network nodes and across network edges, the application of the statistical and mathematical modeling tools to the study of real-world systems, and the development of theory with formal analysis (in the mathematical sense of the term). Notably, each of these subdomains has the potential to inform systems neuroscience, computational neuroscience, theoretical neuroscience, and of course cognitive neuroscience.

Studies utilizing the conceptual framework, the statistical approaches, the mathematical models, or the theories of network science to address open questions in neuroscience fall under the umbrella term of *network neuroscience*. In the early days of network neuroscience, most studies focused on describing neural systems using network- or graph-based statistics. Common goals included descriptions of how brain networks change as children mature or as adults move into old age, descriptions of how brain network architecture reflects cognitive abilities, and descriptions of how brain networks differ in individuals with mental illness. Moreover, the goals of these studies have ranged from those of basic science seeking to characterize and understand the world around us to the goals of clinical medicine seeking to accurately diagnose mental illness or to predict disease progression or recovery. While the descriptive approaches used in these studies have proven very useful and remain a focus of current work, their methods have been described in earlier editions of this same book and in several notable review articles.

Rather than rehashing those same generic topics, here we instead provide three chapters focusing on quite new methodological approaches in the domain of network neuroscience, which collectively press beyond descriptive analyses in an effort to offer predictive models and posit underlying mechanisms. In the first chapter of this set, Vértes offers an introduction to generative modeling, which seeks to posit, exercise, and validate rules for network growth. The approach is particularly relevant for understanding brain network development and for testing hypotheses about altered developmental trajectories consistent with observed network alterations in adults with mental illness thought to be of neurodevelopmental origin. In the second chapter of this set, Bassett and Pasqualetti offer an introduction to network control theory, which combines a model of a system's network architecture with a model of a system's dynamics to infer optimal control points and accessible control strategies. The theory is currently being exercised to better understand cognitive-control capacities in humans, as well as to inform exogenous interventions in the form of brain stimulation and pharmacology. Finally, in the third chapter of this set, Battaglia and Brovelli offer an introduction to models of oscillatory behavior and their roles in flexible information transmission in neural systems. The three chapters share a focus on mathematical modeling, and each addresses dynamics that occur over different spatiotemporal scales.

Current Frontiers

Understanding cognitive processes requires both an understanding of the encoding of information within the brain and the transmission of that information across potentially spatially distributed areas. Recent tools have emerged to address each of these components separately—for example, representational similarity analysis, multivoxel pattern analysis, and similar techniques seek to capture information encoding, while network models and graph-based techniques seek to capture information transmission. Yet little work has sought to formally link these two approaches. In the near future, we envision that the two subfields will come together with integrated approaches for the study of information encoding and transmission.

56 Representational Models and the Feature Fallacy

JÖRN DIEDRICHSEN

ABSTRACT In this chapter I discuss models that specify how neuronal population activity relates to things that happen in the world—that is, models of neuronal representations. Focusing on the application to functional magnetic resonance imaging (fMRI) data, I discuss current approaches to estimate and test such models. Encoding models conceptualize representations in terms of sets of underlying features. We show that these models ultimately test hypotheses about the distribution of activity profiles in the space of the experimental conditions and that the exact choice of feature sets is, to some degree, arbitrary. Overinterpreting the significance of specific feature sets constitutes an intellectual dead-end, termed here a *feature fallacy*. The same representational models can also be tested using a Bayesian approach, pattern component modeling (PCM), which abstracts from the underlying features and allows direct model evaluation. It also provides a powerful means to test more flexible representational models, including models that consist of combinations of different feature sets.

Discovering how the human brain gives rise to intelligent behavior is one of the most daunting scientific challenges of our time. How do neurons represent and process information? In trying to answer this question, many researchers have taken an experimental approach, independently varying stimuli or task conditions and repeatedly measuring the activity of neurons. The responses of each neuron across experimental conditions form a tuning function or activity profile. The goal of the approach is ultimately to discover the mapping between the stimulus characteristics and the neuronal response (Wu, David, & Gallant, 2006).

Of course, neurons do not work in isolation. Therefore, representational models need to be models of population codes, the activity of groups of neurons. Models of population codes usually specify a class of activity profiles. For example, the responses of neurons in M1 can be described as having cosine tuning to movement direction. While each neuron has a different preferred direction (fires maximally for a different movement), the underlying coding principle is the same across neurons. Together, the population codes for movement direction in a distributed fashion (Georgopoulos, Schwartz, & Kettner, 1986). In this chapter I am concerned with such models of population activity and will review current methods that can be used to specify and compare different models.

Measuring Representations with fMRI

While many of the techniques discussed here are inspired by sensory neurophysiology, I will concentrate on their application to fMRI data. The concept of an activity profile can be generalized to the main measurement unit of fMRI, the voxel. A collection of neighboring voxels can be analyzed as a "population code" in the hope that this will reveal something about the underlying neuronal representation. A number of different analysis methods have been proposed to achieve this goal. I refer to these methods collectively as representational fMRI analysis to distinguish them from the classical fMRI approach, which simply asks whether a region increases its activity in response to a specific task.

One might say that analyzing a "population code" across voxels in the hope of learning something about the underlying neuronal population code is simply one step too far. The activity measured in a single voxel combines the activity of millions of neurons; any information encoded in the activity differences within a voxel will be lost (Kriegeskorte & Diedrichsen, 2016). Furthermore, the sluggish hemodynamic response removes nearly all useful temporal information from the signal. Finally, fMRI does not measure neural activity directly but through the indirect lens of vascular responses, with many possible nonlinearities and an irregular spatial spread. Despite these severe limitations, representational fMRI analysis has been successful in uncovering many known characteristics of basic sensory and motor representations (Ejaz, Hamada, & Diedrichsen, 2015; Kay, Naselaris, Prenger, & Gallant, 2008; Norman-Haignere, Kanwisher, & McDermott, 2013). With adequate caution, it is therefore possible to use representational fMRI analysis to study higher-order regions of the human brain—already, the approach has provided insights into the nature of spatial (Kim, Jeffery, & Maguire, 2017), sequential (Yokoi, Arbuckle, & Diedrichsen, 2018), categorical (Kriegeskorte et al., 2008), and semantic (Huth, de Heer, Griffiths, Theunissen, & Gallant, 2016) representations. It is here that the power of representational fMRI analysis lies, as many of these concepts are not easily accessible in animal models.

Describing Activity Profiles Using Features

The goal of representational fMRI analysis is to build models of the activation profiles of groups of voxels across an ideally rich set of experimental conditions. An intuitive way to characterize activity profiles is through a flexible combination of features, an approach taken in so-called encoding models (Naselaris, Kay, Nishimoto, & Gallant, 2011). For example, the responses of voxels in V1 to natural stimuli can be captured using a feature set of Gabor functions with a different location, frequency, and orientation (Kay et al., 2008). The responses of voxels in primary auditory cortex for complex sounds are well predicted by a model that uses the power in specific frequency bands as features (De Angelis et al., 2017).

To understand the relationship between different representational analysis methods, it is useful to take a multilevel modeling view (figure 56.1). On the level of the data, the fMRI data of the pth voxel, $\mathbf{y}_p$, is expressed as a function of a design matrix $\mathbf{Z}$ that captures the temporally delayed and smoothed response to each event or condition and an activity profile $\mathbf{u}_p$, which determines the size of the response to each of the experimental conditions. In encoding models, the activity profile is then modeled by a linear combination of a set of features: Each column of the matrix $\mathbf{M}$ corresponds to a feature, and the feature weights, $\mathbf{w}_p$, determine to what degree voxel p responds to those features. Each feature set spans a specific subspace of the activation profiles that are "allowed" under the model. Any deviation from these permissible profiles would reduce the measure of fit. Early encoding models used multiple regression to estimate the feature weights (Mitchell et al., 2008), followed by either decoding or classical statistical approaches to determine the quality of the fit.

Using standard regression, however, has severe limitations. When the number of features approaches the number of distinct experimental conditions, all models

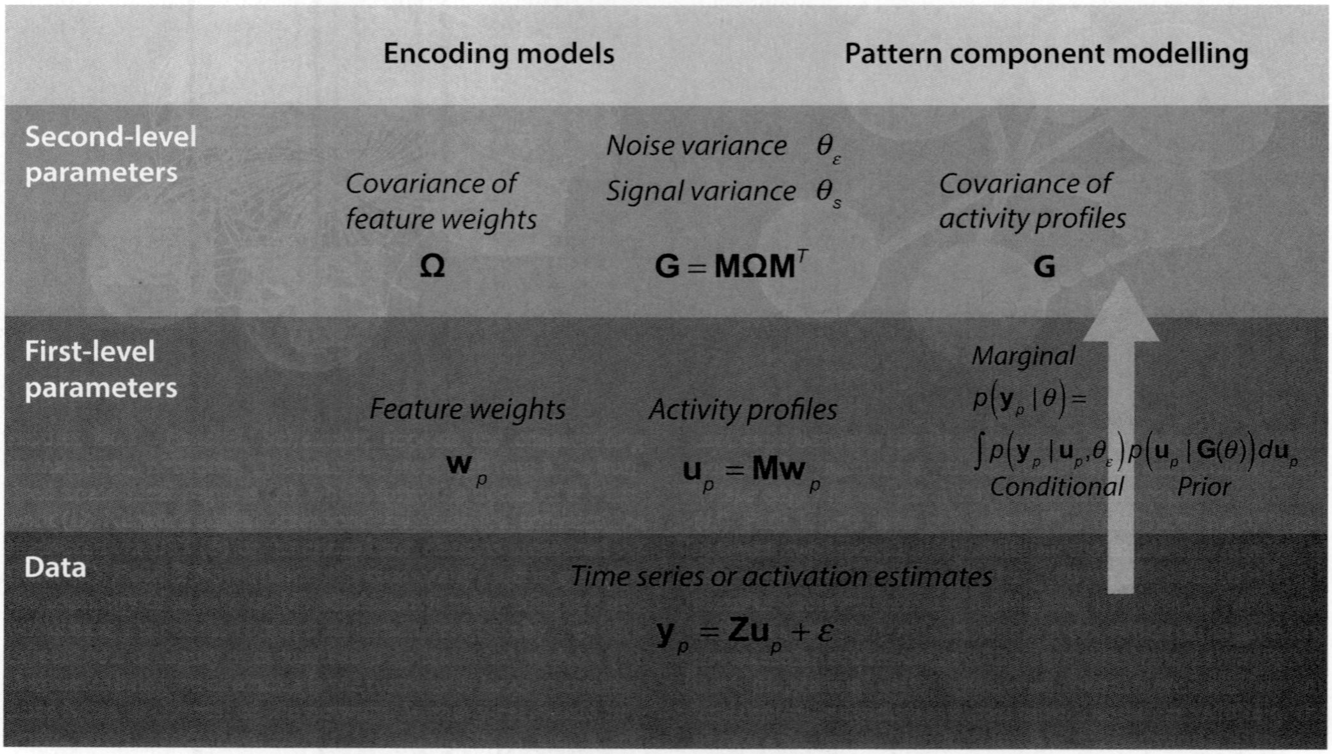

FIGURE 56.1 Multiple levels of a representational model. The data ($\mathbf{y}_p$) are either fMRI time-series or activity estimates from a lower-level analysis for each voxel p. Encoding models (*left column*) model the data using a set of features $\mathbf{M}$ and feature weights $\mathbf{w}_p$. Second-level parameters determine the distribution of feature weights and noise across voxels. In PCM (*right column*), the distribution is directly specified on the activity profiles $\mathbf{u}_p$. The marginal likelihood of the data under a given model can be directly determined by integrating out of first-level parameters (*arrow*).

fit the data equally well. Therefore, it is common practice to introduce a second modeling level that specifies a prior distribution on the voxel weights $p(\mathbf{w}_p)$, often assumed to be Gaussian with mean $\mathbf{0}$ and (co)variance $\Omega\theta_s$. Matrix Ω determines the shape of the distribution, and θ_s simply scales the overall variance of the signal. Together with the feature matrix, this prior specifies how likely activity profiles are under a specific model. If we take the data to be the vector of estimated activity profiles $\hat{\mathbf{u}}_p$, obtained from a first-level time-series model, and if we assume these are estimated with error variance θ_ϵ, then the best linear predictor of $\mathbf{w}_p$ is

$$\hat{\mathbf{w}}_p = \left(\mathbf{M}^T\mathbf{M} + \Omega^{-1}\theta_\epsilon/\theta_s\right)^{-1}\mathbf{M}^T\hat{\mathbf{u}}_p. \tag{56.1}$$

The term $\Omega^{-1}\theta_\epsilon/\theta_s$ shrinks the estimates toward the more likely regions of the prior distribution. In the case of $\Omega = \mathbf{I}$, the above equation simplifies to ridge regression.

The "fit" of such a model is usually compared using cross-validation. Leaving out a small subset of data (~10%), the voxel weights are estimated from the remaining training set. The prediction of the left-out data is then assessed using cross-validated R^2 or correlation between measured and predicted voxel activities. Cross-validation automatically penalizes model complexity, making it possible to compare models with different numbers of features directly. In essence, cross-validation assesses the likelihood of the activity profiles under the hypothesized distribution of the activity profiles, independent of the actual voxel-feature weights. This distribution is Gaussian (implicitly assumed when using equation 56.1), with zero mean and a (co)variance matrix that depends on the features and their prior, $\mathbf{G} = \mathbf{M}\Omega\mathbf{M}^T$. Matrix $\mathbf{G}$ therefore fully specifies the representational model.

The Feature Fallacy

This observation has an important consequence: two representational models are identical if the predicted (co)variance matrix $\mathbf{G}$ is identical. When we find a well-fitting feature model, we need to consider that many other feature sets can predict the same activity profile distribution and therefore describe and predict the data equally well. Thus, feature sets may be useful tools to describe the distributions of activity profiles, but they do not carry a special significance in and of themselves. With the term *feature fallacy*, I am referring to the common mistake of confusing the tools we use to describe the data and the very thing we seek to understand.

As an example, consider the representation of finger movements in M1 and S1 (Ejaz, Hamada, & Diedrichsen, 2015). In this experiment, participants produced isometric finger presses with each finger of the contralateral hand. In a restricted area around the central sulcus, voxels vary their activity systematically with the finger used. Figure 56.2*A* shows the activity of a set of these voxels, plotted into the space of three of the fingers. As we can see, the middle and ring finger activity is highly correlated across voxels, whereas thumb activity is relatively independent. The covariance matrix of the activity profiles is well preserved across participants and coincides with the covariance matrix of finger movements in everyday life. The findings suggest that the finger movement representations are shaped by the everyday structure of such movements.

We can model this distribution of voxels using the five fingers as features. The cross-validated fit will be especially good when the prior variance Ω is set to the covariance matrix of natural movements. The estimated tuning for each voxel for each finger can then be visualized in a winner-take-all map of the cortical surface (figure 56.2*D*), revealing a somatotopic map, an orderly representation of individual digits from the thumb (ventral) to the little finger (dorsal).

Alternatively, we can describe the distribution using the principal components of the covariance matrix of the natural statistics of finger movements (figure 56.2*B*). In the motor control literature, the components underlying natural behaviors are termed *synergies*. In this coordinate system, the weights for the different features are approximately uncorrelated. When mapping the synergy preference of each voxel, an ordered *synergy representation* (Leo et al., 2016) emerges (figure 56.2*E*). Finally, we could equally describe the distribution with random features (figure 56.2*C*). As long as the prior (co)variance matrix (and hence the regularization) is adjusted accordingly, the cross-validated accuracy of this model remains the same. Again, a convincing-looking map of the random features can be produced (figure 56.2*F*).

The deeper point here is that all three feature sets are equally good descriptors of the data, and all would result in the same cross-validated prediction accuracy. Even when we constrain the prior (co)variance matrix to be diagonal, there are an infinite number of feature sets that would lead to the same prediction for left-out data. With equally strong conviction, we can therefore conclude that primary sensorimotor cortex represents fingers, natural synergies, or random features. All three conclusions would be valid to some degree, but none would provide a deeper insight into the underlying neuronal computations. Debates about which feature set is most appropriate therefore miss the point. The main finding is that the distribution of activity profiles is highly structured, that this structure is preserved across individuals, and that it relates in systematic fashion to the correlation of these movements in everyday life. The feature labels we use to describe the distribution are secondary.

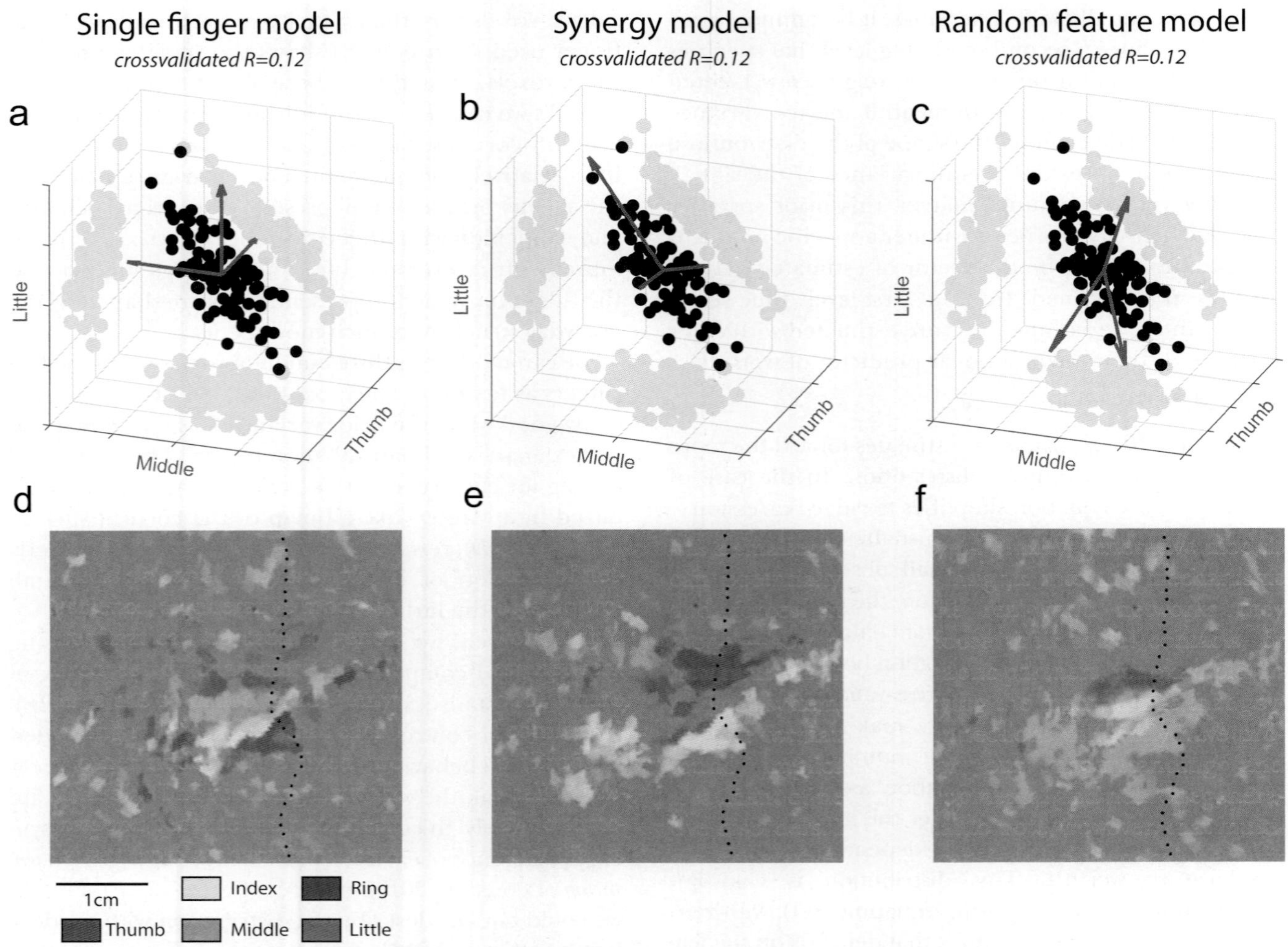

FIGURE 56.2 Three feature models that describe the activity during finger movements in M1 and S1 equally well. *A*, The distribution of voxels in the space of three of the experimental conditions. Each dot represents a voxel's activity profile across the thumb, the middle finger, and the ring finger. The distribution can be described by using individual fingers as features. *B*, Surface map of the human S1 and M1 with voxels colored according to which finger they are most activated. The color saturation reflects the strength of the tuning, with gray areas showing no finger preference. *Dotted line*, Fundus of the central sulcus. The length scale is approximate. *C*, The distribution can also be described using the principal components of the natural statistics of finger movements (synergies). *D*, Surface map of feature weights for the synergy model with voxels either scoring high (+) or low (−) on each PC. *E*, A model with five random feature vectors explains the data equally well, but (*F*) lead to a different feature map. (See color plate 62.)

This is not to say that we shouldn't be allowed to think about neural activity in terms of underlying features. Features can provide a semantically meaningful description of population codes and representational spaces (see below). They are useful tools in constructing representational models, especially in cases in which there are no discrete experimental conditions—for example, when a continuous movie is used as a stimulus. Taking the representation of features too literally, however, constitutes an intellectual dead-end. Neurophysiological research, for decades, has striven to determine whether the firing rate of M1 neurons is better described in terms of muscle activities, extrinsic movement direction, or synergies. The ultimate answer has been that none of these features describes the population especially well. Instead, M1 neurons are commonly found to exhibit *mixed selectivity*. This means that motor cortex represents movement not according to any particular feature set but rather in a latent space that represents the context dependence of complex movement while also producing the dynamics necessary to generate the required patterns of muscle activity (Churchland et al., 2012). Rather than getting stuck in the search for the underlying features, we should compare models that make testable predictions about the distribution of activity profiles.

Pattern component modeling This insight motivated the development of PCM, which seeks to evaluate the likelihood of the observed activity profiles under the hypothesized distribution directly. In the multilevel view of representational models (figure 56.1), PCM is functionally equivalent to an encoding model with a Gaussian prior. Instead of defining features, estimating the feature weights, and then using cross-validation to evaluate their predictive power, it evaluates the marginal likelihood of the data under the model (and the second-level parameters) directly.

$$p(\mathbf{y}_p \mid \theta) = \int p(\mathbf{y}_p \mid \mathbf{u}_p) p(\mathbf{u}_p \mid \mathbf{G}(\theta)) d\mathbf{u}_p \tag{56.2}$$

The first term of the integral is the conditional probability of the data given the activity profiles; the second term, the prior probability of the activity profiles under the model. The evaluation of the integral in analytical form is possible, as both noise and the signal are assumed to be Gaussian. The result is the marginal likelihood, the probability of the data under the assumed distribution, independent of the actual value of individual activity profiles. The marginalization achieves the same as the cross-validation employed in encoding models: it corrects for the complexity (number of features) of the model.

To determine which model provides the most appropriate description of the data, we can therefore simply choose the model with the highest marginal likelihood. The ratio of the likelihoods is the Bayes factor, a measure of the evidence of one model over the other (Kass & Raftery, 1995). This approach is valid for models that predict a fixed representational structure—that is, models based on a single feature set with only one signal variance and noise variance parameter on the second level (θ). Under these circumstances, the marginal likelihood can serve as an approximation of the model evidence—the probability of the data given the model. Using Bayes factors allows for a more powerful model comparison than is possible through encoding models or representational similarity analysis (RSA; Diedrichsen & Kriegeskorte, 2017).

All second-level parameters can be efficiently optimized, as analytical derivatives of the marginal likelihood with respect to these parameters are easily derived. An implementation of the corresponding algorithms is openly available (Diedrichsen, Yokoi, & Arbuckle, 2017). In sum, the core idea behind PCM is to abstract from the actual activity patterns and model features to make direct and powerful inferences about representational models.

Representational similarity analysis and representational spaces The abstraction from activity patterns and features is shared with a number of other representational analysis techniques, first and foremost RSA (Kriegeskorte, Mur, & Bandettini, 2008). A central concept in this approach is the notion of a representational space (Guntupalli et al., 2016; Kriegeskorte et al., 2008). Instead of thinking about voxel activity profiles as points in the space of experimental conditions (figure 56.3*B*), we can think about the conditions as points in the space of voxel activities (figure 56.3*C*). The relationship between the different activity patterns in this space defines the representation. In RSA, the relationship is quantified through a dissimilarity measure, with higher values indicating more distinct activity patterns. The representational geometry can be summarized succinctly by the matrix of all pairwise dissimilarities, the representational dissimilarity matrix (RDM).

Dissimilarity measures have the intuitive appeal of reflecting how strongly the distinction between two conditions is represented in an area. That is, they tell us how well a readout neuron that has access to the whole population code could distinguish between the two conditions. More generally, the representational geometry determines how well any feature that describes the underlying conditions could be read out. An especially useful dissimilarity measure is the cross-validated estimate of the Mahalanobis distance (Diedrichsen & Kriegeskorte, 2017). This distance estimate is unbiased—that is, the expected value of the dissimilarity is zero if two activity patterns only differ by noise. Thus, like cross-validated decoding performance, it can be used to assess whether there is a true difference between two activity patterns.

RSA, PCM, and encoding models are tightly related. This relationship is due to the fact that all three approaches assess model fit by comparing the second-moment matrices of the activity profile distributions. The second moment of P activity profiles is defined as

$$\mathbf{G} = \sum_p \mathbf{u}_p \mathbf{u}_p^T / P. \tag{56.3}$$

Both Euclidean ($d_{1,2} = \sqrt{G_{1,1} - 2G_{2,1} + G_{2,2}}$) and correlation distances ($d_{1,2} = 1 - G_{2,1} / \sqrt{G_{1,1} G_{2,2}}$) are direct functions of this matrix. As the RDM, the second-moment matrix determines the representational geometry completely, with the only difference being that it also specifies the distance of each condition from the resting baseline.

The definition of the second-moment matrix is nearly identical to the definition of the (co)variance matrix of the activity profiles, the central statistics in encoding and PCM models. The only difference is that for a (co)variance matrix, the mean activity profile (across voxels) is subtracted before applying equation 56.3. However, in the context of representational

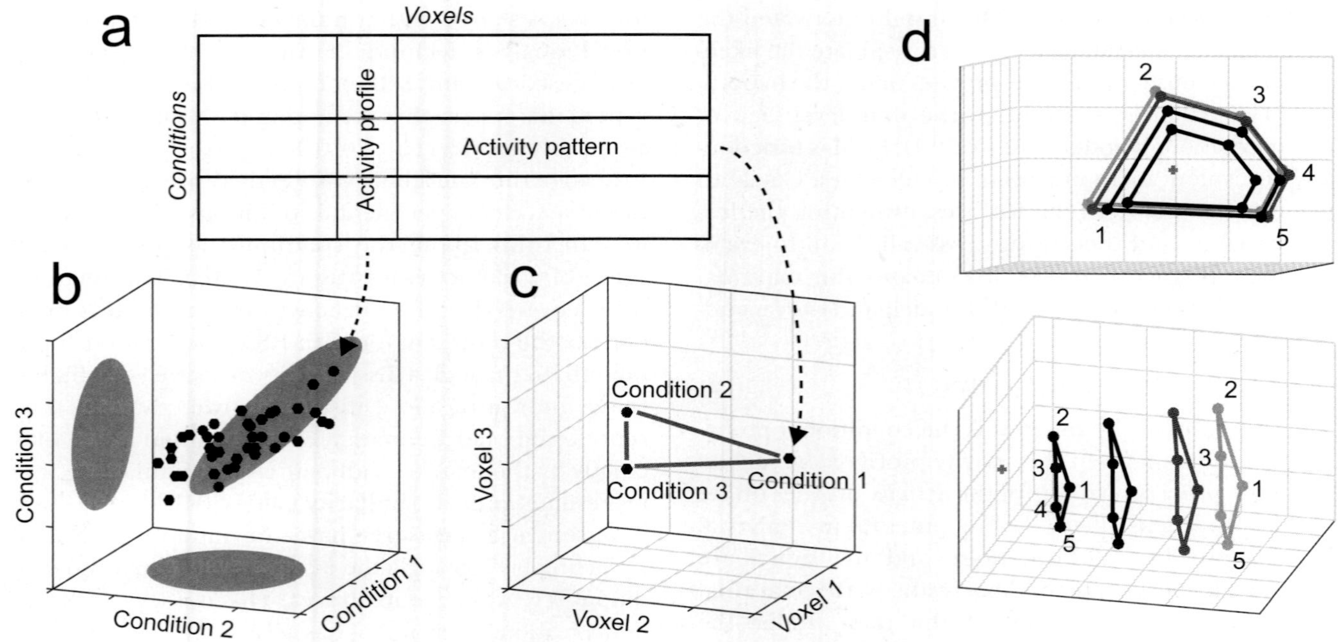

FIGURE 56.3 Representational spaces. *A*, The data consist of repeated measures of the same set of voxels across a range of conditions. Each column of the matrix constitutes an activity profile; each row an activity pattern across voxels. *B*, The activity profiles can be plotted in the space of the experimental conditions. Representational models specify a distribution of activity profiles. *C*, The activity patterns can be plotted in the space spanned by the voxels. The relationships between activity patterns in this high-dimensional space define the representational geometry (*lines*). *D*, Two views of a low-dimensional projection of the representational geometry of individual finger movements in M1 (*1*, thumb–*5*, little finger) at four different movement speeds (*black*, slow–*gray*, fast).

analysis, differences in mean activity between conditions are meaningful—indeed, many important dimensions, such as stimulus intensity, are encoded in the overall activity. Subtracting the mean activity profile (row means in figure 56.3*A*) would remove such difference and severely distort the representational geometry. Although sometimes performed, such subtractions are therefore not meaningful from a representational standpoint. By not performing this subtraction, the PCM and encoding approaches assume the mean of the signal distribution to be zero. Any meaningful mean activity differences are therefore captured in the second-moment matrix. In sum, all three approaches define models by the second moment of the predicted activity profile distribution. The only technical difference is how exactly the mismatch between the empirical and predicted second-moment matrix is measured (Diedrichsen & Kriegeskorte, 2017).

Using representational spaces, one can visualize the representational geometry of a population code without the need to define a model a priori. To be able to generate graphs in two or three dimensions, we typically need to reduce the dimensionality of the space. A common approach here is to use the first three eigenvectors of **G**—that is, the three patterns that best differentiate between conditions. It is, however, also useful to explore other views—for example, by picking dimensions that maximize specific experimental contrasts of interest (Diedrichsen, Yokoi, & Arbuckle, 2017; Kobak et al., 2016). While visualizations of representational spaces can be very powerful (figure 56.3*D*), they need to be treated with caution, as different views of the same data sometimes can tell very different visual stories. This emphasizes the importance of formal model comparison, which should always be conducted in the full, rather than the reduced, visualized representational space.

In summary, representational spaces are an important concept for cognitive neuroscience. Representational spaces abstract from the underlying spatial activity pattern and from the hypothetical features used to describe them. Indeed, it is often observed that the specific activity patterns are quite variable across individuals, whereas the structure of the representation is well preserved (Ejaz, Hamada, & Diedrichsen, 2015; Guntupalli et al., 2016). The concept of a representational space asserts that what matters is the representational content of a population code but not the details of how it is laid out on the cortical sheet.

Flexible representational models The comparison of representational models can be performed using encoding approaches RSA or PCM. While PCM offers slightly more powerful inferences (Diedrichsen & Kriegeskorte, 2017), the small increase in power alone may not constitute an overwhelming argument for its use. The strength of PCM, however, becomes apparent when considering more complicated or flexible representational models.

The representational models considered so far have been "fixed." The cross-validation in encoding models, the calculation of the marginal likelihood in PCM, and the calculation of distances in RSA effectively integrate out the first-level parameters (figure 56.1). That is, even with thousands of voxel-feature weights, encoding models with a single ridge coefficient predict a fixed distribution of activity profiles.

It is, however, a rare stroke of luck if we can fully explain the neural activity patterns in a specific region using a single feature set. From a computational perspective, it may be intuitive at first to think of information processing as a clear sequence of transformations. Speech comprehension, for example, could be conceptualized as starting with a stage that analyzes the spectral features, followed by a stage that detects phonemes and then ending in a stage that identifies the semantics of the entire word (Poeppel et al., 2012).

Unfortunately, nature does not usually accommodate neuroscientists by arranging computations in an anatomically orderly fashion, with each region corresponding to one distinct stage of information processing. Rather, most areas show a mixture of representations from multiple processing stages. For example, coding in the caudal premotor and primary motor cortex exhibits a mixture of extrinsic and muscle-like tuning (Wu & Hatsopoulos, 2007). Similarly, most sensory regions respond to a mixture of basic perceptual features, context information, and attentional signals. Therefore, our default assumption should not be that a representation can be explained by a single feature set but that each region will exhibit a specific mixture of different representations.

This should be reflected in the way we analyze brain data. We therefore require techniques that allow us to test for a mixture of representations. Different approaches have been developed for this purpose. For example, using RSA, Khaligh-Razavi and Kriegeskorte (2014) combined the predictions derived from different layers of a deep neural network with a categorical model to explain object representations in inferior temporal cortex. Similarly, using an encoding approach, de Heer, Huth, Griffiths, Gallant, and Theunissen (2017) considered a mixture of spectral, articulatory, and semantic properties to explain representations in the auditory processing stream. Using PCM, Yokoi, Arbuckle, and Diedrichsen (2018) characterized finger sequence representations in cortical motor areas through a mixture of representations of single fingers and the transition between finger presses.

Methodologically, this is an inference problem that is relatively well studied in the statistical literature (Clyde, 1999). The first step is to be able to fit more complex models to the data, which requires us to find the optimal weighting of each feature set. In the context of PCM, the second-moment matrix of the data can be modeled using a weighted linear combination of the predicted second-moment matrix of $i = 1 \dots I$ components

$$\mathbf{G} = \sum_i \theta_i \mathbf{G}_i, \tag{56.4}$$

where θ_i are the nonnegative component weights. Linear combination here assumes that the features across different components are mutually independent. Because the analytical derivates of the marginal likelihood in respect to each weight can be easily derived, the optimal weights can be estimated very efficiently.

The second step is to find the best feature sets to combine. This can be achieved by fitting all possible combinations of components (i.e., feature sets). Because each of the K component can independently be either present or absent, this amounts to fitting 2^K models, which is easily possible for small K. Once the number of candidate components becomes large, one needs to resort to model-search strategies, such as stepwise or approximate Bayesian approaches (Clyde, 1999). In evaluating each model, we need to take into account the increasing complexity of models with more components. Even though PCM does not require cross-validation for fixed models, as first-level parameters are integrated out in the marginal likelihood, this analytical approach cannot easily be applied to second-level parameters (figure 56.2). Practically, we can resort here to using the Akaike information criterion (AIC), the Bayesian information criterion (BIC), or cross-validation within or across participants (Diedrichsen, Yokoi, & Arbuckle, 2017). Each of these measures gives us an estimate of the model evidence for each combination of features.

The last step is to make inferences about the presence or absence of a specific component. For example, we would like to map how much evidence there is for a semantic representation in secondary auditory areas in the context of the other competing explanations. It has been suggested that the concept of variance partitioning (de Heer et al., 2017) be used to express what proportion of the variability can be explained uniquely by a feature set. This concept from unregularized linear models, however, does not readily generalize to the probabilistic setting. For example, the combination of two feature sets often provides a poorer description of the data than either feature set alone, even if the two feature sets account for different aspects of the data. Thus, a probabilistic approach is needed. One attractive solution is to evaluate the strength of the evidence for each component in terms of a Bayes factor. We can apply Bayesian model averaging and divide the total posterior probability of all models containing the component by the posterior probability of all models not containing the component (Clyde, 1999; Shen & Ma, 2019). These techniques are not restricted to models, in which the second-moment matrix is a linear function of components (equation 56.4), but extend to any situation in which **G** is a differentiable (but otherwise arbitrary) function of the second-level parameters, $\mathbf{G} = F(\theta)$. In PCM, these types of models can be handled with the existing machinery (Diedrichsen, Yokoi, & Arbuckle, 2017). This allows the user to model receptive fields with different widths, arbitrary correlations between feature sets, and the relative weighting of individual features within correlated feature sets.

Conclusions

The structure and analysis of standard representational models are starting to be well understood. Many interesting problems, including group inference, non-Gaussian distributions, and the analysis of the spatial characteristics of activity patterns, still await a thorough analysis. Building representational models using machine-learning techniques, essentially defining the model by the computational problem it has to solve rather than by the represented features, is also a promising direction (see chapters 33 and 59 in this volume).

Acknowledgments

This work was supported by the James S. McDonnell Foundation (Scholar Award), a Natural Sciences and Engineering Research Council Discovery Grant (RGPIN-2016-04890), and the Canada First Research Excellence Fund (BrainsCAN).

REFERENCES

Churchland, M. M., Cunningham, J. P., Kaufman, M. T., Foster, J. D., Nuyujukian, P., Ryu, S. I., & Shenoy, K. V. (2012). Neural population dynamics during reaching. *Nature, 487*(7405), 51–56.

Clyde, M. A. (1999). Bayesian model averaging and model search strategies. In J. M. Bernardo, J. O. Berger, A. P. Dawid, & A. F. M. Smith (Eds.), *Bayesian statistics* (Vol. 6, pp. 157–185). Oxford: Oxford University Press.

De Angelis, V., De Martino, F., Moerel, M., Santoro, R., Hausfeld, L., & Formisano, E. (2018). Cortical processing of pitch: Model-based encoding and decoding of auditory fMRI responses to real-life sounds. *Neuroimage, 180,* 291–300.

de Heer, W. A., Huth, A. G., Griffiths, T. L., Gallant, J. L., & Theunissen, F. E. (2017). The hierarchical cortical organization of human speech processing. *Journal of Neuroscience, 37*(27), 6539–6557.

Diedrichsen, J., & Kriegeskorte, N. (2017). Representational models: A common framework for understanding encoding, pattern-component, and representational-similarity analysis. *PLOS Computational Biology, 13*(4), e1005508.

Diedrichsen, J., Yokoi, A., & Arbuckle, S. A. (2018). Pattern component modeling: A flexible approach for understanding the representational structure of brain activity patterns. *Neuroimage, 180,* 119–133.

Ejaz, N., Hamada, M., & Diedrichsen, J. (2015). Hand use predicts the structure of representations in sensorimotor cortex. *Nature Neuroscience, 18*(7), 1034–1040.

Georgopoulos, A. P., Schwartz, A. B., & Kettner, R. E. (1986). Neuronal population coding of movement direction. *Science, 233*(4771), 1416–1419.

Guntupalli, J. S., Hanke, M., Halchenko, Y. O., Connolly, A. C., Ramadge, P. J., & Haxby, J. V. (2016). A model of representational spaces in human cortex. *Cerebral Cortex, 26*(6), 2919–2934.

Huth, A. G., de Heer, W. A., Griffiths, T. L., Theunissen, F. E., & Gallant, J. L. (2016). Natural speech reveals the semantic maps that tile human cerebral cortex. *Nature, 532*(7600), 453–458.

Kass, R. E., & Raftery, A. E. (1995). Bayes factors. *Journal of the American Statistical Association, 90*(430), 773–795.

Kay, K. N., Naselaris, T., Prenger, R. J., & Gallant, J. L. (2008). Identifying natural images from human brain activity. *Nature, 452*(7185), 352–355.

Khaligh-Razavi, S. M., & Kriegeskorte, N. (2014). Deep supervised, but not unsupervised, models may explain it cortical representation. *PLOS Computational Biology, 10*(11), e1003915.

Kim, M., Jeffery, K. J., & Maguire, E. A. (2017). Multivoxel pattern analysis reveals 3D place information in the human hippocampus. *Journal of Neuroscience, 37*(16), 4270–4279.

Kobak, D., Brendel, W., Constantinidis, C., Feierstein, C. E., Kepecs, A., Mainen, Z. F., Machens, C. K. (2016). Demixed principal component analysis of neural population data. *eLife, 5.*

Kriegeskorte, N., & Diedrichsen, J. (2016). Inferring brain-computational mechanisms with models of activity measurements. *Proceedings of the Royal Society B, 371*(1705), 20160278.

Kriegeskorte, N., Mur, M., & Bandettini, P. (2008). Representational similarity analysis—connecting the branches of systems neuroscience. *Frontiers in Systems Neuroscience, 2,* 4.

Kriegeskorte, N., Mur, M., Ruff, D. A., Kiani, R., Bodurka, J., Esteky, H., … Bandettini, P. A. (2008). Matching categorical object representations in inferior temporal cortex of man and monkey. *Neuron, 60*(6), 1126–1141.

Leo, A., Handjaras, G., Bianchi, M., Marino, H., Gabiccini, M., Guidi, A., …, Ricciardi, E. (2016). A synergy-based hand control is encoded in human motor cortical areas. *eLife, 5.*

Mitchell, T. M., Shinkareva, S. V., Carlson, A., Chang, K. M., Malave, V. L., Mason, R. A., & Just, M. A. (2008). Predicting human brain activity associated with the meanings of nouns. *Science, 320*(5880).

Naselaris, T., Kay, K. N., Nishimoto, S., & Gallant, J. L. (2011). Encoding and decoding in fMRI. *Neuroimage, 56*(2), 400–410.

Norman-Haignere, S., Kanwisher, N., & McDermott, J. H. (2013). Cortical pitch regions in humans respond primarily to resolved harmonics and are located in specific tonotopic regions of anterior auditory cortex. *Journal of Neuroscience, 33*(50), 19451–19469.

Poeppel, D., Emmorey, K., Hickok, G., & Pylkkanen, L. (2012). Towards a new neurobiology of language. *Journal of Neuroscience, 32*(41), 14125–14131.

Shen, S., & Ma, W. J. (2019). Variable precision in visual perception. *Psychological Review, 126*(1), 89–132.

Wu, M. C., David, S. V., & Gallant, J. L. (2006). Complete functional characterization of sensory neurons by system identification. *Annual Review of Neuroscience, 29*, 477–505.

Wu, W., & Hatsopoulos, N. G. (2007). Coordinate system representations of movement direction in the premotor cortex. *Experimental Brain Research, 176*(4), 652–657.

Yokoi, A., Arbuckle, S. A., & Diedrichsen, J. (2018). The role of human primary motor cortex in the production of skilled finger sequences. *Journal of Neuroscience, 38*(6), 1430–1442.

57 An Introduction to Time-Resolved Decoding Analysis for M/EEG

THOMAS A. CARLSON, TIJL GROOTSWAGERS, AND AMANDA K. ROBINSON

ABSTRACT The human brain is constantly processing information in order to make decisions and interact with the world for tasks ranging from recognizing a familiar face to playing a game of tennis. These complex cognitive processes require communication between large populations of neurons. The noninvasive neuroimaging methods of electroencephalography (EEG) and magnetoencephalography (MEG) provide population measures of neural activity with millisecond precision that allow us to study the temporal dynamics of cognitive processes. However, multisensor M/EEG data is inherently high dimensional, making it difficult to parse important signals from noise. Multivariate pattern analysis (MVPA) or *decoding* methods offer vast potential for understanding high-dimensional M/EEG neural data. MVPA can be used to distinguish between different conditions and map the time courses of various neural processes, from basic sensory processing to high-level cognitive processes. In this chapter we discuss the practical aspects of performing decoding analyses on M/EEG data, as well as the limitations of the method, and then review some applications for understanding representational dynamics in the human brain.

One of the most remarkable aspects of the human brain is its speed and processing efficiency. A familiar friend is recognized in an instant, speech communication feels like a natural dynamic exchange, and we can make split-second decisions in life-threatening situations. A crucial question in cognitive neuroscience is how the brain manages such complex tasks with ease. As noninvasive brain-imaging techniques with millisecond resolution, magnetoencephalography and electroencephalography (M/EEG) offer unparalleled potential in capturing the dynamics of human cognition. Multivariate pattern analysis (MVPA or *decoding*) methods, in conjunction with M/EEG data, give insight into the temporal dynamics of information processing in the brain. Such methods yield insight not only into the time course of specific cognitive processes but also into the relative order of different cognitive processes. This can lead to increased understanding of the temporal dynamics of brain representations—for example, how low-level visual representations are transformed into high-level object representations (Contini, Wardle, & Carlson, 2017).

When MVPA was first introduced to functional magnetic resonance imaging (fMRI), it gave cognitive neuroscientists the unprecedented capacity to decode information in the brain (for recent reviews, see Haxby, 2012; Haynes, 2015; Tong & Pratte, 2012). MVPA in combination with fMRI's excellent spatial resolution enabled researchers to target and scrutinize the information represented in different brain areas. MVPA methods for M/EEG data complement fMRI findings by providing valuable insight into the time course of neural processing. Accordingly, M/EEG decoding methods are becoming increasingly popular in the cognitive neuroscience community.

Most cognitive neuroscientists are aware of decoding methods, as they have been actively used in fMRI research for two decades. Interestingly, methods for decoding mental states from the brain predate this work by a quarter century. Vidal first developed these methods for EEG in the context of brain-computer communication, asking the provocative question "Can observable electrical brain signals be put to work as carriers of information in man-computer communication or for the purpose of controlling such external apparatus as prosthetic devices or spaceships?" (Vidal, 1973, p. 157). Vidal's and subsequent work focused on the practical applications of having access to the brain's internal mental states. Many useful applications for decoding methods exist in M/EEG research, including brain-computer interfaces (BCI; e.g., Wolpaw et al., 2002), lie detection (e.g., Davatzikos et al., 2005), and the diagnosis of brain disorders (e.g., Ewers et al., 2011). Practical applications such as these place greater weight on performance over explanation. If diagnostic accuracy for schizophrenia using a particular method is increased by 2%, this represents a significant achievement that can have wide-reaching benefits. In contrast, the cognitive neuroscientist is interested in how the brain works. Here, the priority is to elucidate the neural mechanisms underlying cognitive processes, and metrics like prediction accuracy are of lesser importance (Hebart & Baker, 2017). The cognitive neuroscience application of decoding methods for M/EEG is the topic of this chapter. Henceforth, we will use the term *decoding* to reference the use of this method for understanding the brain.

This chapter focuses explicitly on time-resolved decoding analysis for M/EEG. MVPA can be applied to M/EEG data (frequency analysis, wavelet decomposition, connectivity, and more) in a variety of domains. We will limit the scope of our discussion to time-series decoding, but many of the same principles apply to other domains. This research uses paradigms that closely resemble EEG's event-related potential (ERP) or MEG's event-related field (ERF) research. In these experiments, participants are presented with stimuli (e.g., images, sounds, and others) from different conditions (e.g., faces vs. houses; attended vs. unattended), and the analysis is time locked to the presentation of the stimulus. MVPA and ERP/ERF analyses both seek to determine any differences between experimental conditions. The main difference is that ERP/ERF research uses a univariate approach, such as testing for differences between conditions from a single recording site or from data averaged across multiple sites. MVPA, in contrast, utilizes pattern information across recording sites and thus is often a more sensitive measure. As the topic of this chapter is MVPA, we will not discuss further differences between the approaches except to make illustrative points (for more information, see Grootswagers, Wardle, & Carlson, 2017; Hebart & Baker, 2017). For those interested in ERP/ERF analysis or more general information about the analysis of EEG recordings, we refer the reader to the excellent text on ERP research by Luck (2005).

The fundamental goal of the decoding analysis is to learn what and how information is represented by the brain (see chapters 56 and 58). M/EEG recordings are a measure of brain activity with high temporal resolution, and MVPA is a sensitive measure to determine whether or not any information in the recordings can distinguish between experimental conditions. The combination of the two can tell the researcher if there is information in the recordings of brain activity that can distinguish experimental conditions for each point in time. Going beyond this to address the more critical questions of "what" and "how" information is represented in the brain requires specific knowledge of the field of inquiry, which requires the development of robust experimental paradigms.

The goal of this chapter is to inform readers with little MVPA experience about the mechanics of applying these methods to M/EEG data and guide researchers on using these methods to answer the critical questions of "what" and "how" information is represented in the brain. We assume the reader has a basic knowledge M/EEG and thus will focus on the practical details of decoding analyses rather than aspects of conducting M/EEG experiments. We also note that M/EEG-MVPA is a relatively new field of research. On the one hand, this means that many exciting new theoretical questions can be explored using these powerful analysis methods. On the other hand, it means the techniques themselves are still evolving, and there is currently limited technical guidance (see Grootswagers, Wardle, & Carlson, 2017 for an advanced tutorial). At times, we will give recommendations based on our own experience that lack a scientific reference. We expect many of these recommendations to be followed up empirically in the coming years.

In this chapter, we first provide a foundation by situating pattern analysis methods in the context of M/EEG data. We then discuss some of the intricacies of using decoding methods for analyzing M/EEG data. We then conclude with a section describing advanced methods to familiarize the reader with more sophisticated approaches to time-resolved decoding.

The Basics of Time-Resolved Decoding Analysis for M/EEG

To give intuition to the analysis and a framework for discussion, we describe a hypothetical experiment. Consider a standard EEG system with 64 sensors (figure 57.1*A*), each of which continuously measures local electrical currents on the scalp (figure 57.1*B*). In each trial, participants are shown an image of either an *X* or an *O*, and there are 40 trials for each condition. To simplify the description of the analysis, we start by considering only two electrodes centered on the occipital cortex (e.g., O1 and O2) and data from a single time point (e.g., 100 ms poststimulus onset). This narrow slice of data can be represented as a two-dimensional scatterplot (figure 57.1*C*). In the plot, the two axes are the measurements from the two electrodes in microvolts (μV) for the single time point. The 80 data points are the measurements recorded for the individual trials, denoted by *X*s and *O*s. In this simplified scenario, the EEG data take the form of the familiar diagram given in virtually all texts describing pattern classification analysis (e.g., Duda, Hart, & Stork, 2001).

Pattern classification analysis is used to determine if there is information in the EEG measurements that can predict whether the participant was viewing an *X* or an *O* on a given trial. The analysis is a two-step process. The first step is to train the classifier to find a decision boundary that best separates the *X*s and *O*s using the EEG data. For the purposes of this initial description, we use a linear classifier, which uses a linear decision

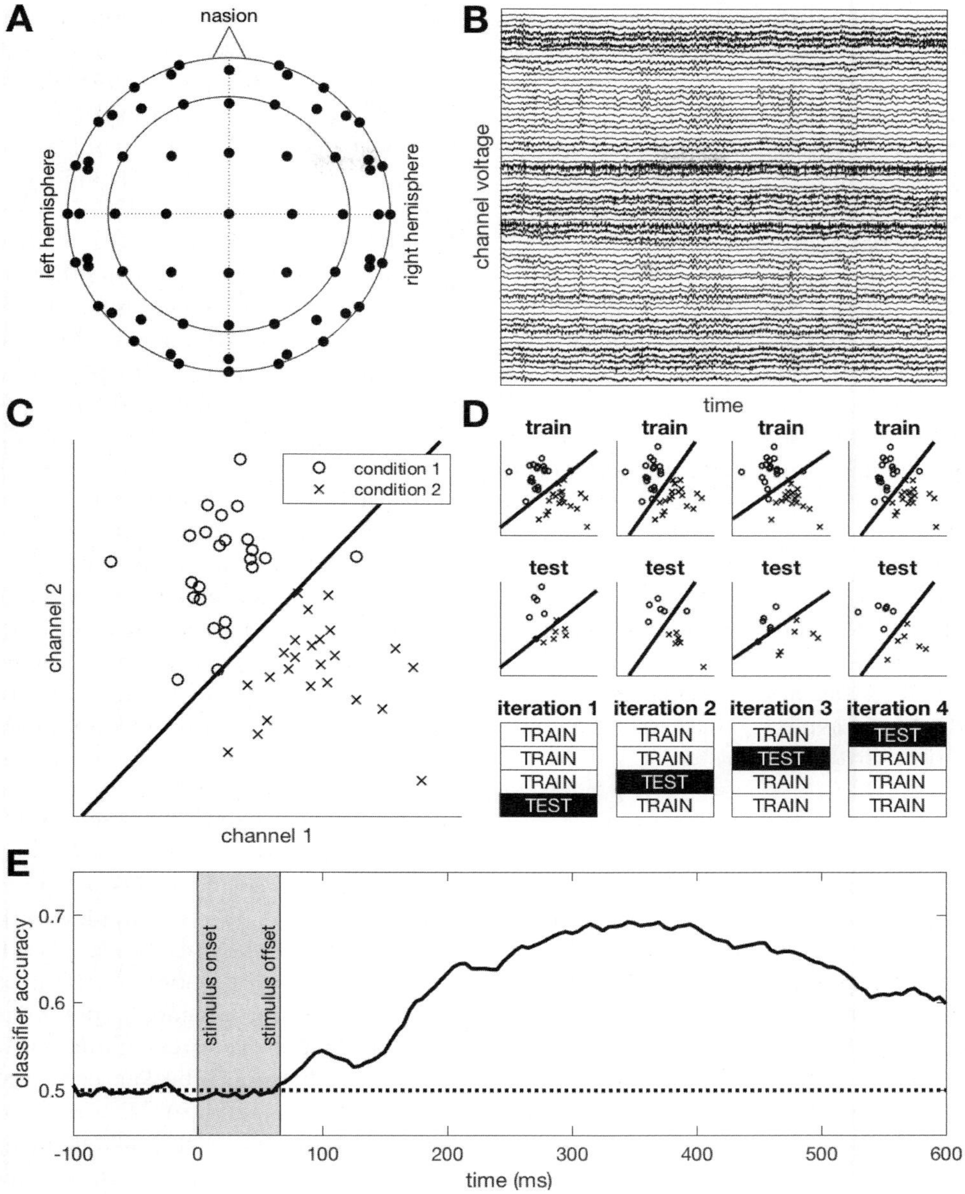

FIGURE 57.1 Time-resolved decoding analysis for M/EEG. *A*, Top-down view of a channel layout for a 64-channel EEG. *B*, Sample recording data from an EEG experiment. *C*, Scatterplot showing hypothetical EEG data from a single time point for a decoding experiment. The two axes are the measured current from two EEG electrodes. The points are individual trials for the *X* and *O* conditions. The line denotes the optimal decision boundary for classifying *X*s and *O*s derived from the generative model. *D*, Four iterations of cross-validation using 75% of the data to train the classifier and 25% to test the classifier. The top row of plots show the training data along with the classification boundary computed using linear discriminant analysis (LDA) for each iteration. The plots below show the test data for each iteration with the decision boundary computed from the training data. *E*, An example of a time-resolved decoding analysis showing classifier accuracy averaged over participants. Data from Grootswagers, Ritchie, Wardle, Heathcote, and Carlson (2017).

boundary for classification. In the plot in figure 57.1*C*, the observations to the left of the boundary are classified as *O*s, and those to the right of the boundary are classified as *X*s.

The second step is testing the classifier, in which a trained classifier is used to predict whether the participant was viewing an *X* or an *O* in a new set of trials. It is essential to use independent data for the two steps (training and test) to show that the classifier can generalize to independent data outside the training sample. For this, we use cross-validation, which involves splitting the data into training and test sets. In this example, we divide the data into four blocks and use 75% of the data to train the classifier and the other 25% to test the classifier. The classifier is trained four times, each time using three blocks of data to train the classifier and the remaining block to test the classifier (figure 57.1*D*). The average prediction accuracy (performance) of the classifier across the four iterations is taken as an estimate of decoding performance for classifying *X*s and *O*s from the EEG data at this time point. Importantly, if the classifier accuracy is above chance (i.e., 50%), this is taken as evidence there were different patterns of activation across the two channels when the participant viewed an *X* or an *O*.

To bring this simplified scenario back to a full M/EEG data set is only a matter of scaling up. First, instead of limiting the analysis to two channels, all of the channels or a selected subset of the channels is used for the analysis. Second, the analysis is repeated for every time point instead of only on a single time point. The result is a time-varying measure of stimulus/condition decodability (figure 57.1*E*). Finally, the experiment is repeated in multiple subjects to enable statistical inference, treating the subject as a random effect (see the section on statistical assessment).

The above provides a broad overview of the critical features of decoding analysis for M/EEG. In the following sections, we go into more detail about the steps in the analysis and discuss specific issues for the analysis of M/EEG data.

Practical Aspects of M/EEG Decoding Analysis

On preprocessing M/EEG data for multivariate pattern analysis M/EEG data is inherently noisy, picking up electrical activity from a variety of nonbrain sources, such as muscle activity, eye movements, and environmental noise. M/EEG research uses many preprocessing procedures to reduce this noise, such as filtering, resampling, and artifact rejection. The goal of preprocessing is to increase the signal-to-noise ratio (SNR), reducing the probability of obtaining a false negative (i.e., failing to observe an effect that is present). Preprocessing pipelines can vary between different laboratories (and even across individual researchers in a laboratory), as there is no consensus on preprocessing pipelines. The variability in preprocessing pipelines has been identified as an important issue for reproducibility in neuroimaging research (Poldrack et al., 2017). In developing a preprocessing pipeline, one thus needs to balance the desire to maximize SNR with minimizing the number of preprocessing steps to ensure the results are robust.

The analysis of ERP/ERF in EEG/MEG, respectively, has been around for decades, while decoding methods are relatively new. A natural starting point for thinking about a preprocessing pipeline for M/EEG decoding analysis might be to adopt established methods from ERP/ERF research (Luck, 2005). However, MVPA is different from ERP/ERF analyses, and thus the preprocessing steps need not be the same. In particular, decoding is more robust to artifacts than these other methods. Standard classifiers, such as linear discriminant analysis (LDA) and support vector machines (SVM), implicitly model the noise in the data. Therefore, classifiers can replace steps that typically require laborious manual inspection, such as bad sensors and artifact rejection. To understand this, we need to look at how machine-learning classifiers like LDA and SVM maximize the prediction.

First, the classifier assigns weights to each sensor. High weights enhance information in the measurement that informs the classification, and low weights suppress uninformative information in the measurement. This aspect is relevant because it potentially makes some preprocessing procedures redundant. In the context of EEG, an example is a broken or very low impendance electrode. In a standard ERP analysis, this electrode would be removed or interpolated during preprocessing. For MVPA, this step can be omitted because the classifier would learn in training that the noisy electrode is not informative for the prediction and would assign it a low weight. Note also that channel interpolation is a linear combination of surrounding electrodes and thus adds no additional information to the classifier. Eyeblink and muscular artifacts are other examples. For ERP analysis, these artifacts can have a significant effect on the quality of the data and thus need to be removed. For MVPA, these artifacts are of lesser concern. If the artifact is not informative for the prediction, the classifier will suppress the artifacts by giving low weight to the component of the signal related to the artifacts. Removing trials with eyeblink artifacts is thus another preprocessing step that potentially could be removed from the MVPA preprocessing pipeline (Grootswagers, Wardle, & Carlson, 2017). Note that artifacts confounded with the conditions of interest are a serious concern for all (MVPA and ERP) analyses!

Second, classifiers can cancel noise in the data to improve the prediction. This is relevant when considering the effect of environmental noise. For ERP analysis, a noisy channel increases variance in the estimate of the average evoked response, and thus it is advisable to interpolate noisy channels or exclude them from the analysis. In contrast, classifiers (e.g., LDA) can use the information about the noise to improve prediction. For example, suppose we have two hypothetical channels. One channel contains brain activity differentiating the two experimental conditions and also environmental noise. The second channel is far from the source of brain activity and only contains the environmental noise. The classifier can use an estimate of the environmental noise from the second channel and subtract it from the first channel so that what remains is the signal that differentiates the two experimental conditions. Including the noisy channel can therefore give the classifier *more* information about the signal of interest than if the channel had been removed (Averbeck, Latham, & Pouget, 2006; Haufe et al., 2014).

In summary, MVPA is a relatively new approach to analyzing M/EEG data that does not have stringent noise-reduction requirements, as do typical ERP analyses. The critical lesson to be taken concerning preprocessing is that MVPA can inherently deal with some noise and artifacts in the data. This is an advantage of the decoding approach because it reduces data-processing steps relying on subjective criteria (e.g., the interpolation of noisy channels identified by visual inspection). While systematic research is still required, M/EEG MVPA research will need to develop its own guidelines for preprocessing. These new standards will need to take into account how classifiers operate to strike a balance between optimizing SNR and minimizing the number of preprocessing steps and options for reproducible research (Poldrack et al., 2017; for early explorations into the effect of preprocessing pipelines on M/EEG decoding, see Grootswagers et al., 2017).

Classifier selection Decoding analysis generally favors linear classifiers. As the name implies, linear classifiers use a linear boundary (e.g., see figure 57.1) or a hyperplane for greater than two dimensions for classification. Compelling arguments have been made for the use of linear classifiers in fMRI research (Kriegeskorte, 2011; Misaki et al., 2010; Muller, Anderson, & Birch, 2003; Mur, Bandettini, & Kriegeskorte, 2009; Pereira, Mitchell, & Botvinick, 2009; Schwarzkopf & Rees, 2011). In contrast, there has been little systematic discussion of this for M/EEG. Below, we give an overview of the different classifiers available for M/EEG decoding research and step into the fMRI community's argument for linear classifiers, discussing them in the context of M/EEG.

Although nonlinear classifiers have the added capacity to fit more complex class boundaries using nonlinear terms, in practice they generally perform equal to or worse than their linear counterparts in neuroimaging decoding studies (Misaki et al., 2010). This is because the increased flexibility of nonlinear classifiers comes at the cost of overfitting, which limits generalization performance. Furthermore, the results from nonlinear classifiers are harder to interpret. As previously noted, it is also essential to keep in mind that cognitive neuroscience applications of MVPA focus on understanding the brain, where improvements in performance are of marginal value (see the introduction to this chapter). A small increase in performance is a hefty price for the loss of interpretability, which we discuss below. Thus, unless there is substantial justification, linear classifiers are the preferred method for cognitive neuroscience applications of MVPA for the analysis of M/EEG data.

Linear classifiers have two interpretive advantages from a cognitive neuroscience perspective. First, linear classifiers are a biologically plausible form of a "read-out" (DiCarlo & Cox, 2007). Specifically, the information used by a linear classifier could also be "read out" by a single downstream neuron (Kriegeskorte, 2011; Misaki et al., 2010). In the context of M/EEG, this is less compelling. Embedded in this reasoning is the fact that the downstream neuron has access to the same information as the classifier (Carlson et al., 2017). A single EEG channel records the aggregated activity of thousands of neurons from different brain regions, and the entire EEG cap has access to most of the neurons on the cortical surface. It is not biologically plausible that a single neuron has access to all the information represented on the cortical surface.

The second interpretive advantage is that linear classifiers produce weight maps that can be visualized to gain insight into the source of decodable information (Kamitani & Tong, 2005). An fMRI-MVPA experiment studying faces, for example, might show that voxels in the fusiform face area (FFA; Kanwisher, McDermott, & Chun, 1997) are given high weights. The weight maps thus can be used to infer that the FFA is a strong candidate source of decodable information. Similarly, for M/EEG, the classifier weights can be projected back to the topographical map of the sensors to gain insight into the source of decodable information (but see Haufe et al. [2014] and the discussion below).

The final point of discussion is the choice of classifier. In a recent study, we compared the performance of five classifiers when decoding minimally preprocessed MEG data: LDA, Gaussian naive Bayes (GNB), linear SVMs, Spearman's rank correlation, and Pearson's correlation (Grootswagers, Wardle, & Carlson, 2017). The study found that SVM, LDA, and GNB performed the best, suggesting that they are all excellent choices for M/EEG decoding studies.

Localizing the source of decodable information in M/EEG Cognitive neuroscientists are often interested in both spatial (i.e., where) and temporal (i.e., when) signatures of neural processing. Recovering the underlying source(s) of neural activity is a long-standing challenge for M/EEG research. This challenge extends to M/EEG decoding research, although advanced decoding methods do offer at least one possible solution (see the section on representational dynamics). In this discussion it is essential to make a distinction between resolution and spatial precision. Resolution is the capacity to resolve two points in space. Using MVPA, MEG studies have shown that it may be possible to resolve activation patterns spanning V1 columns that are just a millimeter in width (Cichy, Ramirez, & Pantazis, 2015; Wardle et al., 2016). These results highlight the remarkable sensitivity of MVPA for detecting subtle differences in patterns of activation. However, localizing the precise source of this neural information (i.e., spatial precision) remains an ongoing challenge.

Broadly, three methods can be used to localize sources in the brain using MVPA. Below, we will discuss two of these methods; the third will be presented later in the chapter. The first approach is weight projection. MVPA returns both a performance metric (e.g., percent correct) and the weights used by the classifier to make the prediction. The more informative the sensor, the higher the weight. One straightforward means of identifying the source of decodable information is to plot the weights on the scalp map. When interpreting these maps, it is important to consider the definition of *informative* (c.f., de-Wit et al., 2016; Haufe et al., 2014). Previously, we discussed the three ways classifiers optimize their performance. In addition to weights being assigned to distinguish condition-specific information, weights are also used by the classifier to suppress noise. When using the weight projection method to interpret the underlying sources, it is thus essential to consider only weights that reflect the differences between conditions (Grootswagers, Wardle, & Carlson, 2017; Haufe et al., 2014). The second approach is to extract multivariate brain activity from a region of interest (ROI) for MVPA. The most straightforward variant of this is to select sensors located above the ROI—for example, using ten sensors over the occipital cortex to study early visual processing. This can also be applied in a sensor-searchlight approach (similar to fMRI; see Haynes et al., 2007; Kriegeskorte, Goebel, & Bandettini, 2006), in which the analysis is repeated on local clusters of sensors, providing a scalp map of decoding accuracies (e.g., see Collins, Robinson, & Behrmann, 2018; Kaiser, Oosterhof, & Peelen, 2016).

These two methods are subject to interpretive criticism arising from the fact that brain activity will propagate outside the M/EEG sensors located above a source (information loss), and brain activity from nearby areas will propagate into the selected sensors (information leakage). Nevertheless, it can be a useful means for gross localization (e.g., left vs. right hemisphere) in some circumstances, and the limitations in its interpretability are transparent. Alternatively, the sensor-level data can first be projected into source space using source reconstruction methods such as minimum norm estimate (MNE; Hamalainen & Ilmoniemi, 1994) and beamforming (Hamalainen & Ilmoniemi, 1994; Van Veen et al., 1997). Multivariate time-series data can be reconstructed using these methods using virtual voxels in an ROI, and decoding analyses can be performed for different ROIs. As with univariate analysis, the quality of the source reconstruction can be improved using anatomical MRI data from the participant and fMRI data to precisely define the ROI. Notably, these more sophisticated methods for localization are also subject to issues of information loss and leakage (Brookes, Woolrich, & Barnes, 2012; Gohel et al., 2018; Hipp et al., 2012; Nolte et al., 2004; Sato et al., 2018), but the algorithms will attempt to minimize their effect.

In summary, MVPA is a sensitive measure for differentiating conditions based on the evoked spatial distribution of activity from M/EEG. The MVPA approach, however, remains limited for localizing sources, which stems from the fundamental inverse problem for M/EEG.

Statistical assessment of information at the group level Time-resolved MVPA tests the presence of information at the group level. That is, we want to know whether across our sample of subjects, classifier performance is higher than would be expected by chance. There are multiple proposed methods to assess this, and there is no consensus about the optimal approach. Group-level classifier accuracies can, for example, be tested against chance using a t-test, or nonparametric tests such as a sign-rank test (Wilcoxon, 1945) or permutation test (Oostenveld et al., 2011). Because these tests are repeated at each time point, they must also be corrected for multiple comparisons—for example, using Bonferroni, false discovery rate (FDR; Nichols & Holmes, 2002), or cluster-based corrections (e.g., Oostenveld et al., 2011;

Smith & Nichols, 2009; Stelzer, Chen, & Turner, 2013). As with the choice of test statistic, there is no consensus on the optimal method for correcting multiple comparisons. Thus, the choice of statistical analysis must be guided by the experimental questions (see Allefeld, Gorgen, & Haynes, 2016; Hebart & Baker, 2017; Thirion et al., 2015).

Advanced Methods for M/EEG Decoding: Representational Dynamics

Exemplar-based decoding methods expand on categorical approaches by studying the structure of how information is represented. Figure 57.2*A, B*, shows the difference between the two approaches. The two parts of the figure show a reconstruction of the brain's representation of 24 objects from MEG data 140 ms after the onset of the stimulus (data from Carlson et al., 2013). The stimulus set included 12 animate objects (e.g., camel and alligator) and 12 inanimate objects (e.g., chair and kiwi fruit). In the category decoding approach (figure 57.2*A*), the stimuli are treated as equivalent class members (labeled *A* = animate or *IA* = inanimate), and the analysis is conducted using standard methods (see the section on the basics of time-resolved decoding). If the analysis shows the classifier can decode animate and inanimate objects from the MEG data, this is evidence that the brain representation of the stimuli is encoding object animacy.

Exemplar-based decoding methods study how the individual exemplars of each category are represented. Here, the decoding analysis is run for all possible pairwise combinations of exemplars—including within-category. The decodability (i.e., classification performance) of each pair is taken as a measure of "distance" between the two items in the neural representation (Walther et al., 2016). Figure 57.2*B* demonstrates this approach to plotting the individual stimuli. A line connecting the object exemplars in the figure indicates the decodability of the connected pair. The length of the line is proportional to decodability—that is, stimuli that are close to one another (a human and a monkey face) are less decodable than stimuli that are far apart (e.g., a monkey and a television) at the given point in time during visual processing.

Representational similarity analysis (RSA; Kriegeskorte & Kievit, 2013; Kriegeskorte, Mur, & Bandettini, 2008) is the standard framework for hypothesis testing for the exemplar-based decoding approach. In the RSA, the pairwise decodability of the stimuli from the brain recordings is encoded in a representational dissimilarity matrix (RDM; figure 57.2*C*). The rows and columns of the matrix correspond to individual exemplars, and each cell is the neural decodability between the two exemplars. Hypotheses are tested by constructing model RDMs that make predictions about the structure of the brain representation. An animacy model, for example, predicts that animate and inanimate objects form separable clusters in the representation. Formally, the animacy model predicts that objects within the animate and inanimate object categories will be close to one another (distance = 0), and objects that span the category boundary will be far apart (distance = 1; see figure 57.2*C*). To test the model, the entries of the observed neural RDM are correlated with the model RDM. For the animacy model, a high correlation between the neural RDM and the model allows us to conclude that animacy is represented in the brain.

The exemplar-based decoding approach taken in RSA makes the structure of the representation and the models more explicit, which has many advantages (Kriegeskorte & Kievit, 2013; Kriegeskorte, Mur, & Bandettini, 2008; Nili et al., 2014). The exemplar-based analysis can be used to study the time-varying structure of brain representations using M/EEG. As before, the analysis and model testing are performed on all of the time points. By examining how brain representations emerge over time, we gain a deeper understanding of how information is dynamically transformed and represented in the brain. Figure 57.2*D* shows the time-varying analysis from a MEG data set with a more extensive stimulus set using several different category models (92 exemplars; Cichy, Pantazis, & Oliva, 2014). The figure shows that category structure for different categories (animacy, human, face) emerge at different time points in the brain's emerging representation of objects. By studying the emergence of different categories in time, studies have shown that basic-level category information emerges first (e.g., human face), and abstract-level category information (e.g., animacy) comes at the later stages of object processing (Carlson et al., 2013; Cichy, Pantazis, & Oliva, 2014; Contini, Wardle, & Carlson, 2017).

Exemplar-based decoding methods and the RSA framework also provide a unique solution to the challenge of acquiring data that have the spatial resolution to study regional brain activity and the temporal resolution to explore the fine-grained temporal dynamics of neural activity (Cichy, Pantazis, & Oliva, 2014, 2016). This novel approach takes advantage of the fine-grained structure of information in brain representations, as indexed by MEG and fMRI, and integrates the data to estimate regional brain activity. The fMRI and MEG experiments are run using identical stimuli; thus, the two imaging methods produce RDMs of equal size that are directly comparable. The MEG RDM describes the brain representation at each time point, which includes activity from multiple brain regions. The fMRI RDM

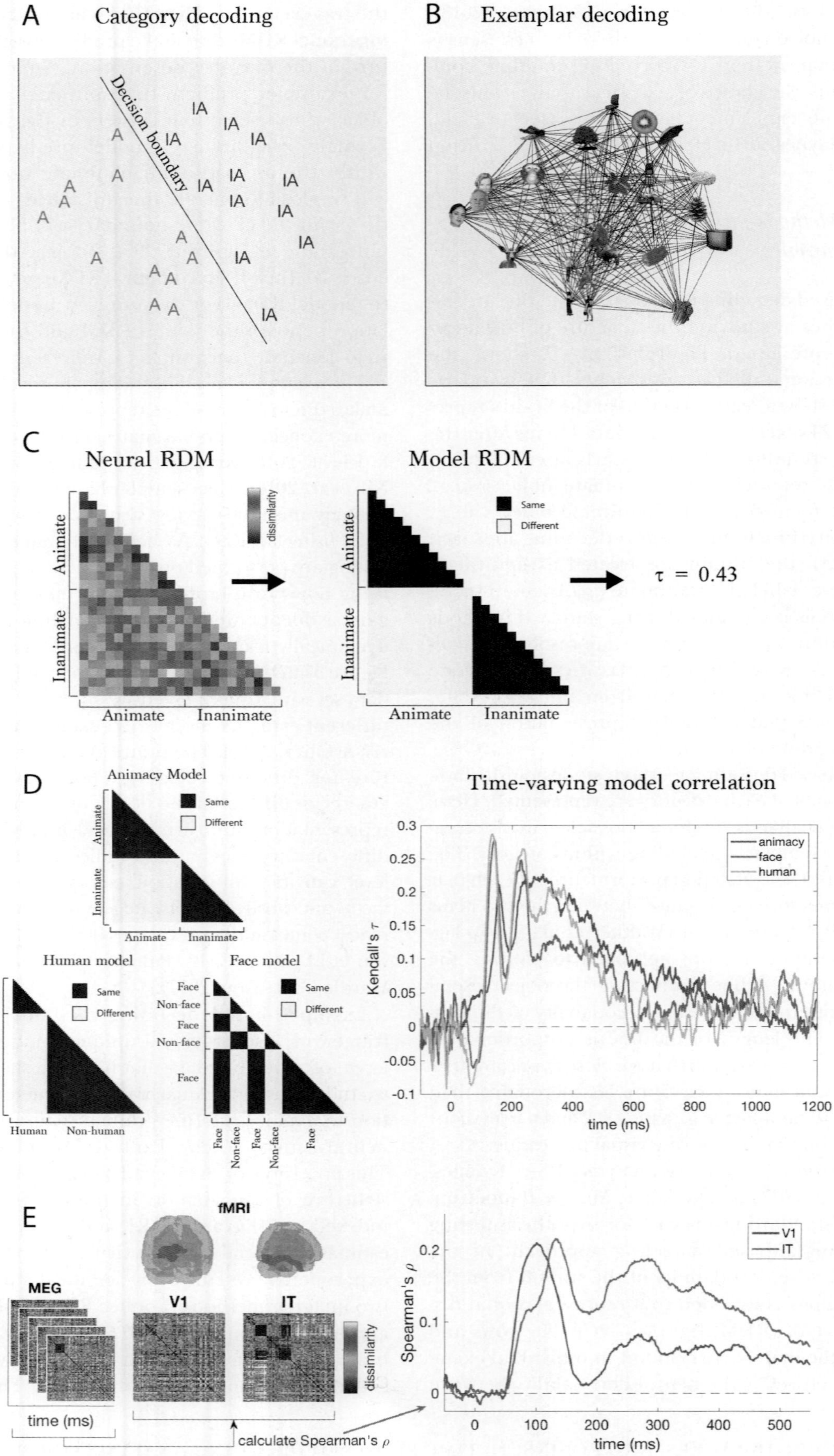

A
Category decoding
Decision boundary
A
IA
B
Exemplar decoding
C
Neural RDM
dissimilarity
Animate
Inanimate
Model RDM
Same
Different
τ = 0.43
D
Animacy Model
Human model
Face model
Human
Non-human
Face
Non-face
Time-varying model correlation
animacy
face
human
Kendall's τ
time (ms)
E
fMRI
MEG
V1
IT
time (ms)
calculate Spearman's ρ
Spearman's ρ

from an ROI describes that brain area's representation. Importantly, different brain areas have unique representations, although some areas will be similar, such as V1 and V2. By correlating a brain area's fMRI RDM with the MEG RDM for each time point, we get a time-varying estimate of that area's contribution to the MEG signal across time. This method was used to study the contribution of early visual cortex (V1) and inferior temporal cortex (IT) to the MEG signal in time for object recognition (figure 57.2*E*; Cichy, Pantazis, & Oliva, 2014).

Conclusion

M/EEG decoding methods provide a powerful set of tools for cognitive neuroscientists to gain insight into perceptual and cognitive functions by revealing the brain's processing dynamics with millisecond resolution. In this chapter we described the fundamental aspects of M/EEG decoding analysis, practical considerations in running these analyses, and advanced methods to study representational dynamics. These methods have been incorporated into a variety of MVPA toolboxes (Bode et al., 2018; Fahrenfort et al., 2018; Gramfort et al., 2014; Hanke et al., 2009; Meyers, 2013; Oostenveld et al., 2011; Oosterhof, Connolly, & Haxby, 2016). We expect M/EEG decoding to have a broad impact on the future of cognitive neuroscience research. For researchers interested in taking the next step and learning more about these methods, we refer the reader to advanced tutorials (Grootswagers, Wardle, & Carlson, 2017; Lemm et al., 2011).

Acknowledgments

We would like to thank the reviewers and Nick McNair, Denise Moerel, and Selene Petit for their feedback and suggestions.

REFERENCES

Allefeld, C., Gorgen, K., & Haynes, J. D. (2016). Valid population inference for information-based imaging: From the second-level t-test to prevalence inference. *Neuroimage, 141*, 378–392.

Averbeck, B. B., Latham, P. E., & Pouget, A. (2006). Neural correlations, population coding and computation. *Nature Reviews Neuroscience, 7*, 358–366.

Bode, S., Feuerriegel, D., Bennett, D., & Alday, P. M. (2018). The decision decoding ToolBOX (DDTBOX)—a multivariate pattern analysis toolbox for event-related potentials. *Neuroinformatics, 17*(1), 27–42.

Brookes, M. J., Woolrich, M. W., & Barnes, G. R. (2012). Measuring functional connectivity in MEG: A multivariate approach insensitive to linear source leakage. *Neuroimage, 63*, 910–920.

Carlson, T., Goddard, E., Kaplan, D. M., Klein, C., & Ritchie, J. B. (2017). Ghosts in machine learning for cognitive neuroscience: Moving from data to theory. *Neuroimage, 180*, 88–100.

Carlson, T., Tovar, D. A., Alink, A., & Kriegeskorte, N. (2013). Representational dynamics of object vision: The first 1000 ms. *Journal of Vision, 13*(10), 1–19.

Cichy, R. M., Pantazis, D., & Oliva, A. (2014). Resolving human object recognition in space and time. *Nature Neuroscience, 17*, 455–462.

Cichy, R. M., Pantazis, D., & Oliva, A. (2016). Similarity-based fusion of MEG and fMRI reveals spatio-temporal dynamics in human cortex during visual object recognition. *Cerebral Cortex, 26*, 3563–3579.

Cichy, R. M., Ramirez, F. M., & Pantazis, D. (2015). Can visual information encoded in cortical columns be decoded from magnetoencephalography data in humans? *Neuroimage, 121*, 193–204.

Collins, E., Robinson, A. K., & Behrmann, M. (2018). Distinct neural processes for the perception of familiar versus unfamiliar faces along the visual hierarchy revealed by EEG. *Neuroimage, 181*, 120–131.

Contini, E. W., Wardle, S. G., & Carlson, T. A. (2017). Decoding the time-course of object recognition in the human brain: From visual features to categorical decisions. *Neuropsychologia, 105*, 165–176.

Davatzikos, C., Ruparel, K., Fan, Y., Shen, D. G., Acharyya, M., Loughead, J. W., Gur, R. C., & Langleben, D. D. (2005). Classifying spatial patterns of brain activity with machine

FIGURE 57.2 Advanced methods for M/EEG decoding. *A, B*, Plots demonstrate the difference between category and exemplar-based decoding methods. Data for the plots come from a single time point (140 ms poststimulus onset) of an MEG experiment studying the brain's representation of objects (Carlson et al., 2013). *A*, Category-decoding approach with stimuli labeled by their category (A=animate object; IA=inanimate object) and the decision boundary from a linear classifier trained to discriminate animate and inanimate objects from the MEG data. *B*, Exemplar-decoding approach with individual stimuli displayed as images. The lines in the plot represent the pairwise decodability of individual exemplars for all possible pairwise comparisons. The distance (*line length*) indicates the relative decodability of exemplar pairs. *C*, Representational similarity (RSA) model testing applied to the MEG data. The entries of the neural RDM are correlated with the animacy model to study whether the brain represents object animacy at 140 ms poststimulus onset. *D*, RSA used to test multiple candidate models from another MEG experiment investigating the brain's representation of objects. Data from Cichy, Pantazis, and Oliva (2014). Plotted is the time correlation between the time-varying RDMs from the MEG and animacy, human, and face models. *E*, Graphical depiction of the MEG-fMRI fusion approach estimating regional time-varying neural activity from fMRI regions of interest (ROI). The time-varying MEG RDMs are correlated with RDMs from ROIs (obtained using fMRI). For each ROI, this gives a time-varying correlation indexing neural activity in the ROI. Shown is the estimated neural activity for V1 and inferior temporal cortex. Data from Cichy, Pantazis, and Oliva (2014). (See color plate 63.)

learning methods: Application to lie detection. *Neuroimage, 28*, 663–668.

de-Wit, L., Alexander, D., Ekroll, V., & Wagemans, J. (2016). Is neuroimaging measuring information in the brain? *Psychonomic Bulletin & Review, 23*, 1415–1428.

DiCarlo, J. J., & Cox, D. D. (2007). Untangling invariant object recognition. *Trends in Cognitive Sciences, 11*, 333–341.

Duda, R. O., Hart, P. E., & Stork, D. G. (2001). *Pattern classification* (2nd ed.). New York: Wiley.

Ewers, M., Sperling, R.A., Klunk, W. E., Weiner, M. W., & Hampel, H. (2011). Neuroimaging markers for the prediction and early diagnosis of Alzheimer's disease dementia. *Trends in Neurosciences, 34*, 430–442.

Fahrenfort, J. J., van Driel, J., van Gaal, S., & Olivers, C. N. L. (2018). From ERPs to MVPA using the Amsterdam decoding and modeling toolbox (ADAM). *Frontiers in Neuroscience, 12*, 368.

Gohel, B., Lim, S., Kim, M. Y., Kwon, H., & Kim, K. (2018). Dynamic pattern decoding of source-reconstructed MEG or EEG data: Perspective of multivariate pattern analysis and signal leakage. *Computers in Biology and Medicine, 93*, 106–116.

Gramfort, A., Luessi, M., Larson, E., Engemann, D. A., Strohmeier, D., Brodbeck, C., Parkkonen, L., & Hamalainen, M. S. (2014). MNE software for processing MEG and EEG data. *Neuroimage, 86*, 446–460.

Grootswagers, T., Ritchie, J. B., Wardle, S. G., Heathcote, A., & Carlson, T. A. (2017). Asymmetric compression of representational space for object animacy categorization under degraded viewing conditions. *Journal of Cognitive Neuroscience, 29*, 1995–2010.

Grootswagers, T., Wardle, S. G., & Carlson, T. A. (2017). Decoding dynamic brain patterns from evoked responses: A tutorial on multivariate pattern analysis applied to time series neuroimaging data. *Journal of Cognitive Neuroscience, 29*, 677–697.

Hamalainen, M. S., & Ilmoniemi, R. J. (1994). Interpreting magnetic fields of the brain: Minimum norm estimates. *Medical & Biological Engineering & Computing, 32*, 35–42.

Hanke, M., Halchenko, Y. O., Sederberg, P. B., Hanson, S. J., Haxby, J. V., & Pollmann, S. (2009). PyMVPA: A python toolbox for multivariate pattern analysis of fMRI data. *Neuroinformatics, 7*, 37–53.

Haufe, S., Meinecke, F., Gorgen, K., Dahne, S., Haynes, J. D., Blankertz, B., & Biessmann, F. (2014). On the interpretation of weight vectors of linear models in multivariate neuroimaging. *Neuroimage, 87*, 96–110.

Haxby, J. V. (2012). Multivariate pattern analysis of fMRI: The early beginnings. *Neuroimage, 62*, 852–855.

Haynes, J. D. (2015). A primer on pattern-based approaches to fMRI: Principles, pitfalls, and perspectives. *Neuron, 87*, 257–270.

Haynes, J. D., Sakai, K., Rees, G., Gilbert, S., Frith, C., & Passingham, R. E. (2007). Reading hidden intentions in the human brain. *Current Biology, 17*, 323–328.

Hebart, M. N., & Baker, C. I. (2017). Deconstructing multivariate decoding for the study of brain function. *Neuroimage, 180*, 4–18.

Hipp, J. F., Hawellek, D. J., Corbetta, M., Siegel, M., & Engel, A. K. (2012). Large-scale cortical correlation structure of spontaneous oscillatory activity. *Nature Neuroscience, 15*, 884–890.

Kaiser, D., Oosterhof, N. N., & Peelen, M. V. (2016). The neural dynamics of attentional selection in natural scenes. *Journal of Neuroscience, 36*, 10522–10528.

Kamitani, Y., & Tong, F. (2005). Decoding the visual and subjective contents of the human brain. *Nature Neuroscience, 8*, 679–685.

Kanwisher, N., McDermott, J., & Chun, M. M. (1997). The fusiform face area: A module in human extrastriate cortex specialized for face perception. *Journal of Neuroscience, 17*, 4302–4311.

Kriegeskorte, N. (2011). Pattern-information analysis: From stimulus decoding to computational-model testing. *Neuroimage, 56*, 411–421.

Kriegeskorte, N., Goebel, R., & Bandettini, P. (2006). Information-based functional brain mapping. *Proceedings of the National Academy of Sciences of the United States of America, 103*, 3863–3868.

Kriegeskorte, N., & Kievit, R. A. (2013). Representational geometry: Integrating cognition, computation, and the brain. *Trends in Cognitive Sciences, 17*, 401–412.

Kriegeskorte, N., Mur, M., & Bandettini, P. (2008). Representational similarity analysis—connecting the branches of systems neuroscience. *Frontiers in Systems Neuroscience, 2*, 4.

Lemm, S., Blankertz, B., Dickhaus, T., & Muller, K. R. (2011). Introduction to machine learning for brain imaging. *Neuroimage, 56*, 387–399.

Luck, S. J. (2005). *An introduction to the event-related potential technique.* Cambridge, MA: MIT Press.

Meyers, E. M. (2013). The neural decoding toolbox. *Frontiers in Neuroinformatics, 7*, 8.

Misaki, M., Kim, Y., Bandettini, P. A., & Kriegeskorte, N. (2010). Comparison of multivariate classifiers and response normalizations for pattern-information fMRI. *Neuroimage, 53*, 103–118.

Muller, K. R., Anderson, C. W., & Birch, G. E. (2003). Linear and nonlinear methods for brain-computer interfaces. *IEEE Transactions on Neural Systems and Rehabilitation Engineering, 11*, 165–169.

Mur, M., Bandettini, P. A., & Kriegeskorte, N. (2009). Revealing representational content with pattern-information fMRI—an introductory guide. *Social Cognitive and Affective Neuroscience, 4*, 101–109.

Nichols, T. E., & Holmes, A. P. (2002). Nonparametric permutation tests for functional neuroimaging: A primer with examples. *Human Brain Mapping, 15*, 1–25.

Nili, H., Wingfield, C., Walther, A., Su, L., Marslen-Wilson, W., & Kriegeskorte, N. (2014). A toolbox for representational similarity analysis. *PLOS Computational Biology, 10*, e1003553.

Nolte, G., Bai, O., Wheaton, L., Mari, Z., Vorbach, S., & Hallett, M. (2004). Identifying true brain interaction from EEG data using the imaginary part of coherency. *Clinical Neurophysiology, 115*, 2292–2307.

Oostenveld, R., Fries, P., Maris, E., & Schoffelen, J. M. (2011). FieldTrip: Open source software for advanced analysis of MEG, EEG, and invasive electrophysiological data. *Computational Intelligence and Neuroscience, 2011*, 156869.

Oosterhof, N. N., Connolly, A. C., & Haxby, J. V. (2016). CoSMoMVPA: Multi-modal multivariate pattern analysis of neuroimaging data in Matlab/GNU octave. *Frontiers in Neuroinformatics, 10*, 27.

Pereira, F., Mitchell, T., & Botvinick, M. (2009). Machine learning classifiers and fMRI: A tutorial overview. *Neuroimage, 45*, S199–209.

Poldrack, R. A., Baker, C. I., Durnez, J., Gorgolewski, K. J., Matthews, P. M., Munafo, M. R., Nichols, T. E., Poline, J. B., Vul, E., & Yarkoni, T. (2017). Scanning the horizon: Towards transparent and reproducible neuroimaging research. *Nature Reviews Neuroscience, 18*, 115–126.

Sato, M., Yamashita, O., Sato, M. A., & Miyawaki, Y. (2018). Information spreading by a combination of MEG source estimation and multivariate pattern classification. *PLoS One, 13*, e0198806.

Schwarzkopf, D. S., & Rees, G. (2011). Interpreting local visual features as a global shape requires awareness. *Proceedings of the Royal Society B: Biological Sciences, 278*, 2207–2215.

Smith, S. M., & Nichols, T. E. (2009). Threshold-free cluster enhancement: Addressing problems of smoothing, threshold dependence and localisation in cluster inference. *Neuroimage, 44*, 83–98.

Stelzer, J., Chen, Y., & Turner, R. (2013). Statistical inference and multiple testing correction in classification-based multi-voxel pattern analysis (MVPA): Random permutations and cluster size control. *Neuroimage, 65*, 69–82.

Thirion, B., Pedregosa, F., Eickenberg, M., & Varoquaux, G. (2015). Correlations of correlations are not reliable statistics: Implications for multivariate pattern analysis. ICML Workshop on Statistics, Machine Learning and Neuroscience (Stamlins), Lille, France.

Tong, F., & Pratte, M. S. (2012). Decoding patterns of human brain activity. *Annual Review of Psychology, 63*, 483–509.

Van Veen, B. D., van Drongelen, W., Yuchtman, M., & Suzuki, A. (1997). Localization of brain electrical activity via linearly constrained minimum variance spatial filtering. *IEEE Transactions on Biomedical Engineering, 44*, 867–880.

Vidal, J. J. (1973). Toward direct brain-computer communication. *Annual Review of Biophysics and Bioengineering, 2*, 157–180.

Walther, A., Nili, H., Ejaz, N., Alink, A., Kriegeskorte, N., & Diedrichsen, J. (2016). Reliability of dissimilarity measures for multi-voxel pattern analysis. *Neuroimage, 137*, 188–200.

Wardle, S. G., Kriegeskorte, N., Grootswagers, T., Khaligh-Razavi, S. M., & Carlson, T. A. (2016). Perceptual similarity of visual patterns predicts dynamic neural activation patterns measured with MEG. *Neuroimage, 132*, 59–70.

Wilcoxon, F. (1945). Individual comparisons by ranking methods. *Biometrics Bulletin, 1*, 80–83.

Wolpaw, J. R., Birbaumer, N., McFarland, D. J., Pfurtscheller, G., & Vaughan, T. M. (2002). Brain-computer interfaces for communication and control. *Clinical Neurophysiology, 113*, 767–791.

58 Encoding and Decoding Framework to Uncover the Algorithms of Cognition

JEAN-RÉMI KING, LAURA GWILLIAMS, CHRIS HOLDGRAF, JONA SASSENHAGEN, ALEXANDRE BARACHANT, DENIS ENGEMANN, ERIC LARSON, AND ALEXANDRE GRAMFORT

ABSTRACT A central challenge to cognitive neuroscience consists of decomposing complex brain signals into an interpretable sequence of operations—an algorithm—that ultimately accounts for intelligent behaviors. Over the past decades, a variety of analytical tools have been developed to (1) isolate each algorithmic step and (2) track their ordering from neuronal activity. In the present chapter, we briefly review the main methods to encode and decode temporally resolved neural recordings, show how these approaches relate to one another, and summarize their main premises and challenges. Throughout, we illustrate with recent findings the increasing role of machine learning, both as a method to extract convoluted patterns of neural activity and as an operational framework to formalize cognitive processes. Overall, we discuss how modern analyses of neural time series help identify the algorithmic bases of cognition.

An algorithm is a sequence of simple computations designed to solve a complex problem. Under this definition, a major goal of cognitive neuroscience therefore consists of uncovering the algorithms of the mind—identifying the nature and the order of computations implemented in the brain to adequately interact with the environment (Marr, 1982).

Over the years, this foundational endeavor has adopted a variety of methods, spanning from the decomposition of reaction times (Donders, 1969; Sternberg, 1998) to modern electrophysiology and neuroimaging paradigms. In the present chapter, we focus on two major pillars necessary to recover an interpretable sequence of operations from neuronal activity. First, we review how individual computations can be isolated by identifying and linking neural codes to mental representations. Second, we review how the analysis of dynamic neural responses can recover the order of these computations. Throughout, we discuss how recent developments in machine learning offer complementary methods to analyze convoluted patterns of neural activity while also helping to formalize the computational foundations of cognition.

Neuronal Activity: Codes and Contents

The triple quest of cognitive neuroscience Three challenges must be addressed to isolate the elementary computations underlying a given cognitive process (figure 58.1). First, we must identify what variables the brain generates and exploits to solve a particular task; in other words, determine the psychologically relevant dimensions. For example, speech has been formally described in terms of phonemes (e.g., /b/, /p/, /k/), but the existence of these psychological variables has been debated given their extensive overlap with low-level acoustic properties (Hickok, 2014). In a recent study, Mesgarani, Cheung, Johnson, and Chang (2014) have shown that activity of the superior temporal gyrus in response to speech is better accounted for by phonetic features than by acoustic ones. This suggests that the brain computes phonetic variables to understand speech. More generally, the search for the relevant mental variables is ubiquitous in cognitive neuroscience. For example, studies have been able to characterize the neural bases of faces (Freiwald & Tsao, 2010; Haxby, 2006; Kanwisher, McDermott, & Chun, 1997), word strings (Dehaene & Cohen, 2007; Price, 2010), and semantics (Huth, de Heer, Griffiths, Theunissen, & Gallant, 2016), to name a few.

Second, we must identify *how* neurons read and communicate such informational content. For example, the fact that neurons usually do not discharge at precise moments has led some to claim that they transmit information through spike rate, rather than through the precise time at which they spike (Shadlen & Newsome, 1998). By contrast, the short duration of certain cognitive processes has led others to argue that spike timing may also carry additional information (Kistler & Gerstner, 2002). More generally, whether neurons and neural populations code information via their firing rates (Shadlen & Newsome, 1998), their oscillatory activity (Buzsaki, 2006; Fries, 2005; Singer & Gray, 1995), or even

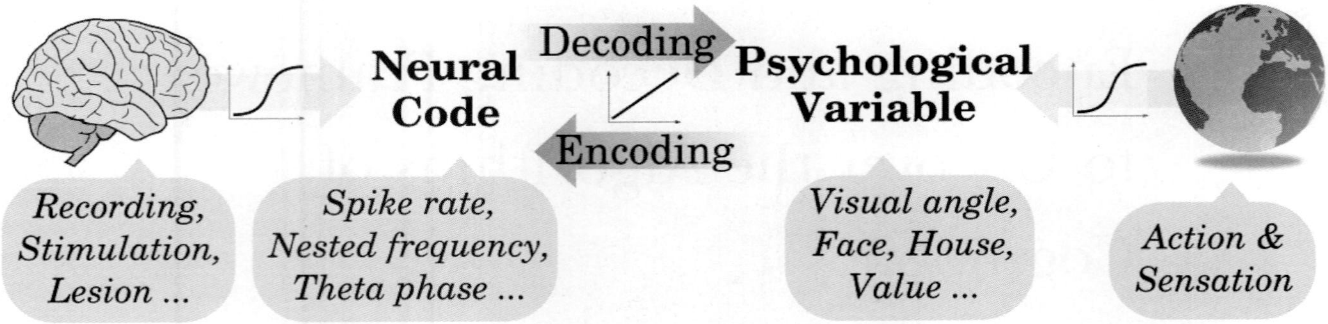

FIGURE 58.1 The representational paradigm consists of three challenges: (1) identifying the content of mental representations, (2) determining how this information is encoded in the brain, and (3) linking these two levels of description. These challenges increasingly benefit from statistical modeling and machine learning. In this view, decoding analyses predict experimental factors (e.g., the speed of a hand movement, the luminance of a flashed image) from specific brain activity features (e.g., spike rate, electric field), whereas encoding analyses predict the reverse. Currently, the relevant features of (1) brain activity and (2) experimental factors are defined a priori by the experimenters and mapped onto one another with linear mapping. However, the capacity of machine learning to find nonlinear structures in large data sets may ultimately help in finding what the brain represents and how. (See color plate 64.)

in the interaction between spikes and the phase of local field potentials (Bose & Recce, 2001; Lisman & Idiart, 1995) remains an outstanding research question.

Third, we must identify how these two levels of description—*what* is coded and *how* it is coded—relate with one another—that is, we must find the patterns of neural activity (e.g., a spike or an oscillation) that are both sensitive and specific to putative variables (e.g., the position of a rat in a maze). In the context of correlational settings, such as neuroimaging and electrophysiological recordings, solving the code-content equation (figure 58.1) is asymmetric. Either the code is assumed and multiple variables are comparatively tested, or vice versa. For example, one can assume a rate code and compare how the position of a rat in a maze predicts spike activity (O'Keefe & Dostrovsky, 1971). Reciprocally, one can assume that spatial locations are coded in the brain and compare how spike rates and the oscillations of the local field potentials predict this variable (Agarwal et al., 2014).

The statistical framework of encoding and decoding analyses The asymmetry of the code-content mapping contributes to the distinction between encoding and decoding analyses. Specifically, encoding consists of predicting neuronal responses from internal (e.g., confidence) or environmental variables (e.g., the presence of an object): P(brain activity pattern | variables). Conversely, decoding consists of predicting variables from neuronal activity (see figure 58.1–58.2): P(variables | brain activity pattern). Generally, encoding and decoding both depend on multivariate models whose objective is univariate, meaning that they fit several parameters to minimize a scalar that results from a loss function (figure 58.3) For example, functional magnetic resonance imaging (fMRI) studies routinely use encoding analyses by fitting a general linear model (GLM) to evaluate the extent to which multiple variables independently contribute to blood oxygen level-dependent (BOLD) measurements. Such variables can be difficult to orthogonalize a priori (1) because of the slow temporal profile of the BOLD response or (2) because the variables of interest can intrinsically covary (e.g., in natural images, the orientation of visual edges correlate with their spatial position; Sigman et al., 2001). Conversely, decoding analyses are predominantly used to maximally predict subjects' behavior or postdict their sensory stimulations. For example, brain-computer interfaces (BCI) studies typically examine several collinear patterns of brain activity in order to maximally predict subjects' actions, intentions (Lebedev & Nicolelis, 2006), or mental state (Zander & Kothe, 2011).

A variety of multivariate linear analyses are routinely used in cognitive neuroscience and range from linear discriminant analysis (LDA) and GLM to ridge and logistic regression and, more recently, to algorithms developed in the field of machine learning such, as linear support vector machines (SVM). Despite their various denominations and historical origins, these analyses can be described within a common statistical framework (figure 58.3, *bottom*). For example, they can be solved via the same convex optimization and identify the linear combination of features that maximally predict a brain response (e.g., encoding a spike) or a putative variable (e.g., decoding the presence of a face). In the context of electrophysiology and neuroimaging data, most of these analyses lead to similar results (Hastie, Tibshirani, & Friedman, 2009; Lebedev & Nicolelis,

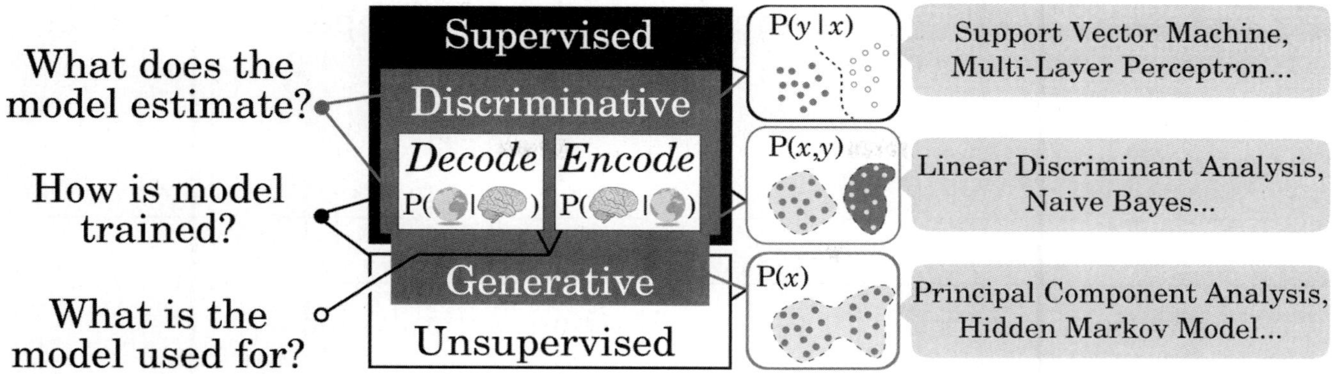

FIGURE 58.2 *Left*, The most common models used in cognitive neuroscience can be distinguished with three main dimensions of the statistical framework. Models that are trained to estimate a conditional probability $P(y \mid x)$ between two sets of variables (e.g., x=psychological variables, y=neuronal recordings, or vice versa) are referred to as *discriminative*. By contrast, models that estimate the joint distribution $P(x, y)$ are considered to be *generative*. All generative models can thus be used to derive $P(y \mid x)$ for prediction. These models are *supervised* in that they are trained to predict one variable (e.g., y) from another (e.g., x). By contrast, *unsupervised* models estimate the distribution of a single (possibly multidimensional) variable (x). Finally, a trained model can ultimately be used for different purposes—for example, decoding or encoding (see figure 58.1). *Right*, Examples of classical supervised and unsupervised models. (See color plate 65.)

2006; Varoquaux et al., 2017). However, distinct multivariate linear analyses assume distinct data distributions (e.g., LDA assumes normal distributions and equal variance-covariance matrices across classes, whereas logistic regression does not). Consequently, the choice of analysis depends on the problem (e.g., regression or classification), the amount of data, and its distribution (e.g., if the data are normally distributed, LDA can outperform logistic regression and vice versa). Interpreting the parameters of an analysis can be particularly challenging because (1) all parameters are simultaneously fitted, which makes the interpretation of individual parameters difficult, (2) some parameters may be related to the noise distribution, and (3) a given parameter need not actually affect the goodness of fit of a given model (Davis et al., 2014; Haufe et al., 2014; Hebart & Baker, 2018; Todd, Nystrom, & Cohen, 2013). For model interpretation, it is thus advised to supplement the inspection of model parameters with an explicit model comparison evaluated on prediction error.

Encoding and decoding models are not always subject to comparable constraints and can thus lead to different conclusions. In particular, decoding can pick up uncontrolled noise or signal structures in brain activity in a way that encoding cannot. For example, if an encoding model predicting the neural response to an image shows that its *luminosity* improves the prediction of brain activity, one can conclude that luminosity causally influences brain activity (provided that a number of assumptions are met; see Weichwald et al., 2015). However, no causal conclusion may actually be drawn from an analogous decoding model: for example, if including parietal neurons in a decoding model improves the decoding performance of image luminosity, parietal activity may not necessarily be caused by luminosity. Instead, parietal activity may simply reflect subjects' vigilance, which itself modulates the representation of luminosity in sensory regions; combining sensory and parietal regions may thus improve the decoding performance of luminosity. In this sensory-based paradigm, decoding can thus be less conclusive than encoding. However, this difference in conclusiveness comes with a benefit: because the decoding model can capture uncontrolled factors (e.g., vigilance), its predictive power may surpass the encoding model's (see Davis et al., 2014 for a related issue).

Overall, encoding and decoding models can therefore be profoundly distinct in their ability to detect and make use of uncontrolled factors and confounds. For a more detailed discussion on the causal inference and on the interpretation of encoding and decoding models, we refer the reader to Haufe et al. (2014), Weichwald and Grosse-Wentrup (2017), and Weichwald et al. (2015).

Where do the linearity assumptions come from? Most encoding and decoding analyses are based on linear modeling. This linear constraint is motivated by two theoretical principles: (1) the general *linear superposition* principle and (2) the neurocentric *linear readout* principle (figure 58.4).

Linear superposition The linear superposition principle is a general assumption based on the notion that measurements derive from a weighted sum of underlying

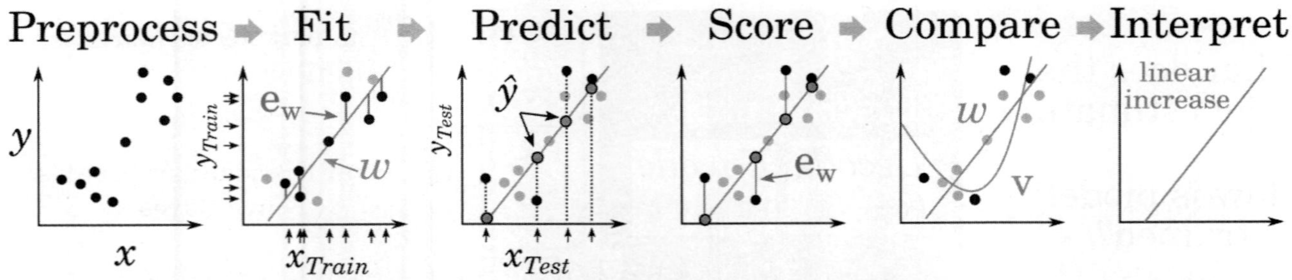

Fitting Objective: Find parameters (w) to best predict targets (y) from input data (x).

max $P(w \mid x, y)$, *where:* $x_i^T w = y_i$

Method: optimization

$P(w \mid x, y) \propto P(x, y \mid w)\, P(w)$

$\min_w Loss(y, x^T w) + Regularization(w)$

Name	Loss	Linear Model	Reg	Optimization
OLS:	l_2	$x_i^T w = \hat{y}_i$	*None*	$\min_w \sum_i (y_i - \hat{y}_i)^2$
Ridge:	l_2		l_2	$\min_w \sum_i (y_i - \hat{y}_i)^2 + \lambda \lvert w \rvert^2$
Logistic:	*log*		l_2	$\min_w \sum_i \log(1 + \exp(-y_i \hat{y}_i)) + \lambda \lvert w \rvert^2$
SVM:	*hinge*		l_2	$\min_w \sum_i \max(0, 1 - y_i \hat{y}_i) + \lambda \lvert w \rvert^2$

FIGURE 58.3 *Top*, Statistical modeling follows a standard multistep pipeline that starts with (1) preprocessing (any transformation that can be defined independently of the data—e.g., filtering) followed by (2) model fitting on a subset of training data, (3) prediction, and (4) scoring of independent and identically distributed held-out test data. The score (a.k.a, *goodness-of-fit* or *prediction error*, here labeled as e_w) can be summarized with a variety of metrics (e.g., accuracy, AUC, R^2, cross-entropy, F-value). Finally, one can subsequently perform a model comparison to interpret results. *Middle*, The fitting stage can make use of a variety of models, such as support vector machines (SVM) or ordinary least squares (OLS). These models can be formalized as maximum a posteriori, which balances the data likelihood (*red*) and the priors on the distribution of the model parameters (*blue*). These model parameters can be found with optimization, which consists of jointly minimizing the loss and regularization function. Note that for categorical models (e.g., logistic regression, SVM), the predicted values ŷ are subsequently transformed into discrete categories. *Bottom*, An example of common linear models used in cognitive neuroscience, together with their corresponding loss, regularization, and optimization functions. (See color plate 66.)

sources. For example, the electric potential measured by an electrode depends on the electric reference and the local field potential, as well as on the pre- and postsynaptic activity of surrounding neurons. Following Maxwell's equations, under the quasi-static approximation the electric fields of these sources linearly sum onto the electrode and do not interact with one another. Similarly, the analysis of hemodynamic responses is often based on an analogous assumption: each voxel contains hundreds of thousands of neurons whose activity is summarized in a unique BOLD measurement. Under the linear superposition assumption, a measurement (from an electrode or from a voxel) linearly covaries with a variable only if one or a combination of sources (the underlying neural responses) linearly covaries with such variables. When multivariate measurements are available (e.g., spatially distributed electrodes), it is possible to separate the independent contribution of several sources, based on physical assumptions (as in magnetoencephalography (MEG) source reconstruction (Hämäläinen, Hari, Ilmoniemi, Knuutila, & Lounasmaa, 1993) or based on statistical assumptions—for example, spike sorting (Quiroga, Nadasdy, & Ben-Shaul, 2004). Note that the linear superposition assumption is generally applicable within a limited range. For example, the BOLD response is known to saturate above certain values, above which the linear superposition assumption fails to hold (Heeger & Ress, 2002).

Linear readout The linear readout principle is mainly relevant to decoding analyses. It builds upon the assumption that individual neurons can be approximated as a nonlinear transformation (e.g., a spike) of a weighted sum of input (e.g., the sum of excitatory and inhibitory presynaptic potentials). Consequently, distributed patterns of simultaneous neuronal activity are thought to

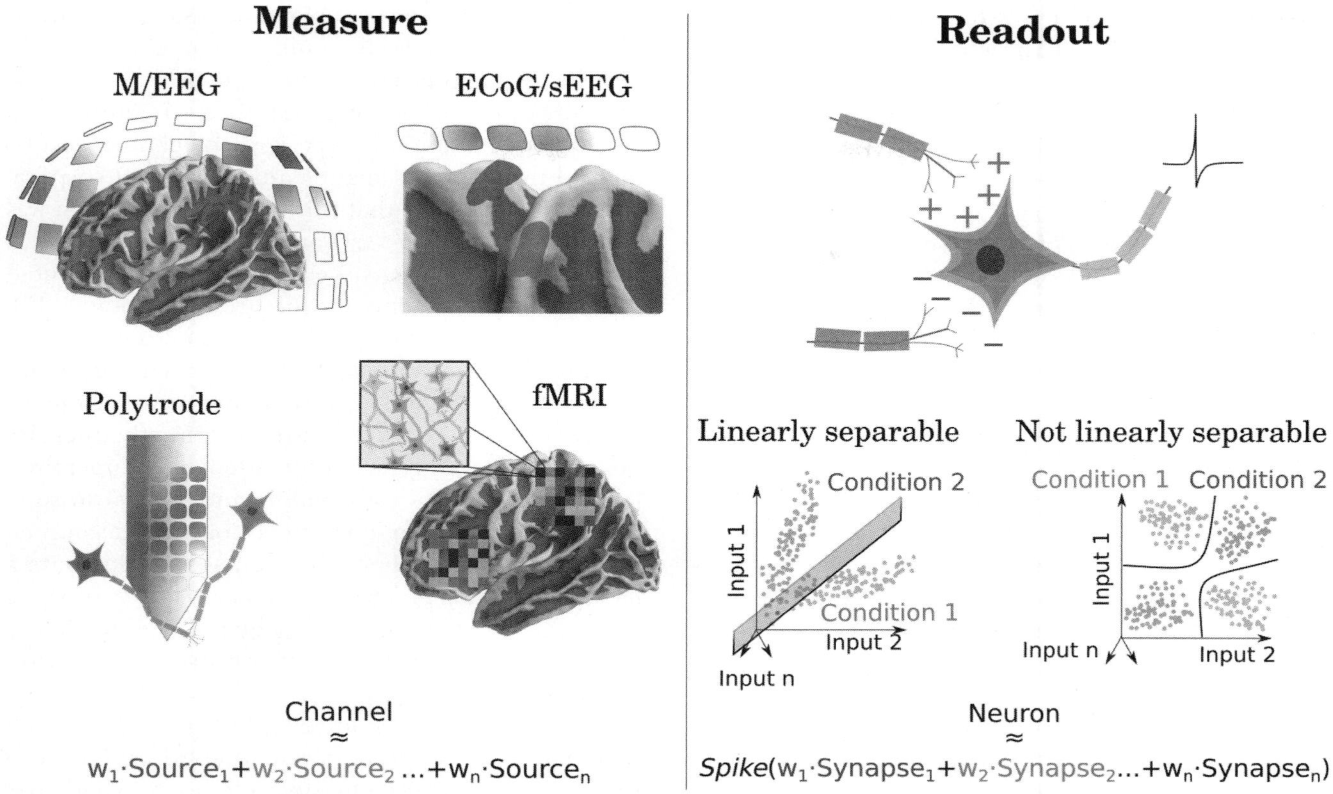

FIGURE 58.4 *Left,* The electromagnetic fields produced by neurons linearly project onto polytrodes, electrocorticography (ECoG), or magnetoelectroencephalography (M/EEG) sensors. A similar concept of linear mixture can be extended to fMRI: each voxel approximates a weighted sum of BOLD responses. Overall, measuring channels can thus be used to infer the underlying neural responses. *Right,* Neuronal computations can be approximated as nonlinear functions (i.e., spike) applied onto weighted combinations of synaptic inputs. Statistically, this means that neurons are sensitive to (i.e., "read") linearly separable inputs. (See color plate 67.)

represent variables because any neuron connected to this distributed population can systematically covary with such variables (Hung et al., 2005; Kamitani & Tong, 2005; King & Dehaene, 2014; Kriegeskorte & Kievit, 2013; Misaki et al., 2010). The linear readout principle clarifies the distinction between information and explicitly represented features. For example, the retina may encode *information* about faces and letter strings but would not explicitly *represent* these categories because faces and letter strings cannot be linearly separated from retinal activations. By contrast, the fusiform face regions (Kanwisher, McDermott, & Chun, 1997; Tsao et al., 2006) and the visual word form area (Dehaene & Cohen, 2011) have been shown to linearly map these two types of visual categories onto their patterns of neuronal activity.

It is important to highlight that encoding and decoding analyses are equally limited in their ability to determine whether a representation de facto constitutes information that the neural system *uses.* For example, one may find a linear relationship between a variable and (1) a spike, (2) an increase in BOLD response, or (3) an oscillation of a linear combination of EEG sensors, without that variable being effectively read and used by any neuron. Similarly to other correlational methods, encoding and decoding should thus be used in conjunction with comparative computational modeling and experimental manipulations in order to identify the causal or epiphenomenal nature of an identified pattern of brain activity.

Challenges of the representational paradigm and the promises of machine learning Constraining the triple quest of cognitive neuroscience (figure 58.1) to linear modeling leads to two main challenges. First, and as discussed elsewhere (Ritchie et al., 2017), the linear readout assumption undermines the nonlinear readout abilities of certain neurons (Brincat & Connor, 2004; Chichilnisky, 2001; Mineault et al., 2012; Sahani & Linden, 2003; Van Steveninck & Bialek, 1988), cortical columns (Bastos et al., 2012), and large neural assemblies (Ritchie et al., 2017). The definition of an explicitly encoded variable is thus likely to change with our improved understanding of the neuronal codes.

Second, linear modeling implies a strong dependence on a priori human insight (Kording et al., 2018). Specifically, linear models only fit the features explicitly provided by the experimenter. They are thus limited in their ability to identify unexpected patterns of neuronal activity or unanticipated mental representations. For example, the discovery of grid cells—hippocampal neurons that fire when an animal is located at regularly interspaced locations in an arena—resulted from human insights from visual data inspection. Indeed, Fyhn, Moser, and their collaborators had to view their electrophysiological recordings in a spatial representation before they could conjecture the grid-coding scheme (Fyhn et al., 2004; Moser, Kropff, & Moser, 2008). Only then did they implement a grid feature in a linear model to formally test and verify the robustness of this hypothesis (Hafting et al., 2005). In other words, a linear model blindly fitting spiking activity to a two-dimensional spatial position variable would have missed the seminal discovery of grid-coding cells.

The rapid development of machine learning may partially roll back this epistemic dependence on human insights. For example, Benjamin et al. (2017) have recently investigated the ability of linear models to predict spiking activity in the macaque motor cortex given conventional variables of the animal's arm movement, such as instantaneous velocity and acceleration. The authors first show that linear-encoding models can accurately predict the macaque's neural responses based on a weighted combination of these variables. However, they then demonstrate that linear models are outperformed by machine-learning models that can efficiently capture nonlinear relationships, such as random forests (Liaw et al., 2002) and long short-term memory neural networks (LSTMs; Hochreiter & Schmidhuber, 1997). In other words, random forests and LSTMs can identify unsuspected features of the arm movements that are represented in the neural activity. More generally, this study illustrates how machine learning may supplement human insights and help to discover unanticipated representations.

Undoubtedly, applying machine-learning algorithms to cognitive neuroscientific data will lead to new challenges (Kording et al., 2018; Poldrack & Farah, 2015; Stevenson & Kording, 2011; Varoquaux & Thirion, 2014). In particular, interpreting a multivariate model, and with greater reason a nonlinear one, can be particularly difficult. For example, in Benjamin et al. (2017)'s study discussed above, machine-learning algorithms proved better at predicting the neural activity associated with arm movements than linear models. However, this came at the cost of diminished interpretability: the exact nature of these captured, unsuspected representations currently remains unclear. Note that interpretability is not a problem specific to nonlinear models. For example, Huth and colleagues predicted the fMRI BOLD responses to spoken stories (Huth et al., 2016) from linear combinations of very large semantics vectors derived from latent semantic analysis of text corpora. The authors showed that this modeling approach was above chance level in a vast number of cortical regions, which thus strengthens the hypothesis of distributed representations of semantic features (Barsalou, 2017). However, to interpret such a model one would need to investigate, for each voxel, the hundreds of coefficients associated with each semantic vector. To make things worse, these vectors are not directly interpretable. In fact, when the authors used an unsupervised linear model (principal component analysis) to summarize the main semantic dimensions that accounted for BOLD activity, they only managed to attribute a meaningful interpretation to a small subset of these principal components. Consequently, even linear modeling does not necessarily ensure a straightforward interpretation.

Overall, these two studies highlight how the interpretability of a neural representation, which has been essential for generating insights and novel hypotheses, runs a risk of becoming increasingly anecdotal as models are (justifiably) increasingly evaluated on the basis of their prediction accuracy.

From Isolated Computations to Algorithms

The above methods isolate the result of individual computations by linking putative variables with patterns of neural activity. However, to uncover the *algorithm* of a given cognitive ability, one must also identify the order in which these computations are performed. In this second section, we will first briefly review a variety of established sequences of neural activity and their algorithmic interpretations. We will then summarize the main methods that (1) isolate specific neural sequences, (2) identify their selective input sequence, and (3) help interpret the computations associated with such neural dynamics.

Sequences of neural responses across spatial scales With the advances in temporally resolved fMRI (Ekman, Kok, & de Lange, 2017) and the increasing ability to simultaneously record multiple neurons (Jun et al., 2017) and brain regions (Boto et al., 2018; Tybrandt et al., 2018), specific sequences of neural activity have been revealed across multiple spatial scales. At the network level, for example, visual stimulations trigger a long cascade of neural responses from occipital to associative cortices

(Gramfort et al., 2011; King, Pescetelli, & Dehaene, 2016; figure 58.5*C*). This long sequence of brain responses has been successfully compared to the deep convolutional networks developed in artificial vision (Cichy et al., 2016; Eickenberg et al., 2017; Gwilliams & King, 2017; Kriegeskorte, 2015; Yamins et al., 2014). At the columnar level, neural activity has been shown to propagate from and to the supra- and infragranular layers of the cortex via frequency-specific traveling waves (van Kerkoerle et al., 2014; figure 58.5*B*) and has been argued to reflect a predictive-coding algorithm (Bastos et al., 2012). Finally, at the cellular level, spatial positions (Girardeau & Zugaro, 2011; Jones & Wilson, 2005) are associated with specific sequences of spikes (figure 58.5*A*) that may reflect the learning and anticipatory simulation of spatial navigation (Girardeau & Zugaro, 2011).

Isolating a sequence of processing stages from mixed neural activity Overall, these sequences of neural activations and their computational interpretations have mainly been manually established. However, a number of signal-processing methods have recently been put forward to automatically detect and characterize specific spatiotemporal patterns of neural activity from high-dimensional recordings.

Analyzing neural recordings is often complicated by the fact that, as discussed above, they usually result from linear mixtures. For example, polytrodes, electrocorticography (ECoG), and MEG all measure the fluctuations of the electromagnetic fields of a linear combination of underlying neural electric sources (figure 58.4, *left*). To assess the spatial dynamics of such linear mixtures, it is thus common to use spatial-filtering techniques. For example, one can decompose the evoked

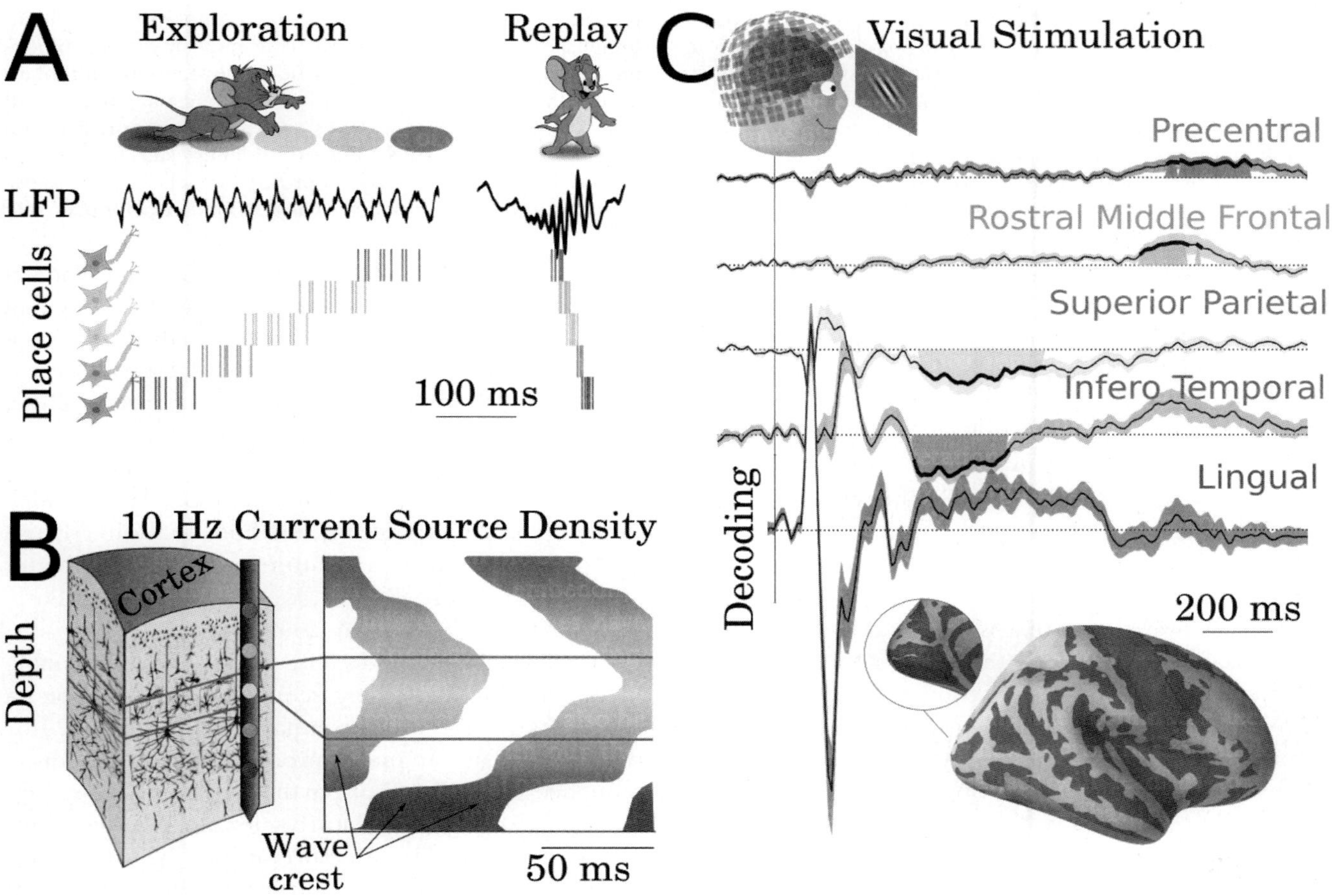

FIGURE 58.5 Sequences of neural responses across multiple spatial scales. Sequences of neural responses have been identified with a variety of methods and across various spatiotemporal scales. *A*, Hippocampal place cells spike when the rodent moves to specific spatial positions. When the rodent rests, the previous spike sequence is replayed in reverse order. Adapted from Girardeau and Zugaro (2011). *B*, The 10 Hz oscillatory activity recorded in the macaque early visual cortex travels from superficial (*red*) and deep layers (*purple*) to the granular layer (*green*). Adapted from van Kerkoerle et al. (2014). *C*, Visual stimulation triggers a long-lasting cascade of macroscopic brain responses, starting from the early visual cortex and ultimately reaching the associative areas (King, Pescetelli, & Dehaene, 2016). (See color plate 68.)

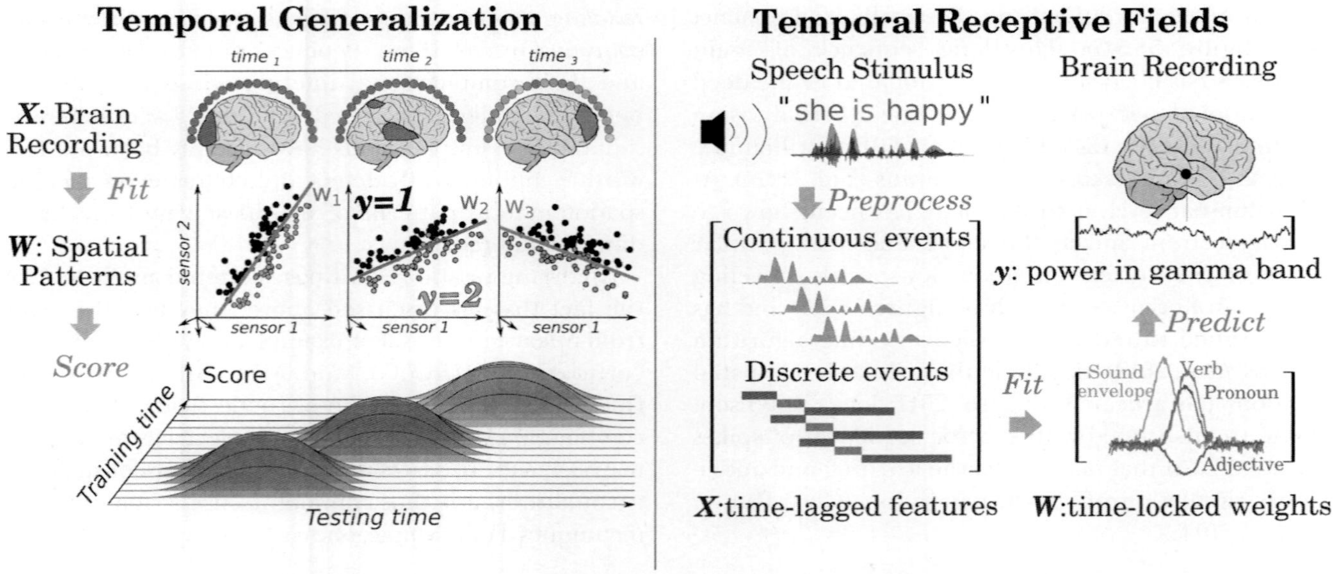

FIGURE 58.6 Decoding and encoding neural sequences. *Left*, Temporal generalization consists of fitting a spatial filter at each time sample locked to an event and testing whether it can be used to decode the brain responses at all time samples. This analysis can be used to determine whether a sustained decoding score results from a stable pattern of neural responses or whether it reflects a series of transient neural responses. Adapted from King and Dehaene (2014). *Right*, Temporal receptive field analyses involve first annotating the latent dimensions that characterize the continuously varying experimental variables—for example, the spectral modulation of an acoustic waveform, the lexical categories of spoken words, and others (*top left*). These features are then transformed into a time-delay matrix whose linear modeling can be used to recover the temporal response profile of each feature (*bottom right*). Adapted from Holdgraf et al. (2017). (See color plate 69.)

response to a visual stimulus by fitting a series of linear models, one at each time sample locked to the stimulus onset, in order to track the rise and fall of a given variable (e.g., contrast: King, Pescetelli, & Dehaene, 2016; or the presence of a face: Cichy, Pantazis, & Oliva, 2014) and test whether a model optimized to decode a variable at time t can generalize to decode at time t' (King & Dehaene, 2014; Meyers et al., 2008; Stokes et al., 2013). If a model generalizes, this suggests that the brain activity pattern it isolated is present at both training and testing time samples. By systematically assessing how each model generalizes over time—that is, by building a temporal generalization matrix, it is thus possible to determine whether the neural dynamics consists of a sequence of transient activations, a recurrent dynamic, or simply a sustained activation directly from a linearly mixed neural time series (figure 58.6*A*; King & Dehaene, 2014).

The above methods focus on "evoked" spatial patterns—that is, neural responses whose phases are consistent across trial repetitions. However, analogous spatial-filtering methods have been developed to specifically capture "induced" spatial patterns—that is, linear mixtures of neural responses, such as oscillatory activity, whose dynamics are not phase locked to an external event. These methods are generally based on the spatial covariance of electrophysiological recordings. For example, the common spatial pattern (CSP) method is a popular spatial-filtering technique to identify the spatial pattern of neural activity that maximizes the discrimination between the induced responses of two external events (e.g., left versus right hand; Koles, Lazar, & Zhou, 1990). Similarly, the source power comodulation (SPoC) method extracts spatial patterns that are modulated by a continuous variable (e.g., hand position; Dähne, Meinecke et al., 2014). Recently, methods that directly use the spatial covariance as a feature proved to reach better results (Barachant, Bonnet, & Congedo, 2012; Barachant et al., 2013; Farquhar, 2009).

These novel techniques have already proved valuable in establishing processing sequences. For example, a recent study (Heikel, Sassenhagen, & Fiebach, 2018) suggested that language processing depends on a sequence of distinct but partially overlapping stages. Such a *cascading* architecture goes against the recurrent or the serial architecture argued by others (see Friederici, 2011 for an overview). Similarly, visual selection has recently been shown to recruit parallel hierarchies of sensory processing, converging into a serial executive-processing stage (Marti & Dehaene, 2017; Marti, King, & Dehaene, 2015). Specifically, MEG recordings show that subjects can specifically attend to an image presented in a rapid stream and represent its content in the associative cortices. However, temporal generalization showed that these late processing stages are systematically delayed when subjects are distracted by auditory stimuli, which suggests a strictly serial architecture. Overall, these studies thus illustrate how the automatic analyses of high-dimensional neural time series can help recover the underlying algorithmic organization of cognitive processes.

Identifying the sequence of inputs detected by a neuronal population A second major challenge consists of identifying the sequence of input detected by one or several neurons. This approach is particularly common in auditory and speech studies. In such cases it is indeed common to model the neural dynamics with multiple "time lags" of continuously varying features (Almon, 1965; Ho & Kálmán, 1966) in order to capture the possibility that some neural responses are time-locked to specific (a combination of) events. For example, to model auditory and language processing, one can (1) extract the continuously changing envelope of a recorded speech waveform and (2) annotate the onsets of individual words (figure 58.6*B*; de Heer et al., 2017; Holdgraf et al., 2017). A temporal "receptive field" model can finally be fitted to isolate the neural responses to these continuous fluctuations and discrete events, respectively. Temporal receptive fields are similar to the predominant approach in fMRI analysis, in which a design matrix is convolved with an impulse waveform that is either assumed (i.e., the canonical hemodynamic response function) or estimated from data, eventually improving fMRI-encoding and -decoding models (Pedregosa et al., 2015). In the case of electrophysiology, the shapes of the impulse response functions are typically not known a priori, so the impulse response must be estimated from the data.

Temporal receptive fields and related methods can be used to (1) encode the average brain responses to categorical events (e.g., de Heer et al., 2017; Smith & Kutas, 2015) and spectrotemporal patterns of sensory input (Theunissen et al., 2001), as well as (2) decode overlapping sequences of neural correlates of discrete (Rivet et al., 2009; Theunissen et al., 2001) and continuously changing events (Dähne, Nikulin, et al., 2014). For example, DiLiberto et al. (2014) have shown that encoding categorical phonemic representations in electroencephalogram EEG signals provides a superior model fit compared to acoustic features. Overall, these methods thus promise to help automatically find the sequence of input that maximally drives each brain response.

Toward directly mapping algorithms onto brain activity The last decade has been marked by an explosion of machine-learning models that efficiently mimic basic cognitive operations. Most remarkably, computer vision models can now detect, locate, or describe objects in static natural images with an accuracy that matches or even exceeds human performance (although, see Lake et al., 2017). One of the most popular approaches, the deep convolutional neural network, takes an image as input and sequentially applies a long series of nonlinear transformations that are optimized to identify objects. Interestingly, this specific sequence of operations has been found to map onto both the spatial (Cichy et al., 2016; Eickenberg et al., 2017; Gwilliams & King, 2017; Kriegeskorte, 2015; Yamins et al., 2014) and the temporal organization (Cichy et al., 2016; Gwilliams & King, 2017; van de Nieuwenhuijzen et al., 2013) of the visual system in the mammalian brain. For example, the activity in the primary visual cortex specifically and linearly correlates with the activation in early layers of the artificial neural network, whereas the later responses of the inferior temporal cortex specifically and linearly correlate with the activation of the deepest layers. This result suggests that the sequence of computations applied by the human brain to solve a given task may be parsed and modeled with deep neural networks trained to solve the same task. This finding thus supports the notion that the visual system is organized as an extended hierarchy (Hubel & Wiesel, 1963; Riesenhuber & Poggio, 1999). More generally, it illustrates how sequences of brain activity patterns may be interpreted with performance-optimized algorithms: the algorithms that (1) can efficiently perform a task while (2) doing so in a way that maps the spatiotemporal characteristics of brain activity are likely to adequately model cognition.

Overall, the rapid development of machine learning provides a threefold promise to cognitive neuroscience. First, these tools support the automation, denoising, and summary of complex electrophysiological and neuroimaging time series (Jas et al., 2017). Second, these tools offer an operational ground to data-driven investigation: unanticipated patterns of data may be automatically identified from large data sets without requiring the preface of human insight (Kording et al., 2018). Finally, machine learning and cognitive neuroscience share the common goal of identifying the elementary components of knowledge acquisition and information processing. The interface between cognitive neuroscience and machine learning is thus mutually beneficial. On the one hand, machine learning can help define, identify, and formalize the computations of the brain. On the other hand, cognitive neuroscience can help provide insights and principled directions to shape the computational architecture of complex cognitive processes (Hassabis et al., 2017; Lake et al., 2017).

Acknowledgments

This project received funding from the European Union's Horizon 2020 Programme for Research and Innovation under grant agreement No. 660086, the Bettencourt-Schueller Foundation, and the Philippe Foundation.

REFERENCES

Agarwal, G., Stevenson, I. H., Berényi, A., Mizuseki, K., Buzsáki, G., & Sommer, F. T. (2014). Spatially distributed local fields in the hippocampus encode rat position. *Science, 344*(6184), 626–630.

Almon, S. (1965). The distributed lag between capital appropriations and expenditures. *Econometrica, 33*(1), 178–196.

Barachant, A., Bonnet, S., & Congedo, M. (2012). Multiclass brain–computer interface classification by Riemannian geometry. *IEEE Transactions on Biomedical Engineering, 59*(4). http://ieeexplore.ieee.org/abstract/document/6046114/.

Barachant, A., Bonnet, S., Congedo, M., & Jutten, C. (2013). Classification of covariance matrices using a Riemannian-based kernel for BCI applications. *Neurocomputing, 112*, 172–178.

Barsalou, L. W. (2017). What does semantic tiling of the cortex tell us about semantics? *Neuropsychologia, 105*, 18–38.

Bastos, A. M., Usrey, W. M., Adams, R. A., Mangun, G. R., Fries, P., & Friston, K. J. (2012). Canonical microcircuits for predictive coding. *Neuron, 76*(4), 695–711.

Benjamin, A. S., Fernandes, H. L., Tomlinson, T., Ramkumar, P., VerSteeg, C., Chowdhury, R., Kording, K. P. (2017). Modern machine learning outperforms GLMs at predicting spikes. BioRxiv, https://doi.org/10.1101/111450

Bose, A., & Recce, M. (2001). Phase precession and phase-locking of hippocampal pyramidal cells. *Hippocampus, 11*(3), 204–215.

Boto, E., Holmes, N., Leggett, J., Roberts, G., Shah, V., Meyer, S. S., ... Brookes, M. J. (2018). Moving magnetoencephalography towards real-world applications with a wearable system. *Nature, 555*(7698), 657–661.

Brincat, S. L., & Connor, C. E. (2004). Underlying principles of visual shape selectivity in posterior inferotemporal cortex. *Nature Neuroscience, 7*(8), 880–886.

Buzsaki, G. (2006). *Rhythms of the brain*. Oxford: Oxford University Press.

Chichilnisky, E. J. (2001). A simple white noise analysis of neuronal light responses. *Network, 12*(2), 199–213.

Cichy, R. M., Khosla, A., Pantazis, D., Torralba, A., & Oliva, A. (2016). Comparison of deep neural networks to spatio-temporal cortical dynamics of human visual object recognition reveals hierarchical correspondence. *Scientific Reports, 6*, 27755.

Cichy, R. M., Pantazis, D., & Oliva, A. (2014). Resolving human object recognition in space and time. *Nature Neuroscience, 17*(3), 455–462.

Dähne, S., Meinecke, F. C., Haufe, S., Höhne, J., Tangermann, M., Müller, K.-R., & Nikulin, V. V. (2014). SPoC: A novel framework for relating the amplitude of neuronal oscillations to behaviorally relevant parameters. *NeuroImage, 86*, 111–122.

Dähne, S., Nikulin, V. V., Ramírez, D., Schreier, P. J., Müller, K.-R., & Haufe, S. (2014). Finding brain oscillations with power dependencies in neuroimaging data. *NeuroImage, 96*, 334–348.

Davis, T., LaRocque, K. F., Mumford, J. A., Norman, K. A., Wagner, A. D., & Poldrack, R. A. (2014). What do differences between multi-voxel and univariate analysis mean? How subject-, voxel-, and trial-level variance impact fMRI analysis. *NeuroImage, 97*, 271–283.

Dehaene, S., & Cohen, L. (2007). Cultural recycling of cortical maps. *Neuron, 56*(2), 384–398.

Dehaene, S., & Cohen, L. (2011). The unique role of the visual word form area in reading. *Trends in Cognitive Sciences, 15*(6), 254–262.

de Heer, W. A., Huth, A. G., Griffiths, T. L., Gallant, J. L., & Theunissen, F. E. (2017). The hierarchical cortical organization of human speech processing. *Journal of Neuroscience, 37*(27), 6539–6557.

Di Liberto, G. M., O'Sullivan, J. A., & Lalor, E. C. (2015). Low-Frequency cortical entrainment to speech reflects phoneme -level processing. *Current Biology, 25*(19), 2457–2465.

Donders, F. C. (1969). On the speed of mental processes. *Acta Psychologica, 30*, 412–431.

Eickenberg, M., Gramfort, A., Varoquaux, G., & Thirion, B. (2017). Seeing it all: Convolutional network layers map the function of the human visual system. *NeuroImage, 152*, 184–194.

Ekman, M., Kok, P., & de Lange, F. P. (2017). Time-compressed preplay of anticipated events in human primary visual cortex. *Nature Communications, 8*, 15276.

Farquhar, J. (2009). A linear feature space for simultaneous learning of spatio-spectral filters in BCI. *Neural Networks, 22*(9), 1278–1285.

Freiwald, W. A., & Tsao, D. Y. (2010). Functional compartmentalization and viewpoint generalization within the macaque face-processing system. *Science, 330*(6005), 845–851.

Friederici, A. D. (2011). The brain basis of language processing: From structure to function. *Physiological Reviews, 91*(4), 1357–1392.

Fries, P. (2005). A mechanism for cognitive dynamics: Neuronal communication through neuronal coherence. *Trends in Cognitive Sciences, 9*(10), 474–480.

Fyhn, M., Molden, S., Witter, M. P., Moser, E. I., & Moser, M.-B. (2004). Spatial representation in the entorhinal cortex. *Science, 305*(5688), 1258–1264.

Girardeau, G., & Zugaro, M. (2011). Hippocampal ripples and memory consolidation. *Current Opinion in Neurobiology, 21*(3), 452–459.

Gramfort, A., Papadopoulo, T., Baillet, S., & Clerc, M. (2011). Tracking cortical activity from M/EEG using graph cuts with spatiotemporal constraints. *NeuroImage, 54*(3), 1930–1941.

Gwilliams, L., & King, J.-R. (2017). Performance-optimized neural network only partially account for the spatio-temporal organization of visual processing in the human brain. *BioRxiv*. https://doi.org/10.1101/221630

Hafting, T., Fyhn, M., Molden, S., Moser, M.-B., & Moser, E. I. (2005). Microstructure of a spatial map in the entorhinal cortex. *Nature, 436*(7052), 801–806.

Hämäläinen, M., Hari, R., Ilmoniemi, R. J., Knuutila, J., & Lounasmaa, O. V. (1993). Magnetoencephalography—theory, instrumentation, and applications to noninvasive studies of the working human brain. *Reviews of Modern Physics, 65*(2), 413–497.

Hassabis, D., Kumaran, D., Summerfield, C., & Botvinick, M. (2017). Neuroscience-inspired artificial intelligence. *Neuron, 95*(2), 245–258.

Hastie, T., Tibshirani, R., & Friedman, J. (2009). Unsupervised learning. In T. Hastie, R. Tibshirani, & J. Friedman (Eds.), *The elements of statistical learning: Data mining, inference, and prediction* (pp. 485–585). New York: Springer.

Haufe, S., Meinecke, F., Görgen, K., Dähne, S., Haynes, J.-D., Blankertz, B., & Biessmann, F. (2014). On the interpretation of weight vectors of linear models in multivariate neuroimaging. *NeuroImage, 87,* 96–110.

Haxby, J. V. (2006). Fine structure in representations of faces and objects. *Nature Neuroscience, 9*(9), 1084–1086.

Haxby, J. V., Gobbini, M. I., Furey, M. L., Ishai, A., Schouten, J. L., & Pietrini, P. (2001). Distributed and overlapping representations of faces and objects in ventral temporal cortex. *Science, 293*(5539), 2425–2430.

Hebart, M. N., & Baker, C. I. (2018). Deconstructing multivariate decoding for the study of brain function. *NeuroImage, 180*(Pt. A), 4–18.

Heeger, D. J., & Ress, D. (2002). What does fMRI tell us about neuronal activity? *Nature Reviews Neuroscience, 3,* 142.

Heikel, E., Sassenhagen, J., & Fiebach, C. J. (2018). Time-generalized multivariate analysis of EEG responses reveals a cascading architecture of semantic mismatch processing. *Brain and Language, 184,* 43–53.

Hickok, G. (2014). The architecture of speech production and the role of the phoneme in speech processing. *Language and Cognitive Processes, 29*(1), 2–20.

Ho, B. L., & Kálmán, R. E. (1966). Effective construction of linear state-variable models from input/output functions. *At-Automatisierungstechnik, 14*(1–12), 545–548.

Hochreiter, S., & Schmidhuber, J. (1997). Long short-term memory. *Neural Computation, 9*(8), 1735–1780.

Holdgraf, C. R., Rieger, J. W., Micheli, C., Martin, S., Knight, R. T., & Theunissen, F. E. (2017). Encoding and decoding models in cognitive electrophysiology. *Frontiers in Systems Neuroscience, 11,* 61.

Hubel, D. H., & Wiesel, T. N. (1963). Shape and arrangement of columns in cat's striate cortex. *Journal of Physiology, 165,* 559–568.

Hung, C. P., Kreiman, G., Poggio, T., & DiCarlo, J. J. (2005). Fast readout of object identity from macaque inferior temporal cortex. *Science, 310*(5749), 863–866.

Huth, A. G., de Heer, W. A., Griffiths, T. L., Theunissen, F. E., & Gallant, J. L. (2016). Natural speech reveals the semantic maps that tile human cerebral cortex. *Nature, 532*(7600), 453–458.

Jas, M., Engemann, D. A., Bekhti, Y., Raimondo, F., & Gramfort, A. (2017). Autoreject: Automated artifact rejection for MEG and EEG data. *NeuroImage, 159,* 417–429.

Jones, M. W., & Wilson, M. A. (2005). Theta rhythms coordinate hippocampal–prefrontal interactions in a spatial memory task. *PLoS Biology, 3*(12), e402.

Jun, J. J., Steinmetz, N. A., Siegle, J. H., Denman, D. J., Bauza, M., Barbarits, B., … Harris, T. D. (2017). Fully integrated silicon probes for high-density recording of neural activity. *Nature, 551*(7679), 232–236.

Kamitani, Y., & Tong, F. (2005). Decoding the visual and subjective contents of the human brain. *Nature Neuroscience, 8*(5), 679–685.

Kanwisher, N., McDermott, J., & Chun, M. M. (1997). The fusiform face area: A module in human extrastriate cortex specialized for face perception. *Journal of Neuroscience, 17*(11), 4302–4311.

King, J.-R., & Dehaene, S. (2014). Characterizing the dynamics of mental representations: The temporal generalization method. *Trends in Cognitive Sciences, 18*(4), 203–210.

King, J.-R., Pescetelli, N., & Dehaene, S. (2016). Brain mechanisms underlying the brief maintenance of seen and unseen sensory information. *Neuron, 92*(5), 1122–1134.

Kistler, W. M., & Gerstner, W. (2002). Stable propagation of activity pulses in populations of spiking neurons. *Neural Computation, 14*(5), 987–997.

Koles, Z. J., Lazar, M. S., & Zhou, S. Z. (1990). Spatial patterns underlying population differences in the background EEG. *Brain Topography, 2*(4), 275–284.

Kording, K. P., Benjamin, A., Farhoodi, R., & Glaser, J. I. (2018). The roles of machine learning in biomedical science. In *Frontiers of engineering: Reports on leading-edge engineering from the 2017 symposium.* Washington, DC: National Academies Press.

Kriegeskorte, N. (2015). Deep neural networks: A new framework for modeling biological vision and brain information processing. *Annual Review of Vision Science, 1,* 417–446.

Kriegeskorte, N., & Kievit, R. A. (2013). Representational geometry: Integrating cognition, computation, and the brain. *Trends in Cognitive Sciences, 17*(8), 401–412.

Lake, B. M., Ullman, T. D., Tenenbaum, J. B., & Gershman, S. J. (2017). Building machines that learn and think like people. *Behavioral and Brain Sciences, 40,* e253.

Lebedev, M. A., & Nicolelis, M. A. L. (2006). Brain-machine interfaces: Past, present and future. *Trends in Neurosciences, 29*(9), 536–546.

Liaw, A., & Wiener, M. (2002). Classification and regression by random forest. *R News, 2*(3), 18–22.

Lisman, J., & Idiart, M. (1995). Storage of 7 /- 2 short-term memories in oscillatory subcycles. *Science, 267*(5203), 1512–1515.

Marr, D. (1982). *Vision: A computational investigation into the human representation and processing of visual information.* San Francisco: W. H. Freeman. https://mitpress.mit.edu/books/978-0-262-29037-1.

Marti, S., & Dehaene, S. (2017). Discrete and continuous mechanisms of temporal selection in rapid visual streams. *Nature Communications, 8*(1), 1955.

Marti, S., King, J.-R., & Dehaene, S. (2015). Time-resolved decoding of two processing chains during dual-task interference. *Neuron, 88*(6), 1297–1307.

Mesgarani, N., Cheung, C., Johnson, K., & Chang, E. F. (2014). Phonetic feature encoding in human superior temporal gyrus. *Science, 343*(6174), 1006–1010.

Meyers, E. M., Freedman, D. J., Kreiman, G., Miller, E. K., & Poggio, T. (2008). Dynamic population coding of category information in inferior temporal and prefrontal cortex. *Journal of Neurophysiology, 100*(3), 1407–1419.

Mineault, P. J., Khawaja, F. A., Butts, D. A., & Pack, C. C. (2012). Hierarchical processing of complex motion along the primate dorsal visual pathway. *Proceedings of the National Academy of Sciences of the United States of America, 109*(16), E972–E980.

Misaki, M., Kim, Y., Bandettini, P. A., & Kriegeskorte, N. (2010). Comparison of multivariate classifiers and response normalizations for pattern-information fMRI. *NeuroImage, 53*(1), 103–118.

Moser, E. I., Kropff, E., & Moser, M.-B. (2008). Place cells, grid cells, and the brain's spatial representation system. *Annual Review of Neuroscience, 31*, 69–89.

O'Keefe, J., & Dostrovsky, J. (1971). The hippocampus as a spatial map: Preliminary evidence from unit activity in the freely-moving rat. *Brain Research, 34*(1), 171–175.

Pedregosa, F., Eickenberg, M., Ciuciu, P., Thirion, B., & Gramfort, A. (2015). Data-driven HRF estimation for encoding and decoding models. *NeuroImage, 104*, 209–220.

Poldrack, R. A., & Farah, M. J. (2015). Progress and challenges in probing the human brain. *Nature, 526*(7573), 371–379.

Price, C. J. (2010). The anatomy of language: A review of 100 fMRI studies published in 2009. *Annals of the New York Academy of Sciences, 1191*, 62–88.

Quiroga, R. Q., Nadasdy, Z., & Ben-Shaul, Y. (2004). Unsupervised spike detection and sorting with wavelets and superparamagnetic clustering. *Neural Computation, 16*(8), 1661–1687.

Riesenhuber, M., & Poggio, T. (1999). Hierarchical models of object recognition in cortex. *Nature Neuroscience, 2*(11), 1019–1025.

Ritchie, J. B., Brendan Ritchie, J., Kaplan, D., & Klein, C. (2017). Decoding the brain: Neural representation and the limits of multivariate pattern analysis in cognitive neuroscience. *The British Journal for the Philosophy of Science, 70*(2), 581–607. https://doi.org/10.1101/127233

Rivet, B., Souloumiac, A., Attina, V., & Gibert, G. (2009). xDAWN algorithm to enhance evoked potentials: Application to brain-computer interface. *IEEE Transactions on Bio-Medical Engineering, 56*(8), 2035–2043.

Sahani, M., & Linden, J. F. (2003). How linear are auditory cortical responses? In S. Becker, S. Thrun, & K. Obermayer (Eds.), *Advances in neural information processing systems* (Vol. 15, pp. 125–132). Cambridge, MA: MIT Press.

Shadlen, M. N., & Newsome, W. T. (1998). The variable discharge of cortical neurons: Implications for connectivity, computation, and information coding. *Journal of Neuroscience, 18*(10), 3870–3896.

Sigman, M., Cecchi, G. A., Gilbert, C. D., & Magnasco, M. O. (2001). On a common circle: Natural scenes and gestalt rules. *Proceedings of the National Academy of Sciences of the United States of America, 98*(4), 1935–1940.

Singer, W., & Gray, C. M. (1995). Visual feature integration and the temporal correlation hypothesis. *Annual Review of Neuroscience, 18*, 555–586.

Smith, N. J., & Kutas, M. (2015). Regression-based estimation of ERP waveforms: II. Nonlinear effects, overlap correction, and practical considerations. *Psychophysiology, 52*(2), 169–181.

Sternberg, S. (1998). Discovering mental processing stages: The method of additive factors. An invitation to cognitive science. Methods, models, and conceptual issues, vol. 4, 703–863. https://doi.org/1999-02657-014

Stevenson, I. H., & Kording, K. P. (2011). How advances in neural recording affect data analysis. *Nature Neuroscience, 14*, 139.

Stokes, M. G., Kusunoki, M., Sigala, N., Nili, H., Gaffan, D., & Duncan, J. (2013). Dynamic coding for cognitive control in prefrontal cortex. *Neuron, 78*(2), 364–375.

Theunissen, F. E., David, S. V., Singh, N. C., Hsu, A., Vinje, W. E., & Gallant, J. L. (2001). Estimating spatio-temporal receptive fields of auditory and visual neurons from their responses to natural stimuli. *Network, 12*(3), 289–316.

Todd, M. T., Nystrom, L. E., & Cohen, J. D. (2013). Confounds in multivariate pattern analysis: Theory and rule representation case study. *NeuroImage, 77*, 157–165.

Tsao, D. Y., Freiwald, W. A., Tootell, R. B. H., & Livingstone, M. S. (2006). A cortical region consisting entirely of face-selective cells. *Science, 311*(5761), 670–674.

Tybrandt, K., Khodagholy, D., Dielacher, B., Stauffer, F., Renz, A. F., Buzsáki, G., & Vörös, J. (2018). High-density stretchable electrode grids for chronic neural recording. *Advanced Materials, 30*(15), e1706520.

van de Nieuwenhuijzen, M. E., Backus, A. R., Bahramisharif, A., Doeller, C. F., Jensen, O., & van Gerven, M. A. J. (2013). MEG-based decoding of the spatiotemporal dynamics of visual category perception. *NeuroImage, 83*, 1063–1073.

van Kerkoerle, T., Self, M. W., Dagnino, B., Gariel-Mathis, M.-A., Poort, J., van der Togt, C., & Roelfsema, P. R. (2014). Alpha and gamma oscillations characterize feedback and feedforward processing in monkey visual cortex. *Proceedings of the National Academy of Sciences of the United States of America, 111*(40), 14332–14341.

Van Steveninck, R., & Bialek, W. (1988). Real-time performance of a movement-sensitive neuron in the blowfly visual system: Coding and information transfer in short spike sequences. *Proceedings of the Royal Society B: Biological Sciences, 234*(1277). Retrieved from http://rspb.royalsocietypublishing.org/content/234/1277/379.short.

Varoquaux, G., Raamana, P. R., Engemann, D. A., Hoyos-Idrobo, A., Schwartz, Y., & Thirion, B. (2017). Assessing and tuning brain decoders: Cross-validation, caveats, and guidelines. *NeuroImage, 145*(Pt. B), 166–179.

Varoquaux, G., & Thirion, B. (2014). How machine learning is shaping cognitive neuroimaging. *GigaScience, 3*, 28.

Weichwald, S., & Grosse-Wentrup, M. (2017). The right tool for the right question—beyond the encoding versus decoding dichotomy. *arXiv.* Retrieved from http://arxiv.org/abs/1704.08851.

Weichwald, S., Meyer, T., Özdenizci, O., Schölkopf, B., Ball, T., & Grosse-Wentrup, M. (2015). Causal interpretation rules for encoding and decoding models in neuroimaging. *NeuroImage, 110*, 48–59.

Yamins, D. L. K., Hong, H., Cadieu, C. F., Solomon, E. A., Seibert, D., & DiCarlo, J. J. (2014). Performance-optimized hierarchical models predict neural responses in higher visual cortex. *Proceedings of the National Academy of Sciences of the United States of America, 111*(23), 8619–8624.

Zander, T. O., & Kothe, C. (2011). Towards passive brain-computer interfaces: Applying brain-computer interface technology to human-machine systems in general. *Journal of Neural Engineering, 8*(2), 025005.

59 Deep Learning for Cognitive Neuroscience

KATHERINE R. STORRS AND NIKOLAUS KRIEGESKORTE

ABSTRACT Neural network models can now recognize images, understand text, translate languages, and play many human games at human or superhuman levels. These systems are highly abstracted but are inspired by biological brains and use only biologically plausible computations. In the coming years, neural networks are likely to become less reliant on learning from massive labeled data sets and more robust and generalizable in their task performance. From their successes and failures, we can learn about the computational requirements of the different tasks at which brains excel. Deep learning also provides the tools for testing cognitive theories. In order to test a theory, we need to realize the proposed information-processing system at scale and assess its feasibility and emergent behaviors. Deep learning allows us to scale up from principles and circuit models to end-to-end trainable models capable of performing complex tasks. Cognitive neuroscientists can use deep learning in their work at many levels, from inspiring theories to serving as full computational models. Ongoing advances in deep learning bring us closer to understanding how cognition and perception may be implemented in the brain—the grand challenge at the core of cognitive neuroscience.

Introduction: Neuro-Inspired Artificial Intelligence and Artificial Intelligence-Inspired Neuroscience

To explain how brains reason, remember, perceive, and act, theories must bridge from biology to behavior. Rigorous explanation requires models that use neurobiologically plausible components to implement cognitive processes (Kriegeskorte & Douglas, 2018; Kriegeskorte & Mok, 2017; Poeppel, 2012). Artificial neural networks, composed of simplified simulated neurons, have long promised such models (McCulloch & Pitts, 1943; Rumelhart & McClelland, 1986). Thanks to computational and methodological advances, neural networks now outperform all other engineering solutions to pattern recognition problems (e.g., Russakovsky et al., 2015), as well as many cognitive challenges, such as language translation (e.g., Wu et al., 2016), visual reasoning (e.g., Santoro et al., 2017), and game-playing (e.g., Jaderberg et al., 2018; Mnih et al., 2015).

Ideas from psychology and neuroscience have inspired engineers and underlie many features in modern networks (Hassabis et al., 2017; Kriegeskorte, 2015). Networks used for visual tasks process images hierarchically with spatially restricted receptive fields, as does mammalian visual cortex (LeCun et al., 1998; Russakovsky et al., 2015). *Attention networks* dynamically select subparts of their inputs to which they sequentially devote processing resources (e.g., Xu et al., 2015). Research on the complementary roles of the hippocampus and neocortex in learning (Kumaran et al., 2016; McLelland, McNaughton, & O'Reilly, 1995) has inspired networks that bootstrap their learning via *experience replay*, with something akin to episodic memory (e.g., Mnih et al., 2015).

Reciprocally, cognitive neuroscientists have been inspired and informed by ideas and results from deep learning. Neural networks provide proofs of concept that help address difficult questions, such as the degree to which visual object recognition requires recurrent processing (O'Reilly et al., 2013; Riesenhuber & Poggio, 1999; Spoerer, McClure, & Kriegeskorte, 2017). Machine learning offers new perspectives, encouraging us to think about cortical and subcortical specialization in terms of the learning objectives present in different brain regions and the prior world knowledge, which may be ingrained in regionally specific cytoarchitectures (Marblestone, Wayne, & Kording, 2016). The humbling experience of trying to teach machines to see and think has impressed upon neuroscientists just how difficult these achievements are computationally. Engineering has also demonstrated how much can be done with relatively simple computational components (e.g., the "GIST" model of Oliva and Torralba [2001]).

Neural network models excel at pattern recognition and pattern generation tasks, whether the patterns are static or dynamic. They do not yet capture the human ability to make comprehensive inferences from tiny details, learn new concepts from single experiences, and generalize successfully to entirely novel domains (Kriegeskorte & Douglas, 2018; Lake et al., 2017). More abstract cognitive computational models, based on symbolic representations and probabilistic inference, currently come closer to matching these amazing human cognitive feats, although they fall short of the human brain in terms of computational efficiency

(Gershman, Horvitz, & Tenenbaum, 2015) and are harder to relate to neurobiology. Understanding human cognition and its implementation in the brain will require both cognitive-level computational models and neural network models. Here we focus on the latter and invite cognitive neuroscientists to incorporate these models into their research.

What Are Deep Neural Networks?

Models of brain computation range from biologically detailed simulations of circuits of spiking neurons to highly abstract cognitive models. The term *neural network model* usually refers to models at an intermediate level of abstraction. Neural network models consist of interconnected units, each computing a weighted sum of its inputs and passing it through a nonlinear activation function (figure 59.1*A*). The resulting scalar output can be thought of as the firing rate of an idealized neuron (called *unit*), with an instantaneous response and no adaptation or refractory period. Networks of such units can implement arbitrarily complex functions between inputs (e.g., photographs) and outputs (e.g., the name of the main object in each photograph) as compositions of linear-nonlinear subfunctions.

Learned connection weights: the flexible knowledge of a network Like biological neural networks, the models don't come into the world knowing everything about their environment or how to perform the tasks that will be required of them. Instead, they learn. The knowledge inside a network is held by the weights associated with connections between units, loosely analogous to synaptic weights in a brain. Weights are usually initially set to small random numbers and incrementally updated via a learning algorithm.

Learning algorithms fall into three broad types: supervised, unsupervised, and reinforcement. In supervised learning, the learning algorithm will adjust the weights to bring the outputs for a large set of input patterns closer to prespecified desired outputs. A cost captures how much the current outputs deviate from the desired outputs. Weights are adjusted in proportion to how strongly their adjustment reduces the cost. This requires computing the derivative of the cost function with respect to each weight. An efficient algorithm for computing these derivatives is backpropagation. The derivative of the cost for a weight tells us in which direction and how much to tweak each connection weight in order to bring the output for a training input closer to the desired output. Many recent engineering achievements—for example, in machine vision and language translation—owe their success to supervised learning on massive ground-truth-labeled data sets (Russakovsky et al., 2015; Wu et al., 2016).

In unsupervised learning, training data are provided without any labels. The network learns to model the statistics of its input data. An autoencoder, for example, learns to compress its input pattern within a lower-dimensional "code" layer. The network maps the input pattern to the lower-dimensional code (encoder component) and then back to the full input pattern (decoder component). Autoencoders learn to encode and decode (i.e., reconstruct) each input pattern (e.g., figure 59.1*B*). They can also learn to remove noise or predict future states of a dynamic input. Unsupervised learning signals are extremely rich (an image autoencoder derives a training signal for every pixel of its reconstruction attempt), enabling the network to better exploit the information in its experiential data. The network attempts to learn all regularities in its data, not only those relevant for a particular task.

Finally, in reinforcement learning (Sutton & Barto, 1998), the network outputs an action (e.g., a movement in a simulated or real environment) and learns to model the expected cumulative reward associated with possible actions. Certain events in the environment are defined as rewarding, and their occurrence drives learning. Reinforcement learning can be combined with deep learning, so as to enable the value function to be represented by a neural network (e.g., deep Q-learning; Mnih et al., 2015). When rewards are few and far between in the environment, reinforcement learning is difficult because it provides fewer direct constraints for adjusting the weights than unsupervised or supervised learning. Rooted in biology and psychology, however, reinforcement learning has high ecological plausibility. It has recently also brought significant advances in engineering, illustrated, for example, by its success in the domain of video game playing (Jaderberg et al., 2018; Mnih et al., 2015).

Networks will overfit their training data to some extent and are therefore always tested on an independent test set of novel inputs (e.g., images) and outputs (e.g., category labels). Unlike in biological brains, learning (the adjustment of connection weights) and perception (the processing sweep of a new input through a network with fixed connection weights) are typically separate processes, and learning generally takes place in a distinct initial phase.

Architecture: the fixed structure of a network Infinitely many network architectures can be created by linking up units in different configurations. In feedforward architectures, units form a single processing hierarchy, with no unit connected to itself or to any previous unit. Feedforward networks have no temporal dynamics beyond the feedforward sweep and compute a static

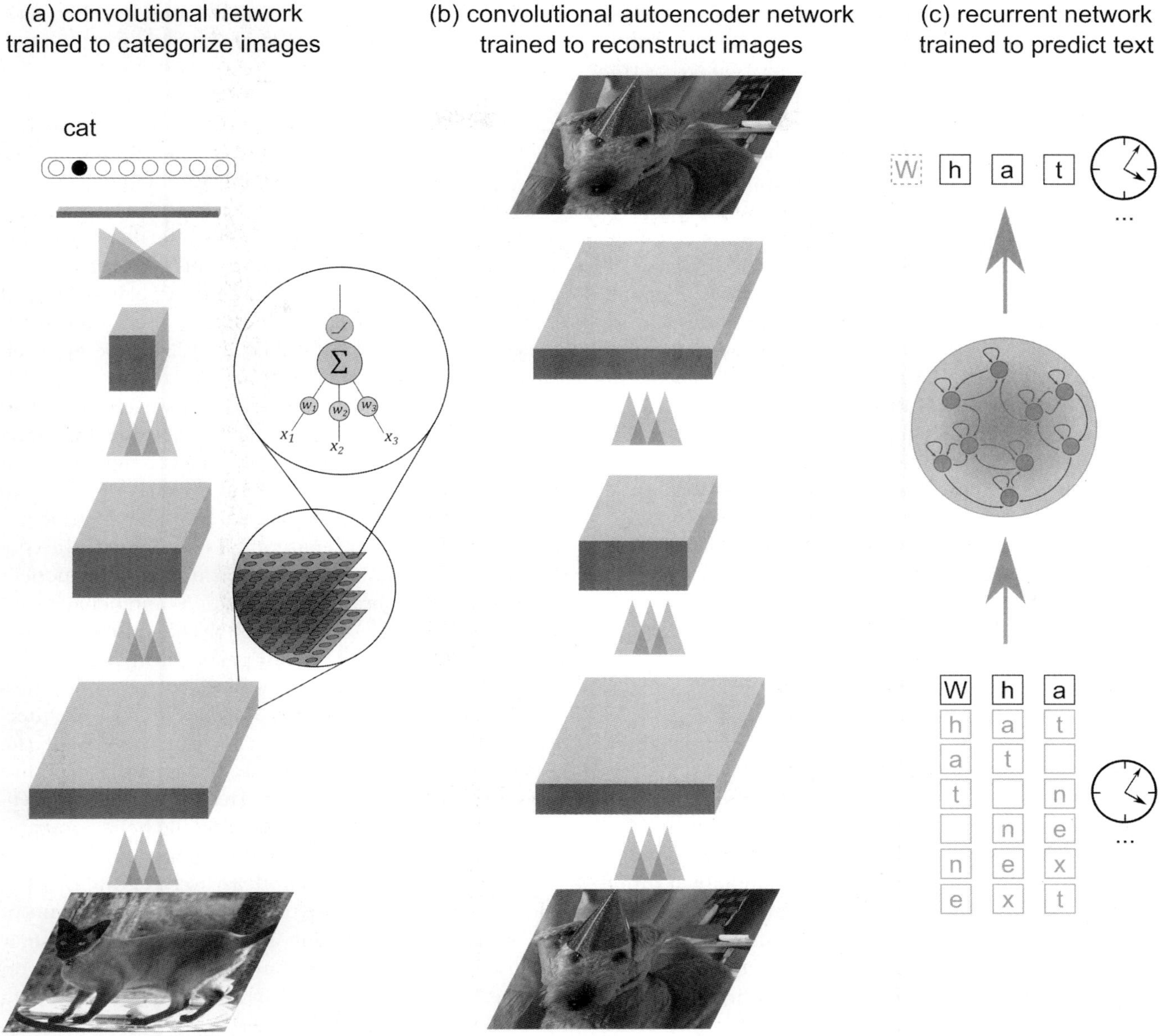

FIGURE 59.1 Diagrams of some of the species of neural network. *A*, A convolutional neural network with four hidden layers (three convolutional and one fully connected) that is able, after supervised training on labeled images, to categorize objects in novel photographs it has not been trained with. The convolutional layers consist of multiple "feature maps," each of which contains units with spatially restricted receptive fields centered on different locations of the input but sharing the same connection weights. This allows each map to represent the presence or absence of a particular feature at each location in the image. *B*, A convolutional autoencoder with three convolutional hidden layers, which is able, after unsupervised training, to compress novel images into a more concise format within the middle "bottleneck" layer and approximately reconstruct them thereafter. *C*, A recurrent neural network that takes as input text, one character at a time, and is able, after unsupervised training, to predict the next character in sentences it has never seen before. (See color plate 70.)

mapping from inputs to outputs. A network that contains one or more loops in its directed connection graph, such as top-down feedback, is recurrent, and its internal state will evolve over time in discrete steps. Recurrent networks are often used to process time-series data such as video or text, with a new frame or character being fed into the network at each time step (figure 59.1*C*).

A feedforward network with only a single "hidden" layer of units between its input data and output responses is known as a *shallow* network. Shallow feedforward networks can already approximate any continuous function with increasing precision as the number of units increases (Hornik, 1991). However, adding more intermediate processing layers can allow the network to express more complex functions with the same number of units by letting later units reuse and recombine features calculated by previous units. A network is called *deep* when it has more than one hidden layer intervening between input and output units. In modern computer vision, deep neural networks capable of near-human image classification performance typically contain ten or more hidden layers and over 1 million individual units (e.g., He et al., 2016; Simonyan & Zisserman, 2015).

Specialized architectures and unit types can exploit prior knowledge about the domain or task. For example, deep *convolutional* neural networks (CNNs) are often used for visual tasks (figure 59.1*A*). CNNs are loosely inspired by mammalian visual cortex and use units with spatially restricted receptive fields and shared weights (LeCun et al., 1998). Weight sharing means that multiple units within each layer use the same template of connection weights, applying this template at different locations on the input. This reduces the number of parameters the network must learn and captures the prior belief that the same feature (e.g., a vertical edge, or the wing of a bird) may appear anywhere within an image. Many recurrent networks use *long short-term memory* (LSTM) units (Hochreiter & Schmidhuber, 1997), which can dynamically retain recent information for as long as it is useful in the current task. Other custom units perform local response normalization or max pooling over their inputs, inspired by the neuroscientific idea of canonical cortical computations (Carandini & Heeger, 2012; Riesenhuber & Poggio, 1999).

Using Neural Networks as Models of Cognition and Perception

Cognitive neuroscientists can engage with neural networks at different levels requiring varying degrees of technical know-how and resource commitment. At one end of the spectrum, we can read and think about networks, using their successes, failures, and principles to seed theories and hypotheses. We can consider task-performing models as proofs of principle for computational mechanisms of perception and cognition. For example, the extent to which visual object recognition can be achieved by feedforward systems has been a matter of historical contention (e.g., Riesenhuber & Poggio, 1999). Modern CNNs demonstrate that object naming and segmentation in natural scenes can be achieved, to a large extent, without lateral or feedback connections. At the same time, recurrent connectivity can substantially improve the recognition performance of neural networks under challenging circumstances (Nayebi et al., 2018; O'Reilly et al., 2013; Spoerer, McClure, & Kriegeskorte, 2017).

At a deeper level of engagement, we can begin to bring DNN models into our empirical work by taking pretrained networks built by other researchers (many of which are available online) and comparing them to human and nonhuman subjects in terms of their behavior and internal representations. To disentangle the contributions of architecture and training to a model's ability to explain brain representations and behavior, we can retrain or fine-tune prebuilt architectures on stimuli and tasks that relate to our particular hypotheses.

Finally, we can develop new models exploring different training data, learning objectives, and architectures. This lets us test how different aspects of the environment, learning process, and neural structure may affect cognitive function. Designing complex neural network models and performing the behavioral and neural experiments to evaluate them requires a wide range of expertise that is not easy to integrate in a single lab. We will therefore need to develop new forms of collaboration and sharing across labs. Some labs may choose to focus more on building models; others, on testing shared models with brain and behavioral data or on developing tasks designed to highlight and quantify the remaining shortcomings of existing models.

Testing neural network models with behavioral and brain-activity data At the behavioral level, a model should be able to perform the task of interest at a similar level to a human or animal subject. However, a model merely being able to perform well in a task in which humans perform well (e.g., generating a verbal description of an image; Xu et al., 2015) does not provide strong evidence that the model performs similar computations or arrives at the output via similar internal representations as humans. Good models will be able to predict detailed patterns of behavioral variation across different instances of the task.

We can also compare each model's internal representations to human perceptual judgments. The similarity of stimuli or conditions in the model's internal activation patterns should predict a perceived similarity to humans (Kubilius, Bracci, & de Beeck, 2016). Stimuli that elicit identical responses within the model should appear identical to humans (Wallis et al., 2017). As stimuli are degraded or distorted, the model's performance should decline in a way quantitatively similar to human performance (contrary to this, Geirhos et al. [2018] found that CNNs were far less robust to image noise than humans). Models should be able to predict patterns of confusion and errors, ideally at the single-stimulus level. For example, in one study CNN classifications accurately predicted object-matching confusion in monkeys and humans at the object-category level but failed to predict which specific images were confused (Rajalingham et al., 2018).

At the level of the internal representation, a model should go through the same sequences of representational transformations across space (brain regions) and time (sequence of processing). Comparing the internal representations between model and brain is complicated by the fact that we may not know the detailed spatial and temporal correspondence mapping between model activity patterns and brain activity patterns. Chapter 56 in this section introduces the framework of representational models, which can be used to test neural network models by comparing their internal representations with brain-activity measurements. Briefly, encoding models predict each measured response channel as a linear combination of the units of the neural network (Dumoulin & Wandell, 2008; Kay et al., 2008; Mitchell et al., 2008; Naselaris, Kay, Nishimoto, & Gallant, 2011). Representational similarity analysis (Kriegeskorte & Diedrichsen, 2016; Kriegeskorte, Mur, & Bandettini, 2008; Nili et al., 2014) and pattern component modeling (Diedrichsen et al., 2011) use a stimulus-by-stimulus matrix of dissimilarities or similarities to characterize the representational space and compare model layers to brain regions. These approaches have complementary pros and cons. For example, encoding models lend themselves to analyzing responses in each voxel separately and mapping these out over the cortex (Eickenberg et al., 2017; Güçlü & van Gerven, 2015; Wen et al., 2017); representational similarity analysis and pattern component modeling obviate the need for fitting thousands of parameters of a linear encoding model while still allowing the flexibility of estimating weights to model the relative prevalence of different features in a brain representation (e.g., Khaligh-Razavi, Henriksson, Kay, & Kriegeskorte, 2017; Khaligh-Razavi & Kriegeskorte, 2014). Encoding models, representational similarity analysis, and pattern component modeling are best thought of as part of a tool box of representational-modeling techniques that can be combined as appropriate to the goals of a study (Diedrichsen & Kriegeskorte, 2016).

Khaligh-Razavi and Kriegeskorte (2014) found that the later layers of a categorization-trained CNN outperformed any of 27 shallow computer vision models in predicting the representation of natural object images in inferior temporal cortex. Categorization-trained CNNs also best predict visual cortical activity whether measured via electrophysiology (Cadieu et al., 2014), fMRI (Eickenberg et al., 2017; Güçlü & van Gerven, 2015; Wen et al., 2017), magnetoencaphalography (MEG) (Cichy et al., 2016, 2017), or electroencephalography (EEG) (Greene & Hansen, 2018), and a DNN trained on speech and music best predicted auditory cortical activity (Kell et al., 2018). Variants of this approach that focus on decoding or reconstructing stimuli from neural activity likewise find that CNNs provide excellent feature spaces for decoding activity in visual areas (Horiwaka & Kamitani, 2017; Wen et al., 2017).

Whatever method is used to evaluate a DNN model, it is crucial to include comparisons to control models and/or to test multiple DNNs that instantiate alternative theories of computation or learning. Even trivial models of stimulus processing explain small but significant variance in sensory brain regions (e.g., the pixel-based control models in Khaligh-Razavi and Kriegeskorte [2014]), as do randomly weighted untrained DNNs (e.g., Cichy et al., 2016).

Interpreting neural network models: doing cognitive neuroscience in silico Once we have a DNN that can perform some task and explain brain and behavioral data (up to the limit determined by noise and intersubject variability), we have reached an important milestone. However, our job is not done. Because the model has been trained on the task and has many parameters, we may not understand its computational mechanism. We therefore need to analyze the representations and dynamics in the model. Like brains, DNNs can be studied at many different levels of resolution and abstraction. Unlike brains, they permit perfect access to the entire system. We can stop and restart a network, have it relearn under different environments or task demands, gather data from it continuously without fatigue or damage, lesion and reinstate any combination of its components, use stimulus optimization techniques to see what features it has learned, and even analytically prove some of its properties.[1]

[1] For example, the mathematics of group theory have been used to explain why successive layers of deep feedforward networks learn increasingly complex features (Paul & Venkatasubramanian, 2014).

Because of our unfettered access, DNNs are far more amenable to "electrophysiological" methods than are real brains. To analyze the feature preferences learned by individual units in a network, experimenters can present thousands or millions of stimuli and record each unit's activation (e.g., Yosinski et al., 2015) or, complementarily, occlude different regions of an image stimulus to find which features are most crucial for each unit's activation (Zhou et al., 2014). Figure 59.2*C* shows how the activation of a single LSTM unit inside a recurrent text-prediction network trained on movie reviews changes as the passage the network is predicting unfolds (Radford, Jozefowicz, & Sutskever, 2017). In this striking example, the visualized unit appears to have learned to represent whether the movie review is expressing a positive or negative sentiment.

DNNs are also amenable to novel interrogation techniques that could never be applied to a brain. Because neural network models are differentiable, it is possible to use gradient-descent optimization to create stimuli that elicit specific patterns of network activation, including generating optimal stimuli for individual units (e.g., Yosinski et al., 2015). Several techniques exist for doing this, as beautifully summarized and illustrated by Olah (2017). Figure 59.2*A* shows a noise image that has been iteratively optimized to strongly activate each of four different layers in two category-trained CNNs using Google's DeepDream algorithm (Mordvintsev, Olah, & Tyka, 2015). The flow of information can be traced through networks to reveal where or when in the stimulus evidence was drawn in order to support a certain output decision (again, beautifully illustrated by Olah [2018]) or to which part of the stimulus a network with dynamic "attention" is currently devoting its resources (figure 59.2*B*).

Direct optimization methods have revealed that the hierarchy of features learned by deep visual networks bears a striking similarity to that in the mammalian ventral stream (see, e.g., Yamins & DiCarlo, 2016). They also allow us to explore how feature preferences develop over the course of training, suggesting tantalizing similarities with human perceptual learning (Wenliang & Seitz, 2018). Task-performing networks may help settle long-standing debates in cognitive neuroscience, such as how sparse or distributed neural codes are likely to be (Agrawal, Girschick, & Malik, 2014; Morcos, Barrett, Rabinowitz, & Botvinick, 2018; Zhuang, Wang, Yamins, & Hu, 2017).

Network representations can also be summarized at higher levels of abstraction. Traditional functional localization methods can identify units or layers that are preferentially selective for particular classes of stimuli, with initially counterintuitive results such as the emergence of object-selective units in a visual network trained only to classify scenes (Zhou et al., 2014). Representational similarity analysis (RSA) can help show how a network transforms stimulus information across the layers (e.g., Khaligh-Razavi & Kriegeskorte, 2014) or time steps (e.g., Cichy et al., 2016) of its processing.

The accessibility of neural network models raises new possibilities for efficient model search and falsification that could substantially change how we design experiments. Currently, experimental conditions are usually chosen "by hand" to differentiate among hypotheses suggested by verbal theories. In the future, conditions could be optimized algorithmically to best differentiate between explicit computational models (e.g., Wang & Simoncelli, 2008).

Neural Networks as Models of Diverse Cognitive Functions

Visual object and face recognition have received a great deal of attention from engineers and cognitive neuroscientists because DCNNs have been able to reach near-human-level performance using static stimuli and a single feedforward sweep of processing (He, Zhang, Ren, & Sun, 2016; Taigman et al., 2014). However, the deep learning revolution has touched almost all domains in cognition and perception, including our higher-level abilities, such as language and reasoning.

Language in neural networks Language-processing DNNs are widely used in automatic translation—for example, of websites and online search results. Such high-level tasks tend to be achieved by more complex compositional network architectures than we have yet encountered. The state-of-the-art Google Neural Machine Translation system (Wu et al., 2016) consists of two eight-layer recurrent neural networks, one that encodes the text from its original language to a latent representation within the network and another that takes the latent representation and decodes it into the target language using *attentional* weights to change which parts of the encoded representation it processes at each time step. The entire dynamic process is learned through backpropagation by training on tens of millions of human-translated sentences, including the form of the latent representation and the allocation of attention during decoding (Wu et al., 2016). A similar approach has been used to generate verbal descriptions of images (Xu et al., 2015). There, an image-processing convolutional network is used to transform the image into a latent representation of high-level image features, and a recurrent network is trained to output a sequence of words based on those features using spatial attention

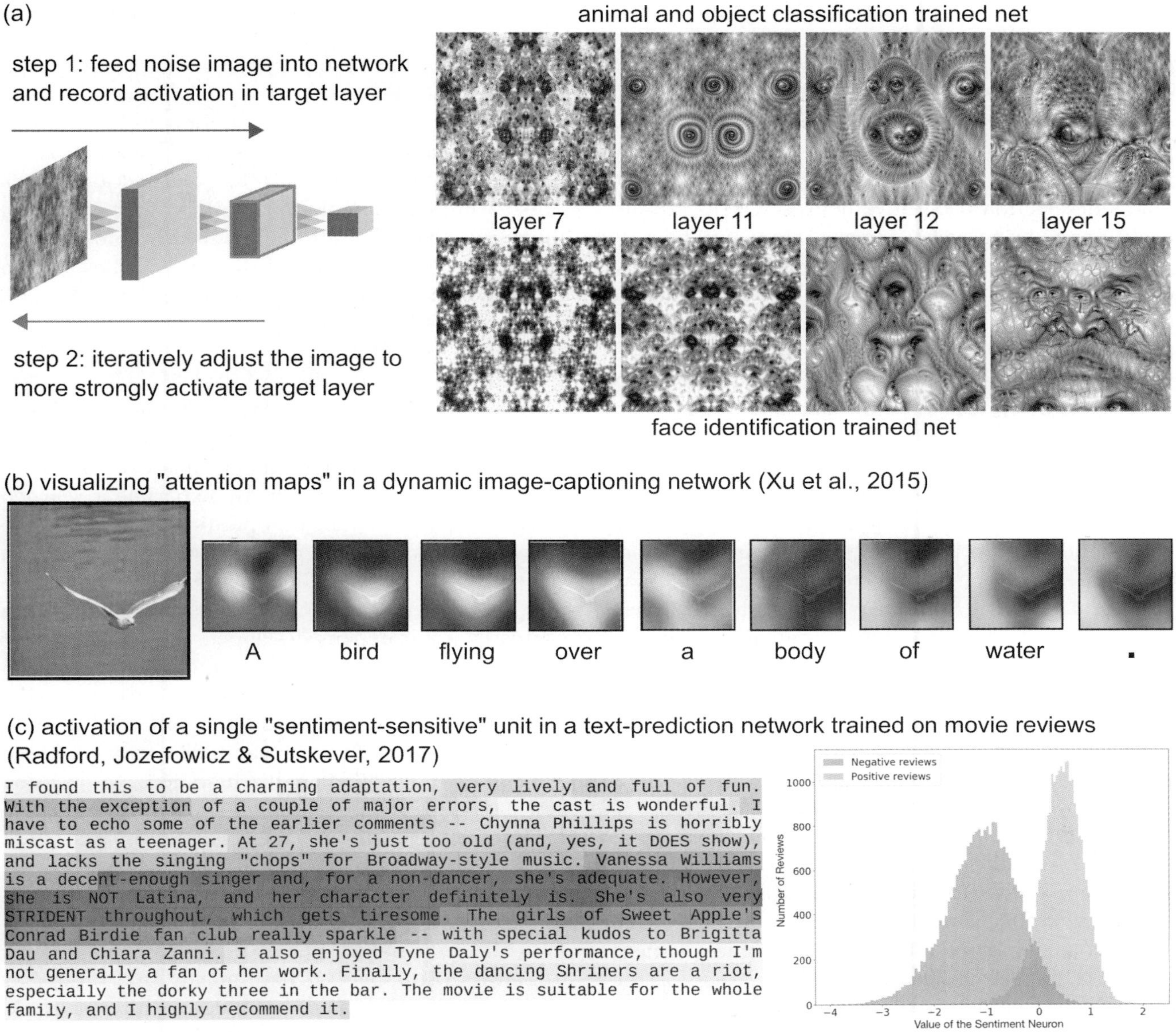

FIGURE 59.2 The relative transparency of deep neural networks. *A*, The backpropagation algorithm used to train networks can also be used to visualize features in trained networks. Here, a noise image has been iteratively optimized to increase how strongly it activates units in each of four layers in a 16-layer DCNN, which has been trained either to categorize objects and animals (*top row*) or identify faces (*bottom row*). Training has a striking impact on the features learned. *B*, *Attention networks* have a spatial attention mask (a multiplicative weighting of the input), which they learn to allocate in ways that improve performance at their training task. Here, the white patches show which regions of this image were attended to at each time step by a recurrent network trained to generate text captions of images, outputting one word per time step. Reproduced with permission from Xu et al. (2015). *C*, Visualization of the output activation of a single LSTM unit in a recurrent network trained to predict text one character at a time, for an example passage the network has never seen before. High activation is indicated by green and low by red. The responses of this unit, when analyzed across a large sample of reviews, predicted whether a review was positive or negative in sentiment. Reproduced with permission from Radford, Jozefowicz, and Sutskever (2017). (See color plate 71.)

to modulate which parts of the image most guide word choice at each time step (figure 59.2*B*).

Although these networks are engineering solutions, it is striking how the composition of multiple pattern recognition DNNs, coupled with attentional guidance (a psychologically inspired feature), can enable such uniquely human skills as language translation and image description. Cognitive neuroscientists are just beginning to bring such principles into neural network models of human cognition and language. Devereux, Clarke, and Tyler (2018) used a modular visual-semantic model to predict fMRI activation patterns as human observers named objects in images. In the model, images were first processed by a feedforward image CNN trained on object classification, and the penultimate-layer features from the CNN were then fed into a single-layer recurrent network, which was trained to encode the presence or absence of a large number of semantic properties of the pictured objects ("is a fruit," "grows on trees," and others). The similarity of representations within the CNN better predicted similarity in human visual cortex, while similarity within the recurrent semantic network better predicted similarity in human perirhinal cortex, an area associated with semantic processing.

Reasoning in neural networks Artificial general intelligence may not yet be in sight, but some deep networks display impressively flexible reasoning. Relation networks (Santoro et al., 2017) achieved superhuman performance on the Compositional Language and Elementary Visual Reasoning (CLEVR) benchmark, a notoriously difficult task involving reasoning about the relations among objects in images. Example questions include "How many objects have the same shape as the green object?" and "Are there any rubber things that have the same size as the yellow metallic cylinder?" (Santoro et al., 2017). A feedforward image CNN extracts high-level visual features from the image, and a recurrent language network encodes the verbal question. The relation network then considers all possible pairs of "objects" in the image (operationalized as pairs of spatial locations), along with the question, and evaluates the likelihood of each possible response. Relation Networks achieved a 95% accuracy rate on the CLEVR task; human performance is around 93% (Santoro et al., 2017). The image, language, and relation-reasoning subnetworks form an end-to-end differentiable system and are simply trained together using backpropagation.

Compositional network architectures have proved to be an important engineering trick for achieving intelligent behavior in DNNs. Recent work by Yang, Song, Newsome, and Wang (2017) begins to unpick how task requirements might determine neural architectures. A recurrent network was trained simultaneously to perform 20 very simple tasks from human and animal cognitive research, such as speeded classification and delayed match-to-sample. The authors were then able to analyze which units were involved in which tasks. They found that the network had learned a compositional representation of the 20 tasks, in that clusters of units specialized in the components of the tasks (e.g., "Remember the direction of the cue" or "Respond in the direction opposite to the cue"), and any given task was performed by a coalition of clusters, according to the composition of its subtasks.

Embodied neural networks DNNs are not necessarily disembodied. For example, a recurrent network trained to simulate the muscle movements made by monkeys reaching toward targets in an experiment learned internal representations that predicted the firing patterns of motor neurons recorded while the monkeys were reaching (Sussilo, Churchland, Kaufman, & Shenoy, 2015). Neural networks trained to navigate in a simulated environment developed units with similar tuning properties to place, grid, border, and head direction cells found in rat entorhinal cortex (Banino et al., 2018; Cueva & Wei, 2018). Embodied models, whether with physical or virtual bodies, may be able to learn more efficiently than disembodied ones by using active learning methods to seek out informative training data (e.g., Haber, Mrowca, Fei-Fei, & Yamins, 2018).

Current Challenges for Deep Neural Networks in Cognitive Neuroscience

Cognitive neuroscientists have much work to do. Any particular current instance of a deep neural network, presented in the machine-learning literature for a particular practical application, will almost certainly not prove to be a fully satisfying model of how similar functions are performed by human brains. We must seek out ways to make cognition in DNNs more flexible and generalizable and learning more ecologically plausible, and turn the predictive power of neural network models into theoretical understanding. Happily, many of these goals align at least in part with those of industrial artificial intelligence researchers.

Modeling more robust and flexible cognition and perception Although DNNs are architecturally deep, they can seem conceptually shallow. Their intelligence is more fragile than that of humans, apparently failing to develop the causal and conceptual understanding that allows us to generalize performance across tasks and environments (Lake, Ullman, Tenenbaum, & Gershman, 2017; Marcus, 2018). Visual object-recognition DNNs can be tricked into drastic misclassifications using minute image perturbations imperceptible to humans (Szegedy et al., 2013). Reading comprehension networks able to answer questions based on passages of text can be similarly tricked by small changes to the text that do not distract human readers (Jia & Liang, 2017). Standard DNNs are poor at estimating the confidence of their decisions and may confidently return seemingly nonsense classifications when shown stimuli with statistics different from those in their training data set (Gharamanhi, 2015; Nguyen, Yosinski, & Clune, 2015). Reinforcement-learning networks can outcompete humans at playing arcade video games (Mnih et al., 2015) but appear to depend on features that humans would consider superficial or irrelevant—changing the shape or position of the player's paddle by just a few pixels can disastrously undermine the network's performance (Kansky et al., 2017).

It is worth bearing in mind that current DNNs are still very limited in computational scale compared to biological brains. A modern DNN might contain hundreds of thousands or millions of units, on the same order as the number of cortical neurons in a handful of fMRI voxels.[2] Importantly, the units in neural network models cannot be equated to biological neurons, which are more complex in structure and dynamics and thus potentially more computationally powerful. As a result, DNNs likely fall short by a far larger factor than counting units and neurons would suggest.

To compound their disadvantage in scale, DNNs so far have tended to live exclusively within a single modality and task. At best this makes them models of specific sensory-processing areas, but even these areas in human brains have extensive interaction with other modalities, executive control, and action. The examples described in the section on reasoning in neural networks hint at some of the possible solutions, involving compositional architectures (Santoro et al., 2017; Xu et al., 2015) as well as hybrid solutions that combine neural networks with symbolic-reasoning models (e.g., Battaglia et al., 2016; Evans & Grefenstette, 2018).

Cognitive neuroscience is well placed to devise benchmark tasks that capture important hallmarks of human cognition. Many existing experimental paradigms may be ideal—for example, *violation of expectation* tasks used to measure the development of intuitive physics concepts in children can be applied to assess models' reasoning about physical interactions (Piloto, 2018). Lake and Baroni (2017) and Marcus (2018) suggest a number of other tasks to evaluate the profundity and generality of machine cognition. Formalizing these as benchmarks would provide concrete goals for the modeling community and drive progress (Kriegeskorte & Douglas, 2018), as object-recognition benchmarks have done in computer vision (Russakovsky et al., 2015).

Building networks that learn in more ecologically and biologically plausible ways Both supervised and unsupervised learning depend on the backpropagation algorithm, which is traditionally considered biologically implausible (Glaser, Benjamin, Farhoodi, & Kording, 2019). Recent theoretical work suggests that deep learning with an algorithm similar to backpropagation might be biologically feasible (Guerguiev, Lillicrap, & Richards, 2017; Hinton & McClelland, 1988; Körding & König, 2001; Lillicrap et al., 2016; Scellier & Bengio, 2016). However, supervised training on millions of labeled stimuli is still an ecologically unrealistic requirement. Animal brains are capable of learning far more data-efficiently, often without explicit supervision and sometimes even from single examples.

Unsupervised, weakly supervised, and one-shot learning are current focuses of the machine-learning community (Hassabis et al., 2017), and we can expect advances in these areas over the next few years. Cognitive neuroscientists may be able to draw from brain theory and data for creative and biologically feasible solutions. One influential idea within neuroscience is that *prediction* may be fundamentally important to how brains learn and perceive (Hawkins & Blakeslee, 2007). Predictive-coding theories, in particular, propose that during perception, neural activity primarily encodes the differences between predicted and actual sensory information (Srinivasan, Laughlin, & Dubs, 1982). Explicit computational models of predictive coding have previously been formulated at the neuronal-circuit level (Rao & Ballard, 1999) and have received some support from electrophysiological (Rao & Ballard, 1999) and fMRI data (Muckli et al., 2015). In machine learning, prediction offers a rich unsupervised-training signal requiring no additional reward or label information. When implemented in a modern deep-learning framework, a predictive-coding network was capable of predicting future frames in video of natural environments (Lotter, Kreiman, & Cox, 2017; 2018). Furthermore, the network spontaneously discovered, in its deeper layers, higher-level properties of the objects depicted in videos,

[2] The successful image recognition network AlexNet has around 658,000 units (Krizhevsky, 2012), and there is a density of around 630,000 neurons per 3 mm isotropic voxel in the human cortex (https://cfn.upenn.edu/aguirre/wiki/public:neurons_in_a_voxel).

such as facial identity and pose (Lotter, Kreiman & Cox, 2017), and its individual units reproduced certain temporal dynamics of primate visual neurons (Lotter, Kreiman & Cox, 2018). *Curiosity-based* learning is another tantalizing method motivated by animal learning, in which networks embodied in simulated environments actively seek out the most informative parts of those environments during learning (Haber, Mrowca, Fei-Fei, & Yamins, 2018).

Illuminating the black box: from prediction to explanation A criticism leveled at neural networks as cognitive models is that success at predicting brain and behavioral data, which DNNs have inarguably enjoyed (Cadieu et al., 2014; Cichy et al., 2016; Eickenberg et al., 2017; Güçlü & van Gerven, 2015; Khaligh-Razavi et al., 2014; Kubilius et al., 2016; Wen et al., 2017), does not constitute success at explaining or understanding the brain or mind (Kay, 2017). If a model exhaustively predicted neural and behavioral data with generalizations to new instances of the task, we might be convinced that it is an accurate model of a certain cognitive function. But unless we can express more concisely how it performs that function, we are unlikely to be fully satisfied as scientists.

We have described some techniques for (*in silico*) electrophysiology and visualization of internal representations to illuminate the black box. An even more satisfying approach is to go from a neural network model to a concise mathematical description. Gonçalves and Welchmann (2017) trained a CNN to judge the relative depth of objects in scenes from pairs of stereo images, requiring it to solve the classic *correspondence problem* (Which points in the two retinal images correspond to the same point on an external object?). By tracing the strength and sign of connection weights in the trained network, they discovered a surprising computational strategy: rather than using positive weights to find positive matches between points in the two images, the network primarily used negative weights to suppress incorrect matches. The authors distilled this strategy into a succinct mathematical image-filtering model. The model displays several unintuitive and previously unexplained phenomena in human depth perception, and substantially changes our understanding of the correspondence problem, which has stood as a computational puzzle for at least 100 years (Gonçalves & Welchmann, 2017). DNN models do not replace intuitive explanations, verbal theories, and concise mathematical descriptions. Instead they make complex hypotheses testable and help us bridge the levels of description between verbally communicable theory and neural implementation.

Conclusion

DNNs are currently at the cutting edge of artificial intelligence and are the most powerful predictive models yet discovered for many aspects of behavior (Gonçalves & Welchmann, 2017; Kubilius, Bracci, & de Beeck, 2016; Wallis et al., 2017), single-neuron activity (e.g., Cadieu et al., 2014), and large-scale cortical activity (Cichy et al., 2016, 2017; Eickenberg et al., 2017; Greene & Hansen, 2018; Güçlü & van Gerven, 2015; Horiwaka & Kamitani, 2017; Kell et al., 2018; Khaligh-Razavi & Kriegeskorte, 2014; Cichy et al., 2016; Cichy et al., 2017; Wen et al., 2017). With biologically plausible components and structures, they are the closest we have yet come to explicit end-to-end models of how perception and cognition might be performed in brains.

Deep learning, therefore, is relevant to everyone interested in how perception and cognition arise from neural activity. There are many levels at which cognitive neuroscientists can use deep learning in their work, from considering the modeling literature in theoretical discussions through testing models built by others to building new models based on cognitive and neuroscientific theories. Conversely, the cognitive neuroscience community may help drive engineering progress, bringing DNNs that learn in more ecologically feasible ways, learn more profound conceptual representations, and display more robust and generalizable task performance. Machine learning of computations capable of real-world tasks in biologically plausible systems will play a major role in understanding how intelligent behavior arises from brains.

Appendix: Resources for Getting Started with Deep Learning

Articles Many excellent review articles are available on the cognitive neuroscientific applications of DNN models. For a recent overview of the complementary exchanges between neuroscience and artificial intelligence, see Hassabis et al. (2017). For a more extensive treatment of neural network architectures and principles and their application as models of brain function, see Kriegeskorte (2015) and Kietzmann, McClure, and Kriegeskorte (2019). For reviews of neural network models in sensory and systems neuroscience, see Yamins and DiCarlo (2016) and Glaser, Benjamin, Farhoodi, and Kording (2019).

Tutorials, courses, and books For accessible introductions to machine learning and DNNs, the free online courses currently offered by Geoff Hinton (https://www.coursera.org/learn/neural-networks) and Andrew Ng (https://www.coursera.org/learn/machine-learning) are invaluable, as is the online book by Michael Nielsen (http://neuralnetworksanddeeplearning.com/). *Deep Learning* (Goodfellow, Bengio, & Courville, 2016) is a comprehensive book written by some of the leaders in the field.

Software Frameworks to implement DNN models are rapidly evolving. At the time of writing, the leading frameworks are free and Python-based: Tensorflow (developed by Google), PyTorch (developed by Facebook), and Caffe (developed by the University of California, Berkeley). Keras and Sonnet are higher-level packages that work atop Tensorflow to provide readier access to many advanced features. MATLAB (MathWorks, Inc.) is less widely supported or used by machine-learning teams in industry but is worth mentioning due to its popularity in psychology labs. The Neural Network Toolbox in versions of MATLAB from 2017 onward allows all core deep-learning operations, such as defining and training feedforward and recurrent models, loading pretrained models, importing models defined in other frameworks (currently, Keras and Caffe are supported), and visualizing units within trained networks.

Acknowledgment

This project has received funding from the European Union's Horizon 2020 Programme for Research and Innovation under the Specific Grant Agreement No. 720270 and 785907 (Human Brain Project SGA1 and SGA2).

REFERENCES

Agrawal, P., Girshick, R., & Malik, J. (2014). Analyzing the performance of multilayer neural networks for object recognition. In *European conference on computer vision* (pp. 329–344). Cham, Switzerland: Springer.

Banino, A., Barry, C., Uria, B., Blundell, C., Lillicrap, T., Mirowski, P., ... Wayne, G. (2018). Vector-based navigation using grid-like representations in artificial agents. *Nature, 557*(7705), 429–433.

Battaglia, P., Pascanu, R., Lai, M., & Rezende, D. J. (2016). Interaction networks for learning about objects, relations and physics. *Advances in Neural Information Processing Systems, 29*, 4502–4510.

Cadieu, C. F., Hong, H., Yamins, D. L., Pinto, N., Ardila, D., Solomon, E. A., ... DiCarlo, J. J. (2014). Deep neural networks rival the representation of primate IT cortex for core visual object recognition. *PLoS Computational Biology, 10*(12), e1003963.

Carandini, M., & Heeger, D. J. (2012). Normalization as a canonical neural computation. *Nature Reviews Neuroscience, 13*(1), 51–62.

Cichy, R. M., Khosla, A., Pantazis, D., & Oliva, A. (2017). Dynamics of scene representations in the human brain revealed by magnetoencephalography and deep neural networks. *NeuroImage, 153*, 346–358.

Cichy, R. M., Khosla, A., Pantazis, D., Torralba, A., & Oliva, A. (2016). Comparison of deep neural networks to spatio-temporal cortical dynamics of human visual object recognition reveals hierarchical correspondence. *Scientific Reports, 6*, 27755.

Cueva, C. J., & Wei, X. X. (2018). Emergence of grid-like representations by training recurrent neural networks to perform spatial localization. *arXiv*. Retrieved from 1803.07770.

Devereux, B. J., Clarke, A. D., & Tyler, L. K. (2018). Integrated deep visual and semantic attractor neural networks predict fMRI pattern-information along the ventral object processing pathway. *Scientific Reports, 8*, 10636.

Diedrichsen, J., & Kriegeskorte, N. (2017). Representational models: A common framework for understanding encoding, pattern-component, and representational-similarity analysis. *PLoS Computational Biology, 13*(4), e1005508.

Diedrichsen, J., Ridgway, G. R., Friston, K. J., & Wiestler, T. (2011). Comparing the similarity and spatial structure of neural representations: A pattern-component model. *NeuroImage, 55*(4), 1665–1678.

Dumoulin, S. O., & Wandell, B. A. (2008). Population receptive field estimates in human visual cortex. *NeuroImage, 39*(2), 647–660.

Eickenberg, M., Gramfort, A., Varoquaux, G., & Thirion, B. (2017). Seeing it all: Convolutional network layers map the function of the human visual system. *NeuroImage, 152*, 184–194.

Evans, R., & Grefenstette, E. (2018). Learning explanatory rules from noisy data. *Journal of Artificial Intelligence Research, 61*, 1–64.

Geirhos, R., Temme, C. R. M., Rauber, J., Schuett, H. H., Bethge, M., & Wichmann, F. A. (2018). Generalisation in humans and deep neural networks. *Advances in Neural Information Processing Systems, 31*, 7538–7550.

Gershman, S. J., Horvitz, E. J., & Tenenbaum, J. B. (2015). Computational rationality: A converging paradigm for intelligence in brains, minds, and machines. *Science, 349*(6245), 273–278.

Ghahramani, Z. (2015). Probabilistic machine learning and artificial intelligence. *Nature, 521*(7553), 452–459.

Glaser, J. I., Benjamin, A. S., Farhoodi, R., & Kording, K. P. (2019). The roles of supervised machine learning in systems neuroscience. *Progress in Neurobiology, 175*, 126–137.

Gonçalves, N. R., & Welchman, A. E. (2017). "What not" detectors help the brain see in depth. *Current Biology, 27*(10), 1403–1412.

Goodfellow, I., Bengio, Y., & Courville, A. (2016). *Deep learning* (Vol. 1). Cambridge, MA: MIT Press.

Greene, M. R., & Hansen, B. C. (2018). Shared spatiotemporal category representations in biological and artificial deep neural networks. *PLoS Computational Biology, 14*(7), e1006327.

Güçlü, U., & van Gerven, M. A. (2015). Deep neural networks reveal a gradient in the complexity of neural representations across the ventral stream. *Journal of Neuroscience, 35*(27), 10005–10014.

Guerguiev, J., Lillicrap, T. P., & Richards, B. A. (2017). Towards deep learning with segregated dendrites. *eLife, 6*, e22901.

Haber, N., Mrowca, D., Fei-Fei, L., & Yamins, D. L. (2018). Emergence of structured behaviors from curiosity-based intrinsic motivation. *arXiv*. Retrieved from 1802.07461.

Hassabis, D., Kumaran, D., Summerfield, C., & Botvinick, M. (2017). Neuroscience-inspired artificial intelligence. *Neuron, 95*(2), 245–258.

Hawkins, J., & Blakeslee, S. (2007). *On intelligence: How a new understanding of the brain will lead to the creation of truly intelligent machines*. New York: Macmillan.

He, K., Zhang, X., Ren, S., & Sun, J. (2016). Deep residual learning for image recognition. *The IEEE Conference on Computer Vision and Pattern Recognition*, 2016, 770–778. http://openaccess.thecvf.com/content_cvpr_2016/html/He_Deep_Residual_Learning_CVPR_2016_paper.html.

Hinton, G. E., & McClelland, J. L. (1988). Learning representations by recirculation. *Advances in Neural Information Processing Systems, 1*, 358–366.

Hochreiter, S. & Schmidhuber, J. (1997). Long short-term memory. *Neural Computation, 9*(8), 1735–1780.

Horikawa, T., & Kamitani, Y. (2017). Generic decoding of seen and imagined objects using hierarchical visual features. *Nature Communications, 8*, 15037.

Hornik, K. (1991). Approximation capabilities of multilayer feedforward networks. *Neural Networks, 4*(2), 251–257.

Huys, Q. J., Maia, T. V., & Frank, M. J. (2016). Computational psychiatry as a bridge from neuroscience to clinical applications. *Nature Neuroscience, 19*(3), 404–413.

Jaderberg, M., Czarnecki, W. M., Dunning, I., Marris, L., Lever, G., Castaneda, A. G., … Sonnerat, N. (2018). Human-level performance in first-person multiplayer games with population-based deep reinforcement learning. *arXiv*. Retrieved from 1807.01281.

Jia, R., & Liang, P. (2017). Adversarial examples for evaluating reading comprehension systems. *arXiv*. Retrieved from 1707.07328.

Kansky, K., Silver, T., Mély, D. A., Eldawy, M., Lázaro-Gredilla, M., Lou, X., … George, D. (2017). Schema networks: Zero-shot transfer with a generative causal model of intuitive physics. *Proceedings of the 34th International Conference on Machine Learning, 70*, 1809–1818.

Kay, K. N. (2017). Principles for models of neural information processing. *NeuroImage, 180*, 101–109.

Kay, K. N., Naselaris, T., Prenger, R. J., & Gallant, J. L. (2008). Identifying natural images from human brain activity. *Nature, 452*(7185), 352–355.

Kell, A. J., Yamins, D. L., Shook, E. N., Norman-Haignere, S. V., & McDermott, J. H. (2018). A task-optimized neural network replicates human auditory behavior, predicts brain responses, and reveals a cortical processing hierarchy. *Neuron, 98*(3), 630–644.

Khaligh-Razavi, S. M., Henriksson, L., Kay, K., & Kriegeskorte, N. (2017). Fixed versus mixed RSA: Explaining visual representations by fixed and mixed feature sets from shallow and deep computational models. *Journal of Mathematical Psychology, 76*, 184–197.

Khaligh-Razavi, S. M., & Kriegeskorte, N. (2014). Deep supervised, but not unsupervised, models may explain IT cortical representation. *PLoS Computational Biology, 10*(11), e1003915.

Kietzmann, T. C., McClure, P., & Kriegeskorte, N. (2019). Deep neural networks in computational neuroscience. *Oxford Research Encyclopedia of Neuroscience*. doi:10.1093/acrefore/9780190264086.013.46

Kording, K. P., & König, P. (2001). Supervised and unsupervised learning with two sites of synaptic integration. *Journal of Computational Neuroscience, 11*(3), 207–215.

Kriegeskorte, N. (2015). Deep neural networks: A new framework for modeling biological vision and brain information processing. *Annual Review of Vision Science, 1*, 417–446.

Kriegeskorte, N., & Diedrichsen, J. (2016). Inferring brain-computational mechanisms with models of activity measurements. *Philosophical Transactions of the Royal Society of London B: Biological Sciences, 371*(1705), 20160278.

Kriegeskorte, N., & Douglas, P. K. (2018). Cognitive computational neuroscience. *Nature Neuroscience, 21*, 1148–1160.

Kriegeskorte, N., & Mok, R. M. (2017). Building machines that adapt and compute like brains. *Behavioral and Brain Sciences, 40*, e269.

Kriegeskorte, N., Mur, M., & Bandettini, P. A. (2008). Representational similarity analysis-connecting the branches of systems neuroscience. *Frontiers in Systems Neuroscience, 2*, 4.

Krizhevsky, A., Sutskever, I., & Hinton, G. E. (2012). Imagenet classification with deep convolutional neural networks. *Advances in Neural Information Processing Systems, 25*, 1097–1105.

Kubilius, J., Bracci, S., & de Beeck, H. P. O. (2016). Deep neural networks as a computational model for human shape sensitivity. *PLoS Computational Biology, 12*(4), e1004896.

Kumaran, D., Hassabis, D., & McClelland, J. L. (2016). What learning systems do intelligent agents need? Complementary learning systems theory updated. *Trends in Cognitive Sciences, 20*(7), 512–534.

Lake, B. M., & Baroni, M. (2017). Still not systematic after all these years: On the compositional skills of sequence-to-sequence recurrent networks. *arXiv*. Retrieved from 1711.00350.

Lake, B. M., Ullman, T. D., Tenenbaum, J. B., & Gershman, S. J. (2017). Building machines that learn and think like people. *Behavioral and Brain Sciences, 40*, e253.

LeCun, Y., Bottou, L., Bengio, Y., & Haffner, P. (1998). Gradient-based learning applied to document recognition. *Proceedings of the IEEE, 86*(11), 2278–2324.

Lillicrap, T. P., Cownden, D., Tweed, D. B., & Akerman, C. J. (2016). Random synaptic feedback weights support error backpropagation for deep learning. *Nature Communications, 7*, 13276.

Lotter, W., Kreiman, G., & Cox, D. (2017). Deep predictive coding networks for video prediction and unsupervised learning. *arXiv*. Retrieved from 1605.08104.

Lotter, W., Kreiman, G., & Cox, D. (2018). A neural network trained to predict future video frames mimics critical properties of biological neuronal responses and perception. *arXiv*. Retrieved from 1805.10734.

Marblestone, A. H., Wayne, G., & Kording, K. P. (2016). Toward an integration of deep learning and neuroscience. *Frontiers in Computational Neuroscience, 10*, 94.

Marcus, G. (2018). Deep learning: A critical appraisal. *arXiv*. Retrieved from 1801.00631.

McClelland, J. L., McNaughton, B. L., and O'Reilly, R. C. (1995). Why there are complementary learning systems in the hippocampus and neocortex: Insights from the successes and failures of connectionist models of learning and memory. *Psychological Review, 102*, 419–457.

McCulloch, W. & Pitts, W. (1943). A logical calculus of the ideas immanent in nervous activity. *Bulletin of Mathematical Biophysics, 5*, 115–133.

Mitchell, T. M., Shinkareva, S. V., Carlson, A., Chang, K. M., Malave, V. L., Mason, R. A., & Just, M. A. (2008). Predicting human brain activity associated with the meanings of nouns. *Science, 320*(5880), 1191–1195.

Mnih, V., Kavukcuoglu, K., Silver, D., Rusu, A. A., Veness, J., Bellemare, M. G., ... Petersen, S. (2015). Human-level control through deep reinforcement learning. *Nature, 518*(7540), 529.

Morcos, A. S., Barrett, D. G., Rabinowitz, N. C., & Botvinick, M. (2018). On the importance of single directions for generalization. *arXiv*. Retrieved from 1803.06959.

Mordvintsev, A., Olah, C., & Tyka, M. (2015) Inceptionism: Going deeper into neural networks. Google technical blog post. https://ai.googleblog.com/2015/06/inceptionism-going-deeper-into-neural.html.

Muckli, L., De Martino, F., Vizioli, L., Petro, L. S., Smith, F. W., Ugurbil, K., ...Yacoub, E. (2015). Contextual feedback to superficial layers of V1. *Current Biology, 25*(20), 2690–2695.

Naselaris, T., Kay, K. N., Nishimoto, S., & Gallant, J. L. (2011). Encoding and decoding in fMRI. *NeuroImage, 56*(2), 400–410.

Nayebi, A., Bear, D., Kubilius, J., Kar, K., Ganguli, S., Sussillo, D., ... Yamins, D. L. (2018). Task-driven convolutional recurrent models of the visual system. *arXiv*. Retrieved from 1807.00053.

Nguyen, A., Yosinski, J., & Clune, J. (2015). Deep neural networks are easily fooled: High confidence predictions for unrecognizable images. *The IEEE Conference on Computer Vision and Pattern Recognition*, 2015, 427–436.

Nili, H., Wingfield, C., Walther, A., Su, L., Marslen-Wilson, W., & Kriegeskorte, N. (2014). A toolbox for representational similarity analysis. *PLoS Computational Biology, 10*(4), e1003553.

Olah, C. (2017). Feature visualization. Distill. Retrieved from https://distill.pub/2017/feature-visualization/.

Olah, C. (2018). The building blocks of interpretability. Distill. Retrieved from https://distill.pub/2018/building-blocks/.

Oliva, A., & Torralba, A. (2001). Modeling the shape of the scene: A holistic representation of the spatial envelope. *International Journal of Computer Vision, 42*(3), 145–175.

O'Reilly, R. C., Wyatte, D., Herd, S., Mingus, B., & Jilk, D. J. (2013). Recurrent processing during object recognition. *Frontiers in Psychology, 4*, 124.

Paul, A., & Venkatasubramanian, S. (2014). Why does deep learning work? A perspective from group theory. *arXiv*. Retrieved from 1412.6621.

Piloto, L., Weinstein, A., Ahuja, A., Mirza, M., Wayne, G., Amos, D., ... Botvinick, M. (2018). Probing physics knowledge using tools from developmental psychology. *arXiv*. Retrieved from 1804.01128.

Poeppel, D. (2012). The maps problem and the mapping problem: Two challenges for a cognitive neuroscience of speech and language. *Cognitive Neuropsychology, 29*(1–2), 34–55.

Radford, A., Jozefowicz, R., & Sutskever, I. (2017). Learning to generate reviews and discovering sentiment. *arXiv*. Retrieved from 1704.01444.

Rajalingham, R., Issa, E. B., Bashivan, P., Kar, K., Schmidt, K., & DiCarlo, J. J. (2018). Large-scale, high-resolution comparison of the core visual object recognition behavior of humans, monkeys, and state-of-the-art deep artificial neural networks. *Journal of Neuroscience, 38*(33), 7255–7269.

Rao, R. P., & Ballard, D. H. (1999). Predictive coding in the visual cortex: A functional interpretation of some extra-classical receptive-field effects. *Nature Neuroscience, 2*(1), 79–87.

Riesenhuber, M., & Poggio, T. (1999). Hierarchical models of object recognition in cortex. *Nature Neuroscience, 2*(11), 1019–1025.

Rumelhart, D. E., & McClelland, J. L. (1986). *Parallel distributed processing: Explorations in the microstructure of cognition*. Cambridge, MA: MIT Press.

Russakovsky, O., Deng, J., Su, H., Krause, J., Satheesh, S., Ma, S., ... Berg, A. C. (2015). Imagenet large scale visual recognition challenge. *International Journal of Computer Vision, 115*(3), 211–252.

Santoro, A., Raposo, D., Barrett, D. G., Malinowski, M., Pascanu, R., Battaglia, P., & Lillicrap, T. (2017). A simple neural network module for relational reasoning. *Advances in Neural Information Processing Systems, 30*, 4967–4976.

Scellier, B., & Bengio, Y. (2016). Towards a biologically plausible backprop. *arXiv*. Retrieved from 1602.05179, 914.

Simonyan, K., & Zisserman, A. (2015). Very deep convolutional networks for large-scale image recognition. *arXiv*. Retrieved from 1409.1556.

Spoerer, C. J., McClure, P., & Kriegeskorte, N. (2017). Recurrent convolutional neural networks: A better model of biological object recognition. *Frontiers in Psychology, 8*, 1551.

Srinivasan, M. V., Laughlin, S. B., & Dubs, A. (1982). Predictive coding: A fresh view of inhibition in the retina. *Proceedings of the Royal Society B: Biological Sciences, 216*(1205), 427–459.

Sussillo, D., Churchland, M. M., Kaufman, M. T., & Shenoy, K. V. (2015). A neural network that finds a naturalistic solution for the production of muscle activity. *Nature Neuroscience, 18*(7), 1025.

Sutton, R. S., & Barto, A. G. (1998). *Introduction to reinforcement learning* (Vol. 135). Cambridge, MA: MIT Press.

Szegedy, C., Zaremba, W., Sutskever, I., Bruna, J., Erhan, D., Goodfellow, I., & Fergus, R. (2013). Intriguing properties of neural networks. *arXiv*. Retrieved from 1312.6199.

Taigman, Y., Yang, M., Ranzato, M. A., & Wolf, L. (2014). Deepface: Closing the gap to human-level performance in face verification. *The IEEE Conference on Computer Vision and Pattern Recognition*, 2014, 1701–1708.

Wallis, T. S., Funke, C. M., Ecker, A. S., Gatys, L. A., Wichmann, F. A., & Bethge, M. (2017). A parametric texture model based on deep convolutional features closely matches texture appearance for humans. *Journal of Vision, 17*(12), 5.

Wang, Z., & Simoncelli, E. P. (2008). Maximum differentiation (MAD) competition: A methodology for comparing computational models of perceptual quantities. *Journal of Vision, 8*(12), 8.

Wen, H., Shi, J., Zhang, Y., Lu, K. H., Cao, J., & Liu, Z. (2017). Neural encoding and decoding with deep learning for dynamic natural vision. *Cerebral Cortex, 28*(12), 4136–4160.

Wenliang, L., & Seitz, A. R. (2018). Deep neural networks for modeling visual perceptual learning. *Journal of Neuroscience, 38*(27), 6028–6044.

Wu, Y., Schuster, M., Chen, Z., Le, Q. V., Norouzi, M., Macherey, W., ... Klingner, J. (2016). Google's neural machine translation system: Bridging the gap between human and machine translation. *arXiv.* Retrieved from 1609.08144.

Xu, K., Ba, J., Kiros, R., Cho, K., Courville, A., Salakhudinov, R., … Bengio, Y. (2015). Show, attend and tell: Neural image caption generation with visual attention. *Proceedings of the 32nd Annual International Conference on Machine Learning, 37,* 2048–2057.

Yamins, D., & DiCarlo, J. J. (2016) Using goal-driven deep learning models to understand sensory cortex. *Nature Neuroscience, 19,* 356–365.

Yang, G. R., Song, H. F., Newsome, W. T., & Wang, X. J. (2017). Clustering and compositionality of task representations in a neural network trained to perform many cognitive tasks. *bioRxiv,* 183632.

Yosinski, J., Clune, J., Nguyen, A., Fuchs, T., & Lipson, H. (2015). Understanding neural networks through deep visualization. *arXiv.* Retrieved from 1506.06579.

Zhou, B., Khosla, A., Lapedriza, A., Oliva, A., & Torralba, A. (2014). Object detectors emerge in deep scene CNNs. *arXiv.* Retrieved from 1412.6856.

Zhuang, C., Wang, Y., Yamins, D., & Hu, X. (2017). Deep learning predicts correlation between a functional signature of higher visual areas and sparse firing of neurons. *Frontiers in Computational Neuroscience, 11,* 100.

60 Connectomes, Generative Models, and Their Implications for Cognition

PETRA E. VÉRTES

ABSTRACT We begin this chapter with a brief overview of how network science has been applied in recent years to advance our understanding of the human brain. We particularly focus on large-scale networks derived from neuroimaging data, which have proven important in describing both healthy brain organization and how it changes with age, cognitive demand, and disease. While a sophisticated set of network measures is now available to describe the architecture of brain networks, the driving forces that shape these networks are only just beginning to be explored. Here we review recent efforts to gain such mechanistic insights using generative modeling of human brain networks, both in health and in disease. We present the detailed process for how a generative model (or set of simple, algorithmic wiring rules) can be designed, fitted to observed data, cross-validated, and further improved. Throughout the chapter we touch on parallel efforts to understand brain organization at the neuronal scale in model organisms, and we highlight both the challenges and opportunities that lie ahead in this fast-moving field.

Networks in Neuroscience

Human brain networks Today the notion that the human brain can be viewed as a network is unlikely to raise any eyebrows. At the macroscopic level, we are used to thinking of white matter tracts connecting different brain regions as highways that connect cities to one another. At the cellular level, we have also grown accustomed to the idea that neurons, connected by gap junctions and synapses, form extensive networks for information processing. However, the sheer number of neurons (10^{11}) and their dense packing still precludes us from observing the individual nodes and edges of this network and from reconstructing the connectome—the wiring diagram of the brain—at the cellular level. While efforts are underway to map out neuronal connectivity in 1 mm^3 of mouse and human cortex (Allen Institute for Brain Science, 2016), it is questionable whether individual cells and synapses are the right scale at which to understand brain organization. Indeed, neighboring neurons tend to have similar characteristics and correlated activity, so it should be possible to coarse grain the network so that nodes represent larger-scale anatomical brain regions as seen, for example, in figure 60.1*A* (Bullmore & Sporns, 2009; Van Essen et al., 2012).

The best way to define these coarse-grained regions is still an active field of research (Glasser et al., 2016). Nevertheless, once the brain has been parcellated into tens or hundreds of regions of interest (ROIs), we can use diffusion tensor imaging (DTI) tractography to measure the white matter tracts connecting these nodes in a large-scale whole-brain network (see figure 60.1*B*). In addition, functional magnetic resonance imaging (fMRI) data can be used to construct so-called functional networks. Here, too, each node corresponds to a coarse-grained brain region, and fMRI is used to measure a blood-oxygen-level-dependent (BOLD) time series—as a proxy of brain activity—in each region (figure 60.1*C*). The brain functional network is then constructed by connecting pairs of nodes whose activity is highly correlated (Bullmore & Sporns, 2009).[1]

Network metrics and why they are useful in neuroimaging The organization and structure of brain networks can be studied using a range of tools and quantitative metrics from network science (see figure 60.1*E–H*). A key insight that has resulted from this approach is that human brain networks share a number of organizational features with many other real-world complex networks, from social networks to transportation networks and the Internet (Bullmore & Sporns, 2009).

---

[1] More recently, other types of data have also been leveraged to construct brain networks. For example, transcriptomic brain networks (TBN) can be constructed by connecting pairs of nodes with correlated gene expression patterns (Romero-Garcia et al., 2017).

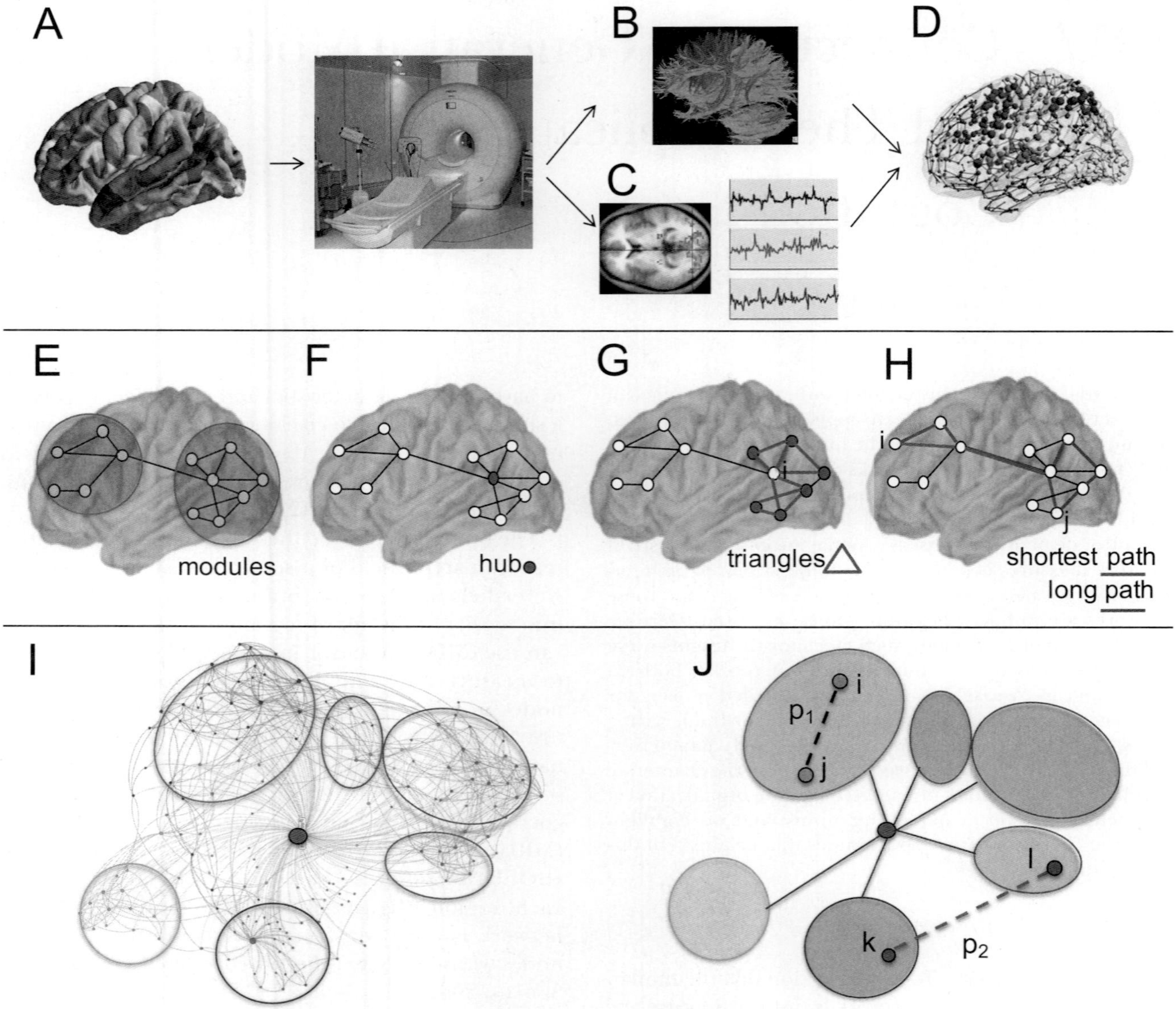

FIGURE 60.1 *Panel 1*, Schematic process for constructing human brain networks. The brain is parcellated into a few hundred anatomical regions (*A*), which form the nodes of the network (*D*). Network edges correspond to anatomical connections measured via DTI (*B*) or functional correlations measured via fMRI (*C*). *Panel 2*, Network metrics commonly used to quantify the topology of networks. (*E*) Modularity: many complex networks have a modular community structure, whereby they contain subsets of highly interconnected nodes called *modules*. (*F*) Degree: the degree of a node is the number of links or connections it has to the rest of the network. In many real-world networks, most nodes have only a few connections, but a few so-called hub nodes are extensively connected. (*G*) Clustering: the clustering coefficient measures the number of connections that exist between the first neighbors of a node i as a proportion of the maximum possible number of such connections. These edges form triangular connections around the node i. (*H*) Path length and efficiency: The minimum path length between two nodes i and j is simply the minimum number of edges that need to be traversed to get from one node to another. Many of these features are highlighted in the brain functional network shown in (*D*). Node size is proportional to degree, and the modular organization is highlighted in color. *Panel 3*, Network diagram of an individual's LinkedIn connections (*I*). The central individual is shown in red, and the modules of the network are highlighted in color. A simplified model of the network is shown in (*J*). Panel 2 was adapted from Vértes and Bullmore (2015), with permission. (See color plate 72.)

For example, all these systems exhibit modules, or clusters of nodes that are more densely connected to one another than to the rest of the network (figure 60.1*E*). This separability of systems into relatively autonomous subunits (such as departments within a company) has well-known operational advantages (Simon, 1962). In a transportation network, modules might represent continents or countries, since international transport connections tend to be sparser than national ones. In the brain, the modules are also anatomically grouped (figure 60.1*D*) and functionally relevant (Meunier, Lambiotte, & Bullmore, 2010).

Another example of common organizational structure is the existence of highly connected nodes, or hubs (figure 60.1*F*). In air transportation networks, for example, these hubs would correspond to large international airports, such as Heathrow, with many incoming and outgoing flights both domestic and international. In the brain, too, key hubs are characterized by high levels of connectivity both locally and globally across the network. Furthermore, hubs tend to have key functional importance (integrating information from various parts of the brain) but are also points of vulnerability, as attacks targeting these nodes will rapidly compromise the system as a whole (van den Heuvel & Sporns, 2013).

A further property shared by human brain networks and many other complex networks is the *small-world* phenomenon (Watts & Strogatz, 1998), whereby the network simultaneously has high clustering (figure 60.1*G*) and short path lengths (figure 60.1*H*). Formally, the clustering coefficient and related network metrics count the number of triangles in the network to quantify the propensity of two nodes connected to one another to share common nearest neighbors. Social networks, for example, are highly clustered because a pair of friends are likely to share a social circle, with many common friends. Despite this clustered connectivity, social networks are also famously easy to traverse—with just six steps needed, on average, to connect any two individuals in the world, according to Milgram's experiment (Milgram, 1967). In network terms this means social networks have a short average path length, or high global efficiency (Latora & Marchiori, 2001).[2]

One obvious benefit of network approaches to understanding brain organization is that they allow us to draw on intuitions from other complex networks, such as social or transportation networks, which are perhaps easier to relate to. An intriguing idea emerging from this interdisciplinary thinking is that diverse complex systems may share broad organizational features because they have been optimized to perform similar tasks (e.g., the transmission of information or traffic) under similar constraints (e.g., minimizing connection costs). This point of view would predict that network features (such as modularity, node degree, clustering, and path length) are carefully tuned in the healthy brain network, according to a trade-off between cost and function (Bullmore & Sporns, 2012).

Consistent with this prediction, human brain networks tend to have predominantly short connections since, as in all spatially embedded networks, the energetic cost of maintaining and forming connections in the brain is proportional to their length (Barthélemy, 2011). In addition, network properties that necessitate long, high-cost connections are expected to underpin high-value adaptive functions (Bullmore & Sporns, 2012). For example, network efficiency was found to correlate with IQ across individuals, both in structural and functional brain networks (Li et al., 2009; van den Heuvel, Stam, Kahn, & Hulshoff Pol, 2009). Similarly, a growing body of literature describes how these network properties are affected in a range of brain disorders, how they are sensitive to pharmacological and other interventions, and how individual brain network architecture may be predictive of treatment response (Bassett, Khambhati, & Grafton, 2017; Fornito, Zalesky, & Breakspear, 2015; Morgan, White, Bullmore, & Vértes, 2018; and references within).

Connectomes in nonhuman animals The network formalism allows comparisons not just to other kinds of complex systems but to other kinds of nervous systems as well. Here, too, we find that brain networks across different scales and species all share general organizational properties such as those highlighted in figure 60.1*E–H*.

To date, the nematode *Caenorhabditis elegans* (comprising around 300 neurons) and the larva of the sea squirt *Ciona intestinalis* (with about 180 neurons) are the only organisms for which a nearly complete wiring diagram at the cellular level is available (Ryan, Lu, & Meinertzhagen, 2016; Varshney, Chen, Paniagua, Hall, & Chklovskii, 2011; White, Southgate, Thomson, & Brenner, 1986). Since the first version of the *C. elegans* connectome has been available for several decades, *C. elegans* has proved a particularly rich test bed for network approaches in complex systems (Chen, Hall, & Chklovskii, 2006; Cherniak, 1994; Kaiser & Hilgetag, 2006; Latora & Marchiori, 2001; Milo et al., 2002, 2004; Nicosia, Vértes, Schafer, Latora, & Bullmore, 2013; Sporns & Kötter, 2004; Towlson, Vértes, Ahnert, Schafer, & Bullmore, 2013; Varshney et al., 2011; Watts & Strogatz, 1998; Yan et al., 2017).

Recent years have seen an unprecedented rate of technological progress in mapping both anatomical

[2] The network metric of global efficiency simply refers to the average of all inverse path lengths in the network.

and functional connectivity in model organisms, including *C. elegans* (Nguyen et al., 2016; Xu et al., 2013), *Drosophila* (Berck et al., 2016; Chiang et al., 2011; Eichler et al., 2017), zebra fish (Naumann et al., 2016), mice (Kashturi et al., 2015; Oh et al., 2014), and monkeys (Felleman & Van Essen, 1991; Stephan, 2013). It is expected, for example, that the complete cellular-scale connectome of the *Drosophila* larva (comprising about 12,000 neurons) will be mapped within the next few years. The increasing speed and ease with which connectomes can now be reconstructed is also paving the way for connectomic studies at the individual level: understanding how circuit structure changes over development, how it relates to behavior, and how it varies between individuals in health and disease.

Generative Models

Why model networks? A toy example In the context of network science, a generative model is a set of rules or procedures that can be used to generate a network with a required set of characteristics. The simplest network model is the so-called Erdös-Rényi (ER) model, where the wiring rule is simply that each node will connect to every other node with a fixed probability p (Erdős & Rényi, 1959). The resulting networks are also called *random networks* because their connections display no structure apart from a characteristic network density that is tuned by the parameter p.

Real-world networks, on the other hand, tend to display some very obvious structure—for example, a network of social interactions will tend to have modules, as shown in figure 60.1*I*, which displays the network of an individual's professional relationships recorded via LinkedIn. Modules are just one example of a property that can never be replicated by an ER random network, and a number of more complex generative models have therefore been developed to capture the existence of a range of structural features. In addition to achieving the desired structure, generative models are also often designed to reflect reasonable assumptions about the forces shaping the observed network properties. For example, one might hypothesize that the modules in the LinkedIn network in figure 60.1*I* correspond to the different institutions or companies at which the central individual has worked or studied. A simple model might therefore define two different connection probabilities: p_1 for connecting two nodes within the same company and a lower probability p_2 for connecting nodes in different companies. This is still an overly simplistic model of the empirically observed network; however, it has the potential to be much more realistic than an ER model because it will, by definition, have modules.

This toy example also illustrates the key reasons why network models can be useful. Looking at figure 60.1, it is immediately obvious that the amount of information required to fully describe the model network in figure 60.1*J* (one company or module label per node and the probabilities p_1 and p_2) is much smaller than that required to capture every detail of the original network in figure 60.1*I* (a list of all edges in the network)—so network models provide a way of compressing information. Also, network models enable generalization from observed to unobserved data. For example, we can generalize from the past to the future and predict where links are likely to appear next in a system. Similarly, we can predict the likelihood of missing links in incomplete data. In figure 60.1*J*, we know based on the simple toy model that the link from node i to node j has a probability p_1, while that from node k to node l has a smaller probability p_2. Finally, generative models can potentially give us insights into the mechanisms and forces shaping the real-world networks they model—how people choose their social contacts, or how neurons choose synaptic partners, for example. In nervous systems, identifying the mechanisms that govern the formation of connections will likely have profound implications for our understanding and treatment of neurodevelopmental and neurodegenerative disorders. It is crucial to note, however, that multiple generative models can lead to identical network features, so even very accurately fitting models must be interpreted cautiously.

How to model a network? Canonical examples As already sketched out in the toy example above, modeling a network typically involves two key steps. First, one must identify a set of *stylized facts* that characterize the structure within the network that one is interested in modeling (e.g., the modules observed in the social network in figure 60.1*I*). Second, one must propose a plausible but parsimonious mechanism that could lead to the emergence of this structure (e.g., preferential connection formation between individuals in the same company).

In practice, both the validity and the utility of a network model will depend on the investigator's flair in selecting appropriate stylized facts and in proposing a mechanism that strikes an appropriate balance between parsimony and plausibility. For example, when selecting a stylized fact to model, it is important to consider whether this apparent fact is (1) real (statistically robust), (2) interesting (linked to a scientific question), and (3) surprising (not a corollary arising naturally from a simpler model). In what follows, we will revisit each of these points in the context of a few canonical examples of stylized facts and associated network models.

Network density and the Erdös-Rényi model One of the most basic properties of a network is the density of connections. The ER model was designed to capture only this very basic characteristic, ignoring all other structures (Erdős & Rényi, 1959). In this model the network has n nodes, and edges are added at random with a fixed probability p of including any given edge, independent of all other edges. The probability p is the model parameter, and it can be tuned such that the model generates networks with a connection density that matches the empirically observed density. The model generates synthetic networks with a binomial degree distribution, such that on average each node will have $(n-1)p$ connections.

Network hubs and the Barabási-Albert model New network models are often born from a discrepancy between the basic properties of an observed network and the synthetic networks generated by simpler, preexisting models. For example, many real-world networks have a heavy-tailed degree distribution, where most nodes have low degree (just a few connections), but certain so-called hub nodes have a very high degree of connectivity (figure 60.1*F*). In 1999 Barabási and colleagues proposed that the degree distribution for many real-world networks follows a power law (Barabási & Albert, 1999). This observation was both surprising (given its sharp deviation from the binomial distribution expected for an ER network) and interesting (given its functional relevance). For example, it is intuitively obvious that the spread of rumors or disease will be heavily influenced by the underlying degree distribution and that hub nodes will have a disproportionate influence on spreading processes. Starting from the power-law degree distribution as the relevant stylized fact, Barabási and colleagues then proposed a simple model (the Barabási-Albert, or BA, model), which was able to produce networks with this property. The key aspects of this model include:

- Growth: nodes are added one by one to the network, and each new node makes connections with preexisting nodes only.
- Preferential attachment: a new node will preferentially attach to preexisting nodes with high degree such that the probability of a new node connecting to an existing node is proportional to k_i, the degree of node i.

This generates synthetic networks with a power-law degree distribution, where the probability of nodes having degree k varies as $P(k) \sim k^{-3}$

Subsequent studies noted that many real-world networks with heavy-tailed degree distributions (i.e., where hubs exist) are not in fact well fitted by this power-law function (Clauset, Shalizi, & Newman, 2009). This is a useful reminder that in addition to being surprising and interesting, the stylized facts of interest also need to be statistically robust. Nevertheless, the BA model provides a parsimonious and plausible explanation for the emergence of heavy-tailed distributions, and in many cases, minor adjustments to the model can capture deviations from the ideal power-law form of the degree distribution (Amaral, Scala, Barthélémy, & Stanley, 2000).

Small-worldness and the Watts-Strogatz model In 1998 Watts and Strogatz published a seminal paper noting that many complex networks exhibit the small-world property of simultaneously having high clustering and short path lengths (Watts & Strogatz, 1998; figure 60.1*H*). In this sense, real-world networks seem to occupy a middle ground between regular lattices (with high clustering and long path lengths) and random networks (with low clustering and short path lengths). The Watts-Strogatz (WS) model therefore starts with a regular lattice and randomly rewires a small proportion β of edges. For intermediate values of the parameter β, the model generates so-called small-world networks. Here, too, the seminal nature of this work can be traced to the choice of stylized fact to model: small-worldness is a widespread, statistically robust, surprising (not predicted by previous models), and interesting (functionally relevant) property. As Watts and Strogatz (1998) point out: "Models of dynamical systems with small-world coupling display enhanced signal-propagation speed, computational power, and synchronizability. In particular, infectious diseases spread more easily in small-world networks than in regular lattices."

Modularity and related network models Modules are also a ubiquitous feature of real-world networks, and their key functional relevance has been understood since Herbert Simon's seminal work in 1962 (Kashtan & Alon, 2005; Simon, 1962). Despite this, modular organization will not spontaneously appear from simple network models such as those described above. In general, modules tend to arise from the preferential linking of nodes with certain shared intrinsic characteristics, so most generative models yielding modular networks require the inclusion of some nodal attributes. For example, in the toy example of a LinkedIn network of professional contacts (figure 60.1*J*), we assumed knowledge of each individual's place of work to define the connection probabilities p_1 and p_2. In stochastic block models (SMB), the relevant nodal characteristics are latent attributes learned from the observed data, along with other parameters of the model such as the matrix of connection probabilities within and between all inferred modules (Snijders & Nowicki, 1997). In other cases the models may use certain nodal characteristics as given and define wiring rules that are dependent on them. One prominent example of this are spatially embedded models, which use the spatial location of each node as an input (Kaiser & Hilgetag, 2006; 2007).

Short connection lengths and spatially embedded network models Many real-world networks such as power grids, road networks, or neural networks are embedded in physical space, and spatial considerations will severely constrain the type of connections that will evolve and be sustained in these networks (Barthélemy, 2011; Klimm, Bassett, Carlson, & Mucha, 2014). Typically, these networks tend to have a predominance of short connections, and a number of generative models have been designed to replicate this observation. Many of these models also lead to other emergent features. These will be discussed more in depth in the next section.

Generative Models in Network Neuroscience

Modeling human brain networks In the previous section, we discussed a number of generative models that were designed to *qualitatively* reproduce an individual network feature, or stylized fact, commonly observed in real-world data. Now we look at how these concepts can be applied to develop generative models that provide *quantitative* fits to a number of network features observed in human brain networks. In such applied settings, the process for designing generative models can be summarized in the following five-step recipe:

1. Make a list of stylized facts observed in the data (making sure they are statistically robust, surprising, and interesting).
2. Propose a plausible but parsimonious mechanism that could lead to the emergence of the stylized facts identified in step 1.
3. Fit the parameters of the model and compare to other models.
4. Validate the model on independent data.
5. Think of what the model doesn't capture.

In what follows we look at each of these steps in turn:

Step 1: Identify the stylized facts of interest in brain networks As described previously, brain networks have hub nodes and a heavy-tail degree distribution, exhibit high modularity, and are small-world networks with high clustering and high efficiency (Bullmore & Sporns, 2009). The brain is also a spatial network, and as such the distribution of connection lengths is heavily skewed toward short connections (Bullmore & Sporns, 2012). In the previous section, we showed that individually, most of the above network properties can be reproduced by simple generative models. Here we seek a simple model that can capture all these properties simultaneously, based on just one or two terms—and therefore just one or two free parameters.

Step 2: Propose a plausible but parsimonious mechanism In practice the choice of terms in the model will be guided by two principles:

- Parsimony: including terms that are as independent of each other as possible is likely to lead to a more parsimonious model.
- Plausibility: including terms that reflect a plausible biological mechanism (such as the minimization of distance/energetic costs or genetic constraints) will lead to more interpretable and more useful models.

Ever since Cajal's work in the 19th century, it has been acknowledged that nervous systems are wired in an economical way that does not waste material or energy on unnecessarily long connections. It therefore seems natural to start with a model implementing this simple spatial constraint first, before examining the other network properties listed above.

In 2004 Kaiser and Hilgetag published an economical network model in which:

- new nodes are added at randomly chosen positions in two-dimensional space,
- connections between a new node i and each preexisting node j are formed with a probability that decreases exponentially with the distance between the two nodes: $P_{ij} = \beta e^{-\alpha d_{ij}}$, and
- new nodes that don't establish connections with any of the preexisting nodes are removed from the system.

As the authors pointed out, the distance penalty in this model is both motivated from an evolutionary perspective, and could easily be implemented biologically—for example, through gradients of guidance molecules, which decay exponentially with distance (Kaiser & Hilgetag, 2004). Despite its simplicity, the model is able to capture not just the predominance of short-range connections but also the high clustering observed empirically in brain networks. The model is also quite flexible, and by tuning the model parameters, the authors were able to generate synthetic networks that approximated the German highway system (for very small β) or the cortical connectivity in the cat and macaque brain (for β approaching 1).

In 2012 Vértes and colleagues revisited this model in the context of human brain functional networks. They found that the model fit could be further improved by including a second term that would promote the formation of hubs and help connect up hubs even if they are far apart (much like large international airports are connected in air transportation networks). In this model the probability of nodes i and j connecting to each other is $P_{ij} \propto d_{ij}^{-\eta} k_{ij}^{\gamma}$, where d_{ij} is the distance between nodes i and j, k_{ij} is the number of common neighbors between nodes i and j, and η and γ are the two parameters of the model. The first term in the model is a distance penalty based on the cost of maintaining long-range connections, and the second term is a bias, reminiscent of Hebbian learning rules, toward clustering or linking regions with a large amount of shared input (Vértes, Alexander-Bloch, & Bullmore, 2014; Vértes et al., 2012). As shown in figure 60.2*A*, this model was able to simultaneously reproduce the degree distribution and distance distribution of connections in human brain functional networks, as well as their observed level of clustering, modularity, and efficiency. Subsequently, the same model was also shown to provide a good fit to anatomical brain networks (Betzel, Avena-Koenigsberger, et al., 2016).

Step 3: Fit the model to the data and compare to other models Fitting the model to data entails tuning the model parameters in order to minimize a so-called objective function, which quantifies the difference between the empirically observed networks and those generated by the model. One popular approach is to estimate model parameters based on maximizing the likelihood that the precise arrangement of edges in the

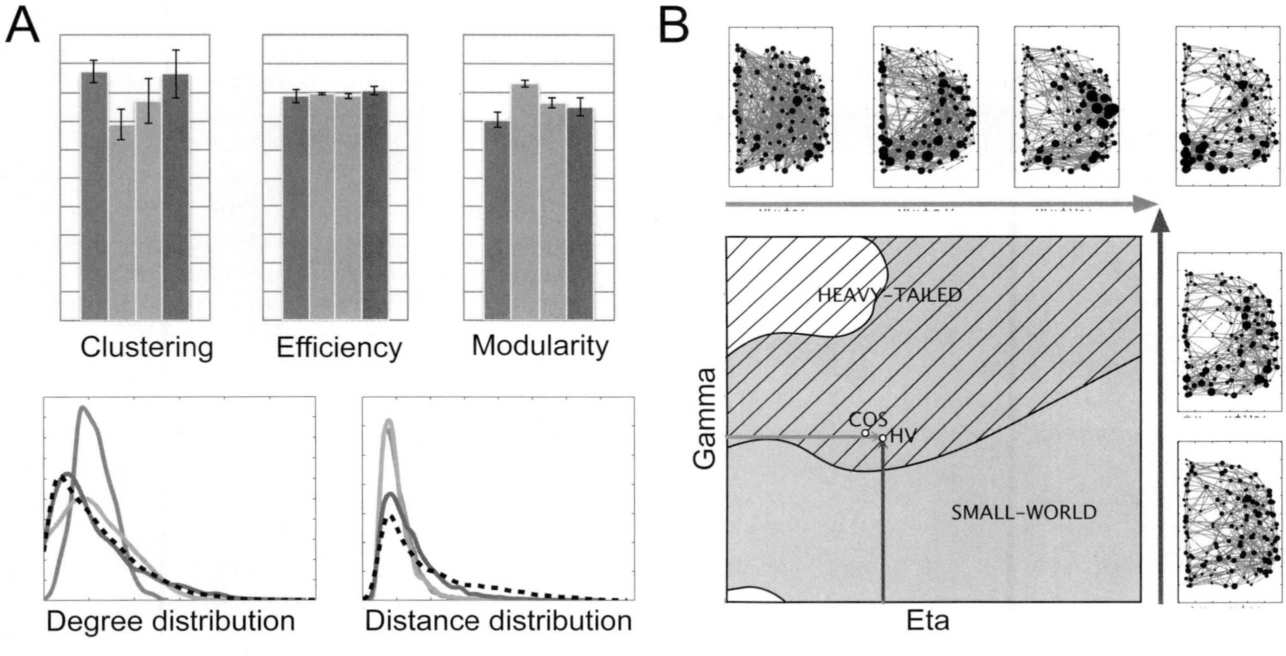

FIGURE 60.2 (*A*) Comparison of networks simulated by three generative models versus brain functional networks derived from experimental fMRI data (*blue*). Both the exponential distance penalty model (*green*, $P_{ij} \propto e^{-\alpha d_{ij}}$) and the economical preferential attachment model (*orange*, $P_{ij} \propto d_{ij}^{-\eta}(k_i k_j)^{\gamma}$) fail to simultaneously capture several topological characteristics of functional brain networks. In contrast, the economical-clustering model (*red*, $P_{ij} \propto d_{ij}^{-\eta} k_{ij}^{\gamma}$) yields significantly more realistic networks by all of the following measures: normalized clustering coefficient, global efficiency, modularity, degree distribution, and distance distribution (*data shown in dashed black lines*). (*B*) Phase diagram of the economical-clustering model. Most values of eta (η) and gamma (γ) yield small-world networks (*gray area*), whereas only high values of γ yield networks with heavy-tailed (skew >1) degree distribution (*hashed area*). The model parameters estimated to minimize mismatch between simulated and experimental fMRI data sets are shown here for both healthy volunteers (HV) and participants with childhood onset schizophrenia (COS). The orange (and purple) arrows show sections through the phase space, varying only η (or γ), respectively, whereas the other parameter is held at its optimal value estimated in healthy volunteers. Schematics of the networks obtained at various points along these sections are also shown (axial view of right hemisphere only). Adapted from Vértes et al. (2012), with permission. (See color plate 73.)

observed network could have been generated by the model in question. However, we have seen that in many cases the functional similarity between two networks is not well captured by the number of overlapping connections. For example, we have seen that in the WS model a small number of rewired edges can produce a dramatic functional difference in terms of the ease with which information might spread on the network. Conversely, two networks that have a large number of differences in the placement of individual connections may still perform quite similarly from a functional point of view. Therefore, it often makes more sense to design an objective function that tries to match the observed networks in terms of the stylized facts chosen in step 1 since these were selected precisely for their anticipated functional relevance. For example, in Vértes et al. (2012) and Betzel, Avena-Koenigsberger, et al. (2016), the objective function is based on the difference between observed and model data in a number of network features, such as clustering, efficiency, modularity, and degree distribution. Simulated annealing or Monte Carlo methods can then be used to find the parameter setting that minimizes this objective function. Importantly, once the parameters are fitted, the same objective function can also be used to compare model fit across different kinds of models. This is crucial because the principle of parsimony requires us to compare any new model to a set of null models—verifying whether the same network features could be explained more simply. The simplest null model is the ER random network, but it is in many ways also a straw man because we already know that it cannot reproduce many observed network features. Whenever a new and more complex model is designed, it therefore makes sense to compare it to the previous best models.

Step 4: Validate the model on independent data When fitting a model to a data set, it is always possible to refine the model with additional parameters to provide a better fit. However, in general we are not interested in producing a perfect fit of the observed network itself (e.g., an individual set of brain networks) but rather in understanding the wiring rules for a class of similar networks (e.g., brain networks in the population at large). It is therefore important to cross-validate the model against independent data. In its simplest form, this requires using an equivalent but independent data set to the one used to fit the model and demonstrating that the same model (with the same parameter settings) still accurately captures these new data. This suggests that the model is not overfitting the original data.

Another approach to help validate the model's general applicability is to test whether the resulting synthetic networks also recapitulate other observed network properties they were not explicitly constrained to possess (Betzel & Bassett, 2017; Vértes et al., 2012).

Finally, it is interesting to explore whether the same model with slightly modified parameters can capture a related family of observations. For example, in Vértes et al. (2012) the authors found that starting from the model for healthy brain networks, slightly detuned parameters could reproduce the pattern of network changes observed in people with schizophrenia (see figure 60.2*B*). Similarly, in Betzel, Avena-Koenigsberger, et al. (2016) the authors fit the model to individual subjects aged 7–85 years and found that increasing age resulted in a gradual shift in parameter toward a weaker distance penalty, as well as a poorer model fit. These kinds of analyses not only help validate the model but also highlight its potential usefulness in understanding individual differences in cognition. Indeed, the stochastic nature of the model allows for individual differences between model instantiations, but additionally, small differences in model parameters could also explain more systematic brain differences between distinct populations (see the section on the implications for cognitive neuroscience).

Step 5: Think of what the model does not capture In the process of fitting and validating the model, it is easy to come across additional network features that the model does not yet adequately capture. These can be seen as additional stylized facts that can be added to the list in step 1, leading to increasingly sophisticated models (Klimm, Bassett, Carlson, & Mucha, 2014).

One key approach to designing new models is to allow for the inclusion of additional domain-specific knowledge, which could help explain more complex or more detailed features of the network (additional stylized facts). For example, the simple models above were designed to capture cortical connectivity within a single hemisphere only, and it is widely accepted that connectivity between hemispheres or in the subcortex and cerebellum may follow different wiring rules. In the next section, we will see that animal models provide a unique opportunity to develop and test increasingly realistic models.

Modeling brain networks in other organisms In addition to a complete wiring diagram, in *C. elegans* we also have access to detailed information about individual nodes of the network. For example, the birth time of individual neurons is known. This allows a shift from generative models (which aim to reproduce a set of network features measured in a given connectome) to growth models (which aim to model the way in which network features emerge over time as the nervous system develops). For example, in Nicosia et al. (2013) the authors sought to reproduce the curve describing how the number of connections in the network grows as neurons are born one by one. They found that a simple model incorporating (1) a distance penalty and (2) a bias for hub nodes to attract new connections is able to reproduce an otherwise surprising, abrupt transition from exponential to linear growth in the number of connections. Crucially, they showed that the success of this model depends on incorporating information on how node locations change over time as the worm elongates over the course of development. Other examples including additional biological information are models of the *Xenopus* tadpole nervous system, which have incorporated information on axon and dendrite geography (Li et al., 2007), neuron type (Sautois, Soffe, Li, & Roberts, 2007), and developmental factors such as chemical gradients and physical constraints (Roberts et al., 2014). Interestingly, this additional information led to a model detailed enough to reproduce observed swimming behaviors. For larger-scale connectomes, the mouse, cat, and macaque have all been used as model organisms to demonstrate the importance of cytoarchitectural features in determining connectivity, with cytoarchitecturally similar regions being more likely to connect to one another (Beul, Barbas, & Hilgetag, 2017; Beul, Grant, & Hilgetag, 2015; Goulas, Uylings, & Hilgetag, 2017).

Emerging research directions: new directions born from better data As in the case of model organisms, the inclusion of additional or more complex data is likely to drive more complete models of human brain networks, with increasing relevance to cognitive function. While the bulk of network neuroscience literature has focused on static brain networks based on a single data modality, it is widely acknowledged that brain networks change over time (with development and ageing) and can also be viewed as multilayer networks (Battiston, Nicosia, Chavez, & Latora, 2017; Bentley et al., 2016), with different types of connections defining distinct network layers (e.g., anatomical connectivity, functional connectivity, gene coexpression similarity). Recent work has begun to build on these additional aspects of brain networks. For example, the inclusion of developmental data enables growth modeling of human brain networks (Betzel, Avena-Koenigsberger, et al., 2016; Betzel & Bassett, 2017; Tang et al., 2017).

New directions born from advances in network science As network science develops, it is likely that new kinds of network features will be found to play a key role in network function and will therefore drive the development of new generative models. For example, recent interest in the use of algebraic topology to quantify non-pairwise relationships between nodes has driven the development of generative models for simplicial complexes[3] (Courtney & Bianconi, 2017) and some preliminary work in applying these tools to network neuroscience (Giusti, Ghrist, & Bassett, 2016; Giusti, Pastalkova, Curto, & Itskov, 2015).

Another recent development has been the application of control theory to illuminate the functional role of diverse nodes in brain networks (Betzel, Gu, et al., 2016; Gu et al., 2015; Tang & Bassett, 2017; Yan et al., 2017; and references therein). This work is based on the key assumption that the brain and nervous system are optimized to solve a control problem, enabling sensory inputs to control particular outputs across the body based on the dynamics of the nervous system, which acts as the control network. This suggests that control theoretical considerations may be key to determining the brain's wiring and vice versa. For instance, Yan et al. (2017) used network control principles to elucidate the role of specific neurons in *C. elegans* locomotor behavior. In Tang et al. (2017), the authors designed a growth model initiated from an observed human brain network where edges were progressively rewired according to a trade-off between average and modal controllability. The range of networks generated over the course of this rewiring procedure were strikingly similar to developmental data in a large data set of 882 children and adolescents aged 8–22 years.

Implications for Cognitive Neuroscience

Until recently, network approaches to understanding neuroimaging and other neuroscientific data have been mostly descriptive, with growing numbers of network metrics being used to characterize how networks change with particular behavioral traits, with age, or with disease. The great promise of generative models is to move beyond description, toward a more mechanistic understanding of the driving forces that shape brain networks in health and disease.

However, as discussed in a previous section, multiple generative models can lead to very similar networks. In other words, even if we design a growth model that accurately matches the development of the nervous system, the model may still fail to represent the true biological mechanisms shaping the network. It is therefore important to select model terms that plausibly embody specific developmental pressures.

[3] Simplicial complexes are composed of simplices where a 0-simplex is a node, a 1-simplex is a dyad, a 2-simplex is a face, a 3-simplex is a tetrahedron, etc.

In such cases, where the parameters of the model are biologically interpretable, it becomes possible to correlate these parameters with behavioral features or to map out how they are affected by age or within disease groups (Betzel, Avena-Koenigsberger, 2016; Betzel & Bassett, 2017; Vértes et al., 2012). As we begin to model both typical and atypical brain development, the developmental trajectories traced out in parameter space may also suggest early interventions to steer network growth away from undesirable states. In other words, network descriptions of human brain organization will need to be coupled with network models of how it is disrupted in disease. In turn, these models have the potential to yield a new generation of network-based biomarkers and therapeutic approaches to mental ill health.

Acknowledgment

Petra E. Vértes is supported by the Medical Research Council (grant no. MR/K020706/1), is a fellow of MQ: Transforming Mental Health (grant no. MQF17_24), and a fellow of the Alan Turing Institute funded under EPSRC (EP/N510129/1).

REFERENCES

Allen Institute for Brain Science. (2016). IARPA awards $18.7 million contract to Allen Institute for Brain Science, as part of project with Baylor College of Medicine and Princeton University, to reconstruct neuronal connections. Retrieved March 27, 2018, from http://www.alleninstitute.org/what-we-do/brain-science/news-press/press-releases/iarpa-awards-187-million-contract-allen-institute-brain-science-part-project-baylor-college-medicine.

Amaral, L. A. N., Scala, A., Barthélémy, M., & Stanley, H. E. (2000). Classes of small-world networks. *Proceedings of the National Academy of Sciences of the United States of America, 97*(21), 11149–11152. doi:10.1073/pnas.200327197

Barabasi, A.-L., & Albert, R. (1999). Emergence of scaling in random networks. *Science, 286*(5439), 509–512.

Barthélemy, M. (2011). Spatial networks. *Physics Reports, 499*, 1–101. doi:10.1016/j.physrep.2010.11.002

Bassett, D. S., Khambhati, A. N., & Grafton, S. T. (2017). Emerging frontiers of neuroengineering: A network science of brain connectivity. *Annual Review of Biomedical Engineering, 19*, 327–352. doi:10.1146/annurev-bioeng-071516-044511

Battiston, F., Nicosia, V., Chavez, M., & Latora, V. (2017). Multilayer motif analysis of brain networks. *Chaos: An Interdisciplinary Journal of Nonlinear Science, 27*, 047404. doi:10.1063/1.4979282

Bentley, B., Branicky, R., Barnes, C. L., Chew, Y. L., Yemini, E., Bullmore, E. T., Vértes, P. E., & Schafer, W. R. (2016). The multilayer connectome of *Caenorhabditis elegans*. *PLoS Computational Biology, 12*, e1005283. doi:10.1371/ journal.pcbi.1005283

Berck, M. E., Khandelwal, A., Claus, L., Hernandez-Nunez, L., Si, G., Tabone, C. J., Li, F., Truman, J. W., Fetter, R. D., Louis, M., Samuel, A. D., & Cardona, A. (2016). The wiring diagram of a glomerular olfactory system. *eLife, 5*, e14859. doi:10.7554/eLife.14859

Betzel, R. F., Avena-Koenigsberger, A., Goni, J., He, Y., de Reus, M. A., Griffa, A., Vértes, P. E., Misic, B., Thiran, J. P., Hagmann, P., van den Heuvel, M., Zuo, X. N., Bullmore, E. T., & Sporns, O. (2016). Generative models of the human connectome. *NeuroImage, 124*, 1054.

Betzel, R. F., & Bassett, D. S. (2017). Generative models for network neuroscience: Prospects and promise. *Journal of the Royal Society Interface, 14*, 20170623. doi:10.1098/rsif.2017.0623

Betzel, R. F., Gu, S., Medaglia, J. D., Pasqualetti, F., & Bassett, D. S. (2016). Optimally controlling the human connectome: The role of network topology. *Scientific Reports, 6*, 30770. doi:10.1038/srep30770

Beul, S. F., Barbas, H., & Hilgetag, C. C. (2017). A predictive structural model of the primate connectome. *Scientific Reports, 7*, 43176. doi:10.1038/srep43176

Beul, S. F., Grant, S., & Hilgetag, C. C. (2015). A predictive model of the cat cortical connectome based on cytoarchitecture and distance. *Brain Structure and Function, 220*, 3167–3184. doi:10.1007/s00429-014-0849-y

Bullmore, E. T., & Sporns, O. (2009). Complex brain networks: Graph theoretical analysis of structural and functional systems. *Nature Reviews Neuroscience, 10*, 186–198. doi:10.1038/nrn2618

Bullmore, E. T., & Sporns, O. (2012). The economy of brain networks. *Nature Reviews Neuroscience, 13*, 336–349. doi:10.1038/nrn3214

Chen, B. L., Hall, D. H., & Chklovskii, D. B. (2006). Wiring optimization can relate neuronal structure and function. *Proceedings of the National Academy of Sciences of the United States of America, 103*(12), 4723–4728. doi:10.1073/pnas.0506806103

Cherniak, C. (1994). Component placement optimization in the brain. *Journal of Neuroscience, 14*(4), 2418–2427.

Chiang, A. S., Lin, C. Y., Chuang, C. C., Chang, H. M., Hsieh, C. H., Yeh, C. W., Shih, C. T., et al. (2011). Three-dimensional reconstruction of brain-wide wiring networks in *Drosophila* at single-cell resolution. *Current Biology, 21*(1), 1–11.

Clauset, A., Shalizi, C. R., & Newman, M. E. J. (2009). Power-law distributions in empirical data. *SIAM Review, 51*, 661–703.

Courtney, O. T., & Bianconi, G. (2017). Weighted growing simplicial complexes. *Physical Review E, 95*, 062301. doi:10.1103/PhysRevE.95.062301

Eichler, K., Li, F., Litwin-Kumar, A., Park, Y., Andrade, I., Schneider-Mizell, C. M., Saumweber, T., et al. (2017). The complete connectome of a learning and memory centre in an insect brain. *Nature, 548*(7666), 175–182. doi:10.1038/nature23455

Erdős, P., & Rényi, A. (1959). On random graphs. I. *Publicationes Mathematicae, 6*, 290–297.

Felleman, D. J., & Van Essen, D. C. (1991). Distributed hierarchical processing in the primate cerebral cortex. *Cerebral Cortex, 1*(1), 1–47.

Fornito, A., Zalesky, A., & Breakspear, M. (2015). The connectomics of brain disorders. *Nature Reviews Neuroscience, 16*(3), 159–172. doi:10.1038/nrn3901

Giusti, C., Ghrist, R., & Bassett, D. S. (2016). Two's company, three (or more) is a simplex: Algebraic-topological tools for understanding higher-order structure in neural data.

Journal of Computational Neuroscience, 41, 1–14. doi:10.1007/s10827-016-0608-6

Giusti, C., Pastalkova, E., Curto, C., & Itskov, V. (2015). Clique topology reveals intrinsic geometric structure in neural correlations. *Proceedings of the National Academy of Sciences of the United States of America, 112*(13), 455–460. doi:10.1073/pnas.1506407112

Glasser, M. F., Coalson, T. S., Robinson, E. C., Hacker, C. D., Harwell, J., Yacoub, E., Ugurbil, K., et al. (2016). A multimodal parcellation of human cerebral cortex. *Nature, 536*, 171–178. doi:10.1038/nature18933

Goulas, A., Uylings, H. B., & Hilgetag, C. C. (2017). Principles of ipsilateral and contralateral cortico-cortical connectivity in the mouse. *Brain Structure and Function, 222*, 1281–1295. doi:10.1007/s00429-016-1277-y

Gu, S., Pasqualetti, F., Cieslak, M., Telesford, Q. K., Yu, A. B., Kahn, A. E., Medaglia, J. D., et al. (2015). Controllability of structural brain networks. *Nature Communications, 6*, 8414. doi:10.1038/ ncomms9414

Kaiser, M., & Hilgetag, C. C. (2004). Spatial growth of real-world networks. *Physical Review E, 69*(3), 036103.

Kaiser, M., & Hilgetag, C. C. (2006). Nonoptimal component placement, but short processing paths, due to long-distance projections in neural systems. *PLoS Computational Biology 2*(7), e95. doi:10.1371/journal.pcbi.0020095

Kaiser, M., & Hilgetag, C. C. (2007). Development of multi-cluster cortical networks by time windows for spatial growth. *Neurocomputing, 70*(10–12), 1829–1832.

Kashtan, N., & Alon, U. (2005). Spontaneous evolution of modularity and network motifs. *Proceedings of the National Academy of Sciences of the United States of America, 102*(39), 13773–13778. doi:10.1073/pnas.0503610102

Kasthuri, N., Hayworth, K. J., Berger, D. R., Schalek, R. L., Conchello, J. A., Knowles-Barley, S., Lee, D., et al. (2015). Saturated reconstruction of a volume of neocortex. *Cell, 162*(3), 648–661.

Klimm, F., Bassett, D. S., Carlson, J. M., & Mucha, P. J. (2014). Resolving structural variability in network models and the brain. *PLoS Computational Biology, 10*(3), e1003491. doi:10.1371/journal.pcbi.1003491

Latora, V., & Marchiori, M. (2001) Efficient behavior of small-world networks. *Physical Review Letters, 87*, 198701.

Li, W. C., Cooke, T., Sautois, B., Soffe, S. R., Borisyuk, R., & Roberts, A. (2007). Axon and dendrite geography predict the specificity of synaptic connections in a functioning spinal cord network. *Neural Development, 2*, 17. doi:10.1186/1749-8104-2-17

Li, Y., Liu, Y., Li, J., Qin, W., Li, K., Yu, C., & Jiang T. (2009) Brain anatomical network and intelligence. *PLoS Computational Biology, 5*(5), e1000395. doi:10.1371/journal.pcbi.1000395

Meunier, D., Lambiotte, R., & Bullmore, E. T. (2010). Modular and hierarchically modular organization of brain networks. *Frontiers in Neuroscience, 4*, 200. doi:10.3389/fnins.2010.00200

Milgram, S. (1967). The small world problem. *Psychology Today, 2*, 60–67.

Milo, R., Itzkovitz, S., Kashtan, N., Levitt, R., Shen-Orr, S., Ayzenshtat, I., Sheffer, M., & Alon, U. (2004). Superfamilies of evolved and designed networks. *Science, 303*, 1538–1542. doi:10.1126/science.1089167

Milo, R., Shen-Orr, S., Itzkovitz, S., Kashtan, N., Chklovskii, D., & Alon, U. (2002). Network motifs: Simple building blocks of complex networks. *Science, 298*, 824–827. doi:10.1126/science.298.5594.824

Morgan, S. E., White, S. R., Bullmore, E. T., & Vértes, P. E. (2018). A network neuroscience approach to typical and atypical brain development. *Biological Psychiatry: Cognitive Neuroscience and Neuroimaging, 3*(9), 754–766. doi:10.1016/j.bpsc.2018.03.003

Naumann, E. A., Fitzgerald, J. E., Dunn, T. W., Rihel, J., Sompolinsky, H., & Engert, F. (2016). From whole-brain data to functional circuit models: The zebrafish optomotor response. *Cell, 167*(4), 947–960.e20.

Nguyen, J. P., Shipley, F. B., Linder, A. N., Plummer, G. S., Liu, M., Setru, S. U., Shaevitz, J. W., & Leifer, A. M. (2016). Whole-brain calcium imaging with cellular resolution in freely behaving *Caenorhabditis elegans*. *Proceedings of the National Academy of Sciences of the United States of America, 113*(8), E1074–E1081. doi:10.1073/pnas.1507110112

Nicosia, V., Vértes, P. E., Schafer, W. R., Latora, V., & Bullmore, E. T. (2013). Phase transition in the economically modeled growth of a cellular nervous system. *Proceedings of the National Academy of Sciences of the United States of America, 110*(19), 7880–7885. doi:10.1073/pnas.1300753110

Oh, S. W., Harris, J. A., Ng, L., Winslow, B., Cain, N., Mihalas, S., Wang, Q., et al. (2014). A mesoscale connectome of the mouse brain. *Nature, 508*(7495), 207–214.

Roberts, A., Conte, D., Hull, M., Merrison-Hort, R., Kalam al Azad, A., Bhul, E., Borisyuk, R., & Soffe, S. R. (2014). Can simple rules control development of a pioneer vertebrate neuronal network generating behaviour? *Journal of Neuroscience, 34*, 608–621. doi:10.1523/JNEUROSCI.3248-13.2014

Romero-Garcia, R., Whitaker, K. J., Váša, F., Seidlitz, J., Shinn, M., Fonagy, P., Dolan, R. J., et al. (2017). Structural covariance networks are coupled to expression of genes enriched in supragranular layers of the human cortex. *NeuroImage, 171*, 256–267. doi:10.1016/j.neuroimage.2017.12.060

Ryan, K., Lu, Z., & Meinertzhagen, I. A. (2016). The CNS connectome of a tadpole larva of *Ciona intestinalis* (L.) highlights sidedness in the brain of a chordate sibling. *eLife*, 5:e16962.

Sautois, B., Soffe, S., Li, W. C., & Roberts, A. (2007). Role of type-specific neuron properties in a spinal cord motor network. *Journal of Computational Neuroscience, 23*, 59–77. doi:10.1007/s10827-006-0019-1

Simon, H. A. (1962). The architecture of complexity. *Proceedings of the American Philosophical Society, 106*, 467–482.

Snijders, T. A. B., & Nowicki, K. (1997). Estimation and prediction for stochastic block models for graphs with latent block structure. *Journal of Classification, 14*(1), 75–100.

Sporns, O., & Kötter, R. (2004). Motifs in brain networks. *PLoS Biology, 2*(11), e369. doi:10.1371/journal.pbio.0020369

Stephan, K. E. (2013). The history of CoCoMac. *NeuroImage, 80*, 46–52.

Tang, E., & Bassett, D. S. (2017). Control of dynamics in brain networks. *Reviews of Modern Physics, 90*(031003). https://journals.aps.org/rmp/abstract/10.1103/RevModPhys.90.031003.

Tang, E., Giusti, C., Baum, G. L., Gu, S., Pollock, E., Kahn, A. E., Roalf, D. R., et al. (2017). Developmental increases in white matter network controllability support a growing diversity of brain dynamics. *Nature Communications, 8*, 1252. doi:10.1038/s41467-017-01254-4

Travers, J., & Stanley, M. (1969). An experimental study of the small world problem. *Sociometry, 32*, 425–443.

Towlson, E. K., Vértes, P. E., Ahnert, S., Schafer, W. R., & Bullmore, E. T. (2013). The rich club of the *C. elegans* neuronal connectome. *Journal of Neuroscience, 33*(15), 6380–6387. doi:10.1523/JNEUROSCI.3784-12.2013 van den Heuvel, M. P., & Sporns, O. (2013). Network hubs in the human brain. *Trends in Cognitive Sciences, 17*, 683–696.

van den Heuvel, M. P., Stam, C. J., Kahn, R. S., & Hulshoff Pol, H. E. (2009). Efficiency of functional brain networks and intellectual performance. *Journal of Neuroscience, 29*(23), 7619–7624. doi:10.1523/JNEUROSCI.1443-09.2009

Van Essen, D. C., Ugurbil, K., Auerbach, E., Barch, D., Behrens, T. E., Bucholz, R., Chang, A., et al. (2012). The human connectome project: A data acquisition perspective. *NeuroImage, 62*(4), 2222–2231. doi:10.1016/j.neuroimage.2012.02.018

Varshney, L. R., Chen, B. L., Paniagua, E., Hall, D. H., & Chklovskii, D. B. (2011). Structural properties of the *Caenorhabditis elegans* neuronal network. *PLoS Computational Biology, 7*(2), e1001066. doi:10.1371/journal.pcbi.1001066

Vértes, P. E., Alexander-Bloch, A. F., & Bullmore, E. T. (2014). Generative models of rich clubs in Hebbian neuronal networks and large-scale human brain networks. *Philosophical Transactions of the Royal Society of London B: Biological Sciences, 369*(1653), 20130531.

Vértes, P. E., Alexander-Bloch, A. F., Gogtay, N., Giedd, J., Rapoport, J. L., & Bullmore, E. T. (2012). Simple models of human brain functional networks. *Proceedings of the National Academy of Sciences of the United States of America, 109*, 5868–5872.

Vértes, P. E., & Bullmore, E. T. (2015). Annual research review: Growth connectomics—the organization and reorganization of brain networks during normal and abnormal development. *Journal of Child Psychology and Psychiatry, 56*(3), 299–320. doi:10.1111/jcpp.12365

Watts, D. J., & Strogatz, S. H. (1998). Collective dynamics of "small-world" networks. *Nature, 393*, 440–442.

White, J. G., Southgate, E., Thomson, J. N., & Brenner, S. (1986). The structure of the nervous system of the nematode *Caenorhabditis elegans. Philosophical Transactions of the Royal Society of London B: Biological Sciences, 314*, 1–340. doi:10.1098/rstb.1986.0056

Xu, M., Jarrell, T. A., Wang, Y., Cook, S. J., Hall, D. H., & Emmons, S. W. (2013). Computer assisted assembly of connectomes from electron micrographs: Application to *Caenorhabditis elegans. PLoS One, 8*(1), e54050. doi:10.1371/journal.pone.0054050

Yan, G., Vértes, P. E., Towlson, E. K., Chew, Y. L., Walker, D. S., Schafer, W. R., & Barabási, A.-L. (2017). Network control principles predict neuron function in the *Caenorhabditis elegans* connectome. *Nature, 550*, 519–523. doi:10.1038/nature24056

61 Network-Based Approaches for Understanding Intrinsic Control Capacities of the Human Brain

DANIELLE BASSETT AND FABIO PASQUALETTI

ABSTRACT The human brain is inherently a networked system, displaying rich connectivity patterns across a broad range of spatial scales. The network architecture of the brain places notable constraints on how activity can flow between regions, how regions can communicate with one another, and what computations can be performed. In this chapter, we review recent work capitalizing on the principles of control theory applied to network systems to better understand the nature of these constraints. We begin with a simple primer on network control, and its applicability to the brain. We then recount evidence that network control offers a possible structural mechanism for cognitive control, and then we consider whether such principles can also inform exogenous interventions, for example via brain stimulation. We aim to provide a simple and therefore particularly accessible introduction to the field, coupled with a broad review of the recent literature, and a few thoughts regarding emerging challenges and opportunities.

The human brain is a beautifully complex organ, replete with rich cellular diversity (Arnatkeviciute, Fulcher, Pocock, & Fornito, 2018; Reimann, Horlemann, Ramaswamy, Muller, & Markram, 2017; Seung & Sumbul, 2014; Sumbul et al., 2014), genetic programming (Bale, 2015; Bock, Wainstock, Braun, & Segal, 2015), and biochemical signaling dynamics (Nishiyama & Yasuda, 2015). But to some, the system's most intriguing characteristic is its intricate wiring pattern, from which emerges computation (McCulloch & Pitts, 1943), communication (Fries, 2015), and information propagation (Betzel & Bassett, 2018). Such intricate wiring spans a broad range of scales, from dendritic spines and their marked spatiotemporal dynamics (Chen, Lu, & Zuo, 2014; Nishiyama & Yasuda, 2015) to macroscopic tracts linking subcortical nuclei and cortical areas (Betzel & Bassett, 2018; Hagmann et al., 2008). For many years, progress in quantitatively describing the statistical properties of these wiring patterns was hampered by the lack of an appropriate mathematical formalism. With the recent development of tools, models, and theories in network science (Newman, 2010), many of the long-standing challenges in understanding the relevance of connectivity for circuit function have been overcome, leading to interdisciplinary investigations under the broad umbrella of *network neuroscience* (Bassett & Sporns, 2017). Concerted efforts in building appropriate network models of neural systems across scales (Scholtens, Schmidt, de Reus, & van den Heuvel, 2014) and species (van den Heuvel, Bullmore, & Sporns, 2016), and in determining their descriptive, explanatory, and predictive validity (Bassett, Zurn, & Gold, 2018), now form important components of contemporary work in cognitive neuroscience (Medaglia, Lynall, & Bassett, 2015; Petersen & Sporns, 2015; Sporns, 2014).

The architecture of cellular, ensemble, or areal networks has important implications for information transmission and circuit function (Kirst, Timme, & Battaglia, 2016; Palmigiano, Geisel, Wolf, & Battaglia, 2017). At the microscale, the pattern of synapses between neurons allows for a wide repertoire of cellular dynamics (Feldt, Bonifazi, & Cossart, 2011), including the rather surprising induction of a synchronized ensemble burst from the activation of a single neuron (Miles & Wong, 1983). At the macroscale, corticothalamic loops display features that are specific to distinct cell types, thereby enriching functional diversity (Guo, Yamawaki, Svoboda, & Shepherd, 2018) while the microstructural integrity of fibers in the corpus callosum allows interhemispheric communication (Berlucchi, 2014; Doron & Gazzaniga, 2008), and projections among the basal ganglia, cerebellum, and cortex produce a topographical organization allowing interconnections between motor, cognitive, and affective territories (Bostan & Strick, 2018). While not yet mapped as exhaustively, corticocortical circuits also have clear relevance for cognitive functions—for example, recently being implicated in the coupling of spatial memory and navigation to diverse aspects of sensorimotor integration and motor control (Yamawaki, Radulovic, & Shepherd, 2016). By using network models, the link between connectivity architecture and function can be made even more explicit, allowing for inferences regarding the types of communication dynamics that a given network topology can support (Avena-Koenigsberger, Misic, & Sporns, 2017). For example, disassortative structures have notable information transmission properties, core-periphery structures support the broadcasting and receiving of information, and assortative structures facilitate the segregation and integration of information (Betzel, Medaglia, & Bassett, 2018).

But while the relevance of network architecture for information transmission and circuit behavior is intuitive, studies of this structure-function link have traditionally remained within the realm of correlative descriptions (Hermundstad et al., 2013; Honey et al., 2009; Reimann et al., 2017), thereby lacking a strong theoretical fundament. Put simply, we have a strikingly exiguous understanding of how a given wiring pattern supports or constrains the process by which an increase (or decrease) in the activity of one neural unit alters the activity of other neural units. This gap in knowledge hampers our ability to pinpoint formal mechanisms of top-down control in executive function, to parameterize homeostatic processes in the resting brain (Deco & Corbetta, 2011; Deco, Jirsa, & McIntosh, 2011), and to deduce the computational capacities of specific projection patterns (Curto, Degeratu, & Itskov, 2012; 2013). Here we summarize a candidate solution in the form of network control theory, an emerging field of physics and engineering (Liu & Barabasi, 2016), which provides theoretical and computational tools to determine whether and how a complex networked system can be driven toward a desired configuration, or state, by influencing specific system components (figure 61.1). As applied to the brain, network control theory builds on formal network models of connectivity between neural units (Bassett, Zurn, & Gold, 2018), models of the dynamics produced by neural units (Breakspear, 2017), and models of control in dynamical systems (Kailath, 1980; Kalman, Ho, & Narendra, 1963). The approach thereby presses beyond descriptive statistics and into the realm of predictive models and theories for how specific cognitive functions can arise from a pattern of interconnections (Tang & Bassett, 2018).

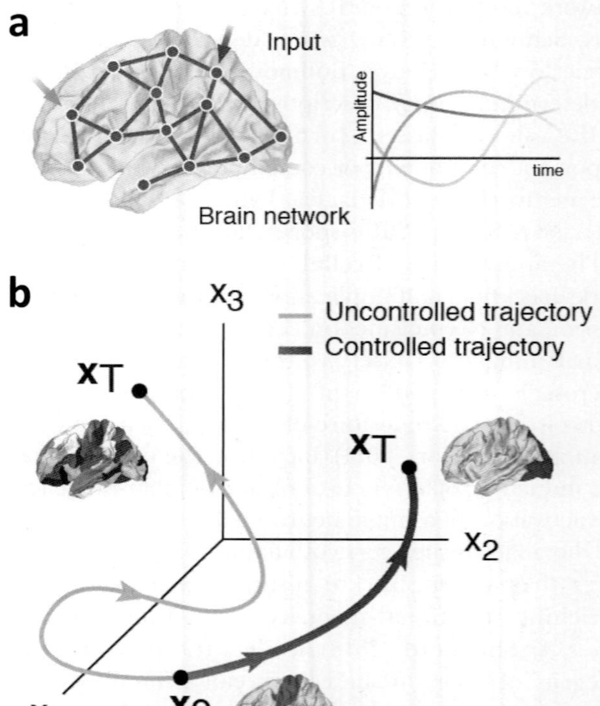

FIGURE 61.1 Controllability of human brain networks. *A*, A set of time-varying inputs are injected into the system at different control points (network nodes, brain regions). The aim is to drive the system from some particular initial state to a target state (e.g., from activation of the somatosensory system to activation of the visual system). *B*, Example trajectory through state space. Without external input (control signals), the system's passive dynamics leads to a state in which random brain regions are more active than others; with input the system is driven into the desired target state. Reproduced with permission from Betzel et al. (2016). (See color plate 74.)

In this chapter we will discuss the methodological advances underpinning recent work in extending the conceptual framework and computational tools of network control theory to neural systems (Tang & Bassett, 2018). We will begin with a brief primer on the mathematical details of the theory and associated models while pointing out other didactic literature in mathematics, physics, and engineering for the interested reader. We then turn to a review of empirical studies that use the theory and associated models to extract controllability statistics from neuroimaging data (Pasqualetti, Zampieri, & Bullo, 2014) and use those statistics to offer candidate explanations for intrinsic human capacities such as cognitive control (Gu et al., 2015). Next, we describe current frontiers in expanding tools from network control theory to enhance their applicability and utility in answering open questions in cognitive neuroscience. In the context of these open questions, we also mention the utility of network control theory for informing exogenous interventions in the form of neurofeedback (Bassett & Khambhati, 2017) and brain stimulation (Tang & Bassett, 2018). Such extensions could prove useful for the treatment of neurological disease or psychiatric disorders that impinge on cognitive capacities (Braun et al., 2018) or for the enhancement of cognition function in healthy individuals (Stiso et al., 2018). Our goal is to offer an accessible introduction to the field, a brief review of the recent literature, and a clear vision for the challenges and opportunities of the near future.

A Primer on Network Control

Networks are fundamental components of many engineering, social, physical, and biological systems. Electrical power grids, mass transportation systems, and cellular networks are instances of modern technological networks, while social networks and nervous systems are sociological and biological examples. Despite arising in different contexts and with diverse purposes, networks are typically characterized by an intricate interconnection of heterogeneous components, which guarantees adaptability to changing environmental conditions, resilience against component failure and perturbations, and complex functionality. *Network controllability* refers to the possibility of changing the network state toward a desired configuration through external stimuli. Understanding network controllability is crucially important in determining how networked systems may be designed (either by man or by evolution), in deducing their functionality, and in inferring the reliability and efficiency of that functionality.

Networks are usually described by a graph representing the interconnections among different parts, a state vector containing characteristic values associated with every component, and a map describing the dynamic evolution of the network state. In a simple setting, the network structure is encoded by a directed graph $G=(V, E)$, where $V=\{1,\ldots,n\}$, and $E \subseteq V \times V$ are the vertex and edge sets, respectively, the state of each node is a real number, the network state is x: $N_{\geq 0} \rightarrow R^n$, and the state dynamics are captured by the linear, discrete time and time-invariant recursion $x(t+1) = Ax(t)$. In the latter equation, A is a weighted adjacency matrix of G, where the (i,j)-th entry is zero if the edge $(i,j) \notin E$, and it equals a real number corresponding to the connection strength otherwise. To ensure network controllability, a subset $K=\{k_1,\ldots,k_m\} \subseteq V$ of control nodes is selected to define the input matrix $B_K=[e_{k1},\ldots,e_{km}]$, where e_i denotes the i-th canonical vector of dimension n. The network dynamics with control nodes K read as

$$x(t+1) = Ax(t) + B_K u_K(t), \qquad (61.1)$$

where u_K: $N_{\geq 0} \rightarrow R^m$ is the control signal injected into the network via the nodes K. The network controllability problem asks for the selection of the control set and the control input such that the network state transitions from rest to any desired state in finite time—that is, the selection of the set K and the sequence u_K such that $x(0) = 0$ and $x(T) = x_d$ for a desired state $x_d \in R^n$ and a final time $T \in N\geq_0$.

From classic systems theory, we know several equivalent conditions ensuring network controllability (Kailath, 1980; Kalman, Ho, & Narendra, 1963). For instance, the network (1) is controllable in time $T \in N\geq_0$ by the control nodes K if and only if the controllability matrix

$$C_{K,T}=[B_K \; AB_K \; A^2B_K \ldots A^{T-1}B_K]$$

is full rank or, equivalently, if and only if the controllability Gramian

$$W_{K,T} = \sum_{\tau=0}^{T-1} A^\tau B_K B_K^T (A^T)^\tau = C_{K,T} C_{K,T}^\tau$$

is invertible. When the duration of the control task, or control horizon, satisfies $T \geq n$, network controllability is also equivalent to the matrix $[\lambda I—A \; B_K]$ being full rank for every eigenvalue λ of A. Finally, when A is (Schur) stable and the control horizon satisfies $T \geq n$, network controllability is ensured by the existence of a unique positive-definite solution X to the Lyapunov equation $A X A^T - X = -B_K B_K^T$ (in which case, the unique solution is $X = W_{K,\infty}$).

The above controllability tests are descriptive, in the sense that they allow us to test the controllability of a network by a set of control nodes, but not prescriptive, in the sense that they do not indicate how to select control nodes to ensure controllability. For the selection of control nodes ensuring controllability, the theory of structured systems provides valuable tools (Reinschke, 1988; Wonham, 1985). In fact, network controllability is a generic property with respect to the specific choices of the network matrix A and, under certain connectivity conditions on the network graph G, network controllability is guaranteed for almost all numerical choices of the network matrix A. For instance, a network is generically controllable if and only if the control nodes can be positioned in a way to decompose the network graph G into a disjoint set of cacti, a specific graph structure (Dion, Commault, & van der Woude, 2003). The structural characterization of network controllability leads to efficient algorithms for the selection of control nodes and the analysis of complex networks based on the network interconnection structure only (e.g., see Liu, Slotine, & Barabasi, 2011; Olshevsky, 2014).

The notion of network controllability presented in the previous paragraphs is only qualitative, and it does not quantify the difficulty of the control task. As a matter of fact, many networks are controllable even with a few control nodes (see the structural analysis above), although their controllability degree may vary significantly as a function of the network parameters and edge weights. One way to measure the degree of controllability of a network is through the energy of the control input needed to transfer the state from rest to a desired state. As a classic result in systems theory, the controllability Gramian contains complete information about the control energy needed to reach a desired target state. In fact, the control energy needed to reach the state x_d in time T equals $x^T_d W^{-1}_{K,T} x_d$. Recent studies have demonstrated connections between the control energy of a network and its structure and parameters. For instance, it has been shown that most complex networks cannot be controlled by a few nodes because the control energy grows exponentially with the network cardinality (Pasqualetti, Zampieri, & Bullo, 2014). This property ensures that existing complex structures are in fact robust to targeted perturbations or failures. On the other hand, there exist network topologies that violate this paradigm, where a few controllers can arbitrarily reprogram large structures with little effort (Pasqualetti & Zampieri, 2014).

It is worth noting that network controllability is an active field of research with broad implications for natural, social, and technological systems (Acemoglu, Ozdaglar, & ParandehGheibi, 2010; Gu et al., 2015; Rajapakse, Groudine, & Mesbahi, 2011; Rahmani, Mesbahi, & Egerstedt, 2009; Skardal & Arenas, 2015). Various controllability measures have been proposed (Cortesi, Summers, & Lygeros, 2014; Kumar, Menolascino, Kafashan, & Ching, 2015), as well as diverse network interpretations (Bof, Baggio, & Zampieri, 2015; Olshevsky, 2015). In this section we simply recount specific relevant advances but do not attempt to be comprehensive.

The Utility of Network Control in Explaining Intrinsic Human Capacities

Now that we have provided a brief introduction to the formalism of network control, we turn to the question of how that formalism, and its associated computational tools, models, and theory, can be used to better understand human cognitive capacities and their neurophysiological basis. We separate our discussion into four main areas, considering studies of cellular scale processes, studies identifying large-scale brain areas relevant for diverse control strategies, studies focusing on a few well-specified brain-state transitions, and studies identifying alterations in control capabilities in neurological disease, psychiatric disorders, and brain injury. Our discussion will lay the groundwork for the next section considering current limitations of the field and emerging frontiers.

Network control at the cellular scale Because network models of the neural system can be built across a range of scales (Betzel & Bassett, 2017), it is useful to consider the applicability of network control in the context of both cellular and areal dynamics. While the initial applications of the theory considered large-scale network architecture in humans (Gu et al., 2015), exercising the theory at the cellular scale in nonhuman animals allows for invasive perturbative experiments for theory validation. In a hallmark study, Yan et al. (2017) used the principles of network control to predict the involvement of specific neurons in the locomotor behaviors of *Caenorhabditis elegans* (Yan et al., 2017). Specifically, based on the same model of linear dynamics stipulated in the previous section and informed by the well-known cellular-level connectome of the nematode (White, Southgate, Thomson, & Brenner, 1986), the authors predicted that muscle control requires 12 distinct neuronal classes, 11 of which had previously been implicated in locomotion by laser ablation. The 12th class was the previously uncharacterized neuron, PDB, which the authors then subjected to laser ablation, subsequently finding a significant loss of dorsoventral polarity in large body bends. The work provides critical support for the utility of network control theory for understanding neural dynamics and associated behaviors. Notably, both code (Towlson et al., 2018) and data (Chew et al., 2017) related to the study have been publicly released. An important open question lies in the degree to which the linear model of dynamics can be used to predict nonlinear dynamics, with preliminary supporting evidence arising in the context of ensemble bursting in the presence of neuronal autapses (Wiles et al., 2017).

Control points: anatomical location supports diverse control strategies Early work in the study of the controllability of complex networks posed the problem of localizing driver nodes or control points in the system that can guide the system's entire dynamics with time-dependent control (Liu, Slotine, & Barabasi, 2011). In the context of the brain, it is interesting to ask not only whether such global controllability is possible (Menara, Gu, Bassett, & Pasqualetti, 2017; Menara, Bassett, & Pasqualetti, 2017) but also whether diverse control strategies could be preferentially implemented by different cortical or subcortical areas. Such a question is motivated by the fact that cognitive neuroscience abounds in examples of specific regions that perform specific functions by exerting influence on (or sharing information with) other regions in a broader circuit. Fortunately, recent technical work has begun defining diverse control strategies and developing statistical metrics to characterize the existence and strength of those strategies in arbitrary networked systems (Pasqualetti, Zampieri, &

Bullo, 2014). In a notable early study using tract-tracing data in macaques and diffusion magnetic resonance-imaging data in humans, Gu et al. (2015) reported evidence that regions of the brain located in frontoparietal cortex and implicated (in the neuroscience literature) in task switching and cognitive control across a wide variety of tasks (Crossley et al., 2013) were also regions that network control theory predicted to be strong *modal controllers*, having the capacity to effectively drive the system into distant states (Gu et al., 2015). In contrast, regions of the brain located in the default mode and implicated (in the neuroscience literature) in baseline dynamics were also regions that network control theory predicted to be strong *average controllers*, having the capacity to effectively drive the system into all reachable states. Both findings underscore a marked similarity between the functions that brain areas are known to perform and the functions that those areas are theoretically predicted to enact effectively, based on their location within the structural white matter network. Interestingly, both average and modal controllability increase as children develop, differ in males and females, and vary across individuals in a manner that tracks cognitive performance (Cornblath et al., 2019; Tang et al., 2017).

Optimal trajectories: implications for brain-state transitions In addition to understanding the capacity for a specific brain region to enact a particular control strategy, it is also of interest to ask how the brain transitions from one state to another via the injection of regional input (Betzel, Gu, Medaglia, Pasqualetti, & Bassett, 2016). Such input can naturally take the form of stimulus-induced activation or information transiently arriving at an area from a distant part of the circuit (Bassett & Khambhati, 2017). Within the linear control setup described in the primer and after applying tractography algorithms to high-resolution diffusion magnetic resonance-imaging data to estimate the human structural connectome, Betzel et al. (2016) calculated the optimal input signals to drive the brain to and from states dominated by different cognitive systems (Betzel et al., 2016). The authors report that *optimal states*—in which the brain should start (and finish) in order to minimize transition energy—display high activity in hub regions (Hagmann et al., 2008), implicating the brain's rich club, including areas of the default mode (van den Heuvel & Sporns, 2011). These inferences are in line with those of a complementary study (Gu et al., 2018) that invokes principles of maximum entropy to suggest that brain states that minimize energy display activation of spatially contiguous sets of brain regions reminiscent of cognitive systems that are coactivated frequently (Crossley et al., 2013). Interestingly, Cornblath et al. (2018) computed the minimum control energy required to maintain each brain state given the underlying white matter architecture and showed that this persistence energy was lowest for the brain state characterized by activation of the default mode regions. Collectively, these studies provide a structural explanation for the brain's baseline dynamics but leave open the question of which control points might be important for some state transitions and not others. In considering transitions from the default mode to the activation of primary sensory, motor, and auditory systems, preliminary evidence suggests that the optimal control points for a given state transition are characterized by high communicability to the target state (Gu et al., 2017). In recent work from Kim et al. (2018), this proposed role of long-distance paths in the propagation of control energy has been derived formally and has also been made more precise, providing evidence that control capacity differs across species. It will be particularly interesting in the future to study the transitions between brain states that support higher-order cognitive functions, such as memory encoding, decision-making, and the inhibition of prepotent responses (Cui et al., 2018), especially in light of the fact that these functions are commonly altered in disorders of mental health (Braun et al., 2018).

Alteration in control capacity in injury and disease In mapping the control capacities of brain networks in healthy humans, it is natural to ask whether those capacities are diminished by injury or altered by disease. In a pioneering recent study, Jeganathan et al. (2018) considered white matter connectivity in 38 young patients with bipolar disorder (BD), 84 healthy relatives of these patients, and 96 age- and gender-matched controls. The authors report that disconnectivity in frontolimbic circuitry leads to impaired network controllability in BD patients and those at high genetic risk, suggesting potential functional consequences of altered brain networks in the disorder. Such relatively localized effects stand in contrast to the reported diffuse alterations in network controllability in mild traumatic brain injury (Gu et al., 2017). An important open question is how such structural changes in the network control capacity of a brain could affect its large-scale functional network dynamics, perhaps in a manner mediated by the biochemical signatures of the specific disease in question. For example, does network controllability offer a structural explanation for the altered functional network dynamics observed during working-memory task performance in schizophrenia (Braun et al., 2015, 2016)? And is the relationship between structural network controllability and functional network dynamics mediated by the alterations in excitatory-inhibitory balance that are characteristic of the disease (Braun et al., 2016)? More generally, it is also interesting to speculate that a better understanding of the network controllability deficits in neurological disease and psychiatric disorders could lead to more targeted interventions (Braun et al., 2018). Future work could consider modeling the effects of cognitive interventions—such as mindfulness and cognitive behavioral therapy—as effectors of network control and test whether the energy required by the intervention task is related to a patient's response to the therapy. Such a possibility is supported by recent work demonstrating that individual differences in network topology, as measured by betweenness centrality, can distinguish nonresponders from responders to transcranial magnetic stimulation in major depression (Downar et al., 2014).

Current Frontiers

While the extension and application of network control theory in the context of understanding human cognition is an extremely exciting recent development in the field, there remain several important limitations of the current work and its associated emerging frontiers. Here we highlight just a few of these pertinent methodological considerations, as well as what we see as particularly important opportunities for future work, and we also point the reader to the primary literature for further information. We separate our comments into three main areas spanning the development of controllability metrics, the extension of current principles and methods to the context of nonlinear control, and the potential to inform experimental paradigms for exogenous brain stimulation.

Development of controllability metrics In earlier sections of this chapter, we mentioned approaches to identify control points (Liu, Slotine, & Barabasi, 2011), estimate the energy required for specific brain-state transitions (Betzel et al., 2016; Gu et al., 2017), and calculate metrics for specific control strategies, such as average and modal controllability (Pasqualetti, Zampieri, & Bullo, 2014). While these specific methods have proven useful in understanding brain structure and its implications for human cognition (Tang & Bassett, 2018), other potentially useful methods also exist, and the further development of controllability metrics is an active and swiftly growing area of inquiry. For example, methods exist to estimate the controllability radius, a measure of the robustness of a network to perturbations of the edges (Bianchin, Frasca, Gasparri, & Pasqualetti, 2017; Menara, Katewa, Bassett, & Pasqualetti, 2018), and it would be interesting to test whether the controllability radius might track with markers of brain or cognitive reserve (Medaglia, Pasqualetti, Hamilton, Thompson-Schill, & Bassett, 2017). One could also consider estimating the controllability of single edges (rather than single nodes; Pang, Wang, Hao, & Lai, 2017), potentially to inform neurofeedback or other intervention approaches to specific connections (Bassett & Khambhati, 2017; Murphy & Bassett, 2017). Further work is also needed in producing analytic results marking the relation between a network's local, mesoscale, and global topology and its capacity for enacting diverse types of control (Kim et al., 2018). Finally, these examples all started from existing network control techniques and asked how they might inform our study of the brain; yet many future advances might instead be enabled by starting with existing neurophysiological or cognitive processes and asking how they might be formulated as network control strategies. For example, processes such as the tuning of sensory gating (Whalley, 2015), the regulation of structural plasticity (Caroni, Donato, & Muller, 2012), and the modulation of gain as a function of arousal (Eldar, Cohen, & Niv, 2013) intuitively appear to be particularly appropriate candidates for which to develop novel controllability statistics tracking their proposed functions.

Extension to nonlinear control Many of the available approaches for network control are built upon the assumption that the system's dynamics can be approximated to follow the linear form stipulated in the primer (Kailath, 1980; Kalman, Ho, & Narendra, 1963). However, extensive evidence points to nontrivial, nonlinear dynamics as a hallmark of the complex functional repertoire characteristic of neural systems (Breakspear, 2017). How and when can linear approximations of such dynamics be useful? Intuitively, linear approximations of nonlinear dynamics hold true over short time horizons and in the vicinity of the system's current operating point (Leith & Leithead, 2000). Additional evidence suggests that time-averaged dynamics and slow fluctuations in the blood oxygen level-dependent signal can also be reasonably modeled with assumptions of linearity (Galan, Ermentrout, & Urban, 2008; Gu et al., 2018; Honey et al., 2009). Moreover, even when the dynamics of a system are truly nonlinear, one can ask whether the predictions of control from the linear model can be used to infer the response of the nonlinear system, either statistically or formally (Coron, 2009; Whalen, Brennan, Sauer, & Schiff, 2015). Initial evidence in neural systems suggests that controllability statistics derived from the linear model of network dynamics can be used to predict transitions into and out of bursting regimes in neuronal ensembles (Wiles et al., 2017) and changes in activity states induced by stimulation in Wilson-Cowan oscillator models of cortical columns (Muldoon et al., 2016). Nevertheless, it remains an important and interesting direction to build on the emerging approaches for nonlinear control in the physics and engineering literature (Motter, 2015) to tackle questions such as the control of oscillations for the purposes of synchronization (Menara, Baggio, Bassett, & Pasqualetti, 2018), attentional gating (Newman & Grace, 1999) or communication (Fries, 2015), or the transfer of information across rhythms at different frequencies (Canolty & Knight, 2010).

Informing exogenous brain stimulation While we have focused this chapter on the utility of network control theory for understanding cognitive function in humans, we would be remiss not to mention the potential utility of network control for exogenous brain stimulation. Since the early work in network neuroscience, it has been evident that the network perspective could have important implications for the effects of brain stimulation, including the mechanisms of its efficacy (McIntyre & Hahn, 2010) and the optimization of its location, duration, intensity, and frequency (Johnson et al., 2013). Although neuromodulation therapies, including deep-brain stimulation, intracranial cortical stimulation, transcranial direct-current stimulation, and transcranial magnetic stimulation, have traditionally targeted specific brain regions, their impact extends far beyond the target location, reaching spatially distributed areas and the tracts leading to them (Johnson et al., 2013; Laxton et al., 2010; Lozano & Lipsman, 2013). Understanding the nonlocal effects of stimulation is critical for the optimization of positive outcomes and the mitigation of any deleterious effects of the stimulation protocol (Medaglia, Zurn, & Bassett, 2017). Initial efforts informing the study of brain stimulation with network control theory include computational studies of the basic mechanisms of stimulation propagation (Muldoon et al., 2016), as well as of the effectiveness of a pseudospectral method for seizure abatement (Taylor et al., 2015). More recent work has complemented such numerical experiments with grid stimulation (Khambhati et al., 2018; Stiso et al., 2018) and transcranial magnetic stimulation (Medaglia et al., 2018) experiments in humans, which provide initial evidence that network controllability statistics can be used to choose targets for stimulation that will affect a specific cognitive outcome, such as memory encoding (Khambhati et al., 2018; Stiso et al., 2018) or language production (Medaglia et al., 2018).

Conclusion

The beauty and richness of human cognition as we know it is supported by an intricate pattern of cellular and regional interconnections. Recent work building formal models of those interconnection patterns as networks has fundamentally changed the types of questions that developmental, cognitive, and systems neuroscience can meaningfully tackle. Yet most network neuroscience studies remain descriptive in nature, limiting their potential to identify underlying mechanisms and to validate the relevance of those mechanisms for various domains of cognitive function. In this chapter we sought to provide a simple and accessible introduction to network control theory, an emerging field of systems engineering that has begun to offer a novel and more mechanistic perspective on human cognition informed by both brain network architecture and dynamics. Our hope is that our account will induce the interested young readers of this textbook to dig deeper into potential network-based explanations of higher-order cognitive function.

Acknowledgments

We would like to thank Jason Z. Kim and Christopher W. Lynn for helpful comments on an earlier version of this chapter. We are grateful to the National Science Foundation (NSF) for a Collaborative Research in Computational Neuroscience grant (BCS-1441502; PO Betty Tuller) to support the initial collaborative efforts of Danielle S. Bassett and Fabio Pasqualetti. We are also grateful for subsequent funding from the NSF (BCS-1430087, BCS-1631550) and an Office of Naval Research Young Investigator grant to Bassett. She would also like to acknowledge support from the Alfred P. Sloan Foundation, the John D. and Catherine T. MacArthur Foundation, the ISI Foundation, and the Paul Allen Foundation.

REFERENCES

Acemoglu, D., Ozdaglar, A., & ParandehGheibi, A. (2010). Spread of (mis)information in social networks. *Games and Economic Behavior, 70*, 194–227.

Arnatkeviciute, A., Fulcher, B. D., Pocock, R., & Fornito, A. (2018). Hub connectivity, neuronal diversity, and gene expression in the *Caenorhabditis elegans* connectome. *PLOS Computational Biology, 14*, e1005989.

Avena-Koenigsberger, A., Misic, B., & Sporns, O. (2017). Communication dynamics in complex brain networks. *Nature Reviews Neuroscience, 19*, 17–33.

Bale, T. L. (2015). Epigenetic and transgenerational reprogramming of brain development. *Nature Reviews Neuroscience, 16*, 332–344.

Bassett, D. S., & Khambhati, A. N. (2017). A network engineering perspective on probing and perturbing cognition with neurofeedback. *Annals of the New York Academy of Sciences, 1396*, 126–143.

Bassett, D. S., & Sporns, O. (2017). Network neuroscience. *Nature Neuroscience, 20*, 353–364.

Bassett, D. S., Zurn, P., & Gold, J. I. (2018). On the nature and use of models in network neuroscience. *Nature Reviews Neuroscience, 19*(9), 566–578. doi:10.1038/s41583-018-0038-8

Berlucchi, G. (2014). Visual interhemispheric communication and callosal connections of the occipital lobes. *Cortex, 56*, 1–13.

Betzel, R. F., & Bassett, D. S. (2017). Multi-scale brain networks. *NeuroImage, 160*, 73–83.

Betzel, R. F., & Bassett, D. S. (2018). Specificity and robustness of long-distance connections in weighted, interareal connectomes. *Proceedings of the National Academy of Sciences of the United States of America, 115*, E4880–E4889.

Betzel, R. F., Gu, S., Medaglia, J. D., Pasqualetti, F., & Bassett, D. S. (2016). Optimally controlling the human connectome: The role of network topology. *Scientific Reports, 6*, 30770.

Betzel, R. F., Medaglia, J. D., & Bassett, D. S. (2018). Diversity of meso-scale architecture in human and non-human connectomes. *Nature Communications, 9*, 346.

Bianchin, G., Frasca, P., Gasparri, A., & Pasqualetti, F. (2017). The observability radius of networks. *IEEE Transactions on Automatic Control, 62*, 3006–3013.

Bock, J., Wainstock, T., Braun, K., & Segal, M. (2015). Stress in utero: Prenatal programming of brain plasticity and cognition. *Biological Psychiatry, 78*, 315–326.

Bof, N., Baggio, G., & Zampieri, S. (2015). On the role of network centrality in the controllability of complex networks. *arXiv*. Retrieved from 1509.04154.

Bostan, A. C., & Strick, P. L. (2018). The basal ganglia and the cerebellum: Nodes in an integrated network. *Nature Reviews Neuroscience, 19*, 338–350.

Braun, U., Schäfer, A., Walter, H., Erk, S., Romanczuk-Seiferth, N., Haddad, L., … Bassett, D. S. (2015). Dynamic reconfiguration of frontal brain networks during executive cognition in humans. *Proceedings of the National Academy of Sciences of the United States of America, 112*, 11678–11683.

Braun, U., Schäfer, A., Bassett, D. S., Rausch, F., Schweiger, J. I., Bilek, E., … Tost, H. (2016). Dynamic reconfiguration of brain networks: A potential schizophrenia genetic risk mechanism modulated by NMDA receptor function. *Proceedings of the National Academy of Sciences of the United States of America, 113*, 12568–12573.

Braun, U., Schaefer, A., Betzel, R. F., Tost, H., Meyer-Lindenberg, A., & Bassett, D. S. (2018). From maps to multi-dimensional network mechanisms of mental disorders. *Neuron, 97*, 14–31.

Breakspear, M. (2017). Dynamic models of large-scale brain activity. *Nature Neuroscience, 20*, 340–352.

Canolty, R. T., & Knight, R. T. (2010). The functional role of cross-frequency coupling. *Trends in Cognitive Sciences, 14*, 506–515.

Caroni, P., Donato, F., & Muller, D. (2012). Structural plasticity upon learning: Regulation and functions. *Nature Reviews Neuroscience, 13*, 478–490.

Chen, C. C., Lu, J., & Zuo, Y. (2014). Spatiotemporal dynamics of dendritic spines in the living brain. *Frontiers in Neuroanatomy, 8*, 28.

Chew, Y. L., Walker, D. S., Towlson, E. K., Vértes, P. E., Yan, G., Barabási, A. L., & Schafer, W. R. (2017). Recordings of *Caenorhabditis elegans* locomotor behaviour following targeted ablation of single motorneurons. *Scientific Data, 4*, 170156.

Cornblath, E. J., Tang, E., Baum, G. L., Moore, T. M., Adebimpe, A., Roalf, D. R., … Bassett, D. S. (2019). Sex differences in network controllability as a predictor of executive function in youth. *NeuroImage, 88*, 122–134.

Cornblath, E. J., Ashourvan, A., Kim, J. Z., Betzel, R. F., Ciric, R., Baum, G. L., … Bassett, D. S. (2018). Context-dependent architecture of brain state dynamics is explained by white matter connectivity and theories of network control. *arXiv*. Retrieved from 1809.02849.

Coron, J.-M. (2009). *Control and nonlinearity*. American Mathematical Society, Providence, RI.

Cortesi, F. L., Summers, T. H., & Lygeros, J. (2014). Submodularity of energy related controllability metrics. *IEEE Conference on Decision and Control*, 2883–2888, Los Angeles, CA.

Crossley, N. A., Mechelli, A., Vértes, P. E., Winton-Brown, T. T., Patel, A. X., Ginestet, C. E., … Bullmore, E. T. et al. (2013). Cognitive relevance of the community structure of the human brain functional coactivation network. *Proceedings of the National Academy of Sciences of the United States of America, 110*, 11583–11588.

Cui, Z., Stiso, J., Baum, G. L., Kim, J. Z., Roalf, D. R., Betzel, R. F., … Satterthwaite, T. D. (2018). Optimization of energy state transition trajectory supports the development of executive function during youth. *bioRxiv, 1101*, 424929.

Curto, C., Degeratu, A., & Itskov, V. (2012). Flexible memory networks. *Bulletin of Mathematical Biology, 74*, 590–614.

Curto, C., Degeratu, A., & Itskov, V. (2013). Encoding binary neural codes in networks of threshold-linear neurons. *Neural Computation, 25*, 2858–2903.

Deco, G., & Corbetta, M. (2011). The dynamical balance of the brain at rest. *Neuroscientist, 17*, 107–123.

Deco, G., Jirsa, V. K., & McIntosh, A. R. (2011). Emerging concepts for the dynamical organization of resting-state activity in the brain. *Nature Reviews Neuroscience, 12*, 43–56.

Dion, J. M., Commault, C., & van der Woude, J. (2003). Generic properties and control of linear structured systems: A survey. *Automatica, 39*, 1125–1144.

Doron, K. W., & Gazzaniga, M. S. (2008). Neuroimaging techniques offer new perspectives on callosal transfer and interhemispheric communication. *Cortex, 44*, 1023–1029.

Downar, J., Geraci, J., Salomons, T. V., Dunlop, K., Wheeler, S., McAndrews, M. P., … Giacobbe, P. (2014). Anhedonia and reward-circuit connectivity distinguish nonresponders from responders to dorsomedial prefrontal repetitive transcranial magnetic stimulation in major depression. *Biological Psychiatry, 76*, 176–185.

Eldar, E., Cohen, J. D., & Niv, Y. (2013). The effects of neural gain on attention and learning. *Nature Neuroscience, 16*(8), 1146–1153.

Feldt, S., Bonifazi, P., & Cossart, R. (2011). Dissecting functional connectivity of neuronal microcircuits: Experimental and theoretical insights. *Trends in Neurosciences, 34*, 225–236.

Fries, P. (2015). Rhythms for cognition: Communication through coherence. *Neuron, 88*, 220–235.

Galan, R. F., Ermentrout, G. B., & Urban, N. N. (2008). *Journal of Neurophysiology, 99*, 277–283.

Gu, S., Pasqualetti, F., Cieslak, M., Telesford, Q. K., Yu, A. B., Kahn, A. E., … Bassett, D. S. (2015). Controllability of structural brain networks. *Nature Communications, 6*, 8414.

Gu, S., Betzel, R. F., Mattar, M. G., Cieslak, M., Delio, P. R., Grafton, S. T., … Bassett, D. S. (2017). Optimal trajectories of brain state transitions. *NeuroImage, 148*, 305–317.

Gu, S., Cieslak, M., Baird, B., Muldoon, S. F., Grafton, S. T., Pasqualetti, F., & Bassett, D. S. (2018). The energy landscape of neurophysiological activity implicit in brain network structure. *Scientific Reports, 8*, 2507.

Guo, K., Yamawaki, N., Svoboda, K., & Shepherd, G. M. G. (2018). Anterolateral motor cortex connects with a medial subdivision of ventromedial thalamus through cell-type-specific circuits, forming an excitatory thalamo-cortico-thalamic loop via layer 1 apical tuft dendrites of layer 5B pyramidal tract type neurons. *Journal of Neuroscience*, Epub Ahead of Print, 1333–1318.

Hagmann, P., Cammoun, L., Gigandet, X., Meuli, R., Honey, C. J., Wedeen, V. J., & Sporns, O. (2008). Mapping the structural core of human cerebral cortex. *PLoS Biology, 6*, e159.

Hermundstad, A. M., Bassett, D. S., Brown, K. S., Aminoff E. M., Clewett, D., Freeman, S., … Carlson, J. M. (2013). Structural foundations of resting-state and task-based functional connectivity in the human brain. *Proceedings of the National Academy of Sciences, 110*, 6169–6174.

Honey, C., Sporns, O., Cammoun, L., Gigandet, X., Thiran, J. P., Meuli, R., & Hagmann, P. (2009). Predicting human resting-state functional connectivity from structural connectivity. *Proceedings of the National Academy of Sciences, 106*, 2035–2040.

Jeganathan, J., Perry, A., Bassett, D. S., Roberts, G., Mitchell, P. B., & Breakspear, M. (2018). Fronto-limbic dysconnectivity leads to impaired brain network controllability in young people with bipolar disorder and those at high genetic risk. *NeuroImage: Clinical, 19*, 71–81.

Johnson, M. D., Lim, H. H., Netoff, T., Connolly, A. T., Johnson, N., Roy, A., … He, B. (2013). Neuromodulation for brain disorders: Challenges and opportunities. *IEEE Transactions on Biomedical Engineering, 60*, 610–624.

Kailath, T. (1980). *Linear systems.* Upper Saddle River, NJ: Prentice-Hall.

Kalman, R. E., Ho, Y. C., & Narendra, S. K. (1963). Controllability of linear dynamical systems. *Contributions to Differential Equations, 1*, 189–213.

Khambhati, A. N., Kahn, A. E., Costantini, J., Ezzyat, Y., Solomon, E. A., Gross, R. E., … Bassett, D. S. (2018). Functional control of electrophysiological network architecture using direct neurostimulation in humans. *Network Neuroscience*, 1–46. https://www.mitpressjournals.org/toc/netn/0/ja.

Kim, J. Z., Soffer, J. M., Kahn, A. E., Vettel, J. M., Pasqualetti, F., & Bassett, D. S. (2018). Role of graph architecture in controlling dynamical networks with applications to neural systems. *Nature Physics, 14*, 91–98.

Kirst, C., Timme, M., & Battaglia, D. (2016). Dynamic information routing in complex networks. *Nature Communications, 7*, 11061.

Kumar, G., Menolascino, D., Kafashan, M., & Ching, S. (2015). Controlling linear networks with minimally novel inputs. In *American Control Conference (ACC)*, 5896–5900. July 1–2, Chicago.

Laxton, A. W., Tang-Wai, D. F., McAndrews, M. P., Zumsteg, D., Wennberg, R., Keren, R., … Lozano, A. M. (2010). A phase I trial of deep brain stimulation of memory circuits in Alzheimer's disease. *Annals of Neurology, 68*, 521–534.

Leith, D. J., & Leithead, W. E. (2000). Survey of gain-scheduling analysis and design. *International Journal of Control, 73*, 1001–1025.

Liu, Y.-Y., & Barabasi, A.-L. (2016). Control principles of complex systems. *Reviews of Modern Physics, 88*, 035006.

Liu, Y.-Y., Slotine, J. J., & Barabasi, A. L. (2011). Controllability of complex networks. *Nature, 473*, 167–173.

Lozano, A. M., & Lipsman, N. (2013). Probing and regulating dysfunctional circuits using deep brain stimulation. *Neuron, 77*, 406–424.

McCulloch, W. S., & Pitts, W. (1943). A logical calculus of the ideas immanent in nervous activity. *Bulletin of Mathematical Biology, 5*, 115–133.

McIntyre, C. C., & Hahn, P. J. (2010). Network perspectives on the mechanisms of deep brain stimulation. *Neurobiology of Disease, 38*, 329–337.

Medaglia, J. D., Lynall, M. E., & Bassett, D. S. (2015). Cognitive network neuroscience. *Journal of Cognitive Neuroscience, 27*, 1471–1491.

Medaglia, J. D., Pasqualetti, F., Hamilton, R. H., Thompson-Schill, S. L., & Bassett, D. S. (2017). Brain and cognitive reserve: Translation via network control theory. *Neuroscience & Biobehavioral Reviews, 75*, 53–64.

Medaglia, J. D., Zurn, P., & Bassett, D. S. (2017). Mind control as a guide for the mind. *Nature Human Behaviour, 1*, 0119.

Medaglia, J. D., Harvey, D. Y., White, N., Kelkar, A., Zimmerman, J., Bassett, D. S., & Hamilton, R. H. (2018). Network controllability in the inferior frontal gyrus relates to controlled language variability and susceptibility to TMS. *Journal of Neuroscience, 38*, 6399–6410.

Menara, T., Baggio, G., Bassett, D. S., & Pasqualetti, F. (2019). Stability conditions for cluster synchronization in networks of Kuramoto oscillators. *IEEE Transactions on Control of Network Systems.* doi:10.1109/TCNS.2019.2903914

Menara, T., Bassett, D. S., & Pasqualetti, F. (2019). Structural controllability of symmetric networks. *IEEE Transactions on Automatic Control, 64*(9), 3740–3747.

Menara, T., Gu, S., Bassett, D. S., & Pasqualetti, F. (2017). On structural controllability of symmetric (brain) networks. *arXiv.* Retrieved from 1706, 05120.

Menara, T., Katewa, V., Bassett, D. S., & Pasqualetti, F. (2018). The structured controllability radius of symmetric (brain) networks, 2802–2807. *Annual American Control Conference (ACC)*, June 27–29, Wisconsin Center, Milwaukee.

Miles, R., & Wong, R. K. (1983). Single neurones can initiate synchronized population discharge in the hippocampus. *Nature, 306*, 371–373.

Motter, A. E. (2015). Networkcontrology. *Chaos, 25*, 097621.

Muldoon, S. F., Pasqualetti, F., Gu, S., Cieslak, M., Grafton, S. T., Vettel, J. M., & Bassett, D. S. (2016). Stimulation-based control of dynamic brain networks. *PLOS Computational Biology, 12*, e1005076.

Murphy, A. C., & Bassett, D. S. (2017). A network neuroscience of neurofeedback for clinical translation. *Current Opinion in Biomedical Engineering, 1*, 63–70.

Newman, J., & Grace, A. A. (1999). Binding across time: The selective gating of frontal and hippocampal systems modulating working memory and attentional states. *Consciousness and Cognition, 8*, 196–212.

Newman, M. E. J. (2010). *Networks: An introduction.* Cambridge, MA: MIT Press.

Nishiyama, J., & Yasuda, R. (2015). Biochemical computation for spine structural plasticity. *Neuron, 87,* 63–75.

Olshevsky, A. (2014). Minimal controllability problems. *IEEE Transactions on Control of Network Systems, 1,* 249–258.

Olshevsky, A. (2015). Eigenvalue clustering, control energy, and logarithmic capacity. *arXiv.* Retrieved from 1511.00205.

Palmigiano, A., Geisel, T., Wolf, F., & Battaglia, D. (2017). Flexible information routing by transient synchrony. *Nature Neuroscience, 20,* 1014–1022.

Pang, S. P., Wang, W. X., Hao, F., & Lai, Y. C. (2017). Universal framework for edge controllability of complex networks. *Scientific Reports, 7,* 4224.

Pasqualetti, F., & Zampieri, S. (2014). On the controllability of isotropic and anisotropic networks. *IEEE Conference on Decision and Control,* 607–612, Los Angeles, CA.

Pasqualetti, F., Zampieri, S., & Bullo, F. (2014). Controllability metrics, limitations and algorithms for complex networks. *IEEE Transactions on Control of Network Systems, 1,* 40–52.

Petersen, S. E., & Sporns, O. (2015). Brain networks and cognitive architectures. *Neuron, 88,* 207–219.

Rahmani, A., Ji, M., Mesbahi, M., & Egerstedt, M. (2009). Controllability of multi-agent systems from a graph-theoretic perspective. *SIAM Journal on Control and Optimization, 48,* 162–186.

Rajapakse, I., Groudine, M., & Mesbahi, M. (2011). Dynamics and control of state-dependent networks for probing genomic organization. *Proceedings of the National Academy of Sciences, 108,* 17257–17262.

Reimann, M. W., Horlemann, A. L., Ramaswamy, S., Muller, E. B., & Markram, H. (2017). Morphological diversity strongly constrains synaptic connectivity and plasticity. *Cerebral Cortex, 27,* 4570–4585.

Reimann, M. W., Nolte, M., Scolamiero, M., Turner, K., Perin, R., Chindemi, G., … Markram, H. (2017). Cliques of neurons bound into cavities provide a missing link between structure and function. *Frontiers in Computational Neuroscience, 11,* 48.

Reinschke, K. J. (1988). *Multivariable control: A graph-theoretic approach.* Berlin: Springer.

Scholtens, L. H., Schmidt, R., de Reus, M. A., & van den Heuvel, M. P. (2014). Linking macroscale graph analytical organization to microscale neuroarchitectonics in the macaque connectome. *Journal of Neuroscience, 34,* 12192–12205.

Seung, H. S., & Sumbul, U. (2014). Neuronal cell types and connectivity: Lessons from the retina. *Neuron, 83,* 1262–1272.

Skardal, P. S., & Arenas, A. (2015). Control of coupled oscillator networks with application to microgrid technologies. *Science Advances, 1*(7), e1500339. doi:10.1126/sciadv.1500339

Sporns, O. (2014). Contributions and challenges for network models in cognitive neuroscience. *Nature Neuroscience, 17,* 652–660.

Stiso, J., Khambhati, A. N., Menara, T., Kahn, A. E., Stein, J. M., Das, S. R., … Bassett, D. S. (2019). White matter network architecture guides direct electrical stimulation through optimal state transitions. *Cell Reports, 28,* 2554–2566.

Sumbul, U., Song, S., McCulloch, K., Becker, M., Lin, B., Sanes, J. R., … Seung, H. S. (2014). A genetic and computational approach to structurally classify neuronal types. *Nature Communications, 5,* 3512.

Tang, E., & Bassett, D. S. (2018). Control of dynamics in brain networks. *Reviews of Modern Physics, 90,* 031003.

Tang, E., Giusti, C., Baum, G. L., Gu, S., Pollock, E., Kahn, A. E., … Bassett, D. S. (2017). Developmental increases in white matter network controllability support a growing diversity of brain dynamics. *Nature Communications, 8,* 1252.

Taylor, P. N., Thomas, J., Sinha, N., Dauwels, J., Kaiser, M., Thesen, T., & Ruths, J. (2015). Optimal control based seizure abatement using patient derived connectivity. *Frontiers in Neuroscience, 9,* 202.

Towlson, E. K., Vértes, P. E., Yan, G., Chew, Y. L., Walker, D. S., Schafer, W. R., & Barabási, A. L. (2018). *Caenorhabditis elegans* and the network control framework—FAQs. *Philosophical Transactions of the Royal Society of London B: Biological Sciences, 373,* 1758.

van den Heuvel, M. P., Bullmore, E. T., & Sporns, O. (2016). Comparative connectomics. *Trends in Cognitive Sciences, 20,* 345–361.

van den Heuvel, M. P., & Sporns, O. (2011). Rich-club organization of the human connectome. *Journal of Neuroscience, 31,* 15775–15786.

Whalen, A. J., Brennan, S. N., Sauer, T. D., & Schiff, S. J. (2015). Observability and controllability of nonlinear networks: The role of symmetry. *Physical Review, X,* 5.

Whalley, K. (2015). Attention: Tuning sensory selection. *Nature Reviews Neuroscience, 16,* 64–65.

White, J. G., Southgate, E., Thomson, J. N., & Brenner, S. (1986). The structure of the nervous system of the nematode *Caenorhabditis elegans. Philosophical Transactions of the Royal Society of London B: Biological Sciences, 314,* 1–340.

Wiles, L., Gu. S., Pasqualetti, F., Parvesse, B., Gabrieli, D., Bassett, D. S., & Meaney, D. F. (2017). Autaptic connections shift network excitability and bursting. *Scientific Reports, 7,* 44006.

Wonham, W. M. (1985). *Linear multivariable control: A geometric approach* (3rd ed.). Berlin: Springer, 3.

Yamawaki, N., Radulovic, J., & Shepherd, G. M. (2016). A corticocortical circuit directly links retrosplenial cortex to M2 in the mouse. *Journal of Neuroscience, 36,* 9365–9374.

Yan, G., Vértes, P. E., Towlson, E. K., Chew, Y. L., Walker, D. S., Schafer, W. R., & Barabási, A. L. (2017). Network control principles predict neuron function in the *Caenorhabditis elegans* connectome. *Nature, 550,* 519–523.

62 Functional Connectivity and Neuronal Dynamics: Insights from Computational Methods

DEMIAN BATTAGLIA AND ANDREA BROVELLI

ABSTRACT Brain function relies on flexible communication between cortical regions. However, the mechanisms underlying flexible information routing are still largely unknown. Here we hypothesize that the emergence of flexible information routing patterns is due to the complex dynamics—often oscillatory—supported by the underlying structural network. Through analyses of computational models of circuits with interacting areas, we show that different dynamic states compatible with a given connectome mechanistically implement different ways of exchanging information. As a result, a fast, network-wide, and self-organized reconfiguration of information-routing patterns—and functional connectivity networks, seen as their proxy—is achieved by inducing transitions between the available intrinsic dynamic states. We present here a survey of theoretical and modeling results, as well as of metrics of functional connectivity that are compliant with the daunting task of characterizing dynamic routing. We thus suggest both a theoretical framework and a tool box for future studies of how neural dynamics serve as a substrate for cognitive algorithms.

Theory: Function Follows Dynamics Rather than Structure

Brain functions in general require the control of distributed networks of interregional communication on fast timescales incompatible with the plasticity of connectivity tracts (Bressler & Kelso, 2001; Varela et al., 2001). This argument has led to the definition of notions of connectivity that are not based on the underlying *structural connectivity* (SC; i.e. anatomic) but rather attempt to capture the exchange of information between neuronal populations. All these *functional connectivities* (FCs) share the key property of being reconfigurable even when the underlying SC is fixed. Yet it is not fully understood which circuit mechanisms allow for flexible FC.

Proposals for circuit mechanisms underlying reconfigurable interregional communication range from hypothetical circuitry dedicated to routing (Vogels & Abbott, 2009; Zylberberg et al., 2010) to conditional signal propagation along interacting synfire chains (Hahn et al., 2014; Kumar, Rotter, & Aertsen, 2008) to the hypothesis that oscillatory rate modulation enables signal multiplexing (Akam & Kullmann, 2014). More in general, dynamic patterns of interregional oscillatory coherence have the potential to orchestrate selective and directed information transfer (Engel, Fries, & Singer, 2001; Varela et al., 2001) over multiple frequency bands (Bastos et al., 2015). According to the influential communication-through-coherence (CTC) hypothesis (see Fries, 2015 for a recent review), neuronal groups oscillating in a suitable phase coherence relation will exchange information more efficiently than neuronal groups that are not synchronized. A growing body of experimental evidence has been accumulated in support of CTC. Yet our understanding of how interareal phase coherence is flexibly regulated is still largely incomplete, especially given how far from ideal *metronomes* are oscillations in vivo (Ray & Maunsell, 2015; Xing et al., 2012).

In this chapter, after shortly reviewing some techniques to estimate flexible FC, we will show how theoretical and computational neuroscience approaches can bring fresh air into the debate on flexible routing and dynamic functional connectivity. Our core theoretical idea is that the relation between SC and FC is not direct but necessarily mediated by emergent collective dynamics at the system's level. More specifically, the anatomy of brain circuits constrains the functional interactions that these circuits can support but cannot determine them fully. Generally, a given structural network will engender a rich repertoire of possible collective dynamic states, also known as the *dynome* (Kopell et al., 2014). On its turn, every dynamic state within the dynome (e.g., different patterns of oscillatory phase coherence between interconnected neuronal populations) will mechanistically implement a different modality of exchanging information among the network nodes, or *information-routing pattern* (Kirst, Timme, & Battaglia, 2016). Thus, streams of input information will propagate through the network (or not) along different pathways, conditionally on the dynamic state in which the system is prepared (figure 62.1*A*). Switching from one information-routing pattern to another can simply be induced by biasing neural circuit dynamics to self-organize collectively into another of its possible intrinsic modes.

Since the dynome level is not accessible to direct experimental observation, computational models of neural dynamics are necessary to investigate it. Time series can be generated from simulations of "virtual brains" of increasing complexity—from toy brains with a few coupled areas (Battaglia et al., 2012; Palmigiano et al., 2017) up to whole-brain networks (Deco, Jirsa, & Mcintosh, 2011)—and FC estimated from simulated time series of activity, using precisely the same metrics used for actual brain recordings. Furthermore, models can be used to interpret functional connectivity dynamics (FCD)—that is, the structured temporal variability of FC networks observed in the resting state (Hutchison et al., 2013)—also referred to as the *chronnectome* (Calhoun et al., 2014)—or through the steps of a task (Brovelli et al., 2017) in terms of the sampling of the available dynome (Hansen et al., 2015).

Measuring Dynamic Routing and Functional Connectivity

We provide here a quick survey of common FC metrics. Despite their different specializations and relative complexity, all these metrics share a fundamental qualitative aspect: their dependence from the underlying dynamic

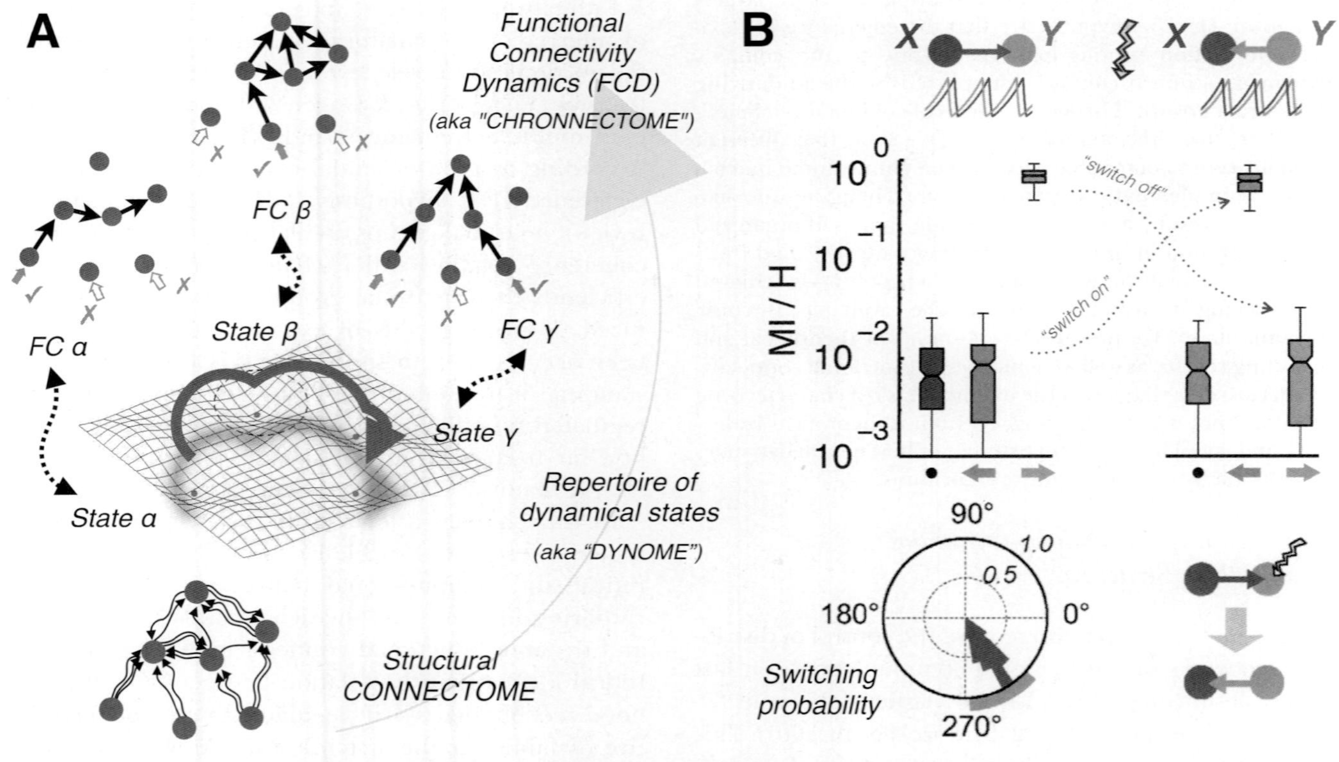

FIGURE 62.1 From structural to functional connectivity via dynamics. *A*, Structural connectivity (SC) of a neuronal circuit shapes but does not fully determine neural dynamics. Even for a fixed connectome, a multiplicity of collective dynamic states can exist—for example, different patterns of oscillatory phase locking between network units. The set of possible dynamic states compatible with a given connectome constitutes its associated *dynome*, or internal repertoire of available dynamic modes. Every dynamic state implements a different way of exchanging information between network units, leading to alternative functional connectivities. As a result of the stochastic sampling of the dynome, switching transitions between these many possible functional connectivity (FC) networks may occur even at rest, giving rise to nontrivial functional connectivity dynamics (FCD), also referred to as the *chronnectome*. *B*, We consider here, for example, a toy brain of two coupled model brain regions *X* and *Y*, undergoing sparsely synchronized oscillations. Two possible interregional phase-locking modes exist, in which either the *X* (*left*) or *Y* (*right*) region is leading in phase, associated to FC motifs with opposite directions of information transfer. In each of the two possible states, information conveyed by spiking code words emitted by source neurons in the phase-leading area can be decoded from code words emitted by target neurons in the phase-laggard area (70% of shared information). However, decoding efficiency does not rise above chance level (•) in the opposite laggard-to-leader direction. Switching between phase-locking modes can be induced by precisely phased pulse perturbations, applied within a specific control phase range (correctly predicted by theory, *highlighted range*). Adapted from Battaglia et al. (2012).

state. Furthermore, all of them can be applied to the analysis of both empirical and simulated time series.

We restrict our presentation to data-driven FC metrics, inferring FC directly from time series of activity without a priori hypotheses on existing couplings, referring the reader to, for example, Friston (2011) for model-driven effective connectivity.

A zoo of functional connectivity metrics The plethora of FC metrics used in cognitive neuroscience can be categorized into undirected and directed measures. FC undirected metrics include various measures based on the covariance notion, such as Pearson's and Spearman's rank *correlation coefficients* (CC). As an extension of linear CC, *mutual information* (MI) provides a more general measure of the dependence between signals by also capturing, in principle, nonlinear relations. MI quantifies shared information between two signals, and it reflects the reduction in uncertainty about one variable given the knowledge of another (MacKay, 2003). When dealing with oscillatory neural signals, their functional coupling can vary as a function of frequency. The most commonly used metric quantifying coupling in the frequency domain is *magnitude-squared coherence* (MSC), which can be seen as the frequency-domain analog of squared CC. The coupling between neural oscillations can also be quantified using *phase synchronization* (Rosenblum, Pikovsky, & Kurths, 1996), defined as the entrainment of phases irrespectively of amplitude correlations, or *phase-locking value* (Lachaux et al., 1999), detecting preferred values of the phase difference at a given frequency between signals. A more general way to establish FC among spectrally complex oscillatory signals relies on cross-frequency coupling (Canolty & Knight, 2010), tracked, for example, by means of *phase-to-amplitude coupling* (Aru et al., 2015).

Directed FC metrics include statistical approaches that can resolve the direction of influence between neural signals and are thus, in principle, better suited to capture dynamic information routing. In the sense of *Granger-Wiener causality* (GC; Granger, 1969; Wiener, 1956), a time series exerts a causal influence on another if the variance of the autoregressive prediction error of the latter is reduced by including the past measurements of the former. Beyond autoregressive modeling, Granger (1980) formalized a general condition of "Granger noncausality" between two time series X and Y as

$$p(Y_{i+1} \mid Y^{(i)}, X^{(i)}) = p(Y_{i+1} \mid Y^{(i)}), \tag{62.1}$$

where the superindex *(i)* refers to the past history of the time series up to and including sample *i*. Accordingly, *causality* can be defined as a deviation from this condition of "noncausality" and quantified by calculating the information-theoretical Kullback-Leibler divergence (MacKay, 2003) between the two conditional probabilities in equation (62.1). In a bivariate context comprising only X and Y, this divergence can be written as follows:

$$\begin{aligned} \mathrm{TE}_{X\to Y} &\equiv H(Y_{i+1} \mid Y^{(i)}) - H(Y_{i+1} \mid X^{(i)}, Y^{(i)}) \\ &= \mathrm{MI}(Y_{i+1}; X^{(i)} \mid Y^{(i)}). \end{aligned} \tag{62.2}$$

The difference of two conditional entropies H on the right-hand side of equation (62.2) quantifies the decrease in uncertainty about future values Y_{i+1} when the past history $X^{(i)}$ is also known. However, even more interesting is the further rewriting of $\mathrm{TE}_{X\to Y}$ as a mutual information term $\mathrm{MI}(Y_{i+1}; X^{(i)} \mid Y^{(i)})$. In layman terms, this term quantifies the amount of information that wasn't already encoded by Y's past history but that can be found in Y's present because it was transferred there from X. Such quantity $\mathrm{TE}_{X\to Y}$ has been named *transfer entropy* (TE; Schreiber, 2000) and represents the most general measure of information transfer capturing any (linear and nonlinear) time-lagged conditional dependence (Wibral, Vicente, & Lizier, 2014).

Directed FC metrics have also been generalized to capture information transfer in the frequency domain, a feature particularly suitable when investigating the role of neural oscillations in establishing interregional interactions at different frequencies. Tools for the nonparametric estimation of *spectrally decomposed Granger causality* directly from Fourier and wavelet transforms of time-series data are available (Dhamala, Rangarajan, & Ding, 2008). However, there is not yet consensus on how to generalize TE to the spectral domain.

Single-trial-based functional connectivity metrics A common strategy to track the temporal dynamics of FC couplings, independently from the used metric, is to assume that experimental trials are realizations of the same stationary stochastic process. In the framework of autoregressive models, this allows the estimation of model coefficients across trials on short time windows for the computation of coherence and Granger causality spectra with high temporal precision (see, e.g., Brovelli et al. [2004, 2015] introducing a powerful hierarchical pipeline). Neural coupling, however, may vary across trials and reflect behavioral modulations occurring during learning and adaptive behaviors (e.g., changes in reaction time across trials). There is therefore a need for FC metrics that can be extracted based on single trials.

A classical approach to estimate single-trial FC is to compute the spectral density matrices over subsegments of time series within a trial, stepped to cover the whole duration of the trial. Such an approach can be used to estimate *single-trial phase synchrony* (Lachaux et al., 2000) and *single-trial Granger causality* using a combination of general linear models and nonparametric spectral techniques (Brovelli, 2012) or covariance-based methods (Brovelli et al., 2015). Alternatively, jackknife approaches have proved to be adequate for single-trial estimates of spectrally resolved FC metrics (Richter et al., 2015).

To conclude, a note of caution should be sounded regarding concerns with the estimation of directed and directed FC metrics, especially when time resolved. The most common factors limiting the correct estimate and interpretability of FC measures are the sample-size bias problem, varying levels of signal-to-noise ratio, volume conduction, and common input or indirect interaction effects (see Bastos and Schoffelen [2015] for a review). Note that the problem of FC estimation is much less severe when dealing with simulated signals, which can be arbitrarily long and artifact-free. We expect, nevertheless, that new techniques first tested in silico will also become applicable to actual data, thanks to the development of improved estimators—for example, as for *time-resolved TE* (Wollstadt et al., 2014).

Functional connectivity dynamics along a task Ultimately, cognition necessarily unrolls in time, and mental operations are built out of successive steps, which assemble into a cognitive architecture mixing serial and massively parallel information processing, also dubbed a *human Turing machine* (Zylberberg et al., 2011). Time-resolved FC analyses can be used to probe how cognitive functions arise from the time-ordered interplay of multiple networks. For instance, in a recent work (Brovelli et al., 2017) we used time-resolved FC analyses of human high-gamma activity to show that visuomotor mapping arises from a sequential *recruitment schedule* of FC networks (figure 62.2): first, a network involving visual and parietal regions coordinated with sensorimotor and premotor areas; second, the dorsal frontoparietal circuit, together with the sensorimotor and associative frontostriatal networks, took the lead; and finally, corticocortical interhemispheric coordination among bilateral sensorimotor regions coupled with the left frontoparietal network and visual areas. These corticocortical and corticosubcortical FC networks—partly overlapping—were interpreted as reflecting the processing of visual information, the emergence of visuomotor plans, and the processing of somatosensory reafference or action outcomes, respectively.

More generally, FCD analyses show that the interdependence between brain regions and networks is nonstationary and displays switching dynamics and areal flexibility over timescales relevant for task performance. FCD approaches thus help elucidate the relation between fast dynamic FC reconfiguration and the algorithmic buildups of executive functions.

Modeling Dynamic Routing and Functional Connectivity

One structural network engenders many functional networks As previously discussed, dynamics on top of a fixed connectome will give rise to an entire repertoire of possible dynamic modes, composing the connectome's dynome. This phenomenon is epitomized by simple toy models involving only a small number of coupled areas. Following Battaglia et al. (2012), we consider in figure 62.1*B* a toy brain of two reciprocally connected brain regions. Such an abstract structural motif serves as a metaphor for canonical cortical circuits in which the relative weights of top-down and bottom-up functional influences must be dynamically adjusted. Every brain region is modeled as a local network of thousands of excitatory and inhibitory spiking neurons, connected by random recurrent connectivity. Parameters

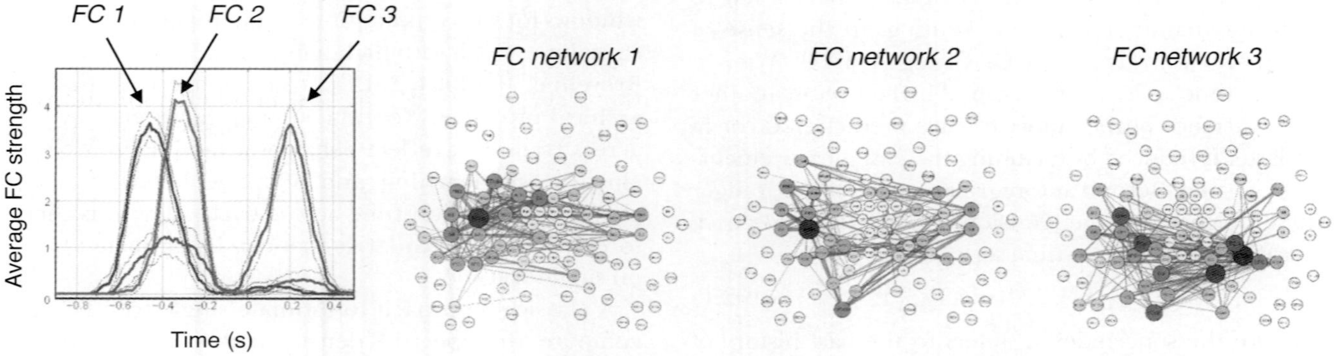

FIGURE 62.2 Functional connectivity dynamics along a task. Time-resolved FC estimated along the performance of a similar task. Three different partially overlapping networks (*right*) activate and deactivate with a characteristic recruitment schedule (*left*). Adapted from Brovelli et al. (2017).

are selected in such a way that each local region generates sparsely synchronized collective oscillations—that is, the firing of individual neurons remains realistically irregular even when the average population activity oscillates periodically at frequencies in the gamma range (40–80 Hz). Since firing is Poisson-like, spike trains have a high entropy, and a large amount of information can be conveyed by the oscillating population within every oscillation cycle. In other words, the oscillations themselves are not likely to encode information but act as carriers for general code words encoded in detailed spiking patterns "surfing on the wave." When coupled with long-range excitation, the oscillating regions will phase lock with preferred phase relations that depend on interareal delays but are also especially influenced by the strength of local inhibition within each region (Battaglia et al., 2012; Palmigiano et al., 2017). For sufficiently strong inhibition, a multiplicity of out-of-phase locking modes tend to emerge, in which one of the two regions leads in phase over the other, despite the reciprocity of coupling.

We quantified the FC associated with different phase-locking modes, using TE as a metric of choice. By evaluating TE between time series of LFP-like signals (average regional activity), we found that for weak interregional coupling, TE was significant only in the direction from the leader to the laggard region, in agreement with physiological intuition from the CTC hypothesis (Fries, 2015). Besides the unidirectional transfer of information, other functional motifs can be implemented by our toy brain (bidirectional, either anisotropic or symmetric; effective disconnection; and so on). Importantly, the directionality of coupling inferred by TE between collective region-level activations also predicts the direction of communication for information encoded at the microscopic level (figure 62.1*B*, *top*). Indeed, the spiking code words of the leader region can be decoded from laggard spiking code words in a matching cycle with approximately 70% accuracy, but not the other way around, as quantified by MI analyses (Battaglia et al., 2012).

Self-organized control of information routing Under the effect of an arbitrary perturbation, the system will be transiently destabilized, but its dynamics will then converge back to one of the available intrinsic modes. If the applied perturbation kicks the system out of the phase-space basin of attraction of the current dynamic state—a valley in an idealized landscape—the system will converge toward a different state within its dynome. As a result, the implemented FC network will also switch to the one associated with the newly recruited state (cf., figure 62.1*A*). Various mechanisms could force the system to leave its current state and then be used for implementing routing control. A first possibility would be to modulate the relative attractiveness of different states (in the landscape metaphor of figure 62.1, this would correspond to make one valley deeper and broader than the others). In the presence of multistability between multiple dynamic configurations, it would be enough to apply a steady input bias to one of the two populations to automatically enhance its probability of becoming phase leader and thus act as an effective information sender (Palmigiano et al., 2017). An unspecific, weak bias would be enough because its role would just be to favor the otherwise self-organized selection of a specific routing state from a preexisting repertoire. Therefore, no additional circuitry for routing control would be required besides the one already responsible for the generation of collective oscillations themselves, in contrast with other proposed mechanisms (e.g., Vogels & Abbott, 2009; Zylberberg et al., 2010). Such a steady bias could be provided by some top-down modulatory signal, neuromodulation, or even stimulus saliency itself.

Furthermore, our theory predicts that if the system's dynamic states are sufficiently stable—as in the case of strong oscillatory power—robust rerouting could even be induced just by precise phase pulse-like inputs, removing the need for a steadily applied bias. Simulations in Battaglia et al. (2012), in agreement with analytical expectations, demonstrate that the reversal of the information transfer direction can be triggered with near-to-one probability by a pulse perturbation delivered to a small fraction of randomly chosen neurons (e.g., in the laggard region), provided that the pulse is applied within a suitable and narrow phase range (but not outside of it; figure 62.1*B*, *bottom*). Such theoretical prediction has not yet been confirmed but could be experimentally validated using, for example, closed-loop optogenetic stimulation (Witt et al., 2013).

We have also generalized our findings to larger networks with arbitrarily complex modular topologies (Kirst, Timme, & Battaglia, 2016). When moving to these larger network models, another nonintuitive—and, in perspective, testable—prediction of our theory is that local perturbations to a target region could induce distributed changes of FC even between distant regions. In our study we indeed predicted that the dominant direction of connectivity between two regions X and Y could be reverted by applying a drive bias to a third "remote controller" region Z. Such a mechanism, emphasizing the nonlocality of the effects of a localized system perturbation, is robust since connectivity patterns would be stable over broad ranges of parameters, and switching would occur—suddenly and "everywhere"—only in the proximity of specific, critical working points of operation.

Self-organized routing with transient and stochastic oscillations The toy models considered in figure 62.1*B* give rise to unrealistically "clock-like" collective oscillations. In reality, oscillatory episodes in vivo are usually transient, arising at stochastic timings and with an inconsistently volatile frequency (Ray & Maunsell, 2015; Xing et al., 2012). In Palmigiano et al. (2017), we have extended the toy model of figure 62.1*B* to a more realistic regime in which asynchronous activity coexists with stochastic oscillatory bursts, as in vivo.

Remarkably, model simulations—as well as experiments (Roberts et al., 2013)—show that the oscillatory bursts of coupled regions continue to be stochastic but that correlations—in both occurrence time and frequency—spontaneously develop between the coupled regions. Furthermore, these co-occurring bursts manifest intrinsically with different sets of favorite phase

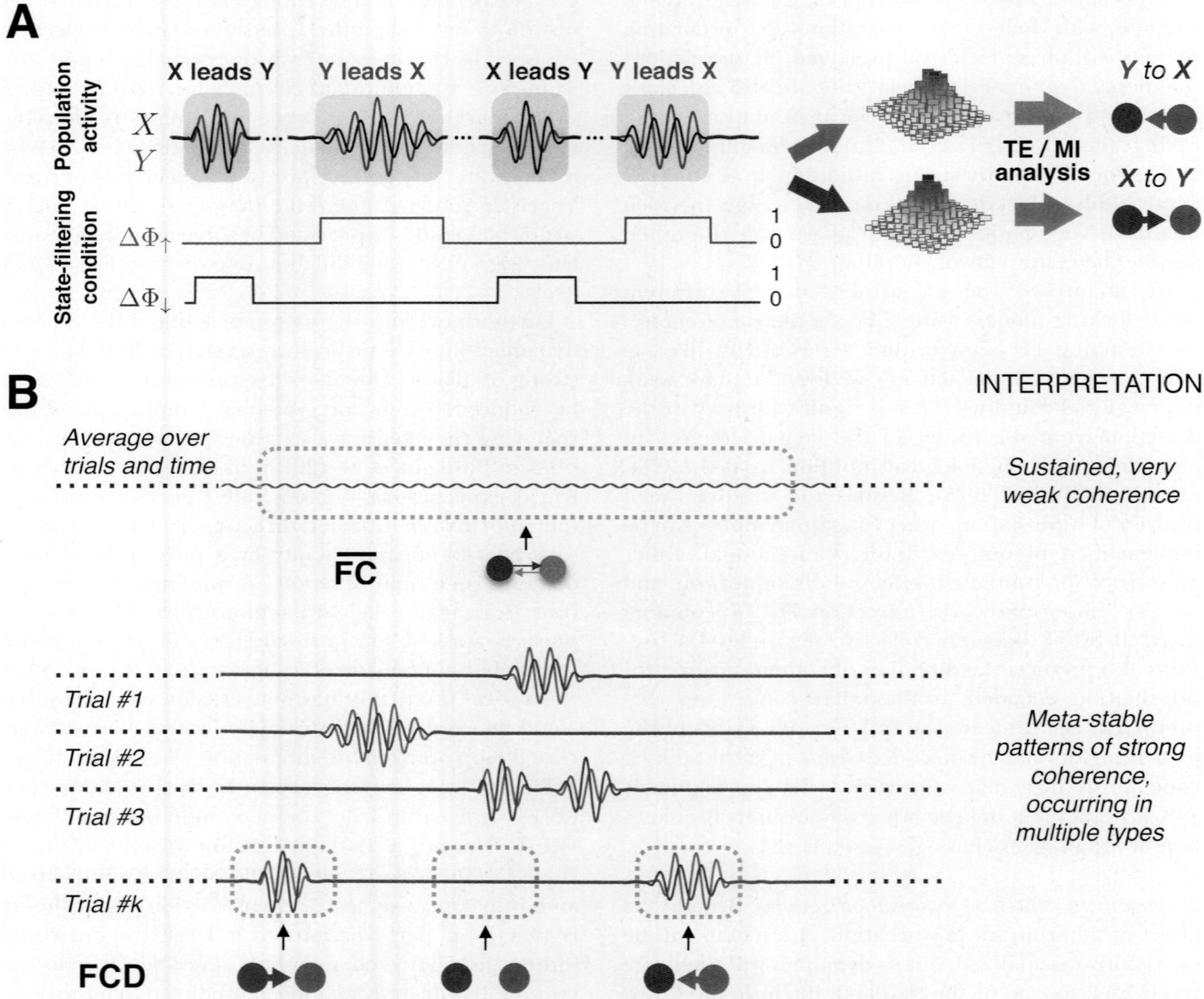

FIGURE 62.3 Transient information-routing patterns. *A*, Oscillatory events in vivo are highly transient and occur at stochastic times. FC analyses can be restricted only to time epochs for which a specific set of state-filtering conditions are fulfilled, such as, for example, instantaneous coherence above a threshold and phase relation within alternative specified ranges (here, $\Delta\Phi_{\uparrow\downarrow}$, corresponding, respectively, to *X* or *Y* as phase-leading regions). Thus, an associated information-routing pattern can be computed for each specific class of metastable oscillatory transients. Adapted from Palmigiano et al. (2017). *B*, The stochasticity of the timing of different routing oscillatory events may lead to spurious interpretations when computing average FC over time-aligned trials, rather than computing FCD along single trials (e.g., weak sustained vs. strong but dynamic coherence patterns).

relations, and each set of phase relations continues to map to a different information-routing pattern, analogous to the case of higher synchrony models but now metastable and transient.

This can be proven by restricting TE analyses to time epochs prelabeled as belonging to a specified target state. In figure 62.3*A*, we defined *state-selecting filters*, tagging an epoch as belonging to a given routing state if instantaneous coherence exceeds a certain threshold, and the interregional phase difference between two coupled regions *X* and *Y* falls within a specified interval. Different filters can be defined to track the stochastic manifestation of different routing states (e.g., *X* phase leading or phase lagging over *Y*). A *state-dependent TE*—or any other FC metric of choice—is then extracted by pooling together activity measurements collected at instants tagged to belong to each given state. Note that state-resolved FC analyses could be seen as a generalization of representational similarity analyses (Kriegeskorte, Mur, & Bandettini, 2008) from activation to connectivity patterns.

Via state-resolved analyses, we can thus conclude that the transient and stochastic nature of oscillations is not an obstacle to the flexible and controllable selective routing of input signals, thanks to collective self-organization. A key prediction of the model—that calls for an experimental confirmation—is that directed information transfer between coupled regions should be intermittent—that is, strongly enhanced during co-occurring oscillatory bursts and reduced to baseline, or even actively suppressed, between these oscillatory events (Palmigiano et al., 2017).

Whole brains? Recently, mean-field whole-brain modeling (Deco, Jirsa, & Mcintosh, 2011) has been used to study the emergence of FC networks from the collective self-organized dynamics of an SC network embedding realistic connectome data. While early analyses have been limited to the rendering of time-averaged resting-state FC, in a recent modeling study (Hansen et al., 2015) we have shown that a chronnectome can also be qualitatively reproduced. In agreement with our theory, a nontrivial FCD arises when the global parameters of the model are tuned to a working point that maximizes the richness of the model's dynome.

However, modeling FCD at the whole-brain level is still in its first steps and currently limited to resting state only (i.e., not yet for task FC schedules, as in figure 62.2). Promising recent developments (e.g., Mejias et al., 2016) nevertheless suggest that mean-field models could in the near future become a valuable tool to study emergent brain-wide networks of flexible multifrequency coherence.

Implications for Functional Connectivity Analyses

We propose that FC networks are a measurable proxy for information-routing patterns implemented by the collective dynamics of neural circuits. According to this vision, the richness of the dynome of a given structural circuit will translate into a parallel variety of possible FC network that can be observed at different moments in time.

A large number of classic analyses of FC are based on averaging FC metrics over very long times or many trials eventually time-aligned to some extrinsic reference event, such as a sensory cue given during a cognitive task. However, if a rich repertoire of states is sampled, either spontaneously as an effect of noise or in a way guided by exogenous (sensory) or endogenous (cognitive) bias, every averaging procedure is going to destroy precious information (Hutchison et al., 2013). This is true even for averaging over time-aligned trials since we cannot a priori guarantee that transitions between internal states are really so tightly linked to task-related events.

Figure 62.3*B* depicts a cartoon situation in which trial averaging would lead to the conclusion that a weak, sustained interareal phase coherence exists between two probed channels. In reality, matching oscillatory bursting events with different phase relations are stochastically occurring along each trial and at different timings for different trials. A more correct interpretation, then, would be that the two regions transiently exchange information with great efficiency (in different possible directions) but only at selected times.

The two interpretations are qualitatively different and lead to radically diverging visions of how information processing works. The static vision conveyed by time and trial averaging may be too strongly influenced by our a priori hypotheses. We thus foster, even in task-based studies, the use of methods able to agnostically detect intrinsic connectivity states and their dynamics.

We foresee that tackling the formidable technical challenge of developing new approaches for single-trial and state-based FC analyses will lead us to find—paraphrasing Haldane (1927)—that the brain is way queerer than we suppose (if not queerer than we can suppose).

Acknowledgments

We acknowledge support from the CNRS Mission pour l'Interdisciplinarité (INFINITI "BrainTime") and from the French program Investissements d'Avenir (through the Institut de Convergence ILCB).

REFERENCES

Akam, T. E., & Kullmann, D. M. (2014). Oscillatory multiplexing of population codes for selective communication in the mammalian brain. *Nature Reviews Neuroscience, 15*(2), 111–122.

Aru, J., et al. (2015). Untangling cross-frequency coupling in neuroscience. *Current Opinion in Neurobiology, 31*, 51–61.

Bastos, A. M., et al. (2015). Visual areas exert feedforward and feedback influences through distinct frequency channels. *Neuron, 85*(2), 390–401.

Bastos, A. M., & Schoffelen, J.-M. (2015). A tutorial review of functional connectivity analysis methods and their interpretational pitfalls. *Frontiers in Systems Neuroscience, 9*, 175.

Battaglia, D., Witt, A., Wolf, F., & Geisel, T. (2012). Dynamic effective connectivity of inter-areal brain circuits. *PLoS Computational Biology, 8*(3), e1002438.

Bressler, S. L., & Kelso, J. A. (2001). Cortical coordination dynamics and cognition. *Trends in Cognitive Sciences, 5*, 26–36.

Brovelli, A. (2012). Statistical analysis of single-trial Granger causality spectra. *Computational and Mathematical Methods in Medicine, 2012*, 697610.

Brovelli, A., Badier, J. M., Bonini, F., Bartolomei, F., Coulon, O., & Auzias, G. (2017). Dynamic reconfiguration of visuomotor-related functional connectivity networks. *Journal of Neuroscience, 37*(4), 839–853.

Brovelli, A., Chicharro, D., Badier, J.-M., Wang, H., & Jirsa, V. (2015). Characterization of cortical networks and cortico-cortical functional connectivity mediating arbitrary visuomotor mapping. *Journal of Neuroscience, 35*, 12643–12658.

Brovelli, A., Ding, M., Ledberg, A., Chen, Y., Nakamura, R., et al. (2004). Beta oscillations in a large-scale sensorimotor cortical network: Directional influences revealed by Granger causality. *Proceedings of the National Academy of Sciences of the United States of America, 101*, 9849–9854.

Calhoun, V. D., Miller, R., Pearlson, G., & Adali, T. (2014). The Chronnectome: Time-varying connectivity networks as the next frontier in fMRI data discovery. *Neuron, 84*(2), 262–274.

Canolty, R. T., & Knight, R. T. (2010). The functional role of cross-frequency coupling. *Trends in Cognitive Sciences, 14*(11), 506–515.

Deco, G., Jirsa, V. K., & Mcintosh, A. R. (2011). Emerging concepts for the dynamical organization of resting-state activity in the brain. *Nature Reviews Neuroscience, 12*(1), 43–56.

Dhamala, M., Rangarajan, G., & Ding, M. (2008). Estimating Granger causality from Fourier and wavelet transforms of time series data. *Physical Review Letters*, 100(1), 018701.

Engel, A. K., Fries, P., & Singer, W. (2001). Dynamic predictions: Oscillations and synchrony in top-down processing. *Nature Reviews Neuroscience, 2*(10), 704–716.

Fries, P. (2015). Rhythms for cognition: Communication through coherence. *Neuron, 88*(1), 220–235.

Friston, K. J. (2011). Functional and effective connectivity: A review. *Brain Connect, 1*, 13–36.

Granger, C. W. J. (1969). Investigating causal relations by econometric models and cross-spectral methods. *Econometrica, 37*, 424.

Granger, C. W. J. (1980). Testing for causality. *Journal of Economic Dynamics and Control, 2*, 329–352.

Hahn, G., Bujan, A. F., Frégnac, Y., Aertsen, A., & Kumar, A. (2014). Communication through resonance in spiking neuronal networks. *PLoS Computational Biology, 10*, e1003811–e1003816.

Haldane, J. B. S. (1927). Possible worlds: and other essays. London: Chatto and Windus.

Hansen, E. C. A., Battaglia, D., Spiegler, A., Deco, G., & Jirsa, V. K. (2015). Functional connectivity dynamics: Modeling the switching behavior of the resting state. *NeuroImage, 105*, 525–535.

Hutchison, R. M., Womelsdorf, T., Alles, E. A., Bandettini, P. A., Calhoun, V. D., Corbetta, M., Della Penna, S., et al. (2013). Dynamic functional connectivity: Promise, issues, and interpretations. *NeuroImage, 80*, 360–378.

Kirst, C., Timme, M., & Battaglia, D. (2016). Dynamic information routing in complex networks. *Nature Communications, 7*, 11061.

Kopell, N. J., Gritton, H. J., Whittington, M. A., & Kramer, M. A. (2014). Beyond the connectome: The dynome. *Neuron, 83*(6), 1319–1328.

Kriegeskorte, N., Mur, M., & Bandettini, P. (2008). Representational similarity analysis—connecting the branches of systems neuroscience. *Frontiers in Systems Neuroscience, 2*, 4.

Kumar, A., Rotter, S., & Aertsen, A. (2008). Conditions for propagating synchronous spiking and asynchronous ring rates in a cortical network model. *Journal of Neuroscience, 28*, 5268–5280.

Lachaux, J.-P., Rodriguez, E., Martinerie, J., & Varela, F. J. (1999). Measuring phase synchrony in brain signals. *Human Brain Mapping, 8*, 194–208.

Lachaux, J.-P., Rodriguez, E., Van Quyen, M. L. E., Lutz, A., Martinerie, J., & Varela, F. J. (2000). Studying single-trials of phase synchronous activity in the brain. *International Journal of Bifurcation and Chaos, 10*, 2429–2439.

MacKay, D. (2003). *Information theory, inference, and learning algorithms*. Cambridge: Cambridge University Press.

Mejias, J. F., Murray, J. D., Kennedy, H., & Wang, X.-J. (2016). Feedforward and feedback frequency-dependent interactions in a large-scale laminar network of the primate cortex. *Science Advances, 2*(11), e1601335–e1601335.

Palmigiano, A., Geisel, T., Wolf, F., & Battaglia, D. (2017). Flexible information routing by transient synchrony. *Nature Neuroscience, 20*(7), 1014–1022.

Ray, S., & Maunsell, J. H. R. (2010). Differences in gamma frequencies across visual cortex restrict their possible use in computation. *Neuron, 67*(5), 885–896.

Ray, S., & Maunsell, J. H. R. (2015). Do gamma oscillations play a role in cerebral cortex? *Trends in Cognitive Sciences, 19*(2), 78–85.

Richter, C. G., Thompson, W. H., Bosman, C. A., & Fries, P. (2015). A jackknife approach to quantifying single-trial correlation between covariance-based metrics undefined on a single-trial basis. *NeuroImage, 114*, 57–70.

Roberts, M. J., Lowet, E., Brunet, N. M., Ter Wal, M., Tiesinga, P., Fries, P., & De Weerd, P. (2013). Robust gamma coherence between macaque V1 and V2 by dynamic frequency matching. *Neuron, 78*(3), 523–536.

Rosenblum, M. G., Pikovsky, A. S., & Kurths, J. (1996). Phase synchronization of chaotic oscillators. *Physical Review Letters, 76*, 1804–1807.

Schreiber, T. (2000). Measuring information transfer. *Physical Review Letters, 85*(2), 461–464.

Varela, F., Lachaux, J. P., Rodriguez, E., & Martinerie, J. (2001). The brainweb: Phase synchronization and large-scale integration. *Nature Reviews Neuroscience, 2*, 229–239.

Vogels, T. P., & Abbott, L. F. (2009). Gating multiple signals through detailed balance of excitation and inhibition in spiking networks. *Nature Neuroscience, 12,* 483–491.

Wibral, M., Vicente, R., & Lizier, J. T. (2014). *Directed information measures in neuroscience.* In M. Wibral, R. Vicente, & J. T. Lizier (Eds.). Berlin: Springer. doi:10.1007/978-3-642-54474-3

Wiener, N. (1956). Nonlinear prediction and dynamics. *In proceedings of the third Berkeley symposium on mathematical statistics and probability* (Vol. 3), 247–252. Berkeley: University of California Press.

Witt, A., Palmigiano, A., Neef, A., El Hady, A., Wolf, F., & Battaglia, D. (2013). Controlling the oscillation phase through precisely timed closed-loop optogenetic stimulation: A computational study. *Frontiers in Neural Circuits, 7,* 49.

Wollstadt, P., Martínez-Zarzuela, M., Vicente, R., Díaz-Pernas, F. J., & Wibral, M. (2014). Efficient transfer entropy analysis of non-stationary neural time series. *PLoS One, 9*(7), e102833.

Xing, D., Shen, Y., Burns, S., Yeh, C.-I., Shapley, R., & Li, W. (2012). Stochastic generation of gamma-band activity in primary visual cortex of awake and anesthetized monkeys. *Journal of Neuroscience, 32*(40), 13873–13880a.

Zylberberg, A., Dehaene, S., Roelfsema, P. R., & Sigman, M. (2011). The human Turing machine: A neural framework for mental programs. *Trends in Cognitive Sciences, 15*(7), 293–300.

Zylberberg, A., Fernández Slezak, D., Roelfsema, P. R., Dehaene, S., & Sigman, M. (2010). The brain's router: A cortical network model of serial processing in the primate brain. *PLoS Computational Biology, 6,* e1000765.

IX
CONCEPTS AND CORE DOMAINS

Introduction

MARINA BEDNY AND ALFONSO CARAMAZZA

Human knowledge covers a wide range of content, from physical objects and how they interact to internal mental states (e.g., goals, beliefs) and abstract numerical quantities (e.g., 3). A key question that has motivated cognitive neuroscience from its inception concerns the degree to which knowledge about different domains is represented and processed by dedicated systems, as opposed to domain-general ones. Neuropsychological disorders provided the earliest insight for an organization of the brain into distinct cognitive systems. Arithmetic deficits (acalculia) dissociate from language deficits (aphasia), and deficits in reasoning about objects dissociate from deficits in action processing. Finer-grained dissociations can be found, as well. The processing of living things can be damaged or spared relative to that of nonliving things (Hillis & Caramazza, 1991; Warrington & Shallice, 1984), raising the question of how the conceptual system for representing objects might be organized so that brain damage can result in *category-specific* deficits. One proposal eschewed the idea that such deficits might reflect an organization by object category and instead suggested they arise from damage to object features that are disproportionately important for one object category over others—for example, the greater role of "visual" features for living things (Warrington & Shallice, 1983). However, the observation that the principal dissociation in object knowledge deficits is between animate and inanimate entities, such as, for example, that the processing of animals (excluding nonanimate living things) can be damaged selectively for all aspects of this object category, suggested that object knowledge might be organized by domain—broad, evolutionarily salient object categories such as animals, conspecifics, and tools

(Caramazza & Shelton, 1998). The neuropsychological results, together with neuroimaging evidence showing that different regions of occipitotemporal cortex respond preferentially to animate versus inanimate objects, reinforced the idea that object knowledge is organized into domain-specific networks of hierarchically organized, distributed representations.

Studies of the neural basis of concrete objects and actions continue to be a test bed for cognitive neuroscience theories of concepts. The identification of distinctions among conceptual domains has made it possible to better understand the overlap and interactions between them. Although it is now well established that object and action processing recruit partially distinct neural networks, the distinction is not absolute: there are aspects of object knowledge that are closely connected to and even overlapping with action knowledge, implying an organization of action and object concepts that cuts across those two major domains. For example, perceptual and conceptual information about the graspability of an object is captured in areas closely linked to manipulation and action (Leshinskaya, Wurm, & Caramazza, chapter 63, this volume; Mahon, chapter 64, this volume) Analogously, brain regions recruited during mechanical reasoning about objects overlap with those involved in action planning and tool use (Fischer, chapter 65, this volume).

Beyond establishing the neural separability of different knowledge domains and the relationship between them, studies of concrete objects and actions have tackled a number of theoretically significant issues: What is the topographic arrangement of the different domains, and why do they take that form? What are the major levels and dimensions of representation within a domain? What are the respective contributions of innately determined structure versus experience (this volume: Bedny, chapter 68; Bi, chapter 66; Clarke & Tyler, chapter 67; Mahon, chapter 64)?

Among these questions, perhaps one of the most difficult has been the very definition of a "concept" itself. There is disagreement in the field regarding what concepts are, what data are relevant to constraining our theories, and what would constitute an adequate explanation. One specific point of contention is how sensorimotor experience relates to concepts. Humans have detailed sensorimotor knowledge of specific objects and actions, and much of the research in this field has been directed at identifying the sensorimotor or appearance-related features associated with objects and actions and identifying how these features are encoded in the brain. For example, a large body of work has examined occipitotemporal representations of objects (Bi, chapter 66, this volume). This research has been highly productive, demonstrating, for example, that even perceptual representations of objects show a major divide between animate and inanimate domains and, within the inanimate domain, a distinction between manipulable objects (tools) and navigation-relevant objects (Bi, chapter 66, this volume; Mahon, chapter 64, this volume). Yet even a complete account of how we represent what objects look like would leave us a long way away from a satisfactory cognitive neuroscience theory of concepts.

Conceptual representations "must" be sufficiently general to encompass the great variety of sensorimotor experiences we consider instances of an object or action. Moreover, even young children override sensorimotor information in favor of inferences and unobservable essences during categorization (e.g., a bird that looks like a bat still lays eggs like a bird; Gelman & Markman, 1986). Cognitive neuroscience theories of concepts must align with these basic facts about human behavior.

Studies with individuals who lack a particular modality of sensory information from birth (e.g., blind, deaf, or amelic individuals) suggest that even seemingly concrete concepts have rich abstract representations that are neurally dissociable from those that are sensory: sensory loss leaves the neural organization of concepts largely unchanged while causing large-scale plasticity in "deprived" sensory systems (Bedny, chapter 68, this volume; Bi, chapter 66, this volume). Some representations that were originally viewed as perceptual/visual have turned out to be modality independent. This is well illustrated by the case of the so-called parahippocampal place area (PPA). While this region was originally identified with sighted subjects viewing pictures of scenes, the PPA is also activated during the tactile exploration of LEGO scenes and when subjects listen to the names of navigable spaces (e.g., meadow, barn) in both sighted and congenitally blind individuals (Bi, chapter 66, this volume). Such evidence raises questions about how innate constraints and experiences give rise to the neural architecture of concepts. If the sensory modality of input is not what determines the neural basis of concepts, what does?

Cognitive neuroscience research on concrete objects and actions has sometimes proceeded in isolation from research on other domains that are part of the mainstay of conceptual research in psychology—for example, spatial reasoning, intuitive physics, and numbers (this volume: Cantlon, chapter 70; Epstein, chapter 69; Fischer, chapter 65). In this respect, one might worry that cognitive neuroscience theories are overly skewed toward explaining object and action domains. For example, while knowing color, shape, and texture could be a part of our concept of *banana*, such sensorimotor features are unlikely to contribute to the concept of *3*. Work in the domains of intuitive physics, spatial reasoning, and numerical concepts is highly interdisciplinary, incorporating insights not only from cognitive neuroscience but also from developmental psychology and evolutionary biology.

Inspired by developmental psychology research on the early emergence of intuitive physical reasoning, Fischer (see chapter 65) reviews research on the cortical networks that support reasoning about the interactions of objects with each other and with human agents (e.g., How likely is a stack of blocks fall?). In the domain of spatial cognition, a network of cortical areas codes information that humans and other animals use to recognize and navigate spatial environments, with the PPA playing a specific role in scene recognition and categorization. Research on the PPA may be of particular interest to theories of concepts since, as noted above, it is activated not only during navigation but also during the comprehension of words that refer to place categories (Epstein, chapter 69, this volume). Finally, the domain of numerical concepts is one of the best examples of integrating research across disciplines. Cantlon (chapter 70, this volume) reviews studies of the cognitive and neural basis of number across the life span and across different species. This work demonstrates that elements of the cognitive and neural capacity for numerical reasoning are present in our evolutionary lineage, while the capacity for exact and symbolic calculation is uniquely human. Language and education transform innately constrained neural systems and enable them to support far more powerful cognitive mechanisms (Cantlon, chapter 70, this volume).

Additionally, this volume covers two important advances in research on concepts. First, until recently, most cognitive neuroscience research examined concepts in isolation, presenting subjects with single words or images. By contrast, in everyday reasoning we combine concepts into structured wholes—"The dog chased the black cat" means more than the sum of its parts. An important goal is to uncover the neural mechanisms of such combinations (Coutanche, Solomon, & Thompson-Schill, chapter 71, this volume). Another important development has been the incorporation of neural network models into the analysis of functional magnetic resonance imaging (fMRI) data. Such models can be trained on data banks of images, human-generated features, or large text corpora, and the similarity structure represented within different layers of the networks can be related to representations in different levels of cortical networks (Clarke & Tyler, chapter 67, this volume).

Since Warrington's (1975) seminal paper on semantic deficits, tremendous strides have been made in charting the neural organization of conceptual and high-level perceptual processing, and the coming years promise to be highly productive. Methods such as multivoxel pattern analysis have made it possible to interrogate the semantic dimensions made explicit by neural population codes within different networks. By applying these tools and working to bridge gaps between cognitive neuroscience and its allied disciplines, we can make progress toward answering the difficult questions: *What* are concepts and *how* are they represented in the brain? The chapters in this section represent efforts toward this goal.

REFERENCES

Caramazza, A., & Shelton, J. R. (1998). Domain-specific knowledge systems in the brain: The animate-inanimate distinction. *Journal of Cognitive Neuroscience, 10*(1), 1–34.

Gelman, S. A., & Markman, E. M. (1986). Categories and induction in young children. *Cognition, 23*, 183–209.

Hillis, A. E., & Caramazza, A. (1991). Category specific naming and comprehension impairment: A double dissociation. *Brain, 114*, 2081–2094.

Sirigu, A., Duhamel, J. R., & Poncet, M. (1991). The role of sensorimotor experience in object recognition. *Brain, 114*, 2555–2573.

Warrington, E. K. (1975). The selective impairment of semantic memory. *Quarterly Journal of Experimental Psychology, 27*, 635–657.

Warrington, E. K., & Shallice, T. (1984). Category specific semantic impairments. *Brain, 107*(3), 829–853.

63 Concepts of Actions and Their Objects

ANNA LESHINSKAYA, MORITZ F. WURM, AND ALFONSO CARAMAZZA

ABSTRACT We take concepts to be mental representations involving stored knowledge with some level of generality and modality invariance. Here we explore the neural organization of action concepts. In the neuropsychological literature on action production and comprehension, a *mechanical reasoning* system diverges from a system based more on object identity, and within the latter system, only rarely is the understanding of action selectively impaired relative to concepts of the object involved in an action. The more frequent co-occurrence of action and tool knowledge deficits reflects the close proximity or even extensive overlap of their corresponding neural representations. Neuroimaging work has identified at least two loci important for (primarily concrete) action concepts: in the posterior middle temporal gyrus (pMTG) and the inferior parietal lobe (IPL). Yet both loci seem equally central to aspects of knowledge about tools. Shared neural territory between action concepts and tools seems to reflect more than the fact that tools cue actions. Rather, we argue that it reflects the fact that possibilities for action are inherent attributes of tools and that action concepts inherently specify their typical instruments as part of their predicate structure.

This chapter is about action concepts, but we begin with the inherent problems of the terms *concepts* and *action*. *Concepts* has different uses in the literature: here, we take concepts to be representations with certain properties, rather than any information retrieved during "conceptual tasks" (Leshinskaya & Caramazza, 2016). Specifically, concepts involve stored knowledge that captures some generality about the world and can be accessed from different modalities of stimuli. Just how general is a theoretical issue. Is a view-invariant representation of a specific chair a concept, or must it span many different chairs? We suspend this issue and take a broader, inclusive view.

What is an *action*? In sensorimotor content, the distinction between static shapes (*objects*) and body movements (*actions*) is clear, but at the conceptual level, different distinctions emerge. Movement is neither necessary nor sufficient in action concepts: we do not have concepts for meaningless movements; meanwhile, mental actions have no physical motion at all. Furthermore, action concepts often specify relations among participating objects as instruments or targets, and likewise, many artifacts have physical features that are imbued with relevance for action. Thus, at the conceptual level, the distinction of object versus action may not be primary.

The evidence we review regarding the neural organization of action concepts reflects this: neural representations of action concepts are entangled with those of objects—specifically, tools. Although content-selective conceptual deficits have long been reported in object domains such as animate and inanimate (Capitani, Laiacona, Mahon, & Caramazza, 2003; Caramazza & Shelton, 1998), they rarely seem to selectively affect action concepts. This raises the question of what organizing principles govern conceptual representations of actions; we describe some possibilities in our review of concepts *for* action and concepts *of* action. Neuroimaging has identified at least two loci important for action concepts; below, we attempt to better understand their representational roles. We find that neither is characterized by pure selectivity to action concepts per se but that both also contain information about tools. Furthermore, both are embedded within complex functional landscapes spanning multiple specialized areas; we suggest that these adjacency relations may be important clues to their broader function.

Dissociations among Action Knowledge Systems

Concepts for action Deficits in knowledge that support action planning are typically probed using *pantomime tasks*. An object is named or shown to a patient, then taken away; patient must demonstrate how they would typically use it with their hands. A deficit in this ability, along with intact basic motor and visual function, is termed *apraxia* (Heilman, Maher, Greenwald, & Rothi, 1997). In these tasks, the object serves as a cue to the relevant stored knowledge about action. The neuropsychological evidence suggests the existence of dissociations among such knowledge into two distinct systems: one based on object identity and the other on *mechanical reasoning*.

One way to solve the pantomime task is to recognize the object, retrieve one's knowledge about how to use this kind of object, and act accordingly (the object identity route). However, it can alternatively be solved by *mechanical reasoning*: computing actions on the basis of information about an object's physical properties—its shape, weight, rigidity, and so on (Goldenberg & Hagmann, 1998; Riddoch, Humphreys, & Price, 1989). Rather than relying on knowledge of the identity of an object, this system enables inferences from the object's physical characteristics available from its visible properties. When patients successfully perform pantomime tasks in response to objects they don't recognize, it is possible they use this mechanical-reasoning system.

A direct way to test the mechanical system is with a *novel tools* task: patients are asked to reason about novel tools whose conventional function is not known, such as a set of unconventional hooks to determine which one can lift another object out of a container (Heilman et al., 1997) or open a box (Hartmann, Goldenberg, Daumüller, & Hermsdörfer, 2005). By requiring only the *selection* of the novel tool, deficits cannot be due to motor execution problems. Such tasks can be solved at ceiling by patients who have deficits in object recognition—that is, who cannot name familiar tools or retrieve other semantic information about them (Bozeat, Lambon Ralph, Patterson, & Hodges, 2002; Hodges, Spatt, & Patterson, 1999; Sirigu, Duhamel, & Poncet, 1991). This even includes patients who cannot pantomime successfully to familiar objects. Conversely, novel tools performance can be impaired in patients with otherwise intact semantic knowledge (Goldenberg & Hagmann, 1998; Goldenberg & Spatt, 2009). Thus, either the object's identity or a mechanical reasoning system can be used to reason about action, and these appear dissociable.

This mechanical-reasoning system is sometimes characterized as *nonconceptual*, but it is not clear that it contains no conceptual content. This content must be independent of the knowledge of the identity of specific objects, but it might well be conceptually rich in other ways. It might contain general, intuitive physics principles relating object properties to inferences about support, containment, propulsion, and other forms of physical interaction. It could also represent how objects can interact with the hand to work as levers or enable reaching. A key direction for future research is to probe what patients with impairments in identifying objects do or do not know about various aspects of intuitive physics (see chapter 65). If their knowledge turns out to be conceptually rich, it would support the idea of a dissociable aspect of the conceptual system that is specifically important for intuitive physics concepts.

There are also cases of deficits to the object identity system that may be selective to action knowledge specifically. Such patients exhibit conceptual errors when using objects with conventional functions, such as brushing the teeth with a spoon (De Renzi & Lucchelli, 1988; Heilman et al., 1997; Ochipa, Rothi, & Heilman, 1989; Sirigu, Duhamel, & Poncet, 1991). These errors appear to result from conceptual confusion about what to do, rather than errors in a mechanical-reasoning system. Ochipa, Rothi, and Heilman's (1989) patient, who made such errors in action, was also poor at describing those objects' typical functions but able to name objects and actions. These cases are suggestive of a specialized conceptual system involved in the knowledge of the conventional functions of objects but distinct from both mechanical reasoning and the ability to name those objects, though the latter part of this dissociation remains tentative (see Bozeat et al., 2002; Daprati & Sirigu, 2006 for discussion).

In summary, a least two varieties of conceptual representations support acting with objects. Mechanical reasoning—the knowledge of intuitive principles linking physical properties of objects to inferences about action—doubly dissociates from other aspects of conceptual knowledge, which in turn allow the use of object-specific action knowledge by identifying the objects and retrieving their conventional functions.

A major limitation is that this work focuses specifically on transitive (object-based) actions. It remains possible that concepts of intransitive (non–object based) actions have different principles of organization. However, from the evidence on hand, it seems difficult to disentangle knowledge about action from that about objects; the mechanical-reasoning system has to make reference to the physical qualities of objects in order to support judgments about acting with them. And while a "concepts for *object-based* action" system is an alluring idea, evidence for it separate from conceptual knowledge regarding nonaction attributes remains tentative.

Concepts of actions Concepts *of* actions enable recognizing and understanding actions that one observes. Action recognition is typically tested by having patients match an action name to a video or picture, and it doubly dissociates from production abilities, as in the pantomime tests described above (Negri et al., 2007; Tarhan, Watson, & Buxbaum, 2016). Action recognition can fail for multiple reasons, however, and not all are due to deficits at the conceptual level. For example, visual agnosia is an impairment specific to the visual modality, leaving intact the ability to make judgments about actions presented as names. Agnosia can selectively affect action or object stimuli (Rothi, Mack, & Heilman, 1986; Tarhan, Watson, & Buxbaum, 2016), suggesting there may be an action-selective component within the visual recognition system but not necessarily in the conceptual system.

Attempts to avert these issues and look for conceptual-level deficits to action concepts per se have failed to provide conclusive evidence. One study (Pillon & D'Honincthun, 2011) reports on a patient with broad, crossmodal conceptual deficits and intact lower-level visual, motor, and lexical abilities. For example, he could discriminate meaningful from meaningless gestures. However, when asked to name pictures, select related pictures, or verify properties of named objects, he showed a consistent pattern of impairment, performing the worst on living things and significantly better on man-made objects and actions, which in turn did not differ from each other. This was the case even for actions that did not involve objects (e.g., between two people). Another study (Vannuscorps & Pillon, 2011) reports a complementary performance profile of a patient with a conceptual-level impairment regarding tools, nontool artifacts, and actions to equal degree, with spared abilities for animals, plants, and famous people and buildings. Thus, rather than a selective semantic system for actions, these findings demonstrate selectivity within the semantic system for actions and artifacts together. The authors argue that a common domain-selective system exists supporting conceptual knowledge for actions and artifacts; collectively, perhaps, it represents concepts that pertain to goals or purposes. This argument relies on the observation that actions and artifacts were damaged to a similar degree across these patients and the premise that this coincidence is not due to damage to adjacent but functionally independent neural structures. Reports do exist of inanimate object impairment without impairment to actions, but these domains were not compared directly (Bi, Han, Shu, & Caramazza, 2007). In a direct comparison of performance in naming the action (sewing) versus the instrument (needle) in an action, there is a report of a patient with a clear dissociation between the two (Shapiro & Caramazza, 2003). The patient performed quite well in naming the objects but very poorly in naming the actions with those objects. Importantly, the difference in performance could not be attributed to differences in the grammatical class of the words (nouns vs. verbs) since the patient showed normal grammatical class (morphosyntactic) processing. Altogether, more evidence is needed to fully resolve whether there is a content-selective system for concepts of actions, and its exact relation to concepts of objects.

Neural Organization of Action Concepts

Dissociations among impairments in action-related tasks, as reviewed above, have shed light on which cognitive components are neurally separable, though leaving many issues unresolved. Neuroimaging and lesion-mapping evidence provide additional insight into cortical organization by demonstrating how action knowledge is spatially arranged in cortex.

The principal findings from this work are centered on areas in lateral temporal and lateral parietal cortex (figure 63.1). It has become clear that parts of these areas represent conceptual content about actions but that these, too, reflect object knowledge, specifically about tools, as would be expected under an account of action concepts as predicates and their arguments. The most compelling facts of these data are that representations about actions and tools are closely entangled in neural space rather than strictly separated, even as the broader roles of those areas—comprising multiple specializations—are best described as serving action planning and understanding.

Concepts of actions A large set of experiments suggests that a relatively anatomically consistent area in the left lateral posterior temporal cortex preferentially responds when participants retrieve action knowledge (Watson, Cardillo, Ianni, & Chatterjee, 2013). We term this area *action-MTG*, to designate a functional area in and around the posterior middle temporal gyrus with this profile. Activation in this area is increased when participants name actions that correspond to pictures or names of tools, relative to naming their typical colors (Martin et al., 1995); effects at nearby coordinates are seen for retrieving action attributes relative to size attributes, for both tools and fruit (Phillips, Noppeney, Humphreys, & Price, 2002) and for semantic judgments about names of actions versus names of objects (Kable, Kan, Wilson, Thompson-Schill, & Chatterjee, 2005; Kable, Lease-Spellmeyer, & Chatterjee, 2002).

Is action-MTG an area specifically involved in action concepts, and what about them does it represent? It could reflect action concepts, or the grammatical category of verbs, or motion imagery. To approach this question, one must describe it in the context of a complex landscape of responses in the broad cortical area surrounding it—we refer to this anatomical region spanning multiple functional areas as the lateral occipitotemporal cortex (LOTC). Essential to this effort is evidence that directly compares functional activations within the same group of subjects, and we rely on

A

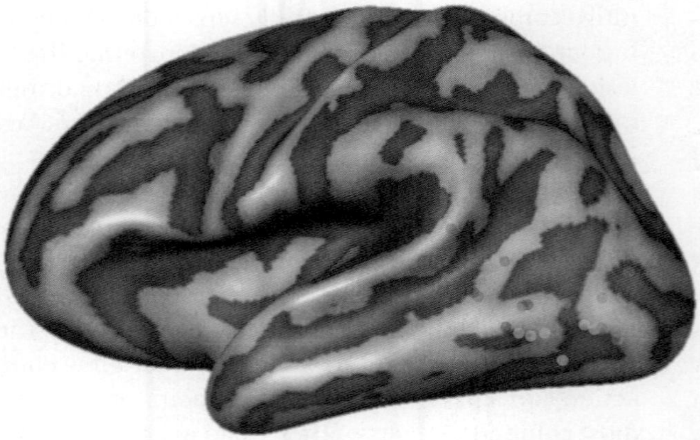

Study	Tal X	Tal Y	Tal Z
Action attribute retrieval			
Martin et al., 1995 (Study 1)	-50	-50	4
Martin et al., 1995 (Study 2)	-54	-62	8
Phillips et al., 2002	-50	-62	5
Kable et al., 2005	-53	-60	-5
Verbs			
Bedny et al., 2008	-53	-41	3
Peelen et al., 2012	-49	-53	12
Shapiro et al., 2006	-57	-40	9
Bedny et al., 2013	-60	-51	11
Hernandez et al., 2014	-45	-43	7
Bedny et al., 2011	-53	-49	6
Tools			
Beauchamp et al., 2002 (Study 1)	-38	-63	-6
Beauchamp et al., 2002 (Study 2)	-46	-70	-4
Valyear et al., 2007	-48	-60	-4
Peelen et al., 2013	-50	-60	-5
Bracci et al., 2011 (Study 1)	-48	-65	-6
Bracci et al., 2011 (Study 2)	-46	-68	-2
Feature-general action representation			
Wurm & Lingnau, 2015	-41	-76	-4
Wurm et al., 2017	-44	-64	3
Oosterhof et al., 2010	-49	-61	2
Wurm & Caramazza, 2018	-54	-61	4
Basic motion			
Bedny et al., 2008	-46	-71	7
Zeki et al. 1991	-38	-74	8
Bracci et al., 2011	-44	-72	-1

B

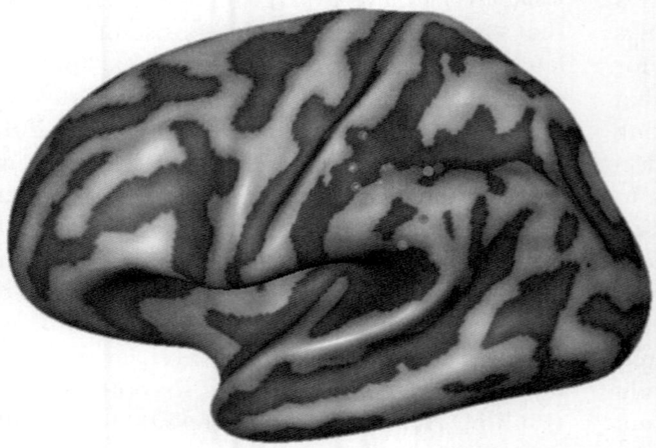

Study	Tal X	Tal Y	Tal Z
Tool experience			
Creem-Regehr et al., 2007	-56	-29	29
Valyear et al., 2012	-43	-39	43
Vingerhoets et al., 2011	-42	-32	42
Weisberg et al., 2007	-42	-43	38
Feature-general action representation			
Oosterhof et al., 2010	-44	-31	44
Oosterhof et al., 2012	-49	-31	42
Hafri et al., 2017	-56	-36	28
Wurm & Lingnau, 2015	-51	-29	36
Wurm et al., 2017	-47	-27	37
Feature-general object function			
Leshinskaya & Caramazza, 2015	-62	-38	38
Tools			
Garcea & Mahon, 2014	-43	-43	41

FIGURE 63.1 Peak coordinates of action-related effects in MTG (*A*) and IPL (*B*) reported in studies discussed in the section on the neural organization of action concepts. The different kinds of effects are based on the following contrasts/classifications: action attribute retrieval (*blue*) = tasks requiring the retrieval of actions or action attributes versus action-unrelated attributes (e.g., color) from pictures or names of actions or manipulable objects; tool experience (*magenta*) = familiar/typical versus unfamiliar/atypical tool use knowledge; verbs (*red*) = verbs versus nouns (various contrast; see the text); basic motion (*orange*) = moving versus static dots; feature-general action representation (*light blue*) = multivoxel pattern classification of action videos across perceptual features; feature-general object function (*green*) = multivoxel pattern classification of abstract categories of functions; tools (*yellow*) = images or videos of tools versus nonmanipulable artifacts or animals. Note that peaks do not reflect the spatial extent or the overlap of effects. (See color plate 75.)

evidence from such comparisons to assess whether different functions are attributable to the same area.

Posteriorly in LOTC is the functional area MT+, which is selective to moving versus static stimuli across content domains (Zeki, Kennard, Watson, Lueck, & Frackowiak, 1991). Anterior to MT+ in left LOTC is another area, which preferentially responds to images of tools relative to human bodies (Beauchamp, Lee, Haxby, & Martin, 2002) and other categories (Valyear, Cavina-Pratesi, Stiglick, & Culham, 2007), and which we term *tool-MTG.* Within-study functional comparisons show that tool-MTG diverges from motion-sensitive MT+ (Beauchamp et al., 2002); the effects of action attribute retrieval—that is, action-MTG—also diverge from MT+ (Kable et al., 2005). Thus, tool and action responses in the MTG do not reflect the retrieval of simple visual motion. One possibility is that they reflect retrieval of more complex kinds of motion. Indeed, tool-MTG responds more strongly to functionally moving tools than to static tools or moving human bodies (Beauchamp et al., 2002). However, tool responsiveness in the MTG is preserved in congenitally blind participants, who have no visual experience (Peelen et al., 2013), suggesting that responses in this area are unlikely due only to the visual imagery of tool motion.

Tool- and action-MTG areas are anatomically nearby; both are reliably anterior to MT+. Critically, a within-subject functional region of interest (ROI) analysis showed that tool-MTG also responds to action attribute retrieval (Perini, Caramazza, & Peelen, 2014). Thus, we suggest that overlapping tool and action responses likely reflect the same functional area (tool/action-MTG hereafter)—one that exhibits preferential responses to tools, particularly moving ones, and the retrieval of action attributes. However, this area is not driven specifically by visual experience, and its content is not reducible to low-level visual or kinematic features. It is thus consistent with being a conceptual-level representation, though not definitively so.

The observation of seemingly shared neural space between responses to actions and tools converges with some of the above-reviewed findings from neuropsychology: that conceptual representations of artifacts and actions sometimes pattern together in semantic impairment (but see Shapiro & Caramazza, 2003). However, there are important differences: tool-MTG is more responsive to tools than to other artifacts (Bracci, Cavina-Pratesi, Ietswaart, Caramazza, & Peelen, 2011; Valyear et al., 2007) and is not the locus of all tool-related knowledge (see chapter 64 on tool concepts). Thus, tool/action-MTG may be just one locus of shared neural territory between action and artifact knowledge.

This shared territory could reflect a common representation accessed by both action attributes and tools; tool images might simply be cues to actions, for example. An alternative is that it reflects something about both tools and actions per se. Recent work finds that tool-MTG represents information about not only the physical uses of tools but also their taxonomic category, such as musical instruments versus garage tools (Bracci, Daniels, & Op de Beeck, 2017), which might support the latter view. Nonetheless, such categories might also reflect action knowledge because playing music versus repairing a house are also distinct categories of actions. Our work also finds that information in and around the MTG represents whether an object or a person is a participant in an action (Wurm & Caramazza, 2018; Wurm, Caramazza, & Lingnau, 2017). In short, what aspects of actions and objects are represented in the MTG remains an open question, but the evidence does not allow the conclusion that the information represented in this area is only about actions and not also about tools.

There is further evidence that responses in and around the anatomical location of tool/action-MTG reflect conceptual similarity among actions, although it is not known whether these responses occur in exactly the same functional area. For example, in posterior parts of the LOTC, videos of opening actions elicit reliably distinct response patterns from videos of closing actions, while kinematically and perceptually different opening actions (opening a bottle vs. a jar) elicit relatively similar patterns (Wurm & Lingnau, 2015). This suggests that regions around tool/action-MTG encode the distinction between meaningfully different actions, generalizing across perceptually different instantiations of an action. Other whole-brain studies report similar effects nearby: actions like "lift" versus "tilt" elicit reliably distinguishable responses while generalizing across visual viewpoints and different hand configurations (Oosterhof, Wiggett, Diedrichsen, Tipper, & Downing, 2010) and the effector used to carry out the action (Vannuscorps, Wurm, Striem-Amit, & Caramazza, 2018). In more anterior LOTC, spanning tool/action-MTG, representations generalize across specific actions and encode more general attributes, such as whether an action involves interaction with manipulable objects or another person (Wurm, Caramazza, & Lingnau, 2017). Moreover, these representations have been shown to generalize across videos and sentences (Wurm & Caramazza, 2019), controlling for the possible effects of verbalization or imagery. In summary, a large set of recent findings shows effects in posterior LOTC that reveal abstract representation of action information.

Action concepts are often, but not necessarily, expressed with a certain grammatical category in language: verbs. Responses to verbs over and above nouns are also found in anterior and superior areas surrounding the MTG, which we term *verb-MTG* (Bedny, Caramazza, Grossman, Pascual-Leone, & Saxe, 2008; Peelen, Romagno, & Caramazza, 2012; Shapiro, Moo, & Caramazza, 2006). Verb-selective responses are preserved in the congenitally blind (Bedny, Dravida, & Saxe, 2013) and cannot be explained by differences in the amount of visual motion they denote (Bedny et al., 2008). Notably, these effects range over a wide range of verb types beyond just action verbs, including those referring to mental states (Bedny et al., 2008; Bedny, Caramazza, Pascual-Leone, & Saxe, 2011), abstract states (*include, exist*; Peelen, Romagno, & Caramazza, 2012), perception (*gaze*), and emission (*clang*; Bedny, Dravida, & Saxe, 2013). They thus reflect more than action concepts. In addition, these responses scale with transitivity, the number of objects a verb requires: *take* requires more arguments than *die* (Hernandez, Fairhall, Lenci, Baroni, & Caramazza, 2014). This suggests that verb-MTG has a role in representing predicate-argument structures, a function that is both grammatical and semantic.

A critical question is whether verb-MTG overlaps with tool/action-MTG. In support of their overlap, action-responsive MTG also responds strongly to names of tools (Kable et al., 2005). On the other hand, preferential responses to verbs over nouns, holding semantics constant (state verbs vs. nouns), are found in a more anterior portion of lateral posterior temporal cortex than preferential responses to action semantics (action vs. state verbs; Peelen, Romagno, & Caramazza, 2012). An analysis of coordinates reported in the work cited here shows a reliable anterior to posterior difference in verb and tool effect coordinates ($M = 18.1$ mm, $t(8) = 6.45$, $p < .001$), with verb effects anterior to tool effects. Thus, representations of verbs as a semantic/grammatical category seem to diverge from those of tools/actions.

Nonetheless, the coincidence of responses to tools, action concepts, and verbs in such close quarters raises the possibility that they reflect the operation of a broader functional area in LOTC. We suggest that this broader function would be about neither "actions" nor "tools" alone but rather relate objects to action concepts and action concepts to predicate-argument structures. Verbs typically denote objects as agents or patients (Pinker, 1989), and objects possess properties that are informative about their action possibilities. Sensitivity to verbs, action attributes, action-relevant objects, and the predicate structures of verbs may be adjacent because of their common participation in the process of understanding actions and speaking about them.

Concepts for actions The visuomotor planning of actions critically relies on parietal cortex (Goodale & Milner, 1992), and lesion mapping shows that selective impairments in mechanical reasoning arise from IPL damage (Goldenberg & Spatt, 2009). Like LOTC, IPL clearly contains multiple adjacent specializations within it: while some components represent visuomotor information, others have signatures of a conceptual role. But although the IPL is functionally heterogeneous, the relationship between different functions is rarely assessed directly, and it is unclear whether all conceptual effects arise from the same area. Below, we consider the possibility that parts of the IPL may "contain" conceptual representations, with their exact location unclear, rather than attributing any singular representational type to the IPL as a whole.

Parts of the IPL are sensitive to factors like action familiarity and conventionality during action planning, all consistent with a conceptual role (Creem-Regehr, Dilda, Vicchrilli, Federer, & Lee, 2007; Valyear, Gallivan, McLean, & Culham, 2012; Vingerhoets, Vandekerckhove, Honoré, Vandemaele, & Achten, 2011; Weisberg, van Turennout, & Martin, 2007). Furthermore, the kinds of representations these areas exhibit during action planning are not only motoric or kinematic, and can also be driven by different kinds of stimulus inputs. For example, Gallivan, McLean, Valyear, and Culham (2013) cued participants to perform reaching and grasping actions with either hands or reverse tongs and found that posterior parts of the IPL contained information discriminating reaching and grasping but common to both hands and tongs, even though these have opposite kinematics. These representations are thus not of motor kinematics. Moreover, IPL representations during action planning can be shared with those during action observation, suggesting they are not tied to specifically visual or motor features of the stimuli: Oosterhof et al. (2010) and Oosterhof, Tipper, and Downing (2012) found that anterior parts of the IPL show similar patterns of responses when the same action type (e.g., lift) is executed and observed while showing distinct patterns of response to different actions (e.g., tilt). These results are consistent with the possibility that conceptual factors are reflected in the IPL.

We reported more direct evidence toward this conclusion (Leshinskaya & Caramazza, 2015). Participants evaluated a set of objects toward one of four functions: keeping her or his body warm; protecting objects from water; decorating a house; and dressing up for a night out. Patterns of response in the anterior IPL were similar between pairs of functions that belonged to similar broader categories, Decorate and Protect, without sharing a similar physical manner of execution (e.g., *decorate house* and *dress up* elicited more similar neural patterns relative to *dress up* and *protect body*). This finding suggests that the IPL contains conceptual-level representations about actions, which are not driven by motoric or visual properties.

Despite its importance in action planning, anterior parts of the IPL respond in a similar way to the MTG in studies of action observation, showing discriminable response patterns for actions like "kick," "push," "open," and "close" while generalizing across specific visual differences among instances of those actions (Hafri, Trueswell, & Epstein, 2017; Oosterhof, Tipper, & Downing, 2012; Oosterhof et al., 2010; Wurm, Caramazza, & Lingnau, 2017; Wurm & Lingnau, 2015). However, there are also notable differences between the MTG and the IPL, in that only the former seems to encode action representations that generalize across observed actions and sentences describing actions (Wurm & Caramazza, 2019).

The conceptual effects we outlined hardly account for the entirety of the IPL, which has a highly diverse set of functions in different parts (e.g., Simon, Mangin, Cohen, Le Bihan, & Dehaene, 2002). Notably, any conceptual representation would seem to reside near specifically visuomotor ones (Gallivan et al., 2013; Tunik, Frey, & Grafton, 2005; Valyear et al., 2007). This is critical evidence against a singular kind of representation in the IPL and for the idea that the IPL may include multiple representations with different levels of abstraction. Thus, like LOTC, this area contains multiple specializations that cannot be reduced to each other. However, the reason they are adjacent may be due to participation in a common broader function.

In combination with its role in mechanical reasoning (Fischer, Mikhael, Tenenbaum, & Kanwisher, 2016; Goldenberg & Spatt, 2009), one possible common function for this set of areas is that they all support "how" reasoning—understanding how objects can be used to accomplish certain outcomes—both when this reasoning is about kinematic or mechanical information or abstract properties like aesthetic quality (Leshinskaya & Caramazza, 2015; Spunt, Satpute, & Lieberman, 2011). This account would predict little specialization in the IPL for knowledge about action to the exclusion of knowledge about objects (and how they support action), reflecting the inherent entanglement between action and object properties supporting action.

In support of this idea, the IPL and the MTG are both core parts of a set of regions responding preferentially to tools (Garcea & Mahon, 2014). Like the MTG, parts of the IPL exhibit representations characterizing the associated motor action and taxonomic categories of objects, albeit in a task-dependent manner (Bracci, Daniels, & de Beeck, 2017). Moreover, transcranial magnetic stimulation to the IPL impairs the ability to name tools more than living items (Pobric, Jefferies, & Lambon Ralph, 2010). This indicates that the IPL contains knowledge about tools beyond their roles in cueing action knowledge. Future research should better characterize the specific roles of the IPL and the MTG in action and tool cognition.

General Conclusion

We have described several specialized neural systems involved in action concepts but have not excluded their also being involved in concepts about objects—specifically, tools. This shared neural territory has been most commonly investigated by concrete, transitive action concepts, so this result might therefore not be too surprising. Nonetheless, we marshaled some evidence that these areas represent information about tools per se, and not only as simple cues to their associated actions, and that relatively abstract knowledge about actions can be represented there. Thus, this neural overlap or adjacency might reflect the fact that possibilities for action are inherent attributes of tools/artifacts and that action concepts inherently specify their typical instruments. Neuropsychological evidence that action and tool concepts dissociate is rare, and such impairments can often pattern together.

We conclude that the neural organization of concepts is not drawn primarily along a clear object versus action boundary, given the entanglement between tools and concrete action concepts in neural territory. Moreover, these entangled conceptual representations are further embedded in functionally diverse landscapes in both temporal and parietal cortex. These diverse areas' ways of interacting, their organizing principles, and their broader roles in cognition are major directions for future work.

REFERENCES

Beauchamp, M. S., Lee, K., Haxby, J. V, & Martin, A. (2002). Parallel visual motion processing streams for manipulable objects and human movements. *Neuron, 34*, 149–159.

Bedny, M., Caramazza, A., Grossman, E., Pascual-Leone, A., & Saxe, R. (2008). Concepts are more than percepts: The case of action verbs. *Journal of Neuroscience, 28*(44), 11347–11353.

Bedny, M., Caramazza, A., Pascual-Leone, A., & Saxe, R. (2011). Typical neural representations of action verbs develop without vision. *Cerebral Cortex, 22*(2), 286–293.

Bedny, M., Dravida, S., & Saxe, R. (2013). Shindigs, brunches, and rodeos: The neural basis of event words. *Cognitive, Affective & Behavioral Neuroscience, 14*(3), 891–901.

Bi, Y., Han, Z., Shu, H., & Caramazza, A. (2007). Nouns, verbs, objects, actions, and the animate/inanimate effect. *Cognitive Neuropsychology, 24*(5), 485–504.

Binder, J. R., Desai, R. H., Graves, W. W., & Conant, L. L. (2009). Where is the semantic system? A critical review and meta-analysis of 120 functional neuroimaging studies. *Cerebral Cortex, 19*(12), 2767–2796.

Bozeat, S., Lambon-Ralph, M. A., Patterson, K., & Hodges, J. R. (2002). When objects lose their meaning: What happens to their use? *Cognitive, Affective & Behavioral Neuroscience, 2*(3), 236–251.

Bracci, S., Cavina-Pratesi, C., Ietswaart, M., Caramazza, A., & Peelen, M. V. (2011). Closely overlapping responses to tools and hands in left lateral occipitotemporal cortex. *Journal of Neurophysiology, 107,* 1443–1456.

Bracci, S., Daniels, N., & Op de Beeck, H. (2017). Task context overrules object- and category-related representational content in the human parietal cortex. *Cerebral Cortex, 27*(1), 310–321.

Capitani, E., Laiacona, M., Mahon, B., & Caramazza, A. (2003). What are the facts of semantic category-specific deficits? A critical review of the clinical evidence. *Cognitive Neuropsychology,* 213–261.

Caramazza, A., & Shelton, J. R. (1998). Domain-specific knowledge systems in the brain: The animate-inanimate distinction. *Journal of Cognitive Neuroscience, 10*(1), 1–34.

Creem-Regehr, S. H., Dilda, V., Vicchrilli, A. E., Federer, F., & Lee, J. N. (2007). The influence of complex action knowledge on representations of novel graspable objects: Evidence from functional magnetic resonance imaging. *Journal of the International Neuropsychological Society, 13,* 1009–1020.

Daprati, E., & Sirigu, A. (2006). How we interact with objects: Learning from brain lesions. *Trends in Cognitive Sciences, 10*(6), 265–270. doi:10.1016/j.tics.2006.04.005

De Renzi, E., & Lucchelli, F. (1988). Ideational apraxia. *Brain, 111,* 1173–1185.

Fischer, J., Mikhael, J., Tenenbaum, J. B., & Kanwisher, N. G. (2016). The functional neuroanatomy of intuitive physical inference. *Proceedings of the National Academy of Sciences, 113*(34), E5072–E5081.

Gallivan, J. P., McLean, D. A., Valyear, K. F., & Culham, J. C. (2013). Decoding the neural mechanisms of human tool use. *eLife, 2,* e00425–e00425.

Garcea, F. E., & Mahon, B. Z. (2014). Parcellation of left parietal tool representations by functional connectivity. *Neuropsychologia, 60,* 131–143.

Goldenberg, G., & Hagmann, S. (1998). Tool use and mechanical problem solving in apraxia. *Neuropsychologia, 36*(7), 581–589.

Goldenberg, G., & Spatt, J. (2009). The neural basis of tool use. *Brain, 132,* 1645–1655.

Goodale, M. A., & Milner, A. D. (1992). Separate visual pathways for perception and action. *Trends in Neurosciences, 15*(1), 20–25.

Hafri, A., Trueswell, J. C., & Epstein, R. A. (2017). Neural representations of observed actions generalize across static and dynamic visual input. *Journal of Neuroscience, 37*(11), 2496–3016.

Hartmann, K., Goldenberg, G., Daumüller, M., & Hermsdörfer, J. (2005). It takes the whole brain to make a cup of coffee: The neuropsychology of naturalistic actions involving technical devices. *Neuropsychologia, 43*(4), 625–637.

Heilman, K. M., Maher, L. M., Greenwald, M. L., & Rothi, L. J. G. (1997). Conceptual apraxia from lateralized lesions. *Neurology, 49,* 457–464.

Hernandez, M., Fairhall, S. L., Lenci, A., Baroni, M., & Caramazza, A. (2014). Predication drives verb cortical signatures. *Journal of Cognitive Neuroscience, 26*(8), 1829–1839.

Hodges, J. R., Spatt, J., & Patterson, K. (1999). "What" and "how": Evidence for the dissociation of object knowledge and mechanical problem-solving skills in the human brain. *Proceedings of the National Academy of Sciences, 96*(16), 9444–9448.

Kable, J. W., Kan, I. P., Wilson, A., Thompson-Schill, S. L., & Chatterjee, A. (2005). Conceptual representations of action in the lateral temporal cortex. *Journal of Cognitive Neuroscience, 17*(12), 1855–1870.

Kable, J. W., Lease-Spellmeyer, J., & Chatterjee, A. (2002). Neural substrates of action event knowledge. *Journal of Cognitive Neuroscience, 14*(5), 795–805.

Leshinskaya, A., & Caramazza, A. (2015). Abstract categories of functions in anterior parietal lobe. *Neuropsychologia, 76,* 1–13.

Leshinskaya, A., & Caramazza, A. (2016). For a cognitive neuroscience of concepts: Moving beyond the grounding issue. *Psychonomic Bulletin & Review, 23*(4), 991–1001.

Martin, A., Haxby, J. V, Lalonde, F. M., Wiggs, C. L., Lalonde, F. M., & Ungerleider, L. G. (1995). Discrete cortical regions associated with knowledge of color and knowledge of action. *Science, 270*(5233), 102–105.

Negri, G. A. L., Rumiati, R. I., Zadini, A., Ukmar, M., Mahon, B. Z., & Caramazza, A. (2007). What is the role of motor simulation in action and object recognition? Evidence from apraxia. *Cognitive Neuropsychology, 24*(8), 795–816.

Nelson, S. M., Cohen, A. L., Power, J. D., Wig, G. S., Miezin, F. M., Wheeler, M. E., … Petersen, S. E. (2010). A parcellation scheme for human left lateral parietal cortex. *Neuron, 67*(1), 156–170.

Ochipa, C., Rothi, L. J. G., & Heilman, K. M. (1989). Ideational apraxia: A deficit in tool selection and use. *Annals of Neurology, 25,* 190–193.

Oosterhof, N. N., Tipper, S. P., & Downing, P. E. (2012). Viewpoint (in)dependence of action representations: An MVPA study. *Journal of Cognitive Neuroscience, 24*(4), 975–989.

Oosterhof, N. N., Wiggett, A. J., Diedrichsen, J., Tipper, S. P., & Downing, P. E. (2010). Surface-based information mapping reveals crossmodal vision-action representations in human parietal and occipitotemporal cortex. *Journal of Neurophysiology, 104*(2), 1077–1089. doi:10.1152/jn.00326.2010

Peelen, M. V., Bracci, S., Lu, X., He, C., Caramazza, A., & Bi, Y. (2013). Tool selectivity in left occipitotemporal cortex develops without vision. *Journal of Cognitive Neuroscience, 25*(8), 1225–1233.

Peelen, M. V., Romagno, D., & Caramazza, A. (2012). Independent representations of verbs and actions in left lateral temporal cortex. *Journal of Cognitive Neuroscience, 24*(10), 2096–2107.

Perini, F., Caramazza, A., & Peelen, M. V. (2014). Left occipitotemporal cortex contributes to the discrimination of tool-associated hand actions: fMRI and TMS evidence. *Frontiers in Human Neuroscience, 8,* 1–10.

Phillips, J. A., Noppeney, U., Humphreys, G. W., & Price, C. J. (2002). Can segregation within the semantic system account for category-specific deficits? *Brain, 125,* 2067–2080.

Pillon, A., & D'Honincthun, P. (2011). A common processing system for the concepts of artifacts and actions? Evidence from a case of a disproportionate conceptual impairment for living things. *Cognitive Neuropsychology, 28*(1), 37–41.

Pinker, S. (1989). *Learnability and cognition: The acquisition of argument structure.* Cambridge, MA: MIT Press.

Pobric, G., Jefferies, E., & Lambon-Ralph, M. A. (2010). Category-specific versus category-general semantic impairment induced by transcranial magnetic stimulation. *Current Biology, 20*(10), 964–968.

Riddoch, M. J., Humphreys, G. W., & Price, C. J. (1989). Routes to action: Evidence from apraxia. *Cognitive Neuropsychology, 6*(5), 437–454.

Rothi, L. J., Mack, L., & Heilman, K. M. (1986). Pantomime agnosia. *Journal of Neurology Neurosurgery, and Psychiatry, 49,* 451–454.

Shapiro, K., & Caramazza, A. (2003). Looming a loom: Evidence for independent access to grammatical and phonological properties in verb retrieval. *Journal of Neurolinguistics, 16,* 85–111.

Shapiro, K., Moo, L. R., & Caramazza, A. (2006). Cortical signatures of noun and verb production. *Proceedings of the National Academy of Sciences, 103*(5), 1644–1649.

Simon, O., Mangin, J. F., Cohen, L., Le Bihan, D., & Dehaene, S. (2002). Topographical layout of hand, eye, calculation, and language-related areas in the human parietal lobe. *Neuron, 33*(3), 475–487.

Sirigu, A., Duhamel, J.-R., & Poncet, M. (1991). The role of sensorimotor experience in object recognition. *Brain, 114*(6), 2555–2573.

Spunt, R. P., Satpute, A. B., & Lieberman, M. D. (2011). Dissociable neural systems support retrieval of how and why action knowledge. *Journal of Cognitive Neuroscience, 23*(1), 63–74.

Tarhan, L. Y., Watson, C. E., & Buxbaum, L. J. (2016). Shared and distinct neuroanatomic regions critical for tool-related action production and recognition: Evidence from 131 left-hemisphere stroke patients. *Journal of Cognitive Neuroscience, 27*(12), 2491–2511.

Tunik, E., Frey, S. H., & Grafton, S. T. (2005). Virtual lesions of the anterior intraparietal area disrupt goal-dependent online adjustments of grasp. *Nature Neuroscience, 8*(4), 505–511.

Valyear, K. F., Cavina-Pratesi, C., Stiglick, A. J., & Culham, J. C. (2007). Does tool-related fMRI activity within the intraparietal sulcus reflect the plan to grasp? *NeuroImage, 36,* T94–T108.

Valyear, K. F., Gallivan, J. P., McLean, D. A., & Culham, J. C. (2012). fMRI repetition suppression for familiar but not arbitrary actions with tools. *Journal of Neuroscience, 32*(12), 4247–4259.

Vannuscorps, G., & Pillon, A. (2011). A domain-specific system for representing knowledge of both man-made objects and human actions: Evidence from a case with an association of deficits. *Neuropsychologia, 49*(9), 2321–2341.

Vannuscorps G., Wurm, M. F., Striem-Amit, E., Caramazza A. (2018). Large-scale organization of the hand action observation network in individuals born without hands. *Cerebral Cortex.* doi:10.1093/cercor/bhy212

Vingerhoets, G., Vandekerckhove, E., Honoré, P., Vandemaele, P., & Achten, E. (2011). Neural correlates of pantomiming familiar and unfamiliar tools: Action semantics versus mechanical problem solving? *Human Brain Mapping, 32*(6), 905–918.

Weisberg, J., van Turennout, M., & Martin, A. (2007). A neural system for learning about object function. *Cerebral Cortex, 17*(3), 513–521.

Watson, C. E., Cardillo, E., Ianni, G., & Chatterjee, A. (2013). Action concepts in the brain: An activation-likelihood estimation meta-analysis. *Journal of Cognitive Neuroscience, 25*(8), 1191–1205.

Wurm, M. F., & Caramazza, A. (2019). Representation of action concepts in left posterior temporal cortex that generalize across vision and language. *Nature Communications, 10*(1), 289.

Wurm, M. F., Caramazza, A., & Lingnau, A. (2017). Action categories in lateral occipitotemporal cortex are organized along sociality and transitivity. *Journal of Neuroscience, 37*(3), 562–575.

Wurm, M. F., & Lingnau, A. (2015). Decoding actions at different levels of abstraction. *Journal of Neuroscience, 35*(20), 7727–7735.

Zeki, S., Kennard, C., Watson, J. D. G., Lueck, C. J., & Frackowiak, R. S. J. (1991). A direct demonstration of functional specialization in human visual cortex. *Cerebral Cortex, 11*(3), 641–649.

64 The Representation of Tools in the Human Brain

BRADFORD Z. MAHON

ABSTRACT A domain-specific brain network spanning multiple regions supports the recognition and use of tools. The "domain" of this network is the translation of propositional attitudes into actions. "Tools" are not the domain; they are the "things in the world" that mediate that translation. This frames a new way to think about the neural systems that support tool recognition and use. One immediate implication of this approach is that a test of the *domain-specificity* of a given brain region is not reducible to a test of which stimulus category elicits the maximal neural response. Whether a region is a component of a broader domain-specific network depends on how the processes supported by that region are integrated with the broader computation of the network. This framework emphasizes the role of connectivity-based constraints in driving neural specificity in any given region. The big challenge that we face is bringing neuropsychological findings demonstrating *dissociations* among cognitive processes into register with functional neuroimaging studies demonstrating *coactivation* of dissociable processes. Progress will depend on computationally explicit proposals about the dynamics of information exchange among anatomically remote brain regions and how these dynamics are modulated by the task or goal states of the system.

The neural representation of tools is widely distributed across many brain regions and pathways, and there is no coherent manner in which a "tool representation" can be said to "be there" in the brain. There are representations of visual form, surface-texture and material properties, object location in various reference frames, action goals, object identity, object function, lexical semantics, linguistic forms, and learned motor competencies and skills. Everyday object use involves the orchestration of that entire diversity of representations (see figure 64.1 for a schematic and explanation).

For the sake of delimiting the scope of a short review, I define *tools* narrowly as objects that are handheld and that mechanically extend the body by transforming a movement of the hand into a goal-directed action. An iPhone in this analysis is not a tool and neither is a book—but a brush, a hammer, and a pen are all tools. It is certainly the case that interactions with "nontool manipulable objects" (e.g., books) involve a subset of processes (and brain regions) that mediate interactions with tools (e.g., the regions involved in object grasping). The advantage of defining *tools* in a strict manner is that it affords a clearer view of what are, by hypothesis, universal cognitive and neural constraints on brain organization that support causal reasoning about objects. Another advantage is that tool *concepts* can be operationally defined as representations that specify the function or purpose of the use of objects. Defined as such, tool concepts bridge propositional attitudes (*I want to eat*) with the concrete actions that satisfy those goals (*Grasp and use* this *fork*). A range of experimental findings collectively suggest that representations of function are a key component of tool concepts (Buxbaum, Veramonti, & Schwartz, 2000; Garcea & Mahon, 2012; Ishibashi, Lambon Ralph, Saito, & Pobric, 2011). Ultimately, these working definitions of *tools* and *tool concepts* are paradigm-level hypotheses, and their value will be weighed in how well they serve to organize extant findings and generate new predictions.

In the course of everyday life, tools are, initially and for the most part, useful things—it is hard to find oneself not already embedded in functional relations with objects. As useful things, tools are not "noticed" until one is needed to accomplish a goal and the object is found to be absent—then the "forkness" of the fork becomes conspicuous in its absence (Heidegger, 1996). This is not unlike how we don't "notice" the words we use to communicate our thoughts until the right word is not available (tip of the tongue). And, just as our minds are "set up" to express thoughts in words, I suggest that our minds are "set up" for causal reasoning about objects. When a fork cannot be found, the mind goes immediately to what might be used in place of a fork—a spoon will do for eating peas, but a knife may be required for eating steak. Making things, using them, recognizing them in the context of using them, and perhaps most importantly—*thinking about how to use objects to satisfy goals*—are core universal human cognitive capacities. Those considerations motivate a theoretical framework in which the design of the central nervous system innately anticipates tools, just as it (by

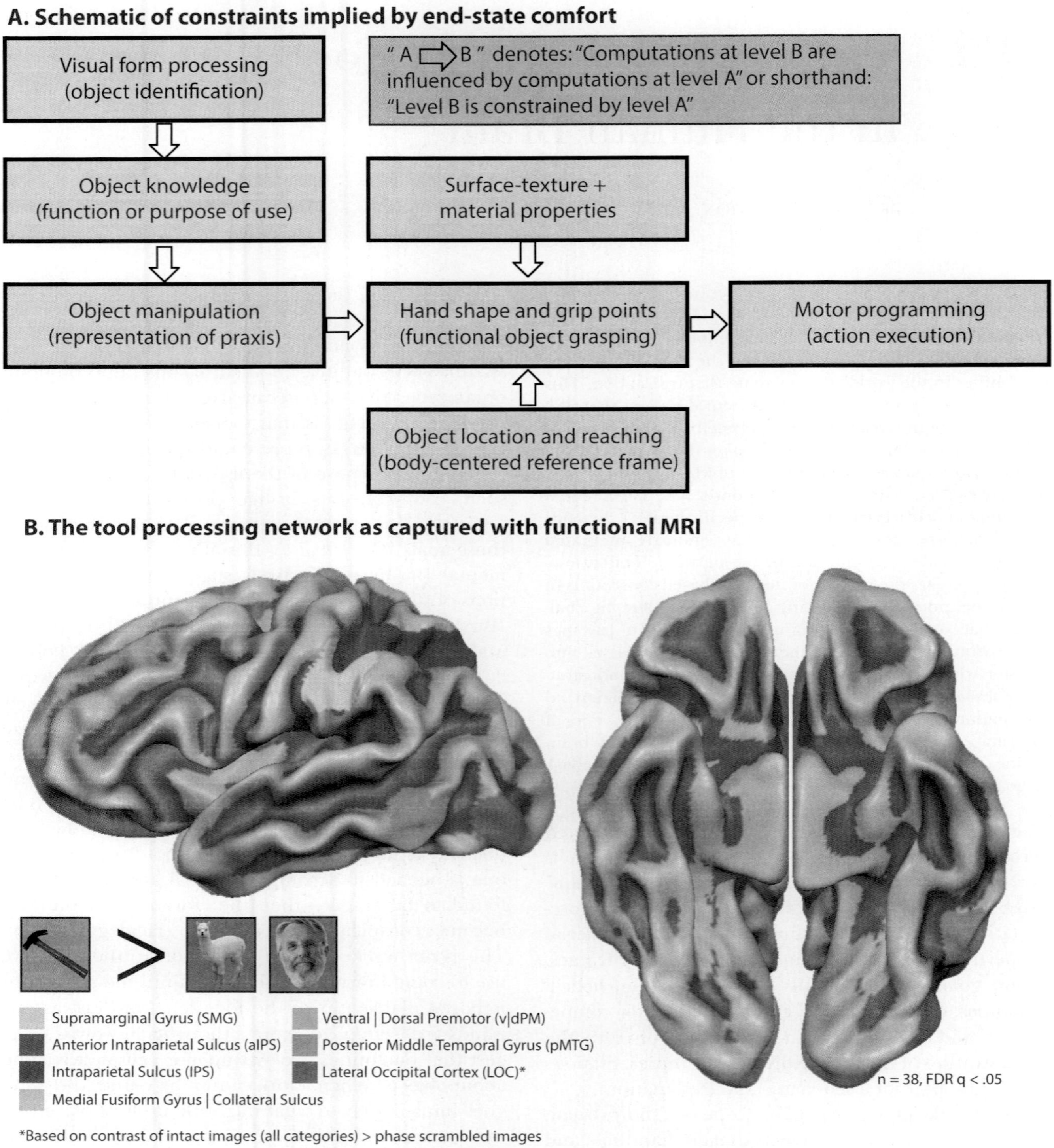
A. Schematic of constraints implied by end-state comfort
Visual form processing (object identification)
" A ⇨ B " denotes: "Computations at level B are influenced by computations at level A" or shorthand: "Level B is constrained by level A"
Object knowledge (function or purpose of use)
Surface-texture + material properties
Object manipulation (representation of praxis)
Hand shape and grip points (functional object grasping)
Motor programming (action execution)
Object location and reaching (body-centered reference frame)
B. The tool processing network as captured with functional MRI
Supramarginal Gyrus (SMG)
Anterior Intraparietal Sulcus (aIPS)
Intraparietal Sulcus (IPS)
Medial Fusiform Gyrus | Collateral Sulcus
Ventral | Dorsal Premotor (v|dPM)
Posterior Middle Temporal Gyrus (pMTG)
Lateral Occipital Cortex (LOC)*
n = 38, FDR q < .05
*Based on contrast of intact images (all categories) > phase scrambled images

hypothesis) anticipates language, human faces, geographic landmarks, biological motion, and other universal stabilities in the human habitat on which successful survival could have depended.

Tools as Instruments of Action: Praxis and Apraxia

Limb apraxia is an impairment for using objects that cannot be attributed to basic sensory or motor deficits (Heilman, 1973; Rumiati, Zanini, Vorano, & Shallice, 2001) and which is classically associated with damage to the left supramarginal gyrus, in the inferior parietal lobule (figure 64.1). Apraxic patients can have greater difficulty with pantomime tasks compared to actual object use (Geschwind, 1965; Heilman, 1973). That asymmetry could be due to the additional cues provided by a real object. In addition, pantomiming involves the re-representation of object properties, whereas actual use involves perception (Goodale, Jakobson, & Keillor, 1994). Some manifestations of apraxia may be due to the disconnection of praxis representations in the inferior parietal lobe from motor structures in the frontal lobe (ideomotor apraxia) or semantic representations in the temporal lobe (ideational apraxia; Geschwind, 1965; see figure 64.2). Additional research will be necessary to tease apart why object-directed actions and pantomimed actions involve partially dissociable neural systems (Freud et al., 2018).

Two theoretically important observations arising from studies of apraxic patients concern what can be *spared* in the setting of apraxia. First, apraxic impairments can occur in the context of spared object naming and spared verbal knowledge of object function (e.g., Buxbaum, Veramonti, & Schwartz, 2000; Garcea, Dombovy, & Mahon, 2013; Negri et al., 2007). While careful testing may yet demonstrate subtle conceptual deficits in some apraxic patients, the basic fact that a range of motor deficits are observed without conceptual deficits indicates that "embodied" theories do not offer a satisfactory account of meaning representation (Mahon, 2015; Mahon & Caramazza, 2008). The second observation is that patients can be impaired at producing actions while having little or no difficulty recognizing actions—both in the domain of manual action (Rapcsak, Ochipa, Anderson, & Poizner, 1995; Rumiati et al., 2001) and in the domain of speech (Rogalsky, Love, Driscoll, Anderson, & Hickok, 2011; Stasenko et al., 2015). Those findings indicate that action production processes are not necessary for action recognition, undermining motor theories of action recognition (Caramazza, Anzellotti, Strnad, & Lingnau, 2014; Hickok, 2009; Mahon & Caramazza, 2005; Negri et al., 2007). Those observations raise the question: Why are motor-relevant processes engaged during conceptual processing and action recognition if, as the patient evidence indicates, those motor processes are not necessary for either? We are left with the inference that access to motor systems is fast and automatic—but *contingent* on access to meaning. Sensorimotor activity during conceptual processing is a reflection of meaning, not meaning itself (Mahon, 2015).

A number of neuroimaging studies converge on the inference, initially supported by patient research, that the left supramarginal gyrus is a key substrate for praxis (see figure 64.1; Boronat et al., 2005; Canessa et al., 2008; Chao & Martin, 2000; Mahon et al., 2007; Orban & Caruana, 2014; Peeters et al., 2009). Simply viewing or naming tools leads to activity in the left supramarginal gyrus, as originally described by Martin and colleagues (Chao & Martin, 2000; Mahon et al., 2007). Furthermore, the same neural representations engaged during

FIGURE 64.1 Overview of constraints among the dissociable processes involved in tool recognition and use. *A*, Consider the everyday act of grasping one's fork to eat. The initial grasp anticipates how the object will be manipulated once it is "in hand." A fork is grasped differently than a knife, even if they have exactly the same handle. A fork is also grasped differently if the goal is to pass it to someone else, rather than to eat. The accommodation of functional object grasps to what the object will be used for once it is in hand, referred to as *end-state comfort* (Rosenbaum, Vaughan, Barnes, Marchak, & Slotta, 1990), implies substantial interaction among what are known to be dissociable representations (Carey, Hargreaves, & Goodale, 1996; Creem & Proffitt, 2001). For instance, the space of *possible* grasps is winnowed down to a space of *functional* grasps, based on representations of what will be done with the object once it is in hand (i.e., praxis; Wu, 2008). Praxis is, in turn, constrained by representations of object function, as objects are manipulated in a manner to accomplish a certain function or purpose of use. Finally, an object (e.g., a fork) is the target of an action only because it has a certain functional role in a broader behavioral goal, and thus the object (prior to any action being directed toward it) must be identified, at some level, for what it is. The schematic in figure 64.1 represents this type of conceptual analysis: the arrows in the figure do not represent *processing direction* but rather (some of) the constraints imposed among dissociable types of representations during functional object grasping and use. *B*, Functional MRI can be used to delineate the neural substrates of the domain-specific system that supports the translation of propositional attitudes into actions. The data shown in the figure were obtained while participants viewed tool stimuli compared to images of animals and faces. Regions are color-coded based on the principal dissociations that have been documented in the neuropsychological literature. The first functional MRI studies describing this set of "tool-preferring" regions were carried out in the laboratory of Alex Martin (Chao, Haxby, & Martin, 1999; Chao & Martin, 2000). (See color plate 76.)

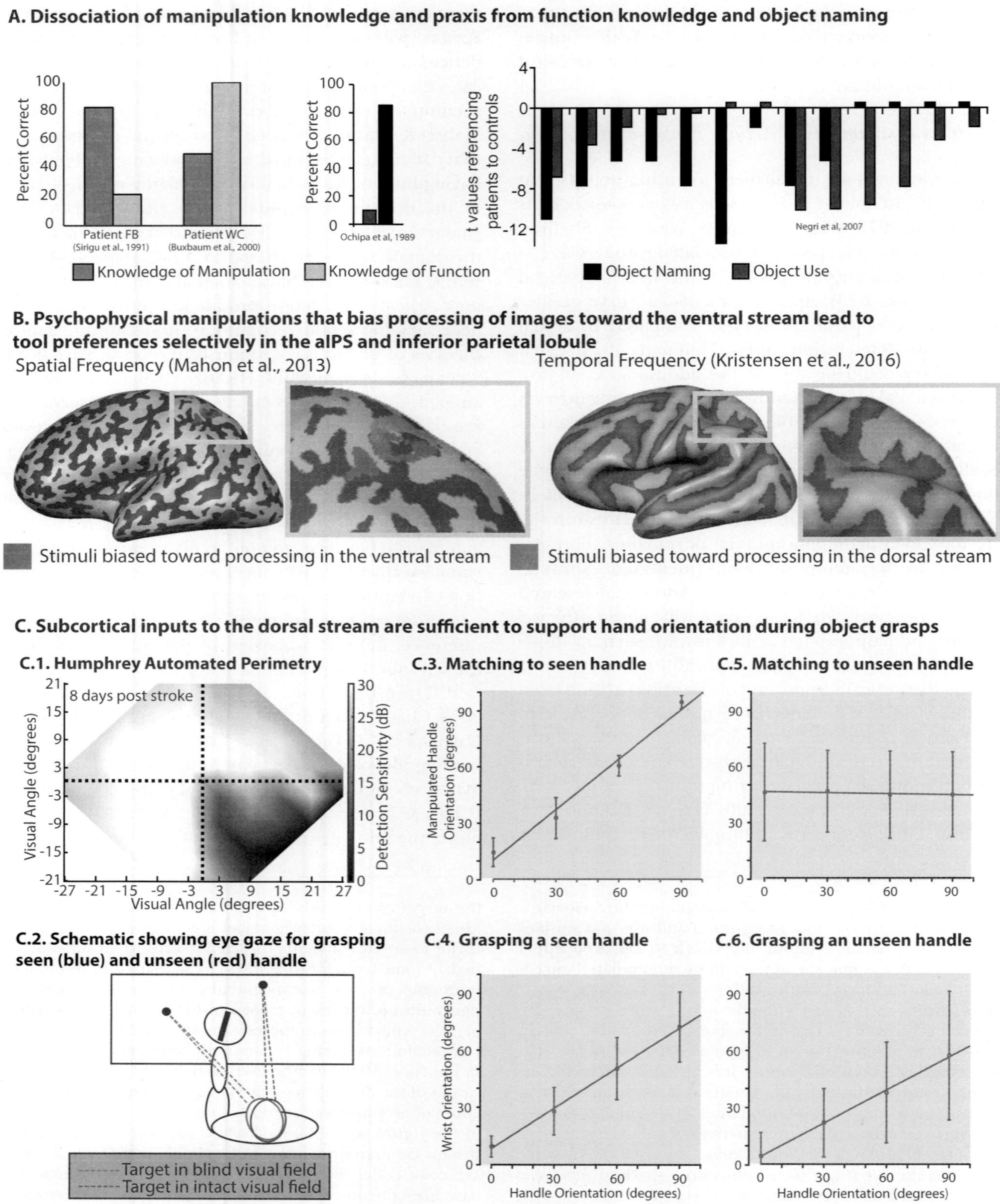
A. Dissociation of manipulation knowledge and praxis from function knowledge and object naming
Percent Correct
Patient FB
(Sirigu et al., 1991)
Patient WC
(Buxbaum et al., 2000)
Knowledge of Manipulation
Knowledge of Function
Ochipa et al, 1989
t values referencing patients to controls
Negri et al, 2007
Object Naming
Object Use
B. Psychophysical manipulations that bias processing of images toward the ventral stream lead to tool preferences selectively in the aIPS and inferior parietal lobule
Spatial Frequency (Mahon et al., 2013)
Temporal Frequency (Kristensen et al., 2016)
Stimuli biased toward processing in the ventral stream
Stimuli biased toward processing in the dorsal stream
C. Subcortical inputs to the dorsal stream are sufficient to support hand orientation during object grasps
C.1. Humphrey Automated Perimetry
8 days post stroke
Visual Angle (degrees)
Detection Sensitivity (dB)
C.3. Matching to seen handle
Manipulated Handle Orientation (degrees)
C.5. Matching to unseen handle
C.2. Schematic showing eye gaze for grasping seen (blue) and unseen (red) handle
Target in blind visual field
Target in intact visual field
C.4. Grasping a seen handle
Wrist Orientation (degrees)
Handle Orientation (degrees)
C.6. Grasping an unseen handle
Handle Orientation (degrees)

object pantomime are engaged during object identification (Chen, Garcea, Jacobs, & Mahon, 2018). But, as noted above, the patient findings indicate that object identification is not necessarily disrupted in the context of apraxic deficits. We are therefore, again, left with the inference that access to praxis information is compulsory and automatic but not necessary for object identification. This means that sensorimotor engagement, in contexts in which no motor response is task relevant, is informative about the connectivity and dynamics of the system.

Tools as Grasp Targets

In a foundational series of papers, Melvyn Goodale, David Milner, and colleagues comprehensively studied patient D. F., who has bilateral lesions to the lateral occipital cortex (LO or LOC, see figure 64.1) caused by anoxic injury. D. F. has intact low-level visual processing, receptive and productive language, executive function, attention, and memory—her principal deficit consists of a dense visual form agnosia. She is unable to make simple judgments about whether a line or object is oriented horizontally or vertically, fails to match visually presented stimuli, and cannot copy simple line drawings (despite being good at drawing from memory). Remarkably, when D. F. reaches to grasp an object or posts a card through a slot, she does so easily, and the parameters of her action accommodate naturally to the target (Goodale, Milner, Jakobson, & Carey, 1991). The dissociation between impaired visual form perception and intact vision-for-action has been observed in subsequent patients, and the reverse pattern has been reported: impaired object-directed reaching and/or grasping in the setting of spared visual form perception (see Goodale, Meenan, et al., 1994; Goodale & Milner, 1992). Goodale and Milner proposed that visual information coming from subcortical structures and early cortical regions is processed in a *dorsal visual pathway* that supports the analysis of object location, orientation, and volumetric properties in the service of action (see also Livingstone & Hubel, 1988; Merigan & Maunsell, 1993). By contrast, the ventral visual pathway supports fine-grained visual analysis in the service of identification and conceptual analysis, and is the substrate of what we experience as phenomenological vision.

Subsequent research has argued that dorsal structures do in fact contribute to perception (Freud, Culham, Plaut, & Behrmann, 2017; Kastner, Chen, Jeong, & Mruczek, 2017; Konen & Kastner, 2008)—thus,

FIGURE 64.2 Functional dissociations among tool representations in neuropsychology and functional neuroimaging. *A*, Limb apraxia is an impairment for using objects correctly that cannot be attributed to elemental sensory or motor disturbance. Variants of limb apraxia are distinguished by the nature of the errors that patients make. A patient with ideomotor apraxia may pantomime the use of a pair of scissors correctly in all ways, except, for instance, he moves the hand backward, opposite the direction of cutting (e.g., Garcea, Dombovy, & Mahon, 2013; for video examples, see www.openbrainproject.org). By contrast, a patient with ideational apraxia may deploy the wrong action for a given object while the action itself is performed correctly (e.g., using a toothbrush to brush one's hair). The distinction between ideomotor apraxia and ideational apraxia is loosely analogous to the distinction between phonological errors in word production (saying "caz" instead of "cat") and semantic errors in speech production (saying "dog" instead of "cat"; Rothi, Ochipa, & Heilman, 1991). The key point is that regardless of the nature of the errors patients make (spatiotemporal, content), the ability to name the same objects or access knowledge about their function can remain intact, indicating that the loss of motor-relevant information does not compromise conceptual processing in a major way. *B*, Laurel Buxbaum and colleagues have synthesized a framework within which to parcellate functional subdivisions within parietal cortex through the lens of everyday actions (Binkofski & Buxbaum, 2013; see also Garcea & Mahon, 2014; Mahon, Kumar, & Almeida, 2013; Peeters et al., 2009; Pisella et al., 2006). Left inferior parietal areas support action planning and praxis and operate over richly interpreted object information, such as that generated through processing in the ventral pathway, while posterior and superior parietal areas support "classic" dorsal stream processing involving online visuomotor control. A recent line of studies sought to determine which tool responses in parietal cortex depend on ventral stream processing by taking advantage of the fact that the dorsal visual pathway receives little parvocellular input (Livingstone & Hubel, 1988; Merigan & Maunsell, 1993). Thus, if images of tools and a baseline category (e.g., animals) are titrated so as to be defined by visual dimensions that are not "seen" by the dorsal pathway (because they require parvocellular processing), one can infer that regions of parietal cortex that continue to exhibit tool preferences receive inputs from the ventral stream. It was found that tool preferences were restricted to the aIPS and the supramarginal gyrus (figure 64.2) when stimuli contained only high spatial frequencies (Mahon, Kumar, & Almeida, 2013), were presented at a low temporal frequency (Kristensen, Garcea, Mahon, & Almeida, 2016), or were defined by red/green isoluminant color contrast (Almeida, Fintzi, & Mahon, 2013). Those findings suggest that neural responses to tools in the left inferior parietal areas are dependent on processing in the ventral visual pathway. *C*, Findings from *action blindsight* indicate that subcortical projections to the dorsal stream can support analysis of basic volumetrics about the shape and orientation of grasp targets. Prentiss, Schneider, Williams, Sahin, and Mahon (2018) described a hemianopic patient who performed at chance when making a perceptual matching judgment about the orientation of a handle presented in the hemianopic field, while he was able to spontaneously and accurately orient his wrist when the handle was the target of a grasp. (See color plate 77.)

perhaps the ventral stream supports perceptual analysis *in the service* of a conceptual interpretation of the input. When the ventral stream is not available—for instance, due to a lesion, such as in patient D. F.—then objects are grasped in a manner that accommodates to the volumetric properties of the object but which is not functional and reflects no understanding of *what* is being grasped. Patient D. F. does not maximize end-state comfort (see figure 64.1) because, by hypothesis, she is not able to access representations of object function and praxis from vision. Once the object is in hand, however, D. F. recognizes it through tactile cues, adjusts her grasp accordingly (Carey et al., 1996), and can demonstrate the correct use. Interestingly, D. F. has great difficulty grasping objects (in a volumetrically appropriate manner) if they do not have a principal axis of elongation (Carey et al., 1996), suggesting that the dorsal stream on its own cannot resolve grasp points on objects that do not have a principal axis of elongation.

A stark demonstration of "grasping without meaning" by the dorsal stream comes from cortically blind patients who can perform visually guided reaches and grasps to stimuli presented in their hemianopic (blind) visual field—*action blindsight* (Danckert & Rossetti, 2005). Perenin and Rossetti (1996) described a patient who could orient her wrist and demonstrate relatively spared grip scaling when grasping objects in the blind field (see figure 64.2). Those findings indicate that subcortical projections bypassing early visual cortex (Lyon, Nassi, & Callaway, 2010; Schmid et al., 2010) are sufficient to support grip scaling and wrist orientation, at least for grasp targets with a principal axis of elongation. Convergent evidence for that inference is provided by studies using the psychophysical technique of continuous flash suppression, a type of interocular suppression in which a stimulus can be rendered "invisible" for prolonged periods of time. Fang and He (2005) showed that Continuous Flash Suppressed (CFS, i.e., invisible) images of elongated tools drive neural responses in dorsal occipital and posterior parietal cortex to the same extent as do visible images of the same stimuli. By comparison, neural responses in the ventral stream were eliminated for CFS-suppressed stimuli. Jorge Almeida and colleagues (Almeida, Mahon, Nakayama, & Caramazza, 2008) showed that when CFS-suppressed images are used as primes in a behavioral priming paradigm, CFS-suppressed images of tools facilitate the subsequent categorization of elongated tool targets, while CFS-suppressed images of vehicles, animals, and faces do not facilitate the categorization of vehicle, animal, or face targets, respectively. Almeida and colleagues subsequently found that *any* elongated CFS-suppressed stimulus was an effective prime for an elongated tool target (e.g., a snake, or even a bar; Almeida et al., 2014; see also Sakuraba, Sakai, Yamanaka, Yokosawa, & Hirayama, 2012). Those findings collectively suggest that "elongation," divorced of any conceptual interpretation, is a visual feature processed by the dorsal visual pathway independent of processing within the ventral stream.

The property of elongation tends to be correlated with "toolness"—rendering it difficult to interpret why certain brain regions exhibit differential neural responses to tools compared to baseline categories, such as animals, faces, and places. To address this, Chen, Snow, Culham, and Goodale (2018) used task-based functional Magnetic Resonance Imaging (fMRI) and effective functional connectivity to distinguish toolness from elongation. The authors found that ventral stream regions process toolness (i.e., as a category) independent of elongation, while mid and posterior IPS regions process elongation independent of toolness. The authors further found that toolness drove connectivity from ventral to dorsal regions, while elongation drove connectivity from dorsal to ventral regions. Other research (Garcea, Almeida, & Mahon, 2012; Garcea, Kristensen, Almeida, & Mahon, 2016; Handy, Grafton, Shroff, Ketay, & Gazzaniga, 2003) may point to asymmetries between the left and right posterior parietal areas in processing object elongation.

Tools as Objects

Alex Martin and colleagues (Chao, Haxby, & Martin, 1999) described two foci of neural specificity for tools in the temporal lobe—one in ventral temporal cortex along the medial fusiform gyrus and adjacent collateral sulcus and one in lateral temporal cortex in the posterior middle (sometimes inferior) temporal gyrus (see figure 64.1). Why is there neural specificity in the visual system for a class of objects defined by motor-relevant properties?

The early literature on category specificity in the ventral stream assumed that the category for which a given subregion exhibits specificity is determined by the stimulus class that elicits the maximal response (Downing, Chan, Peelen, Dodds, & Kanwisher, 2006). Recent work indicates that the maximal univariate response is neither a necessary nor a sufficient empirical criterion for determining neural specificity—stronger indications about representational content are provided by joint analysis of univariate responses, multivoxel pattern analysis, and patterns of functional connectivity with regions outside of the ventral visual pathway. For instance, Yanchao Bi and colleagues (Wang et al., 2017) found a high degree of similarity between congenitally blind and sighted participants in patterns of functional connectivity between ventral-medial occipitotemporal cortex and other brain regions. That ventral-medial occipitotemporal region is likely the same region that exhibits neural specificity for tools. A parallel picture seems to be emerging for lateral occipital cortex, where subregions express neural specificity for tools and hands and also express privileged functional connectivity to regions of somatosensory cortex (Bracci & Peelen, 2013).

The theoretical explanation of how representations of tools are organized in the ventral stream should resonate with how other well-defined classes, such as written words, faces, animals, and geographic places, are recognized and processed. The *uses* to which information is put drive the organization of the system. Geographic landmarks, faces, animals, and tools are all used to very different purposes and project to different systems of the brain. This is a connectivity-constrained account within a domain-specific framework (Bi, Wang, & Caramazza, 2016; Chen, Garcea, Almeida, et al., 2017; Leshinskaya & Caramazza, 2016; Mahon & Caramazza, 2011; Mahon et al., 2007; Martin, 2016; Riesenhuber, 2007). Recent work in the domains of face and printed word recognition (Bouhali et al., 2014; Osher et al., 2016; Saygin et al., 2016) has confirmed core predictions of a connectivity-constrained account and has motivated proof-of-principle computational simulations (Chen & Rogers, 2015).

In the course of everyday action, object grasps are calibrated to *what* is being grasped and to the surface-texture and material properties of the object. The anterior IPS (aIPS) supports hand shaping in the service of object-directed grasping (Binkofski et al., 1998; Culham et al., 2003; Mruczek, von Loga, & Kastner, 2013). In order to grasp an object in a functional manner, by the appropriate part of the object and with the appropriate force, one must take not only visual form information but also the weight distribution and surface-texture of the object into account. The medial fusiform gyrus and collateral sulcus support the extrapolation of surface-texture and object weight from visual cues (Cant & Goodale, 2007; Cavina-Pratesi, Kentridge, Heywood, & Milner, 2010; Gallivan, Cant, Goodale, & Flanagan, 2014). These considerations motivate the stronger hypothesis that neural specificity for tools in the medial fusiform gyrus and collateral sulcus results from two intersecting inputs: inferences about surface-texture and material properties based on analysis of visual information processed in the ventral visual hierarchy and queries from dorsal stream regions that are computing grasp-relevant parameters. On this view, neural specificity for tools in medial ventral temporal cortex is a reflection of the interactions between the ventral and dorsal pathways that allow the system to direct the correct actions to the correct parts of the correct objects. There is no reason to believe that inputs from aIPS to the medial ventral stream regions are "top-down"—that pathway is (by hypothesis) an aspect of how the system initially processes visual information in the service of goal-directed, object-mediated actions (for an analogous proposal, see Bar et al., 2006).

Several expectations follow from the proposal that neural responses to tools in medial ventral stream regions are the result of joint inputs from the ventral visual hierarchy and the dorsal visual pathway. First, there should be privileged connectivity between the medial fusiform gyrus and the aIPS (Chen, Garcea, Almeida, & Mahon, 2017; Gallivan, McLean, Valyear, & Culham, 2013; Garcea, Chen, Vargas, Narayan, & Mahon, 2018; Garcea & Mahon, 2014; Mahon et al., 2007; Stevens, Tessler, Peng, & Martin, 2015). Second, stimulus factors that modulate activity in the aIPS should have echoes of neural activity in the medial fusiform gyrus and the collateral sulcus (Chen et al., 2018; Mahon et al., 2007). An even stronger prediction is that lesions to the aIPS will modulate neural responses to tools in the medial fusiform gyrus (Garcea et al., 2018).

Tools as a Window into Interactions between the Ventral and Dorsal Streams

As anticipated in the earliest formulations of the dorsal/ventral visual pathway hypothesis (Goodale & Milner, 1992), everyday interactions with objects *require* the integration of processing across the ventral and dorsal pathways. Demonstrations that processes supported by the ventral and dorsal streams *can* dissociate are not in conflict with the view that substantial interactions occur between the two streams. The fact that patient D. F. does not display an effect of end-state comfort (see figure 64.1) is what would be expected if the ventral and dorsal streams significantly interacted during object-directed grasps.

The key question is how a conceptual interpretation of visual input (by hypothesis, provenance of the ventral stream) interacts with dorsal stream processing in the service of functional object use. Broadly speaking, two possibilities exist, depending on whether or not it is assumed that there is cognitive penetration of dorsal stream processes (Mahon & Wu, 2015). The first possibility, which does not assume cognitive penetration of the dorsal stream, is that the dorsal stream computes visuomotor parameters blind to what the rest of the brain intends to do with the object—on this view, the dorsal stream precompiles a space of *possible* grasps, while the selection of the final action could be supported by, for instance, frontal regions involved in attentionally mediated selection processes (Jax & Buxbaum, 2010; Kan & Thompson-Schill, 2004; Pisella, Binkofski, Lasek, Toni, & Rossetti, 2006) that have access to ventral stream interpretations of the visual input. The second possibility is that there is true cognitive penetration of the dorsal stream such that the selection of a subset of grasp parameters happens within the dorsal stream on the basis of inputs from ventral stream regions that conceptually interpret the visual structure of the object. On this view, dorsal stream computations "wait" for semantically interpreted information that specifies what the final grasp should be (e.g., to maximize end-state comfort). In either scenario, it is clear that the dorsal stream, on its own, is unable to direct the correct actions to the correct parts of the correct objects—the ventral stream is needed to either set boundaries on what an acceptable action will be or to winnow down the space of possible actions, given a strong prior on what a functionally appropriate action could be. That "prior" is not, by hypothesis, derivable bottom-up from the perceptual input (see figure 64.1).

This chapter's sketches of dorsal-ventral interactions are admittedly cartoonish: for instance, aIPS may "wait" on inputs about visual structure from LO in order to winnow the space of possible grasp parameters while possibly also *driving* detailed analysis in the medial ventral stream areas about aspects of surface texture and material properties that are warranted *because* an action is being planned toward the object. Responses to elongated tools in the aIPS may precede tool-selective responses in the medial fusiform gyrus while a second wave of tool responses in the aIPS could be yoked to outputs of ventral stream regions (LO, medial fusiform gyrus). Similarly, responses in the left supramarginal gyrus that index access to praxis representations may occur relatively late and be contingent on access to object form, identity, and representations of object function, all of which are mediated by processing within the ventral visual pathway. A processing model will likely involve temporally dissociated "waves" of interactions between the dorsal and ventral streams, and the patterning of those interactions will strongly depend on the task (or *goal states* of the system; see figure 64.3).

Toward a Processing Model

The posture of our nervous system is to already always be in a state of interpreting the world in terms of what we might *do* with it—this is reflected in the connectivity of the system and the dynamics of how motor systems are engaged upon the visual presentation of manipulable objects. The empirical evidence reviewed in this chapter is informative about *how* tools are represented. Tool representations are distributed throughout a network that bridges conceptual representations (the units of thought) with sensorimotor systems (the cortical substrates of perception and action). By hypothesis, this network is domain-specific and innately anticipated in the organization of the human brain. What makes this network domain-specific is not that it is about tools, as such—it is domain-specific because it is about translating goals into actions. More generally, what makes a neural network spanning many brain regions domain-specific is not what characterizes the computational scope of the individual regions that form that network. From that a potentially rich methodological precept follows: the test of the domain-specificity of a region is not reducible to a simple test of selectivity to one or another category. By hypothesis, the medial fusiform gyrus and collateral sulcus are part of a domain-specific network—not *because* those regions respond to images of "tools" more than to images of other classes of stimuli but because they carry out certain computations (e.g., surface texture analysis) that are used by a broader system that is itself domain-specific (i.e., translating propositional attitudes into action).

The hypothesis that there is an innately specified domain-specific network focused on the translation of propositional attitudes into actions is a proposal about *why* tool representations have the distribution and organization that they do. Ultimately, the value of this broader proposal will be weighed in its ability to generate new predictions and, at a pragmatic level, its potential to serve as a useful paradigm for studying how the brain represents tools.

Acknowledgments

I am grateful to Frank Garcea for assistance in constructing figure 64.1 and to Jason Gallivan and Jody Culham for making available the graphic used in figure 64.3*B*. Many of the ideas in this chapter grew out of conversations and published collaborations with Alfonso Caramazza over the past 20 years, and I would like to thank Alfonso as well for critical feedback on an earlier version of this chapter. The preparation for this chapter was supported by grants from the National Science Foundation (BCS-134904) and the National Institutes of Health (R01NS089069 and R01EY028535) to Bradford Z. Mahon.

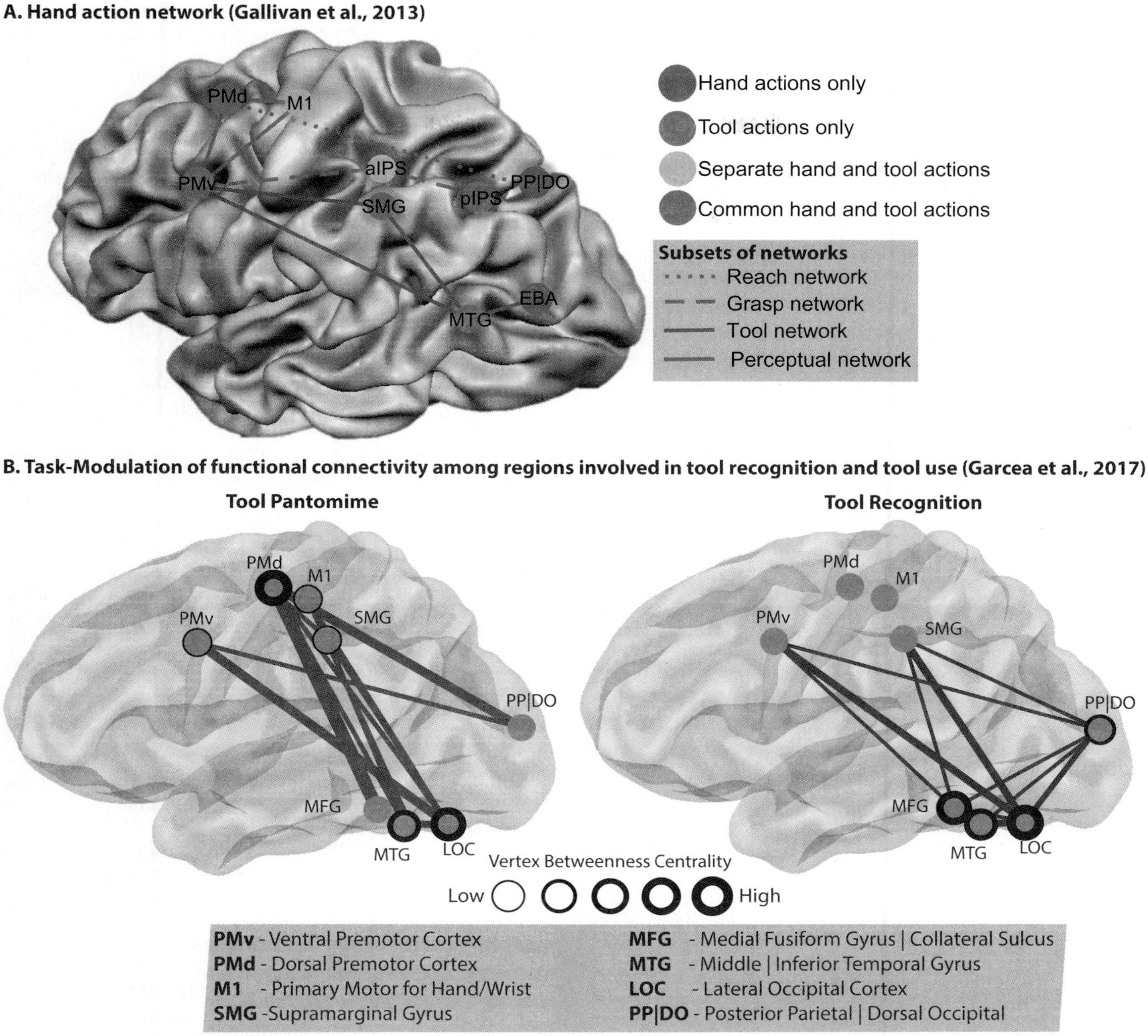

FIGURE 64.3 The next big step is to work toward a processing model that provides an answer to the question: How does the brain translate an abstract goal (eat dinner) into a specific object-directed action (grasp and use *this* fork)? A processing model would specify the types of representations and computations engaged during object recognition and functional object grasping and use, the order in which those computations are engaged, and their neural substrates. The key to developing such a processing model will be a careful analysis of how different tasks modulate connectivity in the system. The stronger suggestion is that it will not be possible to develop generative theories of the computations supported by discrete brain regions without understanding how the connectivity of those regions changes with different "goal states" of the system. Panels *A* and *B* represent two recent attempts using functional MRI to study task-modulated functional connectivity among regions of the brain specialized for translating propositional attitudes into goals (i.e., the "tool-processing network"). Future research with high temporal resolution will be necessary to understand whether there are dissociated "waves" of interactions among overlapping sets of brain regions that unfold in a task-driven manner. (See color plate 78.)

REFERENCES

Almeida, J., Fintzi, A. R., & Mahon, B. Z. (2013). Tool manipulation knowledge is retrieved by way of the ventral visual object processing pathway. *Cortex, 49*(9), 2334–2344. doi:10.1016/j.cortex.2013.05.004

Almeida, J., Mahon, B. Z., Nakayama, K., & Caramazza, A. (2008). Unconscious processing dissociates along categorical lines. *Proceedings of the National Academy of Sciences of the United States of America, 105*(39), 15214–15218. doi:10.1073/pnas.0805867105

Almeida, J., Mahon, B. Z., Zapater-Raberov, V., Dziuba, A., Cabaco, T., Marques, J. F., & Caramazza, A. (2014). Grasping with the eyes: The role of elongation in visual recognition of manipulable objects. *Cognitive Affective & Behavioral Neuroscience, 14*(1), 319–335. doi:10.3758/s13415-013-0208-0

Bar, M., Kassam, K. S., Ghuman, A. S., Boshyan, J., Schmid, A. M., Dale, A. M., … Halgren, E. (2006). Top-down facilitation of visual recognition. *Proceedings of the National Academy of Sciences of the United States of America, 103*(2), 449–454. doi:10.1073/pnas.0507062103

Bi, Y., Wang, X., & Caramazza, A. (2016). Object domain and modality in the ventral visual pathway. *Trends in Cognitive Sciences, 20*(4), 282–290. doi:10.1016/j.tics.2016.02.002

Binkofski, F., & Buxbaum, L. J. (2013). Two action systems in the human brain. *Brain and Language, 127*(2), 222–229. doi:10.1016/j.bandl.2012.07.007

Binkofski, F., Dohle, C., Posse, S., Stephan, K. M., Hefter, H., Seitz, R. J., & Freund, H. J. (1998). Human anterior intraparietal area subserves prehension: A combined lesion and functional MRI activation study. *Neurology, 50*(5), 1253–1259.

Boronat, C. B., Buxbaum, L. J., Coslett, H. B., Tang, K., Saffran, E. M., Kimberg, D. Y., & Detre, J. A. (2005). Distinctions between manipulation and function knowledge of objects: Evidence from functional magnetic resonance imaging. *Brain Research. Cognitive Brain Research, 23*(2–3), 361–373. doi:10.1016/j.cogbrainres.2004.11.001

Bouhali, F., Thiebaut de Schotten, M., Pinel, P., Poupon, C., Mangin, J. F., Dehaene, S., & Cohen, L. (2014). Anatomical connections of the visual word form area. *Journal of Neuroscience, 34*(46), 15402–15414. doi:10.1523/JNEUROSCI.4918-13.2014

Bracci, S., & Peelen, M. V. (2013). Body and object effectors: The organization of object representations in high-level visual cortex reflects body-object interactions. *Journal of Neuroscience, 33*(46), 18247–18258. doi:10.1523/JNEUROSCI.1322-13.2013

Buxbaum, L., Veramonti, T., & Schwartz, M. (2000). Function and manipulation tool knowledge in apraxia: Knowing "what for" but not "how." *Neurocase, 6*, 83–97.

Canessa, N., Borgo, F., Cappa, S. F., Perani, D., Falini, A., Buccino, G., … Shallice, T. (2008). The different neural correlates of action and functional knowledge in semantic memory: An fMRI study. *Cerebral Cortex, 18*(4), 740–751. doi:10.1093/cercor/bhm110

Cant, J. S., & Goodale, M. A. (2007). Attention to form or surface properties modulates different regions of human occipitotemporal cortex. *Cerebral Cortex, 17*(3), 713–731. doi:10.1093/cercor/bhk022

Caramazza, A., Anzellotti, S., Strnad, L., & Lingnau, A. (2014). Embodied cognition and mirror neurons: A critical assessment. *Annual Review of Neuroscience, 37*, 1–15. doi:10.1146/annurev-neuro-071013-013950

Carey, D. P., Hargreaves, E. L., & Goodale, M. A. (1996). Reaching to ipsilateral or contralateral targets: Within-hemisphere visuomotor processing cannot explain hemispatial differences in motor control. *Experimental Brain Research, 112*(3), 496–504.

Cavina-Pratesi, C., Kentridge, R. W., Heywood, C. A., & Milner, A. D. (2010). Separate processing of texture and form in the ventral stream: Evidence from fMRI and visual agnosia. *Cerebral Cortex, 20*(2), 433–446. doi:10.1093/cercor/bhp111

Chao, L. L., Haxby, J. V., & Martin, A. (1999). Attribute-based neural substrates in temporal cortex for perceiving and knowing about objects. *Nature Neuroscience, 2*(10), 913–919. doi:10.1038/13217

Chao, L. L., & Martin, A. (2000). Representation of manipulable man-made objects in the dorsal stream. *NeuroImage, 12*(4), 478–484. doi:10.1006/nimg.2000.0635

Chen, J., Snow, J. C., Culham, J. C., & Goodale, M. A. (2018). What role does "elongation" play in "tool-specific" activation and connectivity in the dorsal and ventral visual streams? *Cerebral Cortex, 28*(4), 1117–1131. doi:10.1093/cercor/bhx017

Chen, L., & Rogers, T. T. (2015). A model of emergent category-specific activation in the posterior fusiform gyrus of sighted and congenitally blind populations. *Journal of Cognitive Neuroscience, 27*(10), 1981–1999. doi:10.1162/jocn_a_00834

Chen, Q., Garcea, F. E., Almeida, J., & Mahon, B. Z. (2017). Connectivity-based constraints on category-specificity in the ventral object processing pathway. *Neuropsychologia, 105*, 184–196. doi:10.1016/j.neuropsychologia.2016.11.014

Chen, Q., Garcea, F. E., Jacobs, R. A., & Mahon, B. Z. (2018). Abstract representations of object-directed action in the left inferior parietal lobule. *Cerebral Cortex, 28*(6), 2162–2174. doi:10.1093/cercor/bhx120

Chen, Q., Garcea, F. E., & Mahon, B. Z. (2016). The representation of object-directed action and function knowledge in the human brain. *Cerebral Cortex, 26*(4), 1609–1618. doi:10.1093/cercor/bhu328

Creem, S. H., & Proffitt, D. R. (2001). Grasping objects by their handles: A necessary interaction between cognition and action. *Journal of Experimental Psychology: Human Perception and Performance, 27*(1), 218–228.

Culham, J. C., Danckert, S. L., DeSouza, J. F., Gati, J. S., Menon, R. S., & Goodale, M. A. (2003). Visually guided grasping produces fMRI activation in dorsal but not ventral stream brain areas. *Experimental Brain Research, 153*(2), 180–189. doi:10.1007/s00221-003-1591-5

Danckert, J., & Rossetti, Y. (2005). Blindsight in action: What can the different sub-types of blindsight tell us about the control of visually guided actions? *Neuroscience and Biobehavioral Reviews, 29*, 1035–1046.

Downing, P. E., Chan, A. W., Peelen, M. V., Dodds, C. M., & Kanwisher, N. (2006). Domain specificity in visual cortex. *Cerebral Cortex, 16*(10), 1453–1461. doi:10.1093/cercor/bhj086

Fang, F., & He, S. (2005). Cortical responses to invisible objects in the human dorsal and ventral pathways. *Nature Neuroscience, 8*(10), 1380–1385. doi:10.1038/nn1537

Freud, E., Culham, J. C., Plaut, D. C., & Behrmann, M. (2017). The large-scale organization of shape processing in the ventral and dorsal pathways. *eLife, 6*. doi:10.7554/eLife.27576

Freud, E., Macdonald, S. N., Chen, J., Quinlan, D. J., Goodale, M. A., & Culham, J. C. (2018). Getting a grip on reality: Grasping movements directed to real objects and images rely on dissociable neural representations. *Cortex, 98*, 34–48. doi:10.1016/j.cortex.2017.02.020

Gallivan, J. P., Cant, J. S., Goodale, M. A., & Flanagan, J. R. (2014). Representation of object weight in human ventral visual cortex. *Current Biology, 24*(16), 1866–1873. doi:10.1016/j.cub.2014.06.046

Gallivan, J. P., McLean, D. A., Valyear, K. F., & Culham, J. C. (2013). Decoding the neural mechanisms of human tool use. *eLife, 2*, e00425. doi:10.7554/eLife.00425

Garcea, F. E., Almeida, J., & Mahon, B. Z. (2012). A right visual field advantage for visual processing of manipulable objects. *Cognitive Affective & Behavioral Neuroscience, 12*(4), 813–825. doi:10.3758/s13415-012-0106-x

Garcea, F. E., Almeida, J., Sims, M., Nunno, A., Meyers, S., Li, Y., ... Mahon, B. (2018). Domain-specific diaschisis: Lesions to parietal action areas modulate neural responses to tools in the ventral stream. *Cerebral Cortex*. doi:10.1093/cercor/bhy183

Garcea, F. E., Chen, Q., Vargas, R., Narayan, D. A., & Mahon, B. Z. (2018). Task- and domain-specific modulation of functional connectivity in the ventral and dorsal object-processing pathways. *Brain Structure and Function, 223*(6), 2589–2607. doi:10.1007/s00429-018-1641-1

Garcea, F. E., Dombovy, M., & Mahon, B. Z. (2013). Preserved tool knowledge in the context of impaired action knowledge: Implications for models of semantic memory. *Frontiers in Human Neuroscience, 7*, 120. doi:10.3389/fnhum.2013.00120

Garcea, F. E., Kristensen, S., Almeida, J., & Mahon, B. Z. (2016). Resilience to the contralateral visual field bias as a window into object representations. *Cortex, 81*, 14–23. doi:10.1016/j.cortex.2016.04.006

Garcea, F. E., & Mahon, B. Z. (2012). What is in a tool concept? Dissociating manipulation knowledge from function knowledge. *Memory & Cognition, 40*(8), 1303–1313. doi:10.3758/s13421-012-0236-y

Garcea, F. E., & Mahon, B. Z. (2014). Parcellation of left parietal tool representations by functional connectivity. *Neuropsychologia, 60*, 131–143. doi:10.1016/j.neuropsychologia.2014.05.018

Geschwind, N. (1965). Disconnexion syndromes in animals and man. II. *Brain, 88*, 585–644.

Goodale, M. A., Jakobson, L. S., & Keillor, J. M. (1994). Differences in the visual control of pantomimed and natural grasping movements. *Neuropsychologia, 32*(10), 1159–1178.

Goodale, M. A., Meenan, J. P., Bulthoff, H. H., Nicolle, D. A., Murphy, K. J., & Racicot, C. I. (1994). Separate neural pathways for the visual analysis of object shape in perception and prehension. *Current Biology, 4*(7), 604–610.

Goodale, M. A., & Milner, A. D. (1992). Separate visual pathways for perception and action. *Trends in Neurosciences, 15*(1), 20–25.

Goodale, M. A., Milner, A. D., Jakobson, L. S., & Carey, D. P. (1991). A neurological dissociation between perceiving objects and grasping them. *Nature, 349*(6305), 154–156. doi:10.1038/349154a0

H. L. (1908). *Drei Aufsatze aus dem Apraxiegebiet*. Berlin: Karger.

Handy, T. C., Grafton, S. T., Shroff, N. M., Ketay, S., & Gazzaniga, M. S. (2003). Graspable objects grab attention when the potential for action is recognized. *Nature Neuroscience, 6*(4), 421–427. doi:10.1038/nn1031

Heidegger, M. (1996). *Being and time: A translation of* Sein und Zeit. Translated by J. Stambaugh. Albany: State University of New York Press.

Heilman, K. M. (1973). Ideational apraxia—a re-definition. *Brain, 96*(4), 861–864.

Hickok, G. (2009). Eight problems for the mirror neuron theory of action understanding in monkeys and humans. *Journal of Cognitive Neuroscience, 21*(7), 1229–1243. doi:10.1162/jocn.2009.21189

Ishibashi, R., Lambon Ralph, M. A., Saito, S., & Pobric, G. (2011). Different roles of lateral anterior temporal lobe and inferior parietal lobule in coding function and manipulation tool knowledge: Evidence from an rTMS study. *Neuropsychologia, 49*(5), 1128–1135. doi:10.1016/j.neuropsychologia.2011.01.004

Jax, S. A., & Buxbaum, L. J. (2010). Response interference between functional and structural actions linked to the same familiar object. *Cognition, 115*(2), 350–355. doi:10.1016/j.cognition.2010.01.004

Kan, I. P., & Thompson-Schill, S. L. (2004). Selection from perceptual and conceptual representations. *Cognitive Affective & Behavioral Neuroscience, 4*(4), 466–482.

Kastner, S., Chen, Q., Jeong, S. K., & Mruczek, R. E. B. (2017). A brief comparative review of primate posterior parietal cortex: A novel hypothesis on the human toolmaker. *Neuropsychologia, 105*, 123–134. doi:10.1016/j.neuropsychologia.2017.01.034

Konen, C. S., & Kastner, S. (2008). Two hierarchically organized neural systems for object information in human visual cortex. *Nature Neuroscience, 11*(2), 224–231. doi:10.1038/nn2036

Kristensen, S., Garcea, F. E., Mahon, B. Z., & Almeida, J. (2016). Temporal frequency tuning reveals interactions between the dorsal and ventral visual streams. *Journal of Cognitive Neuroscience, 28*(9), 1295–1302. doi:10.1162/jocn_a_00969

Leshinskaya, A., & Caramazza, A. (2016). For a cognitive neuroscience of concepts: Moving beyond the grounding issue. *Psychonomic Bulletin & Review, 23*(4), 991–1001. doi:10.3758/s13423-015-0870-z

Livingstone, M., & Hubel, D. (1988). Segregation of form, color, movement, and depth: Anatomy, physiology, and perception. *Science, 240*(4853), 740–749.

Lyon, D. C., Nassi, J. J., & Callaway, E. M. (2010). A disynaptic relay from superior colliculus to dorsal stream visual cortex in macaque monkey. *Neuron, 65*(2), 270–279. doi:10.1016/j.neuron.2010.01.003

Mahon, B. (2015). What is embodied about cognition? *Language Cognition and Neuroscience, 30*(4), 420–429. doi:10.1080/23273798.2014.987791

Mahon, B., Anzellotti, S., Schwarzbach, J., Zampini, M., & Caramazza, A. (2009). Category-specific organization in the human brain does not require visual experience. *Neuron, 63*(3), 397–405. doi:10.1016/j.neuron.2009.07.012

Mahon, B., & Caramazza, A. (2005). The orchestration of the sensory-motor systems: Clues from neuropsychology. *Cognitive Neuropsychology, 22*(3), 480–494. doi:10.1080/02643290442000446

Mahon, B., & Caramazza, A. (2008). A critical look at the embodied cognition hypothesis and a new proposal for grounding conceptual content. *Journal of Physiology, Paris, 102*(1–3), 59–70. doi:10.1016/j.jphysparis.2008.03.004

Mahon, B., & Caramazza, A. (2011). What drives the organization of object knowledge in the brain? *Trends in Cognitive Sciences, 15*(3), 97–103. doi:10.1016/j.tics.2011.01.004

Mahon, B., Kumar, N., & Almeida, J. (2013). Spatial frequency tuning reveals interactions between the dorsal and ventral visual systems. *Journal of Cognitive Neuroscience, 25*(6), 862–871. doi:10.1162/jocn_a_00370

Mahon, B., Milleville, S., Negri, G., Rumiati, R., Caramazza, A., & Martin, A. (2007). Action-related properties shape object representations in the ventral stream. *Neuron, 55*(3), 507–520. doi:10.1016/j.neuron.2007.07.011

Mahon, B., & Wu, W. (2015). Cognitive penetration of the dorsal visual stream? In J. Zeimbekis & A. Raftopoulis (Eds.), *The cognitive penetration of perception: New philosophical perspectives* (pp. 200–217). Oxford: Oxford University Press.

Martin, A. (2016). GRAPES—Grounding representations in action, perception, and emotion systems: How object properties and categories are represented in the human brain. *Psychonomic Bulletin & Review, 23*(4), 979–990. doi:10.3758/s13423-015-0842-3

Merigan, W. H., & Maunsell, J. H. (1993). How parallel are the primate visual pathways? *Annual Review of Neuroscience, 16*, 369–402. doi:10.1146/annurev.ne.16.030193.002101

Mruczek, R. E., von Loga, I. S., & Kastner, S. (2013). The representation of tool and non-tool object information in the human intraparietal sulcus. *Journal of Neurophysiology, 109*(12), 2883–2896. doi:10.1152/jn.00658.2012

Negri, G. A., Rumiati, R. I., Zadini, A., Ukmar, M., Mahon, B. Z., & Caramazza, A. (2007). What is the role of motor simulation in action and object recognition? Evidence from apraxia. *Cognitive Neuropsychology, 24*(8), 795–816. doi:10.1080/02643290701707412

Ochipa, C., Rothi, L. J., & Heilman, K. M. (1989). Ideational apraxia: A deficit in tool selection and use. *Annals of Neurology, 25*(2), 190–193. doi:10.1002/ana.410250214

Orban, G. A., & Caruana, F. (2014). The neural basis of human tool use. *Frontiers in Psychology, 5*, 310. doi:10.3389/fpsyg.2014.00310

Osher, D. E., Saxe, R. R., Koldewyn, K., Gabrieli, J. D., Kanwisher, N., & Saygin, Z. M. (2016). Structural connectivity fingerprints predict cortical selectivity for multiple visual categories across cortex. *Cerebral Cortex, 26*(4), 1668–1683. doi:10.1093/cercor/bhu303

Peeters, R., Simone, L., Nelissen, K., Fabbri-Destro, M., Vanduffel, W., Rizzolatti, G., & Orban, G. A. (2009). The representation of tool use in humans and monkeys: Common and uniquely human features. *Journal of Neuroscience, 29*(37), 11523–11539. doi:10.1523/JNEUROSCI.2040-09.2009

Perenin, M. T., & Rossetti, Y. (1996). Grasping without form discrimination in a hemianopic field. *Neuroreport, 7*, 793–797.

Pisella, L., Binkofski, F., Lasek, K., Toni, I., & Rossetti, Y. (2006). No double-dissociation between optic ataxia and visual agnosia: Multiple sub-streams for multiple visuomanual integrations. *Neuropsychologia, 44*(13), 2734–2748. doi:10.1016/j.neuropsychologia.2006.03.027

Prentiss, E. K., Schneider, C. L., Williams, Z. R., Sahin, B., & Mahon, B. Z. (2018). Spontaneous in-flight accommodation of hand orientation to unseen grasp targets: A case of action blindsight. *Cognitive Neuropsychology, 35*(7), 343–351. doi:10.1080/02643294.2018.1432584

Rapcsak, S. Z., Ochipa, C., Anderson, K. C., & Poizner, H. (1995). Progressive ideomotor apraxia: Evidence for a selective impairment of the action production system. *Brain and Cognition, 27*(2), 213–236. doi:10.1006/brcg.1995.1018

Riesenhuber, M. (2007). Appearance isn't everything: News on object representation in cortex. *Neuron, 55*(3), 341–344. doi:10.1016/j.neuron.2007.07.017

Rogalsky, C., Love, T., Driscoll, D., Anderson, S., & Hickok, G. (2011). Are mirror neurons the basis of speech perception? Evidence from five cases with damage to the purported human mirror system. *Neurocase, 17*, 178–187.

Rosenbaum, D., Vaughan, J., Barnes, H., Marchak, F., & Slotta, J. (1990). Constraints on action selection: Overhand versus underhand grips. In M. Jeannerod (Ed.), *Attention and performance XIII* (pp. 321–342). Hillsdale, NJ: Lawrence Erlbaum.

Rothi, L. J. G., Ochipa, C., & Heilman, K. M. (1991). A cognitive neuropsychological model of limb praxis. *Cognitive Neuropsychology, 8*(6), 443–458. doi:10.1080/02643299108253382

Rumiati, R. I., Zanini, S., Vorano, L., & Shallice, T. (2001). A form of ideational apraxia as a delective deficit of contention scheduling. *Cognitive Neuropsychology, 18*(7), 617–642. doi:10.1080/02643290126375

Sakuraba, S., Sakai, S., Yamanaka, M., Yokosawa, K., & Hirayama, K. (2012). Does the human dorsal stream really process a category for tools? *Journal of Neuroscience, 32*(11), 3949–3953. doi:10.1523/JNEUROSCI.3973-11.2012

Saygin, Z. M., Osher, D. E., Norton, E. S., Youssoufian, D. A., Beach, S. D., Feather, J., … Kanwisher, N. (2016). Connectivity precedes function in the development of the visual word form area. *Nature Neuroscience, 19*(9), 1250–1255. doi:10.1038/nn.4354

Schmid, M. C., Mrowka, S. W., Turchi, J., Saunders, R. C., Wilke, M., Peters, A. J., … Leopold, D. A. (2010). Blindsight depends on the lateral geniculate nucleus. *Nature, 466*(7304), 373–377. doi:10.1038/nature09179

Stasenko, A., Bonn, C., Teghipco, A., Garcea, F. E., Sweet, C., Dombovy, M., … Mahon, B. Z. (2015). A causal test of the motor theory of speech perception: A case of impaired speech production and spared speech perception. *Cognitive Neuropsychology, 32*(2), 38–57. doi:10.1080/02643294.2015.1035702

Stevens, W. D., Tessler, M. H., Peng, C. S., & Martin, A. (2015). Functional connectivity constrains the category-related organization of human ventral occipitotemporal cortex. *Human Brain Mapping, 36*(6), 2187–2206. doi:10.1002/hbm.22764

Wang, X., He, C., Peelen, M. V., Zhong, S., Gong, G., Caramazza, A., & Bi, Y. (2017). Domain selectivity in the parahippocampal gyrus is predicted by the same structural connectivity patterns in blind and sighted individuals. *Journal of Neuroscience, 37*(18), 4705–4716. doi:10.1523/JNEUROSCI.3622-16.2017

Wu, W. (2008). Visual attention, conceptual content and doing it right. *Mind, 117*, 1003–1033.

65 Naïve Physics: Building a Mental Model of How the World Behaves

JASON FISCHER

ABSTRACT To navigate and interact with the world, we must have an intuitive grasp of its physical structure and dynamics. Where should I push to open this door? Can I place this box on top of the others, or will the stack be unstable? Although the natural laws governing physical behavior can be challenging to comprehend in a mathematical sense, we implicitly employ approximate physical models in everyday life to predict objects' physical behaviors and adjust our actions accordingly. Our commonsense understanding of how the world will behave—termed *naïve physics*—emerges early in life and is expanded and refined by experience throughout our development and into adulthood. We draw on naïve physics in nearly all aspects of everyday life, and doing so often feels effortless and automatic. We "see" that a piece of furniture is too heavy to lift or that a surface is too slippery to walk on safely. Just how accurate are our physical intuitions? Do we carry out rich mental simulations of physical dynamics, or do we rely on heuristics that are effective in many scenarios but could break down in others? What brain machinery supports naïve physics? This chapter explores these questions from the vantage points of behavioral and neuroimaging research.

The Development of Physical Cognition in Infancy

Contrary to the once popular Piagetian notion that young infants understand little about the physical structure of the world, research over the past several decades has demonstrated that even in the first months of life, infants have basic expectations about how objects will behave. At just 2.5 months old, infants are surprised when an object seems to jump from one location to another without traversing the space in between, or when one object seems to pass through another. What are the building blocks of these early-emerging physical intuitions? Spelke and colleagues (Spelke, Breinlinger, Macomber, & Jacobson, 1992; Spelke & Kinzler, 2007) argue that we are born with an innate knowledge of some basic principles governing object motion, and this knowledge provides the mental scaffolding for learning more sophisticated physical concepts over the course of development. They propose that the core system of object representation comprises three principles: cohesion (objects move as connected, bounded units), continuity (an object moves along one connected path over space and time), and contact (objects must touch in order to influence each other's motion). Even very young infants apply these principles to individuate objects and predict their motion but initially fail to properly apply other physical principles, such as gravitational and inertial constraints. The emergence of these latter principles appears to hinge on experience—as children learn how particular objects behave in particular circumstances, they acquire piecemeal knowledge that builds upon the core principles. Over the first years of life, children's intuitions regarding gravity and inertia become steadily more adult-like but remain inconsistent across scenarios (Kaiser, Proffitt, & McCloskey, 1985). Likewise, children's sensitivity to the features that discriminate objects (e.g., shape, size, or color) relies on experience with specific events. Young infants fail to make use of such cues to individuate objects (Xu & Carey, 1996), and as infants learn about the attributes relevant for predicting an object's behavior, they often do so in an event-specific fashion that fails to transfer to new scenarios (Wang, Baillargeon, & Paterson, 2005). By contrast, infants rarely display misconceptions about cohesion, continuity, and contact—these principles form the stable core of our physical knowledge that endures throughout development and into adulthood.

How is children's physical knowledge expanded and refined over the course of development? Baillargeon and colleagues have proposed that children's physical representations are enriched through rule learning via explanation-based processes (Baillargeon, 2002; Wang, Zhang, & Baillargeon, 2016). Infants must first notice that two events for which they have similar models have contrastive outcomes that cannot be predicted based on current knowledge. They then search for the conditions that lead to each outcome, engaging in hypothesis-testing behaviors with objects that violated their expectations (Stahl & Feigenson, 2015). Finally, infants attempt to generate an explanation to be incorporated as a new variable that differentiates the outcomes of the two events. This framework supports the learning of event categories (e.g., occlusion, support, collision, and containment) and the relevant variables for interpreting those events (e.g., the shapes and sizes of objects and the spatial relationships between them). Because the same variable can be learned separately and at different times for different events, knowledge about a given variable does not always transfer across event categories. For example, 9-month-old infants attend to the height of an object placed in a container (and are surprised when a tall object fits completely in a short container) but not the height of an object placed in a tube, even when the containment and tube events are visually identical (Wang, Baillargeon, & Paterson, 2005). Hence, most 9-month-olds have not yet identified height as a relevant variable in tube events, even though they have done so for containment events (perhaps because of more experience with containers). After further revision based on experience, infants' rules become sufficiently abstract to unify variables learned under different conditions.

Even before the 1-year mark, infants acquire a broad and diverse catalog of physical knowledge in a systematic fashion. For example, infants incrementally learn increasingly sophisticated notions of support. As early as 3 months old, infants demonstrate an understanding that two objects must be in contact for one to support the other. Infants then come to understand that the spatial arrangement of the objects matters (the supported object must be on top), and ultimately, at about 12 months old, they understand roughly where an object's center of mass must be located relative to a supporting surface in order to be stable (Baillargeon, 1998). Between 5 and 7 months, infants also begin to display expectations about how falling objects will accelerate, and they become sensitive to the causal roles of one object striking and launching another. And infants' learning is not limited to rigid body interactions. By 5 months old, most infants are able to differentiate a liquid from a solid on the basis of movement cues and cohesiveness (Hespos, Ferry, Anderson, Hollenbeck, & Rips, 2016) and have expectations for how nonsolid substances will accumulate when poured (Anderson, Hespos, & Rips, 2018). By about 11 months old, infants can infer the weight of an object based on how much it compresses a soft material (Hauf, Paulus, & Baillargeon, 2012).

The above examples point to a systematic acquisition of physical knowledge during the first years of life, built around a stable core of object-motion principles. Just how sophisticated do our physical inference abilities become in adulthood? Do we ultimately rely on a catalog of situation-specific physical knowledge, or can we employ more generalized processes to predict physical dynamics across a range of scenarios? And what brain machinery supports naïve physics? The remainder of this chapter explores these questions.

Physical Inference Abilities in Adults

In adulthood, the apparent effortlessness with which we predict and reason about object dynamics in daily life belies some striking misconceptions about physical behavior that are revealed upon closer inspection. A classic example comes from McCloskey, Caramazza, and Green (1980), where college students were asked to draw the trajectory of a ball as it exited a curved tube. Many participants drew a curved path, indicating curvilinear motion even in the absence of any external forces. Similarly, many participants indicated that a ball being twirled at the end of a string would follow a curved path when the string was cut. These findings show that people's predictions can be starkly at odds with the physical behaviors they see in the world every day (and in fact, people perceive straight paths to be more natural looking than curved ones when viewing, rather than diagramming, the outcomes of the same scenarios (Kaiser, Proffitt, & Anderson, 1985). In a number of other scenarios, such as when a ball is released from a pendulum (Caramazza, McCloskey, & Green, 1981) or dropped by someone who is walking (McCloskey, Washburn, & Felch, 1983), people draw trajectories that are inconsistent with Newtonian dynamics. People also tend to make systematic errors when predicting how a liquid will be oriented within a tilted container (Vasta & Liben, 1996) or when indicating which of two objects is heavier after observing a collision between them (Gilden & Proffitt, 1989; Todd & Warren, 1982). While this is a surprising pattern of errors to observe in adults, it is consistent with the notion that physical knowledge is acquired in an event-specific fashion. Just as with infants, adults rarely hold misconceptions about the principles of cohesion, continuity, and contact, but judgments of object motion that incorporate gravity and inertia can be highly idiosyncratic. For example, while people tend to make errors regarding the path that a ball will take as it exits a curved tube, they are much more accurate at indicating how water will exit the same tube (Kaiser, Jonides, & Alexander, 1986), perhaps as a result of more experience with the latter scenario. These errors seem to suggest that even in adulthood, we are unable to integrate our learning about various physical scenarios into a unified model of object behavior. Instead, people might construct ad hoc theories of physical behaviors on the fly (Cook & Breedin, 1994) or rely on an incorrect, non-Newtonian model of physics (Clement, 1982; McCloskey, Caramazza, & Green, 1980).

A puzzle remains, though: How are we able to interact so effectively with our everyday environments if our physical predictions draw on idiosyncratic and sometimes incorrect conceptions about object behavior? Recent studies that have tested how people interact with moving objects shed some light on this matter. Using displays like those in Caramazza, McCloskey, and Green (1981), Smith, Battaglia, and Vul (2013) asked people to predict the path a ball would take after it was clipped from a swinging pendulum. Participants' predictions were tested in three ways: (1) drawing the path of the ball, (2) positioning a bin to catch the ball after it was released, and (3) cutting the ball free at the appropriate time so that it would land at a specified location. Results from the first task replicated previous findings that people often make idiosyncratic errors when drawing the path of the ball. However, performance on the latter two tasks revealed a different pattern of errors—participants' biases were less idiosyncratic and more consistent with a correct application of Newtonian mechanics. Other work has shown that in a variety of scenarios, people can be highly accurate and precise when executing actions on falling objects (Zago & Lacquaniti, 2005). People also perform better at judging how a liquid will behave in a container when asked to imagine the action of tilting the container rather than just giving a verbal description (Schwartz & Black, 1999). It may be the case, then, that the implicit physical inferences that support action tap into knowledge separate from that which we use to explicitly describe or diagram the workings of physical systems. When trying to catch the ball cut from the pendulum, people may place the bin in the correct position even without an explicit understanding of why the ball should end up there. Other studies using three-dimensional computer-generated stimuli or videos of object interactions have also found more accurate physical inferences than similar studies that used two-dimensional or schematic stimuli (Flynn, 1994; Hamrick, Battaglia, Griffiths, & Tenenbaum, 2016). The availability of naturalistic cues to the geometry and material properties of objects may be another factor that promotes access to implicit (and more consistently Newtonian) physical knowledge. The errors that people make when explaining the workings of physics nonetheless remain intriguing (Why would implicit and explicit physical predictions draw on distinct knowledge?), but they do not reflect a limit on our ability to make accurate predictions in the real-life scenarios where we use physical inferences to guide behavior.

If we can make accurate, approximately Newtonian physical predictions in at least some circumstances, what mental functions support this ability? One proposal is that we possess a mental "intuitive physics engine" that carries out simulations of physical dynamics (Battaglia, Hamrick, & Tenenbaum, 2013; Ullman, Spelke, Battaglia, & Tenenbaum, 2017). Here, *mental simulation* refers to playing physical dynamics forward in time as a video game physics engine would. Based on an initial scene configuration (e.g., scene layout, object geometry, material properties, and velocities), a mental simulation would step forward through successive states of the scene as physical interactions play out. Such a simulation would likely operate under a number of simplifying assumptions to make efficient simulation tractable, just as video game physics engines do. For example, collision detection may be based on simplified information about an object's three-dimensional shape (e.g., its convex hull) rather than fine-scaled geometry, and objects may only be actively simulated when in motion (akin to "sleep" and "wake" states in a video game physics engine). The end state of a simulation could answer questions such as "Where will the ball land?," and simulating a scenario multiple times over a range of initial parameters could answer questions such as "How should I roll this ball so it will end up in the desired location?" Importantly, this conception of mental simulation does not in itself implicate any particular brain areas or timescales (simulation need not progress in real time) and does not imply that simulation outcomes are always accurate or free of bias. Indeed, recent work has shown that in a number of scenarios both the successes and failures in human judgments are modeled well by probabilistic physics simulations that make similar patterns of errors (Bates, Battaglia, Yildirim, & Tenenbaum, 2015; Battaglia, Hamrick, & Tenenbaum, 2013). Hegarty (2004) has also argued in favor of a mental simulation account of physical inference based on tasks in which participants reason about multicomponent physical systems (e.g., a rope connected to a weight, threaded through a number of pulleys). Participants are slower to make judgments about components that are farther from the beginning of the causal chain, which suggests they step sequentially through the system to determine its behavior rather than simultaneously evaluating the components as a whole.

While probabilistic physics simulations provide good models of human performance under many conditions, there is ample reason to question whether mental simulation is the sole or primary means by which we form physical predictions in many everyday situations. Davis and Marcus (2016) point out that there are many scenarios in which physical outcomes are difficult or inefficient to infer through simulation but are trivial to infer from a rule-based standpoint. For example, to know whether water will spill out of a canteen, it is sufficient simply to know whether the canteen is open or closed. Mental simulation of the water's motion within the canteen would be impractical, and there is no need for the level of detail that a simulation would provide. In scenarios like these, commonsense physical reasoning may be achieved through knowledge-based analysis that relies on a large number of rules, rather than mental simulation (Davis, Marcus, & Frazier-Logue, 2017). Ultimately, it is likely that we draw on some combination of qualitative reasoning and dynamic simulation to form physical predictions. The conditions under which each is used, and the limits of each in terms of precision, processing speed, and adaptability to novel scenarios, will be important to flesh out in future research. Regardless of exactly how precise our naïve physics system is or what algorithms it is built on, there is no doubt we possess some fundamental physical knowledge that allows us to survive and engage with the world. This raises the question of what neural machinery underlies our physical-reasoning abilities.

A Physics Engine in the Brain

Research to identify and characterize the brain regions that support naïve physics is in the early stages, but emerging evidence points to a set of regions in the frontal and parietal cortex. A recent functional magnetic resonance imaging (fMRI) study (Fischer, Mikhael, Tenenbaum, & Kanwisher, 2016) contrasted brain activity from tasks that required physical inference (predicting the direction that an unstable tower of blocks would fall or predicting the trajectory of a bouncing billiard ball) with tasks that did not require physical inference but were otherwise matched on a host of factors. This study revealed a set of brain regions that are reliably engaged when people observe and predict the unfolding of physical events: bilateral frontal regions (dorsal premotor cortex, or PMd, and the supplementary motor area, or SMA), bilateral anterior parietal regions (postcentral sulcus, or PoCS) and the anterior intraparietal sulcus (aIPS), and the left supramarginal gyrus (SMG). Neuroimaging studies using textbook-style tasks have implicated similar regions in more explicit, abstract physical reasoning. A study in which subjects were asked to solve mechanical-reasoning puzzles found that a similar frontoparietal network of regions was engaged (Jack et al., 2013), and another study on the representation of abstract physics concepts (e.g., gravity, potential energy, and wavelength) found information related to these concepts in premotor and anterior parietal areas, among others (Mason & Just, 2016). Thus, although the behavioral work discussed above has established important distinctions between explanation-based physical problem-solving and the implicit physical inferences that we carry out in daily life, these two facets of physical cognition may draw on some common brain machinery.

The brain regions recruited for physical inference appear to largely overlap with those commonly implicated in action planning and tool use (Gallivan & Culham, 2015). This raises the possibility of a close relationship between action planning and naïve physics, and neuropsychological findings from patients with apraxia reinforce this notion. *Apraxia* refers to a pattern of impairments following brain damage that affect the ability to perform meaningful gestures and execute the appropriate actions for particular tools. While apraxia has often been framed as a motor condition, there is evidence that the core impairments in apraxia are in mechanical reasoning and action planning, rather than motor execution per se. When patients with apraxia are presented with novel tools, they show difficulties not only in executing appropriate actions with the tools but also in selecting the appropriate tool for a task based on its geometry (Goldenberg & Hagmann, 1998). The latter task requires mechanical reasoning but not fine-scaled motor execution. Lesions that result in impaired mechanical reasoning in apraxic patients fall in the same frontal and parietal regions as those implicated in physical reasoning in healthy participants (Goldenberg & Spatt, 2009).

The precise degree to which physical inference and action planning engage a common set of brain regions remains to be established by studies that measure both simultaneously. But to the degree that the two functions recruit common brain resources, why might the cortical systems for physical prediction and action planning be closely linked? Perhaps the most fundamental reason is that action planning inherently requires physical prediction. In order to plan appropriate actions, we must have a mental model of how objects will behave when we interact with them, taking into account physical variables such as the objects' shapes, sizes, and material properties. Indeed, there is evidence that many such variables are encoded within the frontal and parietal regions described above. Premotor cortex encodes object mass, both when preparing to lift an object (Gallivan, Cant, Goodale, & Flanagan, 2014) and when observing object interactions in the absence of any intention to perform an action (Schwettmann, Fischer, Tenenbaum, & Kanwisher, 2018). The aIPS encodes visual and somatosensory information about object shape, size, and orientation (Murata, Gallese, Luppino, Kaseda, & Sakata, 2000; Sakata, Taira, Murata, & Mine, 1995). The PMd, the SMA, and the anterior parietal cortex also show tuning to the gravitational constant, responding most strongly when viewing a falling object that accelerates at a rate consistent with natural gravity (Indovina et al., 2005). These variables that are crucial for anticipating objects' behaviors when preparing actions are the same as those we draw on for physical prediction more broadly.

As a result of the interdependence between action planning and physical inference, the two may share cortical machinery in a manner analogous to the relationship between the spatial attention and eye movement systems (Corbetta et al., 1998). Just as covert attention can be deployed off-line from the actual execution of saccades, predictive models in the action-planning system may run off-line from motor execution to simulate the outcomes of physical interactions (Schubotz, 2007). It is critical to note the distinction between this idea and motor simulation theories of perceptual and conceptual processing. Motor simulation theories hold that in a variety of domains, such as object recognition, language processing, and action understanding, covert engagement of the motor system—imaging oneself acting—is required in order to perceive and interpret information in those domains. Theories of this sort have been refuted by empirical evidence showing that disruptions of the motor system do not reliably lead to impairments in perceptual or conceptual processing (Mahon & Caramazza, 2008; Vannuscorps & Caramazza, 2016). The account of physical reasoning presented here does not invoke the notion of imagining one's own actions as a means of understanding physical behavior. The idea is simply that the same physical prediction mechanisms that support action planning may be called upon to subserve physical reasoning more broadly. For example, imagine picking up a bag of tortilla chips and a jar of salsa while grocery shopping. Without much thought, you use a soft grip to handle the chips—any more pressure would crush them—but a firm grip to pick up the salsa so the heavy jar won't slip out of your hand. The same physical inference mechanisms that informed these nuanced actions could alert you to the likelihood of the chips being crushed when you see the checkout attendant pack the salsa on top of the chip bag. Thus, the limits of motor execution need not constrain the kinds of physical behaviors that can be predicted using resources shared with the action-planning system. Interactions between objects that are out of reach may still be understood using the same predictive models that would be applied if the objects were targets of action. A possible reinterpretation of the *mirror neuron* responses implicated in motor simulation is that they reflect predictions regarding the physical outcomes of observed behaviors.

Ventral Stream Contributions to Naïve Physics

While the work discussed above implicates dorsal cortical regions in carrying out physical predictions, the ventral temporal cortex may play a complementary role, computing the object and scene attributes that form the basis for such predictions. In both humans (Cant & Goodale, 2011; Hiramatsu, Goda, & Komatsu, 2011) and monkeys (Goda, Tachibana, Okazawa, & Komatsu, 2014), information about objects' material properties is encoded in the ventral visual pathway. While early visual cortex encodes image-level details that serve as cues to objects' materials, higher-order areas (the posterior inferior temporal (IT) cortex in monkeys; the posterior collateral sulcus/fusiform gyrus in humans) represent more abstract information about dimensions, such as hardness, roughness, and elasticity. The same higher-order ventral regions encode object weight when it can be inferred from surface-texture cues (Gallivan et al., 2014). These material representations can be modified by visuohaptic experience (Goda, Yokoi, Tachibana, Minamimoto, & Komatsu, 2016) and thus may carry supramodal information about objects' material properties to support functions like physical prediction and action planning. Ventral representations of scene elements may also factor importantly into physical prediction—for example, by signaling the orientation of gravity. Humans use visual information (in addition to vestibular input) to infer the direction of gravity (Dichgans, Held, Young, & Brandt, 1972), and Vaziri and Connor (2016) have found that individual neurons in macaque anterior IT cortex are tuned to gravity-aligned scene elements, which may help establish a gravitational reference frame in which to carry out physical predictions.

It remains to be seen whether the object and scene information carried in the ventral visual stream contributes directly to the implicit physical predictions that guide our behavior in everyday life. While a variety of information from the ventral stream would, in principle, be useful for physical prediction, such information may also be present in a more flexible and rapidly accessible format in the dorsal stream (Jeong & Xu, 2017; Vaziri-Pashkam & Xu, 2017). In particular, object representations in posterior parietal cortex that support visually guided action may support physical prediction as well. If these object representations existed solely for the sake of guiding motor behaviors, one might expect them to maintain strict viewpoint specificity (Craighero, Fadiga, Umiltà, & Rizzolatti, 1996) since different object orientations require different actions (James, Humphrey, Gati, Menon, & Goodale,

2002). Instead, these dorsal object representations contain viewpoint-invariant information (Jeong & Xu, 2016; Konen & Kastner, 2008), suggesting they could support a broader range of abilities, such as tracking the stable properties of objects as they move and interact.

Conclusions

Over the past several decades, a flurry of research has led to major strides in understanding the computational and neural basis of our naïve physics abilities. Still, many key questions remain. Beyond allowing us to predict the behavior of objects and plan actions accordingly, how do our physical intuitions shape the way we interpret and engage with the world? Research in computer vision has suggested that naïve physics may have a pervasive role even at the earliest stages of visual processing, helping to segment the surfaces and objects in a scene (Zheng, Zhao, Joey, Ikeuchi, & Zhu, 2013). How does our naïve physics system interact with other aspects of cognition? Recent work has shown that physical cognition is dissociable from social cognition (Kamps et al., 2017), and the two may even be in a mutually inhibitory relationship, limiting our ability to use both in conjunction (Jack et al., 2013). Addressing these broader questions will be key to understanding how our physical intuitions shape our everyday experience.

REFERENCES

Anderson, E. M., Hespos, S. J., & Rips, L. J. (2018). Five-month-old infants have expectations for the accumulation of nonsolid substances. *Cognition, 175*, 1–10.

Baillargeon, R. (1998). Infants' understanding of the physical world. In M. Sabourin, F. Craik, & M. Robert (Eds.), *Advances in psychological science: Biological and cognitive aspects* (Vol. 2, pp. 503–529). Hove, UK: Psychology Press/Erlbaum.

Baillargeon, R. (2002). The acquisition of physical knowledge in infancy: A summary in eight lessons. In U. Goswami (Ed.), *Blackwell handbook of childhood cognitive development* (Vol. 1, 46–83). Oxford, UK: Blackwell.

Bates, C., Yildirim, I., Tenenbaum, J. B., & Battaglia, P. (2015). Humans predict liquid dynamics using probabilistic simulation. In Dale, R., Jennings, C., Maglio, P., Matlock, T., Noelle, D., Warlaumont, A., & Yoshimi, J. (Eds.), *Proceedings of the 37th Annual Conference of the Cognitive Science Society* (pp. 172–178). Austin, TX: Cognitive Science Society.

Battaglia, P. W., Hamrick, J. B., & Tenenbaum, J. B. (2013). Simulation as an engine of physical scene understanding. *Proceedings of the National Academy of Sciences, 110*(45), 18327–18332.

Cant, J. S., & Goodale, M. A. (2011). Scratching beneath the surface: New insights into the functional properties of the lateral occipital area and parahippocampal place area. *Journal of Neuroscience, 31*(22), 8248–8258.

Caramazza, A., McCloskey, M., & Green, B. (1981). Naive beliefs in "sophisticated" subjects: Misconceptions about trajectories of objects. *Cognition, 9*(2), 117–123.

Clement, J. (1982). Students' preconceptions in introductory mechanics. *American Journal of Physics, 50*(1), 66–71.

Cook, N. J., & Breedin, S. D. (1994). Constructing naive theories of motion on the fly. *Memory & Cognition, 22*(4), 474–493.

Corbetta, M., Akbudak, E., Conturo, T. E., Snyder, A. Z., Ollinger, J. M., Drury, H. A., et al. (1998). A common network of functional areas for attention and eye movements. *Neuron, 21*(4), 761–773.

Craighero, L., Fadiga, L., Umiltà, C. A., & Rizzolatti, G. (1996). Evidence for visuomotor priming effect. *Neuroreport, 8*(1), 347–349.

Davis, E., & Marcus, G. (2016). The scope and limits of simulation in automated reasoning. *Artificial Intelligence, 233*, 60–72.

Davis, E., Marcus, G., & Frazier-Logue, N. (2017). Commonsense reasoning about containers using radically incomplete information. *Artificial Intelligence, 248*, 46–84.

Dichgans, J., Held, R., Young, L. R., & Brandt, T. (1972). Moving visual scenes influence the apparent direction of gravity. *Science, 178*(4066), 1217–1219.

Fischer, J., Mikhael, J. G., Tenenbaum, J. B., & Kanwisher, N. (2016). Functional neuroanatomy of intuitive physical inference. *Proceedings of the National Academy of Sciences, 113*(34), E5072–E5081.

Flynn, S. B. (1994). The perception of relative mass in physical collisions. *Ecological Psychology, 6*(3), 185–204.

Gallivan, J. P., Cant, J. S., Goodale, M. A., & Flanagan, J. R. (2014). Representation of object weight in human ventral visual cortex. *Current Biology, 24*(16), 1866–1873.

Gallivan, J. P., & Culham, J. C. (2015). Neural coding within human brain areas involved in actions. *Current Opinion in Neurobiology, 33*, 141–149.

Gilden, D. L., & Proffitt, D. R. (1989). Understanding collision dynamics. *Journal of Experimental Psychology: Human Perception and Performance, 15*(2), 372–383.

Goda, N., Tachibana, A., Okazawa, G., & Komatsu, H. (2014). Representation of the material properties of objects in the visual cortex of nonhuman primates. *Journal of Neuroscience, 34*(7), 2660–2673.

Goda, N., Yokoi, I., Tachibana, A., Minamimoto, T., & Komatsu, H. (2016). Crossmodal association of visual and haptic material properties of objects in the monkey ventral visual cortex. *Current Biology, 26*(7), 928–934.

Goldenberg, G., & Hagmann, S. (1998). Tool use and mechanical problem solving in apraxia. *Neuropsychologia, 36*(7), 581–589.

Goldenberg, G., & Spatt, J. (2009). The neural basis of tool use. *Brain, 132*(Pt. 6), 1645–1655.

Hamrick, J. B., Battaglia, P. W., Griffiths, T. L., & Tenenbaum, J. B. (2016). Inferring mass in complex scenes by mental simulation. *Cognition, 157*, 61–76.

Hauf, P., Paulus, M., & Baillargeon, R. (2012). Infants use compression information to infer objects' weights: Examining cognition, exploration, and prospective action in a preferential-reaching task. *Child Development, 83*(6), 1978–1995.

Hegarty, M. (2004). Mechanical reasoning by mental simulation. *Trends in Cognitive Sciences, 8*(6), 280–285.

Hespos, S. J., Ferry, A. L., Anderson, E. M., Hollenbeck, E. N., & Rips, L. J. (2016). Five-month-old infants have general knowledge of how nonsolid substances behave and interact. *Psychological Science, 27*(2), 244–256.

Hiramatsu, C., Goda, N., & Komatsu, H. (2011). Transformation from image-based to perceptual representation of materials along the human ventral visual pathway. *NeuroImage, 57*(2), 482–494.

Indovina, I., Maffei, V., Bosco, G., Zago, M., Macaluso, E., & Lacquaniti, F. (2005). Representation of visual gravitational motion in the human vestibular cortex. *Science, 308*(5720), 416–419.

Jack, A. I., Dawson, A. J., Begany, K. L., Leckie, R. L., Barry, K. P., Ciccia, A. H., & Snyder, A. Z. (2013). fMRI reveals reciprocal inhibition between social and physical cognitive domains. *NeuroImage, 66,* 385–401.

James, T. W., Humphrey, G. K., Gati, J. S., Menon, R. S., & Goodale, M. A. (2002). Differential effects of viewpoint on object-driven activation in dorsal and ventral streams. *Neuron, 35*(4), 793–801.

Jeong, S. K., & Xu, Y. (2016). Behaviorally relevant abstract object identity representation in the human parietal cortex. *Journal of Neuroscience, 36*(5), 1607–1619.

Jeong, S. K., & Xu, Y. (2017). Task-context-dependent linear representation of multiple visual objects in human parietal cortex. *Journal of Cognitive Neuroscience, 29*(10), 1778–1789.

Kaiser, Mary Kister, Jonides, J., & Alexander, J. (1986). Intuitive reasoning about abstract and familiar physics problems. *Memory & Cognition, 14*(4), 308–312.

Kaiser, Mary Kister, Proffitt, D. R., & Anderson, K. (1985). Judgments of natural and anomalous trajectories in the presence and absence of motion. *Journal of Experimental Psychology: Learning, Memory, and Cognition, 11*(4), 795–803.

Kaiser, Mary Kister, Proffitt, D. R., & McCloskey, M. (1985). The development of beliefs about falling objects. *Perception & Psychophysics, 38*(6), 533–539.

Kamps, F. S., Julian, J. B., Battaglia, P., Landau, B., Kanwisher, N., & Dilks, D. D. (2017). Dissociating intuitive physics from intuitive psychology: Evidence from Williams syndrome. *Cognition, 168,* 146–153.

Konen, C. S., & Kastner, S. (2008). Two hierarchically organized neural systems for object information in human visual cortex. *Nature Neuroscience, 11*(2), 224–231.

Mahon, B. Z., & Caramazza, A. (2008). A critical look at the embodied cognition hypothesis and a new proposal for grounding conceptual content. *Journal of Physiology-Paris, 102*(1–3), 59–70.

Mason, R. A., & Just, M. A. (2016). Neural representations of physics concepts. *Psychological Science, 27*(6), 904–913.

McCloskey, M., Caramazza, A., & Green, B. (1980). Curvilinear motion in the absence of external forces: Naive beliefs about the motion of objects. *Science, 210*(4474), 1139–1141.

McCloskey, M., Washburn, A., & Felch, L. (1983). Intuitive physics: The straight-down belief and its origin. *Journal of Experimental Psychology: Learning, Memory, and Cognition, 9*(4), 636–649.

Murata, A., Gallese, V., Luppino, G., Kaseda, M., & Sakata, H. (2000). Selectivity for the shape, size, and orientation of objects for grasping in neurons of monkey parietal area AIP. *Journal of Neurophysiology, 83*(5), 2580–2601.

Sakata, H., Taira, M., Murata, A., & Mine, S. (1995). Neural mechanisms of visual guidance of hand action in the parietal cortex of the monkey. *Cerebral Cortex, 5*(5), 429–438.

Schubotz, R. I. (2007). Prediction of external events with our motor system: Towards a new framework. *Trends in Cognitive Sciences, 11*(5), 211–218.

Schwartz, D. L., & Black, T. (1999). Inferences through imagined actions: Knowing by simulated doing. *Journal of Experimental Psychology: Learning, Memory, and Cognition, 25*(1), 116–136.

Schwettmann, S., Fischer, J., Tenenbaum, J., & Kanwisher, N. (2018). Neural representation of the intuitive physical dimension of mass. Presented at the *Vision Sciences Society Annual Meeting.* Saint Pete Beach, FL.

Smith, K., Battaglia, P., & Vul, E. (2013). Consistent physics underlying ballistic motion prediction. In Knauff, M., Pauen, M., Sebanz, N., & Wachsmuth, I. (Eds.), *Proceedings of the 35th Conference of the Cognitive Science Society* (pp. 3426–3431). Austin, TX: Cognitive Science Society.

Spelke, E. S., Breinlinger, K., Macomber, J., & Jacobson, K. (1992). Origins of knowledge. *Psychological Review, 99*(4), 605–632.

Spelke, E. S., & Kinzler, K. D. (2007). Core knowledge. *Developmental Science, 10*(1), 89–96.

Stahl, A. E., & Feigenson, L. (2015). Observing the unexpected enhances infants' learning and exploration. *Science, 348*(6230), 91–94.

Todd, J. T., & Warren, W. H. (1982). Visual perception of relative mass in dynamic events. *Perception, 11*(3), 325–335.

Ullman, T. D., Spelke, E., Battaglia, P., & Tenenbaum, J. B. (2017). Mind games: Game engines as an architecture for intuitive physics. *Trends in Cognitive Sciences, 21*(9), 649–665.

Vannuscorps, G., & Caramazza, A. (2016). Typical action perception and interpretation without motor simulation. *Proceedings of the National Academy of Sciences, 113*(1), 86–91.

Vasta, R., & Liben, L. S. (1996). The water-level task: An intriguing puzzle. *Current Directions in Psychological Science, 5,* 171–177.

Vaziri, S., & Connor, C. E. (2016). Representation of gravity-aligned scene structure in ventral pathway visual cortex. *Current Biology, 26*(6), 766–774.

Vaziri-Pashkam, M., & Xu, Y. (2017). Goal-directed visual processing differentially impacts human ventral and dorsal visual representations. *Journal of Neuroscience, 37*(36), 8767–8782.

Wang, S., Baillargeon, R., & Paterson, S. (2005). Detecting continuity violations in infancy: A new account and new evidence from covering and tube events. *Cognition, 95*(2), 129–173.

Wang, S., Zhang, Y., & Baillargeon, R. (2016). Young infants view physically possible support events as unexpected: New evidence for rule learning. *Cognition, 157,* 100–105.

Xu, F., & Carey, S. (1996). Infants' metaphysics: The case of numerical identity. *Cognitive Psychology, 30*(2), 111–153.

Zago, M., & Lacquaniti, F. (2005). Cognitive, perceptual and action-oriented representations of falling objects. *Neuropsychologia, 43*(2), 178–188.

Zheng, B., Zhao, Y., Joey, C. Y., Ikeuchi, K., & Zhu, S.-C. (2013). Beyond point clouds: Scene understanding by reasoning geometry and physics. In *Proceedings of the IEEE Conference on Computer Vision and Pattern Recognition* (pp. 3127–3134). Portland, OR: IEEE.

66 Concepts and Object Domains

YANCHAO BI

ABSTRACT Domain effects have been studied extensively for object perceptual and conceptual processes. Decades of neuroimaging research have identified domain differences in widely distributed brain systems, including various higher-level sensory and motor systems. Investigations of the mechanisms underlying such differences have led to a more detailed understanding of and questions about the computational nature of these regions and their functional roles in object knowledge representation in general. In this chapter, I review recent findings on the variables associated with the response and connectivity profiles of three different domain-preferring clusters in higher ventral visual cortex. The findings reveal the joint effects of visual features and connectivity patterns and an intriguing interaction between input modality and object domain. The available evidence motivates a line of theoretical analyses about the nature of domain-relevant response systems and their relationship with input systems (e.g., vision). A promising hypothesis is that the manner in which bottom-up input information is translated into different response systems for different domains constrains the nature of representation at various object-processing levels.

How does the human brain represent what we know about objects in the world, such as a mouse, a table, or an ax? One hypothesis gleaned from neuropsychological and neuroimaging studies is that the object domains of evolutionary salience, such as animals, tools, and conspecifics, is an important dimension along which object knowledge is organized (Caramazza & Shelton, 1998). Brain lesions may lead to relatively disproportionate deficits in the knowledge of certain domains (Capitani, Laiacona, Mahon, & Caramazza, 2003; Warrington & Shallice, 1984). Stimuli of different domains elicit relatively different strengths of activation in multiple brain regions, including perceptual systems such as higher-order visual cortex, auditory cortex, motor cortex, and so-called higher-order association cortex (see reviews in Brefczynski-Lewis & Lewis, 2017; Martin, 2016).

Objects belonging to various domains systematically differ in many aspects, such as their physical appearance, their movement, the sound they produce, whether and how they can be manipulated, the type of function they serve for humans, and whether and what emotional responses they induce. All of these differences can potentially play a role in accounting for the neuropsychological and functional magnetic resonance imaging (fMRI) findings of domain differences (e.g., Warrington & McCarthy, 1987). In this chapter, I discuss current notions, findings, and the new questions that have emerged.

I first introduce the consensus framework underlying the brain basis of object knowledge representation, which incorporates a domain dimension, focusing on the ventral visual pathway (ventral occipitotemporal cortex, or VOTC); I then discuss how recent empirical patterns pose new challenges for the existing theories. I go on to present a theoretical analysis of the effect of an important domain difference—that is, the manner in which sensory systems map onto the corresponding response systems, on local computations for different domains, and then describe the outstanding questions.

Canonical View of Object Knowledge Representations and the Effects of Object Domains

Decades of neuroimaging studies have consistently localized object-knowledge representations to widely distributed brain regions across the temporal, frontal, and parietal cortices (Binder, Desai, Graves, & Conant, 2009; Mahon & Caramazza, 2011; Martin, 2007). The activations in regions that loosely belong to the sensorimotor cortices are commonly interpreted as representing attributes of corresponding modalities (e.g., form, color, motion, sound, action, and emotion; Lambon Ralph, Jefferies, Patterson, & Rogers, 2017; Martin, 2016). In this distributed-representation framework of object concepts, within each modality, brain subclusters showing a varying degree of sensitivity to objects of different domains have been consistently reported. The higher-order visual cortex includes clusters that show different preferences for pictures of different domains, with a broad animate/inanimate distinction (Chao, Haxby, & Martin, 1999; Grill-Spector & Weiner, 2014; Kanwisher, 2010; Konkle & Caramazza, 2013). In the auditory cortex, clusters have been found that are differentially sensitive to the sounds of people (voices and speech), man-made sounds, and natural sounds (Brefczynski-Lewis & Lewis, 2017). For the action system (prefrontal, inferior frontal, and inferior parietal regions), stronger activation is elicited by small manipulable objects (Lewis, 2006; Martin, Wiggs, Ungerleider, & Haxby, 1996). These domain-preferring nodes distributed in different modality-specific processing streams are linked together by brain connections to form domain-specific networks.

Object Domain Distributions in the Ventral Visual Pathway: Nodal Representations and Connection Structures

Domain organization has been most extensively studied in the VOTC. From the ventral medial to the lateral occipitotemporal cortex, gradients of three clusters showing stronger sensitivity to pictures of three domains of objects have been consistently obtained: the medial-anterior fusiform gyrus/parahippocampal gyrus (medFG/PHG, or the parahippocampal place area, PPA; Epstein & Kanwisher, 1998), which prefers places and large objects; the lateral-posterior fusiform gyrus (latFG; Chao, Haxby, & Martin, 1999), which prefers animals; and the lateral occipitotemporal cortex (LOTC; Bracci, Cavina-Pratesi, Ietswaart, Caramazza, & Peelen, 2012), which prefers tools (figure 66.1; e.g., Konkle & Caramazza, 2013; see reviews in Bi, Wang, & Caramazza, 2016; Bracci, Ritchie, & de Beeck, 2017; Grill-Spector & Weiner, 2014; Peelen & Downing, 2017).

The nature of the domain differences in these regions has been at the heart of discussions about higher-order visual cortex and knowledge representation. The following types of (nonmutually exclusive) hypotheses regarding these differences have been entertained: (1) They compute certain bottom-up visual properties that are correlated with or diagnostic of different domains (e.g., Hasson, Levy, Behrmann, Hendler, & Malach, 2002; Levy, Hasson, Avidan, Hendler, & Malach, 2001; Nasr, Echavarria, & Tootell, 2014; Srihasam, Vincent, & Livingstone, 2014), (2) they are multimodal or amodal (abstract conceptual) domain-specific representations (e.g., Ricciardi, Bonino, Pellegrini, & Pietrini, 2013), and (3) they are driven by the innate brain connections that connect modality-specific representations across different systems for processing a given domain (Mahon & Caramazza, 2011). I will briefly review the following evidence relating to these three notions: whether certain low-level visual features that tend to associate with certain object domains activate these clusters in the absence of object-domain knowledge; whether nonvisual stimuli of the corresponding object domains, even in the case of total visual deprivation (congenitally blind individuals), activate these clusters; and whether they are connected with different brain regions in other sensory/motor systems. The overall findings are summarized in table 66.1 and figure 66.1.

Preference to Navigation-Related Objects in the Medial-Anterior Fusiform Gyrus/Parahippocampal Gyrus

Is this region activated by certain visual features associated with large objects and places? The answer is yes. The lower-level visual properties that have been shown to associate with PPA activation include rectilinear shape (Nasr, Echavarria, & Tootell, 2014), peripheral vision (Levy et al., 2001), and large real-world size (Konkle & Oliva, 2012). Scrambled images of houses, presumably keeping only the low-level visual features and blocking other domain-relevant information that depends on recognition, elicit response patterns similar to those of normal house pictures, with stronger activation in the medFG/PHG areas (Coggan, Liu, Baker, & Andrews, 2016).

Is this region activated by nonvisual stimuli of the corresponding domain and in congenitally blind individuals? The answer is also yes. Compared with various control conditions, this area was more strongly activated when the subjects haptically explored Lego scenes; listened to sounds associated with landmarks, such as the ringing of a church bell; or made semantic judgments on visually presented names of famous sites ("Was the Colosseum constructed before 500 AD?") or size judgments on the auditory names of large nonmanipulable objects (e.g., Adam & Noppeney, 2010; Fairhall & Caramazza, 2013; He et al., 2013; Wolbers, Klatzky, Loomis, Wutte, & Giudice, 2011). In congenitally blind individuals, this region was also more strongly activated when they explored Lego scenes relative to Lego abstract objects (Wolbers et al., 2011) and when they performed size judgment tasks on auditory words of large nonmanipulable objects compared with tools and animals (He et al., 2013).

Brain connectivity pattern Currently, two major types of brain connections are measured noninvasively: white matter structural connectivity, using diffusion tensor imaging (DTI; Le Bihan et al., 2001), and resting-state functional connectivity (rsFC), which is measured by the degree of synchronization (correlation of the activity time course) at rest using functional imaging (Friston, Frith, Liddle, & Frackowiak, 1993; Smith, 2012). The PPA was found to be functionally connected with regions encompassing other scene/large object-sensitive clusters, including the retrosplenial cortex (RSC) and the transverse occipital sulcus (TOS; He et al., 2013). Testing the relationship between the connectivity pattern and domain-preference functional responses, Saygin et al. (2012) showed that a fusiform voxel's domain preference (scenes relative to faces) could be predicted from its structural connectivity patterns with the rest of the brain. Visual experience has minimal influence on the rsFC pattern, the structural connectivity pattern, or the relationship between the structural connectivity pattern and the functional preference for large objects in this area (Wang et al., 2015, 2017). Finally, the properties of the long-range structural connections of the PPA are associated with visual recognition performances of places and large objects (Gomez et al., 2015; Li et al., 2018).

Preference to Small Manipulable Objects (Tools) in the Lateral Occipitotemporal Cortex

Is this region activated by certain visual features associated with tools? The presence of an elongated shape seems sufficient to activate the LOTC (Chen, Snow, Culham, & Goodale, 2017). However, having more elongation features is not necessary to induce preferential activity in this region. It is also activated by items with a very distinct visual shape, such as hands (Bracci et al., 2012; Bracci & Peelen, 2013; Striem-Amit, Vannuscorps, & Caramazza, 2017). Training novel objects to be used as tools results in stronger activation here than pretraining, although the visual properties remain identical before and after training (Weisberg, van Turrennout, & Martin, 2007).

Is this region activated by nonvisual stimuli of the corresponding domain and in congenitally blind individuals? The LOTC's selectivity to tools has been reported when subjects made judgments about or generated names for object sounds, such as the sound of *sawing wood* (Doehrmann, Naumer, Volz, Kaiser, & Altmann, 2008; Lewis, Brefczynski, Phinney, Janik, & DeYoe, 2005; Tranel, Grabowski, Lyon, & Damasio, 2005), or written or spoken tool names (the word *saw*; e.g., Noppeney, Price, Penny, & Friston, 2006; Peelen et al., 2013). For congenitally blind individuals, LOTC's selectivity to tools was reported when the participants performed object-size judgment tasks according to the auditory names of tools compared

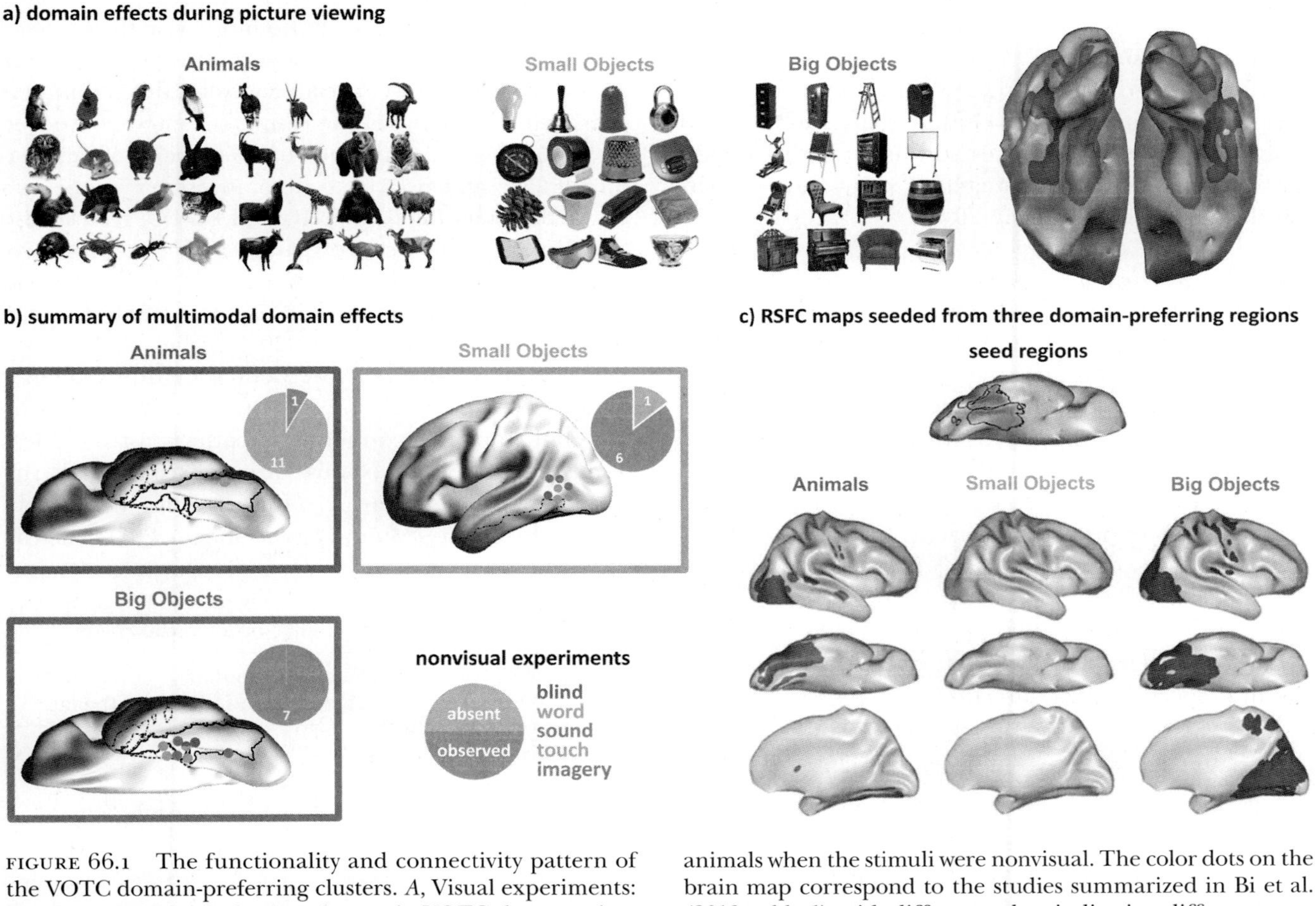

FIGURE 66.1 The functionality and connectivity pattern of the VOTC domain-preferring clusters. *A*, Visual experiments: the three domain-preferring clusters in VOTC that associate with viewing pictures of large objects, small manipulable objects, and animals. Adapted from Konkle and Caramazza (2013). *B*, Nonvisual experiments: The two artifact clusters in (*A*) show consistent domain effects in nonvisual experiments, whereas the animal cluster tended not to show preference to animals when the stimuli were nonvisual. The color dots on the brain map correspond to the studies summarized in Bi et al. (2016, table 1), with different colors indicating different types of nonvisual input. Pie charts show the number of studies in which nonvisual domain effects were observed (*red*) or absent (*blue*). *C*, The resting-state functional connectivity patterns that associate with the three domain-preferring clusters. Adapted from Konkle and Caramazza (2017). (See color plate 79.)

with the names of animals and large nonmanipulable objects (Peelen et al., 2013).

Brain connectivity pattern The results from the rsFC analysis showed that the LOTC is intrinsically linked with the parietal cortex along the intraparietal sulcus and the inferior frontal regions that have been implicated in tool processing (Konkle & Caramazza, 2017; Peelen et al., 2013), which does not seem to be affected by visual deprivation (Wang et al., 2015). DTI studies showed that the pMTG tool region, which roughly corresponds to the LOTC, is structurally connected with the parietal and frontal tool-related regions and lesions affecting the connections between the LOTC and the frontal tool clusters (inferior frontal and ventral premotor cortex) associated with tool conceptual deficits (Bi et al., 2015).

Preference to Animate Items in the LatFG

Is this region activated by certain visual features associated with animate items? Curvature and fovea processing have been suggested to associate with activation in this territory (Hasson et al., 2002; Srihasam, Vincent, & Livingstone, 2014). Nonetheless, after controlling for various visual properties, including shape, texture, and picture size, animal pictures still activate this region more strongly than well-matched man-made objects (Proklova, Kaiser, & Peelen, 2016).

Is this region activated by nonvisual stimuli of the corresponding domain and in congenitally blind individuals? Studies using nonvisual stimuli have failed to observe animal preferences relative to other domains using object sounds (e.g., Adam & Noppeney, 2010; Lewis et al., 2005) or written or spoken animal names (e.g., He et al., 2013; Noppeney, Price, Penny, & Friston, 2006; but see Chao, Haxby, & Martin, 1999). That is, this region is not more strongly activated when subjects listen to animal sounds (e.g., a barking sound) or names (e.g., the word *dog*) relative to nonanimal sounds or words (e.g., a church bell or the word *church*). In congenitally blind participants, listening to animal names does not activate this region more strongly than other objects (He et al., 2013; Wang et al., 2015).

Brain connectivity pattern In sighted individuals, this region is intrinsically functionally connected with the bilateral occipital and posterior ventral temporal cortex, the superior temporal sulcus, and the somatosensory and motor cortex (Konkle & Caramazza, 2017). Visual deprivation has a significant impact on the rsFC pattern of this region; in the congenitally blind, it is additionally connected with the primary and secondary auditory, the bilateral superior parietal, and the inferior frontal regions (Wang et al., 2015).

Support and Challenges Associated with Current Theories

In the first section, I presented three (non–mutually exclusive) notions: the bottom-up visual property account, the amodal domain-specific property account, and the connectivity-constraint account. Each notion is consistent with some of the results reviewed above (see table 66.1). The coexistence of the specific visual feature effects and nonvisual domain effects in the two artifact clusters (medFG/PHG and LOTC) reflects the close interactions between visual and domain representations, which may be optimized for real-world behavior (see discussions in Bracci, Ritchie, & de Beeck, 2017; Proklova et al., 2016). The results showing stronger connectivity between domain-preferring regions across various brain systems and the predictive nature of the connectivity pattern for the local domain-preference response are consistent with the connectivity-constraint account for domain distribution in the VOTC (Mahon & Caramazza, 2011) and the

TABLE 66.1 Summary of the effects of stimuli properties on the domain distribution in the higher-order visual cortex.

		Places and large objects in the medFG/PHG	Small manipulable objects in the LOTC	Animals in the latFG
Visual (view visual features)	Sighted (sufficient, not necessary)	Rectilinear	Elongation	Curvature
Visual (view object pictures)	Sighted	Yes	Yes	Yes
Words (listen to object names)	Sighted	Yes	Yes	Mostly no
	Blind	Yes	Yes	Mostly no
Auditory (listen to object sounds)	Sighted	Yes	Yes	No
	Blind	—	—	—
Tactile (haptic exploration of objects)	Sighted	Yes	—	—
	Blind	Yes	—	—

general notion that connection determines function (Passingham, Stephan, & Kötter, 2002).

None of the accounts, in their current forms, explains the intriguing differences in the input modality effects across domains. When objects are presented in nonvisual modalities, such as haptic or sound, large objects still activate the medFG/PHG and tools LOTC while the latFG no longer has domain preference for animals. Why would hearing the sound of a church bell and the sound of sawing, or hearing the words *church* and *saw*, preferentially activate the two artifact VOTC regions but hearing the barking sound or the word *dog* does not activate the latFG? Does this mean that the nature of representation (format and content) of these three domain-preferring clusters differs, with the animal cluster being more "visual" (representing properties of animals that are primarily sensed through the visual modality), whereas other parts of the VOTC actually represent nonvisual properties (Peelen & Downing, 2017)? If yes, why are there such differences across domains?

Updated Proposal: Further Considerations of Stimulus-Response Mapping

A possible solution for the current empirical package is offered in Bi, Wang, and Caramazza (2016). The central points are that (1) the brain is wired to efficiently map sensory information to response systems that are optimal for survival; (2) the mechanism of mapping is tightly related to the nature of each information system being mapped; (3) different object domains entail mapping sensory information with different types of response systems, and thus the mechanisms of mapping may differ; and (4) the representations that map across systems are more readily accessed from multiple modalities.

Humans engage in different types of responses to different object domains. A typical response to a large, stable object is to go around it (useful for navigation), a response to a tool is to manipulate it in a certain way for a specific function, a response to an animal is to fight or take flight, and a response to other humans would primarily be social. That is, for different object domains, the visual information is primarily mapped onto different nonvisual response systems (figure 66.2; see also figure 1 in Peelen & Downing, 2017). These different target systems may have different types of relationships with the visual system. For instance, the correspondence between manipulation and physical form, such as shape and size, which can be computed through the visual system, may be relatively transparent. Object parts made by humans are of certain shapes and sizes to be manipulated in certain ways using effectors (e.g., elongation for grasping). When mapping visual information onto manipulation information, it can happen at a visual form element level for which corresponding units in the motor system also exist (figure 66.2, midlevel), rather than wait until the object-specific form and manipulation representations, on which mapping can of course also happen based on stored (conceptual) knowledge (figure 66.2, object-specific level). For mapping to the spatial navigation response system, certain shape (e.g., chunky, rectilinear) properties may associate with properties such as "being stable," indicating potential navigation landmarks, and trigger specific navigation actions such as going around or stepping over. Such crossmodal mapping on these midlevel form elements makes them *multimodal*. For animals, however, the type of response (fight or flight) is not associated with specific form features. Being big or small, round or long does not necessarily indicate whether an animal is dangerous or not. Thus, the translation from the visual form information associated with animals to the fight/flight response system does not appear to operate on the same (midlevel) element level as artifacts or through similar mapping mechanisms. The level upon which it operates is unknown—it could be at earlier specific visual detector levels (see below) and/or at later stages (e.g., whole-object [conceptual knowledge] level associations or combinations of multiple types of visual cues, such as shape/motion/color). As a result, in common midlevel "form" elements, the information content could be multimodal for those associated with large objects and small, manipulable objects but not with animate things.

This proposal does not add additional assumptions to the overall framework of object representation. It simply considers the nature of different types of object information and the corresponding crossmodality relationships for major object domains in greater depth. By attributing the VOTC domain effects to the midlevel visual (form) system, this proposal also readily explains why certain low-level visual features might be sufficient to activate these clusters.

Outstanding Questions

This updated proposal highlights the influence of the mapping principles between sensory and response systems in shaping the representation properties in each system. It frames a line of questions to be tested: (1) What is the information content at these domain-preferring regions? Does the "multimodal" domain effect indeed reflect the same types of form representation? (2) The updated proposal argues that the mapping between different object properties may happen on multiple levels and depend on the relationships between the two types of information. What are the mechanisms of

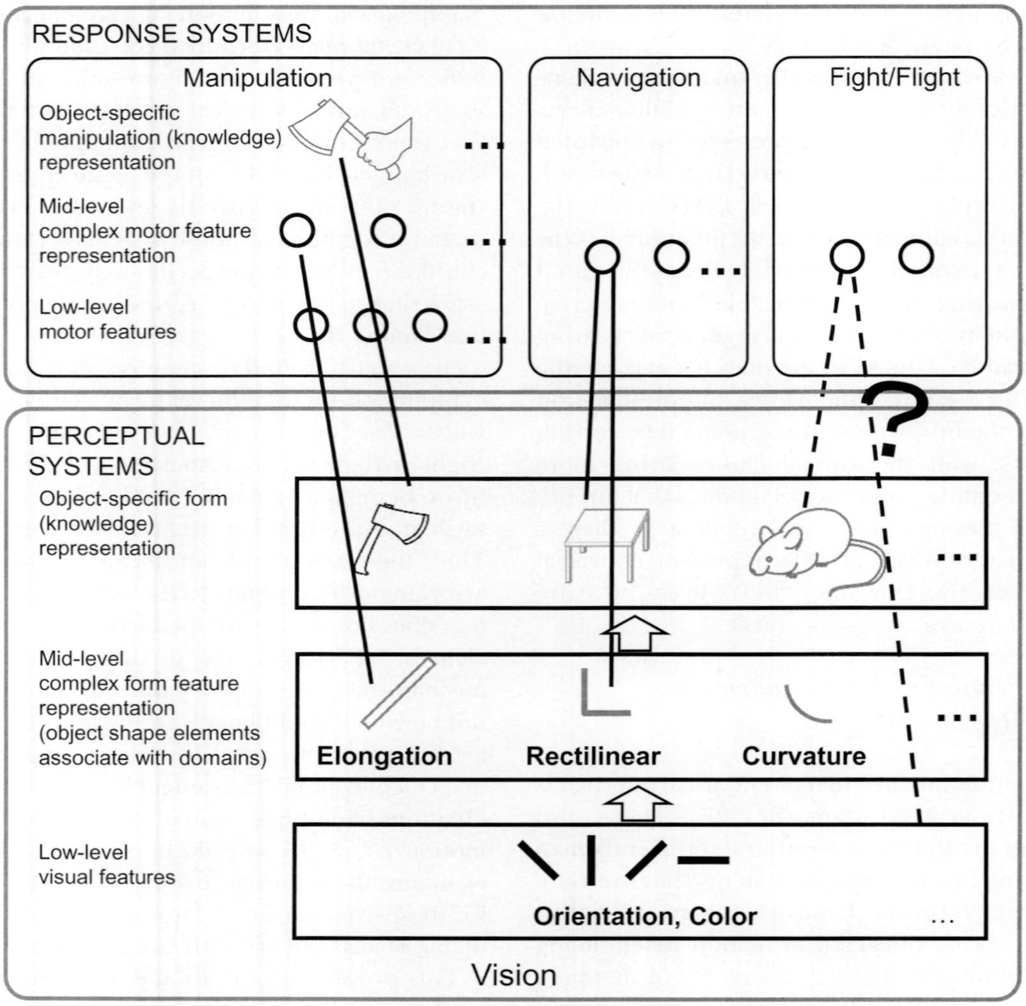

FIGURE 66.2 A schematic sketch of the updated proposal about object-domain representation. Only the example perceptual and response systems are shown. The main point is that the mapping between the perceptual representations and various response systems (corresponding to different object domains) may happen at different levels, depending on the relationships between systems. Note that the representation structures in the navigation and fight/flight response systems are highly simplified.

these mappings (see recent analyses of binding through connection patterns and/or region pattern interactions; Anzellotti & Coutanche, 2018; Fang et al., 2018)? (3) How early is the "domain" influence? Studies of domain representation have focused on the cortical sites where the domain difference is most visible, such as the so-called higher-order cortex. Recent neurophysiological evidence from nonhuman primates has discovered neurons in the primary visual and motor systems that are tuned to features much more complex than previously thought, such as those selective to predators (e.g., snakes) in the pulvinar (Le et al., 2013), curvatures in V1 (Tang et al., 2018), and complex actions in the primary motor cortex (Graziano, 2016). While the complex feature space for objects is large and undetermined (Kourtzi & Connor, 2011), those that are optimized for domain detection and triggering specific stimulus-response mappings might be good candidates for the effective functional units.

Conclusions

For a long time, the field of object processing has aimed to determine whether domain differences originate from bottom-up effects or innate domain-specific circuits. These discussions have led to a more detailed understanding and new questions about the functionalities and connectivity patterns of a range of cortical regions, especially the higher-level visual cortex. I wish to highlight a further dimension: the nature of the interface between different systems. After all, how the brain parses the physical world is driven by the need for optimal responses for survival, which is different for these object domains. How exactly this mapping process affects the regional representations and the connection mechanisms remains to be discovered.

Acknowledgments

I thank Alfonso Caramazza and Xiaoying Wang for the constant discussions about the topic in this chapter. I also thank Xiaosha Wang, Tao Wei, and Wei Wu for comments on earlier drafts and Yuxing Fang for the help in producing figure 66.2. This work has been supported by the Fulbright Visiting Scholar Program.

REFERENCES

Adam, R., & Noppeney, U. (2010). Prior auditory information shapes visual category-selectivity in ventral occipito-temporal cortex. *NeuroImage, 52*(4), 1592–602.

Anzellotti, S., & Coutanche, M. N. (2018). Beyond functional connectivity: Investigating networks of multivariate representations. *Trends in Cognitive Sciences, 22*(3), 258–269.

Bi, Y., Han, Z., Zhong, S., Ma, Y., Gong, G., Huang, R., … Caramazza, A. (2015). The white matter structural network underlying human tool use and tool understanding. *Journal of Neuroscience, 35*(17), 6822–6835.

Bi, Y., Wang, X., & Caramazza, A. (2016). Object domain and modality in the ventral visual pathway. *Trends in Cognitive Sciences, 20*(4), 282–290.

Binder, J. R., Desai, R. H., Graves, W. W., & Conant, L. L. (2009). Where is the semantic system? A critical review and meta-analysis of 120 functional neuroimaging studies. *Cerebral Cortex, 19*(12), 2767–2796.

Bracci, S., Cavina-Pratesi, C., Ietswaart, M., Caramazza, A., & Peelen, M. V. (2012). Closely overlapping responses to tools and hands in left lateral occipitotemporal cortex. *Journal of Neurophysiology, 107*(5), 1443–1456.

Bracci, S., & Peelen, M. V. (2013). Body and object effectors: The organization of object representations in high-level visual cortex reflects body-object interactions. *Journal of Neuroscience, 33*(46), 18247–18258.

Bracci, S., Ritchie, J. B., & de Beeck, H. O. (2017). On the partnership between neural representations of object categories and visual features in the ventral visual pathway. *Neuropsychologia, 105*(June), 153–164.

Brefczynski-Lewis, J. A., & Lewis, J. W. (2017). Auditory object perception: A neurobiological model and prospective review. *Neuropsychologia, 105*, 223–242.

Capitani, E., Laiacona, M., Mahon, B., & Caramazza, A. (2003). What are the facts of semantic category-specific deficits? A critical review of the clinical evidence. *Cognitive Neuropsychology, 20*(3/4/5/6), 213–261.

Caramazza, A., & Shelton, J. R. (1998). Domain-specific knowledge systems in the brain: The animate-inanimate distinction. *Journal of Cognitive Neuroscience, 10*(1), 1–34.

Chao, L. L., Haxby, J. V, & Martin, A. (1999). Attribute-based neural substrates in temporal cortex for perceiving and knowing about objects. *Nature Neuroscience, 2*(10), 913–919.

Chen, J., Snow, J. C., Culham, J. C., & Goodale, M. A. (2017). What role does "elongation" play in "tool-specific" activation and connectivity in the dorsal and ventral visual streams? *Cerebral Cortex,* March, 1–15.

Coggan, D. D., Liu, W., Baker, D. H., & Andrews, T. J. (2016). Category-selective patterns of neural response in the ventral visual pathway in the absence of categorical information. *NeuroImage, 135*, 107–114.

Doehrmann, O., Naumer, M. J., Volz, S., Kaiser, J., & Altmann, C. F. (2008). Probing category selectivity for environmental sounds in the human auditory brain. *Neuropsychologia, 46*(11), 2776–2786.

Epstein, R., & Kanwisher, N. (1998). A cortical representation of the local visual environment. *Nature, 392*(6676), 598–601.

Fairhall, S. L., & Caramazza, A. (2013). Brain regions that represent amodal conceptual knowledge. *Journal of Neuroscience, 33*(25), 10552–10558.

Fang, Y., Wang, X., Zhong, S., Song, L., Han, Z., Gong, G., & Bi, Y. (2018). Semantic representation in the white matter pathway. *PLoS Biology, 16*(4), e2003993.

Friston, K. J., Frith, C. D., Liddle, P. F., & Frackowiak, R. S. (1993). Functional connectivity: The principal-component analysis of large (PET) data sets. *Journal of Cerebral Blood Flow & Metabolism, 13*, 5–14.

Gomez, J., Pestilli, F., Witthoft, N., Golarai, G., Liberman, A., Poltoratski, S., … Grill-Spector, K. (2015). Functionally defined white matter reveals segregated pathways in human ventral temporal cortex associated with category-specific processing. *Neuron, 85*(1), 216–228.

Graziano, M. S. A. (2016). Ethological action maps: A paradigm shift for the motor cortex. *Trends in Cognitive Sciences, 20*(2), 121–132.

Grill-Spector, K., & Weiner, K. S. (2014). The functional architecture of the ventral temporal cortex and its role in categorization. *Nature Reviews Neuroscience, 15*(8), 536–548.

Hasson, U., Levy, I., Behrmann, M., Hendler, T., & Malach, R. (2002). Eccentricity bias as an organizing principle for human high-order object areas. *Neuron, 34*(3), 479–490.

He, C., Peelen, M. V, Han, Z., Lin, N., Caramazza, A., & Bi, Y. (2013). Selectivity for large nonmanipulable objects in scene-selective visual cortex does not require visual experience. *NeuroImage, 79*, 1–9.

Kanwisher, N. (2010). Functional specificity in the human brain: A window into the functional architecture of the mind. *Proceedings of the National Academy of Sciences of the United States of America, 107*(25), 11163–11170.

Konkle, T., & Caramazza, A. (2013). Tripartite organization of the ventral stream by animacy and object size. *Journal of Neuroscience, 33*(25), 10235–10242.

Konkle, T., & Caramazza, A. (2017). The large-scale organization of object-responsive cortex is reflected in resting-state network architecture. *Cerebral Cortex, 27*(10), 4933–4945.

Konkle, T., & Oliva, A. (2012). A real-world size organization of object responses in occipitotemporal cortex. *Neuron, 74*(6), 1114–1124.

Kourtzi, Z., & Connor, C. E. (2011). Neural representations for object perception: Structure, category, and adaptive coding. *Annual Review of Neuroscience, 34*, 45–67.

Lambon Ralph, M. A., Jefferies, E., Patterson, K., & Rogers, T. T. (2017). The neural and computational bases of semantic cognition. *Nature Reviews Neuroscience, 18*, 42–55.

Le, Q. Van, Isbell, L. A., Matsumoto, J., Nguyen, M., Hori, E., Maior, R. S., … Nishijo, H. (2013). Pulvinar neurons reveal neurobiological evidence of past selection for rapid

detection of snakes. *Proceedings of the National Academy of Sciences of the United States of America, 110*(47), 19000–19005.
Le Bihan, D., Mangin, J., Poupon, C., Clark, C., Pappata, S., & Molko, N. (2001). Diffusion tensor imaging: Concepts and applications. *Journal of Magnetic Resonance Imaging, 66*(13), 534–546.
Levy, I., Hasson, U., Avidan, G., Hendler, T., & Malach, R. (2001). Center-periphery organization of human object areas. *Nature Neuroscience, 4*(5), 533–539.
Lewis, J. W. (2006). Cortical networks related to human use of tools. *Neuroscientist, 12*(3), 211–231.
Lewis, J. W., Brefczynski, J. A., Phinney, R. E., Janik, J. J., & DeYoe, E. A. (2005). Distinct cortical pathways for processing tool versus animal sounds. *Journal of Neuroscience, 25*(21), 5148–5158.
Li, Y., Fang, Y., Wang, X., Song, L., Huang, R., Han, Z., ... Bi, Y. (2018). Connectivity of the ventral visual cortex is necessary for object recognition in patients. *Human Brain Mapping, 39*(7), 2786–2799.
Mahon, B. Z., Anzellotti, S., Schwarzbach, J., Zampini, M., & Caramazza, A. (2009). Category-specific organization in the human brain does not require visual experience. *Neuron, 63*(3), 397–405.
Mahon, B. Z., & Caramazza, A. (2009). Concepts and categories: A cognitive neuropsychological perspective. *Annual Review of Psychology, 60*, 27–51.
Mahon, B. Z., & Caramazza, A. (2011). What drives the organization of object knowledge in the brain? *Trends in Cognitive Sciences, 15*(3), 97–103.
Martin, A. (2007). The representation of object concepts in the brain. *Annual Review of Psychology, 58*, 25–45.
Martin, A. (2016). GRAPES—Grounding representations in action, perception, and emotion systems: How object properties and categories are represented in the human brain. *Psychonomic Bulletin and Review, 23*(4), 979–990.
Martin, A., Wiggs, C. L., Ungerleider, L. G., & Haxby, J. V. (1996). Neural correlates of category-specific knowledge. *Nature, 379*(6566), 649–652.
Nasr, S., Echavarria, C. E., & Tootell, R. B. H. (2014). Thinking outside the box: Rectilinear shapes selectively activate scene-selective cortex. *Journal of Neuroscience, 34*(20), 6721–6735.
Noppeney, U., Price, C. J., Penny, W. D., & Friston, K. J. (2006). Two distinct neural mechanisms for category-selective responses. *Cerebral Cortex, 16*(3), 437–445.
Passingham, R. E., Stephan, K. E., & Kötter, R. (2002). The anatomical basis of functional localization in the cortex. *Nature Reviews. Neuroscience, 3*, 606–616.
Peelen, M. V., Bracci, S., Lu, X., He, C., Caramazza, A., & Bi, Y. (2013). Tool selectivity in left occipitotemporal cortex develops without vision. *Journal of Cognitive Neuroscience, 25*(8), 1225–1234.
Peelen, M. V., & Downing, P. E. (2017). Category selectivity in human visual cortex: Beyond visual object recognition. *Neuropsychologia, 105*, 177–183.
Proklova, D., Kaiser, D., & Peelen, M. V. (2016). Disentangling representations of object shape and object category in human visual cortex: The animate-inanimate distinction. *Journal of Cognitive Neuroscience, 28*(5), 680–692.
Ricciardi, E., Bonino, D., Pellegrini, S., & Pietrini, P. (2013). Mind the blind brain to understand the sighted one! Is there a supramodal cortical functional architecture? *Neuroscience and Biobehavioral Reviews, 41*, 64–77.
Saygin, Z. M., Osher, D. E., Koldewyn, K., Reynolds, G., Gabrieli, J. D. E., & Saxe, R. R. (2012). Anatomical connectivity patterns predict face selectivity in the fusiform gyrus. *Nature Neuroscience, 15*(2), 321–327.
Smith, S. M. (2012). The future of fMRI connectivity. *NeuroImage, 62*, 1257–1266.
Srihasam, K., Vincent, J. L., & Livingstone, M. S. (2014). Novel domain formation reveals proto-architecture in inferotemporal cortex. *Nature Neuroscience, 17*(12), 1776–1783.
Striem-Amit, E., Vannuscorps, G., & Caramazza, A. (2017). Sensorimotor-independent development of hands and tools selectivity in the visual cortex. *Proceedings of the National Academy of Sciences, 114*(18), 4787–4792.
Tang, S., Lee, T. S., Li, M., Zhang, Y., Xu, Y., Liu, F., ... Jiang, H. (2018). Complex pattern selectivity in macaque primary visual cortex revealed by large-scale two-photon imaging. *Current Biology, 28*(1), 38–48.
Tranel, D., Grabowski, T. J., Lyon, J., & Damasio, H. (2005). Naming the same entities from visual or from auditory stimulation engages similar regions of left inferotemporal cortices. *Journal of Cognitive Neuroscience, 17*, 1293–1305.
Wang, X., He, C., Peelen, M. V, Zhong, S., Gong, G., Caramazza, A., & Bi, Y. (2017). Domain selectivity in the parahippocampal gyrus is predicted by the same structural connectivity patterns in blind and sighted individuals. *Journal of Neuroscience, 37*(18), 4705–4716.
Wang, X., Peelen, M. V., Han, Z., He, C., Caramazza, A., & Bi, Y. (2015). How visual is the visual cortex? Comparing connectional and functional fingerprints between congenitally blind and sighted individuals. *Journal of Neuroscience, 35*(36), 12545–12559.
Warrington, E. K., & McCarthy, R. A. (1987). Categories of knowledge. *Brain, 110*(5), 1273–1296.
Warrington, E. K., & Shallice, T. (1984). Category specific semantic impairments. *Brain, 107*(3), 829–853.
Weisberg, J., van Turrennout, M., & Martin, A. (2007). A neural system for learning about object function. *Cerebral Cortex, 17*(3), 513–521.
Wolbers, T., Klatzky, R. L., Loomis, J. M., Wutte, M. G., & Giudice, N. A. (2011). Modality-independent coding of spatial layout in the human brain. *Current Biology, 21*, 984–989.

67 Concepts, Models, and Minds

ALEX CLARKE AND LORRAINE K. TYLER

ABSTRACT Conceptual representations form the core of our mental lives, capturing a rich variety of knowledge about the world. Here, different approaches to defining conceptual representations are highlighted, and two prominent approaches to testing these models are discussed—voxel-encoding models and representational similarity analysis. Finally, we show how relating the properties of semantic feature-based models to brain activity from visual objects can explain conceptual processing in the brain, both in terms of the underlying neural architecture and the temporal dynamics.

Conceptual representations of objects and events form the core of our mental lives. They capture a rich variety of knowledge about objects, abstract ideas, mental states, actions, and the relations among them, enabling us to express and understand information about the world. As Murphy (2002) puts it: "Concepts are a kind of mental glue … in that they tie our past experiences to our present interactions with the world, and because the concepts themselves are connected to our larger knowledge structures." Understanding the nature of these representations has long engaged philosophers, linguists, and psychologists and has generated many theoretical accounts and disagreements. Nevertheless, in spite of the difficulties, it is impossible to study the nature of conceptual representations without first defining them. In psychology and cognitive neuroscience, where the goal is to understand how concepts are represented and processed in the mind/brain, there have been a number of attempts to define conceptual representations, resulting in a range of theories, including prototype, exemplar, and theory-theories (for reviews, see Laurence & Margolis, 1999; Murphy, 2002). Here the focus is not to provide a comprehensive review of these different positions but to highlight three prominent paths to defining concrete concepts and show how feature-based accounts in particular can capture the neural representation of concepts, as evoked by visual objects.

What Is a Concept?

Embodied accounts One influential approach defines concepts in terms of their grounding in the neural systems underlying perception and action (Barsalou, 1999; Binder et al., 2016; Martin, 2016; Pulvermüller, 2013). For example, in this embodied view, the motor brain areas involved in producing an action (e.g., kicking) become part of the conceptual representation of the concepts associated with that action (e.g., a soccer ball), such that when we encounter an object or hear its label, its sensorimotor properties and their associated brain regions are reactivated. Different accounts specify different degrees of embodiment, in which conceptual representations are composed of the same substrate required for perceiving and acting (strong embodiment). Others argue for a weaker embodiment by specifying a degree of separation between sensorimotor systems and conceptual representations, but the amodal semantics still directly interact with sensorimotor systems. In this sense, conceptual representations are abstracted away from sensorimotor systems but still depend on them for access to detailed, modality-specific properties. Therefore, the principal dimension upon which grounded accounts seem to vary is the importance they place on having a conceptual representation abstracted from modality-specific information.

While embodied approaches have been very influential, they have also been strongly criticized by those who claim semantics is encapsulated from perception and action systems that are not essential for semantics (Mahon & Caramazza, 2008). It is also unclear how embodied theories account for abstract concepts, which do not have clear sensorimotor relations, but affective information may be important here (Martin, 2016; Vigliocco et al., 2014).

Semantic features Another approach to defining the content of individual concepts assumes that conceptual representations are composed of smaller elements of meaning, called *properties* or *features* (Cree, McNorgan, & McRae, 2006; Farah & McClelland, 1991; McRae & Cree, 2002; Pexman, Holyk, & Monfils, 2003; Taylor, Devereux, & Tyler, 2011; Tyler & Moss, 2001). These semantic features are typically based on the verbal descriptions that participants provide when describing an object (e.g., is green, grows on trees, has a stalk, is round, is tasty; figure 67.1*A*). These labels are neither claimed to be the actual units of the neural representation underpinning object concepts nor claimed to provide a complete account of a concept's meaning. Understanding the nature of the high-level abstract conceptual information that neural populations represent is clearly an issue that needs to be addressed. However, convergent evidence from behavioral studies, computational modeling, functional neuroimaging, and neuropsychology clearly indicates that the statistical regularities captured through semantic features show a good correspondence to the statistical regularities in the brain (see Clarke & Tyler, 2015). Feature-based accounts are not, in principle, incompatible with embodied accounts since the sensorimotor features at the core of embodied approaches could be considered a subtype of features within a broad semantic space that includes many different feature types (e.g., Vigliocco et al., 2014).

Within semantic feature accounts are differences in the ways concepts are defined. For example, in some the notion of semantic "richness" is important, where semantic "richness" is typically operationalized as the number of features (Pexman, Holyk, & Monfils, 2003). Other feature-based accounts, such as the conceptual structure account (Taylor, Devereux, & Tyler, 2011; Tyler & Moss, 2001), have taken a different view, wherein the internal structure of the feature space comprising a concept is important, as are the featural relationships between concepts (figure 67.1*B*). Thus, features both describe the attributes of a concept (e.g., has legs, is tall), which capture its meaning, and also vary in the way they relate to other concepts. That is, features vary in the extent to which they are distinctive of a concept (e.g., the feature *trunk* is distinctive of elephants) or shared by a number of concepts (e.g., *legs* is a feature

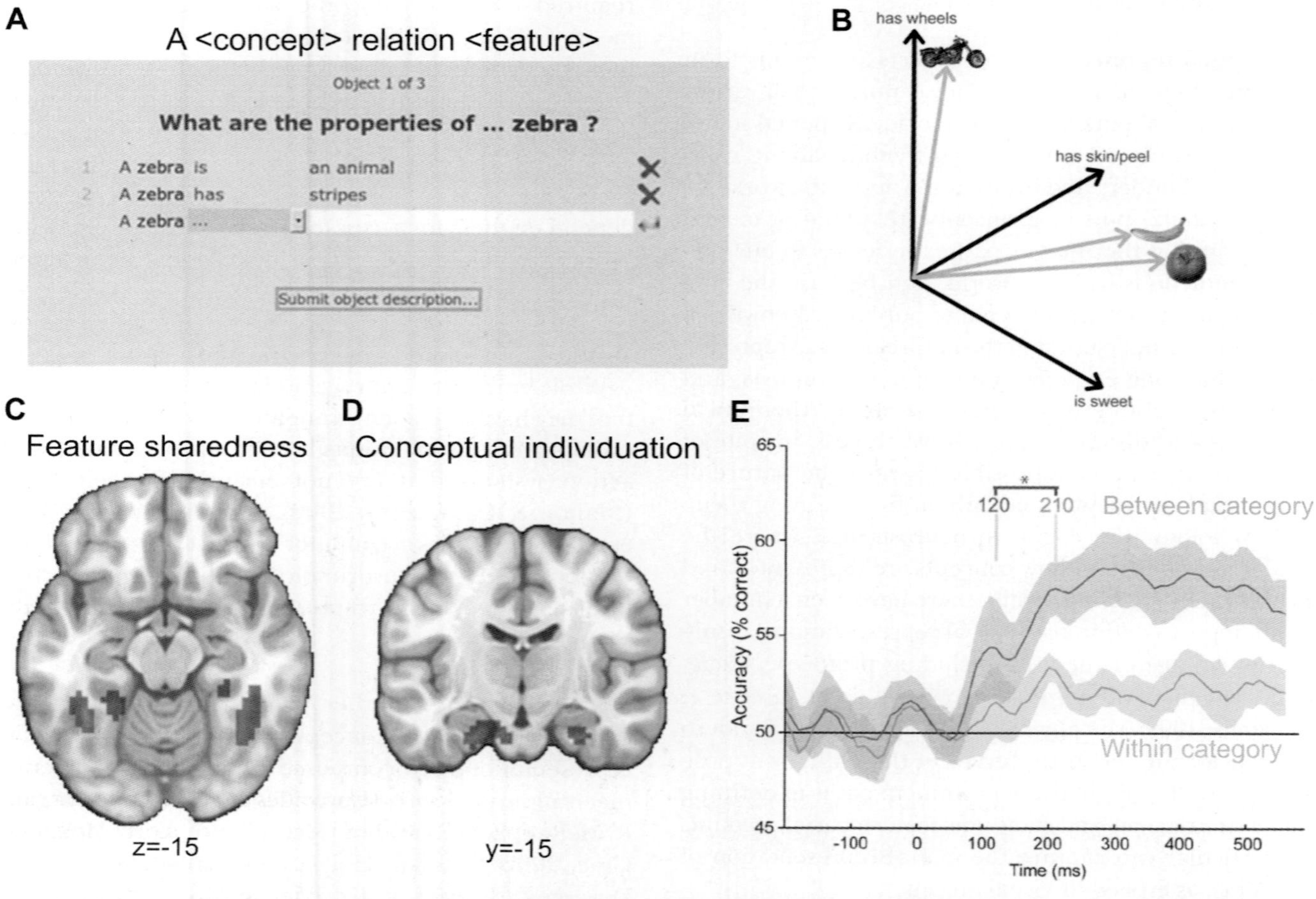

FIGURE 67.1 Semantic features. *A*, Example of collecting features for a given concept in a feature-norming study. *B*, Concepts can be more similar or different based on how similar the feature lists are, meaning they are closer together in a multidimensional feature space (three dimensions shown for clarity). *C*, Regions in the posterior ventral temporal lobe were modulated by feature-based statistics, in which more lateral regions showed increased activity for objects with relatively more shared features, and medial regions showed increased activity for objects with relatively more distinctive. *D*, Bilateral anteromedial temporal cortex (AMTC) activity increases for concepts that are semantically more confusable. *E*, The feature-based model can be used to successfully classify concepts from MEG signals, where between-category information (e.g., animal vs. tool) occurs before within-category information (e.g., lion vs. tiger). Panel (*A*) reproduced from Devereux et al. (2014), panel (*B*) from Devereux et al. (2018), and panel (*E*) from Clarke et al. (2015), all under the Creative Commons License. Panels (*C*) and (*D*) reproduced from Tyler et al. (2013). (See color plate 80.)

shared by many animals). Thus, superordinate category organization is based on the extent to which concepts can be grouped together on the basis of feature similarity, whereas individual concepts can be differentiated from similar concepts by the presence of distinctive features (Taylor, Devereux, & Tyler, 2011; Tyler & Moss, 2001). Feature-based models thus represent the semantics of concrete concepts and enable the quantification of object-specific properties, as well as the similarity between concepts.

Feature-based approaches are readily captured by parallel distributed-processing models. Such models instantiate conceptual knowledge in recurrent neural networks in which simple processing nodes correspond to components of meaning and where individual concepts are captured as patterns of activation over large sets of these microfeatures (Cree, McNorgan, & McRae, 2006; Devereux, Taylor, Randall, Geertzen, & Tyler, 2015; Rogers & McClelland, 2004). These models of the internal structure of concepts have also shown how meaning emerges over time rather than being a punctate event. For example, they show that shared features are activated first, soon followed by distinctive features, generating a gradient of semantic specificity from general to specific over time (Devereux et al., 2015), with a similar temporal trajectory in the brain (Clarke, Taylor, Devereux, Randall, & Tyler, 2013; see the section on neural architecture and the dynamics of accessing meaning from the senses).

When we respond to a concept—be it a written or spoken word or a visual object—the speed of our response and how it varies across different concepts give us insight into the underlying conceptual processes in the brain. This variability in reaction times to different concepts can be explained, in part, by semantic feature statistics. For example, when naming pictures of objects, those with more distinctive features and whose distinctive features are more intercorrelated (such as a typical tool) are named faster than those with more shared features (such as a typical animal; Taylor, Devereux, Acres, Randall, & Tyler, 2012). While it is generally true that tools are named faster than animals, feature-based statistics allow us to understand the variability in naming individual objects—both to explain the origin of distinctions between different superordinate categories and the variability in response times within superordinate categories.

Our responses to concepts are also affected by how we respond. For example, when people name an object at a domain (or superordinate) level, they are generally faster at naming animals than tools. This contrasts with their speed of naming a concept at the basic level (e.g., *hammer, penguin*), when they are slower at naming an animal than a tool. Feature statistics can readily explain this in terms of shared features being relevant for determining that an object is a member of a category while distinctive features are necessary for differentiating between similar objects within a category. Since animals have more shared properties than tools, they are easier to name at a category level, whereas tools have more distinctive properties than animals and are therefore easier to name at the basic level (Taylor et al., 2012). We see similar effects for lexical decisions to words (Devereux et al., 2015). The important point here is that the information captured through feature-based statistics can provide a single framework that captures information about conceptual representations at different levels of description. Equally important, the different feature-based statistical effects when people respond to concepts at different levels of description highlight that we access meaning in a flexible manner, with different kinds of properties taking on importance depending on the goal. This flexibility is also present in a modulation of brain activity, depending on the level of description required (Tyler et al., 2004).

Distributional semantic models Another approach to capturing the semantic content of concepts is known as distributional semantic modeling (DSM). The foundational assumption of distributional semantics is that the meaning of a word or concept can be induced from the contexts in which it occurs since words that occur in similar contexts tend to be related in meaning. This is important in computational linguistics, where meaning can be extracted from large text corpora based on words co-occurring in similar contexts. In the DSM framework, the semantic representation is defined by the distribution as a whole over a vector (Baroni & Lenci, 2010). Compared to feature-based semantic models, this approach has the advantage of being able to characterize different aspects of meaning constrained by the linguistic position of the word in a sentence. For example, the meaning of *lion* as the subject of a verb could be different from its meaning when it is a verb's object. Therefore, DSM provides a better basis for capturing word semantics in naturally occurring language contexts, compared to feature-based accounts. However, DSM, and other accounts in which meaning is defined based on word co-occurrence, still needs to be fully evaluated in terms of its ability to explain how meaning is represented and processed in the brain.

Another example of the DSM approach is topic modeling (Blei, Ng, & Jordan, 2003). In natural language processing, a topic model is a statistical model for obtaining the latent semantics, or "topics," that occur in text. This approach enables mapping between the co-occurring contexts and concepts, capturing various aspects of co-occurrence as a mixture of topics. An important distinction from above is that it fits the model to the data directly in a Bayesian framework, allowing the model to jointly learn the mapping through every observation in the corpus. The learned mapping represents the semantics of contexts and words in terms of the preference of each topic in the form of probability distributions.

Another DSM approach is to characterize meaning in terms of conceptual hierarchical trees. WordNet (Miller, 1995) is a large database that defines such conceptual hierarchies with each node in the tree (called a *synset*) linked to other synsets by means of a small number of conceptual relations. The co-occurrence data in the corpus is propagated through the WordNet hierarchy to obtain a more conceptually based representation of co-occurrence semantics. However, despite the differences across these DSM approaches, they are all informed by the co-occurrence structures of natural language, with a concept defined based on its relations to other concepts. One advantage of these sorts of approaches is that they provide a means of obtaining a rich semantic representation of a concept from its contexts of use.

The Relationship between Semantic Models and the Brain

Given that understanding how cognition is instantiated in neural processes is at the core of cognitive neuroscience, it is essential to have an explicit account of cognition and to determine to what extent such accounts can explain brain activity. The approaches discussed above allow for the creation of explicit definitions of semantics and the semantic relationships between concepts and can often specify semantic structures at different levels of abstraction (e.g., levels of the WordNet tree, such as *animal* or *tiger*). The last decade has seen a great deal of progress in understanding the neural basis of semantic knowledge, with many of these computational approaches to semantics showing robust relationships to brain activity.

A series of studies using functional magnetic resonance imaging (fMRI) by Gallant and colleagues point toward widespread, distributed semantic representations across the cortex (Huth, de Heer, Griffiths, Theunissen, & Gallant, 2016; Huth, Nishimoto, Vu, & Gallant, 2012; Stansbury, Naselaris, & Gallant, 2013), uncovered by voxel-wise encoding models that learn relationships between voxel activity and the presence or absence of thousands of categories. These studies build on the approach taken by Mitchell and colleagues (2008) that modeled semantics based on word co-occurrence probabilities from large text corpora. While these studies demonstrate that the regularities captured through natural language patterns show relationships to how the brain represents semantic knowledge, it has been argued that this type of approach, in the way it has been used, is limited in what it actually tells us about semantic processing. Barsalou (2017), for example, has argued that many (but not all) of the current encoding/decoding applications in fMRI principally establish a relationship between stimuli and brain activity but do so without recourse to an underlying model of the representations and processes involved. In addition, some instances of these approaches point to a neural representation of conceptual knowledge that occupies most of the cortex, arguing against its utility and specificity. What is needed are well-specified cognitive accounts that can begin to bridge this gap between stimulus and neural response, which could explain the nature of the relationships in a more specific manner.

Another approach has been to specify semantic feature dimensions for concepts, based on how different concepts share certain qualities (e.g., have similar visual attributes or a similar function), rather than how conceptual tokens co-occur in language. A number of studies using this type of approach have shown that ventral and medial anterior temporal lobe regions appear to code specifically for the semantic feature relationships between concepts (Bruffaerts et al., 2013; Clarke, Devereux, & Tyler, 2018; Clarke & Tyler, 2014; Devereux, Clarke, & Tyler, 2018; Martin, Douglas, Newsome, Man, & Barense, 2018; Tyler et al., 2013). This work has often used representational similarity analysis (RSA; Kriegeskorte, Mur, & Bandettini, 2008) or other multivariate pattern analysis approaches in which the similarity structure between items based on brain activity is compared to the similarity structure between items based on cognitive measures of semantic similarity (figure 67.1*B*), with a significant relationship showing that the pattern of activity in a specific brain region represents some aspect of semantic feature information. One important distinction between encoding and RSA approaches is that the encoding approach emphasizes that different concepts have more distributed representations, where distant voxels contribute to a semantic representation, while RSA research places more emphasis on brain regions processing specific aspects of all concepts. While many different approaches to univariate and multivariate neuroimaging analyses can address theoretical issues, the contrast between voxel-wise encoding models and RSA provides one example to highlight the different kinds of potential inferences.

Neural Architecture and the Dynamics of Accessing Meaning from the Senses

In this section we discuss research that asks how meaning is accessed by focusing on visual objects. Nonhuman primate research, neuropsychology, and fMRI in humans have provided unequivocal evidence that the visual processing of concepts in the form of visual objects is dependent on a distributed network of regions throughout the occipital, temporal, and parietal lobes (Kravitz, Saleem, Baker, Ungerleider, & Mishkin, 2013). Of particular significance is the ventral visual pathway (VVP) along the axis of the occipital and temporal lobes that acts to transform low-level visual signals into more complex and higher-level visual representations (Bussey, Saksida, & Murray, 2005; Cowell, Bussey, & Saksida, 2010; DiCarlo, Zoccolan, & Rust, 2012; Kravitz et al., 2013; Riesenhuber & Poggio, 1999; Tanaka, 1996). Early visual regions, such as V1, V2, and V4, process the low-level visual properties of the orientation of lines and edges and have small receptive fields (Riesenhuber & Poggio, 1999). Further along the VVP, increasingly complex visual information is coded, such as complex shapes and parts of objects (Tanaka, 1996), while the perirhinal cortex, at the apex of the VVP, codes for the most complex conjunctions of simpler visual information in the posterior temporal regions (Bussey, Saksida, & Murray, 2005; Cowell, Bussey, & Saksida, 2010; Cowell et al., 2010; Miyashita, Okuno, Tokuyama, Ihara, & Nakajima, 1996). Crucially, however, object recognition is more than the visual processing of a visual stimulus and cannot be accomplished without access to object semantics. In this respect, models of semantics need to have explanatory power in relation to behavioral and neural data.

While many theories and accounts of conceptual knowledge in the brain seek to explain superordinate groups of objects, such as animals, tools, and manipulable objects, our approach is to zoom in to a more detailed level and focus on the neural representations of individual, basic-level concepts. When we see an object in the world, we typically understand it at this level as a cat, a hammer, a car, and so on, rather than as an animal. As such, the important questions regarding the neural representation of concepts are perhaps best tackled at the level of individual concepts while also considering how different properties of these concepts can give rise to, or contribute to, superordinate category organization.

Across studies using fMRI, magnetoencephalography (MEG), electroencephalography (EEG), neuropsychology, and intracranial recordings in humans, there is an increasingly clear picture of how the conceptual knowledge of different visual objects is represented and processed in the brain (for reviews, see Bi, Wang, & Caramazza, 2016; Clarke & Tyler, 2015; Lambon Ralph, Jefferies, Patterson, & Rogers, 2017; Martin, 2016). The majority of this evidence comes from visual object concepts, with support also coming from written and spoken words. One recent fMRI study (Tyler et al., 2013) showed clear evidence of the impact that different kinds of conceptual processing have in different regions of the VVP. In this study, regions in the posterior ventral temporal cortex (pVTC) were modulated according to whether the concept had more shared features or relatively more distinguishing semantic features (figure 67.1*C*). A pattern emerged, wherein more lateral regions of the ventral temporal cortex (VTC)—those typically activated more by animals (which tend to have more shared features; Martin, Wiggs, Ungerleider, & Haxby, 1996)—showed higher responses to objects with many shared features, while medial regions of the VTC—those typically activated more by tools and vehicles (which tend to have more distinctive features; Chao & Martin, 2000; Martin et al., 1996)—showed greater responses to objects with more distinguishing properties. The observation of both superordinate category responses (e.g., to animals and tools) and responses modulated by feature sharedness/distinctiveness implies that semantic representations are relatively coarse in the pVTC, where semantic representations of objects from different categories can be distinguished but may not be specific enough to differentiate similar objects from the same category (over and above visual differences).

However, object representations within the VVP must be sufficiently rich and complex to support the recognition of individual objects, not just the category they belong to. Tyler et al. (2013) addressed this issue, showing that the perirhinal cortex (PRC) was modulated by conceptual measures that capture the ease with which concepts are differentiated from one another (figure 67.1*D*). PRC processing was sensitive to how easy it was to differentiate one concept from other similar items (based on measures of conceptual structure). Tyler et al. (2013) provides a clear demonstration of how different regions along the VVP are sensitive to different statistical properties, derived from semantic features, that capture different elements of conceptual processing. While it has been long established that increasingly complex visual information is processed along the VVP, this research highlights a parallel progression of increasingly complex semantic information represented in increasingly anterior regions of the VVP.

Complementary research using semantic features also points to the PRC in the anterior medial temporal lobe as playing a fundamental role in the representation of specific object concepts. Activity patterns in the

PRC relate specifically to semantic-feature information (Bruffaerts et al., 2013; Clarke & Tyler, 2014; Devereux et al., 2018; Martin et al., 2018), suggesting this region has a fundamental part in representing conceptual information at the level of individual items. For example, Clarke and Tyler (2014) calculated the similarity between a large and diverse set of objects based on the overlap of their semantic features. This created a multidimensional map of semantic space, where items close together share many features and are therefore conceptually similar. Using fMRI, they tested across the brain for locations where brain activity patterns elicited by objects showed a matching similarity space. Only PRC activity patterns showed a significant relationship to the semantic similarity space. Similar results have been reported for concepts presented as written words (Bruffaerts et al., 2013; Martin et al., 2018).

The basic architecture of visual object recognition seems, therefore, to be broadly understood and highlights the important relationship between vision and semantics while emphasizing the roles of the pVTC and PRC in object semantics—as captured through semantic-feature models. While this work points to key systems engaged by perceptual and semantic processes, we also need to know *how* visual signals map onto semantic representations. One relevant approach is to create explicit computational models of the visuosemantic processes. For example, Devereux, Clarke, and Tyler (2018) combined a deep convolutional neural network model of visual processing with a recurrent attractor network for semantics. The model produced the expected trajectory from activating shared and visual features prior to distinctive features of objects (given a visual image as input), and further, increasing layers of the visuosemantic model mapped best onto increasingly anterior regions of the VVP. Importantly, this model also allowed us to test the nature of representations in the pVTC and showed that the initial semantic stages of the model, where shared visual features are primarily activated, best explain pVTC representations. This supports the notion that while pVTC undoubtedly represents complex visual object properties, these neural representations also code more abstract semantic details.

The above evidence suggests a view of object recognition in which both neural activity and the complexity of object information progress along the posterior to anterior axis in the VVP. However, this account is fundamentally incomplete, insofar as it implies a feedforward and bottom-up model underpinning cortical processing. The brain's anatomical structure suggests that complex interactions between bottom-up and top-down processes are a key part of object processing, as demonstrated by the abundance of lateral and feedback anatomical connections within the VVP and beyond (Bullier, 2001; Lamme & Roelfsema, 2000). Strictly hierarchical, bottom-up models of recognition can only capture part of the story. More recent work utilizing a time-sensitive methodology has enabled a critical advance by showing that visual and semantic processing are dependent on dynamic interactions within this network, rather than strictly hierarchical processing (Bar et al., 2006; Clarke, Devereux, & Tyler, 2018; Schendan & Ganis, 2012). This research builds on theories of cortical dynamics in which visual signals undergo an initial feedforward phase of processing as signals propagate along the ventral temporal lobe (Bullier, 2001; Lamme & Roelfsema, 2000). Neighboring regions then interact through recurrent interactions, and feedback and recurrent long-range reverberating interactions occur between cortical regions (Bar et al., 2006; Clarke, Devereux, & Tyler, 2018; Schendan & Ganis, 2012).

With regard to how semantic information is accessed from visual inputs, research by Clarke, Tyler, and colleagues (Clarke, Devereux, Randall, & Tyler, 2015; Clarke, Devereux, & Tyler, 2018; Clarke et al., 2013; Clarke, Taylor, & Tyler, 2011) using MEG has sought to determine the timing and dynamic mechanisms with which visual signals become meaningful. Across a series of studies, MEG recordings of brain activity revealed that visual inputs are transformed into coarse semantic representations within the first 150 ms of seeing an object, driven by an initial burst of feedforward processing (Clarke et al., 2013, 2015). This rapid but coarse semantics provides a basis for superordinate category representations in the pVTC, where neural signals initially can dissociate between objects from different semantic categories but do not differentiate between objects within a category (Clarke et al., 2015). Beyond this feedforward procession of signals, long-range recurrent interactions and feedback along the VVP occur. Object representations become more distinct over time, with object-specific semantic information present beyond 200 ms (figure 67.1*E*; Clarke et al., 2013, 2015).

Complementing this temporal transition from vision to semantics, connectivity analysis further suggests that while visual object information is primarily transformed through feedforward activity in the VVP, both feedforward and feedback activity is central to how visual signals relate to the emerging semantic signals (Clarke, Devereux, & Tyler, 2018). Further, the connectivity between anterior and posterior regions in the temporal lobe is modulated according to the level of detail of semantic information required (Clarke, Taylor, & Tyler, 2011). Together, this shows that while feedforward mechanisms can primarily support visual processing, feedback and recurrent dynamics play an important role in accessing semantics.

Concluding Remarks

Given that concepts are central to cognition, cognitive neuroscience must endeavor to understand their representation and processing. Here, we focused on concrete concepts, with research suggesting that the conceptual processing of visual objects is achieved through coordinated activity in the ventral temporal lobe. Both the posterior ventral temporal cortex and the anteromedial temporal cortex engage in feedforward and feedback dynamics to enable specific semantic representations of objects. Increasing our understanding of conceptual processing will enable progress in multiple fields, including language, decision-making, and navigation, all of which rely on first understanding what we perceive.

Acknowledgments

This research was funded by an Advanced Investigator grant to Lorraine K. Tyler from the European Research Council (ERC) under the European Community's Seventh Framework Programme (FP7/2007–2013 ERC grant agreement no. 249640) and an ERC Advanced Investigator grant to Lorraine K. Tyler under the Horizon 2020 Research and Innovation Programme (2014–2020 ERC grant agreement no. 669820).

REFERENCES

Bar, M., Kassam, K. S., Ghuman, A. S., Boshyan, J., Schmid, A. M., Dale, A. M., … Halgren, E. (2006). Top-down facilitation of visual recognition. *Proceedings of the National Academy of Sciences of the United States of America, 103*(2), 449–454.

Baroni, M., & Lenci, A. (2010). Distributional memory: A general framework for corpus-based semantics. *Computational Linguistics, 36*(4), 673–721.

Barsalou, L. W. (1999). Perceptual symbol systems. *Behavioral and Brain Sciences, 22*(4), 577–609.

Barsalou, L. W. (2017). What does semantic tiling of the cortex tell us about semantics? *Neuropsychologia, 105*, 18–38.

Bi, Y., Wang, X., & Caramazza, A. (2016). Object domain and modality in the ventral visual pathway. *Trends in Cognitive Sciences, 20*(4), 282–290.

Binder, J. R., Conant, L. L., Humphries, C. J., Fernandino, L., Simons, S. B., Aguilar, M., & Desai, R. H. (2016). Toward a brain-based componential semantic representation. *Cognitive Neuropsychology, 33*(3–4), 130–174.

Blei, D. M., Ng, A. Y., & Jordan, M. I. (2003). Latent dirichlet allocation. *Journal of Machine Learning Research, 3*(January), 993–1022.

Bruffaerts, R., Dupont, P., Peeters, R., De Deyne, S., Storms, G., & Vandenbergh, R. (2013). Similarity of fMRI activity patterns in left perirhinal cortex reflects semantic similarity between words. *Journal of Neuroscience, 33*(46), 18587–18607.

Bullier, J. (2001). Integrated model of visual processing. *Brain Research Reviews, 36*, 96–107.

Bussey, T. J., Saksida, L. M., & Murray, E. A. (2005). The perceptual-mnemonic/feature conjunction model of perirhinal cortex function. *Quarterly Journal of Experimental Psychology, 58B*, 269–282.

Chao, L. L., & Martin, A. (2000). Representation of manipulable man-made objects in the dorsal stream. *NeuroImage, 12*, 478–484.

Clarke, A., Devereux, B. J., Randall, B., & Tyler, L. K. (2015). Predicting the time course of individual objects with MEG. *Cerebral Cortex, 25*(10), 3602–3612.

Clarke, A., Devereux, B. J., & Tyler, L. K. (2018). Oscillatory dynamics of perceptual to conceptual transformations in the ventral visual pathway. *Journal of Cognitive Neuroscience, 30*(11), 1590–1605.

Clarke, A., Taylor, K. I., Devereux, B., Randall, B., & Tyler, L. K. (2013). From perception to conception: How meaningful objects are processed over time. *Cerebral Cortex, 23*(1), 187–197.

Clarke, A., Taylor, K. I., & Tyler, L. K. (2011). The evolution of meaning: Spatiotemporal dynamics of visual object recognition. *Journal of Cognitive Neuroscience, 23*(8), 1887–1899.

Clarke, A., & Tyler, L. K. (2014). Object-specific semantic coding in human perirhinal cortex. *Journal of Neuroscience, 34*(14), 4766–4775.

Clarke, A., & Tyler, L. K. (2015). Understanding what we see: How we derive meaning from vision. *Trends in Cognitive Sciences, 19*(11), 677–687.

Cowell, R. A., Bussey, T. J., & Saksida, L. M. (2010). Components of recognition memory: Dissociable cognitive processes or just differences in representational complexity? *Hippocampus, 20*, 1245–1262.

Cree, G. S., McNorgan, C., & McRae, K. (2006). Distinctive features hold a privileged status in the computation of word meaning: Implications for theories of semantic memory. *Journal of Experimental Psychology: Learning, Memory, and Cognition, 32*(4), 643–658.

Devereux, B. J., Clarke, A., & Tyler, L. K. (2018). Integrated deep visual and semantic attractor neural networks predict fMRI pattern-information along the ventral object processing pathway. *Scientific Reports, 8*(1), 10636.

Devereux, B. J., Taylor, K3. I., Randall, B., Geertzen, J., & Tyler, L. K. (2015). Feature statistics modulate the activation of meaning during spoken word processing. *Cognitive Science, 40*(2), 325–350.

DiCarlo, J. J., Zoccolan, D., & Rust, N. C. (2012). How does the brain solve visual object recognition? *Neuron, 73*(3), 415–434.

Farah, M. J., & McClelland, J. L. (1991). A computational model of semantic memory impairment: Modality specificity and emergent category specificity. *Journal of Experimental Psychology: General, 120*, 339–357.

Huth, A. G., de Heer, W. A., Griffiths, T. L., Theunissen, F. E., & Gallant, J. L. (2016). Natural speech reveals the semantic maps that tile human cerebral cortex. *Nature, 532*(7600), 453–458.

Huth, A. G., Nishimoto, S., Vu, A. T., & Gallant, J. L. (2012). A continuous semantic space describes the representation of thousands of object and action categories across the human brain. *Neuron, 76*(6), 1210–1224.

Kravitz, D. J., Saleem, K. S., Baker, C. I., Ungerleider, L. G., & Mishkin, M. (2013). The ventral visual pathway: An expanded neural framework for the processing of object quality. *Trends in Cognitive Sciences, 17*(1), 26–49.

Kriegeskorte, N., Mur, M., & Bandettini, P. (2008). Representational similarity analysis—connecting the branches of systems neuroscience. *Frontiers in Systems Neuroscience, 2,* 4. https://doi.org/10.3389/neuro.06.004.2008

Lambon Ralph, M. A., Jefferies, E., Patterson, K., & Rogers, T. T. (2017). The neural and computational bases of semantic cognition. *Nature Reviews Neuroscience, 18*(1), 42–55.

Lamme, V. A., & Roelfsema, P. (2000). The distinct modes of vision offered by feedforward and recurrent processing. *Trends in Neurosciences, 23*(11), 571–579.

Laurence, S., & Margolis, E. (1999). Concepts and cognitive science. In S. Laurence & E. Margolis (Eds.), *Concepts: Core readings* (pp. 3–81). Cambridge, MA: MIT Press.

Mahon, B. Z., & Caramazza, A. (2008). A critical look at the embodied cognition hypothesis and a new proposal for grounding conceptual content. *Journal of Physiology, Paris, 102*(1–3), 59–70.

Martin, A. (2016). GRAPES—Grounding representations in action, perception, and emotion systems: How object properties and categories are represented in the human brain. *Psychonomic Bulletin & Review, 23*(4), 979–990.

Martin, A., Wiggs, C. L., Ungerleider, L., & Haxby, J. V. (1996). Neural correlates of category-specific knowledge. *Nature, 379,* 649–652.

Martin, C. B., Douglas, D., Newsome, R. N., Man, L. L., & Barense, M. D. (2018). Integrative and distinctive coding of visual and conceptual object features in the ventral visual stream. *eLife, 7,* e31873.

McRae, K., & Cree, G. S. (2002). Factors underlying category-specific semantic deficits. In E. M. E. Forde & G. W. Humphreys (Eds.), *Category-specificity in brain and mind* (pp. 211–249). Hove, UK: Psychology Press.

Miller, G. A. (1995). WordNet: A lexical database for English. *Communications of the ACM, 38*(11), 39–41.

Mitchell, T. M., Shinkareva, S. V., Carlson, A., Chang, K., Malave, V. L., Mason, R. A., & Just, M. A. (2008). Predicting human brain activity associated with the meanings of nouns. *Science, 320,* 1191–1195.

Miyashita, Y., Okuno, H., Tokuyama, W., Ihara, T., & Nakajima, K. (1996). Feedback signal from medial temporal lobe mediates visual associative mnemonic codes of inferotemporal neurons. *Cognitive Brain Research, 5,* 81–86.

Murphy, G. L. (2002). *The big book of concepts.* Cambridge, MA: MIT Press.

Pexman, P. M., Holyk, G. G., & Monfils, M. H. (2003). Number-of-features effects and semantic processing. *Memory & Cognition, 31*(6), 842–855.

Pulvermüller, F. (2013). How neurons make meaning: Brain mechanisms for embodied and abstract-symbolic semantics. *Trends in Cognitive Sciences, 17*(9), 458–470.

Riesenhuber, M., & Poggio, T. (1999). Hierarchical models of object recognition. *Nature, 2*(11), 1019–1025.

Rogers, T. T., & McClelland, J. L. (2004). *Semantic cognition: A parallel distributed approach.* Cambridge, MA: MIT Press.

Schendan, H. E., & Ganis, G. (2012). Electrophysiological potentials reveal cortical mechanisms for mental imagery, mental simulation, and grounded (embodied) cognition. *Frontiers in Psychology, 3*(329): doi:10.3389/fpsyg.2012.00329

Stansbury, D. E., Naselaris, T., & Gallant, J. L. (2013). Natural scene statistics account for the representation of scene categories in human visual cortex. *Neuron, 79*(5), 1025–1034.

Tanaka, K. (1996). Inferotemporal cortex and object vision. *Annual Review of Neuroscience, 19,* 109–140.

Taylor, K. I., Devereux, B. J., Acres, K., Randall, B., & Tyler, L. K. (2012). Contrasting effects of feature-based statistics on the categorisation and identification of visual objects. *Cognition, 122*(3), 363–374.

Taylor, K. I., Devereux, B. J., & Tyler, L. K. (2011). Conceptual structure: Towards an integrated neurocognitive account. *Language and Cognitive Processes, 26*(9), 1368–1401.

Tyler, L. K., Chiu, S., Zhuang, J., Randall, B., Devereux, B. J., Wright, P., … Taylor, K. I. (2013). Objects and categories: Feature statistics and object processing in the ventral stream. *Journal of Cognitive Neuroscience, 25*(10), 1723–1735.

Tyler, L. K., & Moss, H. E. (2001). Towards a distributed account of conceptual knowledge. *Trends in Cognitive Sciences, 5*(6), 244–252.

Tyler, L. K., Stamatakis, E. A., Bright, P., Acres, K., Abdallah, S., Rodd, J. M., & Moss, H. E. (2004). Processing objects at different levels of specificity. *Journal of Cognitive Neuroscience,* 16(3), 351–362.

Vigliocco, G., Kousta, S.-T., Rosa, D., Anthony, P., Vinson, D. P., Tettamanti, M., … Cappa, S. F. (2014). The neural representation of abstract words: The role of emotion. *Cerebral Cortex, 24*(7), 1767–1777.

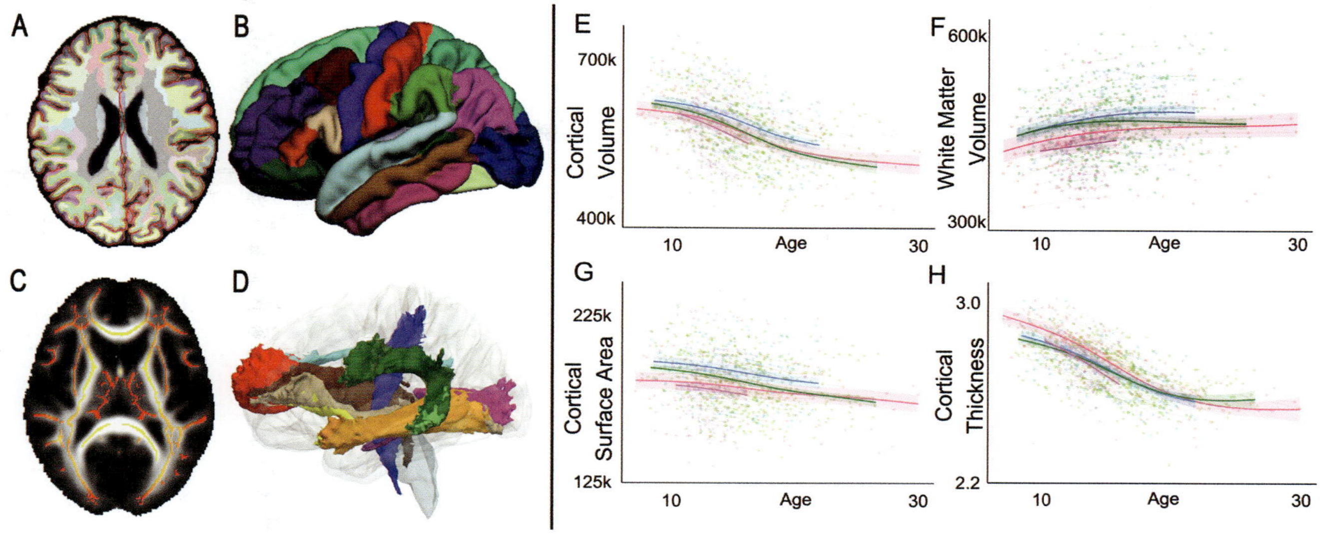

PLATE 1 Illustration of key MRI methods and findings discussed in this review. *A*, Horizontal slice of T1 image showing a whole-brain segmentation used for volumetric analyses. *B*, A left-lateral view of an averaged parcellated cerebral cortex used for surface-based analyses, both from FreeSurfer. *C*, Horizontal slice of Tract-Based Spatial Statistics (TBSS) mean FA white matter skeleton overlaid on a mean FA map. *D*, a left-lateral view of a three-dimensional rendering of probabilistic fiber tracts from the Mori atlas. Developmental trajectories for global cortical measures from four independent samples for cortical volume (*E*), total white matter volume (*F*), cortical surface area (*G*), and mean cortical thickness (*H*). The colored lines represent the generalized additive mixed model (GAMM) fitting while the lighter-colored areas correspond to the 95% confidence intervals. *Note*: *Pink*, Child Psychiatry Branch (CPB); *purple*, Pittsburgh (PIT); *blue*, Neurocognitive Development (NCD); *green*, Braintime (BT). (See figure 2.1.)

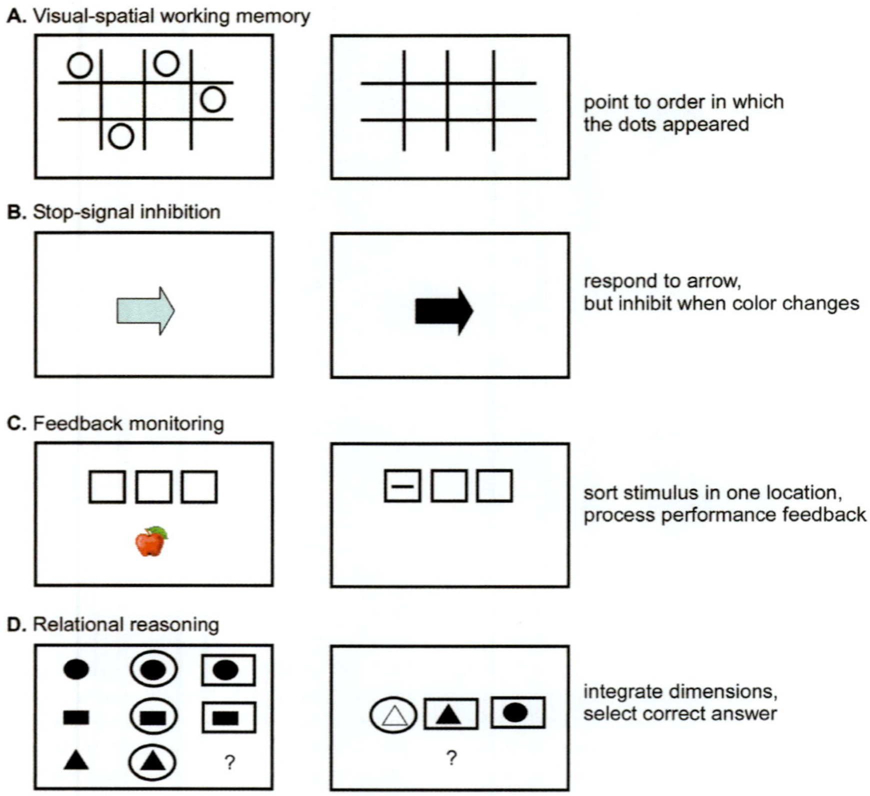

PLATE 2 Examples of cognitive control paradigms. *A*, A visuospatial working memory typically involves the presentation of a grid in which dots are consecutively presented and need to be reproduced on the next trial. More dots will make the task more difficult. *B*, A stop-signal paradigm involves the presentation of a stimulus, which requires a left- or a right-hand response. In some trials, the arrow quickly changes color, which informs the participant that he/she should inhibit responding. *C*, A feedback-learning task typically involves a stimulus, which needs to be sorted in a specific location. The feedback screen informs the participant whether the response was correct or incorrect. *D*, Relational reasoning requires the participant to integrate dimensions of a presented stimulus. One-dimensional trials are those in which only one direction needs to be followed (e.g., a horizontal line), and two-dimensional trials are those in which more dimensions (e.g., a horizontal line and a vertical line) need to be integrated to come to the correct solution. (See figure 3.1.)

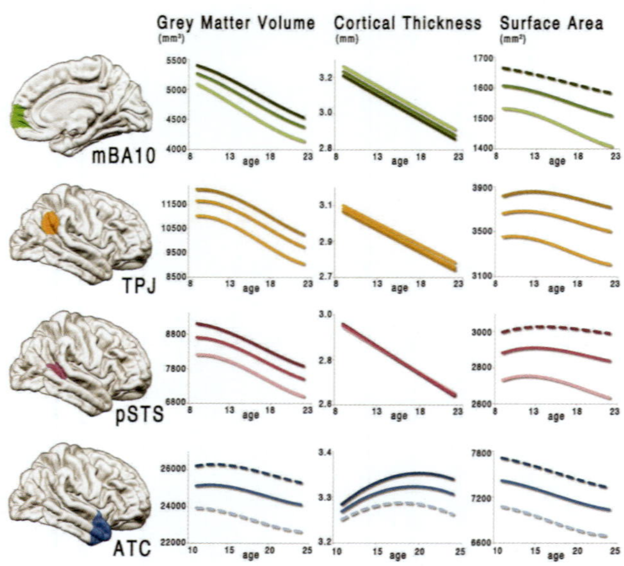

PLATE 3 Structural development of the social brain. Structural developmental trajectories of brain areas associated with mentalizing across adolescence (gray matter volume, cortical thickness, surface area). The best-fitting models for all participants are shown for each region of interest (combined hemispheres). Models are fitted to the middle 80% of the sample (ages 9–22 years for mBA10, TPJ, and pSTS; ages 11–24 years for ATC). The lighter lines show the fitted models applied to females only, and the darker lines show the fitted models applied to males only. Solid lines indicate the fitted model was significant ($p < 0.05$), whereas dashed lines indicate the fitted model was not significant ($p \geq 0.05$). Reproduced from Mills et al. (2014). (See figure 4.1.)

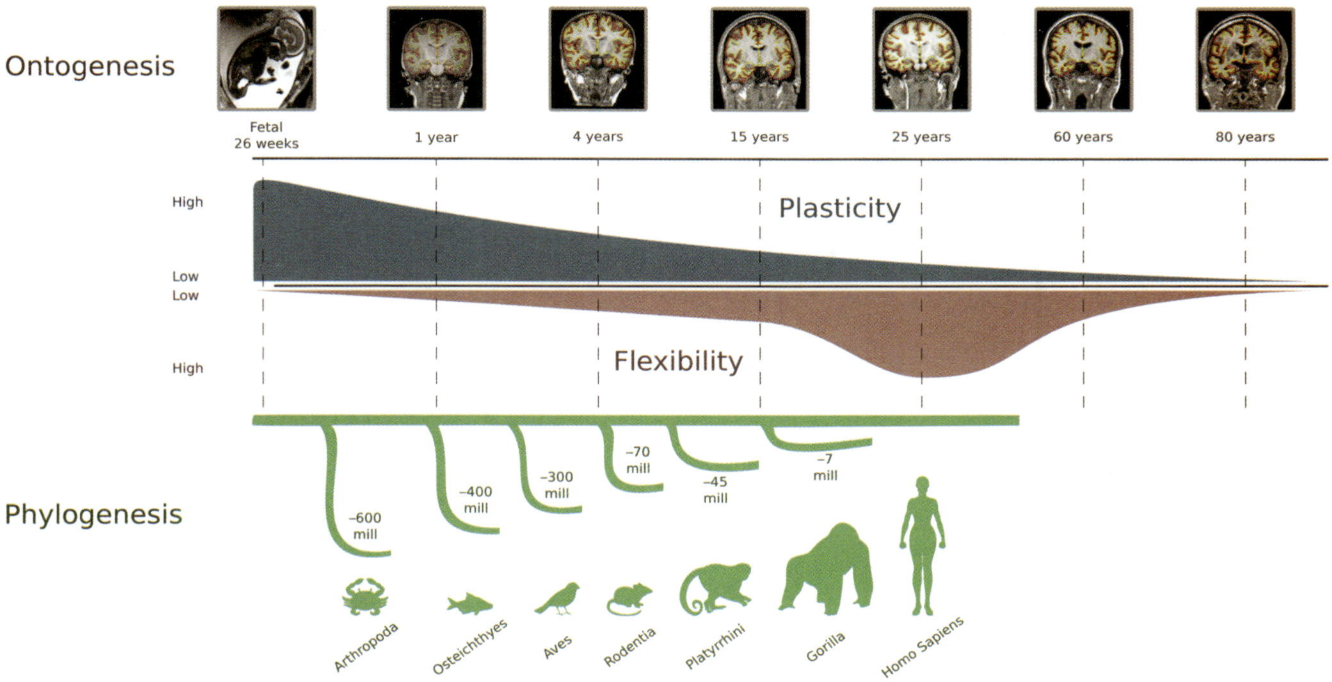

PLATE 4 Schematic figure depicting plasticity (i.e., the capacity for anatomical brain changes induced by environmental demands) and flexibility (i.e., functional capacity) as a function of ontogenesis and phylogenesis. Note that early plasticity is linked to vulnerability, which for a number of functions and their neural substrates appears highest when flexibility is at its lowest in ontogenesis. (See figure 5.1.)

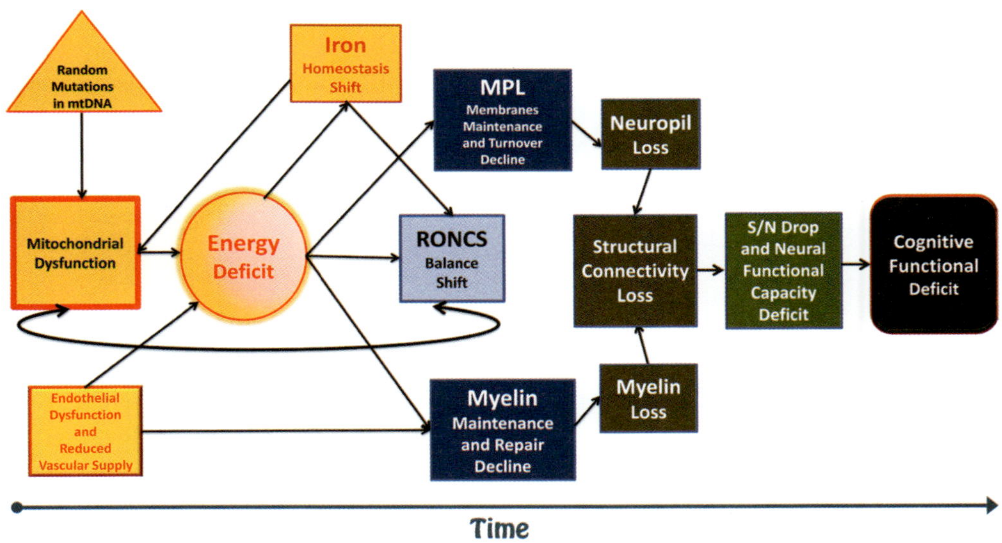

PLATE 5 The FRIENDS model of cognitive aging. Random mutations accumulate in the mitochondrial DNA, and mitochondrial function declines with age. That decline results in the gradual reduction of energy substrates (ATP and NADH). The sequestration and transport of iron are affected, and the accumulation of iron, in turn, exacerbates mitochondrial declines. Endothelial dysfunction impairs the delivery of glucose and oxygen to the end organs in the brain and may directly stunt the regeneration of myelin. An energy crisis and further mitochondrial dysfunction ensue. Impaired mitochondrial respiration and the accumulation of iron lead to a shift in the homeostasis of RONCS and their excess further damages the mitochondria. This cascade of events further reduces the energy available for the maintenance of cell membranes and myelin and for supporting synaptic plasticity. Myelin and neuropil are gradually diminished, and connectivity among the brain regions is lost. That loss impairs the fidelity of neural communication, and the information-processing capabilities of the brain networks are degraded; the system becomes progressively noisier. Noise in the information- processing system affects all cognitive operations, with the more energy-dependent of them faring the worst of all. (See figure 6.1.)

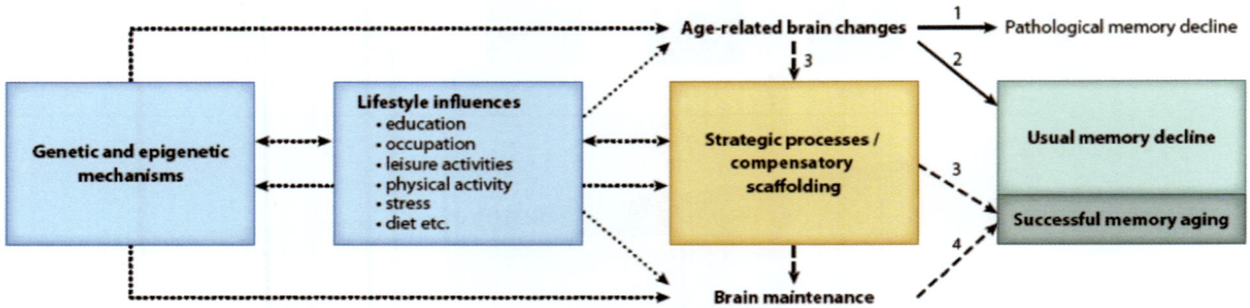

PLATE 6 Model of life course paths to pathological, usual, and successful memory aging. Path 1 (not the focus of this chapter) represents trajectories characterized by pathological memory decline (e.g., dementia). Paths 2, 3, and 4 lead to normal memory aging (usual and successful). Path 2 represents usual memory decline in response to age-related brain changes (e.g., hippocampus atrophy). Path 3 (*dashed line*) is an indirect path to successful memory aging via compensation for age-related brain changes. Path 4 leads directly to successful memory aging via brain maintenance. The dotted lines reflect the influences of genetics, epigenetics, and lifestyle ("what" factors) as well as strategies ("how" factors) on brain integrity (maintenance versus decline) and memory outcomes. From Nyberg & Pudas (2019). (See figure 7.1.)

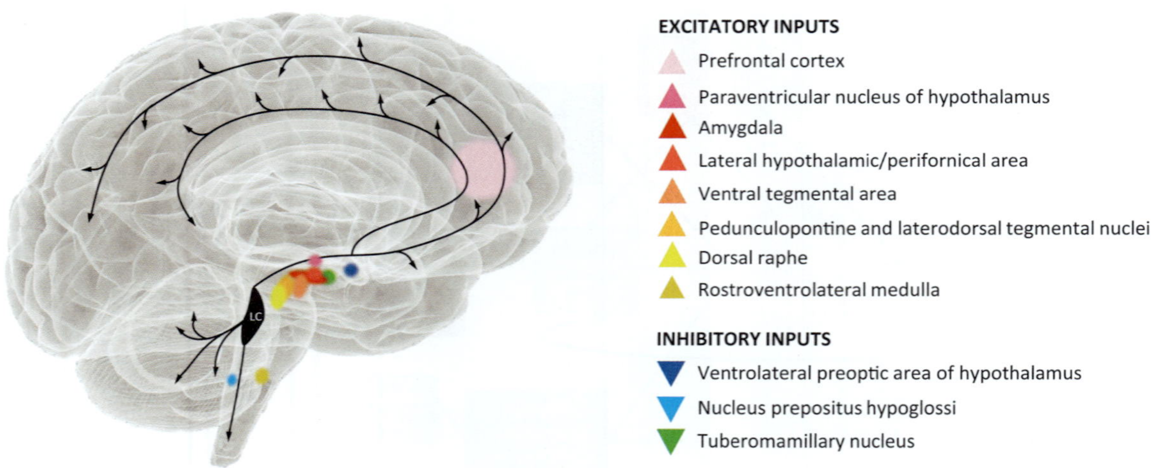

PLATE 7 LC axons project to much of the rest of the brain, but the LC receives direct input from only a subset of these regions. These regions provide information about different aspects of arousal states and include excitatory input from the amygdala, paraventricular nucleus of the hypothalamus, lateral hypothalamus/perifornical area, ventral tegmental area, dorsal raphe, pedunculopontine and laterodorsal tegmental nuclei, and rostroventrolateral medulla and inhibitory input from the ventrolateral preoptic area of the hypothalamus, nucleus prepositus hypoglossi, and tuberomammillary nucleus (for a review, see Samuels & Szabadi, 2008). (See figure 8.1.)

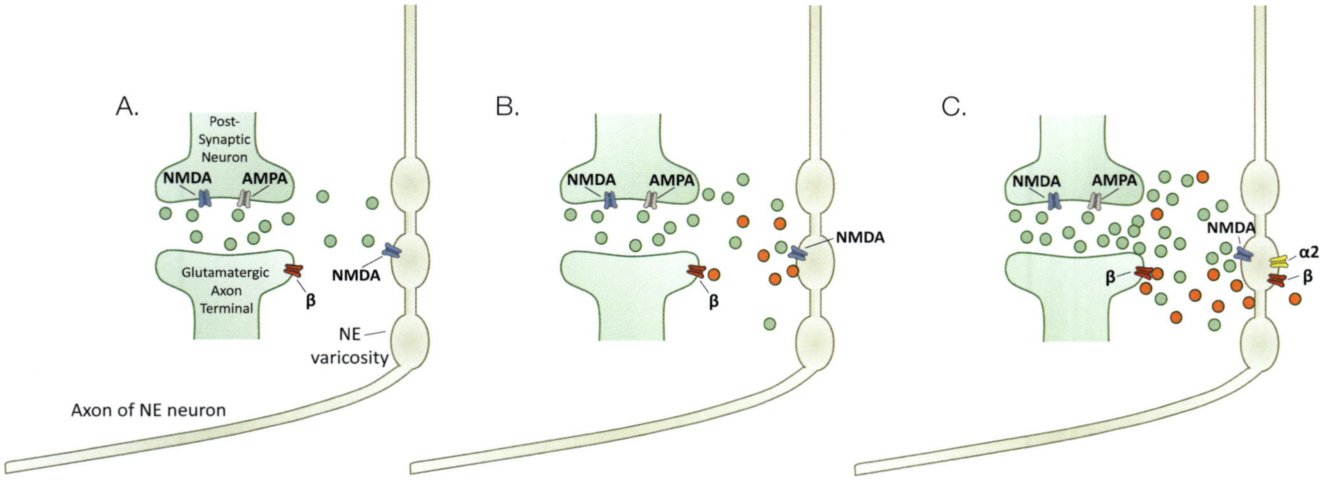

PLATE 8 A graphic depiction of the proposed mechanisms of the GANE model. *A*, Glutamate (*gray circles*) spills over from glutamatergic synapses at the sites of excited representations. *B*, If the LC happens to be activated (that is, depolarized) at the same time as glutamate reaches the N-methyl D-aspartate (NMDA) receptors on the LC axons, this triggers more local release of NE (*black circles*) from those LC varicosities. *C*, Elevated local levels of NE activate β-adrenergic receptors that further stimulate glutamate release, leading to a local hot spot of high excitation. Autoreceptors at LC varicosities also contribute to increasing neural gain by inhibiting the release of noradrenaline when low levels of noradrenaline activate α-adrenergic receptors but increasing the release of noradrenaline when high levels of noradrenaline activate β-adrenergic receptors. Reprinted from Lee et al. (2018). (See figure 8.3.)

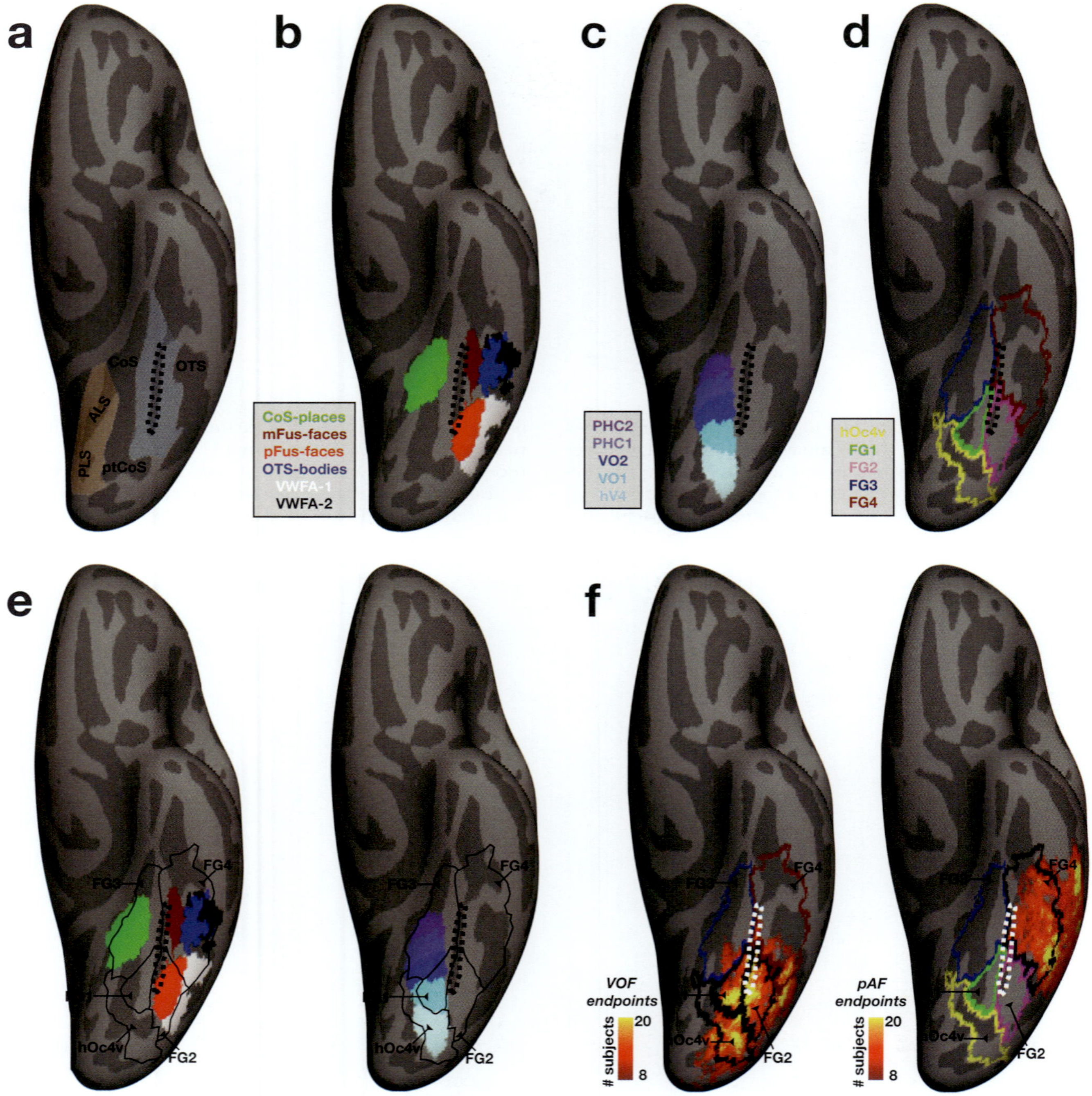

PLATE 9 A correspondence among cortical folding, functional representations, cytoarchitectonic areas, and white matter endpoints in VOTC. Inflated left hemisphere of the FreeSurfer average template (www.freesurfer.net). *Dotted outline (black in A–E; white in F)*: Mid-fusiform sulcus. *A*, *Orange*: lingual gyrus; *blue*: fusiform gyrus; *ALS*: anterior lingual sulcus; *CoS*: collateral sulcus; *OTS*: occipitotemporal sulcus; *PLS*: posterior lingual sulcus; *ptCoS*: posterior transverse collateral sulcus. *B*, Maximum probability maps (MPMs) of domain-specific regions selective for faces, bodies, words, and places (see legend) from 12 individuals (Weiner et al., 2017). *C*, MPMs of retinotopic maps (see legend) defined from 53 individuals (Wang et al., 2015). *D*, MPMs of five cytoarchitectonic areas (see legend) defined from 10 individuals (Rosenke et al., 2018). *E*, Functional parcellations relative to cytoarchitectonic parcellations of VOTC. *Left*: domain-specific regions from (*B*). *Right*: retinotopic maps from (*C*). *F*, Probabilistic maps of cortical endpoints of the VOF (*left*) and pAF (*right*; $N = 37$; Weiner, Yeatman, & Wandell, 2017) relative to cytoarchitectonic areas from (*D*). (See figure 9.1.)

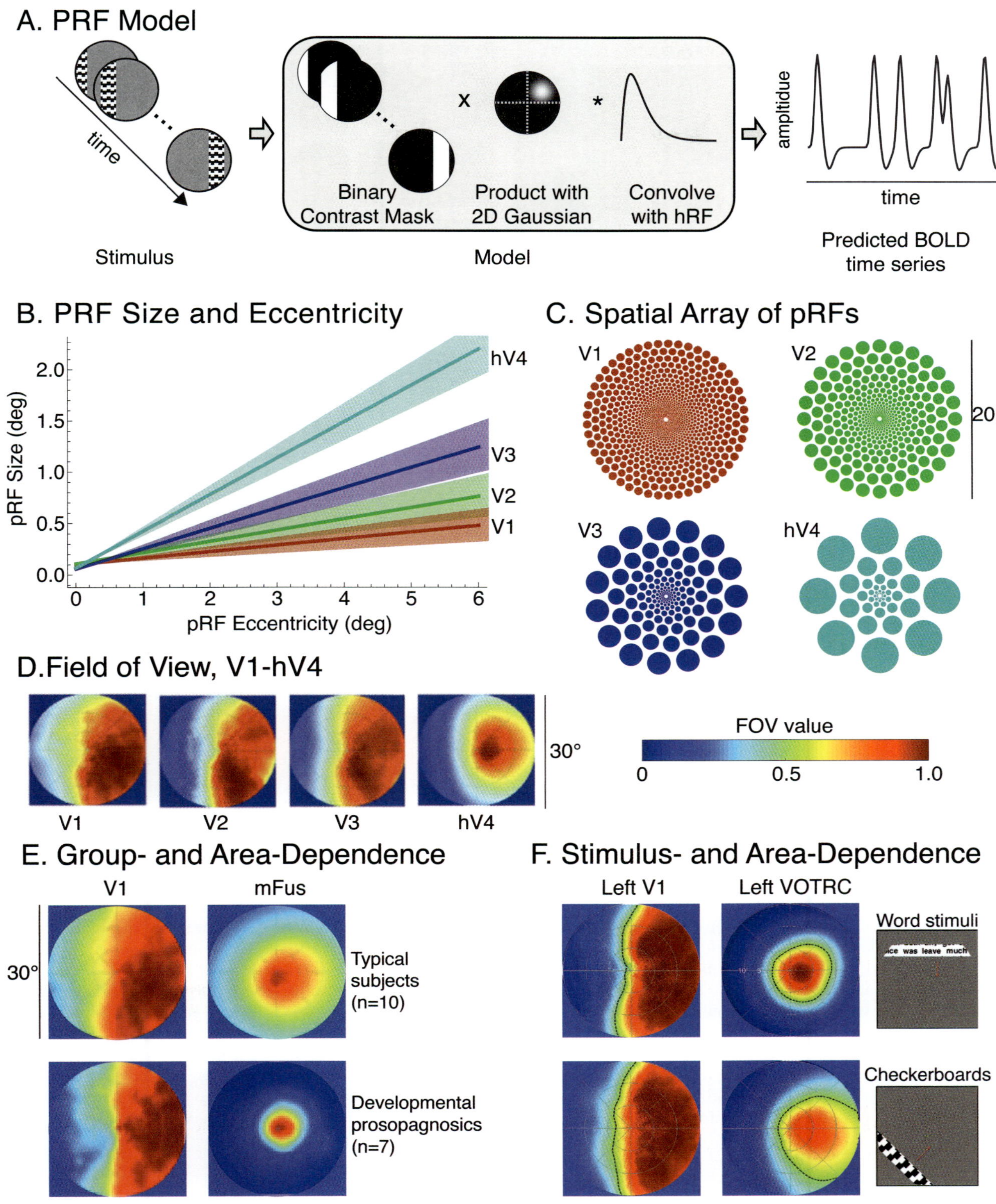

PLATE 10 The fMRI population receptive field. *A*, The pRF model replaces the stimulus at each time point with a binary contrast mask, multiplied point-wise by the pRF, followed by convolution with a hemodynamic response function. *B*, pRF size increases with eccentricity and across visual areas. *C*, A cartoon showing the viewpoint of four visual areas based on pRF models. Circles represent one standard deviation of the pRF at each location (derived from the line plots in *B*). Data in *B* and *C* are from the Human Connectome Project (Benson et al., 2018; Ugurbil et al., 2013; Van Essen et al., 2013). *D*, The FOV for V1–V4, averaged across 10 subjects. pRFs from the right hemisphere are flipped across the y-axis. *E*, The FOV for bilateral V1 and a face-responsive region, mFus, for 10 typical subjects and 7 developmental prosopagnosics. *D* and *E* adapted from Witthoft et al., (2016) as per CC-BY-NC-ND 4.0 International license (https://creativecommons.org/licenses/by-nc-nd/4.0/). *F*, Group average ($n=20$) FOV for left V1 and VOTRC. The pRFs were derived from experiments with word stimuli or checkerboards within the bar apertures. Adapted from Le et al. (2017) as per CC-BY-NC-ND 4.0 International license (https://creativecommons.org/licenses/by-nc-nd/4.0/). (See figure 10.1.)

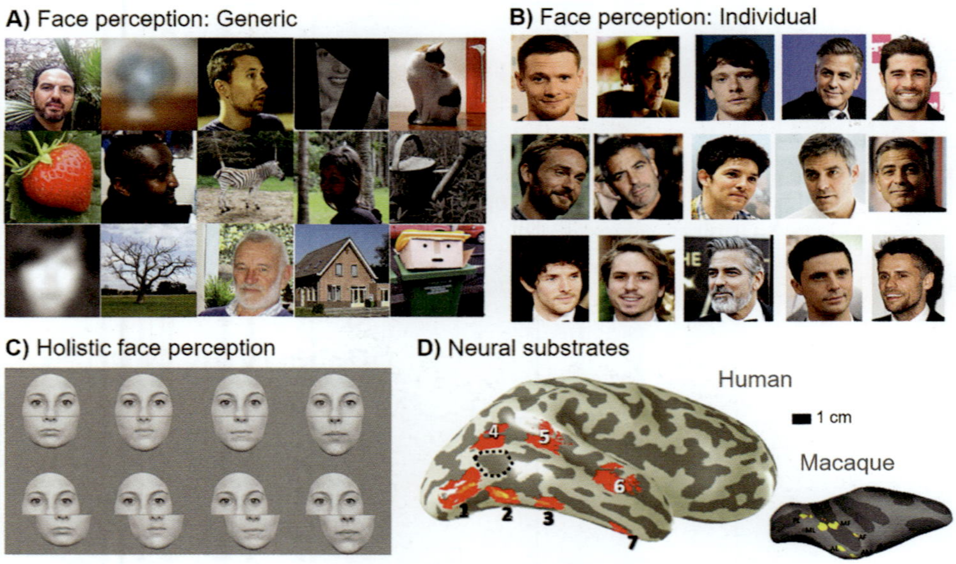

PLATE 11 *A*, The human brain perceives stimuli as faces under a wide variety of viewing conditions, sometimes with little or no shared features. Understanding the nature of such generic face categorization (i.e., discrimination from nonface objects and generalization across variable face exemplars) requires going beyond sensory, quantifiable, low-level information from the visual stimulus and incorporating memory of the category in the process. *B*, Individual face perception can be understood as a more specific function than generic face perception but similarly requiring categorization: discrimination—here from various unfamiliar faces—and generalization across variable viewing conditions. *C*, The composite face illusion (adapted from Rossion, 2013) illustrates how face perception goes beyond the given sensory information, taking advantage of matches to holistic facial memories derived through experience. All four top halves (*above the narrow gap*) are physically identical. Yet aligned (*top row*; compare with misaligned below) with distinct bottom halves (all of different identities), they are perceived as being different. *D*, Inflated segmented brains showing typical locations of face-selective regions in humans (from Weiner & Grill-Spector, 2015) and macaques (from Tsao, Moeller, & Freiwald, 2008), plotted to a common scale. Up to six regions have been labeled in the macaque STS, whereas humans have regions in both the STS (labeled *4*, *5*, and *6*) and the VOTC (*1*, *2*, *3*, *7*). (See figure 11.1.)

PLATE 12 Bayesian causal inference.

A, Generative model. The generative model of Bayesian causal inference explicitly models the potential causal structures that could have generated the sensory signals. For the audiovisual spatial localization example, it models whether the "sight of the bird" and the bird's "singing" are generated by common (C=1) or independent (C=2) sources (Körding et al., 2007). In the case of a common source, the "true" audiovisual location (S_{AV}) is drawn from the prior spatial distribution. In the case of independent sources, the "true" auditory (S_A) and "true" visual (S_V) locations are drawn independently from this prior spatial distribution. Auditory (x_A) and visual (x_V) inputs are generated by adding independent sensory noise.

B, Posterior probability distribution and auditory spatial estimate. (1) The auditory spatial estimate ($\widehat{S_A}$) is shown as a function of audiovisual disparity, with auditory location set to 0° and visual location varying along the x-axis. According to the segregation model, $\widehat{S_A}$ is equal to the auditory location (green line, $\hat{S}_{A,C=2}$). In the visual capture scenario, $\widehat{S_A}$ is equal to the visual location (red line, $\hat{S}_{V,C=2}$). According to the forced-fusion model, auditory and visual inputs are integrated weighted by their relative reliabilities, thereby biasing $\widehat{S_A}$ toward the visual location (yellow line, $\hat{S}_{AV,C=1}$). According to Bayesian causal inference, the bias of $\widehat{S_A}$ depends nonlinearly on audiovisual disparity. For small disparities, the observer integrates auditory and visual inputs approximately weighted by their relative reliabilities. For large spatial disparities, audiovisual biases are attenuated (Rohe & Noppeney, 2015b). *Note*: CMB = crossmodal bias.

2–4. Posterior probability distributions for the segregation $P(S_A|x_A, x_V, C=2)$ (*green*), the forced fusion $P(S_A|x_A, x_V, C=1)$ (*red*), and the Bayesian causal inference model $P(S_A|x_A, x_V)$, with model averaging as a decision strategy (*blue*) for (2) $S_A=0°$, $S_V=10°$, (3) $S_A=0°$, $S_V=20°$, and (4), $S_A=0°$, $S_V=30°$. The auditory location is indicated by the green dot and the visual location by the red dot. For the simulations we set $x_A=S_A$ and $x_V=S_V$ and the parameters for visual noise: $\sigma_V=3°$; auditory noise: $\sigma_A=7°$; spatial prior distribution: $\mu_{prior}=0°$; $\sigma_{prior}=1000°$; and common cause prior: $P_{common}=0.5$.

C, Bayesian causal inference within the cortical hierarchy. Auditory areas represent predominantly the auditory location; early visual areas, the visual location. The posterior intraparietal cortex integrates sensory signals approximately according to forced-fusion principles. The anterior intraparietal sulcus computes the final Bayesian causal inference estimate that takes into account the observer's uncertainty about the world's causal structure (Rohe & Noppeney, 2015a). (See figure 12.1.)

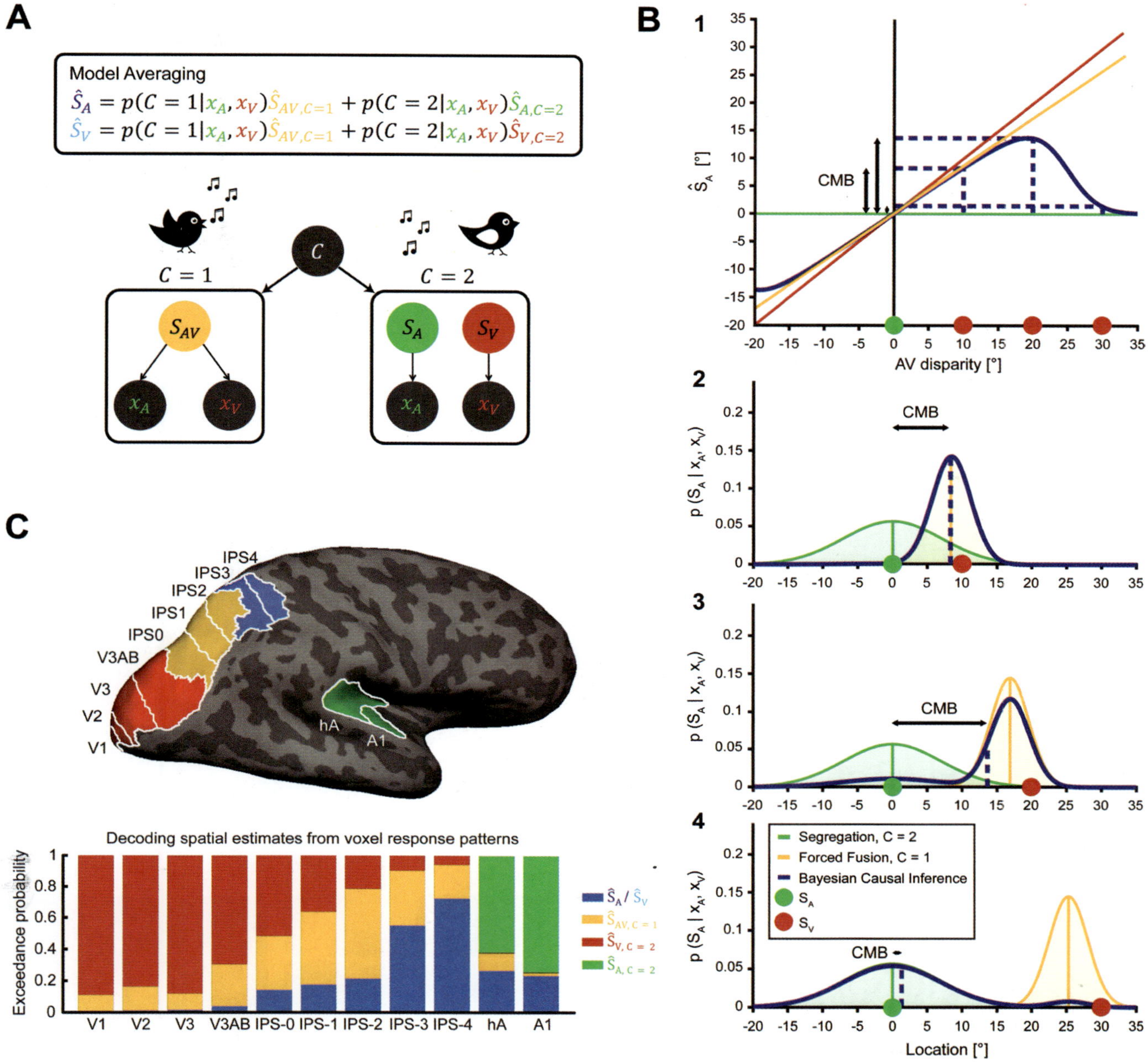
A
Model Averaging
$\hat{S}_A = p(C=1|x_A, x_V)\hat{S}_{AV,C=1} + p(C=2|x_A, x_V)\hat{S}_{A,C=2}$
$\hat{S}_V = p(C=1|x_A, x_V)\hat{S}_{AV,C=1} + p(C=2|x_A, x_V)\hat{S}_{V,C=2}$
C
C = 1
C = 2
S_{AV}
S_A
S_V
x_A
x_V
B
1
CMB
$\hat{S}_A$ [°]
AV disparity [°]
2
CMB
$p(S_A | x_A, x_V)$
3
CMB
4
Segregation, C = 2
Forced Fusion, C = 1
Bayesian Causal Inference
S_A
S_V
CMB
Location [°]
C
IPS4
IPS3
IPS2
IPS1
IPS0
V3AB
V3
V2
V1
hA
A1
Decoding spatial estimates from voxel response patterns
Exceedance probability
$\hat{S}_A$ / $\hat{S}_V$
$\hat{S}_{AV, C=1}$
$\hat{S}_{V, C=2}$
$\hat{S}_{A, C=2}$
V1
V2
V3
V3AB
IPS-0
IPS-1
IPS-2
IPS-3
IPS-4
hA
A1

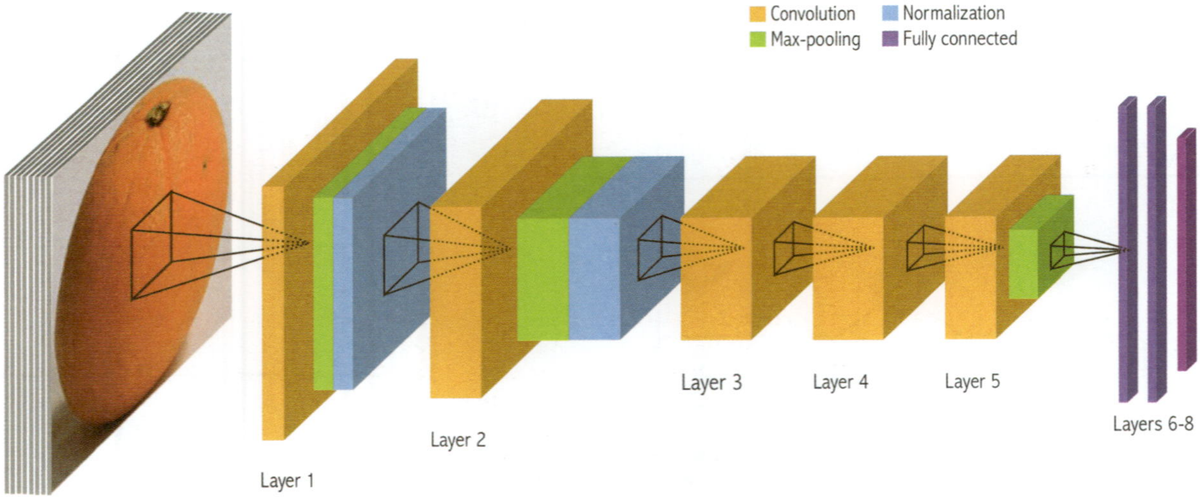

PLATE 13 Deep CNN architecture AlexNet comprises eight layers. Each of layers one through five contains a convolution, some layers with max-pooling and normalization stages. The last three layers are fully connected, with the last layer acting as the output label. The network takes pixel values as inputs and propagates information in a feedforward manner through the layers, activating artificial units with particular weight values successively at each layer. (See figure 13.1.)

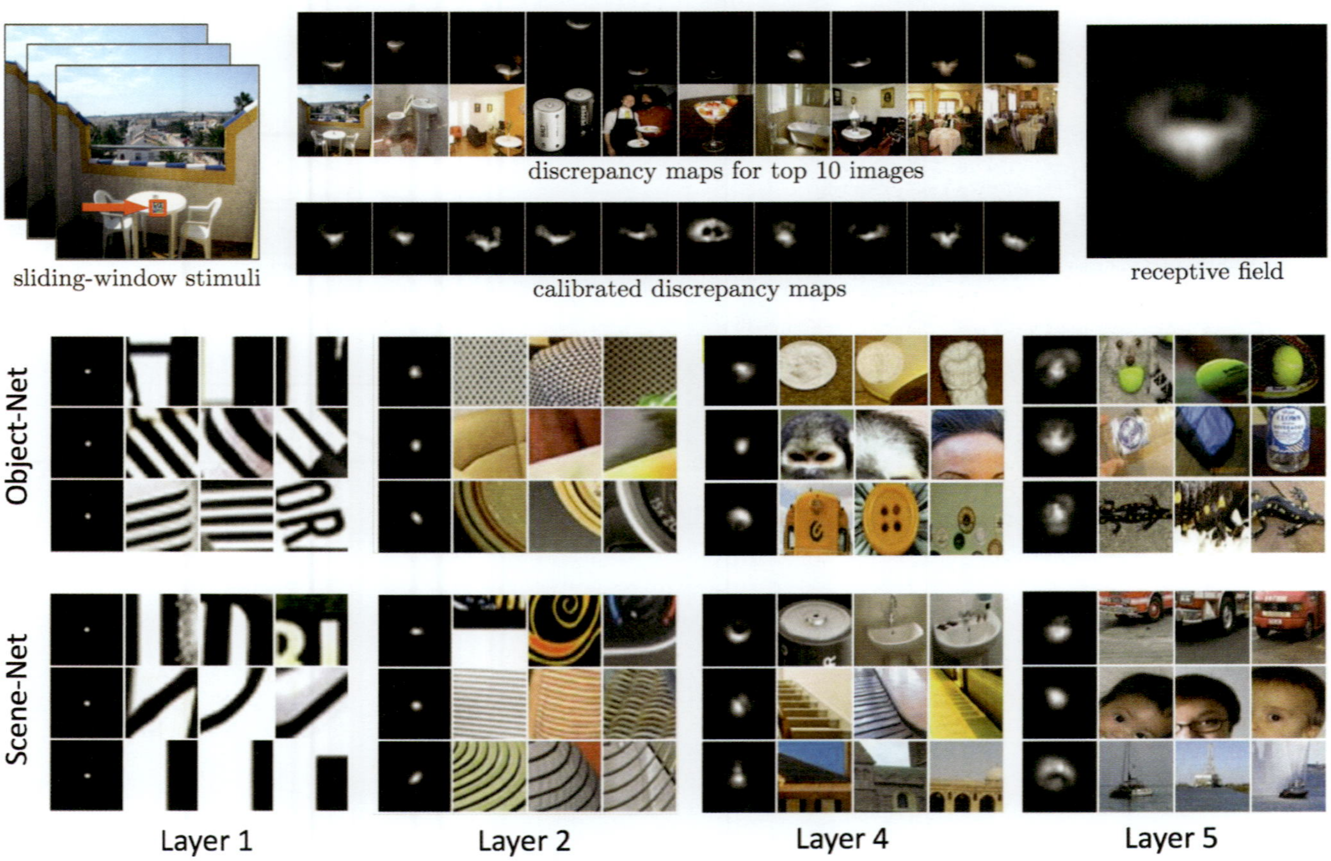

PLATE 14 *A*, The pipeline for estimating the receptive field of each artificial unit of a DNN. Each sliding-window stimulus contains a small, randomized occluding patch (*red arrow*) at different spatial locations. By comparing the activation response of each sliding-window occluder with the activation response of the original image, a discrepancy map for each image can be built (*middle top*). By summing up the calibrated discrepancy maps (*middle bottom*) for the top-ranked images (i.e. 10), we can visualize the receptive field of that unit (*right*). *B*, The receptive fields of three units of pool1, pool2, conv4, and pool5 layers, respectively, for Object-Net (AlexNet trained with the ImageNet data set) and Scene-Net (AlexNet trained with Places data set), along with three images that correspond to particular images that activated these units the most. As the layers go deeper, the receptive field size gradually increases. (See figure 13.2.)

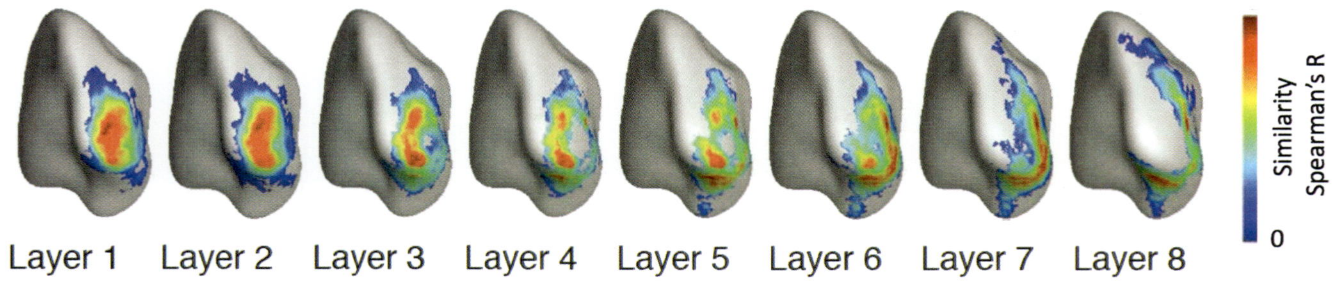

PLATE 15 Spatial maps of visual representations common to human brain and object DNNs. The colors and the color bar show the strength of the brain-DNN similarity (similarity measured as Spearman's correlation, from 0 in *blue*, to a significant value in *red*, the maximum correlation varying between 0.07 and 0.14 depending on the layers). Low layers (one through two) have significant representational similarities with the occipital lobe of the brain—that is, low-level and midlevel visual regions. Higher layers have significant representational similarities with more anterior regions in the temporal and parietal lobe, with layers seven and eight reaching far into the inferior temporal cortex and inferior parietal cortex (from Cichy et al., 2016b). (See figure 13.3.)

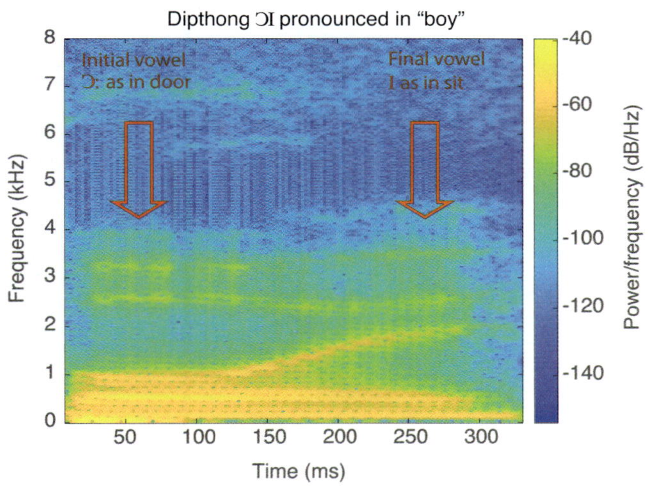

PLATE 16 Spectrogram of the syllable *boy*, showing how dynamic changes in spectral content are nonetheless structured and continuous in time frequency. (See figure 14.2.)

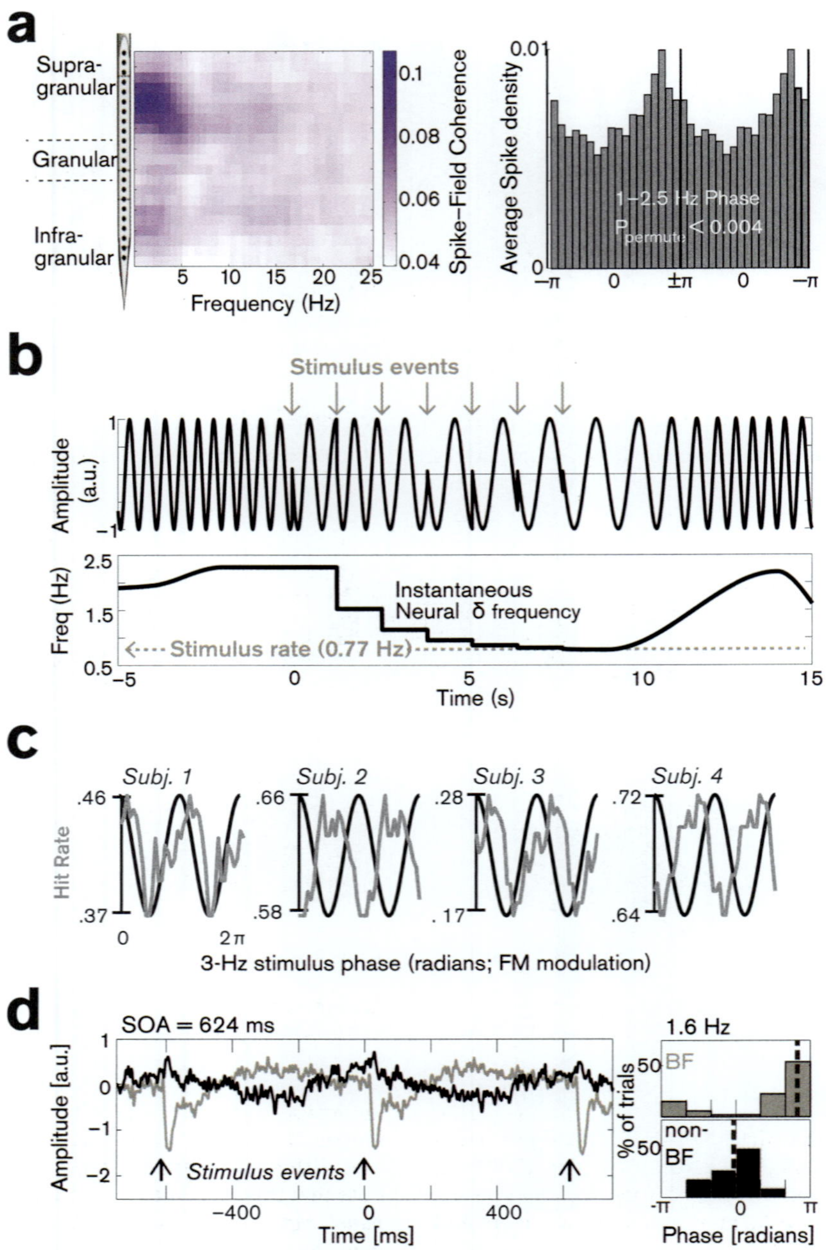

PLATE 17 *A*, Slow neural oscillatory phase in supragranular layers of auditory cortex governs excitability (i.e., spike rate) in a granular layer. Spontaneous data recorded in A1 of an awake macaque. *Left panel*, spike-LFP (local field potential) coherence as a function of frequency (x-axis) and cortical depth (y-axis). *Right panel*, the modulation of granular-layer spike rate by supragranular 1–2.5 Hz (delta) LFP phase. Unpublished data by Jonas Obleser and Peter Lakatos. *B*, Neural oscillatory frequencies can become entrained to stimulus regularities. Here, an approximately 2 Hz neural oscillation adjusts its instantaneous phase (*top panel*; note the "phase slips" in the black waveform when stimulus events, *light gray on top*, occur) and frequency (*black line, bottom panel*) to the entraining 0.77 Hz stimulus rate (*light gray arrow and dashed line, bottom panel*). After stimulus-train offset, the endogenous frequency of approximately 2 Hz is reinstated. Adapted from Obleser, Henry, and Lakatos (2017). *C*, Listening success (target detection) becomes entrained by stimulus regularity. Four individuals from the Henry and Obleser (2012) data set are replotted. Irrespective of average gap-detection performance, detection ("Hit rate," *gray trace*) is a function of the phase of the 3 Hz stimulus modulation (*black trace*) that was employed. Note the individual differences in phase lag: some participants' performance is in phase with the stimulus modulation, while others' is out of phase. *D*, Slow-frequency neural entrainment in A1 as a spectrotemporal filter. Attended (BF, best-frequency, *gray*) and unattended (non-BF, *black*) sites in A1 entrain counterphasic at the stimulation rate (here, 1/624 ms or 1.6 Hz), resulting in amplified, sharpened responses at attended time points. Adapted from Lakatos et al. (2013). (See figure 15.1.)

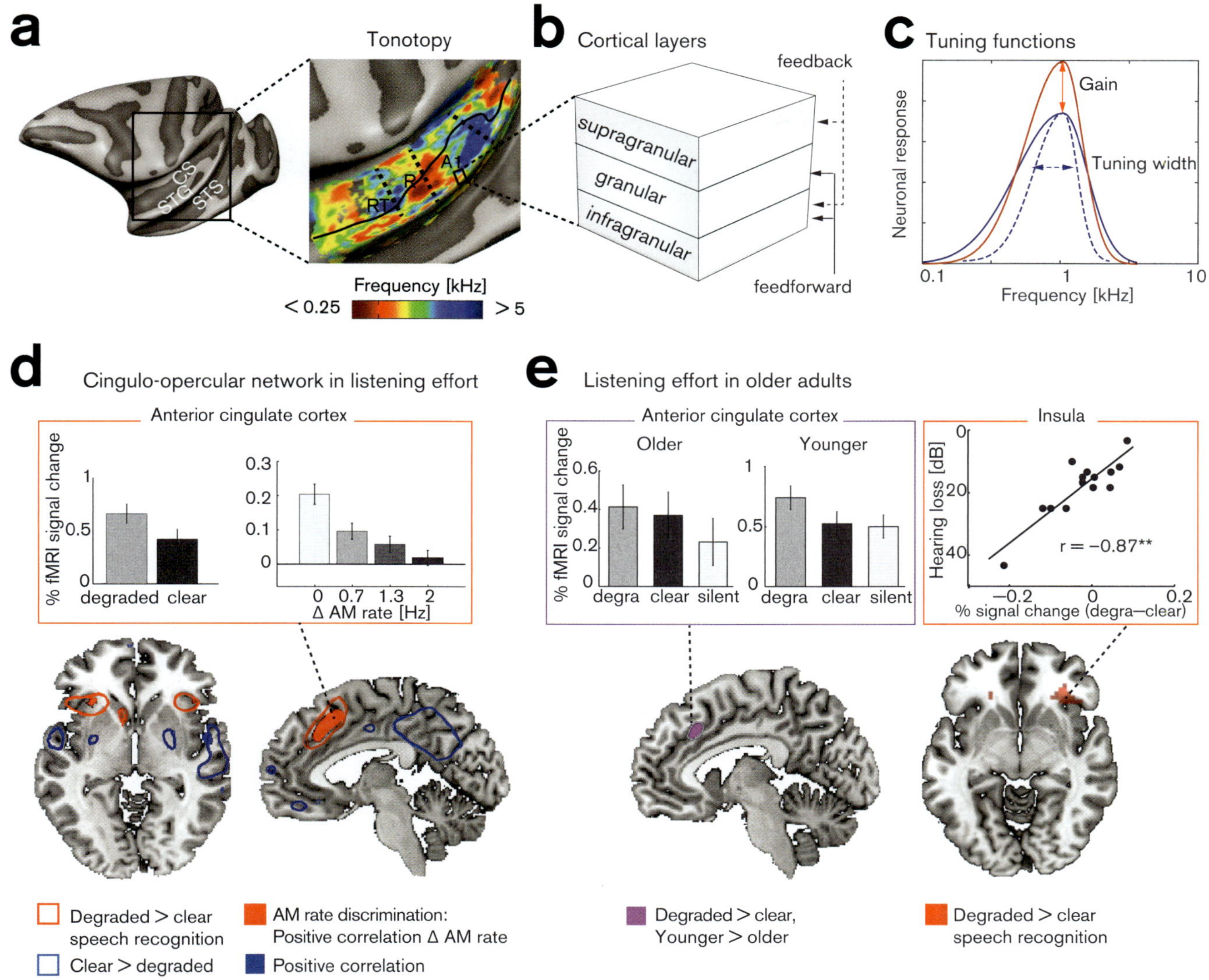

PLATE 18 *A*, Tonotopy mapped with natural sounds. Tonotopic map is shown on the surface of the inflated left hemisphere of one macaque. Modified from Erb et al. (2018). *B*, Schematic of cortical layers in A1 and their inputs: bottom-up sensory feedforward information enters at deep and middle cortical layers; top-down feedback information arrives at superficial and deep layers (see also figure 15.1*A*). *C*, Task demands shape the gain or tuning width of neuronal (population) frequency response functions in a layer-dependent manner (De Martino et al., 2015; O'Connell et al., 2014). *D*, Attentive listening to spectrally degraded compared to clear speech evokes enhanced fMRI responses in insula and anterior cingulate cortex (*top panel, left; bottom panel: contours of the map of the speech degradation effects*). For amplitude modulation (AM) rate discrimination, activity levels parametrically increase in the same areas with decreasing AM rate difference between standard and deviant (Δ AM rate; note that this corresponds to an increasing difficulty level, *top panel, right*). Modified from Erb et al. (2013). *E*, An age-by-degradation interaction in the anterior cingulate cortex is driven by a decreased dynamic range in the older listeners who show an enhanced fMRI signal both in clear and degraded conditions (*left*). Hearing loss correlates with the fMRI signal difference between clear and degraded speech in the insula (*right*). Modified from Erb and Obleser (2013). *Note:* CS: circular sulcus; STG/STS: superior temporal gyrus/sulcus; AM: amplitude modulation; ** $p<0.001$. (See figure 15.3.)

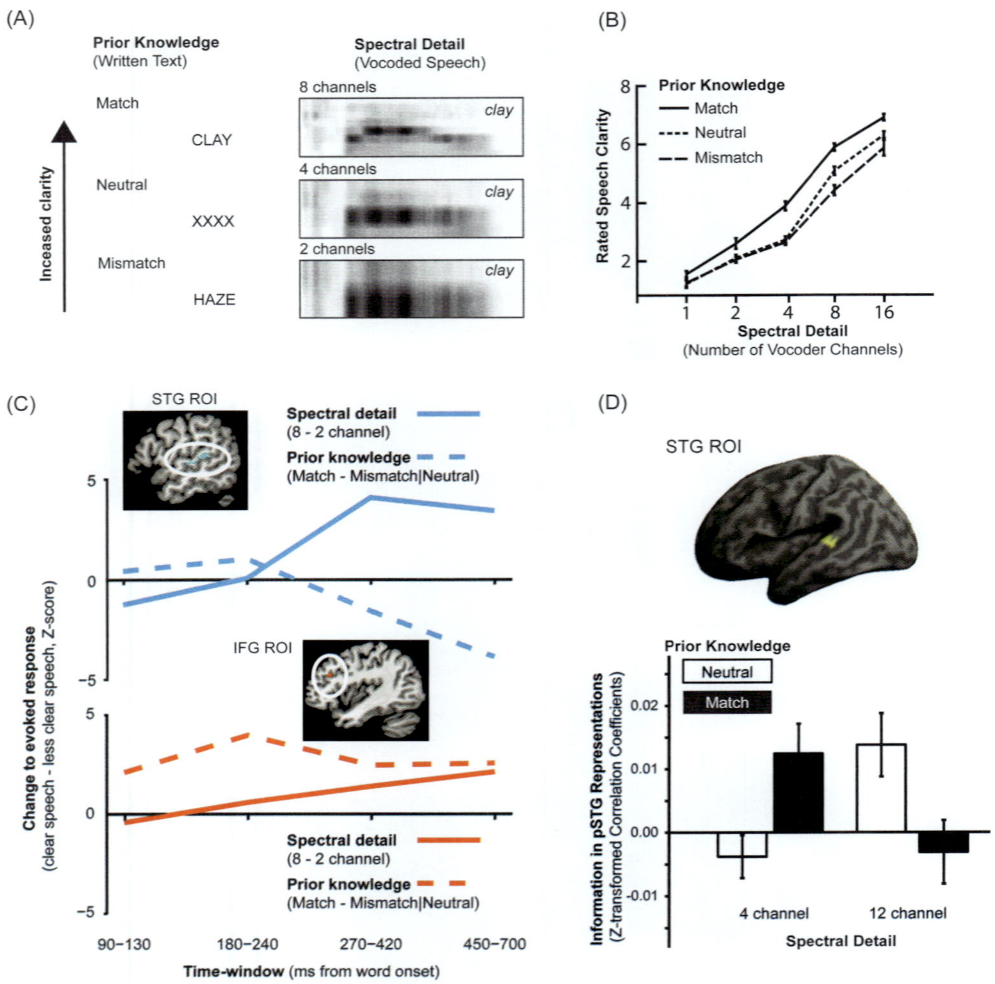

PLATE 19 Experimental findings that illustrate the contrasting behavioral and neural effects of sensory and knowledge-driven manipulations on speech clarity. *A*, Experimental design from Sohoglu et al. (2012, 2014) in which speech clarity is influenced by prior knowledge (matching written text) for vocoded speech that varies in spectral detail (a varying number of channels). *B*, Clarity ratings are enhanced by prior knowledge (matching text) and increased spectral detail (vocoder channels; from Sohoglu et al., 2014, expt. 1). *C*, Differential neural effects of these two manipulations are seen in MEG responses in the superior temporal gyrus (STG, *blue*) and the inferior frontal gyrus (IFG, *red*), consistent with top-down predictions on match trials that reduce STG responses (redrawn from Sohoglu et al., 2012). *D*, Multivariate fMRI responses measured in the STG show interactions between prior knowledge and spectral detail that are in line with prediction error representations. Increased spectral detail enhances STG representations of speech following neutral text but reduces the information content of STG representations after matching text (redrawn from Blank & Davis, 2016). (See figure 16.2.)

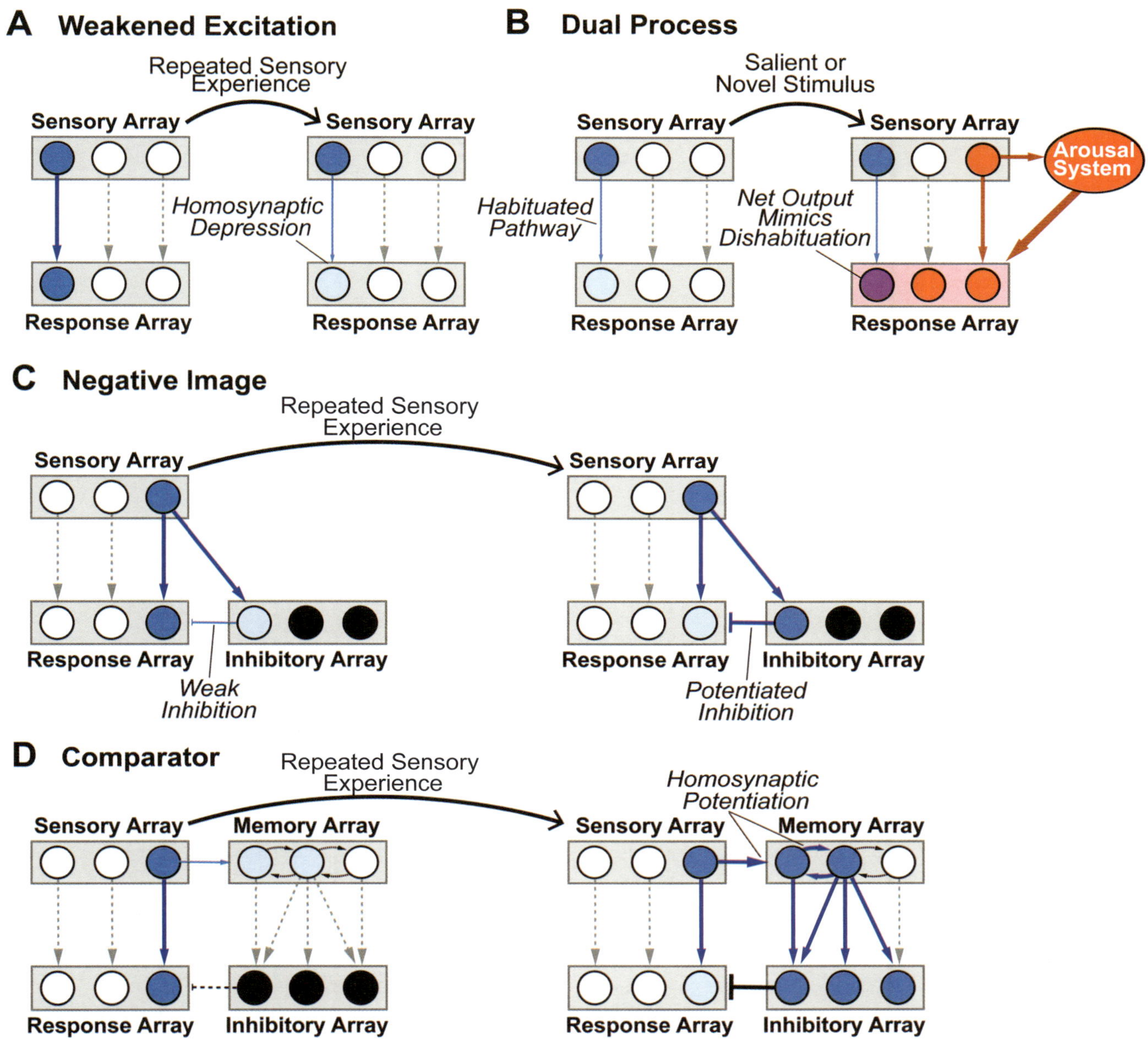

PLATE 20 Models of habituation. Models are presented here with simplified arrays of neurons representing each conceptual stage. Thus, the stimulus-response pathway is modeled as a sensory array connected through one or more intermediates to a response array. *A*, The most parsimonious explanation of behavioral habituation is *weakened excitation* via feedforward synaptic connections between the sensory array, which first responds to sensory input, and an array of unidirectionally connected response neurons, which are responsible for driving behavioral response. Shown on the left of this panel are the arrays prior to habituation, when an innocuous stimulus (*dark blue*) drives a pronounced response through existing feedforward inputs. After repeated presentation of the same innocuous stimulation, the behavioral response is selectively weakened (*light blue*) by homosynaptic depression within these feedforward stimulus-response connections. *B*, An important add-on to this feedforward depression model is included in the *dual process model*, largely to explain the important phenomenon of dishabituation, in which response returns immediately to the habituated stimulus (*blue*) after the presentation of a novel or salient stimulus (*red*). This model regards behavioral output as a net effect (*purple*) of depressed response through habituation, which is highly stimulus-specific, and increased response through sensitization, which is not stimulus-specific and is mediated by a modulatory arousal system. Thus, while stimulus-response synapses remain weakened, a generalized increase in the output of the response array returns the response output to approximately prehabituation levels. *C*, An alternative model contains an added layer of complexity, which is an array of inhibitory interneurons. In this model the primary modification is a selective potentiation of inhibitory neurons that form a *negative image* to suppress the output of the response array. Although not depicted here, dishabituation is proposed to be mediated by disinhibiting the response array, meaning that the previously habituated response truly returns to basal levels after dishabituation. *D*, Finally, *comparator* models have been proposed in which a memory array is an additional intermediary between the stimulus and the response. This memory array, formed by initial experience, is an internal representation of the familiar stimulus. If sensory input subsequently matches this internal representation after comparison, the memory array acts on the response array through the inhibitory intermediaries to suppress its output. If a novel stimulus is presented for which no internal representation exists, then this "top-down" inhibition cannot be applied, leading to behavioral output. The advantages of memory arrays are explained in more detail in figure 17.3. (See figure 17.2.)

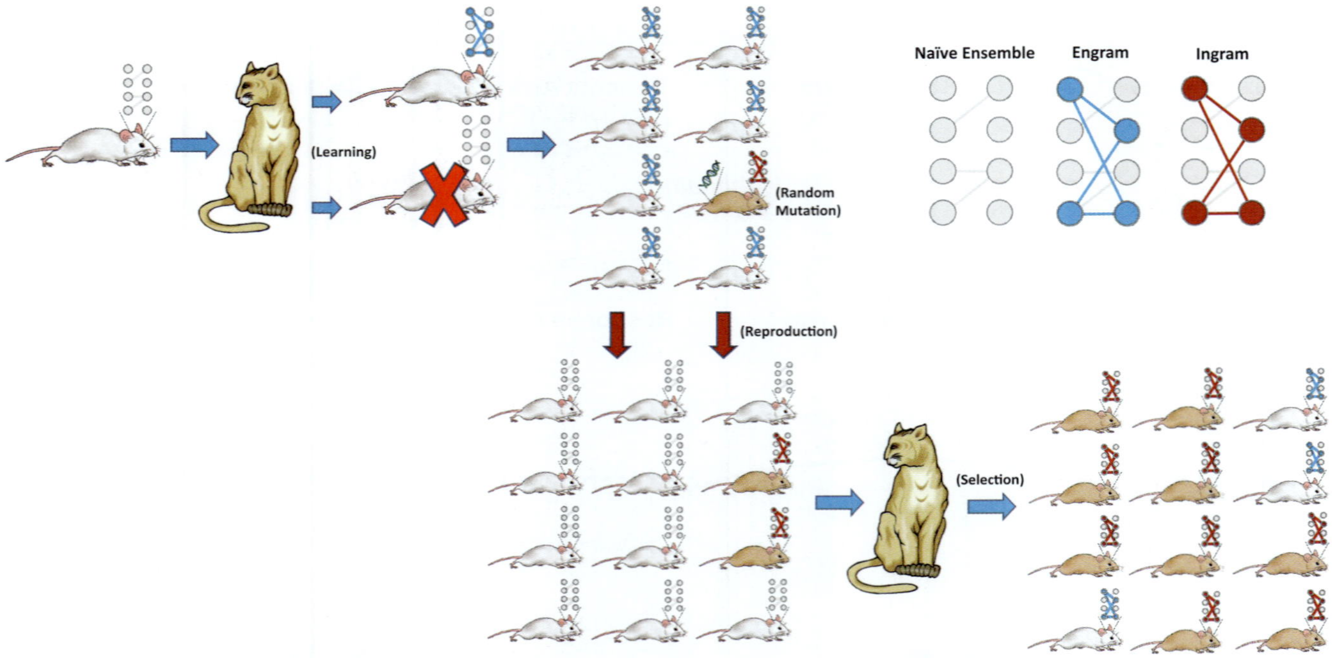

PLATE 21 The innate fear response as an example of the Baldwin effect. *Top left,* Prey animals that encounter a predator must rapidly learn to regard it as a threat and avoid it for survival. All surviving individuals will possess an engram to help negotiate future encounters with the predator. *Top center,* As long as predators are a common part of the ecosystem, a niche will be constructed where information about negotiating the predator will be valuable to individuals. Relevant engrams will become essential for survival in the environment. Within the context of this niche, an individual who acquires a random genetic mutation (originating in the germ cells of one of its parents) experiences a developmental alteration of its brain structure that mimics the engrams of the population. *Bottom center,* The progeny of animals with engrams will begin life as naïve individuals who must learn the crucial memory by experience or social interaction. In contrast, the progeny of animals with ingrams will innately possess information pertaining to the predator. *Bottom right,* During future encounters with predators, animals with innate ingrams will outcompete the naïve individuals who must form engrams to adapt to the environment. Over many generations, animals with ingrams will be selected for, and the ingram will be driven to fixation in the population. *Top right,* Schematic of naïve neuronal ensemble (*gray*), a learning-informed engram ensemble (*blue*), and a genetically informed ingram ensemble (*red*). (See figure 18.2.)

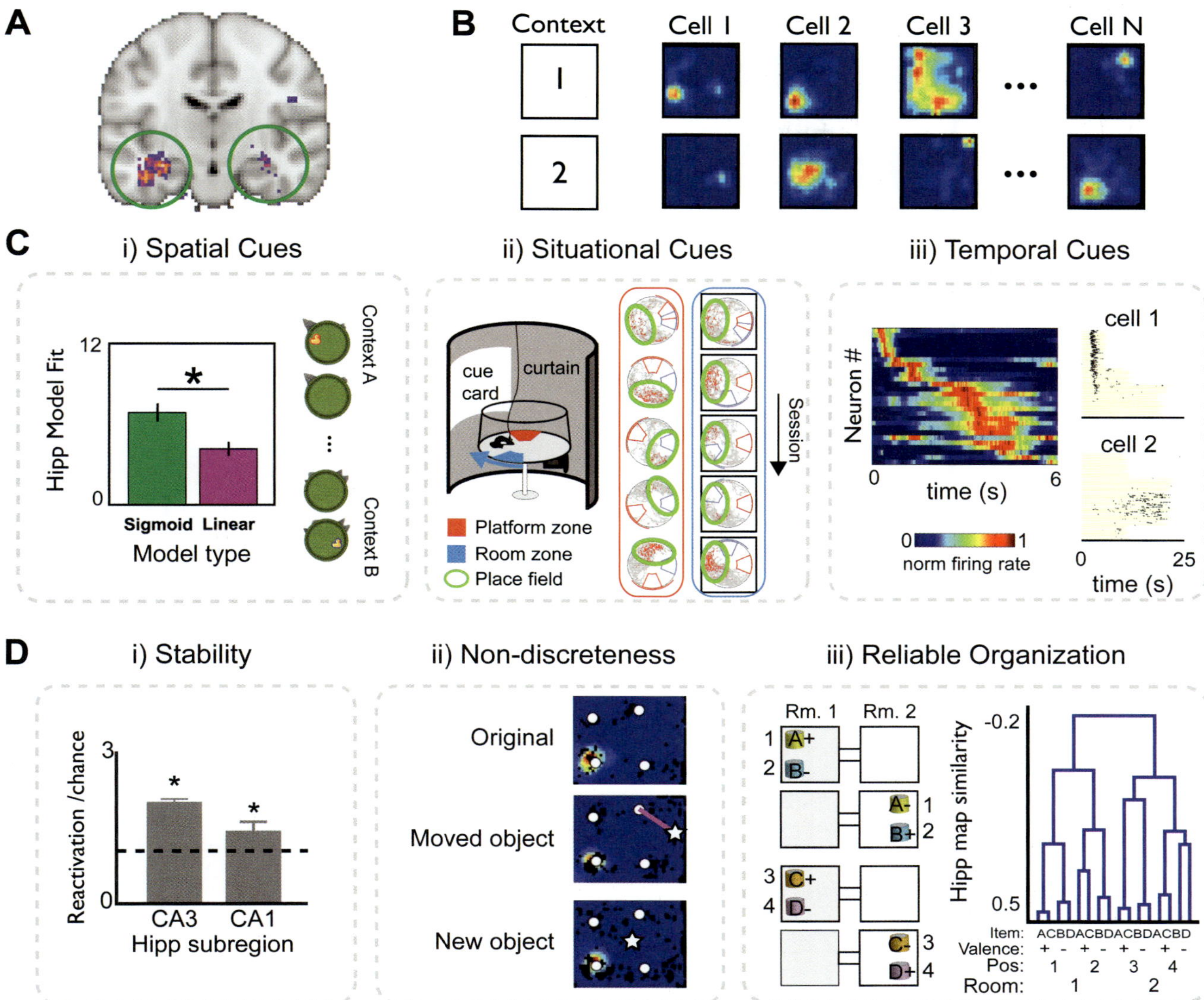

PLATE 22 The hippocampal basis of contextual memory. *A*, Reverse inference meta-analysis (Yarkoni, Poldrack, Nichols, Van Essen, & Wager, 2011) of 36 context-dependent memory human fMRI studies. Overlapping activation across studies was largely localized to the hippocampus (threshold $p<0.01$, false discovery rate (FDR)-corrected). *B*, Contextual memory is indexed by hippocampal remapping, in which all simultaneously recorded neurons alter their firing patterns across contexts (Alme et al., 2015). *C*, Remapping is induced by contextual cue changes: (1) Spatial cues. As visual cues (mountains) were gradually morphed from Context A to B during a spatial memory task, a rapid remapping of fMRI response patterns (*Sigmoidal*) better characterized hippocampal activity than a gradual change (*Linear*) (Steemers et al., 2016); (2) Situational cues. Hippocampal neurons represented locations in two different situational contexts, one relative to a moving platform (*left*) and another relative to the stable room (*right*; Keleman & Fenton, 2010); (3) Temporal cues. *Left*, Hippocampal neurons modulated by time. *Right*, Neurons changed firing patterns when the task's temporal parameters (*yellow bars*) were altered (MacDonald et al., 2011). *D*, Remapping reflects contextual boundary conditions: (1) Stability. The same hippocampal neurons (in subfields CA3 and CA1) reactivated two weeks later after mice were placed in the same context as initial exposure (Tayler et al., 2013); (2) Nondiscreteness. Example hippocampal neuron that did not remap when a discrete object (*white circles*) was moved (*magenta line to star*) or a novel object was added (*star*) (Deshmuch & Knierim, 2013); (3) Reliable organization. When rodents explored two chambers containing objects in different positions associated with different valences, hierarchical cue structure was reflected in hippocampal population activity patterns (McKenzie et al., 2014). (See figure 19.2.)

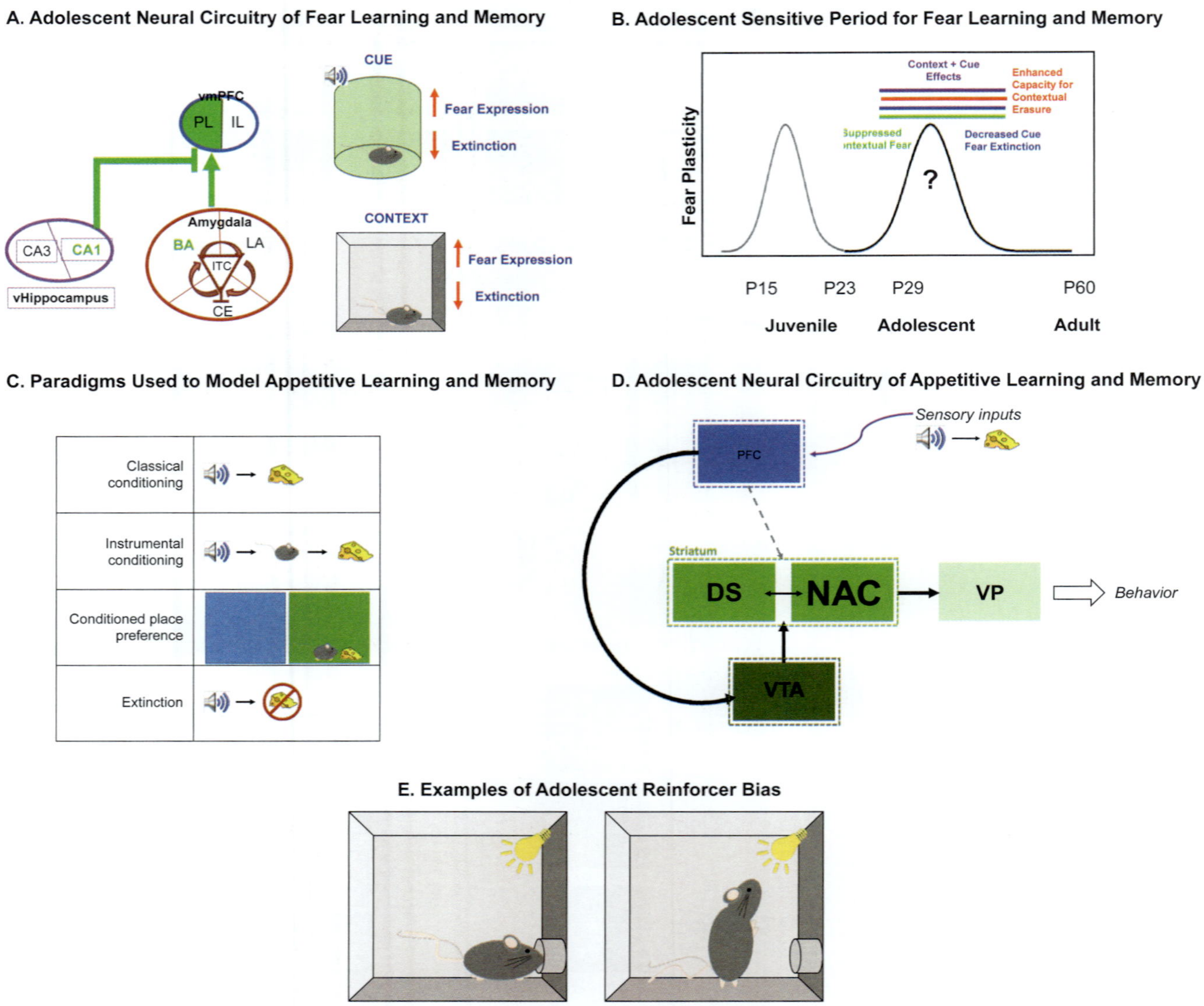

PLATE 23 Emotional memory-formation processes during adolescence. *A*, A schematic of the neural circuitry of adolescent cued fear as simplified from retrograde tracer studies (Pattwell et al., 2016) shows an adolescent surge in connectivity between vCA1 and PL, as well as BA and PL. *Abbreviations*: basal amygdala, BA; central amygdala, CE; infralimbic, IL; intercalated cells, ITC; lateral amygdala, LA; prelimbic, PL. *B*, Developmental sensitive periods for fear learning and memory—insights into adolescent fear memory and behavior. *C*, Paradigms used to model appetitive learning and memory. *D*, A schematic of the neural circuitry of appetitive memory during adolescence. Line weight and font size indicate relative contributions to appetitive memory. Dashed lines indicate notable differences from adult circuitry. *Abbreviations*: dorsal striatum, DS; nucleus accumbens, NAC; prefrontal cortex, PFC; ventral pallidum, VP; ventral tegmental area, VTA. *E*, Appetitive memory strength during adolescence may be influenced by a higher salience of reinforcers, a higher salience of reinforcer-associated cues, or a combination of both. *Left*, An adolescent rodent perseverates on the delivery of a reinforcer by spending more time in the reinforcer receptacle. *Right*, An adolescent rodent perseverates on a visual stimulus associated with reinforcer delivery. (See figure 21.1.)

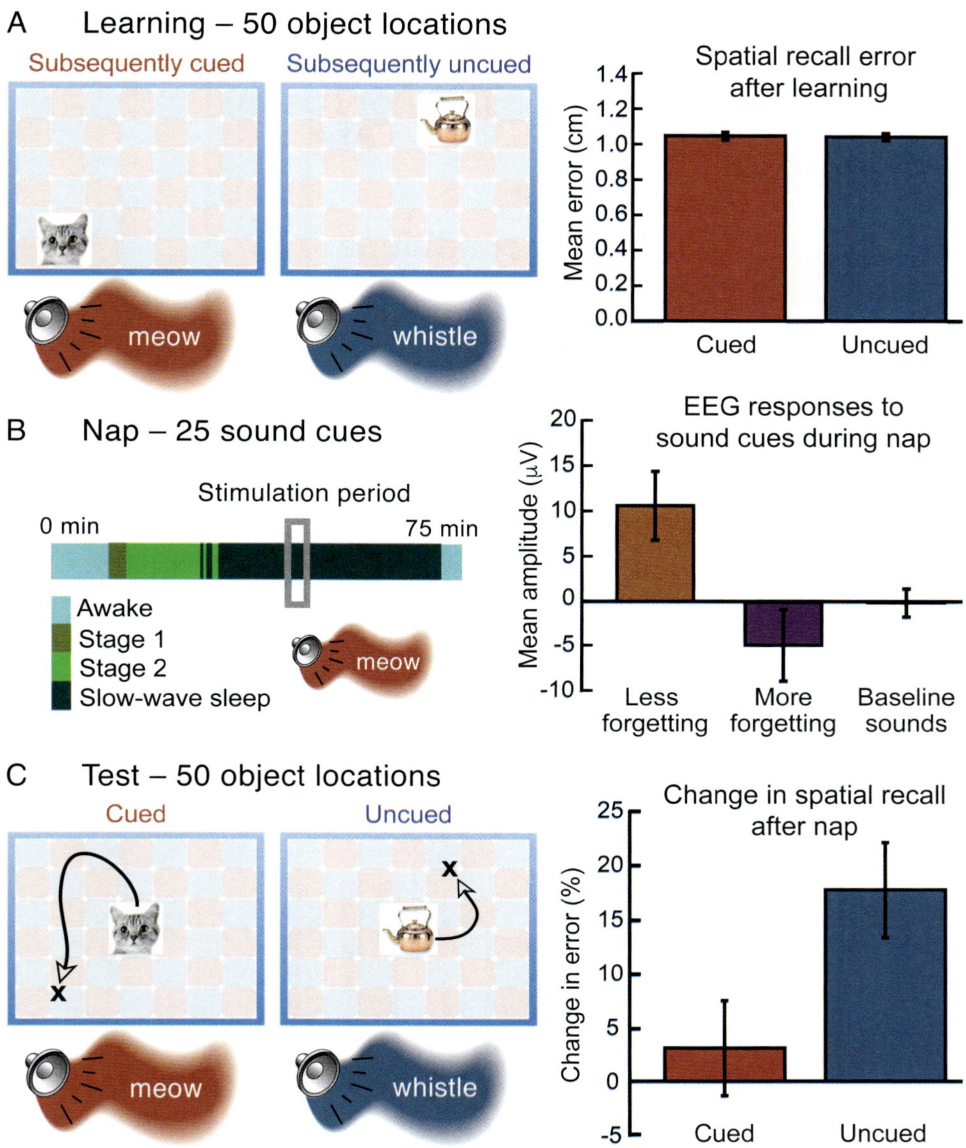

PLATE 24 Targeted memory reactivation (TMR). *A*, Subjects in the study by Rudoy and colleagues (2009) first learned 50 object-location associations. Each object was presented with its characteristic sound. Following an interactive learning procedure, location recall was tested. Half of the objects were assigned to be cued during sleep such that recall accuracy was matched for cued and uncued objects. *B*, Next, subjects slept with EEG monitoring. When signs of SWS were evident, 25 of the sounds were presented at a low intensity. These sounds influenced memory storage without waking people up. *C*, Recall of locations was tested again after the nap. Subjects moved each object from the center to where they thought it belonged (*arrows*). Recall was more accurate for cued versus uncued objects. Mean EEG responses from 400–800 ms following the onset of each sound presented during sleep were found to be more positive for those objects with less decline in recall (*Less forgetting* in *B*) compared to the remaining objects or to baseline sounds. These responses resembled typical event-related potentials predictive of later memory (Dm effects; Paller et al., 1987), suggesting that spatial memory reactivation occurred as a consequence of cue presentation, leading to improved spatial recall after awakening. Reprinted from Rudoy et al. (2009). (See figure 23.1.)

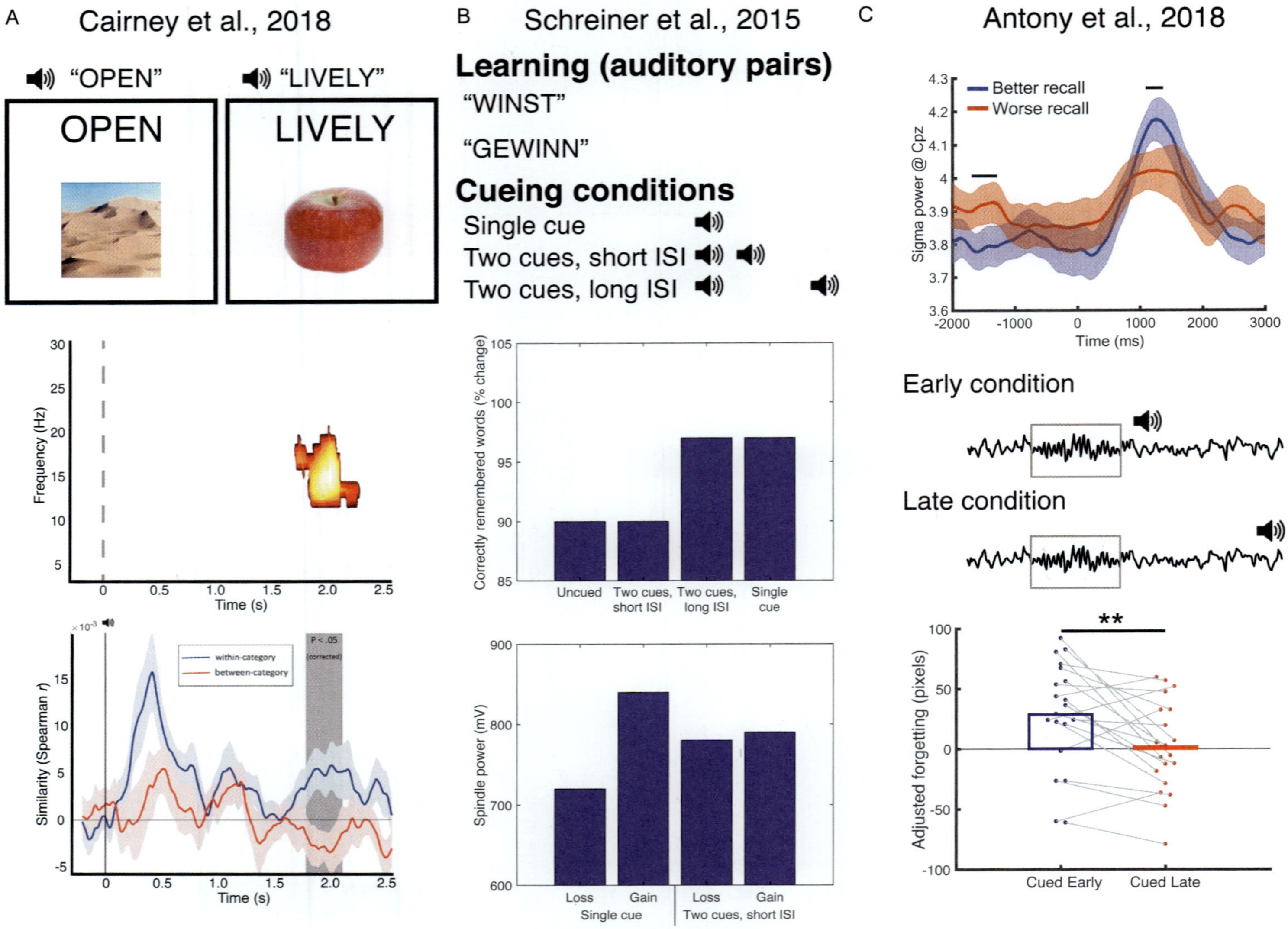

PLATE 25 Sleep spindles and memory as studied in three experiments. *A*, Subjects in Cairney et al. (2018) learned adjective-scene and adjective-object associations. A subset of spoken adjectives were then presented during postlearning sleep. These cues elicited higher EEG power in the spindle band (sigma, ~15 Hz) for learned than for nonlearned words (1.7–2.3 s after cue onset). Additionally, within-category neural similarity (object vs. scene) exceeded between-category similarity at roughly the same time, suggesting that spindles mediate relevant memory reactivation. *B*, Subjects in Schreiner, Lehmann, & Rasch (2015) learned auditory word pairs. Cues presented during sleep included single words, two words separated by 200 ms, or two words separated by 1,500 ms (i.e., a long or short interstimulus interval [ISI]). Subsequent recall was best with single cues or two cues (long ISI), and spindle power within the immediate postcue period predicted memory change with single cues only. *C*, Antony et al. (2018) similarly found that postcue sigma power predicted memory improvement for spatial recall. Additionally, precue sigma power negatively predicted memory, suggesting that precue spindles impede reactivation in that a well-timed postcue spindle is unlikely in these cases. Spindles were found to be most likely to reoccur after about 4–6 s. Using software to track spindles in real time, TMR benefits were better for sounds presented late (long ISI after prior spindle) versus early (short ISI after prior spindle). These results suggest that memory reactivation is linked with spindles, which also means that there may be pauses in reactivation corresponding with the normal pauses between spindles. (See figure 23.2.)

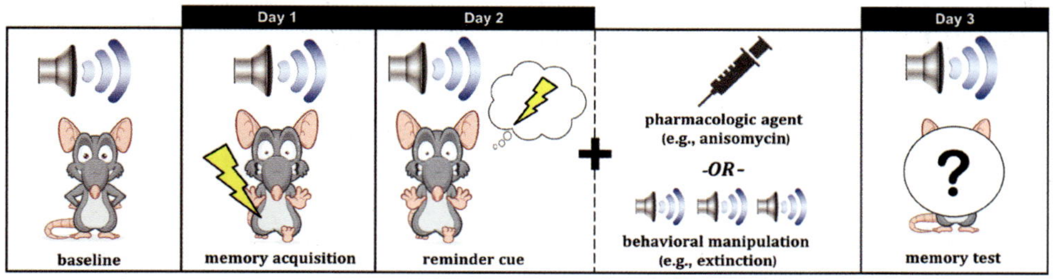

PLATE 26 A standard reconsolidation protocol. A conditioned stimulus (CS; tone) does not originally elicit noteworthy behavioral responding. During conditioning, the animal learns to associate the CS with an unconditioned stimulus (US; shock), and the CS alone then elicits a response. After 24 hours, the CS is presented to reactivate the associative memory. This is followed by a pharmacologic or behavioral manipulation thought to disrupt reconsolidation. Lastly, a memory test is conducted by measuring the behavior indicative of memory retention. (See figure 24.1.)

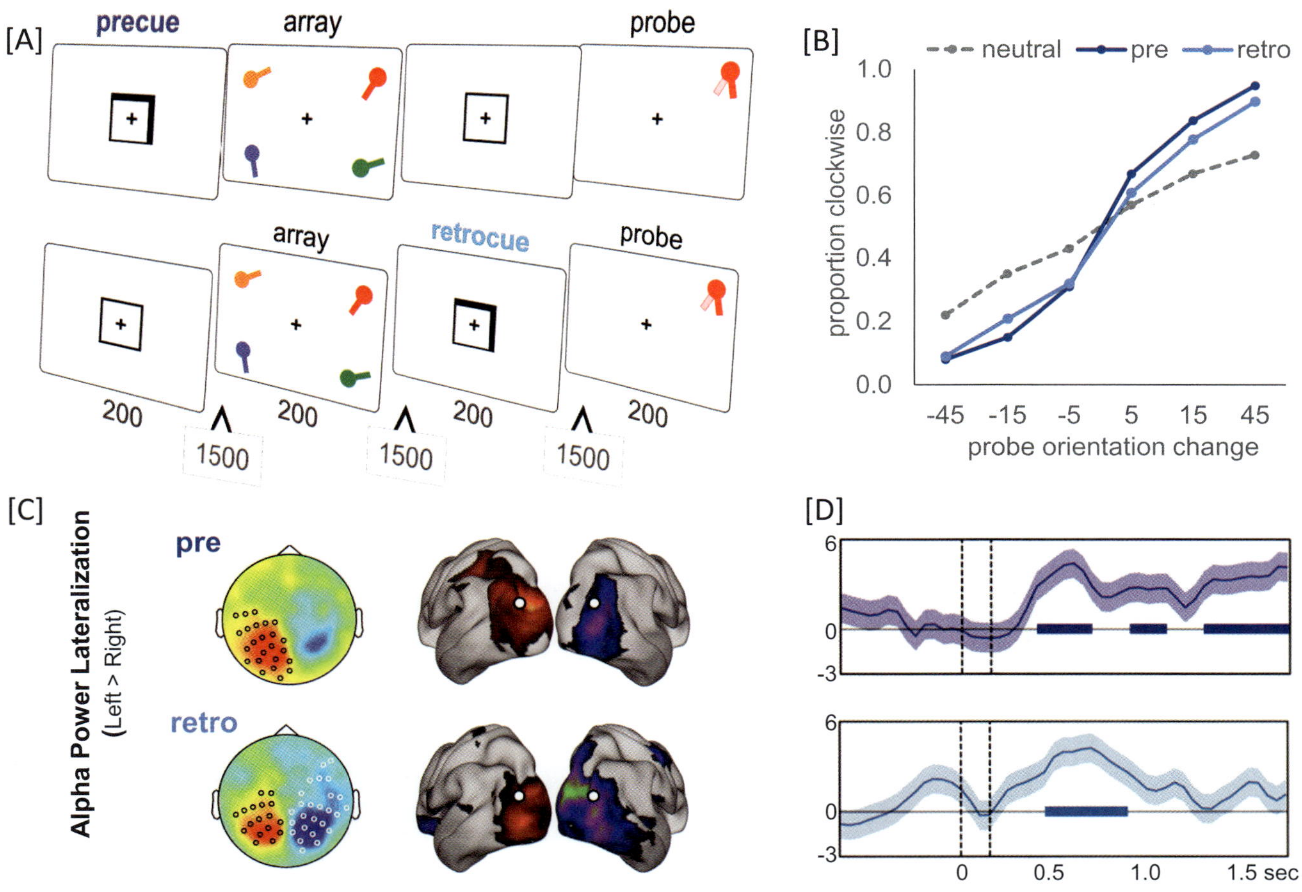

PLATE 27 Attention is also retrospective, operating on the content of working memory. *A*, In this example, Wallis and colleagues (2015) used MEG to compare the neural dynamics underlying prospective and retrospective attention. *B*, Spatial cues presented either before encoding (precue) or during the memory delay (retro-cue) were both effective for optimizing memory performance. *C*, Both cue types also elicited a classic signature of spatial attention (contralateral desynchronization of posterior alpha oscillations; see topological plots). *D*, Time course analysis further showed that preparatory attention involves sustained alpha lateralization, but contralateral desynchronization was relatively transient following the retrocue. (See figure 25.3.)

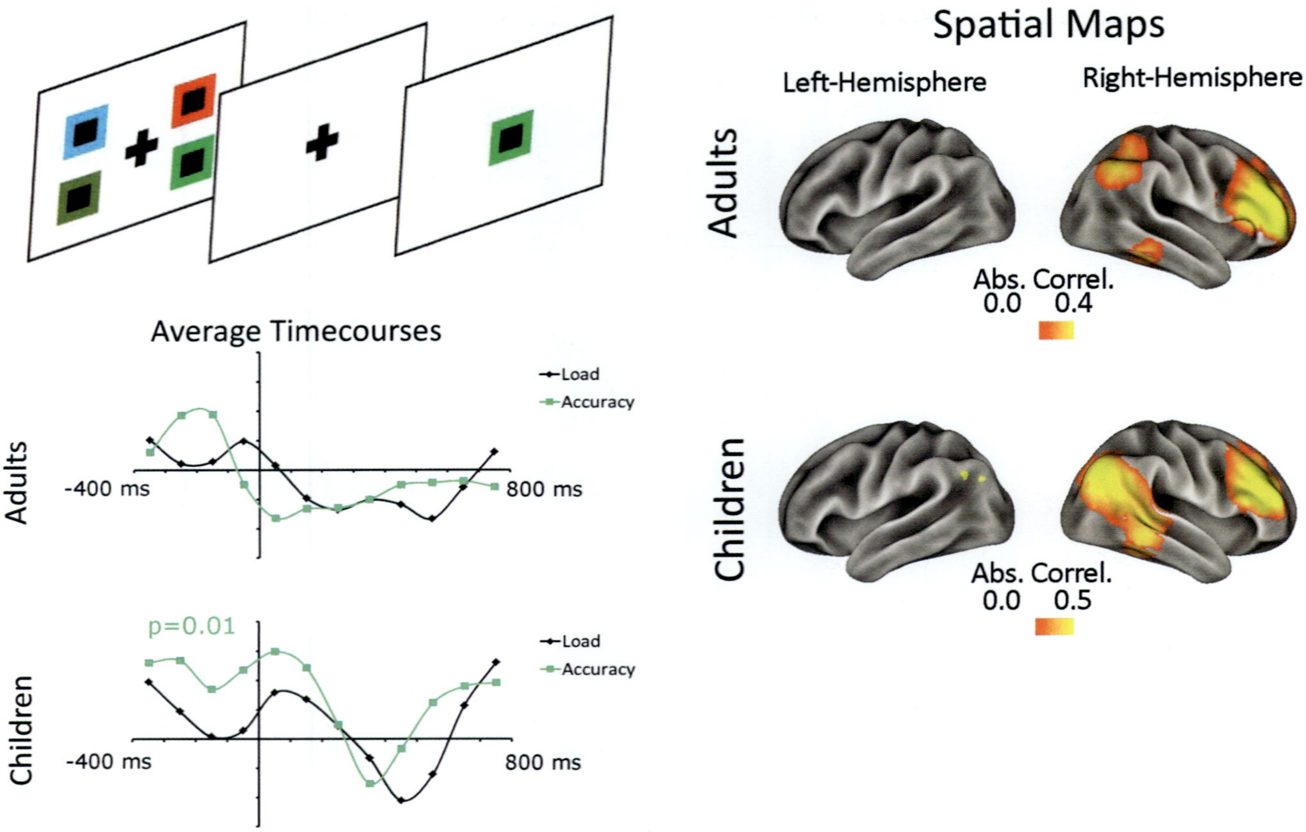

PLATE 28 *A*, Graphical representation of the memory task employed here. *B*, Activity in frontoparietal network (slow frequency theta 4–7 Hz) oscillations predicted accuracy of memory at the end of the trial in children and similarly, but not significantly so, in adults. The map shows the spatial extent of the component networks (in terms of the absolute Pearson correlation values between each brain location and this component). *C*, The time course of the regressor (*black line*) shows that accuracy is predicted by oscillations for this network at the time of encoding of the memoranda. The time course also presents another regressor as a comparison (load—2 vs. 4 items, *cyan line*) to show that this network was not differentially recruited by just any demand, like task difficulty. Adapted with permission from Astle et al. (2014). (See figure 26.1.)

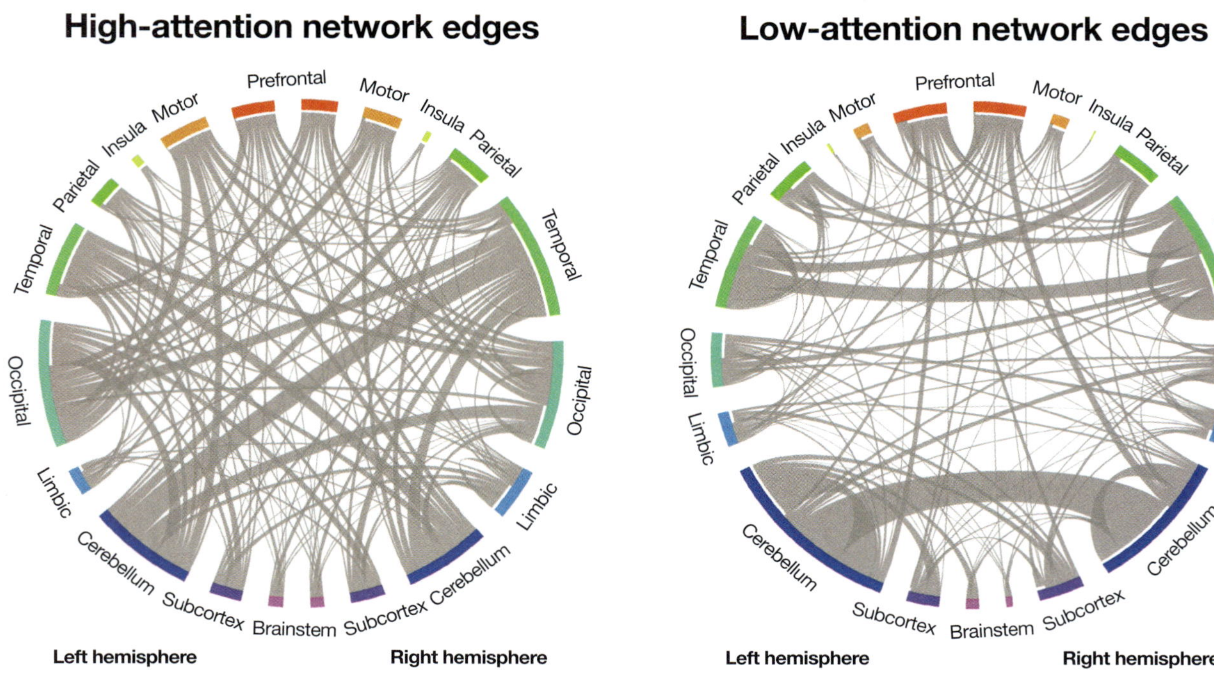

PLATE 29 Functional connections (edges) in the high-attention and low-attention networks (Rosenberg, Finn, et al., 2016). Network nodes are grouped into macroscale brain regions; lines between them represent edges. Line width corresponds to the number of edges between region pairs. (See figure 27.2.)

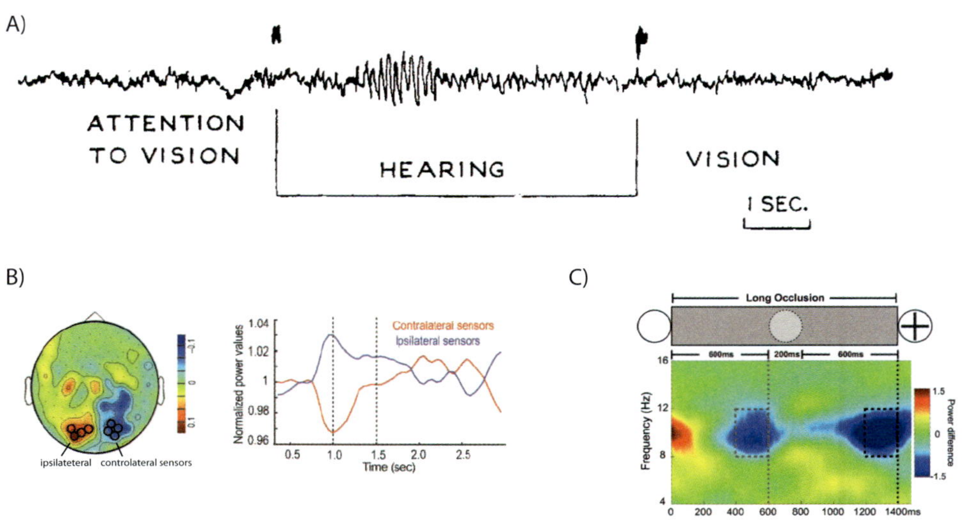

PLATE 30 Alpha and the allocation of selective attention. *A*, In an EEG study, a subject was asked to attend to either visual or auditory input. As attention was allocated to the auditory input stream, the alpha power increased. This observation is consistent with the functional inhibition of visual areas when auditory information is attended to. Reproduced from Adrian (1944). *B*, In a spatial attention task, subjects were asked to attend to items continuously presented in the left or the right visual field. This resulted in an alpha power decrease contralateral to the attended direction, which reflects the engagement of this hemisphere. Importantly, the ipsilateral alpha power prevents the processing of unattended stimuli. Reproduced from Händel, Haarmeier, and Jensen (2011). *C*, In a temporal attention task, an occluded visual item could reappear either at time 800 ms or 1,400 ms. Just prior to these anticipated time points, the alpha power decreased. Reproduced from Rohenkohl and Nobre (2011). (See figure 28.2.)

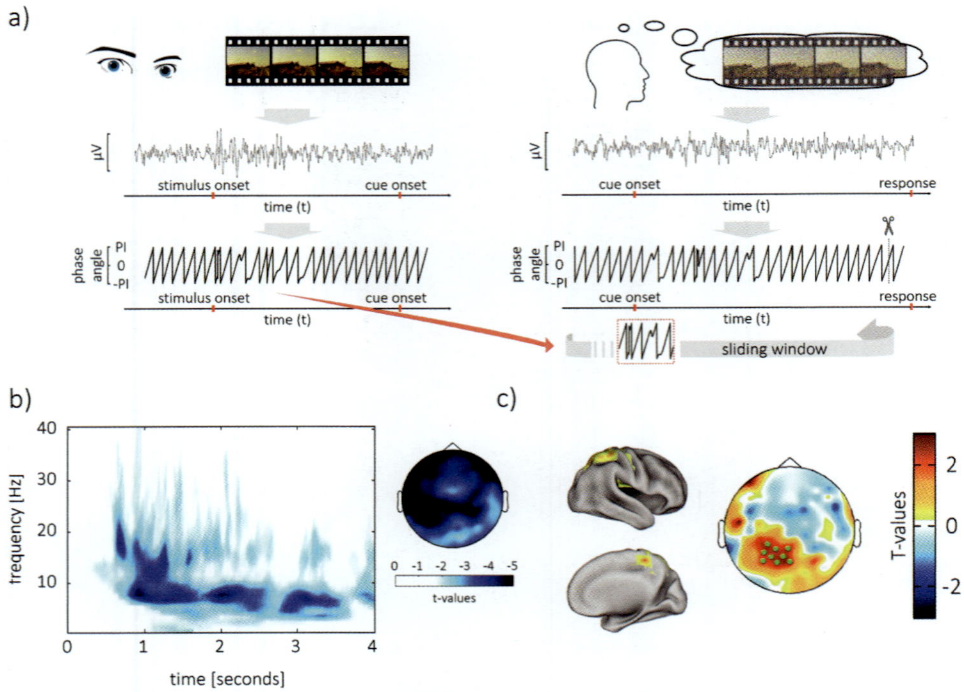

PLATE 31 Alpha power decreases during memory maintenance code stimulus-specific information. *A*, Subjects first encoded a video (*left*) and then maintained a vivid imagination of that video at a later time point (*right*). The phase time course was extracted from the EEG during encoding and retrieval in order to calculate a similarity measure between encoding and retrieval. *B*, During maintenance, strong and sustained alpha power decreases were observed. *C*, Reactivation of stimulus-specific information, as measured with phase similarity, could be detected in the alpha frequency band with a maximum in parietal regions. Reproduced from Michelmann, Bowman, and Hanslmayr (2018). (See figure 28.4.)

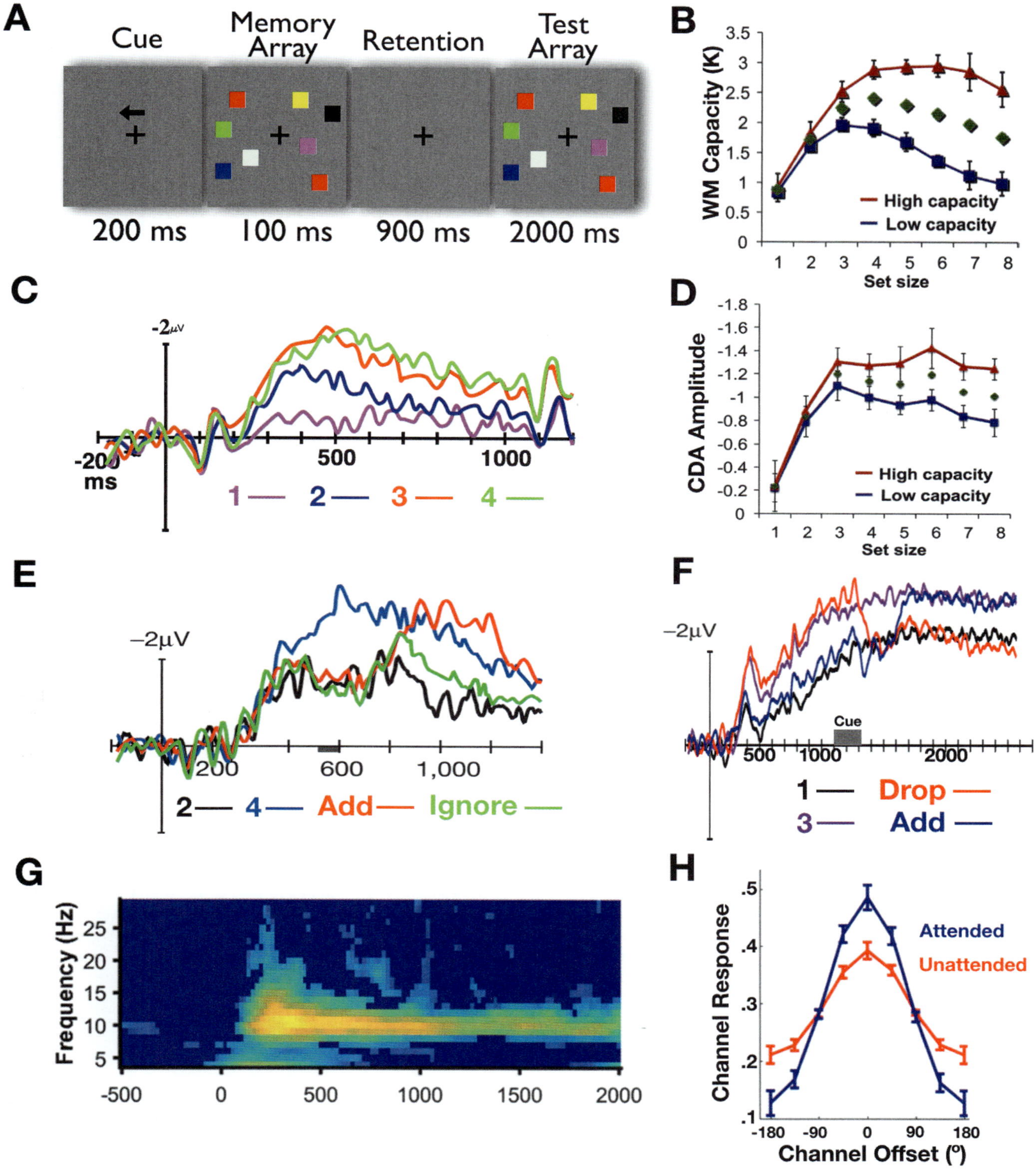

PLATE 32 *A*, Stimuli and procedure for a typical CDA WM paradigm. *B*, Behavioral capacity estimates (K) across set sizes for high- and low-WM-capacity individuals. *C*, CDA amplitudes as a function of the number of memory items. *D*, CDA amplitude across set size for high- and low-WM-capacity individuals. *E*, CDA for sequential displays. In the Add condition, a two-item array is followed by another two-item array that must be stored (i.e., 2+2). In the Ignore condition, a two-item array is followed by another two-item array that must be ignored. *F*, CDA for dynamic load changes. In the Add condition, subjects initially tracked one item and then, following a cue, began tracking two additional items (i.e., 1+3). In the Drop condition, subjects tracked three items but were instructed to drop two of those items (i.e., 3−2). *G*, Sustained location selectivity about remembered position is concentrated in the alpha band (8–12 Hz). *H*, Alpha channel-tuning functions (CTF) show a graded profile of channel activity-tracking locations in WM for attended and unattended items. (See figure 30.1.)

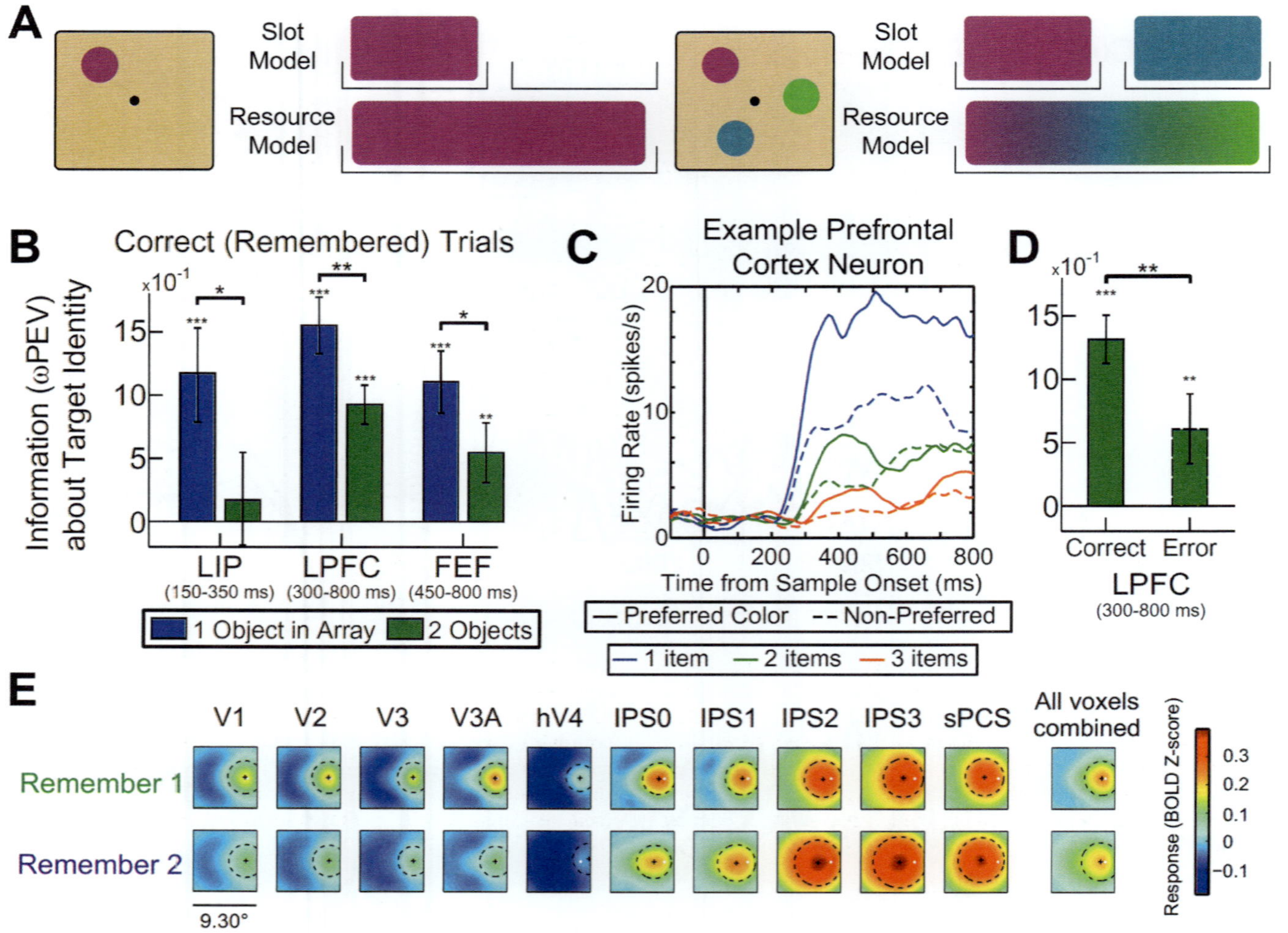

PLATE 33 The slot versus pools models of working-memory capacity limits. *A*, Capacity limits in working memory have been modeled either as the result of a limited number of slots (*top row*) or limited resources (*bottom row*). The slot model predicts that increasing the memory load (*right*) leads to failure to maintain certain memories (e.g., light gray is not stored). In contrast, the resource model predicts that increasing memory load should reduce the information about any single item. *B–C*, Neurophysiological evidence for the resource model. Information about a memory item is reduced in prefrontal and parietal cortex when working-memory load is increased (*B, dark vs. light gray bars*). In addition, prefrontal cortex neurons carry information about a to-be-remembered stimulus even when the animal is unable to report it (i.e., it is forgotten). Adapted from Buschman et al. (2012). *D–E*, Reduced information about a stimulus is thought to be due to the divisive normalization of responses. This is seen at the level of single neurons (*D*) and in the neural population (using blood-oxygen-level dependent activity, BOLD, *E*). Firing rate and decodability are reduced when the number of items to be remembered is increased. Adapted from Buschman et al. (2012) and Sprague, Ester, and Serences (2014). (See figure 31.2.)

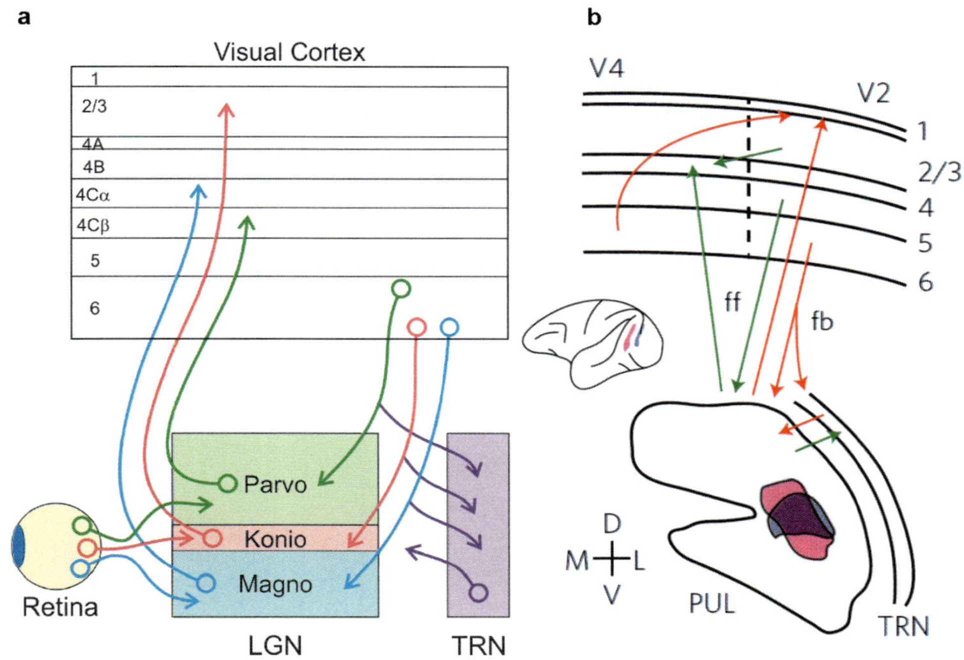

PLATE 34 Thalamocortical connectivity. *A*, The retinogeniculocortical pathway is composed of three distinct streams—the parvocellular, magnocellular, and koniocellular streams—that arise from distinct cell classes in the retina, remain segregated in the LGN, and terminate in different layers of V1. The parallel feedforward streams are matched with similarly specific streams of corticogeniculate feedback. Feedback axons provide monosynaptic excitation to LGN neurons as well as disynaptic inhibition via local interneurons and neurons in the thalamic reticular nucleus. *B*, Pulvinar; direct corticocortical connections (*top*) and indirect corticopulvinocortical loops exemplified by V2-pulvino-V4 circuitry. Tracer injections into V2 (*blue*) and V4 (*pink; inset*) showed overlapping (*purple*) projection zones in the pulvinar (*bottom*). Adapted with permission from Adams et al. (2000). (See figure 32.1.)

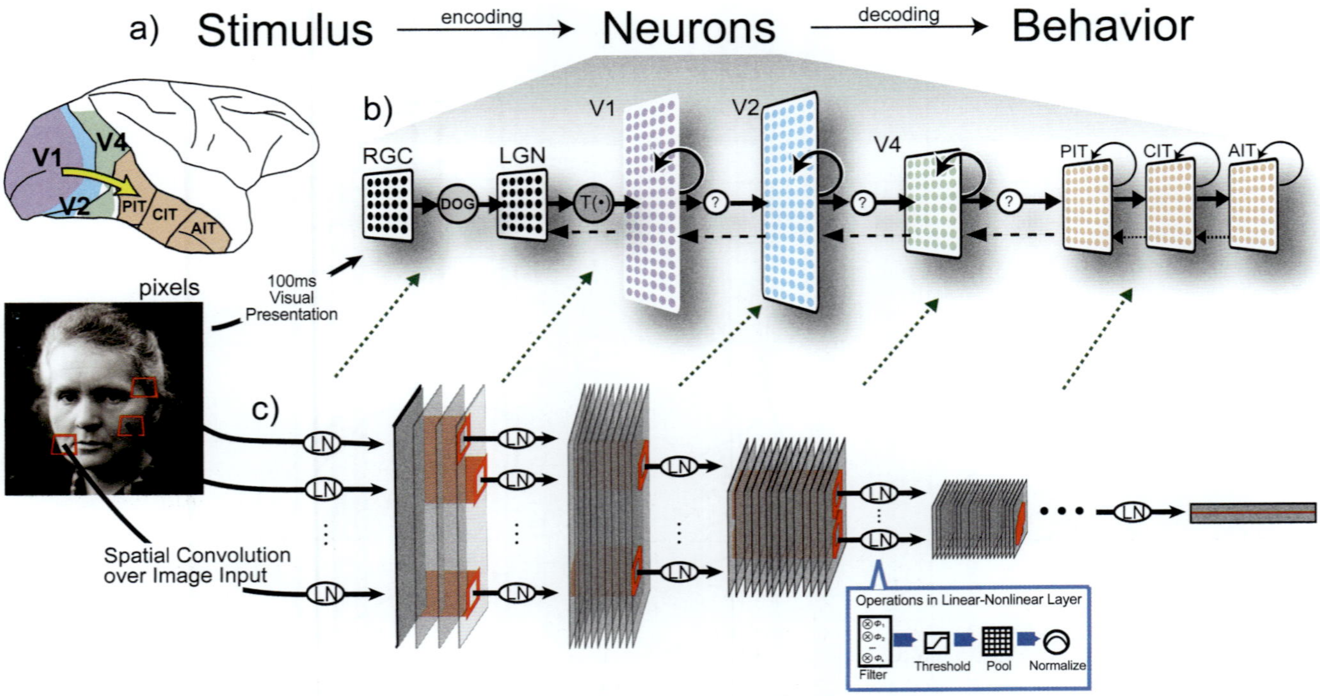

PLATE 35 Hierarchical convolutional neural networks as models of sensory cortex. *A*, The basic framework in which sensory cortex is studied is one of encoding, the process by which stimuli are transformed into patterns of neural activity, and decoding, the process by which neural activity generates behavior. *B*, The ventral visual pathway of humans and nonhuman primates is one of the most comprehensively studied sensory systems in neuroscience. It consists of a series of connected cortical brain areas that are thought to operate in a sensory cascade, from early visual areas such as V1 to later visual areas such as inferior temporal (IT) cortex. Neural responses in the ventral pathway are believed to encode an abstract representation of objects in visual images. *C*, Hierarchical convolutional neural networks (HCNNs) are multilayer neural networks that have been proposed as models of the ventral pathway. Each layer of an HCNN is made up of a linear-nonlinear (LN) combination of simple operations such as filtering, thresholding, pooling, and normalization. The filter bank in each layer consists of a set of weights analogous to synaptic strengths. Each filter in the filter bank corresponds to a distinct template, analogous to Gabor wavelets with different frequencies and orientations (the image shows a model with four filters in layer 1, eight in layer 2, and so on). The operations within a layer are applied locally to spatial patches within the input, corresponding to simple limited-size receptive fields (*red boxes*). The composition of multiple layers leads to a complex nonlinear transform of the original input stimulus. At each layer, retinotopy decreases and effective receptive field size increases. (See figure 33.1.)

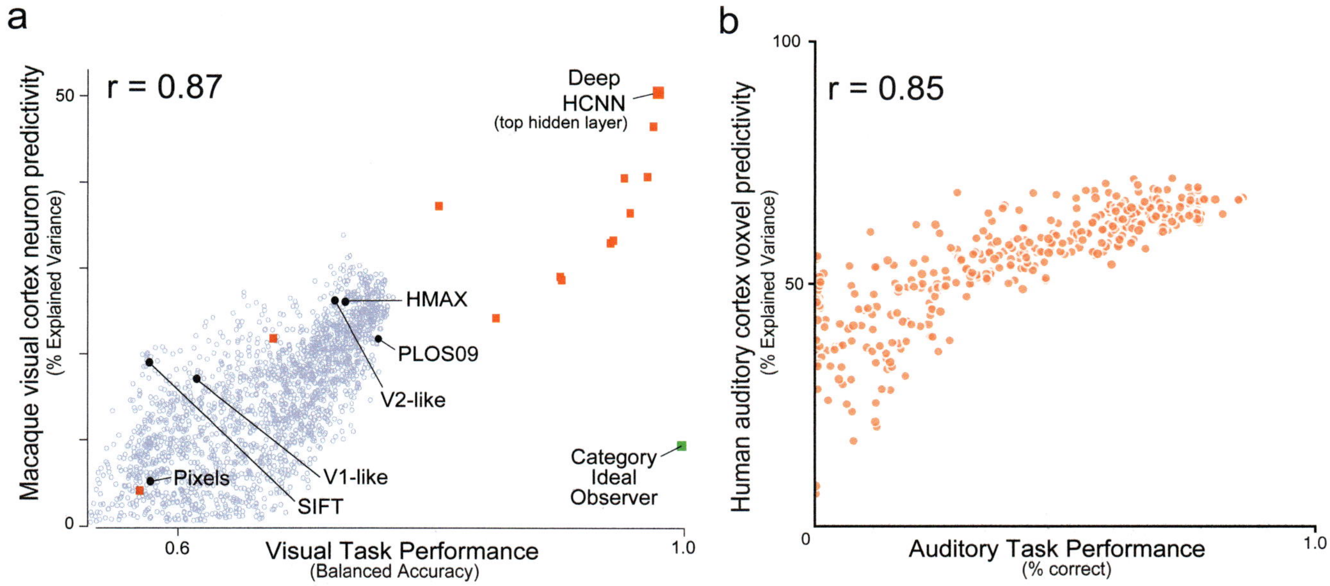

PLATE 36 *A*, Visual object categorization task performance (***x***-axis) is highly correlated with the ability to predict IT cortex neural responses (***y***-axis). Adapted from Yamins et al. (2014). *Blue dots*, Various three-layer (shallow) HCNN models, either with random weights or optimized for categorization performance or to predict IT responses. *Black squares*, A variety of previous models. *Red dots*, Increasing performance and predictivity over time as a deep HCNN is trained. ***r*** value is for red and black points. *Green square*, A category ideal observer with perfect semantic category knowledge, to control for how much neural variance is explained just by categorical features alone. *B*, Analogous result for neural networks optimized for auditory tasks. Adapted from Kell et al. (2018). (See figure 33.2.)

PLATE 37 *A* and *B*, How would you pick up the apples indicated while maintaining a stable arrangement of the other objects? It is easy to see that you will likely need to touch more objects (and probably use two hands) in panel (*A*), while the apple in panel (*B*) can be removed on its own with just one hand. *C* and *D*, What is where? Semantic segmentation maps showing class labels and locations of objects from panels (*A* and *B*). *E*, A child playing with stacking cups. Screenshot from https://www.youtube.com/watch?v=dEnDjyWHN4A. *F*, An orangutan building a tower with large Lego-like blocks. Screenshot from https://www.youtube.com/watch?v=MxRJjzSY_JE&t=21s. (See figure 34.1.)

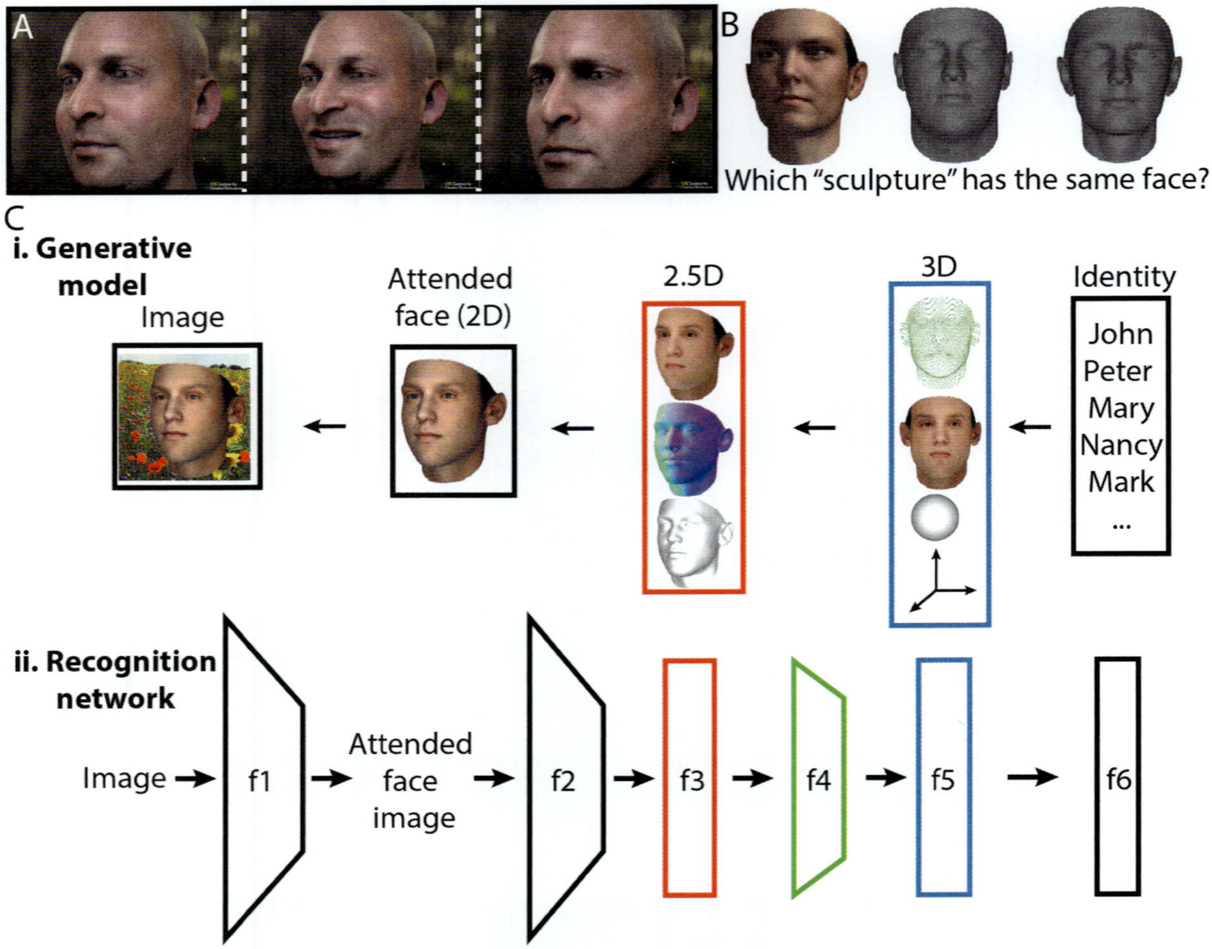

PLATE 38 *A*, Samples from a modern 3-D graphics model of a human face, yielding near photorealistic images (Credit: NVIDIA and University of Southern California Institute for Creative Technologies). Across the three images of this face, in addition to knowing that identity is preserved, we can also appreciate the details of the face's 3-D shape and texture, the subtleties of expression, that vary or remain constant across images. *B*, Despite their unfamiliarity, most observers can match the identity of the naturalistic face on the left to one of the textureless faces ("sculptures"), which must rely on a sense of 3-D shape. *C*, Schematic of the efficient A×S approach, including a probabilistic generative model of face image formation (*panel i*) and the recognition network (*panel ii*). Layers f1 through f6 indicate the different components of the recognition network. Trapezoids show single or multiple layers of transformations where a layer can consist of convolution, normalization, and a nonlinear activation function. Yildirim et al. (2019) found that transformations across the model layers f3, f4, and f5 closely captured the transformations observed in the neural data from ML/MF (middle lateral and middle fundus areas) to AL (anterior lateral area) to AM (anterior medial area; Freiwald & Tsao, 2010). (See figure 34.4.)

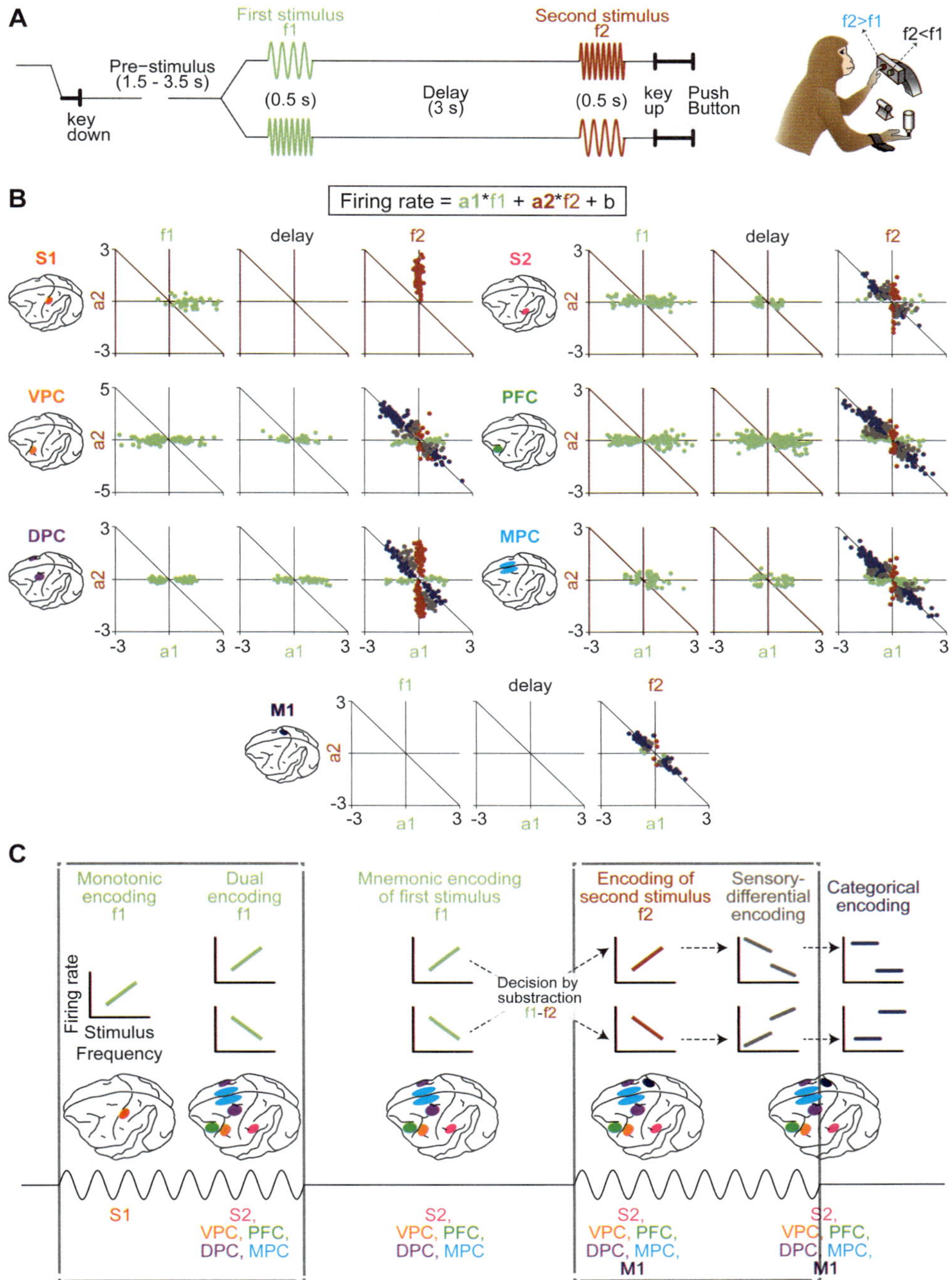

PLATE 39 Vibrotactile discrimination task. *A*, Sequence of events during a frequency discrimination task. As for the detection task (figure 35.1*A*), a trial began when the mechanical probe indented the fingertip of the left restrained hand, triggering the monkey to place its right free hand on an immovable key (key down [kd]). After a variable prestimulus period (1.5–3.5), the probe oscillates vertically at the base stimulus frequency (f1, 10–34 Hz, 0.5 s), and following a delay of 3 s, a second mechanical vibration is delivered at the comparison frequency (f2, 10–34 Hz, 0.5 s). At the end of the second stimulus, the monkey releases the key (key up) and presses either the medial or lateral push-buttons to indicate whether the comparison frequency was lower ($f2 < f1$) or higher ($f2 > f1$) than the base frequency. *B*, Single-neuron dynamics across cortical areas during a discrimination task. For each neuron, responses were fitted to the equation: firing rate $= a1 \times f1 + a2 \times f2 + b$, where f1 is the base stimulus frequency, f2 is the comparison stimulus frequency, and a1, a2, and b are coefficients. Each data point corresponds to one neuron with at least one significant coefficient ($a1 \neq 0$, $a2 \neq 0$, or both are different from zero, $p < 0.05$) evaluated by 200 ms bins. Each panel shows the highest coefficients from each significant neuron coding during three different epochs: the first stimulus period (f1, 0.5 s), the delay between f1 and f2 (delay, 3 s), and the second stimulus period (f2, 0.5 s). Green and red circles correspond to those neurons whose responses depend on f1 only ($a1 \neq 0$, $a2 = 0$; *dots on the abscissa axis*) or on f2 only ($a1 = 0$, $a2 \neq 0$; *red dots on the ordinate axis*), respectively.

(continued)

PLATE 39 *(continued)*

Gray circles correspond to neurons with both significant coefficients of opposite signs (a1 > 0 and a2 < 0; a1 < 0 and a2 > 0) but significantly different magnitudes (a1 ≠ −a2); these responses are classified as partially differential neurons (*dots between the diagonal and the ordinate or abscissa axis*). Blue circles correspond to neurons with both significant coefficients (a1 ≠ 0 and a2 ≠ 0) but opposite signs and statistically equal magnitudes (a1 = −a2); these responses encode f2–f1 in a categorical or fully differential manner (*dots on the diagonal*). *C*, Schematic representation of the coding engagement across cortex during a flutter discrimination task. The macaque brain diagrams depict the activated cortical areas during the three main periods of the task: f1, working memory delay, and f2 (*bottom traces, not drawn to scale*). The primary somatosensory cortex (S1) encodes f1 only through positive monotonic responses (see also panel *B*). This sensory information is distributed to other cortical areas and transformed in such a way that opposite representations (dual encoding) arise in secondary somatosensory cortex (S2), ventral, dorsal, and medial premotor cortices (VPC, DPC, and MPC, respectively) and prefrontal cortex (PFC). Dual representations of the first stimulus (f1) are encoded during the delay period (mnemonic representation) by the sustained activity on the parietal and frontal lobe cortical areas. After the working memory period, the frequency of the second stimulus (f2) is encoded in all recorded areas of the frontal and parietal lobes, including the primary motor cortex (M1). Importantly, at the end of the second stimulus, neurons with decision-related activity in frontal and parietal cortical areas reflected the difference in frequency between the two stimuli frequencies (f2–f1). This decision-related activity could arise by subtracting the firing rates of neurons encoding f1 and f2 with opposite tuning. Note that with the exception of S1, neurons with opposite tuning are observed across all recorded areas. Adapted from Romo and de Lafuente (2013); Romo and Salinas (2003). (See figure 35.2.)

PLATE 40 The proposed human prefrontal function. *A*, Schematic representations of the human cerebral cortex showing the prefrontal cortex (PFC). Compared to monkeys, the human cortex comprises a frontopolar region (poPFC) in the lateral forefront of the PFC with no known homolog regions in monkeys. Additionally, the caudal sector in the lateral PFC witnesses the development of Brodman's area (BA) 44 and 45, yielding to the notion of Broca's area in the left hemisphere. In humans compared to monkeys, task sets thus comprise two nested, abstract levels of chunking involving BA 44 and 45 and playing a major role in language. *B*, Inferential, inhibition, and selection processes forming the human PFC function. Compared to monkeys (see figure 38.2), the human poPFC allows for inferring and monitoring the reliability of a few task sets comprising the actor and three/four counterfactual (i.e., unreliable and not driving current behavior) task sets. This inferential buffer enables the middle lateral PFC to directly select/retrieve a counterfactual task set as an actor when it becomes reliable. *C*, Diagram showing the transitions between exploitation and exploration periods corresponding to creating a new actor task set *p* when no monitored task sets are reliable, to reject and disband newly created actor *p* during exploration periods when one counterfactual task set again becomes reliable to serve as actor, or to confirm and consolidate newly created actor *p* in long-term memory. These transitions realize hypothesis testing has a bearing upon task set creation. See the text for an explanation. (See figure 38.3.)

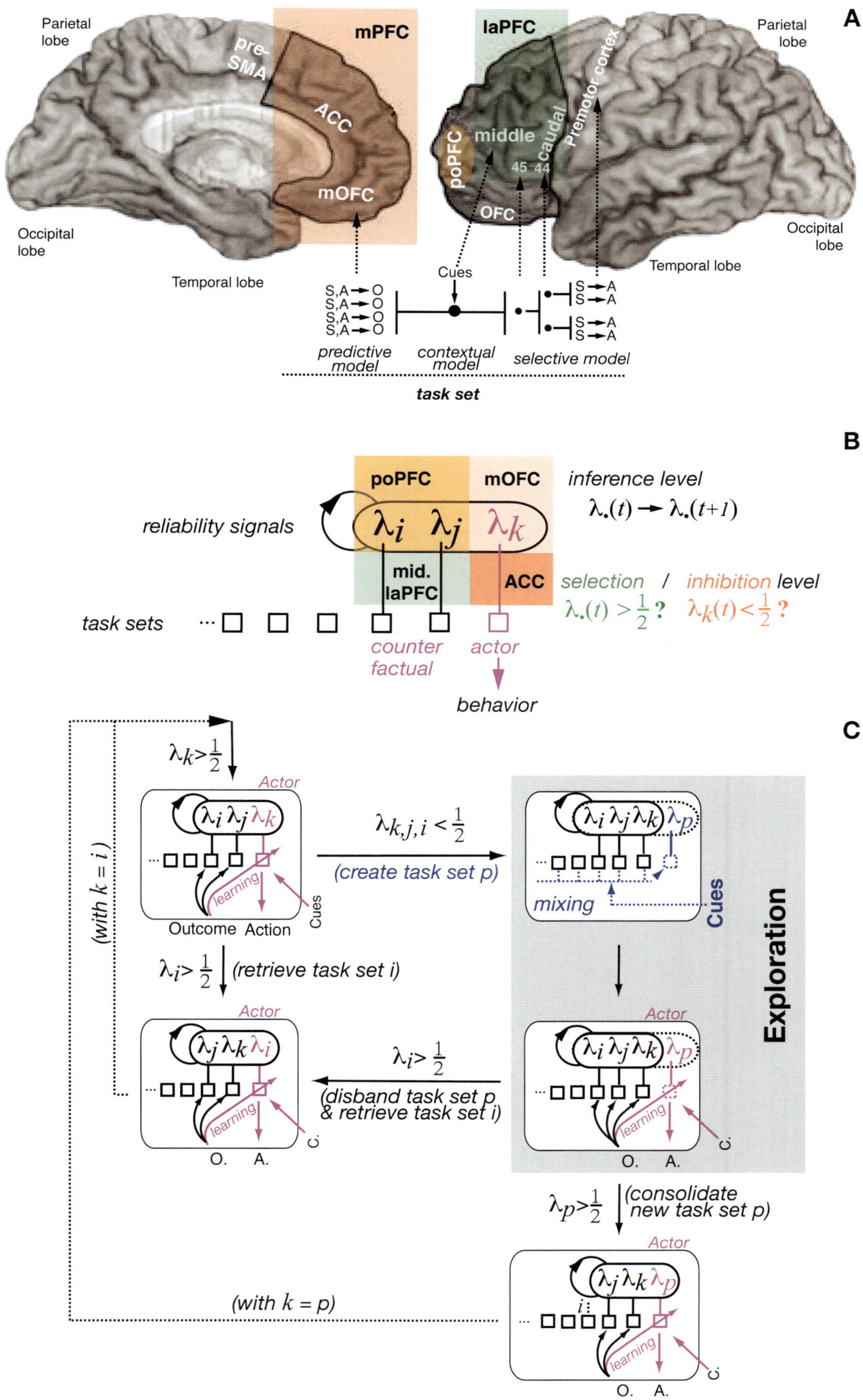

A
Parietal lobe
pre-SMA
mPFC
ACC
mOFC
Occipital lobe
Temporal lobe
laPFC
poPFC
middle
caudal
45
44
OFC
Premotor cortex
Parietal lobe
Occipital lobe
Temporal lobe
Cues
S,A → O
S → A
predictive model
contextual model
selective model
task set
B
poPFC
mOFC
inference level
reliability signals
mid. laPFC
ACC
selection / inhibition level
task sets
counter factual
actor
behavior
C
Actor
(with k = i)
Outcome Action
Cues
(create task set p)
mixing
Cues
Exploration
(retrieve task set i)
(disband task set p & retrieve task set i)
(consolidate new task set p)
(with k = p)
learning
O. A.

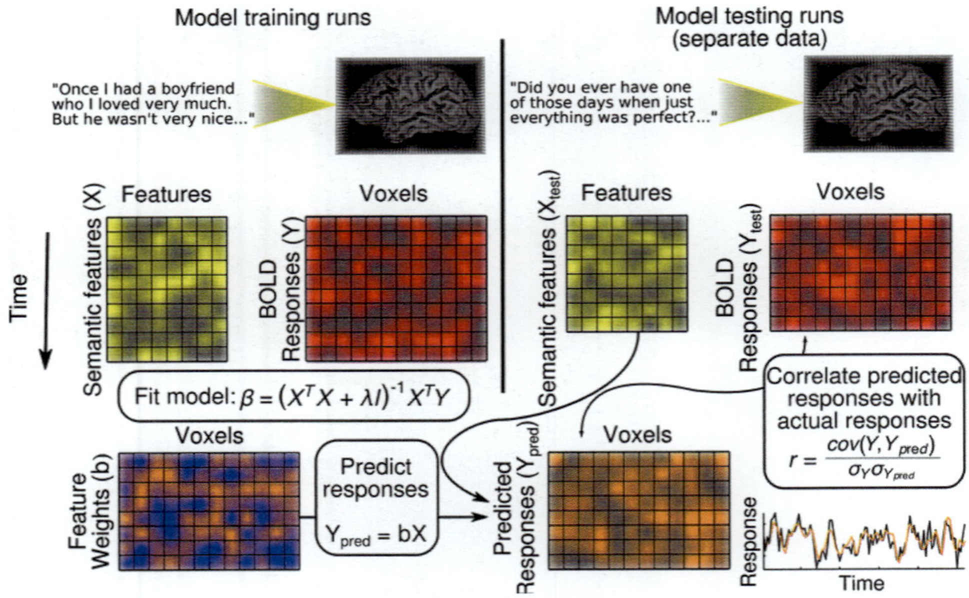

PLATE 41 Voxelwise modeling procedure. Functional MRI data are recorded while subjects listen to natural stories or watch natural movies. These data are separated into two sets: a training set used to fit voxelwise models and a separate test set used to validate the fit models. Semantic features are extracted from the stimuli in each data set. *Left*, For each separate voxel, ridge regression is used to find a model that explains recorded brain activity as a weighted sum of the semantic features in the stories. *Right*, Prediction accuracy of the fit voxelwise models is assessed by using the model weights obtained in the previous step to predict voxel responses to the testing data and then comparing the predictions of the fit models to the obtained brain activity. Statistical significance of predictions and of specific model coefficients is assessed through permutation testing. (See figure 39.1.)

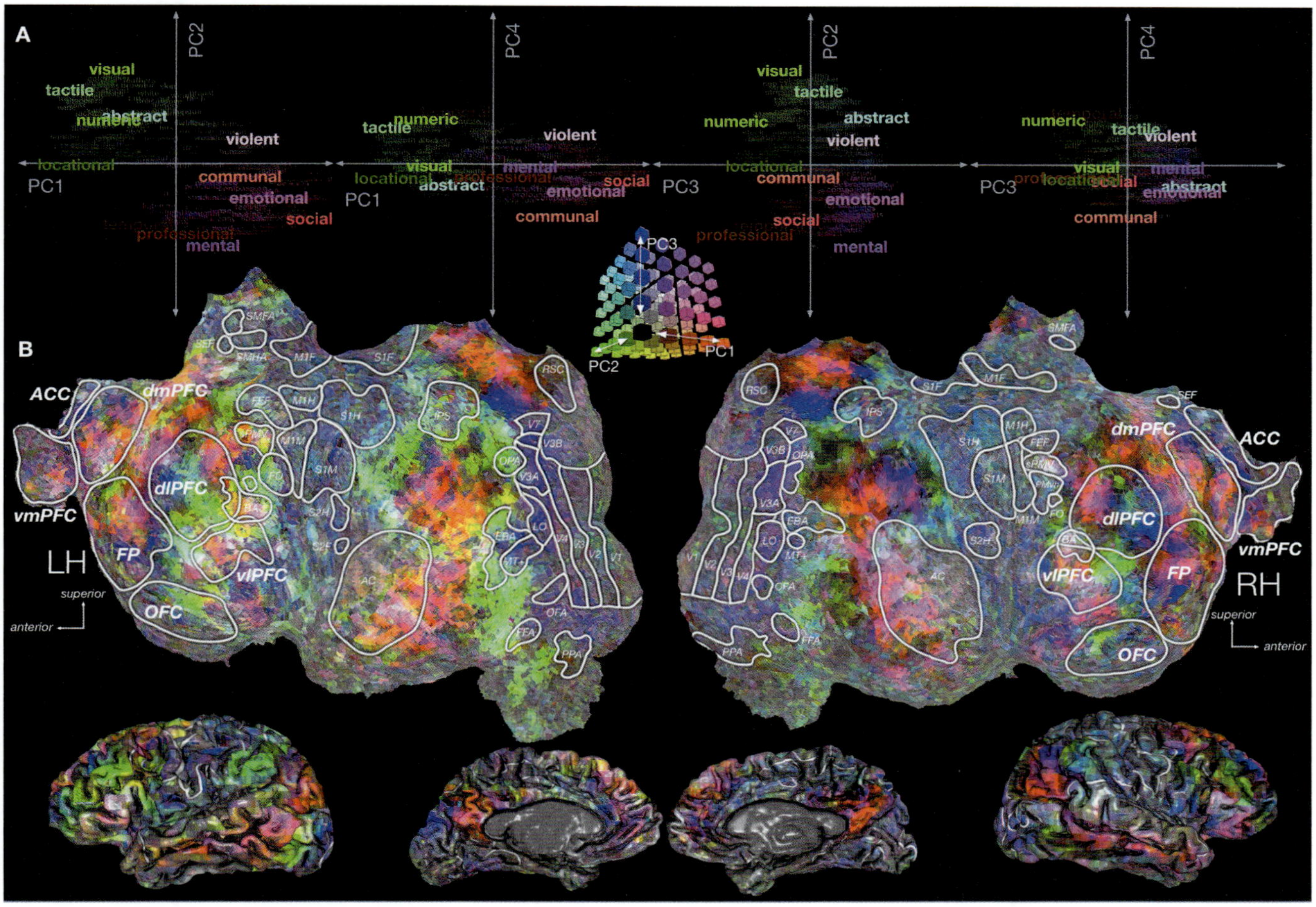

PLATE 42 Semantic maps obtained from subjects who listened to narrative stories. Principal components analysis of voxelwise model weights reveals four important semantic dimensions in the brain. *A*, A Red, Green, Blue (RGB) color map was used to color both words and voxels based on the first three dimensions of the semantic space. Words that best matched the four semantic dimensions were found and then collapsed into 12 categories using k-means clustering. Each category was manually assigned a label. The 12 category labels (large words) and a selection of the 458 best words (small words) are plotted here along four pairs of semantic dimensions. The largest axis of variation lies roughly along the first dimension and separates perceptual and physical categories (tactile, locational) from human-related categories (social, emotional, violent). *B*, Voxelwise model weights were projected onto the semantic dimensions and then colored using the same RGB color map. Projections for one subject (S2) are shown on that subject's cortical surface. Semantic information seems to be represented in intricate patterns across much of the semantic system. White lines show conventional anatomical and/or functional ROIs. Labeled ROIs in prefrontal cortex reflect the typical anatomical parcellation into seven broad regions: dorsolateral prefrontal cortex (dlPFC), ventrolateral prefrontal cortex (vlPFC), dorsomedial prefrontal cortex (dmPFC), ventromedial prefrontal cortex (vmPFC), orbitofrontal cortex (OFC), anterior cingulate cortex (ACC), and the frontal pole (FP). Each of these conventional prefrontal ROIs contains multiple semantic domains, suggesting that the role of prefrontal cortex in semantic comprehension is more complicated than the current cognitive-control view would suggest. Reproduced and modified from Huth et al. (2016). (See figure 39.2.)

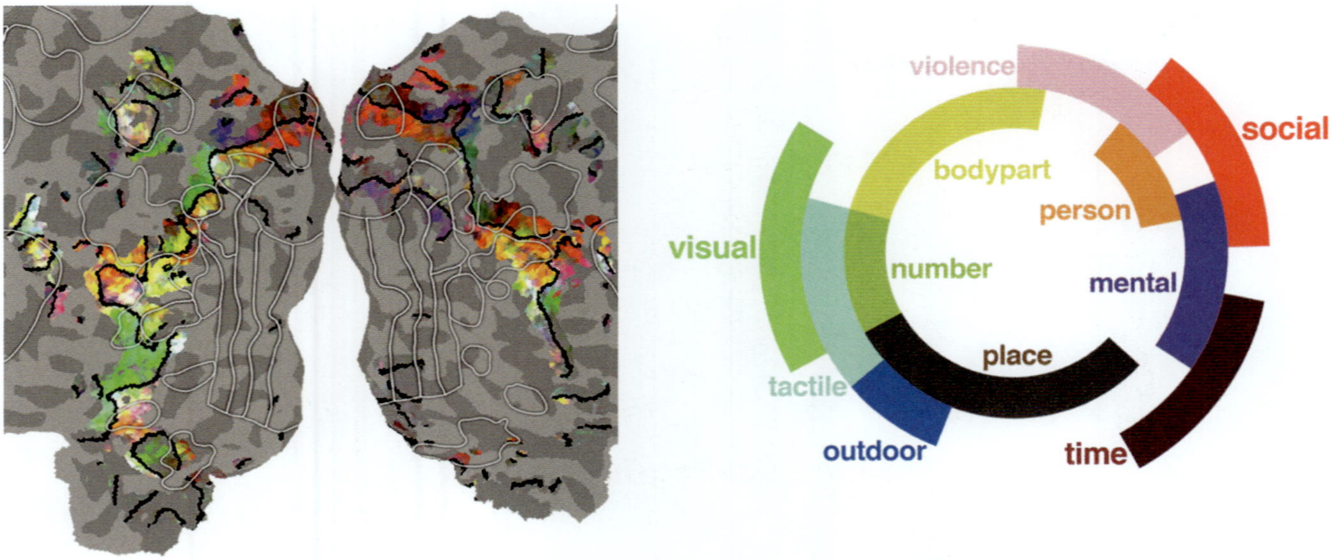

PLATE 43 Relationship between visual and linguistic semantic representations along the boundary of visual cortex. The black boundary indicates the border between cortical regions activated by brief movie clips versus stories. Voxels posterior to the boundary (i.e., nearer the center of the figure) are activated by movie clips but not stories. Voxels anterior to the border are activated by stories but not movie clips. Each of the voxels activated by only one modality is colored based on fit model weights that indicate the semantic category for which it is selective (legend at right; data from Huth et al. [2012] and Huth et al. [2016]). For almost all semantic concepts, the semantic selectivity of voxels posterior to the boundary is similar to the semantic selectivity of voxels anterior to the boundary. The only exception seems to be "mental" concepts (purple voxels located in the dorsal region of the boundary in the right hemisphere), which appear to be represented only in the stories. However, these concepts were not labeled explicitly in the movies and therefore cannot be found in the visual semantic map. (See figure 39.3.)

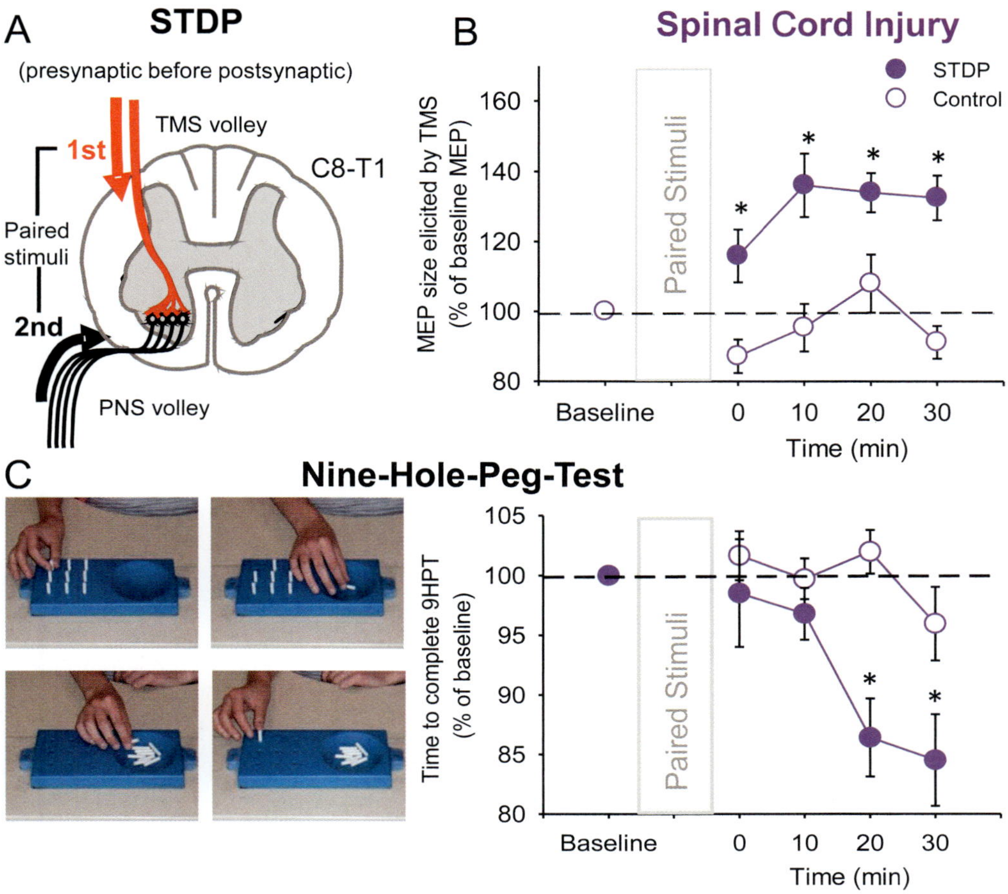

PLATE 44 Paired corticomotoneuronal stimulation (PCMS). *A*, Illustration of the PCMS protocol used to enhance corticospinal function after SCI. Here, corticospinal neurons were activated at a cortical level by using TMS (TMS volley, first) delivered over the hand motor cortex, and spinal motoneurons were activated antidromically by peripheral nerve stimulation (PNS volley, second) delivered to the ulnar nerve. *B*, MEP amplitude increased after PCMS in which postsynaptic pulses were timed to arrive at the synapses 5 ms before presynaptic activation. *C*, Improvement in hand function as assessed with a nine-hold-peg test in 18 participants with chronic cervical SCI. Error bars, SEs, $*p<0.05$. Modified from Bunday and Perez (2012b). (See figure 40.3.)

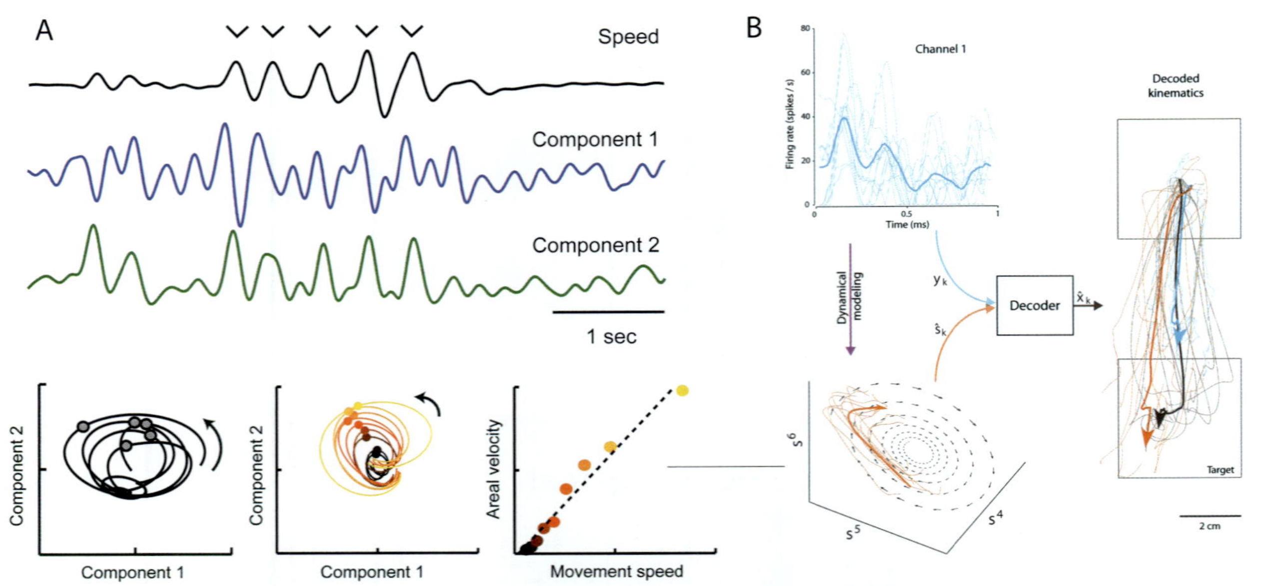

PLATE 45 Neural decoding exploiting dynamic constraints. *A, Top panel,* Movement speed and principal components of associated field potentials recorded from primary motor cortex during an isometric tracking task. Note the regular submovements (indicated by ticks). *Bottom left,* When plotted as a trajectory in principal component space, each submovement is associated with a single cycle with a consistent direction and angular frequency. *Bottom right,* Faster submovements (*yellow traces*) are associated with larger cycles, so cursor speed can be decoded from the areal velocity of trajectories. Adapted from Hall, de Carvalho, and Jackson (2014). *B,* Decoding movement kinematics is improved by incorporating a model of consistent network dynamics. Neural signals (*light blue*) are considered to be noisy observations of a low-dimensional dynamical neural state (*orange*). Movements decoded from this inferred state provide a better match to the movements produced by the subject (*black lines*). Reproduced from Kao et al. (2015). (See figure 41.2.)

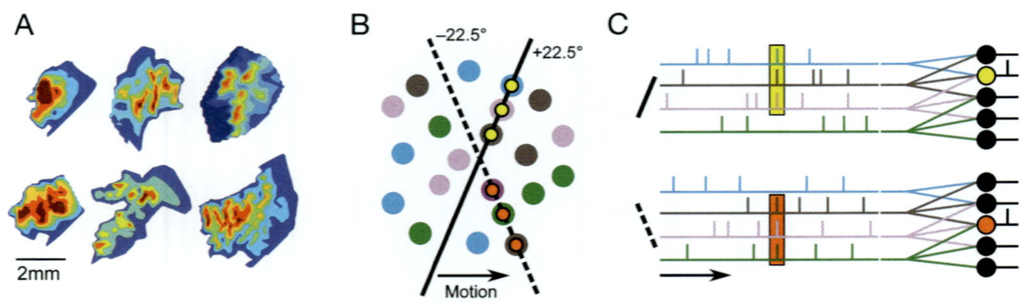

PLATE 46 *A,* Exemplar receptive fields for first-order tactile neurons. The colors reflect sensitivity to a small indentation at that location on the skin. Note that receptive fields are complex, showing many zones of high sensitivity. *B,* Diagram showing the receptive field of four first-order tactile neurons (*color-coded ellipses*), each with multiple highly sensitive zones (*color-coded dots*). Solid and dashed lines illustrate differently oriented edges. *C,* Schematic of how the neurons in (*B*) will respond as the oriented edges move across the skin. Note that edge orientation influences how each neuron responds because of the distribution of its highly sensitive zones. At a given instance, one edge simultaneously stimulates one set of neurons, and the other edge stimulates another set of neurons (*yellow and orange rectangles*). These different sets of coincident action potentials cause responses in different higher-order neurons (*yellow and orange*), propagating edge-orientation information upstream. Reproduced with permission from Pruszynski and Johansson (2014). (See figure 42.3.)

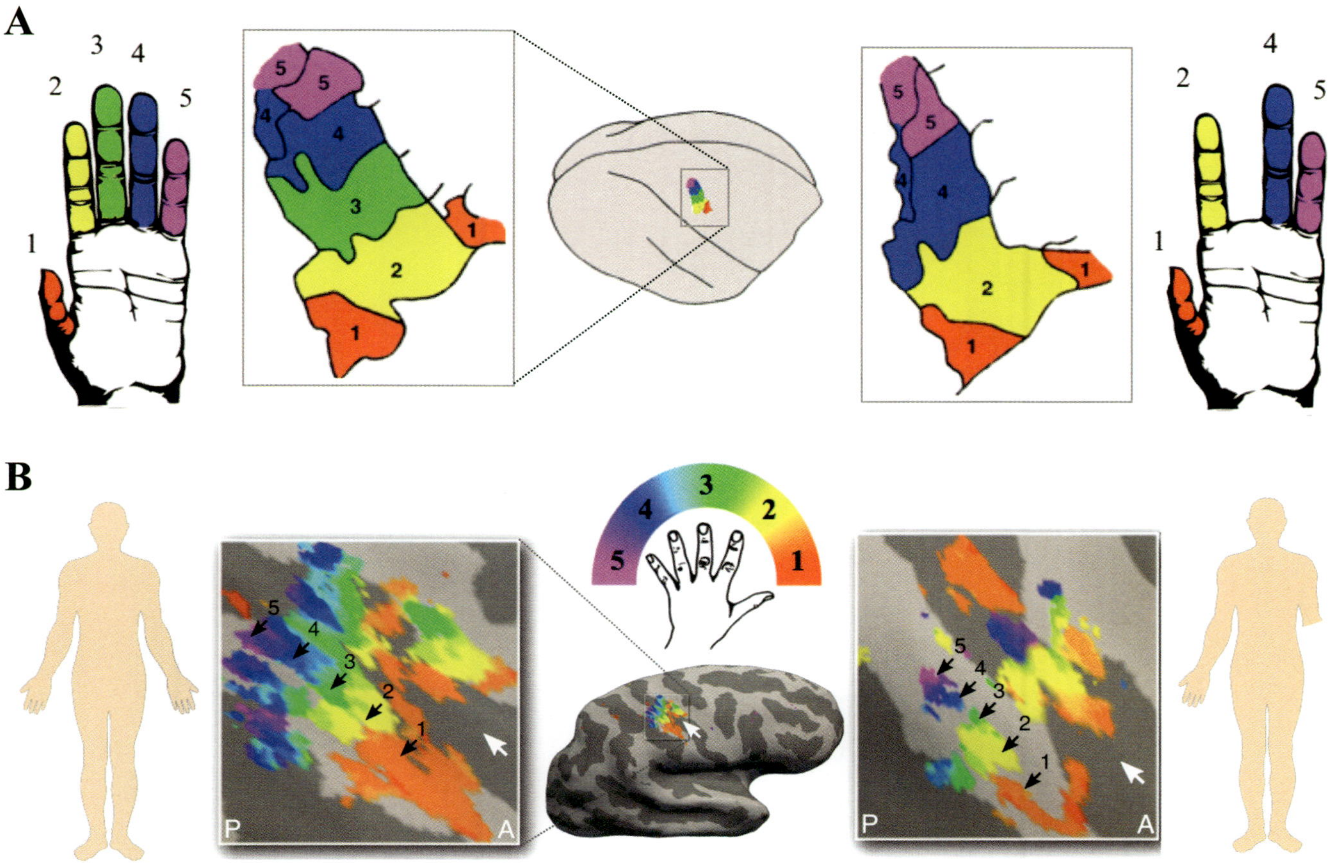

PLATE 47 Digit maps in the monkey (*A*) and human (*B*). The left panels show characteristic maps of intact hands. The right panels show digit maps following the amputation of one digit (*A*) or the entire arm (*B*). *A*, Modified from Merzenich et al. (1984). *B*, Modified from Kikkert et al. (2016). (See figure 43.2.)

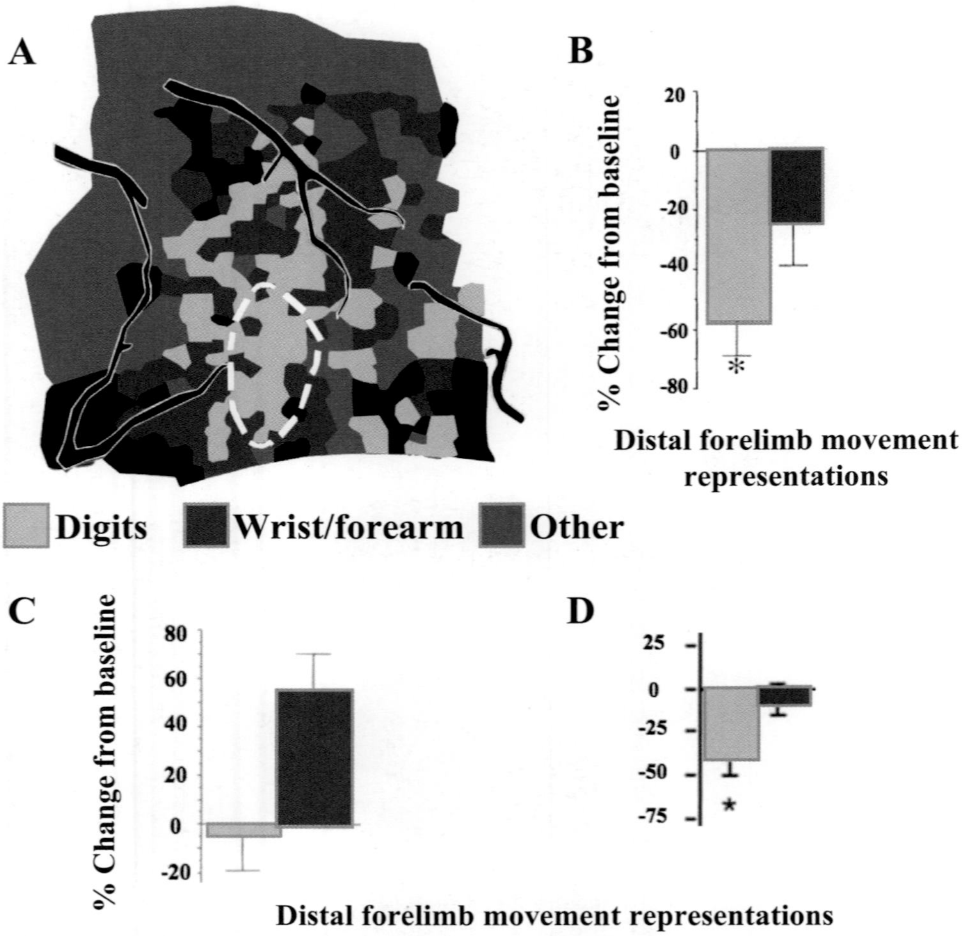

PLATE 48 Remapping of the monkey hand and arm representation following focal M1 infarcts. *A*, Hand representation prior to M1 lesion (*shown by white line*). Remapping of hand representation after lesion to the digit area following (*B*) spontaneous recovery, (*C*) training-based recovery, and (*D*) delayed training. Despite the different remapping profiles, the monkeys regained similar digit function. The y-axis shows the change in motor map area devoted to the distal forelimb following the infarct and training. *A–B*, Modified from Nudo and Milliken (1996). *C*, Modified from Nudo et al. (1996). *D*, Modified from Barbay et al. (2006). (See figure 43.3.)

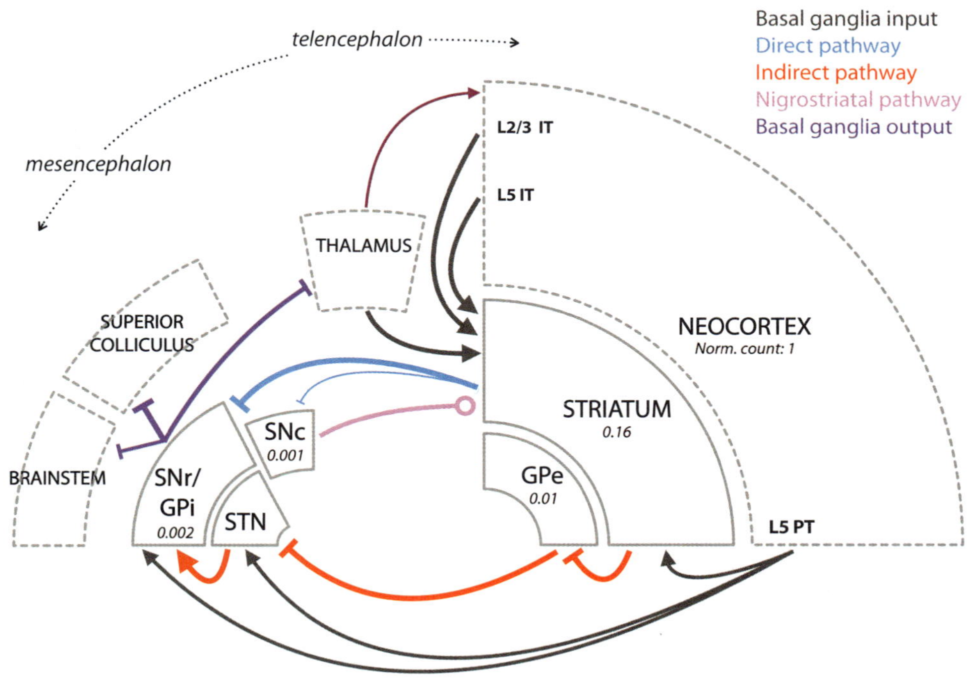

PLATE 49 Functional anatomy of mammalian basal ganglia circuits. The organization of the major pathways of the dorsal basal ganglia is indicated by arrows for predominantly excitatory projections, lines for inhibitory projections, and an open circle for dopaminergic, modulatory projections. Detailed reviews of internal circuitry and topographical organization can be found in Dudman and Gerfen (2015). The weights of arrows reflect the rough density of projection. Numbers in italics reflect approximate normalized cell counts for several key nuclei normalized to the number of cortical inputs and derived from the best stereological counts taken in rats (Oorschot, 1996). *Abbreviations*: globus pallidus external, GPe; globus pallidus internal, GPi; intratelencephalic, IT; layer 5, L5; layer 2/3, L2/3; pyramidal tract, PT; substantia nigra pars compacta, SNc; substantia nigra pars reticulata, SNr; subthalamic nucleus, STN. (See figure 44.1.)

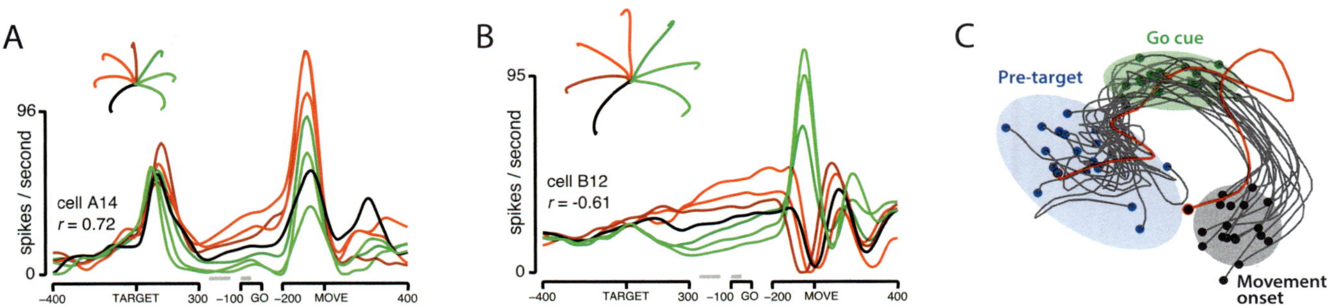

PLATE 50 Neural mechanisms of movement preparation. *A, B*, Preparatory and movement-related activity in two example neurons from motor cortex. Reproduced from Churchland et al. (2010). Different colors indicate activity for different movements (*shown in inset*). *A*, A neuron whose preparatory activity (between TARGET and GO) is positively correlated (r = 0.72) with its activity during movement (MOVE). *B*, A neuron whose analogous correlation is negative (r = −0.61). *C*, At the population level (here, a two-dimensional view of overall population activity), neural activity is initially variable across trials before the target appears (*blue dots*) but converges to a specific state by the time of the GO cue (*green dots*). After the GO cue, neural activity follows a stereotyped activity profile. The thicker red curve indicates data from one outlying trial. Reproduced from Shenoy, Sahani, and Churchland (2013). (See figure 45.2.)

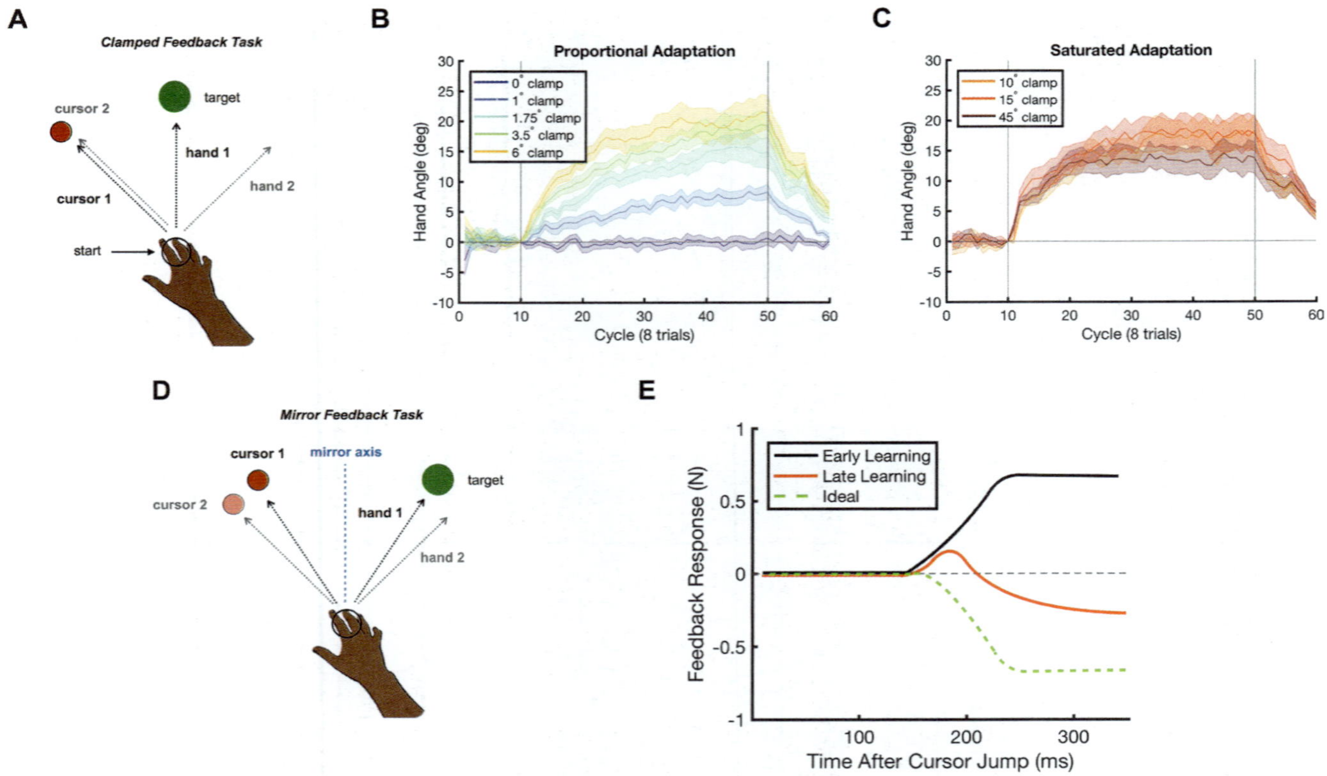

PLATE 51 *A*, Clamped visual feedback used to isolate implicit learning processes. Adapted from Morehead et al. (2017). The hand will adapt in response to a sensory prediction, even though performance error is clamped. *B*, A small range of sensory prediction errors show scaling effects on the rate of learning, but (*C*) large sensory prediction errors do not scale learning rates. Adapted from Kim et al. (2018). *D*, Schematic of mirror task used in Telgen, Parvin, and Diedrichsen (2014). Standard adaptation causes an increase in error, suggesting that the task must be learned in other ways. *E*, Short-latency feedback responses to a displaced cursor do not show ideal adaptation even after prolonged exposure to a mirror perturbation. Adapted from Telgen, Parvin, and Diedrichsen (2014). (See figure 46.2.)

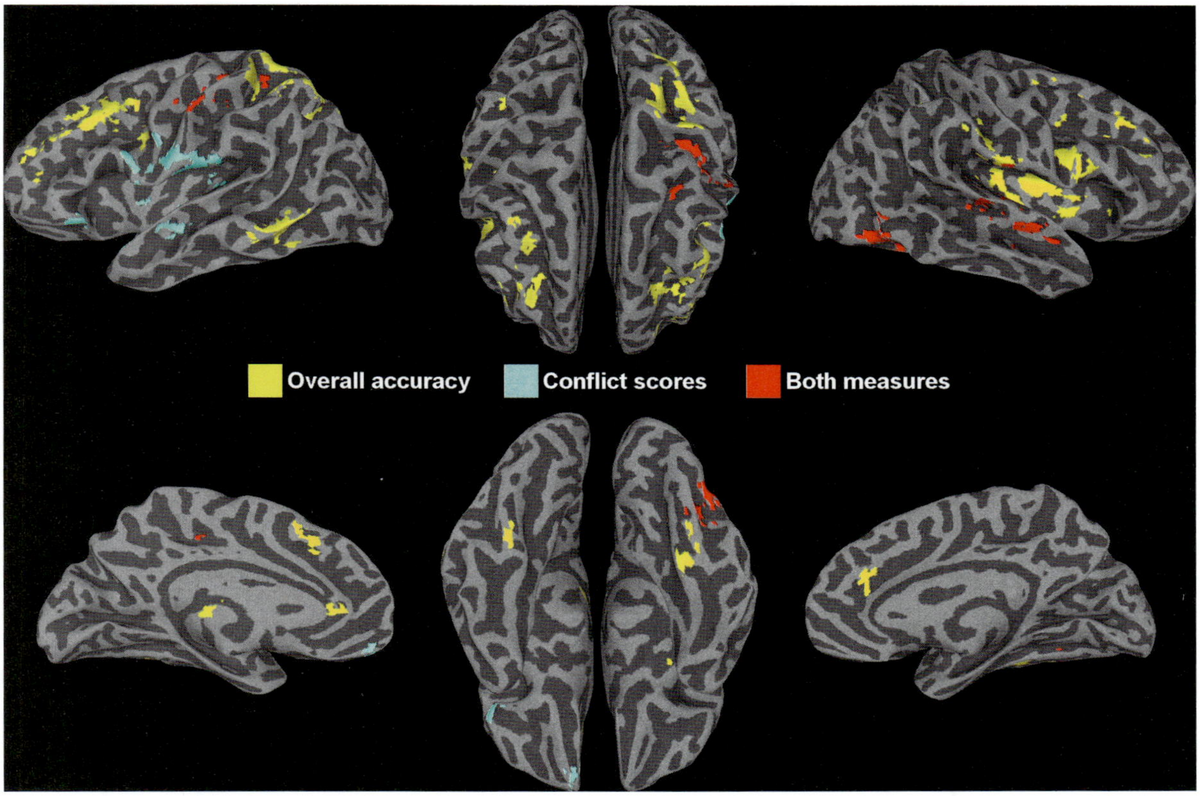

PLATE 52 Brain regions (standardized Talaraich space) whose resting state functional connectivity in left-hemisphere stroke patients is (1) significantly reduced as compared with age-matched control subjects and (2) correlated with performance on tool gesture pantomime (*light gray*), pantomime of "conflict" tools associated with more than one action (*medium gray*), or both measures (*dark gray*). Reproduced from Watson et al. (2019). (See figure 47.1.)

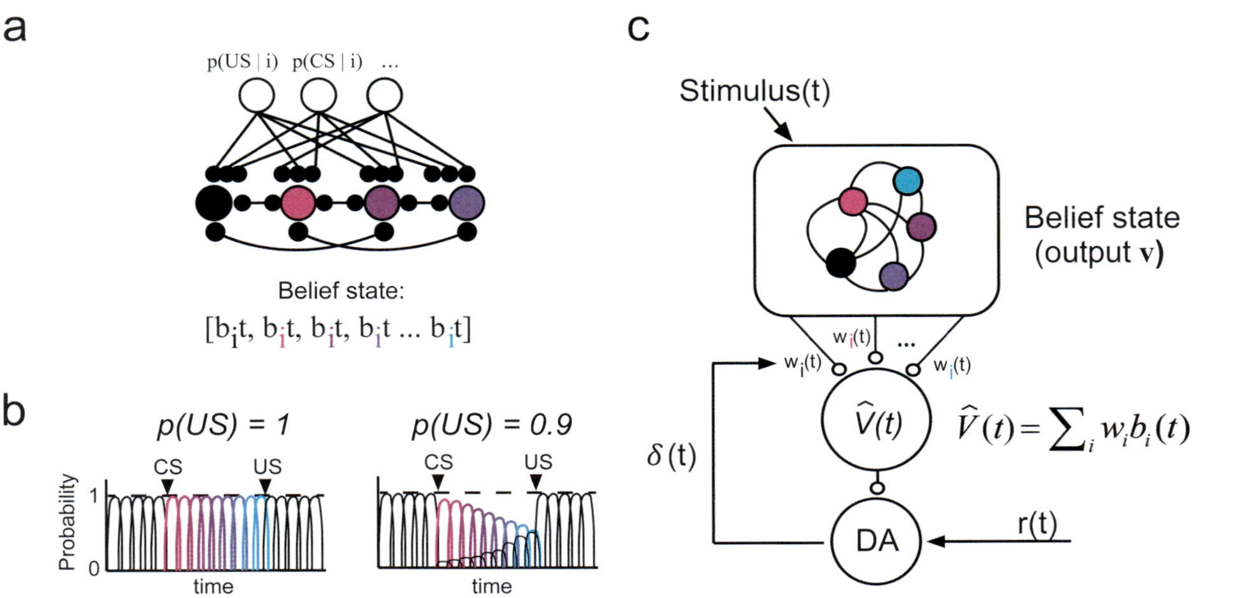

PLATE 53 Recurrent network implementation of a belief state. *A*, A simple network with feedforward connections and a recurrently connected layer, adapted to computing a belief state for the task described in figure 48.2*D*. *B*, Predicted firing rates for output layer neurons in the task described in figure 48.2*D* if these neurons each signal one value within the belief-state vector. *C*, Revised neural circuit hypothesis of TD-learning implementation. Belief-state features could be represented by a recurrent network. The outputs from this recurrent network, multiplied by their weights, would be linearly summed to produce the value estimate. (See figure 48.3.)

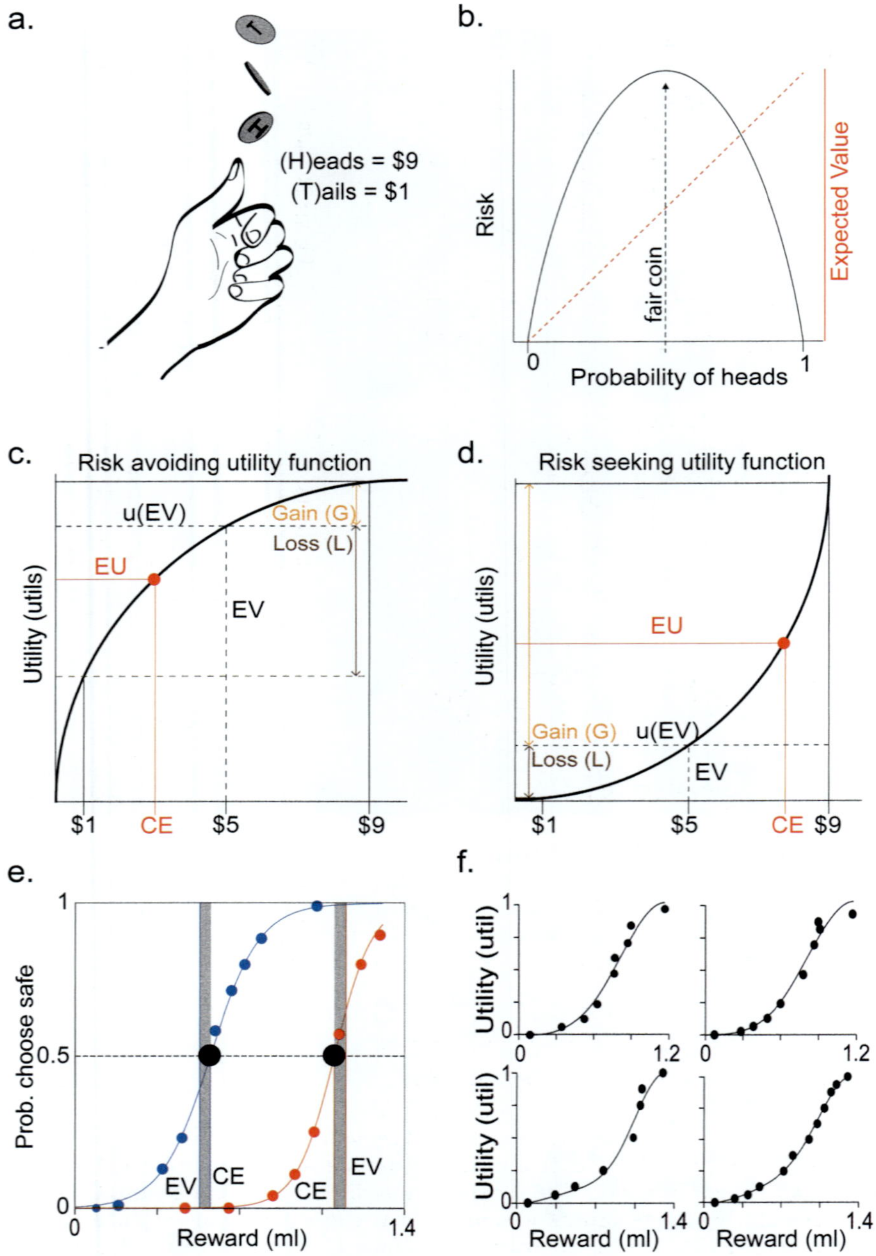

PLATE 54 Risk, value, and utility. *A*, Flipping a coin, where heads pays a large reward ($9) and tails returns a small reward ($1), constitutes an ideal model of the binary gambles we describe. *B*, For a Bernoulli process such as a weighted coin flip, this graph illustrates the relationship between risk, probability, and value. The black curve plateaus at $p=0.5$ (fair coin) and demonstrates the relationship between risk (formally: entropy $\sum_{i=1}^{n} -P(x_i)log_2 P(x_i)$) and the probability that a coin comes up heads. A fair coin has the most unpredictable outcome and therefore has the highest risk. The red dotted line demonstrates that expected value (EV) is a linear function of probability. *C*, A risk-avoiding (concave) utility function describes the behavior of an individual who would prefer a sure $5 to the fair coin flip illustrated in (*A*). The potential utility loss (L) is greater than the potential utility gain (G). Accordingly, the certainty equivalent (CE) is smaller than the gamble EV, and therefore the expected utility (EU) is smaller than the utility of the EV u(EV). *D*, A risk seeking (convex) utility function describes the behavior of an individual who would prefer the gamble described in (*A*) to a sure $5 payout. The potential utility gain (G) outweighs the potential loss (L). The CE is larger than the gamble EV, and therefore EU > u(EV). *E*, Psychometric measurement of monkey risk seeking toward a small EV gamble (*blue*) and risk avoiding for a large EV gamble (*red*). The large black dots represent the certainty equivalents (CE)—the points of choice indifference between the gambles and safe options. Note that the CE is larger than the EV for the small gamble and less than the EV for the large gamble. The red and blue dots represent the probability of choosing a safe reward over the gamble. The red and blue curves represent the fitted psychometric functions. The shaded regions show the risk premiums—the differences between the CE and the gamble EV—for the two gambles. Data adapted with permission from Genest, Stauffer, and Schultz (2016). *F*, Utility functions for four different monkeys reflect risk seeking for small rewards and risk aversion for larger rewards. The black dots represent CE from iterative gambles within the reward range (as in [*E*]). Data adapted with permission from Genest, Stauffer, and Schultz (2016) and Stauffer, Lak, and Schultz (2014). (See figure 49.2.)

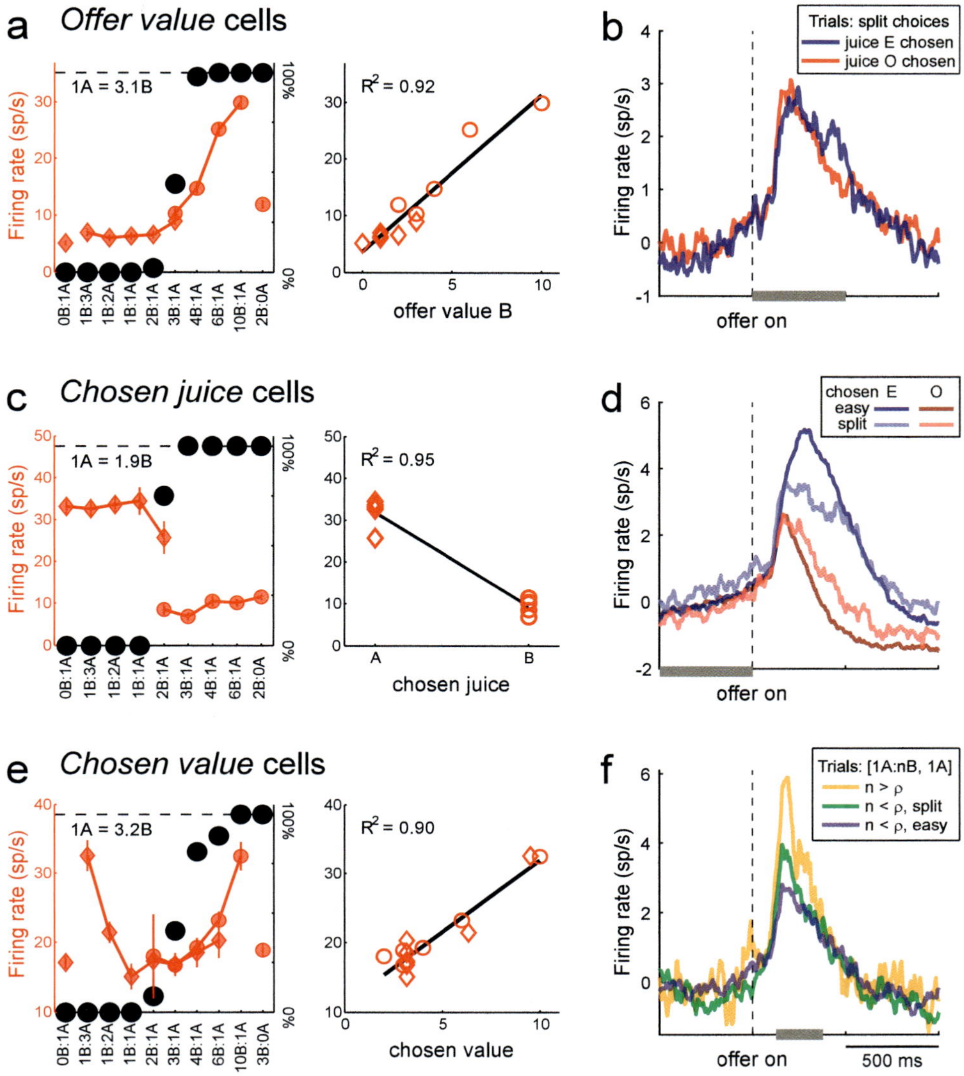

PLATE 55 Three groups of neurons in OFC. *A, C, E,* Example neurons recorded from OFC during a juice choice task. *Left,* Neuronal responses and choice behavior. The x-axis shows the offer types available during the recording session, ranked by the increasing ratio of #B/#A. The black dots represent the proportion of trials for each offer type in which the monkey chose juice B (choice behavior). A sigmoid fit of this data was used to determine the relative value of the two juices. Gray symbols show neuronal activity, with diamonds and circles indicating trials in which the animal chose juice A and juice B, respectively. *Right,* Neuronal response as a function of the encoded variable. *Offer value* and *chosen value* neurons respond to value in a linear way. Neurons shown encode (*A*) offer value A, (*C*) chosen juice A, and (*E*) chosen value. *B, D, F,* The time course of neuronal activity for different choice types. *B,* Activity fluctuations in offer value neurons. Traces show the average baseline-subtracted activity of offer value neurons for offer types in which a monkey's choices were split between juice A and juice B. Traces are separated based on whether the monkey chose the juice encoded by the neuron (juice E) or the other juice (juice O). The juice E is slightly elevated compared to the juice O trace in the time window following the offer presentation. *D,* Predictive activity of *chosen juice* cells. Traces show the average baseline-subtracted activity of chosen juice neurons. Activity was divided into four groups depending on whether the animal chose the encoded juice (juice E) or the other juice (juice O) and whether the decisions were easy (all choices for one of the two juices) or hard (decisions split between the two juices). For offers with split decisions, neuronal activity was slightly elevated before offer onset in trials in which the monkey chose the encoded juice. Separation may reflect residual activity from the previous trial as well as random fluctuations in neuronal activity. *F,* Activity overshooting in chosen value neurons. Traces show the average baseline-subtracted activity of a large number of chosen value cells, including only trials in which the monkey chose 1A. Activity is divided into three groups depending on whether the quantity of the nonchosen juice (n) was greater or less than the relative value of the two juices (ρ). Cases with $n < \rho$ were also separated based on whether the decision was easy or split. During the decision window (~200–450 ms after the offer), chosen value neurons show the greatest peak activity when n is higher, which corresponds to more difficult decisions. Adapted with permission from Padoa-Schioppa (2013). (See figure 50.2.)

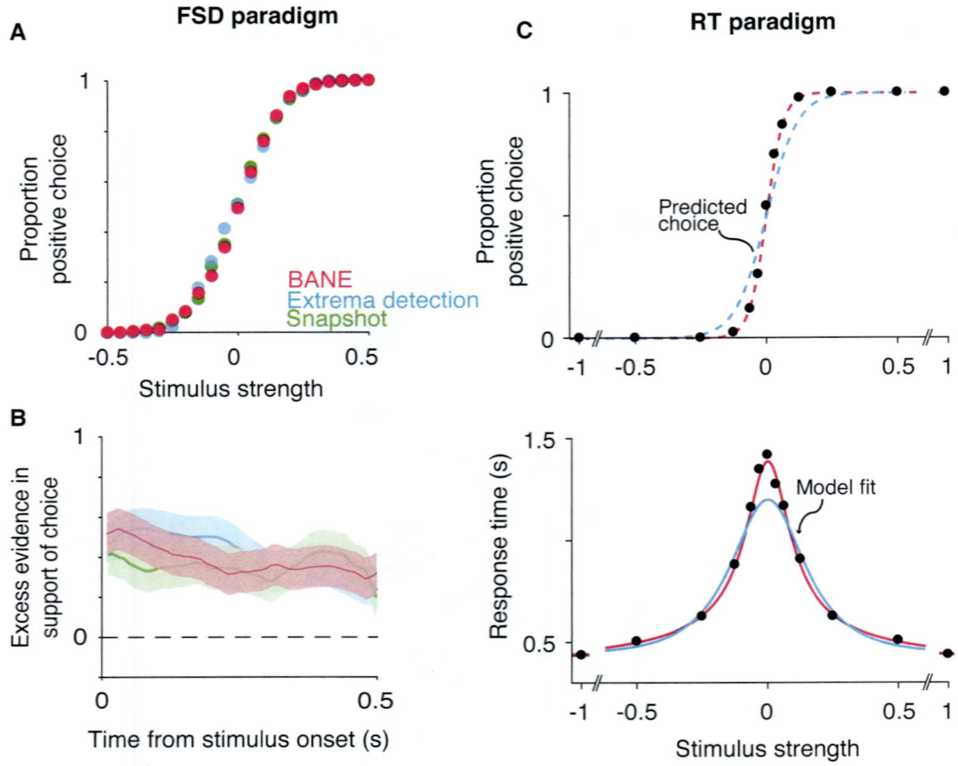

PLATE 56 Alternative strategies to evidence accumulation. *A*, Simulated choice data from three decision-making strategies in a fixed stimulus duration (FSD) paradigm. Colors correspond to different strategies. *B*, Simulated psychophysical kernels using the same parameters as in (*A*). Kernels are calculated by computing the choice-conditioned average of stimulus fluctuations at several time points during stimulus presentation. Note that all three strategies make similar predictions in both (*A*) and (*B*). Stimulus fluctuations affect choice at different times in different trials in FSD and extrema detection. In BANE, stimulus fluctuations affect choices across time in any one trial. The kernel analysis fails to reflect this important distinction. *C*, The choice-response time paradigm can disentangle the models. Choice-RT data (*symbols*) are simulated from a BANE model. RT means are fit with both a BANE model and an extrema detection model (*solid curves*). Snapshot is not shown because it furnishes no explanation of stimulus strength-dependent RTs. The dashed lines show the predicted choice data using the parameters obtained from the RT fits. Note that the predictions and RT fits of the extrema detection strategy are substantially worse than those of BANE. (See figure 51.3.)

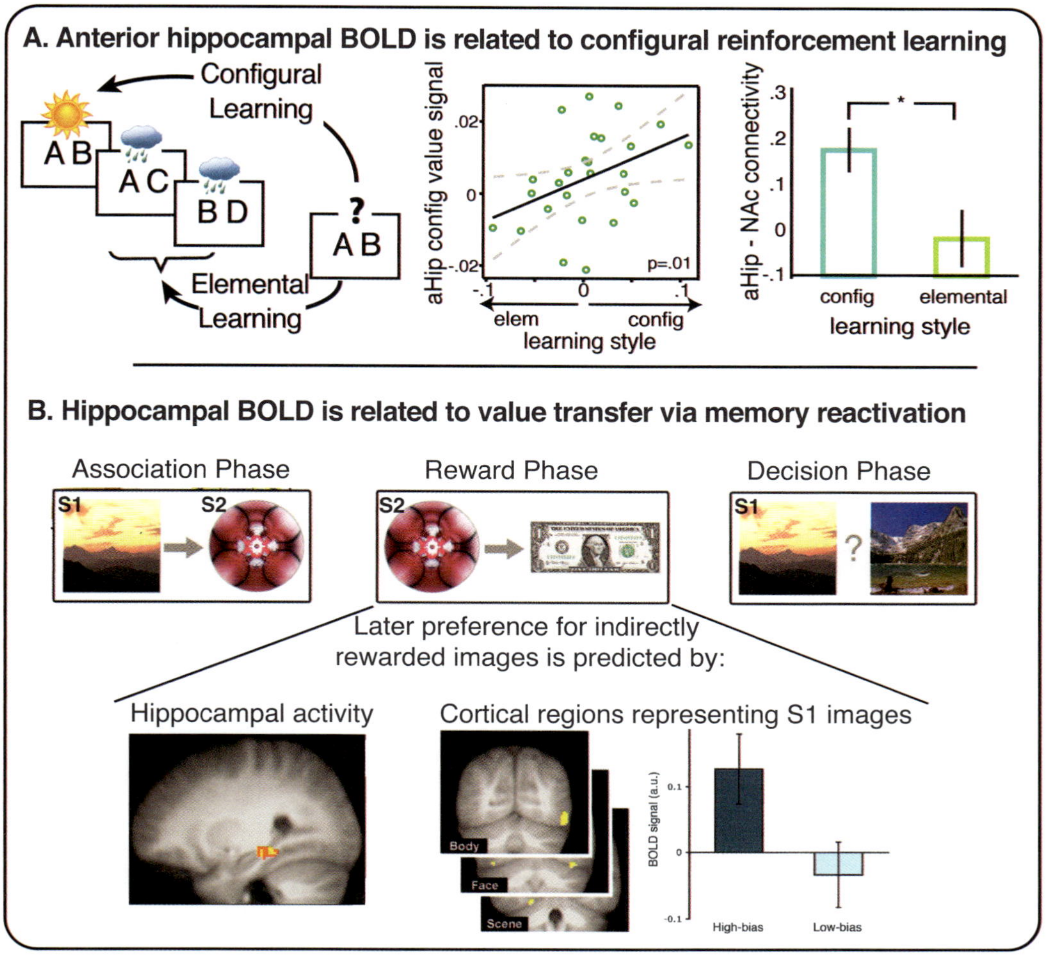

PLATE 57 fMRI evidence for hippocampal involvement in decisions that depend on relational processing. *A*, Participants made weather predictions using pairs of abstract cues in a probabilistic classification task. Reinforcement-learning models quantified the likelihood that choices were made using experience with configurations (e.g., AB) versus individual elements (e.g., A). BOLD responses in the anterior hippocampus (aHip) and functional connectivity between the aHip and the nucleus accumbens tracked the degree to which participants used configural learning. Adapted from Duncan et al. (2018). *B*, Participants performed a sensory-preconditioning task in which multiple S1-S2 pairs were first associated with each other. S2 stimuli were then either rewarded or not rewarded, and preferences for indirectly rewarded S1 stimuli were measured in the final *Decision Phase*. The transfer of value to S1 stimuli was related to BOLD activity during the reward phase in both the hippocampus and the category-specific visual area corresponding to the S1 stimulus' class (scene, face, or body part). Adapted from Wimmer and Shohamy (2012). (See figure 52.2.)

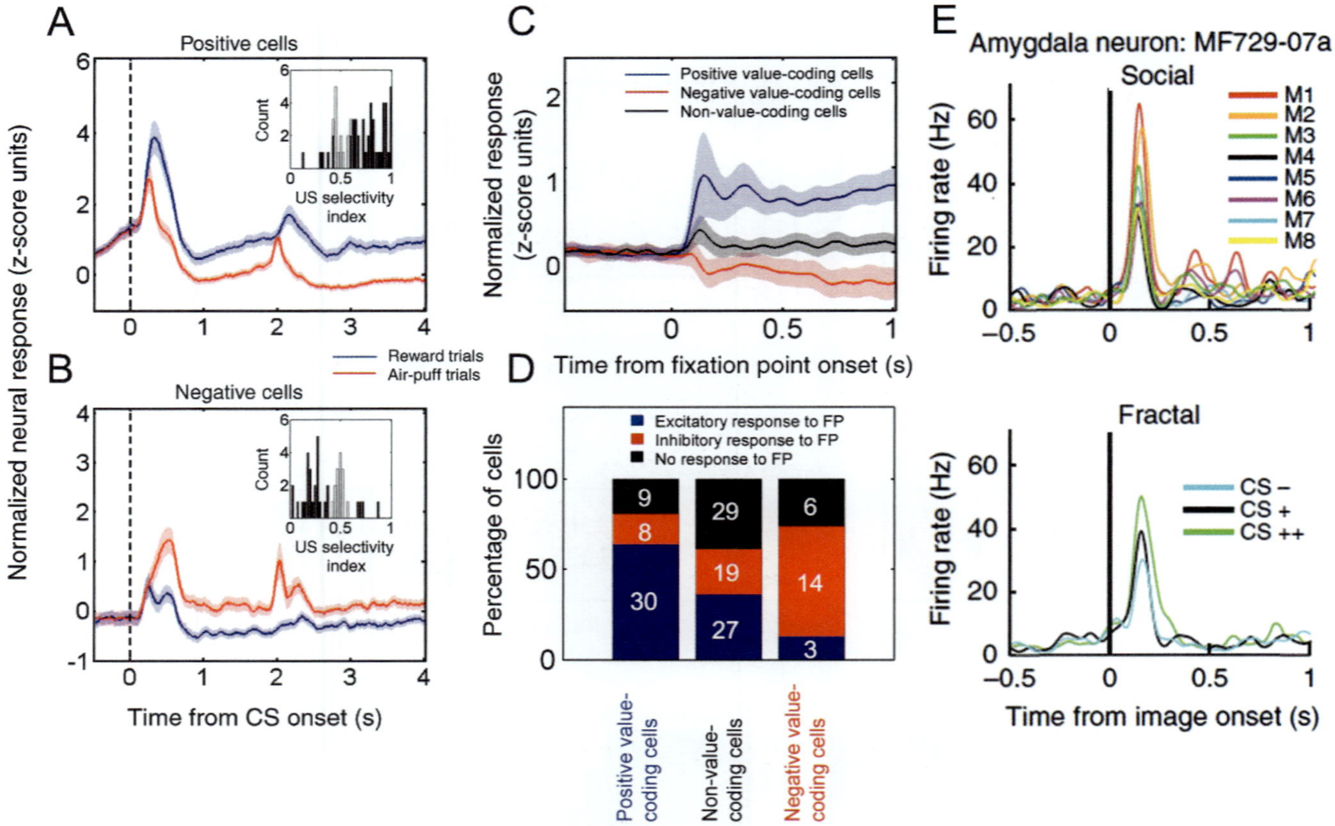

PLATE 58 The amygdala represents the positive and negative value of stimuli. *A, B*, Normalized and averaged neural responses plotted as a function of time relative to CS onset for neurons in the amygdala that respond more strongly to CSs associated with rewards (*A*) or air puffs (*B*). *Blue traces*, Responses in rewarded trials; *red traces*, responses in trials in which subjects received an aversive air puff. *Inset histograms*, Selectivity index characterizing the preference for expected reward or air puff, where values > 0.5 indicate preference for reward and < 0.5 preference for the expected air puff. *C*, Normalized and averaged neural responses to the fixation point presentation for neurons classified according to their selectivity for the valence of CSs (*blue*, cells responding strongest to CSs predicting reward; *red*, cells that responded strongest to CSs predicting air puff; *black*, cells without valence selectivity). *D*, The percentage of cells that responded with increased (*blue*), decreased (*red*), or no change (*black*) activity upon fixation presentation classified according to the preference of the cell for CSs of different valence. *E*, Average activity of a single amygdala neuron recorded during passive viewing of either social images (images of faces of a monkey's social cohort, *top*) or nonsocial images that predicted different reward amounts (*bottom*). Monkeys within the social cohort formed a stable social hierarchy. Neural responses to the face images encoded the hierarchical rank of viewed monkey faces, with the strongest response observed when viewing the most dominant monkey (M1) and weaker responses the more submissive the viewed monkey (e.g., M8). The same neuron responded most strongly to a fractal image that predicted a large reward compared to smaller rewards. Since social rank is a motivationally significant parameter, these data indicate that this amygdala neuron signaled information about the value of both social and nonsocial stimuli. (See figure 53.1.)

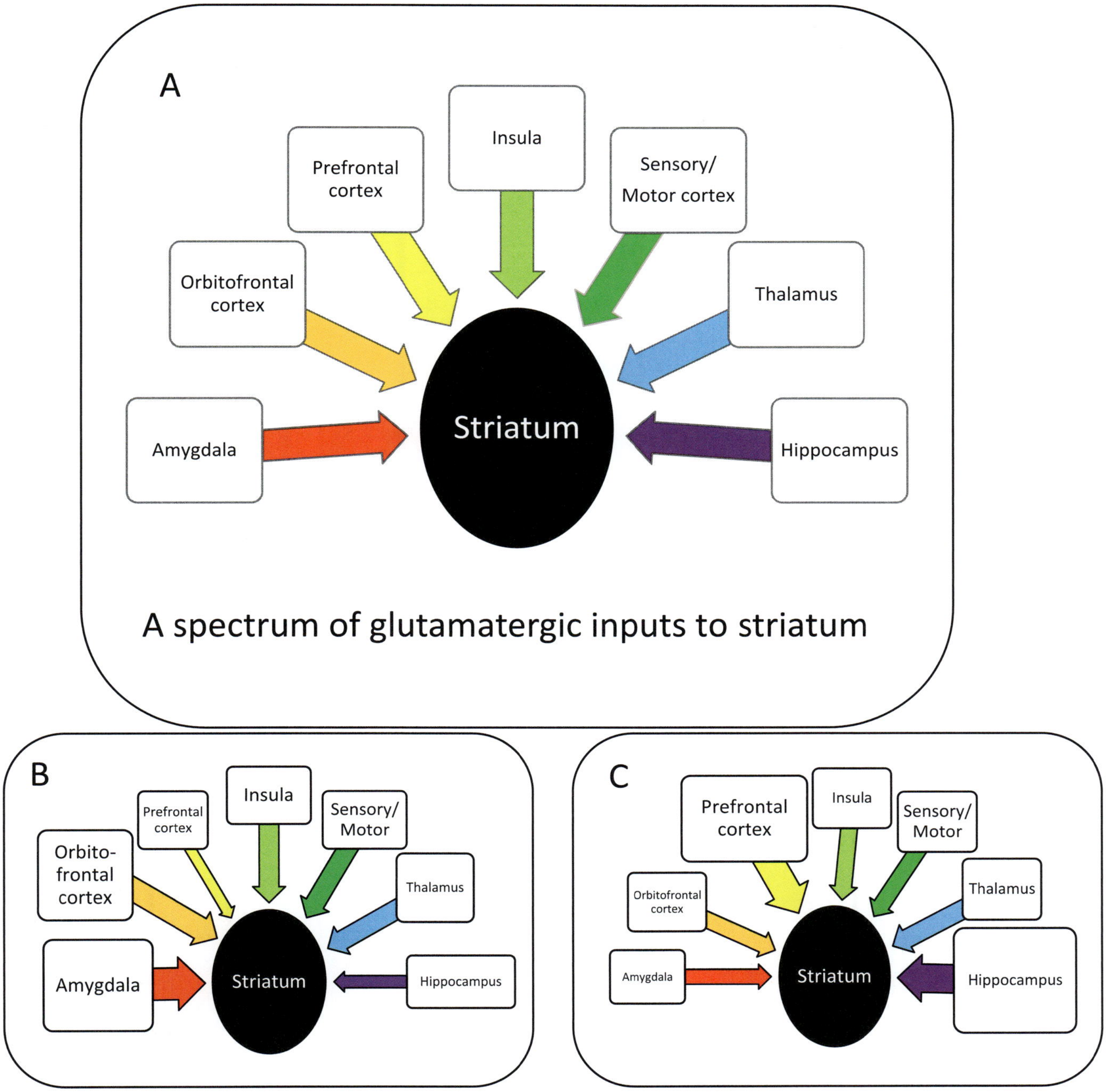

PLATE 59 The striatum as a ballot box that undergoes change in the relative strength of "votes" coming from different regions through development. *A*, A schematic of the spectrum of glutamatergic inputs to the striatum. In theory, glutamatergic inputs all have the capacity to drive excitatory activity in the striatal medium spiny neurons and therefore have a "vote" in the context of the striatal ballot box metaphor (hence in [*A*] all inputs are shown as equal). Note, neuromodulatory inputs from dopamine or other neurotransmitters are not represented. *B*, In a hypothetical model, in early adolescence, inputs from the orbitofrontal cortex, amygdala, and insula to the striatum may dominate relative to other inputs. *C*, With adolescent development, prefrontal cortex inputs to the striatum, particularly from cortical cognitive control networks, may increase their relative strength compared to the amygdala, orbitofrontal cortex, and insula. Hippocampal inputs, although considered "limbic," may also increasingly contribute with age. These changes in relative influence over striatal activity may explain changes in striatal computation and behavior with age. (See figure 54.1.)

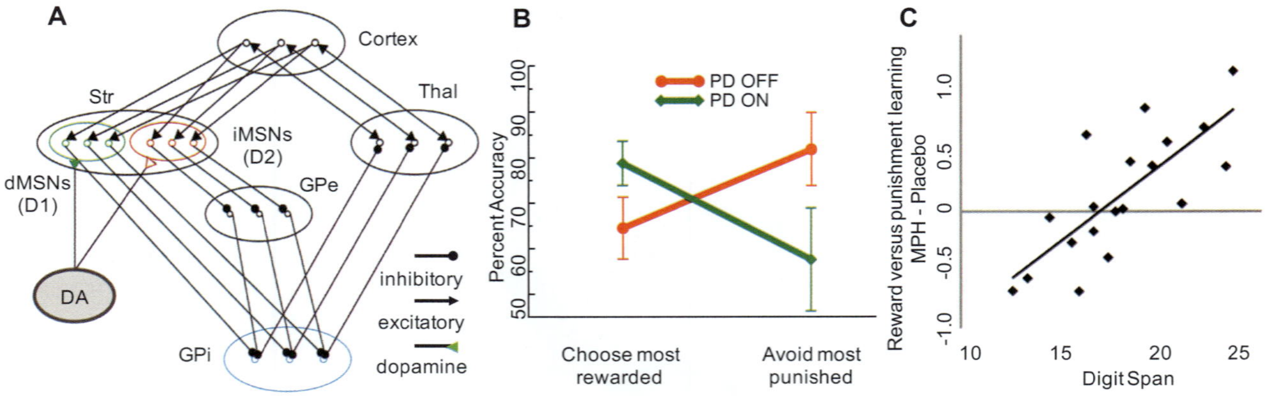

PLATE 60 *A*, Phasic dopamine (DA) signals train corticostriatal synapses to reflect reward and punishment statistics for cortically represented actions. If these synapses facilitate enough activity in the D1-expressing direct versus the D2-expressing indirect pathway, actions are selected via disinhibition of the thalamus (Thal), which recurrently excites the evoking cortical representation. DA: dopamine; GPe: external globus pallidus; GPi: internal globus pallidus; MSN: medium spiny neuron; Str: Striatum. From Collins and Frank (2014). *B*, Medication effects on reinforcement learning in PD. Unmedicated PD patients are relatively impaired at learning to choose the most rewarded stimuli but are relatively better at avoiding the most punished stimuli. The pattern flips when patients are medicated with dopaminergic agents. From Frank, Seeberger, & O'Reilly (2004). Reprinted with permission from the AAAS. *C*, Effects of the dopamine-elevating drug methylphenidate (MPH) on learning vary by digit span, a putative index of individual differences in baseline striatal dopamine function (Cools & D'Esposito, 2011). MPH increases reward versus punishment learning for those with high span while decreasing it for those with low span. From Van der Schaaf et al. (2013). (See figure 55.2.)

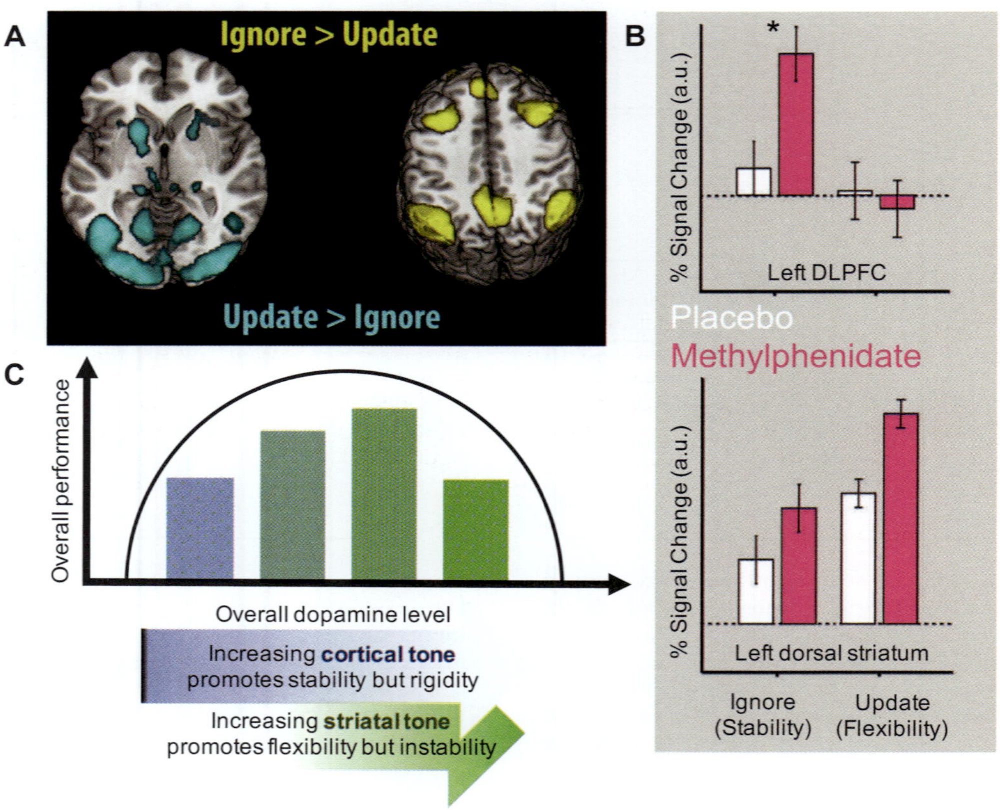

PLATE 61 *A*, Brain regions recruited differentially as a function of demands for cognitive stability versus flexibility in a working memory task. *A*, The frontoparietal cortical network is more active for stable ignore/maintenance versus updating trials, while striatal regions are more active for flexible updating. *B*, Dopamine upregulation with MPH reliably increases the frontal cortical response when demands for stability are higher and numerically increases the striatal response when demands for flexibility are higher. *C*, Globally increased dopamine release has opposing effects in increasing cognitive stability via the cortex and increasing flexibility via the striatum, with optimal overall performance at the midrange of global dopamine function. *A* and *B* adapted from Fallon et al. (2016). (See figure 55.3.)

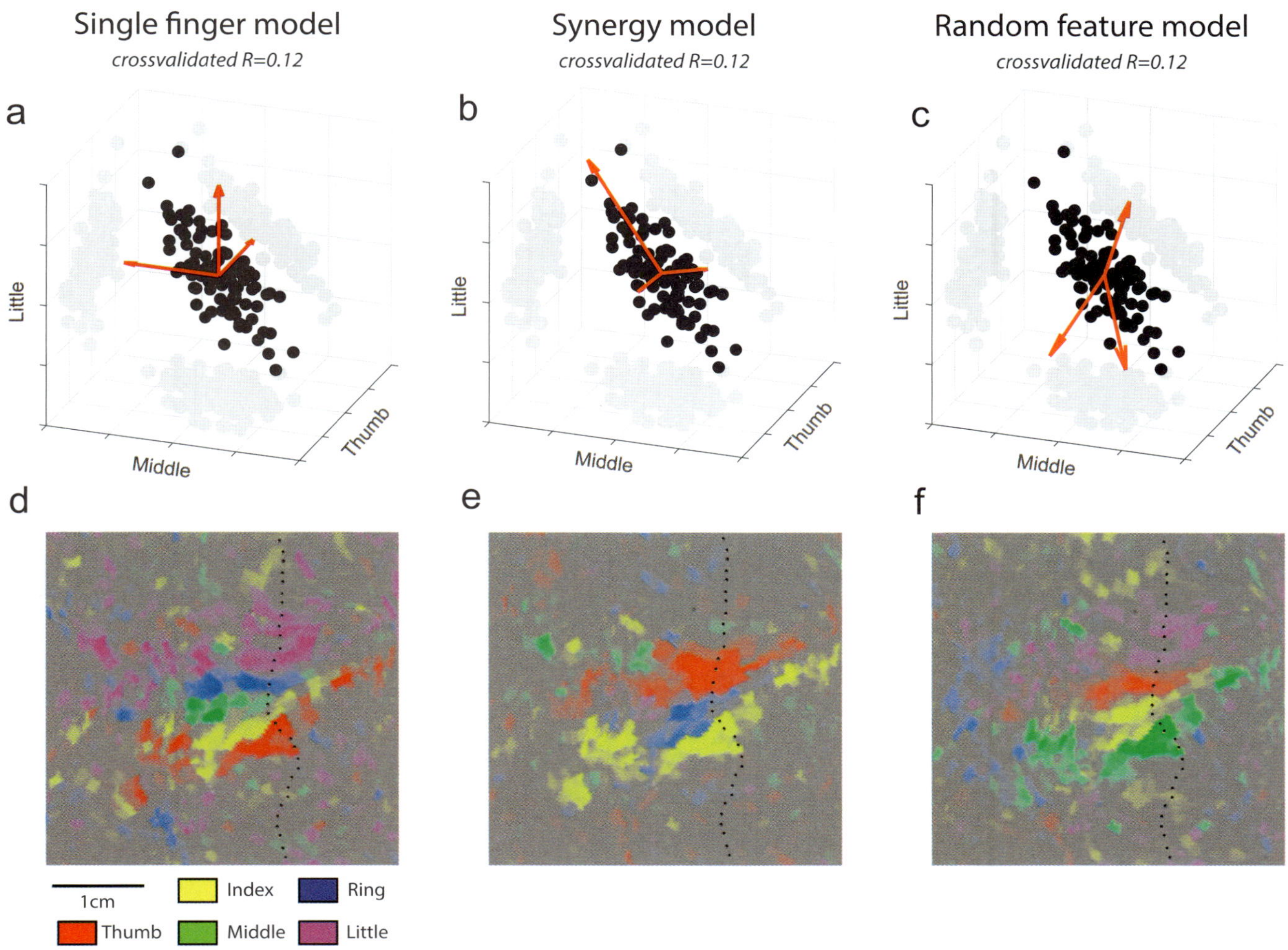

PLATE 62 Three feature models that describe the activity during finger movements in M1 and S1 equally well. *A*, The distribution of voxels in the space of three of the experimental conditions. Each dot represents a voxel's activity profile across the thumb, the middle finger, and the ring finger. The distribution can be described by using individual fingers as features. *B*, Surface map of the human S1 and M1 with voxels colored according to which finger they are most activated. The color saturation reflects the strength of the tuning, with gray areas showing no finger preference. *Dotted line*, Fundus of the central sulcus. The length scale is approximate. *C*, The distribution can also be described using the principal components of the natural statistics of finger movements (synergies). *D*, Surface map of feature weights for the synergy model with voxels either scoring high (+) or low (−) on each PC. *E*, A model with five random feature vectors explains the data equally well, but (*F*) lead to a different feature map. (See figure 56.2.)

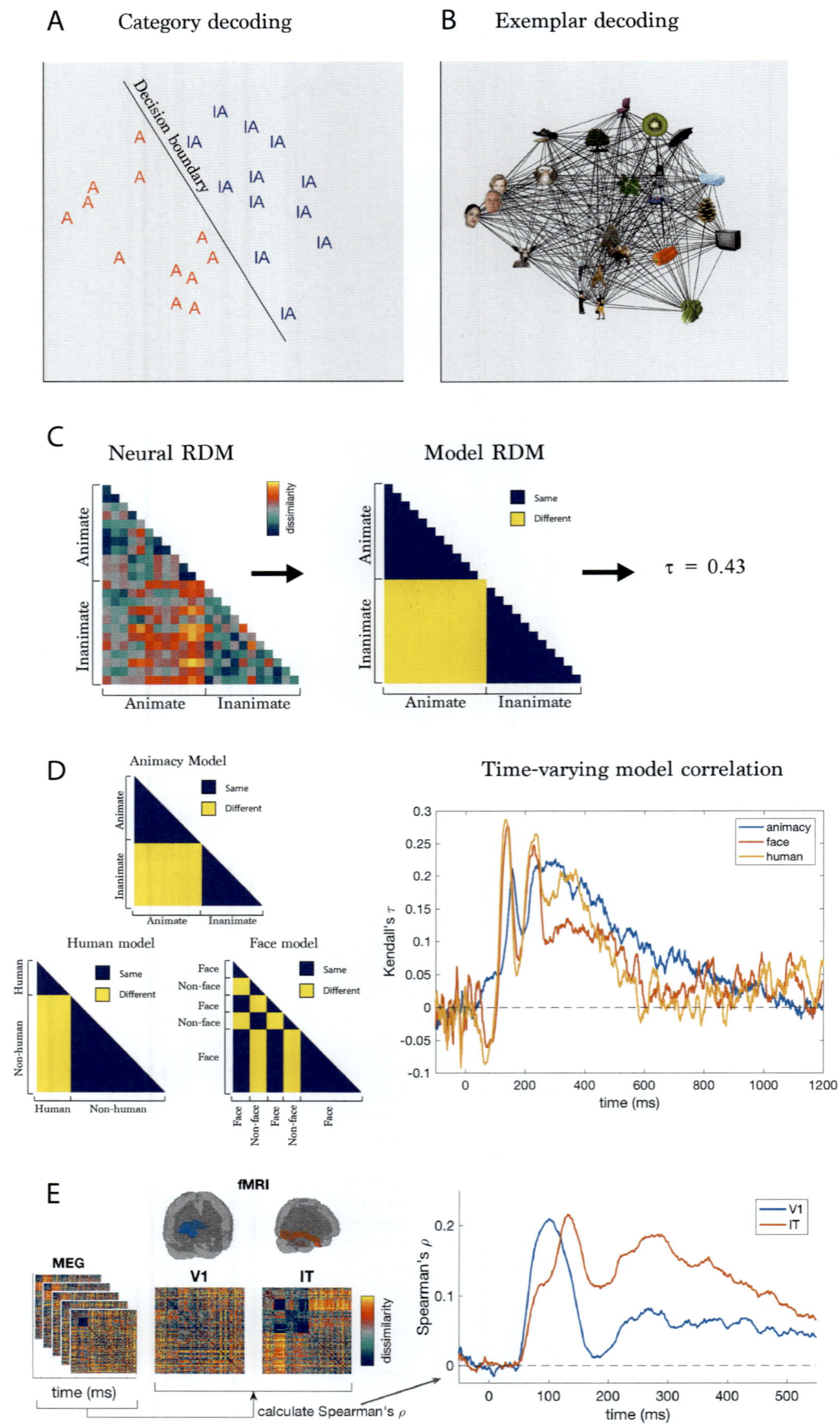

A
Category decoding
Decision boundary
A
IA
B
Exemplar decoding
C
Neural RDM
dissimilarity
Animate
Inanimate
Animate
Inanimate
Model RDM
Same
Different
Animate
Inanimate
Animate
Inanimate
τ = 0.43
D
Animacy Model
Same
Different
Animate
Inanimate
Animate
Inanimate
Human model
Same
Different
Human
Non-human
Human
Non-human
Face model
Same
Different
Face
Non-face
Face
Non-face
Face
Face
Non-face
Face
Non-face
Face
Time-varying model correlation
animacy
face
human
Kendall's τ
time (ms)
0.3
0.25
0.2
0.15
0.1
0.05
0
-0.05
-0.1
0
200
400
600
800
1000
1200
E
fMRI
MEG
V1
IT
dissimilarity
time (ms)
calculate Spearman's ρ
V1
IT
Spearman's ρ
0.2
0.1
0
0
100
200
300
400
500
time (ms)

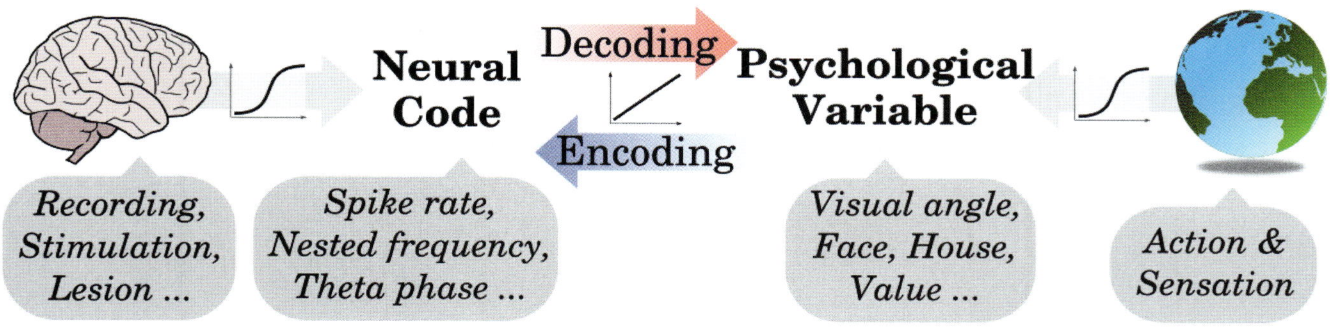

PLATE 64 The representational paradigm consists of three challenges: (1) identifying the content of mental representations, (2) determining how this information is encoded in the brain, and (3) linking these two levels of description. These challenges increasingly benefit from statistical modeling and machine learning. In this view, decoding analyses predict experimental factors (e.g., the speed of a hand movement, the luminance of a flashed image) from specific brain activity features (e.g., spike rate, electric field), whereas encoding analyses predict the reverse. Currently, the relevant features of (1) brain activity and (2) experimental factors are defined a priori by the experimenters and mapped onto one another with linear mapping. However, the capacity of machine learning to find nonlinear structures in large data sets may ultimately help in finding what the brain represents and how. (See figure 58.1.)

PLATE 63 Advanced methods for M/EEG decoding. *A, B,* Plots demonstrate the difference between category and exemplar-based decoding methods. Data for the plots come from a single time point (140 ms poststimulus onset) of an MEG experiment studying the brain's representation of objects (Carlson et al., 2013). *A,* Category-decoding approach with stimuli labeled by their category (A = animate object; IA = inanimate object) and the decision boundary from a linear classifier trained to discriminate animate and inanimate objects from the MEG data. *B,* Exemplar-decoding approach with individual stimuli displayed as images. The lines in the plot represent the pairwise decodability of individual exemplars for all possible pairwise comparisons. The distance (*line length*) indicates the relative decodability of exemplar pairs. *C,* Representational similarity (RSA) model testing applied to the MEG data. The entries of the neural RDM are correlated with the animacy model to study whether the brain represents object animacy at 140 ms poststimulus onset. *D,* RSA used to test multiple candidate models from another MEG experiment investigating the brain's representation of objects. Data from Cichy, Pantazis, and Oliva (2014). Plotted is the time correlation between the time-varying RDMs from the MEG and animacy, human, and face models. *E,* Graphical depiction of the MEG-fMRI fusion approach estimating regional time-varying neural activity from fMRI regions of interest (ROI). The time-varying MEG RDMs are correlated with RDMs from ROIs (obtained using fMRI). For each ROI, this gives a time-varying correlation indexing neural activity in the ROI. Shown is the estimated neural activity for V1 and inferior temporal cortex. Data from Cichy, Pantazis, and Oliva (2014). (See figure 57.2.)

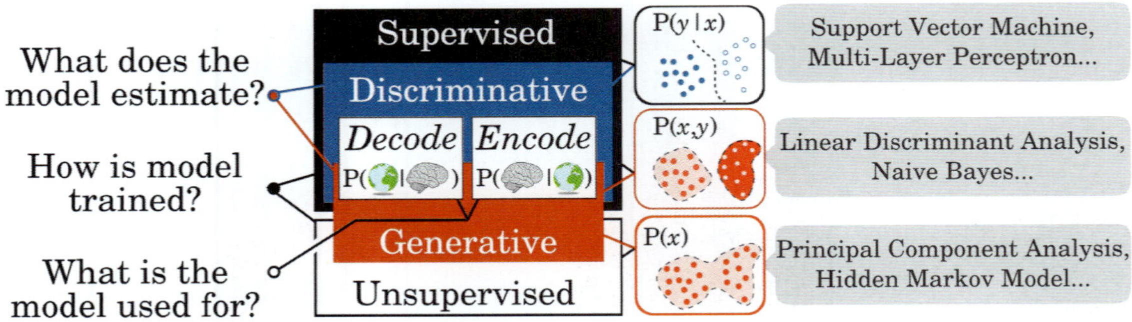

PLATE 65 *Left*, The most common models used in cognitive neuroscience can be distinguished with three main dimensions of the statistical framework. Models that are trained to estimate a conditional probability $P(y\,|\,x)$ between two sets of variables (e.g., x = psychological variables, y = neuronal recordings, or vice versa) are referred to as *discriminative*. By contrast, models that estimate the joint distribution $P(x, y)$ are considered to be *generative*. All generative models can thus be used to derive $P(y\,|\,x)$ for prediction. These models are *supervised* in that they are trained to predict one variable (e.g., y) from another (e.g., x). By contrast, *unsupervised* models estimate the distribution of a single (possibly multidimensional) variable (x). Finally, a trained model can ultimately be used for different purposes—for example, decoding or encoding (see figure 58.1). *Right*, Examples of classical supervised and unsupervised models. (See figure 58.2.)

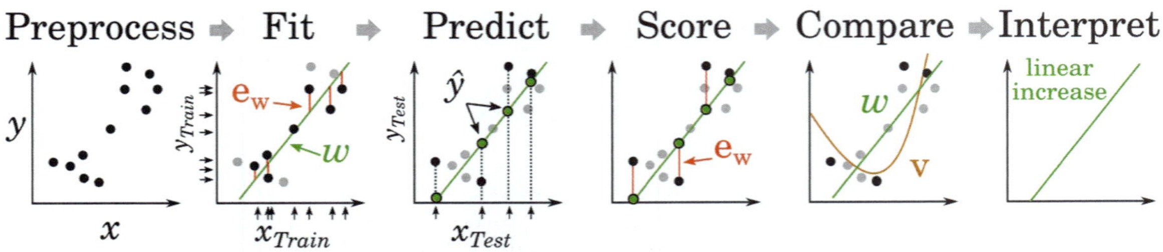

Fitting Objective: Find parameters (w) to best predict targets (y) from input data (x).

max $P(w\,|\,x, y)$, where: $x_i^T w = y_i$

Method: optimization

$$P(w\,|\,x, y) \propto P(x, y\,|\,w)\,P(w)$$

$$\min_w Loss(y, x^T w) + Regularization(w)$$

Name	Loss	Linear Model	Reg	Optimization
OLS:	l_2		None	$\min_w \sum_i (y_i - \hat{y}_i)^2$
Ridge:	l_2	$x_i^T w = \hat{y}_i$	l_2	$\min_w \sum_i (y_i - \hat{y}_i)^2 + \lambda \| w \|^2$
Logistic:	log		l_2	$\min_w \sum_i \log(1 + \exp(-y_i \hat{y}_i)) + \lambda \| w \|^2$
SVM:	hinge		l_2	$\min_w \sum_i \max(0, 1 - y_i \hat{y}_i) + \lambda \| w \|^2$

PLATE 66 *Top*, Statistical modeling follows a standard multistep pipeline that starts with (1) preprocessing (any transformation that can be defined independently of the data—e.g., filtering) followed by (2) model fitting on a subset of training data, (3) prediction, and (4) scoring of independent and identically distributed held-out test data. The score (a.k.a, *goodness-of-fit* or *prediction error*, here labeled as e_w) can be summarized with a variety of metrics (e.g., accuracy, AUC, R^2, cross-entropy, F-value). Finally, one can subsequently perform a model comparison to interpret results. *Middle*, The fitting stage can make use of a variety of models, such as support vector machines (SVM) or ordinary least squares (OLS). These models can be formalized as maximum a posteriori, which balances the data likelihood (*red*) and the priors on the distribution of the model parameters (*blue*). These model parameters can be found with optimization, which consists of jointly minimizing the loss and regularization function. Note that for categorical models (e.g., logistic regression, SVM), the predicted values ŷ are subsequently transformed into discrete categories. *Bottom*, An example of common linear models used in cognitive neuroscience, together with their corresponding loss, regularization, and optimization functions. (See figure 58.3.)

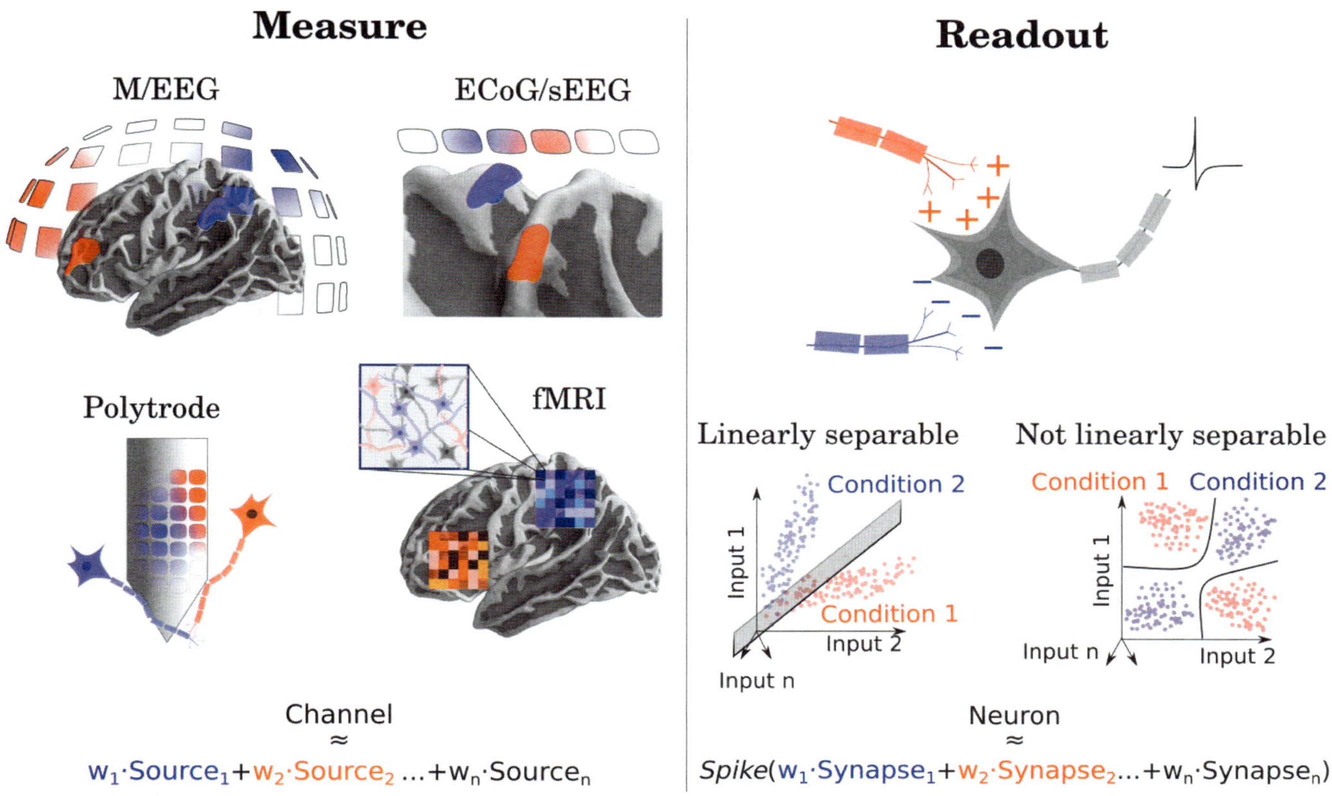

PLATE 67 *Left*, The electromagnetic fields produced by neurons linearly project onto polytrodes, electrocorticography (ECoG), or magnetoelectroencephalography (M/EEG) sensors. A similar concept of linear mixture can be extended to fMRI: each voxel approximates a weighted sum of BOLD responses. Overall, measuring channels can thus be used to infer the underlying neural responses. *Right*, Neuronal computations can be approximated as nonlinear functions (i.e., spike) applied onto weighted combinations of synaptic inputs. Statistically, this means that neurons are sensitive to (i.e., "read") linearly separable inputs. (See figure 58.4.)

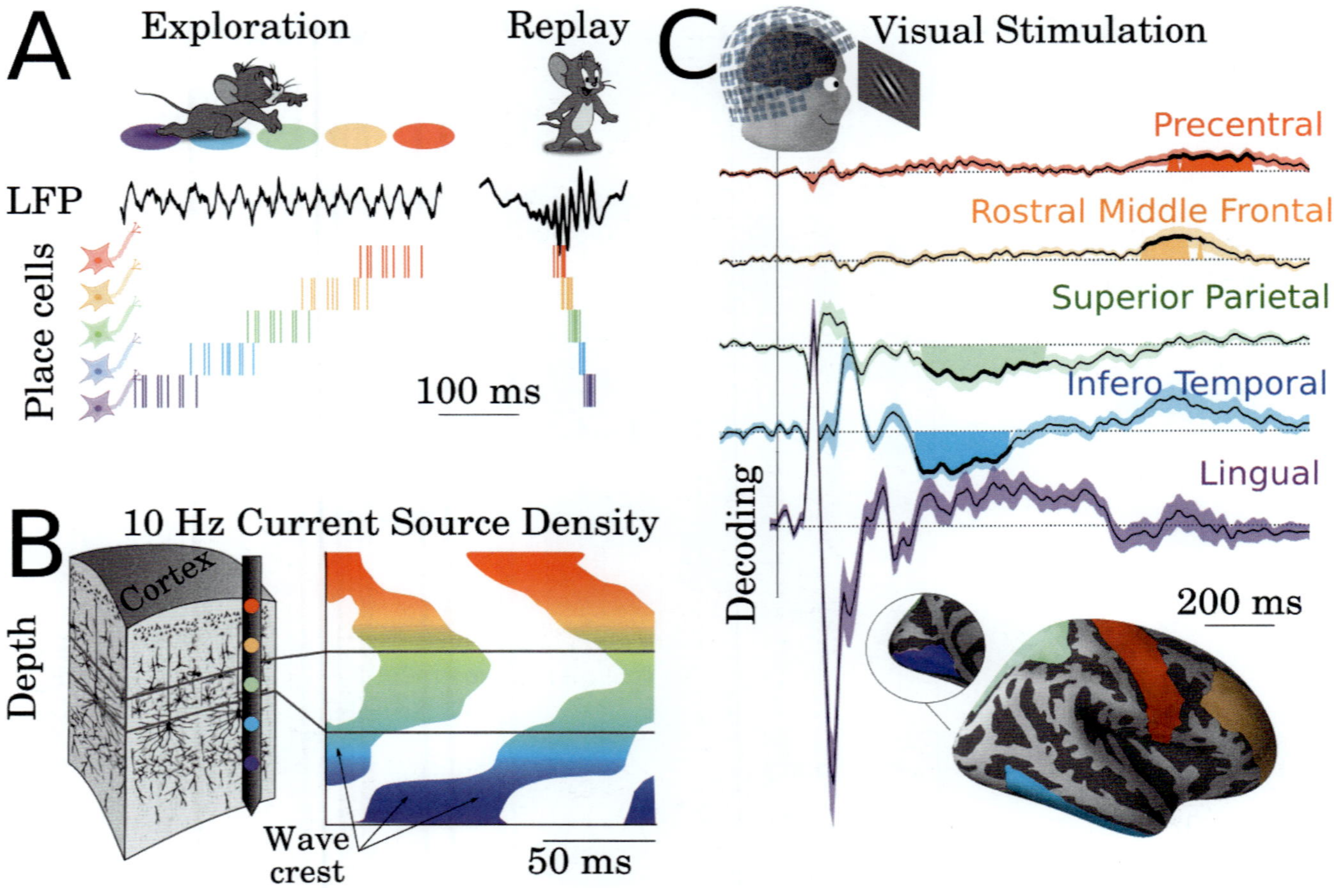

PLATE 68 Sequences of neural responses across multiple spatial scales. Sequences of neural responses have been identified with a variety of methods and across various spatiotemporal scales. *A*, Hippocampal place cells spike when the rodent moves to specific spatial positions. When the rodent rests, the previous spike sequence is replayed in reverse order. Adapted from Girardeau and Zugaro (2011). *B*, The 10 Hz oscillatory activity recorded in the macaque early visual cortex travels from superficial (*red*) and deep layers (*purple*) to the granular layer (*green*). Adapted from van Kerkoerle et al. (2014). *C*, Visual stimulation triggers a long-lasting cascade of macroscopic brain responses, starting from the early visual cortex and ultimately reaching the associative areas (King, Pescetelli, & Dehaene, 2016). (See figure 58.5.)

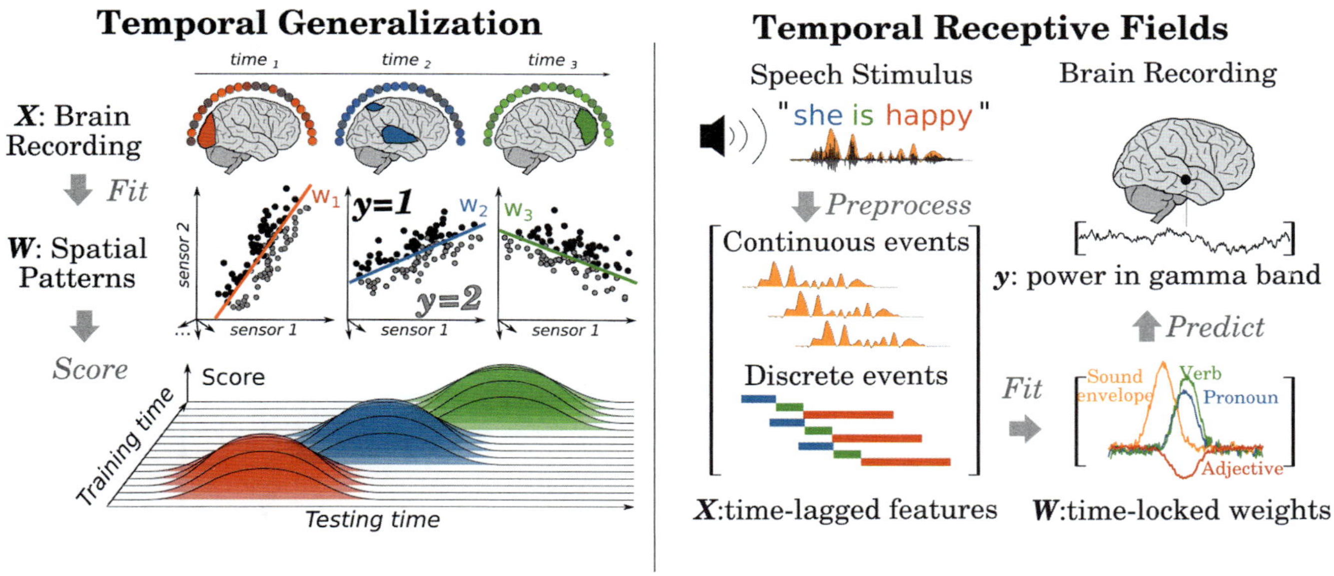

PLATE 69 Decoding and encoding neural sequences. *Left,* Temporal generalization consists of fitting a spatial filter at each time sample locked to an event and testing whether it can be used to decode the brain responses at all time samples. This analysis can be used to determine whether a sustained decoding score results from a stable pattern of neural responses or whether it reflects a series of transient neural responses. Adapted from King and Dehaene (2014). *Right,* Temporal receptive field analyses involve first annotating the latent dimensions that characterize the continuously varying experimental variables—for example, the spectral modulation of an acoustic waveform, the lexical categories of spoken words, and others (*top left*). These features are then transformed into a time-delay matrix whose linear modeling can be used to recover the temporal response profile of each feature (*bottom right*). Adapted from Holdgraf et al. (2017). (See figure 58.6.)

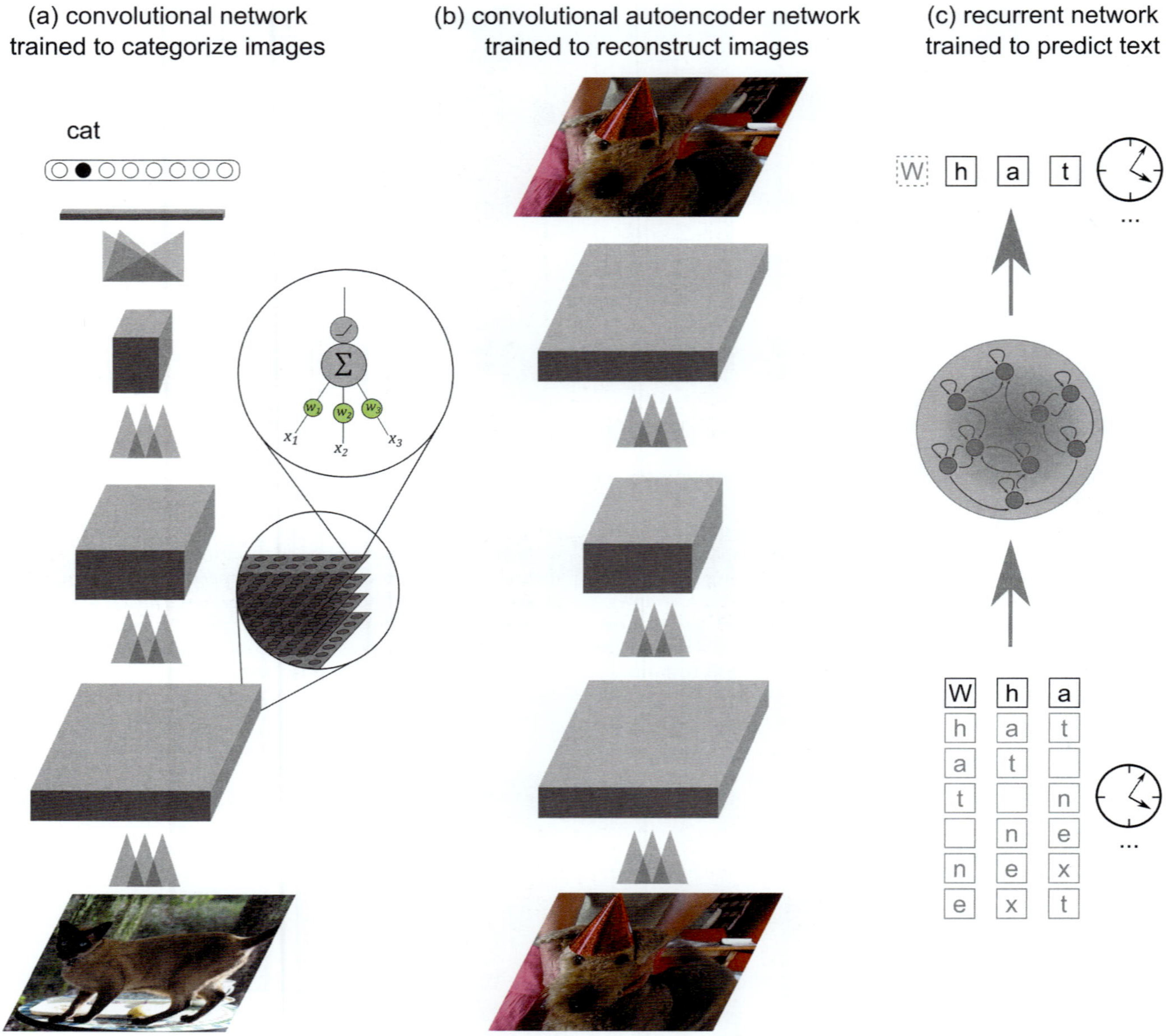

PLATE 70 Diagrams of some of the species of neural network. *A*, A convolutional neural network with four hidden layers (three convolutional and one fully connected) that is able, after supervised training on labeled images, to categorize objects in novel photographs it has not been trained with. The convolutional layers consist of multiple "feature maps," each of which contains units with spatially restricted receptive fields centered on different locations of the input but sharing the same connection weights. This allows each map to represent the presence or absence of a particular feature at each location in the image. *B*, A convolutional autoencoder with three convolutional hidden layers, which is able, after unsupervised training, to compress novel images into a more concise format within the middle "bottleneck" layer and approximately reconstruct them thereafter. *C*, A recurrent neural network that takes as input text, one character at a time, and is able, after unsupervised training, to predict the next character in sentences it has never seen before. (See figure 59.1.)

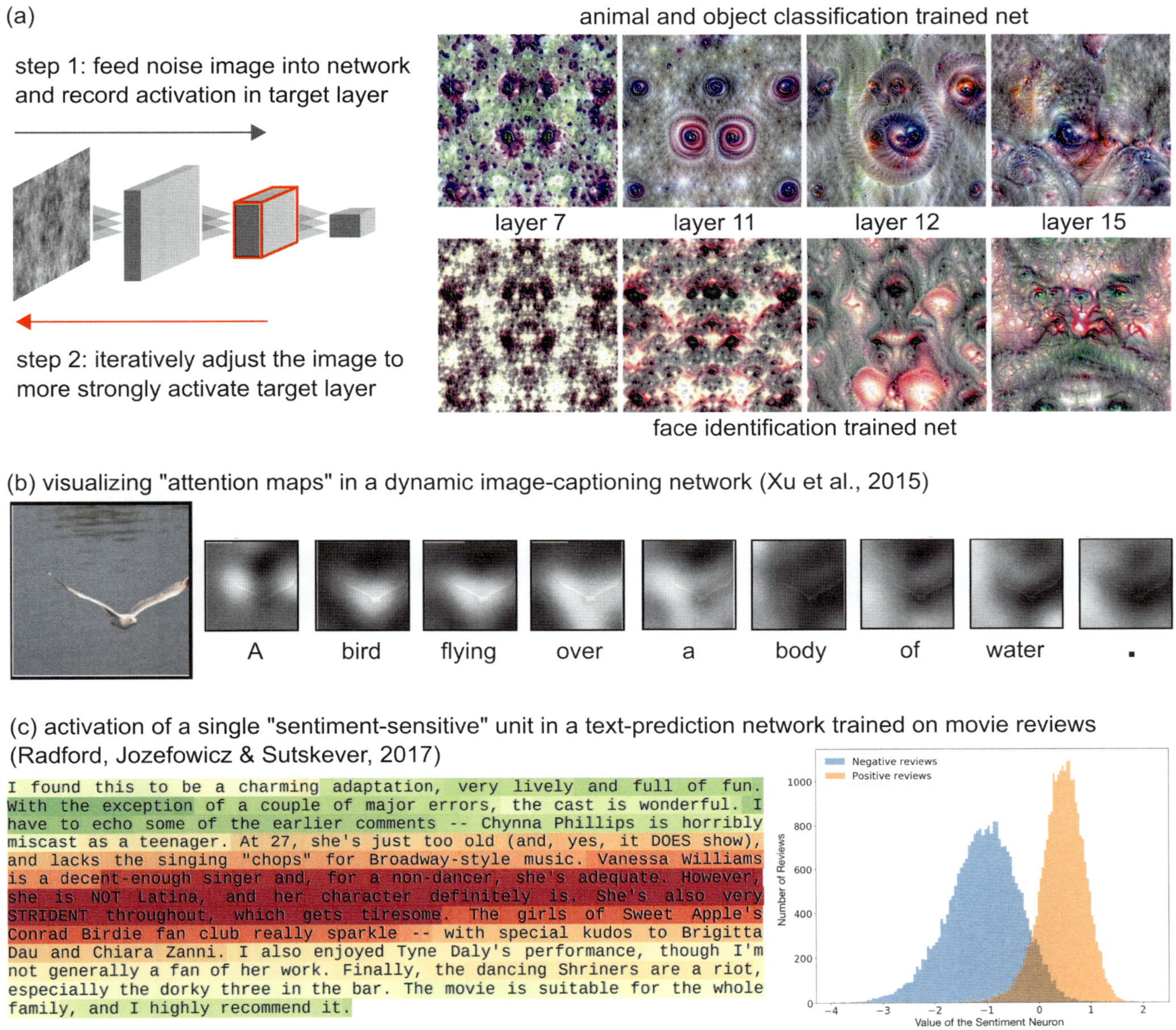

PLATE 71 The relative transparency of deep neural networks. *A*, The backpropagation algorithm used to train networks can also be used to visualize features in trained networks. Here, a noise image has been iteratively optimized to increase how strongly it activates units in each of four layers in a 16-layer DCNN, which has been trained either to categorize objects and animals (*top row*) or identify faces (*bottom row*). Training has a striking impact on the features learned. *B*, *Attention networks* have a spatial attention mask (a multiplicative weighting of the input), which they learn to allocate in ways that improve performance at their training task. Here, the white patches show which regions of this image were attended to at each time step by a recurrent network trained to generate text captions of images, outputting one word per time step. Reproduced with permission from Xu et al. (2015). *C*, Visualization of the output activation of a single LSTM unit in a recurrent network trained to predict text one character at a time, for an example passage the network has never seen before. High activation is indicated by green and low by red. The responses of this unit, when analyzed across a large sample of reviews, predicted whether a review was positive or negative in sentiment. Reproduced with permission from Radford, Jozefowicz, and Sutskever (2017). (See figure 59.2.)

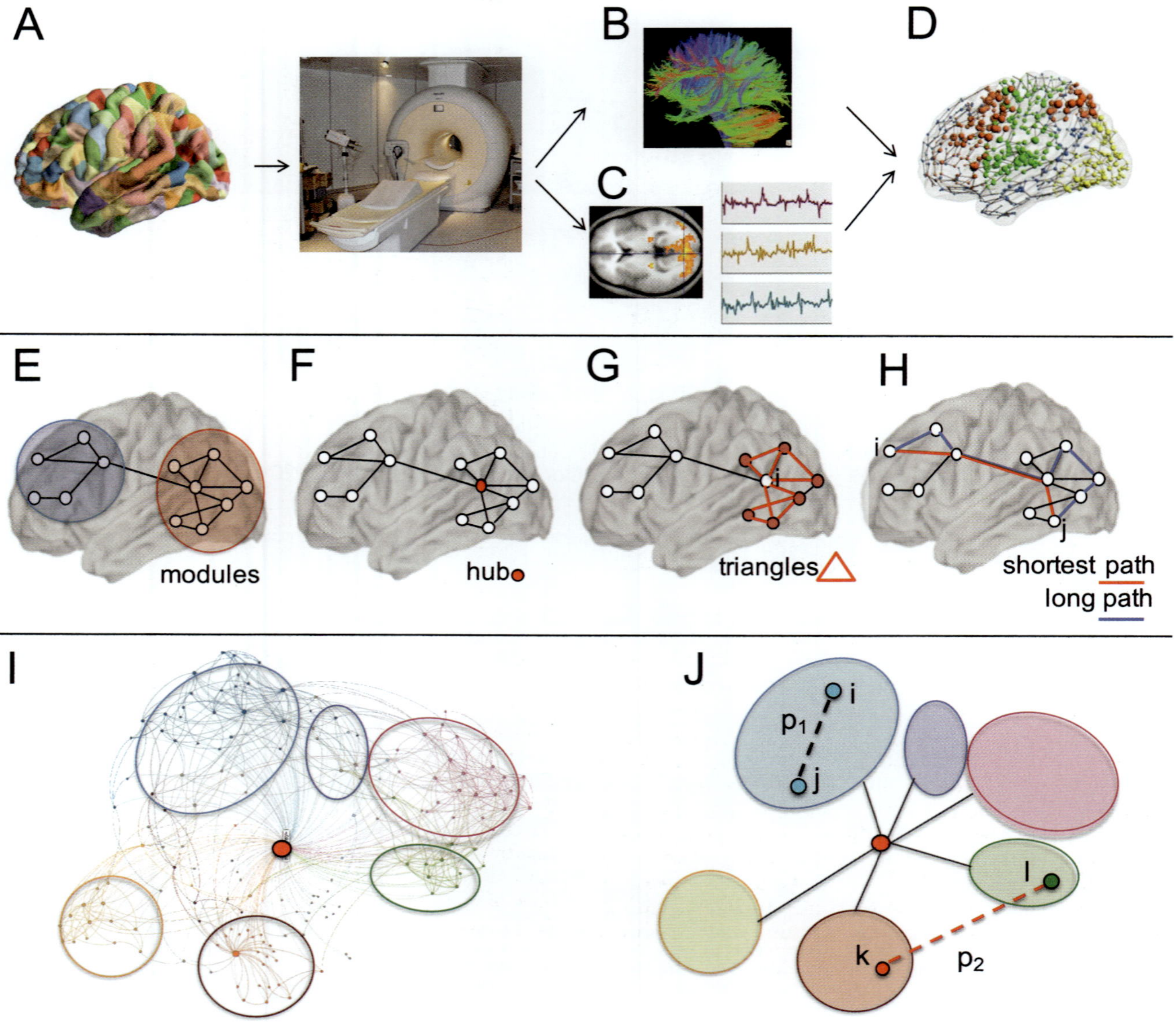

PLATE 72 *Panel 1,* Schematic process for constructing human brain networks. The brain is parcellated into a few hundred anatomical regions (*A*), which form the nodes of the network (*D*). Network edges correspond to anatomical connections measured via DTI (*B*) or functional correlations measured via fMRI (*C*). *Panel 2,* Network metrics commonly used to quantify the topology of networks. (*E*) Modularity: many complex networks have a modular community structure, whereby they contain subsets of highly interconnected nodes called *modules.* (*F*) Degree: the degree of a node is the number of links or connections it has to the rest of the network. In many real-world networks, most nodes have only a few connections, but a few so-called hub nodes are extensively connected. (*G*) Clustering: the clustering coefficient measures the number of connections that exist between the first neighbors of a node i as a proportion of the maximum possible number of such connections. These edges form triangular connections around the node i. (*H*) Path length and efficiency: The minimum path length between two nodes i and j is simply the minimum number of edges that need to be traversed to get from one node to another. Many of these features are highlighted in the brain functional network shown in (*D*). Node size is proportional to degree, and the modular organization is highlighted in color. *Panel 3,* Network diagram of an individual's LinkedIn connections (*I*). The central individual is shown in red, and the modules of the network are highlighted in color. A simplified model of the network is shown in (*J*). Panel 2 was adapted from Vértes and Bullmore (2015), with permission. (See figure 60.1.)

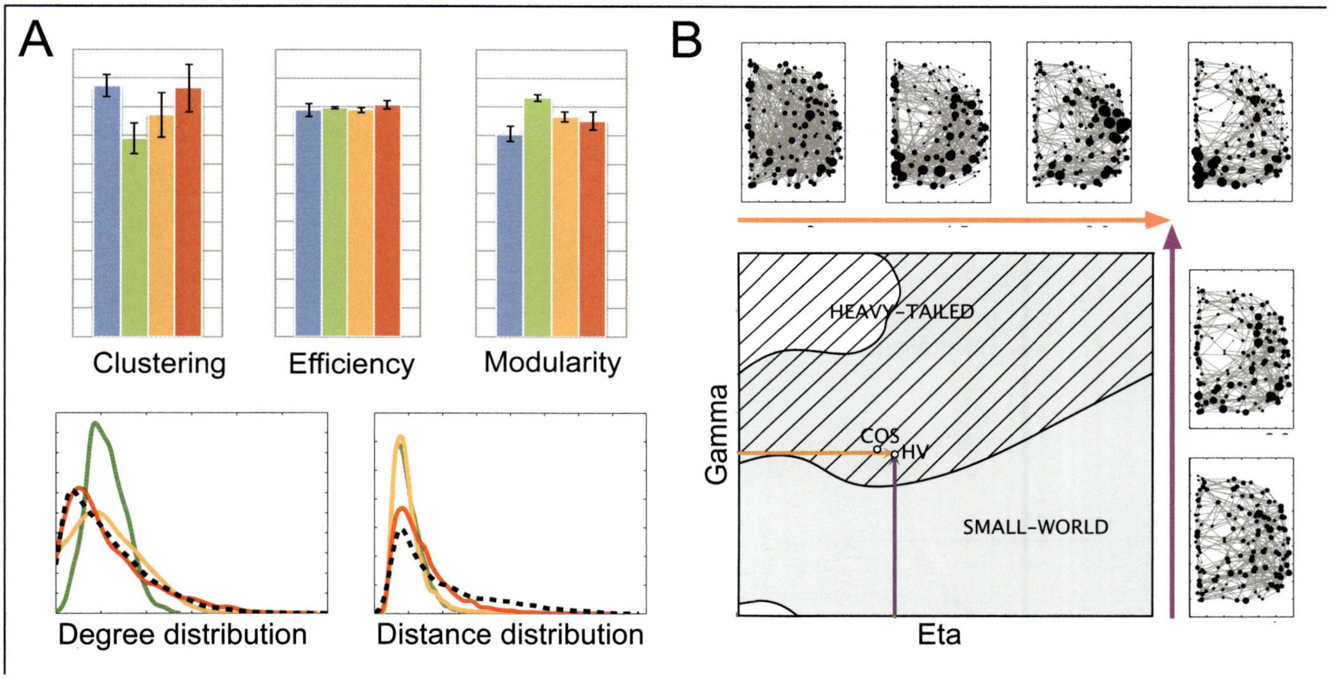

PLATE 73 (*A*) Comparison of networks simulated by three generative models versus brain functional networks derived from experimental fMRI data (*blue*). Both the exponential distance penalty model (*green*, $P_{ij} \propto e^{-\alpha d_{ij}}$) and the economical preferential attachment model (*orange*, $P_{ij} \propto d_{ij}^{-\eta}(k_i k_j)^{\gamma}$) fail to simultaneously capture several topological characteristics of functional brain networks. In contrast, the economical-clustering model (*red*, $P_{ij} \propto d_{ij}^{-\eta} k_{ij}^{\gamma}$) yields significantly more realistic networks by all of the following measures: normalized clustering coefficient, global efficiency, modularity, degree distribution, and distance distribution (*data shown in dashed black lines*). (*B*) Phase diagram of the economical-clustering model. Most values of eta (η) and gamma (γ) yield small-world networks (*gray area*), whereas only high values of γ yield networks with heavy-tailed (skew >1) degree distribution (*hashed area*). The model parameters estimated to minimize mismatch between simulated and experimental fMRI data sets are shown here for both healthy volunteers (HV) and participants with childhood onset schizophrenia (COS). The orange (and purple) arrows show sections through the phase space, varying only η (or γ), respectively, whereas the other parameter is held at its optimal value estimated in healthy volunteers. Schematics of the networks obtained at various points along these sections are also shown (axial view of right hemisphere only). Adapted from Vértes et al. (2012), with permission. (See figure 60.2.)

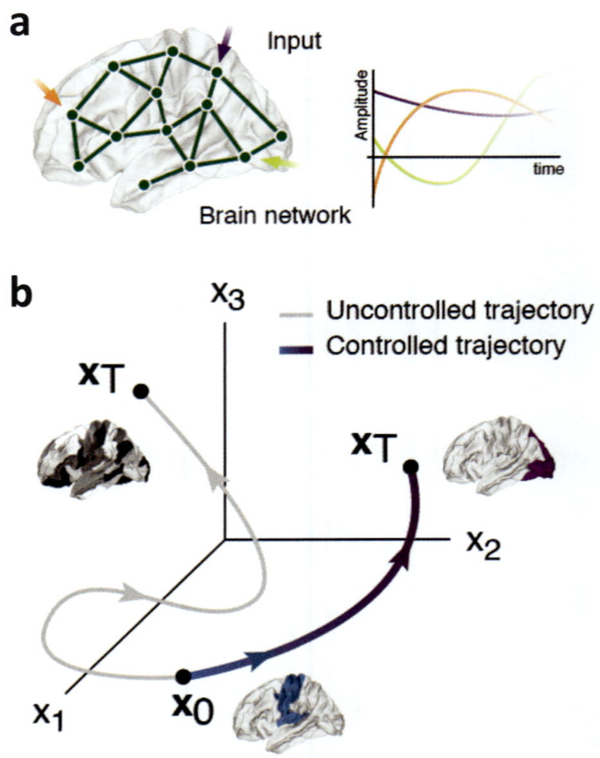

PLATE 74 Controllability of human brain networks. *A*, A set of time-varying inputs are injected into the system at different control points (network nodes, brain regions). The aim is to drive the system from some particular initial state to a target state (e.g., from activation of the somatosensory system to activation of the visual system). *B*, Example trajectory through state space. Without external input (control signals), the system's passive dynamics leads to a state in which random brain regions are more active than others; with input the system is driven into the desired target state. Reproduced with permission from Betzel et al. (2016). (See figure 61.1.)

A

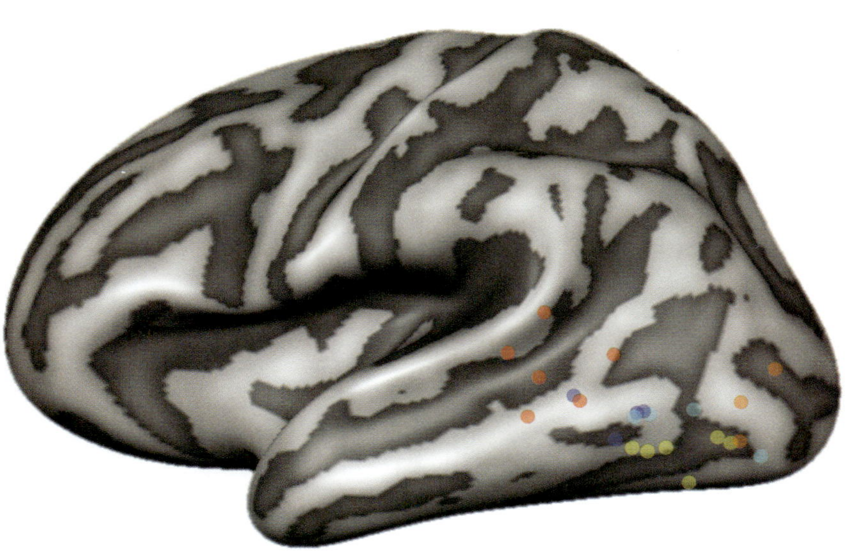

Study	Tal X	Tal Y	Tal Z
Action attribute retrieval			
Martin et al., 1995 (Study 1)	-50	-50	4
Martin et al., 1995 (Study 2)	-54	-62	8
Phillips et al., 2002	-50	-62	5
Kable et al., 2005	-53	-60	-5
Verbs			
Bedny et al., 2008	-53	-41	3
Peelen et al., 2012	-49	-53	12
Shapiro et al., 2006	-57	-40	9
Bedny et al., 2013	-60	-51	11
Hernandez et al., 2014	-45	-43	7
Bedny et al., 2011	-53	-49	6
Tools			
Beauchamp et al., 2002 (Study 1)	-38	-63	-6
Beauchamp et al., 2002 (Study 2)	-46	-70	-4
Valyear et al., 2007	-48	-60	-4
Peelen et al., 2013	-50	-60	-5
Bracci et al., 2011 (Study 1)	-48	-65	-6
Bracci et al., 2011 (Study 2)	-46	-68	-2
Feature-general action representation			
Wurm & Lingnau, 2015	-41	-76	-4
Wurm et al., 2017	-44	-64	3
Oosterhof et al., 2010	-49	-61	2
Wurm & Caramazza, 2018	-54	-61	4
Basic motion			
Bedny et al., 2008	-46	-71	7
Zeki et al. 1991	-38	-74	8
Bracci et al., 2011	-44	-72	-1

B

Study	Tal X	Tal Y	Tal Z
Tool experience			
Creem-Regehr et al., 2007	-56	-29	29
Valyear et al., 2012	-43	-39	43
Vingerhoets et al., 2011	-42	-32	42
Weisberg et al., 2007	-42	-43	38
Feature-general action representation			
Oosterhof et al., 2010	-44	-31	44
Oosterhof et al., 2012	-49	-31	42
Hafri et al., 2017	-56	-36	28
Wurm & Lingnau, 2015	-51	-29	36
Wurm et al., 2017	-47	-27	37
Feature-general object function			
Leshinskaya & Caramazza, 2015	-62	-38	38
Tools			
Garcea & Mahon, 2014	-43	-43	41

PLATE 75 Peak coordinates of action-related effects in MTG (*A*) and IPL (*B*) reported in studies discussed in the section on the neural organization of action concepts. The different kinds of effects are based on the following contrasts/classifications: action attribute retrieval (*blue*) = tasks requiring the retrieval of actions or action attributes versus action-unrelated attributes (e.g., color) from pictures or names of actions or manipulable objects; tool experience (*magenta*) = familiar/typical versus unfamiliar/atypical tool use knowledge; verbs (*red*) = verbs versus nouns (various contrast; see the text); basic motion (*orange*) = moving versus static dots; feature-general action representation (*light blue*) = multivoxel pattern classification of action videos across perceptual features; feature-general object function (*green*) = multivoxel pattern classification of abstract categories of functions; tools (*yellow*) = images or videos of tools versus nonmanipulable artifacts or animals. Note that peaks do not reflect the spatial extent or the overlap of effects. (See figure 63.1.)

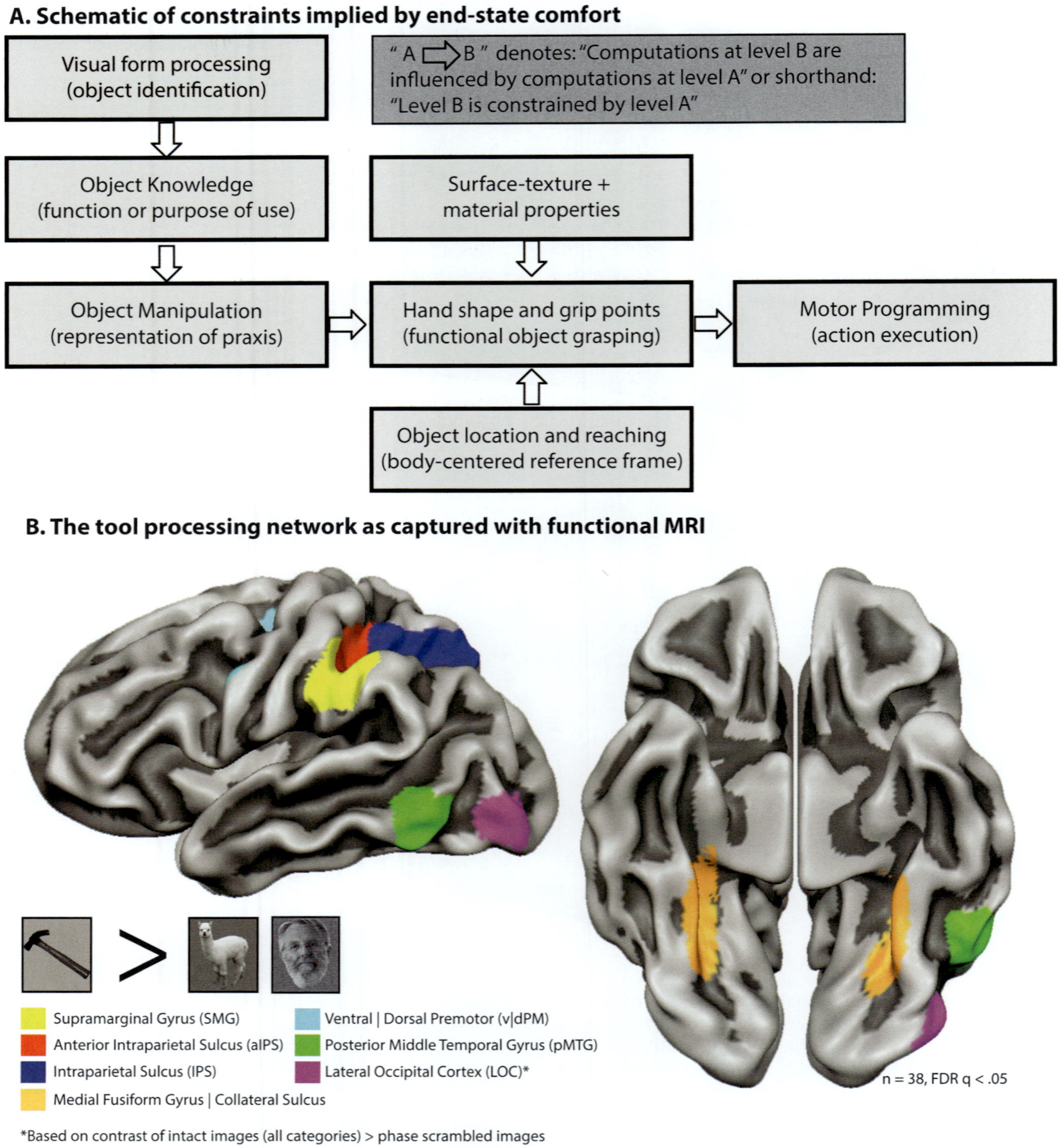
A. Schematic of constraints implied by end-state comfort
Visual form processing (object identification)
" A ⇨ B " denotes: "Computations at level B are influenced by computations at level A" or shorthand: "Level B is constrained by level A"
Object Knowledge (function or purpose of use)
Surface-texture + material properties
Object Manipulation (representation of praxis)
Hand shape and grip points (functional object grasping)
Motor Programming (action execution)
Object location and reaching (body-centered reference frame)
B. The tool processing network as captured with functional MRI
Supramarginal Gyrus (SMG)
Anterior Intraparietal Sulcus (aIPS)
Intraparietal Sulcus (IPS)
Medial Fusiform Gyrus | Collateral Sulcus
Ventral | Dorsal Premotor (v|dPM)
Posterior Middle Temporal Gyrus (pMTG)
Lateral Occipital Cortex (LOC)*
n = 38, FDR q < .05
*Based on contrast of intact images (all categories) > phase scrambled images

PLATE 76 Overview of constraints among the dissociable processes involved in tool recognition and use. *A*, Consider the everyday act of grasping one's fork to eat. The initial grasp anticipates how the object will be manipulated once it is "in hand." A fork is grasped differently than a knife, even if they have exactly the same handle. A fork is also grasped differently if the goal is to pass it to someone else, rather than to eat. The accommodation of functional object grasps to what the object will be used for once it is in hand, referred to as *end-state comfort* (Rosenbaum, Vaughan, Barnes, Marchak, & Slotta, 1990), implies substantial interaction among what are known to be dissociable representations (Carey, Hargreaves, & Goodale, 1996; Creem & Proffitt, 2001). For instance, the space of *possible* grasps is winnowed down to a space of *functional* grasps, based on representations of what will be done with the object once it is in hand (i.e., praxis; Wu, 2008). Praxis is, in turn, constrained by representations of object function, as objects are manipulated in a manner to accomplish a certain function or purpose of use. Finally, an object (e.g., a fork) is the target of an action only because it has a certain functional role in a broader behavioral goal, and thus the object (prior to any action being directed toward it) must be identified, at some level, for what it is. The schematic in figure 64.1 represents this type of conceptual analysis: the arrows in the figure do not represent *processing direction* but rather (some of) the constraints imposed among dissociable types of representations during functional object grasping and use. *B*, Functional MRI can be used to delineate the neural substrates of the domain-specific system that supports the translation of propositional attitudes into actions. The data shown in the figure were obtained while participants viewed tool stimuli compared to images of animals and faces. Regions are color-coded based on the principal dissociations that have been documented in the neuropsychological literature. The first functional MRI studies describing this set of "tool-preferring" regions were carried out in the laboratory of Alex Martin (Chao, Haxby, & Martin, 1999; Chao & Martin, 2000). (See figure 64.1.)

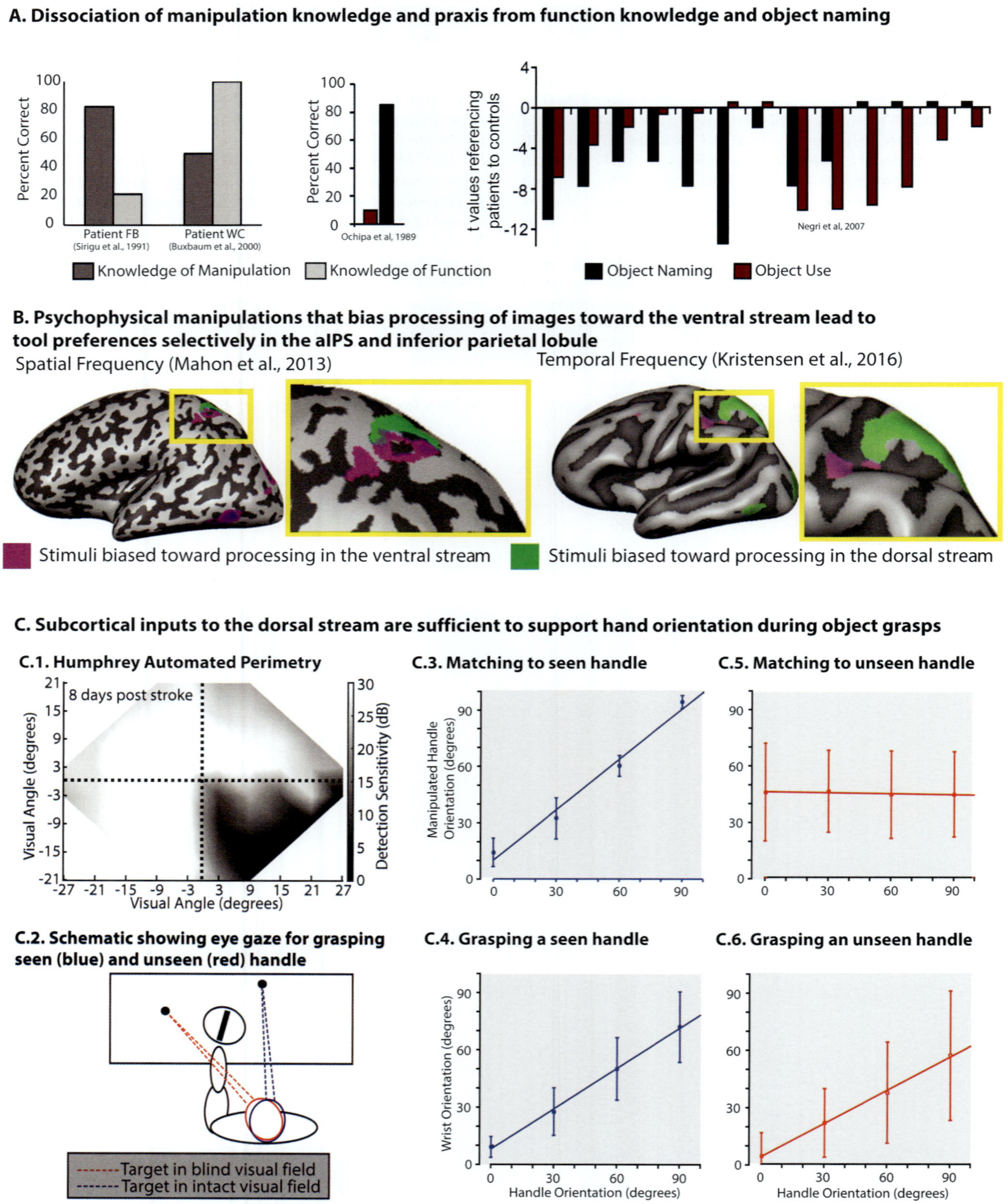
A. Dissociation of manipulation knowledge and praxis from function knowledge and object naming
Percent Correct
Patient FB
(Sirigu et al., 1991)
Patient WC
(Buxbaum et al., 2000)
Ochipa et al, 1989
t values referencing patients to controls
Negri et al, 2007
Knowledge of Manipulation
Knowledge of Function
Object Naming
Object Use
B. Psychophysical manipulations that bias processing of images toward the ventral stream lead to tool preferences selectively in the aIPS and inferior parietal lobule
Spatial Frequency (Mahon et al., 2013)
Temporal Frequency (Kristensen et al., 2016)
Stimuli biased toward processing in the ventral stream
Stimuli biased toward processing in the dorsal stream
C. Subcortical inputs to the dorsal stream are sufficient to support hand orientation during object grasps
C.1. Humphrey Automated Perimetry
8 days post stroke
Visual Angle (degrees)
Detection Sensitivity (dB)
C.2. Schematic showing eye gaze for grasping seen (blue) and unseen (red) handle
Target in blind visual field
Target in intact visual field
C.3. Matching to seen handle
Manipulated Handle Orientation (degrees)
C.4. Grasping a seen handle
Wrist Orientation (degrees)
Handle Orientation (degrees)
C.5. Matching to unseen handle
C.6. Grasping an unseen handle
Handle Orientation (degrees)

PLATE 77 Functional dissociations among tool representations in neuropsychology and functional neuroimaging. *A*, Limb apraxia is an impairment for using objects correctly that cannot be attributed to elemental sensory or motor disturbance. Variants of limb apraxia are distinguished by the nature of the errors that patients make. A patient with ideomotor apraxia may pantomime the use of a pair of scissors correctly in all ways, except, for instance, he moves the hand backward, opposite the direction of cutting (e.g., Garcea, Dombovy, & Mahon, 2013; for video examples, see www.openbrainproject.org). By contrast, a patient with ideational apraxia may deploy the wrong action for a given object while the action itself is performed correctly (e.g., using a toothbrush to brush one's hair). The distinction between ideomotor apraxia and ideational apraxia is loosely analogous to the distinction between phonological errors in word production (saying "caz" instead of "cat") and semantic errors in speech production (saying "dog" instead of "cat"; Rothi, Ochipa, & Heilman, 1991). The key point is that regardless of the nature of the errors patients make (spatiotemporal, content), the ability to name the same objects or access knowledge about their function can remain intact, indicating that the loss of motor-relevant information does not compromise conceptual processing in a major way. *B*, Laurel Buxbaum and colleagues have synthesized a framework within which to parcellate functional subdivisions within parietal cortex through the lens of everyday actions (Binkofski & Buxbaum, 2013; see also Garcea & Mahon, 2014; Mahon, Kumar, & Almeida, 2013; Peeters et al., 2009; Pisella et al., 2006). Left inferior parietal areas support action planning and praxis and operate over richly interpreted object information, such as that generated through processing in the ventral pathway, while posterior and superior parietal areas support "classic" dorsal stream processing involving online visuomotor control. A recent line of studies sought to determine which tool responses in parietal cortex depend on ventral stream processing by taking advantage of the fact that the dorsal visual pathway receives little parvocellular input (Livingstone & Hubel, 1988; Merigan & Maunsell, 1993). Thus, if images of tools and a baseline category (e.g., animals) are titrated so as to be defined by visual dimensions that are not "seen" by the dorsal pathway (because they require parvocellular processing), one can infer that regions of parietal cortex that continue to exhibit tool preferences receive inputs from the ventral stream. It was found that tool preferences were restricted to the aIPS and the supramarginal gyrus (figure 64.2) when stimuli contained only high spatial frequencies (Mahon, Kumar, & Almeida, 2013), were presented at a low temporal frequency (Kristensen, Garcea, Mahon, & Almeida, 2016), or were defined by red/green isoluminant color contrast (Almeida, Fintzi, & Mahon, 2013). Those findings suggest that neural responses to tools in the left inferior parietal areas are dependent on processing in the ventral visual pathway. *C*, Findings from *action blindsight* indicate that subcortical projections to the dorsal stream can support analysis of basic volumetrics about the shape and orientation of grasp targets. Prentiss, Schneider, Williams, Sahin, and Mahon (2018) described a hemianopic patient who by chance performed when making a perceptual matching judgment about the orientation of a handle presented in the hemianopic field, while he was able to spontaneously and accurately orient his wrist when the handle was the target of a grasp. (See figure 64.2.)

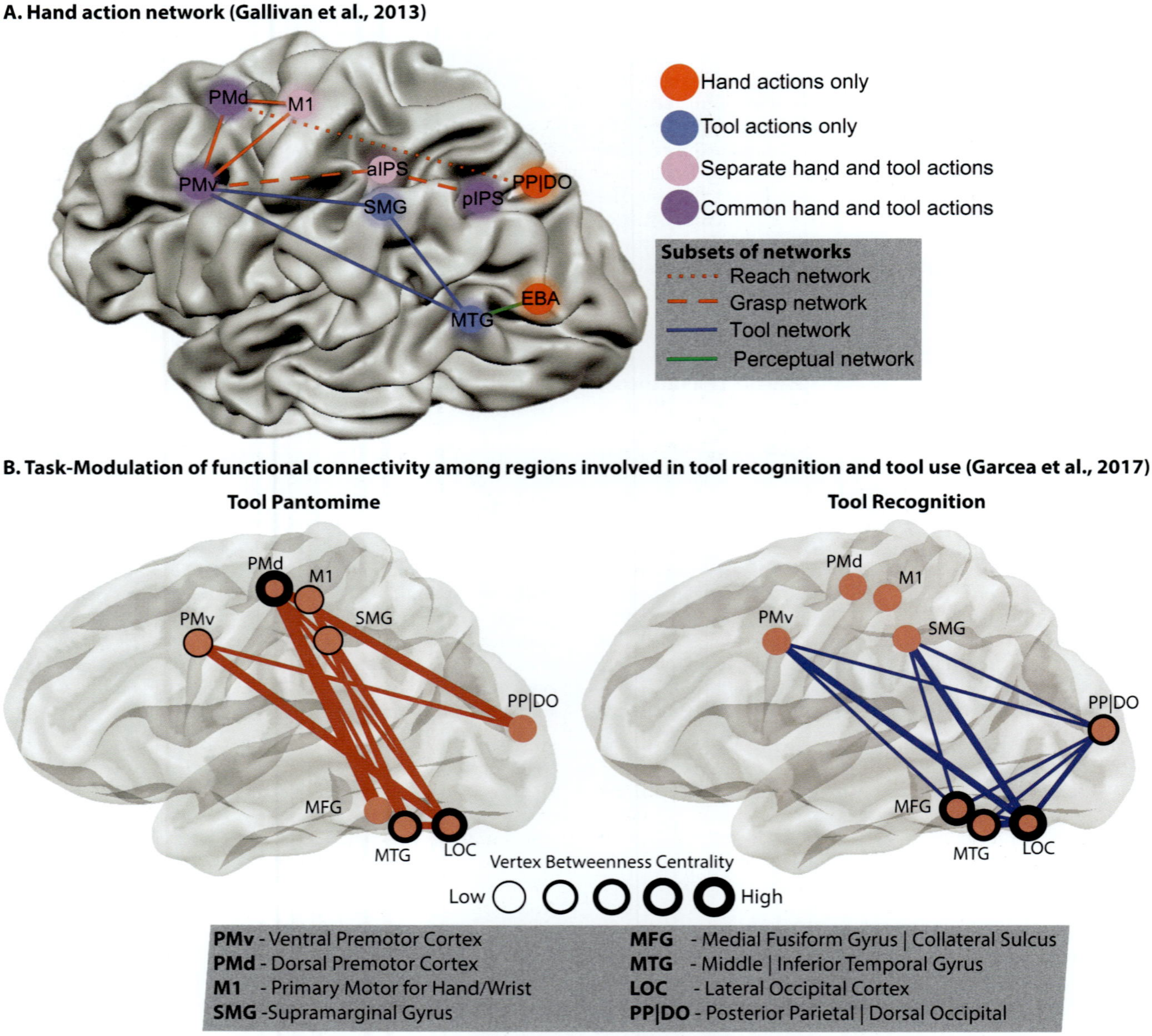

PLATE 78 The next big step is to work toward a processing model that provides an answer to the question: How does the brain translate an abstract goal (eat dinner) into a specific object-directed action (grasp and use *this* fork)? A processing model would specify the types of representations and computations engaged during object recognition and functional object grasping and use, the order in which those computations are engaged, and their neural substrates. The key to developing such a processing model will be a careful analysis of how different tasks modulate connectivity in the system. The stronger suggestion is that it will not be possible to develop generative theories of the computations supported by discrete brain regions without understanding how the connectivity of those regions changes with different "goal states" of the system. Panels *A* and *B* represent two recent attempts using functional MRI to study task-modulated functional connectivity among regions of the brain specialized for translating propositional attitudes into goals (i.e., the "tool-processing network"). Future research with high temporal resolution will be necessary to understand whether there are dissociated "waves" of interactions among overlapping sets of brain regions that unfold in a task-driven manner. (See figure 64.3.)

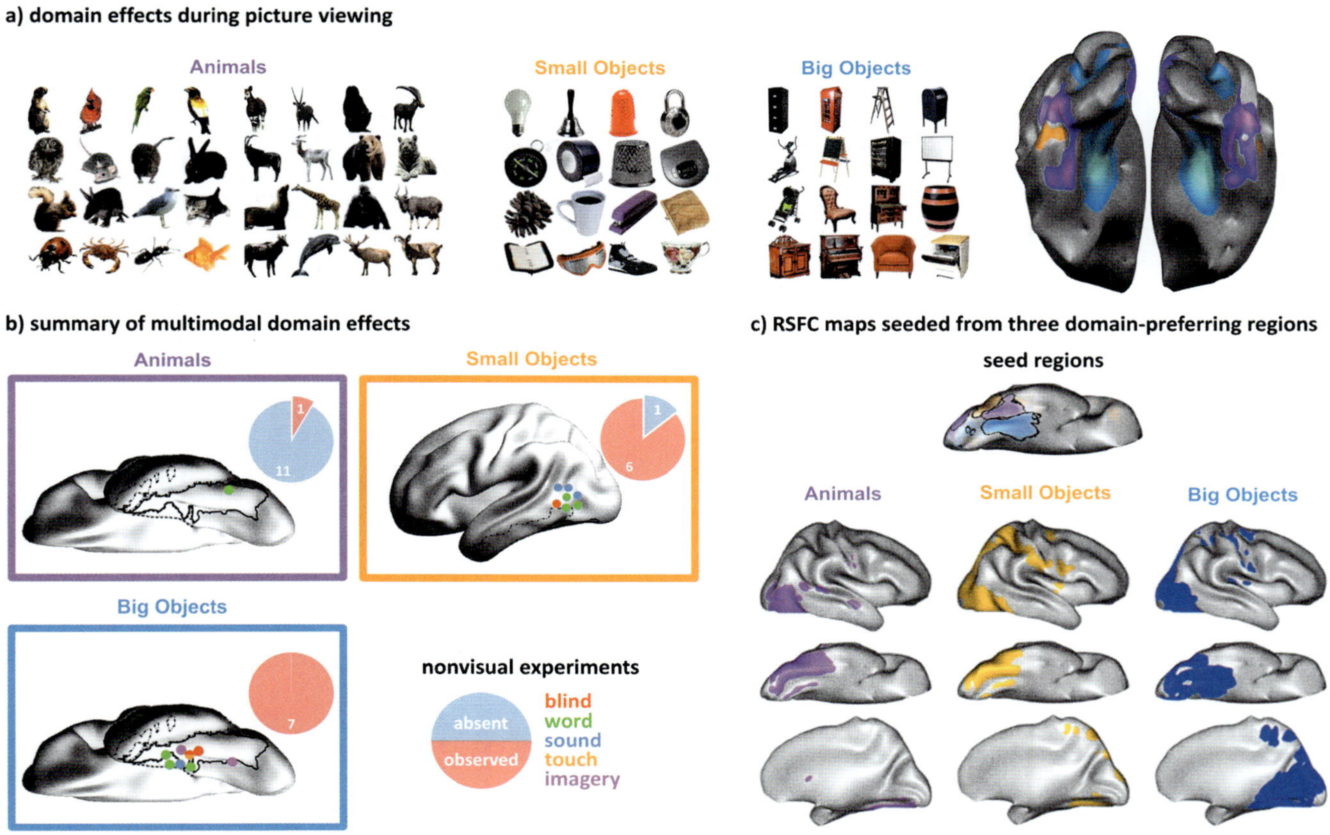

PLATE 79 The functionality and connectivity pattern of the VOTC domain-preferring clusters. *A*, Visual experiments: the three domain-preferring clusters in VOTC that associate with viewing pictures of large objects, small manipulable objects, and animals. Adapted from Konkle and Caramazza (2013). *B*, Nonvisual experiments: The two artifact clusters in (*A*) show consistent domain effects in nonvisual experiments, whereas the animal cluster tended not to show preference to animals when the stimuli were nonvisual. The color dots on the brain map correspond to the studies summarized in Bi et al. (2016, table 1), with different colors indicating different types of nonvisual input. Pie charts show the number of studies in which nonvisual domain effects were observed (*red*) or absent (*blue*). *C*, The resting-state functional connectivity patterns that associate with the three domain-preferring clusters. Adapted from Konkle and Caramazza (2017). (See figure 66.1.)

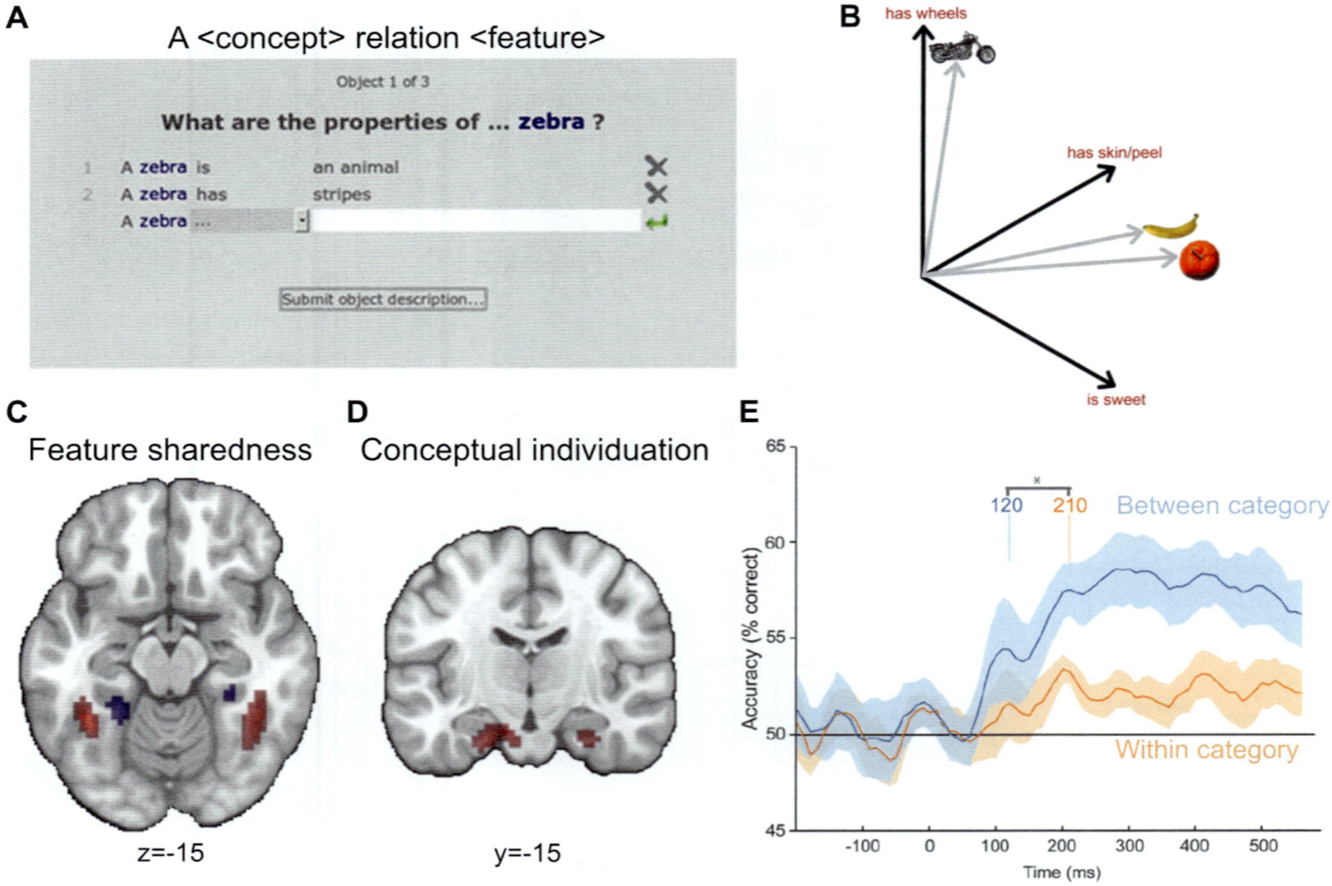

PLATE 80 Semantic features. *A*, Example of collecting features for a given concept in a feature-norming study. *B*, Concepts can be more similar or different based on how similar the feature lists are, meaning they are closer together in a multidimensional feature space (three dimensions shown for clarity). *C*, Regions in the posterior ventral temporal lobe were modulated by feature-based statistics, in which more lateral regions showed increased activity for objects with relatively more shared features, and medial regions showed increased activity for objects with relatively more distinctive. *D*, Bilateral anteromedial temporal cortex (AMTC) activity increases for concepts that are semantically more confusable. *E*, The feature-based model can be used to successfully classify concepts from MEG signals, where between-category information (e.g., animal vs. tool) occurs before within-category information (e.g., lion vs. tiger). Panel (*A*) reproduced from Devereux et al. (2014), panel (*B*) from Devereux et al. (2018), and panel (*E*) from Clarke et al. (2015), all under the Creative Commons License. Panels (*C*) and (*D*) reproduced from Tyler et al. (2013). (See figure 67.1.)

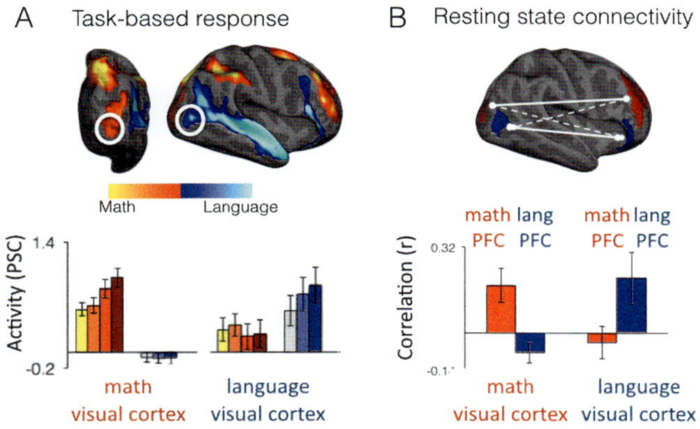

PLATE 81 Responses to language and number in visual cortices of congenitally blind individuals. *A*, Math-responsive "visual" areas (*red*) show an effect of math equation difficulty (*increasingly dark-red bars*). Language-responsive "visual" areas show an effect of grammatical complexity: lists of nonwords (*gray*), grammatically simple sentences (*light blue*), and complex (*dark blue*) sentences. *B*, Stronger resting-state correlations with language-responsive PFC in language-responsive visual cortex and with math-responsive PFC in math-responsive visual cortex. (See figure 68.1.)

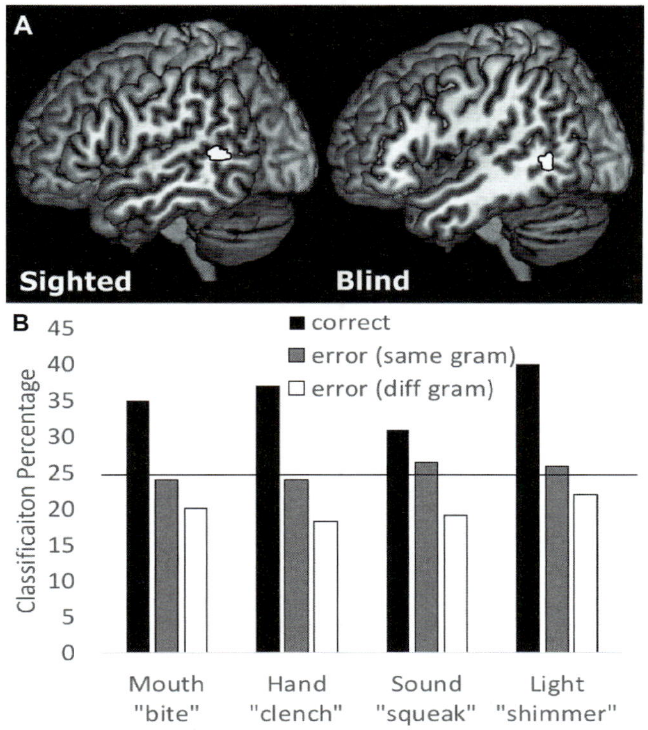

PLATE 82 Representations of verb meanings in the left middle temporal gyrus (LMTG). *A*, Action verbs > object nouns in sighted (*left*) and congenitally blind individuals (*right*). Reprinted from Bedny et al. (2012). *B*, Performance of linear classifier distinguishing among four verb types based on patterns of activity in the LMTG of sighted individuals: transitive mouth and hand actions and intransitive light- and sound-emission events. The classifier successfully distinguished among mouth and hand actions and light- and sound-emission events. Errors across grammatical type (*white bars*; e.g., transitive mouth action mistaken for intransitive light-emission event) are less common than within grammatical type (*gray bars*; e.g., mouth action mistaken for hand action). From Elli, Lane, and Bedny (2019). (See figure 68.2.)

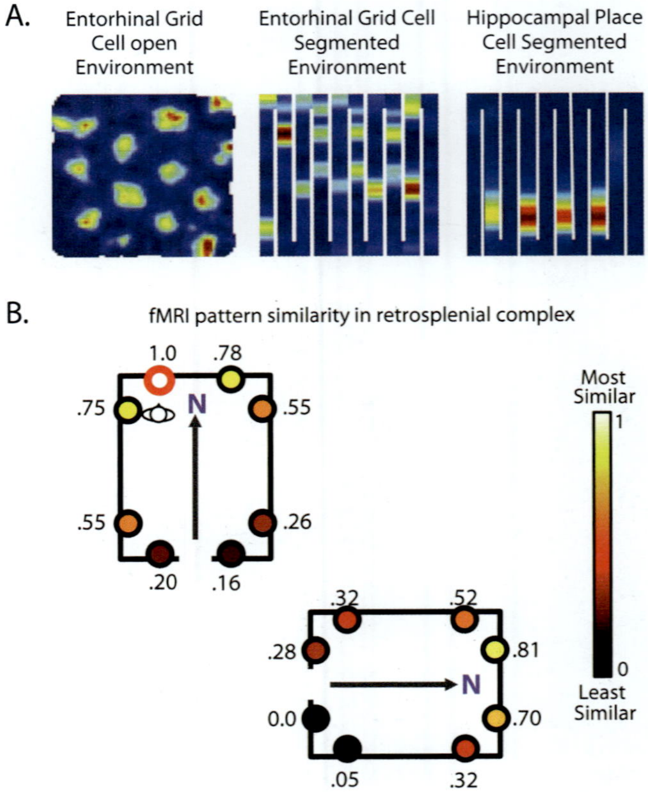

PLATE 83 Spatial representations in structured environments. *A*, Grid cells code a regular triangular grid in open environments, but this pattern fragments into repetitive local fields when the environment is segmented into smaller subchambers (*white lines indicate walls*). A similar effect of pattern fragmentation is observed in hippocampal place cells. *B*, In a multichamber environment, RSC represents local geometric organization. Participants imagined facing an object along the wall at each location indicated by a circle. Colors and numbers indicate the similarity of multivoxel patterns for each view compared to the reference view (*red circle*). There is a high degree of similarity between views facing "local north" (i.e., away from the entrance) in different subchambers. (See figure 69.2.)

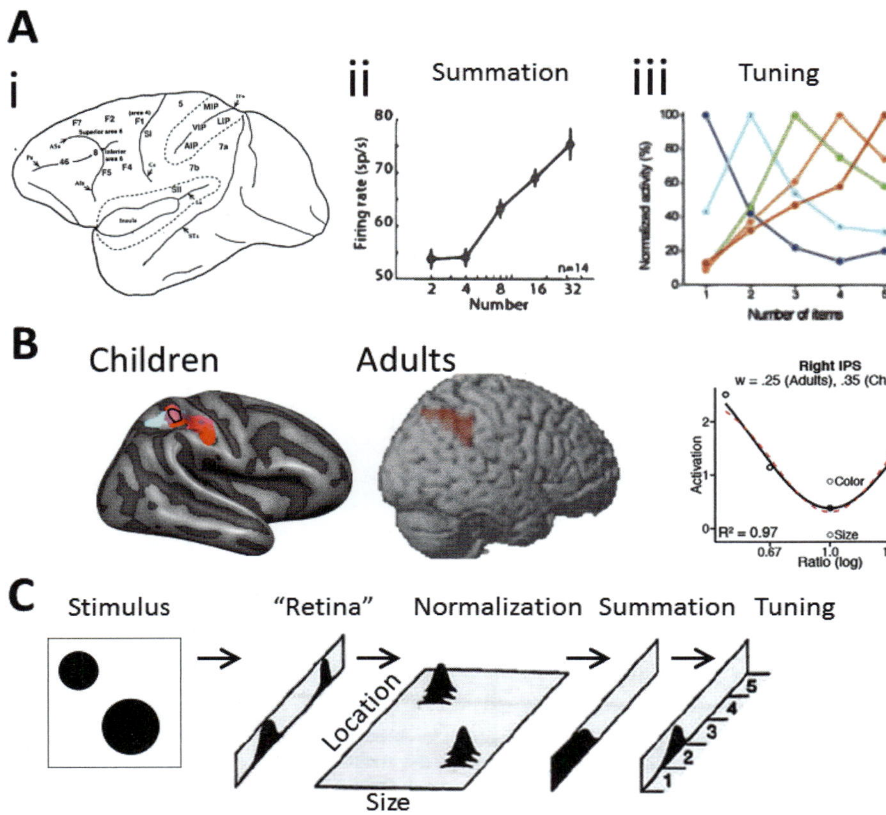

PLATE 84 *A*, Neurons in intraparietal areas VIP and LIP show numerical sensitivity (*1*). In area VIP, neurons respond to numerical stimuli with a monotonic *summation* response (*2*) and in LIP with a *tuning* response (*3*). *B*, Human children and adults show numerical sensitivity in the IPS (*red*). Neural responses in the IPS (*right*) show tuning to numerosity during fMRI adaptation based on the ratio of change in the adaptation stream. Adults show sharper neural tuning to numerosity in the left IPS compared to children. *C*, Dehaene and Changeux (1993) modeled numerical representation in a neural network. Visual objects in an array stimulus are first normalized to a location- and size-independent code. Activation is then summed to yield an estimate of the input numerosity. Numerosity detectors are connected to summation activation, and neural activity is tuned to numerosity in an on-center, off-surround pattern. (See figure 70.1.)

PLATE 85 Two plausible interpretations of the novel concept *robin hawk*. *Top*, A hawk with the red breast of a robin. *Bottom*, A hawk that preys on robins. (See figure 71.1.)

	Combinatory network shows more activity for **sentences** compared to **word lists**	**Simple composition network** shows more activity for **simple phrases** compared to **words**	**Sensitive to semantics** in **syntactically parallel** expressions	Activity **better fit by hierarchical models** compared to **sequence-based** models
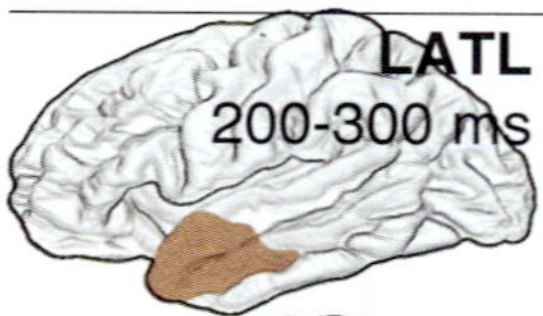 **LATL** 200-300 ms	☒ ☒ ☒	☒ ☒ ☒	☒ ☒ ☒	☒
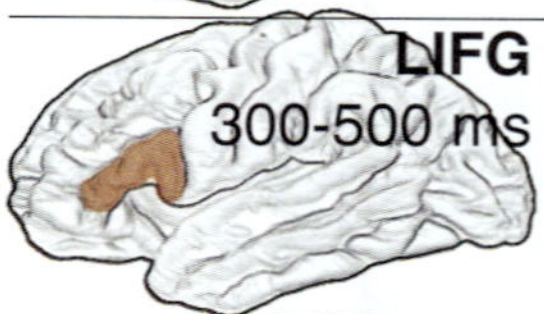 **LIFG** 300-500 ms	☒ ☒ ☐	☒ ☐ ☐	☐ ☐	☐
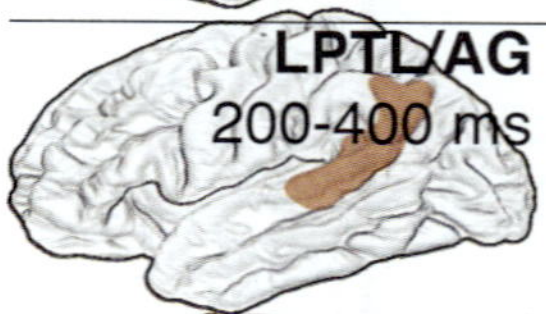 **LPTL/AG** 200-400 ms	☒ ☒ ☐	☒ ☐ ☐	☐ ☐	☒
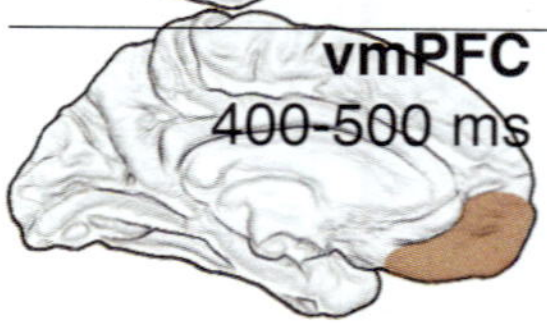 **vmPFC** 400-500 ms	☒ ☐ ☐	☒ ☐	☒ ☒	(no data)

PLATE 86 An informal depiction of our current understanding of the brain regions supporting composition and the extent to which the functional roles of individual network nodes are understood. A lack of understanding can result either from a lack of studies or from a lack of generalizations across studies. Here, the number of boxes in each cell represents the general quantity of studies addressing the role of the region, and the checks inside the boxes represent the amount of positive evidence for the generalization in the first row. Timing estimates primarily reflect results from MEG studies comparing sentence-versus-list activation (e.g., Brennan & Pylkkänen, 2012) or phrase-versus-word activation (e.g., Bemis & Pylkkänen, 2011). The table does not separate results according to method; thus, for example, positive results for the LIFG come primarily from fMRI (and are thus ambiguous as regards timing) and ones for the vmPFC from MEG. Connecting separate findings from different methods is a major goal for future research. In all, the only network node showing a high degree of consistency across the literature is the LATL. (See figure 74.1.)

PLATE 87 *A*, The general topography of the high-level language network. This representation was derived by overlaying 207 individual activation maps for the contrast of reading sentences versus nonword sequences (Fedorenko et al., 2010). *B*, Language activations in six individuals tested in their native languages (using a contrast between listening to passages from *Alice's Adventures in Wonderland* versus the acoustically degraded versions of those passages; Scott, Gallee, & Fedorenko, 2016) that come from distinct language families. *C*, Language activations in three individuals tested across two scanning sessions. *D*, Key functional properties of two sample high-level language regions. The parcels used to define the individual functional regions of interest are shown in gray (each fROI is defined as the top 10% most language-responsive voxels); on the left, we show responses to several linguistic manipulations, and on the right, we show responses to nonlinguistic tasks. (See figure 75.1.)

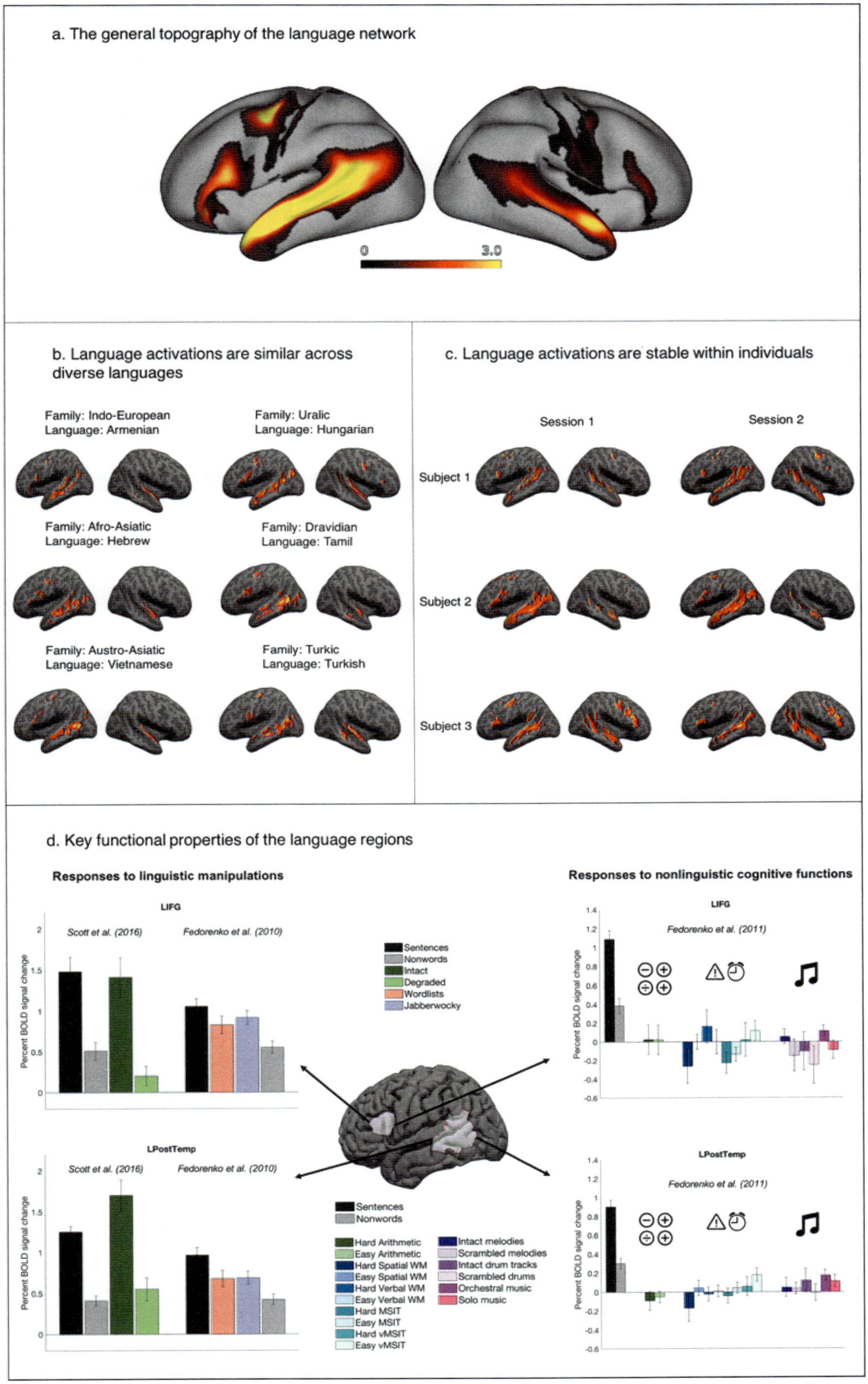

a. The general topography of the language network
0
3.0
b. Language activations are similar across diverse languages
Family: Indo-European
Language: Armenian
Family: Uralic
Language: Hungarian
Family: Afro-Asiatic
Language: Hebrew
Family: Dravidian
Language: Tamil
Family: Austro-Asiatic
Language: Vietnamese
Family: Turkic
Language: Turkish
c. Language activations are stable within individuals
Session 1
Session 2
Subject 1
Subject 2
Subject 3
d. Key functional properties of the language regions
Responses to linguistic manipulations
LIFG
Scott et al. (2016)
Fedorenko et al. (2010)
Percent BOLD signal change
Sentences
Nonwords
Intact
Degraded
Wordlists
Jabberwocky
LPostTemp
Responses to nonlinguistic cognitive functions
Fedorenko et al. (2011)
Hard Arithmetic
Easy Arithmetic
Hard Spatial WM
Easy Spatial WM
Hard Verbal WM
Easy Verbal WM
Hard MSIT
Easy MSIT
Hard vMSIT
Easy vMSIT
Intact melodies
Scrambled melodies
Intact drum tracks
Scrambled drums
Orchestral music
Solo music

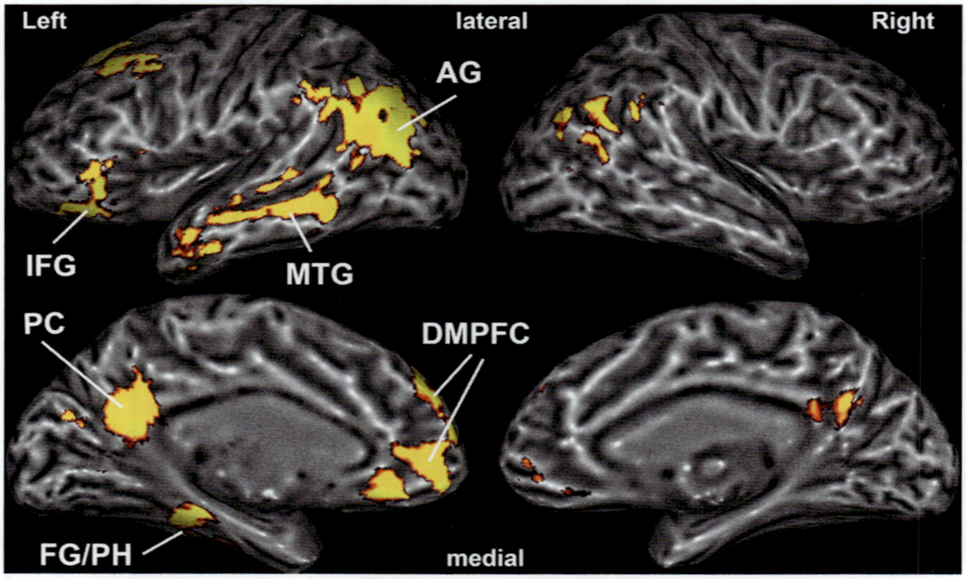

PLATE 88 Results of an activation likelihood estimate meta-analysis of 87 published studies (691 activation foci) using controlled semantic contrasts (Binder et al., 2009). AG = angular gyrus; DMPFC = dorsomedial prefrontal cortex; FG/PH = fusiform and parahippocampal gyri; IFG = inferior frontal gyrus; MTG = middle temporal gyrus; PC = posterior cingulate/precuneus. (See figure 76.1.)

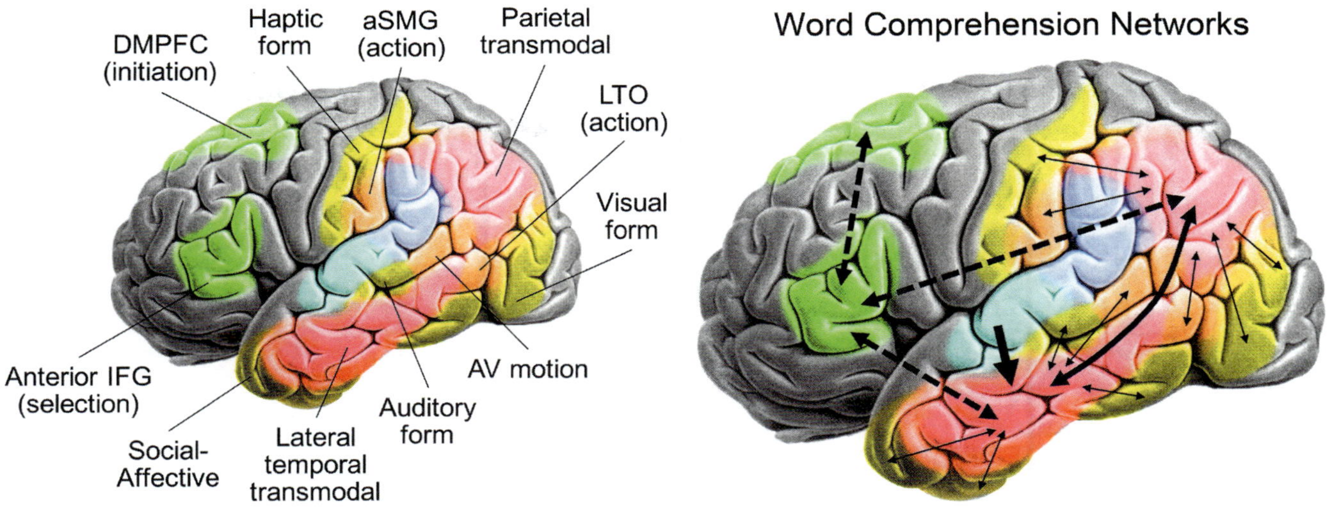

PLATE 89 A schematic model of lexical storage and access networks, showing some principal unimodal (*yellow*), multimodal (*orange*), and transmodal (*red*) conceptual stores; semantic control regions (*green*); and speech perception (*cyan*) and phonological access (*blue*) areas. Spoken-word comprehension (*diagram at right*) involves mapping from auditory speech forms to high-level conceptual representations (*fat arrow*). The subsequent activation of multimodal and unimodal experiential representations (*thin arrows*) enables perceptual grounding and perceptual imagery and likely varies with task demands. Concept selection and information flow (depth of processing) are controlled by initiation and selection mechanisms in dorsomedial and inferolateral prefrontal cortex. (See figure 76.2.)

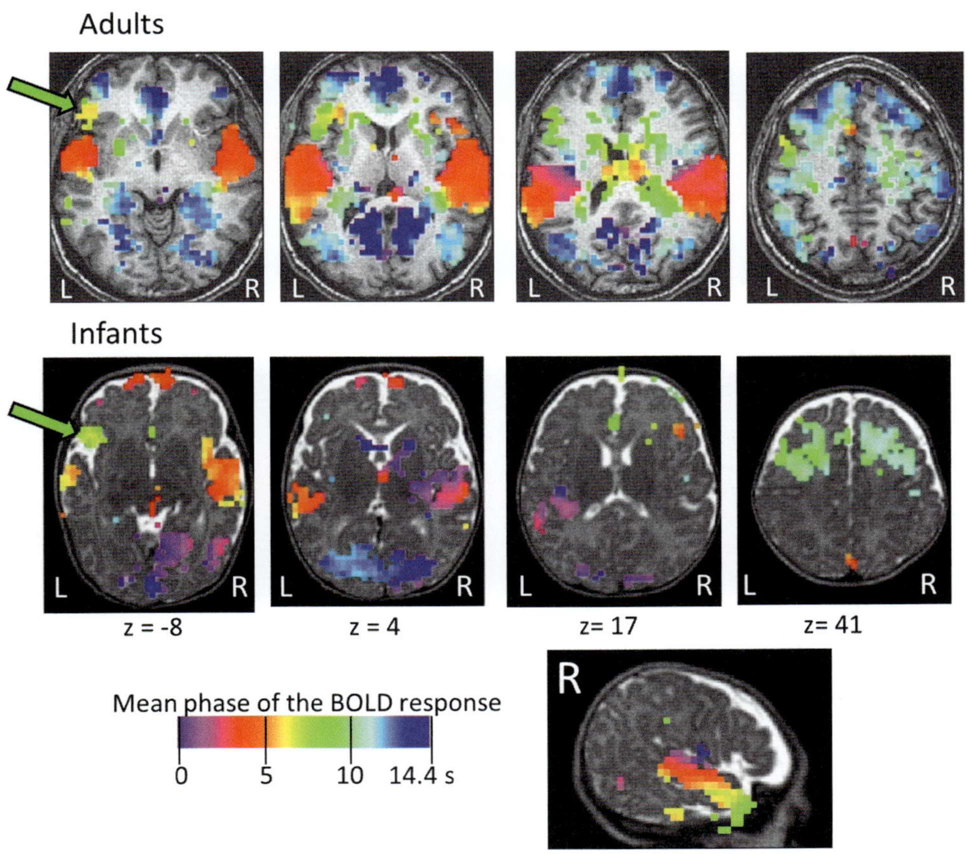

PLATE 90 Hierarchical organization of the perisylvian regions in 3-month-old infants and adults, illustrated by the phase gradient of the BOLD response to a single sentence. The mean phase is presented on axial slices placed at similar locations in the adult (*top row*) and infant (*bottom row*) standard brains and on a sagittal slice in the infant's right hemisphere. Colors encode the circular mean of the phase of the BOLD response, expressed in seconds relative to sentence onset. The same gradient is observed in both groups along the superior temporal region, extending until Broca's area (*arrow*). Blue regions are out of phase with stimulation (Dehaene-Lambertz, Hertz-Pannier, et al., 2006; Dehaene-Lambertz, Dehaene, et al., 2006). (See figure 78.1.)

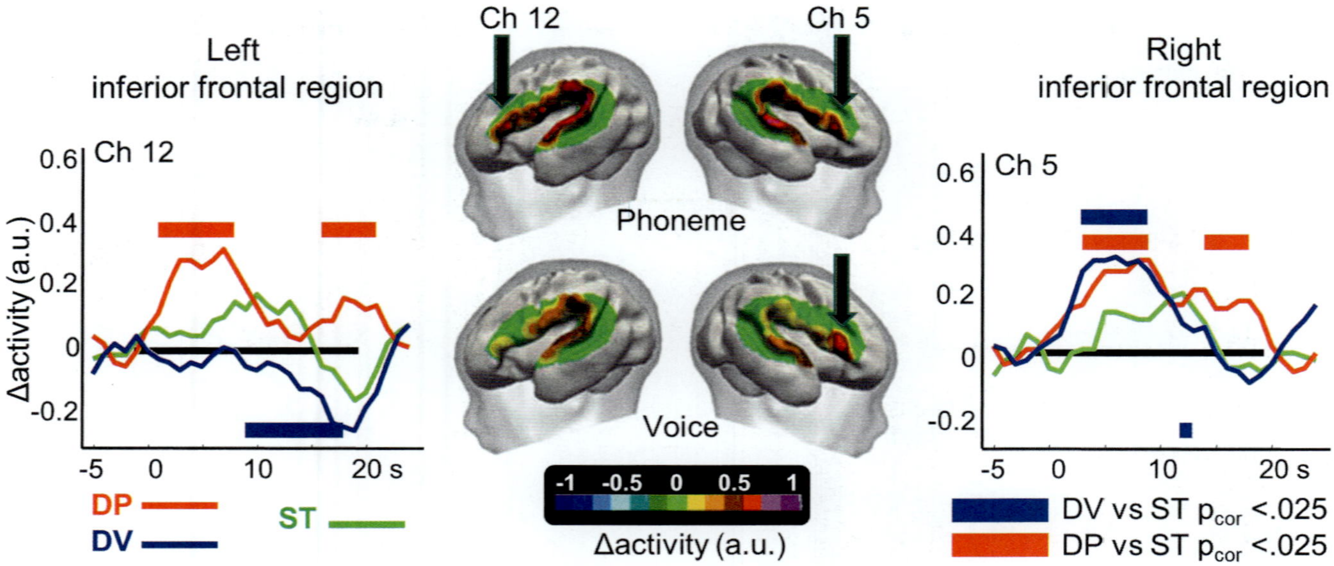

PLATE 91 Parallel pathways in preterms. Oxyhemoglobin responses to a change of phoneme (*/ba/* vs. */ga/*) and a change of voice (male vs. female) measured with NIRS in 30 weeks gestational age—old preterm neonates. A significant increase in the response to a change of phoneme (DP, deviant phoneme) relative to the standard condition (ST) was observed in both temporal and frontal regions, whereas the response to a change of voice (DV, deviant voice) was limited to the right inferior frontal region. The left inferior frontal region responded only to a change of phoneme, whereas the right responded to both changes. The colored rectangles represent the periods of significant differences between the deviant and the standard conditions in the left and right inferior region (*black arrows*; Mahmouzadeh et al., 2013). (See figure 78.2.)

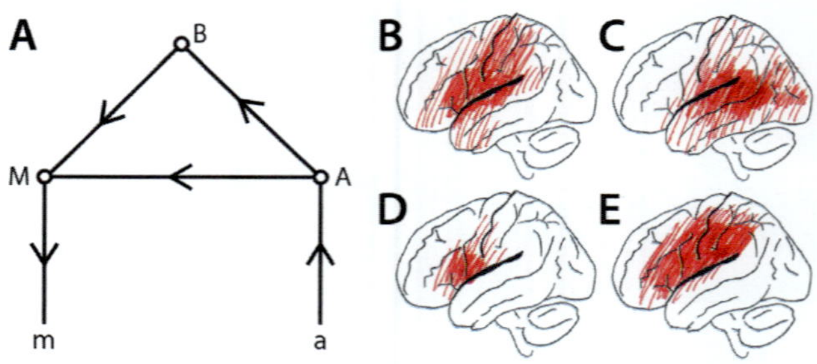

PLATE 92 *A*, The Wernicke-Lichtheim model (Lichtheim, 1885). *B*, Lesion overlay of 14 patients with Broca's aphasia (Kertesz et al., 1977). The intensity of shading indicates the number of patients with lesions. *C*, Lesion overlay of 13 patients with Wernicke's aphasia (Kertesz et al., 1977). *D*, Lesion overlay of 13 patients with infarction restricted to Broca's area (Mohr, 1976). *E*, Lesion overlay of 10 patients with persistent Broca's aphasia (Mohr, 1976). (See figure 79.1.)

PLATE 93 Neural correlates of language deficits in individuals. Voxel-based morphometry revealed distinct regions where atrophy was predictive of speech (*A*), lexical (*B*), or syntactic (*C*) deficits (Wilson et al., 2010). Arrows denote increases or decreases in the prevalence of the phenomena listed. Dorsal and ventral language tracts were identified with diffusion tensor imaging (*D*). ECFS = extreme capsule fiber system; SLF/AF = superior longitudinal fasciculus/arcuate fasciculus. The degeneration of dorsal tracts was associated with deficits in syntactic comprehension (*E*) and production (*F*), while the degeneration of ventral tracts had no effects on syntactic comprehension (*G*) or production (*H*) (Wilson et al., 2011). Functional imaging identified brain regions where recruitment for syntactic processing was predictive of success in syntactic processing in PPA (*I*). In the inferior frontal gyrus (*J, K*) and posterior temporal cortex (*L, M*), modulation of functional signal by syntactic complexity was predictive of accuracy (*J, L*), but nonspecific recruitment for the task was not (*K, M*) (Wilson et al., 2016). (See figure 79.2.)

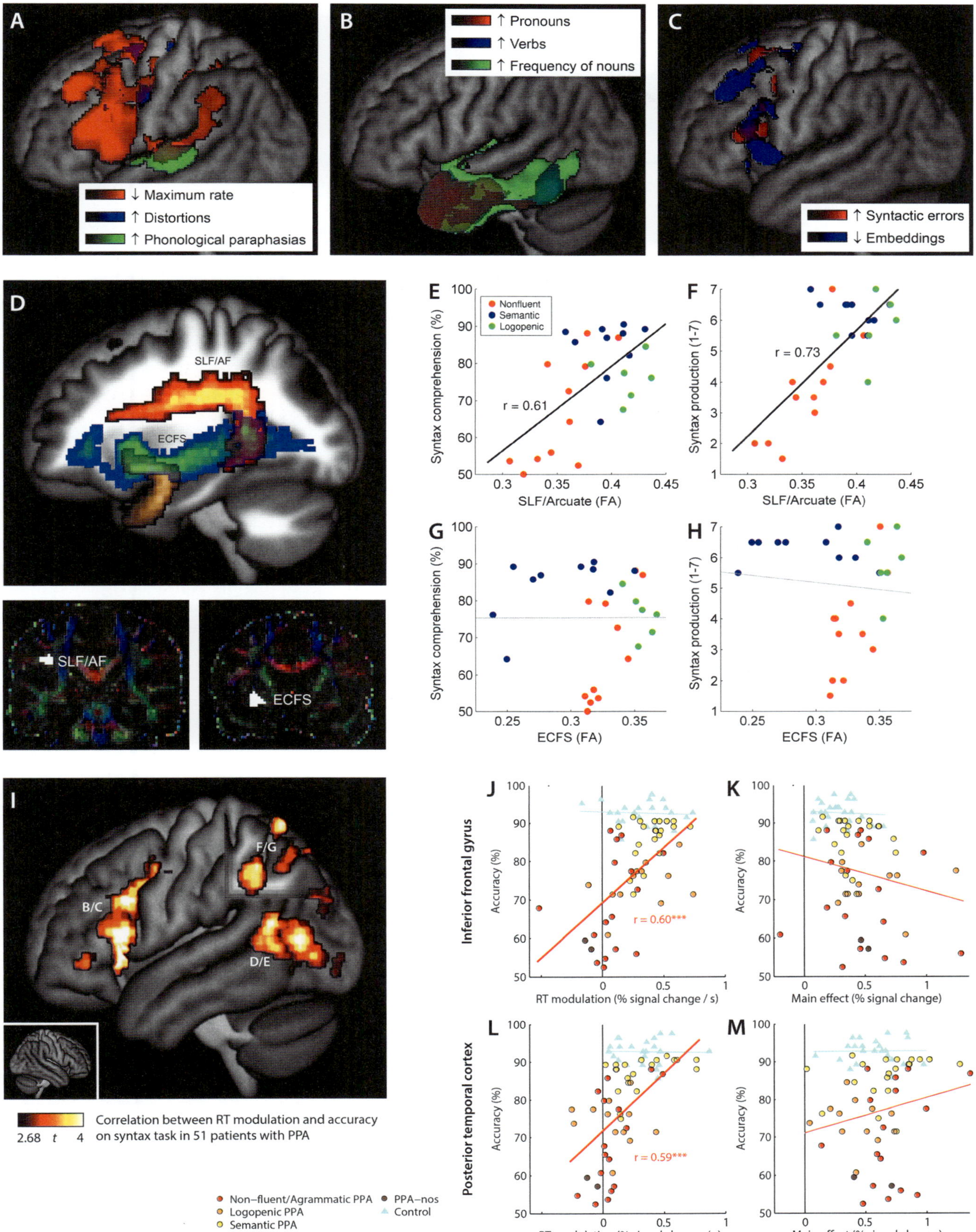
A
↓ Maximum rate
↑ Distortions
↑ Phonological paraphasias
B
↑ Pronouns
↑ Verbs
↑ Frequency of nouns
C
↑ Syntactic errors
↓ Embeddings
D
SLF/AF
ECFS
E
Nonfluent
Semantic
Logopenic
r = 0.61
Syntax comprehension (%)
SLF/Arcuate (FA)
F
r = 0.73
Syntax production (1-7)
SLF/Arcuate (FA)
G
Syntax comprehension (%)
ECFS (FA)
H
Syntax production (1-7)
ECFS (FA)
I
B/C
F/G
D/E
Correlation between RT modulation and accuracy on syntax task in 51 patients with PPA
2.68 t 4
J
K
Inferior frontal gyrus
Accuracy (%)
r = 0.60***
RT modulation (% signal change / s)
Main effect (% signal change)
L
M
Posterior temporal cortex
r = 0.59***
Non–fluent/Agrammatic PPA
Logopenic PPA
Semantic PPA
PPA–nos
Control

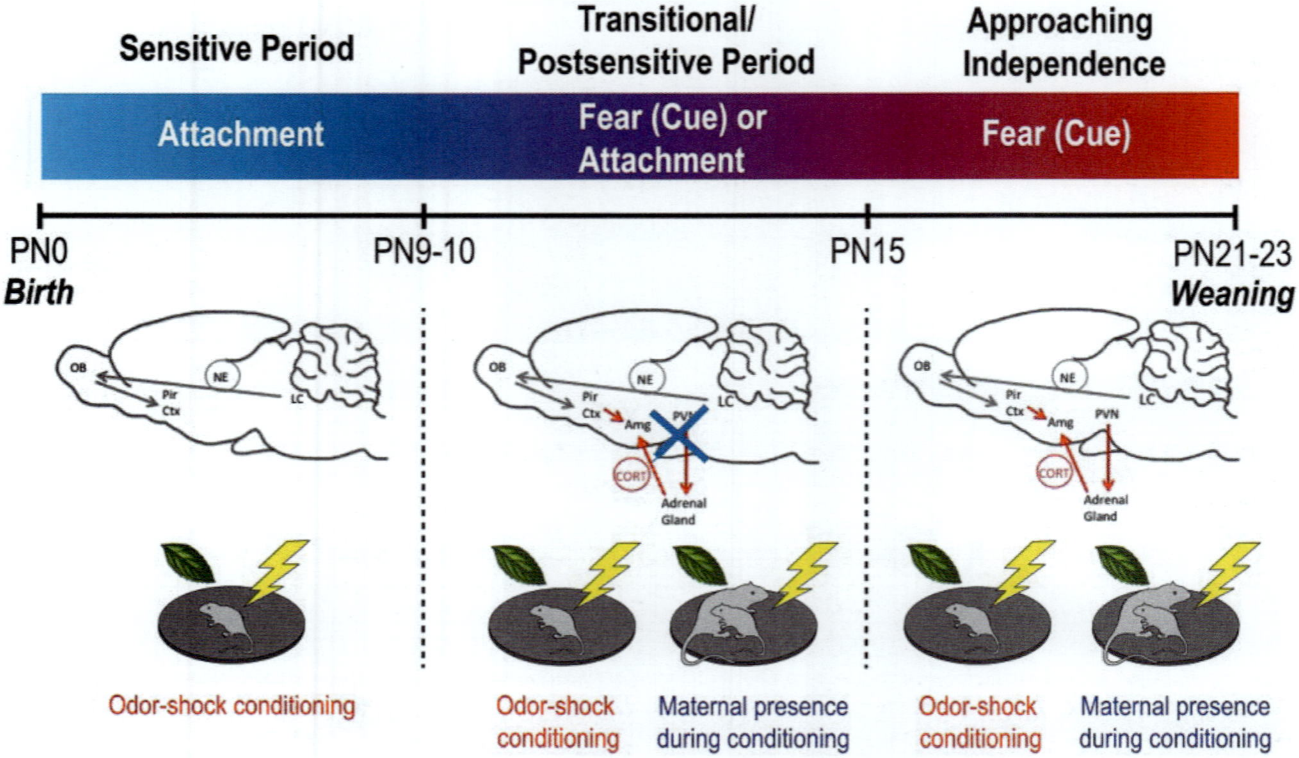

PLATE 94 This schematic represents pups' developmental learning transitions with odor-0.5 mA shock conditioning. Our previous work suggests PN10 is a transitional age for the onset of amygdala-dependent fear conditioning, although until PN15, this learning depends on CORT levels, which can be modulated pharmacologically or by the maternal presence during conditioning. During this transitional period (until PN15), pups conditioned alone learn to avoid an odor paired with shock but will learn attachment when conditioned with lowered levels of CORT. After PN15, conditioning alone or with the maternal presence produces odor avoidance. (See figure 80.2.)

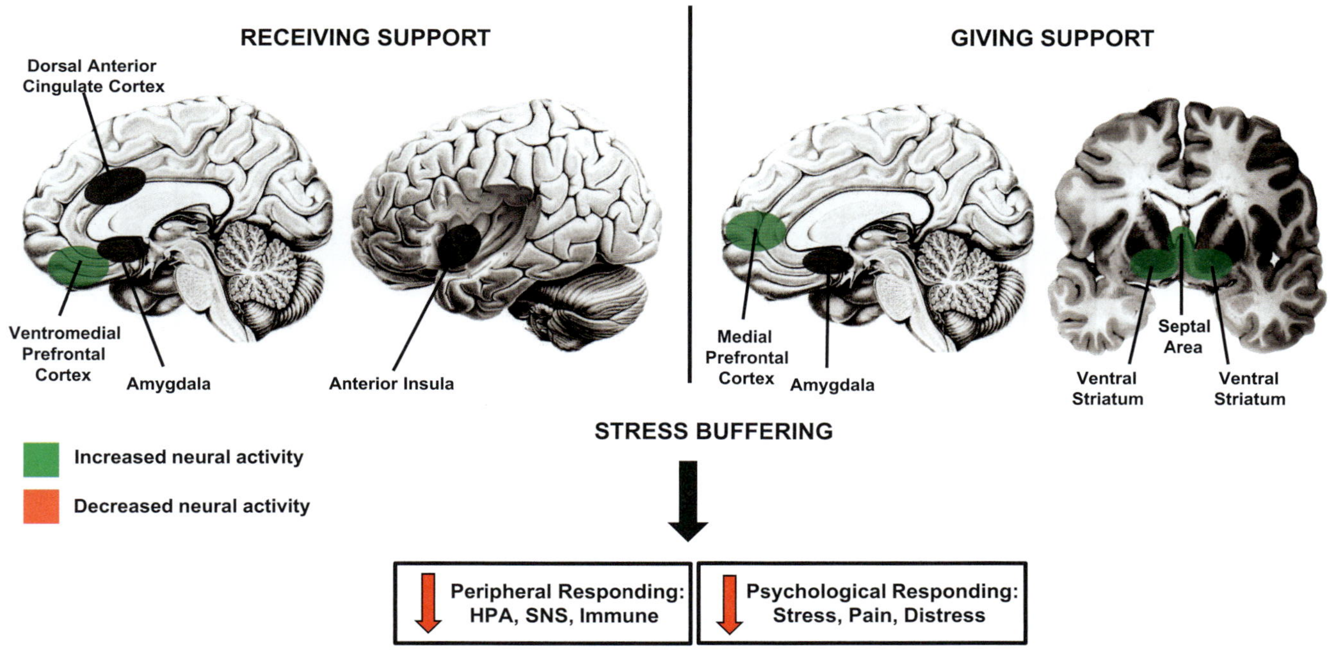

PLATE 95 Neural mechanisms underlying the stress-buffering effects of social support. Receiving support leads to increased activity (*green*) in the ventromedial prefrontal cortex (vmPFC) and decreased activity (*red*) in the dorsal anterior cingulate cortex (dACC) and anterior insula (AI), regions that play a critical role in the distressing experience of pain. Giving support leads to increased activity in the medial prefrontal cortex (mPFC), ventral striatum (VS), and septal area (SA). Given the known inhibitory connections between the vmPFC (active during receiving support) and the SA (active during giving support) with the amygdala, both receiving and giving support may lead to decreased activity in the amygdala, a threat-related region that plays a key role in the stress response, resulting in the reduced activation of peripheral systems (hypothalamic-pituitary-adrenal axis [HPA], sympathetic nervous system [SNS], and immune system) and reduced psychological stress. (See figure 81.1.)

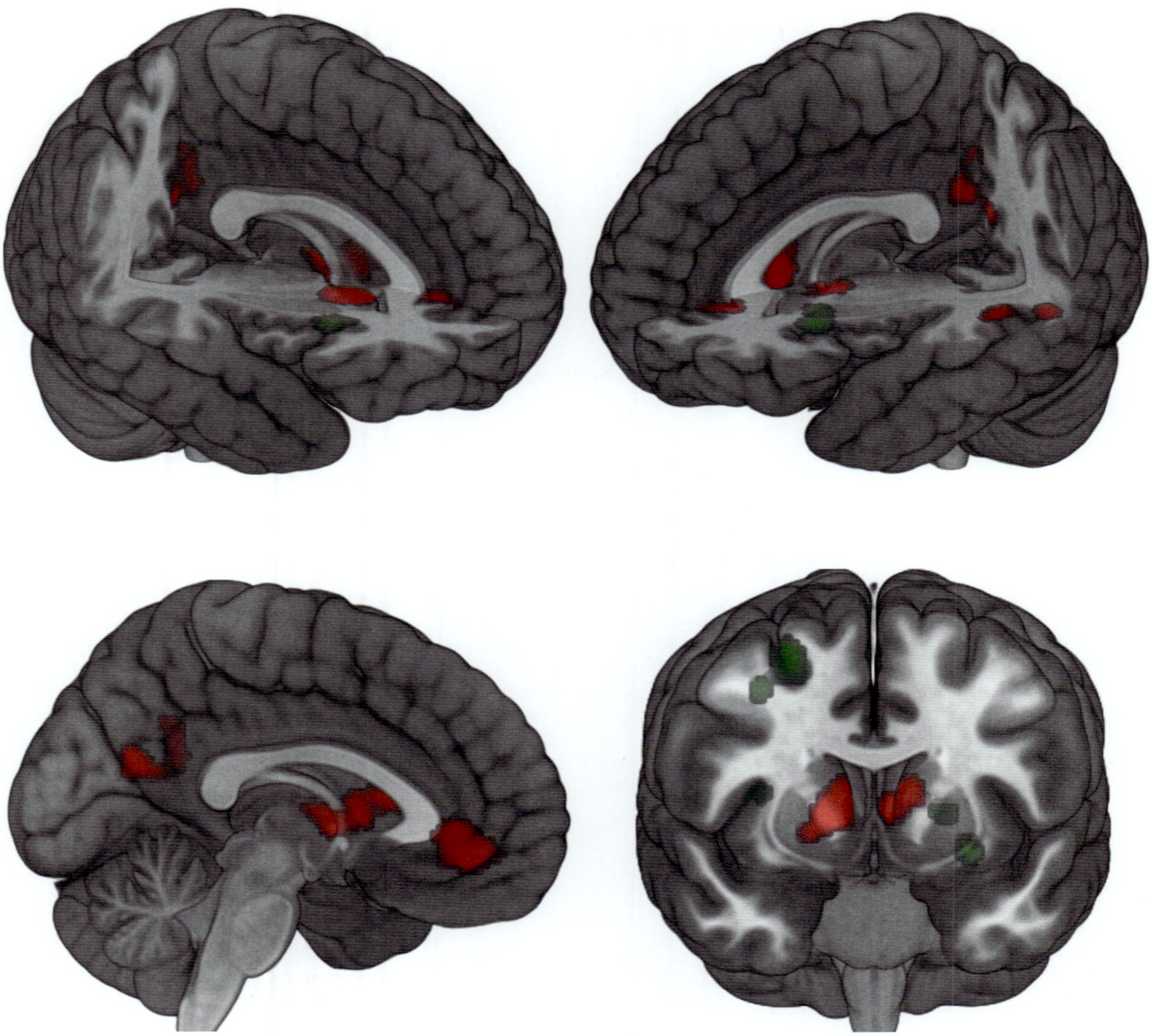

PLATE 96 Activation-likelihood meta-analyses using GingerALE (Eickhoff et al., 2009) were conducted to generate illustrative maps of neural circuitry supporting "learning from" (*green*) and "learning about" (*red*) others. Maps were set to an initial height threshold of $p < .005$ and corrected at the cluster level to $p < .05$. Studies included in these meta-analyses are marked with an * (*learning from*) and a † (*learning about*) in the reference section. (See figure 83.1.)

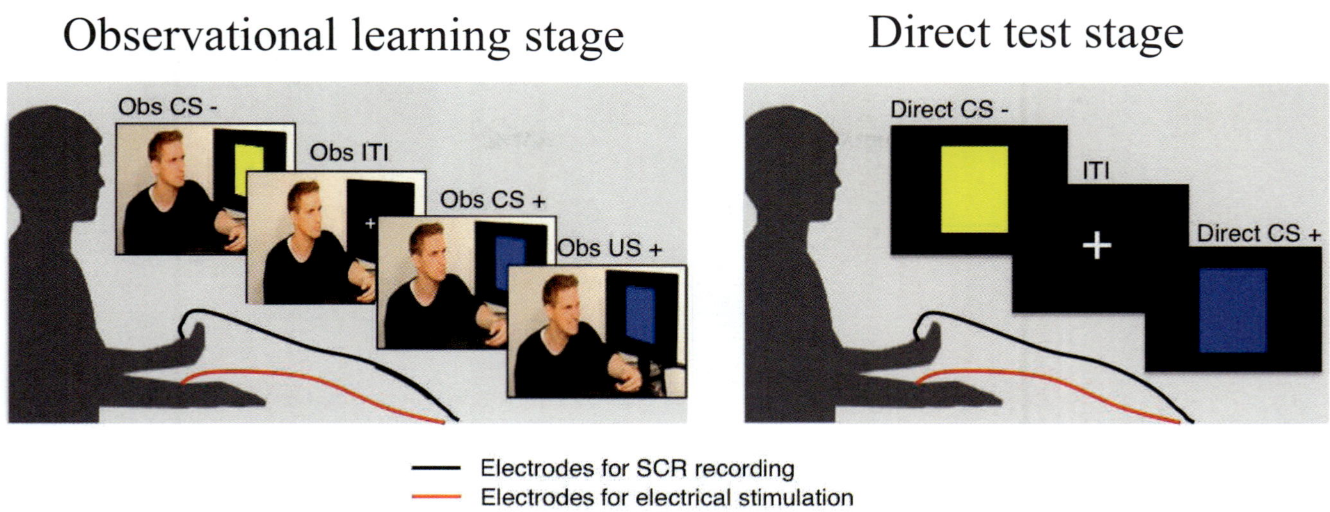

PLATE 97 General design of the observational fear-conditioning protocol depicting the observer (participant; *in shaded gray*), first watching the demonstrator's responses to the CS-US shock pairings (observational learning stage), followed by being exposed to the CS (direct test stage). The observer receives no shocks during the test stage. *Notes*: CS−, conditioned stimulus never paired with shock; CS+, conditioned stimulus paired with shock; ITI, intertrial interval; Obs, observational. Adapted from Haaker, Golkar, et al. (2017). (See figure 84.1.)

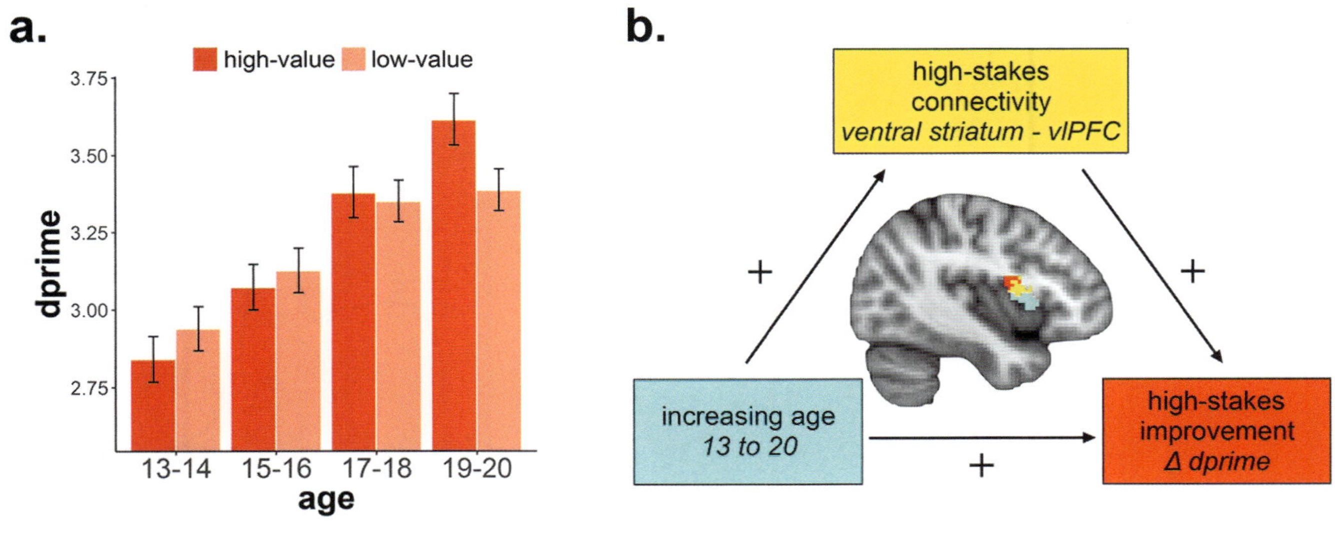

PLATE 98 *A*, When performing a cognitive-control task for low- versus high-value outcomes, older participants selectively improved performance (*dprime* on y-axis) when high-value incentives were at stake, whereas younger participants performed similarly for low-value and high-value conditions. *B*, Functional connectivity analyses seeded in the ventral striatum identified connectivity with ventrolateral prefrontal cortex (VLPFC) that was greater for high-value relative to low-value trials. This pattern of corticostriatal connectivity mediated the relationship between age and value-selective performance. Figure adapted with permission from Insel et al. (2017). (See figure 85.1.)

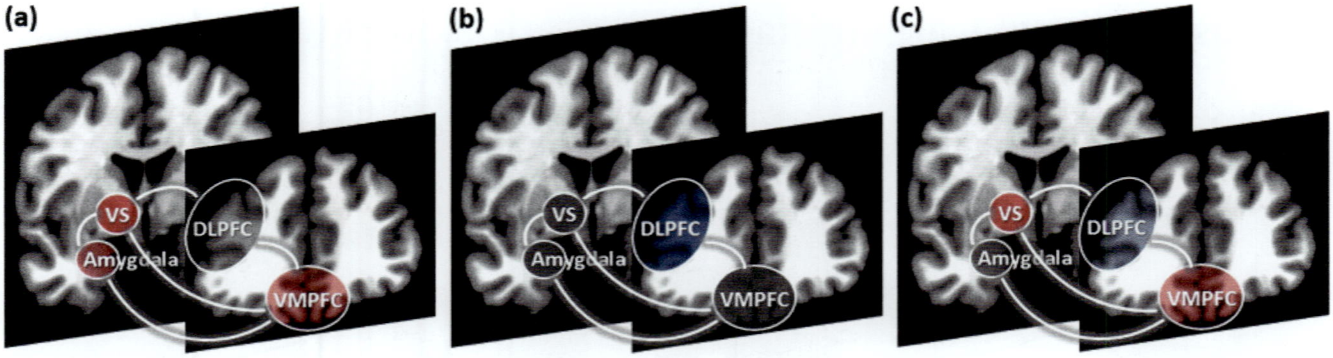

PLATE 99 Candidate neural systems of cooperative decision-making. Dual-process models of prosocial behavior predict cooperation stems from either (*A*) neural regions involved in intuition (*red*) or (*B*) neural regions involved in deliberation (*blue*). Or, (*C*) value-based models predict cooperation should stem from regions typically recruited during decision making (*red*), as well as heightened connectivity between the dlPFC (*blue*) and vmPFC for decisions that require more effort. VS = ventral striatum; vmPFC = ventromedial prefrontal cortex; dlPFC = dorsolateral prefrontal cortex. Graphics adapted from (Phelps, Lempert, & Sokol-Hessner, 2014). (See figure 86.1.)

Stage	Addiction formation	Addiction maintenance
Behavior	Reinforcement learning; Goal-directed behaviors	Habitual response; Compulsive drug taking
Neural candidates	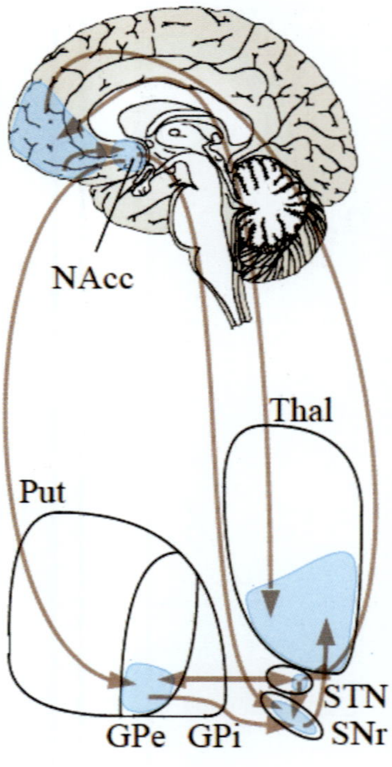	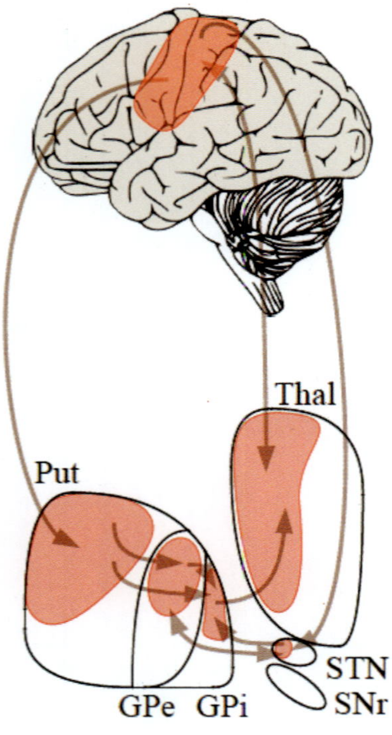

PLATE 100 Behaviors and neural candidates during different stages of addiction targeted by computational models. During the early formation of addiction, individuals are primarily driven by the rewarding effects of substances of abuse. This goal-directed behavior can be nicely quantified by computational RL models and is implemented in the ventral corticostriatal circuit. After the individual has become addicted, the habitual system, primarily implemented through the dorsal corticostriatal circuit, takes over. Images modified from Fiore, Dolan, Strausfeld, and Hirth (2015). (See table 91.1.)

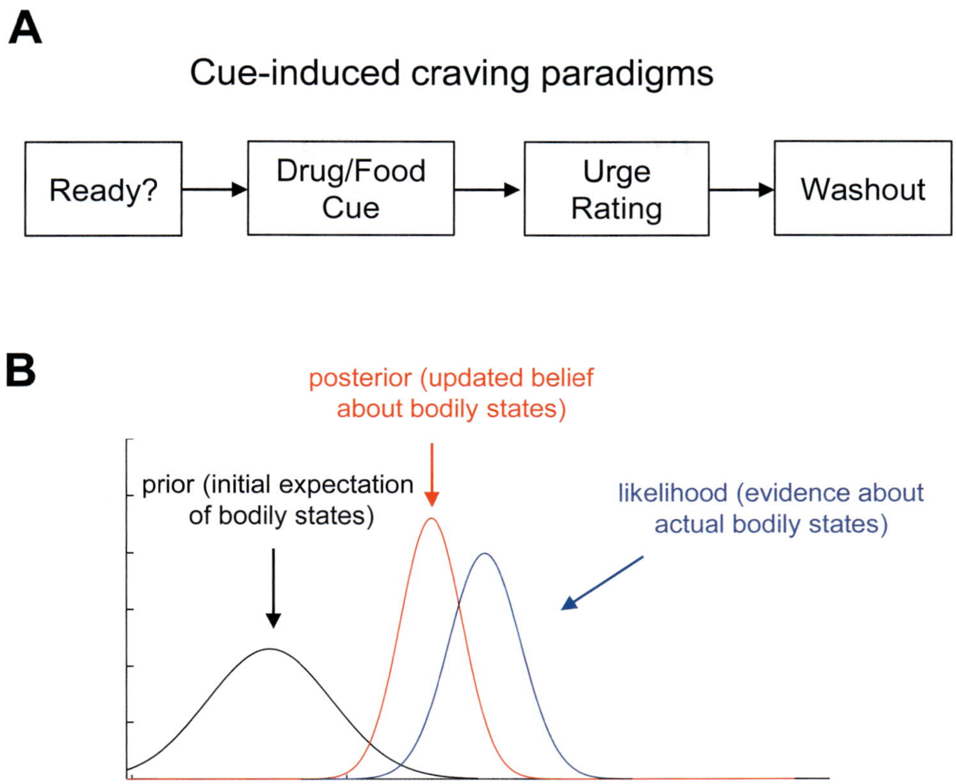

PLATE 101 *A*, Typical cue-induced craving paradigms in the human addiction literature. *B*, A recently proposed Bayesian framework of drug craving (Gu & Filbey, 2017). (See figure 91.2.)

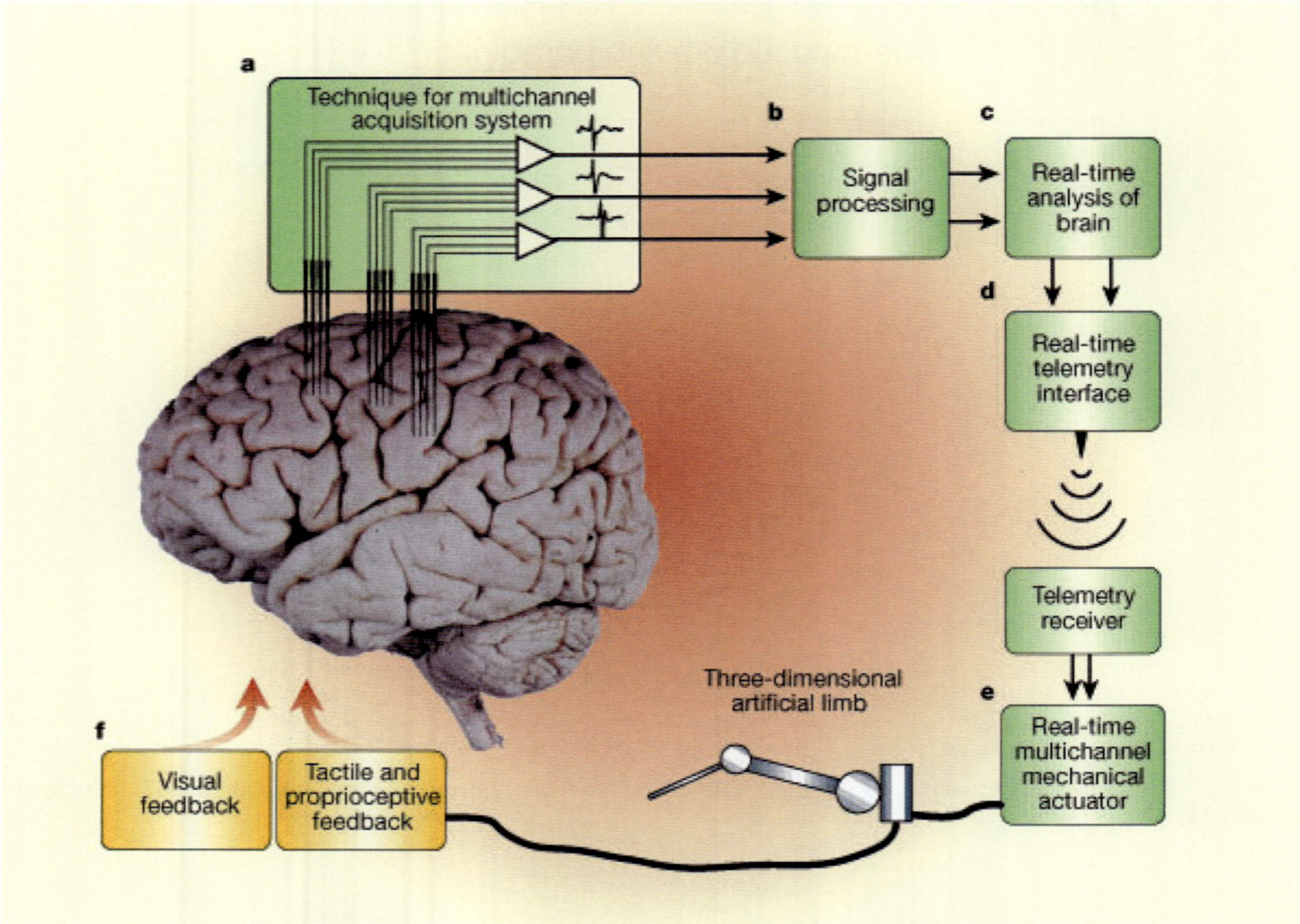

PLATE 102 Classical configuration of a brain-machine interface. Through the employment of multichannel intracranial extracellular recordings, multiple motor commands can be extracted, in real time, from the combined electrical activity of several hundred neurons, distributed across multiple cortical areas. This operation is carried out through the employment of mathematical decoders. Extracted motor commands are then used by subjects to directly control the movements of a variety of artificial devices. Reproduced with permission from Nicolelis (2001). (See figure 94.1.)

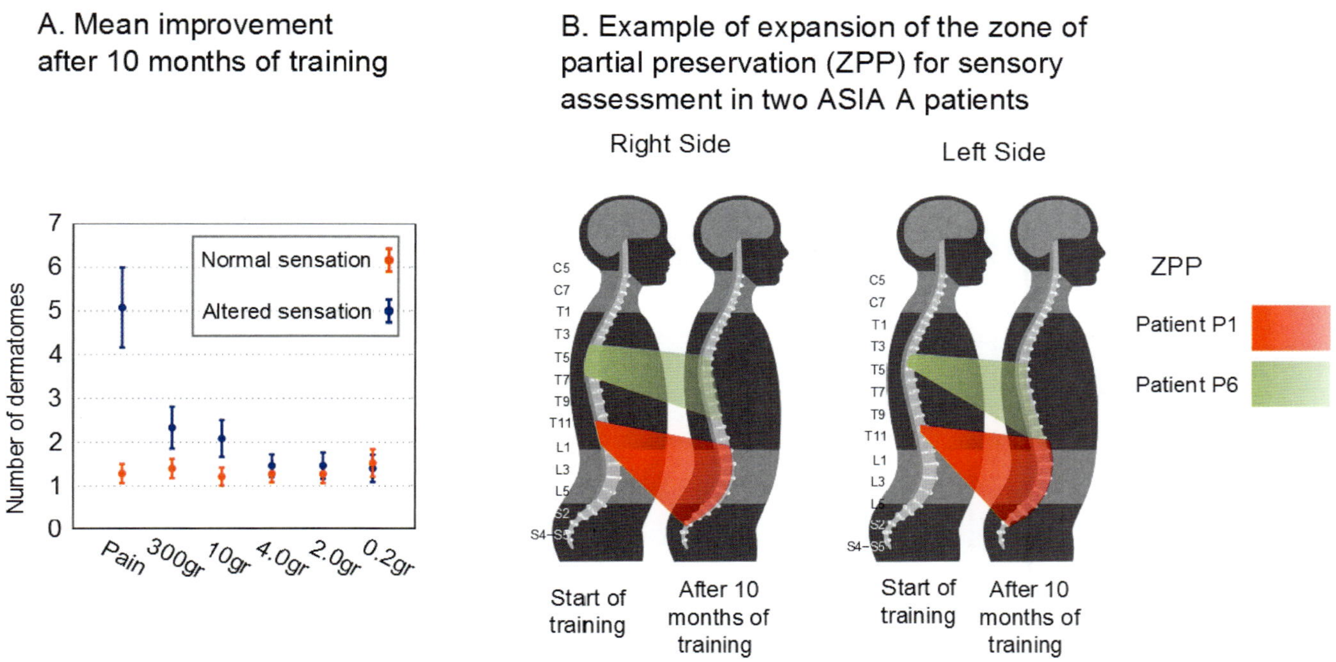

PLATE 103 Partial sensory improvement in chronic SCI patients following training with a BMI protocol. Top shelf: Sensory improvement after neurorehabilitation training. *A*, Average sensory improvement (mean +/− SEM over all patients) after 10 months of training. *B*, Example of improvement in the zone of partial preservation on a sensory evaluation of two patients. Reproduced with permission from Donati et al. (2016). (See figure 94.2.)

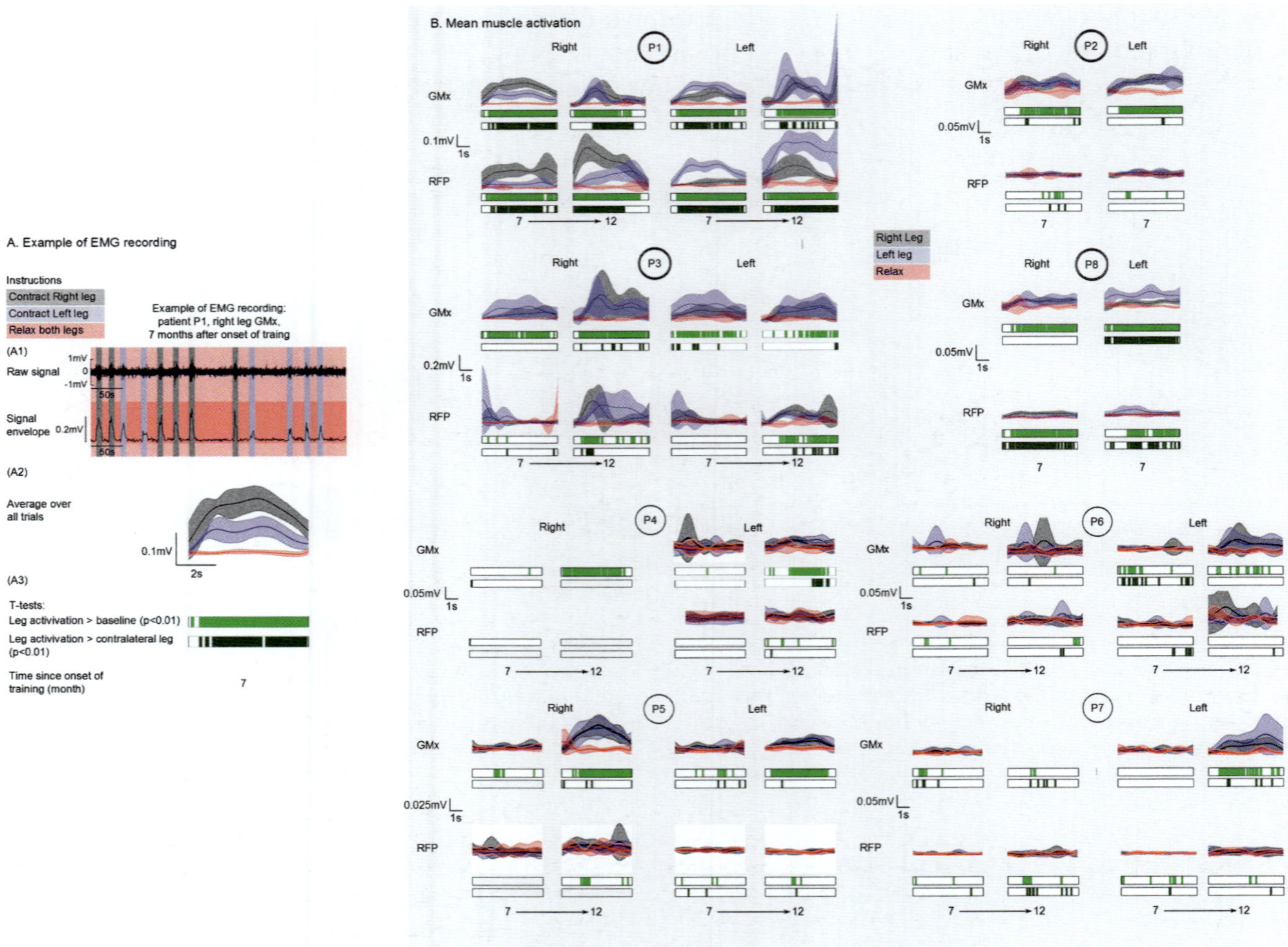

PLATE 104 Lower-limb motor recovery. *A*, Details of the EMG recording procedure in SCI patients. *A1*, Raw EMG for the right gluteus maximus muscle for patient P1 is shown at the top of the topmost graph. The lower part of this graph depicts the envelope of the raw EMG after the signal was rectified and low-pass filtered at 3 Hz. Gray-shaded areas represent periods in which the patient was instructed to move the right leg, while the blue-shaded areas indicate periods of left-leg movement. Red areas indicate periods in which patients were instructed to relax both legs. *A2*, All trials over one session were averaged (mean +/− standard deviation envelopes are shown) and plotted as a function of instruction type (*gray envelope*= contract right leg; *blue*= contract left leg; *red*= relax both legs). *A3*, Below the averaged EMG record, light-green bars indicate instances in which the voluntary muscle contraction (right leg) was significantly different (t-test, $p<0.01$) than the baseline (periods in which she/he was instructed to relax both legs). Dark-green bars depict periods in which there was a significant difference ($p<0.01$) between muscle contraction in the right versus the left leg. *B*, EMG envelopes and t-tests for all recording sessions, involving four muscles, for all eight patients: left and right gluteus maximus (GMx) and rectus femoris proximal (RFP) muscles. Color convention and figure organization follows the one on panel *A*. The data were collected after 7 months of training for all patients and for all but patients P2 and P8 after 12 months. Reproduced with permission from Donati et al. (2016). (See figure 94.3.)

68 The Contribution of Sensorimotor Experience to the Mind and Brain

MARINA BEDNY

ABSTRACT How does sensorimotor experience shape the human mind? This question has been of interest to thinkers for thousands of years, from Plato to the British empiricists. This chapter highlights insights into this puzzle from psychology and cognitive neuroscience. In what ways do knowledge and the functional organization of the cortex arise from sensory experiences? A key source of evidence comes from studies with individuals who have altered sensory experience from birth: those who are congenitally blind, deaf, or missing limbs. Such studies demonstrate that changes in early sensory experience dramatically alters the function of sensory cortices. In congenital blindness, "visual" cortices take on higher cognitive functions, including language and number. This plasticity is believed to occur as a result of top-down input from higher cognitive systems into "visual" cortices. In contrast to these dramatic changes in the "deprived" sensory systems, the neural basis of concepts is largely unchanged in sensory loss. The cognitive and neural basis of concrete objects, events, and properties is similar in congenitally blind and sighted individuals. Insights from developmental psychology further suggest that human concepts are not constructed from sensations. Even seemingly sensory concepts such as "blue" have a rich abstract structure early in life. At the same time, studies of training and expertise show that sensorimotor experience does influence our knowledge of what things look like and how to motorically interact with objects. Semantic knowledge broadly construed includes both abstract conceptual and sensorimotor representations. These different types of information are represented in different cortical systems, each of which is sensitive to different aspects of our experience.

How do sensory experiences contribute to the mind? In what sense do our experiences of seeing, hearing, and touching give rise to concepts such as *tiger, chair,* and *running*? Such questions have puzzled thinkers for thousands of years, dating back to Plato, who held that we are born knowing everything we will ever know, and the role of experience is merely to awaken this knowledge. By contrast, empiricist philosophers such as Locke and Hume proposed that all concepts are built out of sensorimotor experiences and are represented in their terms (Hume, 1748; Locke, 1690; Plato, 1961). Empirically disentangling the contributions of nature and nurture has proven a daunting task since humans share much of their genetic makeup as well as important aspects of experience—for example, vision, audition, motor experience, and the presence of objects, agents, and events in the environment.

A key source of insight comes from studies with individuals who have drastically different sensorimotor histories from birth: individuals who are blind, deaf, or have altered motor experiences. Studies of sensory loss provide a unique window into how the mind and brain responds to alterations in *species typical* or *expected* experiences, that is, experiences that were ubiquitous to the species during our evolutionary history. As a result, the brain may plausibly have evolved to "expect" such experiences (Greenough, Black, & Wallace, 1987). How does the human brain and mind develop when such experiences are absent? This chapter reviews research examining the effects of sensory loss on different cognitive systems. To set the stage, I begin by describing the effects of sensory loss on the cortical systems that typically support sensory perception in the "deprived" modality, focusing on how congenital blindness influences the visual system. Next, I turn to the effect of sensory loss on conceptual representations of objects and events. By comparing how sensorimotor experience affects these different types of representations, we can better understand which experiences are most relevant to which cognitive systems. To complement these findings, I highlight insights from studies of cognitive development. Finally, I discuss findings from studies of sensorimotor expertise and training. Together, these data provide insights into how sensorimotor experience does and does not contribute to conceptual representations. I end by discussing implications for cognitive neuroscience theories of concepts.

Large-Scale Change to the Function of Sensorimotor Systems in Sensory Loss

Early imaging studies with blind and deaf humans provided some of the first demonstrations that early sensory experience changes cortical function. The "visual" cortices of individuals who are blind from birth are highly active during tactile and auditory tasks (Sadato et al., 1996). Analogously, the "auditory" cortices of deaf individuals show robust responses to visual stimuli (Finney, Fine, & Dobkins, 2001). In *crossmodal* plasticity, apart from changing their preferred modality of input, cortices change their sensitivity to information. For example, in blind but not sighted participants, parts of the dorsal "visual" stream respond to moving sounds and are active during sound localization (Collignon et al., 2011). Dorsal "visual" areas thus enhance their sensitivity to auditory information that comes from an analogous domain to the original visual function (i.e., spatial/motion).

In other examples of crossmodal plasticity, the degree of functional reorganization is still more dramatic. Large swaths of "visual" cortices respond to linguistic information in blindness. This includes not only portions of the ventral and lateral occipital cortex but also parts of V1 (Lane, Kanjlia, Omaki, & Bedny, 2015; Röder, Stock, Bien, Neville, & Rösler, 2002). Responses are observed both to spoken and written (Braille) language and occipital activity is sensitive to high-level linguistic content (e.g., the grammar and meaning of sentences). For example, "visual" language areas respond more to sentences than to lists of words, more to jabberwocky than lists of nonwords, and more to grammatically complex sentences than to simple ones (Lane et al., 2015; Röder et al., 2002). There is also some evidence that these responses are behaviorally relevant. Transcranial magnetic stimulation (TMS) to the occipital pole causes blind but not sighted participants to make semantic errors during verb generation (Amedi, Floel, Knecht, Zohary, & Cohen, 2004).

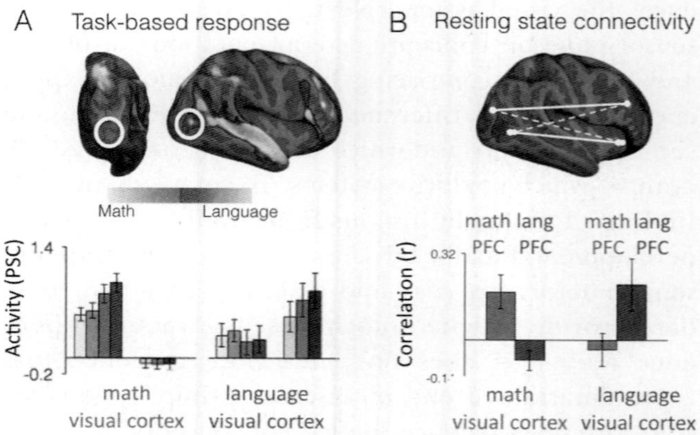

FIGURE 68.1 Responses to language and number in visual cortices of congenitally blind individuals. *A*, Math-responsive "visual" areas (*red*) show an effect of math equation difficulty (*increasingly dark-red bars*). Language-responsive "visual" areas show an effect of grammatical complexity: lists of nonwords (*gray*), grammatically simple sentences (*light blue*), and complex (*dark blue*) sentences. *B*, Stronger resting-state correlations with language-responsive PFC in language-responsive visual cortex and with math-responsive PFC in math-responsive visual cortex. (See color plate 81.)

Language is not the only higher-cognitive function that invades the deafferented visual system. Other parts of "visual" cortices acquire responses to numerical information and still others to executive load in nonverbal tasks (figure 68.1*A*; Kanjlia, Lane, Feigenson, & Bedny, 2016; Loiotile & Bedny, 2018). According to one hypothesis, the invasion of "visual" networks by higher cognitive information in blindness occurs through input from frontoparietal and frontotemporal networks (Amedi, Hofstetter, Maidenbaum, & Heimler, 2017; Bedny, 2017). In the absence of bottom-up information from the retinogeniculate pathway, top-down frontoparietal connectivity takes over "visual" circuits. Consistent with this idea, studies of resting-state connectivity find that in blindness visual areas become more functionally coupled with multiple higher cognitive circuits in frontal and parietal cortices in a functionally specific way (figure 68.1*B*; Deen, Saxe, & Bedny, 2015; Kanjlia et al., 2016). Interestingly, this extreme functional reorganization is curtailed to sensitive periods of development. Although "visual" cortices of adult-onset blind individuals also respond to sound and touch, these responses seem to lack the kind of cognitive specificity observed in congenital blindness (Bedny, Pascual-Leone, Dravida, & Saxe, 2011; Collignon et al., 2013).

The studies reviewed above suggest that early sensory loss has the capacity to profoundly change the function of cortical systems. Even sensory systems believed to be predisposed by evolution for specific sensory processes, undergo substantial functional reorganization when the type of experience they have evolved to "expect" is absent during early development (Greenough, Black, & Wallace, 1987).

The Abstractness of Blue: Resilience of Concepts to Congenital Sensory Loss

Early sensory loss leads to large-scale plasticity in "deprived" sensory cortices. Do these changes carry forward into conceptual systems? Are the cognitive and neural bases of concepts of concrete properties (e.g., blue), entities (e.g., dog), and events (e.g., run) very different in people who are blind from birth? The evidence reviewed below suggests that this is not the case. Even for seemingly purely "visual" concepts, such as *look* and *blue*, blind and sighted people's concepts turn out to have a lot in common. Blind children acquire "visual" words at around the same time as sighted children and use them in appropriate ways, making subtle distinctions between the meanings of words such as *look* and *see*—you can look without seeing. Blind children and adults have a coherent understanding of how color works. By the preschool years, blind children understand that a car can be blue but a thunderstorm and an idea cannot (Landau & Gleitman, 1985). Blind adults know the similarity structure of color space, that orange is more similar to red than to blue—although this knowledge is more variable across blind than sighted subjects (Shepard & Cooper, 1992). Blind people are less likely to know object color pairings (e.g., elephants are grey) and less likely to automatically use object color when sorting fruits and vegetables but nevertheless have preserved understanding of the relationship between object kind (natural kind vs. artifact) and color (Connolly, Gleitman, & Thompson-Schill, 2007; Elli, Lane, & Bedny, 2019; Kim, Elli, & Bedny, 2019).

Analogous evidence comes from studies with individuals who are born without hands. Amelic individuals show typical categorization and perception of hand actions (e.g., typing, playing a guitar). Both reasoning about and the perception of actions is intact. Individuals who themselves have never thrown a ball can nevertheless tell when a basketball throw is likely to hit its mark and are sensitive to whether a hand movement is or isn't awkward to perform (Vannuscorps & Caramazza, 2016). Thus, neither visual nor motor experience is necessary for the development of fine-grained reasoning about seemingly sensorimotor information, such as actions, perceptual experiences, light, and color. Even for concrete concepts, sensory loss does not substantially change what we know.

Consistent with the behavioral literature, the neural basis of concrete concepts is resilient to congenital sensory loss. Many cortical areas that are active during conceptual tasks in the sighted and were once thought to represent "visual" modality-specific information, turn out to be preserved in congenital blindness. When sighted subjects make semantic judgments about concrete objects, they activate a distributed network of regions, including parts of the medial and lateral ventral occipitotemporal cortex (Martin, 2016). One interpretation of this ventral occipitotemporal activation is that it involves the retrieval of modality-specific visual representations of appearance-related knowledge (e.g., of color and shape). However, a number of studies have identified similar ventral occipitotemporal responses in people who are blind. Those parts of the mediate occipitotemporal and parietal cortex that preferentially respond to nonliving entities in sighted participants (medial occipitotemporal and inferior parietal) also prefer inanimate entities in blind participants (Mahon, Anzellotti, Schwarzbach, Zampini, & Caramazza, 2009; Wang, Peelen, Han, Caramazza, & Bi, 2016). When blind individuals listen to the characteristic sounds of entities (e.g., of people or artifacts), patterns of activity in ventral occipitotemporal cortex can be used to decode among the classes of entities (van den Hurk, Van Baelen, & Op de Beeck, 2017). Category-specific responses to concrete objects elsewhere in the brain are also preserved in blindness. For example, a recent study finds that different parts of the anterior temporal lobe (ATL) are involved in retrieving knowledge about concrete (e.g., dog) and abstract entities (e.g., idea) in sighted and blind participants alike although some words, such as "rainbow," appear to activate different parts of the ATL across groups (Striem-Amit, Wang, Bi, & Caramazza, 2018). In sum, a distributed but clearly defined network of cortical areas involved in representing knowledge about entities is shared among sighted and congenitally blind individuals.

An analogous picture of preservation has emerged from studies of concrete events. Secondary motor areas and parts of the frontoparietal cortices are active when subjects reason about actions (Hauk, Johnsrude, & Pulvermüller, 2004; Kemmerer & Gonzalez-Castillo, 2008). Such activations could in principle arise because of prior motor experiences of performing the actions. However, amelic individuals born without hands activate the same action-related neural systems when viewing videos of meaningful hand actions (e.g., taking a tea bag out of a cup, closing a sugar bowl), including regions within the frontoparietal *mirror neuron* system (Gazzola et al., 2007). Individuals who are blind from birth similarly activate frontoparietal circuits when listening to meaningful action sounds (Ricciardi et al., 2009).

Analogously, lateral temporal cortices (left middle temporal gyrus, or LMTG) that were originally thought to code visual motion features relevant to action verbs are active during verb comprehension in blind and sighted individuals alike (Bedny, Caramazza, Pascual-Leone, & Saxe, 2012; Noppeney, 2003; figure 68.2*A*). LMTG representations that are active during verb comprehension have turned out to be neither vision nor motion related, as was originally hypothesized, since even in the sighted the LMTG is equally responsive to abstract verbs that involve no motion at all, such as *believe* and *want* (Bedny, Caramazza, Grossman, Pascual-Leone, & Saxe, 2008). This suggests that the meanings of concrete verbs, such as *run*, are represented alongside the meanings of abstract verbs, such as *believe*. Spatial patterns of activity within the LMTG distinguish between different semantic categories of verbs, including the very types of verbs thought to dissociate within sensorimotor cortical systems. The LMTG distinguishes between hand (e.g., *slap*) and mouth (e.g., *chew*) actions, which in some views are distinguished based on patterns within motor cortex (Hauk,

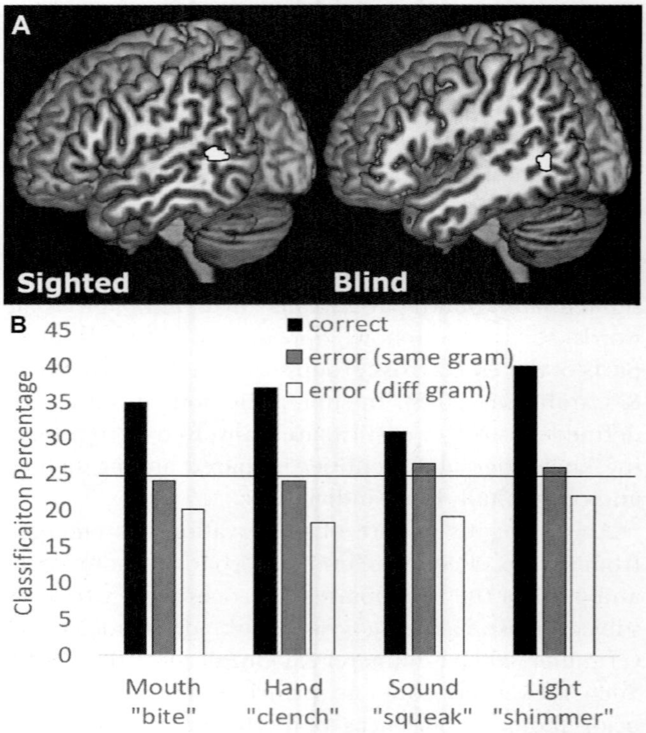

FIGURE 68.2 Representations of verb meanings in the left middle temporal gyrus (LMTG). *A*, Action verbs > object nouns in sighted (*left*) and congenitally blind individuals (*right*). Reprinted from Bedny et al. (2012). *B*, Performance of linear classifier distinguishing among four verb types based on patterns of activity in the LMTG of sighted individuals: transitive mouth and hand actions and intransitive light- and sound-emission events. The classifier successfully distinguished among mouth and hand actions and light- and sound-emission events. Errors across grammatical type (*white bars*; e.g., transitive mouth action mistaken for intransitive light-emission event) are less common than within grammatical type (*gray bars*; e.g., mouth action mistaken for hand action). From Elli, Lane, and Bedny (2019). (See color plate 82.)

Johnsrude, & Pulvermüller, 2004). It also distinguishes between events of light (e.g., *sparkle*) versus those of sound (e.g., *boom*) emission (Elli, Lane, & Bedny, 2019), semantic features previously said to dissociate based on responses in visual and auditory cortices (figure 68.2*B*; e.g., Kiefer, Sim, Herrnberger, Grothe, & Hoenig, 2008). Seemingly "sensory" features are represented in abstract conceptual systems.

Converging evidence for the idea that rich semantic representations develop in the absence of first-person sensory access comes from studies of reasoning about mental states. Neural population codes within the mentalizing network (e.g., the right temporoparietal junction) distinguish between beliefs based on seeing as opposed to hearing experiences (e.g., recognizing someone based on her handwriting versus her voice). And they do so equally in individuals who are sighted and in those who are congenitally blind (Koster-Hale, Bedny, & Saxe, 2014). Similarly, there is evidence that both the cognitive and neural architectures of numerical representations is preserved in blindness (Kanjlia et al., 2016). In sum, across a variety of conceptual domains and cortical systems, early and dramatic changes to sensory experience leave the cognitive and neural basis of concepts largely unchanged. This is true not only for abstract concepts such as *want* and *idea* but also for concrete ones such as *dog*, *run*, *see*, and *sparkle*. Although sensorimotor experience changes sensory systems themselves, many conceptual representations of "sensory" knowledge are unchanged.

Insights into Origins of Concepts from Developmental Psychology

The evidence reviewed above suggests that a rich array of conceptual representations is independent from our sensorimotor experiences. This view is consistent with evidence from developmental psychology. Research with infants suggests that rather than beginning with sensory representations and gradually progressing toward abstract conceptual ones, children think abstractly from the beginning. Within the first few months of life, infants expect entities that look like agents (e.g., have arms or faces) to behave according to goals and intentions, even though goals are not directly observable (Woodward, 1998). Even without any perceptual evidence, preverbal infants infer the presence of intentional agents when things seem to have occurred "on purpose" (Saxe, Tenenbaum, & Carey, 2005). Infants show early sensitivity to the causal structure of events (Leslie & Keeble, 1987) and expect inanimate entities to obey the laws of intuitive physics (e.g., two things cannot be in the same place at once; Baillargeon, Spelke, & Wasserman, 1985; Saxe, Tenenbaum, & Carey, 2005). Children seek an underlying causal structure in the world around them. Preschoolers treat natural things (e.g., tigers and gold) as having an internal, unobservable essence that makes them what they are. A "three-legged, tame, toothless, albino tiger" is still a tiger because it came from a tiger mother (Armstrong, Gleitman, & Gleitman, 1983). Preschoolers recognize that the insides of objects are more important to determining kind than the observable outsides (e.g., pigs are more similar to cows than piggy banks) (Gelman & Wellman, 1991; Keil, Smith, Simons, & Levin, 1998). As noted above, studies with children who are blind further reveal abstract knowledge about seemingly sensory concepts, such *blue* and *see* (Landau & Gleitman, 1985).

The claim that concepts are abstract from early infancy does not imply that concepts are hardwired fully formed into the brain and learning is unimportant. Children use their sensory systems to collect information from the environment, which enables them to elaborate and revise their representations (Carey, 2009). Importantly, learning itself does not appear to involve the gradual binding of sensations. With just a few examples and in some cases no sensory access to the thing being named, children learn labels for new categories and generalize these labels appropriately to novel instances. Children's learning appears to be a problem-solving process that involves hypothesis testing and revising theories (Gopnik & Meltzoff, 1998; Xu & Tenenbaum, 2007). From this perspective, it is not terribly surprising that concepts of people with altered sensory experience are not so different. The sophisticated learning devices that make up the human brain gather conceptually relevant information through various sensory channels (e.g., there are many clues to whether something is animate).

Sensorimotor Knowledge and Semantics: Insights from Studies of Expertise and Training

Not everything that we know about concrete entities and events is independent of the sensorimotor aspects of experience. Studies of expertise and training demonstrate that subtle and specific variation in sensorimotor experience in adulthood changes our long-term knowledge. Hockey experts (both players and fans) show differential priming effects when matching pictures of hockey actions to sentences that describe them ("The hockey player finished the stride"). When the same participants listen to these sentences in the scanner, experts (players and fans) activate left-lateralized secondary motor areas more than novices, and the degree of activation is correlated with priming effects outside the scanner (Beilock, Lyons, Mattarella-Micke, Nusbaum, & Small, 2008). Details of our sensorimotor experiences with objects are stored in long-term memory. When presented with photographs of objects, right-handers are faster at judging whether the object (e.g., a whisk) would be picked up by a "pinch" or a "clench" when its handle is oriented toward their own right hand. This effect reverses in patients who were previously right-handed but are now restricted to using their left hands due to brain injury (Chrysikou, Casasanto, & Thompson-Schill, 2017). Such evidence suggests that we acquire effector-specific information about canonical object-related motor actions and retrieve this information automatically, even when it is not required for the task.

Similar evidence comes from studies of color knowledge. For example, making detailed judgments about object color (e.g., Which is more similar to a school bus in color, egg yolk or butter?) activates cortical areas that partially overlap with those involved in color perception, particularly in people who report having a visual cognitive style (Hsu, Kraemer, Oliver, Schlichting, & Thompson-Schill, 2011). Such responses are influenced by training. Subjects who learn the diagnostic colors of novel objects over the course of a week activate color perception regions during recall, even when color is not relevant to the task (Hsu, Schlichting, & Thompson-Schill, 2014). Sensorimotor experience thus changes our reasoning about the physical world and changes representations in sensorimotor cortices.

At first glance, evidence from studies of sensory loss and sensorimotor expertise might seem contradictory. On the one hand, global and early changes to sensorimotor experience dramatically reorganize perceptual systems while leaving conceptual representations largely unchanged. Yet subtle alternations of sensorimotor experience in adulthood give rise to measurably different neural responses during conceptual tasks. How is it that blind and sighted people have similar representations of color, but the representations of sighted subjects trained on a color task for one week differ from those who have not been trained?

It is tempting to dismiss the findings from one of these literatures as "peripheral." One might argue that the representations retrieved by sighted subjects while making cross-category color judgments and those used by blind individuals when thinking about color are shallow or "verbal" and therefore not truly conceptual. This argument, however, leaves us in the odd position of claiming that much of our linguistic communication and reasoning occurs without using concepts. On the other hand, we might suppose that sensorimotor representations retrieved during conceptual tasks are merely "sensory imagery" and not relevant to cognition and behavior. There is, however, evidence that such representations are behaviorally relevant.

Rather, different tasks engage different types of representations. Sighted people engage color-perception areas only when retrieving detailed information about color hue and saturation, that is, when judging the colors of objects from the same color category (i.e., school buses, egg yolks, and butter.) No such activation is observed when deciding whether a strawberry is more similar in color to a lemon or a cherry (Hsu et al., 2011). This does not imply that the latter judgment is "shallow" or "verbal." It still relies on abstract and detailed information about what color is and how it works (e.g., a physical property perceptible only with the eyes, comes in different types, varies across object types and within an object, e.g., inside vs. outside) and knowledge of the color categories of specific objects (e.g., cherries are red). The within-category judgments additionally tap into perceptual knowledge of object colors (e.g., cherries are darker than strawberries). Even if we consider the perceptual knowledge of the color distinction between cherries and strawberries conceptual, it is a small fraction of conceptual color knowledge.

Implications for Cognitive Neuroscience Theories of Concepts

Where are concepts in the brain? The answer to this question depends on what one means by the term *concept*. If what we mean are the representations that enable us to judge whether something is or is not a dog, then concepts are represented in amodal cortical systems. Such representations enable us to say that a dog that looks like a cat is still a dog, as long as it has dog DNA. These abstract representations play a crucial role in reasoning, even for seemingly "sensory" categories (e.g., *blue*). This is why people who are blind have a similar concept of *blue* to people who are sighted, while those fish, birds, and insects that perceive blue nevertheless do not. If instead by *concept* one means everything we know about a category, then not only amodal representations of what something is but also sensorimotor representations of what it looks like, sounds like, and smells like are included.

Different aspects of our semantic knowledge have distinct developmental origins and are represented in different cortical systems. Experience affects these systems in different ways. Seeing a dog, hearing it bark, and even hearing someone say "dog" are qualitatively different experiences from the perspective of our sensory systems in that they modify different neural circuits (i.e., visual vs. auditory cortices). These experiences are equivalent, however, from the perspective of the abstract conceptual system that represents animate entities: they provide evidence for the existence of an animal of the type dog. Our abstract conceptual knowledge depends on the information the senses convey but not on the modality-specific aspects of experience. This perspective on the origins of knowledge has implications for cognitive neuroscience theories of concepts.

A prominent view is that concepts are distributed across sensorimotor cortical systems (Barsalou, Kyle Simmons, Barbey, & Wilson, 2003). In recent years there has been increasing evidence that modality-independent cortical areas (e.g., the anterior temporal and inferior parietal lobes) play a role in conceptual processes (Binder & Desai, 2011). One construal of this evidence is that the neural basis of human semantic memory consists of sensorimotor features represented in sensorimotor cortices plus the domain-general binding hubs that bind and weigh these features. The evidence reviewed in this chapter does not favor this view. Modality-independent cortical areas represent abstract conceptual information, rather than binding sensory features elsewhere.

Moreover, conceptual modality-independent cortical areas are numerous, heterogeneous among themselves, and, in some cases, organized at the regional scale by cognitive domain (entity vs. event; Leshinskaya & Caramazza, 2016). The list of these areas continues to grow, and multivariate methods are beginning to uncover neural population codes within them (Fairhall & Caramazza, 2013). These population codes make explicit those aspects of objects, events, and properties that are causally central and relevant to category membership (e.g., agent/object, artifact/natural kind, intentional/accidental), including information about seemingly sensory categories (e.g., blue is a physical property perceptible with the eyes). These abstract conceptual systems interact with modality-specific sensory cortical systems when we think, talk about and act on the world (Mahon & Caramazza, 2008).

Conclusions

Evidence from studies of sensory loss demonstrates that the human cortex is functionally flexible early in life. Early changes in experience can alter the representational content of cortical networks dramatically—for example, from low-level vision to linguistic processing (Bedny, 2017). Yet cortical systems are also remarkably specific in the type of experience to which they are sensitive. The same experience that reorganizes sensory systems has little effect on abstract conceptual ones. Innate connectivity patterns constrain which part of experience a given cortical system will be sensitive to (Mahon & Caramazza, 2011; Saygin et al., 2016). Each cortical system can be thought of as a powerful learning device with a particular window onto the world (Gallistel, Brown, Carey, Gelman, & Keil, 1991). Abstract conceptual systems for representing entities, properties, and events are examples of such specialized neural learning devices, each of which only "sees" a particular part of our experience. An important goal for future research is to uncover the physiological properties that make neurocognitive systems so good at learning in general, as well as properties that prepare each system for representing and learning specific types of information. One prediction of such a "specialized learning systems" view is that although abstract conceptual systems do not change much in sensory loss, they would change if information available about objects, entities, and events were altered early in development.

REFERENCES

Amedi, A., Floel, A., Knecht, S., Zohary, E., & Cohen, L. G. (2004). Transcranial magnetic stimulation of the occipital pole interferes with verbal processing in blind subjects. *Nature Neuroscience, 7*(11), 1266–1270.

Amedi, A., Hofstetter, S., Maidenbaum, S., & Heimler, B. (2017). Task selectivity as a comprehensive principle for brain organization. *Trends in Cognitive Sciences, 21*(5), 307–310.

Armstrong, S. L., Gleitman, L. R., & Gleitman, H. (1983). What some concepts might not be. *Cognition, 13*(3), 263–308.

Baillargeon, R., Spelke, E., & Wasserman, S. (1985). Object permanence in five-month-old infants. *Cognition, 20,* 191–208.

Barsalou, L. W., Kyle Simmons, W., Barbey, A. K., & Wilson, C. D. (2003). Grounding conceptual knowledge in modality-specific systems. *Trends in Cognitive Sciences, 7*(2), 84–91.

Bedny, M. (2017). Evidence from blindness for a cognitively pluripotent cortex. *Trends in Cognitive Sciences, 21*(9), 637–648.

Bedny, M., Caramazza, A., Grossman, E., Pascual-Leone, A., & Saxe, R. (2008). Concepts are more than percepts: The case of action verbs. *Journal of Neuroscience, 28*(44), 11347–11353.

Bedny, M., Caramazza, A., Pascual-Leone, A., & Saxe, R. (2012). Typical neural representations of action verbs develop without vision. *Cerebral Cortex, 22*(2), 286–293.

Bedny, M., Pascual-Leone, A., Dravida, S., & Saxe, R. (2012). A sensitive period for language in the visual cortex: Distinct patterns of plasticity in congenitally versus late blind adults. *Brain and Language, 122*(3), 162–170.

Beilock, S. L., Lyons, I. M., Mattarella-Micke, A., Nusbaum, H. C., & Small, S. L. (2008). Sports experience changes the neural processing of action language. *Proceedings of the National Academy of Sciences of the United States of America, 105*(36), 13269–13273.

Binder, J. R., & Desai, R. H. (2011). The neurobiology of semantic memory. *Trends in Cognitive Sciences, 15*(11), 527–536.

Carey, S. (2009). *The origin of concepts: Oxford series in cognitive development.* Oxford: Oxford University Press.

Chrysikou, E. G., Casasanto, D., & Thompson-Schill, S. L. (2017). Motor experience influences object knowledge. *Journal of Experimental Psychology: General, 146*(3), 395–408.

Collignon, O., Dormal, G., Albouy, G., Vandewalle, G., Voss, P., Phillips, C., & Lepore, F. (2013). Impact of blindness onset on the functional organization and the connectivity of the occipital cortex. *Brain, 136*(9), 2769–2783.

Collignon, O., Vandewalle, G., Voss, P., Albouy, G., Charbonneau, G., Lassonde, M., & Lepore, F. (2011). Functional specialization for auditory-spatial processing in the occipital cortex of congenitally blind humans. *Proceedings of the National Academy of Sciences, 108*(11), 4435–4440.

Connolly, A. C., Gleitman, L. R., & Thompson-Schill, S. L. (2007). Effect of congenital blindness on the semantic representation of some everyday concepts. *Proceedings of the National Academy of Sciences, 104*(20), 8241–8246.

Deen, B., Saxe, R., & Bedny, M. (2015). Occipital cortex of blind individuals is functionally coupled with executive control areas of frontal cortex. *Journal of Cognitive Neuroscience, 27*(8), 1633–1647.

Elli, G. V., Lane, C., & Bedny, M. (2019). A double dissociation in sensitivity to verb and noun semantics across cortical networks. *Cerebral Cortex.* doi:10.1093/cercor/bhz014

Fairhall, S. L., & Caramazza, A. (2013). Brain regions that represent amodal conceptual knowledge. *Journal of Neuroscience, 33*(25), 10552–10558.

Finney, E. M., Fine, I., & Dobkins, K. R. (2001). Visual stimuli activate auditory cortex in the deaf. *Nature Neuroscience, 4*(12), 1171–1173.

Gallistel, C. R., Brown, A. L., Carey, S., Gelman, R., & Keil, F. (1991). Lessons from animal learning for the study of cognitive development. In S. Carey and R. Gelman (Eds.), *The epigenesis of mind: Essays on biology and cognition* (pp. 1–36). Hillsdale, NJ: L. Erlbaum

Gazzola, V., van der Worp, H., Mulder, T., Wicker, B., Rizzolatti, G., & Keysers, C. (2007). Aplasics born without hands mirror the goal of hand actions with their feet. *Current Biology, 17*(14), 1235–1240.

Gelman, S. A., & Wellman, H. M. (1991). Insides and essences: Early understandings of the non-obvious. *Cognition, 38*(3), 213–244.

Gopnik, A., & Meltzoff, A. N. (1998). *Words, thoughts, and theories (learning, development, and conceptual change).* Cambridge, MA: MIT Press.

Greenough, W. T., Black, J. E., & Wallace, C. S. (1987). Experience and brain development. *Child Development, 58*(3), 539–559.

Hauk, O., Johnsrude, I., & Pulvermüller, F. (2004). Somatotopic representation of action words in human motor and premotor cortex. *Neuron, 41*(2), 301–307.

Hsu, N. S., Kraemer, D. J. M., Oliver, R. T., Schlichting, M. L., & Thompson-Schill, S. L. (2011). Color, context, and cognitive style: Variations in color knowledge retrieval as a function of task and subject variables. *Journal of Cognitive Neuroscience, 23*(9), 2544–2557.

Hsu, N. S., Schlichting, M. L., & Thompson-Schill, S. L. (2014). Feature diagnosticity affects representations of novel and familiar objects. *Journal of Cognitive Neuroscience, 26*(12), 2735–2749.

Hume, D. (1748). *An enquiry concerning human understanding* (pp. 1–88). Collier & Son.

Kanjlia, S., Lane, C., Feigenson, L., & Bedny, M. (2016). Absence of visual experience modifies the neural basis of numerical thinking. *Proceedings of the National Academy of Sciences, 113*(40), 11172–11177.

Keil, F. C., Smith, W. C., Simons, D. J., & Levin, D. T. (1998). Two dogmas of conceptual empiricism: Implications for hybrid models of the structure of knowledge. *Cognition, 65*(2–3), 103–135.

Kemmerer, D., & Gonzalez-Castillo, J. (2008). The two-level theory of verb meaning: An approach to integrating the semantics of action with the mirror neuron system. *Brain and Language,* 1–23. doi:10.1016/j.bandl.2008.09.010

Kiefer, M., Sim, E. J., Herrnberger, B., Grothe, J., & Hoenig, K. (2008). The sound of concepts: Four markers for a link between auditory and conceptual brain systems. *Journal of Neuroscience, 28*(47), 12224–12230.

Kim, J. S., Elli, G. V., & Bedny, M. (2019). Knowledge of animal appearance among sighted and blind adults. *Proceedings of the National Academy of Sciences, 116*(23), 11213–11222.

Koster-Hale, J., Bedny, M., & Saxe, R. (2014). Thinking about seeing: Perceptual sources of knowledge are encoded in the theory of mind brain regions of sighted and blind adults. *Cognition, 133*(1), 65–78.

Landau, B., & Gleitman, L. R. (1985). *Language and experience: Evidence from the blind child.* Cambridge, MA: Harvard University Press.

Lane, C., Kanjlia, S., Omaki, A., & Bedny, M. (2015). "Visual" cortex of congenitally blind adults responds to syntactic movement. *Journal of Neuroscience, 35*(37), 12859–12868.

Leshinskaya, A., & Caramazza, A. (2016). For a cognitive neuroscience of concepts: Moving beyond the grounding issue. *Psychonomic Bulletin & Review, 23*(4), 991–1001.

Leslie, A. M., & Keeble, S. (1987). Do six-month-old infants perceive causality? *Cognition, 25*(3), 265–288.

Locke, J. (1690). *An essay concerning human understanding.*

Loiotile, R. E., & Bedny, M. (2018). "Visual" cortices of congenitally blind adults respond to executive demands. *bioRxiv.* https://doi.org/10.1101/39045

Mahon, B. Z., Anzellotti, S., Schwarzbach, J., Zampini, M., & Caramazza, A. (2009). Category-specific organization in the human brain does not require visual experience. *Neuron, 63*(3), 397–405.

Mahon, B. Z., & Caramazza, A. (2008). A critical look at the embodied cognition hypothesis and a new proposal for grounding conceptual content. *Journal of Physiology, Paris, 102*(1–3), 59–70.

Mahon, B. Z., & Caramazza, A. (2011). What drives the organization of object knowledge in the brain? *Trends in Cognitive Sciences, 15*(3), 97–103.

Martin, A. (2016). GRAPES—Grounding representations in action, perception, and emotion systems: How object properties and categories are represented in the human brain. *Psychonomic Bulletin & Review, 23*(4), 979–990.

Noppeney, U. (2003). Effects of visual deprivation on the organization of the semantic system. *126*(7), 1620–1627.

Plato, P., Hamilton, E., Cairns, H., & Cooper, L. (1963). *The Collected Dialogues of Plato, Including the Letters.* New York: Pantheon Books.

Ricciardi, E., Bonino, D., Sani, L., Vecchi, T., Guazzelli, M., Haxby, J. V., et al. (2009). Do we really need vision? How blind people "see" the actions of others. *Journal of Neuroscience, 29*(31), 9719–9724.

Röder, B., Stock, O., Bien, S., Neville, H., & Rösler, F. (2002). Speech processing activates visual cortex in congenitally blind humans. *European Journal of Neuroscience, 16*(5), 930–936.

Sadato, N., Pascual-Leone, A., Grafman, J., Ibañez, V., Deiber, M. P., Dold, G., & Hallett, M. (1996). Activation of the primary visual cortex by Braille reading in blind subjects. *Nature, 380*(6574), 526–528.

Saxe, R., Tenenbaum, J., & Carey, S. (2005). Secret agents: 10-and 12-month-old infants' inferences about hidden causes. *Psychological Science, 16*, 995–1001.

Saygin, Z. M., Osher, D. E., Norton, E. S., Youssoufian, D. A., Beach, S. D., Feather, J., et al. (2016). Connectivity precedes function in the development of the visual word form area. *Nature Neuroscience, 19*(9), 1250–1255.

Shepard, R. N., & Cooper, L. A. (1992). Representation of colors in the blind, color-blind, and normally sighted. *Psychological Science, 3*(2), 97–104.

Striem-Amit, E., Wang, X., Bi, Y., & Caramazza, A. (2018). Neural representation of visual concepts in people born blind. *Nature Communications, 9*(1), 5250.

van den Hurk, J., Van Baelen, M., & Op de Beeck, H. P. (2017). Development of visual category selectivity in ventral visual cortex does not require visual experience. *Proceedings of the National Academy of Sciences, 114*(22), E4501–E4510.

Vannuscorps, G., & Caramazza, A. (2016). Typical action perception and interpretation without motor simulation. *Proceedings of the National Academy of Sciences, 113*(1), 86–91.

Wang, X., Peelen, M. V., Han, Z., Caramazza, A., & Bi, Y. (2016). The role of vision in the neural representation of unique entities. *Neuropsychologia, 87*(C), 144–156.

Woodward, A. L. (1998). Infants selectively encode the goal object of an actor's reach. *Cognition, 69*(1), 1–34.

Xu, F., & Tenenbaum, J. B. (2007). Word learning as Bayesian inference. *Psychological Review, 114*(2), 245–272.

69 Spatial Knowledge and Navigation

RUSSELL A. EPSTEIN

ABSTRACT Spatial knowledge is knowledge about where things are in the world and how they are spatially related to each other. One important use of spatial knowledge is to guide navigation from place to place. To accomplish this function, the brain must represent navigationally relevant aspects of the local environment, such as landmarks, scene geometry, and navigational affordances. It must also form representations of the space beyond the current sensory horizon, which might take the form of a cognitive map or graph. Research to date indicates that representations of the local environmental are supported primarily by scene-responsive regions, such as the parahippocampal place area (PPA), occipital place area (OPA), and retrosplenial complex (RSC). Global spatial representations, on the other hand, are supported primarily by the hippocampal formation and the RSC. A key challenge for the field, which this chapter attempts to address, is to understand how the spatial knowledge representations revealed by cognitive behavioral studies are mediated by neural systems.

Space, for a navigator, is structured by both the body and the environment. The body is a point that is distinct from all other points. The body faces a specific direction (its *heading*), which determines which way the organism can move without turning and what it can see. Only the immediate environment (*vista space*) can be sensed; the world beyond the sensory horizon (*environmental space*) must be traveled to or recalled from memory (Montello, 1993). Perception and movement are constrained by barriers and facilitated by openings, passageways, and paths. Some objects in the world are stable and thus likely to maintain their location; others are movable and thus might appear in different locations. As these observations indicate, when considering how spatial knowledge is encoded in the mind/brain, it is essential to consider the spatial organization of the world, and how this organization might facilitate or hinder navigation.

Vista Space: Scenes and Landmarks

A navigating organism must be able to perceive and understand its immediate spatial surroundings (vista space). Of particular importance is the ability to perceive landmarks—items that have a reliable relationship to a location, direction, or point along a path. Landmarks can come in many forms. Some are discrete objects such as buildings, statues, traffic lights, and mailboxes. Others are more distributed entities, such as the arrangement of streets at an intersection, the shape of a room, or the topography of a landscape. Indeed, in many cases the surroundings as a whole (the "local scene") act as a kind of landmark.

Psychological research suggests that several qualities make some items more useful as landmarks than others (Burnett, Smith, & May, 2001; Jansen-Osmann, 2002; Janzen, 2006). First, good landmarks are perceptually salient: they are easy to perceive and easy to distinguish from other landmarks. Second, good landmarks are stable: they are reliably associated with certain locations or bearings. Third, good landmarks are located in navigationally relevant places—for example, an intersection or other decision point. Consider, for example, a church on a town square: this is an ideal landmark because it is distinctive and visible, always in the same location, and in the center of the road network of the town.

Objects that have landmark-suitable qualities appear to hold a special status in the cognitive system of animals and humans. Consider stability. Rats will use an object that is fixed in space as a reference from which to encode the distance and direction to a goal, but they will not use an equivalent object that is not fixed (Biegler & Morris, 1993). Spatial position also has an effect on whether objects are encoded as landmarks. Janzen (2006) asked participants to learn a path through a virtual reality environment. Objects were placed in various locations along the path. After training, participants were presented with the same objects in isolation, intermixed with foils, and asked to report whether each item was familiar or not. Reaction times were faster for objects that had been at navigational decision points than for objects that had been at other locations along the path. This suggests that the decision point objects had obtained a special status in memory.

An especially salient and stable aspect of the perceptible environment is the geometric layout of a local space—for example, the shape of a room or the arrangement of streets at an intersection. A prominent line of research suggests that this geometric information might play a special role in spatial orientation (Cheng, 1986). When rats are trained to dig for a buried food reward in one location in a rectangular chamber and then removed from the chamber, disoriented, and placed back in the chamber, they will search for the reward in either the correct location or the diagonally opposite location. This behavior is notable because these two locations are equivalent in terms of the geometric shape of the chamber. Geometric errors are observed even in chambers that include visual markings on the walls or corners that could, in theory, disambiguate the two conflated locations. Thus, the animals appear to preferentially use the geometry of the chamber to reorient themselves. These results spawned the idea—much debated—that reorientation is mediated by a *geometric module* that is impenetrable to nongeometric cues (see Cheng, Huttenlocher, & Newcombe, 2013). In any case, several lines of evidence suggest that environmental boundaries act as important references for spatial memory (Hartley, Trinkler, & Burgess, 2004; Lee, 2017).

Another important navigational cue is the overall visual appearance of the local scene, which is determined not only by geometric but also by nongeometric features, such as color, texture, and the spatial distribution of visual features. Insects use this kind of raw visual information to identify specific locations (Collett, Chittka, & Collett, 2013), and humans have the ability to use a similar strategy (Gillner, Weiss, & Mallot, 2008). Notably, this viewpoint-dependent "snapshot" appears to differ from representations of the spatial structure of the local environment, with visual appearance used primarily for place recognition and geometry used primarily for spatial orientation (Burgess, Spiers, & Paleologou, 2004; Valiquette & McNamara, 2007; Waller & Hodgson, 2006). Consistent with this idea, in a recent study we found that disoriented rodents use nongeometric visual cues, such as a visual pattern along a wall, to identify their overall navigational context (i.e., the experimental chamber they are in) while using local geometric cues to recover their heading direction within this context (Julian, Keinath, Muzzio, & Epstein, 2015). This suggests the existence of a mechanism for appearance-based place recognition that is behaviorally dissociable from the mechanism for geometry-based reorientation.

Scenes and Landmarks in the Brain

fMRI studies have identified three brain regions that exhibit greater response when subjects view scenes (landscapes, street scenes, rooms, or buildings) than when they view other meaningful visual stimuli, such as artifacts, animals, vehicles, bodies, or faces: the parahippocampal place area (PPA), the retrosplenial complex (RSC), and the occipital place area (OPA; Epstein, 2014). The PPA encodes multiple aspects of the scene that might be useful for identifying it as a particular place or category of place, including the spatial expanse of the scene (Kravitz, Peng, & Baker, 2011; Park, Brady, Greene, & Oliva, 2011), the individual objects within it (Harel, Kravitz, & Baker, 2013), and the scene's 3-D structure (Walther, Chai, Caddigan, Beck, & Fei-Fei, 2011). The RSC shows similar responses but, additionally, codes explicitly spatial quantities such as the implied heading and location of the observer relative to both local scene geometry (Marchette, Vass, Ryan, & Epstein, 2014) and the wider environment (Baumann & Mattingley, 2010; Shine, Valdés-Herrera, Hegarty, & Wolbers, 2016; Vass & Epstein, 2013). Damage to the PPA leads to a deficit in recognizing scenes and landmarks—a syndrome that has been labeled *landmark agnosia*—while damage to the RSC leads to a deficit in the ability to use scenes and landmarks to recover one's heading and orient oneself in space (Aguirre & D'Esposito, 1999).

The OPA may process visual features that are essential for both scene/landmark recognition and spatial perception. When processing in the OPA is disrupted by transcranial magnetic stimulation (TMS), impairments are observed in the ability to visually categorize scenes (Ganaden, Mullin, & Steeves, 2013), discriminate scenes based on their spatial layout (Dilks, Julian, Paunov, & Kanwisher, 2013), and perceive environmental boundaries in scenes (Julian, Ryan, Hamilton, & Epstein, 2016). Complementing this TMS work, a recent fMRI study from our lab suggests that the *navigational affordances* of the local environment might be processed in the OPA (Bonner & Epstein, 2017). Participants in the study viewed artificial rooms or natural scenes, which varied in terms of the direction that one could move to egress the scene. For example, one scene might depict a room with a door on the left wall, while another might depict a room with a door on the right wall. Multivoxel activation patterns within the OPA contained information about these navigational affordances, even when other visual and spatial features of the scenes were strictly controlled. Navigational affordances and environmental boundaries may be complementary aspects of the spatial structure of scenes processed by the OPA: affordances are where one can go in the local environment, and boundaries are where one's movement is blocked.

Beyond their role in processing scenes, several studies suggest that the PPA, RSC, and OPA may play a broader role in processing landmarks, including object-like landmarks. These regions respond more strongly to objects that have intrinsic qualities that make them more useful as landmarks (Troiani, Stigliani, Smith, & Epstein, 2014), such as being large and stable (Auger, Mullally, & Maguire, 2012; Konkle & Oliva, 2012) or distant from the viewer (Amit, Mehoudar, Trope, & Yovel, 2012). This preference for large, stable objects is even observed in blind participants making size judgments in response to auditory cues (He et al., 2013). There is also evidence for a neural correlate of the decision point effect, in the form of greater response to decision point objects compared to non–decision point objects when they are viewed in isolation outside of the navigational context (Janzen & van Turennout, 2004). Multivoxel codes in the PPA, RSC, and OPA contain information about landmarks that generalizes across different views (Marchette, Vass, Ryan, & Epstein, 2015), and all three regions respond during the retrieval of information about specific familiar landmarks even when no picture of the landmark is provided (Fairhall, Anzellotti, Ubaldi, & Caramazza, 2013). Taken as a whole, these results suggest that the PPA, RSC, and OPA may play a role in the processing of landmarks that goes beyond mere visual perception.

Environmental Space: Cognitive Maps and Structured Representations

I now turn to a discussion of environmental space—the space that one can locomote to, typically extending beyond the current sensory horizon. Essential to any discussion of this topic is the concept of a cognitive map. This idea was first proposed by Tolman (1948) to account for aspects of the navigational behaviors of rats that could not be easily explained by behaviorist theories. Tolman observed, for example, that when animals were faced with a situation in which a familiar (but roundabout) path to a goal was blocked, they would often choose an alternative strategy of moving directly toward the goal. Such findings indicated that the animals must have some kind of internal representation of space—akin to a map—that could be flexibly used to guide behavior.

In a later formulation, which has become the "classic" view, O'Keefe and Nadel (1978) argued that the cognitive map is a Euclidean representation of navigational space—that is, a representation of space in terms of spatial coordinates. It is clear, however, that cognitive maps must be more complex than a single sheet of mental graph paper. At a minimum, an organism would need separate maps for different environments: it is highly unlikely that my cognitive map of Philadelphia picks up uninterrupted when I get off the plane in San

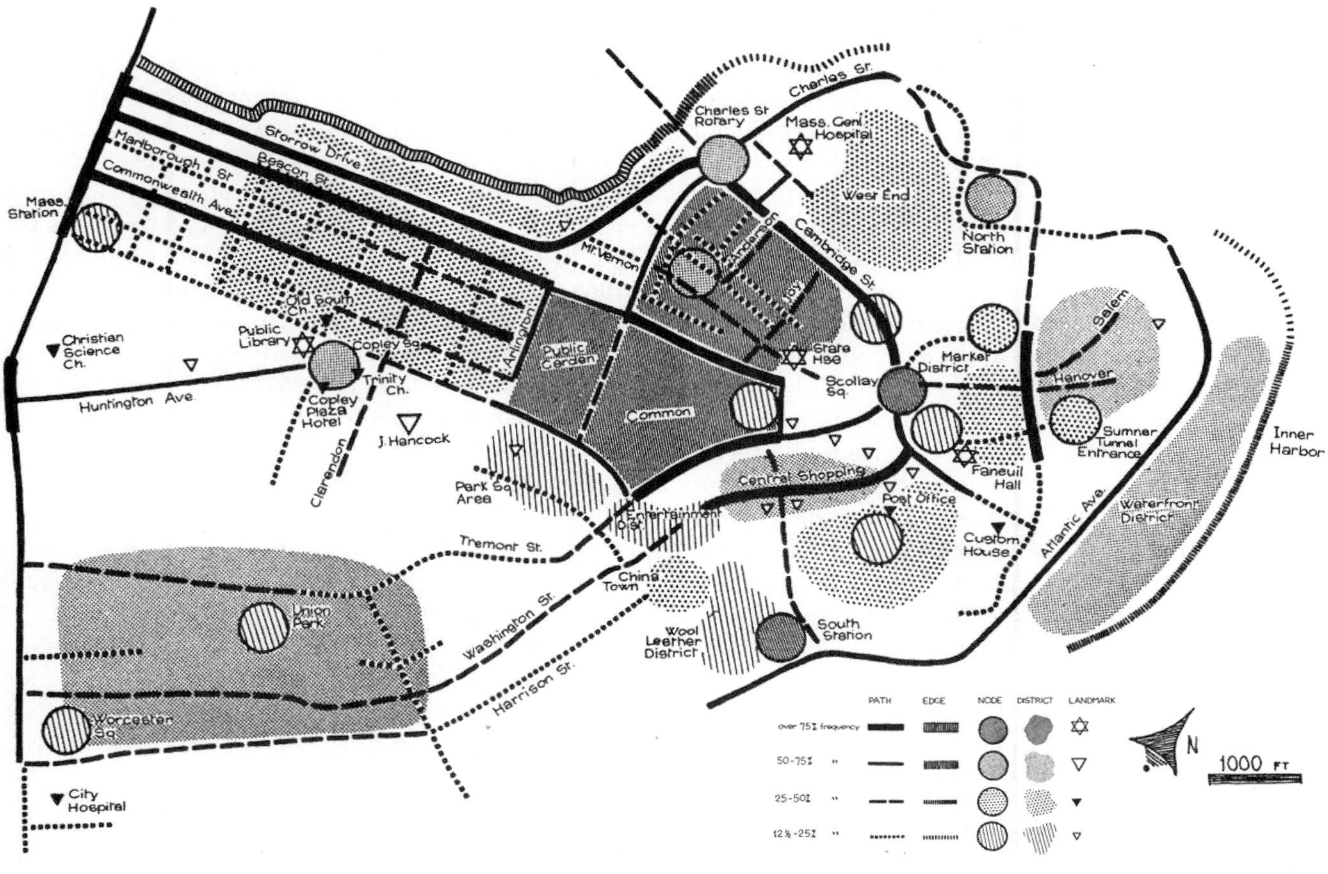

FIGURE 69.1 A cognitive map of Boston, Massachusetts, containing many structural elements (paths, edges, nodes, districts, and landmarks). Compiled by Lynch (1960) from resident reports.

Francisco. Even within the same city or campus, environmental spatial knowledge is structured in multiple ways. As a qualitative illustration of this, Lynch (1960) asked people to describe their experiences of their home cities (figure 69.1). From these accounts he identified five elements that made up their "image" of the city, including paths (streets, highways, bridges), edges (linear boundaries such as a riverbank), districts (regions with geographical and conceptual cohesion), nodes (strategic foci, often junctions of paths), and landmarks. Clearly, their mental map of the environment was more than just a collection of labeled coordinates.

Results from human psychological experiments support the idea that spatial knowledge is structured. Environmental spaces are often represented in a hierarchical manner, with locations grouped together into clusters or regions (Hirtle & Jonides, 1985; McNamara, Hardy, & Hirtle, 1989). For example, Wiener and Mallot (2003) taught subjects a virtual maze containing several objects that were grouped into regions based on conceptual similarity between the objects (e.g., all objects in one region were cars). When asked to navigate through this environment, participants chose paths that minimized the number of regions they had to pass through, even when an equivalent path had the same physical distance. The existence of hierarchical and regional structure may account for long-standing observations that spatial knowledge is distorted relative to metric truth, as evidenced by the fact that people make systematic errors in their estimates of distances and directions between locations (Tversky, 1993).

Relevant to this discussion of spatial structure is the notion of a spatial reference frame. To define coordinates, one must have reference axes. Much of what we know about how these axes are coded comes from studies using the judgment of relative direction (JRD) task. Participants in these experiments first learn an environment containing several objects. Later, after being removed from the environment, they are asked to imagine they are standing at one object while facing a second; from that imagined position and heading they are asked to indicate the remembered bearing to a third object. A consistent result from these experiments is that performance is orientation-dependent; that is, accuracy varies as a function of imagined facing direction (McNamara, Sluzenski, & Rump, 2008). The preferred direction is often aligned with the geometric shape of the environment or with the direction the subject was facing when first entering the environment (Shelton & McNamara, 2001). These results suggest that we assign spatial axes to environments when we first encounter them, which are used to lay down spatial memories. Memory retrieval is more accurate for imagined headings that are aligned rather than misaligned to these spatial axes.

This brings up an important question: If spatial knowledge is hierarchical, what is the relationship between the local reference frame (perhaps encompassing vista space but perhaps extending beyond it) and the groups or regions that constitute the higher level of the hierarchy? One possibility is that local reference frames are connected to each other by stored vectors to make a "network of reference frames" akin to a graph (Meilinger, 2008). Indeed, the idea that spatial knowledge is organized like a graph is one that recurs throughout the literature (Poucet, 1993; Trullier, Wiener, Berthoz, & Meyer, 1997; Warren, Rothman, Schnapp, & Ericson, 2017). Another possibility—not mutually exclusive—is that each local reference frame is a separate "map," which can be retrieved by a separate context recognition mechanism (Julian et al., 2015; Marchette, Ryan, & Epstein, 2017). We consider both of these possibilities in the next section.

Neural Systems for Representing Environmental Space

Some of the strongest evidence for the existence of a cognitive map comes from neuroscience. O'Keefe and Dostrovsky (1971) were the first to report the existence of neurons in the rodent hippocampus that fire when the animal is in specific locations in the world. O'Keefe and Nadel (1978) hypothesized that these *place cells* were the neural instantiation of the cognitive map. Extensive work over the past few decades has fleshed out this picture by showing that place cells are complemented by other classes of spatial cells in the hippocampal formation and related structures that support a neural mechanism for cognitive map-based navigation (Hartley, Lever, Burgess, & O'Keefe, 2014). These include grid cells (which provide a distance metric for the cognitive map), head direction cells (which provide a measure of the animal's orientation), and border/boundary cells (which allow cognitive maps to be anchored to environmental boundaries). Although initially identified in rodents, similar cells have since been found in humans (Ekstrom et al., 2003; Jacobs et al., 2013).

Neuroimaging studies support the idea that the hippocampus and entorhinal cortex play an important role in mediating a cognitive map in humans (see Epstein, Patai, Julian, & Spiers, 2017 for a review). Distances between locations—a key feature of a metric map—are reflected in fMRI adaptation effects (Morgan, Macevoy, Aguirre, & Epstein, 2011) and dissimilarities between multivoxel activation patterns (Deuker, Bellmund, Schröder, & Doeller, 2016; Nielson, Smith, Sreekumar, Dennis, & Sederberg, 2015). Moreover, the size of the right posterior hippocampus predicts participants' abilities to form allocentric representations of the environment (Hartley & Harlow, 2012; Schinazi, Nardi, Newcombe, Shipley, & Epstein, 2013), and this structure increases in volume in London taxi drivers as they acquire "the knowledge" of city streets and landmarks (Woollett & Maguire, 2011). These findings from humans indicate that the hippocampus is involved in memory for large-scale, real-world environmental spaces, not just the small-scale, single-chamber spaces commonly used in rodent-recording experiments.

Neuropsychological studies indicate that spatial memories for premorbidly learned environments are not obliterated by hippocampal damage (Teng & Squire, 1999), though they do become less detailed (Rosenbaum et al., 2000) and more schematic (Maguire, Nannery, & Spiers, 2006). This suggests that neocortical structures may also play a role in mediating environmental spatial knowledge. The retrosplenial/medial parietal region encompassing the RSC may be especially important for this function. This region is highly active in fMRI studies when spatial knowledge is retrieved (Epstein, Parker, & Feiler, 2007; Rosenbaum, Ziegler, Wincour, Grady, & Moscovitch, 2004). Moreover, fMRI activity in this region correlates with the amount of survey knowledge that a navigator has acquired about the environment (Wolbers & Buchel, 2005), and the number of spatially responsive cells in rodent retrosplenial cortex increases as an environment becomes more familiar (Smith, Barredo, & Mizumori, 2012).

But what kind of knowledge is encoded in the hippocampal formation and the RSC? As I noted above, we have many cognitive maps, not just one, and individual cognitive maps might be hierarchical or fragmented. The well-established phenomenon of remapping is likely to be the mechanism by which multiple maps are supported by the hippocampal-entorhinal system (Colgin, Moser, & Moser, 2008). Within any given environment, about a quarter of the hippocampal place cells exhibit place fields, while the remainder are quiescent. When an animal changes its environment—for example, if it is moved from an experimental chamber in one room to a different experimental chamber in another room—the set of active versus quiescent cells changes in an unpredictable manner, and even cells that are active in both environments change their firing locations relative to each other dramatically. Thus, the rodent hippocampal formation appears to have mechanisms for representing multiple maps as distinct pages within a larger "cognitive atlas."

What about hierarchical or fragmented structure within a map? The majority of neurophysiological recording studies are performed in open field environments. Thus, it is not surprising that the responses in these environments—for example, the regular tessellation of grid fields—reflects something that looks very much like a Euclidean map. However, when the environment becomes more structured, the place and grid representations become structured as well (figure 69.2*A*). For example, when an open field is divided by barriers into smaller subchambers, grid fields are observed to reflect the geometry of each subchamber, resetting their phase as the animal moves from one subchamber to another, rather than representing the environment as a whole (Derdikman et al., 2009). A similar effect of field repetition has been observed in hippocampal place cells (Spiers, Hayman, Jovalekic, Marozzi, & Jeffery, 2013). In

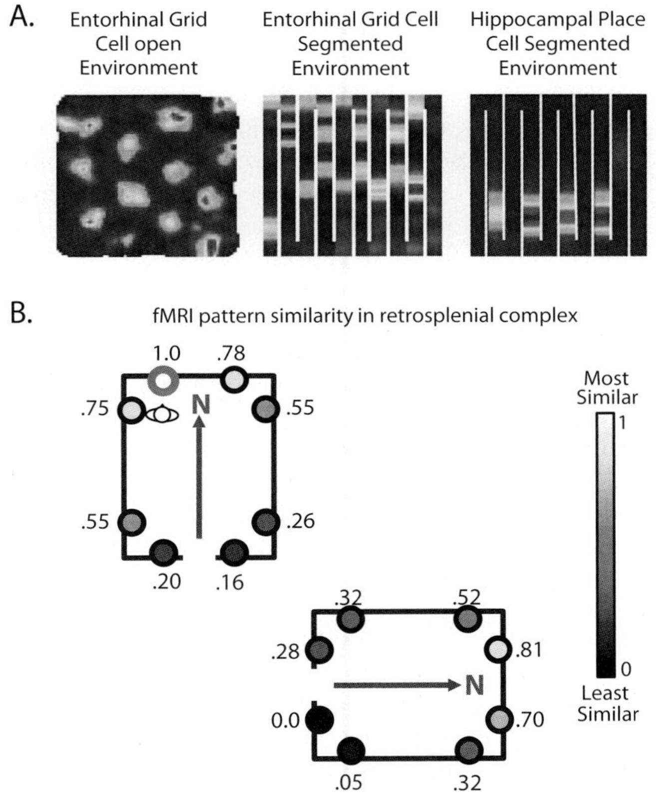

FIGURE 69.2 Spatial representations in structured environments. *A*, Grid cells code a regular triangular grid in open environments, but this pattern fragments into repetitive local fields when the environment is segmented into smaller subchambers (*white lines indicate walls*). A similar effect of pattern fragmentation is observed in hippocampal place cells. *B*, In a multichamber environment, RSC represents local geometric organization. Participants imagined facing an object along the wall at each location indicated by a circle. Colors and numbers indicate the similarity of multivoxel patterns for each view compared to the reference view (*red circle*). There is a high degree of similarity between views facing "local north" (i.e., away from the entrance) in different subchambers. (See color plate 83.)

related fMRI studies in humans, the RSC exhibits repeated use of the same spatial schema across geometrically similar subchambers (Marchette et al., 2014; figure 69.2*B*).

Beyond compartmentalization, two recent studies provide some evidence for the coding of graph-like structure when rats navigate through mazes consisting of constrained paths. In one study the animals navigated in the dark through a maze consisting of 10 path segments, which were connected flexibly to each other so that the angle between them could be varied (Dabaghian, Brandt, & Frank, 2014). Hippocampal place fields reflected the animal's position relative to the topography of the path rather than its position in Euclidean space. In another study, rats were trained to run paths through a maze consisting of three arms connected at a central choice point (Wu & Foster, 2014). When the animals rested at the end of the arms, "replay" activity was observed during sharp-wave-ripple events. Notably, these replay sequences reflected the connectivity structure of the maze, with the direction of replay reversing at the choice point. In humans, activity in the hippocampus has been observed to reflect both Euclidean measures of space (e.g., the total size of the space or the Euclidean distance to a destination) and graph-like measures of space (e.g., the complexity of the space, the path distance to a destination, the global connectivity) (Baumann & Mattingley, 2013; Howard et al., 2014; Javadi et al., 2017). Evidence for the graph-like coding of space has also been observed in rodent retrosplenial cortex (Alexander & Nitz, 2017) and human RSC (Schinazi & Epstein, 2010).

Conclusion

Although there is now a burgeoning cognitive neuroscience literature on spatial navigation, the knowledge structures that underlie navigation are relatively unexplored. As with any topic in cognitive neuroscience, spatial knowledge can be studied in terms of the cognitive representations that underlie it and the neural systems that support it. Until recently, these investigations have largely been the province of different fields: cognitive psychologists and animal behavior researchers on the one hand; neuroimagers and electrophysiologists on the other. In this chapter I have made a preliminary attempt to link these two literatures, but the field is ripe for further exploration.

Acknowledgment

This work was supported by National Institutes of Health grants EY022350 and EY0370470.

REFERENCES

Aguirre, G. K., & D'Esposito, M. (1999). Topographical disorientation: A synthesis and taxonomy. *Brain, 122,* 1613–1628.

Alexander, A. S., & Nitz, D. A. (2017). Spatially periodic activation patterns of retrosplenial cortex encode route sub-spaces and distance traveled. *Current Biology, 27*(11), 1551–1560.

Amit, E., Mehoudar, E., Trope, Y., & Yovel, G. (2012). Do object-category selective regions in the ventral visual stream represent perceived distance information? *Brain and Cognition, 80*(2), 201–213.

Auger, S. D., Mullally, S. L., & Maguire, E. A. (2012). Retrosplenial cortex codes for permanent landmarks. *PLoS One, 7*(8), e43620.

Baumann, O., & Mattingley, J. B. (2010). Medial parietal cortex encodes perceived heading direction in humans. *Journal of Neuroscience, 30*(39), 12897–12901.

Baumann, O., & Mattingley, J. B. (2013). Dissociable representations of environmental size and complexity in the human hippocampus. *Journal of Neuroscience, 33*(25), 10526–10533.

Biegler, R., & Morris, R. G. M. (1993). Landmark stability is a prerequisite for spatial but not discrimination-learning. *Nature, 361*(6413), 631–633.

Bonner, M. F., & Epstein, R. A. (2017). Coding of navigational affordances in the human visual system. *Proceedings of the National Academy of Sciences, 114*(18), 4793–4798.

Burgess, N., Spiers, H. J., & Paleologou, E. (2004). Orientational manoeuvres in the dark: Dissociating allocentric and egocentric influences on spatial memory. *Cognition, 94*(2), 149–166.

Burnett, G., Smith, D., & May, A. (2001). Supporting the navigation task: Characteristics of "good" landmarks. *Contemporary Ergonomics, 1,* 441–446.

Cheng, K. (1986). A purely geometric module in the rats spatial representation. *Cognition, 23*(2), 149–178.

Cheng, K., Huttenlocher, J., & Newcombe, N. S. (2013). Twenty-five years of research on the use of geometry in spatial reorientation: A current theoretical perspective. *Psychonomic Bulletin & Review, 20*(6), 1033–1054.

Colgin, L. L., Moser, E. I., & Moser, M. B. (2008). Understanding memory through hippocampal remapping. *Trends in Neurosciences, 31*(9), 469–477.

Collett, M., Chittka, L., & Collett, T. S. (2013). Spatial memory in insect navigation. *Current Biology, 23*(17), R789–R800.

Dabaghian, Y., Brandt, V. L., & Frank, L. M. (2014). Reconceiving the hippocampal map as a topological template. *eLife, 3.*

Derdikman, D., Whitlock, J. R., Tsao, A., Fyhn, M., Hafting, T., Moser, M.-B., & Moser, E. I. (2009). Fragmentation of grid cell maps in a multicompartment environment. *Nature Neuroscience, 12*(10), 1325.

Deuker, L., Bellmund, J. L., Schröder, T. N., & Doeller, C. F. (2016). An event map of memory space in the hippocampus. *eLife, 5.*

Dilks, D. D., Julian, J. B., Paunov, A. M., & Kanwisher, N. (2013). The occipital place area is causally and selectively involved in scene perception. *Journal of Neuroscience, 33*(4), 1331–1336y.

Ekstrom, A. D., Kahana, M. J., Caplan, J. B., Fields, T. A., Isham, E. A., Newman, E. L., & Fried, I. (2003). Cellular networks underlying human spatial navigation. *Nature, 425*(6954), 184–188.

Epstein, R. A. (2014). Neural systems for visual scene recognition. In M. Bar & K. Kveraga (Eds.), *Scene vision* (pp. 105–134). Cambridge, MA: MIT Press.

Epstein, R. A., Parker, W. E., & Feiler, A. M. (2007). Where am I now? Distinct roles for parahippocampal and retrosplenial cortices in place recognition. *Journal of Neuroscience, 27*(23), 6141–6149.

Epstein, R. A., Patai, E. Z., Julian, J. B., & Spiers, H. J. (2017). The cognitive map in humans: Spatial navigation and beyond. *Nature Neuroscience, 20*(11), 1504.

Fairhall, S. L., Anzellotti, S., Ubaldi, S., & Caramazza, A. (2013). Person-and place-selective neural substrates for entity-specific semantic access. *Cerebral Cortex, 24*(7), 1687–1696.

Ganaden, R. E., Mullin, C. R., & Steeves, J. K. (2013). Transcranial magnetic stimulation to the transverse occipital sulcus affects scene but not object processing. *Journal of Cognitive Neuroscience, 25*(6), 961–968.

Gillner, S., Weiss, A. M., & Mallot, H. A. (2008). Visual homing in the absence of feature-based landmark information. *Cognition, 109*(1), 105–122.

Harel, A., Kravitz, D. J., & Baker, C. I. (2013). Deconstructing visual scenes in cortex: Gradients of object and spatial layout information. *Cerebral Cortex, 23*(4), 947–957.

Hartley, T., & Harlow, R. (2012). An association between human hippocampal volume and topographical memory in healthy young adults. *Frontiers in Human Neuroscience, 6*, 338.

Hartley, T., Lever, C., Burgess, N., & O'Keefe, J. (2014). Space in the brain: How the hippocampal formation supports spatial cognition. *Philosophical Transactions of the Royal Society of London B: Biological Sciences, 369*(1635), 20120510.

Hartley, T., Trinkler, I., & Burgess, N. (2004). Geometric determinants of human spatial memory. *Cognition, 94*(1), 39–75.

He, C., Peelen, M. V., Han, Z., Lin, N., Caramazza, A., & Bi, Y. (2013). Selectivity for large nonmanipulable objects in scene-selective visual cortex does not require visual experience. *NeuroImage, 79*, 1–9.

Hirtle, S. C., & Jonides, J. (1985). Evidence of hierarchies in cognitive maps. *Memory & Cognition, 13*(3), 208–217.

Howard, L. R., Javadi, A. H., Yu, Y. C., Mill, R. D., Morrison, L. C., Knight, R., ... Spiers, H. J. (2014). The hippocampus and entorhinal cortex encode the path and Euclidean distances to goals during navigation. *Current Biology, 24*(12), 1331–1340.

Jacobs, J., Weidemann, C. T., Miller, J. F., Solway, A., Burke, J. F., Wei, X. X., ... Kahana, M. J. (2013). Direct recordings of grid-like neuronal activity in human spatial navigation. *Nature Neuroscience, 16*(9), 1188–1190.

Jansen-Osmann, P. (2002). Using desktop virtual environments to investigate the role of landmarks. *Computers in Human Behavior, 18*(4), 427–436.

Janzen, G. (2006). Memory for object location and route direction in virtual large-scale space. *Quarterly Journal of Experimental Psychology, 59*(3), 493–508.

Janzen, G., & van Turennout, M. (2004). Selective neural representation of objects relevant for navigation. *Nature Neuroscience, 7*(6), 673–677.

Javadi, A.-H., Emo, B., Howard, L. R., Zisch, F. E., Yu, Y., Knight, R., ... Spiers, H. J. (2017). Hippocampal and prefrontal processing of network topology to simulate the future. *Nature Communications, 8*, 14652.

Julian, J. B., Keinath, A. T., Muzzio, I. A., & Epstein, R. A. (2015). Place recognition and heading retrieval are mediated by dissociable cognitive systems in mice. *Proceedings of the National Academy of Sciences of the United States of America, 112*(20), 6503–6508.

Julian, J. B., Ryan, J., Hamilton, R. H., & Epstein, R. A. (2016). The occipital place area is causally involved in representing environmental boundaries during navigation. *Current Biology, 26*(8), 1104–1109.

Konkle, T., & Oliva, A. (2012). A real-world size organization of object responses in occipitotemporal cortex. *Neuron, 74*(6), 1114–1124.

Kravitz, D. J., Peng, C. S., & Baker, C. I. (2011). Real-world scene representations in high-level visual cortex: It's the spaces more than the places. *Journal of Neuroscience, 31*(20), 7322–7333.

Lee, S. A. (2017). The boundary-based view of spatial cognition: A synthesis. *Current Opinion in Behavioral Sciences, 16*, 58–65.

Lynch, K. (1960). *The image of the city*. Cambridge, MA: Technology Press.

Maguire, E. A., Nannery, R., & Spiers, H. J. (2006). Navigation around London by a taxi driver with bilateral hippocampal lesions. *Brain, 129*(Pt. 11), 2894–2907.

Marchette, S. A., Ryan, J., & Epstein, R. A. (2017). Schematic representations of local environmental space guide goal-directed navigation. *Cognition, 158*, 68–80.

Marchette, S. A., Vass, L. K., Ryan, J., & Epstein, R. A. (2014). Anchoring the neural compass: Coding of local spatial reference frames in human medial parietal lobe. *Nature Neuroscience, 17*(11), 1598–1606.

Marchette, S. A., Vass, L. K., Ryan, J., & Epstein, R. A. (2015). Outside looking in: Landmark generalization in the human navigational system. *Journal of Neuroscience, 35*(44), 14896–14908.

McNamara, T. P., Hardy, J. K., & Hirtle, S. C. (1989). Subjective hierarchies in spatial memory. *Journal of Experimental Psychology: Learning, Memory, and Cognition, 15*(2), 211–227.

McNamara, T. P., Sluzenski, J., & Rump, B. (2008). Human spatial memory and navigation. In I. H. L. Roediger (Ed.), *Cognitive psychology of memory* (pp. 157–178). Oxford: Elsevier.

Meilinger, T. (2008). The network of reference frames theory: A synthesis of graphs and cognitive maps. *Spatial cognition VI. Learning, reasoning, and talking about space*, 344–360. New York: Springer.

Montello, D. R. (1993). Scale and multiple psychologies of space. In *European Conference on Spatial Information Theory*, 313–321. Berlin: Springer.

Morgan, L. K., Macevoy, S. P., Aguirre, G. K., & Epstein, R. A. (2011). Distances between real-world locations are represented in the human hippocampus. *Journal of Neuroscience, 31*(4), 1238–1245.

Nielson, D. M., Smith, T. A., Sreekumar, V., Dennis, S., & Sederberg, P. B. (2015). Human hippocampus represents space and time during retrieval of real-world memories. *Proceedings of the National Academy of Sciences of the United States of America, 112*(35), 11078–11083.

O'Keefe, J. & Dostrovsky, J. (1971). The hippocampus as a spatial map: Preliminary evidence from unit activity in the freely-moving rat. *Brain Research, 34* (1), 171–175.

O'Keefe, J., & Nadel, L. (1978). *The hippocampus as a cognitive map*. Oxford: Clarendon Press.

Park, S., Brady, T. F., Greene, M. R., & Oliva, A. (2011). Disentangling scene content from spatial boundary: Complementary roles for the parahippocampal place area and lateral occipital complex in representing real-world scenes. *Journal of Neuroscience, 31*(4), 1333–1340.

Poucet, B. (1993). Spatial cognitive maps in animals: New hypotheses on their structure and neural mechanisms. *Psychological Review, 100*(2), 163.

Rosenbaum, R. S., Priselac, S., Kohler, S., Black, S. E., Gao, F., Nadel, L., & Moscovitch, M. (2000). Remote spatial memory in an amnesic person with extensive bilateral hippocampal lesions. *Nature Neuroscience, 3*(10), 1044–1048.

Rosenbaum, R. S., Ziegler, M., Wincour, G., Grady, C. L., & Moscovitch, M. (2004). "I have often walked down this street before": fMRI studies on the hippocampus and other structures during mental navigation of an old environment. *Hippocampus, 14* (7), 826–835.

Schinazi, V. R., & Epstein, R. A. (2010). Neural correlates of real-world route learning. *NeuroImage, 53* (2), 725–735.

Schinazi, V. R., Nardi, D., Newcombe, N. S., Shipley, T. F., & Epstein, R. A. (2013). Hippocampal size predicts rapid learning of a cognitive map in humans. *Hippocampus, 23*(6), 515–528.

Shelton, A. L., & McNamara, T. P. (2001). Systems of spatial reference in human memory. *Cognitive Psychology, 43*(4), 274–310.

Shine, J. P., Valdés-Herrera, J. P., Hegarty, M., & Wolbers, T. (2016). The human retrosplenial cortex and thalamus code head direction in a global reference frame. *Journal of Neuroscience, 36*(24), 6371–6381.

Smith, D. M., Barredo, J., & Mizumori, S. J. Y. (2012). Complimentary roles of the hippocampus and retrosplenial cortex in behavioral context discrimination. *Hippocampus, 22*(5), 1121–1133.

Spiers, H. J., Hayman, R. M., Jovalekic, A., Marozzi, E., & Jeffery, K. J. (2013). Place field repetition and purely local remapping in a multicompartment environment. *Cerebral Cortex, 25*(1), 10–25.

Teng, E., & Squire, L. R. (1999). Memory for places learned long ago is intact after hippocampal damage. *Nature, 400*(6745), 675–677.

Tolman, E. C. (1948). Cognitive maps in rats and men. *Psychological Review, 55*, 189–208.

Troiani, V., Stigliani, A., Smith, M. E., & Epstein, R. A. (2014). Multiple object properties drive scene-selective regions. *Cerebral Cortex, 24*(4), 883–897.

Trullier, O., Wiener, S. I., Berthoz, A., & Meyer, J. A. (1997). Biologically based artificial navigation systems: Review and prospects. *Progress in Neurobiology, 51*(5), 483–544.

Tversky, B. (1993). Cognitive maps, cognitive collages, and spatial mental models. In *European Conference on Spatial Information Theory*, 14–24. Berlin: Springer.

Valiquette, C., & McNamara, T. P. (2007). Different mental representations for place recognition and goal localization. *Psychonomic Bulletin & Review, 14*(4), 676–680.

Vass, L. K., & Epstein, R. A. (2013). Abstract representations of location and facing direction in the human brain. *Journal of Neuroscience, 33*(14), 6133–6142.

Waller, D., & Hodgson, E. (2006). Transient and enduring spatial representations under disorientation and self-rotation. *Journal of Experimental Psychology: Learning, Memory, and Cognition, 32*(4), 867.

Walther, D. B., Chai, B., Caddigan, E., Beck, D. M., & Fei-Fei, L. (2011). Simple line drawings suffice for functional MRI decoding of natural scene categories. *Proceedings of the National Academy of Sciences of the United States of America, 108*(23), 9661–9666.

Warren, W. H., Rothman, D. B., Schnapp, B. H., & Ericson, J. D. (2017). Wormholes in virtual space: From cognitive maps to cognitive graphs. *Cognition, 166*, 152–163.

Wiener, J. M., & Mallot, H. A. (2003). "Fine-to-coarse" route planning and navigation in regionalized environments. *Spatial Cognition and Computation, 3*(4), 331–358.

Wolbers, T., & Buchel, C. (2005). Dissociable retrosplenial and hippocampal contributions to successful formation of survey representations. *Journal of Neuroscience, 25*(13), 3333–3340.

Woollett, K., & Maguire, E. A. (2011). Acquiring "the knowledge" of London's layout drives structural brain changes. *Current Biology, 21*(24), 2109–2114.

Wu, X., & Foster, D. J. (2014). Hippocampal replay captures the unique topological structure of a novel environment. *Journal of Neuroscience, 34*(19), 6459–6469.

70 The Nature of Human Mathematical Cognition

JESSICA F. CANTLON

ABSTRACT John Locke called the concept of number "the simplest and most universal idea" (1690, p. 127). This is because quantity is central to human rationality, and numerical concepts are the bedrock of all human measurement—number "measures all measurables," as Locke says. Whether measuring sets, time, distance, size, weight, or value, humans primarily use numerical scales to formalize and unitize quantities. Numbers are abstract representations that describe incremental changes in object quantity and that can be logically evaluated and transformed. Simple logical operations on numbers, such as comparison and arithmetic, are the building blocks of human mathematics. Substantial evidence indicates that numerical value can be represented without language, in an analog format, and is cognitively manipulated using nonlinguistic logical operations. This primitive arithmetic exists in modern humans in a psychological and neural format similar to other species. However, human cultures symbolically formalize numerical relations that have a unique impact on human cognition, behavior, and brain activity compared to other species. We present research from the field of numerical cognition across multiple levels of analysis to understand the mutual interactions between its origins and purpose and its computations and biology.

The origins and organization of numerical concepts are studied integratively at multiple levels of analysis. This is important because there are interacting constraints on the mechanisms the brain can implement. Research into the nature of mathematical representations in humans addresses several levels of analysis by comparing species, cultures, and stages of human development (getting at Tinbergen's questions; Tinbergen, 1963) and also across the computational, algorithmic, and neural explanations of representations (getting at Marr's levels; Marr & Poggio, 1976). This approach is necessary because it accounts for different pressures—evolutionary and developmental, neural and functional, environmental, and algorithmic—that limit the mechanisms the brain can or will implement. The field of numerical cognition not only investigates the underlying domain representations but also examines the ways those representations arise from the dynamic interaction between genetic constraints and environmental input. In this review we discuss the different levels of analysis at which numerical cognition is understood. We show comparisons of cognitive and neural processes that reveal numerical cognition's developmental and evolutionary basis, neural and algorithmic properties, computational function, and uniqueness among humans.

Developmental Basis

Studies on human newborns and preverbal infants suggest that domain knowledge about numerical relations establishes the foundation of numerical development in humans. Neonates, just hours after birth, can discriminate the numerical values of sets nonverbally with crude acuity. Numerical discrimination in infants follows Weber's law of psychophysics—it is more difficult to discriminate values that are close together than far apart (i.e., the ratio effect). Izard, Sann, Spelke, and Streri (2009) showed that newborn infants look longer at visual arrays that numerically match the number of sounds they hear in an auditory sequence compared to numerically different visual arrays. Alternative dimensions such as surface area or duration could not explain newborns' looking-time behavior in that study because of its crossmodal design. The study showed that newborn infants represent numerical value at an abstract perceptual level across modalities. Several studies of older infants have produced results that show the early representation of number (Barth et al., 2005; Feron, Gentaz, & Streri, 2006; Jordan & Brannon, 2006; Libertus & Brannon, 2010). The implication is that experience-expectant cognitive processes detect quantitative variation in sets and events at birth.

These studies raise questions about how infants, and humans more generally, disentangle numerical representations from other correlated information in the environment. There are natural correlations between quantitative dimensions in the environment (Cantrell & Smith, 2013; Ferrigno et al., 2017; Gebius & Reynvoet, 2012; Piantadosi & Cantlon, 2017). Infants are sensitive to quantitative dimensions beyond numerical value, including surface area, duration, and density (Clearfield & Mix, 2001; Cordes & Brannon, 2008; Lourenco & Longo, 2010). These dimensions also provide valuable quantitative information about sets and events and they are often correlated with number. For example, a set of six figs often (but not always) has a greater number, cumulative surface area, and volume than a set of three figs. Some have argued that infants are initially "one bit" and only represent a general magnitude value across different dimensions including number, area, and duration (Cantrell & Smith, 2013; Walsh, 2003). Infants are thought to learn to disentangle quantitative dimensions from correlation patterns in the environment. However, how an infant would ever disentangle correlated dimensions without first making some prediction about or interpretation of the underlying components is unclear. For example, in order for infants to detect breeches of correlated structure among dimensions they would have to know that multiple different quantities exist. Thus, it is as yet unclear what algorithm or process might permit infants to develop representations of number from "one bit."

Findings of numerical sensitivity in neonates also raise questions about the nature of innate knowledge in numerical cognition. One proposal for what constitutes innate knowledge, echoing Locke (1690, pp. 127–131), is a representation of the base quantity "one," from which all other numbers can be arithmetically generated by a successor function (Leslie, Gelman, & Gallistel, 2008). A base quantity of one provides a foundation for calculating all integers by adding one to one, and so on, up to any size. Another proposal for innate knowledge is that the algebraic properties of neural codes for numerosity inherently represent arithmetic relations (see Hannagan et al., 2018). Models of number coding are described further in the section on algorithmic models; however, the point here is that some theoretical proposals about the nature of innate numerical knowledge require only simple psychological constraints.

Evolutionary Basis

The extensive literature on numerical abilities in nonhuman animal species converges with developmental data from human infants in an evolutionary interpretation of the origins of numerical cognition (e.g., Agrillo, Piffer, & Bisazza, 2011; Beran, Parrish, & Evans, 2015; Brannon & Terrace, 1998; Cantlon & Brannon, 2006; Emmerton & Renner, 2006; Gallistel & Gelman, 2000; Rugani, Regolin, & Vallortigara, 2010; Scarf, Hayne, & Colombo, 2011).

Like human infants, newborn chicks bear a sensitivity to numerosity from birth. In several studies with newborn chicks, chicks raised in controlled environments imprinted on a set of objects and followed that set as their "mother" (e.g., Rugani, Regolin, & Vallortigara, 2010). Once those chicks imprinted on a set, the experimenters tested them in trials with novel "mother" sets that varied in numerosity. The results showed that the chicks established their imprinting response on numerosity—they were more likely to follow sets with similar numerical values to their original "mother." The representation of numerosity emerged spontaneously at birth in chicks. Birds trained on quantity discrimination tasks in the lab show sensitivity to numerical value when tested with stimuli that are controlled for alternative cues like surface area.

Quantitative abilities in young animals suggest that a core function of animal brains is to compare amounts. Indeed, many species compute amounts of various types—even worms are sensitive to differences in ion concentration (Sambongi et al., 1999). The simple logic of quantity comparison is likely widespread across different nervous systems.

Primates have sophisticated numerical abilities and are likely to share homologous cognitive, neural, and developmental processes with humans. The ability to make numerical choices develops rapidly and spontaneously in nonhuman primates. Infant monkeys are able to make reliable quantitative choices within one year of life (Ferrigno, Hughes, & Cantlon, 2016). Monkeys' ability to make quantitative choices develops three times faster than in humans, which is a ratio similar to other aspects of perceptual and motor development. Numerical development is thus a primitive and rapidly emerging aspect of all primate cognition.

Primates have been shown to engage in a range of logical operations with numerical values. Behavioral research with lemurs, monkeys, and apes shows they possess logical capacities for comparison, incrementing, ordination, proportion, and addition and subtraction with quantities (Beran, Parrish, & Evans, 2015; Cantlon et al., 2015; Nieder, 2016). When monkeys compare visual arrays of dots to determine the smaller quantity, their performance closely resembles that of human subjects who are prevented from counting. Estimation functions for monkeys and humans are parallel and adhere to Weber's law of analog quantity comparison. Some of the more complex arithmetic abilities of primates are proportional reasoning, addition, and subtraction. Monkeys can compare the relative lengths of two pairs of lines to determine whether the proportion relation between pairs is similar (Vallentin & Nieder, 2008). Apes and monkeys can predict the arithmetic outcome of sets combined behind an occluder. For example, if 6 items are covered by an occluder, then 3 more items are added behind the occluder, monkeys will guess that there are 9 items behind the occluder (versus 3, 6, or 12). Monkeys also track the relative values of sets during one-by-one set construction, showing an ability to represent count-like incremental changes in numerical value (Cantlon et al., 2015). Finally, monkeys can make metacognitive judgments about their accuracy during numerosity tasks, and thus their numerical processes are available to internally monitor (Beran, Smith, Redford, & Washburn, 2006). These capacities in nonhuman primates suggest that several logical tools for quantitative cognition emerged many millions of years ago in the human lineage.

Basic numerical reasoning emerges spontaneously in wild primates. For example, baboons make collective troop movements by estimating the number of individual animals in a subgroup that took each of a few possible paths and choosing the greatest number (Strandburg-Peshkin et al., 2015). Wild baboons' troop movements are based on the number of individuals in a subgroup, as opposed to their mass or size (Piantadosi & Cantlon, 2017). As the difference (Weber fraction) between the number of baboons in each subgroup increases, animals are more successful at choosing the larger group. These findings are evidence that numerical comparison is computed naturally by wild primates.

The use of numerical reasoning in the wild is not limited to primates. Numerosity abilities have been observed in so many species that it would be newsworthy to discover a species that lacked it. Fish have been shown to use numerical comparisons during schooling and collective behaviors (e.g., Agrillo & Dadda, 2007). Even insects and other invertebrates are suspected to use numerical representations in their natural behaviors (Chittka & Geiger, 1995; Gallistel, 1990; Wittlinger, Wehner, & Wolf, 2006). However, it is currently unclear whether true numerical reasoning is involved in many of these other cases versus rate, duration, mass, or density perception. A recent study that directly compared spontaneous numerical reasoning in human adults from different cultures, children, and monkeys reported significant qualitative similarities in numerosity perception between groups, but such direct comparisons have not been conducted with nonprimate animals (Ferrigno et al., 2017).

Computational Function

The natural functions of numerical cognition offer clues to its adaptive value as a system of representation and its design—the problems numerical cognition was selected to solve over evolution will constrain the algorithms that are implemented and their neural "wiring."

One domain where numerical reasoning provides adaptive advantages for many species is in foraging (e.g., Gallistel, 1990; Godin & Keenleyside, 1984; Harper, 1982). Wild orangutans, for example, preferentially forage in fig trees with the largest number of ripe figs (Utami et al., 1997). Evolutionary simulations of numerical cognition show a plausible route to numerosity representation through natural selection. Hope, Stoianov, and Zorzi (2010) used artificial life simulations built on the hypothesis that quantity comparison originated from foraging adaptations to maximize food intake. In their model, quantity sensitivity was determined by a naturalistic genetic algorithm (with mutation and crossover) that determined the agent's genome, which in turn determined the connection parameters and size of a hidden layer in a neural network underlying quantitative choice. The model shows that numerical sensitivity could plausibly emerge by genetic selection for foraging efficiency over evolution.

Numerical reasoning could also underlie aspects of social behavior (McComb, Packer, & Pusey, 1994; Wilson, Hauser, & Wrangham, 2001). Social playback experiments show that animals such as chimpanzees and lions use "number of calls" as a cue for deciding intergroup confrontations. When lions were played a small number of foreign lion calls from a hidden speaker, they were more likely to confront the source than if played a large number of calls (McComb, Packer, & Pusey, 1994). Both the social and foraging functions of numerical reasoning show that object-based, crossmodal quantity judgments in natural environments likely shaped the design of numerical mechanisms.

Neural Basis

In humans and nonhuman primates, homologous neural areas within the intraparietal sulcus (IPS), as well as areas of prefrontal cortex (PFC), are engaged during numerical representation. In monkeys, intraparietal areas (ventral, or VIP, and lateral intraparietal, or LIP) contain neurons that are sensitive to numerosity. Neurons in area LIP, called *summation neurons,* show responses that are modulated by the absolute numerical value of a stimulus (figure 70.1*A*; Roitman, Brannon, & Platt, 2007). Neurons in the VIP area show responses that are coarsely tuned to preferred cardinal values and are modulated by the relative numerical value of a stimulus to the preferred numerical value (figure 70.1*B*; Nieder & Miller, 2004; Nieder, 2012). Neurons in monkey VIP, *tuning neurons,* peak at a preferred numerical value (1.0 ratio), and their firing rate to other numerical values diverges from the peak as a function of the ratio between the value presented and its preferred value.

Neural recordings from naïve monkeys that were not trained to discriminate number show that single neurons in LIP represent numerosity monotonically, and VIP and PFC neurons spontaneously tune their firing

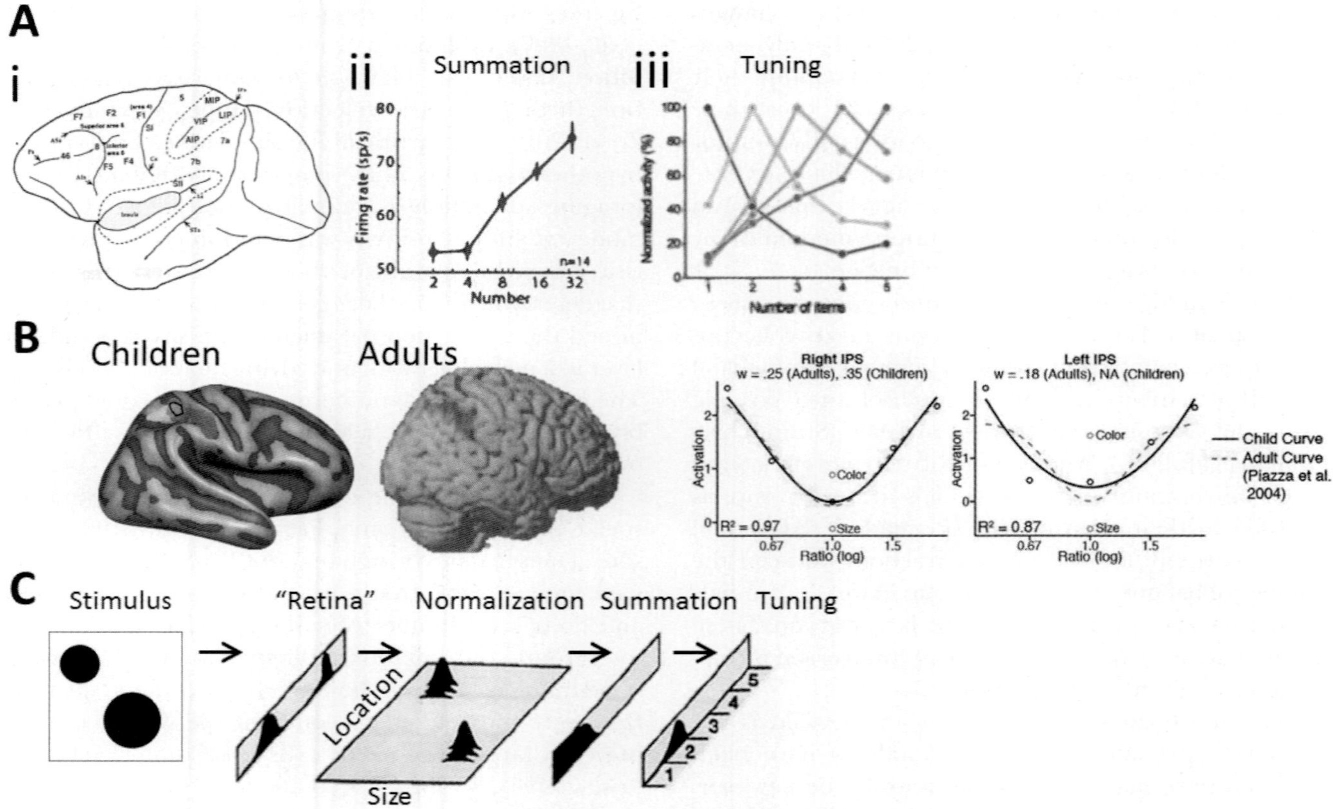

FIGURE 70.1 *A*, Neurons in intraparietal areas VIP and LIP show numerical sensitivity (*1*). In area VIP, neurons respond to numerical stimuli with a monotonic *summation* response (*2*) and in LIP with a *tuning* response (*3*). *B*, Human children and adults show numerical sensitivity in the IPS (*red*). Neural responses in the IPS (*right*) show tuning to numerosity during fMRI adaptation based on the ratio of change in the adaptation stream. Adults show sharper neural tuning to numerosity in the left IPS compared to children. *C*, Dehaene and Changeux (1993) modeled numerical representation in a neural network. Visual objects in an array stimulus are first normalized to a location- and size-independent code. Activation is then summed to yield an estimate of the input numerosity. Numerosity detectors are connected to summation activation, and neural activity is tuned to numerosity in an on-center, off-surround pattern. (See color plate 84.)

patterns to specific numerosities (Nieder & Miller, 2004; Roitman, Brannon, & Platt, 2007). An open question is whether and how summation neurons and tuning neurons work together to represent numerical value (Piazza & Izard, 2009). One possibility is that summation neurons accumulate entities to compute a set representation, and tuning neurons place those sums within the relative context of a number line.

A whole-brain monkey and human comparative functional magnetic resonance imaging (fMRI) study of numerical processing confirmed that the number network includes regions of the IPS and PFC (Wang et al., 2015). Human neuroimaging studies indicate that the representation of numerosity also occurs spontaneously early in child development, in a parallel network of neural regions (figure 70.1*C*). Such findings suggest that evolutionarily primitive and early-developing properties of frontoparietal circuits are responsible for the emergence of number-coding neurons.

Human adults and children show neural tuning to numerosity in functionally overlapping regions of intraparietal cortex (Kersey & Cantlon, 2017; Piazza et al., 2004). Neural responses observed in humans (with fMRI) are modulated by the relative values of numerical stimuli and, like monkeys' responses, follow a ratio-dependent neural tuning curve. Studies using fMRI have shown that parallel regions in the IPS and PFC are activated during number comparisons in adults and young children (Ansari, 2008; Ansari & Dhital, 2006; Bugden et al., 2012; Lussier & Cantlon, 2017). Similarly, near-infrared spectroscopy and electroencephalography studies of infants in the first year show numerical sensitivity in the right parietal cortex (Edwards, Wagner, Simon, & Hyde, 2016; Izard et al., 2008; Libertus, Brannon, & Woldorff, 2011). Tuning neurons have been observed in children and adults with fMRI (Kersey & Cantlon, 2017). Summation neurons have not been observed in humans, but it is unclear whether those neurons would be observed at the population level with fMRI. Observations of summation neurons in humans could require more granular data than those currently available.

The prefrontal and parietal cortices are regions that process stimuli at a high level of perceptual and motor abstraction in primates (Nieder, 2016). Numerical representation requires abstraction across object and event features, including space, time, perspective, and modality, to represent a "set." The demand for abstraction and integration across objects and events is a constraint on neural processing—it limits which neural regions could do the job. The parietofrontal network observed in numerical processing in humans and monkeys is known to meet these demands because those regions take inputs from multiple sensory and perceptual regions, have large spatial and temporal receptive windows, and provide abstract outputs to premotor structures (Cavada & Goldman-Rakic, 1989; Hasson, Yang, Vallines, Heeger, & Rubin, 2008). Parietal regions also show biases toward topographic representation, and numerosities appear to be topographically mapped there, which could be critically related to the ordinality of number (Harvey, Klein, Petridou, & Dumoulin, 2013).

Neural representations of numerosity are multimodal. PFC and IPS neurons are sensitive to numerical quantity from both auditory and visual modalities in monkeys (Nieder, 2016). Similarly, the human IPS exhibits sensitivity to auditory and visual numerical stimuli but not to comparable nonnumeric control stimuli (e.g., Eger et al., 2003).

Intracranial electrocorticography recordings from intraparietal cortex in three epilepsy patients showed number-related neural activity during natural numerical reasoning over the course of 7 to 10 days (Dastjerdi et al., 2013). The subjects were implanted with chronic intracranial electrodes covering lateral parietal cortex while being continuously monitored by video recording. Each electrode captured a signal from a population of around 500,000 parietal neurons. Researchers identified regions in each participant that showed elevated high-frequency broadband (HFB) activity during an experimental arithmetic task versus a control task—those regions included segments of the IPS. They then tested activation patterns from the natural numerical reasoning events identified in participants' video footage. Subjects showed HFB peaks during naturalistic numerical tasks, and even during the mention of number words, in the same IPS region that showed peak activation during the experimental arithmetic task.

Physiological interventions in monkeys suggest that posterior parietal cortex plays a causal role in numerical representation. Brief periods of pharmacological inactivation to posterior parietal cortex (area 5) with muscimol caused monkeys to underestimate the number of items in a sequence of movements (Sawamura, Shima, & Tanji, 2010). The underestimation was not caused by impairment in motor control because the monkeys were able to perform correct movement types in response to an auditory tone—they only failed to produce the correct number of movements. Human neuropsychological data also show that focal lesions to posterior parietal cortex cause number-specific deficits (Dehaene & Cohen, 1997). Together, those data indicate that neural signatures of numerical processing from posterior parietal regions are not simply correlational but causal.

Evidence showing functional homologies between humans and monkeys in the IPS suggests that the numerical functions of the IPS could be homologous among primates. Although there is no homologous structure to the IPS in the avian brain, neural recordings from crows reveal similar neural signatures of numerosity representation within a structure analogous to the primate neocortex, the nidopallium caudolateral (Ditz & Nieder, 2015). Neurons within the nidopallium caudolateral fire with a pattern similar to neural tuning responses in primates; however, the underlying neural anatomy is distinct. These findings from birds show that there are at least two similar yet independently evolved neural implementations of numerical representation in the animal kingdom (Nieder, 2016).

Algorithmic Models

Multiple plausible models of basic numerical representation from different computational approaches and levels of analysis are available. Each model explains some of the underlying algorithm for how the perception of number is encoded at the cognitive or neural level. While there is no comprehensive model of number representation, each model is consistent with some aspect of the behavioral and neural data from humans and animals.

A neural network model by Dehaene and Changeux (1993) takes a set of spatially distributed objects and represents its numerosity as an analog estimate (figure 70.1*D*). The first stage of processing in the model is a location map of the set in which objects' locations are topographically represented. Objects in the location map are normalized for size and location in activity levels on the map—larger objects do not elicit greater activity than smaller objects. Activity in the location map is summed up, with larger numbers of objects causing greater activation than smaller numbers. Finally, summation clusters project to ordered numerosity detectors that respond to preferred numerosities and exhibit the central excitation and lateral inhibition of nonpreferred numerosities. Activation decreases proportionally with increasing numerical distance between the preferred and actual number. This model is supported by neural data from monkeys showing tuning neurons that behave like numerosity detectors (Nieder & Miller, 2004) and summation neurons that are conceptually similar to summation clusters (Roitman, Brannon, & Platt, 2007). Empirical support for the other components of the model, such as the normalized location map, the lateral inhibition, and the processing hierarchy, is currently lacking. This model is further limited because it only accounts for spatially distributed sets, not temporally distributed sets, and it is not crossmodal.

Deep-learning networks engage in unsupervised learning over large amounts of input stimuli to form abstract representations that allow the future prediction of those stimuli in the environment. A deep-learning network by Stoianov and Zorzi (2012) was presented with tens of thousands of images of dot arrays that varied in number, spatial configuration, and size. The model consisted of two hidden layers: one layer (HL2) exhibited properties of summation neurons wherein activity was dependent on the number of dots in the array, and the other layer (HL1) responded like a spatial map of object locations. HL2 represented the numerical value of the objects in the stimuli as opposed to spatial characteristics such as density or surface area. The behavior of the model paralleled numerosity discrimination performance in humans and monkeys, and responses were modulated by Weber's law. The results showed that numerical representations emerge from the abstraction of visual arrays by a process that spontaneously normalizes variability in the spatial features of objects and sets. The current limitations of this model are that it applies only to the narrow case of spatially distributed visual sets, and the model's learning mechanism currently lacks empirical support at the cognitive and neural levels.

Hannagan, Nieder, Viswanathan, and Dehaene (2018) provided a mathematical description of number coding based on the population-coding properties of neurons. In their model, each number is encoded by a sparse, normalized vector, and the vectors for consecutive numbers are iteratively linked because numerical codes are generated through multiplication by a fixed random matrix. Activating a particular number code n requires iterating through the whole sequence of vectors from 0 to n. Number-coding neurons are conceived of as a vector-based population of interrelated codes intrinsically linked by the successor function, $S(n) = n + 1$. This model suggests that ordered numerical representation could emerge spontaneously from simple constraints on neural processes.

The neural network, deep-learning, and mathematical models are not mutually exclusive, as each explains a slightly different part or scale of the processing structure. These models reflect progress in formalizing a description of numerosity representation, but it remains unclear how these distinct explanations will be integrated and elaborated to explain the whole phenomenon of numerical representation.

Human Uniqueness

Humans have a sense of the discrete and logical properties of numbers that goes beyond the nonverbal "numerosity" cognition of nonhuman animals. Significant conceptual change occurs in human children as a consequence of learning verbal counting—qualitative change that could not be achieved simply by mapping words to preverbal representations of numerosities (Carey, 2004). According to Carey (2004), the linguistic form of number, the verbal count list, "transcends the representational power" of any nonlinguistic precursors.

Language appears to play a central role in transforming primitive numeric concepts into a discrete, logical grammar—this is unsurprising because language is generally central to all human concepts. Yet human groups with or without grammatical number (singular/plural) and lexical number (quantity words) can reason about quantities nonverbally, and some human groups communicate concepts of quantity that surpass their lexicon using body parts, gestures, or material representations (Ferrigno et al., 2017; Overmann, 2015; Pica et al., 2004). The concept of discrete, labeled cardinal numbers thus seems somewhat independent of verbal counting in humans. However, since all humans have language, the role of generative labeling (in general) could be a necessary precursor to counting. Precise ordered representations of numbers have not been observed in humans who lack symbolic counting systems, and no nonhuman animal has been trained successfully to count despite multiple attempts, suggesting that uniquely human cognition, possibly generative labeling, is necessary to acquire counting (Matsuzawa, 2009; but see Pepperberg & Carey, 2012).

Some evidence suggests that simple symbolic counting and arithmetic abilities partly draw on nonverbal numerosity estimation mechanisms developmentally (Dillon et al., 2017; Geary & van Marle, 2016; Halbera, Mazzocco, & Feigenson, 2008; Starr, Libertus, & Brannon, 2013), that they share a neural level of computation (Ansari, 2008; Cantlon & Li, 2013; Piazza et al.,

2007; Price et al., 2007), and that numerosity estimation is perhaps a component of ballparking mathematical outcomes during higher math reasoning (Amalric & Dehaene, 2016).

The neural mechanisms recruited to verbally and symbolically reason about numerical values overlap with those used to estimate numerosity in the IPS (Dehaene & Cohen, 2007). The primitive numerosity systems of quantity representation in the IPS seem to ground the evolutionarily recent cultural innovation of verbal counting. Verbal operations like number naming, counting, and arithmetic facts (e.g., multiplication tables) differ, however, from nonverbal numerosity processing at the neural level in that they engage the left perisylvian language areas and the left angular gyrus (Dehaene et al., 1999). The symbolic number code for representing Arabic numerals engages the fusiform and lingual gyri of the ventral stream. The left hemisphere IPS plays a more important role in symbolic numerical development than in numerosity development in children, perhaps due to a proximity effect with the left hemisphere language network (Ansari, 2008; Lussier & Cantlon, 2016). Thus, while there is overlap between numerosity estimation and symbolic mathematics in the brain, particularly in parietal cortex, they are distinct processes. For example, a strong neural predictor of higher mathematical ability in older children is hippocampal volume and the functional connectivity of the hippocampus to the rest of the cortex (Supekar et al., 2013). It seems likely that disparate networks of semantic and logical information are integrated with primitive numerosity representations and domain-general processes in humans to acquire the functions of higher mathematics (Lyons, Ansari, & Bielock, 2012; Bulthé, De Smedt, & Op de Beeck, 2014).

Conclusion

Human numerical cognition at birth includes the perception of object sets in space, time, and across modalities as expressing a numerical quantity. The ability to conceive of quantity to make relative comparisons appears to be evolutionarily primitive across species. The natural functions of this mechanism include foraging efficiency but also comparisons of social group sizes. Processing demands such as crossmodal processing and object-based decision-making could have played an important role in the algorithmic and neural implementation of numerical cognition. The neural basis of numerical cognition appears to be conserved across primates in intraparietal cortex, at least in terms of basic mechanisms like summation neurons and tuning neurons that express relative values. Human symbolic counting and arithmetic are critically associated with primitive numerical cognition throughout the life span, although uniquely human demands on mathematical reasoning require semantic, linguistic, and logical processes that go beyond primitive mechanisms and remain to be explained. Yet whatever unique cognition humans acquire, the study of numerical cognition shows how a mechanism that began with simple set comparisons now grounds human mathematical thinking throughout development and serves as an important anchor to human rationality.

REFERENCES

Agrillo, C., & Dadda, M. (2007). Discrimination of the larger shoal in the Poeciliid fish *Girardinus falcatus. Ethology Ecology & Evolution, 19*(2), 145–157.

Agrillo, C., Piffer, L., & Bisazza, A. (2011). Number versus continuous quantity in numerosity judgments by fish. *Cognition, 119*(2), 281–287.

Amalric, M., & Dehaene, S. (2016). Origins of the brain networks for advanced mathematics in expert mathematicians. *Proceedings of the National Academy of Sciences, 113*(18), 4909–4917.

Ansari, D. (2008). Effects of development and enculturation on number representation in the brain. *Nature Reviews Neuroscience, 9*(4), 278–291.

Ansari, D., & Dhital, B. (2006). Age-related changes in the activation of the intraparietal sulcus during nonsymbolic magnitude processing: An event-related functional magnetic resonance imaging study. *Journal of Cognitive Neuroscience, 18*(11), 1820–1828.

Barth, H., La Mont, K., Lipton, J., & Spelke, E. S. (2005). Abstract number and arithmetic in preschool children. *Proceedings of the National Academy of Sciences of the United States of America, 102*(39), 14116–14121.

Beran, M. J. (2012). Quantity judgments of auditory and visual stimuli by chimpanzees (*Pan troglodytes*). *Journal of Experimental Psychology: Animal Behavioral Processes, 38*(1), 23–29.

Beran, M. J., Parrish, A. E., & Evans, T. A. (2015). Numerical cognition and quantitative abilities in nonhuman primates. In D. C. Geary, D. B. Berch, & K.M. Koepke (Eds.), *Evolutionary origins and early development of basic number processing* (Vol. 1, pp. 91–119). Amsterdam: Elsevier.

Beran, M. J., Smith, J. D., Redford, J. S., & Washburn, D. A. (2006). Rhesus macaques (Macaca mulatta) monitor uncertainty during numerosity judgments. *Journal of Experimental Psychology: Animal Behavior Processes, 32*(2), 111–119.

Brannon, E. M., & Terrace, H. S. (1998). Ordering of the numerosities 1 to 9 by monkeys. *Science, 282*(5389), 746–749.

Bugden, S., Price, G. R., McLean, D. A., & Ansari, D. (2012). The role of the left intraparietal sulcus in the relationship between symbolic number processing and children's arithmetic competence. *Developmental Cognitive Neuroscience, 2*(4), 448–457.

Bulthé, J., De Smedt, B., & Op de Beeck, H. (2014). Format-dependent representations of symbolic and non-symbolic numbers in the human cortex as revealed by multi-voxel pattern analyses. *NeuroImage, 87*, 311–322.

Cantlon, J. F., & Brannon, E. M. (2006). Shared system for ordering small and large numbers in monkeys and humans. *Psychological Science, 17*(5), 401–406.

Cantlon, J. F., & Li, R. (2013). Neural activity during natural viewing of *Sesame Street* statistically predicts test scores in early childhood. *PLoS Biology, 11*(1), e1001462.

Cantlon, J. F., Piantadosi, S. T., Ferrigno, S., Hughes, K. D., & Barnard, A. M. (2015). The origins of counting algorithms. *Psychological Science, 26*(6), 853–865.

Cantrell, L., & Smith, L. B. (2013). Open questions and a proposal: A critical review of the evidence on infant numerical abilities. *Cognition, 128*(3), 331–352.

Carey, S. (2004). Bootstrapping and the origin of concepts. *Daedalus, 133*(1), 59–68.

Cavada, C., & Goldman-Rakic, P. S. (1989). Posterior parietal cortex in rhesus monkey: II. Evidence for segregated corticocortical networks linking sensory and limbic areas with the frontal lobe. *Journal of Comparative Neurology, 287*(4), 422–445.

Chafee, M. V., & Goldman-Rakic, P. S. (2000) Inactivation of parietal and prefrontal cortex reveals interdependence of neural activity during memory-guided saccades. *Journal of Neurophysiology, 83*(3), 1550–1566.

Chittka, L., & Geiger, K. (1995). Can honey bees count landmarks? *Animal Behaviour, 49*(1), 159–164.

Clearfield, M. W., & Mix, K. S. (2001). Amount versus number: Infants' use of area and contour length to discriminate small sets. *Journal of Cognition and Development, 2*(3), 243–260.

Cordes, S., & Brannon, E. M. (2008). Quantitative competencies in infancy. *Developmental Science, 11*(6), 803–808.

Dastjerdi, M., Ozker, M., Foster, B. L., Rangarajan, V., & Parvizi, J. (2013). Numerical processing in the human parietal cortex during experimental and natural conditions. *Nature Communications, 4,* 25–28.

Dehaene, S., & Changeux, J. P. (1993). Development of elementary numerical abilities: A neuronal model. *Journal of Cognitive Neuroscience, 5*(4), 390–407.

Dehaene, S., & Cohen, L. (1997). Cerebral pathways for calculation: Double dissociation between rote verbal and quantitative knowledge of arithmetic. *Cortex, 33*(2), 219–250.

Dehaene, S., & Cohen, L. (2007). Cultural recycling of cortical maps. *Neuron, 56*(2), 384–398.

Dehaene, S., Spelke, E., Pinel, P., Stanescu, R., & Tsivkin, S. (1999). Sources of mathematical thinking: Behavioral and brain-imaging evidence. *Science, 284*(5416), 970–974.

Dillon, M. R., Kannan, H., Dean, J. T., Spelke, E. S., & Duflo, E. (2017). Cognitive science in the field: A preschool intervention durably enhances intuitive but not formal mathematics. *Science, 357*(6346), 47–55.

Ditz, H. M., & Nieder, A. (2015). Neurons selective to the number of visual items in the corvid songbird endbrain. *Proceedings of the National Academy of Sciences, 112*(25), 7827–7832.

Edwards, L. A., Wagner, J. B., Simon, C. E., & Hyde, D. C. (2016). Functional brain organization for number processing in preverbal infants. *Developmental Science, 19*(5), 757–769.

Eger, E., Sterzer, P., Russ, M. O., Giraud, A. L., & Kleinschmidt, A. (2003). A supramodal number representation in human intraparietal cortex. *Neuron, 37*(4), 719–726.

Emmerton, J., & Renner, J. C. (2006). Scalar effects in the visual discrimination of numerosity by pigeons. *Learning & Behavior, 34*(2), 176–192.

Féron, J., Gentaz, E., & Streri, A. (2006). Evidence of amodal representation of small numbers across visuo-tactile modalities in 5-month-old infants. *Cognitive Development, 21*(2), 81–92.

Ferrigno, S., Hughes, K. D., & Cantlon, J. F. (2016). Precocious quantitative cognition in monkeys. *Psychonomic Bulletin & Review, 23*(1), 141–147.

Ferrigno, S., Jara-Ettinger, J., Piantadosi, S. T., & Cantlon, J. F. (2017). Universal and uniquely human factors in spontaneous number perception. *Nature Communications, 8,* 13968.

Gallistel, C. R. (1990). *The organization of learning* (Vol. 336). Cambridge, MA: MIT Press.

Gallistel, C. R., & Gelman, R. (2000). Non-verbal numerical cognition: From reals to integers. *Trends in Cognitive Sciences, 4*(2), 59–65.

Geary, D. C., & Vanmarle, K. (2016). Young children's core symbolic and nonsymbolic quantitative knowledge in the prediction of later mathematics achievement. *Developmental Psychology, 52*(12), 2130–2144.

Gebuis, T., & Reynvoet, B. (2012). The interplay between nonsymbolic number and its continuous visual properties. *Journal of Experimental Psychology: General, 141(4),* 642–648.

Godin, J. G. J., & Keenleyside, M. H. (1984). Foraging on patchily distributed prey by a cichlid fish (Teleostei, *Cichlidae*): A test of the ideal free distribution theory. *Animal Behaviour, 32*(1), 120–131.

Halberda, J., Mazzocco, M. M., & Feigenson, L. (2008). Individual differences in non-verbal number acuity correlate with maths achievement. *Nature, 455*(7213), 665–668.

Hannagan, T., Nieder, A., Viswanathan, P., & Dehaene, S. (2018). A random-matrix theory of the number sense. *Philosophical Transactions of the Royal Society B, 373*(1740), 20170253.

Harper, D. G. C. (1982). Competitive foraging in mallards: "Ideal free" ducks. *Animal Behaviour, 30*(2), 575–584.

Harvey, B. M., Klein, B. P., Petridou, N., & Dumoulin, S. O. (2013). Topographic representation of numerosity in the human parietal cortex. *Science, 341*(6150), 1123–1126.

Hasson, U., Yang, E., Vallines, I., Heeger, D. J., & Rubin, N. (2008). A hierarchy of temporal receptive windows in human cortex. *Journal of Neuroscience, 28*(10), 2539–2550.

Hope, T., Stoianov, I., & Zorzi, M. (2010). Through neural stimulation to behavior manipulation: A novel method for analyzing dynamical cognitive models. *Cognitive Science, 34*(3), 406–433.

Izard, V., Dehaene-Lambertz, G., & Dehaene, S. (2008). Distinct cerebral pathways for object identity and number in human infants. *PLoS Biology, 6*(2), e11.

Izard, V., Sann, C., Spelke, E. S., & Streri, A. (2009). Newborn infants perceive abstract numbers. *Proceedings of the National Academy of Sciences, 106*(25), 10382–10385.

Jordan, K. E., & Brannon, E. M. (2006). The multisensory representation of number in infancy. *Proceedings of the National Academy of Sciences of the United States of America, 103*(9), 3486–3489.

Kersey, A. J., & Cantlon, J. F. (2017). Neural tuning to numerosity relates to perceptual tuning in 3–6-year-old children. *Journal of Neuroscience, 37*(3), 512–522.

Leslie, A. M., Gelman, R., & Gallistel, C. R. (2008). The generative basis of natural number concepts. *Trends in Cognitive Sciences, 12*(6), 213–218.

Libertus, M. E., & Brannon, E. M. (2010). Stable individual differences in number discrimination in infancy. *Developmental Science, 13*(6), 900–906.

Libertus, M. E., Brannon, E. M., & Woldorff, M. G. (2011). Parallels in stimulus-driven oscillatory brain responses to numerosity changes in adults and seven-month-old infants. *Developmental Neuropsychology, 36*(6), 651–667.

Locke, John. (1690). An Essay Concerning Humane Understanding. 1st ed. 1 vols. London: Thomas Basset.

Lourenco, S. F., & Longo, M. R. (2010). General magnitude representation in human infants. *Psychological Science, 21*(6), 873–881.

Lussier, C. A., & Cantlon, J. F. (2017). Developmental bias for number words in the intraparietal sulcus. *Developmental Science, 20*(3), e12385.

Lyons, I. M., Ansari, D., & Beilock, S. L. (2012). Symbolic estrangement: Evidence against a strong association between numerical symbols and the quantities they represent. *Journal of Experimental Psychology: General, 141*(4), 635–641.

Marr, D., & Poggio, T. (1976). From understanding computation to understanding neural circuitry. *Neurosciences Research Program Bulletin, 15*, 470–488.

Matsuzawa, T. (2009). Symbolic representation of number in chimpanzees. *Current Opinion in Neurobiology, 19*(1), 92–98.

McComb, K., Packer, C., & Pusey, A. (1994). Roaring and numerical assessment in contests between groups of female lions, *Panthera leo. Animal Behaviour, 47*(2), 379–387.

Molko, N., Cachia, A., Rivière, D., Mangin, J. F., Bruandet, M., Le Bihan, D., ... & Dehaene, S. (2003). Functional and structural alterations of the intraparietal sulcus in a developmental dyscalculia of genetic origin. *Neuron, 40*(4), 847–858.

Nieder, A. (2012). Supramodal numerosity selectivity of neurons in primate prefrontal and posterior parietal cortices. *Proceedings of the National Academy of Sciences, 109*(29), 11860–11865.

Nieder, A. (2016). The neuronal code for number. *Nature Reviews Neuroscience, 17*(6), 366–382.

Nieder, A., & Miller, E. K. (2004). A parieto-frontal network for visual numerical information in the monkey. *Proceedings of the National Academy of Sciences of the United States of America, 101*(19), 7457–7462.

Overmann, K. A. (2015). Numerosity structures the expression of quantity in lexical numbers and grammatical number. *Current Anthropology, 56*(5), 638–653.

Pepperberg, I. M. (2006). Grey parrot numerical competence: A review. *Animal Cognition, 9*(4), 377–391.

Pepperberg, I. M., & Carey, S. (2012). Grey parrot number acquisition: The inference of cardinal value from ordinal position on the numeral list. *Cognition, 125*(2), 219–232.

Piantadosi, S. T., & Cantlon, J. F. (2017). True numerical cognition in the wild. *Psychological Science, 28*(4), 462–469.

Piazza, M., & Izard, V. (2009). How humans count: Numerosity and the parietal cortex. *Neuroscientist, 15*(3), 261–273.

Piazza, M., Izard, V., Pinel, P., Le Bihan, D., & Dehaene, S. (2004). Tuning curves for approximate numerosity in the human intraparietal sulcus. *Neuron, 44*(3), 547–555.

Piazza, M., Pinel, P., Le Bihan, D., & Dehaene, S. (2007). A magnitude code common to numerosities and number symbols in human intraparietal cortex. *Neuron, 53*(2), 293–305.

Pica, P., Lemer, C., Izard, V., & Dehaene, S. (2004). Exact and approximate arithmetic in an Amazonian indigene group. *Science, 306*(5695), 499–503.

Price, G. R., Holloway, I., Räsänen, P., Vesterinen, M., & Ansari, D. (2007). Impaired parietal magnitude processing in developmental dyscalculia. *Current Biology, 17*(24), R1042–R1043.

Roitman, J. D., Brannon, E. M., & Platt, M. L. (2007). Monotonic coding of numerosity in macaque lateral intraparietal area. *PLoS Biology, 5*(8), e208.

Rugani, R., Regolin, L., & Vallortigara, G. (2010). Imprinted numbers: Newborn chicks' sensitivity to number vs. continuous extent of objects they have been reared with. *Developmental Science, 13*(5), 790–797.

Sambongi, Y., Nagae, T., Liu, Y., Yoshimizu, T., Takeda, K., Wada, Y., & Futai, M. (1999). Sensing of cadmium and copper ions by externally exposed ADL, ASE, and ASH neurons elicits avoidance response in *Caenorhabditis elegans. NeuroReport, 10*(4), 753–757.

Sawamura, H., Shima, K., & Tanji, J. (2010). Deficits in action selection based on numerical information after inactivation of the posterior parietal cortex in monkeys. *Journal of Neurophysiology, 104*(2), 902–910.

Scarf, D., Hayne, H., & Colombo, M. (2011). Pigeons on par with primates in numerical competence. *Science, 334*(6063), 1664.

Sokolowski, H. M., Fias, W., Mousa, A., & Ansari, D. (2017). Common and distinct brain regions in both parietal and frontal cortex support symbolic and nonsymbolic number processing in humans: A functional neuroimaging meta-analysis. *NeuroImage, 146*, 376–394.

Starr, A., Libertus, M. E., & Brannon, E. M. (2013). Number sense in infancy predicts mathematical abilities in childhood. *Proceedings of the National Academy of Sciences, 110*(45), 18116–18120.

Stoianov, I., & Zorzi, M. (2012). Emergence of a "visual number sense" in hierarchical generative models. *Nature Neuroscience, 15*(2), 194–196.

Strandburg-Peshkin, A., Farine, D. R., Couzin, I. D., & Crofoot, M. C. (2015). Shared decision-making drives collective movement in wild baboons. *Science, 348*(6241), 1358–1361.

Supekar, K., Swigart, A. G., Tenison, C., Jolles, D. D., Rosenberg-Lee, M., Fuchs, L., & Menon, V. (2013). Neural predictors of individual differences in response to math tutoring in primary-grade school children. *Proceedings of the National Academy of Sciences, 110*(20), 8230–8235.

Tinbergen, N. (1963). On aims and methods of ethology. *Ethology, 20*(4), 410–433.

Utami, S. S., Wich, S. A., Sterck, E. H., & Van Hooff, J. A. (1997). Food competition between wild orangutans in large fig trees. *International Journal of Primatology, 18*(6), 909–927.

Vallentin, D., & Nieder, A. (2008). Behavioral and prefrontal representation of spatial proportions in the monkey. *Current Biology, 18*(18), 1420–1425.

Walsh, V. (2003). A theory of magnitude: Common cortical metrics of time, space and quantity. *Trends in Cognitive Sciences, 7*(11), 483–488.

Wang, L., Uhrig, L., Jarraya, B., & Dehaene, S. (2015). Representation of numerical and sequential patterns in macaque and human brains. *Current Biology, 25*(15), 1966–1974.

Wilson, M. L., Hauser, M. D., & Wrangham, R. W. (2001). Does participation in intergroup conflict depend on numerical assessment, range location, or rank for wild chimpanzees? *Animal Behaviour, 61*(6), 1203–1216.

Wittlinger, M., Wehner, R., & Wolf, H. (2006). The ant odometer: Stepping on stilts and stumps. *Science, 312*(5782), 1965–1967.

71 Conceptual Combination

MARC N. COUTANCHE, SARAH H. SOLOMON, AND SHARON L. THOMPSON-SCHILL

ABSTRACT Much has been learned about how individual concepts and semantic dimensions are represented in the human brain using methods from the field of cognitive neuroscience; however, the process of conceptual combination, in which a new concept is created from preexisting concepts, has received far less attention. We discuss theories and findings from cognitive science and cognitive neuroscience that shed light on the processing stages and neural systems that allow humans to form new conceptual combinations. We review systematic and creative applications of cognitive neuroscience methods, including neuroimaging, neuropsychological patients, neurostimulation, and behavioral studies, that have yielded fascinating insights into the cognitive nature and neural underpinnings of conceptual combination. Studies have revealed important features of the cognitive processes central to successful conceptual combination. Furthermore, we are beginning to understand how regions of the semantic system, such as the anterior temporal lobe and angular gyrus, integrate features and concepts, and how they evaluate the plausibility of potential resulting combinations, bridging work in linguistics and semantic memory. Despite the relative newness of these questions for cognitive neuroscience, the investigations we review give a very strong foundation for ongoing and future work that seeks to fully understand how the human brain can flexibly integrate existing concepts to form new and never before experienced combinations at will.

Conceptual Combination

Our ability to construct complex concepts from simpler constituents, referred to as *conceptual combination*, is fundamental to many aspects of cognition. One can, often effortlessly, comprehend a novel utterance, event, or idea via the manipulation, integration, or synthesis of other simpler or more familiar concepts; for example, upon hearing a news report that as a result of climate change the Pacific Northwest robin hawk is under threat of extinction, you might construct one of several plausible interpretations of the meaning of *robin hawk* (see figure 71.1).

In order to understand such novel concepts, one must recruit a series of cognitive processes that might include identifying combinable features of the attributing and receiving concepts; selecting which of these features are to be transferred between concepts; integrating the selected features into a unitary conceptual representation; and confirming the plausibility of the resulting concept. Methods of cognitive neuroscience can be deployed to investigate the neural signatures for combined concepts and the subprocesses that create them in order to understand conceptual combination more broadly. Investigating how individuals combine concepts can shed unique light on different aspects of conceptual knowledge, including the cognitive mechanisms that enable the generative and flexible use of language.

One might protest that it is premature to attempt to explain the processes by which simple or familiar concepts are combined to form complex or new concepts, and the representations of the resulting combined concepts, prior to having a more developed understanding of the cognitive and neural architecture of their building blocks. Several of the other chapters in this volume describe the progress—and also the many, many open questions that remain—in our quest to understand the representation of concepts and the processes by which they are learned, stored, and retrieved. Why would one

FIGURE 71.1 Two plausible interpretations of the novel concept *robin hawk*. *Top*, A hawk with the red breast of a robin. *Bottom*, A hawk that preys on robins. (See color plate 85.)

embark on a quest to understand how concepts are combined before we better understand the seemingly more fundamental questions about conceptual representation? We suggest that questions that arise when considering the processes and resulting representations of conceptual combination may help shed light on—or at least, suggest lines of fruitful inquiry into—more basic questions of conceptual representation. For example, what conceptual structures are flexible enough to allow for the decomposition and recomposition of features into novel combinations? In addition, some of the processes that govern the integration of simple concepts (such as *finger* and *lime*) into complex concepts (such as a *finger lime*) might also govern how simple sensory features (such as *round*, *tart*, and *green*) are integrated into so-called simple concepts (such as *lime*). In other words, combination occurs at multiple levels of semantic processing, even for so-called simple concepts. As such, we can potentially advance our understanding of conceptual processing of all sorts by asking questions about how concepts are combined. That is our undertaking in this chapter: How do we construct meaning out of ideas represented in, for example, noun-noun phrases such as *robin hawk*? What neural systems are recruited as we understand these newly combined concepts? We view these questions as critical to understanding not only one of the most fundamental and generative aspects of cognition but also basic questions about conceptual systems.

One note before we begin: Familiar phrases that are now treated as "simple concepts," such as *doorstop* or *straightjacket*, were at one time novel combinations of existing concepts. As certain conceptual combinations fall into common use, they can become integrated into language as unitary lexical entities (*compound words*). When examining the process of conceptual combination, we will not consider these established phrases, which might be treated as a singular word after repeated use. Indeed, some investigations explicitly regress out the natural frequency of combinations to ensure that any identified neural substrate is not driven by the familiarity of a compound word or phrase (e.g., Graves, Binder, Desai, Conant, & Seidenberg, 2010). This is not to say that once a combination becomes familiar, it ceases to act in a combinatorial manner. Though familiar and novel combinations differ in their lexical retrieval, their respective patterns of response times suggest that both undergo similar computations (Estes & Jones, 2008; Gagné & Spalding, 2004). But we have found that studies of novel combinations provide a unique opportunity to explore critical questions about conceptual processing, and for this reason these studies are the focus of this chapter.

The Structure of Conceptual Combinations

The two interpretations of *robin hawk* depicted in figure 71.1 illustrate a potentially useful distinction: certain conceptual combinations (*canary crayon*) are feature-based (or *attributive*), which are understood by selecting a property from a modifier noun (yellow from *canary*), mapping this onto a dimension of the head noun (the color of a crayon), and then integrating them to form the combined concept (a yellow crayon). Schema-based theories of conceptual structure frame this process in terms of each concept containing a set of different dimensions into which alternative properties can be placed (in this case, through a modifier noun; e.g., Murphy, 1988). Comprehending a combined concept involves understanding the transference of a correct property into the other concept's appropriate dimension. Other conceptual combinations (*crayon box*) are *relational*. For these combinations, understanding the relation between items (e.g., containment) is crucial and allows a person to understand that a *crayon box* is a box that contains crayons. The precise relationship between attributive and relational combinatorial processing has been a topic of debate in the field of cognitive science (e.g., Estes, 2003; Gagné & Shoben, 1997). One contribution of cognitive neuroscience has been to shed new light on questions such as this (e.g., Boylan, Trueswell, & Thompson-Schill, 2017, discussed later in this chapter).

A large variety of relations can exist between constituent concepts, and the precise relation between two concepts is extremely significant. *Bird nest* involves one concept (bird) inhabiting another (nest); *flower girl* involves the temporary possession of an object (flower) by an agent (girl). Work in this area shows that we represent the relation between concepts in a relatively precise way. Concept combinations with particular combinatorial relationships can prime other concept compounds that are represented in the same way (e.g., *bird nest* primes *fish pond*; Estes, 2003; Estes & Jones, 2006; Gagné, 2001). Yet the priming combinations can be remarkably specific. For example, *bird nest* does not prime *toy box* (Estes & Jones, 2006). Though the relationships involved in *bird nest* and *toy box* are superficially similar, the presence of a *potential* common relation (such as containment) is not sufficient to induce priming. Instead, *bird nest* is more accurately characterized by the relationship of habitation, allowing it to prime *fish pond* but not *toy box*. This example illustrates the broader idea that conceptual combinations are represented as very particular interactions between composing concepts. The identification of these precise interactions often requires the empirical study of behavioral responses in carefully designed tasks.

Cognitive Processes in Conceptual Combination

The cognitive process of combining concepts appears to be automatic and implicit, without requiring top-down instruction. This is well illustrated by behavioral priming, in which a person's behavioral response speeds up after perceiving two items that share a particular stimulus characteristic in quick succession. The presence of a priming effect based on a particular stimulus dimension informs theories of cognitive and neural architecture (Churchland, 1998); for example, priming based on word phonology or meaning indicates an organization of language and memory systems that reflects those characteristics. A priming effect has also been identified based on the compatibility of two concepts for being combined (Estes & Jones, 2009). Specifically, presenting a word that starts a potential conceptual combination (e.g., *farm*) speeds subsequent judgments of combination-compatible words (*mouse*), with a similar magnitude and prevalence to more common forms of priming, such as that based on semantic meaning (*mouse—rat*; Estes & Jones, 2009). The presence of priming based on conceptual combination suggests that the plausibility of potential combinatorial relationships is automatically calculated during language comprehension.

At what point during word processing does combinatorial processing occur? A number of investigations converge on an important time frame of approximately 400 ms after presenting combining words. This matches the poststimulus delay that has long been associated with the integration of meaning during typical sentence processing, reflected in the N400, an event-related potential (ERP) that is observed 400 ms after a person encounters a word that is unexpected relative to its surrounding sentence (Kutas & Hillyard, 1980). The attenuation of the brain's electroencephalographic signal at 400 ms poststimulus indicates the successful integration of a compound word's meaning (El Yagoubi et al., 2008). Interestingly, a reduction in neural activity is also observed at approximately 400 ms after plausible, compared to less plausible, compounds (Koester, Holle, & Gunter, 2009), suggesting that this stage of conceptual combination is more than a signal of familiarity. Instead, cognitive processes during this time frame include those necessary to calculate a combination's plausibility.

What is the nature of the cognitive processes that underlie conceptual combination? Historically, conceptual combinations have been framed as resulting from amodal operations in predicate-like structures (Fodor & Pylyshyn, 1988; Smith, Osherson, Rips, & Keane, 1988). For example, *red cup* is the result of binding the relevant value (red) to an argument (color) within *cup*, which is composed of many different arguments (color, shape, volume, and more). These arguments are not independent: changing an argument's value can propagate correlated values to other arguments. For example, *large bird* not only affects a bird's expected size but also changes its beak shape from straight to curved (Medin & Shoben, 1988). Connectionist approaches provided an alternative explanatory framework for conceptual combination by replacing predicate-like processes with statistical mechanisms (Pollack, 1990; Smolensky, 1990). A further alternative has drawn on simulation theory to suggest that people combine multimodal simulations (with associated perceptions, beliefs, and emotions) of individual concepts into larger, more complex simulations so that *red* and *cup* simulations are combined to successfully simulate *red cup* (Barsalou, 1999; Wu & Barsalou, 2009).

To test this last idea, Wu and Barsalou (2009) have investigated how the generated characteristics of items change after conceptual combination by having participants read combined concepts (e.g., *rolled-up lawn*) and generate features for each combination. Their results showed that the act of combining concepts shifted the features that participants generated in a way that respected visual occlusion even though the actual features remained largely unchanged in the concept. For example, a lawn has features that include blades, dirt, green, is played on, and more. The features generated by participants to the cue *lawn* shifted from being dominated by external features (blades) to being dominated by internal features (dirt) once presented as a combined concept (*rolled-up lawn*). The generated features were similar to those generated when participants were explicitly told to engage in imagery, which is consistent with the idea that participants were spontaneously deploying perceptual simulation when they processed the combined concepts. The shift in generated features was not simply a function of the modifier: features generated for *rolled-up snake* did not differ from *snake*, suggesting the shift is driven by a recombination of features within the head concept (*lawn*), rather than simply the addition of new features by a modifier (*rolled-up*). Shifts in generated features were observed for known (*convertible car*) and novel (*glass car*) conceptual combinations, suggesting it was not simply a product of how conceptual combinations are represented in memory. Nevertheless, an open question remains of whether simulation plays a role in the construction of combined concepts or is part of a postcombination process. One such process could be the automatic calculation of a combination's plausibility, which (based on priming) appears to occur even in the absence of explicit plausibility judgments (Estes &

Jones, 2009). A fascinating direction for future work will be to characterize the cognitive processes that are unique or shared across different steps toward successful conceptual combination.

The Neural Basis for Conceptual Combination

Cognitive neuroscience theories of semantic knowledge suggest a number of alternatives for how conceptual combination is instantiated in the brain. Some theories represent semantic knowledge as distributed patterns of semantic features across areas of neocortex (Martin, 2007); these theories would suggest that a combined concept is similarly represented across the same neural regions that represent its constituent concepts and corresponding features. On the other hand, theories of semantic knowledge that posit integration sites or semantic "hubs" (Damasio, 1989; Patterson, Nestor, & Rogers, 2007) suggest that processing combined concepts involves additional neural regions involved in the integration or abstraction of conceptual information. Cognitive neuroscience investigations have repeatedly highlighted two cortical sites as being particularly involved in conceptual combination: the anterior temporal lobe (ATL) and the angular gyrus (AG). We explore how these regions (among others) relate to conceptual combination in the remainder of the chapter.

The anterior temporal lobe Classical and recent findings in cognitive neuroscience have established that the properties often combined during conceptual combination, such as color (Zeki et al., 1991), shape (Tanaka, 1996), size (Coutanche & Koch, 2018; Konkle & Oliva, 2012), and manipulation (Buxbaum, Kyle, Tang, & Detre, 2006), are represented across distributed areas of neocortex. How are such features integrated? The process of integration itself is nontrivial, as the same property can vary when combined with different noun concepts: for example, *red* takes on different values when integrated with *face*, *fire*, or *truck* (e.g., Halff, Ortony, & Anderson, 1976). Cognitive neuroscience studies implicate both the ATL and AG in conceptual integration. The ATL is known to be a key brain area underlying semantic knowledge (Patterson et al., 2007). The processing of semantic associations has been linked to ATL activity through multiple neuroimaging methods, including functional magnetic resonance imaging (fMRI) and magnetoencephalography (MEG; Lau, Gramfort, Hämäläinen, & Kuperberg, 2013). Importantly, the ATL appears to play a key role in integrating features to form semantic representations, potentially acting as a "hub" that links and integrates information between feature-specific regions across sensorimotor cortex (Lambon Ralph, Jefferies, Patterson, & Rogers, 2017). For instance, to study how features converge to form object concepts, Coutanche and Thompson-Schill (2015) identified the ATL as a potential *convergence zone* (Damasio, 1989) for the shape and color of known objects. Coutanche and Thompson-Schill scanned participants with fMRI as they held a known object in mind (e.g., tangerine) during a top-down visual detection task. Using multivariate techniques, the specific shape (e.g., sphere) and color (e.g., orange) of the retrieved objects were decodable in regions involved in shape (lateral occipital complex) and color (V4) processing, respectively. In an exploratory analysis, the only region with activity patterns for the identity of the retrieved object (e.g., tangerine) fell within the left ATL. Furthermore, a time series analysis showed that significant ATL object decoding was best predicted by significant feature decoding in shape and color regions. In other words, feature information in visual cortex predicted object representations in the ATL, consistent with the ATL acting as a site for integration.

A current area of debate and investigation is whether the ATL is specialized for the visual modality or if it also acts as an amodal hub for concepts with limited visual relationships (e.g., *truth*; Bonner & Price, 2013). Given its large size, one possibility is that ATL subregions have differential roles for visual and nonvisual combinations. For example, a meta-analysis has suggested that ventral ATL regions are more likely to be recruited during visual object processing, whereas lateral ATL areas are employed in auditory processing (Visser, Jefferies, & Lambon Ralph, 2010).

If the ATL is a semantic hub where conceptual features are integrated to form more complex conceptual representations (Lambon Ralph et al., 2017; Patterson, Nestor, & Rogers, 2007), it should respond more to combinations that involve integration, versus those that do not. In an MEG study, Bemis and Pylkkanen (2011) contrasted integrative and nonintegrative combinations to find brain regions sensitive to property integration. Specifically, they isolated regions whose activity was more strongly modulated by the comprehension of integrative combinations (e.g., *red boat* vs. *xkq boat*) than by nonintegrative combinations (e.g., *cup boat* vs. *xkq boat*). The use of adjective-noun combinations, rather than noun-noun combinations, can be a valuable way to isolate the integration process, independent of the additional processes of property selection (required for noun-noun combinations). The left ATL was specifically sensitive to integrative combinations, with related activity occurring approximately 200–250 ms after stimulus presentation. In a similar MEG study, the left ATL was found to be sensitive to conceptual integration during the basic comprehension of visual and auditory stimuli (Bemis & Pylkkanen, 2013). Furthermore, an integration-sensitive response occurs in the ATL when the task does not explicitly require integration, even if the order of concepts is flipped (e.g., *boat red*), suggesting that ATL integration is automatic and reflects semantic, rather than syntactic, composition (Bemis & Pylkkanen, 2013).

Is the ATL modulated by the form of the interactions between modifiers and object concepts? Westerlund and Pylkkanen (2014) varied the specificity of object concepts and observed whether the ATL integration response was affected. Brain responses were collected as participants processed combinations with low-specificity nouns (e.g., *blue boat*) or combinations with a highly specific counterpart (e.g., *blue canoe*). Other linguistic properties, such as frequency and the transition probability between adjective and noun, were carefully matched. The left ATL responded more strongly (250 ms after the noun presentation) for low-specificity combinations than for high-specificity combinations. This effect indicates that the left ATL's combinatorial response is influenced by semantic properties of the noun, such as conceptual specificity, nicely linking language-focused and semantic-hub accounts of the ATL's role in conceptual combination (Westerlund & Pylkkanen, 2014).

As well as assessing the magnitude of ATL activity during the comprehension of combined concepts, researchers can explore the content of the resulting representations. Baron, Thompson-Schill, Weber, and Osherson (2010) presented fMRI participants with images of faces varying in gender, and age and collected multivoxel patterns that corresponded to each of the target properties (i.e., male, female, young, old) as well as combinations (e.g., young woman). The combined concepts resulted in multivoxel patterns in the left ATL that were predicted by the superimposition of the constituent concepts.

Taken together, these results suggest that the left ATL might represent the conjunction of concepts, in addition to representing the conjunction of basic perceptual features. An open question concerns whether the ATL neural computations that might bind features to form basic concepts overlap completely, or only partially, with neural computations used to bind concepts into conceptual combinations. This will be an important question for the field going forward.

The angular gyrus The AG is another region that has been consistently implicated in studies of conceptual combination. This region of inferior parietal cortex has widespread connections across cortex, including sensory and language networks (Caspers et al., 2011), supporting the idea that the AG lies at the top of a semantic processing hierarchy (Binder, Desai, Graves, & Conant, 2009). The AG has been linked to the combinatorial strength, or plausibility, of conceptual combination. In an fMRI study, Price, Bonner, Peelle, and Grossman (2015) found that the AG responds preferentially to combinations that form meaningful, compared to less meaningful, concepts (e.g., *plaid jacket* vs. *moss pony*) and that this did not depend on the kind of information being integrated (e.g., visual, tactile). Further, right AG cortical thickness predicted how people responded to combined concepts, such as the magnitude of their response-time advantage for phrases with higher combinatorial strengths (Price et al., 2015). Adding to evidence for the region's key role in combinatorial processing, neurological patients with damage to the left AG have shown impairments in combinatorial tasks, with larger impairments experienced by patients with greater AG atrophy (Price et al., 2015).

In a recent neurostimulation study, Price, Peelle, Bonner, Grossman, and Hamilton (2016) stimulated the AG of healthy participants to observe the behavioral consequences for combinatorial processing. Anodal high-definition transcranial direct current stimulation (tDCS) was used to excite left AG cortical sites, which led to faster responses to meaningful (*tiny radish*), compared to nonmeaningful (*fast blueberry*), adjective-noun combinations. The effect of stimulation on response times was correlated with the degree of semantic coherence between the adjective and noun in the combination. In contrast, stimulation of the right AG slowed responses to meaningful (vs. nonmeaningful) combinations.

As the studies discussed thus far illustrate, the relative roles of the left and right AG are not currently clear. Neurological damage and stimulation of the left AG both affect behavioral responses to combinatorial pairings, but cortical thickness in the right (but not left) AG has predicted individual differences in response times to combinable word pairings. In functional studies, AG activity is often lateralized. For instance, Graves et al. (2010) compared meaningful noun-noun combinations (e.g., *lake house*) to less meaningful reversals (*house lake*) during an fMRI scan and found that combinatorial comparisons activate the AG with a large right-sided bias, while lexical processing stimulated the left AG. They proposed that regions of the right hemisphere have larger semantic fields, enabling a broader array of conceptual links to be made during combinatorial processing (Beeman et al., 1994; Graves et al., 2010). Alternatively, the right AG might differ from the left in having access to *implicit* relational content in combinations (Boylan, Trueswell, & Thompson-Schill, 2017). The left AG instead might require the presence of explicit syntactic cues about a relation in order to process the corresponding combination.

The left inferior frontal gyrus Processing less meaningful combinations has been associated with increased activity in left frontal cortex, including the left inferior frontal gyrus (LIFG; Graves et al., 2010), which is implicated in semantic selection (Thompson-Schill, D'Esposito, Aguirre, & Farah, 1997). This LIFG activation might reflect attempts to select the appropriate information to integrate, which is a more effortful process for combinations with a less obvious meaning. The need for a selection process—for successfully comprehending conceptual combinations—is particularly apparent for feature-based combinations, in which a subset of features is selected and applied. For example, the intended referent of *canary crayon* does not involve an actual canary: rather, the person comprehending must select the property *yellow* from a set that includes *small, has wings*, and more. The selected property is then integrated with *crayon*. Similarly, *prune skin* does not involve actual prunes, and the term *piano key teeth* does not involve actual piano keys. These combinations are thus similar to metaphors (e.g., "His teeth are piano keys"), where the processes of selection and integration still apply. In order to study how appropriate features are selected during the comprehension of these attributive metaphors, Solomon and Thompson-Schill (2017) computed a metaphor-specific measure of property selection. They observed the extent to which certain properties became activated after metaphor comprehension by presenting participants with a metaphor (e.g., "Her skin is a prune") and then asking how much faster participants agree that a metaphor-relevant property (e.g., wrinkly) applies to a modifier concept (e.g., *prune*), relative to a metaphor-irrelevant property (e.g., sweet). During an fMRI scan, this property-selection measure predicted activity in the LIFG, suggesting this region is involved in the selection of conceptual properties during metaphor comprehension. This same process might underlie property selection when processing noun-noun conceptual combinations.

Regional interactions and differences How does combinatorial processing in the ATL and AG relate to each other? Processing in the ATL occurs approximately 200 ms after relevant stimuli, followed by processing in the AG 200 ms later (Bemis & Pylkkänen, 2013). Molinaro, Paz-Alonso, Duñabeitia, and Carreiras (2015) examined how regions of lexical and semantic networks, particularly the ATL and AG, respond to differing levels of combinatorial processing. The authors examined concepts and attributes with differing degrees of typicality: prototypical (*wet rain*), contrastive (opposing the typical property: *dry rain*), and noncomposable (*blind rain*). Participants' ATLs were sensitive to the typicality of the perceived word pairings, with greater responses to contrastive, compared to typical, combinations. The ATL also showed particularly strong coupling with the AG during the contrastive combination condition, suggesting coordination between these regions during difficult semantic integrations. This coupling occurred in the context of activation across the broader lexical-semantic network, with activation in the posterior middle temporal gyrus (a region involved in lexical/semantic processing; Lau, Phillips, & Poeppel, 2008) for all conditions and in the LIFG (possibly for the controlled retrieval of lexical-semantic information; Thompson-Schill, Aguirre, D'Esposito, & Farah, 1999) for complex constructions (*dry rain/blind rain*). The ATL was connected with both the medial temporal lobe and the IFG during these complex constructions but only with the AG during the contrastive combination. Coordination between these regions appears to play an important role in combinatorial processing.

The ATL and AG appear to respond differently based on the type of conceptual combination being processed. Boylan, Trueswell, and Thompson-Schill (2017) compared how the regions respond to attributive versus relational nominal compounds. The two regions responded to both types of compound, but the nature of the regions' response differed based on the kind of combination. The AG responded more strongly to relational, compared to attributive, compounds. In contrast, the ATL responded with a similar magnitude to both but had an earlier response to attributive combinations. These findings shed light on a potential greater role for the AG when combinations require more relational processing and suggest that attributive combinations might be processed first in the ATL (Boylan, Trueswell, & Thompson-Schill, 2017).

Summary

As the work described in this chapter indicates, conceptual combination is a multifaceted process, involving feature selection, integration across concepts, and plausibility assessments. The human tendency to engage in conceptual combination is often automatic and implicit, leading to the processing of conceptual combinations 400 ms after combinable items are presented. The ATL and AG appear central to the combinatorial process. Interactions between these regions, and with other areas of the lexical and semantic networks, are crucial to successfully combining concepts. As the methods of cognitive neuroscience continue to be applied to explore how our brains combine and comprehend concepts, we move closer to understanding the place of conceptual combination within the operation of the semantic system more generally.

REFERENCES

Baron, S., Thompson-Schill, S., Weber, M., & Osherson, D. (2010). An early stage of conceptual combination: Superimposition of constituent concepts in left anterolateral temporal lobe. *Cognitive Neuroscience, 1*, 44–51.

Barsalou, L. W. (1999). Perceptual symbol systems. *Behavioral and Brain Sciences, 22*(4), 577–609.

Barsalou, L. W. (2017). What does semantic tiling of the cortex tell us about semantics? *Neuropsychologia, 105*, 18–38.

Beeman, M., Friedman, R. B., Grafman, J., Perez, E., Diamond, S., & Lindsay, M. B. (1994). Summation priming and coarse semantic coding in the right hemisphere. *Journal of Cognitive Neuroscience, 6*(1), 26–45.

Bemis, D. K., & Pylkkänen, L. (2011). Simple composition: A magnetoencephalography investigation into the comprehension of minimal linguistic phrases. *Journal of Neuroscience, 31*(8), 2801–2814.

Bemis, D. K., & Pylkkänen, L. (2013). Combination across domains: An MEG investigation into the relationship between mathematical, pictorial, and linguistic processing. *Frontiers in Psychology, 3*, 1–20.

Binder, J. R., Desai, R. H., Graves, W. W., & Conant, L. L. (2009). Where is the semantic system? A critical review and meta-analysis of 120 functional neuroimaging studies. *Cerebral Cortex, 19*(12), 2767–2796.

Bonner, M. F., & Price, A. R. (2013). Where is the anterior temporal lobe and what does it do? *Journal of Neuroscience, 33*(10), 4213–4215.

Boylan, C., Trueswell, J. C., & Thompson-Schill, S. L. (2017). Relational vs. attributive interpretation of nominal compounds differentially engages angular gyrus and anterior temporal lobe. *Brain and Language, 169*, 8–21.

Buxbaum, L. J., Kyle, K. M., Tang, K., & Detre, J. A. (2006). Neural substrates of knowledge of hand postures for object grasping and functional object use: Evidence from fMRI. *Brain Research, 1117*(1), 175–185.

Caspers, S., Eickhoff, S. B., Rick, T., von Kapri, A., Kuhlen, T., Huang, R., … Zilles, K. (2011). Probabilistic fibre tract analysis of cytoarchitectonically defined human inferior parietal lobule areas reveals similarities to macaques. *NeuroImage, 58*(2), 362–380.

Churchland, P. M. (1998). Conceptual similarity across sensory and neural diversity: The Fodor/Lepore challenge answered. *Journal of Philosophy, 95*, 5–32.

Coutanche, M. N., & Koch, G. E. (2018). Creatures great and small: Real-world size of animals predicts visual cortex representations beyond taxonomic category. *NeuroImage, 183*, 627–634.

Coutanche, M. N., & Thompson-Schill, S. L. (2015). Creating concepts from converging features in human cortex. *Cerebral Cortex, 25*(9), 2584–2593.

Damasio, A. R. (1989). The brain binds entities and events by multiregional activation from convergence zones. *Neural Computation, 1*(1), 123–132.

El Yagoubi, R., Chiarelli, V., Mondini, S., Perrone, G., Danieli, M., & Semenza, C. (2008). Neural correlates of Italian nominal compounds and potential impact of headedness effect: An ERP study. *Cognitive Neuropsychology, 25*(4), 559–581.

Estes, Z. (2003). Attributive and relational processes in nominal combination. *Journal of Memory and Language, 48*(2), 304–319.

Estes, Z., & Jones, L. L. (2006). Priming via relational similarity: A copper horse is faster when seen through a glass eye. *Journal of Memory and Language, 55*(1), 89–101.

Estes, Z., & Jones, L. L. (2008). Relational processing in conceptual combination and analogy. *Behavioral and Brain Sciences, 31*(4), 385–386.

Estes, Z., & Jones, L. L. (2009). Integrative priming occurs rapidly and uncontrollably during lexical processing. *Journal of Experimental Psychology: General, 138*(1), 112–130.

Fodor, J. A., & Pylyshyn, Z. W. (1988). Connectionism and cognitive architecture: A critical analysis. *Cognition, 28*(1), 3–71.

Gagné, C. L. (2001). Relation and lexical priming during the interpretation of noun-noun combinations. *Journal of Experimental Psychology: Learning, Memory, and Cognition, 27*(1), 236–254.

Gagné, C. L., & Shoben, E. J. (1997). Influence of thematic relations on the comprehension of modifier-noun combinations. *Journal of Experimental Psychology: Learning, Memory, and Cognition, 23*(1), 71–87.

Gagné, C. L., & Spalding, T. L. (2004). Effect of discourse context and modifier relation frequency on conceptual combination. *Journal of Memory and Language, 50*(4), 444–455.

Graves, W. W., Binder, J. R., Desai, R. H., Conant, L. L., & Seidenberg, M. S. (2010). Neural correlates of implicit and explicit combinatorial semantic processing. *NeuroImage, 53*(2), 638–646.

Halff, H., Ortony, A., & C. Anderson, R. (1976). A context-sensitive representation of word meanings. *Memory & Cognition, 4*, 378–383.

Koester, D., Holle, H., & Gunter, T. C. (2009). Electrophysiological evidence for incremental lexical-semantic integration in auditory compound comprehension. *Neuropsychologia, 47*(8), 1854–1864.

Konkle, T., & Oliva, A. (2012). A real-world size organization of object responses in occipito-temporal cortex. *Neuron, 74*(6), 1114–1124.

Kutas, M., & Hillyard, S. A. (1980). Reading senseless sentences: Brain potentials reflect semantic incongruity. *Science, 207*(4427), 203–205.

Lambon Ralph, M. A., Jefferies, E., Patterson, K., & Rogers, T. T. (2017). The neural and computational bases of semantic cognition. *Nature Reviews Neuroscience, 18*(1), 42–55.

Lau, E. F., Gramfort, A., Hämäläinen, M. S., & Kuperberg, G. R. (2013). Automatic semantic facilitation in anterior temporal cortex revealed through multimodal neuroimaging. *Journal of Neuroscience, 33*(43), 17174–17181.

Lau, E. F., Phillips, C., & Poeppel, D. (2008). A cortical network for semantics: (De)constructing the N400. *Nature Reviews Neuroscience, 9*(12), 920–933.

Martin, A. (2007). The representation of object concepts in the brain. *Annual Review of Psychology, 58*, 25–45.

Medin, D. L., & Shoben, E. J. (1988). Context and structure in conceptual combination. *Cognitive Psychology, 20*(2), 158–190.

Molinaro, N., Paz-Alonso, P. M., Duñabeitia, J. A., & Carreiras, M. (2015). Combinatorial semantics strengthens angular-anterior temporal coupling. *Cortex, 65*, 113–127.

Murphy, G. L. (1988). Comprehending complex concepts. *Cognitive Science, 12*(4), 529–562.

Patterson, K., Nestor, P. J., & Rogers, T. T. (2007). Where do you know what you know? The representation of semantic knowledge in the human brain. *Nature Reviews Neuroscience, 8*(12), 976–987.

Pollack, J. B. (1990). Recursive distributed representations. *Artificial Intelligence, 46*(1), 77–105.

Price, A. R., Bonner, M. F., Peelle, J. E., & Grossman, M. (2015). Converging evidence for the neuroanatomic basis of combinatorial semantics in the angular gyrus. *Journal of Neuroscience, 35*(7), 3276–3284.

Price, A. R., Peelle, J. E., Bonner, M. F., Grossman, M., & Hamilton, R. H. (2016). Causal evidence for a mechanism of semantic integration in the angular gyrus as revealed by high-definition transcranial direct current stimulation. *Journal of Neuroscience, 36*(13), 3829–3838.

Smith, E. E., Osherson, D. N., Rips, L. J., & Keane, M. (1988). Combining prototypes: A selective modification model. *Cognitive Science, 12*(4), 485–527.

Smolensky, P. (1990). Tensor product variable binding and the representation of symbolic structures in connectionist systems. *Artificial Intelligence, 46*(1), 159–216.

Solomon, S. H., & Thompson-Schill, S. L. (2017). Finding features, figuratively. *Brain and Language, 174*, 61–71.

Tanaka, K. (1996). Inferotemporal cortex and object vision. *Annual Review of Neuroscience, 19*, 109–139.

Thompson-Schill, S. L., Aguirre, G. K., D'Esposito, M., & Farah, M. J. (1999). A neural basis for category and modality specificity of semantic knowledge. *Neuropsychologia, 37*(6), 671–676.

Thompson-Schill, S. L., D'Esposito, M., Aguirre, G. K., & Farah, M. J. (1997). Role of left inferior prefrontal cortex in retrieval of semantic knowledge: A reevaluation. *Proceedings of the National Academy of Sciences, 94*(26), 14792–14797.

Visser, M., Jefferies, E., & Lambon Ralph, M. A. (2010). Semantic processing in the anterior temporal lobes: A meta-analysis of the functional neuroimaging literature. *Journal of Cognitive Neuroscience, 22*(6), 1083–1094.

Westerlund, M., & Pylkkänen, L. (2014). The role of the left anterior temporal lobe in semantic composition vs. semantic memory. *Neuropsychologia, 57*, 59–70.

Wu, L., & Barsalou, L. W. (2009). Perceptual simulation in conceptual combination: Evidence from property generation. *Acta Psychologica, 132*(2), 173–189.

Zeki, S., Watson, J. D., Lueck, C. J., Friston, K. J., Kennard, C., & Frackowiak, R. S. (1991). A direct demonstration of functional specialization in human visual cortex. *Journal of Neuroscience, 11*(3), 641–649.

X
LANGUAGE

Introduction

LIINA PYLKKÄNEN AND KAREN EMMOREY

Language relies on a complex neural system that can be studied through a wide variety of methods, including functional magnetic resonance imaging (fMRI), positron emission tomography (PET), magnetoencephalography (MEG), event-related potentials (ERPs), transcranial magnetic stimulation (TMS), electrocorticography (ECoG), patient studies, and lesion-symptom mapping, among others. In addition, insights into the neurobiology of language can be obtained from investigations into how language systems are established in the developing brain and how language processes break down (and recover) after brain injury. Adding to these rich sources of evidence is the linguistic diversity found across the world's languages (both signed and spoken) that can be tapped to investigate cognitive and linguistic constraints on the neural implementation of language. The chapters in this section draw on these sources of evidence and methodologies to provide a state-of-the-art snapshot of our understanding of the neural bases of language.

The chapter by Bornkessel-Schlesewsky and Schlesewsky focuses on how cross-linguistic diversity affects the neural processing involved in language comprehension. Current evidence indicates that neural tuning to language-specific phonemic distinctions emerges in early auditory cortex and is not observed for subcortical auditory regions that track acoustic changes but are insensitive to phonemic categories. Cross-linguistic category differences may be more difficult to determine at the word level for syntactic categories—for example, neural differences between nouns and verbs appear to be linked to semantic categories (objects, events), and this syntactic distinction between nouns and verbs may not occur in all of the world's languages. At the sentence level, neural responses to implausible semantic role reversals differ across languages depending upon whether or not the language relies heavily on word order for sentence interpretation (i.e., sequence-dependent vs. sequence-independent languages). Bornkessel-Schlesewsky and Schlesewsky propose a neurobiological explanation for these cross-linguistic differences based on how predictive error is neuronally encoded and propagated within the cortex.

MacSweeney and Emmorey capitalize on the perceptual and sensorimotor differences between signed and spoken languages to identify neural systems that are modality-independent and modality-dependent. Their review reveals that a very similar left-lateralized perisylvian network supports both spoken and signed language processing, including classic language regions such as Broca's area and Wernicke's area. Signed and spoken languages differ, however, with respect to the role of parietal cortex. For example, left inferior parietal cortex is engaged to a greater extent for phonological (form) encoding for sign languages, and superior parietal cortex plays a larger role in processing spatial language, most likely because sign languages use locations in signing space to express spatial relationships. These authors point to future studies that use multivoxel pattern analysis as a way to investigate whether the same computations and/or representations occur within shared regions of activation for signed and spoken languages.

Pylkkänen and Brennan summarize our current understanding of the neural mechanisms that compose individual words into phrases and sentences. They highlight the need for systematic, incremental research to unpack the functional roles of various integrative nodes within the brain's *combinatory network*, comprising at least the left anterior temporal cortex, left posterior temporal lobe, left inferior frontal cortex, temporoparietal junction, and ventromedial prefrontal cortex. Among these nodes, our understanding is the most developed for the left anterior temporal lobe, which appears to contribute a conceptually based combinatory operation fairly quickly after stimulus onset (~200 ms), both in comprehension and production. These authors compare and contrast results from traditional factorial experiments and studies comparing model fits for data gathered during narrative comprehension, highlighting the complementary nature of the two methods.

Fedorenko, too, discusses higher-level language processing, with a focus on the domain specificity of the *language network*, on the one hand, and on the possible divisions within this network regarding lexical versus combinatory processing, on the other. While Fedorenko reviews several strands of compelling evidence that language tasks dissociate from various nonlinguistic tasks (such as arithmetic and music processing) in both neuroimaging and deficit-lesion data, she argues for a lack of dissociation in brain regions supporting lexical versus combinatory processing. The core evidence for this lack of dissociation is that although regional specificity can be found in the brain's sensitivity to specific lexical or syntactic variables, it is impossible to find regions that activate (in general) for the presence of syntax and not for the presence of lexical access, and vice versa. Instead, the robust modulator of language activation in Fedorenko's data is the meaningfulness of the stimulus: the richer the meaning, the more strongly the network responds. A similar point is made by Pylkkänen and Brennan, who argue that the extant evidence in combinatory processing is compatible with the hypothesis that all regions within the *combinatory network* perform semantic, as opposed to purely syntactic, combinatory operations.

Binder and Fernandino discuss the neural representation of meaning in more detail, focusing on the word level. They first give a broad overview of different theoretical accounts of concepts and then make a case for a distributed neural architecture that combines aspects of both symbolic and embodied theories of meaning in a three-layer hierarchical model. At the lowest, unimodal level, sensory association areas represent modality-specific aspects of meaning. Information from two or more modalities is then combined in multimodal regions, such as the left posterior temporal cortex and the anterior supramarginal gyrus. The highest-level convergence zones, labeled *transmodal*, combine information from many experiential domains and comprise regions such as the left anterior temporal lobe and the angular gyrus. During processing, activation is proposed to spread "high" to "low," with spoken words first activating transmodal concept representations in lateral temporal cortex and then spreading to more modality-specific features and thematically associated concepts.

Before meaning can be accessed, the brain must decode the speech (or sign) signal, and Adank approaches this problem for speech through the lens of listening under adverse conditions. Her review reveals that partially segregated neural networks are recruited when listening to speech during noise (environmental distortions of speech) versus listening to source-distorted speech (e.g., unfamiliar accents or speech styles)—compared to listening to speech in quiet. The use of TMS has been particularly useful in determining whether a neural region plays a causal role in speech perception under adverse conditions, and current evidence suggests that (pre-) motor cortex is critical for understanding speech in noisy environments.

The chapter by Dehaene-Lambertz and Kabdebon also focuses primarily on speech perception, but their goal is to understand how the developing infant brain is able to accurately analyze aspects of continuous speech (despite poor motor abilities) and whether the functional organization for language in the infant brain parallels that observed for the adult linguistic system. Recent results suggest that the hierarchical organization within perisylvian temporal cortex observed for adults (e.g., neural response gradients that operate on different timescales for parsing speech) is also found in the developing brain. Further, laterality differences in separating voice identity (who is speaking) from linguistic content (vowel identity) are also observed for young infants (with right- and left-lateralized functions, respectively). Dehaene-Lambertz and Kabdebon note further similarities in the function of left inferior frontal cortex between infants and adults, but this region is slower to mature than the auditory cortices. Given that the brain becomes tuned to an individual's native language (i.e., there is an interaction between environmental exposure and neural processing), understanding the maturational changes that occur with learning at both the microstructural and network levels can provide essential insight into the mechanisms that underlie neural plasticity.

Wilson and Fridriksson examine neural plasticity in the adult by exploring the functional reorganization of the language system that can occur after brain injury. They begin their chapter with an extremely insightful and useful review of historical accounts of aphasic disorders, including primary progressive aphasia, and then illustrate how the use of new multivariate analysis methods are enhancing our understanding of the brain bases of aphasic symptoms and subtypes. Wilson and Fridriksson also provide a historical perspective on aphasia recovery before describing modern approaches to aphasia treatment. The mechanisms that underlie aphasia recovery differ in the various stages of recovery. Current evidence indicates that right frontal regions play a compensatory role in the acute stage, whereas increased activation in left perilesional cortex may be more critical for recovery at later stages.

Together with the rest of cognitive neuroscience, the neurobiology of language is experiencing an explosion of new methodological approaches, mostly brought about by rapidly developing computational possibilities in data analysis. In the face of these exciting developments, the contributions in this section highlight the continued importance of theoretically grounded investigations of a system as complex as language. The picture that emerges may be quite different from the theories that motivate the experimentation, but only with a testable theory can we show when it may be wrong (and, perhaps more interestingly, *how* it might be wrong). Moving toward the 2020s, we hope for an explosion of not only exciting new methods but also systematic bodies of research with sufficiently constant methods from study to study to allow robust theoretical generalizations to emerge.

72 The Crosslinguistic Neuroscience of Language

INA BORNKESSEL-SCHLESEWSKY AND MATTHIAS SCHLESEWSKY

ABSTRACT With approximately 7,000 living languages, language is a singularly diverse cognitive ability. Neuroscientific research is only just beginning to illuminate the implications of this diversity for the neurobiological implementation of language. This chapter reviews the current state of the art in this field, focusing primarily on language comprehension. It discusses crosslinguistic variability in the neural categorization of sounds, of words and concepts, and in the information-processing strategies that support linguistic combinatorics. Across all of these domains, the human brain attunes to the input features that are particularly relevant in the language under consideration, thus giving rise to diverse processing patterns. For categorization, this attunement likely draws upon a language-specific refinement of cortical feature detectors. For information processing, it may be based on a language-specific weighting of top-down (feedback) and bottom-up (feedforward) information as part of a hierarchically organized cortical predictive-coding architecture. Finally, the chapter argues that crosslinguistic generalizations can inform the neuroscience of language, as these may reflect cognitive and/or neurobiological constraints on how the human brain learns and processes information.

With approximately 7,000 living languages (Simons & Fennig, 2018), language is one of the most diverse human cognitive abilities. Consequently, information-processing affordances may differ profoundly between languages.

Is There Unity beyond the Diversity?

Scholars disagree about whether, among this diversity, there are universals: properties that hold for all human languages (e.g., Comrie, 1989). However, absolute universals are surprisingly difficult to find (Evans & Levinson, 2009), as even some seemingly simple assumptions (e.g., "all languages have vowels") do not hold in all languages—sign languages being a case in point. Implications for the cognitive (neuro)science of language are potentially far-reaching. Given this diversity, is it feasible to treat language as a single, unified cognitive domain?[1]

A more fruitful perspective may be to examine recurring properties across the world's languages—that is, nonarbitrary skewings in crosslinguistic distributions (*statistical universals*; cf., Bickel, 2015). For example, of the six logically possible basic orders of subject (S), object (O), and verb (V), two are highly dominant (SVO: 35%, SOV: 41%, of the 1,377 language samples in Dryer, 2013). Such typological skewings might be linked to the way in which the brain learns and processes information (Bickel et al., 2015; Bornkessel-Schlesewsky & Schlesewsky, 2016; Bornkessel-Schlesewsky, Schlesewsky, Small, & Rauschecker, 2015; Christiansen & Chater, 2008, 2016). They can thus inform the neuroscience of language by providing targets for explanation in terms of cognitive and/or neurobiological mechanisms (for word order, see Bickel et al., 2015; Bornkessel-Schlesewsky & Schlesewsky, 2009; Kemmerer, 2012).

Given the tendency of most neuroscientific approaches to assume crosslinguistic unity (or only examine a small range of languages), this chapter aims to highlight how the neurobiology of language may be shaped by crosslinguistic diversity. This should, however, not be taken to suggest that brain mechanisms of language processing show no crosslinguistic generalizations at all. Rather, we assume that many basic assumptions laid out in the other chapters of this section hold across the languages of the world. There is no evidence to date to suggest that the basic networks underlying language processing (see chapter 75) differ across languages. Likewise, all languages must draw on basic combinatory mechanisms (see chapter 74) and complex, distributed conceptual representations (see chapter 76). Further compelling evidence for crosslinguistic similarities stems from sign languages, which show a wide range of neurocognitive-processing parallels to spoken languages (see chapter 73). Building on these basic observations, we will examine how the brain *attunes* its linguistic information processing to crosslinguistic diversity.

[1] For counterarguments by proponents of linguistic universals, see, e.g., Hauser, Chomsky, and Fitch (2002); Berwick, Friederici, Chomsky, and Bolhuis (2013). See also Skeide and Friederici (2016) for a detailed discussion of a "universal grammar"–based approach to the neuroscience of language.

Overview and limitations This chapter focuses on language *comprehension*, the area in which crosslinguistic similarities and differences have been studied most extensively from a neuroscientific perspective to date. While recent years have seen a sharp increase in examinations of language production in understudied and typologically diverse languages (cf., Norcliffe, Harris, & Jaeger, 2015), these highly interesting investigations have virtually all been behavioral. However, in view of ongoing research we are hopeful that the next decade will see the emergence of a more complete picture of the crosslinguistic neuroscience of language that integrates production and comprehension.

A second highly relevant topic that is not covered here is bilingualism. We believe that this complex area requires separate treatment, particularly since the implications of crosslinguistic similarities and differences in multilingual language acquisition and processing have not yet been studied systematically.

Finally, the chapter aims to present a framework for the crosslinguistic neuroscience of language based on the evidence currently available. To this end, it draws primarily on domains/phenomena for which systematic crosslinguistic comparisons exist.

The remainder of the chapter is structured as follows. We first review the effects of crosslinguistic diversity on the processing of linguistic categories before addressing information-processing mechanisms and concluding with a discussion of future directions. Throughout the chapter, we focus primarily on mechanisms and, where possible, aim to link observations to neurobiologically plausible explanations (cf., Poeppel, Emmorey, Hickok, & Pylkkänen, 2012; Small, 2008).

Categories

Language provides a powerful means of categorizing perceptual input. Different languages offer different categorization systems at multiple linguistic levels, including sounds, prosody (speech melody), words, and possibly even higher-order combinations of words into phrases and sentences. This section focuses on sounds, prosodic units, words, and concepts. Higher-level units will be discussed in the information-processing section.

Sounds Categorization is crucial for speech-sound processing, as it defines the perception of phonemes: the smallest units that differentiate meaning. For example, the contrast between *l* and *r* is phonemic in English—*lap* and *rap* have distinguishable meanings—but not Japanese. Thus, English speakers perceive a categorical contrast between the syllables *la* and *ra* that transcends acoustic variability (cf., Kuhl, 2004), while speakers of Japanese show near-chance-level discrimination (Miyawaki et al., 1975). Language-specific features for phoneme categorization are learned during the first year of life (Werker & Hensch, 2015) as the brain learns to group input from the same category in a given language, thereby allowing for effective communication in spite of massive acoustic variability (e.g., speaker-dependent differences).

Phoneme categorization emerges at the cortical level—namely, in early auditory areas (Bidelman & Lee, 2015). It relies on feature detectors attuned to relevant (language-specific) cues (Chang et al., 2010). By contrast, subcortical sound processing appears to involve more direct acoustic representations (Bidelman, Moreno, & Alain, 2013): the brain stem frequency following response (FFR; Chandrasekaran & Kraus, 2010) mirrors continuous acoustic changes rather than phoneme categories (Bidelman, Moreno, & Alain, 2013).

While this basic neural architecture appears to be shared across languages, cortical responses attune to language-specific phonemic properties. This has been demonstrated convincingly using the mismatch negativity (MMN; Näätänen, Paavilainen, Rinne, & Alho, 2007), a preattentive event-related brain potential (ERP) component that is observable to infrequent (*deviant*) stimuli within a sequence of common "standards." The MMN is thought to reflect change detection within an auditory scene. Plausibly, this occurs via an integration of top-down predictions with bottom-up input (*predictive coding*; Garrido, Kilner, Stephan, & Friston, 2009), including both the adjustment of the current predictive model (of the sensory memory trace) to the deviant (Näätänen & Winkler, 1999; Winkler, Karmos, & Näätänen, 1996) and the adaptation of auditory cortex activity to the standard (Jääskeläinen et al., 2004).

The MMN is sensitive to language-specific phoneme categories: it is amplified when the difference between standards and deviants crosses a native phoneme category boundary (e.g., Dehaene-Lambertz, 1997; Näätänen et al., 1997). For example, Dehaene-Lambertz (1997) presented French participants with stimuli involving native and nonnative (Hindi) phoneme contrasts and observed an MMN only for deviants that crossed a native phoneme boundary. Importantly, these results cannot be explained via acoustic distance. These findings have been replicated using a range of other language comparisons and have been shown to persist even for acoustically variable standards and deviants (see the review in Näätänen et al., 2007).

Similar observations hold for pitch in tone languages—that is, languages in which pitch has a phonemic status (e.g., Mandarin Chinese). Comparing native speakers of English and Mandarin, Chandrasekaran, Krishnan, and Gandour (2007) observed an MMN effect of similar magnitude for both groups when standards and deviants were acoustically dissimilar Mandarin tones. For acoustically similar tones, by contrast, Mandarin speakers showed a larger MMN than English speakers. Thus, pitch-based MMN amplitude differences vary depending on participants' experience in using specific tonal information for categorizing speech.

While phonemes and tones show similar cortical patterns of linguistic attunement (see also Bidelman and Lee [2015] for evidence regarding the categorical cortical encoding of tone contrasts), tones appear to differ from phonemes in that they also shape responses at the subcortical level. Specifically, the brain stem FFR shows a stronger representation of tones and closer pitch tracking for speakers of tonal as opposed to nontonal languages (Krishnan, Xu, Gandour, & Cariani, 2005; Yu & Zhang, 2018).[2]

In summary, the brain attunes to relevant features for categorization in a given language. This shapes feature detectors at the cortical level, is evident in preattentive sound processing, and feeds into categorical perception. Subcortical responses within the ascending auditory system, by contrast, are more closely tied to the (continuous) acoustic structure of the input. However, when language experience engenders increasing sensitivity to additional features such as pitch, such features may already be tracked at the level of the brain stem.

Words Word categories (e.g., nouns, verbs) have played a crucial role in several prominent debates on the neuroscience of language—for example, whether the brain represents abstract category information (cf., Vigliocco, Vinson, Druks, Barber, & Cappa, 2011) or whether there is a primacy of basic syntactic structure building over other information sources (e.g., Friederici, 2002; Hagoort, 2005).

In language typology, by contrast, the crosslinguistic validity of word categories is controversial. An extreme stance posits that some languages lack word category distinctions altogether (for critical discussion, see Evans & Osada, 2005). While this may be too extreme, many languages show a higher category fluidity than most familiar European languages—that is, have many words with multiple functions (akin to the noun/verb ambiguity of English words such as *cut*). But even assuming that all languages have word categories, it remains controversial whether these categories are, in fact, comparable across languages (Croft, 2001).

In regard to neural representation, Vigliocco et al. (2011) present compelling evidence that apparent differences at the individual word level reflect semantic categories (e.g., events vs. objects) rather than true word category differences (e.g., verbs vs. nouns). True word category differences only emerge in a sentence context. From this observation, Vigliocco et al. (2011) argue for an emergentist view of word categories in the brain—that is, categories emerge from the combination of an individual word with the context in which it is encountered. This is highly compatible with the typological evidence.

Does higher category fluidity (e.g., in Mandarin Chinese; Bisang, 2008) therefore affect sentence processing? Several ERP experiments have compared syntactic (word category), semantic, and combined violations in Mandarin (Ye, Luo, Friederici, & Zhou, 2006; Yu & Zhang, 2008), building on similar studies conducted primarily in German (see Friederici, 2002 for an overview). On the basis of their findings, both groups of authors argued for a more rapid use of semantic information, vis-à-vis word category information in Mandarin as opposed to Western European languages. At first glance, this would appear to suggest that the rigidity (or lack thereof) of category information in a particular language changes the time course of information processing during sentence comprehension. Somewhat problematically, though, these conclusions were partly based on absolute functional interpretations of language-related ERP components—for example, the assumption that N400 effects reflect lexical-semantic processing. It has now been demonstrated repeatedly that there is no one-to-one mapping between components such as the N400 and P600 and particular linguistic domains (for a recent overview, see Bornkessel-Schlesewsky, Staub, & Schlesewsky, 2016; Bornkessel-Schlesewsky & Schlesewsky, 2019). In addition, recent predictive coding–based perspectives on the neurocognition of sentence processing (e.g., Bornkessel-Schlesewsky et al., 2015; Dikker & Pylkkänen, 2011; Dikker, Rabagliati, Farmer, & Pylkkänen, 2010) call for a new perspective on time course–related questions. They highlight the need to consider both the potential specificity of a prediction from the sentence context and the type of evidence in the input that leads to a prediction match or prediction error (see Bornkessel-Schlesewsky, Staub, & Schlesewsky, 2016 for a detailed discussion). It would be illuminating to reexamine the effects of category fluidity on sentence-level combinatorics from the perspective of these approaches.

Concepts Beyond sounds and word categories, languages also provide powerful classification systems for concepts, as revealed by certain concepts receiving similar grammatical treatment, as opposed to others. The

[2] For a comprehensive review of research on the differences between tone and nontone languages, including neuroanatomical differences, see Gandour and Krishnan (2016).

systematic examination of crosslinguistic similarities and differences in conceptual categorization is called *semantic typology* (Evans, 2010).

Kemmerer (2017) presents a comprehensive and compelling overview of possible synergies between semantic typology and concept representation, arguing that "the ways in which categories of object concepts are organised and represented in the brain reflect not only universal tendencies but also language-particular idiosyncrasies" (p. 402). Drawing on results regarding the distributed, categorical representation of objects in ventral temporal cortex, he suggests that crosslinguistic categorization differences along particular semantic parameters (e.g., animacy, spatial aspects of object representation) may tap into neurobiological primitives of how this information is represented. For example, languages with shape-related nominal classifiers (i.e., classificatory words that accompany certain classes of nouns) may assign particular importance to certain shape-based primitives of object recognition in the brain. Conversely, the way in which these properties cluster to form categories in object recognition and conceptualization may depend on language-specific categories. This intriguing hypothesis has not yet been tested systematically but appears highly congruent with existing insights into the brain's language-specific attunement to certain acoustic features in sound categorization.

Information-Processing Strategies

Of course, language is more than just categorization. Indeed, one of its most fascinating properties is its vast combinatory power: words flexibly combine to form sentences and discourses, thus allowing for the expression of ever-new meanings (see chapter 74).

Languages differ with regard to the information sources used for this, as first proposed by the competition model (CM; Bates, Devescovi, & Wulfeck, 2001; MacWhinney, Bates, & Kliegl, 1984). The CM views language processing as a direct form-to-function mapping driven by various information sources (*cues*), such as word order, animacy, case marking, and more. The weighting of individual cues differs from language to language and is governed by *cue validity*. Highly valid cues are both applicable (i.e., often present) and reliable (i.e., unambiguous and not misleading when present). Thus, as for linguistic categorization, the language-processing system attunes to those cues that are the most relevant for sentence interpretation in a given language.

Crosslinguistic diversity in combinatorial strategies In the neuroscience of language, this idea has been generalized to differing *combinatorial strategies*. Specifically, the human brain appears to apply distinct information-processing strategies in languages that rely primarily on word order (*sequence-dependent combinatorics*) compared to languages that rely more strongly on other cues, such as case marking or animacy (*sequence-independent combinatorics*). Supporting evidence stems from ERP studies on semantic reversal anomalies (SRAs; Bornkessel-Schlesewsky et al., 2011): sentences such as "The fries have eaten the boys" (Bourguignon, Drury, Valois, & Steinhauer, 2012), in which the grammatically required interpretation contradicts world knowledge due to an implausible role reversal.

SRAs first piqued the interest of psycholinguists because they engendered only a late positivity in English and Dutch in comparison to plausible control sentences (e.g., Kolk, Chwilla, van Herten, & Oor, 2003; Kuperberg, Sitnikova, Caplan, & Holcomb, 2003), rather than the expected N400 effect for implausible sentence continuations (cf., Kutas & Federmeier, 2011). However, subsequent crosslinguistic research revealed qualitatively different ERP patterns across languages, with SRAs in German, Turkish, and Mandarin Chinese eliciting increased N400 effects (Bornkessel-Schlesewsky et al., 2011). (For a replication of the German vs. English result using another phenomenon, see Tune et al. [2014].) Strikingly, the dissociation cuts across language families as well as subjective language similarities—it forms a *neurotypology*.

The common denominator distinguishing English and Dutch from German, Turkish, and Mandarin is that the former rely heavily on word order for sentence interpretation (sequence-dependent languages), while the latter weigh other sequence-independent cues, such as case marking (German, Turkish) and animacy (Mandarin) more strongly (Bornkessel-Schlesewsky et al., 2011, 2015).

We have proposed that the N400-related SRA dissociation for the two types of languages can be explained in terms of differences in the weighting between top-down and bottom-up information in the context of a predictive-coding framework (Bornkessel-Schlesewsky, Staub, & Schlesewsky, 2016; Bornkessel-Schlesewsky & Schlesewsky, 2019; Tune et al., 2014). In primarily sequence-based languages, word-order regularities permit top-down predictions regarding upcoming categories. In English, for example, the majority of sentences can be processed via a sequential agent-action-object template (Bever, 1970), and sentences not adhering to this template are more likely to be misunderstood (Ferreira, 2003). Consequently, argument and verb features (e.g., animacy, case marking, agreement) are less relevant for sentence interpretation. However, these bottom-up features are considerably more important in sequence-independent languages. The crosslinguistic presence or absence of N400 effects for sentence-level interpretation can thus be explained by differences in the treatment of prediction errors induced by bottom-up features. Feedforward error signals propagated up the cortical hierarchy are weighted by precision (Bastos et al., 2012), which is defined as the inverse of variance (Feldman & Friston, 2010; Kok, Rahnev, Jehee, Lau, & de Lange, 2012). In a linguistic context, low variance in the form-to-meaning mapping characterizes cues with high language-specific validity—that is, highly valid cues induce high-precision prediction errors, and N400 effects only result when a prediction error's precision weighting is sufficiently high (Bornkessel-Schlesewsky & Schlesewsky, 2019). Neurobiologically, this can be modeled by changes in the postsynaptic gain of the pyramidal cells in superficial cortical layers that encode prediction errors and propagate these to higher cortical areas (Bastos et al., 2012).

Neuroanatomically, sequence-dependent and sequence-independent sentence interpretation strategies may be more closely tied to the dorsal and ventral auditory streams, respectively (Bornkessel-Schlesewsky & Schlesewsky, 2013; Bornkessel-Schlesewsky et al., 2015), but more empirical evidence is required to verify this generalization.[3]

Crosslinguistic generalizations Complementing the abovementioned diversity, sentence processing also shows patterns that recur across typologically diverse languages. These include crosslinguistically applicable interpretation strategies related to linguistic actors (i.e., the participants primarily responsible for a linguistically described state of affairs). Across languages, comprehenders prefer (1) actor-initial word orders and (2) sentences with prototypical (i.e., human) actors. Sentences deviating from these preferences engender model updating (N400) responses (Bornkessel-Schlesewsky & Schlesewsky, 2009). The actor-first preference holds even in languages in which an actor interpretation of the initial argument is not the most frequent option (e.g., Hindi: Bickel et al., 2015; Turkish: Demiral, Schlesewsky, & Bornkessel-Schlesewsky, 2008) and in which actors are marked differently depending on sentence transitivity and other factors (ergative languages; Hindi: Bickel et al., 2015; Basque: Erdocia, Laka, Mestres-Missé, & Rodriguez-Fornells, 2009).[4] The preference for animate actors remains observable even in contexts that unambiguously signal the presence of an inanimate actor (Muralikrishnan, Schlesewsky, & Bornkessel-Schlesewsky, 2015). Both preferences can be derived from more general information-processing strategies employed by the brain: the tendency to preferentially attend to potential causers (typically animates) over entities that are less likely to cause events (typically inanimates; New, Cosmides, & Tooby, 2007) and the association between agency and properties related to animacy, such as biological motion (Frith & Frith, 2010). Converging evidence for this view stems from overlapping neuroanatomical correlates for nonlinguistic agency detection (Frith & Frith, 2010) and actor-related language processing (Grewe et al., 2007), both of which engage the posterior superior temporal sulcus (pSTS). The preference for actor-first word orders and actors as a uniform category is further supported by crosslinguistic distributions (Bickel et al., 2015; Dryer, 2013): while counterexamples are attested, they are considerably less frequent. Actor-related preferences in processing and grammar are thus clear sentence-level candidates for which linguistic distributions accord with neurocognitive-processing mechanisms. Note, however, that this assumption needs to be tested more rigorously in languages that constitute clear exceptions to these crosslinguistic generalizations (for a first attempt, see Yasunaga, Yano, Yasugi, & Koizumi, 2015).

Conclusions and Future Directions

We have outlined a framework for how crosslinguistic diversity affects neurobiological mechanisms of language processing. While the underlying neurobiological processing architecture appears similar across languages, it attunes to relevant language-specific features in both categorization and information processing, thus giving rise to diverse processing signatures that may manifest themselves in apparent qualitative differences. Crucially, the neurobiological architecture explicitly *permits* variability as to how its processing goals (e.g., to minimize prediction errors) are fulfilled. Crosslinguistically variable properties can thus be viewed as

[3] Note that the degree of sequence-(in)dependent processing may also vary within languages in accordance with current processing demands. See Bornkessel-Schlesewsky et al. (2011) for Icelandic and Bourgouignon et al. (2012) for English. Thus, the classifications discussed here are language-specific defaults rather than absolutes.

[4] Our discussion of word-order processing only touches on simple sentences rather than more complex cases involving embeddings (e.g., relative clauses). The rich literature on crosslinguistic differences in the processing of relative clauses has hitherto focused exclusively on possible cognitive effects and mechanisms—e.g., the question of whether subject relative clauses are universally easier to process than object relative clauses. This is the case even for studies that have used neuroscientific methods. For a recent review, see Norcliffe, Harris, and Jaeger (2015). Thus, the implications for the neuroscience of language are not yet clear.

alternative solutions to underlying architectural requirements. In some cases, certain solutions may be preferred over others—for example, because, like the actor strategy, they align with processing in other nonlinguistic domains—and this is reflected in skewed linguistic distributions.

We have already outlined how this can be envisaged within the context of a hierarchically organized cortical predictive-coding architecture, in which top-down predictions are integrated with bottom-up input via feedback and feedforward connections, respectively, and in which the brain draws active inferences about the causes of its sensorium (Bastos et al., 2012; e.g., Friston, 2005). Crosslinguistic diversity (via linguistic experience) shapes this process both in terms of how continuous and ambiguous input is mapped onto linguistic categories and in regard to the dynamics of information processing.

A second potential neurobiological processing mechanism for language concerns the temporal "chunking" of information into temporal windows of integration (TWIs) at multiple timescales. TWIs provide a temporal equivalent to receptive fields in the visual domain and are thought to be implemented neurally via oscillatory brain activity (Canolty & Knight, 2010; Fries, 2005). Oscillatory activity entrains to linguistic categories of different sizes (e.g., phonemes, syllables; possibly, words and phrases; Ahissar et al., 2001; Ding et al., 2017; Ding, Melloni, Zhang, Tian, & Poeppel, 2016; Luo & Poeppel, 2007), thereby aligning receptive phases of neuronal information processing with the most informative (i.e., high energy) portions of the speech stream (Giraud & Poeppel, 2012). Initial evidence suggests that this mechanism holds across typologically diverse languages. TWIs could thus provide a crosslinguistically applicable neurobiological basis for linguistic categories. However, potential crosslinguistic differences in speech tracking remain to be explored—for example, in languages in which there is little evidence for syllables (e.g., moraic languages such as Japanese or the Nigerian language Gokana; Hyman, 1983).

We thus suggest that future examinations of the assumed relationship between a shared neurobiological information-processing architecture and crosslinguistic diversity will need to test predictions derived from neurobiological models, particularly for those exceptional languages that do not fit crosslinguistic generalizations.

REFERENCES

Ahissar, E., Nagarajan, S., Ahissar, M., Protopapas, A., Mahncke, H., & Merzenich, M. M. (2001). Speech comprehension is correlated with temporal response patterns recorded from auditory cortex. *Proceedings of the National Academy of Sciences, 98*(23), 13367–13372.

Aust, F., & Barth, M. (2018). *Papaja: Create APA manuscripts with R Markdown.* R package version 0.1.0.9842.

Bastos, A. M., Usrey, W. M., Adams, R. A., Mangun, G. R., Fries, P., & Friston, K. J. (2012). Canonical microcircuits for predictive coding. *Neuron, 76*(4), 695–711. doi:10.1016/j.neuron.2012.10.038

Bates, E., Devescovi, A., & Wulfeck, B. (2001). Psycholinguistics: A cross-language perspective. *Annual Review of Psychology, 52,* 369–396.

Berwick, R., Friederici, A., Chomsky, N., & Bolhuis, J. (2013). Evolution, brain, and the nature of language. *Trends in Cognitive Sciences, 17*(2), 89–98.

Bever, T. G. (1970). The cognitive basis for linguistic structures. In J. Hayes (Ed.), *Cognition and the development of language* (pp. 279–362). New York: Wiley.

Bickel, B. (2015). Distributional typology: Statistical inquiries into the dynamics of linguistic diversity. In B. Heine & H. Narrog (Eds.), *The Oxford handbook of linguistic analysis* (2nd ed., pp. 901–923). Oxford: Oxford University Press.

Bickel, B., Witzlack-Makarevich, A., Choudhary, K. K., Schlesewsky, M., & Bornkessel-Schlesewsky, I. (2015). The neurophysiology of language processing shapes the evolution of grammar: Evidence from case marking. *PLOS One, 10*(8), e0132819. doi:10.1371/journal.pone.0132819

Bidelman, G. M., & Lee, C.-C. (2015). Effects of language experience and stimulus context on the neural organization and categorical perception of speech. *NeuroImage, 120,* 191–200. doi:10.1016/j.neuroimage.2015.06.087

Bidelman, G. M., Moreno, S., & Alain, C. (2013). Tracing the emergence of categorical speech perception in the human auditory system. *NeuroImage, 79,* 201–212. doi:10.1016/j.neuroimage.2013.04.093

Bisang, W. (2008). Precategoriality and syntax-based parts of speech: The case of late archaic Chinese. *Studies in Language, 32,* 568–589.

Bornkessel-Schlesewsky, I., Kretzschmar, F., Tune, S., Wang, L., Gençc, S., Philipp, M., … Schlesewsky, M. (2011). Think globally: Cross-linguistic variation in electrophysiological activity during sentence comprehension. *Brain and Language, 117*(3), 133–152.

Bornkessel-Schlesewsky, I., & Schlesewsky, M. (2009). The role of prominence information in the real-time comprehension of transitive constructions: A cross-linguistic approach. *Linguistics and Language Compass, 3*(1), 19–58. doi:10.1111/j.1749-818X.2008.00099.x

Bornkessel-Schlesewsky, I., & Schlesewsky, M. (2013). Reconciling time, space and function: A new dorsal ventral stream model of sentence comprehension. *Brain and Language, 125*(1), 60–76. doi:10.1016/j.bandl.2013.01.010

Bornkessel-Schlesewsky, I., & Schlesewsky, M. (2016). The importance of linguistic typology for the neurobiology of language. *Linguistic Typology, 20*(3), 615–621.

Bornkessel-Schlesewsky, I., & Schlesewsky, M. (2019). Towards a neurobiologically plausible model of language-related, negative event-related potentials. *Frontiers in Psychology, 10*(298), 1–17. doi:10.3389/fpsyg.2019.00298

Bornkessel-Schlesewsky, I., Schlesewsky, M., Small, S. L., & Rauschecker, J. P. (2015). Neurobiological roots of language in primate audition: Common computational properties. *Trends in Cognitive Sciences, 19*(3), 1–9. doi:10.1016/j.tics.2014.12.008

Bornkessel-Schlesewsky, I., Staub, A., & Schlesewsky, M. (2016). The timecourse of sentence processing in the brain. In G. Hickok & S. L. Small (Eds.), *Neurobiology of language* (pp. 607–620). Amsterdam: Elsevier.

Bourguignon, N., Drury, J., Valois, D., & Steinhauer, K. (2012). Decomposing animacy reversals between agents and experiencers: An ERP study. *Brain and Language, 122*, 179–189.

Canolty, R. T., & Knight, R. T. (2010). The functional role of cross-frequency coupling. *Trends in Cognitive Sciences, 14*(11), 506–515. doi:10.1016/j.tics.2010.09.001

Chandrasekaran, B., & Kraus, N. (2010). The scalp-recorded brainstem response to speech: Neural origins and plasticity. *Psychophysiology, 47*(2), 236–246. doi:10.1111/j.1469-8986.2009.00928.x

Chandrasekaran, B., Krishnan, A., & Gandour, J. T. (2007). Mismatch negativity to pitch contours is influenced by language experience. *Brain Research, 1128*(1), 148–156. doi:10.1016/j.brainres.2006.10.064

Chang, E. F., Rieger, J. W., Johnson, K., Berger, M. S., Barbaro, N. M., & Knight, R. T. (2010). Categorical speech representation in human superior temporal gyrus. *Nature Neuroscience, 13*(11), 1428–1432. doi:10.1038/nn.2641

Christiansen, M. H., & Chater, N. (2008). Language as shaped by the brain. *Behavioral and Brain Sciences, 31*(05). doi:10.1017/S0140525X08004998

Christiansen, M. H., & Chater, N. (2016). *Creating language: Integrating evolution, acquisition, and processing.* Cambridge, MA: MIT Press.

Comrie, B. (1989). *Linguistic universals and language typology.* Oxford: Blackwell.

Croft, W. A. (2001). *Radical construction grammar: Syntactic theory in typological perspective.* Oxford: Oxford University Press.

Dehaene-Lambertz, G. (1997). Electrophysiological correlates of categorical phoneme perception in adults. *NeuroReport, 8*, 919–924.

Demiral, Ş. B., Schlesewsky, M., & Bornkessel-Schlesewsky, I. (2008). On the universality of language comprehension strategies: Evidence from Turkish. *Cognition, 106*(1), 484–500. doi:10.1016/j.cognition.2007.01.008

Dikker, S., & Pylkkänen, L. (2011). Before the N400: Effects of lexical-semantic violations in visual cortex. *Brain and Language, 118*, 23–28.

Dikker, S., Rabagliati, H., Farmer, T. A., & Pylkkänen, L. (2010). Early occipital sensitivity to syntactic category is based on form typicality. *Psychological Science, 21*(5), 629–634. doi:10.1177/0956797610367751

Ding, N., Melloni, L., Yang, A., Wang, Y., Zhang, W., & Poeppel, D. (2017). Characterizing neural entrainment to hierarchical linguistic units using electroencephalography (EEG). *Frontiers in Human Neuroscience, 11.* doi:10.3389/fnhum.2017.00481

Ding, N., Melloni, L., Zhang, H., Tian, X., & Poeppel, D. (2016). Cortical tracking of hierarchical linguistic structures in connected speech. *Nature Neuroscience, 19*(1), 158–164. doi:10.1038/nn.4186

Dryer, M. S. (2013). Order of subject, object and verb. In M. S. Dryer & M. Haspelmath (Eds.), *The world atlas of language structures online.* Leipzig: Max Planck Institute for Evolutionary Anthropology.

Erdocia, K., Laka, I., Mestres-Missé, A., & Rodriguez-Fornells, A. (2009). Syntactic complexity and ambiguity resolution in a free word order language: Behavioral and electrophysiological evidences from Basque. *Brain and Language, 109*(1), 1–17. doi:10.1016/j.bandl.2008.12.003

Evans, N. (2010). Semantic typology. In J. J. Song (Ed.), *The Oxford handbook of linguistic typology* (pp. 504–533). Oxford: Oxford University Press.

Evans, N., & Levinson, S. (2009). The myth of language universals: Language diversity and its importance for cognitive science. *Behavioral and Brain Sciences, 32*, 429–492.

Evans, N., & Osada, T. (2005). Mundari: The myth of a language without word classes. *Linguistic Typology, 9*(3). doi:10.1515/lity.2005.9.3.351

Feldman, H., & Friston, K. J. (2010). Attention, uncertainty, and free-energy. *Frontiers in Human Neuroscience*, article 215.

Ferreira, F. (2003). The misinterpretation of noncanonical sentences. *Cognitive Psychology, 47*, 164–203.

Friederici, A. D. (2002). Towards a neural basis of auditory sentence processing. *Trends in Cognitive Sciences, 6*(2), 78–84.

Fries, P. (2005). A mechanism for cognitive dynamics: Neuronal communication through neuronal coherence. *Trends in Cognitive Sciences, 9*(10), 474–480.

Friston, K. (2005). A theory of cortical responses. *Philosophical Transactions of the Royal Society B: Biological Sciences, 360*, 815–836. doi:10.1098/rstb.2005.1622

Frith, U., & Frith, C. D. (2010). The social brain: Allowing humans to boldly go where no other species has been. *Philosophical Transactions of the Royal Society B: Biological Sciences, 365*, 165–176.

Gandour, J. T., & Krishnan, A. (2016). Processing tone languages. In *Neurobiology of Language* (pp. 1095–1107). New York: Elsevier. doi:10.1016/B978-0-12-407794-2.00087-0

Garrido, M. I., Kilner, J. M., Stephan, K. E., & Friston, K. J. (2009). The mismatch negativity: A review of underlying mechanisms. *Clinical Neurophysiology, 120*(3), 453–463. doi:10.1016/j.clinph.2008.11.029

Giraud, A.-L., & Poeppel, D. (2012). Cortical oscillations and speech processing: Emerging computational principles and operations. *Nature Neuroscience, 15*(4), 511–517. doi:10.1038/nn.3063

Grewe, T., Bornkessel-Schlesewsky, I., Zysset, S., Wiese, R., von Cramon, D. Y., & Schlesewsky, M. (2007). The role of the posterior superior temporal sulcus in the processing of unmarked transitivity. *Neuroimage, 35*, 343–352.

Hagoort, P. (2005). On Broca, brain, and binding: A new framework. *Trends in Cognitive Sciences, 9*(9), 416–423. doi:10.1016/j.tics.2005.07.004

Hauser, M., Chomsky, N., & Fitch, W. T. (2002). The faculty of language: What it is, who has it, and how did it evolve? *Science, 298*, 1569–1579.

Hyman, L. M. (1983). Are there syllables in Gokana. *Current Approaches to African Linguistics, 2*, 171–179.

Jääskeläinen, I. P., Ahveninen, J., Bonmassar, G., Dale, A. M., Ilmoniemi, R. J., Levänen, S., … Belliveau, J. W. (2004). Human posterior auditory cortex gates novel sounds to consciousness. *Proceedings of the National Academy of Sciences, 101*(17), 6809–6814. doi:10.1073/pnas.0303760101

Kemmerer, D. (2012). The cross linguistic prevalence of SOV and SVO word orders reflects the sequential and hierarchical representation of action in Broca's area. *Language and Linguistics Compass, 6*(1), 50–66.

Kemmerer, D. (2017). Categories of object concepts across languages and brains: The relevance of nominal classification systems to cognitive neuroscience. *Language, Cognition and Neuroscience, 32*(4), 401–424. doi:10.1080/23273798.2016.1198819

Kok, P., Rahnev, D., Jehee, J. F. M., Lau, H. C., & de Lange, F. P. (2012). Attention reverses the effect of prediction in silencing sensory signals. *Cerebral Cortex, 22*(9), 2197–2206.

Kolk, H. H., Chwilla, D. J., van Herten, M., & Oor, P. (2003). Structure and limited capacity in verbal working memory: A study with event-related potentials. *Brain and Language, 85*, 1–36.

Krishnan, A., Xu, Y., Gandour, J., & Cariani, P. (2005). Encoding of pitch in the human brainstem is sensitive to language experience. *Cognitive Brain Research, 25*(1), 161–168. doi:10.1016/j.cogbrainres.2005.05.004

Kuhl, P. K. (2004). Early language acquisition: Cracking the speech code. *Nature Reviews Neuroscience, 5*, 831–843.

Kuperberg, G. R., Sitnikova, T., Caplan, D., & Holcomb, P. (2003). Electrophysiological distinctions in processing conceptual relationships within simple sentences. *Cognitive Brain Research, 17*, 117–129.

Kutas, M., & Federmeier, K. D. (2011). Thirty years and counting: Finding meaning in the N400 component of the event-related brain potential (ERP). *Annual Review of Psychology, 62*, 621–647. doi:10.1146/annurev.psych.093008.131123

Luo, H., & Poeppel, D. (2007). Phase patterns of neuronal responses reliably discriminate speech in human auditory cortex. *Neuron, 54*(6), 1001–1010. doi:10.1016/j.neuron.2007.06.004

MacWhinney, B., Bates, E., & Kliegl, R. (1984). Cue validity and sentence interpretation in English, German, and Italian. *Journal of Verbal Learning and Verbal Behavior, 23*, 127–150.

Miyawaki, K., Jenkins, J. J., Strange, W., Liberman, A. M., Verbrugge, R., & Fujimura, O. (1975). An effect of linguistic experience: The discrimination of [r] and [l] by native speakers of Japanese and English. *Perception & Psychophysics, 18*(5), 331–340. doi:10.3758/BF03211209

Muralikrishnan, R., Schlesewsky, M., & Bornkessel-Schlesewsky, I. (2015). Animacy-based predictions in language comprehension are robust: Contextual cues modulate but do not nullify them. *Brain Research, 1608*, 108–137. doi:10.1016/j.brainres.2014.11.046

Näätänen, R., Lehtokoski, A., Lennes, M., Cheour, M., Huotilainen, M., Iivonen, A., ... Alho, K. (1997). *Nature, 385*, 432–434.

Näätänen, R., Paavilainen, P., Rinne, T., & Alho, K. (2007). The mismatch negativity (MMN) in basic research of central auditory processing: A review. *Clinical Neurophysiology, 118*(12), 2544–2590. doi:10.1016/j.clinph.2007.04.026

Näätänen, R., & Winkler, I. (1999). The concept of auditory stimulus representation in cognitive neuroscience. *Psychological Bulletin, 125*(6), 826–859.

New, J., Cosmides, L., & Tooby, J. (2007). Category-specific attention for animals reflects ancestral priorities, not expertise. *Proceedings of the National Academy of Sciences of the United States of America, 104*(42), 16598–16603.

Norcliffe, E., Harris, A. C., & Jaeger, T. F. (2015). Cross-linguistic psycholinguistics and its critical role in theory development: Early beginnings and recent advances. *Language, Cognition and Neuroscience, 30*(9), 1009–1032. doi:10.1080/23273798.2015.1080373

Poeppel, D., Emmorey, K., Hickok, G., & Pylkkänen, L. (2012). Towards a new neurobiology of language. *Journal of Neuroscience, 32*(41), 14125–14131. doi:10.1523/jneurosci.3244-12.2012

Simons, G. F., & Fennig, C. D. (Eds.). (2018). *Ethnologue: Languages of the world* (21st ed.). Dallas: SIL International.

Skeide, M. A., & Friederici, Angela D. (2016). The ontogeny of the cortical language network. *Nature Reviews Neuroscience, 17*, 323–332.

Small, S. L. (2008). The neuroscience of language. *Brain and Language, 106*, 1–3.

Tune, S., Schlesewsky, M., Small, S. L., Sanford, A. J., Bohan, J., Sassenhagen, J., & Bornkessel-Schlesewsky, I. (2014). Cross-linguistic variation in the neurophysiological response to semantic processing: Evidence from anomalies at the borderline of awareness. *Neuropsychologia, 56*, 147–166. doi:10.1016/j.neuropsychologia.2014.01.007

Vigliocco, G., Vinson, D. P., Druks, J., Barber, H., & Cappa, S. F. (2011). Nouns and verbs in the brain: A review of behavioural, electrophysiological, neuropsychological and imaging studies. *Neuroscience and Biobehavioral Reviews, 35*, 407–426.

Werker, J. F., & Hensch, T. K. (2015). Critical periods in speech perception: New directions. *Annual Review of Psychology, 66*(1), 173–196. doi:10.1146/annurev-psych-010814-015104

Winkler, I., Karmos, G., & Näätänen, R. (1996). Adaptive modeling of the unattended acoustic environment reflected in the mismatch negativity event-related potential. *Brain Research, 742*(1–2), 239–252. doi:10.1016/S0006-8993(96)01008-6

Yasunaga, D., Yano, M., Yasugi, Y., & Koizumi, M. (2015). Is the subject-before-object preference universal? An event-related potential study in the Kaqchikel Mayan language. *Language, Cognition and Neuroscience, 30*(9), 1209–1229. doi:10.1080/23273798.2015.1080372

Ye, Z., Luo, Y.-J., Friederici, A. D., & Zhou, X. (2006). Semantic and syntactic processing in Chinese sentence comprehension: Evidence from event-related potentials. *Brain Research, 1071*, 186–196.

Yu, J., & Zhang, Y. (2008). When Chinese semantics meets failed syntax. *NeuroReport, 19*(7), 745–749. doi:10.1097/WNR.0b013e3282fda21d

Yu, L., & Zhang, Y. (2018). Testing native language neural commitment at the brainstem level: A cross-linguistic investigation of the association between frequency-following response and speech perception. *Neuropsychologia, 109*, 140–148. doi:10.1016/j.neuropsychologia.2017.12.022

73 The Neurobiology of Sign Language Processing

MAIRÉAD MACSWEENEY AND KAREN EMMOREY

ABSTRACT By investigating sign languages, which are purely visual and not derived from auditory-vocal processes, we gain unique insight into the neurobiology of language. Sign languages represent a powerful tool with which to test constraints and plasticity of the language system. In this chapter we review the current literature on the neural systems supporting the production and comprehension of signed languages, focusing on native users. The literature clearly shows that the left-lateralized perisylvian language network identified as reliably engaged during spoken language processing, involving the core regions of the inferior frontal gyrus and superior temporal cortex, is recruited during sign language processing. Similarity of processing has also been identified in aspects of the timing of the linguistic processing of sign and speech. However, there are important differences in how the brain processes sign and speech. The left parietal lobe appears to play a particularly important role in sign language production and comprehension. In particular, parietal cortex is involved in processing the linguistic use of space, in phonological encoding (left supramarginal gyrus), and in self-monitoring during sign production (left superior parietal lobule).

Sign languages arise wherever Deaf communities come together, and they differ across countries. For example, American Sign Language (ASL) and British Sign Language (BSL) are mutually unintelligible. Importantly, the grammar of signed languages is not dependent on the surrounding spoken language. Further, studies have clearly shown that deaf (and hearing) children who learn a signed language from birth show the same developmental milestones in their language acquisition as hearing children learning a spoken language (Meier & Newport, 1990). Therefore, we can compare the neural systems established to support language production and comprehension in those who have acquired a signed or a spoken language as their first language.

In this chapter we review the literature to date and show that signed and spoken language processing both recruit modality independent neural circuits (e.g., the perisylvian cortices, including the inferior frontal and superior temporal gyri) and modality-dependent neural regions (e.g., left parietal cortex for sign language processing). Evidence from electroencephalography (EEG) and magnetoencephalography (MEG) indicates that the temporal neural dynamics of language production and comprehension is similar for signed and spoken languages, despite sensorimotor differences. Finally, we explore the role of parietal cortex in supporting spatial-processing demands that are unique to sign languages.

The Neurobiology of Sign Language Production

The primary linguistic articulators for sign language are the hands and arms, which are independent, symmetrical articulators; in contrast, the speech articulators include the larynx, velum, tongue, jaw, and lips, which are all located along the midline of the body. Although much is known about the neural networks involved in speech-motor control, we know very little about the neural systems that control manual sign production. Nonetheless, linguistic and psycholinguistic research has revealed both modality-independent and modality-specific properties of sign and speech production (see Corina, Gutierrez, & Grosvald, 2014 for a review). For example, both sign and speech production require the phonological assembly of sublexical units (handshape, location, and movement for sign language), as evidenced by systematic production errors (slips of the hand; e.g., Hohenberger, Happ, & Leuninger, 2002). Both signed and spoken languages encode syllables and constrain syllable internal structure in a similar manner (e.g., Berent, Dupuis, & Brentari, 2013). Both sign and speech production involve a two-stage process in which lexical semantic representations are retrieved independently of phonological representations, as evidenced by tip-of-the-tongue and tip-of-the-finger states (Thompson, Emmorey, & Gollan, 2005). Syntactic priming in sentence production occurs for both signed and spoken languages (Hall, Ferreira, & Mayberry, 2015). However, language output monitoring likely differs for sign and speech due to differences in perceptual feedback: speakers hear themselves speak, but signers do not see themselves sign (Emmorey, Bosworth, & Kraljic, 2009). Below, we explore the evidence for shared functional neural substrates for sign and speech production, as well as evidence for neural substrates that are specific to sign production.

Modality-independent cortical regions involved in language production Both sign and speech production are strongly lateralized to the left hemisphere. Signers with left, but not right, hemisphere damage produce phonological and semantic paraphasias (Hickok, Bellugi, & Klima, 1996). Phonological paraphasias in sign language involve the substitution of one phonological unit for another, as illustrated in figure 73.1. Recently, Gutierrez and colleagues used functional transcranial Doppler sonography (fTCD) to investigate hemispheric lateralization during natural (nonrestricted) speech and

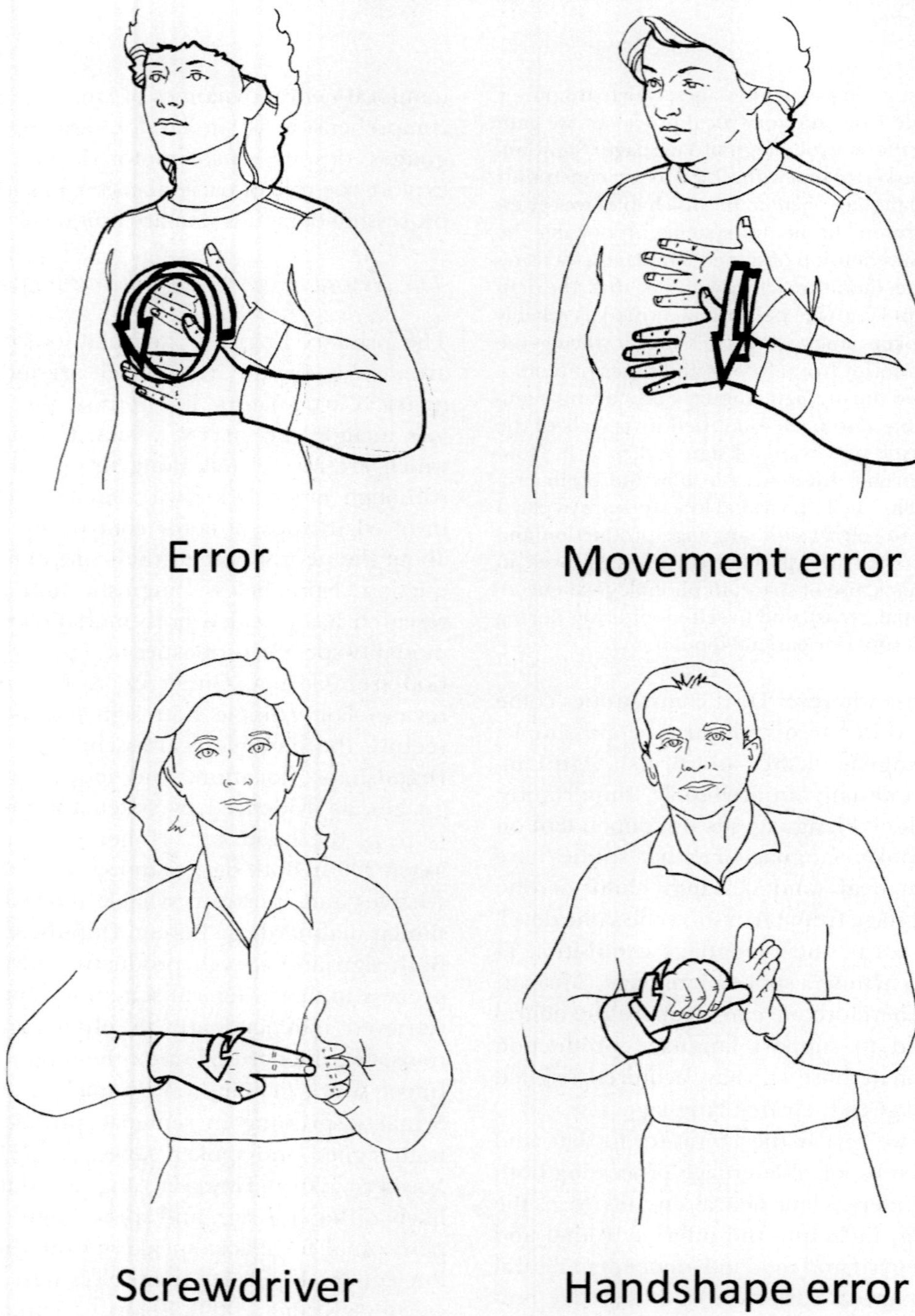

FIGURE 73.1 Examples of phonological paraphasias in ASL created by movement or hand-shape substitutions. Illustrations copyright Ursula Bellugi, Salk Institute for Biological Studies.

sign production in neurotypical adults (Gutierrez-Sigut et al., 2015; Gutierrez-Sigut, Payne, & MacSweeney, 2016). fTCD is a noninvasive technique that measures changes in blood flow velocity within the middle cerebral arteries. Hearing participants who were bilingual in English and British Sign Language (BSL) exhibited stronger left lateralization for sign than speech production when performing verbal fluency tasks (Gutierrez-Sigut et al., 2015). A control experiment with sign-naïve participants indicated that the difference in laterality was not driven by greater motoric demands for manual articulation. Native deaf signers also exhibited stronger left lateralization for both covert and overt sign production in comparison to hearing bilinguals producing speech (Gutierrez-Sigut, Payne, & MacSweeney, 2016). The authors speculate that the increased left lateralization for signing may be due to modality-specific properties of sign production, such as the increased use of proprioceptive self-monitoring mechanisms or the nature of phonological encoding of signs (see below).

Within the left hemisphere, the inferior frontal gyrus (IFG) has been implicated as a key region involved in both sign and speech production. In a positron emission tomography (PET) study, Braun, Guillemin, Hosey, and Vargus (2001) asked hearing ASL-English bilinguals to produce spontaneous narratives in either speech or sign language, and a conjunction analysis that subtracted out oral and manual motor movements revealed a common activation in the left frontal operculum (BA 45, 47) for both languages. Similarly, Emmorey, Mehta, and Grabowski (2007) found that the left IFG (BA 45) was equally engaged for word and sign production when deaf signers and hearing speakers performed a picture-naming task. Horwitz et al. (2003) used probabilistic cytoarchitectonic maps of BA 45 and BA 44 along with the PET data from Braun et al. (2001) to show that BA 45 was involved in higher-level linguistic processes, while BA 44 (and not BA 45) was engaged in the generation of complex oral and manual movements. Consistent with this finding, cortical stimulation of BA 44 during picture naming and sign/pseudosign repetition by a deaf signer resulted in motor execution errors (e.g., lax or imprecise articulation), rather than phonological errors (e.g., handshape substitution; Corina et al., 1999).

Evidence that the left IFG (BA 45, 47) is involved in lexical-semantic processes during sign production comes from PET studies in which signers generated verbs in response to videos of noun signs (Corina et al., 2003; Petitto et al., 2000) or videos of transitive actions (San José-Robertson, Corina, Ackerman, Guillemin, & Braun, 2004). Greater activation was observed in the left IFG for verb generation compared to the passive viewing of nouns or of action videos, regardless of whether the verbs were articulated with the right or left hand (Corina et al., 2003). Thus, engagement of the left IFG during verb generation is not driven by motoric factors related to the use of the dominant right hand in signing. Studies of verb generation in spoken languages have indicated that the left IFG is involved in lexical selection or the strategic control of semantic processing (e.g., Thompson-Schill, D'Esposito, Aguirre, & Farah, 1997).

With respect to higher-level processes involved in language production, a recent MEG study by Blanco-Elorrieta, Kastner, Emmorey, and Pylkännen (2018) investigated whether the same neurobiology underlies the online construction of complex linguistic structures in sign and speech. Two-word compositional phrases and two-word noncompositional "lists" were elicited from signers and speakers using identical pictures. In one condition, participants combined an adjective and a noun to describe the color of the object in the picture (e.g., *white lamp*) and in the control condition, participants named the color of the picture background and then the object (e.g., *white, lamp*). For both signers and speakers, phrase building engaged left anterior temporal and ventromedial cortices, with similar timing. The left anterior temporal lobe may be involved in computing the intersection of semantic features (Poortman & Pylkännen, 2016), while the ventromedial prefrontal cortex may be more specifically involved in constructing combinatorial plans (Pylkkänen, Bemis, & Elorrieta, 2014). Overall, this work indicates that the same frontotemporal network achieves the planning of structured linguistic expressions for both signed and spoken languages.

Modality-specific cortical regions involved in sign language production The supramarginal gyrus (SMG) has been found to be significantly more engaged during sign than word production when deaf signers are compared to hearing speakers (Emmorey, Mehta, & Grabowski, 2007) and when sign and speech production are directly compared within hearing bimodal bilinguals (Braun et al., 2001; Emmorey, McCullough, Mehta, & Grabowski, 2014). The study by Emmorey, Mehta, McCullough, and Grabowski (2016) also implicated the SMG as a key region for sign production. This study elicited the following sign types: one-handed signs (articulated in "neutral" space in front of the signer), two-handed (neutral space) signs, and one-handed body-anchored signs (produced with contact on or near the body). A conjunction analysis comparing each sign type with a baseline task revealed common activation in the SMG bilaterally (greater involvement on the left) for all sign types. Importantly, Corina et al. (1999) found that stimulation to the left SMG resulted in phonological substitutions, rather than motor execution errors. Further, bilateral SMG activation (larger on the left) has been found during the covert rehearsal of pseudosigns but not during the covert rehearsal of pseudowords (Buchsbaum et al., 2005). In addition, Cardin et al. (2016) recently found that linguistic knowledge modulated activation within the SMG in a phonological monitoring task (detecting target handshapes or locations). Specifically, the contrast between illegal nonsigns and real signs was significantly larger for deaf signers than for nonsigners (with increased SMG activation for nonsigns that violated phonological rules in both BSL and Swedish Sign Language). Together, these results suggest that the SMG is likely to be critically involved in the phonological decoding and encoding for sign language.

Emmorey and colleagues also reported that the superior parietal lobule (SPL) was significantly more active during sign than word production (Emmorey, Mehta, & Grabowski, 2007; Emmorey et al., 2014). These authors hypothesized that the SPL may be involved in self-monitoring overt sign output via proprioceptive feedback. Results from Emmorey et al. (2016) provide some support for this hypothesis: the production of body-anchored signs resulted in greater activation in the SPL compared to signs produced in neutral space. Greater engagement of the SPL may reflect the motor control and somatosensory monitoring required to direct the hand toward a specific location on the face or body. It is important to note that signing is not visually guided—signers do not look at their hands when they sign, and visual feedback does not appear to be used to fine-tune sign articulation (Emmorey, Bosworth, & Kraljic, 2009). Thus, the self-monitoring of sign articulation is likely to rely heavily on proprioceptive feedback. The SPL is known to play a role in updating postural representations of the arm and hand when movements are not visually guided (e.g., Parkinson, Condon, & Jackson, 2010). A recent transcranial magnetic stimulation (TMS) study by Vinson et al. (2019) has also implicated the SPL in sign production. While signers named pictures, TMS was administered to the left SPL or a control site. TMS to the SPL had a very specific effect: an increased rate of phonological substitution errors for two-handed signs that required hand contact. However, TMS did not slow or otherwise impair performance. Thus, TMS decreased the likelihood of detecting or correcting phonological errors during otherwise successful bimanual coordination. Interestingly, overt articulation is not required to engage the SPL for sign language production. MacSweeney et al. (2008) reported greater left SPL activation, extending into the superior portion of the SMG, when deaf signers made phonological judgments about the sign names of pictures (Were they produced at the same location?) than in hearing speakers making a phonological decision about words (Do they rhyme?). Although these regions appear to be more involved for signed than spoken language processing, a conjunction analysis by MacSweeney et al. (2008) showed that form-based judgments about both languages recruited the left SPL (extending into the SMG) to a significant degree. This result suggests that regions within parietal cortex may also be involved in phonological processes that are supramodal. The inferior parietal lobule has been implicated in phonological processing during reading and as a component of phonological working memory for speech. Supramodal processes that might be subserved by parietal cortex include sublexical sequencing or assembly processes that are independent of the modality of the to-be-combined phonological units. However, further research is needed to establish the nature and location of shared language-production processes within parietal cortex.

The Neurobiology of Sign Language Comprehension

Although we most often see people when we speak to them—that is, we perceive audiovisual speech—audition is key to speech perception. In contrast, signed languages must be perceived through the visual modality alone. Despite these differences in the modality of perceiving signed and spoken languages, the shared goal is comprehension. As with production, numerous psycholinguistic studies have shown extensive similarities between sign and speech comprehension processes. For example, studies have found evidence for categorical perception (Palmer, Fais, Golinkoff, & Werker, 2012), phonological and semantic priming (Meade, Lee, Midgley, Holcomb, & Emmorey, 2018), Stroop effects (Dupuis & Berent, 2015), incremental processing (Lieberman, Borovsky, & Mayberry, 2018), and many other parallels between the processes involved in comprehending signed and spoken languages (see Emmorey, 2002 for review). Below we explore the evidence for shared functional neural substrates for sign and speech comprehension, as well as the evidence for neural substrates that are specific to sign comprehension.

Modality-independent cortical regions involved in language comprehension As in spoken language users, damage to the left posterior superior temporal cortices and inferior parietal cortices typically leads to problems with sign language comprehension (e.g., Hickok, Love-Geffen, & Klima, 2002; Marshall, Atkinson, Woll, & Thacker,

2005). Neuroimaging studies also indicate a critical role for the left hemisphere during sign language comprehension. The first fMRI study to contrast audiovisual speech perception by hearing speakers with sign language perception in deaf signers used a conjunction analysis to identify regions common to both language modalities (MacSweeney et al., 2002). A primarily left frontotemporal network involving the superior temporal gyrus and sulcus as well as the left inferior frontal gyrus, extending into the prefrontal gyrus, was identified to be involved in processing both sign language and speech (see also Sakai, Tatsuno, Suzuki, Kimura, & Ichida, 2005). Numerous studies of sign language comprehension have also identified a primarily left lateralized frontotemporal network involved in sign language perception when contrasted with nonlinguistic hand movements (MacSweeney et al., 2004), gestures (Newman, Supalla, Fernandez, Newport, & Bavelier, 2015), or transitive actions (Corina et al., 2007). Similarities in subcortical structures supporting sign and speech processing have also been reported (Moreno, Limousin, Dehaene, & Pallier, 2018). Newman, Supalla, Hauser, Newport, and Bavelier (2010a) also demonstrated the recruitment of a predominantly left lateralized network, the components of which were modulated depending on whether the ASL sentences being viewed included inflectional morphology or word order alone to convey grammatical information. Together, these fMRI studies suggest that the classic left-lateralized perisylvian network is resilient to change in the sensory modality of language.

Event-related potential (ERP) studies further suggest that the timing of processing within this network is very similar across sign and speech comprehension. For example, a similar modulation of the N400 is observed for semantic anomalies in signed sentences as in spoken sentences (e.g., Hanel-Faulhaber et al., 2014). Although there is clear evidence for a predominantly left-lateralized network recruited for sign language comprehension, the right hemisphere also plays a supporting role—just as for spoken language processing (e.g., MacSweeney et al., 2002). Newman, Supalla, Hauser, Newport, and Bavelier (2010b) investigated the role of the right hemisphere in sign language comprehension by manipulating the narrative content of ASL sentences. They reported increased activation of the right inferior frontal gyrus and superior temporal cortex in deaf signers watching ASL sentences containing narrative devices, such as affective prosody and role shift, compared to sentences that did not contain these devices. Moreover, these regions included those recruited when hearing people perceive spoken-language sentences that include these narrative features.

Modality-specific cortical regions involved in sign language comprehension Although the overlap between the networks supporting sign and speech processing is extensive, there are some differences. Not surprisingly, direct contrasts have highlighted differences reflecting early sensory processing. Signed languages elicit greater activation than audiovisual speech in biological motion-processing regions of the posterior middle temporal gyri, bilaterally. In contrast, audiovisual speech perception in hearing participants elicits greater activation than sign language perception in deaf participants in auditory-processing regions in the superior temporal cortices (Emmorey et al., 2014; MacSweeney et al., 2002).

It is important to note, however, that although these studies show greater activation in the auditory cortices of hearing people perceiving speech than in deaf people perceiving sign language, these regions *do* respond to visual input in deaf people. This issue of crossmodal plasticity of the auditory cortices in deaf people and the extent to which these regions are involved in sign language comprehension have been topics of much recent research interest. There is mixed evidence regarding whether sign language, or any other visual stimuli, activates the primary auditory cortices in those born deaf (see Cardin et al., 2016; Scott, Karns, Dow, Stevens, & Neville, 2014). However, there are now numerous reports of increased activation in secondary auditory and auditory association cortices in superior temporal cortex (STC) in deaf compared to hearing individuals during sign language perception. This is even the case when deaf native signers are compared to hearing native signers, and sign language experience is therefore similar across groups (Capek et al., 2010; MacSweeney et al., 2004; Twomey et al., 2017).

Sign Language Makes Special Use of Space

As outlined above, the left parietal lobe appears to be particularly involved in sign language production, especially during phonological processing and self-monitoring. In addition, the left parietal lobe appears to be recruited by sign languages when spatial-processing demands are increased.

The use of space for linguistic purposes (e.g., coreference, spatial language) is unique to sign languages. In particular, signers use *classifier constructions* to express spatial relationships, in contrast to speakers, who typically use spatial prepositions or locative affixes. The handshape within a classifier construction is a morpheme that encodes information about the referent object (e.g., its semantic category or size and shape) while the placement and movement of the hands in signing space depict the location and movement of the referent objects. Lesion studies indicate that right hemisphere damage can cause difficulties in both producing and comprehending classifier constructions, but it does not result in sign language aphasia (Atkinson, Marshall, Woll, & Thacker, 2005; Hickok, Pickell, Klima, & Bellugi, 2009). Using a picture-description task and PET imaging, Emmorey, McCullough, Mehta, Ponto, and Grabowski (2013) found that the production of lexical signs and classifier handshape morphemes engaged left inferior frontal and temporal cortices, while the expression of gradient locations and movements engaged the bilateral SPL (extending into the SMG). Emmorey et al. (2013) argued that to express spatial information, signers must transform visual-spatial representations into a body-centered reference frame and reach toward target locations in signing space.

With regard to comprehension, Capek et al. (2009) highlighted the special role of spatial processing in sign language syntax. Using ERPs, they found that syntactic violations in ASL elicited early frontal negativities that varied as a function of how space was used to create the violation. MacSweeney et al. (2004) reported greater activation in the left SMG and SPL when deaf signers viewed BSL sentences that involved classifier constructions than when they viewed sentences that did not (see also Jednorog et al., 2015). McCullough, Saygin, Korpics, and Emmorey (2012) explored this finding further and demonstrated that the left SPL and SMG were particularly engaged during comprehension sentences containing classifier constructions that expressed spatial relations between referents, rather than movement of the referent. Emmorey et al. (2013) also found that the left intraparietal sulcus was more engaged when classifier constructions expressed object location rather than object movement. Sign language processing requires attention to the location and configuration of the hands in space and is likely to explain the enhanced involvement of these regions. The semantic focus on these features when producing and comprehending classifier constructions is likely to increase these processing demands further.

Conclusion

Despite great differences in their surface forms, both signed and spoken language-processing in native users engage very similar, predominantly left-lateralized, networks. This is an important conclusion that should be taken into account in theories of hemispheric specialization for language processing. Some have argued that the left hemisphere shows a predisposition to process certain temporal aspects of auditory information that are critical to speech processing (see McGettigan & Scott, 2012 for discussion). The inference is then made, explicitly or implicitly, that this is the cause of left-hemisphere lateralization for language processing. That signed languages are also predominantly processed in the left hemisphere poses a problem for any purely auditory-based account of language lateralization. It is possible that sign languages recruit the neural infrastructure already established for spoken languages. This proposal is in line with the neuronal recycling hypothesis proposed by Dehaene and Cohen (2007) to account for the preference of the ventral occipitotemporal cortex to process written words. However, we suggest that a recycling hypothesis is unlikely to account for the left lateralization of sign languages. If the left perisylvian cortices are "specialized" for speech, then the use of these regions for sign language processing should come at a cost. That is, native learners of sign languages should show delays/deficits compared to native learners of a spoken language, but this is not the case (Meier & Newport, 1990). Although the research to date with signed languages does not allow us to answer *why* language is predominantly left lateralized in most people, it should prompt the field to generate hypotheses that are modality-independent and can account for the left-hemisphere lateralization of both sign language and speech.

Observing such striking similarities in the neural systems recruited for sign and speech processing has led the field to assume that the same processes are being carried out in these regions for both language types, using similar representations (e.g., MacSweeney et al., 2008). However, this is an assumption based on null findings of no significant differences in activation between languages. Multivoxel pattern analysis (MVPA) has been used in a number of domains to examine *patterns* of activation rather than the overall level of activation. This approach has the potential to identify common neural representations for different modes, inputs, or states. These approaches will also allow us to directly test hypotheses about the similarity of processing and the similarity of representations. Pursuing questions about the computations that occur and the representations used in the regions identified as showing overlap between sign and speech processing is likely to produce novel insights into the neurobiology of language. So, too, is pursuing the small but interesting differences that have to date been identified in the neural systems supporting sign and speech processing. The left inferior and superior parietal lobules, especially, appear to be more involved in sign comprehension, production, memory, and metalinguistic processes compared to spoken language. In sum, the study of sign languages will continue to offer unique insights into the neuroplasticity of the language networks and representations in the brain.

Acknowledgments

We would like to acknowledge the Deaf communities involved in our research for their support. We also thank the following funding agencies that have supported our work: Mairéad MacSweeney, Wellcome Trust (100229/Z/12/Z), Economic and Social Research Council (RES-620-28-0002); Karen Emmorey, National Institutes of Health (R01 DC010997).

REFERENCES

Atkinson, J., Marshall, J., Woll, B., & Thacker, A. (2005). Testing comprehension abilities in users of British Sign Language following CVA. *Brain and Language, 94*(2), 233–248.

Berent, I., Dupuis, A., & Brentari, D. (2013). Amodal aspects of linguistic design. *PLoS One, 8*(4), e60617.

Blanco-Elorrieta, E., Kastner, I., Emmorey, K., & Pylkkänen, L. (2018). Shared neural correlates for building phrases in signed and spoken language. *Scientific Reports, 8*, 5492. doi:10.1038/s41598-018-23915-0

Braun, A. R., Guillemin, A., Hosey, L., & Varga, M. (2001). The neural organization of discourse: An $H_2{}^{15}O$-PET study of narrative production in English and American Sign Language. *Brain, 124*(10), 2028–2044.

Buchsbaum, B., Pickell, B., Love, T., Hatrak, M., Bellugi, U., & Hickok, G. (2005). Neural substrates for verbal working memory in deaf signers: fMRI study and lesion case report. *Brain and Language, 95*(2), 265–272.

Capek, C. M., Grossi, G., Newman, A. J., McBurney, S. L., Corina, D., Roeder, B., & Neville, H. J. (2009). Brain systems mediating semantic and syntactic processing in deaf native signers: Biological invariance and modality specificity. *Proceedings of the National Academy of Sciences, 106*(21), 8784–8789.

Capek, C. M., Woll, B., MacSweeney, M., Waters, D., McGuire, P. K., David, A. S., Brammer, M. J., & Campbell, R. (2010). Superior temporal activation as a function of linguistic knowledge: Insights from deaf native signers who speechread. *Brain and Language, 112*(2), 129–134.

Cardin, V., Orfanidou, E., Kästner, L., Rönnberg, J., Woll, B., Capek, C. M., & Rudner, M. (2016). Monitoring different phonological parameters of sign language engages the same cortical language network but distinctive perceptual ones. *Journal of Cognitive Neuroscience, 28*(1), 20–40.

Cardin, V., Orfanidou, E., Ronnberg, J., Capek, C. M., Rudner, M., & Woll, B. (2013). Dissociating cognitive and sensory neural plasticity in human superior temporal cortex. *Nature Communications, 4*, 1473.

Cardin, V., Smittenaar, R. C., Orfanidou, E., Ronnberg, J., Capek, C. M., Rudner, M., & Woll, B. (2016). Differential activity in Heschl's gyrus between deaf and hearing individuals is due to auditory deprivation rather than language modality. *NeuroImage, 124*, 96–106.

Corina, D. P., Chiu, Y. S., Knapp, H., Greenwald, R., San Jose-Robertson, L., & Braun, A. (2007). Neural correlates of human action observation in hearing and deaf subjects. *Brain Research, 1152*(1), 111–129.

Corina, D. P., Gutierrez, E., & Grosvald, M. (2014). Sign language production: An overview. In M. Goldrick, V. Ferreira, & M. Miozzo (Eds.), *The Oxford handbook of language production* (pp. 393–416). Oxford: Oxford University Press.

Corina, D. P., McBurney, S. L., Dodrill, C., Hinshaw, K., Brinkley, J., & Ojemann, G. (1999). Functional roles of Broca's area and SMG: Evidence from cortical stimulation mapping in a deaf signer. *Neuroimage, 10*(5), 570–581.

Corina, D. P., San José-Robertson, L., Guillemin, A., High, J., & Braun, A. R. (2003). Language lateralization in a bimanual language. *Journal of Cognitive Neuroscience, 15*(5), 718–730.

Dehaene, S., & Cohen, L. (2007). Cultural recycling of cortical maps. *Neuron, 56*(2), 384–398.

Ding, H., Qin, W., Liang, M., Ming, D., Wan, B., Li, Q., & Yu, C. (2015). Cross-modal activation of auditory regions during visuo-spatial working memory in early deafness. *Brain, 138*(9), 2750–2765.

Dupuis, A., & Berent, I. (2015). Signs are symbols: Evidence from the Stroop task. *Language, Cognition and Neuroscience, 30*(10), 1339–1344.

Emmorey, K. (2002). *Language, cognition, and the brain: Insights from sign language research.* Mahwah, NJ: Lawrence Erlbaum.

Emmorey, K., Bosworth, R., & Kraljic, T. (2009). Visual feedback and self-monitoring of sign language. *Journal of Memory and Language, 61*, 398–411.

Emmorey, K., & Corina, D. P. (1990). Lexical recognition in sign language: Effects of phonetic structure and morphology. *Perceptual and Motor Skills, 71*, 1227–1252.

Emmorey, K., McCullough, S., Mehta, S., & Grabowski, T. J. (2014). How sensory-motor systems impact the neural organization for language: Direct contrasts between spoken and signed language. *Frontiers in Psychology, 5*(484). doi:10.3389/fpsyg.2014.00484

Emmorey, K., McCullough, S., Mehta, S., Ponto, L. L., & Grabowski, T. J. (2013). The biology of linguistic expression impacts neural correlates for spatial language. *Journal of Cognitive Neuroscience, 25*(4), 517–533.

Emmorey, K., Mehta, S., & Grabowski, T. J. (2007). The neural correlates of sign and word production. *NeuroImage, 36*, 202–208.

Emmorey, K., Mehta, S., McCullough, S., & Grabowski, T.G. (2016). The neural circuits recruited for the production of signs and fingerspelled words. *Brain and Language, 160*, 30–41. doi.org/10.1016/j.bandl.2016.07.003

Gutierrez-Sigut, E., Daws, R., Payne, H., Blott, J., Marshall, C., & MacSweeney, M. (2015). Language lateralization of hearing native signers: A functional transcranial Doppler sonography (fTCD) study of speech and sign production. *Brain and Language, 151*, 23–34.

Gutierrez-Sigut, E., Payne, H., & MacSweeney, M. (2016). Examining the contribution of motor movement and language dominance to increased left lateralization during sign generation in native signers. *Brain and Language, 159*, 109–117.

Hall, M. L., Ferreira, V. S., & Mayberry, R. I. (2015). Syntactic priming in American Sign Language. *PloS One, 10*(3), e0119611.

Hänel-Faulhaber, B., Skotara, N., Kügow, M., Salden, U., Bottari, D, & Röder, B. (2014). ERP correlates of German Sign Language processing in deaf native signers. *BMC Neuroscience, 10*(15), 62.

Hickok, G., Bellugi, U., & Klima, E. S. (1996). The neurobiology of sign language and its implications for the neural basis of language. *Nature, 381*(6584), 699.

Hickok, G., Love-Geffen, T., & Klima, E. S. (2002). Role of the left hemisphere in sign language comprehension. *Brain and Language, 82*(2), 167–178.

Hickok, G., Pickell, H., Klima, E., & Bellugi, U. (2009). Neural dissociation in the production of lexical versus classifier signs in ASL: Distinct patterns of hemispheric asymmetry. *Neuropsychologia, 47*(2), 382–387.

Hohenberger, A., Happ, D., & Leuninger, H. (2002). Modality-dependent aspects of sign language production: Evidence from slips of the hands and their repairs in German sign language. In R. Meier, K. Cormier, & D. Quinto-Pozos (Eds.), *Modality and structure in signed and spoken languages* (pp. 112–142). Cambridge: Cambridge University Press.

Horwitz, B., Amunts, K., Bhattacharyya, R., Patkin, D., Jeffries, K., Zilles, K., & Braun, A. R. (2003). Activation of Broca's area during the production of spoken and signed language: A combined cytoarchitectonic mapping and PET analysis. *Neuropsychologia, 41*(14), 1868–1876.

Jednoróg, K., Bola, Ł., Mostowski, P., Szwed, M., Boguszewski, P. M., Marchewka, A., & Rutkowski, P. (2015). Three-dimensional grammar in the brain: Dissociating the neural correlates of natural sign language and manually coded spoken language. *Neuropsychologia, 71,* 191–200.

Lieberman, A. M., Borovsky, A., & Mayberry, R. I. (2018). Prediction in a visual language: Real-time sentence processing in American Sign Language across development. *Language, Cognition and Neuroscience, 33*(4), 387–401.

MacSweeney, M., Campbell, R., Woll, B., Giampietro, V., David, A. S., McGuire, P. K., Calvert, G. A., & Brammer, M. J. (2004). Dissociating linguistic and nonlinguistic gestural communication in the brain. *NeuroImage, 22*(4), 1605–1618.

MacSweeney, M., & Cardin, V. (2015). What is the function of auditory cortex without auditory input? *Brain, 138*(Pt. 9), 2468–2470.

MacSweeney, M., Waters, D., Brammer, M. J., Woll, B., & Goswami, U. (2008). Phonological processing in deaf signers and the impact of age of first language acquisition. *Neuroimage, 40*(3), 1369–1379.

MacSweeney, M., Woll, B., Campbell, R., Calvert, G., McGuire, P., David, A., Simmons, A., & Brammer, M. (2002). Neural correlates of British Sign Language comprehension: Spatial processing demands of topographic language. *Journal of Cognitive Neuroscience, 14*(7), 1064–1075.

MacSweeney, M., Woll, B., Campbell, R., McGuire, P. K., David, A. S., Williams, S. C. R., Suckling, J., Calvert, G. A., & Brammer, M. J. (2002). Neural systems underlying British Sign Language and audio-visual English processing in native users. *Brain, 125*(7), 1583–1593.

Marshall, J., Atkinson, J., Woll, B., & Thacker, A. (2005). Aphasia in a bilingual user of British Sign Language and English: Effects of cross-linguistic cues. *Cognitive Neuropsychology, 22*(6), 719–736.

McCullough, S., Saygin, A. P., Korpics, F., & Emmorey, K. (2012). Motion-sensitive cortex and motion semantics in American Sign Language. *NeuroImage, 63,* 111–118.

McGettigan, C., & Scott, S. K. (2012). Cortical asymmetries in speech perception: What's wrong, what's right and what's left? *Trends in Cognitive Science, 16*(5), 269–276.

Meade, G., Lee, B., Midgley, K. J., Holcomb, P. J., & Emmorey, K. (2018). Phonological and semantic priming in American Sign Language: N300 and N400 effects. *Language, Cognition and Neuroscience, 33*(9), 1092–1106. doi.org/10.1080/23273798.2018.1446543

Meier, R. P., & Newport, E. L. (1990). Out of the hands of babes: On a possible sign advantage in language acquisition. *Language, 66*(1), 1–23.

Moreno, A., Limousin, F., Dehaene, S., & Pallier, C. (2018). Brain correlates of constituent structure in sign language comprehension. *Neuroimage, 15*(167), 151–161.

Newman, A. J., Supalla, T., Fernandez, N., Newport, E. L., & Bavelier, D. (2015). Neural systems supporting linguistic structure, linguistic experience, and symbolic communication in sign language and gesture. *Proceedings of the National Academy of Sciences of the United States of America, 112*(37), 11684–11689.

Newman, A. J., Supalla, T., Hauser, P., Newport, E., & Bavelier, D. (2010a). Dissociating neural subsystems for grammar by contrasting word order and inflection. *Proceedings of the National Academy of Sciences of the United States of America, 107*(16), 7539–7544.

Newman, A. J., Supalla, T., Hauser, P., Newport, E., & Bavelier, D. (2010b) Processing narrative content in American Sign Language: An fMRI study. *NeuroImage, 52*(2), 669–676.

Palmer, S. B., Fais, L., Golinkoff, R. M., & Werker, J. F. (2012). Perceptual narrowing of linguistic sign occurs in the 1st year of life. *Child Development, 83*(2), 543–553.

Parkinson, A., Condon, L., & Jackson, S. R. (2010). Parietal cortex coding of limb posture: In search of the body-schema. *Neuropsychologia, 48*(11), 3228–3234.

Petitto, L. A., Zatorre, R. J., Gauna, K., Nikelski, E. J., Dostie, D., & Evans, A. C. (2000). Speech-like cerebral activity in profoundly deaf people processing signed languages: Implications for the neural basis of human language. *Proceedings of the National Academy of Sciences, 97*(25), 13961–13966.

Poortman, E. B., & Pylkkänen, L. (2016). Adjective conjunction as a window into the LATL's contribution to conceptual combination. *Brain and Language, 160,* 50–60.

Pylkkänen, L., Bemis, D. K., & Elorrieta, E. B. (2014). Building phrases in language production: An MEG study of simple composition. *Cognition, 133*(2), 371–384.

Sakai, K. L., Tatsuno, Y., Suzuki, K., Kimura, H., & Ichida, Y. (2005). Sign and speech: Amodal commonality in left hemisphere dominance for comprehension of sentences. *Brain, 128*(6), 1407–1417.

San José-Robertson, L., Corina, D. P., Ackerman, D., Guillemin, A., & Braun, A. R. (2004). Neural systems for sign language production: Mechanisms supporting lexical selection, phonological encoding, and articulation. *Human Brain Mapping, 23*(3), 156–167.

Saygin, A., McCullough, S., Alac, M., & Emmorey, K. (2010). Modulation of BOLD response in motion sensitive lateral temporal cortex by real and fictive motion sentences. *Journal of Cognitive Neuroscience, 22*(11), 2480–2490.

Scott, G. D., Karns, C. M., Dow, M. W., Stevens, C., & Neville, H. J. (2014). Enhanced peripheral visual processing in congenitally deaf humans is supported by multiple brain regions, including primary auditory cortex. *Frontiers in Human Neuroscience, 8,* 17.

Thompson, R., Emmorey, K., & Gollan, T. (2005). Tip-of-the-fingers experiences by ASL signers: Insights into the organization of a sign-based lexicon. *Psychological Science, 16*(11), 856–860.

Thompson-Schill, S. L., D'Esposito, M., Aguirre, G. K., & Farah, M. J. (1997). Role of left inferior prefrontal cortex in retrieval of semantic knowledge: A reevaluation. *Proceedings of the National Academy of Sciences, 94*(26), 14792–14797.

Twomey, T., Waters, D., Price, C. J., Evans, S., & MacSweeney, M. (2017). How auditory experience differentially influences the function of left and right superior temporal cortices. *Journal of Neuroscience, 37*(39), 9564–9673.

Vinson, D., Fox, N., Devlin, J. T., Vigliocco, G., & Emmorey, K. (2019). Transcranial magnetic stimulation during British Sign Language production reveals monitoring of discrete linguistic units in left superior parietal lobule. *BioRxiv.* https://doi.org/10.1101/679340

74 The Neurobiology of Syntactic and Semantic Structure Building

LIINA PYLKKÄNEN AND JONATHAN R. BRENNAN

ABSTRACT Language is a combinatory system able to create an infinite array of complex meanings from memory representations in our mental dictionary. How is this integrative process neurally implemented? Studies on the processing of structured versus unstructured language stimuli have identified a largely left-lateral *combinatory network*, with the anterior temporal lobe as its most consistent integrative node, likely contributing the first stage of a multistage combinatory process. This chapter summarizes our current understanding of the neurobiology of composition, with a focus on three paradigms from extant literature that most directly address the basic process of combining words into phrases and sentences: the so-called sentence-versus-list paradigm, the two-word-phrase paradigm, and approaches using model comparison with natural narratives as stimuli.

What Is Composition? Many Correlated Computations Executed According to Our Grammatical Knowledge

While the retrieval of stored representations is a process shared by many domains of cognition—this is how you distinguish a dog barking from a baby crying or recognize your favorite hat—only in language do the memory representations of elementary building blocks compose into infinite meaningful configurations along an intricate rule system tacitly living in our brains, the grammar. The composition of structured meanings is the essence of language, the source of its expressive power. What do we know about the brain basis of this remarkable ability?

When approaching the neuroscience of composition, one is immediately faced with a principle challenge: the composition of words into complex messages is achieved by a cascade of tightly correlated and possibly simultaneous computations. Thus, understanding this process requires ways to unpack the constituent processes. For example, the comprehension of *cat* in the context *someone fed the* ... will elicit activity reflecting the syntactic combination of *cat* with the article *the* to form a noun phrase, the processes by which this noun phrase completes both the verb phrase *fed the cat*, and the processes by which the verb phrase completes the whole sentence. But of course, the sole purpose of building these syntactic structures is to determine how the words semantically combine with each other. Thus, each step of syntactic structure building is also paired with a potentially multifaceted process of meaning composition. Further, these processes are interdependent: the word *fed* both contributes to building a verb phrase and also allows one to predict semantically compatible words to complete the sentence—namely, things that can eat. Clearly then, careful experimental design resting on a solid theoretical foundation is paramount for the neuroscience of composition. For each possible neural correlate of composition, we must ask: Which of the tightly correlated computations can be ruled out as a functional hypothesis for this particular neural activity? This type of workflow is illustrated in figure 74.1, which first shows a broader network implicated for some type of composition-relevant operations, dubbed the *combinatory network*, and then depicts the extent to which the various network nodes are engaged for more specific computations. In this spirit, our chapter will review the extent to which the brain basis of composition is understood within the current neurobiology of language, with a focus on the ability of extant results to rule out specific functional hypotheses about the activity associated with composition.

In addition to multiple correlated computations, the neural signals elicited by composition also reflect online structure-building computations that conform to the *grammar of some particular language*. Thus, to correctly model this process, we would need not only the right description of the online computational algorithm—that is, exactly the steps by which structure gets built—but also the right description of the grammar—that is, a correct model of the representations that get built. Since both of these are in a sense the "end questions" of entire fields within linguistics, we can never assume that our answers to them are correct; our answers will always be works in progress. In practice, our understanding has progressed along two fronts: (1) by examining the effects

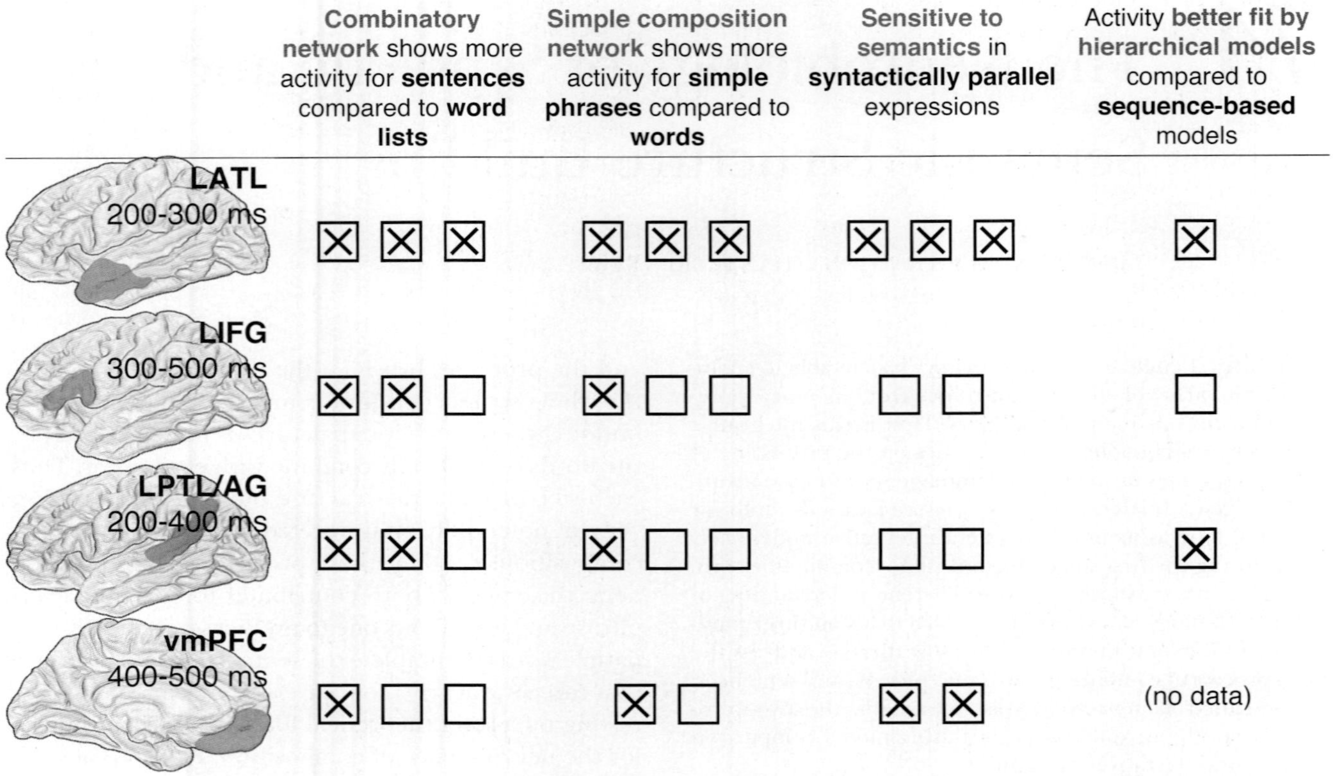

FIGURE 74.1 An informal depiction of our current understanding of the brain regions supporting composition and the extent to which the functional roles of individual network nodes are understood. A lack of understanding can result either from a lack of studies or from a lack of generalizations across studies. Here, the number of boxes in each cell represents the general quantity of studies addressing the role of the region, and the checks inside the boxes represent the amount of positive evidence for the generalization in the first row. Timing estimates primarily reflect results from MEG studies comparing sentence-versus-list activation (e.g., Brennan & Pylkkänen, 2012) or phrase-versus-word activation (e.g., Bemis & Pylkkänen, 2011). The table does not separate results according to method; thus, for example, positive results for the LIFG come primarily from fMRI (and are thus ambiguous as regards timing) and ones for the vmPFC from MEG. Connecting separate findings from different methods is a major goal for future research. In all, the only network node showing a high degree of consistency across the literature is the LATL. (See color plate 86.)

of contrasts that differ in composition under any theory and (2) by comparing the ways in which brain data correlate with different combinations of grammars and algorithms. These two approaches are discussed in the two sections below, respectively.

Composition in Controlled Experiments

Sentence versus list: extracting the combinatory network When stimuli that engage composition in language are contrasted with maximally similar ones that do not, what brain activity is affected? A large literature has addressed this question, yielding descriptions of what could be called the combinatory network. The most common experimental paradigm within this literature contrasts well-formed sentences with random unstructured lists of words. Starting with the classic findings of Mazoyer et al. (1993) and Stowe et al. (1998), the most replicated result from this contrast is that sentences elicit higher activation than lists in the temporal poles, often bilaterally (Humphries et al., 2006; Rogalsky & Hickok, 2009). Other than this finding, the results of these studies have been variable. This is not surprising, given that different studies have used different sentence materials, and thus, to the extent that different sentences may require different interpretive processes, the only consistency we can expect across all studies is activity reflecting processes that are shared across most sentences. However, broadly speaking, the sentence-versus-list literature has also identified the posterior temporal lobe (e.g., Friederici, Meyer, & von Cramon, 2000; Pallier, Devauchelle, & Dehaene, 2011; Snijders et al., 2009; Vandenberghe et al., 2002) and the inferior frontal gyrus (Pallier, Devauchelle, & Dehaene, 2011; Snijders et al., 2009) as relatively common loci of increased activity for sentence stimuli. Finally, some evidence also exists for the middle temporal gyrus

(Brennan & Pylkkänen, 2012; Pallier, Devauchelle, & Dehaene, 2011), the temporoparietal junction (Pallier et al., 2011), and medial parts of ventral prefrontal cortex (Brennan & Pylkkänen, 2012) as part of this combinatory network.

How do all these regions contribute to composing words into complex structures? At the outset, numerous hypotheses are capable of explaining each of the activations: they could reflect any aspect of syntax, semantics or referential processing. Our understanding of the specific contributions of most of these regions is still nascent, but in what follows we summarize the extent to which the hypothesis space has been narrowed down. In particular, we will focus on two questions: (1) Which of these activations survive if we simplify the stimulus to just one instance of composition and (2) to what extent can syntactic or semantic computations be ruled out for any of these activities?

Two-word phrases: identifying the minimal composition network for comprehension and production One way to rule out large classes of hypotheses for the sentence-versus-list results is to get rid of the sentences and ask what activity is elicited by the composition of the smallest possible structures. Results for such minimal combinations would not be reflective of any processes pertaining to the sentence level—such as the formation of long-distance dependencies (as in, e.g., relative clauses in which the object of a verb is expressed outside its canonical object position: *the ball that the dog ate*), agreement, resolution of coreference, and so forth—but rather would be promising correlates of building a single phrase. From such a result, one could scale up to assess what additional activity is recruited for the construction of larger structures.

This approach was taken in Bemis and Pylkkänen (2011), who created a type of two-word version of the sentence-versus-list paradigm to test what neural activity would be sensitive to the presence of just a single step of composition. In this paradigm, magnetoencephalography (MEG) activity is recorded while subjects comprehend pairs of words that either form a phrase or not, as well as single-word controls. The words are followed by pictures that either match or mismatch the verbal description; indicating the match or mismatch serves as a comprehension task. The initial studies used just color-object combinations (*white lamp, red boat,* and so on), which made it possible to run parallel production experiments on the same expressions (Pylkkänen, Bemis, & Blanco Elorrieta, 2014). In the production versions, the same pictures were used as a production prompt, with subjects naming the pictures "green cup" and so forth. MEG activity was analyzed in response to the second word (e.g., *cup* in *blue cup*) in the comprehension versions and in response to the picture in the production versions, where the temporal resolution of MEG allows one to capture the planning of speech production prior to the onset of motion artifacts. Across a series of studies, a stable pattern was observed in the left anterior temporal lobe (LATL) and the ventromedial prefrontal cortex (vmPFC), showing increased activity for phrases relative to noncombinatory controls both in comprehension and production. Unsurprisingly, the time courses of the effects were different between comprehension and production. In comprehension, the LATL increase elicited by composition was early (200–250 ms after the second word) and the vmPFC increase, late (~400 ms), whereas in production, both effects of phrasal structure had an onset that was early and relatively in parallel, starting around 200 ms.

Crosslinguistic generality for this pattern has been demonstrated for Arabic (Westerlund et al., 2015) and American Sign Language (Blanco-Elorrieta et al., 2018), and the LATL effects generalize to predicate-argument configurations as well, such as *eats meat* (Westerlund et al., 2015). This generality is compatible with the hypothesis that the LATL reflects syntactic aspects of composition, but as we discuss in the next section, subsequent research has compellingly ruled out this account.

A thorny question: syntax versus semantics Manipulations of the combinatory structure of sentences are most often described as manipulations of syntax. But in almost all cases, changing structure also changes the semantic combinatorics. What do we know about the processes by which syntactic structures get built, versus the combinatory steps that build complex meanings?

Isolating syntax with "semantics-free" stimuli: jabberwocky and artificial grammar studies Since semantic confounds make the pure study of syntax so difficult with natural language, much research on this question has given up on natural language as stimuli. Instead, many groups have chosen so-called jabberwocky sentences as *semantics-free* expressions (inspired by Lewis Carroll's "Jabberwocky" poem). In these stimuli, the grammatical elements of sentences are preserved, but the conceptual/lexical items are replaced with pseudowords, as in *the solims on a sonting grilloted a yome and a sovir* (Humphries et al., 2006). Although these expressions are standardly labeled as *minus SEMANTICS*, most researchers would acknowledge that these stimuli are not void of meaning—they have a rich relational structure, including tense, argument structure, anaphoric relationships, and so forth, even while lacking actual conceptual labels. Further, exactly what subjects do when comprehending such expressions is unclear: we do not have theoretical or psycholinguistic models that speak to this. Nevertheless, comparing such stimuli to "jabberwocky lists"—as in *rooned the sif inot lilf and the foig aurene to* (Humphries et al., 2006)—has been a common approach to the study of syntactic composition. However, given that what is missing from jabberwocky sentences is specifically the conceptual labels, as opposed to *all* semantics, the hypothesis-killing power of these stimuli is in fact rather narrow: increases for jabberwocky sentences must not reflect the composition of complex conceptual content (since the conceptual labels are missing). But, subjects' strategies for dealing with these unnatural stimuli could substantially complicate things.

Indeed, no consistent result emerges from this literature: while many classic studies on jabberwocky sentences found them to elicit increased activation in anterior temporal cortex (Friederici, Meyer, & von Cramon, 2000; Humphries et al., 2006; Mazoyer et al., 1993), more recent studies have found left ATL increases only in the presence of natural semantics (Matchin, Hammerly, & Lau, 2017; Pallier, Devauchelle, & Dehaene, 2011). We are not aware of aims to reconcile these findings—understanding this pattern would require a within-subjects elicitation of the difference to begin with and then careful hypothesizing about possible contrasts in the stimuli and tasks.

But, the jabberwocky literature has given rise to another candidate for purely structural processing: the pars opercularis (Brodmann Area [BA], 44). For example, increased BA 44 activity was found both for jabberwocky determiner phrases (such as *diese Flirk* in German; Zaccarella & Friederici, 2015) and for natural ones (*this ship* vs. *ship*; Schell et al., 2017), as compared to control stimuli.

However, under the hypothesis that these regions build syntactic structure, their insensitivity to the sentence-versus-list contrast and minimal composition manipulations in numerous studies remains a puzzle. As a possible solution, Zaccarella, Schell, and Friederici (2017) propose that the list conditions of many sentence-versus-list studies may in fact have elicited combinatory processing due to the mixing of content and function words in the lists (e.g., *her eyes during close the she ceremony*). In a meta-analysis, they show that in studies employing only content words or only function words in the lists, BA 44 increases for sentences are systematically elicited. This highlights the importance of experimental design and the challenges in choosing appropriate noncombinatory control conditions.

Posterior superior temporal regions have also emerged as possibly relevant in jabberwocky studies; for example, both Pallier, Devauchelle, and Dehaene (2011) and Matchin, Hammerly, and Lau (2017) found increased pSTS activity for sentences over lists, but while in the Pallier study this effect was shared by jabberwocky stimuli, in the Matchin one, it was not. One account links activity in this region with thematic role assignment (Frankland & Greene, 2015), but numerous specific functions are compatible with existing results. All this suggests that jabberwocky manipulations may not be a fruitful way to approach the syntax-versus-semantics question.

Another literature that has moved away from natural language in order to isolate syntactic processing has used artificial grammars containing no natural language words at all. This literature also points toward an involvement of BAs 44 and 45 in the processing of artificial grammars (Bahlmann, Schubotz, & Friederici, 2008; Friederici et al., 2006; Petersson & Hagoort, 2012; Uddén et al., 2008). Ultimately, viable hypotheses for any activity sensitive to the combinatory properties of language—and language-like artificial stimuli—must explain not only the positive results in the literature but also the negative ones. Given this, the hypothesis that regions within Broca's area perform hierarchical structure building is too general: these regions are often silent when syntactic structures are built (e.g., Bemis & Pylkkänen, 2011, 2012, 2013; Blanco-Elorrieta et al., 2018; Humphries et al., 2006; Mazoyer et al., 1993; Pylkkänen, Bemis, & Blanco Elorrieta, 2014; Stowe et al. 1998; Rogalsky & Hickok, 2009). If neural activity reflects structure building, it in principle should engage every time words are combined into larger structures. Currently, this behavior has not been demonstrated for either the posterior temporal or inferior frontal areas discussed above.

Possible oscillatory reflexes of syntactic phrase building One promising candidate that does show the type of generality required for a correlate of syntactic composition has been recently identified in the frequency domain: Ding et al. (2016) presented subjects with words, phrases, and sentences at predictable rates, with no acoustic cues to structure, and demonstrated that despite the monotonicity of the stimulus, cortical activity as measured by MEG peaked exactly at the frequencies at which the words, phrases, and sentences occurred in their design, specifically at 4, 2, and 1 Hz, respectively. Unpacking this initial result will require elaborate follow-up work that tests how these effects can be eliminated and whether they are semantically sensitive, but as things currently stand, the extant data are compatible with the hypothesis that the 2 Hz peak in some sense reflects syntactic structure building (though the result is also equally compatible with the hypothesis that it reflects semantic composition).

Evidence for a nonsyntactic role for the left anterior temporal lobe and ventromedial prefrontal cortex Given that composition is many correlated computations, each experiment on the brain basis of composition should have the potential to narrow down the hypothesis space in some informative way. The original minimal composition findings of Bemis and Pylkkänen (2011) revealed an early (~200 ms) increase in the LATL and a later one in the vmFPC (~400 ms) in response to a word that completes a two-word phrase. That study was informed by prior work that had showed the vmPFC to be sensitive to the semantic complexity of expressions even when their surface syntax was kept constant (reviewed in Pylkkänen, Brennan, & Bemis, 2011), suggesting that the vmPFC activity elicited by the small phrases most likely reflects semantic aspects of composition.

Since purely semantic manipulations of composition had not engaged the LATL, Bemis and Pylkkänen (2011) conjectured that the LATL activity elicited by adjective-noun combinations might reflect the syntactic side of composition. This hypothesis was compatible with sentence-versus-list findings and also with the deficit/lesion data available at the time (Dronkers et al., 2004). However, subsequent research has compellingly ruled out any purely structural interpretation of the early LATL activity: it does not show the generality required for syntactic composition. For example, although number phrases such as *two boats* require the application of syntactic composition, number phrases do not appear to engage the LATL (e.g., Blanco-Elorrieta & Pylkkänen, 2016). This finding eliminates any straightforward account of the LATL in terms of syntactic composition and instead points toward explanations that perhaps depend more on the conceptual content of the composing elements (cf., Baron & Osherson, 2011): while the modifier *blue* adds a feature to *boats*, the number term *two* enumerates the number of tokens in the boat set but adds no (obvious) conceptual content.

This general idea connects with the prior literature on semantic dementia, which had shown that left ATL atrophy affects the specificity of an individual's conceptual space, leading the person to lose specific conceptual labels such as *poodle* and to resort to more general ones, such as *dog* or *thing*. ATL atrophy has not been linked with deficits in phrase-structure processing (Wilson et al., 2014). This invites the hypothesis that perhaps LATL effects of composition actually reflect not composition but rather the increase in conceptual specificity created by the addition of an adjectival modifier. Subsequent studies have, however, shown that LATL amplitudes do not linearly increase as a function of conceptual specificity, blind to the single word versus phrase distinction. Instead, when lexical frequency is controlled, single-word specificity tends not to affect the LATL reliably, and composition is required to elicit an increase (Westerlund & Pylkkänen, 2014; Zhang & Pylkkänen, 2015). However, the size of the composition effect is modulated by conceptual specificity, with conceptually richer modifiers increasing the amplitudes of their head nouns more. Thus, LATL activity at approximately 200–250 ms appears to act as a specificity-driven conceptual combiner. Additional findings have also shown that its function may be restricted to relatively simple and perhaps mostly intersective conceptual combinations (Poortman & Pylkkänen, 2016; Ziegler & Pylkkänen, 2016). This makes sense, given its early timing: the activity may reflect an early concept combiner that applies whenever the input meanings have been sufficiently accessed—the idea being that more complex input meanings may not have been sufficiently accessed by 200 ms (Pylkkänen, 2016). Thus, for the LATL, we take a purely syntactic explanation to be off the table, leaving us with a much more nuanced, conceptually based explanation (surely to be refined in the future).

Summary Simple manipulations of sentence and phrase structure have yielded a description of the so-called combinatory network covering much of the left temporal cortex (LATL, MTG, pSTS) and potentially extending to areas of prefrontal and temporoparietal cortex (vmPFC, LIFG, TPJ). However, our functional understanding of this network is still rudimentary. A sequence of studies on the LATL shows that activity sensitive to a contrast such as sentence versus list may, in fact, reflect specificity-modulated conceptual composition. The bigger picture is that a more detailed understanding of the functional contributions of the various network nodes requires subtler contrasts in the stimulus materials and systematic efforts to modulate the effects observed for gross contrasts. Only in this way can we understand the computational limits of each type of neural activity.

Modeling Composition for Naturalistic Stimuli

Simple linguistic comparisons allow researchers to target neural composition operations that must be present under any theory. One question we might ask is whether conclusions about localization and timing from such studies generalize to richer linguistic contexts. A second question is whether these neural signals offer insight into the representations and algorithms that are implemented within these neural circuits beyond the coarse-grain labels of syntax versus semantics. Both questions may be answered by leveraging computational models to characterize neural signals recorded from participants as they process natural, ecologically rich, linguistic stimuli.

Computational models of composition A study by Brennan et al. (2012) illustrates the basic idea. They model composition word by word by counting the number of new phrases that have been completed after each word. For example, the sentence "Eleanor's sister pet the dog" includes the phrases *Eleanor's sister*, *the dog*, and *pet the dog*, as well as the trivial phrases made by each word and the entire sentence. Counting phrases that are completely processed at each word yields <*Eleanor's*, 1><*sister*, 2><*pet*, 1><*the*, 1><*dog*, 4> . This is a simple extension of Pallier, Devauchelle, and Dehaene (2011) and Bemis and Pylkkänen (2011), mentioned above, in that each phrase evokes composition operations.

Brennan et al. apply a broad-coverage account of phrase structure (Marcus, Marcinkiewicz, & Santorini, 1993) to annotate every sentence of a selection of real-word language: a chapter from *Alice's Adventures in Wonderland*. Participants listened to this chapter while undergoing functional Magnetic Resonance Imaging (fMRI) scanning. Convolving word-by-word composition steps with the hemodynamic response function yields an estimator for a neural time series that would be recorded from a brain region that was sensitive to the number of phrases completed word by word. A linear regression between this estimator and fMRI data revealed LATL activity that was positively correlated with the number of phrases being composed. Such a correlation was not seen in other regions of the combinatory network, like the LIFG.

The same method can also be applied with electrophysiology (Brennan & Pylkkänen, 2017; Frank et al., 2015; Nelson et al., 2017). For example, Brennan and Pylkkänen (2017) compared the number of phrases posited incrementally with MEG data that were recorded during story reading. Cross-correlating the number of completed phrases with the evoked signal revealed a significant effect in the LATL at 350 ms after word onset. Altogether, there is a promising alignment in terms of both localization and, to some extent, timing between results using rich natural stimuli and results from cases of minimal composition, reviewed above. Interestingly, though, while LATL effects emerge at approximately 200 ms for simple phrases, they occur about 100 ms later within a narrative (time locked to word onset). Given the larger amount of top-down information present in sentences, the opposite could easily be true—that is, effects could be faster in sentence contexts. Alternatively, the later timing could be due to the more complex meanings of sentences as compared to small phases. Careful study is needed to clarify the factors that may affect the timing of composition operations.

One assumption of the previous studies is that hemodynamic signals vary in proportion to the number of phrases composed. This proportionality is a linking hypothesis that connects the cognitive states that are computationally modeled to the neural signals that are experimentally measured. Other linking hypotheses between composition and brain signals can also be explored. For example, Henderson et al. (2016) use the linking hypothesis of surprisal. This quantity, which comes from information theory, reflects the (un)expectedness of a word given the preceding linguistic context (Hale, 2001). Using naturalistic story reading and the same account of phrase structure mentioned above, Henderson et al. report that LATL activity increases for unexpected words. Different linking hypotheses like surprisal and phrase counting tap into different cognitive operations. Ongoing work aims to more clearly specify this mapping—for instance, by probing how expectedness affects structure-building operations (Hale, 2014). Pairing computational models with naturalistic data is, in general, a highly flexible approach: by specifying different kinds of linking hypotheses, or different kinds of linguistic information in the models, multiple aspects of composition can be studied simultaneously (Wehbe et al., 2014).

Comparing alternative accounts of composition Computational models of naturalistic processing also furnish a way to rigorously compare alternative hypotheses of composition operations. The basic logic is to specify a family of models that share the same linking hypothesis but differ on some parameter of interest, say, the grammar rules used to define phrases. Such models can be ranked in terms of their fit to a target neural signal.

Nelson et al. (2017) apply this model comparison logic to probe how predictively the brain composes sentences. A nonpredictive *bottom-up* strategy holds that phrases are only postulated once all of the words that belong to that phrase have already been encountered. A more predictive strategy is also plausible. For example, upon encountering the word *the* at the beginning of a sentence, the composition system might reasonably predict that the next word will be a noun. Subsequently encountering a noun, like *cat*, verifies this prediction and also licenses a new prediction that a verb phrase will come next. This second approach is the *left-corner* strategy. The most eager *top-down* strategy composes syntactic structure prior to encountering any of the words that belong to a particular phrase. Under this strategy, for example, the composition system would postulate a noun phrase and a verb phrase prior to encountering any words in a new sentence.

Nelson et al. (2017) compare these three different strategies by correlating each with electrocorticography (ECoG) signals recorded from patients undergoing monitoring for severe epilepsy. Patients read sentences with a variety of grammatical structures and answered comprehension questions. The goodness of fit between the three models shows that signals recorded from left anterior temporal, left inferior frontal, and midline frontal recording sites are better fit when phrases are incrementally postulated according to either a left-corner or a bottom-up strategy but do not fill well with a top-down strategy.

The strategy for composing words is just one of many parameters needed for a fully explicit computational account. Another parameter is the grammar that guides licit composition. Brennan et al. (2016) probed two different grammars using fMRI. One was a simple phrase-structure grammar that captured major constituent structures in English (Marcus, Marcinkiewicz, & Santorini, 1993). A second grammar, adapted from a generative syntax textbook (Sportiche, Koopman, & Stabler, 2013), incorporated more abstract rules to account for linguistic regularities and dependencies within English and other languages. A model comparison revealed better fits to LATL and left posterior temporal lobe fMRI signals when the more abstract grammar was used. A key takeaway is that composition operations unfold differently under alternative syntactic theories of what is being composed, and neural signals that reflect composition are sensitive to such differences.

Summary Computational models of incremental sentence comprehension offer a tool to tease out neural signals related to composition when participants process rich natural stimuli. Studies have applied these models to hemodynamic and electrophysiological data collected under a variety of tasks. Thus far, the results show general agreement with those from minimal linguistic comparisons: the LATL and the posterior temporal lobes stand out as sensitive to composition (Brennan et al., 2012, 2016; Henderson et al., 2016; Nelson et al., 2017). One question that remains open is whether a continuous predictor of conceptual combination (created based on findings from the two-word phrase literature) would explain LATL activity better than the grammatical predictors used in extant computational-modeling work.

Because computational models are explicit, the fit between any particular model and neural signals can be quantified. Comparing different models in terms of their fit offers a novel way to test claims about the functions that are instantiated across the composition network, such as the nature of the syntactic representations that guide composition (Brennan et al., 2016; Frank et al., 2015; Nelson et al., 2017). But the explicitness required by this approach is also a limitation. To offer quantitative predictions, one must commit to a number of specific assumptions about composition, including the syntactic and semantic grammar, the composition strategy, the navigation of multiple possibilities when the input is ambiguous, and other parameters. The literature has just begun to systematically explore the hypothesis space defined by these many parameters.

Conclusion

Our understanding of the neural basis of sentence processing must rely on a solid characterization of the basic processes by which the brain composes complex structures and meaning from elementary building blocks. The current literature teaches us that the left anterior temporal cortex is the most consistent locus of combinatory effects across many methodologies, likely contributing to an early process of conceptual combination. How more grammatically based composition proceeds is still a mystery, though studies using artificial stimuli that perhaps enhance the subject's awareness of grammatical processing point toward a role for the left inferior frontal cortex (e.g, Petersson & Hagoort, 2012; Zaccarella & Friederici, 2015). Results from cortical stimulation mapping have also shown that disrupting processing in the left inferior frontal gyrus causes errors in grammatical encoding (Chang, Kurteff, & Wilson, 2018), suggesting that the LIFG does in some way interface with grammatical knowledge. But simple studies varying composition and model comparisons using naturalistic stimuli have failed to support a role for left inferior frontal cortex in structure building (Bemis & Pylkkänen, 2011, 2012; Brennan et al., 2012, 2016), creating a tension in the literature that has yet to be resolved. A second area of tension concerns the potential role of posterior temporal regions in composition, as they show mixed sensitivity to simple manipulations of composition. Overall, research using model comparisons in narrative processing has shown that activity within the combinatory network is generally better explained by parsing strategies that take lexical content into account, at least to some extent, and by grammars that are more abstract, as opposed to simpler. Going forward, the naturalistic methodology should include a systematic test of the hypotheses arising from more tightly controlled experiments, both to test the extent to which those findings "scale up" and to help us better connect the two bodies of literature.

Acknowledgments

Support for the writing of this chapter and the research summarized within it was provided by National Science Foundation grants BCS-1221723 (Liina Pylkkänen) and IIS-1607251 (Jonathan R. Brennan) and grant G1001 from the NYUAD Institute, New York University Abu Dhabi (Liina Pylkkänen).

REFERENCES

Bahlmann, J., Schubotz, R. I., & Friederici, A. D. (2008). Hierarchical artificial grammar processing engages Broca's area. *NeuroImage, 42*, 525–534.

Baron, S. G., & Osherson, D. (2011). Evidence for conceptual combination in the left anterior temporal lobe. *NeuroImage, 55*(4), 1847–1852.

Bemis, D. K., & Pylkkänen, L. (2011). Simple composition: A magnetoencephalography investigation into the comprehension of minimal linguistic phrases. *Journal of Neuroscience, 31*(8), 2801–2814.

Bemis, D. K., & Pylkkänen, L. (2012). Combination across domains: An MEG investigation into the relationship between mathematical, pictorial, and linguistic processing. *Frontiers in Psychology, 3*, 583.

Bemis, D. K., & Pylkkänen, L. (2013). Basic linguistic composition recruits the left anterior temporal lobe and left angular gyrus during both listening and reading. *Cerebral Cortex, 23*(8), 1859–1873.

Blanco-Elorrieta, E., Kastner, I., Emmorey, K., & Pylkkänen, L. (2018). Shared neural correlates for building phrases in signed and spoken language. *Scientific Reports*. doi:10.1038/s41598-018-23915-0

Blanco-Elorrieta, E., & Pylkkänen, L. (2016). Composition of complex numbers: Delineating the computational role of the left anterior temporal lobe. *NeuroImage, 124*, 194–203.

Brennan, J. R., Nir, Y., Hasson, U., Malach, R., Heeger, D. J., & Pylkkänen, L. (2012). Syntactic structure building in the anterior temporal lobe during natural story listening. *Brain and Language, 120*, 163–173.

Brennan, J., & Pylkkänen, L. (2012). The time-course and spatial distribution of brain activity associated with sentence processing. *NeuroImage, 60*(2), 1139–1148.

Brennan, J. R., & Pylkkänen, L. (2017). MEG evidence for incremental sentence composition in the anterior temporal lobe. *Cognitive Science, 41*(S6), 1515–1531.

Brennan, J. R., Stabler, E. P., Van Wagenen, S. E., Luh, W.-M., & Hale, J. T. (2016). Abstract linguistic structure correlates with temporal activity during naturalistic comprehension. *Brain and Language, 157–158*, 81–94.

Chang, E. F., Kurteff, G., & Wilson, S. M. (2018). Selective interference with syntactic encoding during sentence production by direct electrocortical stimulation of the inferior frontal gyrus. *Journal of Cognitive Neuroscience, 30*(3), 411–420.

Ding, N., Melloni, L., Zhang, H., Tian, X., & Poeppel, D. (2016). Cortical tracking of hierarchical linguistic structures in connected speech. *Nature Neuroscience, 19*(1), 158.

Dronkers, N. F., Wilkins, D. P., Van Valin, R. D., Redfern, B. B., & Jaeger, J. J. (2004). Lesion analysis of the brain areas involved in language comprehension: Towards a new functional anatomy of language. *Cognition, 92*(1–2), 145–177.

Frank, S. L., Otten, L. J., Galli, G., & Vigliocco, G. (2015). The ERP response to the amount of information conveyed by words in sentences. *Brain and Language, 140*(0), 1–11.

Frankland, S. M., & Greene, J. D. (2015). An architecture for encoding sentence meaning in left mid-superior temporal cortex. *Proceedings of the National Academy of Sciences of the United States of America, 112*(37), 11732–11737.

Friederici, A. D., Bahlmann, J., Heim, S., Schubotz, R. I., & Anwander, A. (2006). The brain differentiates human and non-human grammars: Functional localization and structural connectivity. *Proceedings of the National Academy of Sciences of the United States of America, 103*, 2458–2463.

Friederici, A. D., Meyer, M., & von Cramon, D. Y. (2000). Auditory language comprehension: An event-related fMRI study on the processing of syntactic and lexical information. *Brain and Language, 74*(2), 289–300.

Hale, J. T. (2001). A probabilistic Earley parser as a psycholinguistic model. In *North American Chapter of the Association for Computational Linguistics* (pp. 1–8). Morristown, NJ: Association for Computational Linguistics.

Hale, J. T. (2014). *Automaton theories of human sentence comprehension*. Stanford, CA: CSLI.

Henderson, J. M., Choi, W., Lowder, M. W., & Ferreira, F. (2016). Language structure in the brain: A fixation-related fMRI study of syntactic surprisal in reading. *NeuroImage, 132*, 293–300.

Humphries, C., Binder, J. R., Medler, D. A., & Liebenthal, E. (2006). Syntactic and semantic modulation of neural activity during auditory sentence comprehension. *Journal of Cognitive Neuroscience, 18*(4), 665–679.

Marcus, M., Marcinkiewicz, M., & Santorini, B. (1993). Building a large annotated corpus of English: The Penn treebank. *Computational Linguistics, 19*(2), 313–330.

Matchin, W., Hammerly, C., & Lau, E. F. (2017). The role of the IFG and pSTS in syntactic prediction: Evidence from a parametric study of hierarchical structure in fMRI. *Cortex, 88*, 106–123.

Mazoyer, B. M., Tzourio, N., Frak, V., Syrota, A., Murayama, N., Levrier, O., … & Mehler, J. (1993). The cortical representation of speech. *Journal of Cognitive Neuroscience, 5*(4), 467–479.

Nelson, M. J., El Karoui, I., Giber, K., Yang, X., Cohen, L., Koopman, H., Cash, S. S., Naccache, L., Hale, J. T., Pallier, C., & Dehaene, S. (2017). Neurophysiological dynamics of phrase-structure building during sentence processing. *Proceedings of the National Academy of Sciences of the United States of America, 114*(18), E3669–E3678.

Pallier, C., Devauchelle, A.-D., & Dehaene, S. (2011). Cortical representation of the constituent structure of sentences. *Proceedings of the National Academy of Sciences of the United States of America, 108*(6), 2522–2527.

Petersson, K. M., Folia, V., & Hagoort, P. (2012). What artificial grammar learning reveals about the neurobiology of syntax. *Brain and Language, 120*(2), 83–95.

Petersson, K. M., & Hagoort, P. (2012). The neurobiology of syntax: Beyond string sets. *Philosophical Transactions of the Royal Society of London B: Biological Sciences, 367*(1598), 1971–1983.

Poortman, E. B., & Pylkkänen, L. (2016). Adjective conjunction as a window into the LATL's contribution to conceptual combination. *Brain and Language, 160*, 50–60.

Pylkkänen, L. (2015). Composition of complex meaning: Interdisciplinary perspectives on the left anterior temporal lobe. In G. Hickok & S. Small (Eds.) *Neurobiology of Language* (pp. 621–631). Amsterdam: Academic Press.

Pylkkänen, L., Bemis, D. K., & Blanco Elorrieta, E. (2014). Building phrases in language production: An MEG study of simple composition. *Cognition, 133*(2), 371–384.

Pylkkänen, L., Brennan, J., & Bemis, D. K. (2011). Grounding the cognitive neuroscience of semantics in linguistic theory. *Language and Cognitive Processes, 26*(9), 1317–1337.

Rogalsky, C., & Hickok, G. (2008). Selective attention to semantic and syntactic features modulates sentence processing networks in anterior temporal cortex. *Cerebral Cortex, 19*(4), 786–796.

Schell, M., Zaccarella, E., & Friederici, A. D. (2017). Differential cortical contribution of syntax and semantics: An fMRI study on two-word phrasal processing. *Cortex, 96*, 105–120.

Snijders, T. M., Vosse, T., Kempen, G., Van Berkum, J. J., Petersson, K. M., & Hagoort, P. (2008). Retrieval and unification of syntactic structure in sentence comprehension: an fMRI study using word-category ambiguity. *Cerebral Cortex, 19*(7), 1493–1503.

Sportiche, D., Koopman, H., & Stabler, E. (2013). *An introduction to syntactic analysis and theory*. West Sussex: Wiley -Blackwell.

Stowe, L. A., Broere, C. A., Paans, A. M., Wijers, A. A., Mulder, G., Vaalburg, W., & Zwarts, F. (1998). Localizing components of a complex task: Sentence processing and working memory. *NeuroReport, 9*(13), 2995–2999.

Uddén, J., Folia, V., Forkstam, C., Ingvar, M., Fernández, G., Overeem, S., … Petersson, K. M. (2008). The inferior frontal cortex in artificial syntax processing: An rTMS study. *Brain Research, 1224*, 69–78.

Vandenberghe, R., Nobre, A. C., & Price, C. J. (2002). The response of left temporal cortex to sentences. *Journal of Cognitive Neuroscience, 14*(4):550–560.

Wehbe, L., Murphy, B., Talukdar, P., Fyshe, A., Ramdas, A., & Mitchell, T. (2014). Simultaneously uncovering the patterns of brain regions involved in different story reading subprocesses. *PLoS One, 9*(11), e112575.

Westerlund, M., Kastner, I., Al Kaabi, M., & Pylkkänen, L. (2015). The LATL as locus of composition: MEG evidence from English and Arabic. *Brain and Language, 141*, 124–134.

Westerlund, M., & Pylkkänen, L. (2014). The role of the left anterior temporal lobe in semantic composition vs. semantic memory. *Neuropsychologia, 57*, 59–70.

Wilson, S. M., DeMarco, A. T., Henry, M. L., Gesierich, B., Babiak, M., Mandelli, M. L., Miller, B. L., & Gorno-Tempini, M. L. (2014). What role does the anterior temporal lobe play in sentence-level processing? Neural correlates of syntactic processing in semantic variant primary progressive aphasia. *Journal of Cognitive Neuroscience, 26*(5), 970–985.

Zaccarella, E., & Friederici, A. D. (2015). Merge in the human brain: A sub-region based functional investigation in the left pars opercularis. *Frontiers in Psychology, 6*, 1818.

Zaccarella, E., Schell, M., & Friederici, A. D. (2017). Reviewing the functional basis of the syntactic merge mechanism for language: A coordinate-based activation likelihood estimation meta-analysis. *Neuroscience & Biobehavioral Reviews, 80*, 646–656.

Zhang, L., & Pylkkänen, L. (2015). The interplay of composition and concept specificity in the left anterior temporal lobe: An MEG study. *NeuroImage, 111*, 228–240.

Ziegler, J., & Pylkkänen, L. (2016). Scalar adjectives and the temporal unfolding of semantic composition: An MEG investigation. *Neuropsychologia, 89*, 161–171.

75 The Brain Network That Supports High-Level Language Processing

EVELINA FEDORENKO

ABSTRACT Humans are endowed with a capacity to share complex thoughts with one another via language. Here, I review what we know about the neural substrates of spoken- (cf., sign) language processing. I briefly examine the perceptual and motor brain areas that support speech perception and articulation, respectively, and then—in greater depth—discuss the brain network that supports the higher-level processes of interpretation and generation of linguistic utterances. I summarize this network's basic functional characteristics and then review evidence that informs two big questions about this network. First, do brain regions that support high-level language processing also support nonlinguistic abilities, such as math or music? And second, do different brain regions within this network support different aspects of high-level language processing? In particular, I focus on the distinction between lexicosemantic processing/storage and combinatorial syntactic/semantic processing. I argue that although language-responsive regions are selective for language over diverse nonlinguistic cognitive processes, no language region is selective for lexicosemantic or syntactic processing: any region that responds to individual word meanings also responds to combinatorial processing. Both of these answers importantly constrain our theorizing about the language architecture.

Language is a powerful code through which we can exchange information about the world and form deep interpersonal relationships. What are the knowledge representations and mental computations that underlie this sophisticated capacity? I review what we know about the neural substrates of language processing, with a focus on findings that inform its cognitive architecture.

The Anatomical Scope of Language-Related Brain Areas

Historically, given the early evidence from aphasia (language deficits that arise following brain damage), the focus has been on two perisylvian (adjacent to the sylvian fissure) brain areas: *Broca's area* in the left frontal lobe, linked to language production, and *Wernicke's area* in the left temporoparietal cortex, linked to language comprehension (Geschwind, 1970; see chapter 79 of this volume). However, it is now clear that (1) language processing engages a broader set of brain regions both within and outside perisylvian cortex (Binder et al., 1997; Fedorenko et al., 2010), and (2) the original hypotheses about the functions of Broca's area and Wernicke's area are likely incorrect. In fact, the inconsistency in the definitions and use of the latter terms over the years have recently led Tremblay and Dick (2016) to argue—quite reasonably—for their abolition.

So what regions in the human brain support language processing? If we adopt the broadest possible definition of what it means to "support language processing"—*that is, engagement at some point in the process of understanding or producing linguistic utterances*—then we have a lot of neural machinery that spans both lower-level perceptual and motor areas and higher-level association areas. Although we are still far from mechanistic-level accounts of how these different brain regions contribute to language processing, we have accumulated substantial knowledge about their functional properties, which places constraints on their computations. In the remainder of the chapter, I briefly discuss the perceptual and motor areas that subserve language comprehension and production and then focus on a set of brain regions that support higher-level language processing (i.e., interpreting and generating utterances).

Perceptual and Motor Language-Related Brain Regions

Speech perception Speech perception requires mapping the acoustic stream onto representations that can mediate processes like word recognition. Parts of the auditory cortex in the superior temporal gyrus and sulcus respond robustly to speech (Overath et al., 2015). For these areas, it doesn't matter whether the signal is meaningful: they respond as strongly to speech made up of nonwords or speech in an unfamiliar language as they do to interpretable speech. Although debated at some point (Price, Thierry, and Griffiths 2005), Norman-Haignere, Kanwisher, and McDermott (2015) have established that these regions are selective for speech over many other types of sounds. Overath et al. (2015)

further found that these areas have a preferred temporal window: responses increase with segment length up to approximately 500 ms and then plateau. Thus, speech-responsive auditory areas appear to be tuned to speech-specific spectrotemporal structure and plausibly play a role in encoding phonemes and syllables.

Speech production (articulation) Fluent speech requires the planning of sound sequences, followed by the execution of corresponding motor plans. Portions of the precentral gyrus, supplementary motor area (SMA), inferior frontal cortex, superior temporal cortex, and cerebellum respond robustly during speech production (Basilakos et al., 2018; Bohland & Guenther, 2006). Like the speech perception areas, these regions do not care about the meaning of the articulated sequence, working as hard during the production of a syllable sequence as they do when we produce words or sentences. Furthermore, the articulation-responsive areas in the precentral gyrus and SMA respond as strongly during the production of nonspeech oral-motor movements as during articulation, in line with the somatotopic organization of sensorimotor cortex (Bouchard et al., 2013). In contrast, the articulation-responsive part of the left posterior inferior frontal gyrus (IFG) is relatively selective for speech production and thus plausibly supports speech-specific functions (e.g., preparing an articulatory code to be sent to the motor cortex; Flinker et al., 2015).

Written-language perception and production Speech perception areas have an analog in the visual cortex of literate individuals: a small area on the ventral temporal surface that responds to written linguistic stimuli (McCandliss, Cohen, & Dehaene, 2003). This area's name—the *visual word form area* (Cohen & Dehaene, 2004)—is somewhat of a misnomer because like the speech regions, this area is not sensitive to the meaningfulness of the stimulus: it responds as strongly to letter sequences as it does to real words. Also, like the speech regions, this area is selective for its preferred stimulus (letters in a familiar script) over many other visual stimuli (Baker et al., 2007).

Written (and typed) language production has received relatively little attention in cognitive neuroscience. The few studies that have investigated written production have observed activation in some areas that appear similar to those reported in studies of articulation (within the left IFG, the precentral gyrus, and the cerebellum) and also in some areas that do not typically emerge in studies of articulation (e.g., within the superior frontal gyrus and the superior parietal lobule; Planton et al., 2013). At least some parts of this written production network show selectivity for writing relative to matched movements (Planton et al., 2013) or even for writing letters relative to writing nonletter symbols (Longcamp et al., 2014).

High-Level Language Brain Regions

Basic properties A set of brain regions in the frontal, temporal, and parietal lobes (figure 75.1*A*) appears to support higher-level aspects of language processing. These regions receive input from the perceptual language areas during comprehension and provide input to the motor language areas during production. The goals of these high-level language regions are to derive a representation of the intended meaning in comprehension (decoding) and to convert thoughts into a linguistic format in production (encoding). How these brain regions achieve these goals is what the field of language research aims to understand.

High-level language brain regions exhibit several key properties, including

1. a similar general topography across individuals and languages,
2. left lateralization,
3. input and output modality-independence,
4. functional integration within the network,
5. sensitivity to the meaningfulness of the signal, and
6. a causal role in language processing.

First, the general topography of the frontotemporal language network is similar across individuals, including individuals with vastly different developmental experiences (Bedny et al., 2011; Newman et al., 2010),

FIGURE 75.1 *A*, The general topography of the high-level language network. This representation was derived by overlaying 207 individual activation maps for the contrast of reading sentences versus nonword sequences (Fedorenko et al., 2010). *B*, Language activations in six individuals tested in their native languages (using a contrast between listening to passages from *Alice's Adventures in Wonderland* versus the acoustically degraded versions of those passages; Scott, Gallee, & Fedorenko, 2016) that come from distinct language families. *C*, Language activations in three individuals tested across two scanning sessions. *D*, Key functional properties of two sample high-level language regions. The parcels used to define the individual functional regions of interest are shown in gray (each fROI is defined as the top 10% most language-responsive voxels); on the left, we show responses to several linguistic manipulations, and on the right, we show responses to nonlinguistic tasks. (See color plate 87.)

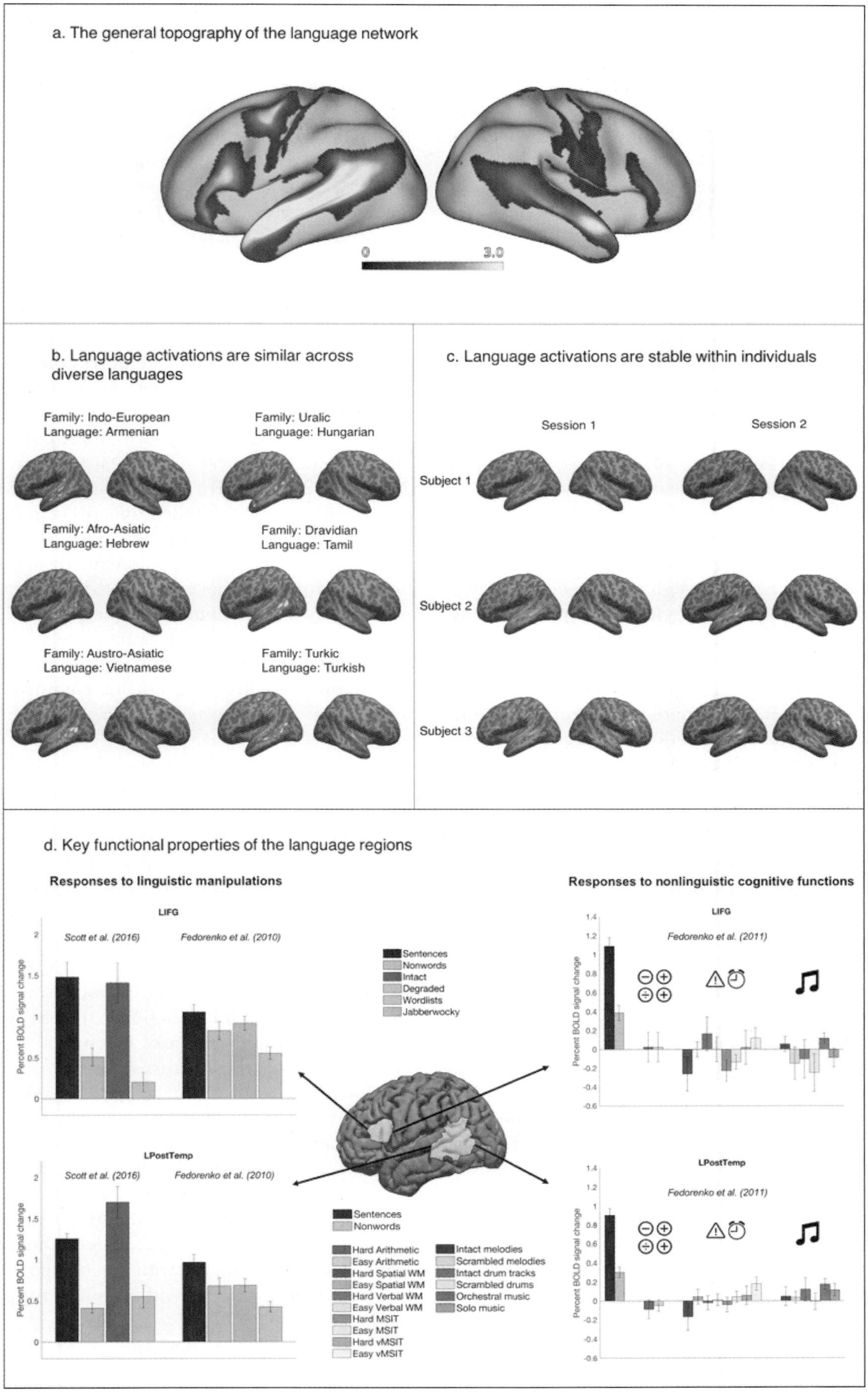
a. The general topography of the language network
0
3.0
b. Language activations are similar across diverse languages
Family: Indo-European
Language: Armenian
Family: Uralic
Language: Hungarian
Family: Afro-Asiatic
Language: Hebrew
Family: Dravidian
Language: Tamil
Family: Austro-Asiatic
Language: Vietnamese
Family: Turkic
Language: Turkish
c. Language activations are stable within individuals
Session 1
Session 2
Subject 1
Subject 2
Subject 3
d. Key functional properties of the language regions
Responses to linguistic manipulations
Responses to nonlinguistic cognitive functions
LIFG
Scott et al. (2016)
Fedorenko et al. (2010)
Percent BOLD signal change
Sentences
Nonwords
Intact
Degraded
Wordlists
Jabberwocky
Fedorenko et al. (2011)
LPostTemp
Hard Arithmetic
Easy Arithmetic
Hard Spatial WM
Easy Spatial WM
Hard Verbal WM
Easy Verbal WM
Hard MSIT
Easy MSIT
Hard vMSIT
Easy vMSIT
Intact melodies
Scrambled melodies
Intact drum tracks
Scrambled drums
Orchestral music
Solo music

and across diverse languages (van Heuven & Dijkstra, 2010; figure 75.1*B*). Nevertheless, the detailed topography varies substantially across individuals (figure 75.1*C*), in line with the well-established anatomical variability (Fedorenko & Kanwisher, 2009). Some have therefore argued for the importance of defining these regions functionally at the individual-subject level instead of attempting to align activations in the common brain space (Demonet, Wise, & Frackowiak, 1993; Fedorenko et al., 2010).

Second, language activations tend to be stronger in the left hemisphere (figure 75.1*A, C*). However, individuals vary—in a stable way—with regard to the amount of right-hemisphere activity (Mahowald & Fedorenko, 2016). Whether the degree of language lateralization has behavioral consequences among the neurotypical population remains debated. However, reduced lateralization has been reported across diverse neurodevelopmental disorders (Lindell & Hudry, 2013), suggesting that it commonly accompanies atypical brain development.

Third, in comprehension, high-level language regions respond to linguistic input regardless of which sensory modality it came through (Braze et al., 2011; Fedorenko et al., 2010). And although not tested extensively, these regions would be expected to respond during language production in a similar way regardless of the eventual output modality (spoken vs. written/typed vs. signed for sign languages; Blanco-Elorrieta et al., 2018).

Fourth, high-level language regions form a functionally integrated system. In addition to similar functional profiles across diverse linguistic manipulations (see the section on the internal architecture), two lines of evidence suggest strong relationships among the language regions. First, the language network emerges robustly from patterns of low-frequency oscillations across the brain during naturalistic-cognition paradigms (Blank, Kanwisher, & Fedorenko, 2014). And second, the cortical thinning patterns in primary progressive aphasia—a neurodegenerative condition that disproportionately affects language (Mesulam, 2001)—bear a striking resemblance to the functional activation pattern that has emerged in neuroimaging work (figure 75.1*A*). According to one proposal, degeneration proceeds along transsynaptic connections (Seeley et al., 2009), in line with strong interconnectivity across the network.

Fifth, in stark contrast to the perceptual and motor language areas, high-level language regions are robustly sensitive to meaning. They respond two or three times more strongly to meaningful phrases and sentences compared to perceptually matched stimuli that lack meaning (Fedorenko et al., 2010; Scott, Gallee, & Fedorenko, 2016; Snijders et al., 2009).

And sixth, at least some parts of the language network are causally important for language: interfering with the activity of these brain regions (through electrical stimulation during surgery: Whitaker & Ojemann, 1977; or via transcranial magnetic stimulation (TMS): Devlin & Watkins, 2006) or their permanent loss in adulthood (due to stroke or degeneration: Bates et al., 2003; Mesulam, 2001; or as a result of surgical resection: Wilson et al., 2015) leads to linguistic deficits. However, research to characterize the relationship between specific brain regions and the resulting linguistic deficits is still ongoing (e.g., chapter 79 in this volume).

Beyond these basic characteristics, how do we go about understanding the precise computations that high-level language regions support? One property that can guide and constrain theorizing is *functional selectivity*: the degree to which a brain region prefers a particular stimulus yields critical information about the nature and scope of its possible function(s) (Mather, Cacioppo, & Kanwisher, 2013). For any domain, selectivity can be assessed with respect to other domains (e.g., does a brain region that responds to language also respond to music?) and with respect to within-domain distinctions (e.g., does a brain region that responds to word meanings also respond when we combine those meanings into complex representations?). Below, I summarize what we know about the selectivity of high-level language regions for (1) language versus nonlinguistic cognitive functions and (2) different aspects of language.

Selectivity for language relative to nonlinguistic functions A relatively recent human invention, language emerged against the backdrop of perceptual, motor, and cognitive machinery. It is therefore reasonable to ask whether language may have co-opted existing neural mechanisms (Anderson, 2010). Furthermore, language itself may have enabled the development of sophisticated cognitive capacities like arithmetic or some aspects of Theory of Mind. One might therefore expect that brain regions that support language processing would also support nonlinguistic functions. Indeed, many have argued that language shares machinery with diverse nonlinguistic cognitive processes, from arithmetic to executive function, to music, to social cognition, to navigation (reviewed in Fedorenko & Varley, 2016). However, brain-imaging studies that have carefully compared activations for linguistic and nonlinguistic tasks (including in individual subjects; Nieto-Castanon & Fedorenko, 2012) have consistently found that the brain's high-level language regions are not engaged by nonlinguistic tasks. Evidence from individuals with aphasia has yielded convergent evidence: such individuals suffer from linguistic deficits, but other aspects of cognition appear unimpaired. Below, I summarize the evidence for nonoverlap between language and two cognitive abilities: arithmetic and music perception. For a more extended discussion of these and other abilities, see Fedorenko and Varley (2016; figure 75.1*D*).

Language versus arithmetic In addition to evolutionarily conserved numerical abilities (magnitude estimation and subitizing), humans have developed means to represent arbitrary exact quantities, using verbal representations (words for numbers). The verbal nature of these representations led to proposals that exact arithmetic relies on the neural system that underlies linguistic processing. For example, Dehaene et al. (1999) observed activation for exact arithmetic within the left IFG. Given that other studies had reported inferior frontal activations for linguistic tasks, Dehaene and colleagues argued that the activation they observed during exact calculations reflects the engagement of the language system.

However, in 2005 Varley and colleagues reported a study of severely aphasic patients, with extensive damage to left-hemisphere language regions, who could nevertheless solve diverse arithmetic problems, suggesting that the brain's language regions are not needed for arithmetic. Recent neuroimaging studies have provided converging evidence: Fedorenko, Behr, and Kanwisher (2011) found no response in the brain's language regions during arithmetic problem-solving, and Monti, Parsons, and Osherson (2012; see also Amalric & Dehaene, 2016) found that linguistic, but not algebraic, syntax produced activations in the IFG.

In summary, brain regions that support linguistic processing are not active when we solve arithmetic problems, and (even extensive) damage to the language network appears to leave our arithmetic abilities intact. Thus, linguistic processing occurs in brain circuits distinct from those that support arithmetic processing.

Language versus music processing Language and music share multiple features, including their structural properties (Jackendoff & Lerdahl, 2006). In both domains, relatively small sets of elements (words in language, notes in music) are used to create a large number of sequential structures (sentences in language, melodies in music). And in both domains, this combinatorial process is rule governed. Inspired by these similarities, researchers have looked for overlap in the processing of structure in language and music.

For example, using a structural-violation paradigm in which participants listen to stimuli that do or do not contain a structurally unexpected element, many studies have observed similar event related potential (ERP) components and similar activations in functional magnetic resonance imaging (fMRI) (Koelsch et al., 2002; Maess et al., 2001; Patel et al., 1998; Tillmann, Janata, & Bharucha, 2003). However, a note or word that is incongruent with the preceding context is a salient event. Thus, the observed responses could reflect a generic mental process such as attentional capture or error detection/correction. Indeed, a meta-analysis of activation peaks from fMRI studies investigating brain responses to unexpected sensory events (Corbetta & Shulman, 2002) has revealed brain regions that closely resemble those activated by structural violations in music (Koelsch et al., 2002).

Later brain-imaging studies that compared neural responses to language and music outside the context of the violation paradigms (Fedorenko, Behr, & Kanwisher, 2011; Rogalsky et al., 2011) found no overlap, in line with the dissociation between linguistic and musical abilities that has been reported in the neuropsychological literature (Peretz & Hyde, 2003). It therefore appears that distinct sets of brain regions support high-level linguistic and music processing.

To summarize more broadly, the available evidence suggests that in a mature human brain, regions that support high-level language processing do so selectively, and damage to these regions affects the ability to understand and produce language but not to engage in many forms of complex thought. The key motivation for investigating the degree of functional specialization in the human mind and brain is that such investigations constrain hypotheses about possible computations (Mather, Cacioppo, & Kanwisher, 2013). For example, had we found a brain region within the high-level language network that responded to both linguistic and musical/arithmetic syntax, we could have hypothesized that this region was sensitive to some abstract features of the structure present in both kinds of stimuli or to the processing of those structural features, such as establishing the dependencies among the relevant elements (Patel, 2003) or the engagement of a recursive operation (Hauser, Chomsky, & Fitch, 2002). The fact that high-level language regions appear to not be active during a wide range of nonlinguistic tasks suggests that these regions respond to some features that are only present in (or some mental operations that only apply to) linguistic stimuli. I discuss what these might be next.

The internal architecture of high-level language processing The high-level language regions span extensive portions of the left frontal, temporal, and parietal lobes (figure 75.1*A*). Is there a meaningful way to divide this network into component parts? And if so, how is linguistic labor shared across those parts in space and time?

A good starting point is the current theorizing about the functional architecture of human language. A core component is a set of knowledge representations, which include knowledge of the sounds, the words and their meanings, and the probabilistic constraints on how sounds can combine to create words and how words can combine to create sentences. During comprehension (decoding) we look for matches between the linguistic signal and these stored knowledge representations, and during production (encoding) we search our knowledge store for the right words/constructions and arrange them in a particular way to express a target idea.

Within this architecture, a distinction has traditionally been drawn between lexicosemantic processing (the knowledge and access of word meanings) and syntactic/combinatorial processing (the knowledge of constraints on combining words into phrases and sentences and inferring or constructing interword dependencies during comprehension and production, respectively) (Chomsky, 1965). Consequently, with the development of neuroimaging techniques, many have searched for, and claimed to have observed, a dissociation between brain regions that selectively—or at least preferentially—support lexicosemantic processing and those that selectively support syntactic, or more general combinatorial, processing (Dapretto & Bookheimer, 1999; Embick et al., 2000; Friederici, Opitz, & von Cramon, 2000). The alleged syntax-selective regions have sparked particular excitement due to claims that (some aspects of) syntax is what makes human language unique (Hauser et al., 2002).

Over the years the distinction between word meanings and grammar has gotten blurry as evidence has accumulated suggesting that our grammatical knowledge goes beyond abstract syntactic rules that operate over categories like nouns and verbs and instead appears to be specific to particular words (Bybee, 2010; Goldberg, 2002; Jackendoff, 2007). Nevertheless, even if our linguistic knowledge representations are characterized by a strong degree of integration between lexical and grammatical knowledge, these representations may be stored in brain regions distinct from those that implement the flexible combination of these representations during comprehension and/or production.

Indeed, most current proposals of the neural architecture of language postulate a distinction between regions that support lexicosemantic storage/processing and those that support syntactic/combinatorial processing (e.g., Baggio & Hagoort, 2011; Friederici, 2012; Tyler et al., 2011). However, the precise regions that are argued to support lexicosemantic versus combinatorial processing, and the construal of these regions' contributions, differ across proposals. Further, much evidence now suggests that any given language region is robustly sensitive to both word meanings (stronger responses to real words than nonwords) and syntactic/combinatorial processing (stronger responses to structured representations, like phrases/sentences, than lists of unconnected words and even to meaningless jabberwocky sentences compared to lists of nonwords; Bedny et al., 2011; Fedorenko et al., 2010; Keller, Carpenter, & Just, 2001; figure 75.1*D*; see Bautista & Wilson, 2016; Blank et al., 2016; Roder et al., 2002 for evidence from other paradigms). This pattern also holds in temporally sensitive methods like ECoG (Fedorenko et al., 2016).

Some studies further suggest a bias toward lexicosemantic and combinatorial semantic processing over syntactic processing. For example, using multivariate analyses, Fedorenko et al. (2012) found that lexicosemantic information is represented more robustly than syntactic information. And Frankland and Greene (2015) found that activation patterns in temporal cortex distinguish thematic roles (agent/patient) but not grammatical positions (subject/object). Thus, the language network may be more strongly concerned with meaning than structure (see chapter 74). This bias is not surprising given that the goal of communication is to transfer *meanings*. However, it is surprising in light of the emphasis that has traditionally been placed on syntax as the core computational capacity of language that emerged in humans and gave us the power to express or understand an infinite number of ideas using a finite number of linguistic signals (Friederici et al., 2006; Hauser et al., 2002).

The tentative working hypothesis is therefore that the frontotemporal language network stores linguistic knowledge representations (whatever their form may be; e.g., Bybee, 2010; Goldberg, 2002; Jackendoff, 2007) in a highly distributed fashion, in line with evidence of the successful decoding of linguistic meanings from neural activity across the frontotemporal cortex (Huth et al., 2016; Pereira et al., 2018). And instead of being localized to a particular brain region, the basic combinatorial operation is either instantiated ubiquitously throughout the language network, allowing for the flexible combination of the relevant representations, or is actually performed by the very same units that also store linguistic knowledge (e.g., Hasson, Chen, & Honey, 2015). It is worth noting that distributed and overlapping lexicosemantic and combinatorial processing is largely consistent with the picture that has emerged from the patient literature (Dick et al., 2001). In particular, (1) damage to different regions within the language network leads to similar syntactic deficits, and (2) syntactic deficits appear to always be accompanied by lexical deficits.

Summary and Open Questions

I have reviewed some of what we know about the brain basis of high-level language processing. I discussed the separability of the neural machinery that supports utterance interpretation and generation from the machinery that supports lower-level perceptual and motor aspects of language. I then reviewed some basic properties of the high-level language regions and discussed two key questions about their functional profiles—namely, whether they are selective for language over nonlinguistic processes that have been argued to share machinery with language and whether different aspects of language recruit distinct regions within the language network. The answers that have emerged so far are as follows. The brain regions that support high-level language processing are selective for language showing little or no response during diverse cognitive tasks, including arithmetic, executive function tasks, music perception, social cognition, and action/gesture observation, among others. However, lexicosemantic and combinatorial processing appear to be implemented in a distributed fashion across the network, contra proposals that postulate regions that selectively support syntactic/combinatorial processing.

These answers importantly constrain our theorizing about the language architecture. However, many questions remain. We strive for a mechanistic-level account (Marr, 1982) of the language network, which would specify the input and output of each brain region, the precise computations that each region performs, the feedforward and feedback connections within the network (and with other networks, such as the domain-general executive network or the network that supports social reasoning), and the time course of the computations that take place as we understand or produce an utterance. This level of understanding would enable us to both (1) develop detailed hypotheses about different kinds of linguistic deficits so that we can improve diagnostics and inform treatments and (2) engineer machines capable of human-like language comprehension and generation in the service of information extraction or automatic translation.

I want to conclude by outlining three open issues/questions that I hope we, as language researchers, can tackle in the coming years.

Comprehension versus production Although most agree that comprehension and production rely on the same linguistic knowledge representations, an important asymmetry exists between them. The goal of comprehension is to infer the intended meaning from the linguistic signal, and abundant evidence now suggests that the representations we extract and maintain during comprehension are probabilistic and noisy (Gibson, Bergen, & Piantadosi, 2013). In contrast, in production the goal is to express a particular meaning, about which we have little or no uncertainty. To do so, we have to utter a precise sequence of words where each word takes a particular morphosyntactic form, and the words appear in a particular order. These pressures for precision and for the linearization of words, morphemes, and sounds may lead to a clearer temporal and/or spatial segregation among the different stages of the production process and, correspondingly, to functional dissociations among the brain regions implicated in production (Indefrey & Levelt, 2004; Fedorenko et al., 2018), compared to comprehension, where the very same brain regions appear to support different aspects of the interpretation, as discussed above. Methods with high spatial and temporal resolution that afford causal inferences are especially promising for this enterprise (Lee et al., 2018).

The relationship between linguistic and conceptual representations As discussed above, high-level language regions appear to be selective for language processing over many nonlinguistic processes. However, given that language is used to convey meanings, our linguistic representations have to be linked to our semantic knowledge. There is at present no consensus about where and how the latter is neurally instantiated or about the relationship between linguistic representations and abstract conceptual ones. More work is needed to develop and evaluate specific proposals about the nature of concepts, the organizing principles of the semantic space, the computations that underlie concept composition, and the relationship between concepts and words/constructions.

White matter tracts that support language processing A number of white matter tracts have been implicated in some aspects of language processing (Dick & Tremblay, 2012), but most current hypotheses are vague and thus difficult to evaluate. As we make progress in deciphering the representations and computations that different brain regions may support, we should start formulating more precise proposals about the role of each relevant tract in language comprehension and/or production.

Building on the foundation of linguistic theorizing, rigorous behavioral experimentation, and computational modeling, we can use cognitive neuroscience approaches to develop a rich and comprehensive understanding of the cognitive and neural architecture of language.

Acknowledgments

I would like to thank Karen Emmorey, Liina Pylkkänen, and the attendees of the 2018 Summer Institute in Cognitive Neuroscience (Lake Tahoe, CA) for their constructive comments and criticisms and Matt Siegelman, Yev Diachek, and Moataz Assem for their help with figure 75.1. The author was supported by National Institutes of Health award R01-DC016607 and R01-DC016950. Finally, due to the strict word limit, I had to omit numerous relevant citations; I refer readers to empirical and review papers from our group, where we review and cite the relevant subliteratures more extensively.

REFERENCES

Anderson, M. L. (2010). Neural reuse: A fundamental organizational principle of the brain. *Behavioral and Brain sciences, 33*(4), 245–266.

Amalric, M., & Dehaene, S. (2016). Origins of the brain networks for advanced mathematics in expert mathematicians. *Proceedings of the National Academy of Sciences, 113*(18), 4909–4917.

Baggio, G., & Hagoort, P. (2011). The balance between memory and unification in semantics: A dynamic account of the N400. *Language and Cognitive Processes, 26*(9), 1338–1367.

Baker, C. I., Liu, J., Wald, L. L., Kwong, K. K., Benner, T., & Kanwisher, N. (2007). Visual word processing and experiential origins of functional selectivity in human extrastriate cortex. *Proceedings of the National Academy of Sciences of the United States of America, 104*(21), 9087–9092.

Basilakos, A., Smith, K., Fillmore, P., Fridriksson, J., & Fedorenko, E. (2018). Functional characterization of the human speech articulation network. *Cerebral Cortex, 28*(5), 1816–1830.

Bates, E., Wilson, S. M., Saygin, A. P., Dick, F., Sereno, M. I., Knight, R. T., & Dronkers, N. F. (2003). Voxel-based lesion-symptom mapping. *Nature Neuroscience, 6*(5), 448–450.

Bautista, A., & Wilson, S. M. (2016). Neural responses to grammatically and lexically degraded speech. *Language Cognition and Neuroscience, 31*(4), 567–574.

Bedny, M., Pascual-Leone, A., Dodell-Feder, D., Fedorenko, E., & Saxe, R. (2011). Language processing in the occipital cortex of congenitally blind adults. *Proceedings of the National Academy of Sciences of the United States of America, 108*(11), 4429–4434.

Binder, J. R., Frost, J. A., Hammeke, T. A., Cox, R. W., Rao, S. M., & Prieto, T. (1997). Human brain language areas identified by functional magnetic resonance imaging. *Journal of Neuroscience, 17*(1), 353–362.

Blanco-Elorrieta, E., Kastner, I., Emmorey, K., & Pylkkänen, L. (2018). Shared neural correlates for building phrases in signed and spoken language. *Scientific Reports, 8*(1), 5492.

Blank, I., Balewski, Z., Mahowald, K., & Fedorenko, E. (2016). Syntactic processing is distributed across the language system. *NeuroImage, 127*, 307–323.

Blank, I., Kanwisher, N., & Fedorenko, E. (2014). A functional dissociation between language and multiple-demand systems revealed in patterns of BOLD signal fluctuations. *Journal of Neurophysiology, 112*(5), 1105–1118.

Bohland, J. W., & Guenther, F. H. (2006). An fMRI investigation of syllable sequence production. *NeuroImage, 32*(2), 821–841.

Bouchard, K. E., Mesgarani, N., Johnson, K., & Chang, E. F. (2013). Functional organization of human sensorimotor cortex for speech articulation. *Nature, 495*(7441), 327–332.

Braze, D., Mencl, W. E., Tabor, W., Pugh, K. R., Constable, R. T., Fulbright, R. K., & Shankweiler, D. P. (2011). Unification of sentence processing via ear and eye: An fMRI study. *Cortex, 47*(4), 416–431.

Bybee, J. (2010). *Language, usage and cognition* (Vol. 98). Cambridge: Cambridge University Press.

Chomsky, N. (1965). *Aspects of the theory of syntax*. Cambridge, MA: MIT Press.

Cohen, L., & Dehaene, S. (2004). Specialization within the ventral stream: The case for the visual word form area. *NeuroImage, 22*(1), 466–476.

Corbetta, M., & Shulman, G. L. (2002). Control of goal-directed and stimulus-driven attention in the brain. *Nature Reviews Neuroscience, 3*(3), 201–215.

Dapretto, M., & Bookheimer, S. Y. (1999). Form and content: Dissociating syntax and semantics in sentence comprehension. *Neuron, 24*(2), 427–432.

Dehaene, S., Spelke, E., Pinel, P., Stanescu, R., & Tsivkin, S. (1999). Sources of mathematical thinking: Behavioral and brain-imaging evidence. *Science, 284*(5416), 970–974.

Demonet, J. F., Wise, R., & Frackowiak, R. S. J. (1993). Language functions explored in normal subjects by positron emission tomography: A critical review. *Human Brain Mapping, 1*, 39–47.

Devlin, J. T., & Watkins, K. E. (2006). Stimulating language: Insights from TMS. *Brain, 130*(3), 610–622.

Dick, A. S., & Tremblay, P. (2012). Beyond the arcuate fasciculus: Consensus and controversy in the connectional anatomy of language. *Brain, 135*(12), 3529–3550.

Dick, F., Bates, E., Wulfeck, B., Utman, J. A., Dronkers, N., & Gernsbacher, M. A. (2001). Language deficits, localization, and grammar: evidence for a distributive model of language breakdown in aphasic patients and neurologically intact individuals. *Psychological Review, 108*(4), 759–788.

Embick, D., Marantz, A., Miyashita, Y., O'Neil, W., & Sakai, K. L. (2000). A syntactic specialization for Broca's area. *Proceedings of the National Academy of Sciences of the United States of America, 97*(11), 6150–6154.

Fedorenko, E., Behr, M. K., & Kanwisher, N. (2011). Functional specificity for high-level linguistic processing in the human brain. *Proceedings of the National Academy of Sciences of the United States of America, 108*(39), 16428–16433.

Fedorenko, E., Hsieh, P. J., Nieto-Castanon, A., Whitfield-Gabrieli, S., & Kanwisher, N. (2010). New method for fMRI investigations of language: Defining ROIs functionally in individual subjects. *Journal of Neurophysiology, 104*(2), 1177–1194.

Fedorenko, E., & Kanwisher, N. (2009). Neuroimaging of language: Why hasn't a clearer picture emerged? *Language and Linguistics Compass, 3*(4), 839–865.

Fedorenko, E., Nieto-Castañon, A. & Kanwisher, N. (2012). Lexical and syntactic representations in the brain: An fMRI investigation with multi-voxel pattern analyses. *Neuropsychologia*, 50(4), 499–513.

Fedorenko, E., Scott, T. L., Brunner, P., Coon, W. G., Pritchett, B., Schalk, G., & Kanwisher, N. (2016). Neural correlate of the construction of sentence meaning. *Proceedings of the*

National Academy of Sciences of the United States of America, 113(41), E6256–E6262.

Fedorenko, E., & Varley, R. (2016). Language and thought are not the same thing: Evidence from neuroimaging and neurological patients. *Annals of the New York Academy of Sciences, 1369*(1), 132–153.

Fedorenko, E., Williams, Z. M., & Ferreira, V. S. (2018). Remaining puzzles about morpheme production in the posterior temporal lobe. *Neuroscience,* 392, 160–163.

Flinker, A., Korzeniewska, A., Shestyuk, A. Y., Franaszczuk, P. J., Dronkers, N. F., Knight, R. T., & Crone, N. E. (2015). Redefining the role of Broca's area in speech. *Proceedings of the National Academy of Sciences of the United States of America, 112*(9), 2871–2875.

Frankland, S. M., & Greene, J. D. (2015). An architecture for encoding sentence meaning in left mid-superior temporal cortex. *Proceedings of the National Academy of Sciences of the United States of America, 112*(37), 11732–11737.

Friederici, A. D. (2012). The cortical language circuit: From auditory perception to sentence comprehension. *Trends in Cognitive Sciences, 16*(5), 262–268.

Friederici, A. D., Bahlmann, J., Heim, S., Schubotz, R. I., & Anwander, A. (2006). The brain differentiates human and non-human grammars: Functional localization and structural connectivity. *Proceedings of the National Academy of Sciences of the United States of America, 103*(7), 2458–2463.

Friederici, A. D., Opitz, B., & von Cramon, D. Y. (2000). Segregating semantic and syntactic aspects of processing in the human brain: An fMRI investigation of different word types. *Cerebral Cortex, 10*(7), 698–705.

Geschwind, N. (1970). The organization of language and the brain. *Science, 170,* 940–944.

Gibson, E., Bergen, L., & Piantadosi, S. T. (2013). Rational integration of noisy evidence and prior semantic expectations in sentence interpretation. *Proceedings of the National Academy of Sciences of the United States of America, 110*(20), 8051–8056.

Goldberg, A. (2002). *Construction grammar: Encyclopedia of cognitive science.* New York: Macmillan Reference Limited Nature.

Hasson, U., Chen, J., & Honey, C. J. (2015). Hierarchical process memory: Memory as an integral component of information processing. *Trends in Cognitive Sciences, 19*(6), 304–313.

Hauser, M. D., Chomsky, N., & Fitch, W. T. (2002). The faculty of language: What is it, who has it, and how did it evolve? *Science, 298*(5598), 1569–1579.

Indefrey, P., & Levelt, W. J. (2004). The spatial and temporal signatures of word production components. *Cognition, 92*(1–2), 101–144.

Jackendoff, R. (2007). A parallel architecture perspective on language processing. *Brain Research, 1146,* 2–22.

Jackendoff, R., & Lerdahl, F. (2006). The capacity for music: What is it, and what's special about it? *Cognition, 100*(1), 33–72.

Keller, T. A., Carpenter, P. A., & Just, M. A. (2001). The neural bases of sentence comprehension: A fMRI examination of syntactic and lexical processing. *Cerebral Cortex, 11*(3), 223–237.

Koelsch, S., Gunter, T. C., von Cramon, D. Y., et al. (2002). Bach speaks: A cortical "language-network" serves the processing of music. *NeuroImage,* 17, 956–966.

Lee, D. K., Fedorenko, E., Simon, M. V., Curry, W. T., Nahed, B., Cahill, D. P., & Williams, Z. M. (2018). Neural encoding and production of functional morphemes in the posterior temporal lobe. *Nature Communications, 9,* 1877.

Lindell, A. K., & Hudry, K. (2013). Atypicalities in cortical structure, handedness, and functional lateralization for language in autism spectrum disorders. *Neuropsychology Review, 23*(3), 257–270.

Longcamp, M., Lagarrigue, A., Nazarian, B., Roth, M., Anton, J. L., Alario, F. X., & Velay, J. L. (2014). Functional specificity in the motor system: Evidence from coupled fMRI and kinematic recordings during letter and digit writing. *Human Brain Mapping, 35*(12), 6077–6087.

Maess, B., Koelsch, S., Gunter, T. C., et al. (2001). Musical syntax is processed in Broca's area: An MEG study. *Nature Neuroscience, 4,* 540–545.

Mahowald, K., & Fedorenko, E. (2016). Reliable individual-level neural markers of high-level language processing: A necessary precursor for relating neural variability to behavioral and genetic variability. *NeuroImage, 139,* 74–93.

Marr, D. (1982). Vision: a computational investigation into the human representation and processing of visual information. New York: Henry Holt.

Mather, M., Cacioppo, J. T., & Kanwisher, N. (2013). How fMRI can inform cognitive theories. *Perspectives on Psychological Science, 8*(1), 108–113.

McCandliss, B. D., Cohen, L., & Dehaene, S. (2003). The visual word form area: Expertise for reading in the fusiform gyrus. *Trends in Cognitive Sciences, 7*(7), 293–299.

Mesulam, M. M. (2001). Primary progressive aphasia. *Annals of Neurology, 49*(4), 425–432.

Monti, M. M., Parsons, L. M., & Osherson, D. N. (2012). Thought beyond language: Neural dissociation of algebra and natural language. *Psychological Science, 23*(8), 914–922.

Newman, A. J., Supalla, T., Hauser, P. C., Newport, E. L., & Bavelier, D. (2010). Prosodic and narrative processing in American Sign Language: An fMRI study. *NeuroImage, 52*(2), 669–676.

Nieto-Castanon, A., & Fedorenko, E. (2012). Subject-specific functional localizers increase sensitivity and functional resolution of multi-subject analyses. *NeuroImage, 63*(3), 1646–1669.

Norman-Haignere, S., Kanwisher, N. G., & McDermott, J. H. (2015). Distinct cortical pathways for music and speech revealed by hypothesis-free voxel decomposition. *Neuron, 88*(6), 1281–1296.

Overath, T., McDermott, J. H., Zarate, J. M., & Poeppel, D. (2015). The cortical analysis of speech-specific temporal structure revealed by responses to sound quilts. *Nature Neuroscience, 18*(6), 903–911.

Patel, A. D. (2003). Language, music, syntax and the brain. *Nature Neuroscience, 6*(7), 674–681.

Patel, A. D., Gibson, E., Ratner, J., Besson, M. & Holcomb, P. (1998). Processing grammatical relations in language and music: An event-related potential study. *Journal of Cognitive Neuroscience,* 10(6), 717–733.

Pereira, F., Lou, B., Pritchett, B., Ritter, S., Gershman, S. J., Kanwisher, N., Fedorenko, E. (2018). Toward a universal decoder of linguistic meaning from brain activation. *Nature Communications, 9,* 963.

Peretz, I., & Hyde, K. L. (2003). What is specific to music processing? Insights from congenital amusia. *Trends in Cognitive Sciences, 7*(8), 362–367.

Planton, S., Jucla, M., Roux, F. E., & Demonet, J. F. (2013). The "handwriting brain": A meta-analysis of neuroimaging studies of motor versus orthographic processes. *Cortex, 49*(10), 2772–2787.

Price, C., Thierry, G., & Griffiths, T. (2005). Speech-specific auditory processing: Where is it? *Trends in Cognitive Sciences, 9*(6), 271–276.

Pylkkänen, L., & Brennan, J. (forthcoming). The neurobiology of syntactic and semantic structure building. In David Poeppel, George R. Mangun, & Michael Gazzaniga (Eds.), *The cognitive neurosciences* (6th ed.). Cambridge, MA: MIT Press.

Roder, B., Stock, O., Neville, H., Bien, S., & Rosler, F. (2002). Brain activation modulated by the comprehension of normal and pseudo-word sentences of different processing demands: A fMRI study. *NeuroImage, 15*(4), 1003–1014.

Rogalsky, C., Rong, F., Saberi, K., & Hickok, G. (2011). Functional anatomy of language and music perception: Temporal and structural factors investigated using fMRI. *Journal of Neuroscience, 31*(10), 3843–3852.

Sandler, W., & Lillo-Martin, D. (2006). *Sign language and linguistic universals.* Cambridge: Cambridge University Press.

Scott, T. L., Gallee, J., & Fedorenko, E. (2016). A new fun and robust version of an fMRI localizer for the frontotemporal language system. *Cognitive Neuroscience, 8*(3), 167–176.

Seeley, W. W., Crawford, R. K., Zhou, J., Miller, B. L., & Greicius, M. D. (2009). Neurodegenerative diseases target large-scale human brain networks. *Neuron, 62*(1), 42–52.

Snijders, T. M., Vosse, T., Kempen, G., Van Berkum, J. J. A., Petersson, K. M., & Hagoort, P. (2009). Retrieval and unification of syntactic structure in sentence comprehension: An fMRI study using word-category ambiguity. *Cerebral Cortex, 19*(7), 1493–1503.

Tillmann, B., Janata, P., & Bharucha, J. (2003). Activation of the inferior frontal cortex in musical priming. *Cognitive Brain Research, 16,* 145–161.

Tremblay, P., & Dick, A. S. (2016). Broca and Wernicke are dead, or moving past the classic model of language neurobiology. *Brain and Language, 162,* 60–71.

Tyler, L. K., Marslen-Wilson, W. D., Randall, B., Wright, P., Devereux, B. J., Zhuang, J., & Stamatakis, E. A. (2011). Left inferior frontal cortex and syntax: Function, structure and behaviour in patients with left hemisphere damage. *Brain, 134,* 415–431.

van Heuven, W. J. B., & Dijkstra, T. (2010). Language comprehension in the bilingual brain: fMRI and ERP support for psycholinguistic models. *Brain Research Reviews, 64*(1), 104–122.

Whitaker, H. A., & Ojemann, G. A. (1977). Graded localisation of naming from electrical stimulation mapping of left cerebral cortex. *Nature, 270*(5632), 50.

Wilson, S. M. & Fridriksson, J. (forthcoming). Aphasia and aphasia recovery. In David Poeppel, George R. Mangun, & Michael Gazzaniga (Eds.), *The cognitive neurosciences* (6th ed.). Cambridge, MA: MIT Press.

Wilson, S. M., Lam, D., Babiak, M. C., Perry, D. W., Shih, T., Hess, C. P., & Chang, E. F. (2015). Transient aphasias after left hemisphere resective surgery. *Journal of Neurosurgery, 123*(3), 581–593.

76 Neural Processing of Word Meaning

JEFFREY R. BINDER AND LEONARDO FERNANDINO

ABSTRACT Accessing word meaning is a core process in language comprehension and production. Neuroimaging and neuropsychological data suggest that lexical semantic knowledge is partly "embodied" in perception, action, and emotion systems but that more abstract crossmodal or amodal representations also play a role. The evidence points to a hierarchical architecture in which modal association cortices converge at multiple levels, culminating in high-level temporal and inferior parietal lobe convergence zones that enable word associations and mapping between abstract semantic codes and phonological forms.

Concepts and Lexical Semantics

Language production and comprehension depend on the ability to connect linguistic (phonological and orthographic) forms with mental representations of concepts. The nature of concepts has been a major concern of philosophers for millennia and remains a central problem in linguistics and psychology. As usually understood, a *concept* is a representation (which may be relatively simple or complex) resulting from generalization over many similar experiences, capturing what is common to these experiences. The concept of a concrete object like *dog*, for example, is an idealized or schematic representation of the characteristics of previously experienced dogs. Concepts thus have defining intrinsic features (e.g., shapes, colors, parts, movements, sounds) but also exist within a complex network of other associated concepts. The concept *dog*, for example, may have associations with concepts like *friend*, *love*, *loyalty*, *leash*, *bone*, *walk*, *cat*, *breed*, *pedigree*, and more, established through co-occurrences in complex verbal and nonverbal experiences. Although concrete object concepts like *dog* have dominated much of the theoretical and empirical work on concepts (especially in the neuroimaging world), it should be obvious that the set of all concepts is as ontologically varied as the set of all content words in a language. Thus, concepts include not just concrete things and their sensory features but also concrete actions and events (represented in language mainly by verbs and sentences but also by nouns like *party* and *explosion*); quantity concepts (number, duration, and size); mental states and events (emotions and thoughts); complex social/behavioral constructs (*honor*, *loyalty*, *democracy*, *justice*); cognitive and scientific domains (*geometry*, *law*, *philosophy*); spatial, temporal, and causal relation concepts; and so on.

When concepts are useful to discuss, they are labeled with arbitrary symbols that are the words of a shared language, though certainly not all concepts are labeled in this way. The shape of a dog's head, for instance, has invariant properties that differ from the shape of a cat's head, but we have no need of labels for these shapes because they are so reliably associated with the more general concepts *dog* and *cat*. Cultures vary in what concepts they choose to label, as vividly demonstrated by the variation of color labeling across languages (Berlin & Kay, 1969) and by borrowed words like *schadenfreude*. *Semantics* refers to the formal study of meaning attached to linguistic signs (words, phrases, discourse). *Semantic memory* was later used to refer to conceptual knowledge stored in the brain (Tulving, 1972). Most theorists make a distinction between semantic memory stores and semantic memory retrieval mechanisms that search the memory store and select context-appropriate information. The general term *semantic processing* refers to the activation of semantic memory stores by either external stimuli or internal retrieval mechanisms.

Theories of Concept Representation

Modern theories of concept representation in the brain fall into three major types. The oldest, dating at least to 18th-century British empiricist philosophy (Locke, 1690/1959), holds that concepts are stored as combinations of sensory and action representations that constitute the content of the concept. The concept *dog*, for example, is equivalent to a set of schematic visual, auditory, tactile, and other representations derived from experiences with dogs. Activating the concept entails activation of this modal sensorimotor information. This account was the default theory among early brain scientists (Freud, 1891/1953; Wernicke, 1874) and has regained popularity in recent decades as the *embodied* or *grounded* view of concept representation (Allport, 1985; Barsalou, 1999). The alternative *symbolic* view arose in the latter half of the 20th century in connection with advancing computer technology, work on artificial intelligence, and the "cognitive revolution" in psychology. This theory holds that concepts are abstract, self-contained representations (Fodor, 1975; Pylyshyn, 1984). Just as retrieving a symbol in a computer system is equivalent to retrieving a meaning, the activation of a symbolic concept representation in the brain is sufficient for, and equivalent to, the activation of the concept it represents: the activation of a concept does not require access to associated perceptual content. Some proponents of this view acknowledge that perceptual content may be activated in the course of concept retrieval but argue that, by definition, modal perceptual content is not conceptual in nature (Mahon, 2015). A third "hybrid" view has gained adherents in recent years (Binder & Desai, 2011; Patterson, Nestor, & Rogers, 2007; Vigliocco, Meteyard, Andrews, & Kousta, 2009). According to this type of theory, concept representation involves both distributed modal (i.e., sensory, motor, affective, and others) content and more abstract or *amodal* representations, the latter viewed as necessary for highly associative processes such as concept learning via language (and word association in general).

A Large-Scale Network for Lexical Semantic Processing

The neuropsychological and neuroimaging literature on lexical concept representation and retrieval is vast and can only be broadly sketched here. In the neurological syndrome known as *transcortical sensory aphasia*, patients show an inability to understand spoken and written words despite normal hearing, vision, and phonological abilities, suggesting either damage to, or an inability to access, concept representations. Lesions causing the syndrome are generally large, involving ventral temporal, posterior parietal, and/or prefrontal cortex in the left hemisphere (Alexander, Hiltbrunner, & Fischer, 1989; Jefferies & Lambon Ralph, 2006; Otsuki et al., 1998; Rapcsak & Rubens, 1994), sparing phonological networks near the sylvian fissure. Beginning around 1990, systematic work on patients with the temporal lobe variant of frontotemporal dementia (*semantic dementia, semantic variant primary progressive aphasia*) showed that bilateral damage focused on the anterior half of the temporal lobe cortex can also produce a profound loss of lexical concept knowledge (Hodges, Patterson, Oxbury, & Funnell, 1992; Snowden, Goulding, & Neary, 1989). The lesion evidence thus suggests a broadly distributed network for semantic processing, involving much of the temporal lobe, as well as large regions of inferior parietal and prefrontal cortex.

fMRI data provide further evidence for this view. Binder, Desai, Conant, and Graves (2009) performed a voxel-wise meta-analysis of 87 neuroimaging studies examining the retrieval of general semantic knowledge. A notable feature of these experiments is that the lexical stimuli were chosen without regard to sensorimotor content or category membership; thus, the results can be interpreted as showing common brain areas involved in semantic processing regardless of specific conceptual content. Each study had to include a nonsemantic comparison task with controls for phonological, orthographic, and cognitive-control demands of the semantic task. The results (figure 76.1) reveal a distributed, left-lateralized network that includes (1) inferior parietal cortex (mainly the angular gyrus), (2) lateral and ventral temporal cortex (middle temporal, inferior temporal, fusiform, and parahippocampal gyri), (3) dorsal and medial prefrontal cortex (mainly the superior frontal gyrus), (4) ventrolateral prefrontal cortex (mainly pars orbitalis of the inferior frontal gyrus), and (5) the posterior cingulate gyrus and precuneus. The parietal and temporal zones are high-level crossmodal association areas distant from primary sensory and motor cortices and positioned at points of convergence across multiple sensory streams (Jones & Powell, 1970; Mesulam, 1985; Sepulcre, Sabuncu, Yeo, Liu, & Johnson, 2012).

The two frontal nodes of this network likely have distinct functions. Neuroimaging and neuropsychological evidence indicates that the ventrolateral prefrontal cortex plays a key role in the top-down activation and selection of conceptual information (Badre, Poldrack, Pare-Blagoev, Insler, & Wagner, 2005; Jefferies & Lambon Ralph, 2006; Thompson-Schill, D'Esposito, Aguirre, & Farah, 1997; Wagner, Pare-Blagoev, Clark, & Poldrack, 2001). Damage to this region impairs the ability to retrieve conceptual information, particularly when the context allows for many salient competing alternatives, the retrieval process is more complex or ambiguous, or retrieved information must be maintained in short-term memory. In contrast, damage to dorsomedial prefrontal cortex (superior frontal gyrus) does not impair concept selection per se but instead the ability to autonomously activate the selection process, manifesting as an inability to spontaneously generate nonformulaic language when no constraining cues are given (Alexander & Benson, 1993; Robinson, Blair, & Cipolotti, 1998). The dorsomedial prefrontal cortex lies between medial prefrontal areas involved in emotion and reward and lateral prefrontal networks involved in cognitive control and may act as a link between these

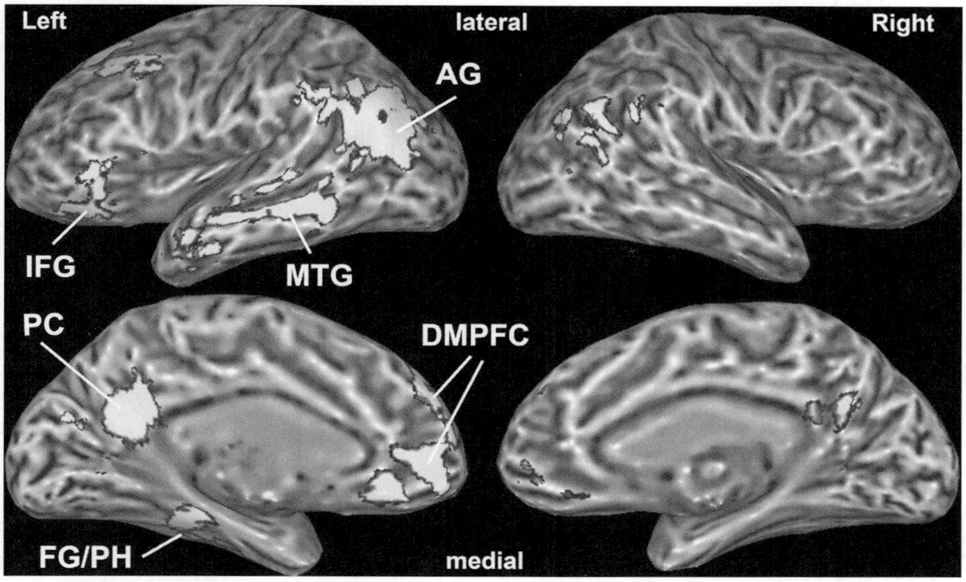

FIGURE 76.1 Results of an activation likelihood estimate meta-analysis of 87 published studies (691 activation foci) using controlled semantic contrasts (Binder et al., 2009). AG = angular gyrus; DMPFC = dorsomedial prefrontal cortex; FG/PH = fusiform and parahippocampal gyri; IFG = inferior frontal gyrus; MTG = middle temporal gyrus; PC = posterior cingulate/precuneus. (See color plate 88.)

processing systems, translating affective drive states into a plan for concept retrieval.

The Case for Experience-Based Concept Representations

As mentioned above, the representational content of conceptual knowledge in the brain has been a matter of intense ongoing debate. The "concepts as abstract symbols" approach, which has had many useful applications in artificial intelligence and cognitive science (e.g., semantic nets, spreading activation models, feature lists, ontologies, schemata), fails to address some core features of human semantic cognition. For example, people can pick out the referent of a word—that is, indicate a thing out in the environment that corresponds to the meaning of the symbol. A conceptual representation composed purely of symbols and their associations with other symbols would not have this capacity, demonstrating that the symbols are not "grounded" in physical reality (Harnad, 1990). They are more like the entries in a monolingual dictionary for a language that one does not know: looking up a word only leads to a set of more words one does not know. People also say they experience feelings, mental images, and other subjective *qualia* (such as the feeling of sadness or the experience of the color red) when they think about concepts. Such experiences are at the core of what distinguishes people from very sophisticated symbol-computing devices. Finally, symbol-based models are largely silent on the question of how conceptual knowledge is acquired.

The view that concepts are acquired as generalizations from everyday experiences provides a natural account of many of these phenomena. Over the course of many real-world experiences with a particular entity, invariant aspects of these sensory, motor, and affective experiences are encoded as increasingly abstract information within modality-specific processing systems. It is easy to see how perceptual "simulation" in these systems during concept retrieval allows people to indicate real-world referents of concepts and experience mental images and other qualia. General aspects of the theory are supported by a broad range of empirical data, including numerous studies showing the activation of modal sensory and motor regions during conceptual tasks (Binder & Desai, 2011; Kiefer & Pulvermüller, 2012; Martin, 2007; Meteyard, Rodriguez Cuadrado, Bahrami, & Vigliocco, 2012). It has been argued that sensorimotor systems are activated merely as a postconceptual epiphenomenon—that is, that this activation is not critical for concept understanding (Mahon & Caramazza, 2008). Countering this assertion is evidence that patients with specific motor (Bak & Hodges, 2004; Boulenger et al., 2008; Buxbaum & Saffran, 2002; Desai, Herter, Riccardi, Rorden, & Fridriksson, 2015; Fernandino et al., 2013; Grossman et al., 2008) or sensory (Bonner & Grossman, 2012; Trumpp, Kliese, Hoenig, Haarmeier, &

Kiefer, 2013) deficits may have specific difficulty processing corresponding conceptual knowledge. Studies using transcranial magnetic stimulation (TMS) to induce transient alterations of the motor system also attest to causal links between neural processing in these regions and action concept retrieval (Buccino et al., 2005; Ishibashi, Lambon Ralph, Saito, & Pobric, 2011).

Underlying some of the ongoing debate about these data are varying interpretations of what *sensorimotor cortex* refers to. Initial theories focused on the involvement (or lack thereof) of primary motor and sensory areas in concept representation (de Zubicaray, Arciuli, & McMahon, 2013; Hauk, Johnsrude, & Pulvermüller, 2004; Postle, McMahon, Ashton, Meredith, & de Zubicaray, 2008; Pulvermüller, 1999), whereas the majority of fMRI results have implicated modal association cortices located some distance away from primary areas (Binder & Desai, 2011; Fernandino, Binder, et al., 2016; Kiefer & Pulvermüller, 2012; Martin, 2007; Meteyard et al., 2012; Thompson-Schill, 2003; Watson, Cardillo, Ianni, & Chatterjee, 2013). An exclusive focus on primary cortices is unwarranted since all sensory and motor systems in the brain are known to be hierarchically organized, with increasingly abstract (i.e., schematic, conjunctive) representational content at higher levels (Simmons & Barsalou, 2003). The fMRI evidence suggests that unimodal conceptual content is stored mainly at higher levels of these hierarchical systems rather than in primary cortices. This explains why patients can have severe motor or perceptual deficits from primary cortex lesions or damage in subcortical white matter pathways but have little or no corresponding impairment of modality-specific concept processing (Mahon & Hickok, 2016).

The neural representation of concepts that do not have simple physical features presents something of a challenge to embodiment theories. Such "abstract" concepts make up a large portion of the lexicon (Recchia & Jones, 2012) and include, for example, products of cognition (*concept, theory, idea*), cognitive states or activities (*believe, ponder, doubt*), abstract situational entities (*criterion, clause, factor*), abstract attributes (*aspect, demeanor, extent*), complex social acts and situations (*cheat, imply, argument*), human mental traits (*honesty, curiosity, wisdom*), and so on. Barsalou (1999) notes that while such concepts do not have physical features, they are nevertheless learned from experiences, albeit often complex experiences that can include purely mental phenomena. Thus, they could be represented in the brain by spatially and temporally complex scenarios involving physical and mental events, rather than by simple sensorimotor information. Certain types of experience might also play a larger role in learning and representing abstract concepts, compared to concrete concepts. Abstract concepts tend to have strong affective and social content (Borghi, Flumini, Cimatti, Marocco, & Scorolli, 2011; Kousta, Vigliocco, Vinson, Andrews, & Del Campo, 2011; Vigliocco et al., 2009), and thus they might be represented to a greater extent in emotion and social cognition networks (Ross & Olson, 2010; Zahn et al., 2007).

Concept Representation in a Hierarchical Convergence Architecture

The model shown in figure 76.2 is a distillation of fMRI studies focused on lexical semantic processing, combined with functional and structural studies of high-level sensory and motor systems. The model distinguishes three levels of representation, referred to here as unimodal, multimodal, and transmodal. At the unimodal level, evidence to date implicates ventral visual areas in retrieving conceptual color knowledge (Fernandino, Binder, et al., 2016; Hsu, Kraemer, Oliver, Schlichting, & Thompson-Schill, 2011; Kellenbach, Brett, & Patterson, 2001; Martin, Haxby, Lalonde, Wiggs, & Ungerleider, 1995; Simmons et al., 2007), auditory association areas in retrieving sound-related knowledge (Fernandino, Binder, et al., 2016; Goldberg, Perfetti, & Schneider, 2006; Kellenbach, Brett, & Patterson, 2001; Kiefer, Sim, Herrnberger, Grothe, & Hoenig, 2008; Kurby & Zacks, 2013), and olfactory and gustatory association areas in retrieving corresponding odor and taste knowledge (Barros-Loscertales et al., 2012; Goldberg, Perfetti, & Schneider, 2006; González et al., 2006). In one recent study (Fernandino, Binder, et al., 2016), shape content (i.e., the degree to which a concept is defined by its shape features) correlated with fMRI activation in both visual areas associated with high-level shape perception (lateral occipital complex, ventral temporal-occipital junction) and somatosensory association areas implicated in tactile shape perception (Miquée et al., 2008). Emotion can also be considered a unimodal experience, and many imaging studies have examined brain activation as a function of the emotional content of words or phrases. There is a clear preponderance of activations in the temporal pole and ventromedial prefrontal cortex (Binder & Desai, 2011), which are areas believed to support cognitive aspects of emotion (Etkin, Egner, & Kalisch, 2011).

Multimodal regions combine information from two or more modalities. Multimodal sensory areas have been extensively documented, yet relatively few studies have addressed multimodal combinations in word meaning. In the most comprehensive study to date on this topic, Fernandino, Binder, et al. (2016) examined

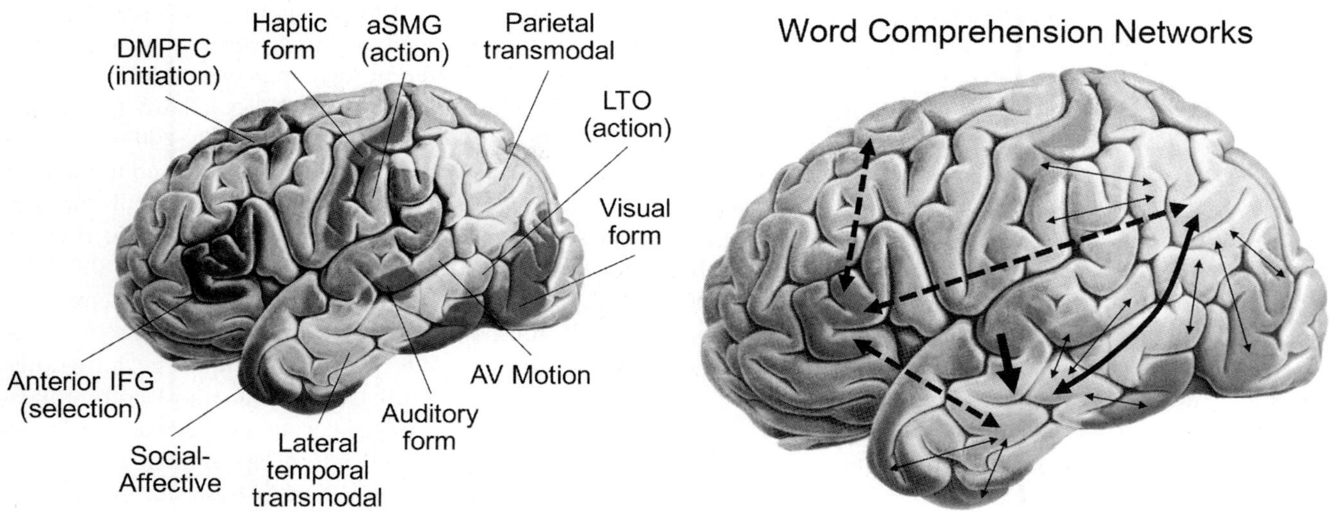

FIGURE 76.2 A schematic model of lexical storage and access networks, showing some principal unimodal (*yellow*), multimodal (*orange*), and transmodal (*red*) conceptual stores; semantic control regions (*green*); and speech perception (*cyan*) and phonological access (*blue*) areas. Spoken-word comprehension (*diagram at right*) involves mapping from auditory speech forms to high-level conceptual representations (*fat arrow*). The subsequent activation of multimodal and unimodal experiential representations (*thin arrows*) enables perceptual grounding and perceptual imagery and likely varies with task demands. Concept selection and information flow (depth of processing) are controlled by initiation and selection mechanisms in dorsomedial and inferolateral prefrontal cortex. (See color plate 89.)

blood-oxygen-level-dependent (BOLD) fMRI responses correlated with variation in the amount of action, color, shape, sound, and visual motion content of 900 noun concepts. A region in the left posterior superior temporal sulcus (STS) and middle temporal gyrus (MTG), known to be involved in audiovisual sensory integration and to receive inputs from both auditory association cortex and from visual motion perception area MT (Beauchamp, 2005), showed sensitivity to both auditory and visual motion semantic content (but not color, shape, or action content), suggesting that this region may store knowledge about correlated dynamic properties of objects in auditory and visual space. Other regions showed sensitivity to both shape and action (but not color, sound, or motion) content. One of these, the anterior supramarginal gyrus, is located near tertiary somatosensory association cortex and probably combines high-level proprioceptive, motor, and haptic shape information. Another, at the junction of the posterior middle temporal gyrus and the anterior occipital lobe (named the *lateral temporal-occipital area* by Fernandino, Binder et al., 2016), partly overlaps the lateral occipital complex and likely combines high-level visual shape and sensorimotor manipulation information. Both areas have been consistently implicated in object-directed action planning, action perception, and action execution (Caspers, Zilles, Laird, & Eickhoff, 2010; Grosbras, Beaton, & Eickhoff, 2012; Lewis, 2006), as well as in prior meta-analyses of tool and action concept processing (Binder et al., 2009; Watson et al., 2013).

At the highest level of convergence are brain regions that combine information from many experiential domains. Debate continues regarding the existence of such regions and the representational content they encode. Strong claims concern the anterior temporal lobe (ATL), where damage in patients with semantic dementia causes a profound multimodal loss of lexical semantic knowledge (Bozeat, Lambon Ralph, Patterson, Garrard, & Hodges, 2000; Rogers et al., 2004). Patterson, Nestor, and Rogers (2007) proposed a hybrid "hub and spoke" neuroanatomical model of concept representation based on this evidence, in which the ATL hub stores amodal concept representations that connect with distributed "spoke" systems, providing perceptual input to the hub for object and word recognition. The exact location of the ATL hub is unclear, however, as the pattern of damage in semantic dementia typically extends into the midportion and even posterior aspects of the temporal lobe (Rohrer et al., 2009), and lesion-behavioral correlation studies in semantic dementia have implicated various ventral and lateral temporal sites (Mion et al., 2010; Rogers et al., 2006). Much of the MTG also seems to play an important role as a hub (Bonner, Peelle, Cook, & Grossman, 2013; Sepulcre et al., 2012; Turken & Dronkers, 2011). Anterior portions of the ATL have been implicated more specifically in processing emotion and social concepts (Kober et al., 2008; Olson, Plotzker, & Ezzyat, 2007; Ross & Olson, 2010; Simmons, Reddish, Bellgowan, & Martin, 2010; Zahn et al., 2007). In addition to the ATL and MTG, there is strong evidence for broad convergence of information streams in the angular gyrus and posterior cingulate region (Bonner et al., 2013; Fernandino, Binder, et al., 2016; Sepulcre et al., 2012). Information encoded at these high-level hubs is variously claimed to be amodal (i.e., containing no modal content) or *heteromodal* (containing many kinds of content with no modal predominance). We use the neutral term *transmodal* to suggest a high level of abstraction arising from broadly multimodal conjunctions. Two fMRI studies suggested the preservation of multimodal information in these hub regions, particularly in the angular gyrus (Bonner et al., 2013; Fernandino, Humphries, Conant, Seidenberg, & Binder, 2016).

Highly abstract, transmodal representations are thought to have several functions in semantic cognition (Binder, 2016). They provide a computationally efficient means of capturing multimodal conceptual similarity (Patterson, Nestor, & Rogers, 2007; Rogers & McClelland, 2004), they provide a mechanism for learning purely thematic (non-feature-based) word associations and word definitions through language (Binder, 2016; Dove, 2011; Hoffman, McClelland, & Lambon Ralph, 2018; Vigliocco et al., 2009), and they facilitate arbitrary mappings between semantic and phonological representations.

Lexical Semantic Access

Given the spatial proximity of the superior temporal lobe phoneme perceptual system to the lateral temporal conceptual hub (figure 76.2), it seems likely that spoken word representations first activate transmodal concept representations in the lateral temporal hub, with activation then spreading to modal concept features and thematically associated concepts represented throughout posterior association cortex. Long-range temporoparietal white matter fasciculi—principally the inferior and middle longitudinal fasciculi (Zhang et al., 2010)—enable the rapid transmission of information across this temporal-parietal-occipital network. Frontal lobe selection mechanisms are engaged to varying degrees during these processes depending on stimulus characteristics and task demands. For example, concept retrieval includes the transient activation of conceptual representations for phonological "neighbors" of the input word (Marslen-Wilson & Welsh, 1978); frontal lobe selection mechanisms likely play a role in inhibiting these activations. Similarly, context-based selection is required to resolve conceptual ambiguity arising from homonymy and polysemy (i.e., words that sound the same but have different meanings or senses; Hino, Lupker, & Pexman, 2002; Hoffman, McClelland, & Lambon Ralph, 2018; Rodd, Davis, & Johnsrude, 2005). Conceptual content activated by a word is not limited to intrinsic attributes of the target concept but also typically includes a network of associated concepts and pragmatic information (Hare, Jones, Thomson, Kelly, & McRae, 2009); frontal lobe mechanisms select (i.e., selectively activate or inhibit) components of this broad conceptual representation to suit the needs of the moment. Large, long-range white matter fasciculi that enable rapid communication between frontal, temporal, and parietal cortices—principally the inferior fronto-occipital fasciculus—likely play a central role in these frontal-posterior interactions (Duffau et al., 2005).

REFERENCES

Alexander, M. P., and D. F. Benson. 1993. The aphasias and related disturbances. In R. J. Joynt (Ed.), *Clinical neurology*. Philadelphia: J. B. Lipincott.

Alexander, M. P., B. Hiltbrunner, and R. S. Fischer. 1989. Distributed anatomy of transcortical sensory aphasia. *Archives of Neurology* 46:885–892.

Allport, D. A. 1985. Distributed memory, modular subsystems and dysphasia. In S. K. Newman and R. Epstein (Eds.), *Current perspectives in dysphasia*. Edinburgh: Churchill Livingstone.

Badre, D., R. A. Poldrack, E. J. Pare-Blagoev, R. Z. Insler, and A. D. Wagner. 2005. Dissociable controlled retrieval and generalized selection mechanisms in ventrolateral prefrontal cortex. *Neuron* 47:907–918.

Bak, T. H., and J. R. Hodges. 2004. The effects of motor neurone disease on language: Further evidence. *Brain and Language* 89:354–361.

Barros-Loscertales, A. , J. Gonzalez, F. Pulvermüller, N. Ventura-Campos, J. C. Bustamante, V. Costumero, M. A. Parcet, and C. Avila. 2012. Reading salt activates gustatory brain regions: fMRI evidence for semantic grounding in a novel sensory modality. *Cerebral Cortex* 22:2554–2563.

Barsalou, L. W. 1999. Perceptual symbol systems. *Behavioral and Brain Sciences* 22:577–660.

Beauchamp, M. S. 2005. See me, hear me, touch me: Multisensory integration in lateral occipital-temporal cortex. *Current Opinion in Neurobiology* 15:145–153.

Berlin, B., and P. Kay. 1969. *Basic color terms: Their universality and evolution*. Berkeley: University of California Press.

Binder, J. R. 2016. In defense of abstract conceptual representations. *Psychonomic Bulletin and Review* 23:1096–1108.

Binder, J. R., and R. H. Desai. 2011. The neurobiology of semantic memory. *Trends in Cognitive Sciences* 15:527–536.

Binder, J. R., R. H. Desai, L. L. Conant, and W. W. Graves. 2009. Where is the semantic system? A critical review and meta-analysis of 120 functional neuroimaging studies. *Cerebral Cortex* 19:2767–2796.

Bonner, M. F., and M. Grossman. 2012. Gray matter density of auditory association cortex relates to knowledge of sound concepts in primary progressive aphasia. *Journal of Neuroscience* 32:7986–7991.

Bonner, M. F., J. E. Peelle, P. A. Cook, and M. Grossman. 2013. Heteromodal conceptual processing in the angular gyrus. *NeuroImage* 71:175–186.

Borghi, A. M., A. Flumini, F. Cimatti, D. Marocco, and C. Scorolli. 2011. Manipulating objects and telling words: A study on concrete and abstract words acquisition. *Frontiers in Psychology* 2:15.

Boulenger, V., L. Mechtouff, S. Thobois, E. Broussolle, M. Jeannerod, and T. A. Nazir. 2008. Word processing in Parkinson's disease is impaired for action verbs but not for concrete nouns. *Neuropsychologia* 46:743–756.

Bozeat, S., M. A. Lambon Ralph, K. Patterson, P. Garrard, and J. R. Hodges. 2000. Nonverbal semantic impairment in semantic dementia. *Neuropsychologia* 38:1207–1215.

Buccino, G., L. Riggio, G. Melli, F. Binkofski, V. Gallese, and G. Rizzolatti. 2005. Listening to action-related sentences modulates the activity of the motor system: A combined TMS and behavioral study. *Brain Research: Cognitive Brain Research* 24:355–363.

Buxbaum, L. J., and E. M. Saffran. 2002. Knowledge of object manipulation and object function: Dissociations in apraxic and nonapraxic subjects. *Brain and Language* 82:179–199.

Caspers, S., K. Zilles, A. R. Laird, and S. B. Eickhoff. 2010. ALE meta-analysis of action observation and imitation in the human brain. *NeuroImage* 50:1148–1167.

Desai, R. H., T. Herter, N. Riccardi, C. Rorden, and J. Fridriksson. 2015. Concepts within reach: Action performance predicts action language processing in stroke. *Neuropsychologia* 7:217–224.

de Zubicaray, G., J. Arciuli, and K. McMahon. 2013. Putting an "end" to the motor cortex representations of action words. *Journal of Cognitive Neuroscience* 25:1957–1974.

Dove, G. 2011. On the need for embodied and dis-embodied cognition. *Frontiers in Psychology* 1:Article 242.

Duffau, H., P. Gatignol, E. Mandonnet, P. Peruzzi, N. Tzourio-Mazoyer, and L. Capelle. 2005. New insights into the anatomo-functional connectivity of the semantic system: A study using cortico-subcortical electrostimulations. *Brain* 128:797–810.

Etkin, A., T. Egner, and R. Kalisch. 2011. Emotional processing in anterior cingulate and medial prefrontal cortex. *Trends in Cognitive Sciences* 15:85–93.

Fernandino, L., J. R. Binder, R. H. Desai, S. L. Pendl, C. J. Humphries, W. Gross, L. L. Conant, and M. S. Seidenberg. 2016. Concept representation reflects multimodal abstraction: A framework for embodied semantics. *Cerebral Cortex* 26:2018–2034.

Fernandino, L., L. L. Conant, J. R. Binder, K. Blindauer, B. Hiner, K. Spangler, and R. H. Desai. 2013. Parkinson's disease disrupts both automatic and controlled processing of action verbs. *Brain and Language* 127:65–74.

Fernandino, L., C. J. Humphries, L. L. Conant, M. S. Seidenberg, and J. R. Binder. 2016. Heteromodal cortical areas encode sensory-motor features of word meaning. *Journal of Neuroscience* 36:9763–9769.

Fodor, J. 1975. *The language of thought.* Cambridge, MA: Harvard University Press.

Freud, S. 1891/1953. *On aphasia: A critical study.* Madison, CT: International Universities Press.

Goldberg, R. F., C. A. Perfetti, and W. Schneider. 2006. Distinct and common cortical activations for multimodal semantic categories. *Cognitive, Affective and Behavioral Neuroscience* 6:214–222.

González, J., A. Barros-Loscertales, F. Pulvermüller, V. Meseguer, A. Sanjuán, V. Belloch, and C. Avila. 2006. Reading cinnamon activates olfactory brain regions. *NeuroImage* 32:906–912.

Grosbras, M.-H., S. Beaton, and S. B. Eickhoff. 2012. Brain regions involved in human movement perception: A quantitative voxel-based meta-analysis. *Human Brain Mapping* 33:431–454.

Grossman, M., C. Anderson, A. Khan, B. Avants, L. Elman, and L. McCluskey. 2008. Impaired action knowledge in amyotrophic lateral sclerosis. *Neurology* 71:1396–1401.

Hare, M., M. N. Jones, C. Thomson, S. Kelly, and K. McRae. 2009. Activating event knowledge. *Cognition* 111:151–167.

Harnad, S. 1990. The symbol grounding problem. *Physica D: Nonlinear Phenomena* 42:335–346.

Hauk, O., I. Johnsrude, and F. Pulvermüller. 2004. Somatotopic representation of action words in human motor and premotor cortex. *Neuron* 41:301–307.

Hino, Y., S. J. Lupker, and P. M. Pexman. 2002. Ambiguity and synonymy effects in lexical decision, naming, and semantic categorization tasks: Interactions between orthography, phonology, and semantics. *Journal of Experimental Psychology: Learning, Memory, and Cognition* 28:686–713.

Hodges, J. R., K. Patterson, S. Oxbury, and E. Funnell. 1992. Semantic dementia: Progressive fluent aphasia with temporal lobe atrophy. *Brain* 115:1783–1806.

Hoffman, P., J. L. McClelland, and M. A. Lambon Ralph. 2018. Concepts, control, and context: A connectionist account of normal and disordered semantic cognition. *Psychological Review* 125:293–328.

Hsu, N. S., D. J. M. Kraemer, R. T. Oliver, M. L. Schlichting, and S. L. Thompson-Schill. 2011. Color, context, and cognitive style: Variations in color knowledge retrieval as a function of task and subject variables. *Journal of Cognitive Neuroscience* 1–14. doi:10.1162/jocn.2011.21619

Ishibashi, R., M. A. Lambon Ralph, S. Saito, and G. Pobric. 2011. Different roles of lateral anterior temporal lobe and inferior parietal lobule in coding function and manipulation tool knowledge: Evidence from an rTMS study. *Neuropsychologia* 49:1128–1135.

Jefferies, E., and M. A. Lambon Ralph. 2006. Semantic impairment in stroke aphasia versus semantic dementia: A case-series comparison. *Brain* 129:2132–2147.

Jones, E. G., and T. S. P. Powell. 1970. An anatomical study of converging sensory pathways within the cerebral cortex of the monkey. *Brain* 93:793–820.

Kellenbach, M. L., M. Brett, and K. Patterson. 2001. Large, colourful or noisy? Attribute- and modality-specific activations during retrieval of perceptual attribute knowledge. *Cognitive, Affective, and Behavioral Neuroscience* 1: 207–221.

Kiefer, M., and F. Pulvermüller. 2012. Conceptual representations in mind and brain: Theoretical developments, current evidence and future directions. *Cortex* 48:805–825.

Kiefer, M., E.-J. Sim, B. Herrnberger, J. Grothe, and K. Hoenig. 2008. The sound of concepts: Four markers for a link between auditory and conceptual brain systems. *Journal of Neuroscience* 28:12224–12230.

Kober, H., L. F. Barrett, J. Joseph, E. Bliss-Moreau, K. Lindquist, and T. D. Wager. 2008. Functional grouping and cortical-subcortical interactions in emotion: A meta-analysis of neuroimaging studies. *NeuroImage* 42:998–1031.

Kousta, S.-T., G. Vigliocco, D. P. Vinson, M. Andrews, and E. Del Campo. 2011. The representation of abstract words: Why emotion matters. *Journal of Experimental Psychology: General* 140:14–34.

Kurby, C. A., and J. M. Zacks. 2013. The activation of modality-specific representations during discourse processing. *Brain and Language* 126:338–349.

Lewis, J. W. 2006. Cortical networks related to human use of tools. *Neuroscientist* 12:211–231.

Locke, J. 1690/1959. *An essay concerning human understanding.* New York: Dover.

Mahon, B. Z. 2015. What is embodied about cognition? *Language, Cognition and Neuroscience* 30:420–429.

Mahon, B. Z., and A. Caramazza. 2008. A critical look at the embodied cognition hypothesis and a new proposal for grounding conceptual content. *Journal of Physiology-Paris* 102:59–70.

Mahon, B. Z., and G. Hickok. 2016. Arguments about the nature of concepts: Symbols, embodiment, and beyond. *Psychonomic Bulletin & Review* 23:941–958.

Marslen-Wilson, W. D., and A. Welsh. 1978. Processing interactions and lexical access during word recognition in continuous speech. *Cognitive Psychology* 10:29–63.

Martin, A. 2007. The representation of object concepts in the brain. *Annual Review of Psychology* 58:25–45.

Martin, A., J. V. Haxby, F. M. Lalonde, C. L. Wiggs, and L. G. Ungerleider. 1995. Discrete cortical regions associated with knowledge of color and knowledge of action. *Science* 270:102–105.

Mesulam, M. 1985. Patterns in behavioral neuroanatomy: Association areas, the limbic system, and hemispheric specialization. In M. Mesulam (Ed.), *Principles of behavioral neurology.* Philadelphia: F. A. Davis.

Meteyard, L., S. Rodriguez Cuadrado, B. Bahrami, and G. Vigliocco. 2012. Coming of age: A review of embodiment and the neuroscience of semantics. *Cortex* 48:788–804.

Mion, M., K. Patterson, J. Acosta-Cabronero, G. Pengas, D. Izquierdo-Garcia, Y. T. Hong, T. D. Fryer, G. B. Williams, J. R. Hodges, and P. J. Nestor. 2010. What the left and right anterior fusiform gyri tell us about semantic memory. *Brain* 133:3256–3268.

Miquée, A., C. Xerri, C. Rainville, J. L. Anton, B. Nazarian, M. Roth, and Y. Zennou-Azogui. 2008. Neuronal substrates of haptic shape encoding and matching: A functional magnetic resonance imaging study. *Neuroscience* 152:29–39.

Olson, I. R., A. Plotzker, and Y. Ezzyat. 2007. The enigmatic temporal pole: A review of findings on social and emotional processing. *Brain* 130:1718–1731.

Otsuki, M., Y. Soma, A. Koyama, N. Yoshimura, H. Furukawa, and S. Tsuji. 1998. Transcortical sensory aphasia following left frontal infarction. *Journal of Neurology* 245:69–76.

Patterson, K., P. J. Nestor, and T. T. Rogers. 2007. Where do you know what you know? The representation of semantic knowledge in the human brain. *Nature Reviews Neuroscience* 8:976–987.

Postle, N., K. L. McMahon, R. Ashton, M. Meredith, and G. I. de Zubicaray. 2008. Action word meaning representations in cytoarchitectonically defined primary and premotor cortices. *NeuroImage* 43:634–644.

Pulvermüller, F. 1999. Words in the brain's language. *Behavioral and Brain Sciences* 22:253–336.

Pylyshyn, Z. W. 1984. *Computation and cognition: Toward a foundation for cognitive science.* Cambridge, MA: MIT Press.

Rapcsak, S. Z., and A. B. Rubens. 1994. Localization of lesions in transcortical aphasia. In A. Kertesz (Ed.), *Localization and neuroimaging in neuropsychology.* San Diego: Academic Press.

Recchia, G., and M. N. Jones. 2012. The semantic richness of abstract concepts. *Frontiers in Human Neuroscience* 6:Article 315.

Robinson, G., J. Blair, and L. Cipolotti. 1998. Dynamic aphasia: An inability to select between competing verbal responses? *Brain* 121:77–89.

Rodd, J. M., M. H. Davis, and I. S. Johnsrude. 2005. The neural mechanisms of speech comprehension: fMRI studies of semantic ambiguity. *Cerebral Cortex* 15:1261–1269.

Rogers, T. T., P. Garrard, J. L. McClelland, M. A. Lambon Ralph, S. Bozeat, J. R. Hodges, and K. Patterson. 2004. Structure and deterioration of semantic memory: A neuropsychological and computational investigation. *Psychological Review* 111:205–235.

Rogers, T. T., J. Hocking, U. Noppeney, A. Mechelli, M. L. Gorno-Tempini, K. Patterson, and C. J. Price. 2006. Anterior temporal cortex and semantic memory: Reconciling findings from neuropsychology and functional imaging. *Cognitive, Affective and Behavioral Neuroscience* 6:201–213.

Rogers, T. T., and J. L. McClelland. 2004. *Semantic cognition: A parallel distributed processing approach.* Cambridge, MA: MIT Press.

Rohrer, J. D., J. D. Warren, M. Modat, G. R. Ridgway, A. Douiri, M. N. Rossor, S. Ourselin, and N. C. Fox. 2009. Patterns of cortical thinning in the language variants of frontotemporal lobar degeneration. *Neurology* 72:1562–1569.

Ross, L. A., and I. R. Olson. 2010. Social cognition and the anterior temporal lobes. *NeuroImage* 49:3452–3462.

Sepulcre, J., M. R. Sabuncu, T. B. Yeo, H. Liu, and K. A. Johnson. 2012. Stepwise connectivity of the modal cortex reveals the multimodal organization of the human brain. *Journal of Neuroscience* 32:10649–10661.

Simmons, W. K., and L. W. Barsalou. 2003. The similarity-in-topography principle: Reconciling theories of conceptual deficits. *Cognitive Neuropsychology* 20:451–486.

Simmons, W. K., V. Ramjee, M. S. Beauchamp, K. McRae, A. Martin, and L. W. Barsalou. 2007. A common neural substrate for perceiving and knowing about color. *Neuropsychologia* 45:2802–2810.

Simmons, W. K., M. Reddish, P. S. Bellgowan, and A. Martin. 2010. The selectivity and functional connectivity of the anterior temporal lobes. *Cerebral Cortex* 20:813–825.

Snowden, J. S., P. J. Goulding, and D. Neary. 1989. Semantic dementia: A form of circumscribed temporal atrophy. *Behavioural Neurology* 2:167–182.

Thompson-Schill, S. L. 2003. Neuroimaging studies of semantic memory: Inferring "how" from "where." *Neuropsychologia* 41:280–292.

Thompson-Schill, S. L., M. D'Esposito, G. K. Aguirre, and M. J. Farah. 1997. Role of left inferior prefrontal cortex in retrieval of semantic knowledge: A reevaluation. *Proceedings of the National Academy of Sciences of the United States of America* 94:14792–14797.

Trumpp, N. M., D. Kliese, K. Hoenig, T. Haarmeier, and M. Kiefer. 2013. Losing the sound of concepts: Damage to auditory association cortex impairs the processing of sound-related concepts. *Cortex* 49:474–486.

Tulving, E. 1972. Episodic and semantic memory. In E. Tulving and W. Donaldson (Eds.), *Organization of memory*. New York: Academic Press.

Turken, A. U., and N. F. Dronkers. 2011. The neural architecture of the language comprehension network: Converging evidence from lesion and connectivity analyses. *Frontiers in Systems Neuroscience* 5:Article 1.

Vigliocco, G., L. Meteyard, M. Andrews, and S. Kousta. 2009. Toward a theory of semantic representation. *Language and Cognition* 1:219–248.

Wagner, A. D., E. J. Pare-Blagoev, J. Clark, and R. A. Poldrack. 2001. Recovering meaning: Left prefrontal cortex guides semantic retrieval. *Neuron* 31:329–338.

Watson, C. E., E. R. Cardillo, G. R. Ianni, and A. Chatterjee. 2013. Action concepts in the brain: An activation likelihood estimation meta-analysis. *Journal of Cognitive Neuroscience* 25:1191–1205.

Wernicke, C. 1874. *Der Aphasische Symptomenkomplex*. Breslau: Cohn & Weigert.

Zahn, R., J. Moll, F. Krueger, E. D. Huey, G. Garrido, and J. Grafman. 2007. Social concepts are represented in the superior anterior temporal cortex. *Proceedings of the National Academy of Sciences of the United States of America* 104: 6430–6435.

Zhang, Y., J. Zhang, K. Oishi, A. V. Faria, H. Jiang, X. Li, K. Akhter, P. Rosa-Neto, G. B. Pike, A. Evans, A. W. Toga, R. Woods, J. C. Mazziotta, M. I. Miller, P. C. M. van Zijl, and S. Mori. 2010. Atlas-guided tract reconstruction for automated and comprehensive examination of the white matter anatomy. *NeuroImage* 52:1289–1301.

77 Neural Mechanisms Governing the Perception of Speech under Adverse Listening Conditions

PATTI ADANK

ABSTRACT Listeners are able to understand each other in a wide variety of adverse listening conditions. Listening conditions that present a challenge to speech perception can be attributed to environmental and/or source-related distortions. Environmental distortions originate from outside the speaker and include background sounds such as noise (energetic masking) or competing speakers (informational masking). For source distortions, degradation originates from the speaker's speech style or voice (e.g., an unfamiliar accent). This chapter integrates results from neuroimaging (e.g., functional magnetic resonance imaging) and neurostimulation (e.g., transcranial magnetic stimulation) studies focusing on the cognitive and neural mechanisms governing listening under adverse listening conditions. Neuroimaging studies indicate that the neural substrates for processing speech in adverse listening conditions compared to speech in quiet conditions are distributed across temporal, frontal, and medial areas. Informational masking tends to recruit a network of areas associated with auditory processing (particularly superior temporal cortex), while energetic masking and source distortions recruit additional areas, including motor and premotor regions. Neurostimulation studies suggest that premotor cortex is crucial for processing speech in energetic maskers. Future studies using a combination of both methods can further elucidate the precise neural mechanisms involved in understanding speech under distinct adverse listening conditions through the systematic scrutiny of areas across temporal as well as (pre)motor regions.

Perceiving speech in everyday situations seems effortless. We are able to understand each other in a wide variety of ecological situations. This ability of human listeners to perceive speech in demanding circumstances demonstrates the robustness and flexibility of the human spoken-language comprehension system. Speech perception is defined here in the broadest sense: as all auditory, cognitive, and neural processes required to classify, understand, and interpret spoken utterances at all linguistic levels, from phoneme to discourse. Most of everyday speech perception in fact occurs in adverse listening conditions, and it is fairly rare that a conversation occurs under ideal listening conditions—that is, in quiet, with our full attention on the conversation and speaking to someone whose voice and speaking style is familiar. Speech perception in adverse listening conditions is often slower and less efficient than under less challenging conditions. Adverse conditions that present a challenge to speech perception can be classified into environmental and source distortions (Mattys, Davis, Bradlow, & Scott, 2012). First, the speech signal can be masked by distortions originating from the speaker's environment, such as background noise or competing speakers (figure 77.1). Second, the distortions can originate from the source—that is, directly from the speaker's speech production—for example, a hoarse voice or an unfamiliar regional or foreign accent.

Environmental distortions can be further classified into two main types: energetic and informational (Mattys et al., 2012). Energetic distortions are defined as variation sources masking the target speech spectrally and temporally—for example, simultaneous background noise. The presence of background noise tends to decrease the intelligibility of the speech signal. It has been possible since the 1950s to predict the relative intelligibility of the speech signal based on its signal-to-noise ratio (SNR). Lower SNRs decrease the intelligibility of the speech signal for speech-shaped noise (i.e., noise with the long-term spectral characteristics of speech). Informational distortions are generally defined as competing speech signals—for example, the presence of one or more background speakers. As energetic maskers, they tend to completely block the target speech spectrotemporally and compete with the speech signal at the level of the cochlea.

Informational masking can be defined as the acoustic consequences of the informational distortion after all acoustic consequences of the energetic masking are accounted for (Mattys et al., 2012). The effects of informational masking on intelligibility are less straightforward to pinpoint than those of energetic masking

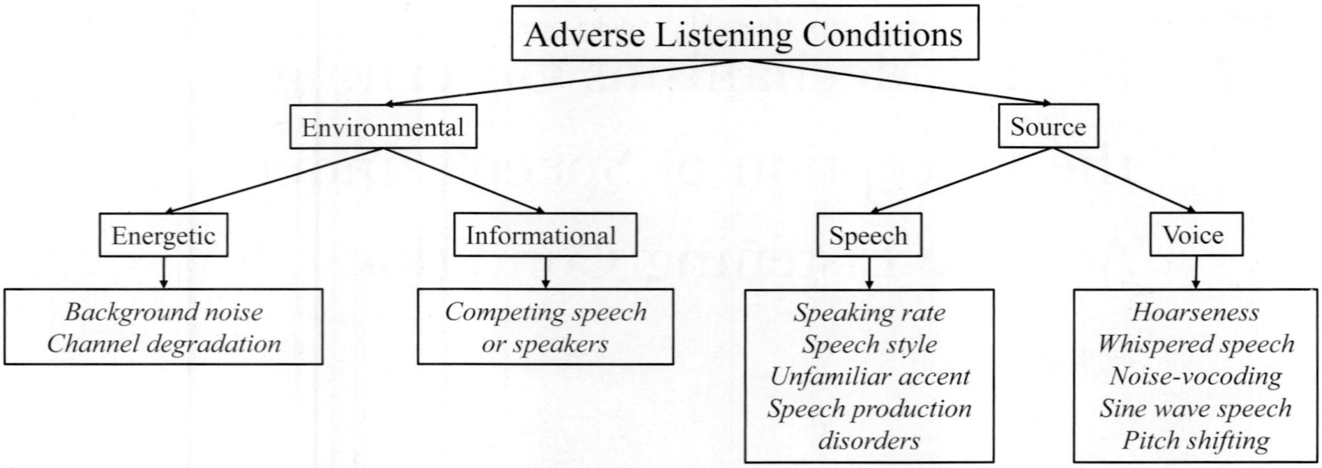

FIGURE 77.1 Overview of the types of adverse listening conditions discussed in this chapter.

because informational masking signals often allow listeners to glimpse parts of the target due to the fluctuating spectral amplitude of the masking signal (Cooke, 2003). Moreover, in contrast with energetic masking, the extent to which informational masking affects speech perception is dependent on the segmental and lexical familiarity of the listener with the masker. Speech perception is more perturbed by informational maskers containing semantically observable information—for example, with babble noise constructed from intelligible speakers.

Source distortions originate from the speaker's style of speech (e.g., regional or foreign accent, fast or slow speech rate, sloppy or formal speaking style) or voice (e.g., hoarse voice, noise-vocoded speech). Listeners tend to show less efficient perception for speech in an unfamiliar regional accent, specifically when combined with an environmental masker (Adank, Evans, Stuart-Smith, & Scott, 2009), for fast speech (Dupoux & Green, 1997), and for noise-vocoded speech (Davis, Johnsrude, Hervais-Adelman, Taylor, & McGettigan, 2005). Fast speech is generally generated using artificial time compression, using a manipulation that reduces the utterance duration without affecting its fundamental frequency. Noise-vocoded speech is created by passing the original speech signal through a channel noise vocoder. Noise-vocoded speech sounds like a harsh, rough whisper yet is largely intelligible (depending on the number of channels used; > six channels is intelligible), but the harmonic structure is no longer intact, so the intonation pattern is disrupted.

This chapter will provide an overview of how the processing of adverse listening conditions has been investigated using functional neuroimaging methods, specifically functional magnetic resonance imaging (fMRI) and positron emission tomography (PET), and brain stimulation methods, such as transcranial magnetic stimulation (TMS). fMRI and PET are ideally suited for outlining the network of brain areas involved in speech processing in adverse listening conditions. However, it remains unclear to which extent any brain areas active during the processing of adverse listening conditions are causally involved, as neuroimaging methods can only establish a correlative link between the activation of a brain area and task performance. Neurostimulation methods, such as TMS, involve the direct stimulation of neural tissues using a pulse delivered noninvasively through the scalp and skull. Specifically, TMS can be used in two main ways: First, unlike neuroimaging methods, TMS can establish causal links by temporarily disrupting neural functioning in a target brain area and measuring task performance before and after stimulation. If task performance is affected poststimulation, then a causal link can be assumed. Second, TMS can be used to determine the extent to which primary motor cortex (M1) is facilitated during task performance or perception by measuring motor evoked potentials (MEPs). MEPs are comparable to fMRI/PET in terms of explanative power, as they are also used to show correlational links between behavior and brain activation (Adank, Nuttall, & Kennedy-Higgins, 2016). This chapter discusses neuroimaging and neurostimulation studies related to environmental and source distortions, with the aim of elucidating the neural mechanisms associated with processing speech in adverse listening conditions in general.

Neuroimaging

Environmental: energetic Several fMRI studies scanned participants while they listened to speech target stimuli in the presence of energetic maskers. Osnes, Hugdahl, and Specht (2011) presented participants with consonant-vowel (CV, /da/ and /ta/) syllables in quiet and in seven SNRs of white noise, in a sparse sampling design. They also presented participants with nonspeech sounds and musical sounds (piano or guitar chords), and participants were to identify the stimuli as speech, noise, or music. Osnes, Hugdahl, and Specht (2011) report a graded increase in activation in the left superior temporal sulcus (STS) for decreasing SNRs. Premotor cortex activity was present at intermediate SNRs, when the syllables were identifiable but still distorted. Premotor activity was not reported for syllables in the most favorable SNRs.

Participants in Du, Buchsbaum, Grady, and Alain (2014) identified the initial phoneme in four CV syllables (/ba/, /ma/, /da/, or /ta/) presented in six SNRs (−2, −9, −6, −2, 8 dB, and in quiet). Du et al. (2014) tested the hypothesis that speech production motor areas contribute to categorical speech perception under adverse, but not quiet, listening conditions. A negative correlation was observed between neural activity and perceptual accuracy in left premotor cortex, which specifically contributed to phoneme categorization at moderate-to-adverse SNRs.

Wong, Uppanda, Parrish, and Dhar (2008) presented participants with words in quiet, in moderately loud noise (+20 dB SNR), and in loud noise (−5 dB SNR). Wong et al. (2008) used a sparse temporal scanning paradigm, thus ensuring that the stimuli were presented in relative silence. The noise was multitalker babble noise, classified here as an energetic masker. They report increased activation in the posterior superior temporal gyrus (STG) and left anterior insula for the words presented in −5 dB SNR noise compared to +20 dB SNR noise.

Adank, Davis, and Hagoort (2012) scanned listeners while they performed a semantic verification task for sentences in quiet and background noise (+2 dB SNR). Compared to sentences in quiet, listening to sentences in noise was associated with increased activation in the left inferior frontal gyrus (IFG) and the left frontal operculum (FO) and medial areas including the anterior cingulate cortex (ACC), parahippocampal gyrus, and caudate nucleus.

Zekveld, Heslenfeld, Festen, and Schoonhoven (2006) presented participants with sentences in increasing noise levels; the SNR was varied in 144 steps between +5 dB and −35 dB SNR. Higher activation was found in the left middle frontal gyrus (MFG), left IFG, and bilateral temporal areas for increasing noise levels.

Finally, Hwang, Wu, Chen, and Liu (2006) measured neural responses while participants heard stories in quiet or mixed with white noise at +5 dB SNR. They report reduced activation in the left superior and middle temporal gyri, parahippocampal gyrus, cuneus, and thalamus for the +5 dB condition relative to speech in quiet. They also report reductions in the right lingual gyrus, anterior and middle STG, uncus, fusiform gyrus, and right IFG.

Environmental: informational Several fMRI and PET studies scanned participants while they listened to speech target stimuli in the presence of informational maskers, or studies directly compared the neural networks associated with processing speech in the presence of informational or energetic maskers. Dole, Meuneir, and Hoen (2014) investigated neural correlates of speech-in-speech perception (informational masking) in neurotypical controls and participants with dyslexia (not discussed here) using fMRI. Listeners performed a subjective intelligibility-rating test with single words played against concurrent maskers consisting of babble noise from four speakers. In the condition designed to maximize informational masking, target words were presented to the right ear, whereas babble noise was presented to the left ear at equal intensity. The authors argue that a second condition maximized energetic masking, as both the target word and noise were presented to the right ear only at an SNR of 0 dB. In this condition, both signals were to be encoded in the same cochlea, thus maximizing energetic masking (albeit using a noise signal that is classified here as an informational masker). The informational masking minus energetic masking contrast showed increases in the blood oxygen level-dependent (BOLD) response in the right STG, while the reverse contrast showed increased activity in the right IFG, left MFG, left STG, and left supplementary motor area (SMA).

Using PET, Scott, Rosen, Beaman, Davis, and Wise (2009) examined the neural effects of masking from speech and two additional maskers derived from the original speech while participants listened passively to sentences. The first additional maskers consisted of spectrally rotated versions of the sentences, while the second consisted of speech-modulated noise. Rotated speech represents a spectral inversion of the original speech signal, in which the spectrum of low-pass-filtered speech is inverted around a center frequency. It has a temporal and spectral structure similar to the original speech signal but is not intelligible. Three sets of stimuli were presented to participants: speech-in-speech, speech-in-rotated-speech, and speech-in-speech-modulated-noise (energetic masking baseline). The speech-in-speech masker was linked to increased bilateral STG activation, compared to the speech-modulated-noise baseline, and masking speech with spectrally rotated speech was related only to right STG activation relative to the baseline. Scott et al. (2009) argue that informational masking links to two main asymmetrically distributed neural loci, one related to linguistic processes engaging the left STS/STG and the other involving the right STG, reflecting signal segregation processes related to separating out the signal and masking signals.

Nakai, Kato, and Natsuo (2005) measured the BOLD response while participants listened to a story narrated by a female speaker that was masked by speech from a male or female speaker (same person as as the narrator). Bilateral increases in the BOLD response were reported in the STG for the male talker blocks compared to the unmasked baseline condition. However, the masked condition with the female (same) speaker resulted in greater activation in a network spanning the bilateral temporal lobes and the prefrontal and parietal lobes. A direct contrast of the same-speaker and different-speaker masked conditions showed increases in the BOLD response in the pre-SMA, left precentral gyrus (PCG) and bilateral IFG, right FO, and right supramarginal gyrus (SMG).

Conclusions Energetic and informational maskers appear to recruit a similar network of cortical areas in frontal, temporal, and medial regions (table 77.1). However, there are subtle differences between the activation patterns associated with both types of maskers, and these may point to different neural strategies. While both types of maskers recruit bilateral areas in the STS/STG, informational maskers seem to recruit these areas more than energetic maskers. Moreover, energetic maskers appear to recruit a wider network of areas—notably, including premotor and motor areas. It has been suggested that processing a speech target that is completely masked spectrally and temporally leads listeners to rely to a greater extent on top-down processes and may be related to an increased reliance on executive processes, including working memory and attention (Mattys et al., 2012). Further studies that directly contrast energetic and informational maskers, ideally using different types of informational maskers (e.g., overlapping in semantic or syntactic content/structure as well as speaker-specific aspects), will further elucidate the question of to which extent the neural mechanisms for both types of maskers are similar.

Source: unfamiliar accent Adank, Noordzij, and Hagoort (2012) presented listeners in an fMRI study with sentences spoken in familiar and unfamiliar accents. Compared to the familiar accent, increased activation was found for the unfamiliar accent in frontal (bilateral FO and insulas), temporal (left middle temporal gyrus [MTG], bilateral STG), and parietal regions (SMG). In Adank, Davis, and Hagoort (2012), listeners were again exposed to sentences spoken in both accents while performing a speeded semantic verification task. Compared to the familiar accent, listening to sentences in the unfamiliar accent was associated with increased activation in the left STG/STS. Yi, Smiljanic, and Chandrasekaran (2014) tested participants in an fMRI study while they listened to native- and Korean-accented English sentences. They report that foreign-accented speech evoked greater activity in the bilateral STG/STS and the IFG.

Source: fast speech Poldrack et al. (2001) presented participants with sentences compressed to 60%, 45%, 30%, and 15% of their original duration. They report compression-related increases in BOLD in the left MFG, right IFG, ACC, and striatum. Peelle, McMillan, Moore, Grossman, and Wingfield (2004) presented listeners with sentences that were time compressed to 80%, 65%, and 50% of their duration. Processing speech at higher compression rates recruited areas in the bilateral ACC, left striatum, and right caudate nucleus but also in bilateral premotor areas. Participants in Adank and Devlin (2010) listened to sentences at their original speech rate and compressed to 45%. Compression-related increases were found in the bilateral anterior and posterior STG/STS, pre-SMA, cingulate sulcus, and bilateral FOs. Processing fast sentences thus seems to recruit a network comprising bilateral temporal areas; midline areas including the anterior cingulate, pre-SMA, striatum, and caudate nucleus; and a set of frontal areas including the left IFG and the bilateral FOs.

Source: noise-vocoded speech Hervais-Adelman, Carlyon, Johnsrude, and Davis (2012) scanned participants while they listened to six-channel noise-vocoded words, clear words, and nonspeech stimuli and performed a nonspeech target-detection task. In comparison with clear words, noise-vocoded words were associated with increases in the BOLD response in frontal areas, including the left IFG, precentral gyrus, and left insula. Erb, Henry, Eisner, and Obleser (2013) presented participants in an fMRI experiment with spoken sentences in three conditions: four-band vocoded sentences, clear (nonvocoded) sentences, and trials lacking any auditory stimulation (silent trials). An increase in the BOLD signal was reported in the left SMA, left ACC, anterior insula, and bilateral caudate nucleus for degraded relative to clear sentences.

TABLE 77.1 Overview of studies contrasting speech perception under adverse listening conditions versus easier listening conditions.

Study	Adverse condition	Stimuli	Method	Areas
Neuroimaging studies				
Osnes et al. (2011)	Environmental (energetic)	Syllables in noise	fMRI	Bilateral STS, left PMv
Du et al. (2014)	Environmental (energetic)	Syllables in noise	fMRI	Left PMv
Wong et al. (2008)	Environmental (energetic)	Words in noise	fMRI	Bilateral STG, left insula
Adank et al. (2012)	Environmental (energetic)	Sentences in noise	fMRI	Left IFG, left FO, ACC, parahippocampal gyrus, caudate nucleus
Zekveld et al. (2006)	Environmental (energetic)	Sentences in noise	fMRI	Left MFG, left IFG, bilateral STG/STS
Hwang et al. (2006)	Environmental (energetic)	Stories in noise	fMRI	Left STG/MTG, parahippocampal gyrus, cuneus, thalamus
Dole et al. (2014)	Environmental (energetic)	Words in monaural, binaural, or dichotic conditions	fMRI	Right IFG, left STG, left MFG, left SMA
Dole et al. (2014)	Environmental (informational)	Words in monaural, binaural, or dichotic conditions	fMRI	Right STG
Scott et al. (2009)	Environmental (informational)	Sentences masked by noise-vocoded or spectrally rotated maskers	PET	Bilateral STG
Nakai et al. (2005)	Environmental (informational)	Stories masked by same > different speaker	fMRI	Pre-SMA, left PCG, bilateral IFG, right FO, right SMG
Adank et al. (2012)	Source (accent)	Sentences in unfamiliar accent	fMRI	Bilateral FO, bilateral insula, left MTG, bilateral STG, SMG
Adank et al. (2012)	Source (accent)	Sentences in unfamiliar accent	fMRI	Left STG/STS
Yi et al. (2014)	Source (accent)	Sentences in unfamiliar accent	fMRI	Bilateral STG/STS, bilateral IFG
Poldrack et al. (2001)	Source (time compression)	Time-compressed sentences	fMRI	Left MFG, right IFG, ACC, striatum
Peelle et al. (2004)	Source (time compression)	Time-compressed sentences	fMRI	Bilateral ACC, left striatum, right caudate nucleus, bilateral premotor areas
Adank & Devlin (2010)	Source (time compression)	Time-compressed sentences	fMRI	Bilateral STG/STS, ACC, pre-SMA, striatum, caudate nucleus, left IFG, bilateral FO
Hervais-Adelman et al. (2012)	Source (noise vocoding)	Noise-vocoded words	fMRI	Left IFG, left PCG, left insula
Erb et al. (2013)	Source (noise vocoding)	Noise-vocoded sentences	fMRI	Left SMA, left ACC, anterior insula, bilateral caudate nuclei
Neurostimulation studies				
D'Ausilio et al. (2009)	Environmental, (energetic)	Syllables in noise	MEP	Tongue M1
Murakami et al. (2011)	Environmental (energetic)	Syllables in noise	MEP	Lip M1
Nuttall et al. (2017)	Environmental (energetic)	Syllables in noise	MEP	Lip M1
Nuttall et al. (2016)	Source (motoric)	Syllables, tongue depressed	MEP	Lip M1
Meister et al. (2007)	Environmental (energetic)	Syllables in noise	TMS	Left PMv
Nuttall et al. (forthcoming)	Source (motoric)	Tongue-depressed sentences	TMS	Right PMv, left lip M1

Notes: ACC: anterior cingulate cortex; fMRI: functional magnetic resonance imaging; FO: frontal operculum; IFG: inferior frontal gyrus; M1: primary motor cortex; MEP: motor evoked potential; MFG: middle frontal gyrus; MTG: middle temporal gyrus; PET: positron emission tomography; PMv: ventral premotor cortex; SMA: supplementary motor area; SMG: supramarginal gyrus; STG: superior temporal gyrus; STS: superior temporal sulcus; TMS: transcranial magnetic stimulation.

Conclusions The extended network for processing source-related distortions recruits areas in the bilateral STG/STS, FO and insula, left MFG, left MTG, bilateral IFG, ACC, anterior insula, bilateral ventral premotor cortex (PMv), striatum, caudate nucleus, pre-SMA and SMA, left SMG, and precentral gyrus (table 77.1). It is not straightforward to determine to which extent the networks associated with processing different source-related distortions differ from each other and how the overall network for source-related distortions differs from the network recruited for environmental distortions. Most studies report strong involvement of the bilateral STS/STG in processing source-distorted speech relative to clear speech, and it seems likely that the neural mechanisms for processing this type of adverse condition are predominantly auditory in nature, as is probably also the case for informational maskers.

Neurostimulation

Motor evoked potentials Several neuroimaging studies assessing speech perception in the adverse listening conditions discussed earlier report the involvement of (pre)motor areas (Du et al., 2014; Nakai, Kato, & Matsuo, 2005; Osnes, Hugdahl, & Specht, 2011). It has been suggested that (pre)motor areas, particularly the lip and tongue areas of M1, play an active role in supporting speech perception. This is thought to be especially the case if the incoming speech signal is distorted or unclear. Articulatory M1 is thought to support speech perception using an analysis-by-synthesis approach, in which articulatory motor patterns are used to "fill in" the missing parts during speech perception (e.g., Skipper, Devlin, & Lametti, 2017). Several MEP studies tested this specific hypothesis by testing if lip M1 is activated to a greater degree when listening to speech in challenging conditions compared to less challenging conditions.

Environmental: energetic Murakami, Restle, and Ziemann (2011) recorded MEPs after stimulation to the lip area of M1 while participants listened to syllables embedded in quiet and in several noise levels. Lip MEPs were enhanced for perceiving syllables in noise relative to perceiving clear syllables (experiment 4). This result was interpreted to reflect the increased excitability of articulatory lip motor representations when listening to speech in noise.

Nuttall, Kennedy-Higgins, Devlin, and Adank (2017) recorded MEPs to test if lip M1 shows differential sensitivity depending on distortion type (motor-distorted or noise, experiment 1) and quantity (two levels of syllables in noise, experiment 2) and if lip M1 excitability relates to individual hearing ability. For experiment 1, larger lip M1 MEPs were reported during the perception of motor-distorted speech produced using a tongue depressor or presented in background noise, relative to natural speech in quiet. However, no difference was reported between both distortion types. Experiment 2 did not find evidence of motor system facilitation when speech was presented in noise at SNRs where speech intelligibility for individual listeners was at 50% (harder) or 75% (easier). However, there was a significant interaction between noise condition and hearing ability, which indicated that when speech stimuli were correctly classified at 50%, speech motor facilitation was observed in individuals with better hearing. Individuals with relatively worse but still normal hearing showed more activation of lip M1 during the perception of clear speech. Taken together, these results indicate that articulatory M1 is activated more during the perception of speech under adverse conditions, thus supporting claims suggesting a role for M1 in processing distorted speech signals (Skipper, Devlin, & Lametti, 2017). Moreover, results from Nuttall et al. (2017) indicate that M1 becomes more activated whenever the speech signal is more difficult to process, irrespective of whether the distortion is environmental or source related.

Environmental and source: energetic, motor distorted Nuttall, Kennedy-Higgins, Hogan, Devlin, and Adank (2016) recorded MEPs from lip and hand (control site) M1 while participants listened to clearly articulated syllables (clear) or syllables articulated while the speaker held a tongue depressor in the mouth (tongue depressed). Participants passively listened to clear and tongue-depressed vowel-consonant-vowel (VCV) syllables (/apa/, /aba/, /ata/, /ada/) in separate blocks while hand and lip MEPs were collected. After MEP collection was completed, participants performed an identification task for the tongue-depressed stimuli. The results showed facilitation for lip MEPs for tongue-depressed compared to clear stimuli. Moreover, this facilitation was increased for stimuli containing a lip-articulated consonant (/apa/ and /aba/) compared to a tongue-articulated consonant (/ata/ and /ada/). Finally, participants who performed best on the identification task showed the greatest amount of facilitation for lip MEPs.

Transcranial magnetic stimulation—environmental: energetic Meister et al. (2007) tested the causal role of the left STG and left PMv in the perception of CV syllables embedded in white noise and of simple tones. Participants received 15 minutes of 1 Hz of repetitive TMS to either target site. The study aimed to establish the role of the left PMv and left STG in processing speech in noise. Participants performed either a phoneme or tone identification task or a color identification control task. Repetitive TMS to the left PMv only impaired phoneme discrimination, thus demonstrating a causal effect of TMS on speech perception, but had no effect on tone or color discrimination tasks. TMS to the left STG impaired tone discrimination but had no effect on phoneme or color discrimination tasks. Meister et al. (2007) argue that the lack of an inhibitory effect of TMS to the left STG during syllable discrimination can be attributed to the recruitment of a more extensive, bilateral neural network for speech processing than for tone perception. Speech perception is arguably a more complex process than tone perception, as it encompasses a basic auditory signal-processing stage as well as higher-level phonetic and phonological processing stages, which tend to recruit areas in bilateral temporal areas.

Participants in D'Ausilio et al. (2009) performed a phoneme identification task for CV syllables in which the consonant was articulated using either the lips (/ pœ/ and /bœ/) or tongue (/tœ/ and /dœ/) embedded in white noise. Participants received TMS pulses to the left lip or tongue area of M1 in an online TMS design. Responses were also collected when no TMS pulse was given (baseline). The results showed a double dissociation between the stimulation site (lip or tongue) and discrimination performance between the primary articulator of the stimuli (lips or tongue). Participants were faster to classify a tongue sound following TMS to tongue M1 and slower to classify a lips sound following a TMS pulse to tongue M1, and vice versa. This pattern in the results was not replicated when the stimuli were presented in quiet, thus showing that the causal role of articulatory M1 was specific to noisy syllables. The results from the virtual lesion TMS studies discussed here demonstrate that articulatory M1 plays a causal role in the perception of speech masked by environmental maskers, thus further supporting the proposed role of M1 in the perception of distorted speech.

Source: Motor-distorted speech Nuttall, Kennedy-Higgins, Devlin, and Adank (forthcoming) examined the connection between left PMv and left lip M1 during challenging speech perception in two experiments that combined the collection of MEPs with virtual lesion TMS. Experiment 1 tested intrahemispheric connectivity between left PMv and left M1 lip perception during the comprehension of speech under clear and distorted listening conditions. TMS was applied to the left PMv. Next, participants performed a speeded sentence-verification task on motor-distorted and clear speech while also undergoing stimulation of left lip M1 to elicit MEPs. Experiment 2 aimed to clarify the role of interhemispheric connectivity between right-hemisphere PMv and the left-hemisphere M1 lip area. Dual-coil TMS was applied to right PMv and left lip M1. The results from both experiments indicated that the disruption of PMv during speech perception affected the comprehension of distorted speech specifically, and listening to distorted speech was found to modulate the balance of intra- and interhemispheric interactions, with a larger sensorimotor network implicated during the comprehension of distorted speech than when speech perception is optimal.

Conclusions Only three TMS studies thus far have examined the causal role of cortical areas in processing speech in adverse listening conditions. The results from the three studies clearly support a causal role for (pre) motor regions in the perception of motor-distorted speech and speech in the presence of an energetic masker, thus supporting accounts that propose a supporting role for speech production substrates in speech perception in challenging listening conditions (Skipper, Devlin, & Lametti, 2017). Note that only a single study (Meister et al., 2007) examined the causal role of an area in the temporal lobe (left STG). Yet Meister et al. (2007) did not report a causal role of this area in processing syllables in noise (but did report a causal role of the left STG in tone discrimination). Due to the inherent limitations of TMS, it is not possible to stimulate more medial target areas, but there is a clear lack of research directly targeting accessible lateral cortical areas, specifically in the STG/STS, while participants process distorted speech at prelexical or lexical levels.

General Conclusions

This chapter discussed neuroimaging (fMRI/PET) and neurostimulation (MEP/TMS) studies aiming to further our understanding of how the brain processes speech under environmental and source-related adverse listening conditions. The overview of neurostimulation studies in table 77.1 displays a different picture from the neuroimaging results. While neuroimaging studies report the involvement of cortical areas in the frontal, temporal, and parietal lobes as well as an extended network of medial areas, neurostimulation studies seem to have mostly focused on frontal areas including articulatory M1 and left PMv. For MEP studies, M1 is the obvious target since it is not straightforward (if not impossible) to elicit MEPs from cortical areas outside (pre)motor areas of the brain. There is a clear lack of neurostimulation studies examining the role of (bilateral) temporal areas in processing speech in adverse conditions. It is not possible to collect MEPs from areas outside the (pre)motor areas, but it is surprising that only a single virtual lesion TMS study (Meister et al., 2007) examined the role of the STS/STG in processing distorted speech signals but did not confirm a causal role for the STG. It may not be straightforward to establish a clear causal effect for temporal regions, presumably due to possible interhemispheric compensation during speech processing. Interhemispheric compensation can especially occur in so-called off-line TMS paradigms, where the application of pulses occurs several minutes before task performance, allowing for online reorganization or compensation by the nontargeted hemisphere. Future TMS studies might therefore explore either the use of online TMS (where the TMS pulse is delivered during stimulus presentation) or target areas in both temporal lobes simultaneously in either an online or off-line paradigm, to limit compensation mechanisms.

This chapter aimed to outline the neural mechanisms associated with processing different types of distortions. The results discussed here can be summarized as the fact that informational maskers tend to recruit a network of areas associated with auditory processing in the STS/STG, while energetic maskers and source distortions also recruit areas outside the STS/STG, including motor and (pre)motor regions. Premotor cortex appears to be crucial for processing speech in energetic maskers. Yet the precise neural mechanisms associated with each type of distortion remain largely unclear, and it is suggested that future studies exploit the respective strengths of neuroimaging and neurostimulation to further elucidate these mechanisms. For example, future studies might systematically link fMRI and TMS by first identifying the relevant nodes and, second, establishing their causal role in processing speech under adverse listening conditions using a variety of speech stimuli and environmental and source distortions.

REFERENCES

Adank, P., Davis, M., & Hagoort, P. (2012). Neural dissociation in processing noise and accent in spoken language comprehension. *Neuropsychologia, 50*(1), 77–84. doi:10.1016/j.neuropsychologia.2011.10.024

Adank, P., & Devlin, J. T. (2010). On-line plasticity in spoken sentence comprehension: Adapting to time-compressed speech. *NeuroImage, 49*(1), 1124–1132. doi:10.1016/j.neuroimage.2009.07.032

Adank, P., Evans, B. G., Stuart-Smith, J., & Scott, S. K. (2009). Comprehension of familiar and unfamiliar native accents under adverse listening conditions. *Journal of Experimental Psychology: Human Perception and Performance, 35*(2), 520–529. doi:10.1037/a0013552

Adank, P., Noordzij, M. L., & Hagoort, P. (2012). The role of planum temporale in processing accent variation in spoken language comprehension. *Human Brain Mapping, 33*(2), 360–372. doi:10.1002/hbm.21218

Adank, P., Nuttall, H. E., & Kennedy-Higgins, D. (2016). Transcranial magnetic stimulation (TMS) and motor evoked potentials (MEPs) in speech perception research. *Language Cognition & Neuroscience, 32*(7), 1–10. doi:10.1080/23273798.2016.1257816

Cooke, M. (2003). A glimpsing model of speech perception in noise. *Journal of the Acoustical Society of America, 119*(3), 1562–1573. doi:10.1121/1.2166600

D'Ausilio, A., Pulvermüller, F., Salmas, P., Bufalari, I., Begliomini, C., & Fadiga, L. (2009). The motor somatotopy of speech perception. *Current Biology, 19*(5), 381–385. doi:10.1016/j.cub.2009.01.017

Davis, M. H., Johnsrude, I. S., Hervais-Adelman, A. G., Taylor, K., & McGettigan, C. (2005). Lexical information drives perceptual learning of distorted speech: Evidence from the comprehension of noise-vocoded sentences. *Journal of Experimental Psychology: General, 134*(2), 222–241.

Dole, M., Meunier, F., & Hoen, M. (2014). Functional correlates of the speech-in-noise perception impairment in dyslexia: An MRI study. *Neuropsychologia, 60*, 103–114. doi:10.1016/j.neuropsychologia.2014.05.016

Du, Y., Buchsbaum, B. R., Grady, C. L., & Alain, C. (2014). Noise differentially impacts phoneme representations in the auditory and speech motor systems. *Proceedings of the National Academy of Sciences, 111*, 7126–7131. doi:10.1073/pnas.1318738111

Dupoux, E., & Green, K. (1997). Perceptual adjustment to highly compressed speech: Effects of talker and rate changes. *Journal of Experimental Psychology: Human Perception and Performance, 23*(3), 914–927.

Erb, J., Henry, M. J., Eisner, F., & Obleser, J. (2013). The brain dynamics of rapid perceptual adaptation to adverse listening conditions. *Journal of Neuroscience, 33*(26), 10688–10697. doi:10.1523/jneurosci.4596-12.2013

Hervais-Adelman, A. G., Carlyon, R. P., Johnsrude, I. S., & Davis, M. H. (2012). Brain regions recruited for the effortful comprehension of noise-vocoded words. *Language and Cognitive Processes, 27*(7–8), 1145–1166. doi:10.1080/01690965.2012.662280

Hwang, J. H., Wu, C. W., Chen, J. H., & Liu, T. C. (2006). The effects of masking on the activation of auditory-associated cortex during speech listening in white noise. *Acta Oto-Laryngologica, 126*(9), 916–920. doi:10.1080/00016480500546375

Mattys, S. L., Davis, M. H., Bradlow, A. R., & Scott, S. K. (2012). Speech recognition in adverse conditions: A review. *Language and Cognitive Processes, 27*(7/8), 953–978. doi:10.1080/01690965.2012.705006

Meister, I. G., Wilson, S. M., Deblieck, C., Wu, A. D., & Iacoboni, M. (2007). The essential role of pre-motor cortex in speech perception. *Current Biology, 17*, 1692–1696. doi:10.1016/j.cub.2007.08.064

Murakami, T., Restle, J., & Ziemann, U. (2011). Observation-execution matching and action inhibition in human primary motor cortex during viewing of speech-related lip movements or listening to speech. *Neuropsychologia, 49*(7), 2045–2054. doi:10.1016/j.neuropsychologia.2011.03.034

Nakai, T., Kato, C., & Matsuo, K. (2005). An fMRI study to investigate auditory attention: A model of the cocktail party phenomenon. *Magnetic Resonance in Medical Sciences, 4*(2), 75–82. doi:10.2463/mrms.4.75

Nuttall, H. E., Kennedy-Higgins, D., Devlin, J. T., & Adank, P. (2017). The role of hearing ability and speech distortion in the facilitation of articulatory motor cortex. *Neuropsychologia, 94*(8), 13–22. doi:10.1016/j.neuropsychologia.2016.11.016

Nuttall, H. E., Kennedy-Higgins, D., Devlin, J. T., & Adank, P. (forthcoming). Modulation of intra- and inter-hemispheric connectivity between primary and premotor cortex during speech perception. *Brain & Language.* (forthcoming). doi:10.1016/j.bandl.2017.12.002

Nuttall, H. E., Kennedy-Higgins, D., Hogan, J., Devlin, J. T., & Adank, P. (2016). The effect of speech distortion on the excitability of articulatory motor cortex. *NeuroImage, 128,* 218–226. doi:10.1016/j.neuroimage.2015.12.038

Osnes, B., Hugdahl, K., & Specht, K. (2011). Effective connectivity analysis demonstrates involvement of premotor cortex during speech perception. *NeuroImage, 54*(3), 2437–2445. doi:10.1016/j.neuroimage.2010.09.078

Peelle, J. E., McMillan, C., Moore, P., Grossman, M., & Wingfield, A. (2004). Dissociable patterns of brain activity during comprehension of rapid and syntactically complex speech: Evidence from fMRI. *Brain and Language, 91,* 315–325. doi:10.1016/j.bandl.2004.05.007

Poldrack, R. A., Temple, E., Protopapas, A., Nagarajan, S., Tallal, P., Merzenich, M., & Gabrieli, J. D. E. (2001). Relations between the neural bases of dynamic auditory processing and phonological processing: Evidence from fMRI. *Journal of Cognitive Neuroscience, 13*(5), 687–697. doi:10.1162/089892901750363235

Scott, S. K., Rosen, S., Beaman, P., Davis, J. P., & Wise, R. J. S. (2009). The neural processing of masked speech: Evidence for different mechanisms in the left and right temporal lobes. *Journal of the Acoustical Society of America, 125*(1737–1743). doi:10.1121/1.3050255

Skipper, J., Devlin, J. T., & Lametti, D. R. (2017). The hearing ear is always found close to the speaking tongue: Review of the role of the motor system in speech perception. *Brain and Language, 164,* 77–105. doi:10.1016/j.bandl.2016.10.004

Wong, P. C. M., Uppanda, A. K., Parrish, T. B., & Dhar, S. (2008). Cortical mechanisms of speech perception in noise. *Journal of Speech, Hearing and Language Research, 51,* 1026–1041. doi:10.1044/1092-4388(2008/075

Yi, H., Smiljanic, R., & Chandrasekaran, B. (2014). The neural processing of foreign-accented speech and its relationship to listener bias. *Frontiers in Human Neuroscience, 8,* 1–12. doi:10.3389/fnhum.2014.00768

Zekveld, A. A., Heslenfeld, D. J., Festen, J. M., & Schoonhoven, R. (2006). Top-down and bottom-up processes in speech comprehension. *NeuroImage, 32*(4), 1826–1836. doi:10.1016/j.neuroimage.2006.04.199

78 The Cerebral Bases of Language Acquisition

GHISLAINE DEHAENE-LAMBERTZ AND CLAIRE KABDEBON

ABSTRACT The development of noninvasive brain-imaging techniques has opened the black box of the infant brain. Instead of postulating theories based on the delayed consequences of, fortunately rare, early lesions, we can now study healthy infant responses to speech. Rather than a brain limited to primary areas or, on the contrary, a poorly specialized brain, brain-imaging studies have revealed a functional architecture in infants that is close to what is described in adults. In particular, a hierarchy of increasingly integrated computations is observed along the superior temporal regions, and the processing of different speech features is already segregated along parallel neural pathways with different hemispheric biases. Yet, although highly structured, the infant brain still differs from the adult brain, with particularly delayed brain responses arising from frontal regions. We can expect that a better understanding of the computational abilities of this early network may provide insight into the mechanisms underlying language acquisition.

Speech is a remarkable communication device whose efficiency to convey information is based on the combination of units (phonemes in words, words in sentences) according to rules. Before the end of the first year of life, human infants display amazing capacities in processing speech. First, they show an extraordinary ability to analyze the auditory content of the speech stream. They learn the repertoire of sounds (or *phonemes*) used by their native language (Kuhl, Williams, Lacerda, Stevens, & Lindblom, 1992; Werker & Tees, 1984) and the rules (*phonotactics*) for combining these sounds within words (Jusczyk, Luce, & Charles-Luce, 1994). They notice the frequent words of stories they have heard a few times (Jusczyk & Hohne, 1997) and that content words are surrounded by recurrent syllables (e.g., *ing*, *the*, *a*) that have a different function in the sentence (Shi, 2014), as they start to figure out sentence organization. This early learning is based on distributional analyses at different levels of the linguistic structure, from the syllabic level (Saffran, Aslin, & Newport, 1996) to a more abstract level, such as word category (Gervain, Nespor, Mazuka, Horie, & Mehler, 2008). Second, they rapidly discover the referential aspect of speech: they know that speech conveys information from at least 4 months of age (Marno et al., 2015), and at 6–9 months of age, they already know the meaning of a few words, such as *mommy*, *hug*, some body parts, and more (Bergelson & Swingley, 2012; Tincoff & Jusczyk, 1999). Third, infants might also rapidly understand that speech is a symbolic system. They can create equivalence between a label and a category (Kabdebon & Dehaene-Lambertz, 2019), which helps them to sort items into named categories (e.g., dinosaur vs. fish pictures; Ferry, Hespos, & Waxman, 2013).

What are the cerebral bases of these impressive competences? Is language acquisition based on a functional organization similar to the adult linguistic network? This question is not trivial, as the development of the human brain is complex and extends over two decades. Its weight increases from 400 g at birth to 1400 g in adults. The organization of cortical layers and large fiber networks is well established at term birth (Dubois & Dehaene-Lambertz, 2015), although neuronal migration is still ongoing in the frontal areas during the first months of life (Paredes et al., 2016). Maturation consists of waves of synaptogenesis followed by pruning with an acceleration of signal transmission speed due to myelination of the tracts. These phenomena are relatively well described, but brain maturation covers many other aspects essential to the effectiveness of neural networks, such as the maturation of glia and various types of neurons, the production of neurotransmitters, changes in receptors, the accumulation of proteoglycan chains, and more, whose maturational sequences are unknown in the human brain. Additionally, depending on the region, maturational spurts occur at different moments and at different rates, generating dynamic shifts within and between regions and adding a dimension of complexity to how networks interact. Although the description of the immature human brain becomes more refined thanks to the development of noninvasive brain-imaging techniques, we are still far from understanding what crucial features of the infant brain allow for this rapid linguistic development. Nevertheless, based on the brain-imaging data acquired from the last trimester of gestation onward, we can start to propose hypotheses on how the functional architecture of the infant brain may explain some of the early linguistic competencies.

The Organization of Perisylvian Regions

In human adults, linguistic and nonlinguistic representations of speech are computed in parallel along distinct hierarchical pathways in the superior temporal lobe, reaching the inferior frontal regions. This hierarchical and parallel functional organization is already observed in infants' perisylvian regions.

A hierarchy of linguistic processes When infants—even neonates—listen to speech, activation occurs along the superior temporal region bilaterally and extends to distant left inferior parietal and frontal regions (Dehaene-Lambertz, Hertz-Pannier, et al., 2006; Pena et al., 2003; Shultz, Vouloumanos, Bennett, & Pelphrey, 2014; Sato et al., 2012). Interestingly, the phase of the Blood Oxygen Level Dependent (BOLD) response progressively slows down as we move away from the primary auditory cortex (Heschl's gyrus) toward the temporal pole and toward the temporoparietal junction (Dehaene-Lambertz, Hertz-Pannier, et al., 2006). Whereas the BOLD response rapidly peaks and decreases in Heschl's gyrus, it becomes more and more delayed and sustained anteriorly in the superior temporal sulcus and even starts at the end of a sentence in the most anterior regions (figure 78.1). This temporal gradient is not related to an immature neurovascular coupling since a similar, although faster, gradient is visible in adults (Dehaene-Lambertz, Dehaene, et al., 2006). Because in infants, as in adults, the superior region of the temporal areas is more sensitive to acoustic features than the more ventral regions involved in the computation of abstract and integrated representations

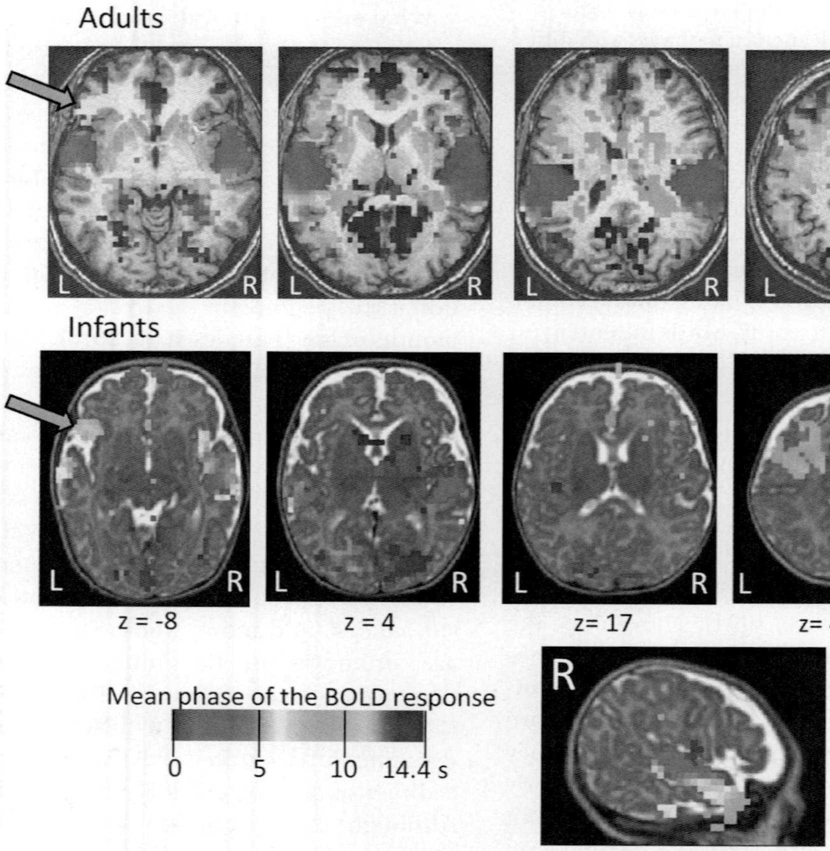

FIGURE 78.1 Hierarchical organization of the perisylvian regions in 3-month-old infants and adults, illustrated by the phase gradient of the BOLD response to a single sentence. The mean phase is presented on axial slices placed at similar locations in the adult (*top row*) and infant (*bottom row*) standard brains and on a sagittal slice in the infant's right hemisphere. Colors encode the circular mean of the phase of the BOLD response, expressed in seconds relative to sentence onset. The same gradient is observed in both groups along the superior temporal region, extending until Broca's area (*arrow*). Blue regions are out of phase with stimulation (Dehaene-Lambertz, Hertz-Pannier, et al., 2006; Dehaene-Lambertz, Dehaene, et al., 2006). (See color plate 90.)

(Bristow et al., 2009; DeWitt & Rauschecker, 2012), we proposed that this gradient might be the consequence of the hierarchical organization of the perisylvian networks: the increasingly delayed and sustained responses would correspond to larger and larger windows for integrating speech chunks, as described in adults (Ding, Melloni, Zhang, Tian, & Poeppel, 2016).

Such hierarchical organization might explain infants' early sensitivity to sentence organization and why they prefer listening to sentences with pauses located at prosodic boundaries rather than within prosodic units (Hirsh-Pasek et al., 1987). With its embedded units, the prosodic hierarchy is a natural input for these regions, helping infants segment the speech stream into coherent chunks. Analyses can then be restricted to each prosodic unit, explaining why the computations of transitional probabilities between syllables—which is the main proposed mechanism in infants for extracting words from a stream of speech (Saffran, Aslin, & Newport, 1996)—cannot occur across a prosodic boundary (Shukla, White, & Aslin, 2011). Finally, as prosody and syntax are tightly related, this hierarchical organization might also secondarily facilitate the learning of native syntax (Christophe, Millotte, Bernal, & Lidz, 2008).

Parallel pathways for voice and phoneme processing Speech conveys information not only about the language but also about the speaker. Both elements are crucial for infants to understand what is said and to identify who is speaking. Thus, they should simultaneously neglect local variations in timbre, pitch, speech rate, and so on to extract the linguistic information and use them to be able to keep track of the speaker's identity, actual emotion, and location in space. Using event-related potentials (ERPs), we showed that these computations are done in parallel: After a series of repeated auditory-visual vowels, a change in vowel identity or the speaker's gender evokes two different mismatch responses, characterized by a different voltage topography on the scalp but within the same time window, in 3-month-olds (Bristow et al., 2009). Although spatial information is coarse with electroencephalography (EEG), a model of brain sources suggests a right-lateralized response for the change of voice, contrasting with a left-lateralized response for a change of vowel.

These hemispheric biases are confirmed with functional magnetic resonance imaging (fMRI) in 2-month-old infants who listened to their mother's voice or to the voice of an unknown mother. In the left posterior temporal region, activations are enhanced in response to the voice of one's own mother, probably because familiarity with the voice allows for better phonetic access. Right-hemisphere differences are also observed in a more anterior temporal region, described as the voice region in adults. This region is also found when nonlinguistic vocal sounds are contrasted with environmental sounds in 3- to 7-month-olds (Blasi et al., 2011). All of these experiments underline a parallel organization from the first months of life channeling voice and phoneme processing along different pathways.

Early lateralization of speech processing The previous studies suggest that adults' left-right functional differences have their roots in early development. Indeed, a larger left-hemispheric response is reported in most studies using speech during the first trimester of life: at the level of the *planum temporale* in fMRI studies (Dehaene-Lambertz, Dehaene, & Hertz-Pannier, 2002; Dehaene-Lambertz, Hertz-Pannier, et al., 2006; Dehaene-Lambertz et al., 2010) and less precisely over the superior temporal region in near-infrared spectroscopy (NIRS) studies (Pena et al., 2003; Sato et al., 2012; Vannasing et al., 2016). Activations in response to one's native language are also more left-lateralized than to music (Dehaene-Lambertz et al., 2010) and to other biological sounds, such as nonspeech vocalization, footsteps, and monkey calls (Shultz et al., 2014), but not compared to a foreign language and backward speech (Dehaene-Lambertz, Dehaene, & Hertz-Pannier, 2002), at least initially. After a few months, however, the difference between native and nonnative speech becomes larger. Five-month-olds, but not 3-month-olds, show larger NIRS activation for their own dialect than for a foreign dialect (Quebecois vs. Parisian French) over only the left, but not the right, temporal region (Cristia et al., 2014). A left-hemispheric advantage to process fast temporal transitions (Zatorre, Belin, & Penhune, 2002) might explain an early left bias for speech-like stimuli that is further reinforced through linguistic experience. However, the fact that sign language is also left lateralized in adults argues for a multifactorial contribution to the robust left lateralization of language in humans (see chapter 73).

A Precise Temporal Encoding since the Fetal Life

This functional organization finds its roots during fetal life. At 6 months gestation, 3 months before term, the subcortical sensory system begins to react to external sounds, and the thalamocortical connections reach the cortical plate, feeding the first cortical circuits with external information (Kostovic & Judas, 2010). Although the local microcircuitry is very different from later ages since most of the neurons are still migrating to reach their final location and dendritic trees are sparse, the brain's general connectivity is already visible at the

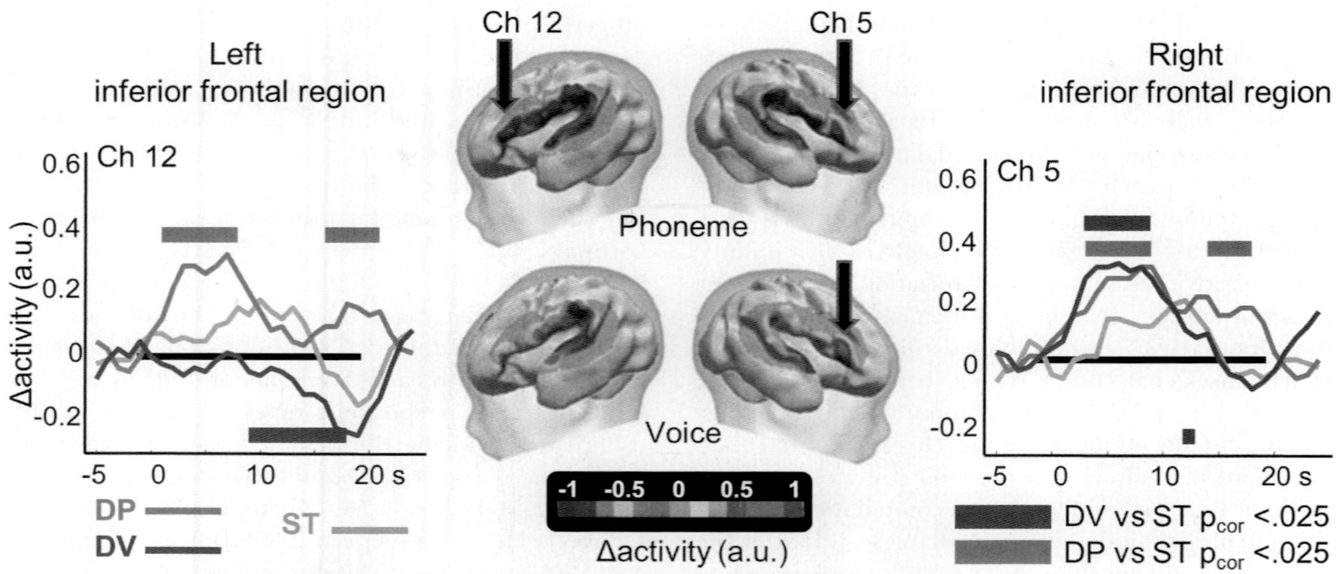

FIGURE 78.2 Parallel pathways in preterms. Oxyhemoglobin responses to a change of phoneme (*/ba/* vs. */ga/*) and a change of voice (male vs. female) measured with NIRS in 30 weeks gestational age—old preterm neonates. A significant increase in the response to a change of phoneme (DP, deviant phoneme) relative to the standard condition (ST) was observed in both temporal and frontal regions, whereas the response to a change of voice (DV, deviant voice) was limited to the right inferior frontal region. The left inferior frontal region responded only to a change of phoneme, whereas the right responded to both changes. The colored rectangles represent the periods of significant differences between the deviant and the standard conditions in the left and right inferior region (*black arrows*; Mahmouzadeh et al., 2013). (See color plate 91.)

structural (Takahashi, Folkerth, Galaburda, & Grant, 2011) and functional level (Fransson et al., 2007; Smyser, Snyder, & Neil, 2011). Already at this age, preterm neonates react to a change of consonant (/ba/ vs. /ga/) and to a change of voice (male vs. female) randomly occurring in a series of repeated syllables (figure 78.2). Furthermore, as in older infants, the temporal and spatial responses generated by both types of changes measured with EEG and NIRS are different, with larger and more mature responses for the change of phoneme than for the change of voice, revealing not only that these two features are processed differently but that the human brain is very sensitive to the temporal dimension of speech from the onset of the thalamocortical circuitry (Mahmoudzadeh et al., 2013; Mahmoudzadeh, Wallois, Kongolo, Goudjil, & Dehaene-Lambertz, 2017).

These results are not trivial since anesthetized rats tested in the same paradigm reacted more strongly to a change in voice than consonant, with a right-lateralized response for both changes (Mahmoudzadeh, Dehaene-Lambertz, & Wallois, 2017). Rats also display a strong reaction to differences in voice, obscuring language discrimination (Toro, Trobalon, & Sebastian-Galles, 2005). By contrast, human adults and infants are commonly better at recovering linguistic content, even for different voices, than at recognizing the same voice for different linguistic content (Dehaene-Lambertz, Dehaene, et al., 2006; Johnson, Westrek, Nazzi, & Cutler, 2011), suggesting a particular human sensitivity to linguistic features beyond general mammal auditory responses. A fine temporal encoding of the auditory world, observed from 30 weeks of gestational age onward, might be one of the important human auditory features.

Several experiments have illustrated the relation between the precision of temporal encoding and better performance in tasks using speech stimuli in normal subjects. For example, Kabdebon et al. (2015) recorded high-density EEG in 8-month-old infants while they were listening to a stream of syllables concatenated according to an AxC structure (i.e., the first syllable (A) predicted the third syllable (C) of successive triplets whereas the middle syllable (x) is variable). The infants were then tested with isolated trisyllabic words that either respected or did not respect the hidden structure of the training stream. The difference between these two conditions at test was significantly correlated with the temporal locking to the syllable frequency during the training stream, as observed with EEG. Similarly in adults, the temporal similarity between auditory cortical activity and speech envelopes predicted speech comprehension (Ahissar et al., 2001). A deficit in temporal encoding has been proposed as one of the mechanisms underlying some oral and written language impairments (Abrams, Nicol, Zecker, & Kraus, 2009; Lehongre, Ramus, Villiermet,

Schwartz, & Giraud, 2011), and the size of the production lexicon can be predicted from the performance of a phonetic discrimination task at 6 months (Tsao, Liu, & Kuhl, 2004). Lexicon size is also correlated with the speed of recognition of auditorily presented words at 18 months (Fernald, Perfors, & Marchman, 2006), demonstrating the interplay between early refined phonetic encoding abilities and later higher-level linguistic abilities.

Immature but Nonetheless Functional Frontal Areas

Activation to speech does not remain limited to auditory areas but extends to higher levels in the parietal and frontal lobes (figures 78.1 and 78.2). Because of their protracted development, frontal areas were classically assumed to function poorly in infants. Many brain-imaging studies have now revealed their involvement in infant cognition: the inferior frontal region reacts to a change in auditory sequences as early as 6 months gestation, on the left for a change of phoneme and on the right for both a change of voice and a change of phoneme (Mahmoudzadeh et al., 2013). At 3 months postterm, an increase in activation in the frontal areas is observed in response to the repetition of a short sentence (Dehaene-Lambertz, Hertz-Pannier, et al., 2006) or in response to repetition of the same vowel across modalities (Bristow et al., 2009). Enhanced frontal activations are also recorded when a complex auditory pattern is violated (Basirat, Dehaene, & Dehaene-Lambertz, 2014). These results reveal the frontal regions' involvement in short-term memory. At the same age, recognition of the prosodic contours of one's native language activates the right dorsolateral prefrontal region in attentive infants (Dehaene-Lambertz, Dehaene, & Hertz-Pannier, 2002), whereas voice familiarity modulates the balance between the median prefrontal regions, sensitive to stimulus familiarity, and the orbitofrontal limbic circuit, involved in stimulus emotional valence (Dehaene-Lambertz et al., 2010). Thus, the frontal lobes in infants are not only activated but are also parceled into different regions distinctively engaged depending on the task, exactly as in older participants.

However, frontal regions react at a slower pace in infancy than later in life. ERP studies have shown that late responses, which depend on higher levels of processing, are disproportionally slower in infants, relative to adults, compared to the infant-adult differences in early sensory regions. Electrical components proposed to be the equivalent of the adult P300 have been recorded after 700 ms, and even around 1 s, until at least the end of the first year (Kouider et al., 2013). By contrast, the latency of the visual P1 reaches adult values around 3 months of age (McCulloch, Orbach, & Skarf, 1999). These time delays should be further studied to analyze whether, and how, they might confer an advantage in learning.

Because maturation improves both local computations and the speed in the connections between regions, the balance between networks may change with development, and patterns of maturation may thus reveal the crucial role of certain circuits at a given moment in acquiring new skills. Adjusting the weights of the different pathways—and thus how they learn—through maturational lags at precise nodes of the perisylvian cortex might be a way to genetically control language development. Combining different techniques makes it possible to study this question—for example, the efficiency of the dorsal and ventral pathway connecting inferior frontal areas and superior temporal areas. A longitudinal study of the functional connectivity over the first 2 years of life reports a rapid increase of connectivity within the left linguistic network between the frontal and posterior temporal areas within the first year of life (Emerson, Gao, & Lin, 2016). At the structural level, the T2 MRI signal component, which is sensitive to free water in the tissues, and diffusion tensor imaging (DTI), which provides measures of the movement of water molecules (measures of diffusivity) and their direction (measure of fractional anisotropy), can be used to study gray and white matter maturation. These markers show that structures belonging to the dorsal pathway (frontal area 44, the posterior superior temporal sulcus, and the arcuate fasciculus) mature in synchrony. While the dorsal pathway displays a delayed maturation relative to the ventral pathway, it starts to catch up after 3 months of age (Dubois et al., 2015; Leroy et al., 2011). This adjustment might be related to the increase in vocalization and progression in the analysis of the segmental part of speech observed at the same age.

The involvement of the inferior frontal regions and the dorsal pathway provides infants with a short-term auditory memory, which seems to be lacking in macaques (Fritz, Mishkin, & Saunders, 2005). A long buffer may favor the discovery of second-order rules by keeping track of segmental elements (Basirat, Dehaene, & Dehaene-Lambertz, 2014; Kovacs & Endress, 2014). Coupled with hierarchical coding along the superior temporal regions, this may favor computations on chunks of chunks and increase sensitivity to deeper hierarchical structures, as well as algebraic rules, as demonstrated in 8-month-olds (Marcus, Vijayan, Bandi Rao, & Vishton, 1999). The early role of the dorsal pathway is confirmed by the observation that fractional anisotropy values measured at term birth in the arcuate fasciculi are correlated with linguistic scores at 2 years of age (Salvan et al., 2017).

When infants listen to speech, activations are not limited to the classical linguistic areas, and the involvement of frontal areas outside the linguistic system may improve infants' focus on speech as a relevant stimulus. Motivation and pleasure, as well as understanding the referential aspect of speech through social cues, have been shown to be important for speech learning (Kuhl, Tsao, & Liu, 2003). The activation in dorsolateral prefrontal regions shown in awake infants recognizing their native language, as well as activation in prefrontal median regions when the voice is familiar, may very well explain these behavioral observations.

Nature versus Nurture

During the first year of life, infants become attuned to the prosody and phonetic repertoire of their native language (Kuhl, Williams, Lacerda, Stevens, & Lindblom, 1992; Werker & Tees, 1984), which can have long-term effects. Chinese adoptees in Quebec, no longer exposed to Chinese after the first year of life, on average, still perceive a tonal contrast and activate the left planum temporale similarly to native Chinese speakers. This contrasts with French-speaking controls never exposed to Chinese, who activate only the right hemisphere (Pierce, Klein, Chen, Delcenserie, & Genesee, 2014). Because preterm infants are exposed earlier than full-term neonates to aerial speech, they can be compared to full-term neonates to study the effects of *ex-utero* exposure versus the brain's developmental age on the sensitivity to foreign speech. In two different studies in preterm infants, Pena and colleagues reported that the decrease in the sensitivity to foreign-language prosody (Pena, Pittaluga, & Mehler, 2010) and foreign phonetic contrasts (Pena, Werker, & Dehaene-Lambertz, 2012) is related to the brain's developmental age rather than the duration of *ex-utero* life. By contrast, learning the phonotactic rules of one's native language is dependent on the duration of exposure to aerial speech (Gonzalez-Gomez & Nazzi, 2012). This discrepancy may point to a critical distinction between a learning mechanism (here, statistical learning allowing for the accumulation of positive evidence on the frequency of phonetic categories and combinations of phonemes) and the critical period during which this learning mechanism is workable. In the mouse visual cortex, it has been proposed that the opening and closing of "critical" windows relies on two thresholds in the accumulation of a homeoprotein, Otx2, in GABAergic parvalbumin interneurons (Hensch, 2004). When the Otx2 level reaches one threshold, learning starts; when it reaches the other, learning then stops or at least becomes more difficult. A similar mechanism might explain how computation of the statistics of the native phonetic environment can only begin after a certain maturational age (probably after 35 weeks gestational age, when the migration and maturation of interneurons is sufficiently advanced, but no study has yet examined this point) and stops around the end of the first year, when the second threshold is reached.

Conclusion

We have emphasized here the early brain organization and its similarities with adult networks and have sought to relate brain-imaging results to behavioral performance. This architecture and its complex maturational calendar have been selected through human evolution as the most efficient in helping infants detect correct cues in the environment in order to learn their native language. A better understanding of brain plasticity and, notably, its changes with age and learning at the microstructural and network levels is a necessary step to refine models of language acquisition.

Acknowledgments

This research was supported by grants from Sodiaal-Fondation Motrice, the Fondation de France, the Fondation NRJ-Institut de France, and the European Research Council (Babylearn project).

REFERENCES

Abrams, D. A., Nicol, T., Zecker, S., & Kraus, N. (2009). Abnormal cortical processing of the syllable rate of speech in poor readers. *Journal of Neuroscience, 29*(24), 7686–7693. doi:10.1523/jneurosci.5242-08.2009

Ahissar, E., Nagarajan, S., Ahissar, M., Protopapas, A., Mahncke, H., & Merzenich, M. M. (2001). Speech comprehension is correlated with temporal response patterns recorded from auditory cortex. *Proceedings of the National Academy of Sciences of the United States of America, 98*(23), 13367–13372.

Basirat, A., Dehaene, S., & Dehaene-Lambertz, G. (2014). A hierarchy of cortical responses to sequence violations in three-month-old infants. *Cognition, 132*(2), 137–150. doi:10.1016/j.cognition.2014.03.013

Bergelson, E., & Swingley, D. (2012). At 6–9 months, human infants know the meanings of many common nouns. *Proceedings of the National Academy of Sciences of the United States of America, 109*(9), 3253–3258. doi:10.1073/pnas.1113380109

Blasi, A., Mercure, E., Lloyd-Fox, S., Thomson, A., Brammer, M., Sauter, D., … Murphy, D. G. (2011). Early specialization for voice and emotion processing in the infant brain. *Current Biology, 21*(14), 1220–1224. doi:10.1016/j.cub.2011.06.009

Bristow, D., Dehaene-Lambertz, G., Mattout, J., Soares, C., Gliga, T., Baillet, S., & Mangin, J. F. (2009). Hearing faces: How the infant brain matches the face it sees with the speech it hears. *Journal of Cognitive Neuroscience, 21*(5), 905–921.

Christophe, A., Millotte, S., Bernal, S., & Lidz, J. (2008). Bootstrapping lexical and syntactic acquisition. *Language and Speech, 51*(Pt. 1–2), 61–75.

Cristia, A., Minagawa-Kawai, Y., Egorova, N., Gervain, J., Filippin, L., Cabrol, D., & Dupoux, E. (2014). Neural correlates of infant accent discrimination: An fNIRS study. *Developmental Science, 17*(4), 628–635. doi:10.1111/desc.12160

Dehaene-Lambertz, G., Dehaene, S., Anton, J. L., Campagne, A., Ciuciu, P., Dehaene, G. P., … Poline, J. B. (2006). Functional segregation of cortical language areas by sentence repetition. *Human Brain Mapping, 27*(5), 360–371. doi:10.1002/hbm.20250

Dehaene-Lambertz, G., Dehaene, S., & Hertz-Pannier, L. (2002). Functional neuroimaging of speech perception in infants. *Science, 298*(5600), 2013–2015. doi:10.1126/science.1077066

Dehaene-Lambertz, G., Hertz-Pannier, L., Dubois, J., Meriaux, S., Roche, A., Sigman, M., & Dehaene, S. (2006). Functional organization of perisylvian activation during presentation of sentences in preverbal infants. *Proceedings of the National Academy of Sciences of the United States of America, 103*(38), 14240–14245. doi:10.1073/pnas.0606302103

Dehaene-Lambertz, G., Montavont, A., Jobert, A., Allirol, L., Dubois, J., Hertz-Pannier, L., & Dehaene, S. (2010). Language or music, mother or Mozart? Structural and environmental influences on infants' language networks. *Brain and Language, 114*(2), 53–65.

DeWitt, I., & Rauschecker, J. (2012). Phoneme and word recognition in the auditory ventral stream. *Proceedings of the National Academy of Sciences of the United States of America, 109*(8), 14.

Ding, N., Melloni, L., Zhang, H., Tian, X., & Poeppel, D. (2016). Cortical tracking of hierarchical linguistic structures in connected speech. *Nature Neuroscience, 19*(1), 158–164. doi:10.1038/nn.4186

Dubois, J., & Dehaene-Lambertz, G. (2015). Fetal and postnatal development of the cortex: MRI and genetics. In A. W. Toga (Ed.), *Brain mapping: An encyclopedic reference* (Vol. 2, pp. 11–19). New York: Elsevier.

Dubois, J., Poupon, C., Thirion, B., Simonnet, H., Kulikova, S., Leroy, F., … Dehaene-Lambertz, G. (2015). Exploring the early organization and maturation of linguistic pathways in the human infant brain. *Cerebral Cortex, 26*(5), 2283–2298. doi:10.1093/cercor/bhv082

Emerson, R. W., Gao, W., & Lin, W. (2016). Longitudinal study of the emerging functional connectivity asymmetry of primary language regions during infancy. *Journal of Neuroscience, 36*(42), 10883–10892. doi:10.1523/jneurosci.3980-15.2016

Fernald, A., Perfors, A., & Marchman, V. A. (2006). Picking up speed in understanding: Speech processing efficiency and vocabulary growth across the 2nd year. *Developmental Psychology, 42*(1), 98–116. doi:10.1037/0012-1649.42.1.98

Ferry, A. L., Hespos, S. J., & Waxman, S. R. (2013). Nonhuman primate vocalizations support categorization in very young human infants. *Proceedings of the National Academy of Sciences of the United States of America, 110*(38), 15231–15235. doi:10.1073/pnas.1221166110

Fransson, P., Skiold, B., Horsch, S., Nordell, A., Blennow, M., Lagercrantz, H., & Aden, U. (2007). Resting-state networks in the infant brain. *Proceedings of the National Academy of Sciences of the United States of America, 104*(39), 15531–15536.

Fritz, J., Mishkin, M., & Saunders, R. C. (2005). In search of an auditory engram. *Proceedings of the National Academy of Sciences of the United States of America, 102*(26), 9359–9364.

Gervain, J., Nespor, M., Mazuka, R., Horie, R., & Mehler, J. (2008). Bootstrapping word order in prelexical infants: A Japanese-Italian cross-linguistic study. *Cognitive Psychology, 57*(1), 56–74. doi:10.1016/j.cogpsych.2007.12.001

Gonzalez-Gomez, N., & Nazzi, T. (2012). Phonotactic acquisition in healthy preterm infants. *Developmental Science, 15*(6), 885–894. doi:10.1111/j.1467-7687.2012.01186.x

Hensch, T. K. (2004). Critical period regulation. *Annual Review of Neuroscience, 27*, 549–579. doi:10.1146/annurev.neuro.27.070203.144327

Hirsh-Pasek, K., Nelson, D. G. K., Jusczyk, P. W., Cassidy, K. W., Druss, B., & Kennedy, L. (1987). Clauses are perceptual units for young infants. *Cognition, 26*, 269–286.

Johnson, E. K., Westrek, E., Nazzi, T., & Cutler, A. (2011). Infant ability to tell voices apart rests on language experience. *Developmental Science, 14*(5), 1002–1011. doi:10.1111/j.1467-7687.2011.01052.x

Jusczyk, P. W., & Hohne, E. A. (1997). Infants' memory for spoken words. *Science, 277*(5334), 1984–1986.

Jusczyk, P. W., Luce, P. A., & Charles-Luce, J. (1994). Infants' sensitivity to phonotactic patterns in the native language. *Journal of Memory and Language, 33*, 630–645.

Kabdebon, C., & Dehaene-Lambertz, G. (2019). Symbolic labelling in 5-month-old human infants. *Proceedings of the National Academy of Sciences of the United States of America, 116*(12), 5805–5810. doi:10.1073/pnas.1809144116

Kabdebon, C., Pena, M., Buiatti, M., & Dehaene-Lambertz, G. (2015). Electrophysiological evidence of statistical learning of long-distance dependencies in 8-month-old preterm and full-term infants. *Brain and Language, 148*, 25–36. doi:10.1016/j.bandl.2015.03.005

Kostovic, I., & Judas, M. (2010). The development of the subplate and thalamocortical connections in the human foetal brain. *Acta Paediatrica, 99*(8), 1119–1127.

Kouider, S., Stahlhut, C., Gelskov, S. V., Barbosa, L. S., Dutat, M., de Gardelle, V., … Dehaene-Lambertz, G. (2013). A neural marker of perceptual consciousness in infants. *Science, 340*(6130), 376–380. doi:10.1126/science.1232509

Kovacs, A. M., & Endress, A. D. (2014). Hierarchical processing in seven-month-old infants. *Infancy, 19*(4), 409–425.

Kuhl, P. K., Tsao, F.-M., & Liu, H.-M. (2003). Foreign-language experience in infancy: Effects of short-term exposure and social interaction on phonetic learning. *Proceedings of the National Academy of Sciences of the United States of America, 100*, 9096–9101.

Kuhl, P. K., Williams, K. A., Lacerda, F., Stevens, K. N., & Lindblom, B. (1992). Linguistic experience alters phonetic perception in infants by 6 months of age. *Science, 255*, 606–608.

Lehongre, K., Ramus, F., Villiermet, N., Schwartz, D., & Giraud, A. L. (2011). Altered low-gamma sampling in auditory cortex accounts for the three main facets of dyslexia. *Neuron, 72*(6), 1080–1090. doi:10.1016/j.neuron.2011.11.002

Leroy, F., Glasel, H., Dubois, J., Hertz-Pannier, L., Thirion, B., Mangin, J. F., & Dehaene-Lambertz, G. (2011). Early maturation of the linguistic dorsal pathway in human infants. *Journal of Neuroscience, 31*(4), 1500–1506.

Mahmoudzadeh, M., Dehaene-Lambertz, G., Fournier, M., Kongolo, G., Goudjil, S., Dubois, J., … Wallois, F. (2013). Syllabic discrimination in premature human infants prior to complete formation of cortical layers. *Proceedings of the National Academy of Sciences of the United States of America, 110*(12), 4846–4851. doi:10.1073/pnas.1212220110

Mahmoudzadeh, M., Dehaene-Lambertz, G., & Wallois, F. (2017). Electrophysiological and hemodynamic mismatch responses in rats listening to human speech syllables. *PLoS One, 12*(3), e0173801. doi:10.1371/journal.pone.0173801

Mahmoudzadeh, M., Wallois, F., Kongolo, G., Goudjil, S., & Dehaene-Lambertz, G. (2017). Functional maps at the onset of auditory inputs in very early preterm human neonates. *Cerebral Cortex, 27*(4), 2500–2512. doi:10.1093/cercor/bhw103

Marcus, G. F., Vijayan, S., Bandi Rao, S., & Vishton, P. M. (1999). Rule learning by seven-month-old infants. *Science, 283*(5398), 77–80.

Marno, H., Farroni, T., Vidal Dos Santos, Y., Ekramnia, M., Nespor, M., & Mehler, J. (2015). Can you see what I am talking about? Human speech triggers referential expectation in four-month-old infants. *Scientific Reports, 5*, 13594. doi:10.1038/srep13594

McCulloch, D. L., Orbach, H., & Skarf, B. (1999). Maturation of the pattern-reversal VEP in human infants: A theoretical framework. *Vision Research, 39*(22), 3673–3680.

Paredes, M. F., James, D., Gil-Perotin, S., Kim, H., Cotter, J. A., Ng, C., … Alvarez-Buylla, A. (2016). Extensive migration of young neurons into the infant human frontal lobe. *Science, 354*(6308). doi:10.1126/science.aaf7073

Pena, M., Maki, A., Kovacic, D., Dehaene-Lambertz, G., Koizumi, H., Bouquet, F., & Mehler, J. (2003). Sounds and silence: An optical topography study of language recognition at birth. *Proceedings of the National Academy of Sciences of the United States of America, 100*(20), 11702–11705.

Pena, M., Pittaluga, E., & Mehler, J. (2010). Language acquisition in premature and full-term infants. *Proceedings of the National Academy of Sciences of the United States of America, 107*(8), 3823–3828. doi:10.1073/pnas.0914326107

Pena, M., Werker, J. F., & Dehaene-Lambertz, G. (2012). Earlier speech exposure does not accelerate speech acquisition. *Journal of Neuroscience, 32*(33), 11159–11163. doi:10.1523/jneurosci.6516-11.2012

Pierce, L. J., Klein, D., Chen, J. K., Delcenserie, A., & Genesee, F. (2014). Mapping the unconscious maintenance of a lost first language. *Proceedings of the National Academy of Sciences of the United States of America, 111*(48), 17314–17319. doi:10.1073/pnas.1409411111

Saffran, J. R., Aslin, R. N., & Newport, E. L. (1996). Statistical learning by 8-month-old infants. *Science, 274*, 1926–1928.

Salvan, P., Tournier, J. D., Batalle, D., Falconer, S., Chew, A., Kennea, N., … Counsell, S. J. (2017). Language ability in preterm children is associated with arcuate fasciculi microstructure at term. *Human Brain Mapping, 38*(8), 3836–3847. doi:10.1002/hbm.23632

Sato, H., Hirabayashi, Y., Tsubokura, H., Kanai, M., Ashida, T., Konishi, I., … Maki, A. (2012). Cerebral hemodynamics in newborn infants exposed to speech sounds: A whole-head optical topography study. *Human Brain Mapping, 33*(9), 2092–2103. doi:10.1002/hbm.21350

Shi, R. (2014). Functional morphemes and early language acquisition. *Child Development Perspectives, 8*(1), 6–11.

Shukla, M., White, K. S., & Aslin, R. N. (2011). Prosody guides the rapid mapping of auditory word forms onto visual objects in 6-mo-old infants. *Proceedings of the National Academy of Sciences of the United States of America, 108*(15), 6038–6043. doi:10.1073/pnas.1017617108

Shultz, S., Vouloumanos, A., Bennett, R. H., & Pelphrey, K. (2014). Neural specialization for speech in the first months of life. *Developmental Science, 17*(5), 766–774. doi:10.1111/desc.12151

Smyser, C. D., Snyder, A. Z., & Neil, J. J. (2011). Functional connectivity MRI in infants: Exploration of the functional organization of the developing brain. *Neuroimage*. doi:10.1016/j.neuroimage.2011.02.073

Takahashi, E., Folkerth, R. D., Galaburda, A. M., & Grant, P. E. (2011). Emerging cerebral connectivity in the human fetal brain: An MR tractography study. *Cerebral Cortex*. doi:10.1093/cercor/bhr126

Tincoff, R., & Jusczyk, P. W. (1999). Some beginnings of word comprehension in 6-month-olds. *Psychological Science, 10*(2), 172–175.

Toro, J. M., Trobalon, J. B., & Sebastian-Galles, N. (2005). Effects of backward speech and speaker variability in language discrimination by rats. *Journal of Experimental Psychology: Animal Behavior Processes, 31*(1), 95–100.

Tsao, F. M., Liu, H. M., & Kuhl, P. K. (2004). Speech perception in infancy predicts language development in the second year of life: A longitudinal study. *Child Development, 75*(4), 1067–1084.

Vannasing, P., Florea, O., Gonzalez-Frankenberger, B., Tremblay, J., Paquette, N., Safi, D., … Gallagher, A. (2016). Distinct hemispheric specializations for native and non-native languages in one-day-old newborns identified by fNIRS. *Neuropsychologia, 84*, 63–69. doi:10.1016/j.neuropsychologia.2016.01.038

Werker, J. F., & Tees, R. C. (1984). Phonemic and phonetic factors in adult cross-language speech perception. *Journal of the Acoustical Society of America, 75*(6), 1866–1878.

Zatorre, R. J., Belin, P., & Penhune, V. B. (2002). Structure and function of auditory cortex: Music and speech. *Trends in Cognitive Sciences, 6*(1), 37–46.

79 Aphasia and Aphasia Recovery

STEPHEN M. WILSON AND JULIUS FRIDRIKSSON

ABSTRACT Aphasia is an acquired impairment of language processing. In this chapter we describe the 19th-century foundations of the classical model of aphasia and how it has been refined over time in response to increasingly sophisticated neuropsychological and neuroimaging studies. In most individuals with aphasia, language function recovers to some extent, suggesting that the language network is not immutable but is capable of functional reorganization. We discuss predictors of aphasia recovery and brain changes that may be associated with a successful recovery.

19th-Century Foundations

Aphasia is an acquired impairment of the production and/or comprehension of language, due to brain injury. The most common etiology is stroke, but any kind of brain injury can cause aphasia, including neurodegeneration, tumors, resective surgery, and traumatic brain injury.

Descriptions of aphasia in the medical literature date back to about 400 BC, but the modern field of aphasia research began in 1861, when Paul Broca (1861) published a report of a patient with expressive aphasia and a lesion centered on the posterior left inferior frontal gyrus, the region now known as Broca's area. The details of the patient's speech impairment and cortical damage were complicated. However, what is important is that Broca proposed the idea that damage to a specific brain region would result in an expressive language deficit because that region has a specific role in speech production. In 1861, Broca did not make anything of the fact that his patient's lesion was in the left hemisphere, but after observing several dozen cases of aphasia over the next few years, all associated with left-hemisphere damage at autopsy, he famously declared, "*Nous parlons avec l'hémisphère gauche*" (we speak with the left hemisphere; Broca, 1865).

Ten years later, Carl Wernicke (1874), a young German neurologist, wrote a remarkable monograph on aphasia. Wernicke not only described a different kind of aphasia—a receptive aphasia we now call *Wernicke's aphasia*—but also derived, from his observations, an insightful model of language processing and the ways in which it can be disrupted by brain damage. Ludwig Lichtheim (1885), a German neurologist, refined and expanded on Wernicke's model, yielding the Wernicke-Lichtheim model (figure 79.1*A*). The model describes input and output transformations: in language comprehension, auditory inputs (a) map onto phonological representations in the posterior superior temporal gyrus (A), which are linked to neurally distributed semantic representations (B), while in language production, these same semantic representations (B) are linked to articulatory representations in Broca's area (M), which project to motor effectors (m). But critically, there is also a link between A and M. Wernicke motivated this link based on his observations that speech production was not intact in his patients with receptive deficits. While their speech was fluent (reflecting the preservation of M and m), it was garbled, with words and sounds misselected; today, we would say *paraphasic*. Wernicke concluded that speech production must not only rely on the pathway from B to M to m but must also depend on the phonological representations that he localized to the superior temporal gyrus (A). This architecture also raised the possibility that the pathway between A and M could be selectively disrupted, in which case language comprehension would be preserved (because a to A to B is intact), while production would be fluent (because M to m is intact) yet paraphasic (because of the disconnection of the phonological representations in A). Wernicke called this syndrome *conduction aphasia*. Similarly,

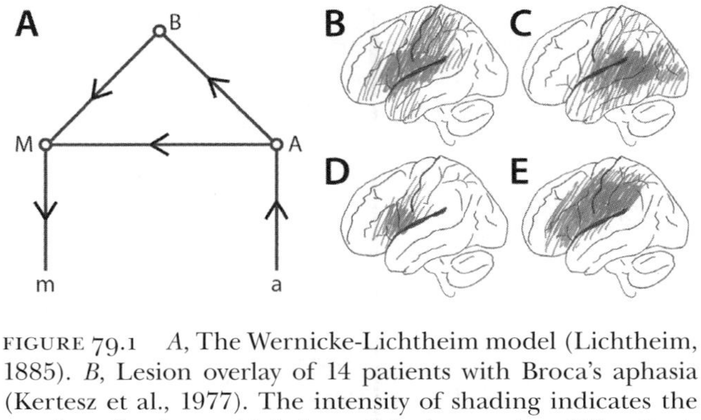

FIGURE 79.1 *A*, The Wernicke-Lichtheim model (Lichtheim, 1885). *B*, Lesion overlay of 14 patients with Broca's aphasia (Kertesz et al., 1977). The intensity of shading indicates the number of patients with lesions. *C*, Lesion overlay of 13 patients with Wernicke's aphasia (Kertesz et al., 1977). *D*, Lesion overlay of 13 patients with infarction restricted to Broca's area (Mohr, 1976). *E*, Lesion overlay of 10 patients with persistent Broca's aphasia (Mohr, 1976). (See color plate 92.)

disconnections of other pathways predict other patterns of deficits; for instance, disruption of the pathway between A and B leads to *transcortical sensory aphasia*, in which comprehension is impaired with relative sparing of repetition (because of the intact link between A and M). From these examples, the predictive nature of the model can be readily appreciated.

Evolving Understanding of the Classic Syndromes

In the 1960s researchers at the Boston Veterans Administration (VA)—Norman Geschwind, Harold Goodglass, Edith Kaplan, Frank Benton, and others—developed a sophisticated, multidisciplinary approach to aphasia, broadly based on the Wernicke-Lichtheim model. Geschwind's (1965) work on disconnection syndromes put the model on a more modern anatomical footing while Goodglass and Kaplan's (1972) *Boston Diagnostic Aphasia Examination* (BDAE) provided a means for diagnosing major aphasic syndromes that are, in most cases, closely based on the syndromes proposed by Wernicke and Lichtheim. The BDAE remains widely used today.

In the 1970s and 1980s, research on the neuroanatomical basis of aphasia was transformed by the development of structural imaging (CT and MRI) and metabolic imaging with PET. Whereas previous generations of researchers had needed to wait potentially decades until autopsy to learn the neural correlates of observed language deficits, this information could now be obtained immediately. It became feasible to study groups of patients and identify general patterns, rather than relying on single cases and their idiosyncrasies. One of the most informative approaches was to create *lesion overlays* of patients sharing an aphasic syndrome or a particular kind of language deficit, so the common neural substrates could be identified. Lesion overlays of classic aphasia syndromes proved to be at least broadly consistent with the Wernicke-Lichtheim model (Basso, Lecours, Moraschini, & Vanier, 1985; Kertesz, Lesk, & McCabe, 1977; Naeser & Hayward, 1978), with Broca's and Wernicke's aphasias associated with relatively anterior and posterior lesion locations (figure 79.1*B, C*), and with transcortical aphasias sparing the perisylvian language network.

Yet there were some striking findings that challenged traditional concepts. Mohr (1976) showed that circumscribed damage to Broca's area (figure 79.1*D*) did not suffice to cause persistent Broca's aphasia, which only followed from much larger lesions (figure 79.1*E*). Basso et al. (1985) found that most patients' lesions were in accordance with the model, but a substantial minority had unexpected lesion localizations. In an elegant series of studies, Metter et al. (1989) showed that regardless of the particulars of structural damage, metabolic abnormalities in left temporoparietal cortex were highly predictive of aphasia severity.

In the new millennium, *dual-stream* models of language have been influential (Bornkessel-Schlesewsky & Schlesewsky, 2013; Hickok & Poeppel, 2007; Wilson et al., 2011). These models propose a ventral stream through the temporal lobes that maps auditory inputs onto meaning and a dorsal stream that maps acoustic or phonological representations onto motor plans for speech production (Hickok & Poeppel, 2007) or may be involved in sequential processing more generally (Bornkessel-Schlesewsky & Schlesewsky, 2013; Wilson et al., 2011). In some respects, this ventral/dorsal dichotomy has supplanted the old posterior/anterior dichotomy of the Wernicke-Lichtheim model (Fridriksson et al., 2016, 2018). While the dual-stream model has introduced some important novel concepts, such as the linguistic capacity of the right-hemisphere ventral stream and the idea that metalinguistic perceptual tasks depend on the dorsal stream, there is also considerable continuity with the classic model: the ventral stream essentially corresponds to the mapping between A and B in the Wernicke-Lichtheim model, while the dorsal stream corresponds to the link between A and M.

Primary Progressive Aphasia

Primary progressive aphasia (PPA) is a clinical syndrome in which the neurodegeneration of dominant-hemisphere language regions leads to progressive language deficits, with relative sparing of other cognitive functions. In contrast to aphasia caused by stroke, its onset is insidious, and language deficits become progressively more severe over time. The study of PPA over the past few decades has contributed greatly to our understanding of the neural architecture of language. One reason for this is that different regions are damaged in PPA than in stroke. For instance, focal damage to the anterior temporal lobe is uncommon in stroke due to vascular anatomy, so the critical role of this region in lexical knowledge was largely unknown until the systematic investigation of semantic dementia in the 1990s (Hodges, Patterson, Oxbury, & Funnell, 1992).

Patients with progressive language deficits have been described for over 100 years (e.g., Imura, 1943; Pick, 1892; Serieux, 1893), but the modern exploration of PPA began in the mid-1970s, when Elizabeth Warrington (1975) described three patients who presented with what she described as a selective impairment of semantic memory. In each case, deficits emerged gradually, and there was no discrete precipitating event like a stroke. The patients demonstrated severe lexical impairments in both production and comprehension. In fact, their deficits were not strictly linguistic: they also demonstrated a loss of object knowledge. Meanwhile, their general cognitive function was well preserved, as were many language domains, including syntax, phonology, and speech production. A few years later, Marsel Mesulam (1982) described six patients with slowly progressive aphasia in the absence of generalized dementia. Imaging findings were generally consistent with left perisylvian atrophy. The selectivity of the language deficits was remarkable in both case series and clearly demonstrated that neurodegenerative processes can be focal in nature and have the potential to affect language areas of the brain.

In the next decade, pioneering research on PPA was carried out by Mesulam and his team and many others, including John Hodges, Karalyn Patterson, and Julie Snowden. It became apparent that PPA patients could be classified into variants based on linguistic features and that each variant was associated with distinct patterns of atrophy (Gorno-Tempini et al., 2004) and different underlying pathologies (Davies et al., 2005; Josephs et al., 2008). Maria Luisa Gorno-Tempini et al. (2004, 2011) defined three specific variants, which are now termed nonfluent/agrammatic variant PPA, semantic variant PPA, and logopenic variant PPA. The nonfluent/agrammatic variant PPA involves deficits in speech production and/or grammar and left-posterior fronto-insular atrophy. The semantic variant is defined by impaired naming, as well as poor comprehension of single words, in association with anterior temporal atrophy. Object knowledge is impaired, except possibly at the earliest stages, and surface dyslexia (reading exception words as they are spelled) is almost invariably present. The patients described by Warrington (1975) would now be diagnosed with semantic variant PPA. The logopenic variant is characterized by impaired retrieval of single words and impaired repetition, with atrophy centered around the left temporoparietal region. Phonemic paraphasias are also common. Most of the patients described by Mesulam (1982) would meet the criteria for the logopenic variant.

Individual Differences and Multivariate Perspectives

Much of our discussion so far has been framed around aphasic syndromes, which are helpful concepts for drawing generalizations and smoothing out the idiosyncrasies of individual cases. However, patients can be classified according to numerous different schemes (e.g., Botha et al., 2015; Goodglass & Kaplan, 1972; Gorno-Tempini et al., 2011; Kertesz, 1982; Schuell, 1965), many patients are classified differently depending on which aphasia battery is used (Wertz, Deal, & Robinson, 1984), and there can be considerable variability among patients diagnosed with the same type of aphasia (Casilio, Rising, Beeson, Bunton, & Wilson, 2019; Kertesz, 1982). These considerations have led many researchers in the new millennium to approach individuals with aphasia not as undifferentiated members of groups but as unique points in a multidimensional symptom space (Bates, Saygin, Moineau, Marangolo, & Pizzamiglio, 2005). In this view, syndromes would reflect regions of this space where patients tend to cluster.

An early example of this approach is a study by Elizabeth Bates et al. (2003) that investigated the neural correlates of fluency and auditory comprehension deficits and quantified each on a continuum. The authors' approach, which they dubbed *voxel-based lesion-symptom mapping*, involved making statistical inferences on the relationship between continuous behavioral measures and damage to each voxel in the brain. A similar approach, *voxel-based morphometry*, was applied to study lexical access in neurodegenerative cohorts (Grossman et al., 2004)

This general approach can be applied to whole batteries of language measures at once—for instance, a set of measures derived from quantitative linguistic analysis of connected speech samples (Wilson et al., 2010; figure 79.2*A–C*). Brain damage can be quantified voxel by voxel, or linguistic deficits can be correlated with damage to specific regions (Caplan et al., 2007) or white matter tracts (Wilson et al., 2011; figure 79.2*D–H*). Linguistic behavioral measures can be considered in relation to one another. Such a study by Myrna Schwartz et al. (2009) identified an anterior temporal region as critical for lemma retrieval in speech production by mapping regions associated with the production of semantic errors, after controlling for semantic function itself by covarying out scores on the Pyramids and Palm Trees Test of semantic association.

The same basic idea can be extended to functional imaging studies, in which language measures in individuals with aphasia can be correlated with functional activation across the brain (Crinion & Price, 2005; Fridriksson, Baker, & Moser, 2009; Griffis, Nenert, Allendorfer, & Szaflarski, 2017; Wilson et al., 2016). For instance, Wilson et al. (2016) showed that in a large cohort of patients with PPA, individuals with spared syntactic processing recruited a left-lateralized frontotemporal-parietal network, whereas those with syntactic processing deficits did not (figure 79.2*I–M*).

In the last few years, researchers have begun to apply multivariate approaches, such as factor analysis and machine-learning methods, to unraveling the complex relationships between patterns of brain damage and

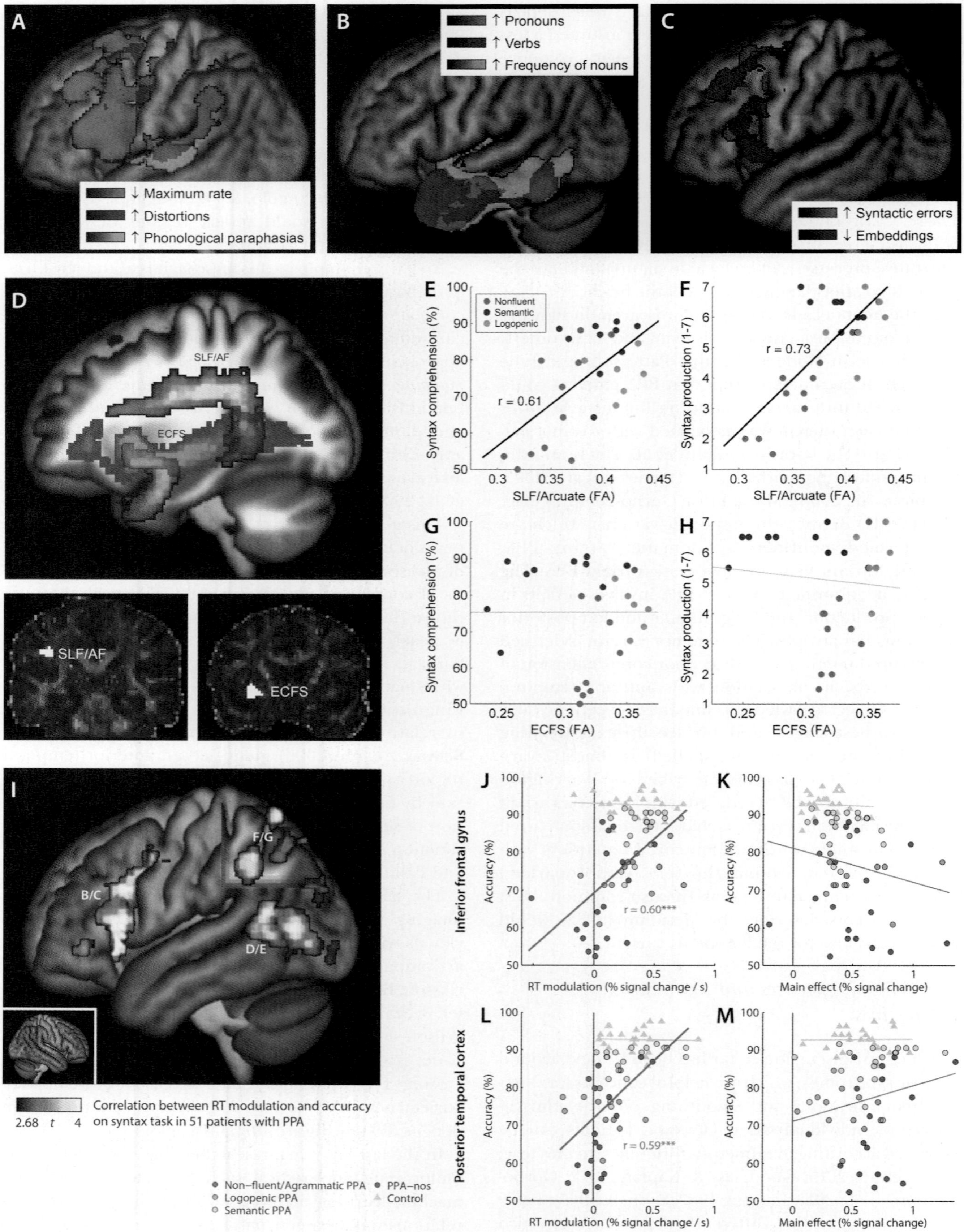

A
↓ Maximum rate
↑ Distortions
↑ Phonological paraphasias
B
↑ Pronouns
↑ Verbs
↑ Frequency of nouns
C
↑ Syntactic errors
↓ Embeddings
D
SLF/AF
ECFS
SLF/AF
ECFS
E
Nonfluent
Semantic
Logopenic
r = 0.61
Syntax comprehension (%)
SLF/Arcuate (FA)
F
r = 0.73
Syntax production (1-7)
SLF/Arcuate (FA)
G
Syntax comprehension (%)
ECFS (FA)
H
Syntax production (1-7)
ECFS (FA)
I
F/G
B/C
D/E
Correlation between RT modulation and accuracy on syntax task in 51 patients with PPA
2.68 t 4
J
K
Inferior frontal gyrus
Accuracy (%)
r = 0.60***
RT modulation (% signal change / s)
Main effect (% signal change)
L
M
Posterior temporal cortex
Accuracy (%)
r = 0.59***
RT modulation (% signal change / s)
Main effect (% signal change)
Non–fluent/Agrammatic PPA
Logopenic PPA
Semantic PPA
PPA–nos
Control

profiles of language deficits. Multivariate analyses of language deficits have shown that panels of linguistic variables can be reduced to smaller numbers of underlying explanatory factors (Butler, Lambon Ralph, & Woollams, 2014; Casilio et al., 2019; Mirman et al., 2015). For instance, Casilio et al. (2019) showed that 79% of the variance in a set of 27 connected speech measures could be explained with reference to just four underlying factors, which they labeled paraphasia, logopenia, agrammatism, and motor speech. The explanatory factors can then be associated with patterns of brain damage. For example, Mirman et al. (2015) showed that speech recognition and speech production factors were associated with damage to adjacent regions in the superior temporal gyrus and supramarginal gyrus, respectively.

Taken together, these kinds of studies have resulted in a fundamental shift in how we think about language and the brain. Traditionally, researchers thought in terms of associations between brain regions and aphasic syndromes. Nowadays, we think in terms of interacting brain networks and the roles they play in specific language domains and processes (Fedorenko & Thompson-Schill, 2014).

Historical Perspectives on Aphasia Recovery

Most research on aphasia has focused on its nature, primarily in relation to specific language and speech impairments and the links between impairment and lesion location. Far less research has been devoted to understanding recovery from aphasia. Nevertheless, aphasia treatment was addressed by some of the early pioneers of aphasiology. Broca (1865) speculated as to whether the right hemisphere could be trained to take on language function in individuals with aphasia with left-hemisphere damage. His premise was that even though the left hemisphere is dominant for language, the right hemisphere may have the potential to learn language much as a child initially learns language. Broca actually administered aphasia therapy to at least one patient who, based on Broca's report, showed improvements in vocabulary and reading. Although Broca did not describe his approach to improve vocabulary, the reading remediation focused on initially relearning the letters of the alphabet. Then, the training moved to putting letters together to form syllables and, finally, to form whole words. However, the transfer to whole words did not proceed as Broca had expected, as the patient relied more on whole-word recognition than letter-by-letter reading. Interestingly, Broca suggested that a main reason why aphasic patients could not relearn language more quickly was because they also tended to have cognitive problems that impaired the learning process. This is one of the earliest accounts of aphasia therapy, and it demonstrates that even 150 years ago, it was recognized that aphasic patients could potentially benefit from therapy.

The era of modern aphasia therapy is typically thought to start with work by Hildred Schuell, a speech-language pathologist at the Minneapolis VA Hospital, who primarily treated soldiers who were aphasic as a result of gunshot wounds suffered during World War II. Schuell's approach was based on engaging the impaired language system using controlled and often repeated auditory stimuli and a hierarchy of treatment steps, many of which are still in use today in clinical aphasia therapy. The premise of the approach was to enable the retrieval of words that, in Schuell's opinion, had not been lost as a result of brain damage but rather were preserved but could not be easily accessed. Schuell's approach improves lexical access while also promoting encouragement and the confidence to transfer treatment gains to real-life communication. Today, many different aphasia treatment approaches are used in clinical practice, and the focus varies from impairment-based approaches that directly target speech and language improvement to more functional approaches that emphasize successful communication over lessening the severity of the language impairment.

Predicting Recovery from Aphasia

Most patients with stroke-induced aphasia experience some improvements in speech and language processing in the weeks and months following onset, regardless of

FIGURE 79.2 Neural correlates of language deficits in individuals. Voxel-based morphometry revealed distinct regions where atrophy was predictive of speech (*A*), lexical (*B*), or syntactic (*C*) deficits (Wilson et al., 2010). Arrows denote increases or decreases in the prevalence of the phenomena listed. Dorsal and ventral language tracts were identified with diffusion tensor imaging (*D*). ECFS = extreme capsule fiber system; SLF/AF = superior longitudinal fasciculus/arcuate fasciculus. The degeneration of dorsal tracts was associated with deficits in syntactic comprehension (*E*) and production (*F*), while the degeneration of ventral tracts had no effects on syntactic comprehension (*G*) or production (*H*) (Wilson et al., 2011). Functional imaging identified brain regions where recruitment for syntactic processing was predictive of success in syntactic processing in PPA (*I*). In the inferior frontal gyrus (*J*, *K*) and posterior temporal cortex (*L*, *M*), modulation of functional signal by syntactic complexity was predictive of accuracy (*J*, *L*), but nonspecific recruitment for the task was not (*K*, *M*) (Wilson et al., 2016). (See color plate 93.)

whether they receive aphasia therapy (Pedersen, Jorgensen, Nakayama, Raaschou, & Olsen, 1995). This is typically referred to as *spontaneous recovery*, and its extent can vary widely across patients. The bulk of spontaneous aphasia recovery occurs within the first 3 months after stroke onset (Enderby & Petheram, 2002; Pedersen et al., 1995), and most patients are considered stable with regard to aphasia severity at 6–12 months poststroke. Although it can be difficult to predict if, and how much, individual patients will recover, some general guidelines exist. One of the strongest predictors of poor outcome is larger lesion size (Kertesz, 1988). This makes sense since patients with more extensive cortical damage have less residual brain tissue to assume whatever language functions were lost as a result of the stroke. Naturally, the patients with the largest lesions also tend to have the most extensive language impairment, which is probably why overall aphasia severity predicts long-term recovery (Kertesz, 1988; Kertesz, Harlock, & Coates, 1979). Lesion location is also important for spontaneous aphasia recovery. Patients with relatively greater damage to perisylvian regions experience less recovery compared to patients with similar lesion size but less perisylvian involvement, and damage to temporal lobe language areas is more likely to result in lasting language deficits than damage to frontal lobe language areas (Metter et al., 1989; Mohr, 1976).

Stroke type matters, as patients with ischemic stroke experience less early recovery compared to those with aphasia as a result of hemorrhagic stroke (Holland, Greenhouse, Fromm, & Swindell, 1989). In the acute stage, the sequelae of hemorrhagic stroke are more complicated than in ischemic stroke, and hemorrhagic patients tend to be sicker than those with ischemic stroke, as indicated by higher mortality rates and longer stays in the hospital. However, a surviving hemorrhagic patient can expect to experience a greater return of function compared to patients with ischemic stroke.

Even though the bulk of aphasia recovery occurs within the first year after stroke, aphasia severity can sometimes be quite dynamic in the chronic phase. In a longitudinal study, Holland, Fromm, Forbes, and MacWhinney (2017) followed individuals with chronic aphasia who were tested twice at least 1 year apart. They found that over half of their participants experienced improvements in overall aphasia severity that were greater than the standard error of measurement, whereas approximately a quarter of the participants were stable, and the remaining participants declined. The mean time poststroke among the participants was 5.5 years, which suggests that individuals can experience considerable aphasia recovery even several years after stroke.

Brain Changes Associated with Aphasia Recovery

What are the neural substrates that underlie recovery from aphasia? This question has been addressed in many functional-imaging studies. It is clear that the mechanisms of recovery are different at different stages of recovery. In the acute poststroke period, reperfusion of the ischemic penumbra appears to be a major determinant of the rapid improvements that are often seen (Hillis et al., 2002). In the early subacute period (the first few weeks after stroke), there is some evidence that right frontal regions may play a compensatory role (Saur et al., 2006; Winhuisen et al., 2005), which is more likely to reflect the recruitment of domain-general cognitive resources than language reorganization (Geranmayeh, Brownsett, & Wise, 2014). However, the recruitment of these regions decreases over time (Winhuisen et al., 2007), with left lateralization returning over time (Heiss & Thiel, 2006; Saur et al., 2006).

Language outcome has been shown to be associated with the extent to which typical left frontal and temporal language regions can be activated by language processing (Griffis et al., 2017). Fridriksson (2010) found a strong association between anomia treatment success and increased cortical activation (as measured using functional magnetic resonance imaging [fMRI]) in the left hemisphere. Specifically, patients who fared well in treatment also experienced a significant increase in left-hemisphere activation, suggesting that recovery from anomia in chronic stroke may be mediated by the left hemisphere. In a follow-up study, Fridriksson, Richardson, Fillmore, and Cai (2012) related change in functional activity in perilesional cortex to change in correct naming. To address the relationship between change in brain activation and improvement in naming, activation was compared between two baseline and two posttreatment fMRI runs in perilesional cortex. A regression analysis revealed that activation change in the perilesional frontal lobe was a predictor of correct naming improvement. Treatment-related change in the production of semantic paraphasias was most robustly predicted by activation change in the temporal lobe, while change in phonemic paraphasias was predicted by activation change involving both the left temporal and parietal lobes. These findings suggest that changes in activation in perilesional regions are associated with treated recovery from anomia.

Other researchers have argued that the right hemisphere plays a major role in aphasia recovery. For example, Weiller et al. (1995) reported that right-hemisphere homotopic areas were activated for language processing in a group of patients who had largely recovered from Wernicke's aphasia. However, it is possible that a group selected for excellent recovery from Wernicke's aphasia may represent a rather exceptional group of individuals. In a larger and more representative group, Crinion and Price (2005) showed that the recruitment of right posterior temporal cortex for narrative comprehension was associated with preserved comprehension in poststroke aphasia. However, this was not interpreted as a finding of reorganization per se because narrative comprehension depends on both temporal lobes in neurologically normal individuals, too.

Whereas localized changes in brain activity may be important for aphasia recovery, it seems plausible that changes in functional network connectivity also play a role. In fact, it could be that changes in connectivity are the primary drivers of aphasia recovery. In a recent well-powered study, Siegel et al. (2018) found that the reemergence of network modularity, a measure comparing the density of connectivity within networks to the density of connectivity between networks, was associated with aphasia recovery in stroke patients at 3 months and 1 year poststroke.

Conclusion

The study of aphasia has provided some groundbreaking findings in regard to the neuroanatomical organization of language. Much of this work has relied on lesion-symptom associations to infer which regions of the brain are crucial for, not just associated with, the execution of given speech or language tasks. Although the technologies and methodologies used in these studies have evolved enormously, especially in the last three decades, the basic premise of the studies has not changed: if a given cortical region or network supports a specific function, then damage to that region should cause an impairment in that same function. The influence of aphasia studies on the neuropsychological understanding of language is perhaps most evident in the current zeitgeist of dual-stream models that have become mainstream in the field. Although much of the work on aphasia has focused on understanding normal brain-behavior relationships, a parallel focus has centered on the clinical manifestations of speech and language impairment to inform clinical practice. Ideally, the study of aphasia will proceed with a united focus where basic science informs clinical research, and vice versa.

Acknowledgment

This work was supported in part by the National Institute on Deafness and Other Communication Disorders (R01 DC013270 and P50 DC014664).

REFERENCES

Basso, A., Lecours, A., Moraschini, S., & Vanier, M. (1985). Anatomoclinical correlations of the aphasias as defined through computerized tomography: Exceptions. *Brain and Language, 26*, 201–229.

Bates, E., Saygin, A. P., Moineau, S., Marangolo, P., & Pizzamiglio, L. (2005). Analyzing aphasia data in a multidimensional symptom space. *Brain and Language, 92*, 106–116.

Bates, E., Wilson, S. M., Saygin, A. P., Dick, F., Sereno, M. I., Knight, R. T., & Dronkers, N. F. (2003). Voxel-based lesion-symptom mapping. *Nature Neuroscience, 6*, 448–450.

Bornkessel-Schlesewsky, I., & Schlesewsky, M. (2013). Reconciling time, space and function: A new dorsal-ventral stream model of sentence comprehension. *Brain and Language, 125*, 60–76.

Botha, H., Duffy, J., Whitwell, J., Strand, E., Machulda, M., Schwarz, C., … Lowe, V. (2015). Classification and clinicoradiologic features of primary progressive aphasia (PPA) and apraxia of speech. *Cortex, 69*, 220–236.

Broca, P. (1861). Remarques sur le siège de la faculté du langage articulé, suivies d'une observation d'aphémie (perte de la parole). In *Bulletins de la Société d'anatomie (Paris), 2e serie* (pp. 330–357).

Broca, P. (1865). Sur le siège de la faculté du langage articulé. In *Bulletins de La Société D'anthropologie de Paris, 6*(1), 377–393.

Butler, R. A., Lambon Ralph, M. A., & Woollams, A. M. (2014). Capturing multidimensionality in stroke aphasia: Mapping principal behavioural components to neural structures. *Brain, 137*, 3248–3266.

Caplan, D., Waters, G., Kennedy, D., Alpert, N., Makris, N., DeDe, G., … Reddy, A. (2007). A study of syntactic processing in aphasia II: Neurological aspects. *Brain and Language, 101*, 151–177.

Casilio, M., Rising, K., Beeson, P. M., Bunton, K., & Wilson, S. M. (2019). Auditory-perceptual rating of connected speech in aphasia. *American Journal of Speech-Language Pathology, 28*, 550–568.

Crinion, J., & Price, C. (2005). Right anterior superior temporal activation predicts auditory sentence comprehension following aphasic stroke. *Brain, 128*, 2858–2871.

Davies, R., Hodges, J., Kril, J., Patterson, K., Halliday, G., & Xuereb, J. (2005). The pathological basis of semantic dementia. *Brain, 128*, 1984–1995.

Enderby, P., & Petheram, B. (2002). Has aphasia therapy been swallowed up? *Clinical Rehabilitation, 16*, 604–608.

Fedorenko, E., & Thompson-Schill, S. L. (2014). Reworking the language network. *Trends in Cognitive Sciences, 18*, 120–126.

Fridriksson, J. (2010). Preservation and modulation of specific left hemisphere regions is vital for treated recovery from anomia in stroke. *Journal of Neuroscience, 30*, 11558–11564.

Fridriksson, J., Baker, J. M., & Moser, D. (2009). Cortical mapping of naming errors in aphasia. *Human Brain Mapping, 30*, 2487–2498.

Fridriksson, J., den Ouden, D.-B., Hillis, A., Hickok, G., Rorden, C., Basilakos, A., … Bonilha, L. (2018). Anatomy of aphasia revisited. *Brain, 141*, 848–862.

Fridriksson, J., Richardson, J. D., Fillmore, P., & Cai, B. (2012). Left hemisphere plasticity and aphasia recovery. *NeuroImage, 60*, 854–863.

Fridriksson, J., Yourganov, G., Bonilha, L., Basilakos, A., Den Ouden, D.-B., & Rorden, C. (2016). Revealing the dual streams of speech processing. *Proceedings of the National Academy of Sciences of the United States of America, 113*, 15108–15113.

Geranmayeh, F., Brownsett, S. L. E., & Wise, R. J. S. (2014). Task-induced brain activity in aphasic stroke patients: What is driving recovery? *Brain, 137*, 2632–2648.

Geschwind, N. (1965). Disconnexion syndromes in animals and man. *Brain, 88*, 237–294.

Goodglass, H., & Kaplan, E. (1972). *Boston Diagnostic Aphasia Examination*. Philadelphia: Lea & Febiger.

Gorno-Tempini, M. L., Dronkers, N., Rankin, K., Ogar, J., La Phengrasamy, B., Rosen, H., … Miller, B. (2004). Cognition and anatomy in three variants of primary progressive aphasia. *Annals of Neurology, 55*, 335–346.

Gorno-Tempini, M. L., Hillis, A., Weintraub, S., Kertesz, A., Mendez, M., Cappa, S., … Grossman, M. (2011). Classification of primary progressive aphasia and its variants. *Neurology, 76*, 1006–1014.

Griffis, J. C., Nenert, R., Allendorfer, J. B., & Szaflarski, J. P. (2017). Linking left hemispheric tissue preservation to fMRI language task activation in chronic stroke patients. *Cortex, 96*, 1–18.

Grossman, M., McMillan, C., Moore, P., Ding, L., Glosser, G., Work, M., & Gee, J. (2004). What's in a name: Voxel-based morphometric analyses of MRI and naming difficulty in Alzheimer's disease, frontotemporal dementia and corticobasal degeneration. *Brain, 127*, 628–649.

Heiss, W.-D., & Thiel, A. (2006). A proposed regional hierarchy in recovery of post-stroke aphasia. *Brain and Language, 98*, 118–123.

Hickok, G., & Poeppel, D. (2007). The cortical organization of speech processing. *Nature Reviews Neuroscience, 8*, 393–402.

Hillis, A. E., Wityk, R. J., Barker, P. B., Beauchamp, N. J., Gailloud, P., Murphy, K., … Metter, E. J. (2002). Subcortical aphasia and neglect in acute stroke: The role of cortical hypoperfusion. *Brain, 125*, 1094–1104.

Hodges, J., Patterson, L., Oxbury, S., & Funnell, E. (1992). Semantic dementia: Progressive fluent aphasia with temporal lobe atrophy. *Brain, 115*, 1783–1806.

Holland, A., Fromm, D., Forbes, M., & MacWhinney, B. (2017). Long-term recovery in stroke accompanied by aphasia: A reconsideration. *Aphasiology, 31*, 152–165.

Holland, A. L., Greenhouse, J. B., Fromm, D., & Swindell, C. S. (1989). Predictors of language restitution following stroke: A multivariate analysis. *Journal of Speech and Hearing Research, 32*, 232–238.

Imura, T. (1943). Aphasia: Characteristics symptoms in Japanese. *Psychiatra et Neurologia Japonica, 47*, 196–218.

Josephs, K., Whitwell, J., Duffy, J., Vanvoorst, W., Strand, E., Hu, W., … Petersen, R. (2008). Progressive aphasia secondary to Alzheimer disease vs FTLD pathology. *Neurology, 70*, 25–34.

Kertesz, A. (1982). *Western Aphasia Battery*. London: Grune and Stratton.

Kertesz, A. (1988). What do we learn from recovery from aphasia? *Advances in Neurology, 47*, 277–292.

Kertesz, A., Harlock, W., & Coates, R. (1979). Computer tomographic localization, lesion size, and prognosis in aphasia and nonverbal impairment. *Brain and Language, 8*, 34–50.

Kertesz, A., Lesk, D., & Mccabe, P. (1977). Isotope localization of infarcts in aphasia. *Archives of Neurology, 34*, 590–601.

Lichtheim, L. (1885). On aphasia. *Brain, 7*, 433–484.

Mesulam, M. (1982). Slowly progressive aphasia without generalized dementia. *Annals of Neurology, 11*, 592–598.

Metter, E. J., Kempler, D., Jackson, C., Hanson, W. R., Mazziotta, J. C., & Phelps, M. E. (1989). Cerebral glucose metabolism in Wernicke's, Broca's, and conduction aphasia. *Archives of Neurology, 46*, 27–34.

Mirman, D., Chen, Q., Zhang, Y., Wang, Z., Faseyitan, O. K., Coslett, H. B., & Schwartz, M. F. (2015). Neural organization of spoken language revealed by lesion-symptom mapping. *Nature Communications, 6*, 6762.

Mohr, J. P. (1976). Broca's area and Broca's aphasia. In H. Whitaker & H. A. Whitaker (Eds.), *Studies in neurolinguistics* (pp. 201–233). New York: Academic Press.

Naeser, M., & Hayward, R. (1978). Lesion localization in aphasia with cranial computed tomography and the Boston Diagnostic Aphasia Exam. *Neurology, 28*, 545–551.

Pedersen, P., Jorgensen, H., Nakayama, H., Raaschou, H., & Olsen, T. (1995). Aphasia in acute stroke: Incidence, determinants and recovery. *Annals of Neurology, 38*, 659–666.

Pick, A. (1892). Ueber die Beziehungen der senile Hirnatrophie zur Aphasie. *Prager Medicinische Wochenschrift, 17*, 165–167.

Saur, D., Lange, R., Baumgaertner, A., Schraknepper, V., Willmes, K., Rijntjes, M., & Weiller, C. (2006). Dynamics of language reorganization after stroke. *Brain, 129*, 1371–1384.

Schuell, H. (1965). *Differential diagnosis of aphasia with the Minnesota test: Administrative manual for the Minnesota Test for Differential Diagnosis of Aphasia* (Vol. 1). Minneapolis: University of Minnesota Press.

Schwartz, M. F., Kimberg, D. Y., Walker, G. M., Faseyitan, O., Brecher, A., Dell, G. S., & Coslett, H. B. (2009). Anterior temporal involvement in semantic word retrieval: Voxel-based lesion-symptom mapping evidence from aphasia. *Brain, 132*, 3411–3427.

Serieux, P. (1893). Sur un cas de surdite verbale pure. *Revue de Médecine, 13*, 733–750.

Siegel, J. S., Seitzman, B. A., Ramsey, L. E., Ortega, M., Gordon, E. M., Dosenbach, N. U. F., … Corbetta, M. (2018). Re-emergence of modular brain networks in stroke recovery. *Cortex, 101*, 44–59.

Warrington, E. (1975). The selective impairment of semantic memory. *Quarterly Journal of Experimental Psychology, 27*, 635–657.

Weiller, C., Isensee, C., Rijntjes, M., Huber, W., Müller, S., Bier, D., … Diener, H. C. (1995). Recovery from Wernicke's aphasia: A positron emission tomographic study. *Annals of Neurology, 37*, 723–732.

Wernicke, C. (1874). *Der Aphasische Symptomencomplex*. Breslau: Cohn and Weigert.

Wertz, R., Deal, J., & Robinson, A. (1984). Classifying the aphasias: A comparison of the Boston Diagnostic Aphasia Examination and the Western Aphasia Battery. In *Clinical aphasiology* (pp. 40–47). London: BBK.

Wilson, S. M., Demarco, A. T., Henry, M. L., Gesierich, B., Babiak, M., Miller, B. L., & Gorno-Tempini, M. L. (2016). Variable disruption of a syntactic processing network in primary progressive aphasia. *Brain, 139*, 2994–3006.

Wilson, S. M., Galantucci, S., Tartaglia, M. C., Rising, K., Patterson, D. K., Henry, M. L., … Gorno-Tempini, M. L. (2011). Syntactic processing depends on dorsal language tracts. *Neuron, 72*, 397–403.

Wilson, S. M., Henry, M. L., Besbris, M., Ogar, J. M., Dronkers, N. F., Jarrold, W., … Gorno-Tempini, M. L. (2010).

Connected speech production in three variants of primary progressive aphasia. *Brain, 133*, 2069–2088.

Winhuisen, L., Thiel, A., Schumacher, B., Kessler, J., Rudolf, J., Haupt, W. F., & Heiss, W. D. (2005). Role of the contralateral inferior frontal gyrus in recovery of language function in poststroke aphasia: A combined repetitive transcranial magnetic stimulation and positron emission tomography study. *Stroke, 36*, 1759–1763.

Winhuisen, L., Thiel, A., Schumacher, B., Kessler, J., Rudolf, J., Haupt, W. F., & Heiss, W. D. (2007). The right inferior frontal gyrus and poststroke aphasia: A follow-up investigation. *Stroke, 38*, 1286–1292.

XI SOCIAL NEUROSCIENCE

Introduction

ELIZABETH PHELPS AND MAURICIO DELGADO

Although this volume is titled *The Cognitive Neurosciences*, in some ways this name is misleading because it has become a quinquennial marker for the status of the field of human neuroscience more broadly. In contrast to the early days of psychology, which led to the parsing of human mental life and behavior into subdisciplines for study (e.g., cognitive, social, clinical, developmental), the introduction of human neuroscience techniques into the study of mind and behavior has taught us that we cannot so easily parse the brain. As successive generations of psychological scientists have recognized the value of using neuroscience techniques to understand the human mind, they have had to grapple with how to (re)connect the science from the subdisciplines of psychology. This is the path that social neuroscience has taken. In the first volume of *The Cognitive Neurosciences*, social behavior was not included as a topic of investigation, as most studies of human brain function at the time focused on behaviors that fell under the domain of cognitive psychology. The second volume was the first to include emotion as a topic, which began to touch on some social psychology topics and techniques, but it was not until the third volume that research on social neuroscience (along with emotion) merited its own section. Perhaps not surprisingly, the researchers contributing these sections come from a range of disciplines and approaches and are pulled together by their shared interest in understanding the neural underpinnings of social and emotional behaviors. The current volume is no exception. The contributors to this section are social, developmental, and cognitive neuroscientists, as well as a neurobiologist, who are using diverse psychological approaches, from laboratory studies in animal models of attachment to the analysis of social networks. They are tied together by the overlapping brain circuits they are investigating to understand these complex social and emotional behaviors.

The first section of chapters focuses on the impact of social connections on one's well-being. In the first chapter, Robinson-Drummer and colleagues examine rodent models of attachment and the influence of the caregiver on the infant brain. This chapter highlights the potential issues associated with poor caregiver quality and the implications for threat learning. Hornstein, Inagaki, and Eisenberger present evidence demonstrating how social connections can act as buffers against the deleterious effects of stress and emphasize how not only receiving but also giving support can contribute to social ties and overall health. In contrast, Cacioppo and Cacioppo focus on the consequences of a lack of social connection and show how social isolation, or loneliness, in the elderly is a risk factor for mortality. This chapter discusses some of the potential pathways through which social isolation can negatively influence the mechanisms that contribute to overall health.

The next section focuses on learning and decision-making in our social world. Fareri, Chang, and Delgado discuss the neural mechanisms involved in learning from and about others that help adjust social expectations and foster social relationships. Olsson and colleagues then consider social learning in the aversive domain, comparing neural mechanisms involved in threat learned via nonsocial versus social means with more empathic processes mediating learning from social observation. With respect to decision-making, Insel, Davidow, and Somerville take a neurodevelopmental approach to explore how value signals can be used to guide goal-directed behavior, in particular discussing cognitive-control capabilities that change during development. Willis and colleagues then explore decision-making in the social domain and review the neural and behavioral mechanisms that underlie cooperative decision-making among individuals.

Finally, the last chapter by Wheatley and Boncz presents the next frontier in social neuroscience and explores social networks. The chapter considers novel efforts that are attempting to go beyond individual brains to understand the social mind by opting to study more complex, yet common, naturalistic social interactions and how they occur in the context of intricate social networks.

Comparing this section to previous sections on social and emotional neuroscience in *The Cognitive Neurosciences*, we can observe the evolution of social neuroscience from an extension of studies on cognition and emotion to a diverse discipline using neuroscience techniques to investigate topics that touch on some of society's most pressing issues. This social neuroscience research takes advantage of what has been learned about brain function from previous studies on affective and cognitive neuroscience and builds on it. However, it is important to remember that although social neuroscience often begins with what has been learned about the human brain from other subdisciplines of neuroscience research, these other topics of investigation will also likely benefit from emerging research on social neuroscience in the future. For example, how can one fully understand the dynamics of language acquisition or use without an appreciation of attachment or social connections? What proportion of our decisions depends on at least some comprehension of the social dynamics of the decision context? And to what extent are our autobiographical memories embedded in our social networks?

As we strive to move beyond the laboratory to use our science to address real-world issues, understanding the impact of social factors becomes increasingly important. The chapters in this section illustrate various facets of our social lives that we are just beginning to investigate using neuroscience techniques in humans, along with several other aspects of our social world still untouched by neuroscience investigations. The increased understanding of the neuroscience of social functions underscored in this section is one more indication that human neuroscience research is changing psychology by forcing us to consider an integrated mind that does not so cleanly separate domains of mental life and behavior.

80 Neurobiology of Infant Threat Processing and Developmental Transitions

PATRESE A. ROBINSON-DRUMMER, TANIA ROTH, CHARLIS RAINEKI, MAYA OPENDAK, AND REGINA M. SULLIVAN

ABSTRACT Early life experiences have the dual purpose of producing adaptive infant behaviors that support attachment to the caregiver while promoting later behavior adaptive for independent living and survival. Adaptive infant behaviors rely on learning to attach to the caregiver and remaining in the caregiver's proximity to receive resources and protection for survival. This sensitive period for attachment learning relies on a unique neural circuitry that includes (1) a hyperfunctioning noradrenergic locus coeruleus that supports rapid olfactory system plasticity for learning, approaching, and remembering the maternal odor; and (2) attenuated amygdala plasticity that ensures pups do not learn to avoid the mother if pain is associated with maternal care. This attachment circuitry constrains the infant to form an attachment to the caretaker regardless of the quality of the care received but minimizes threat and hippocampus-dependent context learning. Poor-quality maternal care, however, profoundly influences brain development, including the early termination of the sensitive period of learning and the accelerated development of threat learning. Overall, these data suggest a strong link between the threat and attachment systems that are concurrently modified as pups experience the natural environment of the mother-infant dyad.

Experiences in early life have the dual purpose of producing adaptive infant behaviors within the attachment system to support interactions with the mother while concurrently programming adaptive, later-life behavior for independent living and survival. Here, we focus on infant adaptive behavior centering on attachment and threat learning and briefly consider the impact of these early experiences on later life.

To survive, the altricial infant must learn and remember the attachment figure and direct social behavior toward the attachment figure to receive the food, protection, and warmth necessary for survival. This phylogenetically conserved attachment system involves rapidly learning and remembering the caregiver and quickly expressing prosocial behaviors. This biologically predisposed (i.e., innate) attachment system supports the caregiver as the target of infant social behavior and facilitates the continued expression of proximity-seeking behaviors toward the caregiver, irrespective of care quality. Learning about the caregiver and the emergence of social behavior directed toward the caregiver, occurs within a temporally limited sensitive period and is referred to as *attachment*, a process of wide phylogenetic representation that includes chicks, rodents, nonhuman primates, and humans and was initially described by Bowlby (1969). Importantly, the infant brain is not an immature version of the adult brain, which is designed for self-care and defense; the infant brain is designed to engage the attachment figure for these necessities.

The robustness of this attachment system is highlighted by the fact that altricial infants form attachments to the caregiver regardless of the quality of care received, even if the caregiver is abusive. However, these experiences profoundly alter behaviors that emerge in later life. That is, the dependent, altricial infant appears somewhat protected from the detrimental effects of compromised caregiving and remains attached; however, infant experiences have an impact on an independent, maturing animal's behaviors, including self-protection against threats and appropriate behavior within a social hierarchy (Opendak, Briones, & Gould, 2016). Here, we focus on the infant attachment and threat system to consider more carefully the role and effects of trauma within the context of attachment.

Early-Life Social Behavior: Attachment Learning

Altricial infants of many species, including humans and rodents, must learn to identify, approach, and prefer their own mother. These developing animals also possess a sensitive period during which learning is rapid and robust due to a specialized learning system (Bowlby, 1969). Once learned, the attachment figure is approached and proximity is actively maintained. In humans, all sensory systems are used, while the neonatally deaf rodent relies heavily on olfaction and somatosensory cues. The maternal cues (e.g., odor) learned during development elicit attachment-associated prosocial behaviors (i.e., approach to the mother) and nipple attachment but also blunt infant stress responding (Hostinar, Sullivan, & Gunnar, 2014). In rodents, maternal odor learning occurs naturally within the nest but can be induced outside the nest; a classically conditioned novel odor rapidly becomes a new maternal odor (Sullivan, Perry, Sloan, Kleinhaus, & Burtchen, 2011).

One of the most striking features of this infant learning is the broad range of stimuli, including, presumably, painful or pleasurable stimuli, able to support odor-approach learning outside of the nest (Camp & Rudy, 1988; Haroutunian & Campbell, 1979; Sullivan, Brake, Hofer, & Williams, 1986). Specifically, paired presentations of odor and unconditioned stimuli (i.e., food, warmth, tactile stimulation, 0.5 mA shock, tail pinch) all support learned odor preferences (i.e., odor approach in a Y-maze test) and nipple attachment, behaviors evoked by maternal odor. Similarly, during a 1 h conditioning procedure in which a novel odor is placed on either an abusive or nurturing mother in the nest, the novel odor becomes a preferred odor, with properties of a new maternal odor (Perry, Al Ain, Raineki, Sullivan, & Wilson, 2016; Roth & Sullivan, 2005; Sullivan, Wilson, Wong, Correa, & Leon, 1990). Together, these results illustrate the robustness of the attachment system under natural- and artificial-learning conditions.

Attachment Learning Circuitry

Considering that neural structures in adult rats, well-documented to support learning, have a protracted development in infancy (e.g., hippocampus, frontal cortex, amygdala), the neural circuitry for sensitive-period attachment learning in the developing rat is relatively unique (Moriceau, Shionoya, Jakubs, & Sullivan, 2009; Morrison, Fontaine, Harley, & Yuan, 2013). Indeed, during the sensitive period, infant attachment odor learning relies heavily on plasticity within the olfactory system, with both anatomical and physiological changes within the olfactory bulb documented to support odor preference learning. Learning-induced olfactory bulb changes are caused by a large influx of norepinephrine (NE), released from the locus coeruleus (LC), which prevents the mitral cells of the olfactory bulb from habituating to repeated olfactory stimulation and plasticity (Wilson & Sullivan, 1994). In the infant, this abundant NE is induced by myriad sensory stimuli (including strong odors, shock, tactile stimulation, and maternal behaviors), and NE is both necessary and sufficient for the infant's learning-induced neurobehavioral changes. Indeed, LC suppression or blocking olfactory bulb NE receptors prevents pup learning, while increasing olfactory bulb NE (via LC or NE microinfusions) supports learning (Sullivan, Landers, Yeaman, & Wilson, 2000; Yuan, Harley, Darby-King, Neve, & McLean, 2003). These data indicate that the contingent events of stimulus-induced NE release from the LC and NE-induced physiological and molecular changes in the olfactory bulb ultimately support the neural plasticity responsible for the acquisition of olfactory-based attachment behavior in the infant rat.

Threat Responding and the Amygdala Are Attenuated in Early Life

In addition to the enhanced approach/attachment learning supported by the neural circuitry discussed above, the infant sensitive period for attachment is also characterized by limitations on aversive learning. For instance, shocking a chick during imprinting actually enhances following of the surrogate caregiver, although shock supports avoidance just hours after the imprinting critical period closes (Hess, 1962; Salzen, 1970). Similarly, shocking an infant dog or rat results in a strong attachment to the caregiver (Camp & Rudy, 1988; Stanley, 1962; Sullivan et al., 2000). Finally, nonhuman primate and human infants exhibit strong proximity-seeking behavior toward an abusive mother (Harlow & Harlow, 1965; Sanchez, Ladd, & Plotsky, 2001; Suomi, 2003).

Rodent models have been used extensively to understand how the infant brain fails to learn to avoid an abusive caregiver. Indeed, the brain area most closely associated with threat learning, the amygdala, is not involved in postnatal day (PN)8 pup odor-shock conditioning, but it is by PN12, when odor-shock conditioning significantly increases amygdala activity and odor aversion (Sullivan et al., 2000). It should be noted that pups younger than PN10 can learn an odor aversion. For instance, Lithium Chloride (LiCl) injection or a very high 1.2 mA shock will induce illness or malaise and malaise learning (Haroutunian & Campbell, 1979; Richardson & McNally, 2003). However, this learning is dependent upon the piriform cortex; the adult-like, amygdala-dependent malaise-learning system does not appear until weaning age (Shionoya et al., 2006). Remarkably, this odor-malaise effect exists even within maternally-controlled constraints; if neonatal rats are nursing during odor-LiCl conditioning, this prevents a learned odor aversion and instead produces a learned odor preference (Shionoya et al., 2006). Together, data indicate that learning to avoid threat is compromised in

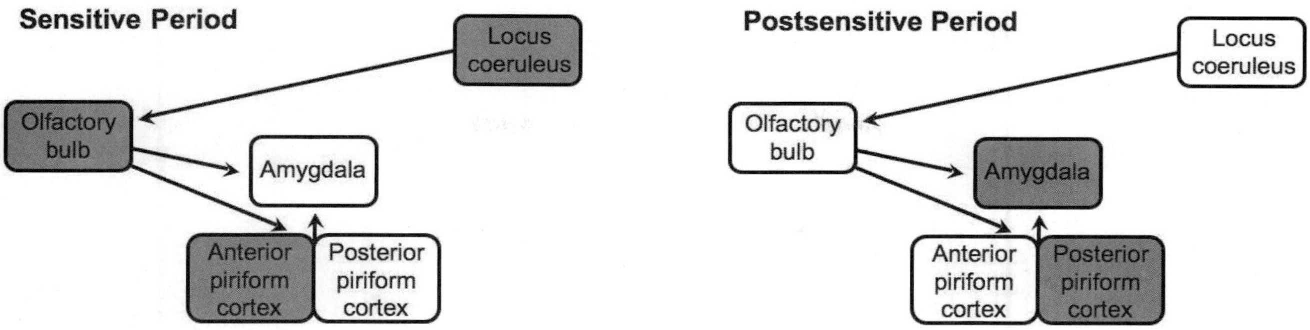

FIGURE 80.1 The neural basis of attachment learning with odor-0.5 mA shock conditioning. During the sensitive period for attachment, presumably noxious and pleasant stimuli support odor-attachment learning, which depends upon the olfactory bulb, anterior piriform cortex, and LC releasing NE into the olfactory bulb. After the sensitive period, learning becomes more specific and odor-pain pairings support amygdala-dependent threat learning in these older pups.

infants. In figure 80.1, we provide a model of our current understanding of this simplistic, early-life social attachment circuit. The figure illustrates how this circuitry changes to transition the developing animal from attachment learning to learning that can accommodate environmental contingencies.

Corticosterone Switches Amygdala Plasticity On/Off to Permit Pup Responses to Cue Threat

We initially reasoned that the delayed functional emergence of amygdala-dependent fear/threat learning was due to an immature amygdala. However, two pieces of evidence suggested an alternative explanation. First, Takahashi showed that threat responding in pups could be precociously induced in younger pups with a systemic injection of the stress hormone corticosterone (CORT; Takahashi, 1994). Second, in infant rats CORT baseline levels are low, and the ability of most stressful stimuli (i.e., restraint, shock) to evoke CORT release is greatly attenuated compared to that in older animals—this is a developmental period termed the stress hyporesponsive period (SHRP; Dallman, 2014; Levine, 1994). Interestingly, maternal sensory stimulation provided during nursing and grooming seems to control both the pups' low CORT levels and pups' failure to mount a stress response during the SHRP (van Oers, Kloet, Whelan, & Levine, 1998).

Since Takahashi's research suggested the amygdala was mature enough to participate in threat responding, but required exogenous CORT, our subsequent research tested whether low CORT levels during the SHRP were responsible for pups' failure to show amygdala-dependent fear/threat learning before PN10. Indeed, young pups still in the sensitive period for attachment (also the SHRP) that had CORT levels increased during odor-0.5 mA shock conditioning readily learned to avoid the odor paired with shock (Barr et al., 2009; Moriceau, Wilson, Levine, & Sullivan, 2006). Specifically, amygdala-dependent odor avoidance typically emerges at PN10 in rat pups, although the age can be reduced to PN5 by increasing CORT systemically or with intra-amygdala infusions during the 0.5 mA odor-shock conditioning. Furthermore, blocking systemic or amygdala CORT activity in older pups (PN10–PN15) caused them to revert to sensitive-period learning with odor-shock conditioning producing an odor preference. Thus, CORT functions as a switch that can turn the amygdala on to support the acquisition of cues associated with threat.

We also showed naturalistic ways in which CORT levels toggle infant attachment/threat learning: rearing pups with a maltreating mother prematurely ends the SHRP, elevates CORT, and permits amygdala-dependent threat learning at PN6 (Moriceau, Shionoya, et al., 2009). This precocious aversion learning can also be produced by exposing pups as young as PN6 to a novel odor paired with the alarm odor of a fearful mother, which acutely increases CORT (Debiec & Sullivan, 2014). In older pups (PN10–15), we capitalized on social buffering, a process by which the maternal presence blocks stressor-induced CORT release (Hostinar, Sullivan, & Gunnar, 2014). We found that a naturalistic blockade of CORT via maternal presence blocked fear/threat learning by preventing the participation of the amygdala in learning (Moriceau & Sullivan, 2006). We verified the causal relationship between maternal presence and the suppression of a shock-induced CORT release in pups' odor aversion learning by systemic and intra-amygdala CORT infusions (Moriceau & Sullivan, 2006; Moriceau et al., 2006). By PN15, the ability of the maternal presence to block amygdala threat learning wanes, and

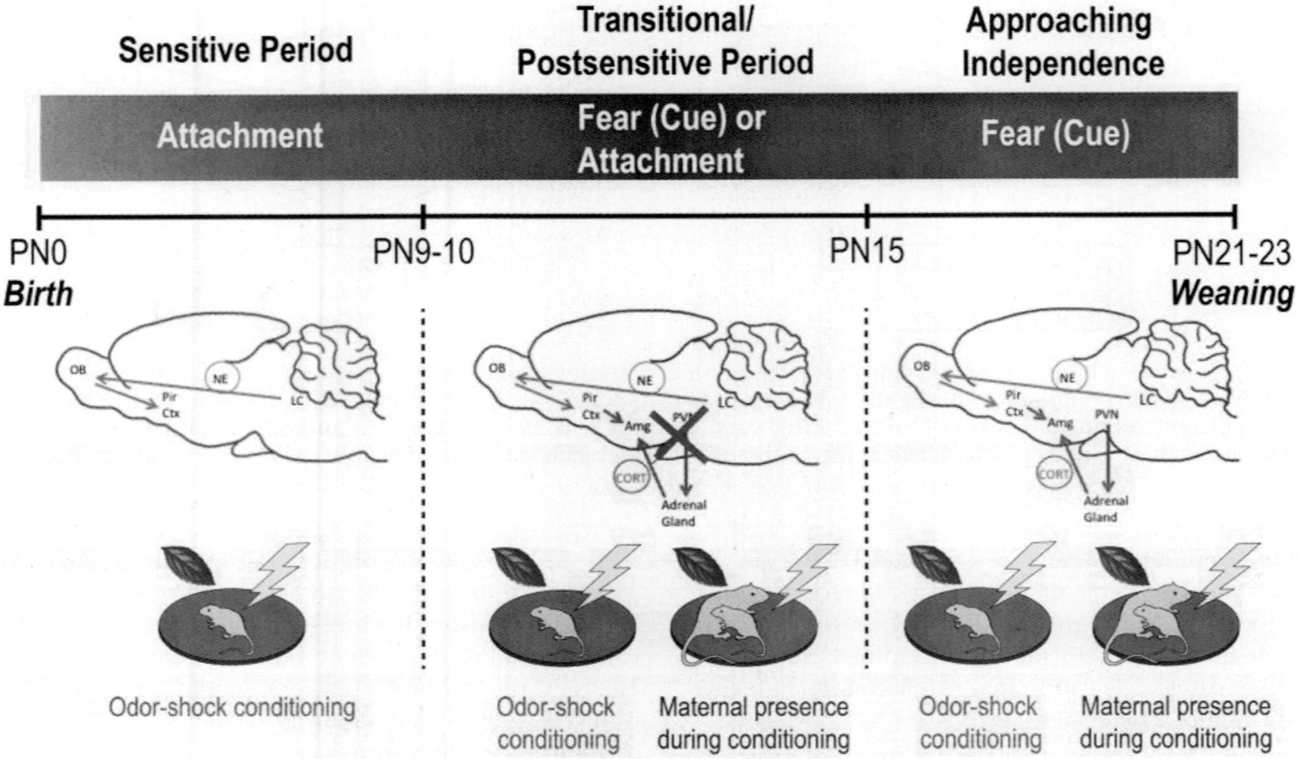

FIGURE 80.2 This schematic represents pups' developmental learning transitions with odor-0.5 mA shock conditioning. Our previous work suggests PN10 is a transitional age for the onset of amygdala-dependent fear conditioning, although until PN15, this learning depends on CORT levels, which can be modulated pharmacologically or by the maternal presence during conditioning. During this transitional period (until PN15), pups conditioned alone learn to avoid an odor paired with shock but will learn attachment when conditioned with lowered levels of CORT. After PN15, conditioning alone or with the maternal presence produces odor avoidance. (See color plate 94.)

pups learn to avoid an odor paired with shock, even if the mother is present (Upton & Sullivan, 2010). The naturalistic modulation of CORT by the mother and our experimental manipulation of CORT are in sharp contrast to the high levels of CORT used in adult learning experiments, where CORT has a modulatory role in conditioning (Corodimas, LeDoux, Gold, & Schulkin, 1994; Roozendaal, Quirarte, & McGaugh, 2002). This specialized role of corticosterone in infant attachment and threat/fear learning is illustrated in figure 80.2.

Development of Contextual Threat Learning

During the acquisition of threat conditioning using odor-shock pairings, fear/threat learning to the physical location and environmental cues (i.e., the context) are also learned. This "background" contextual fear/threat is expressed when the animal is placed back into the context where the conditioning took place without presentation of the discrete cue. Although amygdala input is critical for both cued and context fear, hippocampal activity is specialized for context learning. However, behavioral differences between developing and adult animals in context learning suggest potentially divergent supporting neural activity and circuitry across development.

The earliest assessments of the development of contextual fear/threat learning relied on behavioral assessment and did not include measures of hippocampal function. This landmark study demonstrated that contextual fear conditioning could be behaviorally demonstrated in PN23 rats; however, when tested for long-term memory, no learning was found in PN18 pups, although immediate postshock testing did reveal evidence of some context learning (Rudy, 1993). Interestingly, across several conditioning and retrieval manipulations, learning-induced hippocampal activity was absent in infant (PN17–19) rats, suggesting the aforementioned context learning was hippocampus-independent (Robinson-Drummer, Chakraborty, Heroux, Rosen, & Stanton, 2018; Santarelli, Khan, & Poulos, 2018). Indeed, an assessment of hippocampus c-Fos during contextual fear conditioning suggested the hippocampus was not engaged in the infant (PN17) pups and matured to adult-like activity after weaning (Raineki, Holman, et al., 2010; Santarelli, Khan, & Poulos, 2018). Importantly, weaning (PN24–31), but not infant (PD17) rats, have increased hippocampal Egr-1 (an immediate early gene) during nonreinforced context learning, suggesting that context learning is unable to fully engage the plasticity mechanisms required for learning (Robinson-Drummer et al., 2018). Overall, these results suggest that context learning at PN17 is supported by nonhippocampal structures or an immature hippocampal neural response that matures following weaning (PN23–24) to support long-term context memory.

There is evidence that even without hippocampus-dependent learning, the experience of being conditioned produces enduring effects. Infant conditioning significantly alters glutamatergic function during adolescent contextual fear conditioning in rats (Chan, Baker, & Richardson, 2015). This effect is mediated by the hippocampus in adults, although whether or not this is the case in infant animals is currently unknown. Neurological effects of context learning (i.e., no discrete cue or shock present during conditioning) may also be preserved molecularly (Chan, Baker, & Richardson, 2015), although behaviorally (Robinson-Drummer & Stanton, 2015), there is no evidence of learning. This unexpressed-learning effect extends to other learning phenomena and ages far into adulthood, where infant fear conditioning potentiates negative affective behaviors and sensitizes subsequent context conditioning (Poulos et al., 2014; Quinn, Skipper, & Claflin, 2014). Transient alterations in learning and memory likely have the role of facilitating ecologically relevant behaviors (i.e., attachment learning in the nest or exploration during adolescence) necessary for proper development (Pattwell et al., 2012; Spear, 2000). It is possible that these results reflect the enduring effects of nonlearning experiences in early life.

Can Understanding the Development of the Threat System Provide Insight into the Infant's Abuse-Associated Attachment and Long-Term Outcome?

The previous sections revealed an invaluable use of animal modeling, using the threat system, to inform our understanding of infant-caregiving attachment. Here, we review the detrimental effects of early life trauma on neurobehavioral development using similar models. The developmental and clinical literature suggests that the infant's attachment relationship with the caregiver is of the utmost importance in shaping a child's brain (Callaghan & Tottenham, 2016; Gee, 2016; Gunnar & Quevedo, 2007; Perry et al., 2016; Teicher et al., 2003). The paradoxical attachment of children to caregivers, regardless of care quality, is the product of a robust attachment system designed to ensure strong infant-caregiver bonding.

Early-life adverse experiences can derail long-term neurobehavioral development; long-term effects appear at periadolescence as compromised affective, cognitive, and social behavior (Bremner, 2003; Gunnar & Quevedo, 2007; Luby, Barch, Whalen, Tillman, & Belden, 2017; Nemeroff, 2004; VanTieghem & Tottenham, 2017), as well as long-term modification of neuromolecular function (Doherty, Blaze, Keller, & Roth, 2017; Doherty & Roth, 2016). Importantly, animal and human research has revealed particularly robust disruption of both amygdala and hippocampal development and provides insight into specific structural and functional outcomes of early-life trauma and stress on the developing attachment circuit (Rincòn-Cortès et al., 2015; van Bodegom, Homberg, & Henckens, 2017).

The scarcity-adversity model of maternal maltreatment has been instrumental in accessing the neurobiology of threat and attachment learning. Providing insufficient nest-building materials during the infant sensitive period causes pup maltreatment by the mother. As mentioned in the previous section, when guided by maternal odor, pups will still learn to nipple attach during nursing to both nurturing and abusive mothers, using this paradigm (Raineki, Pickenhagen, et al., 2010). A new odor is readily learned by pups within the abusive context; an abusive mother scented with peppermint supports the learning of that peppermint odor as it takes on the qualities of the maternal odor, a process previously demonstrated in typical nurturing mothers (Galef & Kaner, 1980; Perry, Blair, & Sullivan, 2017; Roth & Sullivan, 2005). This classical conditioning of the novel odor with an abusive mother results in the paradoxical learning of an odor preference.

While attachment is preserved following prolonged maternal abuse, a more careful assessment of neurobiological and behavioral processes suggests some atypical features. Specifically, maltreated pups still approach the maternal odor in a Y-maze but less robustly than controls. Additionally, maltreated pups display atypical social behavior to an anesthetized mother, and there is reduced maternal odor neural network activation (e.g., olfactory bulb, piriform cortex) relative to normally reared pups (Perry et al., 2016). Being reared with an abusive mother also prematurely ends various developmental stages, such as the rodent SHRP and attachment-sensitive period (Moriceau, Raineki, Holman, Holman, & Sullivan, 2009; Moriceau, Shionoya, et al., 2009; Moriceau & Sullivan, 2004; Plotsky & Meaney, 1993), a finding that converges with the human literature (Gunnar, Hostinar, Sanchez, Tottenham, & Sullivan, 2015;

Gunnar & Quevedo, 2007; Hostinar, Sullivan, & Gunnar, 2014). Although unclear in humans, this rodent work suggests that there is a unique role for stress hormones in early life that defines the functioning of brain areas important to threat responding, such as the amygdala.

In children, the infant's primary environment is the caregiver, and while the environment expands as the child becomes more mobile, the caregiver remains an important base of safety to explore the world. When the caregiver is the source of trauma, this safety base can be disrupted to affect interactions with the mother, as well as interactions with the world that could produce further neurobehavioral disruptions. The importance of this relationship for children is further validated by early interventions research that targets the caregiver-infant relationship or the infant's neurobehavioral function and has been shown to have great repair value for neurobehavioral function (Bernard, Lee, & Dozier, 2017; Dozier, Roben, Caron, Hoye, & Bernard, 2018; Theise et al., 2014).

Summary and Implications

The infant attachment system is designed to encourage infant-caregiver interactions and is uniquely equipped to reinforce these interactions. The olfactory system, in conjunction with increased NE activity and reduced threat learning, ensure attachment of the infant to a range of maternal stimuli regardless of maternal care quality. Though it is becoming increasingly clear that disruptions to infant attachment have profound maladaptive effects on adult behavior, the research only hints at *how* trauma and sensory input immediately influence the developing brain to produce individual differences and initiate the pathway to pathology. Different models studying the involvement of the early-life environment and its enduring effect have been developed over the years (i.e., maternal separation/deprivation, rearing environment alteration and CORT manipulation, neonatal handling, low bedding, such as in the scarcity-adversity model and the more stressful fragmentation model). When combined with our rat model of attachment using odor-shock and advanced neuroimaging techniques, a clearer understanding of the link between infant attachment learning and the damaging effects of early trauma on adult behavior is emerging (Bremner, 2003; Gee, 2016; Nemeroff, 2004; Opendak, Gould, & Sullivan, 2017; Teicher et al., 2003; VanTieghem & Tottenham, 2017). These models provide invaluable tools for understanding the long-term effects of early human trauma; however, further research is needed to fully uncover and remedy the neurobehavioral effects of developmental adversity.

Acknowledgments

This work was supported by grants DC009910, MH091451, and HD083217 to Regina M. Sullivan. We thank Dr. Mark E. Stanton for the comments and editing that greatly improved this chapter.

REFERENCES

Barr, G. A., Moriceau, S., Shionoya, K., Muzny, K., Gao, P., Wang, S., & Sullivan, R. M. (2009). Transitions in infant learning are modulated by dopamine in the amygdala. *Nature Neuroscience, 12*(11), 1367–1369. doi:10.1038/nn.2403

Bernard, K., Lee, A. H., & Dozier, M. (2017). Effects of the ABC intervention on foster children's receptive vocabulary: Follow-up results from a randomized clinical trial. *Child Maltreatment, 22*(2), 174–179. doi:10.1177/1077559517691126

Bowlby, J. (1969). *Attachment and loss* (Vol. 1). New York: Basic Books.

Bremner, J. D. (2003). Long-term effects of childhood abuse on brain and neurobiology. *Child and Adolescent Psychiatric Clinics of North America, 12*(2), 271–292.

Callaghan, B. L., & Tottenham, N. (2016). The neuro-environmental loop of plasticity: A cross-species analysis of parental effects on emotion circuitry development following typical and adverse caregiving. *Neuropsychopharmacology, 41*(1), 163–176. doi:10.1038/npp.2015.204

Camp, L. L., & Rudy, J. W. (1988). Changes in the categorization of appetitive and aversive events during postnatal development of the rat. *Developmental Psychobiology, 21*(1), 25–42.

Chan, D., Baker, K. D., & Richardson, R. (2015). Relearning a context-shock association after forgetting is an NMDAr-independent process. *Physiology & Behavior, 148*, 29–35. doi:10.1016/j.physbeh.2014.11.004

Corodimas, K. P., LeDoux, J. E., Gold, P. W., & Schulkin, J. (1994). Corticosterone potentiation of conditioned fear in rats. *Annals of the New York Academy of Sciences, 746*, 392–393.

Dallman, M. F. (2014). Early life stress: Nature and nurture. *Endocrinology, 155*(5), 1569–1572. doi:10.1210/en.2014-1267

Debiec, J., & Sullivan, R. M. (2014). Intergenerational transmission of emotional trauma through amygdala-dependent mother-to-infant transfer of specific fear. *Proceedings of the National Academy of Sciences of the United States of America, 111*(33), 12222–12227. doi:10.1073/pnas.1316740111

Doherty, T. S., Blaze, J., Keller, S. M., & Roth, T. L. (2017). Phenotypic outcomes in adolescence and adulthood in the scarcity-adversity model of low nesting resources outside the home cage. *Developmental Psychobiology, 59*(6), 703–714. doi:10.1002/dev.21547

Doherty, T. S., & Roth, T. L. (2016). Insight from animal models of environmentally driven epigenetic changes in the developing and adult brain. *Development and Psychopathology, 28*(4, pt. 2), 1229–1243. doi:10.1017/s095457941600081x

Dozier, M., Roben, C. K. P., Caron, E., Hoye, J., & Bernard, K. (2018). Attachment and biobehavioral catch-up: An evidence-based intervention for vulnerable infants and their families. *Psychotherapy Research, 28*(1), 18–29. doi:10.1080/10503307.2016.1229873

Galef Jr., B. G., & Kaner, H. C. (1980). Establishment and maintenance of preference for natural and artificial olfactory stimuli in juvenile rats. *Journal of Comparative & Physiological Psychology, 94*(4), 588–595.

Gee, D. G. (2016). Sensitive periods of emotion regulation: Influences of parental care on frontoamygdala circuitry and plasticity. *New Directions for Child and Adolescent Development, 2016*(153), 87–110. doi:10.1002/cad.20166

Gunnar, M. R., Hostinar, C. E., Sanchez, M. M., Tottenham, N., & Sullivan, R. M. (2015). Parental buffering of fear and stress neurobiology: Reviewing parallels across rodent, monkey, and human models. *Society for Neuroscience, 10*(5), 474–478. doi:10.1080/17470919.2015.1070198

Gunnar, M. R., & Quevedo, K. (2007). The neurobiology of stress and development. *Annual Review of Psychology, 58*, 145–173.

Harlow, H., & Harlow, M. (1965). The affectional system. In A. Schrier, H. Harlow, & F. Stollnitz (Eds.), *Behavior of nonhuman primates* (Vol. 2). New York: Academic Press.

Haroutunian, V., & Campbell, B. A. (1979). Emergence of interoceptive and exteroceptive control of behavior in rats. *Science, 205*(4409), 927–929.

Hess, E. (1962). Ethology: An approach to the complete analysis of behavior. In R. Brown, E. Galanter, E. Hess, & G. Mendler (Eds.), *New directions in psychology* (pp. 159–199). New York: Holt, Rinehart and Winston.

Hostinar, C. E., Sullivan, R. M., & Gunnar, M. R. (2014). Psychobiological mechanisms underlying the social buffering of the hypothalamic-pituitary-adrenocortical axis: A review of animal models and human studies across development. *Psychological Bulletin, 140*(1), 256–282. doi:10.1037/a0032671

Levine, S. (1994). The ontogeny of the hypothalamic-pituitary-adrenal axis: The influence of maternal factors. *Annals of the New York Academy of Sciences, 746*, 275–288, discussion, 289–293.

Luby, J. L., Barch, D., Whalen, D., Tillman, R., & Belden, A. (2017). Association between early life adversity and risk for poor emotional and physical health in adolescence: A putative mechanistic neurodevelopmental pathway. *JAMA Pediatrics, 171*(12), 1168–1175. doi:10.1001/jamapediatrics.2017.3009

Moriceau, S., Raineki, C., Holman, J. D., Holman, J. G., & Sullivan, R. M. (2009). Enduring neurobehavioral effects of early life trauma mediated through learning and corticosterone suppression. *Frontiers in Behavioral Neuroscience, 3*, 22. doi:10.3389/neuro.08.022.2009

Moriceau, S., Shionoya, K., Jakubs, K., & Sullivan, R. M. (2009). Early-life stress disrupts attachment learning: The role of amygdala corticosterone, locus ceruleus corticotropin releasing hormone, and olfactory bulb norepinephrine. *Journal of Neuroscience, 29*(50), 15745–15755. doi:10.1523/jneurosci.4106-09.2009

Moriceau, S., & Sullivan, R. M. (2004). Corticosterone influences on mammalian neonatal sensitive-period learning. *Behavioral Neuroscience, 118*(2), 274–281.

Moriceau, S., & Sullivan, R. M. (2006). Maternal presence serves as a switch between learning fear and attraction in infancy. *Nature Neuroscience, 9*(8), 1004–1006.

Moriceau, S., Wilson, D. A., Levine, S., & Sullivan, R. M. (2006). Dual circuitry for odor-shock conditioning during infancy: Corticosterone switches between fear and attraction via amygdala. *Journal of Neuroscience, 26*(25), 6737–6748. doi:10.1523/jneurosci.0499-06.2006

Morrison, G. L., Fontaine, C. J., Harley, C. W., & Yuan, Q. (2013). A role for the anterior piriform cortex in early odor preference learning: Evidence for multiple olfactory learning structures in the rat pup. *Journal of Neurophysiology, 110*(1), 141–152. doi:10.1152/jn.00072.2013

Nemeroff, C. B. (2004). Neurobiological consequences of childhood trauma. *Journal of Clinical Psychiatry, 65*(Suppl 1), 18–28.

Opendak, M., Briones, B. A., & Gould, E. (2016). Social behavior, hormones and adult neurogenesis. *Frontiers in Neuroendocrinology, 41*, 71–86. doi:10.1016/j.yfrne.2016.02.002

Opendak, M., Gould, E., & Sullivan, R. (2017). Early life adversity during the infant sensitive period for attachment: Programming of behavioral neurobiology of threat processing and social behavior. *Developmental Cognitive Neuroscience, 25*, 145–159. doi:10.1016/j.dcn.2017.02.002

Pattwell, S. S., Duhoux, S., Hartley, C. A., Johnson, D. C., Jing, D., Elliott, M. D., ... Lee, F. S. (2012). Altered fear learning across development in both mouse and human. *Proceedings of the National Academy of Sciences of the United States of America, 109*(40), 16318–16323. doi:10.1073/pnas.1206834109

Perry, R. E., Al Ain, S., Raineki, C., Sullivan, R. M., & Wilson, D. A. (2016). Development of odor hedonics: Experience-dependent ontogeny of circuits supporting maternal and predator odor responses in rats. *Journal of Neuroscience, 36*(25), 6634–6650. doi:10.1523/jneurosci.0632-16.2016

Perry, R. E., Blair, C., & Sullivan, R. M. (2017). Neurobiology of infant attachment: Attachment despite adversity and parental programming of emotionality. *Current Opinion in Psychology, 17*, 1–6. doi:10.1016/j.copsyc.2017.04.022

Plotsky, P. M., & Meaney, M. J. (1993). Early postnatal experience alters hypothalamic corticotropin-releasing factor (CRF) mRNA, median eminence CRF content and stress-induced release in adult rats. *Molecular Brain Research, 18*, 195–200.

Poulos, A. M., Reger, M., Mehta, N., Zhuravka, I., Sterlace, S. S., Gannam, C., ... Fanselow, M. S. (2014). Amnesia for early life stress does not preclude the adult development of posttraumatic stress disorder symptoms in rats. *Biological Psychiatry, 76*(4), 306–314. doi:10.1016/j.biopsych.2013.10.007

Quinn, J. J., Skipper, R. A., & Claflin, D. I. (2014). Infant stress exposure produces persistent enhancement of fear learning across development. *Developmental Psychobiology, 56*(5), 1008–1016. doi:10.1002/dev.21181

Raineki, C., Holman, P. J., Debiec, J., Bugg, M., Beasley, A., & Sullivan, R. M. (2010). Functional emergence of the hippocampus in context fear learning in infant rats. *Hippocampus, 20*(9), 1037–1046. doi:10.1002/hipo.20702

Raineki, C., Pickenhagen, A., Roth, T. L., Babstock, D. M., McLean, J. H., Harley, C. W., ... Sullivan, R. M. (2010). The neurobiology of infant maternal odor learning. *Brazilian Journal of Medical and Biological Research, 43*(10), 914–919.

Richardson, R., & McNally, G. P. (2003). Effects of an odor paired with illness on startle, freezing, and analgesia in rats. *Physiology & Behavior, 78*(2), 213–219.

Rincòn-Cortès, M., Barr, G. A., Mouly, A. M., Shionoya, K., Nunez, B. S., & Sullivan, R. M. (2015). Enduring good memories of infant trauma: Rescue of adult neurobehavioral deficits via amygdala serotonin and corticosterone interaction. *Proceedings of the National Academy of Sciences of the United States of America, 112*(3), 881–886. doi:10.1073/pnas.1416065112

Robinson-Drummer, P. A., Chakraborty, T., Heroux, N. A., Rosen, J. B., & Stanton, M. E. (2018). Age and experience dependent changes in Egr-1 expression during the ontogeny of the context preexposure facilitation effect (CPFE). *Neurobiology of Learning and Memory, 150*, 1–12. doi:10.1016/j.nlm.2018.02.008

Robinson-Drummer, P. A., & Stanton, M. E. (2015). Using the context preexposure facilitation effect to study long-term context memory in preweanling, juvenile, adolescent, and adult rats. *Physiology & Behavior, 148*, 22–28. doi:10.1016/j.physbeh.2014.12.033

Roozendaal, B., Quirarte, G. L., & McGaugh, J. L. (2002). Glucocorticoids interact with the basolateral amygdala beta-adrenoceptor—cAMP/cAMP/PKA system in influencing memory consolidation. *European Journal of Neuroscience, 15*(3), 553–560.

Roth, T. L., & Sullivan, R. M. (2005). Memory of early maltreatment: Neonatal behavioral and neural correlates of maternal maltreatment within the context of classical conditioning. *Biological Psychiatry, 57*(8), 823–831.

Rudy, J. W. (1993). Contextual conditioning and auditory cue conditioning dissociate during development. *Behavioral Neuroscience, 107*(5), 887–891.

Salzen, E. (1970). Imprinting and environmental learning. In L. Aronson, E. Tobach, D. Lehrman, & J. Rosenblatt (Eds.), *Development and evolution of behavior.* San Francisco: W. H. Freeman.

Sanchez, M. M., Ladd, C. O., & Plotsky, P. M. (2001). Early adverse experience as a developmental risk factor for later psychopathology: Evidence from rodent and primate models. *Development & Psychopathology, 13*(3), 419–449.

Santarelli, A. J., Khan, A. M., & Poulos, A. M. (2018). Contextual fear retrieval-induced Fos expression across early development in the rat: An analysis using established nervous system nomenclature ontology. *Neurobiology of Learning and Memory, 155*, 42–49. doi:10.1016/j.nlm.2018.05.015

Shionoya, K., Moriceau, S., Lunday, L., Miner, C., Roth, T. L., & Sullivan, R. M. (2006). Development switch in neural circuitry underlying odor-malaise learning. *Learning & Memory, 13*(6), 801–808.

Spear, L. P. (2000). The adolescent brain and age-related behavioral manifestations. *Neuroscience & Biobehavioral Reviews, 24*(4), 417–463. doi:10.1016/s0149-7634(00)00014-2

Stanley, W. (1962). Differential human handling as reinforcing events and as treatments influencing later social behavior in Basenji puppies. *Psychological Reports, 10*, 775–788.

Sullivan, R. M., Brake, S. C., Hofer, M. A., & Williams, C. L. (1986). Huddling and independent feeding of neonatal rats can be facilitated by a conditioned change in behavioral state. *Developmental Psychobiology, 19*(6), 625–635.

Sullivan, R. M., Landers, M., Yeaman, B., & Wilson, D. A. (2000). Good memories of bad events in infancy. *Nature, 407*(6800), 38–39. doi:10.1038/35024156

Sullivan, R. M., Perry, R., Sloan, A., Kleinhaus, K., & Burtchen, N. (2011). Infant bonding and attachment to the caregiver: Insights from basic and clinical science. *Clinics in Perinatology, 38*(4), 643–655. doi:10.1016/j.clp.2011.08.011

Sullivan, R. M., Wilson, D. A., Wong, R., Correa, A., & Leon, M. (1990). Modified behavioral and olfactory bulb responses to maternal odors in preweanling rats. *Brain Research: Developmental Brain Research, 53*(2), 243–247.

Suomi, S. J. (2003). Gene-environment interactions and the neurobiology of social conflict. *Annals of the New York Academy of Sciences, 1008*, 132–139.

Takahashi, L. K. (1994). Organizing action of corticosterone on the development of behavioral inhibition in the preweanling rat. *Brain Research: Developmental Brain Research, 81*(1), 121–127.

Teicher, M. H., Andersen, S. L., Polcari, A., Anderson, C. M., Navalta, C. P., & Kim, D. M. (2003). The neurobiological consequences of early stress and childhood maltreatment. *Neuroscience & Biobehavioral Reviews, 27*(1–2), 33–44.

Theise, R., Huang, K. Y., Kamboukos, D., Doctoroff, G. L., Dawson-McClure, S., Palamar, J. J., & Brotman, L. M. (2014). Moderators of intervention effects on parenting practices in a randomized controlled trial in early childhood. *Journal of Clinical Child and Adolescent Psychology, 43*(3), 501–509. doi:10.1080/15374416.2013.833095

Upton, K. J., & Sullivan, R. M. (2010). Defining age limits of the sensitive period for attachment learning in rat pups. *Developmental Psychobiology, 52*(5), 453–464. doi:10.1002/dev.20448

van Bodegom, M., Homberg, J. R., & Henckens, M. (2017). Modulation of the hypothalamic-pituitary-adrenal axis by early life stress exposure. *Frontiers in Cellular Neuroscience, 11*, 87. doi:10.3389/fncel.2017.00087

van Oers, H., Kloet, E. D., Whelan, T., & Levine, S. (1998). Maternal deprivation effect on the infant's neural stress markers is reversed by tactile stimulation and feeding but not by suppressing corticosterone. *Neuroscience, 18*, 10171–10179.

VanTieghem, M. R., & Tottenham, N. (2017). Neurobiological programming of early life stress: Functional development of amygdala-prefrontal circuitry and vulnerability for stress-related psychopathology. *Current Topics in Behavioral Neurosciences.* doi:10.1007/7854_2016_42

Wilson, D. A., & Sullivan, R. M. (1994). Neurobiology of associative learning in the neonate: Early olfactory learning. *Behavioral & Neural Biology, 61*(1), 1–18.

Yuan, Q., Harley, C. W., Darby-King, A., Neve, R. L., & McLean, J. H. (2003). Early odor preference learning in the rat: Bidirectional effects of cAMP response element-binding protein (CREB) and mutant CREB support a causal role for phosphorylated CREB. *Journal of Neuroscience, 23*(11), 4760–4765.

81 More than Just Friends: An Exploration of the Neurobiological Mechanisms Underlying the Link between Social Support and Health

ERICA A. HORNSTEIN, TRISTEN K. INAGAKI, AND NAOMI I. EISENBERGER

ABSTRACT Although links between social ties and mental and physical health outcomes have been repeatedly demonstrated, the mechanisms underlying this connection are still being determined. One prevailing theory states that social bonds, and the supportive interactions they produce, may act as a buffer against stress responses and their negative downstream consequences. Illuminating the avenues by which social connection might achieve this buffering, recent neuroimaging research suggests that social support has an impact on stress-response and threat-detection systems at both the physiological and neural levels, ultimately reducing stress. Here, we review investigations of the neurobiological buffering effects of the two sides of social interaction, receiving support *from* others and giving support *to* others, and examine how each might contribute to the link between social ties and health.

Our relationships with others have powerful effects on our mental and physical well-being. Research has repeatedly demonstrated that having strong social connections is associated with health, while a lack of connections is associated with various disease outcomes (House, Landis, & Umberson, 1988; Cacioppo, Hawkley, & Thisted, 2010). Yet the psychological and neural pathways underlying these effects are not well understood. One prevailing theory suggests that social ties contribute to health by reducing the physiological stress responses that can ultimately lead to negative health outcomes (Cobb, 1976). Here, we examine two sides of social-support interactions: (1) how receiving support *from* others and (2) how giving support *to* others contribute to the link between social ties and health. By breaking down the impacts of social support along these dimensions—receiving support and the less well-studied effects of giving support—this chapter will focus on how social-support processes affect physiological and neural function and explore potential mechanisms underlying the ability of social support to buffer against stress and ultimately benefit health.

Receiving Support

One route by which our social bonds influence health is through the support we receive or perceive from others. It is hypothesized that the care, resources, and protection we receive from those closest to us, or even just perceiving that such support is available, signal an accessibility of the means necessary to deal with threats in the environment, changing our appraisal of threatening cues or situations (Cohen, 2004). This suggests that both the receipt of social support and perceptions of available social support lead to reduced threat-related responding, ultimately leading to downstream health benefits.

Although adaptive in the face of acute events, bodily systems set in place to facilitate threat or stress responding can be deleterious when chronically activated. Of particular interest are the sympathetic nervous system (SNS) and the hypothalamic-pituitary-adrenal (HPA) axis, which prepare the body for action in the face of stress and result in a myriad of physiological and endocrine outcomes, ranging from increased blood pressure (BP) to altered immune function, but may contribute to negative health if consistently activated over time (Miller, Chen, & Cole, 2009). By reducing appraisals of threat and consequently reducing activity in these stress-response systems, social support may provide a buffer against the experience of stress and its harmful long-term effects.

Evidence for social buffering: received and perceived support reduce stress responding Evidence for the buffering impact of social support on stress-response systems can be found within the animal and human literature. In animals, the presence of a conspecific reduces escape or avoidance behavior to a threatening context (Baum, 1969; Hall, 1955), decreases freezing behavior in the face of threats (Davitz & Mason, 1955), increases tolerance for novel environments (Liddell, 1950, 1954), and mitigates anxious behavior following social defeat (Nakayasu & Ishii, 2008; Nakayasu & Kato, 2011). In addition to effects on behavioral responses to threat, the presence of familiar others ameliorates physiological stress responses to threatening events or contexts. For example, guinea pigs placed in novel environments exhibit dampened HPA axis activity when with a familiar conspecific (Hennessy, Zate, & Maken, 2008; Sachser, Durschlag, & Hirzel, 1998).

Similarly, research in humans demonstrates that receiving social support alleviates stress to threatening or stressful events. For example, receiving social support during a stressful event reduces cortisol, a hormone triggered by the HPA axis that prepares the body to react to acute threats (contact with a close other: Heinrichs, Baumgartner, Kirschbaum, & Ehlert, 2003; verbal encouragement: Roberts, Klatzkin, & Mechlin, 2015). Those who report having more contact with social-support figures also show lower cortisol responses to stressors (Eisenberger, Taylor, Gable, Hilmert, & Lieberman, 2007). Moreover, in addition to receiving support, perceptions of available support also reduce stress in humans. Thus, perceptions of strong social connections are associated with decreased physiological stress responses to acute stressors (for a review, see Hostinar, Sullivan, & Gunnar, 2014) and lower basal levels of cortisol overall (Rosal, King, Ma, & Reed, 2004). Furthermore, the ability of both received and perceived social support to reduce HPA axis activity occurs across the life span, from childhood through adulthood, although different situational or individual factors (e.g., sex of support provider, sex of receiver, early life history) may determine when and if they occur (Hostinar, Sullivan, & Gunnar, 2014).

Neural investigations provide corresponding evidence for the ability of social support to reduce HPA activity; one study found that during experiences of social pain, reminders of social-support figures led to decreased activity in the hypothalamus, a region associated with stress responding and a component of the HPA axis (Karremans, Heslenfeld, van Dillen, & Van Lange, 2011). Perceived and received social support show a similar inhibitory effect on the SNS, leading to a lower heart rate, to lower BP, and to lower skin-conductance responses (SCR) in the presence of acute stressors (peripheral measures of SNS activity: Che, Cash, Fitzgerald, & Fitzgibbon, 2018; Gerin, Pieper, Levy, & Pickering, 1992; Roberts, Klatzkin, & Mechlin, 2015; Thorsteinsson & James, 1999).

Developing a mechanism for social buffering: social support as a safety signal One explanation for these stress-buffering effects is that social support signals safety and consequently mitigates the experience of threat. Neural investigations of social support provide some evidence for this view, indicating that experiencing or being reminded of social support leads to the activation of safety-related neural regions and reduces activation in regions known to be involved in processing pain and distress. Of particular interest is the link between both received and perceived social support and activity in the ventromedial prefrontal cortex (vmPFC), a region associated with processing safety (Delgado, Olsson, & Phelps, 2006; Eisenberger et al., 2011). For example, the vmPFC shows greater activity in response to safety cues (Phelps, Delgado, Nearing, & LeDoux, 2004) and even tracks different types of safety cues (dissociation between cues that were always safe vs. cues that switched from being threatening to safe: Schiller, Levy, Niv, LeDoux, & Phelps, 2008). Importantly, the vmPFC is also known to play a role in inhibiting threat responding and extinguishing learned fear via inhibitory connections with the amygdala (Phelps et al., 2004), a region that is crucial during fear learning and influences downstream fear-related activity in both the SNS and the HPA axis (Adolphs, Tranel, Damasio, & Damasio, 1995; Delgado, Olsson, & Phelps, 2006). Social support has also been linked to decreased activity in the dorsal anterior cingulate (dACC) and anterior insula (AI), regions associated with the distressing experience of both physical (Lieberman & Eisenberger, 2015; Price, 2000) and social pain (Eisenberger, 2012; Eisenberger, Lieberman, & Williams, 2003) (see Figure 81.1, left panel).

Mirroring findings in behavioral research showing that social support can decrease subjective distress during painful experiences (Brown, Sheffield, Leary, & Robinson, 2003; Che et al., 2018; Master et al., 2009), neural investigations have demonstrated that simply viewing pictures of a social-support figure while receiving pain leads to increased activity in the vmPFC and decreased activity in the dACC and AI, suggesting that social support may lead to increased perceptions of safety and a decreased subjective experience of pain (Eisenberger, 2011; Younger, Aron, Parke, Chatterjee, & Mackey, 2010). Given the previously discussed inhibitory connection between the vmPFC and the amygdala, these findings also suggest that social-support figures may be acting as a type of safety signal, leading to reduced perceptions of threat. This link between social-support

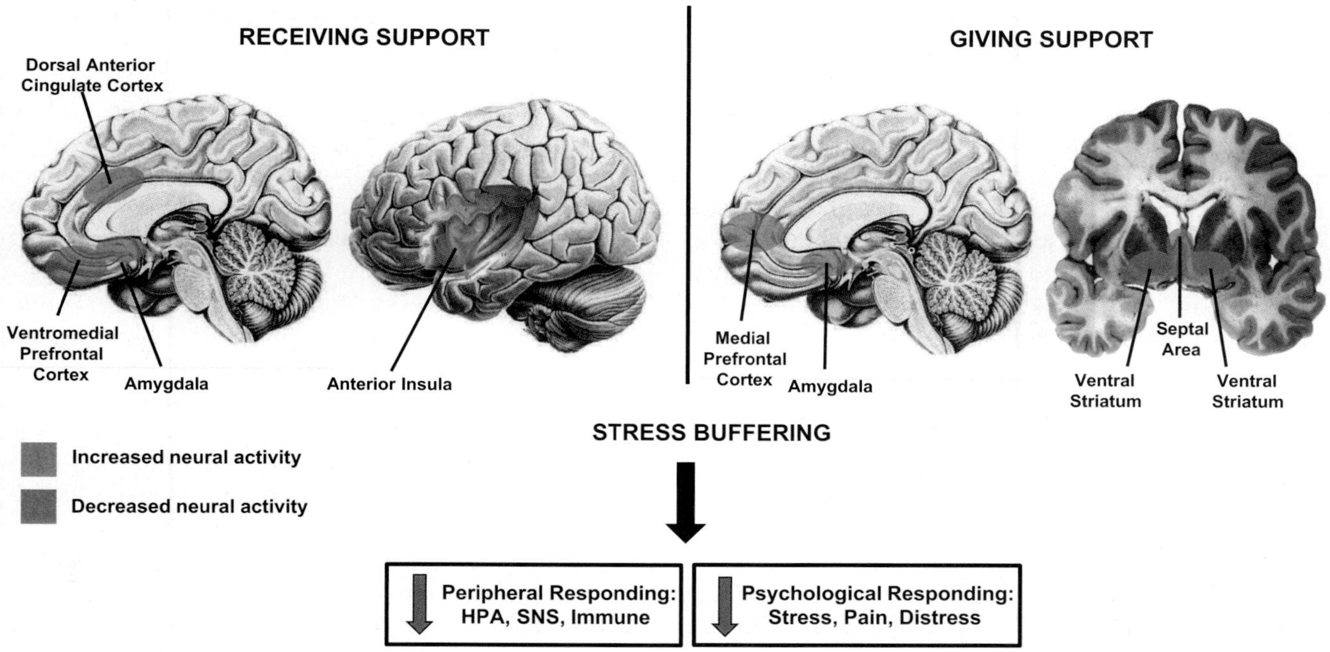

FIGURE 81.1 Neural mechanisms underlying the stress-buffering effects of social support. Receiving support leads to increased activity (*green*) in the ventromedial prefrontal cortex (vmPFC) and decreased activity (*red*) in the dorsal anterior cingulate cortex (dACC) and anterior insula (AI), regions that play a critical role in the distressing experience of pain. Giving support leads to increased activity in the medial prefrontal cortex (mPFC), ventral striatum (VS), and septal area (SA). Given the known inhibitory connections between the vmPFC (active during receiving support) and the SA (active during giving support) with the amygdala, both receiving and giving support may lead to decreased activity in the amygdala, a threat-related region that plays a key role in the stress response, resulting in the reduced activation of peripheral systems (hypothalamic-pituitary-adrenal axis [HPA], sympathetic nervous system [SNS], and immune system) and reduced psychological stress. (See color plate 95.)

processes and safety-related neural activity indicates that social buffering effects may be supported by reductions in both perceptions and responses to threats on a neural level.

Additional evidence that social support acts as a safety signal can be found in recent work examining the unique functions of social-support figures during fear-learning processes. These investigations have revealed that social-support figures are one category of prepared safety stimuli—stimuli that have historically enhanced survival and thus are naturally able to perform the functions of the most powerful learned safety signals (Hornstein & Eisenberger, 2018). Drawing from the tests required of the most powerful learned safety signals—conditioned inhibitors (Rescorla, 1969)—this work investigated whether social-support figures belong in the prepared safety category. Results demonstrated that without undergoing the lab-based, threat-specific safety training essential for learned safety signals to perform their functions, social-support figures: (1) cannot become associated with fear and 2) are able to inhibit fear responses elicited by other cues and thus pass the two tests required of conditioned inhibitors (Hornstein, Fanselow, & Eisenberger, 2016). Specifically, when images were repeatedly paired with a mild electric shock in a fear-acquisition procedure, images of social-support figures did not elicit a fear response, while images of strangers or neutral objects did (Hornstein, Fanselow, & Eisenberger, 2016). Moreover, this effect was not simply due to the familiar or rewarding aspects of social-support figures, as subjects were still able to acquire fear to familiar and rewarding images but not to their social-support figures (Hornstein, Fanselow, & Eisenberger, 2016). In addition to not becoming associated with the fear response, social-support figures inhibited the fear response elicited by other fearful cues. Specifically, pairing a social-support-figure image with a fear cue inhibited the fear response, while pairing images of strangers or neutral objects with fear cues led to no inhibition of the fear response (Hornstein, Fanselow, & Eisenberger, 2016). Because learned safety signals require training to perform these inhibitory functions, these findings indicate that social-support stimuli represent a unique category of safety signals that are distinct in their ability to signal safety universally, without requiring specific training to do so.

Subsequent work has further explored the functions of social-support figures, revealing that social-support figures are uniquely prepared to signal safety and that they also have distinctive effects on fear-learning processes. This work shows that an image of a social-support figure prevents fear acquisition from occurring for other stimuli (no fear becomes associated with other cues: Hornstein & Eisenberger, 2017), an effect that stands in contrast to what is expected for learned safety signals, which are known to enhance fear acquisition (Rescorla, 1971). Further investigations demonstrate that images of social-support figures lead to enhanced fear extinction, such that there is no return of the fear response for fearful cues paired with social-support-figure images even 24 hours postextinction (Hornstein, Haltom, Shirole, & Eisenberger, 2018). This effect is especially surprising, as the current understanding argues that all safety signals are harmful to fear extinction and prevent it from occurring (Lovibond, Davis, & O'Flaherty, 2000; Rescorla, 2003). Together, these results indicate that social support plays a powerful and distinct role in influencing fear-learning outcomes, reducing fear responding by both preventing fear acquisition and enhancing fear extinction and, consequently, reducing threat-related stress.

The ability of social support to not only signal safety from novel threats naturally but also to reduce fear elucidates a previously unexplored route by which social support buffers against threat. These unique safety functions might account for the mitigated threat responding demonstrated in the brain and body when social support is present during experiences of threat.

Giving Support

Although receiving support from others has been assumed to be the primary way in which social connections benefit health, recent thinking proposes that *giving* support may also make a substantial contribution to the social ties-health link via neural regions known to support effective parental care in mammals (Brown & Brown, 2006; Eisenberger, 2013; Inagaki, 2018; Inagaki & Orehek, 2017). Giving support to others is crucial to the survival of human offspring and therefore may rely on mechanisms that ensure parental-giving behavior toward offspring. Similar mechanisms may also extend beyond the parent-infant bond to supportive behavior that is directed toward those other than offspring. In particular, we have proposed that giving support relies on mechanisms that reinforce giving behavior and reduce stress or withdrawal, which might inhibit care-related activities. Giving support may then contribute to long-term health through repeated actions on both mechanisms.

Giving support and health An accumulating body of research shows that giving support to others is associated with health benefits for the giver. For instance, giving more support is related to reductions in self-reported stress (Poulin, Brown, Dillard, & Smith, 2013) and depressive symptoms following the death of a spouse (Brown, Brown, House, & Smith, 2008). Similar associations exist with physiological responding and longevity; giving more support predicts lower BP and heart rate over a 24-hour period (Piferi & Lawler, 2006), and giving support to a close other is associated with lower mortality over a 5-year period, even when controlling for support that is received (Brown, Nesse, Vinokur, & Smith, 2003).

Building on these correlational findings, experimental work has shown that giving support by writing a supportive note to a close other in need (vs. not giving support) leads to reductions in SNS-related responding (systolic BP, salivary alpha-amylase) to a psychosocial stressor (Inagaki & Eisenberger, 2016). Similarly, in interventions outside the lab, random assignment to give support (vs. control conditions) leads to lower resting BP (Whillans et al., 2016), lower proinflammatory gene expression (Nelson-Coffey et al., 2017), lower cholesterol levels (Schreier, Schonert-Reeichl, & Chen, 2013), and fewer physical symptoms in cancer survivors (Rini et al., 2014). Thus, giving support may ultimately have an impact on health via reductions in stress; however, the neural mechanisms that contribute to the health effects of giving are largely unknown.

Giving support may reduce stress through parental-care related neural regions Animal models of parental care provide insight into the neurobiological mechanisms that underlie the broad and multifaceted support-giving behavior of humans. These models show that subcortical neural regions involved in normal parental care contribute to the reinforcing and stress-reducing actions of parenting (Numan, 2007). Thus, behaviors that ensure the development and survival of the litter, such as huddling, licking, and grooming, elicit activity in the ventral striatum (VS), septal area (SA), medial preoptic area (mPOA), ventral bed of the stria terminalis, and ventral tegmental area (VTA) to reinforce effective parental care. Though not an original focus of animal models of parental care, there is also an increasing appreciation for the role of the medial prefrontal cortex (mPFC) in parental care (Febo, Felix-Ortiz, & Johnson, 2010; Pereira & Morrell, 2011). In animals, the VS, SA, and amygdala appear to play a causal role in normal parental behavior. For example, the nucleus accumbens (NAcc), a region of the VS, shows increased activity during parental behavior (Stack, Balakrishnan, Numan, & Numan, 2002). Conversely, lesions to either the VS or SA lead to substantial reductions in typical parenting behaviors (Hansen, 1994; Slotnick & Nigrosh, 1975) (see Figure 81.1, right panel).

Importantly, these regions interact with stress-related regions to inhibit withdrawal or ineffective care. In particular, animal work shows that the SA has an inhibitory connection with the amygdala (Thomas, 1998). Stimulating the amygdala increases cardiovascular responses (BP, heart rate; Tellioğlu, Aker, Oktay, & Onat, 1997), whereas electrical stimulation of the SA decreases the same responses (Covian, Antunes-Rodrigues, & O'Flaherty, 1964; Malmo, 1961). Lesions to the amygdala reduce stressor-evoked cardiovascular responding (Galeno, Van Hoesen, & Brody, 1984; Sanders, Wirtz-Nole, DeFord, & Erling, 1994), but lesions to the SA result in the opposite effect, increasing startle and other stress-related behavior (Melia, Sananes, & Davis, 1992). To the extent that the SA has an inhibitory relationship with stress-related responding, giving support may reduce stress via interactions between the SA and the amygdala. Indeed, it has been hypothesized that the SA contributes to parental behavior by reducing threat responding in caregivers so they can engage in adaptive caregiving responses toward offspring (Stack et al., 2002).

Translation of animal findings to humans Results from imaging studies on human parents largely align with the animal literature. The VS and SA show increased activity to images of one's infant (vs. unknown infants; Lorberbaum et al., 2002), but mothers who show deficits in parenting behavior (characterized as dismissive from the Adult Attachment Interview) show lower VS (Strathearn, Fonagy, Amico, & Montague, 2009) and greater amygdala activity to images of their infant (vs. unknown infants; Atzil, Hendler, & Feldman, 2011). Giving support to those other than infants similarly elicits activity in reward-related regions. The VS is more active when giving financial support to others than when benefitting the self (Harbaugh, Mayr, & Burghart, 2007; Moll et al., 2006). Similarly, SA activity to emotional scenes is related to daily support-giving behavior (Morelli, Rameson, & Lieberman, 2012), and self-reported care for others predicts greater SA activity when listening to biographies of others in need (Ashar, Andrews-Hanna, Dimidjian, & Wager, 2017).

In the first demonstration of the role of parental-care-related neural regions in support giving in humans, giving supportive touch (vs. no support) to a partner who was under threat of electric shock led to increased activity in both the VS and SA (Inagaki & Eisenberger, 2012). Interestingly, VS and SA activity were also greater during support giving than a condition in which the participant simply touched the partner without the threat of shock, suggesting that in this situation it is more reinforcing to give support than engage in physical touch with a close other. Furthermore, greater feelings of social connection were associated with greater VS and SA activity while giving support, and SA activity while giving support was negatively correlated with amygdala activity in the same task, providing initial evidence for the inhibitory connection between the SA and the amygdala during support giving (Inagaki & Eisenberger, 2012). In a separate study, greater SA activity when giving support to a close other in need was associated with less amygdala activity in response to negative emotional faces, suggesting that the stress-reducing effects of giving support may extend to subsequent stressors outside of the context of giving (Inagaki & Ross, 2018).

Individual differences in self-reported support giving similarly relate to activity in parental care-related neural regions, such that greater self-reports are associated with greater VS activity to images of one's own close others and greater VS and SA activity when giving to others (Inagaki et al., 2016). In addition, those with higher trait levels of giving support show less amygdala activity to a social stressor (Inagaki et al., 2016) and to negative emotional faces (Inagaki & Ross, 2018). Finally, patients with basolateral amygdala damage (vs. healthy controls) display more giving behavior but no differences during a nonsocial risk-taking game (van Honk, Eisenegger, Terburg, Stein, & Morgan, 2013), providing some causal evidence that the amygdala might play an inhibitory role in giving support.

Exploring Another Mechanism: The Opioid System

Another lens through which to explore the mechanisms underlying the stress-reducing effects of receiving and giving support is their underlying neurobiology. One potential neurobiological account centers on the opioid system. Opioids are released in response to supportive social interactions and also reduce pain and threat responses (Eisenberger, 2012; Fanselow, 1981). Specifically, opioids attenuate SNS and HPA activity to stressors (Drolet et al., 2001). Thus, the opioid system is a likely route through which social ties may reduce stress responding.

Receiving support and opioids Given that receiving support has been shown to buffer against threat and stress, it is important to note that the opioid system plays a crucial role in both fear acquisition and fear extinction (Fanselow, 1998; Rescorla & Wagner, 1972) and hence may be directly involved in the threat-reducing effects of receiving support. Blocking opioid processes leads to enhanced fear acquisition (Fanselow, 1981) and prevents fear extinction from occurring (McNally & Westbrook, 2003). By triggering a release of endogenous opioids (Eisenberger, 2012; Nelson & Panksepp, 1998), social support may introduce additional opioids into the fear-learning circuit, preventing fear acquisition and enhancing fear extinction. Importantly, social support may do so while continuing to signal safety—a pattern of effects that would be unique in the fear-learning literature. Ultimately, this might suggest that social support not only signals safety and reduces perceptions of threat as they occur, influencing activity at a neurobiological level to mitigate threat responding, but also prevents acquisition of new fears and enhances extinction of ones already held, consequently diminishing the number of threats people perceive in the environment. This would represent a powerful buffering tool with implications for both mental and physical health outcomes.

Giving support and opioids Opioids may also play a critical role in the reinforcing and pleasurable aspects of giving support. Opioids have long been theorized to contribute to parental behavior in animals (Nelson & Panksepp, 1998) and to alter parenting behavior in humans (Slesnick, Feng, Brakenhoff, & Brigham, 2014). Many of the neural regions implicated in parental care in animals, including the VS and amygdala, are densely concentrated with opioid receptors. Thus, opioids may similarly affect support-giving behavior via actions on the neural regions we have proposed are most critical for such behavior. Further research directly measuring or manipulating the opioid system during support giving is needed, but in the context of mammalian parent-infant relationships, opioids appear to affect stress-related responses to parenting. Thus, morphine decreases aggression toward offspring and increases parental behavior (Kendrick & Keverne, 1991), whereas naltrexone increases aggression and reduces parental behavior (Kendrick & Keverne, 1989). These results suggest that opioids may also be involved in the stress-reducing effects seen in human support giving (e.g., Inagaki & Eisenberger, 2016). However, whether the health benefits of giving support *rely* on the opioid system remains open for further inquiry.

Conclusion

Research exploring the neurobiological underpinnings of social-buffering effects suggests that receiving and giving support reduces physiological and neural stress-related responding. Interestingly, this work also suggests that these stress-reducing properties may be a by-product of systems set in place to maintain social ties. Specifically, the mechanisms that have evolved to reinforce and maintain social bonds may have secondary functions that promote health. By mitigating neural responses to threats and even interfering in neural pathways that support fear learning, as in the case of receiving support, and by reducing stress and increasing reward in order to boost parenting and other supportive behavior, as in the case of giving support, these mechanisms may ultimately alleviate the negative consequences of physiological stress. Although much more work is required to elaborate on these processes, the evidence reviewed provides a strong foundation for understanding the link between social ties and health.

Acknowledgments

The authors would like to thank the members of the Social Affective Neuroscience and Social Cognitive Neuroscience labs at the University of California, Los Angeles, and the Social Health and Affective Neuroscience lab at the University of Pittsburgh for their support.

REFERENCES

Adolphs, R., Tranel, D., Damasio, H., & Damasio, A. R. (1995). Fear and the human amygdala. *Journal of Neuroscience, 15*(9), 5879–5891.

Ashar, Y. K., Andrews-Hanna, J. R., Dimidjian, S., & Wager, T. D. (2017). Empathic care and distress: Predictive brain markers and dissociable brain systems. *Neuron, 94*(6), 1263–1273.

Atzil, S., Hendler, T., & Feldman, R. (2011). Specifying the neurobiological basis of human attachment: Brain, hormones, and behavior in synchronous and intrusive mothers. *Neuropsychopharmacology, 36*, 2603.

Baum, M. (1969). Extinction of an avoidance response motivated by intense fear: Social facilitation of the action of response prevention (flooding) in rats. *Behaviour Research and Therapy, 7*(1), 57–62.

Brown, S. L., & Brown, R. M. (2006). Selective investment theory: Recasting the functional significance of close relationships. *Psychological Inquiry, 17*(1), 1–29.

Brown, S. L., Brown, R. M., House, J. S., & Smith, D. M. (2008). Coping with spousal loss: Potential buffering effects of self-reported helping behavior. *Personality and Social Psychology Bulletin, 34*(6), 849–861.

Brown, S. L., Nesse, R. M., Vinokur, A. D., & Smith, D. M. (2003). Providing social support may be more beneficial than receiving it: Results from a prospective study of mortality. *Psychological Science, 14*(4), 320–327.

Brown, J. L., Sheffield, D., Leary, M. R., & Robinson, M. E. (2003). Social support and experimental pain. *Psychosomatic Medicine, 65*(2), 276–283.

Cacioppo, J. T., Hawkley, L. C., & Thisted, R. A. (2010). Perceived social isolation makes me sad: 5-year cross-lagged analyses of loneliness and depressive symptomatology in the Chicago Health, Aging, and Social Relations Study. *Psychology and Aging, 25*(2), 453.

Che, X., Cash, R., Fitzgerald, P., & Fitzgibbon, B. M. (2018). The social regulation of pain: Autonomic and neurophysiological changes associated with perceived threat. *Journal of Pain, 19*(5), 496–505.

Cobb, S. (1976). Social support as a moderator of life stress. *Psychosomatic Medicine, 38(5),* 300–314.

Cohen, S. (2004). Social relationships and health. *American Psychologist, 59*(8), 676.

Covian, M. R., Antunes-Rodrigues, J., & O'Flaherty, J. J. (1964). Effects of stimulation of the septal area upon blood pressure and respiration in the cat. *Journal of Neurophysiology, 27*(3), 394–407.

Davitz, J. R., & Mason, D. J. (1955). Socially facilitated reduction of a fear response in rats. *Journal of Comparative and Physiological Psychology, 48*(3), 149.

Delgado, M. R., Olsson, A., & Phelps, E. A. (2006). Extending animal models of fear conditioning to humans. *Biological Psychology, 73*(1), 39–48.

Drolet, G., Dumont, É. C., Gosselin, I., Kinkead, R., Laforest, S., & Trottier, J. F. (2001). Role of endogenous opioid system in the regulation of the stress response. *Progress in Neuro-Psychopharmacology and Biological Psychiatry, 25*(4), 729–741.

Eisenberger, N. I. (2012). The pain of social disconnection: Examining the shared neural underpinnings of physical and social pain. *Nature Reviews Neuroscience, 13*(6), 421.

Eisenberger, N. I. (2013). An empirical review of the neural underpinnings of receiving and giving social support: Implications for health. *Psychosomatic Medicine, 75*(6), 545.

Eisenberger, N. I., Lieberman, M. D., & Williams, K. D. (2003). Does rejection hurt? An fMRI study of social exclusion. *Science, 302*(5643), 290–292.

Eisenberger, N. I., Master, S. L., Inagaki, T. K., Taylor, S. E., Shirinyan, D., Lieberman, M. D., & Naliboff, B. D. (2011). Attachment figures activate a safety signal-related neural region and reduce pain experience. *Proceedings of the National Academy of Sciences, 108*(28), 11721–11726.

Eisenberger, N. I., Taylor, S. E., Gable, S. L., Hilmert, C. J., & Lieberman, M. D. (2007). Neural pathways link social support to attenuated neuroendocrine stress responses. *Neuroimage, 35*(4), 1601–1612.

Fanselow, M. S. (1981). Naloxone and Pavlovian fear conditioning. *Learning and Motivation, 12*(4), 398–419.

Fanselow, M. S. (1998). Pavlovian conditioning, negative feedback, and blocking: Mechanisms that regulate association formation. *Neuron, 20*(4), 625–627.

Febo, M., Felix-Ortiz, A. C., & Johnson, T. R. (2010). Inactivation or inhibition of neuronal activity in the medial prefrontal cortex largely reduces pup retrieval and grouping in maternal rats. *Brain Research, 1325,* 77–88.

Galeno, T. M., Van Hoesen, G. W., & Brody, M. J. (1984). Central amygdaloid nucleus lesion attenuates exaggerated hemodynamic responses to noise stress in the spontaneously hypertensive rat. *Brain Research, 291*(2), 249–259.

Gerin, W., Pieper, C., Levy, R., & Pickering, T. G. (1992). Social support in social interaction: A moderator of cardiovascular reactivity. *Psychosomatic Medicine, 54*(3), 324–336.

Hall, J. C. (1955). Some conditions of anxiety extinction. *Journal of Abnormal and Social Psychology, 51,* 126–132.

Hansen, S. (1994). Maternal behavior of female rats with 6-OHDA lesions in the ventral striatum: Characterization of the pup retrieval deficit. *Physiology & Behavior, 55*(4), 615–620.

Harbaugh W. T., Mayr, U. & Burghart. D. R. (2007). Neural responses to taxation and voluntary giving reveal motives for charitable donations. *Science, 316,* 1622–1625.

Heinrichs, M., Baumgartner, T., Kirschbaum, C., & Ehlert, U. (2003). Social support and oxytocin interact to suppress cortisol and subjective responses to psychosocial stress. *Biological Psychiatry, 54*(12), 1389–1398.

Hennessy, M. B., Zate, R., & Maken, D. S. (2008). Social buffering of the cortisol response of adult female guinea pigs. *Physiology & Behavior, 93*(4–5), 883–888.

Hornstein, E. A., & Eisenberger, N. I. (2017). Unpacking the buffering effect of social-support figures: Social support attenuates fear acquisition. *PloS One, 12*(5), e0175891.

Hornstein, E. A., & Eisenberger, N. I. (2018). A social safety net: Developing a model of social-support figures as prepared safety stimuli. *Current Directions in Psychological Science, 27*(1), 25–31.

Hornstein, E. A., Fanselow, M. S., & Eisenberger, N. I. (2016). A safe haven: Investigating social-support figures as prepared safety stimuli. *Psychological Science, 27*(8), 1051–1060.

Hornstein, E. A., Haltom, K. E., Shirole, K., & Eisenberger, N. I. (2018). A unique safety signal: Social-support figures enhance rather than protect from fear extinction. *Clinical Psychological Science, 6*(3), 407–415.

Hostinar, C. E., Sullivan, R. M., & Gunnar, M. R. (2014). Psychobiological mechanisms underlying the social buffering of the hypothalamic-pituitary-adrenocortical axis: A review of animal models and human studies across development. *Psychological Bulletin, 140*(1), 256.

House, J. S., Landis, K. R., & Umberson, D. (1988). Social relationships and health. *Science, 241*(4865), 540–545.

Inagaki, T. K. (2018). Neural mechanisms of the link between giving social support and health. *Annals of the New York Academy of Sciences, 1428*(1), 33–50.

Inagaki, T. K., Byrne Haltom, K. E., Suzuki, S., Jevtic, I., Hornstein, E., Bower, J. E., & Eisenberger, N. I. (2016). The neurobiology of giving versus receiving reward: The role of stress-related and social reward-related neural activity. *Psychosomatic Medicine, 78*(4), 443–453.

Inagaki, T. K., & Eisenberger, N. I. (2012). Neural correlates of giving support to a loved one. *Psychosomatic Medicine, 74*(1), 3–7.

Inagaki, T. K., & Eisenberger, N. I. (2016). Giving support to others reduces sympathetic nervous system-related responses to stress. *Psychophysiology, 53*(4), 427–435.

Inagaki, T. K., & Orehek, E. (2017). On the benefits of giving social support: When, why, and how support providers gain by caring for others. *Current Directions in Psychological Science, 26*(2), 109–113.

Inagaki, T. K., & Ross, L. P. (2018). Neural correlates of giving social support: Giving targeted and untargeted support. *Psychosomatic Medicine.* doi:10.1097/PSY.0000000000000623. Advanced online publication

Karremans, J. C., Heslenfeld, D. J., van Dillen, L. F., & Van Lange, P. A. (2011). Secure attachment partners attenuate neural responses to social exclusion: An fMRI investigation. *International Journal of Psychophysiology, 81*(1), 44–50.

Kendrick, K. M., & Keverne, E. B. (1989). Effects of intracerebroventricular infusions of naltrexone and phentolamine on central and peripheral oxytocin release and on maternal behaviour induced by vaginocervical stimulation in the ewe. *Brain Research, 505*(2), 329–332.

Kendrick, K. M., & Keverne, E. B. (1991). Importance of progesterone and estrogen priming for the induction of maternal behavior by vaginocervical stimulation in sheep: Effects of maternal experience. *Physiology & Behavior, 49*(4), 745–750.

Liddell, H. S. (1950). Some specific factors that modify tolerance for environmental stress. In H. G. Wolff, S. G. Wolff Jr., & C. C. Hare (Eds.), *Life stress and bodily disease* (pp. 155–171). Baltimore: Williams and Wilkins.

Liddell, H. S. (1954). Conditioning and emotions. *Scientific American, 190,* 48–57.

Lieberman, M. D., & Eisenberger, N. I. (2015). The dorsal anterior cingulate cortex is selective for pain: Results from large-scale reverse inference. *Proceedings of the National Academy of Sciences, 112*(49), 15250–15255.

Lorberbaum, J. P., Newman, J. D., Horwitz, A. R., Dubno, J. R., Lydiard, R. B., Hamner, M. B., Bohning, D. E., & George, M. S. (2002). A potential role for thalamocingulate circuitry in human maternal behavior. *Biological Psychiatry, 51,* 431–445.

Lovibond, P. F., Davis, N. R., & O'Flaherty, A. S. (2000). Protection from extinction in human fear conditioning. *Behaviour Research and Therapy, 38*(10), 967–983.

Malmo, R. B. (1961). Slowing of heart rate after septal self-stimulation in rats. *Science, 133*(3459), 1128–1130.

Master, S. L., Eisenberger, N. I., Taylor, S. E., Naliboff, B. D., Shirinyan, D., & Lieberman, M. D. (2009). A picture's worth: Partner photographs reduce experimentally induced pain. *Psychological Science, 20*(11), 1316–1318.

McNally, G. P., & Westbrook, R. F. (2003). Opioid receptors regulate the extinction of Pavlovian fear conditioning. *Behavioral Neuroscience, 117*(6), 1292.

Melia, K. R., Sananes, C. B., & Davis, M. (1992). Lesions of the central nucleus of the amygdala block the excitatory effects of septal ablation on the acoustic startle reflex. *Physiology & Behavior, 51*(1), 175–180.

Miller, G., Chen, E., & Cole, S. W. (2009). Health psychology: Developing biologically plausible models linking the social world and physical health. *Annual Review of Psychology, 60,* 501–524.

Moll, J., Krueger, F., Zahn, R., Pardini, M., de Oliveira-Souza, R., & Grafman, J. (2006). Human fronto-mesolimbic networks guide decisions about charitable donation. *Proceedings of the National Academy of Sciences, 103*(42), 15623–15628.

Morelli, S. A., Rameson, L. T., & Lieberman, M. D. (2012). The neural components of empathy: Predicting daily prosocial behavior. *Social Cognitive and Affective Neuroscience, 9*(1), 39–47.

Nakayasu, T., & Ishii, K. (2008). Effects of pair-housing after social defeat experience on elevated plus-maze behavior in rats. *Behavioural Processes, 78*(3), 477–480.

Nakayasu, T., & Kato, K. (2011). Is full physical contact necessary for buffering effects of pair housing on social stress in rats? *Behavioural Processes, 86*(2), 230–235.

Nelson, E. E., & Panksepp, J. (1998). Brain substrates of infant-mother attachment: Contributions of opioids, oxytocin, and norepinephrine. *Neuroscience & Biobehavioral Reviews, 22*(3), 437–452.

Nelson-Coffey, S. K., Fritz, M. M., Lyubomirsky, S., & Cole, S. W. (2017). Kindness in the blood: A randomized controlled trial of the gene regulatory impact of prosocial behavior. *Psychoneuroendocrinology, 81,* 8–13.

Numan, M. (2007). Motivational systems and the neural circuitry of maternal behavior in the rat. *Journal of the International Society for Developmental Psychobiology, 49*(1), 12–21.

Pereira, M., & Morrell, J. I. (2011). Functional mapping of the neural circuitry of rat maternal motivation: Effects of site-specific transient neural inactivation. *Journal of Neuroendocrinology, 23*(11), 1020–1035.

Phelps, E. A., Delgado, M. R., Nearing, K. I., & LeDoux, J. E. (2004). Extinction learning in humans: Role of the amygdala and vmPFC. *Neuron, 43,* 897–905.

Piferi, R. L., & Lawler, K. A. (2006). Social support and ambulatory blood pressure: An examination of both receiving and giving. *International Journal of Psychophysiology, 62*(2), 328–336.

Poulin, M. J., Brown, S. L., Dillard, A. J., & Smith, D. M. (2013). Giving to others and the association between stress and mortality. *American Journal of Public Health, 103*(9), 1649–1655.

Price, D. D. (2000). Psychological and neural mechanisms of the affective dimension of pain. *Science, 288*(5472), 1769–1772.

Rescorla, R. A. (1969). Pavlovian conditioned inhibition. *Psychological Bulletin, 72*(2), 77.

Rescorla, R. A. (1971). Variation in the effectiveness of reinforcement and nonreinforcement following prior inhibitory conditioning. *Learning and Motivation, 2*(2), 113–123.

Rescorla, R. A. (2003). Protection from extinction. *Animal Learning & Behavior, 31*(2), 124–132.

Rescorla, R. A., & Wagner, A. R. (1972). A theory of Pavlovian conditioning: Variations in the effectiveness of reinforcement and nonreinforcement. *Classical Conditioning II: Current Research and Theory, 2,* 64–99.

Rini, C., Austin, J., Wu, L. M., Winkel, G., Valdimarsdottir, H., Stanton, A. L., Isola, L., Rowley, S., & Redd, W. H. (2014). Harnessing benefits of helping others: A randomized controlled trial testing expressive helping to address survivorship problems after hematopoietic stem cell transplant. *Health Psychology, 33*(12), 1541.

Roberts, M. H., Klatzkin, R. R., & Mechlin, B. (2015). Social support attenuates physiological stress responses and experimental pain sensitivity to cold pressor pain. *Annals of Behavioral Medicine, 49*(4), 557–569.

Rosal, M. C., King, J., Ma, Y., & Reed, G. W. (2004). Stress, social support, and cortisol: Inverse associations? *Behavioral Medicine, 30*(1), 11–22.

Sachser, N., Dürschlag, M., & Hirzel, D. (1998). Social relationships and the management of stress. *Psychoneuroendocrinology, 23*(8), 891–904.

Sanders, B. J., Wirtz-Nole, C., DeFord, S. M., & Erling, B. F. (1994). Central amygdaloid lesions attenuate cardiovascular responses to acute stress in rats with borderline hypertension. *Physiology & Behavior, 56*(4), 709–713.

Schiller, D., Levy, I., Niv, Y., LeDoux, J. E., & Phelps, E. A. (2008). From fear to safety and back: Reversal of fear in the human brain. *Journal of Neuroscience, 28*(45), 11517–11525.

Schreier, H. M., Schonert-Reichl, K. A., & Chen, E. (2013). Effect of volunteering on risk factors for cardiovascular disease in adolescents: A randomized controlled trial. *JAMA Pediatrics, 167*(4), 327–332.

Slesnick, N., Feng, X., Brakenhoff, B., & Brigham, G. S. (2014). Parenting under the influence: The effects of

opioids, alcohol and cocaine on mother–child interaction. *Addictive Behaviors, 39*(5), 897–900.
Slotnick, B. M., & Nigrosh, B. J. (1975). Maternal behavior of mice with cingulate cortical, amygdala, or septal lesions. *Journal of Comparative and Physiological Psychology, 88*, 118–127.
Stack, E. C., Balakrishnan, R., Numan, M. J., & Numan, M. (2002). A functional neuroanatomical investigation of the role of the medial preoptic area in neural circuits regulating maternal behavior. *Behavioural Brain Research, 131*(1–2), 17–36.
Strathearn, L., Fonagy, P., Amico, J., & Montague, P. R. (2009). Adult attachment predicts maternal brain and oxytocin response to infant cues. *Neuropsychopharmacology, 34*(13), 2655.
Tellioğlu, T., Aker, R., Oktay, S., & Onat, F. (1997). Effect of brain acetylcholine depletion on bicuculline-induced cardiovascular and locomotor responses. *International Journal of Neuroscience, 89*(3–4), 143–152.
Thomas, E. (1988). Forebrain mechanisms in the relief of fear: The role of the lateral septum. *Psychobiology, 16*(1), 36–44.
Thorsteinsson, E. B., & James, J. E. (1999). A meta-analysis of the effects of experimental manipulations of social support during laboratory stress. *Psychology and Health, 14*(5), 869–886.
van Honk, J., Eisenegger, C., Terburg, D., Stein, D. J., & Morgan, B. (2013). Generous economic investments after basolateral amygdala damage. *Proceedings of the National Academy of Sciences, 110*(7), 2506–2510.
Whillans, A. V., Dunn, E. W., Sandstrom, G. M., Dickerson, S. S., & Madden, K. M. (2016). Is spending money on others good for your heart? *Health Psychology, 35*(6), 574.
Younger, J., Aron, A., Parke, S., Chatterjee, N., & Mackey, S. (2010). Viewing pictures of a romantic partner reduces experimental pain: Involvement of neural reward systems. *PloS One, 5*(10), e13309.

82 Mechanisms of Loneliness

STEPHANIE CACIOPPO AND JOHN T. CACIOPPO

ABSTRACT Loneliness has long been suggested to be a contributing factor to poor mental health and well-being. Only recently, however, has loneliness been recognized as a significant risk factor for morbidity and mortality in older adults, representing a 26% increase in the odds of early mortality even after controlling statistically for demographic factors and objective social isolation. The extant data suggest that there is no single pathway linking loneliness to morbidity or mortality; rather, loneliness is associated with a number of cognitive, neural, hormonal, cellular, and molecular mechanisms that, individually or together, contribute to poor health outcomes. We identified and reviewed the evidence for eight interrelated pathways. Although there may be limited deleterious health effects associated with each pathway and loneliness, the cumulative effects of these pathways over time aggregate to produce significant damage to health and well-being. Given the prevalence of loneliness and the size of the association between loneliness and mortality, it is important to develop inexpensive and accessible interventions to prevent or address chronic loneliness.

Mechanisms of Loneliness

Scientific research on the topic of loneliness (the subjective feeling of being isolated or disconnected from others) was nearly nonexistent in 1959 (Cacioppo & Cacioppo, 2018a, 2018b). The oldest of these scientific papers, by nearly a decade, was a summary of six case studies published by Parfitt (1937) in the *Journal of Neurology and Psychopathology*. Based on these case studies, Parfitt suggested that "loneliness is a potent factor in the development of [paranoid] psychoses" in middle age or early senility and that "cardiovascular degeneration and high blood pressure are the commonest physical findings" (pp. 319, 321) in these cases (Cacioppo & Cacioppo, 2018a, 2018b). The plurality of the remaining articles reflected subjective work on loneliness from a psychiatric perspective and a need for more rigorous scientific research on loneliness.

However, it was not until the 21st century that research in loneliness burgeoned, fueled in part by the rapidly growing number of elderly adults, the rising costs of health care, and concerns about the prevalence of loneliness. A search of Web of Science for the term *loneliness* for the period 2000–2016 produced 4,970 hits (Mn = 292.35 articles/year)—more than an 800% increase in the rate of published work during the previous 40 years. Among the developments during this period were increased interest in the cross-cultural (e.g., Cacioppo et al., 2016) and genetic (cf., Goossens et al., 2015) determinants of loneliness, growing evidence that loneliness may have significant effects on both mental (cf., Cacioppo, Grippo, et al., 2015) *and* physical health (cf., Cacioppo, Cacioppo, Capitanio, & Cole, 2015; Holt-Lunstad, Smith, Baker, Harris, & Stephenson, 2015), and the increased use of prospective designs with population-based samples and animal data to more rigorously assess the potential causal role of loneliness in deleterious physical and mental health outcomes (cf., Cacioppo, Capitanio, & Cacioppo, 2014; Cacioppo, Cacioppo, Cole, et al., 2015). For instance, the associations between loneliness and health and well-being were found to persist after controlling for various potential influences, including objective social isolation, social support, age, gender, ethnicity, income, and marital status.

Prevalence of and Effect Size for Loneliness

Research shows that most individuals do not feel lonely at any given moment, just as most people do not feel hungry, thirsty, or in physical pain at any given moment (Cacioppo & Cacioppo, 2018b). Furthermore, establishing the prevalence of loneliness across time and geographic location is difficult given the differences in the measures of loneliness that have been used, the criteria used for classifying individuals as lonely, the populations and ages of participants, and the sampling procedures and sample sizes (Cacioppo & Cacioppo, 2018a). In the United States, for instance, estimates of its prevalence for adults who are 65 or older was 19.3%, based on a single item from the population-based Health and Retirement Study (HRS; Theeke, 2009), whereas responses to the three-item loneliness scale (Cacioppo & Cacioppo, 2018b) in the HRS indicated that 29% of adults 75 years or older in the HRS reported feeling lonely at least some of the time (Perissinotto, Stojacic Cenzer, & Covinsky, 2012). A recent survey of respondents from North Carolina, Texas, New York, and Ohio using responses from the three-item loneliness scale revealed an even higher prevalence rate: 27% reported moderate levels of loneliness, and 28% reported severe levels of loneliness (Musich, Wang, Hawkins, & Yeh, 2015).

Despite the differences in methods, samples, time periods, and locations, the overall pattern suggests that loneliness ranges from the approximately 20% to 60% who report feeling lonely at least some of the time to the 5% to 10% who report feeling lonely frequently or always. These prevalence rates are similar to those for other modifiable risk factors in industrialized nations. In the United States, for instance, the prevalence rate for (1) hypertension is approximately 29% (NCHS Data Brief, 2013); for (2) extreme obesity (BMI > 39), 6.3% and for obesity (BMI = 30–39), 35.7%; for (3) excessive drinking (15+ drinks/week for men, 8+ drinks/week for women), 6% and for binge drinking (5+ drinks on an occasion for men, 4+ drinks on occasion for women), 17% (www.cdc.gov/alcohol/data-stats.htm); and for (4) smoking, 15.1%.

The prevalence rates for these traditional risk factors are noteworthy because they represent (1) a large and growing number of adults, (2) modifiable targets for improving national health and well-being, and (3) significantly increased odds of premature mortality. A meta-analysis of loneliness as a risk factor for mortality covering data from 70 independent prospective studies involving over 3.4 million participants followed for an average of 7 years revealed that even after accounting for multiple covariates (e.g., objective social isolation), loneliness was associated with a 26% increase in likelihood of death (Holt-Lunstad et al., 2015).

Multiple Pathways

The Cacioppo evolutionary theory of loneliness (ETL, Cacioppo & Cacioppo, 2018b) predicts evolutionary social fitness as a function of the multiple pathways through which chronic loneliness may have deleterious effects (see Cacioppo & Cacioppo, 2018b, for details). These pathways, which are separable but not orthogonal, include (1) decreased sleep quality, (2) heightened activation of the hypothalamic-pituitary-adrenocortical (HPA) axis, (3) selectively elevated sympathetic tonus, (4) altered transcriptome dynamics, (5) decreased viral immunity, (6) increased inflammatory substrate, (7) increased prepotent (e.g., impulsive) responding, and (8) increased depressive symptomatology. Although the deleterious health effects of each pathway may be limited, the cumulative effects of these pathways over time aggregate to produce significant damage to health and well-being (see figure 82.1). We turn next to a review of the extant evidence regarding loneliness and the processes within each pathway.

Decreased sleep quality While it is easy for most individuals to detect signs of loneliness in friends or neighbors, it is more difficult to become aware of our own subjective feelings of loneliness, as loneliness is a condition with deep subconscious roots (Cacioppo, Balogh, & Cacioppo, 2015). Pathways of loneliness are most likely to occur when consciousness is less dominant—that is, during sleep at night (Cacioppo, Hawkley, Berntson, et al., 2002).

The association between loneliness and poor sleep quality has been replicated in middle-aged and older adults in different nations as well as in adolescents and young adults (see Cacioppo & Cacioppo, 2018b for a review). In addition, this association has been replicated in longitudinal investigations even after controlling for various covariates such as sleep quality at baseline (Hawkley, Preacher, & Cacioppo, 2010; McHugh & Lawlor, 2013), and loneliness has been related to poor sleep quality when participants are tested individually (e.g., Cacioppo, Hawkley, Berntson, et al., 2002) and in a population-based study of older adults whether or not participants slept alone (e.g., Hawkley, Preacher, & Cacioppo, 2011).

Heightened activation of the hypothalamic-pituitary-adrenocortical axis The HPA axis regulates physiological functions that include metabolism, digestion, immunity, and energy storage and expenditure and the physiological preparation for and responses to a perceived harmful event, attack, or threat to survival. Among the major hormones produced in the HPA axis are glucocorticoids (e.g., cortisol in humans, corticosterone in rodents). Under normal sleeping conditions, cortisol levels are highest in the morning and lowest shortly after midnight. In addition, cortisol levels increase about 50% approximately 30 minutes after awakening in the morning, a phenomenon termed the cortisol awakening response (CAR).

A robust association between loneliness and HPA activation in humans has been found in studies of the CAR (e.g., Adam et al., 2006; Okamura, Tsuda, & Matsuishi, 2011; Steptoe, Owen, Kunz-Ebrecht, & Brydon, 2004). In addition, studies using biomarkers of glucocorticoid receptor sensitivity indicate that loneliness is associated with decreased glucocorticoid receptor sensitivity (Cole, 2008; Cole et al., 2007, 2015), consistent with an association between loneliness and tonic HPA activation.

There are inconsistencies in the literature, as well. For instance, Steptoe et al. (2004) found loneliness to be associated with larger CARs, but they did not find it to be significantly associated with salivary cortisol levels in the laboratory. Kiecolt-Glaser et al. (1984) found that loneliness was associated with higher urinary cortisol

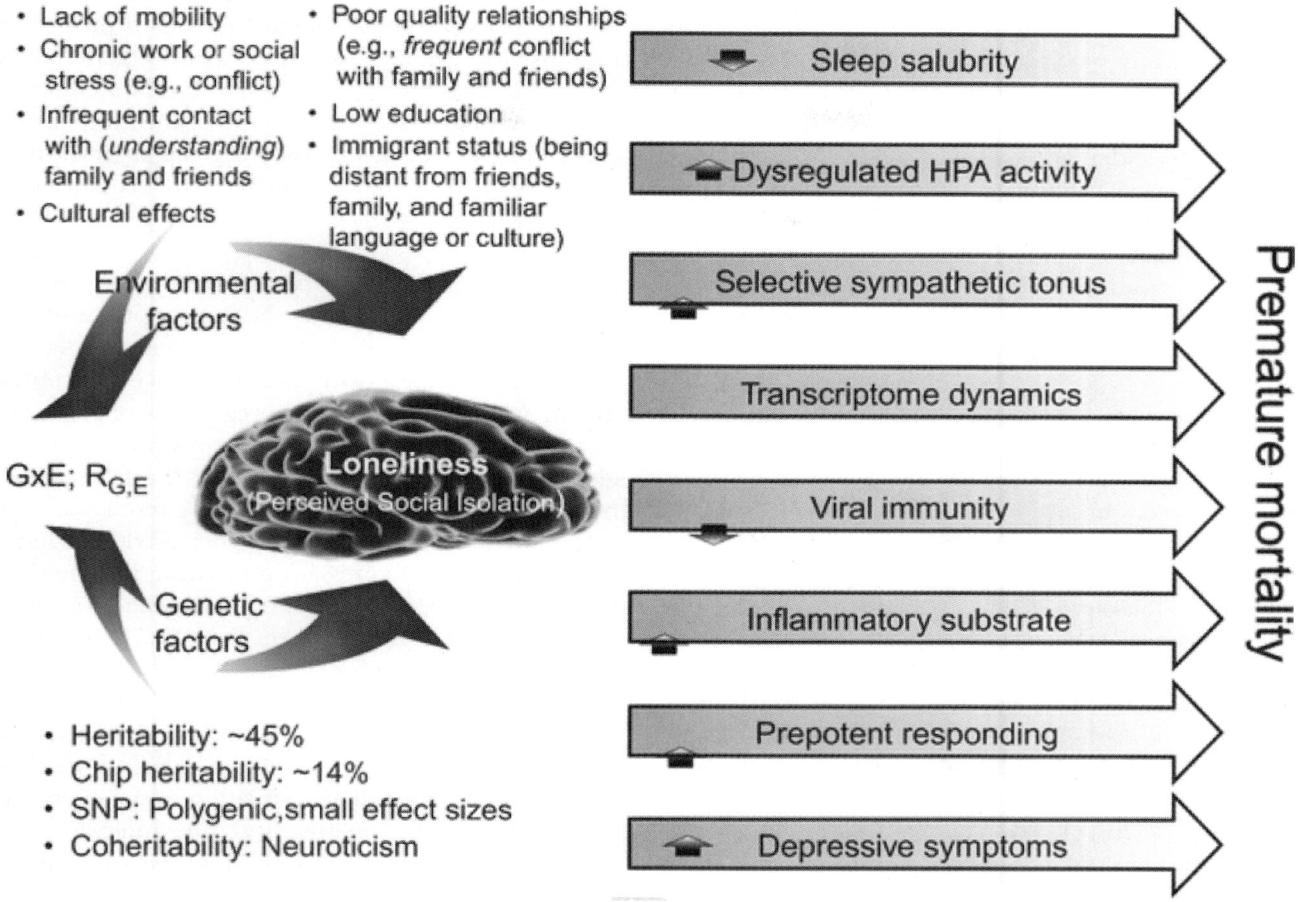

FIGURE 82.1 Eight pathways through which loneliness undermines health and longevity. From Cacioppo and Cacioppo (2018b).

levels in a sample of psychiatric inpatients on the day after admission, whereas Hawkley et al. (2006) discovered that loneliness was not related to urinary cortisol levels measured in overnight urine in a population-based sample of older adults. There is also conflicting evidence regarding the extent to which loneliness is related to cortisol levels over the course of a day, with some studies suggesting a relationship (Adam et al., 2006) and others suggesting no relationship (Sladek & Doane, 2015; Steptoe et al., 2004). Cacioppo et al. (2000) found that University of California Los Angeles (UCLA), loneliness scores and state loneliness scores were positively but nonsignificantly correlated with mean salivary cortisol levels, whereas trait (chronic) loneliness was positively and significantly related to mean cortisol levels, especially in the evening (Cacioppo & Cacioppo, 2018b).

Similar effects have been found in other species, including anthropoid primates (e.g., Cole et al., 2015; Mendoza & Mason, 1986). The increased HPA activation for experimentally isolated animals is not an inevitable consequence of objective social isolation but depends on the organization of the brain and the nature of the relationship of the animal to the conspecific with whom it is separated. For example, following one hour of social isolation from their pair mates, monogamous titi monkeys (for whom behavioral assessment has shown partner preference is high) show a significant increase in plasma cortisol, whereas squirrel monkeys (for whom behavioral assessment has shown partner preference is relatively low) do not (Mendoza & Mason, 1986). In contrast, the squirrel monkey mothers show significant increases in HPA activation when separated from their infant (for whom behavioral assessment has shown pair preference is high), while the titi monkeys (for whom behavioral assessment has shown pair preference is relatively low) do not (cf., Cacioppo, Cacioppo, Capitanio, et al., 2015).

In sum, there is evidence from human and animal studies that loneliness is associated with elevated HPA activation. However, HPA activity is influenced by a number of physiological (e.g., time of day, digestion) and psychological factors (e.g., work stress), and the presence of any such additional influences can make the association between loneliness and the level of HPA activation difficult to discern. In whom (e.g., psychiatric patients, older adults), what (e.g., CAR, cortisol level), where (e.g., lab versus naturalistic settings), how (e.g., levels measured in blood, urine, or saliva), and when (e.g., CAR, midday levels, evening levels) HPA activity is measured are likely to prove to be important considerations in studies of the association between loneliness and HPA activation.

Selectively elevated sympathetic tonus The sympathetic adrenomedullary system (SAM) is involved in the fight-or-flight response to stressors, and there is evidence that increased broad sympathetic contributions to stress reactivity can increase the risk of the development of disease onset or progression (e.g., Cacioppo, Berntson, Malarkey, et al., 1998). Research suggests that the sympathetic nervous system may be affected by or related to loneliness in subtler ways—for instance, by increasing the basal sympathetic tonus to vascular and myeloid tissue, rather than to the viscera more broadly, as part of a fight-or-flight response (cf., Cacioppo, Hawkley, & Berntson, 2003).

Although Parfitt (1937) noted "cardiovascular degeneration" and high blood pressure in his case studies of loneliness, Lynch (1977; Lynch & Convey, 1979) appears to be the first to have pursued the investigation of an association between loneliness and chronic cardiovascular conditions such as high blood pressure and cardiovascular disease. Lynch, however, did not clearly differentiate between the effects of objective versus perceived social isolation. Subsequent research, including prospective studies, has reported a significant association between loneliness and cardiovascular disease even after controlling for various covariates (see Cacioppo & Cacioppo, 2018b for a review).

The research to date on loneliness in humans suggests that it is more closely tied to the tonic activation of the vasculature (hemodynamics) rather than activation of the heart (cardiodynamics). Elevated vascular resistance in young adults is a risk factor for higher blood pressure later in life. In cross-sectional (Cacioppo, Hawkley, Crawford, et al., 2002; Hawkley, Masi, Berry, & Cacioppo, 2006; Ong, Rothstein, & Uchino, 2012) and longitudinal studies (Hawkley, Thisted, Masi, & Cacioppo, 2010; Momtaz et al., 2012) of older adults, loneliness has been associated with elevated basal levels of blood pressure. Some studies have failed to find a statistically significant association between loneliness and blood pressure (Tomaka, Thompson, & Palacios, 2006; Whisman, 2010), and Steptoe et al. (2004) found loneliness to be related to diastolic blood pressure in response to experimental stressors rather than to basal levels. Advances in the diagnosis and treatment of elevated blood pressure may be complicating factors, especially in light of evidence that lonely individuals are more, rather than less, likely to access and use medical services.

Altered transcriptome dynamics In an early investigation, Cole et al. (2007) found that the transcriptome dynamics of leukocytes differed between individuals high versus low in loneliness, with individuals high, in contrast to low, in loneliness showing upregulation of proinflammatory genes and downregulation of genes involved in glucocorticoid receptor signaling and interferon responses (i.e., viral immunity). This result was replicated in subsequent studies of older adults, including upregulated gene expression underlying inflammation (Cole, Capitanio, et al., 2015; Cole, Hawkley, Arevalo, & Cacioppo, 2011; Cole, Levine, et al., 2015). To investigate the potential causal role of loneliness, cross-lagged panel models were calculated (Cole, Capitanio, et al., 2015). Results indicated that increases in loneliness led to an upregulation of the expression of genes underlying inflammation and a downregulation of the expression of genes that defend against viral infections when measured one year later (the CTRA), and the CTRA propagated the feelings of loneliness measured one year later. These results were specific to loneliness and could not be explained in terms of depressive symptomatology or social support.

Together, these studies support a mechanistic model in which chronic loneliness predicts a sympathetically mediated increase in the release of immature monocytes from the bone marrow, a downregulation of glucocorticoid receptor sensitivity and antiviral gene expression, and an upregulation of inflammatory gene expression.

Decreased viral immunity The transcriptome changes associated with loneliness in humans and rhesus monkeys suggested that loneliness may be associated with a reduction in viral immunity. To investigate the potential functional significance of these transcriptome changes, the expression of type I and II interferons were assessed in an additional sample of macaques before and at 2 weeks and 10 weeks following experimental infection with the simian immunodeficiency virus (SIV; Cole, Capitanio, et al., 2015). Measures at baseline again showed that lonely, compared to control, macaques showed lower levels of interferon gene expression. Two weeks after the experimental infection (peak of acute viral replication), interferon gene expression was significantly elevated and did not differ as a function of loneliness. However, 10 weeks after the experimental infection (after establishment of a long-term viral replication set point), lonely macaques showed lower levels of interferon gene expression than control animals. The lonely animals also showed poorer suppression of SIV gene expression between the postinfection measurement periods as well as an elevated SIV viral load and reduced anti-SIV immunoglobulin G (IgG) antibody titers at 10 weeks. These results underscore the importance of the timing of the immune response in studies of loneliness and viral immunity.

Suggestions that loneliness is associated with diminished viral immunity date back more than three decades. Mixed results have also been reported. Jaremka, Fagundes, Glaser, et al. (2013) investigated the association between loneliness and latent herpesvirus reactivation in both cytomegalovirus (CMV) and Epstein-Barr virus (EBV) in breast cancer survivors two months to three years posttreatment. Results showed that loneliness was related to higher CMV antibody titers (suggesting poor viral immunity) but was unrelated to EBV antibody titer levels. Jabaaij et al. (1993) reported that loneliness was unrelated to the antibody response to a low-dose hepatitis B vaccine.

The immune system is highly diversified, so it is not surprising that when, in whom, and how loneliness and immunity are measured may be important considerations. The extant work suggests that loneliness may be associated with diminished immunity, particularly viral immunity, but the details of the underlying mechanism have yet to be delineated.

Increased inflammatory substrate Several studies have failed to find a significant association between loneliness and inflammatory markers (e.g., C-reactive protein) at baseline (Mezuk et al., 2016; O'Luanaigh et al., 2012; Shankar, McMunn, Banks, & Steptoe, 2011). However, the changes in inflammatory biology suggested by the transcriptome differences in circulating leukocytes may be better reflected in the synthesis of proinflammatory cytokines rather than more tonic, indirect markers. In a study bearing on this notion, Hackett, Endrighi, Brydon, and Steptoe (2012) investigated the association between loneliness and inflammatory responses to a laboratory stressor in middle-age adults from the Whitehall cohort. Interleukin-6 (IL-6), interleukin-1 receptor antagonist (IL-1Ra), and the chemokine monocyte chemotactic protein (MCP-1) served as inflammatory markers. Hackett et al. (2012) found that loneliness in women was associated with elevated levels of MCP-1 at baseline and throughout the task and with the IL-6 and IL-1Ra response to the psychological stressor. These associations were not significant for men.

Subsequent studies have found an association between loneliness and the inflammatory response to acute experimental stressors and have not found gender differences (Jaremka, Fagundes, Peng, et al., 2013; Moieni et al., 2015).

Inflammation, like immunity, is a multifarious process. Although investigations of the association between loneliness and inflammation are still relatively new and limited, there is less evidence to support an association between loneliness and circulating markers of chronic inflammation than for an association between loneliness and: (1) the gene expression in leukocytes contributing to the synthesis of pro-inflammatory cytokines or (2) inflammatory responses to an acute stressor.

Increased prepotent responding Early evidence that prepotent responding may be greater in lonely than nonlonely individuals has been observed since 2000 (Cacioppo et al., 2000; Cacioppo & Cacioppo, 2018b for a review). Also, experimental manipulations that lead people to believe they face a future of social isolation have also been shown to decrease self-regulation (Baumeister & DeWall, 2005). Interestingly, subsequent experiments showed that such effects could be eliminated by offering a cash incentive or increasing self-awareness (Baumeister et al., 2005). Finally, in a study contrasting Future Alone and control conditions, Campbell et al. (2006) measured neural activity using magnetoencephalography (MEG) while participants performed moderately difficult math problems. Results indicated that the brains of the future socially isolated participants were less active in the areas involved in the *executive control* of attention, and activation in the parietal and right prefrontal cortex mediated the differences in performance on the math problems (see also Layden et al., 2017).

In sum, the literature is limited, but it suggests that loneliness is associated with increased prepotent responding. The finding that the differences in prepotent responding can be eliminated by offering performance incentives (Baumeister et al., 2005) is consistent with the proposition in the evolutionary model that loneliness increases prepotent responding through its effects on motivation rather than on ability. This result also raises the possibility that the exertion of self-control may play an important role in overcoming prepotent response predispositions. Loneliness has been associated with low perceived self-control, but the extent to which perceived self-control mediates the association between loneliness and prepotent responding has not been investigated.

Increased depressive symptomatology The most common clinical focus on loneliness has been on its association with poor mental health, with an emphasis on depressive symptomatology (e.g., see reviews by Cacioppo & Cacioppo, 2018b; Ernst & Cacioppo, 1998). Numerous studies have reported significant correlations between loneliness and depressive symptomatology (e.g., Cacioppo & Cacioppo, 2018b for a review; Cacioppo, Hawkley, et al., 2006), and for decades many clinicians believed that loneliness was simply an aspect of depression with no distinct concept worthy of study (see Cacioppo & Cacioppo, 2018b for review). Importantly, longitudinal research indicates that loneliness and depression are separable, loneliness predicts increases in depressive symptomatology above and beyond what can be explained by initial levels of depressive symptomatology, and the prospective association between loneliness and depressive symptomatology is reciprocal (e.g., Cacioppo, Hughes, et al., 2006). In addition, experimental manipulations of loneliness have been found to produce higher negative mood, anxiety, anger, and depressive symptomatology (Cacioppo & Cacioppo, 2018a, 2018b), and coheritability analyses in genome-wide association studies further show that loneliness and depression are distinct phenotypes (Abdellaoui et al., 2018).

Although the effects of chronic loneliness on depressive behaviors may prove deleterious, they may be adaptive in the short term. For instance, the depression resulting from loneliness may decrease the likelihood that an individual attempts to force its way back into a group from which it feels excluded and increase the likelihood that an individual will exhibit facial displays, postural displays, and acoustic signals that may serve as a call for others to come to its aid to provide companionship and support (Cacioppo, Cacioppo, & Boomsma, 2014; Cacioppo & Patrick, 2008). Whether this passive strategy succeeds and benefits the individual depends on the social environment, such as the likelihood that a caring conspecific will see and be willing and able to respond to the distress cues before predators or foes take advantage of the vulnerable individual.

Among the early animal models of depression were those based on maternal separation and social isolation in early life (e.g., Sanchez, Ladd, & Plotsky, 2001). Importantly, social separation in adulthood also produces behavioral indicators of depression, anxiety, and/or social withdrawal in a number of species, including the monogamous prairie vole (e.g., Grippo, Cushing, & Carter, 2007), the Sprague-Dawley rat (e.g., Wallace et al., 2009), the Wistar rat (e.g., Evans, Sun, McGregor, & Connor, 2012), the C57BL/6J mouse (Martin & Brown, 2010), and the rhesus monkey (Suomi, Eisele, Grady, & Harlow, 1975). Chronic social isolation in many of these species now serves as an animal model for studying depression and anxiety and treatment responses (e.g., Martin & Brown, 2010; Nin et al., 2011).

In sum, the cumulative research suggests that loneliness contributes to depressive symptomatology, which in turn can have adverse health effects through mechanisms such as autonomic activity, health behaviors, and suicidal behavior. Although the association between loneliness and various other pathways is not mediated by depressive symptomatology, the effects of loneliness on depressive symptomatology represent yet another pathway through which loneliness may contribute to premature mortality.

Conclusion

Loneliness has long been suggested to be a contributing factor to poor mental health and well-being. The fact that loneliness (perceived social isolation) predicts mortality independently of objective social isolation underscores the key role of the brain for (1) forming, monitoring, maintaining, repairing, and replacing salutary connections with others; (2) determining the level of loneliness at any moment in time; and (3) modulating molecular, cellular, hormonal, neural, and behavioral processes to deal with any perceived deficiencies in available social relationships.

Moreover, this body of research emphasizes the fact that no single pathway links loneliness to morbidity or mortality. Instead, the extant data suggest that loneliness is associated with a number of cognitive, neural, hormonal, cellular, and molecular mechanisms that, individually or together, contribute to poor health outcomes. Each of these pathways is influenced by a number of factors in addition to loneliness, and the multiple determined natures of each pathway in everyday life implies that the association between loneliness and each pathway is likely to be small. Additional research is needed to establish the existence and nature of the association between loneliness and specific processes within each pathway.

As more is learned about the specific mechanisms through which loneliness is linked to deleterious health outcomes, new behavioral or pharmacological interventions may be identified to break the chain of events and block the adverse outcomes within one or more pathways. Although there is much yet to be done, our scientific understanding of loneliness and its treatment has increased immensely since it was featured in the first episode of *The Twilight Zone* almost 60 years ago. We may not have solved the problem of loneliness yet, but efforts to understand loneliness, its health effects, and the mechanisms underlying its deleterious effects and interventions to mitigate loneliness have become active and exciting areas of scientific research. Given the rich set of questions that remains, these areas are likely to remain active and exciting for some time to come.

Acknowledgment

This chapter is dedicated to Professor John T. Cacioppo, founder of the field of social neuroscience, pioneer in the neuroscience of loneliness, and extraordinary husband. He will be—is—immensely missed.

REFERENCES

Abdellaoui, A., Chen, H. Y., Willemsen, G., Ehli, E. A., Davies, G. E., Verweij, K. J. H., ... Cacioppo, J. T. (2018). Associations between loneliness and personality are mostly driven by a genetic association with neuroticism. *Journal of Personality, May*, 1–12.

Adam, E. K., Hawkley, L. C., Kudielka, B. M., & Cacioppo, J. T. (2006). Day-to-day dynamics of experience-cortisol associations in a population-based sample of older adults. *Proceedings of the National Academy of Sciences, 103*, 17058–17063.

Baumeister, R. F., & DeWall, C. N. (2005). The inner dimension of social exclusion: Intelligent thought and self-regulation among rejected persons. *The social outcast: Ostracism, social exclusion, rejection, and bullying* (pp. 53–73). New York: Psychology Press.

Baumeister, R. F., DeWall, C. N., Ciarocco, N. J., & Twenge, J. M. (2005). Social exclusion impairs self-regulation. *Journal of Personality and Social Psychology, 88*(4), 589–604.

Cacioppo, J. T., Adler, A. B., Lester, P. B., McGurk, D., Thomas, J. L., Chen, H. Y., & Cacioppo, S. (2015). Building social resilience in soldiers: A double dissociative randomized controlled study. *Journal of Personality and Social Psychology, 109*(1), 90–105.

Cacioppo, J. T., Berntson, G. G., Malarkey, W. B., Kiecolt-Glaser, J. K., Sheridan, J. F., Poehlmann, K. M., ... Glaser, R. (1998). Autonomic, neuroendocrine, and immune responses to psychological stress: The reactivity hypothesis. *Annals of the New York Academy of Sciences, 840*, 664–673.

Cacioppo, J. T., & Cacioppo, S. (2018a). The growing problem of loneliness. *The Lancet, 391*(10119), 426.

Cacioppo, J. T., & Cacioppo, S. (2018b). Loneliness in the modern age: An evolutionary theory of loneliness (ETL). *Advances in Experimental Social Psychology, 58*, 127–197.

Cacioppo, J. T., Cacioppo, S., Adler, A. B., Lester, P. B., McGurk, D., Thomas, J. L., & Chen, H. Y. (2016). The cultural context of loneliness: Risk factors in active duty soldiers. *Journal of Social and Clinical Psychology, 35*, 865–882.

Cacioppo, J. T., Cacioppo, S., & Boomsma, D. (2014). Evolutionary mechanisms for loneliness. *Cognition and Emotion, 28*, 3–21.

Cacioppo, J. T., Cacioppo, S., Capitanio, J. P., & Cole, S. W. (2015). The neuroendocrinology of social isolation. *Annual Review of Psychology, 66*, 733–767.

Cacioppo, J. T., Cacioppo, S., Cole, S. W., Capitanio, J. P., Goossens, L., & Boomsma, D. I. (2015). Loneliness across phylogeny and a call for comparative studies and animal models. *Perspectives on Psychological Science, 10*, 202–212.

Cacioppo, S., Capitanio, J. P., & Cacioppo, J. T. (2014). Toward a neurology of loneliness. *Psychological Bulletin, 140*, 1464–1504.

Cacioppo, J. T., Ernst, J. M., Burleson, M. H., McClintock, M. K., Malarkey, W. B., Hawkley, L. C., ... Berntson, G. G. (2000). Lonely traits and concomitant physiological processes: The MacArthur social neuroscience studies. *International Journal of Psychophysiology, 35*(2–3), 143–154.

Cacioppo, J. T., Hawkley, L. C., & Berntson, G. G. (2003). The anatomy of loneliness. *Current Directions in Psychological Science, 12*, 71–74.

Cacioppo, J. T., Hawkley, L. C., Berntson, G. G., Ernst, J. M., Gibbs, A. C., Stickgold, R., & Hobson, J. A. (2002). Do lonely days invade the nights? Potential social modulation of sleep efficiency. *Psychological Science, 13*(4), 384–387.

Cacioppo, J. T., Hawkley, L. C., Crawford, L. E., Ernst, J. M., Burleson, M. H., Kowalewski, R. B., ... Berntson, G. G. (2002). Loneliness and health: Potential mechanisms. *Psychosomatic Medicine, 64*, 407–417.

Cacioppo, J. T., Hawkley, L. C., Ernst, J. M., Burleson, M., Berntson, G. G., Nouriani, B., & Spiegel, D. (2006). Loneliness within a nomological net: An evolutionary perspective. *Journal of Research in Personality, 40*, 1054–1085.

Cacioppo, J. T., Hawkley, L. C., & Thisted, R. A. (2010). Perceived social isolation makes me sad: 5-year cross-lagged analyses of loneliness and depressive symptomatology in the Chicago Health, Aging, and Social Relations Study. *Psychology and Aging, 25*(2), 453–463.

Cacioppo, J. T., Hughes, M. E., Waite, L. J., Hawkley, L. C., & Thisted, R. A. (2006). Loneliness as a specific risk factor for depressive symptoms: Cross-sectional and longitudinal analyses. *Psychology and Aging, 21*(1), 140–151.

Cacioppo, J. T., Norris, C. J., Decety, J., Monteleone, G., & Nusbaum, H. (2009). In the eye of the beholder: Individual differences in perceived social isolation predict regional brain activation to social stimuli. *Journal of Cognitive Neuroscience, 21*, 83–92.

Cacioppo, J. T., & Patrick, B. (2008). *Loneliness: Human nature and the need for social connection*. New York: W. W. Norton.

Cacioppo, S., Balogh, S., & Cacioppo, J. T. (2015). Implicit attention to negative social, in contrast to nonsocial, words in the Stroop task differs between individuals high and low in loneliness: Evidence from event-related brain microstates. *Cortex, 70*, 213–233.

Cacioppo, S., Bangee, M., Balogh, S., Cardenas-Iniguez, C., Qualter, P., & Cacioppo, J. T. (2016). Loneliness and implicit attention to social threat: A high-performance electrical neuroimaging study. *Journal of Cognitive Neuroscience, 7*(1–4), 138–159.

Cacioppo, S., Grippo, A. J., London, S., Goossens, L., & Cacioppo, J. T. (2015). Loneliness: Clinical import and interventions. *Perspectives on Psychological Science, 10*(2), 238–249.

Campbell, W. K., Krusemark, E. A., Dyckman, K. A., Brunell, A. B., McDowell, J. E., Twenge, J. M., & Clementz, B. A. (2006). A magnetoencephalography investigation of neural correlates for social exclusion and self-control. *Social Neuroscience, 1*(2), 124–134.

Capitanio, J. P., Hawkley, L. C., Cole, S. W., & Cacioppo, J. T. (2014). A behavioral taxonomy of loneliness in monkeys and humans. *PLoS One, 9*(10), e110307.

Cole, S. W. (2008). Social regulation of leukocyte homeostasis: The role of glucocorticoid sensitivity. *Brain, Behavior, & Immunity, 22*, 1049–1065.

Cole, S. W., Capitanio, J. P., Chun, K., Arevalo, J. M. G., Ma, J., & Cacioppo, J. T. (2015). Myeloid differentiation architecture of leukocyte transcriptome dynamics in perceived social isolation. *Proceedings of the National Academy of Sciences, 112*, 15142–15147.

Cole, S. W., Hawkley, L. C., Arevalo, J. M., & Cacioppo, J. T. (2011). Transcript origin analysis identifies antigen presenting cells as primary targets of socially regulated leukocyte gene expression. *Proceedings of the National Academy of Sciences, 108*, 3080–3085.

Cole, S. W., Hawkley, L. C., Arevalo, J. M., Sung, C. Y., Rose, R. M., & Cacioppo, J. T. (2007). Social regulation of gene expression in human leukocytes. *Genome Biology, 8*(9), R189.

Cole, S. W., Levine, M. E., Arevalo, J. M., Ma, J., Weir, D. R., & Crimmins, E. M. (2015). Loneliness, eudaimonia, and the human conserved transcriptional response to adversity. *Psychoneuroendocrinology, 62*, 11–17.

Ernst, J. M., & Cacioppo, J. T. (1998). Lonely hearts: Psychological perspectives on loneliness. *Applied and Preventive Psychology, 8*(1), 1–22.

Evans, J., Sun, Y., McGregor, A., & Connor, B. (2012). Allopregnanolone regulates neurogenesis and depressive/anxiety-like behavior in a social isolation rodent model of chronic stress. *Neuropharmacology, 63*, 1315–1326.

Gao, J., Davis, L. K., Hart, A. B., Sanchez-Roige, S., Han, L., Cacioppo, J. T., & Palmer, A. A. (2017/forthcoming). Genome-wide association study of loneliness demonstrates a role for common variation. *Neuropsychopharmacology, 42*, 811–821. doi:10.1038/npp.2016.197

Goossens, L., van Roekel, E., Verhagen, M., Cacioppo, J. T., Cacioppo, S., Maes, M., & Boomsma, D. I. (2015). The genetics of loneliness: Linking evolutionary theory to genome-wide genetics, epigenomics, and social science. *Perspectives on Psychological Science, 10*, 213–226.

Grippo, A. J., Cushing, B. S., & Carter, C. S. (2007). Depression-like behavior and stressor-induced neuroendocrine activation in female prairie voles exposed to chronic social isolation. *Psychosomatic Medicine, 69*, 149–157.

Grippo, A. J., Gerena, D., Huang, J., Kumar, N., Shah, M., Ughreja, R., & Carter, C. S. (2007). Social isolation induces behavioral and neuroendocrine disturbances relevant to depression in female and male prairie voles. *Psychoneuroendocrinology, 32*, 966–980.

Hackett, R. A., Hamer, M., Endrighi, R., Brydon, L., & Steptoe, A. (2012). Loneliness and stress-related inflammatory and neuroendocrine responses in older men and women. *Psychoneuroendocrinology, 37*(11), 1801–1809.

Hawkley, L. C., Burleson, M. H., Berntson, G. G., & Cacioppo, J. T. (2003). Loneliness in everyday life: Cardiovascular activity, psychosocial context, and health behaviors. *Journal of Personality and Social Psychology, 85*, 105–120.

Hawkley, L. C., Hughes, M. E., Waite, L. J., Masi, C. M., Thisted, R. A., & Cacioppo, J. T. (2008). From social structural factors to perceptions of relationship quality and loneliness: The Chicago health, aging, and social relations study. *Journals of Gerontology Series B: Psychological Sciences and Social Sciences, 63*(6), S375–S384.

Hawkley, L. C., Masi, C. M., Berry, J. D., & Cacioppo, J. T. (2006). Loneliness is a unique predictor of age-related differences in systolic blood pressure. *Psychology and Aging, 21*(1), 152–164.

Hawkley, L., Preacher, K. J., & Cacioppo, J. (2011). As we said, loneliness (not living alone) explains individual differences in sleep quality: Reply. *American Psychological Association, 30*(2), 136.

Hawkley, L. C., Preacher, K. J., & Cacioppo, J. T. (2010). Loneliness impairs daytime functioning but not sleep duration. *Health Psychology, 29*(2), 124–129.

Hawkley, L. C., Thisted, R. A., & Cacioppo, J. T. (2009). Loneliness predicts reduced physical activity: Cross-sectional and longitudinal analyses. *Health Psychology, 28*, 354–363.

Hawkley, L. C., Thisted, R. A., Masi, C. M., & Cacioppo, J. T. (2010). Loneliness predicts increased blood pressure: 5-year cross-lagged analyses in middle-aged and older adults. *Psychology and Aging, 25*(1), 132.

Holt-Lunstad, J., Smith, T. B., Baker, M., Harris, T., & Stephenson, D. (2015). Loneliness and social isolation as risk factors for mortality: A meta-analytic review. *Perspectives on Psychological Science, 10*, 227–237.

Hughes, M. E., Waite, L. J., Hawkley, L. C., & Cacioppo, J. T. (2004). A short scale for measuring loneliness in large surveys: Results from two population-based studies. *Research on Aging, 26*, 655–672.

Jabaaij, L., Grosheide, P. M., Heijtink, R. A., Duivenvoorden, H. J., Ballieux, R. E., & Vingerhoets, A. J. (1993). Influence of perceived psychological stress and distress on antibody response to low dose rDNA hepatitis B vaccine. *Journal of Psychosomatic Research, 37*, 361–369.

Jaremka, L. M., Fagundes, C. P., Glaser, R., Bennett, J. M., Malarkey, W. B., & Kiecolt-Glaser, J. K. (2013). Loneliness predicts pain, depression, and fatigue: Understanding the role of immune dysregulation. *Psychoneuroendocrinology, 38*(8), 1310–1317.

Jaremka, L. M., Fagundes, C. P., Peng, J., Bennett, J. M., Glaser, R., Malarkey, W. B., & Kiecolt-Glaser, J. K. (2013). Loneliness promotes inflammation during acute stress. *Psychological Science, 24*(7), 1089–1097.

Kiecolt-Glaser, J. K., Ricker, D., George, J., Messick, G., Speicher, C. E., Garner, W., & Glaser, R. (1984). Urinary cortisol levels, cellular immunocompetency, and loneliness in psychiatric inpatients. *Psychosomatic Medicine, 46*(1), 15–23.

Layden, E. A., Cacioppo, J. T., Cacioppo, S., Cappa, S. F., Dodich, A., Falini, A., & Canessa, N. (2017). Perceived social isolation is associated with altered functional connectivity in neural networks associated with tonic alertness and executive control. *Neuroimage*, 145, 58–73.

Lynch, J. J. (1977). *The broken heart: The medical consequences of loneliness.* New York: Basic Books.

Lynch, J. J., & Convey, W. H. (1979). Loneliness, disease, and death: Alternative approaches. *Psychosomatics, 20*, 702–708.

Martin, A. L., & Brown, R. E. (2010). The lonely mouse: Verification of a separation-induced model of depression in female mice. *Behavioural Brain Research, 207*, 197–207.

McHugh, J. E., & Lawlor, B. A. (2013). Perceived stress mediates the relationship between emotional loneliness and sleep quality over time in older adults. *British Journal of Health Psychology, 18*(3), 546–555.

Mendoza, S. P., & Mason, W. A. (1986). Contrasting responses to intruders and to involuntary separation by monogamous and polygynous New World monkeys. *Physiology & Behavior, 38*, 795–801.

Mezuk, B., Choi, M., DeSantis, A. S., Rapp, S. R., Roux, A. V. D., & Seeman, T. (2016). Loneliness, depression, and inflammation: Evidence from the multi-ethnic study of atherosclerosis. *PLoS One, 11*(7), doi:10.1371/journal.pone.0158056

Moieni, M., Irwin, M. R., Jevtic, I., Breen, E. C., Cho, H. J., Arevalo, J. M., & Eisenberger, N. I. (2015). Trait sensitivity to social disconnection enhances pro-inflammatory responses to a randomized controlled trial of endotoxin. *Psychoneuroendocrinology, 62*, 336–342.

Momtaz, Y. A., Hamid, T. A., Yusoff, S., Ibrahim, R., Chai, S. T., Yahaya, N., & Abdullah, S. S. (2012). Loneliness as a risk factor for hypertension in later life. *Journal of Aging and Health, 24*(4), 696–710.

Musich, S., Wang, S. S., Hawkins, K., & Yeh, C. S. (2015). The impact of loneliness on quality of life and patient satisfaction among older, sicker adults. *Gerontology & Geriatric Medicine, 1*, 2333721415582119.

Nin, M. S., Martinez, L. A., Pibiri, F., Nelson, M., & Pinna, G. (2011). Neurosteroids reduce social isolation-induced behavior deficits: A proposed link with neurosteroid-mediated upregulation of BDNF expression. *Frontiers in Endocrinology, 2*, 1–12.

Okamura, H., Tsuda, A., & Matsuishi, T. (2011). The relationship between perceived loneliness and cortisol awakening responses on work days and weekends. *Japanese Psychological Research, 53*(2), 113–120.

O'Luanaigh, C., O'Connell, H., Chin, A. V., Hamilton, F., Coen, R., Walsh, C., ... Cunningham, C. J. (2012). Loneliness and vascular biomarkers: The Dublin healthy ageing study. *International Journal of Geriatric Psychiatry, 27*(1), 83–88.

Ong, A. D., Rothstein, J. D., & Uchino, B. N. (2012). Loneliness accentuates age differences in cardiovascular responses to social evaluative threat. *Psychology and Aging, 27*(1), 190–198.

Parfitt, D. N. (1937). Loneliness and the paranoid syndrome. *Journal of Neurology and Psychopathology, 1*(68), 318–321.

Perissinotto, C. M., Stojacic, C. I., & Covinsky, K. E. (2012). Loneliness in older persons: A predictor of functional decline and death. *Archives of Internal Medicine, 172*, 1078–1084.

Sanchez, M. M., Ladd, C. O., & Plotsky, P. M. (2001). Early adverse experience as a developmental risk factor for later psychopathology: Evidence from rodent and primate models. *Development and Psychopathology, 13*(03), 419–449.

Shankar, A., McMunn, A., Banks, J., & Steptoe, A. (2011). Loneliness, social isolation, and behavioral and biological health indicators in older adults. *Health Psychology, 30*(4), 377.

Sladek, M. R., & Doane, L. D. (2015). Daily diary reports of social connection, objective sleep, and the cortisol awakening response during adolescents' first year of college. *Journal of Youth and Adolescence, 44*, 298–316.

Steptoe, A., Owen, N., Kunz-Ebrecht, S. R., & Brydon, L. (2004). Loneliness and neuroendocrine, cardiovascular, and inflammatory stress responses in middle-aged men and women. *Psychoneuroendocrinology, 29*(5), 593–611.

Suomi, S. J., Eisele, C. D., Grady, S. A., & Harlow, H. F. (1975). Depressive behavior in adult monkeys following separation from family environment. *Journal of Abnormal Psychology, 84*, 576–578.

Theeke, L. A. (2009). Predictors of loneliness in US adults over age sixty-five. *Archives of Psychiatric Nursing, 23*(5), 387–396.

Tomaka, J., Thompson, S., & Palacios, R. (2006). The relation of social isolation, loneliness, and social support to disease outcomes among the elderly. *Journal of Aging and Health, 18*(3), 359–384.

Wallace, D. L., Han, M. H., Graham, D. L., Green, T. A., Vialou, V., Iñiguez, S. D., ... Nestler, E. J. (2009). CREB regulation of nucleus accumbens excitability mediated social isolation-induced behavioral deficits. *Nature Neuroscience, 12*, 200–209.

Whisman, M. (2010). Loneliness and the metabolic syndrome in a population-based sample of middle-aged and older adults. *Health Psychology, 29*, 550–554.

Young, L. J., Lim, M. M., Gingrich, B., & Insel, T. R. (2001). Cellular mechanisms of social attachment. *Hormones and Behavior, 40*, 133–138.

83 Neural Mechanisms of Social Learning

DOMINIC S. FARERI, LUKE J. CHANG, AND MAURICIO DELGADO

ABSTRACT Our well-being is contingent upon our ability to navigate challenges and make decisions within a dynamic social environment. Social learning provides unique opportunities to meet such challenges by helping us to reduce uncertainty, update social expectations, and ultimately maximize social gains by developing close relationships. This chapter will review the mechanisms of social learning, focusing on how we can learn from and about others, how we can learn about others' mental states, and how we come to represent social relationships and social distance.

Our days are often spent navigating a complex and dynamic social environment in pursuit of various goals. For example, conducting simple transactions (e.g., buying a meal) often leads to interactions with complete strangers. We typically interact with others on a daily basis who comprise multiple interleaved social networks (e.g., family, friends, professional colleagues). Even when we are ostensibly alone, we can still be immersed in a social world when consuming media through a book, television, or the Internet. Given the preponderance of our lives spent embedded in a social context, a key question is understanding how and what types of information we learn from the social environment.

Humans have strong motivations to approach resources, while avoiding harm for self and others, and reduce uncertainty about the world (Crockett, Kurth-Nelson, Siegel, Dayan, & Dolan, 2014; FeldmanHall & Chang, 2018). We are also intensely driven to form close relationships with others (Baumeister & Leary, 1995). These two overarching goals motivate much of social learning. We can accelerate reducing our uncertainty about the world by learning vicariously *from* others' experiences from both observation and direct communication. Similarly, we can also reduce our uncertainty about others by learning *about* their beliefs, motivations, preferences, and overall character—for example, how does a certain person think about the world? What types of experiences have shaped their beliefs and perspective? What type of moral character do they have and would they be a good colleague? The reduction of social uncertainty can facilitate subsequent social interactions and the development of close relationships. This chapter will review several aspects of social learning, such as how we learn: from and about others, what other people are thinking, and how people are connected to each other. Much of this work is based on basic learning concepts (e.g., Pavlovian, instrumental, goal-directed, habitual), such as forming associations between stimuli and updating beliefs based on feedback, and suggests reliance on neural circuits comprising the amygdala, dorsal and ventral striatum (DS, VS), anterior cingulate cortex (ACC), and ventromedial prefrontal cortex (vmPFC; Delgado, 2007; Haber & Knutson, 2010; O'Doherty, 2004; Phelps & LeDoux, 2005; Yin & Knowlton, 2006). However, rather than simple sensory or affective signals, this information is often gleaned through the lens of social cognition. Thus, much of the literature reviewed involves interactions between neural systems supporting learning, affect, and social reasoning.

Learning from Others

We are motivated at once to both maximize our self-interest and minimize our uncertainty about the world. This requires us to frequently switch between exploiting what we know and exploring the unknown (Cohen, McClure, & Yu, 2007). Social learning offers the advantage of minimizing our uncertainty about the world based on others' experiences without incurring our own costs from exploring. This type of fictive learning (Lohrenz, McCabe, Camerer, & Montague, 2007) could be based on simply observing the outcomes of others' actions (i.e., observational learning). Alternatively, it can be learned from directly communicating these experiences, such as being explicitly told which is the best option.

Observational learning Observing the outcomes of others while minimizing our own costs is vital for survival from the earliest stages of life. This extension of Pavlovian learning can provide key insight into the nature of threats in the environment and how to avoid them, thereby ensuring survival (reviewed in Olsson & Phelps, 2007). The observational learning of stimuli paired with aversive outcomes results in equivalent learning as direct experience. Observationally learned cues are associated with increased physiological arousal and increased activation of the amygdala, anterior cingulate cortex (ACC), and insula (Olsson, Nearing, & Phelps, 2007). Rodent work has demonstrated that neurons projecting from the ACC, the basolateral nucleus of the amygdala (BLA), preferentially fire to cues learned via observing a conspecific undergo fear conditioning, while BLA neurons demonstrate reduced responding to such cues when ACC projections are inhibited (Allsop et al., 2018). Single-cell recordings in epilepsy patients also implicate rostral ACC neurons in the encoding of computational signals of observation, in contrast to amygdala and medial prefrontal cortex (mPFC) neurons, which show stronger involvement during firsthand experience of outcomes (Boorman, Fried, & Hill, 2016). Importantly, the extinction of a learned fear association can transmit vicariously across individuals (Golkar, Selbing, Flygare, Ohman, & Olsson, 2013), suggesting that this method of gleaning information from others aids in reducing uncertainty and avoiding harm.

Observational learning can also help us maximize gain and approach resources. For example, observing a person perform a given task can serve as an anchor (i.e., prior) that we can use to maximize our own performance based on subsequent experience. Similarly, we can make predictions about whether success will come to others and adjust our expectations after observing their outcomes. Such observational prediction error signals (i.e., expected observed outcomes) have been captured in the vmPFC, VS (Burke, Tobler, Baddeley, & Schultz, 2010), and DS (Cooper, Dunne, Furey, & O'Doherty, 2012), regions implicated in functional magnetic resonance imaging (fMRI) studies of associative and instrumental learning (Garrison, Erdeniz, & Done, 2013), as well as the intraparietal sulcus and dorsomedial prefrontal cortex (dmPFC; Dunne, D'Souza, & O'Doherty, 2016). Action prediction errors (i.e., of what others will *do*) are more associated with lateral PFC (Burke et al., 2010). Taken together, observational learning is a powerful social mechanism—through which we learn about the environment while reducing exposure to possible harm—that relies heavily on neural circuits supporting learning from direct experiences.

Social nudges Efforts to reduce uncertainty in the social world are often complicated by considerations of risk. In such situations we may look to others as a guide for whether to be risky or more prudent. Hearing from a friend or colleague who just invested in a stable rather than a more volatile stock may sway or nudge our own investments, with positive or negative consequences. Indeed, participants become risk averse when others are risk averse and become more risk seeking when others are risk seeking (Chung, Christopoulos, King-Casas, Ball, & Chiu, 2015), suggesting a *utility* placed on others' behavior that tracks with changes in vmPFC activity. This pattern of "contagion" is driven by a change in one's own risk attitudes (Suzuki, Jensen, Bossaerts, & O'Doherty, 2016). Relatedly, the vmPFC also appears to track others' *confidence* about their choice, which can influence our own decisions to pursue risk and uncertainty (Campbell-Meiklejohn, Simonsen, Frith, & Daw, 2017). These findings suggest that the overall value of these social and nonsocial signals appears to be integrated in the vmPFC and guides learning in uncertain environments (Behrens, Hunt, Woolrich, & Rushworth, 2008). Social nudges can also arise from evaluative feedback from peers, which is particularly important to consider given the dramatic rise in engagement with social media (Rodman, Powers, & Somerville, 2017). For example, even the mere presence of a peer can have an impact on reward-related neural activation (Fareri, Niznikiewicz, Lee, & Delgado, 2012), influence decisions to take risks (Chein, Albert, O'Brien, Uckert, & Steinberg, 2011), and lead to prosocial decision-making (Izuma, Saito, & Sadato, 2010), in possible anticipation of social approval. In sum, taking cues from others can significantly influence day-to-day decisions, particularly with respect to reducing uncertainty and validating our own choices.

Instructed learning A more explicit way of reducing uncertainty comes through directly receiving rules about environmental contingencies from another person. Learning via instruction is a more top-down and rapid process that can impact the goals of reducing uncertainty and maximizing one's best interest. For example, being provided (incorrect) instructed information about which of two stimuli will most likely lead to a reward will bias choice toward ostensibly more rewarding options, which hold even in the face of inconsistent feedback (i.e., punishment). Thus, explicit instruction may inhibit the appropriate updating of one's expectations (Doll, Jacobs, Sanfey, & Frank, 2009), consistent with prefrontal regulation of instrumental striatal learning processes (Li, Delgado, & Phelps, 2011). Instructions can also impact our ability to learn to avoid harm via corticostriatal circuitry during reversal learning (Atlas, Doll, Li, Daw, & Phelps, 2016). Interestingly, instructions from others concerning the *reliability* of upcoming feedback may moderate these biased processes (Schiffer, Siletti, Waszak, & Yeung, 2017).

Learning about Others

In addition to reducing uncertainty about the world, we are also motivated to build relationships and forge connections with others. This requires building a model of a person that can predict their behavior across a range of contexts (e.g., how good or trustworthy is this person?). We can then update this model based on simple information about a person's social relations and group membership through direct interactions or vicariously through another person's experience. More sophisticated models might incorporate information about an agent's personality, preferences, or how the agent thinks about the world—that is, the agent's beliefs, desires, and intentions (Baker, Jara-Ettinger, Saxe, & Tenenbaum, 2017).

Trait learning and impression updating We often form simple models of others by trying to infer their traits. Upon meeting someone novel, we might make implicit judgments about their level of trustworthiness or approachability based on facial characteristics (Todorov, Baron, & Oosterhof, 2008), assumed knowledge of their affiliations with a particular social group (Stanley, Sokol-Hessner, Banaji, & Phelps, 2011), or their beliefs about the world (i.e., stereotypes; Freeman & Johnson, 2016). These snap judgments contribute to the initial models we construct about others based on social approach and avoidance motives (Willis & Todorov, 2006). Forming first impressions implicates the amygdala (Engell, Haxby, & Todorov, 2007) and posterior cingulate cortex (PCC) in representing valenced social information, as well as the dmPFC in representing more general information about a person (Schiller, Freeman, Mitchell, Uleman, & Phelps, 2009).

Navigating our social landscapes requires constantly updating our initial models of others. We can do this readily when we acquire new information about a person that is perceived to occur with high statistical frequency in the social environment (i.e., more people tend to act trustworthy than not; Mende-Siedlecki, Baron, & Todorov, 2013). The dmPFC, PCC, and superior temporal sulcus (STS), all regions supporting social cognition (Stanley & Adolphs, 2013), are especially important for tracking inconsistencies in diagnostic social information about a target (Mende-Siedlecki, Cai, & Todorov, 2012). Further, positive changes in impressions (based on information about competence) may be mediated by increasing activation in lateral PFC, while negative changes in impressions of competence tend to recruit activation in mPFC, the striatum, and the STS (Bhanji & Beer, 2013).

Social interactions and reputation First impressions serve as a baseline expectation of other individuals that inform the likelihood of future successful interactions with them. Violations of social expectations (e.g., thinking we will be liked, only to find out we are not) tend to recruit regions involved in processing cognitive conflict and error monitoring, such as the dorsal ACC, whereas the ventral ACC discriminates between the valence of social outcomes agnostic to initial expectations (Cooper, Dunne, Furey, & O'Doherty, 2014; Somerville, Heatherton, & Kelley, 2006). The encoding of such signals in the ACC, VS, and mPFC provide neural mechanisms through which we can learn about social targets likely to provide opportunities for social inclusion and affiliation during repeated interactions (Jones et al., 2011).

Repeated interactions with a partner enable learning about *reputation*, which facilitates the development of relationships (Fareri & Delgado, 2014b). Trust underscores learning about one's reputation and can be operationalized as the expectation that someone will reciprocate generosity in situations involving mutual, interdependent risk (Simpson, 2007). Reciprocity serves as a valued social commodity that is consistently represented in corticostriatal reward systems (Bellucci, Chernyak, Goodyear, Eickhoff, & Krueger, 2016; Phan, Sripada, Angstadt, & McCabe, 2010). Experienced reciprocity during repeated interactions with a partner significantly predicts whether we should continue to collaborate with someone, as peak blood oxygen level-dependent (BOLD) activation in the caudate nucleus exhibits a temporal shift from the time at which a partner's choice to reciprocate is revealed to an anticipatory peak prior to the revelation of a partner's response (King-Casas et al., 2005). This pattern of striatal activation is consistent with temporal difference learning models that have been reported in midbrain dopaminergic neurons of nonhuman primates (Hollerman & Schultz, 1998), suggesting a social reward prediction error that can aid in updating social expectations/reputation. Expectations of reciprocity are susceptible to outside influence (i.e., prior instructed information about a partner's moral character): people tend to trust those of positive moral character over those of negative moral character, even when faced with information inconsistent with said priors (Delgado, Frank, & Phelps, 2005). This phenomenon may be driven by the interference of instructed social priors with striatal learning mechanisms to appropriately update social expectations.

Computational mechanisms of impression updating Updating social impressions is thus a dynamic process requiring a comparison of initial expectations/impressions and current experiences (Chang, Doll, van 't Wout, Frank, & Sanfey, 2010), and recent years have seen a steady increase in the incorporation of computational approaches to learning about others. Reinforcement-learning (RL) approaches (Dayan & Daw, 2008; Sutton & Barto, 1998), for example, offer opportunities to apply additional precision to social neuroscientific questions via the mathematical formalization of specific hypotheses regarding social behavior (Cheong, Jolly, Sul, & Chang, 2017). The recent application of RL models to learning about others has delineated neurocomputational mechanisms supporting trait versus reward learning. When faced with the task of choosing between social targets that could share some portion of an endowment, participants appear to use information about outcomes (i.e., amount shared) and generosity (i.e., what was the total amount *available* to be shared by someone) to inform choice and learning (Hackel, Doll, & Amodio, 2015). This study also reported overlapping activation in the VS for learning signals associated with both reward and generosity, consistent with extant research (Garrison, Erdeniz, & Done, 2013), but generosity also recruited a network of putative social regions (PCC, precuneus and right temporoparietal junction [rTPJ]). A related study found that learning about an individual's traits could be described using the same Bayesian model as learning about monetary reward, but the neurocomputational signals supporting social learning are encoded almost exclusively in putative social regions (i.e., precuneus; Stanley, 2016).

RL approaches have also been applied to studies examining trust and reputation learning. Models assuming that trust is a dynamic process posit that initial impressions shape the manner in which new information is incorporated into belief updating about another individual (Chang et al., 2010). Indeed, if initial impressions are strong enough, they can influence how much we subsequently value and use reciprocity/defection to learn about a partner. When priors acquired through direct social experience exist about another person, individuals show higher learning rates for outcomes that are *consistent* with initial impressions than for outcomes that are inconsistent, demonstrating that prior expectations computationally influence impression updating (Fareri, Chang, & Delgado, 2012). Strong instructional priors also modulate the neurocomputational mechanisms of social learning. During violations of trust, connectivity between the striatum and ventrolateral prefrontal regions is enhanced when priors are present, suggesting inhibitory functional interactions that prevent successful impression updating (Fouragnan et al., 2013).

Learning about mental representations Inherent in our ability to use social outcomes to build a model of another's reputation is the idea that we also need to be able to understand what types of goals motivate their behavior (Baker et al., 2017). Being able to represent something about others' mental states and affective experiences (Spunt & Adolphs, 2017)—cornerstones of social cognition—is key to social learning across development, with the dmPFC supporting such computations (Sul, Guroglu, Crone, & Chang, 2017). Multivariate analyses reveal that neural networks that support mentalizing represent information about others' mental states along three key dimensions—rationality (dmPFC, anterior temporal lobe), social impact or relevance (TPJ, precuneus, rostral ACC, dorsal ACC [dACC]), and valence (TPJ, dlPFC, inferior frontal gyrus/insula; Tamir, Thornton, Contreras, & Mitchell, 2016). These dimensions of mental state representation are critically involved in the ability to predict the manner in which individuals will transition between similar/different emotional states, something that overall we tend to be able to predict with high degrees of accuracy (Thornton & Tamir, 2017). In addition, modeling others' mental states requires reasoning about how others will interpret and respond to our actions. Complex computational strategies instantiated in the mPFC and STS (and supported by interactions with the VS) indeed track both another's (e.g., teacher) actions on a trial-by-trial basis and estimations of how one's own behavior will influence the future actions of another (Hampton, Bossaerts, & O'Doherty, 2008). Further, learning about others' preferences for risky behavior (Suzuki et al., 2016) to inform our own choices relies on Bayesian mechanisms and mentalizing circuitry (e.g., dmPFC, dlPFC, inferior parietal lobule [iPL]), such that we use our own baseline preferences as a starting point from which to update beliefs about others.

Learning about social space Social interactions typically occur within rich environments with more than one person. Thus, we can derive important information about people by learning about their place within social space. Indeed, humans develop and immerse themselves in widely interconnected social networks comprised of close others, varying degrees of friends of friends, and other acquaintances. As such, this type of social learning provides information indirectly about traits and the value of others through understanding how people relate to each other within a network of individuals. For example, networks of individuals characterized by empathy tend to be those that involve closer, trusting relationships between individuals (Morelli, Ong, Makati, Jackson, & Zaki, 2017). Interestingly, social network complexity maps on to ventrolateral and medial amygdala functional connectivity (Bickart, Hollenbeck, Barrett, & Dickerson, 2012), and other findings implicate mPFC in distinguishing representations of self and others as a function of similarity and closeness (Krienen, Tu, & Buckner, 2010; Mitchell, Macrae, & Banaji, 2006).

Other work indicates that both reward-related (VS) and social regions (mPFC) differentially integrate information about relationship closeness into value representations of in-network versus out-of-network social experiences (Fareri et al., 2012; Fareri & Delgado, 2014a). For example, collaborative interactions with close others are associated with computational signals of social reward value, represented in the VS and mPFC when experiencing reciprocity, that are contingent upon interpersonal aspects of a close relationship (Fareri, Chang, & Delgado, 2015). Relatedly, people are willing to forgo self-interest (i.e., higher monetary gain) in favor of more equitable splits with another person as a function of social closeness, a pattern that scales with activation in value-related (vmPFC) and social (rTPJ) brain regions (Strombach et al., 2015). Conversely, decisions to trust out-of-network members requires connectivity between regions implicated in cognitive control (i.e., dACC, lateral PFC) and the striatum, presumably to inhibit prepotent responses to distrust such individuals (Hughes, Ambady, & Zaki, 2016).

More recently, there has been growing interest in exploring how we learn the structure of social relationships. Judging social distance within a social network appears to recruit the same regions involved in judging spatial and temporal distance (Parkinson, Liu, & Wheatley, 2014), whereas judging the popularity of various members of a social network appears to recruit activation in reward circuitry (vmPFC, amygdala, VS) and social cognition networks (dmPFC, precuneus, left TPJ) (Zerubavel, Bearman, Weber, & Ochsner, 2015). Patterns within social cognition networks when viewing faces can also predict which members have the highest social value within a social network (i.e., sources of friendship, empathy, and support) (Morelli, Leong, Carlson, Kullar, & Zaki, 2018). Finally, there is intriguing recent evidence of neural homophily that suggests we may have more similar patterns of brain activity to our friends when viewing videos than to more distant others (i.e., friends of friends) (Parkinson, Kleinbaum, and Wheatley, 2018). Taken together, these findings suggest that shared preferences and interpretations of the world may help explain why we become closer to certain individuals than others.

Future Directions and Conclusions

Social learning serves to reduce uncertainty in the environment, maximize gains and avoid harm, and forge close relationships with others. The neural systems across many different types of social learning covered here rely heavily on interactions between corticostriatal circuitry and the cortical regions supporting social processing (Figure 83.1).

We note that the topics covered here are not exhaustive. For instance, social learning can occur via other means, such as through the adherence to and enforcement of social norms (Chang & Sanfey, 2013; Montague & Lohrenz, 2007; Xiang, Lohrenz, & Montague, 2013; Zaki, Schirmer, & Mitchell, 2011; Zhong, Chark, Hsu, & Chew, 2016) or the desire to avoid feelings of guilt for committing social transgressions (Chang, Smith, Dufwenberg, & Sanfey, 2011; Nihonsugi, Ihara, & Haruno, 2015).

With respect to future directions, one exciting path concerns more concrete models of observational learning—that is, does this type of learning occur simply via the simple imitation of an agent or rather through using our observations of others to generate a model about environmental states (i.e., inverse reinforcement learning; Collette, Pauli, Bossaerts, & O'Doherty, 2017)? Better characterizing observational learning mechanisms can foster a deeper understanding of theory of mind processes and how they may break down in clinical samples (i.e., autism). Another interesting direction involves harnessing machine-learning algorithms to facilitate the prediction of psychological states (i.e., negative affect) based on decoding patterns of brain activation (Chang, Gianaros, Manuck, Krishnan, & Wager, 2015). Translating these types of predictive techniques to questions of social appraisals (i.e., reputation, bias) and social decisions (i.e., trust) has implications for understanding breakdowns in representations of others with interpersonal difficulties. Finally, developing stable, long-lasting relationships depends heavily upon the processes reviewed in this chapter. Learning about and from others facilitates the development and maintenance of close, trusting relationships, which supports our overall well-being (Uchino, 2009). Future work can take a more comprehensive approach to characterizing the dynamics of relationships and shared experiences across groups of individuals as they relate to processing, learning, and remembering social information in more naturalistic contexts (Chen et al., 2016) and how this subsequently influences mental health.

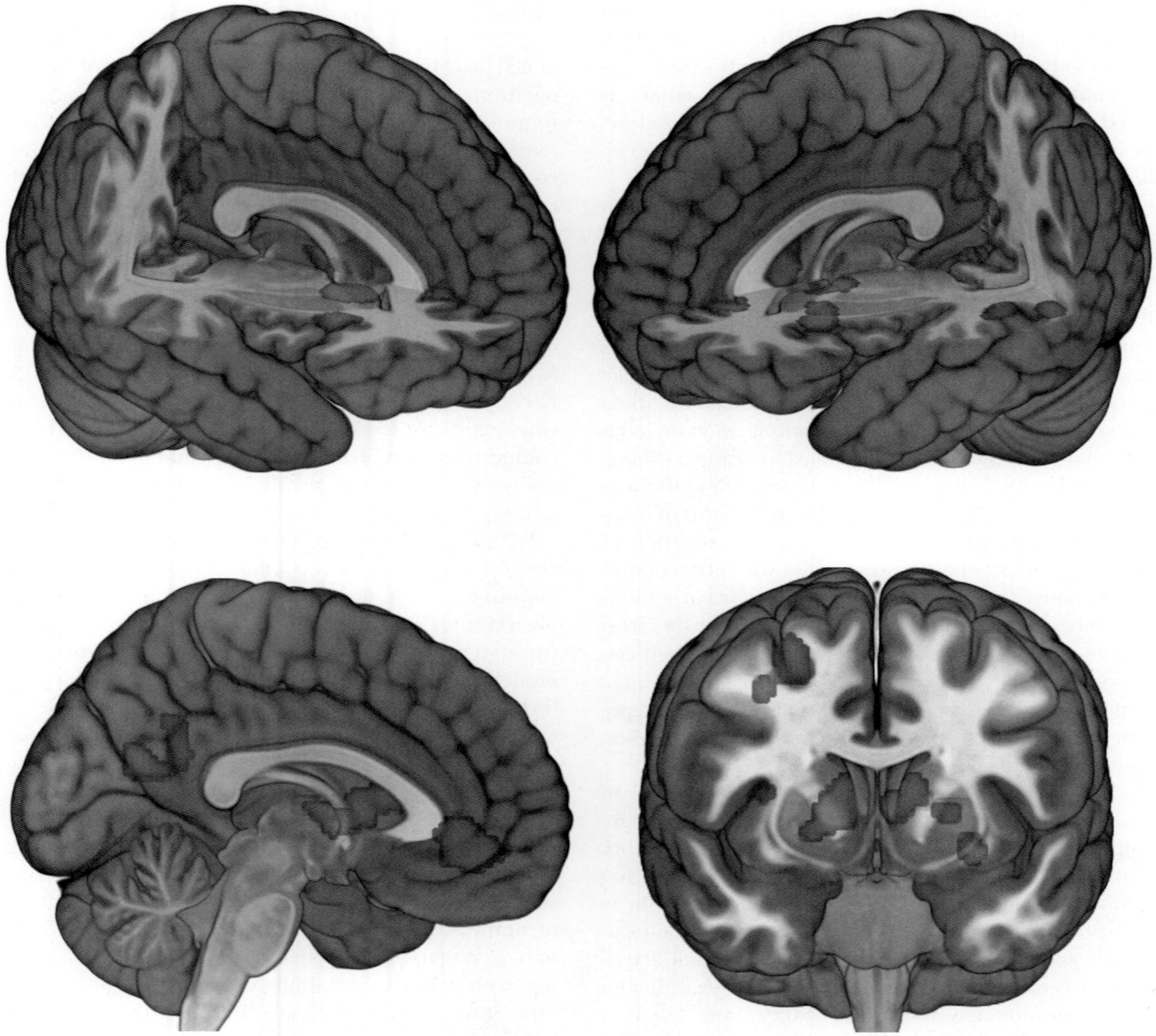

FIGURE 83.1 Activation-likelihood meta-analyses using GingerALE (Eickhoff et al., 2009) were conducted to generate illustrative maps of neural circuitry supporting "learning from" (*green*) and "learning about" (*red*) others. Maps were set to an initial height threshold of $p < .005$ and corrected at the cluster level to $p < .05$. Studies included in these meta-analyses are marked with an * (*learning from*) and a † (*learning about*) in the reference section. (See color plate 96.)

REFERENCES

Achterberg, M., van Duijvenvoorde, A. C. K., Bakermans-Kranenburg, M. J., & Crone, E. A. (2016).* Control your anger! The neural basis of aggression regulation in response to negative social feedback. *Social Cognitive and Affective Neuroscience, 11*(5), 712–720. http://doi.org/10.1093/scan/nsv154

Allsop, S. A., Wichmann, R., Mills, F., Burgos-Robles, A., Chang, C.-J., Felix-Ortiz, A. C., et al. (2018). Corticoamygdala transfer of socially derived information gates observational learning. *Cell, 173*(6), 1–33. http://doi.org/10.1016/j.cell.2018.04.004

Atlas, L. Y., Doll, B. B., Li, J., Daw, N. D., & Phelps, E. A. (2016).* Instructed knowledge shapes feedback-driven aversive learning in striatum and orbitofrontal cortex, but not the amygdala. *eLife, 5*, e15192. http://doi.org/10.7554/eLife.15192

Baker, C. L., Jara-Ettinger, J., Saxe, R., & Tenenbaum, J. B. (2017). Rational quantitative attribution of beliefs, desires and percepts in human mentalizing. *Nature Human Behavior, 1*, 1–10. http://doi.org/10.1038/s41562-017-0064

Baumeister, R., & Leary, M. (1995). The need to belong: Desire for interpersonal attachments as a fundamental human motivation. *Psychological Bulletin, 117*(3), 497–529.

Behrens, T. E. J., Hunt, L. T., Woolrich, M. W., & Rushworth, M. F. S. (2008).* Associative learning of social value. *Nature, 456*(7219), 245–249. http://doi.org/10.1038/nature07538

Bellucci, G., Chernyak, S. V., Goodyear, K., Eickhoff, S. B., & Krueger, F. (2016). Neural signatures of trust in reciprocity: A coordinate-based meta-analysis. *Human Brain Mapping, 38*(3), 1233–1248. http://doi.org/10.1002/hbm.23451

Bhanji, J. P., & Beer, J. S. (2013).† Dissociable neural modulation underlying lasting first impressions, changing your mind for the better, and changing it for the worse. *Journal of Neuroscience, 33*(22), 9337–9344. http://doi.org/10.1523/jneurosci.5634-12.2013

Bickart, K. C., Hollenbeck, M. C., Barrett, L. F., & Dickerson, B. C. (2012). Intrinsic amygdala-cortical functional connectivity predicts social network size in humans. *Journal of Neuroscience, 32*(42), 14729–14741. http://doi.org/10.1523/jneurosci.1599-12.2012

Boorman, E. D., Fried, I., & Hill, M. R. (2016). Observational learning computations in neurons of the human anterior cingulate cortex. *Nature Communications, 7*, 1–12. http://doi.org/10.1038/ncomms12722

Burke, C. J., Tobler, P. N., Baddeley, M., & Schultz, W. (2010).* Neural mechanisms of observational learning. *Proceedings of the National Academy of Sciences of the United States of America, 107*(32). http://doi.org/10.1073/pnas.1003111107

Cacioppo, S., & Cacioppo, J. T. (2019). Mechanisms of loneliness. In M. S. Gazzaniga, G. R. Mangun, & D. Poeppel (Eds.), *The cognitive neurosciences* (6th ed.). Cambridge, MA: MIT Press.

Campbell-Meiklejohn, D., Simonsen, A., Frith, C. D., & Daw, N. D. (2017).* Independent neural computation of value from other people's confidence. *Journal of Neuroscience, 37*(3), 673–684. http://doi.org/10.1523/jneurosci.4490-15.2016

Chang, L. J., Doll, B. B., van 't Wout, M., Frank, M. J., & Sanfey, A. G. (2010). Seeing is believing: Trustworthiness as a dynamic belief. *Cognitive Psychology, 61*(2), 87–105. http://doi.org/10.1016/j.cogpsych.2010.03.001

Chang, L. J., Gianaros, P. J., Manuck, S. B., Krishnan, A., & Wager, T. D. (2015). A sensitive and specific neural signature for picture-induced negative affect. *PLoS Biology, 13*(6), e1002180–28. http://doi.org/10.1371/journal.pbio.1002180

Chang, L. J., & Sanfey, A. G. (2013). Great expectations: Neural computations underlying the use of social norms in decision-making. *Social Cognitive and Affective Neuroscience, 8*(3), 277–284. http://doi.org/10.1093/scan/nsr094

Chang, L. J., Smith, A., Dufwenberg, M., & Sanfey, A. G. (2011). Triangulating the neural, psychological, and economic bases of guilt aversion. *Neuron, 70*(3), 560–572. http://doi.org/10.1016/j.neuron.2011.02.056

Chein, J., Albert, D., O'Brien, L., Uckert, K., & Steinberg, L. (2011). Peers increase adolescent risk taking by enhancing activity in the brain's reward circuitry. *Developmental Science, 14*(2), F1–10. http://doi.org/10.1111/j.1467-7687.2010.01035.x

Chen, J., Leong, Y. C., Honey, C. J., Yong, C. H., Norman, K. A., & Hasson, U. (2016). Shared memories reveal shared structure in neural activity across individuals. *Nature Neuroscience, 20*(1), 115–125. http://doi.org/10.1038/nn.4450

Cheong, J. H., Jolly, E., Sul, S., & Chang, L. J. (2017). Computational models in social neuroscience. In A. A. Moustafa (Ed.), *Computational models of brain and behavior* (pp. 1–16). Hoboken, NJ: John Wiley & Sons.

Chung, D., Christopoulos, G. I., King-Casas, B., Ball, S. B., & Chiu, P. H. (2015).* Social signals of safety and risk confer utility and have asymmetric effects on observers' choices. *Nature Neuroscience, 18*(6), 912–916. http://doi.org/10.1038/nn.4022

Cohen, J. D., McClure, S. M., & Yu, A. J. (2007). Should I stay or should I go? How the human brain manages the trade-off between exploitation and exploration. *Philosophical Transactions of the Royal Society B: Biological Sciences, 362*(1481), 933–942. http://doi.org/10.1098/rstb.2007.2098

Collette, S., Pauli, W. M., Bossaerts, P., & O'Doherty, J. (2017). Neural computations underlying inverse reinforcement learning in the human brain. *eLife, 6*. http://doi.org/10.7554/eLife.29718

Cooper, J. C., Dunne, S., Furey, T., & O'Doherty, J. P. (2012).* Human dorsal striatum encodes prediction errors during observational learning of instrumental actions. *Journal of Cognitive Neuroscience, 24*(1), 106–118. http://doi.org/10.1162/jocn_a_00114

Cooper, J. C., Dunne, S., Furey, T., & O'Doherty, J. P. (2014). The role of the posterior temporal and medial prefrontal cortices in mediating learning from romantic interest and rejection. *Cerebral Cortex, 24*(9), 2502–2511. http://doi.org/10.1093/cercor/bht102

Crockett, M. J., Kurth-Nelson, Z., Siegel, J. Z., Dayan, P., & Dolan, R. J. (2014). Harm to others outweighs harm to self in moral decision making. *Proceedings of the National Academy of Sciences of the United States of America, 111*(48), 17320–17325. http://doi.org/10.1073/pnas.1408988111

Dayan, P., & Daw, N. D. (2008). Decision theory, reinforcement learning, and the brain. *Cognitive, Affective, & Behavioral Neuroscience, 8*(4), 429–453. http://doi.org/10.3758/CABN.8.4.429

Delgado, M. R. (2007). Reward-related responses in the human striatum. *Annals of the New York Academy of Sciences, 1104*, 70–88. http://doi.org/10.1196/annals.1390.002

Delgado, M. R., Frank, R., & Phelps, E. A. (2005).† Perceptions of moral character modulate the neural systems of reward during the trust game. *Nature Neuroscience, 8*, 1611–1618. http://doi.org/doi:10.1038/nn1575

Dewall, C. N., Deckman, T., Gailliot, M. T., & Bushman, B. J. (2011). Sweetened blood cools hot tempers: Physiological self-control and aggression. *Aggressive Behavior, 37*(1), 73–80. http://doi.org/10.1002/ab.20366

Doll, B. B., Jacobs, W. J., Sanfey, A. G., & Frank, M. J. (2009). Instructional control of reinforcement learning: A behavioral and neurocomputational investigation. *Brain Research, 1299*, 1–21. http://doi.org/10.1016/j.brainres.2009.07.007

Dunne, S., D'Souza, A., & O'Doherty, J. P. (2016).* The involvement of model-based but not model-free learning signals during observational reward learning in the absence of choice. *Journal of Neurophysiology, 115*(6), 3195–3203. http://doi.org/10.1152/jn.00046.2016

Eickhoff, S. B., Laird, A. R., Grefkes, C., Wang, L. E., Zilles, K., & Fox, P. T. (2009). Coordinate-based activation likelihood estimation meta-analysis of neuroimaging data: A random-effects approach based on empirical estimates of spatial uncertainty. *Human Brain Mapping, 30*(9), 2907–2926. http://doi.org/10.1002/hbm.20718

Engell, A. D., Haxby, J. V., & Todorov, A. (2007).† Implicit trustworthiness decisions: Automatic coding of face properties in the human amygdala. *Journal of Cognitive Neuroscience, 19*(9), 1508–1519. http://doi.org/10.1162/jocn.2007.19.9.1508

Fareri, D. S., Chang, L. J., & Delgado, M. R. (2012).† Effects of direct social experience on trust decisions and neural reward circuitry. *Frontiers in Neuroscience, 6,* 148–217. http://doi.org/10.3389/fnins.2012.00148

Fareri, D. S., Chang, L. J., & Delgado, M. R. (2015).† Computational substrates of social value in interpersonal collaboration. *Journal of Neuroscience, 35*(21), 8170–8180. http://doi.org/10.1523/jneurosci.4775-14.2015

Fareri, D. S., & Delgado, M. R. (2014a). Differential reward responses during competition against in- and out-of-network others. *Social Cognitive and Affective Neuroscience, 9*(4), 412–420. http://doi.org/10.1093/scan/nst006

Fareri, D. S., & Delgado, M. R. (2014b). The importance of social rewards and social networks in the human brain. *Neuroscientist, 20*(4). http://doi.org/10.1177/1073858414521869

Fareri, D. S., Niznikiewicz, M. A., Lee, V. K., & Delgado, M. R. (2012). Social network modulation of reward-related signals. *Journal of Neuroscience, 32*(26), 9045–9052. http://doi.org/10.1523/jneurosci.0610-12.2012

FeldmanHall, O., & Chang, L. J. (2018). Social learning: Emotions aid in optimizing goal-directed social behavior. In A. M. Bornstein & A. Shenhav (Eds.), *Understanding goal-directed decision-making computations and circuits.* Amsterdam: Elsevier.

Fouragnan, E., Chierchia, G., Greiner, S., Neveu, R., Avesani, P., & Coricelli, G. (2013).† Reputational priors magnify striatal responses to violations of trust. *Journal of Neuroscience, 33*(8), 3602–3611. http://doi.org/10.1523/jneurosci.3086-12.2013

Freeman, J. B., & Johnson, K. L. (2016). More than meets the eye: Split-second social perception. *Trends in Cognitive Sciences, 20*(5), 362–374. http://doi.org/10.1016/j.tics.2016.03.003

Garrison, J., Erdeniz, B., & Done, J. (2013). Prediction error in reinforcement learning: A meta-analysis of neuroimaging studies. *Neuroscience & Biobehavioral Reviews, 37*(7), 1297–1310. http://doi.org/10.1016/j.neubiorev.2013.03.023

Golkar, A., Selbing, I., Flygare, O., Ohman, A., & Olsson, A. (2013). Other people as means to a safe end. *Psychological Science, 24*(11), 2182–2190. http://doi.org/10.1177/0956797613489890

Haber, S. N., & Knutson, B. (2010). The reward circuit: Linking primate anatomy and human imaging. *Neuropsychopharmacology, 35*(1), 4–26. http://doi.org/10.1038/npp.2009.129

Hackel, L. M., Doll, B. B., & Amodio, D. M. (2015).† Instrumental learning of traits versus rewards: Dissociable neural correlates and effects on choice. *Nature Neuroscience, 18*(9), 1233–1235. http://doi.org/10.1038/nn.4080

Hampton, A. N., Bossaerts, P., & O'Doherty, J. P. (2008).† Neural correlates of mentalizing-related computations during strategic interactions in humans. *Proceedings of the National Academy of Sciences of the United States of America, 105*(18), 6741–6746. http://doi.org/10.1073/pnas.0711099105

Hollerman, J., & Schultz, W. (1998). Dopamine neurons report an error in the temporal prediction of reward during learning. *Nature Neuroscience, 1*(4), 304–309.

Hornstein, E. A., Inagaki, T. K., & Eisenberger, N. I. (2019). More than just friends: An exploration of the neurobiological mechanisms underlying the link between social support and health. In M. S. Gazzaniga, G. R. Mangun, & D. Poeppel (Eds.), *The cognitive neurosciences* (6th ed.). Cambridge, MA: MIT Press.

Hughes, B. L., Ambady, N., & Zaki, J. (2016).† Trusting outgroup, but not ingroup members, requires control: Neural and behavioral evidence. *Social Cognitive and Affective Neuroscience, 12*(3), 372–381. http://doi.org/10.1093/scan/nsw139

Izuma, K., Saito, D. N., & Sadato, N. (2010). Processing of the incentive for social approval in the ventral striatum during charitable donation. *Journal of Cognitive Neuroscience, 22*(4), 621–631. http://doi.org/10.1162/jocn.2009.21228

Jones, R. M., Somerville, L. H., Li, J., Ruberry, E. J., Libby, V., Glover, G., et al. (2011).† Behavioral and neural properties of social reinforcement learning. *Journal of Neuroscience, 31*(37), 13039–13045. http://doi.org/10.1523/jneurosci.2972-11.2011

Kahneman, D., & Tversky, A. (1984). Choices, values, and frames. *American Psychologist, 39*(4), 341–350. http://doi.org/10.1037/0003-066X.39.4.341

King-Casas, B., Tomlin, D., Anen, C., Camerer, C., Quartz, S., & Montague, P. (2005).† Getting to know you: Reputation and trust in a two-person economic exchange. *Science, 308*(5718), 78–83. doi:10.1126/science.1108062

Koban, L., Schneider, R., Ashar, Y. K., Andrews-Hanna, J. R., Landy, L., Moscovitch, D. A., et al. (2017). Social anxiety is characterized by biased learning about performance and the self. *Emotion, 17*(8), 1144–1155. http://doi.org/10.1037/emo0000296

Kriegeskorte, N. (2008). Representational similarity analysis—connecting the branches of systems neuroscience. *Frontiers in Systems Neuroscience, 2,* 1–28. http://doi.org/10.3389/neuro.06.004.2008

Krienen, F. M., Tu, P. C., & Buckner, R. L. (2010). Clan mentality: Evidence that the medial prefrontal cortex responds to close others. *Journal of Neuroscience, 30*(41), 13906–13915. http://doi.org/10.1523/jneurosci.2180-10.2010

Li, J., Delgado, M. R., & Phelps, E. A. (2011).* How instructed knowledge modulates the neural systems of reward learning. *Proceedings of the National Academy of Sciences of the United States of America, 108*(1), 1–6. http://doi.org/10.1073/pnas.1014938108

Lohrenz, T., McCabe, K., Camerer, C. F., & Montague, P. R. (2007). Neural signature of fictive learning signals in a sequential investment task. *Proceedings of the National Academy of Sciences of the United States of America, 104*(22), 9493–9498. http://doi.org/10.1073/pnas.0608842104

Mende-Siedlecki, P., Baron, S. G., & Todorov, A. (2013).† Diagnostic value underlies asymmetric updating of impressions in the morality and ability domains. *Journal of Neuroscience, 33*(50), 19406–19415. http://doi.org/10.1523/jneurosci.2334-13.2013

Mende-Siedlecki, P., Cai, Y., & Todorov, A. (2012).† The neural dynamics of updating person impressions. *Social Cognitive and Affective Neuroscience, 8*(6). http://doi.org/10.1093/scan/nss040

Mitchell, J. P., Macrae, C. N., & Banaji, M. R. (2006). Dissociable medial prefrontal contributions to judgments of similar and dissimilar others. *Neuron, 50*(4), 655–663. http://doi.org/10.1016/j.neuron.2006.03.040

Montague, P. R., & Lohrenz, T. (2007). To detect and correct: Norm violations and their enforcement. *Neuron, 56*(1), 14–18.

Morelli, S. A., Leong, Y. C., Carlson, R. W., Kullar, M., & Zaki, J. (2018). Neural detection of socially valued community members. *Proceedings of the National Academy of Sciences, 115*(32), 8149–8154.

Morelli, S. A., Ong, D. C., Makati, R., Jackson, M. O., & Zaki, J. (2017). Empathy and well-being correlate with centrality in different social networks. *Proceedings of the National Academy of Sciences of the United States of America, 114*(37), 9843–9847. http://doi.org/10.1073/pnas.1702155114

Nihonsugi, T., Ihara, A., & Haruno, M. (2015). Selective increase of intention-based economic decisions by noninvasive brain stimulation to the dorsolateral prefrontal cortex. *Journal of Neuroscience, 35*(8), 3412–3419.

O'Doherty, J. P. (2004). Reward representations and reward-related learning in the human brain: Insights from neuroimaging. *Current Opinion in Neurobiology, 14*(6), 769–776. http://doi.org/10.1016/j.conb.2004.10.016

Olsson, A., Nearing, K. I., & Phelps, E. A. (2007).* Learning fears by observing others: The neural systems of social fear transmission. *Social Cognitive and Affective Neuroscience, 2*(1), 3–11. http://doi.org/10.1093/scan/nsm005

Olsson, A., & Phelps, E. A. (2007). Social learning of fear. *Nature Neuroscience, 10*(9), 1095–1102. http://doi.org/10.1038/nn1968

Parkinson, C., Kleinbaum, A. M., & Wheatley, T. (2018). Similar neural responses predict friendship. *Nature Communications, 9*(1), 332.

Parkinson, C., Liu, S., & Wheatley, T. (2014).† A common cortical metric for spatial, temporal, and social distance. *Journal of Neuroscience, 34*(5), 1979–1987. http://doi.org/10.1523/jneurosci.2159-13.2014

Phan, K. L., Sripada, C. S., Angstadt, M., & McCabe, K. (2010).† Reputation for reciprocity engages the brain reward center. *Proceedings of the National Academy of Sciences of the United States of America, 107*(29), 13099–13104. http://doi.org/10.1073/pnas.1008137107

Phelps, E. A., & LeDoux, J. E. (2005). Contributions of the amygdala to emotion processing: From animal models to human behavior. *Neuron, 48*(2), 175–187. http://doi.org/10.1016/j.neuron.2005.09.025

Rodman, A. M., Powers, K. E., & Somerville, L. H. (2017). Development of self-protective biases in response to social evaluative feedback. *Proceedings of the National Academy of Sciences of the United States of America, 114*(50), 13158–13163. http://doi.org/10.1073/pnas.1712398114

Schiffer, A.-M., Siletti, K., Waszak, F., & Yeung, N. (2017). Adaptive behaviour and feedback processing integrate experience and instruction in reinforcement learning. *NeuroImage, 146*, 626–641. http://doi.org/10.1016/j.neuroimage.2016.08.057

Schiller, D., Freeman, J. B., Mitchell, J. P., Uleman, J. S., & Phelps, E. A. (2009).† A neural mechanism of first impressions. *Nature Neuroscience, 12*(4), 508–514. http://doi.org/10.1038/nn.2278

Simpson, J. A. (2007). Psychological foundations of trust. *Current Directions in Psychological Science, 16*(5), 264–268.

Sip, K. E., Smith, D. V., Porcelli, A. J., Kar, K., & Delgado, M. R. (2015). Social closeness and feedback modulate susceptibility to the framing effect. *Social Neuroscience, 10*(1), 35–45. http://doi.org/10.1080/17470919.2014.944316

Somerville, L. H., Heatherton, T. F., & Kelley, W. M. (2006).† Anterior cingulate cortex responds differentially to expectancy violation and social rejection. *Nature Neuroscience, 9*(8), 1007–1008. http://doi.org/10.1038/nn1728

Spunt, R. P., & Adolphs, R. (2017). The neuroscience of understanding the emotions of others. *Neuroscience Letters, 693*, 1–5. http://doi.org/10.1016/j.neulet.2017.06.018

Stanley, D. A. (2016).† Getting to know you: General and specific neural computations for learning about people. *Social Cognitive and Affective Neuroscience, 11*(4), 525–536. http://doi.org/10.1093/scan/nsv145

Stanley, D. A., & Adolphs, R. (2013). Toward a neural basis for social behavior. *Neuron, 80*(3), 816–826. http://doi.org/10.1016/j.neuron.2013.10.038

Stanley, D. A., Sokol-Hessner, P., Banaji, M. R., & Phelps, E. A. (2011). Implicit race attitudes predict trustworthiness judgments and economic trust decisions. *Proceedings of the National Academy of Sciences of the United States of America, 108*(19), 7710–7715. http://doi.org/10.1073/pnas.1014345108

Strombach, T., Weber, B., Hangebrauk, Z., Kenning, P., Karipidis, I. I., Tobler, P. N., & Kalenscher, T. (2015).† Social discounting involves modulation of neural value signals by temporoparietal junction. *Proceedings of the National Academy of Sciences of the United States of America, 112*(5), 201414715. http://doi.org/10.1073/pnas.1414715112

Sul, S., Lu, Güroğlu, B., Crone, E. A., & Chang, L. J. (2017). Medial prefrontal cortical thinning mediates shifts in other-regarding preferences during adolescence. *Scientific Reports, 7*(8510), 1–10. http://doi.org/10.1038/s41598-017-08692-6

Sutton, R., & Barto, A. (1998). *Reinforcement learning*. Cambridge, MA: MIT Press.

Suzuki, S., Jensen, E. L. S., Bossaerts, P., & O'Doherty, J. P. (2016).* Behavioral contagion during learning about another agent's risk-preferences acts on the neural representation of decision-risk. *Proceedings of the National Academy of Sciences of the United States of America, 113*(14), 3755–3760. http://doi.org/10.1073/pnas.1600092113

Tamir, D. I., Thornton, M. A., Contreras, J. M., & Mitchell, J. P. (2016).† Neural evidence that three dimensions organize mental state representation: Rationality, social impact, and valence. *Proceedings of the National Academy of Sciences of the United States of America, 113*(1), 194–199. http://doi.org/10.1073/pnas.1511905112

Thornton, M. A., & Tamir, D. I. (2017). Mental models accurately predict emotion transitions. *Proceedings of the National Academy of Sciences of the United States of America, 114*(23), 5982–5987. http://doi.org/10.1073/pnas.1616056114

Todorov, A., Baron, S. G., & Oosterhof, N. N. (2008).† Evaluating face trustworthiness: A model based approach. *Social Cognitive and Affective Neuroscience, 3*(2), 119–127. http://doi.org/10.1093/scan/nsn009

Uchino, B. N. (2009). Understanding the links between social support and physical health. *Perspectives on Psychological Science, 4*(3), 236–255.

Willis, J., & Todorov, A. (2006). First impressions making up your mind after a 100-ms exposure to a face. *Psychological Science, 17*(7), 592–598. http://doi.org/10.1111/j.1467-9280.2006.01750.x

Xiang, T., Lohrenz, T., & Montague, P. R. (2013). Computational substrates of norms and their violations during social exchange. *Journal of Neuroscience, 33*(3), 1099–1108.

Yin, H. H., & Knowlton, B. J. (2006). The role of the basal ganglia in habit formation. *Nature Reviews Neuroscience, 7*(6), 464–476. http://doi.org/10.1038/nrn1919

Zaki, J., Schirmer, J., & Mitchell, J. P. (2011). Social influence modulates the neural computation of value. *Psychological Science, 22*(7), 894–900.

Zerubavel, N., Bearman, P. S., Weber, J., & Ochsner, K. N. (2015).[†] Neural mechanisms tracking popularity in real-world social networks. *Proceedings of the National Academy of Sciences, 112*(49), 15072–15077. http://doi.org/10.1073/pnas.1511477112

Zhong, S., Chark, R., Hsu, M., & Chew, S. H. (2016). Computational substrates of social norm enforcement by unaffected third parties. *NeuroImage, 129*(C), 95–104. http://doi.org/10.1016/j.neuroimage.2016.01.040

84 Social Learning of Threat and Safety

ANDREAS OLSSON, PHILIP PÄRNAMETS, ERIK C. NOOK, AND BJÖRN LINDSTRÖM

ABSTRACT In rapidly changing environments, humans and other animals often glean information about the value of objects and behaviors through social learning. For example, in humans, observing others' behaviors and their consequences enables the transmission of a wide range of information about what stimuli should be avoided and approached, and what behaviors are useful to that end. We survey important developments in our understanding of the behavioral, computational, and neural aspects of social learning of threat and safety. In particular, we discuss the study of social learning through observation, which has enabled comparisons across species. This research shows that observational threat and safety learning draw on mechanisms partially shared with direct (Pavlovian) threat conditioning and extinction learning. Importantly, however, social and asocial learning also differ from each other, for instance in the role that empathic processes play in observational but not asocial learning. We conclude by underscoring the importance of studying social learning across species using behavioral, computational, and neural measures.

To cope with challenges and opportunities in their environment, animals need to learn the value of stimuli and contexts. In our hypersocial species, such information is rapidly propagated between individuals. For example, in the wake of a natural disaster you might have received alerts and warnings from friends on social media, hopefully followed by status postings indicating that your friends are safe. Research has shown that such emotionally charged information, as well as information about value, whom to trust, and what food to prefer, rapidly spreads between individuals in social networks (Christakis & Fowler, 2009; Nook & Zaki, 2015). Regardless of whether these networks are real or virtual, vicariously experienced information can dramatically influence individuals' thoughts, feelings, and behaviors. Indeed, social learning between peers and across generations has such profound influence on behavior that it has been ascribed a key role in the evolution of culture (Boyd & Richerson, 2009). Social learning is, however, not always adaptive, and there is currently much evidence describing its role in the development of a range of psychological disorders (Askew & Field, 2007; Rachman, 1972). Importantly, because social learning can be studied across species in well-controlled experiments, it provides unique insights into the links between the brain, behavior and society across phylogenetic development. Yet most of what is known about the neural and computational properties of learning derives from the study of learning in a social vacuum (Debiec & Olsson, 2017). The chief reason for this relative lack of understanding social learning is the immense complexity added when multiple individuals are dynamically interconnected. Nonetheless, gaining greater insight into how we learn from others' experiences is a central scientific question relevant to researchers at both basic and applied levels.

In this chapter we will survey important developments in cognitive and affective neuroscience that examine the social learning of threat and safety. We begin by introducing core behavioral, computational, and neural mechanisms of learning, followed by a discussion of a social cognitive process with particular relevance to social learning: empathy. We will then discuss similarities and differences between social and nonsocial forms of learning before concluding by underscoring the importance of studying social learning across species and levels of analyses.

Learning through Direct Experiences

Neural correlates of threat learning Pavlovian conditioning is commonly used in the laboratory to study the neural systems and cellular mechanisms underlying emotional learning and memory. In a Pavlovian threat-conditioning paradigm, a neutral stimulus (conditioned stimulus, or CS) is paired with a directly experienced naturally aversive event (unconditioned stimulus, or US), endowing the CS with an ability to trigger a conditioned (threat) response (CR). This paradigm enables the study of the formation and maintenance of directly learned aversions. Direct threat conditioning critically involves the amygdala, a subcortical structure in the temporal lobe consisting of several interconnected parts with relevance to conditioning, in particular the basolateral part (BL, consisting of the lateral [LA], basal [BA], and basomedial [BM] nuclei) and the central (CeA) nucleus (Pitkänen, Savander, & LeDoux, 1997). Sensory information from the midbrain, thalamus, and cortex converges in the LA where CS-US associations are formed during learning (Rogan, Stäubli, & LeDoux, 1997). The LA and BA neurons also receive information from other brain structures involved in different aspects of threat learning, among them the hippocampus and prefrontal cortex (PFC), which provide contextual information (Maren, Aharonov, & Fanselow, 1997; Morgan, Romanski, & LeDoux, 1993). The LA projects to the CeA, which propagates information downstream to brain regions controlling behavioral, autonomic, and somatic defensive responses, such as freezing and the release of stress hormones (LeDoux, Iwata, Cicchetti, & Reis, 1988).

Direct threat conditioning in patients and healthy human participants using functional magnetic resonance imaging (fMRI) supports the notion that the basic neural and computational mechanisms for acquiring and expressing learned threat are conserved across species (LaBar, Gatenby, Gore, LeDoux, & Phelps, 1998; Phelps & LeDoux, 2005). As in other animals, the human amygdala is interconnected with cortical regions, including the hippocampus and ventromedial PFC necessary for encoding and retrieving contextual information and regulating conditioned threat responses. A major role of the hippocampal-PFC circuitry is thus to disambiguate the meaning of cues in a given context (Maren, Phan, & Liberzon, 2013), be it social or nonsocial. Other cortical regions linked to the amygdala, among them the anterior insula (AI) and the anterior cingulate cortex (ACC), have been implicated in the experiences of receiving, anticipating, and controlling painful or otherwise aversive experiences (Craig, 2009; Shackman et al., 2011). For example, a recent meta-analysis (Fullana et al., 2016) consistently implicated the ACC and AI during human threat learning, supporting the role of these regions in homeostatic autonomic and behavioral regulation. Interestingly, recent research has suggested that some of these regions also serve crucial functions during social learning (Apps, Rushworth, & Chang, 2016; Lindström, Haaker & Olsson, 2018).

Formal theories of learning Several formal theories and computational models have been proposed to account for direct or Pavlovian and instrumental learning in conditioning experiments. Prominent examples are the Rescorla-Wagner (RW; Rescorla & Wagner, 1972), temporal difference (TD; Sutton & Barto, 1998) and Pearce-Hall (PH; Pearce & Hall, 1980) models. All models have in common the assertion that the organism learns by first predicting outcomes based on the available cues and then comparing the actual outcome with what was predicted. This discrepancy (called prediction error), is used to update associations between cues and their predicted outcomes. In the RW model, and the closely related TD model that generalizes RW across multiple time steps, the change in predicted outcome is given by the prediction error multiplied by a constant learning rate determined by the salience of the cue and the reinforcer. These models are sometimes called US-driven learning models, implying that the amount of learning in a trial will depend on the strength of the US received. By contrast, the Pearce-Hall model is CS-driven, meaning that how much is learned from a given cue is determined from recent experiences with that cue. In the PH model, an associability term governs how much is learned from a given reinforcer. Associability evolves as a function of the absolute, or unsigned, prediction error in the previous trial, capturing the animal's surprise at the previous outcome. In the cognitive neurosciences, researchers have recently begun combining ideas from RW and PH models, yielding what is known as the *hybrid model* (Le Pelley, 2004), which has successfully explained both direct and social threat learning.

Neural correlates of formal theories Phasic responses of dopaminergic neurons in the midbrain have been found to be consistent with predictions from RW/TD models—that is, excitatory firing for positive prediction errors and reduced firing for negative prediction errors (Schultz, Dayan, & Montague, 1997). These associations have most frequently been reported for the substantia nigra (SN), ventral tegmental area (VTA), and the ventral striatum (VS; a major projection site for the VTA). While neural correlates of prediction errors are often investigated in the context of reward-learning tasks, prediction errors are also found in aversive-learning tasks featuring punishing reinforcers, such as air puffs and shocks (Matsumoto & Hikosaka, 2009). Such aversive prediction errors have been reported in a range of structures, including the amygdala (McHugh et al., 2014), the striatum (Delgado, Li, Schiller, & Phelps, 2008), and the periaqueductal gray (PAG; Roy et al., 2014). Furthermore, fMRI studies have found correlates of both unsigned prediction errors and the associability term from the hybrid learning model in the amygdala (Boll, Gamer, Gluth, Finsterbusch, & Büchel, 2013; Li, Schiller, Schoenbaum, Phelps, & Daw, 2011). Other work has emphasized the interaction between the amygdala and the dorsal anterior cingulate cortex (dACC), suggesting a critical role for dACC in interpreting the signs of errors and highlighting the complex interplay between various brain systems for learning (Klavir, Genud-Gabai, & Paz, 2013). In sum, there is considerable evidence that the mammalian brain implements and uses the quantities posited by formal learning theory for both rewarding and threatening stimuli. Characterizing how these quantities are processed in social learning is an important topic in current research.

Learning in a Social World

Social learning of threat Traditionally, the study of learning and its underlying processes has focused on the organism in a social vacuum. In the natural ecology of many species, however, much of the learning of new skills and the value of objects and contexts takes place in social situations, where individuals serve as intentional or unintentional demonstrators. *Social learning*, here broadly defined as learning from, or in interaction with, other conspecifics is often adaptive, as it minimizes exposure to threats and gives access to others' innovations. Work in theoretical biology shows, however, that social learning is not always adaptive because information gleaned from others might be error prone. So-called social learning strategies (SLS) specify when and from whom the individual should learn from to optimize its behavior in different situations (Boyd & Richerson, 2009; Kendal et al., 2018). It remains unclear, however, if SLS are distinctly social or emerge from basic asocial-learning mechanisms. Moreover, the neural bases of SLS are so far unexplored.

A well-studied example of social learning across species, including rodents (Jeon et al., 2010; Kavaliers, Choleris, & Colwell, 2001; Knapska et al., 2006) and monkeys (Mineka & Cook, 1993), is observational threat learning. In humans, early work on observational, or *vicarious*, threat learning used experimental confederates serving as demonstrators, who responded with distress to a CS (Berger, 1961; Hygge & Öhman, 1976). Seminal studies by Mineka and her colleagues (Mineka, Davidson, Cook, & Keir, 1984) showed that cage-reared monkeys quickly acquired long-lasting threat responses toward snakes after only one exposure to a conspecific's facial expressions of distress. These studies established that relationships between the demonstrator's expressed distress, the observer's immediate response to the demonstrator's distress, and the subsequent threat expression in the observer were comparable to relationships between the US, UR, and CR in direct threat conditioning. Together with similar findings in other animals, these findings support the view that vicarious threat learning partly relies on the same learning mechanisms as classical conditioning. More recently, these findings have been confirmed and extended to human children (Askew & Field, 2007) and adults (Hooker, Germine, Knight, & D'Esposito, 2006; Lindström, Haaker, & Olsson, 2018; Olsson & Phelps, 2004) using both behavioral and neural measures. The workhorse of much recent experimental research is the video-based paradigm introduced by Olsson and Phelps (2004) and described in detail by Haaker, Golkar, Selbing, and Olsson (2017; see figure 84.1). A "demonstrator" is filmed making stereotypical threat and pain responses when receiving

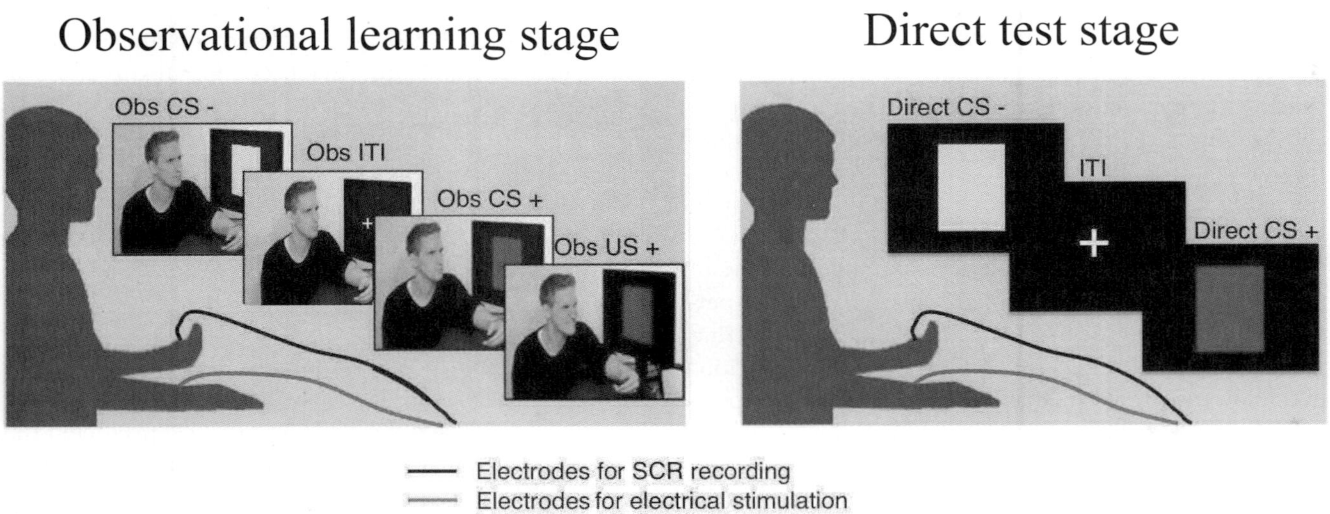

FIGURE 84.1 General design of the observational fear-conditioning protocol depicting the observer (participant; *in shaded gray*), first watching the demonstrator's responses to the CS-US shock pairings (observational learning stage), followed by being exposed to the CS (direct test stage). The observer receives no shocks during the test stage. *Notes*: CS−, conditioned stimulus never paired with shock; CS+, conditioned stimulus paired with shock; ITI, intertrial interval; Obs, observational. Adapted from Haaker, Golkar, et al. (2017). (See color plate 97.)

electric shocks during exposure to a CS. The procedure gives a high degree of control over what the participant is exposed to and can readily be combined with, for example, measures of visual attention (Kleberg, Selbing, Lundqvist, Hofvander, & Olsson, 2015). This paradigm can also be modified to examine the role of observer-demonstrator match in terms of social group and observational safety learning (Golkar et al. 2013; see below). Findings from studies outlined above also suggest that social cognitive processes, such as empathy, are important for the social US to be effective.

The role of empathic processes Research on empathic processes is of special interest to the understanding of social threat learning. Empathy is a collection of psychological processes allowing people to share and understand others' thoughts and feelings. One formulation of empathy subdivides it into three specific processes: affect sharing (resonating with another person's emotions), mentalizing (understanding the contents of another person's mind), and prosocial motivation (wanting to reduce others' suffering; Zaki, 2014). Social learning can depend on both affect sharing and mentalizing processes (Olsson et al., 2016), and prosocial behaviors can be both learned and serve as reinforcers themselves in social interaction (Lockwood, Apps, Valton, Viding, & Roiser, 2016; Nook, Ong, Morelli, Mitchell, & Zaki, 2016).

The last few decades have seen an explosion of interest in charting the neural bases of empathy, with good results. Paradigms for studying affect sharing during fMRI scanning have participants observe photos of people experiencing pain or videos of people discussing emotional experiences. These paradigms have revealed that networks of brain regions thought to underlie affect sharing overlap with those thought to contribute to action, sensation, affect, and interception, such as the premotor cortex, visual cortex, inferior frontal gyrus, inferior parietal lobule, and insula (Zaki & Ochsner, 2012). By contrast, tasks that assess mentalizing involve asking participants to specifically take the perspective of an observer or answer questions about what another person believes about a situation. These studies show that mentalizing is subserved by a network of brain regions that allow people to construct mental models and "project" themselves into other times, places, and perspectives (Mitchell, Banaji, & Macrae, 2005; Saxe & Wexler, 2005). These regions include the medial PFC, temporoparietal junction (TPJ), superior temporal sulcus (STS), precuneus, and temporal poles. It is worth noting that networks underlying affect sharing and mentalizing are separable from each other (Zaki & Ochsner, 2012).

Several lines of work converge to show how empathizing relates to observational threat learning. For example, individuals high in psychopathic traits do not readily acquire conditioned responses, when compared to normal controls (Aniskiewicz, 1979). Other work has demonstrated that observers higher in trait empathy acquire greater conditioned responses in observational fear tasks (Kleberg et al., 2015) and that instructing observers to make empathic appraisals of demonstrators also enhances their learned responses (Olsson et al., 2016). A recent extension of the observational threat-learning paradigm using live, naïve participant pairs tested whether observational threat learning could be modulated by empathic experience sharing. During learning, the degree of synchrony between observers' and demonstrators' electrodermal activity, which indexes autonomic nervous system activity, was found to predict the strength of the observers' later conditioned responses (Pärnamets, Espinosa, & Olsson, 2018). This aligns with work on rodent models that similarly indicates that the experiential sharing of emotional states might underpin observational learning in those species (Meyza, Bartal, Monfils, Panksepp, & Knapska, 2017). It is an intriguing possibility that the functionality of empathic experiences in many cases might be the learning opportunity they offer to the individual.

Neural Mechanisms Involved in Observational Threat Learning

A critical question in the study of social threat and safety learning is to what extent social learning relies on the same neural mechanisms as nonsocial forms of learning. As in Pavlovian learning, the amygdala has been heavily implicated in observational threat learning (Jeon et al., 2010; Knapska et al., 2006; Olsson, Nearing, & Phelps, 2007). In humans, overlapping amygdala activity was found in participants watching a video of the demonstrator receiving shocks paired with a CS and later when encountering the same CS without receiving any shocks, indicating that a specific CS-US association was formed and expressed in the same regions of the amygdala (Olsson, Nearing, & Phelps, 2007). The same study also reported activations in the anterior insula (AI) and ACC in the CS+>CS− contrast during the test stage, and this activity during observation predicted the strength of the CR (electrodermal activity) during the test stage, consistent with the roles these areas play in empathic processing.

The joint role of the ACC and amygdala for observational threat learning has been directly investigated in studies in rodents. For example, Jeon et al. (2010) showed that during observational learning, theta band synchronization increased between the ACC and the basolateral amygdala (BLA), indicating a close interaction between these regions during learning. Selectively deactivating either region impaired observational learning, demonstrating that both regions play a causal role in the formation of threat memories during social learning. These findings have been extended and refined by Allsop et al. (2018) using optogenetic techniques to selectively inhibit cells projecting from the ACC to the BLA (ACC−>BLA). The results showed that the ACC, or, more specifically, its input to the BLA, is critical for learning about the aversive value of a cue predicting the aversive treatment of a demonstrator. These findings suggest that the homologous circuitry in the primate ACC might play a similar role. In support of this, studies tracing the white tract fibers of the primate brain (Vogt & Paxinos, 2014) indicate that the gyrus of the ACC (ACCg) is uniquely connected with the neural circuitry implicated in mentalizing and in simulating others' actions: the medial PFC, TPJ, and the action system.

A recent fMRI study directly investigated the contributions of three of the core brain regions discussed so far—the amygdala, the AI, and the ACC—to both direct and observational threat learning by contrasting the two types of learning within subjects (Lindström, Haaker, & Olsson, 2018). The behavioral expectancy ratings data from both the direct- and observational-learning conditions were best described by the hybrid model, which both provided the first evidence that this model applies to observational learning and suggested overlap in the mechanisms underlying the two types of learning. Furthermore, overlapping activity in the amygdala, the AI, and the ACC in the two types of learning indicated commonalities in the underlying neural systems. The associability term from both direct and observational learning were found in the right AI, in line with earlier findings from direct learning (Li et al., 2011). Finally, dynamic causal modeling (DCM) was used to investigate the flow of information between the amygdala, the AI, and the ACC in response to the UCS (see figure 84.2). The DCM analysis indicated that the US signal likely entered the network through the amygdala for direct learning and through the AI for observational learning, consistent with the notion that the AI (and its empathic functioning) contributes to observational learning.

Like the study by Lindström, Haaker, and Olsson (2017), other work has used formal theories to better understand the contributions of different neural regions to observational threat learning, primarily by investigating the role of prediction errors. Meffert, Brislin, White, and Blair (2015) conducted a study in which participants learned about objects serving as a CS through their pairings with observing happy or angry facial expressions (USs) directed toward the CS. Prediction errors correlated with amygdala activity for both happy and angry emotional expressions, suggesting amygdala involvement in learning about social USs. The role of prediction errors in the amygdala in direct learning is well characterized and involves *N*-methyl-D-aspartate (NMDA) receptors in the lateral amygdala (Johansen, Cain, Ostroff, & LeDoux, 2011). Prediction errors are also downregulated by the involvement of

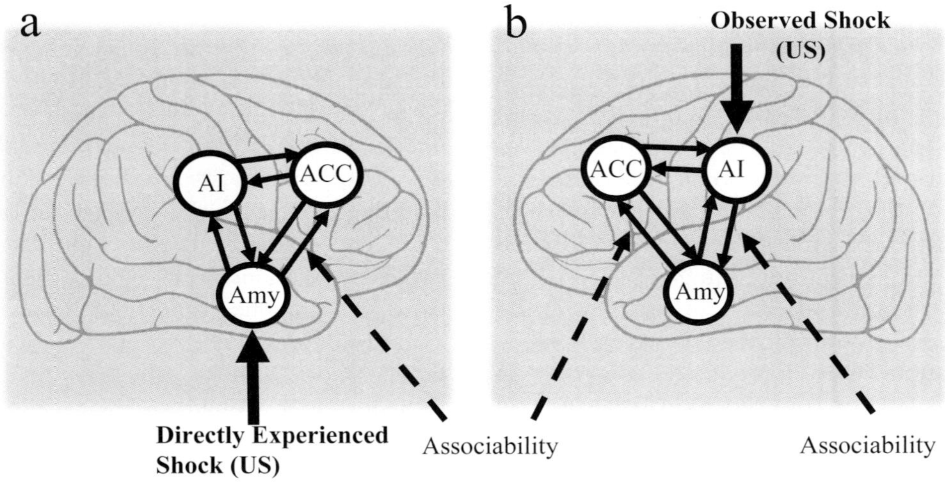

FIGURE 84.2 Dynamic causal modeling (DCM) of (*A*) direct and (*B*) observational threat learning (Lindström, Haaker, & Olsson, 2017). The most likely input region for the US in direct and observational learning was the amygdala and AI, respectively. The dotted arrows show the most likely targets for associability gating. *Notes*: ACC, anterior cingulate cortex; AI, anterior insula; Amy, amygdala; US, unconditioned stimulus.

opioidergic circuits in the periaqueductal gray (PAG; McNally & Cole, 2006), a region projecting to the amygdala and involved in regulating freezing and other defensive behaviors, as well as in analgesia. In an observational threat learning study on humans (Haaker, Yi, Petrovic, & Olsson, 2017), naltrexone, an opioid antagonist, was administered prior to learning. Compared to placebo controls, naltrexone-treated participants exhibited stronger CRs (electrodermal activity) and stronger activation to the US in the amygdala and in the PAG. When comparing naltrexone participants to placebo controls, increased functional connectivity was displayed between the PAG and the STS, a region associated with the integrative processing of social stimuli and mentalizing.

Observational safety learning Equally important to learning what is potentially dangerous is learning when something that was previously dangerous no longer poses a threat. This form of safety learning has traditionally been studied through extinction protocols in which the participant is repeatedly and directly exposed to the CS in the absence of the US (Bouton, 2002). Extinction training has become the standard experimental protocol to understand both the etiology and the treatment of dysfunctional fear and anxieties (Craske, Hermans, & Vervliet, 2018). A growing literature has shown that safety learning through direct extinction involves the ventromedial PFC and its interaction with the amygdala in both rodents (Milad & Quirk, 2002) and humans (Phelps et al., 2004; see Dunsmoor et al., 2015 for a review). A major goal for the study of social safety learning is to understand whether social safety learning involves a change of the CS-US associations (the fear memory) or a strengthening of the inhibitory safety memories formed during extinction.

Observing a demonstrator approach the target of a phobia in a calm and controlled manner has been shown to reduce anxiety and increase approach behavior toward that target (Bandura, Grusec, & Menlove, 1967). Using a modified version of the video-based threat-learning paradigm described above, research has demonstrated that undergoing observational safety learning was more effective in preventing the recovery of directly conditioned threat responses (during a subsequent reinstatement test) compared to direct extinction (Golkar et al., 2013). A first study on observational safety learning using fMRI (Golkar, Haaker, Selbing, & Olsson, 2016) extended these findings and found that the ventromedial PFC activity decreased to an observationally extinguished CS and increased to an observationally reinforced CS during safety learning. The ventromedial PFC activity was interpreted as tracking the relative cue value. More work is needed to fully understand its role in observational safety learning.

Social instrumental learning Social learning is not only passive; crucially, it can also involve actively intervening in the environment to learn how actions can bring about rewarding or punishing consequences—instrumental learning (Balleine & Dickinson, 1998). There has been considerable work on how stimulus-action-outcome contingencies are learned and the computational properties of the underlying neural systems (Dolan & Dayan, 2013; Ruff & Fehr, 2014). However, less is known about the computational and neural mechanisms involved when learning from others.

In one experiment, participants made choices between options that were probabilistically rewarded or punished. Participants made choices without and with social information derived from viewing a demonstrator make choices, as well as seeing the outcome of the observer's choice (Burke, Tobler, Baddeley, & Schultz, 2010). Increased social information monotonically increased the quality of participants' choices. When social information was restricted to the demonstrators' actions, observational action prediction errors (the difference between the observed and predicted action) were expressed in dorsolateral PFC activity, thought to reflect increased uncertainty in selection given the choice of the demonstrator. When social information included both the actions of, and the outcomes for, the demonstrator, observational prediction instead correlated with ventromedial PFC activity and inversely with ventral striatal activity, indicating the full integration of these quantities into the brain's valuation circuits. The behavioral findings from this experiment were replicated and refined in a study using the same conditions but additionally manipulating the skill of the demonstrator (Selbing, Lindström, & Olsson, 2014). Participants performed better when observing both skilled and unskilled demonstrators relative to when learning on their own. The demonstrator's skill level modulated an imitation rate parameter in an RL model, which determined how much the demonstrator's choices affected the participant. Together, these studies show that participants readily and adaptively use observational information from others' choices, and this process can be well described using formal learning theories.

Concluding Remarks and Future Directions

Our understanding of social learning has developed dramatically over recent years thanks to both theoretical and empirical advancements, including the use of experimental models comparable across species. For example, research on observational threat and safety learning has shown that these learning procedures draw on computational and neural mechanisms partially shared with direct (Pavlovian) threat conditioning and extinction learning, respectively. Importantly, however, social learning is distinguished from direct forms of learning by its dependence on social cognition, including empathic processes.

An important task for future research is to continue bringing together work on the functions (phylogenetically and computationally) of social learning with its neural architecture. This will move our understanding closer to the basic workings of social learning, hopefully providing insights into whether, and if so, in what ways, social is distinct from non-social, domain general, learning. Second, and related to the first direction, extended work on non-human animals is needed to uncover the molecular and cellular levels of social learning. This development should benefit both from new animal models, but equally from translating human experimental paradigms to non-human animals. Third, because social learning plays an important role in the development of psychological problems, such as anxiety disorders and post-traumatic stress, more research is needed about both the social etiology of such disorders, and the ways social learning, for example, vicarious safety modelling, can inform new and improved treatments. Finally, future research should continue to examine how social learning scale up from the individual brain to social networks, and even societal. In this chapter, we have argued that the study of social learning provides an ideal experimental model to address these concerns.

Acknowledgments

We thank Tove Hensler and Armita Golkar for comments on an earlier draft and for assistance with this manuscript. This research was supported by an Independent Starting Grant (284366; Emotional Learning in Social Interaction) from the European Research Council the Knut and Alice Wallenberg Foundation (KAW 2014.0237), and a Swedish Research Council Consolidator grant 2018-00877 to Andreas Olsson.

REFERENCES

Allsop, S. A., Wichmann, R., Mills, F., Burgos-Robles, A., Chang, C. J., Felix-Ortiz, A. C., …Tye, K. M. (2018). Corticoamygdala transfer of socially derived information gates observational learning. *Cell, 173*(6), 1329–1342. http://doi.org/10.1016/j.cell.2018.04.004

Aniskiewicz, A. S. (1979). Autonomic components of vicarious conditioning and psychopathy. *Journal of Clinical Psychology, 35*(1), 60–67. http://doi.org/10.1002/1097-4679(197901)35:1<60:AID-JCLP2270350106>3.0.CO;2-R

Apps, M. A. J., Rushworth, M. F. S., & Chang, S. W. C. (2016). The anterior cingulate gyrus and social cognition: Tracking the motivation of others. *Neuron, 90*(4), 692–707. http://doi.org/10.1016/J.NEURON.2016.04.018

Askew, C., & Field, A. P. (2007). Vicarious learning and the development of fears in childhood. *Behaviour Research and Therapy, 45*(11), 2616–2627. http://doi.org/10.1016/j.brat.2007.06.008

Balleine, B. W., & Dickinson, A. (1998). Goal-directed instrumental action: Contingency and incentive learning and their cortical substrates. *Neuropharmacology, 37*(4), 407–419. http://doi.org/https://doi.org/10.1016/S0028-3908(98)00033-1

Bandura, A., Grusec, J. E., & Menlove, F. L. (1967). Vicarious extinction of avoidance behavior. *Journal of Personality and Social Psychology, 5*(1), 16–23. http://doi.org/10.1037/h0024182

Berger, S. M. (1961). Incidental learning through vicarious reinforcement. *Psychological Reports, 9*(3), 477–491. http://doi.org/10.2466/pr0.1961.9.3.477

Boll, S., Gamer, M., Gluth, S., Finsterbusch, J., & Büchel, C. (2013). Separate amygdala subregions signal surprise and predictiveness during associative fear learning in humans. *European Journal of Neuroscience, 37*(5), 758–767. http://doi.org/10.1111/ejn.12094

Bouton, M. E. (2002). Context, ambiguity, and unlearning: Sources of relapse after behavioral extinction. *Biological Psychiatry, 52*(10), 976–986. http://www.ncbi.nlm.nih.gov/pubmed/12437938.

Boyd, R., & Richerson, P. J. (2009). Culture and the evolution of human cooperation. *Philosophical Transactions of the Royal Society of London B: Biological Sciences, 364*(1533), 3281 LP–3288. http://rstb.royalsocietypublishing.org/content/364/1533/3281.abstract.

Burke, C. J., Tobler, P. N., Baddeley, M., & Schultz, W. (2010). Neural mechanisms of observational learning. *Proceedings of the National Academy of Sciences of the United States of America, 107*(32), 14431–14436. http://doi.org/10.1073/pnas.1003111107

Christakis, N. A., & Fowler, J. H. (2009). *Connected: The surprising power of our social networks and how they shape our lives.* New York: Little, Brown.

Craig, A. D. (2009). How do you feel—now? The anterior insula and human awareness. *Nature Reviews Neuroscience, 10*(1), 59–70. http://doi.org/10.1038/nrn2555

Craske, M. G., Hermans, D., & Vervliet, B. (2018). State-of-the-art and future directions for extinction as a translational model for fear and anxiety. *Philosophical Transactions of the Royal Society of London B: Biological Sciences, 373*(1742). http://rstb.royalsocietypublishing.org/content/373/1742/20170025.abstract.

Debiec, J., & Olsson, A. (2017). Social fear learning: From animal models to human function. *Trends in Cognitive Sciences, 21*(7), 546–555. http://doi.org/10.1016/j.tics.2017.04.010

Delgado, M. R., Li, J., Schiller, D., & Phelps, E. A. (2008). The role of the striatum in aversive learning and aversive prediction errors. *Philosophical Transactions of the Royal Society of London B: Biological Sciences, 363*(1511), 3787 LP–3800. http://rstb.royalsocietypublishing.org/content/363/1511/3787.abstract.

Dolan, R. J., & Dayan, P. (2013). Goals and habits in the brain. *Neuron, 80*(2), 312–325. http://doi.org/10.1016/J.NEURON.2013.09.007

Dunsmoor, J. E., Niv, Y., Daw, N., & Phelps, E. A. (2015). Rethinking extinction. *Neuron, 88*(1), 47–63. http://doi.org/10.1016/j.neuron.2015.09.028

Fullana, M. A., Harrison, B. J., Soriano-Mas, C., Vervliet, B., Cardoner, N., Àvila-Parcet, A., & Radua, J. (2016). Neural signatures of human fear conditioning: An updated and extended meta-analysis of fMRI studies. *Molecular Psychiatry, 21*(4), 500. http://doi.org/10.1038/mp.2015.88

Golkar, A., Haaker, J., Selbing, I., & Olsson, A. (2016). Neural signals of vicarious extinction learning. *Social Cognitive and Affective Neuroscience, 11*(10), 1541–1549. http://doi.org/10.1093/scan/nsw068

Golkar, A., Selbing, I., Flygare, O., Öhman, A., & Olsson, A. (2013). Other people as means to a safe end. *Psychological Science, 24*(11), 2182–2190. http://doi.org/10.1177/0956797613489890

Haaker, J., Golkar, A., Selbing, I., & Olsson, A. (2017). Assessment of social transmission of threats in humans using observational fear conditioning. *Nature Protocols, 12,* 1378. http://dx.doi.org/10.1038/nprot.2017.027

Haaker, J., Yi, J., Petrovic, P., & Olsson, A. (2017). Endogenous opioids regulate social threat learning in humans. *Nature Communications, 8,* 15495. http://dx.doi.org/10.1038/ncomms15495

Hooker, C. I., Germine, L. T., Knight, R. T., & D'Esposito, M. (2006). Amygdala response to facial expressions reflects emotional learning. *Journal of Neuroscience, 26*(35), 8915–8922. http://doi.org/10.1523/jneurosci.3048-05.2006

Hygge, S., & Öhman, A. (1976). Conditioning of electrodermal responses through vicarious instigation and through perceived threat to a performer. *Scandinavian Journal of Psychology, 17*(1), 65–72. http://doi.org/10.1111/j.1467-9450.1976.tb00213.x

Jeon, D., Kim, S., Chetana, M., Jo, D., Ruley, H. E., Lin, S.-Y., … Shin, H.-S. (2010). Observational fear learning involves affective pain system and Cav1.2 Ca2+ channels in ACC. *Nature Neuroscience, 13*(4), 482–488. http://doi.org/10.1038/nn.2504

Johansen, J. P., Cain, C. K., Ostroff, L. E., & LeDoux, J. E. (2011). Molecular mechanisms of fear learning and memory. *Cell, 147*(3), 509–524. http://doi.org/10.1016/J.CELL.2011.10.009

Kavaliers, M., Choleris, E., & Colwell, D. D. (2001). Learning from others to cope with biting flies: Social learning of fear-induced conditioned analgesia and active avoidance. *Behavioral Neuroscience, 115*(3), 661–674. http://doi.org/10.1037/0735-7044.115.3.661

Kendal, R. L., Boogert, N. J., Rendell, L., Laland, K. N., Webster, M., & Jones, P. L. (2018). Social learning strategies: Bridge-building between fields. *Trends in Cognitive Sciences, 22*(7), 651–665. http://doi.org/10.1016/j.tics.2018.04.003

Klavir, O., Genud-Gabai, R., & Paz, R. (2013). Functional connectivity between amygdala and cingulate cortex for adaptive aversive learning. *Neuron, 80*(5), 1290–1300. http://doi.org/10.1016/J.NEURON.2013.09.035

Kleberg, J. L., Selbing, I., Lundqvist, D., Hofvander, B., & Olsson, A. (2015). Spontaneous eye movements and trait empathy predict vicarious learning of fear. *International Journal of Psychophysiology, 98*(3), 577–583. http://doi.org/10.1016/j.ijpsycho.2015.04.001

Knapska, E., Nikolaev, E., Boguszewski, P., Walasek, G., Blaszczyk, J., Kaczmarek, L., & Werka, T. (2006). Between-subject transfer of emotional information evokes specific pattern of amygdala activation. *Proceedings of the National Academy of Sciences, 103*(10), 3858–3862. http://doi.org/10.1073/PNAS.0511302103

LaBar, K. S., Gatenby, J. C., Gore, J. C., LeDoux, J. E., & Phelps, E. A. (1998). Human amygdala activation during conditioned fear acquisition and extinction: A mixed-trial fMRI study. *Neuron, 20*(5), 937–945. http://doi.org/10.1016/S0896-6273(00)80475-4

LeDoux, J. E., Iwata, J., Cicchetti, P., & Reis, D. J. (1988). Different projections of the central amygdaloid nucleus mediate autonomic and behavioral correlates of conditioned fear. *Journal of Neuroscience, 8*(7), 2517 LP–2529. http://www.jneurosci.org/content/8/7/2517.abstract.

Le Pelley, M. E. (2004). The role of associative history in models of associative learning: A selective review and a hybrid model. *Quarterly Journal of Experimental Psychology, 57*(3b), 193–243. http://doi.org/10.1080/02724990344000141

Li, J., Schiller, D., Schoenbaum, G., Phelps, E. A., & Daw, N. D. (2011). Differential roles of human striatum and amygdala in associative learning. *Nature Neuroscience, 14*(10), 1250–1252. http://doi.org/10.1038/nn.2904

Lindström, B., Haaker, J., & Olsson, A. (2017). A common neural network differentially mediates direct and social fear learning. *NeuroImage, 167,* 121–129. http://doi.org/10.1016/j.neuroimage.2017.11.039

Lockwood, P. L., Apps, M. A. J., Valton, V., Viding, E., & Roiser, J. P. (2016). Neurocomputational mechanisms of prosocial learning and links to empathy. *Proceedings of the National Academy of Sciences, 113*(35), 9763–9768. http://doi.org/10.1073/pnas.1603198113

Maren, S., Aharonov, G., & Fanselow, M. S. (1997). Neurotoxic lesions of the dorsal hippocampus and Pavlovian fear conditioning in rats. *Behavioural Brain Research, 88*(2), 261–274. http://doi.org/10.1016/S0166-4328(97)00088-0

Maren, S., Phan, K. L., & Liberzon, I. (2013). The contextual brain: Implications for fear conditioning, extinction and psychopathology. *Nature Reviews Neuroscience, 14*(6), 417–428. http://doi.org/10.1038/nrn3492

Matsumoto, M., & Hikosaka, O. (2009). Two types of dopamine neuron distinctly convey positive and negative motivational signals. *Nature, 459,* 837. http://dx.doi.org/10.1038/nature08028

McHugh, S. B., Barkus, C., Huber, A., Capitão, L., Lima, J., Lowry, J. P., & Bannerman, D. M. (2014). Aversive prediction error signals in the amygdala. *Journal of Neuroscience, 34*(27), 9024 LP–9033. http://www.jneurosci.org/content/34/27/9024.abstract.

McNally, G. P., & Cole, S. (2006). Opioid receptors in the midbrain periaqueductal gray regulate prediction errors during Pavlovian fear conditioning. *Behavioral Neuroscience, 120*(2), 313–323. http://doi.org/10.1037/0735-7044.120.2.313

Meffert, H., Brislin, S. J., White, S. F., & Blair, J. R. (2015). Prediction errors to emotional expressions: The roles of the amygdala in social referencing. *Social Cognitive and Affective Neuroscience, 10*(4), 537–544. http://doi.org/10.1093/scan/nsu085

Meyza, K. Z., Bartal, I. B.-A., Monfils, M. H., Panksepp, J. B., & Knapska, E. (2017). The roots of empathy: Through the lens of rodent models. *Neuroscience & Biobehavioral Reviews, 76,* 216–234. http://doi.org/10.1016/j.neubiorev.2016.10.028

Milad, M. R., & Quirk, G. J. (2002). Neurons in medial prefrontal cortex signal memory for fear extinction. *Nature, 420,* 70. http://dx.doi.org/10.1038/nature01138

Mineka, S., & Cook, M. (1993). Mechanisms involved in the observational conditioning of fear. *Journal of Experimental Psychology: General, 122*(1), 23–38. http://doi.org/10.1037/0096-3445.122.1.23

Mineka, S., Davidson, M., Cook, M., & Keir, R. (1984). Observational conditioning of snake fear in rhesus monkeys. *Journal of Abnormal Psychology, 93*(4), 355–372. http://doi.org/10.1037/0021-843X.93.4.355

Mitchell, J. P., Banaji, M. R., & Macrae, N. C. (2005). The link between social cognition and self-referential thought in the medial prefrontal cortex. *Journal of Cognitive Neuroscience, 17*(8), 1306–1315.

Morgan, M. A., Romanski, L. M., & LeDoux, J. E. (1993). Extinction of emotional learning: Contribution of medial prefrontal cortex. *Neuroscience Letters, 163*(1), 109–113. http://doi.org/10.1016/0304-3940(93)90241-C

Nook, E. C., Ong, D. C., Morelli, S. A., Mitchell, J. P., & Zaki, J. (2016). Prosocial conformity: Prosocial norms generalize across behavior and empathy. *Personality and Social Psychology Bulletin, 42*(8), 1045–1062. http://doi.org/10.1177/0146167298248001

Nook, E. C., & Zaki, J. (2015). Social norms shift behavioral and neural responses to foods. *Journal of Cognitive Neuroscience, 27*(7), 1412–1426. http://doi.org/10.1162/jocn

Olsson, A., McMahon, K., Papenberg, G., Zaki, J., Bolger, N., & Ochsner, K. N. (2016). Vicarious fear learning depends on empathic appraisals and trait empathy. *Psychological Science, 27*(1), 25–33. http://doi.org/10.1177/0956797615604124

Olsson, A., Nearing, K. I., & Phelps, E. A. (2007). Learning fears by observing others: The neural systems of social fear transmission. *Social Cognitive and Affective Neuroscience, 2*(1), 3–11. http://doi.org/10.1093/scan/nsm005

Olsson, A., & Phelps, E. A. (2004). Learned fear of "unseen" faces after Pavlovian, observational, and instructed fear. *Psychological Science, 15*(12), 822–828. http://doi.org/10.1111/j.0956-7976.2004.00762.x

Pärnamets, P., Espinosa, L., & Olsson, A. (2018). Physiological synchrony between individuals predicts observational threat learning in humans. *BioRxiv.* https://doi.org/10.1101/454819

Pearce, J. M., & Hall, G. (1980). A model for Pavlovian learning: Variations in the effectiveness of conditioned but not of unconditioned stimuli. *Psychological Review, 87*(6), 532–552. http://doi.org/10.1037/0033-295X.87.6.532

Phelps, E. A., Delgado, M. R., Nearing, K. I., & LeDoux, J. E. (2004). Extinction learning in humans: Role of the amygdala and vmPFC. *Neuron, 43*(6), 897–905. http://doi.org/10.1016/J.NEURON.2004.08.042

Phelps, E. A., & LeDoux, J. E. (2005). Contributions of the amygdala to emotion processing: From animal models to human behavior. *Neuron, 48*(2), 175–187. http://doi.org/10.1016/J.NEURON.2005.09.025

Pitkänen, A., Savander, V., & LeDoux, J. E. (1997). Organization of intra-amygdaloid circuitries in the rat: An emerging framework for understanding functions of the amygdala. *Trends in Neurosciences, 20*(11), 517–523. http://doi.org/10.1016/S0166-2236(97)01125-9

Rachman, S. (1972). Clinical applications of observational learning imitation and modeling. *Behavior Therapy, 3*(3), 379–397. http://doi.org/10.1016/S0005-7894(72)80139-4

Rescorla, R. A., & Wagner, A. R. (1972). A theory of Pavlovian conditioning: Variations in the effectiveness of reinforcement and nonreinforcement. In A. H. Black & W. F. Prokasy (Eds.), *Classical conditioning II: Current research and theory* (pp. 64–99). New York: Appleton-Century-Crofts.

Rogan, M. T., Stäubli, U. V., & LeDoux, J. E. (1997). Fear conditioning induces associative long-term potentiation in the amygdala. *Nature, 390*(6660), 604–607. http://doi.org/10.1038/37601

Roy, M., Shohamy, D., Daw, N., Jepma, M., Wimmer, G. E., & Wager, T. D. (2014). Representation of aversive prediction errors in the human periaqueductal gray. *Nature Neuroscience, 17,* 1607. http://dx.doi.org/10.1038/nn.3832

Ruff, C. C., & Fehr, E. (2014). The neurobiology of rewards and values in social decision making. *Nature Reviews Neuroscience, 15,* 549. http://dx.doi.org/10.1038/nrn3776

Saxe, R., & Wexler, A. (2005). Making sense of another mind: The role of the right temporo-parietal junction. *Neuropsychologia, 43*(10), 1391–1399. http://doi.org/10.1016/j.neuropsychologia.2005.02.013

Schultz, W., Dayan, P., & Montague, P. R. (1997). A neural substrate of prediction and reward. *Science, 275*(5306), 1593–1599. http://www.ncbi.nlm.nih.gov/pubmed/9054347.

Selbing, I., Lindström, B., & Olsson, A. (2014). Demonstrator skill modulates observational aversive learning. *Cognition, 133*(1), 128–139. http://doi.org/10.1016/j.cognition.2014.06.010

Shackman, A. J., Salomons, T. V., Slagter, H. A., Fox, A. S., Winter, J. J., & Davidson, R. J. (2011). The integration of negative affect, pain and cognitive control in the cingulate cortex. *Nature Reviews Neuroscience, 12*(3), 154–167. http://doi.org/10.1038/nrn2994

Sutton, R. S., & Barto, A. G. (1998). *Reinforcement learning: An introduction.* Cambridge, MA: MIT Press.

Vogt, B. A., & Paxinos, G. (2014). Cytoarchitecture of mouse and rat cingulate cortex with human homologies. *Brain Structure & Function, 219*(1), 185–192. http://doi.org/10.1007/s00429-012-0493-3

Zaki, J. (2014). Empathy: A motivated account. *Psychological Bulletin, 140*(6), 1608–1647. http://doi.org/10.1037/a0037679

Zaki, J., & Ochsner, K. (2012). The neuroscience of empathy: Progress, pitfalls and promise. *Nature Neuroscience, 15*(5), 675–680. http://doi.org/10.1038/nn.3085

85 Neurodevelopmental Processes That Shape the Emergence of Value-Guided Goal-Directed Behavior

CATHERINE INSEL, JULIET Y. DAVIDOW, AND LEAH H. SOMERVILLE

ABSTRACT Adolescents are challenged to orchestrate goal-directed actions in increasingly independent and consequential ways. In doing so, it is advantageous to use information about value to select which goals to pursue and how much effort to devote to them. Here, we examine age-related changes in how individuals use value signals to orchestrate goal-directed behavior, with a focus on cognitive control and learning. Emerging research suggests that even young children can detect value signals and use them to guide their goal-directed behaviors, but this process is constrained by ongoing cognitive development. That is, the facilitatory effects of value emerge throughout adolescence for more challenging cognitive demands and are constrained by the ongoing development of striatocortical system interactions.

Signals denoting value pervade contemporary life. From social signals indicating what actions are desirable, to price tags that denote the worth of various goods, to compensation for hours worked, value cues communicate the importance and worth of objects and actions. These kinds of signals also fill the lives of children and adolescents—for example, when children receive money for completing chores, when students must decide what content is the most important to learn on a given day, or when parents communicate the importance of certain activities. Psychological theory and empirical research in adults has underscored the importance of letting value guide goal-directed behaviors (Balleine & O'Doherty, 2010; Braver et al., 2014; Rangel & Hare, 2010; Shenhav et al., 2017). Indeed, we do not devote our energy and cognitive resources randomly—we use cues from the environment about what is valuable and important to guide the prioritization of resources toward actions most relevant to our goals.

In this chapter, we review research demonstrating how value computations change across development from childhood to adulthood, how value influences goal-directed behavior, and how neurodevelopment shapes these processes. We will examine how two domains of complex cognition that support goal-directed behavior—cognitive control and learning—are differently guided by value in children, adolescents, and adults.

Neurodevelopment through Adolescence

The lengthy, complex process of brain development was initially documented by early neuroanatomists such as Peter Huttenlocher and Patricia Goldman-Rakic, whose work revealed progressive and regressive changes in neuronal structure and organization well into the human adolescent years (Huttenlocher, 1984, 1990; Rakic, 1974; Rakic, Bourgeois, Eckenhoff, Zecevic, & Goldman-Rakic, 1986). Since then, the increasingly widespread use of noninvasive brain imaging to study neurodevelopment has generated a strong body of evidence that brain development continues throughout adolescence (Gogtay et al., 2004; Mills, Goddings, Clasen, Giedd, & Blakemore, 2014) and beyond (Somerville, 2016; Sowell, Thompson, Holmes, Jernigan, & Togan, 1999; Tamnes et al., 2010). At the structural level, the developing brain shows reductions in cortical gray matter and increases in the volume and anisotropy of white matter from childhood to adulthood (Giedd et al., 1999; Simmonds, Hallquist, Asato, & Luna, 2014). Although the field continues to refine its understanding of the cellular-molecular mechanisms underlying patterns observable with magnetic resonance imaging (MRI), such changes are broadly thought to reflect synaptic pruning, myelination, and increased connectivity across widely distributed brain circuitry. These progressive and regressive patterns occur at different timelines across the brain, such that some structures lag behind others in neurodevelopment (e.g., Casey, 2015; Somerville & Casey, 2010). Concurrent with structural development is the development of complex brain function, which is thought to be underpinned by increasing the interconnectivity and functional coordination of distributed brain networks (e.g., Chai et al., 2017). Here, we focus on human brain-imaging work that can chart the functioning and coordination of distributed subcortical-cortical pathways that integrate information about value, action, and regulatory demands in the service of cognitive control and learning.

Cognitive control represents a collection of mental processes that allow individuals to select contextually appropriate behavior to pursue superordinate goals (Miller & Cohen, 2001). The maturation of cognitive control follows a protracted developmental trajectory, with continued improvements observed through adolescence. The ongoing refinement of cognitive control through adolescence is paralleled by the continued functional development of brain systems that subserve effortful cognition, including the prefrontal cortex (PFC) and parietal cortices. In addition, age-related differences in PFC recruitment during cognitive control may reflect developmental shifts in cognitive strategy implementation (Crone & Steinbeis, 2017). Older adolescents and young adults are more likely to implement optimal strategies to enhance the precision of control (Church, Bunge, Petersen, & Schlaggar, 2017), such as the engagement of proactive processes that allow individuals to strategically recruit PFC control systems in anticipation of an upcoming cognitive demand, supported by increased connectivity between the striatum and PFC with age (Vink et al., 2014). As an example, trial-by-trial working-memory accuracy and reaction times become more consistent with age, which is supported by age-related increases in the functional recruitment of PFC-centered brain networks (Satterthwaite et al., 2013). Thus, the recruitment of control-related brain systems becomes increasingly stable and strategic with age, and these shifts ultimately promote successful and efficient control performance.

Similar to cognitive control, learning abilities continue to mature throughout adolescence in tandem with active neurodevelopment. Many different forms of learning rely on different brain systems. Associative learning (Kersey & Emberson, 2017) and learning from observation (Hunnius & Bekkering, 2014) are available as early as in infancy and are thought to underlie primary cognitive development. Associative learning is thought to be supported by the hippocampus (Gómez, 2017) and distributed cortical networks (Kersey & Emberson, 2017). Despite the ongoing development and complexity of cognition supported by the PFC, infants and adults alike recruit PFC while generalizing learned information (Gerraty, Davidow, Wimmer, Kahn, & Shohamy, 2014; Werchan, Collins, Frank, & Amso, 2016). In contrast to the rapid learning that relies on the hippocampus, the striatal learning system guides slow learning from repetition and the valence of feedback-based outcomes (Shohamy & Turk-Browne, 2013). Though these complementary learning strategies and the neural systems underpinning them are available in basic forms very early in life, these cognitive processes and brain systems continue to refine and mature to support increasingly complex demands over the course of development (Schlichting, Guarino, Schapiro, Turk-Browne, & Preston, 2017).

Age-Related Change in Value Assignment

A crucial building block to value-guided goal pursuit is the detection and assignment of value to cues in the environment (Rangel, Camerer, & Montague, 2008), which allows an individual to evaluate the potential positive and negative outcomes of their thoughts and actions. Although value cannot be measured directly, higher value can be inferred from indirect assessments of behavior: higher subjective ratings of positive valence and importance, invigoration of physical speed, higher response rate, greater time allocation, and greater effort exertion (Niv, Daw, Joel, & Dayan, 2007).

Based on research in children, adolescents, and adults, individuals across a wide age range are able to detect and assign value to information in the environment. For example, young children (ages 3–6) can readily distinguish between high-value and low-value rewards and can indicate their preference for high-value options (Blake & Rand, 2010; Rodriguez, Mischel, & Shoda, 1989). When asked to provide self-reported subjective value ratings, children, adolescents, and adults similarly rank monetary outcomes according to their relative value (Bjork, Smith, Chen, & Hommer, 2010; Insel, Kastman, Glenn, & Somerville, 2017; Paulsen, Hallquist, Geier, & Luna, 2015). Further, children and adults alike exhibit speeded motor responses to high-reward cues (Galvan et al., 2006). Thus, individuals across a wide developmental window reliably identify value-related cues, and their behavioral responses often reflect value-selective detection in the environment.

Brain-imaging research assesses valuation processes by measuring neural responses to the cued expectation, or receipt, of valued outcomes. Developmental research has shown that even children exhibit robust engagement of these brain regions when receiving rewards (Galvan et al., 2006; Luking, Luby, & Barch, 2014). There is also evidence that the ventral striatum is hyperresponsive to the anticipation and/or receipt of rewards in adolescents (Barkley-Levenson & Galvan,

2014; Braams, van Duijvenvoorde, Peper, & Crone, 2015; Silverman, Jedd, & Luciana, 2015), although it is important to note that this elevation is not always observed (Bjork et al., 2010; Insel et al., 2017; Paulsen et al., 2015), and more research is needed to pinpoint the conditions in which an elevated striatum response is (or is not) observed during adolescence (Sherman, Steinberg, & Chein, 2018). Further, the preponderance of prior research in this area has assessed striatal reactivity to the passive receipt of rewards, rather than examining the differential impact of striatal signaling on goal-directed behavior development. This leaves many unanswered questions regarding the impact of value signaling on goal-directed behavior, the central focus of this chapter.

Using Value to Guide Goal-Directed Behavior

Contemporary neural and computational models of motivation-cognition interactions posit that in adults, value cues shape motor coordination and action selection via interactions among the ventral striatum, dorsal striatum, and PFC (Botvinick & Braver, 2015; Dalley, Everitt, & Robbins, 2011; Haber & Knutson, 2010). Research on cortico-striatal-thalamic-cortical circuits has established that the basal ganglia interacts with the cortex via multiple parallel loops, with distributed connections to and from areas of cortex including lateral PFC, lateral orbitofrontal cortex, anterior cingulate cortex, and ventromedial PFC (Alexander, DeLong, & Strick, 1986). Distinct loops have been associated with varying cognitive functions and computations. As a result, bottom-up and top-down connectivity within these separate and interacting loops is thought to differentially exert influence over cognitive, motor, and motivated behavior (Graybiel, 1990; Haber & Knutson, 2010; Tanaka et al., 2004).

Computational network models propose that the striatum serves a gating function (Botvinick & Braver, 2015; Frank, Loughry, & O'Reilly, 2001), orchestrating goal-directed titration of cognitive and motor control. Dopamine-mediated value signals in the ventral striatum project to the dorsal striatum via indirect, looped connections with the midbrain through nigrostriatal pathways (Aarts, van Holstein, & Cools, 2011; Haber & Knutson, 2010). The dorsal striatum coordinates motor output through connections with PFC and motor cortex. Accordingly, the striatum modulates the active maintenance of goal states in the PFC and motor action selection via output gating (Frank & Badre, 2011). This selective gating determines how goal states influence appropriate action decisions in a context-dependent manner (i.e., selecting the appropriate action in response to a given stimulus; Frank & Badre, 2011). As such, the value of an action can influence its selection and execution in the moment.

Consistent with this model, adults integrate motivational pursuits with cognitive demands through the selective and coordinated recruitment of corticostriatal systems. For example, when incentives are at stake, adults typically improve performance in high-stakes contexts (Botvinick & Braver, 2015). These high-stakes performance improvements are often paralleled by the upregulated functional recruitment of PFC systems (Braver et al., 2014) or increased corticostriatal connectivity (Kinnison, Padmala, Choi, & Pessoa, 2012). If engaging cognitive control is costly (Kool, McGuire, Rosen, & Botvinick, 2010; Westbrook, Kester, & Braver, 2013), individuals may compute cost-benefit analyses to determine whether and when engaging control is worthwhile, given the value of the goal at stake (Boureau, Sokol-Hessner, & Daw, 2015; Shenhav, Botvinick, & Cohen, 2013). Thus, for adults, motivated contexts tune the allocation of cognitive effort and attentional resources by selectively gating prefrontal control systems in a goal-directed fashion.

Value-Guided Goal-Directed Behavior across Development

Cognitive control Do value cues similarly facilitate cognitive control in children and adolescents? There are select circumstances when children and adolescents use value to upregulate control performance; however, task difficulty and increasing cognitive demands constrain this tendency. For example, young children (age 4–5) use value to improve performance when promised a reward for accurate performance on a developmentally appropriate response-inhibition task, but value no longer benefits performance for a more difficult cognitive flexibility task (Qu, Finestone, Qin, & Reena, 2013). If cognitive difficulty is titrated to an individual's ability, children, adolescents, and adults alike improve control accuracy for rewarding versus neutral outcomes (Strang & Pollak, 2014). Finally, if participants can anticipate imminent control demands, such as during an antisaccade task that signals the upcoming need to implement control, children and adolescents can improve control when pursuing performance-contingent rewards (Geier, Terwilliger, Teslovich, Velanova, & Luna, 2010; Padmanabhan, Geier, Ordaz, Teslovich, & Luna, 2011).

However, when cognitive-control demands are particularly challenging, adolescents do not adjust performance in a value-selective fashion. We have recently proposed that the beneficial effects of value on cognitive performance may not stabilize until late adolescence or early adulthood (Davidow, Insel, & Somerville,

2018) and, crucially, emerge along with the capacity to achieve the cognitive challenge at hand. For example, in a visual search task (Stormer, Eppinger, & Li, 2014) that invoked sustained attention and context monitoring (e.g., Chatham et al., 2012) containing low-value or high-value cues (one cent vs. five cents), young adults aged 20–29 exhibited a value-specific enhancement in performance, responding more quickly and consistently for high-value cues. In contrast, the child (age 8–11) and adolescent (age 14–16) groups showed no change in response consistency across low- and high-value trials. Notably, participants of all ages exhibited speeded responses to high-value cues, indicating they detected high-value cues, which invigorated the responses, but this did not translate into better performance for the younger age groups. Similar developmental trends were reported in a study examining the effects of value on selective attention during memory encoding (Castel et al., 2011). Participants encoded word lists, with different words associated with different monetary reward amounts if they recalled them accurately at a later test. Children (age 5–9), adolescents (age 10–17), and young adults (age 18–23) recalled significantly more high-value words. However, this effect was the most pronounced in young adults (age 18–23), indicating value-selective memory continues to become more robust through adolescence.

Recent work has also identified the neurodevelopmental processes that emerge through adolescence to support value-guided behavioral control—namely, the late refinement of corticostriatal network connectivity. In a recent neuroimaging study by our group (Insel et al., 2017), participants aged 13–20 completed a go/no-go task with low-value or high-value payouts for accurate performance. Selective performance improvements for high-value trials emerged in late adolescence (figure 85.1*A*). Individuals who improved performance for high-value incentives exhibited increased functional connectivity between the ventral striatum and ventrolateral prefrontal cortex (VLPFC) during high-value trials (figure 85.1*B*). Moreover, this value-specific connectivity profile mediated age-related increases in value-guided control. Thus, we propose that the late refinement of corticostriatal connectivity sets the stage for successful value-guided cognitive control.

Learning Using value cues to guide when and what to learn is a second key domain underpinning mature goal-directed behavior. To use value cues to guide actions, one must learn the value of particular actions or choice options. Experimentally, learning to associate stimuli or actions with valued outcomes is indexed by choosing the highest-value stimuli or actions based on reinforcement history. Basic forms of value-driven learning are available early in life, including in early childhood (Winter & Sheridan, 2014). Several studies have also shown comparable performance on value-based learning tasks in adolescents and adults (Hauser, Iannaccone, Walitza, Brandeis, & Brem, 2015; van den Bos, Guroglu, van den Bulk, Rombouts, & Crone, 2009). One such study tested adolescents (age 12–16) and adults (age 20–29) in a probabilistic-learning task using monetary gains and losses as reinforcement. Individuals learned to select one of two cues that was reinforced with 80% probability (Hauser et al., 2015), which rendered learning fairly easy and resulted in similar accuracy for adolescents and adults.

Learning demands can be titrated upward by increasing the number of cues to learn, reducing the reinforcement probability, or increasing the complexity of the feedback given. These more complex learning situations challenge adolescents' learning abilities. For example, Palminteri et al. (2016) tested adolescents (age 12–17) and adults (age 18–32) on a probabilistic-learning task presenting gain and loss information along with counterfactual outcome information (i.e., feedback for the chosen and unchosen cue). A comparison of alternative computational models revealed that adults' performance advantage was explained by their tendency to incorporate reinforcement valence (gain/loss) and outcome information for both chosen and unchosen cue options. Adolescents learned according to a simple value-updating rule and did not integrate the complex feedback. Hence, age-related improvements in learning from adolescence to adulthood reveal themselves when learning environments are particularly complex.

Interestingly, there are some learning situations in which adolescents outperform adults. In a probabilistic-learning study, Davidow et al. (2016) demonstrated that adolescents (age 13–17) formed reinforced stimulus-stimulus associations better than adults (age 20–30), suggesting enhanced learning from experience. Relatedly, when presented with a false instruction, adolescents (age 13–17) prioritized learning from actually experienced feedback (resulting in a performance advantage), whereas adults (age 18–34) persisted longer following the false instruction (Decker, Lourenco, Doll, & Hartley, 2015). At a later test, adolescents showed less residual influence from the false instruction than adults, further suggesting that they had prioritized their experienced feedback. Together, these studies suggest that some conditions can be leveraged to reveal key learning advantages during adolescence.

Multiple systems in the brain support the learning and goal-directed implementation of value. In adults, the hippocampus and striatum can functionally couple

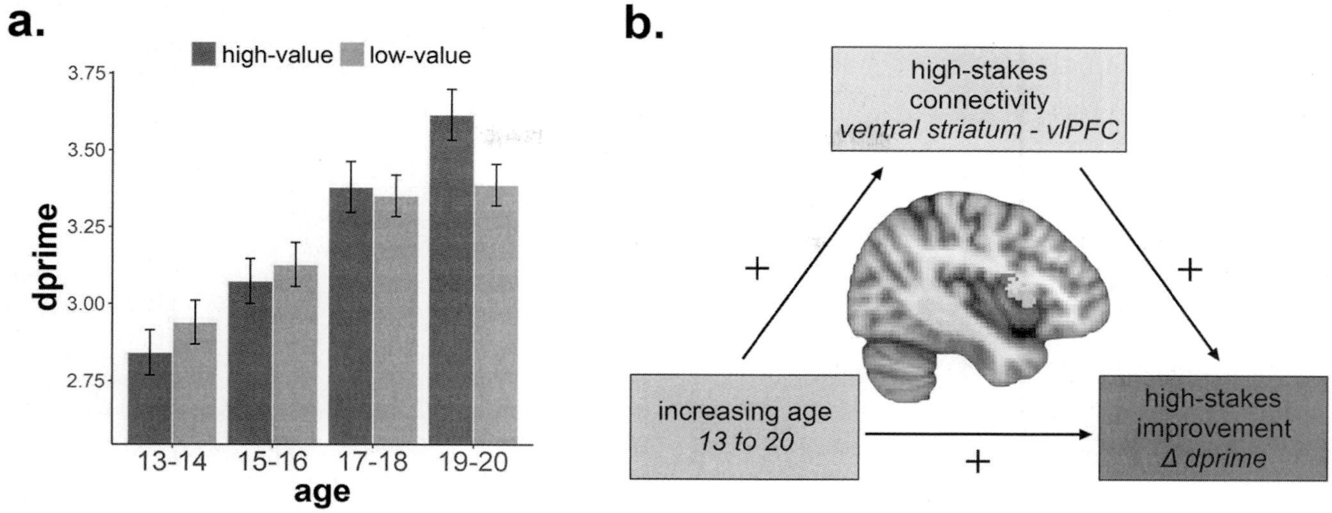

FIGURE 85.1 *A*, When performing a cognitive-control task for low- versus high-value outcomes, older participants selectively improved performance (*dprime* on y-axis) when high-value incentives were at stake, whereas younger participants performed similarly for low-value and high-value conditions. *B*, Functional connectivity analyses seeded in the ventral striatum identified connectivity with ventrolateral prefrontal cortex (VLPFC) that was greater for high-value relative to low-value trials. This pattern of corticostriatal connectivity mediated the relationship between age and value-selective performance. Figure adapted with permission from Insel et al. (2017). (See color plate 98.)

to spread value information (Dickerson, Li, & Delgado, 2011; Kahnt, Park, Burke, & Tobler, 2012; Wimmer & Shohamy, 2012), allowing value learned in one context to transfer into a novel context without requiring relearning. Such generalization informs preferences and supports first-time decision-making (Wimmer & Shohamy, 2012), a tool that could greatly benefit adolescents as they encounter unfamiliar situations. Whether, and when, adolescents benefit from such neural coupling is important for understanding how value can influence goals via alternative routes of learning beyond the corticostriatal value circuit. For example, greater coactivation between the striatum and hippocampus during learning led to stronger learning and memory associations in adolescents (age 13–17) when compared to adults (age 24–30) and may have contributed to adolescents' superior overall learning (Davidow et al., 2016). Recent studies have revealed a shift with age from greater subcortical-subcortical functional connectivity (Davidow et al., 2016; Insel et al., 2017) to increased subcortical-frontal (Insel et al., 2017; Silvers et al., 2016; van den Bos et al., 2012) functional connectivity. The stronger subcortical-frontal connectivity that is observed in adults in these studies (Insel et al., 2017; Silvers et al., 2016; van den Bos et al., 2012) is thought to facilitate sophisticated goal-directed performance.

Age-related shifts in the strategic influence of value parallel the emergence of model-based learning strategies (i.e., the representation of the transitional structure in a decision space acquired through reinforcement experience). Recent work has shown that young adults typically exhibit a "mixture" of model-based and model-free (i.e., purely feedback-driven) learning strategies (Daw, Gershman, Seymour, Dayan, & Dolan, 2011). Moreover, while the representational structure of the environment may emerge in childhood, the strategic implementation of that knowledge, such as selecting the sequences of actions needed to obtain valuable outcomes, may emerge from adolescence into adulthood (Decker et al., 2015; Potter, Bryce, & Hartley, 2017; Stormer, Eppinger, & Li, 2014). Thus, even if younger individuals are capable of using valued feedback to guide learning, the greater complexity of learning demands reveals the continued developmental gains in the strategy and optimization of learning. Recent work suggests that adults adopt model-based strategies when pursuing high-stakes, relative to low-stakes, rewards (Kool, Gershman, & Cushman, 2017). Given that the implementation of model-based learning continues to increase across adolescence, the ability to strategically modulate learning in a value-driven fashion may not emerge until late adolescence and into early adulthood.

Synthesis and Conclusion

More broadly, we propose that the value-based facilitation of goal-directed behaviors such as cognitive control and learning scaffolds on cognitive development, emerging in tandem with the capacity to meet more and more challenging cognitive demands. As such, adolescents may capitalize on value to improve performance when executing relatively easier cognitive tasks, once they have demonstrated stable competence for a given cognitive process. However, when faced with difficult tasks taxing cognitive processes undergoing continued maturation, adolescents face capacity limits that prevent value from bolstering performance. Thus, value may not permeate control performance until the developing capacity for or mastery over a cognitive skill stabilizes.

While we have primarily suggested that this trajectory scaffolds on cognitive development, it is also possible that strategic shifts with age could influence the cost-benefit calculations that guide decisions of when to engage control processes. For example, if a cognitive challenge is more difficult for younger individuals and thus more costly to perform, they may be less likely to choose to engage in a challenging process even if valued outcomes are at stake. Likewise, because cognitive demands are more taxing at younger ages, higher rewards may be required to provoke performance improvements. Future developmental work is needed to identify how these cost-benefit calculations for cognitive effort allocation change with age in tandem with cognitive capabilities.

Acknowledgment

The preparation of this manuscript was supported by a National Science Foundation CAREER award (BCS-1452530) to Leah H. Somerville.

REFERENCES

Aarts, E., van Holstein, M., & Cools, R. (2011). Striatal dopamine and the interface between motivation and cognition. *Frontiers in Psychology, 2*, 163.

Alexander, G. E., DeLong, M. R., & Strick, P. L. (1986). Parallel organization of functionally segregated circuits linking basal ganglia and cortex. *Annual Review of Neuroscience, 9*, 357–381.

Balleine, B. W., & O'Doherty, J. P. (2010). Human and rodent homologies in action control: Corticostriatal determinants of goal-directed and habitual action. *Neuropsychopharmacology, 35*(1), 48–69.

Barkley-Levenson, E., & Galvan, A. (2014). Neural representation of expected value in the adolescent brain. *Proceedings of the National Academy of Sciences, 111*(4), 1646–1651.

Bjork, J. M., Smith, A. R., Chen, G., & Hommer, D. W. (2010). Adolescents, adults and rewards: Comparing motivational neurocircuitry recruitment using fMRI. *PLoS One, 5*(7), e11440.

Blake, P. R., & Rand, D. G. (2010). Currency value moderates equity preference among young children. *Evolution and Human Behavior, 31*(3), 210–218.

Botvinick, M., & Braver, T. (2015). Motivation and cognitive control: From behavior to neural mechanism. *Annual Review of Psychology, 66*, 83–113.

Boureau, Y.-L., Sokol-Hessner, P., & Daw, N. D. (2015). Deciding how to decide: Self-control and meta-decision making. *Trends in Cognitive Sciences, 19*(11), 700–710.

Braams, B. R., van Duijvenvoorde, A. C., Peper, J. S., & Crone, E. A. (2015). Longitudinal changes in adolescent risk-taking: A comprehensive study of neural responses to rewards, pubertal development, and risk-taking behavior. *Journal of Neuroscience, 35*(18), 7226–7238.

Braver, T. S., Krug, M. K., Chiew, K. S., Kool, W., Westbrook, J. A., Clement, N. J., ... Somerville, L. H. (2014). Mechanisms of motivation-cognition interaction: Challenges and opportunities. *Cognitive, Affective, & Behavioral Neuroscience, 14*(2), 443–472.

Casey, B. J. (2015). Beyond simple models of self-control to circuit-based accounts of adolescent behavior. *Annual Review of Psychology, 66*, 295–319.

Castel, A. D., Humphreys, K. L., Lee, S. S., Galvan, A., Balota, D. A., & McCabe, D. P. (2011). The development of memory efficiency and value-directed remembering across the life span: A cross-sectional study of memory and selectivity. *Developmental Psychology, 47*(6), 1553–1564.

Chai, L. R., Khambhati, A. N., Ciric, R., Moore, T. M., Gur, R. C., Gur, R. E., ... Bassett, D. S. (2017). Evolution of brain network dynamics in neurodevelopment. *Network Neuroscience, 1*(1), 14–30.

Chatham, C. H., Claus, E. D., Kim, A., Curran, T., Banich, M. T., & Munakata, Y. (2012). Cognitive control reflects context monitoring, not motoric stopping, in response inhibition. *PLoS One, 7*(2), e31546.

Church, J. A., Bunge, S. A., Petersen, S. E., & Schlaggar, B. L. (2017). Preparatory engagement of cognitive control networks increases late in childhood. *Cerebral Cortex, 27*(3), 2139–2153.

Crone, E. A., & Steinbeis, N. (2017). Neural perspectives on cognitive control development during childhood and adolescence. *Trends in Cognitive Sciences, 21*(3), 205–215.

Dalley, J. W., Everitt, B. J., & Robbins, T. W. (2011). Impulsivity, compulsivity, and top-down cognitive control. *Neuron, 69*(4), 680–694.

Davidow, J. Y., Foerde, K., Galván, A., & Shohamy, D. (2016). An upside to reward sensitivity: The hippocampus supports enhanced reinforcement learning in adolescence. *Neuron, 92*(1), 93–99.

Davidow, J. Y., Insel, C., & Somerville, L. H. (2018). Adolescent development of value-guided goal pursuit. *Trends in Cognitive Sciences, 22*(8), 725–736.

Daw, N. D., Gershman, S. J., Seymour, B., Dayan, P., & Dolan, R. J. (2011). Model-based influences on humans' choices and striatal prediction errors. *Neuron, 69*(6), 1204–1215.

Decker, J. H., Lourenco, F. S., Doll, B. B., & Hartley, C. A. (2015). Experiential reward learning outweighs instruction prior to adulthood. *Cognitive, Affective, & Behavioral Neuroscience, 15*(2), 310–320.

Dickerson, K. C., Li, J., & Delgado, M. R. (2011). Parallel contributions of distinct human memory systems during probabilistic learning. *NeuroImage, 55*(1), 266–276.

Frank, M. J., & Badre, D. (2011). Mechanisms of hierarchical reinforcement learning in corticostriatal circuits 1: Computational analysis. *Cerebral Cortex, 22*(3), 509–526.

Frank, M. J., Loughry, B., & O'Reilly, R. C. (2001). Interactions between frontal cortex and basal ganglia in working memory: A computational model. *Cognitive, Affective, & Behavioral Neuroscience, 1*(2), 137–160.

Galvan, A., Hare, T. A., Parra, C. E., Penn, J., Voss, H., Glover, G., & Casey, B. J. (2006). Earlier development of the accumbens relative to orbitofrontal cortex might underlie risk-taking behavior in adolescents. *Journal of Neuroscience, 26*(25), 6885–6892.

Geier, C. F., Terwilliger, R., Teslovich, T., Velanova, K., & Luna, B. (2010). Immaturities in reward processing and its influence on inhibitory control in adolescence. *Cerebral Cortex,* 20, 1613–1629.

Gerraty, R. T., Davidow, J. Y., Wimmer, G. E., Kahn, I., & Shohamy, D. (2014). Transfer of learning relates to intrinsic connectivity between hippocampus, ventromedial prefrontal cortex, and large-scale networks. *Journal of Neuroscience, 34*(34), 11297–11303.

Giedd, J. N., Blumenthal, J., Jeffries, N. O., Castellanos, F. X., Liu, H., Zijdenbos, A., ... Rapoport, J. (1999). Brain development during childhood and adolescence: A longitudinal MRI study. *Nature Neuroscience, 2,* 861–863.

Gogtay, N., Giedd, J. N., Lusk, L., Hayashi, K. M., Greenstein, D., Vaituzis, A. C., ... Thompson, P. M. (2004). Dynamic mapping of human cortical development during childhood through early adulthood. *Proceedings of the National Academy of Sciences of the United States of America, 101*(21), 8174–8179.

Gómez, R. L. (2017). Do infants retain the statistics of a statistical learning experience? Insights from a developmental cognitive neuroscience perspective. *Philosophical Transactions of the Royal Society of London B, 372*(1711), 20160054.

Graybiel, A. M. (1990). Neurotransmitters and neuromodulators in the basal ganglia. *Trends in Neurosciences, 13*(7), 244–254.

Haber, S. N., & Knutson, B. (2010). The reward circuit: Linking primate anatomy and human imaging. *Neuropsychopharmacology, 1,* 1–23.

Hauser, T. U., Iannaccone, R., Walitza, S., Brandeis, D., & Brem, S. (2015). Cognitive flexibility in adolescence: Neural and behavioral mechanisms of reward prediction error processing in adaptive decision making during development. *NeuroImage, 104,* 347–354.

Hunnius, S., & Bekkering, H. (2014). What are you doing? How active and observational experience shape infants' action understanding. *Philosophical Transactions of the Royal Society of London B, 369*(1644), 20130490.

Huttenlocher, P. R. (1984). Synapse elimination and plasticity in developing human cerebral cortex. *American Journal of Mental Deficiency, 88*(5), 488–496.

Huttenlocher, P. R. (1990). Morphometric study of human cerebral cortex development. *Neuropsychologia, 28*(6), 517–527.

Insel, C., Kastman, E. K., Glenn, C. R., & Somerville, L. H. (2017). Development of corticostriatal connectivity constrains goal directed behavior through adolescence. *Nature Communications, 8,* 1605.

Kahnt, T., Park, S. Q., Burke, C. J., & Tobler, P. N. (2012). How glitter relates to gold: Similarity-dependent reward prediction errors in the human striatum. *Journal of Neuroscience, 32*(46), 16521–16529.

Kersey, A. J., & Emberson, L. L. (2017). Tracing trajectories of audio-visual learning in the infant brain. *Developmental Science, 20*(6), e12480.

Kinnison, J., Padmala, S., Choi, J.-M., & Pessoa, L. (2012). Network analysis reveals increased integration during emotional and motivational processing. *Journal of Neuroscience, 32*(24), 8361–8372.

Kool, W., Gershman, S. J., & Cushman, F. A. (2017). Cost-benefit arbitration between multiple reinforcement-learning systems. *Psychological Science, 28*(9), 1321–1333.

Kool, W., McGuire, J. T., Rosen, Z. B., & Botvinick, M. M. (2010). Decision making and the avoidance of cognitive demand. *Journal of Experimental Psychology: General, 139*(4), 665–682.

Luking, K. R., Luby, J. L., & Barch, D. M. (2014). Kids, candy, brain and behavior: Age differences in responses to candy gains and losses. *Developmental Cognitive Neuroscience, 9,* 82–92.

Miller, E. K., & Cohen, J. D. (2001). An integrative theory of prefrontal cortex function. *Annual Reviews in Neuroscience, 24,* 167–202.

Mills, K. L., Goddings, A.-L., Clasen, L. S., Giedd, J. N., & Blakemore, S.-J. (2014). The developmental mismatch in structural brain maturation during adolescence. *Developmental Cognitive Neuroscience, 36*(3–4), 147–160.

Niv, Y., Daw, N. D., Joel, D., & Dayan, P. (2007). Tonic dopamine: Opportunity costs and the control of response vigor. *Psychopharmacology, 191*(3), 507–520.

Padmanabhan, A., Geier, C. F., Ordaz, S. J., Teslovich, T., & Luna, B. (2011). Developmental changes in brain function underlying the influence of reward processing on inhibitory control. *Developmental Cognitive Neuroscience, 1*(4), 517–529.

Palminteri, S., Kilford, E. J., Coricelli, G., & Blakemore, S. J. (2016). The computational development of reinforcement learning during adolescence. *PLoS Computational Biology, 12*(6), e1004953.

Paulsen, D. J., Hallquist, M. N., Geier, C. F., & Luna, B. (2015). Effects of incentives, age, and behavior on brain activation during inhibitory control: A longitudinal fMRI study. *Developmental Cognitive Neuroscience, 11,* 105–115.

Potter, T. C. S., Bryce, N. V., & Hartley, C. A. (2017). Cognitive components underpinning the development of model-based learning. *Developmental Cognitive Neuroscience, 25,* 272–280.

Qu, L., Finestone, D. L., Qin, L. J., & Reena, L. Z. (2013). Focused but fixed: The impact of expectation of external rewards on inhibitory control and flexibility in preschoolers. *Emotion, 13*(3), 562–572.

Rakic, P. (1974). Neurons in rhesus monkey visual cortex: Systematic relation between time of origin and eventual disposition. *Science, 183,* 425–427.

Rakic, P., Bourgeois, J. P., Eckenhoff, M. F., Zecevic, N., & Goldman-Rakic, P. S. (1986). Concurrent overproduction of synapses in diverse regions of the primate cerebral cortex. *Science, 232,* 232–235.

Rangel, A., Camerer, C., & Montague, P. R. (2008). A framework for studying the neurobiology of value-based decision making. *Nature Reviews Neuroscience, 9,* 545–556.

Rangel, A., & Hare, T. (2010). Neural computations associated with goal-directed choice. *Current Opinion in Neurobiology, 20*(2), 262–270.

Rodriguez, M. L., Mischel, W., & Shoda, Y. (1989). Cognitive person variables in the delay of gratification of older children at risk. *Journal of Personality and Social Psychology, 57*(2), 359–367.

Satterthwaite, T. D., Wolf, D. H., Erus, G., Ruparel, K., Elliott, M. A., Gennatas, E. D., … Bilker, W. B. (2013). Functional maturation of the executive system during adolescence. *Journal of Neuroscience, 33*(41), 16249–16261.

Schlichting, M. L., Guarino, K. F., Schapiro, A. C., Turk-Browne, N. B., & Preston, A. R. (2017). Hippocampal structure predicts statistical learning and associative inference abilities during development. *Journal of Cognitive Neuroscience, 29*(1), 37–51.

Shenhav, A., Botvinick, M. M., & Cohen, J. D. (2013). The expected value of control: An integrative theory of anterior cingulate cortex function. *Neuron, 79*(2), 217–240.

Shenhav, A., Musslick, S., Lieder, F., Kool, W., Griffiths, T. L., Cohen, J. D., & Botvinick, M. M. (2017). Toward a rational and mechanistic account of mental effort. *Annual Review of Neuroscience, 40*, 99–124.

Sherman, L. E., Steinberg, L., & Chein, J. (2018). Connecting brain responsivity and real-world risk taking: Strengths and limitations of current methodological approaches. *Developmental Cognitive Neuroscience, 33*, 27–41.

Shohamy, D., & Turk-Browne, N. B. (2013). Mechanisms for widespread hippocampal involvement in cognition. *Journal of Experimental Psychology: General, 142*(4), 1159–1170.

Silverman, M. H., Jedd, K., & Luciana, M. (2015). Neural networks involved in adolescent reward processing: An activation likelihood estimation meta-analysis of functional neuroimaging studies. *Neuroimage, 122*, 427–439.

Silvers, J. A., Insel, C., Powers, A., Franz, P., Helion, C., Martin, R. E., … Ochsner, K. N. (2016). vlPFC–vmPFC–amygdala interactions underlie age-related differences in cognitive regulation of emotion. *Cerebral Cortex, 27*(7), 3502–3514.

Simmonds, D. J., Hallquist, M. N., Asato, M., & Luna, B. (2014). Developmental stages and sex differences of white matter and behavioral development through adolescence: A longitudinal diffusion tensor imaging (DTI) study. *Neuroimage, 92*, 356–368.

Somerville, L. H. (2016). Searching for signatures of brain maturity: What are we searching for? *Neuron, 92*(6), 1164–1167.

Somerville, L. H., & Casey, B. J. (2010). Developmental neurobiology of cognitive control and motivational systems. *Current Opinion in Neurobiology, 20*(2), 236–241.

Sowell, E. R., Thompson, P. M., Holmes, C. J., Jernigan, T. L., & Togan, A. W. (1999). In vivo evidence for post-adolescent brain maturation in frontal and striatal regions. *Nature Neuroscience, 2*(10), 859–861.

Stormer, V., Eppinger, B., & Li, S. C. (2014). Reward speeds up and increases consistency of visual selective attention: A lifespan comparison. *Cognitive, Affective, and Behavioral Neuroscience, 14*(2), 659–671.

Strang, N. M., & Pollak, S. D. (2014). Developmental continuity in reward-related enhancement of cognitive control. *Developmental Cognitive Neuroscience, 10*, 34–43.

Tamnes, C. K., Østby, Y., Fjell, A. M., Westlye, L. T., Due-Tønnessen, P., & Walhovd, K. B. (2010). Brain maturation in adolescence and young adulthood: Regional age-related changes in cortical thickness and white matter volume and microstructure. *Cerebral Cortex, 20*, 534–548.

Tanaka, S. C., Doya, K., Okada, G., Ueda, K., Okamoto, Y., & Yamawaki, S. (2016). Prediction of immediate and future rewards differentially recruits cortico-basal ganglia loops. In *Behavioral economics of preferences, choices, and happiness* (pp. 593–616). Tokyo: Springer.

van den Bos, W., Guroglu, B., van den Bulk, B. G., Rombouts, S. A., & Crone, E. A. (2009). Better than expected or as bad as you thought? The neurocognitive development of probabilistic feedback processing. *Frontiers in Human Neuroscience, 3*, 52.

van den Bos, W., Cohen, M. X., Kahnt, T., & Crone, E. A. (2012). Striatum–medial prefrontal cortex connectivity predicts developmental changes in reinforcement learning. *Cerebral Cortex, 22*(6), 1247–1255.

Vink, M., Zandbelt, B. B., Gladwin, T., Hillegers, M., Hoogendam, J. M., van den Wildenberg, W. P., … Kahn, R. S. (2014). Frontostriatal activity and connectivity increase during proactive inhibition across adolescence and early adulthood. *Human Brain Mapping, 35*(9), 4415–4427.

Werchan, D. M., Collins, A. G., Frank, M. J., & Amso, D. (2016). Role of prefrontal cortex in learning and generalizing hierarchical rules in 8-month-old infants. *Journal of Neuroscience, 36*(40), 10314–10322.

Westbrook, A., Kester, D., & Braver, T. S. (2013). What is the subjective cost of cognitive effort? Load, trait, and aging effects revealed by economic preference. *PloS One, 8*(7), e68210.

Wimmer, G. E., & Shohamy, D. (2012). Preference by association: How memory mechanisms in the hippocampus bias decisions. *Science, 338*(6104), 270–273.

Winter, W., & Sheridan, M. (2014). Previous reward decreases errors of commission on later No-Go trials in children 4 to 12 years of age: Evidence for a context monitoring account. *Developmental Science, 17*(5), 797–807.

86 The Social Neuroscience of Cooperation

JULIAN A. WILLS, LEOR HACKEL, ORIEL FELDMANHALL, PHILIP PÄRNAMETS, AND JAY J. VAN BAVEL

ABSTRACT Cooperation occurs at all stages of human life and is necessary for large-scale societies to emerge and thrive. We review literature from several fields to characterize cooperative decision-making. Building on work in neuroeconomics, we suggest a value-based account may provide the most powerful understanding of the psychology and neuroscience of group cooperation. We also review the role of individual differences and social context in shaping the mental processes that underlie cooperation and consider gaps in the literature and potential directions for future research on the social neuroscience of cooperation. We suggest that this multi-level approach provides a more comprehensive understanding of the mental and neural processes that underlie the decision to cooperate with others.

The Social Neuroscience of Cooperation

Cooperation occurs at all stages of human life and is necessary for large-scale societies to emerge and thrive: when individuals prioritize themselves over their community, the consequences can damage social communities, scientific institutions, and our planet. Hence, understanding the psychological and neural underpinnings of cooperative behavior is an important goal for social and cognitive neuroscience. Yet extensive research devoted to the mental processes underlying human prosociality has failed to produce a satisfying framework for understanding how the selfish and prosocial impulses unfold in the human brain.

For centuries, philosophers have debated whether prosocial tendencies are rooted in institutions that regulate our selfish impulses (Hobbes, 1650) or emerge through natural intuitions (Rousseau, 1754). These ancient philosophical debates about human nature remain unresolved. Contemporary scientists continue to grapple with the origins of human prosociality. One on hand, models of *prosocial restraint* assert that the better angels of our nature stem from the deliberate restraint of selfish impulses (DeWall, Baumeister, Gailliot, & Maner, 2008; Kocher, Martinsson, Myrseth, & Wollbrant, 2012; Stevens & Hauser, 2004), whereas models of *prosocial intuition* argue that cooperation stems from intuition and is only corrupted by deliberate attempts to maximize self-interest (Rand, 2016; Rand, Greene, & Nowak, 2012). In this chapter we bridge cognitive neuroscience, neuroeconomics, and social psychology to examine the issue of human prosociality and cooperation.

In the first section, we review literature from several fields to describe common experimental tasks used to measure human cooperation. In the second section, we review the dominant theoretical models that have been used to characterize cooperative decision-making, as well as the brain regions implicated in cooperation. Building on work in neuroeconomics, we suggest that a value-based account may provide the most powerful understanding of the psychology and neuroscience of group cooperation. In the third and fourth sections, we review the role of individual differences and social context in shaping the mental processes that underlie cooperation. Finally, we consider gaps in the literature and offer directions for future research on the cognitive neuroscience of cooperation. We suggest that this multilevel approach provides a more comprehensive understanding of the mental and neural processes that underlie the decision to cooperate with others.

Measuring Cooperation

Cooperation involves any action in which one individual incurs a cost in order to benefit others (Rand & Nowak, 2013). These costs and benefits can range from primary reinforcers (e.g., food, drugs, sex) to secondary reinforcers (e.g., wealth, status, fame). Critically, cooperative acts are not always selfless; sometimes we help others at a cost to obtain rewards in the future. For instance, you may be motivated to tip a bartender not only to reward attentive service but to continue receiving excellent service in the future. For this reason, some researchers distinguish between pure or *altruistic cooperation* (i.e., when current or future rewards are ignored) and *strategic cooperation* (i.e., when future rewards motivate the cooperative act; Camerer & Fehr, 2004; Gintis, 2014). Cooperative acts can be pure, strategic, or a mixture of both. As a result, researchers go to great lengths to disambiguate these motives (Camerer & Fehr, 2004).

To better understand the motives that underlie cooperation and how they are studied, we briefly review four measures of cooperation.

Social dilemmas The most common approach to studying cooperation involves the use of social dilemmas, and perhaps the most widely used measure of cooperation is the *prisoner's dilemma* (PD) game.[1] In the PD, two players are each given the choice to either defect (D) or cooperate (C). This game has been popularized on the British game show *Golden Balls* because it creates a tension in which the fates of two players are tied together. In the standard, symmetric version of the game, both players receive payoff R(eward) if both choose C, payoff P(unishment) if both choose D, and payoffs T(emptation) or S(ucker) if one defects and the other cooperates, respectively. Thus, the hierarchical payoff structure is T>R>P>S. As in the legal system, there is a strong temptation not to be a sucker.

In the PD, each player can maximize individual profit by choosing D, regardless of what the other player chooses. In other words, outcome DD is the unique Nash equilibrium of the game and the prediction for fully rational and selfish players. However, the cooperative outcome, CC, maximizes their collective profit. This feature—that the players are always worse off if both defect compared to cooperate, but each is individually better off by defecting—is what makes the PD a social dilemma (Dawes, 1980; Van Lange, Joireman, Parks, & Van Dijk, 2013). Pitting self-interest against collective interest captures the dynamic at play in countless real-world social decisions, from negotiating nuclear arms agreements to sharing research ideas.

Since decisions are typically made simultaneously, anonymous one-shot PDs (i.e., one round only) are used to measure pure cooperation in both players. In contrast, the iterated PD, in which players play multiple rounds with one another, measures strategic cooperation since players' decisions may affect expectations for subsequent choices. In addition, people cooperate strategically when their choices are made public and players can select partners known to be cooperative (Barclay & Willer, 2007; Feinberg, Willer, & Schultz, 2014). Despite understanding that defecting is in one's best self-interest, decades of evidence from both iterated and one-shot versions of the PD reveal that people willingly cooperate—even with complete strangers.

[1] Invented in 1950 by Merrill Flood and Melvin Dresher, while working at the RAND Corporation (no known relation to Dave Rand, who is cited throughout this chapter) as part of the research investigating the use of game theory to inform nuclear strategy.

To understand cooperation in groups with more than two players, researchers employ the *public goods game* (PGG). In this game, players choose between contributing their endowment to a collective pool (i.e., maximizing joint payoffs) or free riding, in which they keep their own endowment while also reaping the benefits of others' contributions (i.e., maximizing individual payoffs in the short term). The PGG has a similar incentive structure to the PD and is sometimes suggested to be a generalization of it (Rand & Nowak, 2013). The PGG inherits many properties of the PD (e.g., anonymous one-shot games index pure cooperation) since contributing and free riding are group-based analogs of cooperating and defecting. Similar to the findings in the PD, evidence reveals that in typical variants of the PGG, people donate, on average, in 60% of the trials. However, because the PGG also inherits properties of group psychology, important differences can emerge (Dawes, 1980). For instance, contributions in iterated PGGs routinely diminish over time (Andreoni, 1988), whereas those in the PD do not. This may be due to the diffusion of responsibility or absence of direct reciprocity in the PGG, in which punishing one free rider equally penalizes the entire group. PGGs may also be particularly sensitive to other aspects of group psychology, such as norms concerning promise keeping (Bicchieri, 2002) and social identity (Kramer & Brewer, 1984). Furthermore, the PGG likely provides superior ecological validity to the PD since the most pressing real-world cooperative dilemmas, like climate change or science reform, involve more than two people (Camerer, 2011).

Social dilemmas sometimes include additional dimensions, such as introducing reinforcement or punishment opportunities (Fehr & Gächter, 2002; Kelley, 2003),[2] confronting reputational concerns (Milinski, Semmann, & Krambeck, 2002), or manipulating the framing of the game (Van Lange et al., 2013). For instance, framing a social dilemma as a "community game" can double rates of cooperation compared to framing it as a "Wall Street game," likely due to activating norms associated with those contexts (Liberman, Samuels, & Ross, 2004). Moreover, introducing opportunities for reward and punishment almost always boosts contributions (Andreoni, Harbaugh, & Vesterlund, 2002; Dreber, Rand, Fudenberg, & Nowak, 2008; Fehr & Gächter, 2002). These factors appear to alter the value people place on the decision to cooperate.

Bargaining games Another measure of cooperation comes from bargaining games in which responsiveness to

[2] This manipulation also provides an opportunity to observe costly punishment.

fairness norms can be assessed. In the *ultimatum game* (UG; Güth, Schmittberger, & Schwarze, 1982), two players take the role of either proposer or responder. The proposer is given some endowment *E* and must offer the responder some amount *O* (which may be zero). The responder can either accept or reject the offer. If the offer is accepted, the responder receives *O*, and the proposer keeps the remainder (*E* minus *O*). If the offer is rejected, neither player receives anything. From an economically rational standpoint, responders should accept any non-zero offer since some money is better than no money. However, it has been repeatedly observed across cultures that responders will reject offers that are considered unfair according to local norms (Camerer & Fehr, 2004; Henrich et al., 2005), which is typically anything below 20% of the endowment. By rejecting the offer, people are signaling their willingness to forgo their own profit to punish a transgressor who violated fairness norms—harming both parties. Thus, a degree of cooperation is normally required to ensure a fair offer is accepted.[3]

To capture pure prosociality, a modified UG is used in which the responder is not given the option to reject the proposer's offer (Kahneman, Knetsch, & Thaler, 1986)—known as the *dictator game* (DG). In this game, the experimenter endows a sum of money to the dictator, who can then decide how much to give to the receiver. True to its name, the receiver has no bargaining power in the DG and has no choice but to accept the initial offer from the dictator. Surprisingly, dictators nevertheless make non-zero offers in these one-sided games, revealing just how altruistic people can be. This is the case even when the experimenter ensures complete anonymity between the two players, providing a measure of pure prosociality for the dictator since there is no opportunity to reciprocate or punish an unfair split. These games provide some evidence for the tendency of humans to cooperate under a wide variety of conditions.

Models of Cooperation

Models of prosocial behavior make assumptions about the underlying mental computations that guide people toward self-interest or cooperation. In the following section, we contrast three such models of cooperation. The first two are based on a dual-process account that casts intuitive and deliberative processes as competing for control in cooperative behavior. The third offers a single-process framework from neuroeconomics that emphasizes the role of valuation circuits. We briefly review each approach and argue that social and cognitive neuroscience might prove fruitful for arbitrating between these different models.

Intuition versus deliberation One of the most ubiquitous frameworks in psychology is the dual-process model, which posits that the mind can be carved into two core systems: *intuition* (i.e., fast, automatic, and unconscious processes) and *deliberation* (i.e., slow, controlled, and rational processes; Chaiken & Trope, 1989; Evans & Stanovich, 2013; Kahneman, 2011). Research in social neuroscience has attempted to map neural systems onto intuition and deliberation (Cohen, 2005; Satpute & Lieberman, 2006). For instance, patients with ventromedial prefrontal cortex (vmPFC) or amygdala damage presented with blunted affective processing (Bechara, 2000), whereas damage to the dorsolateral prefrontal cortex (dlPFC) impaired deliberative processes, like working memory, reasoning, and self-regulation (Barbey, Koenigs, & Grafman, 2013). The dissociations between these systems have been seen by several scholars as further evidence for dual-process models. In psychology, these models have been used to explain a wide range of phenomena, including stereotypes (Devine, 1989), persuasion (Petty & Cacioppo, 1986), and moral judgment (Greene, Sommerville, Nystrom, Darley, & Cohen, 2001). More recently, competing dual-process models of cooperation have proven reminiscent of old philosophical debates regarding humanity's intrinsic benevolence (Rousseau, 1754) versus the need for institutions to restrain our greedy impulses (Hobbes, 1650).

The most prominent dual-process models of cooperation have argued that prosocial decisions stem primarily from intuition (Rand et al., 2014; Zaki & Mitchell, 2013). For instance, the *social heuristics hypothesis* (Rand et al., 2014) makes three core assumptions: (1) rational self-interested agents should never cooperate in anonymous one-shot games; (2) cooperation stems from error-prone intuitions, whereas self-interest stems from more corrective deliberation; and (3) experimentally boosting reliance on intuition (vs. deliberation) should only result in increased or static cooperation. In their words, "Deliberation only ever reduces cooperation in social dilemmas ... or has no effect ... but never increases social-dilemma cooperation" (Bear, Kagan, & Rand, 2017). According to this view, cooperation is frequently rational—but people develop error-prone heuristics to cooperate even when it would be irrational.

Support for the social heuristics hypothesis comes from a mix of behavioral and neural evidence. The most important behavioral evidence comes from experiments showing that people are slower to make

[3] This can be considered a departure from the strict definition of cooperation we introduced above. However, we include it here for completeness since this class of games is used to study prosociality.

self-interested choices compared to cooperative choices in both the one-shot PD and PGG (Everett, Ingbretsen, Cushman, & Cikara, 2017; Rand et al., 2012). Moreover, putting people under time pressure increases cooperation rates (Rand, Greene, & Nowak, 2012). However, a recent international replication effort came up with mixed support for this key finding, suggesting that the behavioral evidence in support of the social heuristics hypothesis may be weaker than previously thought (Bouwmeester et al., 2017; but see also Rand, 2017). Recent functional magnetic resonance imaging (fMRI) studies found that greater dlPFC activity was associated with decisions that prioritize selfish gain over another's pain (FeldmanHall et al., 2013), while reduced dlPFC functional activity and volume were associated with more generosity in a dictator game, which together suggest a link between deliberation and self-interest (Fermin et al., 2016; Yamagishi et al., 2016). Those findings are in line with dual-process models in general and the social heuristics hypothesis in particular.

This perspective has proven particularly provocative and controversial because it contrasts with more traditional *prosocial restraint* models, whereby cooperation primarily stems from the deliberate restraint of our selfish impulses (Achtziger, Alós-Ferrer, & Wagner, 2015; Lohse, 2016; Martinsson, Myrseth, & Wollbrant, 2012). That is, some argue that humans' unique capacity for self-reflection (i.e., compared to other primates) provides a critical avenue to promote prosocial behavior (Stevens & Hauser, 2004). Moreover, prosocial restraint models are supported by evidence that depleting cognitive resources impairs helping behavior (DeWall et al., 2008) and amplifies dishonesty (Mead, Baumeister, Gino, Schweitzer, & Ariely, 2009; but see Saraiva & Marshall, 2015). We recently found that patients with damage to the dlPFC showed impaired cooperation—and reductions in cooperation scaled with the scope of damage in this region (Wills et al., 2017). We found no such decrements for patients with damage to the vmPFC or the amygdala or other brain-damaged control patients. One limitation of this research area is that several preregistered attempts to replicate ego-depletion effects have found null or very small effect sizes—calling many findings in this literature into question. As such, the evidence behind these models has proven unconvincing to opposing camps.

A value-based approach to cooperation A central approach to neuroeconomics has examined how value is represented in the human brain and used to guide decision-making. Instead of conceptualizing cooperation as arising from distinct, competing psychological systems, we argue that cooperation, and social preferences in general, should be situated within such a value-based decision framework. Central to this framework is the assumption, found in most economic and psychological theories of choice, that prior to deciding between one or several alternatives, an organism determines the subjective value of each alternative. Subjective value allows comparisons between complex and qualitatively different alternatives by placing them on a common scale (Bartra, McGuire, & Kable, 2013; Levy & Glimcher, 2012; Rangel, Camerer, & Montague, 2008). Moreover, this approach allows for individual differences and contextual factors to shape the value of these alternatives. We provide an overview of this perspective, examine the underlying neural system involved in value computations, and describe how this might be fruitfully applied to the study of cooperation.

The field of neuroeconomics has focused on understanding how the brain computes the value of alternative actions during decisions, such as when people are forced to decide between engaging in self-interest or cooperation. The decision-making literature across topics has consistently found that brain activation in the orbitofrontal cortex or vmPFC, ventral striatum (VS), and posterior cingulate cortex increases with subjective value during choice tasks and while receiving monetary, primary, or social rewards (Bartra, McGuire, & Kable, 2013; Levy & Glimcher, 2012). This has been taken as evidence that representations of value are computed in these regions and used as a common currency to decide between different options (Grabenhorst & Rolls, 2011; Levy & Glimcher, 2012).

Recent studies suggest that a value-based framework better explains human cooperation than either dual-process accounts mentioned above. Prosocial intuition models argue that intuitive responses are shorter than deliberative ones. But from the perspective of value-based frameworks, response times are a function of the discriminability of alternatives: people make faster choices when deciding between very different values as opposed to similar values (Krajbich, Armel, & Rangel, 2010; Shadlen & Kiani, 2013). Thus, these models make competing predictions about cooperation. In one such experiment, participants played multiple PGGs with varying returns on money contributed (Krajbich, Bartling, Hare, & Fehr, 2015). In one condition, for each monetary unit contributed, each player received 50% back. In the other conditions, the multipliers were 30% (rewarding selfishness) and 90% (rewarding cooperation).[4] Consistent with the value-based approach, the

[4] Recall that in a PGG a player is *always* better off keeping the money rather than cooperating. In other words, the multiplier per monetary unit and player is always strictly less than 1.

relationship between reaction time and cooperation was determined by the reward structure: cooperation was fast when it was rewarded, and selfishness was fast when it was rewarded. In other words, cooperation decisions were fastest when the reward structure made the alternatives clear. These findings also highlight why researchers should be cautious when interpreting reaction time differences as evidence for intuition or deliberation.

A growing body of work in cognitive neuroscience also supports the value-based account of cooperation. Specifically, several studies have found that vmPFC activation relates to value-based quantities during cooperative decisions (FeldmanHall, Dalgeish, Evans, & Mobbs, 2015; Hutcherson, Bushong, & Rangel, 2015; Zaki, Lopez, & Mitchell, 2014). During altruistic decision-making, for instance, the brain forms an overall value signal as a weighted sum of two quantities: the payoffs available for oneself and to a recipient (Hutcherson, Bushong, & Rangel, 2015). Both quantities were associated with activation in the vmPFC during people's choices, supporting the idea that the vmPFC encodes the overall value of prosocial choices.

The notion that the vmPFC encodes the subjective value of cooperation is also supported by findings from a neuroimaging study conducted while people engaged in the PGG (Wills, Hackel, & Van Bavel, 2018). We found that vmPFC activity was greater when participants made choices aligned with their overall social preferences (i.e., when cooperative players made the decision to cooperate and selfish players made the decision to act selfishly). In contrast, dlPFC activity was associated with choices that went against players' social preferences. Moreover, there was increased connectivity between the vmPFC and dlPFC when people made cooperative decisions that violated social norms. In these cases, the dlPFC may be needed to integrate value signals computed in the vmPFC (Domenech, Redoute, Koechling, & Dreher, 2017), as value-related signals in the dlPFC activate after those in the vmPFC (Sokol-Hessner, Hutcherson, Hare, & Rangel, 2012). Clarifying the connectivity between regions will likely be key to further arbitrating between the value-based model and competing frameworks (see figure 86.1).

There is growing research into the various psychological factors that modulate (i.e., suppress or amplify) value. After all, when constructing interventions to promote cooperation, it is vital to understand *when* and *for whom* cooperation is valued. For instance, interventions designed to block "deliberative self-interest" could fail—or even backfire—among those who do not intrinsically value cooperation and need to deliberate longer to fully consider the potential value of cooperation. Similarly, while efforts to deter "intuitive self-interest" could prevail under some circumstances, these same interventions might also reduce cooperation under contexts in which cooperation is strongly valued. Here we review two broad classes of these potential value modulators: (1) contextual factors and (2) individual differences.

Contextual Factors

Several contextual factors can influence cooperative decision-making by shaping social value. For instance, group norms have been known to boost compliance in perceptual judgments (Asch, 1951) and prosocial behavior (Cialdini, Reno, & Kallgren, 1990; Nook, Ong, Morelli, Mitchell, & Zaki, 2016). Evidence for cognitive neuroscience suggests that group norms also modulate

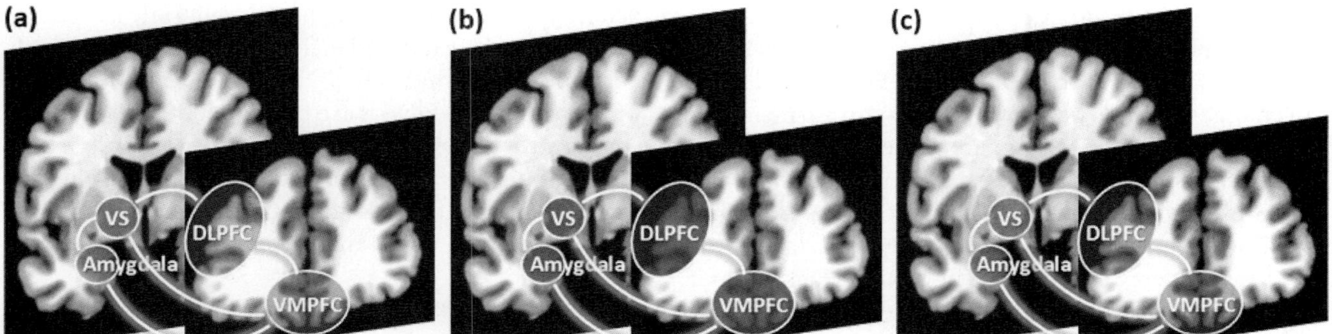

FIGURE 86.1 Candidate neural systems of cooperative decision-making. Dual-process models of prosocial behavior predict cooperation stems from either (*A*) neural regions involved in intuition (*red*) or (*B*) neural regions involved in deliberation (*blue*). Or, (*C*) value-based models predict cooperation should stem from regions typically recruited during decision making (*red*), as well as heightened connectivity between the dlPFC (*blue*) and vmPFC for decisions that require more effort. VS = ventral striatum; vmPFC = ventromedial prefrontal cortex; dlPFC = dorsolateral prefrontal cortex. Graphics adapted from (Phelps, Lempert, & Sokol-Hessner, 2014). (See color plate 99.)

the neural substrates of subjective value (Nook & Zaki, 2015; Wills, Hackel, & Van Bavel, 2018), as well as systems implicated in conflict monitoring (Chang & Sanfey, 2013) and control (Knoch, Pascual-Leone, Meyer, Treyer, & Fehr, 2006; Richeson et al., 2003). For example, disrupting the dlPFC has been shown to disrupt participants' ability to act in accordance with fairness norms and reject unfair offers in ultimatum games (Knoch et al., 2006). Notably, participants still reported accurate valuations of the offers, suggesting a role of the dlPFC in integrating the outputs from valuation circuits.

Social psychologists distinguish between descriptive norms (i.e., how do others typically behave?) and injunctive norms (i.e., how should others behave?). Since there is strong evidence that descriptive norms influence cooperation (Kopelman, Weber, & Messick, 2002), the same is likely true for injunctive norms—especially since cooperation is often characterized as a moral imperative. Consider, for instance, an influential finding in which framing the PGG as "the community game" boosts cooperation significantly more than when it is called the "Wall Street game" (Liberman, Samuels, & Ross, 2004). Even when other players were expected to be selfish, those assigned to the community condition decided to cooperate nonetheless, suggesting that injunctive norms can bias moral behavior.

Social identity—a person's sense of who they are based on their group membership—is another core social psychological construct that drives cooperation and conflict (Tajfel & Turner, 2001). For instance, cooperative decisions can be influenced by existing intergroup conflicts, such as race relations (Kubota, Li, Bar-David, Banaji, & Phelps, 2013) and political partisanship (Iyengar & Westwood, 2015), as well as by artificially created identities (Marcus-Newhall, Miller, Holtz, & Brewer, 1993). Social identity may drive cooperation because it connotes interdependence: people assume in-group members will reciprocate with one another (Yamigishi, 1992). There is also reason to believe that identity can change the value people place on in-group members and their outcomes. For example, one study found greater activation in the ventral striatum when participants observed in-group members receive rewards compared to out-group members, but only for participants who heavily identified with the in-group (Hackel, Zaki, & Van Bavel, 2017). Indeed, simply categorizing faces of in-group members activates the neural circuitry associated with valuation, including the amygdala, orbitofrontal cortex, and dorsal striatum (Van Bavel, Packer, & Cunningham, 2008). Thus, generating a shared group identity can induce cooperation by imbuing in-group members with value or increasing the expectations of future reward due to reciprocity.

Individual Differences

People differ in their tendency to cooperate, and these preferences tend to be stable over time (Volk, Thöni, & Ruigrok, 2012). Within PGGs, for instance, researchers have estimated that a substantial majority of people (50%–55%) are conditional cooperators (i.e., those who only cooperate when others cooperate), a sizable portion (23%–30%) are considered consistent free riders (Fischbacher, Gächter, & Fehr, 2001), and only a small percentage (5%–10%) fall into the category of consistent contributors who always cooperate (Weber & Murnighan, 2008). Some measures, such as the Social Value Orientation measure, are designed to capture these differences (see Van Lange, 1999). *Proselfs* are people who place a high value on their own rewards, whereas *prosocials* are people who place a high value on collective rewards. Research in the past decade has consistently found that prosocials are more inclined to cooperate in both one-shot and iterated games (Balliet, Parks, & Joireman, 2009). Thus, individual differences are robust predictors of cooperative (vs. selfish) behavior.

Critically, individual differences may determine which contextual factors steer cooperative decision-making. Take, for instance, consistent contributors, who are defined by their iconoclastic commitment to cooperating under any circumstance (i.e., even when everyone else in their group is free riding). There is evidence that the mere presence of these consistent contributors can boost cooperation in others by activating moral identities (Gill, Packer, & Van Bavel, 2013). That is, consistent contributors may provide a contextual cue that predominantly boosts cooperation among individuals who consider generosity and fairness to be central features of their identity (Packer, Gill, Chu, & Van Bavel, 2018). In addition, there is evidence that experimentally invoking deliberation promotes cooperation, but only for people exhibiting prosocial tendencies (Mischkowski & Glöckner, 2015). Thus, individual differences can also predict which contextual factors are more or less likely to shape cooperative decision-making. More work should examine this interplay using neuroscientific methods to better understand how individual differences and context are integrated in the brain during decision-making.

Future Directions

Attention A key element of dynamic value-based cognition is the role of attention. By measuring participants' fixations during simple economic choices, researchers have shown that attention to certain options influences decisions (Krajbich, Armel, & Rangel, 2010). These findings have been shown to also hold for more complicated value-based choices, such as those that are moral (Pärnamets, Balkenius, & Richardson, 2014). By tracking participants' fixations and prompting them to make a choice only after sufficiently fixating on one option, researchers were even able to influence what choice participants made (Pärnamets et al., 2015). Moreover, one study found that value signals in the striatum and vmPFC were modulated by the relative value of fixated versus nonfixated food options (Lim, O'Doherty, & Rangel, 2011). Thus, visual attention influences valuation and alters prosocial behavior. In our view, integrating measures of attention and other sensory information into models of cooperative decision-making offers significant opportunities for understanding more about the underlying mental processes and potentially even designing effective interventions for increasing cooperation.

Learning A key element of value-based models is that people learn the value of different actions over time, whether through personal experience (FeldmanHall, Otto, & Phelps, 2018) or social observation (Haaker et al., 2017; Lindström, Haacker, & Olsson, 2018). Understanding this process may offer new insights into how people choose to cooperate.

Canonical models of reciprocity suggest that people form impressions of others' generosity and tend to help those viewed as generous (Wedekind & Milinski, 2000). However, models of value learning in neuroscience suggest another route by which people may learn to cooperate with others. During cooperative interactions, people experience *reward value*—that is, the material benefits of the interaction. When receiving money from an interaction partner, people engage not only the neural regions associated with forming social impressions but also the neural regions associated with reward learning (e.g., the ventral striatum; Hackel, Doll, & Amodio, 2015). As a result, people learn to reciprocate not only with givers who frequently display generosity but also with givers who have greater wealth and thus provide larger rewards (Hackel & Zaki, 2018). Modeling how experience and feedback is integrated into value to guide future decisions is key to fully understanding cooperation. Although the evidence is currently sparse, value learning likely plays a similar role in shaping whether people contribute to collective goods in social dilemmas.

Conclusion

Unlocking the secret to group cooperation is critical for solving social dilemmas ranging from climate change to public resource management to improving science. For this reason, the study of cooperation has attracted an enormous amount of attention in recent years. We believe that a value-based approach holds significant promise for understanding how different people in different contexts make cooperative decisions. This approach not only has an explanatory power that can generate important directions in learning and attention but offers to bridge a number of literatures under a common multilevel framework. This has important implications since models consistent with neural architecture should be privileged over models that are not biologically described, and theories that provide consistent evidence across multiple levels of analysis are most likely to provide a complete and enduring explanation of behavior (Wilson, 1998). If this approach can harness the collective intelligence of scientists and scholars from philosophy to neuroscience, it will allow them to cooperate on solving a long-standing scientific debate as well as some of the most pressing problems facing humanity.

Acknowledgments

This chapter was partially funded by a grant from the National Science Foundation to Jay J. Van Bavel (award #1349089) and from the Swedish Research Council to Philip Pärnamets (2016-06793).

REFERENCES

Achtziger, A., Alós-Ferrer, C., & Wagner, A. K. (2011). *Social preferences and self-control.* Working Paper, University of Constance.

Andreoni, J. (1988). Why free ride? Strategies and learning in public goods experiments. *Journal of Public Economics, 37,* 291–304.

Andreoni, J., Harbaugh, W. T., & Vesterlund, L. (2002). The carrot or the stick: Rewards, punishments and cooperation. University of Oregon Department of Economics Working Paper. Eugene, OR.

Apps, M. A. J., & Sallet, J. (2017). Social learning in the medial prefrontal cortex. *Trends in Cognitive Science, 21,* 151–152.

Asch, S. E. (1951). Effects of group pressure upon the modification and distortion of judgments. In H. Guetzkow (Ed.), *Groups, leadership, and men* (pp. 222–236). Pittsburgh, PA: Carnegie Press.

Balliet, D., Parks, C., & Joireman, J. (2009). Social value orientation and cooperation in social dilemmas: A meta-analysis. *Group Processes & Intergroup Relations, 12,* 533–547.

Barbey, A. K., Koenigs, M., & Grafman, J. (2013). Dorsolateral prefrontal contributions to human working memory. *Cortex, 49,* 1195–1205.

Barclay, P., & Willer, R. (2007). Partner choice creates competitive altruism in humans. *Proceedings of the Royal Society of London B: Biological Sciences, 274,* 749–753.

Bartra, O., McGuire, J. T., & Kable, J. W. (2013). The valuation system: A coordinate-based meta-analysis of BOLD fMRI experiments examining neural correlates of subjective value. *Neuroimage, 76,* 412–427.

Bear, A., Kagan, A., & Rand, D. G. (2017). Co-evolution of cooperation and cognition: The impact of imperfect deliberation and context-sensitive intuition. *Proceedings of the Royal Society B: Biological Sciences, 284*, 20162326.

Bechara, A. (2000). Emotion, decision making and the orbitofrontal cortex. *Cerebral Cortex, 10*, 295–307.

Bicchieri, C. (2002). Covenants without swords: Group identity, norms, and communication in social dilemmas. *Rationality and Society, 14*, 192–228.

Bouwmeester, S., Verkoeijen, P. P., Aczel, B., Barbosa, F., Bègue, L., Brañas-Garza, P., … Evans, A. M. (2017). Registered replication report: Rand, Greene, and Nowak (2012). *Perspectives on Psychological Science, 12*, 527–542.

Camerer, C. (2011). The promise and success of lab-field generalizability in experimental economics: A critical reply to Levitt and List. Available at SSRN 1977749.

Camerer, C. F., & Fehr, E. (2004). Measuring social norms and preferences using experimental games: A guide for social scientists. In J. Henrich, R. Boyd, S. Bowles, C. Camerer, E. Fehr, & H. Ginti (Eds.), *Foundations of human sociality: Economic experiments and ethnographic evidence from fifteen small-scale societies* (pp. 55–95). Oxford: Oxford University Press.

Chaiken, S., & Trope, Y. (Eds.). (1999). *Dual-process theories in social psychology*. New York: Guilford Press.

Chang, L. J., & Sanfey, A. G. (2013). Great expectations: Neural computations underlying the use of social norms in decision-making. *Social Cognitive and Affective Neuroscience, 8*, 277–284.

Cialdini, R. B., Reno, R. R., & Kallgren, C. A. (1990). A focus theory of normative conduct: Recycling the concept of norms to reduce littering in public places. *Journal of Personality and Social Psychology, 58*, 1015–1026.

Cohen, J. D. (2005). The vulcanization of the human brain: A neural perspective on interactions between cognition and emotion. *Journal of Economic Perspectives, 19*, 3–24.

Dawes, R. M. (1980). Social dilemmas. *Annual Review of Psychology, 31*, 169–193.

Devine, P. G. (1989). Stereotypes and prejudice: Their automatic and controlled components. *Journal of Personality and Social Psychology, 56*, 5–18.

DeWall, C. N., Baumeister, R. F., Gailliot, M. T., & Maner, J. K. (2008). Depletion makes the heart grow less helpful: Helping as a function of self-regulatory energy and genetic relatedness. *Personality and Social Psychology Bulletin, 34*(12), 1653–1662. doi:10.1177/0146167208323981

Domenech, P., Redouté, J., Koechlin, E., & Dreher, J. C. (2017). The neuro-computational architecture of value-based selection in the human brain. *Cerebral Cortex, 28*, 585–601.

Dreber, A., Rand, D. G., Fudenberg, D., & Nowak, M. A. (2008). Winners don't punish. *Nature, 452*, 348–351.

Engel, C. (2011). Dictator games: A meta study. *Experimental Economics, 14*, 583–610.

Evans, J. S. B., & Stanovich, K. E. (2013). Dual-process theories of higher cognition: Advancing the debate. *Perspectives on Psychological Science, 8*, 223–241.

Everett, J. A., Ingbretsen, Z., Cushman, F., & Cikara, M. (2017). Deliberation erodes cooperative behavior—even towards competitive out-groups, even when using a control condition, and even when eliminating selection bias. *Journal of Experimental Social Psychology, 73*, 76–81.

Fehr, E., & Gächter, S. (2002). Altruistic punishment in humans. *Nature, 415*, 137–140.

FeldmanHall, O., Dalgleish, T., Evans, D., & Mobbs, D. (2015). Empathic concern drives costly altruism. *Neuroimage, 105*, 347–356.

FeldmanHall, O., Dalgleish, T., Mobbs, D. (2013). Alexithymia decreases altruism in real social decisions. *Cortex, 49*(3), 899–904.

FeldmanHall, O., Dalgleish, T., Thompson, R., Evans, D., Schweizer, S., & Mobbs, D. (2012). Differential neural circuitry and self-interest in real vs hypothetical moral decisions. *Social Cognitive Affective Neuroscience, 7*, 743–751.

FeldmanHall, O., Otto, A. R., & Phelps, E. A. (2018). Learning moral values: Another's desire to punish enhances one's own punitive behavior. *Journal of Experimental Psychology: General, 147*, 1211–1224.

FeldmanHall, O., Son, J., & Heffner, J. (2018). Norms and the flexibility of moral action. *Personality Neuroscience, 1*, 1–14.

Feinberg, M., Willer, R., & Schultz, M. (2014). Gossip and ostracism promote cooperation in groups. *Psychological Science, 25*, 656–664.

Fermin, A. S., Sakagami, M., Kiyonari, T., Li, Y., Matsumoto, Y., & Yamagishi, T. (2016). Representation of economic preferences in the structure and function of the amygdala and prefrontal cortex. *Scientific Reports, 6*, 20982.

Fischbacher, U., Gächter, S., & Fehr, E. (2001). Are people conditionally cooperative? Evidence from a public goods experiment. *Economics Letters, 71*, 397–404.

Gill, M. J., Packer, D. J., & Van Bavel, J. (2013). More to morality than mutualism: Consistent contributors exist and they can inspire costly generosity in others. *Behavioral and Brain Sciences, 36*, 90.

Gintis, H. (2014). *The bounds of reason: Game theory and the unification of the behavioral sciences*. Princeton, NJ: Princeton University Press.

Grabenhorst, F., & Rolls, E. T. (2011). Value, pleasure and choice in the ventral prefrontal cortex. *Trends in Cognitive Sciences, 15*, 56–67.

Greene, J. D., Sommerville, R. B., Nystrom, L. E., Darley, J. M., & Cohen, J. D. (2001). An fMRI investigation of emotional engagement in moral judgment. *Science, 293*, 2105–2108.

Gu, X., Wang, X., Hula, A., Wang, S., Xu, S., Lohrenz, T. M., … Montague, P. R. (2015). Necessary, yet dissociable contributions of the insular and ventromedial prefrontal cortices to norm adaptation: Computational and lesion evidence in humans. *Journal of Neuroscience, 35*, 467–473.

Güth, W., Schmittberger, R., & Schwarze, B. (1982). An experimental analysis of ultimatum bargaining. *Journal of Economic Behavior & Organization, 3*, 367–388.

Haaker, J., Yi, J., Petrovic, P., & Olsson, A. (2017). Endogenous opioids regulate social threat learning in humans. *Nature Communications, 8*, 15495.

Hackel, L. M., Doll, B. B., & Amodio, D. M. (2015). Instrumental learning of traits versus rewards: Dissociable neural correlates and effects on choice. *Nature Neuroscience, 18*, 1233–1235.

Hackel, L. M., & Zaki, J. (2018). Propagation of economic inequality through reciprocity and reputation. *Psychological Science, 29*, 604–613.

Hackel, L. M., Zaki, J., & Van Bavel, J. J. (2017). Social identity shapes social valuation: Evidence from prosocial behavior and vicarious reward. *Social Cognitive and Affective Neuroscience, 12*, 1219–1228.

Henrich, J., Boyd, R., Bowles, S., Camerer, C., Fehr, E., Gintis, H., … Henrich, N. S. (2005). "Economic man" in cross-cultural perspective: Behavioral experiments in 15 small-scale societies. *Behavioral and Brain Sciences, 28*, 795–815.

Hobbes, T. (1650). *Human nature. Leviathan.* England.

Hutcherson, C. A., Bushong, B., & Rangel, A. (2015). A neurocomputational model of altruistic choice and its implications. *Neuron, 87*, 451–462.

Iyengar, S., & Westwood, S. J. (2015). Fear and loathing across party lines: New evidence on group polarization. *American Journal of Political Science, 59*, 690–707.

Kahneman, D. (2011). *Thinking, fast and slow.* New York: Farrar, Straus and Giroux.

Kahneman, D., Knetsch, J. L., & Thaler, R. H. (1986). Fairness and the assumptions of economics. *Journal of Business, 59*(4), S285–S300.

Kelley, H. H. (2003). *An atlas of interpersonal situations.* Cambridge: Cambridge University Press.

Knoch, D., Pascual-Leone, A., Meyer, K., Treyer, V., & Fehr, E. (2006). Diminishing reciprocal fairness by disrupting the right prefrontal cortex. *Science, 314*, 829–832.

Kocher, M. G., Martinsson, P., Myrseth, K. O. R., & Wollbrant, C. E. (2012). Strong, bold, and kind: Self-control and cooperation in social dilemmas. Working Papers in Economics, No. 523, University of Gothenburg, Sweden.

Kopelman, S., Weber, J. M., & Messick, D. M. (2002). Factors influencing cooperation in commons dilemmas: A review of experimental psychological research. In E. Ostrom, T. Dietz, N. Dolsak, P. C. Stern, S. Stonich, & E. U. Weber (Eds.), *The drama of the commons* (pp. 113–156). Washington, DC: National Academy Press.

Krajbich, I., Armel, C., & Rangel, A. (2010). Visual fixations and the computation and comparison of value in simple choice. *Nature Neuroscience, 13*, 1292–1298.

Krajbich, I., Bartling, B., Hare, T., & Fehr, E. (2015). Rethinking fast and slow based on a critique of reaction-time reverse inference. *Nature Communications, 6*, 7455.

Kramer, R. M., & Brewer, M. B. (1984). Effects of group identity on resource use in a simulated commons dilemma. *Journal of personality and social psychology, 46*, 1044–1057.

Kubota, J. T., Li, J., Bar-David, E., Banaji, M. R., & Phelps, E. A. (2013). The price of racial bias: Intergroup negotiations in the ultimatum game. *Psychological Science, 24*, 2498–2504.

Levy, D. J., & Glimcher, P. W. (2012). The root of all value: A neural common currency for choice. *Current Opinion in Neurobiology, 22*, 1027–1038.

Liberman, V., Samuels, S. M., & Ross, L. (2004). The name of the game: Predictive power of reputations versus situational labels in determining prisoner's dilemma game moves. *Personality and Social Psychology Bulletin, 30*, 1175–1185.

Lim, S. L., O'Doherty, J. P., & Rangel, A. (2011). The decision value computations in the vmPFC and striatum use a relative value code that is guided by visual attention. *Journal of Neuroscience, 31*, 13214–13223.

Lindström, B., Haaker, J., & Olsson, A. (2018). A common neural network differentially mediates direct and social fear learning. *Neuroimage, 167*, 121–129.

Lohse, J. (2016). Smart or selfish–When smart guys finish nice. *Journal of Behavioral and Experimental Economics, 64*(C), 28–40.

Marcus-Newhall, A., Miller, N., Holtz, R., & Brewer, M. B. (1993). Cross-cutting category membership with role assignment: A means of reducing intergroup bias. *British Journal of Social Psychology, 32*, 125–146.

Martinsson, P., Myrseth, K. O. R., & Wollbrant, C. (2012). Reconciling pro-social vs. selfish behavior: On the role of self-control. *Judgment and Decision Making, 7*(3), 304.

Mead, N. L., Baumeister, R. F., Gino, F., Schweitzer, M. E., & Ariely, D. (2009). Too tired to tell the truth: Self-control resource depletion and dishonesty. *Journal of Experimental Social Psychology, 45*, 594–597.

Milinski, M., Semmann, D., & Krambeck, H. J. (2002). Reputation helps solve the "tragedy of the commons." *Nature, 415*, 424–426.

Mischkowski, D., & Glöckner, A. (2016). Spontaneous cooperation for prosocials, but not for proselfs: Social value orientation moderates spontaneous cooperation behavior. *Scientific Reports, 6*, 21555.

Nook, E. C., Ong, D. C., Morelli, S. A., Mitchell, J. P., & Zaki, J. (2016). Prosocial conformity prosocial norms generalize across behavior and empathy. *Personality and Social Psychology Bulletin, 42*(8), 1045–1062.

Nook, E. C., & Zaki, J. (2015). Social norms shift behavioral and neural responses to foods. *Journal of Cognitive Neuroscience, 27*, 1412–1426.

Packer, D. J., Gill, M. J., Chu, K., & Van Bavel, J. J. (2018). *How does a person like me behave? On how consistent contributors can inspire generous giving among people with prosocial values.* Unpublished manuscript.

Pärnamets, P., Balkenius, C. & Richardson, D. C. (2014). Modelling moral choice as a diffusion process dependent on visual fixations. In P. Bello, M. Guarini, M. McShane, & B. Scassellati (Eds.), *Proceedings of the 36th Annual Conference of the Cognitive Science Society.* Austin, TX: Cognitive Science Society.

Pärnamets, P., Johansson, P., Balkenius, C., Hall, L., Spivey, M. J. & Richardson, D. C. (2015). Biasing moral choices by exploiting the dynamics of eye gaze. *Proceedings of the National Academy of Sciences, 112*, 4170–4175.

Petty, R. E. & Cacioppo, J. T. (1986). The elaboration likelihood model of persuasion. *Advances in Experimental Social Psychology, 19*, 124–129.

Rand, D. G. (2016). Cooperation, fast and slow: Meta-analytic evidence for a theory of social heuristics and self-interested deliberation. *Psychological Science.* https://doi.org/10.1177/0956797616654455

Rand, D. G. (2017). Reflections on the time-pressure cooperation registered replication report. *Perspectives on Psychological Science.* https://doi.org/10.1177/1745691617693625

Rand, D. G., Greene, J. D., & Nowak, M. A. (2012). Spontaneous giving and calculated greed. *Nature, 489*, 427–430.

Rand, D. G., & Nowak, M. A. (2013). Human cooperation. *Trends in Cognitive Sciences, 17*, 413–425.

Rand, D. G., Peysakhovich, A., Kraft-Todd, G. T., Newman, G. E., Wurzbacher, O., Nowak, M. A., & Greene, J. D. (2014). Social heuristics shape intuitive cooperation. *Nature Communications,* 5, 3677.

Rangel, A., Camerer, C., & Montague, P. R. (2008). A framework for studying the neurobiology of value-based decision making. *Nature Reviews Neuroscience, 9*, 545–556.

Richeson, J. A, Baird, A. A., Gordon, H. L., Heatherton, T. F., Wyland, C. L., Trawalter, S., & Shelton, J. N. (2003). An fMRI investigation of the impact of interracial contact on executive function. *Nature neuroscience, 6*(12), 1323–1328.

Rousseau, Jean Jacques. (1754). A discourse on a subject proposed by the Academy of Dijon: What is the origin of inequality among men, and is it authorised by natural law? *Constitution Society.* Retrieved January 23, 2009, from http://www.constitution.org/jjr/ineq.htm.

Saraiva, A. C., & Marshall, L. (2015). Dorsolateral-ventromedial prefrontal cortex interactions during value-guided choice:

A function of context or difficulty? *Journal of Neuroscience, 35*, 5087–5088.

Satpute, A. B., & Lieberman, M. D. (2006). Integrating automatic and controlled processes into neurocognitive models of social cognition. *Brain Research, 1079*, 86–97.

Shadlen, M. N., & Kiani, R. (2013). Decision-making as a window on cognition. *Neuron, 80*, 791–806.

Sokol-Hessner, P., Hutcherson, C., Hare, T., & Rangel, A. (2012). Decision value computation in DLPFC and VMPFC adjusts to the available decision time. *European Journal of Neuroscience, 35*, 1065–1074.

Stevens, Jeffrey R., & Hauser, Marc D. (2004). Why be nice? Psychological constraints on the evolution of cooperation. *Trends in Cognitive Sciences, 8*(2), 60–65. doi:10.1016/j.tics.2003.12.003

Tajfel, H., & Turner, J. (2001). An integrative theory of intergroup conflict. In M. A. Hogg & D. Abrams (Eds.), *Key readings in social psychology. Intergroup relations: Essential readings* (pp. 94–109). New York: Psychology Press.

Van Bavel, J. J., Packer, D. J., & Cunningham, W. A. (2008). The neural substrates of in-group bias: A functional magnetic resonance imaging investigation. *Psychological Science, 19*, 1131–1139.

Van Lange, P. A. M. (1999). The pursuit of joint outcomes and equality in outcomes: An integrative model of social value orientation. *Journal of personality and social psychology, 77*(2), 337.

Van Lange, P. A. M., Joireman, J., Parks, C. D, & Van Dijk, E. (2013). The psychology of social dilemmas: A review. *Organizational Behavior and Human Decision Processes, 120*, 125–141.

Volk, S., Thöni, C., & Ruigrok, W. (2012). Temporal stability and psychological foundations of cooperation preferences. *Journal of Economic Behavior & Organization, 81*, 664–676.

Weber, J. M., & Murnighan, J. K. (2008). Suckers or saviors? Consistent contributors in social dilemmas. *Journal of Personality and Social Psychology, 95*, 1340–1353.

Wedekind, C., & Milinski, M. (2000). Cooperation through image scoring in humans. *Science, 288*, 850–852.

Wills, J., & FeldmanHall, O., NYU PROSPEC Collaboration, Meager, M. R., & Van Bavel, J. J. (2018). Dissociable contributions of the prefrontal cortex in group-based cooperation. *Social Cognitive and Affective Neuroscience*. doi:10.1093/scan/nsy023

Wills, J. A., Hackel, L. M., & Van Bavel, J. J. (2018). Shifting prosocial intuitions: Neurocognitive evidence for a value based account of group-based cooperation. Unpublished manuscript.

Wilson, E. O. (1998). *Consilience: The unity of knowledge*. New York: Knopf.

Yamagishi, T. (1992). Group size and the provision of a sanctioning system in a social dilemma. In W. B. G. Liebrand, D. M. Messick, & H. A. M. Wilke (Eds.), *Social dilemmas: Theoretical issues and research findings* (pp. 267–287). International Series in Experimental Social Psychology. Elmsford, NY: Pergamon Press.

Yamagishi, T., Takagishi, H., Fermin, A. D. S. R., Kanai, R., Li, Y., & Matsumoto, Y. (2016). Cortical thickness of the dorsolateral prefrontal cortex predicts strategic choices in economic games. *Proceedings of the National Academy of Sciences, 113*, 5582–5587.

Zaki, J., & Mitchell, J. P. (2013). Intuitive prosociality. *Current Directions in Psychological Science, 22*, 466–470.

87 Interpersonal Neuroscience

THALIA WHEATLEY AND ADAM BONCZ

ABSTRACT Social interaction is woven into the fabric of daily life. From one interaction to the next, we share ideas and emotions, form bonds, and create new patterns of thought and behavior that ripple outward through our vast social networks. Despite our social nature, scientific understanding of the human brain rests almost entirely on studying single brains in isolation. As a result, we know a lot about how the isolated brain functions and little about how or why brains interact. This is not a minor omission. The fact that social interaction is universal and ubiquitous despite being metabolically expensive suggests it may have been evolutionarily adaptive. Under this assumption, a deep understanding of the human brain requires understanding how and why this behavior occurs. This chapter reviews recent strides in neuroscience to understand social interaction and concludes by highlighting many of the open questions for this exciting new field.

We think and create in near-constant dialogue. From birth, we learn from caregivers and, later, from teachers and peers. Long after developmental milestones have been reached, interaction continues to be the medium through which we share ideas and experiences, align understanding, forge social ties, and leverage collective expertise. Despite our social nature, the traditional approach in neuroscience has been to examine the human brain in isolation: mapping circuits involved in mental processes one brain at a time. Using this approach, we have learned a great deal about sensory, linguistic, motor, affective, and other neural systems yet little about how these systems achieve, support, and benefit from the collective contexts the brain evolved to solve.

Our limited knowledge about how and why brains interact is understandable. The human brain contains a billion neurons arranged to form local and distributed neural circuits. Studying two or more brains in interaction would appear to increase that complexity exponentially. Others have argued that data-reducing constraints inherent in coupled systems may, in fact, constrain such complexity (Kauffman, 1996; Riley, Richardson, Shockley, & Ramenzoni, 2011). Regardless, studying individual brains can only get us so far. In his famous paper on reductionism in science, the neuroscientist Luria points out that water cannot be studied fruitfully by investigating hydrogen and oxygen separately. Similarly, a coupled dyad such as two people interacting may be the "minimum meaningful unit" (Luria, 1987) for social behavior. If this is true, even the most complete picture of a single brain would yield only an impoverished prediction of what happens when brains interact. With increasing technological advances, neuroscientists are beginning to explore interacting brains—the so-called dark matter of social neuroscience (Przyrembel, Smallwood, Pauen, & Singer, 2012). Here we review these advances, the early discoveries they have enabled, and the future directions they afford.

Brain-to-Brain Alignment

According to Pickering and Garrod (2004), a primary goal of interaction is alignment. In their interactive-alignment account, conversation is successful to the degree that interaction partners align their mental models of the world. Such alignment has been deduced by behavioral signals, such as the convergence of phonetics (Pardo, 2006), speech rate (Giles, Coupland, & Coupland, 1991), syntactic structure (Branigan, Pickering, & Cleland, 2000), eye movements (Dale, Warlaumont, & Richardson, 2011), and motor mimicry between interacting partners; cues that both index and promote cooperation and rapport (Marsh, Johnston, Richardson, & Schmidt, 2009; Ramseyer & Tschacher, 2011; Wiltermuth & Heath, 2009). In an attempt to measure this alignment more directly, neuroscientists have investigated whether the brain activity of two individuals also becomes more synchronous when they share similar mental models (see Hasson & Frith, 2016; Nummenmaa, Lahnakoski, & Glerean, 2018 for reviews).

In a now classic paradigm to investigate neural synchrony, the brain responses of speakers and listeners are compared. Here, speakers tell a story while scanned in functional magnetic resonance imaging (fMRI), and later, listeners are also scanned while hearing the speaker's story. Uri Hasson and colleagues' pioneering work demonstrated that speakers' brain activity while telling their stories is similar to the brain activity of listeners hearing those same stories. Thus, synchronous spatiotemporal fluctuations of blood oxygen levels between brains appear to index shared understanding (Silbert et al., 2014; Stephens et al., 2010). Subsequent studies have demonstrated the utility of this approach to discriminate brain alignment at lower (perceptual) as well as higher (semantic) levels of processing (Honey, Thompson, Lerner, & Hasson, 2012; Yeshurun et al., 2017). Brain-to-brain synchrony has also been observed during nonverbal communication (gestural communication: Schippers et al., 2010; facial communication of affect: Anders et al., 2011).

If interbrain synchrony indexes a common understanding, neural synchrony should be greater among people who share a similar way of seeing the world. Parkinson and colleagues investigated this hypothesis by scanning people from a large social network while they watched political, science, humor, and music videos that they had never seen before. Friends in the network had strikingly similar brain responses to these videos compared to people who were further removed from each other in their social network (Parkinson, Kleinbaum, & Wheatley, 2018). These patterns were widespread across many regions and held even after controlling for shared demographics such as age, gender, and ethnicity. Collectively, these studies demonstrate that synchronous neural activity is a useful index of mental alignment and suggest that synchrony may play a role in social bonding. To be clear, the word *synchrony* in these studies only refers to individuals having similar neural responses to the same stimuli. None of these individuals were scanned at the same time. Elucidating the role of synchrony within actual social interaction requires the simultaneous recording of two or more brains in real time: a technique known as *hyperscanning*.

Synchrony in Real-Time Interaction

Great conversation is often described colloquially as feeling "in sync" or "being on the same wavelength." The work mentioned so far suggests that these metaphors are not merely poetic but echo the very machinery that underpins social connection. Hyperscanning reveals that machinery by allowing scientists to observe interacting minds in real time. Here we highlight a few influential hyperscanning approaches using different imaging modalities and the insights they have afforded thus far. Due to limitations of space, we cannot mention all active branches of research—for example, behavioral economics and decision-making. Interested readers are advised to papers by Astolfi et al. (2011), Ciaramidaro et al. (2018), Jahng, Kralik, Hwang, & Jeong (2017), and Tang et al. (2015). Also, here we focus on the results of recent years; for a comprehensive review of earlier studies, see Babiloni and Astolfi (2014).

Hyperscanning Research: Electroencephalography/Magnetoencephalography

Electroencephalography (EEG) is the most widespread neuroimaging technique for hyperscanning due to its flexibility, low cost, and superior temporal resolution (together with magnetoencephalography [MEG]) for the study of rapidly varying phenomena. Since in many everyday interactions—from speech to holding hands—we rely on fast-paced sensorimotor coordination (Jackson & Decety, 2004), it is no surprise that most hyperscanning experiments to date have utilized dual or even group EEG.

Interpersonal sensorimotor coordination has been an active field of research behaviorally, often under the names joint action (Sebanz & Knoblich, 2009) or coordination dynamics (Schmidt & Richardson, 2008). A number of hyperscanning studies have attempted to describe these and related behavioral results in terms of neural synchrony. For example, Kawasaki, Kitajo, and Yamaguchi (2018) used an alternating tapping task whereby participants were tasked with keeping a steady rhythm of taps with their partner but could only see the effects of the taps, not the actions themselves. Pairs that performed well had greater brain-to-brain amplitude correlations and phase synchronization in the higher alpha band (~12 Hz). These correlations were located in frontocentral regions, while phase connectivity differences were mainly observed in sensorimotor areas.

Other hyperscanning studies have applied larger ecological contexts (e.g., Babiloni et al., 2011; for a review see Acquadro, Congado, & De Riddeer, 2016). For example, in a series of studies Lindenberger and his colleagues investigated phase synchronization and graph-theoretical networks across guitarists playing in duets (Lindenberger, Gruber, & Müller, 2009; Müller, Sänger, & Lindenberger, 2013; Sänger, Müller, & Lindenberger, 2012). They found that phase synchronization across musicians was greater in segments requiring more coordination, manifesting primarily in lower (delta and theta) frequency bands. Further, graph theory–based analyses showed that brain-to-brain networks displayed characteristics of optimal complexity (small-world properties) around moments of larger coordination demands.

The superior temporal resolution of EEG (and also MEG) is also important for verbal interaction. Speech is best understood on multiple timescales (e.g., Giraud et al., 2007), and while the semantic content unfolds over seconds or minutes, other features vary quickly over time. One particularly interesting phenomenon in speech is turn taking. Turn taking is inherently social, with turns representing the moments of coordinated role changes from speaker to listener. Turn taking also happens very quickly (~200 ms for a typical gap; see Stivers et al., 2009)—a duration so short it must rely on predictive models in the brains of listeners (Levinson, 2016) that overlap in time with the end of the speaker's turn. Studying the neural underpinnings of turn taking requires high temporal resolution as well as the ability to measure both participants simultaneously, as their behaviors are interdependent.

Recently, Mandel, Bourguignon, Parkkonen, and Hari (2016) used a dual-MEG setup (described in Zhdanov et al., 2015) to investigate the role of motor-related oscillations (~10 and ~20 Hz) over the motor cortex in turn taking. They asked pairs of participants to engage in free conversation over an audio channel. At the end of speaker turns, they found transient peaks of power in the approximately 10 Hz band over the left primary motor cortex of listeners, preceding the listener's turn by at least one second. These results are consistent with the idea that switches in interactive roles are predicted by power changes in the listener's brain associated with motor (possibly respiratory) preparation. Using a similar paradigm, Ahn et al. (2018) discovered phase synchronization in gamma and alpha bands across participants when participants took turns counting numbers compared to counting individually. In the gamma band, specifically, turn taking evoked strong left frontal and left temporal phase synchronization across participants. In the alpha band, turn taking-associated interbrain phase synchrony arose in frontotemporal and right central-parietal regions. Interestingly, alpha band phase synchrony was also captured between regions, across brains: the left frontotemporal areas of one person synchronized with the right central-parietal regions of the other person. These results demonstrated a tight coupling between interacting partners in turn taking and highlighted the role of a coupled network of putative sensorimotor (frontocentral alpha), auditory (temporal alpha and gamma), and executive control (frontal gamma) processes.

A leap forward in terms of the scope of EEG hyperscanning was achieved by Dikker and her colleagues (2017), who extended simultaneous measurements to a whole classroom of students. They employed portable EEG devices with a small number of electrodes each and studied brain-to-brain synchrony across the entire group as a function of different classroom activities. In general, they found that synchrony on the group level was modulated by shared attention. Both student-to-group and overall group synchrony was higher in group discussion and video segments relative to reading and frontal lectures, a pattern predicted by students' ratings of subjective engagement. Students' individual traits (subjective level of focus, empathy) were also linked to their level of synchrony with the group.

Hyperscanning Research: Functional Magnetic Resonance Imaging

Nonverbal communication Although fMRI has a lower temporal resolution than EEG, its superior spatial resolution enables researchers to localize the effects of interactions in great detail. Research groups have already started utilizing fMRI hyperscanning for the study of social interactions, including communication (see Schoot, Hagoort, & Segaert, 2016 for a review), and the results so far are promising. For example, a recent study captured the gradual development of converging neural activity as participants (who could not speak to each other) worked out a way to communicate over time by moving abstract shapes in particular patterns (Stolk et al, 2014; see also Stolk, Verhagen, & Toni, 2016). They found that the development of a successful communication method was predicted by coherent activity between the partners' right superior temporal cortices. Consistent with the development of shared conceptual structures and mental strategies, this correlated activity was independent of the sensorimotor demands of the task itself.

Joint attention Joint attention—considered the minimal building block of human sociality—has also been studied in fMRI hyperscanning (Bilek et al., 2015; Bilek et al., 2017; Koike et al., 2016; Saito et al., 2010). Saito et al. (2010) used a gaze-cueing task to create conditions of stimulus-driven and gaze-driven attentional shifts. After regressing out task effects, interpersonal correlations of the residual time courses showed synchronization in the right inferior frontal gyrus (rIFG) for real relative to pseudo pairs of participants. The authors concluded that rIFG coupling reflected pair-specific effects of joint attention. Building on this work, Koike et al. (2016) found that coupling across brains during mutual gaze was enhanced in the rIFG after a joint attention task. Enhanced rIFG coupling also correlated with behavioral synchrony as measured by eyeblink synchronization. Conceptually, these studies are interesting as they establish a model of neural synchrony corresponding to a minimal communicative context that is nonetheless an important feature of face-to-face interaction (Kang & Wheatley, 2017).

Bilek et al. (2017) used a similar paradigm to explore how brain-to-brain synchronization during joint attention (gaze cueing) might be disrupted in individuals diagnosed with a clinical disorder (borderline personality disorder, or BPD). They found that pairs that included a member with BPD displayed reduced neural synchrony in the right temporoparietal junction compared to neurotypical control pairs. This finding could not be explained by behavioral accuracy, task-related activity differences, or gray matter differences but was positively associated with childhood maltreatment.

To our knowledge only one study has investigated live verbal communication in dual fMRI (Spiegelhalder et al., 2014). In this study, pairs of participants were shown descriptions of life events (e.g., "being lied to"). In speaker-listener trials, one participant was asked to describe such an event while the other listened. In other trials, both participants imagined such an event independently. By using speakers' motor and premotor activity as a regressor for the listeners' brain, the researchers found evidence of coupling during the speaker-listener trials. Speakers' motor-related activity was correlated with listeners' activity in auditory and medial parietal areas, consistent with predictions from single-brain storytelling studies.

Hyperscanning Research: Functional Near-Infrared Spectroscopy

Among the current stable of noninvasive neuroimaging techniques, functional near-infrared spectroscopy (fNIRS) provides the greatest ecological validity. Like fMRI, fNIRS captures fluctuations in blood oxygen levels but with optodes on the scalp rather than magnetic coils. These optodes direct near-infrared (NIR) light, which scatters nonuniformly through brain tissue because of the distinctive absorption spectrum of hemoglobin due to its oxygenation levels. Thus, fNIRS relies on the same assumption as fMRI that neural activation and vascular responses are tightly coupled. Although the current spatial resolution of fNIRS is coarser than fMRI and NIR cannot reach subcortical regions, the steady progress of fNIRS technology makes it an exciting new tool for the study of interaction. Moreover, it allows participants to not only sit upright in the same room but move around and talk, thereby affording a more ecologically valid context for face-to-face interaction.

Many recent hyperscanning studies have employed fNIRS to test whether features of naturalistic interactions modulate brain-to-brain synchrony (Holper et al., 2013; Jiang et al., 2012, 2015; Liu et al., 2016; Nozawa et al., 2016; Osaka et al., 2015). In general, all of these experiments report brain-to-brain synchrony, but the exact results vary depending on the task and manipulation employed. Jiang et al. (2012) studied face-to-face versus back-to-back dialogue and monologue and found that brain-to-brain wavelet coherence (in the left inferior frontal regions) was only present in the face-to-face condition. Liu et al. (2016) employed a joint Jenga game with cooperative versus obstructive conditions. During the game, participants were encouraged to freely discuss their strategy. Interestingly, their results showed increased coherence in the right prefrontal cortex during both cooperation and obstructive interaction, relative to rest, suggesting that synchronization may not depend on shared goals. Using a group-interaction paradigm, Nozawa et al. (2016) identified wavelet coherence in frontopolar areas during face-to-face communication. Osaka et al. (2015) found interbrain coherence during joint humming and singing, but this coherence was not modulated by whether people faced each other or the wall.

The considerable variability in fNIRS results is due, in part, to a small number of available channels and to the selection of varying regions for those channels. As such, different experiments cast light (literally and metaphorically) on different areas, resulting in an incomplete picture that continues to develop. Importantly, though, Liu et al. (2017) demonstrated convergent brain-to-brain synchrony results in fNIRS and fMRI during the storytelling-listening paradigm. Future studies employing more channels will yield interesting insights about the ecological validity of earlier fMRI findings and will extend them to more naturalistic situations.

The advantages of fNIRS over fMRI also make it possible to investigate the role of interbrain synchrony in parent-child interactions (see also Hasegawa et al., 2016 for a dual-MEG approach). The degree to which caregivers and children synchronize their behaviors has been assumed to be stable within, and specific to, that relationship (Feldman, 2015). Do caregiver-children pairs show the same patterns of neural coupling? Reindl et al. (2018) measured wavelet coherence across parents and children (age 5–9 years) playing a simple synchronization game. They reported stronger coherence at frontopolar and dorsolateral prefrontal cortices in parent-child pairs than in stranger-child pairings in the same task and compared to competitive versions (cf., Cui, Bryant, & Reiss, 2012; Pan, Cheng, Zhang, Li, & Hu, 2017). Related research has revealed that the strength of coupling within an adult-infant pair can be increased by mutual gaze, as well as by infant behavior such as smiling (Piazza et al., 2018). During the smiles, the adult's prefrontal activity lagged behind the infant's, suggesting a dynamic wherein infants provide cues for interaction, and the adult's appropriate and contemporaneous feedback establishes neural synchrony. This finding echoes a similar EEG result in which direct gaze (adult-to-infant) increased neural synchrony, as did increased infant vocalizations. From infancy to adulthood, we employ various strategies to optimize our coupling with other minds.

Hyperstimulation: Transcranial Alternating Current Stimulation

All studies reviewed so far are observational in their nature, as they interpret differences in neural correlates of behavior across conditions but do not directly manipulate neural synchrony itself. In other words, they cannot answer whether neural synchrony is a cause or consequence of alignment. Methods that permute brain activity can help answer this question (e.g., transcranial magnetic stimulation [TMS] and transcranial direct/alternating current stimulation [tDCS/tACS]; see Miniussi, Harris, & Ruzzoli, 2013 for a review). Of these approaches, tACS has recently been used with interacting participants. Novembre, Knoblich, Dunne, and Keller (2017) applied a current to synchronize beta oscillations across members of a dyad (at 20 Hz over the left motor cortex) and observed a boost in behavioral synchrony between participants tapping together, with a metronome as a guide. This effect was specific to beta oscillations administered, in phase, to both participants and could not be explained as a mere by-product of motor entrainment to the metronome. Following a similar logic, Szymanski et al. (2017) applied tACS in the theta frequency range (5–7 Hz in their study) to right frontal and parietal areas of pair members engaged in a synchronous drumming task (cf., Müller, Sänger, & Lindenberger, 2013; Sänger, Müller, & Lindenberger, 2012). However, they did not find the expected positive link between neural stimulation and behavioral synchrony. The method of simultaneous tACS is still a very recent development in this field, but it promises to be a useful tool in elucidating a causal role of neural synchrony between brains.

Future Directions

New computational approaches Investigating neural synchrony is still the cutting edge of interpersonal neuroscience and a tractable starting point. A full accounting of interacting brains, however, will require going beyond synchrony. For example, brains in interaction show not only time-locked synchrony but also leader-follower dynamics (Holper et al. 2013; Jiang et al., 2015). These could be measured more elastically to capture lags that fluctuate in step with the accuracy of each parties' predictive codes. It is also likely that engaging and enduring interactions involve a balance between novelty and synchrony, thereby allowing a conversation to evolve while maintaining shared understanding. Such dynamics would be consistent with other dynamic biological systems that evolve and maintain stability via a mix of new inputs and pressure to maintain order. Many interactions may also involve between-brain complementarity, as in the case of a calm parent consoling an anxious child. And there are likely many other spatiotemporal dependencies between coupled brains yet to be identified (Hasson & Frith, 2016). Recent theoretical accounts of communication aim to accommodate the tension between synchrony and complementarity. For example, the model by Friston and Frith (2015a, 2015b) can derive complementary contributions on the basis of synchronous coupling on hidden levels.

Going beyond synchrony may appear to open up a Pandora's box of mathematical challenges. But complexity in dynamic systems—particularly dynamic systems built from the massive implementation of simple rules—also generates "order for free" (Kauffman, 1996). Such order might not be straightforward to capture, but promising approaches characterizing coupling beyond synchrony (e.g., transfer entropy: see Lizier, Heinzle, Horstmann, Haynes, & Prokopenko, 2011; recurrence quantification: see Fusaroli, Konvalinka, & Wallott, 2014) could help us describe even a multibrain unit. After all, we have good reason to believe that coupled neural systems do not represent an explosion in complexity. The combined neural landscape of two interacting brains operates under the structural constraints enabled by a shared language and shared norms that are themselves the product of social interaction. The possibility space of what one brain can usefully contribute is constrained by what would be fruitfully understood and acted on by the other. Feedback loops, similar neurological scaffolding, and shared priors help two brains self-organize into a unitary coupled system. Separate contributions are blurred together as topics and emotions are coauthored, allowing for "simultaneous mutual access to internal states" (Semin, 2007, p. 631). How brains in interaction create not just dependencies between brains but also emergent patterns across brains is a wide-open and exciting question at the intersection of theoretical biology, neuroscience, and applied mathematics.

Toward more ecological validity Lying supine in a noisy tube or being wired up to 128 scalp electrodes is a poor setting for lively conversation. However, new technologies and inventive paradigms are enabling participants to play games with each other and even have conversations (Schillbach et al., 2013). Ultimately, these paradigms should extend beyond dyads, given the role of groups in adapting efficient communication and social norms (Fay, Garrod, & Roberts, 2008). Hardware and software advances are also increasing the temporal and spatial resolution of neuroimaging data in step with computational approaches that are increasingly allowing these data to reveal their natural patterns (Jack, Crivelli, & Wheatley, 2018; Jolly & Chang, 2018).

Conclusions

Human physical and mental health depends on shared social understanding. Atypical social understanding is a defining feature of several disorders, such as autism spectrum disorders (ASD; see White, Keonig, & Scahill, 2007 for a review) and schizophrenia (see Brune, 2005; Couture, Penn, & Roberts, 2006 for reviews), and contributes to social isolation, with the associated risks of disease and death (Cacioppo et al., 2003; Pantell et al., 2013). Moreover, the underlying neural signatures of these disorders may limit a person's capacity to become coupled with other brains (e.g., ASD: Bolis & Schillbach, 2018; Hasegawa et al., 2016; von der Luhe et al., 2016; schizophrenia: Kupper et al., 2015), thereby creating an upper bound on critical social and cognitive competencies (e.g., joint attention for learning: Yu & Smith, 2016; interpersonal coordination for action prediction: Yin et al., 2016). Challenges with social interaction can further limit social relationships (Soleimani et al., 2014), which in turn reduces opportunities to interact. In the extreme case of solitary confinement, social isolation results in insomnia, confusion, and acute anxiety, as well as delusions and hallucinations. The lack of even minimal social contact exacerbates existing mental illness and disproportionately predicts suicide (for a review, see Smith, 2006). As Charles Dickens famously wrote after witnessing solitary confinement at the Cherry Hill prison in Philadelphia, "I hold this slow and daily tampering with the mysteries of the brain, to be immeasurably worse than any torture of the body" (1842/1985, p. 146). Although tremendous progress has been made over the last 50 years in understanding the processes involved in social perception and cognition within a single brain, we know very little about why interactive, mutual adaptation with other brains is so critical for our cognitive development and mental stability.

Social neuroscientists are beginning to push in this direction, with initial breakthroughs that use neural synchrony as a window on mental alignment. Future methods and analyses will likely uncover more complex mathematical relationships, such as complementary dynamics and patterns that manifest across interacting brains. This exciting endeavor promises a more complete picture of the social challenges associated with different neurological disorders, with implications for intervention. Its scope also includes the age-old questions of what makes people "click," whether we can formalize interpersonal "chemistry," and why people hold different roles in their larger social networks (Parkinson, Kleinbaum, & Wheatley, 2018). Characterizing the dynamic coupling of human minds will also inform the coupling of minds and machines in the form of brain-computer interaction—an endeavor with great promise as well as its own interesting ethical challenges.

Social interaction necessitates dynamic interactions among two or more brains as individuals mutually adapt to reach shared understanding. Elucidating the processes involved requires shifting from a "one-brain" to a "multibrain" frame of reference (Hasson et al., 2012), as well as from artificial laboratory conditions to interactive social contexts. A deep understanding of the human mind cannot be achieved without understanding why our brains expend so much time and energy on being coupled with others.

REFERENCES

Acquadro, M. A., Congedo, M., & De Riddeer, D. (2016). Music performance as an experimental approach to hyperscanning studies. *Frontiers in Human Neuroscience, 10*, 242.

Ahn, S., Cho, H., Kwon, M., Kim, K., Kwon, H., Kim, B. S., … Jun, S. C. (2018). Interbrain phase synchronization during turn-taking verbal interaction—a hyperscanning study using simultaneous EEG/MEG. *Human Brain Mapping*, 39, 171–188.

Anders, S., Heinzle, J., Weiskopf, N., Ethofer, T., & Haynes, J. D. (2011). Flow of affective information between communicating brains. *Neuroimage, 54*, 439–446.

Astolfi, L., De Vico Fallani, F., Toppi, J., Cincotti, F., Salinari, S., Vecchiato, G., … Babiloni, F. (2011). Imaging the social brain by simultaneous hyperscanning of different subjects during their mutual interactions. *IEEE Intelligent Systems, 26*, 38–45.

Babiloni, C., Vecchio, F., Infarinato, F., Buffo, P., Marzano, N., Spada, D., Rossi, S., Rossini, P. M., Bruni, I., & Perani, D. (2011). Simultaneous recording of electroencephalographic data in musicians playing in ensemble. *Cortex, 47*, 1082–1090.

Babiloni, F., & Astolfi, L. (2014). Social neuroscience and hyperscanning techniques: Past, present and future. *Neuroscience & Biobehavioral Reviews, 44*, 76–93.

Bilek, E., Ruf, M., Schäfer, A., Akdeniz, C., Calhoun, V. D., Schmahl, C., … Meyer-Lindenberg, A. (2015). Information flow between interacting human brains: Identification, validation, and relationship to social expertise. *Proceedings of the National Academy of Sciences, 112*, 5207–5212.

Bilek, E., Stößel, G., Schäfer, A., Clement, L., Ruf, M., Robnik, L., … Meyer-Lindenberg, A. (2017). State-dependent cross-brain information flow in borderline personality disorder. *JAMA Psychiatry, 74*, 949–957.

Bolis, D., & Schillbach, L. (2018). Observing and participating in social interactions: Action perception and action control across the autistic spectrum. *Developmental Cognitive Neuroscience, 29,* 168–175.

Branigan, H. P., Pickering, M. J., & Cleland, A. A. (2000). Syntactic co-ordination in dialogue. *Cognition, 75,* B13–B25.

Brune, M. (2005). "Theory of mind" in schizophrenia: A review of the literature. *Schizophrenia Bulletin, 31,* 21–42.

Cacioppo, J. T., & Hawkley, L. C. (2003). Social isolation and health, with an emphasis on underlying mechanisms. *Perspectives in Biology and Medicine,* S39–52.

Ciaramidaro, A., Toppi, J., Casper, C., Freitag, C. M., Siniatchkin, M., & Astolfi, L. (2018). Multiple-brain connectivity during third party punishment: An EEG hyperscanning study. *Scientific Reports, 8,* 6822.

Couture, S. M., Penn, D. L., & Roberts, D. L. (2006). The functional significance of social cognition in schizophrenia: A review. *Schizophrenia Bulletin, 32* (Suppl. 1), S44–63.

Cui, X., Bryant, D. M., & Reiss, A. L. (2012). NIRS-based hyperscanning reveals increased interpersonal coherence in superior frontal cortex during cooperation. *NeuroImage, 59,* 2430–2437.

Dale, R., Warlaumont, A. S., & Richardson, D. C. (2011). Nominal cross recurrence as a generalized lag sequential analysis for behavioral streams. *International Journal of Bifurcation and Chaos, 21,* 1153–1161.

Dickens, C. (1842/1985). *American notes.* London: Penguin.

Dikker, S., Wan, L., Davidesco, I., Kaggen, L., Oostrik, M., McClintock, J., … Poeppel, D. (2017). Brain-to-brain synchrony tracks real-world dynamic group interactions in the classroom. *Current Biology,* 27, 1375–1380.

Fay, N., Garrod, S., & Roberts, L. (2008). The fitness and functionality of culturally evolved communication systems. *Philosophical Transactions of the Royal Society B, 363,* 3553–3561.

Feldman, R. (2015). The adaptive human parental brain: Implications for children's social development. *Trends in Neurosciences, 38,* 387–399.

Friston, K., & Frith, C. (2015a). A duet for one. *Consciousness and Cognition, 36,* 390–405.

Friston, K. J., & Frith, C. D. (2015b). Active inference, communication and hermeneutics. *Cortex, 68,* 129–143.

Fusaroli, R., Konvalinka, I., & Wallot, S. (2014). Analyzing social interactions: The promises and challenges of using cross recurrence quantification analysis. In *Translational recurrences* (pp. 137–155). Cham, Germany: Springer.

Giles, H., Coupland, J., & Coupland, N. (1991). Accommodation theory: Communication, context, and consequence. In H. Giles, J. Coupland, & N. Coupland (Eds.), *Contexts of accommodation: Developments in applied sociolinguistics.* Cambridge: Cambridge University Press.

Giraud, A. L., Kleinschmidt, A., Poeppel, D., Lund, T. E., Frackowiak, R. S., & Laufs, H. (2007). Endogenous cortical rhythms determine cerebral specialization for speech perception and production. *Neuron, 56,* 1127–1134.

Hasegawa, C., Ikeda, T., Yoshimura, Y., Hiraishi, H., Takahashi, T., Furutani, N., Hayashi, N., Minabe, Y., Hirata, M., Asada, M., & Kikuchi, M. (2016). Mu rhythm suppression reflects mother-child face-to-face interactions: A pilot study with simultaneous MEG recording. *Scientific Reports, 6,* 34977.

Hasson, U., & Frith, C. (2016). Mirroring and beyond: Coupled dynamics as a generalized framework for modelling social interactions. *Philosophical Transactions of the Royal Society B, 371,* 1–9.

Hasson, U., Ghazanfar, A. A., Galantucci, B., Garrod, S., & Keysers, C. (2012). Brain-to-brain coupling: A mechanism for creating and sharing a social world. *Trends in Cognitive Sciences, 16,* 114–121.

Holper, L., Goldin, A. P., Shalóm, D. E., Battro, A. M., Wolf, M., & Sigman, M. (2013). The teaching and the learning brain: A cortical hemodynamic marker of teacher-student interactions in the Socratic dialog. *International Journal of Educational Research, 59,* 1–10.

Honey, C. J., Thompson, C. R., Lerner, Y., & Hasson, U. (2012). Not lost in translation: Neural responses shared across languages. *Journal of Neuroscience, 32,* 15277–15283.

Jack, R., Crivelli, C., & Wheatley, T. (2018). Using data-driven methods to diversify knowledge of human psychology. *Trends in Cognitive Sciences, 22,* 1–5.

Jackson, P. L., & Decety, J. (2004). Motor cognition: A new paradigm to study self-other interactions. *Current Opinion in Neurobiology, 14,* 259–263.

Jahng, J., Kralik, J. D., Hwang, D. U., & Jeong, J. (2017). Neural dynamics of two players when using nonverbal cues to gauge intentions to cooperate during the Prisoner's Dilemma Game. *NeuroImage, 157,* 263–274.

Jiang, J., Chen, C., Dai, B., Shi, G., Ding, G., Liu, L., & Lu, C. (2015). Leader emergence through interpersonal neural synchronization. *Proceedings of the National Academy of Sciences, 112,* 4274–4279.

Jiang, J., Dai, B., Peng, D., Zhu, C., Liu, L., & Lu, C. (2012). Neural synchronization during face-to-face communication. *Journal of Neuroscience, 32,* 16064–16069.

Kang, O. E., & Wheatley, T. (2017). Pupil dilation patterns spontaneously synchronize across individuals during shared attention. *Journal of Experimental Psychology: General, 146,* 569–576.

Kauffman, S. A. (1996). *At home in the universe: The search for laws of self-organization and complexity.* London: Penguin Books.

Kawasaki, M., Kitajo, K., & Yamaguchi, Y. (2018). Sensory-motor synchronization in the brain corresponds to behavioral synchronization between individuals. *Neuropsychologia, 119,* 59–67.

Koike, T., Tanabe, H. C., Okazaki, S., Nakagawa, E., Sasaki, A. T., Shimada, K., … Sadato, N. (2016). Neural substrates of shared attention as social memory: A hyperscanning functional magnetic resonance imaging study. *NeuroImage, 125,* 401–412.

Kupper, Z., Ramseyer, F., Hoffmann, H., & Tschacher, W. (2015). Nonverbal synchrony in social interactions of patients with schizophrenia indicates socio-communicative deficits. *PLoS One, 10,* e0145882.

Leong, V., Byrne, E., Clackson, K., Goergieva, S., Lam, S., & Wass, S. (2017). Speaker gaze increases information coupling between infant and adult brains. *Proceedings of the National Academy of Sciences, 114,* 13290–13295.

Levinson, S. C. (2016). Turn-taking in human communication-origins and implications for language processing. *Trends in Cognitive Sciences, 20,* 6–14.

Lindenberger, U., Li, S. C., Gruber, W., & Müller, V. (2009). Brains swinging in concert: Cortical phase synchronization while playing guitar. *BMC Neuroscience, 10,* 22.

Liu, N., Mok, C., Witt, E. E., Pradhan, A. H., Chen, J. E., & Reiss, A. L. (2016). NIRS-based hyperscanning reveals inter-brain neural synchronization during cooperative Jenga game with face-to-face communication. *Frontiers in Human Neuroscience, 10,* 82.

Liu, Y., Piazza, E. A., Simony, E., Shewokis, P. A., Onaral, B., Hasson, U., & Ayaz, H. (2017). Measuring speaker-listener neural coupling with functional near infrared spectroscopy. *Scientific Reports, 7,* srep43293.

Lizier, J. T., Heinzle, J., Horstmann, A., Haynes, J. D., & Prokopenko, M. (2011). Multivariate information-theoretic measures reveal directed information structure and task relevant changes in fMRI connectivity. *Journal of Computational Neuroscience, 30,* 85–107.

Luria, A. R. (1987). *The mind of a mnemonist: A little book about a vast memory.* Cambridge, MA: Harvard University Press.

Mandel, A., Bourguignon, M., Parkkonen, L., & Hari, R. (2016). Sensorimotor activation related to speaker vs. listener role during natural conversation. *Neuroscience Letters, 614,* 99–104.

Marsh, K. L., Richardson, M. J., & Schmidt, R. C. (2009). Social connection through joint action and interpersonal coordination. *Topics in Cognitive Science, 1,* 320–339.

Miniussi, C., Harris, J. A., & Ruzzoli, N. (2013). Non-invasive brain stimulation in cognitive neuroscience. *Clinical Neurophysiology, 124,* e51.

Müller, V., Sänger, J., & Lindenberger, U. (2013). Intra-and inter-brain synchronization during musical improvisation on the guitar. *PloS One, 8,* e73852.

Novembre, G., Knoblich, G., Dunne, L., & Keller, P. E. (2017). Interpersonal synchrony enhanced through 20 Hz phase-coupled dual brain stimulation. *Social Cognitive and Affective Neuroscience, 12,* 662–670.

Nozawa, T., Sasaki, Y., Sakaki, K., Yokoyama, R., & Kawashima, R. (2016). Interpersonal frontopolar neural synchronization in group communication: An exploration toward fNIRS hyperscanning of natural interactions. *NeuroImage, 133,* 484–497.

Nummenmaa, L., Lahnakoski, J. M., & Glerean, E. (2018). Sharing the social world via intersubject neural synchronization. *Current Opinion in Psychology, 24,* 7–14.

Osaka, N., Minamoto, T., Yaoi, K., Azuma, M., Shimada, Y. M., & Osaka, M. (2015). How two brains make one synchronized mind in the inferior frontal cortex: fNIRS-based hyperscanning during cooperative singing. *Frontiers in Psychology, 6,* 1811.

Pan, Y., Cheng, X., Zhang, Z., Li, X., & Hu, Y. (2017). Cooperation in lovers: An fNIRS-based hyperscanning study. *Human Brain Mapping, 38,* 831–841.

Pantell, M., Rehkopf, D., Jutte, D., Syme, S. L., Balmes, J., & Adler, N. (2013). Social isolation: A predictor of mortality comparable to traditional clinical risk factors. *American Journal of Public Health, 103,* 2056–2062.

Pardo, J. S. (2006). On phonetic convergence in speech production. *Frontiers in Psychology: Cognitive Science, 4,* 559.

Parkinson, C., Kleinbaum, A., & Wheatley, T. (2018). Similar neural responses predict friendship. *Nature Communications, 9,* 332.

Parkinson, C., Wheatley, T., & Kleinbaum, A. (forthcoming). *The neuroscience of social networks.* In R. Light & J. Moody (Eds.), *Oxford handbook of social network analysis.* Oxford: Oxford University Press.

Piazza, E. A., Hasenfratz, L., Hasson, U., & Lew-Williams, C. (2018). Infant and adult brains are coupled to the dynamics of natural communication. *BioRxiv,* 359810.

Pickering, M. J., & Garrod, S. (2004). Toward a mechanistic psychology of dialogue. *Behavioral and Brain Sciences, 27,* 169–225.

Przyrembel, M., Smallwood, J., Pauen, M., & Singer, T. (2012). Illuminating the dark matter of social neuroscience: Considering the problem of social interaction from philosophical, psychological, and neuroscientific perspectives. *Frontiers in Human Neuroscience, 6,* 190.

Ramseyer, F., & Tschacher, W. (2011). Nonverbal synchrony in psychotherapy: Coordinated body movement reflects relationship quality and outcome. *Journal of Consulting and Clinical Psychology, 79,* 284–295.

Reindl, V., Gerloff, C., Scharke, W., & Konrad, K. (2018). Brain-to-brain synchrony in parent-child dyads and the relationship with emotion regulation revealed by fNIRS-based hyperscanning. *NeuroImage, 178,* 493–502.

Riley, M. A., Richardson, M. J., Shockly, K., & Ramenzoni, V. C. (2011). Interpersonal synergies. *Frontiers in Psychology, 2,* 38.

Saito, D. N., Tanabe, H. C., Izuma, K., Hayashi, M. J., Morito, Y., Komeda, H., … Sadato, N. (2010). "Stay tuned": Inter-individual neural synchronization during mutual gaze and joint attention. *Frontiers in Integrative Neuroscience, 4,* 127.

Sänger, J., Müller, V., & Lindenberger, U. (2012). Intra-and interbrain synchronization and network properties when playing guitar in duets. *Frontiers in Human Neuroscience, 6,* 312.

Sebanz, N., & Knoblich, G. (2009). Prediction in joint action: What, when, and where. *Topics in Cognitive Science, 1,* 353–367.

Semin, G. R. (2007). Grounding communication: Synchrony. In A. W. Kruglanski & E. T. Higgins (Eds.), *Social psychology: Handbook of basic principles* (2nd ed., pp. 630–649). New York: Guilford Press.

Schilbach, L., Timmermans, B., Reddy, V., Costall, A., Bente, G., Schlicht, T., et al. (2013). Toward a second-person neuroscience. *Behavioral and Brain Sciences, 36,* 393–462.

Schippers, M. B., Roebroeck, A., Renken, R., Nanetti, L., & Keysers, C. (2010). Mapping the information flow from one brain to another during gestural communication. *Proceedings of the National Academy of Sciences, 107,* 9388–9393.

Schmidt, R. C., & Richardson, M. J. (2008). Dynamics of interpersonal coordination. In *Coordination: Neural, behavioral and social dynamics* (pp. 281–308). Berlin: Springer.

Schoot, L., Hagoort, P., & Segaert, K. (2016). What can we learn from a two-brain approach to verbal interaction? *Neuroscience & Biobehavioral Reviews, 68,* 454–459.

Silbert, L., Honey, C., Simony, E., Poeppel, D., & Hasson, U. (2014). Coupled neural systems underlie the production and comprehension of naturalistic narrative speech. *Proceedings of the National Academy of Sciences, 111,* E4687–E4696.

Smith, P. S. (2006). The effects of solitary confinement on prison inmates: A brief history and review of the literature. *Crime and Justice, 43,* 441–528.

Soleimani, M. A., Negarandeh, R., Bastani, F., & Greysen, R. (2014). Disrupted social connectedness in people with Parkinson's disease. *British Journal of Community Nursing, 19,* 136–141.

Spiegelhalder, K., Ohlendorf, S., Regen, W., Feige, B., van Elst, L. T., Weiller, C., … Tüscher, O. (2014). Interindividual synchronization of brain activity during live verbal communication. *Behavioural Brain Research, 258,* 75–79.

Stephens, G., Honey, C., & Hasson, U. (2013). A place for time: The spatiotemporal structure of neural dynamics during natural audition. *Journal of Neurophysiology, 110,* 2019–2026.

Stivers, T., Enfield, N. J., Brown, P., Englert, C., Hayashi, M., Heinemann, T., … Levinson, S. C. (2009). Universals and cultural variation in turn-taking in conversation. *Proceedings of the National Academy of Sciences, 106*, 10587–10592.

Stolk, A., Noordzij, M. L., Verhagen, L., Volman, I., Schoffelen, J. M., Oostenveld, R., … Toni, I. (2014). Cerebral coherence between communicators marks the emergence of meaning. *Proceedings of the National Academy of Sciences, 111*, 18183–18188.

Stolk, A., Verhagen, L., & Toni, I. (2016). Conceptual alignment: How brains achieve mutual understanding. *Trends in Cognitive Sciences, 20*, 180–191.

Szymanski, C., Müller, V., Brick, T. R., Von Oertzen, T., & Lindenberger, U. (2017). Hyper-transcranial alternating current stimulation: Experimental manipulation of inter-brain synchrony. *Frontiers in Human Neuroscience, 11*, 539.

Tang, H., Mai, X., Wang, S., Zhu, C., Krueger, F., & Liu, C. (2015). Interpersonal brain synchronization in the right temporo-parietal junction during face-to-face economic exchange. *Social Cognitive and Affective Neuroscience, 11*, 23–32.

von der Lühe, T., Manera, V., Barisic, I., Becchio, C., Vogeley, K., & Schilbach, L. (2016). Interpersonal predictive coding, not action perception, is impaired in autism. *Philosophical Transactions of the Royal Society B, 371*, 1–8.

White, W. S., Keonig, K., & Scahill, L. (2007). Social skills development in children with autism spectrum disorders: A review of the intervention research. *Journal of Autism Developmental Disorders, 37*, 1858–1868.

Wiltermuth, S. S., & Heath, C. (2009). Synchrony and cooperation. *Psychological Science, 20*, 1–5.

Yeshurun, Y., Swanson, S., Simony, E., Chen, J., Lazaridi, C., Honey, C. J., & Hasson, U. (2017). Same story, different story: The neural representation of interpretive frameworks. *Psychological Science, 28*, 307–319.

Yin, J., Xu, H., Ding, X., Liang, J., Shui, R., & Shen, M. (2016). Social constraints from an observer's perspective: Coordinated actions make an agent's position more predictable. *Cognition, 151*, 10–17.

Yu, C., & Smith, L. B. (2016). The social origins of sustained attention in one-year-old human infants. *Current Biology, 26*, 1235–1240.

Zhdanov, A., Nurminen, J., Baess, P., Hirvenkari, L., Jousmäki, V., Mäkelä, J. P., … Parkkonen, L. (2015). An Internet-based real-time audiovisual link for dual MEG recordings. *PLoS One, 10*, e0128485.

XII NEUROSCIENCE AND SOCIETY

Introduction

ANJAN CHATTERJEE AND ADINA ROSKIES

The cognitive neurosciences have been advancing in leaps and bounds over the last decade. We have seen both technological and theoretical innovations that promise this trajectory will continue. Important questions then arise, such as how do these advances in basic and clinical research affect society, and how do we best employ this knowledge for the greatest benefit? These are concerns for the public in general, as well as for policy-makers and ethicists, and answering these questions will require the understanding of both the vanguard of the science as well as of relevant fields in the social sciences and the humanities. Here we choose several areas and issues in which neuroscience is already having an effect on society. This list is by no means exhaustive. In addition to the chapters that follow—on the brain and morality, law, socioeconomic status (SES), addiction, mind reading, cognitive enhancement, brain-computer interfaces, aesthetics, and music—we might easily have included chapters on marketing, architecture, racial bias, education, and even religious belief and experience.

Society is made possible by the fact that our brains are geared for interaction with others. Moral neuroscience is the science of the cognitive processes and characteristics that undergird value judgments in social interactions. Greene and Young dispel the idea that there are dedicated brain circuits whose domain is moral cognition and argue instead that diverse brain areas involved in representing value, exerting cognitive control, mentalizing about others, reasoning, imagining, and reading and responding to social cues contribute their general-purpose functions to what we identify as moral thought and behavior. Their chapter highlights the role of value representations in moral cognition, reviews what we have learned from people with deficits in moral cognition, and argues that what we have learned is nicely accommodated under a dual-process framework. The chapter draws important connections between philosophical thought about the nature of morality and what we know about the brain.

The law is, to some degree, a codification of moral intuitions and a normative framework for social interaction. Jones and Wagner offer a look at the intersection of neuroscience and law, discussing the sociology of and the progress made by the burgeoning "neurolaw" movement. They chart advances in a number of legally relevant areas of neuroscience in recent years. But how is the relevance of neuroscience to the law to be assessed? Jones and Wagner give us a taxonomy of ways in which neuroscience could influence the law, and they highlight important caveats that temper wild enthusiasm about its potential reach. Although cognitive neuroscience has made great strides and significant effort has gone into exploring how these developments could be harnessed in the law, important limitations have been identified. For instance, although many experiments claim that lie detection using brain measures is effective, the results are confounded by design flaws and fail to provide compelling evidence that brain measures can be used for lie detection in normal contexts. However, despite the limitations, results from cognitive neuroscience are bound to increasingly affect legal proceedings.

Neuroscience has long been aware that the environments animals are raised in affect the development of their cognitive capacities. What is only now being recognized is that the lesson should be broadly applied, not just to lab animals but to humans as well. Increasingly, data demonstrate that low SES is correlated with poor cognitive and mental health outcomes and with structural changes in the brain. The real question is whether this correlation is a result of causation. Farah reviews the large and growing body of data that evidence the correlation and makes compelling arguments that the relation is at least partly causal. It remains to be determined which of the many factors that correlate with low SES, such as high stress, poor food choices and options, or less verbal interaction, are responsible for the detrimental outcomes and how they can best be combatted. The article is mostly forward looking, as this is a relatively nascent area of inquiry, but the policy implications are dramatic. It is possible that significant positive social change could result from social policy guided by cognitive neuroscience.

Addiction is a societal ill, often associated with poverty, that increasingly crosses socioeconomic borders. Deaths from opioid addiction have skyrocketed since the last edition of *The Cognitive Neurosciences*. Although neither neuroscience nor medicine has an answer to the problem of addiction, significant progress has been made in understanding its neurobiology. Gu and Adinoff discuss addiction in light of recent work in what they call *computational psychiatry*. They review literature that integrates computational approaches with biochemical and biophysical models of addiction. They discuss theoretical models of addiction induction, habit formation and maintenance, and craving and their relationship to empirical data. Using machine learning, they also explore data-driven approaches that have been used to characterize addiction phenotypes and to discover cognitive predictors and biomarkers of addiction and treatment outcome. Although still in its infancy, computational approaches to addiction may significantly enhance more traditional approaches to understanding and treating addictive disorders.

One of the primary tools of today's cognitive neuroscience is functional brain imaging. As methods for imaging and analyzing imaging data improve, researchers are able to extract ever-more information about the content of mental states. Some worry that neuroimaging can lay bare the contents of our thoughts and that the end of mental privacy is near. Roskies explores the power of neuroimaging to discern mental content and the limits of this so-called mind reading. While machine learning has radically improved our ability to correlate content with brain states, brute force decoding using machine learning is limited in the absence of a theoretical account of how semantic content is represented, which would enable the construction of generative models. However, significant strides have been made in understanding visual and auditory representations using encoding models. Recent work on semantic representation suggests that generative models of semantics are, in principle, possible. While it is unlikely that this information will infringe mental privacy in forensic contexts, it is nonetheless imperative to better understand the value of mental privacy in order to assess whether and in what ways it might be under threat.

As we learn more about how to treat cognitive and emotional disorders of the brain, people have compelling reasons to want to enhance cognitive abilities in health. Savulich and Sahakian point out that such abilities are increasingly important in a competitive global environment. Demands on attention, memory, and higher-order executive functions push healthy people into using "smart drugs." These cognitive enhancements might offer a range of advantages for individuals and society, including better treatments for patients as well as the possibility of greater productivity within the general populace. However, these benefits need to be weighed against uncertain risks and ethical concerns. Reasons for caution include threats to fairness, peer and parental coercion, and the promotion of societal inequities. As Savulich and Sahakian review, we struggle with how to decide which drugs are acceptable for whom (e.g., soldiers, doctors) and when (e.g., war, shift work) even as we promote their use in patient populations.

Pharmacological enhancement in some form to augment brain functions has been around for a long time. More recently, engineers, roboticists, cognitive scientists, and computer scientists are investigating the uses of direct links between human brains and different mechanical (e.g., robotic prostheses), electronic (e.g., computers), and virtual tools (e.g., limb and body avatars) to enhance function. These biomechanical links are collectively labeled brain-machine interfaces (BMIs). BMI research leverages the dynamic properties of neural circuits as learned from experimental systems to design and deploy novel neurorehabilitation approaches and, more recently, to restore mobility and communication in patients with debilitating brain injury. Nicolelis reviews the major BMI approaches and the rapidly evolving basic and clinical science in this exciting, almost science fiction-like area. His chapter includes a discussion of the potential impact of a new generation of neuroprostheses and concludes with speculations about shared BMIs, a novel way in which Internet-based protocols might be applied to treatment.

Beyond enhancement by drugs and machines, engagement with aesthetics, art, and music is fundamental to human flourishing. The pace of research in the cognitive neuroscience of aesthetics and music has accelerated in the last few decades, and for the first time, this edition of *The Cognitive Neurosciences* includes these domains of scientific inquiry as bearing directly on society. Vartanian and Chatterjee remind us that aesthetic experiences influence our actions in important contexts, such as the selection of mates, principles of design, choices made by consumers, and the appreciation and production of art, many of which are important sources of our well-being. While aesthetics has been a part of psychology for the last 150 years, cognitive neuroscience has only recently added further layers of understanding to these core processes. This chapter, like Greene and Young's view of moral valuation, argues that there are no dedicated brain circuits for aesthetic valuation. The authors review research demonstrating how aesthetic valuation and experiences emerge from interactions within and across a triad of large-scale neural systems that implement emotion-valuation, sensorimotor, and meaning-knowledge understanding.

In a similar vein, Zatorre and Penhume delve into the way that music engages our nervous system, from basic perceptual mechanisms to motor, attentional, memory, cognitive, and emotion systems. They review research from the past three decades that probes the neural basis for musical perception and production, the plasticity associated with expertise, and the mechanisms behind the pleasure we experience from music. They emphasize the role of hierarchical and parallel auditory cortical pathways and situate the extant science within a framework of processes underlying prediction. As they point out, the psychological experience of music is strongly influenced by the generation of expected outcomes derived from the temporal sequence of stimuli that are compared to experienced events. The dynamics of these predictions, which guide learning and behavior, can also undergird the pleasure of music.

This entire volume of *The Cognitive Neurosciences* covers a wide range of remarkable advances in our understanding of how the brain and behavior are related. The chapters in this section go beyond this basic understanding to emphasize the extended impact of these advances when we consider how cognitive neuroscience touches nearly every aspect of our society.

88 The Cognitive Neuroscience of Moral Judgment and Decision-Making

JOSHUA D. GREENE AND LIANE YOUNG

ABSTRACT This article reviews recent history and advances in the cognitive neuroscience of moral judgment and behavior. This field is conceived not as the study of a distinct set of neural functions but as an attempt to understand how the brain's core neural systems coordinate to solve problems that we define, for nonneuroscientific reasons, as "moral." At the heart of moral cognition are representations of value and the ways in which they are encoded, acquired, and modulated. Research dissociates distinct value representations—often within a dual-process framework—and explores the ways in which representations of value are informed or modulated by knowledge of mental states, explicit decision rules, the imagination of distal events, and social cues. Studies illustrating these themes examine the brains of morally pathological individuals, the responses of healthy brains to prototypically immoral actions, and the brain's responses to more complex philosophical and economic dilemmas.

Cognitive neuroscience aims to understand the mind in physical terms. Against this philosophical backdrop, the cognitive neuroscience of moral judgment takes on special significance. Moral judgment is, for many, the quintessential operation of the mind beyond the body, the earthly signature of the soul. Indeed, in many religious traditions it's the quality of a soul's moral judgment that determines where it ends up. Thus, the prospect of understanding morality in physical terms may be especially alluring, or unsettling, depending on your point of view. In this brief review we provide a progress report on these efforts. Here we focus on research using neuroscientific/biological methods, but we regard this as an artificial restriction, useful only for limiting our scope.

The Paradox of the "Moral Brain"

The fundamental problem with the "moral brain" is that it threatens to take over the entire brain and thus ceases to be a meaningful neuroscientific topic. This is not because morality is meaningless but rather because neuroscience is centrally concerned with physical mechanisms, and it's increasingly clear that morality has few, if any, neural mechanisms of its own (Young & Dungan, 2012). By way of analogy, the things we call *vehicles* are bound together, not by their internal mechanics—which include, pedals, sails, and nuclear reactors—but by their common function. So, too, with morality. More specifically, we regard morality as a suite of cognitive mechanisms that enable otherwise selfish individuals to reap the benefits of cooperation (Frank, 1988; Greene, 2013). Humans have psychological features that are straightforwardly moral (such as empathy) and others that are not (such as in-group favoritism) because they enable us to achieve goals that we can't achieve through pure selfishness. We won't defend this controversial thesis here. Instead, our point is that *if* this unified theory of morality is correct, it doesn't bode well for a unified theory of moral neuroscience. Previously, some hoped to find a dedicated "moral organ" in the brain (Hauser, 2006). It's now clear, however, that the "moral brain" is, more or less, the whole brain, applying its computational powers to problems that we, for nonneuroscientific reasons, classify as "moral."

Understanding this is, itself, a kind of progress, but it leaves the cognitive neuroscience of morality—and the authors of a chapter that would summarize it—in an awkward position. To truly understand the neuroscience of morality, we must understand the many neural systems that shape moral thinking, none of which, so far, appears to be specifically moral. At the heart of moral cognition are interlocking systems that represent the value of actions and outcomes (Bartra, McGuire, & Kable, 2013; Craig, 2009; Knutson, Taylor, Kaufman, Peterson, & Glover, 2005). Representations of value are informed and modulated by systems that represent mental states (Frith & Frith, 2006; Koster-Hale et al., 2017) and that orchestrate thought and action in accordance with more abstract knowledge, rules, and goals (Miller & Cohen, 2001). This often gives rise to a dual-process dynamic, whereby automatic processes compete with more controlled processes (Kahneman, 2003).

Other systems enable us to imagine complex distal events (Buckner, Andrews-Hanna, & Schacter, 2008) and keep track of who's who in the social world (Cikara & Van Bavel, 2014). These computational themes recur in lessons learned from abnormally antisocial brains, the responses of healthy brains to basic transgressions, and the ways in which our brains resolve more complex philosophical and economic dilemmas.

Bad Brains

The neuroscience of morality began with the study of brain damage leading to antisocial behavior. Such research accelerated in the 1990s with a series of path-breaking studies of decision-making in patients with damage to ventromedial prefrontal cortex (vmPFC), one of the regions damaged in the famous case of Phineas Gage (Damasio, 1994). Such patients made poor real-life decisions, but their deficits typically evaded detection using conventional measures of executive function (Saver & Damasio, 1991) and moral reasoning (Anderson, Bechara, Damasio, Tranel, & Damasio, 1999). Using a game designed to simulate real-world risky decision-making (the Iowa Gambling Task), Bechara, Tranel, Damasio, and Damasio (1996) documented these behavioral deficits and demonstrated, using autonomic measures, that these deficits are emotional. It seems that such patients make poor decisions because they lack the feelings that guide complex decision-making in healthy individuals. These early studies identified the vmPFC as critical for affectively driven moral choice and underscored the role of learning in moral development, as early-onset vmPFC damage leads not only to poor judgment but to a more psychopathic behavioral profile (Anderson et al., 1999).

Psychopathy is characterized by a pathological degree of callousness, a lack of empathy or emotional depth, a lack of genuine remorse for antisocial actions (Hare, 1991), and a tendency toward instrumental aggression (Blair, 2001). Psychopaths exhibit profound emotional deficits. In clinical and subclinical psychopathy, the amygdala, which plays a central role in emotional learning and memory (Phelps, 2006), exhibits weaker responses to fearful faces (Marsh et al., 2008) and to depictions of moral transgressions (Harenski, Harenski, Shane, & Kiehl, 2010). Critically, these muted affective responses are selective, responding to threats but not distress (Blair, Jones, Clark, & Smith, 1997). This pattern reemerges in more recent work showing that psychopaths, when prompted to imagine painful injuries to themselves and others, exhibit normal neural responses to their own imagined pain but reduced responses in the amygdala and insula, as well as reduced connectivity with the orbitofrontal cortex (OFC) and vmPFC, when imagining the pain of others (Decety, Skelly, & Kiehl, 2013). Likewise, a study of incarcerated psychopaths revealed reduced responses to distress cues in the vmPFC/OFC (Decety, Skelly, & Kiehl, 2013). A similar pattern, featuring the amygdala, has been observed in youths with psychopathic traits (Marsh et al., 2008, 2013).

Consistent with the above, Blair (2007) has proposed that psychopathy arises primarily from dysfunction in the amygdala, which is crucial for stimulus-reinforcement learning (Davis & Whalen, 2001). He argues further that psychopathy involves core deficits in response-outcome learning, which depends critically on the frontostriatal pathway, including the dorsal and ventral striatum as well as the vmPFC (Blair, 2017). This leads to abnormal socialization, such that psychopathic individuals fail to attach negative affective values to socially harmful outcomes and actions. These learning deficits manifest in judgment as well as behavior, such that psychopaths (or a subset thereof: Aharoni, Sinnott-Armstrong, & Kiehl, 2012) fail to distinguish between rules that authorities cannot legitimately change ("moral" rules—e.g., a classroom rule against hitting) from rules that authorities can legitimately change ("conventional" rules—e.g., a rule prohibiting talking out of turn; Blair, 1995).

Psychopaths, in addition to their weak affective responses to harm, tend to be impulsive (Hare, 1991). Psychopaths, compared to other incarcerated criminals, exhibit signs of reduced response conflict when behaving dishonestly (Abe, Greene, & Kiehl, 2018), and related responses to an impulse-control task (go/no-go) predict criminal rearrest (Aharoni et al., 2013). These deficits may ultimately derive from abnormal reward processing: psychopaths who harm impulsively exhibit heightened responses to reward within the frontostriatal pathway (Buckholtz et al., 2010).

An illuminating recent study (Darby et al., 2017) combines lesion data and resting-state functional connectivity data to explain why so many neural regions are implicated in antisocial behavior and why some of these regions appear to be more central than others. They find that the regions most reliably implicated in antisocial behavior are positively functionally connected to the frontostriatal pathway and/or the amygdala/anterior temporal lobe. By contrast, these regions tend to be *negatively* functionally connected to the frontoparietal control network, consistent with a dual-process framework (see below).

Responsive Brains

Consistent with studies of psychopathology, research on how healthy brains respond to moral transgressions and opportunities highlights the importance of the frontostriatal pathway (Decety & Porges, 2011; Moll et al., 2006; Shenhav & Greene, 2010) and the amygdala-vmPFC circuit (Blair, 2007; Decety & Porges, 2011). Bookending their research in psychopaths, Marsh et al. (2014) have shown that extraordinary altruists (who have donated kidneys to strangers) tend to have larger amygdalae that are more sensitive to facial fear expressions. Likewise, several studies highlight the importance of the insula, which represents subjective value and appears to be an expanded somatosensory region (Craig, 2009). The insula's responses reflect the aversiveness of moral transgressions (Baumgartner, Fischbacher, Feierabend, Lutz, & Fehr, 2009; Schaich Borg, Lieberman, & Kiehl, 2008), employing a multimodal code that also reflects pain, vicarious pain, disgust, and unfairness (Corradi-Dell'Acqua, Tusche, Vuilleumier, & Singer, 2016).

As Oliver Wendell Holmes Jr. famously observed, even a dog knows the difference between being tripped over and being kicked. Likewise, the human amygdala distinguishes between depictions of intentional and accidental harm within 200 ms, as revealed by depth electrode recordings (Hesse et al., 2016). The temporoparietal junction (TPJ) is the region most reliably implicated in the representation of morally relevant mental states and mental states more generally (Frith & Frith, 2006). The TPJ is especially sensitive to attempted harms (Koster-Hale, Saxe, Dungan, & Young, 2013; Young, Cushman, Hauser, & Saxe, 2007), which are wrong only because of the agent's mental state. More recent evidence indicates that the TPJ separately encodes information about agents' beliefs and values (Koster-Hale et al., 2017).

Both attempted harms and accidental harms set up a tension between outcome-based and intention-based judgment. This can give rise to a dual-process dynamic (see below), such that an understanding of mental states overrides an impulse to blame, or generates a more abstract reason to blame, despite the absence of harm. Consistent with this, TMS applied to the TPJ results in a childlike (Piaget, 1965), "no harm, no foul" pattern of judgment in which attempted harms are judged less harshly (Young, Camprodon, Hauser, Pascual-Leone, & Saxe, 2010). In addition, a network of brain regions, including the TPJ and dorsal anterior cingulate cortex (ACC), appear to suppress amygdala responses to emotionally salient unintentional transgressions (Treadway et al., 2014). The "no harm, no foul" pattern is also observed in patients with vmPFC damage (Young, Bechara, et al., 2010), connecting the aforementioned effects in the amygdala and TPJ to the frontostriatal pathway. Consistent with this, psychopaths (Young, Koenigs, Kruepke, & Newman, 2012) and patients with alexithymia (Patil & Salani, 2014a), a condition that reduces awareness of one's own emotional states, judge accidental harms to be more acceptable, reflecting reduced affective responses to harmful outcomes. Individuals with high-functioning autism exhibit a complementary pattern, "if harm, then foul," judging accidental harms unusually harshly (Moran et al., 2011). Finally, split-brain patients (Miller et al., 2010), like vmPFC patients, exhibit a "no harm, no foul" pattern, indicating that sensitivity to intention depends on the integration of information across the cerebral hemispheres.

Puzzled Brains

To better understand more complex moral judgments, researchers have used moral dilemmas that capture the tension between competing moral considerations. The research described above emphasizes the role of emotion (Haidt, 2001), while traditional developmental theories emphasize controlled reasoning (Kohlberg, 1969). Greene and colleagues (Greene, 2013; Greene et al., 2001, 2004) have developed a dual-process (Kahneman, 2003) theory of moral judgment that synthesizes these perspectives. More specifically, this theory associates controlled cognition with utilitarian/consequentialist moral judgment aimed at promoting the greater good (Mill, 1861/1998) while associating automatic emotional responses with competing deontological judgments that are naturally justified in terms of rights or duties (Kant, 1785/1959).

This theory was inspired by a long-standing philosophical puzzle known as the *trolley problem* (Foot, 1978; Thomson, 1985). In the *switch* version of the problem, one can save five people who are mortally threatened by a runaway trolley by hitting a switch that will turn the trolley onto a side track, killing one person. Here, most people approve of acting to save more lives. In the contrasting *footbridge* dilemma, the only way to save the five is to push a large person off a footbridge and into the trolley's path. Here, most people disapprove. Why the difference? And what does this tell us about moral judgment?

In short, people say no to the action in the *footbridge* case because that action elicits a relatively strong negative emotional response, and this response tends to override the cost-benefit reasoning that favors pushing. In the *switch* case, the harmful action is less emotionally salient, and therefore cost-benefit reasoning tends to prevail. The first evidence for these conclusions came from a functional magnetic resonance imaging (fMRI) study (Greene et al., 2001) that contrasted sets of "personal" and "impersonal" dilemmas loosely modeled after the *footbridge* and *switch* cases. It found that "personal" dilemmas elicited increased activity in the mPFC, medial parietal cortex, and TPJ. These regions were previously associated with emotion and are now recognized as comprising most of the default mode network (DMN) (Buckner, Andrews-Hanna, & Schacter, 2008). In contrast, the "impersonal" dilemmas elicited relatively greater activity in the frontoparietal control network. A subsequent experiment found increased activity for utilitarian judgment within this network, including regions of DLPFC (Greene et al., 2004). Likewise, a more recent study found increased engagement of the DLPFC when participants were instructed to focus exclusively on utilitarian outcomes (Shenhav & Greene, 2014). Greene et al. (2004) also found increased amygdala responses to "personal" dilemmas. More recent evidence indicates that the DMN's response to "personal" dilemmas is best understood not as an emotional response per se but as the increased engagement of a mechanism that enables the construction and representation of nonpresent episodes such as memories of the past, "prospections" of the future, and hypothetical imaginings (Buckner, Andrews-Hanna, & Schacter, 2008; DeBrigard, Addis, Ford, Schacter, & Giovanello, 2013). Consistent with this, Amit and Greene (2012) found that individuals with more visual cognitive styles tend to make fewer utilitarian judgments in response to high-conflict personal dilemmas and that disrupting visual imagery while contemplating these dilemmas increases utilitarian judgment.

Some of the most compelling evidence for the dual-process theory comes from studies of patients with emotion-related deficits. Mendez, Anderson, and Shapira (2005) found that patients with frontotemporal dementia, who are known for their "emotional blunting," are disproportionately likely to approve of the utilitarian action in the *footbridge* dilemma. Likewise, patients with vmPFC lesions make up to five times as many utilitarian judgments in response to standard high-conflict dilemmas (Ciaramelli, Muccioli, Ladàvas, & di Pellegrino, 2007; Koenigs et al., 2007). Such patients also make more utilitarian judgments in response to dilemmas pitting familial duty against the greater good (e.g., your sister vs. five strangers; Thomas, Croft, & Tranel, 2011). As expected, vmPFC patients exhibit correspondingly weak physiological responses when making utilitarian judgments (Moretto, Ladàvas, Mattioli, & di Pellegrino, 2010), and healthy people who are more physiologically reactive are less likely to make utilitarian judgments (Cushman, Gray, Gaffey, & Mendes, 2012). Paralleling their more lenient responses to accidental harms (see above), low-anxiety psychopaths (Koenigs et al., 2012) and people with alexithymia (Koven, 2011; Patil & Silani, 2014b) are also more approving of utilitarian sacrifices. Critically, these effects depend not only on the disruption of the affective pathway that favors deontological judgment but also on a preserved capacity for cost-benefit reasoning, without which their judgments would simply be disordered, rather than more utilitarian.

Other studies using dilemmas highlight the shared and distinctive functions of the amygdala and vmPFC. Citalopram—a selective serotonin-reuptake inhibitor (SSRI) that increases emotional reactivity in the short term through its influence on the amygdala and vmPFC—increases deontological judgment (Crockett, Clark, Hauser, & Robbins, 2010). By contrast, lorazepam, an antianxiety drug, has the opposite effect (Perkins et al., 2012), as does the administration of testosterone (Chen, Decety, Huang, Chen, & Cheng, 2016). Consistent with this, individuals with psychopathic traits exhibit reduced amygdala responses to personal moral dilemmas (Glenn, Raine, & Schug, 2009). In healthy people, amygdala activity tracks self-reported emotional responses to harmful transgressions and predicts deontological judgments in response to them (Shenhav & Greene, 2014). The same study shows a different pattern for the vmPFC, which is most active when people have to make integrative, "all things considered" judgments, as compared to simply reporting on emotional reactions or assessing options solely in terms of their consequences. This suggests that the amygdala generates an initial negative response to personally harmful actions while the vmPFC weighs that signal against a competing signal reflecting the value of the greater good (see also Hutcherson, Montaser-Kouhsari, Woodward, & Rangel, 2015).

The vmPFC (along with the ventral striatum) also represents *expected moral value*, integrating information concerning the number of lives to be saved and the probability of saving them (Shenhav & Greene, 2010). These findings are consistent with our understanding of the frontostriatal pathway, and the vmPFC more specifically, as a domain-general integrator of decision values (Bartra, McGuire, & Kable, 2013; Knutson et al., 2005). We note that these structures evolved in mammals to evaluate goods, such as food, that tend to exhibit diminishing marginal returns. (The more food you've eaten, the less you need additional food.) This may explain our puzzling tendency to regard the saving of human lives as exhibiting diminishing marginal returns, as if the 100th life to be saved is somehow worth less than the first (Dickert, Västfjäll, Kleber, & Slovic, 2012).

Patients with hippocampal damage, unlike vmPFC patients, are less likely to make utilitarian judgments (McCormick, Rosenthal, Miller, & McGuire, 2016). This result is surprising (cf., Amit & Greene, 2012; Greene et al., 2001) but ultimately consistent with the dual-process theory. The hippocampus is a critical node within the DMN (Buckner, Andrews-Hanna, & Schacter, 2008), which is, once again, essential for the imagination of nonpresent events. The inability of hippocampal patients to fully imagine dilemma scenarios may thus cause them to rely more on emotional responses to the types of actions proposed, as reflected in skin-conductance responses and self-reports (for contrasting null results, however, see Craver et al., 2016).

In an important theoretical development, Cushman (2013) and Crockett (2013) have proposed that the dissociation between deontological and utilitarian/consequentialist judgment reflects a more general dissociation between model-free and model-based learning systems (Daw & Doya, 2006). Model-free learning mechanisms assign values directly to actions based on past experience, while model-based learning attaches values to actions indirectly by attaching values to outcomes and linking outcomes to actions via internal models of causal relations. Thus, an action may seem wrong "in itself" because past experience has associated actions of that type (e.g., pushing people) with negative consequences (e.g., social disapproval), and yet the same action may seem right because it will, according to one's causal world model, produce optimal consequences (saving five lives instead of one). Thus, the fundamental tension in normative ethics, reflected in the competing philosophies of Kant and Mill, may find its origins in a competition between distinct, domain-general mechanisms for assigning values to actions. With respect to the more deontological judgments made by hippocampal patients, McCormick et al. (2016) suggest that their judgments, influenced by a limited capacity for imagination, may be understood as relatively model-free.

Trolley dilemmas are, perhaps, an unlikely tool for scientists, and some researchers have questioned their widespread use. Kahane et al. (2015) have claimed that the utilitarian judgments they elicit are not truly utilitarian and merely reflect antisocial tendencies. This critique is based largely on a misunderstanding about how the term *utilitarian* has been used. The judgments are called *utilitarian* because they are required by utilitarianism and are thought to reflect simple cost-benefit reasoning, not because the judges are thought to be generally committed to utilitarian values (Conway, Goldstein-Greenwood, Polacek, & Greene, 2018). (One can make a utilitarian judgment without being a utilitarian, just as one can make an Italian meal without being an Italian.) Addressing the provocative claim that utilitarian judgments are motivated entirely by antisocial tendencies, a series of studies replicating Kahane et al.'s studies with the addition of process dissociation measures confirms the predictions of the dual-process theory: utilitarian judgments reflect *both* decreased concern about causing harm and increased concern for the greater good (Conway et al., 2018). Conway et al. also examined the judgments of professional philosophers and showed, contra Kahane (2015), that trolley judgments do indeed reflect the fundamental tension between consequentialists and deontologists. Others have challenged the use of hypothetical dilemmas based on concerns about their ecological validity (e.g., Bostyn, Sevenhant, & Roets, 2018). For replies, see Conway et al. (2018) and Plunkett and Greene (in press).

Cooperative Brains

Research on altruism and cooperation, though often considered apart from "morality," could not be more central to our understanding of the moral brain. The most basic question about the cognitive neuroscience of altruism and cooperation is this: What neural processes enable and motivate people to be "nice"—that is, to pay costs to benefit others?

Consistent with our evolving story, the value of helping others, in both unidirectional altruism and bidirectional cooperation, is represented in the frontostriatal pathway and modulated by both economic incentives and social signals (Declerck, Boone, & Emonds, 2013). Activity in this pathway tracks the value of charitable contributions (Moll et al., 2006) and of sharing resources with other individuals (Zaki & Mitchell, 2011). Likewise, it encodes the discounted value of rewards gained at the expense of others (Crockett, Siegel, Kurth-Nelson, Dayan, & Dolan, 2017). Here, signals from the DLPFC appear to modulate striatal signals, resulting in more altruistic behavior. The same pattern is observed in the case of increased altruism following compassion training (Weng et al., 2013). Striatal signals, likewise, track the value of punishing transgressors (Crockett et al., 2013; de Quervain et al., 2004; Singer et al., 2006). And, as above, the DMN appears to have a hand in altruism: TPJ volume (Morishima, Schunk, Bruhin, Ruff, & Fehr, 2012) and medial PFC activity (Waytz, Zaki, & Mitchell, 2012) both predict altruistic behavior, with more dorsal mPFC regions representing the value of rewards for others (Apps & Ramnani, 2014).

As noted above, the brain uses its endogenous carrots—reward signals—to motivate cooperative behavior. It also uses its sticks—negative affective responses to uncooperative behavior. Activity in the insula scales with the unfairness of ultimatum game (UG) offers (Gabay, Radua, Kempton, & Mehta, 2014; Sanfey, Rilling, Aronson, Nystrom, & Cohen, 2003) including offers to third parties (Corradi-Dell'Acqua, Civai, Rumiati, & Fink, 2012). Insula responses also predict aversion to inequality in the distribution of resources (Hsu, Anen, & Quartz, 2008) and egalitarian behavior and attitudes (Dawes et al., 2012). The insula and the amygdala both respond to the punishment of well-behaved people (Singer, Kiebel, Winston, Dolan, & Frith, 2004). Perhaps surprisingly, vmPFC damage leads to increased *rejection* of unfair UG offers (Koenigs & Tranel, 2007), mirroring patterns observed in psychopaths (Koenigs, Kruepke, & Newman, 2010.) This may be because the vmPFC integrates signals responding to material gain as well as unfairness (which compete in the UG) and because, in the absence of such signals, one applies a reciprocity rule.

Honesty is a form of cooperation, and dishonesty is a form of defection. Greene and Paxton (2009) gave people repeated opportunities to gain money by lying about their accuracy in predicting the outcomes of coin flips. Consistently honest subjects appeared to be "gracefully" honest, exhibiting no additional engagement of the frontoparietal control network in forgoing dishonest gains. By contrast, subjects who behaved dishonestly exhibited increased control-related activity, both when lying and when refraining from lying. These individual differences in (dis)honesty are predicted by striatal responses to rewards in an unrelated task (Abe & Greene, 2014). Baumgartner et al. (2009) describe a similar dual-process dynamic in which breaking promises involves increased engagement of the amygdala and the frontoparietal control network.

Cooperation depends on trust, which in turn requires evaluating people's trustworthiness (Delgado, Frank, & Phelps, 2005). We describe the people we trust as "close," and this metaphor is reflected in how the brain represents social relationships: A region of the inferior parietal lobe has been shown to represent spatial, temporal, and social proximity using a common code, as demonstrated by cross-trained pattern classification (Parkinson, Liu, & Wheatley, 2014). Cooperation is more likely with friends than strangers, and the additional social value of cooperation with friends is reflected in ventral-striatal signals and in the mPFC (Fareri, Chang, & Delgado, 2015). Likewise, our brains respond differently to in-group and out-group members, including members of "minimal" groups formed in the lab (Cikara & Van Bavel, 2014). Both neural and behavioral data indicate that cooperation with in-group members is rewarding and relatively effortless, while cooperation with out-group members engages more cognitive control (Hughes, Ambady, & Zaki, 2017), consistent with evolutionarily inspired theories of dual-process cooperation (Bear & Rand, 2016; Greene, 2013; Rand, Greene, & Nowak, 2012. But see Everett, Ingbretsen, Cushman, and Cikara [2017] for evidence of intuitive cooperation with "minimal" out-groups).

Oxytocin is a neuropeptide implicated in social attachment and affiliation across mammals (Insel & Young, 2001). In humans it's been associated with empathy and prosocial behavior (Bartz et al., 2015; Heinrichs, von Dawans, & Domes, 2009). An early and influential study found that intranasally administered oxytocin increases trust among strangers (Kosfeld, Heinrichs, Zak, Fischbacher, & Fehr, 2005), and many studies have associated variation in the oxytocin receptor gene (*OXTR*) with morally relevant phenotypes, including empathic concern (Rodrigues, Saslow, Garcia, John, & Keltner, 2009), generosity (Israel et al., 2009), and psychopathy (Dadds et al., 2014). As with many candidate gene studies, subsequent studies with larger samples have failed to replicate many such effects (Apicella et al., 2010; Bakermans-Kranenburg & van IJzendoorn, 2014), and doubts have been raised about the relation between oxytocin and trust (Nave, Camerer, & McCullough, 2015). A recent study employing separate exploratory and confirmatory samples found an association between an *OXTR* variant and two types of dilemma judgments (Bernhard et al., 2016).

Recent research indicates that the effects of oxytocin are highly variable across personality types (Bartz et al., 2015) and sex (Rilling et al., 2014) and may even include antisocial behavior (Ne'eman, Perach-Barzilay, Fischer-Shofty, Atias, & Shamay-Tsoory, 2016). According to a recent influential theory, the variable effects of oxytocin across individuals, contexts, and relationships are best understood as effects of heightening the salience of social cues, again through modulation of the frontostriatal pathway (Shamay-Tsoory & Abu-Akel, 2016). Most notable of all, there is mounting evidence that the effects of oxytocin are "parochial," biasing judgment and behavior in favor of in-group members (De Dreu et al., 2010; Shalvi & De Dreu, 2014).

Although such results were surprising, given oxytocin's well-established role in affiliative behavior, they make evolutionary sense. Morality evolved, not as a device for universal cooperation but as a competitive weapon—as a system for turning Me into Us, which in turn enables Us to outcompete Them. It does not follow from this, however, that we are doomed to be warring tribalists. Drawing on our ingenuity and flexibility, it's possible to put human values ahead of evolutionary imperatives, as we do when we use birth control.

Looking Back, and Ahead

How does the moral brain work? Answer: exactly the way you'd expect it to work if you understand (1) which cognitive functions morality requires and (2) which cognitive functions are performed by the brain's core neural systems. Our conclusion that human morality depends on the brain's general-purpose machinery for representing value, applying cognitive control, mentalizing, reasoning, imagining, and reading social cues will come as no surprise to today's neuroscientists. But the emergence of morality as a source of tractable neuroscientific problems is itself significant. For the broader sciences and the general public, our increasingly detailed, mechanistic understanding of human morality is radically demystifying, challenging traditional dualistic assumptions about human nature, with important implications for law, public policy, and our collective self-image (Farah, 2012; Greene & Cohen, 2004; Shariff et al., 2014).

From its inception, cognitive neuroscience has focused on structure-function relationships, teaching us which parts of the brain do what. By contrast, we know very little about how ideas move around and interact in the brain. We can track our neural responses to the thought of pushing someone off of a footbridge, but how do our brains even compose such a thought in the first place? We are just beginning to understand how the brain can represent, for example, the morally significant difference between a baby kicking a grandfather and a grandfather kicking a baby (Frankland & Greene, 2015)—a modest step. However, with the confluence of multivariate analysis methods (Kriegeskorte, Goebel, & Bandettini, 2006; Norman, Polyn, Detre, & Haxby, 2006), network approaches (Bullmore & Sporns, 2009), and neurally inspired models of high-level cognition (Graves et al., 2016; Kriete et al., 2013; Lake, Ullman, Tenenbaum, & Gershman, 2017), we may finally be ready to understand how the brain flexibly and precisely manipulates the *contents* of thoughts (Fodor, 1975; Marcus, 2001). And that's a good thing, because understanding moral thinking may require a more general understanding of thinking.

Acknowledgments

Many thanks to Catherine Holland for research assistance. Thanks to Joshua Buckholtz, Joe Paxton, Adina Roskies, and Walter Sinnott-Armstrong for helpful comments.

REFERENCES

Abe, N., & Greene, J. D. (2014). Response to anticipated reward in the nucleus accumbens predicts behavior in an independent test of honesty. *Journal of Neuroscience, 34*(32), 10564–10572.

Abe, N., Greene, J. D., & Kiehl, K. A. (2018). Reduced engagement of the anterior cingulate cortex in the dishonest decision-making of incarcerated psychopaths. *Social Cognitive and Affective Neuroscience*, 797–807.

Aharoni, E., Sinnott-Armstrong, W., & Kiehl, K. A. (2012). Can psychopathic offenders discern moral wrongs? A new look at the moral/conventional distinction. *Journal of Abnormal Psychology, 121*(2), 484.

Aharoni, E., Vincent, G. M., Harenski, C. L., Calhoun, V. D., Sinnott-Armstrong, W., Gazzaniga, M. S., & Kiehl, K. A. (2013). Neuroprediction of future rearrest. *Proceedings of the National Academy of Sciences, 110*(15), 6223–6228.

Amit, E., & Greene, J. D. (2012). You see, the ends don't justify the means: Visual imagery and moral judgment. *Psychological Science, 23*(8), 861–868.

Anderson, S. W., Bechara, A., Damasio, H., Tranel, D., & Damasio, A. R. (1999). Impairment of social and moral behavior related to early damage in human prefrontal cortex. *Nature Neuroscience, 2*, 1032–1037.

Apicella, C. L., Cesarini, D., Johannesson, M., Dawes, C. T., Lichtenstein, P., Wallace, B., ... Westberg, L. (2010). No association between oxytocin receptor (OXTR) gene polymorphisms and experimentally elicited social preferences. *PLoS One, 5*(6), e11153.

Apps, M. A., & Ramnani, N. (2014). The anterior cingulate gyrus signals the net value of others' rewards. *Journal of Neuroscience, 34*(18), 6190–6200.

Bakermans-Kranenburg, M. J., & van IJzendoorn, M. H. (2014). A sociability gene? Meta-analysis of oxytocin receptor genotype effects in humans. *Psychiatric Genetics, 24*(2), 45–51.

Bartra, O., McGuire, J. T., & Kable, J. W. (2013). The valuation system: A coordinate-based meta-analysis of BOLD fMRI experiments examining neural correlates of subjective value. *Neuroimage, 76*, 412–427.

Bartz, J. A., Lydon, J. E., Kolevzon, A., Zaki, J., Hollander, E., Ludwig, N., & Bolger, N. (2015). Differential effects of oxytocin on agency and communion for anxiously and avoidantly attached individuals. *Psychological Science, 26*(8), 1177–1186.

Baumgartner, T., Fischbacher, U., Feierabend, A., Lutz, K., & Fehr, E. (2009). The neural circuitry of a broken promise. *Neuron, 64*(5), 756–770.

Bear, A., & Rand, D. G. (2016). Intuition, deliberation, and the evolution of cooperation. *Proceedings of the National Academy of Sciences, 113*(4), 936–941.

Bechara, A., Tranel, D., Damasio, H., & Damasio, A. R. (1996). Failure to respond autonomically to anticipated future outcomes following damage to prefrontal cortex. *Cerebral Cortex, 6*, 215–225.

Bernhard, R. M., Chaponis, J., Siburian, R., Gallagher, P., Ransohoff, K., Wikler, D., ... Greene, J. D. (2016). Variation in the oxytocin receptor gene (OXTR) is associated with differences in moral judgment. *Social Cognitive and Affective Neuroscience, 11*(12), 1872–1881.

Blair, R. J. (1995). A cognitive developmental approach to mortality: Investigating the psychopath. *Cognition, 57*, 1–29.

Blair, R. J. (2001). Neurocognitive models of aggression, the antisocial personality disorders, and psychopathy. *Journal of Neurology, Neurosurgery, and Psychiatry, 71*, 727–731.

Blair, R. J. (2007). The amygdala and ventromedial prefrontal cortex in morality and psychopathy. *Trends in Cognitive Sciences, 11*, 387–392.

Blair, R. J. (2017). Emotion-based learning systems and the development of morality. *Cognition, 167*, 38–45.

Blair, R. J., Jones, L., Clark, F., & Smith, M. (1997). The psychopathic individual: A lack of responsiveness to distress cues? *Psychophysiology, 34,* 192–198.

Bostyn, D. H., Sevenhant, S., & Roets, A. (2018). Of mice, men, and trolleys: Hypothetical judgment versus real-life behavior in trolley-style moral dilemmas. *Psychological Science*, 0956797617752640.

Buckholtz, J. W., Treadway, M. T., Cowan, R. L., Woodward, N. D., Benning, S. D., Li, R., … Zald, D. H. (2010). Mesolimbicdopamine reward system hypersensitivity in individuals with psychopathic traits. *Nature Neuroscience, 13*(4), 419–421.

Buckner, R. L., Andrews-Hanna, J. R., & Schacter, D. L. (2008). The brain's default network. *Annals of the New York Academy of Sciences, 1124*(1), 1–38.

Bullmore, E., & Sporns, O. (2009). Complex brain networks: Graph theoretical analysis of structural and functional systems. *Nature Reviews Neuroscience, 10*(3), 186.

Chen, C., Decety, J., Huag, P. C., Chen, C. Y., & Cheng, Y. (2016). Testosterone administration in females modulates moral judgment and patterns of brain activation and functional connectivity. *Human Brain Mapping, 37*(10), 3417–3430.

Ciaramelli, E., Muccioli, M., Ladavas, E., & di Pellegrino, G. (2007). Selective deficit in personal moral judgment following damage to ventromedial prefrontal cortex. *Social Cognitive and Affective Neuroscience, 2,* 84–92.

Cikara, M., & Van Bavel, J. J. (2014). The neuroscience of intergroup relations: An integrative review. *Perspectives on Psychological Science, 9*(3), 245–274.

Conway, P., Goldstein-Greenwood, J., Polacek, D., & Greene, J. D. (2018). Sacrificial utilitarian judgments do reflect concern for the greater good: Clarification via process dissociation and the judgments of philosophers. *Cognition, 179,* 241–265.

Corradi-Dell'Acqua, C., Civai, C., Rumiati, R. I., & Fink, G. R. (2012). Disentangling self-and fairness-related neural mechanisms involved in the ultimatum game: An fMRI study. *Social Cognitive and Affective Neuroscience, 8*(4), 424–431.

Corradi-Dell'Acqua, C., Tusche, A., Vuilleumier, P., & Singer, T. (2016). Cross-modal representations of first-hand and vicarious pain, disgust and fairness in insular and cingulate cortex. *Nature Communications, 7,* 10904.

Craig, A. D. (2009). How do you feel—Now? The anterior insula and human awareness. *Nature Reviews Neuroscience, 10*(1), 59–70.

Craver, C. F., Keven, N., Kwan, D., Kurczek, J., Duff, M. C., & Rosenbaum, R. S. (2016). Moral judgment in episodic amnesia. *Hippocampus, 26*(8), 975–979.

Crockett, M. J. (2013). Models of morality. *Trends in Cognitive Sciences, 17*(8), 363–366.

Crockett, M. J., Apergis-Schoute, A., Herrmann, B., Lieberman, M. D., Müller, U., Robbins, T. W., & Clark, L. (2013). Serotonin modulates striatal responses to fairness and retaliation in humans. *Journal of Neuroscience, 33*(8), 3505–3513.

Crockett, M. J., Clark, L., Hauser, M. D., & Robbins, T. W. (2010). Serotonin selectively influences moral judgment and behavior through effects on harm aversion. *Proceedings of the National Academy of Sciences, 107*(40), 17433–17438.

Crockett, M. J., Kurth-Nelson, Z., Siegel, J. Z., Dayan, P., & Dolan, R. J. (2014). Harm to others outweighs harm to self in moral decision making. *Proceedings of the National Academy of Sciences, 111*(48), 17320–17325.

Crockett, M. J., Siegel, J. Z., Kurth-Nelson, Z., Dayan, P., & Dolan, R. J. (2017). Moral transgressions corrupt neural representations of value. *Nature neuroscience, 20*(6), 879.

Cushman, F. (2013). Action, outcome, and value: A dual-system framework for morality. *Personality and Social Psychology Review, 17*(3), 273–292.

Cushman, F., Gray, K., Gaffey, A., & Mendes, W. B. (2012). Simulating murder: The aversion to harmful action. *Emotion, 12*(1), 2.

Dadds, M. R., Moul, C., Cauchi, A., Dobson-Stone, C., Hawes, D. J., Brennan, J., & Ebstein, R. E. (2014). Methylation of the oxytocin receptor gene and oxytocin blood levels in the development of psychopathy. *Development and Psychopathology, 26*(1), 33–40.

Damasio, A. R. (1994). *Descartes' error: Emotion, reason, and the human brain.* New York: G. P. Putnam.

Darby, R. R., Horn, A., Cushman, F., & Fox, M. D. (2017). Lesion network localization of criminal behavior. *Proceedings of the National Academy of Sciences, 115*(3), 601–606.

Davis, M., & Whalen, P. J. (2001). The amygdala: Vigilance and emotion. *Molecular Psychiatry, 6,* 13–34.

Daw, N. D., & Doya, K. (2006). The computational neurobiology of learning and reward. *Current Opinion in Neurobiology, 16*(2), 199–204.

Dawes, C. T., Loewen, P. J., Schreiber, D., Simmons, A. N., Flagan, T., McElreath, R., … Paulus, M. P. (2012). Neural basis of egalitarian behavior. *Proceedings of the National Academy of Sciences, 109*(17), 6479–6483.

De Brigard, F., Addis, D. R., Ford, J. H., Schacter, D. L., & Giovanello, K. S. (2013). Remembering what could have happened: Neural correlates of episodic counterfactual thinking. *Neuropsychologia, 51*(12), 2401–2414.

Decety, J., Chen, C., Harenski, C., & Kiehl, K. A. (2013). An fMRI study of affective perspective taking in individuals with psychopathy: Imagining another in pain does not evoke empathy. *Frontiers in Human Neuroscience, 7,* 489.

Decety, J., & Porges, E. C. (2011). Imagining being the agent of actions that carry different moral consequences: An fMRI study. *Neuropsychologia, 49*(11), 2994–3001.

Decety, J., Skelly, L. R., & Kiehl, K. A. (2013). Brain response to empathy-eliciting scenarios involving pain in incarcerated individuals with psychopathy. *JAMA Psychiatry, 70*(6), 638–645.

Declerck, C. H., Boone, C., & Emonds, G. (2013). When do people cooperate? The neuroeconomics of prosocial decision making. *Brain and Cognition, 81*(1), 95–117.

De Dreu, C. K., Greer, L. L., Handgraaf, M. J., Shalvi, S., Van Kleef, G. A., Baas, M., … Feith, S. W. (2010). The neuropeptide oxytocin regulates parochial altruism in intergroup conflict among humans. *Science, 328*(5984), 1408–1411.

Delgado, M. R., Frank, R., & Phelps, E. A. (2005). Perceptions of moral character modulate the neural systems of reward during the trust game. *Nature Neuroscience, 8,* 1611–1618.

de Quervain, D. J., Fischbacher, U., Treyer, V., Schellhammer, M., Schnyder, U., Buck, A., & Fehr, E. (2004). The neural basis of altruistic punishment. *Science, 305,* 1254–1258.

Dickert, S., Västfjäll, D., Kleber, J., & Slovic, P. (2012). Valuations of human lives: Normative expectations and psychological mechanisms of (ir) rationality. *Synthese, 189*(1), 95–105.

Everett, J. A., Ingbretsen, Z., Cushman, F., & Cikara, M. (2017). Deliberation erodes cooperative behavior—even towards competitive out-groups, even when using a control

condition, and even when eliminating selection bias. *Journal of Experimental Social Psychology, 73,* 76–81.
Farah, M. J. (2012). Neuroethics: The ethical, legal, and societal impact of neuroscience. *Annual Review of Psychology, 63,* 571–591.
Fareri, D. S., Chang, L. J., & Delgado, M. R. (2015). Computational substrates of social value in interpersonal collaboration. *Journal of Neuroscience, 35*(21), 8170–8180.
Fodor, J. A. (1975). *The language of thought* (Vol. 5). Cambridge, MA: Harvard University Press.
Foot, P. (1978). The problem of abortion and the doctrine of double effect. In *Virtues and vices.* Oxford: Blackwell.
Frank, R. H. (1988). *Passions within reason: The strategic role of the emotions.* New York: W. W. Norton.
Frankland, S. M., & Greene, J. D. (2015). An architecture for encoding sentence meaning in left mid-superior temporal cortex. *Proceedings of the National Academy of Sciences, 112*(37), 11732–11737.
Frith, C. D., & Frith, U. (2006). The neural basis of mentalizing. *Neuron, 50*(4), 531–534.
Gabay, A. S., Radua, J., Kempton, M. J., & Mehta, M. A. (2014). The ultimatum game and the brain: A meta-analysis of neuroimaging studies. *Neuroscience & Biobehavioral Reviews, 47,* 549–558.
Glenn, A. L., Raine, A., & Schug, R. A. (2009). The neural correlates of moral decision-making in psychopathy. *Molecular Psychiatry, 14*(1), 5.
Graves, A., Wayne, G., Reynolds, M., Harley, T., Danihelka, I., Grabska-Barwińska, A., … Badia, A. P. (2016). Hybrid computing using a neural network with dynamic external memory. *Nature, 538*(7626), 471.
Greene, J. (2013). *Moral tribes: Emotion, reason, and the gap between us and them.* New York: Penguin Press.
Greene, J., & Cohen, J. (2004). For the law, neuroscience changes nothing and everything. *Philosophical Transactions of the Royal Society B: Biological Sciences, 359*(1451), 1775.
Greene, J. D., Nystrom, L. E., Engell, A. D., Darley, J. M., & Cohen, J. D. (2004). The neural bases of cognitive conflict and control in moral judgment. *Neuron, 44,* 389–400.
Greene, J. D., & Paxton, J. M. (2009). Patterns of neural activity associated with honest and dishonest moral decisions. *Proceedings of the National Academy of Sciences, 106*(30), 12506–12511.
Greene, J. D., Sommerville, R. B., Nystrom, L. E., Darley, J. M., & Cohen, J. D. (2001). An fMRI investigation of emotional engagement in moral judgment. *Science, 293,* 2105–2108.
Haidt, J. (2001). The emotional dog and its rational tail: A social intuitionist approach to moral judgment. *Psychological Review, 108,* 814–834.
Hare, R. D. (1991). *The hare psychopathy checklist—Revised.* Toronto: Multi-Health Systems.
Harenski, C. L., Harenski, K. A., Shane, M. S., & Kiehl, K. A. (2010). Aberrant neural processing of moral violations in criminal psychopaths. *Journal of Abnormal Psychology, 119*(4), 863.
Hauser, M. (2006). The liver and the moral organ. *Social Cognitive and Affective Neuroscience, 1,* 214–220.
Heinrichs, M., von Dawans, B., & Domes, G. (2009). Oxytocin, vasopressin, and human social behavior. *Frontiers in Neuroendocrinology, 30*(4), 548–557.
Hesse, E., Mikulan, E., Decety, J., Sigman, M., Garcia, M. D. C., Silva, W., … Lopez, V. (2015). Early detection of intentional harm in the human amygdala. *Brain, 139*(1), 54–61.
Hsu, M., Anen, C., & Quartz, S. R. (2008). The right and the good: Distributive justice and neural encoding of equity and efficiency. *Science, 320,* 1092–1095.
Hughes, B. L., Ambady, N., & Zaki, J. (2017). Trusting outgroup, but not ingroup members, requires control: Neural and behavioral evidence. *Social Cognitive and Affective Neuroscience, 12*(3), 372–381.
Hutcherson, C. A., Bushong, B., & Rangel, A. (2015). A neurocomputational model of altruistic choice and its implications. *Neuron, 87*(2), 451–462.
Hutcherson, C. A., Montaser-Kouhsari, L., Woodward, J., & Rangel, A. (2015). Emotional and utilitarian appraisals of moral dilemmas are encoded in separate areas and integrated in ventromedial prefrontal cortex. *Journal of Neuroscience, 35*(36), 12593–12605.
Insel, T. R., & Young, L. J. (2001). The neurobiology of attachment. *Nature Reviews Neuroscience, 2,* 129–136.
Israel, S., Lerer, E., Shalev, I., Uzefovsky, F., Riebold, M., Laiba, E., … Ebstein, R. P. (2009). The oxytocin receptor (OXTR) contributes to prosocial fund allocations in the dictator game and the social value orientations task. *PLoS One, 4*(5), e5535.
Kahane, G. (2015). Sidetracked by trolleys: Why sacrificial moral dilemmas tell us little (or nothing) about utilitarian judgment. *Social Neuroscience, 10*(5), 551–560.
Kahane, G., Everett, J. A., Earp, B. D., Caviola, L., Faber, N. S., Crockett, M. J., & Savulescu, J. (2018). Beyond sacrificial harm: A two-dimensional model of utilitarian psychology. *Psychological Review, 125*(2), 131.
Kahane, G., Everett, J. A., Earp, B. D., Farias, M., & Savulescu, J. (2015). "Utilitarian" judgments in sacrificial moral dilemmas do not reflect impartial concern for the greater good. *Cognition, 134,* 193–209.
Kahneman, D. (2003). A perspective on judgment and choice: Mapping bounded rationality. *American Psychologist, 58,* 697–720.
Kant, I. (1785/1959). *Foundation of the metaphysics of morals.* Indianapolis: Bobbs-Merrill.
Knutson, B., Taylor, J., Kaufman, M., Peterson, R., & Glover, G. (2005). Distributed neural representation of expected value. *Journal of Neuroscience, 25*(19), 4806–4812.
Koenigs, M., Kruepke, M., & Newman, J. P. (2010). Economic decision-making in psychopathy: A comparison with ventromedial prefrontal lesion patients. *Neuropsychologia, 48*(7), 2198–2204.
Koenigs, M., Kruepke, M., Zeier, J., & Newman, J. P. (2012). Utilitarian moral judgment in psychopathy. *Social Cognitive and Affective Neuroscience, 7*(6), 708–714.
Koenigs, M., & Tranel, D. (2007). Irrational economic decision-making after ventromedial prefrontal damage: Evidence from the ultimatum game. *Journal of Neuroscience, 27,* 951–956.
Koenigs, M., Young, L., Adolphs, R., Tranel, D., Cushman, F., Hauser, M., & Damasio, A. (2007). Damage to the prefrontal cortex increases utilitarian moral judgements. *Nature, 446,* 908–911.
Kohlberg, L. (1969). Stage and sequence: The cognitive-developmental approach to socialization. In D. A. Goslin (Ed.), *Handbook of socialization theory and research* (pp. 347–480). Chicago: Rand McNally.
Kosfeld, M., Heinrichs, M., Zak, P. J., Fischbacher, U., & Fehr, E. (2005). Oxytocin increases trust in humans. *Nature, 435,* 673–676.

Koster-Hale, J., Richardson, H., Velez, N., Asaba, M., Young, L., & Saxe, R. (2017). Mentalizing regions represent distributed, continuous, and abstract dimensions of others' beliefs. *NeuroImage, 161*, 9–18.

Koster-Hale, J., Saxe, R., Dungan, J., & Young, L. L. (2013). Decoding moral judgments from neural representations of intentions. *Proceedings of the National Academy of Sciences, 110*(14), 5648–5653.

Koven, N. S. (2011). Specificity of meta-emotion effects on moral decision-making. *Emotion, 11*(5), 1255.

Kriegeskorte, N., Goebel, R., & Bandettini, P. (2006). Information-based functional brain mapping. *Proceedings of the National Academy of Sciences of the United States of America, 103*(10), 3863–3868.

Kriete, T., Noelle, D. C., Cohen, J. D., & O'Reilly, R. C. (2013). Indirection and symbol-like processing in the prefrontal cortex and basal ganglia. *Proceedings of the National Academy of Sciences, 110*(41), 16390–16395.

Lake, B. M., Ullman, T. D., Tenenbaum, J. B., & Gershman, S. J. (2017). Building machines that learn and think like people. *Behavioral and Brain Sciences, 40.*

Marcus, G. F. (2001). *The algebraic mind: Integrating connectionism and cognitive science.* Cambridge, MA: MIT Press.

Marsh, A., Finger, E., Mitchell, D., Reid, M., Sims, C., Kosson, D., ... Blair, R. (2008). Reduced amygdala response to fearful expressions in children and adolescents with callous-unemotional traits and disruptive behavior disorders. *American Journal of Psychiatry, 165*(6), 712–720.

Marsh, A. A., Finger, E. C., Fowler, K. A., Adalio, C. J., Jurkowitz, I. T., Schechter, J. C., ... Blair, R. J. R. (2013). Empathic responsiveness in amygdala and anterior cingulate cortex in youths with psychopathic traits. *Journal of child psychology and psychiatry, 54*(8), 900–910.

Marsh, A. A., Stoycos, S. A., Brethel-Haurwitz, K. M., Robinson, P., VanMeter, J. W., & Cardinale, E. M. (2014). Neural and cognitive characteristics of extraordinary altruists. *Proceedings of the National Academy of Sciences, 111*(42), 15036–15041.

McCormick, C., Rosenthal, C. R., Miller, T. D., & Maguire, E. A. (2016). Hippocampal damage increases deontological responses during moral decision making. *Journal of Neuroscience, 36*(48), 12157–12167.

Mendez, M. F., Anderson, E., & Shapira, J. S. (2005). An investigation of moral judgement in frontotemporal dementia. *Cognitive and Behavioral Neurology, 18*, 193–197.

Mill, J. S. (1861/1998). In R. Crisp (Ed.), *Utilitarianism.* New York: Oxford University Press.

Miller, E. K., & Cohen, J. D. (2001). An integrative theory of prefrontal cortex function. *Annual Review of Neuroscience, 24*, 167–202.

Miller, M. B., Sinnott-Armstrong, W., Young, L., King, D., Paggi, A., Fabri, M., ... Gazzaniga, M. S. (2010). Abnormal moral reasoning in complete and partial callosotomy patients. *Neuropsychologia, 48*(7), 2215–2220.

Moll, J., Krueger, F., Zahn, R., Pardini, M., de Oliveira-Souza, R., & Grafman, J. (2006). Human fronto-mesolimbic networks guide decisions about charitable donation. *Proceedings of the National Academy of Sciences of the United States of America, 103*, 15623–15628.

Moran, J. M., Young, L. L., Saxe, R., Lee, S. M., O'Young, D., Mavros, P. L., & Gabrieli, J. D. (2011). Impaired theory of mind for moral judgment in high-functioning autism. *Proceedings of the National Academy of Sciences, 108*(7), 2688–2692.

Moretto, G., Làdavas, E., Mattioli, F., & Di Pellegrino, G. (2010). A psychophysiological investigation of moral judgment after ventromedial prefrontal damage. *Journal of Cognitive Neuroscience, 22*(8), 1888–1899.

Morishima, Y., Schunk, D., Bruhin, A., Ruff, C. C., & Fehr, E. (2012). Linking brain structure and activation in temporoparietal junction to explain the neurobiology of human altruism. *Neuron, 75*(1), 73–79.

Nave, G., Camerer, C., & McCullough, M. (2015). Does oxytocin increase trust in humans? A critical review of research. *Perspectives on Psychological Science, 10*(6), 772–789.

Ne'eman, R., Perach-Barzilay, N., Fischer-Shofty, M., Atias, A., & Shamay-Tsoory, S. G. (2016). Intranasal administration of oxytocin increases human aggressive behavior. *Hormones and Behavior, 80*, 125–131.

Norman, K. A., Polyn, S. M., Detre, G. J., & Haxby, J. V. (2006). Beyond mind-reading: Multi-voxel pattern analysis of fMRI data. *Trends in Cognitive Sciences, 10*(9), 424–430.

Parkinson, C., Liu, S., & Wheatley, T. (2014). A common cortical metric for spatial, temporal, and social distance. *Journal of Neuroscience, 34*(5), 1979–1987.

Patil, I., & Silani, G. (2014a). Alexithymia increases moral acceptability of accidental harms. *Journal of Cognitive Psychology, 26*(5), 597–614.

Patil, I., & Silani, G. (2014b). Reduced empathic concern leads to utilitarian moral judgments in trait alexithymia. *Frontiers in Psychology, 5*, 501.

Perkins, A. M., Leonard, A. M., Weaver, K., Dalton, J. A., Mehta, M. A., Kumari, V., ... Ettinger, U. (2012). A dose of ruthlessness: Interpersonal moral judgment is hardened by the anti-anxiety drug lorazepam. *Journal of Experimental Psychology: General, 142*(3), 612.

Phelps, E. A. (2006). Emotion and cognition: Insights from studies of the human amygdala. *Annual Review of Psychology, 57*, 27–53.

Piaget, J. (1965). *The moral judgement of the child.* New York: Free Press.

Plunkett, D., & Greene, J. D. (in press). Overlooked evidence and a misunderstanding of what trolley dilemmas do best: A comment on Bostyn, Sevenhant, & Roets (2018). *Psychological Science.*

Poulin, M. J., Holman, E. A., & Buffone, A. (2012). The neurogenetics of nice: Receptor genes for oxytocin and vasopressin interact with threat to predict prosocial behavior. *Psychological Science, 23*(5), 446–452.

Rand, D. G., Greene, J. D., & Nowak, M. A. (2012). Spontaneous giving and calculated greed. *Nature, 489*(7416), 427–430.

Ransohoff, K. J. (2011). Patients on the trolley track: The moral cognition of medical practitioners and public health professionals (Undergraduate thesis). Harvard University, Cambridge, MA.

Rilling, J. K., DeMarco, A. C., Hackett, P. D., Chen, X., Gautam, P., Stair, S., ... Pagnoni, G. (2014). Sex differences in the neural and behavioral response to intranasal oxytocin and vasopressin during human social interaction. *Psychoneuroendocrinology, 39*, 237–248.

Rodrigues, S. M., Saslow, L. R., Garcia, N., John, O. P., & Keltner, D. (2009). Oxytocin receptor genetic variation relates to empathy and stress reactivity in humans. *Proceedings of the National Academy of Sciences, 106*(50), 21437–21441.

Sanfey, A. G., Rilling, J. K., Aronson, J. A., Nystrom, L. E., & Cohen, J. D. (2003). The neural basis of economic decision-making in the ultimatum game. *Science, 300*, 1755–1758.

Saver, J., & Damasio, A. (1991). Preserved access and processing of social knowledge in a patient with acquired sociopathy due to ventromedial frontal damage. *Neuropsychologia, 29*, 1241–1249.

Schaich Borg, J., Lieberman, D., & Kiehl, K. A. (2008). Infection, incest, and iniquity: Investigating the neural correlates of disgust and morality. *Journal of Cognitive Neuroscience, 20*, 1–19.

Shalvi, S., & De Dreu, C. K. (2014). Oxytocin promotes group-serving dishonesty. *Proceedings of the National Academy of Sciences, 111*(15), 5503–5507.

Shamay-Tsoory, S. G., & Abu-Akel, A. (2016). The social salience hypothesis of oxytocin. *Biological Psychiatry, 79*(3), 194–202.

Shariff, A. F., Greene, J. D., Karremans, J. C., Luguri, J. B., Clark, C. J., Schooler, J. W., … Vohs, K. D. (2014). Free will and punishment: A mechanistic view of human nature reduces retribution. *Psychological Science, 25*(8), 1563–1570.

Shenhav, A., & Greene, J. D. (2010). Moral judgments recruit domain-general valuation mechanisms to integrate representations of probability and magnitude. *Neuron, 67*(4), 667–677.

Shenhav, A., & Greene, J. D. (2014). Integrative moral judgment: Dissociating the roles of the amygdala and ventromedial prefrontal cortex. *Journal of Neuroscience, 34*(13), 4741–4749.

Singer, T., Kiebel, S. J., Winston, J. S., Dolan, R. J., & Frith, C. D. (2004). Brain responses to the acquired moral status of faces. *Neuron, 41*(4), 653–662.

Singer, T., Seymour, B., O'Doherty, J. P., Stephan, K. E., Dolan, R. J., & Frith, C. D. (2006). Empathic neural responses are modulated by the perceived fairness of others. *Nature, 439*, 466–469.

Thomas, B. C., Croft, K. E., & Tranel, D. (2011). Harming kin to save strangers: Further evidence for abnormally utilitarian moral judgments after ventromedial prefrontal damage. *Journal of Cognitive Neuroscience, 23*(9), 2186–2196.

Thomson, J. (1985). The trolley problem. *Yale Law Journal, 94*, 1395–1415.

Treadway, M. T., Buckholtz, J. W., Martin, J. W., Jan, K., Asplund, C. L., Ginther, M. R., … Marois, R. (2014). Corticolimbic gating of emotion-driven punishment. *Nature Neuroscience, 17*(9), 1270.

Waytz, A., Zaki, J., & Mitchell, J. P. (2012). Response of dorsomedial prefrontal cortex predicts altruistic behavior. *Journal of Neuroscience, 32*(22), 7646–7650.

Weng, H. Y., Fox, A. S., Shackman, A. J., Stodola, D. E., Caldwell, J. Z., Olson, M. C., … Davidson, R. J. (2013). Compassion training alters altruism and neural responses to suffering. *Psychological Science, 24*(7), 1171–1180.

Young, L., Bechara, A., Tranel, D., Damasio, H., Hauser, M., & Damasio, A. (2010). Damage to ventromedial prefrontal cortex impairs judgment of harmful intent. *Neuron, 65*(6), 845–851.

Young, L., Camprodon, J. A., Hauser, M., Pascual-Leone, A., & Saxe, R. (2010). Disruption of the right temporoparietal junction with transcranial magnetic stimulation reduces the role of beliefs in moral judgments. *Proceedings of the National Academy of Sciences, 107*(15), 6753–6758.

Young, L., Cushman, F., Hauser, M., & Saxe, R. (2007). The neural basis of the interaction between theory of mind and moral judgment. *Proceedings of the National Academy of Sciences of the United States of America, 104*(20), 8235–8240.

Young, L., & Dungan, J. (2012). Where in the brain is morality? Everywhere and maybe nowhere. *Social Neuroscience, 7*(1), 1–10.

Young, L., Koenigs, M., Kruepke, M., & Newman, J. P. (2012). Psychopathy increases perceived moral permissibility of accidents. *Journal of abnormal psychology, 121*(3), 659.

Zaki, J., & Mitchell, J. P. (2011). Equitable decision making is associated with neural markers of intrinsic value. *Proceedings of the National Academy of Sciences, 108*(49), 19761–19766.

89 Law and Neuroscience: Progress, Promise, and Pitfalls

OWEN D. JONES AND ANTHONY D. WAGNER

ABSTRACT This chapter provides an overview of new developments at the interface of law and neuroscience. It describes what is happening, explains the promise and potential influences of neuroscientific evidence, and explores the contexts in which neuroscience can be useful to law. Along the way, it considers some of the legal problems on which neuroscientific data are thought, at least by some, to provide potential answers and it highlights some illustrative cases. It also surveys emerging research that documents how interdisciplinary teams of legal scholars, judges, and neuroscientists are yielding progress and identifying potential pitfalls.

Cognitive neuroscientific discoveries about minds and brains not only advance scientific theory but also hold promise to inform, and often directly bear on, real-world problems of the human condition. This is increasingly evident at the intersection of law and neuroscience. The law often concerns itself with making judgments about human behavior, and the cognitive neurosciences aim to explain the psychological and neurobiological mechanisms that give rise to thought and action. The legal system—including legal decision-makers (such as judges and juries) and legal policy-makers (such as legislators)—is frequently charged with making decisions based on limited or noisy evidence. Given the challenges of doing so, the hope has naturally arisen that cognitive neuroscientific advances may yield informative evidence that facilitates fact-based legal decisions and policy.

While neuroscientific evidence, such as the presence of a neural injury or disorder, has long been a staple of tort law (the law of injuries), the remarkable neuroscientific advances made in recent decades have not gone unnoticed by the legal community. Increasingly, legal actors are offering neuroscientific evidence during litigation and citing neuroscientific studies during policy discussions. It appears that such evidence often has some influence on outcomes. In a complementary manner, cognitive neuroscientists are coming to appreciate how their approach can be leveraged to address important problems the law regularly confronts, as well as how their methods and results may be used, for better or worse, by legal actors.

In this review, we provide a high-level summary of recent activities at the interface of law and neuroscience, including overviews of what is happening, of the potential influences of neuroscientific evidence, and of contexts in which neuroscience can be useful to law. Along the way we consider some of the legal problems on which neuroscientific data are thought, at least by some, to provide potential answers, and we highlight some illustrative cases. Throughout, the chapter reflects our view that there is a zone of suitable sense that lies somewhere between being too zealous about the long-term effects of neuroscience on law and being too skeptical that neuroscience has anything useful to offer.

Cross-Field Interactions—the Emergence of "Neurolaw"

We begin by considering some of the key developments that have propelled interactions between neuroscience and law over the last 10 to 15 years.

First, as already noted, lawyers are increasingly offering neuroscientific evidence in the courtroom. In the civil (noncriminal) domain, for example, one core issue of the multidistrict National Football League (NFL) concussion litigation concerns the neurological effects of repetitive impacts to the head (In re: NFL Players' Concussion Injury Litigation, 2015; Grey & Marchant, 2015). Neuroscience also appears in contexts as varied as medical malpractice litigation, on one hand, and suits seeking disability benefits, on the other. In the criminal domain, many defendants now offer evidence of brain abnormalities—such as tumors, cysts, or unusual features—to argue during the sentencing phase of a trial that they should receive a lesser punishment than would someone who acted identically, but with a "normal" brain. Former mayor of San Diego Maureen O'Connor, for instance, claimed that a tumor contributed to her gambling addiction, which in turn led to the embezzlements of which she was convicted (*United States v. O'Connor*, 2013). The past decade has even seen attempts to enter functional brain imaging evidence purported to reveal the veracity of a defendant's testimony, a development to which we return below.

In 2007, the John D. and Catherine T. MacArthur Foundation funded the interdisciplinary Law and Neuroscience Project (under Michael Gazzaniga and, later, Owen D. Jones, directors) to help build direct links between neuroscience, psychological science, academic law, and legal actors such as judges and attorneys. In 2011, the foundation funded the new Research Network on Law and Neuroscience (Owen D. Jones, director). Over 12 years, these efforts propelled exploration of the promise and the limitations of using neuroscientific research to further the goals of criminal justice, building bridges between neuroscientists and legal scholars. Together with leading federal and state judges, teams codesigned and published dozens of legally relevant experiments, as well as many analyses and proposals for ways the legal system could use neuroscience usefully while simultaneously minimizing misuses. (See www.lawneuro.org for details, including members, publications, resources, and more.)

Given the rapid expansion in the types and technical complexity of the neuroscientific evidence available, along with the growth in its submission as evidence, cross-field education is critical. Some of this, of course, will come in the form of expert witnesses, when neuroscientists share knowledge with the legal system, in the context of specific litigation (Jones, Wagner, et al., 2013). But more broadly, this education often takes the form of training sessions and seminars. For example, a number of organizations have offered, and judges are increasingly requesting, some basic exposure for judges in the technologies, vocabularies, capabilities, and limitations of neuroscientific techniques. Over the past decade, more than 1,000 judges—along with many legal scholars, prosecutors, and defense attorneys—have participated in training sessions offered by the American Association for the Advancement of Science, the Federal Judicial Center, the MacArthur Foundation Research Network on Law and Neuroscience, and the MacArthur Law and Neuroscience Project.

Finally, burgeoning activity in law and neuroscience (sometimes called *neurolaw*) is evident along other critical dimensions. To give but a few examples, for context, consider that neurolaw publications numbered barely 100 in 2005 but swelled nineteen fold over the next decade, to over 1,900 today. Across the same time span, over 150 law and neuroscience conferences and symposia were hosted, a variety of law and neuroscience societies formed around the globe, and a number of law schools and other departments started offering neurolaw courses, some using a dedicated textbook on the subject (Jones, Schall, & Shen, 2014). Broader knowledge sharing has taken the forms, for instance, of cover-page articles in the *New York Times Sunday Magazine* (Rosen, 2007) and the *American Bar Association Journal* (Davis, 2012), a multipart television program, various radio documentaries and interviews, a complimentary electronic newsletter (*Neurolaw News*) and more than 50 neurolaw video lectures (at https://www.youtube.com/user/lawneuroorg).

Driving the Interest

There are doubtless many drivers of the increased interest in neurolaw. But at the most basic level, it arises from the intersection of (1) perennial questions that the legal system has been grappling with for generations and (2) the proliferation of new neurotechnological capabilities. Where these overlap springs the hope—or, at the very least, the active curiosity—that neuroscientific tools that can be applied to humans may yield better answers to some legally relevant questions that have historically yielded unsatisfying or uncertain solutions.

For instance: Is this person responsible for his or her behavior? What was this person's likely mental state at the time of the act? How competent is this person? Is this person lying? What does this person remember? How accurate is this person's memory? Is this person really in pain and, if so, how much? How can we improve juror and judge decisions?

And what developments have laid foundation for the hope that cognitive neuroscience can help answer these questions? For one thing, many people—including legal thinkers—increasingly recognize that the brain is not a product of either nature or nurture but rather necessarily exists at the intersection of genes and environments. They increasingly understand that the brain is the product of evolutionary processes, including natural selection, that have shaped it to readily associate various environmental inputs with behavioral outputs that tended (on average, in past environments) to increase the chances of survival and reproduction. And they increasingly understand that human cognition and behavior—including both relatively "automatic," nonconscious phenomena (e.g., implicit racial biases) and more "controlled," conscious phenomena (e.g., planning future acts)—are products of the brain, with some emerging from functionally specialized neural processes and others from large-scale network computations.

Against this background, there has also been increasing awareness of the remarkable rate of technological progress in the neurosciences. This includes awareness of key new tools of cognitive neuroscience that provide unprecedented insights into how human minds and brains work, as well as unique opportunities to try to "read out" from neural signals what a person is perceiving, thinking, or remembering (e.g., Naselaris, Kay, Nishimoto, & Gallant, 2011; Norman, Polyn, Detre, & Haxby, 2006). These cutting-edge tools—including brain imaging methods such as positron emission tomography (PET) and functional magnetic resonance imaging (fMRI) and data analytic methods such as machine learning, as well as the combination of both kinds of methods—have yielded both striking new discoveries as well as overhyped illusory advances. In turn, cognitive neuroscience's many discoveries and advances have, for better or for worse, tantalized the legal system with the prospects of answering some of its most challenging questions and commensurate concerns for associate risks (Aharoni, Funk, Sinnott-Armstrong, & Gazzaniga, 2008; Alces, 2018; Blitz, 2010, 2017; Brown & Murphy, 2010; Denno, 2015; Farahany, 2011; Freeman, 2011; Gazzaniga, 2008; Goodenough & Tucker, 2010; Greeley, 2009, 2013; Moore, 2011; Morse, 2011, 2013; Morse & Roskies, 2013b; Patterson & Pardo, 2016; Zeki & Goodenough, 2006; Slobogin, 2017).

Illustrative Research

In this section we provide a sampling, for general flavor, of some of the legal problems on which neurolaw experiments have been published in the last decade or so. We focus on the works with which we are most familiar, given that we each served on the MacArthur Foundation Research Network on Law and Neuroscience (the "Network"). Readers interested in the broader neurolaw literature can access a sortable and searchable bibliography at http://www.lawneuro.org/bibliography.php.

Brain-based memory detection Behavioral expressions of memory serve as critical evidence for the law, including eyewitness identifications and memory-based statements about an individual's intent or frame of mind during a past act (National Research Council, 2014). Mnemonic evidence is often challenged by the opposing side, leaving the jury to decide whether to believe, and how heavily to weigh, the evidence. Given this longstanding challenge for the law, there is interest in whether neural measures can detect the presence or absence of a memory or distinguish true from false memories (Lacy & Stark, 2013; Nadel & Sinnott-Armstrong, 2012; Schacter & Loftus, 2013). Being able to detect reliable neural signals of memory could be useful in a variety of investigative contexts, including probing the probability of deception (see the next subsection).

To examine whether functional brain imaging can be used to detect real-world memories, one Network working group, led by one of us (Wagner), put cameras that automatically took photos around the necks of undergraduate students as they navigated their lives for a few weeks (Rissman et al., 2016; see related work by St. Jacques et al., 2011; St. Jacques & Schacter, 2013). Subsequently, selected photos from a subject's camera were interleaved with photos from other subjects' cameras and displayed while the subject made memory decisions during fMRI. Machine-learning techniques applied to the fMRI data—here, multivoxel pattern analyses—revealed that activity patterns in numerous cortical regions along with the medial temporal lobe can be used to classify whether subjects are viewing and recognizing photos of their own past (i.e., hits) versus viewing and perceiving as new photos from someone else's camera (i.e., correct rejections). Classifier accuracy was well above chance (approaching ceiling performance in some cases) and, intriguingly, this was the case even when the classifier was applied to brain data from subjects other than the ones on which it was trained. In addition to detecting memories for real-world autobiographical events, a lab-based study revealed high accuracy when classifying brain patterns associated with recognizing studied faces versus correctly rejecting novel faces, as well as discriminating higher confidence versus lower confidence memories (Rissman, Greely, & Wagner, 2010).

While the above findings suggest that under controlled experimental conditions memory states can be detected from fMRI-measured brain patterns, initial studies also point to important boundary conditions. First, while high classification accuracy is possible (under some conditions) when discriminating recognized stimuli from stimuli perceived as novel, classification accuracy was only slightly above chance when attempting to discriminate true versus false recognition of faces (Rissman, Greely, & Wagner, 2010). This finding converges with a wealth of other data highlighting the similarity of brain responses during true and false memory (Schacter & Slotnick, 2004) and suggests that brain-based measures may not solve the law's frequent quandary of knowing when a witness's memory is accurate or mistaken. Second, classification accuracy was essentially at chance when applied to implicit memory—that is, discriminating between old stimuli that a subject failed to recognize (i.e., misses) from new stimuli perceived as novel (i.e., correct rejections; Rissman, Greely, & Wagner, 2010). Finally, the high level of fMRI-based classification of hits versus correct rejections fell to chance when subjects used cognitive countermeasures (shifting how they attended to memory) in an effort to mask their neural patterns of memory (Uncapher et al., 2015). As with the polygraph (National Research Council, 2003) and fMRI-based lie detection (see below), the potential real-world application of brain-based memory detection can be defeated by motivated noncompliant individuals. Thus, while extant data highlight that brain-based memory detection is possible, significant hurdles to real-world application remain.

Brain-based lie detection As noted at the outset, lawyers are increasingly proffering (i.e., "offering into evidence") neuroscientific evidence, both structural and functional. In many cases such evidence is the subject of admissibility hearings, in which a judge determines (according to state or federal law) whether the jury will be allowed to hear and see the evidence. For instance, in the case of *United States v. Semrau* (2010), the defendant Lorne Semrau, who ran a psychiatric group, was prosecuted for Medicare and Medicaid fraud. Although not all criminal statutes require knowledge of wrongdoing to be guilty, it is in fact one element of proving fraud that Dr. Semrau have known that what he was doing was illegal. In his defense, Dr. Semrau sought to introduce a report from the company Cephos purporting to show that an fMRI lie-detection protocol "indicated he is telling the truth in regards to not cheating or defrauding the government." Following 16 hours of hearings before a magistrate judge, the magistrate convincingly recommended to the trial judge that the evidence be excluded from the jury, due to specific flaws in the particular protocol, as well as doubt that the urged inferences could properly be drawn from the results (Shen & Jones, 2011).

With the advent of fMRI, cognitive neuroscientists are examining whether brain-based lie detection is possible. Despite some very promising studies (Greene & Paxton, 2009), the prospects for legal use remain almost entirely speculative (Bizzi et al., 2009; Wagner, 2010; Wagner et al., 2016). Take-home points from the literature (Christ et al., 2009; Farah et al., 2014) include (1) laboratory-based studies predominantly use instructed or permitted lie paradigms and have negligible stakes for failure to successfully deceive (in contrast to the stakes in real-world settings); (2) a set of frontal and parietal lobe regions are often more active during the putative "lie" versus "truth" conditions, and most evidence comes from group-based analyses that average over trials and subjects (c.f., the law requires an assessment of truthfulness about individual facts in individual brains); (3) experimental design limitations raise uncertainty as to whether these neural effects reflect responses associated with deception or whether they reflect attention and memory confounds that are unrelated to deception; and (4) countermeasures appear to alter these neural responses, suggesting that even if associated with deception, it may be possible to mask such responses. These limitations will frequently prevent brain-based techniques from satisfying the legal standards for admissibility of scientific findings. Indeed, some of these limitations and boundary conditions, along with others, were considered in the Semrau case, as well as the handful of other cases in which judges decided not to admit fMRI-based "lie detection" testimony into evidence.

Detection and classification of mental states Generally speaking, the government must prove, in order to get a criminal conviction, both that a defendant performed a prohibited act (*actus reus*) and that he did so in one of several defined states of mind (*mens rea*; for more on this, see Morse & Newsome, 2013). Because most crimes are matters of state law rather than federal law, the mental state definitions can vary. However, the *Model Penal Code*—which itself has no legal force—has been widely influential on the mental state definitions in most states. By its taxonomy, culpable mental states include *purposeful*, *knowing*, *reckless*, and *negligent*—in descending sequence of severity, each with importantly different sentencing results. In Colorado, for instance, the difference between being convicted of a knowing homicide, on the one hand, or a reckless homicide, on the other, could mean the difference between 14 years in prison and incarceration-free probation.

Scholars have long debated whether the knowing-versus-reckless distinction drawn by law actually exists in the brains of defendants, a concern heightened by recent behavioral work strongly suggesting that juror-like subjects have a difficult time distinguishing between the two (Ginther et al., 2014, 2018; Shen et al., 2011). Consequently, another line of research seeks to explore the extent to which coupling fMRI with machine-learning algorithms could shed light on whether there is a real psychological distinction between a "knowing" frame of mind and a "reckless" frame of mind. And one Network working group, led by Gideon Yaffe, found that the combination of fMRI and machine-learning algorithms could (under laboratory conditions) predict with high accuracy whether a subject was in a knowing versus a reckless frame of mind. This arguably suggests that the distinction the law had posited academically actually exists neurologically. And this is the first proof of concept that it is possible to read out a law-relevant mental state of a subject, in a scanner, in real time (Vilares et al., 2017).

Intent and punishment Humans are notoriously prone to various kinds of psychological biases. At the same time, few things are more crucial to the fair administration of criminal justice than trying to ensure that jurors and judges are minimally biased in their decisions about whether or not a defendant is criminally liable (typically a decision for the jury) and, if he is, how much to punish him (typically a decision for the judge). Until recently, nothing was known about how human brains make these important decisions.

Consequently, one line of research explores the extent to which fMRI might illuminate the neural processes underlying these determinations, which could potentially be an important first step in learning how to debias them (through, for instance, more effective training interventions). A first fMRI study found correlations between guilt and punishment decisions and activity in regions commonly associated with analytic, emotional, and theory-of-mind processes (Buckholtz et al., 2008). A subsequent study suggested that theory-of-mind circuitry may either gate or suppress affective neural responses, tempering the effect of emotion on punishment levels when, for instance, a perpetrator's culpability was very low while, at the same time, the harm he caused was very high (Treadway et al., 2014). A third study, using repetitive transcranial magnetic stimulation (rTMS) to test the causal role of right dorsolateral prefrontal cortex, found, as predicted, that compared to sham stimulation rTMS changed the *amount* that subjects punished protagonists in scenarios without altering how much they *blamed* those protagonists (Buckholtz et al., 2015). Breaking liability and punishment decisions down into constituent steps, a Network working group led by Owen Jones recently identified distinct neural responses that separately correlate with four key components of liability/punishment decisions: (1) assessing harms, (2) discerning mental states in others, (3) integrating those two pieces of information, and (4) choosing punishment amounts (Ginther et al., 2016).

Adolescent and young adult brains A constant challenge for legal systems is figuring how best to handle young offenders. While it has always been obvious that the very young are not as culpable for bad behavior as are the mature, legal systems have often struggled to develop juvenile justice regimes that are stable and fair. Several U.S. Supreme Court cases reflect this struggle. In *Roper v. Simmons* (2005), the court held unconstitutional any sentence to death for a crime committed by an adolescent of 16 or 17 years old. In *Graham v. Florida* (2010), the court similarly held it unconstitutional to sentence any juvenile offender, in a nonhomicide crime, to a sentence of life imprisonment without the possibility of parole. In *Miller v. Alabama* (2012), the court went further. It held that *mandatory* life imprisonment without the possibility of parole for those under the age of 18 at the time of their crimes was unconstitutional—even in cases of homicide. (However, the court left open the possibility of such a sentence, if the judge were to make an individualized assessment of the particular juvenile, crime, and surrounding circumstances.) Although the role neuroscientific arguments actually played in the disposition of these cases is debatable (Morse, 2013), it is notable in itself that neuroscientific arguments about adolescent brain development were provided to the court in each case and cited in some of them (Bonnie & Scott, 2013).

Complementing structural data that suggest that full maturation of the human brain may occur as late as into one's 20s (Gogtay et al., 2004; Mills et al., 2014), a wealth of behavioral and functional neural data highlight the context-dependence of developmental trajectories (Albert, Chein, & Steinberg, 2013; Luna, 2012). Importantly, these studies of adolescents and young adults might illuminate issues potentially relevant to juvenile and young adult justice. For example, potentially bearing on the legal system's challenge of deciding when and how to hold juveniles criminally responsible for their behavior, a Network working group led by B. J. Casey is exploring whether it is possible to draw meaningful lines between juveniles and young adults using fMRI and behavioral assays (Casey et al., 2017). In one study (Cohen et al., 2016), fMRI data and behavioral measures from 250 juveniles and young adults examined cognitive control under affectively arousing versus neutral conditions. Among the findings was that the brains and behaviors of 18- to 21-year-olds operate more like older adults under some environmental circumstances—specifically, when arousal and affective states are neutral—and more like juveniles in others—when arousal and affect are elevated (such as when emotion is triggered by stimuli or when performance is under peer observation). These data may have broad implications for the law, as they suggest that the age at which mature behavior may be fully realized is context-dependent.

Categories of Relevance

Neuroscience can be relevant to law in at least seven contexts (Jones, 2013).

Buttressing Most commonly, perhaps, neuroscientific evidence can be used to buttress other, typically behavioral, evidence. For example, suppose a criminal defendant has raised an insanity defense. If there is behavioral evidence consistent with insanity, those data will be the most salient evidence. If it turns out that there is also evidence of an acute abnormality in brain form or function, then the latter will buttress the former. But note that the neuroscientific evidence, no matter how strong, would be insufficient on its own to build a credible insanity defense if there were no behavioral evidence consistent with insanity to accompany it. In such a case, the buttressing effect of neuroscientific evidence would add to the weight of the behavioral evidence, not independently supplant it; that is, the brain data could support a conclusion but not drive it.

Detecting One of the most potent uses of neuroscience, perhaps, is its ability to detect facts that may be legally relevant. For example, in the 1992 New York case *People v. Weinstein,* Mr. Weinstein, an executive in Manhattan, came home one day, strangled his wife, and threw her out of the couple's 12th-floor apartment building. After his arrest Mr. Weinstein complained of headaches, which led to a discovery, through PET, of a very large subarachnoid cyst compressing his prefrontal cortex, which is known to be important for impulse control and executive function (Davis, 2017). Although it is unknown—and perhaps unknowable—how much the cyst contributed to the murder, the possession by the defense of a visually powerful brain image contributed to Mr. Weinstein's plea agreement with the state. And it illustrates the extent to which neuroscientific methods for detecting brain structures and functions may uncover new, legally relevant avenues to pursue. The same is true, for instance, of the extent to which brain imaging might more clearly detect injuries—or even the existence and amount of pain—in torts cases (Davis, 2017; Kolber, 2007; Pustilnik, 2012, 2015). Of course, as noted earlier, some maintain the hope that functional neuroimaging may one day enable the reliable detection of lies or legally relevant memories.

Sorting Neuroscience might also aid the legal system in sorting individuals into different categories, for different purposes. A paradigmatic example, perhaps, would be if neuroscientific measures could reliably identify the criminal addicts most susceptible to rehabilitative interventions. In theory, the legal system could then send such individuals into drug rehabilitation programs instead of into the general prison populations.

Predicting Over time, neuroscience may make important contributions to the law's efforts to predict various kinds of behaviors. For instance, two papers (Aharoni et al., 2013, 2014) provided initial evidence that certain brain-based variations in incarcerated individuals predict some of the variance in the probability of their rearrests after release. It was a small part of the variance, and the magnitude of the effect is debated due to questions about the analytic approach (Poldrack, 2013; Poldrack et al., 2017). Nevertheless, as parole boards, for instance, sometimes expand and revise their actuarial approaches to predicting recidivism (including age, sex, type of crime, and more), such observations raise the possibility that at some point in the future neuroscientific measures may become relevant. A determination of if and when such application emerges will be informed by meaningful debates about how best to interpret and apply neuroprediction (Nadelhoffer et al., 2010; Poldrack, 2013; Poldrack et al., 2017; Singh, Sinnott-Armstrong, & Savulecu, 2013; Slobogin, 2013).

Intervening In theory, neuroscience could aid law through the development and validation of intervention approaches. For example, if certain drug treatments prove to substantially decrease the probability of recidivism, psychopharmacological interventions may be recommended for inclusion as a condition of parole.

Of course, like many aspects of neurolaw, this can raise important ethical considerations about what trade-offs we as a society are willing to make between perceived benefits, attendant risks and costs, and individual rights (Illes, 2017; Morse, 2017).

Explaining Neuroscientific methods are beginning to uncover regions of the brain, neural responses, and interactions within and between regions that subserve the processes by which decisions—key to the functioning of law—are made (Heekeren, Marrett, & Ungerleider, 2008; Shadlen & Kiani, 2013). As discussed above, when considering adolescent brain development, these could provide new insights into why and how individuals transgress the law, in criminal or civil domains (Scott & Steinberg, 2008; Scott, Bonnie, & Steinberg, 2016; Steinberg, 2016). Such discoveries could also provide new insights into the experiences of individuals who have been wronged. And, as noted above, they could provide insights into the processes by which jurors and judges make their decisions. All of these might increase the knowledge base on which new behavioral interventions and legal policy are deployed in furtherance of improving decisions and the legal consequences they create.

Challenging assumptions in the legal system Neuroscience may sometimes challenge assumptions in the legal system. For example, the legal system currently assumes that solitary confinement is insufficiently damaging to the brain to constitute "cruel and unusual punishment," and thus it is not prohibited as unconstitutional. Perhaps that's right. Perhaps it isn't. The tools of neuroscience may eventually help us to know which. If the assumption is wrong, that may provide impetus for law reform.

Similarly, note that the rules of evidence can be thought of as designed to keep certain information from entering the brains of jurors because of assumptions about how that information might affect the decisions of jurors. The evidentiary rules also reflect underlying neuroscientific assumptions about witness brains. For instance, a general rule of evidence, known as the prohibition against *hearsay*, typically operates to prevent person A from testifying as to what person B said they observed at the time of an act relevant to the trial (such as the name of a perpetrator). The logic is that (so long as person B is available) person B's testimony is deemed to be more reliable than person A's. But there are some exceptions. Among them is the exception for *excited utterances*. That exception allows person A to testify as to what person B said—so long as person B was excited and believed to be more or less blurting things out at the time. The explicit assumption underlying this rule is that person B, being in an excited state, would not have time to lie about what she was witnessing. Perhaps that's true. Perhaps it isn't. The tools of cognitive neuroscience might help us to know which. And if the assumption is wrong—with respect to this evidentiary rule or others—neuroscience may again provide the potential foundation for reform.

Two Key Caveats

Of course, many cautions and caveats exist regarding whether neuroscientific information should directly affect legal decisions and policy and, if so, how to carefully, sensibly, and responsibly incorporate such information (Campbell & Eastman, 2013; Faigman, Monahan, & Slobogin, 2014; Jones, Buckholtz, Schall, & Marois, 2009; Morse 2013). For example, we described some of the open questions and potential boundary conditions surrounding brain-based memory and lie detection. In each of the areas of research we briefly considered, as well as others being explored in the field, additional cautions and caveats are warranted.

Here we consider two especially salient, crosscutting caveats.

The long chain of inference First, it is not a simple thing to reason from the presence of a brain feature (a large subarachnoid cyst, for example) to the conclusion that that feature contributed meaningfully to generating or enabling a specific behavior (such as murder). Such a conclusion requires a long chain of inferences, with many potential weak links. What exactly is the brain feature at issue? How long was it there? What is known to correlate with the presence of the brain feature? What are the known causal pathways of influence? In many instances, answers to one or more of these critical questions are unknown, which greatly tempers confidence in any inferences drawn.

Unknown frequencies of predictors and outcomes Second, and relatedly, one key limitation to drawing logical and informed inferences is that the relative frequency of a feature in the population—Mr. Weinstein's cyst, for instance—is often not known. Without that information we have no idea how many people are walking around in the population with the same feature without engaging in the same behavior as the accused. Knowing the relative frequency of a predictor, as well as the frequency of a particular outcome (i.e., the base rate), is necessary to determine the increased likelihood, if any, of engagement in an undesirable behavior given the feature in question (National Research Council, 2003). Without this information, proper inferences are difficult to draw. With what confidence could one say that Mr. Weinstein's cyst meaningfully, and legally, caused him to commit murder?

The issue of unknown predictor frequencies is particularly relevant given the remarkable pace of progress in neuroscientific methods in recent years. Whereas structural imaging of the human brain has been available for a few decades and the detection of a structural abnormality is often relatively straightforward for neuroradiologists, functional imaging is a more recent development, and the machine-learning characterization of functional patterns and their relation to cognition is at an even earlier stage. Thus, whereas some limited information is available on the relative frequencies of structural abnormalities and their relationships to altered behavior, cognitive neuroscience is only just beginning to conduct large-scale individual difference studies of the relationships between functional brain patterns (which themselves vary depending on the particulars of the analytic approach) and cognitive states and behaviors. Early work is focused on characterizing the heterogeneity evident in healthy young adults—we seem far from the point at which we can say anything about the relative frequencies of particular functional patterns in healthy individuals and their associated outcomes, let alone those of atypical patterns and states.

Legal Impact of Neuroscience Evidence

When neuroscientific evidence is admitted, what are its impacts? We know that jurors are, at least sometimes, significantly affected by neuroscience evidence. For example, in the case of *State of Florida v. Grady Nelson* (2010), the defendant was quickly convicted of a murder, leaving the question to the jury by simple majority vote (under Florida death penalty law at the time) as to whether Mr. Nelson should be executed or given life in prison without parole. With Mr. Nelson's life hanging in the balance, the defense introduced qEEG (quantified electroencephalography) evidence in support of the inference that Mr. Nelson's brain was too abnormal to warrant his execution. By the narrowest of possible votes, the jury gave Mr. Nelson life in prison. Afterward, two jurors granted interviews indicating that the brain data had turned their prior inclinations, to vote in favor of execution, completely around.

Some members of the judiciary are increasingly invoking neuroscience in judicial opinions (Farahany, 2015), sometimes drawing colleagues into public debates over its relevance. High profile examples include the U.S. Supreme Court cases of *Graham v. Florida* and *Miller v. Alabama* (mentioned earlier). And Supreme Court Justice Sotomayor recently referred to "a major neurocognitive disorder that compromises [the defendant's] decision-making abilities" in her dissent from the court's refusal to hear the appeal in *Wessinger v. Vannoy* (2018).

Of course, given the complexity of neuroscience, one natural concern is that both judges and jurors may have a hard time understanding where it is—and, equally importantly, is not—relevant. Relatedly, some have expressed worry that jurors may be overawed by the pictorial nature of some brain data and give it more weight than it is due (Weisberg et al., 2008). Two laboratory studies investigating this phenomenon found that the images themselves appear to have no particular biasing effect on subjects—above and beyond nonpictorial neuroscientific testimony—except in the case of death penalty decisions, wherein images decreased the probability of a vote for execution (Saks et al., 2014; Schweitzer & Saks, 2011).

Given the complex interactions between law and neuroscience, there is a need for reasoned consideration of the ethical and legal impacts of neuroscientific evidence (e.g., see Presidential Commission for the Study of Bioethical Issues [2015] and selected recommendations submitted to the commission [Jones, Bonnie, et al., 2014]).

Conclusions

The domains of science and law have very different goals. Painted with a broad brush, these are the attempt to uncover truths, on one hand, and the attempt to fairly and effectively govern the behaviors of large populations, on the other. While truths may inform governance, they don't dictate it. Indeed, most scholars (including ourselves) believe it impossible to argue directly from a description to a prescription without reference to other values. Put another way, explanation isn't justification. And, therefore, we do not expect the law will or should automatically change, or refuse to change, in light of a neuroscientific finding alone.

At the same time, advances in the cognitive neurosciences effectively guarantee a future in which the law increasingly interacts with neuroscientific evidence. Even at this relatively early stage, there is a gradual but discernible shift from nearly exclusive reliance on structural brain evidence (in cases involving any brain evidence) to increasing reliance on functional neural assays. As this shift continues to develop and accelerate, there will be divergent views on whether and when particular types of neural data should be drawn upon to inform legal decisions.

In this review we have highlighted a few illustrative legal problems on which neuroscience research is beginning to yield potentially informative data, as well as others in which the science suggests it is premature to move from the lab to the courtroom (for other overviews, see Jones, Marois, et al., 2013; Jones, Schall, & Shen, 2014). Concurrently, we have considered the categories of potential relevance for neuroscience evidence, along with crosscutting caveats. The growth of neurolaw—which crucially depends on interdisciplinary interactions—has produced significant progress and suggests promise. At the same time, there is ample cause for caution, lest overexuberance pave a path to pitfall.

Acknowledgments

This work was supported in part by a grant to Owen D. Jones from the John D. and Catherine T. MacArthur Foundation and a gift from the Glenn M. Weaver Foundation to Vanderbilt Law School. Its contents do not necessarily represent the official views of the MacArthur Foundation or the MacArthur Foundation Research Network on Law and Neuroscience or the Weaver Foundation. We are grateful to Peter Imrey for helpful comments and to Emily M. Lamm for research assistance.

REFERENCES

Aharoni, E., Funk, C., Sinnott-Armstrong, W., & Gazzaniga, M. (2008). Can neurological evidence help courts assess criminal responsibility? Lessons from law and neuroscience. *Annals of the New York Academy of Sciences, 1124*(1), 145–160.

Aharoni, E., Mallett, J., Vincent, G. M., Harenski, C. L., Calhoun, V. D., Sinnott-Armstrong, W., ... Kiehl, K. A. (2014). Predictive accuracy in the neuroprediction of rearrest. *Social Neuroscience, 9*(4), 332–336.

Aharoni, E., Vincent, G. M., Harenski, C. L., Calhoun, V. D., Sinnott-Armstrong, W., Gazzaniga, M. S., ... Kiehl, K. K. (2013). Neuroprediction of future rearrest. *Proceedings of the National Academy of Sciences of the United States of America, 110*(15), 6223–6228.

Albert, D., Chein, J., & Steinberg, L. (2013). Peer influences on adolescent decision making. *Current Directions in Psychological Science, 22*(2), 114–120.

Alces, P. A. (2018). *The moral conflict of law and neuroscience.* Chicago: University of Chicago Press.

Bizzi, E., Hyman, S. E., Raichle, M. E., Kanwisher, N., Phelps, E. A., Morse, S. J., … Greely, H. T. (2009). *Using imaging to identify deceit: Scientific and ethical questions.* Cambridge, MA: American Academy of Arts & Sciences.

Blitz, M. J. (2010). Freedom of thought for the extended mind: Cognitive enhancement and the Constitution. *Wisconsin Law Review, 2010,* 1049–1118.

Blitz, M. J. (2017). *Searching minds by scanning brains: Neuroscience technology and constitutional privacy protection.* Dordrecht, Switzerland: Springer Nature.

Bonnie, R. J., & Scott, E. S. (2013). The teenage brain: Adolescent brain research and the law. *Current Directions in Psychological Science, 22*(2), 158–161.

Brown, T., & Murphy, E. (2010). Through a scanner darkly: Functional neuroimaging as evidence of a criminal defendant's past mental states. *Stanford Law Review, 62*(4), 1119–1208.

Buckholtz, J. W., Asplund, C. L., Dux, P. E., Zald, D. H., Gore, J. C., Jones, O. D., & Marois, R. (2008). The neural correlates of third-party punishment. *Neuron, 60*(5), 940–950.

Buckholtz, J. W., Martin, J. W., Treadway, M. T., Jan, K., Zald, D. H., Jones, O., & Marois, R. (2015). From blame to punishment: Disrupting prefrontal cortex activity reveals norm enforcement mechanisms. *Neuron, 87*(6), 1369–1380.

Campbell, C., & Eastman, N. (2013). The limits of legal use of neuroscience. In I. Singh, W. P. Sinnott-Armstrong, & J. Savulescu (Eds.), *Bioprediction biomarkers, and bad behavior: Scientific, legal and ethical challenges* (pp. 91–117). New York: Oxford University Press.

Casey, B. J., Bonnie, R. J., Davis, A., Faigman, D. L., Hoffman, M. B., Jones, O. D., … Wagner, A. D. (2017). How should justice policy treat young offenders? A knowledge brief of the MacArthur Foundation Research Network on Law and Neuroscience. MacArthur Foundation Research Network on Law and Neuroscience.

Christ, S. E., Van Essen, D. C., Watson, J. M., Brubaker, L. E., & McDermott, K. B. (2009). The contributions of prefrontal cortex and executive control to deception: Evidence from activation likelihood estimate meta-analyses. *Cerebral Cortex, 19*(7), 1557–1566.

Cohen, A. O., Breiner, K., Steinberg, L., Bonnie, R. J., Scott, E. S., Taylor-Thompson, K., … Casey, B. J. (2016). When is an adolescent an adult? Assessing cognitive control in emotional and nonemotional contexts. *Psychological Science, 27*(4), 549–562.

Davis, K. (2012). Brain trials: Neuroscience is taking a stand in the courtroom. *American Bar Association Journal, 98,* 36–37.

Davis, K. (2017). *The brain defense: Murder in Manhattan and the dawn of neuroscience in America's courtrooms.* New York: Penguin Press.

Davis, K. D., Flor, H., Greely, H. T., Iannetti, G. D., Mackey, S., Ploner, M., … Wager, T. D. (2017). Brain imaging tests for chronic pain: Medical, legal, and ethical issues and recommendations. *Nature Reviews Neurology, 13*(10), 624–638.

Denno, D. W. (2015). The myth of the double-edged sword: An empirical study of neuroscience evidence in criminal cases. *Boston College Law Review, 56*(2), 493–551.

Faigman, D. L., Monahan, J., & Slobogin, C. (2014). Group to individual (G2i) inference in scientific expert testimony. *University of Chicago Law Review, 81*(2), 417–480.

Farah, M. J., Hutchinson, J. B., Phelps, E. A., & Wagner, A. D. (2014). Functional MRI-based lie detection: Scientific and societal challenges. *Nature Reviews Neuroscience, 15*(2), 123–131.

Farahany, N. A. (2011). Incriminating thoughts. *Stanford Law Review, 64,* 351–408.

Farahany, N. A. (2015). Neuroscience and behavioral genetics in US criminal law: An empirical analysis. *Journal of Law & the Biosciences, 2*(3), 485–509.

Freeman, M. (2011). *Law and neuroscience: Current legal issues* (Vol. 13). New York: Oxford University Press.

Gazzaniga, M. S. (2008). The law and neuroscience. *Neuron, 60*(3), 412–415.

Ginther, M. R., Bonnie, R. J., Hoffman, M. B., Shen, F. X., Simons, K. W., Jones, O. D., & Marois, R. (2016). Parsing the behavioral and brain mechanisms of third-party punishment. *Journal of Neuroscience, 36*(36), 9420–9434.

Ginther, M. R., Shen, F. X., Bonnie, R. J., Hoffman, M. B., Jones, O. D., Marois, R., & Simons, K. (2014). The languages of mens rea. *Vanderbilt Law Review, 67,* 1327–1372.

Ginther, M. R., Shen, F. X., Bonnie, R. J., Hoffman, M. B., Jones, O. D., & Simons, K. W. (2018). Decoding guilty minds. *Vanderbilt Law Review, 71,* 241–284.

Gogtay, N., Giedd, J. N., Lusk, L., Hayashi, K. M., Greenstein, D., Vaituzis, A. C., … Thompson, P. M. (2004). Dynamic mapping of human cortical development during childhood through early adulthood. *Proceedings of the National Academy of Sciences of the United States of America, 101*(21), 8174–8179.

Goodenough, O. R., & Tucker, M. (2010). Law and cognitive neuroscience. *Annual Review of Law and Social Science, 6,* 61–92.

Graham v. Florida, 560 U.S. 48 (2010).

Greely, H. T. (2009). Law and the revolution in neuroscience: An early look at the field. *Akron Law Review, 42,* 687–715.

Greely, H. T. (2013). Neuroscience, mindreading, and the law. In S. J. Morse & A. L. Roskies (Eds.), *A primer on criminal law and neuroscience* (pp. 120–149). New York: Oxford University Press.

Greene, J. D., & Paxton, J. M. (2009). Patterns of neural activity associated with honest and dishonest moral decisions. *Proceedings of the National Academy of Science of the United States of America, 106*(30), 12506–12511.

Grey, B. J., & Marchant, G. E. (2015). Biomarkers, concussions, and the duty of care. *Michigan State Law Review, 2015,* 1911–1981.

Heekeren, H. R., Marrett, S., & Ungerleider, L. G. (2008). The neural systems that mediate human perceptual decision making. *Nature Reviews Neuroscience, 9,* 467–479.

Illes, J. (2017). *Neuroethics: Anticipating the future.* New York: Oxford University Press.

In re: Nat'l Football League Players' Concussion Injury Litigation. No. 2:12-md-02323-AB, 2015 WL 12827803 (E.D. Pa. 2015).

Jones, O. D. (2013). Seven ways neuroscience aids law. *Scripta Varia, 121,* 1–14.

Jones, O. D., Bonnie, R. J., Casey, B. J., Davis A., Faigman, D. L., Hoffman, M. B., ... Yaffe, G. (2014). Law and neuroscience: Recommendations submitted to the president's bioethics commission. *Journal of Law & the Biosciences, 1*(2), 224–236.

Jones, O. D., Buckholtz, J., Schall, J., & Marois, R. (2009). Brain imaging for legal thinkers: A guide for the perplexed. *Stanford Technology Law Review, 2009*, 5–91.

Jones, O. D., Marois, R., Farah, M. J., & Greely, H. T. (2013). Law and neuroscience. *Journal of Neuroscience, 33*(45), 17624–17630.

Jones, O. D., Schall, J. D., & Shen, F. X. (2014). *Law and neuroscience*. New York: Wolters Kluwer Law & Business.

Jones, O. D., Wagner, A. D., Faigman, D. L., & Raichle, M. (2013). Neuroscientists in court. *Nature Reviews Neuroscience, 14*, 730–736.

Kolber, A. J. (2007). Pain detection and the privacy of subjective experience. *American Journal of Law & Medicine, 33*, 433–456.

Lacy, J. W., & Stark, C. E. L. (2013). The neuroscience of memory: Implications for the courtroom. *Nature Reviews Neuroscience, 14*, 649–658.

Luna, B. (2012). The relevance of immaturities in the juvenile brain to culpability and rehabilitation. *Hastings Law Journal, 63*(6), 1469–1486.

Miller v. Alabama, 567 U.S. 460 (2012).

Mills, K. L., Goddings, A. L., Clasen, L. S., Giedd, J. N., & Blakemore, S. J. (2014). The developmental mismatch in structural brain maturation during adolescence. *Developmental Neuroscience, 36*(3–4), 147–160.

Moore, M. (2011). Responsible choices, desert-based legal institutions, and the challenges of contemporary neuroscience. *Social Philosophy & Policy, 29*(1), 233–279.

Morse, S. J. (2011). Avoiding irrational exuberance: A plea for neuromodesty. *Mercer Law Review, 62*, 837–859.

Morse, S. J. (2013). Brain overclaim redux. *Law & Inequality, 31*(2), 509–534.

Morse, S. J. (2017). Neuroethics: Neurolaw. *Oxford handbooks online*. Oxford: Oxford University Press.

Morse, S. J., & Newsome, W. T. (2013). Criminal responsibility, criminal competence, and criminal law prediction. In S. J. Morse & A. L. Roskies (Eds.), *A primer on criminal law and neuroscience* (pp. 150–178). New York: Oxford University Press.

Morse, S. J., & Roskies, A. L. (2013a). *A primer on criminal law and neuroscience*. New York: Oxford University Press.

Morse, S. J., & Roskies, A. L. (2013b). The future of law and neuroscience. In S. J. Morse & A. L. Roskies (Eds.), *A primer on criminal law and neuroscience*. New York: Oxford University Press.

Nadel, L., & Sinnott-Armstrong, W. P. (2012). *Memory and law*. New York: Oxford University Press.

Nadelhoffer, T., Bibas, S., Grafton, S., Kiehl, K., Mansfield, A., Sinnott-Armstrong, W., & Gazzaniga, M. (2010). Neuroprediction, violence, and the law: Setting the stage. *Neuroethics, 5*(1), 67–99.

Naselaris, T., Kay, K. N., Nishimoto, S., & Gallant, J. L. (2011). Encoding and decoding in fMRI. *Neuroimage, 56*(2), 400–410.

National Research Council. (2003). *The polygraph and lie detection*. Washington, DC: National Academies Press. https://doi.org/10.17226/10420

National Research Council. (2014). *Identifying the culprit: Assessing eyewitness identification*. Washington, DC: National Academies Press. https://doi.org/10.17226/18891

Neurolaw News. The MacArthur Foundation Research Network on Law and Neuroscience.

Norman, K. A., Polyn, S. M., Detre, G. J., & Haxby, J. V. (2006). Beyond mind-reading: Multi-voxel pattern analysis of fMRI data. *Trends in Cognitive Sciences, 10*(9), 424–430.

Patterson, D., & Pardo, M. S. (2016). *Philosophical foundations of law and neuroscience*. New York: Oxford University Press.

People v. Weinstein, 591 N.Y.S.2d 715 (Sup. Ct. 1992).

Poldrack, R. (2013, April 6). How well can we predict future criminal acts from fMRI data? russpoldrack.org.

Poldrack, R., Monahan, J., Imrey, P., Reyna, V., Raichle, M., Faigman, D., & Buckholtz, J. W. (2017). Predicting violent behavior: What can neuroscience add? *Trends in Cognitive Sciences, 22*(2), 111–123.

Presidential Commission for the Study of Bioethical Issues. (2015). *Gray matters: Topics at the intersection of neuroscience, ethics, and society* (Vol. 2). Washington, DC.

Pustilnik, A. C. (2012). Pain as fact and heuristic: How pain neuroimaging illuminates moral dimensions of law. *Cornell Law Review, 97*(4), 801–847.

Pustilnik A. C. (2015). Imaging brains, changing minds: How pain neuroimaging can inform the law. *Alabama Law Review, 66*(5), 1099–1158.

Rissman, J., Chow, T. E., Reggente, N., & Wagner, A. D. (2016). Decoding fMRI signatures of real-world autobiographical memory retrieval. *Journal of Cognitive Neuroscience, 28*(4), 604–620.

Rissman, J., Greely, H., & Wagner, A. (2010). Detecting individual memories through the neural decoding of memory states and past experience. *Proceedings of the National Academy of Sciences of the United States of America, 107*(21), 9849–9854.

Roper v. Simmons, 543 U.S. 551 (2005).

Rosen, J. (2007, March 11). The brain on the stand. *New York Times Magazine*.

Saks, M. J., Schweitzer, N. J., Aharoni, E., & Kiehl, K. (2014). The impact of neuroimages in the sentencing phase of capital trials. *Journal of Empirical Legal Studies, 11*(1), 105–300.

Schacter, D. L., & Loftus, E. F. (2013). Memory and law: What can cognitive neuroscience contribute? *Nature Neuroscience, 16*, 119–123.

Schacter, D. L., & Slotnick, S. D. (2004). The cognitive neuroscience of memory distortion. *Neuron, 44*(1), 149–160.

Schweitzer, N. J., & Saks, M. J. (2011). Neuroimage evidence and the insanity defense. *Behavioral Science Law, 29*(4), 592–607.

Scott, E. S., Bonnie, R. J., & Steinberg, L. (2016). Young adulthood as a transitional legal category: Science, social change, and justice policy. *Fordham Law Review, 85*(2), 641–666.

Scott, E. S., & Steinberg, L. (2008). *Rethinking juvenile justice*. Cambridge, MA: Harvard University Press.

Shadlen, M. N., & Kiani, R. (2013). Decision making as a window on cognition. *Neuron, 80*(3), 791–806.

Shen, F. X., Hoffman, M. B., Jones, O. D., Greene, J. D., & Marois, R. (2011). Sorting guilty minds. *New York University Law Review, 86*, 1306–1360.

Shen, F. X., & Jones, O. D. (2011). Brain scans as evidence: Truths, proofs, lies, and lessons. *Mercer Law Review, 62*, 861–884.

Singh, I., Sinnott-Armstrong, W. P., & Savulecu, J. (2013). *Bioprediction, biomarkers, and bad behavior: Scientific, legal and ethical challenges.* New York: Oxford University Press.

Slobogin, C. (2013). Bioprediction in criminal cases. In I. Singh, W. P. Sinnott-Armstrong, & J. Savulecu (Eds.), *Bioprediction, biomarkers, and bad behavior: Scientific, legal and ethical challenges* (pp. 77–90). New York: Oxford University Press.

Slobogin, C. (2017). Neuroscience nuance: Dissecting the relevance of neuroscience in adjudicating criminal culpability. *Journal of Law & the Biosciences, 4*(3), 577–593.

St. Jacques, P. L., Conway, M. A., Lowder, M. W., & Cabeza, R. (2011). Watching my mind unfold versus yours: An fMRI study using a novel camera technology to examine neural differences in self-projection of self versus other perspectives. *Journal of Cognitive Neuroscience, 23*(6), 1275–1284.

St. Jacques, P. L., & Schacter, D. L. (2013). Modifying memory: Selectively enhancing and updating personal memories for a museum tour by reactivating them. *Psychological Science, 24,* 537–543.

State of Florida v. Grady Nelson, F05-00846 (11th Fla. Cir. Ct. 2010).

Steinberg, L. (2016). *Age of opportunity: Lessons from the new science of adolescence.* Boston: Houghton Mifflin Harcourt.

Treadway, M. T., Buckholtz, J. W., Martin, J. W., Jan, K., Asplund, C. L., Ginther, M. R., ... Marois, R. (2014). Corticolimbic gating of emotion-driven punishment. *Nature Neuroscience, 17*(9), 1270–1275.

Uncapher, M. R., Boyd-Meredith, J. T., Chow, T. E., Rissman, J., & Wagner, A. D. (2015). Goal-directed modulation of neural memory patterns: Implications for fMRI-based memory detection. *Journal of Neuroscience, 35*(22), 8531–8545.

United States v. O'Connor, 3:13-cr-00537 (S.D. Cal. 2013).

United States v. Semrau, 2010 WL 6845092 (W.D. Tenn. 2010), aff'd, 693 F.3d 510 (6th Cir. 2012).

Vilares, I., Wesley, M., Ahn, W. Y., Bonnie, R. J., Hoffman, M. B., Jones, O. D., ... Montague, P. R. (2017). Predicting the knowledge-recklessness distinction in the human brain. *Proceedings of the National Academy of Sciences of the United States of America, 114*(12), 3222–3227.

Wagner, A. D. (2010). Can neuroscience identify lies? In M. S. Gazzaniga & J. S. Rakoff (Eds.), *A judge's guide to neuroscience: A concise introduction* (pp. 13–25). Santa Barbara: University of California.

Wagner, A. D., Bonnie, R. J., Casey, B. J., Davis, A., Faigman, D. L., Hoffman, M. B., ... Yaffe, G. (2016). fMRI and lie detection. A knowledge brief of the MacArthur Foundation Research Network on Law and Neuroscience. MacArthur Foundation Research Network on Law and Neuroscience.

Weisberg, D. S., Keil, F. C., Goodstein, J., Rawson, E., & Gray, J. R. (2008). The seductive allure of neuroscience explanations. *Journal of Cognitive Neuroscience, 20*(3), 470–477.

Wessinger v. Vannoy, 138 S. Ct. 952 (2018) (Sotomayor, J., dissenting).

Zeki, S., & Goodenough, O. (2006). *Law and the brain.* New York: Oxford University Press.

90 Neuroscience and Socioeconomic Status

MARTHA J. FARAH

ABSTRACT This chapter reviews the concept of socioeconomic status (SES) and the scientific and societal reasons for studying SES in relation to the brain. Epidemiologists have long noted SES gradients in physical and mental health and cognitive capabilities. The research reviewed here is aimed at better understanding these SES disparities using neuroscience and eventually improving well-being for people of low SES. In addition, this research informs the practice of cognitive neuroscience research in general, by clarifying the ways that subjects' SES may influence research findings. The SES research reviewed here includes descriptive studies—aimed at establishing the structural and functional neural correlates of SES—and more explanatory studies—aimed at understanding how and why these correlations arise. At this early stage of development, the neuroscience of SES gives us more questions than answers but has already highlighted SES differences in various cortical and subcortical regions and a number of potential environmental causes.

Early cognitive neuroscience research was focused on understanding the "typical" brain. This was in part an effort to go beyond clinical neuropsychology to address the basic science of how normal brains work. It also reflected the sensible decision to map out the general principles of brain function before grappling with patterns of variation—variation due to pathology or due simply to normal individual differences. Of course, the "typical" brains being studied were usually the brains of people working or studying at universities. These young, educated, middle-class subjects were readily available and understood task instructions easily. And for most of us working in cognitive neuroscience labs, who were also young, educated, and middle-class, they seemed the paradigm of typical, normal humanity.

As cognitive neuroscience has matured, its conception of humanity has broadened. Studies of normal aging pushed the age range of "typical brains" upward and provided an important framework for contrasting with dementia. Studies of early life development, which flourished as the methodological challenges of studying young brains were gradually overcome, pushed the age range downward. Other differences in brain structure and function, related to sex and gender, culture, personality, attitudes, intelligence, and bilingualism, have also been embraced as part of understanding the normal human brain.

One dimension of variation that remains relatively unexplored is socioeconomic status (SES). Low SES afflicts people around the world. In the United States, individuals meeting the government's definition of "low income" face food and housing insecurity and comprise 30% of the population (Kaiser Family Foundation, 2017). This makes low-income people more "typical" than many other subpopulations of interest to cognitive neuroscience. In this chapter I will review what has so far been learned about SES and the brain.

What Is Socioeconomic Status?

SES is a fairly intuitive concept, corresponding to our everyday understanding of wealth, prestige, and power. The epidemiologist Michael Marmot (2004) conveys the idea of SES with a parade analogy (2004). Imagine lining everyone up in order of their income, with the lowest-paid people at the front of the parade and the highest-paid people at the back. As you watch the procession, you notice changes in what Marmot describes as comportment, demeanor, confidence, and signs of physical health, all trending more positive further back in the line. Now, reorganize people in terms of education so that the head of the parade includes those with no formal education, followed by grade school, high school dropouts, and so on, with the postgraduate degree bearers bringing up the rear. Or, organize them in terms of occupational prestige (day laborers and cleaning staff in front, surgeons and judges in back) or in terms of parents' social class (independent of one's own) and watch these parades go by. You will observe the same trends in physical and behavioral signs of well-being as the people file past, and indeed, most people will be at roughly the same point between front and back as in the income parade.

For Marmot, who studies health disparities, the association of these social and economic rankings with health is his key point. However, for understanding the idea of SES, three additional points can be taken away:

First, most people stay in roughly the same part of the parade, regardless of how the parade is organized. In other words, different measures of SES are moderately correlated with one another. Second, the whole length of the parade is sorted, rather than just sorting people into a vanguard with low SES and everyone else crowded together. In other words, while some SES-related phenomena do show a threshold-like pattern, with little difference associated with SES differences between middle and higher SES, most disparities follow a gradient, with differences in SES mattering at all levels. Finally, recall that attributes such as comportment, demeanor, and confidence change as people march by. SES gradients are not confined to health but show up in a wide range of psychological attributes. Emotional health and well-being increase with SES, with progressively less depression, anxiety, and psychosis at higher levels of SES (Kessler et al., 2005; Lorant et al., 2003; McLaughlin et al., 2012). Intelligence and academic achievement also show positive gradients with SES. From the "school readiness" of kindergarteners to performance on standardized achievement and IQ tests throughout life, higher SES is associated with higher performance (McLoyd, 1998; Sirin, 2005). Above and beyond the value for neuroscience of understanding this ubiquitous dimension of human variability, understanding how SES interacts with human development is an important goal from the perspective of public health and human capital.

The most common economic measure of SES involves income. Often, income is measured as an income-to-needs ratio, to take into account the number of mouths to be fed. The US government's "poverty line" is an income-to-needs ratio measure, with the current line equivalent to an income of $25,100 for a family of four (Federal Register, 2018). People living on less than 200% of the poverty line are considered low income, and higher incomes can be expressed as larger percentages of the poverty line. In addition, wealth influences one's economic situation independently of income. Turning to noneconomic measures, the most commonly used is educational attainment. Childhood SES is measured by the educational attainment of the parents. Occupational prestige, for which there are standard ratings (e.g., Hauser & Warren, 1997), is also sometimes used, with a parental occupation standing in for studies of children. Measures of neighborhood socioeconomic characteristics, typically based on census data regarding financial, educational, and other measures of residents' SES, have also been used (Gianaros et al., 2017). Finally, among the commonly used measures of SES is subjective social status, which captures people's sense of where they sit in the status hierarchy of the nation or their community (Adler et al., 2000).

Why Apply Neuroscience to the Study of Socioeconomic Status?

Cognitive neuroscience is viewed by some as overambitious, a roaring young field that does not recognize its own limitations and has been misapplied to problems beyond its reach (e.g., Satel & Lilienfeld, 2013). We should therefore pause and ask: Why apply neuroscience to the study of SES? There are a number of good reasons to pursue SES neuroscience. Although the ultimate success of the enterprise cannot be predicted at present, the pursuit is neither scientifically unrealistic nor the result of an academic bandwagon mentality.

The first reason to study the relationship between SES in the brain was raised at the outset of this chapter: if we want to understand normal brain function, then we need to study a full array of normal brains. The vast majority of cognitive neuroscience research has been carried out with subjects who are middle class, but in the United States, at least a third of our citizens are not in this category, limiting our understanding of what the normal brain is like. People differ in part as a function of SES, and even if it seems reasonable to put value judgments on differences in physical and mental health and cognitive ability, it is not reasonable to classify low SES as a pathology. A complete understanding of human brain function needs to include brain function at all levels of SES.

A related reason for SES neuroscience comes from the emerging applications of neuroscience in everyday life. As we begin to base schooling and educational policy on neuroscience (Gabrieli, 2016), use neural measures as evidence in legal proceedings (Farahany, 2016), or design marketing campaigns to sell or persuade (Lee et al., 2017), the performance of these systems will depend on their validity for all levels of SES.

Another reason to study SES and the brain is to understand why SES is associated with so many important life outcomes, from health to cognitive ability, and how to reduce these disparities. It is obvious that cognition and mental health are dependent on brain function. Physical health, too, is related to the brain, which plays a central role in the endocrine and immune responses linked to stress (McEwen & Gianaros, 2010; Muscatell, 2018; Nusslock & Miller, 2016). Finally, as neuroscience gradually yields insights into SES, it will reveal why low SES is associated with many forms of diminished human potential. Such insights may eventually be a source of practical help (Farah, 2018).

Neural Correlates of Socioeconomic Status

In the past 10 to 15 years, it has become clear that the structure and function of normal human brains depend in part on SES. Using structural and functional magnetic resonance imaging (MRI) and event-related potentials (ERPs), in children and adults, these studies have revealed regional and network-level differences as a function of SES. Reviews of this recent but rapidly growing literature have been provided by Kim et al. (2018); Farah (2017); Johnson, Riis, and Noble (2016); and Lipina and Segretin (2015). Here I will briefly summarize the literature so far and illustrate the summary with representative examples.

Brain structure Structural studies are the easiest to synthesize because, in contrast to fMRI and ERP, their outcomes do not depend on any particular task. Many different studies have examined brain structure as a function of SES in children and adults (see Farah, 2017, for a review). As an illustration of this approach, let us consider one particularly large and rigorously analyzed study. Kim Noble and collaborators used brain images and associated data from the Pediatric Imaging Neurocognition and Genetics (PING) consortium. With over a thousand subjects, ranging in age from 3 to 21, they identified differences in cortical surface area as a function of both family income and parental education, with covariates including genetic ancestry (Noble et al., 2015). Several different analyses found that surface area differences were not uniform over the brain. For example, when parental education was added to the other covariates, income had significant effects on surface area in bilateral inferior frontal, cingulate, insula, and inferior temporal regions and in the right superior frontal cortex and precuneus.

Some additional findings of Noble et al. are worth noting here. First, the relationship between surface area and SES was strongest at the lowest SES levels; SES had a positive relationship with surface area at all levels of income and education, but the difference between poverty and near-poverty mattered most. Second, the SES differences in cortical surface area partially mediated the relation between SES and performance on two tests of executive function. This suggests that the surface area differences index some aspect of brain structure that is relevant to cognitive ability. Third, among subcortical regions, their whole-brain analysis revealed a positive relation between SES and hippocampal volume.

How generalizable are these findings, to other child samples or to people more generally? A definitive answer will require much more research. Most studies of SES and brain structure involve smaller samples than Noble et al. (2015), and many fail to control for race (either in the sense of genetic ancestry, as done by Noble et al. (2015), or as a self-defined social construct, as in Lawson et al. (2013). Many other sources of variance across studies exist and are discussed at the end of the second section.

In general, the relation between SES and cortical structure, when found, is positive. For example, Gianaros et al. (2017) found larger volumes in their midlife adult subjects, and Mackey et al. (2015) found thicker cortex overall in their early adolescent subjects. However, the possibility of more prolonged experience-driven cortical thinning in higher SES young adults (see Piccolo et al., 2016) suggests that the relation of SES to cortical thickness (CT) may not be positive in adulthood; relevant studies have not yet been published. Different dimensions of cortical anatomy index different developmental processes (Raznahan et al., 2011), so we should not expect SES differences in the cortical thickness of a certain area to be "replicated" in surface area or vice versa. Further thwarting the effort to combine results on frontal structure across studies is the multitude of ways that cortical subregions have been defined, including gyral and sulcal divisions, Brodmann areas, other designations such as simply *dorsolateral* or *medial*, or even the whole lobes as regions of interest (ROIs). White matter volume, and integrity as assessed by diffusion tensor imaging, have also been examined and found to relate positively to SES in some cases (e.g., Gianaros et al., 2013; Johnson, Kim, & Gold, 2013; Ursache & Noble, 2016).

The volumes of subcortical structures, especially the hippocampus and amygdala, have also been studied in relation to SES. These structures, central to emotional experience, stress regulation, and learning, might be expected to show differences based on the more stressful nature of life for people of low SES. Hanson et al. (2011) first examined the relation of the hippocampus and amygdala to SES in the National Institutes of Health (NIH) Pediatric MRI Repository. They reported that children from higher-income families have larger hippocampi, along with an additional, more limited, relation between parental education and hippocampal volume (significant only for fathers' education and right hippocampal volume). They found no relation with amygdala volume and indeed reported it as a control region (i.e., predicting a null result). Amygdala volume has sometimes been found to vary with SES, but findings are inconsistent. Merz, Tottenham, and Noble (2018) review this literature and tentatively propose that conditions of low SES may lead to amygdala hypertrophy during early development as a result of chronic activation but amygdala atrophy later as a result of the excitotoxic effects of ongoing hyperactivation.

Hippocampus-SES relations have been widely tested. In the large study of Noble et al. (2015) described earlier, left hippocampal volume was positively related to parental education but not income. Similar to this group's findings with income and cortical surface area, the relation between parental education and hippocampal volume was strongest at the lowest levels of education. Many studies have replicated some form of SES-hippocampus relation in children, while such findings in adults are rarer. Indeed, Yu et al. (2017) contrasted the relations in child and adult samples. They found a positive relation for the children only and further demonstrated a statistical interaction showing that SES had significantly more effect for children than adults. As mentioned earlier, the temporal relations between SES and brain structure remain to be unraveled. They will reflect ongoing environmental influences with different effects on the brain at different stages, alongside genetic effects that may be apparent at some ages and not others (see Papenberg, Lindenberger, & Bäckman, 2015).

Before turning to SES disparities in brain function, it is worth noting that structural correlates of the kind just reviewed have in some cases been tested as mediators of SES-behavior relations. As already mentioned, the cortical surface area differences found by Noble and colleagues (2015) partially mediated the SES-executive function (EF) relation. That is, the effect of SES on EF could be partly accounted for in terms of the relations between SES and surface area and the relations between surface area and EF. Another, more specific, example of anatomical mediation is Romeo et al.'s (2017) finding that cortical thickness in the vicinity of Broca's area fully mediated the effect of SES on vocabulary in children (see Farah, 2017 for other examples).

Brain function: cognition Purely behavioral studies have found SES disparities in standardized measures of general cognitive ability such as IQ tests (Gottfried et al., 2003) and in specific cognitive systems, such as EF (Lawson, Hook, & Farah, 2017), language (Weisleder & Fernald, 2014), and memory (Hermann & Guadagno, 1997). These behavioral correlations accord with findings from a variety of brain function studies, including electroencephalography (EEG), ERP, and fMRI. A few illustrative studies will be summarized here, and readers can find a more complete and granular description of the literature in the review papers cited earlier, as well as the recent review of SES, EF, and language in children by Merz, Wiltshire, and Noble (forthcoming).

In an early study by Kishiyama and colleagues (2009), children's ERPs were recorded in a stimulus-monitoring task involving attending to a sequence of visually presented stimuli for target stimuli. In addition, occasional unrelated novel stimuli were presented. When the authors compared the ERPs evoked by the novel stimuli between groups of lower and higher SES children, they found differences in the ERP waveform that they attributed (on the basis of the prior ERP literature) to prefrontal mechanisms of executive attention. A more familiar way of operationalizing EF, with an N-back task, was used in an fMRI study of child subjects by Finn and colleagues (2016). They found SES differences in regions of the brain engaged in working-memory processes and further found that the relations between task difficulty and brain activity differed between higher and lower SES subjects. SES moderated the relation between brain task demands and activation regions, including prefrontal and parietal cortex, with lower SES subjects activating these classic EF regions more than higher SES at low working-memory loads and the relation reversing at higher loads. In other words, the SES differences in EF observed here are not simply a matter of more or less of the same brain processes but different patterns of brain processes.

Language function has been studied using ERP and fMRI (see Farah, 2017 for a review). For illustration, a study of syntactic parsing by Pakulak and Neville (2010) had adults distinguish between sentences with correct and incorrect syntax and measured the left anterior negativity (LAN), an ERP component that indexes syntactic processing. They found a larger LAN in subjects with higher parental education and occupational status when controlling for language ability and other factors. The scalp localization and timing of the SES difference was limited to the LAN, suggesting that SES differences in syntactic ability are related to neural systems for syntax, as opposed to more general differences in verbal semantics or attention. The earliest fMRI study of SES focused on phonological ability in children of lower and higher SES (Noble et al., 2006). Phonological ability is predictive of early reading ability, but its relation to reading-related brain activity differs by SES. Lower-SES children showed a strong relation between phonological skill and activity in classic reading areas, such as the fusiform visual word form area, whereas this relation was attenuated in higher-SES children. As with the working-memory findings just mentioned, this SES moderation effect suggests qualitative differences in *how*, not just *how well*, higher- and lower-SES children perform cognitive tasks. The authors suggested that the more extensive experience of higher-SES children with books and written language may provide additional, nonphonological processes to support their decoding of print.

Few studies have examined declarative memory and SES with functional methods, and none directly support a simple relation between common measures of SES and hippocampal activation, although more complex relations have been reported (see Farah, 2017 for a summary). For example, Duval et al. (2017) analyzed the association between hippocampal activation and childhood SES in adults while covarying adult SES and found a borderline significant effect of childhood SES on hippocampal activation during recognition. A stronger result concerned moderation: childhood SES significantly moderated the relation between performance and activation—that is, it changed the relation between these two measures. Those who had not been poor as children showed the expected positive relation between recognition accuracy and activation in the hippocampus, whereas those who had been poor showed an opposite effect.

Brain function: social and emotional processes SES is associated with neural-processing differences in social cognition and affect. Clinical and behavioral studies show higher rates of affective disorders among lower SES people (Kessler et al., 2005; Lorant et al., 2003; McLaughlin et al., 2012) and lower levels of self-esteem (Twenge & Campbell, 2002). SES is also related to interpersonal attention, with low-SES individuals allocating relatively more attention to people than to objects (e.g., Dietze & Knowles, 2016). As with the cognitive processes just reviewed, findings from functional imaging are broadly consistent with these SES differences in behavior. As before, a few illustrative examples are summarized here.

Neural responses to negative stimuli tend to be negatively related to SES, and neural responses to positive stimuli tend to follow the opposite pattern (see Farah, 2017 for a review). In the first report of this phenomenon, Gianaros and colleagues (2008) found that amygdala reactivity to threatening faces was higher at lower levels of social status after controlling for a variety of personality factors. Swartz, Hariri, & Williamson (2016) explored the relation of SES to amygdala reactivity in a multimethod longitudinal study of adolescents. They found that the change in methylation of the serotonin transporter gene across time was greater for low-SES subjects and predicted change in amygdala reactivity. For subjects with a positive family history of depression, these changes in turn predicted amount of change in depression.

For positive and rewarding stimuli, lower-SES adults show lower activity in frontal, ACC, and striatal regions (Gianaros et al., 2011; Silverman et al., 2009). Pilyoung Kim et al. (2017) and colleagues have observed similar patterns in first-time mothers exposed to pictures and sounds of infants. For example, amygdala responses to images of happy-looking infants were lower in low SES, whereas amygdala responses to distressed-looking infants were enhanced.

Muscatell and colleagues (2012, 2016) have investigated social cognition as a function of SES in three different tasks. An illustrative example comes from their 2012 article, in which they found that lower-SES individuals spontaneously activated medial prefrontal cortex, an area associated with mentalizing, or thinking about other people's mental states, more than their higher-SES counterparts when performing tasks involving images of people. This is consistent with the greater spontaneous attentional focus on people, compared with objects, noted earlier.

Generalizing, but not overgeneralizing, about socioeconomic status differences in brain structure and function With several dozen studies linking SES to brain structure and function, the time is right to begin seeking generalizations. There is a degree of convergence among these studies—even between structural and functional studies—which is encouraging. The neural substrates of language, EF, memory, and positive and negative emotions have all been implicated by a number of studies each. Of course, not every relevant brain difference will show up in MRI or ERP, and some of the relations linking SES, the brain, and behavior will be complex, as in the examples reviewed here of the moderation of brain-behavior relations by SES.

The integration of findings across studies is also complicated by the many ways in which studies differ. As already mentioned, structural measures such as cortical surface area and thickness reflect the effects of different developmental and experiential processes, and task-activation differences will be comparable only insofar as the tasks can be related to one another. In addition, SES may manifest differently in subjects of different ages, changing not just the asymptotic levels of brain development or decline but, possibly, the trajectory's shape (Noble et al., 2012; Piccolo et al., 2016). Furthermore, the socioeconomic environment is not a one-time "treatment" but a set of factors that impinge on the brain continually, from prenatal life through maturity and senescence. Brain development involves different processes at different stages, and SES may therefore manifest in different ways at these different stages. Finally, SES itself is operationalized differently in different studies, with different dimensions (e.g., education or occupation) used for measurement and different ranges (e.g., encompassing deep poverty or not) represented in samples.

Mechanisms: How Does Socioeconomic Status Get into the Head?

Causal pathways are difficult to pin down for all complex human phenotypes, and SES is no exception. Indeed, for SES and the brain, even the direction of causality is a subject of debate (Farah, 2018): Are the brains of the poor different because of the effects of living in poverty? Or do people live their lives in poverty because they have different brains? While genetic factors may be involved, there are ample candidate environmental factors in low-SES environments capable of impeding healthy brain development and function. These include prenatal and postnatal exposure to environmental toxins, inadequate nutrition, psychosocial stress, and lower levels of cognitive and linguistic stimulation. Neuroscience has provided clues to the role of all of these causal factors in the socioeconomic environment (Hackman, Farah, & Meaney, 2010).

One type of evidence concerning the causes of SES-brain relations comes from studies of the statistical mediation of brain structure or function by aspects of the environment. In many cases, stress or a related factor is found to correlate with both SES and some measured aspect of the brain, and these two relations account for some or all of the relation between SES and the relevant aspect of the brain (Farah, 2017). An example of such a finding comes from the work of Luby et al. (2013), who reported a relation between SES and childhood hippocampal volume. They measured child stress and unsupportive maternal behavior toward children, two factors shown to affect hippocampal development in animals, and found that these factors fully mediated the SES-hippocampus relation. That is, the effect of SES on hippocampal volume could be entirely accounted for in terms of the relations between SES and child stress and maternal behavior and between stress and maternal behavior on hippocampal volume.

Of course, statistical mediation does not imply causal mediation. Perhaps some unmeasured factor associated with the number of stressful events in a child's life and with the mother's behavior is what truly drives relation. It is even logically possible that the arrow of causality goes in the opposite direction. For example, we could imagine that there is a genetic predisposition to small hippocampi, to finding one's way into stressful situations, and to being an unsupportive mother. Perhaps these genetically transmitted traits result in behaviors that cause opportunities for socioeconomic advancement to be lost so that over time SES will drop. This might seem unlikely, but it cannot be ruled out on the basis of Luby et al.'s (2013) findings.

The causal ambiguity of correlational studies, including those that include tests of statistical mediation, is a weakness of all observational research. Only an experiment involving the random assignment of people to levels of SES, ideally for a lifetime, can provide definitive evidence on causality. While this is hardly feasible, two other kinds of evidence do have bearing on the question of what causes SES differences in the brain.

First, animal research allows individuals to be randomly assigned to different environmental conditions. Obviously, these models do not manipulate SES per se because there is no straightforward animal equivalent of SES. Instead, they manipulate candidate causal factors by which SES might affect the brain. Among the SES-linked aspects of human experience, which are candidate causes of brain differences, are the amount and quality of cognitive and linguistic stimulation, psychological stress, and, for children, parenting practices (Farah, 2017). Corresponding animal models have shown pervasive effects on the brain of stimulation (although not linguistic stimulation; van Praag, Kempermann, & Gage, 2000), of stress (McEwen & Gianaros, 2010) and of parenting differences (Francis et al., 1999), the latter being, in part, an effect of stress (Murgatroyd & Nephew, 2013; Rosenblum & Paully, 1984). The environmental effects demonstrated in the aforementioned animal studies show a broad brush-stroke similarity to SES effects.

Second, intervention studies in humans often manipulate environmental factors related to SES in an attempt to ameliorate developmental outcomes in low SES. Well-designed intervention studies take pains to ensure that subjects who receive the intervention do not differ from those in the control group, generally by random assignment. This rules out the possibility that people who signed up for the intervention were especially proactive or otherwise above average and thus clarifies the direction of causality. Although most intervention studies relevant to SES do not measure brain structure or function, a few have. These studies do not manipulate SES per se but instead manipulate aspects of the environment typically correlated with SES, using the random assignment of poor children to the intervention or a control.

The first such study was a parenting intervention carried out by Neville et al. (2013), teaching stress regulation, parental language and responsiveness, methods of discipline, and so on. They found changes in children's attention and language processes reflected in ERPs. With a different parenting intervention, focused on communication, support, safety, and managing racism, Brody et al. (2017) demonstrated less hippocampal volume loss in adulthood, compared to individuals who had been equally poor as children. Comprehensive programs, including early-childhood cognitive enrichment for normal low-SES children, have resulted in changed neuroendocrine function (Blair & Raver, 2014) and later brain structure (Farah et al., 2017). A recently launched study comes the closest yet to manipulating SES using a randomized income intervention (*Economist*, 2018). This study will collect a broad array of behavioral measures in children and mothers and also children's EEGs.

Future Directions

The neuroscience of SES is a young field. Some of the open questions that will propel research in the coming years include the following:

By what mechanisms does SES affect the brain? This fundamental question requires that we understand how genes and environments interact to determine brain function over the life span, beginning in prenatal life. In other words, this question will be answered when the field of neuroscience has been completed, and not before! However, in the coming years we can look forward to partial answers, taking the form of pathways linking specific aspects of the environment to the development, structure, and function of specific brain systems. These pathways will be framed at various levels of description, from molecular to computational and psychological. As already noted, there are numerous SES-linked aspects of the environment, and they are moderately correlated with one another, implying that they will often operate in concert. Provisional clarity on individual pathways will allow us to begin investigating their interactions.

At what ages is the human brain more and less sensitive to the socioeconomic environment? The answer to this question will likely vary, depending on the specific proximal causes in the environment under consideration, as the causal factors discussed above may exert their effects at different times in different ways. Plausible answers include prenatal life, childhood, adolescence, and old age as stages of particular vulnerability to low SES, but of course the many decades of adult life are also filled with opportunities for brain health to be diminished or enhanced by factors linked to SES. To appreciate how daunting this research goal is, consider that development and function at any point in life depends on both the current situation and on earlier formative influences. An environmental challenge to healthy development, such as low levels of parental speech during childhood, may have more impact on a brain with genetically heightened sensitivity to environmental influences or that has been shaped by earlier adversity, such as prenatal exposure to neurotoxic pollutants.

How do the findings of neuroscience relate to actual human lives at different levels of SES? Our interest in the relations between SES and the brain is primarily motivated by the role of the brain in mental and physical health and cognitive capabilities. We therefore want neuroscience to help explain why people of low SES, those leading Marmot's parade, are more depressed, less healthy, and less cognitively capable compared to others and why health, well-being, and ability rise as we look further back in the parade. The mediation findings cited earlier suggest that neuroscience may indeed be a fruitful approach to understanding these facts. However, the causal web linking environment, biology, and life outcome is massively complex. If we think of the explanation as a process of connecting the dots, relating observed phenomena to their causes, then the life outcomes predicted by SES comprise many dots, as do SES-associated features of the environment and the brain. So far we have connected only a tiny subset of the dots, and it is clear that not every barrier to humans flourishing in low-SES environments is located inside the skull. External social and economic impediments work alongside these brain-mediated pathways to constrain life chances.

Finally, how might the neuroscience of SES inform policy related to families, childcare, health care, economics, and education? The answer to this question is closely related to the others, insofar as beneficial policies will be those that prevent or reverse the neural and psychological effects of low SES and extend the corresponding advantages of high SES to more people. The policy implications of our current, rudimentary understanding of these effects is limited, but as the science progresses, so will its potential for translation. At present, neuroscience adds to the weight of evidence for policies already supported by behavioral research, such as the importance of prenatal health, family stress reduction, and conversation with children. Neuroscience has also been used as part of a communication strategy to convey information to policy-makers and the public about the needs of children (Shonkoff & Bales, 2011). In the near future, neuroscience may contribute in more distinctive and consequential ways to guide policy (Farah, 2018). For example, it offers potential biomarkers to screen for the risk of emotional or cognitive difficulties and to facilitate intervention research by providing early predictors of success (Pavlakis et al., 2015).

Ultimately, interventions themselves may be designed based on an understanding of the neural mechanisms of SES disparities. Such interventions, in targeting components of the causal pathways linking SES, the brain, and behavior, will also provide powerful new evidence about these causal pathways, as well as improve life chances for people of low SES.

REFERENCES

Adler, N. E., Epel, E. S., Castellazzo, G., & Ickovics, J. R. (2000). Relationship of subjective and objective social status with psychological and physiological functioning: Preliminary data in healthy, white women. *Health Psychology, 19*(6), 586–592.

Blair, C., & Raver, C. C. (2014). Closing the achievement gap through modification of neurocognitive and neuroendocrine function: Results from a cluster randomized controlled trial of an innovative approach to the education of children in kindergarten. *PLoS One, 9*, e112393.

Brody, G. H., Gray, J. C., Yu, T., Barton, A. W., Beach, S. R., Galván, A., ... Sweet, L. H. (2017). Protective prevention effects on the association of poverty with brain development. *JAMA Pediatrics, 171*(1), 46–52.

Dietze, P., & Knowles, E. D. (2016). Social class and the motivational relevance of other human beings: Evidence from visual attention. *Psychological Science, 27*(11), 1517–1527.

Duval, E. R., Garfinkel, S. N., Swain, J. E., Evans, G. W., Blackburn, E. K., Angstadt, M., Sripada, C. S., & Liberzon, I. (2017). Childhood poverty is associated with altered hippocampal function and *visuospatial memory in adulthood. Developmental Cognitive Neuroscience, 23*, 39–44.

Economist. (May 3, 2018). Mother's money: Does growing up poor harm brain development? https://www.economist.com/united-states/2018/05/03/does-growing-up-poor-harm-brain-development.

Evans, G. W., Otto, S., & Kaiser, F. G. (2018). Childhood origins of young adult environmental behavior. *Psychological Science, 29*(5), 679–687.

Farah, M. J. (2017). The neuroscience of socioeconomic status: Correlates, causes and consequences. *Neuron, 96*(1), 56–71.

Farah, M. J. (2018). Socioeconomic status and the brain: Prospects for neuroscience informed policy. *Nature Reviews Neuroscience, 19*, 428–438.

Farah, M. J., Duda, J. T., Nichols, T. A., Ramey, S. L., Montague, P. R., Lohrenz, T. M., & Ramey, C. T. (2017). Early educational intervention for poor children modifies brain structure in adulthood. Paper presented at the Society for Neuroscience Annual Meeting, Washington, DC.

Farahany, N. A. (2016). Neuroscience and behavioral genetics in US criminal law: An empirical analysis. *Journal of Law and the Biosciences, 2*(3), 485–509.

Federal Register, Vol. 83, No. 12, January 18, 2018, pp. 2642–2644. https://www.federalregister.gov/documents/2018/01/18/2018-00814/annual-update-of-the-hhs-poverty-guidelines.

Finn, A. S., Minas, J. E., Leonard, J. A., Mackey, A. P., Salvatore, J., Goetz, C., ... Gabrieli, J. D. E. (2016). Functional brain organization of working memory in adolescents varies in relation to family income and academic achievement. *Developmental Science, 20*(5), e12450.

Francis, D., Diorio, J., Liu, D., & Meaney, M. J. (1999). Nongenomic transmission across generations of maternal behavior and stress responses in the rat. *Science, 286*(5442), 1155–1158.

Gabrieli, J. D. E. (2016). The promise of educational neuroscience: Comment on Bowers. *American Psychological Association, 123*(5), 613.

Gianaros, P. J., Horenstein, J. A., Hariri, A. R., Sheu, L. K., Manuck, S. B., Matthews, K. A., & Cohen, S. (2008). Potential neural embedding of parental social standing. *Social Cognitive and Affective Neuroscience, 3*(2), 91–96.

Gianaros, P. J., Kuan, D. C., Marsland, A. L., Sheu, L. K., Hackman, D. A., Miller, K. G., & Manuck, S. B. (2017). Community socioeconomic disadvantage in midlife relates to cortical morphology via neuroendocrine and cardiometabolic pathways. *Cerebral Cortex, 27*(1), 460–473.

Gianaros, P. J., Manuck, S. B., Sheu, L. K., Kuan, D. C. H., Votruba-Drzal, E., Craig, A. E., & Hariri, H. R. (2011). Parental education predicts corticostriatal functionality in adulthood. *Cerebral Cortex, 21*(4), 896–910.

Gianaros, P. J., Marsland, A. L., Sheu, L. K, Erikson, K. I., & Verstynen, T. D. (2013). Inflammatory pathways link socioeconomic inequalities to white matter architecture. *Cerebral Cortex, 23*(9), 2058–2071.

Gottfried, A. W., Gottfried, A. E., Bathurst, K., Guerin, D. W., & Parramore, M. M. (2003). Socioeconomic status in children's development and family environment: Infancy through adolescence. In M. H. Bornstein & R. H. Bradley (Eds.), *Monographs in parenting series. Socioeconomic status, parenting, and child development* (pp. 189–207). Mahwah, NJ: Lawrence Erlbaum.

Hackman, D. A., Farah, M. J., & Meaney, M. J. (2010). Socioeconomic status and the brain: Mechanistic insights from human and animal research. *Nature Reviews Neuroscience, 11*, 651–659.

Hanson, J. L., Chandra, A., Wolfe, B. L., & Pollak, S. D. (2011). Association between income and the hippocampus. *PLoS One, 6*(5), e18712.

Hauser, R. M., & Warren, J. R. (1997). Socioeconomic indexes for occupations: A review, update, and critique. *Sociological Methodology, 27*(1), 177–298.

Hermann, D., & Guadagno, M. A. (1997). Memory performance and socioeconomic status. *Applied Cognitive Psychology, 11*, 113–120

Johnson, N. F., Kim, C., & Gold, B. T. (2013). Socioeconomic status is positively correlated with frontal white matter integrity in aging. *Age, 6*, 2045–2056.

Johnson, S. B., Riis, J. L., & Noble, K. G. (2016). State of the art review: Poverty and the developing brain. *Pediatrics, 137*(4), e20153075.

Kaiser Family Foundation. (2017). Distribution of total population by federal poverty level. https://www.kff.org/other/state-indicator/distribution-by-fpl/?currentTimeframe=0&sortModel=%7B%22colId%22:%22Location%22,%22sort%22:%22asc%22%7D#notes.

Kessler, R. C., Berglund, P., Demler, O., Jin, R., Merikangas, K. R., & Walters, E. E. (2005). Lifetime prevalence and age-of-onset distributions of DSM-IV disorders in the National Comorbidity Survey Replication. *Archives of General Psychiatry, 62*(6), 593–602.

Kim, P., Capistrano, C. G., Erhart, A., Gray-Schiff, R., & Xu, N. (2017). Socioeconomic disadvantage, neural responses to infant emotions, and emotional availability among first-time new mothers. *Behavioural Brain Research, 325*, 188–196.

Kim, P., Evans, G. W., Chen, E., Miller, G., & Seeman, T. (2018). How socioeconomic disadvantages get under the skin and into the brain to influence health development across the lifespan. In *Handbook of life course health development* (pp. 463–497). Cham, Switzerland: Springer.

Kishiyama, M. M., Boyce, W. T., Jimenez, A. M., Perry, L. M., & Knight, R. T. (2009). Socioeconomic disparities affect prefrontal function in children. *Journal of Cognitive Neuroscience, 21*(6), 1106–1115.

Lawson, G. M., Duda, J. T., Avants, B. B., Wu, J., & Farah, M. J. (2013). Associations between children's socioeconomic status and prefrontal cortical thickness. *Developmental Science, 16*(5), 641–652.

Lawson, G. M., Hook, C. J., & Farah, M. J. (2017). A meta-analysis of the relationship between socioeconomic status and executive function performance among children. *Developmental Science, 21*(2), e12529.

Lee, N., Brandes, L., Chamberlain L., & Senior, C. (2017). This is your brain on neuromarketing: Reflections on a decade of research. *Journal of Marketing Management, 33*(11–12), 878–892.

Lipina, S. J., & Segretin, M. S. (2015). Strengths and weaknesses of neuroscientific investigations of childhood poverty: Future directions. *Frontiers in Human Neuroscience, 9*, doi:10.3389/fnhum.2015.00053

Lorant, V., Deliège, D., Eaton, W., Robert, A., Philippot, P., & Ansseau, M. (2003). Socioeconomic inequalities in depression: A meta-analysis. *American Journal of Epidemiology, 157*(2), 98–112.

Luby, J., Belden, A., Botteron, K., Marrus, N., Harms, M. P., Babb, C., Nishino, T., & Barch, D. (2013). The effects of poverty on childhood brain development: The mediating effect of caregiving and stressful life events. *JAMA Pediatrics, 167*(12), 1135–1142.

Mackey, A. P., Finn, A. S., Leonard, J. A., Jacoby-Senghor, D. S., West, M. R., Gabrieli, C. F. O., & Gabrieli, J. D. E. (2015). Neuroanatomical correlates of the income-achievement gap. *Psychological Science, 26*(6), 925–933.

Marmot, M. (2004). *Status syndrome*. London: Bloomsbury.

McEwen, B. S., & Gianaros, P. J. (2010). Central role of the brain in stress and adaptation: Links to socioeconomic status, health, and disease. *Annals of the New York Academy of Sciences, 1186*(1), 190–222.

McLaughlin, K. A., Costello, E. J., Leblanc, W., Sampson, N. A., & Kessler, R. C. (2012). Socioeconomic status and adolescent mental disorders. *American Journal of Public Health, 102*(9), 1742–1750.

McLoyd, V. C. (1998). Socioeconomic disadvantage and child development. *American Psychologist, 53*(2), 185.

Merz, E. C., Tottenham, N., & Noble, K. G. (2018). Socioecomomic status, amygdala volume and internalizing symptoms in children and adolescents. *Journal of Clinical Child & Adolescent Psychology, 47*(2), 312–323.

Merz, E. C., Wiltshire, C., & Noble, K. G. (forthcoming). Socioeconomic inequality and the developing brain: Spotlight on language and executive function. *Child Development Perspectives, 13*(1), 15–20.

Murgatroyd, C. A., & Nephew, B. C. (2013). Effects of early life social stress on maternal behavior and neuroendocrinology. *Psychoneuroendocrinology, 38*(2), 219–228.

Muscatell, K. A. (2018). Socioeconomic influences on brain function: Implications for health. *Annals of the New York Academy of Sciences*. doi:10.1111/nyas.13862. Advance online publication

Muscatell, K. A., Dedovic, K., Slavich, G. M., Jarcho, M. R., Breen, E. C., Bower, J. E., Irwin, M. R., & Eisenberger, N. I. (2016). Neural mechanisms linking social status and inflammatory responses to social stress. *Social Cognitive and Affective Neuroscience, 11*(6), 915–922.

Muscatell, K. A., Morelli, S. A., Falk, E. B., Baldwin, M. W., Pfeifer, J. H., Galinsky, A. D., ... Eisenberger, N. I. (2012). Social status modulates neural activity in the mentalizing network. *NeuroImage, 60*, 1771–1777.

Neville, H. J., Stevens, C., Pakulak, E., Bell, T. A., Fanning, J., Klein, S., & Isbell, E. (2013). Family-based training program improves brain function, cognition, and behavior in lower socioeconomic status preschoolers. *Proceedings of the National Academy of Sciences, 110*(29), 12138–12143.

Noble, K. G., Houston, S. M., Brito, N. H., Bartsch, H., Kan, E., Kuperman, J. M., ... Sowell, E. R. (2015). Family income, parental education and brain structure in children and adolescents. *Nature Neuroscience, 18*(5), 773–778.

Noble, K. G., Houston, S. M., Kan, E., & Sowell, E. R. (2012). Neural correlates of socioeconomic status in the developing human brain. *Developmental Science, 15*(4), 516–527.

Noble K. G., Wolmetz, M. E., Ochs, L. G., Farah, M. J., & McCandliss, B. D. (2006). Brain-behavior relationships in reading acquisition are modulated by socioeconomic factors. *Developmental Science, 9*, 642–654.

Nusslock, R., & Miller, G. E. (2016). Early-life adversity and physical and emotional health across the lifespan: A neuroimmune network hypothesis. *Biological Psychiatry, 80*(1), 23–32. doi:10.1016/j.biopsych.2015.05.017

Pakulak, E., & Neville, H. J. (2010). Proficiency differences in syntactic processing of monolingual native speakers indexed by event-related potentials. *Journal of Cognitive Neuroscience, 22*(12), 2728–2744.

Papenberg, G., Lindenberger, U., & Bäckman, L. (2015). Aging-related magnification of genetic effects on cognitive and brain integrity. *Trends in Cognitive Science, 19*(9), 506–514.

Pavlakis, A. E., Noble, K., Pavlakis, S. G., Ali, N., & Frank, Y. (2015). Brain imaging and electrophysiology biomarkers: Is there a role in poverty and education outcome research? *Pediatric Neurology, 52*(4), 383–388.

Piccolo, L. R., Merz, E. C., He, X., Sowell, E. R., & Noble, K. G. (2016). Age-related differences in cortical thickness vary by socioeconomic status. *PLoS One, 11*(9), e0162511.

Raznahan, A., Shaw, P., Lalonde, F., Stockman, M., Wallace, G. L., Greenstein, D., ... Giedd, J. N. (2011). How does your cortex grow? *Journal of Neuroscience, 31*(19), 7174–7177.

Romeo, R. R., Christodoulou, J. A., Halverson, K. K., Murtagh, J., Cyr, A. B., Schimmel, C., ... Gabrieli, J. D. (2017). Socioeconomic status and reading disability: Neuroanatomy and plasticity in response to intervention. *Cerebral Cortex, 28*(7), 1–16.

Rosenblum, L. A., & Paully, G. S. (1984). The effects of varying environmental demands on maternal and infant behavior. *Child Development, 55*(1), 305–314.

Satel, S., & Lillenfield, S. O. (2013). *Brainwashed: The seductive appeal of mindless neuroscience.* New York: Basic Books.

Shonkoff, J. P., & Bales, S. N. (2011). Science does not speak for itself: Translating child development research for the public and its policymakers. *Child Development, 82*(1), 17–32.

Silverman, M. E., Muennig, P., Liu, X., Rosen, Z., & Goldstein, M. A. (2009). The impact of socioeconomic status on the neural substrates associated with pleasure. *Open Neuroimaging Journal, 3,* 58–63.

Sirin, S. R. (2005). Socioeconomic status and academic achievement: A meta-analytic review of research. *Review of Educational Research, 75,* 417–453.

Swartz, J. R., Hariri, A. R., & Williamson, D. E. (2016). An epigenetic mechanism links socioeconomic status to changes in depression-related brain function in high-risk adolescents. *Molecular Psychiatry, 22*(2), 209–214.

Twenge, J. M., & Campbell, K. (2002). Self-esteem and socioeconomic status: A meta-analytic review. *Personality and Social Psychology Review, 6*(1), 59–71.

Ursache, A., & Noble, K. G. (2016). Neurocognitive development in socioeconomic context: Multiple mechanisms and implications for measuring socioeconomic status. *Psychophysiology, 53*(1), 71–82.

van Praag, H., Kempermann, G., & Gage, F. H. (2000). Neural consequences of environmental enrichment. *Nature Reviews Neuroscience, 1,* 191–198.

Weisleder, A., & Fernald, A. (2014). Social environments shape children's language experiences, strengthening language processing and building vocabulary. In I. Arnon, M. Casillas, C. Kurumada, & B. Estigarribia (Eds.), *Language in interaction: Studies in honor of Eve V. Clark.* Amsterdam: John Benjamins.

Yu, Q., Daugherty, A. M., Anderson, D. M., Nishimura, M., Brush, D., Hardwick, A., Lacey, W., Raz, S., & Ofen, N. (2017). Socioeconomic status and hippocampal volume in children and young adults. *Developmental Science, 21*(3), e12561.

91 A Computational Psychiatry Approach toward Addiction

XIAOSI GU AND BRYON ADINOFF

ABSTRACT Addictive behaviors are seen in a wide spectrum of disorders, including substance use disorders, binge eating, and behavioral addictions (e.g., pathological gambling). A mechanistic understanding of addiction is thus crucial for addressing these public health issues. To date, addiction research has made tremendous progress in terms of uncovering the neurobiological and neuropsychological correlates of addiction. However, little is known about the computational principles implemented by the brain (i.e., "software") that underlie addiction, in spite of the wealth of knowledge we have gained regarding its neurobiological mechanisms (i.e., "hardware"). This explanatory gap hinders the understanding of the mechanisms of addiction as well as the development of effective therapeutics. In this chapter we will review recent efforts in the nascent field of computational psychiatry that have started to address this problem. First, we will introduce David Marr's trilevel of analysis as a foundational framework for computational psychiatry, as well as the importance of computational approaches in investigating psychiatric and addictive disorders. Second, we will review studies utilizing theory-driven computational approaches, including reinforcement-learning models, Bayesian models, and biophysical models, that address addiction. Third, we will present recent studies using big data approaches (e.g., machine learning) to reveal new neural and cognitive dimensions of addiction. Last, we will outline a road map for computational work on addiction to move forward.

Why Do We Need a Computational Psychiatry Approach toward Addiction Research?

The explanatory gap Addiction remains one of the most serious threats to public health. Addictive behaviors are observed in a wide spectrum of disorders, including substance use disorders (SUD), binge eating, and behavioral addictions (e.g., pathological gambling). In the United States, SUDs alone cost $740 billion and lead to 640,000 deaths per year. Much effort has been devoted to the study of the neurobiology of addiction at cellular and molecular levels. Despite the progress made in addiction neuroscience, a major explanatory gap still exists between animal models of addiction and human addiction in real life, which prevents the translation of bench work to patient care. One example of this explanatory gap is the finding that the neuropharmacological effects of drugs on the brain can be overridden by cognitive factors such as beliefs and expectancies in humans (Gu et al., 2015, 2016; Robinson et al., 2014); these findings cannot be readily accounted for by even the most detailed mapping of the cellular and molecular pathways associated with substances of abuse using animal models. Thus, a purely neurochemical approach alone is not sufficient to account for the complexity of addiction in humans.

In this chapter, we will review recent endeavors in the nascent field of computational psychiatry that have started to address this disconnection. Computational psychiatry seeks to understand the algorithms underlying mental function and dysfunction using computational approaches and has been used to illustrate the mechanisms of addiction in recent years (see Redish, 2004 for an example). First, we will explain the rationale of using computational psychiatry by introducing David Marr's trilevel analytical framework. Second, we will review studies utilizing theory-driven approaches of computational psychiatry, including reinforcement-learning models, Bayesian models, and biophysical models, that address addiction. Third, we will present recent studies using big data approaches (e.g., machine learning) to reveal new neurocognitive dimensions and phenotypes of addiction. Last, we will outline a roadmap for moving forward with computational work on addiction.

Marr's trilevel analysis David Marr (1945–1980) is best known for his work on vision. By integrating approaches from neuroscience, artificial intelligence, and psychology, Marr's contribution to science goes way beyond vision. In particular, Marr proposed that one must understand any information-processing system (e.g., vision) at three distinct levels (Marr & Poggio, 1976): computational, algorithmic/representational, and implementational/physical (figure 91.1*A*). The computational level addresses the question of "what"—what problems does the system solve? For example, what are the goals of the human brain? The algorithmic/

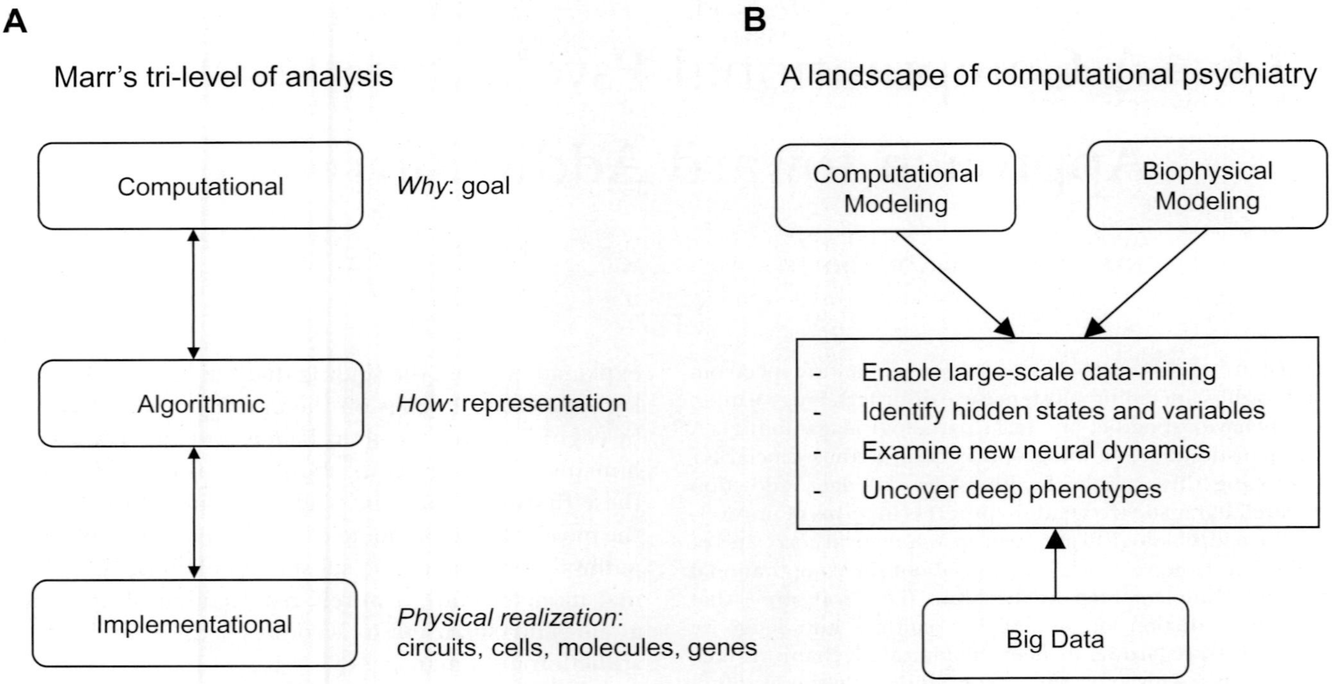

FIGURE 91.1 *A*, David Marr's trilevel analysis framework. Each level addresses its own questions and has its own vocabulary. The ultimate goal is to understand any system from all three levels. *B*, A landscape of computational psychiatry. The combination of top-down approaches (computational modeling and biophysical modeling) and bottom-up approaches (big data analytics) allows the identification of hidden states and variables of behavior and the brain, permits *deep phenotyping*, and enables large-scale data mining.

representational level speaks to the question of "how"—how does the system do what it does? For example, by what processes do addictive substances lead to these aberrant choices? Last, the implementational level deals with the question of which physical substrates are involved. For example, which neurotransmitters, neurons, and brain regions subserve addiction?

Under this framework, the majority of addiction neuroscience work deals with the implementational level ("hardware"). Such work is important, as any account (in the context of human addiction) needs to be biophysically plausible, and we have gained a tremendous amount of knowledge in this domain. However, the "software" problem of *how* the individual exhibits certain behaviors and forms certain beliefs as a result of addictive substance remains a bigger challenge for addiction research. Behavioral addiction, such as gambling, is a unique example in which individuals can develop addiction without the intervention of addictive substances. Thus, the focus of this chapter will be to review literature that bridges biochemical and biophysical models of addiction with computational work.

What is computational psychiatry? Computational psychiatry is a nascent field that seeks to understand mental function and dysfunction across various levels of analysis using computational approaches. In relation to Marr's trilevel analysis, computational psychiatry primarily focuses on the *computational* (what) and *algorithmic* (how) levels—both address the software problem—and how they relate to the *implementational* level (neurobiology or hardware). Scattered efforts had existed before 2010 (for examples, see Braver, Barch, & Cohen, 1999; Chiu, Kayali, et al., 2008; Chiu, Lohrenz, & Montague, 2008; Waltz, Frank, Robinson, & Gold, 2007), but it was not until the 2010s that we saw a systematic push for the growth and acknowledgment of computational psychiatry as a field (Huys, Moutoussis, & Williams, 2011; Kishida, King-Casas, & Montague, 2010; Maia & Frank, 2011; Montague, Dolan, Friston, & Dayan, 2012). Since then, we have seen exponential growth in the application of computational methods to mental illness and addiction research.

Computational psychiatry primarily entails two major classes of methods (figure 91.1*B*). One is theory- or model-driven and is considered a *top-down* approach for hypothesis testing. For example, by having a computational model of dopamine function and reinforcement learning (RL), one can address the question of how these processes are impaired in addicted populations

(Ersche et al., 2012; Goldstein & Volkow, 2002; Naqvi, Rudrauf, Damasio, & Bechara, 2007; Redish, 2004). A second approach is data-driven, usually considered a *bottom-up* approach for data mining—that is, using methods such as machine learning to classify participants (e.g., addicted vs. nonaddicted), to predict certain variables (e.g., treatment outcome), or to identify new phenotypes or dimensions emerging from the data without a prior hypothesis. In addiction research, this data-driven approach is relatively new and has yielded only a handful of empirical findings (see Ahn, Ramesh, Moeller, & Vassileva, 2016; Ahn & Vassileva, 2016; Pariyadath, Stein, & Ross, 2014; and Sakoglu et al, 2019 for examples). Theory-driven and data-driven approaches are complementary to each other, and the integration of both would lead to important new insights into the mechanisms of addiction. In the next sections, we will review studies on addiction that use these approaches separately.

Theory-Driven Approaches in Computational Psychiatry

We will first introduce a set of models and studies that rely on the hypothesis- or theory-driven approach of computational psychiatry. In addiction research, such work has mostly focused on the neural mechanisms underlying choice behavior (table 91.1), as drug taking has been considered the most significant aspect of addiction. The elimination of drug-taking behavior is also considered a main treatment objective. Most of this work is built upon computational models of value-based decision-making and learning. We discuss this in detail below.

Reinforcement-learning models of addiction formation: Goal-oriented drug seeking RL models are a natural candidate to account for the computational mechanisms of addiction formation due to the intertwined relationship

TABLE 91.1

<table>
<tr><th>Stage</th><th>Addiction formation</th><th>Addiction maintenance</th></tr>
<tr><td>Behavior</td><td>Reinforcement learning;
Goal-directed behaviors</td><td>Habitual response;
Compulsive drug taking</td></tr>
<tr><td>Neural candidates</td><td>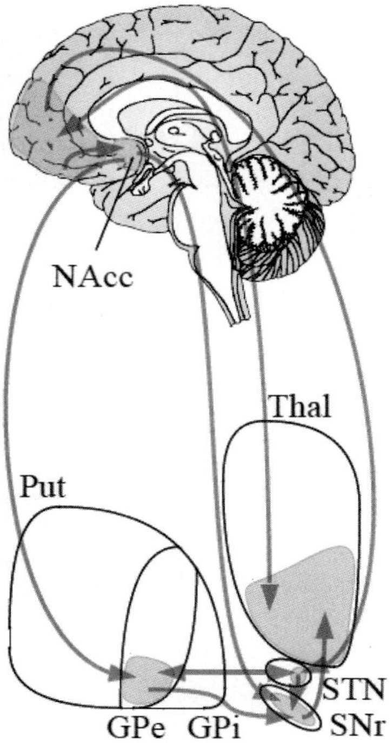
</td><td>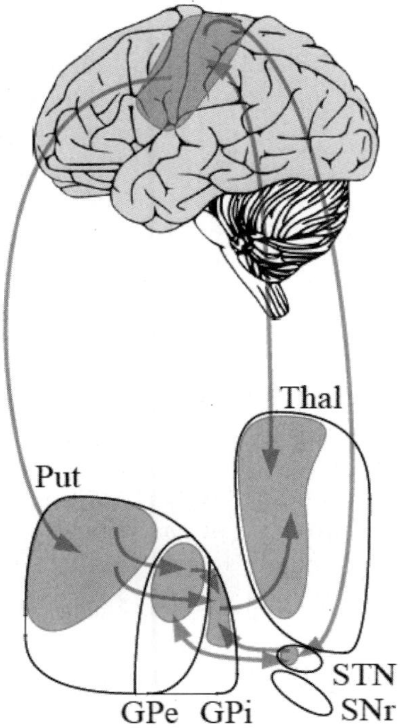
</td></tr>
</table>

Behaviors and neural candidates during different stages of addiction targeted by computational models. During the early formation of addiction, individuals are primarily driven by the rewarding effects of substances of abuse. This goal-directed behavior can be nicely quantified by computational RL models and is implemented in the ventral corticostriatal circuit. After the individual has become addicted, the habitual system, primarily implemented through the dorsal corticostriatal circuit, takes over. Images modified from Fiore, Dolan, Strausfeld, and Hirth (2015). (See color plate 100.)

between addictive substances, the neurotransmitter dopamine, and learning behavior. Naturally, studies driven by the RL hypothesis have mostly focused on choice behaviors (but not subjective states, such as craving) related to addiction. The majority of computational models of addiction are RL models or some reincarnation of the RL model. The basic idea of an RL model is that an agent always seeks to maximize its reward and minimize its punishment. One of the most commonly used RL models is called *temporal difference reinforcement-learning* (TDRL). Under TDRL, for each time point t, the agent is in a certain state s_t and takes an action a_t, among all possible options. This decision is based on the subjective values, called Q-values, assigned to the options based on previous experiences. To learn and update the Q-values to guide future choices, the agent needs to calculate an important signal called the *prediction error* δ_t:

$$\delta_t = \gamma(r_{t+1} + V_{S(t+1)}) - Q(s_t, \alpha_t) \qquad (91.1)$$

Here V_{t+1} is the maximum value of all possible actions in the next state S_{t+1}, and r_{t+1} is the reward received at the next time point ($t + 1$). γ is a discount factor representing how sensitive the agent is to future versus immediate rewards. The Q-value is then updated using

$$Q(s_t, \alpha_t) \leftarrow Q(s_t, \alpha_t) + \alpha\delta_t \qquad (91.2)$$

Here α is the learning rate, a parameter representing how much influence the prediction error δ has and how quickly the agent learns. Converging evidence suggests that the prediction error signal is encoded by the phasic activities of midbrain dopamine neurons (Bayer & Glimcher, 2005; Hollerman & Schultz, 1998; Montague, Dayan, & Sejnowski, 1996; Schultz, Dayan, & Montague, 1997).

If phasic dopamine computes learning signals, then any process that interferes with normal striatal dopamine function would lead to aberrant learning and value-based decision-making. In parallel to the computational work on RL, the animal literature has examined the neurophysiological processes related to the administration of addictive substances for decades (see De Biasi & Dani, 2011; Hyman, 2005; Nestler & Aghajanian, 1997, for reviews). These efforts lead to the conclusion that most addictive substances, including nicotine (Pidoplichko, De Biasi, Williams, & Dani, 1997; Rice & Cragg, 2004), cocaine (Hernandez & Hoebel, 1988), alcohol (Boileau et al., 2003; Weiss, Lorang, Bloom, & Koob, 1993), and cannabis and heroin (Tanda, Pontieri, & Di Chiara, 1997) increase extracellular dopamine release in the nucleus accumbens and interfere with many synaptic and cellular processes involved in dopamine neurotransmission. Thus, RL models become a natural candidate to provide a computational mechanism linking physiological substrates and addiction.

Redish (2004) proposed the first TDRL model to systematically account for addiction. In a standard "healthy" RL model, the prediction error δ eventually becomes 0 if the agent keeps on learning and updating the value function. In other words, once the value function correctly predicts reward, learning stops. In the RL model for addiction, however, due to the surge in dopamine its signaling of prediction error δ is also increased. Under this condition the prediction error δ can no longer be compensated by change in the value.

$$\delta_t = \max\left[\gamma(r_{t+1} + V(S_{(t+1)}) - Q(s_t, \alpha_t) + D(s_t), D(s_t)\right] \qquad (91.3)$$

Here $D(s_t)$ signals a dopamine surge due to the drug effect. When $D(s_t) = 0$, equation 91.3 is exactly the same as equation 91.1. However, under the drug condition, $D(s_t)$ will always be greater than 0, and the prediction error will always be positive. In such a case, the value of the drug state will be infinite. Using this modified model, Redish (2004) was able to simulate several behaviors commonly observed in addiction, including the overselection of drug rewards and the inelasticity to costs under drug states. There have been a few iterated versions of the Redish model, such as the homeostatic RL model developed by Keramati and Gutkin (2014) and Dayan's (2009) actor-critic model. Collectively, these models provide one of the most comprehensive computational frameworks for addiction. The significance of these models lies in their providing a mechanism for how a neurochemical (e.g., dopamine) can actually lead to addictive behaviors based on well-controlled neuroscience studies (particularly animal studies), in contrast to previous studies that simply provide a correlation between the two.

Compared to the elegant theoretical and animal work on the RL model of addiction, empirical findings from humans are much more mixed. For instance, Chiu, Lohrenz, et al. (2008) examined the two different types of RL signals: prediction errors (i.e., "what I actually received vs. what I expected to get") and fictive errors (i.e., "what I could have gotten vs. what I actually received") in nicotine-dependent smokers. The authors found that (1) smokers showed intact neural activations related to fictive errors, but their behaviors were less guided by these learning signals compared to nonsmokers; and (2) overnight deprivation decreased the computation of prediction errors in deprived smokers, compared to nondeprived smokers. Park et al. (2010) examined prediction error representations in alcohol-dependent participants but found no evidence of aberrant choice behavior or striatal activations related to prediction errors, despite the abnormal connectivity between the striatum and dorsolateral prefrontal cortex. Tanabe et al. (2013) examined participants across multiple substance-dependent groups (stimulants, nicotine, alcohol, opioids, cannabis, other) and found reduced neural representation of prediction errors across groups, compared to the non-using group. The reason for the inconsistency between theoretical, animal, and human neuroscience work on addiction is complex and multifaceted. Contributing factors include between-study variability in participant characteristics, type of substance used, homeostatic/deprivation state, and others. For instance, previous results suggest that participants' interoceptive state (deprived/craving or nondeprived/noncraving) can significantly modulate the impact of RL signals on behavior and their neural representations (Chiu, Lohrenz, et al., 2008; Gu et al., 2015, 2016). Thus, it is important for human studies to carefully consider these subtleties related to study design; it is also critical to develop more nuanced computational models to account for the complexity of addiction in humans.

Models of addiction maintenance: habitual drug taking Once an addiction is formed, the rewarding values of drugs and drug-related stimuli drastically decrease. Instead, habitual, automatic, and compulsive drug taking becomes a dominant feature, as suggested by Everitt and Robbins (2005). While this theory has been influential in the addiction literature for over a decade, it is primarily based on animal studies, and empirical evidence from humans supporting this claim remains rare. In one study, Ersche et al. (2016) used an instrumental-learning task to examine participants with cocaine use disorder (CUD). The authors found that compared to control subjects, those with CUD showed impaired goal-directed learning but enhanced habitual responses. Thus, this study provides behavioral evidence suggesting that the reinforcing effects of drugs are no longer sufficient to account for habitual drug taking during the maintenance (addicted) stage of addiction.

There is a rich literature on the neural and computational mechanisms of normative habit learning (stimulus-response association), which involves the dorsolateral striatum, putamen, and cortical motor regions (see Dickinson, 1985; Dolan & Dayan, 2013 for reviews). Several computational theories have been proposed to account for the emergence of habitual behaviors. First, Daw, Niv, and Dayan (2005) used a simulation to show that the competition between the habitual and goal-oriented systems is based on uncertainty. Over repeated training and accumulating experience, the habitual system has less uncertainty, which the brain prefers and consequently chooses. A second theory considers the habitual system more advantageous over time because of its reduced requirement on the computational load of the brain (Moors & De Houwer, 2006). Compared to the goal-oriented system, which requires the deliberative calculation of values and predictions, the habitual system does not involve such high cognitive demand and thus takes control of behavior once the individual has learned the statistics of the environment. Last, FitzGerald, Dolan, and Friston (2014) more recently proposed that the balance between habitual and goal-oriented systems comes from Bayesian model averaging. Specifically, this view suggests that individuals may hold both simpler (e.g., habitual) and complicated (e.g., goal-directed) models of the environment and weigh them based on the evidence supporting each model. Critically, the evidence is calculated as the trade-off between the accuracy and complexity of the model. In other words, the goal system could lose its advantage due to its high complexity, which makes the habitual system, and simpler models, a winner. It remains unclear which of the hypotheses could account better for habitual drug taking in the context of drug addiction, and more empirical evidence is needed to test between these hypotheses. Nevertheless, understanding the computational mechanisms of habits would be crucial, as we could then develop interventions and therapies to break habits.

Drug craving: Bayesian models of subjective states A second important aspect of addiction is craving (Tiffany & Wray, 2012). Since it is a subjective state, craving is difficult to measure objectively and quantitatively. Clinically, although the primary outcome of treatment has typically been the elimination of drug consumption, it has been suggested that craving should be considered a critical clinical target, as it directly relates to the subjective well-being and life quality of the individual and often drives continued substance use (Tiffany, Friedman, Greenfield, Hasin, & Jackson, 2012; Tiffany & Wray, 2012). Unfortunately, craving is more difficult to treat compared to physical dependence symptoms and can persist after drug consumption is reduced or stopped (Nestler, 2002). Despite the association between dopamine and craving shown by numerous studies (Heinz et al., 2004; Volkow et al., 2006; Wong et al., 2006), recent evidence also suggests that the manipulation of the dopamine system (e.g., induced by pharmacological treatments such as nicotine replacement therapy; Waters et al., 2004) is not sufficient to reduce craving by itself in humans. The question now is how do we reconcile these different views and findings?

In humans, craving has been extensively studied using cue-exposure paradigms (figure 91.2*A*; see Chase,

Eickhoff, Laird, & Hogarth, 2011; Engelmann et al., 2012; Jasinska, Stein, Kaiser, Naumer, & Yalachkov, 2014; Tang, Fellows, Small, & Dagher, 2012; Yalachkov, Kaiser, & Naumer, 2012 for reviews and meta-analyses). However, it remains controversial as to what psychological processes are actually elicited by these paradigms and how they relate to real-life craving (Shiffman et al., 2015). For instance, drug cues are inherently valuable to addicted individuals and could thus induce reward processing, along with craving, in the brain. Cue-elicited response studies typically directly contrast brain activities elicited by drug cues with those induced by nondrug cues (e.g., cigarette vs. pencil) and have reported widespread activations in dopaminergic and limbic regions, including the midbrain (ventral tegmental area, or VTA), ventral striatum, insula, anterior cingulate cortex (ACC), ventromedial prefrontal cortex (vmPFC), amygdala, and more (Chase et al., 2011; Engelmann et al., 2012; Jasinska et al., 2014; Tang et al., 2012; Yalachkov, Kaiser, & Naumer, 2012). However, many of these regions are also involved in the general encoding of stimulus and action values (Rangel, Camerer, & Montague, 2008; Rushworth & Behrens, 2008). Although drug cues naturally elicit value encoding as well as craving, these two processes are distinct. Craving has a strong interoceptive basis (i.e., is usually associated with altered bodily signals, such as increased heart rate), whereas value computation is one key cognitive component for learning and decision-making. Thus, it is important that distinct neural mechanisms underlying subjective craving versus those supporting value encoding can be examined in cue-exposure paradigms.

We recently proposed the first Bayesian model of craving (Gu & Filbey, 2017; figure 91.2*B*). Bayesian models have been widely used to account for perception (Knill & Pouget, 2004), beliefs (Brown, Adams, Parees, Edwards, & Friston, 2013; Lawson, Mathys, & Rees, 2017; Powers, Mathys, & Corlett, 2017), and emotional states (Barrett & Simmons, 2015; Seth & Friston, 2016). The Bayesian brain hypothesis suggests that the brain actively infers the causes of sensations, using evidence collected from our external and internal environments, and updates beliefs based on the Bayes rule:

$$\textit{posterior propability} = \frac{\textit{likelihood} \times \textit{prior probability}}{\textit{marginal likelihood}}$$

Importantly, we (Gu & FitzGerald, 2014; Gu, Hof, Friston, & Fan, 2013), among others (Barrett & Simmons,

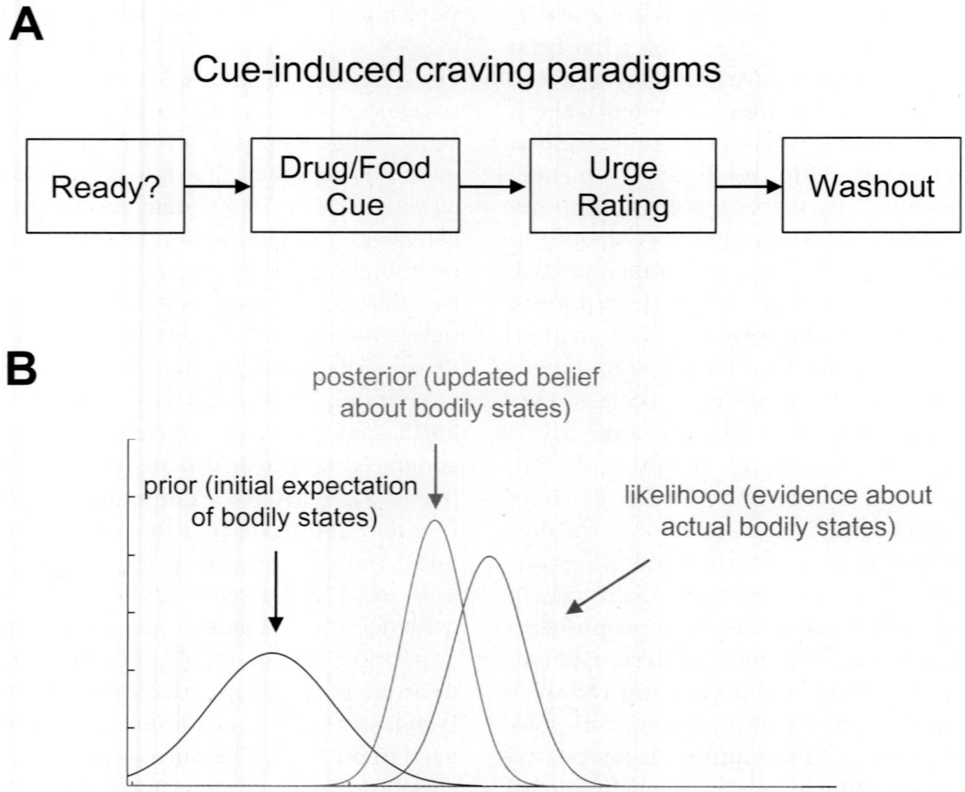

FIGURE 91.2 *A*, Typical cue-induced craving paradigms in the human addiction literature. *B*, A recently proposed Bayesian framework of drug craving (Gu & Filbey, 2017). (See color plate 101.)

2015; Seth & Friston, 2016), previously proposed that the brain also actively predicts bodily and interoceptive states, which forms the basis of subjective feelings. Based on this model, craving can be considered a special case of interoceptive inference—a posterior belief about the bodily states associated with the availability of addictive substances (Gu & Filbey, 2017).

This Bayesian model of craving has proven effective in accounting for several important experimental findings not explained by previous models. In the human addiction neuroscience literature, several studies have shown that nicotine craving depends not only on the availability of the addictive substance in the body but also on smokers' beliefs about the presence of nicotine (Gu et al., 2015; Juliano, Fucito, & Harrell, 2011; Kelemen & Kaighobadi, 2007; McBride, Barrett, Kelly, Aw, & Dagher, 2006). For instance, one study showed that craving was only reduced if smokers had a nicotine cigarette they believed contained nicotine but not when they believed the cigarette contained no nicotine (Gu et al., 2015). These findings contradict previous theories that predict craving should be reduced by the intake of drugs alone. Using a Bayesian framework, we were able to simulate this finding by systematically manipulating the prior beliefs (e.g., whether the smoker expected to receive a cigarette with nicotine or a placebo cigarette) and the likelihood of drug administration (e.g., whether the cigarette has nicotine or not; Gu & Filbey, 2017). Incubation of craving, referring to the effect that craving increases rather than decreases during early abstinence, is another important finding that remained unexplained by any computational framework. In one recent paper, I used the same Bayesian model to further simulate previous experimental findings (Bedi et al., 2011; Conrad et al., 2008; Grimm, Hope, Wise, & Shaham, 2001; Lu et al., 2005; Parvaz, Moeller, & Goldstein, 2016) that craving could increase over time during early abstinence (Gu, 2018). Taken together, this Bayesian framework is powerful in accounting for craving in addiction.

Data-Driven Approaches in Computational Psychiatry

A second focus of the computational psychiatry work on addiction is the application of *big data* analytical tools, such as machine learning, to mine data. The aim of this approach is to uncover hidden features and dimensions in the data that may not be predicted by existing theories and to predict certain characteristics of a new sample or at a future time using existing data. Machine-learning algorithms, in particular, have been widely used. The application of these tools in addiction research has allowed researchers to discover new addiction phenotypes, multivariate cognitive predictors (Ahn et al., 2016; Ahn & Vassileva, 2016), and biomarkers (Ding, Yang, Stein, & Ross, 2015; Pariyadath, Stein, & Ross, 2014) of disease diagnosis (Sakoglu et al., 2019), trajectory (Squeglia et al., 2017), and treatment outcome (Steele, Rao, Calhoun, & Kiehl, 2017).

What is machine learning and why use it for addiction research? Machine-learning techniques comprise two main families: unsupervised learning and supervised learning. Unsupervised learning is mainly used to find hidden structures or dimensions in "unlabeled" data. Cluster analysis (e.g., k-means), for example, tries to regroup the data points in a way that individuals in the same groups share the most similarity and are the most dissimilar to individuals in other groups. In neuropsychiatry research, unsupervised methods have been used to uncover new phenotypes, subphenotypes, or new definitions of patients. For example, one recent study used a hierarchical-clustering method in combination with cognitive and psychiatric assessments and resting-state functional magnetic resonance imaging (fMRI) data to find phenotypic subgroups in a community sample that went beyond *Diagnostic and Statistical Manual of Mental Disorders* (*DSM*) diagnosis (Van Dam et al., 2017). This line of work echoes very nicely with the Research Domain Criteria (RDoC) initiative proposed by the National Institute of Mental Health. Its utility for addiction research, however, remains to be examined.

The majority of machine-learning work on addiction uses supervised learning. In supervised learning, a training data set is required for the algorithm to find *classifiers*. In simple terms, the classifiers refer to inferred functions from the seen training data, which can then be used for mapping new, unseen data. For example, by finding patterns that can classify cigarette smokers versus marijuana users in an existing sample, we can then use these classifiers to predict who is a cigarette smoker and who is a marijuana smoker in a new sample. There are many algorithms to choose from, such as support vector machines (SVM) and neural networks (the latter can be extended into convolutional neural networks [CNN] or deep learning). We review the application of these techniques in the next section in detail.

Machine-learning examples: supervised learning in addiction research SVM has been a popular choice for addiction research due to its simplicity. In SVM the goal is to find hyperplanes that can separate the classes of training-data points with the largest distance. For example, Pariyadath, Stein, and Ross (2014) used SVM in combination with resting-state fMRI data to predict smoking status (i.e., smokers vs. nonsmokers). Structural-imaging data, such as voxel-based morphometry (VBM), has also been used in conjunction with SVM to classify smokers versus nonsmokers (Ding et al., 2015). Single-photon emission computerized tomography (SPECT) is another imaging modality that has been explored in combination with SVM to predict participant's SUD status. For example, using SPECT and SVM, Mete et al. (2016) identified 30 distinct clusters involved in cognitive control, behavioral inhibition, memory, and self-referential processing that classified whether a participant was cocaine-dependent or a healthy control. SVM has also been used in longitudinal studies to predict disease trajectory. For example, Steele et al. (2017) used resting-state fMRI and SVM to predict treatment completion in stimulant- or heroin-dependent incarcerated participants who volunteered for a 12-week substance abuse treatment program. These authors achieved a sensitivity of approximately 80% using resting-state connectivity between networks including the anterior cingulate, insula, and striatum.

Random forest is another supervised-learning technique used in addiction research. Different from SVM, random forest uses decision trees based on the random selection of data points and of variables during training. Each random forest consists of a large number of decision trees (hence forest), and combining these decision trees can reduce the overall variance. One example of the random forest application can be seen in Squeglia et al. (2017), where the authors used both structural MRI and resting-state fMRI data, in combination with demographic and neuropsychological data, to predict the initiation of alcohol use in a large group of adolescents. This analysis identified a multivariate pattern of demographic and behavioral characteristics (e.g., being male, higher socioeconomic status, and so on), worse executive functioning, thinner cortices, and less resting-state activity that predicted early alcohol use.

Ahn and colleagues used another machine-learning algorithm, called elastic net, to predict SUDs (i.e., heroin, amphetamine) using demographic, personality, psychiatric, and neuropsychological data of impulsivity. The authors found distinct multivariate patterns that mark either heroin or amphetamine dependence, challenging the notion that different SUDs could be subserved with the same behavioral and cognitive profiles. In a different study, these authors used the same algorithm to examine the behavioral predictors of cocaine dependence. The authors found that cocaine dependence was predicted by higher motor and cognitive impulsivity, poor response inhibition, and suboptimal decision-making (Ahn et al., 2016).

Taken together, these studies have provided promising new avenues for computational work on addiction. More studies are needed, such as longitudinal studies, for us to be able to truly make individual-level predictions and forecast disease trajectory and treatment outcomes into the future.

Conclusion and Future Directions

Advances in computational cognitive neuroscience have started to benefit psychiatry research in recent years. Yet the awareness and application of these new research paradigms and methods are still scarce in the neuroscience research on addiction. Here we identify the following as major questions for the field to address in the next few years.

- How do the initial reinforcing effects of drugs eventually lead to habitual responses?
- How do the dynamics between ventral and dorsal corticostriatal neural circuits change during different stages of addiction?
- How does drug craving interact with both RL and habitual systems?
- What biomarkers and cognitive markers during adolescence might forecast one's likelihood to develop addiction or relapse in adulthood?
- How can computational psychiatry approaches better inform intervention and treatment?

It is inevitable that addiction research will utilize more and more computational methods and "model thinking." This also means that a dialogue among researchers from different areas—computational modelers, rodent neuroscientists, human neuroscientists, and clinicians—will need to take place. We believe that this cultural shift will not only elucidate the mechanisms of addiction but also help to eventually develop new treatments and interventions.

In addition, a major issue, alluded to earlier, is the limited empirical research to validate computational-derived models and predictions. For example, only a very small number of neuroimaging studies directly quantify RL signals in addicted humans, in contrast to the strong theoretical development in this field based on animal models. Thus, this field urgently needs more "model-driven" empirical studies in humans and an increased integration of computational, clinical, and experimental approaches.

Acknowledgements

We are thankful to Dr. Vincenzo Fiore for his comments on this chapter. XG is supported by NIDA R01DA043695 and the Mental Illness Research, Education, and Clinical Center (MIRECC VISN 2) at the James J. Peter Veterans Affairs Medical Center, Bronx, New York.

REFERENCES

Ahn, W.-Y., Ramesh, D., Moeller, F. G., & Vassileva, J. (2016). Utility of machine-learning approaches to identify behavioral markers for substance use disorders: Impulsivity dimensions as predictors of current cocaine dependence. *Frontiers in Psychiatry, 7*(34). doi:10.3389/fpsyt.2016.00034

Ahn, W.-Y., & Vassileva, J. (2016). Machine-learning identifies substance-specific behavioral markers for opiate and stimulant dependence. *Drug and Alcohol Dependence, 161*, 247–257. doi:10.1016/j.drugalcdep.2016.02.008

Barrett, L. F., & Simmons, W. K. (2015). Interoceptive predictions in the brain. *Nature Reviews, Neuroscience, 16*(7), 419–429. doi:10.1038/nrn3950

Bayer, H. M., & Glimcher, P. W. (2005). Midbrain dopamine neurons encode a quantitative reward prediction error signal. *Neuron, 47*(1), 129–141. doi:10.1016/j.neuron.2005.05.020

Bedi, G., Preston, K. L., Epstein, D. H., Heishman, S. J., Marrone, G. F., Shaham, Y., & de Wit, H. (2011). Incubation of cue-induced cigarette craving during abstinence in human smokers. *Biological Psychiatry, 69*(7), 708–711. doi:10.1016/j.biopsych.2010.07.014

Boileau, I., Assaad, J. M., Pihl, R. O., Benkelfat, C., Leyton, M., Diksic, M., … Dagher, A. (2003). Alcohol promotes dopamine release in the human nucleus accumbens. *Synapse, 49*(4), 226–231. doi:10.1002/syn.10226

Braver, T. S., Barch, D. M., & Cohen, J. D. (1999). Cognition and control in schizophrenia: A computational model of dopamine and prefrontal function. *Biological Psychiatry, 46*(3), 312–328.

Brown, H., Adams, R. A., Parees, I., Edwards, M., & Friston, K. (2013). Active inference, sensory attenuation and illusions. *Cognitive Processing, 14*(4), 411–427. doi:10.1007/s10339-013-0571-3

Chase, H. W., Eickhoff, S. B., Laird, A. R., & Hogarth, L. (2011). The neural basis of drug stimulus processing and craving: An activation likelihood estimation meta-analysis. *Biological Psychiatry, 70*(8), 785–793. doi:10.1016/j.biopsych.2011.05.025

Chiu, P. H., Kayali, M. A., Kishida, K. T., Tomlin, D., Klinger, L. G., Klinger, M. R., & Montague, P. R. (2008). Self responses along cingulate cortex reveal quantitative neural phenotype for high-functioning autism. *Neuron, 57*(3), 463–473. doi:10.1016/j.neuron.2007.12.020

Chiu, P. H., Lohrenz, T. M., & Montague, P. R. (2008). Smokers' brains compute, but ignore, a fictive error signal in a sequential investment task. *Nature Neuroscience, 11*(4), 514–520. doi:10.1038/nn2067

Conrad, K. L., Tseng, K. Y., Uejima, J. L., Reimers, J. M., Heng, L. J., Shaham, Y., … Wolf, M. E. (2008). Formation of accumbens GluR2-lacking AMPA receptors mediates incubation of cocaine craving. *Nature, 454*(7200), 118–121. doi:10.1038/nature06995

Daw, N. D., Niv, Y., & Dayan, P. (2005). Uncertainty-based competition between prefrontal and dorsolateral striatal systems for behavioral control. *Nature Neuroscience, 8*(12), 1704–1711.

Dayan, P. (2009). Dopamine, reinforcement learning, and addiction. *Pharmacopsychiatry, 42* (suppl. 1), S56–65. doi:10.1055/s-0028-1124107

De Biasi, M., & Dani, J. A. (2011). Reward, addiction, withdrawal to nicotine. *Annual Review of Neuroscience, 34*, 105–130. doi:10.1146/annurev-neuro-061010-113734

Dickinson, A. (1985). Actions and habits: The development of behavioural autonomy. *Philosophical Transactions of the Royal Society of London B: Biological Sciences, 308*(1135), 67–78.

Ding, X., Yang, Y., Stein, E. A., & Ross, T. J. (2015). Multivariate classification of smokers and nonsmokers using SVM-RFE on structural MRI images. *Human Brain Mapping, 36*(12), 4869–4879. doi:10.1002/hbm.22956

Dolan, R. J., & Dayan, P. (2013). Goals and habits in the brain. *Neuron, 80*(2), 312–325. doi:10.1016/j.neuron.2013.09.007

Engelmann, J. M., Versace, F., Robinson, J. D., Minnix, J. A., Lam, C. Y., Cui, Y., … Cinciripini, P. M. (2012). Neural substrates of smoking cue reactivity: A meta-analysis of fMRI studies. *Neuroimage, 60*(1), 252–262. doi:10.1016/j.neuroimage.2011.12.024

Ersche, K. D., Gillan, C. M., Jones, P. S., Williams, G. B., Ward, L. H., Luijten, M., … Robbins, T. W. (2016). Carrots and sticks fail to change behavior in cocaine addiction. *Science, 352*(6292), 1468–1471. doi:10.1126/science.aaf3700

Ersche, K. D., Jones, P. S., Williams, G. B., Turton, A. J., Robbins, T. W., & Bullmore, E. T. (2012). Abnormal brain structure implicated in stimulant drug addiction. *Science, 335*(6068), 601–604. doi:10.1126/science.1214463

Everitt, B. J., & Robbins, T. W. (2005). Neural systems of reinforcement for drug addiction: From actions to habits to compulsion. *Nature Neuroscience, 8*(11), 1481–1489. doi:10.1038/Nn1579

Fiore, V. G., Dolan, R. J., Strausfeld, N. J., & Hirth, F. (2015). Evolutionarily conserved mechanisms for the selection and maintenance of behavioural activity. *Philosophical Transactions of the Royal Society of London B: Biological Sciences, 370*(1684). doi:10.1098/rstb.2015.0053

FitzGerald, T. H., Dolan, R. J., & Friston, K. J. (2014). Model averaging, optimal inference, and habit formation. *Frontiers in Human Neuroscience, 8*, 457. doi:10.3389/fnhum.2014.00457

Goldstein, R. Z., & Volkow, N. D. (2002). Drug addiction and its underlying neurobiological basis: Neuroimaging evidence for the involvement of the frontal cortex. *American Journal of Psychiatry, 159*(10), 1642–1652.

Grimm, J. W., Hope, B. T., Wise, R. A., & Shaham, Y. (2001). Neuroadaptation: Incubation of cocaine craving after withdrawal. *Nature, 412*(6843), 141–142. doi:10.1038/35084134

Gu, X. (2018). Incubation of craving: A Bayesian account. *Neuropsychopharmacology, 43*(12), 2337–2339. doi:10.1038/s41386-018-0108-7

Gu, X., & Filbey, F. (2017). A Bayesian observer model of drug craving. *JAMA Psychiatry, 74*(4), 419–420. doi:10.1001/jamapsychiatry.2016.3823

Gu, X., & FitzGerald, T. H. (2014). Interoceptive inference: Homeostasis and decision-making. *Trends in Cognitive Sciences, 18*(6), 269–270. doi:10.1016/j.tics.2014.02.001

Gu, X., Hof, P. R., Friston, K. J., & Fan, J. (2013). Anterior insular cortex and emotional awareness. *Journal of Comparative Neurology, 521*(15), 3371–3388. doi:10.1002/cne.23368

Gu, X., Lohrenz, T., Salas, R., Baldwin, P. R., Soltani, A., Kirk, U., … Montague, P. R. (2015). Belief about nicotine selectively modulates value and reward prediction error signals in smokers. *Proceedings of the National Academy of Sciences of the United States of America, 112*(8), 2539–2544. doi:10.1073/pnas.1416639112

Gu, X., Lohrenz, T., Salas, R., Baldwin, P. R., Soltani, A., Kirk, U., … Montague, P. R. (2016). Belief about nicotine modulates subjective craving and insula activity in deprived smokers. *Front Psychiatry, 7*(126), 126. doi:10.3389/fpsyt.2016.00126

Heinz, A., Siessmeier, T., Wrase, J., Hermann, D., Klein, S., Grusser, S. M., … Bartenstein, P. (2004). Correlation between dopamine D(2) receptors in the ventral striatum and central processing of alcohol cues and craving. *American Journal of Psychiatry, 161*(10), 1783–1789. doi:10.1176/appi.ajp.161.10.1783

Hernandez, L., & Hoebel, B. G. (1988). Food reward and cocaine increase extracellular dopamine in the nucleus accumbens as measured by microdialysis. *Life Sciences, 42*(18), 1705–1712. doi:10.1016/0024-3205(88)90036-7

Hollerman, J. R., & Schultz, W. (1998). Dopamine neurons report an error in the temporal prediction of reward during learning. *Nature Neuroscience, 1*(4), 304–309. doi:10.1038/1124

Huys, Q. J. M., Moutoussis, M., & Williams, J. (2011). Are computational models of any use to psychiatry? *Neural Networks, 24*(6), 544–551. doi:https://doi.org/10.1016/j.neunet.2011.03.001

Hyman, S. E. (2005). Addiction: A disease of learning and memory. *American Journal of Psychiatry, 162*(8), 1414–1422. doi:10.1176/appi.ajp.162.8.1414

Jasinska, A. J., Stein, E. A., Kaiser, J., Naumer, M. J., & Yalachkov, Y. (2014). Factors modulating neural reactivity to drug cues in addiction: A survey of human neuroimaging studies. *Neuroscience & Biobehavioral Reviews, 38*, 1–16. doi:10.1016/j.neubiorev.2013.10.013

Juliano, L. M., Fucito, L. M., & Harrell, P. T. (2011). The influence of nicotine dose and nicotine dose expectancy on the cognitive and subjective effects of cigarette smoking. *Experimental and Clinical Psychopharmacology, 19*(2), 105–115. doi:10.1037/a0022937

Kelemen, W. L., & Kaighobadi, F. (2007). Expectancy and pharmacology influence the subjective effects of nicotine in a balanced-placebo design. *Experimental and Clinical Psychopharmacology, 15*(1), 93–101. doi:10.1037/1064-1297.15.1.93

Keramati, M., & Gutkin, B. (2014). Homeostatic reinforcement learning for integrating reward collection and physiological stability. *Elife, 3*. doi:10.7554/eLife.04811

Kishida, K. T., King-Casas, B., & Montague, P. R. (2010). Neuroeconomic approaches to mental disorders. *Neuron, 67*(4), 543–554. doi:10.1016/j.neuron.2010.07.021

Knill, D. C., & Pouget, A. (2004). The Bayesian brain: The role of uncertainty in neural coding and computation. *Trends in Neurosciences, 27*(12), 712–719.

Lawson, R. P., Mathys, C., & Rees, G. (2017). Adults with autism overestimate the volatility of the sensory environment. *Nature Neuroscience, 20*(9), 1293–1299. doi:10.1038/nn.4615

Lu, L., Hope, B. T., Dempsey, J., Liu, S. Y., Bossert, J. M., & Shaham, Y. (2005). Central amygdala ERK signaling pathway is critical to incubation of cocaine craving. *Nature Neuroscience, 8*(2), 212–219. doi:10.1038/nn1383

Maia, T. V., & Frank, M. J. (2011). From reinforcement learning models to psychiatric and neurological disorders. *Nature Neuroscience, 14*(2), 154–162. doi:10.1038/nn.2723

Marr, D., & Poggio, T. (1976). From understanding computation to understanding neural circuitry. *Neurosciences Research Program Bulletin, 15*, 470–488.

McBride, D., Barrett, S. P., Kelly, J. T., Aw, A., & Dagher, A. (2006). Effects of expectancy and abstinence on the neural response to smoking cues in cigarette smokers: An fMRI study. *Neuropsychopharmacology, 31*(12), 2728–2738. doi:10.1038/sj.npp.1301075

Mete, M., Sakoglu, U., Spence, J. S., Devous, M. D., Harris, T. S., & Adinoff, B. (2016). Successful classification of cocaine dependence using brain imaging: A generalizable machine learning approach. *BMC Bioinformatics, 17*. doi:ARTN 357 10.1186/s12859-016-1218-z

Montague, P. R., Dayan, P., & Sejnowski, T. J. (1996). A framework for mesencephalic dopamine systems based on predictive Hebbian learning. *Journal of Neuroscience, 16*(5), 1936–1947.

Montague, P. R., Dolan, R. J., Friston, K. J., & Dayan, P. (2012). Computational psychiatry. *Trends in Cognitive Sciences, 16*(1), 72–80. doi:10.1016/j.tics.2011.11.018

Moors, A., & De Houwer, J. (2006). Automaticity: A theoretical and conceptual analysis. *Psychological Bulletin, 132*(2), 297–326. doi:10.1037/0033-2909.132.2.297

Naqvi, N. H., Rudrauf, D., Damasio, H., & Bechara, A. (2007). Damage to the insula disrupts addiction to cigarette smoking. *Science, 315*(5811), 531–534. doi:10.1126/science.1135926

Nestler, E. J. (2002). From neurobiology to treatment: Progress against addiction. *Nature Neuroscience, 5* (suppl.), 1076–1079. doi:10.1038/nn945

Nestler, E. J., & Aghajanian, G. K. (1997). Molecular and cellular basis of addiction. *Science, 278*(5335), 58–63.

Pariyadath, V., Stein, E. A., & Ross, T. J. (2014). Machine learning classification of resting state functional connectivity predicts smoking status. *Frontiers in Human Neuroscience, 8*, 425. doi:10.3389/fnhum.2014.00425

Park, S. Q., Kahnt, T., Beck, A., Cohen, M. X., Dolan, R. J., Wrase, J., & Heinz, A. (2010). Prefrontal cortex fails to learn from reward prediction errors in alcohol dependence. *Journal of Neuroscience, 30*(22), 7749–7753. doi:10.1523/jneurosci.5587-09.2010

Parvaz, M. A., Moeller, S. J., & Goldstein, R. Z. (2016). Incubation of cue-induced craving in adults addicted to cocaine measured by electroencephalography. *JAMA Psychiatry, 73*(11), 1127–1134. doi:10.1001/jamapsychiatry.2016.2181

Pidoplichko, V. I., De Biasi, M., Williams, J. T., & Dani, J. A. (1997). Nicotine activates and desensitizes midbrain dopamine neurons. *Nature, 390*(6658), 401–404. doi:10.1038/37120

Powers, A. R., Mathys, C., & Corlett, P. R. (2017). Pavlovian conditioning-induced hallucinations result from overweighting of perceptual priors. *Science, 357*(6351), 596–600. doi:10.1126/science.aan3458

Rangel, A., Camerer, C., & Montague, P. R. (2008). A framework for studying the neurobiology of value-based decision making. *Nature Reviews Neuroscience, 9*(7), 545–556. doi:10.1038/nrn2357

Redish, A. D. (2004). Addiction as a computational process gone awry. *Science, 306*(5703), 1944–1947.

Rice, M. E., & Cragg, S. J. (2004). Nicotine amplifies reward-related dopamine signals in striatum. *Nature Neuroscience, 7*(6), 583–584. doi:10.1038/nn1244

Robinson, J. D., Engelmann, J. M., Cui, Y., Versace, F., Waters, A. J., Gilbert, D. G., … Cinciripini, P. M. (2014). The effects of nicotine dose expectancy and motivationally relevant distracters on vigilance. *Psychology of Addictive Behaviors, 28*(3), 752–760. doi:10.1037/a0035122

Rushworth, M. F., & Behrens, T. E. (2008). Choice, uncertainty and value in prefrontal and cingulate cortex. *Nature Neuroscience, 11*(4), 389–397. doi:10.1038/nn2066

Sakoglu, U., Mete, M., Esquivel, J., Rubia, K., Briggs, R., & Adinoff, B. (2019). Classification of cocaine-dependent participants with dynamic functional connectivity from functional magnetic resonance imaging data. *Journal of Neuroscience Research, 97*(7), 790–803. doi:10.1002/jnr.24421

Schultz, W., Dayan, P., & Montague, P. R. (1997). A neural substrate of prediction and reward. *Science, 275*(5306), 1593–1599.

Seth, A. K., & Friston, K. J. (2016). Active interoceptive inference and the emotional brain. *Philosophical Transactions of the Royal Society of London B: Biological Sciences, 371*(1708). doi:10.1098/rstb.2016.0007

Shiffman, S., Li, X., Dunbar, M. S., Tindle, H. A., Scholl, S. M., & Ferguson, S. G. (2015). Does laboratory cue reactivity correlate with real-world craving and smoking responses to cues? *Drug and Alcohol Dependence, 155*, 163–169. doi:http://dx.doi.org/10.1016/j.drugalcdep.2015.07.673

Squeglia, L. M., Ball, T. M., Jacobus, J., Brumback, T., McKenna, B. S., Nguyen-Louie, T. T., ... Tapert, S. F. (2017). Neural predictors of initiating alcohol use during adolescence. *American Journal of Psychiatry, 174*(2), 172–185. doi:10.1176/appi.ajp.2016.15121587

Steele, V. R., Rao, V., Calhoun, V. D., & Kiehl, K. A. (2017). Machine learning of structural magnetic resonance imaging predicts psychopathic traits in adolescent offenders. *Neuroimage, 145*(Pt. B), 265–273. doi:10.1016/j.neuroimage.2015.12.013

Tanabe, J., Reynolds, J., Krmpotich, T., Claus, E., Thompson, L. L., Du, Y. P., & Banich, M. T. (2013). Reduced neural tracking of prediction error in substance-dependent individuals. *American Journal of Psychiatry, 170*(11), 1356–1363. doi:10.1176/appi.ajp.2013.12091257

Tanda, G., Pontieri, F. E., & Di Chiara, G. (1997). Cannabinoid and heroin activation of mesolimbic dopamine transmission by a common mu1 opioid receptor mechanism. *Science, 276*(5321), 2048–2050.

Tang, D. W., Fellows, L. K., Small, D. M., & Dagher, A. (2012). Food and drug cues activate similar brain regions: A meta-analysis of functional MRI studies. *Physiology & Behavior, 106*(3), 317–324. doi:10.1016/j.physbeh.2012.03.009

Tiffany, S. T., Friedman, L., Greenfield, S. F., Hasin, D. S., & Jackson, R. (2012). Beyond drug use: A systematic consideration of other outcomes in evaluations of treatments for substance use disorders. *Addiction, 107*(4), 709–718.

Tiffany, S. T., & Wray, J. M. (2012). The clinical significance of drug craving. *Annals of the New York Academy of Sciences, 1248*, 1–17. doi:10.1111/j.1749-6632.2011.06298.x

Van Dam, N. T., O'Connor, D., Marcelle, E. T., Ho, E. J., Cameron Craddock, R., Tobe, R. H., ... Milham, M. P. (2017). Data-driven phenotypic categorization for neurobiological analyses: Beyond DSM-5 labels. *Biological Psychiatry, 81*(6), 484–494. doi:10.1016/j.biopsych.2016.06.027

Volkow, N. D., Wang, G. J., Telang, F., Fowler, J. S., Logan, J., Childress, A. R., ... Wong, C. (2006). Cocaine cues and dopamine in dorsal striatum: Mechanism of craving in cocaine addiction. *Journal of Neuroscience, 26*(24), 6583–6588. doi:10.1523/jneurosci.1544-06.2006

Waltz, J. A., Frank, M. J., Robinson, B. M., & Gold, J. M. (2007). Selective reinforcement learning deficits in schizophrenia support predictions from computational models of striatal-cortical dysfunction. *Biological Psychiatry, 62*(7), 756–764. doi:10.1016/j.biopsych.2006.09.042

Waters, A. J., Shiffman, S., Sayette, M. A., Paty, J. A., Gwaltney, C. J., & Balabanis, M. H. (2004). Cue-provoked craving and nicotine replacement therapy in smoking cessation. *Journal of Consulting and Clinical Psychology, 72*(6), 1136–1143. doi:10.1037/0022-006X.72.6.1136

Weiss, F., Lorang, M. T., Bloom, F. E., & Koob, G. F. (1993). Oral alcohol self-administration stimulates dopamine release in the rat nucleus accumbens: Genetic and motivational determinants. *Journal of Pharmacology and Experimental Therapeutics, 267*(1), 250–258.

Wong, D. F., Kuwabara, H., Schretlen, D. J., Bonson, K. R., Zhou, Y., Nandi, A., ... London, E. D. (2006). Increased occupancy of dopamine receptors in human striatum during cue-elicited cocaine craving. *Neuropsychopharmacology, 31*(12), 2716–2727. doi:10.1038/sj.npp.1301194

Yalachkov, Y., Kaiser, J., & Naumer, M. J. (2012). Functional neuroimaging studies in addiction: Multisensory drug stimuli and neural cue reactivity. *Neuroscience & Biobehavioral Reviews, 36*(2), 825–835. doi:10.1016/j.neubiorev.2011.12.004

92 Neurotechnologies for Mind Reading: Prospects for Privacy

ADINA ROSKIES

ABSTRACT Neuroimaging techniques provide unprecedented access to a variety of information about the brain, including, to some extent, the contents of thoughts. This chapter describes the extent to which fMRI allows us to "read minds." As this chapter chronicles, our current abilities to read minds are more limited than many realize, but dramatic progress has been made in the last decade. Even moderate prospects for improvement in mind reading raise ethical and legal questions about how this information relates to privacy rights and the evidential status of imaging data. These are pressing questions for society that are only now beginning to be explored.

American culture values liberty, but privacy, arguably an aspect of liberty, is not so carefully defended. Technology in the form of social media is ubiquitous and is built upon a business model that undermines privacy, yet it is only now coming under scrutiny for clear violations of privacy in the commerce of user data. However, intrusions on privacy by consumer technology do not yet encroach directly upon the space between our ears. The contents of one's thoughts seem to be directly accessible only to the thinker, unless revealed by voluntary disclosure. Mental privacy has been taken for granted, but should it be? Rapid advances in brain-imaging neurotechnologies allow unprecedented access to the brain activity that constitutes our minds. Can neurotechnologies read our thoughts? This chapter explores the prospects for mind reading, its potential for use in legal settings, and the ethical challenges it raises.

Neurotechnologies make accessible to us information that is relevant to a host of phenomena that a person may wish to keep private. For example, brain scans can reveal medically relevant information about a subject's brain, such as early signs of dementia (Salvatore et al., 2015; Teipel et al., 2013), indicators of mental illness (Cetin et al., 2016; Ebisch et al., 2018; Mueller et al., 2012; Tang et al., 2012; Whalley et al., 2013), or incidental findings (Carre et al., 2013; Paulsen et al., 2011; Scott, Murphy, & Illes, 2012). There is evidence that certain relatively stable character traits, such as risk-aversiveness or anxiety, may be inferred on the basis of neuroimaging data (Carre et al., 2013; Malpas et al., 2016; Paulsen et al., 2011), without the subject's knowledge or consent (Farah et al., 2008). Neuroimaging data can also reveal information about a person's unconscious attitudes and biases (Azevedo et al., 2012; Harle et al., 2012; Katsumi & Dolcos, 2018; Liu, Lin, Xu, Zhang, & Luo, 2015; Stanley et al., 2012; Van Bavel, Packer, & Cunningham, 2008), often by way of passive measures or alternative tasks. Protections must be in place to ensure that such information is not obtained or misused. There has been significant discussion and technological advance on this front (Patel, 2018), and many of the provisions designed to protect medical or genetic data may provide a promising model for protecting the privacy of this sort of information.

This chapter focuses upon a different kind of personal information that one might glean from neuroimaging: information about the content of a person's thoughts. Media coverage and scholarship in this area often declare that the sky is falling, virtually taking for granted that the technology for mind reading is all but upon us and that the only significant barrier to state infringement on mental privacy lies with the law. This characterization is far from the case, but it is nonetheless worthwhile to address important ethical and legal issues before the sky does fall.

Brain Reading and Mind Reading

"Reading," at a minimum, involves a mapping of a physical pattern to meaning. I have made a distinction between what I call *brain reading*, on the one hand, and *mind reading*, on the other (Roskies, 2014). Brain reading is supported by rough and brute-force empirical correlations between measurements of the physical state of the brain and mental functions, capacities, and the world. It allows one to infer coarse-grained content from brain data largely on the basis of empirical correlations—for example, inferring emotional reactivity or fear from amygdala activation or perception or imagining a face from activation in the FFA (fusiform face area). Brain reading is here, and although it provides some information regarding mental content, it doesn't distinguish between a large number of semantically different possibilities. Whose face is the subject seeing? Is she experiencing fear or anxiety? What is the object of her fear? Is it present or imagined?

In contrast, the propositional contents of thought are more fine-grained. Mind reading with brain-imaging devices would involve being able to distinguish approximately the same distinctions in content that language enables us to express. What are the relations that constitute their mental contents? Can we distinguish the thought "the baby kicked the grandfather" from the thought "the grandfather kicked the baby"? Can we distinguish "Tom was angry" from "Tom was disappointed"? Practically speaking, what may distinguish mind reading from brain reading is how systematic and generative is the mapping that we establish between mental and physical states. Although a clear distinction between the two may not be sustainable in principle, determining where one is on a brain-reading/mind-reading spectrum may be important in practice, especially when the data are relevant to realms of human interaction that trade in shades of gray, such as the ethical, social, and legal.

A brute-force approach to decoding propositional content from the measurement of brain activity would involve establishing correlations for virtually all the atomic or simple concepts we possess—a dictionary for the mind. But while dictionaries provide translations for individual words or conceptual elements, if we are concerned with content, we must be concerned not only with the elements of thought but also with their relations. After all, "The butler did it" and "The butler did not do it" have contrary meanings, as do "George provoked Harry" and "Harry provoked George." Language and thought are infinitely generative, and a brute-force approach would require establishing an infinite number of correlations. It is therefore practically intractable.

To exhaustively identify mental contents would require understanding the principles of mental representation sufficiently well to develop a generative model of mental representation so that one could reliably and accurately predict patterns of brain activity for novel words, concepts, or propositions. Researchers have made some headway in showing a limited proof of principle for generative models of semantics (discussed below), but the degree to which fine-grained content is encoded systematically rather than fortuitously in the brain is unclear. In addition, there is a need to understand the encoding of relational structure in thought or inner speech in the brain. Whether these aspects of content can be decoded from brain signals is an open question. A similar problem exists for the attitudes one takes to propositions. And, whether concepts combine in terms of their neural signals in a compositional way similar to the way that language is compositional is also unknown (see Reverberi, Görgen, & Haynes, 2011). The problem for mind reading is thus threefold: (1) identifying activity corresponding to individual content elements, (2) identifying activity reflecting conceptual relations, and (3) being able to infer content across subjects. The rest of this chapter will concern these issues.

Can We Read Minds with Neuroimaging Methods?

In an early brain-reading study, O'Craven and Kanwisher (2000) presented subjects with pictures of faces and places and demonstrated the expected changes in BOLD (blood oxygen-level dependent) signal in the FFA (fusiform face area) and PPA (parahippocampal place area) to the perception of faces and places, respectively. They then showed that these same brain regions were active during mental imagery of those same stimulus classes. This turns out to be a common finding: many of the same brain regions involved in processing external stimuli are active during thoughts about the same type of stimuli (see, e.g., Polyn et al., 2005). Having localized the FFA and PPA in their individual subjects, the researchers then showed they could decode whether a subject was imagining a face or a place on the basis of the brain data. Their results demonstrated that specific *classes* of thought content could be determined from brain activation data. Tong et al. (1998) demonstrated that activation in the FFA and PPA during binocular rivalry varied with the *conscious* experience of a face or a place and that the conscious percept could be predicted by the BOLD activation levels.

Importantly, researchers who were "brain reading" in these cases knew that the stimuli fell into one of two broad classes, so they needed only to determine which of two mental state types was more likely than the other. This is common in many "decoding" studies, in which successful decoding is in the context of a number of prespecified possibilities. In addition, the studies were done on types of stimuli for which the brain shows distinct anatomical specificity for processing. The studies reveal nothing about the ability to identify mental states of arbitrary classes of stimuli using functional magnetic resonance imaging (fMRI) or about the possibility of distinguishing particulars within these classes. For example, the studies do not address whether it is possible to distinguish thinking about Bill Clinton from thinking about Robin Williams, whether distinctions can be made among arbitrary numbers or kinds of classes for which separate brain areas are not known to mediate specific types of representations, or whether sense can be made of the propositional content of the subjects' mental states. These and other early studies showed merely that brain reading was possible in principle, in the most clear-cut of cases.

Multivariate analysis of fMRI data and decoding of semantic information Neuroimaging underwent a sea change with the advent of multivariate techniques for data analysis (also called multivoxel pattern analysis, or MVPA). MVPA analyzes patterns of brain activity across many voxels, rather than just the net change of signal in a localized region. In a seminal study, Haxby et al. (2001) presented subjects with pictures of objects from a variety of categories, including faces, shoes, tools, chairs, and cats. Haxby et al. (2001) found that activity for all these categories was widespread across cortex, and in one of the pioneering uses of multivariate techniques in fMRI analysis, he showed that *patterns* of activation differed among brain regions for each stimulus class even when those regions did not show significant net changes in activity between categories. Moreover, even when one eliminated the information from the brain region responding maximally to a class of stimuli (such as ignoring the information from the FFA for face processing), one could still identify the stimulus class to which the item belonged on the basis of activation patterns in other cortical areas. Thus, information encoding the identity of visual categories was widespread throughout cortex.

Early decoding studies used classification-based decoding in which classifiers are trained to discriminate between fMRI data associated with a specified set of stimulus categories and then used to classify novel fMRI data as belonging to one of these categories. More recent work often employs model-based decoding, in which generative representational models of a problem space are constructed based on elementary features, thus allowing, in principle, the identification of arbitrary contents on the basis of neural response data. These methods are becoming more widely used but are limited by the accuracy of the encoding models. While these are well-grounded when it comes to early perception, high-level encoding models are still quite speculative. Other technical advances that have proven powerful have included combining modeling with Bayesian analysis with methods for compensating for individual neuroanatomical variability (Conroy, Singer, Guntupalli, Ramadge, & Haxby, 2013; Haxby et al., 2011) and with methods for probing representational structure (Kriegeskorte, Mur, & Bandettini, 2008).

Reconstructing visual stimuli Vision is perhaps the best understood cortical system. Thirion et al. (2006) have used insights from the organization of the visual system to reconstruct simple visual stimuli from neuroimages, showing proof of principle that one can infer stimuli from knowledge of the transfer function from the visual stimulus to cortex. Drawing on this general approach, Kay et al. (2008) developed methods to reconstruct natural visual scenes from brain data. By examining cortical activation profiles to a large set of images, they constructed a receptive field model for each voxel of the brain (i.e., a model of how various low-level image features at a location of visual space maps to brain activity). Their model described tuning along spatial orientation and spatial frequency domains (Kay et al., 2008). They then presented subjects with a novel image drawn from a large library of images and measured the brain activity. Based on the activation pattern, they could identify the image in the library most likely to have produced that activation pattern. With a library of 120 images, the decoding selected the correct image 92% of the time (chance performance is .8%). With a much larger library (1,000 images), accuracy remained high, falling to 82%. The authors estimate that performance on the entire Google library of images would remain well above chance. The authors also note that decoding remained above chance with single-trial data. This finding suggests the possibility of real-time decoding.

More recent work by the same group improved upon the early visual reconstruction paradigm by combining a visual decoding scheme (like that described above) with a semantic decoder, which relied upon information from anterior brain areas. They also combined this with a Bayesian approach, which used a prior based on the statistics of natural image structure to help with image selection. The three approaches combined allowed them to "reconstruct" images that were structurally and semantically similar to the target image (Naselaris et al., 2009). Importantly, even this method does not do pure bottom-up reconstruction: the reconstructions are always of an image originally sampled in the set of priors. Since for a real-world reconstruction task an infinite number of possible images exist, this method cannot hope to reproduce exactly any arbitrarily viewed image. However, with a large enough database, the thought is that the reconstruction for an arbitrary natural image could still be quite good.

Gallant and colleagues (Nishimoto et al., 2011) extended this approach to new domains, such as the reconstruction of dynamic visual scenes (movie clips) from brain data. Again, they relied upon priors obtained from a large library of video clips. When tested on novel clips not included in the library, the algorithm selected the clip in the library most likely to be the stimulus. The authors report a high degree of similarity between the chosen clip and the novel stimulus clip. Work by Hasson and colleagues (Hasson, Nir, Levy, Fuhrmann, & Malach, 2004; Hasson et al., 2008) indicates that human brains share common activity profiles when viewing dynamic natural scenes, which implies that this method will work relatively well across subjects. Nishimoto et al. (2011) suggest that their method for reconstructing dynamic visual stimuli may also be useful for reconstructing dynamic visual imagery from brain data. They have demonstrated that low-level features in early visual cortex are activated with imagery, and an encoding model can be used for decoding visual imagery (Naselaris, Olman, Stansbury, Ugurbil, & Gallant, 2015).

Auditory reconstruction Just as visual images or scenes can be seen or imagined, so auditory experiences can be heard or imagined. As in vision, the human auditory system follows simple organizational principles in primary cortical areas, with increasing complexity as one ascends the cortical system. This organization has been exploited to enable some aspects of sound to be decoded from fMRI signals. There is evidence, for instance, that different patterns of brain activity encode aspects of the category of acoustic signal (human speech, animal sounds, and more; Formisano et al., 2008). In an early study (Formisano et al., 2008), subjects were asked to listen to repeated presentations of three vowel sounds spoken by three different speakers. Using pattern recognition methods and training on this data set, experimenters were able to determine which of the three sounds was being uttered and by which speaker, even on trials not in the training set. On the basis of their data, Formisano et al. (2008) postulate separate distributed regions of cortex for encoding phonemes and speaker identity. They also found they could train a classifier on vowels from one speaker and correctly classify the vowels spoken by the others. This suggests, in addition, that the speech sound cortical representations are acoustically invariant along certain dimensions. This may be the first demonstration of the feasibility of decoding auditory speech information. It has a number of significant limitations that should make one circumspect about the near-term prospects of decoding speech from brain activity. For one thing, the classifier only discriminated between three vowels, a highly impoverished set of stimuli relative to the approximately 44 in English, and the many more in some other languages. Second, these sounds were presented in isolation, not embedded in a speech stream. Indeed, the temporal order of sounds is a crucial aspect of language—only order disambiguates the phonemic sequences of "super" and "pursue," and grammar is highly dependent on temporal order. Merely distinguishing individual phonemes is a long way from decoding real speech and speech content. In more recent work, Formisano and colleagues constructed an encoding model of spectrotemporal modulation from natural auditory stimuli and showed that this model was above chance in auditory reconstruction in spectral and temporal domains. Reconstructions looked like temporally smoothed versions of the original stimulus, with enough detail to occasionally identify the original source file but insufficient detail to decode speech (Santoro et al., 2017). The authors identified a number of significant theoretical barriers to the development of accurate fMRI-based speech decoders.

Object representation In a groundbreaking series of studies, groups from Carnegie Mellon University showed that brain signatures related to perceiving individual objects could be recognized and that a generative model based upon statistical association could, to a large degree, predict whole-brain fMRI patterns. In an initial study, Shinkareva et al. (2008) showed that they could predict, with high accuracy within and across subjects, which of a set of 10 line drawings a person was looking at, based on his or her fMRI data and, with even more accuracy, which of two object categories the drawing was from. This study suggests that individual objects have unique and discernable neural signals within individuals, raising the possibility that particular objects of mental states could be decoded if classifiers could be trained on a broad array of data from an individual subject.

Perhaps more significantly, it suggests that the overall structure of object encoding and processing is uniform enough across individuals to enable the decoding of some mental states based on information obtained from others. In a landmark paper, Mitchell et al. (2008) trained a classifier to predict the fMRI signatures of 60 objects drawn from 12 categories. The classifier related the statistics of word associations between the objects and common verbs with the MRI results. Then, when presented with a novel object upon which the classifier had not been trained, the classifier predicted an fMRI activation pattern that was very similar to the actual fMRI pattern observed when the subject saw that object. These results suggest that the way our brains encode object information is systematic enough that a reasonably good model of the semantics of object representation could be developed to generalize to novel stimuli. This study was the first indication of the feasibility, in principle, of a generative model of object semantics based on brain data; subsequent work has extended this approach to build predictive models of brain activity and to classify stimuli with respect to their similarity to predicted patterns. More recent work has characterized semantic and visual dimensions of object representation from brain data and has shown that these dimensions can be used to predict significantly above chance activity to novel objects, both within and across subjects (Just, Cherkassky, Aryal, & Mitchell, 2010). If so, general pattern recognition systems could potentially be developed to decode arbitrary kinds of mental content.

Beyond objects An understanding of how brains encode meaning could significantly boost decoding ability. Recent work suggests that there are smooth gradients of semantic representation in the brain that are shared across individuals. Using fMRI data from natural movie viewing, Huth, Nishimoto, Vu, and Gallant (2012) computed semantic selectivity indices for individual voxels. On this basis they constructed a semantic space and characterized several semantic dimensions that varied smoothly across cortex. Similar methods were used to show semantic gradients across cortex with natural speech as input (Huth, Heer, Griffiths, Theunissen, & Gallant, 2016). While the implications of this work are unclear, evidence of systematic variation across cortex supports the possibility of generative mappings. Data-driven approaches thus may allow a more fine-grained characterization of semantic content from neuroimaging data than would be possible by brute correlation methods. If so, arbitrary semantic content should also be recoverable to some degree from imaging data. Along those lines researchers used hierarchical models of semantic relatedness to show that some of the semantic content of naturally viewed movies is decodable from fMRI data.

Cognitive neuroscientific work in a variety of areas emphasizes the importance of understanding the representation of natural stimuli in context. Both language and thought combine representations in context-sensitive ways. How does the brain represent grammatical distinctions? We know that representations of words in context differ from words presented in isolation (Just, Wang, & Cherkassky, 2017). Frankland and Greene (2015) investigated agent-patient relationships and showed that areas of the left lateral superior temporal sulcus are sensitive to agent and patient relationships and distinguish semantically different sentences such as "The baby kicked the grandfather" and "The grandfather kicked the baby." Other work suggests that the angular gyrus may be preferentially involved in verb representation (Boylan, Trueswell, & Thompson-Schill, 2015). Just, Wang, and Cherkassky (2017) used semantic models to explore the possibility of predicting activity to novel protosentences on the basis of seeing words in context, demonstrating the possibility of extracting the components of a proposition from fMRI data. However, here, too, the set of possibilities was quite constrained, with only 36 sentences to choose from.

Work in crosslinguistic encoding has suggested that the semantics of sentences are similarly encoded in individuals across languages. Using semantic features to characterize sentences in L1, researchers were able to rank order sentences in L2 well above chance, on the basis of predicted fMRI signal (Yang, Wang, Bailer, Cherkassky, & Just, 2017b). This is possible even when languages are dissimilar, such as English, Portuguese, and Mandarin. Yang and colleagues showed that decoding was improved when classifiers were trained on sentences in two different languages (Yang, Wang, Bailer, Cherkassky, & Just, 2017a). This suggests that semantic representations in the brain informative at the level of resolution of fMRI are language-independent. If so, we should expect advances in mind reading to generalize universally, rather than only across individuals within a linguistic community.

Identifying memories and lie detection One long-standing goal of memory research has been to understand the way in which memories are encoded in the brain. Such knowledge could potentially be leveraged into a method of decoding memory content, or assessing the veridicality of memory-like signatures. However, despite ongoing advances in understanding memory processes, little progress has been made in understanding content-specific aspects of encoding and retrieval.

With regard to aspects of memory neuroimaging relevant to mind reading, most of the work has focused on proof of possibility. Some memory-related information is accessible by MVPA in temporal lobe structures (Chadwick et al., 2010), but current work does not indicate that anything like the full range of information needed for classifying or reconstructing a remembered stimulus is recoverable from imaging data. Other work shows that fMRI data can indicate how subjectively familiar stimuli are but not whether a putative memory is accurate when subjective status is controlled for (Rissman, Greely, & Wagner, 2010). In addition, it has revealed that decoding is poor in an implicit memory task. Thus, while fMRI might be able to distinguish subjective memory states in forensic contexts, its value will be highly limited in noncooperative contexts or when applied to questions of objective veridicality.

Lie detection, and, more generally, the detection of deception, is related to memory processes, and these techniques have perhaps raised the greatest concerns about privacy in the public sphere. An enormous amount of effort has been directed to adapting neuroimaging techniques to distinguish lying from truth telling. While such measures are relatively effective at distinguishing lies from true responses in the experimental contexts in which they are developed, there are deep problems with external and ecological validity and little insight into content-related aspects that could elevate them into true mind-reading experiments. It is also doubtful that the methods so far developed are robust in the face of countermeasures or otherwise noncompliant subjects. For a critical review of neuroimaging for lie detection, see Farah, Hutchinson, Phelps, and Wagner (2014); Roskies (2015); and Wolpe, Foster, and Langleben (2005).

What Might Neuroscience Be Able to Discern in the Future?

Until very recently we did not understand how the brain represents stimuli beyond early sensory areas. That is now changing, with major advances in understanding the computations underlying specific domains. Face perception provides a good example of a recent advance. Chang and Tsao (2017) demonstrate that single cells in face-selective areas in monkey cortex respond to human faces as projections in a linear multidimensional space onto one of the axes of this space. The firing rates of ensembles of cells thus allowed decoding of and reconstruction of individual face stimuli. If this is representative of a common strategy for neural coding in the brain, we might expect a much deeper understanding of neural representation in the future, such that knowledge of neural firing may permit better encoding models and the reconstruction of mental content viewed more broadly.

However, our understanding of face representation at the single-unit level also allows us to draw some lessons about the limitations of fMRI for mind reading. MVPA of fMRI data of monkey face perception was unable to distinguish face identity in brain areas in which identity was encoded in single-unit data, although face viewpoint was decodable from both methods (Dubois, Berker, & Tsao, 2015). Thus, we can expect that even with an understanding of the neural code, detailed information carried by neural populations may not be fully recoverable by neuroimaging methods, limiting the kind of content that can be discerned from imaging studies.

Big data approaches and advances in machine learning may also illuminate neural coding. For example, a deep neural network (DNN) trained for action recognition has been used to predict brain responses to natural movies, suggesting similarity in representations of DNN layers and dorsal stream areas (Güçlü & van Gerven, 2017). The work also demonstrated that a common representational space underlies neural responses across individuals.

It is likely that with improved technology and methods, our ability to reconstruct the contents of the mental will continue to improve. These efforts at reconstruction will extend to areas so far largely ignored. For example, although little data so far shows that inner speech can be decoded, there is evidence that inner speech leads to the activation of brain structures involved in auditory processing (Shergill et al., 2002) and speech production (Marvel & Desmond, 2012) and thus some promise that at least some aspects of the inner narrative could be decoded. However, no evidence currently exists to suggest that anything like the stream of consciousness of inner speech will be recoverable from brain data.

The experiments described above exhibit both the remarkable progress in neuroimaging in discriminating aspects of mental content and the significant limitations it faces in succeeding as a general mind-reading methodology. Problems remain with spatial and temporal resolution and the mapping to fine-grained mental content, with context effects, with individual variability, and with discriminating between closely related contents. These limitations are likely to persist. It is also doubtful that these methods will work for discerning information that subjects are deliberately withholding.

Ethical and Legal Implications

The value of mental privacy Liberty is a centerpiece of American democracy, and privacy ensures a certain kind of freedom: freedom from the surveillance and intervention of unwanted parties, including the state. As James Moor (1990) has noted, "The concept of privacy seems so obvious, so basic, and so much a part of American values, that there may seem to be little room for any philosophical misgivings about it" (p. 69). However, there is substantial philosophical controversy about both the nature of and the justification for privacy as a right or value. Among the open questions is the value of privacy of the mental. It seems this question has not been a topic of much explicit theorizing, perhaps because, until recently, little seemed to threaten it.

The unsettled nature of the philosophical discourse about privacy is mirrored by the unsettled role of privacy rights in the law. Intimations of the importance of privacy are found in the U.S. Constitution, but nowhere does the Constitution explicitly confer a right to privacy on citizens. The U.S. Supreme Court has variously interpreted the First, Fourth, Fifth, Ninth, and Fourteenth Amendments as grounding protections for privacy—most notably, in its rulings about substantive due process. Considerable disagreement exists about the nature and scope of the privileges to be protected. Both the Fourth and the Fifth Amendments are suggestive of a right to privacy that extends to the mind, but the scope of the right is unclear, as is whether it extends to neuroimaging (Farahany, 2012a, 2012b; Fox, 2009).

Rulings from a series of Supreme Court cases do little to clarify the scope of mental privacy rights. The court affirms a distinction between testimony and physical evidence and holds that the Fifth Amendment protects testimony but not physical evidence (*Schmerber v. California*, 1966). Defendants can thus be compelled to produce physical evidence (such as blood, DNA, and fingerprints) that could be incriminating, but they cannot be compelled to testify (to take communicative action) against themselves. However, neuroimaging techniques call into question the tenability of a physical/testimonial distinction (Farahany, 2012a; Pardo, 2008). Fourth Amendment cases regarding information for which warrants are or are not required distinguish between information that encompasses content (such as the body of an email) and noncontent information (such as the header and address to which the email is sent; *Smith v. Maryland*, 1979). Commentators have argued that this distinction is also untenable given today's technologies (Farahany, 2012b).

No cases have yet ruled on the question of whether neuroimaging infringes on legally protected mental privacy. To date, the main rationales for excluding neuroimaging data as legal evidence have been in the context of lie detection. *United States v. Semrau* (2012) concluded that neuroimaging did not meet the standards for scientific evidence. However, the judge explicitly left the door open for future use of fMRI lie detection in court. The world of social media and changing cultural norms will also have a significant impact on future legal protections for privacy, including mental privacy, for legal doctrines regarding privacy are based on notions of "reasonable expectations" and on cultural standards.

Although techniques for mind reading are improving, it seems unlikely that fine-grained propositional content will be able to be read from brain images any time soon. However, given the lack of clear philosophical grounding and the unsettled nature of the law in this area, it behooves us to devote more effort to articulating the nature and justification for our intuitions about privacy rights and how those intuitions relate to mental privacy. In the law, the ability of the government to forcibly search or seize private property or information is currently governed by a doctrine that requires balancing state interests against reasonable expectations. However, because which expectations are deemed reasonable is culturally dependent, the emergence of technologies that encourage the broad dissemination of personal data threaten the very cultural expectations under which privacy has been enshrined as an inalienable right.

REFERENCES

Azevedo, R. T., Macaluso, E., Avenanti, A., Santangelo, V., Cazzato, V., & Aglioti, S. M. (2012). Their pain is not our pain: Brain and autonomic correlates of empathic resonance with the pain of same and different race individuals. *Human Brain Mapping, 34*, 3168–3181.

Boylan, C., Trueswell, J. C., & Thompson-Schill, S. L. (2015). Compositionality and the angular gyrus: A multi-voxel similarity analysis of the semantic composition of nouns and verbs. *Neuropsychologia, 78*, 130–141. https://doi.org/10.1016/j.neuropsychologia.2015.10.007

Carre, A., Gierski, F., Lemogne, C., Tran, E., Raucher-Chene, D., Bera-Potelle, C., … Limosin, F. (2013). Linear association between social anxiety symptoms and neural activations to angry faces: From subclinical to clinical levels. *Social Cognitive and Affective Neuroscience, 9*(6), 880–886.

Cetin, M. S., Houck, J. M., Rashid, B., Agacoglu, O., Stephen, J. M., Sui, J., … Calhoun, V. D. (2016). Multimodal classification of schizophrenia patients with MEG and fMRI data using static and dynamic connectivity measures. *Frontiers in Neuroscience, 10*. https://doi.org/10.3389/fnins.2016.00466

Chadwick, M., Hassabis, D., Weiskopf, N., & Maguire, E. A. (2010). Decoding individual episodic memory traces in the human hippocampus. *Current Biology, 20*, 544–547.

Chang, L., & Tsao, D. Y. (2017). The code for facial identity in the primate brain. *Cell, 169*(6), 1013–1028.e14. https://doi.org/10.1016/j.cell.2017.05.011

Conroy, B. R., Singer, B. D., Guntupalli, J. S., Ramadge, P. J., & Haxby, J. V. (2013). Inter-subject alignment of human cortical anatomy using functional connectivity. *NeuroImage, 81*, 400–411. https://doi.org/10.1016/j.neuroimage.2013.05.009

Dubois, J., de Berker, A. O., & Tsao, D. Y. (2015). Single-unit recordings in the macaque face patch system reveal limitations of fMRI MVPA. *Journal of Neuroscience, 35*(6), 2791–2802. https://doi.org/10.1523/jneurosci.4037-14.2015

Ebisch, S. J. H., Gallese, V., Salone, A., Martinotti, G., di Iorio, G., Mantini, D., … Northoff, G. (2018). Disrupted relationship between "resting state" connectivity and task-evoked activity during social perception in schizophrenia. *Schizophrenia Research, 193*, 370–376. https://doi.org/10.1016/j.schres.2017.07.020

Farah, M. J., Hutchinson, J. B., Phelps, E. A., & Wagner, A. D. (2014). Functional MRI-based lie detection: Scientific and societal challenges. *Nature Reviews Neuroscience, 15*(2), 123–131. https://doi.org/10.1038/nrn3665

Farah, M. J., Smith, M. E., Gawuga, C., Lindsell, D., & Foster, D. (2008). Brain imaging and brain privacy: A realistic concern? *Journal of Cognitive Neuroscience, 21*, 119–127.

Farahany, N. (2012a). Incriminating thoughts. *Stanford Law Review, 64*, 351–408.

Farahany, N. (2012b). Searching secrets. *University of Pennsylvania Law Review, 160*, 1239–1308.

Formisano, E., De Martino, F., Bonte, M., & Goebel, R. (2008). "Who" is saying "what"? Brain-based decoding of human voice and speech. *Science, 322,* 970–973.

Fox, D. (2009). The right to silence protects mental control. *Akron Law Review, 42,* 763.

Frankland, S. M., & Greene, J. D. (2015). An architecture for encoding sentence meaning in left mid-superior temporal cortex. *Proceedings of the National Academy of Sciences, 112*(37), 11732–11737. https://doi.org/10.1073/pnas.1421236112

Güçlü, U., & van Gerven, M. A. J. (2017). Increasingly complex representations of natural movies across the dorsal stream are shared between subjects. *NeuroImage, 145,* 329–336. https://doi.org/10.1016/j.neuroimage.2015.12.036

Harle, K. M., Chang, L. J., van 't Wout, M., & Sanfey, A. G. (2012). The neural mechanisms of affect infusion in social economic decision-making: A mediating role of the anterior insula. *Neuroimage, 61,* 32–40.

Hasson, U., Landesman, O., Knappmeyer, B., Vallines, U., Rubin, N., & Heeger, D. J. (2008). Neurocinematics: The neuroscience of film. *Projections, 2,* 1–26.

Hasson, U., Nir, Y., Levy, I., Fuhrmann, G., & Malach, R. (2004). Intersubject synchronization of cortical activity during natural vision. *Science, 303,* 1634–1640.

Haxby, J. V., Gobbini, M. I., Ishai, A., Schouten J. L., & Pietrini P. (2001). Distributed and overlapping representations of faces and objects in visual cortex. *Science, 293,* 2425–2430.

Haxby, J. V., Guntupalli, J. S., Connolly, A. C., Halchenko, Y. O., Conroy, B. R., Gobbini, M. I., ... Ramadge, P. J. (2011). A common, high-dimensional model of the representational space in human ventral temporal cortex. *Neuron, 72*(2), 404–416. https://doi.org/10.1016/j.neuron.2011.08.026

Huth, A. G., Heer, W. A., de Griffiths, T. L., Theunissen, F. E., & Gallant, J. L. (2016). Natural speech reveals the semantic maps that tile human cerebral cortex. *Nature, 532*(7600), 453–458. https://doi.org/10.1038/nature17637

Huth, A. G., Nishimoto, S., Vu, A. T., & Gallant, J. L. (2012). A continuous semantic space describes the representation of thousands of object and action categories across the human brain. *Neuron, 76*(6), 1210–1224. https://doi.org/10.1016/j.neuron.2012.10.014

Just, M. A., Cherkassky, V. L., Aryal, S., & Mitchell, T. M. (2010). A neurosemantic theory of concrete noun representation based on the underlying brain codes. *PLoS One, 5*(1), e8622. https://doi.org/10.1371/journal.pone.0008622

Just, M. A., Wang, J., & Cherkassky, V. L. (2017). Neural representations of the concepts in simple sentences: Concept activation prediction and context effects. *NeuroImage, 157,* 511–520. https://doi.org/10.1016/j.neuroimage.2017.06.033

Katsumi, Y., & Dolcos, S. (2018). Neural correlates of racial ingroup bias in observing computer-animated social encounters. *Frontiers in Human Neuroscience, 11.* https://doi.org/10.3389/fnhum.2017.00632

Kay, K. N., Naselaris, T., Prenger, R. J., & Gallant, J. L. (2008). Identifying natural images from human brain activity. *Nature, 452,* 352–355.

Kriegeskorte, N., Mur, M., & Bandettini, P. (2008). Representational similarity analysis—Connecting the branches of systems neuroscience. *Frontiers in Systems Neuroscience, 2.* https://doi.org/10.3389/neuro.06.004.2008

Liu, Y., Lin, W., Xu, P., Zhang, D., & Luo, Y. (2015). Neural basis of disgust perception in racial prejudice. *Human Brain Mapping, 36*(12), 5275–5286. https://doi.org/10.1002/hbm.23010

Malpas, C. B., Genc, S., Saling, M. M., Velakoulis, D., Desmond, P. M., & O'Brien, T. J. (2016). MRI correlates of general intelligence in neurotypical adults. *Journal of Clinical Neuroscience, 24,* 128–134. https://doi.org/10.1016/j.jocn.2015.07.012

Marvel, C. L., & Desmond, J. E. (2012). From storage to manipulation: How the neural correlates of verbal working memory reflect varying demands on inner speech. *Brain and Language, 120,* 42–51.

Mitchell, T. M., Shinkareva, S. V., Carlson, A., Chang, K.-M., Malave, V. L., & Just, M. A. (2008). Predicting human brain activity associated with the meanings of nouns. *Science, 320,* 1191–1195.

Moor, J. (1990). The ethics of privacy protection. *Library Trends, 39.*

Mueller, S., Keeser, D., Reiser, M. F., Teipel, S., & Meindl, T. (2012). Functional and structural MR imaging in neuropsychiatric disorders, part 2: Application in schizophrenia and autism. *American Journal of Neuroradiology, 33,* 2033–2037.

Naselaris, T., Olman, C. A., Stansbury, D. E., Ugurbil, K., & Gallant, J. L. (2015). A voxel-wise encoding model for early visual areas decodes mental images of remembered scenes. *NeuroImage, 105,* 215–228. https://doi.org/10.1016/j.neuroimage.2014.10.018

Naselaris, T., Prenger, R. J., Kay, K. N., Oliver, M., & Gallant, J. L. (2009). Bayesian reconstruction of natural images from human brain activity. *Neuron, 63,* 902–915.

Nishimoto, S., Vu, A. T., Naselaris, T., Benjamini, Y., Yu, B., & Gallant, J. L. (2011). Reconstructing visual experiences from brain activity evoked by natural movies. *Current Biology, 21,* 1641–1646.

O'Craven, K. M., & Kanwisher, N. (2000). Mental imagery of faces and places activates corresponding stimulus-specific brain regions. *Journal of Cognitive Neuroscience, 12,* 1013–1023.

Palko v. Connecticut, 302 U.S. 319 (1937)

Pardo, M. (2008). Self-incrimination and the epistemology of testimony. *Cardozo Law Review, 30,* 1023–1046.

Patel, V. (2018). A framework for secure and decentralized sharing of medical imaging data via blockchain consensus. *Health Informatics Journal.* https://doi.org/10.1177/1460458218769699

Paulsen, D. J., Carter, R. M., Platt, M. L., Huettel, S. A., & Brannon, E. M. (2011). Neurocognitive development of risk aversion from early childhood to adulthood. *Frontiers in Human Neuroscience, 5,* 178.

Polyn, S. M., Natu, V. S., Cohen, J. D., & Norman, K. A. (2005). Category-specific cortical activity precedes retrieval during memory search. *Science, 310,* 1963–1966.

Reverberi, C., Görgen, K., & Haynes, J.-D. (2011). Compositionality of rule representations in human prefrontal cortex. *Cerebral Cortex, 22*(6), 1237–1246.

Rissman, J., Greely, H. T., & Wagner, A. D. (2010). Detecting individual memories through the neural decoding of memory states and past experience. *Proceedings of the National Academy of Sciences of the United States of America, 107,* 9849–9854.

Roskies, A. L. (2014). Mindreading and privacy. In M. S. Gazzaniga& G. R. Mangun (Eds.), *The Cognitive Neurosciences* (5th ed.). Cambridge, MA: MIT Press.

Roskies, A. L. (2015). Mind reading, lie detection, and privacy. In J. Clausen & N. Levy (Eds.), *Handbook of neuroethics* (pp. 679–695). Dordrecht, Netherlands: Springer. http://link.springer.com/referenceworkentry/10.1007/978-94-007-4707-4_123.

Salvatore, C., Cerasa, A., Battista, P., Gilardi, M. C., Quattrone, A., & Castiglioni, I. (2015). Magnetic resonance imaging biomarkers for the early diagnosis of Alzheimer's disease: A machine learning approach. *Frontiers in Neuroscience, 9*. https://doi.org/10.3389/fnins.2015.00307

Santoro, R., Moerel, M., Martino, F. D., Valente, G., Ugurbil, K., Yacoub, E., & Formisano, E. (2017). Reconstructing the spectrotemporal modulations of real-life sounds from fMRI response patterns. *Proceedings of the National Academy of Sciences, 114*(18), 4799–4804. https://doi.org/10.1073/pnas.1617622114

Schmerber v. California, 384 US 757 (1966).

Scott, N. A., Murphy, T. H., & Illes, J. (2012). Incidental findings in neuroimaging research: A framework for anticipating the next frontier. *Journal of Empirical Research on Human Research Ethics, 7*, 53–57.

Shergill, S. S., Brammer, M. J., Fukuda, R., Bullmore, E., Amaro, E., Murray, R. M., & McGuire, P. K. (2002). Modulation of activity in temporal cortex during generation of inner speech. *Human Brain Mapping, 16*, 219–227.

Shinkareva, S. V., Mason, R. A., Malave, V. L., Wang, W., Mitchell, T. M., & Just, M. A. (2008). Using FMRI brain activation to identify cognitive states associated with perception of tools and dwellings. *PloS One, 3*, e1394.

Simanova, I., Hagoort, P., Oostenveld, R., & van Gerven, M. A. J. (2012). Modality-independent decoding of semantic information from the human brain. *Cerebral Cortex, 24*, 426–434.

Smith v. Maryland, 442 U.S. 735 (1979).

Stanley, D. A., Sokol-Hessner, P., Fareri, D. S., Perino, M. T., Delgado, M. R., Banaji, M. R., & Phelps, E. A. (2012). Race and reputation: Perceived racial group trustworthiness influences the neural correlates of trust decisions. *Philosophical Transactions of the Royal Society of London B: Biological Sciences, 367*, 744–753.

Tang, Y., Wang, L., Cao, F., & Tan, L. (2012). Identify schizophrenia using resting-state functional connectivity: An exploratory research and analysis. *Biomedical Engineering Online, 11*, 50.

Teipel, S. J., Grothe, M., Lista, S., Toschi, N., Garaci, F. G., & Hampel, H. (2013). Relevance of magnetic resonance imaging for early detection and diagnosis of Alzheimer disease. *Medical Clinics of North America, 97*, 399–424.

Thirion, B., Duchesnay, E., Hubbard, E., Dubois, J., Poline, J.-B., Lebihan, D., & Dehaene, S. (2006). Inverse retinotopy: Inferring the visual content of images from brain activation patterns. *Neuroimage, 33*, 1104–1116.

Tong, F., Nakayama, K., Vaughan, J. T., & Kanwisher, N. (1998). Binocular rivalry and visual awareness in human extrastriate cortex. *Neuron, 21*, 753–759.

United States v. Semrau, 693 F.3d 510 (2012).

Van Bavel, J. J., Packer, D. J., & Cunningham, W. A. (2008). The neural substrates of in-group bias: A functional magnetic resonance imaging investigation. *Psychological Science, 19*, 1131–1139.

Whalley, H. C., Sussmann, J. E., Romaniuk, L., Stewart, T., Papmeyer, M., Sprooten, E., & McIntosh, A. M. (2013). Prediction of depression in individuals at high familial risk of mood disorders using functional magnetic resonance imaging. *PLoS One, 8*, e57357.

Wolpe, P. R., Foster, K. R., & Langleben, D. D. (2005). Emerging neurotechnologies for lie-detection: Promises and perils. *American Journal of Bioethics, 5*, 39.

Yang, Y., Wang, J., Bailer, C., Cherkassky, V., & Just, M. A. (2017a). Commonalities and differences in the neural representations of English, Portuguese, and Mandarin sentences: When knowledge of the brain-language mappings for two languages is better than one. *Brain and Language, 175*, 77–85. https://doi.org/10.1016/j.bandl.2017.09.007

Yang, Y., Wang, J., Bailer, C., Cherkassky, V., & Just, M. A. (2017b). Commonality of neural representations of sentences across languages: Predicting brain activation during Portuguese sentence comprehension using an English-based model of brain function. *NeuroImage, 146*, 658–666. https://doi.org/10.1016/j.neuroimage.2016.10.029

93 Pharmacological Cognitive Enhancement: Implications for Ethics and Society

GEORGE SAVULICH AND BARBARA J. SAHAKIAN

ABSTRACT Cognitive abilities are becoming more important for successful work performance in a competitive global environment. Increasing demands on everyday cognitive processes such as attention, memory, and higher-order executive functions (e.g., planning, decision-making, and problem-solving) have led to the rise in use of "smart drugs" by healthy people. Pharmacological cognitive enhancement has many advantages for the individual and society, including the potential for better productivity and higher earnings, less fatigue, and a reduced number of accidents. Cognition-enhancing drugs have also been shown to improve functioning and quality of life in patients with neuropsychiatric disorders, thereby reducing the overall economic costs of disease burden. However, the benefits of cognitive enhancement must be considered alongside its associated risks and ethical concerns, particularly in healthy people. These include academic cheating, peer and parental coercion, the globalization of attention deficit hyperactivity disorder, the sharing and selling of medication between students, increasing societal disparity, and the lack of randomized, placebo-controlled trials confirming the safety and efficacy of smart drug use in healthy people. As a society, we need to consider which cognition-enhancing drugs are acceptable for which groups (e.g., military, doctors) and under what conditions (e.g., war, shift work) we wish to improve and flourish.

"Healthy" cognition throughout the life span is critical for everyday functioning, particularly given that we live in a knowledge economy (Beddington et al., 2008). Well-established methods for improving cognition include lifelong learning and education, physical exercise, and lifestyle factors such as diet, sleep, and social interaction (Academy of Medical Sciences, 2012; Frith et al., 2011). None of these methods raise serious, if any, ethical concerns (Maslen, Faulmuller, & Savulescu, 2011). Other nonpharmacological techniques for cognitive enhancement include transcranial magnetic stimulation (TMS), transcranial direct current stimulation (tDCS), and targeted cognitive training (Brühl & Sahakian, 2016; Sahakian et al., 2015; Savulich, Piercy, Brühl, et al., 2017), all of which aim to improve cognition through the stimulation of neural circuits. Drugs with cognition-enhancing potential (so-called smart drugs), such as methylphenidate (Ritalin) and cholinesterase inhibitors, were first developed as treatments for cognitive dysfunction in patients with neuropsychiatric disorders. However, growing evidence indicates the increasing use of smart drugs by healthy people for "lifestyle" rather than medical reasons (d'Angelo, Savulich, & Sahakian, 2017), thus raising ethical and societal concerns surrounding human enhancement.

Recently, a survey with more than 100,000 responders from 15 countries on the use of drugs for the purpose of cognitive enhancement, the largest of its kind, was made public (Maier, Ferris, & Winstock, 2018). Prescription and nontreatment stimulants and modafinil use increased 180% from 2015 to 2017, with rates rising in European countries and remaining consistently high in the United States and Canada. For example, rates rose from 3% to 16% in France and from 5% to 23% in the United Kingdom. In a previous survey of 2,000 students in the United Kingdom, 1 in 10 had reported using modafinil or the peptide nootropic Noopept to help them study, with a quarter of respondents acknowledging that they would consider using them again (The Student Room, 2016). Another survey reported that one in five respondents had used smart drugs for enhancing concentration, memory, or focus (Maher, 2008). Somewhat alarmingly, 34% of respondents had reported obtaining the drug from the Internet. Similarly, 16% of college students and 8% of undergraduate students in the United States admitted to illegally obtaining prescription stimulants (Teter, Falcone, Cranford, Boyd, & McCabe, 2010). With respect to prescription medications, the Care Quality Commission reported a 56% rise in prescriptions of methylphenidate in the United Kingdom from 2007 to 2013. Increases in both the nonmedical and prescription use of substances for cognitive enhancement purposes, even if obtained illegally or through the unsafe purchasing of drugs online from unregulated manufacturing sources, point toward a shift both in the role of drugs in society and in our attitudes toward taking them.

Mechanisms of Action

Pharmacological cognitive enhancement primarily involves the drugs methylphenidate (Ritalin), atomoxetine (Strattera), modafinil (Provigil), and amphetamine. Methylphenidate and dextroamphetamine (Adderall) are potent stimulants of the central nervous system that increase the synaptic concentration of dopamine and noradrenaline by blocking their reuptake in the prefrontal cortex and the cortical and subcortical regions to which they project (Wilens, 2006). Modafinil is a wakefulness-promoting agent used in the treatment of narcolepsy and sleep-related disorders. Its precise mechanism of action in regard to its cognition-enhancing effects is not clear but is known to activate the dopaminergic, glutamatergic, noradrenergic, and serotonergic systems in several regions of the brain, including the prefrontal cortex, hippocampus, hypothalamus, and basal ganglia (Scoriels, Jones, & Sahakian, 2013; Stahl, 2008). Atomoxetine is a relatively selective noradrenaline reuptake inhibitor that blocks the presynaptic norepinephrine transporter (Graf et al., 2011). The classic stimulants, amphetamine and methylphenidate, have abuse potential, whereas atomoxetine does not, and as yet there is no evidence of abuse potential for modafinil at the dose used for enhancing cognition (200 mg; Porsdam Mann & Sahakian, 2015).

Pharmacological Cognitive Enhancement in Healthy People: Motivations and Prevalence of "Smart Drug" Use

Reasons for cognitive enhancement in healthy people are diverse but are mainly driven by two key factors: (1) an increasingly competitive global environment and (2) the desire to maximize performance at work or while in college. Anecdotal evidence points to several benefits of pharmacological cognitive enhancement, mostly using Ritalin, Adderall, and modafinil, including amplified alertness and focus, faster reaction times, feelings of greater possibilities, getting "into the flow," fewer injuries, and more positive well-being. Improving performance affected by a lack of sleep, shift work (longer hours), and jet lag is also cited as a top motivator (Brühl & Sahakian, 2016).

A German survey with 5,017 responders found that those using cognition-enhancing drugs were more worried about their jobs, felt they were already working at their upper limits, or were required to perform activities in which even small mistakes could have serious consequences (Kordt, 2015). They also cited work-related stress (e.g., giving a presentation, completing work successfully within the allotted time, negotiation: 41%), ease of work (35%), attaining goals more easily (32%), more energy and better mood (27%), getting the "competitive edge" (12%), an inability to work otherwise (25%), and fewer requirements for sleep (9%). Smart drug use is also increasingly popular among students wishing to excel in competitive situations and during exam preparation and sessions. Estimates of the prevalence of use vary widely but suggest that somewhere between 13% to 38% of students have used smart drugs to aid memory and concentration (Nicholson, Mayho, & Sharp, 2015; Singh, Bard, & Jackson, 2014; Smith & Farah, 2011). A web-based survey of 2,877 students found that only 65.3% of respondents had decided *not to take* cognition-enhancing drugs (Sattler, Mehlkp, Graeff, & Sauer, 2014).

Although much focus has been given to student populations, many other groups have reportedly used cognition-enhancing drugs, including professional athletes, the military, and the music, entertainment, and tech industries. From the military use of mixed amphetamine salts during World War II to "microdosing" small amounts of psychedelic drugs (e.g., minute quantities of LSD, psilocybin, or mescaline) in Silicon Valley and elsewhere, healthy people have been using psychoactive substances for enhancing not only cognitive processes, such as cognitive flexibility and alertness, but also for serotonin-mediating effects on creativity, euphoria, and well-being (see Sahakian, d'Angelo, & Savulich, 2017). With the rise in the number of novel psychoactive substances surpassing 500 over the last decade (European Monitoring Centre for Drugs and Drug Addiction, 2016), banned hallucinogenic drugs are reemerging for their psychoactive and in some cases antidepressant effects (e.g., ketamine; d'Angelo, Savulich, & Sahakian, 2017).

Effects of Pharmacological Cognitive Enhancement in Healthy People

While the classic stimulants are the most used cognition-enhancing drugs by healthy people in the United States, modafinil is more widely used in the United Kingdom (Maier, Ferris, & Winstock, 2018). In healthy volunteers, methylphenidate has been shown to improve working memory and increase the "efficiency" of the dorsolateral prefrontal cortical network (Elliott et al., 1997; Mehta et al., 2000). Methylphenidate has also been shown to improve sustained attention in those with lower baseline performance (del Campo et al., 2013). Both methylphenidate and amphetamine have been shown to improve inhibitory control in healthy volunteers, but effects are likely to be strongest in individuals with lower baseline performance. Also in healthy volunteers, modafinil has been shown to improve planning and response inhibition (Turner et al., 2003). Of particular interest, modafinil has been shown to improve working memory, planning, decision-making, and cognitive flexibility in sleep-deprived doctors (Sugden, Housden, Aggarwal, Sahakian, & Darzi, 2012). Modafinil has also been shown to improve inhibitory control, working memory, and higher-order executive functions in non–sleep deprived individuals (Battleday & Brem, 2015; Müller et al., 2013; Turner et al., 2003). In chess players, modafinil and methylphenidate enhanced performance in 2,876 games compared to placebo when controlling for game duration and the number of games lost (Franke et al., 2017). In addition to improvements in "cold," or nonemotional, cognition, modafinil has also been shown to improve "hot," or social and emotional cognition, such as the processing of emotional faces (Scoriels et al., 2011). Finally, atomoxetine has been shown to improve response inhibition but not sustained attention or working memory in healthy volunteers, demonstrating more selective effects (Chamberlain et al., 2007).

Thus far, the evidence to date suggests modest effects of cognition-enhancing drugs in healthy people, with modafinil improving higher-order executive functions like planning and decision-making and also mood and methylphenidate and amphetamine improving inhibitory control and memory processes. Nevertheless, not all studies have demonstrated positive effects (Ilieva, Boland, & Farah, 2013), although this might reflect ceiling effects of tests or baseline levels of participants' performance. Furthermore, it has been suggested that some benefits of enhancement are subjective or perceived (Ilieva & Farah, 2013; Vrecko, 2013). However, drugs such as modafinil and methylphenidate have shown at least two separate effects: one as a cognitive enhancer and another on motivational processes. Modafinil in particular has been shown to improve several cognitive tests of planning and working memory, as well as task-related motivation in healthy volunteers (Müller et al., 2013).

Ethical and Safety Concerns

Neuroethics is the study of the ethical, legal, and social questions that arise when scientific findings about the brain are carried out in medical practice, legal interpretations, and health and social policy (Marcus, 2002). In the case of smart drugs, cognitive enhancement can refer to improvement of a cognitive function relative to its previous level or beyond its existing point (Maslen, Faulmuller, & Savulescu, 2011). For example, would taking a cognition-enhancing drug in order to counteract the effects of jet lag or sleep deprivation constitute restoration or enhancement? Similarly, if older adults wish to perform at their earlier peak of cognitive abilities—for example, when they were in their twenties—would this be considered restoration or enhancement (Sahakian et al., 2015)? For patients with neuropsychiatric disorders, impairments in cognitive functions such as attention and episodic memory are clear, and pharmacological cognitive enhancement is used in their treatment ("restoration"). However, in people with subjective cognitive impairment in the absence of a recognizable medical disorder and in healthy people wishing to optimize their already existing cognitive abilities, the use of cognition-enhancing drugs raises both advantages and ethical concerns. From a societal perspective, enhancing cognition could lead to better performance at school and work, which in turn could lead to more productivity and higher earnings. Enhancing cognition would also be particularly advantageous for jobs that require adaptive learning or attentional shifting under high levels of risk or pressure (e.g., surgeons, air traffic controllers, stock traders; Sahakian & Morein-Zamir, 2007).

Despite the benefits of cognitive enhancement, a growing number of societal and ethical concerns have been raised. Concerns around the safety and dangers of using drugs for an unapproved indication remain highly problematic. For example, around 90% of prescriptions for modafinil are being used off-label (Vastag, 2004). Methylphenidate and amphetamine are both Schedule II controlled substances in the United States, indicating high abuse potential. Yet the regulation of these drugs remains difficult given the increase in prescriptions for ADHD in young adults and children, the sharing and selling of medications between students, and the ability to purchase drugs online. In the case of *nootropics*, the umbrella term for drugs, supplements, and other substances purporting cognition-enhancing potential, combinations of "stacks" (individual compounds with claimed benefits when given in combination) are usually sold via the Internet with unknown safety and manufacturing regulations. Often marketed for enhancing a specific cognitive function on an "as needed" basis, stacks are not U.S. Food and Drug Administration (FDA)-approved for this purpose, but the individual compounds might be dietary supplements. Furthermore, anecdotal experiences of nootropics, supplements, and microdosing are largely discussed on Internet forums and social media, which, although prompting open discussion, could also lead to the anonymous misrepresentation of their effects and harms.

Whereas the safety and efficacy of drugs used for the treatment of cognitive dysfunction are regulated and tested using randomized, double-blind, placebo-controlled trials, suppliers of supplements make claims without supporting evidence from rigorous testing.

There are fears of healthy people being coerced into taking cognitive enhancers, either directly by their peers or parents or indirectly through increased workplace competition, particularly in demanding jobs. Concerns of students using smart drugs during exam time has led to some universities banning their use as a form of cheating if not prescribed as a form of a treatment by a doctor. There is also the potential for abuse and dependence, particularly for smart drugs with stimulant effects. Another concern is the exacerbation of societal inequality, with access to drugs depending on having the money to purchase them. It is also possible that attitudes toward drug taking for cognitive enhancement may become normalized at the societal level, with fears that self-improvement through nonpharmacological means will no longer be valued. Finally, concerns of "overenhancement" have been raised, with the suggestion that we run the risk of creating a homogenous society, in which the perception of ourselves could drastically change so that we feel unable to take credit for our achievements and virtues, such as motivation and hard work, become outdated or undervalued.

Overall, the long-term safety, side effects, and efficacy of smart drugs in healthy people remain unknown, particularly on the developing brain, given the lack of large-scale randomized, placebo-controlled trials. With respect to physical health, some negative effects of smart drugs have been reported, including dependence, seizures, cardiovascular problems, and exhaustion due to overworking. There has also been anecdotal evidence of some smart drug users pairing stimulant drugs with alcohol or other "downers" to counteract their effects when no longer required. Although survey studies can indicate patterns of drug use in large numbers of people, they are often informal and subjective. As such, well-designed studies measuring pre- and post-drug changes in cognition and behavior using objective and reliable measurements in large sample sizes are urgently needed.

Pharmacological Cognitive "Restoration" in Neuropsychiatric Disorders

Neuropsychiatric disorders are disorders of cognition, motivation, and their interaction (Sahakian, 2014). They are often of neurodevelopmental origin and disproportionally affect the young, with 75% manifesting before the age of 24 (Kessler et al., 2005). Many affected people do not receive a diagnosis and treatment until much later in the course of the illness (e.g., up to 17 years in individuals with obsessive-compulsive symptoms; Hollander et al., 2016), as stress and other environmental influences continue to have an impact on the developing brain. In contrast, others receive a diagnosis very early in development, for example, at the age of 6 years in a third of children receiving a diagnosis of ADHD in the United States, leading to pharmacological intervention possibly becoming normalized from a young age. In addition to direct costs, the indirect costs of neuropsychiatric disorders are also high when considering poor performance at school, absences from work, early retirement, and other losses in earnings and productivity (Gustavsson et al., 2011).

Neuroscience and mental health policy reports now highlight a shift in focus from attempts to treat chronic relapsing mental health disorders to early detection and intervention (Beddington et al., 2008; Insel et al., 2012, 2013; Sahakian, 2014). Patients with impairments in core cognitive domains such as attention, memory, and executive functions have poorer outcomes, limitations in the activities of daily living, and a reduced quality of life (Savulich, O'Brien, & Sahakian, 2019; Savulich, Piercy, Fox, et al., 2017). Cognition is therefore an important indicator of functional and occupational outcomes across a range of disorders and has been increasingly recognized as an unmet target for treatment (Collins et al., 2011). Drugs with cognition-enhancing potential, such as cholinesterase inhibitors (e.g., donepezil [Aricept], galantamine, rivastigmine) and methylphenidate, are used in the treatment of memory and attentional impairments in Alzheimer's disease and ADHD, respectively, in which cognitive symptoms are the main target of treatment. However, drug treatments available for depression and schizophrenia tend to improve mood and sleep rather than cognitive symptoms and, in the case of schizophrenia, may even exacerbate dose-dependent cognitive impairments (Savulich, Mezquida, Atkinson, Bernardo, & Fernandez-Egea, 2018). Cognitive "restoration" would therefore be beneficial even after the successful remission of these more acute symptoms.

Cognitive symptoms can manifest in a range of other disturbances, including attentional biases toward negative stimuli, aberrant learning, dysfunctional reward systems, and dysregulation in top-down cognitive control by the prefrontal cortex (Sahakian & Savulich, 2019; Sahakian, 2014; Sahakian & Morein-Zamir, 2015). Changes in cognition are often the first or primary characteristic of these disorders. Perhaps most notably, neuropathological changes in the hippocampal formation and temporal neocortex underlie the learning and memory deficits first observed in Alzheimer's disease. Yet cognitive symptoms in other disorders may seem less apparent. Older adults with amnestic mild cognitive impairment (MCI), the so-called transitional stage between "healthy" aging and dementia, experience a subtle but noticeable decline in memory. In addition to persistent low mood, cognitive impairments in depression include difficulties in concentration and decision-making. These disorders are further characterized by problems in motivation, which negatively affect goal-directed behavior, thus representing complex barriers to treatment entry and engagement (Savulich, Piercy, Brühl, et al., 2017). Elderly people with MCI, often the very early stage of Alzheimer's disease, not only have problems with episodic memory but may also have problems of reduced motivation (Savulich, Piercy, Fox, et al., 2017). In children and adolescents with ADHD, problems with inattention, hyperactivity, and impulsivity are highly associated with poor academic performance and increased failure to progress through school (Loe & Feldman, 2007).

Effects of Pharmacological Cognitive Restoration in Neuropsychiatric Disorders

Alzheimer's disease Drugs with cognition-enhancing potential through cholinergic mechanisms show modest benefits for patients with amnestic MCI and Alzheimer's disease but are more likely to be effective for ameliorating attentional rather than memory dysfunction (Sahakian et al., 1993). Studies of cholinesterase inhibitors have shown modest benefits for stabilizing cognitive decline, function, behavior, and global change in Alzheimer's disease (Tan et al., 2014), with continued benefits observed in those with moderate to severe cases still taking Aricept (Howard et al., 2012). However, as yet no drug treatments have been able to modify the underlying disease pathology. In later stages of disease progression, the N-methyl-D-aspartate (NMDA) receptor antagonist memantine, which acts on the glutamate system, is frequently used. The development of more effective symptomatic treatments for the episodic-memory problems in patients with Alzheimer's disease is urgently needed.

Parkinson's disease In patients with Parkinson's disease, weak effects have been found on fatigue using modafinil and methylphenidate (Lou et al., 2009; Mendonca, Menezes, & Jog, 2007). Also in Parkinson's disease, atomoxetine has been shown to reduce impulsivity and risk-taking behavior during a gambling task, with higher levels of plasma concentration showing an association with better problem-solving (Kehagia et al., 2014). Atomoxetine has been further shown to enhance stop-related prefrontal cortical activation and frontostriatal connectivity, suggesting candidate loci for pharmacological intervention in Parkinson's disease (Ye et al., 2015). More recently, modafinil has been considered in the treatment of excessive daytime sleepiness in patients with Parkinson's disease (National Institute for Health and Care Excellence, 2017).

Attention deficit hyperactivity disorder Procognitive effects of methylphenidate have been reported in 60%–70% of adults with ADHD (Spencer & Biederman, 2011), with improvements found in spatial working memory (Turner, Blackwell, Dowson, McLean, & Sahakian, 2005). Methylphenidate has also been shown to normalize and improve stop-signal reaction time in boys age 7–13 (DeVito et al., 2009). Improvements in response inhibition have also been found in ADHD using methylphenidate (Coghill et al., 2014), modafinil (Turner, Clark, Dowson, Robbins, & Sahakian, 2004), and atomoxetine (Chamberlain et al., 2007).

Schizophrenia In first-episode psychosis, modafinil has been shown to selectively enhance spatial working memory and the recognition of emotional facial expressions (Scoriels et al., 2011; Scoriels, Barnett, Soma, Sahakian, & Jones, 2012). In chronic schizophrenia, modafinil has been shown to improve a range of cognitive functions, including working memory (Hunter, Ganesan, Wilkinson, & Spence, 2006), cognitive flexibility (Turner, Clark, Pomarol-Clotet, et al., 2004), episodic memory, and spatial planning (Lees et al., 2017).

Depression Last, modafinil has been shown to improve episodic- and working-memory domains in patients recovering from depression, and crucially, the latter domain is associated with global functioning (Kaser et al., 2017). Combining an antidepressant medication with modafinil also reduces the severity of depression, thus demonstrating the efficacy of augmented therapies (Goss et al., 2013).

Conclusions and Further Considerations

Human cognitive enhancement is a diverse field mainly driven by a global competitive environment and increasing demands to work more productively within it. As such, pharmacological cognitive enhancement has clear benefits for many people, including sleep-deprived doctors, shift workers, air traffic controllers, frequent travelers, and members of the military. The responsible improvement of cognitive functioning in healthy people could also lead to more productivity, higher earnings, fewer accidents, and a better quality of life. Indeed, some authors have argued that it is our moral obligation to cognitively enhance in order to produce the best possible outcomes for future generations (Harris, 2010). However, the advantages of cognitive enhancement must be weighed against its associated risks. Negative factors often driving the desire to enhance, such as stress and increasing demands at school and in the workplace, have implications for severe adverse physical and mental health events. It may be that healthy people are using cognition-enhancing drugs in order to compensate for poor-quality, stressful, or overdemanding work environments. Safety concerns mainly center on the purchasing of unregulated medications online and their potential for misuse, particularly on the developing brain. At the societal level, ethical concerns of unfairness, cheating, coercion, inequality of access, and the potential for discrimination have been raised. The long-term effects of smart drug use, including their side effects and specific effects on cognitive domains and motivation, are unknown.

In terms of mental health, the cognitive symptoms associated with neuropsychiatric disorders can lead to a loss of functioning in everyday life. Neuropsychiatric disorders are also extremely costly, with implications for the government (increasing demands on health care and social services), the economy (loss in productivity and earnings), and the quality of patient life (difficulty living and working independently). Even small increments in cognitive functions in patients (e.g., 1%) could lead to better outcomes and reduce the economic and societal costs of disease burden (Knapp et al., 2007). Novel, more effective drug treatments designed to target cognitive dysfunction are particularly urgent for neuropsychiatric disorders, including for the episodic memory problems in mild cognitive impairment and Alzheimer's disease. In addition, new drugs would be beneficial where cognition is recognized as a target for treatment, such as in schizophrenia, but there are no medications currently licensed by the FDA, European Medicines Agency (EMA), or Medicines and Healthcare products Regulatory Agency (MHRA) for this purpose. Due to advances in physical health care, an increasing number of people will inevitably experience cognitive decline in the later stages of their lives as the population continues to age in the United States, Europe, and elsewhere. Given the high costs of neuropsychiatric disorders, an important next step would be to assess the economic benefits of pharmacological cognitive enhancement at the public health level.

Additional empirical data on the long-term safety and efficacy of cognition-enhancing drugs in healthy people are urgently needed. This could involve public-private partnerships between the government and pharmaceutical industry to conduct well-designed longitudinal studies investigating the safety and the effects of smart drugs in healthy people using objective and reliable tools for assessing cognition, mood, and motivation. More discussion of the impact of the increasing lifestyle use of cognition-enhancing drugs on the individual and society is needed. These discussions should include members of the public, neuroscientists, ethicists, pharmaceutical companies, policy makers, and government regulators. It is important to emphasize that other evidence-based methods can improve cognition or well-being, such as physical exercise, good-quality sleep, mindfulness, social interaction, and lifelong learning (Beddington et al., 2008). Indeed, we have recently focused on cognitive training using game apps in healthy people and in patients with schizophrenia and MCI, demonstrating positive effects on cognition and motivation (Savulich, Thorp, Piercy, et al., 2019; Sahakian et al., 2015; Savulich, Piercy, Fox, et al., 2017). Through research, it is important to continue increasing our knowledge of the effects of pharmacological cognitive enhancement, both in healthy people and for the development of novel cognition-enhancing drugs for the treatment of cognitive dysfunction in patients with neuropsychiatric disorders and brain injury, to ensure the flourishing of the individual and society.

Acknowledgments

George Savulich is funded by Eton College and the Wallitt Foundation. This work was supported by the National Institute for Health Research (NIHR) Cambridge Biomedical Research Centre (BRC) Mental Health Theme. Barbara J. Sahakian receives funding from the NIHR Cambridge BRC Mental Health Theme and the NIHR Brain Injury MedTech and in vitro diagnostic Co-operative (MIC). We thank Alicja Malinowska for assistance with manuscript preparation.

REFERENCES

Academy of Medical Sciences. 2012. Human enhancement and the future of work. Report from a joint workshop hosted by the Academy of Medical Sciences, the British Academy, the Royal Academy of Engineering and the Royal Society, London. https://royalsociety.org/~/media/policy/projects/human-enhancement/2012-11-06-human-enhancement.pdf (15.12.2014).

Battleday, R. M., & Brem, A. K. (2015). Modafinil for cognitive neuroenhancement in healthy non-sleep-deprived subjects: A systematic review. *European Neuropsychopharmacology*, *25*, 1865–1881.

Beddington, J., Cooper, C. L., Field, J., Goswami, U., Huppert, F. A., Jenkins, R., & Thomas, S. M. (2008). The mental wealth of nations. *Nature, 455*, 1057–1060.

Brühl, A. B., & Sahakian, B. J. (2016). Drugs, games, and devices for enhancing cognition: Implications for work and society. *Annals of the New York Academy of Sciences, 1369*, 195–217.

Care Quality Commission. (2013). *The safer management of controlled drugs: Annual report 2012.* http://www.cqc.org.uk/sites/default/files/documents/cdar_2012.pdf.

Chamberlain, S. R., Del Campo, N., Dowson, J., Müller, U., Clark, L., Robbins, T. W., & Sahakian, B. J. (2007). Atomoxetine improved response inhibition in adults with attention deficit/hyperactivity disorder. *Biological Psychiatry, 62*, 977–984.

Coghill, D. R., Seth, S. Pedroso, S., Usala, T., Currie, J., & Gagliano, A. (2014). Effects of methylphenidate on cognitive functions in children and adolescents with attention-deficit/hyperactivity disorder: Evidence from a systematic review and a meta-analysis. *Biological Psychiatry, 76*, 603–615.

Collins, P. Y., Patel, V., Joestel, S. S., March, D., Insel, T. R., Daar, A. S., ... Walport, M. (2011). Grand challenges in global mental health. *Nature, 475*, 27–30.

d'Angelo, C. L-S., Savulich, G., & Sahakian, B. J. (2017). Lifestyle use of drugs by healthy people for enhancing cognition, creativity, motivation and pleasure. *British Journal of Pharmacology, 174*, 3257–3267.

del Campo, N., Fryer, T. D., Hong, Y. T., Smith, R., Brichard, L., Acosta-Cabronero, J., ... Müller, U. (2013). A positron emission tomography study of nigro-striatal dopaminergic mechanisms underlying attention: Implications for ADHD and its treatment. *Brain, 136*, 3252–3270.

Devito, E. E., Blackwell, A. D., Clark, L., Kent, L., Dezersy, A. M., Turner, D. C., ... Sahakian, B. J. (2009). Methylphenidate improves response inhibition but not reflection-impulsivity in children with attention deficit hyperactivity disorder (ADHD). *Psychopharmacology, 202*, 531–539.

Elliott, R., Sahakian, B. J., Matthews, K., Bannerjea, A., Rimmer, J., & Robbins, T. W. (1997). Effects of methylphenidate in adult attention-deficit/hyperactivity disorders. *Psychopharmacology, 178*, 286–295.

European Monitoring Centre for Drugs and Drugs Addiction. (2016). *European drug report 2016: Trends and development.* Luxembourg: European Union. http://www.emcdda.europa.eu/system/files/publications/2637/TDAT16001ENN.PDF.

Franke, A. G., Gränsmark, P., Agricola, A., Schüle, K., Rommel, T., Sebatian, A., ... Lieb, K. (2017). Methylphenidate, modafinil, and caffeine for cognitive enhancement in chess: A double-blind, randomised controlled trial. *European Neuropsychopharmacology, 27*, 248–260.

Frith, U., Bishop, D., Blakemore, C., Blakemore, S.-J., Butterworth, B., Goswami, U., ... Young, C. (2011). *Brain waves module 2: Neuroscience: Implications for education and lifelong learning.* London: The Royal Society.

Goss, A. J, Kaser, M., Costafreda, S. G., Sahakian, B. J., & Fu, C. H. Y. (2013). Modafinil augmentation therapy in unipolar and bipolar depression: A systematic review and meta-analysis of randomized controlled trials. *Journal of Clinical Psychiatry, 74*, 1101–1107.

Graf, H., Abler, B., Freudenmann, R., Beschoner, P., Schaeffeler, E., Spitzer, M., ... Grön, G. (2011). Neural correlates of error monitoring modulated by atomoxetine in healthy volunteers. *Biological Psychiatry, 69*, 890–897.

Gustavsson, A., Svensson, M., Jacobi, F., Allgulander, C., Alonso, J., Beghi, E., ... Olesen, J. (2011). Cost of disorders of the brain in Europe 2010. *European Neuropsychopharmacology, 21*, 718–779.

Harris, J. (2010). Enhancements are a moral obligation. In Julian Savulescu & Nick Bostrom (Eds.), *Human enhancement.* Oxford: Oxford University Press.

Hollander, E., Doernberg, E., Shavitt, R., Waterman, R. J., Soreni, N., Veltman, D. J., & Fineberg, N. A. (2016). The cost and impact of compulsivity: A research perspective. *European Neuropsychopharmacology, 26*, 800–809.

Howard, R., McShane, R., Lindesay, J., Ritchie, C., Baldwin, A., Barber, J., ... Phillips, P. (2012). Donepezil and memantine for moderate-to-severe Alzheimer's disease. *New England Journal of Medicine, 366*, 893–903.

Hunter, M. D., Ganesan, V., Wilkinson, I. D., & Spence, S. A. (2006). Impact of modafinil on prefrontal executive function in schizophrenia. *American Journal of Psychiatry, 163*, 2184–2186.

Ilieva, I., Boland, J., & Farah, M. J. (2013). Objective and subjective cognitive enhancing effects of mixed amphetamine salts in healthy people. *Neuropharmacology, 64*, 496–505.

Ilieva, I., & Farah, M. J. (2013). Enhancement stimulants: Perceived motivational and cognitive advantages. *Frontiers of Neuroscience, 7*, 198.

Insel, T. R., Sahakian, B. J., Voon, V., Nye, J., Brown, V. J., Altevogt, B. M., ... Williams, J. H. (2012). Drug research: A plan for mental illness. *Nature, 483*, 269.

Insel, T. R., Voon, V., Nye, J. S., Brown, V. J., Altevogt, B. M., Bullmore, E. T., ... Sahakian, B. J. (2013). Innovative solutions to novel drug development in mental health. *Neuroscience and Biobehaivoral Reviews, 37*, 2438–2444.

Kaser, M. K., Deakin, J. B., Michael, A., Zapata, C., Bansal, R., Ryan, D., ... Sahakian, B. J. (2017). Modafinil improves episodic memory and working memory cognition in patients with remitted depression: A double-blind, randomized, placebo controlled study. *Psychological Medicine, 2*, 115–122.

Kehagia, A. A., Housden, C. R., Regenthal, R., Barker, R. A., Müller, U., Rowe, J. B., ... Robbins, T. W. (2014). Targeting impulsivity in Parkinson's disease using atomoxetine. *Brain, 137*, 1986–1997.

Kessler, R. C., Berglund, P., Demler, O., Jin, R., Merikangas, K. R., & Walters, E. E. (2005). Lifetime prevalence and age-of-onset distributions of DSM-IV disorders in the National Comorbidity Survey Replication. *Archives of General Psychiatry, 62*, 593–602.

Knapp, M., Prince, M., Albanese, E., Banerjee, S., Dhanasiri, S., Fernandez, J. L., ... Stewart, R. (2007). *Dementia UK.* London: Alzheimer's Society.

Kordt, M. (2015). *DAK-Gesundheitsreport.* Berlin. http://www.dak.de/dak/download/Vollstaendiger_bundesweiter_Gesundheitsreport_2015-1585948.pdf.

Lees, J., Michalopoulou, P. G., Lewis, S. W., Preston, S., Bamford, C., Collier, T., ... Drake, R. J. (2017). Modafinil and cognitive enhancement in schizophrenia and healthy volunteers: The effects of test batter in a randomised controlled trial. *Psychological Medicine, 47*, 2358–2368.

Loe, I. M., & Feldman, H. M. (2007). Academic and educational outcomes of children with ADHD. *Ambulatory Pediatrics, 7*, 82–90.

Lou, J. S., Dimitrova, D. M., Park, B. S., Johnson, S. C., Eaton, R., Arnold, G., & Nutt, J. G. (2009). Using modafinil to treat fatigue in Parkinson disease: A double-blind, placebo-controlled pilot study. *Clinical Neuropharmacology, 32*, 305–310.

Maher, B. (2008). Poll results: Look who's doping. *Nature, 10*, 674–675.

Maier, L. J., Ferris, J. A., & Winstock, A. R. (2018). Pharmacological cognitive enhancement among non-ADHD individuals-a cross-sectional study in 15 countries. *International Journal of Drug Policy, 11*, 104–112.

Marcus, S. J. (Ed.). (2002). *Neuroethics: Mapping the field: Conference proceedings, May 13--14, 2002 San Francisco, California*. New York: Dana Press.

Maslen, H., Faulmuller, N., & Savulescu, J. (2011). Pharmacological cognitive enhancement—How neuroscientific research could advance ethical debate. *Frontiers in Systems Neuroscience, 8*, 107.

Mehta, M. A., Owen, A. M., Sahakian, B. J., Mavaddat, N., Pickard, J. D., & Robbins, T. W. (2000). Methylphendiate enhances working memory by modulating discrete frontal and parietal lobe regions in the human brain. *Journal of Neuroscience, 20*, RC65.

Mendonca, D. A., Menezes, K., & Jog, M. S. (2007). Methylphenidate improves fatigue scores in Parkinson disease: A randomized controlled trial. *Movement Disorders, 22*, 2070–2076.

Müller, U., Rowe, J. B., Rittman, T., Lewis, C., Robbins, T. W., & Sahakian, B. J. (2013). Effects of modafinil on non-verbal cognition, task enjoyment and creative thinking in healthy volunteers. *Neuropharmacology, 64*, 490–495.

National Institute for Health and Care Excellence. (2017). Parkinson's disease in adults: Diagnosis and management. https://bnf.nice.org.uk/treatment-summary/parkinsons-disease.html.

Nicholson, P. J., Mayho, G., & Sharp, C. (2015). Cognitive enhancing drugs and the workplace. British Medical Association, London. https://www.bma.org/uk/advice/employment/occupational-health/cognitive-enhancing-drugs.

Porsdam Mann, S., & Sahakian, B. J. (2015). The increasing lifestyle use of modafinil by healthy people: Safety and ethical issues. *Current Opinion in Behavioral Sciences, 4*, 136–141.

Sahakian, B., d'Angelo, C., & Savulich, G. (2017, February 14). LSD "microdosing" is trending in Silicon Valley—but can it actually make you more creative? *The Conversation*. https://theconversation.com/lsd-microdosing-is-trending-in-silicon-valley-but-can-it-actually-make-you-more-creative-72747.

Sahakian, B. J. (2014). What do the experts think we should do to achieve brain health? *Neuroscience and Biobehavioural Reviews, 43*, 240–258.

Sahakian, B. J., Brühl, A. B., Cook, J., Killikelly, C., Savulich, G., Piercy, T., ... Jones, P. B. (2015). The impact of neuroscience on society: Cognitive enhancement in neuropsychiatric disorders and in healthy people. *Philosophical Transactions of the Royal Society B: Biological Sciences, 370*(1677).

Sahakian, B. J., & Morein-Zamir, S. (2007). Professor's little helper. *Nature, 450*, 1157–1159.

Sahakian, B. J., & Morein-Zamir, S. (2015). Pharmacological cognitive enhancement: Treatment of neuropsychiatric disorders and lifestyle use by healthy people. *Lancet Psychiatry, 2*, 357–362.

Sahakian, B. J., Morris, R. G., Evenden, J. L., Heald, A., Levy, R., Philpot, M., & Robbins, T. W. (1988). A comparative study of visuospatial memory and learning in Alzheimer-type dementia and Parkinson's disease. *Brain, 111*, 695–718.

Sahakian, B. J., Owen, A. M., Morant, N. J., Eagger, S. A., Boddington, S., Crayton, L., ... Levy, R. (1993). Further analysis of the cognitive effects of tetrahydroaminoacridine (THA) in Alzheimer's disease: Assessment of attentional and mnemonic function using CANTAB. *Psychopharmacology, 110*, 395–401.

Sahakian, B.J., & Savulich, G. (2019). Innovative methods for improving cognition, motivation and wellbeing in schizophrenia. *World Psychiatry, 18*, 168–170.

Sattler, S., Mehlkop, G., Graeff, P., & Sauer, C. (2014). Evaluating the drivers of and obstacles to the willingness to use cognitive enhancement drugs: The influence of drug characteristics, social environment, and personal characteristics. *Substance Abuse Treatment, Prevention, and Policy, 9*, 8.

Savulich, G., Mezquida, G., Atkinson, S., Bernardo, M., & Fernandez-Egea, E. (2018). A case study of clozapine and cognition: Friend or foe? *Journal of Clinical Psychopharmacology, 38*, 152–153.

Savulich, G., O'Brien, J. T., & Sahakian, B. J. (2019). Are neuropsychiatric symptoms modifiable risk factors for cognitive decline in Alzheimer's disease and vascular dementia? *British Journal of Psychiatry*, 1–3. doi:10.1192/bjp.2019.98

Savulich, G., Piercy, T., Brühl, A. B., Fox, C., Suckling, J., Rowe, J. B., & Sahakian, B. J. (2017). Focusing the neuroscience and societal implications of cognitive enhancers. *Clinical Pharmacology and Therapeutics, 101*, 170–172.

Savulich, G., Piercy, T., Fox, C., Suckling, J., Rowe, J. B., O'Brien, J. T., & Sahakian, B. J. (2017). Cognitive training using a novel memory game on an iPad in patients with amnestic mild cognitive impairment. *International Journal of Neuropsychopharmacology, 20*, 624–633.

Savulich, G., Thorp, E., Piercy, T., Peterson, K.A., Pickard, J. D., & Sahakian, B. J. (2019). Improvements in attention following cognitive training with the novel 'Decoder' game on an iPad. *Frontiers in Behavioural Neuroscience, 13*, 2.

Scoriels, L., Barnett, J. H., Murray, G. K., Cherukuru, S., Fielding, M., Cheng, F., & Jones, P. B. (2011). Effects of modafinil on emotional processing in first episode psychosis. *Biological Psychiatry, 69*(5), 457–464.

Scoriels, L., Barnett, J. H., Soma, P. K., Sahakian, B. J., & Jones, P. B. (2012). Effects of modafinil on cognitive functions in first episode psychosis. *Psychopharmacology, 22*, 249–258.

Scoriels, L., Jones P. B., & Sahakian, B. J. (2013). Modafinil effects on cognition and emotion in schizophrenia and its neurochemical modulation in the brain. *Neuropharmacology, 64*, 168–184.

Singh, I., Bard, I., & Jackson, J. (2014). Robust resilience and substantial interest: A survey of pharmacological cognitive enhancement among university students in the UK and Ireland. *PLoS One, 9*, e105969.

Smith, M. E., & Farah, M. J. (2011). Are prescription stimulants "smart pills"? The epidemiology and cognitive neuroscience of prescription stimulant use by normal healthy individuals. *Psychologial Bulletin, 137*, 717–741.

Spencer, T., & Biederman, J. (2011). Stimulant treatment of adult ADHD. In K. B. Jan, C. C. Kan, & P. Asherson (Eds.), *ADHD in adults: Characterization, diagnosis, and treatment* (pp.191–197). Cambridge: Cambridge University Press.

Stahl, S. R. (2008). *Stahl's essential psychopharmacology* (3rd ed). Cambridge: Cambridge University Press.

The Student Room. (2016). New research reveals 1 in 10 students have taken study drugs. http://tsrmatters.com/wp-content/uploads/2013/07/New-research-reveals-1-in-10-students-have-taken-study-drugs.pdf.

Sugden, C., Housden, C. R., Aggarwal, R., Sahakian, B. J., & Darzi, A. (2012). Effect of pharmacological enhancement on the cognitive and clinical psychomotor performance of sleep-deprived doctors: A randomized controlled trial. *Annals of Surgery, 255*, 222–227.

Tan, C. C., Yu, J. T., Wang, H. F., Tan, M. S., Meng, X. F., Wang, C., … Tan, L. (2014). Efficacy and safety of donepezil, galantamine, rivastigmine, and memantine for the treatment of Alzheimer's disease: A systematic review and meta-analysis. *Journal of Alzheimer's Disease, 41*, 615–631.

Teter, C. J., Falcone, A. E., Cranford, J. A., Boyd, C. J., & McCabe, S. E. (2010). Nonmedical use of prescription stimulants and depressed mood among college students: Frequency and routes of administration. *Journal of Substance Abuse Treatment, 38*, 292–298.

Turner, D. C., Blackwell, A. D., Dowson, J. H., McLean, A., & Sahakian, B. J. (2005). Neurocognitive effects of methylphenidate in adult attention-deficit/hyperactivity dirods-ers. *Psychopharmacology, 178*, 286–295.

Turner, D. C., Clark, L., Dowson, J., Robbins, T. W., & Sahakian, B. J. (2004). Modafinil improves cognition and response inhibition in adult attention-deficit/hyperactivity disorder. *Biological Psychiatry, 55*, 1031–1040.

Turner, D. C., Clark, L., Pomarol-Clotet, E., Mckenna, P., Robbins, T. W., & Sahakian, B. J. (2004). Modafinil improves cognition and attentional set shifting in patients with chronic schizophrenia. *Neuropsychopharmacology, 29*, 1363–1373.

Turner, D. C., Robbins T. W., Clark, L., Aron, A. R., Dowson, J., & Sahakian, B. J. (2003). Cognitive enhancing effects of modafinil in healthy volunteers. *Psychopharmacology, 165*, 260–269.

Vastag, B. (2004). Poised to challenge need for sleep, "wakefulness enhancer" rouses concerns. *JAMA, 291*, 167–170.

Vrecko, S. (2013). Just how cognitive is "cognitive enhancement"? On the significance of emotions in university students' experiences with study drugs. *Neuroscience, 4*, 4–12.

Wilens, T. E. (2006). Mechanism of action of agents used in attention-deficit/hyperactivity disorder. *Journal of Clinical Psychiatry, 67*, 32–38.

Ye, Z., Altena, E., Nombela, C., Housden, C. R., Maxwell, H., Rittman, T., … Rowe, J. B. (2015). Improving response inhibition in Parkinson's disease with atomoxetine. *Biological Psychiatry, 77*, 740–749.

94 Brain-Machine Interfaces: From Basic Science to Neurorehabilitation

MIGUEL A. L. NICOLELIS

ABSTRACT Over the past two decades, not only neuroscientists, neurologists, and neurosurgeons but also engineers, roboticists, and cognitive and computer scientists alike have investigated the scientific and clinical benefits of establishing direct linkages between living animal or human brains with a large variety of mechanical (e.g., robotic prostheses), electronic (e.g., computers), and even virtual tools (e.g., limb and body avatars). These paradigms are known as brain-machine interfaces (BMIs). BMIs have been employed primarily to either investigate the dynamic properties of neural circuits in experimental animals or to implement novel neurorehabilitation approaches and, more recently, therapies aimed at restoring neurological and cognitive functions such as autonomous mobility and communication in patients suffering from devastating levels of brain injury. As a result of these efforts, BMI research has contributed to the validation of a series of neurophysiological principles and the introduction of novel rehabilitation protocols in spinal cord injury and stroke. This chapter reviews the main BMI paradigms as well as the most significant basic science and clinical findings that resulted from their implementation. It also discusses the potential future impact of BMIs in the development of a new generation of neuroprostheses. The chapter concludes by introducing a potentially disruptive new paradigm, known as Brainets or shared BMIs, which may set the stage in the future for the establishment of Internet-based protocols for neurorehabilitation and treatment.

According to the World Health Organization, about 1 billion people worldwide suffer from some sort of brain disorder. Out of those, hundreds of millions of people have to endure the life-changing effects caused by neurological injuries (Dietz, 2001; Rossignol, Schwab, Schwartz, & Fehlings, 2007; Scivoletto & Di Donna, 2009) and diseases (Calvo et al., 2014). In the United States alone, 5 million people suffer from varying degrees of body paralysis (Paddock, 2009) due to spinal cord injuries (SCIs) only. Worldwide this number increases to about 25 million people. Today, almost 250 million people around the world live with the often devastating, long-term clinical consequences of a stroke, which occurs in 15 million new patients every year. As a result, in 2010 the total global cost of dealing with brain disorders was estimated at $2.5 trillion/year; by 2030 this cost is expected to soar to $6 trillion (Paddock, 2009).

Undoubtedly, the awareness and interest with which most people around the world follow the progress of modern brain research derives from the growing challenges that contemporary societies face in developing new therapies and neurorehabilitation approaches for coping with the tremendous human hardship, and escalating costs, imposed by lesions and diseases of the central nervous system (CNS). Finding novel and cost-efficient solutions to treat and improve the quality of life for such a huge number of people, like the millions suffering from SCIs or stroke, is clearly becoming a high priority for public and private health systems worldwide.

Traditionally, the main therapeutic strategy to cope with the symptoms and debilitation created by brain diseases has focused on the development of new pharmacological agents that could target brain regions, or even particular cell types, compromised by each neurological or psychiatric disorder, in a very specific way. Unfortunately, the development of new CNS drugs is hindered by the immense cost involved in research and clinical translation and the difficulties in mitigating the side effects usually associated with most of these medications. In the last decades, the successful clinical use of medical devices in tens of thousands of people, such as the cochlear implant (Wilson et al., 1991) for treating severely hearing-impaired patients and deep brain stimulation (Benabid, 2003) for the treatment of Parkinson's disease, has raised the hope that a second main strategy to treat CNS disorders—namely, the use of neuroprostheses that interact directly with neuronal tissue—could materialize in the near future. Consistent with this latter view, during the past two decades a new paradigm to interact with the human brain has been gaining considerable attention, not only because it has the clear potential to become a novel neurorehabilitation tool and drive the development of a second generation of neuroprosthetic devices but also because, according to recent clinical findings, it holds promise for leading to potential new therapies for patients severely paralyzed as a result of an SCI or stroke. These paradigms are known as brain-machine interfaces (BMIs; figure 94.1; Nicolelis, 2001). As the name indicates, BMIs establish direct and real-time electronic/computational links between living animal or human brains and a variety of mechanical, electronic, and virtual tools (Nicolelis, 2001). Figure 94.1 illustrates in detail the general basic configuration of the original experimental BMI paradigm introduced in the late 1990s, which today serves as the core concept for the development of a variety of clinical BMI applications. Using this closed-loop control approach, BMIs allow subjects (animals or humans) to use their electrical brain activity to directly control the movements of an artificial device (e.g., a robotic arm or leg exoskeleton) to perform a particular motor task without the overt need of engaging the subject's own body musculature. Essentially, by taking advantage of a combination of neurophysiological recording methods; modern microelectronic instrumentation, which now includes the wireless transmission of hundreds of channels of neuronal data (Schwarz et al., 2014); and a huge library of mathematical and computational decoding algorithms (Li, 2014; Lotte et al., 2018), BMIs allow a series of motor control commands, describing both the kinematic and dynamic parameters of limb movements, to be extracted in real time from a variety of brain-derived electrical signals (e.g., multineuron recordings, local field potentials, the electroencephalogram [EEG], and others) in order to

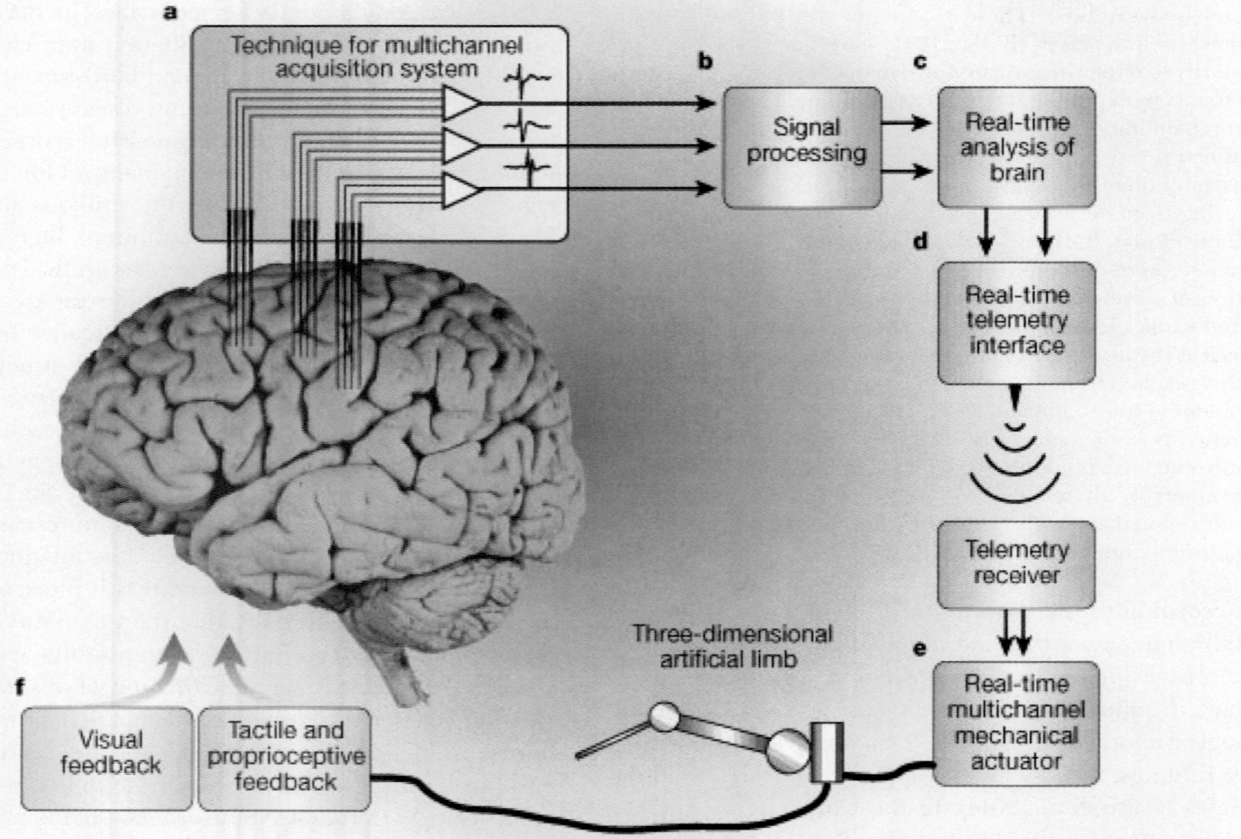

FIGURE 94.1 Classical configuration of a brain-machine interface. Through the employment of multichannel intracranial extracellular recordings, multiple motor commands can be extracted, in real time, from the combined electrical activity of several hundred neurons, distributed across multiple cortical areas. This operation is carried out through the employment of mathematical decoders. Extracted motor commands are then used by subjects to directly control the movements of a variety of artificial devices. Reproduced with permission from Nicolelis (2001). (See color plate 102.)

control a plethora of robotic, electronic, and even virtual tools.

One of the key components of any BMI apparatus, experimental or clinical, resides in the choice of the mathematical decoder and the computational strategy employed to extract in real time the motor commands and features needed to control the movements of an artificial actuator (Li, 2014; Lotte et al., 2018). Beginning with the classic Wiener and Kalman filters; a series of multivariate statistical methods; pattern-recognition techniques, such as artificial neural networks; and even, more recently, machine-learning algorithms have been used to extract motor commands from brain-derived signals in BMI studies (Li, 2014; Lotte et al., 2018; Tseng et al., 2019). As such, the literature on BMI decoders has simply exploded in the past decade and today accounts for a significant part of the published papers in this area.

Historical Background

Because in a BMI setting the experimenter has total control over which motor features are extracted from the recorded brain-derived signals and how they are used to enact the desired movements of a given artificial actuator, as well as the nature of the feedback signals sent back to the user, BMIs have quickly driven a variety of new approaches to investigate how large populations of neurons dynamically encode sensorimotor information. For the same reasons, the growing experience with clinically oriented BMIs has driven the design and implementation of a multitude of neuroprosthetic devices that were considered unfeasible just a few years ago (Lebedev & Nicolelis, 2017).

Yet it took at least 20 years for the BMI concept to entice enough interest in the neuroscience community. I say that because even though the idea of establishing and testing rudimentary versions of BMIs arose in the 1960s with the pioneering experiments of Eberhard Fetz (1969), which involved single-neuronal recordings in nonhuman primates, it was only in the late 1990s that neurophysiologists and clinicians alike were able to demonstrate the feasibility of building BMIs that could be used, albeit in well-controlled laboratory conditions, to either probe the neurophysiological properties of samples of 40–100 cortical neurons in experimental animals or serve as clinical tools for neurology patients. In fact, without knowing of their parallel efforts, two independent groups, one in the United States (Chapin, Moxon, Markowitz, & Nicolelis, 1999; Wessberg et al., 2000) and the other in Germany (Birbaumer et al., 1999), published their pioneer experimental and clinical BMI findings almost simultaneously in 1999. The original experimental BMI was first reported in the United States in a collaboration between the laboratories of John Chapin and Miguel Nicolelis using rats and, soon after, New World monkeys, as experimental animals (Chapin et al., 1999; Wessberg et. al, 2000). A couple of years later, the same group and two other labs reported successful BMIs operated by rhesus monkeys (Carmena et al., 2003). These experimental BMIs became closely associated with the ongoing neurophysiological paradigm shift that shook systems neuroscience in the early 1990s by gradually moving it away from the classic single-neuron recording paradigm to a new electrophysiological technique that allowed, via chronically implanted microelectrode arrays, much larger samples of single cortical and subcortical neurons to be recorded simultaneously in freely behaving animals. Once the first experimental demonstrations of BMIs were published in 1999 and 2000, the widespread dissemination of these findings further enhanced the development of the multielectrode recording approach. Indeed, as early as 2004 a multidisciplinary team from the Duke University Center for Neuroengineering reported that such recordings could be used in human subjects to drive an intraoperative BMI (Patil, Carmena, Nicolelis, & Turner, 2004). It took 2 more years for other groups to report similar results in humans, using another technology for chronic cortical implants (Hochberg et al., 2006). Because of the use of implanted microelectrodes to obtain motor-related electrical brain activity, these BMIs were classified as invasive.

In parallel, the pioneer clinical work on BMIs relied primarily on noninvasive EEG recordings obtained from so-called locked-in patients (Birbaumer et al., 1999): those suffering from advanced stages of the degenerative disorder known as Lou Gehrig's disease (amyotrophic lateral sclerosis, or ALS). Since in advanced ALS patients most, if not all, body musculature is totally paralyzed, they cannot communicate with the external world, their families, and their caregivers. To mitigate this terrible isolating condition, which caused most ALS patients to exist in a state of severe chronic depression, a very ingenious BMI was designed and implemented by researchers at the University of Tubingen (Birbaumer et al., 1999). Led by Niels Birbaumer, this BMI approach enabled ALS patients to use their EEG activity to sequentially select letters displayed on a computer monitor. Through this simple, time-consuming but effective tool, locked-in patients began to write short messages to their families and doctors and even send emails—all by using their own EEGs. Because this system employed EEG recordings to control a computer cursor, this paradigm soon became known as a brain-computer interface (BCI), a more specialized subclass of BMIs since the latter term refers to brain-controlled devices of all sorts, not only computers. Since this original demonstration, many clinical applications have been reported in the literature (Lebedev & Nicolelis, 2017).

The Main Discoveries Associated with Brain-Machine Interface Research

Despite their almost nonoverlapping original aims (to investigate neurophysiological properties of neural circuits in animals or provide severely paralyzed patients with a new communication tool), the two original lines of BMI research categorically demonstrated that both animals and human subjects alike can rather quickly learn to use their raw electrical brain activity to control the movements of artificial devices, even when such tools are unlike the patient's own limbs (e.g., computer cursors, electronic wheelchairs, and other items) or are not positioned next to the subject but lay in remote locations very distant from the BMI operator (Fitzsimmons, Lebedev, Peikon, & Nicolelis, 2009). These early studies also revealed, almost immediately, the importance of providing continuous streams of feedback from the brain-controlled artificial actuators back to the BMI operator in learning to operate these devices (Wessberg et al., 2000).

Paramount to the early BMI studies was the demonstration that interactions with BMIs induced widespread cortical plasticity essential for learning to properly operate a BMI (Carmena et al., 2003; Cramer et al., 2011; Di Pino, Maravita, Zollo, Guglielmelli, & Di Lazzaro, 2014; Dobkin, 2007; Grosse-Wentrup, Mattia, & Oweiss, 2011; Lebedev et al., 2005; Lebedev & Nicolelis, 2006; Nicolelis & Lebedev, 2009; Oweiss & Badreldin, 2015). Basically, what has been repeatedly observed in a series of studies involving different tasks is that motor learning is associated with the gradual improvement in BMI operation observed in animals and humans (Adams, 1987; Bilodeau & Bilodeau, 1961; Doyon et al., 2009; Doyon, Penhune, & Ungerleider, 2003; Hikosaka, Nakamura, Sakai, & Nakahara, 2002; Kleim, Barbay, & Nudo, 1998; Laubach, Wessberg, & Nicolelis, 2000; Mitz, Godschalk, & Wise, 1991; Shadmehr & Wise, 2005).

It is important to stress that more basic neurophysiological findings were corroborated by extensive experimentation with BMIs. For example, an extensive series of BMI studies in rodents and monkeys allowed my laboratory to propose the existence of a series of key neurophysiological principles that govern the operation of large cortical neuronal ensembles in mammals (Nicolelis & Lebedev, 2009). Here, I would like to stress primarily the fact that BMI studies have been instrumental in repeatedly demonstrating the rather distributed and dynamic nature of sensorimotor processing in the primate cortex, particularly regarding the motor and somatosensory cortical areas. For example, one of the first big surprises to emerge from the pioneer BMI experiments was the unequivocal demonstration that one could obtain useful motor control signals to move a robotic arm from pretty much all primary motor and premotor, primary somatosensory, and even posterior parietal cortical areas. Also surprising was the fact that, out of the tens of millions of neurons located in the primary motor cortex (M1) alone, to cite just one example, simultaneous recordings of the electrical activity of populations of a few hundred M1 individual neurons—but not fewer than 10 neurons, as some authors hastily proposed originally (Hochberg et al., 2006)—would suffice to reproduce elaborate three-dimensional (3D) arm movements using a BMI coupled to an industrial robotic arm with multiple degrees of freedom. Moreover, within each of these individual cortical areas, there was no need to target a particular region of the somatotopic maps nor "fish" for a specific cell type. Basically, motor control signals intended to produce the movements of an artificial arm could be obtained throughout the 3D volume of each of these cortical regions through a random sample of 100–700 individual neurons. Since the early days of BMI research, this fundamental finding has been depicted by the now classic neuronal-dropping curves, which relate the mass of neurons recorded simultaneously in a given cortical region with the amount of accuracy in predicting a given motor parameter using a particular BMI decoder. Neuronal dropping curves have become a classic way, therefore, to quantify the amount of predictive information a given mass of cortical neurons contains when using a particular decoder to reproduce a given motor parameter.

Over the years, in addition to these principles of neural ensemble physiology mentioned above, our laboratory has also obtained evidence to support the working hypothesis that the use of BMIs to control the movements of artificial tools is intimately related to the recruitment of the classic frontoparietal cortical mirror neuron system (Fabbri-Destro & Rizzolatti, 2008; Ferrari, Rozzi, & Fogassi, 2005; Ifft, Shokur, Li, Lebedev, & Nicolelis, 2013; Rizzolatti, Cattaneo, Fabbri-Destro, & Rozzi, 2014; Tseng, Rajangam, Lehew, Lebedev, & Nicolelis, 2018). An initial suspicion that this was the case emerged when studies in both animals and human subjects revealed that learning to operate a BMI did not require that individuals generate overt limb movements during the phase required to train the mathematical decoder—that is, the computational model—used to extract motor signals from the combined raw brain activity sampled by the BMI. Instead, if the subject simply observed, on a computer screen, a large library of virtual arm (or leg) trajectories, which the BMI application intended to mimic, one noticed that the subject's performance—and that of the decoder—increased over time, to the point in which both reached the maximum level of accuracy possible for a given neuronal sample. Therefore, based on these data, our theory postulates that when animals and patients learn to operate a BMI system, they are likely to recruit the same mirror neuron cortical circuitry they rely upon to observe, and later mimic, a new motor behavior executed by another member of their species. Accordingly, the learning process involved in becoming a proficient BMI user would be equivalent to what is required, from a neurophysiological point of view, for subjects to learn how to handle a new tool. Since such a process of tool mastery evokes brain plasticity (Berti & Frassinetti, 2000; Di Pino et al., 2014; Iriki, Tanaka, & Iwamura, 1996; Maravita & Iriki, 2004; Maravita, Spence, & Driver, 2003), this provides the basic mechanism through which BMIs could improve patients' neurological functions (see below). On a more basic science level, putting all these observations together and given the fact that we have observed that the number of cortical neurons that become tuned to the artificial actuator—like the robotic arm—tends to increase as subjects learn to operate a BMI (Ifft et al., 2013), one can raise a very interesting corollary from the BMI literature: the number of neurons exhibiting mirror-neuron-like activity may increase over time as subjects learn to operate a new tool or, in the particular case discussed here, a BMI, simply by observing a tutor or computer-screen-generated images. Further experiments will be required to test the full validity of such an interesting possibility.

As mentioned above, the potential parallel between the neurophysiological mechanisms underlying both BMI and tool use is very significant because, essentially, it implies that as users learn to operate an artificial actuator using a BMI they also contribute to an expansion of their own sense of self, or body schema, by incorporating that tool as a true extension of their brain's body representation. Indeed, a series of experiments conducted by the laboratory of Professor Atsushi Iriki in Japan suggest that when these animals learn to use an artificial tool—a rake—to collect objects they could not reach with their own arms and hands, the underlying motor learning triggers the incorporation of the tool as part of the monkeys' body schema, through the process of cortical plasticity (Iriki et al., 1996).

We and others have reported that plastic reorganization of receptive fields and cortical maps takes place when animals learn to interact with a BMI (Carmena et al., 2003; Lebedev & Nicolelis, 2017). For instance, Carmena et al. (2003) reported that as monkeys learned to operate a BMI designed to produce both arm-reaching and hand-grasping movements, changes in both neuronal tuning curves and neuronal firing correlations, within and between motor and somatosensory cortical areas, were observed. Furthermore, Lebedev et al. (2005) and Zacksenhouse et al. (2007) documented that an enhancement in cortical firing modulations, taking place during the learning phase, reduced significantly after monkeys became proficient in BMI operation. Similarly, we have also observed the occurrence of strong but transient enhancements in neuronal correlation while monkeys learned to operate a bimanual BMI (Ifft et al., 2013). In all these studies, the emergence of significant correlations and increased synchrony between the firing of individual neurons, as well as entire neuronal ensembles, to the movements of the artificial actuator being controlled by the BMI was documented. These changes in neuronal tuning were observed even when monkeys continued to perform sporadic arm movements as they used their brain-derived activity to control an artificial actuator. Under these experimental conditions, we observed that neuronal firing in relation to the monkey's own arm movements was reduced, while the same cortical cell's firing became more and more correlated primarily to the artificial actuator (Lebedev et al., 2005). Such newly acquired tuning to the BMI-controlled actuator remained even when monkeys ceased to move their own arms altogether, relying solely on the BMI to achieve only movements of the actuator to solve the motor task (Carmena et al., 2003; Ifft et al., 2013; Lebedev et al., 2005). When patterns of neuronal ensemble firing were analyzed, we observed that such a switch to a phase in which animals only used the BMI to move the actuator, while producing no overt movements of their own limbs or bodies, led to a large increase in neuronal synchrony. This was accompanied by the observation that a large sample of these neurons began to show very similar preferred directions during this "brain-control phase" of the BMI experiment (Carmena et al., 2003; Ifft et al., 2013; Nicolelis & Lebedev, 2009; O'Doherty et al., 2011). Altogether, these and other findings contribute to the proposal that BMI training allows operators to incorporate the artificial actuators, controlled directly by their brain activity, as an extension of their body schema and, in the case of humans, their sense of self (Lebedev & Nicolelis, 2006; Nicolelis, 2011; Shokur et al., 2013).

Another interesting finding that was only observed because of the implementation of a BMI to control bimanual movements—that is, using two virtual arms, each of which had its movements directly controlled by cortical activity generated in one of the monkey's cerebral hemispheres (Ifft et al., 2013)—is that, unexpectedly, bimanual movements cannot be generated by a simple linear summation of the neuronal motor activity produced by cortical areas located in each cerebral hemisphere. Instead, a nonlinear integration of the firing of neurons located bilaterally in homologous premotor and motor cortical areas is required (Ifft et al., 2013).

Emerging Brain-Machine Interface Applications and Technologies

As a result of very fast growth in brain-recording methods, decoding algorithms, and artificial devices controlled by BMIs, the current literature has accumulated a large number of applications designed according to limb movements (Carmena et al., 2003; Collinger et al., 2013; Contreras-Vidal & Grossman, 2013; Hochberg et al., 2012; Kwak, Muller, & Lee, 2015; Lebedev et al., 2005; Lebedev & Nicolelis, 2011; Wang et al., 2015) or even whole-body navigation (Craig & Nguyen, 2007; Long et al., 2012; Moore, 2003; Rajangam et al., 2016; Yin, Tseng, Rajangam, Lebedev, & Nicolelis, 2018; Zhang et al., 2016). For those interested in more details, a recent comprehensive review has covered most applications reported using both noninvasive and invasive BMIs (Lebedev & Nicolelis, 2017).

Heretofore, I focused primarily on the classic BMI design introduced in the late 1990s, which aimed at using brain-derived signals to control the movements of artificial devices, such as a robotic arm or computer cursor. But motor BMIs were not the only ones implemented during the past two decades. To that list we can add BMIs designed to replicate sensations (Bensmaia & Miller, 2014; O'Doherty, Lebedev, Hanson, Fitzsimmons, & Nicolelis, 2009; O'Doherty et al., 2011) and even so-called cognitive BMIs (Andersen, Burdick, Musallam, Pesaran, & Cham, 2004) that seek to reproduce decision-making (Hasegawa, Hasegawa, & Segraves, 2009; Musallam, Corneil, Greger, Scherberger, & Andersen, 2004), memory (Berger et al., 2011) and attention (Fuchs, Birbaumer, Lutzenberger, Gruzelier, & Kaiser, 2003; Lubar, 1995).

During this intense divergence in applications, BMI research has also incorporated into its tool kit a variety of classic neurophysiological techniques. One vital add-on was the method for cortical electrical microstimulation, which in the BMI context was employed to provide a new way to establish a bidirectional interaction between brains and devices that did not require using regular sensory feedback signals, such as visual, auditory, or tactile stimuli. Our lab named this new approach the brain-machine-brain interface (BMBI) since it completely bypassed the sensory periphery to deliver continuous tactile feedback directly into the primary somatosensory cortex of rhesus monkeys (O'Doherty et al., 2011). In a series of experiments, we demonstrated that these monkeys could quickly learn to extract tactile information from electrical pulses delivered through cortical microstimulation of their primary somatosensory cortex (SI) cortex. Bypassing the monkey's skin totally, such cortical microstimulation was the only source of tactile feedback provided to the animals when they used a traditional BMI to control the movements of a virtual hand. In this task the monkeys had to use this BMI-controlled virtual arm to discriminate between the textures of three different objects (O'Doherty et al., 2011). After a few weeks of training, these monkeys not only became proficient in using such a BMBI but reached a tactile discrimination performance level similar to that expected if they were using their own fingertips to touch real objects with the same texture (O'Doherty et al., 2011). These observations raised questions about the potential future clinical relevance of BMBIs by demonstrating that different variations of the original BMI paradigm could help patients who, in addition to exhibiting severe levels of body paralysis, must cope with devastating losses in their ability to process normal tactile stimuli. In line with this idea, a series of studies have implemented the BMBI concept in a clinical setting (Micera & Navarro, 2009).

The potential application of BMIs in sensory replacement was further highlighted recently in a series of studies led by Eric Thomson in my laboratory, when multichannel intracortical microstimulation was employed to allow adult rats to perceive infrared light, as if it were a tactile stimulus (Thomson et al., 2014; Thomson, Carra, & Nicolelis, 2013; Thomson et al., 2017). Using a custom-designed, implantable cortical neuroprosthetic device that converted infrared light beams into trains of electrical pulses delivered, initially, to the whisker representation area of the rat primary somatosensory cortex, and later to the animal's primary visual cortex, Thomson and colleagues were able to demonstrate that adult rats are capable of incorporating a complete new sensory modality, in this case sensing infrared light, despite the fact that mammals—with one single exception—do not contain receptors for detecting the infrared light wavelength in their retinas. Such a demonstration suggests that in the future a similar cortical neuroprosthetic device could be employed in cases of severe blindness. And although the original design of this cortical visual neuroprosthesis does not incorporate the traditional logic of a BMI system, its inception was totally inspired by the successful implementation of the BMBI in monkeys.

At the limit of the expansion that led to the introduction of a variety of new BMI paradigms, the latest innovation came with the somewhat surprising demonstration that multiple subjects—animals or humans—could interact simultaneously in a BMI setup known as a Brainet. Although there are a few other ways in which the term *Brainet* can be used to describe different implementations of a shared BMI—including the so-called brain-to-brain interface already described in both animals and healthy human subjects (Pais-Vieira, Chiuffa, Lebedev, Yadav, & Nicolelis, 2015; Pais-Vieira, Lebedev, Kunicki, Wang, & Nicolelis, 2013; Rao et al., 2014)—for this chapter I am using the term *Brainet* to be synonymous with a shared BMI in which the brain activity of multiple subjects is combined, through mathematical and computational means, to generate a global motor control signal needed to move one or more artificial actuators to complete a social motor task.

In the first experimental animal implementation of such Brainets, two or three rhesus monkeys learned to utilize their combined electrical cortical motor activity to cooperate in the execution of a variety of collective virtual motor tasks, such as producing the 2D and 3D movements of an avatar arm (Ramakrishnan et al., 2015). Interestingly, in this original study pairs or triads of rhesus monkeys acquired a high level of performance in different social motor tasks without being aware that they were in fact part of a social group that interacted via a shared-BMI apparatus. Thus, during the execution of the social task, each individual monkey was isolated in a soundproof chamber, each of which was located in a different room of our laboratory. Despite this arrangement, the subjects were still able to develop the high degree of interbrain cortical synchronization required for the successful completion for each motor task. Attaining such a high level of interbrain cortical synchronization was essential because of the task design, which required that each individual monkey mentally contribute a subset of the control signals needed for the successful completion of the social motor task. For example, in the case in which a monkey pair was used to collectively move the avatar arm in 2D space, monkey 1 was in charge of mentally generating the motor commands to move the avatar arm only in the x-axis, while monkey 2 was in charge of generating the brain-based motor commands for controlling the arm movements on the y-axis. Once animals learned to perform this task, a more complex 3D version of the same social motor task was introduced. Now, instead of animal pairs, a monkey triplet was employed. Moreover, instead of just contributing with one dimension of the movement control, each monkey had to generate brain signals corresponding to two out of the three dimensions required for executing 3D movements of the avatar arm. Thus, while monkey 1 fed the shared BMI with cortically derived signals to control the xy coordinates of the avatar arm movement, monkey 2 was in charge of controlling the arm displacement in the yz-axis while monkey 3 responded by generating the neuronal signals involved in controlling the arm movements in the xz-axes. For such a shared BMI to work properly (i.e., by producing smooth 3D trajectories that allowed the avatar arm to intersect a circular target that appeared randomly, at the beginning of the trial over a particular location of a computer screen), at least two of the three monkeys had to perfectly synchronize their electrical cortical motor activity. Interestingly enough, once animals achieved significant performance in this difficult task, one could detect a large number of trials in which all three brains were highly synchronized (Ramakrishnan et al., 2015). Such a surprising level of interbrain cortical synchrony required just a couple of weeks of training to become very common, despite the fact that the only external signals that could serve as instructions for the monkeys to synchronize their collective brain activity were provided by the visual cues, which each animal received by watching the movements of the avatar arm on a computer screen (each animal only saw the movement dimensions it controlled with its brain), and the fruit juice reward they received at the end of a successful trial. Yet that seemed to be plenty for such monkey Brainets to synchronize and produce coherent 3D arm movements generated by the collective firing of a few hundred cortical neurons recorded simultaneously from three distinct monkey brains.

These initial studies with shared BMIs were followed by a recent demonstration that Brainet-like interbrain cortical motor synchrony may occur naturally when pairs of rhesus monkeys become engaged in a more ethologically meaningful social interaction (Tseng et al., 2018). In the first study of this kind, our laboratory was able to show that when pairs of adult monkeys from the same colony became engaged in a social task that involved the execution of whole-body navigation by one of the subjects (named the Passenger), using an electronic wheelchair while an immobile monkey (named the Observer) attentively observed his companion's displacement through an open room, both animals' motor cortices developed intermittent periods of high neuronal synchronization (Tseng et al., 2018). In this task, for both monkeys to receive a reward the Passenger had to either drive an electronic wheelchair (using a wireless

BMI) or be driven by the experimenter to a location in the opposite corner of the room, where it could collect a fruit reward. The moment the Passenger collected its reward, the Observer also received a juice reward. Therefore, the reward contingency somewhat linked these two animals into participating in such a social interaction.

Wireless multichannel cortical recordings were used to obtain simultaneous brain electrical activity from both monkeys while they interacted socially. A detailed analysis of these periods of interbrain motor cortical synchrony revealed that this combined brain signal can be used to predict not only the spatial position of the Passenger in the room but also the Passenger's proximity to the Observer. More surprisingly, this interbrain-synchronized cortical motor activity can predict the social rank of both animals in their colony (Tseng et al., 2018). Indeed, when the higher-ranking monkey played the role of Passenger, as it neared the lower-ranking monkey, the Observer, the levels of interbrain cortical synchrony were much higher than when these roles were reversed.

As in the case of BMIs, our Brainet experiments raise the likely hypothesis that the mirror-neuron system, activated simultaneously in both animals, is responsible for the development of these strong episodes of interbrain cortical motor synchrony. In addition, by showing that M1 neuronal ensembles are capable of encoding a variety of nonmotor parameters, such as reward value and social rank, these studies suggest that the primate motor cortex is involved in higher cognitive functions and not exclusively related to coding motor programs. At this point it is important to highlight that our experiments with the Passenger-Observer Brainet are somewhat reminiscent of previous studies in which human groups employed an EEG-based shared BMI to collectively control a device, reach a common decision, or plan for a movement together (Eckstein et al., 2012; Poli, Cinel, Matran-Fernandez, Sepulveda, & Stoica, 2013; Poli, Valeriani, & Cinel, 2014; Wang & Jung, 2011; Yuan, Wang, Gao, Jung, & Gao, 2013). As such, I believe that future clinical applications of Brainets may take advantage of the possibility of enhancing interbrain cortical synchrony across subjects to achieve therapeutic effects in neurological patients.

Making the Transition to BMI-Based Neurorehabilitation Tools and Potential Therapies

Following its tremendous impact on systems neuroscience and based on the widespread enthusiasm generated by two decades of preliminary clinical testing in severely paralyzed patients, BMI research is currently being translated into efforts toward a new generation of neurorehabilitation protocols and therapies. Overall, this represents a significant shift in the field's original objectives since the initial central clinical goal proposed for BMIs was to provide new means to restore mobility in severely paralyzed patients, like those suffering from complete spinal cord injuries or strokes. To the surprise of many, however, the introduction of novel prosthetic limbs and orthotics, like exoskeletons, controlled directly by BMIs, as well as the implementation of new multidisciplinary neurorehabilitation paradigms for the long-term BMI training of neurological patients, has collectively produced preliminary clinical findings suggesting that BMIs may be able to evolve from a movement-aiding/movement-restoring technology into a true neurorehabilitation tool (Ang et al., 2015; Dobkin, 2007; Donati et al., 2016; Shokur et al., 2016; Shokur et al., 2018; Silvoni et al., 2011).

The first example that raised the possibility of this major change in clinical focus was the integration of BMIs into protocols for the neurorehabilitation of stroke patients. The rationale behind this idea was that long-term practice with BMIs could allow stroke patients to mentally rehearse limb movements, lost due to the cortical damage caused by the stroke, and use their brain activity to control, for instance, a prosthetic device that not only would help the subjects enact the intended movement but also provide feedback sensory information to the subject's brain. The hope was that this BMI interaction would significantly enhance cortical plasticity in stroke patients, leading to a measurable neurological improvement. In line with this hypothesis, when BMI training was added to regular physical therapy, a significant improvement in motor performance was detected (Broetz et al., 2010; Ramos-Murguialday et al., 2013). Interestingly enough, further analysis using motor evoked potentials (MEPs) indicated that such training was correlated with an enhancement in cortical motor activity in the hemisphere ipsilateral to the stroke side (Brasil et al., 2012). Other studies have also shown that when BMI is combined with other methods, such as robot-assisted physical therapy (Ang et al., 2014, 2015), virtual reality (Bermudez, Garcia Morgade, Samaha, & Verschure, 2013), or even transcranial direct-current stimulation (tDCS; Soekadar, Witkowski, Cossio, Birbaumer, & Cohen, 2014), signs of clear neurological improvement can also be observed.

In addition to stroke, the first long-term assessment of the potential clinical effects of a BMI-based neurorehabilitation protocol for chronic SCI patients was carried out by an international research consortium, the Walk Again Project (WAP; Donati et al., 2016; Shokur et al., 2018). This study, performed by the Associação Alberto Santos Dumont para Apoio à Pesquisa's

(AASDAP) neurorehabilitation laboratory in Brazil, included eight chronic paraplegic patients with no somatic sensations below levels ranging from T4 to T11 of their original spinal cord lesion (occurring 3–13 years earlier). During their neurorehabilitation training, patients learned to operate an EEG-based BMI that allowed them to move a series of artificial devices, from avatar bodies to robotic walkers. The latter included an off-the-shelf gait robotic system (Jezernik, Colombo, Keller, Frueh, & Morari, 2003) and a custom-designed lower-limb exoskeleton, developed by the WAP consortium. One of the important innovations in the BMI apparatus used by these SCI patients was the incorporation of a haptic display to provide users with continuous tactile feedback information while they practiced walking using the BMI system, either in virtual reality (by controlling a body avatar) or through the control of the robotic walkers' movements. A stream of tactile feedback was delivered to the patient's forearm skin surface and was complemented by continuous visual feedback (Donati et al., 2016). Among other effects, such a novel arrangement accounted for the fact that all patients reported experiencing both phantom limb sensations and phantom leg movements during virtual reality training, despite the fact that their real bodies remained totally immobile. By taking advantage of this apparatus, six out of eight patients learned to discriminate above chance level between the three different types of surface upon which the avatar body walked (e.g., sand, grass, and asphalt; Shokur et al., 2016).

Even more stunning, following a 12-month period of interaction with this protocol (twice a week, 1 h per day), all enrolled patients began to exhibit signs and symptoms of a remarkable partial clinical recovery. These included an average expansion, below the original level of the SCI, of five dermatomes in nociceptive sensation, a concurrent one to two dermatome expansion in fine touch, considerable enhancement in vibration and proprioception perception, and, more surprisingly, a partial recovery of voluntary muscle contractions (documented by electromyography [EMG] measurements). Such a partial motor recovery was truly remarkable, given that in some of these patients it was enough to allow them to generate, for the first time in more than a decade, multijoint leg movements resembling walking (while suspended in a weight-support system) under their own volition. But their clinical recovery was not restricted to improvements in sensorimotor functions. In parallel, the patients also underwent a significant improvement in visceral functions, translated by the reappearance/increase of peristaltic and bowel movements, different degrees of bladder

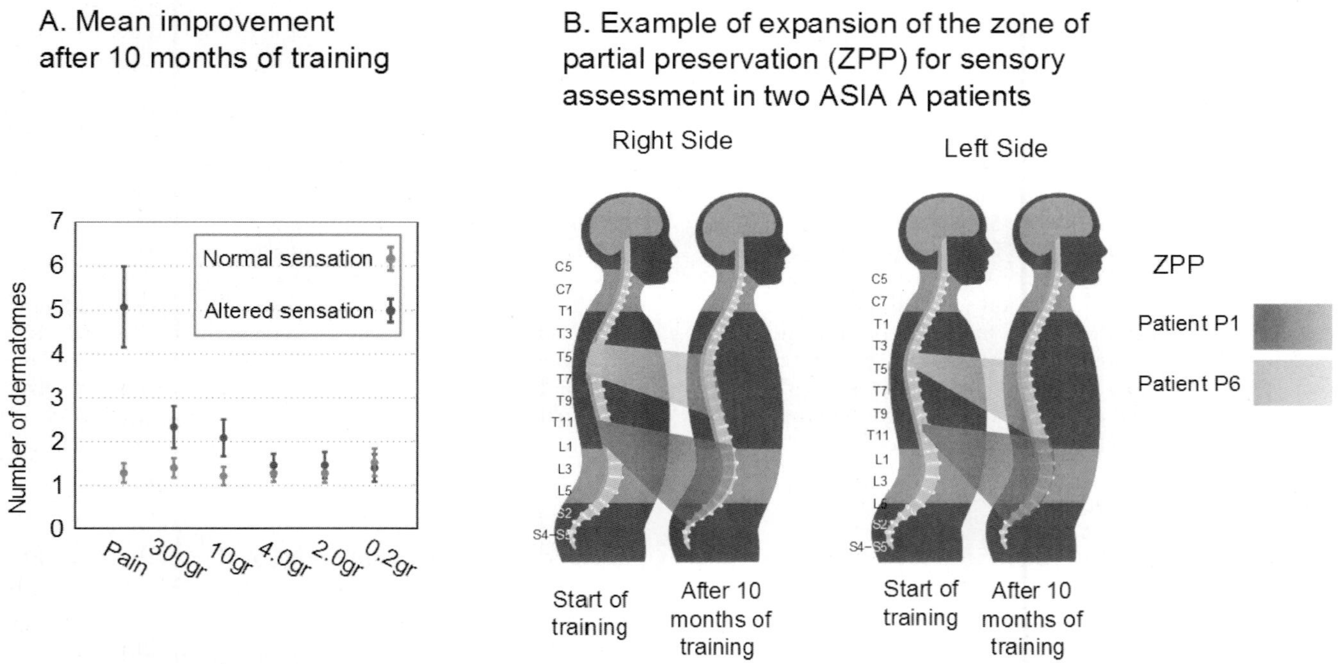

FIGURE 94.2 Partial sensory improvement in chronic SCI patients following training with a BMI protocol. Top shelf: Sensory improvement after neurorehabilitation training. *A*, Average sensory improvement (mean +/− SEM over all patients) after 10 months of training. *B*, Example of improvement in the zone of partial preservation on a sensory evaluation of two patients. Reproduced with permission from Donati et al. (2016). (See color plate 103.)

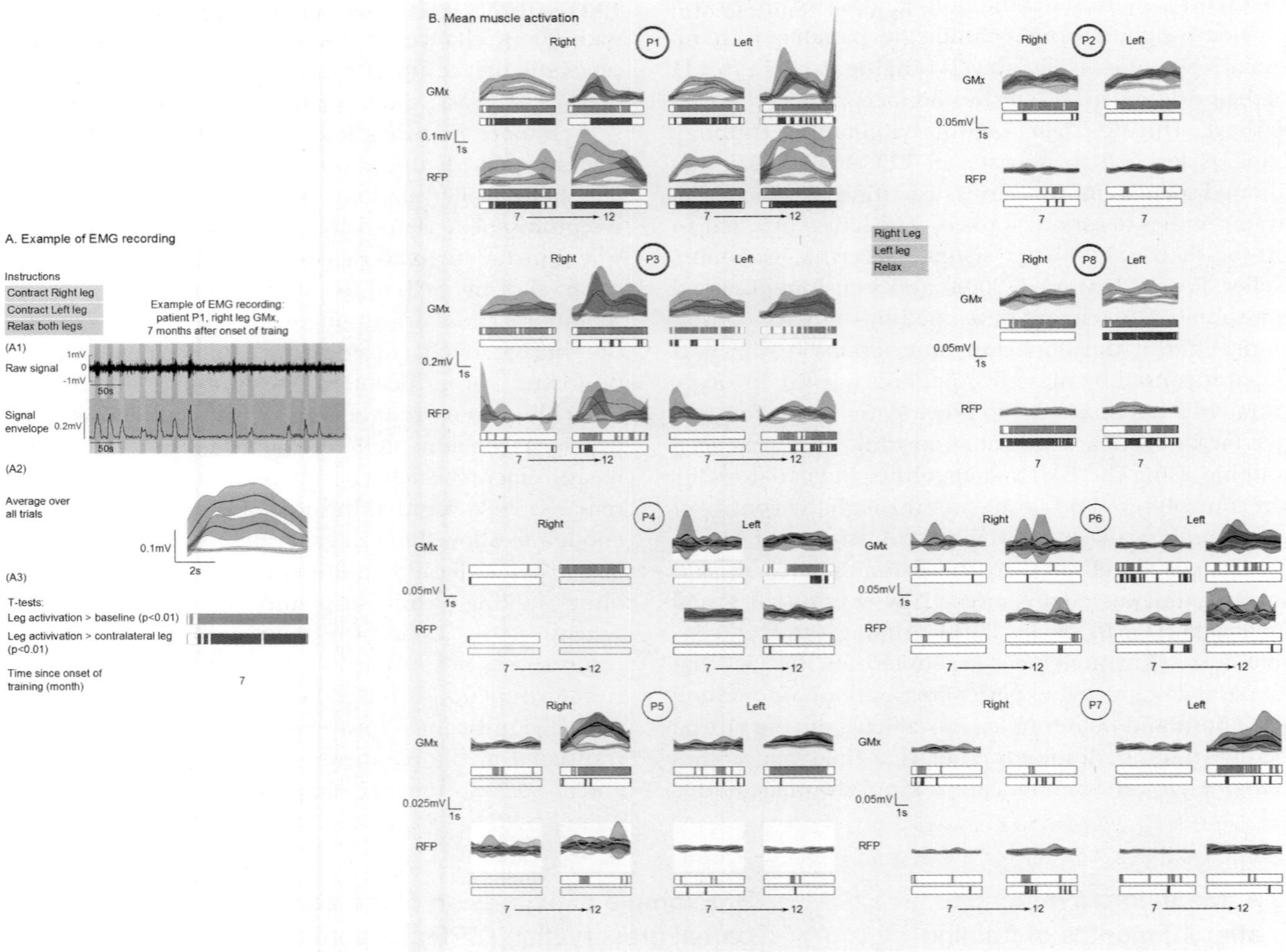

FIGURE 94.3 Lower-limb motor recovery. *A*, Details of the EMG recording procedure in SCI patients. *A1*, Raw EMG for the right gluteus maximus muscle for patient P1 is shown at the top of the topmost graph. The lower part of this graph depicts the envelope of the raw EMG after the signal was rectified and low-pass filtered at 3 Hz. Gray-shaded areas represent periods in which the patient was instructed to move the right leg, while the blue-shaded areas indicate periods of left-leg movement. Red areas indicate periods in which patients were instructed to relax both legs. *A2*, All trials over one session were averaged (mean +/− standard deviation envelopes are shown) and plotted as a function of instruction type (*gray envelope*=contract right leg; *blue*=contract left leg; *red*=relax both legs). *A3*, Below the averaged EMG record, light-green bars indicate instances in which the voluntary muscle contraction (right leg) was significantly different (t-test, $p<0.01$) than the baseline (periods in which she/he was instructed to relax both legs). Dark-green bars depict periods in which there was a significant difference ($p<0.01$) between muscle contraction in the right versus the left leg. *B*, EMG envelopes and t-tests for all recording sessions, involving four muscles, for all eight patients: left and right gluteus maximus (GMx) and rectus femoris proximal (RFP) muscles. Color convention and figure organization follows the one on panel *A*. The data were collected after 7 months of training for all patients and for all but patients P2 and P8 after 12 months. Reproduced with permission from Donati et al. (2016). (See color plate 104.)

control, and an improvement in sexual functions. As a consequence of such a remarkable partial recovery, following 12 months of training, 50% of the patient cohort was upgraded from a complete paraplegia (ASIA A $n=7$, ASIA B $n=1$) to a partial paraplegia classification (ASIA C).

Once again, hints that such a partial clinical recovery was at least in part related to the occurrence of cortical plasticity were also documented by the analysis of EEG records, which depicted an expansion of the lower-limb area in the primary sensorimotor cortex. This suggests that the long-term BMI training, combined with the deliverance of concurrent tactile and visual feedback, may have been the cause of such a clinical improvement. Interestingly enough, in a very recent follow-up paper (Shokur et al., 2018), the same group reported that patients continued to show signs of sensorimotor and visceral improvement as they continued to train for the next 16 months. As a result of this improvement, the average expansion in nociceptive sensation reached 10 dermatomes—twice as many as during the first 12 months—and all 7 patients who remained in the protocol were able to make the transition to the ASIA C category, meaning that they all evolved to become partial paraplegics (Shokur et al., 2018).

Overall, these initial studies applying BMIs to stroke and SCI patients illustrate how tangible is the possibility that a BMI-based neurorehabilitation paradigm may evolve into true therapies aimed at promoting partial neurological recovery. If true, this innovation could mitigate the usual devastating clinical consequences and impact on the quality of life that brain damage can have in patients. In addition to this lofty humanitarian and social goal, because most of the noninvasive BMI applications involve the integration of low-cost technology—perhaps with the exception of exoskeletons, whose price tends to go down with time—one can also envision that BMI-based neurorehabilitation protocols may contribute to a considerable reduction in the overall cost to treat patients with neurological damage. Looking toward that kind of future, I envision that in the next few years there will be a series of major opportunities to combine BMI-based technology with a variety of other therapies, like those derived from stem cell research, or even new medical devices, like neuroprostheses developed to offer new nonpharmacological treatments to other types of neurological disorders. For example, the BMI approach could be combined with the recently introduced method of electrical spinal cord stimulation for improving gait freezing in parkinsonian patients (Yadav & Nicolelis, 2017), a technique that also holds promise for treating chronic drug-resistant epilepsy and even chronic depression (Pais-Vieira et al., 2016).

Among the many potential future BMI-based applications in neurorehabilitation, I also believe that Brainets and other types of shared-BMI applications may open the doors to Internet-based deployment of neurorehabilitation protocols, a revolutionary approach that could potentially affect the lives of tens of millions of severely paralyzed people around the world who have access to neither rehabilitation centers nor caregivers and who cannot afford the costs of most health-care services. According to this view of the future, physical therapists and other rehab professionals could use shared BMIs or Brainet to assist in the training of patients by literally lending their EEG-recorded brain activity so that it can be combined with that of stroke or SCI patients during a shared virtual reality neurorehabilitation protocol. By sharing their EEG activity with a large number of patients simultaneously, these neurorehabilitation professionals could allow populations of patients distributed in many remote locations to engage in a long-term neurorehabilitation training that is capable, via the phenomenon of cortical plasticity, of potentially inducing significant levels of neurological recovery. If such a remarkable future ever materializes, shared-BMI neurorehabilitation protocols could multiply by manyfold the potential social impact of modern brain research, fulfilling, at long last, the enormous hopes and research endeavors invested in this field worldwide.

REFERENCES

Adams, J. A. (1987). Historical review and appraisal of research on the learning, retention, and transfer of human motor-skills. *Psychological Bulletin, 101*(1), 41–74. doi:10.1037//0033-2909.101.1.41

Andersen, R. A., Burdick, J. W., Musallam, S., Pesaran, B., & Cham, J. G. (2004). Cognitive neural prosthetics. *Trends in Cognitive Sciences, 8*(11), 486–493. doi:10.1016/j.tics.2004.09.009

Ang, K. K., Chua, K. S., Phua, K. S., Wang, C., Chin, Z. Y., Kuah, C. W., … Guan, C. (2015). A randomized controlled trial of EEG-based motor imagery brain-computer interface robotic rehabilitation for stroke. *Clinical EEG and Neuroscience, 46*(4), 310–320. doi:10.1177/1550059414522229

Ang, K. K., Guan, C., Phua, K. S., Wang, C., Zhou, L., Tang, K. Y., … Chua, K. S. (2014). Brain-computer interface-based robotic end effector system for wrist and hand rehabilitation: Results of a three-armed randomized controlled trial for chronic stroke. *Frontiers in Neuroengineering, 7*, 30. doi:10.3389/fneng.2014.00030

Benabid, A. L. (2003). Deep brain stimulation for Parkinson's disease. *Current Opinion in Neurobiology, 13*(6), 696–706.

Bensmaia, S. J., & Miller, L. E. (2014). Restoring sensorimotor function through intracortical interfaces: Progress and looming challenges. *Nature Reviews Neuroscience, 15*(5), 313–325. doi:10.1038/nrn3724

Berger, T. W., Hampson, R. E., Song, D., Goonawardena, A., Marmarelis, V. Z., & Deadwyler, S. A. (2011). A cortical neural prosthesis for restoring and enhancing memory. *Journal of Neural Engineering, 8*(4), 046017. doi:10.1088/1741-2560/8/4/046017

Bermudez, i Badia, S., Garcia Morgade, A., Samaha, H., & Verschure, P. F. (2013). Using a hybrid brain computer interface and virtual reality system to monitor and promote cortical reorganization through motor activity and motor imagery training. *IEEE Transactions on Neural Systems and Rehabilitation Engineering, 21*(2), 174–181. doi:10.1109/TNSRE.2012.2229295

Berti, A., & Frassinetti, F. (2000). When far becomes near: Remapping of space by tool use. *Journal of Cognitive Neuroscience, 12*(3), 415–420.

Bilodeau, E. A., & Bilodeau, I. M. (1961). Motor-skills learning. *Annual Review of Psychology, 12*, 243–280. doi:10.1146/annurev.ps.12.020161.001331

Birbaumer, N., Ghanayim, N., Hinterberger, T., Iversen, I., Kotchoubey, B., Kubler, A., … Flor, H. (1999). A spelling device for the paralysed. *Nature, 398*(6725), 297–298. doi:10.1038/18581

Brasil, F., Curado, M., Agostini, M., Liberati, G., Garcia-Cossio, E., Broetz, D., … Soekadar, S. (2012). MEP as predictor of motor recovery in chronic stroke patients after a 4-week daily physical therapy. Paper presented at the Human Brain Mapping Annual Meeting, Beijing, China, June 10–14.

Broetz, D., Braun, C., Weber, C., Soekadar, S. R., Caria, A., & Birbaumer, N. (2010). Combination of brain-computer interface training and goal-directed physical therapy in chronic stroke: A case report. *Neurorehabilitation and Neural Repair, 24*(7), 674–679. doi:10.1177/1545968310368683

Calvo, A. C., Manzano, R., Mendonca, D. M., Munoz, M. J., Zaragoza, P., & Osta, R. (2014). Amyotrophic lateral sclerosis: A focus on disease progression. *BioMed Research International, 2014,* 925101. doi:10.1155/2014/925101

Carmena, J. M., Lebedev, M. A., Crist, R. E., O'Doherty, J. E., Santucci, D. M., Dimitrov, D. F., … Nicolelis, M. A. L. (2003). Learning to control a brain-machine interface for reaching and grasping by primates. *PLoS Biology, 1*(2), E42. doi:10.1371/journal.pbio.0000042

Chapin, J. K., Moxon, K. A., Markowitz, R. S., & Nicolelis, M. A. L. (1999). Real-time control of a robot arm using simultaneously recorded neurons in the motor cortex. *Nature Neuroscience, 2*(7), 664–670. doi:10.1038/10223

Collinger, J. L., Wodlinger, B., Downey, J. E., Wang, W., Tyler-Kabara, E. C., Weber, D. J., … Schwartz, A. B. (2013). High-performance neuroprosthetic control by an individual with tetraplegia. *Lancet, 381*(9866), 557–564. doi:10.1016/S0140-6736(12)61816-9

Contreras-Vidal, J. L., & Grossman, R. G. (2013). NeuroRex: A clinical neural interface roadmap for EEG-based brain machine interfaces to a lower body robotic exoskeleton. Paper presented at the 35th Annual International Conference of the IEEE Engineering in Medicine and Biology Society (EMBC), Osaka, Japan, July 3–7. 10.1109/EMBC.2013.6609816

Craig, D. A., & Nguyen, H. T. (2007). Adaptive EEG thought pattern classifier for advanced wheelchair control. *Conference Proceedings: IEEE Engineering in Medicine and Biology Society*, 2544–2547. doi:10.1109/IEMBS.2007.4352847

Cramer, S. C., Sur, M., Dobkin, B. H., O'Brien, C., Sanger, T. D., Trojanowski, J. Q., … Vinogradov, S. (2011). Harnessing neuroplasticity for clinical applications. *Brain, 134*(Pt. 6), 1591–1609. doi:10.1093/brain/awr039

Dietz, V. (2001). Spinal cord lesion: Effects of and perspectives for treatment. *Neural Plasticity, 8*(1–2), 83–90. doi:10.1155/NP.2001.83

Di Pino, G., Maravita, A., Zollo, L., Guglielmelli, E., & Di Lazzaro, V. (2014). Augmentation-related brain plasticity. *Frontiers in Systems Neuroscience, 8,* 109. doi:10.3389/fnsys.2014.00109

Dobkin, B. H. (2007). Brain-computer interface technology as a tool to augment plasticity and outcomes for neurological rehabilitation. *Journal of Physiology, 579*(Pt. 3), 637–642. doi:10.1113/jphysiol.2006.123067

Donati, A. R., Shokur, S., Morya, E., Campos, D. S., Moioli, R. C., Gitti, C. M., … Nicolelis, M. A. L. (2016). Long-term training with a brain-machine interface-based gait protocol induces partial neurological recovery in paraplegic patients. *Scientific Reports, 6,* 30383. doi:10.1038/srep30383

Doyon, J., Bellec, P., Amsel, R., Penhune, V., Monchi, O., Carrier, J., … Benali, H. (2009). Contributions of the basal ganglia and functionally related brain structures to motor learning. *Behavioural Brain Research, 199*(1), 61–75. doi:10.1016/j.bbr.2008.11.012

Doyon, J., Penhune, V., & Ungerleider, L. G. (2003). Distinct contribution of the cortico-striatal and cortico-cerebellar systems to motor skill learning. *Neuropsychologia, 41*(3), 252–262.

Eckstein, M. P., Das, K., Pham, B. T., Peterson, M. F., Abbey, C. K., Sy, J. L., & Giesbrecht, B. (2012). Neural decoding of collective wisdom with multi-brain computing. *Neuroimage, 59*(1), 94–108. doi:10.1016/j.neuroimage.2011.07.009

Fabbri-Destro, M., & Rizzolatti, G. (2008). Mirror neurons and mirror systems in monkeys and humans. *Physiology (Bethesda), 23,* 171–179. doi:10.1152/physiol.00004.2008

Ferrari, P. F., Rozzi, S., & Fogassi, L. (2005). Mirror neurons responding to observation of actions made with tools in monkey ventral premotor cortex. *Journal of Cognitive Neuroscience, 17*(2), 212–226. doi:10.1162/0898929053124910

Fetz, E. E. (1969). Operant conditioning of cortical unit activity. *Science, 163*(3870), 955–958.

Fitzsimmons, N. A., Lebedev, M. A., Peikon, I. D., & Nicolelis, M. A. L. (2009). Extracting kinematic parameters for monkey bipedal walking from cortical neuronal ensemble activity. *Frontiers in Integrative Neuroscience, 3,* 3. doi:10.3389/neuro.07.003.2009

Fuchs, T., Birbaumer, N., Lutzenberger, W., Gruzelier, J. H., & Kaiser, J. (2003). Neurofeedback treatment for attention-deficit/hyperactivity disorder in children: A comparison with methylphenidate. *Association for Applied Psychophysiology and Biofeedback, 28*(1), 1–12.

Grosse-Wentrup, M., Mattia, D., & Oweiss, K. (2011). Using brain-computer interfaces to induce neural plasticity and restore function. *Journal of Neural Engineering, 8*(2), 025004. doi:10.1088/1741-2560/8/2/025004

Hasegawa, R. P., Hasegawa, Y. T., & Segraves, M. A. (2009). Neural mind reading of multi-dimensional decisions by monkey mid-brain activity. *Neural Networks, 22*(9), 1247–1256. doi:10.1016/j.neunet.2009.07.028

Hikosaka, O., Nakamura, K., Sakai, K., & Nakahara, H. (2002). Central mechanisms of motor skill learning. *Current Opinion in Neurobiology, 12*(2), 217–222.

Hochberg, L. R., Bacher, D., Jarosiewicz, B., Masse, N. Y., Simeral, J. D., Vogel, J., … Donoghue, J. P. (2012). Reach and grasp by people with tetraplegia using a neurally controlled robotic arm. *Nature, 485*(7398), 372–375. doi:10.1038/nature11076

Hochberg, L. R., Serruya, M. D., Friehs, G. M., Mukand, J. A., Saleh, M., Caplan, A. H., … Donoghue, J. P. (2006). Neuronal ensemble control of prosthetic devices by a human with tetraplegia. *Nature, 442*(7099), 164–171. doi:10.1038/nature04970

Ifft, P. J., Shokur, S., Li, Z., Lebedev, M. A., & Nicolelis, M. A. L. (2013). A brain-machine interface enables bimanual arm movements in monkeys. *Science Translational Medicine, 5*(210), 210ra154. doi:10.1126/scitranslmed.3006159

Iriki, A., Tanaka, M., & Iwamura, Y. (1996). Coding of modified body schema during tool use by macaque postcentral neurones. *Neuroreport, 7*(14), 2325–2330.

Jezernik, S., Colombo, G., Keller, T., Frueh, H., & Morari, M. (2003). Robotic orthosis lokomat: A rehabilitation and research tool. *Neuromodulation, 6*(2), 108–115. doi:10.1046/j.1525-1403.2003.03017.x

Kleim, J. A., Barbay, S., & Nudo, R. J. (1998). Functional reorganization of the rat motor cortex following motor skill learning. *Journal of Neurophysiology, 80*(6), 3321–3325. doi:10.1152/jn.1998.80.6.3321

Kwak, N. S., Muller, K. R., & Lee, S. W. (2015). A lower limb exoskeleton control system based on steady state visual evoked potentials. *Journal of Neural Engineering, 12*(5), 056009. doi:10.1088/1741-2560/12/5/056009

Laubach, M., Wessberg, J., & Nicolelis, M. A. L. (2000). Cortical ensemble activity increasingly predicts behaviour outcomes during learning of a motor task. *Nature, 405*(6786), 567–571. doi:10.1038/35014604

Lebedev, M. A., Carmena, J. M., O'Doherty, J. E., Zacksenhouse, M., Henriquez, C. S., Principe, J. C., & Nicolelis, M. A. L. (2005). Cortical ensemble adaptation to represent velocity of an artificial actuator controlled by a brain-machine interface. *Journal of Neuroscience, 25*(19), 4681–4693. doi:10.1523/jneurosci.4088-04.2005

Lebedev, M. A., & Nicolelis, M. A. L. (2006). Brain-machine interfaces: Past, present and future. *Trends in Neurosciences, 29*(9), 536–546. doi:10.1016/j.tins.2006.07.004

Lebedev, M. A., & Nicolelis, M. A. L. (2011). Toward a whole-body neuroprosthetic. *Progress in Brain Research, 194*, 47–60. doi:10.1016/B978-0-444-53815-4.00018-2

Lebedev, M. A., & Nicolelis, M. A. L. (2017). Brain-machine interfaces: From basic science to neuroprostheses and neurorehabilitation. *Physiological Reviews, 97*(2), 767–837. doi:10.1152/physrev.00027.2016

Li, Z. (2014). Decoding methods for neural prostheses: Where have we reached? *Frontiers in Systems Neuroscience, 8*, 129. doi:10.3389/fnsys.2014.00129

Long, J., Li, Y., Wang, H., Yu, T., Pan, J., & Li, F. (2012). A hybrid brain computer interface to control the direction and speed of a simulated or real wheelchair. *IEEE Transactions on Neural Systems and Rehabilitation Engineering, 20*(5), 720–729. doi:10.1109/TNSRE.2012.2197221

Lotte, F., Bougrain, L., Cichocki, A., Clerc, M., Congedo, M., Rakotomamonjy, A., & Yger, F. (2018). A review of classification algorithms for EEG-based brain-computer interfaces: A 10 year update. *Journal of Neural Engineering, 15*(3), 031005. doi:10.1088/1741-2552/aab2f2

Lubar, J. F. (1995). Neurofeedback for the management of attention-deficit/hyperactivity disorders. In M. S. Schwartz (Ed.), *Biofeedback: A practitioner's guide* (pp. 493–522). New York: Guilford Press.

Maravita, A., & Iriki, A. (2004). Tools for the body (schema). *Trends in Cognitive Sciences, 8*(2), 79–86. doi:10.1016/j.tics.2003.12.008

Maravita, A., Spence, C., & Driver, J. (2003). Multisensory integration and the body schema: Close to hand and within reach. *Current Biology, 13*(13), R531–539.

Micera, S., & Navarro, X. (2009). Bidirectional interfaces with the peripheral nervous system. *International Review of Neurobiology, 86*, 23–38. doi:10.1016/S0074-7742(09)86002-9

Mitz, A. R., Godschalk, M., & Wise, S. P. (1991). Learning-dependent neuronal activity in the premotor cortex: Activity during the acquisition of conditional motor associations. *Journal of Neuroscience, 11*(6), 1855–1872.

Moore, M. M. (2003). Real-world applications for brain-computer interface technology. *IEEE Transactions on Neural Systems and Rehabilitation Engineering, 11*(2), 162–165. doi:10.1109/TNSRE.2003.814433

Musallam, S., Corneil, B. D., Greger, B., Scherberger, H., & Andersen, R. A. (2004). Cognitive control signals for neural prosthetics. *Science, 305*(5681), 258–262. doi:10.1126/science.1097938

Nicolelis, M. A. L. (2001). Actions from thoughts. *Nature, 409*(6818), 403–407. doi:10.1038/35053191

Nicolelis, M. A. L. (2011). *Beyond boundaries: The new neuroscience of connecting brains with machines—and how it will change our lives* (1st ed.). New York: Times Books/Henry Holt.

Nicolelis, M. A. L., & Lebedev, M. A. (2009). Principles of neural ensemble physiology underlying the operation of brain-machine interfaces. *Nature Reviews Neuroscience, 10*(7), 530–540. doi:10.1038/nrn2653

O'Doherty, J. E., Lebedev, M. A., Hanson, T. L., Fitzsimmons, N. A., & Nicolelis, M. A. L. (2009). A brain-machine interface instructed by direct intracortical microstimulation. *Frontiers in Integrative Neuroscience, 3*, 20. doi:10.3389/neuro.07.020.2009

O'Doherty, J. E., Lebedev, M. A., Ifft, P. J., Zhuang, K. Z., Shokur, S., Bleuler, H., & Nicolelis, M. A. L. (2011). Active tactile exploration using a brain-machine-brain interface. *Nature, 479*(7372), 228–231. doi:10.1038/nature10489

Oweiss, K. G., & Badreldin, I. S. (2015). Neuroplasticity subserving the operation of brain-machine interfaces. *Neurobiology of Disease, 83*, 161–171. doi:10.1016/j.nbd.2015.05.001

Paddock, C. (2009, April 21). Paralysis affects more Americans than previously thought. *Medical News Today*. https://www.medicalnewstoday.com/articles/146819.php.

Pais-Vieira, M., Chiuffa, G., Lebedev, M., Yadav, A., & Nicolelis, M. A. L. (2015). Building an organic computing device with multiple interconnected brains. *Scientific Reports, 5*, 11869. doi:10.1038/srep11869

Pais-Vieira, M., Lebedev, M., Kunicki, C., Wang, J., & Nicolelis, M. A. L. (2013). A brain-to-brain interface for real-time sharing of sensorimotor information. *Scientific Reports, 3*, 1319. doi:10.1038/srep01319

Pais-Vieira, M., Yadav, A. P., Moreira, D., Guggenmos, D., Santos, A., Lebedev, M., & Nicolelis, M. A. L. (2016). A closed loop brain-machine interface for epilepsy control using dorsal column electrical stimulation. *Scientific Reports, 6*, 32814. doi:10.1038/srep32814

Patil, P. G., Carmena, J. M., Nicolelis, M. A. L., & Turner, D. A. (2004). Ensemble recordings of human subcortical neurons as a source of motor control signals for a brain-machine interface. *Neurosurgery, 55*(1), 27–35, discussion 35–28.

Poli, R., Cinel, C., Matran-Fernandez, A., Sepulveda, F., & Stoica, A. (2013). *Towards cooperative brain-computer interfaces for space navigation*. Paper presented at the Proceedings of the 2013 International Conference on Intelligent User Interfaces, Santa Monica, California, March 19–22.

Poli, R., Valeriani, D., & Cinel, C. (2014). Collaborative brain-computer interface for aiding decision-making. *PLoS One, 9*(7), e102693. doi:10.1371/journal.pone.0102693

Rajangam, S., Tseng, P. H., Yin, A., Lehew, G., Schwarz, D., Lebedev, M. A., & Nicolelis, M. A. L. (2016). Wireless cortical brain-machine interface for whole-body navigation in primates. *Scientific Reports, 6*, 22170. doi:10.1038/srep22170

Ramakrishnan, A., Ifft, P. J., Pais-Vieira, M., Byun, Y. W., Zhuang, K. Z., Lebedev, M. A., & Nicolelis, M. A. L. (2015). Computing arm movements with a monkey brainet. *Scientific Reports, 5*, 10767. doi:10.1038/srep10767

Ramos-Murguialday, A., Broetz, D., Rea, M., Laer, L., Yilmaz, O., Brasil, F. L., … Birbaumer, N. (2013). Brain-machine interface in chronic stroke rehabilitation: A controlled study. *Annals of Neurology, 74*(1), 100–108. doi:10.1002/ana.23879

Rao, R. P., Stocco, A., Bryan, M., Sarma, D., Youngquist, T. M., Wu, J., & Prat, C. S. (2014). A direct brain-to-brain interface in humans. *PLoS One, 9*(11), e111332. doi:10.1371/journal.pone.0111332

Rizzolatti, G., Cattaneo, L., Fabbri-Destro, M., & Rozzi, S. (2014). Cortical mechanisms underlying the organization of goal-directed actions and mirror neuron-based action understanding. *Physiological Reviews, 94*(2), 655–706. doi:10.1152/physrev.00009.2013

Rossignol, S., Schwab, M., Schwartz, M., & Fehlings, M. G. (2007). Spinal cord injury: Time to move? *Journal of Neuroscience, 27*(44), 11782–11792. doi:10.1523/jneurosci.3444-07.2007

Schwarz, D. A., Lebedev, M. A., Hanson, T. L., Dimitrov, D. F., Lehew, G., Meloy, J., … Nicolelis, M. A. L. (2014). Chronic, wireless recordings of large-scale brain activity in freely moving rhesus monkeys. *Nature Methods, 11*(6), 670–676. doi:10.1038/nmeth.2936

Scivoletto, G., & Di Donna, V. (2009). Prediction of walking recovery after spinal cord injury. *Brain Research Bulletin, 78*(1), 43–51. doi:10.1016/j.brainresbull.2008.06.002

Shadmehr, R., & Wise, S. P. (2005). *The computational neurobiology of reaching and pointing: A foundation for motor learning.* Cambridge, MA: MIT Press.

Shokur, S., Donati, A. R., Campos, D. S., Gitti, C. M., Bao, G., Fischer, D., … Nicolelis, M. A. L. (2018). Training with brain-machine interfaces, visuo-tactile feedback and assisted locomotion improves sensorimotor, visceral, and psychological signs in chronic paraplegic patients. *PLoS ONE, 13*(11), e0206464. https://doi.org/10.1371/journal.pone.0206464

Shokur, S., Gallo, S., Moioli, R. C., Donati, A. R., Morya, E., Bleuler, H., & Nicolelis, M. A. L. (2016). Assimilation of virtual legs and perception of floor texture by complete paraplegic patients receiving artificial tactile feedback. *Scientific Reports, 6*, 32293. doi:10.1038/srep32293

Shokur, S., O'Doherty, J. E., Winans, J. A., Bleuler, H., Lebedev, M. A., & Nicolelis, M. A. L. (2013). Expanding the primate body schema in sensorimotor cortex by virtual touches of an avatar. *Proceedings of the National Academy of Sciences of the United States of America, 110*(37), 15121–15126. doi:10.1073/pnas.1308459110

Silvoni, S., Ramos-Murguialday, A., Cavinato, M., Volpato, C., Cisotto, G., Turolla, A., … Birbaumer, N. (2011). Brain-computer interface in stroke: A review of progress. *Clinical EEG and Neuroscience, 42*(4), 245–252. doi:10.1177/155005941104200410

Soekadar, S. R., Witkowski, M., Cossio, E. G., Birbaumer, N., & Cohen, L. G. (2014). Learned EEG-based brain self-regulation of motor-related oscillations during application of transcranial electric brain stimulation: Feasibility and limitations. *Frontiers in Behavioral Neuroscience, 8,* 93. doi:10.3389/fnbeh.2014.00093

Thomson, E. E., Carra, R., & Nicolelis, M. A. L. (2013). Perceiving invisible light through a somatosensory cortical prosthesis. *Nature Communications, 4,* 1482. doi:10.1038/ncomms2497

Thomson, E. E., Lou, J., Sylvester, K., McDonough, A., Tica, S., & Nicolelis, M. A. L. (2014). Basal forebrain dynamics during a tactile discrimination task. *Journal of Neurophysiology, 112*(5), 1179–1191. doi:10.1152/jn.00040.2014

Thomson, E. E., Zea, I., Windham, W., Thenaisie, Y., Walker, C., Pedowitz, J., … Nicolelis, M. A. L. (2017). Cortical neuroprosthesis merges visible and invisible light without impairing native sensory function. *eNeuro, 4*(6). doi:10.1523/ENEURO.0262-17.2017

Tseng, P. H., Rajangam, S., Lehew, G., Lebedev, M. A., & Nicolelis, M. A. L. (2018). Interbrain cortical synchronization encodes multiple aspects of social interactions in monkey pairs. *Scientific Reports, 8*(1), 4699. doi:10.1038/s41598-018-22679-x

Tseng, P. H., Urpí, N. A., Lebedev, M., & Nicolelis, M. (2019). Decoding movements from cortical ensemble activity using a long short-term memory recurrent network. *Neural Computation, 31*(6), 1085–1113. https://doi.org/10.1162/neco_a_01189

Wang, S., Wang, L., Meijneke, C., van Asseldonk, E., Hoellinger, T., Cheron, G., … van der Kooij, H. (2015). Design and control of the MINDWALKER exoskeleton. *IEEE Transactions on Neural Systems and Rehabilitation Engineering, 23*(2), 277–286. doi:10.1109/TNSRE.2014.2365697

Wang, Y., & Jung, T. P. (2011). A collaborative brain-computer interface for improving human performance. *PLoS One, 6*(5), e20422. doi:10.1371/journal.pone.0020422

Wessberg, J., Stambaugh, C. R., Kralik, J. D., Beck, P. D., Laubach, M., Chapin, J. K., … Nicolelis, M. A. L. (2000). Real-time prediction of hand trajectory by ensembles of cortical neurons in primates. *Nature, 408*(6810), 361–365. doi:10.1038/35042582

Wilson, B. S., Finley, C. C., Lawson, D. T., Wolford, R. D., Eddington, D. K., & Rabinowitz, W. M. (1991). Better speech recognition with cochlear implants. *Nature, 352*(6332), 236–238. doi:10.1038/352236a0

Yadav, A. P., & Nicolelis, M. A. L. (2017). Electrical stimulation of the dorsal columns of the spinal cord for Parkinson's disease. *Movement Disorders, 32*(6), 820–832. doi:10.1002/mds.27033

Yin, A., Tseng, P. H., Rajangam, S., Lebedev, M. A., & Nicolelis, M. A. L. (2018). Place cell-like activity in the primary sensorimotor and premotor cortex during monkey whole-body navigation. *Scientific Reports, 8*(1), 9184. doi:10.1038/s41598-018-27472-4

Yuan, P., Wang, Y., Gao, X., Jung, T.-P., & Gao, S. (2013). A collaborative brain-computer interface for accelerating human decision making. In C. Stephanidis & M. Antona (Eds.), *Universal access in human-computer interaction. design methods, tools, and interaction techniques for eInclusion. UAHCI 2013: Lecture notes in computer science* (vol. 8009). Berlin: Springer.

Zacksenhouse, M., Lebedev, M. A., Carmena, J. M., O'Doherty, J. E., Henriquez, C., & Nicolelis, M. A. L. (2007). Cortical modulations increase in early sessions with brain-machine interface. *PLoS One, 2*(7), e619. doi:10.1371/journal.pone.0000619

Zhang, R., Li, Y., Yan, Y., Zhang, H., Wu, S., Yu, T., & Gu, Z. (2016). Control of a wheelchair in an indoor environment based on a brain-computer interface and automated navigation. *IEEE Transactions on Neural Systems and Rehabilitation Engineering, 24*(1), 128–139. doi:10.1109/TNSRE.2015.2439298

95 Aesthetics: From Mind to Brain and Back

OSHIN VARTANIAN AND ANJAN CHATTERJEE

ABSTRACT Aesthetic experiences inform our evaluations of people, places, and things. Exploring the biological basis of aesthetics deepens our understanding of human nature and our actions as relevant to important domains such as the selection of mates, the principles of design, the choices made by consumers, and the appreciation and production of art. Research over the last two decades demonstrates how aesthetic experiences arise from a triad of large-scale systems that involve interaction between emotion-valuation, sensorimotor, and meaning-knowledge neural systems in the brain.

How do people think about the term *aesthetics*? Jacobsen et al. (2004) found that more than 90% of their sample associated the words *beautiful* and 50% associated *ugly* with aesthetics. As such, aesthetics readily evokes the notion of beauty in laypersons. In contrast, contemporary *scientific* treatments of aesthetics expand the denotation of aesthetic judgments more broadly to encompass any *evaluative appraisals of objects* (Brown et al., 2011; Chatterjee & Vartanian, 2014, 2016).

Aesthetic judgments pervade our lives and have a profound impact on our choices. For example, mate preferences are strongly influenced by aesthetics in terms of physical attractiveness (Johnston, 2006; Thornhill & Gangestad, 1999), which also influences our judgments of persons on variables that seem unrelated to physical attractiveness, such as intelligence, personality, and morality (Dion, Berscheid, & Walster, 1972; Tsukiura & Cabeza, 2011). The consumption of goods, choice of where to live, selection of what to wear, and many everyday decisions are influenced by aesthetics.

Gustav Theodor Fechner's (1876) *Vorschule der Aesthetik* (Preschool of aesthetics) marks the beginning of experimental aesthetics. Fechner assumed that the physical properties of stimuli corresponded to the sensations they arouse. Fechner distinguished between *outer psychophysics* and *inner psychophysics*: whereas the former involves the relationship between physical properties of stimuli and their sensations, the latter involves the relationship between those sensations and the neural activities that underlie them. In Fechner's own words, "The stimulus does not immediately arouse a sensation, but in addition, an inner physical activity takes place between it and the sensation" (Robinson, 2010, p. 428). He anticipated a main goal of modern cognitive neuroscience, which is to establish the mediating role of neural processes in relating physical properties of stimuli to their psychological consequences (i.e., mental representations; Ehrenstein & Ehrenstein, 1999; Vartanian, 2015).

Fechner's ideas are present and extended in contemporary experimental aesthetics (Jacobsen, 2006, 2010). First, although he was aware of top-down processes in shaping aesthetic pleasure (Leder, 2015), he advocated strongly for an experimental aesthetics "from below," meaning that he emphasized the search for bottom-up processes that originate in sensation and perception as drivers of aesthetic experiences. Second, since Fechner there has been a strong desire to discover "universal laws" to explain our aesthetic preferences (Berlyne, 1971; Martindale, 1990; McManus, 2013). However, these approaches are complemented by an increased appreciation that top-down processes such as culture, knowledge, and motivation also influence aesthetic experiences (Bullot & Reber, 2013; Nadal, Gallardo, & Marty, 2018; Pelowski, Gerger, et al., 2017).

From Experimental Aesthetics to Neuroaesthetics

Since Fechner, several theoretical advances have been made in our understanding of the processes that underlie aesthetic experiences. More recently, neuroaesthetics has contributed to this knowledge. Neuroaesthetics is motivated to comprehend the neurological underpinnings of aesthetic experiences (Chatterjee, 2011; Nadal & Pearce, 2011; Pearce et al., 2016; Skov & Vartanian, 2009; Zeki, 1999). Specifically, it attempts to discern the proximal and distal *causes* of aesthetic experiences—described by neurologically plausible mechanistic accounts of how they arise. Neuroimaging approaches provide powerful correlational data, which in combination with complementary approaches (e.g., patient studies) can test causal hypotheses to contribute to a richer understanding of aesthetic experiences (Chatterjee, 2004a, 2009, 2011; Chatterjee, Bromberger, Smith, Sternschein, & Widick, 2011; Zaidel, 2005).

Models of Aesthetic Experiences

Several models propose to explain the unfolding of aesthetic experiences in the brain. Chatterjee's (2004b) early model borrowed from visual cognitive neuroscience to propose a framework that rested on two assumptions: First, akin to vision, visual aesthetics has multiple components, each of which contributes to the computation of a different aspect of the overall experience. Second, aesthetic experiences emerge from the organism's response to multiple components of a visual object and their interactions. This framework recognized the contributions of neural structures to processing early and intermediate features of visual objects, as well as downstream processes such as attention and content-specific analyses to the emergence of aesthetic judgments, including the emotions associated with aesthetic experiences.

Consistent with the idea that aesthetic experiences emerge in a progression from processing stimulus-driven properties to "deeper" internally oriented mental states, Cela-Conde et al. (2013) proposed that two different neural systems exert their effects within distinct time frames. Specifically, an initial appraisal of an object as "beautiful" or "not beautiful" emerges rapidly, within 250–750 ms. In turn, a more particular appraisal of the beauty of the object in terms of "the extent to which the stimulus is moving, whether it is interesting or original, how to rate it, and the reasons for considering it attractive" emerges later, around 1,000–1,500 ms. Using magnetoencephalography (MEG), they found that the initial phase engaged structures including the orbitofrontal cortex (OFC), whereas the delayed phase engaged the default mode network (DMN; see also Munar, Nadal, Castellanos, et al., 2012; Munar, Nadal, Roselló, et al., 2012). The delayed engagement of the DMN is consistent with that network's role in internally oriented cognition. Indeed, regions within the DMN are activated during aesthetic experiences (Vartanian, 2018; Vessel, Starr, & Rubin, 2012), suggesting that later stages in aesthetic experiences are experienced as "deeper" and more personally relevant.

Leder et al.'s (2014) important model of aesthetic experience describes five information-processing stages that unfold over an aesthetic episode, eventually leading to aesthetic judgments and emotions (see also Leder & Nadal, 2014). In the Vienna integrated model of top-down and bottom-up processes in art reception (VIMAP), Pelowski, Markey, et al. (2017) updated the original model. Building on Pelowski and Akiba (2011), the effects of top-down as well as bottom-up processes were formally introduced into the framework. The VIMAP also expanded the scope of the model to describe "profound and transformative reactions to artworks." Indeed, there is increasing interest in understanding intense aesthetic episodes and states of "being moved" (Menninghaus et al., 2015; Vessel, Starr, & Rubin, 2012).

Finally, two models introduced recently explicitly aim to describe cognitive mechanisms that lead to the experience of pleasure triggered by negative stimuli, such as sad music or melancholic literature. According to Sachs, Damasio, and Habibi's (2015) homeostatic model, music-evoked sadness leads to pleasure if three conditions are met. First, the sadness evoked by music must be perceived as nonthreatening. Second, it must be experienced as pleasing. Finally, it must produce psychological benefits, including the evocation of memories, empathy, and mood regulation. The realization and interplay of these conditions are influenced by other background factors (e.g., personality). Importantly, the satisfaction of these conditions is not a guarantee that everyone will accrue similar benefits from the experience. Specifically, no pleasure is experienced if the homeostatic imbalance was not present to begin with, or the stimuli failed to correct the imbalance (e.g., no memories were evoked).

According to Menninghaus et al.'s (2017) distancing-embracing model, negative emotions secure attention, elicit intense emotional involvement, and strengthen memories—all of which are claimed to be the core outcomes associated with (good) art. In turn, psychological *distancing* regulates the exposure to the artwork to ensure that negative emotions are not incompatible with expectations of enjoyment, and psychological *embracing* enables one to process the negative emotions to serve deeper aesthetic experiences. Importantly, the model proposes that the simultaneous presence of negative *and* positive emotions is a particularly strong recipe for a deep experience of pleasure.

The Aesthetic Triad

Chatterjee and Vartanian (2014, 2016) proposed that aesthetic experiences emerge from the interaction between emotion-valuation, sensorimotor, and meaning-knowledge neural systems in the brain (figure 95.1). This tripartite model acknowledges that the engagement of all three systems is not necessary for an aesthetic experience to emerge. Specifically, empirical data demonstrate that aesthetic experiences can arise from engagement of any of the three systems alone, or interactions thereof. The three systems can exert their respective influences on aesthetic experiences early or late in the emergence of

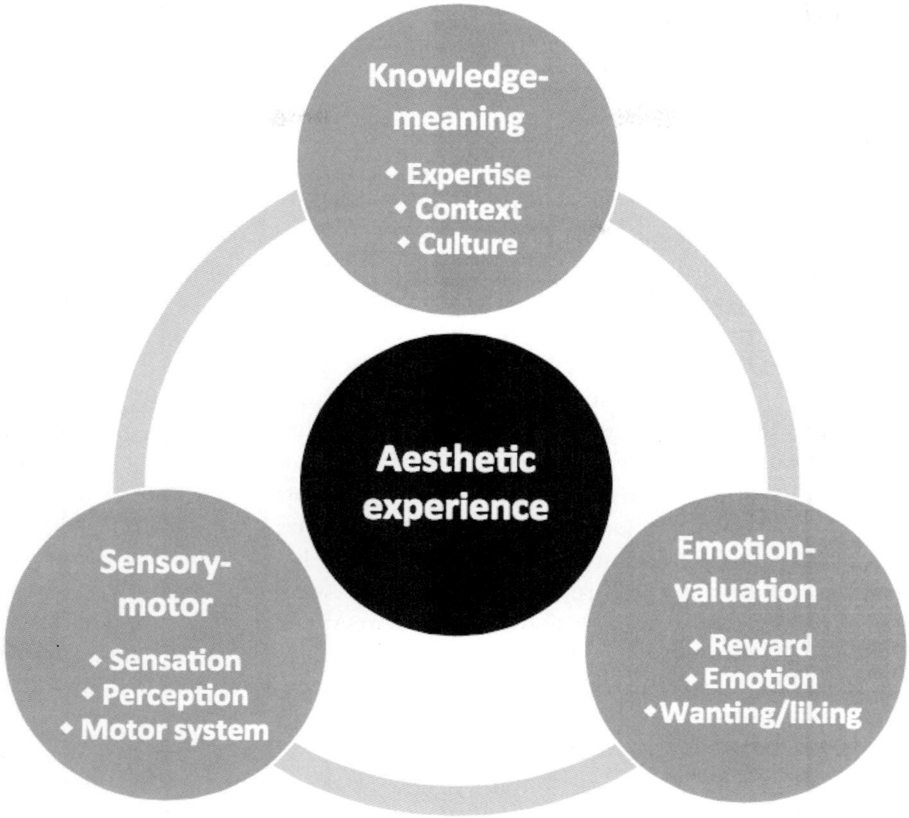

FIGURE 95.1 The aesthetic triad. Aesthetic experiences are an emergent property of the interaction of the sensorimotor, emotion-valuation, and knowledge-meaning neural systems. Reprinted with permission from *Trends in Cognitive Sciences, 18,* © 2014 by Cell Press.

aesthetic experiences. The engagement of the systems is not necessarily sequential but rather interactive. The systems exert their respective influences on the emergence of aesthetic experiences in the larger context, including the person's social and cultural milieu. Finally, the influence of top-down processes need not follow bottom-up processes but can precede them. With these principles in mind, we selectively review neuroaesthetics in support of the *aesthetic triad.*

Emotion-valuation Support for the role of the brain's reward and emotion circuitries in aesthetic experiences emerges from several sources. Prominent are studies of facial attractiveness. Because attractive faces are rewarding, brain regions that are involved in valuation should be activated upon exposure to such faces. Several neuroimaging studies show that assessing faces on attractiveness engages structures implicated in evaluative judgment or reward processing, including the nucleus accumbens, the dorsal striatum, and the OFC (Kampe, Frith, Dolan, & Frith, 2001; Nakamura et al., 1998; O'Doherty et al., 2003; Winston, O'Doherty, Kilner, Perrett, & Dolan, 2007). Consistent with the *common currency* hypothesis, the orbitofrontal and ventral striatal neuronal activity underlies the valuation of rewarding stimuli irrespective of the modality giving rise to the rewarding experience. In this sense, attractive faces could be considered one example of rewarding stimuli that elicit activation in these regions. However, in addition to neural systems common to evaluations across modalities, neural systems may also be tuned to specific modalities and not others. For example, Pegors et al. (2015) instructed participants to rate the attractiveness of faces and places. They found that specific regions of the ventromedial prefrontal cortex and OFC were sensitive to only one type of stimulus category, suggesting that the brain has neural systems for both domain-general and domain-specific aesthetic evaluations.

Interestingly, the pleasure that people derive from looking at beautiful visual objects might *automatically* tap into our reward circuitry (Kühn & Gallinat, 2012). For example, attractive faces activate the fusiform face area (FFA) and parts of the ventral striatum even when people are not thinking explicitly about the attractiveness of the faces (Chatterjee, Thomas, Smith, & Aguirre, 2009; Kim, Adolphs, O'Doherty, & Shimojo, 2007).

Additional evidence for the role of valuation structures in the computation of aesthetic judgments comes from three meta-analyses. Brown et al. (2011) focused on neuroimaging studies in which participants appraised the valence of perceived objects in visual, auditory, gustatory, and olfactory domains. The right anterior insula was activated most consistently across all four modalities. Given that the right anterior insula is part of the brain's core affective system and is strongly associated with visceral perception and the experience of (negative) emotions (Barrett, Mesquita, Ochsner, & Gross, 2007), Brown and colleagues argued that aesthetic processing consists of the appraisal of the valence of perceived objects. In addition, this neural system likely evolved for the appraisal of objects of survival relevance (e.g., food) and was later co-opted to process objects that satisfy social needs, such as artworks.

In turn, Vartanian and Skov (2014) conducted a meta-analysis of studies that involved viewing paintings, regardless of task instructions. This meta-analysis revealed a distributed set of activations in the brain, including the engagement of structures involved in emotion and/or reward processing, such as the anterior insula bilaterally as well as the putamen, a structure reliably activated by the anticipation of rewards (Liu, Hairston, Schrier, & Fan, 2011). Following on the heels of the aforementioned study, Boccia et al. (2016) conducted another meta-analysis of studies in which they focused on visual exposure to a broader set of aesthetically pleasing objects beyond paintings, such as sculptures and scenes. Their results also revealed a distributed set of activations in the brain, including regions within the emotion-valuation system, such as the insula and the amygdala. Together, these quantitative meta-analyses reveal the involvement of structures that underlie the computation of value, affect, and reward in the brain in response to aesthetic experience.

However, structures involved in visual perception also contribute to the computation of value and preferences. For example, the parahippocampal gyrus (PPA) is not only involved in scene perception; its activity while viewing scenes is correlated with pleasure (Yue, Vessel, & Biederman, 2007). Biederman and Vessel (2006) observed that cortical μ-opioid receptor density is greatest in those parts of the ventral visual pathway that process "stimuli that contain a great deal of interpretable information." This suggests that the experience of aesthetic pleasure might arise from the interplay between brain structures that underlie perceptions of specific stimuli (e.g., PPA for scenes) and the distribution of relevant neurotransmitters in the cortex.

Sensorimotor Aesthetic experiences engage brain regions involved in sensation and perception. The meta-analyses by Vartanian and Skov (2014) and Boccia et al. (2015) demonstrated that visual objects activate regions in the visual cortex, including the lingual and the inferior and middle occipital gyrus, as well as the fusiform gyrus. These activations can be attributed to the processing of early, intermediate, and late visual features of the stimuli, including orientation, shape, color, grouping, and categorization. Regarding paintings, specifically, additional activations were observed in the inferior temporal cortex and the precuneus. The inferior temporal cortex has a well-established role in the visual representation of form and color (Gross, 1992) and likely contributes to these processes while a person is viewing paintings as well. The involvement of the precuneus may arise from the visuospatial exploration of pictorial stimuli (Cupchik, Vartanian, Crawley, & Mikulis, 2009; Fairhall & Ishai, 2008.).

Although several studies show that the lateral occipital cortex is engaged in aesthetic experiences involving visual art (Cupchik et al., 2009; Ishizu & Zeki, 2011; Vartanian & Goel, 2004), until recently it was unknown whether this region is causally related to aesthetic experiences. Cattaneo et al. (2015) found that transcranial magnetic stimulation (TMS) applied over the lateral occipital cortex diminished aesthetic evaluations for representational but not abstract artworks, likely because in laypersons an aesthetic preference for representational artworks is driven in large measure by object recognition. TMS applied over the lateral occipital cortex likely affected the perception of objects within those paintings, leading to lower aesthetic evaluations.

The role of visual cortex in aesthetic processing is further supported by additional studies. In one study, activation in visual motion areas MT+ was observed when subjects viewed dynamic paintings that evoke a subjective sense of movement, such as Vincent van Gogh's work (Thakral, Moo, & Slotnick, 2012). Cattaneo et al. (2017) used TMS to test the intriguing possibility that this area might contribute to the aesthetic preference for those artworks. The results demonstrated that TMS not only reduced the perceived sense of motion but also reduced the aesthetic preference for abstract (but not representational) paintings. The authors argued that the reason for a reduced preference for abstract but not representational paintings is that in the former, more attention is allocated to their low-level features, such as motion, color, and orientation (Nadal, 2013), and therefore a disturbance of the sensory-perceptual apparatus has an impact on aesthetic preference for abstract but not representational art.

Viewing paintings that depict actions also engages the motor system. This engagement taps into the extended mirror neuron system. First discovered in a region of the macaque's premotor cortex, mirror neurons respond both when the monkey performs an action or when it observes a similar action being performed by another agent (Gallese, Fadiga, Fogassi, & Rizzolatti, 1996). A similar system has since been discovered in humans (Cattaneo & Rizzolatti, 2009). This system is engaged when people infer the intent of artistic gestures or observe the consequences of actions, such as in the cut canvases of Lucio Fontana. This subtle motor engagement presumably represents an embodied element of our empathetic responses to visual art (Freedberg & Gallese, 2007).

Ticini et al. (2014) asked participants to rate pointillist-style paintings featuring discernible brushstrokes on a Likert scale. Each painting was preceded by either a compatible prime consisting of a static image depicting a hand holding a paintbrush with precision, an incompatible prime depicting a hand holding a power grip, or a control prime depicting a hand resting palm down on a table. The authors hypothesized that if action simulation were causally involved in art appreciation, then participants would like the artwork in the compatible condition the most. This hypothesis was supported, suggesting that involuntary covert movement simulation in perceivers contributes to aesthetic appreciation.

In another study Gerger, Pelowski, and Leder (2018) administered a questionnaire that measures *emotional contagion*—the ability to pick up, mirror, and "feel into" emotions, a propensity correlated positively with empathic abilities. Next, they instructed their subjects to evaluate visual artworks on several aesthetic dimensions while they collected bodily reactions associated with empathic engagement, including facial electromyography (fEMG) and skin conductance responses (SCR). Their results demonstrated that subjects who scored higher on the emotional contagion measure exhibited more intense bodily reactions while viewing artwork and evaluated the artwork higher on all aesthetic dimensions, although the effect was relatively stronger for representational than for abstract artwork.

The evidence presented here is consistent with the argument that our aesthetic experiences often incorporate the role of the mirror neuron system (Gallese & Freedberg, 2007). Empathetic responses to paintings engage our emotional circuitry, mirroring the emotions expressed in art (Freedberg & Gallese, 2007; Gallese & Freedberg, 2007).

Meaning-knowledge A major trend in experimental aesthetics has been the appreciation of top-down processes in aesthetic experience. For example, knowing that a work of art is a forgery reduces our valuation of it (Newman & Bloom, 2012). In addition, when subjects were shown Rembrandt portraits and given information regarding their authenticity, designating a portrait as a "fake" rather than "authentic" was correlated with greater neural activity in the frontopolar cortex and the precuneus (Huang, Bridge, Kemp, & Parker, 2011). The activation in frontopolar cortex could relate to its role in evaluating multiple (potentially conflicting) goals and hypotheses (Koechlin & Hyafil, 2007), whereas the involvement of the precuneus could relate to its role in integrating current input with prior knowledge (Maguire, Frith, & Morris, 1999) or episodic memory retrieval (Krause et al., 1999).

Along similar lines, knowing the title of an artwork facilitates greater engagement and deepening of aesthetic experiences (Leder, Carbon, & Ripsas, 2006), presumably by guiding how the content of the artwork is processed. Gerger and Leder (2015) presented subjects with artworks accompanied by three different title types: semantically matching, semantically nonmatching, and an "untitled" control condition. According to fluency theory (Reber, Schwarz, & Winkielman, 2004), positive aesthetic experiences are driven by processing ease. As such, semantically matching titles in which the titles reference the content of the paintings unambiguously can facilitate processing, whereas the reverse would be true for semantically nonmatching titles. The researchers also obtained fEMG recordings over corrugator supercilii and zygomaticus major muscles to measure subtle changes in emotional and cognitive processing. Likert ratings were higher in the matching and control conditions than in the nonmatching condition. However, M. zygomaticus activated more strongly only in the matching condition. In the matching (fluent) condition, higher aesthetic ratings were associated with positive emotions, suggesting a possible causal role for semantic information on preference.

Another influential study by Kirk et al. (2009) presented their subjects with paintings labeled as sourced from a museum or generated by a computer. Subjects rated the images as aesthetically more pleasing if they were thought to be from the museum. This preference was accompanied by greater neural activity in the medial OFC and ventromedial prefrontal cortex. In addition, believing that a painting came from a museum also produced greater activity in the temporal pole and entorhinal cortex. This suggests that contextual information can activate memories that in turn modulate aesthetic pleasure. The findings of Kirk and colleagues

TABLE 95.1 Outstanding questions.

• What is the nature of the valuation of objects computed in the sensory cortices?
• What is the relationship between aesthetic judgments and approach-avoidance responses?
• Do different parts of the extended reward circuitry play different roles in aesthetic experience?
• How are different aesthetic emotions—including negative feelings such as horror and disgust—implemented in the brain, and how do they give us pleasure?
• How do aesthetic objects evoke moods in viewers that persist after an encounter with an artwork?
• What is the role of the DMN in aesthetic experiences?
• What unique contribution, if any, does each hemisphere make to aesthetic perception and production?
• How does expertise in the visual arts alter the neural structures and functional responses to aesthetic objects?
• What is the relationship between aesthetic judgments and other socially and culturally relevant values, such as judgments of morality and truth?
• Are aesthetic judgments automatic, or do they require deliberation?
• What are the evolutionary underpinnings of the ability of the brain to experience aesthetic pleasure?
• Can art perception and creation be used therapeutically?
• Do art perception and production serve homeostatic needs?

Source: Adapted from Chatterjee and Vartanian (2014).

were replicated by Silveira et al. (2015), who presented subjects with paintings from New York's Museum of Modern Art (MoMA) labeled as being from the MoMA or from an adult education center. The results revealed greater neural activation in the left precuneus and superior and inferior parietal cortex for the MoMA condition. When taking aesthetic preference for a painting into account, the MoMA condition elicited greater involvement of the right precuneus, bilateral anterior cingulate cortex, and temporoparietal junction. Of particular interest is that activation observed in the precuneus that was also observed in the authentic Rembrandts or fake study by Huang et al. (2011), suggesting that the precuneus is sensitive to the labeling of objects. In addition, Lacey et al. (2011) presented their subjects with images easily recognizable as works of art along with other images that were not artworks but matched the art images in terms of content. Their results showed that the ventral striatum, the hypothalamus, and the OFC were activated when subjects viewed images from the former category, suggesting that the status of an image as a work of art, irrespective of its visual content, can activate the reward system in the brain. These studies suggest that aesthetic interactions with artworks are strongly affected by our knowledge of compositional strategies, stylistic conventions, and practices (Seeley, 2013), beyond merely their sensory qualities.

Beyond the Aesthetic Triad

Despite advances in neuroaesthetics, many questions require further study (table 95.1). Perhaps not surprisingly, some humanist scholars question the ability of neuroaesthetics to shed light on our artistic experiences (Croft, 2011; Tallis, 2008) and have criticized the field's focus on aesthetic (rather than artistic) phenomena (Brown & Dissanayake, 2009). However, we believe that the insight provided through biological studies constitutes one component toward developing a complete explanation of how aesthetic experiences arise in our minds. Although in isolation this approach has limitations, we believe that an interdisciplinary framework incorporating neuroscientific approaches will fruitfully advance our understanding of not only aesthetic phenomena but also core aspects of our mind, including our desires and motivations (Chatterjee, 2014).

REFERENCES

Barrett, L. F., Mesquita, B., Ochsner, K. N., & Gross, J. J. (2007). The experience of emotion. *Annual Review of Psychology, 58*, 373–403.

Berlyne, D. E. (1971). *Aesthetics and psychobiology*. New York: Appleton-Century-Crofts.

Biederman, I., & Vessel, E. A. (2006). Perceptual pleasure and the brain. *American Scientist, 94*, 247–253.

Boccia, M., Barbetti, S., Piccardi, L., Guariglia, C., Ferlazzo, F., Giannini, A. M., & Zaidel, D. W. (2016). Where does brain neural activation in aesthetic responses to visual art occur? Meta-analytic evidence from neuroimaging studies. *Neuroscience and Biobehavioral Reviews, 60*, 65–71.

Brown, S., & Dissanayake, E. (2009). The arts are more than aesthetics: Neuroaesthetics as narrow aesthetics. In M. Skov & O. Vartanian (Eds.), *Neuroaesthetics* (pp. 43–57). Amityville, NY: Baywood.

Brown, S., Gao, X., Tisdelle, L., Eickhoff, S. B., & Liotti, M. (2011). Naturalizing aesthetics: Brain areas for aesthetic appraisal across sensory modalities. *Neuroimage, 58*, 250–258.

Bullot, N. J., & Reber, R. (2013). A psycho-historical research program for the integrative science of art. *Behavioral and Brain Sciences, 36*, 163–180.

Cattaneo, Z., Lega, C., Ferrari, C., Vecchi, T., Cela-Conde, C. J., Silvanto, J., & Nadal, M. (2015). The role of the lateral occipital cortex in aesthetic appreciation of representational and abstract paintings: A TMS study. *Brain and Cognition, 95*, 44–53.

Cattaneo, L., & Rizzolatti, G. (2009). The mirror neuron system. *Archives of Neurology, 66*, 557–560.

Cattaneo, Z., Schiavi, S., Silvanto, J., & Nadal, M. (2017). A TMS study on the contribution of visual area V5 to the perception of implied motion in art and its appreciation. *Cognitive Neuroscience, 8*, 59–68.

Cela-Conde, C. J., Garcia-Prieto, J., Ramasco, J., Mirasso, C. R., Bajo, R., Munar, E., ... Maestú, F. (2013). Dynamics of brain networks in the aesthetic appreciation. *Proceedings of the National Academy of Sciences of the United States of America, 110* (suppl. 2), 10454–10461.

Chatterjee, A. (2004a). The neuropsychology of visual artists. *Neuropsychologia, 42*, 1568–1583.

Chatterjee, A. (2004b). Prospects for a cognitive neuroscience of visual aesthetics. *Bulletin of Psychology and the Arts, 4*, 55–60.

Chatterjee, A. (2009). Prospects for a neuropsychology of art. In M. Skov & O. Vartanian (Eds.), *Neuroaesthetics* (pp. 131–143). Amityville, NY: Baywood.

Chatterjee, A. (2011). Neuroaesthetics: A coming of age story. *Journal of Cognitive Neuroscience, 23*, 53–62.

Chatterjee, A. (2014). *The aesthetic brain: How we evolved to desire beauty and enjoy art*. New York: Oxford University Press.

Chatterjee, A., Bromberger, B., Smith II, W. B., Sternschein, R., & Widick, P. (2011). Artistic production following brain damage: A study of three artists. *Leonardo, 44*, 405–410.

Chatterjee, A., Thomas, A., Smith, S. E., & Aguirre, G. K. (2009). The neural response to facial attractiveness. *Neuropsychology, 23*, 135–143.

Chatterjee, A., & Vartanian, O. (2014). Neuroaesthetics. *Trends in Cognitive Sciences, 18*, 370–375.

Chatterjee, A., & Vartanian, O. (2016). Neuroscience of aesthetics. *Annals of the New York Academy of Sciences, 1369*, 172–194.

Croft, J. (2011). The challenges of interdisciplinary epistemology in neuroaesthetics. *Mind, Brain, and Education, 5*, 5–11.

Cupchik, G. C., Vartanian, O., Crawley, A., & Mikulis, D. (2009). Viewing artworks: Contributions of cognitive control and perceptual facilitation to aesthetic experience. *Brain and Cognition, 70*, 84–91.

Dion, K., Berscheid, E., & Walster, E. (1972). What is beautiful is good. *Journal of Personality and Social Psychology, 24*, 285–290.

Ehrenstein, W. H., & Ehrenstein, A. (1999). Psychophysical methods. In U. Windhorst & H. Johansson (Eds.), *Modern techniques in neuroscience research* (pp. 1211–1241). New York: Springer.

Fairhall, S. L., & Ishai, A. (2008). Neural correlates of object indeterminacy in art compositions. *Consciousness and Cognition, 17*, 923–932.

Fechner, G. T. (1876). *Vorschule der Aesthetik* [Experimental aesthetics; "Pre-school" of aesthetics]. Leipzig: Breitkopf & Härtel.

Freedberg, D., & Gallese, V. (2007). Motion, emotion and empathy in esthetic experience. *Trends in Cognitive Science, 11*, 197–203.

Gallese, V., Fadiga, L., Fogassi, L., & Rizzolatti, G. (1996). Action recognition in the premotor cortex. *Brain, 119*, 593–609.

Gallese, V., & Freedberg, D. (2007). Mirror and canonical neurons are crucial elements in esthetic response. *Trends in Cognitive Sciences, 11*, 411.

Gerger, G., & Leder, H. (2015). Titles change the esthetic appreciations of paintings. *Frontiers in Human Neuroscience, 9*, article 464.

Gerger, G., Pelowski, M., & Leder, H. (2018). Empathy, Einfühlung, and aesthetic experience: The effect of emotion contagion on appreciation of representational and abstract art using fEMG and SCR. *Cognitive Processing, 9*, 147–165.

Gross, C. G. (1992). Representation of visual stimuli in inferior temporal cortex. *Philosophical Transactions of the Royal Society B, 335*, 3–10.

Huang, M., Bridge, H., Kemp, M. J., & Parker, A. J. (2011). Human cortical activity evoked by the assignment of authenticity when viewing works of art. *Frontiers in Human Neuroscience, 5*, article 134.

Ishizu, T., & Zeki, S. (2011). Toward a brain-based theory of beauty. *PLoS One, 6*, e21852.

Jacobsen, T. (2006). Bridging the arts and sciences: A framework for the psychology of aesthetics. *Leonardo, 39*, 155–162.

Jacobsen, T. (2010). Beauty and the brain: Culture, history and individual differences in aesthetic appreciation. *Journal of Anatomy, 216*, 184–191.

Jacobsen, T., Buchta, K., Köhler, M., & Schröger, E. (2004). The primacy of beauty in judging the aesthetics of objects. *Psychological Reports, 94*, 1253–1260.

Johnston, V. S. (2006). Mate choice decisions: The role of facial beauty. *Trends in Cognitive Sciences, 10*, 9–13.

Kampe, K. K. W., Frith, C. D., Dolan, R. J., & Frith, U. (2001). Reward value of attractiveness and gaze. *Nature, 413*, 589.

Kim, H., Adolphs, R., O'Doherty, J. P., & Shimojo, S. (2007). Temporal isolation of neural processes underlying face preference decisions. *Proceedings of the National Academy of Sciences of the United States of America, 104*, 18253–18258.

Kirk, U., Skov, M., Hulme, O., Christensen, M. S., & Zeki, S. (2009). Modulation of aesthetic value by semantic context: An fMRI study. *Neuroimage, 44*, 1125–1132.

Koechlin, E., & Hyafil, A. (2007). Anterior prefrontal function and the limits of human decision-making. *Science, 318*, 594–598.

Krause, B. J., Schmidt, D., Mottaghy, F. M., Taylor, J., Halsband, U., Herzog, H., ... Müller-Gärtner, H. W. (1999). Episodic retrieval activates the precuneus irrespective of the imagery content of word pair associates: A PET-study. *Brain, 122*, 255–263.

Kühn, S., & Gallinat, J. (2012). The neural correlates of subjective pleasantness. *Neuroimage, 61*, 289–294.

Lacey, S., Haqtvedt, H., Patrick, V. M., Anderson, A., Stilla, R., Deshpande, G., ... Sathian, K. (2011). Art for reward's sake: Visual art recruits the ventral striatum. *Neuroimage, 55*, 420–433.

Leder, H. (2015). Beyond perception—information processing approaches to art appreciation. In P. L. Tinio & J. K. Smith (Eds.), *The Cambridge handbook of the psychology of aesthetics and the arts* (pp. 115–138). Cambridge: Cambridge University Press.

Leder, H., Belke, B., Oeberst, A., & Augustin, D. (2004). A model of aesthetic appreciation and aesthetic judgments. *British Journal of Psychology, 95,* 489–508.

Leder, H., Carbon, C-C., & Ripsas, A-L. (2006). Entitling art: Influence of title information on understanding and appreciation of paintings. *Acta Psychologica, 121,* 176–198.

Leder, H., & Nadal, M. (2014). Ten years of a model of aesthetic appreciation and aesthetic judgments: The aesthetic episode—developments and challenges in empirical aesthetics. *British Journal of Psychology, 105,* 443–464.

Liu, X., Hairston, J., Schrier, M., & Fan, J. (2011). Common and distinct networks underlying reward valence and processing stages: A meta-analysis of functional neuroimaging studies. *Neuroscience and Biobehavioral Reviews, 35,* 1219–1236.

Maguire, E. A., Frith, C. D., & Morris, R. G. M. (1999). The functional neuroanatomy of comprehension and memory: The importance of prior knowledge. *Brain, 122,* 1839–1850.

Martindale, C. (1990). *The clockwork muse: The predictability of artistic change.* New York: Basic Books.

McManus, C. (2013). The anti-developmental, the anti-narrative, the anti-historical": Mondrian as a paradigmatic artist for empirical aesthetics. *Behavioral and Brain Sciences, 36,* 152–153.

Menninghaus, W., Wagner, V., Hanich, J., Wassiliwizky, E., Jacobsen, T., & Koelsch, S. (2017). The distancing-embracing model of the enjoyment of negative emotions in art reception. *Behavioral and Brain Sciences, 40,* E347.

Menninghaus, W., Wagner, V., Hanich, J., Wassiliwizky, E., Kuehnast, M., & Jacobsen, T. (2015). Towards a psychological construct of being moved. *PLoS One, 10,* e0128451.

Montague, P. R., & Berns, G. S. (2002). Neural economics and the biological substrates of valuation. *Neuron, 36,* 265–284.

Munar E., Nadal, M., Castellanos, N. P., Flexas, A., Maestú, F., Mirasso, C., & Cela-Conde, C. J. (2012). Aesthetic appreciation: Event-related field and time-frequency analyses. *Frontiers in Human Neuroscience, 5,* article 185.

Munar, E., Nadal, M., Roselló, J., Flexas, A., Moratti, S., Maestú, F., ... Cela-Conde, C. J. (2012). Lateral orbitofrontal cortex involvement in initial negative aesthetic impression formation. *PLoS One, 7,* e38152.

Nadal, M. (2013). The experience of art: Insights from neuroimaging. *Progress in Brain Research, 204,* 135–158.

Nadal, M., Gallardo, V., & Marty, G. (2018). Commentary: But is it really art? The classification of images as "art"/"not art" and correlation with appraisal and viewer interpersonal differences. *Frontiers in Psychology, 8,* article 2328.

Nadal, M., & Pearce, M. T. (2011). The Copenhagen Neuroaesthetics Conference: Prospects and pitfalls for an emerging field. *Brain and Cognition, 76,* 172–183.

Nakamura, K., Kawashima, R., Naqumo, S., Ito, K., Suquira, M., Kato, T., ... Kojima, S. (1998). Neuroanatomical correlates of the assessment of facial attractiveness. *Neuroreport, 9,* 753–757.

Newman, G. E., & Bloom, P. (2012). Art and authenticity: The importance of originals in judgments of value. *Journal of Experimental Psychology: General, 141,* 558–569.

O'Doherty, J., Winston, J., Critchley, H., Perrett, D., Burt, D. M., & Dolan, R. J. (2003). Beauty in a smile: The role of medial orbitofrontal cortex in facial attractiveness. *Neuropsychologia, 41,* 147–155.

Pearce, M. T., Zaidel, D. W., Vartanian, O., Skov, M., Leder, H., Chatterjee, A., & Nadal, M. (2016). Neuroaesthetics: The cognitive neuroscience of aesthetic experience. *Perspectives on Psychological Science, 11,* 265–279.

Pegors, T. K., Kable, J., Chatterjee, A., & Epstein, R. (2015). Common and unique representations in pFC for face and place attractiveness. *Journal of Cognitive Neuroscience, 27,* 959–973.

Pelowski, M., & Akiba, F. (2011). A model of art perception, evaluation and emotion in transformative aesthetic experience. *New Ideas in Psychology, 29,* 80–97.

Pelowski, M., Gerger, G., Chetouani, Y., Markey, P. S., & Leder, H. (2017). But is it really art? The classification of images as "Art"/"Not Art" and correlation with appraisal and viewer interpersonal differences. *Frontiers in Psychology, 8,* article 1729.

Pelowski, M., Markey, P. S., Forster, M., Gerger, G., & Leder, H. (2017). Move me, astonish me ... delight my eyes and brain: The Vienna Integrated Model of top-down and bottom-up processes in Art Perception (VIMAP) and corresponding affective, evaluative, and neurophysiological correlates. *Physics of Life Reviews, 21,* 80–125.

Reber, R., Schwarz, N., & Winkielman, P. (2004). Processing fluency and aesthetic pleasure: Is beauty in the perceiver's processing experience? *Personality and Social Psychology Review, 8,* 364–382.

Robinson, D. K. (2010). Fechner's "inner psychophysics." *History of Psychology, 13,* 424–433.

Sachs, M. E., Damasio, A., & Habibi, A. (2015). The pleasures of sad music: A systematic review. *Frontiers in Human Neuroscience, 9,* article 404.

Seeley, W. P. (2013). Art, meaning, and perception: A question of methods for a cognitive neuroscience of art. *British Journal of Aesthetics, 53,* 443–460.

Silveira, S., Fehse, K., Vedder, A., Elvers, K., & Hennig-Fast, K. (2015). Is it the picture or is it the frame? An fMRI study on the neurobiology of framing effects. *Frontiers in Human Neuroscience, 9,* article 528.

Skov, M., & Vartanian, O. (Eds.). (2009). *Neuroaesthetics.* Amityville, NY: Baywood.

Tallis, R. (2008). The limitations of a neurological approach to art. *Lancet, 372,* 19–20.

Thakral, P. P., Moo, L. L., & Slotnick, S. D. (2012). A neural mechanism for aesthetic experience. *Neuroreport, 23,* 310–313.

Thornhill, R., & Gangestad, S. W. (1999). Facial attractiveness. *Trends in Cognitive Sciences, 3,* 452–460.

Ticini, L. F., Rachman, L., Pelletier, J., & Dubal, S. (2014). Enhancing aesthetic appreciation by priming canvases with actions that match the artist's painting style. *Frontiers in Human Neuroscience, 8,* article 391.

Tsukiura, T., & Cabeza, R. (2011). Shared brain activity for aesthetic and moral judgments: Implications for the beauty-is-good stereotype. *Social, Cognitive and Affective Neuroscience, 6,* 138–148.

Vartanian, O. (2015). Empirical aesthetics: Hindsight and foresight. In P. L. Tinio & J. K. Smith (Eds.), *The Cambridge handbook of the psychology of aesthetics and the arts* (pp. 6–34). Cambridge: Cambridge University Press.

Vartanian, O. (2018). Internal orientation in aesthetic experience. In K. C. R. Fox & K. Christoff (Eds.), *The Oxford handbook of spontaneous thought* (pp. 327–336). Oxford: Oxford University Press.

Vartanian, O., & Goel, V. (2004). Neuroanatomical correlates of aesthetic preference for paintings. *Neuroreport, 15*, 893–897.
Vartanian, O., & Skov, M. (2014). Neural correlates of viewing paintings: Evidence from a quantitative meta-analysis of functional magnetic resonance imaging data. *Brain Cognition,* 87, 52–56.
Vessel, E. A., Starr, G. G., & Rubin, N. (2012). The brain on art: Intense aesthetic experience activates the default mode network. *Frontiers in Human Neuroscience, 6,* article 66.
Winston, J. S., O'Doherty, J., Kilner, J. M., Perrett, D. I., & Dolan, R. J. (2007). Brain systems for assessing facial attractiveness. *Neuropsychologia, 45*, 195–206.
Yue, X., Vessel, E. A., & Biederman, I. (2007). The neural basis of scene preferences. *Neuroreport, 18*, 525–529.
Zaidel, D. W. (2005). *Neuropsychology of art: Neurological, cognitive, and evolutionary perspectives.* Hove, UK: Psychology Press.
Zeki, S. (1999). *Inner vision: An exploration of art and the brain.* New York: Oxford University Press.

96 Music: Prediction, Production, Perception, Plasticity, and Pleasure

ROBERT J. ZATORRE AND VIRGINIA B. PENHUNE

ABSTRACT Music exploits every component of our nervous system, from the basic perceptual mechanisms to the motor, attentional, memory, cognitive, and emotion systems. Cognitive neuroscience has sought to understand these mechanisms, providing insight not only into the brain and how it works but also into music itself. In this chapter we focus on the neural basis for musical perception and production, on the plasticity associated with expertise in these domains, and on the mechanisms behind the pleasure that music elicits. We review the literature within the framework of hierarchical and parallel auditory cortical pathways and situate the work within the context of prediction, the concept that neural processing is strongly influenced by the utility of generating expected outcomes in order to compare them to experienced events, which in turn guides learning and behavior.

Perception

Although music is highly varied across cultures and styles, certain elements are commonly shared (Savage, Brown, Sakai, & Currie, 2015). For music to exist, the brain must solve a number of computational problems: it must extract information about patterns of sounds distributed in time, encode relationships among the relevant features of those sounds, generate predictions about upcoming events based both on the immediate context and long-term statistical regularities, and evaluate the outcomes of those predictions on an ongoing basis. These processes may be best understood by considering the functional and structural organization of the auditory nervous system.

Auditory brain stem nuclei transmit information from the cochlea to the cortex but also receive massive input from the cortex (Eggermont, 2001). Electroencephalography (EEG) responses index sound onsets and periodicity via the frequency-following response (FFR; Kraus, Anderson, & White-Schwoch, 2017), which is linked to the perception of pitch (Coffey, Herholz, Chepesiuk, Baillet, & Zatorre, 2016) and consonance and dissonance (Bones, Hopkins, Krishnan, & Plack, 2014). The FFR is generated in brain stem nuclei and cortex (Coffey et al., 2016), thus demonstrating the importance of cortical-subcortical interactions. The auditory cortex consists of multiple subfields that are hierarchically organized from core to belt and parabelt regions, which in turn project to two higher-order pathways (Hickok & Poeppel, 2015; Rauschecker & Scott, 2009). The dorsal pathway connects to parietal, premotor, and frontal cortices and is linked to auditory-motor integration, spatial encoding, and working memory. The ventral pathway connects to the inferior temporal and frontal cortices and is linked to semantic and pattern processing.

Pitch perception is central to most music and is based on the neural encoding of frequency (Saenz & Langers, 2014). However pitch is a higher-order perceptual construct because the same pitch percept can arise from physically different sounds (McDermott & Oxenham, 2008). Based on work in monkeys and humans, pitch is likely computed in lateral auditory regions (Bendor & Wang, 2006; Johnsrude, Penhune, & Zatorre, 2000; Penagos, Melcher, & Oxenham, 2004; Puschmann, Uppenkamp, Kollmeier, & Thiel, 2010). Relationships between individual pitches are encoded in secondary auditory cortices (Gutschalk, Patterson, Scherg, Uppenkamp, & Rupp, 2004), and these regions encode abstract features, such as melodic contour (Lee, Janata, Frost, Hanke, & Granger, 2011) or musical intervals (Klein & Zatorre, 2015), and may even be selective for music as a category (Norman-Haignere, Kanwisher, & McDermott, 2015). Substantial evidence demonstrates a right-hemisphere advantage for pitch-related processing, in keeping with the concept that the right auditory cortex possesses enhanced resolution of spectral information (Cha, Zatorre, & Schonwiesner, 2016; Zatorre, Belin, & Penhune, 2002).

But auditory cortex does more than encode pitch patterns; it also generates predictions about upcoming events, as indicated by EEG studies showing mismatch responses when a change occurs in a pattern of sounds, both for physical properties, such as pitch or timbre (Vuust et al., 2011), and also for more abstract changes, such as melodic contour (Saarinen, Paavilainen, Schoger, Tervaniemi, & Naatanen, 1992). Mismatch responses originating in auditory cortex probably represent relatively automatic change-detection mechanisms, whereas responses from inferior frontal regions likely represent more cognitive processes (Schonwiesner, Krumbholz, Rubsamen, Fink, & von Cramon, 2007).

Amusia

The role of the ventral stream in pattern processing is supported by studies of amusia, a congenital and largely encapsulated disorder of music processing (Peretz, 2016). Although individuals with amusia have preserved auditory cortical responses to pitch deviations (Peretz, Brattico, Jarvenpaa, & Tervaniemi, 2009), they are typically unaware of tonal violations and lack the corresponding physiological response in frontal cortex (Zendel, Lagrois, Robitaille, & Peretz, 2015). They also have deficits in tonal working memory and abnormal auditory cortical responses to tone patterns (Albouy, Cousineau, Caclin, Tillmann, & Peretz, 2016). These deficits are linked to abnormal gray matter (GM) organization in right auditory and inferior frontal regions (Albouy et al., 2013; Hyde et al., 2007) and in the white matter (WM) pathways connecting them (Loui, Alsop, & Schlaug, 2009). Amusia may also be considered a disorder of predictive coding, given that feedback connections seem to be particularly affected, leading to the inability to consciously recognize tonal relationships.

Imagery

Auditory cortical areas engaged for perception also respond to sound imagery (Zatorre & Halpern, 2005). It is even possible to obtain a mismatch response from auditory cortex when a real sound differs from an imagined one (Herholz, Lappe, Knief, & Pantev, 2008), suggesting that imagery serves as an internal model, creating expectancies about upcoming events. Auditory imagery also serves working memory, engaging ventral stream regions when musicians maintain rhythmic or pitch sequences in memory (Brown et al., 2013; Konoike et al., 2012; Schulze, Mueller, & Koelsch, 2011) and engaging the dorsal stream when information must be manipulated (Foster, Halpern, & Zatorre, 2013; Foster & Zatorre, 2010b). In particular, these studies point to the role of the intraparietal region of the dorsal stream in the manipulation of auditory information. Causal evidence for this role comes from a study in which rhythmic transcranial magnetic stimulation (TMS) to the parietal cortex was shown to boost performance during a tonal manipulation task (Albouy, Weiss, Baillet, & Zatorre, 2017). This finding fits with the broader idea that the dorsal stream carries out computations that allow different reference frames to be brought into register and hence participates in diverse processes such as imagery, working memory, and sensorimotor tasks.

Sensorimotor Integration

For musicians, hearing or imagining the notes of a known piece automatically generates activity in motor regions (Bangert et al., 2006; Lahav, Saltzman, & Schlaug, 2007). This phenomenon suggests that there is an integrated representation of sound and action in the brain (Keller & Koch, 2008; Schubotz, 2007). During learning, auditory information and motor responses are linked to form a joint auditory-motor representation, likely encoded in an auditory-parietal-motor network (Brown et al., 2013; Chen, Rae, & Watkins, 2012; Herholz, Coffey, Pantev, & Zatorre, 2016; Lahav, Saltzman, & Schlaug, 2007). Recent brain stimulation studies have shown that dorsal premotor cortex is a crucial node for learning auditory-motor associations (Lega, Stephan, Zatorre, & Penhune, 2016) and that once learned, the predicted or imagined tones activate anticipatory motor representations, even in the absence of movement (Stephan, Lega, & Penhune, 2018). Further, even in naïve participants, listening to musical rhythms elicits activity in motor regions of the brain (Chen, Penhune, & Zatorre, 2008a; Grahn & Brett, 2007). This evidence is consistent with the concept of predictive coding that postulates that the motor system relies on joint representations of sensory targets and their motor implementation to plan upcoming actions (Hickok & Poeppel, 2015; Novembre & Keller, 2014).

Rhythm and Groove

Just as musical pitch and melody generate predictions in tonal space, rhythm generates predictions in a temporal reference frame. *Rhythm* is the pattern of sounds and silences in an auditory stream (Clarke, 1999), and *meter* is their perceived organization into repeating groups, typically based on an isochronous underlying beat (Honig, 2013). Because rhythmic structures are hierarchically organized, they provide the basis for temporal expectations or predictions (Large & Snyder, 2009; Loehr, Large, & Palmer, 2011; Vuust & Witek, 2014). Predictable metrical structures facilitate both perception (Manning & Schutz, 2013; Morillon, Schroeder, & Wyart, 2014) and production (Repp & Su, 2013). Musical "groove" is an excellent example of the power of temporal prediction. Groove is the pleasurable sensation of wanting to move to the musical beat (Janata, Tomic, & Haberman, 2012; Madison, 2006; Witek, Clarke, Wallentin, Kringelbach,

& Vuust, 2014). The sensation of groove is strongest for rhythms that are moderately syncopated (Janata, Tomic, & Haberman, 2012; Matthews, Witek, Heggli, Penhune, & Vuust, 2019; Vuust & Witek, 2014). It has been proposed that more syncopated rhythms, in which predicting the beat is more difficult, require the listener to use movement to assist prediction, thus producing the desire to move, and when that prediction is successful, it produces pleasure (Vuust & Witek, 2014). Thus, metric complexity may contribute to building expectations that lead to pleasurable responses in music.

Rhythm perception and production rely on a network of auditory, motor, and prefrontal regions that are also linked to subcortical structures such as the basal ganglia (BG) and cerebellum. The perception of predictable meter or beat may rely on the BG and its interaction with premotor regions (Grahn & Brett, 2007; Grahn & Rowe, 2009, 2013). The perception and production of more complex meters may rely on more complex premotor and frontal networks (Chen, Penhune, & Zatorre, 2008b; Kung, Chen, Zatorre, & Penhune, 2013), as well as the cerebellum (Teki, Grube, Kumar, & Griffiths, 2011).

A number of related theories propose that the perception of meter and beat may rely on the entrainment of neuronal oscillations in auditory and motor cortex that focus perception and attention, as well as enhance motor accuracy around beat points (e.g., Large & Snyder, 2009; but see Patel & Iversen, 2014). EEG and magnetoencephalography (MEG) studies have found that oscillatory activity in auditory and motor regions tracked the beat structure of musical stimuli (Fujioka, Trainor, Large, & Ross, 2012; Iversen, Repp, & Patel, 2009; Nozaradan, Peretz, & Mouraux, 2012).

Plasticity

Studies of music training have demonstrated the modulation of structure and function, predominantly in the auditory-motor-prefrontal networks underlying perception and production (Brown, Zatorre, & Penhune, 2015; Herholz & Zatorre, 2012). Auditory brain stem and cortical responses are enhanced as a function of years of training (Coffey et al., 2016; Shahin, Bosnyak, Trainor, & Roberts, 2003; Skoe & Kraus, 2013). Similarly, musicians show enhancements in sensorimotor response that are related to the instrument played (Elbert, Pantev, Wienbruch, Rockstroh, & Taub, 1995; Kleber, Veit, Birbaumer, Gruzelier, & Lotze, 2010). Baseline connectivity between the auditory and motor systems is greater for musicians (Palomar-Garcia, Zatorre, Ventura-Campos, Bueicheku, & Avila, 2017), and the two systems are coactivated when pianists play music without auditory feedback or listen to music without playing (Bangert et al., 2006; Baumann et al., 2007; D'Ausilio, Altenmuller, Olivetti Belardinelli, & Lotze, 2006; Haueisen & Knosche, 2001).

Structural-imaging studies have revealed differences between musicians and nonmusicians in both GM and WM architecture in the auditory-motor-prefrontal network. Auditory areas, including primary auditory cortex, are enhanced in musicians and correlate with hours of practice, variations in physiological responses to auditory signals, and performance on musical tasks (Bermudez, Lerch, Evans, & Zatorre, 2009; Foster & Zatorre, 2010a; Gaser & Schlaug, 2003; Schneider et al., 2002).

Trained musicians also have anatomical enhancements in motor regions, including primary motor cortex (Amunts et al., 1997; Bermudez et al., 2009; Gaser & Schlaug, 2003) and the anterior corpus callosum (CC), which connects the motor regions of the two hemispheres (Bengtsson et al., 2005; Schlaug, Jancke, Huang, Staiger, & Steinmetz, 1995). WM integrity in the CC has been shown to be related to hours of practice (Bengtsson et al., 2005), starting age, and sensorimotor synchronization performance (Steele, Bailey, Zatorre, & Penhune, 2013). In addition, measures of WM organization in the corticospinal tract are related to hours of practice in childhood (Bengtsson et al., 2005). Finally, differences have been found in subcortical motor structures, including the putamen (Vaquero et al., 2016) and cerebellum (Abdul-Kareem, Stancak, Parkes, Al-Ameen, et al., 2011; Baer et al., 2015; Gaser & Schlaug, 2003; Hutchinson, Lee, Gaab, & Schlaug, 2003).

Musicians also have greater GM volume in the parietal regions (Foster & Zatorre, 2010a; Gaser & Schlaug, 2003; James et al., 2014), which are engaged in sensorimotor transformations and working memory. Enhanced GM density in area 44/45 has been related to years of music experience (Abdul-Kareem, Stancak, Parkes, & Sluming, 2011; James et al., 2014; Sluming et al., 2002). Importantly, musicians have greater WM integrity in the arcuate fasciculus, the pathway connecting auditory, parietal, and inferior frontal regions in the dorsal stream (Halwani, Loui, Ruber, & Schlaug, 2011). Musicians also have greater cortical thickness in the dorsolateral prefrontal cortex (DLPFC), which covaries with that of auditory regions (Bermudez et al., 2009).

Training

It is difficult to infer causality from cross-sectional studies. However, longitudinal studies indicate that even relatively short-term training can produce changes in behavior, neural response, and brain structure in the auditory-motor-frontal network. For example, children undergoing long-term training show enhanced EEG and MEG responses to musical stimuli (Fujioka, Ross, Kakigi, Pantev, & Trainor, 2006; Putkinen, Tervaniemi, Saarikivi, Ojala, & Huotilainen, 2014; Seither-Preisler, Parncutt, & Schneider, 2014). Longitudinal studies in children also show increased GM in primary and secondary auditory cortex, as well as motor regions correlate with measures of musical skill (Habibi et al., 2017; Hyde et al., 2009; Seither-Preisler, Parncutt, & Schneider, 2014). Studies of short-term training in adults also show structural and functional adaptations in auditory, premotor, and parietal cortex (Chen, Rae, & Watkins, 2012; Herholz et al., 2016; Lahav, Saltzman, & Schlaug, 2007; Woollman, Segado, Charpentier, Penhune, & Zatorre, 2018).

It has long been debated whether predispositions or training makes the biggest contribution to these short- and long-term changes. In two training studies using different instruments, activity in auditory or supplementary motor areas before training predicted subsequent learning (Herholz et al., 2016; Woollman et al., 2018). This finding provides indirect evidence that preexisting neural architecture may mediate the potential to acquire certain musical skills. Individual differences in specific anatomical features, such as the volume of auditory and parietal cortex or WM pathways, also predict performance on musical tasks (Engel et al., 2014; Foster & Zatorre, 2010a; Li et al., 2014; Paquette, Fujii, Li, & Schlaug, 2017; Schneider et al., 2005). Some evidence also suggests that a larger volume of auditory cortex in children may be a preexisting structural feature associated with musical skill (Seither-Preisler, Parncutt, & Schneider, 2014).

Heritability studies show that genetics accounts for a large portion of the variance in music abilities and that the propensity to practice may be genetically transmitted (Mosing, Madison, Pedersen, Kuja-Halkola, & Ullen, 2014). However, a comparison of brain structure in monozygotic twins shows that those who played an instrument had greater cortical thickness in auditory and motor regions, as well as WM enhancements in the CC (de Manzano & Ullen, 2018) than those who did not play. These findings provide the most definitive support yet for the causal effect of music training on brain structure. These authors propose that there are genetic predispositions for both music-specific skills and nonspecific cognitive and personality factors, which interact with the environment to either promote or discourage training (Ullen, Hambrick, & Mosing, 2016). We further propose that the age at which music training begins interacts with genetic predispositions and normative brain maturation to influence long-term behavioral and brain plasticity.

The Interaction between Development and Training

Several early studies showed that M1 and the CC were larger in musicians who began before age 7 (Amunts et al., 1997; Schlaug et al., 1995). These studies did not control for years of training, but when behavior and brain structure were compared in early- (ET, < age 7) and late-trained (LT, > age 7) adult musicians matched for years of experience, ET musicians performed better than LT musicians on rhythm and melody tasks. Additionally, ET musicians had greater GM in the right ventral premotor cortex (Bailey, Zatorre, & Penhune, 2014), enhanced WM integrity in the posterior midbody of the CC (Steele et al., 2013), and reductions in the volume of the cerebellar regions important for motor timing (Baer et al., 2015). These findings demonstrate an interaction between normative maturation and specific training. Both premotor cortex and the CC have their peak period of growth between 6 and 8 years of age (Giedd et al., 1999; Westerhausen et al., 2011). The cerebellum has a later period of peak development between the ages of 12 and 18 (Tiermeier et al., 2010), as well as specialized neural architecture and connections that may result in early training having a different effect on its structure. Taken together, these findings suggest a scaffold, or metaplastic, effect, wherein early training promotes brain plasticity that is sustained or augmented by later practice (Herholz & Zatorre, 2012; Steele et al., 2013).

Why Is Music Such an Effective Driver of Brain Plasticity?

There are four reasons why music may promote brain plasticity. The first is lengthy, consistent training. In the studies reviewed here, musicians trained for an average of 15–20 years, the equivalent of thousands of hours of practice. Second, practice may be thought of as a repeated prediction/feedback/error-correction cycle. Feedback and error correction are key components of motor learning (Wolpert, Diedrichsen, & Flanagan, 2011) and may be particularly effective during periods of high developmental change. Third, music training is inherently multisensory, and the coactivation of circuits from multiple senses may drive plasticity more strongly (Lee & Noppeney, 2011). Finally, we know that learning and plasticity are strongly affected by the reward value of what is learned (Bao, Chan, & Merzenich, 2001; David, Fritz, & Shamma, 2012). Thus, if a piece of music is pleasurable, then playing it should promote plasticity. We turn next, therefore, to music and reward.

Pleasure

Music exists in every human culture (Savage et al., 2015; Trehub, Becker, & Morley, 2015) and is thought to have developed along with other complex cognitive abilities, including language and abstract reasoning (Honig, 2018; Huron, 2001; Patel, 2008). Music serves many distinct social functions: accompanying rituals, underlining cultural identity, and supporting group cohesion (Cross, 2012). At an individual level, music may help regulate mood and provide a safe venue for emotional expression (Zald & Zatorre, 2011). Music generates highly varied and subtle emotional states (Brattico & Pearce, 2013; Juslin & Sloboda, 2011; Zentner, Grandjean, & Scherer, 2008), although its expression differs across cultures (McDermott, Schultz, Undurraga, & Godoy, 2016). That a sequence of abstract sounds should be experienced as beautiful, thrilling, or rewarding seems remarkable. But cognitive neuroscience has recently shed some light on this mystery.

Progress in understanding musical pleasure comes from the reward literature in animals and humans. A key concept derived largely from work in animals using primary reinforcers, such as food or drugs, is that activity in the mesolimbic dopamine system encodes differences between expected and obtained rewards, or reward prediction error (Schultz, 2013). Human neuroimaging research has shown that secondary reinforcers, such as money, also engage the reward system (Sescousse, Caldu, Segura, & Dreher, 2013). These studies converge in showing that a set of common regions—the ventral striatum (VS), ventromedial prefrontal cortex (VMPFC), insula, and thalamus—respond to all rewarding stimuli. Additional cortical regions are also engaged depending on the nature of the stimuli, suggesting that a range of stimuli may access the core reward system via different routes, a concept that may explain why humans experience pleasure and reward from abstract stimuli.

Neuroimaging studies of pleasurable music have consistently shown activity in the striatum and other components of the reward system (Koelsch, 2012) that is linked to the degree of experienced pleasure (Blood & Zatorre, 2001); dopamine binding in the VS is specifically involved in this process (Salimpoor, Benovoy, Larcher, Dagher, & Zatorre, 2011). Further, activity in the VS, and its interaction with auditory and frontal regions, scales with monetary bids for a new tune (Salimpoor et al., 2013). These results support the concept that regions involved in the perceptual and cognitive aspects of music processing interact with regions involved in emotion, value-guided decision-making, and reward to generate the experience of pleasure.

If musical pleasure involves interactions between auditory and reward systems, then the disruption of this network should be associated with a disturbance of pleasure. To test this prediction, Mas-Herrero et al. (2014) identified a cohort of otherwise healthy individuals who do not experience pleasure from music and who do not show the characteristic psychophysiological responses associated with pleasant music. As predicted, they do not engage the VS to music and showed no coupling between the auditory and reward systems (Martinez-Molina, Mas-Herrero, Rodriguez-Fornells, Zatorre, & Marco-Pallares, 2016). Conversely, people who reported higher than average levels of musical enjoyment also showed higher than average coupling across these structures. Direct causal evidence that communication between perceptual and reward systems is key to the experience of musical reward comes from a study that modulated activity in the reward system with TMS (Mas-Herrero, Dagher, & Zatorre, 2017) while listening to music. Stimulation that upregulated dopamine release in the striatum led to increased subjective pleasure, physiological responses, and valuation, whereas stimulation that downregulated dopamine release decreased all these variables.

These results identify the neural mechanisms underlying musical pleasure, but they do not explain why music generates pleasure. To understand this problem, we return to the concept of prediction. Salimpoor et al. (2011) observed that the VS was maximally responsive during the experience of peak pleasure, whereas the dorsal striatum was maximally active during the several seconds leading up to the peak response moment. This anticipatory response may correspond to musical predictions, in which the pattern of sounds currently experienced leads to expectancies about upcoming sounds. This dissociation is consistent with the dorsal striatum's connectivity patterns (Haber & Knutson, 2010) with cortical regions, including the DLPFC and anterior cingulate, involved in cognitive functions such as temporal sequencing, planning, computing expected outcomes, and decision-making. Musicologists and music theorists have formulated models compatible with this idea, in which affective responses are elicited via phenomena such as tension/resolution, delay, anticipation, and surprise (Huron, 2006; Meyer, 1956).

An emerging concept is that musical pleasure may be induced by some aspect of the prediction system, either in the form of expectations and violations thereof (Gebauer, Kringelbach, & Vuust, 2012) and/or more directly in the form of reward prediction errors (Zald & Zatorre, 2011). The latter concept is appealing because it fits with the model that the dopamine mesolimbic system encodes such information, which in the case of music would consist of an abstract evaluation of patterns of sound unfolding in time with a certain, optimal balance of expectedness and surprise (North & Hargreaves, 1995). This computation presumably depends upon the capacity of the cortico-cortical systems previously described to make predictions about outcomes determined both by immediate antecedents (the sounds just experienced), as well as long-term learning (the schematic predictions based on statistical learning through past exposure). Hence, the interactions between reward and perceptual/prediction mechanisms that emerge during musical pleasure would represent the link between cognitive prediction and evaluative/hedonic systems. It remains to be precisely defined how positive or negative reward prediction maps onto an aesthetic context, but computational models (Pearce, Ruiz, Kapasi, Wiggins, & Bhattacharya, 2010) may provide a means to quantify deviation from expectedness to test such ideas.

REFERENCES

Abdul-Kareem, I. A., Stancak, A., Parkes, L. M., Al-Ameen, M., Alghamdi, J., Aldhafeeri, F. M., … Sluming, V. (2011). Plasticity of the superior and middle cerebellar peduncles in musicians revealed by quantitative analysis of volume and number of streamlines based on diffusion tensor tractography. *Cerebellum, 10*(3), 611–623. doi:10.1007/s12311-011-0274-1

Abdul-Kareem, I. A., Stancak, A., Parkes, L. M., & Sluming, V. (2011). Increased gray matter volume of left pars opercularis in male orchestral musicians correlate positively with years of musical performance. *Journal of Magnetic Resonance Imaging, 33*(1), 24–32. doi:10.1002/jmri.22391

Albouy, P., Cousineau, M., Caclin, A., Tillmann, B., & Peretz, I. (2016). Impaired encoding of rapid pitch information underlies perception and memory deficits in congenital amusia. *Scientific Reports, 6*, 18861. doi:10.1038/srep18861

Albouy, P., Mattout, J., Bouet, R., Maby, E., Sanchez, G., Aguera, P. E., … Tillmann, B. (2013). Impaired pitch perception and memory in congenital amusia: The deficit starts in the auditory cortex. *Brain, 136*(Pt. 5), 1639–1661. doi:10.1093/brain/awt082

Albouy, P., Weiss, A., Baillet, S., & Zatorre, R. J. (2017). Selective entrainment of theta oscillations in the dorsal stream causally enhances auditory working memory performance. *Neuron, 94*(1), 193–206 e195. doi:10.1016/j.neuron.2017.03.015

Amunts, K., Schlaug, G., Jancke, L., Steinmetz, H., Schleicher, A., Dabringhaus, A., & Zilles, K. (1997). Motor cortex and hand motor skills: Structural compliance in the human brain. *Human Brain Mapping, 5*(3), 206–215. doi:10.1002/(SICI)1097-0193(1997)5:3<206:AID-HBM5>3.0.CO;2-7

Baer, L., Park, M., Bailey, J., Chakravarty, M., Li, K., & Penhune, V. (2015). Regional cerebellar volumes are related to early musical training and finger tapping performance. *NeuroImage, 109*, 130–139.

Bailey, J. A., Zatorre, R. J., & Penhune, V. B. (2014). Early musical training is linked to gray matter structure in the ventral premotor cortex and auditory-motor rhythm synchronization performance. *Journal of Cognitive Neuroscience, 26*(4), 755–767. doi:10.1162/jocn_a_00527

Bangert, M., Peschel, T., Schlaug, G., Rotte, M., Drescher, D., Hinrichs, H., … Altenmuller, E. (2006). Shared networks for auditory and motor processing in professional pianists: Evidence from fMRI conjunction. *NeuroImage, 30*(3), 917–926. doi:10.1016/j.neuroimage.2005.10.044

Bao, S., Chan, V. T., & Merzenich, M. M. (2001). Cortical remodelling induced by activity of ventral tegmental dopamine neurons. *Nature, 412*(6842), 79–83. doi:10.1038/35083586

Baumann, S., Koeneke, S., Schmidt, C. F., Meyer, M., Lutz, K., & Jancke, L. (2007). A network for audio-motor coordination in skilled pianists and non-musicians. *Brain Research, 1161*, 65–78. doi:10.1016/j.brainres.2007.05.045

Bendor, D., & Wang, X. (2006). Cortical representations of pitch in monkeys and humans. *Current Opinion in Neurobiology, 16*(4), 391–399. doi:10.1016/j.conb.2006.07.001

Bengtsson, S., Nagy, Z., Skare, S., Forsman, L., Forssberg, H., & Ullén, F. (2005). Extensive piano practicing has regionally specific effects on white matter development. *Nature Neuroscience, 8*(9), 1148–1150.

Bermudez, P., Lerch, J. P., Evans, A. C., & Zatorre, R. J. (2009). Neuroanatomical correlates of musicianship as revealed by cortical thickness and voxel-based morphometry. *Cerebral Cortex, 19*(7), 1583–1596. doi:10.1093/cercor/bhn196

Blood, A. J., & Zatorre, R. J. (2001). Intensely pleasurable responses to music correlate with activity in brain regions implicated in reward and emotion. *Proceedings of the National Academy of Sciences of the United States of America, 98*(20), 11818–11823. doi:10.1073/pnas.191355898

Bones, O., Hopkins, K., Krishnan, A., & Plack, C. J. (2014). Phase locked neural activity in the human brainstem predicts preference for musical consonance. *Neuropsychologia, 58*, 23–32. doi:10.1016/j.neuropsychologia.2014.03.011

Brattico, E., & Pearce, M. (2013). The neuroaesthetics of music. *Psychology of Aesthetics, Creativity, and the Arts, 7*(48–61). doi:10.1037/a0031624

Brown, R., Chen, J., Hollinger, A., Penhune, V., Palmer, C., & Zatorre, R. (2013). Repetition suppression in auditory-motor regions to pitch and temporal structure in music. *Journal of Cognitive Neuroscience, 25*(2), 313–328. doi:10.1162/jocn_a_00322

Brown, R., Zatorre, R., & Penhune, V. (2015). Expert music performance: Cognitive, neural, and developmental bases. *Progress in Brain Research, 217*, 57–86. doi:10.1016/bs.pbr.2014.11.021

Cha, K., Zatorre, R. J., & Schonwiesner, M. (2016). Frequency selectivity of voxel-by-voxel functional connectivity in human auditory cortex. *Cerebral Cortex, 26*(1), 211–224. doi:10.1093/cercor/bhu193

Chen, J. L., Penhune, V. B., & Zatorre, R. J. (2008a). Listening to musical rhythms recruits motor regions of the brain. *Cerebral Cortex, 18*(12), 2844–2854. doi:10.1093/cercor/bhn042

Chen, J. L., Penhune, V. B., & Zatorre, R. J. (2008b). Moving on time: Brain network for auditory-motor synchronization is modulated by rhythm complexity and musical training. *Journal of Cognitive Neuroscience, 20*(2), 226–239. doi:10.1162/jocn.2008.20018

Chen, J. L., Rae, C., & Watkins, K. E. (2012). Learning to play a melody: An fMRI study examining the formation of auditory-motor associations. *NeuroImage, 59*(2), 1200–1208. doi:10.1016/j.neuroimage.2011.08.012

Clarke, E. (1999). Rhythm and timing in music. In D. Deutsch (Ed.), *The psychology of music* (2nd ed.). New York: Academic Press.

Coffey, E. B., Herholz, S. C., Chepesiuk, A. M., Baillet, S., & Zatorre, R. J. (2016). Cortical contributions to the auditory frequency-following response revealed by MEG. *Nature Communications, 7,* 11070. doi:10.1038/ncomms11070

Cross, I. (2012). Music as a social and cognitive process. In P. Rebuschat, M. Rohrmeier, J. Hawkins, & I. Cross (Eds.), *Language and music as cognitive systems* (pp. 315–328). Oxford: Oxford University Press.

D'Ausilio, A., Altenmuller, E., Olivetti Belardinelli, M., & Lotze, M. (2006). Cross-modal plasticity of the motor cortex while listening to a rehearsed musical piece. *European Journal of Neuroscience, 24*(3), 955–958. doi:10.1111/j.1460-9568.2006.04960.x

David, S. V., Fritz, J. B., & Shamma, S. A. (2012). Task reward structure shapes rapid receptive field plasticity in auditory cortex. *Proceedings of the National Academy of Sciences of the United States of America, 109*(6), 2144–2149. doi:10.1073/pnas.1117717109

de Manzano, O., & Ullen, F. (2018). Same genes, different brains: Neuroanatomical differences between monozygotic twins discordant for musical training. *Cerebral Cortex, 28*(1), 387–394. doi:10.1093/cercor/bhx299

Eggermont, J. J. (2001). Between sound and perception: Reviewing the search for a neural code. *Hearing Research, 157*(1–2), 1–42.

Elbert, T., Pantev, C., Wienbruch, C., Rockstroh, B., & Taub, E. (1995). Increased cortical representation of the fingers of the left hand in string players. *Science, 270,* 305–307.

Engel, A., Hijmans, B. S., Cerliani, L., Bangert, M., Nanetti, L., Keller, P. E., & Keysers, C. (2014). Inter-individual differences in audio-motor learning of piano melodies and white matter fiber tract architecture. *Human Brain Mapping, 35*(5), 2483–2497. doi:10.1002/hbm.22343

Foster, N. E., Halpern, A. R., & Zatorre, R. J. (2013). Common parietal activation in musical mental transformations across pitch and time. *NeuroImage, 75,* 27–35. doi:10.1016/j.neuroimage.2013.02.044

Foster, N. E., & Zatorre, R. J. (2010a). Cortical structure predicts success in performing musical transformation judgments. *NeuroImage, 53*(1), 26–36. doi:10.1016/j.neuroimage.2010.06.042

Foster, N. E., & Zatorre, R. J. (2010b). A role for the intraparietal sulcus in transforming musical pitch information. *Cerebral Cortex, 20*(6), 1350–1359. doi:10.1093/cercor/bhp199

Fujioka, T., Ross, B., Kakigi, R., Pantev, C., & Trainor, L. J. (2006). One year of musical training affects development of auditory cortical-evoked fields in young children. *Brain, 129*(Pt. 10), 2593–2608. doi:10.1093/brain/awl247

Fujioka, T., Trainor, L. J., Large, E. W., & Ross, B. (2012). Internalized timing of isochronous sounds is represented in neuromagnetic beta oscillations. *Journal of Neuroscience, 32*(5), 1791–1802. doi:10.1523/jneurosci.4107-11.2012

Gaser, C., & Schlaug, G. (2003). Brain structure differences between musicians and non-musicians. *Journal of Neuroscience, 23*(27), 9240–9245.

Gebauer, L., Kringelbach, M., & Vuust, P. (2012). Ever-changing cycles of musical pleasure: The role of dopamine and anticipation. *Psychomusicology: Music, Mind, and Brain, 22*(2), 152.

Giedd, J., Blumenthal, J., Jeffries, N., Castellanos, F., Liu, H., Zijdenbos, A., … Rapoport, J. (1999). Brain development during childhood and adolescence: A longitudinal MRI study. *Nature Neuroscience, 2,* 861–863.

Grahn, J. A., & Brett, M. (2007). Rhythm and beat perception in motor areas of the brain. *Journal of Cognitive Neuroscience, 19*(5), 893–906. doi:10.1162/jocn.2007.19.5.893

Grahn, J. A., & Rowe, J. B. (2009). Feeling the beat: Premotor and striatal interactions in musicians and nonmusicians during beat perception. *Journal of Neuroscience, 29*(23), 7540–7548. doi:10.1523/jneurosci.2018-08.2009

Grahn, J. A., & Rowe, J. B. (2013). Finding and feeling the musical beat: Striatal dissociations between detection and prediction of regularity. *Cerebral Cortex, 23*(4), 913–921. doi:10.1093/cercor/bhs083

Gutschalk, A., Patterson, R. D., Scherg, M., Uppenkamp, S., & Rupp, A. (2004). Temporal dynamics of pitch in human auditory cortex. *NeuroImage, 22*(2), 755–766. doi:10.1016/j.neuroimage.2004.01.025

Haber, S. N., & Knutson, B. (2010). The reward circuit: Linking primate anatomy and human imaging. *Neuropsychopharmacology, 35*(1), 4–26. doi:10.1038/npp.2009.129

Habibi, A., Damasio, A., Ilari, B., Veiga, R., Joshi, A. A., Leahy, R. M., … Damasio, H. (2017). Childhood music training induces change in micro and macroscopic brain structure: Results from a longitudinal study. *Cerebral Cortex, 28*(12), 4336–4347. doi:10.1093/cercor/bhx286

Halwani, G. F., Loui, P., Ruber, T., & Schlaug, G. (2011). Effects of practice and experience on the arcuate fasciculus: Comparing singers, instrumentalists, and non-musicians. *Frontiers in Psychology, 2,* 156. doi:10.3389/fpsyg.2011.00156

Haueisen, J., & Knosche, T. R. (2001). Involuntary motor activity in pianists evoked by music perception. *Journal of Cognitive Neuroscience, 13*(6), 786–792. doi:10.1162/08989290152541449

Herholz, S., Coffey, E., Pantev, C., & Zatorre, R. (2016). Dissociation of neural networks for predisposition and for training-related plasticity in auditory-motor learning. *Cerebral Cortex, 26*(7), 3125–3134. doi:10.1093/cercor/bhv138

Herholz, S., Lappe, C., Knief, A., & Pantev, C. (2008). Neural basis of music imagery and the effect of musical expertise. *European Journal of Neuroscience, 28*(11), 2352–2360. doi:10.1111/j.1460-9568.2008.06515.x

Herholz, S., & Zatorre, R. (2012). Musical training as a framework for brain plasticity: Behavior, function, and structure. *Neuron, 76*(3), 486–502. doi:10.1016/j.neuron.2012.10.011

Hickok, G., & Poeppel, D. (2015). Neural basis of speech perception. *Handbook of Clinical Neurology, 129,* 149–160. doi:10.1016/B978-0-444-62630-1.00008-1

Honig, H. (2013). Structure and interpretation of rhythm in music. In D. Deutsch (Ed.), *The psychology of music* (Vol. 3, pp. 369–404). Amsterdam: AcademicPress.

Honig, H. (2018). On the biological basis of musicality. *Annals of the New York Academy of Sciences. Special Issue: The Neurosciences and Music VI, 1423*(1), 51–56. doi:10.1111/nyas.13638

Huron, D. (2001). Is music an evolutionary adaptation? *Annals of the New York Academy of Sciences, 930*(1), 43–61.

Huron, D. (2006). *Sweet anticipation: Music and the psychology of expectation.* Cambridge, MA: MIT Press.

Hutchinson, S., Lee, L. H., Gaab, N., & Schlaug, G. (2003). Cerebellar volume of musicians. *Cerebral Cortex, 13*(9), 943–949.

Hyde, K. L., Lerch, J., Norton, A., Forgeard, M., Winner, E., Evans, A. C., & Schlaug, G. (2009). Musical training shapes structural brain development. *Journal of Neuroscience, 29*(10), 3019–3025. doi:10.1523/jneurosci.5118-08.2009

Hyde, K. L., Lerch, J. P., Zatorre, R. J., Griffiths, T. D., Evans, A. C., & Peretz, I. (2007). Cortical thickness in congenital amusia: When less is better than more. *Journal of Neuroscience, 27*(47), 13028–13032. doi:10.1523/jneurosci.3039-07.2007

Iversen, J., Repp, B., & Patel, A. (2009). Top-down control of rhythm perception modulates early auditory responses. *New York Academy of Sciences, 1169*, 58–73. doi:10.1111/j.1749-6632.2009.04579.x

James, C. E., Oechslin, M. S., Van De Ville, D., Hauert, C. A., Descloux, C., & Lazeyras, F. (2014). Musical training intensity yields opposite effects on grey matter density in cognitive versus sensorimotor networks. *Brain Structure and Function, 219*(1), 353–366. doi:10.1007/s00429-013-0504-z

Janata, P., Tomic, S. T., & Haberman, J. M. (2012). Sensorimotor coupling in music and the psychology of the groove. *Journal of Experimental Psychology: General, 141*(1), 54–75. doi:10.1037/a0024208

Johnsrude, I. S., Penhune, V. B., & Zatorre, R. J. (2000). Functional specificity in the right human auditory cortex for perceiving pitch direction. *Brain, 123*(Pt. 1), 155–163.

Juslin, P., & Sloboda, J. (2011). *Handbook of music and emotion: Theory, research, applications.* Oxford: Oxford University Press.

Keller, P. E., & Koch, I. (2008). Action planning in sequential skills: Relations to music performance. *Quarterly Journal of Experimental Psychology, 61*(2), 275–291. doi:10.1080/17470210601160864

Kleber, B., Veit, R., Birbaumer, N., Gruzelier, J., & Lotze, M. (2010). The brain of opera singers: Experience-dependent changes in functional activation. *Cerebral Cortex, 20*(5), 1144–1152. doi:10.1093/cercor/bhp177

Klein, M. E., & Zatorre, R. J. (2015). Representations of invariant musical categories are decodable by pattern analysis of locally distributed BOLD responses in superior temporal and intraparietal sulci. *Cerebral Cortex, 25*(7), 1947–1957. doi:10.1093/cercor/bhu003

Koelsch, S. (2012). *Brain and music.* London: Wiley-Blackwell.

Konoike, N., Kotozaki, Y., Miyachi, S., Miyauchi, C. M., Yomogida, Y., Akimoto, Y., ... Nakamura, K. (2012). Rhythm information represented in the fronto-parieto-cerebellar motor system. *NeuroImage, 63*(1), 328–338. doi:10.1016/j.neuroimage.2012.07.002

Kraus, N., Anderson, S., & White-Schwoch, T. (2017). The frequency-following response: A window into human communication. In N. Kraus, S. Anderson, T. White-Schwoch, R. Fay, & A. Popper (Eds.), The frequency-following response. *Springer handbook of auditory research* (Vol. 61). New York: Springer.

Kung, S. J., Chen, J. L., Zatorre, R. J., & Penhune, V. B. (2013). Interacting cortical and basal ganglia networks underlying finding and tapping to the musical beat. *Journal of Cognitive Neuroscience, 25*(3), 401–420. doi:10.1162/jocn_a_00325

Lahav, A., Saltzman, E., & Schlaug, G. (2007). Action representation of sound: Audiomotor recognition network while listening to newly acquired actions. *Journal of Neuroscience, 27*(2), 308–314. doi:10.1523/jneurosci.4822-06.2007

Large, E., & Snyder, J. (2009). Pulse and meter as neural resonance. *Annals of the New York Academy of Sciences, 1169*, 46–57. doi:10.1111/j.1749-6632.2009.04550.x

Lee, H., & Noppeney, U. (2011). Long-term music training tunes how the brain temporally binds signals from multiple senses. *Proceedings of the National Academy of Sciences of the United States of America, 108*(51), E1441–1450. doi:10.1073/pnas.1115267108

Lee, Y., Janata, P., Frost, C., Hanke, M., & Granger, R. (2011). Investigation of melodic contour processing in the brain using multivariate pattern-based fMRI. *NeuroImage, 57*(1), 293–300. doi:10.1016/j.neuroimage.2011.02.006

Lega, C., Stephan, M. A., Zatorre, R. J., & Penhune, V. (2016). Testing the role of dorsal premotor cortex in auditory-motor association learning using transcranical magnetic stimulation (TMS). *PLoS One, 11*(9), e0163380. doi:10.1371/journal.pone.0163380

Li, X., De Beuckelaer, A., Guo, J., Ma, F., Xu, M., & Liu, J. (2014). The gray matter volume of the amygdala is correlated with the perception of melodic intervals: A voxel-based morphometry study. *PLoS One, 9*(6), e99889. doi:10.1371/journal.pone.0099889

Loehr, J. D., Large, E. W., & Palmer, C. (2011). Temporal coordination and adaptation to rate change in music performance. *Journal of Experimental Psychology: Human Perception and Performance, 37*(4), 1292–1309. doi:10.1037/a0023102

Loui, P., Alsop, D., & Schlaug, G. (2009). Tone deafness: A new disconnection syndrome? *Journal of Neuroscience, 29*(33), 10215–10220. doi:10.1523/jneurosci.1701-09.2009

Madison, G. (2006). Experiencing groove induced by music: Consistency and phenomenology. *Music Perception, 24*, 201–208. doi:doi:10.1525/mp.2006.24.2.201

Manning, F., & Schutz, M. (2013). "Moving to the beat" improves timing perception. *Psychonomic Bulletin & Review, 20*(6), 1133–1139. doi:10.3758/s13423-013-0439-7

Martinez-Molina, N., Mas-Herrero, E., Rodriguez-Fornells, A., Zatorre, R. J., & Marco-Pallares, J. (2016). Neural correlates of specific musical anhedonia. *Proceedings of the National Academy of Sciences of the United States of America, 113*(46), E7337–E7345. doi:10.1073/pnas.1611211113

Mas-Herrero, E., Dagher, A., & Zatorre, R. (2017). Modulating musical reward sensitivity up and down with transcranial magnetic stimulation. *Nature Human Behaviour, 2*, 27–32. doi: 10.1038/s41562-017-0241-z

Mas-Herrero, E., Zatorre, R. J., Rodriguez-Fornells, A., & Marco-Pallares, J. (2014). Dissociation between musical and monetary reward responses in specific musical anhedonia. *Current Biology, 24*(6), 699–704. doi:10.1016/j.cub.2014.01.068

Matthews, T., Witek, M., Heggli, O., Penhune, V., & Vuust, P. (2019). The sensation of groove is affected by the interaction of rhythmic and harmonic complexity. *PLoS One, 14*(1): e0204539. https://doi.org/10.1371/journal.pone.0204539

McDermott, J. H., & Oxenham, A. J. (2008). Music perception, pitch, and the auditory system. *Current Opinion in Neurobiology, 18*(4), 452–463. doi:10.1016/j.conb.2008.09.005

McDermott, J. H., Schultz, A. F., Undurraga, E. A., & Godoy, R. A. (2016). Indifference to dissonance in native Amazonians reveals cultural variation in music perception. *Nature, 535*(7613), 547–550. doi:10.1038/nature18635

Meyer, L. (1956). *Emotion and meaning in music.* Chicago: University of Chicago Press.

Morillon, B., Schroeder, C. E., & Wyart, V. (2014). Motor contributions to the temporal precision of auditory attention. *Nature Communications, 5*, 5255. doi:10.1038/ncomms6255

Mosing, M. A., Madison, G., Pedersen, N. L., Kuja-Halkola, R., & Ullen, F. (2014). Practice does not make perfect: No causal effect of music practice on music ability. *Psychological Science, 25*(9), 1795–1803. doi:10.1177/0956797614541990

Norman-Haignere, S., Kanwisher, N. G., & McDermott, J. H. (2015). Distinct cortical pathways for music and speech revealed by hypothesis-free voxel decomposition. *Neuron, 88*(6), 1281–1296. doi:10.1016/j.neuron.2015.11.035

North, A., & Hargreaves, D. (1995). Subjective complexity, familiarity, and liking for popular music. *Psychomusicology: A Journal of Research in Music Cognition, 14*(1–2), 77–93.

Novembre, G., & Keller, P. E. (2014). A conceptual review on action-perception coupling in the musicians' brain: What is it good for? *Frontiers in Human Neuroscience, 8*, 603. doi:10.3389/fnhum.2014.00603

Nozaradan, S., Peretz, I., & Mouraux, A. (2012). Selective neuronal entrainment to the beat and meter embedded in a musical rhythm. *Journal of Neuroscience, 32*(49), 17572–17581. doi:10.1523/jneurosci.3203-12.2012

Palomar-Garcia, M. A., Zatorre, R. J., Ventura-Campos, N., Bueicheku, E., & Avila, C. (2017). Modulation of functional connectivity in auditory-motor networks in musicians compared with nonmusicians. *Cerebral Cortex, 27*(5), 2768–2778. doi:10.1093/cercor/bhw120

Paquette, S., Fujii, S., Li, H. C., & Schlaug, G. (2017). The cerebellum's contribution to beat interval discrimination. *NeuroImage, 163*, 177–182. doi:10.1016/j.neuroimage.2017.09.017

Patel, A. (2008). *Music, language and the brain*. Oxford: Oxford University Press.

Patel, A., & Iversen, J. (2014). The evolutionary neuroscience of musical beat perception: The action simulation for auditory prediction (ASAP) hypothesis. *Frontiers in Systems Neuroscience, 8*, 57. doi:10.3389/fnsys.2014.00057

Pearce, M. T., Ruiz, M. H., Kapasi, S., Wiggins, G. A., & Bhattacharya, J. (2010). Unsupervised statistical learning underpins computational, behavioural, and neural manifestations of musical expectation. *NeuroImage, 50*(1), 302–313. doi:10.1016/j.neuroimage.2009.12.019

Penagos, H., Melcher, J. R., & Oxenham, A. J. (2004). A neural representation of pitch salience in nonprimary human auditory cortex revealed with functional magnetic resonance imaging. *Journal of Neuroscience, 24*(30), 6810–6815. doi:10.1523/jneurosci.0383-04.2004

Peretz, I. (2016). Neurobiology of congenital amusia. *Trends in Cognitive Sciences, 20*(11), 857–867. doi:10.1016/j.tics.2016.09.002

Peretz, I., Brattico, E., Jarvenpaa, M., & Tervaniemi, M. (2009). The amusic brain: In tune, out of key, and unaware. *Brain, 132*(Pt. 5), 1277–1286. doi:10.1093/brain/awp055

Puschmann, S., Uppenkamp, S., Kollmeier, B., & Thiel, C. M. (2010). Dichotic pitch activates pitch processing centre in Heschl's gyrus. *NeuroImage, 49*(2), 1641–1649. doi:10.1016/j.neuroimage.2009.09.045

Putkinen, V., Tervaniemi, M., Saarikivi, K., Ojala, P., & Huotilainen, M. (2014). Enhanced development of auditory change detection in musically trained school-aged children: A longitudinal event-related potential study. *Developmental Science, 17*(2), 282–297. doi:10.1111/desc.12109

Rauschecker, J., & Scott, S. (2009). Maps and streams in the auditory cortex: Nonhuman primates illuminate human speech processing. *Nature Neuroscience, 12*(6), 718–724. doi:10.1038/nn.2331

Repp, B. H., & Su, Y. H. (2013). Sensorimotor synchronization: A review of recent research (2006–2012). *Psychonomic Bulletin & Review, 20*(3), 403–452. doi:10.3758/s13423-012-0371-2

Saarinen, J., Paavilainen, P., Schoger, E., Tervaniemi, M., & Naatanen, R. (1992). Representation of abstract attributes of auditory stimuli in the human brain. *Neuroreport, 3*(12), 1149–1151.

Saenz, M., & Langers, D. R. (2014). Tonotopic mapping of human auditory cortex. *Hearing Research, 307*, 42–52. doi:10.1016/j.heares.2013.07.016

Salimpoor, V. N., Benovoy, M., Larcher, K., Dagher, A., & Zatorre, R. J. (2011). Anatomically distinct dopamine release during anticipation and experience of peak emotion to music. *Nature Neuroscience, 14*(2), 257–262. doi:10.1038/nn.2726

Salimpoor, V. N., van den Bosch, I., Kovacevic, N., McIntosh, A. R., Dagher, A., & Zatorre, R. J. (2013). Interactions between the nucleus accumbens and auditory cortices predict music reward value. *Science, 340*(6129), 216–219. doi:10.1126/science.1231059

Savage, P. E., Brown, S., Sakai, E., & Currie, T. E. (2015). Statistical universals reveal the structures and functions of human music. *Proceedings of the National Academy of Sciences of the United States of America, 112*(29), 8987–8992. doi:10.1073/pnas.1414495112

Schlaug, G., Jancke, L., Huang, Y., Staiger, J. F., & Steinmetz, H. (1995). Increased corpus callosum size in musicians. *Neuropsychologia, 33*(8), 1047–1055.

Schneider, P., Scherg, M., Dosch, H., Specht, H., Gutschalk, A., & Rupp, A. (2002). Morphology of Heschl's gyrus reflects enhanced activation in the auditory cortex of musicians. *Nature Neuroscience, 5*(7), 688–694.

Schneider, P., Sluming, V., Roberts, N., Scherg, M., Goebel, R., Specht, H. J., ... Rupp, A. (2005). Structural and functional asymmetry of lateral Heschl's gyrus reflects pitch perception preference. *Nature Neuroscience, 8*(9), 1241–1247. doi:10.1038/nn1530

Schonwiesner, M., Krumbholz, K., Rubsamen, R., Fink, G. R., & von Cramon, D. Y. (2007). Hemispheric asymmetry for auditory processing in the human auditory brain stem, thalamus, and cortex. *Cerebral Cortex, 17*(2), 492–499. doi:10.1093/cercor/bhj165

Schubotz, R. I. (2007). Prediction of external events with our motor system: Towards a new framework. *Trends in Cognitive Sciences, 11*(5), 211–218. doi:10.1016/j.tics.2007.02.006

Schultz, W. (2013). Updating dopamine reward signals. *Current Opinion in Neurobiology, 23*(2), 229–238. doi:10.1016/j.conb.2012.11.012

Schulze, K., Mueller, K., & Koelsch, S. (2011). Neural correlates of strategy use during auditory working memory in musicians and non-musicians. *European Journal of Neuroscience, 33*(1), 189–196. doi:10.1111/j.1460-9568.2010.07470.x

Seither-Preisler, A., Parncutt, R., & Schneider, P. (2014). Size and synchronization of auditory cortex promotes musical, literacy, and attentional skills in children. *Journal of Neuroscience, 34*(33), 10937–10949. doi:10.1523/JNEUROSCI.5315-13.2014

Sescousse, G., Caldu, X., Segura, B., & Dreher, J. C. (2013). Processing of primary and secondary rewards: A quantitative meta-analysis and review of human functional neuroimaging studies. *Neuroscience & Biobehavioral Reviews, 37*(4), 681–696. doi:10.1016/j.neubiorev.2013.02.002

Shahin, A., Bosnyak, D. J., Trainor, L. J., & Roberts, L. E. (2003). Enhancement of neuroplastic P2 and N1c auditory evoked potentials in musicians. *Journal of Neuroscience, 23*(13), 5545–5552.

Skoe, E., & Kraus, N. (2013). Musical training heightens auditory brainstem function during sensitive periods in development. *Frontiers in Psychology, 4,* 622. doi:10.3389/fpsyg.2013.00622

Sluming, V., Barrick, T., Howard, M., Cezayirli, E., Mayes, A., & Roberts, N. (2002). Voxel-based morphometry reveals increased gray matter density in Broca's area in male symphony orchestra musicians. *NeuroImage, 17*(3), 1613–1622.

Steele, C. J., Bailey, J. A., Zatorre, R. J., & Penhune, V. B. (2013). Early musical training and white-matter plasticity in the corpus callosum: Evidence for a sensitive period. *Journal of Neuroscience, 33*(3), 1282–1290. doi:10.1523/jneurosci.3578-12.2013

Stephan, M. A., Lega, C., & Penhune, V. (2018). Auditory prediction cues motor preparation in the absence of movements. *NeuroImage, 174,* 288–296. doi:10.1016/j.neuroimage.2018.03.044

Teki, S., Grube, M., Kumar, S., & Griffiths, T. D. (2011). Distinct neural substrates of duration-based and beat-based auditory timing. *Journal of Neuroscience, 31*(10), 3805–3812. doi:10.1523/jneurosci.5561-10.2011

Tiemeier, H., Lenroot, R. K., Greenstein, D. K., Tran, L., Pierson, R., & Giedd, J. N. (2010). Cerebellum development during childhood and adolescence: a longitudinal morphometric MRI study. *NeuroImage, 49*(1), 63–70. doi:10.1016/j.neuroimage.2009.08.016

Trehub, S. E., Becker, J., & Morley, I. (2015). Cross-cultural perspectives on music and musicality. *Philosophical Transactions of the Royal Society of London B: Biological Sciences, 370*(1664), 20140096. doi:10.1098/rstb.2014.0096

Ullen, F., Hambrick, D. Z., & Mosing, M. A. (2016). Rethinking expertise: A multifactorial gene-environment interaction model of expert performance. *Psychological Bulletin, 142*(4), 427–446. doi:10.1037/bul0000033

Vaquero, L., Hartmann, K., Ripolles, P., Rojo, N., Sierpowska, J., Francois, C., ... Altenmuller, E. (2016). Structural neuroplasticity in expert pianists depends on the age of musical training onset. *NeuroImage, 126,* 106–119. doi:10.1016/j.neuroimage.2015.11.008

Vuust, P., Brattico, E., Glerean, E., Seppanen, M., Pakarinen, S., Tervaniemi, M., & Naatanen, R. (2011). New fast mismatch negativity paradigm for determining the neural prerequisites for musical ability. *Cortex, 47*(9), 1091–1098. doi:10.1016/j.cortex.2011.04.026

Vuust, P., & Witek, M. A. (2014). Rhythmic complexity and predictive coding: A novel approach to modeling rhythm and meter perception in music. *Frontiers in Psychology, 5,* 1111. doi:10.3389/fpsyg.2014.01111

Westerhausen, R., Luders, E., Specht, K., Ofte, S. H., Toga, A. W., Thompson, P. M., ... Hugdahl, K. (2011). Structural and functional reorganization of the corpus callosum between the age of 6 and 8 years. *Cerebral Cortex, 21*(5), 1012–1017. doi:10.1093/cercor/bhq165

Witek, M. A., Clarke, E. F., Wallentin, M., Kringelbach, M. L., & Vuust, P. (2014). Syncopation, body-movement and pleasure in groove music. *PLoS One, 9*(4), e94446. doi:10.1371/journal.pone.0094446

Wolpert, D. M., Diedrichsen, J., & Flanagan, J. R. (2011). Principles of sensorimotor learning. *Nature Reviews Neuroscience, 12*(12), 739–751. doi:10.1038/nrn3112

Woollman, I., Segado, M., Charpentier, T., Penhune, V., & Zatorre, R. (2018). Neural network retuning and neural predictors of learning success associated with cello training. *Proceedings of the National Academy of Sciences of the United States of America, 115*(26), E6056–E6064.

Zald, D., & Zatorre, R. (2011). Music. In J. Gottfried (Ed.), *Frontiers in neuroscience: Neurobiology of sensation and reward* (pp. 405–428). Boca Raton, FL: CRC Press.

Zatorre, R., Belin, P., & Penhune, V. (2002). Structure and function of auditory cortex: Music and speech. *Trends in Cognitive Sciences, 6*(1), 37–46.

Zatorre, R., & Halpern, A. (2005). Mental concerts: Musical imagery and auditory cortex. *Neuron, 47*(1), 9–12. doi:10.1016/j.neuron.2005.06.013

Zendel, B. R., Lagrois, M. E., Robitaille, N., & Peretz, I. (2015). Attending to pitch information inhibits processing of pitch information: The curious case of amusia. *Journal of Neuroscience, 35*(9), 3815–3824. doi:10.1523/jneurosci.3766-14.2015

Zentner, M., Grandjean, D., & Scherer, K. R. (2008). Emotions evoked by the sound of music: Characterization, classification, and measurement. *Emotion, 8*(4), 494–521. doi:10.1037/1528-3542.8.4.494

XIII

LOOKING AHEAD: CHALLENGES IN ADVANCING COGNITIVE NEUROSCIENCE

97 Toward a Socially Responsible, Transparent, and Reproducible Cognitive Neuroscience

SIKOYA M. ASHBURN, DAVID ABUGABER, JAMES W. ANTONY, KELLY A. BENNION, DAVID BRIDWELL, CARLOS CARDENAS-INIGUEZ, MANOJ DOSS, LUCÍA FERNÁNDEZ, INGE HUIJSMANS, LARA KRISST, REGINA LAPATE, EVAN LAYHER, JOSIAH LEONG, YUANNING LI, FREDDIE MARQUEZ, FELIPE MUNOZ-RUBKE, ELIZABETH MUSZ, TARA K. PATTERSON, JOHN P. POWERS, DARIA PROKLOVA, KRISTINA M. RAPUANO, CHARLES S. H. ROBINSON, JESSICA M. ROSS, JASON SAMAHA, MATTHEW SAZMA, ANDREW X. STEWART, ARIANA STICKEL, ARJEN STOLK, VERONIKA VILGIS, AND MEGAN ZIRNSTEIN

ABSTRACT As members of the next generation of leaders, we aim to promote an ethical, inclusive, humanistic, and curiosity-driven code of conduct within cognitive neuroscience. We critically examine the current state of the field and discuss issues that remain to be addressed. In particular, we highlight the need for cultural change, including a greater focus on integrity and accountability (i.e., addressing sexual harassment), as well as promoting diversity. Further, we highlight current methodological considerations, such as big data and deep data, replicability, and reproducibility. We consider some of the unresolved controversies within cognitive neuroscience and where the research should go from here. Finally, we delve into how neuroscience research is presented to the public and how science communication can benefit society. This chapter presents a call to action not only to further scientific improvements but also to address biases and behaviors that have the potential to hinder scientific advancements.

The previous chapters discussed current directions in cognitive neuroscience. The goal of the current chapter is to address critical issues that limit progress in our field. Scientific progress depends not only on methodological and conceptual improvements, but also on the ethical conduct in science. Therefore, before addressing methodological issues we first discuss two cultural barriers that have impeded the inclusivity of scientific training and contribution. In particular, we highlight the prevalence of sexual harassment and the pervading lack of diversity.

Integrity and Accountability

Whereas ethical issues such as data falsification and plagiarism are typically referenced in discussions on scientific integrity, cultural issues such as sexual harassment and assault are often discussed separately or overlooked entirely. However, more than 50% of female faculty and 20%–50% of female students—with higher rates for women of color and members of the LGBTQ+ community—have reported experiencing gender-based harassment, unwanted sexual attention, and/or sexual coercion (National Academies of Science, Engineering, and Medicine, 2018)—figures that are likely underestimated when considering the fear of retaliation. The innumerable cases of sexual misconduct that have emerged in recent years indicate that our field, like many others, is in need of a critical reevaluation of its ethical practices. Several features inherent to our academic community contribute to an environment where sexual misconduct has become normalized. These include (1) male-dominated work settings, (2) hierarchical power structures, (3) institutional tolerance of sexual harassment and gender-based discrimination, (4) a focus on protecting the liability of institutions, and (5) a lack of informed and proactive leadership. Sexual harassment and assault have an impact on physical, emotional, and professional well-being—and in doing so, obstruct the growth of the field. Thus, we argue that sexual misconduct in academia should be treated as an issue of scientific misconduct (see National Academies of Sciences, Engineering, and Medicine, 2018, pp. 114–118, for further discussion). The responsibility to alleviate these hindrances lies with the scientific community as a whole. The first steps toward this goal include informing the community, changing institutional policies (Melnick, 2018), redefining scientific integrity to include gender-based issues, and reinforcing accountability at all levels.

Equality in Science

The state of diversity and equity within cognitive neuroscience warrants change. Recently, at the 2018 Kavli Summer Institute in Cognitive Neuroscience, a hotel receptionist was asked to contact one of the invited speakers, assumed that Dr. X could not have been the female voice who answered the phone, and asked to speak to her husband. Considering that women make up half of the contributors to this book, including half of the young neuroscientists coauthoring this chapter, it is easy to assume that these unconscious biases will subside. However, the lack of diversity and equity in science is systemic—it is not solely an individual but also a policy failure. The current system favors elite education; disadvantages low-income and first-generation students with few paid opportunities for research experience; reinforces the pay gap; offers poor maternity and paternity leave policies; has a gross underrepresentation of people of color in faculty positions that is often accompanied by greater service expectations; is heteronormative, with diversity initiatives omitting LGBTQ+ scientists (Freeman, 2018); and focuses on experimental results obtained from individuals of higher socioeconomic status (see chapter 90). Achieving participation in a science that represents society as a whole is more than just moral or ethical. It is how we produce the *best* science. Research suggests that groups that are ethnically diverse have higher citation counts on average than homogenous ones (Powell, 2018)—just one demonstration of the value of diversity in high-impact science. Change is necessary. As long as we as scientists remain silent amid this systemic failure, we tacitly propagate it.

Call to Action

We challenge all scientists, especially those in positions of power and influence, to recognize the responsibility of actively promoting a diverse academic and professional environment that supports ethical codes and prevents harassment and discrimination. The failure to take on these challenges perpetuates the sociocultural mechanisms through which sexual harassment and systemic underrepresentation occur. We caution against the unfair enforcement of policies based on individuals' institutional influence or demographics (e.g., gender) and against creating discretionary internal committees in lieu of transparently and publicly acknowledging these issues that affect the quality and integrity of science. Bystanders, potential aggressors, and those benefiting from societal privilege should accept personal responsibility to educate themselves on these issues.

Current Approaches in Cognitive Neuroscience: Big Data and Deep Data

The term *big data* has become a buzzword with multiple definitions; here, we adopt Yarkoni's (2014) definition that refers to data sets of a magnitude considered unusual for the field. Big data can address the lack of replicability, reproducibility, and statistical power from small sample sizes. Larger samples ensure that effect-size estimates and statistical effects can be more accurately captured. Moreover, the practice of validating predictions out of sample, as opposed to within-sample correlations, distinguishes the big data approach from its counterparts and presents an advantage in establishing both the reliability and generalizability of results.

Big data improves the ability to characterize individual differences across participants, which is critical for the successful real-world application of neuroscience research. Individual differences are often examined in models testing mediators and moderators, which require large sample sizes. Hence, big data is ideal for testing such models and reduces the risk of type II errors. However, large-scale studies using cross-sectional designs do not allow for causal inference, which may be better suited for longitudinal analyses (see the next section). Also, since large sample sizes can make trivial effects statistically significant, estimating—and inferring the potential meaning of—effect sizes becomes crucial for the interpretation of results. In the coming years, the collection of big data will continue through initiatives such as the Brain Activity Map (Kandel, Markram, Matthews, Yuste, & Koch, 2013) and the Brain Research through Advancing Innovative Neurotechnologies (BRAIN) Initiative (Choudhury, Fishman, McGowan, & Juengst, 2014).

Additionally, data can be collected across large timescales (e.g., days and years) within single subjects as part of longitudinal or *deep data* collection. This allows for the study of dynamic changes in brain connectivity, gene expression, and metabolites within a single subject (Poldrack et al., 2015).

Application of Deep Data to Development and Neurocognitive Theories

Longitudinal studies involve the collection of behavioral and neuroimaging data at several time points throughout an individual's lifetime and enable researchers to study, track, and predict changes in brain structure and function across the life span. Additionally, the use of longitudinal (vs. cross-sectional) data allows researchers to test for nonlinear trends in how various neuroanatomical measures (e.g., cortical thinning, gray matter volume) change over development. Moreover, longitudinal designs have an increased capacity over cross-sectional designs to distinguish between within- and between-subject variation (Rogosa, Brandt, & Zimowski, 1982). Each additional repeated measure substantially increases both the ways to model change and the number of available models (King et al., 2018).

Recent and ongoing large-scale neuroimaging studies employ longitudinal sampling of individuals. For example, the MyConnectome project demonstrates dynamic changes in brain connectivity, gene expression, and metabolites in a single person over an 18-month period (Poldrack et al., 2015). Similarly, the Adolescent Brain Cognitive Development study (ABCD) plans to collect longitudinal data from over 10,000 individuals, 9–10 years old, across the United States as they progress into early adulthood (Casey et al., 2018). Meanwhile, in Europe, the Lifebrain research project will collect neuroimaging measures at six time points from birth to age 100. The longitudinal effects observed in these decades-long undertakings will help identify and track life span changes in brain structure and function and will inform our understanding of how environmental, genomic, and experiential factors shape brain development.

These studies are influential for developing theories of how brain function and organization change over time, which is also critical for understanding human development. Developmental studies seek to understand how a particular neurocognitive characteristic or process changes throughout an individual's lifetime and to identify genetic and environmental factors that predict individual differences in how such changes vary across people. In addition, they allow us to validate psychological and neurocognitive theories across the life span to address whether these theories can accommodate data beyond those measured in healthy, adult-aged samples.

The life span perspective also helps researchers track the onset and progression of mental disorders and neurodegeneration. In particular, aging has been characterized as an unprogrammed, stochastic wear-and-tear process manifesting in nonlinear and accelerated impairments such as reduced myelination, thinning dendritic spines, and axon deformation (see chapter 6). These quasiperiodic changes occur throughout the life span and must be frequently measured in the same individual in order to chart their progression.

Logistics of Large Data Projects

Future research in cognitive neuroscience will likely require a combination of both big and deep data approaches. These projects often involve collaborations across multiple departments and institutions. Thus, many factors should be considered before undertaking such projects, including determining sample sizes based on a priori power analysis, budget allocations, ethical data-sharing practices, authorship order, preference for theory- versus data-driven approaches, and procedural standardization.

Standardization reduces the risk of confounds related to site differences and should be monitored as these studies progress. Nevertheless, regional differences between participants may remain, making site location an important covariate when examining across-site results. Taken together, these challenges can be time-consuming and require ongoing discussion. However, addressing them will result in better classification of individual differences, as well as more precise translational applications.

Pioneering efforts like the fMRI (functional magnetic resonance imaging) Data Center (Van Horn & Gazzaniga, 2002) have created repositories for neuroimaging data sharing. Although most agree that researchers have an ethical responsibility to share data (Brakewood & Poldrack, 2013), only half report doing so, even though 85% express a willingness to use others' data if easily accessible (Tenopir et al., 2011). The tepid (but growing) consensus to share data is likely related to the providers' confidence in their data sets and results (Wicherts, Bakker, & Molenaar, 2011). However, with over 60,000 practitioners of neuroscience contributing to the field, many will add to the integration of our vast menageries of data scattered across labs around the globe. As Henry Markram predicts, "All the signs indicate that this new generation of neuroscientists will be far more ready to work in teams than the current generation. This is our true hope for the future" (Kandel et al., 2013).

Replicability and Reproducibility

Replication, the ability to obtain similar results when repeating experiments, is a cornerstone of scientific knowledge. In 2015 Brian Nosek and colleagues from the Open Science Collaboration attempted to replicate 100 experiments published in high-impact psychology journals. They found that only 39% of their data sets replicated the original findings at the standard significance level. Standard hypothesis-testing approaches were criticized as early as 1967 by Paul Meehl, and statisticians and research bloggers have voiced concerns about replicability for years. However, Nosek and colleagues highlighted the scope of the problem, and replication has now become a common topic of conversation among scientists and a staple in contemporary research methods courses.

Incentives to report positive rather than negative findings directly contribute to the replication crisis. The current standards arbitrarily define an alpha level of 0.05 and regard findings below this threshold as "significant." These conventions encourage selecting results post hoc with $p<0.05$, often without correcting for multiple statistical comparisons. These problems are referred to as data dredging, *p*-hacking, or, less reproachfully, "the garden of forking paths," reflecting the numerous ways in which one can reasonably preprocess or frame the data analysis until statistically significant results are found (Gelman & Loken, 2013).

At the individual level, we believe the most important solutions to the replication crisis are preregistration and publicly releasing both data and code. In preregistration, authors explicitly and precisely state their methods and hypotheses before running the study. Exploratory analyses are acceptable at later stages but must be acknowledged as such rather than spun into a story about how they were predicted from the outset. Ideally, original findings in exploratory analyses would be followed up with replications using novel data sets.

At the institutional level, journals are often reluctant to publish negative results that are perceived as less exciting or a reflection of poor methodological rigor. Therefore, grant agencies must increase funding for replication efforts of seminal studies with large sample sizes, and journals might agree to publish data before they are collected in order to incentivize researchers to precisely report their findings. Hypothesis testing within these studies may be improved by focusing on Bayes factors or posterior summaries derived from well-defined models (Benjamin et al., 2018), while others have proposed an emphasis on effect sizes rather than null hypothesis significance testing (Szucs & Ioannidis, 2017). Finally, in order to improve the replicability of science, the incentive structure of academia must change. The current incentive structure rewards individual productivity (i.e., number and frequency of publications) over more collaborative "slow science." By shifting incentives away from fast science at the individual level, the field will be better positioned to replicate and extend previous findings through slower and more deliberate collaborative efforts.

Complementary to replicability is reproducibility: the duplication of results given the same data set. In recent years reproducibility has been facilitated by websites like Open Science Framework (https://osf.io) and GitHub (https://github.com) that allow researchers to publicly release their data, experimental and analysis code, stimuli, and other task-related documents. Additionally, the increasing use of open-source languages like Python and R has improved the accessibility of scientific tools. Collectively, these recent efforts should improve the quality of code and the reliability of results prior to publication and ensure that independent labs can reproduce the results.

Improving reproducibility would also influence training for the next generation of cognitive neuroscientists. The classroom is an ideal setting for utilizing these software tools and databases. Research methods courses can directly access data sets, analyses, and experimental protocols from open-access resources to teach new concepts while simultaneously testing the reproducibility of findings in the field. The availability of such materials provides an explicit demonstration of how science (and open science) is conducted, which in turn helps trainees develop into independent researchers who will implement similar techniques.

Choosing Your Method Wisely

Neuroimaging methods have substantially evolved over the past three decades, and the pace at which new methods appear is rapidly increasing. However, rather than being seduced by novel tools and techniques, it is worth reflecting on what new methods can and cannot tell us. Despite much progress, most of the measures we use to infer brain structure and/or function are imperfect proxies for brain activity. For instance, fMRI measures levels of blood oxygenation (BOLD) rather than direct neural activity, and water diffusivity is used to infer white matter microstructure, yet very little is known about how different microstructural properties influence diffusivity. Furthermore, few studies have mapped other measures at the macrostructural level, such as gray matter volume, to microcellular indices (Mills & Tamnes, 2014). Given these shortcomings, when MRI measures are used to build complex network models, every additional analysis step abstracts further from the biological mechanisms of interest.

Researchers estimating brainwide functional and structural connectivity using resting-state fMRI and diffusion tractography should also consider model assumptions and measurement. Functional connectivity first depends on evidence of monosynaptic connection between targeted brain circuits, as in axon-tracing studies of animals. Then, evidence of conditionally dependent activity between circuits, as in studies involving physiological manipulation (e.g., electrophysiology, optogenetics), can further guide researchers' inferences from functional correlation to functional connectivity.

A second consideration in studies of brain connectivity is the operationalization of structural connectivity. Researchers can index structural connectivity with results from diffusion tractography, such as the number of fibers, length of fibers, and water diffusion along fiber trajectories. However, these measures might not estimate structural connectivity with equivalent anatomical precision. The number of fibers traced between brain circuits is a function of manually selected parameters in tractography algorithms, and the length of fibers is a function of the ease of tracking fibers across brain terrain containing simple versus complex fiber crossings. In contrast, measures of water diffusion, such as fractional anisotropy and mean, axial, and radial diffusivity, can be validated with invasive studies of axon myelination and density (e.g., electron microscopy). As modeling approaches to neuroimaging data evolve, so must conceptual clarity about brain connectivity.

Modalities such as fMRI, electroencephalography (EEG), and magnetoencephalography (MEG) are sensitive to different aspects of brain activity with varying spatiotemporal resolutions. Thus, combining these modalities is critical for a complete picture of brain activity. One multimodal approach combines MEG and fMRI under the framework of representational similarity analysis (Cichy, Pantazis, & Oliva, 2014). The latent neuronal space can also be used as the common ground for combining multimodal neural signals, as demonstrated with BOLD fMRI and local field potentials (Hermes, Nguyen, & Winawer, 2017). Moving forward, machine learning (Baltrušaitis, Ahuja, & Morency, 2019) can also be used to develop a rigorous framework of building models that process and combine information from multiple modalities.

Deep neural networks (DNNs) have recently emerged as a new tool for understanding the brain, in particular the visual system. These algorithms are inspired by hierarchical models of the visual system, trained on a large set of images, and, under certain circumstances, can achieve human-level performance in object recognition (Krizhevsky, Sutskever, & Hinton, 2012). The same fusion approach described above for fMRI and MEG has been used to relate neuroimaging data to the output of a DNN, finding similarities between earlier layers of the DNN and early visual areas (fMRI) and earlier time points (MEG) and between later layers and more anterior areas of the ventral visual stream and later time points (Cichy, Khosla, Pantazis, Torralba, & Oliva, 2016). While this suggests that there is similarity between the transformations carried out by the ventral stream and a DNN, this correlation does not mean that the underlying processes are identical. Although DNNs are perhaps the best available models of the visual system to date, they are still imperfect models (e.g., not reaching the noise ceiling) and can only go so far in explaining the brain. One important difference is that a DNN is feedforward (unidirectional), while the visual system has both feedforward and feedback connections. Thus, an important next step will be to use recurrent DNNs to see if they provide a more accurate description of brain activity.

Beyond an understanding of what neural measurements represent *biologically*, we must also consider the ecological validity and real-world relevance of these measurements. Contrived task designs (e.g., Gabor patches, visuomotor rotation) are indispensable for understanding low-level neural mechanisms of systems like vision and movement. However, in assuming similar responses across individuals, such tasks may overlook higher-level factors that capture meaningful sources of variability across participants, such as how people naturally view, navigate, or manipulate their environments. Because real-world tasks can be performed in qualitatively different ways, the study of individual variation in naturalistic settings presents an important avenue for future work. Indeed, human neuroscience has begun to utilize naturalistic task designs (Hasson, Nir, Levy, Fuhrmann, & Malach, 2004) and analyze variability in neural synchrony across individuals during naturalistic movie viewing (Parkinson, Kleinbaum, & Wheatley, 2018). Although these types of unconstrained designs present a different set of challenges (in comparison to standardized, well-validated tasks), these findings provide some insight into the neural signals that support complex, real-world behaviors.

Open Questions in Cognitive Neuroscience

Research in cognitive neuroscience distills complex behaviors and mental phenomena into more parsimonious sets of constructs, which can then be measured and manipulated in the laboratory. The hope is that insights from controlled and reduced experiments will translate meaningfully back to the larger phenomena of interest. This leaves room for both conceptual and methodological errors. For example, when investigating phenomena such as concepts, understanding, embodiment, and consciousness, it is critical to clearly state a working definition of the phenomenon. Clarifying our terms enables us to compare findings when we agree on the phenomenon in question and to diagnose terminological disputes or conceptual differences, thereby enabling us to move toward a scientific consensus. Here, we discuss several areas in cognitive neuroscience where progress in scaling up research beyond the laboratory warrants critical examination.

One of these areas is the way that cognitive neuroscience research and the architecture of computational and robotic systems can mutually inform one another. Bridging these fields is far from trivial, given that our understanding of the brain is limited by neuroimaging measures that capture a subset of brain activity over a narrow range of spatiotemporal scales. In addition, experimental constraints make it difficult to measure complex human behaviors, such as communication in natural environments, face processing in real-life crowds, and cognitive/conscious states that are only available through self-report. Incorporating these realistic contexts within cognitive neuroscience research is a critical next step (Schilbach et. al., 2013). We must also consider how neural processing is shaped by evolutionary pressures to socially evaluate individuals along multiple complex dimensions (e.g., social hierarchies). Resulting biases shape our interpersonal interactions, which in turn form many abstract concepts (e.g., religion, philosophy, morality, government). These social processes are unique and fundamental to human survival, but incorporating them within the field of artificial intelligence (AI) will be a considerable challenge.

Part of the problem here is in identifying the cognitive phenomenon of interest. For instance, having a common language benefits communication, but even commonly used words do not contain fixed meanings that are reliably shared across individuals. In fact, in everyday dialogue, the speaker's intention and a sentence's literal meaning often diverge, and words can take on different meanings based on intonation and other subtle contextual cues. Furthermore, a reply can be a response to an utterance occurring at any time along the interaction's trajectory, irrespective of ordering or syntax. These fleeting conceptual dependencies between utterances cannot be easily grasped by simply analyzing long-term, context-free statistical regularities in speech/text, as do the virtual assistants on our smartphones (Stolk, Verhagen, & Toni, 2016). Therefore, theoretical and empirical approaches that respect the core interpersonal and generative nature of human communication are critical for gaining insights into this remarkable ability and for providing a window into understanding alterations of communication in neurological and neurodevelopmental disorders.

Another controversial topic is that of consciousness. Empirical investigations of consciousness examine the neuronal underpinnings of conscious experience and should aim to account for the subjective, phenomenal aspect of *what it is like* to be that organism. Any viable science of consciousness ought to explain not only the conditions under which conscious perception occurs (correlational claim) but also the necessary and sufficient conditions for subjective experience (causal claim). Two common-sense distinctions between different aspects of consciousness are *conscious states*, referring to the level of consciousness or wakefulness (which can range from coma to sleep to the waking state) and *conscious contents*, referring to specific pieces of information that become accessible to awareness (e.g., the experience of pain or the color blue).

Scaling up consciousness science beyond the laboratory faces several challenges. For one thing, operationalizing consciousness is notoriously tricky. We might assume that if a subject can accurately identify a stimulus, the subject is conscious of it. However, the phenomenon of blindsight, whereby individuals with damage to primary visual cortex are capable of making accurate forced-choice discriminations despite reporting no experiences in the affected visual field, illustrates that the identification of a stimulus can occur in the absence of awareness. Uncovering the neural correlates of consciousness is also uniquely challenging. Neural activity seemingly correlated with specific conscious contents may instead simply precede or follow the true neural correlates of consciousness (Aru, Bachmann, Singer, & Melloni, 2012). Lastly, the emphasis on simplified perceptual tasks (e.g., yes/no detection) allows for careful stimulus control and behavioral modeling but may underestimate the richness of consciousness.

Cognition is largely considered to be a process within the brain, but it is important to consider cognition as it emerges from interactions between the brain, body, and environment. James Gibson (1979) proposed that perception is embedded in our experience and that it cannot be fully understood if we overlook our direct interactions with the environment. Daniel Wolpert's research program suggests that the primary function of the brain is action (i.e., enabling movement of the body through the world). These considerations have prompted a "pragmatic turn" toward dynamic and enactive frameworks grounded in sensorimotor processing (Engel, Friston, & Kragic, 2016) and the notion of *embodied cognition*. Embodied approaches recognize that the brain is part of a broader system that developed to engage with the world around us.

Cognition, perception, and action are mental constructs in the human and brain sciences but are more continuous with each other in implementation (Spivey, 2008). Sensory and motor interactions with cognition have been demonstrated in conceptual metaphor, image schema and prototypes, mental rotation, high-level reasoning, language comprehension, memory, and mathematics, to name just a few. In addition, many higher-level processes (e.g., language, arithmetic) appear to involve cortical areas specialized in lower-level processes that are more directly tied to action and space (Anderson, 2014). Thus, considering the complexity of mind-brain-body-environment interactions will be important for future work in cognitive neuroscience.

Putting Science to Work

While cognitive neuroscience continues to yield discoveries into how mental processes relate to brain structure and function, our success at using these discoveries to benefit society has been less inspiring. A predominant factor in this "translation gap" is the scarce communication between researchers and other professional communities, including policy-makers, educators, and clinicians. These divides are maintained by a complex array of factors, such as cumbersome administrative procedures for translational research; priorities in research funding, publishing, and scientific career metrics; and economic pressures, such as demands for clinical revenue.

Despite these limitations, many avenues with potential for translation are being explored. On the clinical front, new developments suggest that a variety of tools may soon be available for therapeutic application. Some of the most promising approaches include neurofeedback (Marzbani, Marateb, & Mansourian, 2016), brain-machine interfaces (Chaudhary, Birbaumer, & Ramos-Murguialday, 2016), sleep-learning paradigms (Arzi et al., 2014), and neurostimulation techniques (Polanía, Nitsche, & Ruff, 2018).

Another aspect of translation is using research to inform policy. In April 2018, the Trump administration enacted a controversial policy that resulted in the separation of over 2,000 children from their caretakers upon entry into the United States. Many members of the scientific community denounced the policy, citing research on the lasting effects of such experiences on children's development (American Psychological Association, 2018; An open letter to Secretary Nielsen, 2018). The *Washington Post* published an op-ed by two researchers asserting that the actions of the government constituted torture (Juvonen & Silvers, 2018). Only after widespread protest did the administration announce an end to the policy. The family separation crisis highlights the need for vigilance against policies that scientific evidence indicates will cause harm. As cognitive neuroscientists, we have a responsibility to apply our expertise to inform policy-makers and the general public, particularly in cases in which the research is clear and the societal costs are high.

In education, research on school start times is an example of translational potential (Wahlstrom, 2016). Studies going back over 20 years show that later school start times benefit teenagers' academic performance and numerous health outcomes, yet fewer than 15% of U.S. high schools have adopted a later start time. Likewise, reading research supports the necessity of explicitly teaching grapheme-phoneme conversion (Rayner, Foorman, Perfetti, Pesetsky, & Seidenberg, 2001), in contrast to competing "global reading" methods that argue for teaching reading through whole words.

These examples illustrate how cognitive neuroscience can have a positive impact on society. Individual scientists can advance translational efforts by initiating dialogues with other professionals to (1) establish relevant questions from the applied perspective, (2) address how basic research can engage these questions, and (3) relay relevant results to the appropriate professional communities. We can also better connect with lay audiences by utilizing alternative forms of scientific communication, such as op-eds, policy briefs, open letters, and social media. As a community we should aim to create more space for translational work in our training programs and conferences. These efforts are gaining support at the national and institutional levels through programs such as the Mind, Brain, and Education Program at Harvard University, the National Center for Advancing Translational Sciences, and the National Institutes of Health Clinical and Translational Science Award program.

Conclusion

As cognitive neuroscientists, we aim to understand the connections among the brain, psychology, behavior, and their implications for society. To achieve our research goals, we must first address the current cultural and ethical issues that impede our progress as a field, including a lack of integrity, accountability, and diversity; researchers must have equal opportunities to pursue their research goals regardless of their gender, race, sexual identity, or socioeconomic status. The growing popularity of open science and preregistration may help to mitigate the current replicability crisis and may additionally improve the culture of science by allowing it to be more accessible. However, incentive systems must follow, and change its reward structure by explicitly rewarding open-science efforts. In addition, while technological advancements and large-scale data sets have allowed us to explore new avenues of research, it is essential to build the methodologies around the research question and not vice versa. Finally, researchers should strive to improve how they communicate their research with the general public. Cognitive neuroscience is a research area that garners much public interest. Therefore, as cognitive neuroscientists, we have a responsibility to publicly and accurately convey the potential implications of our findings and correct misconceptions. Moving forward, honesty, inclusivity, and transparency will promote future scientific developments with the greatest potential to make a positive impact on our health and society.

REFERENCES

American Psychological Association. (2018, May 29). Statement of APA president regarding the traumatic effects of separating immigrant families [Press release]. https://www.apa.org/news/press/releases/2018/05/separating-immigrant-families.aspx.

Anderson, M. L. (2014). *After phrenology: Neural reuse and the interactive brain.* Cambridge, MA: MIT Press.

Aru, J., Bachmann, T., Singer, W., & Melloni, L. (2012). Distilling the neural correlates of consciousness. *Neuroscience & Biobehavioral Reviews, 36*(2), 737–746.

Arzi, A., Holtzman, Y., Samnon, P., Eshel, N., Harel, E., & Sobel, N. (2014). Olfactory aversive conditioning during sleep reduces cigarette-smoking behavior. *Journal of Neuroscience, 34*(46), 15382–15393.

Baltrušaitis, T., Ahuja, C., & Morency, L.-P. (2019). Multimodal machine learning: A survey and taxonomy. *IEEE Transactions on Pattern Analysis and Machine Intelligence, 41*(2), 423–443.

Benjamin, D. J., Berger, J. O., Johannesson, M., Nosek, B. A., Wagenmakers, E.-J., Berk, R., ... Johnson, V. E. (2018). Redefine statistical significance. *Nature Human Behaviour, 2*(1), 6–10.

Brakewood, B., & Poldrack, R. A. (2013). The ethics of secondary data analysis: Considering the application of Belmont principles to the sharing of neuroimaging data. *Neuroimage, 82,* 671–676.

Casey, B. J., Cannonier, T., Conley, M. I., Cohen, A. O., Barch, D. M., Heitzeg, M. M., Dale, A. M. (2018). The adolescent brain cognitive development (ABCD) study: Imaging acquisition across 21 sites. *Developmental Cognitive Neuroscience, 32,* 43–54.

Chaudhary, U., Birbaumer, N., & Ramos-Murguialday, A. (2016). Brain-computer interfaces for communication and rehabilitation. *Nature Reviews Neurology, 12*(9), 513–525.

Choudhury, S., Fishman, J. R., McGowan, M. L., & Juengst, E. T. (2014). Big data, open science and the brain: Lessons learned from genomics. *Frontiers in Human Neuroscience, 8,* 239.

Cichy, R. M., Khosla, A., Pantazis, D., Torralba, A., & Oliva, A. (2016). Comparison of deep neural networks to spatio-temporal cortical dynamics of human visual object recognition reveals hierarchical correspondence. *Scientific Reports, 6,* 27755.

Cichy, R. M., Pantazis, D., & Oliva, A. (2014). Resolving human object recognition in space and time. *Nature Neuroscience, 17*(3), 455–462.

Engel, A. K., Friston, K. J., & Kragic, D. (Eds.). (2016). *The pragmatic turn: Toward action-oriented views in cognitive science.* Cambridge, MA: MIT Press.

Freeman, J. (2018). LGBTQ scientists are still left out. *Nature, 559*(7712), 27–28.

Gelman, A., & Loken, E. (2013). The garden of forking paths: Why multiple comparisons can be a problem, even when there is no "fishing expedition" or "p-hacking" and the research hypothesis was posited ahead of time [Unpublished paper]. http://www.stat.columbia.edu/~gelman/research/unpublished/p_hacking.pdf.

Gibson, J. (1979). *The ecological approach to visual perception.* Boston: Houghton Mifflin.

Hasson, U., Nir, Y., Levy, I., Fuhrmann, G., & Malach, R. (2004). Intersubject synchronization of cortical activity during natural vision. *Science, 303*(5664), 1634–1640.

Hermes, D., Nguyen, M., & Winawer, J. (2017). Neuronal synchrony and the relation between the blood-oxygen-level dependent response and the local field potential. *PLoS Biology, 15*(7), e2001461.

Juvonen, J., & Silvers, J. (2018, May 15). Separating children from parents at the border isn't just cruel. It's torture. *Washington Post.* https://www.washingtonpost.com.

Kandel, E. R., Markram, H., Matthews, P. M., Yuste, R., & Koch, C. (2013). Neuroscience thinks big (and collaboratively). *Nature Reviews Neuroscience, 14*(9), 659–664.

King, K. M., Littlefield, A. K., McCabe, C. J., Mills, K. L., Flournoy, J., & Chassin, L. (2018). Longitudinal modeling in developmental neuroimaging research: Common challenges, and solutions from developmental psychology. *Developmental Cognitive Neuroscience, 33,* 54–72.

Krizhevsky, A., Sutskever, I., & Hinton, G. E. (2012) ImageNet classification with deep convolutional neural networks. In *Advances in Neural Information Processing Systems, 25,* 1106–1114.

Marzbani, H., Marateb, H. R., & Mansourian, M. (2016). Neurofeedback: A comprehensive review on system design, methodology and clinical applications. *Basic and Clinical Neuroscience, 7*(2), 143–158.

Meehl, P. E. (1967). Theory-testing in psychology and physics: A methodological paradox. *Philosophy of Science, 34*(2), 103–115.

Melnick, R. S. (2018). *The transformation of title IX: Regulating gender equality in education.* Washington, DC: Brookings Institution Press.

Mills, K. L., & Tamnes, C. K. (2014). Methods and considerations for longitudinal structural brain imaging analysis across development. *Developmental Cognitive Neuroscience, 9,* 172–190.

National Academies of Sciences, Engineering, and Medicine. (2018). *Sexual harassment of women: Climate, culture, and consequences in academic sciences, engineering, and medicine.* Washington, DC: National Academies Press. https://doi.org/10.17226/24994

An open letter to Secretary Nielsen. (2018). https://sites.google.com/view/letter-to-secretary-nielsen.

Open Science Collaboration (2015). Estimating the reproducibility of psychological science. *Science, 349*(6251), aac4716.

Parkinson, C., Kleinbaum, A. M., & Wheatley, T. (2018). Similar neural responses predict friendship. *Nature Communications, 9*(1), 332.

Polanía, R., Nitsche, M. A., & Ruff, C. C. (2018). Studying and modifying brain function with non-invasive brain stimulation. *Nature Neuroscience, 21*(2), 174–187.

Poldrack, R. A., Laumann, T. O., Koyejo, O., Gregory, B., Hover, A., Chen, M. Y., ... Mumford, J. A. (2015). Long-term neural and physiological phenotyping of a single human. *Nature Communications, 6,* 8885.

Powell, K. (2018). These labs are remarkably diverse—Here's why they're winning at science. *Nature, 558*(7708), 19–22.

Rayner, K., Foorman, B. R., Perfetti, C. A., Pesetsky, D., & Seidenberg, M. S. (2001). How psychological science informs the teaching of reading. *Psychological Science in the Public Interest, 2*(2), 31–74.

Rogosa, D., Brandt, D., & Zimowski, M. (1982). A growth curve approach to the measurement of change. *Psychological Bulletin, 92*(3), 726–748.

Schilbach, L., Timmermans, B., Reddy, V., Costall, A., Bente, G., Schlicht, T., & Vogeley, K. (2013). Toward a second-person neuroscience. *Behavioral and Brain Sciences, 36*(4), 393–462.

Spivey, M. (2008). *The continuity of mind.* New York: Oxford University Press.

Stolk, A., Verhagen, L., & Toni, I. (2016). Conceptual alignment: How brains achieve mutual understanding. *Trends in Cognitive Sciences, 20*(3), 180–191.

Szucs, D., & Ioannidis, J. P. A. (2017). When null hypothesis significance testing is unsuitable for research: A reassessment. *Frontiers in Human Neuroscience, 11,* 390.

Tenopir, C., Allard, S., Douglass, K., Aydinoglu, A. U., Wu, L., Read, E., ... Frame, M. (2011). Data sharing by scientists: Practices and perceptions. *PloS One, 6*(6), e21101.

Van Horn, J. D., & Gazzaniga, M. S. (2002). Databasing fMRI studies—towards a "discovery science" of brain function. *Nature Reviews Neuroscience, 3*(4), 314–318.

Wahlstrom, K. L. (2016). Later start time for teens improves grades, mood, and safety. *Phi Delta Kappan, 98*(4), 8–14.

Wicherts, J. M., Bakker, M., & Molenaar, D. (2011). Willingness to share research data is related to the strength of the evidence and the quality of reporting of statistical results. *PloS One, 6*(11), e26828.

Yarkoni, T. (2014, May 19). Big data, n. A kind of black magic [Blog post]. https://www.talyarkoni.org/blog/2014/05/19/big-data-n-a-kind-of-black-magic/.

CONTRIBUTORS

David Abugaber University of Illinois at Chicago.

Patti Adank Department of Speech, Hearing and Phonetic Sciences, University College London, London, United Kingdom.

Bryon Adinoff Department of Psychiatry, University of Texas Southwestern Medical Center, Dallas; VA North Texas Health Care System, Dallas.

James W. Antony Department of Psychology and Neuroscience Institute, Princeton University, Princeton, New Jersey.

Sikoya M. Ashburn Center for the Study of Learning, Georgetown University Medical Center, Washington, DC.

Edward Awh Department of Psychology, Institute for Mind & Biology, The University of Chicago, Illinois.

Renée Baillargeon Department of Psychology, University of Illinois at Urbana-Champaign.

Jeffrey R. Binder Medical College of Wisconsin, Milwaukee.

Alexandre Barachant CTRL-Labs, New York.

Danielle Bassett Department of Bioengineering, School of Engineering and Applied Science, University of Pennsylvania, Philadelphia; Department of Physics and Astronomy, College of Arts and Sciences, University of Pennsylvania, Philadelphia; Department of Neurology, Perelman School of Medicine, University of Pennsylvania, Philadelphia; Department of Electrical and Systems Engineering, School of Engineering and Applied Science, University of Pennsylvania, Philadelphia; Department of Psychiatry, Perelman School of Medicine, University of Pennsylvania, Philadelphia.

Demian Battaglia Aix-Marseille Université, Institute for Systems Neuroscience (INS, UMR 1106) and Institute for Language, Communication and the Brain (ILCB), Marseille, France.

Marina Bedny Department of Psychological and Brain Sciences, Johns Hopkins University, Baltimore.

Daniel Bennett Princeton Neuroscience Institute, Princeton University, Princeton, New Jersey.

Kelly A. Bennion California Polytechnic State University, San Luis Obispo.

Noah C. Benson Department of Psychology and Center for Neural Science, New York University, New York.

Sven Bestmann Wellcome Centre for Human Neuroimaging, UCL Institute of Neurology, University College London; Department of Clinical and Movement Neurosciences, UCL Institute of Neurology, University College London, United Kingdom.

Yanchao Bi State Key Laboratory of Cognitive Neuroscience and Learning and IDG/McGovern Institute for Brain Research, Beijing Normal University, China.

Sarah-Jayne Blakemore Professor of Psychology and Cognitive Neuroscience, University of Cambridge, Cambridge, United Kingdom.

Adam Boncz Central European University, Budapest, Hungary.

Ina Bornkessel-Schlesewsky School of Psychology, Social Work and Social Policy, University of South Australia, Adelaide; Centre for Cognitive and Systems Neuroscience, University of South Australia, Adelaide.

David Bridwell The Mind Research Network, Albuquerque, New Mexico.

Jonathan R. Brennan Department of Linguistics, University of Michigan, Ann Arbor.

Andrea Brovelli Aix-Marseille Université, Timone Neuroscience Institute (INT, UMR 7289), Institute for Language, Communication and the Brain (ILCB), Marseille, France.

Timothy J. Buschman Princeton Neuroscience Institute and Department of Psychology, Princeton University, Princeton, New Jersey.

Laurel J. Buxbaum Moss Rehabilitation Research Institute, Medical Arts Building, Elkins Park, Pennsylvania; Thomas Jefferson University, Department of Rehabilitation Medicine, Philadelphia, Pennsylvania.

Melody Buyukozer Dawkins Department of Psychology, University of Illinois at Urbana-Champaign.

John T. Cacioppo Post-mortem.

Stephanie Cacioppo Pritzker School of Medicine, Brain Dynamics Laboratory, Department of Psychiatry and Behavioral Neuroscience, The University of Chicago, Illinois.

JESSICA F. CANTLON Department of Psychology, Carnegie Mellon University, Pittsburgh.

ALFONSO CARAMAZZA Center for Mind/Brain Sciences (CIMeC), University of Trento, Italy, Department of Psychology, Harvard University, Cambridge, Massachusetts.

CARLOS CARDENAS-INIGUEZ Department of Psychology, The University of Chicago, Illinois.

THOMAS A. CARLSON School of Psychology, University of Sydney, Australia.

MARIA CHAIT Ear Institute, University College London, London.

LUKE J. CHANG Department of Psychological and Brain Sciences, Dartmouth College, Hanover, New Hampshire.

ANJAN CHATTERJEE Department of Neurology, Penn Medicine, University of Pennsylvania, Philadelphia.

MARVIN M. CHUN Department of Psychology and Cognitive Science Program, Yale University, New Haven, Connecticut.

ALEX CLARKE Department of Psychology, University of Cambridge, Cambridge, United Kingdom.

KATHERINE E. CONEN Department of Neuroscience, Washington University, Saint Louis, Missouri; Department of Collective Behavior, Max Planck Institute for Animal Behavior, Konstanz, Germany.

SAMUEL F. COOKE Department of Basic and Clinical Neuroscience, King's College London, United Kingdom.

ROSHAN COOLS Donders Institute for Brain, Cognition and Behaviour, Radboud University, Nijmegen, The Netherlands; Department of Psychiatry, Radboud University Medical Centre, Nijmegen, The Netherlands.

MARC N. COUTANCHE Department of Psychology, University of Pittsburgh, Pennsylvania, Learning Research & Development Center, University of Pittsburgh, Pennsylvania, Brain Institute, University of Pittsburgh, Pennsylvania.

EVELINE A. CRONE Department of Psychology, Leiden University, The Netherlands.

JULIET Y. DAVIDOW Department of Psychology and Center for Brain Science, Harvard University, Cambridge, Massachussetts.

MATTHEW H. DAVIS Medical Research Council Cognition and Brain Sciences Unit, University of Cambridge, Cambridge, United Kingdom.

STANISLAS DEHAENE Chair of Experimental Cognitive Psychology, Collège de France, Paris; and NeuroSpin center, INSERM, CEA, Saclay, France.

GHISLAINE DEHAENE-LAMBERTZ INSERM, UMR992, CEA, NeuroSpin Center, University Paris Saclay, Gif-sur-Yvette, France.

MAURICIO DELGADO Department of Psychology, Rutgers University, Newark, New Jersey.

KRISTEN DELEVICH Department of Psychology, University of California, Berkeley.

JÖRN DIEDRICHSEN Brain and Mind Institute, Department for Computer Science, Department for Statistical and Actuarial Sciences, Western University, London, Ontario, Canada.

JOCHEN DITTERICH Center for Neuroscience and Department of Neurobiology, Physiology & Behavior, University of California, Davis.

CHRISTIAN F. DOELLER Kavli Institute for Systems Neuroscience, Centre for Neural Computation, Egil and Pauline Braathen and Fred Kavli Centre for Cortical Microcircuits, NTNU, Norwegian University of Science and Technology, Trondheim; Max Planck Institute for Human Cognitive and Brain Sciences, Leipzig, Germany.

MANOJ DOSS Johns Hopkins University, Baltimore, Maryland.

JOSHUA TATE DUDMAN Janelia Research Campus, Howard Hughes Medical Institute, Ashburn, Virginia.

KATHERINE DUNCAN Department of Psychology, University of Toronto, Toronto, Canada.

NAOMI I. EISENBERGER Department of Psychology, University of California, Los Angeles.

ARNE D. EKSTROM Center for Neuroscience and Department of Psychology, University of California, Davis; Department of Psychology, University of Arizona, Tucson.

KAREN EMMOREY School of Speech, Language, and Hearing Sciences, San Diego State University, San Diego, California.

DENIS ENGEMANN Parietal Team, INRIA, CEA, Université Paris-Saclay, Gif-sur-Yvette, France.

RUSSELL A. EPSTEIN Department of Psychology, University of Pennsylvania, Philadelphia.

JULIA ERB Department of Psychology, University of Lübeck, Germany.

MARTHA J. FARAH Department of Psychology, University of Pennsylvania, Philadelphia.

DOMINIC S. FARERI Gordon F. Derner School of Psychology, Adelphi University, Garden City, New York.

EVELINA FEDORENKO EvLab, Massachusetts Institute of Technology, Cambridge, Massachusetts.

ORIEL FELDMANHALL Brown University, Providence, Rhode Island.

LEONARDO FERNANDINO Medical College of Wisconsin, Milwaukee.

LUCÍA FERNÁNDEZ Centro de Investigación Básica en Psicología, Universidad de la República, Montevideo, Uruguay.

JASON FISCHER Department of Psychological and Brain Sciences, Johns Hopkins University, Baltimore.

JULIUS FRIDRIKSSON Department of Communication Sciences and Disorders, University of South Carolina, Columbia.

MICHAEL J. FRANK Department of Cognitive, Linguistics, and Psychological Sciences, Brown University, Providence, Rhode Island; Carney Institute for Brain Sciences, Brown University, Providence, Rhode Island.

JACK L. GALLANT Helen Wills Neuroscience Institute, University of California, Berkeley; Department of Psychology, University of California, Berkeley.

ADRIANA GALVÁN Department of Psychology, University of California, Los Angeles.

FABIAN GRABENHORST Department of Physiology, Development and Neuroscience, University of Cambridge, Cambridge, United Kingdom.

ALEXANDRE GRAMFORT Parietal Team, INRIA, CEA, Université Paris-Saclay, Gif-sur-Yvette, France.

JOSHUA D. GREENE Department of Psychology, Center for Brain Science, Harvard University, Cambridge, Massachusetts.

KALANT GRILL-SPECTOR Department of Psychology, Wu Tsai Neurosciences Institute, Stanford University, Stanford, California.

MATTHIAS J. GRUBER Cardiff University Brain Research Imaging Centre (CUBRIC), School of Psychology, Cardiff University, United Kingdom.

Tijl Grootswagers School of Psychology, University of Sydney, Australia.

Xiaosi Gu Department of Psychiatry, Icahn School of Medicine at Mount Sinai, New York; Nash Family Department of Neuroscience, Icahn School of Medicine at Mount Sinai, New York; James J. Peter Veterans Affairs Medical Center, Bronx, New York.

Laura Gwilliams Psychology Department, New York University, New York, and New York University Abu Dhabi, United Arab Emirates.

Leor Hackel Stanford University, Stanford, California.

Adrian M. Haith Department of Neurology, Johns Hopkins University School of Medicine, Baltimore, Maryland.

Simon Hanslmayr Centre for Human Brain Health, School of Psychology, University of Birmingham, United Kingdom.

Chris Holdgraf Berkeley Institute for Data Science, Helen Wills Neuroscience Institute, University of California, Berkeley.

Erica A. Hornstein Department of Psychology, University of California, Los Angeles.

Inge Huijsmans Donders Centre for Cognitive Neuroimaging, Radboud University, Nijmegen, The Netherlands.

Tristen K. Inagaki Department of Psychology, University of Pittsburgh, Pennsylvania.

Catherine Insel Department of Psychology and Center for Brain Science, Harvard University, Cambridge, Massachussetts.

Richard B. Ivry Department of Psychology and Helen Wills Neuroscience Institute, University of California, Berkeley.

Andrew Jackson Institute of Neuroscience, Newcastle University, Newcastle upon Tyne, United Kingdom.

Ole Jensen Centre for Human Brain Health, School of Psychology, University of Birmingham, United Kingdom.

Owen D. Jones Glenn M. Weaver, M.D. and Mary Ellen Weaver Chair in Law, Brain, and Behavior; Professor of Law; and Professor of Biological Sciences, Vanderbilt University, Nashville, Tennessee; MacArthur Foundation Research Network on Law and Neuroscience, Vanderbilt University Law School, Nashville, Tennessee.

Donatas Jonikaitis Department of Neurobiology, Stanford University School of Medicine, Stanford, California; Howard Hughes Medical Institute, Stanford, California.

Joshua B. Julian Kavli Institute for Systems Neuroscience, Centre for Neural Computation, Egil and Pauline Braathen and Fred Kavli Centre for Cortical Microcircuits, NTNU, Norwegian University of Science and Technology, Trondheim, Norway.

Claire Kabdebon INSERM, UMR992, CEA, NeuroSpin Center, University Paris Saclay, Gif-sur-Yvette, France.

Solène Kalénine University Lille, Centre national de la recherche scientifique, CHU Lille, UMR 9193, SCALab—Sciences Cognitives et Sciences Affectives, Lille, France.

Sabine Kastner Princeton Neuroscience Institute, Princeton University, Princeton, New Jersey; Department of Psychology, Princeton University, Princeton, New Jersey.

Emma J. Kilford Institute of Cognitive Neuroscience, University College London, United Kingdom.

Jean-Rémi King Psychology Department, New York University, New York, New York University Abu Dhabi, United Arab Emirates, and Frankfurt Institute for Advanced Studies, Frankfurt, Germany.

Etienne Koechlin PSL Research University, Paris; Pierre et Marie Curie University, Paris; Institut National de la Santé et de la Recherche Médicale, Paris.

John W. Krakauer Departments of Neurology and Neuroscience, The Johns Hopkins University School of Medicine, Baltimore, Maryland.

Nikolaus Kriegeskorte Mortimer B. Zuckerman Mind Brain Behavior Institute, Department of Psychology, Department of Neuroscience, and Department of Electrical Engineering (affiliated member), Columbia University, New York.

Lara Krisst Center for Mind and Brain, University of California, Davis.

Regina Lapate Helen Wills Neuroscience Institute, University of California, Berkeley.

Eric Larson Institute for Learning and Brain Sciences, University of Washington, Seattle.

Evan Layher University of California, Santa Barbara.

Anna Leshinskaya Department of Psychology, University of Pennsylvania, Philadelphia.

Josiah Leong Stanford University, Stanford, California.

Yuanning Li Carnegie Mellon University, Pittsburgh, Pennsylvania.

Ulman Lindenberger Max Planck UCL Centre for Computational Psychiatry and Ageing Research, Berlin, Germany; Center for Lifespan Psychology, Max Planck Institute for Human Development, Berlin, Germany; European University Institute, San Domenico di Fiesole (FI), Italy.

Björn Lindström Department of Psychology, University of Amsterdam, The Netherlands.

Martin Lövdén Aging Research Center, Department of Neurobiology, Care Sciences, and Society, Karolinska Institutet and Stockholm University, Sweden.

Steven J. Luck Center for Mind and Brain and Department of Psychology, University of California, Davis.

Mairéad MacSweeney Deafness, Cognition and Language (DCAL) Research Center and Institute of Cognitive Neuroscience, University College London, London, United Kingdom.

Bradford Z. Mahon Department of Psychology and Carnegie Mellon Neuroscience Institute, Carnegie Mellon University, Pittsburgh, Pennsylvania.

Tamar R. Makin Institute of Cognitive Neuroscience, University College London, London, United Kingdom.

Freddie Marquez Department of Neurobiology and Behavior, University of California, Irvine.

Mara Mather University of Southern California, Los Angeles.

Andrew R. Mayes Division of Neuroscience and Experimental Psychology, University of Manchester, United Kingdom.

Josh McDermott Department of Brain and Cognitive Sciences, McGovern Institute of Brain Research, and Center for Brains, Minds and Machines, Massachusetts Institute of Technology; Program in Speech and Hearing Biosciences and Technology, Harvard University, Cambridge Massachusetts.

Samuel D. McDougle Department of Psychology, University of California, Berkeley.

Heidi C. Meyer Department of Psychiatry, Weill Cornell Medicine, New York.

Earl K. Miller The Picower Institute for Learning and Memory, Department of Brain and Cognitive Sciences, Massachusetts Institute of Technology, Cambridge, Massachusetts.

Kathryn L. Mills Department of Psychology, University of Oregon, Eugene.

Tirin Moore Department of Neurobiology, Stanford University School of Medicine, Stanford, California; Howard Hughes Medical Institute, Stanford, California.

Felipe Munoz-Rubke Instituto de Psicología, Universidad Austral de Chile, Puerto Montt, Chile.

Elizabeth Musz Department of Psychological and Brain Sciences, Johns Hopkins University, Baltimore, Maryland.

Miguel A. L. Nicolelis Department of Neurobiology, Department of Neurology, Department of Neurosurgery, Department of Biomedical Engineering, and Department of Psychological Sciences, Center for Neuroengineering, Duke University, Durham, North Carolina.

Yael Niv Princeton Neuroscience Institute and Department of Psychology, Princeton University, Princeton, New Jersey.

A. C. (Kia) Nobre Department of Experimental Psychology, University of Oxford, Oxford; Oxford Centre for Human Brain Activity, Wellcome Centre for Integrative Neuroimaging, Department of Psychiatry, University of Oxford, Oxford, United Kingdom.

Erik C. Nook Department of Psychology, Harvard University, Cambridge, Massachusetts.

Uta Noppeney Computational Neuroscience and Cognitive Robotics Centre, Department of Psychology, University of Birmingham, United Kingdom.

Kenneth A. Norman Department of Psychology and Neuroscience Institute, Princeton University, Princeton, New Jersey.

Lars Nyberg Department of Radiation Sciences, Umeå Center for Functional Brain Imaging (UFBI), and Department of Integrative Medical Biology, Umeå University, Umeå, Sweden.

Jonas Obleser Department of Psychology, University of Lübeck, Germany.

Aude Oliva Computer Science and Artificial Intelligence Laboratory, Massachusetts Institute of Technology, Cambridge.

Temidayo Orederu Department of Psychiatry, Department of Neuroscience, and Friedman Brain Institute, Icahn School of Medicine at Mount Sinai, New York.

Maya Opendak Emotional Brain Institute, Nathan Kline Institute, Child and Adolescent Psychiatry, New York University Langone Medical Center, New York.

Andreas Olsson Division of Psychology, Department of Clinical Neuroscience, Karolinska Institutet, Solna, Sweden.

Camillo Padoa-Schioppa Department of Neuroscience, Department of Economics, and Department of Biomedical Engineering, Washington University, Saint Louis, Missouri.

Ken A. Paller Department of Psychology and Cognitive Neuroscience Program, Northwestern University, Evanston, Illinois.

Philip Pärnamets Division of Psychology, Department of Clinical Neuroscience, Karolinska Institutet, Solna, Sweden; Department of Psychology, New York University, New York.

Fabio Pasqualetti Department of Mechanical Engineering, University of California, Riverside.

Tara K. Patterson Department of Psychology, University of California, Los Angeles.

Siobhan S. Pattwell Human Biology Division, Fred Hutchinson Cancer Research Center, Seattle, Washington.

Virginia B. Penhune Psychology Department, Concordia University, Montreal, Canada.

Monica A. Perez Scientific Chair, Arms+Hands Lab, Shirley Ryan AbilityLab, Chicago; Professor, Department of Physical Medicine and Rehabilitation, Northwestern University, Chicago; Research Scientist, Edward Hines, Jr. VA Hospital, Chicago, Illinois.

Warren Pettine Department of Neurobiology, Stanford University School of Medicine, Stanford, California; Howard Hughes Medical Institute, Stanford, California; Center for Neural Science, New York University, New York.

Elizabeth Phelps Department of Psychology, New York University, New York.

Sara F. Popham Helen Wills Neuroscience Institute, University of California, Berkeley.

John P. Powers Duke University, Durham, North Carolina.

Daria Proklova The Brain and Mind Institute, Western University, London, Ontario, Canada.

J. Andrew Pruszynski Department of Physiology and Pharmacology, Department of Psychology, Robarts Research Institute, and Brain and Mind Institute, Western University, London, Ontario, Canada.

Liina Pylkkänen Department of Linguistics and Department of Psychology, New York University, New York; New York University Abu Dhabi Institute, Abu Dhabi.

Charlis Raineki Cellular and Physiological Sciences, University of British Columbia, Vancouver, Canada.

Mani Ramaswami School of Genetics and Microbiology, Trinity College Institute of Neuroscience, Trinity College, Dublin, Ireland; National Centre for Biological Sciences, Bangalore, India.

Charan Ranganath Center for Neuroscience and Department of Psychology, University of California, Davis.

Kristina M. Rapuano Department of Psychology, Yale University, New Haven, Connecticut.

Naftali Raz Institute of Gerontology and Department of Psychology, Wayne State University, Detroit; Max Planck Institute for Human Development, Lentzeallee 94, Berlin, Germany.

Talia L. Retter Institute of Research in Psychological Science, Institute of Neuroscience, Université de Louvain, Belgium; Department of Psychology, Center for Integrative Neuroscience, University of Nevada, Reno.

Maureen Ritchey Boston College, Department of Psychology, Chestnut Hill, Massachusetts.

David Robbe Aix-Marseille University, Département de Biologie, Parc Scientifique de Luminy, Marseille, France; INSERM-Institut National de la Santé et de la Recherche Médicale, UMR 1249, Marseille, Parc Scientifique de Luminy, Marseille, France; INMED-Institut de Neurobiologie de la Méditerranée, Parc Scientifique de Luminy, Marseille, France.

Amanda K. Robinson School of Psychology, University of Sydney, Australia.

Charles S. H. Robinson The Mind Research Network, Albuquerque, New Mexico.

Patrese A. Robinson-Drummer Emotional Brain Institute, Nathan Kline Institute, Child and Adolescent Psychiatry, New York University Langone Medical Center, New York.

Ranulfo Romo Instituto de Fisiología Celular-Neurociencias, Universidad Nacional Autónoma de México, México City; El Colegio Nacional, Mexico City.

Monica D. Rosenberg Department of Psychology, Yale University, New Haven, Connecticut; Department of Psychology, The University of Chicago, Chicago, Illinois.

Adina Roskies Department of Philosophy, Dartmouth College, Hanover, New Hampshire.

Jessica M. Ross Berenson-Allen Center for Noninvasive Brain Stimulation, Harvard Medical School, Boston, Massachusetts.

Román Rossi-Pool Instituto de Fisiología Celular-Neurociencias, Universidad Nacional Autónoma de México, México City.

Bruno Rossion Université de Lorraine, CNRS, CRAN, Nancy, France; Université de Lorraine, CHRU-Nancy, Service de Neurologie, France; Institute of Research in Psychological Science, Institute of Neuroscience, Université de Louvain, Belgium.

Tania Roth Psychological and Brain Sciences, University of Delaware, Newark.

Tomás J. Ryan School of Biochemistry and Immunology and Trinity College Institute of Neuroscience, Trinity College Dublin, University of Dublin; Florey Institute of Neuroscience and Mental Health, Melbourne Brain Centre, University of Melbourne, Australia.

Barbara J. Sahakian Department of Psychiatry and Behavioural and Clinical Neuroscience Institute, University of Cambridge, School of Clinical Medicine, Cambridge, United Kingdom.

C. Daniel Salzman Departments of Neuroscience and Psychiatry, Columbia University and New York State Psychiatric Institute, New York.

Jason Samaha Department of Psychology, University of California, Santa Cruz.

Jona Sassenhagen Department of Psychology, Goethe-University Frankfurt, Germany.

George Savulich Department of Psychiatry and Behavioral and Clinical Neuroscience Institute, University of Cambridge, School of Clinical Medicine, Cambridge, United Kingdom.

Matthew Sazma Center for Neuroscience, University of California, Davis.

Gaia Scerif Attention, Brain and Cognitive Development Group, Department of Experimental Psychology, University of Oxford, Oxford, United Kingdom.

Daniela Schiller Department of Psychiatry, Department of Neuroscience, and Friedman Brain Institute, Icahn School of Medicine at Mount Sinai, New York.

Matthias Schlesewsky School of Psychology, Social Work and Social Policy, University of South Australia, Adelaide; Centre for Cognitive and Systems Neuroscience, University of South Australia, Adelaide.

Wolfram Schultz Department of Physiology, Development and Neuroscience, University of Cambridge, Cambridge, United Kingdom.

Michael N. Shadlen Department of Neuroscience, Columbia University Medical Center, New York; Zuckerman Mind Brain Behavior Institute, Columbia University, New York; Kavli Institute of Brain Science, Columbia University, New York; Howard Hughes Medical Institute, Columbia University, New York.

Barbara G. Shinn-Cunningham Biomedical Engineering, Boston University, Boston, Massachusetts.

Daphna Shohamy Department of Psychology, Zuckerman Mind, Brain, Behavior Institute and Kavli Institute for Brain Sciences, Columbia University, New York.

Max Siegel Center for Brains, Minds and Machines, Department of Brain & Cognitive Science, Massachusetts Institute of Technology, Cambridge, Massachusetts.

Ediz Sohoglu Medical Research Council Cognition and Brain Sciences Unit, University of Cambridge, Cambridge, United Kingdom.

Sarah H. Solomon Department of Psychology, University of Pennsylvania, Philadelphia.

Leah H. Somerville Department of Psychology and Center for Brain Science, Harvard University, Cambridge, Massachussetts.

Clara K. Starkweather Harvard University, Center for Brain Science, Department of Molecular and Cellular Biology, Cambridge, Massachusetts.

William R. Stauffer Department of Neurobiology, Systems Neuroscience Center, Center for the Neural Basis of Cognition, University of Pittsburgh, Pennsylvania.

Maayan Stavans Department of Psychology, Bar-Ilan University, Ramat Gan, Israel.

Andrew X. Stewart Center for Mind and Brain, University of California, Davis.

Ariana Stickel Department of Psychology, University of Arizona, Tucson.

Gabriel M. Stine Department of Neuroscience, Columbia University Medical Center, New York; Zuckerman Mind Brain Behavior Institute, Columbia University, New York.

M. S. Stokes Department of Experimental Psychology, University of Oxford, Oxford; Oxford Centre for Human Brain Activity, Wellcome Centre for Integrative Neuroimaging, Department of Psychiatry, University of Oxford, Oxford, United Kingdom.

Arjen Stolk University of California, Berkeley.

Katherine R. Storrs Department of Psychology, Justus Liebig University Giessen, Germany.

Regina M. Sullivan Emotional Brain Institute, Nathan Kline Institute, Child and Adolescent Psychiatry, New York University Langone Medical Center, New York.

Christopher Summerfield Department of Experimental Psychology, University of Oxford, Oxford, United Kingdom.

Christian K. Tamnes University of Oslo, Norway.

Jordan A. Taylor Department of Psychology, Princeton University, Princeton, New Jersey.

Joshua Tenenbaum Center for Brains, Minds and Machines, Department of Brain & Cognitive Science, Massachusetts Institute of Technology, Cambridge, Massachusetts.

Sharon L. Thompson-Schill Department of Psychology, University of Pennsylvania, Philadelphia.

Fransisca Ting Department of Psychology, University of Illinois at Urbana-Champaign.

KONSTANTINOS TSETSOS Department of Neurophysiology and Pathophysiology, University Medical Center Hamburg-Eppendorf, Hamburg, Germany.

LORRAINE K. TYLER Department of Psychology, University of Cambridge, Cambridge, United Kingdom.

NAOSHIGE UCHIDA Harvard University, Center for Brain Science, Department of Molecular and Cellular Biology, Cambridge, Massachusetts.

W. MARTIN USREY Center for Neuroscience; Department of Neurobiology, Physiology and Behavior, and Department of Neurology, University of California, Davis.

JAY J. VAN BAVEL Department of Psychology and Neural Science, New York University, New York.

ANNA C. K. VAN DUIJVENVOORDE Department of Psychology, Leiden University, The Netherlands.

OSHIN VARTANIAN Department of Psychology, University of Toronto, Ontario, Canada.

JOSÉ VERGARA Instituto de Fisiología Celular-Neurociencias, Universidad Nacional Autónoma de México, México City.

PETRA E. VÉRTES Behavioural and Clinical Neuroscience Institute, Department of Psychiatry, University of Cambridge, Cambridge; The Alan Turing Institute, London; School of Mathematical Sciences, Queen Mary University of London, United Kingdom.

VERONIKA VILGIS Center for Mind and Brain, University of California, Davis.

EDWARD K. VOGEL Department of Psychology, Institute for Mind & Biology, The University of Chicago, Illinois.

ANTHONY D. WAGNER Professor, Department of Psychology and Wu Tsai Neurosciences Institute, Stanford University, Stanford, California; MacArthur Foundation Research Network on Law and Neuroscience.

KRISTINE BEATE WALHOVD Center for Lifespan Changes in Brain and Cognition, Department of Psychology, University of Oslo, Norway.

JEFFREY WEILER Department of Physiology and Pharmacology, Department of Psychology, Robarts Research Institute, and Brain and Mind Institute, Western University, London, Ontario, Canada.

KEVIN S. WEINER Department of Psychology and Helen Wills Neuroscience Institute, University of California, Berkeley.

ANDREW WESTBROOK Donders Institute for Brain, Cognition and Behaviour, and Department of Psychiatry, Radboud University Medical Centre, Nijmegen, The Netherlands; Department of Cognitive, Linguistics, and Psychological Sciences, Brown University, Providence, Rhode Island.

THALIA WHEATLEY Dartmouth College, Hanover, New Hampshire.

LINDA WILBRECHT Department of Psychology and Helen Wills Neuroscience Institute, University of California, Berkeley.

JULIAN A. WILLS New York University, New York.

STEPHEN M. WILSON Department of Hearing and Speech Sciences, Vanderbilt University Medical Center, Nashville, Tennessee.

JONATHAN WINAWER Department of Psychology and Center for Neural Science, New York University, New York.

MORITZ F. WURM Center for Mind/Brain Sciences (CIMeC), University of Trento, Italy.

DANIEL YAMINS Departments of Psychology and Computer Science, and the Wu Tsai Neurosciences Institute, Stanford University, Stanford, California.

JASON D. YEATMAN Department of Speech and Hearing Sciences and Institute for Learning & Brain Sciences (I-LABS), University of Washington, Seattle.

ILKER YILDIRIM Center for Brains, Minds and Machines, Department of Brain & Cognitive Science, Massachusetts Institute of Technology, Cambridge, Massachusetts.

LIANE YOUNG Department of Psychology, Boston College, Newton, Massachusetts.

ROBERT J. ZATORRE Montreal Neurological Institute, McGill University, Montreal, Canada.

MEGAN ZIRNSTEIN Department of Linguistics and Cognitive Science, Pomona College, Claremont, California.

ARIEL ZYLBERBERG Department of Neuroscience, Columbia University Medical Center, New York; Zuckerman Mind Brain Behavior Institute, Columbia University, New York.

INDEX